SUMMARY OF TECHNICAL CHANGES

For an overview o
edition of the NE(
Changes section

MW00368292

Ground detectors are required to be installed in ungrounded systems to detect such faults. A second insulation failure on a different ungrounded conductor would result in a line-to-line-to-ground fault, with the potential for more extensive damage to electrical equipment.

(B) Ground Detectors. Ground detectors shall be installed in accordance with the following:

(1) Ungrounded ac systems as permitted in 250.21(A)(1) through (A)(4) operating at not less than 120 volts and at 1000 volts or less shall have ground detectors installed on the system.
(2) The ground detection sensing equipment shall be connected as close as practicable to where the system receives its supply.

Ground detectors provide a visual indication, an audible signal, or both to alert system operators and maintenance personnel of a ground-fault condition in the electrical system. The notification of the ground-fault condition, rather than automatic interruption of the circuit, allows the operators of the process to take the necessary steps to initiate an orderly shutdown, to determine where the ground fault is located, and to safely repair the fault.

(C) Marking. Ungrounded systems shall be legibly marked "Caution: Ungrounded System Operating — _____Volts Between Conductors" at the source or first disconnecting means of the system. The marking shall be of sufficient durability to withstand the environment involved.

•

250.24 Grounding of Service-Supplied Alternating-Current Systems.

(A) System Grounding Connections. A premises wiring system supplied by a grounded ac service shall have a grounding electrode conductor connected to the grounded service conductor, at each service, in accordance with 250.24(A)(1) through (A)–(4).

The grounded conductor of an ac service is connected to a grounding electrode system to limit the voltage to ground imposed on the system by lightning, line surges, and unintentional high-voltage crossovers. This connection also stabilizes the voltage to ground during operation, including short circuits.

See also

250.4(A) and **(B)** for the performance requirements for these connections

Δ **(1) General.** The grounding electrode conductor connection shall be made at any accessible point
overhead service conductors, service d
conductors, or service lateral to the ter
grounded service conductor is connec
necting means.

Informational Note: See Article 100 f
Conductors, Overhead; Service Condu
vice Drop; and *Service Lateral.*

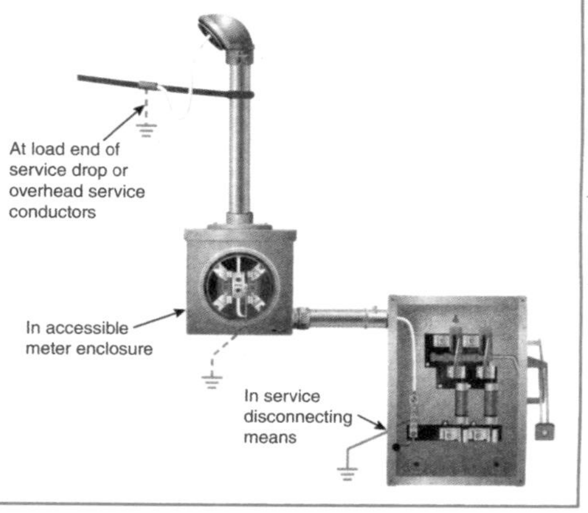

EXHIBIT 250.7 Three locations on an overhead service where the GEC is permitted to be connected to the grounded service conductor.

Allowing various connection locations meets the overall objectives for grounding while permitting a variety of practical options. Exhibit 250.7 illustrates three possible connection points where the GEC is permitted to be connected to the grounded service conductor. The accessibility to the point of connection must be approved by the AHJ based on local conditions such as locked meter socket enclosures.

Δ **(2) Outdoor Transformer.** If the transformer supplying the service is located outside the building, at least one additional grounding connection shall be made from the grounded service conductor to a grounding electrode, either at the transformer or elsewhere outside the building.

Exception: The additional grounding electrode conductor connection shall not be made on impedance grounded systems. Impedance grounded systems shall meet the requirements of 250.36 or 250.187, as applicable.

Outdoor installations are susceptible to lightning as well as accidental primary-to-secondary crossovers. The connection outside a building helps mitigate the effects of these influences on the interior portion of the premises wiring system. Exhibit 250.8 illustrates two grounding electrode connections — one at the service equipment installed inside the building and one
of those connections
equired by 250.24(A)

t are dual fed (double
d together in separate
e, a single grounding
point of the grounded
ll be permitted.

Commentary exhibits are outlined in red to easily differentiate them from code figures.

A bullet indicates where one or more code sections have been removed.

A Δ beside a section number indicates that words within that section were deleted.

(A Δ beside a table or figure number indicates there has been a revision within the table or figure.)

Calculation Example

Calculate the demand load for a 100,000 ft² school building if the calculated load without demand factors is 2.5 MVA (2,500,000 VA).

Solution

Step 1. Determine load per square foot:

$$2.5\ \text{MVA}/100{,}000\ \text{ft}^2 = 25\ \text{VA/ft}^2$$

Step 2. Calculate the demand load using Table 220.86:

First 3 VA/ft² at 100%	3 VA/ft²
Next 17 VA/ft² at 75%	
17 VA/ft² × 0.75	12.75 VA/ft²
Remaining 5 VA/ft² at 25%	
5 VA/ft² × 0.25	1.25 VA/ft²
Overall demand factor	17 VA/ft²

Step 3. Calculate demand load for the school:

$$17\ \text{VA/ft}^2 \times 100{,}000\ \text{ft}^2 = 1{,}700{,}000\ \text{VA} = 1.7\ \text{MVA}$$

Calculation Examples and Application Examples present scenarios outlining how certain requirements can be applied.

National Electrical Code® Handbook

Sixteenth Edition
International Electrical Code Series

Edited by

Dean Austin

Christopher D. Coache

Corey Hannahs

Erik Hohengasser

Jeff Sargent

With the complete text of the 023 edition of NFPA 70®, National Electrical Code®

NATIONAL FIRE PROTECTION ASSOCIATION
The leading information and knowledge resource on fire, electrical and related hazards

Development and Production: Ken Ritchie
Interior Design: Cheryl Langway
Copyediting: Amy Feldman
Composition: KnowledgeWorks Global Ltd.
Printing and Binding: LSC Communications, Inc.

Important Notices and Disclaimers: Publication of this book is for the purpose of circulating information and opinion among those concerned for fire and electrical safet related subjects. While every effort has been made to achieve a work of high quality, neither the NFPA® nor contributors to this handbook guarantee or warrantee the accuracy or completeness of or assume any liability in ction with the information and opinions contained in this handbook. The NFPA and the contributors shall in not be liable for any personal injury, property, or other damages of any nature whatsoever, whether special, indirect onsequential, or compensatory, directly or indirectly resulting from the publication, use of, or reliance upon this book.

This handbook is published with the understanding the NFPA and the contributors to this handbook are supplying information and opinion but are not attempting der engineering or other professional services. If such services are required, the assistance of an appropriate pronal should be sought.

NFPA 70®, National Electrical Code® ("*NFPA 70*"), e all NFPA codes, standards, recommended practices, and guides ("NFPA Standards"), made available for use ct to Important Notices and Legal Disclaimers, which appear at the end of this handbook and can also be view *www.nfpa.org/disclaimers.*

Notice Concerning Code Interpretations: This sixte edition of the *National Electrical Code®* Handbook is based on the 2023 edition of *NFPA 70*. All NFPA cod andards, recommended practices, and guides ("NFPA Standards") are developed in accordance with the pu d procedures of the NFPA by technical committees comprised of volunteers drawn from a broad array of nt interests. The handbook contains the complete text of *NFPA 70* and any applicable Formal Interpretations d by the NFPA at the time of publication. This NFPA Standard is accompanied by explanatory commentary a her supplementary materials.

The commentary and supplementary materials in handbook are not a part of the NFPA Standard and do not constitute Formal Interpretations of the NFPA (wh n be obtained only through requests processed by the responsible technical committees in accordance with th lished procedures of the NFPA). The commentary and supplementary materials, therefore, solely reflect the pel opinions of the editor or other contributors and do not necessarily represent the official position of the NFPA technical committees.

REMINDER: UPDAT DF NFPA STANDARDS

NFPA 70, National Electrical Code, like all NFPA co tandards, recommended practices, and guides ("NFPA Standards"), may be amended from time to time throug issuance of Tentative Interim Amendments or corrected by Errata. An official NFPA Standard at any point in t onsists of the current edition of the document together with any Tentative Interim Amendment and any Erra en in effect. In order to determine whether an NFPA Standard has been amended through the issuance of tive Interim Amendments or corrected by Errata, visit the "Codes & Standards" section on NFPA's website. e, the document information pages located at the "List of NFPA Codes & Standards" provide up-to-date, do nt-specific information, including any issued Tentative Interim Amendments and Errata. To view the document rmation page for a specific NFPA Standard, go to http:// www.nfpa.org/docinfo to choose from the list of NF andards or use the search feature to select the NFPA Standard number (e.g., *NFPA 70*). The document inform page inclues postings of all existing Tentative Interim Amendments and Errata. It also includes the option to ter for an "Alert" feature to receive an automatic email notification when new updates and other information a sted regardir the document.

NFPA No.: 70HB23
ISBN: 978-1-4559-2907-8
ISSN: 0193-7324

Printed in the United States of America
22 23 24 25 26   5 4 3 2 1

Contents

Preface

This handbook contains the 56th edition of the *National Electrical Code®*. More than 125 years have passed since those cold days in March of 1896 (a mere 17 years after the invention of the incandescent light bulb) when a group representing a variety of organizations met at the headquarters of the American Society of Mechanical Engineers in New York City to develop a national code of rules for electrical construction and operation. This was not the first attempt to establish consistent rules for electrical installations, but it was the first national effort. Because the number of electrical fires was increasing, the need for standardization was becoming urgent. By 1881, one insurer had reported electrical fires in 23 of the 65 insured textile mills in New England.

The major problem was the lack of an authoritative, nationwide electrical installation standard. As one of the early participants noted, "We were without standards and inspectors, while manufacturers were without experience and knowledge of real installation needs. The workmen frequently created the standards as they worked, and rarely did two men think and work alike." By 1895, five electrical installation codes were in use in the United States. The manufacture of products that met the requirements of all five codes was difficult, so something had to be done to develop a single national code. The committee that met in 1896 recognized that the five existing codes should be used collectively as the basis for the new code. In the first known instance of international harmonization, the group also referred to the German code, the code of the British Board of Trade, and the Phoenix Rules of England. The importance of industry consensus was immediately recognized — before the committee met again in 1897, the new code was reviewed by 1200 individuals in the United States and Europe. Shortly thereafter, the first standardized US electrical code, the *National Electrical Code*, was published.

The *National Electrical Code* has become the most widely used code in the United States. It is the installation code used in all 50 states and all US territories and is now used in numerous other countries. Use of the *Code* continues to grow because it is a living document, constantly changing to reflect evolving technology. And it continues to offer an open-consensus process. Anyone can submit a public input for change or a public comment, and all public inputs and comments are subject to a rigorous public review process. The *NEC* provides the best and most up-to-date technical information, ensuring the practical safeguarding of persons and property from the hazards arising from the ever-expanding use of electricity.

Throughout its history, the National Electrical Code Committee has been guided by giants in the electrical industry — too many to mention individually. The first chairman, William J. Hammer, provided the leadership necessary to get the *Code* started. More recently, the *Code* has been chaired by outstanding leaders such as Richard G. Biermann, D. Harold Ware, James W. Carpenter, Michael J. Johnston, and Lawrence S. Ayer. Each of these men has devoted many years to the National Electrical Code Committee.

Acknowledgments

Building on our experience from the 2020 edition, the 2023 edition of the *NEC Handbook* has been more of a team effort than any previous edition. The editors are grateful for the efforts of Cheryl Langway for the design of this edition, Amy Feldman for copyediting the revised commentary, and Ken Ritchie for managing the development and production.

The 2023 NEC revision cycle was unusual to say the least. The pandemic forced the NEC Committee to pivot and conduct the First Draft and Second Draft meetings remotely for the first time in the history of the *Code*. Significant preparation and training were necessary to get the work done on schedule. It is testament to professionalism of the code-making panel chairs, the panel members, the NEC correlating committee, and the NFPA staff who facilitated the remote meetings that the 2023 revision cycle was completed without missing a beat.

The 2023 NEC revision cycle also featured significant work on several fronts. Perhaps the most obvious is the relocation of all definitions to Article 100. To do this, the *NEC Style Manual* had to be updated, and this effort was no small task in and of itself. With an updated *NEC Style Manual* specifying that all definitions are to be in Article 100, public inputs were developed for all the Code Making Panels (CMPs) that had definitions is the articles under their purview to initiate the relocation. Additionally, as the definitions were being consolidated to one location, it became necessary to reach common ground where the same term was defined differently by multiple CMPs. At the end of the day, the Usability Task Group, the CMPs, and the Correlation Committee worked as a team to make this huge usability improvement.

The work of the medium-voltage task group was extraordinary. New Articles 235, 245, 305, 315, and 495 were part of their work, along with revisions to several existing articles. Their work makes the current requirements for installations over 1000 volts ac and over 1500 volts dc more accessible and provides a better location for future expansion of requirements on these specialized installations.

Limited-energy systems of the types covered in Chapters 7 and 8 (the back of the book) were another focus of usability for the 2023 revision. As these systems evolve to provide more than their traditional signaling and communication functions, it is necessary to make sure the NEC is keeping pace with the direction the electrical industry is moving in. Class 4 circuits, where limited energy cables are used as a system to power lighting and other utilization equipment, are covered in new Article 726. New Article 722 consolidates many like requirements for limited energy cables that had been repeated in several Chapter 7 articles into a common landing space. Class 1 power-limited circuits have been separated from Class 2 and Class 3 circuits in new Article 724. All these changes position the NEC up for today and the future as the trend of low-voltage, limited-energy ac and dc systems gain more and more momentum in being tomorrow's power distribution systems.

Countless hours were spent on the above initiatives and others that have increased the overall usability of the 2023 NEC. The task group leaders and members took on this work voluntarily and contributed their time and expertise with the singular focus being to make the *Code* a better and more relevant document for all who use and are impacted by it. Thanks to all of you for a job well done.

Richard G. Biermann Award

The Richard G. Biermann Award was created in 2016 to honor the memory of Richard G. Biermann, the former chair of the National Electrical Code Correlating Committee. His service on the NEC included serving as chair of Code-Making Panel No. 16 (CMP 16) and as a member of several other CMPs. Mr. Biermann also served NFPA as a member of the Standards Council and as a member of the Board of Directors. In 1995, he received the Paul C. Lamb Award in recognition of his outstanding service to NFPA.

The Biermann award honors outstanding volunteerism on task groups, code-making panels, the Correlating Committee or on the promotion of adoption and application of the NEC. We are pleased to announce that Michael J. Johnston has been selected as the 2022 recipient.

Michael "Mike" Johnston epitomizes the attributes on which the Richard J. Biermann award was created. A member of the electrical industry for his entire working career, Michael has worked with the tools as a journeyman electrician and project manager, worked as an authority having jurisdiction, served as the director of education for the International Association of Electrical Inspectors (IAEI), and currently holds the position of executive director of codes and standards for the National Electrical Contractors Association (NECA). He is recognized as an electrical industry leader and is one of its most engaging and informative speakers.

Always willing to work toward the greater good of electrical, fire, and building safety, Michael has been a member and chair of Code-Making Panel 5, served as chair of the NEC Correlating Committee for three revision cycles, and is currently serving in his second term as a member of the NFPA Standards Council. He has also served as chair of the NFPA Electrical Section. In addition to his work with NFPA, Michael is a member of ANSI, IBEW, IAEI, NASCLA, SES, and the UL Electrical Council. The fact that the namesake of this award was a NECA contractor and represented NECA on NFPA committees is poignant and is even more special for the 2022 recipient, as Michael recognizes the rich tradition of leadership that NECA has provided to the NEC project. Congratulations Michael, this award is richly deserved.

About the Editors

Dean Austin is a senior electrical content specialist at NFPA, where he serves as an electrical subject matter expert in the development of content and services that support NFPA products and stakeholders. Dean has 30 years of experience in the electrical industry, holding a master electrician license and electrical inspector and electrical plan reviewer registration in the State of Michigan. He spent 10 years teaching in the Associated Builders and Contractor's electrical apprentice program, 11 years as an electrical inspector, and 5 years as Chief of the Electrical Division within the Bureau of Construction Codes of the State of Michigan. Dean was charged with enforcing the laws, rules, and codes governing electrical installations and licensing within the jurisdictional boundaries of the State of Michigan.

Christopher D. Coache is a senior electrical engineer at NFPA. Prior to joining NFPA, he was employed as a compliance engineer in the information technology industry and as an electrical engineer responsible for electrical safety, from the design of equipment to worker safety. Chris has participated in standards development for UL, the IEC, and the Instrument Society of America (ISA). He serves as staff liaison for many NFPA electrical documents, including NFPA 70B, NFPA 70E®, NFPA 73, NFPA 110, NFPA 111, and NFPA 780. Chris is editor of the *Handbook for Electrical Safety in the Workplace* and is a member of IAEI and IEEE.

Corey Hannahs is a senior electrical content specialist at NFPA, where he serves as an electrical subject matter expert in the development of content and services that support NFPA products and stakeholders. Corey is a third-generation electrician, holding licenses as a master electrician, contractor, inspector, and plan reviewer in the state of Michigan. Having held roles as an installer, owner, and executive previously, he has also provided electrical apprenticeship instruction for over 15 years. Corey was twice appointed to the State of Michigan's Electrical Administrative Board, and he received United States Special Congressional Recognition for founding the Building Opportunities for People program, which teaches construction skills to homeless and underprivileged individuals.

Erik Hohengasser is the electrical technical lead for NFPA, directing and advancing the development of content and services that support NFPA products and stakeholders. He is a licensed master electrician and has been involved in various aspects of system installation for over 20 years, including the design and installation of residential, commercial, and photovoltaic electrical systems. In Erik's time at NFPA, he has contributed to the organization in various capacities centering on technical training reviews, strategic business planning, and stakeholder engagement. His industry experience and understanding of how people interact with and apply codes and standards in their day-to-day activities was instrumental in the development of the NFPA LiNK® application.

Jeff Sargent is a principal electrical specialist at NFPA. A member of the electrical industry since 1978, Jeff is a licensed electrician and has been an electrical inspector for municipal and state agencies. He joined the NFPA team in 1997 and has worked in the electrical engineering and field operations divisions. Jeff has contributed to various editions of the *NEC Handbook* and the *Handbook for Electrical Safety in the Workplace* and is the co-author of several other electrical titles, including the *Electrical Inspection Manual, with Checklists*. He is a 30-year member of IAEI.

Summary of Technical Changes

The Summary of Technical Changes table provides an overview of the technical changes from the 2020 to the 2023 edition of NFPA 70®, *National Electrical Code*® (*NEC*®). Purely editorial and formatting changes are not included in this table. For more information about the reason(s) for each change, visit www.nfpa.org/70.

The first revision (FR), first correlating revision (FCR), second revision (SR), second correlating revision (SCR), committee comment (CC), and committee input (CI) numbers are given in the third column of this table for reference to the official documentation of the committees' actions.

For the 2023 edition of the NEC, many common changes were made to improve usability and to align with the updated 2020 *NEC Style Manual*. The following is a summary of some of these changes, which have not been included in the Summary of Technical Changes table due the overall number of common changes:

- *Relocation of definitions.* Definitions previously in the XXX.2 section of articles have been relocated, with some being revised and consolidated, into Article 100.
- *Defined terms.* Titles of defined terms have been modified and grouped by their base term where applicable, followed in parentheses by how the term appears in the *NEC* to assist in electronic searching. Defined terms that only apply to a specific article are followed by the article number in parentheses. All defined terms that do not have a specific article apply throughout the *NEC*.
- *Informational note structure.* The text of informational notes has been revised and restructured for consistency in style throughout the *NEC*, and publication dates have been omitted from referenced standards to indicate that the latest edition should be used.
- *Reference structure.* The language used in referencing other areas of the *NEC* has been revised for consistency and to point users to specific sections or parts, rather than to an entire article.
- *Language efficiency and accuracy.* Editorial revisions have been made throughout the *NEC* toward the objective of being concise and clear in meaning. Examples include but are not limited to eliminating language such as "the provisions of" and "herein," removing the word "Type" where it had preceded the name of a raceway, conduit, or tubing, as well as using "if," "when," and "where" appropriately.

Section	Synopsis	References
90.1, 90.2	Added scope section and combined former requirements with 90.2.	FR-9426, FR-9463
90.4(D)	Added Informational Note.	FR-8401
90.5(C)	Revised to clarify that if a referenced standard does not include a date, the latest edition available should be used.	FR-8379
90.9(B)	Added an exception to accommodate the order in which units are currently displayed in the Annex C tables.	FR-8390
Article 100	Reorganized and restructured article to include all definitions formerly located in XXX.2 sections of articles, to removed Part delineations, and to reorder all defined terms	FR-8496, FR-8497
100, Scope	Revised to address the relocation of all former XXX.2 definitions into Article 100 and that an article number in parentheses following the definition indicates the definition is only applicable in that article.	FCR-352

Section	Synopsis	References
Accessible (as applied to wiring methods)	Revised and expanded definition for clarity.	FR-8399
Air-Conditioning or Comfort-Cooling Equipment	Deleted definition.	SR-8504
Appliance	Revised definition.	SR-8329, SR-8558, SR-8559
Associated Apparatus	Added Informational Note No. 3.	SR-7576
Associated Nonincendive Field Wiring Apparatus	Revised definition and added Informational Note No. 2.	SR-7581
Attachment Fitting, Weight Supporting (WSAF)	Revised definition to distinguish a weight supporting attachment fitting from an attachment plug and added Informational Note No. 2.	FR-7637, SR-8118
Bathroom	Revised definition to remove the term "basin".	FR-8774
Battery	Revised title of defined term from *storage battery* to *battery*, revised the text for consistency with the definition of *battery* in NFPA 855, and removed former *battery system* definition.	FR-9046, FR-9127
Battery, Ffow	Revised to remove the term "fuel cell" from the definition.	FR-9076
Battery, Sealed.	Revised to create separate definitions for *cell, sealed* and *battery, sealed* to differentiate the two.	FR-9629
Berth	Revised definition and added Informational Note.	SR-8582
Branch Circuit, Motor	Added definition to differentiate between a branch circuit and a motor branch circuit.	FR-7980, SR-7587
Branch Circuit, Multiwire	Revised definition to replace the term "grounded conductor" with "neutral conductor."	FR-8779
Building, Floating	Revised definition.	SR-8502
Cable, Abandoned	Revised definition and added an Informational Note.	FR-9570, SCR-112
Cable, Circuit Integrity (CI)	Revised definition and added Informational Note.	FR-9555, SCR-137
Cable, Optical Fiber, Hybrid	Revised definition to replace the term "composite" with "hybrid" to align with industry definitions.	FR-8810
Cable, Portable Power Feeder	Added definition.	FR-8630, SR-8251
Cable Connector [as applied to hazardous (classified) locations]	Added definition and Informational Notes.	FR- 8234, SR-7602
Cable Joint	Added definition.	FR-8622, SR-8234
Cable Management System	Revised definition.	FR-9493, SR-7736
Cable Sheath	Revised definition.	SR-7962
Cable Termination	Added definition.	FR-8623, SR-8239
Cell, Sealed	Revised to create separate definitions for *cell, sealed* and *battery, sealed* to differentiate the two.	FR-9629
Cell Line	Revised definition to remove the reference to only Article 668.	SCR-115
Class 1 Circuit	Revised definition to include Class 1 and deleted the Informational Note.	FR-9556
Class 4 Circuit	Revised definition and added Informational Note.	SCR-97
Class 4 Transmitter	Revised definition and added Informational Note.	SR-8700
Class 4 Tray Cable (CL4TC)	Deleted definition.	SR-8660
Clothes Closet	Revised definition and relocated former Figure 100.1 to 410.16.	SR-8119
Combustible Dust	Revised definition and Informational Note.	SR-7606

Section	Synopsis	References
Commissioning	Revised definition.	FR-8911
Communications Circuit	Revised definition.	FR-9596
Communications Circuit, Premises	Revised definition.	FR-9600
Conductor, Copper-Clad Aluminum	Revised definition.	FR-8430, SR-8347
Conductor, Insulated	Revised definition and Informational Note.	SR-8321, SCR-123
Container (as applied to batteries)	Revised definition and deleted Informational Note.	FR-9042
Control Drawing	Added Informational Note covering ANSI/UL standards.	SR-7608
Control Space	Revised definition and title of defined term.	FR-9284, SR-7575
Controller, Motion	Added Informational Note.	SR-7579
Converter Circuit, DC-to-DC	Revised definition.	FR-9354, SCR-68
Cord, Flexible	Added definition.	FR-8545, SR-8242
Cord Connector	Added definition.	FR-8547, SR-8456
Cord Connector [as applied to hazardous (classified) locations]	Revised definition.	SR-7612
Cord Set	Added definition.	FR-8543
Corrosive Environment	Revised definition.	FR-8765, SCR-138
Counter (Countertop)	Added definition and Informational Notes.	SR-7906
DC Plugging Box	Added definition.	SR-8543
Dead-Front	Revised title of defined term.	SR-7548
Distribution Point (Center Yard Pole) (Meter Pole)	Revised title of defined term, deleted former Informational Note No. 1, and renumbered and revised Informational Note No. 2.	SR-8568
Diversion Controller (Diversion Charge Controller, Diversion Load Controller)	Revised two former definitions and combined into one.	SR-8181, SCR-123
Electric Power Production and Distribution Network	Revised definition.	FR-9125
Electric Vehicle (EV)	Revised definition and Informational Note.	FR-9491
Electric Vehicle Power Export Equipment (EVPE)	Revised Informational Note.	FR-9379
Electric Vehicle Supply Equipment (EVSE)	Revised definition and Informational Note.	FR-9472, SR-7744
Electrical Datum Plane	Revised definition.	SR-8514
Electrical Resistance Trace Heating "60079-30-1"	Revised definition and Informational Note.	SR-7619
Emergency Luminaire, Directly Controlled. (Directly Controlled Emergency Luminaire)	Revised definition and title of defined term.	FR-8777, SR-7899
Emergency Power Supply System (EPSS)	Revised definition.	SR-7897
Emergency Systems	Revised definition and deleted Informational Note.	FR-8775

Section	Synopsis	References
Enclosed-Break	Added definition.	FR-8666, SR-7621
Energized, Likely to Become	Added definition.	FR-8245, SR-8033
Energy Storage System (ESS)	Revised definition and Informational Note.	FR-9075
Equipment Protection Level (EPL)	Added definition and Informational Notes.	FR-8766, SR-7622
Equipotential Plane	Revised definition and title of defined term.	FR-8571, SCR-164
Equipotential Plane (as applied to agriculture buildings)	Revised definition and title of defined term.	SR-8569
Explosionproof Equipment	Revised definition and Informational Notes.	SCR-142
Exposed (to Accidental Contact)	Deleted Informational Note.	SR-7960
Fault-Managed Power	Added definition and Informational Notes.	SCR-98
Feeder Assembly	Revised definition.	FR-7690, SR-8517
Fibers/Flyings, Combustible	Added definition and informational Notes.	SR-7795
Fibers/Flyings, Ignitible	Added definition.	SR-7796
Generating Capacity, Inverter	Revised definition and title of defined term.	FR-9345
Generator (Generator Set)	Added definition.	FR-8978
Ground-Fault	Revised definition.	FR-8243, SR-8026
Ground-Fault Circuit Interrupter, Special Purpose (SPGFCI)	Added definition.	SCR-117
Ground-Fault Current Path, Effective	Revised definition and title of defined term.	FR-8244, SR-8029
Ground-Fault Detector Interrupter, dc (GFDI)	Added definition and Informational Note.	FR-9367, SCR-69
Grounding Conductor, Impedance	Added definition.	FR-8214, FR-8216
Hybrid System	Deleted definition.	FR-9346
Increased Safety "e"	Revised definition and title of defined term.	SR-7773
Insulating End	Revised definition and title of defined term.	SR-8228
Interactive Mode	Revised definition and title of defined term and added Informational Note.	FR-9347, SR-8243
Intrinsically Safe Apparatus	Added Informational Notes.	SR-7626
Intrinsically Safe Circuits, Different	Revised title of defined term and added Informational Note.	SR-7613
Intrinsically Safe System	Added Informational Notes.	SR-7629
Inverter, Interactive	Revised definition.	FR-9126
Inverter, Multimode	Revised definition.	FR-9348
Inverter, Stand-alone	Added definition.	FR-9352
Island Mode	Revised definition.	FR-9349, SR-8245
Legally Required Standby Systems	Relocated Informational Note from definition to Article 701.	FR-9624
Lighting Assembly, Cord-and-Plug Connected	Revised definition and title of defined term.	SR-8360

Section	Synopsis	References
Limited Finishing Workstation	Revised definition and Informational Note.	SCR-146
Load Management	Added definition and Informational Note.	FR-9386, SR-8068
Location, Wet	Revised definition into list format for clarity.	FR-8436
Locations, Hazardous (Classified)	Added definition and revised title of defined term.	FR-8759, SCR-123
Manufactured Home	Revised definition and Informational Note No. 1.	FR-7690, SCR-147
Microgrid	Revised definition.	FR-9360, SCR-119
Microgrid, Health Care (Health Care Microgrid System)	Added definition.	FR-8062, SCR-148
Microgrid Control System (MCS)	Added definition and Informational Note.	FR-9505, SCR-123
Microgrid Interconnect Device (MID)	Revised definition and removed Informational Note.	FR-9359, SR-8252
Mobile Home	Revised definition and added Informational Note.	SCR-149
Mobile Home Park	Revised definition.	SR-8352
Motion Picture Studio Television Studio	Revised definition	SCR-151
Motor Fuel Dispensing Facility	Revised Informational Note.	FR-8671, SR-7571
Multioutlet Assembly	Revised definition.	FR-7638, SR-8117
Nonincendive Field Wiring	Added Informational Note.	SR-7639
Nonsparking	Added definition.	FR-9227, SR-7640
Normal/Emergency Power Source	Added definition.	FR-8781
Normal High-Water Level (as applies to electrical datum plane distances).	Added definition.	FR-7875, SR-8544
Oil Immersion	Revised definition and Informational Note.	SR-7573
Overcurrent Protective Device, Branch-Circuit	Revised definition.	FR-8289, SR-8442
Panelboard	Revised definition.	FR-7946, SR-7505
Panelboard, Enclosed	Added definition.	FR-8470, SR-7508
Passenger Transportation Facilities	Added definition.	SR-8201
Patient Care-Related Electrical Equipment	Added definition.	FR-8064
Patient Care Space Category	Revised definition and category titles.	SR-8665
Photovoltaic Cell (PV) (Solar Cell)	Revised definition.	SR-8291
Point of Entrance	Revised definition.	FR-9572
Pool, Permanently Installed Swimming, Wading, Immersion, and Therapeutic	Revised definition.	FR-8416

Section	Synopsis	References
Pool Lift, Electrically Powered	Revised definition and title of defined term.	SR-8345
Portable	Revised definition.	FR-9396
Power-Supply Cord	Added definition.	FR-8546, SR-8250
Power Outlet, Marina	Revised definition.	SR-8617
Power Source Output Conductors	Revised definition and title of defined term.	FR-9362, SR-8264
Premises	Revised definition.	SR-7955
Pressurized Room "p"	Added definition.	FR-8668, SR-7649
Primary Source	Added definition.	FR-9363
Purpose-Built	Added definition.	SR-8634
PV DC Circuit	Revised definition and title of defined term.	FR-9429, SR-8293
PV DC Circuit, Source	Revised definition.	FR-9431, SR-8293
PV DC Circuit, String	Added definition.	FR-9432, SR-8293
Raceway	Deleted Informational Note.	SR-8148
Receptacle, Weight Supported Ceiling (WSCR)	Added definition.	FR-7639, SR-8130
Reconditioned	Revised definition and moved from purview of CMP-1 to CMP-10.	SR-7745
Recreational Vehicle (RV)	Revised definition reference to NFPA 1192.	FR-7826, SCR-157
Recreational Vehicle Site Supply Equipment	Revised definition.	SR-8550
Relay, Automatic Load Control	Revised definition.	FR-9604
Remote-Control Circuit	Revised definition.	FR-9557, FR-9558, SR-8654
Resistance Heating Element	Revised definition.	SR-8217
Restricted Industrial Establishment [as applied to hazardous (classified) locations]	Added definition.	SR-7700
Safety Circuit	Added definition and Informational Note.	FR-9518, SCR-123
Sealed [as applied to hazardous (classified) locations]	Added definition and Informational Note.	FR-8664, SR-7659
Sealed, Hermetically	Revised definition and Informational Note.	FR-8655, SR-7568
Separately Derived System	Revised definition.	FR-8246
Service-Entrance Conductors	Revised former overhead and underground system definitions and combined into one.	SR-8449
Servicing	Added definition and Informational Note.	FR-8449, SR-7743
Short Circuit	Added definition.	FR-7845
Site-Isolating Device	Revised definition.	FR-7690, SR-8572

Section	Synopsis	References
Slip	Added an Informational Note.	SR-8599
Space	Added definition.	FR-8065
Special Protection "s"	Added definition and Informational Note.	FR-8660, SR-7661
Splash Pad	Revised definition.	FR-8417
Spray Area	Revised definition into list format and added Informational Note No. 2	SR-7665
Stand-Alone System	Revised definition.	FR-9350
Stored-Energy Power Supply System (SEPSS)	Added definition.	FR-9619, SR-7892
Stranding, Compact	Added definition.	FR-8428, SR-8254
Stranding, Compressed (Compressed Stranding)	Added definition.	FR-9024, SR-8255
Supervisory Control and Data Acquisition (SCADA)	Revised definition and added Informational Note.	SR-7903
Support Areas	Added definition.	SR-8632
Switchboard	Revised definition and Informational Note.	SR-7509
Switching Device (as applied to equipment rated over 1000 volts ac, 1500 volts dc, nominal).	Revised title of defined term.	SR-7851
Switching Device, Circuit Breaker	Deleted subheading definition.	FR-8456
Switching Device, Disconnecting Means	Deleted subheading definition.	FR-8410
Switching Device, Disconnecting Switch (or Isolating Switch)	Revised title of defined term.	SR-7846
Switching Device, Interrupter Switch	Revised definition.	SR-7847
Switching Device, Oil-Filled Cutout	Revised title of defined term.	SR-7848
Switching Device, Oil Switch	Revised definition.	SR-7849
Switching Device, Regulator Bypass Switch	Revised definition.	SR-7850
Transfer Switch	Revised title of defined term.	SR-7904
Transfer Switch, Branch-Circuit Emergency Lighting	Revised definition.	FR-8776
Transfer Switch, Meter-Mounted)	Added definition and Informational Note.	FR-9128, SR-7912
Transformer	Revised definition.	FR-7975, SR-7520
Uninterruptable Power Supply (UPS)	Revised definition.	FR-9129
Unit Equipment	Added definition.	FR-9622
Voltage, High	Added Informational Note.	FR-8279, FR-7975, SCR-123
Voltage, Nominal	Revised title and deleted former Informational Note No. 3.	FR-8462, SR-7702, SR-8083

Section	Synopsis	References
Wind Turbine Output Circuit	Revised title of defined term and deleted Informational Note.	FR-9593, SR-8282
Wireless Power Transfer (WPT)	Revised definition.	FR-9400
Wireless Power Transfer Equipment (WPTE)	Revised definition and added two Informational Notes.	FR-9401, SR-7753
Work Surface	Added definition and Informational Notes.	SR-7907
Article 110, Title	Added "General" to title of Article 110 to now read "General Requirements for Electrical Installations."	FR-8494
110.3(A)(8)	Added requirement for cybersecurity for network-connected life safety equipment.	SR-7666
110.3(B)	Revised text to add "identified for a use" and added Informational Note to clarify other options for attaining instructions for installation.	FR-8487, SCR-24
110.4	Revised to remove "throughout this Code" from the text.	SR-7671
110.8	Revised to include "premises wiring system" as an area where wiring can be installed.	FR-8532
110.9	Revised to add "available fault" to clarify the type of current the requirement applies to.	FR-8536
110.12	Revised to replace "neat and workmanlike" with "professional and skillful."	FR-8357
110.12(C)	Deleted former Informational Note No. 1	FR-8552, FR-8358
110.14	Revised to remove example of dissimilar conductors.	SR-7690
110.14(A)	Revised to replace the term "thoroughly good" with "mechanically secure" to aid in enforceability and removed references to unapproved connection methods.	FR-8556, SR-7788
110.16(A)	Revised to add "enclosed" to "panelboard" to read as "enclosed panelboard" to clarify that arc flash labeling be applied to the exterior enclosure of a panelboard and not on the panelboard itself, which is often installed within the cabinet.	FR-8707
110.16(B)	Revised title of section, reduced the ampacity requirements for the label to 1000 amps, removed the list format, and replaced "acceptable" with "applicable."	FR-8772, SR-7704
110.17	Added section that distinguishes between reconditioning and servicing activities and maintenance activities.	FR-8625, SR-7732
110.18	Deleted Informational Note.	FR-8565
110.20	Added section to clarify equipment that can, and cannot, be reconditioned and the process for approval by an AHJ.	FR-3663
110.21(A)(1)	Revised to delete "label" as it is redundant with "marking" and to replace "placed on" with "applied or affixed onto" to clarify marking installation.	FR-8581, SR-7751
110.21(A)(2)	Revised to be in a list format and to provide directions around original markings on the equipment being reconditioned. Deleted Informational Note 3.	FR-8577
110.21(B)	Revised to include labels as a method of marking, added Informational Note with ANSI standard, and revised list item (2) to be consistent with the use of "marking."	FR-8580
110.22(A)	Revised text to require disconnecting means to be marked with both the identification and location of the circuit source that supplies the disconnecting means unless located and arranged so the identification and location are evident.	FR-8583, SR-7762
110.26	Revised to clarify what would be considered as impeding access to equipment.	SCR-25
110.26(A)(1)	Revised subsection (c) to clarify "enclosed" panelboards versus a panelboard that is not installed within a "cabinet, enclosure, or cutout box."	FR-8709
110.26(A)(2)	Relocated requirement for open doors not impeding working space from 110.26(C)(2) and itemized into (a) and (b) subsections.	FR-8635
110.26(A)(3)	Revised to add "enclosed" panelboards to Exception No. 2.	FR-8710
110.26(A)(4)	Revised to clarify that the working depth requirement must be maintained all the way to the floor and that in no case should there be a side reach of more than 6 inches to work within the panel.	FR-8637
110.26(A)(6)	Added requirement to maintain the grade level and flat for the full depth and width of working space required, which correlates with 110.34(A).	FR-8633, FR-7803

Section	Synopsis	References
110.26(C)(2)	Revised to clarify that the combined width is over 6 feet and relocated doors not impeding entry or egress to working space to 110.26(A)(2).	FR-8643, FR-8635
110.26(C)(3)	Revised to require that personnel doors open at least 90 degrees for safe egress.	FR-8363
110.26(D)	Revised to add "enclosed" to "panelboards" for clarity and to correlate with other Article 110 sections.	FR-8711
110.26(E)	Revised to add "service equipment" to clarify that service disconnects also require dedicated equipment space and protection from damage.	FR-8654
110.27(C)	Deleted Informational Note.	SR-7765
110.28	Revised to add "enclosed" to "panelboards" for clarity and to correlate with other Article 110 sections. Revised Informational Notes and added new Informational Notes to Table 110.28 for clarity.	FR-8962, FR-8669
110.29	Added section providing details about equipment "in sight from," "within sight from," or "within sight of."	SR-7706
110.31	Revised Informational Note and deleted redundant Note from Table 110.31.	SR-7809
110.31(A)(4)	Revised to require that personnel doors open at least 90 degrees and added Informational Note for clarity.	FR-8673
110.33(A)	Revised to add the requirement that doors cannot restrict working space or impede entry or egress from working space.	FR-8681, SR-7810
110.33(A)(3)	Revised to require that personnel doors open at least 90 degrees and added Informational Note for clarity.	FR-8771
110.34(A)	Revised to add requirement that working space floors must be as flat and level as practicable for the full depth and width of working space required.	FR-8712, SR-7811
110.53	Revised to remove "metal conduit" as it is also a metal raceway and is already listed.	FR-8714
110.73	Revised to add reference to 110.26(F) as being met by a manhole cover weighing over 100 lbs.	FR-8715
200.2	Revised to into list format and to correlate with the new definitions for *impedance grounded system* and *impedance grounding conductor*.	FR-8223
200.4(A)	Added Informational Note.	FR-8274, SR-7910
200.6(A) & (B)	Revised to clarify that the insulation is required to be identified.	FR-8029
200.6(D)	Revised to add nominal voltage to title and throughout text to clarify different marking requirements.	FR-7984
Article 210	Revised article title to align with voltage limits in Article 210.	SR-8154
210.1	Revised scope to align with voltage limits of Article 210 and added Informational Note directing users to new Article 235 for branch circuits over 1000 volts.	FR-9180, SCR-124
210.2	Relocated from 210.15 and revised to eliminate redundant language.	FR-9095, SR-8038
210.3	Revised to correct references to sections.	FR-8787
210.4(A)	Revised to ensure all conductors of multiwire branch circuits originate from the equipment containing the branch-circuit OCPD or protective devices and added text to exempt column width panelboards.	FR-8788, SR-7916
210.5(C)(1)	Revised to clarify that for premises wiring systems having more than one nominal voltage system the ungrounded conductors must be identified by phase or line and by nominal voltage system at all terminations, connections, and splice points.	FR-8929
210.6(D)	Revised text and title to reflect the new voltage limits found with the creation of new Article 235.	SR-8157
210.6(E)	Deleted section to support the creation of new Article 235.	FR-9208
210.8	Revised to add listing to ensure that the products installed are listed, replaced "appliance" cord with "power" cord, and removed "passing through a window," as a pass-through in a kitchen may not be interpreted as a window but a cord may pass through it. Deleted Informational Note No. 2, 3, and 4.	FR-8804, SR-8140

Section	Synopsis	References
210.8(A)	Revised requirement and exception text, removed Informational Note under list item (5), added new item (7) for sinks, and added Exception No. 4 for factory bath fan receptacles.	SR-7944, SR-7956, SR-7950
210.8(B)	Revised to add list item (4) for serving areas, added item (13) for aquariums and live bait wells, revised other list items for clarify, and relocated exceptions to end of list.	FR-8954, SR-7958, SR-7961
210.8(D)	Revised text, converted into list format, corrected list item (3), and added items (8) through (12).	FR-8865, SR-7966
210.8(F)	Revised to incorporate list item (3) and to require GFCI protection for all dwelling outdoor outlets when they are replaced.	FR-8896, SR-7971
210.11(C)(4)	Revised to clarify that the required receptacle outlets from 210.52(G)(1) must be served from at least one 120-volt, 20-ampere branch circuit. Added Exception No. 2 for a single bay garage to have all outlets served within the garage by the 20-ampere, 120-volt branch circuit.	FR-8970
210.12	Revised to clarify that AFCI devices be listed and to use the acronym AFCI.	FR-9168, FR-9094
210.12(A)	Relocated former 210.12(A) to become 210.12(B), revised title of (A) to "Means of Protection," and combined those means into a list format.	FR-9158
210.12(B)	Revised to add number "10" to the list of ampere ratings due to the introduction and availability of 10-ampere branch circuits for AFCI protection. Added list item (14) for "Similar locations" and Exception No. 2 regarding individual branch circuit supplying arc welding equipment in a dwelling with delayed effective date of January 1, 2025. Relocated former Informational Note No. 1 from 210.12(B) to 210.12(A).	FR-9198, FR-9628, CC-8204, CC-8019, SCR-29
210.12(C)	Revised text for clarity and to align with 210.12(A). Added 10 ampere to the branch circuit requirements to correlate with other sections and clarified patient sleeping rooms in nursing homes.	FR-9093, FR-9176, SR-8205
210.12(D)	Revised text for clarity and to align with 210.12(A). Added 10 ampere to the branch circuit requirements to correlate with other sections and changed "occupancy" to "other occupancies" to clarify that additions or modifications in any area would require the installation of AFCI protection.	FR-9093, FR-9094, SR-8206
210.12(F)	Added section to address sleeping quarters in first responder locations.	FR-9167
210.13	Revised text to require branch circuit disconnects of 1000 amperes or more, on solidly grounded wye systems of more than 150 volts to ground, and 1000-volts phase-to-phase or less be GFCI protected. Changed "disconnect" to "disconnecting means" for correlation with defined term.	FR-9202, SCR-30
210.17	Revised to add assisted living facilities to the list of occupancies that must have branch circuits installed if they have permanent provisions for cooking to meet rules for dwelling units and added Informational Notes for guidance.	FR-9096, SCR-124
210.18	Revised text to read "shall not" and added Exception No. 2 to prohibit 10-ampere branch circuits from supplying receptacle outlets.	FR-9097, SR-8039
210.19	Revised text for clarity on voltage limitation to align with new Article 235 for conductors over 1000 volts.	FR-9298, SR-8042
210.20	Revised text to align with voltage limitations imposed with new Article 235.	SR-8207
210.21(B)(3)	Revised Table 210.21(B)(3) to match existing receptacles that are available.	FR-9099
210.23(A)	Added section to address the introduction of 10-ampere branch circuits and what will be allowed on those branch circuits.	FR-9100
210.23(A)(1)	Revised list item (3) for clarity and usability.	SR-8045
210.24	Added table note to address 10-ampere branch circuits and created Table 210.24(2) specifically for aluminum and copper-clad aluminum conductors.	FR-9190, SCR-43, SR-8154
210.52(A)(2)	Revised text around receptacles along the wall space behind stationary appliances to clarify that it is not included in the wall space.	FR-9110
210.52(B)(1)	Revised Exception No. 1 to clarify that it is permitted and not required.	FR-9111

Section	Synopsis	References
210.52(C)	Revised to retain the 2020 NEC language for multi-outlet assemblies and added Exception No. 2 to address installations where there may be an obstruction such as a window over a counter.	SR-8210, SCR-27
210.52(C)(2)	Revised to clarify that sinks, cooktops, and similar are not included in the calculation of the island or peninsula work surface area to remove confusion around where the perpendicular connection begins for proper measurement.	FR-9130, FR-9131
210.52(C)(3)	Revised to add clarity for the mounting of a receptacle below the countertop or worksurface and for future provisions for receptacles in or on island or peninsular counter spaces.	FR-9138, FR-8209
210.52(D)	Revised to correlate with the change in definition of the term *bathroom*, removing "basin" and using "sink."	FR-8924
210.52(E)	Deleted Informational Note.	SR-8211
210.52(G)	Revised to ensure these receptacles are GFCI protected.	SR-8060
210.62	Revised text for clarity and to appropriately reflect dimensional information.	FR-9143, SR-8061
210.63	Added Informational Note to direct users to the GFCI requirements.	FR-9145
210.65(B)	Revised to replace "or major portion of floor space" with "fraction thereof."	SR-8067
210.70	Revised to add language that if a wall mount switch is battery operated and the battery fails the lights need to be energized and illuminate.	FR-9148
210.70(A)(1)	Revised text to add laundry area to the list for required illumination and revised title to reflect areas covered.	FR-9152, SR-8212
210.70(A)(2)	Revised text to limit use of dimmers on lighting outlets over stairways. Added an exception to list item (2) for dwelling units, attached and detached garages, with electric power for an outdoor grade level bulkhead door with a stairway.	FR-9147, FR-9149, SR-8213
215	Relocated requirements for feeders, outside branch circuits and feeders, and services over 1000 Vac or 1500Vdc to new Article 235.	CC-8472
215.1	Revised scope statement to correlate with new Article 235 and added Informational Note.	FR-7894, FR-7687, SCR-125
215.2(A)(1) and (2)	Revised into list format for clarity and usability.	FR-7684, FR-7685, SR-8472
215.9	Revised to require GFCI protection to be listed.	FR-7686
215.15	Added section for barriers that mirrors 230.62(C).	FR-7688, SR-8216
215.18	Added section for surge protection.	FR-7689, SR-8382, SR-8218
Article 220	Reorganized and restructured various sections of Article 220.	FR-9188
220.1	Revised to include new Parts of Article 220.	SCR-31
220.3	Revised Tale 220.2 for better usability and to align with new Article 235.	FR-9150, SR-8105
220.5(C)	Relocated from former 220.11, changed the floor area calculation to encompass all areas of the dwelling, and revised text around unfinished areas not adaptable for future habitable spaces.	FR-9153, SR-8095
220.10, 220.11	Revised and restructured for better usability.	SR-8214
220.14(C)	Revised language of the requirement to improve clarity.	FR-9155
220.14(H)	Revised to clarify that multioutlet assemblies are not receptacle outlets.	FR-9156
220.14(I) and (K)	Revised to align with other changes made in Article 220.	SR-8111
220.40	Revised section to correlate with other section modifications and split former Informational Note into separate notes.	SCR-42, SCR-28
220.42	Relocated requirements and table from 220.12 and revised.	FR-9154

Section	Synopsis	References
220.43	Relocated from 220.14(K).	FR-9157
220.48	Added section for health care load calculations.	FR-9189
220.50	Revised to add "air-conditioning equipment" and restructured text.	FR-9177
220.51	Revised text to replace the term "might" with "shall be permitted to."	SR-8098
220.53	Revised to not allow demand factors to be applied to the electric vehicle supply equipment.	FR-9182
220.55	Revised Note 4 within Table 220.55 into separate Notes 4, 5, and 6.	FR-9163
220.56	Revised to reflect change in 220.53.	FR-9169
220.57	Added section for electric vehicle supply equipment (EVSE).	FR-9170, SR-8101
220.60	Revised to clarify that an air conditioning load can be considered as a noncoincident load.	FR-9171
220.70	Added section for energy management systems (EMSs).	FR-9172, SR-8106
220.83	Revised text to include the term "feeder" in second sentence for clarity around alternate method for calculating an existing service or feeder capacity.	FR-9173
220.85	Revised to include services.	FR-9174
220.86	Revised Table 220.86 to match the structure of Table 220.88.	FR-9175
220.87	Revised to account for peak load shaving by renewable energy systems.	FR-9179
220, Part VI	Added Part VI, Health Care Facilities.	SCR-36
220.110	Relocated from 220.48.	SCR-36
220, Part VII	Added Part VII, Marinas, Boatyards, Floating Buildings, and Commercial and Noncommercial Docking Facilities.	SCR-32
220.120	Added section for calculating feeders that supply shore power receptacle loads in marinas, boatyards, floating buildings, and docking facilities.	SCR-32
Article 225	Relocated all requirements for outside branch circuits and feeders over 1000 Vac or 1500 Vdc to new Article 235.	CC-8472
225.1	Revised scope statement to reflect the voltage to align with new Article 235.	SCR-26
225.3	Revised Table 225.3 to correlate with changes in and additions to other articles.	FR-7691, SR-8698
225.5	Deleted former section due to redundancy with Articles 210 and 215.	FR-7693
225.7	Deleted former section due to redundancy with Articles 210 and 220.	FR-7694
225.19(A)	Revised measurement for consistency with 230.24(A).	FR-7716
225.27	Revised to provide clarity and correlation with 230.8.	FR-7696
225.41	Added section for emergency disconnects on one- and two-family dwellings.	FR-7706, SCR-41
225.42	Added section requiring surge protective device (SPD) on a feeder.	FR-7707, SR-8379
225, Part III	Relocated former Part III requirements for outside branch circuits and feeders over 1000 Vac or 1500 Vdc to new Article 235.	CC-8472
230	Relocated all requirements for outside branch circuits and feeders over 1000 Vac or 1500 Vdc to new Article 235.	CC-8472
230.1	Revised scope statement to reflect the voltage to align with new Article 235.	SCR-26
230.2(A)	Revised list item (5) to correlate service permissions from Article 705.	FR-7719
230.7	Revised title and text around other conductors in handholes or underground boxes with service conductors other than grounding electrode conductors or supply side bonding jumpers.	FR-7721, SR-8236
230.8	Revised for correlation with 225.27.	FR-7722
230.24(A)	Revised for correlation with 225.19.	FR-7724
230.30(B)	Revised list item (11) to add Type TC-ER cable where identified for direct burial.	FR-7725
230.31	Revised to remove subjective language "shall have adequate mechanical strength."	FR-7728

Section	Synopsis	References
230.33	Revised to add reference to 230.46 to permitted methods for splicing service conductors.	FR-7729
230.43	Revised list item (20) to require that Type TC-ER cable be identified for use as service entrance conductors and added item (21) to allow for flexible bus systems.	FR-7732, SR-8241
230.44	Revised to remove subjective language.	FR-7733
230.46	Revised to remove effective date that has since passed.	FR-8148
230.62(C)	Revised to clarify the intent to protect exposed service conductors.	FR-7736
230.67	Revised to expand the requirement for SPDs beyond dwelling units.	FR-7739
230.67(E)	Added requirement for a nominal discharge rating for SPDs.	FR-8299
230.71(B)	Revised list item (3) to require barriers in switchboard(s), revised item (4) to add transfer switches, added items (5) and (6), and added an exception and Informational Note.	FR-7759, FR-7798, FR-7801, FR-7800, FR-7799
230.75	Revised to replace language removed during 2020 NEC cycle.	FR-7741
230.82	Revised text for clarity and to add the term "energy management system."	FR-8282, SR-8253
230.82	Added list item (12) to protect workers.	FR-7938
230.85(A)	Revised text into subsections and added exception to address 225.41.	FR-7770
230.85(B)	Added section to provide clarity around "suitable for use as service equipment" and "suitable only for use as service equipment" and revised Informational Note.	FR-8300, SCR-66
230.85(C)	Revised to clarify the intent of the section and revised the exception to clarify when equipment replacement triggers the emergency disconnect installation.	FR-8301, SR-8265
230.85(D)	Revised to provide correlation with other isolation disconnect requirements in the code and added Informational Note.	FR-8302, SCR-67
230.85(E)	Relocated marking requirements for clarity.	FR-8303, SR-8722
230.90	Revised to correlate with requirements in 310.12.	FR-7773
230.91	Revised to require the service overcurrent device to be an integral part of the service disconnecting means or be located adjacent thereto.	FR-7778, SR-8453
230, Part VII	Deleted former Part VII and relocated all requirements for outside branch circuits and feeders over 1000 Vac or 1500 Vdc to new Article 235.	SR-8272, CC-8472
Article 235	Created new article for branch circuits, feeders, and services over 1000 Vac, 1500 Vdc, nominal, comprised of requirements from Articles 210, 215, 225, and 230.	CC-8155, CC-8472
240.1	Revised scope to reflect removal of Section IX.	FCR-388
240.2	Relocated and revised reconditioned requirements to consolidate into one section.	SR-8704
240.4(B)	Revised text for clarify and to allow for adjustable OCPD to be used.	FR-7802, SR-8283
240.4(D)(1)	Revised to permit use of Class CF fuses.	FR-7827
240.4(D)(3)	Added section to address copper-clad aluminum.	SR-8285
240.4(H)	Added section for dwelling unit service and feeder conductors.	FR-7803
240.6(A)	Revised to add 10 ampere to standard size breakers and fuses.	FR-7828
240.6(C)	Revised section title to include "Local" and added Informational Note.	FR-7947, SR-8394
240.6(D)	Added section to address remotely accessible adjustable trip circuit breakers.	FR-7947, SR-8391
240.7	Added section for listing requirements.	FR-7805, SR-8295
240.11	Added section for selective coordination.	FR-7807
240.16	Added section around interrupting ratings for branch-circuit overcurrent protective devices.	SR-8440

Section	Synopsis	References
240.24(A)	Revised to remove vague language and to clarify exception application.	FR-7811
240.24(B)(2)	Revised to add sleeping rooms to correlate with 210.60.	FR-7812
240.24(E)	Revised to prohibit OCPDs in bathrooms and similar areas.	FR-7813
240.60(E)	Added section around listing of fuse reducers.	FR-7814
240.67	Revised to remove delayed implementation date.	FR-7834, SR-8323
240.86(A)	Revised to include Informational Note from 240.86(B) for consistency.	FR-7927
240.86(B)	Revised Informational Note to apply to this section only.	FR-7928
240.87	Revised to remove delayed implementation date.	FR-7818, SR-8320
240.89	Added section requiring that replacement trip units be listed for use with the circuit breaker.	FR-7817
240, Part IX	Deleted former sections 240.100, 240.101, and 240.102 on overcurrent protection for systems rated over 1000 Vac, 1500 Vdc and relocated into new Article 245.	SR-8457
242.1	Deleted unnecessary Informational Note.	SR-8331
242.2	Relocated from former 240.7 to address reconditioned equipment and to correlate with other articles.	FR-7956, SR-8705
242, Part II	Deleted Informational Note as it no longer applicable.	FR-7955
242.6	Relocated requirements from former 242.8.	FR-7962
242.8	Relocated requirements from former 242.10.	FR-7962
242.9	Added section to require functional indication on an SPD.	FR-7957
242.12	Relocated requirements from former 242 6.	FR-7962
242.13	Relocated requirements from former 242.12	FR-7962
242.26	Deleted section.	FR-7949
242.28	Revised text to add "SPD line conductors and conductors to ground" for clarity and accuracy.	FR-7934, SR-8334
242.42	Revised text to remove reference to a product no longer produced and consolidated text into one list for clarity.	FR-8283, FR-7952, SR-8335
Article 245	Created new article for overcurrent protection for systems rated over 1000 Vac, 1500 Vdc, comprised of requirements from Article 240.	FCR-387, SR-7853, SCR-87, SCR-88
250.1	Revised to clarity list item (6).	FR-7985
250.3	Deleted section.	FR-8247
250.6(A), (B), and (C)	Revised text to add the term "bonding" as these connections can also create objectionable currents.	FR-7990, SR-7941, SR-7938
250.20	Revised text for clarity and correlation with other sections and added Informational Note.	FR-8159, SCR-37
250.24	Revised text for clarity and to align with defined terms.	FR-8232, SR-7954
250.28(C)	Revised text to ensure methods for bonding jumpers are suitable.	FR-8018
250.30	Revised text to align with defined terms and to eliminate unenforceable terms for clarity.	FR-8233, SR-7965
250.30(A)(3)(b)	Revised text for consistency with 250.24(D)(2).	FR-8233
250.36	Revised text and title to align with defined terms for consistency and clarity.	FR-8235, SCR-38
250.52(A)(3)	Revised text and terms for consistency and clarity.	FR-8066, SR-7969

Section	Synopsis	References
250.53(E)	Revised text and title for clarity and to include copper-clad aluminum.	FR-8072
250.64(B)	Revised to include copper-clad aluminum.	FR-8079
250.64(B)(4)	Revised to add reference to 300.5 to clarify burial depth does not apply to grounding electrode conductors.	FR-8090
250.64(D)	Revised to remove conflict with 250.24(A)(1).	FR-8094
250.64(G)	Added section to address enclosures with ventilation openings.	FR-8075
250.66(A)	Revised to include copper-clad aluminum.	FR-8098
250.70	Revised to restructure into "General" and "Indoor Communications Systems" subsections.	FR-8164
250.92(B)	Revised list item (1) to ensure connections are made with an applicable method in 250.8(A) and revised item (2) to recognize threaded entries in the list of connections.	FR-8165
250.94(A)	Revised text and exceptions for clarity and restructured list item (4).	FR-8257, SR-8087
250.94(B)	Revised text into list format.	FR-8166, SR-7986
250.100	Revised for clarity and to include metal clad cable.	FR-8175
250.102(C)	Revised text and list items for clarity in sizing bonding jumpers in single raceway, cable, or in two or more cables and raceways.	FR-8258
250.104(C)	Revised text and list items for clarity.	FR-8176
250.104(D)	Revised text and list items for clarity.	FR-8225, FR-8226
250.109	Revised text to clarify that metal adapters for metal boxes ensure an effective ground-fault current path and added Informational Note.	FR-8177, SR-7997
250.114	Revised for clarity and added icemakers to list of equipment to require grounding.	FR-8179
250.118(A)	Added stainless steel FMC under list items (5) and (6).	FR-8184, SR-8022
250.118(B)	Revised text with material from 250.121.	FR-8370
250.136	Revised text and title for clarity and to remove unenforceable terms.	FR-8230
250.140	Revised text to provide relief on requirements, to include copper-clad aluminum as a conductor material, and to remove the exceptions.	FR-8231, SR-8053
250.144	Revised text to add compliance with 250.8.	FR-8238
250.148	Revised text to require the connection of all equipment grounding conductors in a box if spliced or terminated.	FR-8240
250.166	Revised text to include copper-clad aluminum as a conductor material.	FR-8100
250.178	Revised text to include copper-clad aluminum as a conductor material.	FR-8104
250.184(B)	Revised to add "covered" to list item (8)b.	FR-8271
250.187	Revised text for clarity and added an exception.	FR-8227
250.190	Revised text to include copper-clad aluminum as a conductor material.	FR-8106
300.2(A)	Revised text to indicate wiring methods in Chapter 3 be used for 100 volts ac, or 1500 volts dc, nominal or less, unless otherwise permitted.	SR-8615
300.3(B)	Revised text to include the "conduit body."	FR-9269
300.4(A)(1)	Revised text to clarify that measurements are from both edges of stud or framing member and revised Exception No. 1 to specify PVC and to include RTRC.	FR-9370, SR-8537
300.4 (E)	Revised title to add "Metal-Corrugated" in front of "Roof Decking", added Exception No. 2 for metal decking covered by concrete, and revised Exception No. 1 to clarify the requirement to use malleable iron fittings and boxes and to make physical protection more generic to allow for other methods.	FR-9375, FR-9382, FR-9408, SR-8541
300.4(F)	Revised Exception No. 1 to specify PVC and to include RTRC.	FR-9377
300.4(G)	Revised text to clarify that protective fittings must be installed before the conductors are installed.	FR-9319

Section	Synopsis	References
300.5(A)	Revised Table 300.5(A) to add "Electrical Metallic Tubing" to Column 3 and added Note 5 to EMT reference section.	FR-9323, SR-8551
300.5(D)	Revised text to replace "Direct-buried conductors" with "Conductors" for clarity.	SR-8553
300.6	Revised text to add "enclosures (other than surrounding fences and walls)" for clarity throughout the section and removed Informational Note from 300.6(A).	FR-9357, SR-8560
300.7(B)	Added Informational Note No. 2.	FR-9364
300.10	Revised section title to add "Cable Armor".	FR-9225
300.11(B)	Revised to clarify that the ceiling grid support wires not be the sole support.	SR-8510
300.11(C)	Revised to add "or Class 3" in list item (2).	FR-9230
300.12	Revised for correlation with other adjoining sections.	SR-8566
300.14	Revised text to clarify that 6 inches of free conductor be permitted to be spliced or unspliced and for consistency with other sections.	FR-9233, SR-8567
300.15	Revised text to add "conductor" to several areas requiring a box or conduit body to be installed.	FR-9241
300.17	Revised text to add "cables" throughout parent text to go along with conductors.	FR-9256, SR-8521
300.18	Revised to add "listed manufactured assemblies as identified in 604.100" and to add "or cables" throughout section text to go along with conductors.	FR-9258
300.19(B)	Revised to replace "Fire Rated" with "Fire Resistive" for consistency with other sections in the NEC.	FR-9262
300.25	Revised text to contain fire resistance rating and to add an exception that permits exterior mounted egress lights to be supplied from the circuit that feeds the exit enclosure lights.	FR-9267, SR-8571
300.26	Added section for remote-control and signaling circuit classification.	SR-8481
300, Part II	Relocated former Part II requirements for wiring methods and materials for systems rated over 1000 Vac, nominal to new Article 305.	FR-9609
Article 305	Created new Article 305, General Requirements for Wiring Methods and Materials for Systems Rated Over 1000 Volts ac, 1500 Volts dc, Nominal.	FR-9609, CC-8485
310.1	Deleted Informational Note.	FR-8373, SCR-44
310.3(B)	Revised into list format, added information for copper-clad aluminum conductors, and added new insulation types XHHN and XHWN.	FR-8372
310.6(C)	Revised text for clarity and added correct section for feeders.	FR-8344
310.10(G)	Revised text to clarify that EGC and supply-side bonding jumpers are not required to be 1/0 or larger.	FR-8314, SR-8420
310.10(G)(4)	Revised text for clarity in applying adjustment factors to parallel runs.	SR-8426
310.15	Added additional notes to Table 310.15(B)(1)(1) and Table 310.15(B)(1)(2).	FR-8336
Article 311	Deleted article and relocated requirements to new Article 315.	FR-8616
312.5	Revised language to accurately cover intent of section.	FR-7704
312.8(A)	Added list item (3) condition for 4 AWG and larger conductors to comply with 314.28(A)(2).	FR-7708
312.8(B)(1)	Revised text to clarify equipment as either an accessory or kit.	FR-7709
312.10	Added section to address specific types of non-factory screws being placed in covers and devices.	FR-7820, SR-7525
312.102	Added section to clearly require doors or covers.	FR-7711
314.5	Added section to address types of screws that enter wiring spaces and how far they can extend into wiring space.	FR-7821, SR-7527
314.16(A)(2)	Revised top row of Table 314.16(A) to correct the number of 8 AWG conductors permitted in a 4 x 1-1/4 in. round/octagon box from 5 to 4 to align with 8 AWG cu. in. capacity.	FR-7727
314.16(B)(2)	Revised to removed second paragraph due to product not coming to market that the section was intended for.	FR-7731

Section	Synopsis	References
314.16(B)(5)	Removed references to "equipment bonding jumpers" from section and replaced "in" with "entering" the box for clarity.	FR-7734, SR-7528
314.16(B)(6)	Added section to address volume allowance for terminal blocks when used within outlet and device boxes.	FR-7868, SCR-1
314.17	Revised to add cables to be included with conductors and change wording for 1/4 in. cable sheath from where it enters box to where it emerges from the clamping mechanism.	FR-7738
314.23(D)(2)	Revised to correct reference to 300.11(B).	FR-7742
314.23(H)(1)	Revised text around cord grips being used in threaded hubs.	FR-7749
314.24	Revised to clarify dimensions of boxes.	FR-7869
314.24(C)	Added section that addresses clearances for side-wiring entrances.	FR-7870, SCR-2
314.25	Revised text to clarify that conduit bodies have covers in place after installation.	FR-7751
314.25(A)	Revised references from "grounding" to "equipment grounding" connections.	FR-7752
314.27(C)	Added requirement for marking of fan boxes internally and clarification on access to structural framing.	FR-7760
314.27(E)	Revised text to use new defined term and to add an Informational Note.	FR-7761, SR-7535
314.29	Revised text for clarity.	FR-7764
314.70	Revised references to reflect changes to Article 314 sections and to change Part IV to Part III.	FR-7777, FR-7772
Article 315	Created new Article 315, Medium Voltage Conductors, Cables, Cable Joints, and Cable Terminations, comprised of requirements from former Article 311.	FR-8616, SCR-65
320.23(A)	Revised text for clarity and to remove vague language.	FR-8453, SCR-45
320.30(A)	Relocated sentence from 320.30(D) stating "Type AC cable fittings shall be used as a means of cable support" as it applies to all sections.	FR-8405
322.56(B)	Revised to replace "color-coded" with "marked" to correlate with terminology in 322.120.	SR-8296
330.10(A)	Revised list item (11) to include Type MC cable being able to be installed in damp locations.	FR-8407
330.30	Relocated sentence stating "Type MC cable fittings shall be used as a means of cable support" from 330.30(D) to 330.30(A) and replaced "provided" with "permitted" in several areas.	FR-8411
330.112(A)	Revised to correlate with the addition of 16 AWG.	SR-8309
334.10	Revised text in list item (2) for use in detached garages in multifamily applications.	FR-8432
334.10(B)	Revised to replace the term "moist" with "wet."	FR-8434
334.15(B) and (B)	Added requirement for abrasion protection when a cable enters or exits a raceway.	FR-8439, FR-8443
334.19	Added section requiring cable sheath to extend at least 1/4 inch beyond any cable clamp or cable entry.	FR-8481
334.24	Added a requirement to use the major diameter (i.e., widest part) of a flat cable when considering bending radius.	FR-8519
334.40(B)	Revised text for clarity and to include new buildings.	FR-8520, SR-8337
334.80	Revised to remove the use of the exception within 310.14(A)(2) when dealing with two or more conductors in contact with thermal insulation.	FR-8521
334.112	Revised Informational Note to remove references to Type NMS and NMS-B cables, which are no longer produced.	FR-8437
Article 335	Relocated former Article 727 into Chapter 3 as new Article 335, Instrumentation Tray Cable: Type ITC.	FCR-457
336.10	Added list item (12) to permit Type TC cable to be used as service-entrance conductors where identified for such use and marked as Type TC-ER.	FR-8524
337.80	Revised to add reference to 310.14(B).	FR-8567

Section	Synopsis	References
337.108	Revised text to state that equipment grounding conductors be provided in multiconductor cable.	FR-8568
337.115	Revised text to allow single conductors without an overall jacket, within enclosures or machinery.	FR-8570
337.116	Revised text to allow for additional armor options.	FR-8572
338.10(B)(4)	Deleted Informational Notes.	FR-8604
338.24	Revised to specify that the major dimension (i.e., widest part) of flat cables is to be used for determining bending radius.	FR-8605
340.24	Revised text to specify that the major dimension (i.e., widest part) of flat cables is to be used for determining bending radius.	FR-8608
342.10(B)	Revised text to allow use of IMC in direct burial applications.	FR-7582
342.20(B)	Revised text to add larger sizes of IMC.	SR-8144
342.24	Revised to combine former 342.24 and 342.26 into one section and to add clarity on total degrees of bends permitted.	FR-7583
342.30(A)(3)	Added an exception that allows fishing of unbroken lengths of IMC within existing walls or panels.	FR-7584
344.10(A)(2)	Revised to remove supplementary corrosion protection requirements that are already part of 344.10(B)(2).	FR-7586
344.10(B)	Revised text to allow for use of RMC in direct burial applications.	FR-7585
344.20(A)	Revised exception text to allow for use of smaller RMC.	FR-7654
344.24	Revised to combine former 344.24 and 344.26 into one section and to add clarity on total degrees of bends permitted.	FR-7588
344.28	Revised text to require manufacturer's instructions be followed when threading conduit to prevent damage to outer coating of PVC-coated RMC and added Informational Note for reference.	FR-7589
348.2	Added section stating that FMC shall not be reconditioned.	FR-7552, SR-8141
348.24	Revised to combine former 348.24 and 348.26 into one section and to add clarity on total degrees of bends permitted.	FR-7553
348.30(A)	Revised Exception No. 4 to correlate with 350.30(A) to clarify use of listed FMC fittings as securing and supporting.	FR-7554
348.60	Revised text into list format and removed redundant requirements found in other applicable articles.	FR-7605
350.2	Added section stating that LFMC shall not be reconditioned.	FR-7556, SR-8142
350.10	Revised to remove reference to 110.14(C).	FR-7623
350.24	Revised to combine former 350.24 and 350.26 into one section and to add clarity on total degrees of bends permitted.	FR-7558
350.60	Revised text into list format and removed redundant requirements found in other applicable articles.	FR-7606
352.1	Revised to delete Informational Note.	FR-7559
352.10(B)	Added section to clearly state that PVC can be encased in concrete.	FR-7561
352.10(G)	Revised to relocate Informational Note to 352.10(K) and to correlate with new sections 352.10(K) and 352.12(C).	FR-7562
352.10(K)	Added section to clearly state that Schedule 80 PVC can be used where subject to physical damage.	FR-7563
352.12(C)	Revised to add clarity and to correlate with new sections 352.10(G) and 352.10(K).	FR-7567
352.24	Revised to combine former 352.24 and 352.26 into one section and to add clarity on total degrees of bends permitted.	FR-7569
352.44(B)	Added section to address frost heave with expansion fittings.	FR-7571

Section	Synopsis	References
353.1	Revised to remove Informational Note.	FR-7517
353.24	Revised to combine former 353.24 and 353.26 into one section and to add clarity on total degrees of bends permitted.	FR-7520
353.60	Revised to remove the second instance of "equipment" from the charging statement to clarify all types of grounding.	FR-7538
354.10	Revised text to align with Table 300.5 and Table 305.15 by replacing "Rigid Nonmetallic Conduit" with "nonmetallic raceways" for clarity.	FR-7539, SR-8135
354.24	Revised to combine former 354.24 and 354.26 into one section and to add clarity on total degrees of bends permitted.	FR-7540
354.60	Revised to remove the second instance of "equipment" from the charging statement to clarify all types of grounding and added two exceptions.	FR-7541
355.1	Revised to delete Informational Note.	FR-7542
355.24	Revised to combine former 355.24 and 355.26 into one section and to add clarity on total degrees of bends permitted.	FR-7545
355.60	Revised to remove the second instance of "equipment" from the charging statement to clarify all types of grounding.	FR-7595
356.10	Revised text in list item (5) for clarity and added permission in item (8) to allow use in corrosive environments to correlate with Article 680.	FR-7502, SR-8131
356.24	Revised to combine former 356.24 and 356.26 into one section and to add clarity on total degrees of bends permitted.	FR-7535
356.60	Revised to remove the second instance of "equipment" from the charging statement to clarify all types of grounding.	FR-7536
358.10(A)	Revised list item (3) for clarification allowing EMT in direct burial applications and added list item (4) for use in manufactured wiring systems.	FR-7591, SR-8159
358.20(A)	Revised the exception to allow for a smaller size EMT.	FR-7592
358.20(B)	Revised text to allow for larger size EMT.	FR-7593
358.24	Revised to combine former 358.24 and 358.26 into one section and to add clarity on total degrees of bends permitted.	FR-7594, SR-8136
362.2	Added section stating that ENT shall not be reconditioned.	FR-7548, SR-8134
362.10	Revised text for permitted uses of ENT when protected with fire sprinkler systems or encased in concrete.	FR-7596, SCR-124
362.24	Revised to combine former 362.24 and 362.26 into one section and to add clarity on total degrees of bends permitted.	FR-7549
362.56	Revised to delete Informational Note.	FR-7550
362.60	Revised to remove the second instance of "equipment" from the charging statement to clarify all types of grounding.	FR-7597
366.10(C)	Added section to permit auxiliary gutters extending up to 30 feet and added exception to permit extension beyond 30 feet to correlate with 620.35.	FR-7609
366.12	Revised to correlate with 366.10 and to delete Informational Note.	FR-7610
Article 369	Created new Article 369, Insulated Bus Pipe (IBP)/Tubular Covered Conductors (TCC) Systems.	FR-7620, SR-8151
370.10	Revised text for clarity and added "indoors" to list item (3) to clarify cablebus can be used indoors.	FR-7612
370.18	Revised text and removed firestop requirements by referencing 300.21.	FR-7613
370.42	Revised text for clarity and removed the term "dead ends" as it is no longer applicable.	FR-7614
370.120	Revised to add subsections for "Nameplates" and "Identification."	FR-7616
Article 371	Created new article 371, Flexible Bus Systems.	FR-7621, SR-8156
374.6	Revised to align with UL 209 in requiring both the floor raceways and fittings to be listed.	FR-7618

Section	Synopsis	References
376.60	Added section that permits a listed metal wireway to be used as an equipment grounding conductor.	FR-7619
378.60	Revised to remove the second instance of "equipment" from the charging statement to clarify all types of grounding.	FR-7515
388.60	Revised to remove the second instance of "equipment" from the charging statement to clarify all types of grounding.	FR-7655
392.10	Revised to add single insulated cables to correlate with 392.10(B)(1).	FR-7573
392.10(A)	Revised Table 392.10(A) to correlate with new Article 722 and changes in Articles 800 and 805.	FR-7574, SR-8180
392.10(B)(1)	Revised to add single insulated conductors to correlate with 392.10.	FR-7575
392.56	Revised to add Type MV cable joints to correlate with Article 311.	FR-7579
393.60	Revised section title to "Equipment Grounding Conductor" and removed load side requirements as they are addressed in 393.61.	FR-7660
393.61	Revised to replace "shall not be grounded" with "shall be permitted to be grounded."	FR-7660
Article 395	Relocated requirements from former Article 399 to create new Article 395, Outdoor Overhead Conductors over 1000 Volts.	FR-8560
398.15(C)	Revised to remove reference to "nonmetallic conduit" and added actual types of conduits that would apply.	FR-8533, SR-8308
398.104	Revised text to specifically reference Table 310.4(1) instead of all of Article 310.	FR-8534
Article 399	Deleted article and relocated requirements to new Article 395.	FR-8560
400.4	Revised Note 2 of Table 400.4 to add an allowance for communications cables, to reflect the need for data transfer in elevator installations, and to correlate with changes in other section.	FR-8612
400.5(A)	Revised to clarify that Table 400.5(A)(1) covers the ampacities for flexible cords and flexible cables.	FR-8471
400, Part III	Revised Part III title to include a maximum of 2000 volts, assisting in establishing new Part IV.	FR-8475
400.35	Revised to include the term "latches."	FR-8468
400, Part IV	Added Part IV to address portable power feeder cables over 2000 volts, nominal.	FR-8465
404.1	Revised scope to clarify that Article 404 does not cover wireless control equipment to which circuit conductors are not connected and added new Informational Note.	FR-7889, SR-7551
404.2(C)	Revised to remove effective date that had passed from the exception and to remove the ability to add a grounded conductor in an accessible switch box after initial installation.	FR-7883
404.8	Revised to convert former exceptions to positive text.	SR-7645
404.14(D)	Added section covering conductors utilized with snap switch terminations.	FR-7881, SR-7859
404.16	Added section covering reconditioned equipment.	FR-7859, SCR-77
404.22	Revised to remove effective date that had passed.	FR-8035
404.30	Added section covering switch enclosures with doors.	FR-7861, SR-7566
406.2	Added section covering reconditioned equipment.	SCR-91
406.3(A)	Revised to delete sentence stating that receptacles shall not be reconditioned, as it is covered by new 406.2.	SCR-89
406.3(D)	Added section covering receptacle terminations.	FR-7601, SCR-13
406.4(B)	Revised title to incorporate equipment grounding conductor.	FR-7565
406.4(D)(5)	Revised text around listing requirements for replacement of nongrounding receptacles and aluminum branch-circuit conductors terminated to tamper-resistant receptacles.	FR-7599, SR-8372
406.4(D)(8)	Added section covering ground-fault protection of equipment (GFPE).	FR-7570

Section	Synopsis	References
406.4(G)	Added section covering protection of floor receptacles using relocated requirements from former 406.9(D).	FR-7667, FR-7668, SR-8147
406.6(D)	Revised to address faceplates with integral night lights and USB chargers and added a new exception covering spring-tensioned contact connections of faceplates.	FR-7564, SR-8143
406.7	Revised to delete sentence stating that attachment plugs, cord connectors, and flanged surface devices shall not be reconditioned, as it is covered by new 406.2	SCR-90
406.9(A)	Revised to require WR ratings for other receptacles.	FR-7598
406.9(A)	Revised to require hinged covers of outlet box hoods to open at least 90 degrees, or fully open if not designed to open 90 degrees, after installation.	SR-8150
406.9(B)	Revised to require other receptacles installed in a wet location to be weather-resistant type and to require hinged covers of outlet box hoods to open at least 90 degrees, or fully open if not designed to open 90 degrees, after installation.	FR-7598, SR-8150
406.9(C)	Revised to clarify receptacle installations within a bathtub and shower space and to add exceptions pertaining to hydromassage bathtub receptacles and weight supporting ceiling receptacles (WSCR).	FR-7578, FR-7608, SR-8152
406.12(1)	Revised list item (1) to include tamper-resistant receptacle requirements in all dwellings, boathouses, mobile homes, and manufactured homes.	FR-7663
406.12(5)	Revised list item (5) to add the term "within" for clarity of what areas require tamper-resistant receptacles and to add lobbies, corridors, waiting spaces, and spaces used for patient sleeping.	FR-7665
406.12(6)	Revised to list item (6) add fitness centers and to clarify that gymnasiums include those within both educational and commercial facilities.	FR-7666
406.12(8) and (9)	Revised list items (8) and 9 to include residential care facilities, rehabilitation facilities, group homes, foster care, nursing homes, and psychiatric hospitals.	FR-7669
406.12(10)	Added list item (10) covering agricultural buildings accessible to the public and any common areas.	FR-7664
406.12, Exception	Revised Item (3) of Exception to (1) through (10) to add clarity.	FR-7662
408.2	Relocated former 408.8 covering reconditioned equipment, revised to remove duplicative requirements with 110.21(A)(2), added the term "corrosive influences", and added "switchboards" to correct inadvertent omission from the 2020 edition.	FR-7951, SCR-78
408.4	Revised for clarity and to remove any potential conflict with the defined term and added "physical location" to the source of supply marking requirement.	FR-7895, SR-7861
408.36(D)	Revised to replace the term "panel" with "panelboard."	FR-7891
408.38	Revised to add an evaluation requirement on systems with fault currents above 10,000 amperes.	FR-7950
408.43	Revised to include mounting of panelboards in the face-down position.	FR-7944
409.3	Deleted Table 409.3 due to redundancy with 90.3.	SR-7662
409.60	Revised section for clarity around grounding and bonding of industrial control panels.	FR-8056, SR-7664
409.70	Added section covering surge protection.	FR-8053, SR-7680
409.102	Revised section title to "Busbars" as that is the only topic addressed in the requirements, with no requirements pertaining to conductors.	FR-8057
409.110	Revised to require permanent markings that are visible after installation, including clarification as to what markings are required to be installed inside and outside of the enclosure.	FR-8058, SR-7742
410.2	Relocated former 410.7 covering reconditioned equipment and revised to add ballasts, LED drivers, and lamps as equipment that cannot be reconditioned.	FR-7504, SCR-79
410.10(D)(1)	Revised to add luminaires (light kits) connected to paddle fans	FR-7505
410.10(F)	Revised to add "where subject to physical damage" and to add exception that recognizes additional protection provided by concrete slab.	FR-7656, SR-8165
410.16	Revised to add text and figure for clothes closet storage space that was formerly a part of the definition that was relocated to Article 100.	SR-8166

Section	Synopsis	References
410.40	Revised section title and removed reference to 250.112.	FR-7507, SR-8167
410.42	Revised into list format for clarity.	FR-7617, SR-8170
410.44	Revised section title and added reference to 410.42.	FR-7510, SCR-15
410.46	Deleted section as content has been relocated into 410.44.	FR-7544
410.62(A)	Revised to remove "pipe size" for clarity.	FR-7512
410.62(C)(1)	Revised to include a new Informational Note referencing 400.10 and 400.12.	FR-7657
410.69	Revised to remove effective date that had passed.	FR-7531
410.71	Added section covering requirements for disconnecting means for fluorescent and LED luminaires that utilize double-ended lamps.	FR-7640, SR-8172
410.130(G)	Deleted section and relocated requirements to 410.71.	FR-7658
410.136(B)	Revised to reference 410.126.	FR-7682, SR-8173
410.155(B)	Revised to incorporate "equipment grounding conductor" and to reference Part V of Article 410.	FR-7524, SR-8174
410.182	Revised to incorporate "equipment grounding conductor" and to reference Part V of Article 410	FR-7525, SR-8175
410.184	Revised to add an exception covering circuits exceeding 150 volts to ground and to add an Informational Note referencing UL 943C.	FR-7526
410.186	Revised to remove "special" from fittings.	FR-7527, SR-8177
410.188	Deleted section as it is already covered by 90.3.	FR-7529
410, Part XVII	Added Part XVII covering special provisions for germicidal irradiation luminaires.	FR-7528, SR-8178, SR-8179
411.1	Revised the scope for clarity.	FR-7646, SCR-16
411.2	Added section covering reconditioned equipment.	SCR-85
411.3	Added section covering voltage limitations.	SR-8186
411.4	Relocated former 411.3 and revised for clarity.	FR-7647, SR-8187, SR-8202
411.4(B)	Deleted section and relocated requirements to 411.2(B).	FR-7648
411.5(B)	Revised to remove reference to Article 680.	FR-7649
411.6(A)	Revised to add an exception covering secondary circuits supplied by a Class 2 power source.	FR-7652
411.7	Revised the exception and removed redundant reference.	SR-8182
411.8	Deleted section.	SR-8188
422.3	Deleted section.	FR-8307
422.4	Deleted section.	FR-8308
422.11(E)	Revised text to include "the overcurrent protection rating."	FR-8343
422.15	Deleted section.	FR-8352
422.16(A)(3)	Relocated requirements from former 422.43.	FR-8725
422.16(B)	Revised section to modify requirements for protecting cords from physical damage and to add "smoothed edge."	FR-8376, SR-8315
422.18	Revised to remove vague and unenforceable language.	FR-8364, SCR-61
422.22	Revised to correlate with terminology in 422.18.	FR-8382

Section	Synopsis	References
422.23	Deleted section.	FR-8387
422.31	Revised language for clarity and usability.	FR-8388
422.31(C)	Revised section to correlate with 422.34.	FR-8389
422.33(B)	Revised section to correlate with 90.5(B).	FR-8392
422.43	Deleted section and relocated requirements to 422.16(A)(3).	FR-8725
422.46	Deleted section.	FR-8394
422.47	Revised section for clarity and deleted Informational Note.	FR-8395
422.50	Deleted section.	FR-8396
424.3	Revised to add the term "additionally" and to add new Table 424.3.	FR-8721, SR-8240
424.4(B)	Revised to clarify that the 125 percent load requirement applies to both the equipment and any associated motor(s).	FR-8617, SR-8222
424.10	Deleted former section 424.10 and renumbered former 424.9 as new 424.10.	FR-8649
424.12(B)	Revised to remove unnecessary language that is already provided in the listing and/or instructions.	SR-8227
424.41(A)	Revised section to distinguish ceiling-rated heating cables from wall-rated heating cables.	FR-8619
424.44(C)	Revised to replace "rigid nonmetallic conduit" with "rigid polyvinyl chloride conduit" for consistency with other sections.	FR-8620
424.48	Added section covering requirements for heating cables installed in walls.	FR-8618, SR-8354
424.66	Revised to replace "sufficient clearance" with "working space."	FR-8621, SCR-62
424.92(A)	Revised to replace panel "finish" with "trim cover."	FR-8638
424.93(A)(2)	Revised to add list item (4) referencing new 424.93(C).	FR-8693
424.93(C)	Added section covering requirements for heating panels installed within walls.	FR-8692
424.98(D)	Revised to replace "rigid nonmetallic conduit" with "rigid polyvinyl chloride conduit" for consistency with other sections.	FR-8641
425.1	Revised scope to only reference items covered by Article 425.	FR-8650
425.10	Deleted former section 425.10 and renumbered former 425.8 as new 425.10.	FR-8651
425.22	Added section title.	FR-8661
426.3	Revised to add the term "additionally" and to add new Table 426.3.	FR-8670, SR-8292
426.14	Deleted section.	FR-8685
426.28	Revised to allow the ground fault trip level to be specified by the manufacturer.	FR-8674
427.3	Revised to add the term "additionally" and to add new Table 427.3.	FR-8708, SR-8358
427.35	Deleted section.	FR-8716
427.48	Deleted Informational Note	FR-8719
430.1	Revised Figure 430.1.	FCR-18
430.2	Added section covering requirements for reconditioned motors.	SCR-81
430.6	Revised section for clarity and usability and added new list item (3) to 430.6(A)(2) for motors exceeding the values in Part XIV.	FR-7986, SCR-3
430.7(A)	Revised section to include design letter A motors.	FR-8031
430.8	Revised to incorporate "motor" with "controller" and "current" with "short-circuit."	FR-7988
430.52(C)	Revised section and Table 430.52(C)(1) to include Design B premium efficiency motors.	FR-8009
430.53	Revised to incorporate "motor" with "controller."	FR-8012, SR-7805
430.61	Revised text to replace "grounds" with "ground-faults" and revised Informational Note.	FR-8013

Section	Synopsis	References
430.72(C)(1)	Revised section title and added references to 724.30 through 724.52 and Part II of Article 725.	FR-8015, SR-7574
430.75, 430.81, 430.82, 430.83	Revised to incorporate "motor" with "controller."	FR-8032, FR-8033, FR-8034
430.83(F)	Added section to provide requirements for the motor controller short-circuit rating.	FR-8034
430.84, 430.85, 430.87, 430.90	Revised to incorporate "motor" with "controller."	FR-8024, FR-8028
430.97(C)	Revised to reference section 312.6.	FR-8036, SR-7536
430.101	Revised to incorporate "motor" with "controller."	FR-8037
430.102	Revised to incorporate "motor" with "controller" and remove "warning" from "warning label."	FR-8037
430.103, 430.109	Revised to incorporate "motor" with "controller."	FR-8037, FR-8043
430.110	Revised to incorporate "motor" with "controller," replaced "ampere" with "current", added a reference to 430.110(C)(1) through (C)(3), and added an exception to 430.110(C)(1) addressing design letter A motors.	FR-8039, SR-7538
430.111	Revised to incorporate "motor" with "controller" and to add "-operable" to "power."	FR-8039
430.113	Revised to add requirement to provide a sign to shut off and also identify the circuits for multiple sources of power.	FR-8040
430.120	Revised to reference requirements for power conversion equipment used in adjustable-speed drive systems.	FR-8122, SR-7544
430.122(A) and (B)	Relocated Informational Note No. 2 formerly in 430.133(A) to 430.133(B).	FR-8124
430.122(C)	Deleted first sentence for clarity.	FR-8126
430.130(A)	Revised to incorporate "motor" with "controller."	FR-8127
430.201	Renumbered Part XI sections and revised to incorporate "motor" with "controller."	FR-8129
430.204	Added section with requirements for wire-bending space in motor controller enclosures.	FR-8044
430.205	Revised to reference requirements for power conversion equipment used in adjustable-speed drive systems.	FR-8045, SR-7555
430.206	Revised to reference requirements for adjustable-speed drive systems.	FR-8046
430.208	Revised to reference requirements for power conversion equipment used in adjustable-speed drive systems, added voltage and current rating requirements, changed the requirement from 115 percent to 100 percent, and replaced "an ampere" with "a current."	FR-8047, SR-7565, SR-7569
430.232, 430.233	Revised to incorporate "motor" with "controller."	FR-8048
430.241	Revised to incorporate "motor" with "controller" and to replace "prevent a voltage above" with "limit voltage to."	FR-8049, SR-7570
430.243, 430.244, 430.245	Revised to incorporate "motor" with "controller."	FR-8050
430, Part XIV	Revised Table 430.249 and Table 430.250 to include 2300 to 2400 volt system voltage ranges.	SR-7559
440.3	Deleted section.	SR-7600
440.8	Revised section to add restrictions on mounting mini-split air conditioning units within bathtub and shower spaces.	FR-8130
440.11	Revised section to add requirements to protect unqualified persons.	FR-8063, SR-7604
440.14	Revised section to reference 110.26(A) for working space requirements and deleted Informational Note No. 2.	FR-8078
440.22(A)	Revised language for clarity and added two exceptions.	FR-8081, SR-7605
440, Part IV	Revised to remove "Branch" from Part IV title.	FR-8091
440.31	Revised section to reference Part III of Article 310.	FR-8082

Section	Synopsis	References
440.32	Revised to incorporate "motor" with "controller."	FR-8083
440.62(B) and (C)	Revised to add "current" in front of "rating."	FR-8088
445.6	Revised section to remove voltage limitations.	FR-8981
445.10	Revised text in Informational Note to align with NFPA 37.	FR-8983
445.11	Revised list items (1) and (3) and added list item (2) requiring that equipment mounted to a generator assembly cannot conceal or obscure the nameplate.	FR-8991, SR-8051
445.18	Revised section to split out generator emergency shutdown and to clarify requirements for generators in parallel.	FR-9028
445.19	Relocated from former 445.18(B) and revised for clarity.	FR-9028, SR-8052
450.2	Added section for interconnection of transformers.	SR-7862
450.3	Revised Informational Note No. 2 to include reference to IEEE standard.	FR-7787
450.3(B)	Revised to delete the exception and reformatted the notes in Table 450.3(A) and Table 450.3(B).	FR-7833
450.10	Revised section title to include "and Bonding."	FR-7790
450.43(C)	Revised section title to "Accessibility" and added an additional requirement for personnel doors to open at least 90 degrees.	FR-7791
460.1	Revised to remove last sentence of scope.	FCR-40
460, Part II	Renumbered former Part I to new Part II and revised Part II title to "1000 Volts, Nominal, or Less."	FR-8132, SR-7596
460, Part III	Renumbered former Part II to new Part III.	FR-8133
460.24(A)	Revised to clarify that switches must be rated for switching of capacitive loads and must open all ungrounded conductors.	FR-8111, SR-7598
460.24(B)(2)	Revised to reference section 495.22.	SR-7717
470.1	Revised to remove the last sentence of the scope.	FCR-42
470, Part II	Renumbered former Part I to new Part II and revised Part II title to "1000 Volts, Nominal, or Less."	FCR-42
470.2	Added a section to cover reconditioned equipment.	SCR-162
470, Part III	Renumbered former Part II to new Part III.	FR-8142
480, title	Revised title of Article 480 to "Stationary Standby Batteries."	FR-9029
480.1	Revised scope of Article 480 to stationary standby batteries having a capacity greater than 3.6 MJ (1 kWh) and added a reference to NFPA 855.	FR-9030, SR-8104
480.7	Revised to replace former terms "stationary battery system" and "battery system" with "stationary standby battery" to align with changes to the scope of the article.	FR-9059, FR-9056, FR-9060, FR-9062
480.10	Revised to replace former terms "stationary battery system" and "battery system" with "stationary standby battery" to align with changes to the scope of the article.	FR-9064, FR-9068, FR-9069, SR-8094
490	Deleted Article 490 and relocated requirements into new Article 495.	SR-7686
495	Relocated requirements from former Article 490 to create new Article 495, Equipment Over 1000 Volts ac, 1500 Volts dc, Nominal.	FR-7941
495.2	Added section to provide requirements for reconditioned equipment.	SR-7755, SCR-84
495.36	Deleted former section.	FR-7963, SR-7755
495.37	Revised section title to "Equipment Grounding Connections" and added a reference to 250.190.	FR-7964, SR-7755
495.48(A)	Revised list item (12) to align with the title of Article 242.	FR-7965

Section	Synopsis	References
495.49	Revised language around reconditioned switchgear.	SCR-83
495, Part IV	Renumbered sections in Part IV.	FR-7969
495.61(B)	Revised section title to "Grounding and Bonding" and added a reference to Part X of Article 250.	FR-7967
495.66	Revised to add references to Parts V, VI, and X of Article 250 and to Part III of Article 400.	FR-7968
495, Part V	Revised title of Part V to "Boilers."	FR-7970
490.71	Revised section to reflect the title change of Part V.	FR-7971
490.72(E)	Revised to change "the electrodes" to "the heating elements."	FR-7972
490.73	Revised to change "the electrodes" to "the heating elements."	FR-7973
495, Part VI	Deleted Part VI.	SR-7755
500.1	Revised scope into a list format and removed scopes of Articles 501, 502, 503, and 504 from this section.	FR-8335, SCR-47
500.4	Revised text and Informational Notes to clarify that documentation needs to be presented to the AHJ.	FR-8955, SR-7506
500.5(A)(1)	Revised to relocate last sentence to 500.1(B)(4).	SR-7513
500.5(B)(1)	Revised to combine former Informational Note No. 1 and 2 into one.	SR-7532
500.5(D)	Revised to correct error in NFPA 499 that was addressed by a TIA and to correlate with changes in Articles 502, 503, and 506.	FR-8522, SR-7836
500.6(A)(2)	Relocated two exceptions to 500.8(B)(1).	FR-8329
500.6(C)	Added section for fibers/flyings to correct error and to correlate with changes in other sections.	FR-8769
500.7	Revised to provide reference to new Table 13 in Chapter 9.	FR-8653
500.7(K)	Revised title and text for correlation and consistency.	SCR-49
500.7(O)	Revised to add the word "locations."	FR-8624
500.7(P)	Added section for protection by electrical resistance trace heating "60079-30-1".	FR-8624
500.7(Q)	Added section for protection by impedance heating "IEEE 844.3".	FR-8624
500.7(R)	Added section for enclosed-break protection.	FR-8624
500.7(S)	Added section for nonsparking protection.	FR-8624
500.7(T)	Added section for sealed protection.	FR-8624
500.7(U)	Revised title to reflect proper context on protection permitted and revised text for clarity.	FR-8624, SR-7542
500.8(A)	Revised Informational Note to remove incorrect reference standard.	SR-7552
500.8(B)	Revised to relocate exceptions from 500.6.	FR-8731
500.8(D)	Revised to correlate with other changes in Articles 502, 503, and 506.	SR-7821
500.8(E)(3)	Revised for clarity on use of close-up plugs and blanking elements.	FR-8384
500.8(F)	Revised for clarity and to replace "composite" with "hybrid".	FR-8476, SR-7561
500.8(G)	Revised to correlate with other sections.	FR-8346, SR-7563
500.9	Deleted section.	FR-8318
501.1	Revised to remove Informational Note and to correlate with 500.5(B).	FR-8347, SR-7775
501.10(A)(1)	Revised for clarity and consistency, added Informational Note, and added cable tray applications.	FR-8647, SR-7709
501.10(A)(2)	Revised for clarity and to correlate with 250.118(5) and (6).	SCR-50
501.10(B)(1)	Revised for clarity, added new Informational Note to correlate with defined term, and relocated last sentence to 501.15(A)(4).	FR-8464, SCR-51

Section	Synopsis	References
501.10(B)(2)	Revised to delete Informational Note, to relocate concepts into list items (2) and (4), and to add Type P cable.	FR-8500
501.15(A)(4)	Relocated text from 501.10(B)(1) to this section and revised for clarity.	FR-8675, SR-7733
501.15(B)(2)	Revised to allow methods from 501.10(B)(1)(1) and 501.10(B)(6) be used and to clarify exceptions.	FR-8680
501.15(D)(1)	Revised and reorganized for clarity and to include Type P and Type TC-ER-HL cable as permissible wiring methods.	FR-8682, SR-7730
501.15(D)(2)	Revised to correct number format and clarify last sentence.	FR-8687, SCR-52
501.15(E)(1)	Revised text and exceptions for clarity and revised Exception No. 2 to align with NFPA 496.	FR-8688, SR-7735
501.30	Revised to separate grounding and bonding into two separate items, which allows for more stringent requirements to be created for bonding.	FR-8391, SR-7852
501.105(B)(6)	Revised to use defined term *restricted industrial establishment [as applied to hazardous (classified) locations]* for clarity and to correlate with other sections and articles.	SR-7747
501.130(B)(4)	Revised exception into two positive enforceable items.	FR-8492, SR-7756
501.135(B)(4)	Added section to correct the omission of luminaires.	FR-8767
501.140(A)	Revised to correlate with definition of new defined term *restricted industrial establishment [as applied to hazardous (classified) locations]*.	SR-7758
501.141	Added section for flexible cables, Class I, Division 2.	FR-9025, SR-7759
501.145	Revised to add listing requirement of attachment plugs and receptacles.	FR-8686
502.10	Added Informational Note referring to Article 100 for new defined term *restricted industrial establishment [as applied to hazardous (classified) locations]*.	SR-7808
502.10(A)	Revised to change references and correlate with definition of new defined term *restricted industrial establishment [as applied to hazardous (classified) locations]*.	SR-7799
502.10(B)	Revised list item (2) to include supplemental corrosion protection coatings and added text to item (3) to allow for listed compression type connectors or couplings for correlation with other articles.	FR-8398, SCR-53
502.15	Revised former Informational Note into enforceable list item (5) and revised other text for conciseness.	FR-8531
502.30	Revised to separate grounding and bonding into two separate items, which allows for additional methods to be created for bonding.	FR-8383, SR-7855
502.140(A)	Revised to use the defined term *restricted industrial establishment [as applied to hazardous (classified) locations* for correlation.	SR-7854
502.150(A)(1)	Revised former exception into positive enforceable text.	FR-8501
502.150(A)(2)	Revised former exception into positive enforceable text.	FR-8502
502.150(B)(1)	Revised former exception into positive enforceable text.	FR-8503
502.150(B)(5)	Revised to use new defined term *restricted industrial establishment [as applied to hazardous (classified) locations]* for correlation.	SR-7807
503.10	Revised Informational Note to refer to new defined term *restricted industrial establishment [as applied to hazardous (classified) locations]* for correlation and to correct references.	SR-7813
503.10(A)(1)	Revised to add Type P cable as acceptable.	FR-8504
503.10(A)(3)	Revised to add Type P cable as acceptable and to delete Informational Note.	FR-8480
503.10(B)	Revised former exception into positive enforceable text.	FR-8485
503.30	Revised to separate grounding and bonding into two separate items, which allows for additional methods to be created for bonding.	FR-8431, SR-7814
504.1	Revised to replace "Articles 500 through 516" with "hazardous (classified) locations."	FR-8745

Section	Synopsis	References
504.30(A)(1)	Revised former exceptions into positive enforceable list items.	FR-8535, SR-7838
504.30(A)(2)	Revised to correlate with 504.30(A)(1).	FR-8538
504.30(A)(3)	Revised former exceptions into positive enforceable list items.	FR-8540
504.60	Revised to correct references to other sections.	SR-7839
505.1	Revised scope text into "Covered" and "Not Covered" subsections and relocated text around polyphoric materials from 505.5(A)(1) into list item (4) of 505.1(B).	FR-8724, SCR-54
505.3	Deleted section due to redundancy with 90.3.	FR-8741
505.4	Revised text and Informational Notes to clarify that documentation needs to be presented to the AHJ.	FR-8689, SR-7766
505.5(A)(1)	Relocated text around pyrophoric materials to 505.1(B)(4).	SR-7767
505.8	Revised Informational Note No. 2 to reference new Table 13 in Chapter 9.	FR-8634
505.8(I)	Revised reference standard in Informational Note and to correlate with Article 100.	SR-7768
505.8(N)	Revised text to add the word "locations" for correlation within the article.	FR-8593
505.8(O)	Added section for protection by impedance heating "IEEE 844.3."	FR-8593
505.8(P)	Added section for pressurized room "p."	FR-8593
505.8(Q)	Added section for special protection "s."	FR-8593, SR-7790
505.9(C)(2)	Deleted Table 505.9(C)(2)(4), relocated table information to new Table 13 in Chapter 9, and revised table notes.	FR-8639, SR-8709
505.9(C)(2)(1)	Revised for clarity for zone equipment marking.	SR-7769
505.9(D)	Revised title for clarity, updated cross references, and consolidated text.	SR-7770
505.9(F)	Revised to correct error and to replace "composite" with "hybrid".	FR-8579, SR-7771
505.15(B)	Revised for correlation and to use defined term *restricted industrial establishment [as applied to hazardous (classified) locations]*.	FR-8414, SCR-55
505.15(B)(2)	Revised for clarity and to use defined term *restricted industrial establishment [as applied to hazardous (classified) locations]*.	FR-8544
505.15(C)(1)	Added Informational Note to reference new defined term *restricted industrial establishment [as applied to hazardous (classified) locations]* and to correct cross references.	FR-8549, SR-7779
505.16(C)(1)	Revised for clarity and to use acronyms for references to RMC and IMC.	FR-8551, SR-7781
505.16(C)(2)	Added enforceable requirements for sealing cables and revised text for consistency.	FR-8555, SR-7783
505.17(A)	Added new Informational Note No. 2 for new defined term and revised protection of cable to be consistent with other sections.	SR-7784
505.20(C)	Added Exception No. 5 and 6 using language from list items (4) and (5) of 501.125(B)(4).	FR-8486, SR-7785
505.30	Relocated requirements from 505.25 and revised text to separate grounding and bonding into two separate items, which allows for additional methods to be created for bonding.	FR-8386, SR-7842
506.1	Revised to correlate with 90.2, to accurately describe what is and is not covered, and to delete Informational Note No. 3.	FR-8642, SCR-56
506.4	Revised text and Informational Notes to clarify that documentation needs to be presented to the AHJ.	FR-8690, SR-7843
506.5	Revised text to correlate with revised definitions and removed selected Informational Notes.	SR-7863
506.6	Revised to correlate with NFPA 499 and added three Informational Notes.	FR-8691, SR-7864
506.7	Revised for clarity and to correlate with revised definitions of *fibers/flyings* and *ignitible fibers/flyings*.	FR-8726, SR-7865
506.8(O)	Added section for pressurized room "p."	FR-8626

Section	Synopsis	References
506.8(P)	Added section for special protection "s."	FR-8626, SR-7857
506.9(C)(2)	Deleted Table 506.9(C)(2)(3), relocated table information to new Table 13 in Chapter 9, and revised table notes.	FR-8644
506.9(E)(3)	Revised text and Informational Note for clarity.	FR-8646
506.9(F)	Revised to replace "composite" with "hybrid."	FR-8647
506.15	Added new Informational Note to reference new defined term *restricted industrial establishment [as applied to hazardous (classified) locations]*.	SCR-58
506.16	Revised for clarity and to correlate with revised definitions of *fibers/flyings* and *ignitible fibers/flyings*.	SR-7866
506.30	Relocated requirements from former 506.25 and revised text to separate grounding and bonding into two separate items, which allows for additional methods to be created for bonding.	FR-8404, SR-7856
Article 510	Deleted article and relocated requirements into appropriate Articles 511–516 to ensure compliance with Articles 500, 505, and 506.	FR-8305
511.2	Added "Other Articles" section and new Table 511.2.	FR-8310, SR-7868
511.3(E)	Revised to correlate with 514.3(B)(1) and to replace "cutoff" with "separated."	FR-8684
511.4(A)	Revised for clarity and updated a reference section.	FR-8733
511.7	Revised to remove redundant language and to correct cross references.	SR-7870
511.8	Revised former exception into positive enforceable language, added requirement of threaded to RMC and IMC, and corrected change in classification term.	FR-8561, SR-7871
512	Created new Article 512, Cannabis Oil Equipment and Cannabis Oil Systems Using Flammable Materials.	FR-8499
513.2	Added "Other Articles" section and new Table 513.2.	FR-8311, SR-7875
513.3(B)	Revised to replace "cut off" with "separated" for clarity.	FR-8698
513.4	Revised for clarity by replacing references to articles with "hazardous (classified) locations."	FR-8736, FR-8737
514.1	Revised to combine Informational Notes into one for clarity.	SR-7876
514.2	Added "Other Articles" section and new Table 514.2.	FR-8312, SR-7877
514.3(C)	Revised to relocate text from former 514.3(D) and (E) under 514.3(C)(2).	FR-8489
514.4	Revised to replace "Class I" with "hazardous (classified)" for clarity.	FR-8564, SR-7878
514.7	Revised to remove redundant language and to correct cross references.	FR-8566, SR-7879
514.8	Revised to replace "and" with "or" for clarity and revised former exception into positive enforceable text.	FR-8569, SR-7880
514.11(A)	Revised to improve readability, use NEC terminology, and correlate with NFPA 30A.	FR-8730
515.2	Added "Other Articles" section and new Table 514.2.	FR-8313, SR-7881
515.4	Revised to replace "Class 1" with "hazardous (classified) locations" for clarity.	FR-8727, SR-7882
515.7	Revised title to replace "Class I" with "Hazardous (Classified)."	SR-7883
515.7(A), (B), (C)	Revised to make wiring methods consistent with other hazardous classified locations.	FR-8738
515.10	Revised to replace "Gasoline" with "Motor Fuel" in reference to dispensers since the text refers to other types of dispensers.	FR-8490, SR-7884
516.1	Revised to combine Informational Note No. 1 and 3.	SR-7885

Section	Synopsis	References
516.2	Added "Other Articles" section and new Table 516.2.	FR-8315, SR-7886
516.4	Revised to correct the dimension in list item (5) and to correct NFPA 33 references.	FR-8747
516.7(A)	Revised text into list format for clarity.	FR-8575, SR-7887
516.38(A)	Revised text into list format for clarity.	FR-8576, SR-7888
517.1	Revised to incorporate former Informational Note from 517.11.	SR-8694
517.4	Deleted section as the text was incorporated into 517.1.	SR-8694
517.6	Added section for patient care-related electrical equipment.	FR-8150
517.10(B)	Revised list item (2) to include that the determination of patient sleeping rooms is done by HCF governing body, replaced "areas" with "spaces" to be consistent with NFPA 99, and added Informational Notes on AFCI and TR devices.	FR-8089, SR-8562
517.11	Deleted section and relocated general information to Part I for consistency.	FR-8151
517.13	Relocated former Exception No. 2 from under 517.13(B)(3).	FR-8170
517.14	Added an exception to allow for termination on ground bus.	FR-8160
517.20	Revised for clarity and added Informational Note No. 2.	FR-8161
517.22	Added section on demand factors to refer users to 220.110.	FR-8222, SCR-35
517.30	Revised to use consistent terms and to correlate with NFPA 99.	FR-8237, SR-8696
517.30(B)(1)	Revised to remove phrase "be permitted to."	SR-8666
517.31(C)(3)	Revised to replace "emergency system" with "EES" and replace "area" with "spaces."	FR-8054
517.31(D)	Removed redundant language in list item (3) for clarity.	FR-8152
517.31(E)	Revised to correct the NFPA 99 reference.	FR-8162
517.34(A)	Revised to replace "selected" with "select" for clarity.	SR-8604
517.40(C)	Deleted Informational Notes.	FR-8107
517.41	Revised to correlate with NFPA 99 and added new Informational Note.	SR-8697
517.44(B)	Revised list item (3) to replace "critical branch" with "equipment branch" to correlate with NFPA 99.	FR-8154
517, Part V	Revised title of Part V to "Diagnostic Imaging and Treatment Equipment" to reflect changes in health care terminology.	FR-8234
517.70	Revised Informational Notes into a list format with examples for clarity and usability.	SR-8606
517.80	Revised to remove requirements for grounding and mechanical protection of Class 2 (PoE) systems and added new Informational Note.	FR-8153
518.1	Relocated Informational Note from 518.2(C) to 518.1.	SR-8644
518.2(A)	Revised list of assembly occupancies to add casinos and gaming facilities.	FR-7981
518.2(C)	Revised to replace "building structure" with "building or structure" for clarity.	FR-8180, SR-8503
518.3	Revised title to reflect context of contained language.	FR-7982
518.4	Revised text into list format and to allow approved Chapter 3 wiring methods.	FR-7983
518.5	Revised text into list format for clarity and usability.	SR-8653
520.5(A) and (B)	Revised text into list format, updated cross references, and clarified communications circuits and those delivering Class 2 power.	FR-7997
520.10	Revised Informational Note to add new standard.	FR-8002
520.53	Revised to differentiate between general purpose switchboards and portable stage switchboards.	SR-8509
520.53(C)	Revised to reference 406.13 for single pole separable connectors.	FR-8004

Section	Synopsis	References
520.54(K)	Revised to replace "qualified personnel" with defined term *qualified persons*.	SR-8515
520.62	Revised to incorporate listing requirements.	FR-8005
520.68(D)	Added section for special-purpose multi-circuit cable systems.	FR-8007, SCR-60
522.22	Revised Note 1 in Table 522.22 with enforceable language.	SR-8715
525.3	Deleted former 525.3(B) and (C) and relocated requirement from former 525.3(D) into 525.3(B).	FR-8017, FR-8019, FR-8021
525.31	Revised to remove non-modifying language.	FR-8023
Article 530	Revised and restructured article extensively to incorporate new technology and to apply to studios in facilities and locations staffed by qualified persons. Revised article title to replace "and Similar Locations" with "and Remote Locations."	FR-8211, SR-8529
540.50	Deleted section.	FR-8220, FR-8221
545.22	Revised to clarify the feeder requirements for relocatable structures to be consistent with Article 550 and to add reference to 590.4(B) in Informational Note.	SR-8603
545.22(D)	Revised to divide into subsections for clarity.	FR-7705,
547.1	Revised article into Part delineations and renumbered sections accordingly.	FR-7816
547.20	Revised text into list format and added Part II of Article 502 to the 547.1 allowable wiring methods.	FR-7746, FR-7816, SR-8605
547.26	Revised to not permit NM cable concealed within walls or above ceilings and added Informational Note.	FR-7748
547.27	Revised to require an insulated equipment grounding conductor if it is separate and not part of a cable assembly.	FR-7789
547.28	Revised list item (1) to replace "having" with "requiring" an equipotential plane.	SR-8619
547.41(B)(2) and (3)	Revised to remove redundant language.	FR-7750, FR-7754
547.44(B)	Revised Informational Note No. 1 and No. 2 references and clarified the bonding termination locations of the equipotential plane conductor.	FR-7771, FR-7756, SCR-124
550.10(I)	Revised to add additional raceway types and use defined terms.	FR-7763, SR-8630
550.13(B)	Revised for clarity and usability for GFCI protection.	SR-8633
550.15	Revised to clarify the intent to avoid terminating dissimilar metals.	FR-7786, SR-8640
550.16(A)(1)	Revised to add "equipment" in front of "grounding."	FR-7792
550.16(C)(1)	Revised to add "solid, insulated, or bare" for applicability.	FR-7794
550.25	Revised to consolidate text and provide reference to Article 100.	FR-7797
550.32(A)	Revised to replace "within 30 ft" with "within sight from the mobile home it serves."	FR-7832, SCR-92
550.33(A)	Added section for feeder equipment to address disconnecting means and renumbered former 550.33(A) to (B).	FR-7874, SR-8651
551.3	Added section around electrical datum plane distances.	FR-7873
551.30(A)	Revised to require the connection of electrical system from a generator be capable of providing an effective ground-fault return path when operational.	FR-7841
551.31(D)	Revised to replace "power supply" with "feeder" to correlate with the revised definition for *feeder assembly*.	SR-8656
551.40(A)	Revised to add the term "nominal" for consistency.	FR-7842
551.40(C)	Revised to replace "power supply" with "feeder."	SR-8658

Section	Synopsis	References
551.40(D)	Deleted former section "Reverse Polarity Device" and added section "Loss of Ground Device" to address the "hot skin" effect that RVs can have and to provide the intended requirement of circuits at detection of a loss of ground.	FR-7849, FR-7851, SR-8708
551.42(C)	Revised to replace "power supply" with "feeder."	SR-8659
551.42(D)	Revised to replace "power supply" with "feeder."	SR-8661
551.44	Revised to replace "power supply" with "feeder."	SR-8662
551.45(C)	Revised to replace "power supply" with "feeder."	SR-8663
551.46(A)	Revised to replace "power supply" with "feeder."	SR-8664
551.47(L)	Revised to correct code references.	FR-7876
551.47(N)	Revised text into list format and to add listing as a requirement for conductors.	FR-7877, SR-8669
551.47(R)	Revised to indicate location of label for generator when a transfer switch is involved.	FR-7878
551.54(C)	Revised to replace "grounded circuit conductor (neutral conductor)" with "neutral conductor" because the neutral in an RV is always floating and insulated from ungrounded conductors.	FR-7879
551.71(B)	Revised to add "weather-resistant" to 30-amp 125-volt receptacle.	FR-7880
551.71(C)	Revised to require weather-resistant receptacles with delayed effective date.	SR-8670
551.71(F)	Revised text into list format and to clarify when GFCI protection is required.	SR-8671
551.72(A)	Revised to clarify that 120/208-volt is not a single-phase system.	FR-7884
551.72(C) and (D)	Revised to relocate neutral conductor language from 551.72(C) to 551.72(D) and to clarify that the RV 50-amp circuit is a feeder.	FR-8168, SR-8672, SR-8673
551.72(E)	Revised to allow auto-transformers in RV parks and revised Informational Note to issue caution.	SR-8667
551.73(A)	Revised text into list format for clarity.	FR-7886
551.76(A)	Revised for clarity around recreational vehicle site supply.	FR-7887
551.77	Revised to add listing requirement to RV site supply equipment.	FR-7871
551.77(D)	Revised to require site supply equipment be 2 feet above electrical datum plane.	FR-7888
552.4	Revised to remove reference to deleted section 552.2 and to correlate language with Article 545.	FR-7892, SR-8680
552.43(C)	Revised to clarify raceway types permitted and to replace "suitable" with the defined term *identified*.	FR-7898, SR-8682
552.44(A)	Revised to replace "power supply" with "feeder" to correlate with the revised definition for *feeder assembly*.	SR-8675
552.44(C)(2)	Revised to replace "power supply" with "feeder."	SR-8676
552.44(D)	Revised to replace "power supply" with "feeder."	SR-8677
552.44(E)	Revised to replace "power supply" with "feeder."	SR-8684
552.45(C)	Revised to replace "power supply" with "feeder."	SR-8678
552.46(A)	Revised to replace "power supply" with "feeder."	SR-8679
552.48(K)	Revised to correct code section references.	FR-7900
552.48(M)	Revised to clarify raceway types permitted and to use the defined term *identified*.	FR-7903
555.3	Added section for electrical datum plane distances.	FR-7873
555.4	Revised to add service equipment location to be adjacent to structure and for the height be measured from electrical datum plane to correlate with Article 682.	FR-7911
555.6	Revised notes in Table 555.6, renumbered table and relocated it to 220.120, and provided reference to 220.120.	FR-7912, SCR-33
555.9	Relocated 555.9 to 555.35(C) with other GFCI requirements.	FR-7917

Section	Synopsis	References
555.13	Revised text indicating that an equipment grounding conductor in a branch circuit or feeder, when extended to circuits or equipment on a dock, does not require an additional bonding conductor.	SR-8686
555.14	Added section for equipotential planes and bonding of equipotential planes.	FR-7925, SR-8687
555.15	Added section for replacement of equipment.	FR-7926, SR-8688
555.30	Revised by consolidating fixed and floating piers under 555.30(A) and renumbering 555.30(C) to (B) to correlate with changes.	FR-7929, TIA 23-5
555.33(B)	Revised to relocate GFCI requirements to 555.35.	FR-7930
555.34(A)	Revised to remove temporary wiring from section, to require an insulated equipment grounding conductor, and to add extra hard usage to portable cables.	FR-7932
555.34(B)(4)	Revised to add IMC as a method of permitted protection from physical damage similar to 300.5.	FR-7932
555.35	Revised text and title around GFPE and GFCI.	FR-7917, TIA 23-6
555.35(A)	Revised to require listed GFPE protection of not more than 100 milliamperes for feeders installed on docking facilities.	FR-7917, TIA 23-6
555.35(B)	Revised to require GFPE protection of not more than 30 milliamperes for receptacles providing shore power and to require GFCI protection for outlets for other than shore power.	FR-7917, TIA 23-6
555.35(C)	Relocated 555.9 to 555.35(C) with other GFCI requirements.	FR-7917
555.35(D)	Revised to add future effective date for listing of leakage current measurement devices to allow for development and implementation of these devices.	SR-8690
555.36(C)	Added section for emergency electrical disconnects for shore power connections.	FR-7936
555.37	Revised to remove redundant language and to align with Article 682.	FR-8169
555.38	Added section for luminaire general requirements and underwater luminaires.	FR-7931, SR-8691
555.53	Revised to add language requiring outdoor outlets, shore power outlets, and boat hoists comply with the GFCI and GFPE protection requirements in 555.35(B) and (C).	FR-7939
590.4(B)	Revised to add the appropriate rules in Article 445 for OCP along with Article 240 to eliminate possible redundant protection.	FR-9459
590.4(F)	Revised to require metal guards for lamps be connected to EGC, if used.	FR-9460
590.6(A)(3)	Revised to update manufacture or re-manufacture date and requirements of GFCI for generators built before the date.	FR-9461
590.8	Revised Informational Note and relocated to 590.8(A).	SR-8581
590.8(B)	Revised text requiring current-limiting OCP where available fault current is greater than 10,000 amperes.	FR-9464
600.4(E)	Deleted section due to redundancy with requirements in 110.21(B).	FR-7625, SR-8194
600.5(A)	Revised to add two exceptions for clarification.	FR-7627
600.5(D)(2)	Revised into list format and to remove redundancies with 90.3.	FR-7628, SCR-17
600.6	Revised for consistency with other sections and to remove redundancies with 90.3.	FR-7629, SCR-18
600.6(A)	Revised to clarify the intent and to replace "permitted to be" with the defined term *accessible*.	FR-7631
600.7(A)	Revised to add "conductor" after "equipment grounding."	FR-7808
600.7(B)(3)	Revised to remove "permitted to be" from text for clarity in not allowing metal parts of a building to be used for bonding.	SR-8200
600.7(B)(7)	Revised to include copper-clad aluminum as a bonding conductor and added "also" to clarify size intent.	FR-7632, SR-8197

Section	Synopsis	References
600.24(B)	Revised title and replaced "grounded by connecting" with "connected."	FR-7634
600.35(B)	Revised to deleted redundant text that is already required by 90.3.	SCR-19
604.100(A)	Revised title to include "tubing" to correlate with content of section.	SR-8623
604.100(A)(1)	Revised to correlate with 110.14 changes around copper-clad aluminum conductors not being dissimilar metals to copper.	FR-7713
604.100(A)(2)	Revised to add EMT to list, add acronyms, and to correlate with 110.14 changes around copper-clad aluminum conductors not being dissimilar metals to copper.	FR-7745, FR-7714, SR-8620
605.3(A) and (B)	Deleted sections.	SCR-20, CC-8193, CC-99
610.13(D)	Deleted section due to redundancy with 90.3.	SR-7777, SR-7584
620.1	Revised to relocate Informational Notes formerly under 620.2 to the scope under Informational Note No. 4 and 5.	FR-9598
620.6	Revised text into subsections for clarity.	FR-9291, SCR-5
620.12(A)(2)	Added section on Class 2 wiring and to provide minimum cable sizes.	FR-9328, SR-7510
620.12(A)(4)	Added section allowing parallel conductors in elevator traveling cables.	FR-9328
620.21	Revised to correlate with other changes in the code and to remove redundant language covered in 90.3.	SCR-6
620.21(A)(1)	Revised to list cable types for Class 2 wiring and to grant permission to use substitute cables.	FR-9327
620.21(B)(2)	Revised to list cable types for Class 2 wiring and to grant permission to use substitute cables.	FR-9309
620.21(C)(2)	Revised to list cable types for Class 2 wiring and to grant permission to use substitute cables.	FR-9304
620.22(A)	Revised to permit additional loads on car lighting branch circuit.	SR-7515
620.22(B)	Revised to add terms to align section with 620.22(A).	FR-9318
620.23	Revised to add "cartop lighting" and "truss interior lighting" for safety and enforceability.	FR-9321, SCR-7
620.25(B)	Revised to correlate with 620.22(A).	FR-9331
620.36	Revised to permit communications limited power cable, shielded pairs, coaxial cables, and communications circuits in traveling cables to accommodate communications and Class 2 applications and for compliance with 800.179.	FR-9333, SR-7545
620.37(A)	Revised text into a list format and added branch circuits identified under 620.24 to be permitted.	FR-9336, SR-7546
620.51(A)	Revised text in Exception No. 2 to identify situations where a cord-and-plug connection is permitted.	FR-9342, SR-7549
620.51(D)(1)	Deleted section since 110.22 covers the identification requirements.	SR-7556
620.51(D)(2)	Revised for clarity and renumbered due to deletion of former subsection (D)(1).	SR-7556
620.51(E)	Revised title for consistency with Article 242 and added text requiring installation of a listed SPD.	FR-9361, SCR-8
620.62	Revised to combine two exceptions to correlate with 700.32 and 701.32.	FR-9368
625.1	Added Informational Note No. 4 and 5.	FR-9373
625.4	Revised to clarify input voltages.	FR-9403
625.6	Renumbered section for consistency and to clarify what equipment must be listed.	FR-9405, SR-7703
625.17(C)(2)	Revised to clarify output cable to primary pad measurement.	FR-9413
625.22	Revised to require a listed personnel protection system (PPS) for WPTE.	FR-9415, SR-7713

Section	Synopsis	References
625.40	Revised to clarify branch circuit requirements and to add an exception.	FR-9416, SR-7714
625.41	Revised to include WPTE in text.	FR-9418
625.42	Revised text into subsections on energy management systems and EVSE with adjustable settings.	FR-9565, SR-7715, SCR-9, SCR-10
625.43	Revised terms for consistency and added language to allow remote disconnecting means locations.	FR-9465, SR-7722
625.44(A)	Revised to add 3-pole, 4-wire configuration and permit 60-amp receptacles.	FR-9466, SR-7725
625.44(B)	Revised to add 125/250 volts as 3-pole, 4-wire configuration and permit 60-amp receptacles.	FR-9467, SR-7727
625.44(C)	Revised to add 125/250 volts as 3-pole, 4-wire configuration.	FR-9467
625.48	Revised to facilitate the integration of WPT into Article 625 and revised Informational Notes.	FR-9469
625.49	Added section for island mode for EVPE and bidirectional EVSE.	FR-9470
625.101	Revised to clarify connection to the circuit equipment grounding conductor.	FR-9478
625.102	Revised to better facilitate the integration WPT into 625.	FR-9479
626.1	Revised to add "nonpropulsion electrical elements" for clarity.	FR-9492
630.8	Added section for ground-fault circuit-interrupter protection for personnel.	FR-9510, SCR-11
630.13	Revised to allow for listed cord-and-plug connector as a disconnecting means.	FR-9506, SR-7589
640.3(C)	Added section for communications cables.	FR-9211
640.3(D)	Revised to include Class 2, Class 3, and Type PLTC cables in cable trays and to correct cross reference.	FR-9211
640.21(B), (C), (D)	Revised for "hybrid" optical fiber cables, to update cross reference to correlate with new Article 722, and to align with current US and international definitions.	FR-9215
640.42(B), (C), (D)	Revised for "hybrid" optical fiber cables and to align with current US and international definitions.	FR-9216
645.1	Deleted Informational Note No. 2.	FR-9530
645.3	Revised to correct cross reference and delete text in subsection (D) that was supposed to be removed during 2020 cycle.	FR-9218
645.3(B)	Revised references from Article 725 to new Article 722.	SR-7669
645.3(F)	Revised references from Article 805 to Article 800.	SR-7673
645.3(G)	Revised references from Article 820 to Article 800.	SR-7675
645.5(B)	Revised Informational Note No. 1, added Informational Note No. 2, and added list item (3).	FR-9554, SR-7677
645.5(D)	Revised terms for consistency throughout article.	FR-9223
645.5(E)(2)	Revised references from Article 725 to new Article 722.	SR-7681
645.5(F)	Revised terms for consistency with those used throughout Article 645.	FR-9231
645.10(B)	Revised references from Article 725 to new Article 722.	SR-7683
645.18	Revised title and text to correlate with new Article 242.	FR-9238, SR-7685
646.1	Relocated Informational Notes from under 646.2 to scope section as Informational Note No. 4 and 5.	FR-9597
646.3(B)	Revised references from Article 725 to new Article 722.	FR-9276, SR-7688
646.3(D)	Revised to correct cross reference and to delete text that was supposed to be removed during 2020 cycle.	FR-9248

Section	Synopsis	References
646.3(F)	Revised references from Article 805 to Article 800.	FR-9251, SR-7691
646.3(G)	Revised references from Article 820 to Article 800.	FR-9252, SR-7693
646.3(H)	Deleted redundant section on storage batteries and replaced with section on surge-protective devices (SPDs).	SR-7694
646.5	Relocated selections of text from Information Notes and revised into enforceable text.	SR-7794
646.19	Revised to require doors open to the full extent of the designed opening.	FR-9255, SR-7699
660.4(A) and (B)	Revised text to add "power-supply cord" to correlate with other similar code sections.	SR-7592, SR-7593
660.7	Revised text to add "power-supply cord" to correlate with other similar code sections.	SR-7594
660.9	Revised references from Article 725 to new Article 724.	SR-7595
660.48	Revised to clarify grounding requirements for battery operated X-ray equipment.	FR-9514
670.1	Revised scope statement to remove the reference to definitions.	SCR-12
670.3(A)	Revised text to align with NFPA 79.	FR-9521
670.4(B)	Added Informational Note to consider application of NFPA 70E before performing maintenance on a machine.	FR-9522, SR-7654
670.5	Revised text into subsections on installation and available short-circuit current field marking for usability.	FR-9525
670.6	Revised title from "Surge" to "Overvoltage" and added "overvoltage" to first sentence of text to clarify protection requirements.	FR-9578, SR-7644
680.5	Revised title and text to place GFCI and special purpose ground-fault circuit interrupter (SPGFCI) protection requirements into one section.	FR-8418, SR-8377
680.5(B)	Revised to add an exception for receptacles and outlets below the low voltage contact limit not requiring GFCI protection.	SR-8381, SR-8714
680.6	Relocated former 680.3 and revised to address listing requirements for correlation with other changes.	FR-8445
680.7	Relocated 680.6 and revised into list format for clarity and usability.	FR-8463
680.9	Relocated Informational Note to 680.9(A) and revised into enforceable language for consistency and usability.	FR-8763
680.9(A)	Revised to replace "an enclosed raceway" with "a raceway" for clarity.	SCR-64
680.9(B)	Revised to reference Chapter 8.	FR-8426, SR-8386
680.10	Revised title and added heat pumps and chillers to address new technology.	FR-8433
680.11(A)	Revised to clarify wiring methods are acceptable when installed complete between outlets, junctions, or splicing points.	FR-8441, SR-8389
680.11(C)	Deleted section.	FR-8442
680.12	Revised into subsections and added text to recognize rated and identified for submersion regarding electrical equipment and GFCI requirements.	FR-8957
680.13	Revised to replace "at least" with "not less than" for consistency.	FR-8444
680.14	Revised text into "Wiring Methods" and "Other Equipment" subsections for clarity within corrosive environments.	FR-8408
680.14(A)	Revised text to add "tubing" as an item not permitted in this section.	SR-8400
680.21	Revised to delete former 680.21(A)(1) and relocated EGC text to 680.7.	FR-8611
680.21(A)(2)	Revised to remove the listed fittings requirement since respective code sections already require listed fittings.	FR-8530
680.21(C)	Revised for proper reference to 680.5 for GFCI and revised exception to indicate ground-fault protection is not required for listed low-voltage motors.	FR-8755

Section	Synopsis	References
680.21(D)	Revised to refer to 680.5 for ground-fault protection requirements and to require ground-fault protection for replaced or repaired motors.	FR-8756, SR-8402
680.22(A)(4)	Revised title for expanded requirements for GFCI and SPGFCI protection.	FR-8757, SR-8411
680.22(A)(5)	Relocated to 680.12(B) and renumbered 680.21(A) accordingly.	FR-8958
680.22(B)(4)	Revised to incorporate GFCI and SPGFCI protection.	FR-8753, SR-8412
680.23(B)(2)	Revised to replace "rigid nonmetallic" with "rigid polyvinyl chloride" conduit and added Informational Note.	FR-8482, SR-8427
680.23(F)(1)	Revised to remove reference to definitions in former 680.2, to relocate LFNMC to 680.14, and to relocate common grounding provisions to 680.7.	FR-9103
680.24(A)	Revised to remove "approved" as it is already required elsewhere in the code.	FR-8506
680.24(B)(1)	Revised list item (3) to correlate with existing requirements in the code.	FR-9104
680.25	Deleted section as most items were relocated to other sections.	FR-8412
680.26(A)	Revised to require equipotential bonding for pools with or without associated electrical equipment related to the pool.	FR-8508, SR-8454
680.26(B)(1)	Revised to clarify reconstructed conductive pool shells require bonding.	FR-9105
680.26(B)(2)	Revised to clarify depth of perimeter bonding conductors in reference to finished grade.	FR-9106, SR-8616
680.26(B)(5)	Added an exception for clarity and usability around metal fittings.	FR-9107
680.26(B)(6)	Revised text regarding equipment that is required to be bonded and to include other equipment typically associated with the pool system.	FR-9108
680.26(B)(7)	Revised former Exception No. 2 and 3 into positive enforceable text.	FR-9109
680.26(C)	Revised to add specific references to code sections.	FR-8515
680.31	Relocated EGC requirements to 680.7(B) for clarity and usability.	FR-8419
680.32	Revised for expanded requirements for GFCI and SPGFCI protection and deleted Informational Note.	FR-8517, FR-8516, SR-8414
680, Part IV	Revised title for clarity and to align with the types of spas and hot tubs.	FR-8760
680.41	Revised text into subsections, added requirements for equipment exceeding low-voltage contact limit, and exempted one-family dwellings from emergency switch.	FR-8751, FR-8758, SR-8433
680.42(C)	Revised title to better reflect section content.	FR-8761
680.43(A)(2)	Revised for expanded requirements for GFCI and SPGFCI protection and deleted Informational Note.	FR-8553, SR-8416
680.44	Revised for expanded requirements for GFCI and SPGFCI protection, deleted Informational Note, and added gas-fired water heaters for consistency.	FR-8554, SR-8614
680.44(B)	Delete former section that was no longer relevant with new ground fault protection technology and replaced with listed units around ground-fault and GFCI protection.	FR-8609
680.45	Relocated equipment grounding requirements to 680.7 and revised for clarity.	FR-8424
680.50	Revised text into subsections, added low voltage contact limit requirements, and added references to proper code sections.	FR-8762, FR-8752
680.54	Revised for clarity, removed Informational Note, and clarified the devices and controls are to be considered equipment.	FR-8557
680.54(C)	Added section for equipotential bonding of splash pads.	FR-8578, SR-8438
680.58	Revised for expanded requirements for GFCI and SPGFCI protection and deleted Informational Note.	FR-8539, SR-8424
680.59	Revised for the requirement of GFCI and SPGFCI protection for permanently installed non-submersible pumps and added exception of listed low-voltage motors not requiring grounding.	FR-8582, SR-8425
680.60	Deleted Informational Note.	FR-8610

Section	Synopsis	References
680.62(D)	Revised into list format for clarity and deleted unnecessary text.	FR-8425
680.62(E)	Revised to refer to ground-fault protection and to correct section reference.	FR-8764
680.74(A)	Added Exception No. 3 to clarify additional items that require bonding and included common terms "mounting strap" and "yoke".	FR-8598
680.83	Revised title and text around equipotential bonding for clarity and to refer to correct sections.	FR-8601
682.11	Revised text into list format and to require all electrical distribution equipment be located at safe measurable distance from water.	FR-8697
682.12	Revised text into subsections for clarity.	FR-8701
682.13	Revised text into subsections and to correlate with 555.34.	FR-8702
682.14	Revised for clarity and better alignment with other sections.	FR-8703
682.15	Revised to add the requirement of listing.	FR-8704
682.30	Revised code references.	FR-8705
682.31(A)	Added section to clarify what must be connected to the EGC and renumbered remaining subsections accordingly.	FR-8706, SR-8319
682.31(E)	Added exception.	FR-8706
690.1	Deleted former Informational Note Figure 690.1(b) and revised Informational Note Figure 690.1 to align with the deletion of PV output circuits and to clarify references to the existing PV source circuit and new PV string circuit.	FR-9476, SR-8300
690.4	Revised to incorporate acronyms related to PV installations.	SR-8312
690.4(B)	Revised to add "electronic power converters" for correlation with new definition in Article 100.	FR-9204
690.4(F)	Revised to deleye the word "exterior" to clarify the installation of inverters.	FR-9210
690.4(G)	Added section for PV equipment floating on bodies of water.	FR-9199
690.7	Revised text into list format for clarity and removed requirements for systems over 1000 volts.	FR-9213, SCR-70
690.7(A)	Revised to recognize the deletion of PV output circuit definition and to add a "PV string circuit."	FR-9433
690.7(B)	Revised title and text to harmonize with other changes made within Article 690.	FR-9434
690.7(C)	Revised text to correlate with the deletion of PV output circuits.	FR-9436
690.7(D)	Relocated from 690.53 and revised for clarity around marking requirements.	FR-9587
690.8(A)(1)	Revised to delete list items for circuit definitions that were deleted elsewhere from this article and added new Informational Note.	FR-9437
690.8(B)	Revised for clarity and to align with Chapter 2 language.	FR-9217, SR-8322
690.8(D)	Revised title and text to correlate with changes made to PV dc circuit terms and to clarify requirements into a list format.	FR-9439, SR-8324
690.9(B)	Revised to constrain the requirement of PV-rated OCPD in a PV system to PV source circuits.	FR-9441
690.9(C)	Revised title and text to harmonize requirements among all PV dc circuit conductors.	FR-9442
690.9(D)	Revised title and text to address proper content and cross references.	FR-9219, SR-8326
690.11	Revised to generalize the exception to all PV system dc circuits not installed on buildings and in metal wiring methods, revised terms in the exception, and deleted Informational Note due to redundancy.	FR-9222, SR-8431
690.12	Added Exception No. 2 to align with existing exception to correlate requirements for firefighter rooftop access.	FR-9278
690.12(A)	Added exception for clarification on exempted control conductors.	FR-9229
690.12(B)	Revised text for clarification on controlled Limits.	FR-9239
690.12(B)(2)	Revised text to simplify option to reduce shock hazard and added Informational Note.	FR-9235, SR-8328

Section	Synopsis	References
690.12(C)	Revised text for clarification and harmonization with Article 230 outdoor location and deleted redundant Informational Note.	FR-9246, SR-8332
690.12(D)	Deleted former section and replaced with revised former section 690.56(C).	FR-9243, FR-9247
690.13(A)	Revised text into subsections for clarity and deleted Informational Note.	FR-9443
690.15(A)	Relocated from 690.15(D) and revised into list format for clarity.	FR-9273, SR-8333
690.15(C)	Revised text into list format for clarity.	FR-9272
690.15(D)	Relocated from former 690.15(A) and revised to align with other parts of the code for clarity.	FR-9268
690.31(A)	Revised and reorganized into subsection for improved usability.	FR-9279
690.31(B)	Revised and reorganized for clarity and added 690.31(B)(1).	FR-9280
690.31(C)(1)	Revised text and deleted former subsection (d).	FR-9282, SR-8336
690.31(C)(2)	Revised text into list format and to address cable tray use.	FR-9283
690.31(D)(1)	Deleted former (D)(1) subsection, renumbered the remaining sections, and revised the exception to incorporate acronyms and remove undefined phrase.	FR-9285, SR-8343
690.31(G)	Added section for over 1000 volts dc.	FR-9286, SR-8338
690.41	Revised to focus on dc grounding and added 690.41(B) for dc ground-fault detector-interrupter (GFDI) protection means.	FR-9288, FR-9287, SR-8344
690.42	Revised and reorganized into subsections and added section on ground-fault detector-interrupter (GFDI) protection.	FR-9289, SR-8349
690.43(A)	Revised Informational Note text and deleted last sentence of parent text for redundancy.	FR-9290
690.43(C)	Revised to clarify that EGCs may be run separately within the array.	FR-9292
690.47	Revised text and reorganized order of statements for clarity.	FR-9500
690.47(B)	Revised to revert to 2020 edition text at the request of the correlating committee; to be further evaluated for 2026 cycle.	SR-8635
690, Part VI	Revised title to "Source Connections."	FR-9487
690.51	Relocated to 690.4.	FR-9586
690.54	Relocated to 705.14.	FR-9588
690.55	Deleted section due to redundancy.	FR-9589
690.56	Revised to better align with reorganization of Part VI.	FR-9613
690, Part VII	Deleted due to other revisions within Article 690.	FR-9490
690.59	Revised to remove incorrect reference.	FR-9494
690, Part VIII	Deleted due to other revisions within the article.	FR-9592
690.71	Deleted section due to redundancy.	FR-9296
691.1	Revised scope to relocate stations with generating capacity not less than 5000 kW to 691.4 and revised Informational Note Figure 691.1 for clarity.	FR-9301, FR-9299
691.4	Revised to ensure 110.31 applies to entire PV system, revised list item (1) to use defined term *qualified person*, added list items (6) and (7), and added Informational Note.	FR-9305, FR-9303, FR-9302, SCR-71
691.9	Revised to refer to any type of disconnection equipment for clarification.	FR-9306
691.10	Revised title to "Fire Mitigation" and added Informational Note.	FR-9308
692.4(B)	Revised to refer to single section for correlation with changes in other articles.	FR-9480
692.4(C)	Revised for correlation with 690.4(C).	FR-9515
692.31	Revised to include recognition of other acceptable wiring methods.	FR-9484

Section	Synopsis	References
692.41, 692.44, 692.45, 692.47	Deleted sections as they did not modify anything in Chapter 2.	FR-9311
692, Part V	Renumbered former Part VI as Part V and renumbered sections within accordingly.	FR-9312
692, Part VI	Renumbered former Part VII as Part VI and deleted former sections 692.60, 692.61, 692.62, 692.64, 692.65 as they did not modify anything in Chapters 1 through 4.	FR-9317
692.60	Added section to refer to specific Parts within Article 705.	FR-9317
692.61	Relocated former 692.59 and revised title and text around transfer switch.	FR-9317
694.1	Deleted former Informational Note Figure 694.1(b) and revised Informational Note accordingly.	FR-9485
694.7	Revised text to indicate any that construction, maintenance, associated wiring, and interconnection be performed by qualified persons	FR-9329
694.7(A)	Revised to list both onshore and offshore installations.	SR-8352
694.7(B)	Revised text to list both onshore and offshore installations and to remove redundant text.	FR-9332, SR-8352
694.7(D)	Revised title to "Overvoltage Protection" and revised text to correlate with Article 242.	FR-9334
694.15(A)	Revised to reference proper sections of the code.	FR-9335
694.24	Revised to delete the term "interactive system" for clarity.	FR-9337
694.30(A)	Revised for clarity and consistency with other articles.	FR-9489
694.50	Relocated former section 705.14.	FR-9340
694.54	Revised to consolidate into one referenced section.	FR-9341
694.62	Revised text align with Articles 690 and 692.	FR-9496
694.68	Deleted section as it was redundant with Article 705.	FR-9344
695.1(B)	Revised to add list item (4) and Informational Note.	FR-9007
695.2	Added section stating that reconditioned equipment is not permitted.	SR-8066
695.3(A)(3)	Revised Informational Note to correlate with NFPA 20.	FR-9072
695.3(C)(2)	Revised text to harmonize with other NFPA documents.	FR-9073
695.3(I)	Revised text to replace "shall not be permitted to be used" with "shall not be used to supply power to."	SR-8063, SR-8468
695.4(B)(1)	Revised text to correct the section reference to NFPA 20.	FR-9008
695.6(B)(1)	Revised terms and converted into list format.	FR-9012
695.6(D)	Revised text for clarity, converted to list format, and added extractions from NFPA 20.	FR-9014, SR-8064
695.6(H)	Revised text of list item (1) to include the word "ceiling."	FR-9016
695.7	Revised to correlate with NFPA 20.	FR-9017
695.7(D)	Revised to correlate with extracted material from NFPA 20.	SR-8054, SR-8469
695.10	Relocated last sentence of section regarding reconditioned equipment to new 695.2 for consistency.	SR-8065
695.15	Revised to identify SPD accurately and to add Informational Note and exception.	FR-9020, SR-8062
700.1	Revised to add Informational Note No. 1 and add a reference to NFPA 111.	FR-8773
700.2	Added section covering reconditioned equipment.	SR-7928
700.3(A)	Revised section title to "Commissioning Witness Test," revised text to add "commissioning," and added Informational Note referencing NECA 90.	FR-8784, SR-7939
700.3(F)	Revised section requirements around listing and interlocking and added new list items on connection points being located outdoors and labeled.	FR-8786

Section	Synopsis	References
700.4(A)	Deleted former section on "rating," relocated text from former 700.4(B), revised to require that the emergency system be sized to accommodate rapid load changes, and added reference to Parts I through IV of Article 220.	FR-8790, SR-7942, SCR-75
700.4(B)	Relocated text from former 700.4(C), revised into list format, and to relocated text around peak load shaving to new section 700.4(C).	FR-8790, FR-8791, SR-7949
700.4(C)	Added section covering parallel operation.	FR-8791
700.5(B)	Revised to include the terms "transfer" and "prevented."	FR-8794, SR-7952
700.5(C)	Revised text to remove reference to reconditioning.	SR-7929
700.5(D)	Added section on redundant transfer equipment.	FR-8796, SR-7952
700.6	Revised to clarify where signal devices are required and replaced "practicable" with "applicable."	FR-8808, SR-7959
700.6(C)	Revised section title and text to focus on battery charging.	FR-8798, SR-7959
700.8	Revised to add switchgear to the equipment requiring a listed SPD.	FR-8802
700.10(B)	Revised to delete references to other Article 700 sections, added new list item (5) addressing traveling cable to an elevator, and added new list item (6)(e) addressing overcurrent protective devices for mixed loads.	FR-8803, SR-8000
700.10(D)(4)	Revised for clarity around requirements for control conductors.	FR-8807, SR-7990
700.11	Added section covering Class 2 powered emergency lighting systems.	FR-8818, SCR-76, SR-7991, SR-7992
700.12(C)	Added section covering supply duration.	FR-8843, SR-7995
700.12(D)	Revised to replace "service" with "power source" and deleted subsection (D)(2).	FR-8843, FR-8844
700.12(E)	Revised section title to "Stored-Energy Power Supply Systems (SEPSS)" and added current available technology types that can be used with SEPSS.	FR-8827
700.12(G)	Deleted former 700.12(G) and relocated requirements into new 700.12(E).	FR-9194,
700.12(H)	Renumbered former 700.12(H) to 700.12(G) and revised title to "Microgrid Systems."	FR-9195
700.16(B)	Revised to replace "control" with "emergency lighting control."	FR-8852
700.17	Revised for clarity around monitoring.	FR-8853
700.24	Revised to clarify the means by which a directly controlled luminaire can be energized to an emergency lighting level.	FR-8857
700.27	Added section to address Class 2 powered emergency lighting systems.	FR-8856
700.32	Revised section to clarify existing requirements where emergency standby system OCPDs are replaced, modified, added, or deleted to ensure selective coordination is maintained.	FR-9113
701.1	Revised section to add two Informational Notes and a reference to NFPA 110.	FR-9623
701.2	Added section covering reconditioned equipment.	SR-8009
701.3	Revised to add "commissioning" and remove "testing."	FR-8867, SR-8005
701.4(B)	Revised to require the system be sized to accommodate rapid load changes, such as large motor startup and corrected cross references.	FR-8869, SR-8006
701.4(C)	Revised title to "Load Management" and removed references to peak load shaving.	FR-8875, SR-8007
701.4(D)	Added section for parallel operation and Informational Note for peak load shaving.	FR-8874
701.5(C)	Revised to remove reconditioning statement from text as it is redundant with 701.2.	SR-8008

Section	Synopsis	References
701.6(C)	Revised section title to "Battery Charging Malfunction."	FR-8877
701.10	Revised text into a list format to add requirements around loads of legally required standby systems and correlate with 700.10(B)(5)(b).	FR-8881
701.12(C)	Revised title to "Supply Duration", relocated 701.12(D)(2) and (3) to correlate with changes in (C), and added new Informational Note to reference NFPA 110.	FR-8967, SR-8012
701.12(D)(1)	Revised to replace "service" with "power source," which allows other types of sources for the normal power.	FR-8965
701.12(D)(2) and (3)	Relocated to 701.12(C)(2) and (3), revised for clarity, and renumbered accordingly.	FR-8967
701.12(E)	Revised title to "Stored-Energy Power Supply Systems (SEPSS)," revised text into list format to add currently available technologies that can be used with SEPSS, and added Informational Note referencing NFPA 853 and 855.	FR-8886, SR-8015
701.12(F)	Revised text to correlate with 700.12(F).	FR-8889
701.12(H)	Relocated from former 701.12(I), revised title to "Microgrid Systems," and removed dc specific references throughout the section.	FR-8969, FR-8891
701.12(I)	Relocated from former 701.12(J), revised to correlate with other sections, and revised title to "Battery-Equipped Emergency Luminaires, Used for Legally Required Standby Systems."	FR-8893
701.32	Revised text into subsections and modified OCPD language to bring awareness to both upstream and downstream applications.	FR-8895
702.1	Relocated Informational Note from 702.2 to 702.1.	FR-9618
702.2	Added section for "Reconditioned Equipment" and to state that it is not permitted.	SR-8024
702.4(A)	Deleted former section 702.4(A) due to redundancy with 110.9 and 110.10 and renumbered sections accordingly.	SR-8020
702.4(A)	Revised text for clarity, revised subsection titles, and added Informational Note.	FR-8898, SR-8021
702.5(A)	Revised to add references to interconnection and removed redundant reconditioning requirements now covered in 702.2.	FR-9614
702.5(B)	Revised title to "Meter-Mounted Transfer Switch" and consolidated text into single section.	SR-8112, SR-8115
702.7(A)	Revised to add requirement for signage to be placed at other than one- and two-family dwellings and corrected cross reference.	FR-8904
702.12(B)	Revised to require flanged inlets and other cord and plug connections to be installed outside of the building or structure.	FR-8907
705.1	Added Informational Note No. 2 and diagrams.	FR-9594
705.1	Revised Informational Note Figure 705.1 to align with terms used in Article 705 and added Informational Note No. 2.	FR-9594, SR-8359
705.5	Relocated from former 705.14, revised text for clarity, and revised title to "Parallel Operation."	FR-9392, SR-8376
705.6	Restructured section for clarity, added Informational Note No. 1 referencing UL 1741, and added Informational Note No. 2 for equipment examples.	FR-9374, SR-8363
705.11	Revised section title to "Source Connections to a Service" and extensively revised subsections around conductor ampacity, disconnecting means, bonding and grounding, and overcurrent protection.	FR-9380, SR-8375
705.12	Revised and reorganized section under new subsections "Feeders and Feeder Taps" and "Busbars."	FR-9389, SR-8364
705.13	Revised section title to "Energy Management Systems (EMS)," condensed into one section, and revised cross reference.	FR-9391, SCR-72
705.16	Deleted section due to unenforceable language and redundancy with 110.9 and 110.10.	FR-9394
705.20	Revised text for specific source disconnecting means, revised cross references, and added Informational Note to reference other applicable code sections.	FR-9395, SCR-73
705.25	Revised text to clarify that section only applies to power source output circuit conductors.	FR-9397, SR-8383

Section	Synopsis	References
705.28	Revised text into list format with revised title to "Power Source Output Maximum Current."	FR-9399, SR-8388
705.28	Revised to add three exceptions for clarity.	FR-9399, SR-8388
705.30	Relocated sections from former 705.12(C), (D), and (E), deleted former 705.30(G), and revised 705.30(F) "Transformers" installation requirements into list format.	FR-9402, SR-8397
705.32	Revised text for ground-fault protection to make it "ac circuit" specific for clarity and alignment with other sections.	FR-9404, SR-8398
705.40	Revised to align with section title being specific to primary source and added word "interactive" for clarity.	FR-9407, SR-8399
705.50	Revised text to better reflect how microgrids operate and added Informational Notes regarding microgrids using multiple sources and for Article 517 for health care facilities.	FR-9411, SR-8403
705.76	Added section for microgrid control systems (MCS) that correlates with the current state of the industry and available standards.	FR-9507, SR-8407
705.80	Added section covering interconnected power production sources that are operating in island mode.	FR-9414, SR-8408
705.81	Added section requiring power sources in island mode be controlled.	FR-9414, SR-8409
705.81	Added section on 120-volt supplies from 710.15(C).	FR-9414.
706.1	Revised text and order of Informational Notes and revised Informational Note No. 1 to align with the scope of Article 480.	FR-9074, SR-8116
706.6	Revised to incorporate applications for outdoor systems and microgrids where there may be multiple ESS on same premises.	FR-9078
706.7	Revised title to add "Commissioning" and revised text into subsections. (A) requires ESS to be commissioned upon installation.	FR-9079
706.7(A)	Revised to require ESS to be commissioned upon installation in other than one- and two-family dwellings.	FR-9079, SR-8086
706.15	Revised text to be able to disconnect ESS from all other wiring and comply with emergency shutdown requirements for one- and two-family dwellings.	FR-9086
706.16(E)	Revised text to correlate with the deletion of Article 712.	SR-8467
706.20(A)	Revised Informational Note No. 1 to include reference to NFPA 855.	FR-9089
706.31(A)	Revised text to align with similar requirements in Article 690 and added Informational Note regarding integral fault protection.	FR-9090, SR-8091
706, Part V	Revised title to use acronym and revised Informational Note to reflect Part V of Article 706.	FR-9617
706.40	Revised to remove reference to Article 692 when dealing with flow batteries and added Informational Note referencing NFPA 855 for installation requirements of flow batteries.	FR-9133
706.50	Revised to remove reference to Part III of Article 705 and added requirement to comply with Parts I, II, III, and IV of Article 706.	FR-9134, SR-8092
706.51	Added section for flywheel ESS (FESS).	FR-9196
708.2	Added section for reconditioned equipment.	SR-8034
708.7	Added section for cybersecurity with COPS.	FR-8914
708.8(A)	Added Informational Note No. 2 referencing 708.7 for cybersecurity assessments.	FR-8917
708.14	Revised text to clarify that a 2-hour listed fire-resistive cable system or a 2-hour electrical circuit protected system meets the requirement.	FR-8931
708.20(F)(1)	Revised to replace "service" with "power source".	FR-8937
708.22(B)	Revised title to remove listed functions and place them into enforceable text.	SR-8037
708.24(A)	Revised to add listing requirements for transfer switches and to clarify cross reference.	FR-8939, SR-8040
708.24(D)	Revised title to "Redundant Transfer Equipment" and revised text to describe the functionality that is needed when emergency loads are supplied by a single feeder.	SR-8041

Section	Synopsis	References
708.54	Added section to correlate with 700.32 on selective coordination to address requirements where critical operations power system OCPDs are replaced, modified, added, or deleted.	FR-8944
710.1	Revised for systems in island mode not connected to utility and deleted Informational Note No. 2 and Informational Note Figure 710.1.	FR-9420, SR-8417
710.6	Revised for clarity and deleted Informational Note.	SCR-74
710.10	Revisions for clarity, to correct cross reference, and to delete exception.	FR-9422
710.15(F)	Deleted former section 710.15(F) as it was redundant with 408.36(D), renumbered rest of section accordingly, and revised text to cover "power sources" for clarity.	FR-9424, FR-9425, SR-8423
Article 712	Deleted Article 712, Direct Current Microgrids, due to redundancy with Article 705 requirements.	SR-8074
Article 720	Deleted Article 720 to remove confusion around applicability of the article throughout the code as it was based on antiquated technology.	FR-9580
Article 722	Created new Article 722, Cables for Power-Limited Circuits and Fault-Managed Power Circuits, covering Class 2 and Class 3 power-limited circuits, power-limited fire alarm (PLFA) circuits, and Class 4 fault-managed power circuits. Common cabling requirements formerly in Articles 725 and 760 have been relocated into new Article 722.	FR-9582, CC-8380
Article 724	Created new Article 724, Class 1 Power-Limited Circuits and Class 1 Power-Limited Remote-Control and Signaling Circuits, covering Class 1 circuits that are not a part of a device or utilization equipment.	FR-9591
Article 725	Revised article title to "Class 2 and Class 3 Power-Limited Circuits."	
725.1	Revised to remove Class I items to correlate with new Article 724, to clarify what is covered, and to add Informational Note 2.	FR-9562, SR-8499
725.3(A)	Relocated former 723.3(A) to new Article 724 renumbered sections accordingly.	FR-9563
725.3(C)	Revised for clarity and to correct cross references.	FR-9575
725.3(E)	Added section on the listing and installation of cables for Class 2 and Class 3 circuits and to correlate with new Article 722.	FR-9621, SR-8473
725.10	Added section covering hazardous (classified) location requirements.	SCR-99
725.24	Revised to add proper cross references and to remove items now covered in Article 722.	SR-8717
725.31	Revised to align Class I items with new Article 724.	FR-9440
725, Part II	Deleted former Part II referencing Class 1 circuits due to creation of new Article 724 and renumbered former Part III as new Part II.	FR-9602
724.60	Relocated section from former 725.121.	SR-8478
725.60(A)	Revised section and Informational Note Figure 725.60 to remove items now covered in Article 722.	SR-8478
725.60(C)	Revised to remove effective date that has passed.	FR-9626, FR-9446
725.127	Revised to remove 20 ampere transformer limitation due to developments in PoE power sources that would exceed that.	FR-9447
725.130	Revised text to correct cross references.	SR-8497
725.130(A)	Revised to permit Class 1 wiring methods with Class 2 and Class 3 circuits.	FR-9449
725.130(B)	Revised title to include "Materials" and revised text to add references to new Article 722 and to correct Informational Note cross reference.	FR-9450
725.135	Deleted former section due to redundancy with Article 722 requirements.	FR-9582
725.144	Revised title to reflect bundling, revised text for alignment with 840.160 and to clarify that 725.144(A) is limited to 4-pair cables.	FR-9458, SR-8479
725.144(B)	Revised to reference section in new Article 722.	FR-9543
725, Part III	Relocated and renumbered former Part IV as new Part III, Listing Requirements.	SR-8477
Article 726	Created new Article 726, Class 4 Fault-Managed Power Systems, for wiring systems and equipment including utilization equipment of Class 4 fault-managed power (FMP) systems.	
Article 727	Deleted former Article 727 and relocated requirements to new Article 335.	FCR-457

Section	Synopsis	References
728.4	Revised text for clarity and deleted Informational Note No. 2.	FR-9477
728.5(B) through (E)	Revised to replace "fire-resistive system" with "fire-resistive cable system" for clarity.	FR-9481, FR-9482
728.5(F)	Revised to delete "systems."	FR-9482
728.5(G)	Revised to add a reference to 300.19.	FR-9483
728.60	Revised to replace "systems" with "cables" and "rated" with "resistive."	FR-9486
750.6	Added section to include listing requirements.	FR-9119, SR-8069
750.30(C)	Revised to add subsections consolidated from other sections.	SR-8070
750.50	Revised to change section title from "Field Marking" to "Directory" and deleted Informational Note.	FR-9120
760.1	Revised Informational Note No. 1 and No. 2	FR-9583, SR-8488
760.3(O)	Added section referencing Part III of Article 760 and Parts I and II of Article 722 for the listing and installation of cable for power-limited fire alarm circuits.	SR-8693
760.10	Added section for hazardous (classified) location requirements.	SCR-104
760.24	Revised to include additional means of support and reference to 300.11.	FR-9523, SR-8594
760.32	Revised to delete Informational Note.	FR-9526
760.33	Added section to require a listed surge-protective device to be installed on the supply side of a fire alarm control panel.	FR-9524
760.41(B)	Revised to delete Informational Note.	SCR-105
760.121(A)	Revised to delete Informational Note.	SCR-106
760.121(B)	Revised into list items and to add a new Informational Note.	FR-9529
760.124	Revised to delete Informational Note.	SR-8595
760.130(A)	Revised to permit some NPLFA wiring methods for PLFA circuits.	FR-9552, SR-8695
760.130(B)	Revised to add subsections for nonconcealed spaces and portable fire alarm systems.	FR-9552, SR-8695
760.135	Deleted former section due to redundancy with Article 722 requirements.	SR-8492
760.136(G)	Added section that permits PLFA circuits to be installed with other circuit conductors when installed using NPLFA wiring methods in accordance with Part II of Article 760 and are protected by an approved method.	FR-9534
760.139(C)	Added section addressing Class 3 and communications circuits installed with PLFA circuits.	FR-9535
760.139(E)	Revised to remove former references and add reference to new 722.135.	SR-8596
760.176	Revised to permit NPLFA cable to contain optical fibers and added an Informational Note.	FR-9538, FR-9536, FR-9553
760.179	Revised to remove former reference to 725.179(J), add references to 725.179(A) through (D), and to add subsections covering listing, voltage and temperature rating, markings, and cable jacket compound.	FR-9541, SR-8602
770.3(C)	Revised to replace "composite" with "hybrid."	FR-8812, SCR-109
770.3(D)	Added section for vertical support of fire-resistive cables.	FR-8812, SCR-109
770.24	Revised to require cable ties to be listed and identified for securement and support and to add a subsection for circuit integrity (CI) cable.	FR-8814, SR-7891
770.27	Added section regarding temperature limitation of optical fiber cables.	FR-8815, SR-7898
770.48(A)	Revised to include "roof" as a potential point of entrance penetration point.	SR-7940

Section	Synopsis	References
770.100(B)(2)	Relocated interior metal water piping from former list item (2) to new list item (7) and renumbered as appropriate.	FR-8885
770.113(B)	Revised to add "Uses Permitted" and "Uses Not Permitted" subsections.	FR-9067
770.114	Revised to add reference to 770.100(A).	FR-8982
770.133	Revised to replace "composite cable" with "hybrid cable" and to replace "composite optical" with "hybrid optical."	FR-8984
770.179	Revised to add Informational Note referencing UL 1651-2015.	FR-8985
800.1	Revised Informational Note No. 2 to add reference to new Article 722.	SR-7919
800.2	Relocated from former 800.3(G).	SCR-86
800.3	Revised parent text for clarity, revised 800.3(A) list item (1) to add reference to Article 722, relocated former 800.3(G) to 800.2, added new 800.3(G) for vertical support for fire-resistive cables and conductors, and added new (I) for bonding and grounding of cable shields.	FR-8800, SR-7920
800.24	Revised to clarify that cable ties can be used for securement and support. and added 800.24(B) for circuit integrity.	FR-8806
800.44	Revised for consolidation of common requirements from Articles 805, 820, 830 and 840.	FR-8842
800.47	Added section for consolidation of common requirements from Articles 805, 820, 830 and 840.	FR-8845
800.48	Added section for consolidation of common requirements from Articles 805, 820, 830 and 840 and revised text to correlate with the definition of *point of entrance*.	FR-8847, SR-7943
800.53	Revised for clarity and for correlation with other related sections.	FR-8849
800.100(A)(3)	Revised to include all metal members.	FR-8890
800.100(A)(4)	Revised Informational Note to focus on reducing lightning damage.	FR-8892
800.100(B)	Revised figure titles of Informational Note Figures 800.100(B)(1) and 800.100(B)(2) to better to differentiate between the two.	FR-8899, FR-9023
800.100(B)(2)	Revised list item (7) to reflect the increased use of nonmetallic water piping systems within buildings and to de-emphasize their use as a grounding connection for communications systems.	FR-8897
800.100(B)(3)	Revised to prohibit steam and hot water pipes as bonding or grounding electrode conductors.	FR-8901
800.106(A)	Revised subsections into list format for clarity.	FR-8903, FR-8906
800.113	Revised for clarity and restructured into "Uses Permitted" and "Uses Not Permitted" subsections for correlation with other sections.	FR-9071, SR-7925
800.133	Added section to consolidate common requirements from 805.133 and 820.133.	FR-8964
800, Part V	Relocated Part V to the head of the listing section.	FR-8960
800.171	Relocated from 805.170 and 840.170 and revised around listing of communications equipment.	SR-8158
800.170	Revised to add listing requirements for plenum cable ties.	FR-9021
800.179	Revised to correct the omission of requirements, to update section title, to correlate with changes in other Chapter 8 articles, and to add and renumber subsections.	FR-9036, SR-7988 SR-7976, SR-7977, SR-7978
805.47	Revised for consolidation of common requirements from other articles into one section.	FR-8850
805.48	Deleted section as part of the consolidation of common requirements from other articles into one section.	FR-8851
805.133	Relocated with 830.133 and consolidated into 800.133.	FR-8966
805.154	Revised for clarity and to correlate with changes in other Chapter 8 articles.	FR-8968
805.170	Revised to relocate listing requirements from this section to Article 800 and to revise section title for accuracy.	FR-9005, SR-7979
805.170(A)	Revised to add listing requirement to primary protector.	FR-9087

Section	Synopsis	References
805.170(C)	Relocated to 800.170.	FR-8837
805.179	Relocated to 800.179.	SR-7980
Article 810	Revised title of article to Antenna Systems.	FR-8952
810.3	Revised to add list item (3) for Chapter 5 requirements.	FR-8912
810.4	Revised for clarity and to reference to the appropriate article of Chapter 8.	FR-8918
810.14	Revised to reference 110.14 (A) and (B).	FR-8925
810.15	Revised title to add "Bonding" and revised text to add "bonded."	FR-8927
810.20(A)	Revised title to "General Requirements," revised exception text into list format, and added "bonded."	FR-8936
810.20(C)	Revised title to add "Bonding" and revised text to add "bonded."	FR-8941
810.21(F)(2)	Revised text in (6) to reflect the increased use of nonmetallic water piping systems within buildings to de-emphasize their use as a grounding connection for antenna systems.	FR-8945
810.52	Revised title to add "conductors."	FR-8948
810.57	Revised for clarity and added "bonded" to within text of Exception No. 2.	FR-8950
820.3	Revised to consolidate text and deleted former 820.3(A) due to redundancy.	SR-7933
820, Part II	Deleted Part II to correlate with changes to other Chapter 8 articles.	FR-8862
820.44	Relocated former 820.44 to 800.44.	FR-8859
820.47	Relocated former 820.47 to 800.47.	FR-8860
820.48	Relocated former 820.48 to 800.48.	FR-8861
820.93(D)	Revised exception to correlate with Chapter 5 sections and the general requirements of Article 800.	SR-7981
820.133	Relocated former 820.133 to 800.133.	FR-8971
820.154	Revised title of Figure 820.154 to add "Coaxial."	FR-8972
820, Part VI	Deleted Part VI to correlate with changes made to other Chapter 8 articles.	SR-7982
820.179	Relocated former 820.179 to 800.179.	FR-9061, SR-7982
830.1	Revised Informational Note to replace "composite" with "hybrid" for optical fiber cable.	FR-8828
830.3	Deleted section due to redundancy with 800.3.	FR-8831, SR-7935
830.24	Deleted section due to redundancy with 800.24.	FR-8833
830.40(B)	Deleted exception as it is not relevant to current installations.	FR-9159
830.47	Revised to correlate with consolidations into Article 800, added note to Table 830.47(A), and removed former note 4.	FR-8866, FR-9026
830.160	Revised to include requirements for bending radius and to add Informational Note.	FR-8976
830.179(D)	Relocated former 830.179(D) to 800.179.	FR-9011, FR-9013
840.1	Revised to remove the term "electrical" from the Informational Note since the signal could be optical.	FR-8816
840.3	Deleted section due to redundancy with 800.3.	SR-7945
840.24, 840.25, 840.26	Deleted sections due to redundancy with Article 800.	FR-8822, FR-8824, FR-8825
840.47	Revised parent text and deleted rest of section based on previous consolidations into Article 800.	FR-8870
840.48	Deleted section based on previous consolidations into Article 800.	FR-8871
840, Part V	Deleted Part V as it is covered under 840.102 in Part IV.	FR-8979
840.133	Deleted section due to redundancy with 800.3.	FR-8977

Section	Synopsis	References
840.160	Revised text to relocate listing requirements for cables to Article 800 and added new Informational Note.	FR-8980, SR-7946
840.170	Revised section to delete redundant requirements that are now in Article 800 and corrected cross references.	SR-7984
Chapter 9, Table 1	Revised text to restricting nipple length to 24 inches without connectors.	FR-7580
Chapter 9, Table 4	Added rows and information to Table 4 for 5-inch and 6-inch EMT.	FR-7626
Chapter 9, Table 4 (Article 342)	Revised to correlate with additional trade sizes for IMC, RMC, and EMT.	SR-8183
Chapter 9, Table 13	Added new table to address protection techniques associated with hazardous location wiring, deleted "v" from level of protection, and corrected cross references in table notes.	FR-8648, SR-7698
Annex A	Expanded and updated Annex A product standard information for NEC requirements specifying the use of listed/certified electrical equipment.	FR-8330, FR-8327, FR-8328, FR-8326, FR-8323, FR-8321, SCR-96
Annex C, Tables C.1 and C.1(A)	Revised to add information into the columns for 5-inch and 6-inch EMT.	FR-7842, FR-7843
Annex C, Table C.14	Revised to correct sections and table numbers for multiconductor MC cable.	FR-7644
Annex C, Table C.8	Revised to add information into the columns for 5-inch and 6-inch IMC.	SR-8190
Annex D, Example D1(b)	Revised to replace "largest motor" with "air-conditioner" and corrected cross reference.	FR-9184
Annex D, Example D3	Revised to correlate with changes in TIA-20-6 from "2-wire, 120/240 V" to "2-wire, 120 V".	FR-9187
Annex D, Examples D5(a) and D5(b)	Removed parentheses under calculations to ensure answer is correct based on mathematical order of operations.	FR-9186, FR-9185
Annex E	Revised text in Type IV construction to address mass and heavy timber construction.	SR-8483
Annex K	Created Informative Annex K to address the increased use of medical equipment in dwellings and residential board-and-care occupancies.	FR-8333, SR-7543

REVISION SYMBOLS IDENTIFYING CHANGES FROM THE PREVIOUS EDITION

Text revisions are shaded. A Δ before a section number indicates that words within that section were deleted and a Δ to the left of a table or figure number indicates a revision to an existing table or figure. When a chapter was heavily revised, the entire chapter is marked throughout with the Δ symbol. Where one or more sections were deleted, a ● is placed between the remaining sections. Chapters, annexes, sections, figures, and tables that are new are indicated with an **N**.

Note that these indicators are a guide. Rearrangement of sections may not be captured in the markup, but users can view complete revision details in the First and Second Draft Reports located in the archived revision information section of each code at www.nfpa.org/docinfo. Any subsequent changes from the NFPA Technical Meeting, Tentative Interim Amendments, and Errata are also located there.

Shaded text = Revisions   Δ = Text deletions and figure/table revisions   ● = Section deletions   **N** = New material

Introduction

ARTICLE
90 Introduction

N 90.1 Scope. This article covers use and application, arrangement, and enforcement of this *Code*. It also covers the expression of mandatory, permissive, and nonmandatory text, provides guidance on the examination of equipment and on wiring planning, and specifies the use and expression of measurements.

90.2 Use and Application.

(A) Practical Safeguarding. The purpose of this *Code* is the practical safeguarding of persons and property from hazards arising from the use of electricity. This *Code* is not intended as a design specification or an instruction manual for untrained persons.

The *National Electrical Code®* (*NEC®*) is prepared by the National Electrical Code Committee, which consists of a correlating committee and 18 code-making panels. The code-making panels have specific technical responsibilities. The scope of the committee is as follows:

> This committee shall have primary responsibility for documents on minimizing the risk of electricity as a source of electric shock and as a potential ignition source of fires and explosions. It shall also be responsible for text to minimize the propagation of fire and explosions due to electrical installations.

> In addition to its overall responsibility for the *NEC*, the Correlating Committee on National Electrical Code is responsible for correlation of the following documents:

1. NFPA 70B, *Recommended Practice for Electrical Equipment Maintenance*
2. NFPA 70E®, *Standard for Electrical Safety in the Workplace®*
3. NFPA 73, *Standard for Electrical Inspections for Existing Dwellings*
4. NFPA 79, *Electrical Standard for Industrial Machinery*
5. NFPA 110, *Standard for Emergency and Standby Power Systems*
6. NFPA 111, *Standard on Stored Electrical Energy Emergency and Standby Power Systems*
7. NFPA 790, *Standard for Competency of Third-Party Field Evaluation Bodies*
8. NFPA 791, *Recommended Practice and Procedures for Unlabeled Electrical Equipment Evaluation*

(B) Adequacy. This *Code* contains provisions that are considered necessary for safety. Compliance therewith and proper maintenance result in an installation that is essentially free from hazard but not necessarily efficient, convenient, or adequate for good service or future expansion of electrical use.

Informational Note: Hazards often occur because of overloading of wiring systems by methods or usage not in conformity with this *Code*. This occurs because initial wiring did not provide for increases in the use of electricity. An initial adequate installation and reasonable provisions for system changes provide for future increases in the use of electricity.

(C) Installations Covered. This *Code* covers the installation and removal of electrical conductors, equipment, and raceways; signaling and communications conductors, equipment, and raceways; and optical fiber cables for the following:

(1) Public and private premises, including buildings, structures, mobile homes, recreational vehicles, and floating buildings
(2) Yards, lots, parking lots, carnivals, and industrial substations
(3) Installations of conductors and equipment that connect to the supply of electricity
(4) Installations used by the electric utility, such as office buildings, warehouses, garages, machine shops, and recreational buildings, that are not an integral part of a generating plant, substation, or control center
(5) Installations supplying shore power to ships and watercraft in marinas and boatyards, including monitoring of leakage current
(6) Installations used to export electric power from vehicles to premises wiring or for bidirectional current flow

Often, the source of supply is the serving electric utility. The point of connection from a premises wiring system to a serving electric utility system is referred to as the *service point*. The conductors on the premises side of the service point are referred to as *service conductors*. See Article 100 for definitions.

The source might be a stand-alone system, such as a generator, a battery system, a photovoltaic system, a fuel cell, a wind turbine, an electric vehicle, or a combination of those sources. Conductors from stand-alone systems are not service conductors, they are feeders. Service conductors are supplied only by a utility source.

See also

Article 230 for service conductor and service equipment requirements if the only source of supply of electricity is from a utility

Article 625 if the source is an electric vehicle arranged for bidirectional current flow

Article 705 if the source of supply includes a utility source(s) in combination with alternate energy sources

Exhibit 90.1 illustrates the distinction between electric utility facilities subject to the *NEC* and those not subject to the *NEC*. The electrical equipment in the generating plant is not governed by the rules of the *NEC*. Office buildings and warehouses of electric utilities are functionally like the facilities owned by other commercial entities. Although the warehouse is owned by the utility, it is a typical commercial facility in which the electrical installation would be governed by the rules of the *NEC*.

Industrial and multibuilding complexes and campus-style complexes often include substations and other installations that employ construction and wiring similar to those of electric utility installations. Because these installations are on the load side of the service point, they are within the purview of the *NEC*. At an increasing number of industrial, institutional, and other campus-style distribution systems, the service point is at an owner-maintained substation, and the conductors extending from that substation to the campus facilities are *feeders* (see Article 100). *NEC* requirements cover these distribution systems in 235.360 and 235.361 and in Article 395. The overhead conductor and live parts clearance requirements in the *NEC* correlate with those in ANSI C2, *National Electrical Safety Code® (NESC®)*, for overhead conductors under the control of an electric utility.

The *NEC* also addresses removal of equipment, such as abandoned conductors. Abandoned conductors and cables from several generations of equipment can increase the fire loading in a building. Abandoned cable is commonly found where communications systems or computer networks have been upgraded. Several places in the *NEC* specify where abandoned cable is required to be removed.

(D) Installations Not Covered. This *Code* does not cover the following:

(1) Installations in ships, watercraft other than floating buildings, railway rolling stock, aircraft, or automotive vehicles other than mobile homes and recreational vehicles

EXHIBIT 90.1 A conventional generating plant, which is governed by the National Electrical Safety Code (*top*); a support facility that is subject to *NEC* requirements (*bottom*).

Informational Note: Although the scope of this *Code* indicates that the *Code* does not cover installations in ships, portions of this *Code* are incorporated by reference into Title 46, Code of Federal Regulations, Parts 110–113.

(2) Installations underground in mines and self-propelled mobile surface mining machinery and its attendant electrical trailing cable

(3) Installations of railways for generation, transformation, transmission, energy storage, or distribution of power used exclusively for operation of rolling stock or installations used exclusively for signaling and communications purposes

(4) Installations of communications equipment under the exclusive control of communications utilities located outdoors or in building spaces used exclusively for such installations

(5) Installations under the exclusive control of an electric utility where such installations

a. Consist of service drops or service laterals, and associated metering, or

b. Are on property owned or leased by the electric utility for the purpose of communications, metering, generation, control, transformation, transmission, energy storage, or distribution of electric energy, or

c. Are located in legally established easements or rights-of-way, or

d. Are located by other written agreements either designated by or recognized by public service commissions, utility commissions, or other regulatory agencies having jurisdiction for such installations. These written agreements shall be limited to installations for the purpose of communications, metering, generation, control, transformation, transmission, energy storage, or distribution of electric energy where legally established easements or rights-of-way cannot be obtained. These installations shall be limited to federal lands, Native American reservations through the U.S. Department of the Interior Bureau of Indian Affairs, military bases, lands controlled by port authorities and state agencies and departments, and lands owned by railroads.

Informational Note to (4) and (5): Examples of utilities may include those entities that are typically designated or recognized by governmental law or regulation by public service/utility commissions and that install, operate, and maintain electric supply (such as generation, transmission, or distribution systems) or communications systems (such as telephone, CATV, Internet, satellite, or data services). Utilities may be subject to compliance with codes and standards covering their regulated activities as adopted under governmental law or regulation. Additional information can be found through consultation with the appropriate governmental bodies, such as state regulatory commissions, the Federal Energy Regulatory Commission, and the Federal Communications Commission.

Section 90.2(B)(5) is not intended to prevent the *NEC* from being used as an installation regulatory document for these types of installations. The *NEC* is fully capable of being utilized for electrical installations in most cases. Rather, 90.2(B)(5) lists specific areas where the nature of the installation requires specialized rules or where other installation rules, standards, and guidelines have been developed for specific uses and industries. For example, the electric utility industry uses ANSI C2, *National Electrical Safety Code (NESC)*, as its primary requirement in the generation, transmission, distribution, and metering of electric energy. See Exhibit 90.1 for examples of electric utility facilities covered or not covered by the *NEC*. In most cases, utility-owned installations are on legally established easements or rights of way. Easements or rights of way might not be available on federally owned lands, Native American reservations, military bases, lands controlled by port authorities or state agencies or departments, and lands owned by railroads. In those limited applications, a written agreement complying with this section can be used to establish the extent of the utility installation.

Utility installations of energy storage are outside the scope of the *NEC*. Utility-owned energy storage would be covered by the *NESC*. Energy storage that is not under the exclusive control of an electric utility is governed by Article 706 of the *NEC*.

Δ **(E) Relation to Other International Standards.** The requirements in this *Code* address the fundamental principles of protection for safety contained in Section 131 of International Electrotechnical Commission Standard 60364-1, *Low-voltage Electrical Installations – Part 1: Fundamental Principles, Assessment of General Characteristics, Definitions*.

Informational Note: See IEC 60364-1, *Low-voltage Electrical Installations – Part 1: Fundamental Principles, Assessment of General Characteristics, Definitions,* Section 131, for fundamental principles of protection for safety that encompass protection against electric shock, protection against thermal effects, protection against overcurrent, protection against fault currents, and protection against overvoltage. All of these potential hazards are addressed by the requirements in this *Code*.

In addition to being an essential part of the safety system of the Americas and the most widely adopted code for the built environment in the United States, the *NEC* has been adopted and is used extensively in many other countries. The *NEC* is compatible with international safety principles, and installations meeting the requirements of the *NEC* are also in compliance with the fundamental principles outlined in International Electrotechnical Commission (IEC) 60364-1, *Electrical Installations of Buildings*, Section 131. Countries that do not have formalized rules for electrical installations can adopt the *NEC* and be fully compatible with the safety principles of IEC 60364-1, Section 131.

(F) Special Permission. The authority having jurisdiction for enforcing this *Code* may grant exception for the installation of conductors and equipment that are not under the exclusive control of the electric utilities and are used to connect the electric utility supply system to the service conductors of the premises served, provided such installations are outside a building or structure, or terminate inside at a readily accessible location nearest the point of entrance of the service conductors.

90.3 Code Arrangement. This *Code* is divided into the introduction and nine chapters, as shown in Figure 90.3. Chapters 1, 2, 3, and 4 apply generally. Chapters 5, 6, and 7 apply to special occupancies, special equipment, or other special conditions and may supplement or modify the requirements in Chapters 1 through 7.

Chapter 8 covers communications systems and is not subject to the requirements of Chapters 1 through 7 except where the requirements are specifically referenced in Chapter 8.

Chapter 9 consists of tables that are applicable as referenced.

Informative annexes are not part of the requirements of this *Code* but are included for informational purposes only.

An example of how the general rules of Chapter 3 are modified is 300.22, which is modified by 725.3(B) and 760.3(B) and is specifically referenced in 800.3(C) and (D). Figure 90.3 is a graphic explanation of the *NEC* arrangement.

The requirements in Chapters 5 through 7 can modify the requirements in Chapters 1 through 7. Prior to the 2017 edition, Chapters 5 through 7 could modify only the requirements in Chapters 1 through 4.

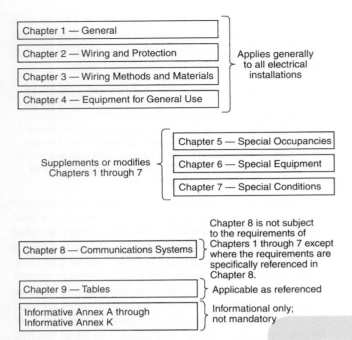

FIGURE 90.3 *Code Arrangement.*

△ **90.4 Enforcement.**

N **(A) Application.** This *Code* is intended to be suitable for mandatory application by governmental bodies that exercise legal jurisdiction over electrical installations, including signaling and communications systems, and for use by insurance inspectors.

All materials and equipment used under the requirements of the *Code* are subject to the approval of the AHJ. Sections 90.7, 110.2, and 110.3 — along with the definitions of the terms *approved*, *identified (as applied to equipment)*, *labeled*, and *listed* — are intended to provide a basis for the AHJ to make judgments about the approval of an installation. The phrase "including signaling and communications systems" emphasizes that those systems are also subject to enforcement.

Some localities do not adopt the most current version of the *NEC*, but in those that do there is sufficient evidence that compliance with the current edition serves as a reference for a safe installation.

N **(B) Interpretations.** The authority having jurisdiction for enforcement of the Code has the responsibility for making interpretations of the rules, for deciding on the approval of equipment and materials, and for granting the special permission contemplated in a number of the rules.

N **(C) Specific Requirements and Alternative Methods.** By special permission, the authority having jurisdiction may waive specific requirements in this *Code* or permit alternative methods where it is assured that equivalent objectives can be achieved by establishing and maintaining effective safety.

The AHJ is responsible for interpreting the *NEC*. Using special permission (written consent), the AHJ may permit alternative methods where specific rules are not established in the *NEC*. For example, the AHJ may waive specific requirements in industrial occupancies, research and testing laboratories, and other occupancies where the specific type of installation is not covered in the *NEC*.

N **(D) New Products, Constructions, or Materials.** This *Code* may require new products, constructions, or materials that may not yet be available at the time the *Code* is adopted. In such event, the authority having jurisdiction may permit the use of the products, constructions, or materials that comply with the most recent previous edition of this *Code* adopted by the jurisdiction.

> Informational Note: See Informative Annex H, Administration and Enforcement, for a model of guidelines that can be used to create an electrical inspection and enforcement program and to adopt *NFPA 70, National Electrical Code.*

The AHJ may waive a new *NEC* requirement during the interim period between acceptance of a new edition of the *NEC* and the availability of a new product, construction, or material redesigned to comply with the level of safety required by the latest edition. Establishing a viable future effective date in each section of the *NEC* is difficult because the time needed to change existing products and standards or to develop new materials and test methods usually is not known at the time the latest edition is adopted.

90.5 Mandatory Rules, Permissive Rules, and Explanatory Material.

(A) Mandatory Rules. Mandatory rules of this *Code* are those that identify actions that are specifically required or prohibited and are characterized by the use of the terms *shall* or *shall not*.

(B) Permissive Rules. Permissive rules of this *Code* are those that identify actions that are allowed but not required, are normally used to describe options or alternative methods, and are characterized by the use of the terms *shall be permitted* or *shall not be required*.

Permissive rules are options or alternative methods of achieving equivalent safety — they are not requirements. Permissive rules are often misinterpreted. For example, the frequently used permissive term *shall be permitted* can be mistaken for a requirement. Substituting "the inspector must allow [item A or method A]" for "[item A or method A] shall be permitted" generally clarifies the interpretation.

(C) Explanatory Material. Explanatory material, such as references to other standards, references to related sections of this *Code,* or information related to a *Code* rule, is included in this *Code* in the form of informational notes or an informative annex. Unless the standard reference includes a date, the reference is to be considered as the latest edition of the standard. Such notes are informational only and are not enforceable as requirements of this *Code*.

Brackets containing section references to another NFPA document are for informational purposes only and are provided as a guide to indicate the source of the extracted text. These bracketed references immediately follow the extracted text.

Several requirements in the *NEC* have been extracted from other NFPA codes and standards. Although *NEC* requirements based on extracted material are under the jurisdiction of the technical committee responsible for the document from which the material was extracted, 90.5(C) clarifies that the *NEC* requirements stand on their own as part of the *NEC*. The extracted material with bracketed references does not indicate that other NFPA documents are adopted through reference. Articles that make use of extracted text include Articles 500, 511, 514, 515, 516, 517, and 695.

The *NEC* contains numerous informational notes. Prior to the 2011 edition, these notes were referred to as *fine print notes*, or *FPNs*. They were renamed *informational notes* to clarify that they do not contain requirements, statements of intent, or recommendations. They present additional supplementary material that aids in the application of the requirement they follow. Because informational notes are not requirements of the *NEC*, they are not enforceable.

Footnotes to tables, although also in fine print, are not explanatory material unless they are identified as informational notes. Table footnotes are part of the tables and are necessary for proper use of the tables. Therefore, they are mandatory and enforceable.

Additional explanatory material is in the informative annexes. The term *informative annex* clarifies that the annexes contain additional information and do not contain recommendations or requirements.

> Informational Note: The format and language used in this *Code* follows guidelines established by NFPA and published in the *NEC Style Manual*. Copies of this manual can be obtained from NFPA.

The *National Electrical Code® Style Manual* can be accessed at nfpa.org/70 by selecting the Current & Prior Editions tab and scrolling down to Additional Information. It can also be downloaded from the NFPA website at nfpa.org/assets/files/AboutTheCodes/70/NEC_Style_Manual_2020.pdf.

(D) Informative Annexes. Nonmandatory information relative to the use of the *NEC* is provided in informative annexes. Informative annexes are not part of the enforceable requirements of the *NEC*, but are included for information purposes only.

90.6 Formal Interpretations. To promote uniformity of interpretation and application of this *Code*, formal interpretation procedures have been established and are found in the *Regulations Governing the Development of NFPA Standards*.

The procedures for Formal Interpretations of the *NEC* are outlined in Section 6 of the NFPA Regulations Governing the Development of NFPA Standards. These regulations are included in the *NFPA Standards Directory*, which is published annually and can be found at nfpa.org/regs.

Δ **90.7 Examination of Equipment for Safety.** For specific items of equipment and materials referred to in this *Code*, examinations for safety made under standard conditions provide a basis for approval where the record is made generally available through promulgation by organizations properly equipped and qualified for experimental testing, inspections of the run of goods at factories, and service-value determination through field inspections. This avoids the necessity for repetition of examinations by different examiners, frequently with inadequate facilities for such work, and the confusion that would result from conflicting reports on the suitability of devices and materials examined for a given purpose.

It is the intent of this *Code* that factory-installed internal wiring or the construction of equipment need not be inspected at the time of installation of the equipment, except to detect alterations or damage, if the equipment has been listed by a qualified electrical testing laboratory that is recognized as having the facilities described in the preceding paragraph and that requires suitability for installation in accordance with this *Code*. Suitability shall be determined by application of requirements that are compatible with this *Code*.

> Informational Note No. 1: See 110.3 for guidance on safety examinations.
> Informational Note No. 2: See Article 100 for definitions of *Listed* and *Reconditioned*.
> Informational Note No. 3: See Informative Annex A for a list of product safety standards that are compatible with this *Code*.

Testing laboratories, inspection agencies, and other organizations concerned with product evaluation publish lists of equipment and materials that have been tested and meet nationally recognized standards or that have been found suitable for use in a specified manner. The *NEC* does not contain detailed information on equipment or materials but refers to products as *listed*, *labeled*, or *identified*. See Article 100 for definitions. Many certification agencies also perform field evaluations of specific installations in order to render an opinion on compliance with *NEC* requirements along with any applicable product standards. NFPA 790, *Standard for Competency of Third-Party Field Evaluation Bodies*, provides requirements for the qualification and competence of a body performing field evaluations on electrical products and assemblies with electrical components. NFPA 791, *Recommended Practice and Procedures for Unlabeled Electrical Equipment Evaluation*, covers recommended procedures for evaluating unlabeled electrical equipment for compliance with nationally recognized standards and with any requirements of the AHJ.

NFPA does not approve, inspect, or certify any installations, procedures, equipment, or materials, nor does it approve or evaluate testing laboratories. In determining the acceptability of installations or procedures, equipment, or materials, the AHJ may base acceptance on compliance with NFPA or other appropriate standards. The AHJ may also refer to the listing or labeling practices of an organization concerned with product evaluations to help determine compliance with appropriate standards.

Informative Annex A contains a list of product safety standards used for product listing. The list includes only product safety standards for which a listing is required by the *NEC*. For example, 344.6 requires that rigid metal conduit, RMC, be listed. By using Informative Annex A, the user finds that the listing standard for rigid metal conduit is UL 6, *Electrical Rigid Metal Conduit — Steel*. Because associated conduit fittings are required to be listed, UL 514B, *Conduit, Tubing, and Cable Fittings*, is found in Informative Annex A also.

90.8 Wiring Planning.

(A) Future Expansion and Convenience. Plans and specifications that provide ample space in raceways, spare raceways, and additional spaces allow for future increases in electric power and communications circuits. Distribution centers located in readily accessible locations provide convenience and safety of operation.

Consideration should always be given to future expansion of the electrical system. Future expansion might be unlikely in some occupancies. In other occupancies, however, it might be cost effective to plan for future additions, alterations, or designs during initial installation.

The requirement for providing the exclusively dedicated equipment space mandated by 110.26(E) supports the intent of 90.8(A) regarding future increases in the use of electricity. Communications circuits are important to consider when planning for future needs. Electrical distribution centers should contain additional space and capacity for future additions and should be located for easy accessibility. In commercial and industrial facilities, a common and cost-effective practice is to purchase switchboards with additional capacity to accommodate future expansion, as shown in Exhibit 90.2.

(B) Number of Circuits in Enclosures. It is elsewhere provided in this *Code* that the number of circuits confined in a single enclosure be varyingly restricted. Limiting the number of circuits in a single enclosure minimizes the effects from a short circuit or ground fault.

90.9 Units of Measurement.

(A) Measurement System of Preference. For the purpose of this *Code*, metric units of measurement are in accordance with the modernized metric system known as the International System of Units (SI).

(B) Dual System of Units. SI units shall appear first, and inch-pound units shall immediately follow in parentheses. Conversion from inch-pound units to SI units shall be based on hard conversion except as provided in 90.9(C).

Exception: The tables located in Informative Annex C shall be permitted to list the trade sizes before SI units.

(C) Permitted Uses of Soft Conversion. The cases given in 90.9(C)(1) through (C)(4) shall not be required to use hard conversion and shall be permitted to use soft conversion.

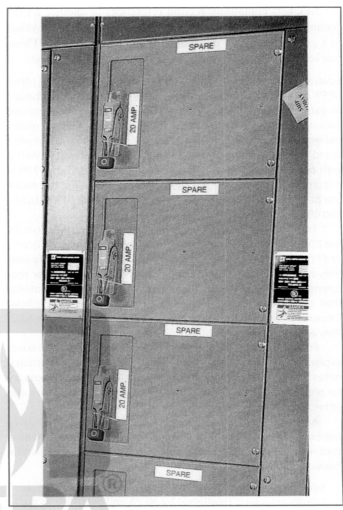

EXHIBIT 90.2 A switchboard with spare circuit breakers installed to facilitate future growth. *(Courtesy of the International Association of Electrical Inspectors)*

(1) Trade Sizes. Where the actual measured size of a product is not the same as the nominal size, trade size designators shall be used rather than dimensions. Trade practices shall be followed in all cases.

(2) Extracted Material. Where material is extracted from another standard, the context of the original material shall not be compromised or violated. Any editing of the extracted text shall be confined to making the style consistent with that of the *NEC*.

(3) Industry Practice. Where industry practice is to express units in inch-pound units, the inclusion of SI units shall not be required.

(4) Safety. Where a negative impact on safety would result, soft conversion shall be used.

(D) Compliance. Conversion from inch-pound units to SI units shall be permitted to be an approximate conversion. Compliance with the numbers shown in either the SI system or the inch-pound system shall constitute compliance with this *Code*.

> Informational Note No. 1: Hard conversion is considered a change in dimensions or properties of an item into new sizes that might or might not be interchangeable with the sizes used in the original measurement. Soft conversion is considered a direct mathematical conversion and involves a change in the description of an existing measurement but not in the actual dimension.
>
> Informational Note No. 2: SI conversions are based on IEEE/ASTM SI 10-1997, *Standard for the Use of the International System of Units (SI): The Modern Metric System*.

Soft conversions are numerically accurate. Hard conversions are rounded or approximate values of the customary unit dimension. Soft conversions are required when an approximation would negatively impact safety. Commentary Table 90.1 offers examples of the hard conversion process. The conversion of the U.S. customary units into metric equivalents shows the small variance between soft conversions and hard conversions.

COMMENTARY TABLE 90.1 Conversions Using the Hard Conversion Method

U.S. Customary Units	Soft Conversions to SI Units	Hard Conversions to SI Units	Equivalent U.S. Customary Units
½ in.	12.7 mm	13 mm	0.51 in.
¾ in.	19.05 mm	19 mm	0.75 in.
1 in.	25.4 mm	25 mm	0.98 in.
4 in.	102 mm	100 mm	3.94 in.
12 in.	305 mm	300 mm	11.81 in.
2 ft	610 mm	600 mm	1.97 ft
3 ft	914 mm	900 mm	2.95 ft
6 ft	1.83 m	1.8 m	5.91 ft
15 ft	4.57 m	4.5 m	14.76 ft

General

Δ

N **Scope.** This article contains only those definitions essential to the application of this *Code*. It is not intended to include commonly defined general terms or commonly defined technical terms from related codes and standards. An article number in parentheses following the definition indicates that the definition only applies to that article.

> Informational Note: A definition that is followed by a reference in brackets has been extracted from one of the following standards. Only editorial changes were made to the extracted text to make it consistent with this *Code*.
>
> (1) NFPA 30A-2021, *Code for Motor Fuel Dispensing Facilities and Repair Garages*
> (2) NFPA 33-2021, *Standard for Spray Application Using Flammable or Combustible Materials*
> (3) NFPA 75-2020, *Standard for the Fire Protection of Information Technology Equipment*
> (4) NFPA 79-2021, *Electrical Standard for Industrial Machinery*
> (5) NFPA 99-2021, *Health Care Facilities Code*
> (6) NFPA *101®*-2022, *Life Safety Code®*
> (7) NFPA 110-2019, *Standard for Emergency and Standby Power Systems*
> (8) NFPA 303-2021, *Fire Protection Standard for Marinas and Boatyards*
> (9) NFPA 307-2021, *Standard for the Construction and Fire Protection of Marine Terminals, Piers, and Wharves*
> (10) NFPA 499-2021, *Recommended Practice for the Classification of Combustible Dusts and of Hazardous (Classified) Locations for Electrical Installations in Chemical Process Areas*
> (11) NFPA 501-2022, *Standard on Manufactured Housing*
> (12) NFPA 790-2021, *Standard for Competency of Third-Party Field Evaluation Bodies*
> (13) NFPA 1192-2021, *Standard on Recreational Vehicles*

The *NEC®* defines a number of technical terms in order to provide a common understanding of their use within the *NEC*. All definitions throughout the *NEC* have been relocated to Article 100, based on revisions within the 2020 *NEC Style Manual*. According to the revised *NEC Style Manual* section 2.2.2, all definitions are required to reside in Article 100, and Article 100 is not permitted to be subdivided. Any definitions that apply in only one article will have the article number in parentheses following the definition. If a specific article does not follow the definition in parentheses, then the definition applies throughout the *NEC*. The Code Making Panel (CMP) responsible for the definition is also listed at the end of the definition.

The *NEC* does not define every term used. General terms are found in nontechnical dictionaries. Some technical terms are used as defined in Institute of Electrical and Electronics Engineers (IEEE) 100, *The Authoritative Dictionary of IEEE Standards Terms*. It is important to note that many terms used in the electrical trades are actually registered trademarks for specific products or lines of products. Trademarked terms are not used in the *NEC* or this handbook.

Accessible (as applied to equipment). Capable of being reached for operation, renewal, and inspection. (CMP-1)

Exhibit 100.1 illustrates equipment considered accessible (as applied to equipment) in accordance with requirements elsewhere in the *NEC*. The requirement for the accessibility of switches and circuit breakers used as switches is shown in (a) and is according to 404.8(A). In (b), the busway installation is according to 368.17(C). The busway switches installation shown in (c), the switch installation adjacent to a motor in (d), and the hookstick-operated isolating switch installation in (e) are according to the exceptions in the list items of 404.8(A).

Accessible (as applied to wiring methods). Capable of being removed or exposed without damaging the building structure or finish or not permanently closed in or blocked by the structure, other electrical equipment, other building systems, or finish of the building. (CMP-1)

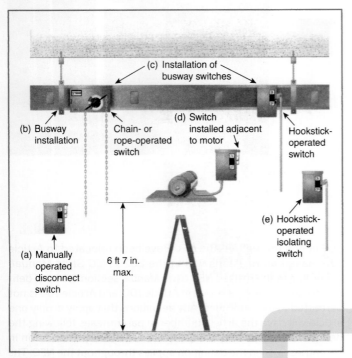

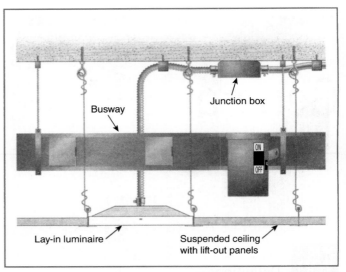

EXHIBIT 100.1 An example of a busway and of switches considered accessible even if located above the 6-feet 7-inch maximum height of the switch handle specified in 404.8(A).

Wiring methods located behind removable panels designed to allow access are not considered permanently enclosed and are considered exposed (as applied to wiring methods). Exhibit 100.2 is an example of wiring methods and equipment that are considered accessible despite being located above a suspended ceiling.

See also

300.4(C) regarding cables located in spaces behind accessible panels

Accessible, Readily. (Readily Accessible) Capable of being reached quickly for operation, renewal, or inspections without requiring those to whom ready access is requisite to take actions such as to use tools (other than keys), to climb over or under, to remove obstacles, or to resort to portable ladders, and so forth. (CMP-1)

Informational Note: Use of keys is a common practice under controlled or supervised conditions and a common alternative to the ready access requirements under such supervised conditions as provided elsewhere in the *NEC*.

Locks are commonly used on electrical equipment and electrical equipment rooms to keep unauthorized persons away from electrical hazards and to prevent tampering. The definition of *readily accessible* does not preclude the use of locks on service equipment doors or doors of rooms containing service equipment, provided a key or lock combination is available to those for whom

ready access is necessary. If a tool is necessary to gain access, the equipment is not readily accessible.

Sections 230.70(A)(1) and 235.352(A) require service disconnecting means to be readily accessible. However, 235.352(A) permits a mechanical linkage or, in multibuilding industrial installations under single management, a remote-control device in a separate building or structure to provide an equivalent to access.

EXHIBIT 100.2 An example of an accessible busway and junction box located above hung ceilings with lift-out panels.

N **Adapter.** A device used to adapt a circuit from one configuration of an attachment plug or receptacle to another configuration with the same current rating. (520) (CMP-15)

Adjustable Speed Drive. Power conversion equipment that provides a means of adjusting the speed of an electric motor. (CMP-11)

Informational Note: A variable frequency drive is one type of electronic adjustable speed drive that controls the rotational speed of an ac electric motor by controlling the frequency and voltage of the electrical power supplied to the motor.

Adjustable Speed Drive System. A combination of an adjustable speed drive, its associated motor(s), and auxiliary equipment. (CMP-11)

Aircraft Painting Hangar. An aircraft hangar constructed for the express purpose of spraying, coating, and/or dipping applications and provided with dedicated ventilation supply and exhaust. (CMP-14)

N **Alternate Power Source.** One or more generator sets, or battery systems where permitted, intended to provide power during the interruption of the normal electrical service; or the public utility electrical service intended to provide power during interruption of service normally provided by the generating facilities on the premises. [**99:**3.3.4] (517) (CMP-15)

See also

517.30(B), which identifies the permitted types of power sources for the essential electrical system (EES) in health care facilities

517.31(B)(2), which permits alternate power sources to serve essential electrical systems of contiguous or multiple building facilities such as a health care campus with a centrally located alternate power plant

N **Ambulatory Health Care Occupancy.** An occupancy used to provide services or treatment simultaneously to four or more patients that provides, on an outpatient basis, one or more of the following:

(1) Treatment for patients that renders the patients incapable of taking action for self-preservation under emergency conditions without the assistance of others.

(2) Anesthesia that renders the patients incapable of taking action for self-preservation under emergency conditions without the assistance of others.

(3) Treatment for patients who, due to the nature of their injury or illness, are incapable of taking action for self-preservation under emergency conditions without the assistance of others.

[*101*:3.3.198.1] (517) (CMP-15)

Ambulatory health care occupancies, including outpatient surgery centers, freestanding emergency medical centers, and hemodialysis units, are subject to the requirements of Part II of Article 517 and 517.45. This definition, which correlates with NFPA 99, *Health Care Facilities Code*, recognizes that some emergency or urgent care is performed at ambulatory health care occupancies.

Ampacity. The maximum current, in amperes, that a conductor can carry continuously under the conditions of use without exceeding its temperature rating. (CMP-6)

"Conditions of use" include factors such as ambient temperature and the number of conductors in the cable or raceway. A conductor with insulation rated at 60°C and installed near a furnace where the ambient temperature is continuously maintained at 60°C has no current-carrying capacity. Any current flowing through the conductor will raise its temperature above the 60°C insulation rating. Therefore, the ampacity of the conductor, regardless of its size, is zero.

Additional conductors in a raceway or cable raise the temperature in the raceway, decreasing the available ampacity. [See 310.15(C)(1).]

See also

Table 310.15(B)(1)(1), Table 310.15(B)(1)(2), and **Informative Annex B** for the ampacity correction factors for temperature

310.15(B)(1) and its commentary for more on the temperature limitation of conductors

N **Amplifier (Audio Amplifier) (Pre-Amplifier).** Electronic equipment that increases the current or voltage, or both, of an audio signal intended for use by another piece of audio equipment. Amplifier is the term used to denote an audio amplifier. (640) (CMP-12)

Appliance. Utilization equipment, generally other than industrial, that is fastened in place, stationary, or portable; is normally built in a standardized size or type; and is installed or connected as a unit to perform one or more functions such as clothes washing, air-conditioning, food mixing, deep frying, and so forth. (CMP-17)

N **Applicator.** The device used to transfer energy between the output circuit and the object or mass to be heated. (665) (CMP-12)

Approved. Acceptable to the authority having jurisdiction. (CMP-1)

Typically, approval of listed equipment is more readily given by an *authority having jurisdiction (AHJ)* if the authority accepts a laboratory's listing mark. Other options may be available for the jurisdiction to approve equipment, including evaluation by the inspection authority or field evaluation by a qualified laboratory or individual. Where an evaluation is conducted on site, industry standards such as NFPA 79, *Electrical Standard for Industrial Machinery,* if applicable, can be used. NFPA 790, *Standard for Competency of Third-Party Field Evaluation Bodies,* can be used to qualify evaluation services. NFPA 791, *Recommended Practice and Procedures for Unlabeled Electrical Equipment Evaluation,* can be used to evaluate unlabeled equipment in accordance with nationally recognized standards and any requirements of the AHJ.

See also

110.2, 110.3, and the definition of *authority having jurisdiction (AHJ)* for more information on the approval process

The definitions of *field evaluation body* and *field labeled* for information on this option for product approval by the AHJ

Arc-Fault Circuit Interrupter (AFCI). A device intended to provide protection from the effects of arc faults by recognizing characteristics unique to arcing and by functioning to de-energize the circuit when an arc fault is detected. (CMP-2)

Arc-fault circuit interrupters (AFCIs) are evaluated in accordance with UL 1699, *Standard for Arc-Fault Circuit-Interrupters,* using testing methods that create or simulate arcing conditions to determine a product's ability to detect and interrupt arcing faults. AFCIs are also tested to verify that arc detection is not inhibited by the presence of loads and circuit characteristics that mask the hazardous arcing condition. In addition, these devices are evaluated to determine resistance to unwanted tripping due to the presence of arcing that occurs in equipment under normal operating conditions or to a loading condition that closely mimics an arcing fault, such as a solid-state electronic ballast or a dimmed load.

N **Array.** A mechanically and electrically integrated grouping of modules with support structure, including any attached system

EXHIBIT 100.3 A PV array support structure that allows for continued use of the walkway. (*Courtesy of Solar Design Associates, LLC*)

components such as inverter(s) or dc-to-dc converter(s) and attached associated wiring. (690) (CMP-4)

An array composed of multiple panels installed on a support structure is illustrated in Exhibit 100.3.

Askarel. A generic term for a group of nonflammable synthetic chlorinated hydrocarbons used as electrical insulating media. (CMP-9)

> Informational Note: Askarels of various compositional types are used. Under arcing conditions, the gases produced, while consisting predominantly of noncombustible hydrogen chloride, can include varying amounts of combustible gases, depending on the askarel type.

Δ **Associated Apparatus.** Apparatus in which the circuits are not necessarily intrinsically safe themselves but that affects the energy in the intrinsically safe circuits and is relied on to maintain intrinsic safety. Such apparatus is one of the following:

(1) Electrical apparatus that has an alternative type of protection for use in the appropriate hazardous (classified) location

(2) Electrical apparatus not so protected that shall not be used within a hazardous (classified) location (CMP-14)

> Informational Note No. 1: Associated apparatus has identified intrinsically safe connections for intrinsically safe apparatus and also might have connections for nonintrinsically safe apparatus.
>
> Informational Note No. 2: An example of associated apparatus is an intrinsic safety barrier, which is a network designed to limit the energy (voltage and current) available to the protected circuit in the hazardous (classified) location under specified fault conditions.
>
> Informational Note No. 3: See ANSI/UL 913, *Intrinsically Safe Apparatus and Associated Apparatus for Use in Class I, II, and III, Division 1, Hazardous (Classified) Locations*; ANSI/ UL 60079-11, *Explosive Atmospheres — Part 11: Equipment Protection by Intrinsic Safety "i"*; and ANSI/ISA RP 12.06.01,

> *Recommended Practice for Wiring Methods for Hazardous (Classified) Locations Instrumentation — Part 1: Intrinsic Safety*, for additional information.

Δ **Associated Nonincendive Field Wiring Apparatus.** Apparatus in which the circuits are not necessarily nonincendive themselves but that affects the energy in nonincendive field wiring circuits and is relied on to maintain nonincendive energy levels. Such apparatus is one of the following:

(1) Electrical apparatus that has an alternative type of protection for use in the appropriate hazardous (classified) location

(2) Electrical apparatus not so protected that shall not be used within a hazardous (classified) location (CMP-14)

> Informational Note No. 1: Associated nonincendive field wiring apparatus has designated associated nonincendive field wiring apparatus connections for nonincendive field wiring apparatus and also might have connections for other electrical apparatus.
>
> Informational Note No. 2: See ANSI/UL 121201, *Nonincendive Electrical Equipment for Use in Class I and II, Division 2 and Class III, Divisions 1 and 2 Hazardous (Classified) Locations*, for additional information.

Attachment Fitting, Weight-Supporting (WSAF). (Weight-Supporting Attachment Fitting) A device that, by insertion into a weight-supporting ceiling receptacle, establishes a connection between the conductors of the attached utilization equipment and the branch-circuit conductors connected to the weight-supporting ceiling receptacle. (CMP-18)

> Informational Note No. 1: A weight-supporting attachment fitting is different from an attachment plug because no cord is associated with the fitting. A weight-supporting attachment fitting in combination with a weight-supporting ceiling receptacle secures the associated utilization equipment in place and supports its weight.
>
> Informational Note No. 2: See ANSI/NEMA WD 6, *American National Standard for Wiring Devices — Dimensional Specifications*, for the standard configuration of weight-supporting attachment fittings and related weight-supporting ceiling receptacles.

The attachment fitting is the mating device that provides both mechanical and electrical connections to the locking support and mounting receptacle. The locking support and mounting receptacle, along with the attachment fitting, have both electrical ratings and mechanical weight ratings.

Attachment Plug (Plug Cap) (Plug). A device that, by insertion in a receptacle, establishes a connection between the conductors of the attached flexible cord and the conductors connected permanently to the receptacle. (CMP-18)

See National Electrical Manufacturers Association Wiring Devices (NEMA WD) 6, *Wiring Devices — Dimensional Specifications*, for the standard on attachment plug configurations.

N **Audio Autotransformer.** A transformer with a single winding and multiple taps intended for use with an amplifier loudspeaker signal output. (640) (CMP-12)

N **Audio Signal Processing Equipment (Audio Equipment).** Electrically operated equipment that produces, processes, or both, electronic signals that, when appropriately amplified and reproduced by a loudspeaker, produce an acoustic signal within the range of normal human hearing (typically 20–20 kHz). Within Article 640, the terms equipment and audio equipment are assumed to be equivalent to audio signal processing equipment. (640) (CMP-12)

> Informational Note: This equipment includes, but is not limited to, loudspeakers; headphones; pre-amplifiers; microphones and their power supplies; mixers; MIDI (musical instrument digital interface) equipment or other digital control systems; equalizers, compressors, and other audio signal processing equipment; and audio media recording and playback equipment, including turntables, tape decks and disk players (audio and multimedia), synthesizers, tone generators, and electronic organs. Electronic organs and synthesizers may have integral or separate amplification and loudspeakers. With the exception of amplifier outputs, virtually all such equipment is used to process signals (using analog or digital techniques) that have nonhazardous levels of voltage or current.

The definition of the term *audio signal processing equipment* clarifies the limits of signal processing (frequency bandwidth), which falls under Article 640.

"MIDI (musical instrument digital interface) equipment or other digital control systems" is mentioned specifically because, while MIDI or similar digital control signals may come from an electronic musical instrument, such signals also can be produced by a computer.

N **Audio System.** The totality of all equipment and interconnecting wiring used to fabricate a fully functional audio signal processing, amplification, and reproduction system. (640) (CMP-12)

N **Audio Transformer.** A transformer with two or more electrically isolated windings and multiple taps intended for use with an amplifier loudspeaker signal output. (640) (CMP-12)

Audio transformers are intended only for use with audio signals, not light and power.

Authority Having Jurisdiction (AHJ). An organization, office, or individual responsible for enforcing the requirements of a code or standard, or for approving equipment, materials, an installation, or a procedure. (CMP-1)

> Informational Note: The phrase "authority having jurisdiction," or its acronym AHJ, is used in NFPA documents in a broad manner, since jurisdictions and approval agencies vary, as do their responsibilities. Where public safety is primary, the authority having jurisdiction may be a federal, state, local, or other regional department or individual such as a fire chief; fire marshal; chief of a fire prevention bureau, labor department, or health department; building official; electrical inspector; or others having statutory authority. For insurance purposes, an insurance inspection department, rating bureau, or other insurance company representative may be the authority having jurisdiction. In many circumstances, the property owner or his or her designated agent assumes the role of the authority having jurisdiction; at government installations,

the commanding officer or departmental official may be the authority having jurisdiction.

In the North American safety system, the importance of the AHJ's role cannot be overstated. The AHJ verifies that an installation complies with the *NEC*. See the definition of *approved*.

> **See also**
>
> **90.4, 90.7, 110.2,** and **110.3** for more information on inspection, approval, and listing

Automatic. Performing a function without the necessity of human intervention. (CMP-1)

Δ **Bathroom.** An area including a sink with one or more of the following: a toilet, a urinal, a tub, a shower, a bidet, or similar plumbing fixtures. (CMP-2)

N **Battery.** A single cell or a group of cells connected together electrically in series, in parallel, or a combination of both. (CMP-13)

N **Battery, Flow. (Flow Battery)** An energy storage component that stores its active materials in the form of one or two electrolytes external to the reactor interface. When in use, the electrolytes are transferred between reactor and storage tanks. (706) (CMP-13)

> Informational Note: Three commercially available flow battery technologies are zinc air, zinc bromine, and vanadium redox, sometimes referred to as *pumped electrolyte ESS*.

N **Battery, Sealed. (Sealed Battery)** A battery that has no provision for the routine addition of water or electrolyte or for external measurement of electrolyte specific gravity and might contain pressure relief venting. (CMP-13)

N **Battery, Stationary Standby. (Stationary Standby Battery)** A battery that spends the majority of the time on continuous float charge or in a high state of charge, in readiness for a discharge event. (CMP-13)

> Informational Note: Uninterruptible Power Supply (UPS) batteries are an example that falls under this definition.

N **Battery-Powered Lighting Units.** Individual unit equipment for backup illumination consisting of a rechargeable battery; a battery-charging means; provisions for one or more lamps mounted on the equipment, or with terminals for remote lamps, or both; and a relaying device arranged to energize the lamps automatically upon failure of the supply to the unit equipment. (517) (CMP-15)

N **Berth.** The water space to be occupied by a boat or other vessel alongside or between bulkheads, piers, piles, fixed and floating docks, or any similar access structure. [**303:**3.3.2] (555) (CMP-7)

> Informational Note: See the definition of *Slip* for additional information.

N **Bipolar Circuit.** A dc circuit that is comprised of two monopole circuits, each having an opposite polarity connected to a common reference point. (CMP-4)

EXHIBIT 100.4 The electrical equipment associated with pumps used to circulate water in this artificial pond is subject to the requirements of Article 682.

N Block. A square or portion of a city, town, or village enclosed by streets and including the alleys so enclosed, but not any street. (800) (CMP-16)

N Boatyard. A facility used for constructing, repairing, servicing, hauling from the water, storing (on land and in water), and launching of boats. [**303**:3.3.3] (555) (CMP-7)

N Bodies of Water, Artificially Made. (Artificially Made Bodies of Water) Bodies of water that have been constructed or modified to fit some decorative or commercial purpose such as, but not limited to, aeration ponds, fish farm ponds, storm retention basins, treatment ponds, and irrigation (channel) facilities. Water depths may vary seasonally or be controlled. (682) (CMP-17)

The term *artificially made bodies of water* includes all bodies of water that are not naturally created and that are not covered by the requirements of Article 680. The uses of artificially made bodies of water include decorative, agricultural, municipal infrastructure, and industrial. The decorative pond shown in Exhibit 100.4 is an example of an artificially made body of water because it was constructed and filled with water and did not occur naturally.

N Bodies of Water, Natural. (Natural Bodies of Water) Bodies of water such as lakes, streams, ponds, rivers, and other naturally occurring bodies of water, which may vary in depth throughout the year. (682) (CMP-17)

Bonded (Bonding). Connected to establish electrical continuity and conductivity. (CMP-5)

Bonding Conductor (Bonding Jumper). A conductor that ensures the required electrical conductivity between metal parts that are required to be electrically connected. (CMP-5)

Either of the two terms, *bonding conductor* or *bonding jumper*, may be used. The term *bonding jumper* is sometimes interpreted to mean a short conductor, although some bonding jumpers may

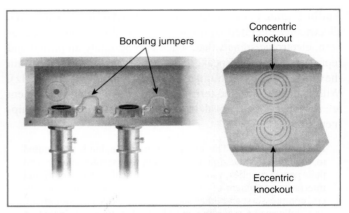

EXHIBIT 100.5 Bonding jumpers installed around concentric or eccentric knockouts.

be several feet in length. The primary purpose of a bonding conductor or jumper is to ensure electrical conductivity between two conductive bodies, such as between a metal box and a metal raceway. Bonding jumpers are particularly important where a box has either concentric- or eccentric-type knockouts. These knockouts can impair the electrical conductivity between metal parts and may actually introduce unnecessary impedance into the grounding path.

Exhibit 100.5 shows the difference between concentric- and eccentric-type knockouts and illustrates one method of applying bonding jumpers at these types of knockouts.

See also

250.92(B) for bonding jumpers at service equipment

250.97 for bonding jumpers at equipment operating over 250 volts

250.98 for bonding jumpers at expansion fittings in metal raceways

Bonding Jumper, Equipment. (Equipment Bonding Jumper) The connection between two or more portions of the equipment grounding conductor. (CMP-5)

Equipment bonding jumpers ensure that an effective ground-fault current path is not compromised by an interruption in mechanical or electrical continuity. For example, conduits entering an open-bottom switchboard usually are not mechanically connected to the switchboard. Expansion fittings may not provide electrical continuity because they are loosely joined raceways. Bonding jumpers are necessary in order to provide electrical continuity.

Exhibit 100.6 shows an external bonding jumper around an expansion joint. Some expansion fittings for metal conduit have an internal bonding jumper that is integral to the fitting. Equipment bonding jumpers are also used to connect the grounding terminal of a receptacle to a metal box that in turn is grounded via an equipment grounding conductor (the raceway system).

Bonding Jumper, Main. (Main Bonding Jumper) The connection between the grounded circuit conductor and the equipment grounding conductor, or the supply-side bonding jumper, or both, at the service. (CMP-5)

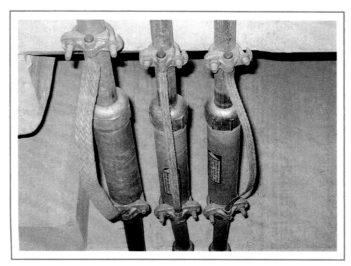

EXHIBIT 100.6 Equipment bonding jumpers installed to maintain electrical continuity around conduit expansion fittings. *(Courtesy of the International Association of Electrical Inspectors)*

Exhibit 100.7 shows a main bonding jumper that provides the connection between the grounded service conductor and the equipment grounding conductor at the service by connecting between the neutral bus and the equipment grounding bus. Bonding jumpers can be located throughout the electrical system, but a main bonding jumper is located only at the service.

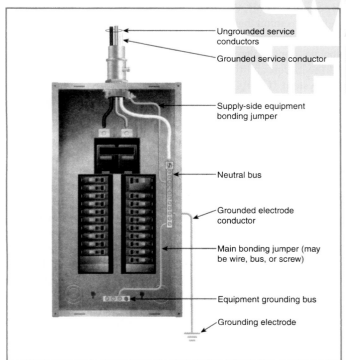

EXHIBIT 100.7 A main bonding jumper installed at the service between the grounded service conductor and the equipment grounding conductor.

See also

250.28 for main bonding jumper requirements

Bonding Jumper, Supply-Side. (Supply-Side Bonding Jumper) A conductor installed on the supply side of a service or within a service equipment enclosure(s), or for a separately derived system, that ensures the required electrical conductivity between metal parts required to be electrically connected. (CMP-5)

Metal equipment enclosures, metal raceways, and metal cable trays are examples of equipment containing supply-side conductors that are required to be bonded. Where bonding jumpers are used, they are required to be installed and sized as specified in 250.102(A), (B), (C), and (E). Bonding jumpers installed on the load side of a service, feeder, or branch-circuit overcurrent protective device (OCPD) are *equipment bonding jumpers.*

Bonding Jumper, System. (System Bonding Jumper) The connection between the grounded circuit conductor and the supply-side bonding jumper, or the equipment grounding conductor, or both, at a separately derived system. (CMP-5)

A system bonding jumper is used to connect the equipment grounding conductor(s) or the supply-side bonding jumper to the grounded conductor of a separately derived system either at the source (see Exhibit 250.14) or at the first system disconnecting means (see Exhibit 250.15). A system bonding jumper is used at the derived system if the derived system contains a grounded conductor.

Like the main bonding jumper at the service equipment, the system bonding jumper provides the necessary link between the equipment grounding conductors and the system grounded conductor in order to establish an effective path for ground-fault current to return to the source.

See also

250.30(A)(1) for system bonding jumper requirements

N **Border Light.** A permanently installed overhead strip light. (520) (CMP-15)

N **Bottom Shield.** A protective layer that is installed between the floor and flat conductor cable (Type FCC) to protect the cable from physical damage and may or may not be incorporated as an integral part of the cable. (324) (CMP-6)

Branch Circuit (Branch-Circuit). The circuit conductors between the final overcurrent device protecting the circuit and the outlet(s). (CMP-2)

Exhibit 100.8 shows the difference between branch circuits and feeders. Conductors between the overcurrent devices in the panelboards and the duplex receptacles are branch-circuit conductors. The overcurrent devices at the panelboards are the *final* OCPD for the circuit and duplex receptacles. Conductors between the service equipment, or source of separately derived systems, and the panelboards are feeders.

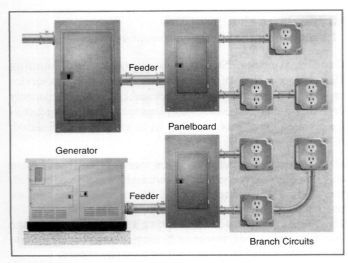

EXHIBIT 100.8 Feeder (circuits) and branch circuits.

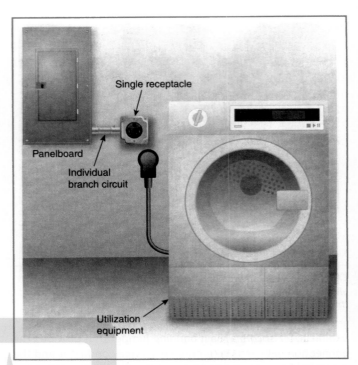

EXHIBIT 100.9 An individual branch circuit supplying only one piece of utilization equipment via a single receptacle.

Branch Circuit, Appliance. (Appliance Branch Circuit) A branch circuit that supplies energy to one or more outlets to which appliances are to be connected and that has no permanently connected luminaires that are not a part of an appliance. (CMP-2)

> **See also**
>
> **210.11(C)(1),** which requires two or more 20-ampere small-appliance branch circuits for dwelling units
>
> **210.52(B)(1),** which requires that small-appliance branch circuits supply receptacle outlets located in rooms such as the kitchen, dining room, and pantry
>
> **210.52(B)(2),** which provides details on small-appliance branch circuits not permitted to supply other outlets

Branch Circuit, General-Purpose. (General-Purpose Branch Circuit) A branch circuit that supplies two or more receptacles or outlets for lighting and appliances. (CMP-2)

Branch Circuit, Individual. (Individual Branch Circuit) A branch circuit that supplies only one utilization equipment. (CMP-2)

Exhibit 100.9 illustrates an individual branch circuit with a single receptacle for connection of one piece of utilization equipment (e.g., one dryer, one range, one space heater, one motor). A branch circuit supplying one duplex receptacle that supplies two cord-and-plug-connected appliances, or similar equipment, is not an individual branch circuit.

> **See also**
>
> **210.21(B)(1),** which requires the single receptacle to have an ampere rating not less than that of the branch circuit
>
> **210.22** for permissible loads on individual branch circuits

N **Branch Circuit, Motor. (Motor Branch Circuit)** The circuit conductors, including equipment, between the motor branch-circuit short-circuit and ground-fault protective device and an individual motor. (CMP-11)

Δ **Branch Circuit, Multiwire. (Multiwire Branch Circuit)** A branch circuit that consists of two or more ungrounded conductors that have a voltage between them, and a neutral conductor that has equal voltage between it and each ungrounded conductor of the circuit and that is connected to the neutral conductor of the system. (CMP-2)

> **See also**
>
> **210.4, 240.15(B)(1),** and **300.13(B)** for specific information about multiwire branch circuits

N **Branch-Circuit Selection Current (BCSC).** The value in amperes to be used instead of the rated-load current in determining the ratings of motor branch-circuit conductors, disconnecting means, controllers, and branch-circuit short-circuit and ground-fault protective devices wherever the running overload protective device permits a sustained current greater than the specified percentage of the rated-load current. The value of branch-circuit selection current will always be equal to or greater than the marked rated-load current. (440) (CMP-11)

N **Breakout Assembly.** An adapter used to connect a multipole connector containing two or more branch circuits to multiple individual branch-circuit connectors. (520) (CMP-15)

N **Broadband.** Wide bandwidth data transmission that transports multiple signals, protocols, and traffic types over various media types. (CMP-16)

Building. A structure that stands alone or that is separated from adjoining structures by fire walls. (CMP-1)

A building is generally considered to be a roofed or walled structure that is intended for supporting or sheltering any use or occupancy. However, a separate structure such as a pole, billboard sign, or water tower may also be considered to be a building. Definitions of the terms *fire walls* and *fire doors* are the responsibility of building codes. Generally, a fire wall may be defined as a wall that separates buildings or subdivides a building to prevent the spread of fire and that has a fire resistance rating and structural stability.

N **Building, Floating. (Floating Building)** A building that floats on water, is moored in a permanent location, and has a premises wiring system served through connection by permanent wiring to an electrical supply system not located on the premises. (CMP-7)

N **Building, Manufactured. (Manufactured Building)** Any building that is of closed construction and is made or assembled in manufacturing facilities on or off the building site for installation, or for assembly and installation on the building site, other than manufactured homes, mobile homes, park trailers, or recreational vehicles. (545) (CMP-7)

N **Building Component.** Any subsystem, subassembly, or other system designed for use in or integral with or as part of a structure, which can include structural, electrical, mechanical, plumbing, and fire protection systems, and other systems affecting health and safety. (545) (CMP-7)

N **Building System.** Plans, specifications, and documentation for a system of manufactured building or for a type or a system of building components, which can include structural, electrical, mechanical, plumbing, and fire protection systems, and other systems affecting health and safety, and including such variations thereof as are specifically permitted by regulation, and which variations are submitted as part of the building system or amendment thereto. (545) (CMP-7)

N **Bulkhead.** A vertical structural wall, usually of stone, timber, metal, concrete, or synthetic material, constructed along, and generally parallel to, the shoreline to retain earth as an extension of the upland, and often to provide suitable water depth at the waterside face. [**303**:3.3.5] (555) (CMP-7)

N **Bull Switch.** An externally operated wall-mounted safety switch that can contain overcurrent protection and is designed for the connection of portable cables and cords. (530) (CMP-15)

N **Bundled.** Cables or conductors that are tied, wrapped, taped, or otherwise periodically bound together. (520) (CMP-15)

N **Busbar.** A noninsulated conductor electrically connected to the source of supply and physically supported on an insulator providing a power rail for connection to utilization equipment, such as sensors, actuators, A/V devices, low-voltage luminaire assemblies, and similar electrical equipment. (393) (CMP-18)

N **Busbar Support.** An insulator that runs the length of a section of suspended ceiling bus rail that serves to support and isolate the busbars from the suspended grid rail. (393) (CMP-18)

N **Busway.** A raceway consisting of a metal enclosure containing factory-mounted, bare or insulated conductors, which are usually copper or aluminum bars, rods, or tubes. (CMP-8)

Cabinet. An enclosure that is designed for either surface mounting or flush mounting and is provided with a frame, mat, or trim in which a swinging door or doors are or can be hung. (CMP-9)

Cabinets are designed for surface or flush mounting with a trim to which a swinging door(s) is hung. A *cutout box* is designed for surface mounting with a swinging door(s) secured directly to the box. A *panelboard* is an electrical assembly designed to be placed in a cabinet or cutout box.

See also

Article 312, which covers both cabinets and cutout boxes

N **Cable, Abandoned. (Abandoned Cable)** Installed cable that is not terminated at equipment other than a termination fitting or a connector and is not identified for future use with a tag. (CMP-16)

> Informational Note: See 640.6(B), 645.5(G), 722.25, 760.25, 770.25, and 800.25 for requirements covering the removal of abandoned cables.

N **Cable, Armored (Type AC). (Armored Cable)** A fabricated assembly of insulated conductors in a flexible interlocked metallic armor. (CMP-6)

Δ **Cable, Circuit Integrity (CI). (Circuit Integrity Cable)** Cable(s) marked with the suffix "-CI" used for remote-control, signaling, power-limited, fire alarm, optical fiber, or communications systems that supply critical circuits to ensure survivability for continued circuit operation for a specified time under fire conditions. (CMP-3)

> Informational Note: See 728.4 for power circuits installed for survivability.

Cable, Coaxial. (Coaxial Cable) A cylindrical assembly composed of a conductor centered inside a metallic tube or shield, separated by a dielectric material, and usually covered by an insulating jacket. (CMP-16)

N **Cable, Festoon. (Festoon Cable)** Single- and multiple-conductor cable intended for use and installation where flexibility is required. (610) (CMP-12)

N **Cable, Flat Conductor (Type FCC). (Flat Conductor Cable)** Three or more separate flat copper conductors placed horizontally edge-to-edge and enclosed within an insulating assembly. (324) (CMP-6)

N **Cable, Instrumentation Tray (Type ITC). (Instrumentation Tray Cable)** A factory assembly of two or more insulated conductors, with or without an equipment grounding conductor(s), enclosed in a nonmetallic sheath. (CMP-3)

N **Cable, Integrated Gas Spacer (Type IGS). (Integrated Gas Spacer Cable)** A factory assembly of one or more conductors, each individually insulated and enclosed in a loose fit, nonmetallic flexible conduit as an integrated gas spacer cable rated 0 volts through 600 volts. (CMP-6)

N **Cable, Limited Use. (Limited-Use Cable)** Cables that are intended to be used with protection such as a raceway or for specific restricted applications. (722) (CMP-3)

N **Cable, Medium Voltage (Type MV). (Medium Voltage Cable)** A single or multiconductor solid dielectric insulated cable rated 2001 volts up to and including 35,000 volts, nominal. (CMP-6)

N **Cable, Metal Clad (Type MC). (Metal Clad Cable)** A factory assembly of one or more insulated circuit conductors with or without optical fiber members enclosed in an armor of interlocking metal tape, or a smooth or corrugated metallic sheath. (CMP-6)

N **Cable, Metallic Conductor. (Metallic Conductor Cable)** A factory assembly of two or more conductors having an overall covering. (CMP-16)

N **Cable, Mineral-Insulated, Metal-Sheathed (Type MI). (Mineral-Insulated, Metal-Sheathed Cable)** A factory assembly of one or more conductors insulated with a highly compressed refractory mineral insulation and enclosed in a liquidtight and gastight continuous copper or alloy steel sheath. (CMP-6)

N **Cable, Nonmetallic-Sheathed.** A factory assembly of two or more insulated conductors enclosed within an overall nonmetallic jacket. (CMP-6)

N **Cable, Nonmetallic-Sheathed (Type NM).** Insulated conductors enclosed within an overall nonmetallic jacket. (CMP-6)

N **Cable, Nonmetallic-Sheathed (Type NMC).** Insulated conductors enclosed within an overall, corrosion resistant, nonmetallic jacket. (CMP-6)

Cable, Optical Fiber. (Optical Fiber Cable) A factory assembly or field assembly of one or more optical fibers having an overall covering. (CMP-16)

> Informational Note: A field-assembled optical fiber cable is an assembly of one or more optical fibers within a jacket. The jacket, without optical fibers, is installed in a manner similar to conduit or raceway. Once the jacket is installed, the optical fibers are inserted into the jacket, completing the cable assembly.

Cable, Optical Fiber, Conductive. (Conductive Optical Fiber Cable) A factory assembly of one or more optical fibers having an overall covering and containing non-current-carrying conductive member(s) such as metallic strength member(s), metallic vapor barrier(s), metallic armor, or metallic sheath. (CMP-16)

Cable, Optical Fiber, Hybrid. (Hybrid Optical Fiber Cable) A cable containing optical fibers and current-carrying electrical conductors. (CMP-16)

Cable, Optical Fiber, Nonconductive. (Nonconductive Optical Fiber Cable) A factory assembly of one or more optical fibers having an overall covering and containing no electrically conductive materials. (CMP-16)

Δ **Cable, Optical Fiber, Protected. (Protected Optical Fiber Cable)** Optical fiber cable protected from releasing optical radiation into the atmosphere during normal operating conditions and foreseeable malfunctions by additional armoring, conduit, cable tray, or raceway. (CMP-14)

> Informational Note: See ANSI/UL 60079-28, *Explosive Atmospheres — Part 28: Protection of Equipment and Transmission Systems Using Optical Radiation*, for additional information.

N **Cable, Portable Power Feeder. (Portable Power Feeder Cable)** One or more flexible shielded insulated power conductors enclosed in a flexible covering rated from 2001 to 25,000 volts. (CMP-6)

N **Cable, Power and Control Tray (Type TC). (Power and Control Tray Cable)** A factory assembly of two or more insulated conductors, with or without associated bare or covered equipment grounding conductors, under a nonmetallic jacket. (CMP-6)

Δ **Cable, Power-Limited Tray (Type PLTC). (Power-Limited Tray Cable)** A factory assembly of two or more insulated conductors rated at 300 volts, with or without associated bare or insulated equipment grounding conductors, under a nonmetallic jacket. (CMP-3)

Cable, Service. (Service Cable) Service conductors made up in the form of a cable. (CMP-10)

N **Cable, Service Entrance. (Service Entrance Cable)** A single conductor or multiconductor cable provided with an overall covering, primarily used for services. (CMP-6)

N **Cable, Service Entrance (Type SE).** Service-entrance cable having a flame-retardant, moisture-resistant covering. (CMP-6)

N **Cable, Service Entrance (Type USE).** Service-entrance cable, identified for underground use, having a moisture-resistant covering, but not required to have a flame-retardant covering. (CMP-6)

N **Cable, Type P.** A factory assembly of one or more insulated flexible tinned copper conductors, with associated equipment grounding conductor(s), with or without a braided metallic armor and with an overall nonmetallic jacket. (CMP-6)

N **Cable, Under Carpet. (Under Carpet Cable)** Cables that are intended to be used under carpeting, floor covering, modular tiles, and planks. (722) (CMP-3)

N **Cable, Underground Feeder and Branch-Circuit (Type UF). (Underground Feeder and Branch-Circuit Cable)** A factory assembly of one or more insulated conductors with an integral or an overall covering of nonmetallic material suitable for direct burial in the earth. (CMP-6)

N **Cable Assembly, Flat (Type FC). (Flat Cable Assembly)** An assembly of parallel conductors formed integrally with an insulating material web specifically designed for field installation in surface metal raceway. (CMP-6)

N **Cable Bundle.** A group of cables that are tied together or in contact with one another in a closely packed configuration for at least 1.0 m (40 in.). (CMP-3)

> Informational Note: Random or loose installation of individual cables can result in less heating. Combing of the cables can result in less heat dissipation and more signal cross talk between cables.

N **Cable Connector.** A connector designed to join flat conductor cables (Type FCC) without using a junction box. (324) (CMP-6)

N **Cable Connector [as applied to hazardous (classified) locations].** An electrical device that is part of a cable assembly and that, by insertion of two mating configurations, establishes a connection between the conductors of the cable assembly and the conductors of a fixed piece of equipment. (CMP-14)

> Informational Note No. 1: See ANSI/UL 121201, *Nonincendive Electrical Equipment for Use in Class I and II, Division 2 and Class III, Divisions 1 and 2 Hazardous (Classified) Locations*, for information on the use of cable connectors.
> Informational Note No. 2: Cable connectors in other than hazardous (classified) locations are referred to as male and female fittings.
> Informational Note No. 3: See ANSI/UL 2238, *Cable Assemblies and Fittings for Industrial Control and Signal Distribution*, and ANSI/UL 2237, *Multi-Point Interconnection Power Cable Assemblies for Industrial Machinery*, for examples of standards on male and female fittings in other than hazardous (classified) locations.

N **Cable Joint.** A connection consisting of an insulation system and a connector where two (or more) medium voltage (Type MV) cables are joined together. (CMP-6)

N **Cable Management System.** An apparatus designed to control and organize lengths of cable or cord. (CMP-12)

Cable Routing Assembly. A single channel or connected multiple channels, as well as associated fittings, forming a structural system that is used to support and route communications wires and cables, optical fiber cables, data cables associated with information technology and communications equipment, Class 2, Class 3, and Type PLTC cables, and power-limited fire alarm cables in plenum, riser, and general-purpose applications. (CMP-16)

N **Cable Sheath.** A single or multiple layers of a protective covering that holds and protects the conductors or optical fibers, or both, contained inside. (CMP-16)

N **Cable System, Fire-Resistive. (Fire-Resistive Cable System)** A cable and components used to ensure survivability of critical circuits for a specified time under fire conditions. (CMP-3)

N **Cable System, Flat Conductor. (Flat Conductor Cable System)** A complete wiring system for branch circuits that is designed for installation under carpet squares. (324) (CMP-6)

> Informational Note: The FCC system includes Type FCC cable and associated shielding, connectors, terminators, adapters, boxes, and receptacles.

N **Cable Termination.** A connection consisting of an insulation system and a connector and installed on a medium voltage (Type MV) cable to connect from a cable to a device, such as equipment. (CMP-6)

N **Cable Tray System.** A unit or assembly of units or sections and associated fittings forming a structural system used to securely fasten or support cables and raceways. (CMP-8)

N **Cablebus.** An assembly of units or sections with insulated conductors having associated fittings forming a structural system to securely fasten or support conductors and conductor terminations in a completely enclosed, ventilated, protective metal housing. This assembly is designed to carry fault current and to withstand the magnetic forces of such current. (CMP-8)

> Informational Note: Cablebus is ordinarily assembled at the point of installation from the components furnished or specified by the manufacturer in accordance with instructions for the specific job.

N **Cell (as applied to batteries).** The basic electrochemical unit, characterized by an anode and a cathode, used to receive, store, and deliver electrical energy. (CMP-13)

N **Cell, Sealed. (Sealed Cell)** A cell that has no provision for the routine addition of water or electrolyte or for external measurement of electrolyte specific gravity and might contain pressure relief venting. (CMP-13)

N **Cell Line.** An assembly of electrically interconnected electrolytic cells supplied by a source of direct-current power. (CMP-12)

N **Cell Line Attachments and Auxiliary Equipment.** A term that includes, but is not limited to, auxiliary tanks; process piping; ductwork; structural supports; exposed cell line conductors; conduits and other raceways; pumps, positioning equipment, and cell cutout or bypass electrical devices. Auxiliary equipment includes tools, welding machines, crucibles, and other portable equipment used for operation and maintenance within the electrolytic cell line working zone. In the cell line working zone, auxiliary equipment includes the exposed conductive surfaces of ungrounded cranes and crane-mounted cell-servicing equipment. (668) (CMP-12)

Charge Controller. Equipment that controls dc voltage or dc current, or both, and that is used to charge a battery or other energy storage device. (CMP-13)

Charger Power Converter. The device used to convert energy from the power grid to a high-frequency output for wireless power transfer. (625) (CMP-12)

N Child Care Facility. A building or structure, or portion thereof, for educational, supervisory, or personal care services for more than four children 7 years old or less. (406) (CMP-18)

Circuit Breaker. A device designed to open and close a circuit by nonautomatic means and to open the circuit automatically on a predetermined overcurrent without damage to itself when properly applied within its rating. (CMP-10)

> Informational Note: The automatic opening means can be integral, direct acting with the circuit breaker, or remote from the circuit breaker.

Circuit Breaker, Adjustable. (Adjustable Circuit Breaker) A qualifying term indicating that the circuit breaker can be set to trip at various values of current, time, or both, within a predetermined range. (CMP-10)

Circuit Breaker, Instantaneous Trip. (Instantaneous Trip Circuit Breaker) A qualifying term indicating that no delay is purposely introduced in the tripping action of the circuit breaker. (CMP-10)

Δ Circuit Breaker, Inverse Time. (Inverse Time Circuit Breaker) A qualifying term indicating that there is a delay purposely introduced in the tripping action of the circuit breaker, and the delay decreases as the magnitude of the current increases. (CMP-10)

Circuit Breaker, Nonadjustable. (Nonadjustable Circuit Breaker) A qualifying term indicating that the circuit breaker does not have any adjustment to alter the value of the current at which it will trip or the time required for its operation. (CMP-10)

Δ Class 1 Circuit. The portion of the wiring system between the load side of the Class 1 power source and the connected equipment. (CMP-3)

Class 2 Circuit. The portion of the wiring system between the load side of a Class 2 power source and the connected equipment. Due to its power limitations, a Class 2 circuit considers safety from a fire initiation standpoint and provides acceptable protection from electric shock. (CMP-3)

Class 3 Circuit. The portion of the wiring system between the load side of a Class 3 power source and the connected equipment. Due to its power limitations, a Class 3 circuit considers safety from a fire initiation standpoint. Since higher levels of voltage and current than for Class 2 are permitted, additional safeguards are specified to provide protection from an electric shock hazard that could be encountered. (CMP-3)

N Class 4 Circuit. The portion of the wiring system between the load side of a Class 4 transmitter and the Class 4 receiver or Class 4 utilization equipment, as appropriate. Due to the active monitoring and control of the voltage and current provided, a Class 4 circuit considers safety from a fire initiation standpoint and provides acceptable protection from electric shock. (726) (CMP-3)

> Informational Note: A Class 4 circuit is also commonly referred to as a fault-managed power circuit.

N Class 4 Device. Any active device connected to the Class 4 circuit; examples include a Class 4 transmitter, a Class 4 receiver, or Class 4 utilization equipment. (CMP-3)

N Class 4 Power System. An actively monitored and controlled system consisting of one or more Class 4 transmitters and one or more Class 4 receivers connected by a cabling system. (CMP-3)

N Class 4 Receiver. A device that accepts Class 4 power and converts it for use by utilization equipment. (CMP-3)

N Class 4 Transmitter. A device that sources Class 4 power. (726) (CMP-3)

> Informational Note: A Class 4 transmitter is different from traditional power sources in that it monitors the line for faults (both line-to-line and line-to-ground) and ceases power transmission if a fault is sensed.

N Class 4 Utilization Equipment. Devices that are directly powered by a Class 4 transmitter without the need for a separate Class 4 receiver (the receiver is integrated into the equipment). (CMP-3)

N Closed Construction. Any building, building component, assembly, or system manufactured in such a manner that all concealed parts of processes of manufacture cannot be inspected after installation at the building site without disassembly, damage, or destruction. (545) (CMP-7)

Clothes Closet. A nonhabitable room or space intended primarily for storage of garments and apparel. (CMP-1)

This definition helps to determine whether the rules of 240.24(D), 410.16, and 550.11(A) apply to an installation. If the definition does not apply, the area may be classified as something other than a clothes closet, such as a bedroom. Other requirements may then be applied, such as 210.52 (Dwelling Unit Receptacle Outlets) and 210.70 (Lighting Outlets Required).

N Clothes Closet Storage Space. The area within a clothes closet in which combustible materials can be kept. (410) (CMP-18)

N Collector Rings. An assembly of slip rings for transferring electric energy from a stationary to a rotating member. (675) (CMP-7)

N Combiner (DC). (dc Combiner) (Direct-Current Combiner) An enclosure that includes devices used to connect two or more PV system dc circuits in parallel. (690) (CMP-4)

Δ Combustible Dust. Solid particles that are 500 μm or smaller (i.e., material passing a U.S. No. 35 Standard Sieve as defined in ASTM E11-17, *Standard Specification for Woven Wire Test Sieve Cloth and Test Sieves*) that can form an explosible mixture when suspended in air at standard atmospheric pressure and temperature. [**499**:3.3.3] (CMP-14)

> Informational Note: See ASTM E1226, *Standard Test Method for Explosibility of Dust Clouds*; ISO 6184-1, *Explosion protection systems — Part 1: Determination of explosion indices of combustible dusts in air*; or ANSI/UL 80079-20-2, *Explosive Atmospheres — Part 20-2: Material Characteristics — Combustible Dusts Test*

Methods, for procedures for determining the explosibility of dusts. Historically, explosibility has been described as presenting a flash fire or explosion hazard. It could be understood that potential hazards due to the formation of an explosible mixture when suspended in air at standard atmospheric pressure and temperature would include ignition.

N **Combustible Gas Detection System.** A protection technique utilizing stationary gas detectors in industrial establishments. (CMP-14)

N **Commissioning.** The process, procedures, and testing used to set up and verify the initial performance, operational controls, safety systems, and sequence of operation of electrical devices and equipment, prior to it being placed into active service. (CMP-13)

N **Communications Circuit.** A metallic, fiber, or wireless circuit that provides voice/data (and associated power) for communications-related services between communications equipment. (CMP-16)

N **Communications Circuit, Network-Powered Broadband. (Network-Powered Broadband Communications Circuit)** The circuit extending from the communications utility's or service provider's serving terminal or tap up to and including the network interface unit (NIU). (830) (CMP-16)

   Informational Note: A typical one-family dwelling network-powered communications circuit consists of a communications drop or communications service cable and an NIU and includes the communications utility's serving terminal or tap where it is not under the exclusive control of the communications utility.

N **Communications Circuit, Premises. (Premises Communications Circuit)** The circuit that extends voice, audio, video, data, interactive services, telegraph (except radio), and outside wiring for fire alarm and burglar alarm from the service provider's network terminal to the customer's communications equipment. (840) (CMP-16)

Communications Equipment. The electronic equipment that performs the telecommunications operations for the transmission of audio, video, and data, and includes power equipment (e.g., dc converters, inverters, and batteries), technical support equipment (e.g., computers), and conductors dedicated solely to the operation of the equipment. (CMP-16)

   Informational Note: As the telecommunications network transitions to a more data-centric network, computers, routers, servers, and their powering equipment, are becoming essential to the transmission of audio, video, and data and are finding increasing application in communications equipment installations.

N **Communications Service Provider.** An organization, business, or individual that offers communications service to others. (CMP-16)

N **Community Antenna Television Circuit (CATV).** The circuit that extends community antenna television systems for audio, video, data, and interactive services from the service provider's network terminal to the appropriate customer equipment. (CMP-16)

N **Concealable Nonmetallic Extension.** A listed assembly of two, three, or four insulated circuit conductors within a nonmetallic jacket, an extruded thermoplastic covering, or a sealed nonmetallic covering. The classification includes surface extensions intended for mounting directly on the surface of walls or ceilings and concealed with paint, texture, joint compound, plaster, wallpaper, tile, wall paneling, or other similar materials. (CMP-6)

Concealed. Rendered inaccessible by the structure or finish of the building. (CMP-1)

   Informational Note: Wires in concealed raceways are considered concealed, even though they may become accessible by withdrawing them.

Raceways and cables supported or located within hollow frames or permanently enclosed by the finish of buildings are considered concealed. Raceways and cables in unfinished basements; in accessible underfloor areas or attics; or behind, above, or below panels designed to allow access and that may be removed without damage to the building structure or finish are not considered concealed. See the definition of *exposed (as applied to wiring methods).*

N **Concealed Knob-and-Tube Wiring.** A wiring method using knobs, tubes, and flexible nonmetallic tubing for the protection and support of single insulated conductors. (CMP-6)

Conductor, Bare. (Bare Conductor) A conductor having no covering or electrical insulation whatsoever. (CMP-6)

Δ **Conductor, Copper-Clad Aluminum. (Copper-Clad Aluminum Conductor)** Conductor drawn from a copper-clad aluminum rod, with the copper metallurgically bonded to an aluminum core. (CMP-6)

Conductor, Covered. (Covered Conductor) A conductor encased within material of composition or thickness that is not recognized by this *Code* as electrical insulation. (CMP-6)

The uninsulated grounded system conductor within the overall exterior jacket of a Type SE cable is an example of a covered conductor. Covered conductors should always be treated as bare conductors for working clearances — because the covering does not have a voltage rating, the conductors are effectively uninsulated. See the definition of *insulated conductor.*

Conductor, Insulated. (Insulated Conductor) A conductor encased within material of composition and thickness that is recognized by this *Code* as electrical insulation. (CMP-6)

N **Conductor, Insulated. (Insulated Conductor)** Overhead service conductor encased in a polymeric material adequate for the applied nominal voltage and any conductor types described in 310.4. (396) (CMP-6)

   Informational Note: See ICEA S-76-474-2011, *Standard for Neutral Supported Power Cable Assemblies with Weather-Resistant*

Extruded Insulation Rated 600 Volts, for information about overhead service conductors.

For the covering on a conductor to be considered insulation, it is generally required to pass minimum testing required by a product standard. One such product standard is UL 83, *Thermoplastic-Insulated Wires and Cables*. To meet the requirements of UL 83, specimens of finished single-conductor wires must pass specified tests that measure (1) resistance to flame propagation, (2) dielectric strength, even while immersed, and (3) resistance to abrasion, cracking, crushing, and impact. Only wires and cables that meet the minimum fire, electrical, and physical properties required by the applicable product standards are permitted to be marked with the letter designations found in Table 310.4(1). Unless a voltage rating is marked on the insulation, a conductor generally should be considered a covered conductor. However, Class 2 conductors are not permitted to have a marked voltage rating.

See also

310.4 for the requirements of insulated conductor construction and applications

N **Conductors, Outdoor Overhead. (Outdoor Overhead Conductors)** Single conductors, insulated, covered, or bare, installed outdoors on support structures in free air. (399) (CMP-6)

N **Conduit, Flexible Metal (FMC). (Flexible Metal Conduit)** A raceway of circular cross section made of helically wound, formed, interlocked metal strip. (CMP-8)

N **Conduit, High Density Polyethylene (HDPE). (High Density Polyethylene Conduit)** A nonmetallic raceway of circular cross section, with associated couplings, connectors, and fittings for the installation of electrical conductors. (CMP-8)

N **Conduit, Intermediate Metal (IMC). (Intermediate Metal Conduit)** A steel threadable raceway of circular cross section designed for the physical protection and routing of conductors and cables and for use as an equipment grounding conductor when installed with its integral or associated coupling and appropriate fittings. (CMP-8)

N **Conduit, Liquidtight Flexible Metal (LFMC). (Liquidtight Flexible Metal Conduit)** A raceway of circular cross section having an outer liquidtight, nonmetallic, sunlight-resistant jacket over an inner flexible metal core with associated couplings, connectors, and fittings for the installation of electric conductors. (CMP-8)

N **Conduit, Liquidtight Flexible Nonmetallic (LFNC). (Liquidtight Flexible Nonmetallic Conduit)** A raceway of circular cross section of various types as follows:

(1) A smooth seamless inner core and cover bonded together and having one or more reinforcement layers between the core and covers, designated as LFNC-A

(2) A smooth inner surface with integral reinforcement within the raceway wall, designated as LFNC-B

(3) A corrugated internal and external surface without integral reinforcement within the raceway wall, designated as LFNC-C (CMP-8)

Informational Note: FNMC is an alternative designation for LFNC.

N **Conduit, Nonmetallic Underground with Conductors (NUCC). (Nonmetallic Underground Conduit with Conductors)** A factory assembly of conductors or cables inside a nonmetallic, smooth wall raceway with a circular cross section. (CMP-8)

N **Conduit, Reinforced Thermosetting Resin (RTRC). (Reinforced Thermosetting Resin Conduit)** A rigid nonmetallic raceway of circular cross section, with integral or associated couplings, connectors, and fittings for the installation of electrical conductors and cables. (CMP-8)

N **Conduit, Rigid Metal (RMC). (Rigid Metal Conduit)** A threadable raceway of circular cross section designed for the physical protection and routing of conductors and cables and for use as an equipment grounding conductor when installed with its integral or associated coupling and appropriate fittings. (CMP-8)

N **Conduit, Rigid Polyvinyl Chloride (PVC). (Rigid Polyvinyl Chloride Conduit)** A rigid nonmetallic raceway of circular cross section, with integral or associated couplings, connectors, and fittings for the installation of electrical conductors and cables. (CMP-8)

Conduit Body. A separate portion of a conduit or tubing system that provides access through a removable cover(s) to the interior of the system at a junction of two or more sections of the system or at a terminal point of the system.

Boxes such as FS and FD or larger cast or sheet metal boxes are not classified as conduit bodies. (CMP-9)

Conduit bodies include the short-radius type as well as capped elbows and service-entrance elbows. Conduit bodies include the LB, LL, LR, C, T, and X designs. A typical conduit body is shown in Exhibit 100.10.

See also

300.15 and **Article 314** for rules on the usage of conduit bodies

Table 314.16(A) for Type FS and Type FD boxes, which are not classified as conduit bodies

EXHIBIT 100.10 Typical conduit bodies. *(Courtesy of Atkore International)*

N **Connector.** An electromechanical fitting. (393) (CMP-18)

N **Connector, Intercell. (Intercell Connector)** An electrically conductive bar or cable used to connect adjacent cells. (CMP-13)

N **Connector, Intertier. (Intertier Connector)** An electrical conductor used to connect two cells on different tiers of the same rack or different shelves of the same rack. (CMP-13)

N **Connector, Load. (Load Connector)** An electromechanical connector used for power from the busbar to utilization equipment. (393) (CMP-18)

N **Connector, Pendant. (Pendant Connector)** An electromechanical or mechanical connector used to suspend low-voltage luminaire or utilization equipment below the grid rail and to supply power to connect from the busbar to utilization equipment. (393) (CMP-18)

N **Connector, Power Feed. (Power Feed Connector)** An electromechanical connector used to connect the power supply to a power distribution cable, to connect directly to the busbar, or to connect from a power distribution cable to the busbar. (393) (CMP-18)

Connector, Pressure (Solderless). (Pressure Connector) A device that establishes a connection between two or more conductors or between one or more conductors and a terminal by means of mechanical pressure and without the use of solder. (CMP-1)

N **Connector, Rail to Rail. (Rail to Rail Connector)** An electromechanical connector used to interconnect busbars from one ceiling grid rail to another grid rail. (393) (CMP-18)

N **Connector Strip.** A metal wireway containing pendant or flush receptacles. (520) (CMP-15)

N **Container (as applied to batteries).** A single-cell or multicell vessel or jar that holds the plates, electrolyte, and other elements of a single unit in a battery. (CMP-13)

Continuous Load. A load where the maximum current is expected to continue for 3 hours or more. (CMP-2)

N **Control.** The predetermined process of connecting, disconnecting, increasing, or reducing electric power. (750) (CMP-13)

Control Circuit. The circuit of a control apparatus or system that carries the electric signals directing the performance of the controller but does not carry the main power current. (CMP-11)

N **Control Circuits, Fault-Tolerant External. (Fault-Tolerant External Control Circuits)** Those control circuits either entering or leaving the fire pump controller enclosure, which if broken, disconnected, or shorted will not prevent the controller from starting the fire pump from all other internal or external means and may cause the controller to start the pump under these conditions. (695) (CMP-13)

N **Control Device, Emergency Lighting. (Emergency Lighting Control Device)** A separate or integral device intended to perform one or more emergency lighting control functions. (700) (CMP-13)

> Informational Note: See UL 924, *Emergency Lighting and Power Equipment*, for information covering emergency lighting control devices.

Control Drawing. A drawing or other document provided by the manufacturer of the intrinsically safe or associated apparatus, or of the nonincendive field wiring apparatus or associated nonincendive field wiring apparatus, that details the allowed interconnections between the intrinsically safe and associated apparatus or between the nonincendive field wiring apparatus or associated nonincendive field wiring apparatus. (CMP-14)

> Informational Note: See the following standards for additional information:
>
> (1) ANSI/ISA/UL 120202, *Recommendations for the Preparation, Content, and Organization of Intrinsic Safety Control Drawings*
> (2) ANSI/UL 913, *Intrinsically Safe Apparatus and Associated Apparatus for Use in Class I, II, and III, Division 1, Hazardous (Classified) Locations*
> (3) ANSI/UL 60079-11, *Explosive Atmospheres — Part 11: Equipment Protection by Intrinsic Safety "i"*
> (4) ANSI/UL 121201, *Nonincendive Electrical Equipment for Use in Class I and II, Division 2 and Class III, Divisions 1 and 2 Hazardous (Classified) Locations*
> (5) ANSI/ISA RP 12.06.01, *Recommended Practice for Wiring Methods for Hazardous (Classified) Locations Instrumentation — Part 1: Intrinsic Safety*

N **Control Room.** An enclosed control space outside the hoistway, intended for full bodily entry, that contains the elevator motor controller. The room could also contain electrical and/or mechanical equipment used directly in connection with the elevator or dumbwaiter but not the electric driving machine or the hydraulic machine. (620) (CMP-12)

N **Control Space.** A space inside or outside the hoistway intended to be accessed with or without full bodily entry that contains the elevator motor controller. This space could also contain electrical and/or mechanical equipment used directly in connection with the elevator, dumbwaiter, escalator, moving walk, or platform lift, but not the electrical driving machine or the hydraulic machine. (620) (CMP-12)

N **Control System.** The overall system governing the starting, stopping, direction of motion, acceleration, speed, and retardation of the moving member. (620) (CMP-12)

Controller. A device or group of devices that serves to govern, in some predetermined manner, the electric power delivered to the apparatus to which it is connected. (CMP-1)

A controller may be a remote-controlled magnetic contactor, switch, circuit breaker, or other device that is normally used to start and stop motors and other apparatus. Stop-and-start stations and similar control circuit components that do not open the power conductors to the motor are not considered controllers.

N **Controller, Motion. (Motion Controller)** The electrical device(s) for that part of the control system that governs the acceleration, speed, retardation, and stopping of the moving member. (620) (CMP-12)

> Informational Note: The motor control function may be integral to the motion controller.

N **Controller, Motor. (Motor Controller)** Any switch or device that is normally used to start and stop a motor by making and breaking the motor circuit current. (CMP-11)

N **Controller, Operation. (Operation Controller)** The electrical device(s) for that part of the control system that initiates the starting, stopping, and direction of motion in response to a signal from an operating device. (620) (CMP-12)

Converter, DC-to-DC. (DC-to-DC Converter) A device that can provide an output dc voltage and current at a higher or lower value than the input dc voltage and current. (CMP-4)

N **Converter Circuit, DC-to-DC. (DC-to-DC Converter Circuit)** The dc circuit conductors connected to the output of a dc-to-dc converter. (CMP-4)

N **Converting Device.** That part of the heating equipment that converts input mechanical or electrical energy to the voltage, current, and frequency used for the heating applicator. A converting device consists of equipment using line frequency, all static multipliers, oscillator-type units using vacuum tubes, inverters using solid-state devices, or motor-generator equipment. (665) (CMP-12)

Cooking Unit, Counter-Mounted. (Counter-Mounted Cooking Unit) A cooking appliance designed for mounting in or on a counter and consisting of one or more heating elements, internal wiring, and built-in or mountable controls. (CMP-2)

Δ **Coordination, Selective. (Selective Coordination)** Localization of an overcurrent condition to restrict outages to the circuit or equipment affected, accomplished by the selection and installation of overcurrent protective devices and their ratings or settings for the full range of available overcurrents, from overload to the available fault current, and for the full range of overcurrent protective device opening times associated with those overcurrents. (CMP-10)

Fuses and circuit breakers have time/current characteristics that determine the time it takes to clear the fault for a given value of fault current. Selectivity occurs when the device closest to the fault opens before the next device upstream operates. Any fault on a branch circuit should open the branch-circuit breaker rather than the feeder overcurrent protection. All faults on a feeder should open the feeder overcurrent protection rather than the service overcurrent protection.

With coordinated overcurrent protection, the faulted or overloaded circuit is isolated by the selective operation of only the OCPD closest to the overcurrent condition. The main goal of selective coordination is to isolate the faulted portion of the electrical circuit quickly while at the same time maintaining power to the remainder of the electrical system. The electrical system overcurrent protection must guard against short circuits and ground faults to ensure that the resulting damage is minimized while other parts of the system not directly involved with the fault are kept operational until other protective devices clear the fault. Where a series-rated system is used, an upstream device in the series will operate to protect a downstream device. For example, a current-limiting fuse will limit the available fault current to the downstream circuit breaker.

See also

700.32, 701.32, and **708.54** for selective coordination requirements for emergency systems, legally required standby systems, and critical operations power systems, respectively

620.62 for selective coordination requirements for elevator feeders

517.31(G), in which coordination is required only for faults that exceed 0.1 second in duration

N **Cord, Flexible. (Flexible Cord)** Two or more flexible insulated conductors enclosed in a flexible covering. (CMP-6)

N **Cord Connector.** A contact device terminated to a flexible cord that accepts an attachment plug or other insertion device. (CMP-6)

Types TC-ER and TC-ER-HL as a wiring method in Articles 501, 502, 503, 505, and 506 require a method of termination that currently is identified as a *cord connector* in the hazardous location (HazLoc) industry. However, the wiring device industry also uses the term *cord connector* to refer to a female electrical connector. The differentiation is addressed by use of the denotation "[as applied to hazardous (classified) locations]."

Cord Connector [as applied to hazardous (classified) locations]. A fitting intended to terminate a cord to a box or similar device and reduce the strain at points of termination and might include an explosionproof, a dust-ignitionproof, or a flameproof seal. (CMP-14)

N **Cord Set.** A length of flexible cord having an attachment plug at one end and a cord connector at the other end. (CMP-6)

N **Corrosive Environment.** Areas or enclosures without adequate ventilation, where electrical equipment is located and pool sanitation chemicals are stored, handled, or dispensed. (680) (CMP-17)

> Informational Note No. 1: See *Advisory: Swimming Pool Chemical: Chlorine*, OSWER 90-008.1, June 1990, available from the EPA National Service Center for Environmental Publications (NSCEP) as sanitation chemicals and pool water are considered to pose a risk of corrosion (gradual damage or destruction of materials) due to the presence of oxidizers (e.g., calcium hypochlorite, sodium hypochlorite, bromine, chlorinated isocyanurates) and chlorinating agents that release chlorine when dissolved in water.

Informational Note No. 2: See ANSI/APSP-11, *Standard for Water Quality in Public Pools and Spas*, ANSI/ASHRAE 62.1, Table 6-4 Minimum Exhaust Rates, and *2021 International Swimming Pool and Spa Code (ISPSC)*, Section 324, including associated definitions and requirements concerning adequate ventilation of indoor spaces such as equipment and chemical storage rooms, which can reduce the likelihood of the accumulation of corrosive vapors. Chemicals such as chlorine cause severe corrosive and deteriorating effects on electrical connections, equipment, and enclosures when stored and kept in the same vicinity.

See also

680.14 for the general requirements covering the permitted wiring methods in a corrosive environment associated with swimming pools, fountains, and similar installations

N **Counter (Countertop).** A fixed or stationary surface typically intended for food preparation and serving, personal lavation, or laundering or a similar surface that presents a routine risk of spillage of larger quantities of liquids upon outlets mounted directly on or in the surface. (CMP-2)

Informational Note No. 1: See UL 498, *Receptacles and Attachment Plugs*, and UL 943, *Ground-Fault Circuit Interrupters*, which establish the performance evaluation criteria and construction criteria.

Informational Note No. 2: See 406.5(E), 406.5(G)(1), and 406.5(H) for information on receptacles for counters and countertops distinguished from receptacles for work surfaces.

N **Crane.** A mechanical device used for lifting or moving boats. [**303**:3.3.6] (555) (CMP-7)

N **Critical Branch.** A system of feeders and branch circuits supplying power for task illumination, fixed equipment, select receptacles, and select power circuits serving areas and functions related to patient care that are automatically connected to alternate power sources by one or more transfer switches during interruption of the normal power source. [**99**:3.3.30] (517) (CMP-15)

N **Critical Operations Areas, Designated (DCOA). (Designated Critical Operations Areas)** Areas within a facility or site designated as requiring critical operations power. (CMP-13)

N **Critical Operations Data System.** An information technology equipment system that requires continuous operation for reasons of public safety, emergency management, national security, or business continuity. (645) (CMP-12)

Similar to the application of Article 708 covering critical operations power systems, the designation of which information technology systems are critical in function is the responsibility of the AHJ, who in many cases may be an emergency management director or similar person rather than the electrical AHJ. Once the system is designated as being "critical," the AHJ responsible for approving the installation ensures compliance with the applicable requirements of Article 645.

N **Critical Operations Power Systems (COPS).** Power systems for facilities or parts of facilities that require continuous operation for the reasons of public safety, emergency management, national security, or business continuity. (CMP-13)

Cutout Box. An enclosure designed for surface mounting that has swinging doors or covers secured directly to and telescoping with the walls of the enclosure. (CMP-9)

N **Data Center, Modular (MDC). (Modular Data Center)** Prefabricated units, rated 1000 volts or less, consisting of an outer enclosure housing multiple racks or cabinets of information technology equipment (ITE) (e.g., servers) and various support equipment, such as electrical service and distribution equipment, HVAC systems, and the like. (646) (CMP-12)

Informational Note: A typical construction may use a standard ISO shipping container or other structure as the outer enclosure, racks or cabinets of ITE, service-entrance equipment and power distribution components, power storage such as a UPS, and an air or liquid cooling system. Modular data centers are intended for fixed installation, either indoors or outdoors, based on their construction and resistance to environmental conditions. MDCs can be configured as an all-in-one system housed in a single equipment enclosure or as a system with the support equipment housed in separate equipment enclosures.

N **DC Plugging Box.** A dc device consisting of one or more 2-pole, 2-wire, nonpolarized, non-grounding-type receptacles intended to be used on dc circuits only. (530) (CMP-15)

Dead-Front. Without live parts exposed to a person on the operating side of the equipment. (CMP-9)

Demand Factor. The ratio of the maximum demand of a system, or part of a system, to the total connected load of a system or the part of the system under consideration. (CMP-2)

N **Dental Office.** A building or part thereof in which the following occur:

(1) Examinations and minor treatments/procedures performed under the continuous supervision of a dental professional;
(2) Use of limited to minimal sedation and treatment or procedures that do not render the patient incapable of self-preservation under emergency conditions; and
(3) No overnight stays for patients or 24-hour operations. [**99**:3.3.38] (CMP-15)

Device. A unit of an electrical system, other than a conductor, that carries or controls electric energy as its principal function. (CMP-1)

Switches, circuit breakers, fuseholders, receptacles, attachment plugs, and lampholders that distribute or control but do not consume electrical energy are considered devices. Devices that consume incidental amounts of electrical energy in the performance of carrying or controlling electricity — such as a switch or a receptacle with an internal pilot light or a magnetic contactor — are considered devices and not utilization equipment. Although conductors are units of the electrical system, they are not devices.

N **Dielectric Heating.** Heating of a nominally insulating material due to its own dielectric losses when the material is placed in a varying electric field. (665) (CMP-12)

Disconnecting Means. A device, or group of devices, or other means by which the conductors of a circuit can be disconnected from their source of supply. (CMP-1)

N **Distribution Point (Center Yard Pole) (Meter Pole).** An electrical supply point from which service drops, service conductors, feeders, or branch circuits to buildings or structures utilized under single management are supplied. (547) (CMP-7)

> Informational Note: The service point is typically located at the distribution point.

N **Diversion Controller (Diversion Charge Controller) (Diversion Load Controller).** Equipment that regulates the output of a source or charging process by diverting power to direct-current or alternating-current loads or to an interconnected utility service. (CMP-13)

N **Diversion Load.** A load connected to a diversion charge controller or diversion load controller, also known as a dump load. (CMP-4)

N **Docking Facility.** A covered or open, fixed or floating structure that provides access to the water and to which boats are secured. [303:3.3.7] (555) (CMP-7)

Dormitory Unit. A building or a space in a building in which group sleeping accommodations are provided for more than 16 persons who are not members of the same family in one room, or a series of closely associated rooms, under joint occupancy and single management, with or without meals, but without individual cooking facilities. (CMP-2)

N **Drop Box.** A box containing pendant- or flush-mounted receptacles attached to a multiconductor cable via strain relief or a multipole connector. (520) (CMP-15)

Δ **Dust-Ignitionproof.** Equipment enclosed in a manner that excludes dusts and does not permit arcs, sparks, or heat otherwise generated or liberated inside of the enclosure to cause ignition of exterior accumulations or atmospheric suspensions of a specified dust on or in the vicinity of the enclosure. (CMP-14)

> Informational Note No. 1: See ANSI/UL 1203, *Explosion-Proof and Dust-Ignition-Proof Electrical Equipment for Use in Hazardous (Classified) Locations*, for additional information on dust-ignitionproof enclosures.
> Informational Note No. 2: See NEMA 250, *Enclosures for Electrical Equipment (1000 Volts Maximum)*, for additional information on dust-ignitionproof enclosures that are sometimes marked additionally marked Type 9.

Δ **Dusttight.** Enclosures constructed so that dust will not enter under specified test conditions. (CMP-14)

> Informational Note No. 1: See ANSI/UL 121201, *Nonincendive Electrical Equipment for Use in Class I and II, Division 2 and Class III, Divisions 1 and 2 Hazardous (Classified) Locations*, for additional information.
> Informational Note No. 2: See NEMA 250, *Enclosures for Electrical Equipment (1000 Volts Maximum)*, and ANSI/UL 50E,

Enclosures for Electrical Equipment, Environmental Considerations, for additional information on enclosure Types 3, 3X, 3S, 3SX, 4, 4X, 5, 6, 6P, 12, 12K, and 13 that are considered dusttight.

Some dusttight constructions are only for wind-blown dust and may not be designed for combustible dusts found in Class II hazardous locations. The basic standard used to investigate dusttight enclosures for Class II, Division 2 locations is UL 1604, *Standard for Safety Electrical Equipment for Use in Class I and II, Division 2, and Class III Hazardous (Classified) Locations*. A dusttight enclosure has been determined to exclude dust under specified test conditions. Combustible dust is 500 microns or smaller. The AHJ determines the suitability of a dusttight enclosure or the acceptance of a specific standard, test, or listing organization.

See also

110.28 and **Table 110.28** for enclosure requirements, as well as the commentary following the definition of *enclosure*

Duty, Continuous. (Continuous Duty) Operation at a substantially constant load for an indefinitely long time. (CMP-1)

Duty, Intermittent. (Intermittent Duty) Operation for alternate intervals of (1) load and no load; or (2) load and rest; or (3) load, no load, and rest. (CMP-1)

Duty, Periodic. (Periodic Duty) Intermittent operation in which the load conditions are regularly recurrent. (CMP-1)

Duty, Short-Time. (Short-Time Duty) Operation at a substantially constant load for a short and definite, specified time. (CMP-1)

Duty, Varying. (Varying Duty) Operation at loads, and for intervals of time, both of which may be subject to wide variation. (CMP-1)

Dwelling, One-Family. (One-Family Dwelling) A building that consists solely of one dwelling unit. (CMP-1)

Dwelling, Two-Family. (Two-Family Dwelling) A building that consists solely of two dwelling units. (CMP-1)

Dwelling, Multifamily. (Multifamily Dwelling) A building that contains three or more dwelling units. (CMP-1)

Dwelling Unit. A single unit, providing complete and independent living facilities for one or more persons, including permanent provisions for living, sleeping, cooking, and sanitation. (CMP-2)

Where dwelling units are referenced throughout the NEC, rooms in motels, hotels, and similar occupancies that have permanent provisions for living, sleeping, cooking, and sanitation are also classified as dwelling units. Exhibit 100.11 illustrates a motel or hotel room that meets this definition. Dorm rooms are not usually considered to be dwelling units because they do not meet the definition.

Electric-Discharge Lighting. Systems of illumination utilizing fluorescent lamps, high-intensity discharge (HID) lamps, or neon tubing. (CMP-18)

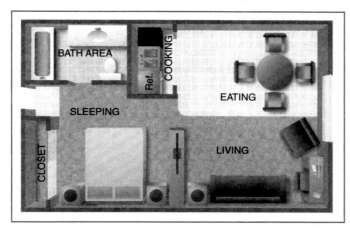

EXHIBIT 100.11 An example of a motel or hotel room considered to be a dwelling unit.

Electric Power Production and Distribution Network. Power production, distribution, and utilization equipment and facilities, such as electric utility systems that are connected to premises wiring and are external to and not controlled by a system that operates in interactive mode. (CMP-13)

Electric Sign. A fixed, stationary, or portable self-contained, electrically operated and/or electrically illuminated utilization equipment with words or symbols designed to convey information or attract attention. (CMP-18)

N **Electric Supply Stations.** Locations containing the generating stations and substations, including their associated generator, storage battery, transformer, and switchgear areas. (CMP-4)

Δ **Electric Vehicle (EV).** An automotive-type vehicle for on-road use, such as passenger automobiles, buses, trucks, vans, neighborhood electric vehicles, and electric motorcycles, primarily powered by an electric motor that draws current from a rechargeable storage battery, fuel cell, photovoltaic array, or other source of electric current. Plug-in hybrid electric vehicles (PHEV) are electric vehicles having a second source of motive power. (CMP-12)

> Informational Note: Off-road, self-propelled electric mobile machines, such as industrial trucks, hoists, lifts, transports, golf carts, airline ground support equipment, tractors, and boats are not considered electric vehicles.

The primary difference between electric vehicles (EVs) as defined in Article 625 and EVs covered by other sections in the *NEC* is in their road and highway worthiness. Automotive EVs are comparable in performance and function to conventional automobiles and light trucks. Automotive EVs must be capable of complying with the Federal Motor Vehicle Safety Standards and other U.S. Department of Transportation (DOT), National Highway Traffic Safety Administration (NHTSA), and U.S. Environmental Protection Agency (EPA) requirements.

A plug-in hybrid electric vehicle can be charged through either its own motive energy storage system or through connection to EV supply equipment located at a home, business, or other location.

The definition of the term *electric vehicle* includes neighborhood electric vehicles (NEVs), which are low-speed, limited-use EVs similar to golf carts but provided with automotive-grade headlights, seat belts, windshields, brakes, and other safety equipment that makes them street legal. Under NHTSA guidelines, the intended use for NEVs is in inner-city areas and planned and retirement communities where the street speed limit is 35 miles per hour or less.

EVs can also be used as a power source for an optional standby system, as covered in Article 702, or as an interconnected electric power production source, as covered in Article 705.

See also

625.48, which covers the transfer of power by the EV supply equipment from the premises to the EV or from the EV to the premises wiring through an interactive system

N **Electric Vehicle Connector.** A device that, when electrically coupled (conductive or inductive) to an electric vehicle inlet, establishes an electrical connection to the electric vehicle for the purpose of power transfer and information exchange. (625) (CMP-12)

> Informational Note: See 625.48 for further information on interactive systems.

N **Electric Vehicle Power Export Equipment (EVPE).** The equipment, including the outlet on the vehicle, that is used to provide electrical power at voltages greater than or equal to 30 Vac or 60 Vdc to loads external to the vehicle, using the vehicle as the source of supply. (625) (CMP-12)

> Informational Note: Electric vehicle power export equipment and electric vehicle supply equipment or wireless power transfer equipment are sometimes contained in one piece of equipment, sometimes referred to as a bidirectional electric vehicle supply equipment (EVSE) or bidirectional wireless power transfer equipment (WPTE).

N **Electric Vehicle Supply Equipment (EVSE).** Equipment for plug-in charging, including the ungrounded, grounded, and equipment grounding conductors, and the electric vehicle connectors, attachment plugs, personnel protection system, and all other fittings, devices, power outlets, or apparatus installed specifically for the purpose of transferring energy between the premises wiring and the electric vehicle. (625) (CMP-12)

> Informational Note: Electric vehicle power export equipment and electric vehicle supply equipment or wireless power transfer equipment (WPTE) are sometimes contained in one piece of equipment, sometimes referred to as a bidirectional EVSE or bidirectional WPTE.

EV supply equipment comprises the components between the skin of the EV and the premises wiring, including any flexible cable, disconnecting means, enclosures, power outlet, and EV connector. The defined term includes all off-vehicle charging

EXHIBIT 100.12 EV supply equipment and a connection to a vehicle.

equipment and does not include charging equipment installed on the vehicle. See Exhibit 100.12 for an EV charging station at a business and a vehicle connector plugged into a vehicle.

Electrical Circuit Protective System. A system consisting of components and materials intended for installation as protection for specific electrical wiring systems with respect to the disruption of electrical circuit integrity upon exterior fire exposure. (CMP-16)

N Electrical Datum Plane. A specified vertical distance above the normal high-water level at which electrical equipment can be installed and electrical connections can be made. (CMP-7)

Throughout Article 555, the physical location of electrical equipment is referenced to the electrical datum plane, which is used as a horizontal benchmark on land and on floating piers. The definition of the term *electrical datum plane* encompasses areas subject to tidal movement and areas in which the water level is affected only by conditions such as climate (rain or snowfall) or by human intervention (the opening or closing of dams and floodgates). In either case, the term covers the normal highest water level, such as astronomical high tides. The term does not cover extremes due to natural or manmade disasters.

N Electrical Ducts. Electrical conduits, or other raceways round in cross section, that are suitable for use underground or embedded in concrete. (CMP-6)

N Electrical Life Support Equipment. Electrically powered equipment whose continuous operation is necessary to maintain a patient's life. [**99**:3.3.45] (517) (CMP-15)

Δ Electrical Resistance Trace Heating "60079-30-1". Type of protection for the purpose of producing heat on the principle of electrical resistance and typically composed of one or more metallic conductors and/or an electrically conductive material, suitably electrically insulated and protected. (CMP-14)

Informational Note: See ANSI/UL 60079-30-1, *Explosive Atmospheres — Part 30-1: Electrical Resistance Trace Heating — General and Testing Requirements*, for additional information.

N Electrically Connected. A connection capable of carrying current as distinguished from connection through electromagnetic induction. (668) (CMP-12)

N Electrified Truck Parking Space. A truck parking space that has been provided with an electrical system that allows truck operators to connect their vehicles while stopped and to use off-board power sources in order to operate on-board systems such as air conditioning, heating, and appliances, without any engine idling. (626) (CMP-12)

Informational Note: An electrified truck parking space also includes dedicated parking areas for heavy-duty trucks at travel plazas, warehouses, shipper and consignee yards, depot facilities, and border crossings. It does not include areas such as the shoulders of highway ramps and access roads, camping and recreational vehicle sites, residential and commercial parking areas used for automotive parking or other areas where ac power is provided solely for the purpose of connecting automotive and other light electrical loads, such as engine block heaters, and at private residences.

N Electrified Truck Parking Space Wiring Systems. All of the electrical wiring, equipment, and appurtenances related to electrical installations within an electrified truck parking space, including the electrified parking space supply equipment. (626) (CMP-12)

N Electrolyte. The medium that provides the ion transport mechanism between the positive and negative electrodes of a cell. (CMP-13)

N Electrolytic Cell. A tank or vat in which electrochemical reactions are caused by applying electric energy for the purpose of refining or producing usable materials. (668) (CMP-12)

N Electrolytic Cell Line Working Zone. The space envelope wherein operation or maintenance is normally performed on or in the vicinity of exposed energized surfaces of electrolytic cell lines or their attachments. (668) (CMP-12)

N Electronic Power Converter. A device that uses power electronics to convert one form of electrical power into another form of electrical power. (CMP-4)

Informational Note: Examples of electronic power converters include, but are not limited to, inverters, dc-to-dc converters, and electronic charge controllers. These devices have limited current capabilities based on the device ratings at continuous rated power.

N Electronically Protected. A motor provided with electronic control that is an integral part of the motor and protects the motor against dangerous overheating due to failure of the electronic control, overload, and failure to start. (430) (CMP-11)

N Emergency Luminaire, Battery-Equipped. (Battery-Equipped Emergency Luminaire) A luminaire with a rechargeable battery, a battery charging means, and an automatic load control relay. (CMP-13)

N Emergency Luminaire, Directly Controlled. (Directly Controlled Emergency Luminaire) A luminaire supplied by the facility emergency power system and with a control input for dimming or switching that provides an emergency illumination level upon loss of normal power. (700) (CMP-13)

> Informational Note: See ANSI/UL 924, *Emergency Lighting and Power Equipment*, for information covering directly controlled emergency luminaires.

N Emergency Power Supply (EPS). The source(s) of electric power of the required capacity and quality for an emergency power supply system (EPSS). (CMP-13)

N Emergency Power Supply System (EPSS). A complete functioning EPS system coupled to a system of conductors, disconnecting means and overcurrent protective devices, transfer switches, and all control, supervisory, and support devices up to and including the load terminals of the transfer equipment needed for the system to operate as a safe and reliable source of electric power. [110:3.3.4] (CMP-13)

N Emergency Systems. Those systems legally required and classed as emergency by municipal, state, federal, or other codes, or by any governmental agency having jurisdiction. These systems are intended to automatically supply illumination, power, or both, to designated areas and equipment in the event of failure of the normal supply or in the event of accident to elements of a system intended to supply, distribute, and control power and illumination essential for safety to human life. (CMP-13)

Δ Encapsulation "m". Type of protection where electrical parts that could ignite an explosive atmosphere by either sparking or heating are enclosed in a compound in such a way that this explosive atmosphere cannot be ignited. (CMP-14)

> Informational Note: See ANSI/UL 60079-18, *Explosive atmospheres — Part 18: Equipment protection by encapsulation "m"*, for additional information.

Enclosed. Surrounded by a case, housing, fence, or wall(s) that prevents persons from accidentally contacting energized parts. (CMP-1)

N Enclosed-Break. Having electrical make-or-break contacts such that, if an internal explosion of the flammable gas or vapor that can enter it occurs, the device will withstand the internal explosion without suffering damage and without communicating the internal explosion to the external flammable gas or vapor. (CMP-14)

> Informational Note: See ANSI/UL 121201, *Nonincendive Electrical Equipment for Use in Class I and II, Division 2 and Class III, Divisions 1 and 2 Hazardous (Classified) Locations*, for additional information.

Enclosure. The case or housing of apparatus, or the fence or walls surrounding an installation to prevent personnel from accidentally contacting energized parts or to protect the equipment from physical damage. (CMP-1)

> Informational Note: See Table 110.28 for examples of enclosure types.

Enclosures are required by 110.28 to be marked with a number that identifies the environmental conditions in which that type of enclosure can be used. Enclosures that comply with the requirements for more than one type of enclosure are marked with multiple designations.

See also

110.28 and its commentary for details on enclosure markings and types

Table 110.28 for a list of the types of enclosures required to be used in specific locations

Energized. Electrically connected to, or is, a source of voltage. (CMP-1)

The term *energized* is not limited to equipment that is "connected to a source of voltage." Equipment such as batteries, capacitors, and conductors with induced voltages must also be considered energized.

N Energized, Likely to Become. (Likely to Become Energized) Conductive material that could become energized because of the failure of electrical insulation or electrical spacing. (CMP-5)

N Energy Management System (EMS). A system consisting of any of the following: a monitor(s), communications equipment, a controller(s), a timer(s), or other device(s) that monitors and/or controls an electrical load or a power production or storage source. (CMP-13)

N Energy Storage System (ESS). One or more devices installed as a system capable of storing energy and providing electrical energy into the premises wiring system or an electric power production and distribution network. (CMP-13)

> Informational Note No. 1: An ESS(s) can include but is not limited to batteries, capacitors, and kinetic energy devices (e.g., flywheels and compressed air). An ESS(s) can include inverters or converters to change voltage levels or to make a change between an ac or a dc system.

> Informational Note No. 2: These systems differ from a stationary standby battery installation where a battery spends the majority of the time on continuous float charge or in a high state of charge, in readiness for a discharge event.

N Entertainment Device. A mechanical or electromechanical device that provides an entertainment experience. (522) (CMP-15)

> Informational Note: These devices can include animated props, show action equipment, animated figures, and special effects, coordinated with audio and lighting to provide an entertainment experience.

Equipment. A general term, including fittings, devices, appliances, luminaires, apparatus, machinery, and the like used as a part of, or in connection with, an electrical installation. (CMP-1)

Equipment, Mobile. (Mobile Equipment) Equipment with electrical components that is suitable to be moved only with mechanical aids or is provided with wheels for movement by a person(s) or powered devices. (513) (CMP-14)

N **Equipment, Portable. (Portable Equipment)** Equipment fed with portable cords or cables intended to be moved from one place to another. (640) (CMP-12)

Equipment, Portable. (Portable Equipment) Equipment with electrical components suitable to be moved by a single person without mechanical aids. (511) (CMP-14)

N **Equipment, Portable. (Portable Equipment)** Equipment fed with portable cords or cables intended to be moved from one place to another. (520) (CMP-15)

N **Equipment, Portable. (Portable Equipment)** Equipment intended to be moved from one place to another. (530) (CMP-15)

N **Equipment, Signal. (Signal Equipment)** Includes audible and visual equipment such as chimes, gongs, lights, and displays that convey information to the user. (620) (CMP-12)

N **Equipment Branch.** A system of feeders and branch circuits arranged for delayed, automatic, or manual connection to the alternate power source and that serves primarily 3-phase power equipment. [**99**:3.3.50] (517) (CMP-15)

N **Equipment Protection Level (EPL).** Level of protection assigned to equipment based on its likelihood of becoming a source of ignition, and distinguishing the differences between explosive gas atmospheres and explosive dust atmospheres. (CMP-14)

> Informational Note: See ANSI/UL 60079-0, *Explosive Atmospheres — Part 0: Equipment — General Requirements*, for additional information.

N **Equipment Rack.** A framework for the support, enclosure, or both, of equipment; can be portable or stationary. (640) (CMP-12)

> Informational Note: See EIA/ECA 310-E-2005, *Cabinets, Racks, Panels and Associated Equipment*, for examples of equipment racks.

Energy Information Administration (EIA)/Energy Analytics Institute (ECA) 310-D, *Cabinets, Racks, Panels and Associated Equipment*, is the standard for commercial equipment racks. Within Article 640, both the terms *equipment rack* and *rack* are used to refer to equipment enclosures that are conceptually similar in intended use to those defined by the American National Standards Institute (ANSI)/EIA standard.

Equipotential Plane. Conductive parts bonded together to reduce voltage gradients in a designated area. (682) (CMP-17)

N **Equipotential Plane.** Conductive elements that are connected together to minimize voltage differences. (CMP-7)

N **Essential Electrical System.** A system comprised of alternate power sources and all connected distribution systems and ancillary equipment, designed to ensure continuity of electrical power

to designated areas and functions of a health care facility during disruption of normal power sources, and also to minimize disruption within the internal wiring system. [**99**:3.3.52] (517) (CMP-15)

Emergency systems in occupancies other than health care occupancies are installed primarily for life safety and building evacuation. The essential system in a hospital supplies power to equipment that provides life safety functions such as egress lighting and exit marking. It also supplies power to equipment directly related to patient care, including task illumination, fixed medical equipment, selected receptacles, and other special power circuits.

Δ **Explosionproof Equipment.** Equipment enclosed in a case that is capable of withstanding an explosion of a specified gas or vapor that might occur within it, that is capable of preventing the ignition of a specified gas or vapor surrounding the enclosure by sparks, flashes, or explosion of the gas or vapor within, and that operates at such an external temperature that a surrounding flammable atmosphere will not be ignited. (CMP-14)

> Informational Note No. 1: See ANSI/UL 1203, *Explosion-Proof and Dust-Ignition-Proof Electrical Equipment for Use in Hazardous (Classified) Locations*, for additional information.
>
> Informational Note No. 2: See NEMA 250, *Enclosures for Electrical Equipment (1000 Volts Maximum)*, for additional information on explosionproof enclosures that are sometimes additionally marked Type 7.

Exposed (as applied to live parts). Capable of being inadvertently touched or approached nearer than a safe distance by a person. (CMP-1)

> Informational Note: This term applies to parts that are not suitably guarded, isolated, or insulated.

Exposed (as applied to wiring methods). On or attached to the surface or behind panels designed to allow access. (CMP-1)

See Exhibit 100.2, which illustrates wiring methods that would be considered exposed because they are located above a suspended ceiling with lift-out panels.

N **Exposed (Optical Fiber Cable Exposed to Accidental Contact).** A conductive optical fiber cable in such a position that, in case of failure of supports or insulation, contact between the cable's non–current-carrying conductive members and an electrical circuit might result. (CMP-16)

N **Exposed (to Accidental Contact).** A circuit in such a position that, in case of failure of supports or insulation, contact with another circuit may result. (CMP-16)

N **Exposed Conductive Surfaces.** Those surfaces that are capable of carrying electric current and that are unprotected, uninsulated, unenclosed, or unguarded, permitting personal contact. [**99**:3.3.54] (517) (CMP-15)

> Informational Note: Paint, anodizing, and similar coatings are not considered suitable insulation, unless they are listed for such use.

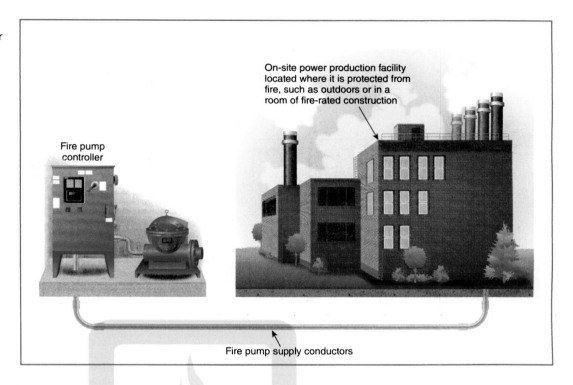

On-site power production facility located where it is protected from fire, such as outdoors or in a room of fire-rated construction

Fire pump controller

Fire pump supply conductors

Externally Operable. Capable of being operated without exposing the operator to contact with live parts. (CMP-1)

N **Facility, On-Site Power Production. (On-Site Power Production Facility)** The normal supply of electric power for the site that is expected to be constantly producing power. (695) (CMP-13)

An on-site power production facility differs in normal application from an on-site standby system such as an engine generator in that it is the normal continuous source of electrical supply for a structure and is not a utility-owned generating facility. Exhibit 100.13 illustrates a cogeneration facility that is the normal source of power for the premises wiring system and meets the definition of *on-site power production facility*. On-site power production is not restricted to generators.

N **Fastened-in-Place.** Mounting means of equipment in which the fastening means are specifically designed to permit removal without the use of a tool. (625) (CMP-12)

N **Fault-Managed Power (FMP).** A powering system that monitors for faults and controls current delivered to ensure fault energy is limited. (726) (CMP-3)

Informational Note No. 1: The monitoring and control systems differentiate fault-managed power from electric light and power circuits; therefore, alternative requirements to those of Chapters 1 through 4 are given regarding minimum wire sizes, ampacity adjustment and correction factors, overcurrent protection, insulation requirements, and wiring methods and materials.
Informational Note No. 2: A fault-managed power circuit is also commonly referred to as a Class 4 circuit.

Fault Current. The current delivered at a point on the system during a short-circuit condition. (CMP-10)

Δ **Fault Current, Available. (Available Fault Current)** The largest amount of current capable of being delivered at a point on the system during a short-circuit condition. (CMP-10)

Informational Note: A short-circuit can occur during abnormal conditions such as a fault between circuit conductors or a ground fault. See Informational Note Figure 100.1.

N **Fault Protection Device.** An electronic device that is intended for the protection of personnel and functions under fault conditions, such as network-powered broadband communications cable short or open circuit, to limit the current or voltage, or both, for a low-power network-powered broadband communications circuit and provide acceptable protection from electric shock. (830) (CMP-16)

Feeder. All circuit conductors between the service equipment, the source of a separately derived system, or other power supply source and the final branch-circuit overcurrent device. (CMP-10)

See the commentary following the definition of *branch circuit*, including Exhibit 100.8, which illustrates the difference between branch circuits and feeders.

N **Feeder Assembly.** The overhead or under-chassis feeder conductors, including the equipment grounding conductor, together with the necessary fittings and equipment; or the power-supply cord assembly for a mobile home, recreational vehicle, or park trailer, identified for the delivery of energy from the source of electrical supply to the panelboard within the mobile home, recreational vehicle, or park trailer. (CMP-7)

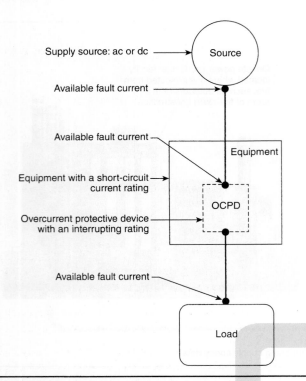

Supply source: ac or dc → Source

Available fault current

Available fault current

Equipment

Equipment with a short-circuit
current rating →

OCPD

Overcurrent protective device
with an interrupting rating →

Available fault current

Load

INFORMATIONAL NOTE FIGURE 100.1
Available Fault Current.

Festoon Lighting. A string of outdoor lights that is suspended between two points. (CMP-18)

Δ **Fibers/Flyings, Combustible. (Combustible Fibers/Flyings)** Fibers/flyings, where any dimension is greater than 500 μm in nominal size, which can form an explosible mixture when suspended in air at standard atmospheric pressure and temperature. [**499**:3.3.4.1] (CMP-14)

> Informational Note No. 1: This definition and Informational Notes No. 2 and No. 3 have been extracted from NFPA 499-2021, *Recommended Practice for the Classification of Combustible Dusts and of Hazardous (Classified) Locations for Electrical Installations in Chemical Process Areas*. The NFPA 499 reference is in brackets. Only editorial changes were made to the extracted text to make it consistent with this *Code*.
>
> Informational Note No. 2: Section 500.5(D) defines a Class III location. Combustible fibers/flyings can be similar in physical form to ignitible fibers/flyings and protected using the same electrical equipment installation methods. Examples of fibers/flyings include flat platelet-shaped particulate, such as metal flake, and fibrous particulate, such as particle board core material. If the smallest dimension of a combustible material is greater than 500 μm, it is unlikely that the material would be combustible fibers/flyings, as determined by test. Finely divided solids with lengths that are large compared to their diameter or thickness usually do not pass through a 500 μm sieve, yet when tested could potentially be determined to be explosible. [**499**:A.3.3.4.1]
>
> Informational Note No. 3: See ASTM E1226, *Standard Test Method for Explosibility of Dust Clouds*, ISO 6184-1, *Explosion protection systems — Part 1: Determination of explosion indices of combustible dusts in air*, or ISO/IEC/UL 80079-20-2,

Explosive atmospheres — Part 20-2: Material characteristics — Combustible dusts test methods, for procedures for determining the explosibility of dusts. A material that is found to not present an explosible mixture could still be an ignitible fiber/flying, as defined in this article. Historically, the explosibility condition has been described as presenting a flash fire or explosion hazard. It could be understood that the potential hazard due to the formation of an explosible mixture when suspended in air at standard atmospheric pressure and temperature would include ignition. [**499**:A.3.3.4.1]

Fibers/Flyings, Ignitible. (Ignitible Fibers/Flyings) Fibers/flyings where any dimension is greater than 500 μm in nominal size, which are not likely to be in suspension in quantities to produce an explosible mixture, but could produce an ignitible layer fire hazard. [**499**:3.3.4.2] (CMP-14)

> Informational Note No. 1: This definition and Informational Note No. 2 have been extracted from NFPA 499-2021, *Recommended Practice for the Classification of Combustible Dusts and of Hazardous (Classified) Locations for Electrical Installations in Chemical Process Areas*. The NFPA 499 reference is in brackets. Only editorial changes were made to the extracted text to make it consistent with this *Code*.
>
> Informational Note No. 2: Section 500.5 of this *Code* prescribes a Class III location as one where ignitible fibers/flyings are present, but not likely to be in suspension in the air in quantities sufficient to produce ignitible mixtures. This description addresses fibers/flyings that do not present a flash-fire hazard or explosion hazard by test. This could be because those fibers/flyings are too large or too agglomerated to be suspended in air in sufficient concentration, or at all, under typical test conditions. Alternatively, this could be because they burn so slowly that, when suspended in air, they do not propagate combustion at any concentration. In this document the zone classification system includes ignitible fibers/flyings as a fire hazard in a layer, which is not addressed in the IEC zone system (see IEC 60079-10-2, *Explosive atmospheres — Part 10-2: Classification of areas — Explosive dust atmospheres*). Where these are present, the user could also consider installation in accordance with Article 503 of this *Code*. [**499**:A.3.3.4.2]

Field Evaluation Body (FEB). An organization or part of an organization that performs field evaluations of electrical or other equipment. [**790**:3.3.4] (CMP-1)

> Informational Note: See NFPA 790-2021, *Standard for Competency of Third-Party Field Evaluation Bodies*, provides guidelines for establishing the qualification and competency of a body performing field evaluations of electrical products and assembles with electrical components.

Some laboratories that evaluate listed products also provide field labeling services. However, this definition does not limit field evaluation bodies to those that list products. A field evaluation usually does not extend beyond the specific equipment seen by the evaluator.

See also

NFPA 790, *Standard for Competency of Third-Party Field Evaluation Bodies*, and **NFPA 791,** *Recommended Practice and Procedures for Unlabeled Electrical Equipment Evaluation*

Field Labeled (as applied to evaluated products). Equipment or materials to which has been attached a label, symbol, or other identifying mark of an FEB indicating the equipment or materials were evaluated and found to comply with requirements as described in an accompanying field evaluation report. [790:3.3.6] (CMP-1)

N **Fire Alarm Circuit.** The portion of the wiring system between the load side of the overcurrent device or the power-limited supply and the connected equipment of all circuits powered and controlled by the fire alarm system. Fire alarm circuits are classified as either non–power-limited or power-limited. (CMP-3)

N **Fire Alarm Circuit, Non-Power-Limited (NPLFA). (Non-Power-Limited Fire Alarm Circuit)** A fire alarm circuit powered by a source that is not power limited. (CMP-3)

> Informational Note: See 760.41 and 760.43 for requirements for non-power-limited fire alarm circuits.

N **Fire Alarm Circuit, Power-Limited (PLFA). (Power-Limited Fire Alarm Circuit)** A fire alarm circuit powered by a power-limited source. (CMP-3)

> Informational Note: See 760.121 for requirements on power-limited fire alarm circuits.

Fitting. An accessory such as a locknut, bushing, or other part of a wiring system that is intended primarily to perform a mechanical rather than an electrical function. (CMP-1)

N **Fixed (as applied to equipment).** Equipment that is fastened or otherwise secured at a specific location. (680) (CMP-17)

N **Fixed-in-Place.** Mounting means of equipment using fasteners that require a tool for removal. (625) (CMP-12)

Δ **Flameproof "d".** Type of protection where the enclosure will withstand an internal explosion of a flammable mixture that has penetrated into the interior, without suffering damage and without causing ignition, through any joints or structural openings in the enclosure of an external explosive gas atmosphere consisting of one or more of the gases or vapors for which it is designed. (CMP-14)

> Informational Note: See ANSI/UL 60079-1, *Explosive Atmospheres — Part 1: Equipment Protection by Flameproof Enclosures "d"*, for additional information.

N **Flammable Anesthetics.** Gases or vapors, such as fluroxene, cyclopropane, divinyl ether, ethyl chloride, ethyl ether, and ethylene, that could form flammable or explosive mixtures with air, oxygen, or reducing gases such as nitrous oxide. (517) (CMP-15)

N **Flexible Bus Systems.** An assembly of flexible insulated bus, with a system of associated fittings used to secure, support, and terminate the bus. (CMP-8)

> Informational Note: Flexible bus systems are engineered systems for a specific site location and are ordinarily assembled at the point of installation from the components furnished or specified by the manufacturer.

N **Flexible Insulated Bus.** A flexible rectangular conductor with an overall insulation. (CMP-8)

N **Flywheel ESS (FESS).** A mechanical ESS composed of a spinning mass referred to as a rotor and an energy conversion mechanism such as a motor-generator that converts the mechanical energy to electrical energy. (706) (CMP-13)

> Informational Note: There are primarily two types of rotor constructions, solid metal mass design and composite fiber design.

N **Footlight.** A border light installed on or in the stage. (520) (CMP-15)

N **Forming Shell.** A structure designed to support a wet-niche luminaire assembly and intended for mounting in a pool or fountain structure. (680) (CMP-17)

N **Fountain.** An ornamental structure or recreational water feature from which one or more jets or streams of water are discharged into the air, including splash pads, ornamental pools, display pools, and reflection pools. The definition does not include drinking water fountains or water coolers. (680) (CMP-17)

N **Frame.** Chassis rail and any welded addition thereto of metal thickness of 1.35 mm (0.053 in.) or greater. (551) (CMP-7)

Free Air (as applied to conductors). Open or ventilated environment that allows for heat dissipation and air flow around an installed conductor. (CMP-6)

Fuel Cell. An electrochemical system that consumes fuel to produce an electric current. In such cells, the main chemical reaction used for producing electric power is not combustion. However, there may be sources of combustion used within the overall cell system, such as reformers/fuel processors. (CMP-4)

Fuel Cell System. The complete aggregate of equipment used to convert chemical fuel into usable electricity and typically consisting of a reformer, stack, power inverter, and auxiliary equipment. (CMP-4)

Fuse. An overcurrent protective device with a circuit-opening fusible part that is heated and severed by the passage of overcurrent through it. (CMP-10)

> Informational Note: A fuse comprises all the parts that form a unit capable of performing the prescribed functions. It may or may not be the complete device necessary to connect it into an electrical circuit.

Fuse, Electronically Actuated. (Electronically Actuated Fuse) An overcurrent protective device that generally consists of a control module that provides current-sensing, electronically derived time–current characteristics, energy to initiate tripping, and an interrupting module that interrupts current when an overcurrent occurs. Such fuses may or may not operate in a current-limiting fashion, depending on the type of control selected. (CMP-10)

Although they are called fuses because they interrupt current by melting a fusible element, electronically actuated fuses respond to a signal from an electronic control rather than from the heat generated by actual current passing through a fusible element. Electronically actuated fuses have controls similar to those of electronic circuit breakers.

△ **Fuse, Expulsion. (Expulsion Fuse)** A vented fuse unit in which the expulsion effect of gases produced by the arc and lining of the fuseholder, either alone or aided by a spring, extinguishes the arc. (CMP-10)

Fuse, Nonvented Power. (Nonvented Power Fuse) A fuse without intentional provision for the escape of arc gases, liquids, or solid particles to the atmosphere during circuit interruption. (CMP-10)

Fuse, Power. (Power Fuse) A vented, nonvented, or controlled vented fuse unit in which the arc is extinguished by being drawn through solid material, granular material, or liquid, either alone or aided by a spring. (CMP-10)

Fuse, Vented Power. (Vented Power Fuse) A fuse with provision for the escape of arc gases, liquids, or solid particles to the surrounding atmosphere during circuit interruption. (CMP-10)

△ **Garage.** A building or portion of a building in which one or more self-propelled vehicles can be kept for use, sale, storage, rental, repair, exhibition, or demonstration purposes. (CMP-1)

> Informational Note: See 511.1 for commercial garages, repair and storage.

Garage, Major Repair. (Major Repair Garage) A building or portions of a building where major repairs, such as engine overhauls, painting, body and fender work, welding or grinding, and repairs that require draining or emptying of the motor vehicle fuel tank are performed on motor vehicles, including associated floor space used for offices, parking, or showrooms. [30A:3.3.12.1] (CMP-14)

Garage, Minor Repair. (Minor Repair Garage) A building or portions of a building used for lubrication, inspection, and minor automotive maintenance work, such as engine tune-ups, replacement of parts, fluid changes (e.g., oil, antifreeze, transmission fluid, brake fluid, air-conditioning refrigerants), brake system repairs, tire rotation, and similar routine maintenance work, including the associated floor space used for offices, parking, or showrooms. [30A:3.3.12.2] (CMP-14)

N **General-Purpose Cables, Cable Routing Assemblies, and Raceways.** Cables, cable routing assemblies, and raceways are suitable for general-purpose applications and are resistant to the spread of fire. (722) (CMP-3)

Generating Capacity, Inverter. (Inverter Generating Capacity) The sum of parallel-connected inverter maximum continuous output power at 40°C in watts, kilowatts, volt-amperes, or kilovolt-amperes. (CMP-4)

N **Generating Station.** A plant wherein electric energy is produced by conversion from some other form of energy (e.g., chemical, nuclear, solar, wind, mechanical, or hydraulic) by means of suitable apparatus. (CMP-4)

N **Generator (Generator Set).** A machine that converts mechanical energy into electrical energy by means of a prime mover and alternator and/or inverter. (CMP-13)

N **Generator, On-Site Standby. (On-Site Standby Generator)** A facility producing electric power on site as the alternate supply of electric power. It differs from an on-site power production facility in that it is not constantly producing power. (695) (CMP-13)

N **Grid Bus Rail.** A combination of the busbar, the busbar support, and the structural suspended ceiling grid system. (393) (CMP-18)

Ground. The earth. (CMP-5)

Ground Fault. An unintentional, electrically conductive connection between an ungrounded conductor of an electrical circuit and the normally non-current-carrying conductors, metal enclosures, metal raceways, metal equipment, or earth. (CMP-5)

△ **Ground-Fault Circuit Interrupter (GFCI).** A device intended for the protection of personnel that functions to de-energize a circuit or portion thereof within an established period of time when a ground-fault current exceeds the values established for a Class A device. (CMP-2)

> Informational Note: See UL 943, *Standard for Ground-Fault Circuit Interrupters*, for further information. Class A ground-fault circuit interrupters trip when the ground-fault current is 6 mA or higher and do not trip when the ground-fault current is less than 4 mA.

See also

210.8 commentary, which includes a table that lists references to other GFCI requirements in the *NEC* as well as exhibits that provide further information on, and examples of, GFCIs

N **Ground-Fault Circuit Interrupter, Special Purpose (SPGFCI). (Special Purpose Ground-Fault Circuit Interrupter)** A device intended for the detection of ground-fault currents, used in circuits with voltage to ground greater than 150 volts, that functions to de-energize a circuit or portion of a circuit within an established period of time when a ground-fault current exceeds the values established for Class C, D, or E devices. (CMP-2)

> Informational Note: See UL 943C, *Outline of Investigation for Special Purpose Ground-Fault Circuit Interrupters*, for information on Classes C, D, or E special purpose ground-fault circuit interrupters.

Ground-Fault Current Path. An electrically conductive path from the point of a ground fault on a wiring system through normally non–current-carrying conductors, grounded conductors, equipment, or the earth to the electrical supply source. (CMP-5)

> Informational Note: Examples of ground-fault current paths are any combination of equipment grounding conductors, metallic raceways, metallic cable sheaths, electrical equipment, and any other electrically conductive material such as metal, water, and gas piping; steel framing members; stucco mesh; metal ducting; reinforcing steel; shields of communications cables; grounded conductors; and the earth itself.

Ground-Fault Current Path, Effective. (Effective Ground-Fault Current Path) An intentionally constructed, low-impedance electrically conductive path designed and intended to carry current during ground-fault events from the point of a ground fault on a wiring system to the electrical supply source and that facilitates the operation of the overcurrent protective device or ground-fault detectors. (CMP-5)

N Ground-Fault Detector-Interrupter, dc (GFDI). A device that provides protection for PV system dc circuits by detecting a ground fault and could interrupt the fault path in the dc circuit. (690) (CMP-4)

> Informational Note: See UL 1741, *Standard for Inverters, Converters, Controllers and Interconnection System Equipment for Use with Distributed Energy Resources*, and UL 62109, *Standard for Power Converters for use in Photovoltaic Power Systems*, for further information on GFDI equipment.

Ground-Fault Protection of Equipment (GFPE). A system intended to provide protection of equipment from damaging line-to-ground fault currents by operating to cause a disconnecting means to open all ungrounded conductors of the faulted circuit. This protection is provided at current levels less than those required to protect conductors from damage through the operation of a supply circuit overcurrent device. (CMP-5)

See also

230.95, 426.28, and **427.22** commentary for more information on ground-fault protection of equipment

Grounded (Grounding). Connected (connecting) to ground or to a conductive body that extends the ground connection. (CMP-5)

N Grounded, Functionally. (Functionally Grounded) A system that has an electrical ground reference for operational purposes that is not solidly grounded. (CMP-4)

> Informational Note: A functionally grounded system is often connected to ground through an electronic means internal to an inverter or charge controller that provides ground-fault protection. Examples of operational purposes for functionally grounded systems include ground-fault detection and performance-related issues for some power sources.

Grounded, Solidly. (Solidly Grounded) Connected to ground without inserting any resistor or impedance device. (CMP-5)

Grounded Conductor. A system or circuit conductor that is intentionally grounded. (CMP-5)

> Informational Note: Although an equipment grounding conductor is grounded, it is not considered a grounded conductor.

N Grounded System, Impedance. (Impedance Grounded System) An electrical system that is grounded by intentionally connecting the system neutral point to ground through an impedance device. (CMP-5)

Grounding Conductor, Equipment (EGC). (Equipment Grounding Conductor) A conductive path(s) that is part of an effective ground-fault current path and connects normally non-current-carrying metal parts of equipment together and to the system grounded conductor or to the grounding electrode conductor, or both. (CMP-5)

> Informational Note No. 1: It is recognized that the equipment grounding conductor also performs bonding.
> Informational Note No. 2: See 250.118 for a list of acceptable equipment grounding conductors.

N Grounding Conductor, Impedance. (Impedance Grounding Conductor) A conductor that connects the system neutral point to the impedance device in an impedance grounded system. (CMP-5)

Grounding Electrode. A conducting object through which a direct connection to earth is established. (CMP-5)

Grounding Electrode Conductor (GEC). A conductor used to connect the system grounded conductor or the equipment to a grounding electrode or to a point on the grounding electrode system. (CMP-5)

Refer to Exhibit 100.7 and Exhibit 250.1, which show a grounding electrode conductor in a typical grounding system for a single-phase, 3-wire service.

N Grouped. Cables or conductors positioned adjacent to one another but not in continuous contact with each other. (520) (CMP-15)

Guarded. Covered, shielded, fenced, enclosed, or otherwise protected by means of suitable covers, casings, barriers, rails, screens, mats, or platforms to remove the likelihood of approach or contact by persons or objects to a point of danger. (CMP-1)

Guest Room. An accommodation combining living, sleeping, sanitary, and storage facilities within a compartment. (CMP-2)

Guest Suite. An accommodation with two or more contiguous rooms comprising a compartment, with or without doors between such rooms, that provides living, sleeping, sanitary, and storage facilities. (CMP-2)

If a guest room or guest suite contains permanent provisions for cooking, dwelling unit requirements apply.

N Gutter, Metal Auxiliary. (Metal Auxiliary Gutter) A sheet metal enclosure used to supplement wiring spaces at meter centers, distribution centers, switchgear, switchboards, and similar points of wiring systems. The enclosure has hinged or removable covers for housing and protecting electrical wires, cable, and busbars. The enclosure is designed for conductors to be laid or set in place after the enclosures have been installed as a complete system. (CMP-8)

N Gutter, Nonmetallic Auxiliary. (Nonmetallic Auxiliary Gutter) A flame-retardant, nonmetallic enclosure used to supplement wiring spaces at meter centers, distribution centers, switchgear, switchboards, and similar points of wiring systems. The enclosure has hinged or removable covers for housing and protecting electrical wires, cable, and busbars. The enclosure is designed for conductors to be laid or set in place after the enclosures have been installed as a complete system. (CMP-8)

Habitable Room. A room in a building for living, sleeping, eating, or cooking, but excluding bathrooms, toilet rooms, closets, hallways, storage or utility spaces, and similar areas. (CMP-2)

EXHIBIT 100.14 An example of a handhole enclosure. *[Courtesy of Quazite® (Hubbell Lenoir City, Inc.)]*

Handhole Enclosure. An enclosure for use in underground systems, provided with an open or closed bottom, and sized to allow personnel to reach into, but not enter, for the purpose of installing, operating, or maintaining equipment or wiring or both. (CMP-9)

Handhole enclosures are required by 314.30 to be "identified" for use in underground systems. Exhibit 100.14 shows the installation of one type of handhole enclosure.

N **Hazard Current.** For a given set of connections in an isolated power system, the total current that would flow through a low impedance if it were connected between either isolated conductor and ground. [**99**:3.3.72] (517) (CMP-15)

N *Hazard Current, Fault. (Fault Hazard Current)* The hazard current of a given isolated power system with all devices connected except the line isolation monitor. [**99**:3.3.72.1] (517) (CMP-15)

N *Monitor Hazard Current.* The hazard current of the line isolation monitor alone. [**99**:3.3.72.2] (517) (CMP-15)

N *Total Hazard Current.* The hazard current of a given isolated system with all devices, including the line isolation monitor, connected. [**99**:3.3.72.3] (517) (CMP-15)

N **Header.** Transverse metal raceways for electrical conductors, providing access to predetermined cells of a precast cellular concrete floor, thereby permitting the installation of electrical conductors from a distribution center to the floor cells. (CMP-8)

N **Health Care Facilities.** Buildings, portions of buildings, or mobile enclosures in which human medical, dental, psychiatric, nursing, obstetrical, or surgical care is provided. [**99**:3.3.73] (CMP-15)

> Informational Note: Examples of health care facilities include, but are not limited to, hospitals, nursing homes, limited care facilities, clinics, medical and dental offices, and ambulatory care centers, whether permanent or movable.

The term *health care facility* should not be confused with the term *health care occupancy*. All health care occupancies, including ambulatory health care occupancies, are considered health care facilities; however, not all health care facilities are considered health care occupancies. A medical office building can be a medical facility, but under NFPA *101, Life Safety Code,* it typically would be a business occupancy. A hospital is a health care facility, but it can be several different occupancies (health care, assembly, business, ambulatory health care, and so forth).

The *NEC* does not designate the type of facility or level of care provided. The governing body of the facility determines the type of services being provided at a facility. For example, depending on the jurisdiction, a chiropractic office might or might not be a facility covered under Article 517.

N **Health Care Facility's Governing Body.** The person or persons who have the overall legal responsibility for the operation of a health care facility. [**99**:3.3.74] (517) (CMP-15)

N **Heating Equipment.** Any equipment that is used for heating purposes and whose heat is generated by induction or dielectric methods. (665) (CMP-12)

N **Heating Panel.** A complete assembly provided with a junction box or a length of flexible conduit for connection to a branch circuit. (CMP-17)

N **Heating Panel Set.** A rigid or nonrigid assembly provided with nonheating leads or a terminal junction assembly identified as being suitable for connection to a wiring system. (CMP-17)

N **Heating System.** A complete system consisting of components such as heating elements, fastening devices, nonheating circuit wiring, leads, temperature controllers, safety signs, junction boxes, raceways, and fittings. (426) (CMP-17)

N **Heating System, Impedance. (Impedance Heating System)** A system in which heat is generated in an object, such as a pipe, rod, or combination of such objects serving as a heating element, by causing current to flow through such objects by direct connection to an ac voltage source from an isolating transformer. In some installations the object is embedded in the surface to be heated or constitutes the exposed component to be heated. (CMP-17)

N **Heating System, Induction. (Induction Heating System)** A system in which heat is generated in a pipeline or vessel wall by inducing current in the pipeline or vessel wall from an external isolated ac field source. (CMP-17)

N **Heating System, Skin Effect. (Skin-Effect Heating System)** A system in which heat is generated on the inner surface of a ferromagnetic envelope embedded in or fastened to the surface to be heated.

> Informational Note: Typically, an electrically insulated conductor is routed through and connected to the envelope at the other end. The envelope and the electrically insulated conductor are connected to an ac voltage source from an isolating transformer. (CMP-17)

Hermetic Refrigerant Motor-Compressor. A combination consisting of a compressor and motor, both of which are enclosed in the same housing, with no external shaft or shaft seals, with the motor operating in the refrigerant. (CMP-11)

Hoistway. Any shaftway, hatchway, well hole, or other vertical opening or space in which an elevator or dumbwaiter is designed to operate. (CMP-12)

N **Hospital.** A building or portion thereof used on a 24-hour basis for the medical, psychiatric, obstetrical, or surgical care of four or more inpatients. [*101*:3.3.152] (CMP-15)

N **Host Sign.** A sign or outline lighting system already installed in the field that is designated for field conversion of the illumination system with a retrofit kit. (600) (CMP-18)

An existing electric sign that is being updated with new lighting components is referred to as the *host sign* in the installation instructions associated with a sign-specific retrofit kit. This definition correlates with the use of the term *host sign* in ANSI/UL 879A, *Standard for LED Sign and Sign Retrofit Kits*.

N **Hydromassage Bathtub.** A permanently installed bathtub equipped with a recirculating piping system, pump, and associated equipment. It is designed so it can accept, circulate, and discharge water upon each use. (680) (CMP-17)

Identified (as applied to equipment). Recognizable as suitable for the specific purpose, function, use, environment, application, and so forth, where described in a particular *Code* requirement. (CMP-1)

> Informational Note: Some examples of ways to determine suitability of equipment for a specific purpose, environment, or application include investigations by a qualified testing laboratory (listing and labeling), an inspection agency, or other organizations concerned with product evaluation.

In Sight From (Within Sight From) (Within Sight). Equipment that is visible and not more than 15 m (50 ft) distant from other equipment is *in sight from* that other equipment. (CMP-1)

> Informational Note: See 110.29 for additional information.

Δ **Increased Safety "e".** Type of protection applied to electrical equipment that does not produce arcs or sparks in normal service and under specified abnormal conditions, in which additional measures are applied to give increased security against the possibility of excessive temperatures and of the occurrence of arcs and sparks. (CMP-14)

> Informational Note: See ANSI/UL 60079-7, *Explosive Atmospheres — Part 7: Equipment Protection by Increased Safety "e"*, for additional information.

N **Induction Heating (Induction Melting) (Induction Welding).** The heating, melting, or welding of a nominally conductive material due to its own I2R losses when the material is placed in a varying electromagnetic field. (665) (CMP-12)

Induction and dielectric heating methods are used for ovens, furnaces, and industrial equipment where pieces of material are heated by a rapidly alternating magnetic or electric field.

Δ **Industrial Control Panel.** An assembly of two or more components consisting of one of the following: (1) power circuit components only, such as motor controllers, overload relays, fused disconnect switches, and circuit breakers; (2) control circuit components only, such as push buttons, pilot lights, selector switches, timers, switches, and control relays; (3) a combination of power and control circuit components. These components, with associated wiring and terminals, are mounted on, or contained within, an enclosure or mounted on a subpanel. (CMP-11)

> Informational Note: The industrial control panel does not include the controlled equipment.

N **Industrial Installation, Supervised. (Supervised Industrial Installation)** The industrial portions of a facility where all of the following conditions are met:

(1) Conditions of maintenance and engineering supervision ensure that only qualified persons monitor and service the system.
(2) The premises wiring system has 2500 kVA or greater of load used in industrial process(es), manufacturing activities, or both, as calculated in accordance with Article 220.
(3) The premises has at least one service or feeder that is more than 150 volts to ground and more than 300 volts phase-to-phase.

This definition excludes installations in buildings used by the industrial facility for offices, warehouses, garages, machine shops, and recreational facilities that are not an integral part of the industrial plant, substation, or control center. (240) (CMP-10)

All process or manufacturing loads from each low-, medium-, and high-voltage system can be added together to satisfy the load requirement. Loads not associated with manufacturing or processing cannot be used to meet the 2500-kilovolt-ampere minimum requirement. Loads are calculated in accordance with Article 220. Part VIII of Article 240 does not apply to electrical systems operating at over 1000 volts, nominal, or to electrical systems that serve separate facilities — such as offices, warehouses, garages, machine shops, or recreational buildings — that are not a part of the manufacturing or industrial process.

Δ **Information Technology Equipment (ITE).** Equipment and systems rated 1000 volts or less, normally found in offices or other business establishments and similar environments classified as ordinary locations, that are used for creation and manipulation of data, voice, video, and similar signals that are not communications equipment and do not process communications circuits. (CMP-12)

> Informational Note: See UL 60950-1, *Information Technology Equipment — Safety — Part 1: General Requirements*, or UL 62368-1, *Audio/Video Information and Communication Technology Equipment Part 1: Safety Requirements*, for information on listing requirements for both information technology equipment and communications equipment.

Information Technology Equipment Room. A room within the information technology equipment area that contains the information technology equipment. [*75*:3.3.15] (CMP-12)

Innerduct. A nonmetallic raceway placed within a larger raceway. (CMP-16)

N **Insulated Bus Pipe (IBP).** A cylindrical solid or hollow conductor with a solid insulation system, having conductive grading layers and a grounding layer imbedded in the insulation, and provided with an overall covering of insulating or metallic material. IBP is also referred to as tubular covered conductor (TCC). (CMP-8)

N **Insulated Bus Pipe System.** An assembly that includes bus pipe, connectors, fittings, mounting structures, and other fittings and accessories. (CMP-8)

N **Insulating End.** An insulator designed to electrically insulate the end of a flat conductor cable (Type FCC). (324) (CMP-6)

Interactive Mode. The operating mode for power production equipment or microgrids that operate in parallel with and are capable of delivering energy to an electric power production and distribution network or other primary source. (CMP-4)

> Informational Note: Interactive mode is an operational mode of both interactive systems and of equipment such as interactive inverters.

Interrupting Rating. The highest current at rated voltage that a device is identified to interrupt under standard test conditions. (CMP-10)

> Informational Note: Equipment intended to interrupt current at other than fault levels may have its interrupting rating implied in other ratings, such as horsepower or locked rotor current.

Interrupting ratings are essential for the coordination of electrical systems. Exhibit 100.15 depicts the label of a 225-ampere frame circuit breaker, showing the interrupting capacity ratings. See the definition of *short-circuit current rating.*

See also

110.9, 240.60(C), 240.83(C), and **240.86,** which deal specifically with interrupting ratings

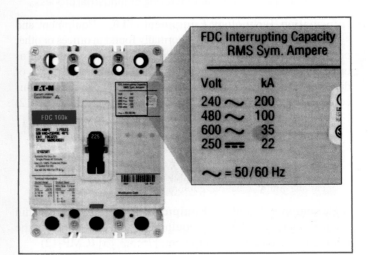

EXHIBIT 100.15 Interrupting ratings information on the label of a 225-ampere frame circuit breaker. *(Courtesy of Eaton)*

Intersystem Bonding Termination (IBT). A device that provides a means for connecting intersystem bonding conductors for communications systems to the grounding electrode system. (CMP-16)

An intersystem bonding termination is a dedicated location for terminating the equipment grounding conductors required in Chapter 8 and 770.93 to the service, building, or structure's grounding electrode system.

Δ **Intrinsic Safety "i".** Type of protection where any spark or thermal effect is incapable of causing ignition of a mixture of flammable or combustible material in air under prescribed test conditions. (CMP-14)

> Informational Note: See ANSI/UL 913, *Intrinsically Safe Apparatus and Associated Apparatus for Use in Class I, II, and III, Division 1, Hazardous (Classified) Locations*; and ANSI/UL 60079-11, *Explosive Atmospheres — Part 11: Equipment Protection by Intrinsic Safety "i"*, for additional information.

Intrinsically Safe Apparatus. Apparatus in which all the circuits are intrinsically safe. (CMP-14)

> Informational Note No. 1: See ANSI/UL 913, *Intrinsically Safe Apparatus and Associated Apparatus for Use in Class I, II, and III, Division 1, Hazardous (Classified) Locations*, and ANSI/UL 60079-11, *Explosive Atmospheres — Part 11: Equipment Protection by Intrinsic Safety "i"*, for additional information.
> Informational Note No. 2: See ANSI/ISA RP 12.06.01, *Recommended Practice for Wiring Methods for Hazardous (Classified) Locations Instrumentation — Part 1: Intrinsic Safety*, for installation information.

Δ **Intrinsically Safe Circuit.** A circuit in which any spark or thermal effect is incapable of causing ignition of a mixture of flammable or combustible material in air under prescribed test conditions. (CMP-14)

> Informational Note: See ANSI/UL 913, *Intrinsically Safe Apparatus and Associated Apparatus for Use in Class I, II, and III, Division 1, Hazardous (Classified) Locations*, and ANSI/UL 60079-11, *Explosive Atmospheres — Part 11: Equipment Protection by Intrinsic Safety "i"*, for test conditions.

Due to its physical and electrical characteristics, an intrinsically safe (IS) circuit does not develop, under normal or abnormal operating conditions, sufficient electrical energy (millijoules) in an arc or spark to cause ignition or sufficient thermal energy resulting from an overload condition to cause the temperature of the installed circuit to exceed the ignition temperature of a specified gas or vapor.

An abnormal condition can occur due to damage, failure of electrical components, excessive voltage, or improper adjustment or maintenance of the equipment. Abnormal conditions are mitigated by associated apparatus such as an IS barrier, an example of which is shown in Exhibit 504.1.

Intrinsically Safe Circuits, Different. (Different Intrinsically Safe Circuits) Intrinsically safe circuits in which the possible interconnections have not been evaluated and identified as intrinsically safe. (CMP-14)

Informational Note: See ANSI/UL 913, *Intrinsically Safe Apparatus and Associated Apparatus for Use in Class I, II, and III, Division 1, Hazardous (Classified) Locations*, and ANSI/UL 60079-11, *Explosive Atmospheres — Part 11: Equipment Protection by Intrinsic Safety "i"*, for additional information.

Intrinsically Safe System. An assembly of interconnected intrinsically safe apparatus, associated apparatus, and interconnecting cables, in which those parts of the system that might be used in hazardous (classified) locations are intrinsically safe circuits. (CMP-14)

Informational Note No. 1: An intrinsically safe system might include more than one intrinsically safe circuit.

Informational Note No. 2: See ANSI/UL 913, *Intrinsically Safe Apparatus and Associated Apparatus for Use in Class I, II, and III, Division 1, Hazardous (Classified) Locations*; ANSI/UL 60079-11, *Explosive Atmospheres — Part 11: Equipment Protection by Intrinsic Safety "i"*; and ANSI/UL 60079-25, *Explosive Atmospheres — Part 25: Intrinsically Safe Electrical Systems*, for additional information.

Informational Note No. 3: See ANSI/ISA RP 12.06.01, *Recommended Practice for Wiring Methods for Hazardous (Classified) Locations Instrumentation — Part 1: Intrinsic Safety*, for installation information.

N **Invasive Procedure.** Any procedure that penetrates the protective surfaces of a patient's body (i.e., skin, mucous membrane, cornea) and that is performed with an aseptic field (procedural site). [Not included in this category are placement of peripheral intravenous needles or catheters used to administer fluids and/or medications, gastrointestinal endoscopies (i.e., sigmoidoscopies), insertion of urethral catheters, and other similar procedures.] [**99**:3.3.91] (517) (CMP-15)

Inverter. Equipment that changes dc to ac. (CMP-4)

Inverter, Interactive. (Interactive Inverter) Inverter equipment having the capability to operate only in interactive mode. (CMP-13)

The most common interactive inverters are string inverters and microinverters. String inverters typically are connected to a number of photovoltaic (PV) modules. Microinverters usually are connected to a single module. If microinverters are used in an installation, each module will have its own microinverter. Both string inverters and microinverters can be interactive.

Inverter, Multimode. (Multimode Inverter) Inverter equipment capable of operating in both interactive and island modes. (CMP-4)

N **Inverter, Stand-alone. (Stand-alone Inverter)** Inverter equipment having the capabilities to operate only in island mode. (CMP-4)

Inverter Input Circuit. Conductors connected to the dc input of an inverter. (CMP-13)

Inverter Output Circuit. Conductors connected to the ac output of an inverter. (CMP-13)

N **Inverter Utilization Output Circuit.** Conductors between the multimode or stand-alone inverter and utilization equipment. (706) (CMP-13)

N **Irrigation Machine.** An electrically driven or controlled machine, with one or more motors, not hand-portable, and used primarily to transport and distribute water for agricultural purposes. (675) (CMP-7)

N **Irrigation Machine, Center Pivot. (Center Pivot Irrigation Machine)** A multimotored irrigation machine that revolves around a central pivot and employs alignment switches or similar devices to control individual motors. (675) (CMP-7)

Δ **Island Mode.** The operating mode for power production equipment or microgrids that allows energy to be supplied to loads that are disconnected from an electric power production and distribution network or other primary power source. (CMP-4)

Isolated (as applied to location). Not readily accessible to persons unless special means for access are used. (CMP-1)

N **Isolated Power System.** A system comprising an isolation transformer or its equivalent, a line isolation monitor, and its ungrounded circuit conductors. [**99**:3.3.93] (517) (CMP-15)

N **Isolation Transformer.** A transformer of the multiple-winding type, with the primary and secondary windings physically separated, that inductively couples its ungrounded secondary winding to the grounded feeder system that energizes its primary winding. [**99**:3.3.94] (517) (CMP-15)

Kitchen. An area with a sink and permanent provisions for food preparation and cooking. (CMP-2)

Labeled. Equipment or materials to which has been attached a label, symbol, or other identifying mark of an organization that is acceptable to the authority having jurisdiction and concerned with product evaluation, that maintains periodic inspection of production of labeled equipment or materials, and by whose labeling the manufacturer indicates compliance with appropriate standards or performance in a specified manner. (CMP-1)

Informational Note: If a listed product is of such a size, shape, material, or surface texture that it is not possible to apply legibly the complete label to the product, the complete label may appear on the smallest unit container in which the product is packaged.

Equipment and conductors required or permitted by the *NEC* are acceptable only if they have been approved for a specific environment or application by the AHJ as stated in 110.2. Listing or labeling by a qualified testing laboratory provides a basis for approval.

See also

90.7 regarding the examination of equipment for safety

Laundry Area. An area containing or designed to contain a laundry tray, clothes washer, or clothes dryer. (CMP-2)

N **Leakage-Current Detector-Interrupter (LCDI).** A device provided in a power supply cord or cord set that senses leakage current flowing between or from the cord conductors and interrupts the circuit at a predetermined level of leakage current. (440) (CMP-11)

Opening of the circuit is accomplished through the use of electronic switching or by "air-break" contacts. The circuit remains open until the cause of the leakage current is eliminated or the protection device is manually reset. Leakage-current detector-interrupter protection is one of the permitted methods for the power supply cord or cord set of a room air conditioner specified in 440.65.

N **LED Sign Illumination System.** A complete lighting system for use in signs and outline lighting consisting of light-emitting diode (LED) light sources, power supplies, wire, and connectors to complete the installation. (600) (CMP-18)

N **Legally Required Standby Systems.** Those systems required and so classed as legally required standby by municipal, state, federal, or other codes or by any governmental agency having jurisdiction. These systems are intended to automatically supply power to selected loads (other than those classed as emergency systems) in the event of failure of the normal source. (CMP-13)

N **Life Safety Branch.** A system of feeders and branch circuits supplying power for lighting, receptacles, and equipment essential for life safety that is automatically connected to alternate power sources by one or more transfer switches during interruption of the normal power source. [*99:*3.3.97] (517) (CMP-15)

N **Lighting Assembly, Cord-and-Plug-Connected. (Cord-and-Plug-Connected Lighting Assembly)** A lighting assembly consisting of a luminaire intended for installation in the wall of a spa, hot tub, or storable pool, and a cord-and-plug-connected transformer or power supply. (680) (CMP-17)

N **Lighting Assembly, Through-Wall. (Through-Wall Lighting Assembly)** A lighting assembly intended for installation above grade, on or through the wall of a pool, consisting of two interconnected groups of components separated by the pool wall. (680) (CMP-17)

Lighting Outlet. An outlet intended for the direct connection of a lampholder or luminaire. (CMP-18)

Δ **Lighting Track. (Track Lighting)** A manufactured assembly designed to support and energize luminaires that are capable of being readily repositioned on the track. Its length can be altered by the addition or subtraction of sections of track. (CMP-18)

N **Limited Care Facility.** A building or portion of a building used on a 24-hour basis for the housing of four or more persons who are incapable of self-preservation because of age; physical limitation due to accident or illness; or limitations such as intellectual disability/developmental disability, mental illness, or chemical dependency. [*101:*3.3.93.2] (CMP-15)

Δ **Limited Finishing Workstation.** A power-ventilated apparatus that is capable of confining the vapors, mists, residues, dusts, or deposits that are generated by a limited spray application process. Such apparatus is not a spray booth or spray room, as herein defined. [**33:**3.3.23.1] (CMP-14)

Informational Note: See NFPA 33, *Standard for Spray Application Using Flammable or Combustible Materials*, Section 14.3, for information on limited finishing workstations.

N **Line Isolation Monitor.** A test instrument designed to continually check the balanced and unbalanced impedance from each line of an isolated circuit to ground and equipped with a built-in test circuit to exercise the alarm without adding to the leakage current hazard. [**99:**3.3.99] (517) (CMP-15)

Δ **Liquid Immersion "o".** Type of protection where electrical equipment is immersed in a protective liquid so that an explosive atmosphere that might be above the liquid or outside the enclosure cannot be ignited. (CMP-14)

Informational Note: See ANSI/UL 60079-6, *Explosive Atmospheres — Part 6: Equipment Protection by Liquid Immersion "o"*, for additional information.

Listed. Equipment, materials, or services included in a list published by an organization that is acceptable to the authority having jurisdiction and concerned with evaluation of products or services, that maintains periodic inspection of production of listed equipment or materials or periodic evaluation of services, and whose listing states that either the equipment, material, or service meets appropriate designated standards or has been tested and found suitable for a specified purpose. (CMP-1)

Informational Note: The means for identifying listed equipment may vary for each organization concerned with product evaluation, some of which do not recognize equipment as listed unless it is also labeled. Use of the system employed by the listing organization allows the authority having jurisdiction to identify a listed product.

Live Parts. Energized conductive components. (CMP-1)

N **Load Management.** The process within an energy management system that limits the total electrical load on an electrical supply system to a set value by adjusting or controlling the individual loads. (625) (CMP-12)

Informational Note: Load management is sometimes called *demand-side management* (DSM).

N **Location, Anesthetizing. (Anesthetizing Location)** Any space within a facility that has been designated for the administration of any flammable or nonflammable inhalation anesthetic agent during examination or treatment, including the use of such agents for relative analgesia. (517) (CMP-15)

In an emergency, it may be necessary to administer an anesthetic almost anywhere in a health care facility. Only those areas specifically set aside for administering anesthetics are required to meet the requirements of Part IV of Article 517. The requirements of Part IV do not apply to administering analgesic or local anesthetics, such as might be used in minor medical or dental procedures.

The definition of the term *anesthetizing location* applies to areas where inhalation anesthetics are used for relative analgesia. The term *relative analgesia* is sometimes referred to as "conscious sedation" and is a state of sedation in which the perception of pain is partially blocked but the patient does not lose consciousness. This form of anesthesia is commonly used by oral surgeons.

N Location, Anesthetizing, Flammable. (Flammable Anesthetizing Location) Any area of the facility that has been designated to be used for the administration of any flammable inhalation anesthetic agents in the normal course of examination or treatment. (517) (CMP-15)

Location, Damp. (Damp Location) Locations protected from weather and not subject to saturation with water or other liquids but subject to moderate degrees of moisture. (CMP-1)

> Informational Note: Examples of such locations include partially protected locations under canopies, marquees, roofed open porches, and like locations, and interior locations subject to moderate degrees of moisture, such as some basements, some barns, and some cold-storage warehouses.

Location, Dry. (Dry Location) A location not normally subject to dampness or wetness. A location classified as dry may be temporarily subject to dampness or wetness, as in the case of a building under construction. (CMP-1)

N Location, Remote. (Remote Location) A location, other than a motion picture or television studio, where a production is filmed or recorded. (530) (CMP-15)

Location, Wet. (Wet Location) A location that is one or more of the following:

(1) Unprotected and exposed to weather
(2) Subject to saturation with water and other liquids
(3) Underground
(4) In concrete slabs or masonry in direct contact with the earth
(CMP-1)

> Informational Note: A vehicle washing area is an example of a wet location saturated with water or other liquids.

Both the inside of a raceway in a wet location and a raceway installed underground are considered wet locations. Therefore, any conductors inside the raceway would be required to be suitable for wet locations.

See also

300.6(D) for the general requirement for protection against corrosion and deterioration, which also identifies several examples of wet locations

410.10(A) for requirements for luminaires installed in wet locations

Table 110.28 for a list of suitable enclosures for wet locations

N Location, Wet Procedure. (Wet Procedure Location) The area in a patient care space where a procedure is performed that is normally subject to wet conditions while patients are present, including standing fluids on the floor or drenching of the work area, either of which condition is intimate to the patient or staff. [**99**:3.3.187] (517) (CMP-15)

> Informational Note: Routine housekeeping procedures and incidental spillage of liquids do not define a wet procedure location. [**99**:A.3.3.187]

Wet procedure locations are designated by the governing body of the facility. Examples of *wet procedure locations* include hydrotherapy areas, dialysis laboratories, and certain wet laboratories. Unless otherwise determined through a risk assessment, operating rooms are considered to be wet procedure locations. The definition excludes lavatories or bathrooms within a health care facility. For infection control purposes, many patient and treatment areas have a sink for hand washing, which also is not a wet procedure location.

N Locations, Hazardous (Classified). [Hazardous (Classified) Locations] Locations where fire or explosion hazards might exist due to flammable gases, flammable liquid–produced vapors, combustible liquid–produced vapors, combustible dusts, combustible fiber/flyings, or ignitible fibers/flyings. (CMP-14)

Locations, Unclassified. (Unclassified Locations) Locations determined to be neither Class I, Division 1; Class I, Division 2; Zone 0; Zone 1; Zone 2; Class II, Division 1; Class II, Division 2; Class III, Division 1; Class III, Division 2; Zone 20; Zone 21; Zone 22; nor any combination thereof. (CMP-14)

N Long-Time Rating. A rating based on an operating interval of 5 minutes or longer. (660) (CMP-12)

N Long-Time Rating (Standby Power). A rating based on an operating interval of 5 minutes or longer. (517) (CMP-15)

N Loudspeaker (Speaker). Equipment that converts an ac electric signal into an acoustic signal. (640) (CMP-12)

N Low-Voltage Contact Limit. A voltage not exceeding the following values:

(1) 15 volts (RMS) for sinusoidal ac
(2) 21.2 volts peak for nonsinusoidal ac
(3) 30 volts for continuous dc
(4) 12.4 volts peak for dc that is interrupted at a rate of 10 to 200 Hz (680) (CMP-17)

The low voltage contact limits are based on the wet contact limits specified in Tables 11(A) and 11(B) in Chapter 9.

N Low-Voltage Suspended Ceiling Power Distribution System. A system that serves as a support for a finished ceiling surface and consists of a busbar and busbar support system to distribute power to utilization equipment supplied by a Class 2 power supply. (393) (CMP-18)

Luminaire. A complete lighting unit consisting of a light source such as a lamp or lamps, together with the parts designed to position the light source and connect it to the power supply. It may also include parts to protect the light source or the ballast or to distribute the light. A lampholder itself is not a luminaire. (CMP-18)

Light pipes and glass fiber optics are sometimes referred to as "lighting systems." The definition of *luminaire* includes such systems because light pipes and fiber optics are actually "parts designed to distribute the light."

Luminaire is the term specified by the Illuminating Engineering Society of North America (IESNA), the Underwriters Laboratories (ANSI/UL) safety standards, and the National Electrical Manufacturers Associations (ANSI/NEMA) performance standards for lighting products previously referred to as "light fixtures." *Luminaire* is also the term used in International Electrotechnical Commission (IEC) standards and accepted globally.

N **Luminaire, Dry-Niche. (Dry-Niche Luminaire)** A luminaire intended for installation in the floor or wall of a pool, spa, or fountain in a niche that is sealed against the entry of water. (680) (CMP-17)

N **Luminaire, No-Niche. (No-Niche Luminaire)** A luminaire intended for installation above or below the water without a niche. (680) (CMP-17)

N **Luminaire, Wet-Niche. (Wet-Niche Luminaire)** A luminaire intended for installation in a forming shell mounted in a pool or fountain structure where the luminaire will be completely surrounded by water. (680) (CMP-17)

N **Machine Room.** An enclosed machinery space outside the hoistway, intended for full bodily entry, that contains the electrical driving machine or the hydraulic machine. The room could also contain electrical and/or mechanical equipment used directly in connection with the elevator or dumbwaiter. (620) (CMP-12)

N **Machine Room and Control Room, Remote. (Remote Machine Room and Control Room)** A machine room or control room that is not attached to the outside perimeter or surface of the walls, ceiling, or floor of the hoistway. (620) (CMP-12)

N **Machinery, Industrial (Industrial Machine). (Industrial Machinery)** A power-driven machine (or a group of machines working together in a coordinated manner), not portable by hand while working, that is used to process material by cutting; forming; pressure; electrical, thermal, or optical techniques; lamination; or a combination of these processes. It can include associated equipment used to transfer material or tooling, including fixtures, to assemble/disassemble, to inspect or test, or to package. The associated electrical equipment, including the logic controller(s) and associated software or logic together with the machine actuators and sensors, are considered as part of the industrial machine. (CMP-12)

This definition permits the inclusion of other types of industrial machines without the need to continuously modify the scope of Article 670 and NFPA 79. Also, the scope and the definition are in harmony with IEC 60204-1, *Safety of Machinery — Electrical Equipment of Machines — Part 1: General Requirements*. Exhibit 100.16 shows an example of an industrial machine.

EXHIBIT 100.16   Thread-spooling machine used in the textile industry. *(Courtesy of the International Association of Electrical Inspectors)*

N **Machinery Space.** A space inside or outside the hoistway, intended to be accessed with or without full bodily entry, that contains the elevator, dumbwaiter, platform lift, or stairway chairlift equipment and could also contain equipment used directly in connection with the elevator, dumbwaiter, platform lift, or stairway chairlift. (620) (CMP-12)

N **Machinery Space and Control Space, Remote. (Remote Machinery Space and Control Space)** A machinery space or control space that is not within the hoistway, machine room, or control room and that is not attached to the outside perimeter or surface of the walls, ceiling, or floor of the hoistway. (620) (CMP-12)

Definitions of the terms *remote machine room and control room (for elevator, dumbwaiter)* and *remote machinery space and control space (for elevator, dumbwaiter)* correlate with their use in ASME A17.1, *Safety Code for Elevators and Escalators*.

N **Manufactured Home.** A structure, transportable in one or more sections, which in the traveling mode is 2.4 m (8 ft) or more in width or 12.2 m (40 ft) or more in length, or when erected on site is 29.77 m² (320 ft²) or more is built on a permanent chassis and is designed to be used as a dwelling with or without a permanent foundation, whether or not connected to the utilities, and includes plumbing, heating, air conditioning, and electrical systems contained therein. The term includes any structure that meets all the requirements of this paragraph except the size requirements and with respect to which the manufacturer voluntarily files a certification required by the regulatory agency. Calculations used to determine the number of square meters (square feet) in a structure are based on the structure's exterior dimensions and include all expandable rooms, cabinets, and other projections containing interior space, but do not include bay windows. [**501**:1.2.12] (CMP-7)

Informational Note No. 1: Unless otherwise indicated, the term *mobile home* includes manufactured home and excludes park trailers.

Informational Note No. 2: See the applicable building code for definition of the term *permanent foundation.*

Informational Note No. 3: See 24 CFR Part 3280, *Manufactured Home Construction and Safety Standards, of the Federal Department of Housing and Urban Development,* for additional information on the definition.

N **Manufactured Wiring System.** A system containing component parts that are assembled in the process of manufacture and cannot be inspected at the building site without damage or destruction to the assembly and used for the connection of luminaires, utilization equipment, continuous plug-in type busways, and other devices. (604) (CMP-7)

N **Marina.** A facility, generally on the waterfront, that stores and services boats in berths, on moorings, and in dry storage or dry stack storage. [**303:**3.3.13] (555) (CMP-7)

N **Maximum Output Power.** The maximum power delivered by an amplifier into its rated load as determined under specified test conditions. (640) (CMP-12)

Informational Note: The maximum output power can exceed the manufacturer's rated output power for the same amplifier.

N **Maximum Output Power.** The maximum 1 minute average power output a wind turbine produces in normal steady-state operation (instantaneous power output can be higher). (694) (CMP-4)

N **Maximum Voltage.** The greatest difference in potential produced between any two conductors of a wind turbine circuit. (694) (CMP-4)

N **Maximum Water Level.** The highest level that water can reach before it spills out. (680) (CMP-17)

Exhibit 100.17 illustrates the difference between the normal water level (typically controlled by the pool filtration system) and the maximum water level.

N **Medical Office.** A building or part thereof in which the following occur:

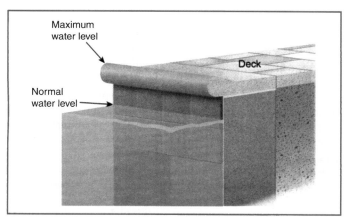

EXHIBIT 100.17 Cutaway view of a pool edge, showing the difference between normal and maximum water levels.

(1) Examinations and minor treatments/procedures performed under the continuous supervision of a medical professional;

(2) The use of limited to minimal sedation and treatment or procedures that do not render the patient incapable of self-preservation under emergency conditions; and

(3) No overnight stays for patients or 24-hour operations. [**99:**3.3.110] (CMP-15)

Δ **Membrane Enclosure.** A temporary enclosure used for the spraying of workpieces that cannot be moved into a spray booth where open spraying is not practical due to proximity to other operations, finish quality, or concerns such as the collection of overspray. (CMP-14)

Informational Note: See NFPA 33, *Standard for Spray Application Using Flammable or Combustible Materials,* Chapter 18, for information on the construction and use of membrane enclosures.

N **Messenger-Supported Wiring.** An exposed wiring support system using a messenger wire to support insulated conductors by any one of the following:

(1) A messenger with rings and saddles for conductor support

(2) A messenger with a field-installed lashing material for conductor support

(3) Factory-assembled aerial cable

(4) Multiplex cables utilizing a bare conductor, factory assembled and twisted with one or more insulated conductors, such as duplex, triplex, or quadruplex type of construction (CMP-6)

Messenger Wire (Messenger). A wire that is run along with or integral with a cable or conductor to provide mechanical support for the cable or conductor. (CMP-6)

N **Metal Shield Connections.** Means of connection for flat conductor cables (Type FCC) designed to electrically and mechanically connect a metal shield to another metal shield, to a receptacle housing or self-contained device, or to a transition assembly. (324) (CMP-6)

N **Microgrid.** An electric power system capable of operating in island mode and capable of being interconnected to an electric power production and distribution network or other primary source while operating in interactive mode, which includes the ability to disconnect from and reconnect to a primary source and operate in island mode. (CMP-4)

Informational Note No. 1: See IEEE 1547, *IEEE Standard for Interconnection and Interoperability of Distributed Energy Resources with Associated Electric Power Systems Interface;* IEEE 2030.7, *IEEE Standard for the Specification of Microgrid Controllers;* IEEE 2030.8, *IEEE Standard for the Testing of Microgrid Controllers;* and UL1008B, *Outline for Source Interconnection,* for additional information about microgrids.

Informational Note No. 2: Examples of power sources in microgrids include such items as photovoltaic systems, generators, fuel cell systems, wind electric systems, energy storage systems, electric vehicles that are used as a source of supply, and electrical power conversion from other energy sources.

N Microgrid, Health Care (Health Care Microgrid System). (Health Care Microgrid) A group of interconnected loads and distributed energy resources within clearly defined boundaries that acts as a single controllable entity with respect to the utility. [**99**:3.3.75] (517) (CMP-15)

N Microgrid Control System (MCS). A structured control system that manages microgrid operations, functionalities for utility interoperability, islanded operations, and transitions. (CMP-4)

> Informational Note: MCS differ from multiple standby generators or uninterruptible power supplies that are evaluated and rated to operate as a single source of backup power upon loss of the primary power source. MCS functions include coordination, transitions, and interoperability between multiple power sources.

N Microgrid Interconnect Device (MID). A device that enables a microgrid system to separate from and reconnect to an interconnected primary power source. (CMP-4)

N Mixer. Equipment used to combine and level match a multiplicity of electronic signals, such as from microphones, electronic instruments, and recorded audio. (640) (CMP-12)

N Mobile. X-ray equipment mounted on a permanent base with wheels and/or casters for moving while completely assembled. (660) (CMP-12)

N Mobile Home. A factory-assembled structure or structures transportable in one or more sections that are built on a permanent chassis and designed to be used as a dwelling without a permanent foundation where connected to the required utilities and that include the plumbing, heating, air-conditioning, and electrical systems contained therein. (CMP-7)

> Informational Note: Unless otherwise indicated, the term *mobile home* includes manufactured home and excludes park trailers.

N Mobile Home Lot. A designated portion of a mobile home park designed for the accommodation of one mobile home and its accessory buildings or structures for the exclusive use of its occupants. (550) (CMP-7)

N Mobile Home Park. A contiguous parcel of land that is used for the accommodation of mobile homes that are intended to be occupied. (550) (CMP-7)

N Module, AC. (AC Module) A complete, environmentally protected unit consisting of solar cells, inverter, and other components, designed to produce ac power. (690) (CMP-4)

N Module System, AC. (AC Module System) An assembly of ac modules, wiring methods, materials, and subassemblies that are evaluated, identified, and defined as a system. (690) (CMP-4)

N Momentary Rating. A rating based on an operating interval that does not exceed 5 seconds. (660) (CMP-12)

N Momentary Rating (Maximum Power). A rating based on an operating interval that does not exceed 5 seconds. (517) (CMP-15)

N Monitor. An electrical or electronic means to observe, record, or detect the operation or condition of the electric power system or apparatus. (750) (CMP-13)

N Monopole Circuit. An electrical subset of a PV system that has two conductors in the output circuit, one positive (+) and one negative (−). (690) (CMP-4)

N Monorail. Overhead track and hoist system for moving material around the boatyard or moving and launching boats. [**303**:3.3.16] (555) (CMP-7)

N Mooring(s). Any place where a boat is wet stored or berthed. [**303**:3.3.17] (555) (CMP-7)

N Motion Picture Studio (Television Studio). A building, group of buildings, other structures, and outdoor areas designed, constructed, permanently altered, designated, or approved for the purpose of motion picture or television production. (530) (CMP-15)

Motor Control Center. An assembly of one or more enclosed sections having a common power bus and principally containing motor control units. (CMP-11)

Δ Motor Fuel Dispensing Facility. That portion of a property where motor fuels are stored and dispensed from fixed equipment into the fuel tanks of motor vehicles or marine craft or into approved containers, including all equipment used in connection therewith. [**30A**:3.3.11] (CMP-14)

> Informational Note: See 511.1 with respect to electrical wiring and equipment for other areas used as lubritoriums, service rooms, repair rooms, offices, salesrooms, compressor rooms, and similar locations.

N Multi-Circuit Cable Outlet Enclosure. An enclosure containing one or more multi-circuit plugs, receptacles, or both. (520) (CMP-15)

Δ Multioutlet Assembly. A surface, flush, or freestanding assemblage with a raceway and fittings or other enclosure provided with one or more receptacles, for the purpose of supplying power to utilization equipment. (CMP-18)

Multioutlet assemblies include freestanding assemblies with multiple receptacles, commonly called power poles. In dry locations, metal and nonmetallic multioutlet assemblies are permitted; however, they are not permitted to be installed if concealed.

Exhibit 100.18 shows a multioutlet assembly used for countertop appliances. Portable assemblies, often called "power strips" or "plug strips," are not multioutlet assemblies but are "relocatable power taps."

See also

Article 380 for details on recessing multioutlet assemblies

N Nacelle. An enclosure housing the alternator and other parts of a wind turbine. (694) (CMP-4)

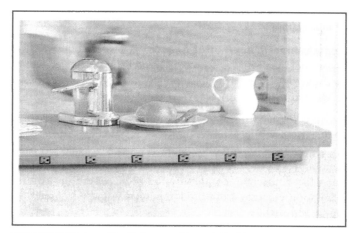

EXHIBIT 100.18 A multioutlet assembly installed to serve countertop appliances. *(Courtesy of Legrand)*

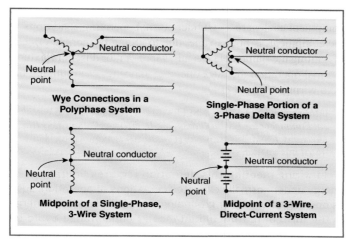

EXHIBIT 100.19 Four examples of a neutral point.

N **Neon Tubing.** Electric-discharge luminous tubing, including cold cathode luminous tubing, that is manufactured into shapes to illuminate signs, form letters, parts of letters, skeleton tubing, outline lighting, other decorative elements, or art forms and filled with various inert gases. (600) (CMP-18)

N **Network Interface Unit (NIU).** A device that converts a broadband signal into component voice, audio, video, data, and interactive services signals and provides isolation between the network power and the premises signal circuits. These devices often contain primary and secondary protectors. (CMP-16)

N **Network Terminal.** A device that converts network-provided signals (optical, electrical, or wireless) into component signals, including voice, audio, video, data, wireless, optical, and interactive services, and is considered a network device on the premises that is connected to a communications service provider and is powered at the premises. (CMP-16)

Neutral Conductor. The conductor connected to the neutral point of a system that is intended to carry current under normal conditions. (CMP-5)

Neutral Point. The common point on a wye-connection in a polyphase system or midpoint on a single-phase, 3-wire system, or midpoint of a single-phase portion of a 3-phase delta system, or a midpoint of a 3-wire, direct-current system. (CMP-5)

> Informational Note: At the neutral point of the system, the vectorial sum of the nominal voltages from all other phases within the system that utilize the neutral, with respect to the neutral point, is zero potential.

Exhibit 100.19 illustrates four examples of a neutral point in a system.

Nonautomatic. Requiring human intervention to perform a function. (CMP-1)

Δ **Nonincendive Circuit.** A circuit, other than field wiring, in which any arc or thermal effect produced under intended operating conditions of the equipment, is not capable, under specified test conditions, of igniting the flammable gas–air, vapor–air, or dust–air mixture. (CMP-14)

> Informational Note: See ANSI/UL 121201, *Nonincendive Electrical Equipment for Use in Class I and II, Division 2 and Class III, Divisions 1 and 2 Hazardous (Classified) Locations,* for additional information.

A nonincendive circuit employs a protection technique that prevents electrical circuits from causing a fire or explosion in a hazardous location under normal conditions. This is in contrast to an intrinsically safe circuit, whose evaluation is conducted under abnormal conditions. Because of its definition, a nonincendive circuit is a low-energy circuit, but many low-voltage, low-energy circuits, including some communications circuits and thermocouple circuits (or Class 2 or 3 circuits as defined in Article 725), are not necessarily nonincendive.

Δ **Nonincendive Component.** A component having contacts for making or breaking an incendive circuit and the contacting mechanism is constructed so that the component is incapable of igniting the specified flammable gas–air or vapor–air mixture. The housing of such a component is not intended to exclude the flammable atmosphere or contain an explosion. (CMP-14)

> Informational Note: See ANSI/UL 121201, *Nonincendive Electrical Equipment for Use in Class I and II, Division 2 and Class III, Divisions 1 and 2 Hazardous (Classified) Locations,* for additional information.

Δ **Nonincendive Equipment.** Equipment having electrical/electronic circuitry that is incapable, under normal operating conditions, of causing ignition of a specified flammable gas–air, vapor–air, or dust–air mixture due to arcing or thermal means. (CMP-14)

> Informational Note: See ANSI/UL 121201, *Nonincendive Electrical Equipment for Use in Class I and II, Division 2 and Class III, Divisions 1 and 2 Hazardous (Classified) Locations,* for additional information.

Nonincendive Field Wiring. Wiring that enters or leaves an equipment enclosure and, under normal operating conditions of the equipment, is not capable, due to arcing or thermal effects, of igniting the flammable gas–air, vapor–air, or dust–air mixture. Normal operation includes opening, shorting, or grounding the field wiring. (CMP-14)

> Informational Note: See ANSI/UL 121201, *Nonincendive Electrical Equipment for Use in Class I and II, Division 2 and Class III, Divisions 1 and 2 Hazardous (Classified) Locations*, for additional information.

Field wiring meeting this definition requires limitations of energy under normally expected conditions of operation, such as opening, shorting, or grounding. For example, stored energy in the form of mutual inductance or capacitance could be released during an opening, shorting, or grounding of nonincendive field wiring, which defeats the purpose of this protection technique.

Δ **Nonincendive Field Wiring Apparatus.** Apparatus intended to be connected to nonincendive field wiring. (CMP-14)

> Informational Note: See ANSI/UL 121201, *Nonincendive Electrical Equipment for Use in Class I and II, Division 2 and Class III, Divisions 1 and 2 Hazardous (Classified) Locations*, for additional information.

Nonlinear Load. A load where the wave shape of the steady-state current does not follow the wave shape of the applied voltage. (CMP-1)

> Informational Note: Electronic equipment, electronic/electric-discharge lighting, adjustable-speed drive systems, and similar equipment may be nonlinear loads.

Nonlinear loads are a major cause of harmonic currents. Additional conductor heating is just one of the undesirable operational effects associated with harmonic currents. Informational Note No. 1(2) following 310.14(A)(3) points out that harmonic current is one of the factors that must be considered when determining the heat generated internally in a conductor.

Actual circuit measurements of current for nonlinear loads should be made using only true root-mean-square (rms) measuring ammeter instruments. Averaging ammeters produces inaccurate values if used to measure nonlinear loads.

See also

310.15(E)(3) and its commentary for more on nonlinear loads

N **Nonmetallic Extension.** An assembly of two insulated conductors within a nonmetallic jacket or an extruded thermoplastic covering. The classification includes surface extensions intended for mounting directly on the surface of walls or ceilings. (CMP-6)

N **Nonsparking.** Constructed to minimize the risk of arcs or sparks capable of creating an ignition hazard during conditions of normal operation. (CMP-14)

> Informational Note: See ANSI/UL 121201, *Nonincendive Electrical Equipment for Use in Class I and II, Division 2 and Class III,*

Divisions 1 and 2 Hazardous (Classified) Locations, for additional information.

N **Normal/Emergency Power Source.** A power source on the output side of a transfer switch or uninterruptible power supply that is automatically available upon loss of normal power. (700) (CMP-13)

N **Normal High-Water Level (as applies to electrical datum plane distances).** Natural or Artificially Made Shorelines: An elevation delineating the highest water level that has been maintained for a sufficient period of time to leave evidence upon the landscape, commonly the point where the natural vegetation changes from predominantly aquatic to predominantly terrestrial.

Rivers and Streams: The elevation of the top of the bank of the channel. Streams, rivers, and tributaries that are prone to flooding and effects of water runoff shall consider the "bankfull stage" where an established gauge height at a given location along a river or stream, above which a rise in water surface will cause the river or stream to overflow the lowest natural stream bank somewhere in the corresponding reach.

Flood Control Bodies of Water: The flood pool maximum water surface elevation of a reservoir, equal to the elevation of the spillway.

Nonflood Control Bodies of Water: The flowage easement boundary in which the highest water surface elevation defined by the area existing between governmental-owned property line(s) and a contour line with perpetual rights to flood the area in connection with the operation of the reservoir. (CMP-7)

N **Nurses' Station.** A space intended to provide a center of nursing activity for a group of nurses serving bed patients, where patient calls are received, nurses dispatched, nurses' notes written, inpatient charts prepared, and medications prepared for distribution to patients. Where such activities are carried on in more than one location within a nursing unit, all such separate spaces are considered a to be parts of the nurses' station. (517) (CMP-15)

N **Nursing Home.** A building or portion of a building used on a 24-hour basis for the housing and nursing care of four or more persons who, because of mental or physical incapacity, might be unable to provide for their own needs and safety without the assistance of another person. [*101*:3.3.150.2] (CMP-15)

N **Office Furnishing.** Cubicle panels, partitions, study carrels, workstations, desks, shelving systems, and storage units that may be mechanically and electrically interconnected to form an office furnishing system. (CMP-18)

Oil Immersion. Electrical equipment immersed in a protective liquid so that an explosive atmosphere that might be above the liquid or outside the enclosure cannot be ignited. (CMP-14)

> Informational Note: See ANSI/UL 121201, *Nonincendive Electrical Equipment for Use in Class I and II, Division 2 and Class III,*

Divisions 1 and 2 Hazardous (Classified) Locations, for additional information.

N **Open Wiring on Insulators.** An exposed wiring method using cleats, knobs, tubes, and flexible tubing for the protection and support of single insulated conductors run in or on buildings. (CMP-6)

N **Operating Device.** The car switch, pushbuttons, key or toggle switch(s), or other devices used to activate the operation controller. (620) (CMP-12)

N **Operator.** The individual responsible for starting, stopping, and controlling an amusement ride or supervising a concession. (525) (CMP-15)

Δ **Optical Radiation.** Electromagnetic radiation at wavelengths in vacuum between the region of transition to X-rays and the region of transition to radio waves that is approximately between 1 nm and 1000 μm. (CMP-14)

> Informational Note: See ANSI/UL 60079-28, *Explosive Atmospheres — Part 28: Protection of Equipment and Transmission Systems Using Optical Radiation*, for information on types of protection that can be applied to minimize the risk of ignition in explosive atmospheres from optical radiation in the wavelength range from 380 nm to 10 μm.

Δ **Optical Radiation, Inherently Safe "op is". (Inherently Safe Optical Radiation "op is")** Type of protection to minimize the risk of ignition in explosive atmospheres from optical radiation where visible or infrared radiation is incapable of producing sufficient energy under normal or specified fault conditions to ignite a specific explosive atmosphere. (CMP-14)

> Informational Note: See ANSI/UL 60079-28, *Explosive Atmospheres — Part 28: Protection of Equipment and Transmission Systems Using Optical Radiation*, for additional information.

Δ **Optical Radiation, Protected "op pr". (Protected Optical Radiation "op pr")** Type of protection to minimize the risk of ignition in explosive atmospheres from optical radiation where visible or infrared radiation is confined inside optical fiber or other transmission medium under normal constructions or constructions with additional mechanical protection based on the assumption that there is no escape of radiation from the confinement. (CMP-14)

> Informational Note: See ANSI/UL 60079-28, *Explosive Atmospheres — Part 28: Protection of Equipment and Transmission Systems Using Optical Radiation*, for additional information.

Δ **Optical System With Interlock "op sh".** Type of protection to minimize the risk of ignition in explosive atmospheres from optical radiation where visible or infrared radiation is confined inside optical fiber or other transmission medium with interlock cutoff provided to reliably reduce the unconfined beam strength to safe levels within a specified time in case the confinement fails and the radiation becomes unconfined. (CMP-14)

> Informational Note: See ANSI/UL 60079-28, *Explosive Atmospheres — Part 28: Protection of Equipment and Transmission Systems Using Optical Radiation*, for additional information.

N **Optional Standby Systems.** Those systems intended to supply power to public or private facilities or property where life safety does not depend on the performance of the system. These systems are intended to supply on-site generated or stored power to selected loads either automatically or manually. (CMP-13)

N **Organ, Electronic. (Electronic Organ)** A musical instrument that imitates the sound of a pipe organ by producing sound electronically. (CMP-12)

> Informational Note: Most new electronic organs produce sound digitally and are called digital organs.

N **Organ, Pipe. (Pipe Organ)** A musical instrument that produces sound by driving pressurized air (called *wind*) through pipes selected via a keyboard. (CMP-12)

N **Organ, Pipe Sounding Apparatus. (Pipe Organ Sounding Apparatus) (Pipe Organ Chamber).** The sound-producing part of a pipe organ, including, but not limited to, pipes, chimes, bells, the pressurized air- (wind-) producing equipment (blower), associated controls, and power equipment. (CMP-12)

Outlet. A point on the wiring system at which current is taken to supply utilization equipment. (CMP-1)

The term *outlet* is frequently misused to refer to receptacles. Although receptacle outlets are outlets, not all outlets are receptacle outlets. Other common examples of outlets include appliance outlets, lighting outlets, and smoke alarm outlets.

N **Outlet Box Hood.** A housing shield intended to fit over a faceplate for flush-mounted wiring devices, or an integral component of an outlet box or of a faceplate for flush-mounted wiring devices. The hood does not serve to complete the electrical enclosure; it reduces the risk of water coming in contact with electrical components within the hood, such as attachment plugs, current taps, surge protective devices, direct plug-in transformer units, or wiring devices. (CMP-18)

Outline Lighting. An arrangement of incandescent lamps, electric-discharge lighting, or other electrically powered light sources to outline or call attention to certain features such as the shape of a building or the decoration of a window. (CMP-18)

Outline lighting includes low-voltage, light-emitting diodes (LEDs), as well as other luminaires installed to form various shapes.

See also

Article 600 for outline lighting requirements

N **Output Cable to the Electric Vehicle.** An assembly consisting of a length of flexible EV cable and an electric vehicle connector (supplying power to the electric vehicle). (625) (CMP-12)

N **Output Cable to the Primary Pad.** A multiconductor, shielded cable assembly consisting of conductors to carry the high-frequency energy and any status signals between the charger power converter and the primary pad. (625) (CMP-12)

Overcurrent. Any current in excess of the rated current of equipment or the ampacity of a conductor. It may result from overload, short circuit, or ground fault. (CMP-10)

> Informational Note: A current in excess of rating may be accommodated by certain equipment and conductors for a given set of conditions. Therefore, the rules for overcurrent protection are specific for particular situations.

Δ **Overcurrent Protective Device, Branch-Circuit. (Branch-Circuit Overcurrent Protective Device)** A device capable of providing protection for service, feeder, and branch circuits and equipment over the full range of overcurrents between its rated current and its interrupting rating. (CMP-10)

The protection provided may be overload, short-circuit, or ground-fault, or a combination, depending on the application.

N **Overcurrent Protective Device, Current-Limiting. (Current-Limiting Overcurrent Protective Device)** A device that, when interrupting currents in its current-limiting range, reduces the current flowing in the faulted circuit to a magnitude substantially less than that obtainable in the same circuit if the device were replaced with a solid conductor having comparable impedance. (240) (CMP-10)

A current-limiting protective device is one that cuts off a fault current in less than one-half cycle, thus preventing short-circuit currents from building up to their full available values. Most electrical distribution systems can deliver high ground-fault or short-circuit currents to components such as conductors and service equipment. Those components can be damaged or destroyed by high fault currents, resulting in serious burndowns and fires. Properly selected current-limiting OCPDs, such as the ones shown in Exhibit 100.20, limit the let-through energy to an amount that does not exceed the rating of the components in spite of high available short-circuit currents.

EXHIBIT 100.20 Class R current-limiting fuses with rejection feature to prohibit the installation of non–current-limiting fuses. *(Courtesy of Eaton, Bussmann Division)*

Overcurrent Protective Device, Supplementary. (Supplementary Overcurrent Protective Device) A device intended to provide limited overcurrent protection for specific applications and utilization equipment such as luminaires and appliances. This limited protection is in addition to the protection provided in the required branch circuit by the branch-circuit overcurrent protective device. (CMP-10)

The definition of *supplementary overcurrent protective device* makes two important distinctions between OCPDs. First, the use of a supplementary device is specifically limited to a few applications. Second, where it is used, the supplementary device must be in addition to and protected by the more robust branch-circuit OCPD.

The devices used to provide overcurrent protection are different, and the differences are found in the product standards UL 489, *Molded-Case Circuit Breakers, Molded-Case Switches, and Circuit-Breaker Enclosures,* and UL 1077, *Supplementary Protectors for Use in Electrical Equipment.*

N **Overhead Gantry.** A structure consisting of horizontal framework, supported by vertical columns spanning above electrified truck parking spaces, that supports equipment, appliances, raceway, and other necessary components for the purpose of supplying electrical, HVAC, internet, communications, and other services to the spaces. (626) (CMP-12)

Overload. Operation of equipment in excess of normal, full-load rating, or of a conductor in excess of its ampacity that, when it persists for a sufficient length of time, would cause damage or dangerous overheating. A fault, such as a short circuit or ground fault, is not an overload. (CMP-10)

N **Packaged Therapeutic Tub or Hydrotherapeutic Tank Equipment Assembly.** A factory-fabricated unit consisting of water-circulating, heating, and control equipment mounted on a common base, intended to operate a therapeutic tub or hydrotherapeutic tank. Equipment can include pumps, air blowers, heaters, lights, controls, sanitizer generators, and so forth. (680) (CMP-17)

Panelboard. A single panel or group of panel units designed for assembly in the form of a single panel, including buses and automatic overcurrent devices, and equipped with or without switches for the control of light, heat, or power circuits; designed to be placed in a cabinet, enclosure, or cutout box placed in or against a wall, partition, or other support; and accessible only from the front. (CMP-9)

N **Panelboard, Enclosed. (Enclosed Panelboard)** An assembly of buses and connections, overcurrent devices, and control apparatus with or without switches or other equipment, installed in a cabinet, cutout box, or enclosure suitable for a panelboard application. (CMP-9)

N **Park Electrical Wiring Systems.** All of the electrical wiring, luminaires, equipment, and appurtenances related to electrical installments within a mobile home park, including the mobile home service equipment. (550) (CMP-7)

N **Park Trailer.** A unit that is built on a single chassis mounted on wheels and has a gross trailer area not exceeding 37 m² (400 ft²) in the set-up mode. (552) (CMP-7)

N **Part-Winding Motors.** A part-winding start induction or synchronous motor is one that is arranged for starting by first energizing part of its primary (armature) winding and, subsequently, energizing the remainder of this winding in one or more steps. A standard part-winding start induction motor is arranged so that one-half of its primary winding can be energized initially, and, subsequently, the remaining half can be energized, both halves then carrying equal current. (CMP-11)

> Informational Note: A hermetic refrigerant motor-compressor is not considered a standard part-winding start induction motor.

N **Passenger Transportation Facilities.** Any area open to the public associated with passenger transportation such as an airport, bus terminal, highway rest stop and service area, marina, seaport, ferry slip, subway station, train station, or port of entry. (CMP-18)

N **Patient Bed Location.** The location of a patient sleeping bed, or the bed or procedure table of a Category 1 space. [**99:**3.3.138] (CMP-15)

N **Patient Care–Related Electrical Equipment.** Electrical equipment appliance that is intended to be used for diagnostic, therapeutic, or monitoring purposes in a patient care vicinity. [**99:**3.3.139] (517) (CMP-15)

N **Patient Care Space Category.** Any space of a health care facility wherein patients are intended to be examined or treated. [**99:**3.3.140] (517) (CMP-15)

> Informational Note No. 1: The health care facility's governing body designates patient care space in accordance with the type of patient care anticipated.
> Informational Note No. 2: Business offices, corridors, lounges, day rooms, dining rooms, or similar areas typically are not classified as patient care spaces. [**99:**A.3.3.140]

The word *space* conveys that the patient care area is often not defined by the four walls of a room. Formerly called patient care areas, patient care spaces are defined by four varying degrees of possible injury to a patient or caregiver (Categories 1–4) that an electrical system failure could cause. Category 1 poses the most risk to a patient or caregiver, and Category 4 provides the least amount of risk to the patient or caregiver.

An operating room, where patients are subject to invasive procedures, is a Category 1 space. See Exhibit 100.21.

N *Category 1 Space (Category 1).* Space in which failure of equipment or a system is likely to cause major injury or death of patients, staff, or visitors. [**99:**3.3.140.1] (CMP-15)

> Informational Note: These spaces, formerly known as critical care rooms, are typically where patients are intended to be subjected to invasive procedures and connected to line-operated, patient

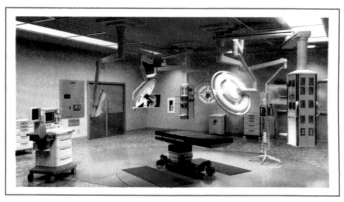

EXHIBIT 100.21 An operating room where patients are connected to line-operated electromedical devices while undergoing invasive procedures. *(From NFPA LiNK®)*

care–related appliances. Examples include, but are not limited to, special care patient rooms used for critical care, intensive care, and special care treatment rooms such as angiography laboratories, cardiac catheterization laboratories, delivery rooms, operating rooms, post-anesthesia care units, trauma rooms, and other similar rooms. [**99:**A.3.3.140.1]

N *Category 2 Space (Category 2).* Space in which failure of equipment or a system is likely to cause minor injury to patients, staff, or visitors. [**99:**3.3.140.2] (CMP-15)

> Informational Note: These spaces were formerly known as general care rooms. Examples include, but are not limited to, inpatient bedrooms, dialysis rooms, in vitro fertilization rooms, procedural rooms, and similar rooms. [**99:**A.3.3.140.2]

N *Category 3 Space (Category 3).* Space in which the failure of equipment or a system is not likely to cause injury to patients, staff, or visitors but can cause discomfort. [**99:**3.3.140.3] (517) (CMP-15)

> Informational Note: These spaces, formerly known as basic care rooms, are typically where basic medical or dental care, treatment, or examinations are performed. Examples include, but are not limited to, examination or treatment rooms in clinics, medical and dental offices, nursing homes, and limited care facilities. [**99:**A.3.3.140.3]

N *Category 4 Space (Category 4).* Space in which failure of equipment or a system is not likely to have a physical impact on patient care. [**99:**3.3.140.4] (517) (CMP-15)

> Informational Note: These spaces were formerly known as support rooms. Examples of support spaces include, but are not limited to, anesthesia work rooms, sterile supply, laboratories, morgues, waiting rooms, utility rooms, and lounges. [**99:**A.3.3.140.4]

N **Patient Care Vicinity.** A space, within a location intended for the examination and treatment of patients, extending 1.8 m (6 ft) beyond the normal location of the bed, chair, table, treadmill, or other device that supports the patient during examination and treatment and extending vertically to 2.3 m (7 ft 6 in.) above the floor. [**99:**3.3.141] (517) (CMP-15)

The patient care vicinity is defined not only by the location of a patient bed but also by other equipment that supports a patient during examination or treatment. The vicinity is also determined by equipment in its normal location used for treatment or in the architect's plans, rather than the temporary location of equipment subject to movement by housekeeping staff or for the convenience of the medical staff.

N **Patient Equipment Grounding Point.** A jack or terminal that serves as the collection point for redundant grounding of electric appliances serving a patient care vicinity or for grounding other items in order to eliminate electromagnetic interference problems. [**99:**3.3.142] (517) (CMP-15)

N **Performance Area.** The stage and audience seating area associated with a temporary stage structure, whether indoors or outdoors, constructed of scaffolding, truss, platforms, or similar devices, that is used for the presentation of theatrical or musical productions or for public presentations. (520) (CMP-15)

N **Permanent Amusement Attraction.** A ride device, entertainment device, or a combination of both that is installed such that portability or relocation is impractical. (522) (CMP-15)

N **Permanently Installed Decorative Fountains and Reflection Pools.** Those that are constructed in the ground, on the ground, or in a building in such a manner that the fountain cannot be readily disassembled for storage, whether or not served by electrical circuits of any nature. These units are primarily constructed for their aesthetic value and are not intended for swimming or wading. (680) (CMP-17)

N **Personnel Protection System (as applied to EVSE).** A system of personnel protection devices and constructional features that when used together provide protection against electric shock of personnel. (625) (CMP-12)

N **Phase, Manufactured. (Manufactured Phase)** The phase that originates at the phase converter and is not solidly connected to either of the single-phase input conductors. (CMP-13)

N **Phase Converter.** An electrical device that converts single-phase power to 3-phase electric power. (CMP-13)

> Informational Note: Phase converters have characteristics that modify the starting torque and locked-rotor current of motors served, and consideration is required in selecting a phase converter for a specific load.

N **Phase Converter, Rotary. (Rotary-Phase Converter)** A device that consists of a rotary transformer and capacitor panel(s) that permits the operation of 3-phase loads from a single-phase supply. (455) (CMP-13)

N **Phase Converter, Static. (Static-Phase Converter)** A device without rotating parts, sized for a given 3-phase load to permit operation from a single-phase supply. (455) (CMP-13)

N **Photovoltaic Cell (PV). (Solar Cell).** The basic photovoltaic device that generates dc electricity when exposed to light. (CMP-4)

Pier. A structure extending over the water and supported on a fixed foundation (fixed pier), or on flotation (floating pier), that provides access to the water. [**303:**3.3.18] (CMP-7)

Pier, Fixed. (Fixed Pier) Pier constructed on a permanent, fixed foundation, such as on piles, that permanently establishes the elevation of the structure deck with respect to land. [**303:**3.3.18.2] (CMP-7)

Pier, Floating. (Floating Pier) Pier designed with inherent flotation capability that allows the structure to float on the water surface and rise and fall with water level changes. [**303:**3.3.18.3] (CMP-7)

N **Pipeline.** A length of pipe including pumps, valves, flanges, control devices, strainers, and/or similar equipment for conveying fluids. (CMP-17)

Plenum. A compartment or chamber to which one or more air ducts are connected and that forms part of the air distribution system. (CMP-3)

Because of concerns about the transfer of products of combustion through environmental air systems, the *NEC* provides specific requirements — in 300.22(B), (C), and (D) and in Articles 725, 760, 770, 800, 820, 830, and 840 — for the installation of wiring methods that are subject to the direct flow of environmental air. The *NEC* definition of the term *plenum* is similar to the definition of *plenum* contained in NFPA 90A, *Standard for the Installation of Air-Conditioning and Ventilating Systems*. The definition is used in conjunction with the requirements for the installation of wiring methods in spaces used for air transfer that are not specifically fabricated as ducts for environmental air.

See also

Article 645 for requirements specific to the air-handling space under a computer room floor

N **Plenum Cable, Cable Routing Assemblies, and Raceways.** Cables, cable routing assemblies, and raceways that have adequate fire-resistant and low smoke-producing characteristics and are suitable for use in ducts, plenums, and other spaces used for environmental air. (722) (CMP-3)

N **Point of Entrance.** The point within a building at which the wire or cable emerges from an external wall, from the roof, or from a concrete floor slab. (CMP-16)

N **Pool.** Manufactured or field-constructed equipment designed to contain water on a permanent or semipermanent basis and used for swimming, wading, immersion, or therapeutic purposes. (680) (CMP-17)

N **Pool, Immersion. (Immersion Pool)** A pool for ceremonial or ritual immersion of users, which is designed and intended to have its contents drained or discharged. (680) (CMP-17)

Immersion pools used for baptisms or other ceremonial rituals that are permanently installed inside or outside a building must comply with the requirements in Parts I and IV of Article 680;

portable immersion pools are covered by the requirements in Parts I and III.

N **Pool, Permanently Installed Swimming, Wading, Immersion, and Therapeutic. (Permanently Installed Swimming, Wading, Immersion, and Therapeutic Pools)** Those that are constructed or installed in the ground or partially in the ground, and all pools installed inside of a building, whether or not served by electrical circuits of any nature. (680) (CMP-17)

Swimming pools without electrical equipment, such as underwater lighting or filtration pumps, are still required to comply with applicable requirements in Article 680 such as those covering overhead clearances, the proximity of underground wiring, and equipotential bonding. This underscores the concern for electrical influences that not only could originate from the premises wiring, but it also addresses the concern about stray currents in the earth resulting from the required grounding and bonding of premises and utility electrical systems or a failure of a premises system at a neighboring property that introduces stray current into the earth. In brook- or stream-fed "natural" swimming pools, water circulation is accomplished by the water flowing through the pool instead of by an electric motor-driven filtration pump.

N **Pool, Storable; used for Swimming, Wading, or Immersion (Storable Immersion Pool). (Storable Pool)** Pools installed entirely on or above the ground that are intended to be stored when not in use and are designed for ease of relocation, regardless of water depth. (680) (CMP-17)

Storable pools are intended to be temporary structures without the need for special wiring or modification to the pool site. They are usually sold as a complete package, consisting of the pool walls, vinyl liner, plumbing kit, and pump/filter device. A storable pool is often disassembled and stored during the winter months.

N **Pool Cover, Electrically Operated. (Electrically Operated Pool Cover)** Motor-driven equipment designed to cover and uncover the water surface of a pool by means of a flexible sheet or rigid frame. (680) (CMP-17)

N **Pool Lift, Electrically Powered. (Electrically Powered Pool Lift)** An electrically powered lift that provides accessibility for people with disabilities to and from a pool or spa. (680) (CMP-17)

N **Portable.** A device intended for indoor or outdoor use that is designed to be hand-carried from location to location, or easily transported without the use of other devices or equipment. (625) (CMP-12)

N **Portable.** X-ray equipment designed to be hand-carried. (660) (CMP-12)

N **Portable (as applied to equipment).** Equipment that is actually moved or can easily be moved from one place to another in normal use. (680) (CMP-17)

N **Portable Power Distribution Unit.** A power distribution box containing receptacles and overcurrent devices. (520) (CMP-15)

Informational Note: See ANSI/UL 1640, *Portable Power-Distribution Equipment*, for information on portable power distribution units.

N **Portable Structures.** Units designed to be moved including, but not limited to, amusement rides, attractions, concessions, tents, trailers, trucks, and similar units. (525) (CMP-15)

N **Portable Substation.** A portable assembly, usually mounted on a trailer, containing primary and secondary switchgear and a transformer. (530) (CMP-15)

Δ **Powder Filling "q".** Type of protection where electrical parts capable of igniting an explosive atmosphere are fixed in position and completely surrounded by filling material (glass or quartz powder) to prevent the ignition of an external explosive atmosphere. (CMP-14)

Informational Note: See ANSI/UL 60079-5, *Explosive Atmospheres — Part 5: Equipment protection by powder filling "q"*, for additional information.

Power Outlet. An enclosed assembly that may include receptacles, circuit breakers, fuseholders, fused switches, buses, and watt-hour meter mounting means; intended to supply and control power to mobile homes, recreational vehicles, park trailers, or boats or to serve as a means for distributing power required to operate mobile or temporarily installed equipment. (CMP-7)

N **Power Outlet, Marina. (Marina Power Outlet)** An enclosed assembly that can include equipment such as receptacles, circuit breakers, fused switches, fuses, watt-hour meters, panelboards, and monitoring means identified for marina use. (555) (CMP-7)

Power Production Equipment. Electrical generating equipment supplied by any source other than a utility service, up to the source system disconnecting means. (CMP-4)

Informational Note: Examples of power production equipment include such items as generators, solar photovoltaic systems, and fuel cell systems.

N **Power Source Output Conductors.** The conductors between power production equipment and the service or other premises wiring. (CMP-4)

N **Power Supply.** A Class 2 power supply connected between the branch-circuit power distribution system and the busbar low-voltage suspended ceiling power distribution system. (393) (CMP-18)

N **Power-Supply Cord.** An assembly consisting of an attachment plug and a length of flexible cord connected to utilization equipment. (CMP-6)

N **Premises.** The land and buildings located on the user's side of the point of demarcation between the communications service provider and the user. (800) (CMP-16)

N **Premises-Powered.** Using power provided locally from the premises. (CMP-16)

Premises Wiring (System). Interior and exterior wiring, including power, lighting, control, and signal circuit wiring together with all their associated hardware, fittings, and wiring devices, both permanently and temporarily installed. This includes one of the following:

(1) Wiring from the service point or power source to the outlets
(2) Wiring from and including the power source to the outlets where there is no service point

Such wiring does not include wiring internal to appliances, luminaires, motors, controllers, motor control centers, and similar equipment. (CMP-1)

Informational Note: Power sources include, but are not limited to, interconnected or stand-alone batteries, solar photovoltaic systems, other distributed generation systems, or generators.

A premises wiring system does not have to be supplied by an electric utility. For example, portable generators and stand-alone PV systems can supply premises wiring systems. If there is no service point, there are no service conductors. The supply conductors are feeder conductors.

Δ **Pressurized.** The process of supplying an enclosure with a protective gas with or without continuous flow at sufficient pressure to prevent the entrance of combustible dust or ignitible fibers/flyings. (CMP-14)

Δ **Pressurized Enclosure "p".** Type of protection for electrical equipment that uses the technique of guarding against the ingress of the external atmosphere, which might be explosive, into an enclosure by maintaining a protective gas therein at a pressure above that of the external atmosphere. (CMP-14)

Informational Note: See ANSI/UL-60079-2, *Explosive Atmospheres — Part 2: Equipment protection by pressurized enclosures "p"*, for additional information.

N **Pressurized Room "p".** A room volume protected by pressurization and of sufficient size to permit the entry of a person who might occupy the room. (CMP-14)

Informational Note: See ANSI/UL 60079-13, *Explosive Atmospheres — Part 13: Equipment protection by pressurized room "p" and artificially ventilated room "v"*, for information on the requirements for rooms intended for human entry where pressurization is used as a means of reducing the risk of explosion.

N **Primary Pad.** A device external to the EV that transfers power via the contactless coupling as part of a wireless power transfer system. (625) (CMP-12)

N **Primary Source.** An electric utility or another source of power that acts as the main forming and stabilizing source in an electric power system. (CMP-4)

Prime Mover. The machine that supplies the mechanical horsepower to a generator. (CMP-13)

Process Seal. A seal between electrical systems and flammable or combustible process fluids where a failure could allow the migration of process fluids into the premises' wiring system. (CMP-14)

Informational Note: See ANSI/UL 122701, *Requirements for Process Sealing Between Electrical Systems and Flammable or Combustible Process Fluids*, for additional information.

N **Production Areas.** Areas where portable electrical equipment is used to implement the capture of images. (530) (CMP-15)

N **Projector, Nonprofessional. (Nonprofessional Projector)** Those types of projectors that do not comply with the definition of *Professional-Type Projector*. (540) (CMP-15)

N **Projector, Professional-Type. (Professional-Type Projector)** A type of projector using 35- or 70-mm film that has a minimum width of 35 mm (1⅜ in.) and has on each edge 212 perforations per meter (5.4 perforations per inch), or a type using carbon arc, xenon, or other light source equipment that develops hazardous gases, dust, or radiation. (540) (CMP-15)

N **Proscenium.** The wall and arch that separates the stage from the auditorium (i.e., house). (520) (CMP-15)

Δ **Protection by Enclosure "t".** Type of protection for explosive dust atmospheres where electrical equipment is provided with an enclosure providing dust ingress protection and a means to limit surface temperatures. (CMP-14)

Informational Note: See ANSI/UL 60079-31, *Explosive Atmospheres — Part 31: Equipment Dust Ignition Protection by Enclosure "t"*, for additional information.

N **Psychiatric Hospital.** A building used exclusively for the psychiatric care, on a 24-hour basis, of four or more inpatients. (517) (CMP-15)

Δ **Purged and Pressurized.** The process of (1) purging, supplying an enclosure with a protective gas at a sufficient flow and positive pressure to reduce the concentration of any flammable gas or vapor initially present to an acceptable level; and (2) pressurization, supplying an enclosure with a protective gas with or without continuous flow at sufficient pressure to prevent the entrance of a flammable gas or vapor, a combustible dust, or an ignitible fiber. (CMP-14)

Informational Note: See NFPA 496, *Standard for Purged and Pressurized Enclosures for Electrical Equipment*, for additional information.

N **Purpose-Built.** A custom luminaire, a piece of lighting equipment, or an effect that is constructed for a specific purpose and is not serially manufactured or available for general sale. (530) (CMP-15)

N **PV DC Circuit (PV System DC Circuit).** Any dc conductor in PV source circuits, PV string circuits, and PV dc-to-dc converter circuits. (690) (CMP-4)

N *PV DC Circuit, Source. (PV Source Circuit)* The PV dc circuit conductors between modules in a PV string circuit, and from PV string circuits or dc combiners, to dc combiners, electronic power converters, or a dc PV system disconnecting means. (690) (CMP-4)

N *PV DC Circuit, String. (PV String Circuit)* The PV source circuit conductors of one or more series-connected PV modules. (690) (CMP-4)

N **PV Module (Module).** A complete, environmentally protected unit consisting of solar cells and other components designed to produce dc power. (CMP-4)

An alternating current (ac) PV module consists of a single integrated unit. Because there is no accessible, field-installed dc wiring in this single unit, the direct current (dc) PV source-circuit requirements in the *NEC* are not applicable to the dc wiring in an ac PV module.

PV (Photovoltaic) System (PV System) (Photovoltaic System). The total components, circuits, and equipment up to and including the PV system disconnecting means that, in combination, convert solar energy into electric energy. (CMP-4)

Qualified Person. One who has skills and knowledge related to the construction and operation of the electrical equipment and installations and has received safety training to recognize and avoid the hazards involved. (CMP-1)

> Informational Note: See NFPA *70E*-2021, *Standard for Electrical Safety in the Workplace*, for electrical safety training requirements.

Section 110.6(A) of the 2021 edition of *NFPA 70E®*, *Standard for Electrical Safety in the Workplace®*, provides training requirements for qualified and unqualified persons who might be exposed to electrical hazards.

Δ **Raceway.** An enclosed channel designed expressly for holding wires, cables, or busbars, with additional functions as permitted in this *Code*. (CMP-8)

Cable trays are support systems for wiring methods and are not considered to be raceways.

> **See also**
>
> **Article 392** for cable tray requirements

N **Raceway Cell.** A single enclosed tubular space in a cellular metal or concrete floor member, the axis of the cell being parallel to the axis of the floor member. (CMP-8)

N **Raceway, Cellular Metal Floor. (Cellular Metal Floor Raceway)** The hollow spaces of cellular metal floors, together with suitable fittings, that may be approved as enclosed channel for electrical conductors. (CMP-8)

Raceway, Communications. (Communications Raceway) An enclosed channel of nonmetallic materials designed expressly for holding communications wires and cables; optical fiber cables; data cables associated with information technology and communications equipment; Class 2, Class 3, and Type PLTC cables; and power-limited fire alarm cables in plenum, riser, and general-purpose applications. (CMP-16)

N **Raceway, Strut-Type Channel. (Strut-Type Channel Raceway)** A metal raceway that is intended to be mounted to the surface of or suspended from a structure, with associated accessories for the installation of electrical conductors and cables. (CMP-8)

N **Raceway, Surface Metal. (Surface Metal Raceway)** A metal raceway that is intended to be mounted to the surface of a structure, with associated couplings, connectors, boxes, and fittings for the installation of electrical conductors. (CMP-8)

N **Raceway, Surface Nonmetallic. (Surface Nonmetallic Raceway)** A nonmetallic raceway that is intended to be mounted to the surface of a structure, with associated couplings, connectors, boxes, and fittings for the installation of electrical conductors. (CMP-8)

N **Raceway, Underfloor. (Underfloor Raceway)** A raceway and associated components designed and intended for installation beneath or flush with the surface of a floor for the installation of cables and electrical conductors. (CMP-8)

N **Rail.** The structural support for the suspended ceiling system typically forming the ceiling grid supporting the ceiling tile and listed utilization equipment, such as sensors, actuators, A/V devices, and low-voltage luminaires and similar electrical equipment. (393) (CMP-18)

Rainproof. Constructed, protected, or treated so as to prevent rain from interfering with the successful operation of the apparatus under specified test conditions. (CMP-1)

> **See also**
>
> **110.28** and its commentary for more information on enclosures considered to be rainproof

Raintight. Constructed or protected so that exposure to a beating rain will not result in the entrance of water under specified test conditions. (CMP-1)

> **See also**
>
> **Table 110.28** for information on enclosure types that are considered to be raintight
>
> **300.6(A)(2)** for related requirements for raintight boxes and cabinets

N **Rated-Load Current (RLC).** The current of a hermetic refrigerant motor-compressor resulting when it is operated at the rated load, rated voltage, and rated frequency of the equipment it serves. (440) (CMP-11)

N **Rated Output Power.** The amplifier manufacturer's stated or marked output power capability into its rated load. (640) (CMP-12)

N **Rated Power.** The output power of a wind turbine at its rated wind speed. (694) (CMP-4)

> Informational Note: See IEC 61400-12-1, *Power Performance Measurements of Electricity Producing Wind Turbines*, for the method for measuring wind turbine power output.

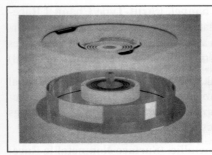

EXHIBIT 100.22 Components of the listed locking support and mounting receptacle and attachment fitting for a luminaire installation. *(Courtesy of Safety Quick Lighting and Fans Corp.)*

Receptacle. A contact device installed at the outlet for the connection of an attachment plug, or for the direct connection of electrical utilization equipment designed to mate with the corresponding contact device. A single receptacle is a single contact device with no other contact device on the same yoke or strap. A multiple receptacle is two or more contact devices on the same yoke or strap. (CMP-18)

> Informational Note: A duplex receptacle is an example of a multiple receptacle that has two receptacles on the same yoke or strap.

Not all receptacles are for cord-and-plug-connected equipment. A receptacle can be used for the connection of specific utilization equipment, such as luminaires. Exhibit 100.22 illustrates a typical locking support and mounting receptacle used for the connection of luminaires.

N **Receptacle, Weight-Supporting Ceiling (WSCR). (Weight-Supporting Ceiling Receptacle)** A contact device installed at an outlet box for the connection and support of luminaries or ceiling-suspended (paddle) fans using a weight-supporting attachment fitting (WSAF). (CMP-18)

> Informational Note: See ANSI/NEMA WD 6, *American National Standard for Wiring Devices — Dimensional Specifications*, for the standard configuration of weight-supporting ceiling receptacles and related weight-supporting attachment fittings.

Receptacle Outlet. An outlet where one or more receptacles are installed. (CMP-18)

Reconditioned. Electromechanical systems, equipment, apparatus, or components that are restored to operating conditions. This process differs from normal servicing of equipment that remains within a facility, or replacement of listed equipment on a one-to-one basis. (CMP-1)

> Informational Note: The term *reconditioned* is frequently referred to as *rebuilt*, *refurbished*, or *remanufactured*.

N **Recreational Vehicle (RV) (Camping Trailer) (Motor Home) (Travel Trailer) (Truck Camper).** A vehicle or slide-in camper that is primarily designed as temporary living quarters for recreational, camping, or seasonal use; has its own motive power or is mounted on or towed by another vehicle; is regulated by the National Highway Traffic Safety Administration as a vehicle or vehicle equipment; does not require a special highway use permit for operation on the highways; and can be easily transported and set up on a daily basis by an individual. [**1192**:3.3.52] (551) (CMP-7)

> Informational Note: See NFPA 1192, *Standard on Recreational Vehicles*, Informative Annex A, for product types and definitions for motor homes and towable recreational vehicles.

N **Recreational Vehicle Park.** Any parcel or tract of land under the control of any person, organization, or governmental entity wherein two or more recreational vehicle, recreational park trailer, and/or other camping sites are offered for use by the public or members of an organization for overnight stays. (551) (CMP-7)

N **Recreational Vehicle Site.** A specific area within a recreational vehicle park or campground that is set aside for use by a camping unit. (551) (CMP-7)

N **Recreational Vehicle Site Supply Equipment.** A power outlet assembly located near the point of entrance of supply conductors to a recreational vehicle site and intended to constitute the disconnecting means for connected recreational vehicles. (551) (CMP-7)

N **Recreational Vehicle Stand.** That area of a recreational vehicle site intended for the placement of a recreational vehicle. (551) (CMP-7)

N **Reference Grounding Point.** The ground bus of the panelboard or isolated power system panel supplying the patient care room. [**99**:3.3.158] (517) (CMP-15)

N **Relative Analgesia.** A state of sedation and partial block of pain perception produced in a patient by the inhalation of concentrations of nitrous oxide insufficient to produce loss of consciousness (conscious sedation). (517) (CMP-15)

N **Relay, Automatic Load Control. (Automatic Load Control Relay)** An emergency lighting control device used to set normally dimmed or normally-off switched emergency lighting equipment to full power illumination levels in the event of a loss of the normal supply by bypassing the dimming/switching controls, and to return the emergency lighting equipment to normal status when the device senses the normal supply has been restored. (700) (CMP-13)

> Informational Note: See ANSI/UL 924, *Emergency Lighting and Power Equipment*, for the requirements covering automatic load control relays.

One use of automatic load control relays is in a lighting branch circuit supplied by the emergency system where the load is controlled by an energy management system. The automatic load control relay functions to restore the required level of emergency lighting where the lighting either has been dimmed or has been completely turned off by an energy management system. When the emergency loads are transferred from the normal source to the alternate source, the relay overrides the energy management mode and provides full power to the load. When the normal source is restored, the relay returns the load to the normal operating mode that is controlled by the energy management system.

Remote-Control Circuit. Any electrical circuit that controls any other circuit through a relay or an equivalent device. (CMP-3)

N **Remote Disconnect Control.** An electric device and circuit that controls a disconnecting means through a relay or equivalent device. (645) (CMP-12)

N **Resistance Heating Element.** A specific separate element to generate heat that is stand-alone, externally attached to, embedded in, integrated with, or internal to the object to be heated. (CMP-17)

> Informational Note: Tubular heaters, strip heaters, heating cable, heating tape, heating blankets, immersion heaters, and heating panels are examples of resistance heaters.

N **Restricted Industrial Establishment [as applied to hazardous (classified) locations].** Establishment with restricted public access, where the conditions of maintenance and supervision ensure that only qualified persons service the installation. (CMP-14)

Δ **Retrofit Kit.** A complete subassembly of parts and devices for field conversion of utilization equipment. (CMP-18)

N **Retrofit Kit, General Use. (General Use Retrofit Kit)** A kit consisting of primary parts, which does not include all the parts for a complete subassembly but includes a list of required parts and installation instructions to complete the subassembly in the field. (600) (CMP-18)

N **Retrofit Kit, Sign Specific. (Sign Specific Retrofit Kit)** A kit consisting of the necessary parts and hardware to allow for field installation in a host sign, based on the included installation instructions. (600) (CMP-18)

N **Reverse Polarity Protection (Backfeed Protection).** A system that prevents two interconnected power supplies, connected positive to negative, from passing current from one power source into a second power source. (393) (CMP-18)

N **Ride Device.** A device or combination of devices that carry, convey, or direct a person(s) over or through a fixed or restricted course within a defined area for the primary purpose of amusement or entertainment. (522) (CMP-15)

N **Riser Cable, Cable Routing Assemblies, and Raceways.** Cables, cable routing assemblies, and raceways that have fire-resistant characteristics capable of preventing the carrying of fire from floor to floor and are suitable for use in a vertical run in a shaft or from floor to floor. (722) (CMP-3)

N **Safe Zone.** Low probability of damage other than a slight swelling of the capacitor case, as identified by the case rupture curve of the capacitor. (460) (CMP-11)

N **Safety Circuit.** The part of a control system containing one or more devices that perform a safety-related function. [79:3.3.95] (CMP-12)

> Informational Note: See NFPA 79-2021, *Electrical Standard for Industrial Machinery. Safety-related control system* and *safety interlock circuit* are common terms that can be used to refer to the safety circuit in other standards. The safety circuit can include hard-wired, communication, and software-related components.

Sealable Equipment. Equipment enclosed in a case or cabinet that is provided with a means of sealing or locking so that live parts cannot be made accessible without opening the enclosure. (CMP-1)

> Informational Note: The equipment may or may not be operable without opening the enclosure.

N **Sealed [as applied to hazardous (classified) locations].** Constructed such that equipment is sealed effectively against entry of an external atmosphere and is not opened during normal operation or for any maintenance activities. (CMP-14)

> Informational Note: See ANSI/UL 121201, *Nonincendive Electrical Equipment for Use in Class I and II, Division 2 and Class III, Divisions 1 and 2 Hazardous (Classified) Locations*, for additional information.

Δ **Sealed, Hermetically. (Hermetically Sealed)** Sealed against the entrance of an external atmosphere, such that the seal is made by fusion of metal to metal, ceramic to metal, or glass to metal. (CMP-14)

> Informational Note: See ANSI/UL 121201, *Nonincendive Electrical Equipment for Use in Class I and II, Division 2 and Class III, Divisions 1 and 2 Hazardous (Classified) Locations*, for additional information.

N **Section Sign.** A sign or outline lighting system, shipped as subassemblies, that requires field-installed wiring between the subassemblies to complete the overall sign. The subassemblies are either physically joined to form a single sign unit or are installed as separate remote parts of an overall sign. (600) (CMP-18)

The definition of the term *section sign* clarifies that the multiple parts are referred to as subassemblies, and the only field wiring involved is the connection between subassemblies and connection of the subassemblies to the power source.

The power source may be a branch circuit or the secondary wiring from a sign power supply. In accordance with 600.3, section signs are required to be listed. In accordance with ANSI/UL 48, *Standard for Electric Signs*, each subassembly is provided with installation instructions containing detailed information on the mechanical and electrical connections to be performed when the subassemblies are installed to form the completed section sign (see Exhibit 100.23).

N **Selected Receptacles.** A minimal number of receptacles selected by the health care facility's governing body as necessary to provide essential patient care and facility services during loss of normal power. [99:3.3.164] (517) (CMP-15)

EXHIBIT 100.23 Example of a section sign, in which the separate remote subassemblies are field installed to form an overall single sign. *(Courtesy of Kieffer-Starlite)*

N Self-Contained Therapeutic Tubs or Hydrotherapeutic Tanks. A factory-fabricated unit consisting of a therapeutic tub or hydrotherapeutic tank with all water-circulating, heating, and control equipment integral to the unit. Equipment may include pumps, air blowers, heaters, light controls, sanitizer generators, and so forth. (680) (CMP-17)

N Separable Power Supply Cable Assembly. A flexible cord or cable, including ungrounded, grounded, and equipment grounding conductors, provided with a cord connector, an attachment plug, and all other fittings, grommets, or devices installed for the purpose of delivering energy from the source of electrical supply to the truck or transport refrigerated unit (TRU) flanged surface inlet. (626) (CMP-12)

Separately Derived System. An electrical power supply output, other than a service, having no direct connection(s) to circuit conductors of any other electrical source other than those established by grounding and bonding connections. (CMP-5)

Examples of separately derived systems include generators, batteries, converter windings, transformers, and solar photovoltaic systems, provided they have no direct electrical connection to another source. The earth, metal enclosures, metal raceways, and equipment grounding conductors may provide incidental connection between systems. This definition clarifies that those systems can still be considered to be separately derived systems as long as the separately derived systems have no direct electrical connection to service-derived systems.

See also

250.30(A)(6), which permits a common grounding electrode conductor to be installed for multiple separately derived systems

Service. The conductors and equipment connecting the serving utility to the wiring system of the premises served. (CMP-10)

A service can be supplied only by the serving utility. If electric energy is supplied by something other than the serving utility, the supplied conductors and equipment are considered feeders and not a service.

Service Conductors. The conductors from the service point to the service disconnecting means. (CMP-10)

The term *service conductors* is broad and may include overhead service conductors, underground service conductors, and service-entrance conductors. The term specifically excludes any wiring on the supply side (serving utility side) of the service point. The service conductors originate at the service point (where the serving utility ends) and end at the service disconnect. Service conductors originate only from the serving utility. The definition no longer includes service drops and service laterals; those terms now apply only to conductors that are under the control of the serving utility.

If the utility has specified that the service point is at the utility pole, the service conductors from an overhead distribution system originate at the utility pole and terminate at the service disconnecting means. See Exhibit 100.24 for overhead service conductors.

If the utility has specified that the service point is at the utility manhole, the service conductors from an underground distribution system originate at the utility manhole and terminate at the service disconnecting means. Where utility-owned primary conductors are extended to outdoor pad-mounted transformers on private property, the service conductors originate at the secondary connections of the transformers only if the utility has specified that the service point is at the secondary connections.

See also

Article 235, Part V, and the commentary following **235.401** for information on service conductors exceeding 1000 volts, nominal

Service Conductors, Overhead. (Overhead Service Conductors) The overhead conductors between the service point and the first point of connection to the service-entrance conductors at the building or other structure. (CMP-10)

Service Conductors, Underground. (Underground Service Conductors) The underground conductors between the service point and the first point of connection to the service-entrance conductors in a terminal box, meter, or other enclosure, inside or outside the building wall. (CMP-10)

> Informational Note: Where there is no terminal box, meter, or other enclosure, the point of connection is considered to be the point of entrance of the service conductors into the building.

Service Drop. The overhead conductors between the serving utility and the service point. (CMP-10)

This definition correlates with the definition of the term *service lateral*. Service-drop and service-lateral conductors are conductors on the line side of the service point and are not subject to the *NEC*. Overhead conductors on the load side of the service point are overhead service conductors.

EXHIBIT 100.24 Conductors in an overhead service where the service point is the connection to the terminals on the load side of a utility transformer.

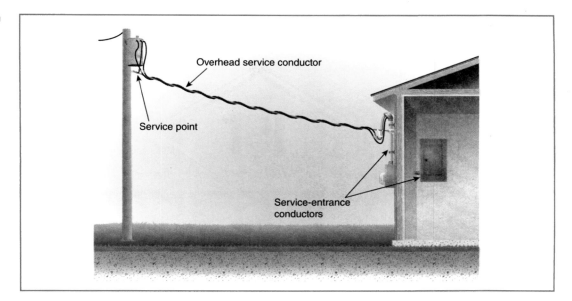

Overhead service conductor

Service point

Service-entrance conductors

EXHIBIT 100.25 Conductors in an overhead service where the service point is the point of attachment at the building structure where the service-drop conductors terminate to the service-entrance conductors.

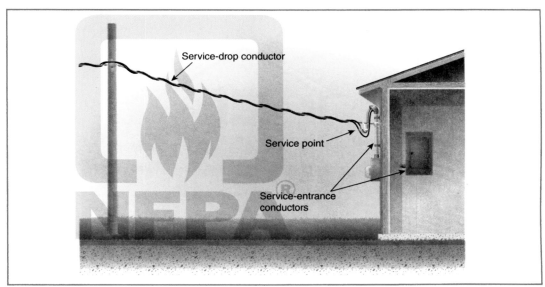

Service-drop conductor

Service point

Service-entrance conductors

In Exhibit 100.25, the service-drop conductors run from the utility pole and connect to the service-entrance conductors at the service point. Conductors on the utility side of the service point are not covered by the *NEC*. The utility specifies the location of the service point. Exact locations of the service point can vary from utility to utility as well as from occupancy to occupancy.

N **Service-Entrance Conductor Assembly.** Multiple single-insulated conductors twisted together without an overall covering, other than an optional binder intended only to keep the conductors together. (CMP-6)

Δ **Service-Entrance Conductors.** The service conductors between the terminals of the service equipment to the service drop, overhead service conductors, service lateral, or underground service conductors. (CMP-10)

Informational Note: Where service equipment is located outside the building walls, there could be no service-entrance conductors or they might be entirely outside the building.

See Exhibit 100.25 for an illustration of service-entrance conductors in an overhead system. The system shows a service drop from a utility pole to attachment on a house and service-entrance conductors from point of attachment (spliced to service-drop conductors), down the side of the house, through the meter socket, and terminating in the service equipment. In this instance, the service point is at the drip loop. The conductors on the line side of the service point are under the control of the utility.

See Exhibit 100.26 for an illustration of service-entrance conductors in an underground system. The illustration on the top shows underground service lateral conductors run from a pole to a service point underground. The conductors from the service point

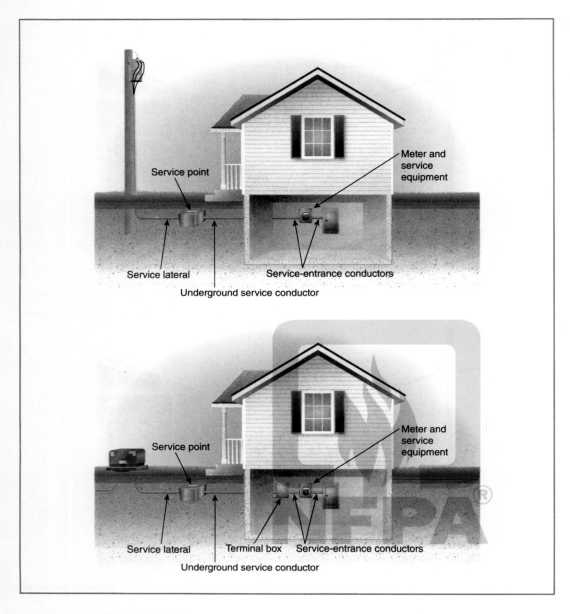

EXHIBIT 100.26 Underground systems showing service laterals run from a pole *(top)* and from a transformer *(bottom)*.

into the building are underground service conductors. The illustration on the bottom shows service lateral conductors run from a utility transformer.

Service Equipment. The necessary equipment, consisting of a circuit breaker(s) or switch(es) and fuse(s) and their accessories, connected to the serving utility and intended to constitute the main control and disconnect of the serving utility. (CMP-10)

Service equipment may consist of circuit breakers or fused switches that are provided to disconnect all ungrounded conductors in a building or other structure from the service-entrance conductors. Individual meter socket enclosures are not considered service equipment according to 230.66(B) but must be listed and rated for the voltage and current rating of the service.

The disconnecting means at any one location in a building or other structure is not allowed to consist of more than six circuit breakers or six switches. The disconnecting means is required to be readily accessible either outside or inside nearest the point of entrance of the service-entrance conductors.

See also

Article 230, Part VI, for service conductor disconnecting means requirements

230.6 for service conductor requirements outside the building

N **Service Equipment, Mobile Home. (Mobile Home Service Equipment)** The equipment containing the disconnecting means, overcurrent protective devices, and receptacles or other means for connecting a mobile home feeder assembly. (550) (CMP-7)

Service Lateral. The underground conductors between the utility electric supply system and the service point. (CMP-10)

Service-drop and service-lateral conductors are conductors on the line side of the service point and are not subject to the requirements of the *NEC*. In other words, these conductors are under the exclusive control of the utility. Underground conductors on the load side of the service point are underground service conductors.

As Exhibit 100.26 shows, the underground service laterals may be run from poles or from transformers, with or without terminal boxes, provided they terminate at the service point. The next transition would be to the underground service conductors, which would connect to the service-entrance conductors, or they may terminate in a terminal box, meter, or some other enclosure, which may be inside or outside the building. Conductors on the utility side of the service point are not covered by the *NEC*. The utility specifies the location of the service point. Exact locations of the service point can vary from utility to utility as well as from occupancy to occupancy.

Service Point. The point of connection between the facilities of the serving utility and the premises wiring. (CMP-10)

> Informational Note: The service point can be described as the point of demarcation between where the serving utility ends and the premises wiring begins. The serving utility generally specifies the location of the service point based on the conditions of service.

The exact location for a service point is generally determined by the utility and can vary from utility to utility. Only those conductors that are located on the premises wiring side of the service point are covered by the *NEC*.

Conductors on the serving utility side of the service point generally are not covered by the *NEC*. For example, a typical suburban residence has overhead conductors from the utility pole to the house. If the utility specifies that the service point is at the point of attachment of the overhead conductors to the house, the overhead conductors are service-drop conductors and are not covered by the *NEC* because the conductors are not on the premises wiring side of the service point. Alternatively, if the utility specifies that the service point is at the pole, the overhead conductors are considered overhead service conductors, and the *NEC* would apply to those conductors.

N **Servicing.** The process of following a manufacturer's set of instructions or applicable industry standards to analyze, adjust, or perform prescribed actions upon equipment with the intention to preserve or restore the operational performance of the equipment. (CMP-1)

> Informational Note: Servicing often encompasses maintenance and repair activities.

N **Shore Power.** The electrical equipment required to power a floating vessel including, but not limited to, the receptacle and cords. (555) (CMP-7)

N **Shoreline.** The farthest extent of standing water under the applicable conditions that determine the electrical datum plane for the specified body of water. (682) (CMP-17)

N **Short Circuit.** An abnormal connection (including an arc) of relatively low impedance, whether made accidentally or intentionally, between two or more points of different potential. (CMP-10)

Short-Circuit Current Rating. The prospective symmetrical fault current at a nominal voltage to which an apparatus or system is able to be connected without sustaining damage exceeding defined acceptance criteria. (CMP-10)

The short-circuit current rating is marked on the equipment nameplate as shown in Exhibit 100.27. The available input current must not exceed this rating. Otherwise, the equipment can be damaged by short-circuit currents, posing a hazard to personnel and property. See also the definition of *interrupting rating*.

Show Window. Any window, including windows above doors, used or designed to be used for the display of goods or advertising material, whether it is fully or partly enclosed or entirely open at the rear and whether or not it has a platform raised higher than the street floor level. (CMP-2)

N **Sign, Photovoltaic (PV) Powered (PV Powered Sign). [Photovoltaic (PV) Powered Sign]** A complete sign powered by solar energy consisting of all components and subassemblies for installation either as an off-grid stand-alone, on-grid interactive, or non-grid interactive system. (600) (CMP-18)

N **Sign Body.** A portion of a sign that may provide protection from the weather but is not an electrical enclosure. (600) (CMP-18)

Signaling Circuit. Any electrical circuit that energizes signaling equipment. (CMP-3)

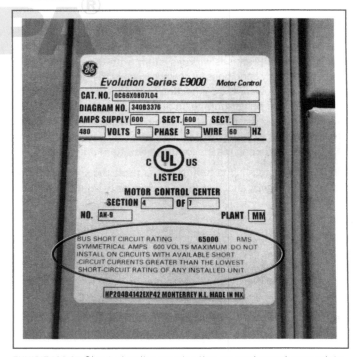

EXHIBIT 100.27 Short-circuit current rating on equipment nameplate.

Δ **Simple Apparatus.** An electrical component or combination of components of simple construction with well-defined electrical parameters that does not generate more than 1.5 volts, 100 mA, and 25 mW, or a passive component that does not dissipate more than 1.3 watts and is compatible with the intrinsic safety of the circuit in which it is used. (CMP-14)

> Informational Note No. 1: The following are examples of simple apparatus:
>
> (1) Passive components; for example, switches, instrument connectors, plugs and sockets, junction boxes, resistance temperature devices, and simple semiconductor devices such as LEDs
> (2) Sources of stored energy consisting of single components in simple circuits with well-defined parameters; for example, capacitors or inductors, whose values are considered when determining the overall safety of the system
> (3) Sources of generated energy; for example, thermocouples and photocells, that do not generate more than 1.5 volts, 100 mA, and 25 mW

> Informational Note No. 2: See ANSI/UL 913, *Intrinsically Safe Apparatus and Associated Apparatus for Use in Class I, II, and III, Division 1, Hazardous (Classified) Locations*, and ANSI/UL 60079-11, *Explosive Atmospheres — Part 11: Equipment Protection by Intrinsic Safety "i"*, for additional information.

Single-Pole Separable Connector. A device that is installed at the ends of portable, flexible, single-conductor cable that is used to establish connection or disconnection between two cables or one cable and a single-pole, panel-mounted separable connector. (CMP-18)

N **Site-Isolating Device.** A pole-mounted disconnecting means installed at the distribution point for the purposes of isolation, system maintenance, emergency disconnection, or connection of optional standby systems. (547) (CMP-7)

The site-isolating device provides a means to disconnect and isolate the agricultural premises wiring system from the serving utility under emergency conditions for maintenance of the load-side wiring system or to allow for the connection of an alternate power source when there is a power outage. In accordance with 547.41(A)(2), the site-isolating device must be pole-mounted. And, as its name implies, it is an isolating switch and is not considered to be the service disconnecting means for the agricultural premises.

N **Skeleton Tubing.** Neon tubing that is itself the sign or outline lighting and is not attached to an enclosure or sign body. (600) (CMP-18)

N **Slip.** A berthing space between or adjacent to piers, wharves, or docks; the water areas associated with boat occupation. [**303**:3.3.21] (555) (CMP-7)

> Informational Note: See the definition of *Berth* for additional information.

N **Solid-State Phase-Control Dimmer.** A solid-state dimmer where the wave shape of the steady-state current does not follow the wave shape of the applied voltage such that the wave shape is nonlinear. (CMP-15)

Solid-state phase-control dimmers are nonlinear devices. Where they are used, nonlinear loading of the neutral conductor occurs, which could necessitate increasing the size of the neutral conductor. See 310.15(E).

N **Solid-State Sine Wave Dimmer.** A solid-state dimmer where the wave shape of the steady-state current follows the wave shape of the applied voltage such that the wave shape is linear. (CMP-15)

Some professional performance lighting systems use a solid-state, 3-phase, 4-wire dimming system whose wave shape varies with the amplitude of the applied voltage wave shape without any of the nonlinear switching found in phase-control solid-state dimmers. Because solid-state sine wave dimmers are linear loads, they do not require the neutral to be considered a current-carrying conductor.

N **Spa or Hot Tub.** A hydromassage pool, or tub for recreational or therapeutic use, not located in health care facilities, designed for immersion of users, and usually having a filter, heater, and motor-driven blower. It may be installed indoors or outdoors, on the ground or supporting structure, or in the ground or supporting structure. Generally, they are not designed or intended to have its contents drained or discharged after each use. (680) (CMP-17)

N **Spa or Hot Tub, Packaged Equipment Assembly. (Packaged Spa or Hot Tub Equipment Assembly)** A factory-fabricated unit consisting of water-circulating, heating, and control equipment mounted on a common base, intended to operate a spa or hot tub. Equipment can include pumps, air blowers, heaters, lights, controls, sanitizer generators, and so forth. (680) (CMP-17)

N **Spa or Hot Tub, Self-Contained. (Self-Contained Spa or Hot Tub)** Factory-fabricated unit consisting of a spa or hot tub vessel with all water-circulating, heating, and control equipment integral to the unit. Equipment can include pumps, air blowers, heaters, lights, controls, sanitizer generators, and so forth. (680) (CMP-17)

N **Spa or Hot Tub, Storable. (Storable Spa or Hot Tub)** Spas or hot tubs installed entirely on or above the ground that are intended to be stored when not in use and are designed for ease of relocation. (680) (CMP-17)

N **Space.** A portion of the health care facility designated by the health care facility's governing body that serves a specific purpose. [**99**:3.3.171] (517) (CMP-15)

Special Permission. The written consent of the authority having jurisdiction. (CMP-1)

The AHJ for enforcement of the *NEC* is responsible for making interpretations and granting special permission as outlined in a number of the rules as stated in 90.4.

110.26(A)(1)(b) and **230.2(B)** for requirements governing special permission situations

N **Special Protection "s".** Type of protection that permits design, assessment, and testing of equipment that cannot be fully assessed within a recognized type of protection or combination of recognized types of protection because of functional or operational limitations, but that can be demonstrated to provide the necessary equipment protection level (EPL). (CMP-14)

> Informational Note: See ANSI/UL 60079-33, *Explosive Atmospheres — Part 33: Equipment Protection by Special Protection "s"*, for additional information.

N **Special-Purpose Multi-Circuit Cable System.** A portable branch-circuit distribution system consisting of one or more trunk cables and optional breakout assemblies or multi-circuit outlet enclosures. (520) (CMP-15)

N **Spider (Cable Splicing Block).** A device that contains busbars that are insulated from each other for the purpose of splicing or distributing power to portable cables and cords that are terminated with single-pole busbar connectors. (530) (CMP-15)

N **Spin Down.** A shutdown condition of the FESS, where energy is being dissipated and the flywheel rotor is slowing down to a stop. (706) (CMP-13)

> Informational Note: A complete stop of a flywheel rotor cannot occur instantaneously because of the high kinetic energy of the rotor, but rather occurs over time as a result of friction forces acting on the rotor.

N **Splash Pad.** A fountain intended for recreational use by pedestrians and designed to contain no more than 25 mm (1 in.) of water depth. This definition does not include showers intended for hygienic rinsing prior to use of a pool, spa, or other water feature. (680) (CMP-17)

Splash pads do not completely immerse users in water; nevertheless, they are wet, their body resistances are lowered, and nothing precludes someone from lying on the splash pad surface where a voltage gradient could exist. Ohm's law tells us that a greater difference of potential across a resistance (a person laying on the splash pad) results in higher current. For all the same reasons that electrical installations associated with a permanently installed swimming pool, including equipotential bonding, exist, splash pads, even though considered to be fountains and subject to the requirements of Parts I and V of Article 680, are also required to comply with Part II for permanently installed swimming pools.

680.50 for more information on the applicability of various parts of Article 680 to splash pads

Δ **Spray Area.** Any fully enclosed, partly enclosed, or unenclosed area in which flammable or combustible vapors, mists, residues, dusts, or deposits are present due to the operation of spray processes, including:

(1) any area in the direct path of a spray application process;
(2) the interior of a spray booth, spray room, or limited finishing workstation, as herein defined;
(3) the interior of any exhaust plenum, eliminator section, or scrubber section;
(4) the interior of any exhaust duct or exhaust stack leading from a spray application process;
(5) the interior of any air recirculation path up to and including recirculation particulate filters;
(6) any solvent concentrator (pollution abatement) unit or solvent recovery (distillation) unit; and
(7) the inside of a membrane enclosure.

The following are not part of the spray area:

(1) fresh air make-up units;
(2) air supply ducts and air supply plenums;
(3) recirculation air supply ducts downstream of recirculation particulate filters; and
(4) exhaust ducts from solvent concentrator (pollution abatement) units. [33:3.3.2.3] (CMP-14)

> Informational Note No. 1: Unenclosed spray areas are locations outside of buildings or are localized operations within a larger room or space. Such areas are normally provided with some local vapor extraction/ventilation system. In automated operations, the area limits are the maximum area in the direct path of spray operations. In manual operations, the area limits are the maximum area of spray when aimed at 90 degrees to the application surface.
>
> Informational Note No. 2: See definitions for *limited finishing workstation* and *membrane enclosure* for additional information.

Spray Area, Outdoor. (Outdoor Spray Area) A spray area that is outside the confines of a building or that has a canopy or roof that does not limit the dissipation of the heat of a fire or dispersion of flammable vapors and does not restrict fire-fighting access and control. For the purpose of this standard, an outdoor spray area can be treated as an unenclosed spray area as defined in this *Code*. [33:3.3.2.3.1] (CMP-14)

Spray Area, Unenclosed. (Unenclosed Spray Area) Any spray area that is not confined by a limited finishing workstation, spray booth, or spray room, as herein defined. [33:3.3.2.3.2] (CMP-14)

Spray Booth. A power-ventilated enclosure for a spray application operation or process that confines and limits the escape of the material being sprayed, including vapors, mists, dusts, and residues that are produced by the spraying operation and conducts or directs these materials to an exhaust system. [33:3.3.19] (CMP-14)

> Informational Note: A spray booth is an enclosure or insert within a larger room used for spraying, coating, and/or dipping applications. A spray booth can be fully enclosed or have open front or face and can include a separate conveyor entrance and exit. The spray booth is provided with a dedicated ventilation exhaust with supply air from the larger room or from a dedicated air supply.

Spray Room. A power-ventilated fully enclosed room with a specified fire resistance rating used exclusively for open spraying of flammable or combustible materials. [33:3.3.20] (CMP-14)

N **Stage Effect (Special Effect).** An electrical or electromechanical piece of equipment used to simulate a distinctive visual or audible effect, such as a wind machine, lightning simulator, or sunset projector. (CMP-15)

N **Stage Equipment.** Equipment at any location on the premises integral to the stage production including, but not limited to, equipment for lighting, audio, special effects, rigging, motion control, projection, or video. (520) (CMP-15)

N **Stage Lighting Hoist.** A motorized lifting device that contains a mounting position for one or more luminaires, with wiring devices for connection of luminaires to branch circuits, and integral flexible cables to allow the luminaires to travel over the lifting range of the hoist while energized. (520) (CMP-15)

N **Stage Property.** An article or object used as a visual element in a motion picture or television production, except painted backgrounds (scenery) and costumes. (530) (CMP-15)

N **Stage Set.** A specific area set up with temporary scenery and properties designed and arranged for a particular scene in a motion picture or television production. (CMP-15)

N **Stage Switchboard, Fixed. (Fixed Stage Switchboard)** A permanently installed switchboard, panelboard, or rack containing dimmers or relays with associated overcurrent protective devices, or overcurrent protective devices alone, used primarily to feed stage equipment. (CMP-15)

N **Stage Switchboard, Portable. (Portable Stage Switchboard)** A portable rack or pack containing dimmers or relays with associated overcurrent protective devices, or overcurrent protective devices alone, used to feed stage equipment. (520) (CMP-15)

N **Stand Lamp.** A portable stand that contains a general-purpose luminaire or lampholder with guard for the purpose of providing general illumination on a stage, in an auditorium, or in a studio. (520) (CMP-15)

Stand-Alone System. A system that is not connected to an electric power production and distribution network. (CMP-4)

N **Stationary (as applied to equipment).** Equipment that is not moved from one place to another in normal use. (680) (CMP-17)

N **Storage, Dry Stack. (Dry Stack Storage)** A facility, either covered or uncovered, constructed of horizontal and vertical structural members designed to allow placement of small boats in defined slots arranged both horizontally and vertically. [303:3.3.24.2] (555) (CMP-7)

N **Stored-Energy Power Supply System (SEPSS).** A complete functioning EPSS powered by a stored-energy electrical source. (CMP-13)

N **Stranding, Compact. (Compact Stranding)** A conductor stranding method in which each layer of strands is pressed together to minimize the gaps between the strands so the overall diameter of the finished conductor is less than a concentric stranded conductor and less than a compressed stranded conductor. (CMP-6)

N **Stranding, Compressed. (Compressed Stranding)** A conductor stranding method in which the outer layer of strands is pressed together so the overall diameter of the finished conductor is less than a concentric stranded conductor but greater than a compact stranded conductor. (CMP-6)

N **Stranding, Concentric. (Concentric Stranding)** A conductor consisting of a straight central strand surrounded by one or more layers of strands, helically laid in a geometric pattern. (CMP-6)

N **Strip Light.** A luminaire with multiple lamps arranged in a row. (520) (CMP-15)

Structure. That which is built or constructed, other than equipment. (CMP-1)

N **Structure, Relocatable. (Relocatable Structure)** A factory-assembled structure or structures transportable in one or more sections that are built on a permanent chassis and designed to be used as other than a dwelling unit without a permanent foundation. (545) (CMP-7)

Informational Note: Examples of relocatable structures are those units that are equipped for sleeping purposes only, contractor's and other on-site offices, construction job dormitories, studio dressing rooms, banks, clinics, stores, shower facilities and restrooms, training centers, or for the display or demonstration of merchandise or machines.

The addition of the reference to relocatable structures and the conductors that connect them to the supply of electricity are necessary to properly address structures such as mobile office units, mobile classrooms, mobile restrooms and shower facilities, mobile break rooms, mobile training centers, and other non-dwelling types of relocatable structures.

N **Subassembly.** Component parts or a segment of a sign, retrofit kit, or outline lighting system that, when assembled, forms a complete unit or product. (600) (CMP-18)

Substation. An assemblage of equipment (e.g., switches, interrupting devices, circuit breakers, buses, and transformers) through which electric energy is passed for the purpose of distribution, switching, or modifying its characteristics. (CMP-9)

N **Supervisory Control and Data Acquisition (SCADA).** An electronic system that provides monitoring and controls for the operation of the critical operations power system. (CMP-13)

Informational Note: This can include the fire alarm system, security system, control of the HVAC, the start/stop/monitoring of the power supplies and electrical distribution system, annunciation and communications equipment to emergency personnel, facility occupants, and remote operators.

Support Areas. Areas, other than fixed production offices, intended to support production and where image capture will not take place. Such areas include, but are not limited to, mobile production offices, storage, and workspaces; vehicles and trailers for cast, makeup, hair, lighting, grip, wardrobe, props, catering, and craft services; and portable restrooms. (530) (CMP-15)

Surge Arrester. A protective device for limiting surge voltages by discharging or bypassing surge current; it also prevents continued flow of follow current while remaining capable of repeating these functions. (CMP-10)

Surge-Protective Device (SPD). A protective device for limiting transient voltages by diverting or limiting surge current; it also prevents continued flow of follow current while remaining capable of repeating these functions and is designated as follows:

(1) Type 1: Permanently connected SPDs intended for installation between the secondary of the service transformer and the line side of the service disconnect overcurrent device

(2) Type 2: Permanently connected SPDs intended for installation on the load side of the service disconnect overcurrent device, including SPDs located at the branch panel

(3) Type 3: Point of utilization SPDs

(4) Type 4: Component SPDs, including discrete components, as well as assemblies. (CMP-10)

Informational Note: See UL 1449, *Standard for Surge Protective Devices*, for further information on SPDs.

Suspended Ceiling Grid. A system that serves as a support for a finished ceiling surface and other utilization equipment. (393) (CMP-18)

Switch, General-Use. (General-Use Switch) A switch intended for use in general distribution and branch circuits. It is rated in amperes, and it is capable of interrupting its rated current at its rated voltage. (CMP-9)

Switch, General-Use Snap. (General-Use Snap Switch) A form of general-use switch constructed so that it can be installed in device boxes or on box covers, or otherwise used in conjunction with wiring systems recognized by this *Code*. (CMP-9)

Switch, Isolating. (Isolating Switch) A switch intended for isolating an electrical circuit from the source of power. It has no interrupting rating, and it is intended to be operated only after the circuit has been opened by some other means. (CMP-9)

Switch, Motor-Circuit. (Motor-Circuit Switch) A switch rated in horsepower that is capable of interrupting the maximum operating overload current of a motor of the same horsepower rating as the switch at the rated voltage. (CMP-11)

Switchboard. A large single panel, frame, or assembly of panels on which are mounted on the face, back, or both, switches, overcurrent and other protective devices, buses, and usually instruments. (CMP-9)

Informational Note: These assemblies can be accessible from the rear or side as well as from the front and are not intended to be installed in cabinets.

Switchgear. An assembly completely enclosed on all sides and top with sheet metal (except for ventilating openings and inspection windows) and containing primary power circuit switching, interrupting devices, or both, with buses and connections. The assembly may include control and auxiliary devices. Access to the interior of the enclosure is provided by doors, removable covers, or both. (CMP-9)

Informational Note: All switchgear subject to *NEC* requirements is metal enclosed. Switchgear rated below 1000 V or less may be identified as "low-voltage power circuit breaker switchgear." Switchgear rated over 1000 V may be identified as "metal-enclosed switchgear" or "metal-clad switchgear." Switchgear is available in non–arc-resistant or arc-resistant constructions.

Switching Device (as applied to equipment rated over 1000 volts ac, 1500 volts dc, nominal). A device designed to close, open, or both, one or more electrical circuits. (CMP-9)

Cutout. An assembly of a fuse support with either a fuseholder, fuse carrier, or disconnecting blade. The fuseholder or fuse carrier may include a conducting element (fuse link) or may act as the disconnecting blade by the inclusion of a nonfusible member.

Disconnecting Switch (or Isolating Switch). A mechanical switching device used for isolating a circuit or equipment from a source of power.

Interrupter Switch. A switching device capable of making, carrying, and interrupting specified currents.

Oil-Filled Cutout. A cutout in which all or part of the fuse support and its fuse link or disconnecting blade is mounted in oil with complete immersion of the contacts and the fusible portion of the conducting element (fuse link) so that arc interruption by severing of the fuse link or by opening of the contacts will occur under oil.

Oil Switch. A switching device having contacts that operate under oil (or askarel or other suitable liquid).

Regulator Bypass Switch. A switching device or combination of switching devices designed to bypass equipment used to control voltage levels or related circuit characteristics.

System Isolation Equipment. A redundantly monitored, remotely operated contactor-isolating system, packaged to provide the disconnection/isolation function, capable of verifiable operation from multiple remote locations by means of lockout switches, each having the capability of being padlocked in the "off" (open) position. (430) (CMP-11)

Tap Conductor. A conductor, other than a service conductor, that has overcurrent protection ahead of its point of supply that exceeds the value permitted for similar conductors that are protected as described elsewhere in 240.4. (240) (CMP-10)

Tap conductors are branch-circuit and feeder conductors subject to the conditional overcurrent protection requirements specified in 240.21. The general requirement is for conductor overcurrent protection to be provided at the point a conductor is supplied. Section 240.21 allows for the installation of branch-circuit and feeder conductors that are protected against overcurrent using an OCPD downstream of the point of supply. Short-circuit and ground-fault protection is provided upstream of the tap.

Exhibit 100.28 illustrates a 1/0 AWG, Type THW (thermoplastic heat and water-resistant wire) copper conductor (150 amperes, from Table 310.16) connected to a 3/0 AWG, Type THW copper feeder conductor with an ampacity of 200 amperes (increased in size to compensate for voltage drop) that is protected by a 150-ampere OCPD. Because the ampacity of the 1/0 AWG conductor is not exceeded by the rating of the overcurrent device, the 1/0 AWG conductor is not considered to be a tap conductor. The overcurrent device protects both the 1/0 AWG and the 3/0 AWG conductors in accordance with the basic rule of 240.4, and additional overcurrent protection is not required at the supply point or the termination point of the 1/0 AWG conductors.

N **Task Illumination.** Provisions for the minimum lighting required to carry out necessary tasks in the areas described in 517.34(A), including safe access to supplies and equipment and access to exits. [**99**:3.3.177] (517) (CMP-15)

N **Technical Power System.** An electrical distribution system where the equipment grounding conductor is isolated from the premises grounded conductor and the premises equipment grounding conductor except at a single grounded termination point within a branch-circuit panelboard, at the originating (main breaker) branch-circuit panelboard or at the premises grounding electrode. (640) (CMP-12)

N **Temporary Equipment.** Portable wiring and equipment intended for use with events of a transient or temporary nature

where all equipment is presumed to be removed at the conclusion of the event. (640) (CMP-12)

Temporary equipment may be used in permanent or temporary facilities or in areas with no services other than a source of electrical power. Temporary equipment may be used in indoor and outdoor areas such as athletic facilities, halls, auditoriums, concert shells, athletic fields, beaches, and other places designated for public assembly.

N **Terminal (as applied to batteries).** That part of a cell, container, or battery to which an external connection is made (commonly identified as post, pillar, pole, or terminal post). (CMP-13)

Thermal Protector (as applied to motors). A protective device for assembly as an integral part of a motor or motor-compressor that, when properly applied, protects the motor against dangerous overheating due to overload and failure to start. (CMP-11)

> Informational Note: The thermal protector may consist of one or more sensing elements integral with the motor or motor-compressor and an external control device.

N **Thermal Resistivity.** The heat transfer capability through a substance by conduction. (CMP-6)

> Informational Note: Thermal resistivity is the reciprocal of thermal conductivity and is designated Rho, which is expressed in the units °C-cm/W.

Thermally Protected (as applied to motors). A motor or motor-compressor that is provided with a thermal protector. (CMP-11)

N **Top Shield.** A grounded metal shield covering under-carpet components of the flat conductor cable (Type FCC) system for the purposes of providing protection against physical damage. (324) (CMP-6)

N **Tower.** A pole or other structure that supports a wind turbine. (694) (CMP-4)

Transfer Switch. An automatic or nonautomatic device for transferring one or more load conductor connections from one power source to another. (CMP-13)

N **Transfer Switch, Branch-Circuit Emergency Lighting. (Branch-Circuit Emergency Lighting Transfer Switch)** A device connected on the load side of a branch-circuit overcurrent protective device that transfers only emergency lighting loads from the normal power source to an emergency power source. (700) (CMP-13)

> Informational Note: See ANSI/UL 1008, *Transfer Switch Equipment,* for information covering branch-circuit emergency lighting transfer switches.

Transfer Switch, Bypass Isolation. (Bypass Isolation Transfer Switch) A manual, nonautomatic, or automatic operated device used in conjunction with a transfer switch to provide a means of directly connecting load conductors to a power source and of disconnecting the transfer switch. (CMP-13)

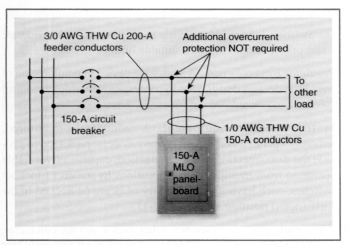

EXHIBIT 100.28 An example in which the sets of 1/0 AWG and the 3/0 AWG conductors are both protected by the 150-ampere circuit breaker, making the 1/0 AWG "tap conductors."

N **Transfer Switch, Meter-Mounted. (Meter-Mounted Transfer Switch)** A transfer switch connected between the utility meter and the meter base. (CMP-13)

> Informational Note: Meter-mounted transfer switches can plug into the meter base. Transfer switches that incorporate the meter base in the transfer equipment assembly are not considered meter-mounted transfer switches.

N **Transformer.** Equipment, either single-phase or polyphase, that uses electromagnetic induction to convert current and voltage in a primary circuit into current and voltage in a secondary circuit. (CMP-9)

N **Transition Assembly.** An assembly to facilitate connection of the flat conductor cable (Type FCC) system to other wiring systems, incorporating (1) a means of electrical interconnection and (2) a suitable box or covering for providing electrical safety and protection against physical damage. (324) (CMP-6)

N **Transport Refrigerated Unit (TRU).** A trailer or container, with integrated cooling or heating, or both, used for the purpose of maintaining the desired environment of temperature-sensitive goods or products. (626) (CMP-12)

N **Transportable.** X-ray equipment that is to be installed in a vehicle or that may be readily disassembled for transport in a vehicle. (660) (CMP-12)

N **Truck.** A motor vehicle designed for the transportation of goods, services, and equipment. (626) (CMP-12)

N **Truck Coupler.** A truck flanged surface inlet and mating cord connector. (626) (CMP-12)

N **Truck Flanged Surface Inlet.** The device(s) on the truck into which the connector(s) is inserted to provide electric energy and other services. This device is part of the truck coupler. For the purposes of this article, the truck flanged surface inlet is considered to be part of the truck and not part of the electrified truck parking space supply equipment. (626) (CMP-12)

N **Trunk Cable.** A portable extension cable containing six or more branch circuits, a male multipole plug, and a female multipole receptacle. (520) (CMP-15)

N **Tubing, Electrical Metallic (EMT). (Electrical Metallic Tubing)** An unthreaded thinwall raceway of circular cross section designed for the physical protection and routing of conductors and cables and for use as an equipment grounding conductor when installed utilizing appropriate fittings. (CMP-8)

N **Tubing, Electrical Nonmetallic (ENT). (Electrical Nonmetallic Tubing)** A nonmetallic, pliable, corrugated raceway of circular cross section with integral or associated couplings, connectors, and fittings for the installation of electrical conductors. It is composed of a material that is resistant to moisture and chemical atmospheres and is flame retardant.

A pliable raceway is a raceway that can be bent by hand with a reasonable force but without other assistance. (CMP-8)

N **Tubing, Flexible Metallic (FMT). (Flexible Metallic Tubing)** A metal raceway that is circular in cross section, flexible, and liquidtight without a nonmetallic jacket. (CMP-8)

N **Two-Fer.** An assembly containing one male plug and two female cord connectors used to connect two loads to one branch circuit. (520) (CMP-15)

A two-fer, as shown in Exhibit 100.29, consists of two cord connectors on separate cords connected to a single supply cord.

Δ **Type of Protection "n".** Type of protection where electrical equipment, in normal operation, is not capable of igniting a surrounding explosive gas atmosphere and a fault capable of causing ignition is not likely to occur. (CMP-14)

> Informational Note: See ANSI/UL 60079-15, *Explosive Atmospheres — Part 15: Equipment Protection by Type of Protection "n"*, for additional information.

Ungrounded. Not connected to ground or to a conductive body that extends the ground connection. (CMP-5)

Δ **Uninterruptible Power Supply (UPS).** A device or system that provides quality and continuity of ac power through the use of a stored-energy device as the backup power source for a period of time when the normal power supply is incapable of performing acceptably. (CMP-13)

N **Unit Equipment.** A battery-equipped emergency luminaire that illuminates only as part of the emergency illumination system and is not illuminated when the normal supply is available. (CMP-13)

Utilization Equipment. Equipment that utilizes electric energy for electronic, electromechanical, chemical, heating, lighting, or similar purposes. (CMP-1)

N **Valve Actuator Motor (VAM) Assemblies.** A manufactured assembly, used to operate a valve, consisting of an actuator motor and other components such as motor controllers, torque switches, limit switches, and overload protection. (430) (CMP-11)

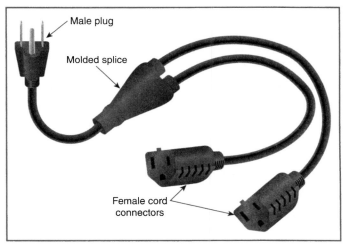

EXHIBIT 100.29 A two-fer.

Informational Note: VAMs typically have short-time duty and high-torque characteristics.

Ventilated. Provided with a means to permit circulation of air sufficient to remove an excess of heat, fumes, or vapors. (CMP-14)

> **See also**
>
> **110.13(B)** and its commentary for information on cooling of equipment

N **Vessel.** A container such as a barrel, drum, or tank for holding fluids or other material. (CMP-17)

Volatile Flammable Liquid. A flammable liquid having a flash point below 38°C (100°F), or a flammable liquid whose temperature is above its flash point, or a Class II combustible liquid that has a vapor pressure not exceeding 276 kPa (40 psia) at 38°C (100°F) and whose temperature is above its flash point. (CMP-14)

The flash point is defined as the minimum temperature of a liquid at which sufficient vapor is given off to form an ignitible mixture with the air, near the surface of the liquid or within the vessel used to contain the liquid. An ignitible mixture is defined as a mixture within the explosive or flammable range (between upper and lower limits) that is capable of the propagation of flame away from the source of ignition when ignited. Some emission of vapors takes place below the flash point but not enough to form an ignitible mixture.

Voltage (of a circuit). The greatest root-mean-square (rms) (effective) difference of potential between any two conductors of the circuit concerned. (CMP-1)

> Informational Note: Some systems, such as 3-phase 4-wire, single-phase 3-wire, and 3-wire direct current, may have various circuits of various voltages.

Common 3-phase, 4-wire wye systems are 480/277 volts and 208/120 volts. The voltage of the circuit is the higher voltage between any two-phase conductors (i.e., 480 volts or 208 volts). The voltage of a single-phase, 2-wire feeder or branch-circuit (with a grounded conductor) derived from these systems would be the voltage between the two conductors (i.e., 277 volts or 120 volts). The same applies to dc or single-phase, 3-wire systems where there are two voltages.

N **Voltage, High. (High Voltage)** A potential difference of more than 1000 volts, nominal. (CMP-9)

> Informational Note: Circuits and equipment rated at potential differences of more than 1000 volts and up to 52 kV are also commonly referred to as medium voltage.

N **Voltage, Low. (Low Voltage)** An electromotive force rated 24 volts, nominal, or less. (551) (CMP-7)

Δ **Voltage, Nominal. (Nominal Voltage)** A nominal value assigned to a circuit or system for the purpose of conveniently

designating its voltage class (e.g., 120/240 volts, 480Y/277 volts, 600 volts). (CMP-1)

> Informational Note No. 1: The actual voltage at which a circuit operates can vary from the nominal within a range that permits satisfactory operation of equipment.
> Informational Note No. 2: See ANSI C84.1-2011, *Voltage Ratings for Electric Power Systems and Equipment (60 Hz).*

> **See also**
>
> **220.5(A),** which provides nominal voltages for use in computing branch-circuit and feeder loads

N **Voltage, Nominal (as applied to battery or cell). (Nominal Voltage)** The value assigned to a cell or battery of a given voltage class for the purpose of convenient designation. The operating voltage of the cell or battery may vary above or below this value. (CMP-13)

> Informational Note: The most common nominal cell voltages are 2 volts per cell for the lead-acid batteries, 1.2 volts per cell for alkali batteries, and 3.2 to 3.8 volts per cell for Li-ion batteries. Nominal voltages might vary with different chemistries.

Voltage to Ground. For grounded circuits, the voltage between the given conductor and that point or conductor of the circuit that is grounded; for ungrounded circuits, the greatest voltage between the given conductor and any other conductor of the circuit. (CMP-1)

This definition can be best illustrated using examples.

System	Voltage to Ground
480/277-volt wye	277
208/120-volt wye	120
3-phase, 3-wire ungrounded 480-volt	480

For a 3-phase, 4-wire delta system with the center of one leg grounded, there are three voltages to ground — that is, on a 240-volt system, two legs would each have 120 volts to ground, and the third, commonly referred to as the "high leg," would have 208 volts to ground.

> **See also**
>
> **110.15, 230.56,** and **408.3(E)** for requirements pertaining to special markings and arrangements on such circuit conductors

Watertight. Constructed so that moisture will not enter the enclosure under specified test conditions. (CMP-1)

Weatherproof. Constructed or protected so that exposure to the weather will not interfere with successful operation. (CMP-1)

> Informational Note: Rainproof, raintight, or watertight equipment can fulfill the requirements for weatherproof where varying weather conditions other than wetness, such as snow, ice, dust, or temperature extremes, are not a factor.

Wharf. A structure at the shoreline that has a platform built along and parallel to a body of water with either an open deck or a superstructure. [307:3.3.28] (555) (CMP-7)

Wind Turbine. A mechanical device that converts wind energy to electrical energy. (CMP-4)

Wind Turbine Output Circuit. (Turbine Output Circuit) The circuit conductors between the internal components of a wind turbine (which might include an alternator, integrated rectifier, controller, and/or inverter) and other equipment. (694) (CMP-4)

Wire. A factory assembly of one or more insulated conductors without an overall covering. (805) (CMP-16)

Wireless Power Transfer (WPT). The transfer of electrical energy from a power source to an electrical load via magnetic fields by a contactless means between a primary device and a secondary device. (625) (CMP-12)

Wireless Power Transfer Equipment (WPTE). Equipment installed specifically for the purpose of transferring energy between the premises wiring and the electric vehicle without physical electrical contact. (625) (CMP-12)

> Informational Note No. 1: The general form of WPTE consists of two physical packages: a control box and a primary pad.
> Informational Note No. 2: Electric vehicle power export equipment and wireless power transfer equipment are sometimes contained in one set of equipment, sometimes referred to as a bidirectional WPTE.

Wireways, Metal. (Metal Wireways) Sheet metal troughs with hinged or removable covers for housing and protecting electrical wires and cable and in which conductors are laid in place after the raceway has been installed as a complete system. (CMP-8)

Wireways, Nonmetallic. (Nonmetallic Wireways) Flame-retardant, nonmetallic troughs with removable covers for housing and protecting electrical wires and cables in which conductors are laid in place after the raceway has been installed as a complete system. (CMP-8)

Work Surface. A fixed, stationary, or portable surface typically intended for dry use and for tasks other than food preparation, personal lavation, or laundering that presents an incidental risk of spillage of smaller quantities of beverages and other liquids upon outlets mounted directly on or recessed in the surface. (CMP-2)

> Informational Note No. 1: See UL 111, *Outline of Investigation for Multioutlet Assemblies*, and UL 962A, *Furniture Power Distribution Units*, which establish the performance evaluation criteria and construction criteria.

> Informational Note No. 2: See 406.5(F), 406.5(G)(1), and 406.5(H) for information on receptacles for work surfaces distinguished from receptacles for counters and countertops.

Zone. A physically identifiable area (such as barriers or separation by distance) within an information technology equipment room, with dedicated power and cooling systems for the information technology equipment or systems. (645) (CMP-12)

ARTICLE 110 General Requirements for Electrical Installations

Part I. General

110.1 Scope. This article covers general requirements for the examination and approval, installation and use, access to and spaces about electrical conductors and equipment; enclosures intended for personnel entry; and tunnel installations.

> Informational Note: See Informative Annex J for information regarding ADA accessibility design.

110.2 Approval. The conductors and equipment required or permitted by this *Code* shall be acceptable only if approved.

> Informational Note: See 90.7, Examination of Equipment for Safety, and 110.3, Examination, Identification, Installation, and Use of Equipment. See definitions of *Approved, Identified, Labeled,* and *Listed.*

Approval of electrical equipment is the responsibility of the authority having jurisdiction (AHJ). Many approvals are based on tests and listings of testing laboratories. Unique equipment is often approved following a field evaluation by a qualified third-party laboratory or qualified individual.

110.3 Examination, Identification, Installation, Use, and Listing (Product Certification) of Equipment.

Listing or labeling is the most common method of establishing suitability. This section does not require listing or labeling of equipment, but it does require considerable evaluation for approval. Before approving the installation, the AHJ may require evidence of compliance with 110.3.

(A) Examination. In judging equipment, considerations such as the following shall be evaluated:

(1) Suitability for installation and use in conformity with this *Code*

> Informational Note No. 1: Equipment may be new, reconditioned, refurbished, or remanufactured.

> Informational Note No. 2: Suitability of equipment use may be identified by a description marked on or provided with a product to identify the suitability of the product for a specific purpose, environment, or application. Special conditions of use or other limitations and other pertinent information may be marked on

the equipment, included in the product instructions, or included in the appropriate listing and labeling information. Suitability of equipment may be evidenced by listing or labeling.

Examples of "special conditions of use" include elevated or reduced ambient temperatures, special environmental limitations, stringent power quality requirements, and specific types of overcurrent protective devices (OCPDs). The additional information needed for such special cases might be marked on the equipment, included as part of the listing information in a listing directory, or included in the information furnished with the equipment.

(2) Mechanical strength and durability, including, for parts designed to enclose and protect other equipment, the adequacy of the protection thus provided

(3) Wire-bending and connection space

(4) Electrical insulation

(5) Heating effects under normal conditions of use and also under abnormal conditions likely to arise in service

(6) Arcing effects

(7) Classification by type, size, voltage, current capacity, and specific use

(8) Cybersecurity for network-connected life safety equipment to address its ability to withstand unauthorized updates and malicious attacks while continuing to perform its intended safety functionality

Informational Note No. 3: See the ANSI/ISA 62443 series of standards for industrial automation and control systems, the UL 2900 series of standards for software cybersecurity for network-connectable products, and UL 5500, *Standard for Remote Software Updates*, which are standards that provide frameworks to mitigate current and future security cybersecurity vulnerabilities and address software integrity in systems of electrical equipment.

(9) Other factors that contribute to the practical safeguarding of persons using or likely to come in contact with the equipment

(B) Installation and Use. Equipment that is listed, labeled, or both, or identified for a use shall be installed and used in accordance with any instructions included in the listing, labeling, or identification.

Informational Note: The installation and use instructions may be provided in the form of printed material, quick response (QR) code, or the address on the internet where users can download the required instructions.

Manufacturers' listing and labeling installation instructions must be followed, even if the equipment itself is not required to be listed. For example, 210.52 permits permanently installed electric baseboard heaters to be equipped with receptacle outlets that meet the requirements for the wall space utilized by such heaters. The installation instructions for permanent baseboard heaters indicate that the heaters should not be mounted beneath a receptacle. In dwelling units, the use of low-density heating units more than 12 feet in length is common. Therefore, to meet the requirements of 210.52(A) and the installation instructions, a receptacle must either be part of the heating unit or be installed

in the floor close to the wall but not above the heating unit. (Exhibit 210.25 and the Informational Note to 210.52 provide more specific details.)

Listing and labeling are the most common methods of establishing suitability. This section does not require listing or labeling of equipment. Before approving the installation, the AHJ may require evidence of compliance with 110.3.

Some sections do require listed or labeled equipment. For example, 250.8 specifies "listed pressure connectors . . . pressure connectors listed as grounding and bonding equipment [or] . . . other listed means" as connection methods for grounding and bonding conductors.

Listing organizations typically require the use of their listing mark on the equipment as the means of determining if the product is listed. Where it is impractical to have the listing mark on the equipment, the listing organization usually requires that the listing mark be on the smallest unit container in which the product is packaged. Since 110.3(B) requires compliance with the listing requirements, the appropriate certification mark is required if that is a requirement of the listing entity.

(C) Listing. Product testing, evaluation, and listing (product certification) shall be performed by recognized qualified electrical testing laboratories and shall be in accordance with applicable product standards recognized as achieving equivalent and effective safety for equipment installed to comply with this *Code*.

Informational Note: The Occupational Safety and Health Administration (OSHA) recognizes qualified electrical testing laboratories that perform evaluations, testing, and certification of certain products to ensure that they meet the requirements of both the construction and general industry OSHA electrical standards. If the listing (product certification) is done under a qualified electrical testing laboratory program, this listing mark signifies that the tested and certified product complies with the requirements of one or more appropriate product safety test standards.

110.4 Voltages. The voltage considered shall be that at which the circuit operates. The voltage rating of electrical equipment shall not be less than the nominal voltage of a circuit to which it is connected.

110.5 Conductors. Conductors used to carry current shall be of copper, aluminum, or copper-clad aluminum unless otherwise provided in this *Code*. If the conductor material is not specified, the sizes given in this *Code* shall apply to copper conductors. If other materials are used, the size shall be changed accordingly.

See also

310.3(B), which specifies the alloy for aluminum conductors and for copper-clad aluminum conductors

110.6 Conductor Sizes. Conductor sizes are expressed in American Wire Gauge (AWG) or in circular mils.

For copper, aluminum, or copper-clad aluminum conductors up to size 4/0 AWG, the *NEC®* uses the American Wire Gage (AWG) for size identification, which is the same as the Brown and Sharpe

(BS) Wire Gage. Wire sizes up to size 4/0 AWG are expressed as XX AWG, with XX being the wire size.

Conductors larger than 4/0 AWG are sized in circular mils, beginning with 250,000 circular mils. Prior to the 1990 edition, a 250,000-circular mil conductor was labeled 250 MCM. The designation *MCM* was defined as 1000 circular mils (the first *M* being the roman numeral 1000). Beginning in the 1990 edition, the notation was changed to *kcmil* to recognize the accepted convention that *k* indicates 1000. UL standards and IEEE standards also use the notation kcmil rather than MCM.

Table 8 in Chapter 9 and its associated commentary provide the circular mil area for AWG-sized conductors. Where stranded conductors are used, the circular mil area of each strand must be multiplied by the number of strands to determine the circular mil area of the conductor.

110.7 Wiring Integrity. Completed wiring installations shall be free from short circuits, ground faults, or any connections to ground other than as required or permitted elsewhere in this *Code*.

Failure of the insulation system is one of the most common causes of problems in electrical installations. The principal causes of insulation failures are heat, moisture, dirt, and physical damage (abrasion or nicks) occurring during and after installation. Some hazards are exposure to chemical attack, excessive voltage stresses, saltwater environments or flood waters, or even sunlight (UV degradation).

Overcurrent protective devices must be selected and coordinated using tables of insulation thermal-withstand ability to ensure that the damage point of an insulated conductor is never reached. Those tables, entitled "Allowable Short-Circuit Currents for Insulated Copper (or Aluminum) Conductors," are contained in the Insulated Cable Engineers Association publication ICEA P-32-382, *Short-Circuit Characteristics of Insulated Cable*.

See also

110.10 for selection criteria for other circuit components

Insulation tests are performed on new or existing installations to determine the quality or condition of the insulation of conductors and equipment. In an insulation resistance test, a voltage ranging from 100 to 5000 (usually 500 to 1000 volts for systems of 1000 volts or less), supplied from a source of constant potential, is applied across the insulation. A megohmmeter is usually the potential source, and it indicates the insulation resistance directly on a scale calibrated in megohms (MΩ). The quality of the insulation is evaluated based on the level of the insulation resistance.

The field data obtained should be corrected to the standard temperature for the class of equipment being tested because insulation resistance of many types of insulation varies with temperature. The megohm value of insulation resistance obtained is inversely proportional to the volume of insulation tested. For example, a cable 1000 feet long would be expected to have one-tenth the insulation resistance of a cable 100 feet long if all other conditions are identical. Exhibit 110.1 shows a typical megohm-meter insulation tester. Information on specific test methods is available from instrument manufacturers.

See also

NFPA 70B, *Recommended Practice for Electrical Equipment Maintenance*, which provides useful information on test methods and on establishing a preventive maintenance program

110.8 Wiring Methods. Only wiring methods recognized as suitable are included in this *Code*. The recognized methods of wiring shall be permitted to be installed in any type of building, occupancy, or premises wiring system, except as otherwise provided in this *Code*.

110.9 Interrupting Rating. Equipment intended to interrupt current at fault levels shall have an interrupting rating at nominal circuit voltage at least equal to the available fault current at the line terminals of the equipment.

Equipment intended to interrupt current at other than fault levels shall have an interrupting rating at nominal circuit voltage at least equal to the current that must be interrupted.

Fuses or circuit breakers that do not have adequate interrupting ratings could rupture while attempting to clear a short circuit. The interrupting rating of an overcurrent protective device is determined under standard test conditions. The rating should meet or exceed the actual installation needs. Interrupting ratings should not be confused with short-circuit current ratings.

110.10 Circuit Impedance, Short-Circuit Current Ratings, and Other Characteristics. The overcurrent protective devices, the total impedance, the equipment short-circuit current ratings, and other characteristics of the circuit to be protected shall be selected

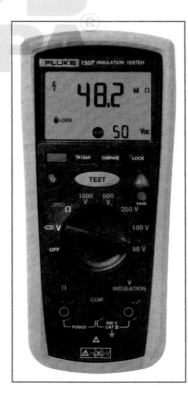

EXHIBIT 110.1 A multivoltage, multirange insulation tester. (*Reproduced with permission, Fluke Corporation*)

and coordinated to permit the circuit protective devices used to clear a fault to do so without extensive damage to the electrical equipment of the circuit. This fault shall be assumed to be either between two or more of the circuit conductors or between any circuit conductor and the equipment grounding conductor(s) permitted in 250.118. Listed equipment applied in accordance with their listing shall be considered to meet the requirements of this section.

Short-circuit current ratings are marked on equipment such as panelboards, switchboards, switchgear, busways, contactors, and starters. Listed products are subjected to rigorous testing as part of their evaluation, which includes tests under fault conditions. Therefore, listed products used within their ratings are considered to have met the requirements of 110.10.

The purpose of overcurrent protection is to open the circuit before conductors or conductor insulation is damaged when an overcurrent condition occurs. An overcurrent condition can be the result of an overload, a ground fault, or a short circuit.

OCPDs (such as fuses and circuit breakers) should be selected to ensure that the short-circuit current rating of the system components is not exceeded if a short circuit or high-level ground fault occurs.

Wire, bus structures, switching, protection and disconnect devices, and distribution equipment all have limited short-circuit ratings and would be damaged or destroyed if those short-circuit ratings were exceeded. Merely providing OCPDs with sufficient interrupting ratings would not ensure adequate short-circuit protection for the system components. When the available short-circuit current exceeds the short-circuit current rating of an electrical component, the OCPD must limit the let-through energy to within the rating of that electrical component.

Utility companies often determine and provide information on available short-circuit current levels to the service equipment.

Application Example

The short-circuit current rating of utilization equipment located and connected near the service equipment should be known because the available short-circuit currents near the service equipment are higher than elsewhere in the system. For example, HVAC equipment is tested at 3500 amperes through a 40-ampere load rating and at 5000 amperes for loads rated more than 40 amperes.

For a typical one-family dwelling with a 100-ampere service using 2 AWG aluminum supplied by a 37½ kilovolt-ampere transformer with 1.72 percent impedance located at a distance of 25 feet, the available short-circuit current would be approximately 6000 amperes.

Where pad-mounted transformers are located close to a multimetering location of a multifamily building, available short-circuit current can be relatively high. For example, the line-to-line fault current values close to a low-impedance transformer could exceed 22,000 amperes. At the secondary of a single-phase, center-tapped transformer, the line-to-neutral fault current is approximately one and one-half times that of the line-to-line fault current.

110.11 Deteriorating Agents. Unless identified for use in the operating environment, no conductors or equipment shall be located in damp or wet locations; where exposed to gases, fumes, vapors, liquids, or other agents that have a deteriorating effect on the conductors or equipment; or where exposed to excessive temperatures.

> Informational Note No. 1: See 300.6 for protection against corrosion.
>
> Informational Note No. 2: Some cleaning and lubricating compounds can cause severe deterioration of many plastic materials used for insulating and structural applications in equipment.

Equipment not identified for outdoor use and equipment identified only for indoor use, such as "dry locations," "indoor use only," "damp locations," or enclosure Types 1, 2, 5, 12, 12K, and/or 13, shall be protected against damage from the weather during construction.

> Informational Note No. 3: See Table 110.28 for appropriate enclosure-type designations.
>
> Informational Note No. 4: See *NFPA 5000*-2015, *Building Construction and Safety Code*, the *International Building Code (IBC)*, and the *International Residential Code for One- and Two-Family Dwellings (IRC)*, for information for minimum flood provisions.

The flood-related requirements contained in Chapter 39 and elsewhere in *NFPA 5000®*, *Building Construction and Safety Code®*, provide methods to reduce flood damage to buildings and attendant systems. Chapter 32 of NFPA 70B, *Recommended Practice for Electrical Equipment Maintenance*, provides guidance to facilities, institutions, and businesses in the event of a disaster to help limit the recovery time, therefore getting them up and running sooner. The requirements in *NFPA 5000* are based on years of collective experience and are consistent with the minimum requirements of the National Flood Insurance Program (NFIP). Fortunately, flood hazard information — which is necessary to implement the flood-related requirements — is readily available for most communities.

While flood hazards have been mapped by a federal agency on a nationwide basis, floodplain management regulations are adopted and decisions are made at the local level. Communities that choose to participate in the NFIP must adopt local floodplain management regulations, which are no less stringent than minimum NFIP regulations.

The flood-related requirements call for a number of actions, including the following:

1. Using available flood hazard information for the siting, design, and construction of buildings
2. Elevating buildings above the design flood elevation (DFE)
3. Designing and constructing structures to withstand anticipated flood loads during the design flood event
4. Using flood damage-resistant materials below the design flood elevation
5. In coastal areas, elevating buildings on a pile or column foundation and keeping areas below the elevated buildings free of obstructions, which could transfer flood loads to or otherwise damage the elevated portions of buildings during a flood

Other aspects of flood-resistant design and construction covered in Chapter 39 of *NFPA 5000* include determining the lowest floor elevation, designing foundations, selecting materials and systems, bringing existing buildings into compliance, and designing and building manufactured homes and temporary structures.

110.12 Mechanical Execution of Work. Electrical equipment shall be installed in a professional and skillful manner.

> Informational Note: See ANSI/NECA 1-2015, *Standard for Good Workmanship in Electrical Construction,* and other ANSI-approved installation standards for information on accepted industry practices.

The requirement for installations to be done in a "professional and skillful manner" is representative of pride in one's work and is emphasized by persons involved in the training of apprentice electricians. A professional and skillful installation is also one that is easy to troubleshoot and is unlikely to have problems properly functioning.

Installations that do not qualify as "professional and skillful" include exposed runs of cables or raceways that are sagging between supports or supported by improper methods; cables that are kinked, flattened, or incorrectly sized; or cabinets, cutout boxes, and enclosures that are not plumb or not properly secured.

(A) Unused Openings. Unused openings, other than those intended for the operation of equipment, those intended for mounting purposes, or those permitted as part of the design for listed equipment, shall be closed to afford protection substantially equivalent to the wall of the equipment. Where metallic plugs or plates are used with nonmetallic enclosures, they shall be recessed at least 6 mm (¼ in.) from the outer surface of the enclosure.

(B) Integrity of Electrical Equipment and Connections. Internal parts of electrical equipment, including busbars, wiring terminals, insulators, and other surfaces, shall not be damaged or contaminated by foreign materials such as paint, plaster, cleaners, abrasives, or corrosive residues. There shall be no damaged parts that may adversely affect safe operation or mechanical strength of the equipment such as parts that are broken; bent; cut; or deteriorated by corrosion, chemical action, or overheating.

Δ **(C) Cables and Conductors.** Cables and conductors installed exposed on the surfaces of ceilings and sidewalls shall be supported by the building structure in such a manner that the cables and conductors will not be damaged by normal building use. Such cables and conductors shall be secured by hardware including straps, staples, cable ties, hangers, or similar fittings designed and installed so as not to damage the cable. The installation shall also conform with 300.4 and 300.11. Nonmetallic cable ties and other nonmetallic cable accessories used to secure and support cables in other spaces used for environmental air (plenums) shall be listed as having low smoke and heat release properties.

> Informational Note No. 1: See NFPA 90A-2021, *Standard for the Installation of Air-Conditioning and Ventilating Systems,* 4.3.11.2.6.5 and 4.3.11.5.5.6, for discrete combustible components installed in accordance with 300.22(C).
>
> Informational Note No. 2: Paint, plaster, cleaners, abrasives, corrosive residues, or other contaminants may result in an undetermined alteration of optical fiber cable properties.

110.13 Mounting and Cooling of Equipment.

(A) Mounting. Electrical equipment shall be firmly secured to the surface on which it is mounted. Wooden plugs driven into holes in masonry, concrete, plaster, or similar materials shall not be used.

(B) Cooling. Electrical equipment that depends on the natural circulation of air and convection principles for cooling of exposed surfaces shall be installed so that room airflow over such surfaces is not prevented by walls or by adjacent installed equipment. For equipment designed for floor mounting, clearance between top surfaces and adjacent surfaces shall be provided to dissipate rising warm air.

Electrical equipment provided with ventilating openings shall be installed so that walls or other obstructions do not prevent the free circulation of air through the equipment.

The term *ventilated* is defined in Article 100. Ventilating openings in equipment are provided to allow the circulation of room air around internal equipment components. Blocking these openings can cause dangerous overheating. For example, a ventilated busway must be located where there are no walls or other obstructions that could interfere with the natural circulation of air and convection principles for cooling. The surfaces of some enclosures, such as panelboards and transformers, may also require normal room air circulation to prevent overheating. Proper placement of equipment requiring ventilation becomes enforceable using the requirements of 110.3(B).

See also

430.14(A) and **430.16** for ventilation requirements for motors
450.9 and **450.45** for ventilation requirements for transformers

Δ **110.14 Electrical Connections.** Because of different characteristics of dissimilar metals, devices such as pressure terminal or pressure splicing connectors and soldering lugs shall be identified for the material of the conductor and shall be properly installed and used. Conductors of dissimilar metals shall not be intermixed in a terminal or splicing connector where physical contact occurs between dissimilar conductors unless the device is identified for the purpose and conditions of use. Materials such as solder, fluxes, inhibitors, and compounds, where employed, shall be suitable for the use and shall be of a type that will not adversely affect the conductors, installation, or equipment.

Connectors and terminals for conductors more finely stranded than Class B and Class C stranding as shown in Chapter 9, Table 10, shall be identified for the specific conductor class or classes.

(A) Terminals. Connection of conductors to terminal parts shall ensure a mechanically secure electrical connection without damaging the conductors and shall be made by means of pressure connectors (including set-screw type), solder lugs, or splices to flexible leads. Connection by means of wire-binding screws or studs and nuts that have upturned lugs or the equivalent shall be permitted for 10 AWG or smaller conductors.

Terminals for more than one conductor and terminals used to connect aluminum shall be so identified.

(B) Splices. Conductors shall be spliced or joined with splicing devices identified for the use or by brazing, welding, or soldering with a fusible metal or alloy. Soldered splices shall first be spliced or joined so as to be mechanically and electrically secure without solder and then be soldered. All splices and joints and the free ends of conductors shall be covered with an insulation equivalent to that of the conductors or with an identified insulating device.

Wire connectors or splicing means installed on conductors for direct burial shall be listed for such use.

Electrical connection failures are the cause of many equipment burnouts and fires. Many of these failures are attributable to improper terminations, poor workmanship, the differing characteristics of dissimilar metals, and improper binding screws or splicing devices.

UL's requirements for listing solid aluminum conductors in 12 AWG and 10 AWG and its requirements for listing snap switches and receptacles for use on 15- and 20-ampere branch circuits incorporate stringent tests that are intended to prevent termination failures. Screwless pressure terminal connectors of the conductor push-in type are for use only with 14 AWG solid copper conductors.

See also

404.14(C) and **406.3(C)** for further information regarding receptacles and switches using copper-aluminum, revised (CO/ALR) rated terminals

The electrical industry has developed new product and material designs that provide increased levels of safety for aluminum wire terminations. See the accompanying Aluminum Wire Terminations feature.

CLOSER LOOK: Aluminum Wire Terminations

For New Installations

The following commentary is based on a report prepared by the Ad Hoc Committee on Aluminum Terminations prior to publication of the 1975 *NEC*. This information is still pertinent today and is necessary for compliance with 110.14(A) when aluminum wire is used in new installations. New installation of aluminum conductors on 15- and 20-ampere branch circuits is not common, but many of these circuits continue to be in use.

New Materials and Devices

For direct connection, only 15- and 20-ampere receptacles and switches marked "CO/ALR" and connected as follows should be used. The "CO/ALR" marking on the device mounting yoke or strap means the devices have been tested to stringent heat-cycling requirements to determine their suitability for use with UL-labeled aluminum, copper, or copper-clad aluminum wire.

Listed solid aluminum wire, 12 AWG or 10 AWG, marked with the aluminum insulated wire label should be used. The installation instructions that are packaged with the wire should be followed.

Installation Method

Exhibit 110.2 illustrates the following correct method of connection:

1. The freshly stripped end of the wire is wrapped two-thirds to three-quarters of the distance around the wire-binding screw post as shown in Step A. The loop is made so that rotation of the screw during tightening will tend to wrap the wire around the post rather than unwrap it.
2. The screw is tightened until the wire is snugly in contact with the underside of the screw head and with the contact plate on the wiring device as shown in Step B.

3. The screw is tightened an additional half-turn to provide a firm connection as shown in Step C.
4. The wires should be positioned behind the wiring device to decrease the likelihood of the terminal screws loosening when the device is positioned into the outlet box.

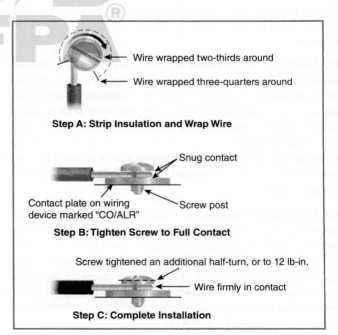

Step A: Strip Insulation and Wrap Wire
— Wire wrapped two-thirds around
— Wire wrapped three-quarters around

Step B: Tighten Screw to Full Contact
— Snug contact
— Screw post
Contact plate on wiring device marked "CO/ALR"

Step C: Complete Installation
Screw tightened an additional half-turn, or to 12 lb-in.
— Wire firmly in contact

EXHIBIT 110.2 Correct method of terminating aluminum wire at wire-binding screw terminals of receptacles and snap switches.

Exhibit 110.3 illustrates incorrect methods of connection. These methods should *not* be used.

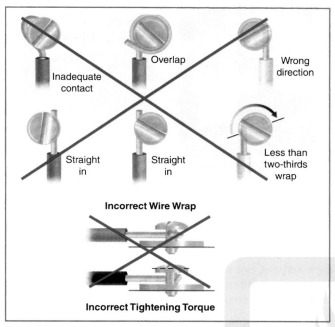

EXHIBIT 110.3 Incorrect methods of terminating aluminum wire at wire-binding screw terminals of receptacles and snap switches.

Existing Inventory

Labeled 12 AWG or 10 AWG solid aluminum wire that does not bear the new aluminum wire label should be used with wiring devices marked "CO/ALR" and connected as described under the heading Installation Method. This is the preferred and recommended method for using such wire.

For the following types of devices, the terminals should not be directly connected to aluminum conductors but may be used with labeled copper or copper-clad conductors:

1. Receptacles and snap switches marked "AL-CU"
2. Receptacles and snap switches having no conductor marking
3. Receptacles and snap switches that have back-wired terminals or screwless terminals of the push-in type

For Existing Installations

If examination discloses overheating or loose connections, the recommendations described under the heading Existing Inventory should be followed.

Splicing Wire Connectors

Splicing wire connectors are required to be marked for the material of the conductor and for their suitability where intermixed. Splicing wire connectors, such as twist-on wire connectors, are not suitable for splicing aluminum conductors or copper-clad aluminum to copper conductors, unless it is so stated and marked as such on the unit container or an information sheet supplied with the unit container. The required marking is "AL-CU (intermixed-dry locations)" where intermixing (direct contact) occurs. Other types of listed splicing wire connectors that are not rated for intermixing between the copper and the aluminum may also be used, as long as the conductors are not in direct physical contact. These connectors are just marked "AL-CU."

UL lists twist-on wire connectors that are suitable for use with aluminum-to-copper conductors, in accordance with UL 486C, *Splicing Wire Connectors*. The UL listing does *not* cover aluminum-to-aluminum combinations. However, more than one aluminum or copper conductor is allowed where used in combination. Suitable wire combinations are marked on the unit container or supplied on the information sheet with the unit container. These listed splicing wire-connecting devices are available for pigtailing short lengths of copper conductors directly to the original aluminum branch-circuit conductors as shown in Exhibit 110.4. Also depicted is a similarly rated crimp splicing device that is also suitable for intermixing (direct contact). Primarily, these pigtailed conductors supply 15- and 20-ampere wiring devices. Pigtailing is permitted, provided suitable space exists within the enclosure.

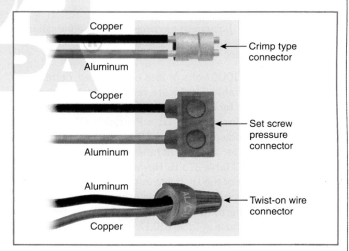

EXHIBIT 110.4 Pigtailing copper-to-aluminum conductors using two listed devices.

(C) Temperature Limitations. The temperature rating associated with the ampacity of a conductor shall be selected and coordinated so as not to exceed the lowest temperature rating of any connected termination, conductor, or device. Conductors with temperature ratings higher than specified for terminations shall be permitted to be used for ampacity adjustment, correction, or both.

(1) Equipment Provisions. The determination of termination provisions of equipment shall be based on 110.14(C)(1)(a) or (C)(1)(b). Unless the equipment is listed and marked otherwise, conductor ampacities used in determining equipment termination provisions shall be based on Table 310.16 as appropriately modified by 310.12.

(a) Termination provisions of equipment for circuits rated 100 amperes or less, or marked for 14 AWG through 1 AWG conductors, shall be used only for one of the following:

(1) Conductors rated 60°C (140°F).
(2) Conductors with higher temperature ratings, provided the ampacity of such conductors is determined based on the 60°C (140°F) ampacity of the conductor size used.
(3) Conductors with higher temperature ratings if the equipment is listed and identified for use with such conductors.
(4) For motors marked with design letters B, C, or D, conductors having an insulation rating of 75°C (167°F) or higher shall be permitted to be used, provided the ampacity of such conductors does not exceed the 75°C (167°F) ampacity.

(b) Termination provisions of equipment for circuits rated over 100 amperes, or marked for conductors larger than 1 AWG, shall be used only for one of the following:

(1) Conductors rated 75°C (167°F)
(2) Conductors with higher temperature ratings, provided the ampacity of such conductors does not exceed the 75°C (167°F) ampacity of the conductor size used, or up to their ampacity if the equipment is listed and identified for use with such conductors

Δ **(2) Separate Connector Provisions.** Separately installed pressure connectors shall be used with conductors at the ampacities not exceeding the ampacity at the listed and identified temperature rating of the connector.

> Informational Note: Equipment markings or listing information may additionally restrict the sizing and temperature ratings of connected conductors.

When equipment of 1000 volts or less is evaluated, conductors sized according to Table 310.16 are required to be used. The UL *Guide Information for Electrical Equipment for Use in Ordinary Locations (AALZ)* located on UL's Product iQ™ database (productiq.ul.com) indicates that the 60°C and 75°C termination temperature ratings for equipment have been determined using conductors from *NEC* Table 310.16. However, installers or designers who are unaware of the UL guide information might attempt to select conductors based on a table other than Table 310.16, especially if a wiring method is used that allows the use of ampacities such as those in Table 310.17, which can result in overheated equipment terminations. The ampacities shown in other tables could be used for various conditions to which the wiring method is subject (such as ambient or ampacity correction conditions), but the conductor size at the equipment termination must be based on ampacities from Table 310.16.

Conductor terminations, as well as conductors, must be rated for the operating temperature of the circuit. For example, the load on an 8 AWG THHN, 90°C copper conductor is limited to 40 amperes where connected to a disconnect switch with terminals rated at 60°C. The same conductor is limited to 50 amperes where connected to a fusible switch with terminals rated at 75°C.

EXHIBIT 110.5 An example of termination temperature marking on a main circuit breaker. *(Courtesy of Siemens Industry, Inc.)*

Not only do termination temperature ratings apply to conductor terminations, but the equipment enclosure marking must also permit terminations above 60°C. Exhibit 110.5 shows an example of termination temperature marking.

Δ **(D) Terminal Connection Torque.** Tightening torque values for terminal connections shall be as indicated on equipment or in installation instructions provided by the manufacturer. An approved means shall be used to achieve the indicated torque value.

> Informational Note No. 1: Examples of approved means of achieving the indicated torque values include torque tools or devices such as shear bolts or breakaway-style devices with visual indicators that demonstrate that the proper torque has been applied.
>
> Informational Note No. 2: See UL Standard 486A-486B, *Standard for Safety-Wire Connectors*, Informative Annex I for torque values in the absence of manufacturer's recommendations. The equipment manufacturer can be contacted if numeric torque values are not indicated on the equipment or if the installation instructions are not available.
>
> Informational Note No. 3: See NFPA 70B-2019, *Recommended Practice for Electrical Equipment Maintenance*, Section 8.11 for additional information for torquing threaded connections and terminations.

Because the reliability and safety of terminations depend on proper connection, the use of an approved means is essential. Approved means include the use of a torque wrench, torque screwdriver, shear bolts, and breakaway devices. For wire connectors for which the manufacturer has not assigned a value appropriate for the design, Informative Annex I provides information on the tightening torques from UL 468A-486B, *Wire Connectors*. These tables should be used for guidance only if no tightening information on a specific wire connector is available. They should not be used to replace the manufacturer's instructions, which should always be followed.

UL 486A-486B refers to conductor stranding by class. Terminals and connectors for conductors that are more finely stranded than Class B and C stranding are required to be identified for the class or classes of conductor stranding and the number of strands. Table 10 in Chapter 9 provides information for the application of this information in the field.

110.15 High-Leg Marking. On a 4-wire, delta-connected system where the midpoint of one phase winding is grounded, only the conductor or busbar having the higher phase voltage to ground shall be durably and permanently marked by an outer finish that is orange in color or by other effective means. Such identification shall be placed at each point on the system where a connection is made if the grounded conductor is also present.

The high leg is common on a 240/120-volt, 3-phase, 4-wire delta system. It is typically designated as "B phase." The high-leg marking, which is required to be the color orange or other similar effective means, is intended to prevent problems caused by the lack of standardization where metered and nonmetered equipment are installed in the same installation. See Exhibit 110.6. Electricians should always test each phase relative to ground with suitable equipment to determine exactly where the high leg is located within the system.

110.16 Arc-Flash Hazard Warning.

(A) General. Electrical equipment, such as switchboards, switchgear, enclosed panelboards, industrial control panels, meter socket enclosures, and motor control centers, that is in other than dwelling units, and is likely to require examination, adjustment, servicing, or maintenance while energized, shall be field or factory marked to warn qualified persons of potential electric arc flash hazards. The marking shall meet the requirements in 110.21(B) and shall be located so as to be clearly visible to qualified persons before examination, adjustment, servicing, or maintenance of the equipment.

Proper warning labels raise the level of awareness of electrical arc-flash hazards and decrease the number of accidents that result when electricians do not wear the proper type of protective clothing while working on "hot" (energized) equipment. Exhibit 110.7 is one example of an equipment warning sign as required by this section.

How to choose suitable personal protective equipment (PPE) appropriate to a particular hazard is described in *NFPA 70E®, Standard for Electrical Safety in the Workplace®*. See Exhibit 110.8.

Accident reports confirm the fact that workers responsible for the installation or maintenance of electrical equipment often do not establish an electrically safe working condition, as described in NFPA *70E*, before working on the equipment.

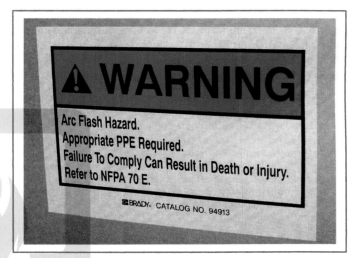

EXHIBIT 110.7 One example of an arc-flash warning sign. *(Courtesy of the International Association of Electrical Inspectors)*

EXHIBIT 110.8 Worker clothed in PPE appropriate for the hazard involved. *(Courtesy of Enespro)*

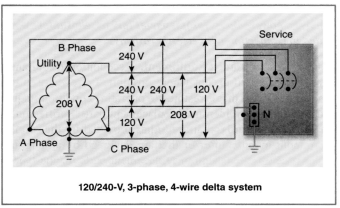

120/240-V, 3-phase, 4-wire delta system

EXHIBIT 110.6 A 240/120-volt, 3-phase, 4-wire delta system.

Working on electrical equipment that is energized is a major safety concern in the electrical industry. This requirement alerts electrical contractors, electricians, facility owners and managers, and other interested parties to some of the hazards present when personnel are exposed to energized electrical conductors or circuit parts. It emphasizes the importance of turning off the power before working on electrical circuits. This section does not apply to equipment in dwelling units. However, multifamily dwellings, which include multiple dwelling units, could have the same electric service as a commercial office building. The intent is to provide warnings to electricians working on these larger services.

Δ **(B) Service Equipment and Feeder Supplied Equipment.** In other than dwelling units, in addition to the requirements in 110.16(A), a permanent arc flash label shall be field or factory applied to service equipment and feeder supplied equipment rated 1000 amperes or more. The arc flash label shall be in accordance with applicable industry practice and include the date the label was applied. The label shall meet the requirements of 110.21(B).

> Informational Note No. 1: See ANSI Z535.4-2011 (R2017), *Product Safety Signs and Labels*, for guidelines for the design of safety signs and labels for application to products.
>
> Informational Note No. 2: See NFPA *70E*-2021, *Standard for Electrical Safety in the Workplace*, for applicable industry practices for equipment labeling. This standard provides specific criteria for developing arc-flash labels for equipment that provides nominal system voltage, incident energy levels, arc-flash boundaries, minimum required levels of personal protective equipment, and so forth.

The available short-circuit current must be known at the time of installation to comply with the interrupting requirements of 110.9 and 110.10. See the Application Example below 110.10. This information is necessary to determine the incident energy and working distance for compliance with *NFPA 70E* for future work on the service equipment. Section 110.16(B) specifies the information that a label is required to provide. The date that the label was applied is important because, over time, available fault current can change. Exhibit 110.9 is an example of the label required by this section.

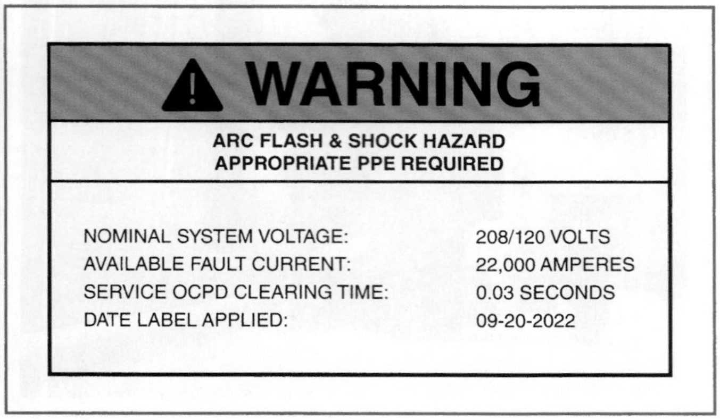

EXHIBIT 110.9 A label that meets the requirements of 110.16(B). *(Courtesy of the National Electrical Contractors Association)*

N **110.17 Servicing and Maintenance of Equipment.** Servicing and electrical preventive maintenance shall be performed by qualified persons trained in servicing and maintenance of equipment and shall comply with the following:

(1) The servicing and electrical preventive maintenance shall be performed in accordance with the original equipment manufacturer's instructions and information included in the listing information, applicable industry standards, or as approved by the authority having jurisdiction.

(2) The servicing and electrical preventive maintenance shall be performed using identified replacement parts that are verified under applicable product standards. The replacement parts shall comply with at least one of the following:
   a. Be provided by the original equipment manufacturer
   b. Be designed by an engineer experienced in the design of replacement parts for the type of equipment being serviced or maintained
   c. Be approved by the authority having jurisdiction

> Informational Note No. 1: For equipment that is not listed or field labeled, or for which components are no longer available from the original equipment manufacturer, one way to determine suitability is to review the documentation that accompanies the replacement parts.
>
> Informational Note No. 2: See NFPA 70B, *Recommended Practice for Electrical Equipment Maintenance*, for information related to preventive maintenance for electrical, electronic, and communication systems and equipment.

Δ **110.18 Arcing Parts.** Parts of electrical equipment that in ordinary operation produce arcs, sparks, flames, or molten metal shall be enclosed or separated and isolated from all combustible material.

Examples of electrical equipment that can produce sparks during ordinary operation include open motors that have centrifugal starting switches, open motors with commutators, and collector rings. Adequate separation from combustible material is essential if open motors with those features are used.

110.19 Light and Power from Railway Conductors. Circuits for lighting and power shall not be connected to any system that contains trolley wires with a ground return.

Exception: Such circuit connections shall be permitted in car houses, power houses, or passenger and freight stations operated in connection with electric railways.

N **110.20 Reconditioned Equipment.** Reconditioned equipment shall be permitted except where prohibited elsewhere in this *Code*. Equipment that is restored to operating condition shall be reconditioned with identified replacement parts, verified under applicable standards, that are either provided by the original equipment manufacturer or that are designed by an engineer experienced in the design of replacement parts for the type of equipment being reconditioned.

N **(A) Equipment Required to Be Listed.** Equipment that is reconditioned and required by this *Code* to be listed shall be listed or field labeled as reconditioned using available instructions from the original equipment manufacturer.

N (B) Equipment Not Required to Be Listed. Equipment that is reconditioned and not required by this *Code* to be listed shall comply with one of the following:

(1) Be listed or field labeled as reconditioned
(2) Have the reconditioning performed in accordance with the original equipment manufacturer instructions

N (C) Approved Equipment. If the options specified in 110.20(A) or (B) are not available, the authority having jurisdiction shall be permitted to approve reconditioned equipment, and the reconditioner shall provide the authority having jurisdiction with documentation of the changes to the product.

110.21 Marking.

(A) Equipment Markings.

Δ **(1) General.** The manufacturer's name, trademark, or other descriptive marking by which the organization responsible for the product can be identified shall be applied or affixed onto all electrical equipment. Other markings that indicate voltage, current, wattage, or other ratings shall be provided as specified elsewhere in this *Code*. The marking shall be of sufficient durability to withstand the environment involved.

See Exhibit 110.10 for an example of a short-circuit current rating marking.

Δ **(2) Reconditioned Equipment.** Reconditioned equipment shall be marked with the following:

(1) Name, trademark, or other descriptive marking of the organization that performed the reconditioning
(2) The date of the reconditioning
(3) The term *reconditioned* or other approved wording or symbol indicating that the equipment has been reconditioned

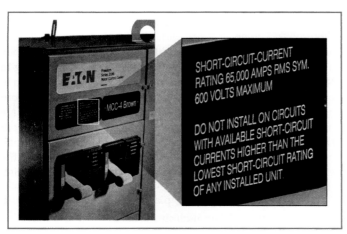

EXHIBIT 110.10 A short-circuit current rating marking. (*Courtesy of Eaton*)

The original listing mark shall be removed or made permanently illegible. The equipment nameplate shall not be required to be removed or made permanently illegible, only the part of the nameplate that includes the listing mark, if applicable. Approval of the reconditioned equipment shall not be based solely on the equipment's original listing.

Exception: In industrial occupancies, where conditions of maintenance and supervision ensure that only qualified persons service the equipment, the markings indicated in 110.21(A)(2) shall not be required for equipment that is reconditioned by the owner or operator as part of a regular equipment maintenance program.

Informational Note No. 1: ANSI-approved standards are available for application of reconditioned and refurbished equipment.
Informational Note No. 2: The term *reconditioned* may be interchangeable with the terms *rebuilt, refurbished,* or *remanufactured* even though these are sometimes different processes.

Δ **(B) Field-Applied Hazard Markings.** Where caution, warning, or danger hazard markings such as labels or signs are required by this *Code*, the markings shall meet the following requirements:

(1) The marking shall be of sufficient durability to withstand the environment involved and warn of the hazards using effective words, colors, symbols, or any combination thereof.

Informational Note No. 1: See ANSI Z535.2-2011 (R2017), *Environmental and Facility Safety Signs*, which describes the design, application, and use of safety signs in facilities and in the environment.
Informational Note No. 2: See ANSI Z535.4-2011 (R2017), *Product Safety Signs and Labels*, which details the design, application, use, and placement of safety signs and labels on a wide variety of products.

(2) The marking shall be permanently affixed to the equipment or wiring method and shall not be handwritten.

Exception to (2): Portions of the markings that are variable, or that could be subject to changes, shall be permitted to be handwritten and shall be legible.

110.22 Identification of Disconnecting Means.

(A) General. Each disconnecting means shall be legibly marked to indicate its purpose unless located and arranged so the purpose is evident. In other than one- or two-family dwellings, the marking shall include the identification and location of the circuit source that supplies the disconnecting means unless located and arranged so the identification and location of the circuit source is evident. The marking shall be of sufficient durability to withstand the environment involved.

Marking of disconnecting means must specifically identify the purpose of each piece of equipment — that is, the marking should not, for example, indicate simply "motor" but rather "motor, water pump," or not simply "lights" but rather "lights, front lobby." The markings must remain legible and not fade or wear off.

See also

408.4 and its commentary for additional requirements and information on circuit directories for switchboards and panelboards

(B) Engineered Series Combination Systems. Equipment enclosures for circuit breakers or fuses applied in compliance with series combination ratings selected under engineering supervision in accordance with 240.86(A) shall be legibly marked in the field as directed by the engineer to indicate the equipment has been applied with a series combination rating. The marking shall meet the requirements in 110.21(B) and shall be readily visible and state the following:

> CAUTION — ENGINEERED SERIES
> COMBINATION SYSTEM RATED _____
> AMPERES. IDENTIFIED REPLACEMENT
> COMPONENTS REQUIRED.

The warnings on replacement components are important to follow in order to maintain the level of protection provided by the design. Likewise, when components are replaced, new or updated warning labels with information based on the new component may be necessary.

(C) Tested Series Combination Systems. Equipment enclosures for circuit breakers or fuses applied in compliance with the series combination ratings marked on the equipment by the manufacturer in accordance with 240.86(B) shall be legibly marked in the field to indicate the equipment has been applied with a series combination rating. The marking shall meet the requirements in 110.21(B) and shall be readily visible and state the following:

> CAUTION — SERIES COMBINATION
> SYSTEM RATED _____ AMPERES. IDENTIFIED
> REPLACEMENT COMPONENTS REQUIRED.

> Informational Note: See IEEE 3004.5-2014 *Recommended Practice for the Application of Low-Voltage Circuit Breakers in Industrial and Commercial Power Systems*, for further information on series tested systems.

110.23 Current Transformers. Unused current transformers associated with potentially energized circuits shall be short-circuited.

110.24 Available Fault Current.

Δ **(A) Field Marking.** Service equipment at other than dwelling units shall be legibly marked in the field with the available fault current. The field marking(s) shall include the date the fault-current calculation was performed and be of sufficient durability to withstand the environment involved. The calculation shall be documented and made available to those authorized to design, install, inspect, maintain, or operate the system.

> Informational Note No. 1: See NFPA *70E*-2021, *Standard for Electrical Safety in the Workplace,* for assistance in determining

the severity of potential exposure, planning safe work practices, and selecting personal protective equipment. The available fault-current marking(s) addressed in 110.24 is related to required short-circuit current and interrupting ratings of equipment.
> Informational Note No. 2: Values of available fault current for use in determining appropriate minimum short-circuit current and interrupting ratings of service equipment are available from electric utilities in published or other forms.

(B) Modifications. When modifications to the electrical installation occur that affect the available fault current at the service, the available fault current shall be verified or recalculated as necessary to ensure the service equipment ratings are sufficient for the available fault current at the line terminals of the equipment. The required field marking(s) in 110.24(A) shall be adjusted to reflect the new level of available fault current.

Exception: The field marking requirements in 110.24(A) and 110.24(B) shall not be required in industrial installations where conditions of maintenance and supervision ensure that only qualified persons service the equipment.

To be used safely, equipment must have an interrupting rating or short-circuit current rating equal to or greater than the available fault current. Any equipment operating with ratings less than the available fault current is potentially unsafe.

Existing electrical distribution systems often experience change over the life of the system. As the system ages, the supply network to which it is connected is affected by growth and is forced to increase capacity or increase efficiency by reducing transformer impedance. In some cases, alternative energy systems are added to existing installations. Such changes to the electrical distribution system can result in an increase of the available fault current. That increase in available fault current can exceed the ratings of the originally installed equipment, violating 110.10 and 110.11 and creating an unsafe condition. This section requires an initial marking of maximum available fault current as well as the requirement to update the information when the system is modified.

The available fault current must be known to complete an arc-flash risk assessment per Section 130.5 of *NFPA 70E, Standard for Electrical Safety in the Workplace*. This assessment is used to determine the arc-flash protection boundary and required PPE in accordance with *NFPA 70E*, Sections 130.5(E) and 130.5(F). The equipment must then be marked with the incident energy or required level of PPE per 130.5(H). According to NFPA *70E*, Article 130, an arc-flash risk assessment is required to be conducted any time work is performed where electrical conductors or circuit parts operate at 50 volts and greater and are not placed in an electrically safe work condition. Additionally, per *NFPA 70E* 130.5(H) the assessment must be reviewed periodically — at least every 5 years — to account for changes in the electrical distribution system that could affect the original arc-flash assessment.

110.25 Lockable Disconnecting Means. If a disconnecting means is required to be lockable open elsewhere in this *Code*, it shall be capable of being locked in the open position. The provisions for locking shall remain in place with or without the lock installed.

Exception: Locking provisions for a cord-and-plug connection shall not be required to remain in place without the lock installed.

The means to lock the switch or circuit breaker in the open position must be an integral part of the enclosure or be an accessory that is not readily removed from the switch or circuit breaker. Portable locking mechanisms that are intended for temporary applications are not acceptable means of compliance.

Part II. 1000 Volts, Nominal, or Less

110.26 Spaces About Electrical Equipment. Working space, and access to and egress from working space, shall be provided and maintained about all electrical equipment to permit ready and safe operation and maintenance of such equipment. Open equipment doors shall not impede access to and egress from the working space. Access or egress is impeded if one or more simultaneously opened equipment doors restrict working space access to be less than 610 mm (24 in.) wide and 2.0 m (6½ ft) high.

Spaces about electrical equipment are divided into two separate and distinct categories: working space and dedicated equipment space. The term *working space* generally applies to the protection of the worker, and *dedicated equipment space* applies to the space reserved for future access to electrical equipment and to protection of the equipment from intrusion by nonelectrical equipment. Storage of material that blocks access or prevents safe work practices must always be avoided. The performance requirements for all spaces about electrical equipment are set forth in this section.

(A) Working Space. Working space for equipment operating at 1000 volts, nominal, or less to ground and likely to require examination, adjustment, servicing, or maintenance while energized shall comply with the dimensions of 110.26(A)(1), (A)(2), (A)(3), and (A)(4) or as required or permitted elsewhere in this *Code*.

Informational Note: See NFPA *70E-2021*, *Standard for Electrical Safety in the Workplace*, for guidance, such as determining severity of potential exposure, planning safe work practices including establishing an electrically safe work condition, arc flash labeling, and selecting personal protective equipment.

The intent is to provide enough space for the performance of any of the operations listed without jeopardizing workers. Minimum working clearances are not required if the equipment is not likely to require examination, adjustment, servicing, or maintenance while energized. However, access and working space are still required by the opening paragraph of 110.26.

Examples of such equipment include panelboards, switches, circuit breakers, controllers, and controls on heating and air-conditioning equipment. Note that the word "examination" includes tasks such as checking for the presence of voltage using a portable voltmeter.

Δ **(1) Depth of Working Space.** The depth of the working space in the direction of live parts shall not be less than that specified in Table 110.26(A)(1) unless the requirements of 110.26(A)(1)(a), (A)(1)(b), or (A)(1)(c) are met. Distances shall be measured from the exposed live parts or from the enclosure or opening if the live parts are enclosed.

TABLE 110.26(A)(1) *Working Spaces*

Nominal Voltage to Ground	Minimum Clear Distance		
	Condition 1	Condition 2	Condition 3
0–150	900 mm (3 ft)	900 mm (3 ft)	900 mm (3 ft)
151–600	900 mm (3 ft)	1.0 m (3 ft 6 in.)	1.2 m (4 ft)
601–1000	900 mm (3 ft)	1.2 m (4 ft)	1.5 m (5 ft)

Note: Where the conditions are as follows:

Condition 1 — Exposed live parts on one side of the working space and no live or grounded parts on the other side of the working space, or exposed live parts on both sides of the working space that are effectively guarded by insulating materials.

Condition 2 — Exposed live parts on one side of the working space and grounded parts on the other side of the working space. Concrete, brick, or tile walls shall be considered as grounded.

Condition 3 — Exposed live parts on both sides of the working space.

Table 110.26(A)(1) provides requirements for clearances from equipment to grounded or ungrounded objects and exposed live parts based on the circuit voltage to ground. See Exhibit 110.11 for the general working clearance requirements for each of the three conditions listed in Table 110.26(A)(1).

Examples of common electrical supply systems covered in the 0 to 150 volts-to-ground group are 120/240-volt, single-phase, 3-wire and 208Y/120-volt, 3-phase, 4-wire systems. Examples of common electrical supply systems covered in the 151 to 1000 volts-to-ground group are 240-volt, 3-phase, 3-wire; 480Y/277-volt, 3-phase, 4-wire; and 480-volt, 3-phase, 3-wire (ungrounded and corner grounded) systems. Where an ungrounded system is used, the voltage to ground (by definition) is the greatest voltage between the given conductor and any other conductor of the circuit. For example, the voltage to ground for a 480-volt ungrounded delta system is 480 volts. For assemblies such as switchboards, switchgear, or motor-control centers that are accessible from the back and expose live parts, the working clearance dimensions are required at the rear of the equipment. For Condition 3, where enclosures are on opposite sides of the working space, the clearance for only one working space is required.

(a) *Dead-Front Assemblies.* Working space shall not be required in the back or sides of assemblies, such as dead-front switchboards, switchgear, or motor control centers, where all connections and all renewable or adjustable parts, such as fuses or switches, are accessible from locations other than the back or sides. Where rear access is required to work on nonelectrical parts on the back of enclosed equipment, a minimum horizontal working space of 762 mm (30 in.) shall be provided.

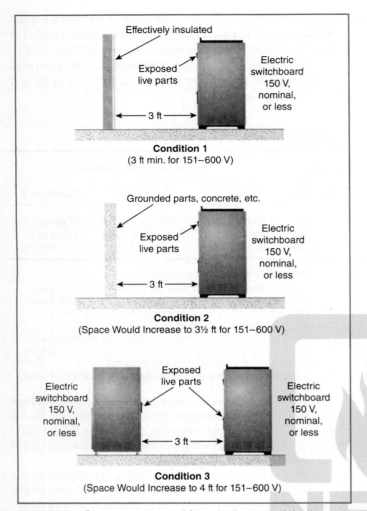

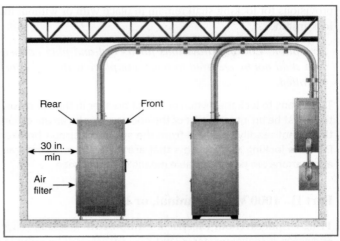

EXHIBIT 110.12 An example of the 30-inch minimum working space at the rear of equipment to allow work on nonelectrical parts, such as the replacement of an air filter.

EXHIBIT 110.11 Distances measured from the live parts if the live parts are exposed or from the enclosure front if the live parts are enclosed.

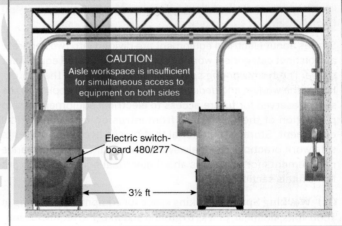

EXHIBIT 110.13 A permitted reduction from a Condition 3 to a Condition 2 clearance.

In many cases, equipment of "dead-front" assemblies requires only front access. The workspace described in 110.26(A)(1) is required for any side of the equipment from which access to live parts or other maintenance is required. For equipment that requires rear access for nonelectrical maintenance, such as replacing a filter, a reduced working space of at least 30 inches must be provided. Exhibit 110.12 shows a reduced working space of 30 inches at the rear of equipment to allow work on nonelectrical parts.

(b) *Low Voltage.* By special permission, smaller working spaces shall be permitted where all exposed live parts operate at not greater than 30 volts rms, 42 volts peak, or 60 volts dc.

(c) *Existing Buildings.* In existing buildings where electrical equipment is being replaced, Condition 2 working clearance shall be permitted between dead-front switchboards, switchgear, enclosed panelboards, or motor control centers located across

the aisle from each other where conditions of maintenance and supervision ensure that written procedures have been adopted to prohibit equipment on both sides of the aisle from being open at the same time and qualified persons who are authorized will service the installation.

This section permits some relief for existing equipment being replaced if it is not practical to increase the existing work space. Exhibit 110.13 illustrates this relief for existing buildings.

(2) Width of Working Space. The width of the working space in front of the electrical equipment shall be the width of the equipment or 762 mm (30 in.), whichever is greater. In all cases, the work space shall permit at least a 90-degree opening of equipment doors or hinged panels.

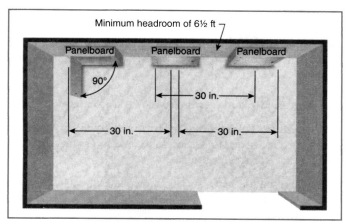

EXHIBIT 110.14 The 30-inch-wide front working space, which is not required to be directly centered on the electrical equipment and can overlap other electrical equipment.

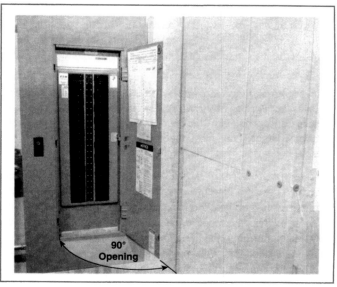

EXHIBIT 110.15 A full 90-degree opening of an equipment door in order to ensure a safe working approach.

Regardless of the width of the electrical equipment, the working space cannot be less than 30 inches wide. This space allows an individual to have at least shoulder-width space in front of the equipment. The 30-inch measurement can be made from either the left or the right edge of the equipment and can overlap other electrical equipment, provided the equipment does not extend into the working space of the other equipment. If the equipment is wider than 30 inches, the space must be equal to the width of the equipment. Exhibit 110.14 illustrates the 30-inch width requirement.

Sufficient depth in the working space is also required to allow a panel or a door to open at least 90 degrees. If doors or hinged panels are wider than 3 feet, more than a 3-foot-deep working space must be provided to allow a full 90-degree opening. (See Exhibit 110.15.) Doors are often part of bolted equipment covers. The 90-degree requirement applies only to the hinged door. It does not require that the bolted cover be capable of swinging 90 degrees.

(3) Height of Working Space. The work space shall be clear and extend from the grade, floor, or platform to a height of 2.0 m (6½ ft) or the height of the equipment, whichever is greater. Within the height requirements of this section, other equipment or support structures, such as concrete pads, associated with the electrical installation and located above or below the electrical equipment shall be permitted to extend not more than 150 mm (6 in.) beyond the front of the electrical equipment.

Exception No. 1: On battery systems mounted on open racks, the top clearance shall comply with 480.10(D).

Exception No. 2: In existing dwelling units, service equipment or enclosed panelboards that do not exceed 200 amperes shall be permitted in spaces where the height of the working space is less than 2.0 m (6½ ft).

Exception No. 3: Meters that are installed in meter sockets shall be permitted to extend beyond the other equipment. The meter socket shall be required to follow the rules of this section.

This requirement allows the placement of a 12-inch × 12-inch wireway on the wall directly above or below a 6-inch-deep panelboard without encroaching on the working space. The allowable 6-inch extension into working space also applies to support structures or housekeeping pads that the equipment may be affixed to. The requirement prohibits large differences in depth of equipment below or above other equipment that specifically requires working space. Freestanding, dry-type transformers are not permitted to be installed where they would extend into the work space for a wall-mounted panelboard, compromising clear access to the panelboard.

(4) Limited Access. Where equipment operating at 1000 volts, nominal, or less to ground and likely to require examination, adjustment, servicing, or maintenance while energized is required by installation instructions or function to be located in a space with limited access, all of the following shall apply:

(1) Where equipment is installed above a lay-in ceiling, there shall be an opening not smaller than 559 mm × 559 mm (22 in. × 22 in.), or in a crawl space, there shall be an accessible opening not smaller than 559 mm × 762 mm (22 in. × 30 in.).

(2) The width of the working space shall be the width of the equipment enclosure or a minimum of 762 mm (30 in.), whichever is greater.

(3) All enclosure doors or hinged panels shall be capable of opening a minimum of 90 degrees.

(4) The space in front of the enclosure shall comply with the depth requirements of Table 110.26(A)(1) and shall be unobstructed to the floor by fixed cabinets, walls, or partitions. Space reductions in accordance with 110.26(A)(1)(b) shall be permitted. The maximum height of the working space shall be the height necessary to install the equipment

in the limited space. A horizontal ceiling structural member or access panel shall be permitted in this space provided the location of weight-bearing structural members does not result in a side reach of more than 150 mm (6 in.) to work within the enclosure.

This requirement was located in 424.66 prior to the 2017 edition. A typical application of this requirement is the installation of duct heaters and other ventilation equipment located above suspended ceilings. The workspace is usually limited, and workers are usually performing maintenance from ladders.

(5) Separation from High-Voltage Equipment. Where switches, cutouts, or other equipment operating at 1000 volts, nominal, or less are installed in a vault, room, or enclosure where there are exposed live parts or exposed wiring operating over 1000 volts, nominal, the high-voltage equipment shall be effectively separated from the space occupied by the low-voltage equipment by a suitable partition, fence, or screen.

N (6) Grade, Floor, or Working Platform. The grade, floor, or platform in the required working space shall be kept clear, and the floor, grade, or platform in the working space shall be as level and flat as practical for the entire required depth and width of the working space.

(B) Clear Spaces. Working space required by this section shall not be used for storage. When normally enclosed live parts are exposed for inspection or servicing, the working space, if in a passageway or general open space, shall be suitably guarded.

Prohibited storage includes portable equipment on rollers. Exhibit 110.16 shows an equipment location that is free of storage. This section and the rest of 110.26 do not prohibit the placement of panelboards in corridors or passageways. When the covers of corridor-mounted panelboards are removed for

EXHIBIT 110.16 An equipment location that is free of storage to allow the equipment to be worked on safely.

authorized energized work, access to the area around the panelboard should be guarded or limited to prevent injury to unqualified persons using the corridor in accordance with NFPA *70E, Standard for Electrical Safety in the Workplace,* Article 130.

(C) Entrance to and Egress from Working Space.

(1) Minimum Required. At least one entrance of sufficient area shall be provided to give access to and egress from working space about electrical equipment.

The requirements in this section provide access to and egress from electrical equipment. However, the primary intent is to provide egress from the area so that workers can escape if an arc-flash incident occurs.

Δ **(2) Large Equipment.** For large equipment that contains overcurrent devices, switching devices, or control devices, there shall be one entrance to and egress from the required working space not less than 610 mm (24 in.) wide and 2.0 m (6½ ft) high at each end of the working space. This requirement shall apply to either of the following conditions:

(1) For equipment rated 1200 amperes or more and over 1.8 m (6 ft) wide
(2) For service disconnecting means installed in accordance with 230.71(B) where the combined ampere rating is 1200 amperes or more and where the combined width is over 1.8 m (6 ft)

A single entrance to and egress from the required working space shall be permitted where either of the conditions in 110.26(C)(2)(a) or (C)(2)(b) is met.

(a) *Unobstructed Egress.* Where the location permits a continuous and unobstructed way of egress travel, a single entrance to the working space shall be permitted.

(b) *Extra Working Space.* Where the depth of the working space is twice that required by 110.26(A)(1), a single entrance shall be permitted. It shall be located such that the distance from the equipment to the nearest edge of the entrance is not less than the minimum clear distance specified in Table 110.26(A)(1) for equipment operating at that voltage and in that condition.

Open equipment doors must not impede access to or egress from the work space. This requirement is intended to prevent workers from being entrapped between equipment doors and walls or other equipment facing the installation. Exhibits 110.17 and 110.18 illustrate access and entrance requirements for working spaces. Exhibit 110.19 shows an unacceptable and hazardous work space arrangement. See Exhibits 110.20 and 110.21 for representations of the single egress requirements for large equipment. Large equipment is equipment rated 1200 amperes or more and over 6 feet wide. Equipment consisting of multiple disconnecting means, in accordance with 230.71, where the combined rating is 1200 amperes or more and over 6 feet wide, is also considered large equipment.

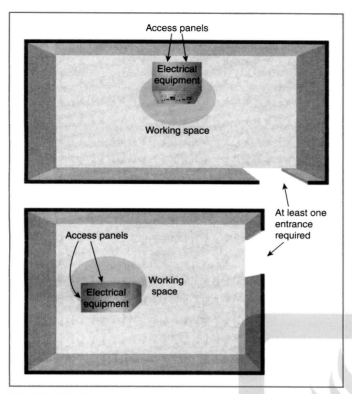

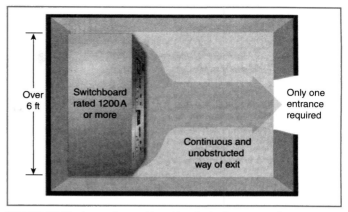

EXHIBIT 110.20 An equipment location that allows a continuous and unobstructed way of exit travel.

EXHIBIT 110.17 At least one entrance is required to provide access to the working space around electrical equipment.

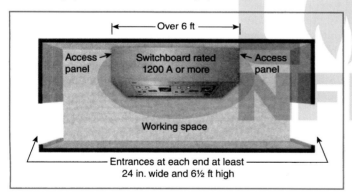

EXHIBIT 110.18 For large equipment, one entrance not less than 24 inches wide and 6½ feet high is required at each end.

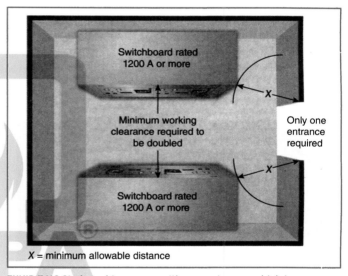

EXHIBIT 110.21 A working space with one entrance, which is permitted if the working space required by 110.26(A) is doubled [see Table 110.26(A)(1) for permitted dimensions of X].

(3) Personnel Doors. Where equipment rated 800 amperes or more that contains overcurrent devices, switching devices, or control devices is installed and there is a personnel door(s) intended for entrance to and egress from the working space less than 7.6 m (25 ft) from the nearest edge of the working space, the door(s) shall open at least 90 degrees in the direction of egress and be equipped with listed panic hardware or listed fire exit hardware.

Informational Note: See UL 305, *Standard For Panic Hardware*, for additional information on panic hardware, and see UL 10C, *Standard for Safety for Positive Pressure Fire Tests of Door Assemblies*, for additional information.

The requirements in this section are based on equipment rated 800 amperes or more and not on the width of the equipment. The 25-foot measurement for the personnel door(s) is made from the nearest edge of the working space. It applies to the personnel door(s) that provide access to and egress from the room containing the working space.

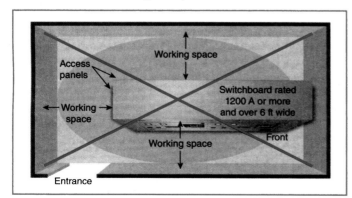

EXHIBIT 110.19 An unacceptable arrangement of a large switchboard in which a worker could be trapped behind arcing electrical equipment.

Not every electrical installation is in an equipment room. Where electrical installations are in a room, personnel doors that are up to 25 feet from the working space are required to have listed panic hardware or listed fire exit hardware, and they must open in the direction of egress from the area. The requirement for listed panic hardware is independent of the need for two exits from the working space. Exhibit 110.22 shows an exit with panic hardware.

Door-opening hardware must not require the turning of a doorknob or similar action that could preclude quick exit from the area in the event of an emergency. This requirement allows an injured worker to exit from the equipment room without having to turn knobs or pull doors open. Exhibit 110.23 illustrates egress doors in relation to the working space around equipment with listed panic hardware that open in the direction of egress.

(D) Illumination. Illumination shall be provided for all working spaces about service equipment, switchboards, switchgear, enclosed panelboards, or motor control centers installed indoors. Control by automatic means shall not be permitted to control all illumination within the working space. Additional lighting outlets shall not be required where the work space is illuminated by an adjacent light source or as permitted by 210.70(A)(1), Exception No. 1, for switched receptacles.

EXHIBIT 110.22 An installation of large equipment showing one of the required exits.

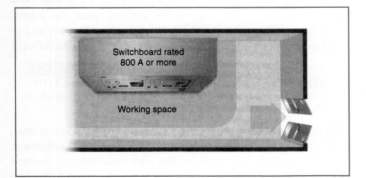

EXHIBIT 110.23 Egress doors from required working space with listed panic hardware that open in the direction of egress.

Automatic lighting control through devices such as timers, occupancy sensors, and similar devices are not prohibited, but a manual means to bypass the automatic control is required.

(E) Dedicated Equipment Space. All service equipment, switchboards, switchgear, panelboards, and motor control centers shall be located in dedicated spaces and protected from damage.

Exception: Control equipment that by its very nature or because of other rules of the Code must be adjacent to or within sight of its operating machinery shall be permitted in those locations.

(1) Indoor. Indoor installations shall comply with 110.26(E)(1)(a) through (E)(1)(d).

(a) *Dedicated Electrical Space.* The space equal to the width and depth of the equipment and extending from the floor to a height of 1.8 m (6 ft) above the equipment or to the structural ceiling, whichever is lower, shall be dedicated to the electrical installation. No piping, ducts, leak protection apparatus, or other equipment foreign to the electrical installation shall be located in this zone.

Exception: Suspended ceilings with removable panels shall be permitted within the 1.8 m (6 ft) zone.

(b) *Foreign Systems.* The area above the dedicated space required by 110.26(E)(1)(a) shall be permitted to contain foreign systems, provided protection is installed to avoid damage to the electrical equipment from condensation, leaks, or breaks in such foreign systems.

(c) *Sprinkler Protection.* Sprinkler protection shall be permitted for the dedicated space where the piping complies with this section.

(d) *Suspended Ceilings.* A dropped, suspended, or similar ceiling that does not add strength to the building structure shall not be considered a structural ceiling.

This requirement applies only to switchboards, panelboards, switchgear, or motor control centers. The dedicated electrical space extends the footprint of the equipment from the floor to a height of 6 feet above the height of the equipment or to the structural ceiling (whichever is lower). The dedicated space is required to be clear of piping, ducts, leak protection apparatus, or equipment foreign to the electrical installation. Plumbing, heating, ventilation, and air-conditioning piping, ducts, and equipment must be installed outside this space. Busways, conduits, raceways, and cables are permitted to enter equipment through this zone.

Foreign systems installed directly above the dedicated space reserved for electrical equipment are required to include protective equipment that ensures that occurrences such as leaks, condensation, and even breaks do not damage the electrical equipment located below.

Sprinkler protection is permitted for the dedicated space. The sprinkler or other suppression system piping must comply with 110.26(E)(1).

A dropped, suspended, or similar ceiling is permitted to be located directly in the dedicated space, because they are not considered structural ceilings. Building structural members are also permitted in this space.

Exhibits 110.24, 110.25, and 110.26 illustrate the two distinct indoor installation spaces required by 110.26(A) and 110.26(E), that is, the working space and the dedicated electrical space.

In Exhibit 110.24, the dedicated electrical space is reserved for the installation of electrical equipment and for the installation of conduits, cable trays, and so forth, entering or exiting that equipment. The outlined area in front of the electrical equipment in Exhibit 110.24 is the working space required by 110.26(A). Note that sprinkler protection is afforded the entire dedicated electrical space and working space without actually entering either space. Also note that the exhaust duct is not located in or directly above the dedicated electrical space.

Exhibit 110.25 illustrates the working space required in front of the panelboard by 110.26(A). No equipment, electrical or otherwise, is allowed in the working space. Exhibit 110.26 illustrates the dedicated electrical space above and below the panelboard required by 110.26(E)(1).

(2) Outdoor. Outdoor installations shall comply with 110.26(E)(2)(a) through (E)(2)(c).

(a) *Installation Requirements.* Outdoor electrical equipment shall be the following:

(1) Installed in identified enclosures
(2) Protected from accidental contact by unauthorized personnel or by vehicular traffic
(3) Protected from accidental spillage or leakage from piping systems

(b) *Work Space.* The working clearance space shall include the zone described in 110.26(A). No architectural appurtenance or other equipment shall be located in this zone.

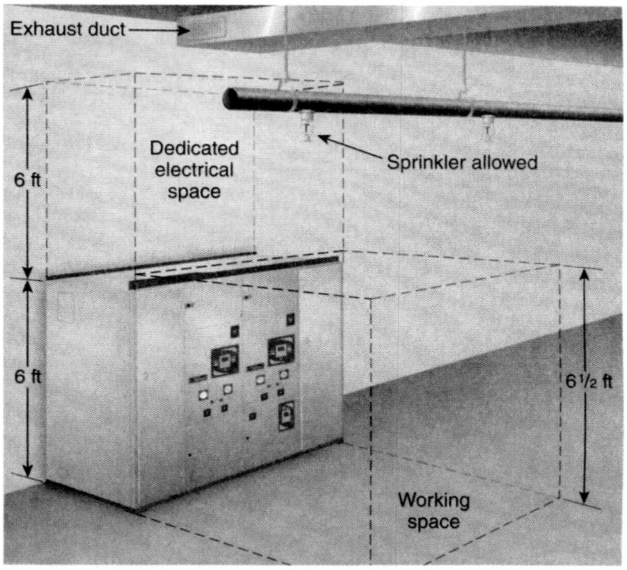

EXHIBIT 110.24 The two distinct indoor installation spaces: the working space and the dedicated electrical space.

If protection of the outdoor installation is requested, consideration should be taken for any underground wiring before any excavation or driving any steel into the ground for the placement of fencing, vehicle stops, or bollards.

(c) *Dedicated Equipment Space.* The space equal to the width and depth of the equipment, and extending from grade to a height of 1.8 m (6 ft) above the equipment, shall be dedicated to the electrical installation. No piping or other equipment foreign to the electrical installation shall be located in this zone.

EXHIBIT 110.25 The working space in front of a panelboard. (This illustration supplements the dedicated electrical space shown in Exhibit 110.24.)

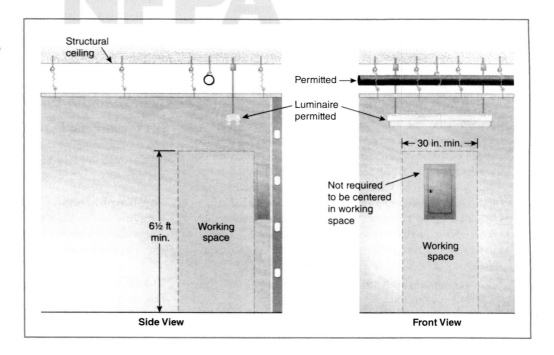

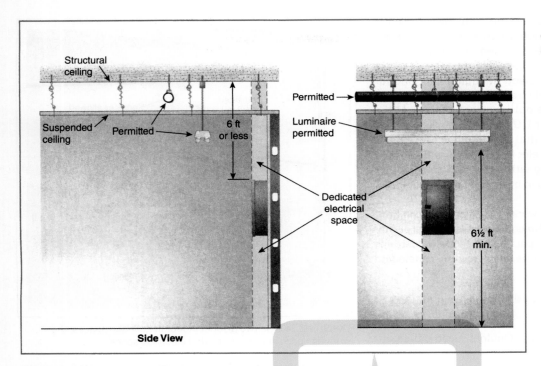

Side View

EXHIBIT 110.26 The dedicated electrical space above and below a panelboard.

Exception: Structural overhangs or roof extensions shall be permitted in this zone.

(F) Locked Electrical Equipment Rooms or Enclosures. Electrical equipment rooms or enclosures housing electrical apparatus that are controlled by a lock(s) shall be considered accessible to qualified persons.

110.27 Guarding of Live Parts.

(A) Live Parts Guarded Against Accidental Contact. Except as elsewhere required or permitted by this *Code,* live parts of electrical equipment operating at 50 to 1000 volts, nominal shall be guarded against accidental contact by approved enclosures or by any of the following means:

(1) By location in a room, vault, or similar enclosure that is accessible only to qualified persons.
(2) By permanent, substantial partitions or screens arranged so that only qualified persons have access to the space within reach of the live parts. Any openings in such partitions or screens shall be sized and located so that persons are not likely to come into accidental contact with the live parts or to bring conducting objects into contact with them.
(3) By location on a balcony, gallery, or platform elevated and arranged so as to exclude unqualified persons.
(4) By elevation above the floor or other working surface as follows:

    a. A minimum of 2.5 m (8 ft) for 50 volts to 300 volts between ungrounded conductors
    b. A minimum of 2.6 m (8 ft 6 in.) for 301 volts to 600 volts between ungrounded conductors

    c. A minimum of 2.62 m (8 ft 7 in.) for 601 volts to 1000 volts between ungrounded conductors

Live parts of electrical equipment should be covered, shielded, enclosed, or otherwise protected by covers, barriers, or platforms to prevent contact by persons or objects.

Contact conductors used for traveling cranes are permitted by 610.13(B) and 610.21(A) to be bare because they are protected from contact by elevation. Although contact conductors obviously must be bare for contact shoes on the moving member to make contact with the conductor, guards can be placed near the conductor to prevent accidental contact with persons and still have slots or spaces through which the moving contacts can operate.

The *NEC* also recognizes the guarding of live parts by elevation. The elevation levels correlate with requirements in ANSI/IEEE C2, *National Electrical Safety Code® (NESC®).*

(B) Prevent Physical Damage. In locations where electrical equipment is likely to be exposed to physical damage, enclosures or guards shall be so arranged and of such strength as to prevent such damage.

Δ **(C) Warning Signs.** Entrances to rooms and other guarded locations that contain exposed live parts shall be marked with conspicuous warning signs forbidding unqualified persons to enter. The marking shall meet the requirements in 110.21(B).

110.28 Enclosure Types. Enclosures (other than surrounding fences or walls covered in 110.31) of switchboards, switchgear, enclosed panelboards, industrial control panels, motor control centers, meter sockets, enclosed switches, transfer switches, power outlets, circuit breakers, adjustable-speed drive systems, pullout switches, portable power distribution equipment, termination boxes, general-purpose transformers, fire pump controllers, fire pump motors, and motor controllers, rated not over 1000 volts nominal and intended for such locations, shall be marked with an enclosure-type number as shown in Table 110.28.

Table 110.28 shall be used for selecting these enclosures for use in specific locations other than hazardous (classified) locations. The enclosures are not intended to protect against conditions such as condensation, icing, corrosion, or contamination that may occur within the enclosure or enter via the raceway or unsealed openings.

Δ **TABLE 110.28** *Enclosure Selection*

Provides a Degree of Protection Against the Following Environmental Conditions	For Outdoor Use									
	Enclosure Type Number									
	3	3R	3S	3X	3RX	3SX	4	4X	6	6P
Incidental contact with the enclosed equipment	X	X	X	X	X	X	X	X	X	X
Rain, snow, and sleet	X	X	X	X	X	X	X	X	X	X
Sleet*	—	—	X	—	—	X	—	—	—	—
Windblown dust	X	—	X	X	—	X	X	X	X	X
Hosedown	—	—	—	—	—	—	X	X	X	X
Corrosive agents	—	—	—	X	X	X	—	X	—	X
Temporary submersion	—	—	—	—	—	—	—	—	X	X
Prolonged submersion	—	—	—	—	—	—	—	—	—	X

Provides a Degree of Protection Against the Following Environmental Conditions	For Indoor Use									
	Enclosure Type Number									
	1	2	4	4X	5	6	6P	12	12K	13
Incidental contact with the enclosed equipment	X	X	X	X	X	X	X	X	X	X
Falling dirt	X	X	X	X	X	X	X	X	X	X
Falling liquids and light splashing	—	X	X	X	X	X	X	X	X	X
Circulating dust, lint, fibers, and flyings	—	—	X	X	—	X	X	X	X	X
Settling airborne dust, lint, fibers, and flyings	—	—	X	X	X	X	X	X	X	X
Hosedown and splashing water	—	—	X	X	—	X	X	—	—	—
Oil and coolant seepage	—	—	—	—	—	—	—	X	X	X
Oil or coolant spraying and splashing	—	—	—	—	—	—	—	—	—	X
Corrosive agents	—	—	—	X	—	—	X	—	—	—
Temporary submersion	—	—	—	—	—	X	X	—	—	—
Prolonged submersion	—	—	—	—	—	—	X	—	—	—

*The mechanism shall be operable when ice covered.

Informational Note No. 1: The term *raintight* is typically used in conjunction with Enclosure Types 3, 3S, 3SX, 3X, 4, 4X, 6, and 6P. The term *rainproof* is typically used in conjunction with Enclosure Types 3R and 3RX. The term *watertight* is typically used in conjunction with Enclosure Types 4, 4X, 6, and 6P. The term *driptight* is typically used in conjunction with Enclosure Types 2, 5, 12, 12K, and 13. The term *dusttight* is typically used in conjunction with Enclosure Types 3, 3S, 3SX, 3X, 4, 4X, 5, 6, 6P, 12, 12K, and 13.

Informational Note No. 2: See ANSI/IEC 60529, *Degrees of Protection Provided by Enclosures*, for ingress protection (IP) ratings.

Informational Note No. 3: See 502.10(A)(3), 502.10(B)(4), 503.10(A)(2), and 506.15(C)(9) for information on the use of dusttight enclosures in hazardous locations.

Informational Note No. 4: Some enclosure types, such as 12, 12K, or 13 enclosures, may be marked with an ancillary "-XH" for corrosive and hosedown capable indoor enclosure.

Informational Note No. 5: Some type 4X enclosures may be marked "indoor only."

Informational Note No. 6: See UL 508A, *Standard for Industrial Control Panels*, for information on determining applicable requirements for evaluating type 4, 4X, and 12 ventilated enclosures.

Informational Note No. 7: See NEMA 250, *Enclosures for Electrical Equipment (1000 Volts Maximum)*, for the description of the "Enclosure Type Rating: Ancillary — PW for Pressure Wash."

Enclosures that comply with the requirements for more than one type of enclosure may be marked with multiple designations. Enclosures marked with a type may also be marked as follows:

Type 1 may be marked "Indoor Use Only."

Type 3, 3S, 4, 4X, 6, or 6P may be marked "Raintight."

Type 3R may be marked "Rainproof."

Type 4, 4X, 6, or 6P may be marked "Watertight."

Type 4X or 6P may be marked "Corrosion Resistant."

Type 2, 5, 12, 12K, or 13 may be marked "Driptight."

Type 3, 3S, 5, 12K, or 13 may be marked "Dusttight."

For equipment designated "raintight," testing designed to simulate exposure to a beating rain will not result in entrance of water. For equipment designated "rainproof," testing designed to simulate exposure to a beating rain will not interfere with the operation of the apparatus or result in wetting of live parts and wiring within the enclosure. "Watertight" equipment is constructed so that water does not enter the enclosure when subjected to a stream of water. "Corrosion-resistant" equipment is constructed so that it provides a degree of protection against exposure to corrosive agents such as salt spray. "Driptight" equipment is constructed so that falling moisture or dirt does not enter the enclosure. "Dusttight" equipment is constructed so that circulating or airborne dust does not enter the enclosure.

Article 100 defines the term *weatherproof* as "constructed or protected so that exposure to the weather will not interfere with successful operation." Rainproof, raintight, or watertight equipment can fulfill the requirements of this definition where varying weather conditions other than wetness, such as snow, ice, dust, or temperature extremes, are not a factor.

N **110.29 In Sight From (Within Sight From, Within Sight).** Where this *Code* specifies that one equipment shall be "in sight from," "within sight from," or "within sight of" another equipment, the specified equipment shall be visible and not more than 15 m (50 ft) distant from the other.

Part III. Over 1000 Volts, Nominal

110.30 General. Conductors and equipment used on circuits over 1000 volts, nominal, shall comply with Part I of this article and with 110.30 through 110.41, which supplement or modify Part I. In no case shall this part apply to equipment on the supply side of the service point.

Δ **110.31 Enclosure for Electrical Installations.** Electrical installations in a vault, room, or closet or in an area surrounded by a wall, screen, or fence, access to which is controlled by a lock(s) or other approved means, shall be considered to be accessible to qualified persons only. The type of enclosure used in a given case shall be designed and constructed according to the nature and degree of the hazard(s) associated with the installation.

For installations other than equipment as described in 110.31(D), a wall, screen, or fence shall be used to enclose an outdoor electrical installation to deter access by persons who are not qualified. A fence shall not be less than 2.1 m (7 ft) in height or a combination of 1.8 m (6 ft) or more of fence fabric and a 300 mm (1 ft) or more extension utilizing three or more strands of barbed wire or equivalent. The distance from the fence to live parts shall be not less than given in Table 110.31.

Informational Note: See ANSI/IEEE C2-2017, *National Electrical Safety Code*, for clearances of conductors for specific system voltages and typical BIL ratings.

Δ **TABLE 110.31** *Minimum Distance from Fence to Live Parts*

Nominal Voltage	Minimum Distance to Live Parts	
	m	ft
1001–13,799	3.05	10
13,800–230,000	4.57	15
Over 230,000	5.49	18

(A) Electrical Vaults. Where an electrical vault is required or specified for conductors and equipment 110.31(A)(1) to (A)(5) shall apply.

(1) Walls and Roof. The walls and roof shall be constructed of materials that have adequate structural strength for the conditions, with a minimum fire rating of 3 hours. For the purpose of this section, studs and wallboard construction shall not be permitted.

(2) Floors. The floors of vaults in contact with the earth shall be of concrete that is not less than 102 mm (4 in.) thick, but where the vault is constructed with a vacant space or other stories below it, the floor shall have adequate structural strength for the load imposed on it and a minimum fire resistance of 3 hours.

(3) Doors. Each doorway leading into a vault from the building interior shall be provided with a tight-fitting door that has a minimum fire rating of 3 hours. The authority having jurisdiction shall be permitted to require such a door for an exterior wall opening where conditions warrant.

Exception to (1), (2), and (3): Where the vault is protected with automatic sprinkler, water spray, carbon dioxide, or halon, construction with a 1-hour rating shall be permitted.

(4) Locks. Doors shall be equipped with locks, and doors shall be kept locked, with access allowed only to qualified persons. Personnel doors shall open at least 90 degrees in the direction of egress and be equipped with listed panic hardware or listed fire exit hardware.

Informational Note: See UL 305, *Standard for Panic Hardware*, for additional information, and UL10C, *Standard for Safety for Positive Pressure Fire Tests of Door Assemblies*.

(5) Transformers. Where a transformer is installed in a vault as required by Part II of Article 450, the vault shall be constructed in accordance with Part III of Article 450.

> Informational Note No. 1: See ANSI/ASTM E119-2018a, *Method for Fire Tests of Building Construction and Materials*, for additional information, and see NFPA 80-2019, *Standard for Fire Doors and Other Opening Protectives*.
>
> Informational Note No. 2: A typical 3-hour construction is 150 mm (6 in.) thick reinforced concrete.

(B) Indoor Installations.

(1) In Places Accessible to Unqualified Persons. Indoor electrical installations that are accessible to unqualified persons shall be made with metal-enclosed equipment. Switchgear, transformers, pull boxes, connection boxes, and other similar associated equipment shall be marked with appropriate caution signs. Openings in ventilated dry-type transformers or similar openings in other equipment shall be designed so that foreign objects inserted through these openings are deflected from energized parts.

(2) In Places Accessible to Qualified Persons Only. Indoor electrical installations considered accessible only to qualified persons in accordance with this section shall comply with 110.34, 110.36, and 495.24.

(C) Outdoor Installations.

(1) In Places Accessible to Unqualified Persons. Outdoor electrical installations that are open to unqualified persons shall comply with Part III of Article 225.

(2) In Places Accessible to Qualified Persons Only. Outdoor electrical installations that have exposed live parts shall be accessible to qualified persons only in accordance with the first paragraph of this section and shall comply with 110.34, 110.36, and 495.24.

(D) Enclosed Equipment Accessible to Unqualified Persons. Ventilating or similar openings in equipment shall be designed such that foreign objects inserted through these openings are deflected from energized parts. Where exposed to physical damage from vehicular traffic, suitable guards shall be provided. Equipment located outdoors and accessible to unqualified persons shall be designed such that exposed nuts or bolts cannot be readily removed, permitting access to live parts. Where equipment is accessible to unqualified persons and the bottom of the enclosure is less than 2.5 m (8 ft) above the floor or grade level, the enclosure door or hinged cover shall be kept locked. Doors and covers of enclosures used solely as pull boxes, splice boxes, or junction boxes shall be locked, bolted, or screwed on. Underground box covers that weigh over 45.4 kg (100 lb) shall be considered as meeting this requirement.

110.32 Work Space About Equipment. Sufficient space shall be provided and maintained about electrical equipment to permit ready and safe operation and maintenance of such equipment. Where energized parts are exposed, the minimum clear work space shall be not less than 2.0 m (6½ ft) high (measured vertically from the floor or platform) and the width of the equipment or 914 mm (3 ft) wide (measured parallel to the equipment), whichever is greater. The depth shall be as required in 110.34(A). In all cases, the work space shall permit at least a 90-degree opening of doors or hinged panels. Within the height requirements of this section, other equipment that is associated with the electrical installation and is located above or below the electrical equipment shall be permitted to extend not more than 150 mm (6 in.) beyond the front of the electrical equipment. Working space required by this section shall not be used for storage. When normally enclosed live parts are exposed for inspection or servicing, the working space, if in a passageway or general open space, shall be suitably guarded.

This requirement allows the placement of an 18-inch × 18-inch wireway on the wall directly above or below a 12-inch-deep panelboard without encroaching on the working space. The allowable 6-inch extension into working space also applies to support structures or housekeeping pads that the equipment may be affixed to. The requirement prohibits large differences in depth of equipment below or above other equipment that specifically requires working space.

110.33 Entrance to Enclosures and Access to Working Space.

(A) Entrance. At least one entrance to enclosures for electrical installations as described in 110.31 not less than 610 mm (24 in.) wide and 2.0 m (6½ ft) high shall be provided to give access to the working space about electrical equipment.

Open equipment doors shall not impede access to and egress from the working space. Access or egress is impeded if one or more simultaneously opened equipment doors restrict working space access to be less than 610 mm (24 in.) wide and 2.0 m (6½ ft) high.

(1) Large Equipment. On switchgear and control panels exceeding 1.8 m (6 ft) in width, there shall be one entrance at each end of the equipment. A single entrance to the required working space shall be permitted where either of the conditions in 110.33(A)(1)(a) or (A)(1)(b) is met.

(a) *Unobstructed Exit.* Where the location permits a continuous and unobstructed way of exit travel, a single entrance to the working space shall be permitted.

(b) *Extra Working Space.* Where the depth of the working space is twice that required by 110.34(A), a single entrance shall be permitted. It shall be located so that the distance from the equipment to the nearest edge of the entrance is not less than the minimum clear distance specified in Table 110.34(A) for equipment operating at that voltage and in that condition.

(2) Guarding. Where bare energized parts at any voltage or insulated energized parts above 1000 volts, nominal, are located adjacent to such entrance, they shall be suitably guarded.

(3) Personnel Doors. Where there are personnel doors intended for entrance to and egress from the working space less than 7.6 m (25 ft) from the nearest edge of the working space, the doors shall open at least 90 degrees in the direction of egress and be equipped with listed panic hardware or listed fire exit hardware.

Informational Note: See UL 305, *Standard for Panic Hardware*, for additional information, and UL 10C, *Standard for Safety for Positive Pressure Fire Tests of Door Assemblies.*

Not every electrical installation is in an equipment room. This section also does not mandate that the installation be enclosed in a room. If there are personnel doors within 25 feet from the working space, they must have listed panic hardware or listed fire exit hardware and they must open in the direction of egress from the area.

(B) Access. Permanent ladders or stairways shall be provided to give safe access to the working space around electrical equipment installed on platforms, balconies, or mezzanine floors or in attic or roof rooms or spaces.

110.34 Work Space and Guarding.

(A) Working Space. Except as elsewhere required or permitted in this *Code,* equipment likely to require examination, adjustment, servicing, or maintenance while energized shall have clear working space in the direction of access to live parts of the electrical equipment and shall be not less than specified in Table 110.34(A). Distances shall be measured from the live parts, if such are exposed, or from the enclosure front or opening if such are enclosed. The grade, floor, or platform in the required working space shall be kept clear, and the floor, grade, or platform in the working space shall be as level and flat as practical for the entire depth and width of the working space.

Exception: Working space shall not be required in back of equipment such as switchgear or control assemblies where there are no renewable or adjustable parts (such as fuses or switches) on the back and where all connections are accessible from locations other than the back. Where rear access is required to work on nonelectrical parts on the back of enclosed equipment, a minimum working space of 762 mm (30 in.) horizontally shall be provided.

(B) Separation from Low-Voltage Equipment. Where switches, cutouts, or other equipment operating at 1000 volts, nominal, or less are installed in a vault, room, or enclosure where there are exposed live parts or exposed wiring operating at over 1000 volts, nominal, the high-voltage equipment shall be effectively separated from the space occupied by the low-voltage equipment by a suitable partition, fence, or screen.

Exception: Switches or other equipment operating at 1000 volts, nominal, or less and serving only equipment within the high-voltage vault, room, or enclosure shall be permitted to be installed in the high-voltage vault, room, or enclosure without a partition, fence, or screen if accessible to qualified persons only.

Δ **TABLE 110.34(A)** *Minimum Depth of Clear Working Space at Electrical Equipment*

Nominal Voltage to Ground	Minimum Clear Distance		
	Condition 1	Condition 2	Condition 3
1001–2500 V	900 mm (3 ft)	1.2 m (4 ft)	1.5 m (5 ft)
2501–9000 V	1.2 m (4 ft)	1.5 m (5 ft)	1.8 m (6 ft)
9001–25,000 V	1.5 m (5 ft)	1.8 m (6 ft)	2.8 m (9 ft)
25,001 V–75 kV	1.8 m (6 ft)	2.5 m (8 ft)	3.0 m (10 ft)
Above 75 kV	2.5 m (8 ft)	3.0 m (10 ft)	3.7 m (12 ft)

Note: Where the conditions are as follows:

Condition 1 — Exposed live parts on one side of the working space and no live or grounded parts on the other side of the working space, or exposed live parts on both sides of the working space that are effectively guarded by insulating materials.

Condition 2 — Exposed live parts on one side of the working space and grounded parts on the other side of the working space. Concrete, brick, or tile walls shall be considered as grounded.

Condition 3 — Exposed live parts on both sides of the working space.

(C) Locked Rooms or Enclosures. The entrance to all buildings, vaults, rooms, or enclosures containing exposed live parts or exposed conductors operating at over 1000 volts, nominal, shall be kept locked unless such entrances are under the observation of a qualified person at all times.

Permanent and conspicuous danger signs shall be provided. The danger sign shall meet the requirements in 110.21(B) and shall read as follows:

DANGER — HIGH VOLTAGE — KEEP OUT

(D) Illumination. Illumination shall be provided for all working spaces about electrical equipment. Control by automatic means only shall not be permitted. The lighting outlets shall be arranged so that persons changing lamps or making repairs on the lighting system are not endangered by live parts or other equipment.

The points of control shall be located so that persons are not likely to come in contact with any live part or moving part of the equipment while turning on the lights.

(E) Elevation of Unguarded Live Parts. Unguarded live parts above working space shall be maintained at elevations not less than required by Table 110.34(E).

TABLE 110.34(E) *Elevation of Unguarded Live Parts Above Working Space*

Nominal Voltage Between Phases	Elevation	
	m	ft
1001–7500 V	2.7	9
7501–35,000 V	2.9	9 ft 6 in.
Over 35 kV	Add 9.5 mm per kV above 35 kV	Add 0.37 in. per kV above 35 kV

(F) Protection of Service Equipment, Switchgear, and Industrial Control Assemblies. Pipes or ducts foreign to the electrical installation and requiring periodic maintenance or whose malfunction would endanger the operation of the electrical system shall not be located in the vicinity of the service equipment, switchgear, or industrial control assemblies. Protection shall be provided where necessary to avoid damage from condensation leaks and breaks in such foreign systems. Piping and other facilities shall not be considered foreign if provided for fire protection of the electrical installation.

110.36 Circuit Conductors. Circuit conductors shall be permitted to be installed in raceways; in cable trays; as metal-clad cable Type MC; as bare wire, cable, and busbars; or as Type MV cables or conductors as provided in 305.3, 305.9, 305.10, and 305.15. Bare live conductors shall comply with 495.24.

Insulators, together with their mounting and conductor attachments, where used as supports for wires, single-conductor cables, or busbars, shall be capable of safely withstanding the maximum magnetic forces that would prevail if two or more conductors of a circuit were subjected to short-circuit current.

Exposed runs of insulated wires and cables that have a bare lead sheath or a braided outer covering shall be supported in a manner designed to prevent physical damage to the braid or sheath. Supports for lead-covered cables shall be designed to prevent electrolysis of the sheath.

110.40 Temperature Limitations at Terminations. Conductors shall be permitted to be terminated based on the 90°C (194°F) temperature rating and ampacity as given in Table 315.60(C)(1) through Table 315.60(C)(20), unless otherwise identified.

110.41 Inspections and Tests.
(A) Pre-energization and Operating Tests. Where required elsewhere in this *Code*, the complete electrical system design, including settings for protective, switching, and control circuits, shall be prepared in advance and made available on request to the authority having jurisdiction and shall be tested when first installed on-site.

(B) Test Report. A test report covering the results of the tests required in 110.41(A) shall be available to the authority having jurisdiction prior to energization and made available to those authorized to install, operate, test, and maintain the system.

Part IV. Tunnel Installations over 1000 Volts, Nominal

110.51 General.

(A) Covered. This part shall apply to the installation and use of high-voltage power distribution and utilization equipment that is portable, mobile, or both, such as substations, trailers, cars, mobile shovels, draglines, hoists, drills, dredges, compressors, pumps, conveyors, underground excavators, and the like.

(B) Protection Against Physical Damage. Conductors and cables in tunnels shall be located above the tunnel floor and so placed or guarded to protect them from physical damage.

110.52 Overcurrent Protection. Motor-operated equipment shall be protected from overcurrent in accordance with Parts III, IV, and V of Article 430. Transformers shall be protected from overcurrent in accordance with 450.3.

Δ **110.53 Conductors.** High-voltage conductors in tunnels shall be installed in metal raceway, Type MC cable, or other approved multiconductor cable. Multiconductor portable cable shall be permitted to supply mobile equipment.

110.54 Bonding and Equipment Grounding Conductors.

(A) Grounded and Bonded. All non–current-carrying metal parts of electrical equipment and all metal raceways and cable sheaths shall be solidly grounded and bonded to all metal pipes and rails at the portal and at intervals not exceeding 300 m (1000 ft) throughout the tunnel.

(B) Equipment Grounding Conductors. An equipment grounding conductor shall be run with circuit conductors inside the metal raceway or inside the multiconductor cable jacket. The equipment grounding conductor shall be permitted to be insulated or bare.

110.55 Transformers, Switches, and Electrical Equipment. All transformers, switches, motor controllers, motors, rectifiers, and other equipment installed belowground shall be protected from physical damage by location or guarding.

110.56 Energized Parts. Bare terminals of transformers, switches, motor controllers, and other equipment shall be enclosed to prevent accidental contact with energized parts.

110.57 Ventilation System Controls. Electrical controls for the ventilation system shall be arranged so that the airflow can be reversed.

110.58 Disconnecting Means. A switch or circuit breaker that simultaneously opens all ungrounded conductors of the circuit shall be installed within sight of each transformer or motor location for disconnecting the transformer or motor. The switch or circuit breaker for a transformer shall have an ampere rating not less than the ampacity of the transformer supply conductors. The switch or circuit breaker for a motor shall comply with the applicable requirements of Part IX of Article 430.

110.59 Enclosures. Enclosures for use in tunnels shall be dripproof, weatherproof, or submersible as required by the environmental conditions. Switch or contactor enclosures shall not be used as junction boxes or as raceways for conductors feeding through or tapping off to other switches, unless the enclosures comply with 312.8.

Part V. Manholes and Other Electrical Enclosures Intended for Personnel Entry

110.70 General. Electrical enclosures intended for personnel entry and specifically fabricated for this purpose shall be of sufficient size to provide safe work space about electrical equipment with live parts that is likely to require examination, adjustment, servicing, or maintenance while energized. Such enclosures shall have sufficient size to permit ready installation or withdrawal of the conductors employed without damage to the conductors or to their insulation. They shall comply with this part.

Exception: Where electrical enclosures covered by Part V of this article are part of an industrial wiring system operating under conditions of maintenance and supervision that ensure that only qualified persons monitor and supervise the system, they shall be permitted to be designed and installed in accordance with appropriate engineering practice. If required by the authority having jurisdiction, design documentation shall be provided.

110.71 Strength. Manholes, vaults, and their means of access shall be designed under qualified engineering supervision and shall withstand all loads likely to be imposed on the structures.

Informational Note: See ANSI C2-2007, *National Electrical Safety Code,* for additional information on the loading that can be expected to bear on underground enclosures.

Δ **110.72 Cabling Work Space.** A clear work space not less than 900 mm (3 ft) wide shall be provided where cables are located on both sides, and not less than 750 mm (2½ ft) where cables are only on one side. The vertical headroom shall be not less than 1.8 m (6 ft) unless the opening is within 300 mm (1 ft), measured horizontally, of the adjacent interior side wall of the enclosure.

Exception: A manhole containing only one or more of the following shall be permitted to have one of the horizontal work space dimensions reduced to 600 mm (2 ft) where the other horizontal clear work space is increased so the sum of the two dimensions is not less than 1.8 m (6 ft):

(1) Optical fiber cables
(2) Power-limited fire alarm circuits supplied in accordance with 760.121
(3) Class 2 or Class 3 remote-control and signaling circuits, or both, supplied in accordance with 725.60

110.73 Equipment Work Space. Where electrical equipment with live parts that is likely to require examination, adjustment, servicing, or maintenance while energized is installed in a manhole, vault, or other enclosure designed for personnel access, the work space and associated requirements in 110.26 shall be met for installations operating at 1000 volts or less. Where the installation is over 1000 volts, the work space and associated requirements in 110.34 shall be met. A manhole access cover that weighs over 45.4 kg (100 lb) shall be considered as meeting the requirements of 110.26(F) and 110.34(C).

110.74 Conductor Installation. Conductors installed in manholes and other enclosures intended for personnel entry shall be cabled, racked up, or arranged in an approved manner that provides ready and safe access for persons to enter for installation and maintenance. The installation shall comply with 110.74(A) or 110.74(B), as applicable.

Δ **(A) 1000 Volts, Nominal, or Less.** Wire bending space for conductors operating at 1000 volts or less shall be provided in accordance with 314.28.

(B) Over 1000 Volts, Nominal. Conductors operating at over 1000 volts shall be provided with bending space in accordance with 314.71(A) and (B), as applicable.

Exception: Where 314.71(B) applies, each row or column of ducts on one wall of the enclosure shall be calculated individually, and the single row or column that provides the maximum distance shall be used.

110.75 Access to Manholes.

Δ **(A) Dimensions.** Rectangular access openings shall not be less than 650 mm × 550 mm (26 in. × 22 in.). Round access openings in a manhole shall be not less than 650 mm (26 in.) in diameter.

Exception: A manhole that has a fixed ladder that does not obstruct the opening or that contains only one or more of the following shall be permitted to reduce the minimum cover diameter to 600 mm (2 ft):

(1) Optical fiber cables
(2) Power-limited fire alarm circuits supplied in accordance with 760.121
(3) Class 2 or Class 3 remote-control and signaling circuits, or both, supplied in accordance with 725.60

(B) Obstructions. Manhole openings shall be free of protrusions that could injure personnel or prevent ready egress.

(C) Location. Manhole openings for personnel shall be located where they are not directly above electrical equipment or conductors in the enclosure. Where this is not practicable, either a protective barrier or a fixed ladder shall be provided.

(D) Covers. Covers shall be over 45 kg (100 lb) or otherwise designed to require the use of tools to open. They shall be designed or restrained so they cannot fall into the manhole or protrude sufficiently to contact electrical conductors or equipment within the manhole.

(E) Marking. Manhole covers shall have an identifying mark or logo that prominently indicates their function, such as "electric."

110.76 Access to Vaults and Tunnels.

(A) Location. Access openings for personnel shall be located where they are not directly above electrical equipment or conductors in the enclosure. Other openings shall be permitted over equipment to facilitate installation, maintenance, or replacement of equipment.

(B) Locks. In addition to compliance with the requirements of 110.34, if applicable, access openings for personnel shall be arranged such that a person on the inside can exit when the access door is locked from the outside, or in the case of normally locking by padlock, the locking arrangement shall be such that the padlock can be closed on the locking system to prevent locking from the outside.

110.77 Ventilation. Where manholes, tunnels, and vaults have communicating openings into enclosed areas used by the public, ventilation to open air shall be provided wherever practicable.

110.78 Guarding. Where conductors or equipment, or both, could be contacted by objects falling or being pushed through a ventilating grating, both conductors and live parts shall be protected in accordance with the requirements of 110.27(A)(2) or 110.31(B)(1), depending on the voltage.

110.79 Fixed Ladders. Fixed ladders shall be corrosion resistant.

Wiring and Protection

CHAPTER 2

ARTICLE 200 — Use and Identification of Grounded Conductors

200.1 Scope. This article provides requirements for the following:

(1) Identification of terminals
(2) Grounded conductors in premises wiring systems
(3) Identification of grounded conductors

Informational Note: See Article 100 for definitions of *Grounded Conductor, Equipment Grounding Conductor,* and *Grounding Electrode Conductor.*

200.2 General. Grounded conductors shall comply with 200.2(A) and (B).

(A) Insulation. The grounded conductor, if insulated, shall have insulation that complies with either one of the following:

(1) Is suitably rated, other than color, for any ungrounded conductor of the same circuit for systems of 1000 volts or less.
(2) Is rated not less than 600 volts for solidly grounded neutral systems of over 1000 volts in accordance with 250.184(A)

(B) Continuity. The continuity of a grounded conductor shall not depend on a connection to a metal enclosure, raceway, or cable armor.

Informational Note: See 300.13(B) for the continuity of grounded conductors used in multiwire branch circuits.

Grounded conductors are required to be connected to a terminal or busbar that is specifically intended and identified for connection of grounded or neutral conductors. Because grounded conductors are current carrying, connecting them to a separate equipment grounding terminal or bar (i.e., directly connected to a metal cabinet or enclosure) results in the enclosure becoming a neutral conductor between the equipment grounding terminal and the point of connection for the grounded conductor.

See also

300.13(B), which does not permit the wiring terminals of a device, such as a receptacle, to be the means of maintaining the continuity of the grounded conductor in a multiwire branch circuit

200.3 Connection to Grounded System. Grounded conductors of premises wiring systems shall be electrically connected to the supply system grounded conductor to ensure a common, continuous grounded system. For the purpose of this section, *electrically connected* shall mean making a direct electrical connection capable of carrying current, as distinguished from induced currents.

Exception: Listed interactive inverters identified for use in distributed resource generation systems such as photovoltaic and fuel cell power systems shall be permitted to be connected to premises wiring without a grounded conductor if the connected premises wiring or utility system includes a grounded conductor.

200.4 Neutral Conductors. Neutral conductors shall be installed in accordance with 200.4(A) and (B).

The grounded conductor is often, but not always, the neutral conductor. A neutral conductor is one that is connected to the neutral point of an electrical system. Some electrical systems do not have a neutral conductor. For example, in a 3-phase, corner-grounded delta system, the intentionally grounded conductor is not a neutral conductor, because it is not connected to a system neutral point. Because the conductor is connected to the same grounding electrode system as the non–current-carrying metal parts of the electrical equipment, generally no potential difference exists between the grounded conductor and those grounded metal parts. However, unlike an equipment grounding conductor, the grounded conductor is a current-carrying circuit conductor. Whether it is referred to as the grounded conductor or as the neutral conductor, the white or gray marking on a circuit conductor indicates that it is intentionally connected to the earth.

Electric shock injuries and electrocutions have occurred as a result of working on the grounded conductor while the circuit is energized. Extreme caution must be exercised where the grounded (neutral) conductor is part of a multiwire branch circuit.

See also

210.4(B) for disconnecting means for multiwire branch circuits

Article 250 for more on the use and installation of grounded conductors

(A) Installation. Neutral conductors shall not be used for more than one branch circuit, for more than one multiwire branch circuit, or for more than one set of ungrounded feeder conductors unless specifically permitted elsewhere in this *Code*.

> Informational Note: See 215.4 for information on common neutrals.

(B) Multiple Circuits. Where more than one neutral conductor associated with different circuits is in an enclosure, grounded circuit conductors of each circuit shall be identified or grouped to correspond with the ungrounded circuit conductor(s) by wire markers, cable ties, or similar means in at least one location within the enclosure.

Exception No. 1: The requirement for grouping or identifying shall not apply if the branch-circuit or feeder conductors enter from a cable or a raceway unique to the circuit that makes the grouping obvious.

Exception No. 2: The requirement for grouping or identifying shall not apply where branch-circuit conductors pass through a box or conduit body without a loop as described in 314.16(B)(1) or without a splice or termination.

A neutral conductor is not required to be installed with the ungrounded circuit conductors if line-to-neutral loads are not supplied by the branch circuit or feeder. However, each multiwire circuit must have its own neutral conductor.

The concern with multiple circuits sharing a neutral conductor is the possibility of overloading the neutral conductor. Typically, an overcurrent protective device is not present, nor is it required to be installed in series with the neutral conductor. Therefore, an overload condition can exist without being detected or responded to, and insulation damage can occur. Because of the electrical relationship between the ungrounded conductors and the neutral conductor in multiwire single-phase and 3-phase circuits, the possibility of overload does not exist except where the load characteristics result in additive harmonic currents in the neutral conductor.

Where multiple circuits enter an enclosure, each circuit is required to have all its conductors grouped together in at least one location to make it obvious which conductors are associated with the circuit. This grouping is not necessary if the conductors enter the enclosure as a single cable.

200.6 Means of Identifying Grounded Conductors.

Δ **(A) Sizes 6 AWG or Smaller.** The insulation of grounded conductors of 6 AWG or smaller shall be identified by one of the following means:

(1) A continuous white outer finish.

(2) A continuous gray outer finish.

(3) Three continuous white or gray stripes along the conductor's entire length on other than green insulation.

(4) Conductors with white or gray insulation and colored tracer threads in the braid identifying the source of manufacture.

(5) A single-conductor, sunlight-resistant, outdoor-rated cable used as a solidly grounded conductor in photovoltaic power systems, as permitted by 690.31(C)(1), shall be identified at the time of installation by markings at terminations in accordance with 200.6(A)(1) through (A)(4).

(6) The grounded conductor of a mineral-insulated, metal-sheathed cable (Type MI) shall be identified at the time of installation by a distinctive white or gray marking at its terminations. The marking shall encircle the conductor insulation.

(7) Fixture wire shall comply with the requirements for grounded conductor identification in accordance with 402.8.

(8) For aerial cable, the identification shall comply with one of the methods in 200.6(A)(1) through (A)(5), or by means of a ridge located on the exterior of the cable so as to identify it.

Δ **(B) Sizes 4 AWG or Larger.** An insulated grounded conductor 4 AWG or larger shall be identified by one of the following means:

(1) A continuous white outer finish.

(2) A continuous gray outer finish.

(3) Three continuous white or gray stripes along the entire length on other than green insulation.

(4) At the time of installation, be identified by a distinctive white or gray marking at its terminations. This marking shall encircle the conductor insulation.

The most common method used to identify a single conductor as a grounded conductor is a white or gray marking to the insulation at all termination points at the time of installation. This field-applied white or gray marking must completely encircle the conductor insulation so that it is clearly visible. The marking can be applied by using tape or by painting the insulation. See Exhibit 200.1.

(C) Flexible Cords. An insulated conductor that is intended for use as a grounded conductor, where contained within a flexible cord, shall be identified by a white or gray outer finish or by methods permitted by 400.22.

Δ **(D) Grounded Conductors of Different Nominal Voltage Systems.** If grounded conductors of different nominal voltage systems are installed in the same raceway, cable, box, auxiliary

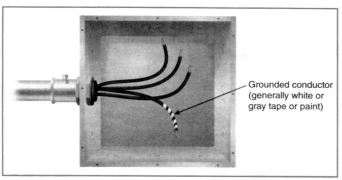

EXHIBIT 200.1 Field-applied identification of a 4 AWG conductor to identify it as the grounded conductor.

Grounded conductor (generally white or gray tape or paint)

gutter, or other type of enclosure, each grounded conductor shall be identified by nominal voltage system. Identification that distinguishes each nominal voltage system grounded conductor shall be permitted by one of the following means:

(1) One nominal voltage system grounded conductor shall have an outer covering conforming to 200.6(A) or (B).
(2) The grounded conductor(s) of other nominal voltage systems shall have a different outer covering conforming to 200.6(A) or (B) or by an outer covering of white or gray with a distinguishable colored stripe other than green running along the insulation.
(3) Other and different means of identification allowed by 200.6(A) or (B) shall distinguish each nominal voltage system grounded conductor.

The means of identification shall be documented in a manner that is readily available or shall be permanently posted where the conductors of different nominal voltage systems originate.

This requirement applies only where grounded conductors of different systems are installed in a common enclosure, such as a junction or pull box or a wireway. Exhibit 200.2 illustrates an enclosure containing grounded conductors of two different systems

that are distinguished from each other by color, which is one of the methods specified by 200.6(D). Gray and white colored insulation on conductors is recognized by the *NEC*® as two separate means of identifying grounded conductors. The industry practice of using white for lower-voltage systems and gray for higher-voltage systems is permitted but not mandated by the *NEC*.

Identifying the grounded conductors associated with one system from the grounded conductors of another system helps to ensure proper system connections. As more systems with grounded conductors are installed in a common enclosure, more means to distinguish each system grounded conductor from the other system grounded conductor in that enclosure become necessary. The only acceptable means of identifying grounded conductors are those specified in 200.6(A) and (B).

The means of identification must be included in readily accessible documentation or be posted where the conductors originate. Exhibit 200.3 is an example of a permanent label posted at electrical distribution equipment, indicating the identification scheme for the grounded conductors of each nominal voltage system.

(E) Grounded Conductors of Multiconductor Cables. The insulated grounded conductor(s) in a multiconductor cable shall be identified by a continuous white or gray outer finish or by three continuous white or gray stripes on other than green insulation along its entire length. For conductors that are 4 AWG or larger in cables, identification of the grounded conductor shall be permitted to comply with 200.6(B). For multiconductor flat cable with conductors that are 4 AWG or larger, an external ridge shall be permitted to identify the grounded conductor.

Exception No. 1: Conductors within multiconductor cables shall be permitted to be re-identified at their terminations at the time of installation by a distinctive white or gray marking or other equally effective means.

Exception No. 2: The grounded conductor of a multiconductor varnished-cloth-insulated cable shall be permitted to be identified at its terminations at the time of installation by a distinctive white marking or other equally effective means.

Informational Note: The color gray may have been used in the past as an ungrounded conductor. Care should be taken when working on existing systems.

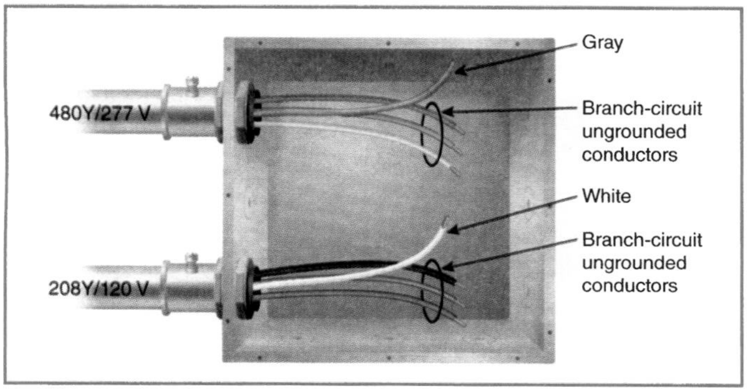

EXHIBIT 200.2 Grounded conductors of different systems in the same enclosure.

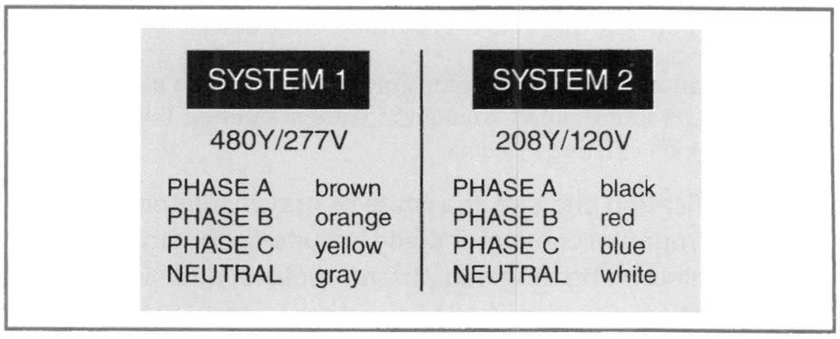

480Y/277 V
208Y/120 V

Gray
Branch-circuit ungrounded conductors
White
Branch-circuit ungrounded conductors

SYSTEM 1		SYSTEM 2	
480Y/277V		208Y/120V	
PHASE A	brown	PHASE A	black
PHASE B	orange	PHASE B	red
PHASE C	yellow	PHASE C	blue
NEUTRAL	gray	NEUTRAL	white

EXHIBIT 200.3 A label on a piece of equipment containing conductors of two nominal voltage systems that provides the required identification of the two systems.

All shades of gray insulation and marking are reserved for grounded conductors. This Informational Note warns that some existing systems have used gray to identify ungrounded conductors in those systems.

200.7 Use of Insulation of a White or Gray Color or with Three Continuous White or Gray Stripes.

(A) General. The following shall be used only for the grounded circuit conductor, unless otherwise permitted in 200.7(B) and (C):

(1) A conductor with continuous white or gray covering

(2) A conductor with three continuous white or gray stripes on other than green insulation

(3) A marking of white or gray color at the termination

(B) Circuits of Less Than 50 Volts. A conductor with white or gray color insulation or three continuous white stripes or having a marking of white or gray at the termination for circuits of less than 50 volts shall be required to be grounded only in accordance with 250.20(A).

(C) Circuits of 50 Volts or More. The use of insulation that is white or gray or that has three continuous white or gray stripes for other than a grounded conductor for circuits of 50 volts or more shall be permitted only as in the following:

(1) If part of a cable assembly that has the insulation permanently reidentified to indicate its use as an ungrounded conductor by marking tape, painting, or other effective means at its termination and at each location where the conductor is visible and accessible. Identification shall encircle the insulation and shall be a color other than white, gray, or green. If used for single-pole, 3-way or 4-way switch loops, the reidentified conductor with white or gray insulation or three continuous white or gray stripes shall be used only for the supply to the switch, but not as a return conductor from the switch to the outlet.

(2) A flexible cord having one conductor identified by a white or gray outer finish or three continuous white or gray stripes, or by any other means in accordance with 400.22, that is used for connecting an appliance or equipment in accordance with 400.10. This shall apply to flexible cords connected to outlets whether or not the outlet is supplied by a circuit that has a grounded conductor.

> Informational Note: The color gray may have been used in the past as an ungrounded conductor. Care should be taken when working on existing systems.

Re-identification and use of a white or gray insulated conductor as an ungrounded conductor could include conductors for water heaters, electric heating equipment, motors, and switch loops. The insulation is re-identified to avoid confusing it with grounded conductors having white or gray insulation or neutral conductors at switch and outlet points and at other points in the wiring system where the conductors are accessible.

In previous editions, the *NEC* permitted switch loops using a white insulated conductor to serve as an ungrounded conductor supplying the switch but not as the return ungrounded conductor to supply the lighting outlet. Re-identification of a white conductor used for this purpose was not required. However, electronic switching devices with small power supplies are available that can be installed at switch locations. These devices require a grounded conductor in order to power the internal components. Although in switch loops the conductor with white or gray insulation is re-identified at all accessible locations to indicate that it is an ungrounded conductor, it is permitted to supply only the switch and cannot be the switched ungrounded conductor to the controlled outlet.

Exhibit 200.4 shows a switch location where the conductor with the white insulation is re-identified with red tape at the termination to indicate that it is the ungrounded supply conductor run to the switch.

△ **200.9 Means of Identification of Terminals.** In devices or utilization equipment with polarized connections, identification of terminals to which a grounded conductor is to be connected shall be white or silver in color. The identification of other terminals shall be of a distinguishable different color.

Exception: If the conditions of maintenance and supervision ensure that only qualified persons service the installations, terminals for grounded conductors shall be permitted to be permanently identified at the time of installation by a distinctive white marking or other equally effective means.

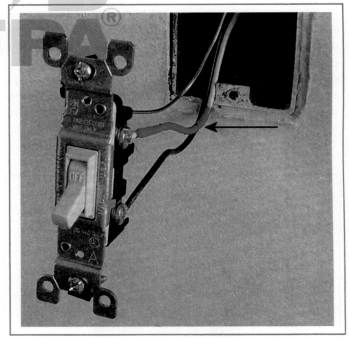

EXHIBIT 200.4 The re-identified conductor at a switch location, limited to use only as the ungrounded conductor that supplies the switch.

200.10 Identification of Terminals.

△ **(A) Device Terminals.** All devices, excluding panelboards, provided with terminals for the attachment of conductors and intended for connection to more than one side of the circuit shall have terminals marked for identification, unless the electrical connection of the terminal intended to be connected to the grounded conductor is clearly evident.

Exception: Terminal identification shall not be required for devices that have a current rating of over 30 amperes, other than polarized attachment plugs and polarized receptacles for attachment plugs in accordance with 200.10(B).

△ **(B) Receptacles, Plugs, and Connectors.** Receptacles, polarized attachment plugs, and cord connectors for plugs and polarized plugs shall have the terminal intended for connection to the grounded conductor identified as follows:

(1) Identification shall be by a metal or metal coating that is white or silver in color or by the word "white" or the letter "W" located adjacent to the identified terminal.
(2) If the terminal is not visible, the conductor entrance hole for the connection shall be colored white or marked with the word "white" or the letter "W."

Informational Note: See 250.126 for identification of wiring device equipment grounding conductor terminals.

(C) Screw Shells. For devices with screw shells, the terminal for the grounded conductor shall be the one connected to the screw shell.

(D) Screw Shell Devices with Leads. For screw shell devices with attached leads, the conductor attached to the screw shell shall have a white or gray finish. The outer finish of the other conductor shall be of a solid color that will not be confused with the white or gray finish used to identify the grounded conductor.

Informational Note: The color gray may have been used in the past as an ungrounded conductor. Caution should be taken when working on existing systems.

(E) Appliances. Appliances that have a single-pole switch or a single-pole overcurrent device in the line or any line-connected screw shell lampholders, and that are to be connected by (1) a permanent wiring method or (2) field-installed attachment plugs and cords with three or more wires (including the equipment grounding conductor), shall have means to identify the terminal for the grounded circuit conductor (if any).

200.11 Polarity of Connections. No grounded conductor shall be attached to any terminal or lead so as to reverse the designated polarity.

ARTICLE 210 — Branch Circuits Not Over 1000 Volts ac, 1500 Volts dc, Nominal

△ **Part I. General**

210.1 Scope. This article provides the general requirements for branch circuits not over 1000 volts ac, 1500 volts dc, nominal.

Informational Note: See Part II of Article 235 for requirements for branch circuits over 1000 volts ac, 1500 volts dc, nominal.

△ **210.2 Reconditioned Equipment.** The following shall not be reconditioned:

(1) Equipment that provides ground-fault circuit-interrupter protection for personnel
(2) Equipment that provides arc-fault circuit-interrupter protection

210.3 Other Articles for Specific-Purpose Branch Circuits. Table 210.3 lists references for specific equipment and applications not located in Chapters 5, 6, and 7 that amend or supplement the requirements of this article.

△ **TABLE 210.3** *Specific-Purpose Branch Circuits*

Equipment	Article	Section
Air-conditioning and refrigerating equipment		440.6, 440.31, and 440.32
Busways		368.17
Central heating equipment other than fixed electric space-heating equipment		422.12
Fixed electric heating equipment for pipelines and vessels		427.4
Fixed electric space-heating equipment		424.4
Fixed outdoor electrical deicing and snow-melting equipment		426.4
Infrared lamp industrial heating equipment		422.48 and 424.3
Motors, motor circuits, and controllers	430	
Switchboards and panelboards		408.52

210.4 Multiwire Branch Circuits.

(A) General. Branch circuits recognized by this article shall be permitted as multiwire circuits. A multiwire circuit shall be permitted to be considered as multiple circuits. Except as permitted in 300.3(B)(4), all conductors of a multiwire branch circuit shall originate from the equipment containing the branch-circuit overcurrent protective device or protective devices.

Using a multiwire branch circuit as multiple circuits satisfies the requirement for providing two small appliance branch circuits for countertop receptacle outlets in a dwelling unit kitchen.

> Informational Note No. 1: A 3-phase, 4-wire, wye-connected power system used to supply power to nonlinear loads might necessitate that the power system design allow for the possibility of high harmonic currents on the neutral conductor.
> Informational Note No. 2: See 300.13(B) for continuity of grounded conductors on multiwire circuits.

Power supplies for equipment such as computers, printers, electric-discharge lighting systems, and adjustable-speed motor drives can introduce harmonic currents in the system neutral conductor. The resulting total harmonic distortion current could exceed the load current of the device itself.

See also

310.15(E) commentary for information on neutral conductor ampacity

(B) Disconnecting Means. Each multiwire branch circuit shall be provided with a means that will simultaneously disconnect all ungrounded conductors at the point where the branch circuit originates.

> Informational Note: See 240.15(B) for information on the use of single-pole circuit breakers as the disconnecting means.

A simultaneous disconnecting means reduces the risk of shock to personnel working on equipment supplied by the multiwire branch circuit. In former editions of the *NEC®*, this requirement applied only where the multiwire branch circuit supplied equipment mounted to a common yoke or strap.

For a single-phase installation, the disconnecting means could be two single-pole circuit breakers with an identified handle tie or a 2-pole circuit breaker, as shown in Exhibit 210.1. For a 3-phase installation, a 3-pole circuit breaker, three single-pole circuit breakers with an identified handle tie, or a 3-pole switch

with branch circuit overcurrent protection provides the required simultaneous opening of the ungrounded conductors.

(C) Line-to-Neutral Loads. Multiwire branch circuits shall supply only line-to-neutral loads.

Exception No. 1: A multiwire branch circuit that supplies only one utilization equipment shall be permitted to supply line-to-line loads.

Exception No. 2: A multiwire branch circuit shall be permitted to supply line-to-line loads if all ungrounded conductors of the multiwire branch circuit are opened simultaneously by the branch-circuit overcurrent device.

A multiwire branch circuit is a branch circuit that consists of two or more ungrounded conductors that have a voltage between them, and a grounded conductor that has equal voltage between it and each ungrounded conductor of the circuit and that is connected to the neutral or grounded conductor of the system. The most commonly used multiwire branch circuit consists of two ungrounded conductors and one grounded conductor supplied from a 120/240-volt, single-phase, 3-wire system. Such multiwire circuits supply appliances that have both line-to-line and line-to-neutral connected loads, such as electric ranges and clothes dryers, or supply loads that are line-to-neutral connected only, such as the split-wired combination device shown in Exhibit 210.1 (bottom). A single-phase, 120/240-volt multiwire branch circuit is also permitted to supply a device with a 250-volt rated receptacle (line-to-line) and a 125-volt rated receptacle (line-to-neutral), as shown in Exhibit 210.2, provided the branch-circuit overcurrent device simultaneously opens both of the ungrounded conductors.

Multiwire branch circuits have many advantages, including using three wires to do the work of four (in place of two 2-wire circuits), less raceway fill, easier balancing and phasing of a system, and less voltage drop. Multiwire branch circuits can be derived from a 120/240-volt, single-phase system; a 208Y/120-volt and 480Y/277-volt, 3-phase, 4-wire system; or a 240/120-volt, 3-phase, 4-wire delta system. If two ungrounded conductors and a common neutral of a multiwire branch circuit are supplied from a 208Y/120-volt, 3-phase, 4-wire system, the neutral carries the same current as the phase conductor with the highest current and, therefore, should be the same size.

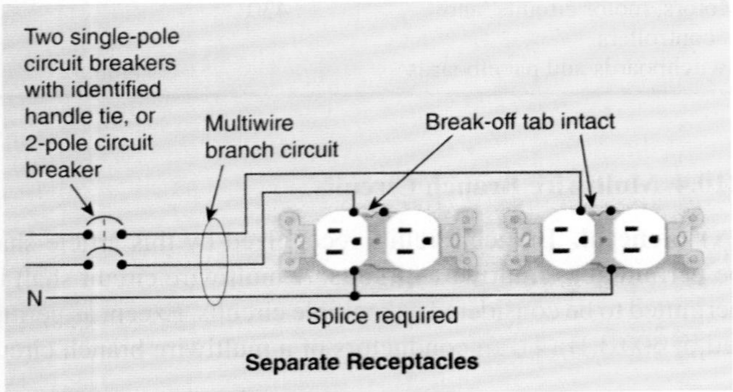

EXHIBIT 210.1 An example where 210.4(B) requires the simultaneous disconnection of all ungrounded conductors in a multiwire branch circuit.

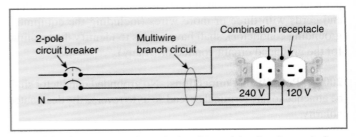

EXHIBIT 210.2 A multiwire branch circuit supplying line-to-neutral and line-to-line connected loads, provided the ungrounded conductors can be opened simultaneously by the branch-circuit overcurrent device.

If loads are connected line-to-line (i.e., utilization equipment connected between 2 or 3 phases), 2-pole or 3-pole circuit breakers are required to disconnect all ungrounded conductors simultaneously. Use of single-pole circuit breakers with handle ties for 240-volt equipment might not open all ungrounded conductors.

See also

210.19 and its commentary for more information on voltage drop for branch circuits

210.4(A) and its commentary for more information on 3-phase, 4-wire system neutral conductors

240.15(B)(1) for more information on circuit breaker overcurrent protection of ungrounded conductors on multiwire branch circuits

300.13(B) and its commentary for hazards associated with device removal on multiwire branch circuits

(D) Grouping. The ungrounded and grounded circuit conductors of each multiwire branch circuit shall be grouped in accordance with 200.4(B).

210.5 Identification for Branch Circuits.

(A) Grounded Conductor. The grounded conductor of a branch circuit shall be identified in accordance with 200.6.

(B) Equipment Grounding Conductor. The equipment grounding conductor shall be identified in accordance with 250.119.

(C) Identification of Ungrounded Conductors. Ungrounded conductors shall be identified in accordance with 210.5(C)(1) or (2), as applicable.

Δ **(1) Branch Circuits Supplied from More Than One Nominal Voltage System.** Where the premises wiring system has branch circuits supplied from more than one nominal voltage system, each ungrounded conductor of a branch circuit shall be identified by phase or line and by nominal voltage system at all termination, connection, and splice points in accordance with 210.5(C)(1)(a) and (C)(1)(b). Different systems within the same premises that have the same nominal voltage shall be permitted to use the same identification.

The requirement for the identification of ungrounded branch-circuit conductors covers all branch-circuit configurations. The identification requirement applies only where more than one nominal voltage system supplies branch circuits (e.g., a 208Y/120-volt system and a 480Y/277-volt system) and requires that the conductors be identified by system and phase.

Different systems within the same premises having the same system voltage class do not have to be identified differently. For example, in a building with five transformers on different floors supplying multiple 208Y/120-volt, three-phase, 4-wire separately derived systems to supply lighting and receptacle loads, the branch circuits conductors supplied from the five different systems having the same voltage characteristics can be identified using a common conductor identification system or

scheme throughout the building. Unlike the requirement of 200.6(D) for identifying the grounded conductors supplied from different voltage systems, application of this requirement is not predicated on the different system conductors sharing a common raceway, cabinet, or enclosure.

(a) *Means of Identification.* The means of identification shall be permitted to be by separate color coding, marking tape, tagging, or other approved means.

(b) *Posting of Identification Means.* The method used for conductors originating within each branch-circuit panelboard or similar branch-circuit distribution equipment shall be documented in a manner that is readily available or shall be permanently posted at each branch-circuit panelboard or similar branch-circuit distribution equipment. The label shall be of sufficient durability to withstand the environment involved and shall not be handwritten.

Although color coding is a popular method, other types of marking or tagging are acceptable alternatives. Whatever identification method is used, it must be consistent throughout the premises. This affords a higher level of safety for personnel working on premises electrical systems having ungrounded conductors supplied from multiple nominal voltage systems. If the identification legend is posted at electrical distribution equipment, the marking has to describe only the identification scheme for the ungrounded conductors supplied from that particular equipment.

Exhibit 210.3 shows an example of two different nominal voltage systems in a building. Each ungrounded system conductor is identified by color-coded marking tape. A label indicating the means of the identification is permanently located at each

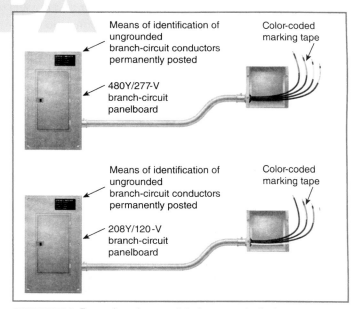

EXHIBIT 210.3 Examples of accessible (ungrounded) phase conductors identified by marking tape at a junction or outlet location where the conductors will be spliced or terminated.

panelboard. Although there are commonly employed color schemes used in the electrical industry, the *NEC* only specifies color designations of ungrounded conductors for a few applications including heating cables, intrinsically safe circuits, isolated systems in health care facilities, or 4-wire delta (high-leg) systems.

See also

210.5(C)(2) and its commentary for identification of direct current (dc) systems

Exception: In existing installations where a voltage system(s) already exists and a different voltage system is being added, it shall be permissible to mark only the new system voltage. Existing unidentified systems shall not be required to be identified at each termination, connection, and splice point in accordance with 210.5(C)(1)(a) and (C)(1)(b). Labeling shall be required at each voltage system distribution equipment to identify that only one voltage system has been marked for a new system(s). The new system label(s) shall include the words "other unidentified systems exist on the premises."

Systems installed prior to the adoption of the 2005 *NEC* were not required to be identified. The *NEC* is not retroactive, so only new installations would require identification. However, if existing installations are not marked, this section requires the new installation labeling to include a warning indicating there are existing unidentified systems.

(2) Branch Circuits Supplied from Direct-Current Systems. Where a branch circuit is supplied from a dc system operating at more than 60 volts, each ungrounded conductor of 4 AWG or larger shall be identified by polarity at all termination, connection, and splice points by marking tape, tagging, or other approved means; each ungrounded conductor of 6 AWG or smaller shall be identified by polarity at all termination, connection, and splice points in compliance with 210.5(C)(2)(a) and (b). The identification methods utilized for conductors originating within each branch-circuit panelboard or similar branch-circuit distribution equipment shall be documented in a manner that is readily available or shall be permanently posted at each branch-circuit panelboard or similar branch-circuit distribution equipment.

(a) *Positive Polarity, Sizes 6 AWG or Smaller.* Where the positive polarity of a dc system does not serve as the connection point for the grounded conductor, each positive ungrounded conductor shall be identified by one of the following means:

(1) A continuous red outer finish
(2) A continuous red stripe durably marked along the conductor's entire length on insulation of a color other than green, white, gray, or black
(3) Imprinted plus signs (+) or the word POSITIVE or POS durably marked on insulation of a color other than green, white, gray, or black and repeated at intervals not exceeding 610 mm (24 in.) in accordance with 310.8(B)

(4) An approved permanent marking means such as sleeving or shrink-tubing that is suitable for the conductor size, at all termination, connection, and splice points, with imprinted plus signs (+) or the word POSITIVE or POS durably marked on insulation of a color other than green, white, gray, or black

(b) *Negative Polarity, Sizes 6 AWG or Smaller.* Where the negative polarity of a dc system does not serve as the connection point for the grounded conductor, each negative ungrounded conductor shall be identified by one of the following means:

(1) A continuous black outer finish
(2) A continuous black stripe durably marked along the conductor's entire length on insulation of a color other than green, white, gray, or red
(3) Imprinted minus signs (–) or the word NEGATIVE or NEG durably marked on insulation of a color other than green, white, gray, or red and repeated at intervals not exceeding 610 mm (24 in.) in accordance with 310.8(B)
(4) An approved permanent marking means such as sleeving or shrink-tubing that is suitable for the conductor size, at all termination, connection, and splice points, with imprinted minus signs (–) or the word NEGATIVE or NEG durably marked on insulation of a color other than green, white, gray, or red

The requirements for the identification of ungrounded conductors are based on whether the system is negatively or positively grounded. Negatively grounded and positively grounded 2-wire dc systems must employ a grounded conductor identified in accordance with 200.6. Grounded conductors of negatively grounded and positively grounded 2-wire dc systems are also required to be identifiable and distinguishable from ungrounded conductors of 3-wire dc systems and of 2-wire dc systems employing high-impedance references to ground. The identification method must be documented and readily available, or it must be posted at each branch-circuit panelboard or branch-circuit distribution equipment.

210.6 Branch-Circuit Voltage Limitations. The nominal voltage of branch circuits shall not exceed the values permitted by 210.6(A) through (D).

(A) Occupancy Limitation. In dwelling units and guest rooms or guest suites of hotels, motels, and similar occupancies, the voltage shall not exceed 120 volts, nominal, between conductors that supply the terminals of the following:

(1) Luminaires
(2) Cord-and-plug-connected loads 1440 volt-amperes, nominal, or less or less than ¼ hp

The term *similar occupancies* in 210.6(A) refers to sleeping rooms in dormitories, fraternities, sororities, nursing homes, and other such facilities. This requirement is intended to reduce the exposure of residents in these types of occupancies to electric shock hazards when using or servicing permanently installed luminaires and cord-and-plug-connected portable lamps and appliances.

Small loads, such as those of 1440 volt-amperes or less and motors of less than 1/4 horsepower, are limited to 120-volt circuits. High-wattage cord-and-plug-connected loads, such as electric ranges, clothes dryers, and some window air conditioners, could be connected to a 208-volt or 240-volt circuit.

(B) 120 Volts Between Conductors. Circuits not exceeding 120 volts, nominal, between conductors shall be permitted to supply the following:

(1) The terminals of lampholders applied within their voltage ratings
(2) Auxiliary equipment of electric-discharge lamps

Informational Note: See 410.137 for auxiliary equipment limitations.

(3) Cord-and-plug-connected or permanently connected utilization equipment

Auxiliary equipment includes ballasts and starting devices for fluorescent and high-intensity-discharge (e.g., mercury vapor, metal halide, and sodium) lamps.

(C) 277 Volts to Ground. Circuits exceeding 120 volts, nominal, between conductors and not exceeding 277 volts, nominal, to ground shall be permitted to supply cord-and-plug-connected or permanently connected utilization equipment, or the following types of listed luminaires:

(1) Electric-discharge luminaires with integral ballasts
(2) LED luminaires with LED drivers between the branch circuit and the lampholders
(3) Incandescent or LED luminaires, equipped with medium-base or smaller screw shell lampholders, where the lampholders are supplied at 120 volts or less from the output of a stepdown autotransformer, LED driver, or other type of power supply that is an integral component of the luminaire

Informational Note No. 1: See 410.90 for requirements regarding the connection of screw shell lampholders to grounded conductors.

An incandescent luminaire is permitted on a 277-volt circuit only if it is a listed luminaire with an integral autotransformer and an output to the lampholder that does not exceed 120 volts. In this application, the autotransformer supplies 120 volts to the lampholder, and the grounded conductor is connected to the screw shell of the lampholder. Listed light-emitting diode (LED) luminaires supplied by an integral LED driver with 120-volt output can be supplied by 277-volt circuits. This application is similar to a branch circuit derived from an autotransformer, except that the 120-volt circuit is the internal wiring of the luminaire.

(4) Luminaires equipped with mogul-base screw shell lampholders
(5) Luminaires equipped with lampholders, other than the screw shell type, when used within the voltage ratings of their lampholders
(6) Luminaires without lampholders

Informational Note No. 2: Luminaires with nonserviceable LEDs are examples of luminaires without lampholders.

(7) Auxiliary equipment of electric-discharge or LED-type lamps

Informational Note No. 3: See 410.137 for auxiliary equipment limitations.

(8) Luminaires converted with listed retrofit kits incorporating integral LED light sources or accepting LED lamps that also conform with 210.6(C)(1), (C)(2), (C)(3), (C)(4), or (C)(5)

Exhibit 210.4 shows some examples of luminaires permitted to be connected to branch circuits. Medium-base screw shell lampholders must not be directly connected to 277-volt branch circuits. Other types of lampholders may be connected to 277-volt circuits but only if the lampholders have a 277-volt rating. A 277-volt branch circuit may be connected to a listed electric-discharge luminaire or to a listed autotransformer-type incandescent luminaire with a medium-base screw shell lampholder.

Typical examples of cord-and-plug-connected equipment include through-the-wall heating and air-conditioning units and

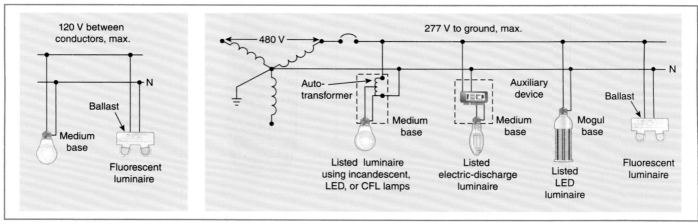

EXHIBIT 210.4 Examples of luminaires permitted to be connected to branch circuits.

restaurant deep fat fryers that operate at 480 volts, 3 phase, from a grounded wye system.

Luminaires listed for and connected to a 480-volt source may be used in applications permitted by 210.6(C), provided the 480-volt system is in fact a grounded wye system that contains a grounded conductor (thus limiting the system "voltage to ground" to the 277-volt level).

(D) 1000 Volts ac or 1500 Volts dc Between Conductors. Circuits exceeding 277 volts, nominal, to ground and not exceeding 1000 volts ac or 1500 volts dc, nominal, between conductors shall be permitted to supply the following:

(1) The auxiliary equipment of electric-discharge lamps mounted in permanently installed luminaires where the luminaires are mounted in accordance with one of the following:
   a. Not less than a height of 6.7 m (22 ft) on poles or similar structures for the illumination of outdoor areas such as highways, roads, bridges, athletic fields, or parking lots
   b. Not less than a height of 5.5 m (18 ft) on other structures such as tunnels

Informational Note: See 410.137 for auxiliary equipment limitations.

(2) Cord-and-plug-connected or permanently connected utilization equipment other than luminaires
(3) Luminaires powered from direct-current systems where either of the following apply:
   a. The luminaire contains a listed, dc-rated ballast that provides isolation between the dc power source and the lamp circuit and protection from electric shock when changing lamps.
   b. The luminaire contains a listed, dc-rated ballast and has no provision for changing lamps.

Exception No. 1 to (B), (C), and (D): For lampholders of infrared industrial heating appliances as provided in 425.14.

Exception No. 2 to (B), (C), and (D): For railway properties as described in 110.19.

210.7 Multiple Branch Circuits. If two or more branch circuits supply devices or equipment on the same yoke or mounting strap, a means to simultaneously disconnect the ungrounded supply conductors shall be provided at the point at which the branch circuits originate.

Δ **210.8 Ground-Fault Circuit-Interrupter Protection for Personnel.** A listed Class A GFCI shall provide protection in accordance with 210.8(A) through (F). The GFCI shall be installed in a readily accessible location.

Informational Note: See 215.9 for GFCI protection on feeders.

For the purposes of this section, the distance from receptacles shall be measured as the shortest path the power supply cord connected to the receptacle would follow without piercing a floor, wall, ceiling, or fixed barrier.

The Consumer Product Safety Commission (CPSC) has reported a decrease in the number of electrocutions in the United States since the introduction of ground-fault circuit-interrupter (GFCI) devices in the 1971 *NEC*. Most safety experts agree that GFCIs are directly responsible for saving numerous lives and preventing countless injuries.

Exhibit 210.5 shows a typical circuit arrangement of a GFCI. The line conductors are passed through a sensor and are connected to a shunt-trip device. As long as the current in the conductors is equal, the device remains in a closed position. If one of the conductors comes in contact with a grounded object, either directly or through a person's body, some of the current returns by the alternative path, resulting in an unbalanced current. The toroidal coil senses the unbalanced current, and the shunt-trip mechanism reacts to open the circuit. The circuit design does not require the presence of an equipment grounding conductor, which is the reason 406.4(D)(2)(c) permits the use of GFCIs as replacements for receptacles where a grounding means does not exist.

A variety of GFCIs are available, including portable and plug-in types and circuit-breaker types, types built into attachment plug caps, and receptacle types. Each type has a test switch so that units can be checked periodically to ensure proper operation. See Exhibits 210.6 and 210.7.

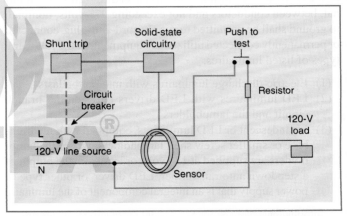

EXHIBIT 210.5 The circuitry and components of a typical GFCI.

EXHIBIT 210.6 A GFCI circuit breaker, which protects all outlets supplied by the branch circuit. *(Courtesy of Siemens)*

EXHIBIT 210.7 Two 20-ampere duplex receptacles with integral GFCI that may also protect downstream loads.

The manufacturer's installation and use instructions specify monthly testing. To facilitate this important ongoing safety check, GFCIs installed to protect the receptacles covered in 210.8(A) and (B) are required to be readily accessible.

GFCIs operate on fault currents of 4 to 6 milliamperes. At trip levels of 5 milliamperes (the instantaneous current could be much higher), a shock can be felt during the fault. The shock can lead to an involuntary reaction that could cause a secondary accident such as a fall. GFCIs do not protect persons from shock hazards where contact is made with the ungrounded (hot) and grounded (neutral) conductors or between different ungrounded (hot) conductors.

Although 210.7 is the main rule for GFCIs, other specific applications that require the use of GFCIs are listed in Commentary Table 210.1.

Δ **(A) Dwelling Units.** All 125-volt through 250-volt receptacles installed in the following locations and supplied by single-phase branch circuits rated 150 volts or less to ground shall have ground-fault circuit-interrupter protection for personnel:

(1) Bathrooms

All 125- through 250-volt, single-phase receptacles for cord-and-plug-connected equipment in bathrooms must have GFCI protection, including receptacles that are integral with luminaires. See commentary following 210.8, Exception to (1) through (4) for other types of receptacle, circuit, and equipment GFCI requirements.

The term *bathroom* (see Article 100) applies to the entire area, whether or not a separating door, as illustrated in Exhibit 210.8, is present. If they are adjacent and in close proximity, one receptacle

COMMENTARY TABLE 210.1 *Additional Requirements for the Application of GFCI Protection*

Location	Applicable Section(s)
Aircraft hangars	513.12
Appliances	422.5(A)
Audio system equipment	640.10(A)
Automotive vacuum machines	422.5(A)
Boathouses	555.35
Carnivals, circuses, fairs, and similar events	525.23(A)
Commercial garages	511.12
Dishwashers	422.5(A)
Drinking fountains	422.5(A)
Electrically operated pool covers	680.27(B)(2)
Elevators, escalators, and moving walkways	620.6
Feeders	215.9
Floating buildings	555.53
Fountains	680.51(A), 680.56(A)
Health care facilities	517.21
High-pressure spray washers	422.5(A)
Hydromassage bathtubs	680.71
Marinas and boatyards	555.33 (B)(1)
Mobile and manufactured homes	550.13(B), 550.13(E), 550.32(E)
Natural and artificially made bodies of water	682.15
Park trailers	552.41(C), 552.41(D)
Pools, permanently installed	680.22(A)(2), 680.22(B)(3), 680.23(A)(3)
Pools, storable	680.32
Sensitive electronic equipment	647.7(A)
Signs within fountains	680.57(B)
Signs, portable or mobile	600.10(C)(2)
Space heating embedded in floor	424.44(E)
Spas and hot tubs	680.42(A)(2). 680.43, 680.44
Sump pumps	422.5(A)
Recreational vehicles	551.40(C), 551.41(C)
Recreational vehicle parks	551.71(F)
Replacement receptacles	406.4(D)(3)
Temporary installations	590.6
Tire inflation machines	422.5(A)
Tubs, therapeutic	680.62(A)
Vending machines	422.5(A)

outlet, meeting the proximity requirement of 210.52(D) for each basin, can be used to meet the receptacle outlet location requirement as shown in Exhibit 210.8 (top). Exhibit 210.8 (bottom) illustrates the requirements of 210.11(C)(3), which provides two acceptable supply circuit arrangements for the bathroom receptacle outlet(s).

(2) Garages and also accessory buildings that have a floor located at or below grade level not intended as habitable rooms and limited to storage areas, work areas, and areas of similar use

The requirement for GFCI-protected receptacles in garages and sheds, as illustrated in Exhibit 210.9, improves safety for persons using portable handheld tools, string trimmers, snow blowers, and similar tools that might be connected to the receptacles.

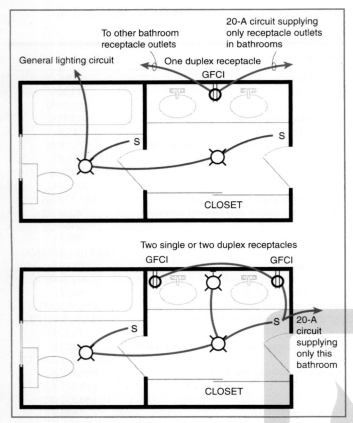

EXHIBIT 210.8 GFCI-protected receptacles in bathrooms.

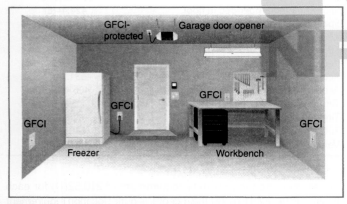

EXHIBIT 210.9 Examples of receptacles in a garage required to have GFCI protection.

GFCI protection is also required because auto repair work and general workshop electrical tools are often used.

There are no exceptions. All 125- through 250-volt, single-phase receptacles installed in garages must provide GFCI protection for the user of appliances or other equipment regardless of where the receptacle is located in the garage. Appliance leakage currents permitted by today's product standards are far less than the operational threshold of a GFCI, so unwanted tripping is unlikely.

(3) Outdoors

The dwelling unit shown in Exhibit 210.10, which has four outdoor receptacles, illustrates the requirement of 210.8(A)(3). Three receptacles must be provided with GFCI protection. The fourth receptacle, located adjacent to the gutter for the roof-mounted snow-melting cable, is not readily accessible and, therefore, is exempt from the GFCI requirements because of its dedicated function to supply the deicing equipment. This receptacle is, however, covered by the equipment ground-fault protection requirements of 426.28.

See also

210.52(E) and **406.9(B)(1)** commentary regarding the installation of outdoor receptacles in wet and damp locations

(4) Crawl spaces — at or below grade level

(5) Basements

All 125- through 250-volt, single-phase receptacles in a basement, as shown in Exhibit 210.11, must have GFCI protection. Conductive floor surfaces usually exist in finished and unfinished basements. Basements, whether finished or unfinished, are prone to moisture, including flooding. GFCI protection is now required in all basements of a dwelling unit. There is an exception for locking support-type receptacles used for luminaires and for paddle fans because these receptacles are not used for cord-and-plug-connected equipment. There is also an exception for receptacles for fire alarm and burglar systems. This correlates with the performance requirements covering fire alarm power supplies contained in *NFPA 72*, *National Fire Alarm and Signaling Code*.

(6) Kitchens

Due to the presence of water and multiple grounded surfaces in a kitchen, there is a higher risk of electric shock. A damaged cord,

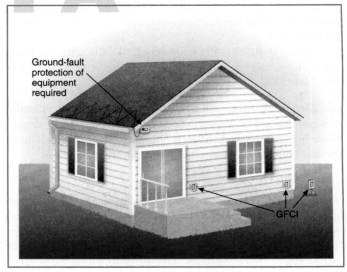

EXHIBIT 210.10 A dwelling unit with three outdoor receptacles that are required to have GFCI protection. It also has one outdoor receptacle that supplies rooftop snow melting equipment, which is protected by ground-fault protection of equipment (GFPE).

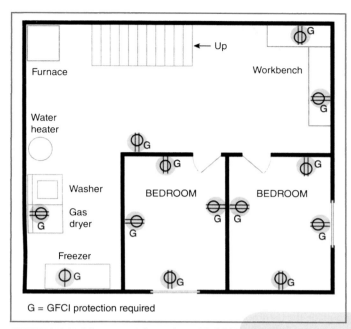

G = GFCI protection required

EXHIBIT 210.11 A basement floor plan with GFCI-protected receptacles in the work area and in the finished areas.

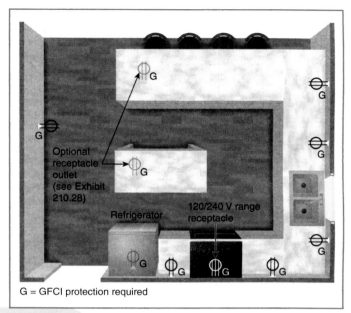

G = GFCI protection required

EXHIBIT 210.12 GFCI protection is required for all receptacles in kitchens.

or faulty or mis-wired appliance, could result in exposed surfaces becoming energized and therefore creating a shock hazard. GFCI protection of all receptacles in a kitchen addresses this potentially dangerous condition. Consumer Product Safety Commission (CPSC) data between 2011 and 2020 reflect 104 electrocutions, 81 percent of which were associated with working on an appliance or equipment. All kitchen receptacles rated 125-volt through 250-volt and supplied by single-phase branch circuits rated 150 volts or less to ground are required to have GFCI protection, provided through the use of a GFCI circuit breaker or GFCI receptacle. See Exhibit 210.12.

(7) Areas with sinks and permanent provisions for food preparation, beverage preparation, or cooking

Receptacles that meet the voltage requirements listed in 210.8(A) are required to have GFCI protection within an area that has both a sink and permanent provisions for food preparation, beverage preparation, or cooking. For example, all receptacles within a wet bar area would require GFCI protection based on having provisions for beverage preparation. Exhibit 210.13 illustrates receptacles installed within a wet bar area with a sink with permanent provisions for beverage preparation, therefore requiring GFCI protection.

(8) Sinks — where receptacles are installed within 1.8 m (6 ft) from the top inside edge of the bowl of the sink

Sinks in kitchens are not the only sinks where a ground-fault shock hazard exists; therefore, this requirement covers all sinks in a dwelling. This GFCI requirement is not limited to receptacles serving countertop surfaces; rather, it covers all 125- through 250-volt, single-phase receptacles within 6 feet of any point along the inside edge at the top of the sink bowl. Many

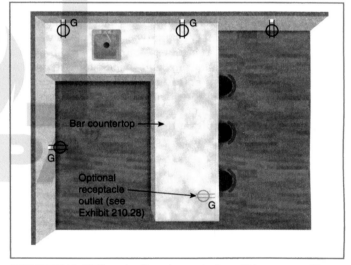

EXHIBIT 210.13 GFCI protection of receptacles within an area with a sink and permanent provisions for beverage preparation.

appliances used in these locations are ungrounded, and the presence of water and grounded surfaces contributes to a condition of increased risk of electric shock.

(9) Boathouses

(10) Bathtubs or shower stalls — where receptacles are installed within 1.8 m (6 ft) of the outside edge of the bathtub or shower stall

(11) Laundry areas

In some instances, bathtubs and shower stalls are installed in areas that might not meet the NEC definition of a bathroom. Many of these areas have tile or other conductive, possibly grounded, floors; therefore, a shock hazard could be present.

GFCI protection of receptacles in laundry areas of dwelling units applies regardless of what the area or room is called. Wet clothes and puddles of water can pose a shock hazard to anyone using any appliance in a laundry area.

(12) Indoor damp and wet locations

Exception No. 1: Receptacles that are not readily accessible and are supplied by a branch circuit dedicated to electric snow-melting, deicing, or pipeline and vessel heating equipment shall be permitted to be installed in accordance with 426.28 or 427.22, as applicable.

Exception No. 2: A receptacle supplying only a permanently installed premises security system shall be permitted to omit ground-fault circuit-interrupter protection.

Exception No. 3: Listed weight-supporting ceiling receptacles (WSCR) utilized in combination with compatible weight-supporting attachment fittings (WSAF) installed for the purpose of supporting a ceiling luminaire or ceiling-suspended fan shall be permitted to omit ground-fault circuit-interrupter protection. If a general-purpose convenience receptacle is integral to the ceiling luminaire or ceiling-suspended fan, GFCI protection shall be provided.

The locking support and mounting receptacles generally are located on the ceiling, or high on the wall, and generally are not readily accessible. These receptacles are used for luminaires, paddle fans, or combination luminaire/paddle fan fixtures. The receptacle is not designed for use with flexible cords and attachment plugs. GFCI protection is required if the assembly supplied by the locking support and mounting device includes a receptacle to supply cord-and-plug-connected equipment. Exhibit 100.14 is an example of a receptacle with a locking support mechanism.

Exception No. 4: Factory-installed receptacles that are not readily accessible and are mounted internally to bathroom exhaust fan assemblies shall not require GFCI protection unless required by the installation instructions or listing.

Informational Note: See 760.41(B) and 760.121(B) for power supply requirements for fire alarm systems.

△ **(B) Other Than Dwelling Units.** All 125-volt through 250-volt receptacles supplied by single-phase branch circuits rated 150 volts or less to ground, 50 amperes or less, and all receptacles supplied by three-phase branch circuits rated 150 volts or less to ground, 100 amperes or less, installed in the following locations shall be provided with GFCI protection:

This requirement now applies to most single-phase receptacles in the areas listed in this section, including receptacles on 240-volt circuits. Most three-phase receptacles on 280Y/120-volt three-phase circuits in those areas would also require GFCI protection. Equipment connected to receptacles of the higher voltage and current ratings presents the same shock hazards as equipment connected to receptacles with lower voltage and current ratings.

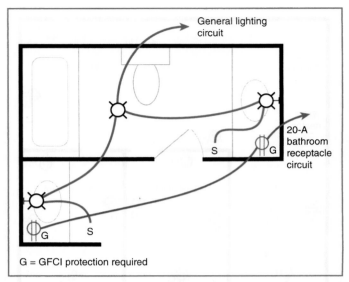

EXHIBIT 210.14 Room plan view showing GFCI protection of receptacles in a motel/hotel bathroom where one basin is located outside the bathroom door.

(1) Bathrooms

Some motel and hotel bathrooms, like the one shown in Exhibit 210.14, have the basin located outside the door to the room containing the tub, toilet, or shower. The *NEC* definition of *bathroom* uses the word "area," so that the sink shown in the lower left section of Exhibit 210.14 is considered as being in the bathroom and is, therefore, subject to the GFCI requirement. Exhibit 210.14 also depicts the supply circuit requirement of 210.11(C)(3) for bathroom receptacle outlets.

(2) Kitchens

This requirement applies to single-phase and three-phase receptacles in kitchens, regardless of whether the receptacle serves countertop areas.

Electrical accident data indicate that there are many electrical hazards in nondwelling kitchens, including poorly maintained electrical equipment, damaged cords, and wet floors. Requiring GFCI protection in those kitchens protects personnel who might be exposed to electric shock. The *NEC* definition of *kitchen* is "an area with a sink and permanent facilities for food preparation and cooking." A portable cooking appliance (e.g., cord-and-plug-connected microwave oven or hot plate) does not constitute a permanent cooking facility. Kitchens in restaurants, hotels, schools, churches, dining halls, and similar facilities are covered by this requirement.

(3) Areas with sinks and permanent provisions for food preparation, beverage preparation, or cooking
(4) Buffet serving areas with permanent provisions for food serving, beverage serving, or cooking
(5) Rooftops
(6) Outdoors
(7) Sinks where receptacles or cord-and-plug-connected fixed or stationary appliances are installed within 1.8 m (6 ft) from the top inside edge of the bowl of the sink

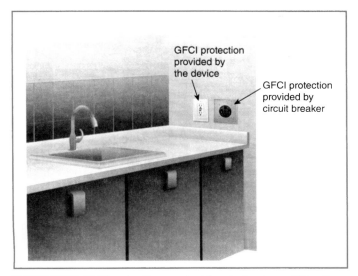

GFCI protection
provided by
the device

GFCI protection
provided by
circuit breaker

EXHIBIT 210.15 GFCI protection required for single-phase, 125-volt, and 250-volt receptacles installed within 6 feet of a sink if the receptacles are rated 50 amperes or less.

Item (7) covers receptacles installed near sinks in lunchrooms, janitors' closets, classrooms, and all other areas not covered by the bathroom and kitchen requirements. Exhibit 210.15 shows one single-phase, 125-volt, 15-ampere receptacle installed within 6 feet of the top inside edge of the sink bowl that is covered by the GFCI requirement. The single-phase, 240-volt, 30-ampere receptacle to the right of the duplex receptacle is also subject to the GFCI requirement. GFCI protection would also be required for any 3-phase, 208Y/120-volt receptacles rated 100 amperes or less.

  (8) Indoor damp or wet locations
  (9) Locker rooms with associated showering facilities
 (10) Garages, accessory buildings, service bays, and similar areas other than vehicle exhibition halls and showrooms

This requirement covers all garages, not just those in which vehicle maintenance is expected to take place. Many commercial garages have receptacles installed for purposes other than the use of hand tools. For instance, winter temperatures in some areas necessitate the use of engine block heaters. Cord-and-plug-connected engine block heaters that are not listed could exceed the maximum leakage current requirements for appliances. If the receptacles are not GFCI protected, the frame of the vehicle could become energized, posing a shock hazard.

 (11) Crawl spaces at or below grade level
 (12) Unfinished areas of basements
 (13) Aquariums, bait wells, and similar open aquatic vessels or containers, such as tanks or bowls, where receptacles are installed within 1.8 m (6 ft.) from the top inside edge or rim or from the conductive support framing of the vessel or container
 (14) Laundry areas
 (15) Bathtubs and shower stalls where receptacles are installed within 1.8 m (6 ft) of the outside edge of the bathtub or shower stall

Exception No. 1: Receptacles that are not readily accessible and are supplied by a branch circuit dedicated to electric snow-melting, deicing, or pipeline and vessel heating equipment shall be permitted to be installed in accordance with 426.28 or 427.22, as applicable.

Exception No. 2: Receptacles on rooftops shall not be required to be readily accessible other than from the rooftop.

Exception No. 3: Receptacles or cord-and-plug-connected fixed and stationary appliances installed within 1.8 m (6 ft) from the top inside edge of a bowl of a sink shall not be required to be GFCI protected in industrial establishments where the conditions of maintenance and supervision ensure that only qualified personnel are involved, an assured equipment grounding conductor program in accordance with 590.6(B)(2) shall be permitted for only those receptacle outlets used to supply equipment that would create a greater hazard if power is interrupted or that has a design not compatible with GFCI protection.

Exception No. 4: Receptacles or cord-and-plug-connected fixed and stationary appliances installed within 1.8 m (6 ft) from the top inside edge of a bowl of a sink shall not be required to be GFCI protected in industrial laboratories where the receptacles are used to supply equipment if removal of power would introduce a greater hazard.

Exception No. 5: Receptacles located in patient bed locations of Category 2 (general care) or Category 1 (critical care) spaces of health care facilities shall be permitted to comply with 517.21.

Exception No. 6: Listed weight-supporting ceiling receptacles (WSCR) utilized in combination with compatible weight-supporting attachment fittings (WSAF) installed for the purpose of serving a ceiling luminaire or ceiling-suspended fan shall be permitted to omit GFCI protection. If a general-purpose convenience receptacle is integral to the ceiling luminaire or ceiling-suspended fan, GFCI protection shall be provided.

(C) Crawl Space Lighting Outlets. GFCI protection shall be provided for lighting outlets not exceeding 120 volts installed in crawl spaces.

Δ **(D) Specific Appliances.** GFCI protection shall be provided for the branch circuit or outlet supplying the following appliances rated 150 volts or less to ground and 60 amperes or less, single- or 3-phase:

  (1) Automotive vacuum machines
  (2) Drinking water coolers and bottle fill stations
  (3) High-pressure spray washing machines
  (4) Tire inflation machines
  (5) Vending machines
  (6) Sump pumps
  (7) Dishwashers
  (8) Electric ranges
  (9) Wall-mounted ovens

(10) Counter-mounted cooking units
(11) Clothes dryers
(12) Microwave ovens

See also

422.5(A) and its commentary for more information on GFCI protection of specific appliance types

(E) Equipment Requiring Servicing. GFCI protection shall be provided for the receptacles required by 210.63.

Δ **(F) Outdoor Outlets.** For dwellings, all outdoor outlets, other than those covered in 210.8(A), Exception No. 1, including outlets installed in the following locations, and supplied by single-phase branch circuits rated 150 volts or less to ground, 50 amperes or less, shall be provided with GFCI protection:

(1) Garages that have floors located at or below grade level
(2) Accessory buildings
(3) Boathouses

If equipment supplied by an outlet covered under the requirements of this section is replaced, the outlet shall be supplied with GFCI protection.

Exception No. 1: GFCI protection shall not be required on lighting outlets other than those covered in 210.8(C).

Exception No. 2: GFCI protection shall not be required for listed HVAC equipment. This exception shall expire September 1, 2026.

See also

Article 100 for the definition of the term *outlet*

210.9 Circuits Derived from Autotransformers. Branch circuits shall not be derived from autotransformers unless the circuit supplied has a grounded conductor that is electrically connected to a grounded conductor of the system supplying the autotransformer.

Exception No. 1: An autotransformer shall be permitted without the connection to a grounded conductor where transforming from a nominal 208 volts to a nominal 240-volt supply or similarly from 240 volts to 208 volts.

Exception No. 2: In industrial occupancies, where conditions of maintenance and supervision ensure that only qualified persons service the installation, autotransformers shall be permitted to supply nominal 600-volt loads from nominal 480-volt systems, and 480-volt loads from nominal 600-volt systems, without the connection to a similar grounded conductor.

A buck-boost autotransformer provides a means of lowering (bucking) or raising (boosting) a supply line voltage by a small amount (usually no more than 20 percent). A buck-boost transformer has two primary windings (H_1–H_2 and H_3–H_4) connected to two secondary windings (X_1–X_2 and X_3–X_4). A single unit is used to buck/boost a single-phase voltage, but two or three units are used to buck/boost a 3-phase voltage.

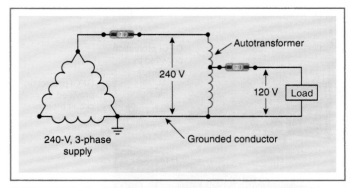

EXHIBIT 210.16 Circuitry for an autotransformer used to derive a 2-wire, 120-volt system for lighting or convenience receptacles from a 240-volt corner-grounded delta system.

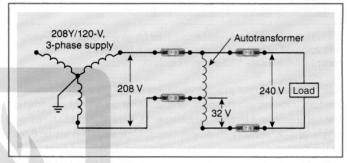

EXHIBIT 210.17 Circuitry for an autotransformer used to derive a 240-volt system for appliances from a 208Y/120-volt source.

In Exhibit 210.16, a 120-volt supply is derived from a 240-volt system. The grounded conductor of the primary system is electrically connected to the grounded conductor of the secondary system.

Exhibit 210.17 illustrates a common application that utilizes Exception No. 1. This circuit derives a single-phase, 240-volt supply system for ranges, air conditioners, heating elements, and motors from a 3-phase, 208Y/120-volt source system. The boosted leg should not be used to supply line-to-neutral loads, because the boosted line-to-neutral voltage will be higher than 120 volts.

Other common applications include increasing a single-phase, 240-volt source to a single-phase, 277-volt supply for lighting systems and transforming 240 volts to 208 volts for use with 208-volt appliances. See Exhibit 210.18. Exhibit 210.19 illustrates typical 3-phase buck-boost transformers connected to change 240 volts to 208 volts and vice versa.

210.10 Ungrounded Conductors Tapped from Grounded Systems. Two-wire dc circuits and ac circuits of two or more ungrounded conductors shall be permitted to be tapped from the ungrounded conductors of circuits that have a grounded neutral conductor. Switching devices in each tapped circuit shall have a pole in each ungrounded conductor. All poles of multipole switching devices shall manually switch together where such switching devices also serve as a disconnecting means as required by the following:

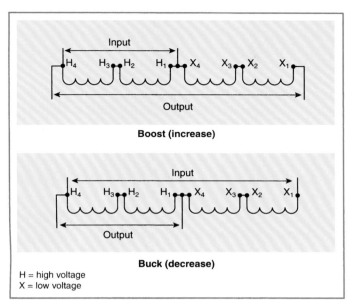

EXHIBIT 210.18 Typical single-phase connection diagrams for buck-boost transformers connected as autotransformers to change 240 volts single-phase to 208 volts and vice versa.

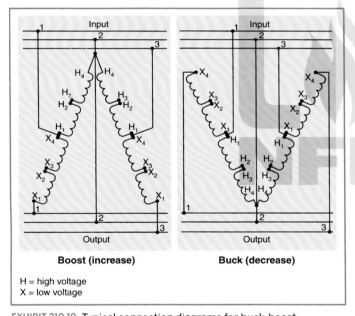

EXHIBIT 210.19 Typical connection diagrams for buck-boost transformers connected in 3-phase open delta as autotransformers to change 240 volts to 208 volts and vice versa.

(1) 410.93 for double-pole switched lampholders
(2) 410.104(B) for electric-discharge lamp auxiliary equipment switching devices
(3) 422.31(B) for an appliance
(4) 424.20 for a fixed electric space-heating unit
(5) 426.51 for electric deicing and snow-melting equipment
(6) 430.85 for a motor controller
(7) 430.103 for a motor

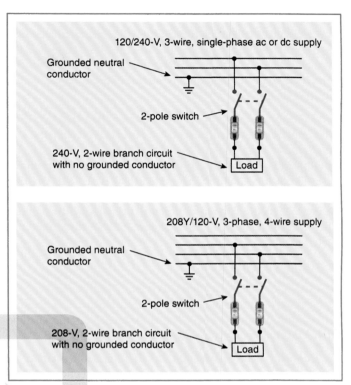

EXHIBIT 210.20 Branch circuits tapped from ungrounded conductors of multiwire systems.

Exhibit 210.20 (top) illustrates an ungrounded 2-wire branch circuit tapped from the ungrounded conductors of a dc or single-phase system to supply a small motor. Exhibit 210.20 (bottom) illustrates a 3-phase, 4-wire wye system. Based on the characteristics of the depicted loads (line-to-line connected) supplied by the tapped circuits, a neutral or grounded conductor is not required to be installed with the ungrounded conductors.

Circuit breakers or switches used as the disconnecting means for the tapped branch circuit must open all poles simultaneously using only the manual operation of the disconnecting means. If switches and fuses are used and one fuse blows, or if circuit breakers (two single-pole circuit breakers with a handle tie) are used and one breaker trips, one pole could possibly remain closed. The intention is not to require a common trip of fuses or circuit breakers but rather to disconnect the ungrounded conductors of the branch circuit with one manual operation.

See also

240.15(B) for information on the use of identified handle ties with single-pole circuit breakers

210.11 Branch Circuits Required. Branch circuits for lighting and for appliances, including motor-operated appliances, shall be provided to supply the loads calculated in accordance with 220.10. In addition, branch circuits shall be provided for specific loads not covered by 220.10 where required elsewhere in this *Code* and for dwelling unit loads as specified in 210.11(C).

(A) Number of Branch Circuits. The minimum number of branch circuits shall be determined from the total calculated load and the size or rating of the circuits used. In all installations, the number of circuits shall be sufficient to supply the load served. In no case shall the load on any circuit exceed the maximum specified by 220.11.

(B) Load Evenly Proportioned Among Branch Circuits. Where the load is calculated on the basis of volt-amperes per square meter or per square foot, the wiring system up to and including the branch-circuit panelboard(s) shall be provided to serve not less than the calculated load. This load shall be evenly proportioned among multioutlet branch circuits within the panelboard(s). Branch-circuit overcurrent devices and circuits shall be required to be installed only to serve the connected load.

(C) Dwelling Units.

(1) Small-Appliance Branch Circuits. In addition to the number of branch circuits required by other parts of this section, two or more 20-ampere small-appliance branch circuits shall be provided for all receptacle outlets specified by 210.52(B).

(2) Laundry Branch Circuits. In addition to the number of branch circuits required by other parts of this section, at least one additional 20-ampere branch circuit shall be provided to supply the laundry receptacle outlet(s) required by 210.52(F). This circuit shall have no other outlets.

(3) Bathroom Branch Circuits. In addition to the number of branch circuits required by other parts of this section, one or more 120-volt, 20-ampere branch circuit shall be provided to supply bathroom(s) receptacle outlet(s) required by 210.52(D) and any countertop and similar work surface receptacle outlets. Such circuits shall have no other outlets.

Exception: Where the 20-ampere circuit supplies a single bathroom, outlets for other equipment within the same bathroom shall be permitted to be supplied in accordance with 210.23(B)(1) and (B)(2).

The installation of a 20-ampere branch for the receptacles in each individual bathroom is not necessary. A single circuit can supply more than one bathroom.

△ **(4) Garage Branch Circuits.** In addition to the number of branch circuits required by other parts of this section, at least one 120-volt, 20-ampere branch circuit shall be installed to supply receptacle outlets, including those required by 210.52(G)(1) for attached garages and in detached garages with electric power. This circuit shall have no other outlets.

Additional branch circuits rated 15 amperes or greater shall be permitted to serve receptacle outlets other than those required by 210.52(G)(1).

Exception No. 1: This circuit shall be permitted to supply outdoor receptacle outlets.

Exception No. 2: Where the 20-ampere circuit supplies a single vehicle bay garage, outlets for other equipment within the same garage shall be permitted to be supplied in accordance with 210.23(B)(1) and (B)(2).

210.12 Arc-Fault Circuit-Interrupter Protection. Arc-fault circuit-interrupter (AFCI) protection shall be installed in accordance with 210.12(B) through (E) by any of the means described in 210.12(A)(1) through (A)(6). The AFCI shall be listed and installed in a readily accessible location.

△ **(A) Means of Protection.** AFCI protection shall be provided by any of the following means:

(1) A listed combination-type AFCI installed to provide protection of the entire branch circuit.

(2) A listed branch/feeder-type AFCI installed at the origin of the branch circuit in combination with a listed outlet branch-circuit-type AFCI installed on the branch circuit at the first outlet box, which shall be marked to indicate that it is the first outlet of the branch circuit.

(3) A listed supplemental arc protection circuit breaker installed at the origin of the branch circuit in combination with a listed outlet branch-circuit-type AFCI installed on the branch circuit at the first outlet box if all of the following conditions are met:

a. The branch-circuit wiring shall be continuous from the branch-circuit overcurrent device to the outlet branch-circuit AFCI.

b. The maximum length of the branch-circuit wiring from the branch-circuit overcurrent device to the first outlet shall not exceed 15.2 m (50 ft) for a 14 AWG conductor or 21.3 m (70 ft) for a 12 AWG conductor.

c. The first outlet box shall be marked to indicate that it is the first outlet of the branch circuit.

Listed arc-fault circuit-interrupter (AFCI) devices are evaluated in accordance with UL 1699, *Standard for Arc-Fault Circuit-Interrupters*. Testing methods create or simulate arcing conditions to determine a product's ability to detect and interrupt arcing faults. The devices are also tested to verify that arc detection is not unduly inhibited by the presence of loads and circuit characteristics that may mask the hazardous arcing condition. The devices are also evaluated to determine resistance to unwanted tripping due to the presence of arcing that occurs in control and utilization equipment under normal operating conditions or to a loading condition that closely mimics an arcing fault, such as that found in a solid-state electronic ballast or a dimmed load.

AFCI devices may also be capable of performing other functions such as overcurrent protection, ground-fault circuit interruption, and surge suppression. UL 1699 currently recognizes four types of AFCIs: branch/feeder, cord, outlet circuit, and portable. (See Exhibit 210.21 for an example of the required marking indicating the type of AFCI protection.) AFCI devices have a maximum rating of 20 amperes and are intended for use in 120 volts alternating current (ac), 60-hertz circuits. Cord AFCIs can be rated up to 30 amperes. Placement of the device in the circuit must be considered when complying with 210.12. Six possible configurations using listed equipment are permitted by the *NEC*. The objective of the *NEC* is to provide protection of the entire branch circuit. The configurations may use a single combination-type AFCI at the origin of the circuit, a combination of devices, or a combination of physical protection for part of the circuit and a device-type AFCI located downstream of the origin of the circuit. Where the AFCI is not located at the origin of the circuit, a higher level of physical protection must be provided for branch-circuit conductors from the origin of the branch circuit to the device-type AFCI. Some configurations have length restrictions to the first outlet, based on the size of the conductors (50 feet for 14 AWG and 70 feet for 12 AWG). Commentary Table 210.2 summarizes the permitted methods of providing AFCI protection.

Branch-circuit/feeder-type AFCI devices provide arcing protection against parallel faults. An example of a parallel arcing fault is a cable stapled to a wooden stud where the staple has been driven deeply into the cable jacket, damaging the conductor insulation. Combination-type AFCIs provide parallel arcing protection as well as protection against series arcing, such as could occur in a cord set.

AFCI protection is required to protect receptacle outlets, lighting outlets, smoke alarm outlets, and other outlets. See the definition of *outlet* in Article 100. Because circuits are often shared between a bedroom and other areas such as closets and hallways, providing AFCI protection on the complete circuit would comply with 210.12. Providing AFCI protection on other circuits, or locations other than those specified in 210.12(B) through (D), is not prohibited.

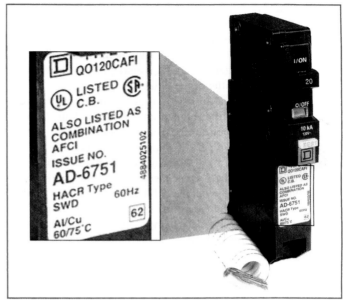

EXHIBIT 210.21 A circuit breaker with the required marking indicating the type of AFCI protection. *(Courtesy of Square D by Schneider Electric)*

(4) A listed outlet branch-circuit-type AFCI installed on the branch circuit at the first outlet in combination with a listed branch-circuit overcurrent protective device if all of the following conditions are met:

a. The branch-circuit wiring shall be continuous from the branch-circuit overcurrent device to the outlet branch-circuit AFCI.

COMMENTARY TABLE 210.2 *AFCI Protection Methods*

210.12(A) Reference	AFCI Protection Method	Additional Installation Requirements
210.12(A)(1)	• Combination-type AFCI circuit breaker installed at origin of branch circuit.	• No additional requirements
210.12(A)(2)	• Branch/feeder-type AFCI circuit breaker installed at origin of branch circuit, plus • Outlet branch-circuit-type AFCI device installed at first outlet in branch circuit.	• Marking of first outlet box in branch circuit
210.12(A)(3)	• Supplemental arc protection-type circuit breaker installed at origin of branch circuit, plus • Outlet branch-circuit-type AFCI device installed at first outlet in branch circuit.	• Continuous branch-circuit wiring • "Home run" conductor length restricted • Marking of first outlet box in branch circuit
210.12(A)(4)	• Branch-circuit overcurrent protective device, plus • Outlet branch-circuit-type AFCI device installed at first outlet in branch circuit. The combination of devices must be listed and identified to provide "system combination-type" arc-fault protection for the "home run" conductors.	• Continuous branch-circuit wiring • "Home run" conductor length restricted (50 ft for 14 AWG, 70 ft for 12 AWG) • Marking of first outlet box in branch circuit
210.12(A)(5)	• Outlet branch-circuit-type AFCI device installed at first outlet in branch circuit.	• Branch-circuit conductors installed in specific types of metal raceways or metal cables and metal boxes from origin of branch circuit to the first outlet
210.12(A)(6)	• Outlet branch-circuit-type AFCI device installed at first outlet in branch circuit.	• Branch-circuit conduit, tubing, or cable encased in 2 inches of concrete from origin of branch circuit to the first outlet

b. The maximum length of the branch-circuit wiring from the branch-circuit overcurrent device to the first outlet shall not exceed 15.2 m (50 ft) for a 14 AWG conductor or 21.3 m (70 ft) for a 12 AWG conductor.

c. The first outlet box shall be marked to indicate that it is the first outlet of the branch circuit.

d. The combination of the branch-circuit overcurrent device and outlet branch-circuit AFCI shall be identified as meeting the requirements for a system combination-type AFCI and listed as such.

(5) If metal raceway, metal wireways, metal auxiliary gutters, or Type MC or Type AC cable meeting the applicable requirements of 250.118, with metal boxes, metal conduit bodies, and metal enclosures are installed for the portion of the branch circuit between the branch-circuit overcurrent device and the first outlet, it shall be permitted to install a listed outlet branch-circuit-type AFCI at the first outlet to provide protection for the remaining portion of the branch circuit.

(6) Where a listed metal or nonmetallic conduit or tubing or Type MC cable is encased in not less than 50 mm (2 in.) of concrete for the portion of the branch circuit between the branch-circuit overcurrent device and the first outlet, it shall be permitted to install a listed outlet branch-circuit-type AFCI at the first outlet to provide protection for the remaining portion of the branch circuit.

Informational Note: See UL 1699-2011, *Standard for Arc-Fault Circuit-Interrupters*, for information on combination-type and branch/feeder-type AFCI devices. See UL Subject 1699A, *Outline of Investigation for Outlet Branch Circuit Arc-Fault Circuit-Interrupters*, for information on outlet branch-circuit type AFCI devices. See UL Subject 1699C, *Outline of Investigation for System Combination Arc-Fault Circuit Interrupters*, for information on system combination AFCIs.

N **(B) Dwelling Units.** All 120-volt, single-phase, 10-, 15-, and 20-ampere branch circuits supplying outlets or devices installed in the following locations shall be protected by any of the means described in 210.12(A)(1) through (A)(6):

(1) Kitchens
(2) Family rooms
(3) Dining rooms
(4) Living rooms
(5) Parlors
(6) Libraries
(7) Dens
(8) Bedrooms
(9) Sunrooms
(10) Recreation rooms
(11) Closets
(12) Hallways
(13) Laundry areas
(14) Similar areas

Exception No. 1: AFCI protection shall not be required for an individual branch circuit supplying a fire alarm system installed

in accordance with 760.41(B) or 760.121(B). The branch circuit shall be installed in a metal raceway, metal auxiliary gutter, steel-armored cable, or Type MC or Type AC cable meeting the applicable requirements of 250.118, with metal boxes, conduit bodies, and enclosures.

Exception No. 2: AFCI protection shall not be required for the individual branch circuit supplying an outlet for arc welding equipment in a dwelling unit until January 1, 2025.

Informational Note No. 1: See *NFPA 72-2022, National Fire Alarm and Signaling Code,* 29.9.4(5), for information on secondary power source requirements for smoke alarms installed in dwelling units.

Informational Note No. 2: See 760.41(B) and 760.121(B) for power source requirements for fire alarm systems.

(C) Dormitory Units. All 120-volt, single-phase, 10-, 15-, and 20-ampere branch circuits supplying outlets or devices installed in the following locations shall be protected by any of the means described in 210.12(A)(1) through (A)(6):

(1) Bedrooms
(2) Living rooms
(3) Hallways
(4) Closets
(5) Bathrooms
(6) Similar rooms

(D) Other Occupancies. All 120-volt, single-phase, 10-, 15-, and 20-ampere branch circuits supplying outlets or devices installed in the following locations shall be protected by any of the means described in 210.12(A)(1) through (A)(6):

(1) Guest rooms and guest suites of hotels and motels
(2) Areas used exclusively as patient sleeping rooms in nursing homes and limited-care facilities
(3) Areas designed for use exclusively as sleeping quarters in fire stations, police stations, ambulance stations, rescue stations, ranger stations, and similar locations

Δ **(E) Branch Circuit Wiring Extensions, Modifications, or Replacements.** If branch-circuit wiring for any of the areas specified in 210.12(B), (C), or (D) is modified, replaced, or extended, the branch circuit shall be protected by one of the following:

(1) By any of the means described in 210.12(A)(1) through (A)(6)
(2) A listed outlet branch-circuit-type AFCI located at the first receptacle outlet of the existing branch circuit

Exception: AFCI protection shall not be required where the extension of the existing branch-circuit conductors is not more than 1.8 m (6 ft) and does not include any additional outlets or devices, other than splicing devices. This measurement shall not include the conductors inside an enclosure, cabinet, or junction box.

Δ **210.13 Ground-Fault Protection of Equipment.** Each branch-circuit disconnecting means rated 1000 amperes or more and installed on solidly grounded wye electrical systems of more than 150 volts to ground, but not exceeding 1000 volts phase-to-phase, shall be provided with ground-fault protection of equipment in accordance with 230.95.

> Informational Note: See 517.17 for requirements on buildings that contain health care occupancies.

> *Exception No. 1: This section shall not apply to a disconnecting means for a continuous industrial process where a nonorderly shutdown will introduce additional or increased hazards.*

> *Exception No. 2: This section shall not apply if ground-fault protection of equipment is provided on the supply side of the branch circuit and on the load side of any transformer supplying the branch circuit.*

Exceptions are provided for continuous industrial processes where shutdown could introduce additional or increased hazards and for installations where ground-fault protection is provided upstream of the branch circuit on the load side of a transformer supplying the branch circuit.

See also

230.95 and its commentary for more information on ground-fault protection of equipment

210.17 Guest Rooms and Guest Suites. Guest rooms and guest suites in the following occupancies that are provided with permanent provisions for cooking shall have branch circuits installed to meet the rules for dwelling units:

(1) Hotels
(2) Motels
(3) Assisted living facilities

> Informational Note No. 1: See 210.11(C)(2) and 210.52(F), Exception No. 2, for information on laundry branch circuits and receptacle outlets.

> Informational Note No. 2: See NFPA *101*-2021, *Life Safety Code,* 3.3.198.12 and A.3.3.198.12(5), for the definition of *assisted living facilities.*

The guest suite configuration shown in Exhibit 210.22 triggers the requirement to install the branch-circuit wiring in the unit using all the branch-circuit requirements that apply to dwelling units.

Part II. Branch-Circuit Ratings

210.18 Rating. Branch circuits recognized by this article shall be rated in accordance with the maximum permitted ampere rating or setting of the overcurrent device. The rating for other than individual branch circuits shall be 10, 15, 20, 30, 40, and 50 amperes. Where conductors of higher ampacity are used for any reason, the ampere rating or setting of the specified overcurrent device shall determine the circuit rating.

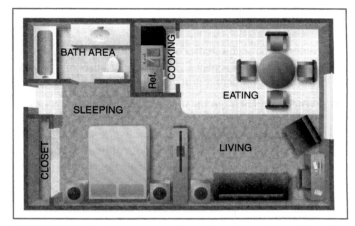

EXHIBIT 210.22 Guest rooms or suites with permanent provisions for cooking in which the installation of branch circuits must follow all the requirements in Article 210 covering dwelling units.

The rating of a branch circuit is determined by the rating of the overcurrent protective device (OCPD), not by the ampacity of the circuit conductors. An increase in conductor ampacity from 20 amperes (12 AWG) to 30 amperes (10 AWG) for any reason does not change the rating of the circuit. If the rating of the branch-circuit OCPD is 20 amperes, then the rating of the circuit is 20 amperes.

> *Exception No. 1: Multioutlet branch circuits greater than 50 amperes shall be permitted to supply nonlighting outlet loads in locations where conditions of maintenance and supervision ensure that only qualified persons service the equipment.*

> *Exception No. 2: Branch circuits rated 10 amperes shall not supply receptacle outlets.*

A common practice at industrial facilities is to provide several single receptacles with ratings of 50 amperes or higher on a single branch circuit to allow quick relocation of equipment (such as electric welders) for production or maintenance. Generally, only one piece of equipment at a time is supplied from this type of receptacle circuit. The type of receptacle used in this situation is generally a pin-and-sleeve receptacle.

210.19 Conductors — Minimum Ampacity and Size. Branch-circuit conductors for circuits not exceeding 1000 volts ac or 1500 volts dc shall be sized in accordance with 210.19(A) through (D).

> Informational Note: Conductors for branch circuits as defined in Article 100, sized to prevent a voltage drop exceeding 3 percent at the farthest outlet of power, heating, and lighting loads, or combinations of such loads, and where the maximum total voltage drop on both feeders and branch circuits to the farthest outlet does not exceed 5 percent, provide reasonable efficiency of operation. See 215.2(A)(2), Informational Note No. 2, for information on voltage drop on feeder conductors.

Excessive voltage drop in supply conductors can cause inefficient operation or malfunction of electrical equipment. Under-voltage conditions reduce the capability and reliability of motors, lighting sources, heaters, and solid-state equipment. It may be necessary to compensate for this with an installation beyond the minimum requirements.

Δ **(A) General.** Branch-circuit conductors shall have an ampacity not less than the larger of the following and comply with 110.14(C) for equipment terminations:

(1) Where a branch circuit supplies continuous loads or any combination of continuous and noncontinuous loads, the minimum branch-circuit conductor size shall have an ampacity not less than the noncontinuous load plus 125 percent of the continuous load in accordance with 310.14.

> *Exception to (1): If the assembly, including the overcurrent devices protecting the branch circuits, is listed for operation at 100 percent of its rating, the ampacity of the branch-circuit conductors shall be permitted to be not less than the sum of the continuous load plus the noncontinuous load in accordance with 110.14(C).*

(2) The minimum branch-circuit conductor size shall have an ampacity not less than the maximum load to be served after the application of any adjustment or correction factors in accordance with 310.15.

Exception to (1) and (2): Where a portion of a branch circuit is connected at both its supply and load ends to separately installed pressure connections as covered in 110.14(C)(2), an allowable ampacity in accordance with 310.15 not less than the sum of the continuous load plus the noncontinuous load shall be permitted. No portion of a branch circuit installed under this exception shall extend into an enclosure containing either the branch-circuit supply or the branch-circuit load terminations.

Conductors of branch circuits must be able to supply power to loads without overheating. The minimum conductor size and ampacity must be based on the larger of the following criteria:

* The noncontinuous load plus 125 percent of the continuous load
* Not less than the maximum load to be served after the application of any adjustment or correction factors

It is not necessary to apply the additional 25 percent continuous load to the adjustment and correction factor calculations of 210.19(A)(2).

See also

220.61 for sizing of the neutral conductor load

(B) Branch Circuits with More than One Receptacle. Conductors of branch circuits supplying more than one receptacle for cord-and-plug-connected portable loads shall have an ampacity of not less than the rating of the branch circuit.

The loading of branch-circuit conductors that supply receptacles for cord-and-plug-connected portable loads is unpredictable. The circuit conductors are required to have an ampacity that is not less than the rating of the branch circuit. The rating of the

branch circuit is the rating of the overcurrent device according to 210.18.

(C) Household Ranges and Cooking Appliances. Branch-circuit conductors supplying household ranges, wall-mounted ovens, counter-mounted cooking units, and other household cooking appliances shall have an ampacity not less than the rating of the branch circuit and not less than the maximum load to be served. For ranges of 8¾ kW or more rating, the minimum branch-circuit rating shall be 40 amperes.

Exception No. 1: Conductors tapped from a branch circuit not exceeding 50 amperes supplying electric ranges, wall-mounted electric ovens, and counter-mounted electric cooking units shall have an ampacity of not less than 20 amperes and shall be sufficient for the load to be served. These tap conductors include any conductors that are a part of the leads supplied with the appliance that are smaller than the branch-circuit conductors. The taps shall not be longer than necessary for servicing the appliance.

Both factory-installed pigtails and field-installed conductors are considered to be tap conductors in the application of Exception No. 1. This exception permits a 20-ampere tap conductor from a range, oven, or cooking unit to be connected to a 50-ampere branch circuit as illustrated in Exhibit 210.23.

Exception No. 2: The neutral conductor of a 3-wire branch circuit supplying a household electric range, a wall-mounted oven, or a counter-mounted cooking unit shall be permitted to

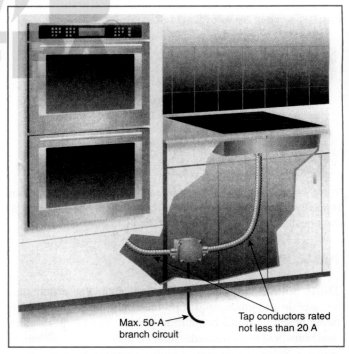

Max. 50-A branch circuit

Tap conductors rated not less than 20 A

EXHIBIT 210.23 Tap conductors sized smaller than the branch-circuit conductors and not longer than necessary for servicing the appliances.

be smaller than the ungrounded conductors where the maximum demand of a range of 8¾ kW or more rating has been calculated according to Column C of Table 220.55, but such conductor shall have an ampacity of not less than 70 percent of the branch-circuit rating and shall not be smaller than 10 AWG.

Column C of Table 220.55 indicates that the maximum demand for one range (not over 12-kilowatt rating) is 8 kilowatts (8000 VA; 8000 VA ÷ 240 V = 33.3 A). The ampacity of an 8 AWG copper conductor from the 60°C column of Table 310.16 is 40 amperes, and this conductor may be used for the range branch circuit. According to Exception No. 2, the neutral of this 3-wire circuit can be smaller than 8 AWG but not smaller than 10 AWG. A 10 AWG conductor has an ampacity of 30 amperes (30 amperes is more than 70 percent of 40 amperes). The maximum demand for the neutral of an 8-kilowatt range circuit seldom exceeds 25 amperes because the only line-to-neutral connected loads are lights, clocks, timers, and the heating elements of some ranges when the control is adjusted to the low-heat setting.

(D) Other Loads. Branch-circuit conductors that supply loads other than those specified in 210.3 and other than cooking appliances as covered in 210.19(C) shall have an ampacity sufficient for the loads served and shall not be smaller than 14 AWG.

Exception No. 1: Tap conductors shall have an ampacity sufficient for the load served. In addition, they shall have an ampacity of not less than 15 for circuits rated less than 40 amperes and not less than 20 for circuits rated at 40 or 50 amperes and only where these tap conductors supply any of the following loads:

(1) Individual lampholders or luminaires with taps extending not longer than 450 mm (18 in.) beyond any portion of the lampholder or luminaire

(2) A luminaire having tap conductors in accordance with 410.117

(3) Individual outlets, other than receptacle outlets, with taps not over 450 mm (18 in.) long

(4) Infrared lamp industrial heating appliances

(5) Nonheating leads of deicing and snow-melting cables and mats

Exception No. 2: Fixture wires and flexible cords shall be permitted to be smaller than 14 AWG as permitted by 240.5.

210.20 Overcurrent Protection. Branch-circuit conductors and equipment for circuits not exceeding 1000 volts ac or 1500 volts dc shall be protected by overcurrent protective devices that have a rating or setting that complies with 210.20(A) through (D).

(A) Continuous and Noncontinuous Loads. Where a branch circuit supplies continuous loads or any combination of continuous and noncontinuous loads, the rating of the overcurrent device shall not be less than the noncontinuous load plus 125 percent of the continuous load.

Exhibit 210.24 is an example of a calculation of a minimum branch-circuit rating for a continuous load (store lighting). The

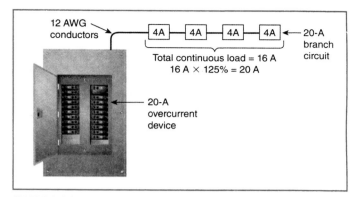

EXHIBIT 210.24 A continuous load calculated at 125 percent.

125 percent requirement from 210.20(A) has been applied to the 16 amperes of continuous load, making it permissible to use a 20-ampere rated overcurrent device in this application.

Exception: Where the assembly, including the overcurrent devices protecting the branch circuit(s), is listed for operation at 100 percent of its rating, the ampere rating of the overcurrent device shall be permitted to be not less than the sum of the continuous load plus the noncontinuous load.

Section 210.19(A)(1) requires that the circuit conductors have an ampacity based on the larger of these two criteria:

- Not less than the sum of the noncontinuous load plus 125 percent of the continuous load
- Not less than the maximum load to be served after the application of any adjustment or correction factors

However, the rating of the overcurrent device cannot exceed the sum of the noncontinuous load plus 125 percent of the continuous load.

See also

220.61 for sizing of the neutral conductor load

(B) Conductor Protection. Conductors shall be protected in accordance with 240.4. Flexible cords and fixture wires shall be protected in accordance with 240.5.

(C) Equipment. The rating or setting of the overcurrent protective device shall not exceed that specified in the applicable articles referenced in Table 240.3 for equipment.

(D) Outlet Devices. The rating or setting shall not exceed that specified in 210.21 for outlet devices.

210.21 Outlet Devices. Outlet devices shall have an ampere rating that is not less than the load to be served and shall comply with 210.21(A) and (B).

(A) Lampholders. Where connected to a branch circuit having a rating in excess of 20 amperes, lampholders shall be of the heavy-duty type. A heavy-duty lampholder shall have a rating of not less than 660 watts if of the admedium type, or not less than 750 watts if of any other type.

(B) Receptacles.

Δ **(1) Single Receptacle on an Individual Branch Circuit.** A single receptacle installed on an individual branch circuit shall have an ampere rating not less than that of the branch circuit.

Exception No. 1: A receptacle installed in accordance with 430.81(B).

Exception No. 2: A receptacle installed exclusively for the use of a cord-and-plug-connected arc welder shall be permitted to have an ampere rating not less than the minimum branch-circuit conductor ampacity determined by 630.11(A) for arc welders.

Informational Note: See Article 100 for the definition of *receptacle.*

A single receptacle installed on an individual branch circuit must have an ampere rating not less than that of the branch circuit. For example, a single receptacle on a 20-ampere individual branch circuit must be rated at 20 amperes in accordance with 210.21(B)(1); however, two or more 15-ampere single receptacles or a 15-ampere duplex receptacle are permitted on a 20-ampere branch circuit in accordance with 210.21(B)(3).

(2) Total Cord-and-Plug-Connected Load. Where connected to a branch circuit supplying two or more receptacles or outlets, a receptacle shall not supply a total cord-and-plug-connected load in excess of the maximum specified in Table 210.21(B)(2).

TABLE 210.21(B)(2) *Maximum Cord-and-Plug-Connected Load to Receptacle*

Circuit Rating (Amperes)	Receptacle Rating (Amperes)	Maximum Load (Amperes)
15 or 20	15	12
20	20	16
30	30	24

Δ **(3) Receptacle Ratings.** Where connected to a branch circuit supplying two or more receptacles or outlets, receptacle ratings shall not be less than the values listed in Table 210.21(B)(3), or, where rated higher than 50 amperes, the receptacle rating shall not be less than the branch-circuit rating.

Δ **TABLE 210.21(B)(3)** *Receptacle Ratings for Circuits Serving More Than One Receptacle or Receptacle Outlet*

Circuit Rating (Amperes)	Receptacle Rating (Amperes)
15	15
20	15 or 20
30	30
40	40 or 50
50	50

Exception No. 1: Receptacles installed exclusively for the use of cord-and-plug-connected arc welders shall be permitted to have ampere ratings not less than the minimum branch-circuit conductor ampacity determined by 630.11(A) or (B) for arc welders.

Exception No. 2: The ampere rating of a receptacle installed for electric discharge lighting shall be permitted to be based on 410.62(C).

(4) Range Receptacle Rating. The ampere rating of a range receptacle shall be permitted to be based on a single range demand load as specified in Table 220.55.

210.22 Permissible Loads, Individual Branch Circuits. An individual branch circuit shall be permitted to supply any load for which it is rated, but in no case shall the load exceed the branch-circuit ampere rating.

Δ **210.23 Permissible Loads, Multiple-Outlet Branch Circuits.** In no case shall the load exceed the branch-circuit ampere rating. A branch circuit supplying two or more outlets or receptacles shall supply only the loads specified according to its size in accordance with 210.23(A) through (E) and as summarized in 210.24.

N **(A) 10-Ampere Branch Circuits.** A 10-ampere branch circuit shall comply with the requirements of 210.23(A)(1) and (A)(2).

N **(1) Loads Permitted for 10-Ampere Branch Circuits.** A 10-ampere branch circuit shall be permitted to supply one or more of the following:

(1) Lighting outlets
(2) Dwelling unit exhaust fans on bathroom or laundry room lighting circuits
(3) A gas fireplace unit supplied by an individual branch circuit

N **(2) Loads Not Permitted for 10-Ampere Branch Circuits.** A 10-ampere branch circuit shall not supply any of the following:

(1) Receptacle outlets
(2) Fixed appliances, except as permitted for individual branch circuits
(3) Garage door openers
(4) Laundry equipment

(B) 15- and 20-Ampere Branch Circuits. A 15- or 20-ampere branch circuit shall be permitted to supply lighting outlets, lighting units, or other utilization equipment, or any combination of them, and shall comply with 210.23(B)(1) and (B)(2).

Exception: The small-appliance branch circuits, laundry branch circuits, and bathroom branch circuits required in a dwelling unit(s) by 210.11(C)(1), (C)(2), and (C)(3) shall supply only the receptacle outlets specified in that section.

(1) Cord-and-Plug-Connected Equipment Not Fastened in Place. The rating of any one cord-and-plug-connected utilization equipment not fastened in place shall not exceed 80 percent of the branch-circuit ampere rating.

(2) Utilization Equipment Fastened in Place. The total rating of utilization equipment fastened in place, other than luminaires, shall not exceed 50 percent of the branch-circuit ampere rating where lighting units, cord-and-plug-connected utilization equipment not fastened in place, or both, are also supplied.

A 15- or 20-ampere branch circuit supplying lighting outlets may also supply utilization equipment fastened in place, such as a dishwasher. The utilization equipment load, whether direct wired or cord-and-plug-connected, must not exceed 50 percent of the branch-circuit ampere rating (7.5 amperes on a 15-ampere circuit and 10 amperes on a 20-ampere circuit).

The requirement does not apply to a branch circuit that supplies only fastened-in-place utilization equipment. For example, on a 20-ampere branch circuit that supplies a waste disposer and a dishwasher, neither appliance is restricted to 50 percent (10 amperes) of the branch-circuit rating. The combined load of the two appliances cannot exceed 20 amperes.

(C) 30-Ampere Branch Circuits. A 30-ampere branch circuit shall be permitted to supply fixed lighting units with heavy-duty lampholders in other than a dwelling unit(s) or utilization equipment in any occupancy. The rating of any one cord-and-plug-connected utilization equipment shall not exceed 80 percent of the branch-circuit ampere rating.

(D) 40- and 50-Ampere Branch Circuits. A 40- or 50-ampere branch circuit shall be permitted to supply cooking appliances that are fastened in place in any occupancy. In other than dwelling units, such circuits shall be permitted to supply fixed lighting units with heavy-duty lampholders, infrared heating units, or other utilization equipment.

(E) Branch Circuits Larger Than 50 Amperes. Branch circuits larger than 50 amperes shall supply only nonlighting outlet loads.

See also

210.18, Exception No. 1, and the commentary regarding multioutlet branch circuits greater than 50 amperes that are permitted to supply nonlighting outlet loads at industrial premises

210.24 Branch-Circuit Requirements — Summary. The requirements for circuits that have two or more outlets or receptacles, other than the receptacle circuits of 210.11(C)(1), (C)(2), and (C)(3), are summarized in Table 210.24(1) for copper conductors and Table 210.24(2) for aluminum and copper-clad aluminum conductors. Table 210.24(1) and Table 210.24(2) provide only a summary of minimum requirements. See 210.19, 210.20, and 210.21 for the specific requirements applying to branch circuits.

Table 210.24(1) summarizes the general branch-circuit requirements where two or more outlets are supplied by copper conductors. Table 210.24(2) summarizes the general branch-circuit requirements where two or more outlets are supplied by aluminum or copper-clad conductors. Small appliance, laundry, and bathroom circuits supplying receptacles are not included, and no allowance is made for conditions of use or adjustments.

The circuit rating is determined by the rating of the overcurrent device based on 210.18. The tap conductor ampacity is governed by Exception No. 1 of 210.19(D). The circuit conductor and tap conductor size are directly from the 60°C column of Table 310.16. The receptacle ratings are from Table 210.21(B)(3).

210.25 Branch Circuits in Buildings with More Than One Occupancy.

(A) Dwelling Unit Branch Circuits. Branch circuits in each dwelling unit shall supply only loads within that dwelling unit or loads associated only with that dwelling unit.

(B) Common Area Branch Circuits. Branch circuits installed for lighting, central alarm, signal, communications, or other purposes for public or common areas of a two-family dwelling, a multifamily dwelling, or a multi-occupancy building shall not be supplied from equipment that supplies an individual dwelling unit or tenant space.

Informational Note: Examples of public or common areas include, but are not limited to, lobbies, corridors, stairways, laundry rooms, roofs, elevators, washrooms, store rooms, driveways (parking), and mechanical rooms.

Δ **TABLE 210.24(1)** *Summary of Branch-Circuit Requirements — Copper Conductors*

Circuit Rating	10 A	15 A	20 A	30 A	40 A	50 A
Conductors (min. size):						
Circuit wires	14	14	12	10	8	6
Taps	14	14	14	14	12	12
Fixture wires and cords			See 240.5.			
Overcurrent Protection	**10 A**	**15 A**	**20 A**	**30 A**	**40 A**	**50 A**
Outlet devices:						
Lampholders permitted	Any type	Any type	Any type	Heavy duty	Heavy duty	Heavy duty
Receptacle rating[1]	Not applicable[2]	15 max. A	15 A or 20 A	30 A	40 A or 50 A	50 A
Maximum Load	**10 A**	**15 A**	**20 A**	**30 A**	**40 A**	**50 A**
Permissible load	See 210.23(A).	See 210.23(B).	See 210.23(B).	See 210.23(C).	See 210.23(D).	See 210.23(D).

[1]For receptacle rating of cord-connected electric-discharge luminaires, see 410.62(C).
[2]Branch circuits rated 10-amperes shall not supply receptacle outlets.

N **TABLE 210.24(2)** *Summary of Branch-Circuit Requirements — Aluminum and Copper-Clad Aluminum Conductors*

Circuit Rating	10 A	15 A	20 A	30 A	40 A	50 A
Conductors (min. size):						
Circuit wires	12	12	10	8	6	4
Taps	12	12	12	12	10	10
Fixture wires and cords			See 240.5.			
Overcurrent Protection	**10 A**	**15 A**	**20 A**	**30 A**	**40 A**	**50 A**
Outlet devices:						
Lampholders permitted	Any type	Any type	Any type	Heavy duty	Heavy duty	Heavy duty
Receptacle rating[1]	Not applicable[2]	15 max. A	15 A or 20 A	30 A	40 A or 50 A	50 A
Maximum Load	**10 A**	**15 A**	**20 A**	**30 A**	**40 A**	**50 A**
Permissible load	See 210.23(A).	See 210.23(B).	See 210.23(B).	See 210.23(C).	See 210.23(D).	See 210.23(D).

[1]For receptacle rating of cord-connected electric-discharge luminaires, see 410.62(C).

[2]Branch circuits rated 10-amperes shall not supply receptacle outlets.

In addition to prohibiting branch circuits from feeding more than one dwelling unit, 210.25 also prohibits an individual dwelling unit branch circuit from supplying shared systems, equipment, or common lighting. Common area circuits in occupancies other than dwelling units are subject to this requirement. "House load" branch circuits must be supplied from equipment that does not directly supply branch circuits for an individual occupancy or tenant space. This requirement permits access to the branch-circuit disconnecting means without the need to enter the space of any tenants. The requirement also prevents a tenant from turning off important circuits that could affect other tenants.

Part III. Required Outlets

210.50 Receptacle Outlets. Receptacle outlets shall be installed as specified in 210.52 through 210.65.

Informational Note: See Informative Annex J for information regarding ADA accessibility design.

(A) Cord Pendants. A cord connector that is supplied by a permanently connected cord pendant shall be considered a receptacle outlet.

(B) Cord Connections. A receptacle outlet shall be installed wherever flexible cords with attachment plugs are used. Where flexible cords are permitted to be permanently connected, receptacles shall be permitted to be omitted for such cords.

(C) Appliance Receptacle Outlets. Appliance receptacle outlets installed in a dwelling unit for specific appliances, such as laundry equipment, shall be installed within 1.8 m (6 ft) of the intended location of the appliance.

> **See also**
>
> **210.52(F)** and **210.11(C)(2)** for requirements regarding laundry receptacle outlets and branch circuits

210.52 Dwelling Unit Receptacle Outlets. This section provides requirements for 125-volt, 15- and 20-ampere receptacle outlets. The receptacles required by this section shall be in addition to any receptacle that is as follows:

(1) Part of a luminaire or appliance, or
(2) Controlled by a listed wall-mounted control device in accordance with 210.70(A)(1), Exception No. 1, or
(3) Located within cabinets or cupboards, or
(4) Located more than 1.7 m (5½ ft) above the floor

Permanently installed electric baseboard heaters equipped with factory-installed receptacle outlets or outlets provided as a separate assembly by the manufacturer shall be permitted as the required outlet or outlets for the wall space utilized by such permanently installed heaters. Such receptacle outlets shall not be connected to the heater circuits.

Informational Note: Listed baseboard heaters include instructions that may not permit their installation below receptacle outlets.

An outlet containing a duplex receptacle that is wired so that only one of the receptacles is controlled by a wall switch can be used to meet the receptacle outlet spacing requirement. However, if both halves are controlled by a wall switch or other wall-mounted control device, such as an occupancy sensor, an additional unswitched receptacle has to be installed to meet the receptacle outlet spacing requirement. Where both halves of the duplex receptacle are controlled by a wall switch, the occupant could run an extension cord from a receptacle that is not controlled by a switch to an appliance that requires continuous power.

(A) General Provisions. In every kitchen, family room, dining room, living room, parlor, library, den, sunroom, bedroom, recreation room, or similar room or area of dwelling units, receptacle outlets shall be installed in accordance with the general provisions specified in 210.52(A)(1) through (A)(4).

(1) Spacing. Receptacles shall be installed such that no point measured horizontally along the floor line of any wall space is more than 1.8 m (6 ft) from a receptacle outlet.

Receptacle outlets are to be installed so that an appliance or lamp with an attached flexible cord can be placed anywhere in the room near a wall and be within 6 feet of a receptacle, minimizing the need for occupants to use extension cords. The receptacle layout can be designed for intended utilization equipment or practical room use. For example, receptacles in a family room that are intended to serve home entertainment equipment can be grouped or placed for convenience. Receptacle outlets intended for window-type holiday lighting can be placed under windows. No point in any wall space is permitted to be more than 6 feet from a receptacle outlet, even where more receptacle outlets than the minimum required are installed in the room.

(2) Wall Space. As used in this section, a wall space shall include the following:

(1) Any space 600 mm (2 ft) or more in width (including space measured around corners) and unbroken along the floor line by doorways and similar openings, fireplaces, stationary appliances, and fixed cabinets that do not have countertops or similar work surfaces

(2) The space occupied by fixed panels in walls, excluding sliding panels

(3) The space afforded by fixed room dividers, such as free-standing bar-type counters or railings

(3) Floor Receptacles. Receptacle outlets in or on floors shall not be counted as part of the required number of receptacle outlets unless located within 450 mm (18 in.) of the wall.

Any wall space that is unbroken along the floor line by doorways, fireplaces, archways, and similar openings must be included in the measurement. The wall space may include two or more walls of a room (around corners) as illustrated in Exhibit 210.25.

Fixed room dividers, such as bar-type counters and railings, are required to be included in the 6-foot measurement. Fixed glass panels in exterior walls are counted as wall space, and a floor-type receptacle close to the wall can be used to meet the required spacing. Isolated, individual wall spaces 2 feet or more in width, which are often used for small pieces of furniture on which a lamp or an appliance might be placed, are required to have a receptacle outlet to preclude the use of an extension cord to supply equipment in such an isolated space.

The word *usable* does not appear at all in 210.52(A)(2) as a condition for determining compliance with the receptacle-spacing requirements. For example, to correctly determine the dimension of the wall line in a room, the wall space behind the swing of a door is included in the measurement. That does not mean that the receptacle outlet has to be located in that space, only that the space is included in the wall-line measurement.

(4) Countertop and Similar Work Surface Receptacle Outlets. Receptacles installed for countertop and similar work surfaces as specified in 210.52(C) shall not be considered as the receptacle outlets required by 210.52(A).

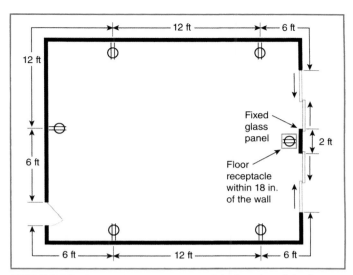

EXHIBIT 210.25 A typical room plan view of the location of dwelling unit receptacles that meet NEC requirements.

Because of the need to provide a sufficient number of receptacles for the appliances used at the kitchen counter area, receptacle outlets installed to serve kitchen or dining area counters cannot also be used as the required receptacle outlet for an adjacent wall space that is subject to the requirements of 210.52(A)(1) and (A)(2).

(B) Small Appliances.

Δ **(1) Receptacle Outlets Served.** In the kitchen, pantry, breakfast room, dining room, or similar area of a dwelling unit, the two or more 20-ampere small-appliance branch circuits required by 210.11(C)(1) shall serve all wall and floor receptacle outlets covered by 210.52(A), all countertop outlets covered by 210.52(C), and receptacle outlets for refrigeration equipment.

The limited exceptions to 210.52(B)(2) keep loads for specific equipment to a minimum so that the majority of the circuit capacity is dedicated to supplying cord-and-plug-connected portable appliance loads. The *NEC* restricts the loads supplied by these receptacle circuits because the number of cord-and-plug-connected portable appliances used by occupants is generally undetermined. Receptacles and other types of outlets in other locations, such as in cabinets, are not permitted to be connected to the small-appliance branch circuit. Such outlets reduce the capacity to supply portable appliances used at a kitchen counter.

No restriction is placed on the number of outlets connected to a general-lighting or small-appliance branch circuit. The minimum number of receptacle outlets in a room is determined by 210.52(A) based on the room perimeter and on 210.52(C) for counter spaces. Installing more than the required minimum number of receptacle outlets can also help reduce the need for extension cords and cords lying across counters.

Exhibit 210.26 illustrates the application of the requirements of 210.52(B)(1), (B)(2), and (B)(3). Only the counter area is required to be supplied by both small-appliance branch circuits.

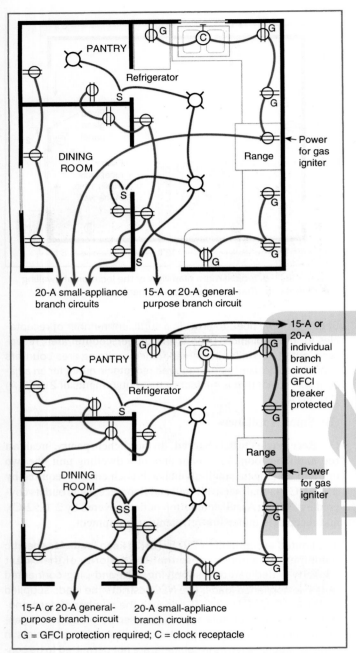

20-A small-appliance branch circuits

15-A or 20-A general-purpose branch circuit

15-A or 20-A individual branch circuit GFCI breaker protected

15-A or 20-A general-purpose branch circuit

20-A small-appliance branch circuits

G = GFCI protection required; C = clock receptacle

EXHIBIT 210.26 Room plan view showing small-appliance branch circuits as required for all receptacle outlets in the kitchen (including refrigerator), pantry, and dining room.

The wall receptacle outlets in the kitchen and dining room are permitted to be supplied by one or both of the circuits as shown in the two diagrams.

Exception No. 1: In addition to the required receptacles specified by 210.52, switched receptacles supplied from a general-purpose 15- or 20-ampere branch circuit shall be permitted in accordance with 210.70(A)(1), Exception No. 1.

Switched receptacles supplied from general-purpose 15-ampere branch circuits may be located in kitchens, pantries, breakfast rooms, and similar areas. See 210.70(A) and Exhibit 210.27 for details.

Exception No. 2: In addition to the required receptacles specified by 210.52, a receptacle outlet to serve a specific appliance shall be permitted to be supplied from an individual branch circuit rated 15 amperes or greater.

This exception provides a choice for installing receptacle outlets to supply refrigeration equipment or other specific appliances in a kitchen or similar area. An individual 15-ampere or larger branch circuit can serve this equipment, or it can be included in the 20-ampere small-appliance branch circuit. A receptacle supplying refrigeration equipment is exempt from the GFCI requirements of 210.8 where the receptacle is greater than 6 feet from the edge of any sink and where it is located so that it cannot be used to serve countertop surfaces, as shown in Exhibit 210.27.

(2) No Other Outlets. The two or more small-appliance branch circuits specified in 210.52(B)(1) shall have no other outlets.

Exception No. 1: A receptacle installed solely for the electrical supply to and support of an electric clock in any of the rooms specified in 210.52(B)(1) shall be permitted to be served by a small-appliance branch circuit.

Exception No. 2: Receptacles installed to provide power for supplemental equipment and lighting on gas-fired ranges, ovens, or counter-mounted cooking units shall be permitted to be served by a small-appliance branch circuit.

Because of the comparatively small load associated with ignition controls and other electronics, gas-fired appliances are permitted to be supplied by a receptacle connected to one of the small-appliance branch circuits. See Exhibit 210.26 for an illustration.

(3) Kitchen Receptacle Requirements. Receptacles installed in a kitchen to serve countertop surfaces shall be supplied by not fewer than two small-appliance branch circuits, either or both of which shall also be permitted to supply receptacle outlets in the same kitchen and in other rooms specified in 210.52(B)(1). Additional small-appliance branch circuits shall be permitted to supply receptacle outlets in the kitchen and other rooms specified in 210.52(B)(1). No small-appliance branch circuit shall serve more than one kitchen.

In most dwellings, the countertop receptacle outlets supply more of the portable cooking appliances than the wall receptacles in the kitchen and dining areas, hence the requirement for the counter areas to be supplied by no fewer than two small-appliance branch circuits. The *NEC* does not specify that both circuits be installed to serve the receptacle outlet(s) at each separate counter area in a kitchen, but rather that the total counter area of a kitchen must be supplied by no fewer than two circuits; the arrangement of the circuits is determined by the designer or installer.

The concept of evenly proportioning the load as specified in 210.11(B) (for loads calculated on the basis of volt-amperes per square foot) can be used as a best practice in distributing the number of receptacle outlets to be supplied by each of the small-appliance branch circuits. If additional small-appliance branch circuits are installed, they are subject to all the requirements that apply to the minimum two required circuits.

The two circuits that supply the countertop receptacle outlets can also supply receptacle outlets in the pantry, dining room, and breakfast room, as well as an electric clock receptacle and electric loads associated with gas-fired appliances. However, these circuits are to supply no other outlets/loads such as outdoor lighting or an outdoor receptacle.

(C) Countertops and Work Surfaces. In kitchens, pantries, breakfast rooms, dining rooms, and similar areas of dwelling units, receptacle outlets for countertop and work surfaces that are 300 mm (12 in.) or wider shall be installed in accordance with 210.52(C)(1) through (C)(3) and shall not be considered as the receptacle outlets required by 210.52(A).

For the purposes of this section, where using multioutlet assemblies, each 300 mm (12 in.) of multioutlet assembly containing two or more receptacles installed in individual or continuous lengths shall be considered to be one receptacle outlet.

(1) Wall Spaces. Receptacle outlets shall be installed so that no point along the wall line is more than 600 mm (24 in.) measured horizontally from a receptacle outlet in that space. The location of the receptacles shall be in accordance with 210.52(C)(3).

Dwelling unit receptacles within kitchens, dining areas, and similar rooms, as illustrated in Exhibit 210.27, are required to be installed as follows:

1. In each wall space wider than 12 inches and spaced so that no point along the wall line is more than 24 inches from a receptacle
2. For countertop islands and peninsular countertops, according to 210.52(C)(2) and 210.52(C)(3) [See commentary following 210.52(C)(3).]
3. Accessible for use and not blocked by appliances occupying dedicated space or fastened in place
4. Supplied by the two (more are permitted) required 20-ampere small-appliance branch circuits and GFCI protected according to 210.8(A)(6)
5. Not more than 20 inches above the countertop (According to 406.5, receptacles cannot be installed in a face-up position, unless in specifically listed countertop or work surface applications. Receptacles installed in a face-up position in a countertop could collect crumbs, liquids, and other debris, resulting in a potential fire or shock hazard.)

Exception No. 1: Receptacle outlets shall not be required directly behind a range, counter-mounted cooking unit, or sink in the installation described in Figure 210.52(C)(1).

This exception and Figure 210.52(C)(1) define the wall space behind a sink, range, or counter-mounted cooking unit that is not

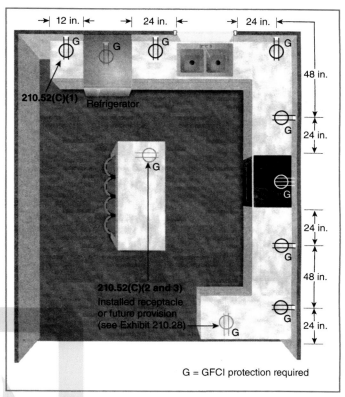

G = GFCI protection required

EXHIBIT 210.27 Dwelling unit receptacles in a kitchen and arranged in accordance with 210.52(C)(1) through (C)(3).

required to be provided with a receptacle outlet. Figure 210.52(C)(1) shows the wall space behind the sink or range that is exempt from the wall line measurement. Where the space behind a sink, range, or counter-mounted cooking unit is 12 inches or more (or 18 inches for corner-mounted configurations), the space must be included in measuring the wall counter space. Receptacle outlets are not prohibited from being installed in this space.

Exception No. 2: Where a required receptacle outlet cannot be installed in the wall areas shown in Figure 210.52(C)(1), the receptacle outlet shall be permitted to be installed as close as practicable to the countertop area to be served. The total number of receptacle outlets serving the countertop shall not be less than the number needed to satisfy 210.52(C)(1). These outlets shall be located in accordance with 210.52(C)(3).

Δ **(2) Island and Peninsular Countertops and Work Surfaces.** Receptacle outlets, if installed to serve an island or peninsular countertop or work surface, shall be installed in accordance with 210.52(C)(3). If a receptacle outlet is not provided to serve an island or peninsular countertop or work surface, provisions shall be provided at the island or peninsula for future addition of a receptacle outlet to serve the island or peninsular countertop or work surface.

Δ **(3) Receptacle Outlet Location.** Receptacle outlets shall be located in one or more of the following:

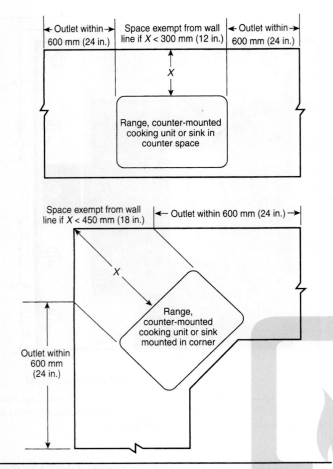

△ **FIGURE 210.52(C)(1)** *Determination of Area Behind a Range, Counter-Mounted Cooking Unit, or Sink.*

(1) On or above, but not more than 500 mm (20 in.) above, a countertop or work surface
(2) In a countertop using receptacle outlet assemblies listed for use in countertops
(3) In a work surface using receptacle outlet assemblies listed for use in work surfaces or listed for use in countertops

Receptacle outlets rendered not readily accessible by appliances fastened in place, appliance garages, sinks, or rangetops as covered in 210.52(C)(1), Exception No. 1, or appliances occupying assigned spaces shall not be considered as these required outlets.

Informational Note No. 1: See 406.5(E) for installation of receptacles in countertops and 406.5(F) for installation of receptacles in work surfaces. See 380.10 for installation of multioutlet assemblies.

Informational Note No. 2: See Informative Annex J and ANSI/ICC A117.1-2009, *Standard on Accessible and Usable Buildings and Facilities*, for additional information.

If a receptacle is installed to serve an island or peninsular countertop, it must be GFCI protected and installed based on Option 1 or Option 2 as depicted in Exhibit 210.28. Option 1 permits the receptacle to be installed above the countertop within an elevated backsplash space, but not more than 20 inches above the countertop. Option 2 allows the receptacle to be installed within the countertop surface within a listed assembly as specified in 210.52(C)(3)(1) and (2). If Option 1 or Option 2 is not selected at the time of installation, the *NEC* requires a default to Option 3, which requires a provision to be added at the island so a receptacle could be installed in the future. One means of accomplishing Option 3 could be installing a junction box, with the appropriate *NEC*-required small-appliance branch circuit, located within the island or peninsular cabinet space.

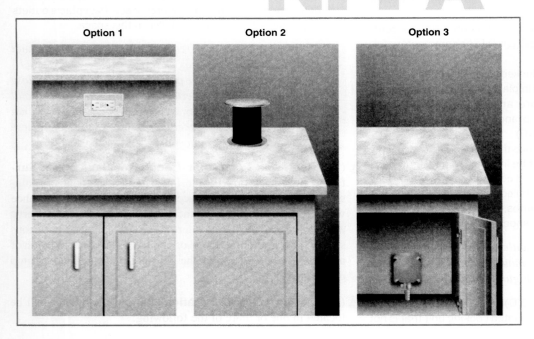

EXHIBIT 210.28 Three separate options for installation, or non-installation, of island and peninsula receptacles: (1) above the counter, (2) in the countertop as part of a listed assembly, or (3) a future provision to install a receptacle.

(D) Bathrooms. At least one receptacle outlet shall be installed in bathrooms within 900 mm (3 ft) of the outside edge of each sink. The receptacle outlet shall be located on a wall or partition that is adjacent to the sink or sink countertop, located on the countertop, or installed on the side or face of the sink cabinet. In no case shall the receptacle be located more than 300 mm (12 in.) below the top of the sink or sink countertop. Receptacle outlet assemblies listed for use in countertops shall be permitted to be installed in the countertop.

> Informational Note: See 406.5(E) and 406.5(G) for requirements on installation of receptacles in countertops.

The permission to install a receptacle outlet on the basin cabinet is not contingent on the adjacent wall location being infeasible. This receptacle is also required to be GFCI protected in accordance with 210.8(A)(1). This receptacle is required in addition to any receptacle that is part of any luminaire or medicine cabinet. If there is more than one basin, a receptacle outlet is required adjacent to each basin location. One receptacle outlet within 36 inches off both basins can be used to satisfy this requirement.

Section 210.11(C)(3) requires that receptacle outlets be supplied from a 20-ampere branch circuit with no other outlets. However, this circuit is permitted to supply the required receptacles in more than one bathroom. If the circuit supplies the required receptacle outlet in only one bathroom, it is also allowed to supply lighting and an exhaust fan in that bathroom, provided the lighting and fan load does not exceed that permitted by 210.23(B)(2). Receptacle outlets installed in addition to the one(s) required by 210.52 (D) to serve countertop or other work surfaces are also permitted to be supplied by the required 20-ampere branch circuit.

> **See also**
>
> **406.9(C),** which prohibits installation of a receptacle over a bathtub or inside a shower stall
>
> **Exhibit 210.8** for a sample electrical layout of a bathroom

△ **(E) Outdoor Outlets.** Outdoor receptacle outlets shall be installed in accordance with 210.52(E)(1) through (E)(3).

Outdoor receptacles must be installed so that the receptacle faceplate rests securely on the supporting surface to prevent moisture from entering the enclosure. On uneven surfaces such as brick, stone, or stucco, it might be necessary to use mounting blocks or to close openings with caulking compound or mastic.

> **See also**
>
> **406.9** for more information on receptacles installed in damp or wet locations

(1) One-Family and Two-Family Dwellings. For a one-family dwelling and each unit of a two-family dwelling that is at grade level, at least one receptacle outlet readily accessible from grade and not more than 2.0 m (6½ ft) above grade level shall be installed at the front and back of the dwelling.

Two outdoor receptacle outlets are required for each dwelling unit. One receptacle outlet is required at the front and one at the back as shown in Exhibit 210.29. The two required receptacle outlets are to be available to a person standing on the ground (at grade level). Where outdoor heating, air-conditioning, and refrigeration equipment is located at grade level, the receptacle outlets required by this section can be used to comply with the requirement of 210.63, provided that one of the outlets is located within 25 feet of the equipment. Outdoor receptacle outlets on decks, porches, and similar structures can be used to meet 210.52(E) as long as the receptacle outlet is not more than 6½ feet above grade and can be accessed by a person standing at grade.

(2) Multifamily Dwellings. For each dwelling unit of a multifamily dwelling where the dwelling unit is located at grade level and provided with individual exterior entrance/egress, at least one receptacle outlet readily accessible from grade and not more than 2.0 m (6½ ft) above grade level shall be installed.

EXHIBIT 210.29 Three one-family dwellings (separated from each other by fire walls) with GFCI-protected receptacles located at the front and the back of each dwelling.

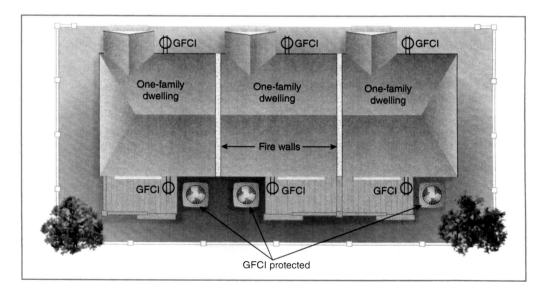

This requirement applies to the dwelling units that are located at grade level and have a doorway that leads directly to the exterior of the structure. The unauthorized use of the outdoor receptacle outlet by other than the dwelling occupant(s) can be allayed by a switch inside the dwelling unit that controls the receptacle. Due to the spatial constraints often associated with the construction of multifamily units, the required receptacle outlet is permitted to be accessible "from grade" rather than "while standing at grade," as is the case for one- and two-family dwellings. An outdoor receptacle outlet located 6½ feet or less above grade and accessible by walking up a set of deck or porch steps can be used to meet this requirement.

(3) Balconies, Decks, and Porches. Balconies, decks, and porches that are within 102 mm (4 in.) horizontally of the dwelling unit shall have at least one receptacle outlet accessible from the balcony, deck, or porch. The receptacle outlet shall not be located more than 2.0 m (6½ ft) above the balcony, deck, or porch walking surface.

Regardless of area, a porch, balcony, or deck must have at least one receptacle outlet installed within its perimeter. This requirement applies only to porches, balconies, or decks that are attached to or within 4 inches horizontally from the unit. The porch, balcony, or deck does not need to be accessible from inside the unit for the requirement to apply. Because it is an outdoor receptacle, GFCI protection is required. Depending on the location of the porch, balcony, or deck, the receptacle outlet can also be used to meet the receptacle requirements of 210.52(E)(1) and (E)(2). The receptacle must be not more than 6½ feet above the balcony, deck, or porch walking surface.

(F) Laundry Areas. In dwelling units, at least one receptacle outlet shall be installed in areas designated for the installation of laundry equipment.

Exception No. 1: A receptacle for laundry equipment shall not be required in a dwelling unit of a multifamily building where laundry facilities are provided on the premises for use by all building occupants.

Exception No. 2: A receptacle for laundry equipment shall not be required in other than one-family dwellings where laundry facilities are not to be installed or permitted.

A laundry receptacle outlet(s) is supplied by a 20-ampere branch circuit that can have no other outlets.

See also

210.11(C)(2) for more information on laundry branch circuits

(G) Basements, Garages, and Accessory Buildings. For one- and two-family dwellings, and multifamily dwellings, at least one receptacle outlet shall be installed in the areas specified in 210.52(G)(1) through (G)(3). These receptacles shall be in addition to receptacles required for specific equipment. Receptacles supplying only a permanently installed premises security system shall not be considered as meeting these requirements.

(1) Garages. In each attached garage and in each detached garage with electric power, at least one receptacle outlet shall be installed in each vehicle bay and not more than 1.7 m (5½ ft) above the floor.

Exception: Garage spaces not attached to an individual dwelling unit of a multifamily dwelling shall not require a receptacle outlet in each vehicle bay.

(2) Accessory Buildings. In each accessory building with electric power.

(3) Basements. In each separate unfinished portion of a basement.

Receptacle outlets are not required for a detached garage if it is not supplied with electricity. Receptacle outlets are not required for individual garages of multifamily dwellings where the garage is not attached to the dwelling unit. An example of how this applies is a multifamily dwelling with a separate detached garage structure. Each of the dwelling units is provided with a garage space but the space is not attached to the dwelling unit, therefore a receptacle outlet is not required in each of the detached garage spaces.

GFCI protection is required by 210.8(A)(5) for receptacles in finished and unfinished basements and by 210.8(A)(2) for receptacles installed in garages. Circuits that supply receptacle outlets in a garage can supply receptacle outlets only in the garage and readily accessible outdoor receptacle outlets.

See also

210.11(C)(4) for more information on garage branch circuits

(H) Hallways. In dwelling units, hallways of 3.0 m (10 ft) or more in length shall have at least one receptacle outlet.
    As used in this subsection, the hallway length shall be considered the length along the centerline of the hallway without passing through a doorway.

This requirement is intended to minimize strain or damage to cords and receptacles for dwelling unit receptacles. The requirement does not apply to common hallways of hotels, motels, apartment buildings, condominiums, and similar occupancies.

(I) Foyers. Foyers that are not part of a hallway in accordance with 210.52(H) and that have an area that is greater than 5.6 m^2 (60 ft^2) shall have a receptacle(s) located in each wall space 900 mm (3 ft) or more in width. Doorways, door-side windows that extend to the floor, and similar openings shall not be considered wall space.

Foyers are not included in the living spaces covered by 210.52(A) even though they may have comparable dimensions in some cases. This requirement provides for receptacle outlets in these spaces to accommodate lamps or other utilization equipment. The primary objective of the requirements covering any of the dwelling areas included in 210.52 is to minimize the need to use extension cords to supply utilization equipment.

210.60 Guest Rooms, Guest Suites, Dormitory Units, and Similar Occupancies.

(A) General. Guest rooms or guest suites in hotels or motels, sleeping rooms in dormitory units, and similar occupancies shall have receptacle outlets installed in accordance with 210.52(A) and (D). Guest rooms or guest suites provided with permanent provisions for cooking shall have receptacle outlets installed in accordance with all of the applicable rules in 210.52.

(B) Receptacle Placement. The total number of receptacle outlets shall not be less than required in 210.52(A). These receptacle outlets shall be permitted to be located conveniently for permanent furniture layout. At least two receptacle outlets shall be readily accessible. Where receptacles are installed behind the bed, the receptacle shall be located to prevent the bed from contacting any attachment plug that may be installed or the receptacle shall be provided with a suitable guard.

The receptacles in guest rooms and guest suites of hotels and motels and in dormitories are permitted to be placed in accessible locations that are compatible with permanent furniture. However, the minimum number of receptacles required by 210.52 is not permitted to be reduced and should be determined by assuming furniture is not in the room. The practical locations of that minimum number of receptacles are then determined based on the permanent furniture layout.

Hotel and motel rooms and suites are commonly used as remote offices for businesspeople who use laptop computers and other plug-in devices. In dormitories, the use of electrical and electronic equipment can be significant, necessitating accessible receptacle outlets. The *NEC* requires that two receptacle outlets be readily accessible without requiring the movement of furniture to access those receptacles. To reduce the risk of bedding material fires, receptacles located behind beds must include guards if attachment plugs could contact the bed to protect against cords being pinched and damaged by bedframes, mattresses, or box springs.

Extended-stay hotels and motels are often equipped with permanent cooking equipment and countertop areas. All applicable receptacle spacing and supply requirements in 210.52 apply to guest rooms or suites that contain such provisions. A portable microwave oven is not considered to be a permanently installed cooking appliance.

Exhibit 210.30 shows a hotel guest room in which the receptacles are located based on the permanent furniture layout. Because of this rule, some spaces that are 2 feet or more in width have no receptacle outlets. However, the receptacle locations are compatible with the permanent furniture layout. In Exhibit 210.30, the receptacle outlet adjacent to the permanent dresser is needed because 210.60(B) applies only to the location of receptacle outlets, not to the minimum number of receptacle outlets.

> **See also**
>
> **210.17** and its commentary for more information on hotel and motel guest rooms and guest suites that are equipped with permanent provisions for cooking

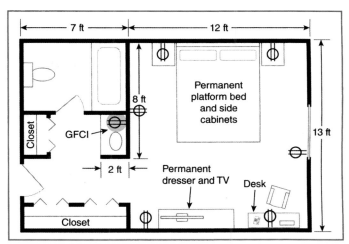

EXHIBIT 210.30 Floor plan of a hotel guest room with receptacles located with respect to permanent furniture.

210.62 Show Windows. At least one 125-volt, single-phase, 15- or 20-ampere-rated receptacle outlet shall be installed within 450 mm (18 in.) of the top of each show window. No point along the top of the window shall be farther than 1.8 m (6 ft) from a receptacle outlet.

Some show windows extend from floor to ceiling for maximum display. To discourage floor receptacles and the use of extension cords, receptacles must be installed directly above the show window, within 18 inches of the top, and no point along the top of the window is permitted to be more than 6 feet from a receptacle. This requirement requires the use of 125-volt, 15- or 20-ampere receptacles. Such a receptacle would still be required if a circuit were provided for a 24-volt lighting system.

> **See also**
>
> **220.14(G)** and **220.46(A)** for information regarding load calculations for show windows

210.63 Equipment Requiring Servicing. A 125-volt, single-phase, 15- or 20-ampere-rated receptacle outlet shall be installed at an accessible location within 7.5 m (25 ft) of the equipment as specified in 210.63(A) and (B).

> Informational Note: See 210.8(E) for requirements on GFCI protection.

(A) Heating, Air-Conditioning, and Refrigeration Equipment. The required receptacle outlet shall be located on the same level as the heating, air-conditioning, and refrigeration equipment. The receptacle outlet shall not be connected to the load side of the equipment's branch-circuit disconnecting means.

Exception: A receptacle outlet shall not be required at one- and two-family dwellings for the service of evaporative coolers.

Requiring a permanently installed receptacle within 25 feet of equipment that requires servicing improves worker safety by eliminating the need to employ makeshift methods of obtaining 120-volt power for servicing and troubleshooting. The exception exempts evaporative coolers (commonly referred to as swamp coolers) from the receptacle requirement where the cooler is installed at a one- or two-family dwelling. Although this type of cooling equipment is exempt from 210.63, one- and two-family dwellings are required to have outdoor receptacle outlets at the front and the back of the structure in accordance with 210.52(E).

(B) Other Electrical Equipment. In other than one- and two-family dwellings, a receptacle outlet shall be located as specified in 210.63(B)(1) and (B)(2).

(1) Indoor Service Equipment. The required receptacle outlet shall be located within the same room or area as the service equipment.

Δ **(2) Indoor Equipment Requiring Dedicated Equipment Spaces.** Where equipment, other than service equipment, requires dedicated equipment space as specified in 110.26(E), the required receptacle outlet shall be located within the same room or area as the electrical equipment and shall not be connected to the load side of the equipment's disconnecting means.

The required receptacle is intended to facilitate the use of portable test and diagnostic equipment that requires a 120-volt power source. The receptacle is not permitted to be on the load side of the equipment's branch-circuit disconnecting means so that the receptacle can be used while the disconnecting means is open. The receptacle is not required in one- or two-family dwellings.

210.65 Meeting Rooms.

(A) General. Each meeting room of not more than 93 m² (1000 ft²) in other than dwelling units shall have outlets for nonlocking-type, 125-volt, 15- or 20-ampere receptacles. The outlets shall be installed in accordance with 210.65(B). Where a room or space is provided with movable partition(s), each room size shall be determined with the partition in the position that results in the smallest size meeting room.

> Informational Note No. 1: For the purposes of this section, meeting rooms are typically designed or intended for the gathering of seated occupants for such purposes as conferences, deliberations, or similar purposes, where portable electronic equipment such as computers, projectors, or similar equipment is likely to be used.
> Informational Note No. 2: Examples of rooms that are not meeting rooms include auditoriums, schoolrooms, and coffee shops.

(B) Receptacle Outlets Required. The total number of receptacle outlets, including floor outlets and receptacle outlets in fixed furniture, shall not be less than as determined in 210.65(B)(1) and (B)(2).

(1) Receptacle Outlets in Fixed Walls. The required number of receptacle outlets shall be determined in accordance with 210.52(A)(1) through (A)(4). These receptacle outlets shall be permitted to be located as determined by the installer, designer, or building owner.

Δ **(2) Floor Outlets.** A meeting room with any floor dimension that is 3.7 m (12 ft) or greater in any direction and that has a floor area of at least 20 m² (215 ft²) shall have at least one floor receptacle outlet, or at least one floor outlet to serve a receptacle(s), located at a distance not less than 1.8 m (6 ft) from any fixed wall for each 20 m² (215 ft²) or fraction thereof.

> Informational Note No. 1: See 314.27(B) for requirements on floor boxes used for receptacles located in the floor.
> Informational Note No. 2: See 518.1 for requirements on assembly occupancies designed for 100 or more persons.

These requirements apply to meeting rooms in a variety of occupancies, including office buildings and hotels. For most rooms in commercial buildings, there are no *NEC* requirements for spacing of wall and countertop receptacles. However, 210.65 recognizes that there is a need to provide receptacles to facilitate meetings in which attendees will be using computers. Means of compliance include floor receptacle outlets and floor outlets supplying hard-wired furniture that contains receptacles. The 6-foot distance from fixed walls is to allow for emergency egress without occupants having to cross over flexible cords.

210.70 Lighting Outlets Required. Lighting outlets shall be installed where specified in 210.70(A), (B), and (C). The switch or wall-mounted control device shall not rely exclusively on a battery unless a means is provided for automatically energizing the lighting outlets upon battery failure.

Δ **(A) Dwelling Units.** In dwelling units, lighting outlets shall be installed in accordance with 210.70(A)(1) and (A)(2).

(1) Habitable Rooms, Kitchens, Laundry Areas, and Bathrooms. At least one lighting outlet controlled by a listed wall-mounted control device shall be installed in every habitable room, kitchen, laundry area, and bathroom. The wall-mounted control device shall be located near an entrance to the room on a wall.

Exception No. 1: In other than kitchens, laundry areas, and bathrooms, one or more receptacles controlled by a listed wall-mounted control device shall be permitted in lieu of lighting outlets.

Exception No. 2: Lighting outlets shall be permitted to be controlled by occupancy sensors that are (1) in addition to listed wall-mounted control devices or (2) located at a customary wall switch location and equipped with a manual override that will allow the sensor to function as a wall switch.

Lighting outlets are permitted to be controlled by wall switches and other wall-mounted control devices, including occupancy sensors. A controlled receptacle outlet is not permitted to serve as the required lighting outlet in kitchens, laundry areas, or bathrooms. Exhibit 210.26 shows a switched receptacle supplied by a 15-ampere general-purpose branch circuit in a dining room. A switched receptacle is not considered one of the receptacle outlets required by 210.52.

Occupancy sensors are permitted to be used for switching lighting outlets in habitable rooms, kitchens, and bathrooms, provided they are equipped with a manual override or are used in addition to standard switches.

Δ **(2) Additional Locations.** Additional lighting outlets shall be installed in accordance with the following:

(1) At least one lighting outlet controlled by a listed wall-mounted control device shall be installed in hallways, stairways, attached garages, detached garages, and accessory buildings with electric power.

(2) For dwelling units, attached garages, and detached garages with electric power, at least one exterior lighting outlet controlled by a listed wall-mounted control device shall be installed to provide illumination on the exterior side of outdoor entrances or exits with grade-level access. A vehicle door in a garage shall not be considered as an outdoor entrance or exit.

Exception to (2): For an outdoor, grade-level bulkhead door with stairway access to a sub-grade-level basement, the required lighting outlet that provides illumination on the stairway steps shall be permitted to be located in the basement interior within 1.5 m (5 ft) horizontally of the bottommost stairway riser. This interior lighting outlet shall be permitted to be controlled by a listed wall-mounted control device or by a unit switch of the interior luminaire or interior lampholder.

(3) Where lighting outlets are installed for an interior stairway with six or more risers between floor levels, there shall be a listed wall-mounted control device at each floor level and at each landing level that includes a stairway entry to control the lighting outlets.

Exception to (1), (2), and (3): Remote, central, or automatic control of lighting shall be permitted in hallways, in stairways, and at outdoor entrances.

(4) Dimmer control of lighting outlets installed in accordance with 210.70(A)(2)(3) shall not be permitted unless the listed control devices can provide dimming control to maximum brightness at each control location for the interior stairway illumination.

(B) Guest Rooms or Guest Suites. In hotels, motels, or similar occupancies, guest rooms or guest suites shall have at least one lighting outlet controlled by a listed wall-mounted control device installed in every habitable room and bathroom.

Exception No. 1: In other than bathrooms and kitchens where provided, one or more receptacles controlled by a listed wall-mounted control device shall be permitted in lieu of lighting outlets.

Exception No. 2: Lighting outlets shall be permitted to be controlled by occupancy sensors that are (1) in addition to listed wall-mounted control devices or (2) located at a customary wall switch location and equipped with a manual override that allows the sensor to function as a wall switch.

A lighting outlet is required in every habitable room (the hotel room or rooms in a suite) and in bathrooms. The lighting outlet must be operated through a wall-mounted control device. Kitchens are required to have at least one lighting outlet controlled by a wall-mounted control device. Rooms other than bathrooms and kitchens can have switched receptacles to meet this lighting requirement. Exception No. 2 permits the use of occupancy sensors to control the lighting outlet, provided it is in the typical switch location and can be manually controlled.

(C) All Occupancies. For attics and underfloor spaces, utility rooms, and basements, at least one lighting outlet containing a switch or controlled by a wall switch or listed wall-mounted control device shall be installed where these spaces are used for storage or contain equipment requiring servicing. A point of control shall be at each entry that permits access to the attic and underfloor space, utility room, or basement. Where a lighting outlet is installed for equipment requiring service, the lighting outlet shall be installed at or near the equipment.

ARTICLE 215 Feeders

Δ **215.1 Scope.** This article covers the installation requirements, overcurrent protection requirements, minimum size, and ampacity of conductors for feeders not over 1000 volts ac or 1500 volts dc, nominal.

Informational Note: See Part III of Article 235 for feeders over 1000 volts ac or 1500 volts dc.

Δ **215.2 Minimum Rating and Size.**

Δ **(A) General.** Feeder conductors shall have an ampacity not less than the larger of 215.2(A)(1) or (A)(2) and shall comply with 110.14(C).

N **(1) Continuous and Noncontinuous Loads.** Where a feeder supplies continuous loads or any combination of continuous and noncontinuous loads, the minimum feeder conductor size shall have an ampacity not less than the noncontinuous load plus 125 percent of the continuous load.

Exception No. 1: If the assembly, including the overcurrent devices protecting the feeder(s), is listed for operation at 100 percent of its rating, the ampacity of the feeder conductors shall be permitted to be not less than the sum of the continuous load plus the noncontinuous load.

Exception No. 2: Where a portion of a feeder is connected at both its supply and load ends to separately installed pressure connections as covered in 110.14(C)(2), it shall be permitted to have an ampacity not less than the sum of the continuous load plus the noncontinuous load. No portion of a feeder installed under this exception shall extend into an enclosure containing either the feeder supply or the feeder load terminations, as covered in 110.14(C)(1).

Exception No. 2 to 215.2(A)(1) addresses installations where feeders are installed in multiple segments. Feeders installed under this scenario comprise three segments: two segments at the terminations — supply and load — and an intervening segment. The allowable ampacity of the intervening segment is permitted to be calculated without applying 125 percent to the continuous load.

The two segments at the terminations of the feeder are sized to accommodate the effects of continuous loading on the termination devices. Their ampacities are selected and coordinated to not exceed the lowest temperature rating of any connected termination, conductor, or device, as covered in 110.14(C)(1).

The intervening segment must terminate in enclosures at the feeder source and destination. This segment is sized in accordance with the ampacity requirements for the conductor specifically to provide wiring that will accommodate the maximum current, whether or not any portion of that current is continuous. The ampacity of this intervening segment is limited only by the ampacity parameters that apply over the length of the intervening segment and the ampacity at the listed and identified temperature rating of the pressure connector, as covered in 110.14(C)(2).

Exhibit 215.1 shows a 400-ampere load supplied by a feeder comprising three segments. The segments at the overcurrent device and load terminate on terminals rated at 75°C and connect to the intervening load with pressure connectors rated at 90°C. A 600-kcmil, Cu THHN, Type MC cable is used at the supply and load ends. A 500-kcmil, Cu THHN, Type MC cable is used for the intervening segment.

Exception No. 3: Grounded conductors that are not connected to an overcurrent device shall be permitted to be sized at 100 percent of the continuous and noncontinuous load.

Feeder grounded/neutral conductors that do not connect to the terminals of an overcurrent protective device are not required to be sized based on 125 percent of the continuous load. For example, if the maximum unbalanced load on a feeder neutral is calculated per 220.61 to be 200 amperes and the load is considered to be continuous, the use of a 3/0 AWG, Type THW copper conductor is permitted as long as the conductor terminates at a neutral bus or terminal bar within the electrical distribution equipment.

N **(2) Ampacity Adjustment or Correction Factors.** The minimum feeder conductor size shall have an ampacity not less than the maximum load to be served after the application of any adjustment or correction factors in accordance with 310.14.

> Informational Note No. 1: See Informative Annex D for Examples D1 through D11.
>
> Informational Note No. 2: Conductors for feeders, as defined in Article 100, sized to prevent a voltage drop exceeding 3 percent at the farthest outlet of power, heating, and lighting loads, or combinations of such loads, and where the maximum total voltage drop on both feeders and branch circuits to the farthest outlet does not exceed 5 percent, will provide reasonable efficiency of operation.
>
> Informational Note No. 3: See 210.19, Informational Note for voltage drop for branch circuits.

Informational notes are not mandatory, as outlined in 90.5(C). Where circuit conductors are increased due to voltage drop, 250.122(B) requires an increase in circular mil area for the associated equipment grounding conductors.

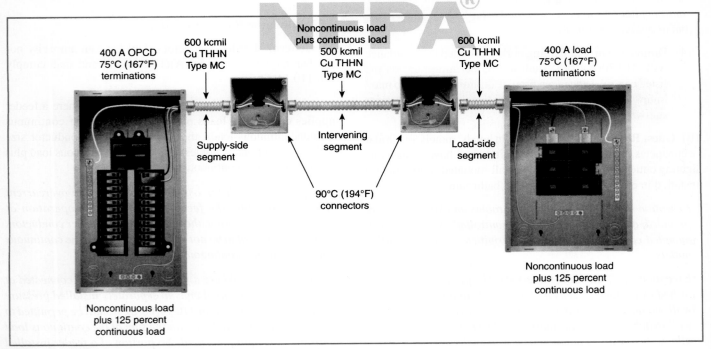

EXHIBIT 215.1 A feeder installed in accordance with 215.2(A)(1) Exception No. 2.

The resistance or impedance of conductors can cause a substantial difference between voltage at service equipment and voltage at the point-of-utilization equipment. Excessive voltage drop impairs the starting and the operation of electrical equipment. Undervoltage can result in inefficient operation of heating, lighting, and motor loads. An applied voltage of 10 percent below rating can result in a decrease in efficiency of substantially more than 10 percent — for example, fluorescent light output would be reduced by 15 percent, and incandescent light output would be reduced by 30 percent. Induction motors would run hotter and produce less torque. With an applied voltage of 10 percent below rating, the running current would increase 11 percent, and the operating temperature would increase 12 percent. At the same time, torque would be reduced 19 percent.

In addition to resistance or impedance, the type of raceway or cable enclosure, the type of circuit [alternating current (ac), direct current (dc), single-phase, 3-phase], and the power factor should be considered to determine voltage drop.

This basic formula can be used to determine the voltage drop in a 2-wire dc circuit, a 2-wire ac circuit, or a 3-wire ac single-phase circuit, all with a balanced load at 100 percent power factor and where reactance can be neglected:

$$VD = \frac{2 \times L \times R \times I}{1000}$$

where:

VD = voltage drop (based on conductor temperature of 75°C)
L = one-way length of circuit (ft)
R = conductor resistance in ohms (Ω) per 1000 ft (from Chapter 9, Table 8)
I = load current (amperes)

For 3-phase circuits (at 100 percent power factor), the voltage drop between any two phase conductors is 0.866 times the voltage drop calculated by the formula. Voltage-drop tables and calculations are available from various manufacturers.

See also

Chapter 9, Table 9, and its commentary for an example of voltage-drop calculation using ac reactance and resistance

(B) Grounded Conductor. The size of the feeder circuit grounded conductor shall not be smaller than the equipment grounding conductor size required by 250.122, except that 250.122(F) shall not apply where grounded conductors are run in parallel.

Additional minimum sizes shall be as specified in 215.2(C) under the conditions stipulated.

Using 250.122 to establish the minimum size grounded conductor in a feeder circuit provides a relationship between the grounded conductor and the feeder circuit overcurrent protective device that is the same as that used for sizing equipment-grounding conductors. It provides an adequate fault current path in the event of a fault between a phase conductor and the grounded conductor.

See also

235.202 for sizing requirements of the grounded feeder conductor for feeder circuits over 1000 volts ac or 1500 volts dc, nominal

220.61 and **310.10(G)** for determining the minimum grounded conductor size for feeder circuits installed in parallel in separate raceways or cables

(C) Ampacity Relative to Service Conductors. The feeder conductor ampacity shall not be less than that of the service conductors where the feeder conductors carry the total load supplied by service conductors with an ampacity of 55 amperes or less.

According to Table 310.16, a 3/0 AWG, Type THW copper conductor has an ampacity of 200 amperes. However, for a 3-wire, single-phase dwelling unit service, as shown in Exhibit 215.2, 310.12 permits a service conductor (or a main power feeder conductor) with an ampacity of 83 percent of the service rating. This permits a minimum of 2/0 AWG, Type THW copper conductors or 4/0 AWG, Type THW aluminum conductors for services or a main power feeder rated at 200 amperes.

215.3 Overcurrent Protection. Feeders shall be protected against overcurrent in accordance with Part I of Article 240. Where a feeder supplies continuous loads or any combination of continuous and noncontinuous loads, the rating of the overcurrent device shall not be less than the noncontinuous load plus 125 percent of the continuous load.

Exception: Where the assembly, including the overcurrent devices protecting the feeder(s), is listed for operation at 100 percent of its rating, the ampere rating of the overcurrent device shall be permitted to be not less than the sum of the continuous load plus the noncontinuous load.

215.4 Feeders with Common Neutral Conductor.

(A) Feeders with Common Neutral. Up to three sets of 3-wire feeders or two sets of 4-wire or 5-wire feeders shall be permitted to utilize a common neutral.

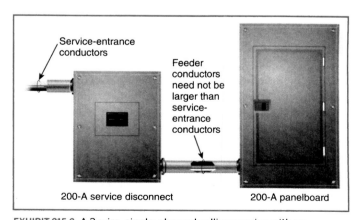

EXHIBIT 215.2 A 3-wire, single-phase dwelling service with an ampacity of 200 amperes for 2/0 AWG copper or 4/0 AWG aluminum conductors used as service-entrance conductors and feeder conductors.

(B) In Metal Raceway or Enclosure. Where installed in a metal raceway or other metal enclosure, all conductors of all feeders using a common neutral conductor shall be enclosed within the same raceway or other enclosure as required in 300.20.

If ac feeder conductors, including the neutral conductor, are installed in metal raceways, the conductors are required to be grouped together to avoid induction heating of the surrounding metal. If it is necessary to run parallel conductors through multiple metal raceways, conductors from each phase plus the neutral must be run in each raceway.

> **See also**
>
> **250.102(E), 250.134, 300.3, 300.5(I),** and **300.20** for requirements associated with conductor grouping of feeder circuits

A 3-phase, 4-wire (208Y/120 or 480Y/277) system is often used to supply both lighting and motor loads. The 3-phase motor loads typically are not connected to the neutral and, thus, will not cause current in the neutral conductor. The maximum current on the neutral is due to lighting loads or circuits where the neutral is used. On this type of system (3-phase, 4-wire), a demand factor of 70 percent is permitted by 220.61(B) for that portion of the neutral load in excess of 200 amperes.

For example, if the maximum possible unbalanced load is 500 amperes, the neutral would have to be large enough to carry 410 amperes (200 amperes plus 70 percent of 300 amperes, or 410 amperes). No reduction of the neutral capacity for that portion of the load consisting of electric-discharge lighting is permitted.

Section 310.15(E) points out that a neutral conductor must be counted as a current-carrying conductor if the load it serves consists of harmonic currents. The maximum unbalanced load for feeders supplying clothes dryers, household ranges, wall-mounted ovens, and counter-mounted cooking units is required to be considered 70 percent of the load on the ungrounded conductors.

> **See also**
>
> **220.61(B)** for other systems in which the 70-percent demand factor can be applied

215.5 Diagrams of Feeders. If required by the authority having jurisdiction, a diagram showing feeder details shall be provided prior to the installation of the feeders. Such a diagram shall show the area in square feet of the building or other structure supplied by each feeder, the total calculated load before applying demand factors, the demand factors used, the calculated load after applying demand factors, and the size and type of conductors to be used.

215.6 Feeder Equipment Grounding Conductor. Where a feeder supplies branch circuits in which equipment grounding conductors are required, the feeder shall include or provide an equipment grounding conductor, to which the equipment grounding conductors of the branch circuits shall be connected.

Where the feeder supplies a separate building or structure, the requirements of 250.32 shall apply.

215.7 Ungrounded Conductors Tapped from Grounded Systems. Two-wire dc circuits and ac circuits of two or more ungrounded conductors shall be permitted to be tapped from the ungrounded conductors of circuits having a grounded neutral conductor. Switching devices in each tapped circuit shall have a pole in each ungrounded conductor.

215.9 Ground-Fault Circuit-Interrupter Protection for Personnel. Feeders shall be permitted to be protected by a listed ground-fault circuit interrupter installed in a readily accessible location in lieu of the provisions for such interrupters as specified in 210.8 and 590.6(A).

Ground-fault circuit-interrupter (GFCI) protection of the feeder circuit protects all branch circuits supplied by that feeder. This type of GFCI installation is permitted in lieu of the branch-circuit GFCI requirements of 210.8. GFCI protection in the feeder can also be used to protect construction site receptacles, as covered in 590.6(A), provided the feeder supplies no lighting branch circuits.

Although it may be more economical or convenient to install GFCIs for feeders, consideration should be given to the possibility that a GFCI might be monitoring several branch circuits and will de-energize all branch circuits in response to a line-to-ground fault from one branch circuit.

Δ **215.10 Ground-Fault Protection of Equipment.** Each feeder disconnect rated 1000 amperes or more and installed on solidly grounded wye electrical systems of more than 150 volts to ground, but not exceeding 1000 volts phase-to-phase, shall be provided with ground-fault protection of equipment in accordance with 230.95.

> Informational Note: See 517.17 for buildings that contain health care occupancies.

Exception No. 1: This section shall not apply to a disconnecting means for a continuous industrial process where a nonorderly shutdown will introduce additional or increased hazards.

Prior to this requirement, an unusually high number of burn-downs on feeders and services operating in this voltage range were reported. Solidly grounded systems operating at 480Y/277 volts were the primary focus of this requirement when it was first introduced in the *NEC*®, but other solidly grounded, wye-connected systems operating over 150 volts to ground and not more than 600 volts phase-to-phase are covered by this requirement. Each ground-fault protection system must be performance tested and documented according to the requirements of 230.95(C) before being put into service.

Ground-fault protection of feeder equipment is not required if protection is provided on an upstream feeder or at the service. However, additional levels of ground-fault protection for feeders might be preferred so that a single ground fault does not de-energize the whole electrical system.

230.95 for further commentary on ground-fault protection of services

517.17, which requires an additional level of ground-fault protection for health care facilities

700.31 for the different ground-fault protection requirements for emergency feeders covered within the scope of Article 700

Exception No. 2: This section shall not apply if ground-fault protection of equipment is provided on the supply side of the feeder and on the load side of any transformer supplying the feeder.

Ground-fault protection installed in equipment supplying the primary of a transformer will not function to protect equipment supplied by the secondary of the transformer. If the equipment supplied by the secondary of the transformer meets the parameters under which ground-fault protection of equipment (GFPE) is required by 215.10, protection must be installed to protect the equipment supplied by the secondary of the transformer.

Exception No. 3: If temporary feeder conductors are used to connect a generator to a facility for repair, maintenance, or emergencies, ground-fault protection of equipment shall not be required. Temporary feeders without ground-fault protection shall be permitted for the time period necessary but shall not exceed 90 days.

215.11 Circuits Derived from Autotransformers. Feeders shall not be derived from autotransformers unless the system supplied has a grounded conductor that is electrically connected to a grounded conductor of the system supplying the autotransformer.

Exception No. 1: An autotransformer shall be permitted without the connection to a grounded conductor where transforming from a nominal 208 volts to a nominal 240-volt supply or similarly from 240 volts to 208 volts.

Exception No. 2: In industrial occupancies, where conditions of maintenance and supervision ensure that only qualified persons service the installation, autotransformers shall be permitted to supply nominal 600-volt loads from nominal 480-volt systems, and 480-volt loads from nominal 600-volt systems, without the connection to a similar grounded conductor.

215.12 Identification for Feeders.

(A) Grounded Conductor. The grounded conductor of a feeder, if insulated, shall be identified in accordance with 200.6.

(B) Equipment Grounding Conductor. The equipment grounding conductor shall be identified in accordance with 250.119.

(C) Identification of Ungrounded Conductors. Ungrounded conductors shall be identified in accordance with 215.12(C)(1) or (C)(2), as applicable.

EXHIBIT 215.3 The different colors identify each ungrounded line or phase conductor of a nominal voltage system. Marking of the armor can also be applied. *(Courtesy of AFC Cable Systems, a Part of Atkore International)*

Parallel with the requirement for ungrounded branch-circuit conductors in 210.5(C), 215.12(C) requires identification of ungrounded feeder conductors by system and phase where there is more than one nominal voltage supply system to a building, structure, or other premises.

For ac circuits, the identification scheme is not specified, but whatever is used is required to be consistent throughout the premises. A permanent legend or directory indicating the feeder identification system for the premises is required to be posted at each point in the distribution system from which feeder circuits are supplied, or the identification scheme is to be described in a facility log or other documentation and made readily available. Exhibit 215.3 is an example of the use of different colors to identify each ungrounded line or phase of a nominal voltage system.

(1) Feeders Supplied from More Than One Nominal Voltage System. Where the premises wiring system has feeders supplied from more than one nominal voltage system, each ungrounded conductor of a feeder shall be identified by phase or line and system at all termination, connection, and splice points in compliance with 215.12(C)(1)(a) and (C)(1)(b).

(a) *Means of Identification.* The means of identification shall be permitted to be by separate color coding, marking tape, tagging, or other approved means.

(b) *Posting of Identification Means.* The method utilized for conductors originating within each feeder panelboard or similar feeder distribution equipment shall be documented in a manner that is readily available or shall be permanently posted at each feeder panelboard or similar feeder distribution equipment.

(2) Feeders Supplied from Direct-Current Systems. Where a feeder is supplied from a dc system operating at more than 60 volts, each ungrounded conductor of 4 AWG or larger shall be identified by polarity at all termination, connection, and splice points by marking tape, tagging, or other approved means; each ungrounded conductor of 6 AWG or smaller shall be identified by polarity at all termination, connection, and splice points in compliance with 215.12(C)(2)(a) and (C)(2)(b). The identification methods utilized for conductors originating within each feeder panelboard or similar feeder distribution equipment shall be documented in a manner that is readily available or shall be permanently posted at each feeder panelboard or similar feeder distribution equipment.

(a) *Positive Polarity, Sizes 6 AWG or Smaller.* Where the positive polarity of a dc system does not serve as the connection for the grounded conductor, each positive ungrounded conductor shall be identified by one of the following means:

(1) A continuous red outer finish
(2) A continuous red stripe durably marked along the conductor's entire length on insulation of a color other than green, white, gray, or black
(3) Imprinted plus signs (+) or the word POSITIVE or POS durably marked on insulation of a color other than green, white, gray, or black, and repeated at intervals not exceeding 610 mm (24 in.) in accordance with 310.8(B)
(4) An approved permanent marking means such as sleeving or shrink-tubing that is suitable for the conductor size, at all termination, connection, and splice points, with imprinted plus signs (+) or the word POSITIVE or POS durably marked on insulation of a color other than green, white, gray, or black

(b) *Negative Polarity, Sizes 6 AWG or Smaller.* Where the negative polarity of a dc system does not serve as the connection for the grounded conductor, each negative ungrounded conductor shall be identified by one of the following means:

(1) A continuous black outer finish
(2) A continuous black stripe durably marked along the conductor's entire length on insulation of a color other than green, white, gray, or red
(3) Imprinted minus signs (–) or the word NEGATIVE or NEG durably marked on insulation of a color other than green, white, gray, or red, and repeated at intervals not exceeding 610 mm (24 in.) in accordance with 310.8(B)
(4) An approved permanent marking means such as sleeving or shrink-tubing that is suitable for the conductor size, at all termination, connection, and splice points, with imprinted minus signs (–) or the word NEGATIVE or NEG durably

marked on insulation of a color other than green, white, gray, or red

N **215.15 Barriers.** Barriers shall be placed such that no energized, uninsulated, ungrounded busbar or terminal is exposed to inadvertent contact by persons or maintenance equipment while servicing load terminations in panelboards, switchboards, switchgear, or motor control centers supplied by feeder taps in 240.21(B) or transformer secondary conductors in 240.21(C) when the disconnecting device, to which the tap conductors are terminated, is in the open position.

N **215.18 Surge Protection.**

N **(A) Surge-Protective Device.** Where a feeder supplies any of the following, a surge-protective device (SPD) shall be installed:

(1) Dwelling units
(2) Dormitory units
(3) Guest rooms and guest suites of hotels and motels
(4) Areas of nursing homes and limited-care facilities used exclusively as patient sleeping rooms

N **(B) Location.** The SPD shall be installed in or adjacent to distribution equipment, connected to the load side of the feeder, that contains branch circuit overcurrent protective device(s) that supply the locations specified in 215.18(A).

Informational Note: Surge protection is most effective when closest to the branch circuit. Surges can be generated from multiple sources including, but not limited to, lightning, the electric utility, or utilization equipment.

N **(C) Type.** The SPD shall be a Type 1 or Type 2 SPD.

N **(D) Replacement.** Where the distribution equipment supplied by the feeder is replaced, all of the requirements of this section shall apply.

N **(E) Ratings.** SPDs shall have a nominal discharge current rating (In) of not less than 10kA.

ARTICLE 220   Branch-Circuit, Feeder, and Service Load Calculations

Part I. General

Δ **220.1 Scope.** This article provides requirements for calculating branch-circuit, feeder, and service loads. Part I provides general requirements for calculation methods. Part II provides calculation methods for branch-circuit loads. Part III and Part IV provide calculation methods for feeder and service loads. Part V provides calculation methods for farm loads. Part VI provides calculation methods for health care facilities. Part VII provides

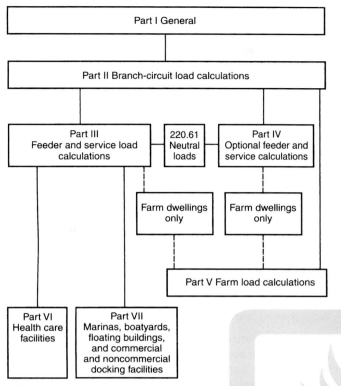

△ **INFORMATIONAL NOTE FIGURE 220.1** *Branch-Circuit, Feeder, and Service Load Calculation Methods.*

△ **TABLE 220.3** *Specific-Purpose Calculation References*

Calculation	Article	Section (or Part)
Air-conditioning and refrigerating equipment, branch-circuit conductor sizing	440	Part IV
Capacitors	460	460.8
Fixed electric heating equipment for pipelines and vessels, branch-circuit sizing	427	427.4
Fixed electric space-heating equipment, branch-circuit sizing	424	424.3
Fixed outdoor electric deicing and snow-melting equipment, branch-circuit sizing	426	426.4
Fixed resistance and electrode industrial process heating equipment	425	425.4
Motors, feeder demand factor	430	430.26
Motors, multimotor and combination-load equipment	430	430.25
Motors, several motors or a motor(s) and other load(s)	430	430.24
Over 1000-volt ac and 1500-volt dc branch-circuit calculations	235	235.19
Over 1000-volt feeder calculations	215	215.2(B)
Phase converters, conductors	455	455.6
Storage-type water heaters	422	422.11

calculation methods for marinas, boatyards, floating buildings, and commercial and noncommercial docking facilities.

Informational Note No. 1: See Informative Annex D for examples.
Informational Note No. 2: See Informational Note Figure 220.1 for information on the organization of this article.

Although this article does not contain the requirements for determining the minimum number of branch circuits, the loads calculated in accordance with Article 220 are used in conjunction with the rules of 210.11 to determine how many branch circuits are needed at a premises.

See also

210.11(A) for the general requirement on determining the minimum number of branch circuits for all occupancy types

220.3 Other Articles for Specific-Purpose Calculations. Table 220.3 shall provide references for specific-purpose calculation requirements not located in Chapters 5, 6, or 7 that amend or supplement the requirements of this article.

220.5 Calculations.

(A) Voltages. Unless other voltages are specified, for purposes of calculating branch-circuit and feeder loads, nominal system voltages of 120, 120/240, 208Y/120, 240, 347, 480Y/277, 480, 600Y/347, and 600 volts shall be used.

(B) Fractions of an Ampere. Calculations shall be permitted to be rounded to the nearest whole ampere, with decimal fractions smaller than 0.5 dropped.

For uniform calculation of load, nominal voltages, as listed in 220.5(A), are required to be used in computing the ampere load on the conductors.

Loads are calculated on the basis of volt-amperes (VA) or kilovolt-amperes (kVA), rather than watts or kilowatts (kW), to calculate the true ampere values. However, the rating of equipment is given in watts or kilowatts for resistive loads. Such ratings are considered to be the equivalent of the same rating in volt-amperes or kilovolt-amperes. This concept recognizes that load calculations determine conductor and circuit sizes, that the power factor of the load is often unknown, and that the conductor "sees" the circuit volt-amperes only, not the circuit power (watts).

See also

310.15 to select conductor sizes using the ampacity tables
Informative Annex D for examples of load calculations

N **(C) Floor Area.** The floor area for each floor shall be calculated from the outside dimensions of the building, dwelling unit, or other area involved. For dwelling units, the calculated floor area shall not include open porches or unfinished areas not adaptable for future use as a habitable room or occupiable space.

Part II. Branch-Circuit Load Calculations

△ **220.10 General.** Branch-circuit loads shall be calculated in accordance with the following sections:

(1) 220.14 for other loads — all occupancies
(2) 220.16 for additions to existing installations
(3) 220.41 for dwelling units
(4) 220.42 for lighting loads for non-dwelling occupancies
(5) 220.44 for hotel and motel occupancies

N 220.11 Maximum Load. The total load on a branch circuit shall not exceed the rating of the branch circuit nor the maximum loads specified in 220.11(A) through (C) under the conditions specified therein.

N (A) Motor-Operated and Combination Loads. Where a circuit supplies only motor-operated loads, the conductor sizing requirement specified in 430.22 shall apply. Where a circuit supplies only air-conditioning equipment, refrigerating equipment, or both, the requirements of 440.6 shall apply. For circuits supplying loads consisting of motor-operated utilization equipment that is fastened in place and has a motor larger than ⅛ hp in combination with other loads, the total calculated load shall be based on 125 percent of the largest motor load plus the sum of the other loads in accordance with 430.24.

N (B) Inductive and LED Lighting Loads. For circuits supplying lighting units that have ballasts, transformers, autotransformers, or LED drivers, the calculated load shall be based on the total ampere ratings of such units and not on the total watts of the lamps.

N (C) Electric Cooking Appliances. Applying demand factors for ranges, wall-mounted ovens, counter-mounted cooking units, and other household cooking appliance loads in excess of 1¾ kW shall be permitted in accordance with Table 220.55, including Notes 4, 5, and 6.

220.14 Other Loads — All Occupancies. Branch-circuit load calculations shall include calculation of a minimum load on each outlet as calculated in 220.14(A) through (K) and then summed to establish the load on the branch circuit.

In all occupancies, the minimum load for each outlet for general-use receptacles and outlets not used for general illumination shall not be less than that calculated in 220.14(A) through (K), with the loads shown being based on nominal branch-circuit voltages.

Exception: The loads of outlets serving switchboards and switching frames in telephone exchanges shall be waived from the calculations.

(A) Specific Appliances or Loads. An outlet for a specific appliance or other load not covered in 220.14(B) through (K) shall be calculated based on the ampere rating of the appliance or load served.

(B) Electric Dryers and Electric Cooking Appliances in Dwellings and Household Cooking Appliances Used in Instructional Programs. Load calculations shall be permitted as specified in 220.54 for electric dryers and in 220.55 for electric ranges and other cooking appliances.

(C) Motor Outlets. The conductor sizing requirements specified in 430.22, 430.24, and 440.6 shall be used to determine the loads for motor outlets.

(D) Luminaires. An outlet supplying a luminaire(s) shall be calculated based on the maximum volt-ampere rating of the equipment and lamps for which the luminaire(s) is rated.

Where the rating of the luminaires installed for general lighting exceeds the minimum load provided for in Table 220.42(A), the minimum general lighting load for that premises must be based on the installed luminaires. In general, no additional calculation is required for luminaires (recessed and surface mounted) installed in or on a dwelling unit, because the load of such luminaires is covered in the 3 volt-amperes per square foot calculation specified in 220.41.

(E) Heavy-Duty Lampholders. Outlets for heavy-duty lampholders shall be calculated at a minimum of 600 volt-amperes.

(F) Sign and Outline Lighting. Sign and outline lighting outlets shall be calculated at a minimum of 1200 volt-amperes for each required branch circuit specified in 600.5(A).

The load calculation must be based on a minimum of 1200 volt-amperes or the actual load of the sign and outline lighting outlets, whichever is greater.

(G) Show Windows. Show windows shall be calculated in accordance with either of the following:

(1) The unit load per outlet as required in other provisions of this section
(2) At 200 volt-amperes per linear 300 mm (1 ft) of show window

As shown in Exhibit 220.1, the linear-foot calculation method is permitted in lieu of the specified unit load per outlet for branch circuits serving show windows.

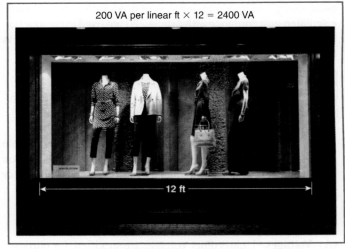

200 VA per linear ft × 12 = 2400 VA

12 ft

EXHIBIT 220.1 An example of the volt-ampere per linear-foot load calculation for branch circuits serving a show window.

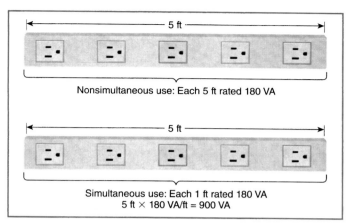

EXHIBIT 220.2 The requirements of 220.14(H)(1) and (H)(2) as applied to fixed multioutlet assemblies.

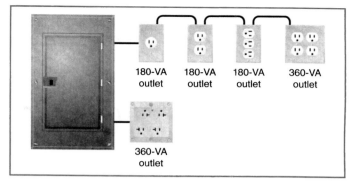

EXHIBIT 220.3 The load requirement of 180 volt-amperes per 220.14(I) applied to single- and multiple-receptacle outlets on single straps and the load of 360 volt-amperes applied to a multiple device consisting of four receptacles.

(H) Fixed Multioutlet Assemblies. Fixed multioutlet assemblies used in other than dwelling units or the guest rooms or guest suites of hotels or motels shall be calculated in accordance with the following:

(1) Where appliances are unlikely to be used simultaneously, each 1.5 m (5 ft) or fraction thereof of each separate and continuous length shall be considered as one outlet of not less than 180 volt-amperes.

(2) Where appliances are likely to be used simultaneously, each 300 mm (1 ft) or fraction thereof shall be considered as an outlet of not less than 180 volt-amperes.

For the purposes of this section, the calculation shall be permitted to be based on the portion that contains receptacles.

In light-use commercial and industrial applications, not all the cord-connected equipment is expected to be used at the same time. An example of light use is a workbench area where one worker uses one electrical tool at a time. In heavy-use applications, all the cord-connected equipment generally is operating at the same time. An example of heavy use is a retail store displaying television sets, where most or all sets are operating simultaneously. Exhibit 220.2 illustrates the difference between 220.14(H)(1) and (H)(2).

See also

220.47 for demand factors that apply to the load calculation in 220.14(H) and (I)

Δ **(I) Receptacle Outlets.** Except as covered in 220.41 and 220.14(J), receptacle outlets shall be calculated at not less than 180 volt-amperes for each single or for each multiple receptacle on one yoke. A single piece of equipment consisting of a multiple receptacle comprised of four or more receptacles shall be calculated at not less than 90 volt-amperes per receptacle. This provision shall not be applicable to the receptacle outlets specified in 210.11(C)(1) and (C)(2).

Section 220.47 contains demand factors that apply to the load calculation in 220.14(H) and (I). In dwelling units, the minimum unit load for receptacles supplied by general-purpose branch circuits is specified in 220.41 and is based on floor area rather than an assigned load per receptacle outlet. The load requirements for the laundry and small-appliance branch circuits are 1500 volt-amperes per circuit, as described in 220.52.

Exhibit 220.3 shows the load of 180 volt-amperes applied to single and multiple receptacles mounted on a single yoke or strap, and the load of 360 volt-amperes applied to the outlet containing two duplex receptacles and to the outlet containing the device with four receptacles. Note that the last outlet of the top circuit consists of two duplex receptacles on separate straps. The multiple receptacle supplied from the bottom circuit in the exhibit consists of four receptacles. For example, single-strap and multiple-receptacle devices are calculated as follows:

Device	Calculated Load
Duplex receptacle	180 VA
Triplex receptacle	180 VA
Double duplex receptacle	360 VA (180×2)
Quadplex-type receptacle	360 VA (90×4)

In Exhibit 220.4, the maximum number of receptacle outlets permitted on 15- and 20-ampere branch circuits is 10 and 13 outlets, respectively, based on the load assigned for each outlet by 220.14(I). This restriction does not apply to receptacle outlets in dwelling occupancies.

N **(J) Receptacle Outlets in Office Buildings.** In office buildings, the receptacle loads shall be calculated to be the larger of the following:

(1) The calculated load from 220.14(I)

(2) 11 volt-amperes/m^2 (1 volt-ampere/ft^2)

(K) Other Outlets. Other outlets not covered in 220.14(A) through (J) shall be calculated based on 180 volt-amperes per outlet.

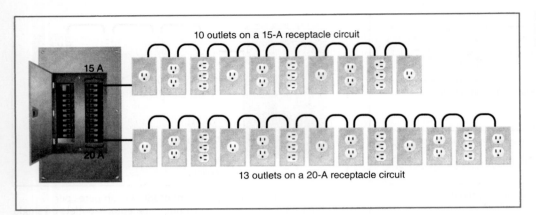

10 outlets on a 15-A receptacle circuit

13 outlets on a 20-A receptacle circuit

EXHIBIT 220.4 Maximum number of receptacle outlets permitted on 15- and 20-ampere branch circuits.

220.16 Loads for Additions to Existing Installations.

(A) Dwelling Units. Loads added to an existing dwelling unit(s) shall comply with the following as applicable:

(1) Loads for structural additions to an existing dwelling unit or for a previously unwired portion of an existing dwelling unit shall be calculated in accordance with 220.14.

(2) Loads for new circuits or extended circuits in previously wired dwelling units shall be calculated in accordance with 220.14.

(B) Other Than Dwelling Units. Loads for new circuits or extended circuits in other than dwelling units shall be calculated in accordance with either 220.42 or 220.14, as applicable.

Part III. Feeder and Service Load Calculations

△ **220.40 General.** The calculated load of a feeder or service shall not be less than the sum of the loads on the branch circuits supplied, as determined by Part II of this article, after any applicable demand factors permitted or required by Part III, IV, V, VI, or VII have been applied.

> Informational Note No. 1: See Informative Annex D, Examples D1(a) through D10, for examples of feeder and service load calculations.
>
> Informational Note No. 2: See 220.11(B) for the maximum load in amperes permitted for lighting units operating at less than 100 percent power factor.

In the example shown in Exhibit 220.5, each panelboard supplies a calculated load of 80 amperes. The main set of service conductors is sized to carry the total calculated load of 240 amperes (3 × 80 A). The service conductors from the meter enclosure to each panelboard (2 AWG Cu = 95 A per 60°C column of Table 310.16) are sized to supply a calculated load of 80 amperes and to meet the requirement of 230.90 relative to overcurrent (overload) protection of service conductors terminating in a single-service overcurrent protective device (OCPD). The main set of service conductors (250 kcmil THWN Cu = 255 A per 75°C column of Table 310.16) is not required to be sized to carry 300

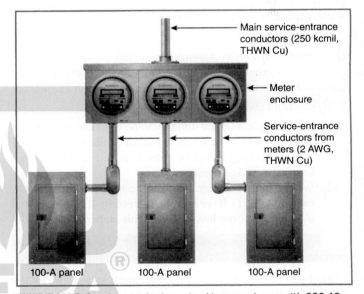

Main service-entrance conductors (250 kcmil, THWN Cu)

Meter enclosure

Service-entrance conductors from meters (2 AWG, THWN Cu)

100-A panel   100-A panel   100-A panel

EXHIBIT 220.5 Service conductors sized in accordance with 220.40.

amperes based on the combined rating of the panelboards. The individual service-entrance conductors to each panelboard (2 AWG THWN) meet the requirement of 230.90.

See also

230.23, 230.31, and **230.42** for specifics on size and rating of service conductors

△ **220.41 Dwelling Units, Minimum Unit Load.** In one-family, two-family, and multifamily dwellings, the minimum unit load shall be not less than 33 volt-amperes/m² (3 volt-amperes/ft²).

Unit loads include the following lighting and receptacle outlets, and no additional load calculations shall be required:

(1) All general-use receptacle outlets of 20-ampere rating or less, including receptacles connected to the circuits specified in 210.11(C)(3) and (C)(4)

(2) The receptacle outlets specified in 210.52(E) and (G)

(3) The lighting outlets specified in 210.70

The minimum lighting load shall be determined using the minimum unit load and the floor area as determined in 220.5(C) for dwelling occupancies. Motors rated less than ⅛ hp and connected to a lighting circuit shall be considered part of the minimum lighting load.

220.42 Lighting Load for Non-Dwelling Occupancies.

(A) General. A unit load of not less than that specified in Table 220.42(A) for non-dwelling occupancies and the floor area determined in 220.5(C) shall be used to calculate the minimum lighting load. Motors rated less than ⅛ HP and connected to a lighting circuit shall be considered general lighting load.

> Informational Note: The unit values of Table 220.42(A) are based on minimum load conditions and 80 percent power factor and might not provide sufficient capacity for the installation contemplated.

(B) Energy Code. Where the building is designed and constructed to comply with an energy code adopted by the local authority, the lighting load shall be permitted to be calculated using the unit values specified in the energy code where the following conditions are met:

(1) A power monitoring system is installed that will provide continuous information regarding the total general lighting load of the building.

(2) The power monitoring system will be set with alarm values to alert the building owner or manager if the lighting load exceeds the values set by the energy code. Automatic means to take action to reduce the connected load shall be permitted.

(3) The demand factors specified in 220.45 are not applied to the general lighting load.

(4) The continuous load multiplier of 125 percent shall be applied.

△ **220.43 Office Buildings.** In office buildings, the receptacle loads shall be calculated to be the larger of the following:

(1) The calculated load from 220.14(I) after Table 220.47 demand factors have been applied

(2) 11 volt-amperes/m^2 or 1 volt-ampere/ft^2

△ **220.44 Hotel and Motel Occupancies.** In guest rooms or guest suites of hotels and motels, the following lighting and receptacle outlets are included in the minimum unit load in Table 220.42(A), and no additional load calculations shall be required for such outlets:

(1) All general-use receptacle outlets of 20-ampere rating or less, including receptacles connected to the circuits in 210.11(C)(3) and (C)(4)

(2) The receptacle outlets specified in 210.52(E)(3)

(3) The lighting outlets specified in 210.70

△ **TABLE 220.42(A)** *General Lighting Loads by Non-Dwelling Occupancy*

Type of Occupancy	Unit Load	
	Volt-amperes/ m^2	Volt-amperes/ ft^2
Automotive facility	16	1.5
Convention center	15	1.4
Courthouse	15	1.4
Dormitory	16	1.5
Exercise center	15	1.4
Fire station	14	1.3
Gymnasium[1]	18	1.7
Health care clinic	17	1.6
Hospital	17	1.6
Hotel or motel, or apartment house without provisions for cooking by tenants[2]	18	1.7
Library	16	1.5
Manufacturing facility[3]	24	2.2
Motion picture theater	17	1.6
Museum	17	1.6
Office[4]	14	1.3
Parking garage[5]	3	0.3
Penitentiary	13	1.2
Performing arts theater	16	1.5
Police station	14	1.3
Post office	17	1.6
Religious facility	24	2.2
Restaurant[6]	16	1.5
Retail[7, 8]	20	1.9
School/university	16	1.5
Sports arena	16	1.5
Town hall	15	1.4
Transportation	13	1.2
Warehouse	13	1.2
Workshop	18	1.7

Note: The 125 percent multiplier for a continuous load as specified in 210.20(A) is included, therefore no additional multiplier shall be required when using the unit loads in this table for calculating the minimum lighting load for a specified occupancy.

[1]Armories and auditoriums are considered gymnasium-type occupancies.

[2]Lodge rooms are similar to hotels and motels.

[3]Industrial commercial loft buildings are considered manufacturing-type occupancies.

[4]Banks are office-type occupancies.

[5]Commercial (storage) garages are considered parking garage occupancies.

[6]Clubs are considered restaurant occupancies.

[7]Barber shops and beauty parlors are considered retail occupancies.

[8]Stores are considered retail occupancies.

220.45 General Lighting. The demand factors specified in Table 220.45 shall apply to that portion of the total branch-circuit load calculated for general illumination. They shall not be applied in determining the number of branch circuits for general illumination.

TABLE 220.45 *Lighting Load Demand Factors*

Type of Occupancy	Portion of Lighting Load to Which Demand Factor Applies (Volt-Amperes)	Demand Factor (%)
Dwelling units	First 3000 at	100
	From 3001 to 120,000 at	35
	Remainder over 120,000 at	25
Hotels and motels, including apartment houses without provision for cooking by tenants*	First 20,000 or less at	60
	From 20,001 to 100,000 at	50
	Remainder over 100,000 at	35
Warehouses (storage)	First 12,500 or less at	100
	Remainder over 12,500 at	50
All others	Total volt-amperes	100

*The demand factors of this table shall not apply to the calculated load of feeders or services supplying areas in hotels and motels where the entire lighting is likely to be used at one time, as in ballrooms or dining rooms.

220.46 Show-Window and Track Lighting.

(A) Show Windows. For show-window lighting, a load of not less than 660 volt-amperes/linear meter or 200 volt-amperes/linear foot shall be included for a show window, measured horizontally along its base.

> Informational Note: See 220.14(G) for branch circuits supplying show windows.

(B) Track Lighting. For track lighting in other than dwelling units or guest rooms or guest suites of hotels or motels, an additional load of 150 volt-amperes shall be included for every 600 mm (2 ft) of lighting track or fraction thereof. Where multicircuit track is installed, the load shall be considered to be divided equally between the track circuits.

Calculation Example

A lighting plan shows 63 linear feet of single-circuit track lighting for a small clothing store. Because the actual track lighting fixtures are owner supplied, neither the quantity of track lighting fixtures nor the lamp size is specified. What is the minimum calculated load associated with the track lighting that must be added to the service or feeder supplying this store?

Solution

According to 220.46(B), the minimum calculated load to be added to the service or feeder supplying this track light installation is calculated as follows:

$$\frac{63\ ft}{2\ ft} = 31.50, \text{ rounded up to } 32$$

$$32 \times 150\ VA = 4800\ VA$$

It is important to note that the branch circuits supplying this installation are covered in 410.150(B). The maximum load on the track must not exceed the rating of the branch circuit supplying the track. Also, the track must be supplied by a branch circuit that has a rating not exceeding the rating of the track.

Section 220.46(B) is not intended to limit the number of feet of track on a single branch circuit, nor is it intended to limit the number of fixtures on an individual track. It is meant to be used solely for load calculations of feeders and services.

Exception: If the track lighting is supplied through a device that limits the current to the track, the load shall be permitted to be calculated based on the rating of the device used to limit the current.

The rating of the branch-circuit OCPD can be used for this calculation, or the device used to limit current can be a supplementary OCPD. This exception enables load calculations to be more in line with the limitations placed on building lighting loads by energy codes.

Calculation Example

A lighting plan shows 63 linear feet of single-circuit track lighting for a small clothing store. Because the actual track lighting fixtures are owner supplied, neither the quantity of track lighting fixtures nor the lamp size is specified in the plan. Because the amount of track is to facilitate easy relocation of the luminaires to accommodate changes in the display of merchandise rather than to accommodate a large number of luminaires, the entire length will be supplied by a single 20-A, 120-V branch circuit. What is the minimum calculated load associated with the track lighting that must be added to the service or feeder supplying this store?

Solution

According to the exception to 220.46(B), the minimum calculated load to be added to the service or feeder supplying this track light installation is calculated as follows:

$$20\ A \times 120\ V = 2400\ VA$$

This calculation could be further reduced where the track is supplied through a supplementary OCPD(s) having a current rating less than 20 A.

220.47 Receptacle Loads — Other Than Dwelling Units.
Receptacle loads calculated in accordance with 220.14(H) and (I) shall be permitted to be made subject to the demand factors given in Table 220.45 or Table 220.47.

TABLE 220.47 *Demand Factors for Non-Dwelling Receptacle Loads*

Portion of Receptacle Load to Which Demand Factor Applies (Volt-Amperes)	Demand Factor (%)
First 10 kVA or less at	100
Remainder over 10 kVA at	50

Δ **220.50 Motors and Air-Conditioning Equipment.**

N **(A) Motors.** The conductor sizing requirements specified in 430.24 and 430.25 and the feeder demand factor calculation method specified in 430.26 shall be used to determine motor loads.

N **(B) Air-Conditioning Equipment.** The conductor sizing requirements specified in Part IV of Article 440 shall be used to determine air-conditioning loads for hermetic refrigerant motor-compressors.

220.51 Fixed Electric Space Heating. Fixed electric space-heating loads shall be calculated at 100 percent of the total connected load. However, in no case shall a feeder or service load current rating be less than the rating of the largest branch circuit supplied.

Exception: If reduced loading of the conductors results from units operating on duty-cycle or intermittently, or from all units not operating at the same time, the authority having jurisdiction shall be permitted to grant permission for feeder and service conductors to have an ampacity less than 100 percent if the conductors have an ampacity for the load so determined.

220.52 Small-Appliance and Laundry Loads — Dwelling Unit.

(A) Small-Appliance Circuit Load. In each dwelling unit, the load shall be calculated at 1500 volt-amperes for each 2-wire small-appliance branch circuit as covered by 210.11(C)(1). Where the load is subdivided through two or more feeders, the calculated load for each shall include not less than 1500 volt-amperes for each 2-wire small-appliance branch circuit. These loads shall be permitted to be included with the general lighting load and subjected to the demand factors provided in Table 220.45.

Exception: The individual branch circuit permitted by 210.52(B)(1), Exception No. 2, shall be permitted to be excluded from the calculation required by 220.52.

(B) Laundry Circuit Load. A load of not less than 1500 volt-amperes shall be included for each 2-wire laundry branch circuit installed as covered by 210.11(C)(2). This load shall be permitted to be included with the general lighting load and shall be subjected to the demand factors provided in Table 220.45.

Where additional small-appliance and laundry branch circuits are provided, they also are calculated at 1500 volt-amperes per circuit. All small appliance and laundry loads are combined with the general lighting load, and the demand factors in Table 220.45 can be applied.

220.53 Appliance Load — Dwelling Unit(s). Applying a demand factor of 75 percent to the nameplate rating load of four or more appliances rated ¼ hp or greater, or 500 watts or greater, that are fastened in place, and that are served by the same feeder or service in a one-family, two-family, or multifamily dwelling shall be permitted. This demand factor shall not apply to the following:

(1) Household electric cooking equipment that is fastened in place
(2) Clothes dryers
(3) Space heating equipment
(4) Air-conditioning equipment
(5) Electric vehicle supply equipment (EVSE)

For appliances fastened in place (other than ranges and other cooking appliances, clothes dryers, space-heating, air-conditioning equipment, and electric vehicle supply equipment), feeder capacity must be provided for the sum of these loads. If the total load includes four or more such appliances, a demand factor of 75 percent is permitted. To reduce the number of small motor-operated appliance loads (e.g., bathroom ventilating fans) that are typically supplied by general-purpose branch circuits from having two demand factors applied (220.53 plus Table 220.45), the minimum appliance rating that can be considered in applying the 75 percent demand of 220.53 is either 500 watts or greater or ¼ horsepower (hp).

See also

Table 430.248 for the full-load current, in amperes, for single-phase ac motors in accordance with 220.50

220.54 Electric Clothes Dryers — Dwelling Unit(s). The load for household electric clothes dryers in a dwelling unit(s) shall be either 5000 watts (volt-amperes) or the nameplate rating, whichever is larger, for each dryer served. The use of the demand factors in Table 220.54 shall be permitted. Where two or more single-phase dryers are supplied by a 3-phase, 4-wire feeder or service, the total load shall be calculated on the basis of twice the maximum number connected between any two phases. Kilovolt-amperes (kVA) shall be considered equivalent to kilowatts (kW) for loads calculated in this section.

TABLE 220.54 *Demand Factors for Household Electric Clothes Dryers*

Number of Dryers	Demand Factor (%)
1–4	100
5	85
6	75
7	65
8	60
9	55
10	50
11	47
12–23	47% minus 1% for each dryer exceeding 11
24–42	35% minus 0.5% for each dryer exceeding 23
43 and over	25%

The use of demand factors is permitted but not required, because the *NEC®* does not prohibit applying the full load of all dryers to a service and/or feeder calculation. However, this method is not necessary or practical, and experience has demonstrated that the use of the Table 220.54 demand factors provides sufficient capacity in the service for dryer loads. It is unlikely that all dryers will be in operation simultaneously, and Table 220.54 was developed based on utility demand that proves this point.

The minimum load to be used is the larger of 5000 volt-amperes or the nameplate rating of the dryer. In the following calculation example, the nameplate rating of 5500 watts (VA) is used. In addition, because of the supply system characteristics, calculation is a little different from simply adding the total rating of all dryers and applying the appropriate demand factor. The load of 10 single-phase dryers will be distributed across a 3-phase supply system.

Calculation Example

Assuming the load of 10 single-phase dryers is connected as evenly as possible to the 3-phase system (3 dryers connected between phases A and B, 3 dryers connected between phases B and C, and 4 dryers connected between phases A and C), the maximum number of dryers connected between any two phases is 4.

Solution

Step 1. Twice the maximum number of dryers connected between any two phases is used as the basis for calculating the demand load:

$$4 \text{ dryers} \times 2 = 8 \text{ dryers}$$

60% demand from Table 220.54:

$$5500 \text{ VA/dryer} \times 8 \text{ dryers} \times 0.6 = 26,400 \text{ VA}$$

(connected between two phases of the 3-phase system)

$$26,400 \text{ VA} \div 2 \text{ phase loads} = 13,200 \text{ VA/phase load}$$

3-phase dryer load on service:

$$13,200 \text{ VA/phase load} \times 3 \text{ phase loads} = 39,600 \text{ VA}$$

Dryer load on each of the ungrounded service or feeder conductors:

$$39,600 \text{ VA} \div (208 \text{ V} \times 1.732) = 110 \text{ A}$$

Step 2. For the grounded (neutral) service or feeder conductor, 220.61(B)(1) permits a 70-percent demand to be applied to the demand load of the ungrounded conductors. However, since this is a 3-phase feeder, 220.61(C)(1) prohibits the application of the demand factor on the neutral conductor load.

If the service and feeder load for this multifamily dwelling is calculated using the optional calculation for multifamily dwellings from Part IV of Article 220, the dryer load is not subject to an individual demand factor as required by the Part III calculation.

The total connected load is subjected to a single demand factor from Table 220.84(B) based on the total number of dwelling units served by a set of service or feeder conductors. For a set of conductors serving 10 dwelling units, the demand factor is 43 percent.

The optional calculation can be used if all the conditions specified in 220.84(A)(1) through (A)(3) are met. Otherwise, the calculation for the multifamily dwelling must be performed in accordance with the requirements of Part III of Article 220.

△ **220.55 Electric Cooking Appliances in Dwelling Units and Household Cooking Appliances Used in Instructional Programs.** The load for household electric ranges, wall-mounted ovens, counter-mounted cooking units, and other household cooking appliances individually rated in excess of 1¾ kW shall be permitted to be calculated in accordance with Table 220.55. Kilovolt-amperes (kVA) shall be considered equivalent to kilowatts (kW) for loads calculated under this section.

Where two or more single-phase ranges are supplied by a 3-phase, 4-wire feeder or service, the total load shall be calculated on the basis of twice the maximum number connected between any two phases.

Informational Note No. 1: See Informative Annex D for examples.
Informational Note No. 2: See Table 220.56 for demand factors for commercial cooking equipment.

The demand factors are based on the diversified use of household appliances, because it is unlikely that all appliances will be used simultaneously or that all cooking elements and the oven of a range will be at maximum heat for any length of time.

The values in Column C are applicable to installations where all ranges in the group have the same rating. Note 1 applies where the ranges are rated greater than 12 kilowatts, and Note 2 applies where the ranges in a group have different ratings. Column C must be used unless Note 3 is applicable. Notes 3 and 4 cover installations where the circuit supplies multiple cooking components, which are combined and treated as a single range.

Table Note 1. For household electric ranges, the size of the conductors must be determined by the rating of the range. According to Table 220.55, for one range rated 12 kilowatts or less, the maximum demand load is 8 kilowatts. Note that 210.19(C) does not permit the branch-circuit rating of a circuit supplying household ranges with a nameplate rating of 8¾ kilowatts to be less than 40 amperes.

The demand in Column C must be increased if an individual range is rated over 12 kilowatts but not more than 27 kilowatts. All ranges in the group must have the same rating to apply Note 1. If the ratings are different, refer to Note 2.

Δ **TABLE 220.55** *Demand Factors and Loads for Household Electric Ranges, Wall-Mounted Ovens, Counter-Mounted Cooking Units, and Other Household Cooking Appliances over 1¾ kW Rating (Column C to be used in all cases except as otherwise permitted in Note 3.)*

Number of Appliances	Demand Factor (%) (See Notes)		Column C Maximum Demand (kW) (See Notes) (Not over 12 kW Rating)
	Column A (Less than 3½ kW Rating)	Column B (3½ kW through 8¾ kW Rating)	
1	80	80	8
2	75	65	11
3	70	55	14
4	66	50	17
5	62	45	20
6	59	43	21
7	56	40	22
8	53	36	23
9	51	35	24
10	49	34	25
11	47	32	26
12	45	32	27
13	43	32	28
14	41	32	29
15	40	32	30
16	39	28	31
17	38	28	32
18	37	28	33
19	36	28	34
20	35	28	35
21	34	26	36
22	33	26	37
23	32	26	38
24	31	26	39
25	30	26	40
26–30	30	24	15 kW + 1 kW for each range
31–40	30	22	
41–50	30	20	25 kW + ¾ kW for each range
51–60	30	18	
61 and over	30	16	

Notes:

1. *Over 12 kW through 27 kW ranges all of same rating.* For ranges individually rated more than 12 kW but not more than 27 kW, the maximum demand in Column C shall be increased 5 percent for each additional kilowatt of rating or major fraction thereof by which the rating of individual ranges exceeds 12 kW.

2. *Over 8¾ kW through 27 kW ranges of unequal ratings.* For ranges individually rated more than 8¾ kW and of different ratings, but none exceeding 27 kW, an average value of rating shall be calculated by adding together the ratings of all ranges to obtain the total connected load (using 12 kW for any range rated less than 12 kW) and dividing by the total number of ranges. Then the maximum demand in Column C shall be increased 5 percent for each kilowatt or major fraction thereof by which this average value exceeds 12 kW.

3. *Over 1¾ kW through 8¾ kW.* In lieu of the method provided in Column C, adding the nameplate ratings of all household cooking appliances rated more than 1¾ kW but not more than 8¾ kW and multiplying the sum by the demand factors specified in Column A or Column B for the given number of appliances shall be permitted. Where the rating of cooking appliances falls under both Column A and Column B, the demand factors for each column shall be applied to the appliances for that column, and the results added together.

4. Calculating the branch-circuit load for one range in accordance with Table 220.55 shall be permitted.

5. The branch-circuit load for one wall-mounted oven or one counter-mounted cooking unit shall be the nameplate rating of the appliance.

6. The branch-circuit load for a counter-mounted cooking unit and not more than two wall-mounted ovens, all supplied from a single branch circuit and located in the same room, shall be calculated by adding the nameplate rating of the individual appliances and treating this total as equivalent to one range.

7. This table shall also apply to household cooking appliances rated over 1¾ kW and used in instructional programs.

Calculation Example

Determine the maximum demand for four ranges each rated at 15 kW.

Solution

Step 1. Determine amount of rating above 12 kW:

$$15 \text{ kW} - 12 \text{ kW} = 3 \text{ kW}$$

Step 2. Calculate the required demand increase:

$$5\% \text{ per kW} \times 3 \text{ kW} = 15\%$$

Step 3. Calculate the maximum demand:

$$17 \text{ kW (Column C demand value for 4 ranges)} \times 115\% = 19.55 \text{ kW}$$

Table Note 2. If ranges in a group installation have different ratings over 8¾ kilowatts through 27 kilowatts, the ratings are added together to determine the average rating and the maximum demand. The demand in Column C must be increased if the average rating is over 12 kilowatts. If the ratings are the same, refer to Note 1.

Table Note 3. For a group installation of ranges with ratings over 1¾ kilowatts through 8¾ kilowatts, the ratings are permitted to be added together for determining a demand factor. Ranges rated below 3½ kilowatts should be grouped independently of those rated 3½ kilowatts and above. The appropriate column (A or B) is used rather than Column C.

Calculation Example

Determine the maximum demand for four ranges rated at 8 kW, 10 kW, 15 kW, and 18 kW.

Solution

Step 1. Total the range ratings:

$$12 \text{ kW (minimum value for 8 kW range)} + 12 \text{ kW (minimum value for 10 kW range)} + 15 \text{ kW} + 18 \text{ kW} = 57 \text{ kW}$$

Step 2. Calculate the average rating:

$$57 \text{ kW} \div 4 \text{ ranges} = 14.25 \text{ kW}$$

Step 3. Determine amount of rating above 12 kW:

$$14.25 \text{ kW} - 12 \text{ kW} = 2.25 \text{ kW}$$

Step 4. Calculate the required demand increase:

$$5\% \text{ per kW} \times 2 \text{ kW (0.25 kW is not a major fraction)} = 10\%$$

Step 5. Calculate the maximum demand:

$$17 \text{ kW (Column C demand value for 4 ranges)} \times 110\% = 18.7 \text{ kW}$$

Calculation Example

Determine the maximum demand for four ranges rated at 4.5 kW, 5 kW, 5 kW, and 8.5 kW.

Solution

Step 1. Combine ratings into single value:

$$4.5 \text{ kW} + 5 \text{ kW} + 5 \text{ kW} + 8.5 \text{ kW} = 23 \text{ kW}$$

Step 2. Determine demand factor:

$$50\% \text{ (Column B, 4 ranges)}$$

Step 3. Calculate the maximum demand:

$$23 \text{ kW} \times 50\% = 11.5 \text{ kW}$$

Table Note 4. The branch-circuit load for one range is permitted to be computed by using either the nameplate rating of the appliance or Table 220.55. A counter-mounted cooking appliance has a smaller load rating than does a full-sized range with an oven. If a single branch circuit supplies a counter-mounted cooking unit and not more than two wall-mounted ovens, all of which are located in the same room, the nameplate ratings of the appliances can be added, and the total treated as the equivalent of one range. For feeder demand factors other than dwelling units (commercial electric cooking equipment, dishwasher booster heaters, water heaters, etc.), see Table 220.56.

Where a counter-mounted cooking appliance like the one illustrated in Exhibit 210.23 is used with a separate wall oven, it is permissible to run a single branch circuit to the kitchen and supply each with branch-circuit tap conductors installed as specified in 210.19(C), Exception No. 1.

Calculation Example

Calculate the load for a single branch circuit that supplies the following cooking appliances:

- One counter-mounted cooking unit with rating of 8 kW
- One wall-mounted oven with rating of 7 kW
- A second wall-mounted oven with rating of 6 kW

Solution

Step 1. Combine the ratings of the cooking appliances:

$$8 \text{ kW} + 7 \text{ kW} + 6 \text{ kW} = 21 \text{ kW}$$

Step 2. Determine amount of rating above 12 kW (see Table 220.55, Note 1):

$$21 \text{ kW} - 12 \text{ kW} = 9 \text{ kW}$$

Step 3. Calculate the demand increase required by Table 220.55, Note 1:

$$5\% \text{ per kW} \times 9 \text{ kW} = 45\%$$

Step 4. Calculate the maximum demand:

8 kW (Column C demand value for single range)
× 145% = 11.6 kW = 11.6 VA

Step 5. Calculate the branch circuit load:

Branch circuit load = 11,600 VA ÷ 240 A
= 48.3 A, rounded down to 48 A

220.56 Kitchen Equipment — Other Than Dwelling Unit(s).

Calculating the load for commercial electric cooking equipment, dishwasher booster heaters, water heaters, and other kitchen equipment in accordance with Table 220.56 shall be permitted. Other kitchen equipment shall include equipment that is fastened in place and rated ¼ hp or greater, or 500 watts or greater. These demand factors shall be applied to all equipment that has either thermostatic control or intermittent use as kitchen equipment. These demand factors shall not apply to space-heating, ventilating, or air-conditioning equipment.

However, in no case shall the feeder or service calculated load be less than the sum of the largest two kitchen equipment loads.

TABLE 220.56 *Demand Factors for Kitchen Equipment — Other Than Dwelling Unit(s)*

Number of Units of Equipment	Demand Factor (%)
1	100
2	100
3	90
4	80
5	70
6 and over	65

The demand factors in Table 220.56 can be applied to a combination of cooking, dishwashing, water heating, and other types of commercial kitchen appliances, and the combined total allows a deeper demand factor to be applied. Thermostatically controlled electric cooking appliances are permitted, but not required, to have the demand applied to the appliance nameplate rating.

Because commercial electric cooking appliances are used more intensively and for longer periods, the demand factors for commercial applications are not as generous as those permitted for dwelling units. The applications of the demand factors in 220.55 are permitted to be used for household cooking equipment used in instructional programs, such as in culinary schools, because the equipment is not subjected to the continuous demands that would exist in a commercial restaurant.

N **220.57 Electric Vehicle Supply Equipment (EVSE) Load.** The EVSE load shall be calculated at either 7200 watts (volt-amperes) or the nameplate rating of the equipment, whichever is larger.

Δ **220.60 Noncoincident Loads.** If it is unlikely that two or more noncoincident loads will be in use simultaneously, using only the largest load(s) that will be used at one time for calculating the total load of a feeder or service shall be permitted. If a motor or air-conditioning load is part of the noncoincident load and is not the largest of the noncoincident loads, 125 percent of either the motor load or air-conditioning load, whichever is larger, shall be used in the calculation.

220.61 Feeder or Service Neutral Load.

(A) Basic Calculation. The feeder or service neutral load shall be the maximum unbalance of the load determined by this article. The maximum unbalanced load shall be the maximum net calculated load between the neutral conductor and any one ungrounded conductor.

Exception: For 3-wire, 2-phase or 5-wire, 2-phase systems, the maximum unbalanced load shall be the maximum net calculated load between the neutral conductor and any one ungrounded conductor multiplied by 140 percent.

Δ **(B) Permitted Reductions.** A service or feeder supplying the following loads shall be permitted to have an additional demand factor of 70 percent applied to the amount in 220.61(B)(1) and a portion of the amount in 220.61(B)(2).

N **(1) Household Electric Ranges, Wall-Mounted Ovens, Counter-Mounted Cooking Units, and Dryers.** A feeder or service supplying household electric ranges, wall-mounted ovens, counter-mounted cooking units, and electric dryers, where the maximum unbalanced load has been determined in accordance with Table 220.55 for ranges and Table 220.54 for dryers.

N **(2) Unbalanced Load in Excess of 200 Amperes.** That portion of the unbalanced load in excess of 200 amperes where the feeder or service is supplied from a 3-wire dc or single-phase ac system; a 4-wire, 3-phase system; a 3-wire, 2-phase system; or a 5-wire, 2-phase system.

Informational Note: See Informative Annex D, Examples D1(a), D1(b), D2(b), D4(a), and D5(a) for examples of unbalanced feeder or service neutral loads.

(C) Prohibited Reductions. There shall be no reduction of the neutral or grounded conductor capacity applied to the amount in 220.61(C)(1), or portion of the amount in (C)(2), from that determined by the basic calculation:

(1) Any portion of a 3-wire circuit consisting of 2 ungrounded conductors and the neutral conductor of a 4-wire, 3-phase, wye-connected system
(2) That portion consisting of nonlinear loads supplied from a 4-wire, wye-connected, 3-phase system

Informational Note: A 3-phase, 4-wire, wye-connected power system used to supply power to nonlinear loads might necessitate that the power system design allows for the possibility of high harmonic neutral conductor currents.

Section 220.61 describes the basis for calculating the neutral load of feeders or services as the maximum unbalanced load that can occur between the neutral and any ungrounded conductor. For a household electric range or clothes dryer, the maximum unbalanced load for a single-phase feeder or service can be assumed to be 70 percent, so the neutral can be sized on that basis. If the unbalanced load exceeds 200 amperes, 220.61(B) permits the reduction of the feeder neutral conductor size under specific conditions of use. However, 220.61(C)(1) and (C)(2) cite a circuit arrangement and a load characteristic as applications where the capacity of a neutral or grounded conductor of a feeder or service is not permitted to be reduced.

The neutral is considered a current-carrying conductor if the load of the electric-discharge lighting, data-processing, or similar equipment on the feeder neutral consists of more than half the total load, in accordance with 310.15(E)(3). Electric-discharge lighting and data-processing equipment could have harmonic currents in the neutral that might exceed the load current in the ungrounded conductors. The Informational Note to item (2) cautions designers and installers to be aware of harmonic contribution and to design the electrical installation to accommodate the harmonic load imposed on neutral conductors. In some instances, the neutral current could exceed the current in the phase conductors.

A 3-phase, 4-wire (208Y/120-volt or 480Y/277-volt) system is often used to supply both lighting and motor loads. The 3-phase motor loads are balanced loads that are not connected to the neutral and, thus, would not be considered when determining the maximum unbalanced load. The maximum current on the neutral is due to lighting loads or circuits that are connected line-to-neutral. On this type of system (3-phase, 4-wire), a demand factor of 70 percent is permitted by 220.61(B) for that portion of the neutral load more than 200 amperes.

For example, if the maximum possible unbalanced load is 500 amperes, the neutral would have to be large enough to carry 410 amperes (200 amperes plus 70 percent of 300 amperes, or 410 amperes). The 70 percent demand cannot be applied for that portion of the line-to-neutral load consisting of electric-discharge lighting.

Section 310.15(C)(1) specifies that a neutral conductor must be counted as a current-carrying conductor if the load it serves consists of harmonic currents. The maximum unbalanced load for feeders supplying clothes dryers, household ranges, wall-mounted ovens, and counter-mounted cooking units is required to be considered 70 percent of the load on the ungrounded conductors.

> **See also**
>
> **220.61(B)** for other systems in which the 70-percent demand factor can be applied
>
> **Informative Annex D,** Examples D1(a) through D2(c), which provide one-family dwelling calculation examples

N 220.70 Energy Management Systems (EMSs). If an energy management system (EMS) is used to limit the current to a feeder or service in accordance with 750.30, a single value equal to the maximum ampere setpoint of the EMS shall be permitted to be used in load calculations for the feeder or service.

The setpoint value of the EMS shall be considered a continuous load for the purposes of load calculations.

Part IV. Optional Feeder and Service Load Calculations

220.80 General. Optional feeder and service load calculations shall be permitted in accordance with Part IV.

220.82 Dwelling Unit.

(A) Feeder and Service Load. This section applies to a dwelling unit having the total connected load served by a single 120/240-volt or 208Y/120-volt set of 3-wire service or feeder conductors with an ampacity of 100 or greater. It shall be permissible to calculate the feeder and service loads in accordance with this section instead of the method specified in Part III of this article. The calculated load shall be the result of adding the loads from 220.82(B) and (C). Feeder and service-entrance conductors whose calculated load is determined by this optional calculation shall be permitted to have the neutral load determined by 220.61.

This optional calculation method applies to a single dwelling unit, whether a separate building or located in a multifamily dwelling. See Article 100 for the definition of *dwelling unit*.

> **See also**
>
> **Informative Annex D,** Examples D2(a), D2(b), D2(c), and D4(b), for examples of the optional calculations for a dwelling unit

(B) General Loads. The general calculated load shall be not less than 100 percent of the first 10 kVA plus 40 percent of the remainder of the following loads:

(1) 33 volt-amperes/m^2 or 3 volt-amperes/ft^2 for general lighting and general-use receptacles. The floor area for each floor shall be calculated from the outside dimensions of the dwelling unit. The calculated floor area shall not include open porches, garages, or unused or unfinished spaces not adaptable for future use.

(2) 1500 volt-amperes for each 2-wire, 20-ampere small-appliance branch circuit and each laundry branch circuit covered in 210.11(C)(1) and (C)(2).

(3) The nameplate rating of the following:
   a. All appliances that are fastened in place, permanently connected, or located to be on a specific circuit
   b. Ranges, wall-mounted ovens, counter-mounted cooking units
   c. Clothes dryers that are not connected to the laundry branch circuit specified in 220.82(B)(2)
   d. Water heaters

(4) The nameplate ampere or kVA rating of all permanently connected motors not included in 220.82(B)(3).

(C) Heating and Air-Conditioning Load. The largest of the following six selections (load in kVA) shall be included:

(1) 100 percent of the nameplate rating(s) of the air conditioning and cooling.

(2) 100 percent of the nameplate rating(s) of the heat pump when the heat pump is used without any supplemental electric heating.

(3) 100 percent of the nameplate rating(s) of the heat pump compressor and 65 percent of the supplemental electric heating for central electric space-heating systems. If the heat pump compressor is prevented from operating at the same time as the supplementary heat, it does not need to be added to the supplementary heat for the total central space heating load.

(4) 65 percent of the nameplate rating(s) of electric space heating if less than four separately controlled units.

(5) 40 percent of the nameplate rating(s) of electric space heating if four or more separately controlled units.

(6) 100 percent of the nameplate ratings of electric thermal storage and other heating systems where the usual load is expected to be continuous at the full nameplate value. Systems qualifying under this selection shall not be calculated under any other selection in 220.82(C).

Section 220.82(C) requires that only the largest of the six choices needs to be included in the feeder or service calculation.

See also

Informative Annex D, Examples D2(a), D2(b), and D2(c), for examples of calculations using air conditioning and heating

220.83 Existing Dwelling Unit. This section shall be permitted to be used to determine if the existing service or feeder is of sufficient capacity to serve additional loads. Where the dwelling unit is served by a 120/240-volt or 208Y/120-volt, 3-wire service or feeder, calculating the total load in accordance with 220.83(A) or (B) shall be permitted.

Δ **(A) Where Additional Air-Conditioning Equipment or Electric Space-Heating Equipment Is Not to Be Installed.** The percentages listed in Table 220.83(A) shall be used for existing and additional new loads.

TABLE 220.83(A) *Without Additional Air-Conditioning or Electric Space-Heating Equipment*

Load (kVA)	Percent of Load
First 8 kVA of load at	100
Remainder of load at	40

Load calculations shall include the following:

(1) General lighting and general-use receptacles at 33 volt-amperes/m^2 or 3 volt-amperes/ft^2 as determined by 220.42

(2) 1500 volt-amperes for each 2-wire, 20-ampere small-appliance branch circuit and each laundry branch circuit covered in 210.11(C)(1) and (C)(2)

(3) The nameplate rating of the following:

a. All appliances that are fastened in place, permanently connected, or located to be on a specific circuit

b. Ranges, wall-mounted ovens, counter-mounted cooking units

c. Clothes dryers that are not connected to the laundry branch circuit specified in item (2)

d. Water heaters

Δ **(B) Where Additional Air-Conditioning Equipment or Electric Space-Heating Equipment Is to Be Installed.** The percentages listed in Table 220.83(B) shall be used for existing and additional new loads. The larger connected load of air conditioning or space heating, but not both, shall be used.

TABLE 220.83(B) *With Additional Air-Conditioning or Electric Space-Heating Equipment*

Load	Percent of Load
Air-conditioning equipment	100
Central electric space heating	100
Less than four separately controlled space-heating units	100
First 8 kVA of all other loads	100
Remainder of all other loads	40

Other loads shall include the following:

(1) General lighting and general-use receptacles at 33 volt-amperes/m^2 or 3 volt-amperes/ft^2 as determined by 220.42

(2) 1500 volt-amperes for each 2-wire, 20-ampere small-appliance branch circuit and each laundry branch circuit covered in 210.11(C)(1) and (C)(2)

(3) The nameplate rating of the following:

a. All appliances that are fastened in place, permanently connected, or located to be on a specific circuit

b. Ranges, wall-mounted ovens, counter-mounted cooking units

c. Clothes dryers that are not connected to the laundry branch circuit specified in item (2)

d. Water heaters

The optional methods described in 220.83(A) and (B) allow an additional load to be supplied by an existing service.

Calculation Example 1

220.83(A). An existing dwelling unit is served by a 100-A service. A 20 ft × 20 ft addition with a second laundry room is to be added to the dwelling. Determine if the service is properly sized to handle the increased load.

Solution

Step 1. Following the requirements of 220.83(A), calculate the existing dwelling unit load before the addition:

General lighting:	
24 ft × 40 ft = 960 ft² × 3 VA per ft²	2,880 VA
Small-appliance circuits: 3 × 1500 VA	4,500 VA
Laundry circuit	1,500 VA
Electric range rated 10.5 kW	10,500 VA
Electric water heater rated 3.0 kW	3,000 VA
Total existing load	22,380 VA

Step 2. Determine new load of the dwelling unit.

Added 20 ft × 20 ft floor area (400 ft² × 3 VA per ft²)	1,200 VA
Added laundry circuit	1,500 VA
Total new load	2,700 VA

Step 3. Following the requirements in 220.83(A), calculate the dwelling unit total load after the new addition:

First 8 kVA of other load at 100%	8,000 VA
Remainder of other load at 40%:	
22,380 VA + 2700 VA – 8000 VA = 17,080 VA × 40%	6,832 VA
Total load	14,832 VA

Step 4. Determine if service is properly rated to handle additional load:

$$14{,}832 \text{ VA} \div 240 \text{ V} = 61.8 \text{ A, rounded up to 62 A}$$

The additional load contributed by the added floor space with a laundry room does not exceed the allowable load permitted on a 100-A service.

Calculation Example 2

220.83(B). In addition to the added floor area and laundry circuit, a single 5-kVA, 240-V air-conditioning unit is to be installed. Determine if the service is properly sized to handle the increased load.

Solution

Step 1. Determine existing load. Total existing load = 22,380 VA from Step 1 of Calculation Example 1, above.
Step 2. Determine new loads of the dwelling unit.

Added 20 ft × 20 ft floor area (400 ft² × 3 VA per ft²)	1,200 VA
Added laundry circuit	1,500 VA
Added air-conditioning unit	5,000 VA

Step 3. Following the requirements in 220.83(B), calculate the dwelling unit total load after the new addition and equipment:

100% of added air-conditioning equipment	5,000 VA
First 8 kVA of other load at 100%	8,000 VA
Remainder of other load at 40%:	
22,380 VA + 2,700 VA – 8,000 VA = 17,080 VA × 40%	6,832 VA
Total load	19,832 VA

Step 4. Determine if service is properly rated to handle additional load:

$$19{,}832 \text{ VA} \div 240 \text{ V} = 82.6 \text{ A, rounded up to 83 A}$$

The additional load contributed by the added floor space with a laundry room and 5-kVA air-conditioning equipment does not exceed the allowable load permitted on a 100-A service.

220.84 Multifamily Dwelling.

(A) Feeder or Service Load. It shall be permissible to calculate the load of a feeder or service that supplies three or more dwelling units of a multifamily dwelling in accordance with Table 220.84(B) instead of Part III of this article if all the following conditions are met:

(1) No dwelling unit is supplied by more than one feeder.
(2) Each dwelling unit is equipped with electric cooking equipment.

Exception: When the calculated load for multifamily dwellings without electric cooking in Part III of this article exceeds that calculated under Part IV for the identical load plus electric cooking (based on 8 kW per unit), the lesser of the two loads shall be permitted to be used.

(3) Each dwelling unit is equipped with either electric space heating or air conditioning, or both. Feeders and service conductors whose calculated load is determined by this optional calculation shall be permitted to have the neutral load determined by 220.61.

(B) House Loads. House loads shall be calculated in accordance with Part III of this article and shall be in addition to the dwelling unit loads calculated in accordance with Table 220.84(B).

(C) Calculated Loads. The calculated load to which the demand factors of Table 220.84(B) apply shall include the following:

(1) 33 volt-amperes/m² or 3 volt-amperes/ft² for general lighting and general-use receptacles
(2) 1500 volt-amperes for each 2-wire, 20-ampere small-appliance branch circuit and each laundry branch circuit covered in 210.11(C)(1) and (C)(2)
(3) The nameplate rating of the following:
    a. All appliances that are fastened in place, permanently connected, or located to be on a specific circuit
    b. Ranges, wall-mounted ovens, counter-mounted cooking units
    c. Clothes dryers that are not connected to the laundry branch circuit specified in item (2)
    d. Water heaters
(4) The nameplate ampere or kVA rating of all permanently connected motors not included in item (3)
(5) The larger of the air-conditioning load or the fixed electric space-heating load

TABLE 220.84(B) *Optional Calculations — Demand Factors for Three or More Multifamily Dwelling Units*

Number of Dwelling Units	Demand Factor (%)
3–5	45
6–7	44
8–10	43
11	42
12–13	41
14–15	40
16–17	39
18–20	38
21	37
22–23	36
24–25	35
26–27	34
28–30	33
31	32
32–33	31
34–36	30
37–38	29
39–42	28
43–45	27
46–50	26
51–55	25
56–61	24
62 and over	23

220.85 Two Dwelling Units. Where two dwelling units are supplied by a single feeder or service and the calculated load under Part III of this article exceeds that for three identical units calculated under 220.84, the lesser of the two loads shall be permitted to be used.

220.86 Schools. The calculation of a feeder or service load for schools shall be permitted in accordance with Table 220.86 in lieu of Part III of this article where equipped with electric space heating, air conditioning, or both. The connected load to which the demand factors of Table 220.86 apply shall include all of the interior and exterior lighting, power, water heating, cooking, other loads, and the larger of the air-conditioning load or space-heating load within the building or structure.

Feeders and service conductors whose calculated load is determined by this optional calculation shall be permitted to have the neutral load determined by 220.61. Where the building

N   **TABLE 220.86** *Optional Method — Demand Factors for Feeders and Service Conductors for Schools*

Connected Load		Demand Factor (%)	Calculated Loads (VA)
Total VA/m²	Total VA/ft²		
0–33	0–3	100	Amount × 100%
Over 33–220	Over 3–20	75	(Amount × 75%) + 3
Remainder over 220	Remainder over 20	25	(Amount × 25%) + 15.75

or structure load is calculated by this optional method, feeders within the building or structure shall have ampacity as permitted in Part III of this article; however, the ampacity of an individual feeder shall not be required to be larger than the ampacity for the entire building.

This section shall not apply to portable classroom buildings.

The air-conditioning load in portable classrooms must comply with Article 440, and the lighting load must be considered continuous. The demand factors in Table 220.86 do not apply to portable classrooms, because those demand factors would decrease the feeder or service size to below that required for the connected continuous load.

Table 220.86 provides a series of demand increments that are permitted to be applied to the initial calculated load for service or feeder conductors supplying the total load of a school building in lieu of the "standard calculation" covered in Part III of Article 220. The incremental steps of Table 220.86 are based on more significant reductions of the total load as the load per square foot increases. Any portion of the load exceeding 20 volt-amperes per square foot is permitted to have a 25-percent demand factor applied. This approach is similar in concept to that applied to dwellings in Table 220.45, except that the demands in Table 220.86 apply to the entire load of the building, not just the general lighting load.

Feeder conductors that do not supply the entire load of the building or structure are to be calculated in accordance with Part III of Article 220. Feeder conductors supplying subdivided building loads are not required to be larger than the service or feeder conductors that supply the entire building or structure load.

Calculation Example

Calculate the demand load for a 100,000 ft² school building if the calculated load without demand factors is 2.5 MVA (2,500,000 VA).

Solution

Step 1. Determine load per square foot:

$$2.5 \text{ MVA}/100,000 \text{ ft}^2 = 25 \text{ VA/ft}^2$$

Step 2. Calculate the demand load using Table 220.86:

First 3 VA/ft² at 100%	3 VA/ft²
Next 17 VA/ft² at 75%	
17 VA/ft² × 0.75	12.75 VA/ft²
Remaining 5 VA/ft² at 25%	
5 VA/ft² × 0.25	1.25 VA/ft²
Overall demand factor	17 VA/ft²

Step 3. Calculate demand load for the school:

$$17 \text{ VA/ft}^2 \times 100,000 \text{ ft}^2 = 1,700,000 \text{ VA} = 1.7 \text{ MVA}$$

Δ **220.87 Determining Existing Loads.** The calculation of a feeder or service load for existing installations shall be permitted to use actual maximum demand to determine the existing load under all of the following conditions:

(1) The maximum demand data is available for a 1-year period.

Exception: If the maximum demand data for a 1-year period is not available, the calculated load shall be permitted to be based on the maximum demand (the highest average kilowatts reached and maintained for a 15-minute interval) continuously recorded over a minimum 30-day period using a recording ammeter or power meter connected to the highest loaded phase of the feeder or service, based on the initial loading at the start of the recording. The recording shall reflect the maximum demand of the feeder or service by being taken when the building or space is occupied and shall include by measurement or calculation the larger of the heating or cooling equipment load, and other loads that might be periodic in nature due to seasonal or similar conditions. This exception shall not be permitted if the feeder or service has a renewable energy system (i.e., solar photovoltaic or wind electric) or employs any form of peak load shaving.

(2) The maximum demand at 125 percent plus the new load does not exceed the ampacity of the feeder or rating of the service.
(3) The feeder has overcurrent protection in accordance with 240.4, and the service has overload protection in accordance with 230.90.

220.88 New Restaurants. Calculation of a service or feeder load, where the feeder serves the total load, for a new restaurant shall be permitted in accordance with Table 220.88 in lieu of Part III of this article.

The overload protection of the service conductors shall be in accordance with 230.90 and 240.4.

Feeder conductors shall not be required to be of greater ampacity than the service conductors.

Service or feeder conductors whose calculated load is determined by this optional calculation shall be permitted to have the neutral load determined by 220.61.

The demand factors in 220.88 recognize the effects of load diversity that are typical of restaurants. It also recognizes the number of continuous loads as a percentage of the total connected load.

The National Restaurant Association, the Edison Electric Institute, and the Electric Power Research Institute based the data for 220.88 on load studies of 262 restaurants. Those studies showed that the demand factors were lower for restaurants with larger connected loads. The service or feeder size is calculated by applying the appropriate demand factor from Table 220.88 to the total load (kVA).

Calculation Example 1

A new, all-electric restaurant has a total connected load of 348 kVA at 208Y/120 V. Calculate the demand load using Table 220.88. Then determine the size of the service-entrance conductors and the maximum-size overcurrent device for the service.

Solution

Step 1. Calculate the demand load using the value in Table 220.88 (Row 3, Column 2):

$$\text{Demand load} = 50\% \text{ of amount over } 325 \text{ kVA} + 172.5 \text{ kVA}$$
$$= [0.50 \times (348 \text{ kVA} - 325 \text{ kVA})] + 172.5 \text{ kVA}$$
$$= 184 \text{ kVA}_{\text{Demand load}}$$

Step 2. Calculate the service size:

$$\text{Service size} = \frac{\text{kVA}_{\text{Demand load}} \times 1000}{\text{voltage } \sqrt{3}}$$
$$= \frac{184 \text{ kVA} \times 1000}{208 \text{ V} \times \sqrt{3}}$$
$$= 510.7 \text{ A, rounded up to } 511 \text{ A}$$

Step 3. Determine the size of the overcurrent device. The next higher standard-size overcurrent device permitted by 240.4(B) is 600 A.

Step 4. Determine minimum conductor size from Table 310.16. The minimum copper conductor size using the 75°C column is a single 900 AWG (allowable ampacity 520 A). Parallel conductors, such a two 300 kcmil copper conductors (allowable ampacity 285 A × 2), can also be used.

Calculation Example 2

A new restaurant has gas cooking appliances plus a total connected electrical load of 348 kVA at 208Y/120 V. Calculate the demand load using Table 220.88. Then determine the size of the service-entrance conductors and the maximum-size overcurrent device for the service.

TABLE 220.88 Optional Method — Permitted Load Calculations for Service and Feeder Conductors for New Restaurants

Total Connected Load (kVA)	All Electric Restaurant Calculated Loads (kVA)	Not All Electric Restaurant Calculated Loads (kVA)
0–200	80%	100%
201–325	10% (amount over 200) + 160.0	50% (amount over 200) + 200.0
326–800	50% (amount over 325) + 172.5	45% (amount over 325) + 262.5
Over 800	50% (amount over 800) + 410.0	20% (amount over 800) + 476.3

Note: Add all electrical loads, including both heating and cooling loads, to calculate the total connected load. Select the one demand factor that applies from the table, then multiply the total connected load by this single demand factor.

Solution

Step 1. Calculate the demand load using the value in Table 220.88 (Row 3, Column 3):

Demand load = 45% of amount over 325 kVA + 262.5 kVA
$$= [0.45 \times (348\ kVA - 325\ kVA)] + 262.5\ kVA$$
$$= 272.85\ kVA_{Demand\ load}$$

Step 2. Calculate the service size:

$$Service\ size = \frac{kVA_{Demand\ load} \times 1000}{voltage\ \sqrt{3}}$$

$$= \frac{272.85\ kVA \times 1000}{208\ V \times \sqrt{3}}$$

$$= 757.36\ A,\ rounded\ down\ to\ 757\ A$$

Step 3. Determine the size of the overcurrent device. The next higher standard-size overcurrent device permitted by 240.6(A) is 800 A.

Step 4. Determine minimum conductor size from Table 310.16. The required conductor ampacity is above those provided on the table for a single conductor. Parallel conductors, such as two 500 kcmil copper conductors (allowable ampacity 285 A × 2), are necessary.

Part V. Farm Load Calculations

220.100 General. Farm loads shall be calculated in accordance with Part V.

220.102 Farm Loads — Buildings and Other Loads.

(A) Dwelling Unit. The feeder or service load of a farm dwelling unit shall be calculated in accordance with the provisions for dwellings in Part III or IV of this article. Where the dwelling has electric heat and the farm has electric grain-drying systems, Part IV of this article shall not be used to calculate the dwelling load where the dwelling and farm loads are supplied by a common service.

(B) Other Than Dwelling Unit. Where a feeder or service supplies a farm building or other load having two or more separate branch circuits, the load for feeders, service conductors, and service equipment shall be calculated in accordance with demand factors not less than indicated in Table 220.102(B).

220.103 Farm Loads — Total. Where supplied by a common service, the total load of the farm for service conductors and service equipment shall be calculated in accordance with the farm dwelling unit load and demand factors specified in Table 220.103. Where there is equipment in two or more farm equipment buildings or for loads having the same function, such loads shall be calculated in accordance with Table 220.102(B) and shall be permitted to be combined as a single load in Table 220.103 for calculating the total load.

TABLE 220.102(B) *Method for Calculating Farm Loads for Other Than Dwelling Unit*

Ampere Load at 240 Volts Maximum	Demand Factor (%)
The greater of the following:	
All loads that are expected to operate simultaneously, or 125 percent of the full load current of the largest motor, or First 60 amperes of the load	100
Next 60 amperes of all other loads	50
Remainder of other loads	25

TABLE 220.103 *Method for Calculating Total Farm Load*

Individual Loads Calculated in Accordance with Table 220.102	Demand Factor (%)
Largest load	100
Second largest load	75
Third largest load	65
Remaining loads	50

Note: To this total load, add the load of the farm dwelling unit calculated in accordance with Part III or IV of this article. Where the dwelling has electric heat and the farm has electric grain-drying systems, Part IV of this article shall not be used to calculate the dwelling load.

N Part VI. Health Care Facilities

N 220.110 Receptacle Loads. Receptacle loads calculated in accordance with 220.14(H) and (I) and supplied by branch circuits not exceeding 150 volts to ground shall be permitted to be subjected to the demand factors provided in Table 220.110(1) and Table 220.110(2) for health care facilities.

N TABLE 220.110(1) *Demand Factors for Receptacles Supplied by General-Purpose Branch Circuits in Category 1 and Category 2 Patient Care Spaces*

Portion of Receptacle Load to Which Demand Factor Applies (Volt-Amperes)	Demand Factor (%)
First 5000 or less	100
From 5001 to 10,000	50
Remainder over 10,000	25

N TABLE 220.110(2) *Demand Factors for Receptacles Supplied by General-Purpose Branch Circuits in Category 3 and Category 4 Patient Care Spaces*

Portion of Receptacle Load to Which Demand Factor Applies (Volt-Amperes)	Demand Factor (%)
First 10,000 or less	100
Remainder over 10,000	50

Informational Note No. 1: See Article 100 for the definitions of patient care space categories.

Informational Note No. 2: See 220.14(I) for the calculation of receptacle outlet loads.

N Part VII. Marinas, Boatyards, Floating Buildings, and Commercial and Noncommercial Docking Facilities

N 220.120 Receptacle Loads. General lighting and other loads in marinas, boatyards, floating buildings, and commercial and noncommercial docking facilities shall be calculated in accordance with Part III of this article and, in addition, the demand factors set forth in Table 220.120 shall be permitted for each service or feeder circuit supplying receptacles that provide shore power for boats. These calculations shall be permitted to be modified as indicated in Notes (1) and (2) of Table 220.120. Where demand factors of Table 220.120 are applied, the demand factor specified in 220.61(B) shall not be permitted.

Informational Note: These demand factors could be inadequate in areas of extreme hot or cold temperatures with loaded circuits for heating, air-conditioning, or refrigerating equipment.

N TABLE 220.120 *Demand Factors for Shore Power Receptacle Loads*

Number of Shore Power Receptacles	Sum of the Rating of the Receptacles (%)
1–4	100
5–8	90
9–14	80
15–30	70
31–40	60
41–50	50
51–70	40
≥71	30

Notes:

1. Where shore power accommodations provide two receptacles specifically for an individual boat slip and these receptacles have different voltages (e.g., one 30-ampere, 125-volt and one 50-ampere, 125/250-volt), only the receptacle with the larger kilowatt demand shall be required to be calculated.

2. For each shore powered pedestal being installed that includes an individual kilowatt-hour submeters for each slip and is being calculated using the criteria listed in Table 220.120, the total demand amperes shall be permitted to be multiplied by 0.9 to achieve the final demand amperes of the facility.

3. If a circuit feeding a boat hoist and shore power for the same boat slip is shared, only the load with the larger kilowatt demand shall be required to be counted in the load calculation.

ARTICLE 225 Outside Branch Circuits and Feeders

N Part I. General

225.1 Scope. This article covers requirements for outside branch circuits and feeders not over 1000 volts ac or 1500 volts dc, nominal, run on or between buildings, structures, or poles on the premises; and electrical equipment and wiring for the supply of utilization equipment that is located on or attached to the outside of buildings, structures, or poles.

Informational Note: See Part IV of Article 235 for outside branch circuits and feeders over 1000 volts ac or 1500 volts dc.

Article 225 provides requirements unique to the installation of feeders and branch circuits outside (overhead and underground, not over 1000 volts ac or 1500 volts dc) of buildings and structures. These circuits may supply specific items of electrical equipment, or they may be the power supply to another building or structure. Examples of outside feeders and branch circuits include the following:

- Conductors supplying the buildings of a multibuilding industrial complex or institutional campus
- Outdoor supply conductors from an emergency system, a standby system, an alternative energy system, or on-site power generation
- Supply conductors between a dwelling unit and a detached garage or other structure

These requirements are in addition to the general requirements for branch circuits and feeders in Articles 210 and 215.

225.3 Other Articles. Application of other articles, including additional requirements to specific cases of equipment and conductors, is shown in Table 225.3.

225.4 Conductor Insulation. Where within 3.0 m (10 ft) of any building or structure other than supporting poles or towers, open individual (aerial) overhead conductors shall be insulated for the nominal voltage. The insulation of conductors in cables or raceways, except Type MI cable, shall be of thermoset or thermoplastic type and, in wet locations, shall comply with 310.10(C). The insulation of conductors for festoon lighting shall be of the thermoset or thermoplastic type.

Exception: Equipment grounding conductors and grounded circuit conductors shall be permitted to be bare or covered as specifically permitted elsewhere in this Code.

This exception and 250.184(A)(1), Exception No. 2, correlate to permit the use of the bare messenger wire of an overhead cable assembly as the grounded (neutral) conductor of an outdoor feeder circuit.

Δ **TABLE 225.3** *Other Articles*

Equipment/Conductors	Article
Branch circuits	210
Class 1 power-limited circuits and Class 1 power-limited remote-control and signaling circuits	724
Class 2 and Class 3 remote-control, signaling, and power-limited circuits	725
Conductors for general wiring	310
Electrically driven or controlled irrigation machines	675
Electric signs and outline lighting	600
Feeders	215
Fire alarm systems	760
Fixed outdoor electric deicing and snow-melting equipment	426
Grounding and bonding	250
Hazardous (classified) locations	500
Hazardous (classified) locations — specific	510
Marinas and boatyards	555
Medium-voltage conductors and cable	311
Messenger-supported wiring	396
Mobile homes, manufactured homes, and mobile home parks	550
Open wiring on insulators	398
Over 1000 volts, general	495
Overcurrent protection	240
Overcurrent protection for systems rated over 1000 volts ac, 1500 volts dc	245
Services	230
Services, feeders, and branch circuits over 1000 volts ac, 1500 volts dc	235
Solar photovoltaic systems	690
Swimming pools, fountains, and similar installations	680
Use and identification of grounded conductors	200

225.6 Conductor Size and Support.

Δ **(A) Overhead Spans.** Open individual conductors shall not be smaller than 10 AWG copper or 8 AWG aluminum for spans up to 15 m (50 ft) in length, and 8 AWG copper or 6 AWG aluminum for a longer span unless supported by a messenger wire.

The size limitation of copper and aluminum conductors for overhead spans is based on the need for adequate mechanical strength to support the weight of the conductors and to withstand wind, ice, and other similar conditions. Exhibit 225.1 illustrates overhead spans that are not messenger supported, are run between buildings, and are 1000 volts or less. The messenger cable provides the necessary mechanical strength rather than relying on the conductors.

See also

396.10 for cable types permitted to be messenger supported

(B) Festoon Lighting. Overhead conductors for festoon lighting shall not be smaller than 12 AWG unless the conductors are supported by messenger wires. In all spans exceeding 12 m (40 ft), the conductors shall be supported by messenger wire. The messenger wire shall be supported by strain insulators.

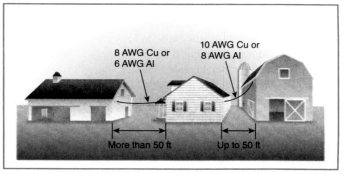

EXHIBIT 225.1 Minimum sizes of conductors in overhead cable spans for 1000 volts, nominal, or less.

EXHIBIT 225.2 Messenger wire required for festoon lighting conductors in a span exceeding 40 feet.

Conductors or messenger wires shall not be attached to any fire escape, downspout, or plumbing equipment.

Article 100 defines *festoon lighting* as "a string of outdoor lights that is suspended between two points." Exhibit 225.2 illustrates an installation of festoon lighting that complies with the minimum clearance above-grade requirement specified in 225.18(1). The maximum distance between conductor supports is 40 feet unless a messenger wire is installed to support the current-carrying conductors.

Attachment of festoon lighting to fire escapes, plumbing equipment, or metal drain spouts is prohibited, because the attachment could provide a path to ground. Such methods of attachment could not be relied on for a permanent or secure means of support.

Δ **225.10 Wiring on Buildings (or Other Structures).** The installation of outside wiring on surfaces of buildings (or other structures) shall be permitted for circuits not exceeding 1000 volts, nominal, as the following:

(1) Auxiliary gutters
(2) Busways

(3) Cable trays
(4) Cablebus
(5) Electrical metallic tubing (EMT)
(6) Flexible metal conduit (FMC)
(7) Intermediate metal conduit (IMC)
(8) Liquidtight flexible metal conduit (LFMC)
(9) Liquidtight flexible nonmetallic conduit (LFNC)
(10) Messenger-supported wiring
(11) Open wiring on insulators
(12) Reinforced thermosetting resin conduit (RTRC)
(13) Rigid metal conduit (RMC)
(14) Rigid polyvinyl chloride conduit (PVC)
(15) Type MC cable
(16) Type MI cable
(17) Type SE cable
(18) Type TC-ER cable
(19) Type UF cable
(20) Wireways

Δ **225.11 Feeder and Branch-Circuit Conductors Entering, Exiting, or Attached to Buildings or Structures.** Feeder and branch-circuit conductors entering or exiting buildings or structures shall be installed in accordance with 230.52. Overhead branch circuits and feeders attached to buildings or structures shall be installed in accordance with 230.54.

225.12 Open-Conductor Supports. Open conductors shall be supported on knobs, racks, brackets, or strain insulators, that are made of glass, porcelain, or other approved materials.

225.14 Open-Conductor Spacings. Conductors shall comply with the spacings provided in Table 230.51(C).

(A) Separation from Other Circuits. Open conductors shall be separated from open conductors of other circuits or systems by not less than 100 mm (4 in.).

(B) Conductors on Poles. Conductors on poles shall have a separation of not less than 300 mm (1 ft) where not placed on racks or brackets. Conductors supported on poles shall provide a horizontal climbing space not less than the following:

(1) Power conductors below communications conductors — 750 mm (30 in.)
(2) Power conductors alone or above communications conductors:
   a. 300 volts or less — 600 mm (24 in.)
   b. Over 300 volts — 750 mm (30 in.)
(3) Communications conductors below power conductors — same as power conductors
(4) Communications conductors alone — no requirement

Sufficient space is required for personnel to be able to climb safely over or through conductors to work with conductors on the pole.

225.15 Supports over Buildings. Outside branch-circuit and feeder conductors passing over a building shall be securely supported.

225.16 Attachment to Buildings.

(A) Point of Attachment. The point of attachment to a building shall be in accordance with 230.26.

(B) Means of Attachment. The means of attachment to a building shall be in accordance with 230.27.

225.17 Masts as Supports. Only feeder or branch-circuit conductors specified within this section shall be permitted to be attached to the feeder and/or branch-circuit mast. Masts used for the support of final spans of feeders or branch circuits shall be installed in accordance with 225.17(A) and (B).

Conductors that are attached to the exterior of the mast are not permitted to be attached between the weatherhead and any coupling that is above a point of securement to the building. Attaching conductors in that space can put additional strain on the mast, which could cause it to fail.

A mast supporting an overhead branch circuit or feeder span is not permitted to support conductors of other systems, such as overhead conductor spans for signaling, communications, or CATV systems.

See also

230.28, which provides similar requirements for masts supporting service drops

(A) Strength. The mast shall have adequate strength or be supported by braces or guy wires to safely withstand the strain imposed by the overhead feeder or branch-circuit conductors. Hubs intended for use with a conduit serving as a mast for support of feeder or branch-circuit conductors shall be identified for use with a mast.

(B) Attachment. Feeder and/or branch-circuit conductors shall not be attached to a mast where the connection is between a weatherhead or the end of the conduit and a coupling where the coupling is located above the last point of securement to the building or other structure, or where the coupling is located above the building or other structure.

225.18 Clearance for Overhead Conductors and Cables. Overhead spans of open conductors and open multiconductor cables of not over 1000 volts, nominal, shall have a clearance of not less than the following:

(1) 3.0 m (10 ft) — above finished grade, sidewalks, or from any platform or projection that will permit personal contact where the voltage does not exceed 150 volts to ground and accessible to pedestrians only
(2) 3.7 m (12 ft) — over residential property and driveways, and those commercial areas not subject to truck traffic where the voltage does not exceed 300 volts to ground

(3) 4.5 m (15 ft) — for those areas listed in the 3.7 m (12 ft) classification where the voltage exceeds 300 volts to ground

(4) 5.5 m (18 ft) — over public streets, alleys, roads, parking areas subject to truck traffic, driveways on other than residential property, and other land traversed by vehicles, such as cultivated, grazing, forest, and orchard

(5) 7.5 m (24½ ft) — over track rails of railroads

This section covers the requirements for clearances from ground; 225.19 covers the requirements for clearances from buildings for conductors not over 1000 volts. Section 225.19(D) provides final span clearances from windows, doors, fire escapes, and similar areas where overhead conductors are subject to contact by persons. Exhibit 225.3 shows an example of where those clearance requirements apply.

225.19 Clearances from Buildings for Conductors of Not over 1000 Volts, Nominal. Overhead spans of open conductors and open multiconductor cables shall comply with 225.19(A), (B), (C), and (D).

(A) Above Roofs. Overhead spans of open conductors and open multiconductor cables shall have a vertical clearance of not less than 2.6 m (8 ft 6 in.) above the roof surface. The vertical clearance above the roof level shall be maintained for a distance not less than 900 mm (3 ft) in all directions from the edge of the roof.

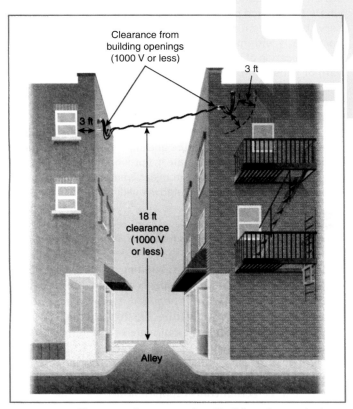

EXHIBIT 225.3 Clearances from ground and buildings for conductors not over 1000 volts.

Exception No. 1: The area above a roof surface subject to pedestrian or vehicular traffic shall have a vertical clearance from the roof surface in accordance with the clearance requirements of 225.18.

Exception No. 2: Where the voltage between conductors does not exceed 300, and the roof has a slope of 100 mm in 300 mm (4 in. in 12 in.) or greater, a reduction in clearance to 900 mm (3 ft) shall be permitted.

Exception No. 3: Where the voltage between conductors does not exceed 300, a reduction in clearance above only the overhanging portion of the roof to not less than 450 mm (18 in.) shall be permitted if (1) not more than 1.8 m (6 ft) of the conductors, 1.2 m (4 ft) horizontally, pass above the roof overhang, and (2) they are terminated at a through-the-roof raceway or approved support.

Exception No. 4: The requirement for maintaining the vertical clearance 900 mm (3 ft) from the edge of the roof shall not apply to the final conductor span where the conductors are attached to the side of a building.

(B) From Nonbuilding or Nonbridge Structures. From signs, chimneys, radio and television antennas, tanks, and other nonbuilding or nonbridge structures, clearances — vertical, diagonal, and horizontal — shall not be less than 900 mm (3 ft).

(C) Horizontal Clearances. Clearances shall not be less than 900 mm (3 ft).

(D) Final Spans. Final spans of feeders or branch circuits shall comply with 225.19(D)(1), (D)(2), and (D)(3).

(1) Clearance from Windows. Final spans to the building they supply, or from which they are fed, shall be permitted to be attached to the building, but they shall be kept not less than 900 mm (3 ft) from windows that are designed to be opened, and from doors, porches, balconies, ladders, stairs, fire escapes, or similar locations.

Exception: Conductors run above the top level of a window shall be permitted to be less than the 900 mm (3 ft) requirement.

(2) Vertical Clearance. The vertical clearance of final spans above or within 900 mm (3 ft) measured horizontally of platforms, projections, or surfaces that will permit personal contact shall be maintained in accordance with 225.18.

(3) Building Openings. The overhead branch-circuit and feeder conductors shall not be installed beneath openings through which materials may be moved, such as openings in farm and commercial buildings, and shall not be installed where they obstruct entrance to these openings.

(E) Zone for Fire Ladders. Where buildings exceed three stories or 15 m (50 ft) in height, overhead lines shall be arranged, where practicable, so that a clear space (or zone) at least 1.8 m (6 ft) wide will be left either adjacent to the buildings or beginning not over 2.5 m (8 ft) from them to facilitate the raising of ladders when necessary for fire fighting.

225.20 Protection Against Physical Damage. Conductors installed on buildings, structures, or poles shall be protected against physical damage as provided for services in 230.50.

225.21 Multiconductor Cables on Exterior Surfaces of Buildings (or Other Structures). Supports for multiconductor cables on exterior surfaces of buildings (or other structures) shall be as provided in 230.51.

225.22 Raceways on Exterior Surfaces of Buildings or Other Structures. Raceways on exteriors of buildings or other structures shall be arranged to drain and shall be listed or approved for use in wet locations.

If raceways are exposed to weather or temperature changes, condensation is likely to occur, causing moisture to accumulate within raceways at low points of the installation and in junction boxes. Therefore, raceways are required to be installed so as to allow moisture to drain from the raceway through drain holes or other means provided at appropriate locations.

225.24 Outdoor Lampholders. Where outdoor lampholders are attached as pendants, the connections to the circuit wires shall be staggered. Where such lampholders have terminals of a type that puncture the insulation and make contact with the conductors, they shall be attached only to conductors of the stranded type.

225.25 Location of Outdoor Lamps. Locations of lamps for outdoor lighting shall be below all energized conductors, transformers, or other electric utilization equipment, unless either of the following apply:

(1) Clearances or other safeguards are provided for relamping operations.
(2) Equipment is controlled by a disconnecting means that is lockable open in accordance with 110.25.

The objective is to protect personnel during relamping of outdoor luminaires. Section 225.18 requires a minimum clearance of 10 feet above grade or platforms for open conductors. In some cases, it may be difficult to keep all electrical equipment above the lamps. Section 225.25(1) allows other clearances or safeguards to permit safe relamping, while the reference to 110.25 provides another alternative for safe relamping through the use of a disconnecting means that can be locked in the open or off position.

225.26 Vegetation as Support. Vegetation such as trees shall not be used for support of overhead conductor spans.

Overhead conductor spans attached to a tree are subject to damage over the course of time because normal tree growth around the attachment device causes the mounting insulators to break. Normal growth can also cause tree bark to grow around

the insulation. This requirement reduces the likelihood of chafing or degradation of the conductor insulation, which can create a shock hazard for tree trimmers and tree climbers.

Outdoor luminaires and associated equipment are permitted by 410.36(G) to be supported by trees. To prevent the chafing damage, conductors are run up the tree from an underground wiring method.

> **See also**
>
> **300.5(D)(1)** for requirements on the protection of direct-buried conductors emerging from below grade

225.27 Raceway Seal. Where a raceway enters a building or structure from outside, it shall be sealed in accordance with 300.5(G) and 300.7(A). Spare or unused raceways shall also be sealed. Sealants shall be identified for use with cable insulation, conductor insulation, bare conductor, shield, or other components.

Part II. Buildings or Other Structures Supplied by a Feeder(s) or Branch Circuit(s)

Part II covers outside branch circuits and feeders as the source of electrical supply for buildings and structures. Understanding the definitions of *service point*, *service*, *service equipment*, *feeder*, and *branch circuit* is important. Determining what constitutes a set of feeder or branch-circuit conductors versus a set of service conductors requires a clear understanding of where the service point is located and where the service and service equipment for a premises are located. In some cases, particularly with medium- and high-voltage distribution, the service location of a campus or multibuilding facility is a switchyard or substation. With the location of the service point and service equipment established, the requirements for outside branch circuits and feeders from Part II (and Article 235 if over 1000 volts) can be properly applied.

The feeders and branch circuits covered in Part II might originate in one building or structure and supply another building or structure, or they might originate in outdoor equipment such as freestanding switchboards, switchgear, transformers, or generators and supply equipment located in buildings or structures.

> **See also**
>
> **Article 230,** which contains similar requirements for services

225.30 Number of Supplies. A building or other structure that is served by a branch circuit or feeder on the load side of a service disconnecting means shall be supplied by only one feeder or branch circuit unless permitted in 225.30(A) through (F). For the purpose of this section, a multiwire branch circuit shall be considered a single circuit.

Where a branch circuit or feeder originates in these additional buildings or other structures, only one feeder or branch circuit shall be permitted to supply power back to the original building or structure, unless permitted in 225.30(A) through (F).

(A) Special Conditions. Additional feeders or branch circuits shall be permitted to supply the following:

(1) Fire pumps
(2) Emergency systems
(3) Legally required standby systems
(4) Optional standby systems
(5) Parallel power production systems
(6) Systems designed for connection to multiple sources of supply for the purpose of enhanced reliability
(7) Electric vehicle power transfer systems listed, labeled, and identified for more than a single branch circuit or feeder
(8) Docking facilities and piers

(B) Common Supply Equipment. Where feeder conductors originate in the same panelboard, switchboard, or other distribution equipment, and each feeder terminates in a single disconnecting means, not more than six feeders shall be permitted. Where more than one feeder is installed in accordance with this section, all feeder disconnects supplying the building or structure shall be grouped in the same location, and the requirements of 225.33 shall not apply. Each disconnect shall be marked to indicate the load served.

(C) Special Occupancies. By special permission, additional feeders or branch circuits shall be permitted for either of the following:

(1) Multiple-occupancy buildings where there is no space available for supply equipment accessible to all occupants
(2) A single building or other structure sufficiently large to make two or more supplies necessary

(D) Capacity Requirements. Additional feeders or branch circuits shall be permitted where the capacity requirements are in excess of 2000 amperes at a supply voltage of 1000 volts or less.

(E) Different Characteristics. Additional feeders or branch circuits shall be permitted for different voltages, frequencies, or phases, or for different uses such as control of outside lighting from multiple locations.

(F) Documented Switching Procedures. Additional feeders or branch circuits shall be permitted to supply installations under single management where documented safe switching procedures are established and maintained.

Buildings on college campuses, multibuilding industrial facilities, and multibuilding commercial facilities are permitted to be supplied by secondary loop supply (secondary selective) networks if documented switching procedures are in place. The switching procedures must establish a method to safely operate switches for the facility during maintenance and during alternative supply and emergency supply conditions. Keyed interlock systems are often used to reduce the likelihood of inappropriate switching procedures that could result in hazardous conditions.

225.31 Disconnecting Means.

N (A) General. Means shall be provided for disconnecting all ungrounded conductors that supply or pass through the building or structure.

(B) Location. The disconnecting means shall be installed either inside or outside of the building or structure served or where the conductors pass through the building or structure. The disconnecting means shall be at a readily accessible location nearest the point of entrance of the conductors. For the purposes of this section, the requirements in 230.6 shall apply.

Exception No. 1: For installations under single management, where documented safe switching procedures are established and maintained, and where the installation is monitored by qualified individuals, the disconnecting means shall be permitted to be located elsewhere on the premises.

Exception No. 2: For buildings or other structures qualifying under 685.1, the disconnecting means shall be permitted to be located elsewhere on the premises.

Exception No. 3: For towers or poles used as lighting standards, the disconnecting means shall be permitted to be located elsewhere on the premises.

Exception No. 4: For poles or similar structures used only for support of signs installed in accordance with 600.1, the disconnecting means shall be permitted to be located elsewhere on the premises.

Although the requirement for locating the disconnecting means for a feeder or branch circuit supplying a structure is essentially the same as that specified for services in 230.70(A), there is an important difference. Where a building or structure is supplied by a feeder or branch circuit, the disconnecting means must always be nearest the point of entrance of the conductors whether inside or outside the building or structure supplied, unless one of the exceptions can be applied.

Many campus-style facilities are supplied by a single utility service, in which the service disconnecting means is remote from the buildings or structures supplied. The supply conductors to the campus buildings are feeders or branch circuits, and this requirement applies to conductors that supply a building as well as conductors that pass through a building.

Exhibit 225.4 shows an example of a disconnecting means for a generator feeder that can be used to meet the requirements of Article 225, Article 445, and Articles 700, 701, or 702 as applicable. The disconnecting means is required to be "within sight" of the building or structure supplied by the generator. An exception to 700.12(D)(4) permits the disconnecting means for an outdoor generator supplying an emergency system to be located other than within sight of the building(s) or structure(s) supplied by the generator. That exception contains the same conditions as used in 225.31(B), Exception No. 1.

EXHIBIT 225.4 Disconnecting means for alternative source feeder supplying an emergency load, a legally required standby load, an optional standby load, or a combination of the three.

225.33 Maximum Number of Disconnects.

(A) General. The disconnecting means for each supply permitted by 225.30 shall consist of not more than six switches or six circuit breakers mounted in a single enclosure, in a group of separate enclosures, or in or on a switchboard or switchgear. There shall be no more than six disconnects per supply grouped in any one location.

Exception: For the purposes of this section, disconnecting means used solely for the control circuit of the ground-fault protection system, or the control circuit of the power-operated supply disconnecting means, installed as part of the listed equipment, shall not be considered a supply disconnecting means.

(B) Single-Pole Units. Two or three single-pole switches or breakers capable of individual operation shall be permitted on multiwire circuits, one pole for each ungrounded conductor, as one multipole disconnect, provided they are equipped with identified handle ties or a master handle to disconnect all ungrounded conductors with no more than six operations of the hand.

225.34 Grouping of Disconnects.

(A) General. The two to six disconnects as permitted in 225.33 shall be grouped. Each disconnect shall be marked to indicate the load served.

Exception: One of the two to six disconnecting means permitted in 225.33, where used only for a water pump also intended to provide fire protection, shall be permitted to be located remote from the other disconnecting means.

(B) Additional Disconnecting Means. The one or more additional disconnecting means for fire pumps or for emergency, legally required standby or optional standby system permitted by 225.30 shall be installed sufficiently remote from the one to six disconnecting means for normal supply to minimize the possibility of simultaneous interruption of supply.

225.35 Access to Occupants. In a multiple-occupancy building, each occupant shall have access to the occupant's supply disconnecting means.

Exception: In a multiple-occupancy building where electric supply and electrical maintenance are provided by the building management and where these are under continuous building management supervision, the supply disconnecting means supplying more than one occupancy shall be permitted to be accessible to authorized management personnel only.

Δ **225.36 Type of Disconnecting Means.** The disconnecting means specified in 225.31 shall be a circuit breaker, molded case switch, general-use switch, snap switch, or other approved means. Where applied in accordance with 250.32(B)(1), Exception No. 1, the disconnecting means shall be suitable for use as service equipment.

The feeder or branch-circuit disconnecting means is required to be suitable for use as service equipment only where the feeder grounded conductor is also used as the return path for ground-fault current per 250.32(B)(1), Exception No. 1. A three- or four-way snap switch cannot be used as a disconnecting means for an outside branch circuit or feeder because it does not provide a positive indication that the circuit is disconnected.

225.37 Identification. Where a building or structure has any combination of feeders, branch circuits, or services passing through it or supplying it, a permanent plaque or directory shall be installed at each feeder and branch-circuit disconnect location denoting all other services, feeders, or branch circuits supplying that building or structure or passing through that building or structure and the area served by each.

Exception No. 1: A plaque or directory shall not be required for large-capacity multibuilding industrial installations under single management, where it is ensured that disconnection can be accomplished by establishing and maintaining safe switching procedures.

Exception No. 2: This identification shall not be required for branch circuits installed from a dwelling unit to a second building or structure.

This requirement correlates with 230.2(E) in that, if a building has multiple sources of supply, permanent identification at each supply (service, feeder, and branch circuit) disconnecting means is required. Permanent identification must have long-term durability. Such identification is an important safety feature during an emergency, because in many cases first responders will not be familiar with the electrical distribution system of the facility.

See also

700.7(A) for emergency power sources
701.7(A) for legally required standby sources
702.7(A) for optional standby sources
705.10 for parallel power production sources
706.21 for energy storage systems (ESS)
710.10 for stand-alone systems

225.38 Disconnect Construction. Disconnecting means shall meet the requirements of 225.38(A) through (D).

(A) Manually or Power Operable. The disconnecting means shall consist of either (1) a manually operable switch or a circuit breaker equipped with a handle or other suitable operating means or (2) a power-operable switch or circuit breaker, provided the switch or circuit breaker can be opened by hand in the event of a power failure.

(B) Simultaneous Opening of Poles. Each building or structure disconnecting means shall simultaneously disconnect all ungrounded supply conductors that it controls from the building or structure wiring system.

(C) Disconnection of Grounded Conductor. Where the building or structure disconnecting means does not disconnect the grounded conductor from the grounded conductors in the building or structure wiring, other means shall be provided for this purpose at the location of the disconnecting means. A terminal or bus to which all grounded conductors can be attached by means of pressure connectors shall be permitted for this purpose.

In a multisection switchboard or switchgear, disconnects for the grounded conductor shall be permitted to be in any section of the switchboard or switchgear, if the switchboard section or switchgear section is marked to indicate a grounded conductor disconnect is contained within the equipment.

(D) Indicating. The building or structure disconnecting means shall plainly indicate whether it is in the open or closed position.

225.39 Rating of Disconnect. The feeder or branch-circuit disconnecting means shall have a rating of not less than the calculated load to be supplied, determined in accordance with Parts I and II of Article 220 for branch circuits, Part III or IV of Article 220 for feeders, or Part V of Article 220 for farm loads. Where the branch circuit or feeder disconnecting means consists of more than one switch or circuit breaker, as permitted by 225.33, combining the ratings of all the switches or circuit breakers for determining the rating of the disconnecting means shall be permitted. In no case shall the rating be lower than specified in 225.39(A), (B), (C), or (D).

(A) One-Circuit Installation. For installations to supply only limited loads of a single branch circuit, the branch circuit disconnecting means shall have a rating of not less than 15 amperes.

(B) Two-Circuit Installations. For installations consisting of not more than two 2-wire branch circuits, the feeder or branch-circuit disconnecting means shall have a rating of not less than 30 amperes.

(C) One-Family Dwelling. For a one-family dwelling, the feeder disconnecting means shall have a rating of not less than 100 amperes, 3-wire.

(D) All Others. For all other installations, the feeder or branch-circuit disconnecting means shall have a rating of not less than 60 amperes.

225.40 Access to Overcurrent Protective Devices. Where a feeder overcurrent device is not readily accessible, branch-circuit overcurrent devices shall be installed on the load side, shall be mounted in a readily accessible location, and shall be of a lower ampere rating than the feeder overcurrent device.

N **225.41 Emergency Disconnects.** For one- and two-family dwelling units, an emergency disconnecting means shall be installed.

N **(A) General.**

N **(1) Location.** The disconnecting means shall be installed in a readily accessible outdoor location on or within sight of the dwelling unit.

N **(2) Rating.** The disconnecting means shall have a short-circuit current rating equal to or greater than the available fault current.

N **(3) Grouping.** If more than one disconnecting means is provided, they shall be grouped.

N **(B) Identification of Other Isolation Disconnects.** Where equipment for isolation of other energy source systems is not located adjacent to the emergency disconnect required by this section, a plaque or directory identifying the location of all equipment for isolation of other energy sources shall be located adjacent to the disconnecting means required by this section.

Informational Note: See 445.18, 480.7, 705.20, and 706.15 for examples of other energy source system isolation means.

N **(C) Marking.** The disconnecting means shall be marked as EMERGENCY DISCONNECT.

Markings shall comply with 110.21(B) and all of the following:

(1) The marking or labels shall be located on the outside front of the disconnect enclosure with red background and white text.
(2) The letters shall be least 13 mm (½ in.) high.

N **225.42 Surge Protection.**

N **(A) Surge-Protective Device.** Where a feeder supplies any of the following, a surge-protective device (SPD) shall be installed:

(1) Dwelling units
(2) Dormitory units
(3) Guest rooms and guest suites of hotels and motels
(4) Areas of nursing homes and limited-care facilities used exclusively as patient sleeping rooms

N **(B) Location.** The SPD shall be installed in or adjacent to the distribution equipment that is connected to the load side of the feeder and contains branch circuit overcurrent protective device(s) that supply the location specified in 225.42(A).

> Informational Note: Surge protection is most effective when closest to the branch circuit. Surges can be generated from multiple sources including, but not limited to, lightning, the electric utility, or utilization equipment.

N **(C) Type.** The SPD shall be a Type 1 or Type 2 SPD.

N **(D) Replacement.** Where the distribution equipment supplied by the feeder is replaced, all of the requirements of this section shall apply.

N **(E) Ratings.** SPDs shall have a nominal discharge current rating (I_n) of not less than 10kA.

> Informational Note: Lead lengths of conductors to the SPD should be kept as short as possible to reduce let-through voltages.

ARTICLE 230 Services

Part I. General

Δ **230.1 Scope.** This article covers service conductors and equipment for control and protection of services not over 1000 volts ac or 1500 volts dc, nominal and their installation requirements.

> Informational Note No. 1: See Informational Note Figure 230.1.
> Informational Note No. 2: See Part V of Article 235 for services over 1000 volts ac or 1500 volts dc, nominal.

230.2 Number of Services. A building or other structure served shall be supplied by only one service unless permitted in 230.2(A) through (D). For the purpose of 230.40, Exception No. 2 only, underground sets of conductors, 1/0 AWG and larger, running to the same location and connected together at their supply end but not connected together at their load end shall be considered to be supplying one service.

The general requirement is for a building or structure to be supplied by only one service. However, under some conditions, a single service might not be adequate. Therefore, the installation

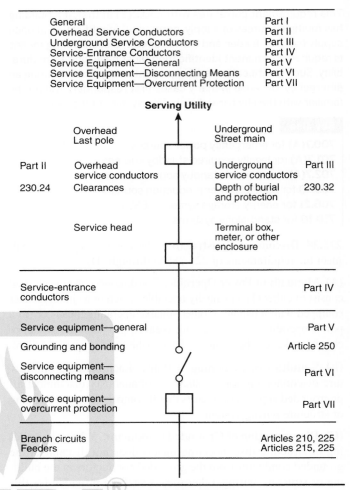

General	Part I
Overhead Service Conductors	Part II
Underground Service Conductors	Part III
Service-Entrance Conductors	Part IV
Service Equipment—General	Part V
Service Equipment—Disconnecting Means	Part VI
Service Equipment—Overcurrent Protection	Part VII

Δ **INFORMATIONAL NOTE FIGURE 230.1** *Services.*

of additional services is permitted as specified in conditions covered by 230.2(A) through (D). Where more than one service (or combination of service, feeder, and branch circuit) is installed, 230.2(E) requires that a permanent plaque or directory with the pertinent information on the multiple sources of supply be located at each supply source disconnecting means. Exhibits 230.1 through 230.6 illustrate examples of the general requirement that a building or structure be supplied by only one service.

(A) Special Conditions. Additional services shall be permitted to supply the following:

(1) Fire pumps
(2) Emergency systems
(3) Legally required standby systems
(4) Optional standby systems
(5) Interconnected electric power production sources
(6) Systems designed for connection to multiple sources of supply for the purpose of enhanced reliability

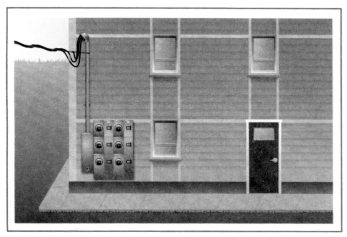

EXHIBIT 230.1 One service drop or one set of overhead service conductors supplying six separate service disconnects in separate compartments (optional arrangement to Exhibit 230.4). The building has more than one occupancy with separate loads by the single service, in accordance with 230.2 and 230.71.

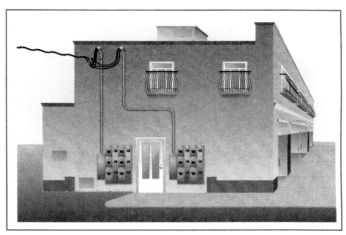

EXHIBIT 230.3 One service drop or one set of overhead service conductors supplying two service equipment enclosures installed at separate locations. The building has two groups of occupancies and two service equipment enclosures. The disconnecting means are in separate compartments within the enclosures. See 230.40, Exception No. 1, and 230.71.

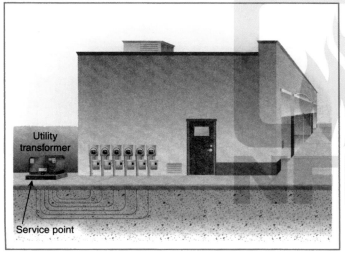

EXHIBIT 230.2 One set of underground service conductors consisting of six sets of conductors 1/0 AWG or larger (connected at their supply ends), terminating in six separate service equipment enclosures. The building has more than one occupancy. The separate loads are fed by one service lateral or one set of underground service conductors, in accordance with 230.2; 230.40, Exception No. 2; and 230.71.

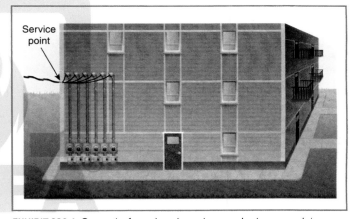

EXHIBIT 230.4 One set of overhead service conductors supplying a maximum of six separate service disconnecting means enclosures for a building having more than one occupancy. Separate loads are fed by one service, in accordance with 230.40, Exception No. 2, and 230.71.

Completely separate services to these systems and equipment increases the overall reliability of the power supply to these systems by negating the impact of a disruption of the main building service. Unless the separate service is supplied from a different utility circuit, no protection is provided against an overall utility system outage. However, the effect of an outage resulting from a problem between the service point and the "normal," or main, building service equipment will be limited to the normal service.

(B) Special Occupancies. By special permission, additional services shall be permitted for either of the following:

(1) Multiple-occupancy buildings where there is no available space for service equipment accessible to all occupants
(2) A single building or other structure sufficiently large to make two or more services necessary

Additional services for certain occupancies are allowed by special permission, that is, written consent of the AHJ. See the definition of *authority having jurisdiction* in Article 100. The expansion of buildings often necessitates the addition of one or more services. It might be impractical or even impossible to install one service for an industrial plant with sufficient capacity for any and all future loads. It is also impractical to run extremely long

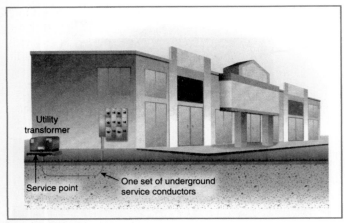

EXHIBIT 230.5 One set of underground service conductors supplying multiple service disconnecting means in separate compartments for a multi-occupancy building. Separate loads are supplied by one service, in accordance with 230.40, Exception No. 1, and 230.71.

EXHIBIT 230.7 One service drop or two sets of overhead service conductors supplying two services, installed at separate locations for a building with more than one occupancy where there is no available space for service equipment accessible to all occupants, in accordance with 230.2(B)(1) and 230.72(C).

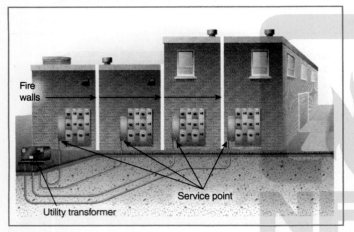

EXHIBIT 230.6 Four sets of service laterals supplying four service equipment enclosures installed at separate locations on a contiguous structure. Note the presence of firewalls, which make the structures separate buildings.

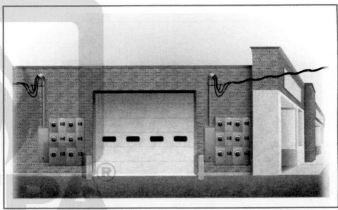

EXHIBIT 230.8 Two service drops or two sets of overhead service conductors supplying two services, installed at separate locations for a building with capacity requirements exceeding 2000 amperes, in accordance with 230.2(C)(1) and 230.71.

feeders. The AHJ should be consulted early in the planning stages to ascertain whether special permission can be obtained.

(C) Capacity Requirements. Additional services shall be permitted under any of the following:

(1) Where the capacity requirements are in excess of 2000 amperes at a supply voltage of 1000 volts or less
(2) Where the load requirements of a single-phase installation are greater than the serving agency normally supplies through one service
(3) By special permission

If two or more services are needed, each service is not required be rated 2000 amperes or that there be one service rated 2000 amperes and the additional service(s) rated for the calculated load in excess of 2000 amperes. For example, a building

with a calculated load of 2300 amperes could have two 1200-ampere services. Additional services for lesser loads also are allowed by special permission.

Many electric utilities have specifications for, and enforce special regulations covering, certain types of electrical loads and service equipment. Before electrical services for large buildings and facilities are designed, the serving utility should be consulted to determine line and transformer capacities.

Exhibits 230.7 through 230.10 illustrate examples of service configurations permitted by 230.2(B) and (C); 230.40, Exceptions No. 1 and No. 2; 230.71; and 230.72. The exhibits depict a number of service arrangements where more than a single service or multiple sets of service-entrance conductors are used to supply a building with one or more than one occupancy. The exhibits also illustrate the 230.71 requirement that not more than six service disconnecting means be grouped in

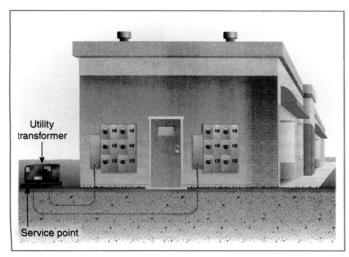

EXHIBIT 230.9 Two sets of underground service conductors, terminating in two service equipment enclosures installed at separate locations. Separate loads are supplied by a separate service, in accordance with 230.2(B)(1) and 230.71.

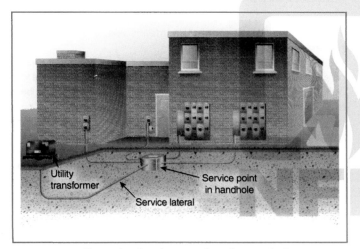

EXHIBIT 230.10 Four sets of underground service conductors supplying four service equipment enclosures installed at different locations. There is more than one occupancy with separate loads supplied by four sets of underground service conductors, in accordance with 230.40, Exception No.1, and 230.71.

one location; defining and determining the "one location" is a matter for the AHJ.

Additionally, a building can also be supplied by a feeder. This commonly occurs where the building is supplied with emergency, legally required standby, and/or optional standby power. Such installations are subject to the requirements of this article as well as Articles 225, 700, 701, and 702 as applicable.

(D) Different Characteristics. Additional services shall be permitted for different voltages, frequencies, or phases, or for different uses, such as for different rate schedules.

An example of different service characteristics is a facility served by a 3-wire, 120/240-volt, single-phase service and a 3-phase, 4-wire, 480Y/277-volt service. Where rate schedules are different, a second service is permitted for supplying a second meter on a different rate. Curtailable loads, interruptible loads, electric heating, and electric water heating are examples of loads that may be on a different rate schedule.

(E) Identification. Where a building or structure is supplied by more than one service, or any combination of branch circuits, feeders, and services, a permanent plaque or directory shall be installed at each service disconnect location denoting all other services, feeders, and branch circuits supplying that building or structure and the area served by each. See 225.37.

The permanent plaque or directory must indicate where the other disconnects that supply the building are located, as illustrated in Exhibit 230.11. All the other services on or in the building or structure and the area served by each also must be noted on the plaques or directories. This information is critically important to first responders, maintenance staff, and other persons who need to disconnect the building or structure from all of its supplies. The plaques or directories must be of sufficient durability to withstand the environment and remain legible.

230.3 One Building or Other Structure Not to Be Supplied Through Another. Service conductors supplying a building or other structure shall not pass through the interior of another building or other structure.

Although service conductors that supply one building or structure are prohibited from being run through the interior of another building or structure, service conductors are permitted to be installed along the exterior of one building to supply another building. Each building served in this manner is required to be provided with a disconnecting means for all ungrounded conductors, in accordance with Part VI.

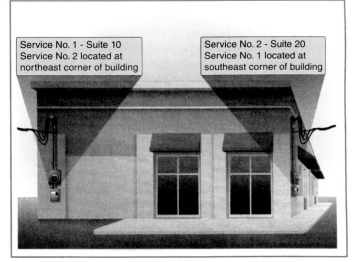

EXHIBIT 230.11 An example of two separate services installed at one building with permanent plaques or directories at each service disconnecting means location containing information describing all other services and the area served by each.

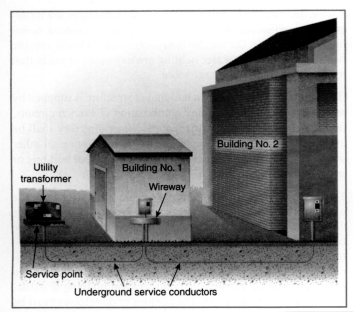

EXHIBIT 230.12 Service conductors installed so as not to pass through the interior of Building No. 1 to supply Building No. 2.

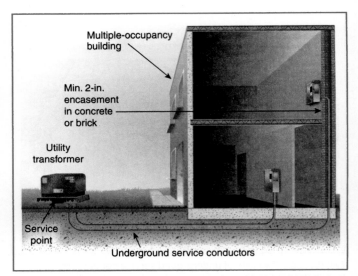

EXHIBIT 230.13 Service conductors considered outside a building where installed under not less than 2 inches of concrete beneath the building or in a raceway encased by not less than 2 inches of concrete or brick within the building.

For example, in Exhibit 230.12, the disconnecting means suitable for use as service equipment is shown on the exterior walls of Building No. 1 and Building No. 2. The prohibition against passing through one building to another applies only to service conductors, because feeders and branch circuits are provided with overcurrent protection at the point they receive their supply unless otherwise permitted by 240.21.

Δ **230.6 Conductors Considered Outside the Building.** Conductors shall be considered outside of a building or other structure under any of the following conditions:

(1) Where installed under not less than 50 mm (2 in.) of concrete beneath a building or other structure
(2) Where installed within a building or other structure in a raceway that is encased in concrete or brick not less than 50 mm (2 in.) thick
(3) Where installed in any vault that meets the construction requirements of Part III of Article 450
(4) Where installed in conduit and under not less than 450 mm (18 in.) of earth beneath a building or other structure
(5) Where installed within rigid metal conduit (RMC) or intermediate metal conduit (IMC) used to accommodate the clearance requirements in 230.24 and routed directly through an eave but not a wall of a building

Exhibit 230.13 illustrates two of the conditions that permit service conductors to be considered outside a building.

Service conductors installed in an interior vault complying with the construction requirements of Part III of Article 450 are considered to be outside a building regardless of whether or not the vault contains a transformer. Once the conductors leave the vault, the service disconnecting means has to be installed as required by 230.70(A)(1).

Service conductors installed in conduit and under 18 inches of earth beneath the building are also considered outside the building according to 230.6(4). An example of that would be a building or structure built on piers with service conductors buried beneath and running to a readily accessible service disconnecting means located within the interior of the building or structure.

Service conductors passing through a roof overhang as covered in 230.24(A), Exception No. 3, and depicted in Exhibit 230.17 are considered to be outside the building per 230.6(5). This requirement would not allow horizontal runs through eaves.

Δ **230.7 Other Conductors.** Circuit conductors other than service conductors, shall not be installed in the same raceway, cable, handhole enclosure, or underground box as the service conductors.

Service conductors are not provided with overcurrent protection where they receive their supply; they are protected against overload conditions at their load end by the service disconnect fuses or circuit breakers. If feeder or branch-circuit conductors are in the same raceway with service conductors during a fault, the fault current that could flow through them would be much higher than the ampacity of the feeder or branch-circuit conductors.

The gutter space of a panelboard cabinet or other electrical equipment enclosure is not a *raceway* (see Article 100 for the definition of *raceway*) and is not subject to the requirement of 230.7. Service conductors, feeder conductors, and branch-circuit conductors can share the same gutter space. The panelboard cabinet gutter space accommodates a set of service conductors terminating on the 200-ampere main breaker, a set of feeder conductors supplying the adjacent panelboard, and several sets of branch-circuit conductors entering the bottom of the cabinet that will connect to overcurrent protective devices (OCPDs) installed on the panelboard.

Exception No. 1: Grounding electrode conductors or supply side bonding jumpers or conductors shall be permitted within service raceways.

Exception No. 2: Load management control conductors having overcurrent protection shall be permitted within service raceways.

Δ **230.8 Raceway Seal.** Where a service raceway enters a building or structure, it shall be sealed in accordance with 300.5(G) and 300.7(A). Spare or unused raceways shall also be sealed. Sealants shall be identified for use with the cable insulation, conductor insulation, bare conductor, shield, or other components.

Sealant, such as duct seal or a bushing incorporating the physical characteristics of a seal, is required to be used to seal the ends of service raceways. The intent is to prevent water — usually the result of condensation due to temperature differences — from entering the service equipment via the raceway. The sealant material should be compatible with the conductor insulation and should not cause deterioration of the insulation over time. See Exhibit 300.9 for an example of a sealing bushing.

230.9 Clearances on Buildings. Service conductors and final spans shall comply with 230.9(A), (B), and (C).

(A) Clearances. Service conductors installed as open conductors or multiconductor cable without an overall outer jacket shall have a clearance of not less than 900 mm (3 ft) from windows that are designed to be opened, doors, porches, balconies, ladders, stairs, fire escapes, or similar locations.

Exception: Conductors run above the top level of a window shall be permitted to be less than the 900 mm (3 ft) requirement.

The 3-foot clearance applies to open conductors, not to a raceway or to a cable assembly that has an overall outer jacket, such as Types SE, MC, and MI cables. This distance protects the conductors from physical damage and persons from accidental contact with the conductors. The exception permits service conductors, including service-entrance conductors, overhead service conductors, and service-drop conductors, to be located just above window openings, because they are considered out of reach, as illustrated in Exhibit 230.14.

(B) Vertical Clearance. The vertical clearance of final spans above, or within 900 mm (3 ft) measured horizontally of platforms, projections, or surfaces that will permit personal contact shall be maintained in accordance with 230.24(B).

Service conductors must not be located where a person could reach and touch them. The amount of vertical clearance required depends on the location and voltage of the overhead conductors.

Exhibit 230.15 illustrates an installation where the 10-foot clearance is based on the conditions described in 230.24(B)(1).

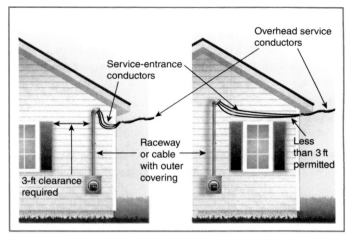

EXHIBIT 230.14 Required dimensions for service conductors located alongside a window *(left)* and overhead service conductors above the top edge of a window designed to be opened *(right)*.

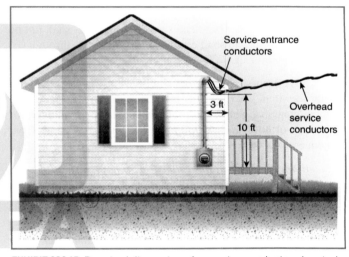

EXHIBIT 230.15 Required dimensions for service conductors located above a stair landing.

(C) Building Openings. Overhead service conductors shall not be installed beneath openings through which materials may be moved, such as openings in farm and commercial buildings, and shall not be installed where they obstruct entrance to these building openings.

230.10 Vegetation as Support. Vegetation such as trees shall not be used for support of overhead service conductors or service equipment.

Part II. Overhead Service Conductors

230.22 Insulation or Covering. Individual conductors shall be insulated or covered.

Exception: The grounded conductor of a multiconductor cable shall be permitted to be bare.

230.23 Size and Ampacity.

(A) General. Conductors shall have sufficient ampacity to carry the current for the load as calculated in accordance with Parts II through V of Article 220 and shall have adequate mechanical strength.

When a significant load is added, such as electric vehicle power transfer equipment, the serving electric utility should be notified to ensure that adequate capacity is available.

(B) Minimum Size. The conductors shall not be smaller than 8 AWG copper or 6 AWG aluminum or copper-clad aluminum.

Exception: Conductors supplying only limited loads of a single branch circuit — such as small polyphase power, controlled water heaters, and similar loads — shall not be smaller than 12 AWG hard-drawn copper or equivalent.

(C) Grounded Conductors. The grounded conductor shall not be less than the minimum size as required by 250.24(D).

230.24 Clearances. Overhead service conductors shall not be readily accessible and shall comply with 230.24(A) through (E) for services not over 1000 volts, nominal.

(A) Above Roofs. Conductors shall have a vertical clearance of not less than 2.6 m (8 ft 6 in.) above the roof surface. The vertical clearance above the roof level shall be maintained for a distance of not less than 900 mm (3 ft) in all directions from the edge of the roof.

Exception No. 1: The area above a roof surface subject to pedestrian or vehicular traffic shall have a vertical clearance from the roof surface in accordance with the clearance requirements of 230.24(B).

Exception No. 2: Where the voltage between conductors does not exceed 300 and the roof has a slope of 100 mm in 300 mm (4 in. in 12 in.) or greater, a reduction in clearance to 900 mm (3 ft) shall be permitted.

A steeply sloped roof (not less than 4 inches vertically in 12 inches horizontally) as illustrated in Exhibit 230.16 is less likely to be walked on. The conductors' length over the roof is not restricted.

Exception No. 3: Where the voltage between conductors does not exceed 300, a reduction in clearance above only the overhanging portion of the roof to not less than 450 mm (18 in.) shall be permitted if (1) not more than 1.8 m (6 ft) of overhead service conductors, 1.2 m (4 ft) horizontally, pass above the roof overhang, and (2) they are terminated at a through-the-roof raceway or approved support.

Informational Note: See 230.28 for mast supports.

This reduction is for service-mast (through-the-roof) installations where the mast is located within 4 feet of the edge of the roof, measured horizontally, as illustrated in Exhibit 230.17.

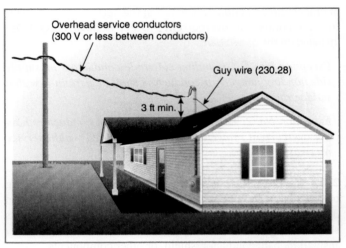

EXHIBIT 230.16 Permitted reduction to 3-foot clearance above a sloped roof.

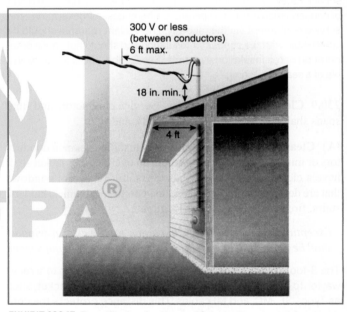

EXHIBIT 230.17 Permitted reduction to 18-inch clearance above an overhang roof penetration.

Exception No. 3 applies to the overhanging portion of sloped and flat roofs. The length of conductors that pass over the roof is not permitted to exceed 6 feet.

Exception No. 4: The requirement for maintaining the vertical clearance 900 mm (3 ft) from the edge of the roof shall not apply to the final conductor span where the service drop or overhead service conductors are attached to the side of a building.

The final span of service drop or overhead service conductors attached to the side of a building is exempt from the 8-foot and 3-foot clearance requirements to allow the service conductors to be attached to the building, as illustrated in Exhibit 230.18.

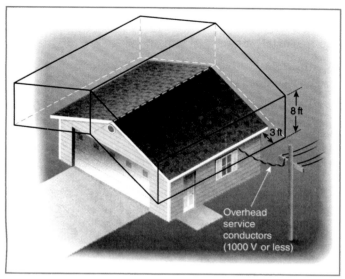

EXHIBIT 230.18 Permitted clearance of the final span of overhead service conductors.

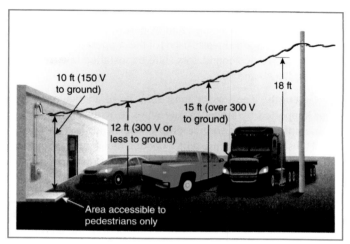

EXHIBIT 230.19 Clearances from grade for overhead service conductors.

Exception No. 5: Where the voltage between conductors does not exceed 300 and the roof area is guarded or isolated, a reduction in clearance to 900 mm (3 ft) shall be permitted.

(B) Vertical Clearance for Overhead Service Conductors. Overhead service conductors, where not in excess of 1000 volts, nominal, shall have the following minimum clearance from final grade:

(1) 3.0 m (10 ft) — at the electrical service entrance to buildings, also at the lowest point of the drip loop of the building electrical entrance, and above areas or sidewalks accessible only to pedestrians, measured from final grade or other accessible surface only for overhead service conductors supported on and cabled together with a grounded bare messenger where the voltage does not exceed 150 volts to ground

(2) 3.7 m (12 ft) — over residential property and driveways, and those commercial areas not subject to truck traffic where the voltage does not exceed 300 volts to ground

(3) 4.5 m (15 ft) — for those areas listed in the 3.7 m (12 ft) classification where the voltage exceeds 300 volts to ground

(4) 5.5 m (18 ft) — over public streets, alleys, roads, parking areas subject to truck traffic, driveways on other than residential property, and other land such as cultivated, grazing, forest, and orchard

(5) 7.5 m (24½ ft) over tracks of railroads

Exhibit 230.19 illustrates the required 10-foot, 12-foot, 15-foot, and 18-foot vertical clearances from ground for overhead service conductors up to 1000 volts. The voltages given are nominal voltages to ground, not the nominal voltage between circuit conductors specified in 230.24(A), Exceptions No. 2 and No. 3. A 480Y/277-volt system (277 volts to ground) is covered by the 12-foot clearance requirement in 230.24(B)(2), but overhead service conductors supplied by an ungrounded 480-volt system (considered to be 480 volts to ground) are required to have a 15-foot clearance over commercial areas not subject to truck traffic, in accordance with 230.24(B)(3).

(C) Clearance from Building Openings. Clearances from building openings shall comply with 230.9(C).

(D) Clearance from Swimming Pools, Fountains, and Similar Installations. Clearances from swimming pools, fountains, and similar installations shall comply with 680.9.

(E) Clearance from Communication Wires and Cables. Clearance from communication wires and cables shall be in accordance with 800.44(A)(4).

230.26 Point of Attachment. The point of attachment of the overhead service conductors to a building or other structure shall provide the minimum clearances as specified in 230.9 and 230.24. In no case shall this point of attachment be less than 3.0 m (10 ft) above finished grade.

230.27 Means of Attachment. Multiconductor cables used for overhead service conductors shall be attached to buildings or other structures by fittings identified for use with service conductors. Open conductors shall be attached to fittings identified for use with service conductors or to noncombustible, nonabsorbent insulators securely attached to the building or other structure.

See also

230.51 for mounting and supporting of service cables and individual open service conductors

230.54 for connections at service heads

230.28 Service Masts as Supports. Only power service-drop or overhead service conductors shall be permitted to be attached to a service mast. Service masts used for the support of service-drop or overhead service conductors shall be installed in accordance with 230.28(A) and (B).

(A) Strength. The service mast shall be of adequate strength or be supported by braces or guy wires to withstand safely the strain imposed by the service-drop or overhead service conductors. Hubs intended for use with a conduit that serves as a service mast shall be identified for use with service-entrance equipment.

(B) Attachment. Service-drop or overhead service conductors shall not be attached to a service mast between a weatherhead or the end of the conduit and a coupling, where the coupling is located above the last point of securement to the building or other structure or is located above the building or other structure.

If the service drop is secured to the mast, a guy wire may be needed to support the mast and provide the required mechanical strength to support the service drop. The service drop is not permitted to be secured to the mast between the weatherhead or end of the conduit and a coupling installed above the last point where the conduit is secured to a building or structure. Securing the conductors above the fitting could put undue stress on the assembly. See Exhibit 230.20. Communications conductors such as those for cable television or telephone service are not permitted to be attached to the service mast.

230.29 Supports over Buildings. Service conductors passing over a roof shall be securely supported by substantial structures. For a grounded system, where the substantial structure is metal, it shall be bonded by means of a bonding jumper and listed connector to the grounded overhead service conductor. Where practicable, such supports shall be independent of the building.

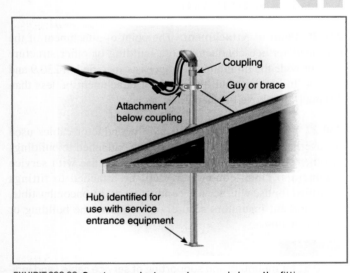

EXHIBIT 230.20 Service conductors not secured above the fitting because of stress above the fitting possibly weakening the assembly.

Part III. Underground Service Conductors

230.30 Installation.

(A) Insulation. Underground service conductors shall be insulated for the applied voltage.

Exception: A grounded conductor shall be permitted to be uninsulated as follows:

(1) Bare copper used in a raceway
(2) Bare copper for direct burial where bare copper is approved for the soil conditions
(3) Bare copper for direct burial without regard to soil conditions where part of a cable assembly identified for underground use
(4) Aluminum or copper-clad aluminum without individual insulation or covering where part of a cable assembly identified for underground use in a raceway or for direct burial

Δ **(B) Wiring Methods.** Underground service conductors shall be installed in accordance with the applicable requirements of this *Code* covering the type of wiring method used and shall be limited to the following methods:

(1) RMC conduit
(2) IMC conduit
(3) Type NUCC conduit
(4) HDPE conduit
(5) PVC conduit
(6) RTRC conduit
(7) Type IGS cable
(8) Type USE conductors or cables
(9) Type MV or Type MC cable identified for direct burial applications
(10) Type MI cable where suitably protected against physical damage and corrosive conditions
(11) Type TC-ER cable where identified for service entrance use and direct burial applications

230.31 Size and Ampacity.

(A) General. Underground service conductors shall have sufficient ampacity to carry the current for the load as calculated in accordance with Parts II through V of Article 220.

(B) Minimum Size. The conductors shall not be smaller than 8 AWG copper or 6 AWG aluminum or copper-clad aluminum.

Exception: Conductors supplying only limited loads of a single branch circuit — such as small polyphase power, controlled water heaters, and similar loads — shall not be smaller than 12 AWG copper or 10 AWG aluminum or copper-clad aluminum.

(C) Grounded Conductors. The grounded conductor shall not be smaller than the minimum size required by 250.24(D).

See also

310.15(D) for the allowable ampacity of bare and covered conductors

310.12(D) for further information on sizing the grounded conductor for dwelling services

230.32 Protection Against Damage. Underground service conductors shall be protected against damage in accordance with 300.5. Service conductors entering a building or other structure shall be installed in accordance with 230.6 or protected by a raceway wiring method identified in 230.43.

230.33 Spliced Conductors. Service conductors shall be permitted to be spliced or tapped in accordance with 110.14, 230.46, 300.5(E), 300.13, and 300.15.

Part IV. Service-Entrance Conductors

230.40 Number of Service-Entrance Conductor Sets. Each service drop, set of overhead service conductors, set of underground service conductors, or service lateral shall supply only one set of service-entrance conductors.

Exception No. 1: A building with more than one occupancy shall be permitted to have one set of service-entrance conductors for each service, as permitted in 230.2, run to each occupancy or group of occupancies. If the number of service disconnect locations for any given classification of service does not exceed six, the requirements of 230.2(E) shall apply at each location. If the number of service disconnect locations exceeds six for any given supply classification, the following conditions shall apply:

(1) All service disconnect locations for all supply characteristics, together with any branch circuit or feeder supply sources, shall be clearly described using graphics or text, or both, on one or more plaques

(2) The plaques shall be located in an approved, readily accessible location(s) on the building or structure served and as near as practicable to the point(s) of attachment or entry(ies) for each service drop or service lateral and for each set of overhead or underground service conductors.

If a building has more than one occupancy — such as multifamily dwellings, strip malls, and office buildings — each service drop, set of overhead service conductors, set of underground service conductors, or service lateral is allowed to supply more than one set of service-entrance conductors, provided they are run to each occupancy or group of occupancies. This requirement allows for multiple service disconnecting means locations for each service that supplies the building or structure. Based on this exception, one service can be arranged similarly to a building or structure that is supplied by multiple services. The exception does not limit the number of disconnecting means locations supplied by each service.

Because this exception permits more than one service equipment location on a single building or structure, a plaque or directory with information about the multiple equipment locations is required to be provided at each location.

If the number of service equipment locations for any class of service exceeds six, a master plaque(s) or directory(s) is required near the point where service conductors attach to or enter a building or structure. The information must describe the multiple service equipment locations using text, a graphic, or a combination. At the individual service equipment locations, the general marking required in 230.70(B) must be provided.

For example, if a mercantile building has eight storefronts and the building is supplied by a single 208Y/120-volt service, eight sets of service-entrance conductors can be installed with one set run to each occupancy. The service equipment at each occupancy can have up to six service disconnecting means in accordance with 230.71. Because the number of service disconnecting means locations for this service exceeds six, a permanent plaque(s) identifying the multiple supply equipment locations is required. The location of the plaque must be acceptable to the AHJ.

Exception No. 2: Where two to six service disconnecting means in separate enclosures are grouped at one location and supply separate loads from one service drop, set of overhead service conductors, set of underground service conductors, or service lateral, one set of service-entrance conductors shall be permitted to supply each or several such service equipment enclosures.

Exception No. 3: A one-family dwelling unit and its accessory structures shall be permitted to have one set of service-entrance conductors run to each from a single service drop, set of overhead service conductors, set of underground service conductors, or service lateral.

A second set of service-entrance conductors supplied by a single service drop or lateral at a single-family dwelling unit is permitted to supply another building on the premises, such as a garage or storage shed. The utility meters may be grouped at one location, but in this application, the service disconnecting means are not required to be grouped at one location.

Exception No. 4: Two-family dwellings, multifamily dwellings, and multiple occupancy buildings shall be permitted to have one set of service-entrance conductors installed to supply the circuits covered in 210.25.

Exception No. 5: One set of service-entrance conductors connected to the supply side of the normal service disconnecting means shall be permitted to supply each or several systems covered by 230.82(5) or 230.82(6).

230.41 Insulation of Service-Entrance Conductors. Service-entrance conductors entering or on the exterior of buildings or other structures shall be insulated.

Exception: A grounded conductor shall be permitted to be unin-sulated as follows:

(1) *Bare copper used in a raceway or part of a service cable assembly*

(2) *Bare copper for direct burial where bare copper is approved for the soil conditions*

(3) *Bare copper for direct burial without regard to soil condi-tions where part of a cable assembly identified for under-ground use*

(4) *Aluminum or copper-clad aluminum without individual insulation or covering where part of a cable assembly or identified for underground use in a raceway, or for direct burial*

(5) *Bare conductors used in an auxiliary gutter*

230.42 Minimum Size and Ampacity.

Δ **(A) General.** Service-entrance conductors shall have an ampac-ity of not less than the maximum load to be served. Conductors shall be sized not less than the largest of 230.42(A)(1) or (A)(2). Loads shall be determined in accordance with Part III, IV, or V of Article 220, as applicable. Ampacity shall be determined from 310.14 and shall comply with 110.14(C). The maximum current of busways shall be that value for which the busway has been listed or labeled.

> Informational Note: See UL 857, *Standard for Safety for Bus-ways*, for information on busways.

N **(1) Continuous and Noncontinuous Loads.** Where the service-entrance conductors supply continuous loads or any combination of noncontinuous and continuous loads, the minimum service-entrance conductor size shall have an ampacity not less than the sum of the noncontinuous loads plus 125 percent of continuous loads.

> *Exception No. 1: Grounded conductors that are not connected to an overcurrent device shall be permitted to be sized at 100 percent of the sum of the continuous and noncontinuous load.*

> *Exception No. 2: The sum of the noncontinuous load and the continuous load if the service-entrance conductors terminate in an overcurrent device where both the overcurrent device and its assembly are listed for operation at 100 percent of their rating shall be permitted.*

N **(2) Application of Adjustment or Correction Factors.** The minimum service-entrance conductor size shall have an ampacity not less than the maximum load to be served after the application of any adjustment or correction factors.

(B) Specific Installations. In addition to the requirements of 230.42(A), the minimum ampacity for ungrounded conductors for specific installations shall not be less than the rating of the service disconnecting means specified in 230.79(A) through (D).

(C) Grounded Conductors. The grounded conductor shall not be smaller than the minimum size as required by 250.24(D).

230.43 Wiring Methods for 1000 Volts, Nominal, or Less. Service-entrance conductors shall be installed in accordance with the applicable requirements of this *Code* covering the type of wiring method used and shall be limited to the following methods:

(1) Open wiring on insulators
(2) Type IGS cable
(3) Rigid metal conduit (RMC)
(4) Intermediate metal conduit (IMC)
(5) Electrical metallic tubing (EMT)
(6) Electrical nonmetallic tubing
(7) Service-entrance cables
(8) Wireways
(9) Busways
(10) Auxiliary gutters
(11) Rigid polyvinyl chloride conduit (PVC)
(12) Cablebus
(13) Type MC cable
(14) Mineral-insulated, metal-sheathed cable, Type MI
(15) Flexible metal conduit (FMC) not over 1.8 m (6 ft) long or liquidtight flexible metal conduit (LFMC) not over 1.8 m (6 ft) long between a raceway, or between a race-way and service equipment, with a supply-side bonding jumper routed with the flexible metal conduit (FMC) or the liquidtight flexible metal conduit (LFMC) according to 250.102(A), (B), (C), and (E)
(16) Liquidtight flexible nonmetallic conduit (LFNC)
(17) High density polyethylene conduit (HDPE)
(18) Nonmetallic underground conduit with conductors (NUCC)
(19) Reinforced thermosetting resin conduit (RTRC)
(20) Type TC-ER cable where identified for use as service entrance conductors
(21) Flexible bus systems

Section 230.43(15) permits no more than 6 feet of flexible metal conduit or liquidtight flexible metal conduit to be used as a ser-vice wiring method. Because of the high levels of fault energy available on the line side of the service disconnecting means, a bonding jumper must be installed where these raceway types are used for service conductors. The bonding jumper is allowed to be installed inside or outside the raceway, but when installed out-side the raceway it must follow the path of the raceway and can-not exceed 6 feet in length. To minimize the impedance of the ground-fault current return path, the bonding jumper must not be wrapped or spiraled around the flexible conduit.

230.44 Cable Trays. Cable tray systems shall be permitted to support service-entrance conductors. Cable trays used to support service-entrance conductors shall contain only service-entrance conductors and shall be limited to the following methods:

(1) Type SE cable
(2) Type MC cable
(3) Type MI cable
(4) Type IGS cable
(5) Single conductors 1/0 and larger that are listed for use in cable tray
(6) Type TC-ER cable

Such cable trays shall be identified with permanently affixed labels with the wording "Service-Entrance Conductors." The labels shall be located so as to be visible after installation with a spacing not to exceed 3 m (10 ft) so that the service-entrance conductors are able to be readily traced through the entire length of the cable tray.

Exception: Conductors, other than service-entrance conductors, shall be permitted to be installed in a cable tray with service-entrance conductors, provided a solid fixed barrier identified for use with the cable tray is installed to separate the service-entrance conductors from other conductors installed in the cable tray.

230.46 Spliced and Tapped Conductors. Service-entrance conductors shall be permitted to be spliced or tapped in accordance with 110.14, 300.5(E), 300.13, and 300.15. Power distribution blocks, pressure connectors, and devices for splices and taps shall be listed. Power distribution blocks installed on service conductors shall be marked "suitable for use on the line side of the service equipment" or equivalent.

Pressure connectors and devices for splices and taps installed on service conductors shall be marked "suitable for use on the line side of the service equipment" or equivalent.

Splices must be in an enclosure or be direct buried using a listed underground splice kit. Splices are permitted where the cable enters a terminal box and a different wiring method, such as conduit, continues to the service equipment. Tapped sets of conductors are commonly used to supply multiple service disconnecting means installed in separate enclosures as permitted by 230.71.

230.50 Protection Against Physical Damage.

(A) Underground Service-Entrance Conductors. Underground service-entrance conductors shall be protected against physical damage in accordance with 300.5.

(B) All Other Service-Entrance Conductors. All other service-entrance conductors, other than underground service entrance conductors, shall be protected against physical damage as specified in 230.50(B)(1) or (B)(2).

(1) Service-Entrance Cables. Service-entrance cables, where subject to physical damage, shall be protected by any of the following:

(1) Rigid metal conduit (RMC)
(2) Intermediate metal conduit (IMC)
(3) Schedule 80 PVC conduit
(4) Electrical metallic tubing (EMT)
(5) Reinforced thermosetting resin conduit (RTRC)
(6) Other approved means

(2) Other Than Service-Entrance Cables. Individual open conductors and cables, other than service-entrance cables, shall not be installed within 3.0 m (10 ft) of grade level or where exposed to physical damage.

Exception: Type MI and Type MC cable shall be permitted within 3.0 m (10 ft) of grade level where not exposed to physical damage or where protected in accordance with 300.5(D).

230.51 Mounting Supports. Service-entrance cables or individual open service-entrance conductors shall be supported as specified in 230.51(A), (B), or (C).

(A) Service-Entrance Cables. Service-entrance cables shall be supported by straps or other approved means within 300 mm (12 in.) of every service head, gooseneck, or connection to a raceway or enclosure and at intervals not exceeding 750 mm (30 in.).

(B) Other Cables. Cables that are not approved for mounting in contact with a building or other structure shall be mounted on insulating supports installed at intervals not exceeding 4.5 m (15 ft) and in a manner that maintains a clearance of not less than 50 mm (2 in.) from the surface over which they pass.

(C) Individual Open Conductors. Individual open conductors shall be installed in accordance with Table 230.51(C). Where exposed to the weather, the conductors shall be mounted on insulators or on insulating supports attached to racks, brackets, or other approved means. Where not exposed to the weather, the conductors shall be mounted on glass or porcelain knobs.

TABLE 230.51(C) *Supports*

| Maximum Volts | Maximum Distance Between Supports | | Minimum Clearance | | | |
| | | | Between Conductors | | From Surface | |
	m	ft	mm	in.	mm	in.
1000	2.7	9	150	6	50	2
1000	4.5	15	300	12	50	2
300	1.4	4½	75	3	50	2
1000*	1.4*	4½*	65*	2½*	25*	1*

*Where not exposed to weather.

230.52 Individual Conductors Entering Buildings or Other Structures. Where individual open conductors enter a building or other structure, they shall enter through roof bushings or through the wall in an upward slant through individual, noncombustible, nonabsorbent insulating tubes. Drip loops shall be formed on the conductors before they enter the tubes.

230.53 Raceways to Drain. Where exposed to the weather, raceways enclosing service-entrance conductors shall be listed or approved for use in wet locations and arranged to drain. Where embedded in masonry, raceways shall be arranged to drain.

230.54 Overhead Service Locations.

(A) Service Head. Service raceways shall be equipped with a service head at the point of connection to service-drop or overhead service conductors. The service head shall be listed for use in wet locations.

(B) Service-Entrance Cables Equipped with Service Head or Gooseneck. Service-entrance cables shall be equipped with a service head. The service head shall be listed for use in wet locations.

Exception: Type SE cable shall be permitted to be formed in a gooseneck and taped with a self-sealing weather-resistant thermoplastic.

Type SE service-entrance cables can be installed without a service head (weatherhead) if they are run continuously from a utility pole to metering or service equipment, or if they are shaped in a downward direction (forming a "gooseneck") and sealed with self-sealing weather-resistant tape, as shown in Exhibit 230.21.

(C) Service Heads and Goosenecks Above Service-Drop or Overhead Service Attachment. Service heads on raceways or service-entrance cables and goosenecks in service-entrance cables shall be located above the point of attachment of the service-drop or overhead service conductors to the building or other structure.

Exception: Where it is impracticable to locate the service head or gooseneck above the point of attachment, the service head or gooseneck location shall be permitted not farther than 600 mm (24 in.) from the point of attachment.

(D) Secured. Service-entrance cables shall be held securely in place.

(E) Separately Bushed Openings. Service heads shall have conductors of different potential brought out through separately bushed openings.

Exception: For jacketed multiconductor service-entrance cable without splice.

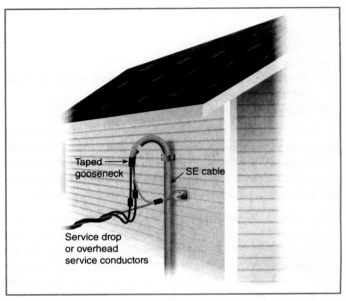

EXHIBIT 230.21 A service-entrance cable that terminates in a gooseneck without a raintight service head (weatherhead).

(F) Drip Loops. Drip loops shall be formed on individual conductors. To prevent the entrance of moisture, service-entrance conductors shall be connected to the service-drop or overhead service conductors either (1) below the level of the service head or (2) below the level of the termination of the service-entrance cable sheath.

(G) Arranged That Water Will Not Enter Service Raceway or Equipment. Service-entrance and overhead service conductors shall be arranged so that water will not enter service raceway or equipment.

230.56 Service Conductor with the Higher Voltage to Ground. On a 4-wire, delta-connected service where the midpoint of one phase winding is grounded, the service conductor having the higher phase voltage to ground shall be durably and permanently marked by an outer finish that is orange in color, or by other effective means, at each termination or junction point.

The unique marking of the service conductor with the higher voltage to ground provides a warning of a potential hazard; connection to this higher voltage (208 volts to ground) conductor can damage equipment or injure personnel. The marking should be at both the point of connection to the service-drop or lateral conductors and the point of connection to the service disconnecting means.

See also

408.3(E) and (F) for phase arrangement requirements and for equipment labeling reflecting the presence of a high leg

Part V. Service Equipment — General

230.62 Service Equipment — Enclosed or Guarded. Energized parts of service equipment shall be enclosed as specified in 230.62(A) or guarded as specified in 230.62(B).

(A) Enclosed. Energized parts shall be enclosed so that they will not be exposed to accidental contact or shall be guarded as in 230.62(B).

(B) Guarded. Energized parts that are not enclosed shall be installed on a switchboard, panelboard, or control board and guarded in accordance with 110.18 and 110.27. Where energized parts are guarded as provided in 110.27(A)(1) and (A)(2), a means for locking or sealing doors providing access to energized parts shall be provided.

(C) Barriers. Barriers shall be placed in service equipment such that no uninsulated, ungrounded service busbar or service terminal is exposed to inadvertent contact by persons or maintenance equipment while servicing load terminations with the service disconnect in the open position.

If disconnecting or de-energizing the service conductors supplying a service panelboard, switchboard, or switchgear is infeasible, it might be necessary for a qualified person (according to NFPA 70E®, Standard for Electrical Safety in the Workplace®) to work on that equipment with the load terminals de-energized but with the service bus still energized. Barriers provide physical separation (adequate distance or an obstacle) between load terminals and the service busbars and terminals. This provides some measure of safety against inadvertent contact with line-energized parts during maintenance and installation of new feeders or branch circuits. Barriers are not necessary in most multi-section switchboards and switchgear because the line-side conductors and busbars are not in the same section as the load terminals.

230.66 Marking.

(A) General. Service equipment rated at 1000 volts or less shall be marked to identify it as being suitable for use as service equipment. All service equipment shall be listed or field evaluated.

"Suitable for Use as Service Equipment (SUSE)" is a common marking found on equipment that can be used at the service location. The marking indicates that the equipment meets service equipment suitability requirements in the applicable product standard (i.e., panelboard, switchboard, enclosed switch, or other equipment product standard). The marking "Suitable Only for Use as Service Equipment" indicates that the grounded conductor or neutral terminal bus is not able to be electrically isolated from the metal equipment enclosure. That inability precludes most feeder applications for this equipment where the equipment grounding terminals and the grounded conductor terminals are required to be electrically isolated.

(B) Meter Sockets. Meter sockets shall not be considered service equipment but shall be listed and rated for the voltage and current rating of the service.

Exception: Meter sockets supplied by and under the exclusive control of an electric utility shall not be required to be listed.

230.67 Surge Protection.

(A) Surge-Protective Device. All services supplying the following occupancies shall be provided with a surge-protective device (SPD):

(1) Dwelling units
(2) Dormitory units
(3) Guest rooms and guest suites of hotels and motels
(4) Areas of nursing homes and limited-care facilities used exclusively as patient sleeping rooms

Informational Note: See 517.10(B)(2).

Sensitive electronics and systems found in modern appliances, safety devices [e.g., arc-fault circuit interrupters (AFCIs), ground-fault circuit interrupters (GFCIs), and smoke alarms], and other equipment used in dwellings warrant protection by surge-protective devices. Additionally, the increased use of distributed energy resources within electrical systems creates more opportunity for the introduction of or greater exposure to surges in the system.

Δ **(B) Location.** The SPD shall be an integral part of the service equipment or shall be located immediately adjacent thereto.

Exception: The SPD shall not be required to be located at the service equipment as required in 230.67(B) if located at each next level distribution equipment downstream toward the load.

(C) Type. The SPD shall be a Type 1 or Type 2 SPD.

(D) Replacement. Where service equipment is replaced, all of the requirements of this section shall apply.

N **(E) Ratings.** SPDs shall have a nominal discharge current rating (In) of not less than 10kA.

Part VI. Service Equipment — Disconnecting Means

230.70 General. Means shall be provided to disconnect all ungrounded conductors in a building or other structure from the service conductors.

(A) Location. The service disconnecting means shall be installed in accordance with 230.70(A)(1), (A)(2), and (A)(3).

(1) Readily Accessible Location. The service disconnecting means shall be installed at a readily accessible location either outside of a building or structure or inside nearest the point of entrance of the service conductors.

No maximum distance between the point of entrance of service conductors to a readily accessible location for the installation of a service disconnecting means is specified. The authority enforcing the NEC® is responsible for the decision on how far inside the building the service-entrance conductors are allowed to travel to the service disconnecting means. The length of service-entrance conductors should be kept to a minimum inside buildings. There is an increased fire hazard because power utilities provide limited overcurrent protection. Some local jurisdictions specify a maximum length that service-entrance conductors may be run within the building before terminating at the disconnecting means.

If the AHJ determines the distance to be excessive, the disconnecting means may be required to be located on the outside of the building or near the building at a readily accessible location.

(2) Bathrooms. Service disconnecting means shall not be installed in bathrooms.

(3) Remote Control. Where a remote control device(s) is used to actuate the service disconnecting means, the service disconnecting means shall be located in accordance with 230.70(A)(1).

(B) Marking. Each service disconnect shall be permanently marked to identify it as a service disconnect.

Δ **(C) Suitable for Use.** Each service disconnecting means shall be suitable for the prevailing conditions. Service equipment installed in hazardous (classified) locations shall comply with the hazardous location requirements.

230.71 Maximum Number of Disconnects. Each service shall have only one disconnecting means unless the requirements of 230.71(B) are met.

(A) General. For the purpose of this section, disconnecting means installed as part of listed equipment and used solely for the following shall not be considered a service disconnecting means:

(1) Power monitoring equipment
(2) Surge-protective device(s)
(3) Control circuit of the ground-fault protection system
(4) Power-operable service disconnecting means

Δ **(B) Two to Six Service Disconnecting Means.** Two to six service disconnects shall be permitted for each service permitted by 230.2 or for each set of service-entrance conductors permitted by 230.40, Exception No. 1, 3, 4, or 5. The two to six service disconnecting means shall be permitted to consist of a combination of any of the following:

(1) Separate enclosures with a main service disconnecting means in each enclosure
(2) Panelboards with a main service disconnecting means in each panelboard enclosure
(3) Switchboard(s) where there is only one service disconnect in each separate vertical section with barriers provided between each vertical section to maintain the inadvertent contact protection required in 230.62 based on access from the adjacent section(s)
(4) Service disconnects in switchgear, transfer switches, or metering centers where each disconnect is located in a separate compartment

(5) Metering centers with a main service disconnecting means in each metering center
(6) Motor control center(s) where there is only one service disconnect in a motor control center unit and a maximum of two service disconnects provided in a single motor control center with barriers provided between each motor control center unit or compartment containing a service disconnect to maintain the inadvertent contact protection required in 230.62 based on access from adjacent motor control center unit(s) or compartment(s)

Exception to (2), (3), (4), (5), and (6): Existing service equipment, installed in compliance with previous editions of this Code that permitted multiple service disconnecting means in a single enclosure, section, or compartment, shall be permitted to contain a maximum of six service disconnecting means.

Informational Note No. 1: See UL 67, *Standard for Panelboards,* for information on metering centers.
Informational Note No. 2: Examples of separate enclosures with a main service disconnecting means in each enclosure include but are not limited to motor control centers, fused disconnects, and circuit breaker enclosures.
Informational Note No. 3: Transfer switches are provided with one service disconnect or multiple service disconnects in separate compartments.

In general, a service should have a single disconnecting means. However, one set of service-entrance conductors is permitted to supply a group of two to six separate service disconnecting means in lieu of a single main disconnect. Both single-occupancy and multiple-occupancy buildings can be provided with one main service disconnect or up to six separate main disconnects for each set of service-entrance conductors. Conductors from renewable energy sources, such as photovoltaic systems and wind generators, are service conductors. Where alternative energy systems are run in parallel with a utility source, see Article 705.

Multiple-occupancy buildings may have service-entrance conductors run to each occupancy, and each such set of service-entrance conductors may have from one to six separate disconnects (see 230.40, Exception No. 1). Exhibit 230.22 shows a single enclosure for grouping service equipment that consists of separate compartments for six circuit breakers or six fused switches. This arrangement does not require a single main service disconnecting means if they are in separate compartments. Some factory-installed switches that disconnect power to ancillary devices that are included as part of listed equipment do *not* count as one of the six service disconnecting means.

Revising the permission for up to six service disconnects to require each disconnecting means to be installed in separate enclosures or separate compartments allows an installer/maintainer to operate a single disconnect that de-energizes conductors and circuit parts except for the line side of the disconnecting means. This enhances safety by reducing the likelihood of an incident with energized conductors or circuit parts in the service equipment.

EXHIBIT 230.22 A service equipment enclosure that groups six service disconnecting means in separate compartments.

230.72 Grouping of Disconnects.

(A) General. The two to six disconnects, if permitted in 230.71, shall be grouped. Each disconnect shall be marked to indicate the load served.

Exception: One of the two to six service disconnecting means permitted in 230.71, where used only for a water pump also intended to provide fire protection, shall be permitted to be located remote from the other disconnecting means. If remotely installed in accordance with this exception, a plaque shall be posted at the location of the remaining grouped disconnects denoting its location.

(B) Additional Service Disconnecting Means. The one or more additional service disconnecting means for fire pumps, emergency systems, legally required standby, or optional standby services permitted by 230.2 shall be installed remote from the one to six service disconnecting means for normal service to minimize the possibility of simultaneous interruption of supply.

The AHJ is responsible for determining what is a sufficiently remote location for the additional service disconnecting means.

(C) Access to Occupants. In a multiple-occupancy building, each occupant shall have access to the occupant's service disconnecting means.

Exception: In a multiple-occupancy building where electric service and electrical maintenance are provided by the building

management and where these are under continuous building management supervision, the service disconnecting means supplying more than one occupancy shall be permitted to be accessible to authorized management personnel only.

230.74 Simultaneous Opening of Poles. Each service disconnect shall simultaneously disconnect all ungrounded service conductors that it controls from the premises wiring system.

230.75 Disconnection of Grounded Conductor. Where the service disconnecting means does not disconnect the grounded conductor from the premises wiring, other means shall be provided for this purpose in the service equipment. A terminal or bus to which all grounded conductors can be attached by means of pressure connectors shall be permitted for this purpose. In a multisection switchboard or switchgear, disconnects for the grounded conductor shall be permitted to be in any section of the switchboard or switchgear, if the switchboard or switchgear section is marked to indicate a grounded conductor disconnect is located within.

> Informational Note: In switchgear or multisection switchboards, the disconnecting means provided for the grounded conductor is typically identified as a neutral disconnect link and is typically located in the bus to which the service grounded conductor is connected.

230.76 Manually or Power Operable. The service disconnecting means for ungrounded service conductors shall consist of one of the following:

(1) A manually operable switch or circuit breaker equipped with a handle or other suitable operating means
(2) A power-operated switch or circuit breaker, provided the switch or circuit breaker can be opened by hand in the event of a power supply failure

230.77 Indicating. The service disconnecting means shall plainly indicate whether it is in the open (off) or closed (on) position.

230.79 Rating of Service Disconnecting Means. The service disconnecting means shall have a rating not less than the calculated load to be carried, determined in accordance with Part III, IV, or V of Article 220, as applicable. In no case shall the rating be lower than specified in 230.79(A), (B), (C), or (D).

(A) One-Circuit Installations. For installations to supply only limited loads of a single branch circuit, the service disconnecting means shall have a rating of not less than 15 amperes.

(B) Two-Circuit Installations. For installations consisting of not more than two 2-wire branch circuits, the service disconnecting means shall have a rating of not less than 30 amperes.

(C) One-Family Dwellings. For a one-family dwelling, the service disconnecting means shall have a rating of not less than 100 amperes, 3-wire.

(D) All Others. For all other installations, the service disconnecting means shall have a rating of not less than 60 amperes.

230.80 Combined Rating of Disconnects. Where the service disconnecting means consists of more than one switch or circuit breaker, as permitted by 230.71, the combined ratings of all the switches or circuit breakers used shall not be less than the rating required by 230.79.

230.81 Connection to Terminals. The service conductors shall be connected to the service disconnecting means by pressure connectors, clamps, or other approved means. Connections that depend on solder shall not be used.

Δ **230.82 Equipment Connected to the Supply Side of Service Disconnect.** Only the following equipment shall be permitted to be connected to the supply side of the service disconnecting means:

(1) Cable limiters.

Cable limiters or other current-limiting devices are often applied ahead of the service disconnecting means for the following reasons:

1. To individually isolate faulted cable(s) from the remainder of the circuit or paralleled set of conductors
2. To maintain continuity of service even though one or more cables are faulted
3. To reduce the possibility of severe equipment damage or burndown as a result of a fault on the service conductors
4. To provide protection against high short-circuit currents for services and to provide compliance with 110.10

(2) Meters and meter sockets nominally rated not in excess of 1000 volts, if all metal housings and service enclosures are grounded in accordance with Part VII and bonded in accordance with Part V of Article 250.

(3) Meter disconnect switches nominally rated not in excess of 1000 volts that have a short-circuit current rating equal to or greater than the available fault current, if all metal housings and service enclosures are grounded in accordance with Part VII and bonded in accordance with Part V of Article 250. A meter disconnect switch shall be capable of interrupting the load served. A meter disconnect shall be legibly field marked on its exterior in a manner suitable for the environment as follows:

METER DISCONNECT
NOT SERVICE EQUIPMENT

The meter disconnect is not the service disconnecting means. It is a load-break disconnect switch designed to interrupt the service load. The purpose of the meter disconnect switch is to facilitate meter change, maintenance, or disconnecting of the service. A meter disconnect switch must have a short-circuit current rating that is not less than the available short-circuit current at the line terminals of the meter disconnect switch.

(4) Instrument transformers (current and voltage), impedance shunts, load management devices, surge arresters, and Type 1 surge-protective devices.

(5) Conductors used to supply energy management systems, circuits for standby power systems, fire pump equipment, and fire and sprinkler alarms, if provided with service equipment and installed in accordance with requirements for service-entrance conductors.

(6) Solar photovoltaic systems, fuel cell systems, wind electric systems, energy storage systems, or interconnected electric power production sources, if provided with a disconnecting means listed as suitable for use as service equipment, and overcurrent protection as specified in Part VII of Article 230.

(7) Control circuits for power-operable service disconnecting means, if suitable overcurrent protection and disconnecting means are provided.

(8) Ground-fault protection systems or Type 2 surge-protective devices, where installed as part of listed equipment, if suitable overcurrent protection and disconnecting means are provided.

(9) Connections used only to supply listed communications equipment under the exclusive control of the serving electric utility, if suitable overcurrent protection and disconnecting means are provided. For installations of equipment by the serving electric utility, a disconnecting means is not required if the supply is installed as part of a meter socket, such that access can only be gained with the meter removed.

(10) Emergency disconnects in accordance with 230.85(B)(2) and (B)(3), if all metal housings and enclosures are grounded in accordance with Part VII and bonded in accordance with Part V of Article 250.

(11) Meter-mounted transfer switches nominally rated not in excess of 1000 volts that have a short-circuit current rating equal to or greater than the available fault current. A meter-mounted transfer switch shall be listed and be capable of transferring the load served. A meter-mounted transfer switch shall be marked on its exterior with both of the following:
    a. Meter-mounted transfer switch
    b. Not service equipment

(12) Control power circuits for protective relays where installed as part of listed equipment, if overcurrent protection and disconnecting means are provided.

Δ **230.85 Emergency Disconnects.** For one- and two-family dwelling units, an emergency disconnecting means shall be installed.

This section recognizes the need for an outdoor disconnect for first responders. Prior to the 2020 *NEC*, first responders and utility personnel have not had a way to safely remove power from a structure. This new requirement mandates that a means to disconnect the electric utility be located in a readily accessible, outdoor location. Mounting the emergency disconnect in a readily accessible location does not preclude locking the disconnect in the "on" position. Replacement of the service equipment will now require the addition of the emergency disconnect, but not repair of service equipment or wiring. Additional labeling for emergency disconnects is required where alternate sources of power are not located adjacent to the disconnect, which will help ensure all sources of power to the structure are removed.

N **(A) General.**

N **(1) Location.** The disconnecting means shall be installed in a readily accessible outdoor location on or within sight of the dwelling unit.

Exception: Where the requirements of 225.41 are met, this section shall not apply.

N **(2) Rating.** The disconnecting means shall have a short-circuit current rating equal to or greater than the available fault current.

N **(3) Grouping.** If more than one disconnecting means is provided, they shall be grouped.

N **(B) Disconnects.** Each disconnect shall be one of the following:

(1) Service disconnect
(2) A meter disconnect integral to the meter mounting equipment not marked as suitable only for use as service equipment installed in accordance with 230.82
(3) Other listed disconnect switch or circuit breaker that is marked suitable for use as service equipment, but not marked as suitable only for use as service equipment, installed on the supply side of each service disconnect

Informational Note 1: Conductors between the emergency disconnect and the service disconnect in 230.85(2) and 230.85(3) are service conductors.

Informational Note 2: Equipment marked "Suitable only for use as service equipment" includes the factory marking "Service Disconnect".

N **(C) Replacement.** Where service equipment is replaced, all of the requirements of this section shall apply.

Exception: Where only meter sockets, service entrance conductors, or related raceways and fittings are replaced, the requirements of this section shall not apply.

N **(D) Identification of Other Isolation Disconnects.** Where equipment for isolation of other energy source systems is not located adjacent to the emergency disconnect required by this section, a plaque or directory identifying the location of all equipment for isolation of other energy sources shall be located adjacent to the disconnecting means required by this section.

Informational Note: See 445.18, 480.7, 705.20, and 706.15 for examples of other energy source system isolation means.

N **(E) Marking.**

N **(1) Marking Text.** The disconnecting means shall marked as follows:

(1) Service disconnect

EMERGENCY DISCONNECT, SERVICE DISCONNECT

(2) Meter disconnects installed in accordance with 230.82(3) and marked as follows:

EMERGENCY DISCONNECT, METER DISCONNECT,
NOT SERVICE EQUIPMENT

(3) Other listed disconnect switches or circuit breakers on the supply side of each service disconnect that are marked suitable for use as service equipment and marked as follows:

EMERGENCY DISCONNECT,
NOT SERVICE EQUIPMENT

N **(2) Marking Location and Size.** Markings shall comply with 110.21(B) and both of the following:

(1) The marking or labels shall be located on the outside front of the disconnect enclosure with red background and white text.
(2) The letters shall be at least 13 mm (½ in.) high.

Part VII. Service Equipment — Overcurrent Protection

230.90 Where Required. Each ungrounded service conductor shall have overload protection.

Service equipment is the main control and means of cutoff of the electrical supply to the premises wiring system. It is usually an overcurrent device, such as a circuit breaker or a fuse, installed in series with each ungrounded service conductor to provide overload protection only.

The service overcurrent device does not protect the service conductors under short-circuit or ground-fault conditions on the line side of the disconnect. Protection against ground faults and short circuits is provided by the special requirements for service conductor protection and the location of the conductors.

Δ **(A) Ungrounded Conductor.** Such protection shall be provided by an overcurrent device in series with each ungrounded service conductor that has a rating or setting not higher than the ampacity of the conductor. A set of fuses shall be considered all the fuses required to protect all the ungrounded conductors of a circuit. Single-pole circuit breakers, grouped in accordance with 230.71(B), shall be considered as one protective device.

Exception No. 1: For motor-starting currents, ratings that comply with 430.52, 430.62, and 430.63 shall be permitted.

Exception No. 2: Fuses and circuit breakers with a rating or setting that complies with 240.4(B) or (C) and 240.6 shall be permitted.

Where the conductor ampacity does not correspond to the standard ampere rating of a circuit breaker or fuse, this exception permits the next-larger standard size circuit breaker or fuse to be installed. The permission to "round up" is limited by 240.4(B)(3) to ratings not exceeding 800 amperes. This exception permits rounding up only to the next standard size fuse or circuit breaker rating and does not permit the load to exceed the allowable ampacity of the service conductors.

See also

240.6 for standard ampere ratings of fuses and circuit breakers

Exception No. 3: Two to six circuit breakers or sets of fuses shall be permitted as the overcurrent device to provide the overload protection. The sum of the ratings of the circuit breakers or fuses shall be permitted to exceed the ampacity of the service conductors, provided the calculated load does not exceed the ampacity of the service conductors.

Section 230.90 requires an overcurrent device to provide overload protection in each ungrounded service conductor. However, Exception No. 3 allows two to six circuit breakers or sets of fuses to be considered as the overcurrent device. None of the individual overcurrent devices can have a rating or setting higher than the ampacity of the service conductors.

In complying with these rules, it is possible for the total of the six overcurrent devices to be greater than the rating of the service-entrance conductors. However, the size of the service-entrance conductors is required to be adequate for the computed load only, and each individual service disconnecting means is required to be large enough for the individual load it supplies.

The combined ratings of the five service disconnecting means OCPDs (350 amperes) shown in Exhibit 230.23 exceed the ampacity of the service-entrance conductors (310 amperes)

permitted by this exception. As specified, the ampacity of the service-entrance conductors is sufficient to carry the calculated load. The combined rating of the five service disconnecting means also complies with 230.80, which requires that the combined rating (350 amperes) not be less than the calculated load (305 amperes), the minimum size required for the service OCPD specified by 240.4 and 230.80.

Exception No. 4: Overload protection for fire pump supply conductors shall comply with 695.4(B)(2)(a).

Exception No. 5: Overload protection in accordance with the conductor ampacities of 310.12 shall be permitted for single-phase dwelling services.

(B) Not in Grounded Conductor. No overcurrent device shall be inserted in a grounded service conductor except a circuit breaker that simultaneously opens all conductors of the circuit.

230.91 Location. The service overcurrent device shall be an integral part of the service disconnecting means or shall be located immediately adjacent thereto. Where fuses are used as the service overcurrent device, the disconnecting means shall be located ahead of the supply side of the fuses.

230.92 Locked Service Overcurrent Devices. Where the service overcurrent devices are locked or sealed or are not readily accessible to the occupant, branch-circuit or feeder overcurrent devices shall be installed on the load side, shall be mounted in a readily accessible location, and shall be of lower ampere rating than the service overcurrent device.

230.93 Protection of Specific Circuits. Where necessary to prevent tampering, an automatic overcurrent device that protects service conductors supplying only a specific load, such as a water

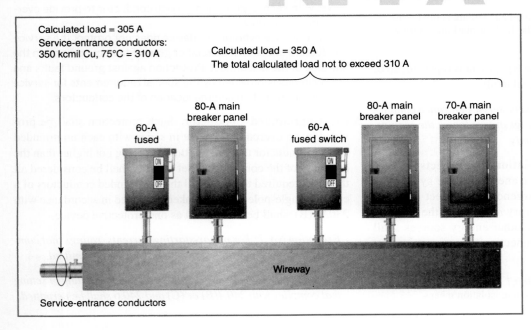

Calculated load = 305 A
Service-entrance conductors:
350 kcmil Cu, 75°C = 310 A

Calculated load = 350 A
The total calculated load not to exceed 310 A

60-A fused

80-A main breaker panel

60-A fused switch

80-A main breaker panel

70-A main breaker panel

Wireway

Service-entrance conductors

EXHIBIT 230.23 An example in which the combined ratings of the five service disconnecting means OCPDs are permitted to exceed the ampacity of the service conductors.

heater, shall be permitted to be locked or sealed where located so as to be accessible.

230.94 Relative Location of Overcurrent Device and Other Service Equipment.
The overcurrent device shall protect all circuits and devices.

Exception No. 1: The service switch shall be permitted on the supply side.

Exception No. 2: High-impedance shunt circuits, surge arresters, Type 1 surge-protective devices, surge-protective capacitors, and instrument transformers (current and voltage) shall be permitted to be connected and installed on the supply side of the service disconnecting means as permitted by 230.82.

Exception No. 3: Circuits for load management devices shall be permitted to be connected on the supply side of the service overcurrent device where separately provided with overcurrent protection.

Exception No. 4: Circuits used only for the operation of fire alarm, other protective signaling systems, or the supply to fire pump equipment shall be permitted to be connected on the supply side of the service overcurrent device where separately provided with overcurrent protection.

Exception No. 5: Meters nominally rated not in excess of 600 volts shall be permitted, provided all metal housings and service enclosures are grounded.

Exception No. 6: Where service equipment is power operable, the control circuit shall be permitted to be connected ahead of the service equipment if suitable overcurrent protection and disconnecting means are provided.

230.95 Ground-Fault Protection of Equipment.
Ground-fault protection of equipment shall be provided for solidly grounded wye electric services of more than 150 volts to ground but not exceeding 1000 volts phase-to-phase for each service disconnect rated 1000 amperes or more. The grounded conductor for the solidly grounded wye system shall be connected directly to ground through a grounding electrode system, as specified in 250.50, without inserting any resistor or impedance device.

The rating of the service disconnect shall be considered to be the rating of the largest fuse that can be installed or the highest continuous current trip setting for which the actual overcurrent device installed in a circuit breaker is rated or can be adjusted.

Ground-fault protection of equipment (GFPE) for service disconnecting means was first required in the 1971 edition of the *NEC* due to the unusually high number of equipment burndowns reported on large capacity 480Y/277-volt solidly grounded services. Other solidly grounded, wye-connected systems (e.g., 600Y/347 volts) are covered within the parameters specified by this section. The requirement does not apply to systems where the grounded conductor is not solidly grounded, as is the case with impedance grounded systems covered in 250.36.

Ground-fault protection of services does not protect the conductors on the supply side of the service disconnecting means, but it is designed to provide protection from line-to-ground faults that occur on the load side of the service disconnecting means.

The two basic types of ground-fault equipment protectors are illustrated in Exhibits 230.24 and 230.25. In Exhibit 230.24, the ground-fault sensor is installed around all the circuit conductors, and a stray current on a line-to-ground fault sets up an imbalance of the currents flowing in individual conductors installed through the ground-fault sensor. When this current exceeds the setting of the ground-fault sensor, the shunt trip operates and opens the circuit breakers.

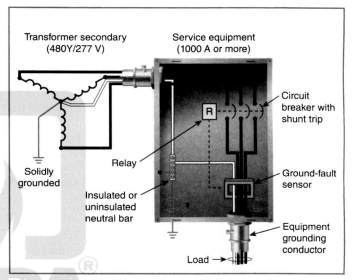

EXHIBIT 230.24 A ground-fault sensor encircling all circuit conductors, including the neutral.

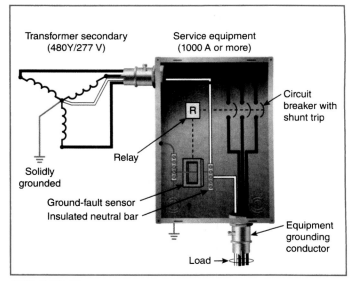

EXHIBIT 230.25 A ground-fault sensor encircling only the bonding jumper conductor.

The ground-fault sensor illustrated in Exhibit 230.25 is installed around the bonding jumper only. When an unbalanced current from a line-to-ground fault occurs, the current flows through the bonding jumper and the shunt trip causes the circuit breaker to operate, removing the load from the line.

Exception: The ground-fault protection provisions of this section shall not apply to a service disconnect for a continuous industrial process where a nonorderly shutdown will introduce additional or increased hazards.

(A) Setting. The ground-fault protection system shall operate to cause the service disconnect to open all ungrounded conductors of the faulted circuit. The maximum setting of the ground-fault protection shall be 1200 amperes, and the maximum time delay shall be one second for ground-fault currents equal to or greater than 3000 amperes.

Ground-fault sensor settings at low levels can increase the likelihood of unwanted shutdowns. The duration of the fault is limited to not more than 1 second for fault currents greater than 3000 amperes. This restriction minimizes the amount of damage done by an arcing fault, which is directly proportional to the time the arcing fault is allowed to burn.

Care should be taken to ensure that interconnecting of multiple supply systems does not interfere with proper sensing by the GFPE. An engineering study should be made to ensure that fault currents do not take parallel paths to the supply system, thereby bypassing the ground-fault detection device.

(B) Fuses. If a switch and fuse combination is used, the fuses employed shall be capable of interrupting any current higher than the interrupting capacity of the switch during a time that the ground-fault protective system will not cause the switch to open.

(C) Performance Testing. The ground-fault protection system shall be performance tested when first installed on site. This testing shall be conducted by a qualified person(s) using a test process of primary current injection, in accordance with instructions that shall be provided with the equipment. A written record of this testing shall be made and shall be available to the authority having jurisdiction.

Informational Note No. 1: Ground-fault protection that functions to open the service disconnect affords no protection from faults on the line side of the protective element. It serves only to limit damage to conductors and equipment on the load side in the event of an arcing ground fault on the load side of the protective element.

Informational Note No. 2: This added protective equipment at the service equipment could make it necessary to review the overall wiring system for proper selective overcurrent protection coordination. Additional installations of ground-fault protective equipment might be needed on feeders and branch circuits where maximum continuity of electric service is necessary.

Informational Note No. 3: Where ground-fault protection is provided for the service disconnect and interconnection is made with another supply system by a transfer device, means or devices could be needed to ensure proper ground-fault sensing by the ground-fault protection equipment.

Informational Note No. 4: See 517.17(A) for information on where an additional step of ground-fault protection is required for hospitals and other buildings with critical areas or life-support equipment.

N | **ARTICLE 235** Branch Circuits, Feeders, and Services Over 1000 Volts ac, 1500 Volts dc, Nominal

N Part I. General

N 235.1 Scope. This article provides the general requirements for branch circuits, feeders, and services over 1000 volts ac or 1500 volts dc, nominal.

Informational Note: See ANSI/IEEE C2-2017, *National Electrical Safety Code*, for additional information on wiring over 1000 volts, nominal.

New Article 235 has been created to cover branch circuits, feeders, and services for installations over 1000 volts ac, 1500 volts dc, nominal. To aid in usability of the *NEC®*, much of the information within Article 235 has been pulled in from long-standing, medium-voltage requirements within other code articles, such as Articles 210, 215, 225, and 230.

N Part II. Branch Circuits

N 235.3 Other Articles for Specific-Purpose Branch Circuits. Table 235.3 lists references for specific equipment and applications not located in Chapters 5, 6, and 7 that amend or supplement the requirements of this article.

N TABLE 235.3 *References for Specific Equipment and Applications Not Located in Chapters 5, 6, and 7*

Equipment	Article	Section
Air-conditioning and refrigerating equipment		440.6, 440.31, and 440.32
Busways		368.17
Central heating equipment other than fixed electric space-heating equipment		422.12
Fixed electric heating equipment for pipelines and vessels		427.4
Fixed electric space-heating equipment		424.4
Fixed outdoor electrical deicing and snow-melting equipment		426.4
Infrared lamp industrial heating equipment		422.48 and 424.3
Motors, motor circuits, and controllers	430	

N 235.5 Conductor Identification for Branch Circuits.

N (A) Grounded Conductor. The grounded conductor of a branch circuit shall be identified in accordance with 200.6.

N (B) Equipment Grounding Conductor. The equipment grounding conductor shall be identified in accordance with 250.119.

N (C) Ungrounded Conductors. Ungrounded conductors shall be identified in accordance with 235.5(C)(1) or (C)(2), as applicable.

N (1) Branch Circuits Supplied from More Than One Nominal Voltage System. Where the premises wiring system has branch circuits supplied from more than one nominal voltage system, each ungrounded conductor of a branch circuit shall be identified by phase or line and by nominal system voltage at all termination, connection, and splice points in accordance with 235.5(C)(1)(a) and (C)(1)(b). Different systems within the same premises that have the same nominal voltage shall be permitted to use the same identification.

(a) *Means of Identification.* The means of identification shall be permitted to be by separate color coding, marking tape, tagging, or other approved means.

(b) *Posting of Identification Means.* The method used for conductors originating within each branch-circuit panelboard or similar branch-circuit distribution equipment shall be documented in a manner that is readily available or shall be permanently posted at each branch-circuit panelboard or similar branch-circuit distribution equipment. The label shall be of sufficient durability to withstand the environment involved and shall not be handwritten.

Exception: In existing installations where a voltage system(s) already exists and a different voltage system is being added, it shall be permissible to mark only the new system voltage. Existing unidentified systems shall not be required to be identified at each termination, connection, and splice point in accordance with 235.5(C)(1)(a) and (C)(1)(b). Labeling shall be required at each voltage system distribution equipment to identify that only one voltage system has been marked for a new system(s). The new system label(s) shall include the words "other unidentified systems exist on the premises."

N (2) Branch Circuits Supplied from Direct-Current Systems. Where a branch circuit is supplied from a dc system operating at more than 1500 volts, each ungrounded conductor of 4 AWG or larger shall be identified by polarity at all termination, connection, and splice points by marking tape, tagging, or other approved means and each ungrounded conductor of 6 AWG or smaller shall be identified by polarity at all termination, connection, and splice points in compliance with 235.5(C)(2)(a) and (C)(2)(b). The identification methods used for conductors originating within each branch-circuit panelboard or similar branch-circuit distribution equipment shall be documented in a

manner that is readily available or be permanently posted at each branch-circuit panelboard or similar branch-circuit distribution equipment.

(a) *Positive Polarity, Sizes 6 AWG or Smaller.* Where the positive polarity of a dc system does not serve as the connection point for the grounded conductor, each positive ungrounded conductor shall be identified by one of the following means:

(1) A continuous red outer finish
(2) A continuous red stripe durably marked along the conductor's entire length on insulation of a color other than green, white, gray, or black
(3) Imprinted plus signs (+) or the word POSITIVE or POS durably marked on insulation of a color other than green, white, gray, or black and repeated at intervals not exceeding 610 mm (24 in.) in accordance with 310.8(B)
(4) An approved permanent marking means such as sleeving or shrink-tubing that is suitable for the conductor size, at all termination, connection, and splice points, with imprinted plus signs (+) or the word POSITIVE or POS durably marked on insulation of a color other than green, white, gray, or black

(b) *Negative Polarity, Sizes 6 AWG or Smaller.* Where the negative polarity of a dc system does not serve as the connection point for the grounded conductor, each negative ungrounded conductor shall be identified by one of the following means:

(1) A continuous black outer finish
(2) A continuous black stripe durably marked along the conductor's entire length on insulation of a color other than green, white, gray, or red
(3) Imprinted minus signs (–) or the word NEGATIVE or NEG durably marked on insulation of a color other than green, white, gray, or red and repeated at intervals not exceeding 610 mm (24 in.) in accordance with 310.8(B)
(4) An approved permanent marking means such as sleeving or shrink-tubing that is suitable for the conductor size, at all termination, connection, and splice points, with imprinted minus signs (–) or the word NEGATIVE or NEG durably marked on insulation of a color other than green, white, gray, or red

N 235.6 Branch-Circuit Voltage Limitations Over 1000 volts ac or 1500 volts dc, Nominal, Between Conductors. Circuits exceeding 1000 volts ac or 1500 volts dc, nominal, between conductors shall be permitted to supply utilization equipment in installations where conditions of maintenance and supervision ensure that only qualified persons service the installation.

N 235.9 Circuits Derived from Autotransformers. Branch circuits shall not be derived from autotransformers unless the circuit supplied has a grounded conductor that is electrically connected to a grounded conductor of the system supplying the autotransformer.

N **235.10 Ungrounded Conductors Tapped from Grounded Systems.** Two-wire dc circuits and ac circuits of two or more ungrounded conductors shall be permitted to be tapped from the ungrounded conductors of circuits that have a grounded neutral conductor. Switching devices in each tapped circuit shall have a pole in each ungrounded conductor. All poles of multipole switching devices shall manually switch together where such switching devices also serve as a disconnecting means as required by the following sections:

(1) 410.93 for double-pole switched lampholders
(2) 410.104(B) for electric-discharge lamp auxiliary equipment switching devices
(3) 422.31(B) for an appliance
(4) 424.20 for a fixed electric space-heating unit
(5) 426.51 for electric deicing and snow-melting equipment
(6) 430.85 for a motor controller
(7) 430.103 for a motor

N **235.11 Branch Circuits Required.** The minimum number of branch circuits shall be determined from the total calculated load and the size or rating of the circuits used. In all installations, the number of circuits shall be sufficient to supply the load served.

N **235.18 Rating.** Branch circuits recognized by this article shall be rated in accordance with the maximum permitted ampere rating or setting of the overcurrent device. Where conductors of higher ampacity are used for any reason, the ampere rating or setting of the specified overcurrent device shall determine the circuit rating.

N **235.19 Conductors — Minimum Ampacity and Size.** The ampacity of conductors shall be in accordance with 310.14 and 315.60, as applicable. Branch-circuit conductors shall be sized in accordance with 235.19(A) or (B).

N **(A) General.** The ampacity of branch-circuit conductors shall not be less than 125 percent of the designed potential load of utilization equipment that will be operated simultaneously.

N **(B) Supervised Installations.** For supervised installations, branch-circuit conductor sizing shall be permitted to be determined by qualified persons under engineering supervision. Supervised installations are defined as those portions of a facility where both of the following conditions are met:

(1) Conditions of design and installation are provided under engineering supervision.

(2) Qualified persons with documented training and experience in over 1000-volt ac or 1500-volt dc systems provide maintenance, monitoring, and servicing of the system.

N **235.20 Overcurrent Protection.** Branch-circuit conductors and equipment shall be protected by overcurrent protective devices that have a rating or setting that complies with 235.20(A) through (C).

N **(A) Continuous and Noncontinuous Loads.** Where a branch circuit supplies continuous loads or any combination of continuous and noncontinuous loads, the rating of the overcurrent device shall not be less than the noncontinuous load plus 125 percent of the continuous load.

Exception: Where the assembly, including the overcurrent devices protecting the branch circuit(s), is listed for operation at 100 percent of its rating, the ampere rating of the overcurrent device shall be permitted to be not less than the sum of the continuous load plus the noncontinuous load.

N **(B) Conductor Protection.** Conductors shall be protected in accordance with the ampacities specified in 310.14 or 315.60, as applicable.

N **(C) Equipment.** The rating or setting of the overcurrent protective device shall not exceed that specified in the applicable articles referenced in Table 240.3 for equipment.

N **235.22 Permissible Loads, Individual Branch Circuits.** An individual branch circuit shall be permitted to supply any load for which it is rated, but in no case shall the load exceed the branch-circuit ampere rating.

N **235.23 Permissible Loads, Multiple-Outlet Branch Circuits.** A branch circuit supplying two or more outlets or receptacles shall supply only the loads specified according to its size in accordance with 210.23(A) through (E) and as summarized in 210.24, and in no case shall the load exceed the branch-circuit ampere rating.

N **(A) 15- and 20-Ampere Branch Circuits.** A 15- or 20-ampere branch circuit shall be permitted to supply lighting outlets, lighting units, or other utilization equipment, or any combination of them, and shall comply with 235.23(A)(1) and (A)(2).

N **(1) Cord-and-Plug-Connected Equipment Not Fastened in Place.** The rating of any one cord-and-plug-connected utilization equipment not fastened in place shall not exceed 80 percent of the branch-circuit ampere rating.

N **(2) Utilization Equipment Fastened in Place.** The total rating of utilization equipment fastened in place, other than luminaires, shall not exceed 50 percent of the branch-circuit ampere rating where lighting units, cord-and-plug-connected utilization equipment not fastened in place, or both, are also supplied.

N **(B) 30-Ampere Branch Circuits.** A 30-ampere branch circuit shall be permitted to supply fixed lighting units with heavy-duty lampholders in other than a dwelling unit(s) or utilization equipment in any occupancy. The rating of any one cord-and-plug-connected utilization equipment shall not exceed 80 percent of the branch-circuit ampere rating.

N **(C) 40- and 50-Ampere Branch Circuits.** A 40- or 50-ampere branch circuit shall be permitted to supply cooking appliances that are fastened in place in any occupancy. In other than dwelling units, such circuits shall be permitted to supply fixed lighting units with heavy-duty lampholders, infrared heating units, or other utilization equipment.

N **(D) Branch Circuits Larger Than 50 Amperes.** Branch circuits larger than 50 amperes shall supply only nonlighting outlet loads.

N **235.63 Equipment Requiring Servicing.** A 125-volt, single-phase, 15- or 20-ampere-rated receptacle outlet shall be installed at an accessible location within 7.5 m (25 ft) of the equipment as specified in 210.63(A) and (B).

> Informational Note: See 210.8(E) for requirements on GFCI protection.

N **(A) Heating, Air-Conditioning, and Refrigeration Equipment.** The required receptacle outlet shall be located on the same level as the heating, air-conditioning, and refrigeration equipment. The receptacle outlet shall not be connected to the load side of the equipment's branch-circuit disconnecting means.

> *Exception: A receptacle outlet shall not be required at one- and two-family dwellings for the service of evaporative coolers.*

N **(B) Other Electrical Equipment.** In other than one- and two-family dwellings, a receptacle outlet shall be located as specified in 210.63(B)(1) and (B)(2).

N **(1) Indoor Service Equipment.** The required receptacle outlet shall be located within the same room or area as the service equipment.

N **(2) Indoor Equipment Requiring Dedicated Equipment Spaces.** Where equipment, other than service equipment, requires dedicated equipment space as specified in 110.26(E), the required receptacle outlet shall be located within the same room or area as the electrical equipment and shall not be connected to the load side of the equipment's disconnecting means.

N **Part III. Feeders**

N **235.201 General.** Part III covers the installation requirements, overcurrent protection requirements, minimum size, and ampacity of conductors for feeders over 1000 volts ac or 1500 volts dc, nominal.

N **235.202 Minimum Rating and Size.** The ampacity of conductors shall be in accordance with 310.14 and 315.60 as applicable. Where installed, the size of the feeder-circuit grounded conductor shall not be smaller than that required by 250.122, except that 250.122(F) shall not apply where grounded conductors are run in parallel. Feeder conductors over 1000 volts shall be sized in accordance with 235.202(A), (B), or (C).

N **(A) Feeders Supplying Transformers.** The ampacity of feeder conductors shall not be less than the sum of the nameplate ratings of the transformers supplied when only transformers are supplied.

N **(B) Feeders Supplying Transformers and Utilization Equipment.** The ampacity of feeders supplying a combination of transformers and utilization equipment shall not be less than the sum of the nameplate ratings of the transformers and 125 percent of the designed potential load of the utilization equipment that will be operated simultaneously.

N **(C) Supervised Installations.** For supervised installations, feeder conductor sizing shall be permitted to be determined by qualified persons under engineering supervision in accordance with 310.14(B) or 315.60(B). Supervised installations are defined as those portions of a facility where all of the following conditions are met:

(1) Conditions of design and installation are provided under engineering supervision.
(2) Qualified persons with documented training and experience in over 1000-volt systems provide maintenance, monitoring, and servicing of the system.

N **235.203 Overcurrent Protection.** Feeders shall be protected against overcurrent.

N **235.205 Diagrams of Feeders.** If required by the authority having jurisdiction, a diagram showing feeder details shall be provided prior to the installation of the feeders. Such a diagram shall show the area in square feet of the building or other structure supplied by each feeder, the total calculated load before applying demand factors, the demand factors used, the calculated load after applying demand factors, and the size and type of conductors to be used.

N **235.206 Feeder Equipment Grounding Conductor.** Where a feeder supplies branch circuits in which equipment grounding conductors are required, the feeder shall include or provide an equipment grounding conductor, to which the equipment grounding conductors of the branch circuits shall be connected. Where the feeder supplies a separate building or structure, the requirements of 250.32 shall apply.

N **235.212 Identification for Feeders.**

N **(A) Grounded Conductor.** The grounded conductor of a feeder, if insulated, shall be identified in accordance with 200.6.

N **(B) Equipment Grounding Conductor.** The equipment grounding conductor shall be identified in accordance with 250.119.

N **(C) Identification of Ungrounded Conductors.** Ungrounded conductors shall be identified in accordance with 235.212(C)(1) or (C)(2), as applicable.

N **(1) Feeders Supplied from More Than One Nominal Voltage System.** Where the premises wiring system has feeders supplied from more than one nominal voltage system, each ungrounded conductor of a feeder shall be identified by phase or line and system at all termination, connection, and splice points in compliance with 235.212(C)(1)(a) and (C)(1)(b).

(a) *Means of Identification.* The means of identification shall be permitted to be by separate color coding, marking tape, tagging, or other approved means.

(b) *Posting of Identification Means.* The method utilized for conductors originating within each feeder panelboard or similar feeder distribution equipment shall be documented in a manner that is readily available or shall be permanently posted at each feeder panelboard or similar feeder distribution equipment.

N **(2) Feeders Supplied from Direct-Current Systems.** Where a feeder is supplied from a dc system operating at more than 1500 volts, each ungrounded conductor of 4 AWG or larger shall be identified by polarity at all termination, connection, and splice points by marking tape, tagging, or other approved means; each ungrounded conductor of 6 AWG or smaller shall be identified by polarity at all termination, connection, and splice points in compliance with 235.212(C)(2)(a) and (C)(2)(b). The identification methods utilized for conductors originating within each feeder panelboard or similar feeder distribution equipment shall be documented in a manner that is readily available or shall be permanently posted at each feeder panelboard or similar feeder distribution equipment.

(a) *Positive Polarity, Sizes 6 AWG or Smaller.* Where the positive polarity of a dc system does not serve as the connection for the grounded conductor, each positive ungrounded conductor shall be identified by one of the following means:

(1) A continuous red outer finish
(2) A continuous red stripe durably marked along the conductor's entire length on insulation of a color other than green, white, gray, or black
(3) Imprinted plus signs (+) or the word POSITIVE or POS durably marked on insulation of a color other than green, white, gray, or black, and repeated at intervals not exceeding 610 mm (24 in.) in accordance with 310.8(B)
(4) An approved permanent marking means such as sleeving or shrink-tubing that is suitable for the conductor size, at all termination, connection, and splice points, with imprinted plus signs (+) or the word POSITIVE or POS durably marked on insulation of a color other than green, white, gray, or black

(b) *Negative Polarity, Sizes 6 AWG or Smaller.* Where the negative polarity of a dc system does not serve as the connection for the grounded conductor, each negative ungrounded conductor shall be identified by one of the following means:

(1) A continuous black outer finish
(2) A continuous black stripe durably marked along the conductor's entire length on insulation of a color other than green, white, gray, or red

(3) Imprinted minus signs (–) or the word NEGATIVE or NEG durably marked on insulation of a color other than green, white, gray, or red, and repeated at intervals not exceeding 610 mm (24 in.) in accordance with 310.8(B)
(4) An approved permanent marking means such as sleeving or shrink-tubing that is suitable for the conductor size, at all termination, connection, and splice points, with imprinted minus signs (–) or the word NEGATIVE or NEG durably marked on insulation of a color other than green, white, gray, or red

N ## Part IV. Outside Branch Circuits and Feeders

N **235.301 General.** Part IV covers requirements for outside branch circuits and feeders over 1000 volts ac or 1500 volts dc, nominal, that are run on or between buildings, structures, or poles on the premises; and electrical equipment and wiring for the supply of utilization equipment that is located on or attached to the outside of buildings, structures, or poles. Outside branch circuits and feeders over 1000 volts ac or 1500 volts dc, nominal, shall comply with the applicable requirements in Parts I and II of Article 225 and with Part IV of this article, which supplements or modifies those requirements.

N **235.306 Conductor Size and Support.** For overhead spans, open individual conductors shall not be smaller than 6 AWG copper or 4 AWG aluminum where open individual conductors and 8 AWG copper or 6 AWG aluminum where in cable.

N **235.310 Wiring on Buildings (or Other Structures).** The installation of outside wiring on surfaces of buildings (or other structures) shall be installed as provided in 305.3.

N **235.314 Open-Conductor Spacings.** Conductors shall comply with the spacings provided in 110.36 and 495.24.

N **235.339 Rating of Disconnect.** The feeder or branch-circuit disconnecting means shall have a rating of not less than the calculated load to be supplied, determined in accordance with Parts I and II of Article 220 for branch circuits, Part III or IV of Article 220 for feeders, or Part V of Article 220 for farm loads.

N **235.350 Sizing of Conductors.** The sizing of conductors over 1000 volts shall be in accordance with 235.19(A) for branch circuits and 235.19(B) for feeders.

N **235.351 Isolating Switches.** Where oil switches or air, oil, vacuum, or sulfur hexafluoride circuit breakers constitute a building disconnecting means, an isolating switch with visible break contacts and meeting the requirements of 235.404(B), (C), and (D) shall be installed on the supply side of the disconnecting means and all associated equipment.

Exception: The isolating switch shall not be required where the disconnecting means is mounted on removable truck panels or switchgear units that cannot be opened unless the circuit is disconnected and that, when removed from the normal operating position, automatically disconnect the circuit breaker or switch from all energized parts.

N 235.352 Disconnecting Means.

N (A) Location. A building or structure disconnecting means shall be located in accordance with 225.31(B), or, if not readily accessible, it shall be operable by mechanical linkage from a readily accessible point. For multibuilding industrial installations under single management, it shall be permitted to be electrically operated by a readily accessible, remote-control device in a separate building or structure.

N (B) Type. Each building or structure disconnect shall simultaneously disconnect all ungrounded supply conductors it controls and shall have a fault-closing rating not less than the available fault current at its supply terminals.

Exception: Where the individual disconnecting means consists of fused cutouts, the simultaneous disconnection of all ungrounded supply conductors shall not be required if there is a means to disconnect the load before opening the cutouts. A permanent legible sign shall be installed adjacent to the fused cutouts and shall read DISCONNECT LOAD BEFORE OPENING CUTOUTS.

Where fused switches or separately mounted fuses are installed, the fuse characteristics shall be permitted to contribute to the fault-closing rating of the disconnecting means.

Where a switch is used, fuses are permitted to help with the fault-closing capability of the switch. Using fused load-break cutouts to switch sections of overhead lines and load-break elbows to switch sections of underground lines is permitted. However, the building disconnecting means must be gang-operated to simultaneously open and close all ungrounded supply conductors. Load-break elbows and fused cutouts cannot be used as the building disconnecting means.

> **See also**
>
> **235.405(B),** which contains a similar requirement for services

N (C) Locking. Disconnecting means shall be lockable open in accordance with 110.25.

Exception: Where an individual disconnecting means consists of fused cutouts, a suitable enclosure capable of being locked and sized to contain all cutout fuse holders shall be installed at a convenient location to the fused cutouts

N (D) Indicating. Disconnecting means shall clearly indicate whether they are in the open "off" or closed "on" position.

N (E) Uniform Position. Where disconnecting means handles are operated vertically, the "up" position of the handle shall be the "on" position.

Exception: A switching device having more than one "on" position, such as a double throw switch, shall not be required to comply with this requirement.

N (F) Identification. Where a building or structure has any combination of feeders, branch circuits, or services passing through or supplying it, a permanent plaque or directory shall be installed at each feeder and branch-circuit disconnect location that denotes all other services, feeders, or branch circuits supplying that building or structure or passing through that building or structure and the area served by each.

N 235.356 Inspections and Tests.

N (A) Pre-Energization and Operating Tests. The complete electrical system design, including settings for protective, switching, and control circuits, shall be prepared in advance and made available on request to the authority having jurisdiction and shall be performance tested when first installed on-site. Each protective, switching, and control circuit shall be adjusted in accordance with the system design and tested by actual operation using current injection or equivalent methods as necessary to ensure that each and every such circuit operates correctly to the satisfaction of the authority having jurisdiction.

The AHJ must be satisfied that the performance tests demonstrate proper operation. Section 235.356(B) requires that a report of all tests performed in accordance with 235.356(A) be provided to the AHJ prior to energizing the system. Adjustments of settings must be in accordance with the electrical system design.

N (1) Instrument Transformers. All instrument transformers shall be tested to verify correct polarity and burden.

N (2) Protective Relays. Each protective relay shall be demonstrated to operate by injecting current or voltage, or both, at the associated instrument transformer output terminal and observing that the associated switching and signaling functions occur correctly and in proper time and sequence to accomplish the protective function intended.

N (3) Switching Circuits. Each switching circuit shall be observed to operate the associated equipment being switched.

N (4) Control and Signal Circuits. Each control or signal circuit shall be observed to perform its proper control function or produce a correct signal output.

N (5) Metering Circuits. All metering circuits shall be verified to operate correctly from voltage and current sources in a similar manner to protective relay circuits.

N (6) Acceptance Tests. Complete acceptance tests shall be performed, after the substation installation is completed, on all assemblies, equipment, conductors, and control and protective systems, as applicable, to verify the integrity of all the systems.

N (7) Relays and Metering Utilizing Phase Differences. All relays and metering that use phase differences for operation shall be verified by measuring phase angles at the relay under actual load conditions after operation commences.

N (B) Test Report. A test report covering the results of the tests required in 235.356(A) shall be delivered to the authority having jurisdiction prior to energization.

Informational Note: See ANSI/NETA ATS, *Acceptance Testing Specifications for Electrical Power Distribution Equipment and Systems*, for an example of acceptance specifications.

N 235.360 Clearances over Roadways, Walkways, Rail, Water, and Open Land.

N (A) 22 kV or Less to Ground. The clearances over roadways, walkways, rail, water, and open land for conductors and live parts up to 22 kV or less to ground shall be not less than the values shown in Table 235.360(A).

N TABLE 235.360(A) *Clearances over Roadways, Walkways, Rail, Water, and Open Land*

| Location | Clearance | |
	m	ft
Open land subject to vehicles, cultivation, or grazing	5.6	18.5
Roadways, driveways, parking lots, and alleys	5.6	18.5
Walkways	4.1	13.5
Rails	8.1	26.5
Spaces and ways for pedestrians and restricted traffic	4.4	14.5
Water areas not suitable for boating	5.2	17.0

N (B) More Than 22 kV to Ground. Clearances for the categories shown in Table 235.360(A) shall be increased by 10 mm (0.4 in.) per kV, or major fraction thereof, more than 22 kV.

N (C) Special Cases. For special cases, such as where crossings will be made over lakes, rivers, or areas using large vehicles such as mining operations, specific designs shall be engineered considering the special circumstances and shall be approved by the authority having jurisdiction.

Informational Note: See ANSI/IEEE C2-2017, *National Electrical Safety Code*, for additional information.

N 235.361 Clearances over Buildings and Other Structures.

N (A) 22 kV or Less to Ground. The clearances over buildings and other structures for conductors and live parts up to 22 kV or less to ground shall be not less than the values shown in Table 235.361(A).

N TABLE 235.361(A) *Clearances over Buildings and Other Structures*

| Clearance from Conductors or Live Parts from: | Horizontal | | Vertical | |
	m	ft	m	ft
Building walls, projections, and windows	2.3	7.5	—	—
Balconies, catwalks, and similar areas accessible to people	2.3	7.5	4.1	13.5
Over or under roofs or projections not readily accessible to people	—	—	3.8	12.5
Over roofs accessible to vehicles but not trucks	—	—	4.1	13.5
Over roofs accessible to trucks	—	—	5.6	18.5
Other structures	2.3	7.5	—	—

N (B) More Than 22 kV to Ground. Clearances for the categories shown in Table 235.361(A) shall be increased by 10 mm (0.4 in.) per kV, or major fraction thereof, more than 22 kV.

Informational Note: See ANSI/IEEE C2-2017, *National Electrical Safety Code*, for additional information.

These clearance requirements and specific distances over buildings and structures correlate with requirements in the *National Electrical Safety Code® (NESC®).*

N Part V. Services

N 235.401 General. Part V covers requirements for service conductors and equipment used on circuits over 1000 volts ac and 1500 volts dc, nominal, shall comply with all of the applicable requirements in Parts I through VII of Article 230 and with Part V of this article, which supplements or modifies those requirements. In no case shall the provisions of Part V apply to equipment on the supply side of the service point.

Only those conductors on the load side of the service-point connection are subject to the requirements of the *NEC*. The service point is a specific location where the supply conductors of the electric utility and the customer-owned (premises wiring) conductors connect.

Exhibit 235.1 depicts an installation where the transformer and service lateral to the service point are owned by the electric utility. The transformer secondary conductors between the service point (separate connection point outside the transformer in this scenario) and the service disconnecting means at the building are *underground service conductors* until the point at which they enter the building. From that point to the termination in the service equipment, they are *service-entrance conductors.*

In the installation depicted in Exhibit 235.2, the service disconnecting means is located at the customer-owned transformer primary. The conductors that connect the service point (at the top of the pole), the service disconnecting means, and the service/transformer overcurrent protective device (OCPD) are underground service conductors. The conductors between the transformer secondary and the line side of the building disconnecting means are *feeders* and are subject to the requirements in

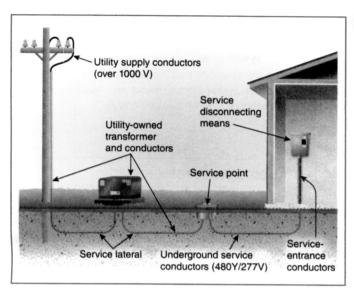

EXHIBIT 235.1 Service rated over 1000 volts supplied by a utility-owned transformer.

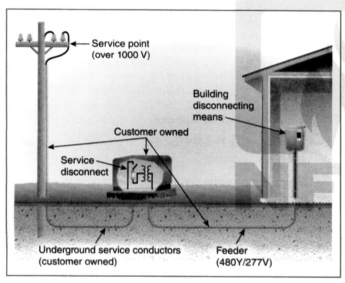

EXHIBIT 235.2 Service rated over 1000 volts supplying a customer-owned transformer.

240.21(C)(4) for outside conductors connected to a transformer secondary. Conductors (not shown) on the load side of the building disconnecting means are also feeders. Each building or structure is required to have a disconnecting means, in accordance with 225.31.

N **235.402 Service-Entrance Conductors.** Service-entrance conductors to buildings or enclosures shall be installed to conform to 235.402(A) and (B).

N **(A) Conductor Size.** Service-entrance conductors shall not be smaller than 6 AWG unless in multiconductor cable. Multiconductor cable shall not be smaller than 8 AWG.

N **(B) Wiring Methods.** Service-entrance conductors shall be installed by one of the wiring methods covered in 305.3 and 305.15.

N **235.404 Isolating Switches.**

N **(A) Where Required.** Where oil switches or air, oil, vacuum, or sulfur hexafluoride circuit breakers constitute the service disconnecting means, an isolating switch with visible break contacts shall be installed on the supply side of the disconnecting means and all associated service equipment.

Exception: An isolating switch shall not be required where the circuit breaker or switch is mounted on removable truck panels or switchgear units where both of the following conditions apply:

(1) Cannot be opened unless the circuit is disconnected
(2) Where all energized parts are automatically disconnected when the circuit breaker or switch is removed from the normal operating position

N **(B) Fuses as Isolating Switch.** Where fuses are of the type that can be operated as a disconnecting switch, a set of such fuses shall be permitted as the isolating switch.

N **(C) Accessible to Qualified Persons Only.** The isolating switch shall be accessible to qualified persons only.

N **(D) Connection to Ground.** Isolating switches shall be provided with a means for readily connecting the load side conductors to a grounding electrode system, equipment ground busbar, or grounded steel structure when disconnected from the source of supply.

A means for grounding the load side conductors to a grounding electrode system, equipment grounding busbar, or grounded structural steel shall not be required for any duplicate isolating switch installed and maintained by the electric supply company.

N **235.405 Disconnecting Means.**

N **(A) Location.** The service disconnecting means shall be located in accordance with 230.70.

For either overhead or underground primary distribution systems on private property, the service disconnect shall be permitted to be located in a location that is not readily accessible, if the disconnecting means can be operated by mechanical linkage from a readily accessible point, or electronically in accordance with 235.405(C), where applicable.

The general requirement for service disconnecting means in systems rated over 1000 volts is that it be readily accessible. Because ready access might not be possible, or desirable (if it could put employees at risk), the use of a readily accessible operating mechanism or an electronic switching device is an acceptable alternative.

N **(B) Type.** Each service disconnect shall simultaneously disconnect all ungrounded service conductors that it controls and shall have a fault-closing rating that is not less than the available fault current at its supply terminals.

Where fused switches or separately mounted fuses are installed, the fuse characteristics shall be permitted to contribute to the fault-closing rating of the disconnecting means.

N **(C) Remote Control.** For multibuilding, industrial installations under single management, the service disconnecting means shall be permitted to be located at a separate building or structure. In such cases, the service disconnecting means shall be permitted to be electrically operated by a readily accessible, remote-control device.

N **235.406 Overcurrent Devices as Disconnecting Means.** Where the circuit breaker or alternative for it, as specified in 235.408 for service overcurrent devices, meets the requirements specified in 235.405, it shall constitute the service disconnecting means.

N **235.408 Protection Requirements.** A short-circuit protective device shall be provided on the load side of, or as an integral part of, the service disconnect, and shall protect all ungrounded conductors that it supplies. The protective device shall be capable of detecting and interrupting all values of current, in excess of its trip setting or melting point, that can occur at its location. A fuse rated in continuous amperes not to exceed three times the ampacity of the conductor, or a circuit breaker with a trip setting of not more than six times the ampacity of the conductors, shall be considered as providing the required short-circuit protection.

> Informational Note: See Table 315.60(C)(1) through Table 315.60(C)(20) for ampacities of conductors rated 2001 volts to 35,000 volts.

Overcurrent devices shall conform to 235.408(A) and (B).

N **(A) Equipment Type.** Equipment used to protect service-entrance conductors shall meet the requirements of Article 495, Part II.

N **(B) Enclosed Overcurrent Devices.** The restriction to 80 percent of the rating for an enclosed overcurrent device for continuous loads shall not apply to overcurrent devices installed in systems operating at over 1000 volts.

N **235.409 Surge Arresters.** Surge arresters installed in accordance with the requirements of Parts II and III of Article 242 shall be permitted on each ungrounded overhead service conductor.

> Informational Note: Surge arresters may be referred to as lightning arresters in older documents.

N **235.410 Service Equipment — General.** Service equipment, including instrument transformers, shall conform to Part I of Article 495.

N **235.411 Switchgear.** Switchgear shall consist of a substantial metal structure and a sheet metal enclosure. Where installed over a combustible floor, suitable protection thereto shall be provided.

A vault in accordance with 235.412 can be used in lieu of switchgear for voltages below 35,000 volts.

N **235.412 Over 35,000 Volts.** Where the voltage exceeds 35,000 volts between conductors that enter a building, they shall terminate in a switchgear compartment or a vault conforming to the requirements of 450.41 through 450.48.

ARTICLE
240 Overcurrent Protection

Part I. General

Δ **240.1 Scope.** Parts I through VII of this article provide the general requirements for overcurrent protection and overcurrent protective devices not more than 1000 volts, nominal. Part VIII covers overcurrent protection for those portions of supervised industrial installations operating at voltages of not more than 1000 volts, nominal.

> Informational Note No. 1: Overcurrent protection for conductors and equipment is provided to open the circuit if the current reaches a value that will cause an excessive or dangerous temperature in conductors or conductor insulation.
>
> Informational Note No. 2: See 110.9 for requirements for interrupting ratings and 110.10 for requirements for protection against fault currents.

N **240.2 Reconditioned Equipment.**

N **(A) Reconditioning Not Permitted.** The following equipment shall not be reconditioned:

(1) Equipment providing ground-fault protection of equipment
(2) Ground-fault circuit interrupters
(3) Low-voltage fuseholders and low-voltage nonrenewable fuses
(4) Molded-case circuit breakers
(5) Low-voltage power circuit breaker electronic trip units.

N **(B) Reconditioning Permitted.** The following equipment shall be permitted to be reconditioned:

(1) Low-voltage power circuit breakers
(2) Electromechanical protective relays and current transformers

Reconditioned equipment shall be listed as reconditioned and comply with 110.21(A)(2).

240.3 Other Articles. Equipment shall be protected against overcurrent in accordance with the article in this *Code* that covers the type of equipment specified in Table 240.3.

240.4 Protection of Conductors. Conductors, other than flexible cords, flexible cables, and fixture wires, shall be protected

△ **TABLE 240.3** Other Articles

Equipment	Article
Air-conditioning and refrigerating equipment	440
Appliances	422
Assembly occupancies	518
Audio signal processing, amplification, and reproduction equipment	640
Branch circuits	210
Busways	368
Capacitors	460
Class 1 power-limited circuits and Class 1 power-limited remote-control and signaling circuits	724
Class 2 and Class 3 remote-control, signaling, and power-limited circuits	725
Cranes and hoists	610
Electric signs and outline lighting	600
Electric welders	630
Electrolytic cells	668
Elevators, dumbwaiters, escalators, moving walks, wheelchair lifts, and stairway chairlifts	620
Emergency systems	700
Fire alarm systems	760
Fire pumps	695
Fixed electric heating equipment for pipelines and vessels	427
Fixed electric space-heating equipment	424
Fixed outdoor electric deicing and snow-melting equipment	426
Generators	445
Health care facilities	517
Induction and dielectric heating equipment	665
Industrial machinery	670
Luminaires, lampholders, and lamps	410
Motion picture and television studios and similar locations	530
Motors, motor circuits, and controllers	430
Phase converters	455
Pipe organs	650
Receptacles	406
Services	230
Solar photovoltaic systems	690
Switchboards, switchgear, and panelboards	408
Theaters, audience areas of motion picture and television studios, and similar locations	520
Transformers and transformer vaults	450
X-ray equipment	660

against overcurrent in accordance with their ampacities specified in 310.14, unless otherwise permitted or required in 240.4(A) through (H).

> Informational Note: See ICEA P-32-382-2018, *Short Circuit Characteristics of Insulated Cables*, for information on allowable short-circuit currents for insulated copper and aluminum conductors.

(A) Power Loss Hazard. Conductor overload protection shall not be required where the interruption of the circuit would create a hazard, such as in a material-handling magnet circuit or fire pump circuit. Short-circuit protection shall be provided.

> Informational Note: See NFPA 20-2019, *Standard for the Installation of Stationary Pumps for Fire Protection*.

(B) Overcurrent Devices Rated 800 Amperes or Less. The next higher standard overcurrent device rating (above the ampacity of the conductors being protected) shall be permitted to be used, provided all of the following conditions are met:

(1) The conductors being protected are not part of a branch circuit supplying more than one receptacle for cord-and-plug-connected portable loads.

(2) The ampacity of the conductors does not correspond with the standard ampere rating of a fuse or a circuit breaker without overload trip adjustments above its rating (but that shall be permitted to have other trip or rating adjustments).

(3) The next higher standard rating selected does not exceed 800 amperes.

If the overcurrent protective device is an adjustable trip device installed in accordance with 240.4(B)(1), (B)(2), and (B)(3), it shall be permitted to be set to a value that does not exceed the next higher standard value above the ampacity of the conductors being protected as shown in Table 240.6(A) where restricted access in accordance with 240.6(C) is provided.

Table 240.6(A) lists the standard ratings of overcurrent devices. Where the ampacity of the conductor specified in Article 310 does not match the rating of the standard overcurrent device, 240.4(B) permits the use of the next larger standard overcurrent device. All three conditions must be met for this permission to apply. For example, in Table 310.16, the ampacity for 3 AWG, 75°C copper, Type THWN, is listed as 100 amperes. That conductor would have to be protected by an overcurrent protective device (OCPD) rated not more than 100 amperes unless otherwise permitted in 240.4(E), (F), or (G).

Section 240.4(B) does not modify or change the allowable ampacity of the conductor — it serves only to provide a reasonable increase in the permitted OCPD rating where the allowable ampacity and the standard OCPD ratings do not correspond.

For example, a 500-kcmil THWN copper conductor has an allowable ampacity of 380 amperes, specified in Table 310.16. This conductor can supply a load not exceeding 380 amperes and, in accordance with 240.4(B), can be protected by a 400-ampere OCPD.

Application of 310.12 is permitted only for single-phase, 120/240-volt, residential services and feeder conductors that supply the entire load of the dwelling unit. The service and main power feeder loads permitted to be supplied by the conductor types and sizes exceed the conductor ampacities for the same conductor types and sizes specified in Table 310.16. The overcurrent protection for these residential supply conductors is also permitted to be based on the increased rating allowed by 310.12. The increased ratings are based on the significant diversity inherent to most dwelling unit loads and the fact that only the two ungrounded service or feeder conductors are considered to be current carrying.

See also

Example D7 in Annex D for a table of conductor sizes for feeders and services based on the application of the equation in 310.12

(C) Overcurrent Devices Rated over 800 Amperes. Where the overcurrent device is rated over 800 amperes, the ampacity of the conductors it protects shall be equal to or greater than the rating of the overcurrent device defined in 240.6.

(D) Small Conductors. Unless specifically permitted in 240.4(E) or (G), the overcurrent protection shall not exceed that required by 240.4(D)(1) through (D)(8) after any correction factors for ambient temperature and number of conductors have been applied.

(1) 18 AWG Copper. 7 amperes, provided all the following conditions are met:

(1) Continuous loads do not exceed 5.6 amperes.
(2) Overcurrent protection is provided by one of the following:
  a. Branch-circuit-rated circuit breakers listed and marked for use with 18 AWG copper conductor
  b. Branch-circuit-rated fuses listed and marked for use with 18 AWG copper conductor
  c. Class CC, Class CF, Class J, or Class T fuses

(2) 16 AWG Copper. 10 amperes, provided all the following conditions are met:

(1) Continuous loads do not exceed 8 amperes.
(2) Overcurrent protection is provided by one of the following:
  a. Branch-circuit-rated circuit breakers listed and marked for use with 16 AWG copper conductor
  b. Branch-circuit-rated fuses listed and marked for use with 16 AWG copper conductor
  c. Class CC, Class CF, Class J, or Class T fuses

N (3) 14 AWG Copper-Clad Aluminum. 10 amperes, provided all the following conditions are met:

(1) Continuous loads do not exceed 8 amperes
(2) Overcurrent protection is provided by one of the following:
  a. Branch-circuit-rated circuit breakers are listed and marked for use with 14 AWG copper-clad aluminum conductor.
  b. Branch-circuit-rated fuses are listed and marked for use with 14 AWG copper-clad aluminum conductor.

(4) 14 AWG Copper. 15 amperes

(5) 12 AWG Aluminum and Copper-Clad Aluminum. 15 amperes

(6) 12 AWG Copper. 20 amperes

(7) 10 AWG Aluminum and Copper-Clad Aluminum. 25 amperes

(8) 10 AWG Copper. 30 amperes

(E) Tap Conductors. Tap conductors shall be permitted to be protected against overcurrent in accordance with the following:

(1) 210.19(C) and (D), Household Ranges and Cooking Appliances and Other Loads
(2) 240.5(B)(2), Fixture Wire
(3) 240.21, Location in Circuit
(4) 368.17(B), Reduction in Ampacity Size of Busway
(5) 368.17(C), Feeder or Branch Circuits (busway taps)
(6) 430.53(D), Single Motor Taps

(F) Transformer Secondary Conductors. Single-phase (other than 2-wire) and multiphase (other than delta-delta, 3-wire) transformer secondary conductors shall not be considered to be protected by the primary overcurrent protective device. Conductors supplied by the secondary side of a single-phase transformer having a 2-wire (single-voltage) secondary, or a three-phase, delta-delta connected transformer having a 3-wire (single-voltage) secondary, shall be permitted to be protected by overcurrent protection provided on the primary (supply) side of the transformer, provided this protection is in accordance with 450.3 and does not exceed the value determined by multiplying the secondary conductor ampacity by the secondary-to-primary transformer voltage ratio.

(G) Overcurrent Protection for Specific Conductor Applications. Overcurrent protection for the specific conductors shall be permitted to be provided as referenced in Table 240.4(G).

Δ **TABLE 240.4(G)** *Specific Conductor Applications*

Conductor	Article	Section
Air-conditioning and refrigeration equipment circuit conductors	440, Parts III, IV, VI	
Capacitor circuit conductors	460	460.8(B) and 460.25
Control and instrumentation circuit conductors (Type ITC)	335	335.9
Electric welder circuit conductors	630	630.12 and 630.32
Fire alarm system circuit conductors	760	760.43, 760.45, 760.121, and Chapter 9, Tables 12(A) and 12(B)
Motor-operated appliance circuit conductors	422, Part II	
Motor and motor-control circuit conductors	430, Parts II, III, IV, V, VI, VII	
Phase converter supply conductors	455	455.7
Remote-control, signaling, and power-limited circuit conductors	725	724.43, 724.45, 725.60, and Chapter 9, Tables 11(A) and 11(B)
Secondary tie conductors	450	450.6

N (H) Dwelling Unit Service and Feeder Conductors. Dwelling unit service and feeder conductors shall be permitted to be protected against overcurrent at the ampacity values in 310.12.

240.5 Protection of Flexible Cords, Flexible Cables, and Fixture Wires. Flexible cord and flexible cable, including tinsel cord and extension cords, and fixture wires shall be protected against overcurrent by either 240.5(A) or (B).

(A) Ampacities. Flexible cord and flexible cable shall be protected by an overcurrent device in accordance with their ampacity as specified in Table 400.5(A)(1) and Table 400.5(A)(2). Fixture wire shall be protected against overcurrent in accordance with its ampacity as specified in Table 402.5. Supplementary overcurrent protection, as covered in 240.10, shall be permitted to be an acceptable means for providing this protection.

(B) Branch-Circuit Overcurrent Device. Flexible cord shall be protected, where supplied by a branch circuit, in accordance with one of the methods described in 240.5(B)(1), (B)(3), or (B)(4). Fixture wire shall be protected, where supplied by a branch circuit, in accordance with 240.5(B)(2).

(1) Supply Cord of Listed Appliance or Luminaire. Where flexible cord or tinsel cord is approved for and used with a specific listed appliance or luminaire, it shall be considered to be protected when applied within the appliance or luminaire listing requirements. For the purposes of this section, a luminaire may be either portable or permanent.

(2) Fixture Wire. Fixture wire shall be permitted to be tapped to the branch-circuit conductor of a branch circuit in accordance with the following:

(1) 15- or 20-ampere circuits — 18 AWG, up to 15 m (50 ft) of run length
(2) 15- or 20-ampere circuits — 16 AWG, up to 30 m (100 ft) of run length
(3) 20-ampere circuits — 14 AWG and larger
(4) 30-ampere circuits — 14 AWG and larger
(5) 40-ampere circuits — 12 AWG and larger
(6) 50-ampere circuits — 12 AWG and larger

(3) Extension Cord Sets. Flexible cord used in listed extension cord sets shall be considered to be protected when applied within the extension cord listing requirements.

(4) Field Assembled Extension Cord Sets. Flexible cord used in extension cords made with separately listed and installed components shall be permitted to be supplied by a branch circuit in accordance with the following:

20-ampere circuits — 16 AWG and larger

240.6 Standard Ampere Ratings.

Δ **(A) Fuses and Fixed-Trip Circuit Breakers.** The standard ampere ratings for fuses and inverse time circuit breakers shall be considered as shown in Table 240.6(A). Additional standard ampere ratings for fuses shall be 1, 3, 6, and 601. The use of fuses and inverse time circuit breakers with nonstandard ampere ratings shall be permitted.

N **TABLE 240.6(A)** *Standard Ampere Ratings for Fuses and Inverse Time Circuit Breakers*

Standard Ampere Ratings				
10	15	20	25	30
35	40	45	50	60
70	80	90	100	110
125	150	175	200	225
250	300	350	400	450
500	600	700	800	1000
1200	1600	2000	2500	3000
4000	5000	6000	—	—

(B) Adjustable-Trip Circuit Breakers. The rating of adjustable-trip circuit breakers having external means for adjusting the current setting (long-time pickup setting), not meeting the requirements of 240.6(C), shall be the maximum setting possible.

(C) Local Restricted Access Adjustable-Trip Circuit Breakers. A circuit breaker(s) that has restricted access to the adjusting means shall be permitted to have an ampere rating(s) that is equal to the adjusted current setting (long-time pickup setting). Restricted access shall be achieved by one of the following methods:

(1) Located behind removable and sealable covers over the adjusting means
(2) Located behind bolted equipment enclosure doors
(3) Located behind locked doors accessible only to qualified personnel
(4) Password protected, with password accessible only to qualified personnel

Informational Note: See NFPA 730, *Guide for Premises Security*, and ANSI/TIA-5017, *Telecommunications Physical Network Security Standard*, for information regarding physical security.

The set long-time pickup rating of an adjustable-trip circuit breaker is permitted to be considered the circuit-breaker rating where access to the adjustment means is limited. Access limitation can be provided by locating the adjustment means behind sealable covers, as shown in Exhibit 240.1. The purpose of limiting access to the adjustment prevents tampering or readjustment by unqualified personnel.

N **(D) Remotely Accessible Adjustable-Trip Circuit Breakers.** A circuit breaker(s) that can be adjusted remotely to modify the adjusting means shall be permitted to have an ampere rating(s) that is equal to the adjusted current setting (long-time pickup setting). Remote access shall be achieved by one of the following methods:

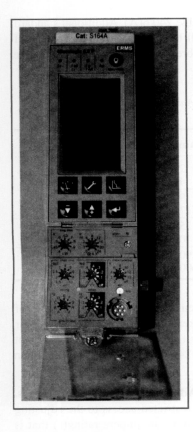

EXHIBIT 240.1 An adjustable-trip circuit breaker with a transparent, removable, and sealable cover. *(Courtesy of Square D by Schneider Electric)*

(1) Connected directly through a local nonnetworked interface.
(2) Connected through a networked interface complying with one of the following methods:

   a. The circuit breaker and associated software for adjusting the settings are identified as being evaluated for cybersecurity.

   b. A cybersecurity assessment of the network is completed. Documentation of the assessment and certification shall be made available to those authorized to inspect, operate, and maintain the system.

Informational Note No. 1: See ANSI/ISA 62443, *Cybersecurity Standards series*, UL 2900 *Cybersecurity Standard series*, or the NIST *Framework for Improving Critical Infrastructure Cybersecurity*, Version 1.1 for assessment requirements.

Informational Note No. 2: Examples of the commissioning certification used to demonstrate the system has been investigated for cybersecurity vulnerabilities could be one of the following:

(1) The ISA Security Compliance Institute (ISCI) conformity assessment program
(2) Certification of compliance by a nationally recognized test laboratory
(3) Manufacturer certification for the specific type and brand of system provided

Informational Note No. 3: Cybersecurity is a specialized field requiring constant, vigilant attention to security vulnerabilities that could arise due to software defects, system configuration

changes, or user interactions. Installation of devices that can be secured is an important first step but not sufficient to guarantee a secure system.

N 240.7 Listing Requirements. The following shall be listed:

(1) Branch-circuit overcurrent protective devices
(2) Relays and circuit breakers providing ground-fault protection of equipment
(3) Ground-fault circuit interrupter devices

240.8 Fuses or Circuit Breakers in Parallel. Fuses and circuit breakers shall be permitted to be connected in parallel where they are factory assembled in parallel and listed as a unit. Individual fuses, circuit breakers, or combinations thereof shall not otherwise be connected in parallel.

240.9 Thermal Devices. Thermal relays and other devices not designed to open short circuits or ground faults shall not be used for the protection of conductors against overcurrent due to short circuits or ground faults, but the use of such devices shall be permitted to protect motor branch-circuit conductors from overload if protected in accordance with 430.40.

240.10 Supplementary Overcurrent Protection. Where supplementary overcurrent protection is used for luminaires, appliances, and other equipment or for internal circuits and components of equipment, it shall not be used as a substitute for required branch-circuit overcurrent devices or in place of the required branch-circuit protection. Supplementary overcurrent devices shall not be required to be readily accessible.

N 240.11 Selective Coordination. If one or more feeder overcurrent protective devices are required to be selectively coordinated with a service overcurrent protective device by other requirements in this *Code*, all feeder overcurrent protective devices supplied directly by the service overcurrent protective device shall be selectively coordinated with the service overcurrent protective device.

240.12 Orderly Shutdown. Where an orderly shutdown is required to minimize the hazard(s) to personnel or equipment, a system of coordination based on the following two conditions shall be permitted:

(1) Coordinated short-circuit protection
(2) Overload indication based on monitoring systems or devices

Informational Note: The monitoring system may cause the condition to go to alarm, allowing corrective action or an orderly shutdown, thereby minimizing personnel hazard and equipment damage.

With coordinated overcurrent protection, the faulted or overloaded circuit is isolated by the selective operation of only the OCPD closest to the overcurrent condition. This selective

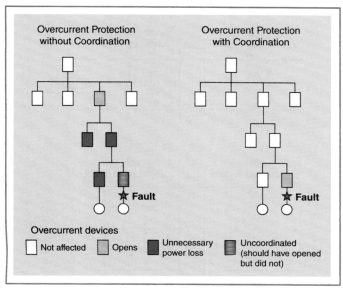

Overcurrent Protection without Coordination

Overcurrent Protection with Coordination

★ Fault

★ Fault

Overcurrent devices

☐ Not affected ▨ Opens ■ Unnecessary power loss ▨ Uncoordinated (should have opened but did not)

EXHIBIT 240.2 Overcurrent protection schemes without system coordination and with system coordination.

operation prevents power loss to unaffected loads. Coordinated short-circuit protection automatically opens the circuit by localizing and de-energizing the faulted portion of the circuit, but an overload condition is not required to result in automatic opening of a protective device. Instead, an alarm can be used to warn of the overload condition, and remedial action can be taken.

In some circumstances, an orderly shutdown of a system or process is more critical to personnel and equipment safety than is the automatic operation of the OCPD in response to an overload. Examples of overcurrent protection with and without coordinated protection are illustrated in Exhibit 240.2.

See also

620.62, 645.27, 695.3(C)(3), 700.32, 701.32, and **708.54** for selective coordination requirements for specific systems or equipment

240.13 Ground-Fault Protection of Equipment. Ground-fault protection of equipment shall be provided in accordance with 230.95 for solidly grounded wye electrical systems of more than 150 volts to ground but not exceeding 1000 volts phase-to-phase for each individual device used as a building or structure main disconnecting means rated 1000 amperes or more.

This section shall not apply to the disconnecting means for the following:

(1) Continuous industrial processes where a nonorderly shutdown will introduce additional or increased hazards
(2) Installations where ground-fault protection is provided by other requirements for services or feeders
(3) Fire pumps

Section 240.13 extends the requirement of 230.95 to building disconnects, regardless of whether they are classified as service

disconnects, building disconnects for feeders, or even branch circuits.

Where ground-fault protection for equipment is installed at the service equipment, and feeders or branch circuits are installed from that service to supply other buildings or structures, the disconnecting means at any subsequent building is not required to be provided with ground-fault protection if the service device provides the required protection. Installations performed prior to the 2008 edition of the *NEC*® permitted "re-grounding" of the grounded conductor at the separate building if an equipment grounding conductor was not included with the supply circuit. Where re-grounding of the neutral occurs downstream from the service, the re-grounding could nullify the ground-fault protection (or result in unwanted operation of the protection) because the neutral current has parallel paths on which to return to the source. Feeders or branch circuits supplying other buildings or structures might have to be isolated to allow for proper operation of the service ground-fault protection, and separate ground-fault protection installed at the building disconnecting means is then necessary to meet the requirements of 240.13.

See also

210.13, 215.10, and **Article 225, Part II,** for the requirements for building disconnects not on the utility service

240.15 Ungrounded Conductors.

(A) Overcurrent Device Required. A fuse or an overcurrent trip unit of a circuit breaker shall be connected in series with each ungrounded conductor. A combination of a current transformer and overcurrent relay shall be considered equivalent to an overcurrent trip unit.

Informational Note: See Parts III, IV, V, and XI of Article 430 for motor circuits.

(B) Circuit Breaker as Overcurrent Device. Circuit breakers shall open all ungrounded conductors of the circuit both manually and automatically unless otherwise permitted in 240.15(B)(1), (B)(2), (B)(3), and (B)(4).

(1) Multiwire Branch Circuits. Individual single-pole circuit breakers, with identified handle ties, shall be permitted as the protection for each ungrounded conductor of multiwire branch circuits that serve only single-phase line-to-neutral loads.

(2) Grounded Single-Phase Alternating-Current Circuits. In grounded systems, individual single-pole circuit breakers rated 120/240 volts ac, with identified handle ties, shall be permitted as the protection for each ungrounded conductor for line-to-line connected loads for single-phase circuits.

(3) 3-Phase and 2-Phase Systems. For line-to-line loads in 4-wire, 3-phase systems or 5-wire, 2-phase systems, individual single-pole circuit breakers rated 120/240 volts ac with identified handle ties shall be permitted as the protection for each ungrounded conductor, if the systems have a grounded neutral point and the voltage to ground does not exceed 120 volts.

(4) 3-Wire Direct-Current Circuits. Individual single-pole circuit breakers rated 125/250 volts dc with identified handle ties shall be permitted as the protection for each ungrounded conductor for line-to-line connected loads for 3-wire, direct-current circuits supplied from a system with a grounded neutral where the voltage to ground does not exceed 125 volts.

N 240.16 Interrupting Ratings. Branch-circuit overcurrent protective devices shall have an interrupting rating no less than 5000 amperes.

The basic rule is that circuit breakers open all ungrounded conductors of the circuit when they trip (automatic operation in response to overcurrent) or are manually operated as a disconnecting means. For 2-wire circuits with one conductor grounded, this rule is simple and needs no further explanation. For branch circuits comprised of multipole ungrounded conductors that do not meet the conditions of B(1) through B(4), multipole common trip circuit breakers are required. Exhibit 240.3 shows where multiple common trip circuit breakers are required. Handle ties are not permitted because the circuits are supplied from ungrounded systems.

Multiwire branch circuits are permitted to supply line-to-line connected loads where the loads are associated with a single piece of utilization equipment or where all the ungrounded conductors are opened simultaneously by the branch-circuit overcurrent device (automatic opening in response to overcurrent).

The most widely used method is to install a multipole circuit breaker with an internal common trip mechanism. The use of such multipole devices provides overcurrent protection and the required disconnecting means. The breaker is operated by an external single lever internally attached to the two or three poles of the circuit breaker, or the external lever can be attached to multiple handles operated as one.

The second method permitted for multiwire branch circuits is to use two or three single-pole circuit breakers and add an identified handle tie to function as a common operating handle. This multipole circuit breaker is field assembled by externally attaching an identified common lever (handle tie) onto the two or three individual circuit breakers. The circuit breakers can be used as the protection of line-to-line connected loads in grounded single-phase and grounded 3-phase alternating current systems, but this requirement is limited to circuit breakers rated 120/240 volts. Handle ties do not cause the circuit breaker to function as a common trip device; rather, they only allow common operation as a disconnecting means. The term *identified* requires the use of hardware designed specifically to perform this common disconnecting means function. The use of homemade hardware such as nails or pieces of wire to perform this function is not an identified means.

Exhibit 240.4 shows where handle ties are required for single-pole circuit breakers on multiwire branch circuits serving line-to-neutral loads. Exhibit 240.5 shows where handle ties are required for single-pole circuit breakers on multiwire branch circuits serving line-to-line loads. The multiwire branch circuits shown in the top and middle diagrams supply a single piece of utilization

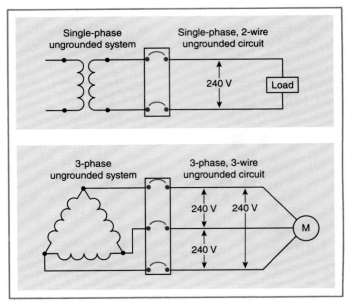

EXHIBIT 240.3 Examples of circuits that require multipole common trip circuit breakers.

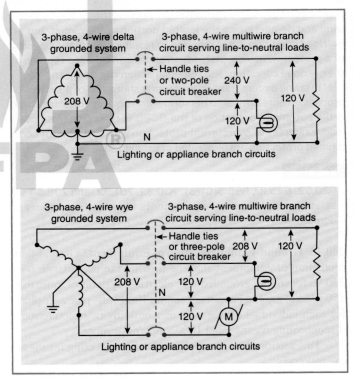

EXHIBIT 240.4 Examples of circuits in which single-pole circuit breakers are permitted because they open the ungrounded conductor of the circuit.

equipment in accordance with 210.4(C), Exception No. 1. For those branch circuits, common trip operation is permitted but is not required.

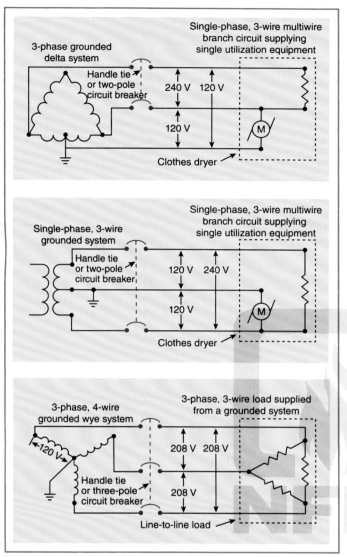

EXHIBIT 240.5 Examples of circuits in which identified handle ties are permitted to provide the simultaneous disconnecting function.

Section 240.15(B)(4) covers protection of direct-current circuits and permits the use of two single-pole circuit breakers (rated for dc application) with identified handle ties to be used for the overcurrent protection of line-to-line connected loads. The 3-wire circuits covered by this requirement are *multiwire branch circuits* per the definition in Article 100 and as such are subject to all of the requirements specified in 210.4.

Part II. Location

240.21 Location in Circuit. Overcurrent protection shall be provided in each ungrounded circuit conductor and shall be located at the point where the conductors receive their supply except as specified in 240.21(A) through (H). Conductors supplied under 240.21(A) through (H) shall not supply another

conductor except through an overcurrent protective device meeting the requirements of 240.4.

(A) Branch-Circuit Conductors. Branch-circuit tap conductors meeting the requirements specified in 210.19 shall be permitted to have overcurrent protection as specified in 210.20.

(B) Feeder Taps. Conductors shall be permitted to be tapped, without overcurrent protection at the tap, to a feeder as specified in 240.21(B)(1) through (B)(5). The tap shall be permitted at any point on the load side of the feeder overcurrent protective device. Section 240.4(B) shall not be permitted for tap conductors.

An OCPD is not permitted to be supplied by a tap conductor having an ampacity less than its rating. For example, the use of a 500-kcmil THWN copper conductor (380 amperes, per Table 310.16) as a tap conductor to supply a 400-ampere rated device is not permitted.

Note that the lengths specified in 240.21(B) and (C) apply to the conductors, not to a raceway enclosing the conductors or to the distance between the enclosures in which the tap conductors originate and terminate.

Δ **(1) Taps Not over 3 m (10 ft) Long.** If the length of the tap conductors does not exceed 3 m (10 ft) and the tap conductors comply with all of the following:

(1) The ampacity of the tap conductors is as follows:

    a. Not less than the combined calculated loads on the circuits supplied by the tap conductors

    b. Not less than the rating of the equipment containing an overcurrent device(s) supplied by the tap conductors or not less than the rating of the overcurrent protective device at the termination of the tap conductors

Exception to b: Where listed equipment, such as a surge-protective device(s) [SPD(s)], is provided with specific instructions on minimum conductor sizing, the ampacity of the tap conductors supplying that equipment shall be permitted to be determined based on the manufacturer's instructions.

(2) The tap conductors do not extend beyond the switchboard, switchgear, panelboard, disconnecting means, or control devices they supply.

(3) Except at the point of connection to the feeder, the tap conductors are enclosed in a raceway, which extends from the tap to the enclosure of an enclosed switchboard, switchgear, a panelboard, or control devices, or to the back of an open switchboard.

(4) For field installations, if the tap conductors leave the enclosure or vault in which the tap is made, the ampacity of the tap conductors is not less than one-tenth of the rating of the overcurrent device protecting the feeder conductors.

Informational Note: See 408.36 for overcurrent protection requirements for panelboards.

(2) Taps Not over 7.5 m (25 ft) Long. Where the length of the tap conductors does not exceed 7.5 m (25 ft) and the tap conductors comply with all the following:

(1) The ampacity of the tap conductors is not less than one-third of the rating of the overcurrent device protecting the feeder conductors.

(2) The tap conductors terminate in a single circuit breaker or a single set of fuses that limit the load to the ampacity of the tap conductors. This device shall be permitted to supply any number of additional overcurrent devices on its load side.

(3) The tap conductors are protected from physical damage by being enclosed in an approved raceway or by other approved means.

Exhibit 240.6 illustrates an installation that meets the feeder tap requirements of 240.21(B)(2). The ampacity of the 3/0 AWG, Type THW copper conductor (200 amperes) is more than one-third the rating of the overcurrent device (400 amperes) protecting the feeder circuit.

> **See also**
>
> **Table 310.16** for the ampacity of copper conductors in a raceway

(3) Taps Supplying a Transformer [Primary Plus Secondary Not over 7.5 m (25 ft) Long]. Where the tap conductors supply a transformer and comply with all the following conditions:

(1) The conductors supplying the primary of a transformer have an ampacity at least one-third the rating of the overcurrent device protecting the feeder conductors.

(2) The conductors supplied by the secondary of the transformer shall have an ampacity that is not less than the value of the primary-to-secondary voltage ratio multiplied by one-third of the rating of the overcurrent device protecting the feeder conductors.

(3) The total length of one primary plus one secondary conductor, excluding any portion of the primary conductor that is protected at its ampacity, is not over 7.5 m (25 ft).

(4) The primary and secondary conductors are protected from physical damage by being enclosed in an approved raceway or by other approved means.

(5) The secondary conductors terminate in a single circuit breaker or set of fuses that limit the load current to not more than the conductor ampacity that is permitted by 310.14.

This section covers applications where the conductor length of 25 feet is applied using the length of one primary conductor plus the length of one secondary conductor for the measurement. The transformer primary conductors are tapped from a feeder, and the secondary conductors are required to terminate in a single OCPD. Exhibit 240.7 illustrates the conditions of 240.21(B)(3)(1) through (5). The overcurrent protection requirements of 408.36 for panelboards and 450.3(B) for transformers also apply.

> **See also**
>
> **240.21(C)(6)** if the primary conductors are protected in accordance with their ampacity

(4) Taps over 7.5 m (25 ft) Long. Where the feeder is in a high bay manufacturing building over 11 m (35 ft) high at walls and the installation complies with all the following conditions:

(1) Conditions of maintenance and supervision ensure that only qualified persons service the systems.

(2) The tap conductors are not over 7.5 m (25 ft) long horizontally and not over 30 m (100 ft) total length.

(3) The ampacity of the tap conductors is not less than one-third the rating of the overcurrent device protecting the feeder conductors.

(4) The tap conductors terminate at a single circuit breaker or a single set of fuses that limit the load to the ampacity of the tap conductors. This single overcurrent device shall be permitted to supply any number of additional overcurrent devices on its load side.

(5) The tap conductors are protected from physical damage by being enclosed in an approved raceway or by other approved means.

(6) The tap conductors are continuous from end-to-end and contain no splices.

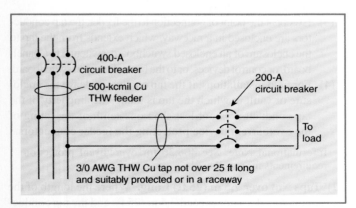

EXHIBIT 240.6 An example in which the feeder taps terminate in a single circuit breaker.

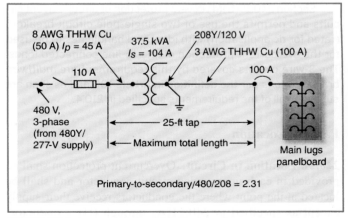

EXHIBIT 240.7 An example in which the transformer feeder taps (primary plus secondary) are not over 25 feet long.

(7) The tap conductors are sized 6 AWG copper or 4 AWG aluminum or larger.
(8) The tap conductors do not penetrate walls, floors, or ceilings.
(9) The tap is made no less than 9 m (30 ft) from the floor.

Exhibit 240.8 illustrates an installation complying with 240.21(B)(4).

(5) Outside Taps of Unlimited Length. Where the conductors are located outside of a building or structure, except at the point of load termination, and comply with all of the following conditions:

(1) The tap conductors are protected from physical damage in an approved manner.
(2) The tap conductors terminate at a single circuit breaker or a single set of fuses that limits the load to the ampacity of the tap conductors. This single overcurrent device shall be permitted to supply any number of additional overcurrent devices on its load side.
(3) The overcurrent device for the tap conductors is an integral part of a disconnecting means or shall be located immediately adjacent thereto.
(4) The disconnecting means for the tap conductors is installed at a readily accessible location complying with one of the following:
  a. Outside of a building or structure
  b. Inside, nearest the point of entrance of the tap conductors
  c. Where installed in accordance with 230.6, nearest the point of entrance of the tap conductors

This tap conductor requirement is similar in some respects to an installation of service conductors. The conductors are supplied from a feeder at an outdoor location and run to a building or structure without limitations on the tap conductor length. The OCPD provides overload protection for the tap conductors. The fused switch or circuit breaker is installed at a readily accessible location either inside or outside a building or structure at a point nearest to where the tap conductors enter the building or structure. This disconnect is subject to the applicable requirements covering feeder disconnecting means in Part II of Article 225.

Δ **(C) Transformer Secondary Conductors.** A set of conductors feeding a single load, or each set of conductors feeding separate loads, shall be permitted to be connected to a transformer secondary, without overcurrent protection at the secondary, as specified in 240.21(C)(1) through (C)(6). Section 240.4(B) shall not be permitted for transformer secondary conductors.

> Informational Note: See 450.3 for overcurrent protection requirements for transformers.

Section 240.21(C) prohibits using the next higher standard size OCPD to protect transformer secondary conductors.

The secondary terminals of a transformer are permitted to supply one or more than one set of secondary conductors. For example, the secondary terminals could supply two separate sets of secondary conductors that feed two panelboards. One set of conductors could be installed using the 25-foot secondary conductor rule of 240.21(C)(6), while the other set of conductors could be installed using the 10-foot secondary conductor rule of 240.21(C)(2). Each set is treated individually in applying the applicable secondary conductor requirement.

The *NEC* requires the protection of both conductors and transformers. Article 240 contains requirements for protection of conductors, and Article 450 contains requirements for the protection of transformers. It is possible to protect both conductors and transformers with the same device, if the device meets the requirements of both articles.

(1) Protection by Primary Overcurrent Device. Conductors supplied by the secondary side of a single-phase transformer having a 2-wire (single-voltage) secondary, or a three-phase, delta-delta connected transformer having a 3-wire (single-voltage) secondary, shall be permitted to be protected by overcurrent protection provided on the primary (supply) side of the transformer, provided this protection is in accordance with 450.3 and does not exceed the value determined by multiplying the secondary conductor ampacity by the secondary-to-primary transformer voltage ratio.

Single-phase (other than 2-wire) and multiphase (other than delta-delta, 3-wire) transformer secondary conductors are not considered to be protected by the primary overcurrent protective device.

Δ **(2) Transformer Secondary Conductors Not over 3 m (10 ft) Long.** If the length of secondary conductor does not exceed 3 m (10 ft) and complies with all of the following:

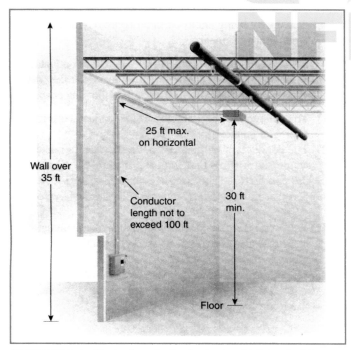

Wall over 35 ft

25 ft max. on horizontal

Conductor length not to exceed 100 ft

30 ft min.

Floor

EXHIBIT 240.8 An illustration of a feeder tap in a high bay building.

(1) The ampacity of the secondary conductors is as follows:
   a. Not less than the combined calculated loads on the circuits supplied by the secondary conductors
   b. Not less than the rating of the equipment containing an overcurrent device(s) supplied by the secondary conductors or not less than the rating of the overcurrent protective device at the termination of the secondary conductors

Exception: Where listed equipment, such as a surge protective device(s) [SPD(s)], is provided with specific instructions on minimum conductor sizing, the ampacity of the tap conductors supplying that equipment shall be permitted to be determined based on the manufacturer's instructions.

(2) The secondary conductors do not extend beyond the switchboard, switchgear, panelboard, disconnecting means, or control devices they supply.

(3) The secondary conductors are enclosed in a raceway, which shall extend from the transformer to the enclosure of an enclosed switchboard, switchgear, a panelboard, or control devices or to the back of an open switchboard.

(4) For field installations where the secondary conductors leave the enclosure or vault in which the supply connection is made, the secondary conductors shall have an ampacity that is not less than the value of the primary-to-secondary voltage ratio multiplied by one-tenth of the rating of the overcurrent device protecting the primary of the transformer.

> Informational Note: See 408.36 for overcurrent protection requirements for panelboards.

The size-rating relationship between the primary OCPD and the secondary conductors is necessary because the transformer primary device provides short-circuit ground-fault protection for the transformer secondary conductors. The ampacity of the conductors must be adequate for the calculated load and not less than the rating of the device or OCPD in which the conductors terminate. The following example illustrates application of this requirement.

Calculation Example

Apply the 10-ft secondary conductor protection criteria of 240.21(C)(2) to a transformer rated 75 kVA, 3-phase, 480 V primary to 208Y/120 V secondary. The transformer primary OCPD is rated 125 A. Determine the minimum secondary conductor size for the installation of one of the sets of secondary conductors.

Solution

Step 1. Determine $1/10$ of the primary OCPD rating using the following calculation:

$$125 \text{ A} \div 10 = 12.5 \text{ A}$$

Step 2. Determine the line-to-line primary-to-secondary voltage ratio:

$$\left(\frac{480}{208}\right) = 2.31$$

Step 3. Determine the minimum ampacity for ungrounded transformer secondary conductor:

$$12.5 \text{ A} \times 2.31 = 29 \text{ A} \rightarrow 10 \text{ AWG copper THWN}$$
(Table 310.16): 30 A from 60°C column

A 10 AWG copper conductor is permitted to be supplied from the secondary of this transformer with primary overcurrent protection rated 125 A. The load supplied by this secondary conductor cannot exceed the conductor's allowable ampacity from Table 310.16 coordinated with the temperature rating of the conductor terminations in accordance with 110.14(C)(1)(a).

---

(3) Industrial Installation Secondary Conductors Not over 7.5 m (25 ft) Long. For the supply of switchgear or switchboards in industrial installations only, where the length of the secondary conductors does not exceed 7.5 m (25 ft) and complies with all of the following:

(1) Conditions of maintenance and supervision ensure that only qualified persons service the systems.
(2) The ampacity of the secondary conductors is not less than the secondary current rating of the transformer, and the sum of the ratings of the overcurrent devices does not exceed the ampacity of the secondary conductors.
(3) All overcurrent devices are grouped.
(4) The secondary conductors are protected from physical damage by being enclosed in an approved raceway or by other approved means.

(4) Outside Secondary Conductors. Where the conductors are located outside of a building or structure, except at the point of load termination, and comply with all of the following conditions:

(1) The conductors are protected from physical damage in an approved manner.
(2) The conductors terminate at a single circuit breaker or a single set of fuses that limit the load to the ampacity of the conductors. This single overcurrent device shall be permitted to supply any number of additional overcurrent devices on its load side.
(3) The overcurrent device for the conductors is an integral part of a disconnecting means or shall be located immediately adjacent thereto.
(4) The disconnecting means for the conductors is installed at a readily accessible location complying with one of the following:

   a. Outside of a building or structure
   b. Inside, nearest the point of entrance of the conductors

c. Where installed in accordance with 230.6, nearest the point of entrance of the conductors

(5) Secondary Conductors from a Feeder Tapped Transformer. Transformer secondary conductors installed in accordance with 240.21(B)(3) shall be permitted to have overcurrent protection as specified in that section.

(6) Secondary Conductors Not over 7.5 m (25 ft) Long. Where the length of secondary conductor does not exceed 7.5 m (25 ft) and complies with all of the following:

(1) The secondary conductors shall have an ampacity that is not less than the value of the primary-to-secondary voltage ratio multiplied by one-third of the rating of the overcurrent device protecting the primary of the transformer.
(2) The secondary conductors terminate in a single circuit breaker or set of fuses that limit the load current to not more than the conductor ampacity that is permitted by 310.14.
(3) The secondary conductors are protected from physical damage by being enclosed in an approved raceway or by other approved means.

(D) Service Conductors. Service conductors shall be permitted to be protected by overcurrent devices in accordance with 230.91.

(E) Busway Taps. Busways and busway taps shall be permitted to be protected against overcurrent in accordance with 368.17.

(F) Motor Circuit Taps. Motor-feeder and branch-circuit conductors shall be permitted to be protected against overcurrent in accordance with 430.28 and 430.53, respectively.

(G) Conductors from Generator Terminals. Conductors from generator terminals that meet the size requirement in 445.13 shall be permitted to be protected against overload by the generator overload protective device(s) required by 445.12.

(H) Battery Conductors. Overcurrent protection shall be permitted to be installed as close as practicable to the storage battery terminals in an unclassified location. Installation of the overcurrent protection within a hazardous (classified) location shall also be permitted.

240.22 Grounded Conductor. No overcurrent device shall be connected in series with any conductor that is intentionally grounded, unless one of the following two conditions is met:

(1) The overcurrent device opens all conductors of the circuit, including the grounded conductor, and is designed so that no pole can operate independently.
(2) Where required by 430.36 or 430.37 for motor overload protection.

240.24 Location in or on Premises.

(A) Accessibility. Circuit breakers and switches containing fuses shall be readily accessible and installed so that the center of the grip of the operating handle of the switch or circuit breaker,

when in its highest position, is not more than 2.0 m (6 ft 7 in.) above the floor or working platform, unless one of the following applies:

(1) For busways, as provided in 368.17(C).
(2) For supplementary overcurrent protection, as described in 240.10.
(3) For overcurrent protective devices, as described in 225.40 and 230.92.
(4) For overcurrent protective devices adjacent to utilization equipment that they supply, access shall be permitted to be by portable means.

Exception: The use of a tool shall be permitted to access overcurrent protective devices located within listed industrial control panels, within enclosures designed for hazardous (classified) locations or enclosures to protect against environmental conditions. An enclosure within the scope of this exception, and all overcurrent protective device(s) within such enclosures as judged with the enclosure open, shall comply with the accessibility provisions of 240.24(A).

For the purposes of this requirement, ready access to the operating handle of a fusible switch or circuit breaker is considered to be not more than 6 feet 7 inches above the finished floor or working platform. The measurement is made from the center of the device operating handle where the handle is at its highest position. This text parallels the requirement of 404.8(A), which applies to all switches and circuit breakers used as switches.

Overcurrent protection devices may be necessary in locations that are not readily accessible, such as above suspended ceilings. Overcurrent devices are permitted to be located so that they are not readily accessible, as long as they are located next to the appliance, motor, or other equipment they supply and can be reached by using a ladder.

See also

240.10 for information regarding the accessibility of supplementary overcurrent devices

(B) Occupancy. Each occupant shall have ready access to all overcurrent devices protecting the conductors supplying that occupancy, unless otherwise permitted in 240.24(B)(1) and (B)(2).

(1) Service and Feeder Overcurrent Protective Devices. Where electric service and electrical maintenance are provided by the building management and where these are under continuous building management supervision, the service overcurrent protective devices and feeder overcurrent protective devices supplying more than one occupancy shall be permitted to be accessible only to authorized management personnel in the following:

(1) Multiple-occupancy buildings
(2) Guest rooms or guest suites

(2) Branch-Circuit Overcurrent Protective Devices. Where electric service and electrical maintenance are provided by the building management and where these are under continuous building management supervision, the branch-circuit overcurrent protective devices supplying any guest rooms, guest suites, or sleeping rooms in dormitory units without permanent provisions for cooking shall be permitted to be accessible only to authorized management personnel.

(C) Not Exposed to Physical Damage. Overcurrent protective devices shall be located where they will not be exposed to physical damage.

> Informational Note: See 110.11 for information on deteriorating agents that could cause physical damage.

(D) Not in Vicinity of Easily Ignitible Material. Overcurrent protective devices shall not be located in the vicinity of easily ignitible material, such as in clothes closets.

(E) Not Located in Bathrooms. Overcurrent protective devices, other than supplementary overcurrent protection, shall not be located in bathrooms, showering facilities, or locker rooms with showering facilities.

(F) Not Located over Steps. Overcurrent protective devices shall not be located over steps of a stairway.

Part III. Enclosures

240.30 General.

(A) Protection from Physical Damage. Overcurrent devices shall be protected from physical damage by one of the following:

(1) Installation in enclosures, cabinets, cutout boxes, or equipment assemblies
(2) Mounting on open-type switchboards, panelboards, or control boards that are in rooms or enclosures free from dampness and easily ignitible material and are accessible only to qualified personnel

(B) Operating Handle. The operating handle of a circuit breaker shall be permitted to be accessible without opening a door or cover.

240.32 Damp or Wet Locations.
Enclosures for overcurrent devices in damp or wet locations shall comply with 312.2.

240.33 Vertical Position.
Enclosures for overcurrent devices shall be mounted in a vertical position. Circuit breaker enclosures shall be permitted to be installed horizontally where the circuit breaker is installed in accordance with 240.81. Listed busway plug-in units shall be permitted to be mounted in orientations corresponding to the busway mounting position.

A wall-mounted vertical position for enclosures for overcurrent devices affords easier access, natural hand operation, normal swinging or closing of doors or covers, and legibility of the manufacturer's markings. Compliance with the requirement that the up position of the handle is on or closed, and the down position of the handle is off or open, in accordance with 240.81, limits the number of pole spaces available in a panelboard where the cabinet is mounted in a horizontal position. Because of the orientation of the buses in the panelboard, many of the circuit breakers would have their "on" and "off" operation upside down, which would not comply with 240.81.

Part IV. Disconnecting and Guarding

240.40 Disconnecting Means for Fuses.
Cartridge fuses in circuits of any voltage, and all fuses in circuits over 150 volts to ground, shall be provided with a disconnecting means on their supply side so that each circuit containing fuses can be independently disconnected from the source of power. A cable limiter without a disconnecting means shall be permitted on the supply side of the service disconnecting means as permitted by 230.82. A single disconnecting means shall be permitted on the supply side of more than one set of fuses as permitted by 430.112, Exception, for group operation of motors, 424.22(C) for fixed electric space-heating equipment, and 425.22(C) for fixed resistance and electrode industrial process heating equipment, or where specifically permitted elsewhere in this *Code*.

240.41 Arcing or Suddenly Moving Parts.
Arcing or suddenly moving parts shall comply with 240.41(A) and (B).

(A) Location. Fuses and circuit breakers shall be located or shielded so that persons will not be burned or otherwise injured by their operation.

(B) Suddenly Moving Parts. Handles or levers of circuit breakers, and similar parts that may move suddenly in such a way that persons in the vicinity are likely to be injured by being struck by them, shall be guarded or isolated.

Part V. Plug Fuses, Fuseholders, and Adapters

240.50 General.

(A) Maximum Voltage. Plug fuses shall be permitted to be used in the following circuits:

(1) Circuits not exceeding 125 volts between conductors
(2) Circuits supplied by a system having a grounded neutral point where the line-to-neutral voltage does not exceed 150 volts

(B) Marking. Each fuse, fuseholder, and adapter shall be marked with its ampere rating.

(C) Hexagonal Configuration. Plug fuses of 15-ampere and lower rating shall be identified by a hexagonal configuration of the window, cap, or other prominent part to distinguish them from fuses of higher ampere ratings.

The 10- and 15-ampere plug fuses shown in Exhibit 240.9 are equipped with the hexagonal window required by 240.50(C), where the T-type 25-ampere plug fuse is not. The 10-ampere fuse

EXHIBIT 240.9 Two plug fuses and a Type S fuse. *(Courtesy of Eaton, Bussmann Division)*

is a Type S plug fuse, which cannot be interchanged with a higher rated fuse when used with its corresponding adapter.

(D) No Energized Parts. Plug fuses, fuseholders, and adapters shall have no exposed energized parts after fuses or fuses and adapters have been installed.

(E) Screw Shell. The screw shell of a plug-type fuseholder shall be connected to the load side of the circuit.

240.51 Edison-Base Fuses.

(A) Classification. Plug fuses of the Edison-base type shall be classified at not over 125 volts and 30 amperes and below.

(B) Replacement Only. Plug fuses of the Edison-base type shall be used only for replacements in existing installations where there is no evidence of overfusing or tampering.

240.52 Edison-Base Fuseholders. Fuseholders of the Edison-base type shall be installed only where they are made to accept Type S fuses by the use of adapters.

240.53 Type S Fuses. Type S fuses shall be of the plug type and shall comply with 240.53(A) and (B).

(A) Classification. Type S fuses shall be classified at not over 125 volts and 0 to 15 amperes, 16 to 20 amperes, and 21 to 30 amperes.

(B) Noninterchangeable. Type S fuses of an ampere classification as specified in 240.53(A) shall not be interchangeable with a lower ampere classification. They shall be designed so that they cannot be used in any fuseholder other than a Type S fuseholder or a fuseholder with a Type S adapter inserted.

240.54 Type S Fuses, Adapters, and Fuseholders.

(A) To Fit Edison-Base Fuseholders. Type S adapters shall fit Edison-base fuseholders.

EXHIBIT 240.10 Type S nonrenewable plug fuse and fuse adapter. *(Courtesy of Eaton, Bussmann Division)*

(B) To Fit Type S Fuses Only. Type S fuseholders and adapters shall be designed so that either the fuseholder itself or the fuseholder with a Type S adapter inserted cannot be used for any fuse other than a Type S fuse.

(C) Nonremovable. Type S adapters shall be designed so that once inserted in a fuseholder, they cannot be removed.

(D) Nontamperable. Type S fuses, fuseholders, and adapters shall be designed so that tampering or shunting (bridging) would be difficult.

(E) Interchangeability. Dimensions of Type S fuses, fuseholders, and adapters shall be standardized to permit interchangeability regardless of the manufacturer.

Exhibit 240.10 shows a Type S nonrenewable plug fuse adapter designed to prevent installation of a fuse exceeding the ampacity of the conductor being protected. The adapter will not accept Type S plug fuses having an ampere rating other than that for which it is specifically designed. Type S fuses have various size screw-in and thread dimensions for various ampere ratings.

Part VI. Cartridge Fuses and Fuseholders

240.60 General.

(A) Maximum Voltage — 300-Volt Type. Cartridge fuses and fuseholders of the 300-volt type shall be permitted to be used in the following circuits:

(1) Circuits not exceeding 300 volts between conductors
(2) Single-phase line-to-neutral circuits supplied from a 3-phase, 4-wire, solidly grounded neutral source where the line-to-neutral voltage does not exceed 300 volts

(B) Noninterchangeable — 0–6000-Ampere Cartridge Fuseholders. Fuseholders shall be designed so that it will be difficult to put a fuse of any given class into a fuseholder that is designed for a current lower, or voltage higher, than that of the class to which the fuse belongs. Fuseholders for current-limiting fuses shall not permit insertion of fuses that are not current-limiting.

See Exhibit 240.11 for an example of a current-limiting fuse and fuseholder with rejection features.

(C) Marking. Fuses shall be plainly marked, either by printing on the fuse barrel or by a label attached to the barrel showing the following:

(1) Ampere rating
(2) Voltage rating
(3) Interrupting rating where other than 10,000 amperes
(4) Current limiting where applicable
(5) The name or trademark of the manufacturer

The interrupting rating shall not be required to be marked on fuses used for supplementary protection.

Exhibit 240.12 shows two examples of Class G fuses rated 300 volts that bear the markings required by 240.60(C). Class H-type cartridge fuses have an interrupting capacity (IC) rating of 10,000 amperes, which does not need to be marked on the fuse. However, Class CC, G, J, K, L, R, and T cartridge fuses exceed the 10,000-ampere IC rating and must be marked with the IC rating.

EXHIBIT 240.11 A fuse block rejection pin and notch in the fuse blade (red arrows) ensure that circuit protection is provided by fuses with a specific level of overcurrent protection. *(Courtesy of Eaton, Bussmann Division)*

EXHIBIT 240.12 Two fuses rated 300 volts with marking to indicate they are Class G. *(Courtesy of Eaton, Bussmann Division)*

(D) Renewable Fuses. Class H cartridge fuses of the renewable type shall be permitted to be used only for replacement in existing installations where there is no evidence of overfusing or tampering.

N (E) Fuse Reducers. Fuse reducers shall be listed.

240.61 Classification. Cartridge fuses and fuseholders shall be classified according to voltage and amperage ranges. Fuses rated 1000 volts, nominal, or less shall be permitted to be used for voltages at or below their ratings.

Δ **240.67 Arc Energy Reduction.** Where fuses rated 1200 amperes or higher are installed, 240.67(A), (B), and (C) shall apply.

(A) Documentation. Documentation shall be available to those authorized to design, install, operate, or inspect the installation as to the location of the fuses.

Documentation shall also be provided to demonstrate that the method chosen to reduce clearing time is set to operate at a value below the available arcing current.

Δ **(B) Method to Reduce Clearing Time.** A fuse shall have a clearing time of 0.07 seconds or less at the available arcing current, or one of the following means shall be provided and shall be set to operate at less than the available arcing current:

(1) Differential relaying
(2) Energy-reducing maintenance switching with local status indicator
(3) Energy-reducing active arc-flash mitigation system
(4) Current-limiting, electronically actuated fuses
(5) An approved equivalent means

Informational Note No. 1: An energy-reducing maintenance switch allows a worker to set a disconnect switch to reduce the clearing time while the worker is working within an arc-flash boundary as defined in *NFPA 70E-2021, Standard for Electrical Safety in the Workplace*, and then to set the disconnect switch back to a normal setting after the potentially hazardous work is complete.

Informational Note No. 2: An energy-reducing active arc-flash mitigation system helps in reducing arcing duration in the electrical distribution system. No change in the disconnect switch or the settings of other devices is required during maintenance when a worker is working within an arc-flash boundary as defined in *NFPA 70E-2021, Standard for Electrical Safety in the Workplace*.

Informational Note No. 3: IEEE 1584-2018, *IEEE Guide for Performing Arc Flash Hazard Calculations*, provides guidance in determining arcing current.

(C) Performance Testing. The arc energy reduction protection system shall be performance tested by primary current injection testing or another approved method when first installed on site. This testing shall be conducted by a qualified person(s) in accordance with the manufacturer's instructions.

A written record of this testing shall be made and shall be available to the authority having jurisdiction.

Informational Note: Some energy reduction protection systems cannot be tested using a test process of primary current injection due to either the protection method being damaged such as with the use of fuse technology or because current is not the primary method of arc detection.

Part VII. Circuit Breakers

240.80 Method of Operation. Circuit breakers shall be trip free and capable of being closed and opened by manual operation. Their normal method of operation by other than manual means, such as electrical or pneumatic, shall be permitted if means for manual operation are also provided.

240.81 Indicating. Circuit breakers shall clearly indicate whether they are in the open "off" or closed "on" position.

Where circuit breaker handles are operated vertically rather than rotationally or horizontally, the "up" position of the handle shall be the "on" position.

If the panelboard enclosure is mounted in a horizontal position, compliance with this requirement that the up position of the handle is on or closed and the down position of the handle is off or open limits the number of pole spaces available in a panelboard. This results in some of the circuit breakers having their "on" and "off" operation upside down.

Opening or turning off a circuit breaker or switch is one of the steps necessary to achieve an electrically safe work condition for servicing or maintenance. *NFPA 70E®, Standard for Electrical Safety in the Workplace®*, Article 120 contains seven additional steps that must be followed before electrical conductors or circuit parts are considered to be in an electrically safe work condition.

240.82 Nontamperable. A circuit breaker shall be of such design that any alteration of its trip point (calibration) or the time required for its operation requires dismantling of the device or breaking of a seal for other than intended adjustments.

240.83 Marking.

(A) Durable and Visible. Circuit breakers shall be marked with their ampere rating in a manner that will be durable and visible after installation. Such marking shall be permitted to be made visible by removal of a trim or cover.

(B) Location. Circuit breakers rated at 100 amperes or less and 1000 volts or less shall have the ampere rating molded, stamped, etched, or similarly marked into their handles or escutcheon areas.

(C) Interrupting Rating. Every circuit breaker having an interrupting rating other than 5000 amperes shall have its interrupting rating shown on the circuit breaker. The interrupting rating shall not be required to be marked on circuit breakers used for supplementary protection.

(D) Used as Switches. Circuit breakers used as switches in 120-volt and 277-volt fluorescent lighting circuits shall be listed and shall be marked SWD or HID. Circuit breakers used as switches in high-intensity discharge lighting circuits shall be listed and shall be marked as HID.

Circuit breakers marked switch duty (SWD) have been subjected to additional endurance and temperature testing to assess their ability for use as the regular control device for fluorescent lighting circuits. Circuit breakers marked HID are also acceptable for use as the regular switching device to control high-intensity discharge (HID) lighting such as mercury vapor, high-pressure or low-pressure sodium, or metal halide lighting. Circuit breakers marked HID can be used for switching both HID and fluorescent lighting loads; however, a circuit breaker marked SWD can be used only as a switching device for fluorescent lighting loads.

(E) Voltage Marking. Circuit breakers shall be marked with a voltage rating not less than the nominal system voltage that is indicative of their capability to interrupt fault currents between phases or phase to ground.

240.85 Applications. A circuit breaker with a straight voltage rating, such as 240V or 480V, shall be permitted to be applied in a circuit in which the nominal voltage between any two conductors does not exceed the circuit breaker's voltage rating. A two-pole circuit breaker shall not be used for protecting a 3-phase, corner-grounded delta circuit unless the circuit breaker is marked 1ϕ–3ϕ to indicate such suitability.

A circuit breaker with a slash rating, such as 120/240V or 480Y/277V, shall be permitted to be applied in a solidly grounded circuit where the nominal voltage of any conductor to ground does not exceed the lower of the two values of the circuit breaker's voltage rating and the nominal voltage between any two conductors does not exceed the higher value of the circuit breaker's voltage rating.

Informational Note: Proper application of molded case circuit breakers on 3-phase systems, other than solidly grounded wye, particularly on corner grounded delta systems, considers the circuit breakers' individual pole-interrupting capability.

A circuit breaker marked 480Y/277 volts is not intended for use on a 480-volt system with up to 480 volts to ground, such as a 480-volt circuit derived from a corner-grounded, delta-connected system. A circuit breaker marked either 480 volts or 600 volts should be used on such a system. Similarly, a circuit breaker marked 120/240 volts is not intended for use on a delta-connected 240-volt circuit. A 240-volt, 480-volt, or 600-volt circuit breaker should be used on such a circuit. The slash (/) between the lower and higher voltage ratings in the marking indicates that the circuit breaker has been tested for use on a circuit with the higher voltage between phases and with the lower voltage to ground.

240.86 Series Ratings. Where a circuit breaker is used on a circuit having an available fault current higher than the marked interrupting rating by being connected on the load side of an approved overcurrent protective device having a higher rating, the circuit breaker shall meet the requirements specified in 240.86(A) or (B), and (C).

A series-rated system is a combination of circuit breakers or fuses and circuit breakers that can be applied at available short-circuit levels above the interrupting rating of the load-side circuit breakers but not above that of the main or line-side device. Series-rated systems can consist of fuses that protect circuit breakers or of circuit breakers that protect circuit breakers. The arrangement of protective components in a series-rated system can be as specified in 240.86(A) for engineered systems applied to existing installations or in 240.86(B) for tested combinations that can be applied in any new or existing installation.

(A) Selected Under Engineering Supervision in Existing Installations. The series rated combination devices shall be selected by a licensed professional engineer engaged primarily in the design or maintenance of electrical installations. The selection shall be documented and stamped by the professional engineer. This documentation shall be available to those authorized to design, install, inspect, maintain, and operate the system. This series combination rating, including identification of the upstream device, shall be field marked on the end use equipment.

For calculated applications, the engineer shall ensure that the downstream circuit breaker(s) that are part of the series combination remain passive during the interruption period of the line side fully rated, current-limiting device.

Informational Note: See 110.22 for marking of series combination systems.

An engineering solution is available for existing facilities where the available fault current is increased beyond the interrupting rating of the existing circuit overcurrent protection equipment. Examples of this increase could include changes in transformer size, transformer impedances, motor replacements, new motor installations, or utility distribution systems. An engineered system essentially redesigns the overcurrent protection scheme to accommodate the increase in available fault current. This eliminates the need for a wholesale replacement of electrical distribution equipment. Where the increase in fault current causes

existing equipment to be "underrated," the engineering approach is to provide upstream protection that functions in concert with the existing protective devices to safely open the circuit under fault conditions. The requirement specifies that the design of such systems is to be performed only by licensed professional engineers whose credentials substantiate their ability to perform this type of engineering. Documentation in the form of stamped drawings and field marking of end-use equipment to indicate it is a component of a series-rated system is required.

Designing a series-rated system requires consideration of the fault-clearing characteristics of the existing protective devices and of their ability to interact with the newly installed upstream protective device(s) under fault conditions. An engineered series-rated system can be applied to all existing installations. The operating parameters of the existing overcurrent protection equipment dictate what can be done in a field-engineered protection scheme.

The circuit breakers that are most likely to be compatible with series-rated systems are those that will remain closed during the interruption period of the fully rated OCPD installed on their line side and those that have an interrupting rating not less than the let-through current of an upstream protective device (such as a current-limiting fuse). In cases where the opening of a circuit breaker (under any level of fault current) begins in less than one-half cycle, the use of a field-engineered series-rated system is likely to be contrary to acceptable application practices specified by the circuit breaker manufacturer. In an engineered system, the upstream device must operate before the downstream device for any fault that exceeds the interrupting rating of the downstream device in the series.

Where there is doubt about the proper application of existing downstream circuit breakers with new upstream OCPDs, the manufacturers of the existing circuit breakers and the new upstream OCPDs must be consulted.

The safety objective of any overcurrent protection scheme is to ensure compliance with 110.9.

(B) Tested Combinations. The combination of line-side overcurrent device and load-side circuit breaker(s) is tested and marked on the end use equipment, such as switchboards and panelboards.

Informational Note: See 110.22 for marking of series combination systems.

The enclosures must have a label affixed by the equipment manufacturer that provides the series rating of the combination(s). Because the equipment often does not have enough room to show all the legitimate series-rated combinations, UL 67, *Standard for Panelboards*, allows a bulletin to be referenced and supplied with the panelboard. These bulletins typically provide all the acceptable combinations. The installer of a series-rated system must also provide the additional labeling on equipment enclosures required by 110.22(C), indicating that the equipment has been applied in a series-rated system.

Δ **(C) Motor Contribution.** Series ratings shall not be used in the following situations:

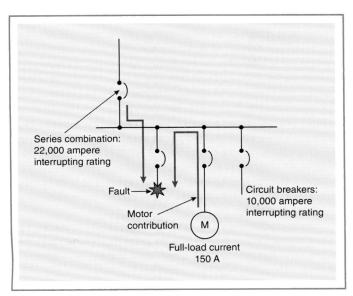

EXHIBIT 240.13 Example of an installation where the level of motor contribution exceeds 1 percent of the interrupting rating for the lowest-rated circuit breaker in this series-rated system.

(1) Where motor circuits are connected between the higher-rated overcurrent device of a series-rated combination and on the lower-rated circuit breaker
(2) Where the sum of these motor full-load currents exceeds 1 percent of the interrupting rating of the lower-rated circuit breaker

These requirements limit the use of series-rated systems in which motors are connected between the line-side (protecting) device and the load-side (protected) circuit breaker. Exhibit 240.13 illustrates the conditions where a series rating is not permitted.

240.87 Arc Energy Reduction. Where the highest continuous current trip setting for which the actual overcurrent device installed in a circuit breaker is rated or can be adjusted is 1200 amperes or higher, 240.87(A), (B), and (C) shall apply.

(A) Documentation. Documentation shall be available to those authorized to design, install, operate, or inspect the installation as to the location of the circuit breaker(s). Documentation shall also be provided to demonstrate that the method chosen to reduce clearing time is set to operate at a value below the available arcing current.

(B) Method to Reduce Clearing Time. One of the following means shall be provided and shall be set to operate at less than the available arcing current:

(1) Zone-selective interlocking
(2) Differential relaying
(3) Energy-reducing maintenance switching with local status indicator
(4) Energy-reducing active arc flash mitigation system

(5) An instantaneous trip setting. Temporary adjustment of the instantaneous trip setting to achieve arc energy reduction shall not be permitted.
(6) An instantaneous override
(7) An approved equivalent means

Informational Note No. 1: An energy-reducing maintenance switch allows a worker to set a circuit breaker trip unit to "no intentional delay" to reduce the clearing time while the worker is working within an arc-flash boundary as defined in *NFPA 70E-2021, Standard for Electrical Safety in the Workplace*, and then to set the trip unit back to a normal setting after the potentially hazardous work is complete.

Informational Note No. 2: An energy-reducing active arc-flash mitigation system helps in reducing arcing duration in the electrical distribution system. No change in the circuit breaker or the settings of other devices is required during maintenance when a worker is working within an arc-flash boundary as defined in *NFPA 70E-2021, Standard for Electrical Safety in the Workplace*.

Informational Note No. 3: An instantaneous trip is a function that causes a circuit breaker to trip with no intentional delay when currents exceed the instantaneous trip setting or current level. If arcing currents are above the instantaneous trip level, the circuit breaker will trip in the minimum possible time.

Informational Note No. 4: See IEEE 1584-2018, *IEEE Guide for Performing Arc Flash Hazard Calculations*, for guidance in determining arcing current.

The requirement to reduce clearing times does not rely on the circuit breaker having or not having an instantaneous trip setting. The continuous current trip setting is the rating of the installed circuit breaker or trip plug installed in the circuit breaker. For devices with adjustable set points, it is the maximum rating of the installed device that is to be used, not the rating set for the specific circuit. If either of these is rated 1200 amperes or higher, then one of the methods to reduce clearing time must be employed.

Circuit breakers without the capability of an instantaneous response can be used to selectively coordinate with other OCPDs where required by the *NEC* or where incorporated as part of an electrical system design. However, a longer protective device opening time can expose the person(s) who may have initiated the fault while working on energized electrical equipment to more incident energy, thus increasing the severity of the hazard.

This section identifies six specific methods that can be used to comply with this requirement and also permits use of an equivalent that is acceptable to the AHJ. Zone-selective interlocking and differential relaying are methods that can be implemented as part of the system design and do not require any manual intervention.

An energy-reducing maintenance switch is a means by which an intentional delay in the opening of a circuit breaker can be overridden while maintenance, service, or diagnostic tasks are being performed. The circuit breaker can then be restored upon completion of the tasks to enable the system to be selectively coordinated or for other system design purposes.

An instantaneous trip setting that is below the available arcing current is a permitted method of achieving arc energy reduction. It is critical that the available arcing current exceeds the instantaneous trip or instantaneous override for the circuit breaker to open as quickly as possible during an arcing fault. It is not enough to just require an instantaneous trip or an instantaneous override, because the arcing current could be low enough that it takes the circuit breaker from many cycles to many seconds to open. If the arcing current is greater than the instantaneous trip or instantaneous override, most circuit breakers will clear somewhere between ½ cycle (smaller molded case type) and 3 cycles (power or "air-frame" type).

The final setting of the instantaneous trip determines whether additional arc energy reduction techniques are required. It is not the minimum setting of the instantaneous trip, as typically shipped from the factory, that is the determining factor of whether additional arc energy reduction is necessary, but rather the final setting as determined by the electrical system requirements such as inrush characteristics or selective coordination. The use of circuit breakers having an instantaneous capability that is adjusted to allow for a longer clearing time is not prohibited. That is a design consideration.

Exhibit 240.14 is an example of an electronic trip unit for a circuit breaker employing an energy-reducing maintenance switch in addition to means for adjusting other settings. The function of the energy reduction maintenance switch is to override any settings that intentionally delay the opening time of the circuit breaker. A selectively coordinated electrical system is one example of where the response time adjustments available on some circuit breakers are set to allow for localization of a fault in the circuit. When these settings are

overridden, the circuit breaker responds faster to a downstream fault in the circuit. The faster response time provides the important benefit of reducing the level of incident energy to which personnel may be exposed where tasks are being performed within the arc-flash boundary. An example of a local status indicator is an LED signifying "maintenance mode," which provides qualified persons with the indication that trip settings are adjusted to reduce incident energy should a fault occur anywhere electrically downstream of the device. See the definition of *arc-flash boundary* in *NFPA 70E*; see the definition of *qualified person* in Article 100.

Once the task is completed, any settings that previously introduced an intentional delay in the circuit-breaker clearing time can be restored.

The instantaneous trip setting adjusted to a lower setting while a worker is working on the equipment and then adjusted back to the desired setting after work is complete does not determine the arc energy reduction.

(C) Performance Testing. The arc energy reduction protection system shall be performance tested by primary current injection testing or another approved method when first installed on site. This testing shall be conducted by a qualified person(s) in accordance with the manufacturer's instructions.

A written record of this testing shall be made and shall be available to the authority having jurisdiction.

Informational Note: Some energy reduction protection systems cannot be tested using a test process of primary current injection due to either the protection method being damaged such as with the use of fuse technology or because current is not the primary method of arc detection.

N **240.89 Replacement Trip Units.** Replacement trip units shall be listed for use with the circuit breaker type in which it is installed.

Informational Note: The replacement trip unit can be a listed unit identical to the original or a different trip unit listed for use with the specific circuit breaker.

Part VIII. Supervised Industrial Installations

240.90 General. Overcurrent protection in areas of supervised industrial installations shall comply with all of the other applicable provisions of this article, except as provided in Part VIII. Part VIII shall be permitted to apply only to those portions of the electrical system in the supervised industrial installation used exclusively for manufacturing or process control activities.

240.91 Protection of Conductors. Conductors shall be protected in accordance with 240.91(A) or 240.91(B).

(A) General. Conductors shall be protected in accordance with 240.4.

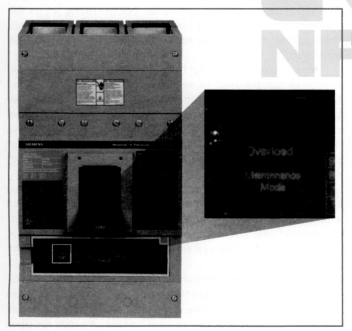

EXHIBIT 240.14 An electronic trip unit for a circuit breaker that has a maintenance mode setting, which is an energy-reducing setting. *(Courtesy of Siemens Industry, Inc.)*

(B) Devices Rated Over 800 Amperes. Where the overcurrent device is rated over 800 amperes, the ampacity of the conductors it protects shall be equal to or greater than 95 percent of the rating of the overcurrent device in accordance with the following:

(1) The conductors are protected within recognized time vs. current limits for short-circuit currents

Informational Note: Table 240.92(B) provides time vs. short-circuit current formulas to determine limits for copper and aluminum conductors.

(2) All equipment in which the conductors terminate is listed and marked for the application.

Contrary to 240.4(C), in supervised industrial installations conductors protected by OCPDs rated more than 800 amperes must have an ampacity that is 95 percent of the rating of the OCPD. For example, if the OCPD is rated 1000 amperes, the conductor ampacity cannot be less than 950 amperes. Calculations must be made to verify that the conductors are adequately protected during the time it takes for the OCPD to open due to a short circuit. In addition, the device and equipment in which the conductors terminate are required to be evaluated and listed by a qualified electrical testing laboratory for this specific application. The requirements in Articles 210, 215, and 230 that conductors have an ampacity not less than the calculated load being supplied still apply.

240.92 Location in Circuit. An overcurrent device shall be connected in each ungrounded circuit conductor as required in 240.92(A) through (E).

(A) Feeder and Branch-Circuit Conductors. Feeder and branch-circuit conductors shall be protected at the point the conductors receive their supply as permitted in 240.21 or as otherwise permitted in 240.92(B), (C), (D), or (E).

(B) Feeder Taps. For feeder taps specified in 240.21(B)(2), (B)(3), and (B)(4), the tap conductors shall be permitted to be sized in accordance with Table 240.92(B).

Table 240.92(B) does not modify the maximum length of the feeder taps specified in 240.21(B)(2), (B)(3), and (B)(4). It only provides an alternative method of sizing the feeder conductors.

(C) Transformer Secondary Conductors of Separately Derived Systems. Conductors shall be permitted to be connected to a transformer secondary of a separately derived system, without overcurrent protection at the connection, where the conditions of 240.92(C)(1), (C)(2), and (C)(3) are met.

(1) Short-Circuit and Ground-Fault Protection. The conductors shall be protected from short-circuit and ground-fault conditions by complying with one of the following conditions:

(1) The length of the secondary conductors does not exceed 30 m (100 ft), and the transformer primary overcurrent device has a rating or setting that does not exceed 150 percent of the value determined by multiplying the secondary

TABLE 240.92(B) *Tap Conductor Short-Circuit Current Ratings*

Tap conductors are considered to be protected under short-circuit conditions when their short-circuit temperature limit is not exceeded. Conductor heating under short-circuit conditions is determined by (1) or (2):

(1) *Short-Circuit Formula for Copper Conductors*
$$(I^2/A^2)t = 0.0297 \log_{10}[(T_2 + 234)/(T_1 + 234)]$$
(2) *Short-Circuit Formula for Aluminum Conductors*
$$(I^2/A^2)t = 0.0125 \log_{10}[(T_2 + 228)/(T_1 + 228)]$$

where:

I = short-circuit current in amperes

A = conductor area in circular mils

t = time of short circuit in seconds (for times less than or equal to 10 seconds)

T_1 = initial conductor temperature in degrees Celsius

T_2 = final conductor temperature in degrees Celsius

Copper conductor with paper, rubber, varnished cloth insulation, $T_2 = 200$

Copper conductor with thermoplastic insulation, $T_2 = 150$

Copper conductor with cross-linked polyethylene insulation, $T_2 = 250$

Copper conductor with ethylene propylene rubber insulation, $T_2 = 250$

Aluminum conductor with paper, rubber, varnished cloth insulation, $T_2 = 200$

Aluminum conductor with thermoplastic insulation, $T_2 = 150$

Aluminum conductor with cross-linked polyethylene insulation, $T_2 = 250$

Aluminum conductor with ethylene propylene rubber insulation, $T_2 = 250$

conductor ampacity by the secondary-to-primary transformer voltage ratio.

(2) The conductors are protected by a differential relay with a trip setting equal to or less than the conductor ampacity.

Informational Note: A differential relay is connected to be sensitive only to short-circuit or fault currents within the protected zone and is normally set much lower than the conductor ampacity. The differential relay is connected to trip protective devices that de-energize the protected conductors if a short-circuit condition occurs.

(3) The conductors shall be considered to be protected if calculations, made under engineering supervision, determine that the system overcurrent devices will protect the conductors within recognized time vs. current limits for all short-circuit and ground-fault conditions.

(2) Overload Protection. The conductors shall be protected against overload conditions by complying with one of the following:

(1) The conductors terminate in a single overcurrent device that will limit the load to the conductor ampacity.

(2) The sum of the overcurrent devices at the conductor termination limits the load to the conductor ampacity. The overcurrent devices shall consist of not more than six circuit

breakers or sets of fuses mounted in a single enclosure, in a group of separate enclosures, or in or on a switchboard or switchgear. There shall be no more than six overcurrent devices grouped in any one location.

(3) Overcurrent relaying is connected [with a current transformer(s), if needed] to sense all of the secondary conductor current and limit the load to the conductor ampacity by opening upstream or downstream devices.

(4) Conductors shall be considered to be protected if calculations, made under engineering supervision, determine that the system overcurrent devices will protect the conductors from overload conditions.

(3) Physical Protection. The secondary conductors are protected from physical damage by being enclosed in an approved raceway or by other approved means.

(D) Outside Feeder Taps. Outside conductors shall be permitted to be tapped to a feeder or to be connected at a transformer secondary, without overcurrent protection at the tap or connection, where all the following conditions are met:

(1) The conductors are protected from physical damage in an approved manner.

(2) The sum of the overcurrent devices at the conductor termination limits the load to the conductor ampacity. The overcurrent devices shall consist of not more than six circuit breakers or sets of fuses mounted in a single enclosure, in a group of separate enclosures, or in or on a switchboard or switchgear. There shall be no more than six overcurrent devices grouped in any one location.

(3) The tap conductors are installed outdoors of a building or structure except at the point of load termination.

(4) The overcurrent device for the conductors is an integral part of a disconnecting means or is located immediately adjacent thereto.

(5) The disconnecting means for the conductors are installed at a readily accessible location complying with one of the following:
    a. Outside of a building or structure
    b. Inside, nearest the point of entrance of the conductors
    c. Where installed in accordance with 230.6, nearest the point of entrance of the conductors

(E) Protection by Primary Overcurrent Device. Conductors supplied by the secondary side of a transformer shall be permitted to be protected by overcurrent protection provided on the primary (supply) side of the transformer, provided the primary device time–current protection characteristic, multiplied by the maximum effective primary-to-secondary transformer voltage ratio, effectively protects the secondary conductors.

ARTICLE 242 Overvoltage Protection

Part I. General

Δ **242.1 Scope.** This article provides the general requirements, installation requirements, and connection requirements for overvoltage protection and overvoltage protective devices. Part II covers surge-protective devices (SPDs) permanently installed on premises wiring systems of not more than 1000 volts, nominal, while Part III covers surge arresters permanently installed on premises wiring systems over 1000 volts, nominal.

The delineation between surge-protective devices (SPDs) covered by Part II and those covered by Part III is the voltage rating of the supply system. The designations of SPDs are varied, depending on their location in the premises wiring system. For instance, a Type 1 SPD is permitted to be connected on the supply side of the service or building disconnecting means. Type 2 and Type 3 SPDs must be installed on the load side of OCPDs and are the devices referred to in previous editions as transient voltage surge suppressors (TVSSs). Two examples of SPDs are shown in Exhibit 242.1.

N **242.2 Reconditioned Equipment.** SPDs and surge arresters shall not be reconditioned.

242.3 Other Articles. Equipment shall be protected against overvoltage in accordance with the article in this *Code* that covers the type of equipment or location specified in Table 242.3.

EXHIBIT 242.1 Two SPDs suitable for service-entrance installation — one for direct connection to panelboard busbars and one for mounting in a cabinet or enclosure knockout. (*Courtesy of Eaton*)

△ **TABLE 242.3** *Other Articles*

Equipment	Article
Class I locations	501
Class II locations	502
Community antenna television and radio distribution systems	820
Critical operations power systems	708
Elevators, dumbwaiters, escalators, moving walks, platform lifts, and stairway chairlifts	620
Emergency systems	700
Equipment over 1000 volts, nominal	495
Fire pumps	695
Industrial machinery	670
Information technology equipment	645
Modular data centers	646
Outdoor overhead conductors over 1000 volts	395
Radio and television equipment	810
Receptacles, cord connectors, and attachment plugs (caps)	406
Wind electric systems	694

EXHIBIT 242.2 An SPD as an integral component of a receptacle, providing local point-of-use protection of equipment when transient events occur within the facility. *(Courtesy of Legrand®)*

△ **Part II. Surge-Protective Devices (SPDs), 1000 Volts or Less**

242.6 Listing. An SPD shall be a listed device.

UL 1449, *Standard for Surge Protective Devices*, covers Types 1, 2, 3, and 4 devices. SPDs are permitted to be installed on ungrounded systems, impedance grounded systems, and corner-grounded systems where the device is listed for the specific characteristic of the system per 242.12(2).

242.8 Short-Circuit Current Rating. The SPD shall be marked with a short-circuit current rating and shall not be installed at a point on the system where the available fault current is in excess of that rating. This marking requirement shall not apply to receptacles.

In residential and small commercial electrical systems, the first SPD is commonly installed either as an integral component of or near to the service-entrance equipment.

Depending on the system voltage, surge protection in larger commercial and industrial electrical systems can be provided by installing a Type 1 SPD (a surge arrester for systems 1000 volts and less) or surge arrester (the devices covered in Part III for systems over 1000 volts) on the line side of the service equipment. Subsequent levels of SPDs are then provided at intermediate points in the distribution system (such as at panelboards that serve loads susceptible to transients) and at the point where utilization equipment connects to the electrical system.

Point-of-use SPDs such as receptacles and relocatable power taps (power strips) can be installed at the equipment (such as computers or equipment with electronic controls). The function of a point-of-use SPD is to remove small transients that pass through the more robust surge devices located at the service. Point-of-use SPDs are also useful in removing small transients that have been generated within the building. See Exhibit 242.2 for a point-of-use, or Type 3 SPD.

N **242.9 Indicating.** An SPD shall provide indication that it is functioning properly.

242.12 Uses Not Permitted. An SPD device shall not be installed in the following:

(1) Circuits over 1000 volts
(2) On ungrounded systems, impedance grounded systems, or corner grounded delta systems unless listed specifically for use on these systems
(3) Where the rating of the SPD is less than the maximum continuous phase-to-ground voltage at the power frequency available at the point of application

242.13 Type 1 SPDs. Type 1 SPDs shall be installed in accordance with 242.13(A) and (B).

(A) Installation. Type 1 SPDs shall be permitted to be connected in accordance with one of the following:

(1) To the supply side of the service disconnect as permitted in 230.82(4)
(2) As specified in 242.14

(B) At the Service. When installed at services, Type 1 SPDs shall be connected to one of the following:

(1) Grounded service conductor
(2) Grounding electrode conductor
(3) Grounding electrode for the service
(4) Equipment grounding terminal in the service equipment

Although four locations for connecting the SPD grounding lead are acceptable, the requirement in 242.24 covering the length and physical routing of the conductor must be followed.

242.14 Type 2 SPDs. Type 2 SPDs shall be installed in accordance with 242.14(A) through (C).

(A) Service-Supplied Building or Structure. Type 2 SPDs shall be connected anywhere on the load side of a service disconnect overcurrent device required in 230.91 unless installed in accordance with 230.82(8).

(B) Feeder-Supplied Building or Structure. Type 2 SPDs shall be connected at the building or structure anywhere on the load side of the first overcurrent device at the building or structure.

(C) Separately Derived System. The SPD shall be connected on the load side of the first overcurrent device in a separately derived system.

242.16 Type 3 SPDs. Type 3 SPDs shall be permitted to be installed on the load side of branch-circuit overcurrent protection up to the equipment served. If included in the manufacturer's instructions, the Type 3 SPD connection shall be a minimum 10 m (30 ft) of conductor distance from the service or separately derived system disconnect.

The point in the electrical system where SPDs are connected is dependent on the type of SPD. UL 1449, *Standard for Surge Protection Devices*, is the product standard used to evaluate safe performance of SPDs. A Type 2 SPD must be installed on the load side of the service-disconnect overcurrent protection unless installed in accordance with 230.82(8), and a Type 3 SPD must be installed on the load side of a branch-circuit OCPD. The requirement for Type 2 SPDs to be connected on the load side of the first OCPD in a feeder-supplied structure is necessary due to the exposure of external feeder conductors to a more hostile surge environment, such as lightning.

    Two requirements — 230.82(4) for Type 1 surge-protection devices and 230.82(8) for Type 2 surge-protection devices — permit the installation of SPDs on the line side of the service disconnecting means. Section 230.71(A)(2) permits an additional disconnecting means at the service equipment for SPDs installed as part of listed equipment. The disconnecting means for the SPD does not count as one of the six permitted by 230.71(A)(2) where the SPD and its disconnecting means are provided in the listed equipment by the manufacturer. Type 2 SPDs must be installed in listed equipment where the SPD is provided with a disconnecting means and overcurrent protection.

242.18 Type 4 and Other Component Type SPDs. Type 4 component assemblies and other component type SPDs shall only be installed by the equipment manufacturer.

Type 4 and other component-type SPDs are incomplete devices that are acceptable only if installed as part of listed equipment.

242.20 Number Required. Where used at a point on a circuit, the SPD shall be connected to each ungrounded conductor.

242.22 Location. SPDs shall be permitted to be located indoors or outdoors and shall be made inaccessible to unqualified persons unless listed for installation in accessible locations.

242.24 Routing of Conductors. The conductors used to connect the SPD to the line or bus and to ground shall not be any longer than necessary and shall avoid unnecessary bends.

High-frequency currents, such as those common to lightning discharges, tend to reduce the effectiveness of a conductor that connects the device to ground. To optimize performance of SPDs, the length of the conductor that connects the device to ground is limited. Short conductors with few bends will have a lower impedance to surge current. Higher impedance drives the clamping voltage higher and reduces the protection provided by the SPD unit. Maximum protection is achieved where the SPD is located as close as practicable to the equipment being protected as shown in Exhibit 242.3.

242.28 Conductor Size. SPD line conductors and conductors to ground shall not be smaller than 14 AWG copper or 12 AWG aluminum.

242.30 Connection Between Conductors. An SPD shall be permitted to be connected between any two conductors —

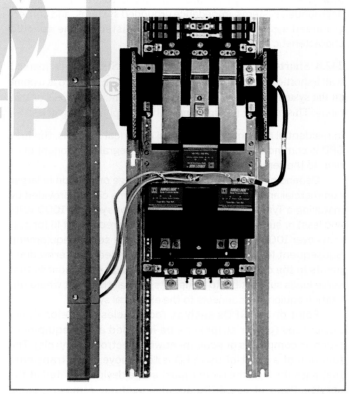

EXHIBIT 242.3 A Type 2 SPD mounted as an integral component of a panelboard, which minimizes conductor length between the electrical system and the SPD. *(Courtesy of Schneider Electric)*

ungrounded conductor(s), grounded conductor, equipment grounding conductor, or grounding electrode conductor. The grounded conductor and the equipment grounding conductor shall be interconnected only by the normal operation of the SPD during a surge.

242.32 Grounding Electrode Conductor Connections and Enclosures. Except as indicated in this article, SPD grounding connections shall be made as specified in Article 250, Part III. Grounding electrode conductors installed in metal enclosures shall comply with 250.64(E).

Part III. Surge Arresters, Over 1000 Volts

Voltage surges with peaks of several thousand volts, even on 120-volt circuits, are not uncommon. Surges occur because of induced voltages in power and transmission lines resulting from lightning strikes in the vicinity of the line. Surges also occur as a result of switching inductive circuits. The standard on surge arresters is IEEE C62.11, *Standard for Metal-Oxide Surge Arresters for Alternating-Current Power Circuits (> 1 kV)*.

242.40 Uses Not Permitted. A surge arrester shall not be installed where the rating of the surge arrester is less than the maximum continuous phase-to-ground voltage at the power frequency available at the point of application.

242.42 Surge Arrester Rating. The duty cycle rating of a surge arrester shall be not less than 125 percent of the maximum continuous operating voltage available at the point of application.

For solidly grounded systems, the maximum continuous operating voltage shall be the phase-to-ground voltage of the system.

For impedance or ungrounded systems, the maximum continuous operating voltage shall be the phase-to-phase voltage of the system.

> Informational Note No. 1: See IEEE C62.11-2020, *Standard for Metal-Oxide Surge Arresters for Alternating-Current Power Circuits (>1 kV)*, and IEEE C62.22-2009, *Guide for the Application of Metal-Oxide Surge Arresters for Alternating-Current Systems*, for further information on surge arresters.
>
> Informational Note No. 2: The selection of a properly rated metal oxide arrester is based on considerations of maximum continuous operating voltage and the magnitude and duration of overvoltages at the arrester location as affected by phase-to-ground faults, system grounding techniques, switching surges, and other causes. See the manufacturer's application rules for selection of the specific arrester to be used at a particular location.

242.44 Number Required. Where used at a point on a circuit, a surge arrester shall be connected to each ungrounded conductor. A single installation of such surge arresters shall be permitted to protect a number of interconnected circuits if no circuit is exposed to surges while disconnected from the surge arresters.

242.46 Location. Surge arresters shall be permitted to be located indoors or outdoors. Surge arresters shall be made inaccessible to unqualified persons unless listed for installation in accessible locations.

Maximum protection is achieved where the SPD is located as close as practicable to the equipment being protected. When a surge passes through an arrester, a wave is reflected in both directions on the conductors connected to the surge arrester. The magnitude of the reflected wave increases as the distance from the arrester increases. If the length of the conductor between the protected equipment and the surge arrester is short, the magnitude of the wave reflected through the equipment is minimized.

242.48 Routing of Surge Arrester Equipment Grounding Conductors. The conductor used to connect the surge arrester to line, bus, or equipment and to an equipment grounding conductor or grounding electrode connection point as provided in 242.50 shall not be any longer than necessary and shall avoid unnecessary bends.

242.50 Connection. The arrester shall be connected to one of the following:

(1) Grounded service conductor
(2) Grounding electrode conductor
(3) Grounding electrode for the service
(4) Equipment grounding terminal in the service equipment

242.52 Surge-Arrester Conductors. The conductor between the surge arrester and the line, and the surge arrester and the grounding connection, shall not be smaller than 6 AWG copper or aluminum.

242.54 Interconnections. The surge arrester protecting a transformer that supplies a secondary distribution system shall be interconnected as specified in 242.54(A), (B), or (C).

(A) Metal Interconnections. A metal interconnection shall be made to the secondary grounded circuit conductor or the secondary circuit grounding electrode conductor, if, in addition to the direct grounding connection at the surge arrester, the connection complies with 242.54(A)(1) or (A)(2).

(1) Additional Grounding Connection. The grounded conductor of the secondary has a grounding connection elsewhere to a continuous metal underground water piping system. In urban water-pipe areas where there are at least four water-pipe connections on the neutral conductor and not fewer than four such connections in each mile of neutral conductor, the metal interconnection shall be permitted to be made to the secondary neutral conductor with omission of the direct grounding connection at the surge arrester.

(2) Multigrounded Neutral System Connection. The grounded conductor of the secondary system is part of a multigrounded neutral system or static wire of which the primary neutral conductor or static wire has at least four grounding connections in each 1.6 km (1 mile) of line in addition to a grounding connection at each service.

Maximum protection is achieved where the SPD protecting a transformer supplying a secondary system has the grounded conductor of the secondary system connected to at least four grounding locations for every one mile of primary neutral conductor.

(B) Through Spark Gap or Device. Where the surge arrester grounding electrode conductor is not connected as in 242.54(A), or where the secondary is not grounded as in 242.54(A) but is otherwise grounded as in 250.52, an interconnection shall be made through a spark gap or listed device as required by 242.54(B)(1) or (B)(2).

A spark gap device has two conductors separated by a gap often filled with a gas such as air. This permits an arc to pass between the conductors when the voltage difference between them exceeds the breakdown voltage of the gas within the gap. The arc forms, ionizing the gas, and reduces its electrical resistance, allowing the high voltage surge to pass to ground.

(1) Ungrounded or Unigrounded Primary System. For ungrounded or unigrounded primary systems, the spark gap or a listed device shall have a 60-Hz breakdown voltage of at least twice the primary circuit voltage but not necessarily more than 10 kV, and there shall be at least one other ground on the grounded conductor of the secondary that is not less than 6.0 m (20 ft) distant from the surge-arrester grounding electrode.

(2) Multigrounded Neutral Primary System. For multigrounded neutral primary systems, the spark gap or listed device shall have a 60-Hz breakdown of not more than 3 kV, and there shall be at least one other ground on the grounded conductor of the secondary that is not less than 6.0 m (20 ft) distant from the surge-arrester grounding electrode.

(C) By Special Permission. An interconnection of the surge-arrester ground and the secondary neutral conductor, other than as provided in 242.54(A) or (B), shall be permitted to be made only by special permission.

242.56 Grounding Electrode Conductor Connections and Enclosures. Except as indicated in this article, surge-arrester grounding electrode conductor connections shall be made as specified in Article 250, Parts III and X. Grounding electrode conductors installed in metal enclosures shall comply with 250.64(E).

N

ARTICLE 245 Overcurrent Protection for Systems Rated Over 1000 Volts ac, 1500 Volts dc

N 245.1 Scope. This article covers overcurrent protection requirements for systems over 1000 volts ac, 1500 volts dc, nominal.

New Article 245 has been created to cover overcurrent protection for systems rated over 1000 volts ac, 1500 volts dc. Former Article 240 Part IX has been relocated here as the basis for new Article 245, which was created to aid in the usability of the *NEC®* when locating medium-voltage overcurrent protection requirements.

N 245.2 Reconditioned Equipment.

N (A) Reconditioned Equipment Permitted. The following reconditioned equipment shall be permitted:

   (1) Medium- and high-voltage circuit breakers
   (2) Electromechanical protective relays and current transformers

N (B) Reconditioned Equipment Not Permitted. Medium-voltage fuseholders and medium-voltage nonrenewable fuses shall not be permitted.

N 245.21 Circuit-Interrupting Devices.

N (A) Circuit Breakers.

N (1) Location.

   (a) Circuit breakers installed indoors shall be mounted either in metal-enclosed units or fire-resistant cell-mounted units, or they shall be permitted to be open-mounted in locations accessible to qualified persons only.
   (b) Circuit breakers used to control oil-filled transformers in a vault shall either be located outside the transformer vault or be capable of operation from outside the vault.
   (c) Oil circuit breakers shall be arranged or located so that adjacent readily combustible structures or materials are safeguarded in an approved manner.

N (2) Operating Characteristics. Circuit breakers shall have the following equipment or operating characteristics:

   (1) An accessible mechanical or other identified means for manual tripping, independent of control power
   (2) Be release free (trip free)
   (3) If capable of being opened or closed manually while energized, main contacts that operate independently of the speed of the manual operation
   (4) A mechanical position indicator at the circuit breaker to show the open or closed position of the main contacts
   (5) A means of indicating the open and closed position of the breaker at the point(s) from which they may be operated

N (3) Nameplate. A circuit breaker shall have a permanent and legible nameplate showing the manufacturer's name or trademark, manufacturer's type or identification number, continuous current rating, interrupting rating in megavolt-amperes (MVA) or amperes, and maximum voltage rating. Modification of a circuit breaker affecting its rating(s) shall be accompanied by an appropriate change of nameplate information.

N **(4) Rating.** Circuit breakers shall have the following ratings:

(1) The continuous current rating of a circuit breaker shall not be less than the maximum continuous current through the circuit breaker.

(2) The interrupting rating of a circuit breaker shall not be less than the available fault current the circuit breaker will be required to interrupt, including contributions from all connected sources of energy.

(3) The closing rating of a circuit breaker shall not be less than the maximum asymmetrical fault current into which the circuit breaker can be closed.

(4) The momentary rating of a circuit breaker shall not be less than the maximum asymmetrical fault current at the point of installation.

(5) The rated maximum voltage of a circuit breaker shall not be less than the maximum circuit voltage.

N **(5) Retrofit Trip Units.** Retrofit trip units shall be listed for use with the specific circuit breaker with which it is installed.

N **(B) Power Fuses and Fuseholders.**

N **(1) Use.** Where fuses are used to protect conductors and equipment, a fuse shall be placed in each ungrounded conductor. Two power fuses shall be permitted to be used in parallel to protect the same load if both fuses have identical ratings and both fuses are installed in an identified common mounting with electrical connections that divide the current equally. Power fuses of the vented type shall not be used indoors, underground, or in metal enclosures unless identified for the use.

N **(2) Interrupting Rating.** The interrupting rating of power fuses shall not be less than the available fault current the fuse is required to interrupt, including contributions from all connected sources of energy.

N **(3) Voltage Rating.** The maximum voltage rating of power fuses shall not be less than the maximum circuit voltage. Fuses having a minimum recommended operating voltage shall not be applied below this voltage.

N **(4) Identification of Fuse Mountings and Fuse Units.** Fuse mountings and fuse units shall have permanent and legible nameplates showing the manufacturer's type or designation, continuous current rating, interrupting current rating, and maximum voltage rating.

N **(5) Fuses.** Fuses that expel flame in opening the circuit shall be designed or arranged so that they function properly without hazard to persons or property.

N **(6) Fuseholders.** Fuseholders shall be designed or installed so that they are de-energized while a fuse is being replaced. A field-applied permanent and legible sign, in accordance with 110.21(B), shall be installed immediately adjacent to the fuseholders and shall be worded as follows:

DANGER — DISCONNECT CIRCUIT
BEFORE REPLACING FUSES.

Exception: Fuses and fuseholders designed to permit fuse replacement by qualified persons using identified equipment without de-energizing the fuseholder shall be permitted.

N **(7) High-Voltage Fuses.** Switchgear and substations that use high-voltage fuses shall be provided with a gang-operated disconnecting switch. Isolation of the fuses from the circuit shall be provided by either connecting a switch between the source and the fuses or providing roll-out switch and fuse-type construction. The switch shall be of the load-interrupter type, unless mechanically or electrically interlocked with a load-interrupting device arranged to reduce the load to the interrupting capability of the switch.

Exception: More than one switch shall be permitted as the disconnecting means for one set of fuses where the switches are installed to provide connection to more than one set of supply conductors. The switches shall be mechanically or electrically interlocked to permit access to the fuses only when all switches are open. A conspicuous sign shall be placed at the fuses identifying the presence of more than one source.

N **(C) Distribution Cutouts and Fuse Links — Expulsion Type.**

N **(1) Installation.** Cutouts shall be located so that they may be readily and safely operated and re-fused, and so that the exhaust of the fuses does not endanger persons. Distribution cutouts shall not be used indoors, underground, or in metal enclosures.

N **(2) Operation.** Where fused cutouts are not suitable to interrupt the circuit manually while carrying full load, an approved means shall be installed to interrupt the entire load. Unless the fused cutouts are interlocked with the switch to prevent opening of the cutouts under load, a conspicuous sign shall be placed at such cutouts identifying that they shall not be operated under load.

N **(3) Interrupting Rating.** The interrupting rating of distribution cutouts shall not be less than the available fault current the cutout is required to interrupt, including contributions from all connected sources of energy.

N **(4) Voltage Rating.** The maximum voltage rating of cutouts shall not be less than the maximum circuit voltage.

N **(5) Identification.** Distribution cutouts shall have on their body, door, or fuse tube a permanent and legible nameplate or identification showing the manufacturer's type or designation, continuous current rating, maximum voltage rating, and interrupting rating.

N **(6) Fuse Links.** Fuse links shall have a permanent and legible identification showing continuous current rating and type.

N **(7) Structure Mounted Outdoors.** The height of cutouts mounted outdoors on structures shall provide safe clearance between lowest energized parts (open or closed position) and standing surfaces, in accordance with 110.34(E).

N **(D) Oil-Filled Cutouts.**

N **(1) Continuous Current Rating.** The continuous current rating of oil-filled cutouts shall not be less than the maximum continuous current through the cutout.

N **(2) Interrupting Rating.** The interrupting rating of oil-filled cutouts shall not be less than the available fault current the oil-filled cutout is required to interrupt, including contributions from all connected sources of energy.

N **(3) Voltage Rating.** The maximum voltage rating of oil-filled cutouts shall not be less than the maximum circuit voltage.

N **(4) Fault Closing Rating.** Oil-filled cutouts shall have a fault closing rating not less than the maximum asymmetrical fault current that can occur at the cutout location, unless suitable interlocks or operating procedures preclude the possibility of closing into a fault.

N **(5) Identification.** Oil-filled cutouts shall have a permanent and legible nameplate showing the rated continuous current, rated maximum voltage, and rated interrupting current.

N **(6) Fuse Links.** Fuse links shall have a permanent and legible identification showing the rated continuous current.

N **(7) Location.** Cutouts shall be located so that they are readily and safely accessible for re-fusing, with the top of the cutout not over 1.5 m (5 ft) above the floor or platform.

N **(8) Enclosure.** Suitable barriers or enclosures shall be provided to prevent contact with nonshielded cables or energized parts of oil-filled cutouts.

N **(E) Load Interrupters.** Load-interrupter switches shall be permitted if suitable fuses or circuit breakers are used in conjunction with these devices to interrupt available fault currents. Where these devices are used in combination, they shall be coordinated electrically so that they will safely withstand the effects of closing, carrying, or interrupting all possible currents up to the assigned maximum short-circuit rating.

Where more than one switch is installed with interconnected load terminals to provide for alternate connection to different supply conductors, each switch shall be provided with a warning sign identifying the presence of more than one source. Each warning sign or label shall comply with 110.21.

N **(1) Continuous Current Rating.** The continuous current rating of interrupter switches shall equal or exceed the maximum continuous current at the point of installation.

N **(2) Voltage Rating.** The maximum voltage rating of interrupter switches shall equal or exceed the maximum circuit voltage.

N **(3) Identification.** Interrupter switches shall have a permanent and legible nameplate, including the following information: manufacturer's type or designation, continuous current rating, interrupting current rating, fault closing rating, maximum voltage rating.

N **(4) Switching of Conductors.** The switching mechanism shall be arranged to be operated from a location where the operator is not exposed to energized parts and shall be arranged to open all ungrounded conductors of the circuit simultaneously with one operation. Switches shall be arranged to be locked in the open position. Metal-enclosed switches shall be operable from outside the enclosure.

N **(5) Stored Energy for Opening.** The stored-energy operator shall be permitted to be left in the uncharged position after the switch has been closed if a single movement of the operating handle charges the operator and opens the switch.

N **(6) Supply Terminals.** The supply terminals of fused interrupter switches shall be installed at the top of the switch enclosure, or, if the terminals are located elsewhere, the equipment shall have barriers installed to prevent persons from accidentally contacting energized parts or dropping tools or fuses into energized parts.

Exhibits 245.1 and 245.2 are examples of a fused interrupter switch and the fuseholder components. The components shown include the spring and cable assembly, refill unit, holder, and snuffler.

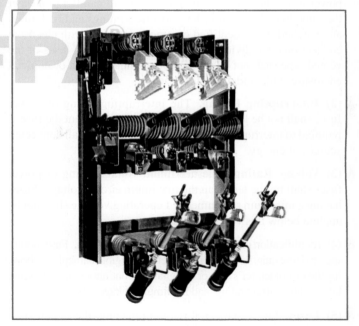

EXHIBIT 245.1 Group-operated interrupter switch and power fuse combination rated at 13.8 kilovolts, 600 amperes continuous and interrupting, 40,000 amperes momentary, 40,000 amperes fault closing. *(Courtesy of S&C Electric Co.)*

EXHIBIT 245.2 Components of the indoor solid-material (SM) power fuseholder (boric-acid arc-extinguishing type) with a 14.4-kilovolt, 400E-ampere maximum, 40,000-ampere rms asymmetrical interrupting rating. (Courtesy of S&C Electric Co.)

N **245.26 Feeders and Branch Circuits.**

N **(A) Location and Type of Protection.** Feeder and branch-circuit conductors shall have overcurrent protection in each ungrounded conductor located at the point where the conductor receives its supply or at an alternative location in the circuit when designed under engineering supervision that includes but is not limited to considering the appropriate fault studies and time–current coordination analysis of the protective devices and the conductor damage curves. The overcurrent protection shall be permitted to be provided by either 250.184(B) or (A)(2).

N **(1) Overcurrent Relays and Current Transformers.** Circuit breakers used for overcurrent protection of 3-phase circuits shall have a minimum of three overcurrent relay elements operated from three current transformers. The separate overcurrent relay elements (or protective functions) shall be permitted to be part of a single electronic protective relay unit.

On 3-phase, 3-wire circuits, an overcurrent relay element in the residual circuit of the current transformers shall be permitted to replace one of the phase relay elements.

An overcurrent relay element, operated from a current transformer that links all phases of a 3-phase, 3-wire circuit, shall be permitted to replace the residual relay element and one of the phase-conductor current transformers. Where the neutral conductor is not regrounded on the load side of the circuit as permitted in 250.184(B), the current transformer shall be permitted to link all 3-phase conductors and the grounded circuit conductor (neutral).

N **(2) Fuses.** A fuse shall be connected in series with each ungrounded conductor.

N **(B) Protective Devices.** The protective device(s) shall be capable of detecting and interrupting all values of current that can occur at their location in excess of their trip-setting or melting point.

N **(C) Conductor Protection.** The operating time of the protective device, the available short-circuit current, and the conductor used shall be coordinated to prevent damaging or dangerous temperatures in conductors or conductor insulation under short-circuit conditions.

N **245.27 Additional Requirements for Feeders.**

N **(A) Rating or Setting of Overcurrent Protective Devices.** The continuous ampere rating of a fuse shall not exceed three times the ampacity of the conductors. The long-time trip element setting of a breaker or the minimum trip setting of an electronically actuated fuse shall not exceed six times the ampacity of the conductor. For fire pumps, conductors shall be permitted to be protected for overcurrent in accordance with 695.4(B)(2).

N **(B) Feeder Taps.** Conductors tapped to a feeder shall be permitted to be protected by the feeder overcurrent device where that overcurrent device also protects the tap conductor.

ARTICLE
250 Grounding and Bonding

Part I. General

Δ **250.1 Scope.** This article covers general requirements for grounding and bonding of electrical installations and the following specific requirements:

(1) Systems, circuits, and equipment required, permitted, or not permitted to be grounded
(2) Circuit conductor to be grounded on grounded systems
(3) Location of grounding connections
(4) Types and sizes of grounding and bonding conductors and electrodes
(5) Methods of grounding and bonding
(6) Conditions under which isolation, insulation, or guards are permitted to be substituted for grounding

Informational Note: See Informational Note Figure 250.1 for information on the organization of this article covering grounding and bonding requirements.

Although frequently used interchangeably in the field, the terms *grounding* and *bonding* are separate concepts that have different outcomes based on the requirements of Article 250. The two concepts are not mutually exclusive though, and in many cases a single physical action, such as connecting the equipment grounding conductor (EGC) to the grounding terminal of a duplex receptacle, results in a bonding, as well as a grounding, connection.

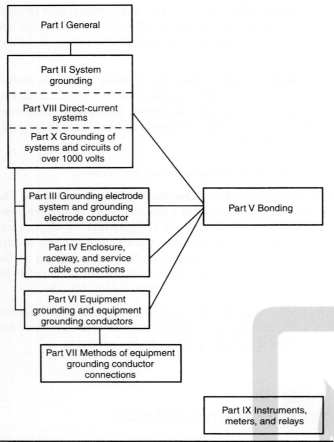

INFORMATIONAL NOTE FIGURE 250.1 *Grounding and Bonding.*

250.4 General Requirements for Grounding and Bonding. The following general requirements identify what grounding and bonding of electrical systems are required to accomplish. The prescriptive methods contained in this article shall be followed to comply with the performance requirements of this section.

The requirements of 250.4(A) and (B) specify the overall performance objectives for grounding and bonding of electrical systems and equipment but do not provide specific details or prescriptive rules for things like where to connect or how to size grounding and bonding conductors. The stated performance goals of 250.4 provide the user with a clear direction of what the intended outcomes are when the prescriptive rules contained in the remainder of Article 250 are applied.

(A) Grounded Systems.

(1) Electrical System Grounding. Electrical systems that are grounded shall be connected to earth in a manner that will limit the voltage imposed by lightning, line surges, or unintentional contact with higher-voltage lines and that will stabilize the voltage to earth during normal operation.

Informational Note No. 1: An important consideration for limiting the imposed voltage is the routing of bonding and grounding electrode conductors so that they are not any longer than necessary to complete the connection without disturbing the permanent parts of the installation and so that unnecessary bends and loops are avoided.

Informational Note No. 2: See NFPA 780-2020, *Standard for the Installation of Lightning Protection Systems*, for information on installation of grounding and bonding for lightning protection systems.

(2) Grounding of Electrical Equipment. Normally non-current-carrying conductive materials enclosing electrical conductors or equipment, or forming part of such equipment, shall be connected to earth so as to limit the voltage to ground on these materials.

(3) Bonding of Electrical Equipment. Normally non-current-carrying conductive materials enclosing electrical conductors or equipment, or forming part of such equipment, shall be connected together and to the electrical supply source in a manner that establishes an effective ground-fault current path.

(4) Bonding of Electrically Conductive Materials and Other Equipment. Normally non-current-carrying electrically conductive materials that are likely to become energized shall be connected together and to the electrical supply source in a manner that establishes an effective ground-fault current path.

Δ **(5) Effective Ground-Fault Current Path.** Electrical equipment and wiring and other electrically conductive material likely to become energized shall be installed in a manner that creates a low-impedance circuit facilitating the operation of the overcurrent device or ground detector for impedance grounded systems. It shall be capable of safely carrying the maximum ground-fault current likely to be imposed on it from any point on the wiring system where a ground fault occurs to the electrical supply source. The earth shall not be considered as an effective ground-fault current path.

The performance objective for the effective ground-fault current path is not always to facilitate operation of an overcurrent protective device (OCPD). For high-impedance grounded systems, for example, the performance objective is to ensure operation of the required ground detector in order to activate some type of an alarm or other signal indicating the existence of a ground-fault condition.

(B) Ungrounded Systems.

(1) Grounding Electrical Equipment. Non-current-carrying conductive materials enclosing electrical conductors or equipment, or forming part of such equipment, shall be connected to earth in a manner that will limit the voltage imposed by lightning or unintentional contact with higher-voltage lines and limit the voltage to ground on these materials.

Informational Note: See NFPA 780-2020, *Standard for the Installation of Lightning Protection Systems*, for information on installation of grounding and bonding for lightning protection systems.

(2) Bonding of Electrical Equipment. Non-current-carrying conductive materials enclosing electrical conductors or equipment, or forming part of such equipment, shall be connected together and to the supply system grounded equipment in a manner that creates a low-impedance path for ground-fault current that is capable of carrying the maximum fault current likely to be imposed on it.

(3) Bonding of Electrically Conductive Materials and Other Equipment. Electrically conductive materials that are likely to become energized shall be connected together and to the supply system grounded equipment in a manner that creates a low-impedance path for ground-fault current that is capable of carrying the maximum fault current likely to be imposed on it.

(4) Path for Fault Current. Electrical equipment, wiring, and other electrically conductive material likely to become energized shall be installed in a manner that creates a low-impedance circuit from any point on the wiring system to the electrical supply source to facilitate the operation of overcurrent devices should a second ground fault from a different phase occur on the wiring system. The earth shall not be considered as an effective fault-current path.

The performance requirements for grounding in both grounded and ungrounded systems can be categorized into two functions: system grounding and equipment grounding. The two functions are kept separate except at the point of supply, such as at the service equipment or at a separately derived system.

Grounding is the intentional connection of a circuit conductor to ground or to something that serves in place of ground. In most instances, the connection is made at the supply source, such as a transformer, and at the main service disconnecting means of the premises using the energy. Where a system operates "ungrounded," it does not have an intentionally grounded circuit conductor, but equipment is required to be grounded with an EGC that connects to a grounding electrode system.

There are two reasons for grounding:

1. To limit the voltages caused by lightning or by accidental contact of the supply conductors with conductors of higher voltage
2. To stabilize the voltage under normal operating conditions (which maintains the voltage at one level relative to ground so that any equipment connected to the system will be subject only to that potential difference)

Exhibit 250.1 shows a grounded single-phase, 3-wire service supplied from a utility transformer. Inside the service disconnecting means enclosure, the grounded conductor of the system is intentionally connected to a grounding electrode via the grounding electrode conductor (GEC). Bonding the equipment grounding bus to the grounded or neutral bus via the main bonding jumper within the enclosure provides a ground reference for

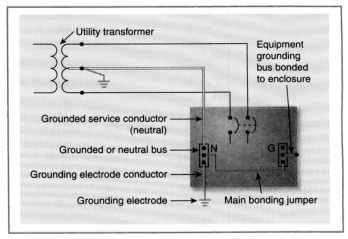

EXHIBIT 250.1 Grounding and bonding arrangement for a single-phase, 3-wire service.

exposed non-current-carrying parts of the electrical system. It also provides a circuit for ground-fault current through the grounded service conductor back to the utility transformer (source of supply). At the utility transformer, often an additional connection is made from the grounded conductor to a separate grounding electrode. It is the bonding of the EGC bus to the neutral bus that facilitates the operation of OCPDs or relays under a ground-fault condition, and not the grounding connection of the system to earth via the grounding electrode system.

250.6 Objectionable Current.

(A) Arrangement to Prevent Objectionable Current. The grounding and bonding of electrical systems, circuit conductors, surge arresters, surge-protective devices, and conductive normally non-current-carrying metal parts of equipment shall be installed and arranged in a manner that will prevent objectionable current.

Δ **(B) Alterations to Stop Objectionable Current.** If the use of multiple grounding or bonding connections results in objectionable current and the requirements of 250.4(A)(5) or (B)(4) are met, one or more of the following alterations shall be permitted:

(1) Discontinue one or more but not all of such grounding or bonding connections.
(2) Change the locations of the grounding or bonding connections.
(3) Interrupt the continuity of the conductor or conductive path causing the objectionable current.
(4) Take other remedial and approved action.

Some electronic equipment is sensitive to stray currents. Circulating currents on EGCs, metal raceways, and building steel develop potential differences between ground and the neutral of electronic equipment. Installation designers must look for ways to isolate electronic equipment from the effects of such stray circulating currents.

Isolating the electronic equipment from all other power equipment by disconnecting it from the power equipment ground is not the correct solution, nor is removing the equipment grounding means or adding nonmetallic spacers in the metallic raceway system. Those solutions are contrary to fundamental safety grounding principles covered in the requirements of Article 250. Furthermore, if the electronic equipment is grounded to an earth ground that is isolated from the common power system ground, a potential difference can be created, which is a shock hazard. The error is compounded because such isolation does not establish a low-impedance ground-fault return path to the power source, which is necessary to actuate the OCPD. Section 250.6(B) does not permit disconnecting all safety grounding and bonding connections to the electronic equipment in order to reduce an electromagnetic interference (EMI) problem.

See also

250.6(D) and its commentary for limitations to alterations

Δ **(C) Currents Not Classified as Objectionable Currents.** Currents resulting from abnormal conditions such as ground faults, and from currents resulting from required grounding and bonding connections shall not be classified as objectionable current for the purposes specified in 250.6(A) and (B).

(D) Limitations to Permissible Alterations. This section shall not be considered as permitting electronic equipment to be operated on ac systems or branch circuits that are not connected to an equipment grounding conductor as required by this article. Currents that introduce electromagnetic interference or data errors in electronic equipment shall not be considered the objectionable currents addressed in this section.

Section 250.6(D) indicates that currents that result in noise or data errors in electronic equipment are not considered to be the objectionable currents referred to in 250.6, which limits the alterations permitted by 250.6(B).

See also

250.96(B) and **250.146(D)** for requirements that provide safe bonding and grounding methods to minimize noise and data errors

(E) Isolation of Objectionable Direct-Current from Cathodic Protection Systems. If isolation of objectionable direct currents from a cathodic protection system is required, a listed isolator device shall be permitted in the equipment grounding conductor path to provide an effective return path for ac ground-fault current while blocking the flow of direct currents.

The listed ac coupling/dc isolating device allowed by this section blocks the dc current on grounding and bonding conductors and allows the ground-fault return path to function properly. These devices are evaluated by a recognized qualified electrical testing laboratory for proper performance under ground-fault conditions.

Where cathodic protection for the piping system is provided, the required grounding and bonding connections associated with metal piping systems allow dc current to be imposed on grounding and bonding conductors.

250.8 Connection of Grounding and Bonding Equipment.

(A) Permitted Methods. Equipment grounding conductors, grounding electrode conductors, and bonding jumpers shall be connected by one or more of the following means:

(1) Listed pressure connectors
(2) Terminal bars
(3) Pressure connectors listed as grounding and bonding equipment
(4) Exothermic welding process
(5) Machine screw-type fasteners that engage not less than two threads or are secured with a nut
(6) Thread-forming machine screws that engage not less than two threads in the enclosure
(7) Connections that are part of a listed assembly
(8) Other listed means

By specifically identifying machine screws and thread-forming machine screws as acceptable connection methods, this section does not permit other types of screws, such as sheet metal screws or drywall screws, to be used for the connection of grounding and bonding conductors or terminals. Coarse-threaded screws do not satisfy product certification standard requirements for grounding and bonding connections to engage two full threads of the screw into a metal box or other metal enclosure.

Listed pressure connectors, such as twist-on wire connectors, do not have to be specifically listed for grounding and bonding. The use of listed pressure connectors other than those that are green in color is permitted for the connection of grounding and bonding conductors. Exhibits 250.2 and 250.3 illustrate two acceptable methods of attaching an equipment bonding jumper to a grounded metal box.

(B) Methods Not Permitted. Connection devices or fittings that depend solely on solder shall not be used.

250.10 Protection of Ground Clamps and Fittings. Ground clamps or other fittings exposed to physical damage shall be enclosed in metal, wood, or equivalent protective covering.

Δ **250.12 Clean Surfaces.** Nonconductive coatings (such as paint, lacquer, and enamel) on equipment to be grounded or bonded shall be removed from threads and other contact surfaces to ensure electrical continuity or shall be connected by means of fittings designed to make such removal unnecessary.

Certain fittings, such as locknuts and star washers, can be designed to ensure good electrical continuity to the contact surface through nonconductive coatings. The key is for the installer to make the connection up tight to ensure that the locknut provides that good electrical connection.

Part II. System Grounding

Δ **250.20 Alternating-Current Systems to Be Grounded.** Alternating-current systems shall be grounded in accordance

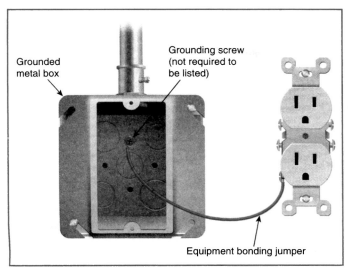

EXHIBIT 250.2 Equipment bonding jumper installed from a grounded metal box to the receptacle equipment grounding terminal.

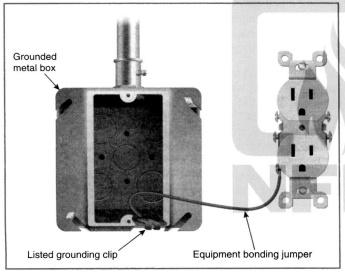

EXHIBIT 250.3 Use of a listed grounding clip to attach an equipment bonding jumper to a metal box.

with 250.20(A), (B), (C), or (D), unless prohibited elsewhere in this *Code*. Other systems shall be permitted to be grounded. If such systems are grounded, they shall comply with the applicable provisions of this article.

> Informational Note No. 1: An example of a system permitted to be grounded is a corner-grounded delta transformer connection.
>
> Informational Note No. 2: See 503.155, 517.61, 517.160, 668.10, and 680.23(A)(2) for examples of circuits prohibited to be grounded.

(A) Alternating-Current Systems of Less Than 50 Volts. Alternating-current systems of less than 50 volts shall be grounded under any of the following conditions:

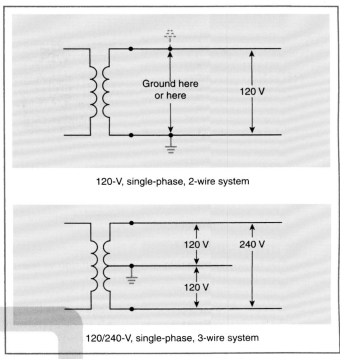

EXHIBIT 250.4 Typical grounding requirements for single-phase 120-volt, 2-wire and 120/240-volt, 3-wire systems.

(1) If supplied by transformers, if the transformer supply system exceeds 150 volts to ground
(2) If supplied by transformers, if the transformer supply system is ungrounded
(3) If installed outside as overhead conductors

(B) Alternating-Current Systems of 50 Volts to 1000 Volts. Alternating-current systems of 50 volts to 1000 volts that supply premises wiring and premises wiring systems shall be grounded under any of the following conditions:

(1) If the system can be grounded so that the maximum voltage to ground on the ungrounded conductors does not exceed 150 volts

Exhibit 250.4 illustrates the grounding requirements of 250.20(B)(1) for a 120-volt, single-phase, 2-wire system and for a 120/240-volt, single-phase, 3-wire system. The selection of which conductor to be grounded is covered in 250.26.

(2) If the system is 3-phase, 4-wire, wye connected in which the neutral conductor is used as a circuit conductor
(3) If the system is 3-phase, 4-wire, delta connected in which the midpoint of one phase winding is used as a circuit conductor

> Informational Note: See NFPA *70E*-2021, *Standard for Electrical Safety in the Workplace*, Annex O, for information on impedance grounding to reduce arc-flash hazards.

Exhibit 250.5 illustrates a 3-phase, 4-wire, wye-connected system covered by 250.20(B)(2). Because this system

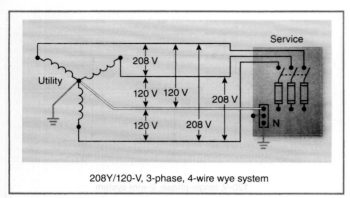

208Y/120-V, 3-phase, 4-wire wye system

EXHIBIT 250.5 A typical system supplying line-to-neutral connected loads in addition to line-to-line connected loads in which the neutral conductor is required to be grounded.

supplies line-to-neutral connected loads in addition to line-to-line connected loads, the neutral conductor is required to be grounded.

Exhibit 250.6 shows a 3-phase, 4-wire, delta-connected system covered by 250.20(B)(3). A connection is made at the midpoint of one phase to enable line-to-neutral loads to be supplied. The same voltage is developed between this grounded conductor and the two phase conductors connected at either end of the winding that is midpoint grounded. The voltage between the third phase conductor and the grounded conductor is higher, and other requirements in the *NEC®* cover the arrangement and identification of the "high leg" in this system.

See also

250.26 for requirements covering which conductor is to be grounded

(C) Alternating-Current Systems of over 1000 Volts. Alternating-current systems supplying mobile or portable equipment shall be grounded in accordance with 250.188. If supplying other than mobile or portable equipment, such systems shall be permitted to be grounded.

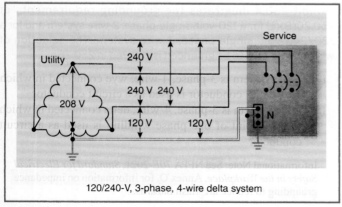

120/240-V, 3-phase, 4-wire delta system

EXHIBIT 250.6 A typical system in which the midpoint of one phase winding is grounded to provide a system grounded circuit conductor.

Δ **(D) Impedance Grounded Systems.** Impedance grounded systems shall be grounded in accordance with 250.36 or 250.187, as applicable.

250.21 Alternating-Current Systems of 50 Volts to 1000 Volts Not Required to Be Grounded.

Δ **(A) General.** The following ac systems of 50 volts to 1000 volts shall be permitted to be grounded but shall not be required to be grounded:

(1) Electrical systems used exclusively to supply industrial electric furnaces used for applications such as melting, refining, or tempering

(2) Separately derived systems used exclusively for rectifiers that supply only adjustable-speed industrial drives

(3) Separately derived systems supplied by transformers that have a primary voltage rating of 1000 volts or less if all the following conditions are met:

    a. The system is used exclusively for control circuits.

    b. The conditions of maintenance and supervision ensure that only qualified persons service the installation.

    c. Continuity of control power is required.

(4) Other systems that are not required to be grounded in accordance with 250.20(B)

Ungrounded systems are systems without an intentionally grounded circuit conductor that is used for normal circuit operation. Systems that normally operate as grounded systems — such as 120/240-volt, single-phase, 3-wire; 208Y/120-volt, 3-phase, 4-wire; and 480Y/277-volt, 3-phase, 4-wire systems — can be operated as ungrounded systems where specifically permitted by the *NEC*. A system that operates without a grounded conductor is not exempt from complying with the requirements for equipment grounding and for the required electrode grounding system.

Delta-connected, 3-phase, 3-wire, 240-volt and 480-volt systems are examples of common electrical distribution systems that are permitted but are not required to have a circuit conductor that is intentionally grounded. The operational advantage of using an ungrounded electrical system is continuity of operation, which in some processes might be a safer condition than that created by the automatic and unplanned opening of the supply circuit. The disadvantage of operating systems ungrounded is increased susceptibility to high transient voltages that can hasten insulation deterioration. As stated in 250.4(A)(1), limiting voltage impressed on the system due to lightning or line surges is a primary function of system grounding. In some limited applications, it may be desirable to not establish a voltage to ground. However, the consensus has been that for general applications, grounded systems provide a higher level of safety.

Unlike solidly grounded systems, in which the first line-to-ground fault causes the OCPD to automatically open the circuit, the same line-to-ground fault in an ungrounded system results in the faulted circuit conductor becoming a grounded conductor until the damaged conductor insulation can be repaired. However, this ground fault may become a hazard if it is undetected.

Ground detectors are required to be installed in ungrounded systems to detect such faults. A second insulation failure on a different ungrounded conductor would result in a line-to-line-to-ground fault, with the potential for more extensive damage to electrical equipment.

(B) Ground Detectors. Ground detectors shall be installed in accordance with the following:

(1) Ungrounded ac systems as permitted in 250.21(A)(1) through (A)(4) operating at not less than 120 volts and at 1000 volts or less shall have ground detectors installed on the system.
(2) The ground detection sensing equipment shall be connected as close as practicable to where the system receives its supply.

Ground detectors provide a visual indication, an audible signal, or both to alert system operators and maintenance personnel of a ground-fault condition in the electrical system. The notification of the ground-fault condition, rather than automatic interruption of the circuit, allows the operators of the process to take the necessary steps to initiate an orderly shutdown, to determine where the ground fault is located, and to safely repair the fault.

(C) Marking. Ungrounded systems shall be legibly marked "Caution: Ungrounded System Operating — _____Volts Between Conductors" at the source or first disconnecting means of the system. The marking shall be of sufficient durability to withstand the environment involved.

250.24 Grounding of Service-Supplied Alternating-Current Systems.

(A) System Grounding Connections. A premises wiring system supplied by a grounded ac service shall have a grounding electrode conductor connected to the grounded service conductor, at each service, in accordance with 250.24(A)(1) through (A)–(4).

The grounded conductor of an ac service is connected to a grounding electrode system to limit the voltage to ground imposed on the system by lightning, line surges, and unintentional high-voltage crossovers. This connection also stabilizes the voltage to ground during operation, including short circuits.

See also

250.4(A) and **(B)** for the performance requirements for these connections

Δ **(1) General.** The grounding electrode conductor connection shall be made at any accessible point from the load end of the overhead service conductors, service drop, underground service conductors, or service lateral to the terminal or bus to which the grounded service conductor is connected at the service disconnecting means.

> Informational Note: See Article 100 for definitions of *Service Conductors, Overhead; Service Conductors, Underground; Service Drop;* and *Service Lateral.*

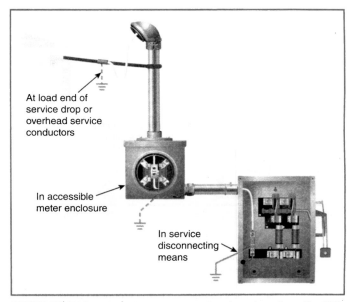

At load end of service drop or overhead service conductors

In accessible meter enclosure

In service disconnecting means

EXHIBIT 250.7 Three locations on an overhead service where the GEC is permitted to be connected to the grounded service conductor.

Allowing various connection locations meets the overall objectives for grounding while permitting a variety of practical options. Exhibit 250.7 illustrates three possible connection points where the GEC is permitted to be connected to the grounded service conductor. The accessibility to the point of connection must be approved by the AHJ based on local conditions such as locked meter socket enclosures.

Δ **(2) Outdoor Transformer.** If the transformer supplying the service is located outside the building, at least one additional grounding connection shall be made from the grounded service conductor to a grounding electrode, either at the transformer or elsewhere outside the building.

> *Exception: The additional grounding electrode conductor connection shall not be made on impedance grounded systems. Impedance grounded systems shall meet the requirements of 250.36 or 250.187, as applicable.*

Outdoor installations are susceptible to lightning as well as accidental primary-to-secondary crossovers. The connection outside a building helps mitigate the effects of these influences on the interior portion of the premises wiring system. Exhibit 250.8 illustrates two grounding electrode connections — one at the service equipment installed inside the building and one installed at the transformer. At least one of those connections must be located outside the building as required by 250.24(A)(2).

(3) Dual-Fed Services. For services that are dual fed (double ended) in a common enclosure or grouped together in separate enclosures and employing a secondary tie, a single grounding electrode conductor connection to the tie point of the grounded conductor(s) from each power source shall be permitted.

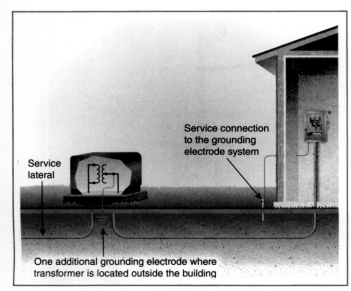

EXHIBIT 250.8 Grounding connection to the grounded conductor at the transformer and at the service equipment.

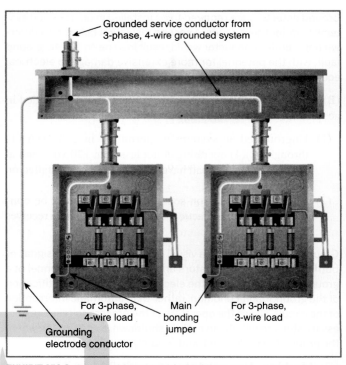

EXHIBIT 250.9 A grounded system in which the grounded service conductor is bonded to the enclosure supplying 3-phase, 4-wire service loads and to the enclosure supplying 3-phase, 3-wire loads.

(4) Main Bonding Jumper as Wire or Busbar. If the main bonding jumper specified in 250.28 is a wire or busbar and is installed from the grounded conductor terminal bar or bus to the equipment grounding terminal bar or bus in the service equipment, the grounding electrode conductor shall be permitted to be connected to the equipment grounding terminal, bar, or bus to which the main bonding jumper is connected.

Ṅ (B) Load-Side Grounding Connections. A grounded conductor shall not be connected to normally non-current-carrying metal parts of equipment, to equipment grounding conductor(s), or be reconnected to ground on the load side of the service disconnecting means except as otherwise permitted in this article.

> Informational Note: See 250.30 for separately derived systems, 250.32 for connections at separate buildings or structures, and 250.142 for use of the grounded circuit conductor for grounding equipment.

Except under limited conditions, 250.24(B) prohibits connecting the grounded conductor of an electrical system to a grounding electrode or grounding electrode system anywhere on the load side of the service disconnecting means. This prohibition correlates with the requirement of 250.142(B), which is a general prohibition on the use of the grounded conductor for grounding equipment. This prevents parallel paths for neutral current on the load side of the service disconnecting means. Parallel paths could include metal raceways, metal piping systems, metal ductwork, structural steel, and other continuous metal paths that are not intended to be current-carrying conductors under normal conditions.

Δ (C) Main Bonding Jumper. For a grounded system, an unspliced main bonding jumper shall be used to connect the equipment grounding conductor(s) and the service-disconnect

enclosure to the grounded conductor within the enclosure for each service disconnect in accordance with 250.28.

Where the service equipment of a grounded system consists of multiple disconnecting means in separate enclosures, a main bonding jumper for each separate service disconnecting means is required to connect the grounded service conductor, the EGC, and the service equipment enclosure. The size of the main bonding jumper in each enclosure is determined in accordance with 250.28(D)(1). See Exhibit 250.9.

> *Exception No. 1: If more than one service disconnecting means is located in an assembly listed for use as service equipment, an unspliced main bonding jumper shall bond the grounded conductor(s) to the assembly enclosure.*

Where multiple service disconnecting means are part of an assembly listed as service equipment, all grounded service conductors are required to be run to and bonded to the assembly. However, only one section of the assembly is required to have the main bonding jumper connection. See Exhibit 250.10.

> *Exception No. 2: Impedance grounded systems shall be permitted to be connected in accordance with 250.36 and 250.187.*

Δ (D) Grounded Conductor Brought to Service Equipment. If an ac system operating at 1000 volts or less is grounded at any point, the grounded conductor(s) shall be routed with the ungrounded conductors to each service disconnecting means and shall be connected to each disconnecting means grounded conductor(s) terminal or bus. A main bonding jumper shall connect the grounded conductor(s) to each service disconnecting means enclosure. The grounded conductor(s) shall be installed in accordance with 250.24(C)(1) and 250.24(D)(1) through (D)(4).

Exception: If two or more service disconnecting means are located in a single assembly listed for use as service equipment, it shall be permitted to connect the grounded conductor(s) to the assembly common grounded conductor(s) terminal or bus. The assembly shall include a main bonding jumper for connecting the grounded conductor(s) to the assembly enclosure.

If the utility service supplying the premises wiring system is grounded, the grounded conductor must be run to the service equipment, regardless of whether line-to-neutral loads are supplied. The grounded conductor must be bonded to the equipment and be connected to a grounding electrode system. Exhibit 250.9 shows an example of the main rule in 250.24(D), which requires the grounded service conductor to be installed and bonded to each service disconnecting means enclosure. On the line side of the service disconnecting means, the grounded conductor is used to complete the ground-fault current path between the service equipment and the utility source. The exception to 250.24(D) permits a single connection of the grounded service conductor to a listed service assembly (such as a switchboard) that contains more than one service disconnecting means, as shown in Exhibit 250.10.

(1) Sizing for a Single Raceway or Cable. The grounded conductor shall not be smaller than specified in Table 250.102(C)(1).

Δ **(2) Conductors Connected in Parallel in Two or More Raceways or Cables.** If the ungrounded service-entrance conductors are connected in parallel in two or more raceways or cables, the grounded conductors shall also be installed in each raceway or cable and shall be connected in parallel. The size of each grounded conductor(s) in each raceway or cable shall not be smaller than 1/0 AWG and shall be sized in accordance with 250.24(D)(2)(a) or (D)(2)(b) in accordance with 250.24(D)(1).

(a) Shall be based on the largest ungrounded conductor in each raceway or cable.

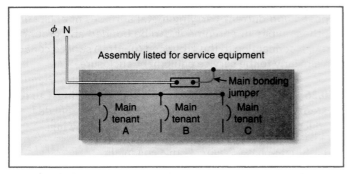

EXHIBIT 250.10 One connection of the grounded service conductor to a listed service assembly containing multiple service disconnecting means.

(b) Shall be based on the sum of the circular mil areas of the largest ungrounded conductors from each set connected in parallel in each raceway or cable.

Informational Note: See 310.10(G) for grounded conductors connected in parallel.

(3) Delta-Connected Service. The grounded conductor of a 3-phase, 3-wire delta service shall have an ampacity not less than that of the ungrounded conductors.

Δ **(4) Impedance Grounded Service.** The impedance grounding conductor on an impedance grounded system shall be connected in accordance with 250.36 or 250.187, as applicable.

(E) Grounding Electrode Conductor. A grounding electrode conductor shall be used to connect the equipment grounding conductors, the service-equipment enclosures, and, if the system is grounded, the grounded service conductor to the grounding electrode(s) required by Part III of this article. This conductor shall be sized in accordance with 250.66.

Impedance grounded system connections shall be made in accordance with 250.36 or 250.187, as applicable.

(F) Ungrounded System Grounding Connections. A premises wiring system that is supplied by an ac service that is ungrounded shall have, at each service, a grounding electrode conductor connected to the grounding electrode(s) required by Part III of this article. The grounding electrode conductor shall be connected to a metal enclosure of the service conductors at any accessible point from the load end of the overhead service conductors, service drop, underground service conductors, or service lateral to the service disconnecting means.

250.25 Grounding of Systems Permitted to Be Connected on the Supply Side of the Service Disconnect. The grounding of systems connected on the supply side of the service disconnect, in accordance with 230.82, that are in enclosures separate from the service equipment enclosure shall comply with 250.25(A) or (B).

(A) Grounded System. If the utility supply system is grounded, the grounding of systems permitted to be connected on the supply side of the service disconnect and are installed in one or more separate enclosures from the service equipment enclosure shall comply with the requirements of 250.24(A) through (D).

(B) Ungrounded Systems. If the utility supply system is ungrounded, the grounding of systems permitted to be connected on the supply side of the service disconnect and are installed in one or more separate enclosures from the service equipment enclosure shall comply with the requirements of 250.24(F).

This section applies to any equipment permitted to be connected on the supply or line side of the service disconnecting means as permitted by 230.82. However, if the system that is connected on the supply side is intended to operate in parallel or simultaneously with the utility service, such systems fall under the purview of Article 705, Interconnected Electric Power Production Sources.

CLOSER LOOK: Sizing Main and System Bonding Jumpers

General

In a grounded system, the primary function of the main bonding jumper and of the system bonding jumper is to create the link for ground-fault current between the EGCs and the grounded conductor. Table 250.102(C)(1) is used to establish the minimum size of main and system bonding jumpers. Unlike the GEC, which carries current to the ground (via connection to a grounding electrode), the main and system bonding jumpers are placed directly in the supply-side ground-fault current return path.

Where the largest ungrounded supply conductor exceeds the parameters of Table 250.102(C)(1), Note 1 in the table requires a proportional relationship between the ungrounded conductor and the main or system bonding jumper. Where the service-entrance conductors are larger than 1100 kcmil copper or 1750 kcmil aluminum, the bonding jumper must have a cross-sectional area of not less than 12½ percent of the cross-sectional area of the largest phase conductor or largest phase conductor set. In equipment such as panelboards or switchboards that are listed for use as service equipment, the manufacturer provides a bonding jumper that can be installed as the main or system bonding jumper. It is not necessary to provide an additional bonding jumper.

Main Bonding Jumper for Service with More Than One Enclosure

Where a service consists of more than one disconnecting means in separate enclosures, each enclosure is treated separately, as depicted in Exhibit 250.11. Based on the 3/0 AWG ungrounded service conductors and Table 250.102(C)(1), the minimum-size main bonding jumper for the enclosure on the left is 4 AWG copper. Similarly, the 1/0 AWG main bonding jumper for the enclosure on the right is derived from Table 250.102(C)(1) using the 500 kcmil ungrounded service conductors. The bonding jumper provided by the manufacturer for listed service equipment provides the equivalent current capacity to that of a field fabricated bonding jumper sized per Table 250.102(C)(1).

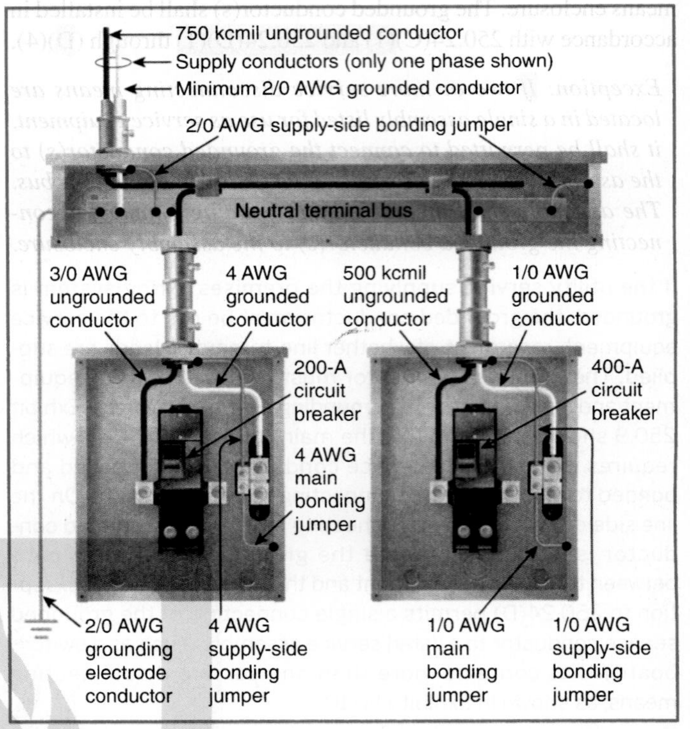

EXHIBIT 250.11 An example of the bonding requirements for service equipment.

Separately Derived System with More Than One Enclosure

To prevent parallel neutral current paths on raceways and enclosures, the system bonding jumper can either be internal to the panelboards or be installed at the separately derived system enclosure so that it connects any supply-side bonding jumpers to the system grounded conductor terminal. The system bonding jumper is not permitted to be installed at both locations.

Energy production systems such as photovoltaic, wind, fuel cells, or engine generators are permitted to make supply-side connections and operate in parallel with the utility service. The disconnecting means required in Article 705 for such systems are service disconnecting means, and the requirements of 250.24 apply to those disconnecting means.

Equipment that is covered by 250.25 includes equipment connected in series with the service conductors on the supply side of the disconnecting means, but are not service equipment or service disconnecting means. Examples of such equipment include meters, meter enclosures, meter disconnect switches, meter-mounted transfer switches, and emergency disconnects per 230.82(2), (3), (10) and (11). The exposed non-current-carrying metal surfaces of enclosures for these types of equipment can be bonded and grounded through connection to the grounded (neutral) service conductor, similarly to the service equipment enclosure.

250.26 Conductor to Be Grounded — Alternating-Current Systems. If an ac premises wiring system is grounded, the conductor to be grounded shall be one of the following:

(1) Single-phase, 2-wire — one conductor
(2) Single-phase, 3-wire — the neutral conductor

Application Example 1

A service is supplied by four 500 kcmil conductors in parallel for each phase. The minimum cross-sectional area of the bonding jumper is calculated as follows:

$$4 \times 500 \text{ kcmil} = 2000 \text{ kcmil}$$

Therefore, the main or system bonding jumper cannot be less than 12½ percent of 2000 kcmil, which results in a 250 kcmil copper conductor. The copper GEC for this set of conductors, based on Table 250.66, is not required to be larger than 3/0 AWG.

Application Example 2

System Bonding Jumpers at Panelboards
A 225-kVA transformer supplies three 3-phase, 208Y/120-V secondary feeders. Two sets of 3/0 AWG copper, THHW, ungrounded conductors terminate in panelboards with 200-A main breakers. One set of 3 AWG copper, THHW, ungrounded conductors terminates in a panelboard with a 100-A main breaker. A system bonding jumper is installed at each of the panelboards. Determine the size of the system bonding jumpers at each panelboard enclosure.

200-A panelboards:
Size of largest ungrounded conductor supplying the two 200-A panelboards:

3/0 AWG copper

System bonding jumper [from Table 250.102(C)(1)]:

3/0 AWG copper ungrounded conductors → 4 AWG copper

or 2 AWG aluminum system bonding jumper

100-A panelboard:
Size of largest ungrounded conductor supplying the 100-A panelboard:

3 AWG copper

System bonding jumper [from Table 250.102(C)(1)]:

3 AWG copper ungrounded conductors → 8 AWG copper

or 6 AWG aluminum system bonding jumper

Application Example 3

System Bonding Jumper at the Transformer
The same electrical equipment arrangement as in the previous example applies. In this case, however, the system bonding jumper is installed at the transformer and connects the transformer neutral terminal (XO) to individual supply-side bonding jumpers that are installed between the panelboard EGC terminals and a terminal bus attached to the transformer enclosure. Determine the size of the system bonding jumper.

Size the system bonding jumper from the terminal bus in the transformer to the transformer neutral terminal (XO). There are two secondary feeder circuits with 3/0 AWG ungrounded conductors and one with 3 AWG ungrounded conductors. Find the cumulative circular mil area of all ungrounded conductors of a phase.

From Chapter 9, Table 8:

3/0 AWG = 167,800 circular mils × 2 (number of sets of secondary conductors) = 335,600 circular mils

3 AWG = 52,620 circular mils

Total = 388,220 circular mils

From Table 250.102(C)(1):

388,220 circular mils copper ungrounded conductors → 2 AWG copper or 1/0 AWG aluminum system bonding jumper

---

(3) Multiphase systems having one wire common to all phases — the neutral conductor
(4) Multiphase systems if one phase is grounded — that phase conductor
(5) Multiphase systems in which one phase is used as in (2) — the neutral conductor

250.28 Main Bonding Jumper and System Bonding Jumper. For a grounded system, main bonding jumpers and system bonding jumpers shall be installed as follows:

The system bonding jumper performs the same electrical function as the main bonding jumper in a grounded ac system by connecting the EGCs to the grounded circuit conductor either at the source of a separately derived system or at the first disconnecting means supplied by the source. The term *system bonding jumper* is used to distinguish it from the main bonding jumper, which is installed in service equipment. See the commentary following the definition of *bonding jumper, system* in Article 100.

(A) Material. Main bonding jumpers and system bonding jumpers shall be of copper, aluminum, copper-clad aluminum, or other corrosion-resistant material. A main bonding jumper and a system bonding jumper shall be a wire, bus, screw, or similar suitable conductor.

(B) Construction. If a main bonding jumper or a system bonding jumper is a screw only, the screw shall be identified with a green finish that shall be visible with the screw installed.

(C) Attachment. Main bonding jumpers and system bonding jumpers shall be connected by one or more of the methods in 250.8 that is suitable for the material of the bonding jumper and enclosure.

(D) Size. Main bonding jumpers and system bonding jumpers shall be sized in accordance with 250.28(D)(1) through (D)(3).

(1) General. Main bonding jumpers and system bonding jumpers shall not be smaller than specified in Table 250.102(C)(1).

(2) Main Bonding Jumper for Service with More Than One Enclosure. If a service consists of more than a single enclosure as permitted in 230.71(B), the main bonding jumper for each enclosure shall be sized in accordance with 250.28(D)(1) based on the largest ungrounded service conductor serving that enclosure.

(3) Separately Derived System with More Than One Enclosure. If a separately derived system supplies more than a single enclosure, the system bonding jumper for each enclosure shall be sized in accordance with 250.28(D)(1) based on the largest ungrounded feeder conductor serving that enclosure, or a single system bonding jumper shall be installed at the source and sized in accordance with 250.28(D)(1) based on the equivalent size of the largest supply conductor determined by the largest sum of the areas of the corresponding conductors of each set.

For more information, see the accompanying Closer Look feature about sizing main and system bonding jumpers.

Δ **250.30 Grounding Separately Derived Alternating-Current Systems.** In addition to complying with 250.30(A) for grounded systems, or as provided in 250.30(B) for ungrounded systems, separately derived systems shall comply with 250.20, 250.21, or 250.26, as applicable. Multiple power sources of the same type that are connected in parallel to form one system that supplies premises wiring shall be treated as a single separately derived system and shall be installed in accordance with 250.30.

An example of multiple power sources of the same type connected in parallel is an emergency or standby (legally required or optional) power system supplied by multiple generators connected to paralleling bus or equipment where the transfer arrangement switches the ungrounded and grounded conductors of the normal and alternate power supplies.

> Informational Note No. 1: An alternate ac power source, such as an on-site generator, is not a separately derived system if the grounded conductor is solidly interconnected to a service-supplied system grounded conductor. An example of such a situation is if the alternate source transfer equipment does not include a switching action in the grounded conductor and allows it to remain solidly connected to the service-supplied grounded conductor when the alternate source is operational and supplying the load served.
>
> Informational Note No. 2: See 445.13 for the minimum size of conductors that carry fault current.

In Exhibit 250.12, the neutral conductor from the generator to the load is not disconnected by the transfer switch. The system has a direct electrical connection between the normal grounded

system conductor (neutral) and the generator neutral through the neutral bus in the transfer switch, thereby grounding the generator neutral. Because the grounded circuit conductor is connected to the normal system grounded conductor, it is not a separately derived system and there are no requirements for grounding the neutral at the generator (see Informational Note No. 1 to 250.30).

In Exhibit 250.13, the grounded conductor (neutral) is connected to the switching contacts of a 4-pole transfer switch. The generator system does not have a direct electrical connection to

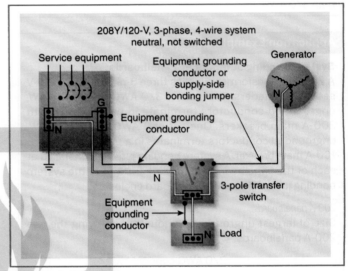

EXHIBIT 250.12 A 208Y/120-volt, 3-phase, 4-wire system that has a direct electrical connection of the grounded circuit conductor (neutral) to the generator and is therefore not considered a separately derived system.

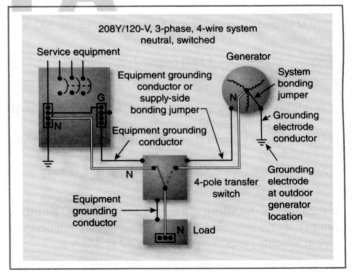

EXHIBIT 250.13 A 208Y/120-volt, 3-phase, 4-wire system that does not have a direct electrical connection of the grounded circuit conductor (neutral) to the generator and is therefore considered a separately derived system.

the other supply system grounded conductor (neutral), other than the bonding and equipment grounding conductors (EGCs). Therefore, the system supplied by the generator is considered separately derived.

Δ **(A) Grounded Systems.** A separately derived ac system that is grounded shall comply with 250.30(A)(1) through (A)(8). Except as otherwise permitted in this article, a grounded conductor shall not be connected to normally non-current-carrying metal parts of equipment, be connected to equipment grounding conductors, or be reconnected to ground on the load side of the system bonding jumper.

> Informational Note: See 250.32 for connections at separate buildings or structures and 250.142 for use of the grounded circuit conductor for grounding equipment.

Exception: Impedance grounded system grounding connections shall be made in accordance with 250.36 or 250.187, as applicable.

(1) System Bonding Jumper. An unspliced system bonding jumper shall comply with 250.28(A) through (D). This connection shall be made at any single point on the separately derived system from the source to the first system disconnecting means or overcurrent device, or it shall be made at the source of a separately derived system that has no disconnecting means or overcurrent devices, in accordance with 250.30(A)(1)(a) or (A)(1)(b). The system bonding jumper shall remain within the enclosure where it originates. If the source is located outside the building or structure supplied, a system bonding jumper shall be installed at the grounding electrode connection in compliance with 250.30(C).

Exception No. 1: For systems installed in accordance with 450.6, a single system bonding jumper connection to the tie point of the grounded circuit conductors from each power source shall be permitted.

Exception No. 2: If a building or structure is supplied by a feeder from an outdoor separately derived system, a system bonding jumper at both the source and the first disconnecting means shall be permitted if doing so does not establish a parallel path for the grounded conductor. If a grounded conductor is used in this manner, it shall not be smaller than the size specified for the system bonding jumper but shall not be required to be larger than the ungrounded conductor(s). For the purposes of this exception, connection through the earth shall not be considered as providing a parallel path.

Exception No. 3: The size of the system bonding jumper for a system that supplies a Class 1, Class 2, or Class 3 circuit, and is derived from a transformer rated not more than 1000 volt-amperes, shall not be smaller than the derived ungrounded conductors and shall not be smaller than 14 AWG copper or 12 AWG aluminum.

(a) *Installed at the Source.* The system bonding jumper shall connect the grounded conductor to the supply-side bonding jumper and the normally non-current-carrying metal enclosure.

(b) *Installed at the First Disconnecting Means.* The system bonding jumper shall connect the grounded conductor to the supply-side bonding jumper, the disconnecting means enclosure, and the equipment grounding conductor(s).

Separately derived systems are required to have a system bonding jumper to connect the grounded circuit conductor (neutral) to the supply-side bonding jumper, the EGC, or both.

The system bonding jumper can be installed in several ways. For example, if a multi-barrel lug is connected to the XO terminal of a transformer, the system bonding jumper, GEC, grounded conductor, and supply-side jumper can be connected at that connector. If a multi-barrel lug is connected to the transformer or generator enclosure, a common practice is to connect the system bonding jumper, the GEC, and the bonding jumper or conductor to that connector. The grounded conductor should always connect directly to the XO terminal.

See also

250.28(D) and its commentary for more information on sizing the system bonding jumper

Exception: Separately derived systems consisting of multiple sources of the same type that are connected in parallel shall be permitted to have the system bonding jumper installed at the paralleling switchgear, switchboard, or other paralleling connection point instead of at the disconnecting means located at each separate source.

The sizing requirements for system bonding jumpers in 250.28(D) and Table 250.102(C)(1) use ungrounded conductor size as the basis. In large-capacity parallel generator standby power systems, the connections from the generators to a collector bus can be done exclusively with busbars. To size a system bonding jumper in that case, it is necessary to convert the busbar ampere rating to an equivalent ungrounded conductor size in order to be able to use Table 250.102(C)(1).

For example, if the busbar rating (per ungrounded phase) for a paralleled installation is 2500 amperes, it is necessary to establish the minimum copper or aluminum wire size to supply 2500 amperes and, using that conversion, then apply Table 250.102(C)(1) to size a single system bonding jumper for the paralleled installation. Because of the system size in this example, the equivalent wire size will exceed 1100 kcmil copper or 1750 kcmil aluminum. Therefore, an additional calculation using the 12½ percent rule will have to be performed to size the system bonding jumper.

(2) Supply-Side Bonding Jumper. If the source of a separately derived system and the first disconnecting means are located in separate enclosures, a supply-side bonding jumper shall be installed with the circuit conductors from the source enclosure to the first disconnecting means enclosure. A supply-side bonding jumper shall not be required to be larger than the derived ungrounded conductors. The supply-side bonding jumper shall be permitted to be of nonflexible metal raceway type or of the wire or bus type as follows:

(1) A supply-side bonding jumper of the wire type shall comply with 250.102(C), based on the size of the derived ungrounded conductors.

(2) A supply-side bonding jumper of the bus type shall have a cross-sectional area not smaller than a supply-side bonding jumper of the wire type as determined in 250.102(C).

Exception: A supply-side bonding jumper shall not be required between enclosures for installations made in compliance with 250.30(A)(1), Exception No. 2.

The supply-side bonding jumper installed between the separately derived system enclosure and the enclosure of the first system disconnecting means provides the path for the circuit ground-fault current, whether the system bonding jumper is installed at the source enclosure or at the disconnecting means enclosure. The supply-side bonding jumper can be a bare, covered, or insulated wire that is sized per 250.102(C), or it can be rigid metal conduit, intermediate metal conduit, or electrical metallic tubing installed between the two enclosures.

Application Example

The source of a separately derived system is a 75-kVA dry-type transformer. Liquidtight flexible metal conduit is used as the wiring method between the transformer and the 200-A fusible safety switch that is the first system disconnecting means. The system bonding jumper is installed in the safety switch enclosure. The ungrounded conductors are 3/0 AWG copper. The wiring method necessitates the installation of a wire-type supply-side bonding jumper. What is the minimum size for this bonding jumper?

Solution

The requirement covering the minimum size for the supply-side bonding jumper is 250.102(C). That section refers to Table 250.102(C)(1) for sizing supply-side bonding jumpers where the ungrounded supply conductors are not greater than 1100 kcmil copper or 1750 kcmil aluminum. From Table 250.102(C)(1), 3/0 AWG ungrounded conductors = 4 AWG copper or 2 AWG aluminum supply-side bonding jumper. If the length of the bonding jumper does not exceed 6 ft, 250.102(E)(2) permits the jumper to be installed outside the raceway and routed with it.

Δ **(3) Grounded Conductor.** If a grounded conductor is installed and the system bonding jumper connection is not located at the source, 250.30(A)(3)(a) through (A)(3)(d) shall apply. The grounded conductor shall not be required to be larger than the derived ungrounded conductors.

(a) *Sizing for a Single Raceway.* The grounded conductor shall not be smaller than specified in Table 250.102(C)(1).

(b) *Conductors Connected in Parallel in Two or More Raceways or Cables.* If the ungrounded conductors are connected in parallel in two or more raceways or cables, the grounded conductors shall also be installed in each raceway or cable and shall be connected in parallel. The size of the grounded conductor(s) in each raceway or cable shall be based on the largest derived ungrounded conductor in each raceway or cable, or the sum of the circular mil areas of the largest derived ungrounded conductors from each set connected in parallel in each raceway or cable, in accordance with 250.30(A)(3)(a), but not smaller than 1/0 AWG.

Informational Note: See 310.10(G) for grounded conductors connected in parallel.

(c) *Delta-Connected System.* The grounded conductor of a 3-phase, 3-wire delta system shall have an ampacity not less than that of the ungrounded conductors.

(d) *Impedance Grounded System.* The impedance grounding conductor of an impedance grounded system shall be installed in accordance with 250.36 or 250.187, as applicable.

Grounded and neutral conductors that carry current as part of normal circuit operation are required to be sized in accordance with 220.61. Grounded conductors that are intended to carry fault current must also meet the sizing requirements of 250.102(C).

For example, if the system bonding jumper is not installed at the source of a grounded separately derived system that has line-to-neutral connected loads, the grounded conductor of a system, such as a single-phase 3-wire or three-phase 4-wire, provides two circuit functions. Under normal operating conditions, it is the circuit conductor for line-to-neutral connected loads. Under the abnormal conditions of a line-to-ground fault, it is also a path for the ground-fault current to return to the source. Based on its dual functions, the grounded (neutral) conductor is sized based on a comparison between 220.61 (normal circuit operation) and 250.102(C) (fault conditions), with the larger conductor resulting from that comparison being the minimum required size.

(4) Grounding Electrode. The building or structure grounding electrode system shall be used as the grounding electrode for the separately derived system. If located outdoors, the grounding electrode shall be in accordance with 250.30(C).

Exception: If a separately derived system originates in equipment that is listed and identified as suitable for use as service equipment, the grounding electrode used for the service or feeder equipment shall be permitted to be used as the grounding electrode for the separately derived system.

Informational Note No. 1: See 250.104(D) for bonding requirements for interior metal water piping in the area served by separately derived systems.

Informational Note No. 2: See 250.50 and 250.58 for requirements for bonding all electrodes together if located at the same building or structure.

As described in 250.50, all electrodes described in 250.52(A)(1) through (7) that are present are required to be bonded together to form the grounding electrode system for the building or structure. Separately derived systems are required to utilize the entire building or structure grounding electrode system as the grounding electrode.

(5) Grounding Electrode Conductor, Single Separately Derived System. A grounding electrode conductor for a single separately derived system shall be sized in accordance with 250.66 for the derived ungrounded conductors. It shall be used to connect the grounded conductor of the derived system to the grounding electrode in accordance with 250.30(A)(4), or as permitted in 250.68(C)(1) and (C)(2). This connection shall be made at the same point on the separately derived system where the system bonding jumper is connected.

Exception No. 1: If the system bonding jumper specified in 250.30(A)(1) is a wire or busbar, it shall be permitted to connect the grounding electrode conductor to the equipment grounding terminal, bar, or bus if the equipment grounding terminal, bar, or bus is of sufficient size for the separately derived system.

Exception No. 2: If the source of a separately derived system is located within equipment listed and identified as suitable for use as service equipment, the grounding electrode conductor from the service or feeder equipment to the grounding electrode shall be permitted as the grounding electrode conductor for the separately derived system, if the grounding electrode conductor is of sufficient size for the separately derived system. If the equipment grounding bus internal to the equipment is not smaller than the required grounding electrode conductor for the separately derived system, the grounding electrode connection for the separately derived system shall be permitted to be made to the bus.

Exception No. 3: A grounding electrode conductor shall not be required for a system that supplies a Class 1, Class 2, or Class 3 circuit and is derived from a transformer rated not more than 1000 volt-amperes, provided the grounded conductor is bonded to the transformer frame or enclosure by a jumper sized in accordance with 250.30(A)(1), Exception No. 3, and the transformer frame or enclosure is grounded by one of the means specified in 250.134.

The location of the GEC connection to the grounded conductor must be at the point at which the system bonding jumper is connected to the grounded conductor. This allows the normal neutral current to be carried on only the system grounded conductor and not be imposed on parallel paths. Exhibits 250.14 and 250.15 illustrate two acceptable locations for connecting the GEC to the grounded conductor of a separately derived system. The exhibits depict typical wiring arrangements for dry-type transformers supplied from a 480-volt, 3-phase feeder to derive a 208Y/120-volt or 480Y/277-volt secondary.

In Exhibit 250.14, the GEC connection is made at the source of the separately derived system (in the transformer on a terminal bar as required by 450.10 where the system bonding jumper is also installed). In Exhibit 250.15, the GEC connection is made at the first disconnecting means where the system bonding jumper is installed. With the GEC, the bonding jumper, and the bonding of the grounded circuit conductor (neutral) connected as shown, line-to-ground fault currents are able to return to the

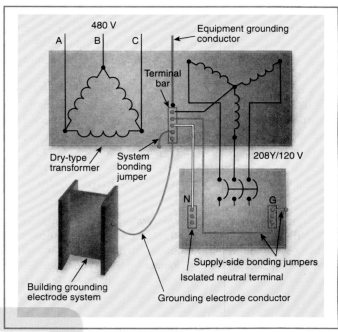

EXHIBIT 250.14 A grounding arrangement for a separately derived system in which the GEC connection is made at the source of the separately derived system (transformer).

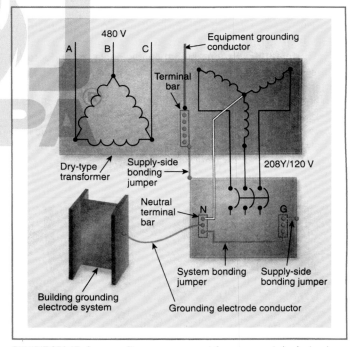

EXHIBIT 250.15 A grounding arrangement for a separately derived system in which the GEC connection is made at the first system disconnecting means.

supply source through a short, low-impedance path. A path of lower impedance is provided that facilitates the operation of overcurrent devices, in accordance with 250.4(A)(5).

(6) Grounding Electrode Conductor, Multiple Separately Derived Systems. A common grounding electrode conductor for multiple separately derived systems shall be permitted. If installed, the common grounding electrode conductor shall be used to connect the grounded conductor of each separately derived system to the grounding electrode as specified in 250.30(A)(4). A grounding electrode conductor tap shall then be installed from each separately derived system to the common grounding electrode conductor. Each tap conductor shall connect the grounded conductor of the separately derived system to the common grounding electrode conductor. This connection shall be made at the same point on the separately derived system where the system bonding jumper is connected.

A common GEC serving several separately derived systems is an alternative to installing individual GECs from each separately derived system to the grounding electrode system. In such an arrangement, a tapped GEC is installed from the common GEC to the point of connection to the individual separately derived system grounding conductor. This tap is sized from Table 250.66 based on the size of the ungrounded conductors for that individual separately derived system.

The minimum size for this conductor is 3/0 AWG copper or 250 kcmil aluminum so that the GEC always is of sufficient size to accommodate the multiple separately derived systems it serves. The sizing requirement for the common GEC is specified in 250.30(A)(6)(a), and the sizing requirement for the individual taps to the common GEC is specified in 250.30(A)(6)(b).

The methods of connecting the individual tap conductor(s) to the common GEC are specified in 250.30(A)(6)(c). The permitted methods include the use of a busbar as a point of connection between the taps and the common GEC. The connections to the busbar must be made using a listed means. Exhibit 250.16 shows a copper busbar used as a connection point for the individual taps from multiple separately derived systems to be connected to the common grounding electrode.

EXHIBIT 250.16 Listed connectors used to connect the common GEC and individual taps to a centrally located copper busbar with a minimum dimension of ¼ inch thick by 2 inches wide.

Exception No. 1: If the system bonding jumper specified in 250.30(A)(1) is a wire or busbar, it shall be permitted to connect the grounding electrode conductor tap to the equipment grounding terminal, bar, or bus, provided the equipment grounding terminal, bar, or bus is of sufficient size for the separately derived system.

Exception No. 2: A grounding electrode conductor shall not be required for a system that supplies a Class 1, Class 2, or Class 3 circuit and is derived from a transformer rated not more than 1000 volt-amperes, provided the system grounded conductor is bonded to the transformer frame or enclosure by a jumper sized in accordance with 250.30(A)(1), Exception No. 3, and the transformer frame or enclosure is grounded by one of the means specified in 250.134.

Exception No. 3: If the source of a separately derived system is located within equipment listed and identified as suitable for use as service equipment, the grounding electrode conductor from the service or feeder equipment to the grounding electrode shall be permitted as the grounding electrode conductor for the separately derived system, if the grounding electrode conductor is of sufficient size for the separately derived system. If the equipment grounding bus internal to the equipment is not smaller than the required grounding electrode conductor for the separately derived system, the grounding electrode connection for the separately derived system shall be permitted to be made to the bus.

(a) *Common Grounding Electrode Conductor.* The common grounding electrode conductor shall be permitted to be one of the following:

(1) A conductor of the wire type not smaller than 3/0 AWG copper or 250 kcmil aluminum
(2) A metal water pipe in accordance with 250.68(C)(1)
(3) The metal structural frame of the building or structure in accordance with 250.68(C)(2) or is connected to the grounding electrode system by a conductor not smaller than 3/0 AWG copper or 250 kcmil aluminum

(b) *Tap Conductor Size.* Each tap conductor shall be sized in accordance with 250.66 based on the derived ungrounded conductors of the separately derived system it serves.

Exception to (a)(1) and (b): If the only electrodes that are present are of the types in 250.66(A), (B), or (C), the size of the common grounding electrode conductor shall not be required to be larger than the largest conductor required by 250.66(A), (B), or (C) for the type of electrode that is present.

(c) *Connections.* All tap connections to the common grounding electrode conductor shall be made at an accessible location by one of the following methods:

(1) A connector listed as grounding and bonding equipment.
(2) Listed connections to aluminum or copper busbars not smaller than 6 mm thick × 50 mm wide (¼ in. thick × 2 in. wide) and of a length to accommodate the number

of terminations necessary for the installation. If aluminum busbars are used, the installation shall also be in accordance with 250.64(A).

(3) The exothermic welding process.

Tap conductors shall be connected to the common grounding electrode conductor in such a manner that the common grounding electrode conductor remains without a splice or joint.

(7) Installation. The installation of all grounding electrode conductors shall comply with 250.64(A), (B), (C), and (E).

(8) Bonding. Structural steel and metal piping shall be connected to the grounded conductor of a separately derived system in accordance with 250.104(D).

(B) Ungrounded Systems. The equipment of an ungrounded separately derived system shall be grounded and bonded as specified in 250.30(B)(1) through (B)(3).

(1) Grounding Electrode Conductor. A grounding electrode conductor, sized in accordance with 250.66 for the largest derived ungrounded conductor(s) or set of derived ungrounded conductors, shall be used to connect the metal enclosures of the derived system to the grounding electrode as specified in 250.30(A)(5) or (A)(6), as applicable. This connection shall be made at any point on the separately derived system from the source to the first system disconnecting means. If the source is located outside the building or structure supplied, a grounding electrode connection shall be made in compliance with 250.30(C).

For ungrounded separately derived systems, a GEC is required to be connected to the metal enclosure of the system disconnecting means. The GEC is sized from Table 250.66 based on the largest ungrounded supply conductor. This connection establishes a reference to ground for all exposed non-current-carrying metal equipment supplied from the ungrounded system. The EGCs of circuits supplied from the ungrounded system are connected to ground via the GEC connection.

(2) Grounding Electrode. Except as permitted by 250.34 for portable and vehicle-mounted generators, the grounding electrode shall comply with 250.30(A)(4).

(3) Bonding Path and Conductor. A supply-side bonding jumper shall be installed from the source of a separately derived system to the first disconnecting means in compliance with 250.30(A)(2).

△ **(C) Outdoor Source.** If the source of the separately derived system is located outside the building or structure supplied, a grounding electrode connection shall be made at the source location to one or more grounding electrodes in accordance with 250.50. In addition, the installation shall be in accordance with 250.30(A) for grounded systems or with 250.30(B) for ungrounded systems.

Exception: The grounding electrode conductor connection for impedance grounded systems shall be in accordance with 250.36 or 250.187, as applicable.

This exception is similar in function to the requirement of 250.24(A)(2) in that it allows an outdoor grounding connection at the source of a separately derived system. This connection provides a first line of defense against the effects of overvoltages due to lightning, transients, or accidental contact between conductors of systems operating at different voltages.

250.32 Buildings or Structures Supplied by a Feeder(s) or Branch Circuit(s).

△ **(A) Grounding Electrode System and Grounding Electrode Conductor.** A building(s) or structure(s) supplied by a feeder(s) or branch circuit(s) shall have a grounding electrode system and grounding electrode conductor installed in accordance with Part III of Article 250.

The equipment grounding bus must be bonded to the grounding electrode system as is shown for Buildings 2 and 3 in Exhibit 250.17. Building 1 is supplied by a service and is grounded in accordance with 250.24(A) through (D), and the disconnecting means enclosure, building steel, and interior metal water piping are also required to be bonded to the grounding electrode system. All exposed non-current-carrying metal parts of electrical equipment are required to be grounded through EGC connections to the equipment grounding bus at the building disconnecting means. The grounded conductor of the feeder supplying Building 2 is permitted to be re-grounded per 250.32(B)(1), Exception No. 1. An EGC is run with the feeder to Building 3 as specified in the general requirement of 250.32(B).

Exception: A grounding electrode system and grounding electrode conductor shall not be required if only a single branch circuit, including a multiwire branch circuit, supplies the building or structure and the branch circuit includes an equipment grounding conductor for grounding the normally non-current-carrying metal parts of equipment.

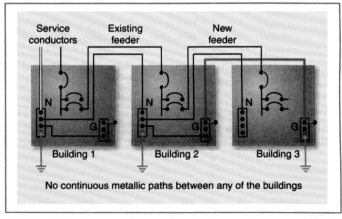

Service conductors | Existing feeder | New feeder

Building 1 | Building 2 | Building 3

No continuous metallic paths between any of the buildings

EXHIBIT 250.17 Example of grounding electrode systems required at feeder-supplied Building 2 and Building 3.

(B) Grounded Systems.

(1) Supplied by a Feeder or Branch Circuit. An equipment grounding conductor, as described in 250.118, shall be run with the supply conductors and be connected to the building or structure disconnecting means and to the grounding electrode(s). The equipment grounding conductor shall be used for grounding or bonding of equipment, structures, or frames required to be grounded or bonded. The equipment grounding conductor shall be sized in accordance with 250.122. Any installed grounded conductor shall not be connected to the equipment grounding conductor or to the grounding electrode(s).

Exception No. 1: For installations made in compliance with previous editions of this Code that permitted such connection, the grounded conductor run with the supply to the building or structure shall be permitted to serve as the ground-fault return path if all of the following requirements continue to be met:

(1) An equipment grounding conductor is not run with the supply to the building or structure.
(2) There are no continuous metallic paths bonded to the grounding system in each building or structure involved.
(3) Ground-fault protection of equipment has not been installed on the supply side of the feeder(s).

If the grounded conductor is used for grounding in accordance with the provision of this exception, the size of the grounded conductor shall not be smaller than the larger of either of the following:

(1) The calculated neutral load in accordance with 220.61
(2) The minimum equipment grounding conductor sized in accordance with 250.122

Exception No. 2: If system bonding jumpers are installed in accordance with 250.30(A)(1), Exception No. 2, the feeder grounded circuit conductor at the building or structure served shall be connected to the equipment grounding conductors, grounding electrode conductor, and the enclosure for the first disconnecting means.

Using the grounded conductor to ground equipment, in lieu of installing a separate EGC, creates parallel paths for normal neutral current along metal raceways, metal piping, metal cable sheaths or shields, and other metal structures such as ductwork. Installing an EGC with the supply circuit conductors and not using the grounded conductor as an EGC helps ensure that normal circuit current is not imposed on continuous metal paths other than the insulated grounded or neutral conductor. At a building or structure supplied by a feeder or branch circuit, the EGC is connected to the grounding electrode system [unless the installation complies with 250.32(A), Exception] in the equipment supplied by the feeder or branch circuit. Where installed, the grounded or neutral conductor is electrically isolated from the EGC and any grounding electrodes at the building or structure supplied by the feeder or branch circuit, which is illustrated in Exhibit 250.18.

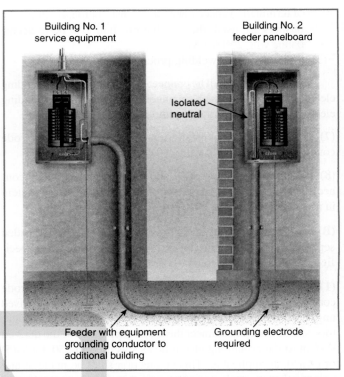

EXHIBIT 250.18 An installation in which a connection between the grounded conductor (neutral) and equipment grounding terminal bar is not permitted. A connection from the equipment grounding terminal bus to the grounding electrode is required.

Because "re-grounding" the neutral or grounded conductor was permitted prior to the 2002 edition of the *NEC*, Exception No. 1 to 250.32(B)(1) permits limited applications where the grounded conductor is used for grounding and bonding of equipment and systems, but only for circuits that were installed in compliance with the *NEC* when re-grounding of the neutral or grounded conductor was permitted and where the three conditions specified in the current exception are met.

(2) Supplied by Separately Derived System.

(a) *With Overcurrent Protection.* If overcurrent protection is provided where the conductors originate, the installation shall comply with 250.32(B)(1).

(b) *Without Overcurrent Protection.* If overcurrent protection is not provided where the conductors originate, the installation shall comply with 250.30(A). If installed, the supply-side bonding jumper shall be connected to the building or structure disconnecting means and to the grounding electrode(s).

The grounding and bonding requirements for a building supplied by a separately derived system are dependent on the location of the OCPD. For example, if a separately derived system originates in a generator and an OCPD is installed at the generator to protect the feeder, an EGC is required to be installed with the feeder conductors. The separately derived system is grounded as specified in 250.30(C) for outdoor sources. Where the building has a disconnecting means [700.12(D)(4), 701.12(D)(3), and 702.12 amend the general disconnecting means requirement in 225.31], the EGC is connected to the disconnecting means enclosure and to the building's grounding electrode system. This installation is no different from what is required for feeders originating in another building. Where an OCPD is not located at the source (as is the case with many outdoor transformer installations), the grounding and bonding requirements of 250.30(A) apply, and a supply-side bonding jumper is used to complete the ground-fault current path between the source and the building or structure supplied.

(C) Ungrounded Systems.

(1) Supplied by a Feeder or Branch Circuit. An equipment grounding conductor, as described in 250.118, shall be installed with the supply conductors and be connected to the building or structure disconnecting means and to the grounding electrode(s). The grounding electrode(s) shall also be connected to the building or structure disconnecting means.

(2) Supplied by a Separately Derived System.

(a) *With Overcurrent Protection.* If overcurrent protection is provided where the conductors originate, the installation shall comply with 250.32(C)(1).

(b) *Without Overcurrent Protection.* If overcurrent protection is not provided where the conductors originate, the installation shall comply with 250.30(B). If installed, the supply-side bonding jumper shall be connected to the building or structure disconnecting means and to the grounding electrode(s).

(D) Disconnecting Means Located in Separate Building or Structure on the Same Premises. If one or more disconnecting means supply one or more additional buildings or structures under single management, and where these disconnecting means are located remote from those buildings or structures in accordance with 225.31(B), Exception No. 1 and No. 2, 700.12(D)(4), 701.12(D)(3), or 702.12, all of the following conditions shall be met:

(1) The connection of the grounded conductor to the grounding electrode, to normally non-current-carrying metal parts of equipment, or to the equipment grounding conductor at a separate building or structure shall not be made.

(2) An equipment grounding conductor for grounding and bonding any normally non-current-carrying metal parts of equipment, interior metal piping systems, and building or structural metal frames is run with the circuit conductors to a separate building or structure and connected to existing grounding electrode(s) required in Part III of this article, or, if there are no existing electrodes, the grounding electrode(s) required in Part III of this article shall be installed if a separate building or structure is supplied by more than one branch circuit.

(3) The connection between the equipment grounding conductor and the grounding electrode at a separate building or structure shall be made in a junction box, panelboard, or

similar enclosure located immediately inside or outside the separate building or structure.

Exceptions to 225.31(B) and requirements within 700.12(D)(4), 701.12(D)(3), and 702.12 permit the disconnecting means to be located elsewhere from the building or structure being supplied. The requirement to make a connection to a grounding electrode system at these buildings or structures still applies, but no disconnecting means is available in which the connection can be made. Sections 250.32(D)(1) through (D)(3) allow for the required connection to be made in a panelboard, junction box, or similar enclosure that is located either inside or outside the building or structure being supplied. The enclosure must be located at a point nearest to where the supply conductors enter the building or structure. An EGC must be run with the supply conductors. The grounded conductor (where installed) must not be bonded to the enclosure or equipment grounding bus. The equipment grounding bus must be connected to a new or existing grounding electrode system at the second building. All non-current-carrying metal parts of equipment, building steel, and interior metal piping systems must be connected to the grounding electrode system. Exhibit 250.19 illustrates an installation in which Building 1 houses the disconnecting means for Building 2.

(E) Grounding Electrode Conductor. The size of the grounding electrode conductor to the grounding electrode(s) shall not be smaller than given in 250.66, based on the largest ungrounded supply conductor. The installation shall comply with Part III of this article.

250.34 Portable, Vehicle-Mounted, and Trailer-Mounted Generators.

(A) Portable Generators. The frame of a portable generator shall not be required to be connected to a grounding electrode as defined in 250.52 for a system supplied by the generator under both of the following conditions:

(1) The generator supplies only equipment mounted on the generator, cord-and-plug-connected equipment through receptacles mounted on the generator, or both.

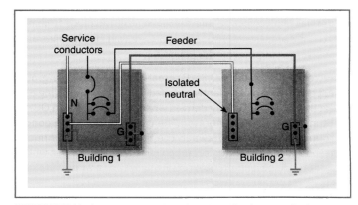

EXHIBIT 250.19 Grounding and bonding for a feeder-supplied separate building under single management with the disconnecting means located remote from the building.

(2) The normally non-current-carrying metal parts of equipment and the equipment grounding conductor terminals of the receptacles are connected to the generator frame.

(B) Vehicle-Mounted and Trailer-Mounted Generators. The frame of a vehicle or trailer shall not be required to be connected to a grounding electrode as defined in 250.52 for a system supplied by a generator located on this vehicle or trailer under all of the following conditions:

(1) The frame of the generator is bonded to the vehicle or trailer frame.
(2) The generator supplies only equipment located on the vehicle or trailer; cord-and-plug-connected equipment through receptacles mounted on the vehicle; or both equipment located on the vehicle or trailer and cord-and-plug-connected equipment through receptacles mounted on the vehicle, trailer, or on the generator.
(3) The normally non-current-carrying metal parts of equipment and the equipment grounding conductor terminals of the receptacles are connected to the generator frame.

Δ **(C) Grounded Conductor Bonding.** A conductor that is required to be grounded by 250.26 shall be connected to the generator frame if the generator is a component of a separately derived system.

> Informational Note: See 250.30 for grounding portable generators supplying fixed wiring systems.

250.35 Permanently Installed Generators. A conductor that provides an effective ground-fault current path shall be installed with the supply conductors from a permanently installed generator(s) to the first disconnecting mean(s) in accordance with 250.35(A) or (B).

(A) Separately Derived System. If the generator is installed as a separately derived system, the requirements in 250.30 shall apply.

(B) Nonseparately Derived System. If the generator is installed as a nonseparately derived system, and overcurrent protection is not integral with the generator assembly, a supply-side bonding jumper shall be installed between the generator equipment grounding terminal and the equipment grounding terminal, bar, or bus of the disconnecting mean(s). It shall be sized in accordance with 250.102(C) based on the size of the conductors supplied by the generator.

The requirements of 250.35(B) create a return path for ground-fault current for permanently installed generators supplying a system that is not separately derived where the first system OCPD is not installed at the generator. The conductor used to conduct ground-fault current between the first system disconnecting means and the generator is a supply-side bonding jumper, which is permitted to be a nonflexible metal raceway or a wire.

> **See also**
>
> **250.102(C)** for sizing of wire-type supply-side bonding jumpers

Δ **250.36 Impedance Grounded Systems — 480 Volts to 1000 Volts.** Impedance grounded systems in which a grounding impedance device, typically a resistor, limits the ground-fault current shall be permitted for 3-phase ac systems of 480 volts to 1000 volts if all the following conditions are met:

(1) The conditions of maintenance and supervision ensure that only qualified persons service the installation.
(2) Ground detectors are installed on the system.
(3) Line-to-neutral loads are not served.

Impedance grounded systems shall comply with 250.36(A) through (G).

> Informational Note: See NFPA *70E*-2021, *Standard for Electrical Safety in the Workplace,* Annex O, for information on impedance grounding to reduce arc-flash hazards.

Exhibit 250.20 shows the location of the grounding impedance device (in this example, a resistor) in a high-impedance grounded neutral system, which limits the amount of ground-fault current in the neutral conductor when a line-to-ground fault occurs. The grounding impedance is selected to limit fault current to a value that is slightly greater than or equal to the capacitive charging current. This system is used where continuity of power is required. Therefore, a ground fault results in an alarm condition rather than in the tripping of a circuit breaker. The alarm allows for the safe and orderly shutdown of a process for which a non-orderly shutdown could introduce additional or increased hazards.

> **See also**
>
> **250.187** for requirements for impedance grounded neutral systems rated over 1000 volts

(A) Location. The grounding impedance device shall be installed between the grounding electrode conductor and the impedance grounding conductor connected to the system neutral

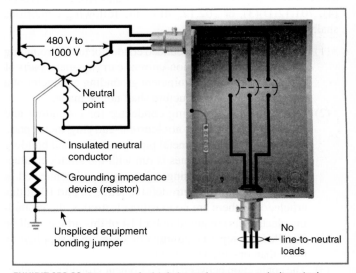

EXHIBIT 250.20 Diagram of a high-impedance grounded neutral system.

point. If a neutral point is not available, the grounding impedance shall be installed between the grounding electrode conductor and the impedance grounding conductor connected to the neutral point derived from a grounding transformer.

(B) Impedance Grounding Conductor Insulation and Ampacity. The impedance grounding conductor from the neutral point of the transformer or generator to its connection point to the grounding impedance shall be fully insulated.

The impedance grounding conductor shall have an ampacity of not less than the maximum current rating of the grounding impedance but in no case shall the impedance grounding conductor be smaller than 8 AWG copper or 6 AWG aluminum or copper-clad aluminum.

(C) System Grounding Connection. The system shall not be connected to ground except through the grounding impedance device.

> Informational Note: The impedance is normally selected to limit the ground-fault current to a value slightly greater than or equal to the capacitive charging current of the system. This value of impedance will also limit transient overvoltages to safe values. For guidance, refer to criteria for limiting transient overvoltages in IEEE 3003.1-2019, *Recommended Practice for System Grounding of Industrial and Commercial Power Systems.*

(D) Impedance Grounding Conductor Routing. The impedance grounding conductor shall be permitted to be installed in a separate raceway from the ungrounded conductors. It shall not be required to run this conductor with the phase conductors to the first system disconnecting means or overcurrent device.

(E) Impedance Bonding Jumper. The impedance bonding jumper (the connection between the equipment grounding conductors and the grounding impedance device) shall be an unspliced conductor run from the first system disconnecting means or overcurrent device to the grounded side of the grounding impedance device.

(F) Grounding Electrode Conductor Connection Location. For services or separately derived systems, the grounding electrode conductor shall be connected at any point from the grounded side of the grounding impedance device to the equipment grounding connection at the service equipment or the first system disconnecting means of a separately derived system.

(G) Impedance Bonding Jumper Size. The impedance bonding jumper shall be sized in accordance with either of the following:

(1) If the grounding electrode conductor connection is made at the grounding impedance device, the equipment bonding jumper shall be sized in accordance with 250.66, based on the size of the service entrance conductors for a service or the derived phase conductors for a separately derived system.

(2) If the grounding electrode conductor is connected at the first system disconnecting means or overcurrent device, the impedance bonding jumper shall be sized the same as the impedance grounding conductor in 250.36(B).

Part III. Grounding Electrode System and Grounding Electrode Conductor

250.50 Grounding Electrode System. All grounding electrodes as described in 250.52(A)(1) through (A)(7) that are present at each building or structure served shall be bonded together to form the grounding electrode system. If none of these grounding electrodes exist, one or more of the grounding electrodes specified in 250.52(A)(4) through (A)(8) shall be installed and used.

The formation of a system of electrodes is required where multiple grounding electrodes are at the building being served. Exhibit 250.21 illustrates a grounding electrode system consisting of multiple electrodes, rather than reliance on a single grounding electrode. Metal in-ground support structures, metal underground water pipe, and concrete footings or foundations are required to be integrated into the grounding electrode system if they qualify under the conditions specified in 250.52(A). The *NEC* does not specify that metal underground water pipe, an in-ground structural metal frame, or concrete-encased-type electrodes have to be installed, only that where they have been installed as part of the building construction, they are to be used as components of the grounding electrode system.

Exception: Concrete-encased electrodes of existing buildings or structures shall not be required to be part of the grounding electrode system if the rebar is not accessible for use without disturbing the concrete.

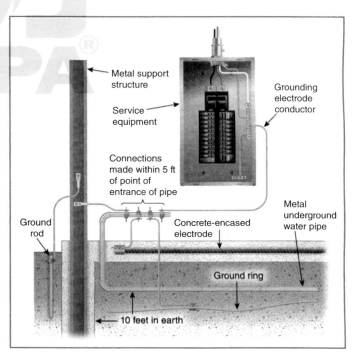

EXHIBIT 250.21 A grounding electrode system comprised of the metal frame of a building, a ground ring, a concrete-encased electrode, a metal underground water pipe, and a ground rod.

Because the installation of the footings and foundation is one of the first elements of a construction project, one that in most cases has long been completed by the time the electric service is installed, compliance with 250.50 requires an awareness and coordinated effort on the part of designers and the construction trades to make sure that the concrete-encased electrode is incorporated into the grounding electrode system. The exception to 250.50 provides a practical solution where modifications are made to an electrical system in an existing building that necessitates connection to the building grounding electrode system.

250.52 Grounding Electrodes.

(A) Electrodes Permitted for Grounding.

(1) Metal Underground Water Pipe. A metal underground water pipe in direct contact with the earth for 3.0 m (10 ft) or more (including any metal well casing bonded to the pipe) and electrically continuous (or made electrically continuous by bonding around insulating joints or insulating pipe) to the points of connection of the grounding electrode conductor and the bonding conductor(s) or jumper(s), if installed.

(2) Metal In-ground Support Structure(s). One or more metal in-ground support structure(s) in direct contact with the earth vertically for 3.0 m (10 ft) or more, with or without concrete encasement. If multiple metal in-ground support structures are present at a building or a structure, it shall be permissible to bond only one into the grounding electrode system.

> Informational Note: Metal in-ground support structures include, but are not limited to, pilings, casings, and other structural metal.

Metal in-ground support structure electrodes extend the metal building frame into the ground. Connection to more than one in-ground structural member is not necessary. Metal frames of buildings that are secured to hold down bolts connected to a concrete-encased electrode are permitted by 250.68(C)(2) to serve as the GEC.

△ **(3) Concrete-Encased Electrode.** A concrete-encased electrode shall consist of at least 6.0 m (20 ft) of either of the following:

   (1) One or more bare or zinc galvanized or other electrically conductive coated rebar of not less than 13 mm (½ in.) in diameter, installed in one continuous 6.0 m (20 ft) length, or if in multiple pieces, the rebar shall be connected together by steel tie wires, exothermic welding, welding, or other effective means to create a 6.0 m (20 ft) or greater length
   (2) Bare copper conductor not smaller than 4 AWG

Metal components shall be encased by at least 50 mm (2 in.) of concrete and shall be located horizontally within that portion of a concrete foundation or footing that is in direct contact with the earth or within vertical foundations or structural components or members that are in direct contact with the earth. If multiple

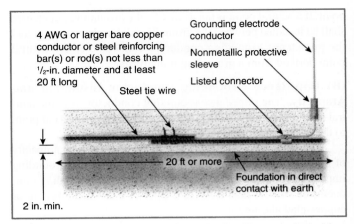

EXHIBIT 250.22 An example of a concrete-encased electrode that is required to be incorporated into the grounding electrode system.

concrete-encased electrodes are present at a building or structure, it shall be permissible to bond only one into the grounding electrode system.

> Informational Note: Concrete installed with insulation, vapor barriers, films, or similar items separating the concrete from the earth is not considered to be in "direct contact" with the earth.

To qualify as a grounding electrode, the horizontal or vertical installation of the steel reinforcing rod or the 4 AWG bare copper conductor within the concrete encasement is required to be in contact with the earth for one continuous 20-foot length. Shorter lengths of reinforcing rod can be connected together to form an electrode of at least 20 feet using the connection methods identified in this requirement. Some buildings or structures may have discontinuous segments of a footing or foundation that individually qualify as grounding electrodes per this section, and once one has been bonded to the grounding electrode system, the remaining ones are exempt from any bonding or grounding requirements. Exhibit 250.22 shows an example of a concrete-encased electrode embedded horizontally. As indicated in the informational note, direct contact with the earth means that no medium is between the concrete and the earth that impedes the grounding connection or insulates the concrete from being in direct contact with the earth.

(4) Ground Ring. A ground ring encircling the building or structure, in direct contact with the earth, consisting of at least 6.0 m (20 ft) of bare copper conductor not smaller than 2 AWG.

△ **(5) Rod and Pipe Electrodes.** Rod and pipe electrodes shall not be less than 2.44 m (8 ft) in length and consist of the following materials.

   (1) Grounding electrodes of pipe or conduit shall not be smaller than metric designator 21 (trade size ¾) and, where of steel, shall have the outer surface galvanized or otherwise metal-coated for corrosion protection.
   (2) Rod-type grounding electrodes of stainless steel and copper or zinc-coated steel shall be at least 15.87 mm (⅝ in.) in diameter, unless listed.

(6) Other Listed Electrodes. Other listed grounding electrodes shall be permitted.

(7) Plate Electrodes. Each plate electrode shall expose not less than 0.186 m² (2 ft²) of surface to exterior soil. Electrodes of bare or electrically conductive coated iron or steel plates shall be at least 6.4 mm (¼ in.) in thickness. Solid, uncoated electrodes of nonferrous metal shall be at least 1.5 mm (0.06 in.) in thickness.

(8) Other Local Metal Underground Systems or Structures. Other local metal underground systems or structures such as piping systems, underground tanks, and underground metal well casings that are not bonded to a metal water pipe.

(B) Not Permitted for Use as Grounding Electrodes. The following systems and materials shall not be used as grounding electrodes:

(1) Metal underground gas piping systems
(2) Aluminum
(3) The structures and structural rebar described in 680.26(B)(1) and (B)(2)

Informational Note: See 250.104(B) for bonding requirements of gas piping.

250.53 Grounding Electrode System Installation.

(A) Rod, Pipe, and Plate Electrodes. Rod, pipe, and plate electrodes shall be free from nonconductive coatings such as paint or enamel. Rod, pipe, and plate electrodes shall meet the requirements of 250.53(A)(1) through (A)(3).

Δ **(1) Below Permanent Moisture Level.** If practicable, rod, pipe, and plate electrodes shall be embedded below permanent moisture level.

(2) Supplemental Electrode Required. A single rod, pipe, or plate electrode shall be supplemented by an additional electrode of a type specified in 250.52(A)(2) through (A)(8). The supplemental electrode shall be permitted to be bonded to one of the following:

(1) Rod, pipe, or plate electrode
(2) Grounding electrode conductor
(3) Grounded service-entrance conductor
(4) Nonflexible grounded service raceway
(5) Any grounded service enclosure

Exception: If a single rod, pipe, or plate grounding electrode has a resistance to earth of 25 ohms or less, the supplemental electrode shall not be required.

There are several methods to measure the resistance to ground of a rod, pipe, or plate electrode. Exhibit 250.23 illustrates one method of determining the ground resistance of a rod-type electrode in which a ground tester is used to measure the fall of potential between the rod being tested and the reference rod (stake) connected to the P1 or P2 terminal of the tester. The clamp-on ground resistance tester is another method of testing

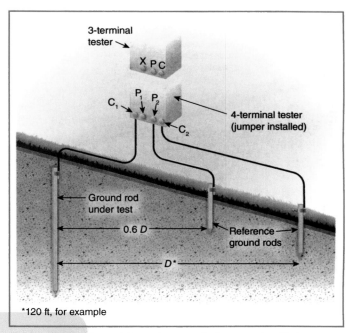

EXHIBIT 250.23 The resistance to ground of a ground rod being measured using a fall-of-potential ground tester. The clamp-on tester is another method of determining the earth resistance of a grounding electrode.

the earth resistance of a grounding electrode. It does not involve the use of reference electrodes to perform the earth resistance test.

(3) Supplemental Electrode. If multiple rod, pipe, or plate electrodes are installed to meet the requirements of this section, they shall not be less than 1.8 m (6 ft) apart.

Informational Note: The paralleling efficiency of rods is increased by spacing them twice the length of the longest rod.

The spacing requirement in 250.53(A)(3) for multiple auxiliary rod, pipe, or plate electrodes is illustrated in Exhibit 250.24. The direct-buried clamps shown are required to be listed for underground installation per 250.70.

(4) Rod and Pipe Electrodes. The electrode shall be installed such that at least 2.44 m (8 ft) of length is in contact with the soil. It shall be driven to a depth of not less than 2.44 m (8 ft) except that, where rock bottom is encountered, the electrode shall be driven at an oblique angle not to exceed 45 degrees from the vertical or, where rock bottom is encountered at an angle up to 45 degrees, the electrode shall be permitted to be buried in a trench that is at least 750 mm (30 in.) deep. The upper end of the electrode shall be flush with or below ground level unless the aboveground end and the grounding electrode conductor attachment are protected against physical damage as specified in 250.10.

Where rock bottom is encountered, the electrodes must be either driven at not more than a 45-degree angle or buried in a

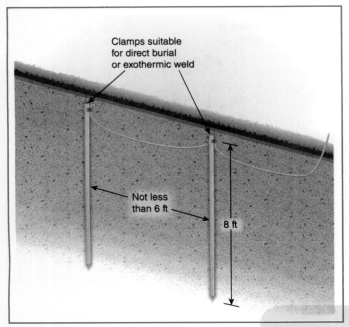

EXHIBIT 250.24 The 6-foot spacing required between multiple rod-type electrodes.

2½-foot-deep trench. Driving the rod at an angle is permitted only if it is impossible to drive the rod vertically to obtain at least 8 feet of earth contact. Burying the ground rod is permitted only if driving the rod vertically or at an angle is not possible. Exhibit 250.25 illustrates these requirements.

Section 250.70 requires ground clamps used on buried electrodes to be listed for direct earth burial. Section 250.10 requires

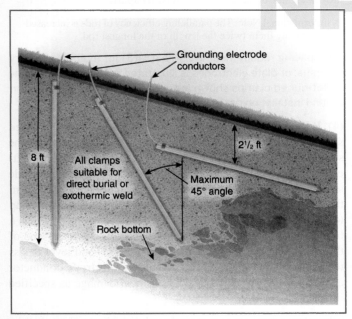

EXHIBIT 250.25 Installation requirements for rod and pipe electrodes shown in specified order of installation from left to right.

ground clamps installed above ground to be protected where subject to physical damage.

(5) Plate Electrode. Plate electrodes shall be installed not less than 750 mm (30 in.) below the surface of the earth.

Δ **(B) Electrode Spacing.** If more than one of the electrodes of the type specified in 250.52(A)(5) or (A)(7) are used, each electrode of one grounding system (including that used for strike termination devices) shall not be less than 1.83 m (6 ft) from any other electrode of another grounding system.

(C) Bonding Jumper. The bonding jumper(s) used to connect the grounding electrodes together to form the grounding electrode system shall be installed in accordance with 250.64(A), (B), and (E), shall be sized in accordance with 250.66, and shall be connected in the manner specified in 250.70. Rebar shall not be used as a conductor to interconnect the electrodes of grounding electrode systems.

> **See also**
>
> **250.68(C)(3)** and its commentary for more information on the use of rebar as a conductor in the grounding electrode system

(D) Metal Underground Water Pipe. If used as a grounding electrode, metal underground water pipe shall meet the requirements of 250.53(D)(1) and (D)(2).

(1) Continuity. Continuity of the grounding path or the bonding connection to interior piping shall not rely on water meters or filtering devices and similar equipment.

(2) Supplemental Electrode Required. A metal underground water pipe shall be supplemented by an additional electrode of a type specified in 250.52(A)(2) through (A)(8). If the supplemental electrode is of the rod, pipe, or plate type, it shall comply with 250.53(A). The supplemental electrode shall be bonded to one of the following:

(1) Grounding electrode conductor
(2) Grounded service-entrance conductor
(3) Nonflexible grounded service raceway
(4) Any grounded service enclosure
(5) As provided by 250.32(B)

Exception: The supplemental electrode shall be permitted to be bonded to the interior metal water piping as specified in 250.68(C)(1).

This requirement clarifies that the supplemental electrode system must be installed as if it were the sole grounding electrode for the system. As specified in the exception to 250.53(A)(2), if a single rod, pipe, or plate electrode has a resistance to earth of 25 ohms or less, it is not necessary to supplement that electrode with one of the types from 250.52(A)(2) through (A)(8). In other words, a single rod, pipe, or plate electrode being used to supplement a metal underground water pipe-type electrode is itself required to be provided with a supplemental electrode unless the condition of 250.53(A)(2), Exception, can be met. One of the permitted methods of bonding a supplemental GEC to the grounding electrode system is to connect it to the grounded service enclosure.

The need for supplemental electrodes for metal water pipe is due to the common practice of using a plastic pipe for replacement when the original metal water pipe fails. Plastic replacement pipe leaves the system without a grounding electrode unless a supplemental electrode is provided.

(E) Supplemental Grounding Electrode Bonding Jumper Size. If the supplemental electrode is a rod, pipe, or plate electrode, that portion of the bonding jumper that is the sole connection to the supplemental grounding electrode shall not be required to be larger than 6 AWG copper wire or 4 AWG aluminum or copper-clad aluminum wire.

(F) Ground Ring. The ground ring shall be installed not less than 750 mm (30 in.) below the surface of the earth.

250.54 Auxiliary Grounding Electrodes. One or more grounding electrodes shall be permitted to be connected to the equipment grounding conductors specified in 250.118 and shall not be required to comply with the electrode bonding requirements of 250.50 or 250.53(C) or the resistance requirements of 250.53(A)(2) Exception, but the earth shall not be used as an effective ground-fault current path as specified in 250.4(A)(5) and (B)(4).

Grounding electrodes, such as ground rods, that are connected to equipment are not permitted to be used in lieu of the EGC, but grounding electrodes may be used to establish a reference to ground in the area of electrically operated equipment. The earth may not be used as the EGC or effective (ground) fault current path. Auxiliary grounding electrodes are not required to be incorporated into the grounding electrode system for the service or other source of electrical supply.

Δ **250.58 Common Grounding Electrode.** If an ac system is connected to a grounding electrode in or at a building or structure, the same electrode shall be used to ground conductor enclosures and equipment in or on that building or structure. If separate services, feeders, or branch circuits supply a building and are required to be connected to a grounding electrode(s), the same grounding electrode(s) shall be used.

250.60 Use of Strike Termination Devices. Conductors and driven pipes, rods, or plate electrodes used for grounding strike termination devices shall not be used in lieu of the grounding electrodes required by 250.50 for grounding wiring systems and equipment. This provision shall not prohibit the required bonding together of grounding electrodes of different systems.

> Informational Note No. 1: See 250.106 for the bonding requirement of the lightning protection system components to the building or structure grounding electrode system.

Informational Note No. 2: Bonding together of all separate grounding electrodes will limit voltage differences between them and between their associated wiring systems.

250.62 Grounding Electrode Conductor Material. The grounding electrode conductor shall be of copper, aluminum, copper-clad aluminum, or the items as permitted in 250.68(C). The material selected shall be resistant to any corrosive condition existing at the installation or shall be protected against corrosion. Conductors of the wire type shall be solid or stranded, insulated, covered, or bare.

250.64 Grounding Electrode Conductor Installation. Grounding electrode conductors at the service, at each building or structure where supplied by a feeder(s) or branch circuit(s), or at a separately derived system shall be installed as specified in 250.64(A) through (G).

(A) Aluminum or Copper-Clad Aluminum Conductors. Grounding electrode conductors of bare, covered, or insulated aluminum or copper-clad aluminum shall comply with the following:

(1) Bare or covered conductors without an extruded polymeric covering shall not be installed where subject to corrosive conditions or be installed in direct contact with concrete.

(2) Terminations made within outdoor enclosures that are listed and identified for the environment shall be permitted within 450 mm (18 in.) of the bottom of the enclosure.

(3) Aluminum or copper-clad aluminum conductors external to buildings or equipment enclosures shall not be terminated within 450 mm (18 in.) of the earth.

Concern over accelerated conductor degradation due to excessive moisture underlies the restrictions on the use of aluminum and copper-clad aluminum GECs outdoors in close proximity to the earth and in direct contact with concrete. If protected against environmental conditions by being installed within an enclosure listed for outdoor locations (such as a panelboard cabinet, switchboard enclosure, or switchgear enclosure), aluminum GECs are permitted to be installed within 18 inches of the enclosure bottom. Where the bottom of the enclosure is concrete, as is the case with open-bottom equipment such as outdoor switchboards and switchgear, covered aluminum or copper-clad aluminum GECs with an extruded polymeric covering and insulated conductors are permitted to be in direct contact with the concrete.

(B) Securing and Protection Against Physical Damage. If exposed, a grounding electrode conductor or its enclosure shall be securely fastened to the surface on which it is carried. Grounding electrode conductors shall be permitted to be installed on or through framing members.

See also

250.64(E) and its commentary for additional information on GECs enclosed in metal raceways

(1) Not Exposed to Physical Damage. A 6 AWG or larger copper, copper-clad aluminum, or aluminum grounding electrode conductor not exposed to physical damage shall be permitted to be run along the surface of the building construction without metal covering or protection.

(2) Exposed to Physical Damage. A 6 AWG or larger copper, copper-clad aluminum, or aluminum grounding electrode conductor exposed to physical damage shall be protected in rigid metal conduit (RMC), intermediate metal conduit (IMC), Schedule 80 rigid polyvinyl chloride conduit (PVC), reinforced thermosetting resin conduit Type XW (RTRC-XW), electrical metallic tubing (EMT), or cable armor.

(3) Smaller Than 6 AWG. Grounding electrode conductors smaller than 6 AWG shall be protected in RMC, IMC, Schedule 80 PVC, RTRC-XW, EMT, or cable armor.

(4) In Contact with the Earth. Grounding electrode conductors and grounding electrode bonding jumpers in contact with the earth shall not be required to comply with 300.5 or 305.15, but shall be buried or otherwise protected if subject to physical damage.

(C) Continuous. Except as provided in 250.30(A)(5) and (A)(6), 250.30(B)(1), and 250.68(C), grounding electrode conductor(s) shall be installed in one continuous length without a splice or joint. If necessary, splices or connections shall be made as permitted in the following:

(1) Splicing of the wire-type grounding electrode conductor shall be permitted only by irreversible compression-type connectors listed as grounding and bonding equipment or by the exothermic welding process.
(2) Sections of busbars shall be permitted to be connected together to form a grounding electrode conductor.
(3) Bolted, riveted, or welded connections of structural metal frames of buildings or structures.
(4) Threaded, welded, brazed, soldered or bolted-flange connections of metal water piping.

A building remodeling project or the replacement of existing electrical equipment might necessitate splicing of the GEC. Section 250.64(C)(1) contains the permitted means by which the splice can be considered a permanent connection that provides the intended GEC continuity.

(D) Building or Structure with Multiple Disconnecting Means in Separate Enclosures. If a building or structure is supplied by a service or feeder with two or more disconnecting means in separate enclosures, the grounding electrode connections shall be made in accordance with 250.64(D)(1), (D)(2), or (D)(3).

(1) Common Grounding Electrode Conductor and Taps. A common grounding electrode conductor and grounding electrode conductor taps shall be installed. The common grounding electrode conductor shall be sized in accordance with 250.66, based

on the sum of the circular mil area of the largest ungrounded conductor(s) of each set of conductors that supplies the disconnecting means. If the service-entrance conductors connect directly to the overhead service conductors, service drop, underground service conductors, or service lateral, the common grounding electrode conductor shall be sized in accordance with Table 250.66, note 1.

A grounding electrode conductor tap shall extend to the inside of each disconnecting means enclosure. The grounding electrode conductor taps shall be sized in accordance with 250.66 for the largest service-entrance or feeder conductor serving the individual enclosure. The tap conductors shall be connected to the common grounding electrode conductor by one of the following methods in such a manner that the common grounding electrode conductor remains without a splice or joint:

(1) Exothermic welding.
(2) Connectors listed as grounding and bonding equipment.
(3) Connections to an aluminum or copper busbar not less than 6 mm thick × 50 mm wide (¼ in. thick × 2 in. wide) and of a length to accommodate the number of terminations necessary for the installation. The busbar shall be securely fastened and shall be installed in an accessible location. Connections shall be made by a listed connector or by the exothermic welding process. If aluminum busbars are used, the installation shall comply with 250.64(A).

The common GEC from which the taps are made must be sized from Table 250.66 based on the sum of the circular mil area of the largest ungrounded conductor(s) of each set of conductors that supply the disconnecting means or the sum of the equivalent cross-sectional area for parallel conductors.

As illustrated in Exhibit 250.26, the tap method permitted by this section eliminates the difficulties found in looping GECs from one enclosure to another. The 2 AWG GEC (based on the 350 kcmil ungrounded conductor) shown in Exhibit 250.26 is required to be installed without a splice or joint, except as permitted in 250.64(C), and the 8 AWG and 4 AWG taps are sized from Table 250.66 based on the size of the ungrounded conductor serving the respective service disconnecting means.

As is the case with connectors used for splicing the GEC [see 250.64(C)(1)], the tap connection must have a high degree of reliability under any abnormal conditions of system operation to which it may be exposed. Therefore, if pressure connectors are used for making the tap to the GEC, this requirement specifies not only that the connector be listed but also that it be listed specifically as grounding and bonding equipment, meaning that it has been submitted for certification in accordance with UL 467, *Grounding and Bonding Equipment*. This high a bar is not required for all grounding and bonding connections covered by the requirements of Article 250.

(2) Individual Grounding Electrode Conductors. A grounding electrode conductor shall be connected between the grounding electrode system and one or more of the following, as applicable:

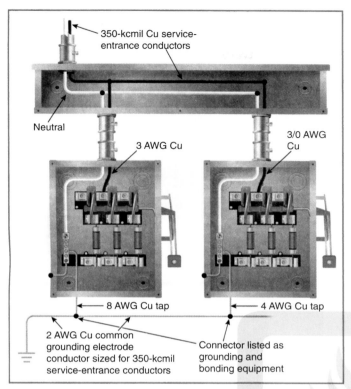

EXHIBIT 250.26 The tap method of connecting the GEC to multiple service disconnecting means enclosures.

(1) Grounded conductor in each service equipment disconnecting means enclosure
(2) Equipment grounding conductor installed with the feeder(s) or branch circuit(s) for other than services
(3) Supply-side bonding jumper

Each grounding electrode conductor shall be sized in accordance with 250.66 based on the service-entrance or feeder conductor(s) supplying the individual disconnecting means.

(3) Common Location. A grounding electrode conductor shall be connected in a wireway or other accessible enclosure on the supply side of the disconnecting means to one or more of the following, as applicable:

(1) Grounded service conductor(s)
(2) Equipment grounding conductor installed with the feeder
(3) Supply-side bonding jumper

The connection shall be made with exothermic welding or a connector listed as grounding and bonding equipment. The grounding electrode conductor shall be sized in accordance with 250.66 based on the service-entrance or feeder conductor(s) at the common location where the connection is made.

(E) Raceways, Cable Armor, and Enclosures for Grounding Electrode Conductors.

(1) General. Ferrous metal raceways, enclosures, and cable armor for grounding electrode conductors shall be electrically

continuous from the point of attachment to cabinets or equipment to the grounding electrode and shall be securely fastened to the ground clamp or fitting. Ferrous metal raceways, enclosures, and cable armor shall be bonded at each end of the raceway or enclosure to the grounding electrode or grounding electrode conductor to create an electrically parallel path. Nonferrous metal raceways, enclosures, and cable armor shall not be required to be electrically continuous.

(2) Methods. Bonding shall be in compliance with 250.92(B) and ensured by one of the methods in 250.92(B)(2) through (B)(4).

(3) Size. The bonding jumper for a grounding electrode conductor(s), raceway(s), enclosure(s), or cable armor shall be the same size as, or larger than, the largest enclosed grounding electrode conductor.

(4) Wiring Methods. If a raceway is used as protection for a grounding electrode conductor, the installation shall comply with the requirements of the applicable raceway article.

These bonding connections are necessary so that the ferrous raceway or cable armor used to protect the GEC does not create an inductive choke on the GEC. Exhibit 250.27 shows the required bonding of a ferrous metal raceway to the GEC at both ends of the raceway to ensure that the raceway and the conductor are in parallel. There are a number of listed fittings available for making these bonding connections, including ones that are designed to connect the raceway or cable armor plus the GEC directly to the grounding electrode.

(F) Installation to Electrode(s). Grounding electrode conductor(s) and bonding jumpers interconnecting grounding

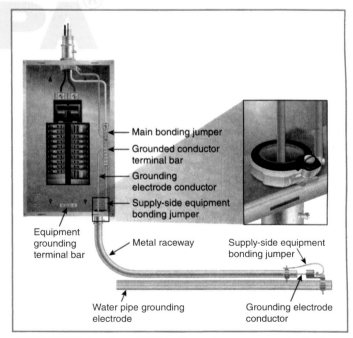

EXHIBIT 250.27 Bonding of GEC to the metal raceway enclosing it at both ends.

electrodes shall be installed in accordance with one of the following. The grounding electrode conductor shall be sized for the largest grounding electrode conductor required among all the electrodes connected to it.

(1) The grounding electrode conductor shall be permitted to be run to any convenient grounding electrode available in the grounding electrode system where the other electrode(s), if any, is connected by bonding jumpers that are installed in accordance with 250.53(C).

(2) Grounding electrode conductor(s) shall be permitted to be run to one or more grounding electrode(s) individually.

(3) Bonding jumper(s) from grounding electrode(s) shall be permitted to be connected to an aluminum or copper busbar not less than 6 mm thick × 50 mm wide (¼ in. thick × 2 in wide.) and of sufficient length to accommodate the number of terminations necessary for the installation. The busbar shall be securely fastened and shall be installed in an accessible location. Connections shall be made by a listed connector or by the exothermic welding process. The grounding electrode conductor shall be permitted to be run to the busbar. Where aluminum busbars are used, the installation shall comply with 250.64(A).

Exhibit 250.28 illustrates a grounding electrode system. The single GEC is permitted to run to any convenient grounding electrode available. The other electrodes are connected together using bonding jumpers sized in accordance with 250.53(C), which references 250.66. A permitted alternative is to run the GEC to a busbar that can be used as a connection point for bonding jumpers from multiple electrodes that form the grounding electrode system. A fully sized GEC, sized in accordance with 250.66 for the largest electrode used, is permitted to be run to the busbar from the point at which it is connected to the grounded or ungrounded system equipment.

N (G) **Enclosures with Ventilation Openings.** Grounding electrode conductors shall not be installed through a ventilation opening of an enclosure.

250.66 Size of Alternating-Current Grounding Electrode Conductor. The size of the grounding electrode conductor and bonding jumper(s) for connection of grounding electrodes shall not be smaller than given in Table 250.66, except as permitted in 250.66(A) through (C).

(A) **Connections to a Rod, Pipe, or Plate Electrode(s).** If the grounding electrode conductor or bonding jumper connected to a single or multiple rod, pipe, or plate electrode(s), or any combination thereof, as described in 250.52(A)(5) or (A)(7), does not extend on to other types of electrodes that require a larger size conductor, the grounding electrode conductor shall not be required to be larger than 6 AWG copper wire or 4 AWG aluminum or copper-clad aluminum wire.

(B) **Connections to Concrete-Encased Electrodes.** If the grounding electrode conductor or bonding jumper connected to

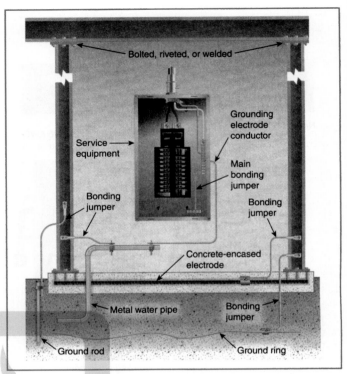

EXHIBIT 250.28 An example of running the GEC to any convenient electrode available as well as bonding together all electrodes that are present at a building or structure to form the required grounding electrode system.

Δ **TABLE 250.66** *Grounding Electrode Conductor for Alternating-Current Systems*

Size of Largest Ungrounded Conductor or Equivalent Area for Parallel Conductors (AWG/kcmil)		Size of Grounding Electrode Conductor (AWG/kcmil)	
Copper	Aluminum or Copper-Clad Aluminum	Copper	Aluminum or Copper-Clad Aluminum
2 or smaller	1/0 or smaller	8	6
1 or 1/0	2/0 or 3/0	6	4
2/0 or 3/0	4/0 or 250	4	2
Over 3/0 through 350	Over 250 through 500	2	1/0
Over 350 through 600	Over 500 through 900	1/0	3/0
Over 600 through 1100	Over 900 through 1750	2/0	4/0
Over 1100	Over 1750	3/0	250

Notes:

1. If multiple sets of service-entrance conductors connect directly to a service drop, set of overhead service conductors, set of underground service conductors, or service lateral, the equivalent size of the largest service-entrance conductor shall be determined by the largest sum of the areas of the corresponding conductors of each set.

2. If there are no service-entrance conductors, the grounding electrode conductor size shall be determined by the equivalent size of the largest service-entrance conductor required for the load to be served.

3. See installation restrictions in 250.64.

a single or multiple concrete-encased electrode(s), as described in 250.52(A)(3), does not extend on to other types of electrodes that require a larger size of conductor, the grounding electrode conductor shall not be required to be larger than 4 AWG copper wire.

(C) Connections to Ground Rings. If the grounding electrode conductor or bonding jumper connected to a ground ring, as described in 250.52(A)(4), does not extend on to other types of electrodes that require a larger size of conductor, the grounding electrode conductor shall not be required to be larger than the conductor used for the ground ring.

Exhibit 250.29 illustrates a GEC installed from service equipment or from a separately derived system to a water pipe grounding electrode. The GEC is required by 250.66 to be sized based on the size of the ungrounded supply conductors. The supply conductors could be service conductors or, in the case of separately derived systems, feeder conductors from a generator or other power source or transformer secondary conductors. The bonding jumpers that connect the other grounding electrodes together are sized using 250.53(C), which refers to 250.66. GECs and bonding jumpers are permitted to be sized based on the electrodes to which they connect, as specified in 250.66(A), (B), or (C). However, if the GEC or bonding jumper extends from this connection to an electrode that is not specified in 250.66(A), (B), or (C), it must be sized per Table 250.66.

In Exhibit 250.29, the size of the GEC and the bonding jumpers is dependent on the electrode to which they are connected. The illustration is not intended to show a mandatory physical routing and connection order of the bonding jumpers and the GEC, since the *NEC* does not specify an order or hierarchy for these connections. The sizes for the bonding jumpers to the ground rod and the concrete-encased electrode shown in Exhibit 250.29 are the maximum sizes required by the *NEC* based on 250.66(A) and (B). If the GEC from the service equipment is run to the ground rod first and then to the water pipe, the GEC to the ground rod is required to be sized based on Table 250.66 as if it were run to the water pipe electrode. The use of bonding jumpers or GECs larger than required by 250.66 is not prohibited.

250.68 Grounding Electrode Conductor and Bonding Jumper Connection to Grounding Electrodes. The connection of a grounding electrode conductor at the service, at each building or structure where supplied by a feeder(s) or branch circuit(s), or at a separately derived system and associated bonding jumper(s) shall be made as specified 250.68(A) through (C).

(A) Accessibility. All mechanical elements used to terminate a grounding electrode conductor or bonding jumper to a grounding electrode shall be accessible.

Exception No. 1: An encased or buried connection to a concrete-encased, driven, or buried grounding electrode shall not be required to be accessible.

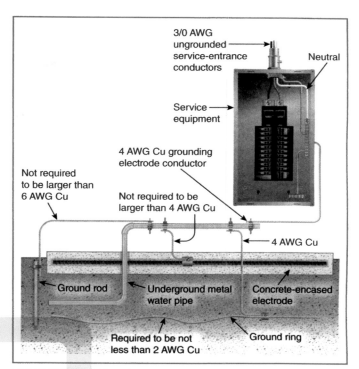

EXHIBIT 250.29 GEC and bonding jumpers sized in accordance with 250.66 for a service supplied by 3/0 AWG copper ungrounded conductors.

Exception No. 2: Exothermic or irreversible compression connections used at terminations, together with the mechanical means used to attach such terminations to fireproofed structural metal whether or not the mechanical means is reversible, shall not be required to be accessible.

Ground clamps and other connectors suitable for use where buried in earth or embedded in concrete must be listed for such use per 250.70. Indication of this listing is either by a marking on the connector or by a tag attached to the connector. See Exhibits 250.22 and 250.25 for illustrations of encased and buried electrodes.

Connections, including the mechanical attachment of a compression lug to structural steel, are permitted to be encapsulated by fireproofing material and are not required to be accessible. This allowance recognizes the importance of maintaining the integrity of the structural fireproofing.

(B) Effective Grounding Path. The connection of a grounding electrode conductor or bonding jumper to a grounding electrode shall be made in a manner that will ensure an effective grounding path. Where necessary to ensure the grounding path for a metal piping system used as a grounding electrode, bonding shall be provided around insulated joints and around any equipment likely to be disconnected for repairs or replacement. Bonding jumpers shall be of sufficient length to permit removal of such equipment while retaining the integrity of the grounding path.

Water meters and water filter systems are examples of equipment likely to be disconnected for repairs or replacement. Shorter bonding jumpers are more likely to be disconnected to facilitate removal or reinstallation of a water meter or filter cartridge. The bonding jumper must be long enough to permit removal of such equipment without disconnecting or otherwise interrupting the bonding jumper.

Δ **(C) Grounding Electrode Conductor Connections.** Grounding electrode conductors and bonding jumpers shall be permitted to be connected at the following locations and used to extend the connection to an electrode(s):

(1) Interior metal water piping that is electrically continuous with a metal underground water pipe electrode and is located not more than 1.52 m (5 ft) from the point of entrance to the building, as measured along the water piping, shall be permitted to extend the connection to an electrode(s). Interior metal water piping located more than 1.52 m (5 ft) from the point of entrance to the building, as measured along the water piping, shall not be used as a conductor to interconnect electrodes of the grounding electrode system.

Exception: In industrial, commercial, and institutional buildings or structures, if conditions of maintenance and supervision ensure that only qualified persons service the installation, interior metal water piping located more than 1.52 m (5 ft) from the point of entrance to the building, as measured along the water piping, shall be permitted as a bonding conductor to interconnect electrodes that are part of the grounding electrode system, or as a grounding electrode conductor, if the entire length, other than short sections passing perpendicularly through walls, floors, or ceilings, of the interior metal water pipe that is being used for the conductor is exposed.

The piping at this point is not a grounding electrode [only the underground portion is an electrode per 250.52(A)(1)]. Rather, it is used to extend grounding and bonding conductor connections to the grounding electrode. The exception permits connections beyond the first 5 feet, and at that point the water piping is considered either a conductor used for bonding grounding electrodes together or the actual GEC.

(2) The metal structural frame of a building shall be permitted to be used as a conductor to interconnect electrodes that are part of the grounding electrode system, or as a grounding electrode conductor. Hold-down bolts securing the structural steel column that are connected to a concrete-encased electrode complying with 250.52(A)(3) and located in the support footing or foundation shall be permitted to connect the metal structural frame of a building or structure to the concrete-encased grounding electrode. The hold-down bolts shall be connected to the concrete-encased electrode by welding, exothermic welding, steel tie wires, or other approved means.

The structural metal frame of a building is not a grounding electrode unless it extends into the earth for 10 feet or more, as specified in 250.52(A)(2).

See also

250.104(A) and **(C)** for the bonding requirements of metal water piping systems and structural metal frames

(3) A rebar-type concrete-encased electrode installed in accordance with 250.52(A)(3) with an additional rebar section extended from its location within the concrete foundation or footing to an accessible location that is not subject to corrosion shall be permitted for connection of grounding electrode conductors and bonding jumpers in accordance with the following:
   a. The additional rebar section shall be continuous with the grounding electrode rebar or shall be connected to the grounding electrode rebar and connected together by steel tie wires, exothermic welding, welding, or other effective means.
   b. The rebar extension shall not be exposed to contact with the earth without corrosion protection.
   c. Rebar shall not be used as a conductor to interconnect the electrodes of grounding electrode systems.

If rebar is used to form the concrete-encased electrode, it is permitted to extend out of the concrete to an accessible location for connection to a GEC or bonding jumper.

The accessible section of rebar extending out of the concrete can be either a continuous extension of the rebar that is in the concrete or a separate piece that is bonded to the rebar in the concrete. Unlike metal water piping and structural steel, the rebar network in the concrete is not permitted as a means to bond different types of electrodes together to form the grounding electrode system. No testing has been done to validate that ½ inch rebar is equivalent to metal water piping or the structural metal frame of a building, both of which are permitted to act as a common bus for interconnecting multiple grounding electrodes.

Contact between unprotected rebar and earth can cause corrosion and is prohibited. Over time, such corrosion will deteriorate the rebar and eliminate the connection to the electrode in the concrete.

Δ **250.70 Methods of Grounding and Bonding Conductor Connection to Electrodes.**

N **(A) General.** The grounding or bonding conductor shall be connected to the grounding electrode by exothermic welding, listed lugs, listed pressure connectors, listed clamps, or other listed means. Connections depending on solder shall not be used. Ground clamps shall be listed for the materials of the grounding electrode and the grounding electrode conductor and, if used on pipe, rod, or other buried electrodes, shall also be listed for direct soil burial or concrete encasement. Not more than one conductor shall be connected to the grounding electrode by a single clamp or fitting unless the clamp or fitting is listed for multiple conductors.

EXHIBIT 250.30 An application of a listed pipe-type ground clamp.

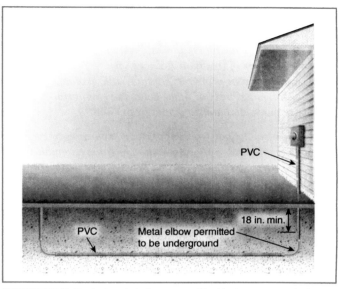

EXHIBIT 250.31 Buried metal elbow permitted to be ungrounded, provided it is isolated from contact by a minimum cover of 18 inches to any part of the elbow.

N (B) Indoor Communications Systems. For indoor communications purposes only, a listed sheet metal strap-type ground clamp having a rigid metal base that seats on the electrode and having a strap of such material and dimensions that it is not likely to stretch during or after installation shall be permitted.

> Informational Note: Listed ground clamps that are identified for direct burial are also suitable for concrete encasement.

If a ground clamp is connected to a galvanized water pipe electrode, the clamp must be of a material compatible with steel to prevent galvanic corrosion. This also applies to ground clamps used with grounding electrodes made of other materials such as copper or steel reinforcing rods or bars.

Exhibit 250.30 shows a listed water pipe ground clamp generally used with 8 AWG through 4 AWG GECs.

Part IV. Enclosure, Raceway, and Service Cable Connections

Δ **250.80 Service Raceways and Enclosures.** Metal enclosures and raceways for service conductors and equipment shall be connected to the grounded conductor if the electrical system is grounded or to the grounding electrode conductor for electrical systems that are not grounded.

> *Exception: Metal components that are installed in a run of underground nonmetallic raceway(s) and are isolated from possible contact by a minimum cover of 450 mm (18 in.) to all parts of the metal components shall not be required to be connected to the grounded conductor, supply-side bonding jumper, or grounding electrode conductor.*

Metal elbows are often installed because nonmetallic elbows can be damaged during conductor installation from the friction caused by pull lines or ropes rubbing against the interior radius of the elbow. The elbows can be isolated from physical contact by

burying all metal components at a depth of 18 inches or more below grade. This exception applies where isolated metal components are used in a service raceway installation. Installations of raceways containing other than service conductors can be found in 250.86, Exception No. 3. See Exhibit 250.31 for an example of this application.

250.84 Underground Service Cable or Raceway.

Δ **(A) Underground Service Cable.** The sheath or armor of a continuous underground metal-sheathed or armored service cable system that is connected to the grounded conductor on the supply side shall not be required to be connected to the grounded conductor at the building or structure. The sheath or armor shall be permitted to be insulated from the interior metal raceway or piping.

Δ **(B) Underground Service Raceway Containing Cable.** An underground metal service raceway that contains a metal-sheathed or armored cable connected to the grounded conductor shall not be required to be connected to the grounded conductor at the building or structure. The sheath or armor shall be permitted to be insulated from the interior metal raceway or piping.

250.86 Other Conductor Enclosures and Raceways. Except as permitted by 250.112(I), metal enclosures and raceways for other than service conductors shall be connected to the equipment grounding conductor.

> *Exception No. 1: Metal enclosures and raceways for conductors added to existing installations of open wire, knob-and-tube wiring, and nonmetallic-sheathed cable shall not be required to be connected to the equipment grounding*

conductor if these enclosures or wiring methods comply with all the following:

(1) Do not provide an equipment ground

(2) Are in runs of less than 7.5 m (25 ft)

(3) Are free from probable contact with ground, grounded metal, metal lath, or other conductive material

(4) Are guarded against contact by persons

Exception No. 2: Short sections of metal enclosures or raceways used to provide support or protection of cable assemblies from physical damage shall not be required to be connected to the equipment grounding conductor.

Exception No. 3: Metal components shall not be required to be connected to the equipment grounding conductor or supply-side bonding jumper if either of the following conditions exist:

(1) The metal components are installed in a run of nonmetallic raceway(s) and isolated from possible contact by a minimum cover of 450 mm (18 in.) to any part of the metal components.

(2) The metal components are part of an installation of nonmetallic raceway(s) and are isolated from possible contact to any part of the metal components by being encased in not less than 50 mm (2 in.) of concrete.

Connectors, couplings, or other similar fittings that perform mechanical and electrical functions must ensure bonding and grounding continuity between the fitting, the metal raceway, and the enclosure. Metal enclosures must be grounded so that when a fault occurs between an ungrounded (hot) conductor and ground, the potential difference between the non-current-carrying parts of the electrical installation is minimized, thereby reducing the shock hazard risk.

Part V. Bonding

250.90 General. Bonding shall be provided if necessary to ensure electrical continuity and the capacity to conduct safely any fault current likely to be imposed.

250.92 Services.

(A) Bonding of Equipment for Services. The normally non-current-carrying metal parts of equipment indicated in the following shall be bonded together:

(1) All raceways, cable trays, cablebus framework, auxiliary gutters, or service cable armor or sheath that enclose, contain, or support service conductors, except as permitted in 250.80

(2) All enclosures containing service conductors, including meter fittings, boxes, or the like, interposed in the service raceway or armor

Δ **(B) Method of Bonding at the Service.** Bonding jumpers meeting the requirements of this article shall be used around impaired connections, such as reducing washers or oversized, concentric,

or eccentric knockouts. Standard locknuts or bushings shall not be the only means for the bonding required by this section but shall be permitted to be installed to make a mechanical connection of the raceway(s).

Electrical continuity at service equipment, service raceways, and service conductor enclosures shall be ensured by one or more of the following methods:

Standard locknuts, sealing locknuts, and metal bushings are not acceptable as the sole means for bonding a raceway or cable to an enclosure on the line side of the service disconnecting means. For concentric, eccentric, or oversized knockouts, electrical continuity must be ensured through a supply-side bonding jumper that connects the raceway to the enclosure through an approved grounding fitting. If these knockouts were in service enclosures, they would impede the bonding connections. Bonding jumpers are required in those situations. They would also be required if reducing washers are used to provide a suitable path for the high level of ground-fault current that is available on the line side of the service disconnecting means. For further information on concentric and eccentric knockouts, see the commentary following the definition of *bonding conductor (bonding jumper)* in Article 100 and the example shown in Exhibit 100.5.

(1) Bonding equipment to the grounded service conductor by an applicable method in 250.8(A)

Exhibit 250.32 illustrates an acceptable grounding and bonding arrangement at a service that has one disconnecting means. Exhibit 250.33 illustrates an acceptable grounding and bonding arrangement for a service that has three disconnecting means as permitted by 230.71.

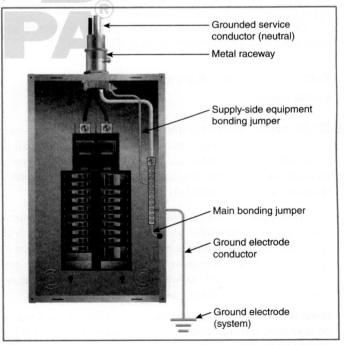

EXHIBIT 250.32 Grounding and bonding arrangement for a service with one disconnecting means.

EXHIBIT 250.33 Grounding and bonding arrangement for a service with three disconnecting means.

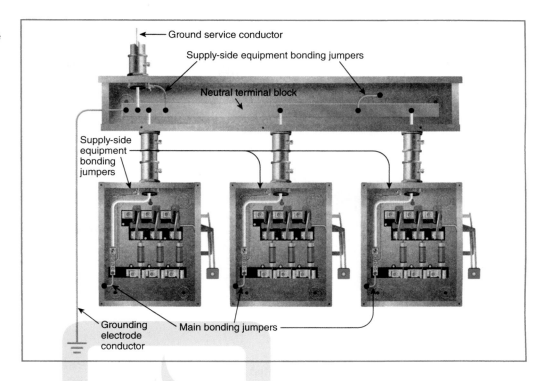

See also

250.24(C), which specifies that the grounded service conductor must be run to each service disconnecting means and be bonded to the disconnecting means enclosure

(2) Connections made up wrenchtight using threaded couplings, threaded entries, or listed threaded hubs on enclosures

(3) Threadless couplings and connectors if made up tight for metal raceways and metal-clad cables

(4) Other listed devices, such as bonding-type locknuts, bushings, or bushings with bonding jumpers

Bonding-type locknuts and grounding and bonding bushings for use with rigid or intermediate metal conduit are provided with means (usually one or more set screws that make positive contact with the conduit) for reliably bonding the bushing and the conduit on which it is threaded to the metal equipment enclosure or box.

Grounding bushings used with fittings for rigid or intermediate metal conduit or with electrical metallic tubing (EMT) have means for connecting a bonding jumper or have means provided by the manufacturer to attach a wire connector (see Exhibit 250.34). This type of bushing may also have one or more set screws to reliably bond the bushing to the conduit.

Threaded hubs used to connect rigid metal and intermediate metal conduits to service equipment and other line-side enclosures are suitable for bonding the conduit only if they have been listed as grounding and bonding equipment per UL 467, *Standard for Grounding and Bonding Equipment.* Such hubs are available with gaskets and are suitable for use in wet locations

so that a conduit connection can be made to an enclosure above the live parts. Grounding and bonding hubs are provided with a grounding and bonding locknut to ensure that the connection between the hub and the conduit is suitable for the higher levels of fault current available at the service location. Although many threaded hubs are listed as conduit fittings, not all threaded hubs have additionally been listed as grounding and bonding equipment. Conduit hubs listed as grounding and bonding equipment are also required (see 250.97) where the

EXHIBIT 250.34 Grounding bushings used to connect a copper bonding conductor to the conduits. *(Courtesy of Thomas and Betts, a Member of the ABB Group)*

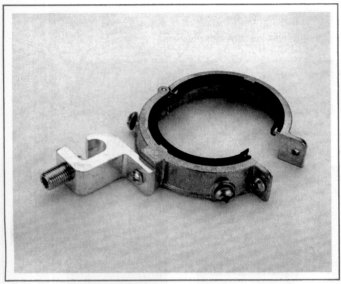

EXHIBIT 250.35 A split-type bond bushing used to provide a bonding connection between a conduit and an enclosure. *(Courtesy of Bridgeport Fittings, Inc.)*

EXHIBIT 250.36 A listed intersystem bonding termination providing the required number of terminals (minimum of three) for connecting building communications systems to the grounding system of the electrical power supply.

conduit contains feeders or branch circuits rated above 250 volts to ground.

Exhibit 250.35 shows a listed split-type bond bushing used to connect a threaded conduit to an enclosure and provide the bonding connection required by 250.92(B).

250.94 Bonding for Communications Systems. Communications system bonding conductor terminations shall be connected in accordance with 250.94(A) or (B).

Δ **(A) The Intersystem Bonding Termination Device.** An intersystem bonding termination (IBT) for connecting intersystem bonding conductors shall be provided external to enclosures at the service equipment or metering equipment enclosure and at the disconnecting means for any buildings or structures that are supplied by a feeder or branch circuit. If an IBT is used, it shall comply with the following:

(1) Be accessible for connection and inspection
(2) Consist of a set of terminals with the capacity for connection of not less than three intersystem bonding conductors
(3) Not interfere with opening the enclosure for a service, building or structure disconnecting means, or metering equipment

Intersystem means that the electrical system and other systems such as communications systems found in Chapter 8 and optical fiber cables (Article 770) are bonded together to minimize the occurrence of potential differences between equipment of different systems. Exhibit 250.36 is an example of an intersystem bonding termination that is to be installed and connected as specified in 250.94(A)(1) through (5).

(4) Be securely mounted as follows:

a. At the service equipment, to a metal enclosure for the service equipment, to a metal meter enclosure, or to an exposed nonflexible metal service raceway, or be connected to the metal enclosure for the grounding electrode conductor with a minimum 6 AWG copper conductor
b. At the disconnecting means for a building or structure that is supplied by a feeder or branch circuit, be electrically connected to the metal enclosure for the building or structure disconnecting means, or be connected to the metal enclosure for the grounding electrode conductor with a minimum 6 AWG copper conductor

(5) Be listed as grounding and bonding equipment

Exception: In existing buildings or structures, if any of the intersystem bonding and grounding electrode conductors required by 770.100(B)(2), 800.100(B)(2), 810.21(F)(2), and 820.100 exist, installation of an IBT shall not be required. An accessible means external to enclosures for connecting intersystem bonding and grounding electrode conductors shall be permitted at the service equipment and at the disconnecting means for any buildings or structures that are supplied by a feeder or branch circuit by at least one of the following means:

(1) Exposed nonflexible metal raceways
(2) An exposed grounding electrode conductor
(3) Approved means for the external connection of a copper or other corrosion-resistant bonding or grounding electrode conductor to the grounded raceway or equipment

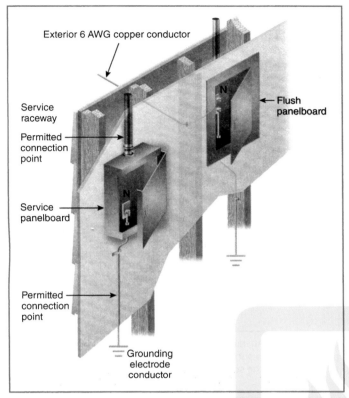

EXHIBIT 250.37 Permitted location to make intersystem bonding connections at an existing building or structure.

The intersystem bonding points at an existing building or structure can be installed with the connection points as shown in Exhibit 250.37. An intersystem bonding termination of the type shown in Exhibit 250.36 is not required to be installed under the conditions of this exception.

> Informational Note: See 770.100, 800.100, 810.21, and 820.100 for intersystem bonding and grounding requirements for conductive optical fiber cables, communications circuits, radio and television equipment, CATV circuits, and network-powered broadband communications systems, respectively.

Lightning protection systems, communications, radio and television, and community antenna television (CATV) systems are required to be bonded together to minimize the potential differences between the systems. Lack of interconnection can result in a severe shock and fire hazard due to differences in potential. Exhibit 250.38 illustrates a CATV cable that is connected to the cable decoder and the tuner of a television set. Also connected to the decoder and the television is the 120-volt supply, with one conductor grounded at the service (the power ground). In each case, resistance to ground is present at the grounding electrode. The resistance to ground varies depending on soil conditions and the type of grounding electrode. For more information, see the accompanying Closer Look feature.

The bonding requirement of 250.94 addresses the difficulties sometimes encountered by communications and CATV installers trying to properly bond their respective systems

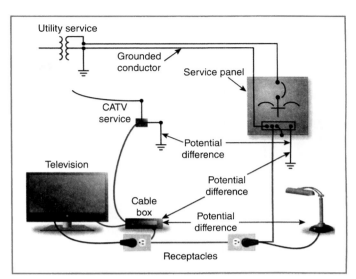

EXHIBIT 250.38 A CATV installation that does not comply with the *NEC*, illustrating why bonding between different systems is necessary.

together and to the electrical supply system. In the past, bonding between communications, CATV, and power systems was usually achieved by connecting the communications protector grounds or cable shield to an interior metal water pipe because the pipe was often used as the power grounding electrode. Thus, the requirement that the power, communications, CATV cable shield, and metal water piping systems be bonded together was easily satisfied. If the power system was grounded to one of the other electrodes permitted by the *NEC* (usually by a made electrode such as a ground rod), the bond was connected to the power GEC or to a metal service raceway, since at least one of these was usually accessible.

With the increased use of nonmetallic water pipe and the service equipment sometimes flush-mounted in finished areas, the GEC is typically concealed. With the increased use of nonmetallic service-entrance conduit, communications and CATV installers often do not have access to a suitable point for connecting bonding jumpers or GECs. In such situations, an intersystem bonding termination device is needed.

See also

800.100 for more information on bonding and grounding of CATV systems

(B) Other Means. Connections to an aluminum or copper busbar not less than 6 mm thick × 50 mm wide (¼ in. thick × 2 in. wide) and of a length to accommodate at least three terminations for communication systems in addition to other connections. The busbar shall be securely fastened and shall be installed in an accessible location. Connections shall be made by a listed connector. If aluminum busbars are used, the installation shall also comply with 250.64(A). The busbar shall be connected to the grounding electrode system by a conductor that is the larger of the following:

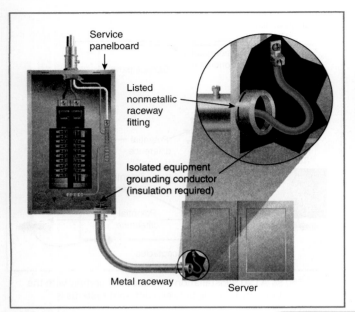

EXHIBIT 250.39 An installation in which the electronic equipment enclosure is grounded through the isolated EGC.

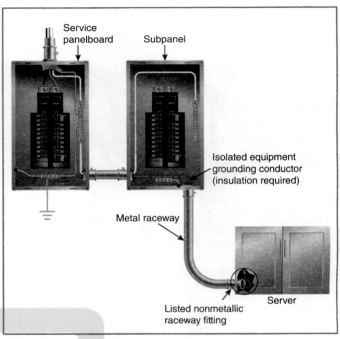

EXHIBIT 250.40 An installation in which the isolated EGC is allowed to pass through the subpanel without connecting to the grounding bus to terminate at the service grounding bus.

(1) The largest grounding electrode conductor that is connected to the busbar

(2) As required or permitted in 250.94(A)

Exception to (A) and (B): Means for connecting intersystem bonding conductors are not required if communications systems are not likely to be used in or on the building or structure.

> Informational Note: The use of an IBT can reduce electrical noise on communication systems.

Exhibits 250.39 and 250.40 illustrate installations with isolated EGCs. In Exhibit 250.39, the metal raceway is bonded through its attachment to the grounded service equipment enclosure, which provides equipment grounding for the raceway. In Exhibit 250.40, the isolated EGC (which is required to be insulated to prevent inadvertent contact with grounded enclosures and raceways) passes through the downstream feeder panelboard and terminates in the service equipment, as permitted by 408.40, Exception. To meet the performance objectives for the grounding and bonding of electrical equipment specified in 250.4(A)(5), the insulated EGC must be connected in a manner that creates an effective path for ground-fault current. The connection point is permitted to be at any location on the distribution system to minimize the electrical noise to an acceptable level.

250.96 Bonding Other Enclosures.

Δ **(A) General.** Metal raceways, cable trays, cable armor, cable sheath, enclosures, frames, fittings, and other metal non-current-carrying parts that are to serve as equipment grounding conductors, with or without the use of wire-type supplementary equipment grounding conductors, shall be bonded if necessary to ensure electrical continuity and the capacity to conduct fault current likely to be imposed on them. Any nonconductive paint, enamel, or similar coating shall be removed at threads, contact points, and contact surfaces or shall be connected by means of fittings designed so as to make such removal unnecessary.

(B) Isolated Grounding Circuits. If installed for the reduction of electromagnetic interference on the grounding circuit, an equipment enclosure supplied by a branch circuit shall be permitted to be isolated from a raceway containing circuits supplying only that equipment by one or more listed nonmetallic raceway fittings located at the point of attachment of the raceway to the equipment enclosure. The metal raceway shall comply with this article and shall be supplemented by an internal insulated equipment grounding conductor installed in accordance with 250.146(D) to ground the equipment enclosure.

> Informational Note: Use of an isolated equipment grounding conductor does not relieve the requirement for grounding the raceway system.

250.97 Bonding for Over 250 Volts to Ground. For circuits of over 250 volts to ground, the electrical continuity of metal raceways and cables with metal sheaths that contain any conductor other than service conductors shall be ensured by one or more of the methods specified for services in 250.92(B), except for (B)(1).

Exception: If oversized, concentric, or eccentric knockouts are not encountered, or if a box or enclosure with concentric or eccentric knockouts is listed to provide a reliable bonding connection, the following methods shall be permitted:

(1) *Threadless couplings and connectors for cables with metal sheaths*
(2) *Two locknuts, on rigid metal conduit or intermediate metal conduit, one inside and one outside of boxes and cabinets*
(3) *Fittings with shoulders that seat tightly against the box or cabinet, such as electrical metallic tubing connectors, flexible metal conduit connectors, and cable connectors, with one locknut on the inside of boxes and cabinets*
(4) *Listed fittings*

Bonding around pre-punched concentric or eccentric knockouts is not required if the enclosure containing the knockouts is listed as suitable for bonding.

UL Guide information indicates that concentric and eccentric knockouts of all metal outlet boxes evaluated in accordance with UL 514A, *Metallic Outlet Boxes,* are suitable for bonding without the use of additional bonding equipment, such as bonding-type locknuts or bonding bushings. The guide information further indicates that metal outlet boxes may be marked to indicate this condition of use.

250.98 Bonding Loosely Jointed Metal Raceways. Expansion, expansion-deflection, or deflection fittings and telescoping sections of metal raceways shall be made electrically continuous by equipment bonding jumpers or other means.

250.100 Bonding in Hazardous (Classified) Locations. Regardless of the voltage of the electrical system, the electrical continuity of normally non-current-carrying metal parts of electrical equipment, raceways, metal-clad cable, and metal enclosures containing electrical equipment in any hazardous (classified) location, as defined in 500.5, 505.5, and 506.5, shall be ensured by any of the bonding methods specified in 250.92(B)(2) through (B)(4). One or more of these bonding methods shall be used whether or not equipment grounding conductors of the wire type are installed in the raceway or in a multiconductor cable assembly.

Informational Note: See 501.30, 502.30, 503.30, 505.30, or 506.30 for specific bonding requirements.

250.102 Grounded Conductor, Bonding Conductors, and Jumpers.

(A) Material. Bonding jumpers shall be of copper, aluminum, copper-clad aluminum, or other corrosion-resistant material. A bonding jumper shall be a wire, bus, screw, or similar suitable conductor.

(B) Attachment. Bonding jumpers shall be attached in the manner specified in 250.8 for circuits and equipment and in 250.70 for grounding electrodes.

(C) Size — Supply-Side Bonding Jumper.

(1) Size for Supply Conductors in a Single Raceway or Cable. The supply-side bonding jumper shall not be smaller than specified in Table 250.102(C)(1).

Δ **(2) Size for Parallel Conductor Installations in Two or More Raceways or Cables.** If the ungrounded supply conductors are connected in parallel in two or more raceways or cables, the supply-side bonding jumper shall be sized in accordance with either of the following:

(1) An individual bonding jumper for each raceway or cable shall be selected from Table 250.102(C)(1) based on the size of the largest ungrounded supply conductor in each raceway or cable.
(2) A single bonding jumper installed for bonding two or more raceways or cables shall be sized in accordance with Table 250.102(C)(1) based on the sum of the circular mil areas of the largest ungrounded conductors from each set connected in parallel in each raceway or cable. The size of the grounded conductor(s) in each raceway or cable shall be based on the largest ungrounded conductor in each raceway or cable, or the sum of the circular mil areas of the largest ungrounded conductors from each set connected in parallel in each raceway or cable.

Informational Note No. 1: The term *supply conductors* includes ungrounded conductors that do not have overcurrent protection on their supply side and terminate at service equipment or the first disconnecting means of a separately derived system.

Informational Note No. 2: See Chapter 9, Table 8, for the circular mil area of conductors 18 AWG through 4/0 AWG.

(D) Size — Equipment Bonding Jumper on Load Side of an Overcurrent Device. The equipment bonding jumper on the load side of an overcurrent device(s) shall be sized in accordance with 250.122.

A single common continuous equipment bonding jumper shall be permitted to connect two or more raceways or cables if the bonding jumper is sized in accordance with 250.122 for the largest overcurrent device supplying circuits therein.

(E) Installation. Bonding jumpers or conductors and equipment bonding jumpers shall be permitted to be installed inside or outside of a raceway or an enclosure.

(1) Inside a Raceway or an Enclosure. If installed inside a raceway, equipment bonding jumpers and bonding jumpers or conductors shall comply with the requirements of 250.119 and 250.148.

(2) Outside a Raceway or an Enclosure. If installed on the outside, the length of the bonding jumper or conductor or equipment bonding jumper shall not exceed 1.8 m (6 ft) and shall be routed with the raceway or enclosure.

Exception: An equipment bonding jumper or supply-side bonding jumper longer than 1.8 m (6 ft) shall be permitted at outside pole locations for the purpose of bonding or grounding isolated sections of metal raceways or elbows installed in exposed risers of metal conduit or other metal raceway, and for bonding grounding electrodes, and shall not be required to be routed with a raceway or enclosure.

Δ **TABLE 250.102(C)(1)** *Grounded Conductor, Main Bonding Jumper, System Bonding Jumper, and Supply-Side Bonding Jumper for Alternating-Current Systems*

Size of Largest Ungrounded Conductor or Equivalent Area for Parallel Conductors (AWG/kcmil)		Size of Grounded Conductor or Bonding Jumper (AWG/kcmil)	
Copper	Aluminum or Copper-Clad Aluminum	Copper	Aluminum or Copper-Clad Aluminum
2 or smaller	1/0 or smaller	8	6
1 or 1/0	2/0 or 3/0	6	4
2/0 or 3/0	4/0 or 250	4	2
Over 3/0 through 350	Over 250 through 500	2	1/0
Over 350 through 600	Over 500 through 900	1/0	3/0
Over 600 through 1100	Over 900 through 1750	2/0	4/0
Over 1100	Over 1750	See Notes 1 and 2.	

Notes:

1. If the circular mil area of ungrounded supply conductors that are connected in parallel is larger than 1100 kcmil copper or 1750 kcmil aluminum, the grounded conductor or bonding jumper shall have an area not less than 12½ percent of the area of the largest ungrounded supply conductor or equivalent area for parallel supply conductors. The grounded conductor or bonding jumper shall not be required to be larger than the largest ungrounded conductor or set of ungrounded conductors.

2. If the circular mil area of ungrounded supply conductors that are connected in parallel is larger than 1100 kcmil copper or 1750 kcmil aluminum and if the ungrounded supply conductors and the bonding jumper are of different materials (copper, aluminum, or copper-clad aluminum), the minimum size of the grounded conductor or bonding jumper shall be based on the assumed use of ungrounded supply conductors of the same material as the grounded conductor or bonding jumper that has an ampacity equivalent to that of the installed ungrounded supply conductors.

3. If there are no service-entrance conductors, the supply conductor size shall be determined by the equivalent size of the largest service-entrance conductor required for the load to be served.

For some metal raceway and rigid conduit systems, installing the bonding jumper where it is visible and accessible for inspection and maintenance is desirable. Exhibit 250.41 illustrates a bonding jumper run outside a length of flexible metal conduit.

(3) Protection. Bonding jumpers or conductors and equipment bonding jumpers shall be installed in accordance with 250.64(A) and (B).

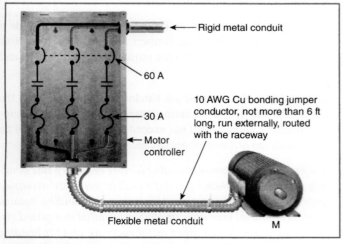

EXHIBIT 250.41 A bonding jumper installed on the outside of a flexible metal conduit.

250.104 Bonding of Piping Systems and Exposed Structural Metal.

(A) Metal Water Piping. The metal water piping system shall be bonded as required in 250.104(A)(1), (A)(2), or (A)(3).

(1) General. Metal water piping system(s) installed in or attached to a building or structure shall be bonded to any of the following:

(1) Service equipment enclosure
(2) Grounded conductor at the service
(3) Grounding electrode conductor, if of sufficient size
(4) One or more grounding electrodes used, if the grounding electrode conductor or bonding jumper to the grounding electrode is of sufficient size

The bonding jumper(s) shall be installed in accordance with 250.64(A), (B), and (E). The points of attachment of the bonding jumper(s) shall be accessible. The bonding jumper(s) shall be sized in accordance with Table 250.102(C)(1) except that it shall not be required to be larger than 3/0 copper or 250 kcmil aluminum or copper-clad aluminum and except as permitted in 250.104(A)(2) and (A)(3).

Bonding a metal water piping system is not the same as using the metal water piping system as a grounding electrode. Bonding to the grounding electrode system places the bonded components at the same voltage level.

CLOSER LOOK: Sizing Supply-Side Bonding Jumpers

Similar to the way that the main bonding jumper is sized per 250.28(D), supply-side bonding jumpers are sized based on the size of the ungrounded conductors with which they are associated. If the ungrounded conductors are 1100 kcmil copper or 1750 kcmil aluminum or smaller, the supply-side bonding jumper is selected from Table 250.102(C)(1) based on the size of the largest ungrounded supply conductor. If the ungrounded conductors are larger than 1100 kcmil copper or 1750 kcmil aluminum, the size of the supply-side bonding jumper(s) is calculated based on 12.5 percent of the area of the largest ungrounded supply conductor or the equivalent area of the parallel supply conductors. Where an installation consists of multiple raceways for parallel conductors, an individual supply-side bonding jumper can be installed for each raceway; the jumper is sized based on the size of the ungrounded conductors in that raceway.

Application Example

A 3-phase, 1600-A service is supplied using five 350 kcmil THWN conductors per phase. The parallel conductors are installed in five separate runs of rigid metal conduit. In accordance with 300.12, Exception No. 2, a supply-side bonding jumper is needed for each raceway at the point it enters the open-bottom switchboard. Using 250.102(C)(2), determine the size of the supply-side bonding jumper.

Multiple supply-side bonding jumpers
Step 1. Determine the size of the largest ungrounded conductor in each raceway.

Step 2. Determine the size of the supply-side bonding jumper for each raceway using Table 250.102(C)(1).

Solution
For 350 kcmil supply conductors, use the "Over 3/0 through 350" row of Table 250.102(C)(1). This results in a 2 AWG copper or 1/0 AWG aluminum supply-side bonding jumper.

Single supply-side bonding jumper
Step 1. Determine the equivalent area of parallel supply conductors.

Step 2. Determine the size of the supply-side bonding jumper using Table 250.102(C)(1), Note 1.

Solution
Five parallel 350 kcmil conductors:

Total equivalent area = 5 × 350 = 1750 kcmil

This total exceeds 1100 kcmil; therefore, the total is multiplied by 0.125 (12.5 percent).

1750 kcmil × 0.125 = 218.75 kcmil

From Table 8 of Chapter 9, the next standard conductor size is 250 kcmil copper.

Isolated sections of metal water piping (such as might be used for a plumbing fixture connection) that are connected to an overall nonmetallic water piping system are not subject to the requirements of 250.104(A). The isolated sections are not a metal water piping system.

Similar to the maximum size required for GECs sized from Table 250.66, the bonding jumper to a metal water piping system, although sized based on Table 250.102(C)(1), is not required to be larger than 3/0 AWG copper or 250 kcmil aluminum or copper-clad aluminum.

See also

250.64(A), (B), and **(E)** for installation requirements that apply to the water piping bonding jumper

(2) Buildings of Multiple Occupancy. In buildings of multiple occupancy where the metal water piping system(s) installed in or attached to a building or structure for the individual occupancies is metallically isolated from all other occupancies by use of nonmetallic water piping, the metal water piping system(s) for each occupancy shall be permitted to be bonded to the equipment grounding terminal of the switchgear, switchboard, or panelboard enclosure (other than service equipment) supplying that occupancy. The bonding jumper shall be sized in accordance with 250.102(D).

In some cases, metal water piping in multiple-occupancy buildings is isolated by nonmetallic fittings and pipe. Therefore, the metal water pipe of an occupancy is permitted to be bonded to the equipment grounding terminal of a feeder-supplied panelboard or switchboard supplying only that individual occupancy instead of the bonding jumper being run to the service equipment location. The bonding jumper, in this case, is sized according to 250.102(D) and Table 250.122, based on the rating of the feeder overcurrent device supplying the occupancy. If the occupancy is supplied by a separate service, its metal water piping system is required to be bonded to any of the locations specified in 250.104(A)(1) for services.

(3) Buildings or Structures Supplied by a Feeder(s) or Branch Circuit(s). The metal water piping system(s) installed in or attached to a building or structure shall be bonded to any of the following:

(1) Building or structure disconnecting means enclosure where located at the building or structure
(2) Equipment grounding conductor run with the supply conductors
(3) One or more grounding electrodes used

The bonding jumper(s) shall be sized in accordance with 250.102(D). The bonding jumper shall not be required to be larger than the largest ungrounded feeder or branch-circuit conductor supplying the building or structure.

Δ **(B) Other Metal Piping.** If installed in or attached to a building or structure, a metal piping system(s), including gas piping, that is likely to become energized shall be bonded to any of the following:

(1) Equipment grounding conductor for the circuit that is likely to energize the piping system
(2) Service equipment enclosure
(3) Grounded conductor at the service
(4) Grounding electrode conductor, if of sufficient size
(5) One or more grounding electrodes used, if the grounding electrode conductor or bonding jumper to the grounding electrode is of sufficient size

The bonding conductor(s) or jumper(s) shall be sized in accordance with Table 250.122, and equipment grounding conductors shall be sized in accordance with Table 250.122 using the rating of the circuit that is likely to energize the piping system(s). The points of attachment of the bonding jumper(s) shall be accessible.

> Informational Note No. 1: Bonding all piping and metal air ducts within the premises will provide additional safety.
> Informational Note No. 2: See NFPA 54, *National Fuel Gas Code*, and NFPA 780, *Standard for the Installation of Lightning Protection Systems*, for information on gas piping systems.

Where the phrase "likely to become energized" is used in the *NEC*, it means that the failure of electrical insulation on a conductor can cause normally non-current-carrying metal parts to become energized. Where mechanical and electrical connections are within equipment, a failure of electrical insulation can result in the connected piping system(s) becoming energized. For example, an insulation failure in an electrical circuit of a gas range could energize the metal gas piping.

The use of an additional bonding jumper is not necessary to comply with 250.104(B), because the equipment grounding connection to the non-current-carrying metal parts of the appliance also provides a bonding connection to the metal piping attached to the appliance. Section 250.104(B) requires the gas piping system to be bonded. It does not make the gas piping part of the grounding electrode system. Therefore, this requirement does not conflict with 250.52(B)(1), which prohibits the use of metal underground gas piping as a grounding electrode. To prevent the underground gas piping from inadvertently becoming a grounding electrode, gas utilities usually provide an isolating fitting. Gas utility companies also often provide cathodic protection of their underground metal piping system.

(C) Structural Metal. Exposed structural metal that is interconnected to form a metal building frame, is not intentionally grounded or bonded, and is likely to become energized shall be bonded to any of the following:

(1) Service equipment enclosure
(2) Grounded conductor at the service
(3) Disconnecting means for buildings or structures supplied by a feeder or branch circuit
(4) Grounding electrode conductor, if not smaller than a conductor sized in accordance with Table 250.102(C)(1)
(5) One or more grounding electrodes used, if the grounding electrode conductor or bonding jumper to the grounding electrode is not smaller than a conductor sized in accordance with Table 250.102(C)(1)

The bonding conductor(s) or jumper(s) shall be sized in accordance with Table 250.102(C)(1), except that it shall not be required to be larger than 3/0 AWG copper or 250 kcmil aluminum or copper-clad aluminum, and installed in accordance with 250.64(A), (B), and (E). The points of attachment of the bonding jumper(s) shall be accessible unless installed in compliance with 250.68(A), Exception No. 2.

(D) Separately Derived Systems. Metal water piping systems and structural metal that is interconnected to form a building frame shall be bonded to separately derived systems in accordance with 250.104(D)(1) through (D)(3).

(1) Metal Water Piping System(s). The grounded conductor of each separately derived system shall be bonded to the nearest accessible point of the metal water piping system(s) in the area served by each separately derived system. This connection shall be made at the same point on the separately derived system where the grounding electrode conductor is connected. Each bonding jumper shall be sized in accordance with Table 250.102(C)(1) based on the largest ungrounded conductor of the separately derived system except that it shall not be required to be larger than 3/0 AWG copper or 250 kcmil aluminum or copper-clad aluminum.

Exception No. 1: A separate bonding jumper to the metal water piping system shall not be required if the metal water piping system is used as the grounding electrode or grounding electrode conductor for the separately derived system and the connection to the water piping system is in the area served by the separately derived system.

Exception No. 2: A separate bonding jumper to the metal water piping system shall not be required if the metal in-ground support structure is used as a grounding electrode or the metal frame of a building or structure is used as the grounding electrode conductor for a separately derived system and is bonded to the metal water piping system in the area served by the separately derived system.

(2) Structural Metal. If exposed structural metal that is interconnected to form the building frame exists in the area served by the separately derived system, it shall be bonded to the grounded conductor of each separately derived system. This connection shall be made at the same point on the separately derived system where the grounding electrode conductor is connected. Each bonding jumper shall be sized in accordance with Table 250.102(C)(1) based on the largest ungrounded conductor of the separately derived system except that it shall not be required to be larger than 3/0 AWG copper or 250 kcmil aluminum or copper-clad aluminum.

Exception No. 1: A separate bonding jumper to the building structural metal shall not be required if the metal in-ground support structure is used as a grounding electrode or the metal frame of a building or structure is used as the grounding electrode conductor for the separately derived system.

Exception No. 2: A separate bonding jumper to the building structural metal shall not be required if the water piping system of a building or structure is used as the grounding electrode or grounding electrode conductor for a separately derived system and is bonded to the building structural metal in the area served by the separately derived system.

(3) Common Grounding Electrode Conductor. If a common grounding electrode conductor is installed for multiple separately derived systems as permitted by 250.30(A)(6), and exposed structural metal that is interconnected to form the building frame or interior metal water piping exists in the area served by the separately derived system, the metal water piping and the structural metal member shall be bonded to the common grounding electrode conductor in the area served by the separately derived system.

Exception: A separate bonding jumper from each derived system to metal water piping and to structural metal members shall not be required if the metal water piping and the structural metal members in the area served by the separately derived system are bonded to the common grounding electrode conductor.

Δ **250.106 Lightning Protection Systems.** The lightning protection system ground terminals shall be bonded to the building or structure grounding electrode system.

> Informational Note No. 1: See 250.60 for use of strike termination devices.
> Informational Note No. 2: See NFPA 780, *Standard for the Installation of Lightning Protection Systems*, which contains detailed information on grounding, bonding, and sideflash distance from lightning protection systems.

Exhibit 250.42 illustrates the bonding of the lightning protection system to the electrical service grounding electrode system. A similar requirement is found in Section 4.14.3 of NFPA 780, *Standard for the Installation of Lightning Protection Systems*. Additional bonding between the lightning protection system and the electrical system might be necessary based on proximity and whether separation between the systems is through air or through building materials. Paragraphs 4.15.2.5.1 and 4.15.2.6.1 of NFPA 780 (2023 edition) describe a method for calculating

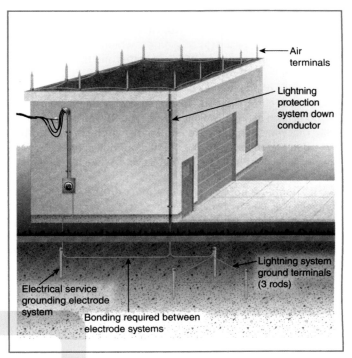

EXHIBIT 250.42 Bonding between the lightning system ground terminals and the electrical service grounding electrode system.

bonding distances between grounded metal bodies and the lightning protection system.

Part VI. Equipment Grounding and Equipment Grounding Conductors

250.109 Metal Enclosures. Metal enclosures shall be permitted to be used to connect bonding jumpers or equipment grounding conductors, or both, together to become a part of an effective ground-fault current path. If installed, metal covers, plaster rings, extension rings, and metal fittings shall be attached to these metal enclosures to ensure an effective ground-fault current path or shall be connected with bonding jumpers or equipment grounding conductors, or both.

> Informational Note: See 250.97 for bonding requirements for over 250 volts to ground.

Metal raceways and cables connect to electrical enclosures (such as cabinets, boxes, and conduit bodies) and fittings at their supply end, at junction or pull points, and at outlets and equipment connections. Metal wiring methods that are recognized as EGCs in 250.118 and wire-type EGCs are required to terminate at and within such enclosures, making the enclosures part of the effective ground-fault current path. The requirements of 250.109 and 250.148 are complementary in purpose.

This requirement, along with 300.10, creates the effective ground-fault current path from the enclosure cover or attached device or equipment, to the connected metal raceway and cables and any wire-type EGCs in those raceways and cables, to the point where the circuit(s) originates, which ultimately is the source of power so that an OCPD operates and the ground fault is indicated.

The use of a metal box or enclosure as part of the effective ground-fault current path is also the case with nonmetallic wiring methods such as Type NM cable and Type SE cable and nonmetallic raceways. An example is a snap switch connected to a grounded metal box using the metal mounting screws to create the electrical continuity to the metal switch yoke and any metal faceplates, as covered in 404.9.

It should be noted that, depending on the supply circuit characteristics, a mechanical connection to an enclosure such as conduit with locknuts, tubing with a fitting and locknut, or metal cable fittings might not provide a sufficient electrical connection to the enclosure. See 250.97 for additional requirements where the supply voltage exceeds 250 volts to ground.

Δ **250.110 Equipment Fastened in Place (Fixed) or Connected by Permanent Wiring Methods.** Exposed, normally non-current-carrying metal parts of fixed equipment supplied by or enclosing conductors or components that are likely to become energized shall be connected to an equipment grounding conductor under any of the following conditions:

(1) If within 2.5 m (8 ft) vertically or 1.5 m (5 ft) horizontally of ground or grounded metal objects and subject to contact by persons
(2) If located in a wet or damp location and not isolated
(3) If in electrical contact with metal
(4) If in a hazardous (classified) location
(5) If supplied by a wiring method that provides an equipment grounding conductor, except as permitted by 250.86, Exception No. 2, for short sections of metal enclosures
(6) If equipment operates with any terminal at over 150 volts to ground

Exception No. 1: If exempted by special permission, the metal frame of electrically heated appliances that have the frame permanently and effectively insulated from ground shall not be required to be grounded.

Exception No. 2: Distribution apparatus, such as transformer and capacitor cases, mounted on wooden poles at a height exceeding 2.5 m (8 ft) above ground or grade level shall not be required to be grounded.

Exception No. 3: Listed equipment protected by a system of double insulation, or its equivalent, shall not be required to be connected to the equipment grounding conductor. If such a system is employed, the equipment shall be distinctively marked.

250.112 Specific Equipment Fastened in Place (Fixed) or Connected by Permanent Wiring Methods. Except as permitted in 250.112(F) and (I), exposed, normally non-current-

carrying metal parts of equipment described in 250.112(A) through (K), and normally non-current-carrying metal parts of equipment and enclosures described in 250.112(L) and (M), shall be connected to an equipment grounding conductor, regardless of voltage.

(A) Switchgear and Switchboard Frames and Structures. Switchgear or switchboard frames and structures supporting switching equipment, except frames of 2-wire dc switchgear or switchboards if effectively insulated from ground.

(B) Pipe Organs. Generator and motor frames in an electrically operated pipe organ, unless effectively insulated from ground and the motor driving it.

(C) Motor Frames. Motor frames, as provided by 430.242.

(D) Enclosures for Motor Controllers. Enclosures for motor controllers unless attached to ungrounded portable equipment.

(E) Elevators and Cranes. Electrical equipment for elevators and cranes.

(F) Garages, Theaters, and Motion Picture Studios. Electrical equipment in commercial garages, theaters, and motion picture studios, except pendant lampholders supplied by circuits not over 150 volts to ground.

(G) Electric Signs. Electric signs, outline lighting, and associated equipment as provided in 600.7.

(H) Motion Picture Projection Equipment. Motion picture projection equipment.

(I) Remote-Control, Signaling, and Fire Alarm Circuits. Equipment supplied by Class 1 circuits shall be grounded unless operating at less than 50 volts. Equipment supplied by Class 1 power-limited circuits, by Class 2 and Class 3 remote-control and signaling circuits, and by fire alarm circuits shall be grounded if system grounding is required by Part II or Part VIII of this article.

(J) Luminaires. Luminaires as provided in Part V of Article 410.

(K) Skid-Mounted Equipment. Permanently mounted electrical equipment and skids shall be connected to the equipment grounding conductor. Wire-type equipment grounding conductors shall be sized as required by 250.122.

(L) Motor-Operated Water Pumps. Motor-operated water pumps, including the submersible type.

(M) Metal Well Casings. If a submersible pump is used in a metal well casing, the well casing shall be connected to the pump circuit equipment grounding conductor.

Δ **250.114 Equipment Connected by Cord and Plug.** Exposed, normally non-current-carrying metal parts of cord-and-plug-connected equipment shall be connected to the equipment grounding conductor under any of the following conditions:

Exception: Listed tools, listed appliances, and listed equipment covered in 250.114, list items 2 through 4, shall not be required to be connected to an equipment grounding conductor if protected by a system of double insulation or its equivalent. Double-insulated equipment shall be distinctively marked.

(1) In hazardous (classified) locations
(2) If operated at over 150 volts to ground

Exception No. 1 to (2): Motors, if guarded, shall not be required to be connected to an equipment grounding conductor.

Exception No. 2 to (2): Metal frames of electrically heated appliances, exempted by special permission, shall not be required to be connected to an equipment grounding conductor, in which case the frames shall be permanently and effectively insulated from ground.

(3) In residential occupancies:

    a. Refrigerators, freezers, icemakers, and air conditioners
    b. Clothes-washing, clothes-drying, and dish-washing machines; ranges; kitchen waste disposers; information technology equipment; sump pumps; and electrical aquarium equipment
    c. Hand-held motor-operated tools, stationary and fixed motor-operated tools, and light industrial motor-operated tools
    d. Motor-operated appliances of the following types: hedge clippers, lawn mowers, snow blowers, and wet scrubbers
    e. Portable handlamps

(4) In other than residential occupancies:

    a. Refrigerators, freezers, icemakers, and air conditioners
    b. Clothes-washing, clothes-drying, and dish-washing machines; information technology equipment; sump pumps; and electrical aquarium equipment
    c. Hand-held motor-operated tools, stationary and fixed motor-operated tools, and light industrial motor-operated tools
    d. Motor-operated appliances of the following types: hedge clippers, lawn mowers, snow blowers, and wet scrubbers
    e. Portable handlamps
    f. Cord-and-plug-connected appliances used in damp or wet locations or by persons standing on the ground, standing on metal floors, or working inside of metal tanks or boilers
    g. Tools likely to be used in wet or conductive locations

Exception: Tools and portable handlamps and portable luminaires likely to be used in wet or conductive locations shall not be required to be connected to an equipment grounding conductor if supplied through an isolating transformer with an ungrounded secondary of not over 50 volts.

250.116 Nonelectrical Equipment. The metal parts of the following nonelectrical equipment described in this section shall be connected to the equipment grounding conductor:

(1) Frames and tracks of electrically operated cranes and hoists
(2) Frames of nonelectrically driven elevator cars to which electrical conductors are attached
(3) Hand-operated metal shifting ropes or cables of electric elevators

> Informational Note: If extensive metal in or on buildings or structures may become energized and is subject to personal contact, adequate bonding and grounding will provide additional safety.

Δ **250.118 Types of Equipment Grounding Conductors.**

As a general rule, the *NEC* requires only one EGC, and it can be any of the types specified in this section. However, some specific requirements, such as 517.13(B), require an additional wire-type EGC regardless of the wiring method. If a wire-type EGC is installed in a metal raceway and is not isolated per 250.96(B), the raceway is in parallel with the wire, and the combination of the metal raceway and the wire is the EGC for enclosed circuit(s).

See also

250.122 for sizing of wire-type EGCs

Δ **(A) Permitted.** Each equipment grounding conductor run with or enclosing the circuit conductors shall be one or more or a combination of the following:

(1) A copper, aluminum, or copper-clad aluminum conductor. This conductor shall be solid or stranded; insulated, covered, or bare; and in the form of a wire or a busbar of any shape.
(2) Rigid metal conduit.
(3) Intermediate metal conduit.
(4) Electrical metallic tubing.
(5) Listed flexible metal conduit meeting all the following conditions:

    a. The conduit is terminated in listed fittings.
    b. The circuit conductors contained in the conduit are protected by overcurrent devices rated at 20 amperes or less.
    c. The size of the conduit does not exceed metric designator 35 (trade size 1¼).
    d. The combined length of flexible metal conduit, flexible metallic tubing, and liquidtight flexible metal conduit in the same effective ground-fault current path does not exceed 1.8 m (6 ft).
    e. If flexibility is necessary to minimize the transmission of vibration from equipment or to provide flexibility for equipment that requires movement after installation, a wire-type equipment grounding conductor or a bonding jumper in accordance with 250.102(E)(2) shall be installed.
    f. If flexible metal conduit is constructed of stainless steel, a wire-type equipment grounding conductor or bonding jumper in accordance with 250.102(E)(2) shall be installed.

(6) Listed liquidtight flexible metal conduit meeting all the following conditions:

  a. The conduit is terminated in listed fittings.

  b. For metric designators 12 through 16 (trade sizes ⅜ through ½), the circuit conductors contained in the conduit are protected by overcurrent devices rated at 20 amperes or less.

  c. For metric designators 21 through 35 (trade sizes ¾ through 1¼), the circuit conductors contained in the conduit are protected by overcurrent devices rated not more than 60 amperes and there is no flexible metal conduit, flexible metallic tubing, or liquidtight flexible metal conduit in metric designators 12 through 16 (trade sizes ⅜ through ½) in the effective ground-fault current path.

  d. The combined length of flexible metal conduit, flexible metallic tubing, and liquidtight flexible metal conduit in the same effective ground-fault current path does not exceed 1.8 m (6 ft).

  e. If flexibility is necessary to minimize the transmission of vibration from equipment or to provide flexibility for equipment that requires movement after installation, a wire-type equipment grounding conductor or a bonding jumper in accordance with 250.102(E)(2) shall be installed.

  f. If liquidtight flexible metal conduit contains a stainless steel core, a wire-type equipment grounding conductor or a bonding jumper in accordance with 250.102(E)(2) shall be installed.

(7) Flexible metallic tubing if the tubing is terminated in listed fittings and meeting the following conditions:

  a. The circuit conductors contained in the tubing are protected by overcurrent devices rated at 20 amperes or less.

  b. The combined length of flexible metal conduit, flexible metallic tubing, and liquidtight flexible metal conduit in the same effective ground-fault current path does not exceed 1.8 m (6 ft).

(8) Armor of Type AC cable as provided in 320.108.

(9) The copper sheath of mineral-insulated, metal-sheathed cable Type MI.

(10) Type MC cable that provides an effective ground-fault current path in accordance with one or more of the following:

  a. It contains an insulated or uninsulated equipment grounding conductor in compliance with 250.118(1).

  b. The combined metallic sheath and uninsulated equipment grounding/bonding conductor of interlocked metal tape–type MC cable that is listed and identified as an equipment grounding conductor

  c. The metallic sheath or the combined metallic sheath and equipment grounding conductors of the smooth or corrugated tube-type MC cable that is listed and identified as an equipment grounding conductor

(11) Cable trays as permitted in 392.10 and 392.60.

(12) Cablebus framework as permitted in 370.60(1).

(13) Other listed electrically continuous metal raceways and listed auxiliary gutters.

(14) Surface metal raceways listed for grounding.

Informational Note: See Article 100 for a definition of *effective ground-fault current path.*

(B) Not Permitted. The following shall not be used as equipment grounding conductors.

(1) Grounding electrode conductors

Exception: A wire-type equipment grounding conductor installed in compliance with 250.6(A) and the applicable requirements for both the equipment grounding conductor and the grounding electrode conductor in Parts II, III, and VI of this article shall be permitted to serve as both an equipment grounding conductor and a grounding electrode conductor.

Equipment grounding conductors (EGCs) and grounding electrode conductors (GECs) have specific functions. GECs are required to be a wire- or busbar-type conductor in accordance with 250.62, whereas EGCs are permitted to be any of the types listed in 250.118. This exception permits a wire-type conductor to be used for both purposes if it satisfies all applicable requirements for both the EGC and the GEC and it does not carry current during normal operating conditions.

(2) Structural metal frame of a building or structure

Δ **250.119 Identification of Wire-Type Equipment Grounding Conductors.**

N **(A) General.** Unless required elsewhere in this *Code,* equipment grounding conductors shall be permitted to be bare, covered, or insulated. Individually covered or insulated equipment grounding conductors of the wire type shall have a continuous outer finish that is either green or green with one or more yellow stripes except as permitted in this section. Conductors with insulation or individual covering that is green, green with one or more yellow stripes, or otherwise identified as permitted by this section shall not be used for ungrounded or grounded circuit conductors.

Exception No. 1: Power-limited Class 2 or Class 3 cables, power-limited fire alarm cables, or communications cables containing only circuits operating at less than 50 volts ac or 60 volts dc if connected to equipment not required to be grounded shall be permitted to use a conductor with green insulation or green with one or more yellow stripes for other than equipment grounding purposes.

In general, most limited-energy ac systems are not required to be grounded per 250.20(A). Similarly, 250.162 does not require grounding of dc systems operating at 60 volts or less. Per 250.112 (I), if the ac or dc limited-energy system is not required to be grounded, the equipment associated with circuits supplied from such systems is not required to be connected to an equipment grounding conductor.

Exception No. 2: Flexible cords having an integral insulation and jacket without an equipment grounding conductor shall be permitted to have a continuous outer finish that is green.

Informational Note: An example of a flexible cord with integral-type insulation is Type SPT-2, 2 conductor.

Exception No. 3: Conductors with green insulation shall be permitted to be used as ungrounded signal conductors where installed between the output terminations of traffic signal control and traffic signal indicating heads. Signaling circuits installed in accordance with this exception shall include an equipment grounding conductor in accordance with 250.118. Wire-type equipment grounding conductors shall be bare or have insulation or covering that is green with one or more yellow stripes.

(B) Conductors 4 AWG and Larger. Equipment grounding conductors 4 AWG and larger shall comply with the following:

(1) At the time of installation, if the insulation does not comply with 250.119(A), it shall be permanently identified as an equipment grounding conductor at each end and at every point where the conductor is accessible.

Exception: Conductors 4 AWG and larger shall not be required to be marked in conduit bodies that contain no splices or unused hubs.

(2) Identification shall encircle the conductor and shall be accomplished by one of the following:
   a. Stripping the insulation or covering from the entire exposed length
   b. Coloring the insulation or covering green at the termination
   c. Marking the insulation or covering with green tape or green adhesive labels at the termination

(C) Multiconductor Cable. One or more insulated conductors in a multiconductor cable, at the time of installation, shall be permitted to be permanently identified as equipment grounding conductors at each end and at every point where the conductors are accessible by one of the following means:

(1) Stripping the insulation from the entire exposed length.
(2) Coloring the exposed insulation green.
(3) Marking the exposed insulation with green tape or green adhesive labels. Identification shall encircle the conductor.

(D) Flexible Cord. Equipment grounding conductors in flexible cords shall be insulated and shall have a continuous outer finish that is either green or green with one or more yellow stripes.

250.120 Equipment Grounding Conductor Installation. An equipment grounding conductor shall be installed in accordance with 250.120(A), (B), and (C).

(A) Raceway, Cable Trays, Cable Armor, Cablebus, or Cable Sheaths. If it consists of a raceway, cable tray, cable armor, cablebus framework, or cable sheath or if it is a wire within a raceway or cable, it shall be installed in accordance with the applicable provisions in this *Code* using fittings for joints and terminations approved for use with the type of raceway or cable used. All connections, joints, and fittings shall be made tight using suitable tools.

Informational Note: See the UL Guide Information for Electrical Circuit Integrity Systems (FHIT) for equipment grounding conductors installed in a raceway that are part of a listed electrical circuit protective system or a listed fire-resistive cable system.

Δ **(B) Aluminum and Copper-Clad Aluminum Conductors.** Equipment grounding conductors of bare, covered, or insulated aluminum or copper-clad aluminum shall comply with the following:

(1) Unless part of an applicable cable wiring method, bare or covered conductors shall not be installed if subject to corrosive conditions or be installed in direct contact with concrete, masonry, or the earth.
(2) Terminations made within outdoor enclosures that are listed and identified for the environment shall be permitted within 450 mm (18 in.) of the bottom of the enclosure.
(3) Aluminum or copper-clad aluminum conductors external to buildings or enclosures shall not be terminated within 450 mm (18 in.) of the earth, unless terminated within a listed wire connector system.

Δ **(C) Equipment Grounding Conductors Smaller Than 6 AWG.** If not routed with circuit conductors as permitted in 250.130(C) and 250.134, Exception No. 2, equipment grounding conductors smaller than 6 AWG shall be protected from physical damage by an identified raceway or cable armor unless installed within hollow spaces of the framing members of buildings or structures and if not subject to physical damage.

250.122 Size of Equipment Grounding Conductors.

(A) General. Copper, aluminum, or copper-clad aluminum equipment grounding conductors of the wire type shall not be smaller than shown in Table 250.122. The equipment grounding conductor shall not be required to be larger than the circuit conductors supplying the equipment. If a cable tray, a raceway, or a cable armor or sheath is used as the equipment grounding conductor, as provided in 250.118 and 250.134(1), it shall comply with 250.4(A)(5) or (B)(4).

Equipment grounding conductors shall be permitted to be sectioned within a multiconductor cable, provided the combined circular mil area complies with Table 250.122.

Δ **(B) Increased in Size.** If ungrounded conductors are increased in size for any reason other than as required in 310.15(B) or 310.15(C), wire-type equipment grounding conductors, if installed, shall be increased in size proportionately to the increase in circular mil area of the ungrounded conductors.

Exception: Equipment grounding conductors shall be permitted to be sized by a qualified person to provide an effective ground fault current path in accordance with 250.4(A)(5) or (B)(4)

Generally, the minimum-sized EGC is selected from Table 250.122 based on the rating or setting of the feeder or branch-circuit OCPD(s). Where the ungrounded circuit conductors are increased in size, the EGCs must be increased proportionately. This will lower the overall impedance of the ground-fault current return path, which will facilitate operation of the OCPD in the event of a line-to-ground fault.

Calculation Example

A 240-V, single-phase, 250-A load is supplied from a 300-A breaker located in a panelboard 500 ft away. The conductors are 250 kcmil copper, installed in rigid nonmetallic conduit, with a 4 AWG copper EGC. If the conductors are increased to 350 kcmil, what is the minimum size for the EGC based on the proportional increase requirement?

Solution

Step 1. Calculate the size ratio of the new conductors to the existing conductors:

$$\text{Size ratio} = \frac{350{,}000 \text{ circular mils}}{250{,}000 \text{ circular mils}} = 1.4$$

Step 2. Calculate the cross-sectional area of the new EGC:

$$41{,}740 \text{ circular mils} \times 1.4 = 58{,}436 \text{ circular mils}$$

According to Table 8 of Chapter 9, the size of the existing grounding conductor, 4 AWG, has a cross-sectional area of 41,740 circular mils.

Step 3. Determine the size of the new EGC. Again, referring to Chapter 9, Table 8, we find that 58,436 circular mils is larger than 3 AWG. The next larger size is 66,360 circular mils, which converts to a 2 AWG copper EGC.

(C) Multiple Circuits. A single equipment grounding conductor shall be permitted to be installed for multiple circuits that are installed in the same raceway, cable, trench, or cable tray. It shall be sized from Table 250.122 for the largest overcurrent device protecting circuit conductors in the raceway, cable, trench, or cable tray. Equipment grounding conductors installed in cable trays shall meet the minimum requirements of 392.10(B)(1)(c).

A single EGC serving multiple circuits is required to be sized based on the size of the largest circuit OCPD supplied; it is not likely that all circuits will develop faults at the same time. For example, three 3-phase circuits in the same raceway, protected by overcurrent devices rated 30, 60, and 100 amperes, would require only one EGC, sized according to the largest overcurrent device (in this case, 100 amperes). Therefore, according to Table 250.122, an 8 AWG copper or 6 AWG aluminum conductor or copper-clad aluminum conductor is required.

(D) Motor Circuits. Equipment grounding conductors for motor circuits shall be sized in accordance with 250.122(D)(1) or (D)(2).

(1) General. The equipment grounding conductor size shall not be smaller than determined by 250.122(A) based on the rating of the branch-circuit short-circuit and ground-fault protective device.

(2) Instantaneous-Trip Circuit Breaker and Motor Short-Circuit Protector. If the overcurrent device is an instantaneous-trip circuit breaker or a motor short-circuit protector, the equipment grounding conductor shall be sized not smaller than that given by 250.122(A) using the maximum permitted rating of a dual element time-delay fuse selected for branch-circuit short-circuit and ground-fault protection in accordance with 430.52(C)(1), Exception No. 1.

(E) Flexible Cord and Fixture Wire. The equipment grounding conductor in a flexible cord with the largest circuit conductor 10 AWG or smaller, and the equipment grounding conductor used with fixture wires of any size in accordance with 240.5, shall not be smaller than 18 AWG copper and shall not be smaller than the circuit conductors. The equipment grounding conductor in a flexible cord with a circuit conductor larger than 10 AWG shall be sized in accordance with Table 250.122.

(F) Conductors in Parallel. For circuits of parallel conductors as permitted in 310.10(G), the equipment grounding conductor shall be installed in accordance with 250.122(F)(1) or (F)(2).

Δ **(1) Conductor Installations in Raceways, Auxiliary Gutters, or Cable Trays.**

(a) *Single Raceway or Cable Tray, Auxiliary Gutter, or Cable Tray.* If circuit conductors are connected in parallel in the same raceway, auxiliary gutter, or cable tray, a single wire-type conductor shall be permitted as the equipment grounding conductor. The wire-type equipment grounding conductor shall be sized in accordance with 250.122 based on the overcurrent protective device for the feeder or branch circuit.

(b) *Multiple Raceways.* If conductors are installed in multiple raceways and are connected in parallel, a wire-type equipment grounding conductor, if used, shall be installed in each raceway and shall be connected in parallel. The equipment grounding conductor installed in each raceway shall be sized in accordance with 250.122 based on the rating of the overcurrent protective device for the feeder or branch circuit.

(c) *Wire-Type Equipment Grounding Conductors in Cable Trays.* Wire-type equipment grounding conductors installed in cable trays shall meet the minimum requirements of 392.10(B)(1)(c).

(d) *Metal Raceways, Auxiliary Gutters, or Cable Trays.* Metal raceways or auxiliary gutters in accordance with 250.118 or cable trays complying with 392.60(B) shall be permitted as the equipment grounding conductor.

(2) Multiconductor Cables.

(a) Except as provided in 250.122(F)(2)(c) for raceway or cable tray installations, the equipment grounding conductor in each multiconductor cable shall be sized in accordance with 250.122 based on the overcurrent protective device for the feeder or branch circuit.

(b) If circuit conductors of multiconductor cables are connected in parallel, the equipment grounding conductor(s) in each cable shall be connected in parallel.

(c) If multiconductor cables are paralleled in the same raceway, auxiliary gutter, or cable tray, a single equipment grounding conductor that is sized in accordance with 250.122 shall be permitted in combination with the equipment grounding conductors provided within the multiconductor cables and shall all be connected together.

(d) Equipment grounding conductors installed in cable trays shall meet the minimum requirements of 392.10(B)(1)(c). Cable trays complying with 392.60(B), metal raceways in accordance with 250.118, or auxiliary gutters shall be permitted as the equipment grounding conductor.

A full-sized EGC is required to prevent overloading and possible burnout of the conductor should a ground fault occur along one of the parallel branches. The installation conditions for paralleled conductors prescribed in 310.10(G) result in proportional distribution of the current-time duty among the several paralleled grounding conductors only for overcurrent conditions downstream of the paralleled set of circuit conductors.

Exhibit 250.43 shows a parallel arrangement with two nonmetallic conduits installed underground. A ground fault at the enclosure will cause the EGC in the top conduit to carry more than its proportional share of fault current. The fault is fed by two different conductors of the same phase, one from the left and one from the right. The shortest and lowest-impedance path to ground from the fault to the supply panelboard is through the EGC in the top conduit. The grounding path from the fault through the bottom conduit is longer and of higher impedance. Therefore, the EGC in each raceway must be capable of carrying a major portion of the fault current without burning open.

Generally, if cables are used in parallel, the EGC in each cable is required to be sized in accordance with 250.122 based on the rating or setting of the OCPD that is protecting the conductors connected in parallel. The EGC might have to be larger than the ungrounded conductor in an individual cable that is used as part of a larger parallel set. The cross-sectional areas of sectioned EGCs in an individual cable are permitted by 250.122(A) to be added together to meet the requirement of a fully sized EGC in each cable. For multiconductor cables installed in a raceway, cable tray, or auxiliary gutter, a separate bare, covered, or insulated full-sized EGC that is based on the rating of the OCPD protecting the parallel cable circuit can be installed as an alternative to requiring that each cable contain a full-sized EGC. However, this separate, full-sized EGC and any EGCs contained in the multiconductor cables are required to be connected together at each end of the circuit.

(G) Feeder Taps. Equipment grounding conductors installed with feeder taps shall not be smaller than shown in Table 250.122 based on the rating of the overcurrent device ahead of the feeder on the supply side ahead of the tap but shall not be required to be larger than the tap conductors.

TABLE 250.122 *Minimum Size Equipment Grounding Conductors for Grounding Raceway and Equipment*

Rating or Setting of Automatic Overcurrent Device in Circuit Ahead of Equipment, Conduit, etc., Not Exceeding (Amperes)	Size (AWG or kcmil)	
	Copper	Aluminum or Copper-Clad Aluminum*
15	14	12
20	12	10
60	10	8
100	8	6
200	6	4
300	4	2
400	3	1
500	2	1/0
600	1	2/0
800	1/0	3/0
1000	2/0	4/0
1200	3/0	250
1600	4/0	350
2000	250	400
2500	350	600
3000	400	600
4000	500	750
5000	700	1250
6000	800	1250

Note: Where necessary to comply with 250.4(A)(5) or (B)(4), the equipment grounding conductor shall be sized larger than given in this table.

*See installation restrictions in 250.120.

250.124 Equipment Grounding Conductor Continuity.

(A) Separable Connections. Separable connections such as those provided in drawout equipment or attachment plugs and

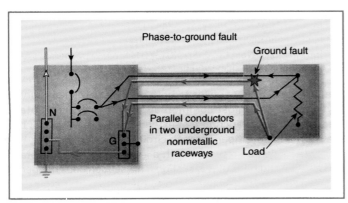

EXHIBIT 250.43 Grounding paths for ground fault at the load supplied by parallel conductors in two underground nonmetallic raceways, illustrating the purpose of 250.122(F).

mating connectors and receptacles shall provide for first-make, last-break of the equipment grounding conductor. First-make, last-break shall not be required if interlocked equipment, plugs, receptacles, and connectors preclude energization without grounding continuity.

(B) Switches. No automatic cutout or switch shall be placed in the equipment grounding conductor of a premises wiring system unless the opening of the cutout or switch disconnects all sources of energy.

250.126 Identification of Wiring Device Terminals. The terminal for the connection of the equipment grounding conductor shall be identified by one of the following:

(1) A green, not readily removable terminal screw with a hexagonal head.
(2) A green, hexagonal, not readily removable terminal nut.
(3) A green pressure wire connector. If the terminal for the equipment grounding conductor is not visible, the conductor entrance hole shall be marked with the word *green* or *ground*, the letters *G* or *GR*, a grounding symbol, or otherwise identified by a distinctive green color. If the terminal for the equipment grounding conductor is readily removable, the area adjacent to the terminal shall be similarly marked.

Informational Note: See Informational Note Figure 250.126.

INFORMATIONAL NOTE FIGURE 250.126 One Example of a Symbol Used to Identify the Grounding Termination Point for an Equipment Grounding Conductor.

Part VII. Methods of Equipment Grounding Conductor Connections

250.130 Equipment Grounding Conductor Connections. Equipment grounding conductor connections at the source of separately derived systems shall be made in accordance with 250.30(A)(1). Equipment grounding conductor connections at service equipment shall be made as indicated in 250.130(A) or (B). For replacement of non-grounding-type receptacles with grounding-type receptacles, or snap switches without an equipment grounding terminal with snap switches with an equipment grounding terminal, and for branch-circuit extensions only in existing installations that do not have an equipment grounding conductor in the branch circuit, connections shall be permitted as indicated in 250.130(C).

(A) For Grounded Systems. The connection shall be made by bonding the equipment grounding conductor to the grounded service conductor and the grounding electrode conductor.

The grounding and bonding arrangement required by 250.130(A) for a grounded system is illustrated in Exhibit 250.44.

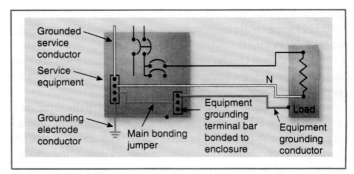

EXHIBIT 250.44 Grounding and bonding arrangement for grounded systems, per 250.130(A), illustrating connection of the EGC (bus) to the enclosures and to the grounded service conductor.

(B) For Ungrounded Systems. The connection shall be made by bonding the equipment grounding conductor to the grounding electrode conductor.

(C) Replacement of Nongrounding Receptacle or Snap Switch and Branch Circuit Extensions. The equipment grounding conductor that is connected to a grounding-type receptacle, a snap switch with an equipment grounding terminal, or a branch-circuit extension shall be permitted to be connected to any of the following:

(1) Any accessible point on the grounding electrode system as described in 250.50
(2) Any accessible point on the grounding electrode conductor
(3) The equipment grounding terminal bar within the enclosure where the branch circuit for the receptacle or branch circuit originates
(4) An equipment grounding conductor that is part of another branch circuit that originates from the enclosure where the branch circuit for the receptacle, snap switch, or branch circuit originates
(5) For grounded systems, the grounded service conductor within the service equipment enclosure
(6) For ungrounded systems, the grounding terminal bar within the service equipment enclosure

Informational Note No. 1: See 406.4(D) for the use of a ground-fault circuit-interrupting type of receptacle.
Informational Note No. 2: See 404.9(B) for requirements regarding grounding of snap switches.

Section 250.130(C) applies to both ungrounded and grounded systems, but its most common application is for receptacle replacement or branch-circuit extensions in single-phase, 120-volt, 15- and 20-ampere branch circuits. A non-grounding-type receptacle can be replaced with a grounding-type receptacle under the conditions of this section. This section does not permit a separate EGC to be connected to the metal water piping of a building beyond the first 5 feet because it is not considered a GEC [unless the conditions of 250.68(C)(1) Exception apply] or part of the electrode system.

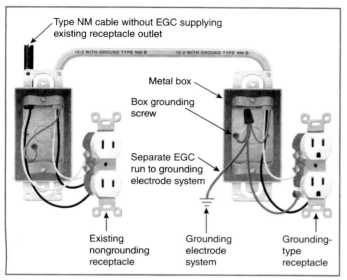

EXHIBIT 250.45 Branch-circuit extension from an existing installation, per 250.130(C), illustrating a separate EGC connected to the grounding electrode system.

Exhibit 250.45 shows a branch-circuit extension made from an existing receptacle outlet without an EGC to a new receptacle outlet. The new receptacle is required to be a grounding-type device. The EGC in the new Type NM cable used to extend the branch circuit is connected to the separate EGC installed per 250.130(C) at the new metal box enclosing the new receptacle. Based on 250.148(C), the EGC is required to be connected to the new metal box installed to enclose the new receptacle and to the existing metal box where the branch-circuit extension originates. The existing non-grounding-type receptacle can remain. If the installer chooses, a new grounding-type receptacle can be installed as a replacement for the non-grounding-type device.

250.132 Short Sections of Raceway or Cable Armor. Isolated sections of metal raceway or cable armor, if required to be connected to an equipment grounding conductor, shall be connected in accordance with 250.134.

250.134 Equipment Fastened in Place or Connected by Permanent Wiring Methods (Fixed). Unless connected to the grounded circuit conductor as permitted by 250.32, 250.140, and 250.142, non-current-carrying metal parts of equipment, raceways, and other enclosures, if grounded, shall be connected to an equipment grounding conductor by one of the following methods:

(1) By connecting to any of the equipment grounding conductors permitted by 250.118(2) through (14)
(2) By connecting to an equipment grounding conductor of the wire type that is contained within the same raceway, contained within the same cable, or otherwise run with the circuit conductors

The EGC run in the same raceway or cable as the circuit conductor(s) allows the magnetic field developed by the circuit conductor and the EGC to cancel, reducing their impedance.

Magnetic flux strength is inversely proportional to the square of the distance between the two conductors. By placing an EGC away from the conductor delivering the fault current, the magnetic flux cancellation decreases. This increases the impedance of the fault path and delays operation of the protective device contrary to the performance requirements specified in 250.4(A)(5) and (B)(4) for the ground-fault current return path.

Exception No. 1: As provided in 250.130(C), the equipment grounding conductor shall be permitted to be run separately from the circuit conductors.

Exception No. 2: For dc circuits, the equipment grounding conductor shall be permitted to be run separately from the circuit conductors

Informational Note No. 1: See 250.102 and 250.168 for equipment bonding jumper requirements.
Informational Note No. 2: See 400.10 for use of flexible cords and flexible cables for fixed equipment.

250.136 Equipment Secured to a Metal Rack or Structure. If a metal rack or structure is connected to an equipment grounding conductor in accordance with 250.134, it shall be permitted to serve as the equipment grounding conductor for electrical equipment secured to and in electrical contact with the metal rack or structure.

Equipment bolted or securely clamped to the rack will typically provide the necessary electrical contact to ensure a low impedance connection between the rack and the equipment. If the rack has been painted, 250.12 requires the paint to be removed to ensure good electrical continuity.

Exhibit 250.46 is an example of electrical equipment secured to and in electrical contact with a metal rack. It is effectively grounded by the equipment bonding jumper installed between the safety switch and the rack. Building steel is not permitted as the equipment grounding conductor by 250.134 and 300.3(B).

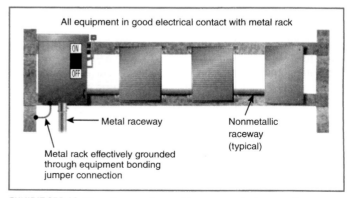

EXHIBIT 250.46 Mounting equipment to a grounded support structure or rack is an acceptable means of providing the required equipment grounding connection.

250.138 Cord-and-Plug-Connected Equipment. Non–current-carrying metal parts of cord-and-plug-connected equipment, if required to be connected to an equipment grounding conductor, shall be connected by one of the methods in 250.138(A) or (B).

(A) By Means of an Equipment Grounding Conductor. By means of an equipment grounding conductor run with the power supply conductors in a cable assembly or flexible cord properly terminated in a grounding-type attachment plug with one fixed grounding contact.

Exception: The grounding contacting pole of grounding-type plug-in ground-fault circuit interrupters shall be permitted to be of the movable, self-restoring type on circuits operating at not over 150 volts between any two conductors or over 150 volts between any conductor and ground.

(B) By Means of a Separate Flexible Wire or Strap. By means of a separate flexible wire or strap, insulated or bare, connected to an equipment grounding conductor, and protected as well as practicable against physical damage, if part of equipment.

△ 250.140 Frames of Ranges and Clothes Dryers. Frames of electric ranges, wall-mounted ovens, counter-mounted cooking units, clothes dryers, and outlet or junction boxes that are part of the circuit shall be connected to the equipment grounding conductor in accordance with 250.140(A) or the grounded conductor in accordance with 250.140(B).

Prior to the 1996 edition of the *NEC,* use of the grounded circuit conductor as a grounding conductor was permitted for range installations. In many instances, the wiring method was service-entrance cable with an uninsulated neutral conductor covered by the cable jacket. Where Type SE cable was used to supply ranges and dryers, the branch circuit was required to originate at the service equipment to avoid neutral current from downstream panelboards being imposed on metal objects, such as pipes or ducts. The grounded conductor (neutral) of newly installed branch circuits supplying ranges and clothes dryers is not permitted to be used for grounding the non-current-carrying metal parts of the appliances. Branch circuits for new appliance installations are required to provide an EGC sized in accordance with 250.122 for grounding the non-current-carrying metal parts.

An older appliance connected to a new branch circuit must have its 3-wire cord and plug replaced with a 4-conductor cord, with one of those conductors being an EGC. The appliance bonding jumper between the neutral and the frame of the appliance must be removed. Where a new range or clothes dryer is connected to an existing branch circuit without an EGC, an appliance bonding jumper must be connected between the neutral terminal and the frame of the appliance.

The grounded circuit conductor of an existing branch circuit is permitted to be used to ground the frame of an electric range, wall-mounted oven, or counter-mounted cooking unit, provided the conditions in 250.140(B)(1), (2), and (3) are all met, and one of either (4) or (5) is met. In addition, the grounded circuit

conductor of the existing branch circuits is also permitted to be used to ground any junction boxes in the circuit supplying the appliance, and a 3-wire pigtail and range receptacle are permitted to be used.

N (A) Equipment Grounding Conductor Connections. The circuit supplying the appliance shall include an equipment grounding conductor. The frame of the appliance shall be connected to the equipment grounding conductor in the manner specified by 250.134 or 250.138.

N (B) Grounded Conductor Connections. For existing branch-circuit installations only, if an equipment grounding conductor is not present in the outlet or junction box the frame of the appliance shall be permitted to be connected to the grounded conductor if all the conditions in the following list items (1), (2), and (3) are met and the grounded conductor complies with either list item (4) or (5):

(1) The supply circuit is 120/240-volt, single-phase, 3-wire; or 208Y/120-volt derived from a 3-phase, 4-wire, wye-connected system.
(2) The grounded conductor is not smaller than 10 AWG copper or 8 AWG aluminum or copper-clad aluminum.
(3) Grounding contacts of receptacles furnished as part of the equipment are bonded to the equipment.
(4) The grounded conductor is insulated, or the grounded conductor is uninsulated and part of a Type SE service-entrance cable and the branch circuit originates at the service equipment.
(5) The grounded conductor is part of a Type SE service-entrance cable that originates in equipment other than a service. The grounded conductor shall be insulated or field covered within the supply enclosure with listed insulating material, such as tape or sleeving to prevent contact of the uninsulated conductor with any normally non-current-carrying metal parts.

250.142 Use of Grounded Circuit Conductor for Grounding Equipment.

(A) Supply-Side Equipment. A grounded circuit conductor shall be permitted to be connected to non-current-carrying metal parts of equipment, raceways, and other enclosures at any of the following locations:

(1) On the supply side or within the enclosure of the ac service disconnecting means
(2) On the supply side or within the enclosure of the main disconnecting means for separate buildings as provided in 250.32(B)(1) Exception No. 1
(3) On the supply side or within the enclosure of the main disconnecting means or overcurrent devices of a separately derived system where permitted by 250.30(A)(1)

△ (B) Load-Side Equipment. Except as permitted in 250.30(A)(1), 250.32(B)(1), Exception No. 1, and Part X of Article 250, a grounded circuit conductor shall not be connected to non–current-carrying metal parts of equipment on the load side of the service disconnecting means or on the load side of a separately derived system disconnecting means or the overcurrent devices for a separately derived system not having a main disconnecting means.

Exception No. 1: The frames of ranges, wall-mounted ovens, counter-mounted cooking units, and clothes dryers under the conditions permitted for existing installations by 250.140 shall be permitted to be connected to the grounded circuit conductor.

Exception No. 2: It shall be permissible to connect meter enclosures to the grounded circuit conductor on the load side of the service disconnect if all of the following conditions apply:

(1) Ground-fault protection of equipment is not installed.
(2) All meter enclosures are located immediately adjacent to the service disconnecting means.
(3) The size of the grounded circuit conductor is not smaller than the size specified in Table 250.122 for equipment grounding conductors.

Exception No. 3: Electrode-type boilers operating at over 1000 volts shall be grounded as required in 495.72(E)(1) and 495.74.

If the grounded circuit conductor is re-grounded on the load side of the service and the grounded conductor is disconnected at any point on the line side of the service, the EGC and all conductive parts connected to it would become energized. Under this condition, the potential to ground of any exposed metal parts could be raised to line voltage. This rise in potential on non–current-carrying conductive parts could result in arcing in concealed spaces and could pose a severe shock hazard, particularly if contact is made with metal piping or ductwork.

Even without an open grounded conductor, a connection between the grounded conductor and the EGC on the load side of the service places the EGC in a parallel circuit path with the grounded conductor. Exposed and concealed non-current-carrying metal parts could also be hazardous.

See also

250.30(A)(3) and **250.32(B),** which prohibit the creation of parallel paths for normal neutral current

Δ **250.144 Multiple Circuit Connections.** If equipment is required to be grounded and is supplied by more than one circuit containing an equipment grounding conductor, a means to terminate each equipment grounding conductor meeting the requirements of 250.8 shall be provided as specified in 250.134 and 250.138.

250.146 Connecting Receptacle Grounding Terminal to an Equipment Grounding Conductor. An equipment bonding jumper shall be used to connect the grounding terminal of a grounding-type receptacle to a metal box that is connected to an equipment grounding conductor, except as permitted in 250.146(A) through (D). The equipment bonding jumper shall be sized in accordance with Table 250.122.

Δ **(A) Surface-Mounted Box.** If a metal box is mounted on the surface, the direct metal-to-metal contact between the device yoke or strap to the box shall be permitted to provide the required effective ground-fault current path. At least one of the insulating washers shall be removed from receptacles that do not have a contact yoke or device to ensure direct metal-to-metal contact. Direct metal-to-metal contact for providing continuity applies to cover-mounted receptacles if the box and cover combination are listed as providing continuity between the box and the receptacle. A listed exposed work cover shall be permitted to be the grounding and bonding means under both of the following conditions:

(1) The device is attached to the cover with at least two fasteners that are permanent (such as a rivet) or have a thread-locking or screw- or nut-locking means.
(2) The cover mounting holes are located on a flat nonraised portion of the cover.

Section 250.146(A) permits the equipment bonding jumper to be omitted where the metal yoke of the device is in direct metal-to-metal contact with the metal device box and at least one of the fiber retention washers for the receptacle mounting screws is removed, as illustrated in Exhibit 250.47.

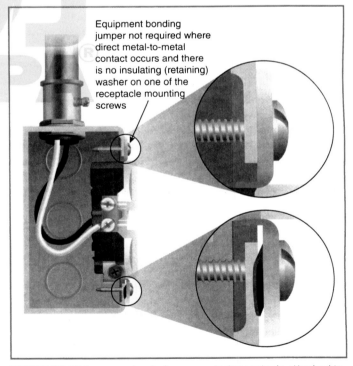

Equipment bonding jumper not required where direct metal-to-metal contact occurs and there is no insulating (retaining) washer on one of the receptacle mounting screws

EXHIBIT 250.47 An example of a box-mounted receptacle attached to a surface box where an equipment bonding jumper from the grounded metal box to the receptacle is not required, provided at least one of the insulating washers is removed.

EXHIBIT 250.48 An example of a cover-mounted receptacle attached to a surface box where an equipment bonding jumper from the grounded metal box to the receptacle is not required. Note that there are two points of attachment between the receptacle and the metal cover.

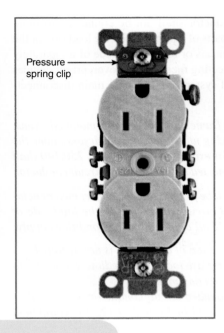

EXHIBIT 250.49 A listed self-grounding receptacle provided with a pressure spring clip attached to the upper device mounting screw. *(Photo courtesy of Leviton Manufacturing Co., Inc.)*

Pressure spring clip

Box-cover and device combinations listed as providing grounding continuity are permitted, as illustrated in Exhibit 250.48. The mounting holes for the cover must be located on a flat, non-raised portion of the cover to provide the best possible surface-to-surface contact, and the receptacle must be secured to the cover using not less than two rivets or locking means for threaded attachment means.

(B) Contact Devices or Yokes. Contact devices or yokes designed and listed as self-grounding shall be permitted in conjunction with the supporting screws to establish equipment bonding between the device yoke and flush-type boxes.

Exhibit 250.49 shows a listed self-grounding receptacle designed with a pressure spring clip that secures the top upper device mounting screw to establish a bonding connection between the receptacle and a grounded metal box. This connection eliminates the need for a wire-type equipment bonding jumper installed from the grounded metal box to the receptacle grounding terminal. Such devices are listed as "self-grounding" and are permitted to be used for bonding a flush-mounted receptacle to a grounded metal box.

Δ **(C) Floor Boxes.** Floor boxes designed for and listed as providing continuity between the box and the device shall be permitted.

(D) Isolated Ground Receptacles. If installed for the reduction of electromagnetic interference on the equipment grounding conductor, a receptacle in which the grounding terminal is purposely insulated from the receptacle mounting means shall be permitted. The receptacle grounding terminal shall be connected to an insulated equipment grounding conductor run with the circuit conductors. This equipment grounding conductor shall be permitted to pass through one or more panelboards without a connection to the panelboard grounding terminal bar as permitted

in 408.40, Exception, so as to terminate within the same building or structure directly at an equipment grounding conductor terminal of the applicable derived system or service. If installed in accordance with this section, this equipment grounding conductor shall also be permitted to pass through boxes, wireways, or other enclosures without being connected to such enclosures.

Informational Note: Use of an isolated equipment grounding conductor does not relieve the requirement for connecting the raceway system and outlet box to an equipment grounding conductor.

An isolated-ground-type receptacle is permitted to be installed without a bonding jumper between the metal device box and the receptacle grounding terminal. However, the isolated EGC must provide an effective path for ground-fault current between the receptacle grounding terminal and the source of the branch circuit supplying the receptacle. An insulated EGC, as shown in Exhibit 250.50, is installed with the branch-circuit conductors. This conductor can originate in the service panel, pass through any number of subpanels without being connected to the equipment grounding bus, and terminate at the isolated-ground-type receptacle ground terminal. Termination of the isolated EGC at the service is not necessary. It can be terminated at any of the intervening panelboards. The objective is to terminate it where the noise is eliminated.

This isolated EGC arrangement does not exempt the metal device box from being grounded. The metal device box must be grounded either by an EGC run with the circuit conductors or by a wiring method that serves as an EGC.

According to 250.146(D), where isolated-ground-type receptacles are used, the isolated EGC can terminate at an equipment grounding terminal of the applicable service or derived system in the same building as the receptacle. If the isolated EGC terminates at a separate building, a large voltage difference could exist between buildings during lightning transients. Such transients

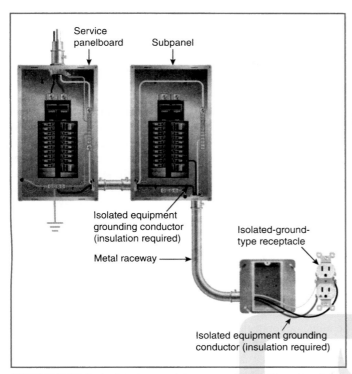

EXHIBIT 250.50 An isolated-ground-type receptacle with an insulated EGC and the device box and receptacle mounting yoke grounded through the metal raceway.

could cause damage to equipment connected to an isolated-ground-type receptacle and present a shock hazard between the isolated equipment frame and other grounded surfaces.

Δ **250.148 Continuity of Equipment Grounding Conductors and Attachment in Boxes.** If circuit conductors are spliced within a box or terminated on equipment within or supported by a box, the installation shall comply with 250.148(A) through (D).

If multiple circuits are spliced within the box or terminated to equipment within the box, all the EGCs must be connected together and to the box unless exempted under 250.148. If a metal box is grounded by a metal raceway system and the circuit conductors are not spliced or terminated to equipment in the metal box, the wire-type EGC is not required to be connected to the box. Wire-type EGCs that are not spliced or terminated within a box do not have to be connected to the box; however, the box (if metal) is required to be connected to the EGC of the circuit with the highest rating that is contained in the box.

Exception: The equipment grounding conductor permitted in 250.146(D) shall not be required to be connected to the other equipment grounding conductors or to the box.

(A) Connections and Splices. All equipment grounding conductors that are spliced or terminated within the box shall be connected together. Connections and splices shall be made in accordance with 110.14(B) and 250.8 except that insulation shall not be required.

(B) Equipment Grounding Conductor Continuity. The arrangement of grounding connections shall be such that the disconnection or the removal of a luminaire, receptacle, or other device fed from the box does not interrupt the electrical continuity of the equipment grounding conductor(s) providing an effective ground-fault current path.

Δ **(C) Metal Boxes.** A connection used for no other purpose shall be made between the metal box and the equipment grounding conductor(s). The equipment bonding jumper or equipment grounding conductor shall be sized from Table 250.122 based on the largest overcurrent device protecting circuit conductors in the box.

Δ **(D) Nonmetallic Boxes.** One or more equipment grounding conductors brought into a nonmetallic outlet box shall be arranged to provide a connection to any fitting or device in that box requiring connection to an equipment grounding conductor.

Part VIII. Direct-Current Systems

250.160 General. Direct-current systems shall comply with Part VIII and other sections of Article 250 not specifically intended for ac systems.

250.162 Direct-Current Circuits and Systems to Be Grounded. Direct-current circuits and systems shall be grounded as provided for in 250.162(A) and (B).

(A) Two-Wire, Direct-Current Systems. A 2-wire, dc system supplying premises wiring and operating at greater than 60 volts but not greater than 300 volts shall be grounded.

Exception No. 1: A system equipped with a ground detector and supplying only industrial equipment in limited areas shall not be required to be grounded if installed immediately adjacent to, or integral with, the source of supply.

Exception No. 2: A rectifier-derived dc system supplied from an ac system complying with 250.20 shall not be required to be grounded.

Exception No. 3: Direct-current fire alarm circuits having a maximum current of 0.030 ampere as specified in Article 760, Part III, shall not be required to be grounded.

(B) Three-Wire, Direct-Current Systems. The neutral conductor of all 3-wire, dc systems supplying premises wiring shall be grounded.

250.164 Point of Connection for Direct-Current Systems.

(A) Off-Premises Source. Direct-current systems to be grounded and supplied from an off-premises source shall have the grounding connection made at one or more supply stations. A grounding connection shall not be made at individual services or at any point on the premises wiring.

Δ **(B) On-Premises Source.** If the dc system source is located on the premises, a grounding connection shall be made at one of the following:

(1) The source
(2) The first system disconnection means or overcurrent device
(3) By other means that accomplish system protection and that use equipment listed and identified for the use

250.166 Size of the Direct-Current Grounding Electrode Conductor.

The size of the grounding electrode conductor for a dc system shall be as specified in 250.166(A) and (B), except as permitted by 250.166(C) through (E). The grounding electrode conductor for a dc system shall meet the sizing requirements in this section but shall not be required to be larger than 3/0 copper or 250 kcmil aluminum or copper-clad aluminum.

(A) Not Smaller Than the Neutral Conductor. If the dc system consists of a 3-wire balancer set or a balancer winding with overcurrent protection as provided in 445.12(D), the grounding electrode conductor shall not be smaller than the neutral conductor and not smaller than 8 AWG copper or 6 AWG aluminum or copper-clad aluminum.

Δ **(B) Not Smaller Than the Largest Conductor.** If the dc system is other than as in 250.166(A), the grounding electrode conductor shall not be smaller than the largest conductor supplied by the system and not smaller than 8 AWG copper or 6 AWG aluminum or copper-clad aluminum.

(C) Connected to Rod, Pipe, or Plate Electrodes. If connected to rod, pipe, or plate electrodes as in 250.52(A)(5) or (A)(7), that portion of the grounding electrode conductor that is the sole connection to the grounding electrode shall not be required to be larger than 6 AWG copper wire or 4 AWG aluminum or copper-clad aluminum wire.

(D) Connected to a Concrete-Encased Electrode. If connected to a concrete-encased electrode as in 250.52(A)(3), that portion of the grounding electrode conductor that is the sole connection to the grounding electrode shall not be required to be larger than 4 AWG copper wire.

(E) Connected to a Ground Ring. If connected to a ground ring as in 250.52(A)(4), that portion of the grounding electrode conductor that is the sole connection to the grounding electrode shall not be required to be larger than the conductor used for the ground ring.

250.167 Direct-Current Ground-Fault Detection.

(A) Ungrounded Systems. Ground-fault detection systems shall be required for ungrounded systems.

(B) Grounded Systems. Ground-fault detection shall be permitted for grounded systems.

Δ **(C) Marking.** Direct-current systems shall be legibly marked to indicate the grounding type at the dc source or the first disconnecting means of the system. The marking shall be of sufficient durability to withstand the environment involved.

> Informational Note: See NFPA 70E-2018, *Standard for Electrical Safety in the Workplace*, which identifies four dc grounding types in detail.

250.168 Direct-Current System Bonding Jumper.

For direct-current systems that are to be grounded, an unspliced bonding jumper shall be used to connect the equipment grounding conductor(s) to the grounded conductor at the source or to the first system disconnecting means where the system is grounded. The size of the bonding jumper shall not be smaller than the system grounding electrode conductor specified in 250.166 and shall comply with 250.28(A), (B), and (C).

250.169 Ungrounded Direct-Current Separately Derived Systems.

Except as otherwise permitted in 250.34 for portable and vehicle-mounted generators, an ungrounded dc separately derived system supplied from a stand-alone power source (such as an engine–generator set) shall have a grounding electrode conductor connected to an electrode that complies with Part III of this article to provide for grounding of metal enclosures, raceways, cables, and exposed non–current-carrying metal parts of equipment. The grounding electrode conductor connection shall be to the metal enclosure at any point on the separately derived system from the source to the first system disconnecting means or overcurrent device, or it shall be made at the source of a separately derived system that has no disconnecting means or overcurrent devices.

The size of the grounding electrode conductor shall be in accordance with 250.166.

Part IX. Instruments, Meters, and Relays

250.170 Instrument Transformer Circuits.

Secondary circuits of current and potential instrument transformers shall be grounded if the primary windings are connected to circuits of 300 volts or more to ground and, if installed on or in switchgear and on switchboards, shall be grounded irrespective of voltage.

Exception No. 1: Circuits where the primary windings are connected to circuits of 1000 volts or less with no live parts or wiring exposed or accessible to other than qualified persons.

Exception No. 2: Current transformer secondaries connected in a three-phase delta configuration shall not be required to be grounded.

250.172 Instrument Transformer Cases.

Cases or frames of instrument transformers shall be connected to the equipment grounding conductor if accessible to other than qualified persons.

Exception: Cases or frames of current transformers, the primaries of which are not over 150 volts to ground and that are used exclusively to supply current to meters.

250.174 Cases of Instruments, Meters, and Relays Operating at 1000 Volts or Less. Instruments, meters, and relays operating with windings or working parts at 1000 volts or less shall be connected to the equipment grounding conductor as specified in 250.174(A), (B), or (C).

(A) Not on Switchgear or Switchboards. Instruments, meters, and relays not located on switchgear or switchboards operating with windings or working parts at 300 volts or more to ground, and accessible to other than qualified persons, shall have the cases and other exposed metal parts connected to the equipment grounding conductor.

(B) On Switchgear or Dead-Front Switchboards. Instruments, meters, and relays (whether operated from current and potential transformers or connected directly in the circuit) on switchgear or switchboards having no live parts on the front of the panels shall have the cases connected to the equipment grounding conductor.

(C) On Live-Front Switchboards. Instruments, meters, and relays (whether operated from current and potential transformers or connected directly in the circuit) on switchboards having exposed live parts on the front of panels shall not have their cases connected to the equipment grounding conductor. Mats of insulating rubber or other approved means of floor insulation shall be provided for the operator where the voltage to ground exceeds 150 volts.

250.176 Cases of Instruments, Meters, and Relays — Operating at Over 1000 Volts. If instruments, meters, and relays have current-carrying parts of over 1000 volts to ground, they shall be isolated by elevation or protected by a barrier(s), grounded metal, or insulating covers or guards. Their cases shall not be connected to the equipment grounding conductor.

Exception: Cases of electrostatic ground detectors shall be permitted to be connected to an equipment grounding conductor if the internal ground segments of the instrument are connected to the instrument case and grounded and the ground detector is isolated by elevation.

Δ **250.178 Instrument Equipment Grounding Conductor.** The equipment grounding conductor for secondary circuits of instrument transformers and for instrument cases shall not be smaller than 12 AWG copper or 10 AWG aluminum or copper-clad aluminum. Cases of instrument transformers, instruments, meters, and relays that are mounted directly on grounded metal surfaces of enclosures or grounded metal of switchgear or switchboard panels shall not be required to be connected to an additional equipment grounding conductor.

Part X. Grounding of Systems and Circuits of over 1000 Volts

250.180 General. If systems over 1000 volts are grounded, they shall comply with all applicable requirements of 250.1 through 250.178 and with 250.182 through 250.194, which supplement and modify the preceding sections.

As a general rule, Table 250.66, for sizing GECs, and Table 250.122, for sizing EGCs, apply to systems operating over 1 kilovolt. There are, however, special requirements in Part X for EGCs for electrical systems utilizing shielded solid dielectric insulated cables rated 2001 to 35,000 volts. Those electrical systems are commonly referred to as medium- and high-voltage systems.

250.182 Derived Neutral Systems. A system neutral point derived from a grounding transformer shall be permitted to be used for grounding systems over 1 kV.

250.184 Solidly Grounded Neutral Systems. Solidly grounded neutral systems shall be permitted to be either single point grounded or multigrounded neutral.

(A) Neutral Conductor.

(1) Insulation Level. The minimum insulation level for neutral conductors of solidly grounded systems shall be 600 volts.

Exception No. 1: For multigrounded neutral systems as permitted in 250.184(C), bare copper conductors shall be permitted to be used for the neutral conductor of the following:

(1) Service-entrance conductors
(2) Service laterals or underground service conductors
(3) Direct-buried portions of feeders

Exception No. 2: Bare conductors shall be permitted for the neutral conductor of overhead portions installed outdoors.

Exception No. 3: The grounded neutral conductor shall be permitted to be a bare conductor if isolated from phase conductors and protected from physical damage.

Informational Note: See 225.4 for conductor covering where within 3.0 m (10 ft) of any building or other structure.

(2) Ampacity. The neutral conductor shall have an ampacity that is not less than the load imposed and be not less than 33⅓ percent of the ampacity of the phase conductors.

Exception: In industrial and commercial premises under engineering supervision, it shall be permissible to size the ampacity of the neutral conductor to not less than 20 percent of the ampacity of the phase conductor.

Δ **(B) Single-Point Grounded Neutral System.** If a single-point grounded neutral system is used, the following shall apply:

(1) A single-point grounded neutral system shall be permitted to be supplied from one of the following:

  a. A separately derived system
  b. A multigrounded neutral system with an equipment grounding conductor connected to the multigrounded neutral conductor at the source of the single-point grounded neutral system

(2) A grounding electrode shall be provided for the system.

(3) A grounding electrode conductor shall connect the grounding electrode to the system neutral conductor.

(4) A bonding jumper shall connect the equipment grounding conductor to the grounding electrode conductor.

(5) An equipment grounding conductor shall be provided to each building, structure, and equipment enclosure.

(6) A neutral conductor shall only be required if phase-to-neutral loads are supplied.

(7) The neutral conductor, if provided, shall be insulated and isolated from earth except at one location.

(8) An equipment grounding conductor shall be run with the phase conductors and shall comply with all of the following:

     a. Shall not carry continuous load

     b. Shall be bare, covered, or insulated

     c. Shall have ampacity for fault current duty

(C) Multigrounded Neutral Systems. If a multigrounded neutral system is used, the following shall apply:

(1) The neutral conductor of a solidly grounded neutral system shall be permitted to be grounded at more than one point. Grounding shall be permitted at one or more of the following locations:

     a. Transformers supplying conductors to a building or other structure

     b. Underground circuits if the neutral conductor is exposed

     c. Overhead circuits installed outdoors

(2) The multigrounded neutral conductor shall be grounded at each transformer and at other additional locations by connection to a grounding electrode.

(3) At least one grounding electrode shall be installed and connected to the multigrounded neutral conductor every 400 m (1300 ft).

(4) The maximum distance between any two adjacent electrodes shall not be more than 400 m (1300 ft).

(5) In a multigrounded shielded cable system, the shielding shall be grounded at each cable joint that is exposed to personnel contact.

Exception: In a multipoint grounded system, a grounding electrode shall not be required to bond the neutral conductor in an uninterrupted conductor exceeding 400 m (1300 ft) if the only purpose for removing the cable jacket is for bonding the neutral conductor to a grounding electrode.

Large-scale photovoltaic electric supply systems (solar farms) and wind electric systems (wind farms) of the types covered in Articles 691 and 694 encompass large tracts of land, and long conductor runs are the norm. Connecting neutral conductors to a grounding electrode at specific intervals is impractical, but, more importantly, it introduces potential operational problems. Recent changes to ANSI/IEEE C2, *National Electrical Safety Code® (NESC®)*, permit intervals greater than 1300 feet

between grounding electrode connections to minimize the number of locations where the cable jacket is removed. Multiple points of grounding necessitate removing the cable jacket to make the connection, which compromises the overall cable integrity and increases the possibility of premature cable failure. Because the cables involved in these installations are typically shielded, the shield is also included in the grounding connection, which has the potential of exposing the shield to undesirable EMI.

250.186 Grounding Service-Supplied Alternating-Current Systems.

(A) Systems with a Grounded Conductor at the Service Point. If an ac system is grounded at any point and is provided with a grounded conductor at the service point, a grounded conductor(s) shall be installed and routed with the ungrounded conductors to each service disconnecting means and shall be connected to each disconnecting means grounded conductor(s) terminal or bus. A main bonding jumper shall connect the grounded conductor(s) to each service disconnecting means enclosure. The grounded conductor(s) shall be installed in accordance with 250.186(A)(1) through (A)(4). The size of the solidly grounded circuit conductor(s) shall be the larger of that determined by 250.184 or 250.186(A)(1) or (A)(2).

Exception: If two or more service disconnecting means are located in a single assembly listed for use as service equipment, it shall be permitted to connect the grounded conductor(s) to the assembly common grounded conductor(s) terminal or bus. The assembly shall include a main bonding jumper for connecting the grounded conductor(s) to the assembly enclosure.

(1) Sizing for a Single Raceway or Overhead Conductor. The grounded conductor shall not be smaller than the required grounding electrode conductor specified in Table 250.102(C)(1) but shall not be required to be larger than the largest ungrounded service-entrance conductor(s).

(2) Parallel Conductors in Two or More Raceways or Overhead Conductors. If the ungrounded service-entrance conductors are installed in parallel in two or more raceways or as overhead parallel conductors, the grounded conductors shall also be installed in parallel. The size of the grounded conductor in each raceway or overhead shall be based on the total circular mil area of the parallel ungrounded conductors in the raceway or overhead, as indicated in 250.186(A)(1), but not smaller than 1/0 AWG.

> Informational Note: See 310.10(G) for grounded conductors connected in parallel.

(3) Delta-Connected Service. The grounded conductor of a 3-phase, 3-wire delta service shall have an ampacity not less than that of the ungrounded conductors.

Δ **(4) Impedance Grounded Systems.** Impedance grounded systems shall be installed in accordance with 250.187.

(B) Systems Without a Grounded Conductor at the Service Point. If an ac system is grounded at any point and is not provided with a grounded conductor at the service point, a supply-side bonding jumper shall be installed and routed with the ungrounded conductors to each service disconnecting means and shall be connected to each disconnecting means equipment grounding conductor terminal or bus. The supply-side bonding jumper shall be installed in accordance with 250.186(B)(1) through (B)(3).

Exception: If two or more service disconnecting means are located in a single assembly listed for use as service equipment, it shall be permitted to connect the supply-side bonding jumper to the assembly common equipment grounding terminal or bus.

(1) Sizing for a Single Raceway or Overhead Conductor. The supply-side bonding jumper shall not be smaller than the required grounding electrode conductor specified in Table 250.102(C)(1) but shall not be required to be larger than the largest ungrounded service-entrance conductor(s).

(2) Parallel Conductors in Two or More Raceways or Overhead Conductors. If the ungrounded service-entrance conductors are installed in parallel in two or more raceways or overhead conductors, the supply-side bonding jumper shall also be installed in parallel. The size of the supply-side bonding jumper in each raceway or overhead shall be based on the total circular mil area of the parallel ungrounded conductors in the raceway or overhead, as indicated in 250.186(A)(1), but not smaller than 1/0 AWG.

Δ **(3) Impedance Grounded Systems.** Impedance grounded systems shall be installed in accordance with 250.187.

250.187 Impedance Grounded Systems. Impedance grounded systems in which a grounding impedance device, typically a resistor, limits the ground-fault current shall be permitted if all of the following conditions are met:

(1) The conditions of maintenance and supervision ensure that only qualified persons service the installation.
(2) Ground detectors are installed on the system.
(3) Line-to-neutral loads are not served.

Impedance grounded systems shall comply with 250.187(A) through (D).

(A) Location. The grounding impedance device shall be installed between the grounding electrode conductor and the impedance grounding conductor connected to the system neutral point.

(B) Insulated. The impedance grounding conductor shall be insulated for the maximum neutral voltage.

Exception: A bare impedance grounding conductor shall be permitted if the bare portion of the grounding impedance device and conductor are not in a readily accessible location and securely separated from the ungrounded conductors.

Informational Note: The maximum neutral voltage in a 3-phase wye system is 57.7 percent of the phase-to-phase voltage.

(C) System Neutral Point Connection. The system neutral point shall not be connected to ground, except through the grounding impedance device.

(D) Equipment Grounding Conductors. Equipment grounding conductors shall be permitted to be bare and shall be electrically connected to the ground bus and grounding electrode conductor.

250.188 Grounding of Systems Supplying Portable or Mobile Equipment. Systems supplying portable or mobile equipment over 1000 volts, other than substations installed on a temporary basis, shall comply with 250.188(A) through (F).

(A) Portable or Mobile Equipment. Portable or mobile equipment over 1000 volts shall be supplied from a system having its neutral conductor grounded through an impedance. If a delta-connected system over 1000 volts is used to supply portable or mobile equipment, a system neutral point and associated neutral conductor shall be derived.

The term *portable* describes equipment that is easily carried from one location to another. The term *mobile* describes equipment that is easily moved on wheels, treads, skids, or similar means.

(B) Exposed Non-Current-Carrying Metal Parts. Exposed non-current-carrying metal parts of portable or mobile equipment shall be connected by an equipment grounding conductor to the point at which the system neutral impedance is grounded.

(C) Ground-Fault Current. The voltage developed between the portable or mobile equipment frame and ground by the flow of maximum ground-fault current shall not exceed 100 volts.

(D) Ground-Fault Detection and Relaying. Ground-fault detection and relaying shall be provided to automatically de-energize any component of a system over 1000 volts that has developed a ground fault. The continuity of the equipment grounding conductor shall be continuously monitored so as to automatically de-energize the circuit of the system over 1000 volts to the portable or mobile equipment upon loss of continuity of the equipment grounding conductor.

(E) Isolation. The grounding electrode to which the portable or mobile equipment system neutral impedance is connected shall be isolated from and separated in the ground by at least 6.0 m (20 ft) from any other system or equipment grounding electrode, and there shall be no direct connection between the grounding electrodes, such as buried pipe and fence, and so forth.

(F) Trailing Cable and Couplers. Trailing cable and couplers of systems over 1000 volts for interconnection of portable or mobile equipment shall meet the requirements of Part III of Article 400 for cables and 495.65 for couplers.

250.190 Grounding of Equipment.

(A) Equipment Grounding. All non-current-carrying metal parts of fixed, portable, and mobile equipment and associated fences, housings, enclosures, and supporting structures shall be grounded.

Exception: If isolated from ground and located such that any person in contact with ground cannot contact such metal parts when the equipment is energized, the metal parts shall not be required to be grounded.

> Informational Note: See 250.110, Exception No. 2, for pole-mounted distribution apparatus.

(B) Grounding Electrode Conductor. If a grounding electrode conductor connects non-current-carrying metal parts to ground, the grounding electrode conductor shall be sized in accordance with Table 250.66, based on the size of the largest ungrounded service, feeder, or branch-circuit conductors supplying the equipment. The grounding electrode conductor shall not be smaller than 6 AWG copper or 4 AWG aluminum or copper-clad aluminum.

(C) Equipment Grounding Conductor. Equipment grounding conductors shall comply with 250.190(C)(1) through (C)(3).

(1) General. Equipment grounding conductors that are not an integral part of a cable assembly shall not be smaller than 6 AWG copper or 4 AWG aluminum or copper-clad aluminum.

(2) Shielded Cables. The metallic insulation shield encircling the current-carrying conductors shall be permitted to be used as an equipment grounding conductor, if it is rated for clearing time of ground-fault current protective device operation without damaging the metallic shield. The metallic tape insulation shield and drain wire insulation shield shall not be used as an equipment grounding conductor for solidly grounded systems.

Shields comprised of copper tape and drain wires cannot be used as the EGC in solidly grounded systems. The use of shielded cables is specified in 315.44. Those requirements are based on the system voltage, installation conditions, and cable construction. As explained in the informational note, the primary purposes of the shielding are to confine voltage stresses on the insulation, dissipate insulation leakage current, and drain off the capacitive charging current.

Exhibit 250.51 shows three different types of single-conductor shielded cable construction. Shielded cables are also available in multiconductor configurations such as Type MV (Article 315) and Type MC (Article 330) cables. Where the ground-fault current is relatively low (as in impedance grounded neutral systems), the metallic shield of any of the cable types pictured in Exhibit 250.51 is permitted to serve as the EGC if it is rated for clearing time of ground-fault current without being damaged. Cable manufacturers can provide permissible short-circuit currents for a metallic shield based on the fault clearing time of the OCPD.

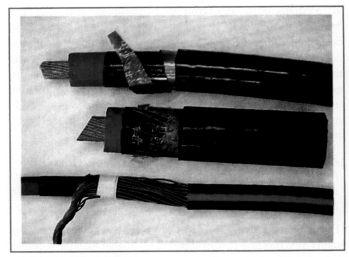

EXHIBIT 250.51 Three different types of single-conductor shielded cables. *(Courtesy of Chuck Mello)*

Where the system is solidly grounded, neither the metallic tape insulation shield (top cable in Exhibit 250.51) nor the drain wire insulation shield (middle cable) can be used as the EGC, because they do not have sufficient circular mil area to provide the effective ground-fault return path required by 250.4(A)(5). A metallic insulation shield encircling the conductor (bottom cable), commonly called concentric neutral cable, has larger conductor strands in the concentric wrap. Because of its larger overall circular mil area, this concentric neutral cable is permitted to be used as an EGC in a solidly grounded system if the metallic shield will not be damaged during the time it takes to open the circuit OCPD. The copper metallic tape is typically 5 mils thick and is helically applied with a 12.5 percent or larger overlap over the insulation shield. Drain wires are typically 24 AWG bare copper wires.

Where the cable shield cannot carry ground-fault current without damage, a separate EGC must be installed. A separate EGC would also be required where tape or drain-wire-type shields are not permitted to carry fault current. The EGC can be integral to a cable assembly, can be run as a separate conductor in a raceway or cable tray, or can be one of the types of EGCs specified in 250.118, such as rigid metal conduit or intermediate metal conduit. Wire-type EGCs are sized in accordance with 250.122.

See also

250.190(C)(3), Informational Note, for more information on how the rating of OCPDs used in systems operating over 1000 volts is determined

(3) Sizing. Equipment grounding conductors shall be sized in accordance with Table 250.122 based on the current rating of the fuse or the overcurrent setting of the protective relay.

> Informational Note: The overcurrent rating for a circuit breaker is the combination of the current transformer ratio and the current pickup setting of the protective relay.

This requirement applies to EGCs that are separately installed as specified in 250.190(C)(1). It also applies to a conductor installed in a cable assembly, other than the cable shield, that is used as an EGC. An EGC contained within a cable assembly can be a single conductor, or it can be sectioned (comprising multiple conductors within the cable jacket or sheath to form a single EGC) as permitted by 310.10(G)(5).

250.191 Grounding System at Alternating-Current Substations. For ac substations, the grounding system shall be in accordance with Part III of this article.

> Informational Note: See IEEE 80, *IEEE Guide for Safety in AC Substation Grounding,* for further information on outdoor ac substation grounding.

250.194 Grounding and Bonding of Fences and Other Metal Structures. Metal fences enclosing, and other metal structures in or surrounding, a substation with exposed electrical conductors and equipment shall be grounded and bonded to limit step, touch, and transfer voltages.

Δ **(A) Metal Fences.** If metal fences are located within 5 m (16 ft) of the exposed electrical conductors or equipment, the fence shall be bonded to the grounding electrode system with wire-type bonding jumpers as follows:

(1) Bonding jumpers shall be installed at each fence corner and at maximum 50 m (160 ft) intervals along the fence.
(2) If bare overhead conductors cross the fence, bonding jumpers shall be installed on each side of the crossing.

(3) Gates shall be bonded to the gate support post, and each gate support post shall be bonded to the grounding electrode system.
(4) Any gate or other opening in the fence shall be bonded across the opening by a buried bonding jumper.
(5) The grounding grid or grounding electrode systems shall be extended to cover the swing of all gates.
(6) The barbed wire strands above the fence shall be bonded to the grounding electrode system.

Alternate designs performed under engineering supervision shall be permitted for grounding or bonding of metal fences.

> Informational Note No. 1: A nonconducting fence or section may provide isolation for transfer of voltage to other areas.
> Informational Note No. 2: See IEEE 80, *IEEE Guide for Safety In AC Substation Grounding,* for design and installation of fence grounding.

(B) Metal Structures. All exposed conductive metal structures, including guy wires within 2.5 m (8 ft) vertically or 5 m (16 ft) horizontally of exposed conductors or equipment and subject to contact by persons, shall be bonded to the grounding electrode systems in the area.

Metal fences around substations must be grounded to limit the rise of hazardous voltage on the fence. For situations where considerations of step potential and touch potential indicate that additional grounding and bonding design is required, alternative designs performed under engineering supervision are permitted.

Wiring Methods and Materials

ARTICLE 300 — General Requirements for Wiring Methods and Materials

300.1 Scope.

(A) All Wiring Installations. This article covers general requirements for wiring methods and materials for all wiring installations unless modified by other articles in Chapter 3.

(B) Integral Parts of Equipment. The requirements of this article are not intended to apply to the conductors that form an integral part of equipment, such as motors, controllers, motor control centers, or factory-assembled control equipment or listed utilization equipment.

Requirements for specific wiring methods can be found in the Chapter 3 article governing that particular wiring method, but the overarching requirements for wiring methods are covered in this first article of Chapter 3. Chapters 5 through 7 modify some of the requirements of Article 300. Chapter 8 is not subject to the requirements of Article 300, except where specifically referenced. Article 300 also covers wiring requirements within boxes, conduit bodies, and fittings.

(C) Metric Designators and Trade Sizes. Metric designators and trade sizes for conduit, tubing, and associated fittings and accessories shall be in accordance with Table 300.1(C).

Metric designators are used for traditional trade size threaded conduit. They do not change the physical dimensions or the traditional "NPT-type" threads of the conduit. Metric designators are simply another method of identifying the size of a circular raceway. Table 300.1(C) identifies a distinct metric designator for each circular raceway trade size. The unit of measure has not been included because it reflects a modular or relative measure rather than an exact dimension. As stated in the table footnote, the metric designators and trade sizes are not actual dimensions.

Each metric designator-sized circular raceway is identical in dimension (including manufacturing tolerances) to its trade size counterpart in Table 4 of Chapter 9. Therefore, the Informative

TABLE 300.1(C) *Metric Designators and Trade Sizes*

Metric Designator	Trade Size
12	⅜
16	½
21	¾
27	1
35	1¼
41	1½
53	2
63	2½
78	3
91	3½
103	4
129	5
155	6

Note: The metric designators and trade sizes are for identification purposes only and are not actual dimensions.

Annex C wire fill tables are applicable to both metric designator and trade size circular raceways.

Threaded joints on circular raceways are a concern. For example, 344.6 requires rigid metal conduit (RMC) to be listed, and the appropriate product standard is ANSI/UL 6, *Electrical Rigid Metal Conduit — Steel*. Listed conduit must be threaded in accordance with ANSI/ASME B.1.20.1-1983, *Pipe Threads, General Purpose (Inch)*. Therefore, only conduit threaded to the traditional dimension of ¾-inch taper per foot is acceptable. Simply stated, although conduit with a metric designator is permitted, metric-threaded conduit is not permitted by the *NEC®*. This aligns with 500.8(E)(2), for example, which states that although metric threads are permitted on equipment, an adapter must be used for connection to conduit.

300.2 Limitations.

(A) Voltage. Wiring methods specified in Chapter 3 shall be used for 1000 volts ac, 1500 volts dc, nominal, or less where not specifically limited elsewhere in Chapter 3. They shall be permitted for over 1000 volts ac, 1500 volts dc, nominal, where specifically permitted elsewhere in this *Code*.

Where an installation involves requirements using different voltage range limits, the limits within each applicable article must be observed.

(B) Temperature. Temperature limitation of conductors shall be in accordance with 310.14(A)(3).

300.3 Conductors.

(A) Single Conductors. Single conductors specified in Table 310.4(1) shall only be permitted where installed as part of a recognized wiring method specified in Chapter 3.

Exception: Individual conductors shall be permitted where installed as separate overhead conductors in accordance with 225.6.

Individual insulated conductors, such as THHN, are prohibited from use in other than recognized wiring methods. The exception points out two long-time permissions that allow individual conductors as overhead spans and as festoon lighting.

Δ **(B) Conductors of the Same Circuit.** All conductors of the same circuit and, where used, the grounded conductor and all equipment grounding conductors and bonding conductors shall be contained within the same raceway, conduit body, auxiliary gutter, cable tray, cablebus assembly, trench, cable, or cord unless otherwise permitted in accordance with 300.3(B)(1) through (B)(4).

All conductors of an individual circuit must be grouped in order to reduce inductive heating and to avoid increases in overall circuit impedance.

> **See also**
>
> **300.5(I)** for a similar requirement pertaining to underground installations

Δ **(1) Paralleled Installations.** Conductors shall be permitted to be run in parallel in accordance with 310.10(G). The requirement to run all circuit conductors within the same raceway, auxiliary gutter, cable tray, trench, cable, or cord shall apply separately to each portion of the paralleled installation, and the equipment grounding conductors shall comply with 250.122. Connections, taps, or extensions made from paralleled conductors shall connect to all conductors of the paralleled set, grounded and ungrounded, as applicable. Parallel runs in cable trays shall comply with 392.20(C).

Exception: Conductors installed in nonmetallic raceways run underground shall be permitted to be arranged as isolated phase, neutral, and grounded conductor installations. The raceways shall be installed in close proximity, and the isolated phase, neutral, and grounded conductors shall comply with 300.20(B).

Δ **(2) Grounding and Bonding Conductors.** Equipment grounding conductors shall be permitted to be installed outside a raceway or cable assembly in accordance with 250.130(C) for certain existing installations or in accordance with 250.134, Exception No. 2, for dc circuits. Equipment bonding conductors shall be permitted to be installed on the outside of raceways in accordance with 250.102(E).

(3) Nonferrous Wiring Methods. Conductors in wiring methods with a nonmetallic or other nonmagnetic sheath, where run in different raceways, auxiliary gutters, cable trays, trenches, cables, or cords, shall comply with 300.20(B). Conductors in single-conductor Type MI cable with a nonmagnetic sheath shall comply with 332.31. Conductors of single-conductor Type MC cable with a nonmagnetic sheath shall comply with 330.31, 330.116, and 300.20(B).

(4) Column-Width Panelboard Enclosures. Where an auxiliary gutter runs between a column-width panelboard and a pull box, and the pull box includes neutral terminations, the neutral conductors of circuits supplied from the panelboard shall be permitted to originate in the pull box.

(C) Conductors of Different Systems.

Δ **(1) 1000 Volts ac, 1500 volts dc, Nominal, or Less.** Conductors of ac and dc circuits rated 1000 volts ac, 1500 volts dc, nominal, or less shall be permitted to occupy the same equipment wiring enclosure, cable, or raceway. All conductors shall have an insulation rating equal to at least the maximum circuit voltage applied to any conductor within the enclosure, cable, or raceway.

Secondary wiring to electric-discharge lamps of 1000 volts ac, 1500 volts dc, or less, if insulated for the secondary voltage involved, shall be permitted to occupy the same luminaire, sign, or outline lighting enclosure as the branch-circuit conductors.

Informational Note No. 1: See 725.136(A) for Class 2 and Class 3 circuit conductors.
Informational Note No. 2: See 690.31(B) for photovoltaic source and output circuits.

For systems of 1000 volts or less, the maximum circuit voltage in the raceway is what determines the minimum voltage rating required for the insulation of conductors, not the maximum insulation voltage rating of the conductors in the raceway. Specific systems with requirements elsewhere in the *NEC* have modifications to this allowance.

> **See also**
>
> **690.31(B),** which prohibits the location of solar photovoltaic dc circuits within the same enclosure as conductors of other systems unless separated by a partition
>
> **700.10(B),** which requires that circuit wiring for emergency systems be kept entirely independent of all other wiring and equipment

Δ **(2) Over 1000 Volts ac, 1500 Volts dc, Nominal.** Conductors of circuits rated over 1000 volts ac, 1500 volts dc, nominal, shall not occupy the same equipment wiring enclosure, cable, or raceway with conductors of circuits rated 1000 volts ac, 1500 volts dc, nominal, or less unless permitted in accordance with 305.4.

300.4 Protection Against Physical Damage. Where subject to physical damage, conductors, raceways, and cables shall be protected.

(A) Cables and Raceways Through Wood Members.

Δ **(1) Bored Holes.** In both exposed and concealed locations, where a cable- or raceway-type wiring method is installed through bored holes in joists, rafters, or wood members, holes shall be bored so that the edge of the hole is not less than 32 mm (1¼ in.) from the edges of the wood member. Where this distance cannot be maintained, the cable or raceway shall be protected from penetration by screws or nails by a steel plate(s) or bushing(s) at least 1.6 mm (¹⁄₁₆ in.) thick, and of appropriate length and width, installed to cover the area of the wiring.

> *Exception No. 1: Steel plates shall not be required to protect rigid metal conduit, intermediate metal conduit, rigid PVC conduit, RTRC, or electrical metallic tubing.*

> *Exception No. 2: A listed and marked steel plate less than 1.6 mm (¹⁄₁₆ in.) thick that provides equal or better protection against nail or screw penetration shall be permitted.*

The intent is to prevent nails and screws from being driven into conductors, cables, and raceways. As shown in Exhibit 300.1, if the edge of a drilled hole is less than 1¼ inch from the nearest edge of a wood stud, a plate is required to prevent screws or nails from penetrating the stud far enough to injure a cable. Building codes limit the maximum size of bored or notched holes in studs, and 300.4(A)(2) indicates that consideration should be given to the size of notches in studs, so as not to affect the strength of the structure.

(2) Notches in Wood. Where there is no objection because of weakening the building structure, in both exposed and concealed

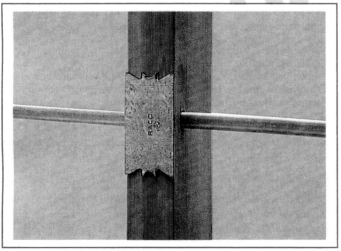

EXHIBIT 300.1 A listed and marked steel plate, permitted to be less than ¹⁄₁₆ inch only if it is listed and marked, and providing equal or better protection, used to protect a nonmetallic-sheathed cable less than 1¼ inches from the edge of a wood stud. *(Courtesy of Hubbell Incorporated)*

locations, cables or raceways shall be permitted to be laid in notches in wood studs, joists, rafters, or other wood members where the cable or raceway at those points is protected from penetration by nails or screws by a steel plate at least 1.6 mm (¹⁄₁₆ in.) thick, and of appropriate length and width, installed to cover the area of the wiring. The steel plate shall be installed before the building finish is applied.

> *Exception No. 1: Steel plates shall not be required to protect rigid metal conduit, intermediate metal conduit, rigid nonmetallic conduit, or electrical metallic tubing.*

> *Exception No. 2: A listed and marked steel plate less than 1.6 mm (¹⁄₁₆ in.) thick that provides equal or better protection against nail or screw penetration shall be permitted.*

(B) Nonmetallic-Sheathed Cables and Electrical Nonmetallic Tubing Through Metal Framing Members.

(1) Nonmetallic-Sheathed Cable. In both exposed and concealed locations where nonmetallic-sheathed cables pass through either factory- or field-punched, cut, or drilled slots or holes in metal members, the cable shall be protected by listed bushings or listed grommets covering all metal edges that are securely fastened in the opening prior to installation of the cable.

(2) Nonmetallic-Sheathed Cable and Electrical Nonmetallic Tubing. Where nails or screws are likely to penetrate nonmetallic-sheathed cable or electrical nonmetallic tubing, a steel sleeve, steel plate, or steel clip not less than 1.6 mm (¹⁄₁₆ in.) in thickness shall be used to protect the cable or tubing.

> *Exception: A listed and marked steel plate less than 1.6 mm (¹⁄₁₆ in.) thick that provides equal or better protection against nail or screw penetration shall be permitted.*

Δ **(C) Cables Through Spaces Behind Panels Designed to Allow Access.** Cables or raceway-type wiring methods, installed behind panels designed to allow access shall be supported according to their applicable articles.

Cable- or raceway-type wiring installed above suspended ceilings with lift-up panels must not be laid on the suspended ceiling, which would inhibit access. Such wiring is required to be supported according to 300.11(B), 300.23, and the requirements of the Chapter 3 article applicable to the particular wiring method.

Similarly, low-voltage, optical fiber, broadband, and communications cables also are not permitted to block access to equipment above a suspended ceiling.

(D) Cables and Raceways Parallel to Framing Members and Furring Strips. In both exposed and concealed locations, where a cable- or raceway-type wiring method is installed parallel to framing members, such as joists, rafters, or studs, or is installed parallel to furring strips, the cable or raceway shall be installed and supported so that the nearest outside surface of the cable or raceway is not less than 32 mm (1¼ in.) from the nearest edge of the framing member or furring strips where nails or screws are likely to penetrate. Where this distance cannot be maintained,

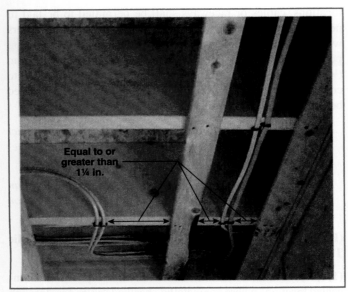

EXHIBIT 300.2 Nonmetallic-sheathed cables adjacent to furring strips in a wood frame structure.

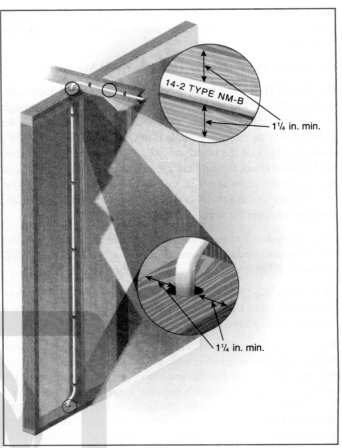

EXHIBIT 300.3 Nonmetallic-sheathed cable installed parallel to framing members.

the cable or raceway shall be protected from penetration by nails or screws by a steel plate, sleeve, or equivalent at least 1.6 mm (¹⁄₁₆ in.) thick.

Exception No. 1: Steel plates, sleeves, or the equivalent shall not be required to protect rigid metal conduit, intermediate metal conduit, rigid nonmetallic conduit, or electrical metallic tubing.

Exception No. 2: For concealed work in finished buildings, or finished panels for prefabricated buildings where such supporting is impracticable, it shall be permissible to fish the cables between access points.

Exception No. 3: A listed and marked steel plate less than 1.6 mm (¹⁄₁₆ in.) thick that provides equal or better protection against nail or screw penetration shall be permitted.

To prevent mechanical damage to cables and raceways from nails and screws, the raceways need the same level of physical protection at furring strips as they do at framing members. Two means of protection that generally apply to exposed or concealed work are described. The first method, shown in Exhibit 300.2, is to position the nonmetallic-sheathed cables a clear distance from the nearest edge of the furring strip. Exhibit 300.3 illustrates the NM cable installed parallel to framing members and fastened to maintain clearance from the edge. The second method requires physical protection, such as a steel plate or a sleeve, if the 1¼-inch clearance cannot be maintained. (A steel plate is illustrated in Exhibit 300.1.)

Δ **(E) Cables, Raceways, or Boxes Installed in or Under Metal-Corrugated Roof Decking.** A cable, raceway, or box, installed in exposed or concealed locations under metal-corrugated sheet roof decking, shall be installed and supported so there is not less than 38 mm (1½ in.) measured from the lowest surface of the roof decking to the top of the cable, raceway, or box. A cable, raceway, or box shall not be installed in concealed locations in metal-corrugated, sheet decking–type roof.

Informational Note: Roof decking material is often repaired or replaced after the initial raceway or cabling and roofing installation and might be penetrated by screws or other mechanical devices designed to provide "hold down" strength of the waterproof membrane or roof insulating material.

Exception No. 1: Rigid metal conduit and intermediate metal conduit, with listed steel or malleable iron fittings and boxes, shall not be required to comply with 300.4(E).

Exception No. 2: The 38 mm (1½ in.) spacing is not required where metal-corrugated sheet roof decking is covered with a minimum thickness 50 mm (2 in.) concrete slab, measured from the top of the corrugated roofing.

For cables, raceways, and boxes installed below a metal-corrugated sheet roof decking installation, 300.4(E) requires at least 1½ inches of separation from any of the roof decking surface, as illustrated in Exhibit 300.4. The 1½-inch dimension is measured from the lowest point of the roof deck to the top surface of the

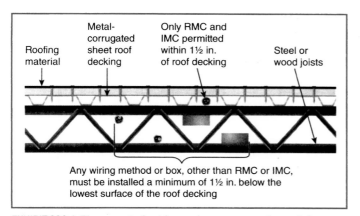

EXHIBIT 300.4 Placement of cables and raceways under metal-corrugated sheet roof decking to avoid physical damage.

cable, raceway, or box. This section prohibits cables, raceways, or boxes from being installed in the space between the metal deck and the roofing material.

Roof replacement materials are fastened in place with long screws, which penetrate the roof decking installation and could continue into cables and raceways installed below. Adequate space between the corrugated sheet metal roof deck installation and the cables and raceways below the deck will prevent future damage to the cable or raceway installation. The exception clarifies that this requirement does not apply to more robust metal raceways such as RMC and IMC.

See also

410.10(F) for similar requirements for luminaires

(F) Cables and Raceways Installed in Shallow Grooves. Cable- or raceway-type wiring methods installed in a groove, to be covered by wallboard, siding, paneling, carpeting, or similar finish, shall be protected by 1.6 mm (¹⁄₁₆ in.) thick steel plate, sleeve, or equivalent or by not less than 32-mm (1¼-in.) free space for the full length of the groove in which the cable or raceway is installed.

Exception No. 1: Steel plates, sleeves, or the equivalent shall not be required to protect rigid metal conduit, intermediate metal conduit, rigid PVC conduit, RTRC, or electrical metallic tubing.

Exception No. 2: A listed and marked steel plate less than 1.6 mm (¹⁄₁₆ in.) thick that provides equal or better protection against nail or screw penetration shall be permitted.

(G) Fittings. Where raceways contain 4 AWG or larger insulated circuit conductors, and these conductors enter a cabinet, a box, an enclosure, or a raceway, prior to the installation of conductors, the conductors shall be protected in accordance with any of the following:

(1) An identified fitting providing a smoothly rounded insulating surface
(2) A listed metal fitting that has smoothly rounded edges

(3) Separation from the fitting or raceway using an identified insulating material that is securely fastened in place
(4) Threaded hubs or bosses that are an integral part of a cabinet, box, enclosure, or raceway providing a smoothly rounded or flared entry for conductors

Conduit bushings constructed wholly of insulating material shall not be used to secure a fitting or raceway. The insulating fitting or insulating material shall have a temperature rating not less than the insulation temperature rating of the installed conductors.

Heavy conductors and cables tend to stress the conductor insulation at raceway terminating points. Care must be taken at raceway and cable terminations to reduce the risk of insulation failure at conductor insulation stress points. This is often done by using insulating bushings or fittings with rounded edges; however, there are multiple options to achieve the desired protection of the insulation. The temperature ratings of the insulating bushing must coordinate with the insulation of the conductor to ensure that the conductor is protected for its entire life cycle.

Whether provided separately or as part of a fitting, listed insulating bushings are colored black or brown to indicate a temperature rating of 150°C and any other color to indicate a rating of 90°C unless specifically marked for a higher temperature. Exhibit 300.5 shows an insulated thermoplastic or fiber bushing used to protect the conductors from chafing against a metal conduit fitting. Note the use of a double locknut, with one on the inside and one on the outside of the enclosure. The locknuts are necessary because the raceway cannot be secured by the fiber or plastic bushing.

See also

342.46, 344.46, and 352.46 for further information relating to bushings

(H) Structural Joints. A listed expansion/deflection fitting or other approved means shall be used where a raceway crosses a structural joint intended for expansion, contraction, or deflection, used in buildings, bridges, parking garages, or other structures.

Raceways can be damaged if improperly installed in structural construction joints, leaving conductors or cables exposed.

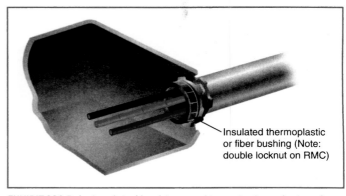

EXHIBIT 300.5 An insulated bushing used to protect conductors from chafing against a metal conduit fitting.

Structural construction joints can experience shear and lateral loads due to gravity, expansion and contraction, and movement of the structure. Listed expansion/deflection fittings are available for use at structural joints. However, other approved means of protecting the integrity of the raceways system are permitted.

300.5 Underground Installations.

(A) Minimum Cover Requirements. Direct-buried cable, conduit, or other raceways shall be installed to meet the minimum cover requirements of Table 300.5(A).

Table 300.5(A) requires conductors under residential driveways to be at least 18 inches below grade. However, if the conductors are protected by an overcurrent device rated at not more than 20 amperes and provided with ground-fault circuit-interpreter (GFCI) protection for personnel, column 4 permits the burial depth to be reduced to 12 inches. Exhibits 300.6 and 300.7 show examples of underground installations of 18 inches and 12 inches, respectively. Footnote 1 in column 5 permits wiring at depths less than the minimum if it is specified by the manufacturer's instructions. Footnote 2 permits column 5 to apply to circuits for pool, spa, and fountain lighting provided that the circuit voltage is 30 volts or less and it is installed in a nonmetallic raceway.

(B) Wet Locations. The interior of enclosures or raceways installed underground shall be considered to be a wet location. Insulated conductors and cables installed in these enclosures or raceways in underground installations shall comply with 310.10(C).

The inside of all raceways and enclosures installed underground is classified as a wet location. Conductors installed

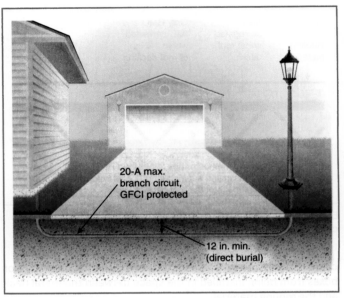

EXHIBIT 300.7 A 20-ampere, GFCI-protected residential branch circuit installed with a minimum burial depth of 12 inches beneath a residential driveway.

in such underground locations must meet one of the following criteria:

1. Be moisture-impervious, metal-sheathed
2. Be Type MTW, RHW, RHW-2, TW, THW, THW-2, THHW, THWN, THWN-2, XHHW, XHHW-2, or ZW
3. Be of a type listed for use in wet locations

(C) Underground Cables and Conductors Under Buildings. Underground cable and conductors installed under a building shall be in a raceway.

Exception No. 1: Type MI cable shall be permitted under a building without installation in a raceway where embedded in concrete, fill, or other masonry in accordance with 332.10(6) or in underground runs where suitably protected against physical damage and corrosive conditions in accordance with 332.10(10).

Exception No. 2: Type MC cable listed for direct burial or concrete encasement shall be permitted under a building without installation in a raceway in accordance with 330.10(A)(5) and in wet locations in accordance with 330.10(A)(11).

(D) Protection from Damage. Conductors and cables shall be protected from damage in accordance with 300.5(D)(1) through (D)(4).

(1) Emerging from Grade. Direct-buried conductors and cables emerging from grade and specified in Columns 1 and 4 of Table 300.5(A) shall be protected by enclosures or raceways extending from the minimum cover distance below grade required by 300.5(A) to a point at least 2.5 m (8 ft) above finished grade. In no case shall the protection be required to exceed 450 mm (18 in.) below finished grade.

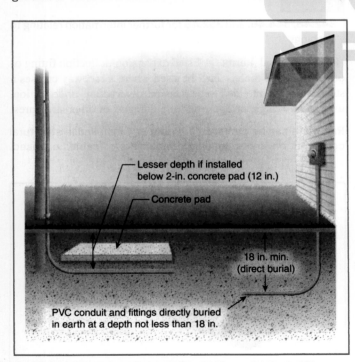

EXHIBIT 300.6 Rigid polyvinyl chloride (PVC) conduit buried 18 inches below grade and also at 12 inches as permitted when installed in a trench below a minimum of a 2-inch concrete pad.

△ **TABLE 300.5(A)** *Minimum Cover Requirements, 0 to 1000 Volts ac, 1500 Volts dc, Nominal, Burial in Millimeters (Inches)*

Location of Wiring Method or Circuit	Column 1 Direct Burial Cables or Conductors		Column 2 Rigid Metal Conduit or Intermediate Metal Conduit		Column 3 Electrical Metallic Tubing, Nonmetallic Raceways Listed for Direct Burial Without Concrete Encasement, or Other Approved Raceways		Column 4 Residential Branch Circuits Rated 120 Volts or Less with GFCI Protection and Maximum Overcurrent Protection of 20 Amperes		Column 5 Circuits for Control of Irrigation and Landscape Lighting Limited to Not More Than 30 Volts and Installed with Type UF or in Other Identified Cable or Raceway	
	mm	in.	mm	in.	mm	in.	mm	in.	mm	in.
All locations not specified below	600	24	150	6	450	18	300	12	150[1,2]	6[1,2]
In trench below 50 mm (2 in.) thick concrete or equivalent	450	18	150	6	300	12	150	6	150	6
Under a building	0 (in raceway or Type MC or Type MI cable identified for direct burial)	0	0	0	0	0	0 (in raceway or Type MC or Type MI cable identified for direct burial)	0	0 (in raceway or Type MC or Type MI cable identified for direct burial)	0
Under minimum of 102 mm (4 in.) thick concrete exterior slab with no vehicular traffic and the slab extending not less than 152 mm (6 in.) beyond the underground installation	450	18	100	4	100	4	150 (direct burial) 100 (in raceway)	6 4	150 (direct burial) 100 (in raceway)	6 4
Under streets, highways, roads, alleys, driveways, and parking lots	600	24	600	24	600	24	600	24	600	24
One- and two-family dwelling driveways and outdoor parking areas, and used only for dwelling-related purposes	450	18	450	18	450	18	300	12	450	18
In or under airport runways, including adjacent areas where trespassing is prohibited	450	18	450	18	450	18	450	18	450	18

[1] A lesser depth shall be permitted where specified in the installation instructions of a listed low-voltage lighting system.

[2] A depth of 150 mm (6 in.) shall be permitted for pool, spa, and fountain lighting, installed in a nonmetallic raceway, limited to not more than 30 volts where part of a listed low-voltage lighting system.

Notes:

1. Cover shall be defined as the shortest distance in mm (in.) measured between a point on the top surface of any direct-buried conductor, cable, conduit, or other raceway and the top surface of finished grade, concrete, or similar cover.

2. Raceways approved for burial only where concrete encased shall require a concrete envelope not less than 50 mm (2 in.) thick.

3. Lesser depths shall be permitted where cables and conductors rise for terminations or splices or where access is otherwise required.

4. Where one of the wiring method types listed in Columns 1 through 3 is used for one of the circuit types in Columns 4 and 5, the shallowest depth of burial shall be permitted.

5. Where solid rock prevents compliance with the cover depths specified in this table, the wiring shall be installed in a metal raceway, or a nonmetallic raceway permitted for direct burial. The raceways shall be covered by a minimum of 50 mm (2 in.) of concrete extending down to rock.

6. Directly buried electrical metallic tubing (EMT) shall comply with 358.10.

(2) Conductors Entering Buildings. Conductors entering a building shall be protected to the point of entrance.

(3) Service Conductors. Underground service conductors that are not encased in concrete and that are buried 450 mm (18 in.) or more below grade shall have their location identified by a warning ribbon that is placed in the trench at least 300 mm (12 in.) above the underground installation.

The warning ribbon reduces the risk of an accident, such as an electrocution or an arc-flash incident, during excavation near underground service conductors that are not encased in concrete, because these circuits are not protected from short circuit and overload.

(4) Enclosure or Raceway Damage. Where the enclosure or raceway is subject to physical damage, the conductors shall be installed in electrical metallic tubing, rigid metal conduit, intermediate metal conduit, RTRC-XW, Schedule 80 PVC conduit, or equivalent.

(E) Splices and Taps. Direct-buried conductors or cables shall be permitted to be spliced or tapped without the use of splice boxes. The splices or taps shall be made in accordance with 110.14(B).

Underground splices are not required to be in a box if they are made in accordance with 110.14(B), which requires the splicing means to be listed for underground use. Listed sealed wire connector systems restore the insulation integrity of the spliced conductors in a permanent joint. Sealed wire connectors are used where future access to the splices will not be necessary.

The difference between multiconductor cables labeled for direct burial and single conductors labeled for direct burial is that direct-burial multiconductor cables might or might not contain individual conductors labeled for direct burial. The overall cable jacket might be the only underground protection technique for the contained conductors. Although the direct-burial splicing techniques used on multiconductor cables can differ widely from the techniques used on direct-burial single-conductor cables, the *NEC* requirements are generally the same. The splicing technique must be listed for the cable type and listed for direct burial. Exhibit 300.8 shows a listed heat-shrink underground splice kit being utilized with a multiconductor cable.

(F) Backfill. Backfill that contains large rocks, paving materials, cinders, large or sharply angular substances, or corrosive material shall not be placed in an excavation where materials might damage raceways, cables, conductors, or other substructures or prevent adequate compaction of fill or contribute to corrosion of raceways, cables, or other substructures.

Where necessary to prevent physical damage to the raceway, cable, or conductor, protection shall be provided in the form of granular or selected material, suitable running boards, suitable sleeves, or other approved means.

(G) Raceway Seals. Conduits or raceways through which moisture might contact live parts shall be sealed or plugged

EXHIBIT 300.8 A listed underground splicing method. *(Courtesy of Ideal Industries, Inc.)*

at either or both ends. Spare or unused raceways shall also be sealed. Sealants shall be identified for use with the cable insulation, conductor insulation, bare conductor, shield, or other components.

Informational Note: Presence of hazardous gases or vapors might also necessitate the sealing of underground conduits or raceways entering buildings.

A conduit sealing bushing, shown in Exhibit 300.9, is one method that prevents the entrance of gas or moisture. See 230.8 for sealing service raceways.

(H) Bushing. A bushing, or terminal fitting, with an integral bushed opening shall be used at the end of a conduit or other raceway that terminates underground where the conductors or cables emerge as a direct burial wiring method. A seal incorporating the physical protection characteristics of a bushing shall be permitted to be used in lieu of a bushing.

(I) Conductors of the Same Circuit. All conductors of the same circuit and, where used, the grounded conductor and all equipment grounding conductors shall be installed in the same raceway or cable or shall be installed in close proximity in the same trench.

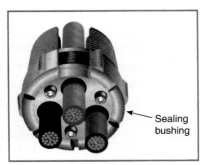

EXHIBIT 300.9 A conduit sealing bushing used to prevent the entrance of moisture.

Sealing bushing

Exception No. 1: Conductors shall be permitted to be installed in parallel in raceways, multiconductor cables, or direct-buried single conductor cables. Each raceway or multiconductor cable shall contain all conductors of the same circuit, including equipment grounding conductors. Each direct-buried single conductor cable shall be located in close proximity in the trench to the other single conductor cables in the same parallel set of conductors in the circuit, including equipment grounding conductors.

Exception No. 2: Isolated phase, polarity, grounded conductor, and equipment grounding and bonding conductor installations shall be permitted in nonmetallic raceways or cables with a nonmetallic covering or nonmagnetic sheath in close proximity where conductors are paralleled as permitted in 310.10(G), and where the conditions of 300.20(B) are met.

Isolated phase installations contain only one phase per raceway or cable. In an ac circuit installation, the spacing between isolated phase raceways and cables should be as small as possible and the length of the run limited to avoid increased circuit impedance and the resulting increase in voltage drop. Isolated phase installations can be used in ac circuits to limit available fault current at downstream equipment.

Isolated phase installations present an inherent hazard of overheating, which is a risk that must be understood and carefully controlled. This hazard results from induced currents in metal raceways that contain only one phase conductor. The surrounding metal acts as a shorted transformer turn. In underground installations, a single conductor is unlikely to be installed in a metal raceway; if it is installed, it is unlikely to present a fire hazard. This is not true for aboveground raceways, which is the reason isolated phase installations have limited application for aboveground installations.

See also

300.20(A) and **(B)** for more information on induced currents in raceways

(J) Earth Movement. Where direct-buried conductors, raceways, or cables are subject to movement by settlement or frost, direct-buried conductors, raceways, or cables shall be arranged so as to prevent damage to the enclosed conductors or to equipment connected to the raceways.

Informational Note: This section recognizes "S" loops in underground direct burial cables and conductors to raceway transitions, expansion fittings in raceway risers to fixed equipment, and, generally, the provision of flexible connections to equipment subject to settlement or frost heaves.

(K) Directional Boring. Cables or raceways installed using directional boring equipment shall be approved for the purpose.

300.6 Protection Against Corrosion and Deterioration. Raceways, cable trays, cablebus, auxiliary gutters, cable armor, boxes, cable sheathing, cabinets, enclosures (other than surrounding fences and walls); elbows, couplings, fittings,

supports, and support hardware shall be of materials suitable for the environment in which they are to be installed.

Δ **(A) Ferrous Metal Equipment.** Ferrous metal raceways, cable trays, cablebus, auxiliary gutters, cable armor, boxes, cable sheathing, cabinets, enclosures (other than surrounding fences and walls), elbows, couplings, nipples, fittings, supports, and support hardware shall be suitably protected against corrosion inside and outside (except threads at joints) by a coating of approved corrosion-resistant material. Where corrosion protection is necessary and the conduit is threaded anywhere other than at the factory where the product is listed, the threads shall be coated with an approved electrically conductive, corrosion-resistant compound.

Exception: Stainless steel shall not be required to have protective coatings.

Ferrous metal equipment must be protected from corrosion with an approved anti-corrosion compound, such as that shown in Exhibit 300.10.

(1) Protected from Corrosion Solely by Enamel. Where protected from corrosion solely by enamel, ferrous metal raceways, cable trays, cablebus, auxiliary gutters, cable armor, boxes, cable sheathing, cabinets, enclosures (other than surrounding fences and walls), elbows, couplings, nipples, fittings, supports, and support hardware shall not be used outdoors or in wet locations as described in 300.6(D).

(2) Organic Coatings on Boxes or Cabinets. Where boxes, cabinets, or enclosures (other than surrounding fences and walls) have an approved system of organic coatings and are marked "Raintight," "Rainproof," or "Outdoor Type," they shall be permitted outdoors.

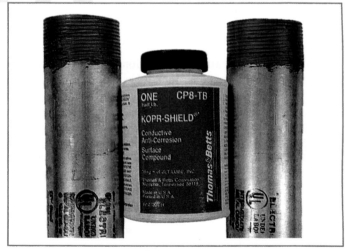

EXHIBIT 300.10 KOPR-Shield® (a registered trademark of Jet Lube) is a conductive anti-corrosion surface compound suitable for application on field-cut conduit threads where protection from corrosion is necessary. *(Courtesy of Thomas and Betts, A Member of the ABB Group)*

△ **(3) In Concrete or in Direct Contact with the Earth.** Ferrous metal raceways, cable armor, boxes, cable sheathing, cabinets, enclosures (other than surrounding fences and walls), elbows, couplings, nipples, fittings, supports, and support hardware shall be permitted to be installed in concrete, in direct contact with the earth, or in areas subject to severe corrosive influences where made of material approved for the condition or where provided with corrosion protection approved for the condition.

Metal raceways installed in the earth can be coated with an asphalt compound, plastic sheath, or other equivalent protection to help prevent deterioration. Also, metal raceways are available with a bonded PVC coating. Galvanized steel RMC and steel IMC generally do not require supplementary corrosion protection.

(B) Aluminum Metal Equipment. Aluminum raceways, cable trays, cablebus, auxiliary gutters, cable armor, boxes, cable sheathing, cabinets, enclosures (other than surrounding fences and walls), elbows, couplings, nipples, fittings, supports, and support hardware embedded or encased in concrete or in direct contact with the earth shall be provided with supplementary corrosion protection.

(C) Nonmetallic Equipment. Nonmetallic raceways, cable trays, cablebus, auxiliary gutters, boxes, cables with a nonmetallic outer jacket and internal metal armor or jacket, cable sheathing, cabinets, enclosures (other than surrounding fences and walls), elbows, couplings, nipples, fittings, supports, and support hardware shall be made of material approved for the condition and shall comply with 300.6(C)(1) and (C)(2) as applicable to the specific installation.

(1) Exposed to Sunlight. Where exposed to sunlight, the materials shall be listed as sunlight resistant or shall be identified as sunlight resistant.

(2) Chemical Exposure. Where subject to exposure to chemical solvents, vapors, splashing, or immersion, materials or coatings shall either be inherently resistant to chemicals based on their listing or be identified for the specific chemical reagent.

(D) Indoor Wet Locations. In portions of dairy processing facilities, laundries, canneries, and other indoor wet locations, and in locations where walls are frequently washed or where there are surfaces of absorbent materials, such as damp paper or wood, the entire wiring system, where installed exposed, including all boxes, cabinets, enclosures (other than surrounding fences and walls), fittings, raceways, and cable used therewith, shall be mounted so that there is at least a 6 mm (¼ in.) airspace between it and the wall or supporting surface.

Exception: Nonmetallic raceways, boxes, and fittings shall be permitted to be installed without the airspace on a concrete, masonry, tile, or similar surface.

Informational Note: In general, areas where acids and alkali chemicals are handled and stored might present such corrosive conditions, particularly when wet or damp. Severe corrosive conditions might also be present in portions of meatpacking plants, tanneries, glue houses, and some stables; in installations immediately adjacent to a seashore and swimming pool areas; in areas where chemical deicers are used; and in storage cellars or rooms for hides, casings, fertilizer, salt, and bulk chemicals.

300.7 Raceways Exposed to Different Temperatures.

(A) Sealing. Where portions of a raceway or sleeve are known to be subjected to different temperatures, and where condensation is known to be a problem, as in cold storage areas of buildings or where passing from the interior to the exterior of a building, the raceway or sleeve shall be sealed to prevent the circulation of warm air to a colder section of the raceway or sleeve. Sealants shall be identified for use with cable insulation, conductor insulation, a bare conductor, a shield, or other components. An explosionproof seal shall not be required for this purpose.

Condensation can form in raceways or sleeves that are subjected to temperature differences as a result of air circulating through the raceway from a warmer to a colder section. Condensation could accumulate, for example, in a raceway used to supply lighting or branch-circuit conductors within a walk-in refrigerator or freezer. Circulation of air can be prevented by sealing the raceway with a suitable pliable compound at a conduit body or junction box, usually installed in the raceway before it enters the colder section. Special sealing fittings such as those used in hazardous (classified) locations are not necessary.

(B) Expansion, Expansion-Deflection, and Deflection Fittings. Raceways shall be provided with expansion, expansion-deflection, or deflection fittings where necessary to compensate for thermal expansion, deflection, and contraction.

Informational Note No. 1: Table 352.44(A) and Table 355.44 provide the expansion information for polyvinyl chloride (PVC) and for reinforced thermosetting resin conduit (RTRC), respectively. A nominal number for steel conduit can be determined by multiplying the expansion length in Table 352.44(A) by 0.20. The coefficient of expansion for steel electrical metallic tubing, intermediate metal conduit, and rigid metal conduit is 1.170×10^{-5} (0.0000117 mm per mm of conduit for each °C in temperature change) [0.650×10^{-5} (0.0000065 in. per in. of conduit for each °F in temperature change)].

A nominal number for aluminum conduit and aluminum electrical metallic tubing can be determined by multiplying the expansion length in Table 352.44(A) by 0.40. The coefficient of expansion for aluminum electrical metallic tubing and aluminum rigid metal conduit is 2.34×10^{-5} (0.0000234 mm per mm of conduit for each °C in temperature change) [1.30×10^{-5} (0.000013 in. per in. of conduit for each °F in temperature change)].

Informational Note No. 2: See NEMA FB 2.40-2019, *Installation Guidelines for Expansion and Expansion/Deflection Fittings*, for further information on expansion and expansion deflection fittings.

Substantial changes in temperature cause destructive amounts of expansion, deflection, and contraction in a raceway system. Properly designed and installed expansion and deflection fittings allow expansion, deflection, and contraction without damage to enclosures, raceways, and their conductors.

COMMENTARY TABLE 300.1 Expansion Coefficients with a 55°F Change

Raceway Type	Referenced Table	Expansion Coefficient (in./100 ft)	Expansion of 40 ft of raceway (in.)
PVC	352.44(A)	2.23	0.892
RTRC	355.44	0.99	0.396
Steel RMC, IMC, or EMT	352.44(A) [as referenced in 300.7(B) Informational Note No. 1]	0.446	0.178

The informational note addresses expansion of aluminum RMC and reinforced thermosetting resin conduit (RTRC). The note also provides a few simple relationships (or ratios) of linear expansion in length of PVC conduit to other types of conduit.

The key to determining any temperature-related change in conduit length is understanding that the change in length is equal to the coefficient of expansion of the conduit material times the change in temperature times the initial length of the conduit.

Commentary Table 300.1 provides the coefficient of expansion for various conduit types. This illustrates the variances of different raceway types due to thermal expansion with a 55°F change.

300.8 Installation of Conductors With Other Systems. Raceways or cable trays containing electrical conductors shall not contain any pipe, tube, or equal for steam, water, air, gas, drainage, or any service other than electrical.

300.9 Raceways in Wet Locations Above Grade. Where raceways are installed in wet locations above grade, the interior of these raceways shall be considered to be a wet location. Insulated conductors and cables installed in raceways in wet locations above grade shall comply with 310.10(C).

Δ **300.10 Electrical Continuity of Metal Raceways, Cable Armor, and Enclosures.** Metal raceways, cable armor, and other metal enclosures for conductors shall be metallically joined together into a continuous electrical conductor and shall be connected to all boxes, fittings, and cabinets to provide effective electrical continuity. Unless specifically permitted elsewhere in this *Code*, raceways and cable assemblies shall be mechanically secured to boxes, fittings, cabinets, and other enclosures.

Exception No. 1: Short sections of raceways used to provide support or protection of cable assemblies from physical damage shall not be required to be made electrically continuous.

Exception No. 2: Equipment enclosures to be isolated, as permitted by 250.96(B), shall not be required to be metallically joined to the metal raceway.

Sections 250.4(A) and (B) require that the metal parts form an effective low-impedance path to ground. This safely conducts any fault current and facilitates the operation of overcurrent devices protecting the enclosed circuit conductors.

300.11 Securing and Supporting.

(A) Secured in Place. Raceways, cable assemblies, boxes, cabinets, and fittings shall be securely fastened in place.

Δ **(B) Wiring Systems Installed Above Suspended Ceilings.** Support wires that do not provide secure support shall not be the sole support. Support wires and associated fittings that provide secure support and that are installed in addition to the ceiling grid support wires shall be permitted as the sole support. Where independent support wires are used, they shall be secured at both ends. Cables and raceways shall not be supported by ceiling grids.

(1) Fire-Rated Assemblies. Wiring located within the cavity of a fire-rated floor–ceiling or roof–ceiling assembly shall not be secured to, or supported by, the ceiling assembly, including the ceiling support wires. An independent means of secure support shall be provided and shall be permitted to be attached to the assembly. Where independent support wires are used, they shall be distinguishable by color, tagging, or other effective means from those that are part of the fire-rated design.

Exception: The ceiling support system shall be permitted to support wiring and equipment that have been tested as part of the fire-rated assembly.

Informational Note: See ASTM E119, *Standard Test Methods for Fire Tests of Building Construction and Materials*, for one method of testing to determine fire rating.

Wiring methods and luminaires are not allowed to be supported or secured to the support wires or T-bars of a fire-rated ceiling assembly unless the assembly has been tested and listed for that use. If wire is selected as the supporting means for the electrical system within a fire-rated ceiling cavity, it must be distinguishable from the ceiling support wires by color, tagging, or other effective means and must be secured at both ends. Independent support and securing both ends of the support wire do preclude a connection to the ceiling grid on one end.

Generally, the rule for supporting electrical equipment is that the equipment must be "securely fastened in place." This phrase means not only that vertical support for the weight of the equipment must be provided but also that the equipment must be secured to prevent horizontal movement or sway. The intention is to prevent the loss of grounding continuity provided by the raceway that could result from horizontal movement.

Sections 300.11(B)(1) and (B)(2) are quite similar. Unless the exceptions apply, these sections prohibit all types of wiring from being attached in any way to the support wires of a ceiling assembly or ceiling grid not part of the building structure.

Refer to the appropriate wiring method article in Chapter 3 for cable and raceway supporting requirements.

See also

Article 410, Part IV, for the proper support of luminaires

314.23 for the support of outlet boxes

760.24 and **770.24** for various low-voltage fire alarm and optical fiber cable supports

800.110(C) for requirements on communications systems cable supports

(2) Non-Fire-Rated Assemblies. Wiring located within the cavity of a non-fire-rated floor–ceiling or roof–ceiling assembly shall not be secured to, or supported by, the ceiling assembly, including the ceiling support wires. An independent means of secure support shall be provided and shall be permitted to be attached to the assembly. Where independent support wires are used, they shall be distinguishable by color, tagging, or other effective means.

Exception: The ceiling support system shall be permitted to support branch-circuit wiring and associated equipment where installed in accordance with the ceiling system manufacturer's instructions.

(C) Raceways Used as Means of Support. Raceways shall be used only as a means of support for other raceways, cables, or nonelectrical equipment under any of the following conditions:

(1) Where the raceway or means of support is identified as a means of support

(2) Where the raceway contains power supply conductors for electrically controlled equipment and is used to support Class 2 or Class 3 circuit conductors or cables that are solely for the purpose of connection to the equipment control circuits

(3) Where the raceway is used to support boxes or conduit bodies in accordance with 314.23 or to support luminaires in accordance with 410.36(E)

As a general rule, this section prohibits supporting cables by securing them to the exterior of a raceway. Electrical, telephone, and data cables wrapped around a raceway can impede dissipation of heat from the raceway, thus affecting the temperature of the conductors contained in the raceway. The weight from large bundles of cables can compromise the mechanical integrity of the raceway system. For those reasons, this section also prohibits the use of a raceway as a means of support for nonelectrical equipment, such as suspended ceilings, water pipes, and signs.

Class 2 conductors or cables are allowed to be supported by a raceway as long as the power supply conductors are inside the raceway or functionally associated with the attached Class 2 circuit conductors. For example, the thermostat conductors for heating or air-conditioning units are permitted to be supported by the conduit supplying power to the unit.

(D) Cables Not Used as Means of Support. Cable wiring methods shall not be used as a means of support for other cables, raceways, or nonelectrical equipment.

300.12 Mechanical Continuity — Raceways and Cables. Raceways, cable armors, and cable sheaths shall be continuous between cabinets, boxes, conduit bodies, fittings, or other enclosures or outlets.

Exception No. 1: Short sections of raceways used to provide support or protection of cable assemblies from physical damage shall not be required to be mechanically continuous.

Exception No. 2: Raceways and cables installed into the bottom of open bottom equipment, such as switchboards, motor control centers, and floor or pad-mounted transformers, shall not be required to be mechanically secured to the equipment.

300.13 Mechanical and Electrical Continuity — Conductors.

(A) General. Conductors in raceways shall be continuous between outlets, boxes, devices, and so forth. There shall be no splice or tap within a raceway unless permitted by 300.15, 368.56(A), 376.56, 378.56, 384.56, 386.56, 388.56, or 390.56.

Δ **(B) Device Removal.** In multiwire branch circuits, the continuity of a grounded conductor shall not depend on device connections such as lampholders, receptacles, and so forth where the removal of such devices would interrupt the continuity.

Grounded conductors (neutrals) of multiwire branch circuits supplying receptacles, lampholders, or other devices are not permitted to depend on terminal connections for continuity between devices. For multiwire installations, a splice can be made with a jumper connected to the terminal or the neutral can be looped. This allows a receptacle or device to be replaced without interrupting the continuity of energized downstream line-to-neutral loads. Opening the neutral could cause unbalanced voltages, and a considerably higher voltage would be impressed on one part of a multiwire branch circuit, especially if the downstream line-to-neutral loads were appreciably unbalanced.

300.14 Length of Free Conductors at Outlets, Junctions, and Switch Points. At least 150 mm (6 in.) of free conductor, measured from the point in the box where it emerges from its raceway or cable sheath, shall be left at each outlet, junction, and switch point for splices or the connection of luminaires or devices. The 150 mm (6 in.) free conductor shall be permitted to be spliced or unspliced. Where the opening to an outlet, junction, or switch point is less than 200 mm (8 in.) in any dimension, each conductor shall be long enough to extend at least 75 mm (3 in.) outside the opening.

Exception: Conductors that are not spliced or terminated at the outlet, junction, or switch point shall not be required to comply with 300.14.

A conductor must have enough slack for the connections to be made easily. The length of slack (free conductor) required for the box size is shown in Exhibit 300.11. The exception excludes conductors running through a box, which should have sufficient slack to prevent physical damage from the insertion of devices or from the use of luminaire studs, hickeys, or other luminaire supports within the box.

△ **300.15 Boxes, Conduit Bodies, or Fittings — Where Required.** A box shall be installed at each outlet and switch point for concealed knob-and-tube wiring.

Fittings and connectors shall be used only with the specific wiring methods for which they are designed and listed.

Where the wiring method is conduit, tubing, Type AC cable, Type MC cable, Type MI cable, nonmetallic-sheathed cable, or other cables, a box or conduit body shall be installed at each outlet point, switch point, conductor splice point, conductor junction point, conductor termination point, wiring method transition point, or conductor pull point, unless otherwise permitted in 300.15(A) through (L).

(A) Wiring Methods with Interior Access. A box or conduit body shall not be required for each splice, junction, switch, pull, termination, or outlet points in wiring methods with removable covers, such as wireways, multioutlet assemblies, auxiliary gutters, and surface raceways. The covers shall be accessible after installation.

(B) Equipment. An integral junction box or wiring compartment as part of approved equipment shall be permitted in lieu of a box.

(C) Protection. A box or conduit body shall not be required where cables enter or exit from conduit or tubing that is used to provide cable support or protection against physical damage. A fitting shall be provided on the end(s) of the conduit or tubing to protect the cable from abrasion.

(D) Type MI Cable. A box or conduit body shall not be required where accessible fittings are used for straight-through splices in mineral-insulated metal-sheathed cable.

(E) Integral Enclosure. A wiring device with integral enclosure identified for the use, having brackets that securely fasten the device to walls or ceilings of conventional on-site frame construction, for use with nonmetallic-sheathed cable, shall be permitted in lieu of a box or conduit body.

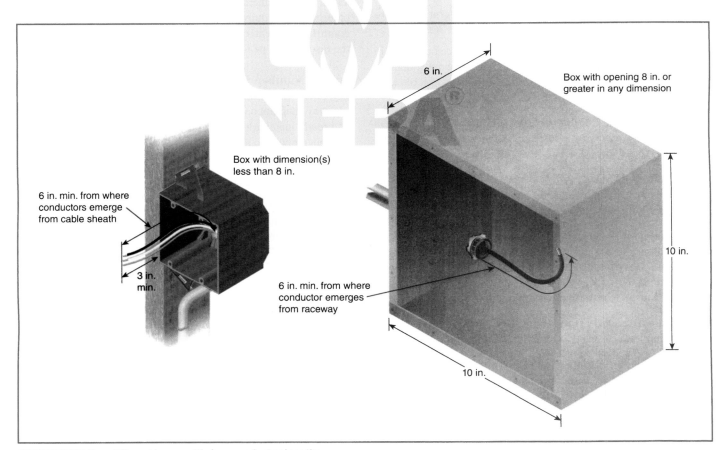

Box with dimension(s) less than 8 in.

6 in. min. from where conductors emerge from cable sheath

3 in. min.

6 in. min. from where conductor emerges from raceway

6 in.

Box with opening 8 in. or greater in any dimension

10 in.

10 in.

EXHIBIT 300.11 Two different boxes with free conductor lengths.

EXHIBIT 300.12 Examples of self-contained, boxless device receptacles. *(Left, courtesy of Legrand®; right, courtesy of Sillites®)*

Informational Note: See 334.30(C); 545.10; 550.15(I); 551.47(E), Exception No. 1; and 552.48(E), Exception No. 1.

This requirement applies to a device with an integral enclosure (boxless device) such as the ones shown in Exhibit 300.12.

(F) Fitting. A fitting identified for the use shall be permitted in lieu of a box or conduit body where conductors are not spliced or terminated within the fitting. The fitting shall be accessible after installation, unless listed for concealed installation.

(G) Direct-Buried Conductors and Cables. As permitted in 300.5(E), a box or conduit body shall not be required for splices and taps in direct-buried conductors and cables.

(H) Insulated Devices. As permitted in 334.40(B), a box or conduit body shall not be required for insulated devices supplied by nonmetallic-sheathed cable.

(I) Enclosures. A box or conduit body shall not be required where a splice, switch, terminal, or pull point is in a cabinet or cutout box, in an enclosure for a switch or overcurrent device as permitted in 312.8, in a motor controller as permitted in 430.10(A), or in a motor control center.

(J) Luminaires. A box or conduit body shall not be required where a luminaire is used as a raceway as permitted in 410.64.

(K) Embedded. A box or conduit body shall not be required for splices where conductors are embedded as permitted in 424.40, 424.41(D), 426.22(C), 426.24(A), and 427.19(A).

△ **(L) Manholes and Handhole Enclosures.** A box or conduit body shall not be required for conductors in manholes or handhole enclosures, except where connecting to electrical equipment.

The installation shall comply with Part V of Article 110 for manholes, and 314.30 for handhole enclosures.

300.16 Raceway or Cable to Open or Concealed Wiring.

△ **(A) Box, Conduit Body, or Fitting.** A box, conduit body, or terminal fitting having a separately bushed hole for each conductor shall be used wherever a change is made from conduit, electrical metallic tubing, electrical nonmetallic tubing, nonmetallic-sheathed cable, Type AC cable, Type MC cable, or mineral-insulated, metal-sheathed cable and surface raceway wiring to open wiring or to concealed knob-and-tube wiring. A fitting used for this purpose shall contain no taps or splices and shall not be used at luminaire outlets. A conduit body used for this purpose shall contain no taps or splices unless it complies with 314.16(C)(2).

A fitting is permitted instead of a box where a cable system transitions to a raceway to protect it against physical damage. For example, when a nonmetallic-sheathed cable runs overhead on floor joists, then needs protection where it drops down on a masonry wall to supply a receptacle, a short length of raceway is installed to the outlet device box. The cable is then inserted into the raceway and secured by a combination fitting that is fastened to the end of the raceway.

(B) Bushing. A bushing shall be permitted in lieu of a box or terminal where the conductors emerge from a raceway and enter or terminate at equipment, such as open switchboards, unenclosed control equipment, or similar equipment. The bushing shall be of the insulating type for other than lead-sheathed conductors.

△ **300.17 Number and Size of Conductors and Cables in Raceway.** The number and size of conductors and cables in any raceway shall not be more than will permit dissipation of the heat and ready installation or withdrawal of the conductors or cables without damage to the conductors or cables, or to their insulation.

Informational Note: See the following sections of this *Code*: intermediate metal conduit, 342.22; rigid metal conduit, 344.22; flexible metal conduit, 348.22; liquidtight flexible metal conduit, 350.22; PVC conduit, 352.22; HDPE conduit, 353.22; RTRC, 355.22; liquidtight nonmetallic flexible conduit, 356.22; electrical metallic tubing, 358.22; flexible metallic tubing, 360.22; electrical nonmetallic tubing, 362.22; cellular concrete floor raceways, 372.22; cellular metal floor raceways, 374.22; metal wireways, 376.22; nonmetallic wireways, 378.22; surface metal raceways, 386.22; surface nonmetallic raceways, 388.22; underfloor raceways, 390.22; fixture wire, 402.7; theaters, 520.6; signs, 600.31(C); elevators, 620.33; audio signal processing, amplification, and reproduction equipment, 640.23(A) and 640.24; Class 1 circuits, 724.3(A); Class 2, Class 3, Class 4, and power-limited fire alarm (PLFA) circuits, 722.3(A); non-power-limited fire alarm (NPLFA) circuits, 760.3(H); and optical fiber cables and raceways, 770.110(B).

300.18 Raceway Installations.

△ **(A) Complete Runs.** Raceways other than busways, listed manufactured assemblies in accordance with 604.100, or exposed raceways having hinged or removable covers shall be installed complete between outlet, junction, or splicing points prior to the installation of conductors or cables. Where required to facilitate the installation of utilization equipment, the raceway shall be permitted to be initially installed without a terminating connection at the equipment. Prewired raceway assemblies shall be permitted only where specifically permitted in this *Code* for the applicable wiring method.

Exception: Short sections of raceways used to contain conductors or cable assemblies for protection from physical damage shall not be required to be installed complete between outlet, junction, or splicing points.

One of the primary functions of a raceway is to provide physical protection for conductors. If raceways are incomplete at the time of conductor installation, a greater possibility of damage to the conductors exists.

The installation of conductors in an incomplete raceway is permitted for connection of utilization equipment. The motor installation shown in Exhibit 300.13 is a typical example, in which the motor will be supplied through liquidtight flexible metal conduit (LFMC) that terminates in the motor terminal box through a 90-degree angle connector. Wiring a luminaire whip prior to connecting a luminaire is also permitted by this section.

(B) Welding. Metal raceways shall not be supported, terminated, or connected by welding to the raceway unless specifically designed to be or otherwise specifically permitted to be in this *Code*.

300.19 Supporting Conductors in Vertical Raceways.

(A) Spacing Intervals — Maximum. Conductors in vertical raceways shall be supported if the vertical rise exceeds the values in Table 300.19(A). At least one support method shall be provided for each conductor at the top of the vertical raceway or as close to the top as practical. Intermediate supports shall be provided as necessary to limit supported conductor lengths to not greater than those values specified in Table 300.19(A).

Exception: Steel wire armor cable shall be supported at the top of the riser with a cable support that clamps the steel wire armor. A safety device shall be permitted at the lower end of the riser to hold the cable in the event there is slippage of the cable in the wire-armored cable support. Additional wedge-type supports shall be permitted to relieve the strain on the equipment terminals caused by expansion of the cable under load.

This requirement prevents the weight of the conductors from damaging the insulation where they leave the conduit and prevents the conductors from being pulled out of the terminals. Support bushings or cleats such as those shown in Exhibits 300.14 and 300.15 may be used, in addition to many other types of grips manufactured for this purpose.

EXHIBIT 300.14 A support bushing, located at the top of a vertical conduit at a cabinet or pull box, used to prevent the weight of the conductors from damaging insulation or placing strain on termination points.

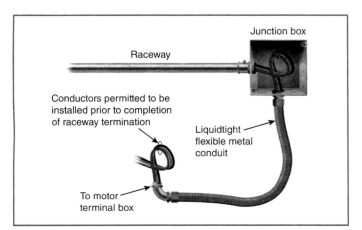

EXHIBIT 300.13 The conductors supplying a motor through LFMC are permitted to be installed prior to the connection of the raceway to the motor terminal box.

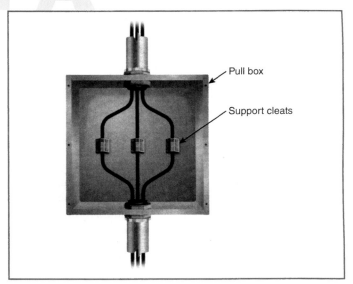

EXHIBIT 300.15 Support cleats used to prevent the weight of vertical conductors from damaging insulation or placing strain on termination points.

TABLE 300.19(A) *Spacings for Conductor Supports*

Conductor Size	Support of Conductors in Vertical Raceways	Aluminum or Copper-Clad Aluminum		Copper	
		m	ft	m	ft
18 AWG through 8 AWG	Not greater than	30	100	30	100
6 AWG through 1/0 AWG	Not greater than	60	200	30	100
2/0 AWG through 4/0 AWG	Not greater than	55	180	25	80
Over 4/0 AWG through 350 kcmil	Not greater than	41	135	18	60
Over 350 kcmil through 500 kcmil	Not greater than	36	120	15	50
Over 500 kcmil through 750 kcmil	Not greater than	28	95	12	40
Over 750 kcmil	Not greater than	26	85	11	35

(B) Fire-Resistive Cables and Conductors. Support methods and spacing intervals for fire-resistive cables and conductors shall comply with any restrictions provided in the listing of the electrical circuit protective system or fire-resistive cable system used and in no case shall exceed the values in Table 300.19(A).

(C) Support Methods. One of the following methods of support shall be used:

(1) Clamping devices constructed of or employing insulating wedges inserted in the ends of the raceways. Where clamping of insulation does not adequately support the cable, the conductor also shall be clamped.

(2) Inserting boxes at the required intervals in which insulating supports are installed and secured in an approved manner to withstand the weight of the conductors attached thereto, the boxes being provided with covers.

(3) In junction boxes, deflecting the cables not less than 90 degrees and carrying them horizontally to a distance not less than twice the diameter of the cable, with the cables being carried on two or more insulating supports and additionally secured thereto by tie wires, if desired. Where this method is used, cables shall be supported at intervals not greater than 20 percent of the support spacing in Table 300.19(A).

(4) Other approved means.

300.20 Induced Currents in Ferrous Metal Enclosures or Ferrous Metal Raceways.

Δ **(A) Conductors Grouped Together.** Where conductors carrying alternating current are installed in ferrous metal enclosures or ferrous metal raceways, they shall be arranged so as to avoid heating the surrounding ferrous metal by induction. To accomplish this, all phase conductors and, where used, the grounded conductor and all equipment grounding conductors shall be grouped together.

Exception No. 1: Equipment grounding conductors for certain existing installations shall be permitted to be installed separate

from their associated circuit conductors where run in accordance with 250.130(C).

Exception No. 2: A single conductor shall be permitted to be installed in a ferromagnetic enclosure and used for skin-effect heating in accordance with 426.42 and 427.47.

Nonferrous metals are defined as those with little or no iron in their composition. Some of the more common nonferrous (nonmagnetic) metals include aluminum, brass, bronze, copper, lead, tin, and zinc. Section 300.20(A) addresses the problem of induction from ac conductors into ferrous (magnetic) metal enclosures and ferrous raceways. Induction into raceways and enclosures can lead to overheating and is also a shock hazard.

Δ **(B) Individual Conductors.** Where a single conductor carrying alternating current passes through metal with magnetic properties, the inductive effect shall be minimized by either cutting slots in the metal between the individual holes through which the individual conductors pass or passing all the conductors in the circuit through an insulating wall sufficiently large for all of the conductors of the circuit.

Exception: In the case of circuits supplying vacuum or electric-discharge lighting systems or signs or X-ray apparatus, the currents carried by the conductors are so small that the inductive heating effect can be ignored where these conductors are placed in metal enclosures or pass through metal.

Informational Note: Because aluminum is not a magnetic metal, there will be no heating due to hysteresis; however, induced currents will be present. They will not be of sufficient magnitude to require grouping of conductors or special treatment in passing conductors through aluminum wall sections.

300.21 Spread of Fire or Products of Combustion. Electrical installations in hollow spaces, vertical shafts, and ventilation or air-handling ducts shall be made so that the possible spread of fire or products of combustion will not be substantially increased. Openings around electrical penetrations into or through fire-resistant-rated walls, partitions, floors, or ceilings shall be firestopped using approved methods to maintain the fire resistance rating.

Informational Note: Directories of electrical construction materials published by qualified testing laboratories contain many listing installation restrictions necessary to maintain the fire-resistive rating of assemblies where penetrations or openings are made. Building codes also contain restrictions on membrane penetrations on opposite sides of a fire-resistance-rated wall assembly. An example is the 600-mm (24-in.) minimum horizontal separation that usually applies between boxes installed on opposite sides of the wall. Assistance in complying with the requirements of 300.21 can be found in building codes, fire resistance directories, and product listings.

Cables, cable trays, and raceways must be installed through fire-rated walls, floors, and ceiling assemblies using an approved firestop method so they do not contribute to the spread of fire or the products of combustion. In the UL *Guide Information for Electrical Equipment,* which can be found at productspec.ul.com, Category XHEZ covers through-penetration firestop systems, and Category XHJI covers firestop devices. These two category sections provide valuable information concerning application, installation, and use of firestop systems and firestop devices. A firestop system, the seals for which are shown in Exhibit 300.16, meets the requirements of 300.21.

Using the proper protection techniques is crucial in limiting or stopping flame, excessive temperature, and smoke from passing through fire-rated construction. Sleeves for the protection of cables passing through fire-rated construction must be firestopped to the original rating of the assembly both around the sleeve and around the cables inside the sleeve.

The structural integrity of the floor or wall assembly needs to be evaluated when openings for the penetrating items are being provided. The rating of the building assembly being penetrated by electrical cables or conduits must also be determined. This information is available from construction documents or building codes. Once the fire resistance rating has been determined, the properties and types of electrical penetration required [whether metal or nonmetallic cable(s), conduit(s), or even cable trays] must be determined. Next, the selected method of firestopping must be approved by the AHJ. Building codes generally

require that a firestop system or device be tested. ASTM E814-2013, *Standard Test Method for Fire Tests of Penetration Firestop Systems,* and ANSI/UL 1479-2015, *Fire Tests of Penetration Firestops,* are two examples of such test methods. Certain construction practices can be used in lieu of tested systems or devices. NFPA 221, *Standard for High Challenge Fire Walls, Fire Walls, and Fire Barrier Walls,* as well as some building codes, contains a statement somewhat similar to the following:

Where concrete, grout, or mortar has been used to fill the annular spaces around . . . steel conduit or tubing that penetrates one or more concrete or masonry walls, the nominal diameter of each penetrating item shall not exceed 6 in., the opening size shall not exceed 144 in.^{2}, and the thickness of the concrete, grout, or mortar shall be the full thickness of the assembly.

300.22 Wiring in Ducts Not Used for Air Handling, Fabricated Ducts for Environmental Air, and Other Spaces for Environmental Air (Plenums). The requirements of this section shall apply to the installation and uses of electrical wiring and equipment in ducts used for dust, loose stock, or vapor removal; ducts specifically fabricated for environmental air; and other spaces used for environmental air (plenums).

Informational Note: See Part VI of Article 424 for requirements on duct heaters.

Other codes and standards refer to all spaces that move air as plenums, including spaces that the *NEC* identifies as "other spaces." Some of the terms used to describe this space include *plenum,* *ceiling cavity plenum,* and *raised floor plenum.* Products are often required to be marked as "suitable for use in other spaces for environmental air" or equivalent language to comply with requirements in the *NEC.*

Section 300.22(B) addresses ducts and spaces used solely for the movement of environmental air. In 300.22(C), the term *plenum* appears with *other spaces used for environmental air* to make it clear that it applies to structures that are not fabricated specifically to handle environmental air as the primary purpose but that handle the air, nonetheless.

(A) Ducts for Dust, Loose Stock, or Vapor Removal. No wiring systems of any type shall be installed in ducts used to transport dust, loose stock, or flammable vapors. No wiring system of any type shall be installed in any duct, or shaft containing only such ducts, used for vapor removal or for ventilation of commercial-type cooking equipment.

(B) Ducts Specifically Fabricated for Environmental Air. Equipment, devices, and the wiring methods specified in this section shall be permitted within such ducts only if necessary for the direct action upon, or sensing of, the contained air. Where equipment or devices are installed and illumination is necessary to facilitate maintenance and repair, enclosed gasketed-type luminaires shall be permitted.

Only wiring methods consisting of Type MI cable without an overall nonmetallic covering, Type MC cable employing a smooth or corrugated impervious metal sheath without an overall

EXHIBIT 300.16 Fire seals used in a through-penetration firestop system to maintain the fire resistance rating of the wall. *(Courtesy of O-Z/Gedney, a division of EGS Electrical Group)*

nonmetallic covering, electrical metallic tubing, flexible metallic tubing, intermediate metal conduit, or rigid metal conduit without an overall nonmetallic covering shall be installed in ducts specifically fabricated to transport environmental air. Flexible metal conduit shall be permitted, in lengths not to exceed 1.2 m (4 ft), to connect physically adjustable equipment and devices permitted to be in these fabricated ducts. The connectors used with flexible metal conduit shall effectively close any openings in the connection.

Exception: Wiring methods and cabling systems, listed for use in other spaces used for environmental air (plenums), shall be permitted to be installed in ducts specifically fabricated for environmental air-handling purposes under both of the following conditions:

(1) The wiring methods or cabling systems shall be permitted only if necessary to connect to equipment or devices associated with the direct action upon or sensing of the contained air.

(2) The total length of such wiring methods or cabling systems shall not exceed 1.2 m (4 ft).

The use of wiring methods within ducts is limited to minimize the contribution of smoke and products of combustion during a fire in an area that handles environmental air.

Δ **(C) Other Spaces Used for Environmental Air (Plenums).** This section shall apply to spaces not specifically fabricated for environmental air-handling purposes but used for air-handling purposes as a plenum. This section shall not apply to habitable rooms or areas of buildings, the prime purpose of which is not air handling.

> Informational Note No. 1: The space over a hung ceiling used for environmental air-handling purposes is an example of the type of other space to which this section applies.
>
> Informational Note No. 2: See NFPA 90A, *Standard for the Installation of Air-Conditioning and Ventilating Systems*, and other mechanical codes for information on how the term *other spaces used for environmental air (plenum)*, as used in this section, correlates with the use of the term *plenum* where the plenum is used for return air purposes, as well as some other air-handling spaces.

Other spaces or plenums — such as the space or cavity between a structural floor or roof and a suspended (hung) ceiling — are used to transport environmental air and are not specifically manufactured as ducts. Many spaces above suspended ceilings are intended to transport return air. Informational Note No. 2 correlates the phrase *other spaces used for environmental air* with the term *plenum* as used in NFPA 90A, *Standard for the Installation of Air-Conditioning and Ventilating Systems.*

Exhibit 300.17 illustrates the distinction between a habitable room and one used for air handling. If the prime purpose of the room or space is air handling, the restrictions in 300.22(C) apply.

Exception: This section shall not apply to the joist or stud spaces of dwelling units where the wiring passes through such spaces perpendicular to the long dimension of such spaces.

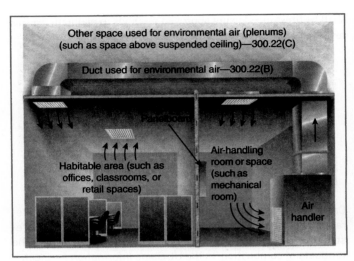

EXHIBIT 300.17 An example of spaces used for environmental air.

This exception permits cable to pass through joist or stud spaces of a dwelling unit where the joist space is used as a return for a forced-air central heating or air-conditioning system. As shown in Exhibit 300.18, the joist space is covered with appropriate material, and the cable passes through the space perpendicular to the vertical run.

Δ **(1) Wiring Methods.** The wiring methods for other spaces used for environmental air shall be limited to totally enclosed, nonventilated, insulated busway having no provisions for plug-in connections, Type MI cable without an overall nonmetallic covering, Type MC cable without an overall nonmetallic covering, Type AC cable, or other factory-assembled multiconductor control or power cable that is specifically listed for use within an air-handling space, or listed prefabricated cable assemblies of metallic manufactured wiring systems without nonmetallic sheath. Other types of cables, conductors, and raceways shall be permitted to be installed in electrical metallic tubing, flexible metallic tubing, intermediate metal conduit, rigid metal conduit

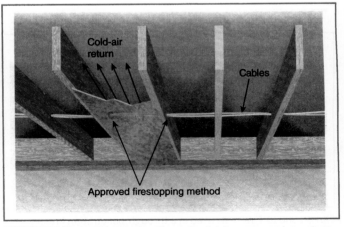

EXHIBIT 300.18 A cable passing through a joist space used as return air for a forced-air system in a dwelling unit.

without an overall nonmetallic covering, flexible metal conduit, or, where accessible, surface metal raceway or metal wireway with metal covers.

Nonmetallic cable ties and other nonmetallic cable accessories used to secure and support cables shall be listed as having low smoke and heat release properties.

> Informational Note: See UL 2043, *Fire Test for Heat and Visible Smoke Release for Discrete Products and Their Accessories Installed in Air-Handling Spaces*, for one method of testing low smoke and heat release properties for nonmetallic cable ties and other nonmetallic cable accessories to determine a maximum peak optical density of 0.50 or less, an average optical density of 0.15 or less, and a peak heat release rate of 100 kW or less.

Δ **(2) Cable Tray Systems.** The requirements in 300.22(C)(2)(a) or (C)(2)(b) shall apply to the use of metallic cable tray systems in other spaces used for environmental air (plenums), where accessible.

(a) *Metal Cable Tray Systems.* Metal cable tray systems shall be permitted to support the wiring methods specified in 300.22(C)(1).

(b) *Solid Side and Bottom Metal Cable Tray Systems.* Solid side and bottom metal cable tray systems with solid metal covers shall be permitted to enclose wiring methods and cables not already covered in 300.22(C)(1) in accordance with 392.10(A) and (B).

Δ **(3) Equipment.** Electrical equipment with a metal enclosure, or electrical equipment with a nonmetallic enclosure listed for use within an air-handling space and having low smoke and heat release properties, and associated wiring material suitable for the ambient temperature shall be permitted to be installed in such other spaces unless prohibited elsewhere in this *Code*.

> Informational Note: See UL 2043, *Fire Test for Heat and Visible Smoke Release for Discrete Products and Their Accessories Installed in Air-Handling Spaces*, for one method of testing low smoke and heat release properties to determine that the equipment exhibits a maximum peak optical density of 0.50 or less, an average optical density of 0.15 or less, and a peak heat release rate of 100 kW or less.

Exception: Integral fan systems shall be permitted where specifically identified for use within an air-handling space.

(D) Information Technology Equipment. Where the installation complies with the special requirements specified in 645.4, electrical wiring in air-handling areas beneath raised floors for information technology equipment shall be permitted in accordance with 645.5(E).

Δ **300.23 Panels Designed to Allow Access.** Cables, raceways, and equipment installed behind panels designed to allow access, including suspended ceiling panels, shall be arranged and secured to allow the removal of panels and access to the equipment.

300.25 Exit Enclosures (Stair Towers). Where an exit enclosure is required to have a fire resistance rating, only electrical wiring methods serving equipment permitted by the authority having jurisdiction in the exit enclosure shall be installed within the exit enclosure.

Exception: Where egress lighting is required on outside exterior doorways from the exit enclosure, luminaires shall be permitted to be supplied from the inside of the exit enclosure.

> Informational Note: See NFPA *101*-2021, *Life Safety Code*, 7.1.3.2.1(10)(b), for more information.

NFPA *101*®, *Life Safety Code*® requires certain stairways, those defined as "exit enclosures," to be separated from the building by fire-rated walls or other means. A critical function of these stairways is to provide safe passage out of the building in the event of an emergency. It is the intent of the requirements in NFPA *101* to limit the materials contained within an exit enclosure to only those needed to serve the intended function of the exit enclosure.

N **300.26 Remote-Control and Signaling Circuits Classification.** Remote-control and signaling circuits shall be classified as either power-limited or non-power-limited and comply with the following:

(1) Class 1 power-limited remote-control and signaling circuits shall comply with 724.3.
(2) Class 2 and Class 3 power-limited remote-control and signaling circuits shall comply with 725.3.
(3) Non-power-limited remote-control and signaling circuits shall be installed in accordance with 300.2 through 300.25.

ARTICLE

N **305** General Requirements for Wiring Methods and Materials for Systems Rated Over 1000 Volts ac, 1500 Volts dc, Nominal

---

N **305.1 Scope.** This article covers wiring methods and materials for systems rated over 1000 volts ac, 1500 volts dc, nominal.

N **305.3 Other Articles.** Conductors shall be permitted to be installed in accordance with any of the wiring methods identified in Table 305.3.

Exposed runs of Type MV cables, bare conductors, and bare busbars shall be permitted in locations accessible only to qualified persons. Busbars shall be permitted to be either copper or aluminum.

Exception: Airfield lighting cable used in series circuits that are powered by regulators and installed in restricted airport lighting vaults shall be permitted as exposed cable installations.

> Informational Note: An example of a common application is FAA L-824 cables installed as exposed runs within a restricted vault area.

N **TABLE 305.3** *Wiring Methods Permitted for Use in Systems Rated Over 1000 Volts ac, 1500 Volts dc, Nominal*

Wiring Methods Permitted for Use Above 1000 Volts ac, 1500 Volts dc	Voltage Levels	Reference
Pull and junction boxes, conduit bodies, and handhole enclosures	Over 1000	Article 314, Part IV
Metal-clad cable (Type MC)	1000–35,000	Article 330
Type IM cable	1000–2000	Article 337
Intermediate metal conduit (IMC)	Over 1000	Article 342
Rigid metal conduit (RMC)	Over 1000	Article 344
Rigid polyvinyl chloride conduit (PVC)	Over 1000	Article 352
Reinforced thermosetting resin conduit (RTRC)	Over 1000	Article 355
Electrical metallic tubing (EMT)	Over 1000	Article 358
Auxiliary gutters	Over 1000	Article 366
Busway	Over 1000	Article 368, Part IV
Cablebus	1000–35,000	Article 370
Cable trays	1000–35,000	Article 392
Messenger-supported wiring	1000–35,000	Article 396
Outdoor overhead conductors	Over 1000	Article 395
Insulated bus pipe (IBP)	1000–35,000 ac	Article 369

N **305.4 Conductors of Different Systems.** Conductors of circuits rated over 1000 volts ac, 1500 volts dc, nominal, shall not occupy the same equipment wiring enclosure, cable, or raceway with conductors of circuits rated 1000 volts ac, 1500 volts dc, nominal, or less unless otherwise permitted as follows:

(1) Where contained within the individual wiring enclosure, primary leads of electric-discharge lamp ballasts insulated for the primary voltage of the ballast shall be permitted to occupy the same luminaire, sign, or outline lighting enclosure as the branch-circuit conductors.

(2) Excitation, control, relay, and ammeter conductors used in connection with any individual motor or starter shall be permitted to occupy the same enclosure as the motor-circuit conductors.

(3) Conductors of different voltage ratings shall be permitted in motors, transformers, switchgear, switchboards, control assemblies, and similar equipment.

(4) If the conductors of each system in a manhole are permanently and effectively separated from the conductors of the other systems and securely fastened to racks, insulators, or other approved supports, conductors of different voltage ratings shall be permitted.

Conductors having nonshielded insulation and operating at different voltage levels shall not occupy the same enclosure, cable, or raceway.

305.5 Conductor Bending Radius. The conductor shall not be bent to a radius less than 8 times the overall diameter for nonshielded conductors or 12 times the overall diameter for shielded or lead-covered conductors during or after installation. For multiconductor or multiplexed single-conductor cables having individually shielded conductors, the minimum bending radius shall be 12 times the diameter of the individually shielded conductors or 7 times the overall diameter, whichever is greater.

Δ **305.6 Protection Against Induction Heating.** Metallic raceways and associated conductors shall be arranged to avoid heating of the raceway in accordance with 300.20.

305.7 Covers Required. Suitable covers shall be installed on all boxes, fittings, and similar enclosures to prevent accidental contact with energized parts or physical damage to parts or insulation.

305.8 Raceways in Wet Locations Above Grade. Where raceways are installed in wet locations above grade, the interior of these raceways shall be considered to be a wet location. Insulated conductors and cables installed in raceways in wet locations above grade shall be either moisture-impervious metal-sheathed or of a type listed for use in wet locations.

305.9 Braid-Covered Insulated Conductors — Exposed Installation. Exposed runs of braid-covered insulated conductors shall have a flame-retardant braid. If the conductors used do not have this protection, a flame-retardant saturant shall be applied to the braid covering after installation. This treated braid covering shall be stripped back a safe distance at conductor terminals, according to the operating voltage. Where practicable, this distance shall not be less than 25 mm (1 in.) for each kilovolt of the conductor-to-ground voltage of the circuit.

305.10 Insulation Shielding. Metallic and semiconducting insulation shielding components of shielded cables shall be removed for a distance dependent on the circuit voltage and insulation. Stress reduction means shall be provided at all terminations of factory-applied shielding.

Metallic shielding components such as tapes, wires, or braids, or combinations of them, shall be connected to an equipment grounding conductor, an equipment grounding busbar, or a grounding electrode.

305.11 Moisture or Mechanical Protection for Metal-Sheathed Cables. Where cable conductors emerge from a metal sheath and where protection against moisture or physical damage is necessary, the insulation of the conductors shall be protected by a cable sheath terminating device.

305.12 Danger Signs. Danger signs shall be conspicuously posted at points of access to conductors in all raceway systems and cable systems. The sign(s) shall meet the requirements in 110.21(B), shall be readily visible, and shall state the following:

DANGER—HIGH VOLTAGE—KEEP OUT

305.15 Underground Installations.

(A) General. Underground conductors shall be identified for the voltage and conditions under which they are installed. Conductors used for direct-burial applications shall be of a type identified for such use. Underground cables shall be installed in accordance with 305.15(A)(1), (A)(2), or (A)(3), and the installation shall meet the depth requirements of Table 305.15(A).

Prior to backfilling a ditch or trench, a warning ribbon must be placed near underground direct-buried conductors over 1000 volts, nominal, that are not encased in concrete, in accordance with Table 305.15(A), Footnote 1. This requirement is intended to reduce the risk of an accident, electrocution, or arc-flash incident during excavation.

Δ **(1) Shielded Cables and Nonshielded Cables in Metal-Sheathed Cable Assemblies.** Underground cables, including nonshielded, Type MC and moisture-impervious metal sheath cables, shall have those sheaths grounded through an effective grounding path meeting the requirements of 250.4(A)(5) or 250.4(B)(4). They shall be direct buried or installed in raceways identified for the use.

(2) Industrial Establishments. In industrial establishments, where conditions of maintenance and supervision ensure that only qualified persons service the installed cable, nonshielded single-conductor cables with insulation types up to 2000 volts that are listed for direct burial shall be permitted to be directly buried.

(3) Other Nonshielded Cables. Other nonshielded cables not covered in 305.15(A)(1) or (A)(2) shall be installed in rigid metal conduit, intermediate metal conduit, or rigid nonmetallic conduit encased in not less than 75 mm (3 in.) of concrete.

(B) Wet Locations. The interior of enclosures or raceways installed underground shall be considered to be a wet location. Insulated conductors and cables installed in these enclosures or raceways in underground installations shall be listed for use in wet locations and shall be either moisture-impervious metal-sheathed or of a type listed for use in wet locations. Any

Δ **TABLE 305.15(A)** *Minimum Cover Requirements*

	General Conditions (not otherwise specified)						Special Conditions (use if applicable)					
	Column 1		Column 2		Column 3		Column 4		Column 5		Column 6	
	Direct-Buried Cables[1]		Electrical Metallic Tubing, RTRC, PVC, and HDPE Conduit[2]		Rigid Metal Conduit and Intermediate Metal Conduit		Raceways Under Buildings or Exterior Concrete Slabs, 100 mm (4 in.) Minimum Thickness[3]		Cables in Airport Runways or Adjacent Areas Where Trespass Is Prohibited		Areas Subject to Vehicular Traffic, Such as Thoroughfares and Commercial Parking Areas	
Circuit Voltage	mm	in.	mm	in.	mm	in.	mm	in.	mm	in.	mm	in.
Over 1000 V ac, 1500 V dc, through 22 kV	750	30	450	18	150	6	100	4	450	18	600	24
Over 22 kV through 40 kV	900	36	600	24	150	6	100	4	450	18	600	24
Over 40 kV	1000	42	750	30	150	6	100	4	450	18	600	24

Notes:

1. Cover shall be defined as the shortest distance in millimeters (inches) measured between a point on the top surface of any direct-buried conductor, cable, conduit, or other raceway and the top surface of finished grade, concrete, or similar cover.

2. Lesser depths shall be permitted where cables and conductors rise for terminations or splices or where access is otherwise required.

3. Where solid rock prevents compliance with the cover depths specified in this table, the wiring shall be installed in a metal or nonmetallic raceway permitted for direct burial. The raceways shall be covered by a minimum of 50 mm (2 in.) of concrete extending down to rock.

4. In industrial establishments, where conditions of maintenance and supervision ensure that qualified persons will service the installation, the minimum cover requirements for other than rigid metal conduit and intermediate metal conduit shall be permitted to be reduced 150 mm (6 in.) for each 50 mm (2 in.) of concrete or equivalent placed entirely within the trench over the underground installation.

[1]Underground direct-buried cables that are not encased or protected by concrete and are buried 750 mm (30 in.) or more below grade shall have their location identified by a warning ribbon that is placed in the trench at least 300 mm (12 in.) above the cables.

[2]Listed by a qualified testing agency as suitable for direct burial without encasement. All other nonmetallic systems shall require 50 mm (2 in.) of concrete or equivalent above conduit in addition to the table depth.

[3]The slab shall extend a minimum of 150 mm (6 in.) beyond the underground installation, and a warning ribbon or other effective means suitable for the conditions shall be placed above the underground installation.

connections or splices in an underground installation shall be approved for wet locations.

The inside of all raceways and enclosures installed underground is classified as a wet location. Conductors installed in such underground locations must be listed for use in wet locations and comply with 310.10(C).

(C) Protection from Damage. Conductors emerging from the ground shall be enclosed in listed raceways. Raceways installed on poles shall be of rigid metal conduit, intermediate metal conduit, RTRC-XW, Schedule 80 PVC conduit, or equivalent, extending from the minimum cover depth specified in Table 305.15(A) to a point 2.5 m (8 ft) above finished grade. Conductors entering a building shall be protected by an approved enclosure or raceway from the minimum cover depth to the point of entrance. Where direct-buried conductors, raceways, or cables are subject to movement by settlement or frost, they shall be installed to prevent damage to the enclosed conductors or to the equipment connected to the raceways. Metallic enclosures shall be grounded.

(D) Splices. Direct burial cables shall be permitted to be spliced or tapped without the use of splice boxes if they are installed using materials suitable for the application. The taps and splices shall be watertight and protected from mechanical damage. Where cables are shielded, the shielding shall be continuous across the splice or tap.

Exception: At splices of an engineered cabling system, metallic shields of direct-buried single-conductor cables with maintained spacing between phases shall be permitted to be interrupted and overlapped. Where shields are interrupted and overlapped, each shield section shall be grounded at one point.

(E) Backfill. Backfill containing large rocks, paving materials, cinders, large or sharply angular substances, or corrosive materials shall not be placed in an excavation where materials can damage or contribute to the corrosion of raceways, cables, or other substructures or where it might prevent adequate compaction of fill.

Protection in the form of granular or selected material or suitable sleeves shall be provided to prevent physical damage to the raceway or cable.

△ **(F) Raceway Seal.** Where a raceway enters from an underground system, the end within the building shall be sealed with an identified compound to prevent the entrance of moisture.

Informational Note: Presence of hazardous gases or vapors might also necessitate sealing of underground conduits or raceways entering buildings.

ARTICLE
310 Conductors for General Wiring

Conductors rated over 2000 volts are now located in Article 315, Medium Voltage Conductors, Cable, Cable Joints, and Cable Terminations.

Part I. General

△ **310.1 Scope.** This article covers general requirements for conductors rated up to and including 2000 volts and their type designations, insulations, markings, mechanical strengths, ampacity ratings, and uses. These requirements do not apply to conductors that form an integral part of equipment, such as motors, motor controllers, and similar equipment, or to conductors specifically provided for elsewhere in this *Code*.

310.3 Conductors.

(A) Minimum Size of Conductors. The minimum size of conductors for voltage ratings up to and including 2000 volts shall be 14 AWG copper or 12 AWG aluminum or copper-clad aluminum, except as permitted elsewhere in this *Code*.

△ **(B) Conductor Material.** Conductors in this article shall be of copper, aluminum, or copper-clad aluminum, unless otherwise specified. Aluminum and copper-clad aluminum shall comply with the following:

(1) Solid aluminum conductors 8, 10, and 12 AWG shall be made of an AA-8000 series electrical grade aluminum alloy conductor material.

(2) Stranded aluminum conductors 8 AWG through 1000 kcmil marked as Type RHH, RHW, XHHW, XHHN, XHWN, THW, THHW, THWN, THHN, service-entrance Type SE Style U, and SE Style R shall be made of an AA-8000 series electrical grade aluminum alloy conductor material.

(3) For copper-clad aluminum conductors, the copper shall form a minimum 10 percent of the cross-sectional area of a solid conductor or each strand of a stranded conductor. The aluminum core of a copper-clad aluminum conductor shall be made of an AA-8000 series electrical grade aluminum alloy conductor material.

(4) Copper-clad aluminum conductor material shall be listed.

This section correlates with the UL listing requirements for testing terminations — such as copper-aluminum, revised (CO/ALR) devices and other connectors — suitable for use with aluminum conductors. The electrical industry has developed AA-8000 series aluminum alloy materials and the connectors suitable for use with aluminum conductors to provide for safe and stable connections. Connections suitable for use with aluminum conductors are also generally listed as suitable for use with copper conductors and are marked accordingly, such as AL7CU or

AL9CU. Numbers 7 and 9 identify the temperature ratings of 75°C and 90°C, respectively, for these terminations.

Copper-clad aluminum conductors are manufactured so that copper forms a minimum of 10 percent of the cross-sectional area of a solid conductor or of each strand of a stranded conductor.

(C) Stranded Conductors. Where installed in raceways, conductors 8 AWG and larger shall be stranded, unless specifically permitted or required elsewhere in this *Code* to be solid.

Large-size conductors are required to be stranded for greater flexibility. This requirement does not apply to conductors outside of raceways, such as busbars and the conductors of Type MI metal-sheathed cable. Special applications elsewhere in the *NEC*® may require or permit different requirements for stranded conductors. For example, the bonding conductors of a permanently installed swimming pool are required to be solid copper conductors of 8 AWG or larger, according to 680.26(B).

(D) Insulated. Conductors not specifically permitted elsewhere in this *Code* to be covered or bare shall be insulated.

Informational Note: See 250.184 for insulation of neutral conductors of a solidly grounded high-voltage system.

Part II. Construction Specifications

310.4 Conductor Constructions and Applications. Insulated conductors shall comply with Table 310.4(1) and Table 310.4(2).

Informational Note: Thermoplastic insulation may stiffen at temperatures lower than −10°C (+14°F). Thermoplastic insulation may also be deformed at normal temperatures where subjected to pressure, such as at points of support.

Table 310.4(1) includes conductor applications and maximum operating temperatures for insulations rated 600 volts. Some conductors with dual ratings are listed for dry, damp, and wet locations. Type XHHW is rated 90°C for dry and damp locations and 75°C for wet locations; Type THW is rated 75°C for dry and wet locations and 90°C for special applications within electric-discharge lighting equipment. Types RHW-2, XHHW-2, and other types identified by the suffix "2" are rated 90°C for dry and wet locations.

Additional detailed wire classification information for sizes 14 AWG through 2000 kcmil is available in standards and directories such as those published by Underwriters Laboratories Inc.

Δ **TABLE 310.4(1)** *Conductor Applications and Insulations Rated 600 Volts*

Trade Name	Type Letter	Maximum Operating Temperature	Application Provisions	Insulation	Thickness of Insulation AWG or kcmil	mm	mils	Outer Covering[1]
Fluorinated ethylene propylene	FEP or FEPB	90°C (194°F)	Dry and damp locations	Fluorinated ethylene propylene	14–10 / 8–2	0.51 / 0.76	20 / 30	None
		200°C (392°F)	Dry locations — special applications[2]	Fluorinated ethylene propylene	14–8	0.36	14	Glass braid
					6–2	0.36	14	Glass or other suitable braid material
Mineral insulation (metal sheathed)	MI	90°C (194°F)	Dry and wet locations	Magnesium oxide	18–16[3]	0.58	23	Copper or alloy steel
		250°C (482°F)	For special applications[2]		16–10	0.91	36	
					9–4	1.27	50	
					3–500	1.40	55	

Trade Name	Type Letter	Maximum Operating Temperature	Application Provisions	Insulation	AWG or kcmil	mm (A)	mm (B)	mils (A)	mils (B)	Outer Covering[1]
Moisture-, heat-, and oil-resistant thermoplastic	MTW	60°C (140°F)	Machine tool wiring in wet locations	Flame-retardant, moisture-, heat-, and oil-resistant thermoplastic						(A) None (B) Nylon jacket or equivalent
		90°C (194°F)	Machine tool wiring in dry locations.		22–12	0.76	0.38	30	15	
					10	0.76	0.51	30	20	
			Informational Note: See NFPA 79-2021, *Electrical Standard for Industrial Machinery.*		8	1.14	0.76	45	30	
					6	1.52	0.76	60	30	
					4–2	1.52	1.02	60	40	
					1–4/0	2.03	1.27	80	50	
					213–500	2.41	1.52	95	60	
					501–1000	2.79	1.78	110	70	
Paper		85°C (185°F)	For underground service conductors, or by special permission	Paper						Lead sheath

(continues)

Δ **TABLE 310.4(1)** *Continued*

Trade Name	Type Letter	Maximum Operating Temperature	Application Provisions	Insulation	Thickness of Insulation			Outer Covering[1]
					AWG or kcmil	mm	mils	
Perfluoro-alkoxy	PFA	90°C (194°F)	Dry and damp locations	Perfluoro-alkoxy	14–10	0.51	20	None
					8–2	0.76	30	
		200°C (392°F)	Dry locations — special applications[2]		1–4/0	1.14	45	
Perfluoro-alkoxy	PFAH	250°C (482°F)	Dry locations only. Only for leads within apparatus or within raceways connected to apparatus (nickel or nickel-coated copper only)	Perfluoro-alkoxy	14–10	0.51	20	None
					8–2	0.76	30	
					1–4/0	1.14	45	
Thermoset	RHH	90°C (194°F)	Dry and damp locations		14–10	1.14	45	Moisture-resistant, flame-retardant, nonmetallic covering[1]
					8–2	1.52	60	
					1–4/0	2.03	80	
					213–500	2.41	95	
					501–1000	2.79	110	
					1001–2000	3.18	125	
Moisture-resistant thermoset	RHW	75°C (167°F)	Dry and wet locations	Flame-retardant, moisture-resistant thermoset	14–10	1.14	45	Moisture-resistant, flame-retardant, nonmetallic covering
					8–2	1.52	60	
					1–4/0	2.03	80	
	RHW-2	90°C (194°F)			213–500	2.41	95	
					501–1000	2.79	110	
					1001–2000	3.18	125	
Silicone	SA	90°C (194°F)	Dry and damp locations	Silicone rubber	14–10	1.14	45	Glass or other suitable braid material
					8–2	1.52	60	
					1–4/0	2.03	80	
		200°C (392°F)	For special application[2]		213–500	2.41	95	
					501–1000	2.79	110	
					1001–2000	3.18	125	
Thermoset	SIS	90°C (194°F)	Switchboard and switchgear wiring only	Flame-retardant thermoset	14–10	0.76	30	None
					8–2	1.14	45	
					1–4/0	1.40	55	
Thermoplastic and fibrous outer braid	TBS	90°C (194°F)	Switchboard and switchgear wiring only	Thermoplastic	14–10	0.76	30	Flame-retardant, nonmetallic covering
					8	1.14	45	
					6–2	1.52	60	
					1–4/0	2.03	80	
Extended polytetra-fluoro-ethylene	TFE	250°C (482°F)	Dry locations only. Only for leads within apparatus or within raceways connected to apparatus, or as open wiring (nickel or nickel-coated copper only)	Extruded polytetra-fluoroethylene	14–10	0.51	20	None
					8–2	0.76	30	
					1–4/0	1.14	45	
Heat-resistant thermoplastic	THHN	90°C (194°F)	Dry and damp locations	Flame-retardant, heat-resistant thermoplastic	14–12	0.38	15	Nylon jacket or equivalent
					10	0.51	20	
					8–6	0.76	30	
					4–2	1.02	40	
					1–4/0	1.27	50	
					250–500	1.52	60	
					501–1000	1.78	70	

Δ **TABLE 310.4(1)** *Continued*

Trade Name	Type Letter	Maximum Operating Temperature	Application Provisions	Insulation	Thickness of Insulation			Outer Covering[1]
					AWG or kcmil	mm	mils	
Moisture- and heat-resistant thermoplastic	THHW	75°C (167°F)	Wet location	Flame-retardant, moisture- and heat-resistant thermoplastic	14–10	0.76	30	None
					8	1.14	45	
					6–2	1.52	60	
		90°C (194°F)	Dry location		1–4/0	2.03	80	
					213–500	2.41	95	
					501–1000	2.79	110	
					1001–2000	3.18	125	
Moisture- and heat-resistant thermoplastic	THW	75°C (167°F)	Dry and wet locations	Flame-retardant, moisture- and heat-resistant thermoplastic	14–10	0.76	30	None
		90°C (194°F)	Special applications within electric discharge lighting equipment. Limited to 1000 open-circuit volts or less. (Size 14-8 only as permitted in 410.68.)		8	1.14	45	
					6–2	1.52	60	
					1–4/0	2.03	80	
					213–500	2.41	95	
					501–1000	2.79	110	
					1001–2000	3.18	125	
	THW-2	90°C (194°F)	Dry and wet locations					
Moisture- and heat-resistant thermoplastic	THWN	75°C (167°F)	Dry and wet locations	Flame-retardant, moisture- and heat-resistant thermoplastic	14–12	0.38	15	Nylon jacket or equivalent
					10	0.51	20	
					8–6	0.76	30	
					4–2	1.02	40	
	THWN-2	90°C (194°F)			1–4/0	1.27	50	
					250–500	1.52	60	
					501–1000	1.78	70	
Moisture-resistant thermoplastic	TW	60°C (140°F)	Dry and wet locations	Flame-retardant, moisture-resistant thermoplastic	14–10	0.76	30	None
					8	1.14	45	
					6–2	1.52	60	
					1–4/0	2.03	80	
					213–500	2.41	95	
					501–1000	2.79	110	
					1001–2000	3.18	125	
Underground feeder and branch-circuit cable — single conductor (for Type UF cable employing more than one conductor, see Part II of Article 340).	UF	60°C (140°F)	See Part II of Article 340.	Moisture-resistant	14–10	1.52	60[5]	Integral with insulation
					8–2	2.03	80[5]	
					1–4/0	2.41	95[5]	
		75°C (167°F)[4]		Moisture- and heat-resistant				
Underground service-entrance cable — single conductor (for Type USE cable employing more than one conductor, see Part II of Article 338).	USE	75°C (167°F)[4]	See Part II of Article 338.	Heat- and moisture-resistant	14–10	1.14	45	Moisture-resistant nonmetallic covering (See 338.2.)
					8–2	1.52	60	
					1–4/0	2.03	80	
					213–500	2.41	95[6]	
					501–1000	2.79	110	
					1001–2000	3.18	125	
	USE-2	90°C (194°F)	Dry and wet locations					

(continues)

Δ **TABLE 310.4(1)** *Continued*

Trade Name	Type Letter	Maximum Operating Temperature	Application Provisions	Insulation	Thickness of Insulation			Outer Covering[1]
					AWG or kcmil	mm	mils	
Thermoset	XHH	90°C (194°F)	Dry and damp locations	Flame-retardant thermoset	14–10	0.76	30	None
					8–2	1.14	45	
					1–4/0	1.40	55	
					213–500	1.65	65	
					501–1000	2.03	80	
					1001–2000	2.41	95	
Thermoset	XHHN	90°C (194°F)	Dry and damp locations	Flame-retardant thermoset	14–12	0.38	15	Nylon jacket or equivalent
					10	0.51	20	
					8–6	0.76	30	
					4–2	1.02	40	
					1–4/0	1.27	50	
					250–500	1.52	60	
					501–1000	1.78	70	
Moisture-resistant thermoset	XHHW	90°C (194°F)	Dry and damp locations	Flame-retardant, moisture-resistant thermoset	14–10	0.76	30	None
					8–2	1.14	45	
		75°C (167°F)	Wet locations		1–4/0	1.40	55	
					213–500	1.65	65	
					501–1000	2.03	80	
					1001–2000	2.41	95	
Moisture-resistant thermoset	XHHW-2	90°C (194°F)	Dry and wet locations	Flame-retardant, moisture-resistant thermoset	14–10	0.76	30	None
					8–2	1.14	45	
					1–4/0	1.40	55	
					213–500	1.65	65	
					501–1000	2.03	80	
					1001–2000	2.41	95	
Moisture-resistant thermoset	XHWN	75°C (167°F)	Dry and wet locations	Flame-retardant, moisture-resistant thermoset	14–12	0.38	15	Nylon jacket or equivalent
					10	0.51	20	
					8–6	0.76	30	
	XHWN-2	90°C (194°F)			4–2	1.02	40	
					1–4/0	1.27	50	
					250–500	1.52	60	
					501–1000	1.78	70	
Modified ethylene tetrafluoro-ethylene	Z	90°C (194°F)	Dry and damp locations	Modified ethylene tetrafluoro-ethylene	14–12	0.38	15	None
					10	0.51	20	
		150°C (302°F)	Dry locations — special applications[2]		8–4	0.64	25	
					3–1	0.89	35	
					1/0–4/0	1.14	45	
Modified ethylene tetrafluoro-ethylene	ZW	75°C (167°F)	Wet locations	Modified ethylene tetrafluoro-ethylene	14–10	0.76	30	None
		90°C (194°F)	Dry and damp locations		8–2	1.14	45	
		150°C (302°F)	Dry locations — special applications[2]					
	ZW-2	90°C (194°F)	Dry and wet locations					

Note: Conductors in Table 310.4(1) shall be permitted to be rated up to 1000 volts if listed and marked.

[1]Outer coverings shall not be required where listed without a covering.

[2]Higher temperature rated constructions shall be permitted where design conditions require maximum conductor operating temperatures above 90°C (194°F).

[3]Conductor sizes shall be permitted for signaling circuits permitting 300-volt insulation.

[4]The ampacity of Type UF cable shall be limited in accordance with 340.80.

[5]Type UF insulation thickness shall include the integral jacket.

[6]Insulation thickness shall be permitted to be 2.03 mm (80 mils) for listed Type USE conductors that have been subjected to special investigations. The nonmetallic covering over individual rubber-covered conductors of aluminum-sheathed cable and of lead-sheathed or multiconductor cable shall not be required to be flame retardant.

TABLE 310.4(2) *Thickness of Insulation for Nonshielded Types RHH and RHW Solid Dielectric Insulated Conductors Rated 2000 Volts*

Conductor Size (AWG or kcmil)	Column A[1]		Column B[2]	
	mm	mils	mm	mils
14–10	2.03	80	1.52	60
8	2.03	80	1.78	70
6–2	2.41	95	1.78	70
1–2/0	2.79	110	2.29	90
3/0–4/0	2.79	110	2.29	90
213–500	3.18	125	2.67	105
501–1000	3.56	140	3.05	120
1001–2000	3.56	140	3.56	140

[1]Column A insulations shall be limited to natural, SBR, and butyl rubbers.

[2]Column B insulations shall be materials such as cross-linked polyethylene, ethylene propylene rubber, and composites thereof.

310.6 Conductor Identification.

(A) Grounded Conductors. Insulated or covered grounded conductors shall be identified in accordance with 200.6.

(B) Equipment Grounding Conductors. Equipment grounding conductors shall be identified in accordance with 250.119.

Δ **(C) Ungrounded Conductors.** Conductors that are intended for use as ungrounded conductors, whether used as a single conductor or in multiconductor cables, shall be finished to be clearly distinguishable from grounded conductors and equipment grounding conductors. Distinguishing markings shall not conflict in any manner with the surface markings required by 310.8(B)(1). Branch-circuit ungrounded conductors shall be identified in accordance with 210.5(C). Feeders shall be identified in accordance with 215.12(C).

Exception: Conductor identification shall be permitted in accordance with 200.7.

Ungrounded conductors with white or gray insulation are permitted if the conductors are permanently reidentified at termination points and are visible and accessible. Common methods of reidentification include colored tape, tagging, or paint. Other applications where white conductors are permitted include flexible cords and circuits less than 50 volts. A white conductor used in single-pole, 3-way and 4-way switch loops also requires reidentification (a color other than white, gray, or green) if it is used as an ungrounded conductor.

See also

200.7(C)(2) for further information about reidentification of grounded conductors

310.8 Marking.

(A) Required Information. All conductors and cables shall be marked to indicate the following information, using the applicable method described in 310.8(B):

(1) The maximum rated voltage.
(2) The proper type letter or letters for the type of wire or cable as specified elsewhere in this *Code*.
(3) The manufacturer's name, trademark, or other distinctive marking by which the organization responsible for the product can be readily identified.
(4) The AWG size or circular mil area.

Informational Note: See Chapter 9, Table 8, Conductor Properties, for conductor area expressed in SI units for conductor sizes specified in AWG or circular mil area.

(5) Cable assemblies where the neutral conductor is smaller than the ungrounded conductors shall be so marked.

(B) Method of Marking.

(1) Surface Marking. The following conductors and cables shall be durably marked on the surface:

(1) Single-conductor and multiconductor thermoset and thermoplastic-insulated wire and cable
(2) Nonmetallic-sheathed cable
(3) Service-entrance cable
(4) Underground feeder and branch-circuit cable
(5) Tray cable
(6) Irrigation cable
(7) Power-limited tray cable
(8) Instrumentation tray cable

The AWG size or circular mil area shall be repeated at intervals not exceeding 610 mm (24 in.). All other markings shall be repeated at intervals not exceeding 1.0 m (40 in.).

Δ **(2) Marker Tape.** Metal-covered multiconductor cables shall employ a marker tape located within the cable and running for its complete length.

Exception No. 1: Type MI cable shall not require a marker tape.

Exception No. 2: Type AC cable shall not require a marker tape.

Exception No. 3: The information required in 310.8(A) shall be permitted to be durably marked on the outer nonmetallic covering of Type MC, Type ITC, or Type PLTC cables at intervals not exceeding 1.0 m (40 in.).

Exception No. 4: The information required in 310.8(A) shall be permitted to be durably marked on a nonmetallic covering under the metallic sheath of Type ITC or Type PLTC cable at intervals not exceeding 1.0 m (40 in.).

Informational Note: Included in the group of metal-covered cables are Type AC cable, Type MC cable, and lead-sheathed cable.

(3) Tag Marking. The following conductors and cables shall be marked by means of a printed tag attached to the coil, reel, or carton:

(1) Type MI cable
(2) Switchboard wires
(3) Metal-covered, single-conductor cables
(4) Type AC cable

(4) Optional Marking of Wire Size. The information required in 310.8(A)(4) shall be permitted to be marked on the surface of the individual insulated conductors for the following multiconductor cables:

(1) Type MC cable
(2) Tray cable
(3) Irrigation cable
(4) Power-limited tray cable
(5) Power-limited fire alarm cable
(6) Instrumentation tray cable

(C) Suffixes to Designate Number of Conductors. A type letter or letters used alone shall indicate a single insulated conductor. The letter suffixes shall be indicated as follows:

(1) D — For two insulated conductors laid parallel within an outer nonmetallic covering
(2) M — For an assembly of two or more insulated conductors twisted spirally within an outer nonmetallic covering

(D) Optional Markings. All conductors and cables contained in Chapter 3 shall be permitted to be surface marked to indicate special characteristics of the cable materials. These markings include, but are not limited to, markings for limited smoke, sunlight resistant, and so forth.

Cable insulations that have special characteristics are permitted to carry surface markings that indicate their characteristics. For example, limited-smoke cables are permitted to be marked "LS" or "ST1." Other characteristics permitted to be marked include sunlight resistance and low corrosiveness. For a detailed list of optional wire and cable marking, see the UL *Wire Marking Guide*.

Part III. Installation

310.10 Uses Permitted. The conductors described in 310.4 shall be permitted for use in any of the wiring methods covered in Chapter 3 and as specified in their respective tables or as permitted elsewhere in this *Code*.

(A) Dry Locations. Insulated conductors and cables used in dry locations shall be any of the types identified in this *Code*.

(B) Dry and Damp Locations. Insulated conductors and cables used in dry and damp locations shall be Types FEP, FEPB, MTW, PFA, RHH, RHW, RHW-2, SA, THHN, THW, THW-2, THHW, THWN, THWN-2, TW, XHH, XHHW, XHHW-2, XHHN, XHWN, XHWN-2, Z, or ZW.

(C) Wet Locations. Insulated conductors and cables used in wet locations shall comply with one of the following:

(1) Be moisture-impervious metal-sheathed
(2) Be types MTW, RHW, RHW-2, TW, THW, THW-2, THHW, THWN, THWN-2, XHHW, XHHW-2, XHWN, XHWN-2 or ZW
(3) Be of a type listed for use in wet locations

(D) Locations Exposed to Direct Sunlight. Insulated conductors or cables used where exposed to direct rays of the sun shall comply with one of the following:

(1) Conductors and cables shall be listed as being sunlight resistant.
(2) Conductors and cables shall be covered with insulating material, such as tape or sleeving, that is listed as being sunlight resistant.

(E) Direct-Burial Conductors. Conductors used for direct-burial applications shall be of a type identified for such use.

(F) Corrosive Conditions. Conductors exposed to oils, greases, vapors, gases, fumes, liquids, or other substances having a deleterious effect on the conductor or insulation shall be of a type suitable for the application.

Nylon-jacketed conductors, such as Type THWN, are suitable for use where exposed to gasoline. The UL *Guide Information for Electrical Equipment* states in the category for Thermoplastic-Insulated Wire (ZLGR) in part:

> THWN — wire that is suitable for exposure to mineral oil, and to liquid gasoline and gasoline vapors at ordinary ambient temperature, is marked "Gasoline and Oil Resistant I" if suitable for exposure to mineral oil at 60°C, or "Gasoline and Oil Resistant II" if the compound is suitable for exposure to mineral oil at 75°C. Gasoline-resistant wire has been tested at 23°C when immersed in gasoline and is considered inherently resistant to gasoline vapors within the limits of the temperature rating of the wire type.

Before a wire-pulling compound is used, it should first be investigated to determine compliance with 310.10(F).

(G) Conductors in Parallel.

(1) General. Aluminum, copper-clad aluminum, or copper circuit conductors for each ungrounded conductor, grounded conductor, or neutral conductor shall be permitted to be connected in parallel (electrically joined at both ends) only in sizes 1/0 AWG and larger and shall be installed in accordance with 310.10(G)(2) through (G)(4).

Conductors connected in parallel are treated by the *NEC* as a single conductor with a total cross-sectional area of all conductors in parallel. The use of parallel conductors is a practical and cost-effective means of installing large-capacity feeders or services. Using conductors larger than 1000 kcmil in raceways is neither economical nor practical unless the conductor size is governed by voltage drop. The ampacity of larger sizes of conductors would increase very little in proportion to the increase in the size of the conductor. Where the cross-sectional area of a conductor increases 50 percent (e.g., from 1000 to 1500 kcmil), a Type THW conductor ampacity increases only 80 amperes (less than 15 percent). A 100-percent increase (from 1000 to 2000 kcmil) causes an increase of only 120 amperes (approximately 22 percent). Generally, where cost is a factor, installation of two (or more) paralleled conductors per phase could be beneficial.

The parallel connection of two or more conductors in place of using one large conductor depends on compliance with 310.10(G)(2) to ensure equal current division to prevent overloading any of the individual paralleled conductors.

Where individual conductors are tapped from conductors in parallel, the tap connection must include *all* of the conductors in parallel for that phase. Tapping into only one of the parallel conductors would result in unbalanced distribution of tap load current between parallel conductors, resulting in one of the conductors carrying more than its share of the load, which could cause overheating and conductor insulation failure. For example, if a 250-kcmil conductor is tapped from a set of two 500-kcmil conductors in parallel, the splicing device must include both 500-kcmil conductors and the single 250-kcmil tap conductor.

Exception No. 1: Conductors in sizes smaller than 1/0 AWG shall be permitted to be run in parallel to supply control power to indicating instruments, contactors, relays, solenoids, and similar control devices, or for frequencies of 360 Hz and higher, provided all of the following apply:

(1) They are contained within the same raceway or cable.
(2) The ampacity of each individual conductor is sufficient to carry the entire load current shared by the parallel conductors.
(3) The overcurrent protection is such that the ampacity of each individual conductor will not be exceeded if one or more of the parallel conductors become inadvertently disconnected.

Exception No. 2: Under engineering supervision, 2 AWG and 1 AWG grounded neutral conductors shall be permitted to be installed in parallel for existing installations.

Informational Note: Exception No. 2 can be used to alleviate overheating of neutral conductors in existing installations due to high content of triplen harmonic currents.

The word *triplen* refers to a third-order harmonic current, such as the third, sixth, ninth, and so on. The concern is limited to odd-number triplen harmonic currents, such as the third, ninth, and fifteenth, since they are additive currents in the neutral conductor and do not cancel.

See also

Chapter 10 of NFPA 70B, *Recommended Practice for Electrical Equipment Maintenance*, for additional information on power quality and harmonics

Δ **(2) Conductor and Installation Characteristics.** The paralleled conductors that comprise each ungrounded conductor, grounded conductor, neutral conductor, equipment grounding conductor, equipment bonding jumper, or supply-side bonding jumper shall comply with all of the following:

(1) Be the same length
(2) Consist of the same conductor material
(3) Be the same size in circular mil area
(4) Have the same insulation type
(5) Be terminated in the same manner

To avoid excessive voltage drop and to ensure equal division of current, different phase conductors must be located close together. Each phase conductor, grounded conductor, and the grounding conductor (if used) must also be grouped together in each raceway or cable. However, isolated phase installations are permitted underground where the phase conductors are run in nonmetallic raceways that are in close proximity.

All conductors of the same phase or neutral are required by 310.10(G)(2) to be of the same conductor material. For example, if 12 conductors are paralleled for a 3-phase, 4-wire, 480Y/277-volt ac circuit, four conductors could be installed in each of three raceways. The *NEC* does not intend that all 12 conductors be copper or aluminum but does intend that the individual conductors in parallel for each phase, grounded conductor, and neutral be the same material, insulation type, length, and so forth. For example, the conductors in phases A and B might be copper, and those in phase C might be aluminum. Also, the three raceways are intended to have the same physical characteristics (e.g., three rigid aluminum conduits, three steel IMCs, three EMTs, or three nonmetallic conduits), not a mixture (e.g., two rigid aluminum conduits and one rigid steel conduit).

See also

300.3(B)(1) for more information on paralleled installations

(3) Separate Cables or Raceways. Where run in separate cables or raceways, the cables or raceways with conductors shall have the same number of conductors and shall have the same electrical characteristics. Conductors composing one paralleled set shall not be required to have the same physical characteristics as those of another paralleled set.

All parallel raceways or cables for a circuit are required to be of the same size, material, and length. In this case, "cables" means wiring method-type cables such as Type MC. The impedance of a circuit in an aluminum raceway or aluminum-sheathed cable differs from the impedance of the same circuit in a steel raceway or steel-sheathed cable; therefore, separate raceways and cables must have the same physical characteristics. Also, the same number of conductors must be used in each raceway or cable.

See also

300.20 regarding induced currents in metal enclosures or raceways

Δ **(4) Ampacity Correction or Adjustment.** Conductors installed in parallel shall comply with 310.15(B)and(C).

(5) Equipment Grounding Conductors. Where parallel equipment grounding conductors are used, they shall be sized in accordance with 250.122. Sectioned equipment grounding conductors smaller than 1/0 AWG shall be permitted in multiconductor cables, if the combined circular mil area of the sectioned equipment grounding conductors in each cable complies with 250.122.

(6) Bonding Jumpers. Where parallel equipment bonding jumpers or supply-side bonding jumpers are installed in raceways, they shall be sized and installed in accordance with 250.102.

The equipment bonding jumper size requirements might be different from the requirements for equipment grounding conductors (EGCs). On the supply side of the service, the size of the bonding jumper is based on 250.102(C)(1), which is the same as the requirement in 250.66 for grounding electrode conductors (GECs). On the load side of the service, the size is based on 250.122, which is also the size requirement for EGCs. The 1/0 AWG minimum size limitation on paralleled conductors does not apply to the equipment bonding jumper.

310.12 Single-Phase Dwelling Services and Feeders. For one-family dwellings and the individual dwelling units of two-family and multifamily dwellings, service and feeder conductors supplied by a single-phase, 120/240-volt system shall be permitted to be sized in accordance with 310.12(A) through (D).

For one-family dwellings and the individual dwelling units of two-family and multifamily dwellings, single-phase feeder conductors consisting of two ungrounded conductors and the neutral conductor from a 208Y/120 volt system shall be permitted to be sized in accordance with 310.12(A) through (C).

(A) Services. For a service rated 100 amperes through 400 amperes, the service conductors supplying the entire load associated with a one-family dwelling, or the service conductors supplying the entire load associated with an individual dwelling unit in a two-family or multifamily dwelling, shall be permitted to have an ampacity not less than 83 percent of the service rating. If no adjustment or correction factors are required, Table 310.12(A) shall be permitted to be applied.

(B) Feeders. For a feeder rated 100 amperes through 400 amperes, the feeder conductors supplying the entire load associated with a one-family dwelling, or the feeder conductors supplying the entire load associated with an individual dwelling

TABLE 310.12(A) *Single-Phase Dwelling Services and Feeders*

Service or Feeder Rating (Amperes)	Conductor (AWG or kcmil)	
	Copper	Aluminum or Copper-Clad Aluminum
100	4	2
110	3	1
125	2	1/0
150	1	2/0
175	1/0	3/0
200	2/0	4/0
225	3/0	250
250	4/0	300
300	250	350
350	350	500
400	400	600

Note: If no adjustment or correction factors are required, this table shall be permitted to be applied.

unit in a two-family or multifamily dwelling, shall be permitted to have an ampacity not less than 83 percent of the feeder rating. If no adjustment or correction factors are required, Table 310.12(A) shall be permitted to be applied.

(C) Feeder Ampacities. In no case shall a feeder for an individual dwelling unit be required to have an ampacity greater than that specified in 310.12(A) or (B).

Δ **(D) Grounded Conductors.** Grounded conductors shall be permitted to be sized smaller than the ungrounded conductors, if the requirements of 220.61 and 230.42 for service conductors or the requirements of 215.2 and 220.61 for feeder conductors are met.

Where correction or adjustment factors are required by 310.15(B) or (C), they shall be permitted to be applied to the ampacity associated with the temperature rating of the conductor.

Informational Note No. 1: See 240.6(A) for standard ampere ratings for fuses and inverse time circuit breakers.
Informational Note No. 2: See Informative Annex D, Example D7.

The main service or feeder to a dwelling unit is permitted to be sized at 83 percent of the disconnect rating. The calculation is not based on the rating of the overcurrent device protecting the main feeder. The minimum disconnect rating for a dwelling unit is 100 amperes according to 225.39 and 230.79. This calculation applies only to conductors carrying 100 percent of the dwelling unit's diversified load.

If a 120/240-volt single-phase service supplies a one-family dwelling or an individual unit of a two-family or multifamily dwelling, the reduced conductor size is applicable to the service-entrance conductors or feeder conductors that supply the dwelling unit. The feeder conductors to a dwelling unit are not required to be larger than its service-entrance conductors.

Exhibit 310.1 illustrates where Table 310.12(A) could be applied. The reduced conductor size permitted is applicable only to the service-entrance conductors run to each apartment from

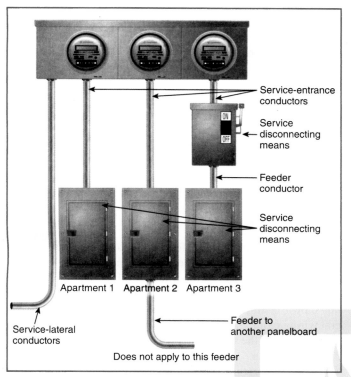

EXHIBIT 310.1 One application where the reduced conductor size is applicable to the service-entrance conductors.

the meters. The reduced conductor size permitted is also applicable to the feeder conductors run to Apartment 3 in the exhibit from the service disconnecting means because that feeder carries the entire load.

310.14 Ampacities for Conductors Rated 0 Volts – 2000 Volts.

(A) General.

Δ **(1) Tables or Engineering Supervision.** Ampacities for conductors shall be permitted to be determined by tables as provided in 310.15 or under engineering supervision as provided in 310.14(B).

> Informational Note No. 1: See 210.19, Informational Note, for voltage drop on branch circuits that this section does not take into consideration. See 215.2(A)(2), Informational Note No. 2, for voltage drop on feeders that this section does not take into consideration.
>
> Informational Note No. 2: See NFPA 79-2021, *Electrical Standard for Industrial Machinery*, Table 12.5.1, for the allowable ampacities of Type MTW wire.

(2) Selection of Ampacity. Where more than one ampacity applies for a given circuit length, the lowest value shall be used.

Exception: Where different ampacities apply to portions of a circuit, the higher ampacity shall be permitted to be used if the total portion(s) of the circuit with lower ampacity does not exceed the lesser of 3.0 m (10 ft) or 10 percent of the total circuit.

Informational Note: See 110.14(C) for conductor temperature limitations due to termination provisions.

Calculation Example

Three 500-kcmil THW conductors in RMC) are run from a motor control center for 12 ft past a heat-treating furnace to a pump motor located 150 ft from the motor control center. Where run in a 78°F to 86°F ambient temperature, the conductors have an ampacity of 380 A, per Table 310.16. The ambient temperature near the furnace, where the conduit is run, is found to be 113°F, and the length of this part of the run is greater than 10 ft and more than 10 percent of the total length of the run at the 78°F to 86°F ambient temperature. Determine the ampacity of the total run, in accordance with 310.14(A)(2).

Solution
Using the ambient temperature correction factors in Table 310.15(B)(1)(1) for 113°F, calculate the ampacity:

$$0.82 \times 380 \text{ A} = 311.6 \text{ A}$$

which is the ampacity of the total run, in accordance with 310.14(A)(2).

If the run near the furnace at the 113°F ambient temperature was 10 ft or less in length, then the ampacity of the entire run would have been 380 A, according to the exception to 310.14(A)(2). The heat-sinking effect of the run at the lower ambient temperature is sufficient to reduce the conductor temperature near the furnace.

---

(3) Temperature Limitation of Conductors. No conductor shall be used in such a manner that its operating temperature exceeds that designated for the type of insulated conductor involved. In no case shall conductors be associated together in such a way, with respect to type of circuit, the wiring method employed, or the number of conductors, that the limiting temperature of any conductor is exceeded.

Most terminations are designed for 60°C or 75°C maximum temperatures. The higher-rated ampacities for conductors of 90°C, 105°C, and so forth cannot be used unless the terminals at which the conductors terminate have comparable ratings.

Ambient temperature must be considered in determining the allowable ampacity of conductors. Conductors should have a rating above the anticipated maximum ambient temperature. The operating temperature of conductors should be controlled at or below the conductor rating by coordinating conductor size, number of associated conductors, and ampacity for the conductor rating and ambient temperature. Tables 310.16 through 310.20 have ampacities based on a 30°C or 40°C ambient temperature, as indicated in the table heading. Where the ambient temperature is different, Table 310.15(B)(1)(1) or Table 310.15(B)(1)(2) is used to correct the ampacity. If more than three conductors are installed without spacing to allow the adequate dissipation of heat, the additional adjustment shown in 310.15(C)(1) must also be applied.

Calculation Example

Determine the ampacity of 2 AWG THHN copper conductors to be installed in a raceway in an ambient temperature of 50°C (122°F).

Solution
Table 310.16 shows that the allowable ampacity of the conductor at 30°C is 130 A, which is multiplied by 0.82 [the ambient temperature correction factor in Table 310.15(B)(1)(1)].

$$130\ A \times 0.82 = 106.6\ A$$

Thus, the allowable ampacity of the 2 AWG conductor at 50°C is reduced to 106.6 A. For six of the conductors run in the raceway, 310.15(C)(1) requires the allowable ampacity to be further reduced to 80 percent:

$$106.6\ A \times 0.8 = 85.28\ A$$

Under these conditions, the 2 AWG conductors would be suitable for an 80-A circuit, based on the standard ampere ratings of circuit breaker and fuses in 240.6(A).

The basis for determining the ampacities of conductors for Tables 310.16 and 310.17 was the NEMA *Report of Determination of Maximum Permissible Current-Carrying Capacity of Code Insulated Wires and Cables for Building Purposes*, dated June 27, 1938. The basis for determining the ampacities of conductors for Tables 310.18 and 310.19 and the ampacity tables in Informative Annex B was the Neher–McGrath method.

Informational Note No. 1: See Table 310.4(1) and Table 315.10(A) for the temperature rating of a conductor that is the maximum temperature, at any location along its length, that the conductor can withstand over a prolonged time period without serious degradation. The ampacity tables of Article 310 and the ampacity tables of Informative Annex B, the ambient temperature correction factors in 310.15(B), and the notes to the tables provide guidance for coordinating conductor sizes, types, ampacities, ambient temperatures, and number of associated conductors. The principal determinants of operating temperature are as follows:

(1) Ambient temperature — ambient temperature may vary along the conductor length as well as from time to time.
(2) Heat generated internally in the conductor as the result of load current flow, including fundamental and harmonic currents.
(3) The rate at which generated heat dissipates into the ambient medium. Thermal insulation that covers or surrounds conductors affects the rate of heat dissipation.
(4) Adjacent load-carrying conductors — adjacent conductors have the dual effect of raising the ambient temperature and impeding heat dissipation.

Informational Note No. 2: Refer to 110.14(C) for the temperature limitation of terminations.

(B) Engineering Supervision. Under engineering supervision, conductor ampacities shall be permitted to be calculated by means of Equation 310.14(B).

$$I = \sqrt{\dfrac{T_c - T_a}{R_{dc}(1 + Y_c)R_{ca}}} \times 10^3 \text{ amperes} \quad \text{[310.14(B)]}$$

where:

T_c = conductor temperature in degrees Celsius (°C)
T_a = ambient temperature in degrees Celsius (°C)
R_{dc} = dc resistance of 305 mm (1 ft) of conductor in microohms at temperature, T_c
Y_c = component ac resistance resulting from skin effect and proximity effect
R_{ca} = effective thermal resistance between conductor and surrounding ambient

310.15 Ampacity Tables.

Δ **(A) General.** Ampacities for conductors rated 0 volts to 2000 volts shall be as specified in the Ampacity Table 310.16 through Table 310.21, as modified by 310.15(A) through (F) and 310.12. Under engineering supervision, ampacities of sizes not shown in ampacity tables for conductors meeting the general wiring requirements shall be permitted to be determined by interpolation of the adjacent conductors based on the conductor's circular-mil area.

The temperature correction and adjustment factors shall be permitted to be applied to the ampacity for the temperature rating of the conductor, if the corrected and adjusted ampacity does not exceed the ampacity for the temperature rating of the termination in accordance with 110.14(C).

Informational Note No. 1: Table 310.16 through Table 310.19 are application tables for use in determining conductor sizes on loads calculated in accordance with Part II, Part III, Part IV, or Part V of Article 220. Ampacities result from consideration of one or more of the following:

(1) Temperature compatibility with connected equipment, especially the connection points
(2) Coordination with circuit and system overcurrent protection
(3) Compliance with the requirements of product listings or certifications
(4) Preservation of the safety benefits of established industry practices and standardized procedures

Informational Note No. 2: See Chapter 9, Table 8, Conductor Properties, for conductor area. Interpolation is based on the conductor circular-mil area and not the conductor overall area.

Informational Note No. 3: See 400.5 for the ampacities of flexible cords and cables. See 402.5 for the ampacities of fixture wires.

Informational Note No. 4: See Table 310.4(1) and Table 310.4(2) for explanation of type letters used in tables and for recognized sizes of conductors for the various conductor insulations. See 310.1 through 310.14 and the various articles of this *Code* for installation requirements. See Table 400.4, Table 400.5(A)(1), and Table 400.5(A)(2) for flexible cords.

Ampacity tables, particularly Table 310.16, do not consider all factors affecting ampacity. However, experience has proven the table values to be adequate for loads calculated in accordance with Article 220 because not all the load diversity occurring in most installations is specifically provided for in Article 220. If loads are not calculated in accordance with the requirements of Article 220, the table ampacities, even if corrected in accordance with ambient correction factors and the notes to the tables, might be higher than needed for the actual load.

See also

310.14(B) and **Informative Annex B** for more information

(B) Ambient Temperature Correction Factors.

(1) General. Ampacities for ambient temperatures other than those shown in the ampacity tables shall be corrected in accordance with Table 310.15(B)(1)(1) or Table 310.15(B)(1)(2), or shall be permitted to be calculated using Equation 310.15(B)(1).

$$I' = I\sqrt{\frac{T_c - T_a'}{T_c - T_a}} \qquad [310.15(B)(1)]$$

where:

I' = ampacity corrected for ambient temperature
I = ampacity shown in the tables
T_c = temperature rating of conductor (°C)
T_a' = new ambient temperature (°C)
T_a = ambient temperature used in the table (°C)

Δ **(2) Rooftop.** For raceways or cables exposed to direct sunlight on or above rooftops where the distance above the roof to the bottom of the raceway or cable is less than 19 mm (¾ in.), a temperature adder of 33°C (60°F) shall be added to the outdoor temperature to determine the applicable ambient temperature for application of the correction factors in Table 310.15(B)(1)(1) or Table 310.15(B)(1)(2).

Exception: Type XHHW-2 insulated conductors shall not be subject to this ampacity adjustment.

> Informational Note: The *ASHRAE Handbook — Fundamentals* is one source for the ambient temperatures in various locations.

Conductors in outdoor conduits or cables installed less than ¾ inches above the surface of the roof are subject to a significant increase in temperature when the roof is exposed to direct sunlight.

Calculation Example

A feeder installed in IMC runs across the top of a commercial building in St. Louis, Missouri (MO). The calculated load on the feeder is 175 A. The lateral portion of the raceway is exposed to sunlight and secured to supports elevated less than ¾ in. above the finished rooftop surface. Determine the minimum size circuit conductor using aluminum 90°C THWN-2 insulation, taking into consideration only the exposure to sunlight. None of the loads are continuous, and the neutral is not considered a current-carrying conductor. The design temperature is based on the averaged June, July, and August 2-percent design temperature from the 2009 *ASHRAE Handbook — Fundamentals*.

Solution

Step 1. Determine the ambient temperature (compensated for proximity of conduit to the rooftop exposure to sunlight):

a. Compensated ambient temperature = design temperature + 60°F as required by 310.15(B)(2)

TABLE 310.15(B)(1)(1) *Ambient Temperature Correction Factors Based on 30°C (86°F)*

For ambient temperatures other than 30°C (86°F), multiply the ampacities specified in the ampacity tables by the appropriate correction factor shown below.

Ambient Temperature (°C)	Temperature Rating of Conductor			Ambient Temperature (°F)
	60°C	75°C	90°C	
10 or less	1.29	1.20	1.15	50 or less
11–15	1.22	1.15	1.12	51–59
16–20	1.15	1.11	1.08	60–68
21–25	1.08	1.05	1.04	69–77
26–30	1.00	1.00	1.00	78–86
31–35	0.91	0.94	0.96	87–95
36–40	0.82	0.88	0.91	96–104
41–45	0.71	0.82	0.87	105–113
46–50	0.58	0.75	0.82	114–122
51–55	0.41	0.67	0.76	123–131
56–60	—	0.58	0.71	132–140
61–65	—	0.47	0.65	141–149
66–70	—	0.33	0.58	150–158
71–75	—	—	0.50	159–167
76–80	—	—	0.41	168–176
81–85	—	—	0.29	177–185

Note: Table 310.15(B)(1)(1) shall be used with Table 310.16 and Table 310.17 as required.

b. Design temperature for St. Louis area = 95°F
c. Temperature adjustment for a raceway elevated less than ¾ in. above rooftop = 60°F
d. Compensated ambient temperature: 95°F + 60°F = 155°F

Step 2. Determine the temperature correction factor for this application from Table 310.15(B)(1)(1). Using the 90°C column and the temperature correction factor row for 155°F, the temperature correction factor is 0.58.

Step 3. Determine the proper conductor size to supply the 175-A load.

a. Because the load is calculated at 175 A noncontinuous, and the neutral conductor is not considered to be a current-carrying conductor, the conductor ampacity is calculated as follows:

$$175 \text{ A} \div 0.58 = 302 \text{ A}$$

b. Select a conductor not less than 302 A from the aluminum 90°C column of Table 310.16:

400 kcmil aluminum THWN-2

c. Verify that the conductor ampacity at 75°C is sufficient for the calculated load to comply with terminal temperature requirements of 110.14(C): The 75°C aluminum column of Table 310.16 ampacity equals 270 A, which is greater than the 175-A calculated load.

Based on a very specific set of circumstances, this example demonstrates that roughly a 42-percent loss of usable conductor ampacity occurred. This loss was due solely to high ambient heat present where the raceway was subjected to sunlight and installed within ¾ in. to the rooftop.

(C) Adjustment Factors.

Δ **(1) More than Three Current-Carrying Conductors.** The ampacity of each conductor shall be reduced as shown in Table 310.15(C)(1) where the number of current-carrying conductors in a raceway or cable exceeds three, or where single conductors or multiconductor cables not installed in raceways are installed without maintaining spacing for a continuous length longer than 600 mm (24 in.). Each current-carrying conductor of a paralleled set of conductors shall be counted as a current-carrying conductor.

TABLE 310.15(B)(1)(2) *Ambient Temperature Correction Factors Based on 40°C (104°F)*

For ambient temperatures other than 40°C (104°F), multiply the ampacities specified in the ampacity tables by the appropriate correction factor shown below.

Ambient Temperature (°C)	Temperature Rating of Conductor						Ambient Temperature (°F)
	60°C	75°C	90°C	150°C	200°C	250°C	
10 or less	1.58	1.36	1.26	1.13	1.09	1.07	50 or less
11–15	1.50	1.31	1.22	1.11	1.08	1.06	51–59
16–20	1.41	1.25	1.18	1.09	1.06	1.05	60–68
21–25	1.32	1.2	1.14	1.07	1.05	1.04	69–77
26–30	1.22	1.13	1.10	1.04	1.03	1.02	78–86
31–35	1.12	1.07	1.05	1.02	1.02	1.01	87–95
36–40	1.00	1.00	1.00	1.00	1.00	1.00	96–104
41–45	0.87	0.93	0.95	0.98	0.98	0.99	105–113
46–50	0.71	0.85	0.89	0.95	0.97	0.98	114–122
51–55	0.50	0.76	0.84	0.93	0.95	0.96	123–131
56–60	—	0.65	0.77	0.90	0.94	0.95	132–140
61–65	—	0.53	0.71	0.88	0.92	0.94	141–149
66–70	—	0.38	0.63	0.85	0.90	0.93	150–158
71–75	—	—	0.55	0.83	0.88	0.91	159–167
76–80	—	—	0.45	0.80	0.87	0.90	168–176
81–90	—	—	—	0.74	0.83	0.87	177–194
91–100	—	—	—	0.67	0.79	0.85	195–212
101–110	—	—	—	0.60	0.75	0.82	213–230
111–120	—	—	—	0.52	0.71	0.79	231–248
121–130	—	—	—	0.43	0.66	0.76	249–266
131–140	—	—	—	0.30	0.61	0.72	267–284
141–160	—	—	—	—	0.50	0.65	285–320
161–180	—	—	—	—	0.35	0.58	321–356
181–200	—	—	—	—	—	0.49	357–392
201–225	—	—	—	—	—	0.35	393–437

Note: Table 310.15(B)(1)(2) shall be used with Table 310.18, Table 310.19, Table 310.20, and Table 310.21 as required.

Where conductors of different systems, as provided in 300.3, are installed in a common raceway or cable, the adjustment factors shown in Table 310.15(C)(1) shall apply only to the number of power and lighting conductors.

The factors in Table 310.15(C)(1) are based on no diversity, meaning that all conductors in the raceway or cable are loaded to their maximum rated load. The table values can be used even if there is diversity because the table values would represent a worst-case scenario. If there is diversity and a more favorable adjustment factor is desired, a calculation can be performed in accordance with Informative Annex B.

The basis for the last paragraph of 310.15(C)(1) is the assumption that the watt loss (heating) from any control and signal conductors in the same raceway or cable will not be enough to significantly increase the temperature of the power and lighting conductors.

See also

724.48, 725.136, and **725.139** for limitations on the installation of control and signal conductors in the same raceway or cable as power and lighting conductors

300.17, Informational Note, for specific cross references for raceway fill and adjustment factors of 310.15(C)

724.51 for fill requirements and adjustment factors for Class 1 conductors

760.51 and **760.130** for fire alarm systems

770.110 for optical fiber cables and raceways

800.110 for communications wires and cables within buildings

Informational Note No. 1: See Informative Annex B for adjustment factors for more than three current-carrying conductors in a raceway or cable with load diversity.

Informational Note No. 2: See 366.23 for adjustment factors for conductors and ampacity for bare copper and aluminum bars in auxiliary gutters and 376.22(B) for adjustment factors for conductors in metal wireways.

The conditions under which derating factors do not apply are specified in 310.15(C)(1)(c). Exhibit 310.2 illustrates the conditions specified in 310.15(C)(1)(c).

TABLE 310.15(C)(1) *Adjustment Factors for More Than Three Current-Carrying Conductors*

Number of Conductors*	Percent of Values in Table 310.16 Through Table 310.19 as Adjusted for Ambient Temperature if Necessary
4–6	80
7–9	70
10–20	50
21–30	45
31–40	40
41 and above	35

*Number of conductors is the total number of conductors in the raceway or cable, including spare conductors. The count shall be adjusted in accordance with 310.15(E) and (F). The count shall not include conductors that are connected to electrical components that cannot be simultaneously energized.

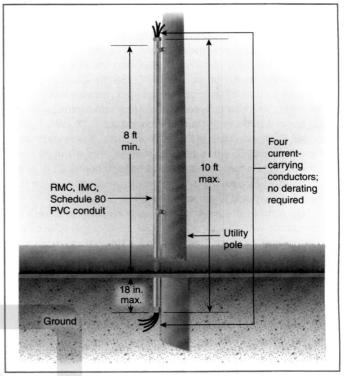

EXHIBIT 310.2 Conditions under which derating factors do not apply.

(a) Where conductors are installed in cable trays, 392.80 shall apply.

(b) Adjustment factors shall not apply to conductors in raceways having a length not exceeding 600 mm (24 in.).

(c) Adjustment factors shall not apply to underground conductors entering or leaving an outdoor trench if those conductors have physical protection in the form of rigid metal conduit, intermediate metal conduit, rigid polyvinyl chloride conduit (PVC), or reinforced thermosetting resin conduit (RTRC) having a length not exceeding 3.05 m (10 ft), and if the number of conductors does not exceed four.

(d) Adjustment factors shall not apply to Type AC cable or to Type MC cable under the following conditions:

(1) The cables do not have an overall outer jacket.
(2) Each cable has not more than three current-carrying conductors.
(3) The conductors are 12 AWG copper.
(4) Not more than 20 current-carrying conductors are installed without maintaining spacing, are stacked, or are supported on "bridle rings."

Exception to (4): If cables meeting the requirements in 310.15(C)(1)(d)(1) through (C)(1)(d)(3) with more than 20 current-carrying conductors are installed longer than 600 mm (24 in.) without maintaining spacing, are stacked, or are supported on bridle rings, a 60 percent adjustment factor shall be applied.

Calculation Example

A commercial office space requires fourteen 277-V fluorescent lighting circuits to serve the office area. The office area lighting is assumed to be a continuous load, and the office ambient temperature does not exceed 30°C (86°F). Each circuit is arranged so that it has a calculated load not exceeding 16 A. The wiring method is Type MC cable, 3-conductor (with an additional equipment grounding conductor), 12 AWG THHN copper. Each individual MC cable contains a 3-wire multiwire branch circuit. To serve the entire area, this arrangement requires a total of seven cables bundled together for 25 ft, without maintaining spacing between them where they leave the electrical room and enter the office area.

Determine the ampacity of each circuit conductor in accordance with 310.15, applying 310.15(C)(1)(d) to account for the bundled cables. Then determine the maximum permitted branch-circuit overcurrent protection.

Solution

Step 1. Determine the quantity of current-carrying conductors. According to 310.15(F), equipment grounding conductors (EGCs) are not counted as current-carrying conductors. Fluorescent lighting is considered a nonlinear load, and, according to 310.15(E)(3), the grounded conductor of each Type MC cable must be counted as a current-carrying conductor:

7 cables × 3 conductors each = 21 current-carrying conductors

Step 2. Determine the ampacity of each current-carrying conductor. Because the quantity of current-carrying conductors being bundled exceeds 20, a 60-percent adjustment factor is required by 310.15(C)(1)(c), Exception to (4). From Table 310.16:

12 AWG THHN = 30 A
30 A × 0.60 = 18 A

Because the actual calculated load is 16 A of continuous load, 210.19(A)(1) is applicable. The conductors must have an ampacity equal to or greater than the load before the adjustment factor is applied. Because the ampacity of the conductors after the adjustment factor is applied is 18 A, the conductors are suitable for this installation.

Step 3. Finally, determine the maximum size of the overcurrent protective device (OCPD) permitted for these bundled cables. Section 240.4(B) permits the use of the next higher standard rating of OCPD. Therefore, although the conductors have a calculated ampacity of 18 A, a 20-A OCPD is permitted. In addition, and of significance, the 20-A OCPD complies with 210.20(A), given that the actual 16-A continuous load would require a 20-A OCPD.

(2) Raceway Spacing. Spacing between raceways shall be maintained.

(D) Bare or Covered Conductors. Where bare or covered conductors are installed with insulated conductors, the temperature rating of the bare or covered conductor shall be equal to the lowest temperature rating of the insulated conductors for the purpose of determining ampacity.

Δ **(E) Neutral Conductor.** Neutral conductors shall be considered current carrying in accordance with any of the following:

(1) A neutral conductor that carries only the unbalanced current from other conductors of the same circuit shall not be required to be counted when applying the provisions of 310.15(C)(1).

(2) In a 3-wire circuit consisting of two phase conductors and the neutral conductor of a 4-wire, 3-phase, wye-connected system, the neutral conductor carries approximately the same current as the line-to-neutral load currents of the other conductors and shall be counted when applying 310.15(C)(1).

(3) On a 4-wire, 3-phase wye circuit where the major portion of the load consists of nonlinear loads, harmonic currents are present in the neutral conductor; the neutral conductor shall therefore be considered a current-carrying conductor.

Nonlinear loads on 3-phase circuits can cause an increase in neutral conductor current. Exhibit 310.3 shows an example of a portable diagnostic analyzer used for more sophisticated power measurements, including measuring harmonic distortion.

(F) Grounding or Bonding Conductor. A grounding or bonding conductor shall not be counted when applying the provisions of 310.15(C)(1).

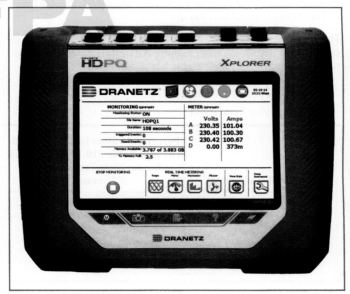

EXHIBIT 310.3 A portable tool for tasks such as diagnostic power analysis, including harmonic distortion, where accurate measurement and thorough circuit analysis is desirable. *(Courtesy of Dranetz)*

310.16 Ampacities of Insulated Conductors in Raceway, Cable, or Earth (Directly Buried). The ampacities shall be as specified in Table 310.16 where all of the following conditions apply:

(1) Conductors are rated 0 volts through 2000 volts.

(2) Conductors are rated 60°C (140°F), 75°C (167°F), or 90°C (194°F).

(3) Wiring is installed in a 30°C (86°F) ambient temperature.

(4) There are not more than three current-carrying conductors.

△ **TABLE 310.16** *Ampacities of Insulated Conductors with Not More Than Three Current-Carrying Conductors in Raceway, Cable, or Earth (Directly Buried)*

	Temperature Rating of Conductor [See Table 310.4(1)]						
	60°C (140°F)	75°C (167°F)	90°C (194°F)	60°C (140°F)	75°C (167°F)	90°C (194°F)	
Size AWG or kcmil	Types TW, UF	Types RHW, THHW, THW, THWN, XHHW, XHWN, USE, ZW	Types TBS, SA, SIS, FEP, FEPB, MI, PFA, RHH, RHW-2, THHN, THHW, THW-2, THWN-2, USE-2, XHH, XHHW, XHHW-2, XHWN, XHWN-2, XHHN, Z, ZW-2	Types TW, UF	Types RHW, THHW, THW, THWN, XHHW, XHWN, USE	Types TBS, SA, SIS, THHN, THHW, THW-2, THWN-2, RHH, RHW-2, USE-2, XHH, XHHW, XHHW-2, XHWN, XHWN-2, XHHN	Size AWG or kcmil
	COPPER			ALUMINUM OR COPPER-CLAD ALUMINUM			
18*	—	—	14	—	—	—	—
16*	—	—	18	—	—	—	—
14*	15	20	25	—	—	—	—
12*	20	25	30	15	20	25	12*
10*	30	35	40	25	30	35	10*
8	40	50	55	35	40	45	8
6	55	65	75	40	50	55	6
4	70	85	95	55	65	75	4
3	85	100	115	65	75	85	3
2	95	115	130	75	90	100	2
1	110	130	145	85	100	115	1
1/0	125	150	170	100	120	135	1/0
2/0	145	175	195	115	135	150	2/0
3/0	165	200	225	130	155	175	3/0
4/0	195	230	260	150	180	205	4/0
250	215	255	290	170	205	230	250
300	240	285	320	195	230	260	300
350	260	310	350	210	250	280	350
400	280	335	380	225	270	305	400
500	320	380	430	260	310	350	500
600	350	420	475	285	340	385	600
700	385	460	520	315	375	425	700
750	400	475	535	320	385	435	750
800	410	490	555	330	395	445	800
900	435	520	585	355	425	480	900
1000	455	545	615	375	445	500	1000
1250	495	590	665	405	485	545	1250
1500	525	625	705	435	520	585	1500
1750	545	650	735	455	545	615	1750
2000	555	665	750	470	560	630	2000

Notes:

1. Section 310.15(B) shall be referenced for ampacity correction factors where the ambient temperature is other than 30°C (86°F).

2. Section 310.15(C)(1) shall be referenced for more than three current-carrying conductors.

3. Section 310.16 shall be referenced for conditions of use.

*Section 240.4(D) shall be referenced for conductor overcurrent protection limitations, except as modified elsewhere in the *Code*.

310.17 Ampacities of Single-Insulated Conductors in Free Air. The ampacities shall be as specified in Table 310.17 where all of the following conditions apply:

(1) Conductors are rated 0 volts through 2000 volts.
(2) Conductors are rated 60°C (140°F), 75°C (167°F), or 90°C (194°F).
(3) Wiring is installed in a 30°C (86°F) ambient temperature.

310.18 Ampacities of Insulated Conductors in Raceway or Cable. The ampacities shall be as specified in Table 310.18 where all of the following conditions apply:

(1) Conductors are rated 0 volts through 2000 volts.
(2) Conductors are rated 150°C (302°F), 200°C (392°F), or 250°C (482°F).
(3) Wiring is installed in a 40°C (104°F) ambient temperature.
(4) There are not more than three current-carrying conductors.

Δ **TABLE 310.17** *Ampacities of Single-Insulated Conductors in Free Air*

Size AWG or kcmil	Temperature Rating of Conductor [See Table 310.4(1)]						Size AWG or kcmil
	60°C (140°F)	75°C (167°F)	90°C (194°F)	60°C (140°F)	75°C (167°F)	90°C (194°F)	
	Types TW, UF	Types RHW, THHW, THW, THWN, XHHW, XHWN, ZW	Types TBS, SA, SIS, FEP, FEPB, MI, PFA, RHH, RHW-2, THHN, THHW, THW-2, THWN-2, USE-2, XHH, XHHW, XHHW-2, XHWN, XHWN-2, XHHN, Z, ZW-2	Types TW, UF	Types RHW, THHW, THW, THWN, XHHW, XHWN	Types TBS, SA, SIS, THHN, THHW, THW-2, THWN-2, RHH, RHW-2, USE-2, XHH, XHHW, XHHW-2, XHWN, XHWN-2, XHHN	
	COPPER			ALUMINUM OR COPPER-CLAD ALUMINUM			
18	—	—	18	—	—	—	—
16	—	—	24	—	—	—	—
14*	25	30	35	—	—	—	—
12*	30	35	40	25	30	35	12*
10*	40	50	55	35	40	45	10*
8	60	70	80	45	55	60	8
6	80	95	105	60	75	85	6
4	105	125	140	80	100	115	4
3	120	145	165	95	115	130	3
2	140	170	190	110	135	150	2
1	165	195	220	130	155	175	1
1/0	195	230	260	150	180	205	1/0
2/0	225	265	300	175	210	235	2/0
3/0	260	310	350	200	240	270	3/0
4/0	300	360	405	235	280	315	4/0
250	340	405	455	265	315	355	250
300	375	445	500	290	350	395	300
350	420	505	570	330	395	445	350
400	455	545	615	355	425	480	400
500	515	620	700	405	485	545	500
600	575	690	780	455	545	615	600
700	630	755	850	500	595	670	700
750	655	785	885	515	620	700	750
800	680	815	920	535	645	725	800
900	730	870	980	580	700	790	900
1000	780	935	1055	625	750	845	1000
1250	890	1065	1200	710	855	965	1250
1500	980	1175	1325	795	950	1070	1500
1750	1070	1280	1445	875	1050	1185	1750
2000	1155	1385	1560	960	1150	1295	2000

Notes:

1. Section 310.15(B) shall be referenced for ampacity correction factors where the ambient temperature is other than 30°C (86°F).

2. Section 310.17 shall be referenced for conditions of use.

*Section 240.4(D) shall be referenced for conductor overcurrent protection limitations, except as modified elsewhere in the *Code.*

△ **TABLE 310.18** *Ampacities of Insulated Conductors with Not More Than Three Current-Carrying Conductors in Raceway or Cable*

	Temperature Rating of Conductor [See Table 310.4(1)]				
	150°C (302°F)	200°C (392°F)	250°C (482°F)	150°C (302°F)	
	Type Z	Types FEP, FEPB, PFA, SA	Types PFAH, TFE	Type Z	
Size AWG or kcmil	COPPER		NICKEL OR NICKEL-COATED COPPER	ALUMINUM OR COPPER-CLAD ALUMINUM	Size AWG or kcmil
14	34	36	39	—	14
12	43	45	54	30	12
10	55	60	73	44	10
8	76	83	93	57	8
6	96	110	117	75	6
4	120	125	148	94	4
3	143	152	166	109	3
2	160	171	191	124	2
1	186	197	215	145	1
1/0	215	229	244	169	1/0
2/0	251	260	273	198	2/0
3/0	288	297	308	227	3/0
4/0	332	346	361	260	4/0

Notes:

1. Section 310.15(B) shall be referenced for ampacity correction factors where the ambient temperature is other than 40°C (104°F).

2. Section 310.15(C)(1) shall be referenced for more than three current-carrying conductors.

3. Section 310.18 shall be referenced for conditions of use.

310.19 Ampacities of Single-Insulated Conductors in Free Air. The ampacities shall be as specified in Table 310.19 where all of the following conditions apply:

(1) Conductors are rated 0 volts through 2000 volts.
(2) Conductors are rated up to 250°C (482°F).
(3) Wiring is installed in a 40°C (104°F) ambient temperature.

TABLE 310.19 *Ampacities of Single-Insulated Conductors in Free Air*

	Temperature Rating of Conductor [See Table 310.4(1)]				
	150°C (302°F)	200°C (392°F)	250°C (482°F)	150°C (302°F)	
	Type Z	Types FEP, FEPB, PFA, SA	Types PFAH, TFE	Type Z	
Size AWG or kcmil	COPPER		NICKEL, OR NICKEL-COATED COPPER	ALUMINUM OR COPPER-CLAD ALUMINUM	Size AWG or kcmil
14	46	54	59	—	14
12	60	68	78	47	12
10	80	90	107	63	10
8	106	124	142	83	8
6	155	165	205	112	6
4	190	220	278	148	4
3	214	252	327	170	3
2	255	293	381	198	2
1	293	344	440	228	1
1/0	339	399	532	263	1/0
2/0	390	467	591	305	2/0
3/0	451	546	708	351	3/0
4/0	529	629	830	411	4/0

Notes:

1. Section 310.15(B) shall be referenced for ampacity correction factors where the ambient temperature is other than 40°C (104°F).

2. Section 310.19 shall be referenced for conditions of use.

310.20 Ampacities of Conductors Supported on a Messenger. The ampacities shall be as specified in Table 310.20 where all of the following conditions apply:

(1) Conductors are rated 0 volts through 2000 volts.
(2) Conductors are rated 75°C (167°F) or 90°C (194°F).
(3) Wiring is installed in a 40°C (104°F) ambient temperature.
(4) There are not more than three single-insulated conductors.

310.21 Ampacities of Bare or Covered Conductors in Free Air. The ampacities shall be as specified in Table 310.21 where all of the following conditions apply:

(1) Wind velocity is 610 mm/sec (2 ft/sec).
(2) Conductors are 80°C (176°F) total conductor temperature.
(3) Wiring is installed in a 40°C (104°F) ambient temperature.

Δ   **TABLE 310.20** *Ampacities of Conductors on a Messenger*

Size AWG or kcmil	Temperature Rating of Conductor [See Table 310.4(1)]				Size AWG or kcmil
	75°C (167°F)	90°C (194°F)	75°C (167°F)	90°C (194°F)	
	Types RHW, THHW, THW, THWN, XHHW, XHWN, ZW	Types MI, THHN, THHW, THW-2, THWN-2, RHH, RHW-2, USE-2, XHHW, XHHW-2, XHWN-2, ZW-2	Types RHW, THW, THWN, THHW, XHHW, XHWN	Types THHN, THHW, RHH, XHHW, RHW-2, XHHW-2, THW-2, THWN-2, XHWN-2, USE-2, ZW-2	
	COPPER		ALUMINUM OR COPPER-CLAD ALUMINUM		
8	57	66	44	51	8
6	76	89	59	69	6
4	101	117	78	91	4
3	118	138	92	107	3
2	135	158	106	123	2
1	158	185	123	144	1
1/0	183	214	143	167	1/0
2/0	212	247	165	193	2/0
3/0	245	287	192	224	3/0
4/0	287	335	224	262	4/0
250	320	374	251	292	250
300	359	419	282	328	300
350	397	464	312	364	350
400	430	503	339	395	400
500	496	580	392	458	500
600	553	647	440	514	600
700	610	714	488	570	700
750	638	747	512	598	750
800	660	773	532	622	800
900	704	826	572	669	900
1000	748	879	612	716	1000

Notes:

1. Section 310.15(B) shall be referenced for ampacity correction factors where the ambient temperature is other than 40°C (104°F).

2. Section 310.15(C)(1) shall be referenced for more than three current-carrying conductors.

3. Section 310.20 shall be referenced for conditions of use.

TABLE 310.21 *Ampacities of Bare or Covered Conductors in Free Air*

| Copper Conductors | | | | AAC Aluminum Conductors | | | |
| Bare | | Covered | | Bare | | Covered | |
AWG or kcmil	Amperes	AWG or kcmil	Amperes	AWG or kcmil	Amperes	AWG or kcmil	Amperes
8	98	8	103	8	76	8	80
6	124	6	130	6	96	6	101
4	155	4	163	4	121	4	127
2	209	2	219	2	163	2	171
1/0	282	1/0	297	1/0	220	1/0	231
2/0	329	2/0	344	2/0	255	2/0	268
3/0	382	3/0	401	3/0	297	3/0	312
4/0	444	4/0	466	4/0	346	4/0	364
250	494	250	519	266.8	403	266.8	423
300	556	300	584	336.4	468	336.4	492
500	773	500	812	397.5	522	397.5	548
750	1000	750	1050	477.0	588	477.0	617
1000	1193	1000	1253	556.5	650	556.5	682
—	—	—	—	636.0	709	636.0	744
—	—	—	—	795.0	819	795.0	860
—	—	—	—	954.0	920	—	—
—	—	—	—	1033.5	968	1033.5	1017
—	—	—	—	1272	1103	1272	1201
—	—	—	—	1590	1267	1590	1381
—	—	—	—	2000	1454	2000	1527

Note: Section 310.21 shall be referenced for conditions of use.

ARTICLE 312 Cabinets, Cutout Boxes, and Meter Socket Enclosures

Part I. General

312.1 Scope. This article covers the installation and construction specifications of cabinets, cutout boxes, and meter socket enclosures. It does not apply to equipment operating at over 1000 volts, except as specifically referenced elsewhere in the *Code*.

Cabinets and cutout boxes are designed with a swinging door(s) to enclose potential transformers, current transformers, switches, overcurrent devices, meters, or control equipment. Cabinets and cutout boxes are required to be of sufficient size to accommodate all devices and conductors without overcrowding or jamming. Additional space is often provided through auxiliary gutters (see Article 366).

See also

Article 100 for definitions of the terms *cabinet* and *cutout box*

The serving electric utility often has equipment specifications or service requirements beyond the *NEC®* for meter sockets, metering cabinets, and metering compartments within switchgear, switchboards, and panelboards. Consulting with the local electric utility on such requirements helps identify suitable equipment for an installation.

Δ **312.2 Damp or Wet Locations.** In damp or wet locations, surface-type enclosures within the scope of this article shall be placed or equipped so as to prevent moisture or water from entering and accumulating within the cabinet or cutout box, and shall be mounted so there is at least 6-mm (¼-in.) airspace between the enclosure and the wall or other supporting surface. Enclosures installed in wet locations shall be weatherproof. For enclosures in wet locations, raceways or cables entering above the level of uninsulated live parts shall use fittings listed for wet locations.

Exception: Nonmetallic enclosures shall be permitted to be installed without the airspace on a concrete, masonry, tile, or similar surface.

Informational Note: See 300.6 for protection against corrosion.

312.3 Position in Wall. In walls of concrete, tile, or other noncombustible material, cabinets shall be installed so that the front edge of the cabinet is not set back of the finished surface more than 6 mm (¼ in.). In walls constructed of wood or other combustible material, cabinets shall be flush with the finished surface or project therefrom.

312.4 Repairing Noncombustible Surfaces. Noncombustible surfaces that are broken or incomplete shall be repaired so there will be no gaps or open spaces greater than 3 mm (⅛ in.) at the edge of the cabinet or cutout box employing a flush-type cover.

The repair of noncombustible surfaces is not limited to plaster or drywall types of construction. This requirement for cabinets and cutout boxes is similar to the outlet box requirement found in 314.21.

312.5 Cabinets, Cutout Boxes, and Meter Socket Enclosures. Cable assemblies and insulated conductors entering enclosures within the scope of this article shall be protected from abrasion and shall comply with 312.5(A) through (C).

(A) Openings to Be Closed. Openings through which conductors enter shall be closed in an approved manner.

(B) Metal Cabinets, Cutout Boxes, and Meter Socket Enclosures. Where metal enclosures within the scope of this article are installed with messenger-supported wiring, open wiring on insulators, or concealed knob-and-tube wiring, conductors shall enter through insulating bushings or, in dry locations, through flexible tubing extending from the last insulating support and firmly secured to the enclosure.

Δ **(C) Cables.** Where cable is used, each cable shall be secured to the cabinet, cutout box, or meter socket enclosure.

The installation of several cables bunched together and run through a knockout or chase nipple is prohibited. Individual cable clamps or connectors are required to be used with only one cable per clamp or connector unless the clamp or connector is identified for more than a single cable.

Exception No. 1: Cables with entirely nonmetallic sheaths shall be permitted to enter the top of a surface-mounted enclosure through one or more nonflexible raceways not less than 450 mm (18 in.) and not more than 3.0 m (10 ft) in length, provided all of the following conditions are met:

(1) Each cable is fastened within 300 mm (12 in.), measured along the sheath, of the outer end of the raceway.

(2) The raceway extends directly above the enclosure and does not penetrate a structural ceiling.

(3) A fitting is provided on each end of the raceway to protect the cable(s) from abrasion and the fittings remain accessible after installation.

(4) The raceway is sealed or plugged at the outer end using approved means so as to prevent access to the enclosure through the raceway.

(5) The cable sheath is continuous through the raceway and extends into the enclosure beyond the fitting not less than 6 mm (¼ in.).

(6) The raceway is fastened at its outer end and at other points in accordance with the applicable article.

(7) Where installed as conduit or tubing, the cable fill does not exceed the amount that would be permitted for complete

conduit or tubing systems by Table 1 of Chapter 9 of this Code and all applicable notes thereto. Note 2 to the tables in Chapter 9 does not apply to this condition.

Informational Note: See Chapter 9, Table 1, including Note 9, for allowable cable fill in circular raceways. See 310.15(C)(1) for required ampacity reductions for multiple cables installed in a common raceway.

Exception No. 2: Single conductors and multiconductor cables shall be permitted to enter enclosures in accordance with 392.46(A) or (B).

This exception correlates with 392.46(A) and (B), which permit single or multiconductor cables entering an enclosure. The term *multiconductor cables* has been added to coordinate with the common practice of transitioning from cable trays with Type TC and Type TC-ER cables. This clarifies and correlates the permitted methods for individual conductors and cables to transition and enter equipment from cable trays.

312.6 Deflection of Conductors. Conductors at terminals or conductors entering or leaving cabinets, cutout boxes, and meter socket enclosures shall comply with 312.6(A) through (C).

Exception: Wire-bending space in enclosures for motor controllers with provisions for one or two wires per terminal shall comply with 430.10(B).

(A) Width of Wiring Gutters. Conductors shall not be deflected within a cabinet or cutout box unless a gutter having a width in accordance with Table 312.6(A) is provided. Conductors in parallel in accordance with 310.10(G) shall be judged on the basis of the number of conductors in parallel.

If Table 312.6(A) is used, bending space is measured in the direction in which the wire leaves the terminal.

Exhibit 312.1 illustrates the requirements of wiring gutter widths as specified in 312.6(A) and Table 312.6(A). The table determines the following required gutter widths, where *G* stands for the dimension of the gutter, *M* stands for the main conductors, and *BC* stands for the branch-circuit conductors:

- Gutter width G_1 is required to be 6 inches based on conductors *M*, which are parallel 4/0 conductors.
- Gutter width G_2 is required to be 1½ inches based on conductors BC_2, which are 6 AWG.
- Gutter width G_3 is required to be 3 inches based on conductors BC_3, which are 1 AWG.
- Gutter width G_4 is required to be 4 inches, based on conductor *N*, which is 4/0 AWG.

As a practical matter, the available gutter space on either side is likely to be equal. Therefore, the allowable space determined for G_3 would typically apply to G_2 as well.

(B) Wire-Bending Space at Terminals. Wire-bending space at each terminal shall be provided in accordance with 312.6(B)(1) or (B)(2).

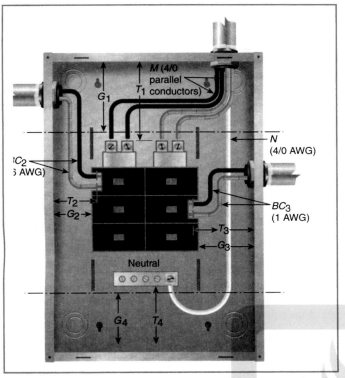

EXHIBIT 312.1 Table 312.6(A) and Table 312.6(B) applied to gutter and wire-bending space within enclosures.

(1) Conductors Not Entering or Leaving Opposite Wall. Table 312.6(A) shall apply where the conductor does not enter or leave the enclosure through the wall opposite its terminal.

(2) Conductors Entering or Leaving Opposite Wall. Table 312.6(B)(2) shall apply where the conductor does enter or leave the enclosure through the wall opposite its terminal.

Exception No. 1: Where the distance between the wall and its terminal is in accordance with Table 312.6(A), a conductor shall be permitted to enter or leave an enclosure through the wall opposite its terminal, provided the conductor enters or leaves the enclosure where the gutter joins an adjacent gutter that has a width that conforms to Table 312.6(B)(2) for the conductor.

Exception No. 2: A conductor not larger than 350 kcmil shall be permitted to enter or leave an enclosure containing only a meter socket(s) through the wall opposite its terminal, provided the distance between the terminal and the opposite wall is not less than that specified in Table 312.6(A) and the terminal is a lay-in type or removable lug with integral mounting tang, where the terminal is either of the following:

(1) Directed toward the opening in the enclosure and within a 45-degree angle of directly facing the enclosure wall

(2) Directly facing the enclosure wall and offset not greater than 50 percent of the bending space specified in Table 312.6(A)

Informational Note: *Offset* is the distance measured along the enclosure wall from the axis of the centerline of the terminal to a line passing through the center of the opening in the enclosure.

TABLE 312.6(A) *Minimum Wire-Bending Space at Terminals and Minimum Width of Wiring Gutters*

Wire Size (AWG or kcmil)		Wires per Terminal									
		1		2		3		4		5	
All Other Conductors	Compact Stranded AA-8000 Aluminum Alloy Conductors (see Note 2)	mm	in.	mm	in.	mm	in.	mm	in.	mm	in.
14–10	12–8	Not specified		—	—	—	—	—	—	—	—
8–6	6–4	38.1	1½	—	—	—	—	—	—	—	—
4–3	2–1	50.8	2	—	—	—	—	—	—	—	—
2	1/0	63.5	2½	—	—	—	—	—	—	—	—
1	2/0	76.2	3	—	—	—	—	—	—	—	—
1/0–2/0	3/0–4/0	88.9	3½	127	5	178	7	—	—	—	—
3/0–4/0	250–300	102	4	152	6	203	8	—	—	—	—
250	350	114	4½	152	6	203	8	254	10	—	—
300–350	400–500	127	5	203	8	254	10	305	12	—	—
400–500	600–750	152	6	203	8	254	10	305	12	356	14
600–700	800–1000	203	8	254	10	305	12	356	14	406	16
750–900	—	203	8	305	12	356	14	406	16	457	18
1000–1250	—	254	10	—	—	—	—	—	—	—	—
1500–2000	—	305	12	—	—	—	—	—	—	—	—

Notes:

1. Bending space at terminals shall be measured in a straight line from the end of the lug or wire connector (in the direction that the wire leaves the terminal) to the wall, barrier, or obstruction.

2. This column shall be permitted to be used to determine the minimum wire-bending space for compact stranded aluminum conductors in sizes up to 1000 kcmil and manufactured using AA-8000 series electrical grade aluminum alloy conductor material in accordance with 310.3(B). The minimum width of the wire gutter space shall be determined using the all other conductors value in this table.

Section 312.6(B)(2) and Table 312.6(B) provide the requirements for wire-bending space where straight-in wiring or offset is employed at terminals. Table 312.6(A) is used when the conductors do not enter or leave the wall opposite their terminal.

If Table 312.6(A) is used, bending space is measured in the direction in which the wire leaves the terminal. If Table 312.6(B) is used, it is measured in a direction perpendicular to the enclosure wall.

Exhibit 312.1 applies the rules of 312.6(B)(1), 312.6(B)(2), and Tables 312.6(A) and 312.6(B). The tables determine the following required gutter widths, where T stands for the wire-bending space, M stands for the main conductors, N is the neutral conductor, and BC stands for the branch-circuit conductors:

- T_1 is required to be 7½ inches based on conductors M, which are parallel 4/0 conductors.
- T_2 is required to be 1½ inches based on conductors BC_2, which are 6 AWG.
- T_3 is required to be 4½ inches, based on conductors BC_3, which are 1 AWG.
- T_4 is required to be 7 inches, based on conductor N, which is 4/0 AWG.

Table 312.6(B), Note 2, permits a reduction in required bending space for removable and lay-in wire terminals. The removable terminal wire connectors can be either the compression type or the set-screw type. However, connectors are required to be of the type intended for a single conductor (single barrel). Removable connectors designed for multiple wires are not permitted to have a reduction in bending space.

A lay-in wire terminal is a pressure wire connector in which part of the connector is removable or swings away so that the stripped end of the conductor can be laid into the fixed portion of the connector. The removable or swing-away portion is then put back in place and the connector tightened down on the conductor.

Exhibit 312.2 illustrates the conditions under which 312.6(B)(2), Exception No. 2, is applicable. The terminal on the left has an offset not greater than 50 percent of bending space, per condition (2) of Exception No. 2. The terminal on the right is within a 45-degree angle of the enclosure, per condition (1) of Exception No. 2.

(C) Conductors 4 AWG or Larger. Installation shall comply with 300.4(G).

312.7 Space in Enclosures. Cabinets and cutout boxes shall have approved space to accommodate all conductors installed in them without crowding.

312.8 Switch and Overcurrent Device Enclosures. The wiring space within enclosures for switches and overcurrent devices shall be permitted for other wiring and equipment subject to limitations for specific equipment as provided in 312.8(A) and (B).

(A) Splices, Taps, and Feed-Through Conductors. The wiring space of enclosures for switches or overcurrent devices shall be permitted for conductors feeding through, spliced, or tapping off to other enclosures, switches, or overcurrent devices where all of the following conditions are met:

(1) The total of all conductors installed at any cross section of the wiring space does not exceed 40 percent of the cross-sectional area of that space.
(2) The total area of all conductors, splices, and taps installed at any cross section of the wiring space does not exceed 75 percent of the cross-sectional area of that space.
(3) The bending space for conductors 4 AWG and larger complies with 314.28(A)(2).
(4) A warning label complying with 110.21(B) is applied to the enclosure that identifies the closest disconnecting means for any feed-through conductors.

The following application example shows a volume calculation used for splices, taps, or feed-through conductors.

Application Example

An enclosure having a wiring space of 4 in. wide by 3 in. deep has a 12 in.² cross-sectional area. Thus, the total conductor fill (see Chapter 9, Table 5, for dimensions of conductors) at any cross section cannot exceed 4.8 in.² (40 percent of 12 in.²), and the maximum space for conductors and splices or taps at any cross section cannot exceed 9 in.² (75 percent of 12 in.²).

(B) Power Monitoring or Energy Management Equipment. The wiring space of enclosures for switches or overcurrent devices shall be permitted to contain power monitoring or energy management equipment in accordance with 312.8(B)(1) through (B)(3).

Many new devices are used for power monitoring that are intended to be installed in enclosures containing panelboards, which are often not supplied by the panelboard manufacturer.

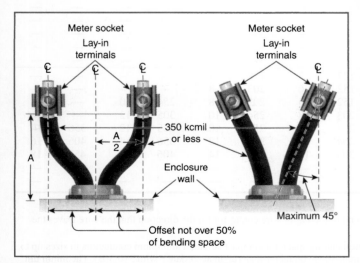

EXHIBIT 312.2 Wiring in a meter socket with lay-in wire terminals.

TABLE 312.6(B)(2) *Minimum Wire-Bending Space at Terminals*

Wire Size (AWG or kcmil)		Wires per Terminal							
		1		2		3		4 or More	
All Other Conductors	Compact Stranded AA-8000 Aluminum Alloy Conductors (See Note 3.)	mm	in.	mm	in.	mm	in.	mm	in.
14–10	12–8	Not specified		—	—	—	—	—	—
8	6	38.1	1½	—	—	—	—	—	—
6	4	50.8	2	—	—	—	—	—	—
4	2	76.2	3	—	—	—	—	—	—
3	1	76.2	3	—	—	—	—	—	—
2	1/0	88.9	3½	—	—	—	—	—	—
1	2/0	114	4½	—	—	—	—	—	—
1/0	3/0	140	5½	140	5½	178	7	—	—
2/0	4/0	152	6	152	6	190	7½	—	—
3/0	250	165[a]	6½[a]	165[a]	6½[a]	203	8	—	—
4/0	300	178[b]	7[b]	190[c]	7½[c]	216[a]	8½[a]	—	—
250	350	216[d]	8½[d]	216[d]	8½[d]	229[b]	9[b]	254	10
300	400	254[e]	10[e]	254[d]	10[d]	279[b]	11[b]	305	12
350	500	305[e]	12[e]	305[e]	12[e]	330[e]	13[e]	356[d]	14[d]
400	600	330[e]	13[e]	330[e]	13[e]	356[e]	14[e]	381[e]	15[e]
500	700–750	356[e]	14[e]	356[e]	14[e]	381[e]	15[e]	406[e]	16[e]
600	800–900	381[e]	15[e]	406[e]	16[e]	457[e]	18[e]	483[e]	19[e]
700	1000	406[e]	16[e]	457[e]	18[e]	508[e]	20[e]	559[e]	22[e]
750	—	432[e]	17[e]	483[e]	19[e]	559[e]	22[e]	610[e]	24[e]
800	—	457	18	508	20	559	22	610	24
900	—	483	19	559	22	610	24	610	24
1000	—	508	20	—	—	—	—	—	
1250	—	559	22	—	—	—	—	—	
1500	—	610	24	—	—	—	—	—	
1750	—	610	24	—	—	—	—	—	
2000	—	610	24	—	—	—	—	—	

Notes:

1. Bending space at terminals shall be measured in a straight line from the end of the lug or wire connector in a direction perpendicular to the enclosure wall.

2. For removable and lay-in wire terminals intended for only one wire, bending space shall be permitted to be reduced by the following number of millimeters (inches):

[a]12.7 mm (½ in.)

[b]25.4 mm (1 in.)

[c]38.1 mm (1½ in.)

[d]50.8 mm (2 in.)

[e]76.2 mm (3 in.)

3. This column shall be permitted to determine the required wire-bending space for compact stranded aluminum conductors in sizes up to 1000 kcmil and manufactured using AA-8000 series electrical grade aluminum alloy conductor material in accordance with 310.3(B).

Much of this equipment is used for load monitoring and energy management. This section provides requirements used to determine if the addition of devices or equipment is acceptable. The inclusion of devices and equipment in a wiring space is limited to those that are identified as field installable accessories as part of the listed equipment or as a listed kit evaluated for field installation in the specific equipment.

(1) Identification. Power monitoring or energy management equipment shall either be identified as a field installable accessory as part of the listed equipment or be a listed kit evaluated for field installation in switch or overcurrent device enclosures.

(2) Area. The total area of all conductors, splices, taps, and equipment at any cross section of the wiring space shall not exceed 75 percent of the cross-sectional area of that space.

(3) Conductors. Conductors used exclusively for control or instrumentation circuits shall comply with either 312.8(B)(3)(a) or (B)(3)(b).

(a) Conductors shall comply with 724.49.

(b) Conductors smaller than 18 AWG, but not smaller than 22 AWG for a single conductor and 26 AWG for a multiconductor cable, shall be permitted to be used where the conductors and cable assemblies meet all of the following conditions:

(1) Are enclosed within raceways or routed along one or more walls of the enclosure and secured at intervals that do not exceed 250 mm (10 in.)
(2) Are secured within 250 mm (10 in.) of terminations
(3) Are secured to prevent contact with current carrying components within the enclosure
(4) Are rated for the system voltage and not less than 600 volts
(5) Have a minimum insulation temperature rating of 90°C

312.9 Side or Back Wiring Spaces or Gutters. Cabinets and cutout boxes shall be provided with back-wiring spaces, gutters, or wiring compartments as required by 312.101(C) and (D).

N **312.10 Screws or Other Fasteners.** Screws or other fasteners installed in the field that enter wiring spaces shall be as provided by or specified by the manufacturer or shall comply with the following as applicable:

(1) Screws shall be machine type with blunt ends.
(2) Other fasteners shall have blunt ends.
(3) Screws or other fasteners shall extend into the enclosure no more than 6 mm (¼ in.) unless the end is protected with an approved means.

Exception to (3): Screws or other fasteners shall be permitted to extend into the enclosure not more than 11 mm (⁷⁄₁₆ in.) if located within 10 mm (³⁄₈ in.) of an enclosure wall.

Part II. Construction Specifications

312.100 Material. Cabinets, cutout boxes, and meter socket enclosures shall comply with 312.100(A) through (C).

(A) Metal Cabinets and Cutout Boxes. Metal enclosures within the scope of this article shall be protected both inside and outside against corrosion.

(B) Strength. The design and construction of enclosures within the scope of this article shall be such as to secure ample strength and rigidity. If constructed of sheet steel, the metal thickness shall not be less than 1.35 mm (0.053 in.) uncoated.

(C) Nonmetallic Cabinets. Nonmetallic cabinets shall be listed, or they shall be submitted for approval prior to installation.

312.101 Spacing. The spacing within cabinets and cutout boxes shall comply with 312.101(A) through (D).

(A) General. Spacing within cabinets and cutout boxes shall provide approved spacing for the distribution of wires and cables placed in them and for a separation between metal parts of devices and apparatus mounted within them in accordance with 312.101(A)(1), (A)(2), and (A)(3).

(1) Base. Other than at points of support, there shall be an airspace of at least 1.59 mm (0.0625 in.) between the base of the device and the wall of any metal cabinet or cutout box in which the device is mounted.

(2) Doors. There shall be an airspace of at least 25.4 mm (1.00 in.) between any live metal part, including live metal parts of enclosed fuses, and the door.

Exception: Where the door is lined with an approved insulating material or is of a thickness of metal not less than 2.36 mm (0.093 in.) uncoated, the airspace shall not be less than 12.7 mm (0.500 in.).

(3) Live Parts. There shall be an airspace of at least 12.7 mm (0.500 in.) between the walls, back, gutter partition, if of metal, or door of any cabinet or cutout box and the nearest exposed current-carrying part of devices mounted within the cabinet where the voltage does not exceed 250. This spacing shall be increased to at least 25.4 mm (1.00 in.) for voltages of 251 to 1000, nominal.

Exception: Where the conditions in 312.101(A)(2), Exception, are met, the airspace for nominal voltages from 251 to 600 shall be permitted to be not less than 12.7 mm (0.500 in.).

(B) Switch Clearance. Cabinets and cutout boxes shall be deep enough to allow the closing of the doors when 30-ampere branch-circuit panelboard switches are in any position, when combination cutout switches are in any position, or when other single-throw switches are opened as far as their construction permits.

(C) Wiring Space. Cabinets and cutout boxes that contain devices or apparatus connected within the cabinet or box to more than eight conductors, including those of branch circuits, meter loops, feeder circuits, power circuits, and similar circuits, but not including the supply circuit or a continuation thereof, shall have back-wiring spaces or one or more side-wiring spaces, side gutters, or wiring compartments.

(D) Wiring Space — Enclosure. Side-wiring spaces, side gutters, or side-wiring compartments of cabinets and cutout boxes shall be made tight enclosures by means of covers, barriers, or partitions extending from the bases of the devices contained in the cabinet, to the door, frame, or sides of the cabinet.

Exception: Side-wiring spaces, side gutters, and side-wiring compartments of cabinets shall not be required to be made tight enclosures where those side spaces contain only conductors that enter the cabinet directly opposite to the devices where they terminate.

Partially enclosed back-wiring spaces shall be provided with covers to complete the enclosure. Wiring spaces that are required by 312.101(C) and are exposed when doors are open shall be provided with covers to complete the enclosure. Where space is provided for feed-through conductors and for splices as required in 312.8, additional barriers shall not be required.

N **312.102 Doors or Covers.** Cabinets, cutout boxes, and meter socket enclosures shall be equipped with doors or covers.

ARTICLE 314 Outlet, Device, Pull, and Junction Boxes; Conduit Bodies; Fittings; and Handhole Enclosures

Δ **Part I. General**

314.1 Scope. This article covers the installation and use of all boxes and conduit bodies used as outlet, device, junction, or pull boxes, depending on their use, and handhole enclosures. Cast metal, sheet metal, nonmetallic, and other boxes such as FS, FD, and larger boxes are not classified as conduit bodies. This article also includes installation requirements for fittings used to join raceways and to connect raceways and cables to boxes and conduit bodies.

314.2 Round Boxes. Round boxes shall not be used where conduits or connectors requiring the use of locknuts or bushings are to be connected to the side of the box.

314.3 Nonmetallic Boxes. Nonmetallic boxes shall be permitted only with open wiring on insulators, concealed knob-and-tube wiring, cabled wiring methods with entirely nonmetallic sheaths, flexible cords, and nonmetallic raceways.

Exception No. 1: Where internal bonding means are provided between all entries, nonmetallic boxes shall be permitted to be used with metal raceways or metal-armored cables.

Exception No. 2: Where integral bonding means with a provision for attaching an equipment bonding jumper inside the box are provided between all threaded entries in nonmetallic boxes listed for the purpose, nonmetallic boxes shall be permitted to be used with metal raceways or metal-armored cables.

314.4 Metal Boxes. Metal boxes shall be grounded and bonded in accordance with Parts I, IV, V, VI, VII, and X of Article 250 as applicable, except as permitted in 250.112(I).

N **314.5 Screws or Other Fasteners.** Screws or other fasteners installed in the field that enter wiring spaces shall be as provided by or specified by the manufacturer or shall comply with the following as applicable:

(1) Screws shall be machine type with blunt ends.
(2) Other fasteners shall have blunt ends.
(3) Screws attaching a cover shall extend no more than 10 mm (⅜ in.).
(4) Screws or other fasteners, other than in (3), penetrating a cover shall extend no more than 8 mm (⁵⁄₁₆ in.).
(5) Screws or other fasteners penetrating a wall of a box exceeding 1650 cm³ (100 in.³) shall extend no more than 6 mm (¼ in.), or more than 11 mm (⁷⁄₁₆ in.) if located within 10 mm (⅜ in.) of an adjacent box wall.

(6) Screws or other fasteners penetrating the wall of a box not exceeding 1650 cm³ (100 in.³) and not covered in 314.23(B)(1) shall be made flush with the box interior.
(7) Screws or other fasteners penetrating the wall of a conduit body shall be made flush with the conduit body interior.

Exception to (3) through (6): A screw shall be permitted to be longer if the end of the screw is protected with an approved means.

Part II. Installation

Δ **314.15 Damp or Wet Locations.** In damp or wet locations, boxes, conduit bodies, outlet box hoods, and fittings shall be placed or equipped so as to prevent moisture from entering or accumulating within the box, conduit body, or fitting. Boxes, conduit bodies, outlet box hoods, and fittings installed in wet locations shall be listed for use in wet locations. Approved drainage openings not smaller than 3 mm (⅛ in.) and not larger than 6 mm (¼ in.) in diameter shall be permitted to be installed in the field in boxes or conduit bodies listed for use in damp or wet locations. For installation of listed drain fittings, larger openings are permitted to be installed in the field in accordance with manufacturer's instructions.

Informational Note No. 1: See 314.27(B) for boxes in floors.
Informational Note No. 2: See 300.6 for protection against corrosion.

Δ **314.16 Number of Conductors in Outlet, Device, and Junction Boxes, and Conduit Bodies.** Boxes and conduit bodies shall be of an approved size to provide free space for all enclosed conductors. In no case shall the volume of the box, as calculated in 314.16(A), be less than the fill calculation as calculated in 314.16(B). The minimum volume for conduit bodies shall be as calculated in 314.16(C).

This section shall not apply to terminal housings supplied with motors or generators.

Informational Note: See 430.12 for volume requirements of motor or generator terminal housings.

Boxes and conduit bodies enclosing conductors 4 AWG or larger shall also comply with 314.28. Outlet and device boxes shall also comply with 314.24.

(A) Box Volume Calculations. The volume of a wiring enclosure (box) shall be the total volume of the assembled sections and, where used, the space provided by plaster rings, domed covers, extension rings, and so forth, that are marked with their volume or are made from boxes the dimensions of which are listed in Table 314.16(A). Where a box is provided with one or more securely installed barriers, the volume shall be apportioned to each of the resulting spaces. Each barrier, if not marked with its volume, shall be considered to take up 8.2 cm³ (½ in.³) if metal, and 16.4 cm³ (1.0 in.³) if nonmetallic.

△ **TABLE 314.16(A)** *Metal Boxes*

Box Trade Size			Minimum Volume		Maximum Number of Conductors* (arranged by AWG size)						
mm	in.		cm³	in.³	18	16	14	12	10	8	6
100 × 32	(4 × 1¼)	round/octagonal	205	12.5	8	7	6	5	5	4	2
100 × 38	(4 × 1½)	round/octagonal	254	15.5	10	8	7	6	6	5	3
100 × 54	(4 × 2⅛)	round/octagonal	353	21.5	14	12	10	9	8	7	4
100 × 32	(4× 1¼)	square	295	18.0	12	10	9	8	7	6	3
100 × 38	(4 × 1½)	square	344	21.0	14	12	10	9	8	7	4
100 × 54	(4 × 2⅛)	square	497	30.3	20	17	15	13	12	10	6
120 × 32	(4¹¹⁄₁₆ × 1¼)	square	418	25.5	17	14	12	11	10	8	5
120 × 38	(4¹¹⁄₁₆ × 1½)	square	484	29.5	19	16	14	13	11	9	5
120 × 54	(4¹¹⁄₁₆ × 2⅛)	square	689	42.0	28	24	21	18	16	14	8
75 × 50 × 38	(3 × 2 × 1½)	device	123	7.5	5	4	3	3	3	2	1
75 × 50 × 50	(3 × 2 × 2)	device	164	10.0	6	5	5	4	4	3	2
75× 50 × 57	(3 × 2 × 2¼)	device	172	10.5	7	6	5	4	4	3	2
75 × 50 × 65	(3 × 2 × 2½)	device	205	12.5	8	7	6	5	5	4	2
75 × 50 × 70	(3 × 2 × 2¾)	device	230	14.0	9	8	7	6	5	4	2
75 × 50 × 90	(3 × 2 × 3½)	device	295	18.0	12	10	9	8	7	6	3
100 × 54 × 38	(4 × 2⅛ × 1½)	device	169	10.3	6	5	5	4	4	3	2
100 × 54 × 48	(4 × 2⅛ × 1⅞)	device	213	13.0	8	7	6	5	5	4	2
100 × 54 × 54	(4 × 2⅛ × 2⅛)	device	238	14.5	9	8	7	6	5	4	2
95 × 50 × 65	(3¾ × 2 × 2½)	masonry box/gang	230	14.0	9	8	7	6	5	4	2
95 × 50 × 90	(3¾ × 2 × 3½)	masonry box/gang	344	21.0	14	12	10	9	8	7	4
min. 44.5 depth	FS — single cover/gang (1¾)		221	13.5	9	7	6	6	5	4	2
min. 60.3 depth	FD — single cover/gang (2⅜)		295	18.0	12	10	9	8	7	6	3
min. 44.5 depth	FS — multiple cover/gang (1¾)		295	18.0	12	10	9	8	7	6	3
min. 60.3 depth	FD — multiple cover/gang (2⅜)		395	24.0	16	13	12	10	9	8	4

*Where no volume allowances are required by 314.16(B)(2) through (B)(6).

(1) Standard Boxes. The volumes of standard boxes that are not marked with their volume shall be as given in Table 314.16(A).

(2) Other Boxes. Boxes 1650 cm³ (100 in.³) or less, other than those described in Table 314.16(A), and nonmetallic boxes shall be durably and legibly marked by the manufacturer with their volume(s). Boxes described in Table 314.16(A) that have a volume larger than is designated in the table shall be permitted to have their volume marked as required by this section.

△ **(B) Box Fill Calculations.** The volumes in 314.16(B)(1) through (B)(6), as applicable, shall be added together. No allowance shall be required for small fittings such as locknuts and bushings. Each space within a box installed with a barrier shall be calculated separately.

(1) Conductor Fill. Each conductor that originates outside the box and terminates or is spliced within the box shall be counted once, and each conductor that passes through the box without splice or termination shall be counted once. Each loop or coil of unbroken conductor not less than twice the minimum length required for free conductors in 300.14 shall be counted twice. The conductor fill shall be calculated using Table 314.16(B)(1). A conductor, no part of which leaves the box, shall not be counted.

Exception: An equipment grounding conductor or conductors or not over four fixture wires smaller than 14 AWG, or both, shall be permitted to be omitted from the calculations where they enter a box from a domed luminaire or similar canopy and terminate within that box.

△ **(2) Clamp Fill.** Where one or more internal cable clamps, whether factory or field supplied, are present in the box, a single volume allowance in accordance with Table 314.16(B)(1) shall be made based on the largest conductor present in the box. No allowance shall be required for a cable connector with its clamping mechanism outside the box.

(3) Support Fittings Fill. Where one or more luminaire studs or hickeys are present in the box, a single volume allowance in accordance with Table 314.16(B)(1) shall be made for each type of fitting based on the largest conductor present in the box.

(4) Device or Equipment Fill. For each yoke or strap containing one or more devices or equipment, a double volume allowance in accordance with Table 314.16(B)(1) shall be made for each yoke or strap based on the largest conductor connected to a device(s) or equipment supported by that yoke or strap. A device or utilization equipment wider than a single

TABLE 314.16(B)(1) *Volume Allowance Required per Conductor*

Size of Conductor (AWG)	Free Space Within Box for Each Conductor	
	cm³	in.³
18	24.6	1.50
16	28.7	1.75
14	32.8	2.00
12	36.9	2.25
10	41.0	2.50
8	49.2	3.00
6	81.9	5.00

50 mm (2 in.) device box as described in Table 314.16(A) shall have double volume allowances provided for each gang required for mounting.

Δ **(5) Equipment Grounding Conductor Fill.** Where up to four equipment grounding conductors enter a box, a single volume allowance in accordance with Table 314.16(B)(1) shall be made based on the largest equipment grounding conductor entering the box. A ¼ volume allowance shall be made for each additional equipment grounding conductor that enters the box, based on the largest equipment grounding conductor entering the box.

N **(6) Terminal Block Fill.** Where a terminal block is present in a box, a single volume allowance in accordance with Table 314.16(B)(1) shall be made for each terminal block assembly based on the largest conductor(s) terminated to the assembly.

This section requires that the total box "volume" be equal to or greater than the total box "fill." The total box volume is determined by adding the individual volumes of the box components. The components include the box itself plus any attachments to it, such as a plaster ring, an extension ring, or a dome cover. The volume of each component is determined either from the volume marking on the component itself or from the standard volumes listed in Table 314.16(A). If a box is marked with a volume larger than listed in Table 314.16(A), the larger volume can be used instead of the table value. If a box contains one or more securely installed dividers, the volume would be apportioned among the resulting spaces. However, the space occupied by the divider must be considered in calculation of the box fill.

Adding all the volume allowances for all items contributing to box fill determines the total box fill. The volume allowance for each fill item is based on the volume listed in Table 314.16(B)(1) for the conductor size indicated. Commentary Table 314.1 summarizes the components contributing to box fill.

Up to four equipment grounding conductors entering a box will be considered as a single volume allowance. A ¼ volume allowance must be added for equipment grounding conductors and equipment bonding conductors where the number of conductors exceeds four.

COMMENTARY TABLE 314.1 Summary of Items Contributing to Box Fill

Items Contained Within Box	Volume Allowance	Based on [see Table 314.16(B)(1)]
Conductors that originate outside box	One for each conductor	Actual conductor size
Conductors that pass through box without splice or connection (less than 12 in. in total length)	One for each conductor	Actual conductor size
Conductors 12 in. or greater that are looped (or coiled) and unbroken (see 300.14 for exact measurement)	Two for a single (entire) unbroken conductor	Actual conductor size
Conductors that originate within box and do not leave box	None (these conductors not counted)	N.A.
Fixture wires [per 314.16(B)(1), Exception]	None (these conductors not counted)	N.A.
Internal cable clamps (one or more)	One only	Largest-sized conductor present
Support fittings (such as luminaire studs or hickeys)	One for each type of support fitting	Largest-sized conductor present
Devices (such as receptacles or switches) or utilization equipment (such as timers, dimmers, AFCI receptacles, GFCI receptacles, TVSS receptacles)	Two for each yoke or mounting strap	Largest-sized conductor connected to device or utilization equipment
Equipment grounding conductor (up to four)	One only	Largest equipment grounding conductor present; ¼ volume more per conductor over four
Isolated equipment grounding conductor (see equipment grounding conductor) [see 250.146(D)]	One only	Largest isolated and insulated equipment grounding conductor present

N.A. = not applicable.

Calculation Example 1

For this example, as shown in Exhibit 314.1, the box does not contain any cable clamps, support fittings, devices, or equipment grounding conductors (EGCs).

Solution

To determine the minimum standard-sized, square, metal box for the number of 12 AWG conductors being installed, count the conductors and compare the total to the maximum number of conductors permitted by Table 314.16(A). Each conductor running through the box without a splice is counted as one conductor, and each other conductor is counted as one conductor. Therefore, the total conductor count is nine. Table 314.16(A) indicates that the maximum fill for a standard 4 in. × 1½ in. square box is nine 12 AWG conductors. The minimum standard square box size is 21 in.³

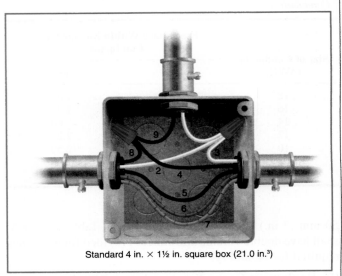

Standard 4 in. × 1½ in. square box (21.0 in.³)

EXHIBIT 314.1 A standard-sized square box containing no fittings or devices, such as luminaire studs, cable clamps, switches, receptacles, or EGCs.

Calculation Example 2

The standard method for determining adequate box size calculates the total box volume first and then subtracts the total box fill to ensure compliance. Using this method, determine whether the box in Exhibit 314.2 is adequately sized.

Solution

Table 314.16(A) shows the minimum volume of a standard 3 in. × 2 in. × 3½ in. device box to be 18 in.³ The box fill for this situation as given in Commentary Table 314.2 is 16 in.³ Because the total box fill of 16 in.³ is less than the 18 in.³ total box volume, the box is adequately sized.

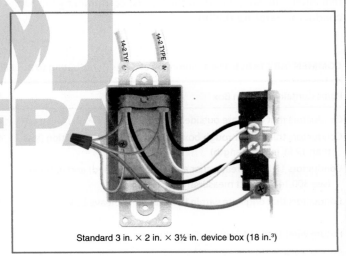

Standard 3 in. × 2 in. × 3½ in. device box (18 in.³)

EXHIBIT 314.2 A standard-sized device box containing a device and conductors requiring deductions.

COMMENTARY TABLE 314.2 Total Box Fill for Calculation Example 2

Items Contained Within Box	Volume Allowance	Unit Volume Based on Table 314.16(B)(1) (in.³)	Total Box Fill (in.³)
4 conductors	4 volume allowances for 14 AWG conductors	2.00	8.00
1 clamp	1 volume allowance (based on 14 AWG conductors)	2.00	2.00
1 device	2 volume allowances (based on 14 AWG conductors)	2.00	4.00
EGCs (not over four)	1 volume allowance (based on 14 AWG conductors)	2.00	2.00
Total			16.00

Calculation Example 3

Using the standard method, determine the adequacy of the device box illustrated in Exhibit 314.3, in which two standard-sized 3 in. × 2 in. × 3½ in. device boxes are ganged to form a single box.

Solution

Table 314.16(A) shows that the minimum volume for a single box is 18 in.^{3} Thus, the total box volume for the ganged box is 36 in.^{3} The total box fill, based on Table 314.16(B)(1), is determined as given in Commentary Table 314.3. With only 26 in.^{3} of the 36 in.^{3} filled, the box is adequately sized.

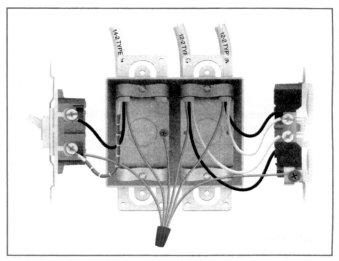

EXHIBIT 314.3 Two standard gangable device boxes containing conductors of different sizes.

COMMENTARY TABLE 314.3 Total Box Fill for Calculation Example 3

Items Contained Within Box	Volume Allowance	Unit Volume Based on Table 314.16(B)(1) (in.^{3})	Total Box Fill (in.^{3})
6 conductors	2 volume allowances for 14 AWG conductors	2.00	4.00
	4 volume allowances for 12 AWG conductors	2.25	9.00
2 clamps	1 volume allowance (based on 12 AWG conductors)	2.25	2.25
2 devices	2 volume allowances (based on 14 AWG conductors)	2.00	4.00
	2 volume allowances (based on 12 AWG conductors)	2.25	4.50
EGCs (up to four)	1 volume allowance (based on 12 AWG conductors)	2.25	2.25
Total			26.00

(C) Conduit Bodies.

(1) General. Conduit bodies enclosing 6 AWG conductors or smaller, other than short-radius conduit bodies as described in 314.16(C)(3), shall have a cross-sectional area not less than twice the cross-sectional area of the largest conduit or tubing to which they can be attached. The maximum number of conductors permitted shall be the maximum number permitted by Table 1 of Chapter 9 for the conduit or tubing to which it is attached.

(2) With Splices, Taps, or Devices. Only those conduit bodies that are durably and legibly marked by the manufacturer with their volume shall be permitted to contain splices, taps, or devices. The maximum number of conductors shall be calculated in accordance with 314.16(B). Conduit bodies shall be supported in a rigid and secure manner.

As illustrated in Exhibit 314.4, conduit bodies are required to have a cross-sectional area not less than twice that of the conduit to which they are attached and are not permitted to contain more

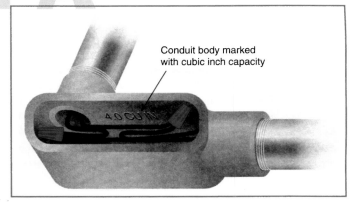

EXHIBIT 314.4 An example of permitted splices in a raceway-supported conduit body.

conductors than the attached raceway. Conduit bodies must be rigidly supported.

See also

314.23 for support requirements

(3) Short Radius Conduit Bodies. Conduit bodies such as capped elbows and service-entrance elbows that enclose conductors 6 AWG or smaller, and are only intended to enable the installation of the raceway and the contained conductors, shall not contain splices, taps, or devices and shall be of an approved size to provide free space for all conductors enclosed in the conduit body.

314.17 Conductors and Cables Entering Boxes, Conduit Bodies, or Fittings. Conductors entering boxes, conduit bodies, or fittings shall be protected from abrasion. Conductors and cables shall comply with 314.17(A) through (C).

(A) Openings to Be Closed. Openings through which conductors enter shall be closed in an approved manner.

(B) Boxes and Conduit Bodies. The installation of the conductors and cables in boxes and conduit bodies shall comply with 314.17(B)(1) through (B)(4).

(1) Conductors Entering Through Individual Holes or Through Flexible Tubing. For messenger-supported wiring, open wiring on insulators, or concealed knob-and-tube wiring, the conductors shall enter the box through individual holes. In installations where metal boxes or conduit bodies are used with conductors unprotected by flexible tubing, the individual openings shall be provided with insulating bushings. Where flexible tubing is used to enclose the conductors, the tubing shall extend from the last insulating support to not less than 6 mm (¼ in.) inside the box or conduit body and 6 mm (¼ in.) beyond the end of any cable clamp. The wiring method shall be secured to the box or conduit body.

(2) Cables Entering Through Cable Clamps. Where cable assemblies with nonmetallic sheaths are used, the sheath shall extend not less than 6 mm (¼ in.) inside the box and 6 mm (¼ in.) beyond the end of any cable clamp. Except as covered in 300.15(C), the wiring method shall be secured to the box or conduit body.

Exception: Where nonmetallic-sheathed cable is used with single gang nonmetallic boxes not larger than a nominal size 57 mm × 100 mm (2¼ in. × 4 in.) mounted in walls or ceilings, and where the cable is fastened within 200 mm (8 in.) of the box measured along the sheath and where the sheath extends through a cable knockout not less than 6 mm (¼ in.), securing the cable to the box shall not be required. Multiple cable entries shall be permitted in a single cable knockout opening.

Exhibit 314.5 is an example of an installation that complies with 314.17(B)(2), Exception.

(3) Conductors and Cables Entering Through Raceways. Where the raceway is complete between boxes, conduit bodies, or both and encloses individual conductors or nonmetallic cable assemblies or both, the conductors or cable assemblies shall not be required to be additionally secured. Where raceways enclose cable assemblies as covered in 300.15(C), the cable

EXHIBIT 314.5 Installation of a single gang outlet with cables securely fastened within 8 inches of the box. *(Courtesy of the International Association of Electrical Inspectors)*

assembly shall not be required to be additionally secured within the box or conduit body.

(4) Temperature Limitation. Nonmetallic boxes and conduit bodies shall be suitable for the lowest temperature-rated conductor entering the box or conduit body.

(C) Conductors 4 AWG or Larger. Installation shall comply with 300.4(G).

Informational Note: See 110.12(A) for requirements on closing unused cable and raceway knockout openings.

314.19 Boxes Enclosing Flush Devices or Flush Equipment. Boxes used to enclose flush devices or flush equipment shall be of such design that the devices or equipment will be completely enclosed on the back and sides, and substantial support for the devices or equipment will be provided. Screws for supporting the box shall not also be used to attach a device or equipment.

314.20 Flush-Mounted Installations. Installations within or behind a surface of concrete, tile, gypsum, plaster, or other noncombustible material, including boxes employing a flush-type cover or faceplate, shall be made so that the front edge of the box, plaster ring, extension ring, or listed extender will not be set back of the finished surface more than 6 mm (¼ in.).

Installations within a surface of wood or other combustible surface material, boxes, plaster rings, extension rings, or listed extenders shall extend to the finished surface or project therefrom.

This section applies only to the construction of the finished surface of the wall or ceiling and not to its structure or subsurface. It is important that the final finished surface, such as a tiled kitchen backsplash, be considered before outlet boxes are installed. This often requires the cooperation of other construction trades (drywall installers, plasterers, and carpenters) and the building designers.

314.21 Repairing Noncombustible Surfaces. Noncombustible surfaces that are broken or incomplete around boxes employing a flush-type cover or faceplate shall be repaired so there will be no gaps or open spaces greater than 3 mm (⅛ in.) at the edge of the box.

314.22 Surface Extensions. Surface extensions shall be made by mounting and mechanically securing an extension ring over the box. Equipment grounding shall be in accordance with Part VI of Article 250.

Exception: A surface extension shall be permitted to be made from the cover of a box where the cover is designed so it is unlikely to fall off or be removed if its securing means becomes loose. The wiring method shall be flexible for an approved length that permits removal of the cover and provides access to the box interior and shall be arranged so that any grounding continuity is independent of the connection between the box and cover.

△ **314.23 Supports.** Enclosures within the scope of this article shall be supported in accordance with 314.23(A) through (H) as applicable.

△ **(A) Surface Mounting.** An enclosure mounted on a building or other surface shall be rigidly and securely fastened in place. If the surface does not provide rigid and secure support, additional support in accordance with this section shall be provided.

(B) Structural Mounting. An enclosure supported from a structural member or from grade shall be rigidly supported either directly or by using a metal, polymeric, or wood brace.

(1) Nails and Screws. Nails and screws, where used as a fastening means, shall secure boxes by using brackets on the outside of the enclosure, or by using mounting holes in the back or in one or more sides of the enclosure, or they shall pass through the interior within 6 mm (¼ in.) of the back or ends of the enclosure. Screws shall not be permitted to pass through the box unless exposed threads in the box are protected using approved means to avoid abrasion of conductor insulation. Mounting holes made in the field shall be approved.

(2) Braces. Metal braces shall be protected against corrosion and formed from metal that is not less than 0.51 mm (0.020 in.) thick uncoated. Wood braces shall have a cross section not less than nominal 25 mm × 50 mm (1 in. × 2 in.). Wood braces in wet locations shall be treated for the conditions. Polymeric braces shall be identified as being suitable for the use.

(C) Mounting in Finished Surfaces. An enclosure mounted in a finished surface shall be rigidly secured thereto by clamps, anchors, or fittings identified for the application.

Where structural members are lacking or where boxes are cut into existing walls, boxes are permitted to be secured by clamps or anchors. Exhibit 314.6 shows several examples of "old work" boxes that are designed to be installed into a finished wall or

EXHIBIT 314.6 Metal and nonmetallic outlet boxes designed for mounting in finished wall and ceiling surfaces.

ceiling. The three nonmetallic boxes have mounting tabs that secure to the back of the finished surface as the screws on the front of the box are tightened. The metal boxes have mounting tabs that are designed to accept screws that secure the box to a surface such as wood paneling or wainscoting. The metal brackets (box hold-its) shown can be used to secure a box to a wall surface such as gypsum wall board.

(D) Suspended Ceilings. An enclosure mounted to structural or supporting elements of a suspended ceiling shall be not more than 1650 cm³ (100 in.³) in size and shall be securely fastened in place in accordance with either 314.23(D)(1) or (D)(2).

(1) Framing Members. An enclosure shall be fastened to the framing members by mechanical means such as bolts, screws, or rivets, or by the use of clips or other securing means identified for use with the type of ceiling framing member(s) and enclosure(s) employed. The framing members shall be supported in an approved manner and securely fastened to each other and to the building structure.

(2) Support Wires. The installation shall comply with 300.11(B). The enclosure shall be secured, using identified methods, to ceiling support wire(s), including any additional support wire(s) installed for ceiling support. Support wire(s) used for enclosure support shall be fastened at each end so as to be taut within the ceiling cavity.

△ **(E) Raceway-Supported Enclosure, Without Devices, Luminaires, or Lampholders.** An enclosure that does not contain a device(s), other than splicing devices, or supports a luminaire(s), a lampholder, or other equipment and is supported by entering raceways shall not exceed 1650 cm³ (100 in.³) in size. It shall have threaded entries or identified hubs. It shall be supported by two or more conduits threaded wrenchtight into the enclosure or hubs. Each conduit shall be secured within 900 mm (3 ft) of the enclosure, or within 450 mm (18 in.) of the enclosure if all conduit entries are on the same side.

Boxes are not permitted to be supported by rigid raceways using locknuts and bushings. A box is not permitted to be supported by a single raceway, but a conduit body is, where it is being supported by the trade size raceway it is designed to accept and a reducing bushing is not used.

Exception: The following wiring methods shall be permitted to support a conduit body of any size, including a conduit body constructed with only one conduit entry, provided that the trade size of the conduit body is not larger than the largest trade size of the conduit or tubing:

(1) Intermediate metal conduit, IMC
(2) Rigid metal conduit, RMC
(3) Rigid polyvinyl chloride conduit, PVC
(4) Reinforced thermosetting resin conduit, RTRC
(5) Electrical metallic tubing, EMT

(F) Raceway-Supported Enclosures, with Devices, Luminaires, or Lampholders. An enclosure that contains a device(s), other than splicing devices, or supports a luminaire(s), a lampholder, or other equipment and is supported by entering raceways shall not exceed 1650 cm³ (100 in.³) in size. It shall have threaded entries or identified hubs. It shall be supported by two or more conduits threaded wrenchtight into the enclosure or hubs. Each conduit shall be secured within 450 mm (18 in.) of the enclosure.

Exception No. 1: Rigid metal or intermediate metal conduit shall be permitted to support a conduit body of any size, including a conduit body constructed with only one conduit entry, provided the trade size of the conduit body is not larger than the largest trade size of the conduit.

Exception No. 2: An unbroken length(s) of rigid or intermediate metal conduit shall be permitted to support a box used for luminaire or lampholder support, or to support a wiring enclosure that is an integral part of a luminaire and used in lieu of a box in accordance with 300.15(B), where all of the following conditions are met:

(1) The conduit is securely fastened at a point so that the length of conduit beyond the last point of conduit support does not exceed 900 mm (3 ft).
(2) The unbroken conduit length before the last point of conduit support is 300 mm (12 in.) or greater, and that portion of the conduit is securely fastened at some point not less than 300 mm (12 in.) from its last point of support.
(3) Where accessible to unqualified persons, the luminaire or lampholder, measured to its lowest point, is at least 2.5 m (8 ft) above grade or standing area and at least 900 mm (3 ft) measured horizontally to the 2.5 m (8 ft) elevation from windows, doors, porches, fire escapes, or similar locations.
(4) A luminaire supported by a single conduit does not exceed 300 mm (12 in.) in any direction from the point of conduit entry.
(5) The weight supported by any single conduit does not exceed 9 kg (20 lb).
(6) At the luminaire or lampholder end, the conduit(s) is threaded wrenchtight into the box, conduit body, integral

wiring enclosure, or identified hubs. Where a box or conduit body is used for support, the luminaire shall be secured directly to the box or conduit body, or through a threaded conduit nipple not over 75 mm (3 in.) long.

(G) Enclosures in Concrete or Masonry. An enclosure supported by embedment shall be identified as suitably protected from corrosion and securely embedded in concrete or masonry.

Boxes are permitted to be embedded in masonry or concrete, provided they are rigid and secure. Corrosion protection is required due to the corrosive effects of concrete. Exhibit 314.7 shows a galvanized box installed in a brick wall. Additional support is not required.

(H) Pendant Boxes. An enclosure supported by a pendant shall comply with 314.23(H)(1) or (H)(2).

(1) Flexible Cord. A box shall be supported from a multiconductor cord or cable in an approved manner that protects the conductors against strain. A connection to a box equipped with a hub shall be made with a listed cord grip attachment fitting marked for use with a threaded hub.

(2) Conduit. A box supporting lampholders or luminaires, or wiring enclosures within luminaires used in lieu of boxes in accordance with 300.15(B), shall be supported by rigid or intermediate metal conduit stems. For stems longer than 450 mm (18 in.), the stems shall be connected to the wiring system with listed swivel hangers suitable for the location. At the luminaire end, the conduit(s) shall be threaded wrenchtight into the box, wiring enclosure, or identified hubs.

Where supported by only a single conduit, the threaded joints shall be prevented from loosening by the use of set-screws or other effective means, or the luminaire, at any point, shall be at least 2.5 m (8 ft) above grade or standing area and at least

EXHIBIT 314.7 A galvanized box installed in a brick wall. *(Courtesy of the International Association of Electrical Inspectors)*

900 mm (3 ft) measured horizontally to the 2.5 m (8 ft) elevation from windows, doors, porches, fire escapes, or similar locations. A luminaire supported by a single conduit shall not exceed 300 mm (12 in.) in any horizontal direction from the point of conduit entry.

These requirements address field installation of pendants mounted on boxes. Flexible fittings, such as a ball and socket supporting a pendant, as shown in Exhibit 314.8, must be installed on the box for stems longer than 18 inches. Swivel hangers for stems longer than 18 inches must be listed.

△ 314.24 Dimensions of Boxes. Outlet and device boxes shall have approved dimensions to allow equipment installed within them to be mounted properly and without likelihood of damage to conductors within the box.

(A) Depth of Outlet Boxes Without Enclosed Devices or Utilization Equipment. Outlet boxes that do not enclose devices or utilization equipment shall have a minimum internal depth of 12.7 mm (½ in.).

△ (B) Depth of Outlet and Device Boxes with Enclosed Devices or Utilization Equipment. Outlet and device boxes that enclose devices or utilization equipment shall have a minimum internal depth that accommodates the rearward projection of the equipment and the size of the conductors that supply the equipment. The internal depth shall include, where used, that of any extension boxes, plaster rings, or raised covers. The internal depth shall comply with 314.24(B)(1) through (B)(5) as applicable.

The box selection must be based on its having sufficient cubic-inch capacity for the device, equipment, and conductors.

(1) Large Equipment. Boxes that enclose devices or utilization equipment that projects more than 48 mm (1⅞ in.) rearward from

EXHIBIT 314.8 Flexible fitting for stems longer than 18 inches. *(Courtesy of Hubbell Incorporated)*

the mounting plane of the box shall have a depth that is not less than the depth of the equipment plus 6 mm (¼ in.).

(2) Conductors Larger Than 4 AWG. Boxes that enclose devices or utilization equipment supplied by conductors larger than 4 AWG shall be identified for their specific function.

Exception: Devices or utilization equipment supplied by conductors larger than 4 AWG shall be permitted to be mounted on or in junction and pull boxes larger than 1650 cm³ (100 in.³) if the spacing at the terminals meets the requirements of 312.6.

(3) Conductors 8, 6, or 4 AWG. Boxes that enclose devices or utilization equipment supplied by 8, 6, or 4 AWG conductors shall have an internal depth that is not less than 52.4 mm (2 1/16 in.).

(4) Conductors 12 or 10 AWG. Boxes that enclose devices or utilization equipment supplied by 12 or 10 AWG conductors shall have an internal depth that is not less than 30.2 mm (1 3/16 in.). Where the equipment projects rearward from the mounting plane of the box by more than 25 mm (1 in.), the box shall have a depth not less than that of the equipment plus 6 mm (¼ in.). Where wiring enters the center portion of the rear of a box opposite to the equipment, the minimum clearance shall be increased to 13 mm (½ in.).

(5) Conductors 14 AWG and Smaller. Boxes that enclose devices or utilization equipment supplied by 14 AWG or smaller conductors shall have a depth that is not less than 23.8 mm (15/16 in.).

Exception: Under any of the conditions specified in 314.24(B)(1) through (B)(5), devices or utilization equipment that is listed to be installed with specified boxes shall be permitted.

The intent here is to prevent damage to conductors that can occur where clearance from the device or utilization equipment to the back of the box is insufficient. Where clearance is insufficient, conductors are often pinched or the insulation is damaged as the device or utilization equipment is pushed into the box.

Examples of utilization equipment often located within an outlet or device box include speakers, timers, motion detectors, alarms, and video and audio surveillance equipment. Typically, minimum box dimensions are included in product installation instructions for listed utilization equipment.

N (C) Clearances for Side-Wiring Entrances. Where devices or equipment are mounted in boxes having side-wiring entries, the conductors entering from the side shall be protected as covered in (1) or (2), as follows. The term *side* applies to any wall of a box other than the one opposite to the opening.

(1) The rearward projection of the device or equipment shall not extend beyond the centerline of the wiring knockout or other entry.

(2) The clearance from the box wall to the installed device or equipment shall be not less than 13 mm (½ in.).

314.25 Covers and Canopies. In completed installations, each box shall have a cover, faceplate, lampholder, or luminaire canopy, except where the installation complies with 410.24(B). Conduit body enclosures shall be installed with a cover, lampholder, or device. Screws used for the purpose of attaching covers, or other equipment, to the box shall be either machine screws matching the thread gauge and size that is integral to the box or shall be in accordance with the manufacturer's instructions.

Δ **(A) Nonmetallic or Metal Covers and Plates.** Nonmetallic or metal covers and plates shall be permitted. Where metal covers or plates are used, they shall be connected to the equipment grounding conductor in accordance with 250.110.

> Informational Note: See 410.42 for metal luminaire canopies and 404.12 and 406.6(B) for metal faceplates for additional grounding requirements.

(B) Exposed Combustible Wall or Ceiling Finish. Where a luminaire canopy or pan is used, any combustible wall or ceiling finish exposed between the edge of the canopy or pan and the outlet box shall be covered with noncombustible material if required by 410.23.

Because heat from a short circuit or a ground fault or heat due to overlamping could create a fire hazard within a luminaire canopy or pan, any exposed combustible wall or ceiling space exceeding 180 square inches between the edge of the outlet box and the perimeter of the luminaire is required to be covered with noncombustible material. The noncombustible material need not be metal. Glass fiber pads, commonly provided as thermal barriers within the ceiling pan of luminaires, are an essential thermal barrier, but they do not serve in place of noncombustible wall coverings. Where the wall or ceiling finish is concrete, tile, gypsum, plaster, or other noncombustible material, the requirements of this section do not apply.

(C) Flexible Cord Pendants. Covers of outlet boxes and conduit bodies having holes through which flexible cord pendants pass shall be provided with identified bushings or shall have smooth, well-rounded surfaces on which the cords can bear. So-called hard rubber or composition bushings shall not be used.

314.27 Outlet Boxes.

(A) Boxes at Luminaire or Lampholder Outlets. Outlet boxes or fittings designed for the support of luminaires and lampholders, and installed as required by 314.23, shall be permitted to support a luminaire or lampholder.

(1) Vertical Surface Outlets. Boxes used at luminaire or lampholder outlets in or on a vertical surface shall be identified and marked on the interior of the box to indicate the maximum weight of the luminaire that is permitted to be supported by the box if other than 23 kg (50 lb).

Exception: A vertically mounted luminaire or lampholder weighing not more than 3 kg (6 lb) shall be permitted to be supported on other boxes or plaster rings that are secured to other boxes, provided that the luminaire or its supporting yoke, or the lampholder, is secured to the box with no fewer than two No. 6 or larger screws.

Device boxes designed for the mounting of snap switches, receptacles, and other devices are usually provided with 6-32 screws. They usually are not suitable for supporting other than lightweight wall-mounted luminaires. Altering the screw holes to increase their size does not increase the weight capacity of the box and may violate the listing.

(2) Ceiling Outlets. At every outlet used exclusively for lighting, the box shall be designed or installed so that a luminaire or lampholder can be attached. Boxes shall be required to support a luminaire weighing a minimum of 23 kg (50 lb). A luminaire that weighs more than 23 kg (50 lb) shall be supported independently of the outlet box, unless the outlet box is listed for not less than the weight to be supported. The interior of the box shall be marked by the manufacturer to indicate the maximum weight the box shall be permitted to support.

Whether a luminaire is attached to an outlet box or is supported independently of the outlet box, the supporting means of the luminaire should be securely fastened. Boxes designed as the sole support of luminaires weighing more than 50 pounds must be listed and marked with the maximum weight.

(B) Floor Boxes. Boxes listed specifically for this application shall be used for receptacles located in the floor.

Exception: Where the authority having jurisdiction judges them free from likely exposure to physical damage, moisture, and dirt, boxes located in elevated floors of show windows and similar locations shall be permitted to be other than those listed for floor applications. Receptacles and covers shall be listed as an assembly for this type of location.

Δ **(C) Boxes at Ceiling-Suspended (Paddle) Fan Outlets.** Outlet boxes or outlet box systems used as the sole support of a ceiling-suspended (paddle) fan shall be listed, shall be marked by their manufacturer on the interior of the box as suitable for this purpose, and shall not support ceiling-suspended (paddle) fans that weigh more than 32 kg (70 lb). For outlet boxes or outlet box systems designed to support ceiling-suspended (paddle) fans that weigh more than 16 kg (35 lb), the required marking shall include the maximum weight to be supported.

Outlet boxes mounted in the ceilings of habitable rooms of dwelling occupancies in a location acceptable for the installation of a ceiling-suspended (paddle) fan shall comply with one of the following:

(1) Listed for the sole support of ceiling-suspended (paddle) fans

(2) Installed so as to allow direct access through the box to structural framing capable of supporting a ceiling-suspended (paddle) fan without removing the box

Outlet boxes mounted in ceilings of habitable rooms in dwelling units must be a box listed for ceiling (paddle) fan support or comply with 314.27(C) where a ceiling (paddle) fan either will be installed initially or might be installed in the future. Outlet boxes specifically listed to support ceiling-mounted paddle fans are available, as are several alternative and retrofit methods that can provide suitable support for a paddle fan. Exhibit 314.9 illustrates two methods of supporting a fan from an outlet box listed for fan support. In new residential construction, it is common to provide a wall-mounted switch(s) with wiring to allow for the future installation of paddle fans or paddle fans with light kits. Such installations are required to have an outlet box or outlet box system that is listed for the sole support of a fan.

(D) Utilization Equipment. Boxes used for the support of utilization equipment other than ceiling-suspended (paddle) fans shall meet the requirements of 314.27(A) for the support of a luminaire that is the same size and weight.

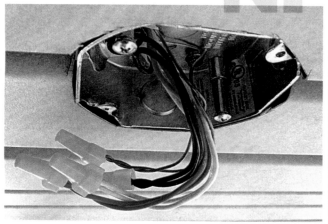

EXHIBIT 314.9 Two methods of supporting a ceiling fan from a listed outlet box. *(Top)* A listed outlet box used as the sole support of a ceiling-suspended (paddle) fan. Note the two clamps on the perimeter of the box that will support a fixture. *(Courtesy of the International Association of Electrical Inspectors). (Bottom)* A box listed for supporting a ceiling-suspended (paddle) fan (35 pounds or less). *(Courtesy of Hubbell Incorporated)*

Exception: Utilization equipment weighing not more than 3 kg (6 lb) shall be permitted to be supported on other boxes or plaster rings that are secured to other boxes, provided the equipment or its supporting yoke is secured to the box with no fewer than two No. 6 or larger screws.

Δ **(E) Weight-Supporting Ceiling Receptacles (WSCR) and Weight-Supporting Attachment Fittings (WSAF).** Outlet boxes required in 314.27 shall be permitted to support listed weight-supporting ceiling receptacles (WSCR). A WSCR shall be used in combination with compatible weight-supporting attachment fittings (WSAF) that are identified for the support of equipment within the weight and mounting orientation limits of the listing. Where the WSCR is installed, it shall be included in the box fill calculation covered in 314.16(B)(4).

Listed WSCR used in combination with compatible WSAF shall be permitted to be installed in outlet boxes for the sole support of ceiling-suspended (paddle) fans, in accordance with 314.27(C).

Informational Note: See ANSI/NEMA WD-6, *American National Standard for Wiring Devices–Dimensional Specifications*, for standard configurations of weight-supporting ceiling receptacles and weight-supporting attachment fittings.

This section permits a combination of a weight-supporting ceiling receptacle and a weight-supporting attachment fitting mounted to an outlet box. It allows a quick connect/disconnect for a luminaire or ceiling fans. The quick connection combination locks the device in place and supports its weight as well as acts as a load make-and-break mechanism.

314.28 Pull and Junction Boxes and Conduit Bodies. Boxes and conduit bodies used as pull or junction boxes shall comply with 314.28(A) through (E).

Exception: Terminal housings supplied with motors shall comply with the provisions of 430.12.

(A) Minimum Size. For raceways containing conductors of 4 AWG or larger that are required to be insulated, and for cables containing conductors of 4 AWG or larger, the minimum dimensions of pull or junction boxes installed in a raceway or cable run shall comply with 314.28(A)(1) through (A)(3). Where an enclosure dimension is to be calculated based on the diameter of entering raceways, the diameter shall be the metric designator (trade size) expressed in the units of measurement employed.

Section 314.28(A) addresses raceways or cables that contain conductors 4 AWG or larger and are required to be insulated. One conductor that is not required to be insulated is a grounding electrode conductor (GEC). Thus, where a GEC is installed as the sole conductor within a raceway, conduit bodies used as part of that raceway system are not required to comply with enclosure dimensions in 314.28(A).

(1) Straight Pulls. In straight pulls, the length of the box or conduit body shall be not less than eight times the metric designator (trade size) of the largest raceway.

For example, a straight pull with trade size 2 conduit would require a 16-inch-long pull box (8 × 2 in. = 16 in.). Although 16 inches is the required minimum length, a longer pull box might be desired for ease in handling conductors.

(2) Angle or U Pulls, or Splices. Where splices or where angle or U pulls are made, the distance between each raceway entry inside the box or conduit body and the opposite wall of the box or conduit body shall not be less than six times the metric designator (trade size) of the largest raceway in a row. This distance shall be increased for additional entries by the amount of the sum of the diameters of all other raceway entries in the same row on the same wall of the box. Each row shall be calculated individually, and the single row that provides the maximum distance shall be used.

Exception: Where a raceway or cable entry is in the wall of a box or conduit body opposite a removable cover, the distance from that wall to the cover shall be permitted to comply with the distance required for one wire per terminal in Table 312.6(A).

The distance between raceway entries enclosing the same conductor shall not be less than six times the metric designator (trade size) of the larger raceway.

When transposing cable size into raceway size in 314.28(A)(1) and (A)(2), the minimum metric designator (trade size) raceway required for the number and size of conductors in the cable shall be used.

The "six times" rule of 314.28(A)(2) applies to straight-through conduit entries if the conductors are spliced as part of the straight-through wiring. Adjusting the example in the preceding paragraph of trade size 2 conduit, if the conductors were spliced within the enclosure, the required pull box dimension could be reduced to a 12-inch-long pull box (6 × 2 in. = 12 in.).

In boxes where splices, angle pulls, or U pulls are made, the distance between each raceway entry inside the box and the opposite wall of the box must be not less than six times the trade diameter of the largest raceway, plus the distance for additional raceway entries (see Exhibit 314.10). The additional distance is calculated by adding the diameters of the other raceway entries in one row on the same side of the box. Raceway entries enclosing the same conductor are required to have a minimum separation between them (see Exhibit 314.11). The intent is to provide adequate space for the conductor to make the bend.

(3) Smaller Dimensions. Listed boxes or listed conduit bodies of dimensions less than those required in 314.28(A)(1) and (A)(2) shall be permitted for installations of combinations of conductors that are less than the maximum conduit or tubing fill (of conduits or tubing being used) permitted by Table 1 of Chapter 9.

Listed conduit bodies of dimensions less than those required in 314.28(A)(2), and having a radius of the curve to the centerline not less than that indicated in Table 2 of Chapter 9 for one-shot and full-shoe benders, shall be permitted for installations of combinations of conductors permitted by Table 1 of Chapter 9. These conduit bodies shall be marked to

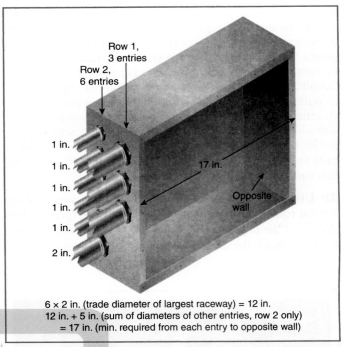

6 × 2 in. (trade diameter of largest raceway) = 12 in.
12 in. + 5 in. (sum of diameters of other entries, row 2 only)
= 17 in. (min. required from each entry to opposite wall)

EXHIBIT 314.10 An example showing calculations for splices, angle pulls, or U pulls.

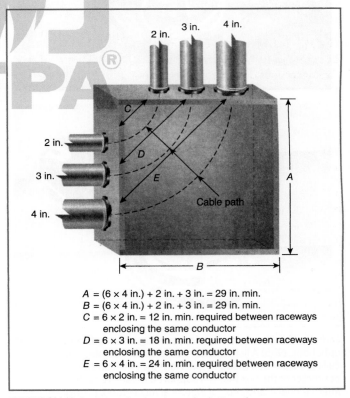

A = (6 × 4 in.) + 2 in. + 3 in. = 29 in. min.
B = (6 × 4 in.) + 2 in. + 3 in. = 29 in. min.
C = 6 × 2 in. = 12 in. min. required between raceways enclosing the same conductor
D = 6 × 3 in. = 18 in. min. required between raceways enclosing the same conductor
E = 6 × 4 in. = 24 in. min. required between raceways enclosing the same conductor

EXHIBIT 314.11 An example showing calculations for raceways enclosing the same conductor.

show they have been specifically evaluated in accordance with this provision.

Where the permitted combinations of conductors for which the box or conduit body has been listed are less than the maximum conduit or tubing fill permitted by Table 1 of Chapter 9, the box or conduit body shall be permanently marked with the maximum number and maximum size of conductors permitted. For other conductor sizes and combinations, the total cross-sectional area of the fill shall not exceed the cross-sectional area of the conductors specified in the marking, based on the type of conductor identified as part of the product listing.

> Informational Note: Unless otherwise specified, the applicable product standards evaluate the fill markings covered here based on conductors with Type XHHW insulation.

(B) Conductors in Pull or Junction Boxes. In pull boxes or junction boxes having any dimension over 1.8 m (6 ft), all conductors shall be cabled or racked up in an approved manner.

(C) Covers. All pull boxes, junction boxes, and conduit bodies shall be provided with covers compatible with the box or conduit body construction and suitable for the conditions of use. Where used, metal covers shall comply with the grounding requirements of 250.110.

(D) Permanent Barriers. Where permanent barriers are installed in a box, each section shall be considered as a separate box.

(E) Power Distribution Blocks. Power distribution blocks shall be permitted in pull and junction boxes over 1650 cm³ (100 in.³) for connections of conductors where installed in boxes and where the installation complies with 314.28(E)(1) through (E)(5).

> *Exception: Equipment grounding terminal bars shall be permitted in smaller enclosures.*

(1) Installation. Power distribution blocks installed in boxes shall be listed.

(2) Size. In addition to the overall size requirement in the first sentence of 314.28(A)(2), the power distribution block shall be installed in a box with dimensions not smaller than specified in the installation instructions of the power distribution block.

(3) Wire Bending Space. Wire bending space at the terminals of power distribution blocks shall comply with 312.6.

(4) Live Parts. Power distribution blocks shall not have uninsulated live parts exposed within a box, whether or not the box cover is installed.

(5) Through Conductors. Where the pull or junction boxes are used for conductors that do not terminate on the power distribution block(s), the through conductors shall be arranged so the power distribution block terminals are unobstructed following installation.

314.29 Boxes, Conduit Bodies, and Handhole Enclosures to Be Accessible. Boxes, conduit bodies, and handhole enclosures shall be installed so that wiring and devices contained in the boxes, conduit bodies, or handhole enclosures can be rendered accessible in accordance with 314.29(A) and (B).

(A) In Buildings and Other Structures. Boxes and conduit bodies shall be installed so the contained wiring and devices are accessible.

A junction box installed on a structural ceiling above a suspended ceiling is permitted, as long as it remains accessible. A suspended ceiling is not considered part of the structure.

(B) Underground. Underground boxes and handhole enclosures shall be installed so they are accessible without excavating sidewalks, paving, earth, or other substance that is to be used to establish the finished grade.

> *Exception: Listed boxes and handhole enclosures shall be permitted where covered by gravel, light aggregate, or noncohesive granulated soil if their location is effectively described and accessible for excavation. The location description shall be available to those authorized to access, maintain, or inspect the wiring.*

314.30 Handhole Enclosures. Handhole enclosures shall be designed and installed to withstand all loads likely to be imposed on them. They shall be identified for use in underground systems.

> Informational Note: See ANSI/SCTE 77-2013, *Specification for Underground Enclosure Integrity*, for additional information on deliberate and nondeliberate traffic loading that can be expected to bear on underground enclosures.

The loads referred to in this requirement is the weight or force of traffic loads on handhole enclosures, not electrical loads.

(A) Size. Handhole enclosures shall be sized in accordance with 314.28(A) for conductors operating at 1000 volts or below, and in accordance with 314.71 for conductors operating at over 1000 volts. For handhole enclosures without bottoms where the provisions of 314.28(A)(2), Exception, or 314.71(B)(1), Exception No. 1, apply, the measurement to the removable cover shall be taken from the end of the conduit or cable assembly.

(B) Wiring Entries. Underground raceways and cable assemblies entering a handhole enclosure shall extend into the enclosure, but they shall not be required to be mechanically connected to the enclosure.

(C) Enclosed Wiring. All enclosed conductors and any splices or terminations, if present, shall be listed as suitable for wet locations.

(D) Covers. Handhole enclosure covers shall have an identifying mark or logo that prominently identifies the function of the enclosure, such as "electric." Handhole enclosure covers shall require the use of tools to open, or they shall weigh over 45 kg (100 lb). Metal covers and other exposed conductive surfaces shall be bonded in accordance with 250.92 if the conductors in the handhole are service conductors, or in accordance with 250.96(A) if the conductors in the handhole are feeder or branch-circuit conductors.

Part III. Pull and Junction Boxes, Conduit Bodies, and Handhole Enclosures for Use on Systems over 1000 Volts, Nominal

314.70 General.

Δ **(A) Pull and Junction Boxes.** Where pull and junction boxes are used on systems over 1000 volts, the installation shall comply with Part III and with the following general provisions of this article:

(1) Part I, 314.2, 314.3, 314.4, and 314.5
(2) Part II, 314.15; 314.17; 314.20; 314.23(A), (B), or (G); 314.28(B); and 314.29
(3) Part III, 314.100(A) and (C); and 314.101

Δ **(B) Conduit Bodies.** Where conduit bodies are used on systems over 1000 volts, the installation shall comply with Part III and with the following general provisions of this article:

(1) Part I, 314.4, and 314.5
(2) Part II, 314.15; 314.17; 314.23(A), (E), or (G); 314.28(A)(3); and 314.29
(3) Part III, 314.100(A) and 314.101

Δ **(C) Handhole Enclosures.** Where handhole enclosures are used on systems over 1000 volts, the installation shall comply with Part III and with the following general provisions of this article:

(1) Part I, 314.3, 314.4, and 314.5
(2) Part II, 314.15, 314.17, 314.23(G), 314.28(B), 314.29, and 314.30

314.71 Size of Pull and Junction Boxes, Conduit Bodies, and Handhole Enclosures.
Pull and junction boxes and handhole enclosures shall provide approved space and dimensions for the installation of conductors, and they shall comply with the specific requirements of this section. Conduit bodies shall be permitted if they meet the dimensional requirements for boxes.

(A) For Straight Pulls. The length of the box shall not be less than 48 times the outside diameter, over sheath, of the largest shielded or lead-covered conductor or cable entering the box. The length shall not be less than 32 times the outside diameter of the largest nonshielded conductor or cable.

(B) For Angle or U Pulls.

(1) Distance to Opposite Wall. The distance between each cable or conductor entry inside the box and the opposite wall of the box shall not be less than 36 times the outside diameter, over sheath, of the largest cable or conductor. This distance shall be increased for additional entries by the amount of the sum of the outside diameters, over sheath, of all other cables or conductor entries through the same wall of the box.

Exception No. 1: Where a conductor or cable entry is in the wall of a box opposite a removable cover, the distance from that wall to the cover shall be permitted to be not less than the bending radius for the conductors as provided in 305.5.

Exception No. 2: Where cables are nonshielded and not lead covered, the distance of 36 times the outside diameter shall be permitted to be reduced to 24 times the outside diameter.

(2) Distance Between Entry and Exit. The distance between a cable or conductor entry and its exit from the box shall not be less than 36 times the outside diameter, over sheath, of that cable or conductor.

Exception: Where cables are nonshielded and not lead covered, the distance of 36 times the outside diameter shall be permitted to be reduced to 24 times the outside diameter.

(C) Removable Sides. One or more sides of any pull box shall be removable.

314.72 Construction and Installation Requirements.

(A) Corrosion Protection. Boxes shall be made of material inherently resistant to corrosion or shall be suitably protected, both internally and externally, by enameling, galvanizing, plating, or other means.

(B) Passing Through Partitions. Suitable bushings, shields, or fittings having smooth, rounded edges shall be provided where conductors or cables pass through partitions and at other locations where necessary.

(C) Complete Enclosure. Boxes shall provide a complete enclosure for the contained conductors or cables.

(D) Wiring Is Accessible. Boxes and conduit bodies shall be installed so that the conductors are accessible without removing any fixed part of the building or structure. Working space shall be provided in accordance with 110.34.

(E) Suitable Covers. Boxes shall be closed by suitable covers securely fastened in place. Underground box covers that weigh over 45 kg (100 lb) shall be considered meeting this requirement. Covers for boxes shall be permanently marked "DANGER — HIGH VOLTAGE — KEEP OUT." The marking shall be on the outside of the box cover and shall be readily visible. Letters shall be block type and at least 13 mm (½ in.) in height.

(F) Suitable for Expected Handling. Boxes and their covers shall be capable of withstanding the handling to which they are likely to be subjected.

Part IV. Construction Specifications

314.100 Metal Boxes, Conduit Bodies, and Fittings.

(A) Corrosion Resistant. Metal boxes, conduit bodies, and fittings shall be corrosion resistant or shall be well-galvanized, enameled, or otherwise properly coated inside and out to prevent corrosion.

> Informational Note: See 300.6 for limitation in the use of boxes and fittings protected from corrosion solely by enamel.

(B) Thickness of Metal. Sheet steel boxes not over 1650 cm³ (100 in.³) in size shall be made from steel not less than 1.59 mm (0.0625 in.) thick. The wall of a malleable iron box or conduit body and a die-cast or permanent-mold cast aluminum, brass, bronze, or zinc box or conduit body shall not be less than 2.38 mm (³⁄₃₂ in.) thick. Other cast metal boxes or conduit bodies shall have a wall thickness not less than 3.17 mm (⅛ in.).

Exception No. 1: Listed boxes and conduit bodies shown to have equivalent strength and characteristics shall be permitted to be made of thinner or other metals.

Exception No. 2: The walls of listed short radius conduit bodies, as covered in 314.16(C)(2), shall be permitted to be made of thinner metal.

(C) Metal Boxes Over 1650 cm³ (100 in.³). Metal boxes over 1650 cm³ (100 in.³) in size shall be constructed so as to be of ample strength and rigidity. If of sheet steel, the metal thickness shall not be less than 1.35 mm (0.053 in.) uncoated.

(D) Equipment Grounding Conductor Provisions. A means shall be provided in each metal box for the connection of an equipment grounding conductor. The means shall be permitted to be a tapped hole or equivalent.

For device boxes and other standard outlet boxes, the means for connecting the equipment grounding conductor is usually provided by the box manufacturer in the form of a 10-32 tapped hole marked "GR" or "GRD," or the equivalent symbol (⏚), next to the hole.

314.101 Covers. Metal covers shall be of the same material as the box or conduit body with which they are used, or they shall be lined with firmly attached insulating material that is not less than 0.79 mm (¹⁄₃₂ in.) thick, or they shall be listed for the purpose. Metal covers shall be the same thickness as the boxes or conduit body for which they are used, or they shall be listed for the purpose. Covers of porcelain or other approved insulating materials shall be permitted if of such form and thickness as to afford the required protection and strength.

△ **314.102 Bushings.** Covers of outlet boxes and conduit bodies having holes through which flexible cord pendants pass shall be provided with approved bushings or shall have smooth, well-rounded surfaces on which the cord will bear. Where individual conductors pass through a metal cover, a separate hole equipped with a bushing of suitable insulating material shall be provided for each conductor. Such separate holes shall be connected by a slot as required by 300.20.

314.103 Nonmetallic Boxes. Provisions for supports or other mounting means for nonmetallic boxes shall be outside of the box, or the box shall be constructed so as to prevent contact between the conductors in the box and the supporting screws.

314.104 Marking. All boxes and conduit bodies, covers, extension rings, plaster rings, and the like shall be durably and legibly marked with the manufacturer's name or trademark.

ARTICLE
315 Medium Voltage Conductors, Cable, Cable Joints, and Cable Terminations

Part I. General

315.1 Scope. This article covers the use, installation, construction specifications, and ampacities for Type MV medium voltage conductors, cable, cable joints, and cable terminations. This article includes voltages from 2001 volts to 35,000 volts ac, nominal and 2001 volts to 2500 volts dc, nominal.

This article is new for the 2023 edition. It consolidates medium voltage requirements that previously appeared in Article 311. Type MV cables are rated 2001 to 35,000 volts. MV cables installed in underground installations must comply with 305.15.

See also

315.36 and **315.44** for shielding requirements

315.6 Listing Requirements. Type MV cables, type MV cable joints, type MV cable terminations, connectors, and associated fittings shall be listed. The listing requirement for Type MV cable joints, cable terminations, and connectors shall be effective January 1, 2026.

Part II. Construction Specifications

315.10 Constructions and Applications. Type MV cables shall comply with the applicable provisions in 315.10(A) through (C).

(A) Conductor Application and Insulation. Conductor application and insulation shall comply with Table 315.10(A).

(B) Thickness of Insulation and Jacket for Nonshielded Insulated Conductors. Thickness of insulation and jacket for nonshielded solid dielectric insulated conductors rated 2001 volts to 5000 volts shall comply with Table 315.10(B).

(C) Thickness of Insulation for Shielded Insulated Conductors. Thickness of insulation for shielded solid dielectric insulated conductors rated 2001 volts to 35,000 volts shall comply with Table 315.10(C) and 315.10(C)(1) through (C)(3).

(1) 100 Percent Insulation Level. Cables shall be permitted to be applied where the system is provided with relay protection such that ground faults will be cleared as rapidly as possible but, in any case, within 1 minute. These cables are applicable to cable installations that are on grounded systems and shall be permitted to be used on other systems provided the above clearing requirements are met in completely de-energizing the faulted section.

(2) 133 Percent Insulation Level. Cables shall be permitted to be applied in situations where the clearing time requirements of the 100 percent level category cannot be met and the faulted section will be de-energized in a time not exceeding 1 hour. Cable shall be permitted to be used in 100 percent insulation level applications where the installation requires additional insulation.

(3) 173 Percent Insulation Level. Cables shall be permitted to be applied under all of the following conditions:

(1) In industrial establishments where the conditions of maintenance and supervision ensure only qualified persons service the installation
(2) Where the fault clearing time requirements of the 133 percent level category cannot be met
(3) Where an orderly shutdown is required to protect equipment and personnel
(4) Where the faulted section will be de-energized in an orderly shutdown

Cables shall be permitted to be used in 100 percent or 133 percent insulation level applications where the installation requires additional insulation.

> Informational Note: See UL 1072, *Medium-Voltage Power Cable*, ANSI/ICEA S-93-639, *American National Standard for 5-46kV Shielded Power Cable for Use in the Transmission and Distribution of Electric Energy*, and ICEA S-94-649 2013, *Standard for Concentric Neutral Cables Rated 5 through 46 kV for Medium Voltage Cables*.

315.12 Conductors.

(A) Minimum Size of Conductors. The minimum size of conductors shall be as shown in Table 315.12(A), except as permitted elsewhere in this *Code*.

(B) Conductor Material. Conductors shall be of aluminum, copper-clad aluminum, or copper unless otherwise specified.

Copper-clad aluminum conductors are manufactured so that copper forms a minimum of 10 percent of the cross-sectional area of a solid conductor or of each strand of a stranded conductor.

(C) Stranded Conductors. Where installed in raceways, conductors not specifically permitted or required elsewhere in this *Code* to be solid shall be stranded.

TABLE 315.10(A) *Conductor Application and Insulation Rated 2001 Volts and Higher*

Trade Name	Type Letter	Maximum Operating Temperature	Application Provision	Insulation	Outer Covering
Medium voltage solid dielectric	MV-90 MV-105*	90°C 105°C	Dry or wet locations	Thermoplastic or thermosetting	Jacket, sheath, or armor

*Where design conditions require maximum conductor temperatures above 90°C.

TABLE 315.10(B) *Thickness of Insulation and Jacket for Nonshielded Solid Dielectric Insulated Conductors Rated 2001 Volts to 5000 Volts*

Conductor Size (AWG or kcmil)	Dry Locations, Single Conductor						Wet or Dry Locations					
	Without Jacket Insulation		With Jacket				Single Conductor				Multiconductor Insulation*	
			Insulation		Jacket		Insulation		Jacket			
	mm	mils	mm	mils	mm	mils	mm	mils	mm	mils	mm	mils
8	2.79	110	2.29	90	0.76	30	3.18	125	2.03	80	2.29	90
6	2.79	110	2.29	90	0.76	30	3.18	125	2.03	80	2.29	90
4–2	2.79	110	2.29	90	1.14	45	3.18	125	2.03	80	2.29	90
1–2/0	2.79	110	2.29	90	1.14	45	3.18	125	2.03	80	2.29	90
3/0–4/0	2.79	110	2.29	90	1.65	65	3.18	125	2.41	95	2.29	90
213–500	3.05	120	2.29	90	1.65	65	3.56	140	2.79	110	2.29	90
501–750	3.30	130	2.29	90	1.65	65	3.94	155	3.18	125	2.29	90
751–1000	3.30	130	2.29	90	1.65	65	3.94	155	3.18	125	2.29	90
1001–1250	3.56	140	2.92	115	1.65	65	4.32	170	3.56	140	2.92	115
1251–1500	3.56	140	2.92	115	2.03	80	4.32	170	3.56	140	2.92	115
1501–2000	3.56	140	2.92	115	2.03	80	4.32	170	3.94	155	3.56	140

*Under a common overall covering such as a jacket, sheath, or armor.

TABLE 315.10(C) *Thickness of Insulation for Shielded Solid Dielectric Insulated Conductors Rated 2001 Volts to 35,000 Volts*

Conductor Size (AWG or kcmil)	2001–5000 Volts		5001–8000 Volts								8001–15,000 Volts						15,001–25,000 Volts					
	100 Percent Insulation Level		100 Percent Insulation Level		133 Percent Insulation Level		173 Percent Insulation Level		100 Percent Insulation Level		133 Percent Insulation Level		173 Percent Insulation Level		100 Percent Insulation Level		133 Percent Insulation Level		173 Percent Insulation Level	
	mm	mils	mm	mils	mm	mils	mm	mils	mm	mils	mm	mils	mm	mils	mm	mils	mm	mils	mm	mils
8	2.29	90	—	—	—	—	—	—	—	—	—	—	—	—	—	—	—	—	—	—
6–4	2.29	90	2.92	115	3.56	140	4.45	175	—	—	—	—	—	—	—	—	—	—	—	—
2	2.29	90	2.92	115	3.56	140	4.45	175	4.45	175	5.59	220	6.60	260	—	—	—	—	—	—
1	2.29	90	2.92	115	3.56	140	4.45	175	4.45	175	5.59	220	6.60	260	6.60	260	8.13	320	10.67	420
1/0–2000	2.29	90	2.92	115	3.56	140	4.45	175	4.45	175	5.59	220	6.60	260	6.60	260	8.13	320	10.67	420

Conductor Size (AWG or kcmil)	25,001–28,000 Volts						28,001–35,000 Volts					
	100 Percent Insulation Level		133 Percent Insulation Level		173 Percent Insulation Level		100 Percent Insulation Level		133 Percent Insulation Level		173 Percent Insulation Level	
	mm	mils	mm	mils	mm	mils	mm	mils	mm	mils	mm	mils
1	7.11	280	8.76	345	11.30	445	—	—	—	—	—	—
1/0–2000	7.11	280	8.76	345	11.30	445	8.76	345	10.67	420	14.73	580

TABLE 315.12(A) *Minimum Size of Conductors*

| Conductor Voltage Rating (Volts) | Minimum Conductor Size (AWG) |
	Copper, Aluminum, or Copper-Clad Aluminum
2001–5000	8
5001–8000	6
8001–15,000	2
15,001–28,000	1
28,001–35,000	1/0

Large-size conductors are required to be stranded for greater flexibility. This requirement does not apply to conductors outside of raceways, such as busbars.

315.14 Conductor Identification. Conductors that are intended for use as ungrounded conductors, whether used as a single conductor or in multiconductor cables, shall be finished to be clearly distinguishable from grounded and grounding conductors. Distinguishing markings shall not conflict in any manner with the surface markings required by 315.16(B)(1). Branch-circuit ungrounded conductors shall be identified in accordance with 210.5(C). Feeders shall be identified in accordance with 215.12.

315.16 Marking for Type MV Cables and Conductors.

(A) Required Information for Type MV Cables and Conductors. All conductors and cables shall be marked to indicate the following information, using the applicable method described in 315.16(B):

(1) The maximum rated voltage

(2) The proper type letter or letters for the type of wire or cable as specified elsewhere in this *Code*

(3) The manufacturer's name, trademark, or other distinctive marking by which the organization responsible for the product can be readily identified

(4) The AWG size or circular mil area

Informational Note: See Chapter 9, Table 8, Conductor Properties, for conductor area expressed in SI units for conductor sizes specified in AWG or circular mil area.

(B) Method of Marking for Type MV Cables and Conductors. One or more of the methods in 315.16(B)(1) through (B)(4) shall be used for marking of cable.

(1) Surface Marking. Cables shall be durably marked on the surface. The AWG size or circular mil area shall be repeated at intervals not exceeding 610 mm (24 in.). All other markings shall be repeated at intervals not exceeding 1.0 m (40 in.).

(2) Marker Tape. Metal-covered multiconductor cables shall employ a marker tape located within the cable and along its complete length.

(3) Tag Marking. Metal-covered, single-conductor cables shall be marked by means of a printed tag attached to the reel.

(4) Optional Marking of Wire Size. The information required in 315.16(A)(4) shall be permitted to be marked on the surface of the individual insulated conductors for multiconductor Type MC cable.

(C) Optional Markings. Cables shall be permitted to be marked to indicate special characteristics of the cable materials, such as limited smoke and sunlight resistance.

N **315.17 Marking for Type MV Cable Joints and Terminations.**

N **(A) Required Information for Type MV Cable Joints, Terminations, and Connectors.** All Type MV cable joints, cable terminations, and connectors shall be marked to indicate the following information, using one or more of the methods described in 315.17(B)(1) or (B)(2), and shall be permitted to be optionally marked as described in 315.17(C):

(1) The maximum rated voltage.
(2) The proper type letter or letters for the type of wire or cable as specified elsewhere in this *Code* that the cable joint or cable terminations is listed for use with.
(3) The manufacturer's name, trademark, or other distinctive marking by which the organizations responsible for the product can be readily identified.
(4) The conductor AWG size or circular mil area size, or range of sizes, that the cable joint or cable terminations is listed for use with.
(5) The cable outer diameter size, or size range, that the cable joint or cable termination is listed for use with.
(6) Connectors shall be marked with the following information; the marking shall also be on the unit container (the smallest container in which the connector is packaged):
    a. The manufacturer's name, trademark, or other distinctive marking by which the organization responsible for the product can be readily identified
    b. The manufacturer's catalog number
    c. The conductor AWG size or circular mil use range, and die number if applicable
    d. The type of conductor material(s) the connector is for use with

N **(B) Method of Marking for Type MV Cable Joints, Terminations, and Connectors.** One or both of the methods in 315.17(B)(1) or (B)(2) shall be used for the marking of cable joints, terminations, or connectors.

N **(1) Surface Marking.** Type MV cable joints, terminations, or connectors shall be durably marked on the surface.

N **(2) Tag Marking.** Type MV cable joints, terminations, or connectors shall be marked by means of a durably printed tag or label attached to joint or termination.

N **(C) Optional Markings.** Type MV cable joints and cable terminations shall be permitted to be marked to indicate special characteristics, such as limited smoke and sunlight resistance.

Part III. Installation

Δ **315.30 Installation.** A qualified person(s) with documented training and experience shall perform the installation and testing of Type MV cable. A qualified person(s) with documented training and experience in the installation of Type MV cable joints shall perform the installation of Type MV cable joints. A qualified person(s) with documented training and experience in the installation of Type MV cable terminations shall perform the installation of Type MV cable terminations.

> Informational Note No. 1: See ANSI/NECA/NCSCB 600-2020, *Standard for Installing and Maintaining Medium-Voltage Cable*, and IEEE 576, *Recommended Practice for Installation, Termination, and Testing of Insulated Power Cables as Used in Industrial and Commercial Applications*, for information about accepted industry practices and installation procedures for medium-voltage cable.
>
> Informational Note No. 2: Where medium-voltage cable is used for dc circuits, low frequency polarization can create hazardous voltages. When handling the cable these voltages could be present or could develop on dc stressed cable while the circuit is energized. Solidly grounding the cable prior to contacting, cutting or disconnecting cables in dc circuits is a method to discharge these voltages.

Δ **315.32 Uses Permitted.**

N **(A) Type MV Cable.** Type MV cable shall be permitted for use on power systems rated up to and including 35,000 volts, nominal, as follows:

(1) In wet or dry locations.
(2) In raceways.
(3) In cable trays, where identified for the use, in accordance with 392.10, 392.20(B), (C), and (D), 392.22(C), 392.30(B)(1), 392.46, 392.56, and 392.60. Type MV cable that has an overall metallic sheath or armor, complies with the requirements for Type MC cable, and is identified as "MV or MC" shall be permitted to be installed in cable trays in accordance with 392.10(B)(2).
(4) In messenger-supported wiring in accordance with Part II of Article 396.
(5) As exposed runs in accordance with 305.3. Type MV cable that has an overall metallic sheath or armor, complies with the requirements for Type MC cable, and is identified as "MV or MC" shall be permitted to be installed as exposed runs of metal-clad cable in accordance with 305.3.
(6) Corrosive conditions where exposed to oils, greases, vapors, gases, fumes, liquids, or other substances having a deleterious effect on the conductor or insulation shall be of a type suitable for the application.
(7) Conductors in parallel in accordance with 310.10(G).
(8) Type MV cable used where exposed to direct sunlight shall be identified for the use.
(9) Direct buried in accordance with 315.36.

N **(B) Type MV Cable Joints and Terminations.** Type MV cable joints and terminations shall be permitted for use on power systems rated up to and including 35,000 volts, nominal, as follows:

(1) Type MV cable joints and terminations, used where exposed to direct sunlight, shall be identified for the use.
(2) Direct buried.
(3) Where used intermittently or continuously submerged in water at a depth not exceeding 7 m (23 ft) type MV cable joints and terminations shall be identified for the use.

(4) The environmental operating temperature range shall be identified.

(5) Where used in one or more of the following conditions Type MV cable joints and terminations shall be identified for the use:
   a. Underground chambers
   b. Tunnels
   c. Conduits
   d. Manholes
   e. Vaults

(6) Corrosive conditions where exposed to oils, greases, vapors, gases, fumes, liquids, or other substances having a deleterious effect on the joint or termination shall be of a type suitable for the application.

(7) In cable trays, where identified for use, in accordance with 392.10, 392.20(B), (C) and (D), 392.22(C), 392.30(B)(1), 392.46, 392.56, and 392.60.

Informational Note No. 1: The "uses permitted" is not an all-inclusive list.

Informational Note No. 2: See IEEE-404, *IEEE Standard for Extruded and Laminated Dielectric Shielded Cable Joints Rated 2.5kV to 500kV*, for more information on cable joints. Cable joints are often referred to as splices. However, the term *splice* includes many other applications not included in the definition of a cable joint.

Informational Note No. 3: See IEEE-48, *IEEE Standard for Test Procedures and Requirements for Alternating-Current Cable Terminations Used on Shielded Cables Having Laminated Insulation Rated 2.5 kV through 765 kV or Extruded Insulation Rated 2.5 kV through 500 kV*, for information on terminations. Type MV cable terminations include terminations used to connect directly to equipment or insulators.

Informational Note No. 4: See IEEE-386, *IEEE Standard for Separable Insulated Connector Systems for Power Distribution Systems Rated 2.5kV through 35 kV*, and IEEE-1215, *IEEE Guide for the Application of Separable Insulated Connectors*, for more information on separable insulated connectors. Type MV cable terminations also include separable insulated connectors, which are a type of pluggable cable termination and can be used for connection to equipment, such as switchgear or transformers. A separable connector has a matching interface that the separable connector plugs into on the equipment, such as switchgear or transformers. Separable connectors can also be ganged together to form a distribution junction using specialized junction brackets.

Type MV cables intended for installation in cable trays in accordance with Article 392 are marked "For CT Use" or "For Use in Cable Trays." Where marked "MV or MC," the cable complies with the crush and impact rating associated with MC cable. Cable marked "MV or MC" is permitted to be installed in accordance with Article 330 as well as Article 392.

315.36 Direct-Burial Conductors. Type MV conductors and cables used for direct burial applications shall be shielded, identified for such use, and installed in accordance with 305.15.

Exception No. 1: Nonshielded multiconductor cables rated 2001 volts to 2400 volts shall be permitted if the cable has an overall metallic sheath or armor.

The metallic shield, sheath, or armor shall be connected to a grounding electrode conductor, a grounding busbar, or a grounding electrode.

Exception No. 2: Airfield lighting cable used in series circuits that are rated up to 5000 volts and are powered by regulators shall be permitted to be nonshielded.

Informational Note to Exception No. 2: Federal Aviation Administration (FAA) Advisory Circulars (ACs) provide additional practices and methods for airport lighting.

315.40 Support. Type MV cable terminated in equipment or installed in pull boxes or vaults shall be secured and supported by metallic or nonmetallic supports suitable to withstand the weight by cable ties listed and identified for securement and support, or other approved means, at intervals not exceeding 1.5 m (5 ft) from terminations or a maximum of 1.8 m (6 ft) between supports.

Δ **315.44 Shielding.** Nonshielded, ozone-resistant insulated conductors with a maximum phase-to-phase voltage of 5000 volts shall be permitted in Type MC cables in industrial establishments where the conditions of maintenance and supervision ensure that only qualified persons service the installation. For other establishments, solid dielectric insulated conductors operated above 2000 volts in permanent installations shall have ozone-resistant insulation and shall be shielded. All metallic insulation shields shall be connected to a grounding electrode conductor, a grounding busbar, an equipment grounding conductor, or a grounding electrode. Equipment grounding conductors installed with circuits using medium voltage cables shall be sized according to 250.190(C).

The construction of metal-armored cable provides enhanced reliability because the conductors have a concentric lay orientation and their insulation is protected from damage during installation. Nonshielded conductors within metal raceways do not provide the same level of reliability. Conductors-into-conduit installation is inconsistent and cannot guarantee that insulation will not be damaged nor that cables will be in concentric lay orientation.

Solid dielectric insulated conductors that are permanently installed and that operate at greater than 2000 volts are required to have ozone-resistant insulation and must be shielded with a grounded metallic shield. Shielding is accomplished by applying a metal tape or nonmetallic semiconducting tape around the conductor surface to prevent corona from forming and to reduce high-voltage stresses.

Corona is a faint glow adjacent to the surface of the electrical conductor at high voltage. If high-voltage stresses and a charging current are flowing between the conductor and ground (usually due to moisture), the surrounding atmosphere is ionized, and ozone — generated by an electric discharge in ordinary oxygen or air — is formed and will attack the conductor jacket and insulation, eventually breaking them down. The shield is at ground potential; therefore, no voltage above ground is present on the jacket outside the shield, thus preventing a discharge from the jacket and the subsequent formation of ozone.

Exhibits 315.1 and 315.2 illustrate shielded cable installations: a three-conductor cable of the shielded type, a stress-relief cone for an indoor cable terminator, and a stress cone on a single-conductor shielded cable terminating inside a pothead. In Exhibit 315.2 (right), a clamping ring provides a grounding connection between the copper shielding tape and the shield to the metallic base of the pothead.

Informational Note: The primary purposes of shielding are to confine the voltage stresses to the insulation, dissipate insulation leakage current, and drain off the capacitive charging current.

Exception No. 1: Nonshielded insulated conductors listed by a qualified testing laboratory shall be permitted for use up to 2400 volts under the following conditions:

(1) Conductors shall have insulation resistant to electric discharge and surface tracking, or the insulated conductor(s) shall be covered with a material resistant to ozone, electric discharge, and surface tracking.

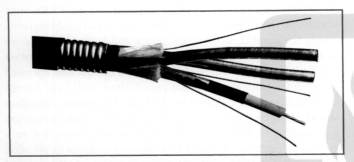

EXHIBIT 315.1 A three-conductor cable of the shielded type.

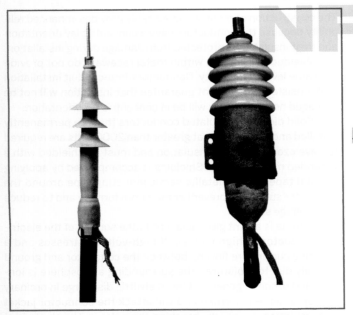

EXHIBIT 315.2 *(Left)* A one-piece, premolded stress-relief cone for indoor cable terminations of up to 35 kilovolts phase-to-phase. *(Right)* A stress cone on a single-conductor shielded cable terminating inside a pothead.

(2) Where used in wet locations, the insulated conductor(s) shall have an overall nonmetallic jacket or a continuous metallic sheath.

(3) Insulation and jacket thicknesses shall be in accordance with Table 315.10(B).

Exception No. 2: Nonshielded insulated conductors listed by a qualified testing laboratory shall be permitted for use up to 5000 volts to replace existing nonshielded conductors, on existing equipment in industrial establishments only, under the following conditions:

(1) Where the condition of maintenance and supervision ensures that only qualified personnel install and service the installation.

(2) Conductors shall have insulation resistant to electric discharge and surface tracking, or the insulated conductor(s) shall be covered with a material resistant to ozone, electric discharge, and surface tracking.

(3) Where used in wet locations, the insulated conductor(s) shall have an overall nonmetallic jacket or a continuous metallic sheath.

(4) Insulation and jacket thicknesses shall be in accordance with Table 315.10(B).

Informational Note: Relocation or replacement of equipment may not comply with the term *existing* as related to this exception.

Where cable in an existing installation requires replacement, it might be preferable to replace a nonshielded cable with another nonshielded cable where existing raceways and termination enclosures do not provide adequate space for shielded conductors and their associated terminations.

Federal Aviation Administration (FAA) Advisory Circulars for airfield lighting cable permit certain circuits up to 5 kilovolts to be unshielded. Permitted uses of nonshielded cable in airfield lighting are provided in 315.36, Exception No. 2.

Specialized training and close adherence to manufacturers' instructions are essential for high-voltage cable installations.

Exception No. 3: Where permitted in 315.36, Exception No. 2.

N **315.45 Shielding at Type MV Cable Joints and Terminations.** Type MV cable joints and terminations shall be provided with means to connect the metallic insulation shield to ground if required.

Part IV. Ampacities

315.60 Ampacities of Conductors.

(A) General.

(1) Tables or Engineering Supervision. Ampacities for solid dielectric-insulated conductors shall be permitted to be determined by tables or under engineering supervision, as provided in 315.60(B) and (C).

The ampacity of Type MV cable installed in cable tray shall be determined in accordance with 392.80(B).

(2) Selection of Ampacity. Where more than one calculated or tabulated ampacity could apply for a given circuit length, the lowest value shall be used.

Exception: Where different ampacities apply to portions of a circuit, the higher ampacity shall be permitted to be used if the total portion(s) of the circuit with the lower ampacity does not exceed the lesser of 3.0 m (10 ft) or 10 percent of the total circuit.

Informational Note: See 110.40 for conductor temperature limitations due to termination provisions.

(B) Engineering Supervision. Under engineering supervision, conductor ampacities shall be permitted to be calculated by using the following general equation:

$$I = \sqrt{\frac{T_c - (T_a + \Delta T_d)}{R_{dc}(1 + Y_c)R_{ca}}} \times 10^3 \text{ amperes} \qquad \textbf{[315.60(B)]}$$

where:

T_c = conductor temperature (°C)
T_a = ambient temperature (°C)
ΔT_d = dielectric loss temperature rise
R_{dc} = dc resistance of conductor at temperature, T_c
Y_c = component ac resistance resulting from skin effect and proximity effect
R_{ca} = effective thermal resistance between conductor and surrounding ambient

Informational Note: The dielectric loss temperature rise (ΔT_d) is negligible for single circuit extruded dielectric cables rated below 46 kV.

(C) Tables. Ampacities for conductors rated 2001 volts to 35,000 volts shall be as specified in Table 315.60(C)(1) through Table 315.60(C)(20). Ampacities for ambient temperatures other than those specified in the ampacity tables shall be corrected in accordance with 315.60(D)(4).

Informational Note No. 1: See IEEE 835, *Standard Power Cable Ampacity Tables*, and the references therein for availability of all factors and constants for ampacities calculated in accordance with 315.60(A).

Informational Note No. 2: See 210.19, Informational Note, for voltage drop on branch circuits that this section does not take into consideration. See 215.2(A)(2), Informational Note No. 2, for voltage drop on feeders that this section does not take into consideration.

Δ **TABLE 315.60(C)(1)** *Ampacities of Insulated Single Copper Conductor Cables Triplexed in Air*

Conductor Size (AWG or kcmil)	Temperature Rating of Conductor			
	2001–5000 Volts Ampacity		5001–35,000 Volts Ampacity	
	90°C (194°F) Type MV-90	105°C (221°F) Type MV-105	90°C (194°F) Type MV-90	105°C (221°F) Type MV-105
8	65	74	—	—
6	90	99	100	110
4	120	130	130	140
2	160	175	170	195
1	185	205	195	225
1/0	215	240	225	255
2/0	250	275	260	295
3/0	290	320	300	340
4/0	335	375	345	390
250	375	415	380	430
350	465	515	470	525
500	580	645	580	650
750	750	835	730	820
1000	880	980	850	950

Note: Refer to 315.60(E) for the basis of ampacities, 315.10(A) for conductor maximum operating temperature and application, and 315.60(D)(4) for the ampacity correction factors where the ambient air temperature is other than 40°C (104°F).

Δ **TABLE 315.60(C)(2)** *Ampacities of Insulated Single Aluminum Conductor Cables Triplexed in Air*

Conductor Size (AWG or kcmil)	Temperature Rating of Conductor			
	2001–5000 Volts Ampacity		5001–35,000 Volts Ampacity	
	90°C (194°F) Type MV-90	105°C (221°F) Type MV-105	90°C (194°F) Type MV-90	105°C (221°F) Type MV-105
8	50	57	—	—
6	70	77	75	84
4	90	100	100	110
2	125	135	130	150
1	145	160	150	175
1/0	170	185	175	200
2/0	195	215	200	230
3/0	225	250	230	265
4/0	265	290	270	305
250	295	325	300	335
350	365	405	370	415
500	460	510	460	515
750	600	665	590	660
1000	715	800	700	780

Note: Refer to 315.60(E) for basis of ampacities, 315.10(A) for conductor maximum operating temperature and application, and 315.60(D)(4) for the ampacity correction factors where the ambient air temperature is other than 40°C (104°F).

Δ   **TABLE 315.60(C)(3)** *Ampacities of Insulated Single Copper Conductor Isolated in Air*

Conductor Size (AWG or kcmil)	Temperature Rating of Conductor					
	2001–5000 Volts Ampacity		5001–15,000 Volts Ampacity		15,001–35,000 Volts Ampacity	
	90°C (194°F) Type MV-90	105°C (221°F) Type MV-105	90°C (194°F) Type MV-90	105°C (221°F) Type MV-105	90°C (194°F) Type MV-90	105°C (221°F) Type MV-105
8	83	93	—	—	—	—
6	110	120	110	125	—	—
4	145	160	150	165	—	—
2	190	215	195	215	—	—
1	225	250	225	250	225	250
1/0	260	290	260	290	260	290
2/0	300	330	300	335	300	330
3/0	345	385	345	385	345	380
4/0	400	445	400	445	395	445
250	445	495	445	495	440	490
350	550	615	550	610	545	605
500	695	775	685	765	680	755
750	900	1000	885	990	870	970
1000	1075	1200	1060	1185	1040	1160
1250	1230	1370	1210	1350	1185	1320
1500	1365	1525	1345	1500	1315	1465
1750	1495	1665	1470	1640	1430	1595
2000	1605	1790	1575	1755	1535	1710

Note: Refer to 315.60(E) for the basis of ampacities, 315.10(A) for conductor maximum operating temperature and application, and 315.60(D)(4) for the ampacity correction factors where the ambient air temperature is other than 40°C (104°F).

Δ   **TABLE 315.60(C)(4)** *Ampacities of Insulated Single Aluminum Conductor Isolated in Air*

Conductor Size (AWG or kcmil)	Temperature Rating of Conductor					
	2001–5000 Volts Ampacity		5001–15,000 Volts Ampacity		15,001–35,000 Volts Ampacity	
	90°C (194°F) Type MV-90	105°C (221°F) Type MV-105	90°C (194°F) Type MV-90	105°C (221°F) Type MV-105	90°C (194°F) Type MV-90	105°C (221°F) Type MV-105
8	64	71	—	—	—	—
6	85	95	87	97	—	—
4	115	125	115	130	—	—
2	150	165	150	170	—	—
1	175	195	175	195	175	195
1/0	200	225	200	225	200	225
2/0	230	260	235	260	230	260
3/0	270	300	270	300	270	300
4/0	310	350	310	350	310	345
250	345	385	345	385	345	380
350	430	480	430	480	430	475
500	545	605	535	600	530	590
750	710	790	700	780	685	765
1000	855	950	840	940	825	920
1250	980	1095	970	1080	950	1055
1500	1105	1230	1085	1215	1060	1180
1750	1215	1355	1195	1335	1165	1300
2000	1320	1475	1295	1445	1265	1410

Note: Refer to 315.60(E) for the basis of ampacities, 315.10(A) for conductor maximum operating temperature and application, and 315.60(D)(4) for the ampacity correction factors where the ambient air temperature is other than 40°C (104°F).

Δ **TABLE 315.60(C)(5)** *Ampacities of an Insulated Three-Conductor Copper Cable Isolated in Air*

| Conductor Size (AWG or kcmil) | Temperature Rating of Conductor | | | |
| | 2001–5000 Volts Ampacity | | 5001–35,000 Volts Ampacity | |
	90°C (194°F) Type MV-90	105°C (221°F) Type MV-105	90°C (194°F) Type MV-90	105°C (221°F) Type MV-105
8	59	66	—	—
6	79	88	93	105
4	105	115	120	135
2	140	154	165	185
1	160	180	185	210
1/0	185	205	215	240
2/0	215	240	245	275
3/0	250	280	285	315
4/0	285	320	325	360
250	320	355	360	400
350	395	440	435	490
500	485	545	535	600
750	615	685	670	745
1000	705	790	770	860

Note: Refer to 315.60(E) for the basis of ampacities, 315.10(A) for conductor maximum operating temperature and application, and 315.60(D)(4) for the ampacity correction factors where the ambient air temperature is other than 40°C (104°F).

Δ **TABLE 315.60(C)(7)** *Ampacities of an Insulated Triplexed or Three Single-Conductor Copper Cables in Isolated Conduit in Air*

| Conductor Size (AWG or kcmil) | Temperature Rating of Conductor | | | |
| | 2001–5000 Volts Ampacity | | 5001–35,000 Volts Ampacity | |
	90°C (194°F) Type MV-90	105°C (221°F) Type MV-105	90°C (194°F) Type MV-90	105°C (221°F) Type MV-105
8	55	61	—	—
6	75	84	83	93
4	97	110	110	120
2	130	145	150	165
1	155	175	170	190
1/0	180	200	195	215
2/0	205	225	225	255
3/0	240	270	260	290
4/0	280	305	295	330
250	315	355	330	365
350	385	430	395	440
500	475	530	480	535
750	600	665	585	655
1000	690	770	675	755

Note: Refer to 315.60(E) for the basis of ampacities, 315.10(A) for conductor maximum operating temperature and application, and 315.60(D)(4) for the ampacity correction factors where the ambient air temperature is other than 40°C (104°F).

Δ **TABLE 315.60(C)(6)** *Ampacities of an Insulated Three-Conductor Aluminum Cable Isolated in Air*

| Conductor Size (AWG or kcmil) | Temperature Rating of Conductor | | | |
| | 2001–5000 Volts Ampacity | | 5001–35,000 Volts Ampacity | |
	90°C (194°F) Type MV-90	105°C (221°F) Type MV-105	90°C (194°F) Type MV-90	105°C (221°F) Type MV-105
8	46	51	—	—
6	61	68	72	80
4	81	90	95	105
2	110	120	125	145
1	125	140	145	165
1/0	145	160	170	185
2/0	170	185	190	215
3/0	195	215	220	245
4/0	225	250	255	285
250	250	280	280	315
350	310	345	345	385
500	385	430	425	475
750	495	550	540	600
1000	585	650	635	705

Note: Refer to 315.60(E) for the basis of ampacities, 315.10(A) for conductor maximum operating temperature and application, and 315.60(D)(4) for the ampacity correction factors where the ambient air temperature is other than 40°C (104°F).

Δ **TABLE 315.60(C)(8)** *Ampacities of an Insulated Triplexed or Three Single-Conductor Aluminum Cables in Isolated Conduit in Air*

| Conductor Size (AWG or kcmil) | Temperature Rating of Conductor | | | |
| | 2001–5000 Volts Ampacity | | 5001–35,000 Volts Ampacity | |
	90°C (194°F) Type MV-90	105°C (221°F) Type MV-105	90°C (194°F) Type MV-90	105°C (221°F) Type MV-105
8	43	48	—	—
6	58	65	65	72
4	76	85	84	94
2	100	115	115	130
1	120	135	130	150
1/0	140	155	150	170
2/0	160	175	175	200
3/0	190	210	200	225
4/0	215	240	230	260
250	250	280	255	290
350	305	340	310	350
500	380	425	385	430
750	490	545	485	540
1000	580	645	565	640

Note: Refer to 315.60(E) for the basis of ampacities, 315.10(A) for conductor maximum operating temperature and application, and 315.60(D)(4) for the ampacity correction factors where the ambient air temperature is other than 40°C (104°F).

Δ **TABLE 315.60(C)(9)** *Ampacities of an Insulated Three-Conductor Copper Cable in Isolated Conduit in Air*

| Conductor Size (AWG or kcmil) | Temperature Rating of Conductor | | | |
| | 2001–5000 Volts Ampacity | | 5001–35,000 Volts Ampacity | |
	90°C (194°F) Type MV-90	105°C (221°F) Type MV-105	90°C (194°F) Type MV-90	105°C (221°F) Type MV-105
8	52	58	—	—
6	69	77	83	92
4	91	100	105	120
2	125	135	145	165
1	140	155	165	185
1/0	165	185	195	215
2/0	190	210	220	245
3/0	220	245	250	280
4/0	255	285	290	320
250	280	315	315	350
350	350	390	385	430
500	425	475	470	525
750	525	585	570	635
1000	590	660	650	725

Note: Refer to 315.60(E) for the basis of ampacities, 315.10(A) for conductor maximum operating temperature and application, and 315.60(D)(4) for the ampacity correction factors where the ambient air temperature is other than 40°C (104°F).

Δ **TABLE 315.60(C)(10)** *Ampacities of an Insulated Three-Conductor Aluminum Cable in Isolated Conduit in Air*

| Conductor Size (AWG or kcmil) | Temperature Rating of Conductor | | | |
| | 2001–5000 Volts Ampacity | | 5001–35,000 Volts Ampacity | |
	90°C (194°F) Type MV-90	105°C (221°F) Type MV-105	90°C (194°F) Type MV-90	105°C (221°F) Type MV-105
8	41	46	—	—
6	53	59	64	71
4	71	79	84	94
2	96	105	115	125
1	110	125	130	145
1/0	130	145	150	170
2/0	150	165	170	190
3/0	170	190	195	220
4/0	200	225	225	255
250	220	245	250	280
350	275	305	305	340
500	340	380	380	425
750	430	480	470	520
1000	505	560	550	615

Note: Refer to 315.60(E) for the basis of ampacities, 315.10(A) for conductor maximum operating temperature and application, and 315.60(D)(4) for the ampacity correction factors where the ambient air temperature is other than 40°C (104°F).

Δ **TABLE 315.60(C)(11)** *Ampacities of Three Single-Insulated Copper Conductors in Underground Electrical Ducts (Three Conductors per Electrical Duct)*

| Conductor Size (AWG or kcmil) | Temperature Rating of Conductor | | | |
| | 2001–5000 Volts Ampacity | | 5001–35,000 Volts Ampacity | |
	90°C (194°F) Type MV-90	105°C (221°F) Type MV-105	90°C (194°F) Type MV-90	105°C (221°F) Type MV-105
One Circuit [See Figure 315.60(D)(3), Detail 1.]				
8	64	69	—	—
6	85	92	90	97
4	110	120	115	125
2	145	155	155	165
1	170	180	175	185
1/0	195	210	200	215
2/0	220	235	230	245
3/0	250	270	260	275
4/0	290	310	295	315
250	320	345	325	345
350	385	415	390	415
500	470	505	465	500
750	585	630	565	610
1000	670	720	640	690
Three Circuits [See Figure 315.60(D)(3), Detail 2.]				
8	56	60	—	—
6	73	79	77	83
4	95	100	99	105
2	125	130	130	135
1	140	150	145	155
1/0	160	175	165	175
2/0	185	195	185	200
3/0	210	225	210	225
4/0	235	255	240	255
250	260	280	260	280
350	315	335	310	330
500	375	405	370	395
750	460	495	440	475
1000	525	565	495	535
Six Circuits [See Figure 315.60(D)(3), Detail 3.]				
8	48	52	—	—
6	62	67	64	68
4	80	86	82	88
2	105	110	105	115
1	115	125	120	125
1/0	135	145	135	145
2/0	150	160	150	165
3/0	170	185	170	185
4/0	195	210	190	205
250	210	225	210	225
350	250	270	245	265
500	300	325	290	310
750	365	395	350	375
1000	410	445	390	415

Note: Refer to 315.60(F) for basis of ampacities and Table 315.10(A) for the temperature rating of the conductor.

△ **TABLE 315.60(C)(12)** *Ampacities of Three Single-Insulated Aluminum Conductors in Underground Electrical Ducts (Three Conductors per Electrical Duct)*

Conductor Size (AWG or kcmil)	2001–5000 Volts Ampacity		5001–35,000 Volts Ampacity	
	90°C (194°F) Type MV-90	105°C (221°F) Type MV-105	90°C (194°F) Type MV-90	105°C (221°F) Type MV-105
One Circuit [See Figure 315.60(D)(3), Detail 1.]				
8	50	54	—	—
6	66	71	70	75
4	86	93	91	98
2	115	125	120	130
1	130	140	135	145
1/0	150	160	155	165
2/0	170	185	175	190
3/0	195	210	200	215
4/0	225	245	230	245
250	250	270	250	270
350	305	325	305	330
500	370	400	370	400
750	470	505	455	490
1000	545	590	525	565
Three Circuits [See Figure 315.60(D)(3), Detail 2.]				
8	44	47	—	—
6	57	61	60	65
4	74	80	77	83
2	96	105	100	105
1	110	120	110	120
1/0	125	135	125	140
2/0	145	155	145	155
3/0	160	175	165	175
4/0	185	200	185	200
250	205	220	200	220
350	245	265	245	260
500	295	320	290	315
750	370	395	355	385
1000	425	460	405	440
Six Circuits [See Figure 315.60(D)(3), Detail 3.]				
8	38	41	—	—
6	48	52	50	54
4	62	67	64	69
2	80	86	80	88
1	91	98	90	99
1/0	105	110	105	110
2/0	115	125	115	125
3/0	135	145	130	145
4/0	150	165	150	160
250	165	180	165	175
350	195	210	195	210
500	240	255	230	250
750	290	315	280	305
1000	335	360	320	345

Note: Refer to 315.60(F) for basis of ampacities and Table 315.10(A) for the temperature rating of the conductor.

△ **TABLE 315.60(C)(13)** *Ampacities of Three Insulated Copper Conductors Cabled Within an Overall Covering (Three-Conductor Cable) in Underground Electrical Ducts (One Cable per Electrical Duct)*

Conductor Size (AWG or kcmil)	2001–5000 Volts Ampacity		5001–35,000 Volts Ampacity	
	90°C (194°F) Type MV-90	105°C (221°F) Type MV-105	90°C (194°F) Type MV-90	105°C (221°F) Type MV-105
One Circuit [See Figure 315.60(D)(3), Detail 1.]				
8	59	64	—	—
6	78	84	88	95
4	100	110	115	125
2	135	145	150	160
1	155	165	170	185
1/0	175	190	195	210
2/0	200	220	220	235
3/0	230	250	250	270
4/0	265	285	285	305
250	290	315	310	335
350	355	380	375	400
500	430	460	450	485
750	530	570	545	585
1000	600	645	615	660
Three Circuits [See Figure 315.60(D)(3), Detail 2.]				
8	53	57	—	—
6	69	74	75	81
4	89	96	97	105
2	115	125	125	135
1	135	145	140	155
1/0	150	165	160	175
2/0	170	185	185	195
3/0	195	210	205	220
4/0	225	240	230	250
250	245	265	255	270
350	295	315	305	325
500	355	380	360	385
750	430	465	430	465
1000	485	520	485	515
Six Circuits [See Figure 315.60(D)(3), Detail 3.]				
8	46	50	—	—
6	60	65	63	68
4	77	83	81	87
2	98	105	105	110
1	110	120	115	125
1/0	125	135	130	145
2/0	145	155	150	160
3/0	165	175	170	180
4/0	185	200	190	200
250	200	220	205	220
350	240	270	245	275
500	290	310	290	305
750	350	375	340	365
1000	390	420	380	405

Note: Refer to 315.60(F) for basis of ampacities and Table 315.10(A) for the temperature rating of the conductor.

Δ **TABLE 315.60(C)(14)** *Ampacities of Three Insulated Aluminum Conductors Cabled Within an Overall Covering (Three-Conductor Cable) in Underground Electrical Ducts (One Cable per Electrical Duct)*

	Temperature Rating of Conductor			
	2001–5000 Volts Ampacity		5001–35,000 Volts Ampacity	
Conductor Size (AWG or kcmil)	90°C (194°F) Type MV-90	105°C (221°F) Type MV-105	90°C (194°F) Type MV-90	105°C (221°F) Type MV-105
One Circuit [See Figure 315.60(D)(3), Detail 1.]				
8	46	50	—	—
6	61	66	69	74
4	80	86	89	96
2	105	110	115	125
1	120	130	135	145
1/0	140	150	150	165
2/0	160	170	170	185
3/0	180	195	195	210
4/0	205	220	220	240
250	230	245	245	265
350	280	310	295	315
500	340	365	355	385
750	425	460	440	475
1000	495	535	510	545
Three Circuits [See Figure 315.60(D)(3), Detail 2.]				
8	41	44	—	—
6	54	58	59	64
4	70	75	75	81
2	90	97	100	105
1	105	110	110	120
1/0	120	125	125	135
2/0	135	145	140	155
3/0	155	165	160	175
4/0	175	185	180	195
250	190	205	200	215
350	230	250	240	255
500	280	300	285	305
750	345	375	350	375
1000	400	430	400	430
Six Circuits [See Figure 315.60(D)(3), Detail 3.]				
8	36	39	—	—
6	46	50	49	53
4	60	65	63	68
2	77	83	80	86
1	87	94	90	98
1/0	99	105	105	110
2/0	110	120	115	125
3/0	130	140	130	140
4/0	145	155	150	160
250	160	170	160	170
350	190	205	190	205
500	230	245	230	245
750	280	305	275	295
1000	320	345	315	335

Note: Refer to 315.60(F) for basis of ampacities and Table 315.10(A) for the temperature rating of the conductor.

Δ **TABLE 315.60(C)(15)** *Ampacities of Single Insulated Copper Conductors Directly Buried in Earth*

	Temperature Rating of Conductor			
	2001–5000 Volts Ampacity		5001–35,000 Volts Ampacity	
Conductor Size (AWG or kcmil)	90°C (194°F) Type MV-90	105°C (221°F) Type MV-105	90°C (194°F) Type MV-90	105°C (221°F) Type MV-105
One Circuit, Three Conductors [See Figure 315.60(D)(3), Detail 9.]				
8	110	115	—	—
6	140	150	130	140
4	180	195	170	180
2	230	250	210	225
1	260	280	240	260
1/0	295	320	275	295
2/0	335	365	310	335
3/0	385	415	355	380
4/0	435	465	405	435
250	470	510	440	475
350	570	615	535	575
500	690	745	650	700
750	845	910	805	865
1000	980	1055	930	1005
Two Circuits, Six Conductors [See Figure 315.60(D)(3), Detail 10.]				
8	100	110	—	—
6	130	140	120	130
4	165	180	160	170
2	215	230	195	210
1	240	260	225	240
1/0	275	295	255	275
2/0	310	335	290	315
3/0	355	380	330	355
4/0	400	430	375	405
250	435	470	410	440
350	520	560	495	530
500	630	680	600	645
750	775	835	740	795
1000	890	960	855	920

Note: Refer to 315.60(F) for basis of ampacities and Table 315.10(A) for the temperature rating of the conductor.

Δ **TABLE 315.60(C)(16)** *Ampacities of Single Insulated Aluminum Conductors Directly Buried in Earth*

Conductor Size (AWG or kcmil)	Temperature Rating of Conductor			
	2001–5000 Volts Ampacity		5001–35,000 Volts Ampacity	
	90°C (194°F) Type MV-90	105°C (221°F) Type MV-105	90°C (194°F) Type MV-90	105°C (221°F) Type MV-105
One Circuit, Three Conductors [See Figure 315.60(D)(3), Detail 9.]				
8	85	90	—	—
6	110	115	100	110
4	140	150	130	140
2	180	195	165	175
1	205	220	185	200
1/0	230	250	215	230
2/0	265	285	245	260
3/0	300	320	275	295
4/0	340	365	315	340
250	370	395	345	370
350	445	480	415	450
500	540	580	510	545
750	665	720	635	680
1000	780	840	740	795
Two Circuits, Six Conductors [See Figure 315.60(D)(3), Detail 10.]				
8	80	85	—	—
6	100	110	95	100
4	130	140	125	130
2	165	180	155	165
1	190	200	175	190
1/0	215	230	200	215
2/0	245	260	225	245
3/0	275	295	255	275
4/0	310	335	290	315
250	340	365	320	345
350	410	440	385	415
500	495	530	470	505
750	610	655	580	625
1000	710	765	680	730

Note: Refer to 315.60(F) for basis of ampacities and Table 315.10(A) for the temperature rating of the conductor.

Δ **TABLE 315.60(C)(17)** *Ampacities of Three Insulated Copper Conductors Cabled Within an Overall Covering (Three-Conductor Cable), Directly Buried in Earth*

Conductor Size (AWG or kcmil)	Temperature Rating of Conductor			
	2001–5000 Volts Ampacity		5001–35,000 Volts Ampacity	
	90°C (194°F) Type MV-90	105°C (221°F) Type MV-105	90°C (194°F) Type MV-90	105°C (221°F) Type MV-105
One Circuit [See Figure 315.60(D)(3), Detail 5.]				
8	85	89	—	—
6	105	115	115	120
4	135	150	145	155
2	180	190	185	200
1	200	215	210	225
1/0	230	245	240	255
2/0	260	280	270	290
3/0	295	320	305	330
4/0	335	360	350	375
250	365	395	380	410
350	440	475	460	495
500	530	570	550	590
750	650	700	665	720
1000	730	785	750	810
Two Circuits [See Figure 315.60(D)(3), Detail 6.]				
8	80	84	—	—
6	100	105	105	115
4	130	140	135	145
2	165	180	170	185
1	185	200	195	210
1/0	215	230	220	235
2/0	240	260	250	270
3/0	275	295	280	305
4/0	310	335	320	345
250	340	365	350	375
350	410	440	420	450
500	490	525	500	535
750	595	640	605	650
1000	665	715	675	730

Note: Refer to 315.60(F) for basis of ampacities and Table 315.10(A) for the temperature rating of the conductor.

△ **TABLE 315.60(C)(18)** *Ampacities of Three Insulated Aluminum Conductors Cabled Within an Overall Covering (Three-Conductor Cable), Directly Buried in Earth*

	Temperature Rating of Conductor			
	2001–5000 Volts Ampacity		5001–35,000 Volts Ampacity	
Conductor Size (AWG or kcmil)	90°C (194°F) Type MV-90	105°C (221°F) Type MV-105	90°C (194°F) Type MV-90	105°C (221°F) Type MV-105
One Circuit [See Figure 315.60(D)(3), Detail 5.]				
8	65	70	—	—
6	80	88	90	95
4	105	115	115	125
2	140	150	145	155
1	155	170	165	175
1/0	180	190	185	200
2/0	205	220	210	225
3/0	230	250	240	260
4/0	260	280	270	295
250	285	310	300	320
350	345	375	360	390
500	420	450	435	470
750	520	560	540	580
1000	600	650	620	665
Two Circuits [See Figure 315.60(D)(3), Detail 6.]				
8	60	66	—	—
6	75	83	80	95
4	100	110	105	115
2	130	140	135	145
1	145	155	150	165
1/0	165	180	170	185
2/0	190	205	195	210
3/0	215	230	220	240
4/0	245	260	250	270
250	265	285	275	295
350	320	345	330	355
500	385	415	395	425
750	480	515	485	525
1000	550	590	560	600

Note: Refer to 315.60(F) for basis of ampacities and Table 315.10(A) for the temperature rating of the conductor.

△ **TABLE 315.60(C)(19)** *Ampacities of Three Triplexed Single Insulated Copper Conductors Directly Buried in Earth*

	Temperature Rating of Conductor			
	2001–5000 Volts Ampacity		5001–35,000 Volts Ampacity	
Conductor Size (AWG or kcmil)	90°C (194°F) Type MV-90	105°C (221°F) Type MV-105	90°C (194°F) Type MV-90	105°C (221°F) Type MV-105
One Circuit, Three Conductors [See Figure 315.60(D)(3), Detail 7.]				
8	90	95	—	—
6	120	130	115	120
4	150	165	150	160
2	195	205	190	205
1	225	240	215	230
1/0	255	270	245	260
2/0	290	310	275	295
3/0	330	360	315	340
4/0	375	405	360	385
250	410	445	390	410
350	490	580	470	505
500	590	635	565	605
750	725	780	685	740
1000	825	885	770	830
Two Circuits, Six Conductors [See Figure 315.60(D)(3), Detail 8.]				
8	85	90	—	—
6	110	115	105	115
4	140	150	140	150
2	180	195	175	190
1	205	220	200	215
1/0	235	250	225	240
2/0	265	285	255	275
3/0	300	320	290	315
4/0	340	365	325	350
250	370	395	355	380
350	445	480	425	455
500	535	575	510	545
750	650	700	615	660
1000	740	795	690	745

Note: Refer to 315.60(F) for basis of ampacities and Table 315.10(A) for the temperature rating of the conductor.

Δ **TABLE 315.60(C)(20)** *Ampacities of Three Triplexed Single Insulated Aluminum Conductors Directly Buried in Earth*

Conductor Size (AWG or kcmil)	Temperature Rating of Conductor			
	2001–5000 Volts Ampacity		5001–35,000 Volts Ampacity	
	90°C (194°F) Type MV-90	105°C (221°F) Type MV-105	90°C (194°F) Type MV-90	105°C (221°F) Type MV-105
One Circuit, Three Conductors [See Figure 315.60(D)(3), Detail 7.]				
8	70	75	—	—
6	90	100	90	95
4	120	130	115	125
2	155	165	145	155
1	175	190	165	175
1/0	200	210	190	205
2/0	225	240	215	230
3/0	255	275	245	265
4/0	290	310	280	305
250	320	350	305	325
350	385	420	370	400
500	465	500	445	480
750	580	625	550	590
1000	670	725	635	680
Two Circuits, Six Conductors [See Figure 315.60(D)(3), Detail 8.]				
8	65	70	—	—
6	85	95	85	90
4	110	120	105	115
2	140	150	135	145
1	160	170	155	170
1/0	180	195	175	190
2/0	205	220	200	215
3/0	235	250	225	245
4/0	265	285	255	275
250	290	310	280	300
350	350	375	335	360
500	420	455	405	435
750	520	560	485	525
1000	600	645	565	605

Note: Refer to 315.60(F) for basis of ampacities and Table 315.10(A) for the temperature rating of the conductor.

(D) Ampacity Adjustment.

(1) Grounded Shields. Ampacities shown in Table 315.60(C)(3), Table 315.60(C)(4), Table 315.60(C)(15), and Table 315.60(C)(16) shall apply for cables with shields grounded at one point only. Where shields for these cables are grounded at more than one point, ampacities shall be adjusted to take into consideration the heating due to shield currents.

Informational Note: Tables other than those listed contain the ampacity of cables with shields grounded at multiple points.

(2) Burial Depth. Where the burial depth of direct burial or electrical duct bank circuits is modified from the values shown in a figure or table, ampacities shall be permitted to

be modified as indicated in 315.60(D)(2)(a) and (D)(2)(b). No ampacity adjustments shall be required where the burial depth is decreased.

(a) Where burial depths are increased in part(s) of an electrical duct run, a decrease in ampacity of the conductors shall not be required, provided the total length of parts of the duct run increased in depth is less than 25 percent of the total run length.

(b) Where burial depths are deeper than shown in a specific underground ampacity table or figure, an ampacity derating factor of 6 percent per 300 mm (1 ft) increase in depth for all values of rho shall be permitted.

(3) Electrical Ducts Entering Equipment Enclosures. At locations where electrical ducts enter equipment enclosures from underground, spacing between such ducts, as shown in Figure 315.60(D)(3), shall be permitted to be reduced without requiring the ampacity of conductors therein to be reduced.

The term *electrical ducts*, as defined in Article 100, is used to differentiate them from other ducts, such as those used for air handling. The term is intended to include nonmetallic electrical ducts commonly used for underground wiring, as well as other raceways [such as rigid metal conduit (RMC), intermediate metal conduit (IMC), polyvinyl chloride (PVC) conduit, and high density polyethylene (HDPE) conduit] listed for use underground in earth or concrete.

(4) Ambient Temperature Correction. Ampacities for ambient temperatures other than those specified in the ampacity tables shall be corrected in accordance with Table 315.60(D)(4) or shall be permitted to be calculated using the following equation:

$$I' = I \sqrt{\frac{T_c - T_a'}{T_c - T_a}} \qquad [315.60(D)(4)]$$

where:

I' = ampacity corrected for ambient temperature
I = ampacity shown in the table for T_c and T_a
T_c = temperature rating of conductor (°C)
T_a' = new ambient temperature (°C)
T_a = ambient temperature used in the table (°C)

Informational Note: See 110.40 for ambient temperature adjustments for terminals.

(E) Ampacity in Air. Ampacities for conductors and cables in air shall be as specified in Table 315.60(C)(1) through Table 315.60(C)(10). Ampacities shall be based on the following:

(1) Conductor temperatures of 90°C (194°F) and 105°C (221°F)
(2) Ambient air temperature of 40°C (104°F)

Informational Note: See 315.60(D)(4) where the ambient air temperature is other than 40°C (104°F).

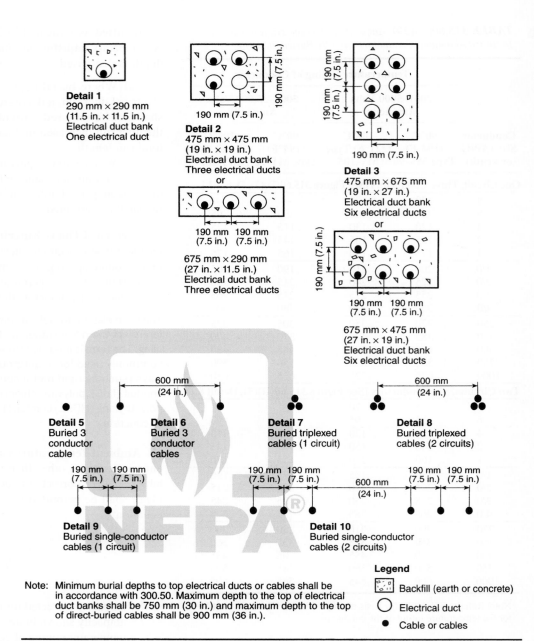

Note: Minimum burial depths to top electrical ducts or cables shall be in accordance with 300.50. Maximum depth to the top of electrical duct banks shall be 750 mm (30 in.) and maximum depth to the top of direct-buried cables shall be 900 mm (36 in.).

Legend

Backfill (earth or concrete)

○ Electrical duct

● Cable or cables

△ **FIGURE 315.60(D)(3)** *Cable Installation Dimensions for Use with Table 315.60(C)(11) Through Table 315.60(C)(20).*

TABLE 315.60(D)(4) *Ambient Temperature Correction Factors*

For ambient temperatures other than 40°C (104°F), multiply the allowable ampacities specified in the ampacity tables by the appropriate factor shown below.

Ambient Temperature (°C)	Temperature Rating of Conductor		Ambient Temperature (°F)
	90°C	105°C	
10 or less	1.26	1.21	50 or less
11–15	1.22	1.18	51–59
16–20	1.18	1.14	60–68
21–25	1.14	1.11	69–77
26–30	1.10	1.07	78–86
31–35	1.05	1.04	87–95
36–40	1.00	1.00	96–104
41–45	0.95	0.96	105–113
46–50	0.89	0.92	114–122
51–55	0.84	0.88	123–131
56–60	0.77	0.83	132–140
61–65	0.71	0.78	141–149
66–70	0.63	0.73	150–158
71–75	0.55	0.68	159–167
76–80	0.45	0.62	168–176
81–85	0.32	0.55	177–185
86–90	—	0.48	186–194
91–95	—	0.39	195–203
96–100	—	0.28	204–212

(F) Ampacity in Underground Electrical Ducts and Direct Buried in Earth. Ampacities for conductors and cables in underground electrical ducts and direct buried in earth shall be as specified in Table 315.60(C)(11) through Table 315.60(C)(20). Ampacities shall be based on the following:

(1) Ambient earth temperature of 20°C (68°F)
(2) Arrangement in accordance with Figure 315.60(D)(3)
(3) 100 percent load factor
(4) Thermal resistance (Rho) of 90
(5) Conductor temperatures 90°C (194°F) and 105°C (221°F)
(6) Minimum burial depths to the top electrical ducts or cables shall be in accordance with 305.15.
(7) Maximum depth to the top of electrical duct banks shall be 750 mm (30 in.), and maximum depth to the top of direct-buried cables shall be 900 mm (36 in.).

ARTICLE 320 Armored Cable: Type AC

Part I. General

320.1 Scope. This article covers the use, installation, and construction specifications for armored cable, Type AC.

Type AC cable is listed in sizes 14 AWG through 1 AWG copper and 12 AWG through 1 AWG aluminum or copper-clad aluminum and is rated at 600 volts or less. Exhibit 320.1 provides two examples of type AC cable. Note the paper wrap on the individual insulated conductors and the 16 AWG aluminum bonding strip that is in contact with the metal cable armor for its entire length. The bonding strip is not required to be brought into any box or enclosure with the circuit conductors and can be terminated at the cable fitting. The paper wrap and bonding strip are two features that distinguish Type AC cable from other metal jacketed cables.

At a glance, Type AC cable and interlocked armor Type MC cable look very similar, particularly in the sizes used for general lighting and receptacle branch circuits. However, there are construction differences between the two based on the respective product standards (UL 4, *Standard for Armored Cable,* for Type AC and UL1569, *Standard for Metal-Clad Cables,* for Type MC). More important, though, are the *NEC*® applications of the two cables in which the requirement calls for certain construction features depending on which of the cables is being used. Examples of such requirements include 517.13(A) and (B), 518.4(A), and 530.11.

Notwithstanding requirements that call for construction features available only in Type MC cable (such as an impervious nonmetallic outer covering), there are also requirements such as 516.7, 516.38, and 690.31(D)(1) that specifically call for the use of Type MC cable. Although many electricians generically refer to cables with an outer metal jacket as "BX," it must be understood that the cable is either a Type AC or a Type MC cable and recognizing the difference between them is important to getting the installation correct. See 310.8(B) for information on how Types AC and MC cables are required to be marked.

320.6 Listing Requirements. Type AC cable and associated fittings shall be listed.

Part II. Installation

320.10 Uses Permitted. Type AC cable shall be permitted as follows:

(1) For feeders and branch circuits in both exposed and concealed installations
(2) In cable trays
(3) In dry locations
(4) Embedded in plaster finish on brick or other masonry, except in damp or wet locations
(5) To be run or fished in the air voids of masonry block or tile walls where such walls are not exposed or subject to excessive moisture or dampness

Informational Note: The "Uses Permitted" is not an all-inclusive list.

320.12 Uses Not Permitted. Type AC cable shall not be used as follows:

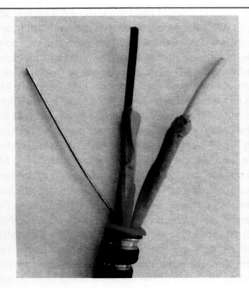

EXHIBIT 320.1 Two examples of Type AC cable. *(Courtesy of AFC Cable Systems, a Part of Atkore International)*

Type AC cable with insulated circuit conductors in a metal armor that also serves as the EGC

Type AC cable with insulated circuit conductors in a metal armor that also serves as one EGC, plus a second insulated wire-type EGC

(1) Where subject to physical damage

(2) In damp or wet locations

(3) In air voids of masonry block or tile walls where such walls are exposed or subject to excessive moisture or dampness

(4) Where exposed to corrosive conditions

(5) Embedded in plaster finish on brick or other masonry in damp or wet locations

320.15 Exposed Work. Exposed runs of cable, except as provided in 300.11(B), shall closely follow the surface of the building finish or of running boards. Exposed runs shall also be permitted to be installed on the underside of joists where supported at each joist and located so as not to be subject to physical damage.

320.17 Through or Parallel to Framing Members. Type AC cable shall be protected in accordance with 300.4(A), (C), and (D) where installed through or parallel to framing members.

320.23 In Accessible Attics. Type AC cables in accessible attics or roof spaces shall be installed as specified in 320.23(A) and (B).

(A) Cables Run Across the Top of Framing Members. Where run across the top of framing members, or across the face of rafters or studding within 2.1 m (7 ft) of the floor or horizontal surface, the cable shall be protected by guard strips that are at least as high as the cable. Where this space is not accessible by permanently installed stairs or ladders, protection shall only be required within 1.8 m (6 ft) of the nearest edge of the scuttle hole or attic entrance.

A permanently installed stairway is typically one that is built in place and if constructed new is built in compliance with the applicable building code. A permanently installed ladder is simply a ladder that is installed so that access to the attic can be accomplished without having to bring a portable ladder to the attic access location. Permanently installed stairs and ladders promote use of the attic space for storage, which can lead to damaged cables if not protected as specified in this section.

(B) Cable Installed Parallel to Framing Members. Where the cable is installed parallel to the sides of rafters, studs, or ceiling or floor joists, neither guard strips nor running boards shall be required, and the installation shall also comply with 300.4(D).

320.24 Bending Radius. Bends in Type AC cable shall be made such that the cable is not damaged. The radius of the curve of the inner edge of any bend shall not be less than five times the diameter of the Type AC cable.

320.30 Securing and Supporting.

(A) General. Type AC cable shall be supported and secured by staples; cable ties listed and identified for securement and support; straps, hangers, or similar fittings; or other approved means designed and installed so as not to damage the cable.

Type AC cable fittings shall be permitted as a means of cable support.

(B) Securing. Unless otherwise permitted, Type AC cable shall be secured within 300 mm (12 in.) of every outlet box, junction box, cabinet, or fitting and at intervals not exceeding 1.4 m (4½ ft).

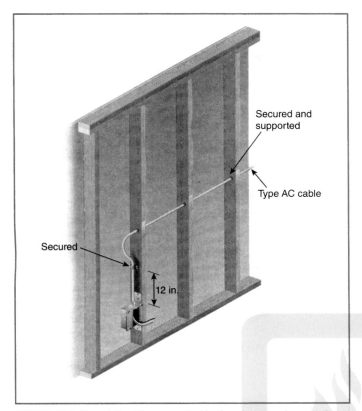

EXHBIT 320.2 Type AC cable supported by framing members and secured within 12 inches of the outlet box.

(C) Supporting. Unless otherwise permitted, Type AC cable shall be supported at intervals not exceeding 1.4 m (4½ ft).

Horizontal runs of Type AC cable installed in wooden or metal framing members or similar supporting means shall be considered supported and secured where such support does not exceed 1.4 m (4½ ft) intervals.

Bored or punched holes in framing members support and secure horizontal runs of Type AC cable. Additional securing and supporting are not needed, provided the cable is secured within 12 inches of the outlet and the framing members are less than 4½ feet apart. Exhibit 320.2 illustrates the difference between securing and supporting.

Δ **(D) Unsupported Cables.** Type AC cable shall be permitted to be unsupported and unsecured where the cable complies with any of the following:

(1) Is fished between access points through concealed spaces in finished buildings or structures and supporting is impracticable
(2) Is not more than 600 mm (2 ft) in length at terminals where flexibility is necessary
(3) Is not more than 1.8 m (6 ft) in length from the last point of cable support to the point of connection to a luminaire(s) or other electrical equipment and the cable and point of connection are within an accessible ceiling

EXHIBIT 320.3 An anti-short bushing designed to protect insulated conductors from abrasion. *(Courtesy of Eaton, Crouse-Hinds Division)*

320.40 Boxes and Fittings. At all points where the armor of AC cable terminates, a fitting shall be provided to protect wires from abrasion, unless the design of the outlet boxes or fittings is such as to afford equivalent protection, and, in addition, an insulating bushing or its equivalent protection shall be provided between the conductors and the armor. The connector or clamp by which the Type AC cable is fastened to boxes or cabinets shall be of such design that the insulating bushing or its equivalent will be visible for inspection. Where change is made from Type AC cable to other cable or raceway wiring methods, a box, fitting, or conduit body shall be installed at junction points as required in 300.15.

An anti-short bushing (sometimes referred to as a "red head") is shown in Exhibit 320.3. It is a plastic insert placed between the metal jacket of a Type AC cable and the insulated conductors at the point where the conductors emerge from the metal jacket of the Type AC cable. The anti-short bushing provides the insulated conductors with an additional level of short-circuit and ground-fault protection where they are most vulnerable.

Armored cable connectors are considered suitable for equipment grounding if installed in accordance with 300.10.

320.80 Ampacity. The ampacity shall be determined in accordance with 310.14.

(A) Thermal Insulation. Armored cable installed in thermal insulation shall have conductors rated at 90°C (194°F). The ampacity of cable installed in these applications shall not exceed that of a 60°C (140°F) rated conductor. The 90°C (194°F) rating shall be permitted to be used for ampacity adjustment and correction calculations; however, the ampacity shall not exceed that of a 60°C (140°F) rated conductor.

Where more than two Type AC cables containing two or more current-carrying conductors in each cable are installed in contact with thermal insulation, caulk, or sealing foam without maintaining spacing between cables, the ampacity of each conductor shall be adjusted in accordance with Table 310.15(C)(1).

Armored cable installed in thermal insulation has a decreased heat dissipation capacity. Cable marked "ACTH" indicates an armored cable rated 75°C and employing conductors having thermoplastic insulation. Cable marked "ACTHH" indicates an armored cable rated 90°C and employing conductors having thermoplastic insulation. Cable marked "ACHH" indicates armored cable rated 90°C and employing conductors having thermosetting insulation. Where conductors are rated 90°C, the 90°C ampacity can be used for derating (ampacity adjustment, correction, or both).

Thermal insulation and similar materials impede the dissipation of heat from the cables. No space between the cables further reduces the heat dissipation, and ampacity adjustment is required. This rule covering the bundles in contact with insulation, caulking, or foam offsets the permission in 310.15(C)(1)(d) to install bundled AC or MC cable without ampacity adjustment factors having to be applied.

(B) Cable Tray. The ampacity of Type AC cable installed in cable tray shall be determined in accordance with 392.80(A).

Part III. Construction Specifications

320.100 Construction. Type AC cable shall have an armor of flexible metal tape and shall have an internal bonding strip of copper or aluminum in intimate contact with the armor for its entire length.

The armor of Type AC cable is recognized as an equipment grounding conductor (EGC) by 250.118. This internal bonding strip is not required to be connected to an equipment grounding terminal and can be cut off at the termination of the armored cable or be bent back on the armor. Its purpose is to reduce the inductive reactance of the spiral armor and increase the armor's effectiveness as an equipment ground. Many installers use this strip to help prevent the insulating (anti-short) bushing required by 320.40 (also known as the "red head") from falling out during rough wiring.

320.104 Conductors. Insulated conductors shall be of a type listed in Table 310.4(1) or those identified for use in this cable. In addition, the conductors shall have an overall moisture-resistant and fire-retardant fibrous covering. For Type ACT, a moisture-resistant fibrous covering shall be required only on the individual conductors.

320.108 Equipment Grounding Conductor. Type AC cable shall provide an adequate path for fault current as required by 250.4(A)(5) or (B)(4) to act as an equipment grounding conductor.

320.120 Marking. The cable shall be marked in accordance with 310.8, except that Type AC shall have ready identification of the manufacturer by distinctive external markings on the cable armor throughout its entire length.

ARTICLE
322 — Flat Cable Assemblies: Type FC

Part I. General

322.1 Scope. This article covers the use, installation, and construction specifications for flat cable assemblies, Type FC.

End cap/puller   Channel coupling

Cable   Channel

Fixture hanger   Channel hanger

Cable pull-in guide

EXHIBIT 322.1 Basic components and accessories that may be used for an installation of Type FC cable. *(Courtesy of Legrand®)*

Type FC cable is an assembly of three or four parallel 10 AWG special stranded copper wires formed integrally with an insulating material web. The cable is marked with the size of the maximum branch circuit to which it can be connected, the cable type designation, the manufacturer's identification, the maximum working voltage, the conductor size, and the temperature rating. A marking accompanying the cable on a tag or reel indicates the special metal raceways and specific Type FC cable fittings with which the cable is intended to be used. Exhibits 322.1 and 322.2 show the basic components of this wiring method.

322.6 Listing Requirements. Type FC and associated fittings shall be listed.

Part II. Installation

322.10 Uses Permitted. Flat cable assemblies shall be permitted only as follows:

EXHIBIT 322.2 A luminaire hanger used with Type FC cable. *(Courtesy of Legrand®)*

(1) As branch circuits to supply suitable tap devices for lighting, small appliances, or small power loads. The rating of the branch circuit shall not exceed 30 amperes.

(2) Where installed for exposed work.

(3) In locations where they will not be subjected to physical damage. Where a flat cable assembly is installed less than 2.5 m (8 ft) above the floor or fixed working platform, it shall be protected by a cover identified for the use.

(4) In surface metal raceways identified for the use. The channel portion of the surface metal raceway systems shall be installed as complete systems before the flat cable assemblies are pulled into the raceways.

322.12 Uses Not Permitted. Flat cable assemblies shall not be used as follows:

(1) Where exposed to corrosive conditions, unless suitable for the application

(2) In hoistways or on elevators or escalators

(3) In any hazardous (classified) location, except as specifically permitted by other articles in this *Code*

(4) Outdoors or in wet or damp locations unless identified for the use

322.30 Securing and Supporting. The flat cable assemblies shall be supported by means of their special design features, within the surface metal raceways.

The surface metal raceways shall be supported as required for the specific raceway to be installed.

322.40 Boxes and Fittings.

(A) Dead Ends. Each flat cable assembly dead end shall be terminated in an end-cap device identified for the use.

The dead-end fitting for the enclosing surface metal raceway shall be identified for the use.

(B) Luminaire Hangers. Luminaire hangers installed with the flat cable assemblies shall be identified for the use.

(C) Fittings. Fittings to be installed with flat cable assemblies shall be designed and installed to prevent physical damage to the cable assemblies.

(D) Extensions. All extensions from flat cable assemblies shall be made by approved wiring methods, within the junction boxes, installed at either end of the flat cable assembly runs.

322.56 Splices and Taps.

(A) Splices. Splices shall be made in listed junction boxes.

Δ **(B) Taps.** Taps shall be made between any phase conductor and the grounded conductor or any other phase conductor by means of devices and fittings identified for the use. Tap devices shall be rated at not less than 15 amperes, or more than 300 volts to ground, and shall be marked in accordance with 322.120(C).

Part III. Construction Specifications

322.100 Construction. Flat cable assemblies shall consist of two, three, four, or five conductors.

322.104 Conductors. Flat cable assemblies shall have conductors of 10 AWG special stranded copper wires.

322.112 Insulation. The entire flat cable assembly shall be formed to provide a suitable insulation covering all the conductors and using one of the materials recognized in Table 310.4(1) for general branch-circuit wiring.

322.120 Marking.

(A) Temperature Rating. In addition to the provisions of 310.8, Type FC cable shall have the temperature rating durably marked on the surface at intervals not exceeding 600 mm (24 in.).

(B) Identification of Grounded Conductor. The grounded conductor shall be identified throughout its length by means of a distinctive and durable white or gray marking.

Informational Note: The color gray may have been used in the past as an ungrounded conductor. Care should be taken when working on existing systems.

(C) Terminal Block Identification. Terminal blocks identified for the use shall have distinctive and durable markings for color or word coding. The grounded conductor section shall have a white marking or other suitable designation. The next adjacent section of the terminal block shall have a black marking or other suitable designation. The next section shall have a red marking or other suitable designation. The final or outer section, opposite the grounded conductor section of the terminal block, shall have a blue marking or other suitable designation.

ARTICLE 324 Flat Conductor Cable: Type FCC

Part I. General

324.1 Scope. This article covers a field-installed wiring system for branch circuits incorporating Type FCC cable and associated accessories as defined by the article. The wiring system is designed for installation under carpet squares.

A Type FCC system is designed to provide a completely accessible, flexible power system. As shown in Exhibit 324.1, it also provides an easy method for reworking obsolete wiring systems currently in use in many office facilities. The carpet squares are not permitted to be larger than 1.0 meter by 1.0 meter, to comply with 324.41. This limitation provides ready access to the cable by lifting a carpet square. It also reduces the likelihood of an individual cutting through the carpet above the cable with a knife or razor blade and possibly penetrating the top shield of the cable.

324.6 Listing Requirements. Type FCC cable and associated fittings shall be listed.

Part II. Installation

324.10 Uses Permitted.

(A) Branch Circuits. Use of FCC systems shall be permitted both for general-purpose and appliance branch circuits and for individual branch circuits.

(B) Branch-Circuit Ratings.

(1) Voltage. Voltage between ungrounded conductors shall not exceed 300 volts. Voltage between ungrounded conductors and the grounded conductor shall not exceed 150 volts.

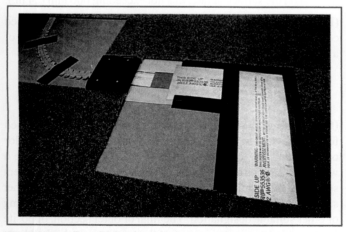

EXHIBIT 324.1 Type FCC cable installed beneath carpet squares. *(Courtesy of Tyco Electronics)*

(2) Current. General-purpose and appliance branch circuits shall have ratings not exceeding 20 amperes. Individual branch circuits shall have ratings not exceeding 30 amperes.

(C) Floors. Use of FCC systems shall be permitted on hard, sound, smooth, continuous floor surfaces made of concrete, ceramic, or composition flooring, wood, and similar materials.

(D) Walls. Use of FCC systems shall be permitted on wall surfaces in surface metal raceways.

(E) Damp Locations. Use of FCC systems in damp locations shall be permitted.

(F) Heated Floors. Materials used for floors heated in excess of 30°C (86°F) shall be identified as suitable for use at these temperatures.

(G) System Height. Any portion of an FCC system with a height above floor level exceeding 2.3 mm (0.090 in.) shall be tapered or feathered at the edges to floor level.

324.12 Uses Not Permitted. FCC systems shall not be used in the following locations:

(1) Outdoors or in wet locations
(2) Where subject to corrosive vapors
(3) In any hazardous (classified) location
(4) In residential buildings
(5) In school and hospital buildings, other than administrative office areas

Type FCC wiring systems are not permitted throughout school and hospital buildings except in parts of those buildings that are administrative office spaces.

324.18 Crossings. Crossings of more than two Type FCC cable runs shall not be permitted at any one point. Crossings of a Type FCC cable over or under a flat communications or signal cable shall be permitted. In each case, a grounded layer of metal shielding shall separate the two cables, and crossings of more than two flat cables shall not be permitted at any one point.

324.30 Securing and Supporting. All FCC system components shall be firmly anchored to the floor or wall using an adhesive or mechanical anchoring system identified for this use. Floors shall be prepared to ensure adherence of the FCC system to the floor until the carpet squares are placed.

324.40 Boxes and Fittings.

(A) Cable Connections and Insulating Ends. All Type FCC cable connections shall use connectors identified for their use, installed such that electrical continuity, insulation, and sealing against dampness and liquid spillage are provided. All bare cable ends shall be insulated and sealed against dampness and liquid spillage using listed insulating ends.

(B) Polarization of Connections. All receptacles and connections shall be constructed and installed so as to maintain proper polarization of the system.

(C) Shields.

(1) Top Shield. A metal top shield shall be installed over all floor-mounted Type FCC cable, connectors, and insulating ends. The top shield shall completely cover all cable runs, corners, connectors, and ends.

(2) Bottom Shield. A bottom shield shall be installed beneath all Type FCC cable, connectors, and insulating ends.

(D) Connection to Other Systems. Power feed, grounding connection, and shield system connection between the FCC system and other wiring systems shall be accomplished in a transition assembly identified for this use.

(E) Metal-Shield Connectors. Metal shields shall be connected to each other and to boxes, receptacle housings, self-contained devices, and transition assemblies using metal-shield connectors.

324.41 Floor Coverings. Floor-mounted Type FCC cable, cable connectors, and insulating ends shall be covered with carpet squares not larger than 1.0 m (39.37 in.) square. Carpet squares that are adhered to the floor shall be attached with release-type adhesives.

324.42 Devices.

(A) Receptacles. All receptacles, receptacle housings, and self-contained devices used with the FCC system shall be identified for this use and shall be connected to the Type FCC cable and metal shields. Connection from any equipment grounding conductor of the Type FCC cable shall be made to the shield system at each receptacle.

Δ **(B) Receptacles and Housings.** Receptacle housings and self-contained devices designed either for floor mounting or for in-wall or on-wall mounting shall be permitted for use with the FCC system. Receptacle housings and self-contained devices shall incorporate means for facilitating entry and termination of Type FCC cable and for electrically connecting the housing or device with the metal shield. Receptacles and self-contained devices shall comply with 406.4. Power and communications outlets installed together in common housing shall be permitted in accordance with 800.133(A)(3), Exception No. 2.

324.56 Splices and Taps.

(A) FCC Systems Alterations. Alterations to FCC systems shall be permitted. New cable connectors shall be used at new connection points to make alterations. It shall be permitted to leave unused cable runs and associated cable connectors in place and energized. All cable ends shall be covered with insulating ends.

(B) Transition Assemblies. All transition assemblies shall be identified for their use. Each assembly shall incorporate means

for facilitating entry of the Type FCC cable into the assembly, for connecting the Type FCC cable to grounded conductors, and for electrically connecting the assembly to the metal cable shields and to equipment grounding conductors.

324.60 Grounding and Bonding. All metal shields, boxes, receptacle housings, and self-contained devices shall be electrically continuous to the equipment grounding conductor of the supplying branch circuit. All such electrical connections shall be made with connectors identified for this use. The electrical resistivity of such shield system shall not be more than that of one conductor of the Type FCC cable used in the installation.

Part III. Construction Specifications

324.100 Construction.

(A) Type FCC Cable. Type FCC cable shall be listed for use with the FCC system and shall consist of three, four, or five flat copper conductors, one of which shall be an equipment grounding conductor.

(B) Shields.

(1) Materials and Dimensions. All top and bottom shields shall be of designs and materials identified for their use. Top shields shall be metal. Both metallic and nonmetallic materials shall be permitted for bottom shields.

(2) Resistivity. Metal shields shall have cross-sectional areas that provide for electrical resistivity of not more than that of one conductor of the Type FCC cable used in the installation.

324.101 Corrosion Resistance. Metal components of the system shall be either corrosion resistant, coated with corrosion-resistant materials, or insulated from contact with corrosive substances.

324.112 Insulation. The insulating material of the cable shall be moisture resistant and flame retardant. All insulating materials in the FCC systems shall be identified for their use.

324.120 Markings.

(A) Cable Marking. Type FCC cable shall be clearly and durably marked on both sides at intervals of not more than 610 mm (24 in.) with the information required by 310.8(A) and with the following additional information:

(1) Material of conductors
(2) Maximum temperature rating
(3) Ampacity

(B) Conductor Identification. Conductors shall be clearly and durably identified on both sides throughout their length as specified in 310.6.

ARTICLE 326 Integrated Gas Spacer Cable: Type IGS

Part I. General

326.1 Scope. This article covers the use, installation, and construction specifications for integrated gas spacer cable, Type IGS.

As illustrated in Exhibit 326.1, IGS cable consists of solid aluminum rod conductors, 250 kcmil minimum size. These conductors are insulated with dry kraft paper and are installed in a medium-density polyethylene gas pipe, minimum trade size 2, which is then filled with sulfur hexafluoride (SF$_6$) gas at a pressure of approximately 20 pounds per square inch.

Part II. Installation

326.10 Uses Permitted. Type IGS cable shall be permitted for use underground, including direct burial in the earth, as the following:

(1) Service-entrance conductors
(2) Feeder or branch-circuit conductors
(3) Service conductors, underground

326.12 Uses Not Permitted. Type IGS cable shall not be used as interior wiring or be exposed in contact with buildings.

326.24 Bending Radius. Where the coilable nonmetallic conduit and cable are bent for installation purposes or are flexed or bent during shipment or installation, the radius of the curve of the inner edge measured to the inside of the bend shall not be less than specified in Table 326.24.

326.26 Bends. A run of Type IGS cable between pull boxes or terminations shall not contain more than the equivalent of four quarter bends (360 degrees total), including those bends located immediately at the pull box or terminations.

326.40 Fittings. Terminations and splices for Type IGS cable shall be identified as a type that is suitable for maintaining the gas pressure within the conduit. A valve and cap shall be provided for each length of the cable and conduit to check the gas pressure or to inject gas into the conduit.

326.80 Ampacity. The ampacity of Type IGS cable shall not exceed the values shown in Table 326.80.

TABLE 326.24 *Minimum Radii of Bends*

Conduit Size		Minimum Radii	
Metric Designator	Trade Size	mm	in.
53	2	600	24
78	3	900	35
103	4	1150	45

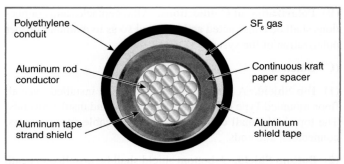

EXHIBIT 326.1 Cross section of single-conductor, 4750-kcmil Type IGS cable.

TABLE 326.80 *Ampacity of Type IGS Cable*

Size (kcmil)	Amperes	Size (kcmil)	Amperes
250	119	2500	376
500	168	3000	412
750	206	3250	429
1000	238	3500	445
1250	266	3750	461
1500	292	4000	476
1750	315	4250	491
2000	336	4500	505
2250	357	4750	519

Part III. Construction Specifications

326.104 Conductors. The conductors shall be solid aluminum rods, laid parallel, consisting of one to nineteen 12.7 mm (½ in.) diameter rods. The minimum conductor size shall be 250 kcmil, and the maximum size shall be 4750 kcmil.

326.112 Insulation. The insulation shall be dry kraft paper tapes and a pressurized sulfur hexafluoride gas (SF$_6$), both approved for electrical use. The nominal gas pressure shall be 138 kPa gauge (20 lb/in.^{2} gauge). The thickness of the paper spacer shall be as specified in Table 326.112.

TABLE 326.112 *Paper Spacer Thickness*

Size (kcmil)	Thickness	
	mm	in.
250–1000	1.02	0.040
1250–4750	1.52	0.060

326.116 Conduit. The conduit shall be a medium density polyethylene identified as suitable for use with natural gas rated pipe in metric designator 53, 78, or 103 (trade size 2, 3, or 4). The percent fill dimensions for the conduit are shown in Table 326.116.

The size of the conduit permitted for each conductor size shall be calculated for a percent fill not to exceed those found in Table 1, Chapter 9.

TABLE 326.116 *Conduit Dimensions*

Conduit Size		Actual Outside Diameter		Actual Inside Diameter	
Metric Designator	Trade Size	mm	in.	mm	in.
53	2	60	2.375	49.46	1.947
78	3	89	3.500	73.30	2.886
103	4	114	4.500	94.23	3.710

326.120 Marking. The cable shall be marked in accordance with 310.8(A), (B)(1), and (D).

ARTICLE
330 Metal-Clad Cable: Type MC

Part I. General

330.1 Scope. This article covers the use, installation, and construction specifications of metal-clad cable, Type MC.

Type MC cable is rated up to 2000 volts in sizes 18 AWG and larger for copper or nickel-coated copper and 12 AWG and larger for aluminum or copper-clad aluminum. Type MC cable rated 2400 to 35,000 volts is classified as medium-voltage cable, is marked "Type MV or MC," and is covered by Article 315. Type MC-HL cable rated up to 35,000 volts is suitable for hazardous locations and is provided with a gas/vapor tight continuous sheath. Composite electrical MC and optical fiber cables are classified as Type MC cable and are marked "Type MC-OF."

Type MC cable is available in three designs: interlocked metal tape, corrugated metal tube, and smooth metal tube. A nonmetallic jacket may be provided over the metal sheath. Cable construction must comply with 250.118(10) for the cable to be used as an equipment grounding conductor (EGC). It must be marked with the maximum rated voltage, the proper insulation-type letter or letters, and the AWG size or circular mil area.

Exhibit 330.1 shows some examples of Type MC cable. The basic standard to investigate cable in this category is UL 1569, *Standard for Metal-Clad Cables.* Summary information regarding listed metal-clad cable may be found in the UL *Guide Information for Electrical Equipment,* under category PJAZ.

330.6 Listing Requirements. Type MC cable shall be listed. Fittings used for connecting Type MC cable to boxes, cabinets, or other equipment shall be listed and identified for such use.

The sheath of Type MC cable is constructed of aluminum, copper, or steel and can also have a supplemental protective polyvinyl

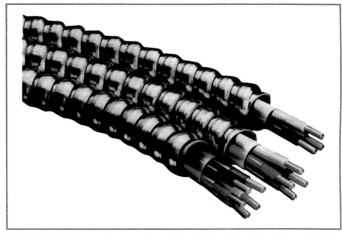

EXHIBIT 330.1 Several examples of Type MC cable. *(Courtesy of AFC Cable Systems, Inc.)*

chloride (PVC) jacket. Connectors should be selected in accordance with the size and type of cable for which they are designated. Bronze connectors are intended for use only with cable employing corrugated copper armor. Fittings with direct-bearing screws are suitable for steel armor only. Type AC cable connectors are also acceptable and listed for use with MC cable if specifically indicated on the fitting or the shipping carton.

Part II. Installation

330.10 Uses Permitted.

(A) General Uses. Type MC cable shall be permitted as follows:

(1) For services, feeders, and branch circuits.
(2) For power, lighting, control, and signal circuits.
(3) Indoors or outdoors.
(4) Exposed or concealed.
(5) To be direct buried where identified for such use.
(6) In cable tray where identified for such use.
(7) In any raceway.
(8) As aerial cable on a messenger.
(9) In hazardous (classified) locations where specifically permitted by other articles in this *Code.*
(10) In dry locations and embedded in plaster finish on brick or other masonry except in damp or wet locations.
(11) In damp or wet locations where a corrosion-resistant jacket is provided over the metallic covering and any of the following conditions are met:
　　a. The metallic covering is impervious to moisture.
　　b. A jacket resistant to moisture is provided under the metal covering.
　　c. The insulated conductors under the metallic covering are listed for use in wet locations.
(12) Where single-conductor cables are used, all phase conductors and, where used, the grounded conductor shall be grouped together to minimize induced voltage on the sheath.

(B) Specific Uses. Type MC cable shall be permitted to be installed in compliance with Parts II and III of Article 725 and 770.133 as applicable and in accordance with 330.10(B)(1) through (B)(4).

Informational Note: The "Uses Permitted" is not an all-inclusive list.

(1) Cable Tray. Type MC cable installed in cable tray shall comply with 392.10, 392.12, 392.18, 392.20, 392.22, 392.30, 392.46, 392.56, 392.60(C), and 392.80.

(2) Direct Buried. Direct-buried cable shall comply with 300.5 or 305.15, as appropriate.

(3) Installed as Service-Entrance Cable. Type MC cable installed as service-entrance cable shall be permitted in accordance with 230.43.

(4) Installed Outside of Buildings or Structures or as Aerial Cable. Type MC cable installed outside of buildings or structures or as aerial cable shall comply with 225.10, 396.10, and 396.12.

330.12 Uses Not Permitted. Type MC cable shall not be used under either of the following conditions:

(1) Where subject to physical damage
(2) Where exposed to any of the destructive corrosive conditions in (a) or (b), unless the metallic sheath or armor is resistant to the conditions or is protected by material resistant to the conditions:
   a. Direct buried in the earth or embedded in concrete unless identified for direct burial
   b. Exposed to cinder fills, strong chlorides, caustic alkalis, or vapors of chlorine or of hydrochloric acids

330.15 Exposed Work. Exposed runs of cable, except as provided in 300.11(B), shall closely follow the surface of the building finish or of running boards. Exposed runs shall also be permitted to be installed on the underside of joists where supported at each joist and located so as not to be subject to physical damage.

330.17 Through or Parallel to Framing Members. Type MC cable shall be protected in accordance with 300.4(A), (C), and (D) where installed through or parallel to framing members.

330.23 In Accessible Attics. The installation of Type MC cable in accessible attics or roof spaces shall also comply with 320.23.

In accessible attics, cable installed across the top of floor joists or within 7 feet of the floor or floor joists across the face of rafters or studs must be protected by guard strips. Where the attic is not accessible by a permanent ladder or stairs, guard strips are required only within 6 feet of the scuttle hole or opening.

330.24 Bending Radius. Bends in Type MC cable shall be so made that the cable will not be damaged. The radius of the curve of the inner edge of any bend shall not be less than required in 330.24(A) through (C).

(A) Smooth Sheath.

(1) Ten times the external diameter of the metallic sheath for cable not more than 19 mm (¾ in.) in external diameter
(2) Twelve times the external diameter of the metallic sheath for cable more than 19 mm (¾ in.) but not more than 38 mm (1½ in.) in external diameter
(3) Fifteen times the external diameter of the metallic sheath for cable more than 38 mm (1½ in.) in external diameter

(B) Interlocked-Type Armor or Corrugated Sheath. Seven times the external diameter of the metallic sheath.

(C) Shielded Conductors. Twelve times the overall diameter of one of the individual conductors or seven times the overall diameter of the multiconductor cable, whichever is greater.

330.30 Securing and Supporting.

(A) General. Type MC cable shall be supported and secured by staples; cable ties listed and identified for securement and support; straps, hangers, or similar fittings; or other approved means designed and installed so as not to damage the cable. Type MC cable fittings shall be permitted as a means of cable support.

A difference exists between securing and supporting. Cable that runs horizontally through or on framing members or racks (spaced less than 6 feet apart) without additional securing is considered supported. Staples or cable ties are not required as the cable passes through or on these members. However, the cable must be secured (fastened in place) within 12 inches of the outlet box. Staples, cable ties, or clamps would be necessary to secure a cable. Both requirements are illustrated in Exhibit 330.2.

(B) Securing. Unless otherwise permitted in this *Code*, cables shall be secured at intervals not exceeding 1.8 m (6 ft). Cables containing four or fewer conductors sized no larger than 10 AWG shall be secured within 300 mm (12 in.) of every box, cabinet, fitting, or other cable termination. In vertical installations, listed cables with ungrounded conductors 250 kcmil and larger shall be permitted to be secured at intervals not exceeding 3 m (10 ft).

(C) Supporting. Unless otherwise permitted in this *Code*, cables shall be supported at intervals not exceeding 1.8 m (6 ft).

Horizontal runs of Type MC cable installed in wooden or metal framing members or similar supporting means shall be considered supported and secured where such support does not exceed 1.8 m (6 ft) intervals.

Δ **(D) Unsupported Cables.** Type MC cable shall be permitted to be unsupported and unsecured where the cable complies with any of the following:

EXHIBIT 330.2 Type MC cable supported and secured at intervals not exceeding 6 feet and within 12 inches of the box [per 330.30(B)] and Type MC cable to be fished in walls, floors, or ceilings [per 330.30(D)(1)].

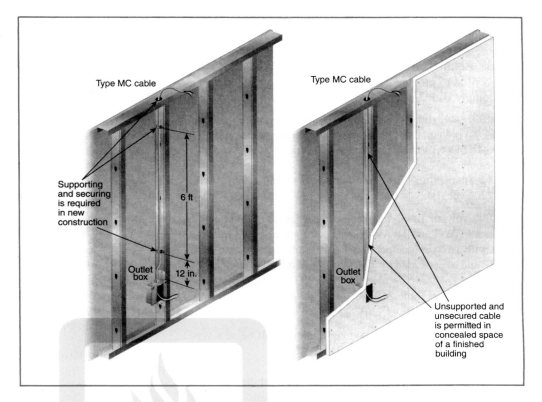

(1) Is fished between access points through concealed spaces in finished buildings or structures and supporting is impractical

(2) Is not more than 1.8 m (6 ft) in length from the last point of cable support to the point of connection to luminaires or other electrical equipment and the cable and point of connection are within an accessible ceiling

(3) Is Type MC of the interlocked armor type in lengths not exceeding 900 mm (3 ft) from the last point where it is securely fastened and is used to connect equipment where flexibility is necessary to minimize the transmission of vibration from equipment or to provide flexibility for equipment that requires movement after installation

330.31 Single Conductors. Where single-conductor cables with a nonferrous armor or sheath are used, the installation shall comply with 300.20.

330.80 Ampacity. The ampacity of Type MC cable shall be determined in accordance with 310.14 or 315.60 for 14 AWG and larger conductors and in accordance with Table 402.5 for 18 AWG and 16 AWG conductors. The installation shall not exceed the temperature ratings of terminations and equipment.

(A) Type MC Cable Installed in Cable Tray. The ampacities for Type MC cable installed in cable tray shall be determined in accordance with 392.80.

(B) Single Type MC Conductors Grouped Together. Where single Type MC conductors are grouped together in a triangular or square configuration and installed on a messenger or exposed with a maintained free airspace of not less than 2.15 times one conductor diameter (2.15 × O.D.) of the largest conductor contained within the configuration and adjacent conductor configurations or cables, the ampacity of the conductors shall not exceed the allowable ampacities in the following tables:

(1) Table 310.20 for conductors rated 0 volts through 2000 volts
(2) Table 315.60(C)(1) and Table 315.60(C)(2) for conductors rated over 2000 volts

(C) Thermal Insulation. Where more than two Type MC cables containing two or more current-carrying conductors in each cable are installed in contact with thermal insulation, caulk, or sealing foam without maintaining spacing between cables, the ampacity of each conductor shall be adjusted in accordance with Table 310.15(C)(1).

Part III. Construction Specifications

330.104 Conductors. For ungrounded, grounded, and equipment grounding conductors, the minimum conductor sizes shall be 14 AWG copper, nickel, or nickel-coated copper and 12 AWG aluminum or copper-clad aluminum.

For control and signal conductors, minimum conductor sizes shall be 18 AWG copper, nickel, or nickel-coated copper, 14 AWG copper-clad aluminum, and 12 AWG aluminum.

Nickel and nickel-coated copper are used as conductors in some fire-rated Type MC cables and are listed in Table 310.4(1) for insulation Types PFAH and TFE (used in high-temperature applications).

330.108 Equipment Grounding Conductor. Where Type MC cable is used to provide an equipment grounding conductor, it shall comply with 250.118(A)(10) and 250.122.

The interlocked sheath cable construction is not recognized as an EGC. An uninsulated grounding/bonding conductor must be integral to this type of MC cable to qualify the sheath as an EGC.

330.112 Insulation. Insulated conductors shall comply with 330.112(A) or (B).

(A) 1000 Volts or Less. Insulated control and signal conductors in sizes 18 AWG and 16 AWG shall be of a type listed in Table 402.3, with a maximum operating temperature not less than 90°C (194°F) and as permitted by 724.49. Ungrounded, grounded, and equipment grounding conductors 16 AWG and larger shall be of a type listed in Table 310.4(1) or of a type identified for use in Type MC cable.

(B) Over 1000 Volts. Insulated conductors shall be of a type listed in Table 310.4(2) and Table 315.10(A).

330.116 Sheath. Metallic covering shall be one of the following types: smooth metallic sheath, corrugated metallic sheath, or interlocking metal tape armor. The metallic sheath shall be continuous and close fitting. A nonmagnetic sheath or armor shall be used on single conductor Type MC. Supplemental protection of an outer covering of corrosion-resistant material shall be permitted and shall be required where such protection is needed. The sheath shall not be used as a current-carrying conductor.

Informational Note: See 300.6 for protection against corrosion.

330.130 Hazardous (Classified) Locations. Where required to be marked MC-HL, the cable shall be listed and shall have a gas/vapor tight continuous corrugated metallic sheath, an overall jacket of suitable polymeric material, and a separate equipment grounding conductor.

ARTICLE 332 Mineral-Insulated, Metal-Sheathed Cable: Type MI

Part I. General

332.1 Scope. This article covers the use, installation, and construction specifications for mineral-insulated, metal-sheathed cable, Type MI.

Type MI cable consists of one or more solid conductors insulated with highly compressed magnesium oxide and enclosed in a continuous copper or alloy steel (e.g., stainless steel) sheath with or

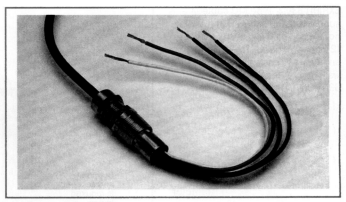

EXHIBIT 332.1 Mineral-insulated (MI) cable with four conductors internal to the cable. *[Courtesy of Tyco Thermal Controls (Canada) Ltd./Pyrotenax]*

without a nonmetallic jacket. (See Exhibit 332.1.) Type MI cable is rated 300 or 600 volts and is available in size 16 AWG and larger with one to seven conductors per cable. It is available in 18 AWG and larger in two conductor cables.

332.6 Listing Requirements. Type MI cable and associated fittings shall be listed.

Part II. Installation

332.10 Uses Permitted. Type MI cable shall be permitted as follows:

(1) For services, feeders, and branch circuits
(2) For power, lighting, control, and signal circuits
(3) In dry, wet, or continuously moist locations
(4) Indoors or outdoors
(5) Where exposed or concealed
(6) Where embedded in plaster, concrete, fill, or other masonry, whether above or below grade
(7) In hazardous (classified) locations where specifically permitted by other articles in this *Code*
(8) Where exposed to oil and gasoline
(9) Where exposed to corrosive conditions not deteriorating to its sheath
(10) In underground runs where suitably protected against physical damage and corrosive conditions
(11) In or attached to cable tray

Informational Note: The "Uses Permitted" is not an all-inclusive list.

332.12 Uses Not Permitted. Type MI cable shall not be used under the following conditions or in the following locations:

(1) In underground runs unless protected from physical damage, where necessary
(2) Where exposed to conditions that are destructive and corrosive to the metallic sheath, unless additional protection is provided

332.17 Through or Parallel to Framing Members. Type MI cable shall be protected in accordance with 300.4 where installed through or parallel to framing members.

332.24 Bending Radius. Bends in Type MI cable shall be so made that the cable will not be damaged. The radius of the inner edge of any bend shall not be less than required as follows:

(1) Five times the external diameter of the metallic sheath for cable not more than 19 mm (¾ in.) in external diameter
(2) Ten times the external diameter of the metallic sheath for cable greater than 19 mm (¾ in.) but not more than 25 mm (1 in.) in external diameter

The minimum bending radius is to prevent mechanical damage to the conductor insulation or the sheath that could result in cracking, a hot spot at the point of damage, or both. Exhibit 332.2 illustrates the minimum bending radius for Type MI cables based upon the external or outside diameter (OD).

332.30 Securing and Supporting. Type MI cable shall be supported and secured by staples, straps, hangers, or similar fittings, designed and installed so as not to damage the cable, at intervals not exceeding 1.8 m (6 ft).

Type MI cable used as part of a fire-rated assembly typically requires more support than do non-fire-rated installations. The manufacturer's instructions will specify the support intervals to achieve the desired fire rating.

See also

300.19(B) for support requirements of fire-resistive cables and conductors

(A) Horizontal Runs Through Holes and Notches. In other than vertical runs, cables installed in accordance with 300.4 shall be considered supported and secured where such support does not exceed 1.8 m (6 ft) intervals.

(B) Unsupported Cable. Type MI cable shall be permitted to be unsupported where the cable is fished between access points through concealed spaces in finished buildings or structures and supporting is impractical.

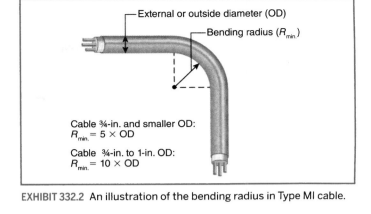

EXHIBIT 332.2 An illustration of the bending radius in Type MI cable.

(C) Cable Trays. All MI cable installed in cable trays shall comply with 392.30(A).

332.31 Single Conductors. Where single-conductor cables are used, all phase conductors and, where used, the neutral conductor shall be grouped together to minimize induced voltage on the sheath.

Because single conductors in a metal sheath can induce voltage on the sheath, all conductors of the circuit must be grouped together to minimize the voltage on the sheath.

Where single conductors enter a ferrous enclosure, inductive heating can occur due to hysteresis loss caused by the magnetic flux occurring in ferrous metals and I^2R losses from the currents induced by the conductor. To minimize this magnetic heating of enclosures, 300.20 requires additional measures, including cutting slots in the metal between the individual holes for each conductor connector. Cable manufacturers offer nonferrous connecting plates that accept individual threaded connections of all circuit conductors, thereby eliminating circulating currents and fully complying with 300.20.

332.40 Boxes and Fittings.

(A) Fittings. Fittings used for connecting Type MI cable to boxes, cabinets, or other equipment shall be identified for such use.

Fittings are required by 332.6 to be listed for use with Type MI cable. As shown in Exhibit 332.3, a complete box connector

EXHIBIT 332.3 A Type MI cable fitting used for terminating cable to an enclosure, to a box, or directly to equipment. *(Courtesy of Tyco Thermal Controls)*

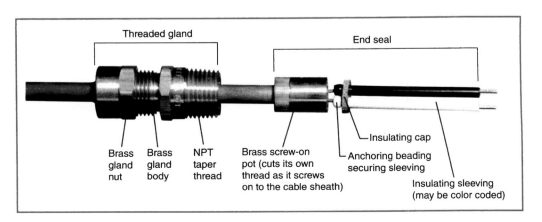

consists of a connector body and a screw-on potting fitting. The fitting is assembled with a special tool and consists of a screw-on pot, insulating cap, insulating sleeving, anchoring bead, and sealing compound.

(B) Terminal Seals. Where Type MI cable terminates, an end seal fitting shall be installed immediately after stripping to prevent the entrance of moisture into the insulation. The conductors extending beyond the sheath shall be individually provided with an insulating material.

332.80 Ampacity. The ampacity of Type MI cable shall be determined in accordance with 310.14. The conductor temperature at the end seal fitting shall not exceed the temperature rating of the listed end seal fitting, and the installation shall not exceed the temperature ratings of terminations or equipment.

(A) Type MI Cable Installed in Cable Tray. The ampacities for Type MI cable installed in cable tray shall be determined in accordance with 392.80(A).

(B) Single Type MI Conductors Grouped Together. Where single Type MI conductors are grouped together in a triangular or square configuration, as required by 332.31, and installed on a messenger or exposed with a maintained free air space of not less than 2.15 times one conductor diameter (2.15 × O.D.) of the largest conductor contained within the configuration and adjacent conductor configurations or cables, the ampacity of the conductors shall not exceed the allowable ampacities of Table 310.17.

Part III. Construction Specifications

332.104 Conductors. Type MI cable conductors shall be of solid copper, nickel, or nickel-coated copper with a resistance corresponding to standard AWG and kcmil sizes.

332.108 Equipment Grounding Conductor. Where the outer sheath is made of copper, it shall provide an adequate path to serve as an equipment grounding conductor. Where the outer sheath is made of steel, a separate equipment grounding conductor shall be provided.

The copper sheath of Type MI cable must be constructed as an equipment grounding conductor (EGC) and is permitted to be used as an EGC according to 250.118(A)(9), but an alloy steel outer sheath is not. A separate EGC is required to be used with steel-sheathed cable.

332.112 Insulation. The conductor insulation in Type MI cable shall be a highly compressed refractory mineral that provides proper spacing for all conductors.

332.116 Sheath. The outer sheath shall be of a continuous construction to provide mechanical protection and moisture seal.

ARTICLE 334 Nonmetallic-Sheathed Cable: Types NM and NMC

Part I. General

334.1 Scope. This article covers the use, installation, and construction specifications of nonmetallic-sheathed cable.

Nonmetallic-sheathed, Type NM and Type NMC, cable was first recognized in the 1928 *NEC®* as a substitute for concealed knob-and-tube wiring (Article 394) and open wiring on insulators (Article 398). The original advantages of nonmetallic-sheathed cable over knob-and-tube wiring were that the outer sheath provided continuous protection in addition to the insulation applied to the conductors; the cable was easily fished in partitions of finished buildings; no insulating supports were required; and only one hole needed to be bored, and that hole could accommodate more than one cable passing through a wood cross member. Type NMS cable has been removed from Article 334 because it has not been available for many years.

334.6 Listing Requirements. Type NM and Type NMC cables and associated fittings shall be listed.

ANSI/UL 719, *Standard for Nonmetallic-Sheathed Cables*, requires a construction and performance evaluation, including testing related to flammability, dielectric voltage-withstand, unwinding at low temperatures, pulling through joists, conductor pullout, crushing, and abrasion.

Part II. Installation

Δ **334.10 Uses Permitted.** Type NM and Type NMC cables shall be permitted to be used in the following, except as prohibited in 334.12:

(1) One- and two-family dwellings and their attached or detached garages, and their storage buildings.

(2) Multi-family dwellings and their detached garages permitted to be of Types III, IV, and V construction.

(3) Other structures permitted to be of Types III, IV, and V construction. Cables shall be concealed within walls, floors, or ceilings that provide a thermal barrier of material that has at least a 15-minute finish rating as identified in listings of fire-rated assemblies.

Informational Note No. 1: See NFPA 220-2021, *Standard on Types of Building Construction*, or the applicable building code, or both for types of building construction and occupancy classification definitions.

Informational Note No. 2: See Informative Annex E for determination of building types.

(4) Cable trays in structures permitted to be Types III, IV, or V where the cables are identified for the use.

Informational Note No. 3: See 310.14(A)(3) for temperature limitation of conductors.

(5) Types I and II construction where installed within raceways permitted to be installed in Types I and II construction.

A well-established means of codifying fire protection and fire safety requirements is to classify buildings by types of construction, based on materials used for the structural elements and the degree of fire resistance afforded by each element. The five fundamental construction types used by the model building codes are as follows:

Type I (fire resistive)

Type II (noncombustible)

Type III (combination of combustible and noncombustible)

Type IV (heavy timber)

Type V (wood frame)

Types I and II basically require all structural elements to be noncombustible, whereas Types III, IV, and V allow some or all of the structural elements to be combustible (wood).

The selection of building construction types is regulated by the local building code, based on the occupancy, height, and area of the building. If a building of a selected height (in feet or stories above grade) and area is permitted to be built of combustible construction (i.e., Types III, IV, or V), the installation of nonmetallic-sheathed cable is permitted. Common areas (corridors) and incidental and subordinate uses (such as laundry rooms and lounge rooms) that serve a multifamily dwelling occupancy are also considered part of the multifamily occupancy. Type NM cable is allowed in such areas.

If a building is of noncombustible construction (Type I or II) by the owner's choice, even though the building code would permit combustible construction, the building is allowed to be wired with Type NM cable. In such an instance, Type NM cable may be installed in the noncombustible building because the building code would have permitted the building to be of combustible construction.

If Type I or Type II construction is required, Section 334.10(5) permits Type NM cables to be installed if the cables are in a raceway. A raceway is permitted to be used only if it complies with the article for the raceway and its use does not violate another article in the *NEC*.

See also

Informative Annex E for information on the types of construction as well as a table that cross-references the five construction types to the types described in the model building codes

(A) Type NM. Type NM cable shall be permitted as follows:

(1) For both exposed and concealed work in normally dry locations except as prohibited in 334.10(3)
(2) To be installed or fished in air voids in masonry block or tile walls

For concealed work, cable should be installed where it is protected from physical damage often caused by nails or screws. Where practical, care should be taken to avoid areas where trim, door and window casings, baseboards, moldings, and so forth are likely to be nailed.

See also

300.4 for details on protection against physical damage

(B) Type NMC. Type NMC cable shall be permitted as follows:

(1) For both exposed and concealed work in dry, wet, damp, or corrosive locations, except as prohibited by 334.10(3)
(2) In outside and inside walls of masonry block or tile
(3) In a shallow chase in masonry, concrete, or adobe protected against nails or screws by a steel plate at least 1.59 mm (1/16 in.) thick and covered with plaster, adobe, or similar finish

If Type NM cable is used in dairy barns and similar agricultural buildings (see Article 547), it must be Type NMC (corrosion resistant). The cable will be exposed to fumes, vapors, or liquids such as ammonia and barnyard acids. Under such circumstances, ordinary Type NM cable can deteriorate rapidly due to ammonia fumes or the growth of fungus or mold.

334.12 Uses Not Permitted.

Restrictions of use of Type NM cable also exist elsewhere in the *NEC*. For example, Type NM cables are not permitted to be installed in ducts, plenums, and other air-handling spaces.

See also

300.22, which limits the use of materials in ducts, plenums, and other air-handling spaces that can contribute smoke and products of combustion during a fire

Δ **(A) Types NM and NMC.** Types NM and NMC cables shall not be permitted as follows:

(1) In any dwelling or structure not specifically permitted in 334.10(1), 334.10(2), 334.10(3), and 334.10(5)
(2) Exposed within a dropped or suspended ceiling cavity in other than one- and two-family and multifamily dwellings

Nonmetallic-sheathed cables are prohibited in the space above hung ceilings that allow access. This requirement does not apply to dwelling-type occupancies. The term *exposed*, as used in this requirement, meets the definition of *exposed (as applied to wiring methods)* found in Article 100, which states "on or attached to the surface or behind panels designed to allow access."

For example, cables installed above a gypsum board ceiling or soffit would not be considered exposed cable, if the area above the ceiling is not accessible (does not have removable tiles or does not contain an access panel). Because hung or dropped ceilings are often accessible, cables installed above those types of ceilings would be considered exposed cables if the cables do not have additional physical protection.

A simple change to an architectural finish schedule during construction could change the acceptability of the wiring method. For example, if a corridor ceiling in an occupancy (other than a dwelling type) calls for a painted gypsum board ceiling and the finish schedule changes the ceiling construction to a suspended ceiling, Type NM cable would not be permitted unless the cable is installed using additional protection.

See also

334.15(B) for examples of additional protection

(3) As service-entrance cable

(4) In commercial garages having hazardous (classified) locations as defined in 511.3

(5) In theaters and similar locations, except where permitted in 518.4(C)

(6) In motion picture studios

(7) In storage battery rooms

(8) In hoistways or on elevators or escalators

(9) Embedded in poured cement, concrete, or aggregate

(10) In hazardous (classified) locations, except where specifically permitted by other articles in this *Code*

(B) Type NM. Type NM cables shall not be used under the following conditions or in the following locations:

(1) Where exposed to corrosive fumes or vapors

(2) Where embedded in masonry, concrete, adobe, fill, or plaster

(3) In a shallow chase in masonry, concrete, or adobe and covered with plaster, adobe, or similar finish

(4) In wet or damp locations

334.15 Exposed Work. In exposed work, except as provided in 300.11(B), cable shall be installed as specified in 334.15(A) through (C).

(A) To Follow Surface. Cable shall closely follow the surface of the building finish or of running boards.

Δ **(B) Protection from Physical Damage.** Cable shall be protected from physical damage where necessary by rigid metal conduit, intermediate metal conduit, electrical metallic tubing, Schedule 80 PVC conduit, RTRC marked with the suffix -XW, or other approved means. Where passing through a floor, the cable shall be enclosed in rigid metal conduit, intermediate metal conduit, electrical metallic tubing, Schedule 80 PVC conduit, RTRC marked with the suffix -XW, or other approved means extending at least 150 mm (6 in.) above the floor. Conduit or tubing shall be provided with a bushing or adapter that provides protection from abrasion at the point the cable enters and exits the raceway.

Type NMC cable installed in shallow chases or grooves in masonry, concrete, or adobe shall be protected in accordance with the requirements in 300.4(F) and covered with plaster, adobe, or similar finish.

Δ **(C) In Unfinished Basements and Crawl Spaces.** Where cable is run at angles with joists in unfinished basements and crawl spaces, it shall be permissible to secure cables not smaller than two 6 AWG or three 8 AWG conductors directly to the lower edges of the joists. Smaller cables shall be run either through bored holes in joists or on running boards. Nonmetallic-sheathed cable installed on the wall of an unfinished basement shall be permitted to be installed in a listed conduit or tubing or shall be protected in accordance with 300.4. Conduit or tubing shall be provided with a bushing or adapter that provides protection from abrasion at the point the cable enters and exits the raceway. The sheath of the nonmetallic-sheathed cable shall extend through the conduit or tubing and into the outlet, device, or junction box not less than 6 mm (¼ in.). The cable shall be secured within 300 mm (12 in.) of the point where the cable enters the conduit or tubing. Metal conduit, tubing, and metal outlet boxes shall be connected to an equipment grounding conductor complying with 250.86 and 250.148.

Type NM cable installed in crawl spaces is more susceptible to physical damage due to the limited height of the space. Specific means of physical protection in crawl spaces and unfinished basements can include the use of raceways, guard strips, and so forth.

Nonmetallic-sheathed cables installed in an unfinished basement or crawl space can be run through holes in joists, attached to the side of joists or beams, and installed on running boards, as shown on the left side in Exhibit 334.1.

See also

300.4(D), which requires cables that run parallel to framing members be installed at least 1¼ inches from the nearest edge of studs, joists, or rafters

334.17 Through or Parallel to Framing Members. Types NM and NMC cable shall be protected in accordance with 300.4 where installed through or parallel to framing members. Grommets used as required in 300.4(B)(1) shall remain in place and be listed for the purpose of cable protection.

Cable that passes through factory- or field-punched holes in metal studs or similar members is required to be protected in accordance with 300.4(B)(1). Listed bushings or listed grommets covering all metal edges must be securely fastened in the opening before the cable is installed.

N **334.19 Cable Entries.** The sheath on nonmetallic-sheathed cable shall extend not less than 6 mm (¼ in.) beyond any cable clamp or cable entry.

334.23 In Accessible Attics. The installation of cable in accessible attics or roof spaces shall also comply with 320.23.

334.24 Bending Radius. Bends in Types NM and NMC cable shall be so made that the cable will not be damaged. The radius of

EXHIBIT 334.1 Nonmetallic-sheathed cables (on plywood running board) and Type SE cables (installed directly to bottom of joists) in an unfinished basement or crawl space.

the curve of the inner edge of any bend during or after installation shall not be less than five times the diameter of the cable. For flat cables, the major diameter dimension of the cable shall be used to determine the bending radius.

334.30 Securing and Supporting. Nonmetallic-sheathed cable shall be supported and secured by staples, cable ties listed and identified for securement and support, or straps, hangers, or similar fittings designed and installed so as not to damage the cable, at intervals not exceeding 1.4 m (4½ ft) and within 300 mm (12 in.) of every cable entry into enclosures such as outlet boxes, junction boxes, cabinets, or fittings. The cable length between the cable entry and the closest cable support shall not exceed 450 mm (18 in.). Flat cables shall not be stapled on edge.

Sections of cable protected from physical damage by raceway shall not be required to be secured within the raceway.

Draping the cable over air ducts, rafters, timbers, joists, pipes, and ceiling grid members without securing the cable with approved means is not permitted, except where the cable is fished, as allowed in 334.30(B)(1).

Two-conductor Type NM cable (or other flat configurations) is prohibited from being stapled on edge (i.e., with its short dimension against a wood joist). If the staple is driven too far into the stud, damage to the insulation and conductors could occur.

See also

300.4(C) for support requirements of cables through spaces behind panels designed to allow access

(A) Horizontal Runs Through Holes and Notches. In other than vertical runs, cables installed in accordance with 300.4 shall be considered to be supported and secured where such support does not exceed 1.4-m (4½-ft) intervals and the nonmetallic-sheathed cable is securely fastened in place by an approved means within 300 mm (12 in.) of each box, cabinet, conduit body, or other nonmetallic-sheathed cable termination.

Informational Note: See 314.17(B)(1) for support where nonmetallic boxes are used.

Cable running horizontally through framing members (spaced less than 54 inches apart) and passing through bored or punched holes in the framing members is considered to be supported by the framing members. Additional securing or fastening with cable ties is not required. However, the cable must be secured within 12 inches of the outlet box. Where the cable terminates at a single gang nonmetallic box that does not contain a cable clamping device, the cable may be secured (fastened in place) within 8 inches of the outlet box, according to 314.17(B)(2), Exception.

(B) Unsupported Cables. Nonmetallic-sheathed cable shall be permitted to be unsupported where the cable:

(1) Is fished between access points through concealed spaces in finished buildings or structures and supporting is impracticable.
(2) Is not more than 1.4 m (4½ ft) from the last point of cable support to the point of connection to a luminaire or other piece of electrical equipment and the cable and point of connection are within an accessible ceiling in one-, two-, or multifamily dwellings.

(C) Wiring Device Without a Separate Outlet Box. A wiring device identified for the use, without a separate outlet box, and incorporating an integral cable clamp shall be permitted where the cable is secured in place at intervals not exceeding 1.4 m (4½ ft) and within 300 mm (12 in.) from the wiring device wall opening, and there shall be at least a 300 mm (12 in.) loop of unbroken cable or 150 mm (6 in.) of a cable end available on the interior side of the finished wall to permit replacement.

334.40 Boxes and Fittings.

(A) Boxes of Insulating Material. Nonmetallic outlet boxes shall be permitted as provided by 314.3.

(B) Devices of Insulating Material. Self-contained switches, self-contained receptacles, and listed nonmetallic-sheathed cable interconnector devices of insulating material that are listed for use without a box shall be permitted to be used without boxes in exposed or concealed installations. Openings in such devices shall form a close fit around the outer covering of the cable, and the device shall fully enclose the part of the cable from which any part of the covering has been removed. Where connections to conductors are by binding-screw terminals, there shall be available as many terminals as conductors.

(C) Devices with Integral Enclosures. Wiring devices with integral enclosures identified for such use shall be permitted as provided by 300.15(E).

△ **334.80 Ampacity.** The ampacity of Types NM and NMC cable shall be determined in accordance with 310.14. The ampacity shall not exceed that of a 60°C (140°F) rated conductor. The 90°C (194°F) rating shall be permitted to be used for ampacity adjustment and correction calculations, provided the final calculated ampacity does not exceed that of a 60°C (140°F) rated conductor. The ampacity of Types NM and NMC cable installed in cable trays shall be determined in accordance with 392.80(A).

Where more than two NM cables containing two or more current-carrying conductors are installed, without maintaining spacing between the cables, through the same opening in wood framing that is to be sealed with thermal insulation, caulk, or sealing foam, the ampacity of each conductor shall be adjusted in accordance with Table 310.15(C)(1) and 310.14(A)(2), Exception, shall not apply.

Where more than two NM cables containing two or more current-carrying conductors are installed in contact with thermal insulation without maintaining spacing between cables, the ampacity of each conductor shall be adjusted in accordance with Table 310.15(C)(1) and 310.14(A)(2), Exception shall not apply.

As stated in 310.15(C)(1), "The ampacity of each conductor shall be reduced as shown in Table 310.15(C)(1), where the number of current-carrying conductors in a raceway or cable exceeds three, or where single conductors or multiconductor cables not installed in raceways are installed without maintaining spacing for a continuous length longer than 600 mm (24 in.)."

Failure to apply the appropriate ampacity adjustment factor called for by Table 310.15(C)(1), where Type NM cables are stacked or bundled without maintaining spacing, can lead to overheating of conductors. The ampacity adjustment requirements prevent overheating of the conductors where they pass though wood-framed draft- and fire-stopping material or are in direct contact with thermal insulation. Not only is thermal insulation provided within structures to reduce heat loss or heat gain, the same thermal insulation material can be used to control sound within structures as well.

Calculation Example

Four 2-conductor, size 12 AWG, copper with ground, Type NM cables are installed in direct contact with thermal insulation without maintaining spacing. Calculate the ampacity of the conductors according to the requirements of 334.80 and determine the maximum overcurrent protection permitted for the four circuits.

Solution

Step 1. Determine the number of current-carrying conductors:

4 cables × 2 conductors per cable =
8 current-carrying conductors

Step 2. Determine the initial conductor ampacity. Using the 90°C copper ampacity from Table 310.16 for derating purposes, the initial ampacity of 12 AWG is 30 A.

Step 3. Determine the adjusted conductor ampacity. Due to the direct contact with thermal insulation, use Table 310.15(C)(1). Eight current-carrying conductors require an adjustment factor of 70 percent.

30 A × 0.7 = 21 A (adjusted)

Section 334.80 does not allow an ampacity greater than that given in the 60°C column of Table 310.16; therefore, the conductor ampacity is limited to 20 A.

Step 4. Determine the maximum permitted overcurrent device for each circuit. According to the footnote in Table 310.16, conductor sizes 14 AWG through 10 AWG must also comply with 240.4(D), which limits protection of a 12 AWG copper conductor to a maximum of 20 A.

Conclusion. The final ampacity for each current-carrying conductor is 20 A, and the maximum overcurrent device permitted for each of the four circuits is 20 A.

This example points out that the 14 AWG to 10 AWG Type NM cable typically used for branch circuits can be installed without spacing and placed within thermal insulation with little impact on most installations. For similar installations, as long as the bundle is limited to not more than nine current-carrying conductors, the adjusted ampacity will not be below the overcurrent protection set in 240.4(D).

Part III. Construction Specifications

334.100 Construction. The outer cable sheath of nonmetallic-sheathed cable shall be a nonmetallic material.

334.104 Conductors. The 600-volt insulated power conductors shall be sizes 14 AWG through 2 AWG copper conductors or sizes 12 AWG through 2 AWG aluminum or copper-clad aluminum conductors. Control and signaling conductors shall be no smaller than 18 AWG copper.

334.108 Equipment Grounding Conductor. In addition to the insulated conductors, the cable shall have an insulated, covered, or bare equipment grounding conductor.

334.112 Insulation. The insulated power conductors shall be one of the types listed in Table 310.4(1) that are suitable for branch-circuit wiring or one that is identified for use in these cables. Conductor insulation shall be rated at 90°C (194°F).

Informational Note: Types NM and NMC cable identified by the markings NM-B and NMC-B meet this requirement.

334.116 Sheath. The outer sheath of nonmetallic-sheathed cable shall comply with 334.116(A) and (B).

(A) Type NM. The overall covering shall be flame retardant and moisture resistant.

(B) Type NMC. The overall covering shall be flame retardant, moisture resistant, fungus resistant, and corrosion resistant.

N | ARTICLE **335** | Instrumentation Tray Cable: Type ITC

N 335.1 Scope. This article covers the use, installation, and construction specifications of instrumentation tray cable (Type ITC) for application to instrumentation and control circuits operating at 150 volts or less and 5 amperes or less.

Former Article 727 has been relocated to new Article 335, Instrumentation Tray Cable: Type ITC. The committee felt this article covers a wiring method type that is better suited to be in Chapter 3, which covers wiring methods and materials. Article 335 permits an alternate wiring method for circuits that do not exceed 5 amperes and 150 volts. Instrumentation tray cable is particularly suited for instrumentation circuits in industrial establishments where qualified persons perform service and maintenance.

N 335.3 Other Articles. In addition to the provisions of this article, installation of Type ITC cable shall comply with other applicable articles of this *Code*.

N 335.4 Uses Permitted. Type ITC cable shall be permitted to be used as follows in industrial establishments where the conditions of maintenance and supervision ensure that only qualified persons service the installation:

(1) In cable trays.
(2) In raceways.
(3) In hazardous locations as permitted in 501.10, 502.10, 503.10, 504.20, 504.30, 504.80, and 505.15.
(4) Enclosed in a smooth metallic sheath, continuous corrugated metallic sheath, or interlocking tape armor applied over the nonmetallic sheath in accordance with 335.6. The cable shall be supported and secured at intervals not exceeding 1.8 m (6 ft).
(5) Cable, without a metallic sheath or armor, that complies with the crush and impact requirements of Type MC cable and is identified for such use with the marking *ITC-ER* shall be permitted to be installed exposed. The cable shall be continuously supported and protected against physical damage using mechanical protection such as dedicated struts, angles, or channels. The cable shall be secured at intervals not exceeding 1.8 m (6 ft).

Exception to (5): Where not subject to physical damage, Type ITC-ER shall be permitted to transition between cable trays and between cable trays and utilization equipment or devices for a distance not to exceed 1.8 m (6 ft) without continuous support. The cable shall be mechanically supported where exiting the cable tray to ensure that the minimum bending radius is not exceeded.

(6) As aerial cable on a messenger.
(7) Direct buried where identified for the use.
(8) Under raised floors in rooms containing industrial process control equipment and rack rooms where arranged to prevent damage to the cable.
(9) Under raised floors in information technology equipment rooms in accordance with 645.5(E)(2).

N 335.5 Uses Not Permitted. Type ITC cable shall not be installed on circuits operating at more than 150 volts or more than 5 amperes.

Installation of Type ITC cable with other cables shall be subject to the stated requirements of the specific articles for the other cables. Where the governing articles do not contain stated requirements for installation with Type ITC cable, the installation of Type ITC cable with the other cables shall not be permitted.

Type ITC cable shall not be installed with power, lighting, Class 1 circuits that are not power limited, or non–power-limited circuits.

Exception No. 1: Where terminated within equipment or junction boxes and separations are maintained by insulating barriers or other means.

Exception No. 2: Where a metallic sheath or armor is applied over the nonmetallic sheath of the Type ITC cable.

N 335.6 Construction. The insulated conductors of Type ITC cable shall be in sizes 22 AWG through 12 AWG. The conductor material shall be copper or thermocouple alloy. Insulation on the conductors shall be rated for 300 volts. Shielding shall be permitted.

The cable shall be listed as being resistant to the spread of fire. The outer jacket shall be sunlight and moisture resistant.

Where a smooth metallic sheath, continuous corrugated metallic sheath, or interlocking tape armor is applied over the nonmetallic sheath, an overall nonmetallic jacket shall not be required.

Informational Note: One method of defining *resistant to the spread of fire* is that the cables do not spread fire to the top of the tray in the UL flame exposure, vertical tray flame test in ANSI/UL 1685-2010, *Vertical-Tray Fire-Propagation and Smoke-Release Test for Electrical and Optical-Fiber Cables*. The smoke measurements in the test method are not applicable.

Another method of defining *resistant to the spread of fire* is for the damage (char length) not to exceed 1.5 m (4 ft 11 in.) when performing the CSA vertical flame test — cables in cable trays, as described in CSA C22.2 No. 0.3-M-2001, *Test Methods for Electrical Wires and Cables*.

N **335.7 Marking.** The cable shall be marked in accordance with 310.8(A)(2) through (A)(5). Voltage ratings shall not be marked on the cable.

N **335.8 Ampacity.** The ampacity of the conductors shall be 5 amperes, except for 22 AWG conductors, which shall have an ampacity of 3 amperes.

N **335.9 Overcurrent Protection.** Overcurrent protection shall not exceed 5 amperes for 20 AWG and larger conductors, and 3 amperes for 22 AWG conductors.

N **335.10 Bends.** Bends in Type ITC cables shall be made so as not to damage the cable.

ARTICLE 336   Power and Control Tray Cable: Type TC

Part I. General

336.1 Scope. This article covers the use, installation, and construction specifications for power and control tray cable, Type TC.

The basic standard to investigate power and control tray cable is ANSI/UL 1277, *Electrical Power and Control Tray Cables with Optional Optical-Fiber Members*. Summary information regarding listed power and control tray cable may be found in the UL *Guide Information for Electrical Equipment*, under category QPOR.

336.6 Listing Requirements. Type TC cables and associated fittings shall be listed.

Part II. Installation

336.10 Uses Permitted. Type TC cable shall be permitted to be used as follows:

(1) For power, lighting, control, and signal circuits.
(2) In cable trays, including those with mechanically discontinuous segments up to 300 mm (1 ft).
(3) In raceways.
(4) In outdoor locations supported by a messenger wire.
(5) For Class 1 circuits as permitted in Parts II and III of Article 725.
(6) For non-power-limited fire alarm circuits if conductors comply with the requirements of 760.49.
(7) Between a cable tray and the utilization equipment or device(s), provided all of the following apply:

    a. The cable is Type TC-ER.
    b. The cable is installed in industrial establishments where the conditions of maintenance and supervision ensure that only qualified persons service the installation.

    c. The cable is continuously supported and protected against physical damage using mechanical protection such as struts, angles, or channels.
    d. The cable complies with the crush and impact requirements of Type MC cable and is identified with the marking "TC–ER."
    e. The cable is secured at intervals not exceeding 1.8 m (6 ft).
    f. Equipment grounding for the utilization equipment is provided by an equipment grounding conductor within the cable. In cables containing conductors sized 6 AWG or smaller, the equipment grounding conductor shall be provided within the cable or, at the time of installation, one or more insulated conductors shall be permanently identified as an equipment grounding conductor in accordance with 250.119(C).

Exception to (7): Where not subject to physical damage, Type TC-ER shall be permitted to transition between cable trays and between cable trays and equipment or devices for a distance not to exceed 1.8 m (6 ft) without continuous support. The cable shall be mechanically supported where exiting the cable tray to ensure that the minimum bending radius is not exceeded.

Specific types of tray cable are permitted to extend from a cable tray to a piece of equipment without the use of conduit. According to ANSI/UL 1277, *Electrical Power and Control Tray Cables with Optional Optical-Fiber Members*, cables suitable for use as exposed wiring between cable tray and the utilization equipment are surface-marked "Type TC-ER" (tray cable for exposed runs).

The exception permits TC-ER cable to exit a cable tray without continuous support for a maximum of 6 feet. However, extension from cable tray to cable tray or from cable tray to equipment is not allowed beyond 6 feet without continuous support.

(8) Type TC cable shall be resistant to moisture and corrosive agents where installed in wet locations.
(9) For one- and two-family dwelling units, Type TC-ER-JP cable containing conductors for both power and control circuits shall be permitted for branch circuits and feeders. Type TC-ER-JP cable used as interior wiring shall be installed per the requirements of Part II of Article 334 and where installed as exterior wiring shall be installed per the requirements of Part II of Article 340.

Exception: Where used to connect a generator and associated equipment having terminals rated 75°C (140°F) or higher, the cable shall not be limited in ampacity by 334.80 or 340.80.

Informational Note No. 1: See 725.136 for limitations on Class 2 or 3 circuits contained within the same cable with conductors of electric light, power, or Class 1 circuits.

Power conductors and control conductors are often used to run between a generator and transfer switch, or between equipment of an HVAC system. Type TC-ER cable allows the power and control circuit conductors to occupy the same cable.

(10) Direct buried, where identified for such use.

(11) In hazardous (classified) locations where specifically permitted by other articles in this *Code*.

(12) For service-entrance conductors where identified for such use and marked Type TC-ER.

Informational Note No. 2: See 310.14(A)(3) for temperature limitation of conductors.

336.12 Uses Not Permitted. Type TC tray cable shall not be installed or used as follows:

(1) Installed where it will be exposed to physical damage

(2) Installed outside a raceway or cable tray system, except as permitted in 336.10(4), 336.10(7), 336.10(9), and 336.10(10)

(3) Used where exposed to direct rays of the sun, unless identified as sunlight resistant

336.24 Bending Radius. Bends in Type TC cable shall be made so as not to damage the cable. For Type TC cable without metal shielding, the minimum bending radius shall be as follows:

(1) Four times the overall diameter for cables 25 mm (1 in.) or less in diameter

(2) Five times the overall diameter for cables larger than 25 mm (1 in.) but not more than 50 mm (2 in.) in diameter

(3) Six times the overall diameter for cables larger than 50 mm (2 in.) in diameter

Type TC cables with metallic shielding shall have a minimum bending radius of not less than 12 times the cable overall diameter.

336.80 Ampacity. The ampacity of Type TC tray cable shall be determined in accordance with 392.80(A) for 14 AWG and larger conductors, in accordance with 402.5 for 18 AWG through 16 AWG conductors where installed in cable trays, and in accordance with 310.14 where installed outside of cable trays, where permitted.

Part III. Construction Specifications

336.100 Construction. A metallic sheath or armor as defined in 330.116 shall not be permitted either under or over the nonmetallic jacket. Metallic shield(s) shall be permitted over groups of conductors, under the outer jacket, or both.

336.104 Conductors. For ungrounded, grounded, and equipment grounding conductors, the conductor sizes shall be 14 AWG through 1000 kcmil copper, nickel, or nickel-coated copper and 12 AWG through 1000 kcmil aluminum or copper-clad aluminum. Insulation types shall be one of the types listed in Table 310.4(1) or Table 310.4(2) that is suitable for branch circuit and feeder circuits or one that is identified for such use.

For control and signal conductors, the minimum conductor sizes shall be 18 AWG copper, nickel, or nickel-coated copper, 14 AWG copper-clad aluminum, and 12 AWG aluminum.

(A) Fire Alarm Systems. Where used for fire alarm systems, conductors shall also be in accordance with 760.49.

(B) Thermocouple Circuits. Conductors in Type TC cable used for thermocouple circuits in accordance with Part III of Article 724 shall also be permitted to be any of the materials used for thermocouple extension wire.

(C) Class 1 Circuit Conductors. Insulated conductors of 18 AWG and 16 AWG copper shall also be in accordance with 724.49.

336.116 Jacket. The outer jacket shall be a flame-retardant, nonmetallic material.

336.120 Marking. There shall be no voltage marking on a Type TC cable employing thermocouple extension wire.

336.130 Hazardous (Classified) Location Cable. Cable listed and marked Type TC-ER-HL shall comply with the following:

(1) The overall nonmetallic jacket shall be suitable for the environment.

(2) The overall cable construction shall be essentially circular in cross-section.

(3) The overall nonmetallic jacket shall be continuous and gas/vapor tight.

(4) For construction greater than 25.4 mm (1 in.) in diameter, the following shall apply:

   a. The equipment grounding conductor shall be bare.

   b. A metallic shield shall be included over all conductors under the outer jacket.

ARTICLE
337 Type P Cable

Part I. General

337.1 Scope. This article covers the use, installation, and construction specifications for up through 2000 volt Type P cable. (armored and unarmored).

337.6 Listing Requirements. Type P cables and associated fittings shall be listed.

Part II. Installation

337.10 Uses Permitted. Type P cable shall be permitted to be used:

(1) Under engineering supervision in industrial installations where conditions of maintenance and supervision ensure that only qualified persons monitor and service the system.

(2) In hazardous (classified) locations where specifically permitted by other articles in this *Code*.

337.12 Uses Not Permitted. Type P cable shall not be installed or used:

(1) Where it will be exposed to physical damage
(2) Where not specifically permitted by other articles in the *Code*

337.24 Bending Radius. The minimum bending radii during installations and handling in service shall be adequate to prevent damage to the cable.

337.30 Securing and Supporting. Type P cable shall be supported and secured by cable ties listed and identified for securement and support; straps, hangers, or similar fittings; or other approved means designed and installed so as not to damage the cable.

337.31 Single Conductors. Where single-conductor cables are used, the installation shall comply with 300.20.

337.80 Ampacity. The ampacity of Type P cable shall be determined in accordance with 310.14(A) or (B) for 14 AWG and larger conductors. For 18 AWG and 16 AWG conductors, the ampacities shall be determined in accordance with Table 402.5 or 310.14(B). When installed in cable tray, the ampacities shall be permitted to be determined in accordance with 392.80. The installation shall not exceed the temperature ratings of terminations and equipment.

Part III. Construction Specifications

337.104 Conductors. Conductors shall be of tinned copper. Conductors shall employ flexible stranding. The minimum conductor size shall be 18 AWG.

337.108 Equipment Grounding Conductor. An equipment grounding conductor complying with 250.122 shall be provided within multiconductor Type P cable.

337.112 Insulation. Insulated conductors shall be a thermoset type identified for use in Type P cable. All conductors shall be suitable for wet locations. The minimum wall thickness shall be 0.76 mm (30 mils).

337.114 Shield. Metallic shield(s) shall be permitted over a single conductor or groups of conductors.

337.115 Jacket. Multiconductor cables shall have an overall nonmetallic jacket that is impervious to moisture, corrosion resistant, and sunlight resistant. When installed external to an enclosure or industrial machinery, single conductor cables shall have an overall nonmetallic jacket that is impervious to moisture, corrosion resistant, and sunlight resistant. Single conductor cables rated 2000 volts with conductor sizes equal to or larger than 4/0 AWG shall be permitted to use an increased insulation thickness in lieu of using a separate cable jacket. When the increased insulation thickness is used, the insulation material shall be sunlight resistant.

337.116 Armor. Armor shall be permitted over the jacket. If provided, the armor or metallic covering shall be a braided basket weave type consisting of wire laid closely together, flat and parallel, and forming a basket weave that shall firmly grip the cable. The wire shall be commercial bronze, tinned copper, stainless steel, or aluminum. The armor shall not be used as a current-carrying conductor or as an equipment grounding conductor. A nonmetallic jacket that conforms to 337.115 shall be provided over the armor.

337.120 Marking. Type P cable shall be marked in accordance with 310.8. When an armor is provided, the cable shall be marked accordingly.

ARTICLE
338   Service-Entrance Cable: Types SE and USE

Part I. General

338.1 Scope. This article covers the use, installation, and construction specifications of service-entrance cable.

According to the UL *Guide Information for Electrical Equipment*, category TYLZ cable (service-entrance cable rated 600 volts) is listed in sizes 14 AWG and larger for copper and 12 AWG and larger for aluminum or copper-clad aluminum. Type SE cable contains Types RHW, RHW-2, XHHW, XHHW-2, THWN, and THWN-2 conductors. Type USE cable contains conductors with insulation equivalent to RHW or XHHW. Type USE-2 contains insulation equivalent to RHW-2 or XHHW-2 and is rated 90°C, wet or dry.

The type designation of the conductors may be marked on the cable surface. If used, the marking indicates the temperature rating for the cable corresponding to the temperature rating of the conductors. If this marking does not appear, the temperature rating of the cable is 75°C. The cables are designated as Type SE, Type USE or USE-2, and submersible water pump cable.

Type SE — Cable suitable for aboveground installations. Both the insulated conductors and the outer jacket are suitable for use where exposed to sunlight.

Type USE or USE-2 — Cable suitable for underground installations, including direct burial. Although both the conductor insulation and the outer covering are suitable for use where exposed to sunlight, the cables are not suitable inside the premises or aboveground other than to terminate at service or metering equipment.

Submersible water pump cable — A multiconductor cable containing two, three, or four single-conductor, Type USE or USE-2 cables in a flat or twisted assembly. The cable is tag-marked "For use within the well casing for wiring deep-well water pumps where the cable is not subject to repetitive handling caused by frequent servicing of the pump units."

338.6 Listing Requirements. Type SE and USE cables and associated fittings shall be listed.

Part II. Installation

338.10 Uses Permitted.

(A) Service-Entrance Conductors. Service-entrance cable shall be permitted to be used as service-entrance conductors and shall be installed in accordance with 230.6, 230.7, and Parts II, III, and IV of Article 230.

(B) Branch Circuits or Feeders.

(1) Grounded Conductor Insulated. Type SE service-entrance cables shall be permitted in wiring systems where all of the circuit conductors of the cable are of the thermoset or thermoplastic type.

Branch circuits using Type SE cable as a wiring method are permitted only if all circuit conductors within the cable are insulated. The equipment grounding conductor (EGC) is the only conductor permitted to be bare or covered within Type SE cable used for branch circuits.

(2) Use of Uninsulated Conductor. Type SE service-entrance cable shall be permitted for use where the insulated conductors are used for circuit wiring and the uninsulated conductor is used only for equipment grounding purposes.

Exception: In existing installations, uninsulated conductors shall be permitted as a grounded conductor in accordance with 250.32 and 250.140, where the uninsulated grounded conductor of the cable originates in service equipment, and with 225.30 through 225.40.

Service-entrance cable containing a bare grounded (neutral) conductor is not permitted for new installations where it is used as a branch circuit to supply appliances such as ranges, wall-mounted ovens, counter-mounted cooking units, or clothes dryers. The exception permits a bare neutral service-entrance cable for existing installations only.

(3) Temperature Limitations. Type SE service-entrance cable used to supply appliances shall not be subject to conductor temperatures in excess of the temperature specified for the type of insulation involved.

Δ **(4) Installation Methods for Branch Circuits and Feeders.**

(a) *Interior Installations.* Interior installations shall comply with the following:

(1) In addition to the provisions of this article, Type SE service-entrance cable used for interior wiring shall comply with the installation requirements of Part II of Article 334, excluding 334.80.

(2) Where more than two Type SE cables containing two or more current-carrying conductors in each cable are installed in contact with thermal insulation, caulk, or sealing foam without maintaining spacing between cables, the ampacity of each conductor shall be adjusted in accordance with Table 310.15(C)(1).

(3) For Type SE cable with ungrounded conductor sizes 10 AWG and smaller, where installed in contact with thermal insulation, the ampacity shall be in accordance with 60°C (140°F) conductor temperature rating. The maximum conductor temperature rating shall be permitted to be used for ampacity adjustment and correction purposes, if the final ampacity does not exceed that for a 60°C (140°F) rated conductor.

Type SE cable is used for a variety of interior circuits, including ranges, clothes dryers, heating, and air-conditioning equipment and as feeders to supply panelboards that are not the service equipment.

While all conductors in nonmetallic-sheathed cable are required to have an insulation temperature rating of 90°C, 334.80 limits the operating (calculated load) ampacity to those contained in the 60°C column of Table 310.16. This limitation applies to all uses of NM cable. In contrast to this restriction, Type SE cable is limited to operating at a 60°C ampacity only if the ungrounded conductor sizes are 10 AWG and smaller and it is installed in thermal insulation. Type SE cable is permitted to have conductors with either 75°C or 90°C insulation. If the cable surface does not have a temperature marking, the conductors have a 75°C insulation temperature rating, and ampacity adjustment or correction is based on that rating. If the cable is marked with a conductor insulation temperature rating, ampacity adjustment or correction of the conductor can be made based on the conductor temperature marked on the cable.

Where Type SE cable with ungrounded conductor sizes 10 AWG and smaller is installed in thermal insulation, the adjusted and/or corrected operating ampacity of the conductors cannot exceed those contained in the 60°C column of Table 310.16. If the cable is not installed in thermal insulation, the limiting factor on conductor ampacity is the requirement, as stated in 110.14(C), for coordinating the operating ampacity of the conductor with the terminal temperature ratings of the electrical equipment.

(b) *Exterior Installations.* Exterior installations shall comply with the following:

(1) In addition to the provisions of this article, service-entrance cable used for feeders or branch circuits, where installed as exterior wiring, shall be installed in accordance with Part I of Article 225. The cable shall be supported in accordance with 334.30.

(2) Type USE cable installed as underground feeder and branch circuit cable shall comply with Part II of Article 340.

Exception: Single-conductor Type USE and multi-rated USE conductors shall not be subject to the ampacity limitations of Part II of Article 340.

338.12 Uses Not Permitted.

Δ **(A) Service-Entrance Cable.** Type SE cable shall not be used under the following conditions or in the following locations:

(1) Where subject to physical damage unless protected in accordance with 230.50(B)

(2) Underground with or without a raceway

(3) For exterior branch circuits and feeder wiring unless the installation complies with Part I of Article 225 and is supported in accordance with 334.30 or is used as messenger-supported wiring as permitted in Part II of Article 396

(B) Underground Service-Entrance Cable. Type USE cable shall not be used under the following conditions or in the following locations:

(1) For interior wiring

(2) For aboveground installations except where USE cable emerges from the ground and is terminated in an enclosure at an outdoor location and the cable is protected in accordance with 300.5(D)

(3) As aerial cable unless it is a multiconductor cable identified for use aboveground and installed as messenger-supported wiring in accordance with 225.10 and Part II of Article 396

Cables marked only as "Type USE service-entrance cable" are not required to have a flame-retardant covering, according to the definition of *cable, service entrance (Type USE)* in Article 100.

338.24 Bending Radius. Bends in Types USE and SE cable shall be so made that the cable will not be damaged. The radius of the curve of the inner edge of any bend, during or after installation, shall not be less than five times the diameter of the cable. For flat cables, the major diameter dimension of the cable shall be used to determine the bending radius.

Part III. Construction Specifications

338.100 Construction.

(A) Assemblies. Cabled assemblies of multiple single-conductor Type USE conductors shall be permitted for direct burial. All conductors shall be insulated.

Informational Note: The term "cabled" refers to a manufacturing process of twisting single conductors together and may also be referred to as "plexed."

(B) Uninsulated Conductor. Type SE or USE cable with an overall covering containing two or more conductors shall be permitted to have one conductor uninsulated.

338.120 Marking. Service-entrance cable shall be marked as required in 310.8. Cable with the neutral conductor smaller than the ungrounded conductors shall be so marked.

ARTICLE 340

Underground Feeder and Branch-Circuit Cable: Type UF

Part I. General

340.1 Scope. This article covers the use, installation, and construction specifications for underground feeder and branch-circuit cable, Type UF.

340.6 Listing Requirements. Type UF cable and associated fittings shall be listed.

Part II. Installation

340.10 Uses Permitted. Type UF cable shall be permitted as follows:

(1) For use underground, including direct burial in the earth.

(2) As single-conductor cables. Where installed as single-conductor cables, all conductors of the feeder or branch circuit, including the grounded conductor and equipment grounding conductor, if any, shall be installed in accordance with 300.3.

(3) For wiring in wet, dry, or corrosive locations.

(4) Installed as nonmetallic-sheathed cable. Where so installed, the installation and conductor requirements shall comply with Parts II and III of Article 334, except for 334.12(B), and shall be of the multiconductor type.

(5) As single-conductor cables as the nonheating leads for heating cables as provided in 424.43.

(6) Supported by cable trays. Type UF cable supported by cable trays shall be of the multiconductor type.

Informational Note: See 310.14(A)(3) for temperature limitation of conductors.

340.12 Uses Not Permitted. Type UF cable shall not be used as follows:

(1) As service-entrance cable

(2) In commercial garages

(3) In theaters and similar locations

(4) In motion picture studios

(5) In storage battery rooms

(6) In hoistways or on elevators or escalators

(7) In hazardous (classified) locations, except as specifically permitted by other articles in this *Code*

(8) Embedded in poured cement, concrete, or aggregate, except where embedded in plaster as nonheating leads where permitted in 424.43

(9) Where exposed to direct rays of the sun, unless identified as sunlight resistant

Informational Note: The sunlight-resistant marking on the jacket does not apply to the individual conductors.

(10) Where subject to physical damage

(11) As overhead cable, except where installed as messenger-supported wiring in accordance with Part II of Article 396

Type UF cable suitable for exposure to the direct rays of the sun is tagged and marked with the designation "Sunlight Resistant." This physical protection requirement ensures that Type UF cable, as it emerges from underground, is protected from ultraviolet damage.

340.24 Bending Radius. Bends in Type UF cable shall be so made that the cable is not damaged. The radius of the curve of the inner edge of any bend shall not be less than five times the diameter of the cable. For flat cables, the major diameter dimension of the cable shall be used to determine the bending radius.

340.80 Ampacity. The ampacity of Type UF cable shall be that of 60°C (140°F) conductors in accordance with 310.14.

If Type UF cable is installed as nonmetallic-sheathed cable, the ampacity of the cable is determined according to rules for Type NM cable in 334.80.

Part III. Construction Specifications

340.104 Conductors. The conductors shall be sizes 14 AWG copper or 12 AWG aluminum or copper-clad aluminum through 4/0 AWG.

340.108 Equipment Grounding Conductor. In addition to the insulated conductors, the cable shall be permitted to have an insulated or bare equipment grounding conductor.

340.112 Insulation. The conductors of Type UF shall be one of the moisture-resistant types listed in Table 310.4(1) that is suitable for branch-circuit wiring or one that is identified for such use. Where installed as a substitute wiring method for NM cable, the conductor insulation shall be rated 90°C (194°F).

340.116 Sheath. The overall covering shall be flame retardant; moisture, fungus, and corrosion resistant; and suitable for direct burial in the earth.

Part I. General

342.1 Scope. This article covers the use, installation, and construction specifications for intermediate metal conduit (IMC) and associated fittings.

IMC is thinner-walled and lighter in weight than rigid metal conduit (RMC) and is satisfactory for uses in all locations where RMC is permitted to be used. Threaded fittings, couplings, connectors, and so forth are interchangeable between IMC and RMC. Threadless fittings for IMC are suitable only for the type of conduit indicated by the carton marking.

342.6 Listing Requirements. IMC, factory elbows and couplings, and associated fittings shall be listed.

Part II. Installation

342.10 Uses Permitted.

(A) All Atmospheric Conditions and Occupancies. Use of IMC shall be permitted under all atmospheric conditions and occupancies.

(B) Corrosion Environments. IMC, elbows, couplings, and fittings shall be permitted to be installed in concrete, in direct contact with the earth, in direct burial applications, or in areas subject to severe corrosive influences where protected by corrosion protection approved for the condition.

Other documents, such as the Steel Tube Institute's 2015 *Guidelines for Installing Steel Conduit/Tubing*, and ANSI/NECA 101-2013, *Standard for Installing Steel Conduits (Rigid, IMC, EMT)*, should be consulted for approval guidance of corrosion-resistant materials or for requirements prior to the installation of nonferrous metal (aluminum) conduit in concrete, since chloride additives in the concrete mix can cause corrosion.

(C) Cinder Fill. IMC shall be permitted to be installed in or under cinder fill where subject to permanent moisture where protected on all sides by a layer of noncinder concrete not less than 50 mm (2 in.) thick; where the conduit is not less than 450 mm (18 in.) under the fill; or where protected by corrosion protection approved for the condition.

(D) Wet Locations. All supports, bolts, straps, screws, and so forth, shall be of corrosion-resistant materials or protected against corrosion by corrosion-resistant materials.

Informational Note: See 300.6 for protection against corrosion.

Galvanized IMC installed in concrete does not require supplementary corrosion protection. Similarly, galvanized IMC installed in contact with soil does not generally require supplementary corrosion protection. As a guide in the absence of experience with the corrosive effects of soil in a specific location, soils producing severe corrosive effects are generally characterized by low resistivity of less than 2000 ohm-centimeter.

(E) Severe Physical Damage. IMC shall be permitted to be installed where subject to severe physical damage.

342.14 Dissimilar Metals. Where practicable, dissimilar metals in contact anywhere in the system shall be avoided to eliminate the possibility of galvanic action.

Stainless steel and aluminum fittings and enclosures shall be permitted to be used with galvanized steel IMC where not subject to severe corrosive influences.

Stainless steel IMC shall only be used with the following:

(1) Stainless steel fittings
(2) Stainless steel boxes and enclosures
(3) Steel (galvanized, painted, powder or PVC coated, and so forth) boxes and enclosures when not subject to severe corrosive influences
(4) Stainless steel, nonmetallic, or approved accessories

342.20 Size.

(A) Minimum. IMC smaller than metric designator 16 (trade size ½) shall not be used.

(B) Maximum. IMC larger than metric designator 155 (trade size 6) shall not be used.

> Informational Note: See 300.1(C) for the metric designators and trade sizes. These are for identification purposes only and do not relate to actual dimensions.

342.22 Number of Conductors. The number of conductors shall not exceed that permitted by the percentage fill specified in Table 1, Chapter 9.

Cables shall be permitted to be installed where such use is not prohibited by the respective cable articles. The number of cables shall not exceed the allowable percentage fill specified in Table 1, Chapter 9.

Table 4 of Chapter 9 provides the usable area within the selected conduit or tubing, and Table 5 provides the required area for each conductor. Examples using these tables to calculate a conduit or tubing size are provided in the commentary following Chapter 9, Notes to Tables, Note 6.

To select the proper trade size of IMC, see the appropriate sub-table for Article 342, Intermediate Metal Conduit (IMC), in Table 4 of Chapter 9. If the conductors are of the same wire size and insulation type, Tables C.4 and C.4(A) for IMC in Informative Annex C can be used instead of performing the calculations.

Δ **342.24 Bends.**

N **(A) How Made.** Bends of IMC shall be so made that the conduit will not be damaged and the internal diameter of the conduit will not be effectively reduced. The radius of the curve of any field bend to the centerline of the conduit shall not be less than indicated in Table 2, Chapter 9.

N **(B) Number in One Run.** The total degrees of bends in a conduit run shall not exceed 360 degrees between pull points.

The number of bends in one run is limited, to reduce pulling tension on conductors. It also helps ensure easy insertion or removal of conductors during later phases of construction, when the conduit may be permanently enclosed by the building's finish.

342.28 Reaming and Threading. All cut ends shall be reamed or otherwise finished to remove rough edges. Where conduit is threaded in the field, a standard cutting die with a taper of 1 in 16 (¾ in. taper per foot) shall be used.

> Informational Note: See ANSI/ASME B1.20.1-2013, *Standard for Pipe Threads, General Purpose (Inch)*.

Conduit is cut using a saw or a roll cutter (pipe cutter). Care should be taken to ensure a straight cut, given that crooked threads result from a die not started on the pipe squarely. After the cut is made, the conduit must be reamed. Proper reaming removes burrs from the interior of the cut conduit so that as wires and cables are pulled through the conduit, no chafing of the insulation or cable jacket occurs. Finally, the conduit is threaded. The number of threads is important, because cutting too many (or not enough) threads could result in improper assembly of the conduit system. If a threaded ring gauge is not available, the same number of threads should be cut on the conduit as are present on the factory (threaded) end of the conduit.

342.30 Securing and Supporting. IMC shall be installed as a complete system in accordance with 300.18 and shall be securely fastened in place and supported in accordance with 342.30(A) and (B).

(A) Securely Fastened. IMC shall be secured in accordance with one of the following:

(1) IMC shall be securely fastened within 900 mm (3 ft) of each outlet box, junction box, device box, cabinet, conduit body, or other conduit termination.
(2) Where structural members do not readily permit fastening within 900 mm (3 ft), fastening shall be permitted to be increased to a distance of 1.5 m (5 ft).
(3) Where approved, conduit shall not be required to be securely fastened within 900 mm (3 ft) of the service head for above-the-roof termination of a mast.

> *Exception: For concealed work in finished buildings or pre-finished wall panels where such securing is impracticable, unbroken lengths (without coupling) of IMC shall be permitted to be fished.*

As illustrated in Exhibit 342.1, IMC is required to be securely fastened within 3 feet of outlet boxes, junction boxes, cabinets, conduit bodies, or other conduit terminations. Couplings are not considered conduit terminations. However, where structural

EXHIBIT 342.1 Minimum IMC fastening requirements.

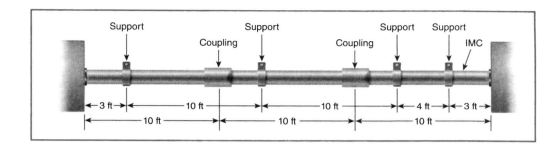

support members do not permit fastening within 3 feet, the support may be located up to 5 feet away. In addition, IMC is required to be supported at least every 10 feet unless permitted otherwise by 342.30(B).

Δ **(B) Supports.** IMC shall be supported in accordance with one of the following:

(1) Conduit shall be supported at intervals not exceeding 3 m (10 ft).
(2) The distance between supports for straight runs of conduit shall be permitted in accordance with Table 344.30(B), provided the conduit is made up with threaded couplings and supports that prevent transmission of stresses to termination where conduit is deflected between supports.
(3) Exposed vertical risers from industrial machinery or fixed equipment shall be permitted to be supported at intervals not exceeding 6 m (20 ft) if the conduit is made up with threaded couplings, the conduit is supported and securely fastened at the top and bottom of the riser, and no other means of intermediate support is readily available.
(4) Horizontal runs of IMC supported by openings through framing members at intervals not exceeding 3 m (10 ft) and securely fastened within 900 mm (3 ft) of termination points shall be permitted.

Lengths of IMC are permitted to be supported (but not necessarily secured) by framing members at 10-foot intervals, provided the IMC is secured and supported at least 3 feet from the box or enclosure. Exhibit 342.2 illustrates an installation where the IMC is installed through the bar joists.

342.42 Couplings and Connectors.

(A) Threadless. Threadless couplings and connectors used with conduit shall be made tight. Where buried in masonry or concrete, they shall be the concretetight type. Where installed in wet locations, they shall comply with 314.15. Threadless couplings and connectors shall not be used on threaded conduit ends unless listed for the purpose.

(B) Running Threads. Running threads shall not be used on conduit for connection at couplings.

342.46 Bushings. Where a conduit enters a box, fitting, or other enclosure, a bushing shall be provided to protect the wires

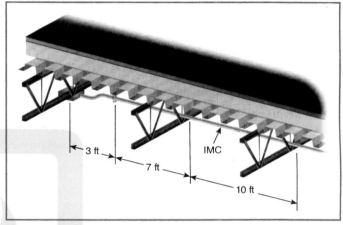

EXHIBIT 342.2 An example of IMC supported by framing members and securely fastened 3 feet from the box.

from abrasion unless the box, fitting, or enclosure is designed to provide such protection.

Informational Note: See 300.4(G) for the protection of conductors 4 AWG and larger at bushings.

342.56 Splices and Taps. Splices and taps shall be made in accordance with 300.15.

342.60 Grounding. IMC shall be permitted as an equipment grounding conductor.

Part III. Construction Specifications

342.100 Construction. IMC shall be made of one of the following:

(1) Steel, with protective coatings
(2) Stainless steel

342.120 Marking. Each length shall be clearly and durably marked at least every 1.5 m (5 ft) with the letters IMC. Each length shall be marked as required in the first sentence of 110.21(A).

ARTICLE 344 Rigid Metal Conduit (RMC)

Part I. General

344.1 Scope. This article covers the use, installation, and construction specifications for rigid metal conduit (RMC) and associated fittings.

344.6 Listing Requirements. RMC, factory elbows and couplings, and associated fittings shall be listed.

Part II. Installation

344.10 Uses Permitted.

(A) Atmospheric Conditions and Occupancies.

(1) Galvanized Steel, Stainless Steel, and Red Brass RMC. Galvanized steel, stainless steel, and red brass RMC shall be permitted under all atmospheric conditions and occupancies.

(2) Aluminum RMC. Aluminum RMC shall be permitted to be installed where approved for the environment.

(3) Ferrous Raceways and Fittings. Ferrous raceways and fittings protected from corrosion solely by enamel shall be permitted only indoors and in occupancies not subject to severe corrosive influences.

(B) Corrosive Environments.

(1) Galvanized Steel, Stainless Steel, and Red Brass RMC, Elbows, Couplings, and Fittings. Galvanized steel, stainless steel, and red brass RMC, elbows, couplings, and fittings shall be permitted to be installed in concrete, in direct contact with the earth, in direct burial applications, or in areas subject to severe corrosive influences where protected by corrosion protection approved for the condition.

(2) Supplementary Protection of Aluminum RMC. Aluminum RMC shall be provided with approved supplementary corrosion protection where encased in concrete or in direct contact with the earth, or in direct burial applications where identified for the application.

Other documents, such as the Steel Tube Institute's 2015 *Guidelines for Installing Steel Conduit/Tubing*, and ANSI/NECA 101-2013, *Standard for Installing Steel Conduits (Rigid, IMC, EMT)*, should be consulted for approval guidance of corrosion-resistant materials or for requirements prior to the installation of nonferrous metal (aluminum) conduit in concrete, since chloride additives in the concrete mix can cause corrosion.

(C) Cinder Fill. Galvanized steel, stainless steel, and red brass RMC shall be permitted to be installed in or under cinder fill where subject to permanent moisture where protected on all sides by a layer of noncinder concrete not less than 50 mm (2 in.)

thick; where the conduit is not less than 450 mm (18 in.) under the fill; or where protected by corrosion protection approved for the condition.

Although cinder fill is not commonly used in modern construction, it is still encountered at older building sites. Cinders used as fill may contain sulfur and, when combined with moisture, form sulfuric acid, which can corrode metal raceways.

(D) Wet Locations. All supports, bolts, straps, screws, and so forth, shall be of corrosion-resistant materials or protected against corrosion by corrosion-resistant materials.

Informational Note: See 300.6 for protection against corrosion.

(E) Severe Physical Damage. RMC shall be permitted to be installed where subject to severe physical damage.

344.14 Dissimilar Metals. Where practicable, dissimilar metals in contact anywhere in the system shall be avoided to eliminate the possibility of galvanic action. Stainless steel and aluminum fittings and enclosures shall be permitted to be used with galvanized steel RMC, and galvanized steel fittings and enclosures shall be permitted to be used with aluminum RMC where not subject to severe corrosive influences. Stainless steel rigid conduit shall only be used with the following:

(1) Stainless steel fittings
(2) Stainless steel boxes and enclosures
(3) Steel (galvanized, painted, powder or PVC coated, and so forth) boxes and enclosures when not subject to severe corrosive influences
(4) Stainless steel, nonmetallic, or approved accessories

Aluminum rigid conduit can be used with steel fittings and enclosures, as can aluminum fittings and enclosures with steel rigid conduit. Tests show that the galvanic corrosion at steel and aluminum interfaces is minor compared to the natural corrosion on the combination of steel and steel or of aluminum and aluminum.

344.20 Size.

(A) Minimum. RMC smaller than metric designator 16 (trade size ½) shall not be used.

Exception: Metric designator 12 (trade size ⅜) shall be permitted for enclosing the leads of motors as permitted in 430.245(B).

(B) Maximum. RMC larger than metric designator 155 (trade size 6) shall not be used.

Informational Note: See 300.1(C) for the metric designators and trade sizes. These are for identification purposes only and do not relate to actual dimensions.

344.22 Number of Conductors. The number of conductors shall not exceed that permitted by the percentage fill specified in Table 1, Chapter 9.

Cables shall be permitted to be installed where such use is not prohibited by the respective cable articles. The number of cables shall not exceed the allowable percentage fill specified in Table 1, Chapter 9.

Table 4 of Chapter 9 provides the usable area within the selected conduit or tubing, and Table 5 provides the required area for each conductor. Examples using these tables to calculate a conduit or tubing size are provided in the commentary following Chapter 9, Notes to Tables, Note 6.

To select the proper trade size of RMC, see the appropriate sub-table for Article 344, Rigid Metal Conduit (RMC), in Table 4 of Chapter 9. If the conductors are of the same wire size and insulation type, Tables C.9 and C.9(A) for RMC in Informative Annex C can be used instead of performing the calculations.

Δ **344.24 Bends.**

N **(A) How Made.** Bends of RMC shall be so made that the conduit will not be damaged and so that the internal diameter of the conduit will not be effectively reduced. The radius of the curve of any field bend to the centerline of the conduit shall not be less than indicated in Table 2, Chapter 9.

N **(B) Number in One Run.** The total degrees of bends in a conduit run shall not exceed 360 degrees between pull points.

Limiting the number of bends in a conduit run reduces pulling tension on conductors. It also helps ensure easy insertion or removal of conductors during later phases of construction when the conduit may be permanently enclosed by the finish of the building.

344.28 Reaming and Threading. All cut ends shall be reamed or otherwise finished to remove rough edges. Where conduit is threaded in the field, a standard cutting die with a 1 in 16 taper (¾ in. taper per foot) shall be used. PVC-coated RMC shall be threaded in accordance with manufacturer's instructions to prevent damage to the exterior coating.

> Informational Note No. 1: See ANSI/ASME B1.20.1-2013, *Standard for Pipe Threads, General Purpose (Inch)*.

> Informational Note No. 2: See NECA 101-2013, *Standard for Installing Steel Conduits (RMC, IMC, EMT)*, for information on threading and clamping methods for RMC and PVC-coated RMC.

Conduit is cut using a saw or a roll cutter (pipe cutter). Crooked threads result from a die not started on the pipe squarely. After the cut is made, the conduit must be reamed. Proper reaming removes burrs from the interior of the cut conduit so that as wires and cables are pulled through the conduit, chafing of the insulation or cable jacket does not occur. Finally, the conduit is threaded. The number of threads is important. When determining the correct number of threads for a conduit end, the same number of threads should be cut on the conduit as are present on the factory (threaded) end of the conduit. Where excessive threads are cut on the conduit and threaded couplings are installed, the conduits within the coupling will butt, resulting in a weak mechanical joint and poor grounding continuity.

To avoid damage to the exterior coating on PVC-coated conduits, proper threading and clamping tools specifically designed for PVC-coated conduit must be used. Standard threading and clamping tools, which have not been modified, can cause damage to the coatings, therefore exposing the conduit to unintended corrosion and potential unsafe conditions. Manufacturer's instructions for threading and bending should also be followed, in order to prevent any additional damage.

344.30 Securing and Supporting. RMC shall be installed as a complete system in accordance with 300.18 and shall be securely fastened in place and supported in accordance with 344.30(A) and (B).

(A) Securely Fastened. RMC shall be secured in accordance with one of the following:

(1) RMC shall be securely fastened within 900 mm (3 ft) of each outlet box, junction box, device box, cabinet, conduit body, or other conduit termination.

(2) Fastening shall be permitted to be increased to a distance of 1.5 m (5 ft) where structural members do not readily permit fastening within 900 mm (3 ft).

(3) Where approved, conduit shall not be required to be securely fastened within 900 mm (3 ft) of the service head for above-the-roof termination of a mast.

Exception: For concealed work in finished buildings or prefinished wall panels where such securing is impracticable, unbroken lengths (without coupling) of RMC shall be permitted to be fished.

As illustrated in Exhibit 344.1, RMC is required to be securely fastened within 3 feet of outlet boxes, junction boxes, cabinets, conduit bodies, or other conduit terminations. Couplings are not considered conduit terminations. However, where structural support members do not permit fastening within 3 feet, secure fastening can be located up to 5 feet away. In addition, RMC is required to be supported at least every 10 feet unless permitted otherwise by 344.30(B).

Δ **(B) Supports.** RMC shall be supported in accordance with one of the following:

(1) Conduit shall be supported at intervals not exceeding 3 m (10 ft).

(2) The distance between supports for straight runs of conduit shall be permitted in accordance with Table 344.30(B), provided the conduit is made up with threaded couplings and supports that prevent transmission of stresses to termination where conduit is deflected between supports.

(3) Exposed vertical risers from industrial machinery or fixed equipment shall be permitted to be supported at intervals not exceeding 6 m (20 ft) if the conduit is made up with threaded couplings, the conduit is supported and securely fastened at the top and bottom of the riser, and no other means of intermediate support is readily available.

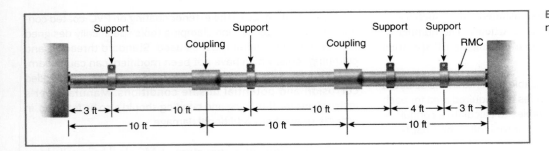

EXHIBIT 344.1 Minimum support required for RMC.

TABLE 344.30(B) *Supports for Rigid Metal Conduit*

Conduit Size		Maximum Distance Between Rigid Metal Conduit Supports	
Metric Designator	Trade Size	m	ft
16–21	½–¾	3.0	10
27	1	3.7	12
35–41	1¼–1½	4.3	14
53–63	2–2½	4.9	16
78 and larger	3 and larger	6.1	20

(4) Horizontal runs of RMC supported by openings through framing members at intervals not exceeding 3 m (10 ft) and securely fastened within 900 mm (3 ft) of termination points shall be permitted.

Lengths of RMC are permitted to be supported (but not necessarily secured) by framing members at 10-foot intervals, provided the RMC is secured and supported at least 3 feet from the box or enclosure. Installations where the RMC is installed through bar joists are just one example, as illustrated in Exhibit 344.2.

344.42 Couplings and Connectors.

(A) Threadless. Threadless couplings and connectors used with conduit shall be made tight. Where buried in masonry or concrete,

they shall be the concrete tight type. Where installed in wet locations, they shall comply with 314.15. Threadless couplings and connectors shall not be used on threaded conduit ends unless listed for the purpose.

Exhibit 344.3 illustrates two different conduit fittings: a threadless conduit coupling and a threadless conduit connector. Threadless fittings might be suitable for other applications, such as in raintight or concretetight applications, provided the product itself, or the product packaging, is marked as such.

In general, threadless fittings are not intended for use over threads, because the fitting will not seat properly. The threaded end of the conduit should be cut off and the conduit reamed before installation.

Exhibit 344.4 illustrates a three-piece threaded coupling (the electrical equivalent of a pipe union), which is used to join two lengths of conduit where turning either length is impossible, such as in underground or concrete slab construction. Another fitting for joining conduit is a bolted split coupling.

(B) Running Threads. Running threads shall not be used on conduit for connection at couplings.

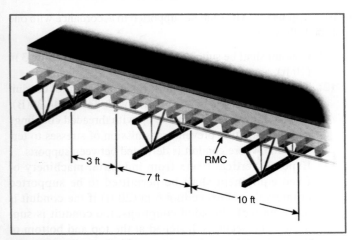

EXHIBIT 344.2 An example of RMC supported by framing members and securely fastened 3 feet from the box.

EXHIBIT 344.3 An example of RMC threadless connector *(bottom)* and a threadless coupling *(top). (Courtesy of Eaton, Crouse-Hinds Division)*

EXHIBIT 344.4 A three-piece (union-type) coupling. *(Courtesy of Appleton™, Emerson Electric Co.)*

344.46 Bushings. Where a conduit enters a box, fitting, or other enclosure, a bushing shall be provided to protect the wires from abrasion unless the box, fitting, or enclosure is designed to provide such protection.

> Informational Note: See 300.4(G) for the protection of conductors sizes 4 AWG and larger at bushings.

344.56 Splices and Taps. Splices and taps shall be made in accordance with 300.15.

344.60 Grounding. RMC shall be permitted as an equipment grounding conductor.

Part III. Construction Specifications

344.100 Construction. RMC shall be made of one of the following:

(1) Steel with protective coatings
(2) Aluminum
(3) Red brass
(4) Stainless steel

344.120 Marking. Each length shall be clearly and durably identified in every 3 m (10 ft) as required in the first sentence of 110.21(A). Nonferrous conduit of corrosion-resistant material shall have suitable markings.

ARTICLE 348 Flexible Metal Conduit (FMC)

Part I. General

348.1 Scope. This article covers the use, installation, and construction specifications for flexible metal conduit (FMC) and associated fittings.

N **348.2 Reconditioned Equipment.** FMC shall not be reconditioned.

348.6 Listing Requirements. FMC and associated fittings shall be listed.

Part II. Installation

348.10 Uses Permitted. FMC shall be permitted to be used in exposed and concealed locations.

FMC ½ inch and larger may be installed in unlimited lengths, provided an equipment grounding conductor (EGC) is installed with the circuit conductors.

See also

250.118(A)(5) and **348.60(A)** for specific requirements related to the use of FMC as an EGC

348.12 Uses Not Permitted. FMC shall not be used in the following:

(1) In wet locations
(2) In hoistways, other than as permitted in 620.21(A)(1)
(3) In storage battery rooms
(4) In any hazardous (classified) location except as permitted by other articles in this *Code*
(5) Where exposed to materials having a deteriorating effect on the installed conductors, such as oil or gasoline
(6) Underground or embedded in poured concrete or aggregate
(7) Where subject to physical damage

348.20 Size.

(A) Minimum. FMC less than metric designator 16 (trade size ½) shall not be used unless permitted in 348.20(A)(1) through (A)(5) for metric designator 12 (trade size ⅜).

(1) For enclosing the leads of motors as permitted in 430.245(B)
(2) In lengths not in excess of 1.8 m (6 ft) for any of the following uses:
  a. For utilization equipment
  b. As part of a listed assembly
  c. For tap connections to luminaires as permitted in 410.117(C)

Trade size ⅜ FMC is permitted to be used as the manufactured or field-installed metal raceway (1½ feet to 6 feet in length) to enclose tap conductors between the outlet box and the terminal housing of recessed luminaires. FMC is also permitted to be used as a 6-foot luminaire whip from an outlet box to a luminaire.

(3) For manufactured wiring systems as permitted in 604.100(A)

A smaller minimum size for manufactured wiring systems is permitted [see Exception No. 3 to 604.100(A)(2)] because conductors assembled under factory-controlled conditions are not as prone to physical damage.

(4) In hoistways as permitted in 620.21(A)(1)
(5) As part of a listed assembly to connect wired luminaire sections as permitted in 410.137(C)

(B) Maximum. FMC larger than metric designator 103 (trade size 4) shall not be used.

> Informational Note: See 300.1(C) for the metric designators and trade sizes. These are for identification purposes only and do not relate to actual dimensions.

348.22 Number of Conductors. The number of conductors shall not exceed that permitted by the percentage fill specified in Table 1, Chapter 9, or as permitted in Table 348.22, or for metric designator 12 (trade size ⅜).

Cables shall be permitted to be installed where such use is not prohibited by the respective cable articles. The number of cables shall not exceed the allowable percentage fill specified in Table 1, Chapter 9.

Table 4 of Chapter 9 provides the usable area within the selected conduit or tubing, and Table 5 provides the required area for each conductor. Examples using these tables to calculate a conduit or tubing size are provided in the commentary following Chapter 9, Notes to Tables, Note 6.

To select the proper trade size of FMC, see the appropriate sub-table for Article 348, Flexible Metal Conduit (FMC), in Table 4 of Chapter 9. If the conductors are of the same wire size and insulation type, Tables C.3 and C.3(A) for FMC in Informative Annex C can be used instead of performing the calculations.

Δ **348.24 Bends.**

N **(A) How Made.** Bends in conduit shall be made so that the conduit is not damaged and the internal diameter of the conduit is not effectively reduced. Bends shall be permitted to be made manually without auxiliary equipment. The radius of the curve to the centerline of any bend shall not be less than shown in Table 2, Chapter 9 using the column "Other Bends."

N **(B) Number in One Run.** The total degrees of bends in a conduit run shall not exceed 360 degrees between pull points.

348.28 Trimming. All cut ends shall be trimmed or otherwise finished to remove rough edges, except where fittings that thread into the convolutions are used.

348.30 Securing and Supporting. FMC shall be securely fastened in place and supported in accordance with 348.30(A) and (B).

(A) Securely Fastened. FMC shall be securely fastened in place by an approved means within 300 mm (12 in.) of each box, cabinet, conduit body, or other conduit termination and shall be supported and secured at intervals not to exceed 1.4 m (4½ ft). Where used, cable ties shall be listed and be identified for securement and support.

Listing of cable ties for securement and support of FMC is necessary because the standard requires markings that identify critical performance characteristics. These characteristics can affect their suitability for the conditions of use, including minimum and maximum operating temperatures and resistance to ultraviolet light for outdoor installations.

Exception No. 1: Where FMC is fished between access points through concealed spaces in finished buildings or structures and supporting is impracticable.

Exception No. 2: Where flexibility is necessary after installation, lengths from the last point where the raceway is securely fastened shall not exceed the following:

(1) 900 mm (3 ft) for metric designators 16 through 35 (trade sizes ½ through 1¼)
(2) 1200 mm (4 ft) for metric designators 41 through 53 (trade sizes 1½ through 2)
(3) 1500 mm (5 ft) for metric designators 63 (trade size 2½) and larger

An example of the phrase "where flexibility is necessary after installation" is an installation of FMC to a motor mounted on an adjustable or sliding frame, where the frame is required to be movable for drive belt maintenance. The length that the exception addresses is the length from the last point where the FMC is securely fastened.

Exception No. 3: Lengths not exceeding 1.8 m (6 ft) from a luminaire terminal connection for tap connections to luminaires as permitted in 410.117(C).

TABLE 348.22 *Maximum Number of Insulated Conductors in Metric Designator 12 (Trade Size ⅜) Flexible Metal Conduit (FMC)*[*]

Size (AWG)	Types RFH-2, SF-2		Types TF, XHHW, TW		Types TFN, THHN, THWN		Types FEP, FEBP, PF, PGF	
	Fittings Inside Conduit	Fittings Outside Conduit	Fittings Inside Conduit	Fittings Outside Conduit	Fittings Inside Conduit	Fittings Outside Conduit	Fittings Inside Conduit	Fittings Outside Conduit
18	2	3	3	5	5	8	5	8
16	1	2	3	4	4	6	4	6
14	1	2	2	3	3	4	3	4
12	—	—	1	2	2	3	2	3
10	—	—	1	1	1	1	1	2

*In addition, one insulated, covered, or bare equipment grounding conductor of the same size shall be permitted.

Exception No. 4: Lengths not exceeding 1.8 m (6 ft) from the last point where the raceway is securely fastened for connections within an accessible ceiling to a luminaire(s) or other equipment. For the purposes of the exceptions, listed FMC fittings shall be permitted as a means of securement and support.

Securing a raceway can be different from supporting the raceway. Specifying that the listed FMC fitting provides the securement also required by this section clarifies that the listed fitting provides both securement and support of the FMC.

(B) Supports. Horizontal runs of FMC supported by openings through framing members at intervals not greater than 1.4 m (4½ ft) and securely fastened within 300 mm (12 in.) of termination points shall be permitted.

348.42 Couplings and Connectors. Angle connectors shall not be concealed.

348.56 Splices and Taps. Splices and taps shall be made in accordance with 300.15.

Δ **348.60 Grounding and Bonding.**

N **(A) Fixed Installation.** FMC shall be permitted to be used as an equipment grounding conductor when installed in accordance

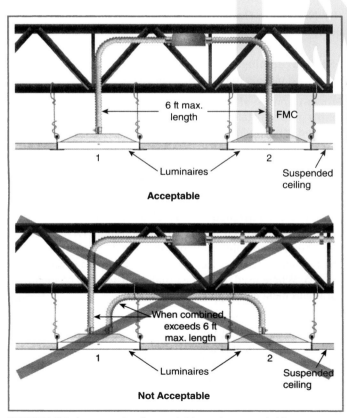

EXHIBIT 348.1 An example of acceptable and unacceptable applications of FMC without separate EGCs used as a luminaire whip, in accordance with 250.118(5)(d).

with 250.118(A)(5) where flexibility is not required after installation.

N **(B) Flexible Installation.** An equipment grounding conductor shall be installed where flexibility is necessary to minimize the transmission of vibration from equipment or to provide flexibility for equipment that requires movement after installation.

An additional EGC is always required where FMC is used for flexibility. Examples of such installations include using FMC to minimize the transmission of equipment vibration such as motors or to provide flexibility for floodlights, spotlights, or other equipment that require adjustment after installation.

According to ANSI/UL 1, *Standard for Flexible Metal Conduit*, FMC longer than 6 feet has not been judged to be suitable for grounding purposes. If the length of the total ground-fault return path exceeds 6 feet or the circuit overcurrent protection exceeds 20 amperes, a separate EGC must be installed with the circuit conductors according to 250.118(5). The top figure in Exhibit 348.1 shows an acceptable application of FMC, where the total length of any ground return path is limited to 6 feet. The bottom figure shows an application that is unacceptable because the grounding return path for Luminaire 2 exceeds the permitted maximum of 6 feet to the box.

Where FMC is used in hazardous (classified) locations, a bonding jumper is required. Section 250.102(E) permits the routing of equipment bonding jumpers on the outside of the raceway in lengths that are no longer than 6 feet and bonded at each end.

N **(C) Equipment Grounding Conductors.** Where required or installed, equipment grounding conductors shall be installed in accordance with 250.134.

N **(D) Equipment Bonding Jumpers.** Where required or installed, equipment bonding jumpers shall be installed in accordance with 250.102.

ARTICLE
350 Liquidtight Flexible Metal Conduit (LFMC)

Part I. General

350.1 Scope. This article covers the use, installation, and construction specifications for liquidtight flexible metal conduit (LFMC) and associated fittings.

LFMC is intended for use in wet locations or where exposed to oil or coolants, at a maximum temperature of 140°F. LFMC is not intended for use where exposed to gasoline or similar light petroleum solvents unless so marked on the product. If properly marked for the application, LFMC is permitted for direct burial in the earth. LFMC is on the permitted list of wiring methods for services (see 230.43), provided the length does not exceed 6 feet and an equipment bonding jumper is installed in accordance with 250.102. LFMC may be installed in unlimited lengths, provided it meets the other requirements of Article 350 and a separate equipment grounding conductor (EGC) is installed with the circuit conductors.

N **350.2 Reconditioned Equipment.** LFMC shall not be reconditioned.

350.6 Listing Requirements. LFMC and associated fittings shall be listed.

Part II. Installation

Δ **350.10 Uses Permitted.** LFMC shall be permitted to be used in exposed or concealed locations as follows:

(1) Where conditions of installation, operation, or maintenance require flexibility or protection from machine oils, liquids, vapors, or solids.
(2) In hazardous (classified) locations where specifically permitted by Chapter 5.
(3) For direct burial where listed and marked for the purpose.
(4) Conductors or cables rated at a temperature higher than the listed temperature rating of LFMC shall be permitted to be installed in LFMC, provided the conductors or cables are not operated at a temperature higher than the listed temperature rating of the LFMC.

350.12 Uses Not Permitted. LFMC shall not be used where subject to physical damage.

350.20 Size.

(A) Minimum. LFMC smaller than metric designator 16 (trade size ½) shall not be used.

Exception: LFMC of metric designator 12 (trade size ⅜) shall be permitted as covered in 348.20(A).

(B) Maximum. The maximum size of LFMC shall be metric designator 103 (trade size 4).

> Informational Note: See 300.1(C) for the metric designators and trade sizes. These are for identification purposes only and do not relate to actual dimensions.

350.22 Number of Conductors or Cables.

(A) Metric Designators 16 through 103 (Trade Sizes ½ through 4). The number of conductors shall not exceed that permitted by the percentage fill specified in Table 1, Chapter 9.

Cables shall be permitted to be installed where such use is not prohibited by the respective cable articles. The number of cables shall not exceed the allowable percentage fill specified in Table 1, Chapter 9.

(B) Metric Designator 12 (Trade Size ⅜). The number of conductors shall not exceed that permitted in Table 348.22, "Fittings Outside Conduit" columns.

Table 4 of Chapter 9 provides the usable area within the selected conduit or tubing, and Table 5 provides the required area for each conductor. Examples using these tables to calculate a conduit or tubing size are provided in the commentary following Chapter 9, Notes to Tables, Note 6.

To select the proper trade size of LFMC, see the appropriate sub-table for Article 350, Liquidtight Flexible Metal Conduit (LFMC), in Table 4 of Chapter 9. If the conductors are of the same wire size, Tables C.8 and C.8(A) for LFMC in Informative Annex C can be used instead of performing the calculations.

The exception to 350.20(A) permits the use of trade size ⅜ LFMC under the limited conditions specified for flexible metal conduit (FMC) in 348.20(A).

See also

Table 348.22 for the number of conductors permitted in trade size ⅜ LFMC

Δ **350.24 Bends.**

N (A) How Made. Bends in conduit shall be so made that the conduit will not be damaged and the internal diameter of the conduit will not be effectively reduced. Bends shall be permitted to be made manually without auxiliary equipment. The radius of the curve to the centerline of any bend shall not be less than required in Table 2, Chapter 9 using the column "Other Bends."

N (B) Number in One Run. The total degrees of bends in a conduit run shall not exceed 360 degrees between pull points.

350.28 Trimming. All cut ends of conduit shall be trimmed inside and outside to remove rough edges.

Proper trimming of the cut ends of LFMC is necessary to allow for the proper installation of the steel grounding ferrule, which is required to maintain ground continuity of the steel sheath.

350.30 Securing and Supporting. LFMC shall be securely fastened in place and supported in accordance with 350.30(A) and (B).

(A) Securely Fastened. LFMC shall be securely fastened in place by an approved means within 300 mm (12 in.) of each box, cabinet, conduit body, or other conduit termination and shall be supported and secured at intervals not to exceed 1.4 m (4½ ft). Where used, cable ties shall be listed and be identified for securement and support.

Listing of cable ties for securement and support of LFMC is necessary because the standard requires markings that identify critical performance characteristics. These characteristics can affect their suitability for the conditions of use, including minimum and maximum operating temperatures, and resistance to ultraviolet light for outdoor installations.

Exception No. 1: Where LFMC is fished between access points through concealed spaces in finished buildings or structures and supporting is impractical.

Exception No. 2: Where flexibility is necessary after installation, lengths from the last point where the raceway is securely fastened shall not exceed the following:

(1) 900 mm (3 ft) for metric designators 16 through 35 (trade sizes ½ through 1¼)

(2) 1200 mm (4 ft) for metric designators 41 through 53 (trade sizes 1½ through 2)

(3) 1500 mm (5 ft) for metric designators 63 (trade size 2½) and larger

Exception No. 3: Lengths not exceeding 1.8 m (6 ft) from a luminaire terminal connection for tap conductors to luminaires, as permitted in 410.117(C).

Exception No. 4: Lengths not exceeding 1.8 m (6 ft) from the last point where the raceway is securely fastened for connections within an accessible ceiling to luminaire(s) or other equipment.

Securing LFMC can be different from supporting it. The listed fittings are now recognized to provide the securement as well as the support required by this section.

For the purposes of the exceptions, listed LFMC fittings shall be permitted as a means of securement and support.

(B) Supports. Horizontal runs of LFMC supported by openings through framing members at intervals not greater than 1.4 m (4½ ft) and securely fastened within 300 mm (12 in.) of termination points shall be permitted.

350.42 Couplings and Connectors. Only fittings listed for use with LFMC shall be used. Angle connectors shall not be concealed. Straight LFMC fittings shall be permitted for direct burial where marked.

350.56 Splices and Taps. Splices and taps shall be made in accordance with 300.15.

Δ **350.60 Grounding and Bonding.**

N **(A) Fixed Installation.** LFMC shall be permitted to be used as an equipment grounding conductor when installed in accordance with 250.118(A)(6) where flexibility is not required after installation.

N **(B) Flexible Installation.** An equipment grounding conductor shall be installed where flexibility is necessary to minimize the transmission of vibration from equipment or to provide flexibility for equipment that requires movement after installation.

N **(C) Equipment Grounding Conductor.** Where required or installed, equipment grounding conductors shall be installed in accordance with 250.134.

N **(D) Equipment Bonding Jumpers.** Where required or installed, equipment bonding jumpers shall be installed in accordance with 250.102.

Informational Note: See 501.30(B)(2), 502.30(B)(2), 503.30(B)(2), 505.30(B)(2), and 506.30(B)(2) for types of equipment grounding conductors.

Part III. Construction Specifications

350.120 Marking. LFMC shall be marked according to 110.21. The trade size and other information required by the listing shall also be marked on the conduit. Conduit suitable for direct burial shall be so marked.

ARTICLE
352 Rigid Polyvinyl Chloride Conduit (PVC)

Part I. General

Δ **352.1 Scope.** This article covers the use, installation, and construction specifications for rigid polyvinyl chloride conduit (PVC) and associated fittings.

The UL *Guide Information for Electrical Equipment* describes rigid PVC conduit, Type PVC, for use in accordance with Article 352. Schedule 40 is suitable for locations not subject to physical damage for underground, aboveground, indoor, and outdoor locations. Schedule 80 is suitable for locations where the conduit will be subject to damage. Types A and EB are intended for underground installations.

352.6 Listing Requirements. PVC conduit, factory elbows, and associated fittings shall be listed.

Part II. Installation

352.10 Uses Permitted. The use of PVC conduit shall be permitted in accordance with 352.10(A) through (K).

Informational Note: Extreme cold may cause some nonmetallic conduits to become brittle and, therefore, more susceptible to damage from physical contact.

(A) Concealed. PVC conduit shall be permitted in walls, floors, and ceilings.

Δ **(B) Encased in Concrete.** PVC conduit shall be permitted to be encased in concrete.

(C) Corrosive Influences. PVC conduit shall be permitted in locations subject to severe corrosive influences as covered in 300.6 and where subject to chemicals for which the materials are specifically approved.

(D) Cinders. PVC conduit shall be permitted in cinder fill.

(E) Wet Locations. PVC conduit shall be permitted in portions of dairies, laundries, canneries, or other wet locations, and in locations where walls are frequently washed, the entire conduit system, including boxes and fittings used therewith, shall be installed and equipped so as to prevent water from entering the conduit. All supports, bolts, straps, screws, and so forth, shall be of corrosion-resistant materials or be protected against corrosion by approved corrosion-resistant materials.

(F) Dry and Damp Locations. PVC conduit shall be permitted for use in dry and damp locations not prohibited by 352.12.

Δ **(G) Exposed.** PVC conduit shall be permitted for exposed work.

(H) Underground Installations. For underground installations, PVC shall be permitted for direct burial and underground encased in concrete. See 300.5 and 305.15.

Schedule 40 and Schedule 80 PVC are both permitted for underground installations, such as under driveways, provided that the required burial depth is met.

> **See also**
>
> **Table 300.5(A)** for burial depth minimum cover requirements for 1000 V ac or less underground installations
>
> **Table 305.15(A)** for burial depth minimum cover requirements for over 1000 V ac underground installations

(I) Support of Conduit Bodies. PVC conduit shall be permitted to support nonmetallic conduit bodies not larger than the largest trade size of an entering raceway. These conduit bodies shall not support luminaires or other equipment and shall not contain devices other than splicing devices as permitted by 110.14(B) and 314.16(C)(2).

(J) Insulation Temperature Limitations. Conductors or cables rated at a temperature higher than the listed temperature rating of PVC conduit shall be permitted to be installed in PVC conduit, provided the conductors or cables are not operated at a temperature higher than the listed temperature rating of the PVC conduit.

Conductors marked with a rated temperature higher than that of the raceway can be used if the conductors are to be operated within the raceway temperature rating. One application is the use of 105°C-rated medium-voltage cables, Type MV, where the cable ampacity at the 105°C rating is reduced to the cable ampacity at 75°C or 90°C to match the listed operating temperature rating of the PVC conduit (75°C or 90°C).

N **(K) Physical Damage.** Where subject to physical damage, Schedule 80 PVC conduit, Schedule 80 PVC elbows, and listed fittings for PVC conduit shall be used.

> Informational Note: All listed PVC conduit fittings are suitable for connection to both Schedule 40 and Schedule 80 PVC conduit.

352.12 Uses Not Permitted. PVC conduit shall not be used under the conditions specified in 352.12(A) through (E).

(A) Hazardous (Classified) Locations. In any hazardous (classified) location, except as permitted by other articles of this *Code*.

(B) Support of Luminaires. For the support of luminaires or other equipment not described in 352.10(I).

(C) Physical Damage. Where subject to physical damage, except as permitted in 352.10(K).

(D) Ambient Temperatures. Where subject to ambient temperatures in excess of 50°C (122°F) unless listed otherwise.

(E) Theaters and Similar Locations. In theaters and similar locations, except as provided in 518.4 and 520.5.

In addition to the conditions in 352.12(A) through (E), PVC conduits are not permitted to be installed in ducts, plenums, and other air-handling spaces.

> **See also**
>
> **300.22** for limitations of the use of materials in ducts, plenums, and other air-handling spaces, which could contribute smoke and products of combustion during a fire

352.20 Size.

(A) Minimum. PVC conduit smaller than metric designator 16 (trade size ½) shall not be used.

Δ **(B) Maximum.** PVC conduit larger than metric designator 155 (trade size 6) shall not be used.

> Informational Note: See 300.1(C) for the trade sizes and metric designators that are for identification purposes only and do not relate to actual dimensions.

352.22 Number of Conductors. The number of conductors shall not exceed that permitted by the percentage fill specified in Table 1, Chapter 9.

Cables shall be permitted to be installed where such use is not prohibited by the respective cable articles. The number of cables shall not exceed the allowable percentage fill specified in Table 1, Chapter 9.

Table 4 of Chapter 9 provides the usable area within the selected conduit or tubing, and Table 5 provides the required area for each conductor. Examples using these tables to calculate a conduit or tubing size are provided in the commentary following Chapter 9, Notes to Tables, Note 6.

To select the proper trade size of PVC conduit, see the appropriate sub-table for Article 352, Rigid PVC Conduit (PVC), in Table 4 of Chapter 9. If the conductors are of the same wire size and insulation type, Tables C.10 and C.10(A) through Tables C.13 and C.13(A) can be used instead of performing the calculations, provided the appropriate table for the given type of PVC conduit is used.

Δ **352.24 Bends.**

N **(A) How Made.** Bends shall be so made that the conduit will not be damaged and the internal diameter of the conduit will not be effectively reduced. Field bends shall be made only with identified bending equipment. The radius of the curve to the centerline of such bends shall not be less than shown in Table 2, Chapter 9.

Pulling conductors in underground conduit can damage nonmetallic elbows. Metal elbows are often used to ensure the raceway's integrity. Metal elbows in runs of PVC conduit that are buried at least 18 inches are not required to be bonded to the system grounded conductor or the grounding electrode conductor (GEC).

See also

250.80, Exception, for more information on metal elbows in service raceways and enclosures

Chapter 9, Table 2, for common raceway field bend measurements

N **(B) Number in One Run.** The total degrees of bends in a conduit run shall not exceed 360 degrees between pull points.

The number of bends in a conduit run is limited, to reduce pulling tension on the conductors and to help ensure easy insertion or removal of conductors during later phases of construction, when the conduit may be permanently enclosed by the building finish. The *NEC®* does not limit the pull points to conduit bodies and boxes, which are only examples of pull points.

352.28 Trimming. All cut ends shall be trimmed inside and outside to remove rough edges.

352.30 Securing and Supporting. PVC conduit shall be installed as a complete system as provided in 300.18 and shall be fastened so that movement from thermal expansion or contraction is permitted. PVC conduit shall be securely fastened and supported in accordance with 352.30(A) and (B).

(A) Securely Fastened. PVC conduit shall be securely fastened within 900 mm (3 ft) of each outlet box, junction box, device box, conduit body, or other conduit termination. Conduit listed for securing at other than 900 mm (3 ft) shall be permitted to be installed in accordance with the listing.

(B) Supports. PVC conduit shall be supported as required in Table 352.30(B). Conduit listed for support at spacings other than as shown in Table 352.30(B) shall be permitted to be installed in accordance with the listing. Horizontal runs of PVC conduit

TABLE 352.30(B) *Support of Rigid Polyvinyl Chloride Conduit (PVC)*

Conduit Size		Maximum Spacing Between Supports	
Metric Designator	Trade Size	mm or m	ft
16–27	½–1	900 mm	3
35–53	1¼–2	1.5 m	5
63–78	2½–3	1.8 m	6
91–129	3½–5	2.1 m	7
155	6	2.5 m	8

supported by openings through framing members at intervals not exceeding those in Table 352.30(B) and securely fastened within 900 mm (3 ft) of termination points shall be permitted.

Δ **352.44 Expansion Fittings.**

N **(A) Thermal Expansion and Contraction.** Expansion fittings for PVC conduit shall be provided to compensate for thermal expansion and contraction where the length change, in accordance with Table 352.44(A), is expected to be 6 mm (¼ in.) or greater in a straight run between securely mounted items such as boxes, cabinets, elbows, or other conduit terminations.

Because PVC conduit exhibits a considerable change in length during a change in temperature, expansion fittings are required for specific temperature variations. According to Table 352.44(A), a 100-foot run of PVC conduit will change 4.06 inches in length if the temperature change is 100°F.

The allowable range of expansion for many PVC conduit expansion couplings is generally 6 inches. Information concerning installation and application of this type of coupling can be obtained from manufacturers' instructions.

Expansion fittings are seldom used underground, where temperatures are relatively constant. If PVC conduit is buried or covered immediately, expansion and contraction are not considered a problem.

See also

300.7(B) and its commentary regarding the expansion of PVC

N **(B) Earth Movement.** Expansion fittings for underground runs of direct buried PVC conduit emerging from the ground shall be provided above grade when required to compensate for earth settling or movement, including frost heave.

Informational Note: See 300.5(J).

352.46 Bushings. Where a conduit enters a box, fitting, or other enclosure, a bushing or adapter shall be provided to protect the wire from abrasion unless the box, fitting, or enclosure design provides equivalent protection.

Informational Note: See 300.4(G) for the protection of conductors 4 AWG and larger at bushings.

352.48 Joints. All joints between lengths of conduit and between conduit and couplings, fittings, and boxes, shall be made by an approved method.

352.56 Splices and Taps. Splices and taps shall be made in accordance with 300.15.

Δ **352.60 Grounding.** Where equipment grounding is required, separate grounding conductor shall be installed in the conduit.

Exception No. 1: The equipment grounding conductor shall be permitted to be run separately from the circuit conductors as permitted in 250.134, Exception No. 2, for dc circuits and

△ **TABLE 352.44(A)** *Expansion Characteristics of PVC Rigid Nonmetallic Conduit Coefficient of Thermal Expansion = 6.084 × 10⁻⁵ mm/ mm/°C (3.38 × 10⁻⁵ in./in./°F)*

Temperature Change (°C)	Length Change of PVC Conduit (mm/m)	Temperature Change (°F)	Length Change of PVC Conduit (in./100 ft)	Temperature Change (°F)	Length Change of PVC Conduit (in./100 ft)
5	0.30	5	0.20	105	4.26
10	0.61	10	0.41	110	4.46
15	0.91	15	0.61	115	4.66
20	1.22	20	0.81	120	4.87
25	1.52	25	1.01	125	5.07
30	1.83	30	1.22	130	5.27
35	2.13	35	1.42	135	5.48
40	2.43	40	1.62	140	5.68
45	2.74	45	1.83	145	5.88
50	3.04	50	2.03	150	6.08
55	3.35	55	2.23	155	6.29
60	3.65	60	2.43	160	6.49
65	3.95	65	2.64	165	6.69
70	4.26	70	2.84	170	6.90
75	4.56	75	3.04	175	7.10
80	4.87	80	3.24	180	7.30
85	5.17	85	3.45	185	7.50
90	5.48	90	3.65	190	7.71
95	5.78	95	3.85	195	7.91
100	6.08	100	4.06	200	8.11

250.134, Exception No. 1, for separately run equipment grounding conductors.

Exception No. 2: The equipment grounding conductor shall not be required where the grounded conductor is used to ground equipment as permitted in 250.142.

Part III. Construction Specifications

352.100 Construction. PVC conduit shall be made of rigid (nonplasticized) polyvinyl chloride (PVC). PVC conduit and fittings shall be composed of suitable nonmetallic material that is resistant to moisture and chemical atmospheres. For use aboveground, it shall also be flame retardant, resistant to impact and crushing, resistant to distortion from heat under conditions likely to be encountered in service, and resistant to low temperature and sunlight effects. For use underground, the material shall be acceptably resistant to moisture and corrosive agents and shall be of sufficient strength to withstand abuse, such as by impact and crushing, in handling and during installation. Where intended for direct burial, without encasement in concrete, the material shall also be capable of withstanding continued loading that is likely to be encountered after installation.

352.120 Marking. Each length of PVC conduit shall be clearly and durably marked at least every 3 m (10 ft) as required in the first sentence of 110.21(A). The type of material shall also be included in the marking unless it is visually identifiable. For conduit recognized for use aboveground, these markings shall be permanent. For conduit limited to underground use only, these markings shall be sufficiently durable to remain legible until the material is installed. Conduit shall be permitted to be surface marked to indicate special characteristics of the material.

Informational Note: Examples of these markings include but are not limited to "limited smoke" and "sunlight resistant."

ARTICLE
353

High Density Polyethylene Conduit: (HDPE Conduit)

Part I. General

△ **353.1 Scope.** This article covers the use, installation, and construction specifications for high density polyethylene (HDPE) conduit and associated fittings.

353.6 Listing Requirements. HDPE conduit and associated fittings shall be listed.

Part II. Installation

353.10 Uses Permitted. The use of HDPE conduit shall be permitted under the following conditions:

(1) In discrete lengths or in continuous lengths from a reel

(2) In locations subject to severe corrosive influences as covered in 300.6 and where subject to chemicals for which the conduit is listed

(3) In cinder fill

(4) In direct burial installations in earth or concrete

Informational Note to (4): See 300.5 and 305.15 for underground installations.

(5) Above ground, except as prohibited in 353.12, where encased in not less than 50 mm (2 in.) of concrete.

(6) Conductors or cables rated at a temperature higher than the listed temperature rating of HDPE conduit shall be permitted to be installed in HDPE conduit, provided the conductors or cables are not operated at a temperature higher than the listed temperature rating of the HDPE conduit.

Conductors marked with a rated temperature higher than that of the raceway can be used if the conductors are to be operated within the raceway temperature rating.

An example of 353.10(6) is the use of 105°C-rated medium-voltage cables, Type MV, where the cable ampacity at the 105°C rating is reduced to the cable ampacity at 75°C or 90°C to match the listed operating temperature rating of HDPE (75°C or 90°C).

353.12 Uses Not Permitted. HDPE conduit shall not be used under the following conditions:

(1) Where exposed

(2) Within a building

(3) In any hazardous (classified) location, except as permitted by other articles in this *Code*

(4) Where subject to ambient temperatures in excess of 50°C (122°F) unless listed otherwise

353.20 Size.

(A) Minimum. HDPE conduit smaller than metric designator 16 (trade size ½) shall not be used.

Δ **(B) Maximum.** HDPE conduit larger than metric designator 155 (trade size 6) shall not be used.

Informational Note: See 300.1(C) for the trade sizes and metric designators that are for identification purposes only and do not relate to actual dimensions.

353.22 Number of Conductors. The number of conductors shall not exceed that permitted by the percentage fill specified in Table 1, Chapter 9.

Cables shall be permitted to be installed where such use is not prohibited by the respective cable articles. The number of cables shall not exceed the allowable percentage fill specified in Table 1, Chapter 9.

Table 4 of Chapter 9 provides the usable area within the selected conduit or tubing, and Table 5 provides the required area for each conductor. Examples using these tables to calculate a conduit or tubing size are provided in the commentary following Chapter 9, Notes to Tables, Note 6.

To select the proper trade size of HDPE, see the appropriate sub-tables for Articles 352 and 353, Rigid PVC Conduit (PVC), Schedule 40, and HDPE Conduit (HDPE) in Table 4 of Chapter 9. If the conductors are of the same wire size and insulation type, Tables C.11 and C.11(A) in Informative Annex C can be used instead of performing the calculations.

Δ **353.24 Bends.**

N **(A) How Made.** Bends shall be so made that the conduit will not be damaged and the internal diameter of the conduit will not be effectively reduced. Bends shall be permitted to be made manually without auxiliary equipment, and the radius of the curve to the centerline of such bends shall not be less than shown in Table 354.24(A). For conduits of metric designators 129 and 155 (trade sizes 5 and 6), the allowable radii of bends shall be in accordance with specifications provided by the manufacturer.

N **(B) Number in One Run.** The total degrees of bends in a conduit run shall not exceed 360 degrees between pull points.

353.28 Trimming. All cut ends shall be trimmed inside and outside to remove rough edges.

353.46 Bushings. Where a conduit enters a box, fitting, or other enclosure, a bushing or adapter shall be provided to protect the wire from abrasion unless the box, fitting, or enclosure design provides equivalent protection.

Informational Note: See 300.4(G) for the protection of conductors 4 AWG and larger at bushings.

Δ **353.48 Joints.** All joints between lengths of conduit and between conduit and couplings, fittings, and boxes shall be made by an approved method.

Informational Note: HDPE conduit can be joined using either heat fusion, electrofusion, or mechanical fittings.

353.56 Splices and Taps. Splices and taps shall be made in accordance with 300.15.

Δ **353.60 Grounding.** Where equipment grounding is required, a separate grounding conductor shall be installed in the conduit.

Exception No. 1: The equipment grounding conductor shall be permitted to be run separately from the conduit where used for grounding dc circuits as permitted in 250.134, Exception No. 2.

Exception No. 2: The equipment grounding conductor shall not be required where the grounded conductor is used to ground equipment as permitted in 250.142.

Part III. Construction Specifications

353.100 Construction. HDPE conduit shall be composed of high density polyethylene that is resistant to moisture and chemical atmospheres. The material shall be resistant to moisture and corrosive agents and shall be of sufficient strength to withstand abuse, such as by impact and crushing, in handling and during installation. Where intended for direct burial, without encasement in concrete, the material shall also be capable of withstanding continued loading that is likely to be encountered after installation.

353.120 Marking. Each length of HDPE shall be clearly and durably marked at least every 3 m (10 ft) as required in 110.21. The type of material shall also be included in the marking.

ARTICLE 354 — Nonmetallic Underground Conduit with Conductors (NUCC)

Part I. General

354.1 Scope. This article covers the use, installation, and construction specifications for nonmetallic underground conduit with conductors (NUCC).

NUCC (preassembled conductors in conduit) has been used by electric utilities for outdoor lighting for several years. It is supplied in continuous lengths on coils or reels or in cartons. NUCC consists of nonmetallic conduit with the conductors pre-installed by the manufacturer. The product is designed to allow conductors to be removed and reinserted.

354.6 Listing Requirements. NUCC and associated fittings shall be listed.

Part II. Installation

Δ **354.10 Uses Permitted.** The use of NUCC and fittings shall be permitted in the following:

(1) For direct burial underground installation (For minimum cover requirements, see Table 300.5(A) and Table 305.15(A).)
(2) Encased or embedded in concrete
(3) In cinder fill
(4) In underground locations subject to severe corrosive influences as covered in 300.6 and where subject to chemicals for which the assembly is specifically approved
(5) Above ground, except as prohibited in 354.12, where encased in not less than 50 mm (2 in.) of concrete

354.12 Uses Not Permitted. NUCC shall not be used in the following:

(1) In exposed locations
(2) Inside buildings

Exception: The conductor or the cable portion of the assembly, where suitable, shall be permitted to extend within the building for termination purposes in accordance with 300.3.

(3) In any hazardous (classified) location, except as permitted by other articles of this *Code*

354.20 Size.

(A) Minimum. NUCC smaller than metric designator 16 (trade size ½) shall not be used.

(B) Maximum. NUCC larger than metric designator 103 (trade size 4) shall not be used.

> Informational Note: See 300.1(C) for the metric designators and trade sizes. These are for identification purposes only and do not relate to actual dimensions.

354.22 Number of Conductors. The number of conductors or cables shall not exceed that permitted by the percentage fill in Table 1, Chapter 9.

Δ **354.24 Bends.**

N **(A) How Made.** Bends shall be manually made so that the conduit will not be damaged and the internal diameter of the conduit will not be effectively reduced. The radius of the curve of the centerline of such bends shall not be less than shown in Table 354.24(A).

The bending radius for NUCC does not follow Chapter 9, Table 2, but rather must conform to Table 354.24(A).

N **(B) Number in One Run.** The total degrees of bends in a conduit run shall not exceed 360 degrees between pull points.

354.28 Trimming. For termination, the conduit shall be trimmed away from the conductors or cables using an approved method that will not damage the conductor or cable insulation or jacket. All conduit ends shall be trimmed inside and out to remove rough edges.

354.46 Bushings. Where the NUCC enters a box, fitting, or other enclosure, a bushing or adapter shall be provided to protect the conductor or cable from abrasion unless the design of the box, fitting, or enclosure provides equivalent protection.

TABLE 354.24(A) *Minimum Bending Radius for Nonmetallic Underground Conduit with Conductors (NUCC)*

Conduit Size		Minimum Bending Radius	
Metric Designator	Trade Size	mm	in.
16	½	250	10
21	¾	300	12
27	1	350	14
35	1¼	450	18
41	1½	500	20
53	2	650	26
63	2½	900	36
78	3	1200	48
103	4	1500	60

Informational Note: See 300.4(G) for the protection of conductors size 4 AWG or larger.

354.48 Joints. All joints between conduit, fittings, and boxes shall be made by an approved method.

354.50 Conductor Terminations. All terminations between the conductors or cables and equipment shall be made by an approved method for that type of conductor or cable.

354.56 Splices and Taps. Splices and taps shall be made in junction boxes or other enclosures.

Δ **354.60 Grounding.** Where equipment grounding is required, an assembly containing a separate grounding conductor shall be used.

> *Exception No. 1: The equipment grounding conductor shall be permitted to be run separately from the conduit where used for grounding dc circuits as permitted in 250.134, Exception No. 2.*
>
> *Exception No. 2: The equipment grounding conductor shall not be required where the grounded conductor is used to ground equipment as permitted in 250.142.*

Part III. Construction Specifications

354.100 Construction.

(A) General. NUCC is an assembly that is provided in continuous lengths shipped in a coil, reel, or carton.

(B) Nonmetallic Underground Conduit. The nonmetallic underground conduit shall be listed and composed of a material that is resistant to moisture and corrosive agents. It shall also be capable of being supplied on reels without damage or distortion and shall be of sufficient strength to withstand abuse, such as impact or crushing, in handling and during installation without damage to conduit or conductors.

(C) Conductors and Cables. Conductors and cables used in NUCC shall be listed and shall comply with 310.10(C). Conductors of different systems shall be installed in accordance with 300.3(C).

(D) Conductor Fill. The maximum number of conductors or cables in NUCC shall not exceed that permitted by the percentage fill in Table 1, Chapter 9.

354.120 Marking. NUCC shall be clearly and durably marked at least every 3.05 m (10 ft) as required by 110.21. The type of conduit material shall also be included in the marking.

Identification of conductors or cables used in the assembly shall be provided on a tag attached to each end of the assembly or to the side of a reel. Enclosed conductors or cables shall be marked in accordance with 310.8.

ARTICLE 355 Reinforced Thermosetting Resin Conduit (RTRC)

Part I. General

Δ **355.1 Scope.** This article covers the use, installation, and construction specification for reinforced thermosetting resin conduit (RTRC) and associated fittings.

355.6 Listing Requirements. RTRC, factory elbows, and associated fittings shall be listed.

Part II. Installation

355.10 Uses Permitted. The use of RTRC shall be permitted in accordance with 355.10(A) through (I).

(A) Concealed. RTRC shall be permitted in walls, floors, and ceilings.

(B) Corrosive Influences. RTRC shall be permitted in locations subject to severe corrosive influences as covered in 300.6 and where subject to chemicals for which the materials are specifically approved.

(C) Cinders. RTRC shall be permitted in cinder fill.

(D) Wet Locations. RTRC shall be permitted in portions of dairies, laundries, canneries, or other wet locations, and in locations where walls are frequently washed, the entire conduit system, including boxes and fittings used therewith, shall be installed and equipped so as to prevent water from entering the conduit. All supports, bolts, straps, screws, and so forth, shall be of corrosion-resistant materials or be protected against corrosion by approved corrosion-resistant materials.

(E) Dry and Damp Locations. RTRC shall be permitted for use in dry and damp locations not prohibited by 355.12.

(F) Exposed. RTRC shall be permitted for exposed work if identified for such use.

> Informational Note: RTRC, Type XW, is identified for areas of physical damage.

(G) Underground Installations. For underground installations, see 300.5 and 305.15.

(H) Support of Conduit Bodies. RTRC shall be permitted to support nonmetallic conduit bodies not larger than the largest trade size of an entering raceway. These conduit bodies shall not support luminaires or other equipment and shall not contain devices other than splicing devices as permitted by 110.14(B) and 314.16(C)(2).

(I) Insulation Temperature Limitations. Conductors or cables rated at a temperature higher than the listed temperature rating of RTRC conduit shall be permitted to be installed in RTRC conduit, if the conductors or cables are not operated at a temperature higher than the listed temperature rating of the RTRC conduit.

355.12 Uses Not Permitted. RTRC shall not be used under the following conditions.

(A) Hazardous (Classified) Locations.

(1) In any hazardous (classified) location, except as permitted by other articles in this *Code*
(2) In Class I, Division 2 locations, except as permitted in 501.10(B)(1)(6)

(B) Support of Luminaires. For the support of luminaires or other equipment not described in 355.10(H).

(C) Physical Damage. Where subject to physical damage unless identified for such use.

Type RTRC installed in a location where the raceway is subject to physical damage must be marked "XW." If the location is above ground and exposed to physical damage, the conduit must be marked "AG XW RTRC."

> **See also**
>
> **305.15(A), 334.15(B), 501.10(B)(1)(6),** and **551.80(B)** for examples of requirements specifying the use of RTRC-XW

(D) Ambient Temperatures. Where subject to ambient temperatures in excess of 50°C (122°F) unless listed otherwise.

(E) Theaters and Similar Locations. In theaters and similar locations, except as provided in 518.4 and 520.5.

In addition to the conditions in 355.12(A) through (E), nonmetallic conduits are not permitted to be installed in ducts, plenums, and other air-handling spaces.

> **See also**
>
> **300.22,** which limits the use of materials in ducts, plenums, and other air-handling spaces, which could contribute smoke and products of combustion during a fire

355.20 Size.

(A) Minimum. RTRC smaller than metric designator 16 (trade size ½) shall not be used.

Δ **(B) Maximum.** RTRC larger than metric designator 155 (trade size 6) shall not be used.

> Informational Note: See 300.1(C) for the trade sizes and metric designators that are for identification purposes only and do not relate to actual dimensions.

355.22 Number of Conductors. The number of conductors shall not exceed that permitted by the percentage fill specified in Table 1, Chapter 9. Cables shall be permitted to be installed where such use is not prohibited by the respective cable articles.

The number of cables shall not exceed the allowable percentage fill specified in Table 1, Chapter 9.

Table 1 of Chapter 9 specifies the maximum percent fill of conduit or tubing. No internal dimensions for Type RTRC are given in the *NEC*® for calculating the allowable number of conductors. Conductor fill calculation can be in accordance with provided dimensions marked on the conduit. For the exact dimensions for fill calculations of RTRC, refer to the product standard or to the manufacturers' product information.

Δ **355.24 Bends.**

N **(A) How Made.** Bends shall be so made that the conduit will not be damaged and the internal diameter of the conduit will not be effectively reduced. Field bends shall be made only with identified bending equipment. The radius of the curve to the centerline of such bends shall not be less than shown in Table 2, Chapter 9.

N **(B) Number in One Run.** The total degrees of bends in a conduit run shall not exceed 360 degrees between pull points.

355.28 Trimming. All cut ends shall be trimmed inside and outside to remove rough edges.

355.30 Securing and Supporting. RTRC shall be installed as a complete system in accordance with 300.18 and shall be securely fastened in place and supported in accordance with 355.30(A) and (B).

(A) Securely Fastened. RTRC shall be securely fastened within 900 mm (3 ft) of each outlet box, junction box, device box, conduit body, or other conduit termination. Conduit listed for securing at other than 900 mm (3 ft) shall be permitted to be installed in accordance with the listing.

(B) Supports. RTRC shall be supported as required in Table 355.30(B). Conduit listed for support at spacing other than as shown in Table 355.30(B) shall be permitted to be installed in accordance with the listing. Horizontal runs of RTRC supported by openings through framing members at intervals not exceeding

TABLE 355.30(B) *Support of Reinforced Thermosetting Resin Conduit (RTRC)*

Conduit Size		Maximum Spacing Between Supports	
Metric Designator	Trade Size	mm or m	ft
16–27	½–1	900 mm	3
35–53	1¼–2	1.5 m	5
63–78	2½–3	1.8 m	6
91–129	3½–5	2.1 m	7
155	6	2.5 m	8

TABLE 355.44 *Expansion Characteristics of Reinforced Thermosetting Resin Conduit (RTRC) Coefficient of Thermal Expansion =*
2.7×10^{-5} *mm/mm/°C* $(1.5 \times 10^{-5}$ *in./in./°F)*

Temperature Change (°C)	Length Change of RTRC Conduit (mm/m)	Temperature Change (°F)	Length Change of RTRC Conduit (in./100 ft)	Temperature Change (°F)	Length Change of RTRC Conduit (in./100 ft)
5	0.14	5	0.09	105	1.89
10	0.27	10	0.18	110	1.98
15	0.41	15	0.27	115	2.07
20	0.54	20	0.36	120	2.16
25	0.68	25	0.45	125	2.25
30	0.81	30	0.54	130	2.34
35	0.95	35	0.63	135	2.43
40	1.08	40	0.72	140	2.52
45	1.22	45	0.81	145	2.61
50	1.35	50	0.90	150	2.70
55	1.49	55	0.99	155	2.79
60	1.62	60	1.08	160	2.88
65	1.76	65	1.17	165	2.97
70	1.89	70	1.26	170	3.06
75	2.03	75	1.35	175	3.15
80	2.16	80	1.44	180	3.24
85	2.30	85	1.53	185	3.33
90	2.43	90	1.62	190	3.42
95	2.57	95	1.71	195	3.51
100	2.70	100	1.80	200	3.60

those in Table 355.30(B) and securely fastened within 900 mm (3 ft) of termination points shall be permitted.

355.44 Expansion Fittings. Expansion fittings for RTRC shall be provided to compensate for thermal expansion and contraction where the length change, in accordance with Table 355.44, is expected to be 6 mm (¼ in.) or greater in a straight run between securely mounted items such as boxes, cabinets, elbows, or other conduit terminations.

Because RTRC exhibits a considerable change in length per degree change in temperature, expansion fittings are required for specific variations in temperature. In some areas, outdoor temperature variations of over 100°F are common. According to Table 355.44, a 100-foot run of RTRC will change 1.80 inches in length if the temperature change is 100°F. Information concerning installation and application of this type of coupling can be obtained from manufacturers' instructions. Expansion fittings are seldom used underground, where temperatures are relatively constant.

See also

300.7(B) and its commentary regarding the expansion of RTRC

355.46 Bushings. Where a conduit enters a box, fitting, or other enclosure, a bushing or adapter shall be provided to protect the wire from abrasion unless the box, fitting, or enclosure design provides equivalent protection.

Informational Note: See 300.4(G) for the protection of conductors 4 AWG and larger at bushings.

355.48 Joints. All joints between lengths of conduit, and between conduit and couplings, fitting, and boxes, shall be made by an approved method.

355.56 Splices and Taps. Splices and taps shall be made in accordance with 300.15.

Δ **355.60 Grounding.** Where equipment grounding is required, a separate grounding conductor shall be installed in the conduit.

Exception No. 1: The equipment grounding conductor shall be permitted to be run separately from the circuit conductors as permitted in 250.134, Exception No. 2, for dc circuits and 250.134, Exception No. 1, for separately run equipment grounding conductors.

Exception No. 2: An equipment grounding conductor shall not be required where the grounded conductor is used to ground equipment as in 250.142(A).

Part III. Construction Specifications

355.100 Construction. RTRC and fittings shall be composed of suitable nonmetallic material that is resistant to moisture and chemical atmospheres. For use aboveground, it shall also be flame retardant, resistant to impact and crushing, resistant to distortion from heat under conditions likely to be encountered in service, and resistant to low temperature and sunlight effects. For use underground, the material shall be acceptably resistant to moisture and corrosive agents and shall be of sufficient strength to withstand abuse, such as by impact and crushing, in handling and during installation. Where intended for direct burial, without encasement in concrete, the material shall also be capable of withstanding continued loading that is likely to be encountered after installation.

355.120 Marking. Each length of RTRC shall be clearly and durably marked at least every 3 m (10 ft) as required in the first sentence of 110.21(A). The type of material shall also be included in the marking unless it is visually identifiable. For conduit recognized for use aboveground, these markings shall be permanent. For conduit limited to underground use only, these markings shall be sufficiently durable to remain legible until the material is installed. Conduit shall be permitted to be surface marked to indicate special characteristics of the material.

Informational Note: Examples of these markings include but are not limited to "limited smoke" and "sunlight resistant."

ARTICLE 356 Liquidtight Flexible Nonmetallic Conduit (LFNC)

Part I. General

356.1 Scope. This article covers the use, installation, and construction specifications for liquidtight flexible nonmetallic conduit (LFNC) and associated fittings.

356.6 Listing Requirements. LFNC and associated fittings shall be listed.

Part II. Installation

Δ **356.10 Uses Permitted.** LFNC shall be permitted to be used in exposed or concealed locations for the following purposes:

(1) Where flexibility is required for installation, operation, or maintenance.
(2) Where protection of the contained conductors is required from vapors, machine oils, liquids, or solids.
(3) For outdoor locations where listed and marked as suitable for the purpose.
(4) For direct burial where listed and marked for the purpose.
(5) Installed in lengths longer than 1.8 m (6 ft) where secured in accordance with 356.30.
(6) LFNC-B as a listed manufactured prewired assembly, metric designator 16 through 27 (trade size ½ through 1) conduit.
(7) For encasement in concrete where listed for direct burial and installed in compliance with 356.42.

(8) In locations subject to severe corrosive influences as covered in 300.6 and where subject to chemicals for which the materials are specifically approved.
(9) Conductors or cables rated at a temperature higher than the listed temperature rating of LFNC shall be permitted to be installed in LFNC, provided the conductors or cables are not operated at a temperature higher than the listed temperature rating of the LFNC.

Informational Note: Extreme cold can cause some types of nonmetallic conduits to become brittle and therefore more susceptible to damage from physical contact.

356.12 Uses Not Permitted. LFNC shall not be used as follows:

(1) Where subject to physical damage
(2) Where any combination of ambient and conductor temperatures is in excess of that for which it is listed
(3) In lengths longer than 1.8 m (6 ft), except as permitted by 356.10(5) or where a longer length is approved as essential for a required degree of flexibility
(4) In any hazardous (classified) location, except as permitted by other articles in this *Code*

356.20 Size.

Δ **(A) Minimum.** LFNC smaller than metric designator 16 (trade size ½) shall not be used unless permitted for metric designator 12 (trade size ⅜) as follows:

(1) For enclosing the leads of motors as permitted in 430.245(B)
(2) In lengths not exceeding 1.8 m (6 ft) as part of a listed assembly for tap connections to luminaires as required in 410.117(C), or for utilization equipment

(B) Maximum. LFNC larger than metric designator 103 (trade size 4) shall not be used.

Informational Note: See 300.1(C) for the metric designators and trade sizes. These are for identification purposes only and do not relate to actual dimensions.

356.22 Number of Conductors. The number of conductors shall not exceed that permitted by the percentage fill specified in Table 1, Chapter 9.

Cables shall be permitted to be installed where such use is not prohibited by the respective cable articles. The number of cables shall not exceed the allowable percentage fill specified in Table 1, Chapter 9.

Table 4 of Chapter 9 provides the usable area within the selected conduit or tubing, and Table 5 provides the required area for each conductor. Examples using these tables to calculate a conduit or tubing size are provided in the commentary following Chapter 9, Notes to Tables, Note 6.

To select the proper trade size of LFNC, see the appropriate sub-table for Article 356, Liquidtight Flexible Nonmetallic

Conduit (LFNC-A), Liquidtight Flexible Nonmetallic Conduit (LFNC-B), or Liquidtight Nonmetallic Conduit (LFNC-C), in Table 4 of Chapter 9. If the conductors are of the same wire size and insulation type, Tables C.5 and C.5(A) (LFNC-A), Tables C.6 and C.6(A) (LFNC-B), or Tables C.7 and C.7(A) (LFNC-C) in Informative Annex C can be used instead of performing the calculations.

Δ **356.24 Bends.**

N **(A) How Made.** Bends in conduit shall be so made that the conduit is not damaged and the internal diameter of the conduit is not effectively reduced. Bends shall be permitted to be made manually without auxiliary equipment. The radius of the curve to the centerline of any bend shall not be less than shown in Table 2, Chapter 9 using the column "Other Bends."

N **(B) Number in One Run.** The total degrees of bends in a conduit run shall not exceed 360 degrees between pull points.

356.28 Trimming. All cut ends of conduit shall be trimmed inside and outside to remove rough edges.

Δ **356.30 Securing and Supporting.** LFNC shall be securely fastened and supported in accordance with one of the following:

(1) Where installed in lengths exceeding 1.8 m (6 ft), the conduit shall be securely fastened at intervals not exceeding 900 mm (3 ft) and within 300 mm (12 in.) on each side of every outlet box, junction box, cabinet, or fitting. Where used, cable ties shall be listed for the application and for securing and supporting.

Listing of cable ties for securement and support of flexible metal conduits is necessary because the standard requires markings that identify critical performance characteristics. These characteristics can affect their suitability for the conditions of use, including minimum and maximum operating temperatures and resistance to ultraviolet light for outdoor installations.

(2) Securing or supporting of the conduit shall not be required where it is fished, installed in lengths not exceeding 900 mm (3 ft) at terminals where flexibility is required, or installed in lengths not exceeding 1.8 m (6 ft) from a luminaire terminal connection for tap conductors to luminaires permitted in 410.117(C).

(3) Horizontal runs of LFNC supported by openings through framing members at intervals not exceeding 900 mm (3 ft) and securely fastened within 300 mm (12 in.) of termination points shall be permitted.

(4) Securing or supporting of LFNC shall not be required where installed in lengths not exceeding 1.8 m (6 ft) from the last point where the raceway is securely fastened for connections within an accessible ceiling to a luminaire(s) or other equipment. For the purpose of 356.30, listed liquidtight flexible nonmetallic conduit fittings shall be permitted as a means of support.

356.42 Couplings and Connectors. Only fittings listed for use with LFNC shall be used. Angle connectors shall not be used for concealed raceway installations. Straight LFNC fittings are permitted for direct burial or encasement in concrete.

356.56 Splices and Taps. Splices and taps shall be made in accordance with 300.15.

Δ **356.60 Grounding.** Where equipment grounding is required, a separate grounding conductor shall be installed in the conduit.

Exception No. 1: The equipment grounding conductor shall be permitted to be run separately from the circuit conductors as permitted in 250.134, Exception No. 2, for dc circuits and 250.134, Exception No. 1, for separately run equipment grounding conductors.

Exception No. 2: The equipment grounding conductor shall not be required where the grounded conductor is used to ground equipment as permitted in 250.142.

Part III. Construction Specifications

356.100 Construction. LFNC-B as a prewired manufactured assembly shall be provided in continuous lengths capable of being shipped in a coil, reel, or carton without damage.

356.120 Marking. LFNC shall be marked at least every 600 mm (2 ft) in accordance with 110.21. The marking shall include a type designation in accordance with the definition of *Conduit, Liquidtight Flexible Nonmetallic (LFNC)* in Article 100 and the trade size. Conduit that is intended for outdoor use or direct burial shall be marked.

The type, size, and quantity of conductors used in prewired manufactured assemblies shall be identified by means of a printed tag or label attached to each end of the manufactured assembly and either the carton, coil, or reel. The enclosed conductors shall be marked in accordance with 310.8.

ARTICLE
358 Electrical Metallic Tubing (EMT)

Part I. General

358.1 Scope. This article covers the use, installation, and construction specifications for electrical metallic tubing (EMT) and associated fittings.

358.6 Listing Requirements. EMT, factory elbows, and associated fittings shall be listed.

Part II. Installation

358.10 Uses Permitted.

(A) Exposed and Concealed. The use of EMT shall be permitted for both exposed and concealed work for the following:

(1) In concrete, in direct contact with the earth, in direct burial applications with fittings identified for direct burial, or in areas subject to severe corrosive influences where installed in accordance with 358.10(B)
(2) In dry, damp, and wet locations
(3) In any hazardous (classified) location as permitted by other articles in this *Code*
(4) For manufactured wiring systems as permitted in 604.100(A)(2)

(B) Corrosive Environments.

(1) Galvanized Steel and Stainless Steel EMT, Elbows, and Fittings. Galvanized steel and stainless steel EMT, elbows, and fittings shall be permitted to be installed in concrete, in direct contact with the earth, or in areas subject to severe corrosive influences where protected by corrosion protection and approved as suitable for the condition.

(2) Supplementary Protection of Aluminum EMT. Aluminum EMT shall be provided with approved supplementary corrosion protection where encased in concrete or in direct contact with the earth.

According to the UL *Guide Information for Electrical Equipment*, category FJMX, galvanized steel EMT installed in concrete, on grade or above, generally requires no supplementary corrosion protection. However, galvanized and stainless steel EMT in concrete slab below grade level or in direct contact with earth is required to be protected from corrosion. Where galvanized steel EMT without supplementary corrosion protection extends directly from concrete encasement to soil burial, severe corrosive effects are likely to occur on the metal in contact with the soil.

Other documents, such as the Steel Tube Institute's 2021 *Guidelines for Installing Steel Conduit/Tubing*, and ANSI/NECA 101-2020, *Standard for Installing Steel Conduits (Rigid, IMC, EMT)*, should be consulted for approval guidance of corrosion-resistant materials or for requirements prior to the installation of nonferrous metal (aluminum) conduit in concrete, since chloride additives in the concrete mix can cause corrosion.

(C) Cinder Fill. Galvanized steel and stainless steel EMT shall be permitted to be installed in cinder concrete or cinder fill where subject to permanent moisture when protected on all sides by a layer of noncinder concrete at least 50 mm (2 in.) thick or when the tubing is installed at least 450 mm (18 in.) under the fill.

Although cinder fill is not commonly used in modern construction, it is still encountered at older building sites. Cinders used as fill may contain sulfur and, when combined with moisture, form sulfuric acid, which can corrode metal raceways.

(D) Wet Locations. All supports, bolts, straps, screws, and so forth shall be of corrosion-resistant materials or protected against corrosion by corrosion-resistant materials.

Informational Note: See 300.6 for protection against corrosion.

(E) Physical Damage. Steel and stainless steel EMT shall be permitted to be installed where subject to physical damage.

358.12 Uses Not Permitted. EMT shall not be used under the following conditions:

(1) Where subject to severe physical damage
(2) For the support of luminaires or other equipment except conduit bodies no larger than the largest trade size of the tubing

358.14 Dissimilar Metals. Where practicable, dissimilar metals in contact anywhere in the system shall be avoided to eliminate the possibility of galvanic action.

Stainless steel and aluminum fittings and enclosures shall be permitted to be used with galvanized steel EMT, and galvanized steel fittings and enclosures shall be permitted to be used with aluminum EMT where not subject to severe corrosive influences.

Stainless steel EMT shall only be used with the following:

(1) Stainless steel fittings
(2) Stainless steel boxes and enclosures
(3) Steel (galvanized, painted, powder or PVC coated, and so forth) boxes and enclosures when not subject to severe corrosive influences
(4) Stainless steel, nonmetallic, or approved accessories

358.20 Size.

(A) Minimum. EMT smaller than metric designator 16 (trade size ½) shall not be used.

Exception: Metric designator 12 (trade size ⅜) shall be permitted for enclosing the leads of motors as permitted in 430.245(B).

(B) Maximum. The maximum size of EMT shall be metric designator 155 (trade size 6).

Informational Note: See 300.1(C) for the metric designators and trade sizes. These are for identification purposes only and do not relate to actual dimensions.

358.22 Number of Conductors. The number of conductors shall not exceed that permitted by the percentage fill specified in Table 1, Chapter 9.

Cables shall be permitted to be installed where such use is not prohibited by the respective cable articles. The number of cables shall not exceed the allowable percentage fill specified in Table 1, Chapter 9.

Table 4 of Chapter 9 provides the usable area within the selected conduit or tubing, and Table 5 provides the required area for each conductor. Examples using these tables to calculate a conduit or tubing size are provided in the commentary following Chapter 9, Notes to Tables, Note 6.

To select the proper trade size of EMT, see the appropriate sub-table for Article 358, Electrical Metallic Tubing (EMT), in Table 4 of Chapter 9. If the conductors are of the same wire size and insulation type, Tables C.1 and C.1(A) for EMT in Informative Annex C can be used instead of performing the calculations.

Δ **358.24 Bends.**

N **(A) How Made.** Bends shall be made so that the tubing is not damaged and the internal diameter of the tubing is not effectively reduced. The radius of the curve of any field bend to the centerline of the tubing shall not be less than shown in Table 2, Chapter 9 for one-shot and full shoe benders.

N **(B) Number in One Run.** The total degrees of bends in a tubing run shall not exceed 360 degrees between pull points.

358.28 Reaming and Threading.

(A) Reaming. All cut ends of EMT shall be reamed or otherwise finished to remove rough edges.

In addition to a reamer, a half-round file has proved practical for removing rough edges from the interior and exterior of the conduit or tubing. The steel handle of a pair of pump pliers, the nose of side-cutting pliers, or an electrician's knife can be effective on the smaller sizes of EMT as well.

(B) Threading. EMT shall not be threaded.

Exception: EMT with factory threaded integral couplings complying with 358.100 shall be permitted.

358.30 Securing and Supporting. EMT shall be installed as a complete system in accordance with 300.18 and shall be securely fastened in place and supported in accordance with 358.30(A) and (B).

(A) Securely Fastened. EMT shall be securely fastened in place in accordance with the following:

(1) At intervals not to exceed 3 m (10 ft)
(2) Within 900 mm (3 ft) of each outlet box, junction box, device box, cabinet, conduit body, or other tubing termination

Type EMT is required to be "securely fastened" at the prescribed intervals as illustrated in Exhibit 358.1.

See also

344.30(A) and its commentary for more information on secure fastening

Exception No. 1: Fastening of unbroken lengths shall be permitted to be increased to a distance of 1.5 m (5 ft) where structural members do not readily permit fastening within 900 mm (3 ft).

Exception No. 2: For concealed work in finished buildings or prefinished wall panels where such securing is impracticable, unbroken lengths (without coupling) of EMT shall be permitted to be fished.

As illustrated in Exhibit 358.2, boxes are permitted to be secured to ceiling or roof support structural members that are spaced not more than 5 feet apart to serve as support for runs of EMT perpendicular to the axis of the ceiling or roof support members.

(B) Supports. Horizontal runs of EMT supported by openings through framing members at intervals not greater than 3 m (10 ft) and securely fastened within 900 mm (3 ft) of termination points shall be permitted.

Horizontal runs of EMT are permitted to be supported (but not necessarily secured) by framing members at 10-foot intervals, provided the EMT is secured at least 3 feet from the box or enclosure. See Exhibit 342.2 in the commentary following 342.30(B)(4) for an example.

358.42 Couplings and Connectors. Couplings and connectors used with EMT shall be made up tight. Where buried in masonry or concrete, they shall be concretetight type. Where installed in wet locations, they shall comply with 314.15.

Only listed fittings are permitted per 358.6, and 314.15 specifically requires that fittings for use in wet locations be listed for such use. According to ANSI/UL 797, *Electrical Metallic Tubing — Steel*, listed fittings suitable for use in poured concrete or where exposed to rain are indicated on the fitting or carton. The term *concretetight* or equivalent on the carton indicates suitability for use in poured concrete. The term *raintight* or the equivalent on the carton indicates suitability for use where directly exposed to rain.

EXHIBIT 358.1 Minimum requirements for securely fastening EMT unless an exception applies.

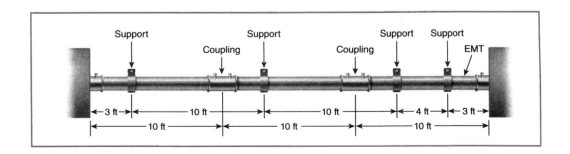

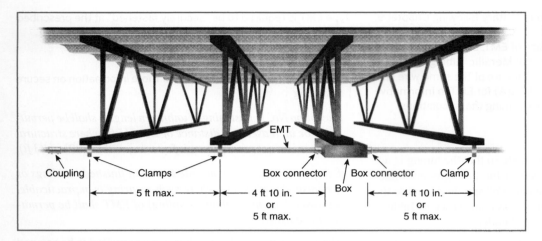

EXHIBIT 358.2 An EMT installation in which the fastening spacing is increased to a maximum of 5 feet.

Coupling | Clamps | Box connector | Box connector | Clamp

5 ft max. — 4 ft 10 in. or 5 ft max. — Box — 4 ft 10 in. or 5 ft max.

EMT

See also

225.22 and **230.54(A)** for raintight requirements as applied to raceways on exterior surfaces of buildings and to service raceways

Indenter-type fittings, utilized only with metallic-coated EMT, require a special tool supplied by the manufacturer for proper installation. See Exhibit 358.3. Fittings are tested for use only with steel EMT, unless specific marking on the device or carton indicates the fittings are suitable for use with aluminum or other material. Fittings are designed for dry-locations or concrete tight when installed per the manufacturer's installation instructions.

358.56 Splices and Taps. Splices and taps shall be made in accordance with 300.15.

358.60 Grounding. EMT shall be permitted as an equipment grounding conductor.

Part III. Construction Specifications

358.100 Construction. EMT shall be made of one of the following:

(1) Steel with protective coatings
(2) Aluminum
(3) Stainless steel

358.120 Marking. EMT shall be clearly and durably marked at least every 3 m (10 ft) as required in the first sentence of 110.21(A).

EXHIBIT 358.3 Indenter-type EMT coupling for use with steel EMT (*top*); Indenter forged crimping tool for proper installation of indenter-type couplings and connectors (*bottom*). *(Courtesy of Hubbell/RACO)*

ARTICLE 360 Flexible Metallic Tubing (FMT)

Part I. General

360.1 Scope. This article covers the use, installation, and construction specifications for flexible metallic tubing (FMT) and associated fittings.

FMT is a type of raceway used for certain specific applications, particularly under the requirements of 300.22(B) and (C) for wiring in ducts and other air-handling spaces. Initially intended for use in such locations, FMT is an effective barrier to the gases and products of combustion. It is very flexible and rarely affected by vibration or other movement.

360.6 Listing Requirements. FMT and associated fittings shall be listed.

Part II. Installation

360.10 Uses Permitted. FMT shall be permitted to be used for branch circuits as follows:

(1) In dry locations
(2) Where concealed
(3) In accessible locations
(4) For system voltages of 1000 volts maximum

The 1000-volt limitation prohibits the use of FMT for the secondary circuits of sign ballasts, sign transformers, electronic sign power supplies, or oil burner ignition transformers unless those circuits are less than 1000 volts.

360.12 Uses Not Permitted. FMT shall not be used as follows:

(1) In hoistways
(2) In storage battery rooms
(3) In hazardous (classified) locations unless otherwise permitted under other articles in this *Code*
(4) Underground for direct earth burial, or embedded in poured concrete or aggregate
(5) Where subject to physical damage
(6) In lengths over 1.8 m (6 ft)

360.20 Size.

(A) Minimum. FMT smaller than metric designator 16 (trade size ½) shall not be used.

Exception No. 1: FMT of metric designator 12 (trade size ⅜) shall be permitted to be installed in accordance with 300.22(B) and (C).

Exception No. 2: FMT of metric designator 12 (trade size ⅜) shall be permitted in lengths not in excess of 1.8 m (6 ft) as part of a listed assembly or for luminaires. See 410.117(C).

(B) Maximum. The maximum size of FMT shall be metric designator 21 (trade size ¾).

Informational Note: See 300.1(C) for the metric designators and trade sizes. These are for identification purposes only and do not relate to actual dimensions.

360.22 Number of Conductors.

(A) FMT — Metric Designators 16 and 21 (Trade Sizes ½ and ¾). The number of conductors in metric designators 16 (trade size ½) and 21 (trade size ¾) shall not exceed that permitted by the percentage fill specified in Table 1, Chapter 9.

Cables shall be permitted to be installed where such use is not prohibited by the respective cable articles. The number of cables shall not exceed the allowable percentage fill specified in Table 1, Chapter 9.

Table 4 of Chapter 9 provides the usable area within the selected conduit or tubing, and Table 5 provides the required area for each conductor. Examples using these tables to calculate a conduit or tubing size are provided in the commentary following Chapter 9, Notes to Tables, Note 6.

To select the proper trade size of FMT, see the appropriate sub-table for Article 348, Flexible Metal Conduit (FMC), in Table 4 of Chapter 9. If the conductors are of the same wire size and insulation type, Tables C.3 and C.3(A) for FMC can be used for FMT sizes ½ inch and ¾ inch in Informative Annex C instead of performing the calculations.

(B) FMT — Metric Designator 12 (Trade Size ⅜). The number of conductors in metric designator 12 (trade size ⅜) shall not exceed that permitted in Table 348.22.

360.24 Bends.

(A) Infrequent Flexing Use. When FMT is infrequently flexed in service after installation, the radii of bends measured to the inside of the bend shall not be less than specified in Table 360.24(A).

TABLE 360.24(A) *Minimum Radii for Flexing Use*

Metric Designator	Trade Size	Minimum Radii for Flexing Use	
		mm	in.
12	⅜	254.0	10
16	½	317.5	12½
21	¾	444.5	17½

(B) Fixed Bends. Where FMT is bent for installation purposes and is not flexed or bent as required by use after installation, the radii of bends measured to the inside of the bend shall not be less than specified in Table 360.24(B).

TABLE 360.24(B) *Minimum Radii for Fixed Bends*

Metric Designator	Trade Size	Minimum Radii for Fixed Bends	
		mm	in.
12	⅜	88.9	3½
16	½	101.6	4
21	¾	127.0	5

360.56 Splices and Taps. Splices and taps shall be made in accordance with 300.15.

360.60 Grounding. FMT shall be permitted as an equipment grounding conductor where installed in accordance with 250.118(A)(7).

Part III. Construction Specifications

360.120 Marking. FMT shall be marked according to 110.21.

ARTICLE 362 Electrical Nonmetallic Tubing (ENT)

Part I. General

362.1 Scope. This article covers the use, installation, and construction specifications for electrical nonmetallic tubing (ENT) and associated fittings.

ENT is made of the same material used for polyvinyl chloride (PVC) conduit. The outside diameters of ENT (½-inch through 2-inch trade sizes only) are such that standard couplings and other fittings for rigid PVC conduit can be used.

Because of the corrugations, the raceway can be bent by hand and has some degree of flexibility. ENT is not intended for use where flexibility is necessary, such as at motor terminations to prevent transmission of noise and vibration, or for connection of adjustable luminaires or moving parts. ENT is suitable for the installation of conductors having a temperature rating as indicated on the product. The maximum allowable ambient temperature is 122°F. Exhibit 362.1 shows an example of ENT.

N 362.2 Reconditioned Equipment. ENT shall not be reconditioned.

362.6 Listing Requirements. ENT and associated fittings shall be listed.

Part II. Installation

Δ 362.10 Uses Permitted. For the purpose of this article, the first floor of a building shall be that floor that has 50 percent or more of the exterior wall surface area level with or above finished grade. One additional level that is the first level and not designed for human habitation and used only for vehicle parking, storage, or similar use shall be permitted. The use of ENT and fittings shall be permitted in the following:

(1) In any building not exceeding three floors above grade as follows:

   a. For exposed work, where not prohibited by 362.12
   b. Concealed within walls, floors, and ceilings

Where exposed and subject to physical damage, ENT is required to be protected and is limited to use in buildings not exceeding three floors above grade. Where concealed or above a suspended ceiling (exposed), ENT is permitted to be installed within walls, floors, or ceilings in buildings of three floors or fewer without the need for fire-rated construction. The three-floor limitation is based on the likelihood that only a small quantity of ENT would be exposed to fire and that the occupants would have adequate time to exit the building before the products of combustion made the building untenable. Exhibit 362.2 illustrates permitted uses of ENT in a building of three floors or fewer.

(2) In any building exceeding three floors above grade concealed within combustible or noncombustible walls, floors, and ceilings where the walls, floors, and ceilings provide a thermal barrier of material that has at least a 15-minute finish rating as identified in listings of fire-rated assemblies.

Exception to (2): Where an approved automatic fire protective system(s) is installed on all floors, ENT shall be permitted to be

EXHIBIT 362.1 Various sizes of ENT suitable for use. (*Courtesy of Thomas and Betts, A Member of the ABB Group*)

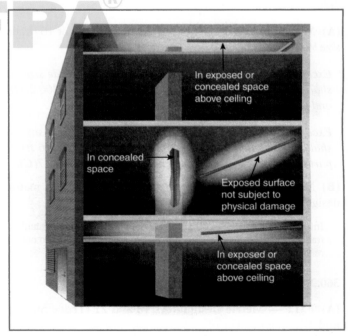

EXHIBIT 362.2 Examples of permitted uses of ENT in a building not exceeding three floors.

used within walls, floors, and ceilings, exposed or concealed, in buildings exceeding three floors above grade.

Informational Note No. 1: A finish rating is established for assemblies containing combustible (wood) supports. The finish rating is defined as the time at which the wood stud or wood joist reaches an average temperature rise of 121°C (250°F) or an individual temperature of 163°C (325°F) as measured on the plane of the wood nearest the fire. A finish rating is not intended to represent a rating for a membrane ceiling.

Informational Note No. 2: See NFPA 13-2022, *Standard for the Installation of Sprinkler Systems*, a recognized fire sprinkler system(s) standard.

ENT is permitted to be installed within the walls, floors, or ceilings of a building of any height where the walls, floors, or ceilings provide a thermal barrier of material that has at least a 15-minute finish rating. Exposed ENT in the first three floors of a building that exceeds three floors is not permitted or intended except as permitted in 362.10(5). Where installed in a building exceeding three floors, ENT must be installed behind the 15-minute thermal barrier on all floors. Exhibit 362.3 illustrates areas in a building exceeding three floors in which ENT is permitted to be used. In accordance with the exception, fire sprinkler systems can also be used as a construction condition under which an expanded use of ENT is allowed.

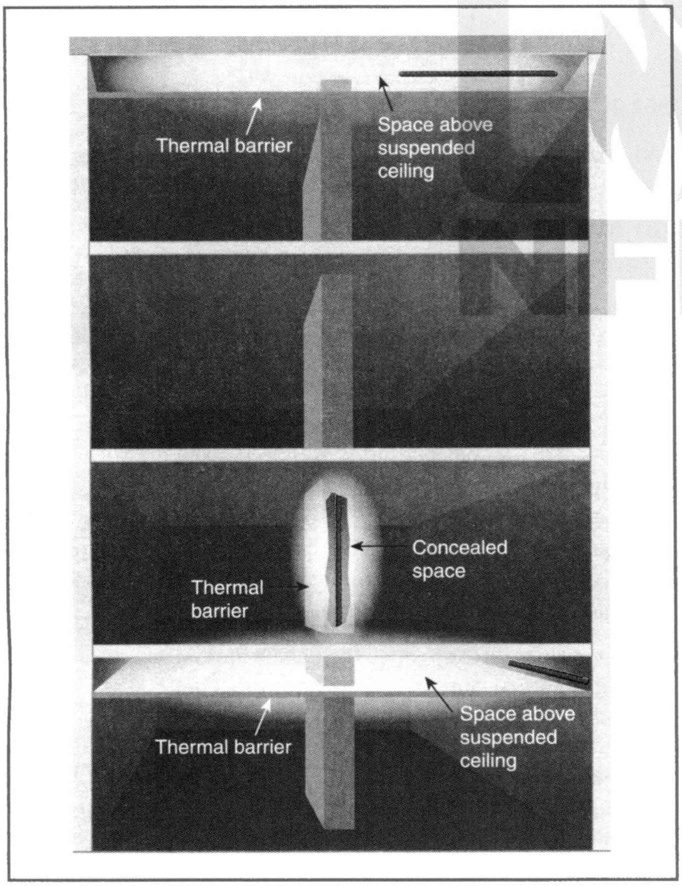

EXHIBIT 362.3 Examples of permitted uses of ENT in a building exceeding three floors.

Interior finish is generally considered to consist of those materials or combinations of materials that form the exposed interior surface of walls and ceilings in a building. Common interior finish materials include plaster, gypsum wallboard, wood, plywood paneling, fibrous ceiling tiles, and a variety of wall coverings. Ordinary paint, wallpaper, or other similar wall coverings not exceeding 1/28 inch in thickness are generally considered incidental to interior finish, except where the AHJ deems them a hazard.

The finish rating of a wall or ceiling finish material is the time required for the unexposed surface of the finish membrane to reach an average temperature rise of 250°F above ambient or an individual temperature rise at any one point not exceeding 325°F when the assembly is tested in accordance with ANSI/UL 263, *Standard for Fire Tests of Building Construction and Materials*, or

COMMENTARY TABLE 362.1 Various Finishes over Wood Framing, One Side (Combustible) with Exposure on Finish Side

Material	Fire Resistance Rating[a] (min.)
Fiberboard, ½ in. thick	5
Fiberboard, flameproofed, ½ in. thick	10
Fiberboard, ½ in. thick, with ½ in.-1:2, 1:2 gypsum-sand plaster	15
Gypsum wallboard, ⅜ in. thick	10
Gypsum wallboard, ½ in. thick	15
Gypsum wallboard, ⅝ in. thick	20
Gypsum wallboard, laminated, two ⅜ in.	28
Gypsum wallboard, laminated, one ⅜ in. plus one ½ in. thick	37
Gypsum wallboard, laminated, two ½ in. thick	47
Gypsum wallboard, laminated, two ⅝ in. thick	60
Gypsum lath, plain or indented, ⅜ in. thick, with ½ in.-1:2, 1:2 gypsum-sand plaster	20
Gypsum lath, perforated, ⅜ in. thick, with ½ in.-1:2, 1:2 gypsum-sand plaster	30
Gypsum-sand plaster, 1:2, 1:3, ½ in. thick, on wood lath	15
Lime-sand plaster, 1:5, 1:7.5, ½ in. thick, on wood lath	15
Gypsum-sand plaster, 1:2, 1:2, ¾ in. thick, on metal lath (no paper backing)	15
Neat gypsum plaster, ¾ in. thick, on metal lath (no paper backing)[b]	30
Neat gypsum plaster, 1 in. thick, on metal lath (no paper backing)[b]	35
Lime-sand plaster, 1:5, 1:7.5, ¾ in. thick, on metal lath (no paper backing)	10
Portland cement plaster, ¾ in. thick, on metal lath (no paper backing)	10
Gypsum-sand plaster, 1:2, 1:3, ¾ in. thick, on paper-backed metal lath	20

Note: For SI units, 1 in. = 25.4 mm.

[a]From National Institute for Standards and Technology, BMS-92.

[b]Unsanded wood-fiber plaster.

ASTM E119, *Standard Test Methods for Fire Tests of Building Construction and Materials.*

The finish rating of wall and ceiling finish materials tested and rated by UL as part of wall and ceiling assemblies can be found in the UL *Fire Resistance Directory*, immediately following the assembly rating and just below the design number. Only assemblies containing combustible support members, however, have published finish ratings. Obviously, limiting ENT to constructions consisting of combustible support members is not the intent. This section is intended to provide a 15-minute thermal barrier as a minimum threshold of acceptability.

Commentary Table 362.1, reproduced from the NFPA *Fire Protection Handbook*, 20th edition (Volume 2, Section 19, Chapter 2, Table 19.2.13), provides ratings for common finish materials. If the finish rating concealing the ENT is unknown or is less than 15 minutes, the ENT can still be used if the installation meets the criteria in 362.10, including the three-floor limitation, where required, and the installation is not prohibited by 362.12. For finish materials not tested and rated in the UL *Fire Resistance Directory*, use Commentary Table 362.1.

See also

10.2.1 of NFPA *101*®, *Life Safety Code*®, for more information regarding classification of interior finish material

(3) In locations subject to severe corrosive influences as covered in 300.6 and where subject to chemicals for which the materials are specifically approved.

(4) In concealed, dry, and damp locations not prohibited by 362.12.

(5) Above suspended ceilings where the suspended ceilings provide a thermal barrier of material that has at least a 15-minute finish rating as identified in listings of fire-rated assemblies, except as permitted in 362.10(1)a.

Exception to (5): ENT shall be permitted to be used above suspended ceilings in buildings exceeding three floors above grade where the building is protected throughout by an approved automatic fire protective system.

Informational Note No. 3: See NFPA 13-2022, *Standard for the Installation of Sprinkler Systems*, a recognized fire sprinkler system(s) standard.

(6) Encased in poured concrete floors, ceilings, walls, and slabs.

(7) Embedded in a concrete slab on grade where ENT is placed on sand or approved screenings, provided fittings identified for this purpose are used for connections.

(8) For wet locations as permitted in this section or in a concrete slab on or belowgrade, with fittings listed for the purpose.

(9) Metric designator 16 through 27 (trade size ½ through 1) as listed manufactured prewired assembly.

Prewired ENT is a listed assembly whose conductors must be installed at the manufacturing facility, where controlled conditions prevent damage to the conductor insulation. Special tools are required for cutting prewired ENT to prevent nicking of the conductor insulation.

(10) With conductors or cables rated at a temperature higher than the listed temperature rating of ENT if the conductors or cables are not operated at a temperature higher than the listed temperature rating of the ENT.

362.12 Uses Not Permitted. ENT shall not be used in the following:

(1) In any hazardous (classified) location, except as permitted by other articles in this *Code*

(2) For the support of luminaires and other equipment

(3) Where subject to ambient temperatures in excess of 50°C (122°F) unless listed otherwise

(4) For direct earth burial

(5) In exposed locations, except as permitted by 362.10(1), 362.10(5), and 362.10(8)

(6) In theaters and similar locations, except as provided in 518.4 and 520.5

(7) Where exposed to the direct rays of the sun, unless identified as sunlight resistant

(8) Where subject to physical damage

Informational Note: Extreme cold may cause some types of nonmetallic conduits to become brittle and therefore more susceptible to damage from physical contact.

362.20 Size.

(A) Minimum. ENT smaller than metric designator 16 (trade size ½) shall not be used.

(B) Maximum. ENT larger than metric designator 63 (trade size 2½) shall not be used.

Informational Note: See 300.1(C) for the metric designators and trade sizes. These are for identification purposes only and do not relate to actual dimensions.

362.22 Number of Conductors. The number of conductors shall not exceed that permitted by the percentage fill in Table 1, Chapter 9.

Cables shall be permitted to be installed where such use is not prohibited by the respective cable articles. The number of cables shall not exceed the allowable percentage fill specified in Table 1, Chapter 9.

Table 4 of Chapter 9 provides the usable area within the selected conduit or tubing, and Table 5 provides the required area for each conductor. Examples using these tables to calculate a conduit or tubing size are provided in the commentary following Chapter 9, Notes to Tables, Note 6.

To select the proper trade size of ENT, see the appropriate sub-table for Article 362, Electrical Nonmetallic Tubing (ENT), in Table 4 of Chapter 9. If the conductors are of the same wire size and insulation type, Tables C.2 and C.2(A) for ENT in Informative Annex C can be used instead of performing the calculations.

Δ **362.24 Bends.**

N **(A) How Made.** Bends shall be so made that the tubing will not be damaged and the internal diameter of the tubing will not be effectively reduced. Bends shall be permitted to be made manually without auxiliary equipment, and the radius of the curve to the centerline of such bends shall not be less than shown in Table 2, Chapter 9 using the column "Other Bends."

N **(B) Number in One Run.** The total degrees of bends in a tubing run shall not exceed 360 degrees between pull points.

362.28 Trimming. All cut ends shall be trimmed inside and outside to remove rough edges.

362.30 Securing and Supporting. ENT shall be installed as a complete system in accordance with 300.18 and shall be securely fastened in place by an approved means and supported in accordance with 362.30(A) and (B).

(A) Securely Fastened. ENT shall be securely fastened at intervals not exceeding 900 mm (3 ft). In addition, ENT shall be securely fastened in place within 900 mm (3 ft) of each outlet box, device box, junction box, cabinet, or fitting where it terminates. Where used, cable ties shall be listed for the application and for securing and supporting.

Exception No. 1: Lengths not exceeding a distance of 1.8 m (6 ft) from a luminaire terminal connection for tap connections to lighting luminaires shall be permitted without being secured.

Exception No. 2: Lengths not exceeding 1.8 m (6 ft) from the last point where the raceway is securely fastened for connections within an accessible ceiling to luminaire(s) or other equipment.

Exception No. 3: For concealed work in finished buildings or prefinished wall panels where such securing is impracticable, unbroken lengths (without coupling) of ENT shall be permitted to be fished.

As illustrated in Exhibit 362.4, where run on the surface of framing members, ENT is required to be fastened to the framing member every 3 feet and within 3 feet of every box.

As illustrated in Exhibit 362.5, ENT is permitted by Exception No. 1 to be used as luminaire whip without support for lengths not exceeding 6 feet.

See also

300.4(D) for requirements for protection against physical damage

410.117(C) for details on tap conductor wiring

(B) Supports. Horizontal runs of ENT supported by openings in framing members at intervals not exceeding 900 mm (3 ft) and securely fastened within 900 mm (3 ft) of termination points shall be permitted.

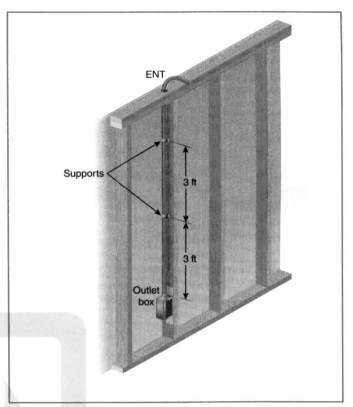

EXHIBIT 362.4 An example showing ENT supported every 3 feet and within 3 feet of the outlet box.

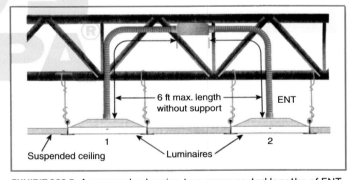

EXHIBIT 362.5 An example showing two unsupported lengths of ENT, each permitted to be installed in a length not to exceed 6 feet.

362.46 Bushings. Where a tubing enters a box, fitting, or other enclosure, a bushing or adapter shall be provided to protect the wire from abrasion unless the box, fitting, or enclosure design provides equivalent protection.

Informational Note: See 300.4(G) for the protection of conductors size 4 AWG or larger.

362.48 Joints. All joints between lengths of tubing and between tubing and couplings, fittings, and boxes shall be by an approved method.

△ **362.56 Splices and Taps.** Splices and taps shall be made only in accordance with 300.15.

△ **362.60 Grounding.** Where equipment grounding is required, a separate grounding conductor shall be installed in the raceway in compliance with Article 250, Part VI.

> *Exception No. 1: The equipment grounding conductor shall be permitted to be run separately from the raceway where used for grounding dc circuits as permitted in 250.134, Exception No. 2.*
>
> *Exception No. 2: The equipment grounding conductor shall not be required where the grounded conductor is used as part of the effective ground-fault path as permitted in 250.142.*

Part III. Construction Specifications

362.100 Construction. ENT shall be made of material that does not exceed the ignitibility, flammability, smoke generation, and toxicity characteristics of rigid (nonplasticized) polyvinyl chloride.

ENT, as a prewired manufactured assembly, shall be provided in continuous lengths capable of being shipped in a coil, reel, or carton without damage.

362.120 Marking. ENT shall be clearly and durably marked at least every 3 m (10 ft) as required in the first sentence of 110.21(A). The type of material shall also be included in the marking. Marking for limited smoke shall be permitted on the tubing that has limited smoke-producing characteristics.

The type, size, and quantity of conductors used in prewired manufactured assemblies shall be identified by means of a printed tag or label attached to each end of the manufactured assembly and either the carton, coil, or reel. The enclosed conductors shall be marked in accordance with 310.8.

ARTICLE 366 Auxiliary Gutters

Part I. General

366.1 Scope. This article covers the use, installation, and construction requirements of metal auxiliary gutters and nonmetallic auxiliary gutters and associated fittings.

An auxiliary gutter provides additional gutter space for wiring in various types of electrical enclosures and equipment. This additional gutter space may be necessary to provide sufficient room for the number of conductors in an enclosure or to provide adequate wiring bending/deflection space where conductors connect to a terminal. Although the construction of an auxiliary gutter is no different from that of a wireway, the field application of this equipment differentiates an auxiliary gutter from a wireway.

A wireway is a raceway in accordance with the definition of *raceway* in Article 100. Auxiliary gutters supplement enclosure wiring spaces and are not encompassed by the definition of raceway. Therefore, *NEC®* requirements that apply only to raceways do not apply to auxiliary gutters. An example of such a requirement is 230.7, which prohibits service conductors from being installed in a raceway with conductors that are not service conductors. This rule applies to wireways installed in accordance with Articles 376 and 378. However, an auxiliary gutter installed to supplement the wiring space of a service equipment enclosure is not a wireway; therefore, service conductors, feeder conductors, and branch-circuit conductors can occupy the same wiring space of an auxiliary gutter. Unlike wireways, which are covered in Articles 376 and 378, auxiliary gutters are permitted to contain bare and insulated copper and aluminum busbars. See 366.23(A) for the ampere rating of bare copper and aluminum busbars installed in auxiliary gutters.

366.6 Listing Requirements.

(A) Outdoors. Nonmetallic auxiliary gutters installed outdoors shall be listed for all of the following conditions:

(1) Exposure to sunlight
(2) Use in wet locations
(3) Maximum ambient temperature of the installation

(B) Indoors. Nonmetallic auxiliary gutters installed indoors shall be listed for the maximum ambient temperature of the installation.

The sections and associated fittings of auxiliary gutters are identical to those of wireways. They differ only in their intended use. If listed, they may be marked as "Wireway," "Auxiliary Gutter," or "Wireway or Auxiliary Gutter," depending on the intended application.

Part II. Installation

366.10 Uses Permitted.

(A) Sheet Metal Auxiliary Gutters.

(1) Indoor and Outdoor Use. Sheet metal auxiliary gutters shall be permitted for indoor and outdoor use.

(2) Wet Locations. Sheet metal auxiliary gutters installed in wet locations shall be suitable for such locations.

(B) Nonmetallic Auxiliary Gutters. Nonmetallic auxiliary gutters shall be listed for the maximum ambient temperature of the installation and marked for the installed conductor insulation temperature rating.

> Informational Note: Extreme cold may cause nonmetallic auxiliary gutters to become brittle and therefore more susceptible to damage from physical contact.

(1) Outdoors. Nonmetallic auxiliary gutters shall be permitted to be installed outdoors where listed and marked as suitable for the purpose.

(2) Indoors. Nonmetallic auxiliary gutters shall be permitted to be installed indoors.

N (C) Extended Distance of Auxiliary Gutters. Auxiliary gutters shall be permitted to extend a distance not greater than 9 m (30 ft) beyond the equipment that it supplements.

Exception: Where used in accordance with 620.35 for elevators, an auxiliary gutter shall be permitted to extend a distance greater than 9 m (30 ft) beyond the equipment it supplements.

Δ 366.12 Uses Not Permitted. Auxiliary gutters shall not be used to enclose switches, overcurrent devices, appliances, or other similar equipment.

366.20 Conductors Connected in Parallel. Where single conductor cables comprising each phase, neutral, or grounded conductor of an alternating-current circuit are connected in parallel as permitted in 310.10(G), the conductors shall be installed in groups consisting of not more than one conductor per phase, neutral, or grounded conductor to prevent current imbalance in the paralleled conductors due to inductive reactance.

366.22 Number of Conductors.

No limit is placed on the size of conductors that can be installed in an auxiliary gutter. If the auxiliary gutter is being used to supplement the wiring "gutter" space of a panelboard cabinet, the wire-bending requirements of 312.6(A) and (B) apply total wiring space created by adding the width of the auxiliary gutter to the width of the wire-bending space of the panelboard cabinet. The dimensions of insulated conductors, found in Tables 5 and 5A of Chapter 9, can be used to calculate the size of auxiliary gutters.

(A) Sheet Metal Auxiliary Gutters. The sum of the cross-sectional areas of all contained conductors and cables at any cross section of a sheet metal auxiliary gutter shall not exceed 20 percent of the interior cross-sectional area of the sheet metal auxiliary gutter.

(B) Nonmetallic Auxiliary Gutters. The sum of cross-sectional areas of all contained conductors and cables at any cross section of the nonmetallic auxiliary gutter shall not exceed 20 percent of the interior cross-sectional area of the nonmetallic auxiliary gutter.

366.23 Ampacity of Conductors.

(A) Sheet Metal Auxiliary Gutters. The adjustment factors in 310.15(C)(1) shall be applied only where the number of current-carrying conductors, including neutral conductors classified as current-carrying under 310.15(E), exceeds 30 at any cross section of the sheet metal auxiliary gutter. Conductors for signaling circuits or controller conductors between a motor and its starter and used only for starting duty shall not be considered as current-carrying conductors. The current carried continuously in bare copper bars in sheet metal auxiliary gutters shall not exceed

1.55 amperes/mm² (1000 amperes/in.²) of cross section of the conductor. For aluminum bars, the current carried continuously shall not exceed 1.09 amperes/mm² (700 amperes/in.²) of cross section of the conductor.

Where sheet metal auxiliary gutters contain 30 or fewer current-carrying conductors, the correction factors in 310.15(B)(2) do not apply. However, if more than 30 conductors are installed in the same cross-sectional area of a sheet metal auxiliary gutter, the ampacity adjustment factors of 310.15(C)(1) apply, and the number of current-carrying conductors is not limited up to the 20-percent fill.

(B) Nonmetallic Auxiliary Gutters. The adjustment factors specified in 310.15(C)(1) shall be applicable to the current-carrying conductors up to and including the 20 percent fill specified in 366.22(B).

The requirements for nonmetallic auxiliary gutters limit the cross-sectional area of all conductors to 20 percent. There is no 30-conductor allowance. The derating factors specified in 310.15(C)(1) must be applied as stated in 366.23(B).

366.30 Securing and Supporting.

(A) Sheet Metal Auxiliary Gutters. Sheet metal auxiliary gutters shall be supported and secured throughout their entire length at intervals not exceeding 1.5 m (5 ft).

(B) Nonmetallic Auxiliary Gutters. Nonmetallic auxiliary gutters shall be supported and secured at intervals not to exceed 900 mm (3 ft) and at each end or joint, unless listed for other support intervals. In no case shall the distance between supports exceed 3 m (10 ft).

366.44 Expansion Fittings. Expansion fittings shall be installed where expected length change, due to expansion and contraction due to temperature change, is more than 6 mm (0.25 in.).

See also

Table 352.44(A) for expansion characteristics of nonmetallic gutters similar to those of polyvinyl chloride (PVC) conduit

300.7(B) commentary for more on expansion requirements

366.56 Splices and Taps. Splices and taps shall comply with 366.56(A) through (D).

(A) Within Gutters. Splices or taps shall be permitted within gutters where they are accessible by means of removable covers or doors. The conductors, including splices and taps, shall not fill the gutter to more than 75 percent of its area.

(B) Bare Conductors. Taps from bare conductors shall leave the gutter opposite their terminal connections, and conductors shall not be brought in contact with uninsulated current-carrying parts of different voltages.

(C) Suitably Identified. All taps shall be suitably identified at the gutter as to the circuit or equipment that they supply.

(D) Overcurrent Protection. Tap connections from conductors in auxiliary gutters shall be provided with overcurrent protection as required in 240.21.

366.58 Insulated Conductors.

(A) Deflected Insulated Conductors. Where insulated conductors are deflected within an auxiliary gutter, either at the ends or where conduits, fittings, or other raceways or cables enter or leave the gutter, or where the direction of the gutter is deflected greater than 30 degrees, dimensions corresponding to one wire per terminal in Table 312.6(A) shall apply.

(B) Auxiliary Gutters Used as Pull Boxes. Where insulated conductors 4 AWG or larger are pulled through an auxiliary gutter, the distance between raceway and cable entries enclosing the same conductor shall not be less than that required in 314.28(A)(1) for straight pulls and 314.28(A)(2) for angle pulls.

366.60 Grounding.
Metal auxiliary gutters shall be connected to an equipment grounding conductor(s), to an equipment bonding jumper, or to the grounded conductor where permitted or required by 250.92(B)(1) or 250.142.

Part III. Construction Specifications

366.100 Construction.

(A) Electrical and Mechanical Continuity. Gutters shall be constructed and installed so that adequate electrical and mechanical continuity of the complete system is secured.

(B) Substantial Construction. Gutters shall be of substantial construction and shall provide a complete enclosure for the contained conductors. All surfaces, both interior and exterior, shall be suitably protected from corrosion. Corner joints shall be made tight, and where the assembly is held together by rivets, bolts, or screws, such fasteners shall be spaced not more than 300 mm (12 in.) apart.

(C) Smooth Rounded Edges. Suitable bushings, shields, or fittings having smooth, rounded edges shall be provided where conductors pass between gutters, through partitions, around bends, between gutters and cabinets or junction boxes, and at other locations where necessary to prevent abrasion of the insulation of the conductors.

(D) Covers. Covers shall be securely fastened to the gutter.

(E) Clearance of Bare Live Parts. Bare conductors shall be securely and rigidly supported so that the minimum clearance between bare current-carrying metal parts of different voltages mounted on the same surface will not be less than 50 mm (2 in.), nor less than 25 mm (1 in.) for parts that are held free in the air. A clearance not less than 25 mm (1 in.) shall be secured between bare current-carrying metal parts and any metal surface. Adequate provisions shall be made for the expansion and contraction of busbars.

366.120 Marking.

(A) Outdoors. Nonmetallic auxiliary gutters installed outdoors shall have the following markings:

(1) Suitable for exposure to sunlight
(2) Suitable for use in wet locations
(3) Installed conductor insulation temperature rating

(B) Indoors. Nonmetallic auxiliary gutters installed indoors shall be marked with the installed conductor insulation temperature rating.

ARTICLE 368 Busways

Part I. General

368.1 Scope. This article covers service-entrance, feeder, and branch-circuit busways and associated fittings.

Exhibit 368.1 illustrates a section of busway with a plug-in device covered by this article. Busway of this type typically is used as a feeder to supply other feeders or to supply branch circuits for utilization equipment. The plug-in devices provide the means to connect feeders and branch circuits to the busway and contain the overcurrent protection for the connected feeder or branch circuit.

In addition to the power distribution busway illustrated in Exhibit 368.1, there are four special application busway designs covered by UL 857, *Standard for Safety for Busways*:

1. Lighting busway, with a maximum current rating of 50 amperes, supplies and supports industrial and commercial luminaires.
2. Trolley busway allows continuous contact with a trolley through a slot in the enclosure and might be marked as "Lighting Busway" if intended for use with industrial and commercial luminaires.
3. Continuous plug-in busway allows for the insertion of plug-in devices at any point along its length. This busway, limited to a maximum current rating of 225 amperes, is intended for general use and is permitted to be installed within reach of persons.
4. Short-run busway is intended primarily to feed switchboards and is limited to a run of 30 feet horizontal or 10 feet vertical.

Part II. Installation

368.10 Uses Permitted. Busways shall be permitted to be installed where they are located in accordance with 368.10(A) through (C).

Informational Note: See 300.21 for information concerning the spread of fire or products of combustion.

Busways are commonly used as feeders. They either are installed horizontally or, where used for power distribution in high-rise

EXHIBIT 368.1 A section of feeder busway with a plug-in tap device, one of the busway types used for power distribution covered by Article 368. *(Courtesy of Square D™ by Schneider Electric)*

buildings, are run vertically from floor to floor to supply transformers and other distribution equipment in "stacked" electric rooms or closets. See Exhibit 368.2 for an illustration of a busway installed horizontally in the space above a dropped or hung ceiling. The space is not being used as an "other space for environmental air (plenums)" covered by 300.22(C) and can contain the plug-in devices for supplying electrical equipment installed above, within, or below the ceiling.

Unless marked to indicate otherwise, busways and associated fittings containing a vapor seal have not been evaluated for passage through a fire-rated wall or floor. Listed firestop systems are available for making busway penetrations through fire-rated walls and floors.

See also

300.21, which provides requirements that are essential to limiting the spread of a fire and products of combustion

(A) Exposed. Busways shall be permitted to be located in the open where visible, except as permitted in 368.10(C).

(B) Behind Access Panels. Busways shall be permitted to be installed behind access panels, provided the busways are totally enclosed, of nonventilating-type construction, and installed so that the joints between sections and at fittings are accessible for maintenance purposes. Where installed behind access panels, means of access shall be provided, and either of the following conditions shall be met:

(1) The space behind the access panels shall not be used for air-handling purposes.
(2) Where the space behind the access panels is used for environmental air, other than ducts and plenums, there shall be no provisions for plug-in connections, and the conductors shall be insulated.

(C) Through Walls and Floors. Busways shall be permitted to be installed through walls or floors in accordance with 368.10(C)(1) and (C)(2).

(1) Walls. Unbroken lengths of busway shall be permitted to be extended through dry walls.

(2) Floors. Floor penetrations shall comply with 368.10(C)(2)(a) and (C)(2)(b).

(a) Busways shall be permitted to be extended vertically through dry floors if totally enclosed (unventilated) where

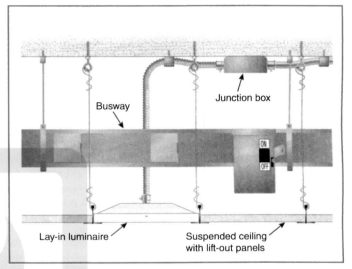

Busway   Junction box

Lay-in luminaire   Suspended ceiling with lift-out panels

EXHIBIT 368.2 An example of a busway mounted horizontally in the space above a hung ceiling.

passing through and for a minimum distance of 1.8 m (6 ft) above the floor to provide adequate protection from physical damage.

(b) In other than industrial establishments, where a vertical riser penetrates two or more dry floors, a minimum 100-mm (4-in.) high curb shall be installed around all floor openings for riser busways to prevent liquids from entering the opening. The curb shall be installed within 300 mm (12 in.) of the floor opening. Electrical equipment shall be located so that it will not be damaged by liquids that are retained by the curb.

368.12 Uses Not Permitted.

(A) Physical Damage. Busways shall not be installed where subject to severe physical damage or corrosive vapors.

(B) Hoistways. Busways shall not be installed in hoistways.

(C) Hazardous Locations. Busways shall not be installed in any hazardous (classified) location, unless specifically approved for such use.

Informational Note: See 501.10(B).

(D) Wet Locations. Busways shall not be installed outdoors or in wet or damp locations unless identified for such use.

(E) Working Platform. Lighting busway and trolley busway shall not be installed less than 2.5 m (8 ft) above the floor or working platform unless provided with an identified cover.

368.17 Overcurrent Protection. Overcurrent protection shall be provided in accordance with 368.17(A) through (D).

> **See also**
>
> **Section 240.21(E),** which specifies that overcurrent protection for busways and busway taps is covered by the rules in 368.17

(A) Rating of Overcurrent Protection — Feeders. A busway shall be protected against overcurrent in accordance with the current rating of the busway.

Exception No. 1: The applicable provisions of 240.4 shall be permitted.

Exception No. 2: Where used as transformer secondary ties, 450.6(A)(3) shall be permitted.

Busways not intended for use on the line (supply) side of service equipment are marked with the maximum rating of overcurrent protection required on the supply side. The current rating of a busway can be determined in the field only by reference to the nameplate data. The applicable sections of 240.4 referenced in Exception No. 1 are 240.4(B) and (C).

(B) Reduction in Ampacity Size of Busway. Overcurrent protection shall be required where busways are reduced in ampacity.

Exception: For industrial establishments only, omission of overcurrent protection shall be permitted at points where busways are reduced in ampacity, provided that the length of the busway having the smaller ampacity does not exceed 15 m (50 ft) and has an ampacity at least equal to one-third the rating or setting of the overcurrent device next back on the line, and provided that such busway is free from contact with combustible material.

In industrial establishments, where the size of a smaller busway is kept within the specified limits, providing overcurrent protection at the point where the size is changed is not required. For example, a busway protected by a 1200-ampere overcurrent device may be reduced in size, provided the smaller busway has a current rating of at least 400 amperes ($\frac{1}{3}$ of 1200 amperes) and does not extend more than 50 feet. Restricting the length of smaller busway and establishing a relationship between the busway current rating and the rating of the feeder overcurrent protective device (OCPD) provides a level of protection for the smaller section of busway in the event of a short circuit or a ground fault. These restrictions are analogous to those required by 240.21(B) for feeder taps.

(C) Feeder or Branch Circuits. Where a busway is used as a feeder, devices or plug-in connections for tapping off feeder or branch circuits from the busway shall contain the overcurrent devices required for the protection of the feeder or branch circuits. The plug-in device shall consist of an externally operable circuit breaker or an externally operable fusible switch. Where

such devices are mounted out of reach and contain disconnecting means, suitable means such as ropes, chains, or sticks shall be provided for operating the disconnecting means from the floor.

> **See also**
>
> **Exhibit 100.1,** which illustrates methods of operating elevated busway plug-in devices from floor level

Exception No. 1: As permitted in 240.21.

Exception No. 2: For fixed or semifixed luminaires, where the branch-circuit overcurrent device is part of the luminaire cord plug on cord-connected luminaires.

Exception No. 3: Where luminaires without cords are plugged directly into the busway and the overcurrent device is mounted on the luminaire.

Exception No. 4: Where the branch-circuit overcurrent plug-in device is directly supplying a readily accessible disconnect, a method of floor operation shall not be required.

Exception No. 4 allows alternative methods of providing ready access to disconnects with an equivalent level of safety. Receptacles may be used as a disconnecting means where permitted.

(D) Rating of Overcurrent Protection — Branch Circuits. A busway used as a branch circuit shall be protected against overcurrent in accordance with 210.20.

368.30 Support. Busways shall be securely supported at intervals not exceeding 1.5 m (5 ft) unless otherwise designed and marked.

Busways are marked if suitable for installation in a specified position, for use in vertical runs, or for support at intervals greater than 5 feet.

368.56 Branches from Busways. Branches from busways shall be permitted to be made in accordance with 368.56(A), (B), and (C).

Δ **(A) General.** Branches from busways shall be permitted to use any of the following wiring methods:

(1) Type AC armored cable
(2) Type MC metal-clad cable
(3) Type MI mineral-insulated, metal-sheathed cable
(4) IMC intermediate metal conduit
(5) RMC rigid metal conduit
(6) FMC flexible metal conduit
(7) LFMC liquidtight flexible metal conduit
(8) PVC rigid polyvinyl chloride conduit
(9) RTRC reinforced thermosetting resin conduit
(10) LFNC liquidtight flexible nonmetallic conduit
(11) EMT electrical metallic tubing
(12) ENT electrical nonmetallic tubing
(13) Busways
(14) Strut-type channel raceway
(15) Surface metal raceway
(16) Surface nonmetallic raceway

Where a separate equipment grounding conductor is used, connection of the equipment grounding conductor to the busway shall comply with 250.8 and 250.12.

(B) Cord and Cable Assemblies. Suitable cord and cable assemblies identified for extra-hard usage or hard usage and listed bus drop cable shall be permitted as branches from busways for the connection of portable equipment or the connection of stationary equipment to facilitate their interchange in accordance with 400.10 and 400.12 and the following conditions:

(1) The cord or cable shall be attached to the building by an approved means.
(2) The length of the cord or cable from a busway plug-in device to a suitable tension take-up support device shall not exceed 1.8 m (6 ft).
(3) The cord and cable shall be installed as a vertical riser from the tension take-up support device to the equipment served.
(4) Strain relief cable grips shall be provided for the cord or cable at the busway plug-in device and equipment terminations.

Exception to (B)(2): In industrial establishments only, where the conditions of maintenance and supervision ensure that only qualified persons service the installation, lengths exceeding 1.8 m (6 ft) shall be permitted between the busway plug-in device and the tension take-up support device where the cord or cable is supported at intervals not exceeding 2.5 m (8 ft).

Exhibit 368.3 illustrates a flexible cable or cord branch from a busway installed according to the requirements of 368.56(B)(2). Exhibit 368.4 illustrates a flexible cable or cord branch from a busway installed according to 368.56(B)(2), Exception.

(C) Branches from Trolley-Type Busways. Suitable cord and cable assemblies identified for extra-hard usage or hard usage and listed bus drop cable shall be permitted as branches from trolley-type busways for the connection of movable equipment in accordance with 400.10 and 400.12.

368.58 Dead Ends. A dead end of a busway shall be closed.

368.60 Grounding. Busway shall be connected to an equipment grounding conductor(s), to an equipment bonding jumper, or to the grounded conductor where permitted or required by 250.92(B)(1) or 250.142.

The metal enclosure of a listed busway is intended for use as an equipment grounding conductor (EGC). In some cases, an additional grounding bus may also act as an EGC.

Part III. Construction

368.120 Marking. Busways shall be marked with the voltage and current rating for which they are designed, and with the manufacturer's name or trademark in such a manner as to be visible after installation.

EXHIBIT 368.3 An example of a flexible cable or cord branch from busway installed according to 368.56(B).

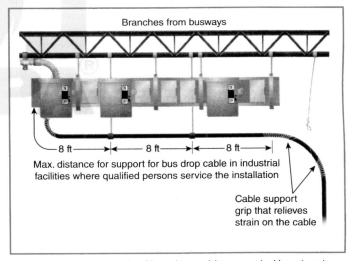

EXHIBIT 368.4 An example of bus drop cable supported by a tension take-up support device attached to the structure of an industrial building.

Part IV. Requirements for Over 1000 Volts, Nominal

368.214 Adjacent and Supporting Structures. Metal-enclosed busways shall be installed so that temperature rise from induced circulating currents in adjacent ferrous metal parts will not be hazardous to personnel or constitute a fire hazard.

368.234 Barriers and Seals.

(A) Vapor Seals. Busway runs that have sections located both inside and outside of buildings shall have a vapor seal at the building wall to prevent interchange of air between indoor and outdoor sections.

Exception: Vapor seals shall not be required in forced-cooled bus.

(B) Fire Barriers. Fire barriers shall be provided where fire walls, floors, or ceilings are penetrated.

Informational Note: See 300.21 for information concerning the spread of fire or products of combustion.

368.236 Drain Facilities. Drain plugs, filter drains, or similar methods shall be provided to remove condensed moisture from low points in busway run.

368.237 Ventilated Bus Enclosures. Ventilated busway enclosures shall be installed in accordance with Article 110, Part III, and 495.24.

368.238 Terminations and Connections. Where bus enclosures terminate at machines cooled by flammable gas, seal-off bushings, baffles, or other means shall be provided to prevent accumulation of flammable gas in the busway enclosures.

All conductor termination and connection hardware shall be accessible for installation, connection, and maintenance.

368.239 Switches. Switching devices or disconnecting links provided in the busway run shall have the same momentary rating as the busway. Disconnecting links shall be plainly marked to be removable only when bus is de-energized. Switching devices that are not load-break shall be interlocked to prevent operation under load, and disconnecting link enclosures shall be interlocked to prevent access to energized parts.

368.240 Wiring 1000 Volts or Less, Nominal. Secondary control devices and wiring that are provided as part of the metal-enclosed bus run shall be insulated by fire-retardant barriers from all primary circuit elements with the exception of short lengths of wire, such as at instrument transformer terminals.

368.244 Expansion Fittings. Flexible or expansion connections shall be provided in long, straight runs of bus to allow for temperature expansion or contraction, or where the busway run crosses building vibration insulation joints.

368.258 Neutral Conductor. Neutral bus, where required, shall be sized to carry all neutral load current, including harmonic currents, and shall have adequate momentary and short-circuit current rating consistent with system requirements.

368.260 Grounding. Metal-enclosed busway shall be grounded.

368.320 Marking. Each busway run shall be provided with a permanent nameplate on which the following information shall be provided:

(1) Rated voltage.
(2) Rated continuous current; if bus is forced-cooled, both the normal forced-cooled rating and the self-cooled (not forced-cooled) rating for the same temperature rise shall be given.
(3) Rated frequency.
(4) Rated impulse withstand voltage.
(5) Rated 60-Hz withstand voltage (dry).
(6) Rated momentary current.
(7) Manufacturer's name or trademark.

Informational Note: See IEEE C37.23-2015, *IEEE Standard for Metal-Enclosed Bus*, for construction and testing requirements for metal-enclosed bus assemblies.

N **369** Insulated Bus Pipe (IBP)/Tubular Covered Conductors (TCC) Systems

N **Part I. General**

N **369.1 Scope.** This article covers the use, installation, and construction specifications for insulated bus pipe (IBP) systems.

New Article 369 has been created to establish requirements for the installation of insulated bus pipe (IBP), which is also known as tubular covered conductor (TCC). This article requires IBP to be listed and installed by qualified persons. A manufacturer supplied terminating means must be utilized for all system connections or terminations. All documentation must be available to the AHJ.

N **369.2 Reconditioned Equipment.** IBP and IBP systems shall not be reconditioned.

N **369.6 Listing Requirements.** IBP and IBP systems shall be listed.

N **Part II. Installation**

N **369.10 Uses Permitted.** IBP systems shall be permitted for use on power systems in accordance with the following:

(1) As exposed runs in accordance with 305.3.
(2) In wet or damp locations only when listed for such use.
(3) Installed through walls, in unbroken lengths. Where IBP penetrates an exterior wall, the entire length that penetrates the wall shall be listed for outdoor use, and the opening in the wall shall be sealed by an approved method.

(4) Extended vertically through dry floors if totally enclosed in metal where passing through the floor and for a minimum distance of 1.8 m (6 ft) above the floor to provide protection from physical damage.

(5) For voltages up to and including 35,000 volts ac nominal.

N **369.12 Uses Not Permitted.** IBP systems shall not be used under the following conditions:

(1) In any hazardous (classified) location except as permitted by other articles in this *Code*
(2) For the support of luminaires or other equipment
(3) Where concealed by the building structure
(4) Where accessible to other than qualified person(s)

N **369.14 Installation.** IBP systems shall be installed by qualified persons. All documentation shall be available to the authority having jurisdiction.

N **369.20 Termination or Connections.** Manufacturer's supplied terminating means shall be used for IBP system connections or terminations. Connections employing dissimilar metals shall be avoided to eliminate the possibility of galvanic action.

> Informational Note No. 1: See 110.14(C) for conductor temperature limitations due to termination provisions for installations up to and including 2000 volts.
>
> Informational Note No. 2: See 110.40 for conductor temperature limitations due to termination provisions for installations 2001 volts to 35,000 volts.

N **369.80 Ampacity.** IBP systems shall be used within the marked ampacity of the IBP.

N **369.90 Temperature Rating.** IBP systems shall be used within the maximum rated conductor temperature.

N **Part III. Construction Specifications**

N **369.100 Construction.** The IBP conductor shall be aluminum or copper. The bus pipe shall be permitted to be solid or hollow.

N **369.110 Barriers.** Fire barriers shall be provided where fire walls, floors, or ceilings are penetrated.

> Informational Note: See 300.21 for information concerning the spread of fire or products of combustion.

N **369.120 Marking.** All IBP shall be marked to indicate the following information:

(1) The maximum rated voltage phase-to-phase or phase-to-ground
(2) The maximum rated ampacity
(3) The manufacturer's name, trademark, or other distinctive marking by which the organization responsible for the product can be readily identified

(4) The equivalent AWG size or circular mil area of the conductor
(5) The maximum rated conductor temperature
(6) The rated peak withstand current rating in rms symmetrical amperes or kA
(7) Enclosure type designation, if other than Type 1
(8) Rated short-time withstand current and duration if greater than 2 seconds

370 Cablebus

Part I. General

370.1 Scope. This article covers the use and installation requirements of cablebus and associated fittings.

Cablebus consists of a metal structure or framework installed in a manner similar to that of a cable tray support system. As illustrated in Exhibit 370.1, continuous runs of insulated conductors of 1/0 AWG or larger are field installed within the framework on special insulating blocks at specified intervals to provide controlled spacing between conductors. A ventilated top cover is attached to the framework to completely enclose the conductors.

Part II. Installation

Δ **370.10 Uses Permitted.** Cablebus shall be permitted as follows:

(1) At any voltage or current for which spaced conductors are rated and where installed only for exposed work, except as permitted in 370.18
(2) For branch circuits, feeders, and services
(3) To be installed indoors, outdoors, or in corrosive, wet, or damp locations where identified for the use

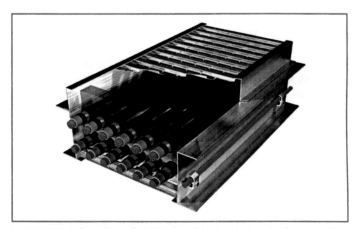

EXHIBIT 370.1 A section of cablebus with conductors in place and the ventilated top cover ready to be attached to the busway frame. *(Courtesy of MP Husky Cable Bus & Cable Tray)*

370.12 Uses Not Permitted. Cablebus shall not be permitted to be installed in the following:

(1) Hoistways
(2) Hazardous (classified) locations, unless specifically permitted in Chapter 5

370.18 Cablebus Installation. Cablebus shall be permitted to extend transversely through partitions and walls or vertically through platforms and floors in wet or dry locations where the installation, complete with the installed cables, is made in accordance with 300.21.

370.20 Conductor Size and Termination.

Δ **(A) Conductors.** The current-carrying conductors in cablebus shall comply with the following:

(1) Have an insulation rating of 75°C (167°F) or higher and be of an approved type suitable for the application
(2) Be sized in accordance with the design of the cablebus but in no case be smaller than 1/0

(B) Termination. Approved terminating means shall be used for connections to cablebus conductors.

Informational Note No. 1: See 110.14(C) for conductor temperature limitations due to termination provisions for installations up to and including 2000 volts.

Informational Note No. 2: See 110.40 for conductor temperature limitations due to termination provisions for installations 2001 volts to 35,000 volts.

The construction of and cable arrangement within cablebus allows for free air circulation around the insulated conductors. Therefore, the ampacity of the conductors can be selected from the applicable "free air (as applied to conductors)" tables in Articles 310 (0–2000 volts) and 315 (2001–35,000 volts) in accordance with 370.80. See the definition of *free air (as applied to conductors)* in Article 100. However, the new informational notes to 370.20 alert the user of cablebus to the fact that the terminations within the supply-end and load-end equipment are based on ampacities for 60°C or 75°C insulated conductors in Table 310.16 for conductors rated 0–2000 volts and for 90°C insulated conductors in Tables 315.60(C)(3) and (C)(4) for conductors rated 2001–35,000 volts.

This restriction may necessitate transitioning to larger or more conductors in parallel at the supply end and the load end of the circuit in order to coordinate the equipment termination temperature limitations with the higher permitted conductor ampacities within the cablebus run. Because cablebus systems are most typically built for a specific installation, the necessary transition connectors or enclosures can be provided by the manufacturer to facilitate connecting the cablebus conductors to conductors that provide compliance with the equipment terminal temperature limitations. This same condition is found with cable tray installations where conductors in cable tray can carry more current than is permitted by the conductor ampacity tables referenced in 110.14(C) and 110.40 that are used to control heating at connections within supply-end and load-end equipment.

370.22 Number of Conductors. The number of conductors shall be that for which the cablebus is designed.

370.23 Overcurrent Protection. Cablebus shall be protected against overcurrent in accordance with the ampacity of the cablebus conductors in accordance with 240.4.

Exception: Overcurrent protection shall be permitted in accordance with 245.26 and 245.27.

370.30 Securing and Supporting.

(A) Cablebus Supports. Cablebus shall be securely supported at intervals not exceeding 3.7 m (12 ft). Where spans longer than 3.7 m (12 ft) are required, the structure shall be specifically designed for the required span length.

(B) Conductor Supports. The insulated conductors shall be supported on blocks or other identified mounting means.

The individual conductors in a cablebus shall be supported at intervals not greater than 900 mm (3 ft) for horizontal runs and 450 mm (1½ ft) for vertical runs. Vertical and horizontal spacing between supported conductors shall be not less than one conductor diameter at the points of support.

Δ **370.42 Fittings.** A cablebus system shall include fittings for the following:

(1) Changes in horizontal or vertical direction of the run
(2) Terminations in or on connected apparatus or equipment or the enclosures for such equipment
(3) Additional physical protection where required, such as guards where subject to severe physical damage

370.60 Grounding. A cablebus system shall be grounded and/or bonded as applicable:

(1) Cablebus framework, where bonded, shall be permitted to be used as the equipment grounding conductor for branch circuits and feeders.
(2) A cablebus installation shall be grounded and bonded in accordance with Part V and Part VI of Article 250, excluding 250.86, Exception No. 2.

Cablebus framework is recognized as an equipment grounding conductor (EGC) by 250.118(A)(12). Similar to cable tray, the product evaluation organizations can test the cablebus framework for ground-fault current-carrying performance and "classify" it for the specific purpose. The classification program tests the cablebus only for equipment grounding performance and is not an overall listing of the product. As is the case with cable tray in Article 392, the overall cablebus product is a cable support system and is not required by the *NEC®* to be a listed product. Specific information relating to the rating of the framework as an EGC is available from the cablebus manufacturer.

370.80 Ampacity of Conductors. The ampacity of conductors in cablebus shall be in accordance with 310.17 and 310.19 for installations up to and including 2000 volts, or with Table 315.60(C)(3) and Table 315.60(C)(4) for installations 2001 volts to 35,000 volts.

Where conductors in cablebus terminate at circuit breakers, distribution switchgear, and similar electrical equipment, the temperature limitations of the electrical equipment terminals should be coordinated with the conductor ampacity. See commentary for 370.20.

Part III. Construction Specifications

Δ 370.120 Marking.

N **(A) Nameplates.** Each cablebus system shall include a nameplate at each terminating end of the system with the manufacturer's name or trade designation and the maximum diameter, number, voltage rating, and ampacity of the conductors to be installed. Nameplates shall be visible after installation.

N **(B) Identification.** Each section and fitting of a cablebus system shall be identified with a marking that corresponds to the manufacturer's installation instructions.

N ARTICLE
371 Flexible Bus Systems

N Part I. General

N **371.1 Scope.** This article covers the use and installation requirements of flexible bus systems and associated fittings.

New Article 371 has been created to recognize flexible bus systems as a wiring method. Flexible bus systems are required to be listed and must be designed and specified by a qualified engineer based upon site-specific installation applications.

N **371.6 Listing Requirements.** Flexible bus systems shall be listed.

N Part II. Installation

N **371.10 Uses Permitted.** Flexible bus systems shall be permitted for the following:

(1) Services, feeders, and branch circuits
(2) Indoors
(3) Outdoors where identified for outdoor use
(4) Installed in corrosive, wet, or damp locations where identified for use
(5) Exposed
(6) Behind access panels where the space behind the access panel is not used for air-handling purposes
(7) To penetrate through walls and floors in accordance with 371.18

N **371.12 Uses Not Permitted.** Flexible bus systems shall not be permitted to be installed in the following:

(1) Hoistways
(2) Where exposed to severe physical damage
(3) Hazardous (classified) locations, unless specifically permitted in Chapter 5
(4) Air-handling spaces

N **371.14 Installation Design.** Flexible bus systems shall be designed and specified for specific installation site applications by a qualified engineer within the limits of the listing and manufacturer's installation instructions. All documentation shall be available to the authority having jurisdiction.

N **371.17 Overcurrent Protection.** Overcurrent protection shall be provided in accordance with 371.17(A) through (G).

N **(A) Rating of Overcurrent Protection — Services.** Flexible bus systems installed for services shall be protected against overcurrent in accordance with 230.90.

N **(B) Rating of Overcurrent Protection — Feeders.** Flexible bus systems installed as feeders shall be protected against overcurrent in accordance with 215.3.

Exception: The applicable requirements of 240.4 shall be permitted.

N **(C) Rating of Overcurrent Protection — Branch Circuits.** Flexible bus systems installed as branch circuits shall be protected against overcurrent in accordance with 210.20.

Exception: The applicable requirements of 240.4 shall be permitted.

N **(D) Transformer Secondary Flexible Bus Systems.** Flexible bus systems installed on a transformer secondary to the disconnect and overcurrent protection device shall be protected from overcurrent in accordance with 240.21(C).

N **(E) Flexible Bus Systems from Generator Terminals.** Flexible bus systems installed from generator terminals that meet the size requirement in 445.13 shall be permitted to be protected against overload by the generator overload protective device(s) required by 445.12.

N **(F) Flexible Bus Systems from Battery Terminals.** Flexible bus systems installed for battery systems shall be protected from overcurrent in accordance with 240.21(H).

N **(G) Reduction in Size of Flexible Bus Systems.** Overcurrent protection shall be required at the point where flexible bus systems are reduced in size.

Exception: For industrial establishments only, omission of overcurrent protection shall be permitted at points where a flexible bus system is reduced in size, provided that the length of the flexible bus system having a reduced size does not exceed 15 m

(50 ft) and has a current rating at least equal to one-third the rating or setting of the overcurrent device ahead of the point of connection and provided that such a flexible bus system is free from contact with combustible material.

N 371.18 Flexible Bus Systems Installation. Installation of flexible bus systems shall comply with 371.18(A) through (E).

N (A) Manufacturer's Installation Instructions. Flexible bus systems shall be installed under design engineering supervision and in accordance with the manufacturer's instructions, including supporting and securing. All documentation shall be available to the authority having jurisdiction.

N (B) Physical Damage. Flexible bus systems subject to physical damage shall have approved protective means installed.

> Informational Note: Typical methods of protecting flexible bus systems from physical damage include suitable barriers, guards, or elevation.

N (C) Transversely Routed. Flexible bus systems shall be permitted to extend transversely through partitions or walls if the section within the wall is continuous and protected against physical damage. Where the flexible bus systems penetrate a fire-resistant-rated wall or partition, the installation shall be made in accordance with 300.21.

N (D) Through Dry Floors and Platforms. Flexible bus systems shall be permitted to extend vertically through dry floors and platforms. Where the flexible bus systems penetrate a fire-resistant-rated floor or ceiling, the installation shall be made in accordance with 300.21.

N (E) Through Floors and Platforms in Wet Locations. Flexible bus systems shall be permitted to extend vertically through floors and platforms in wet locations as follows:

(1) Where there are curbs or other suitable means to prevent waterflow through the floor or platform opening
(2) Where the flexible bus system provides a means to seal the floor penetration

Where the flexible bus systems penetrate a fire-resistant-rated floor or ceiling, the installation shall be made in accordance with 300.21.

N 371.20 Terminations. Flexible bus system terminations shall comply with 371.20(A) and (B).

N (A) Termination Fittings or Connectors. Flexible bus systems shall be terminated with fittings or connectors listed for flexible bus systems.

> Informational Note: See 110.14(C) for conductor temperature limitations due to termination provisions.

N (B) Connection to Equipment. The connection of a flexible bus system to the distribution equipment shall comply with one of the following:

(1) Be listed and marked by the specific distribution equipment as an acceptable means of termination
(2) Be listed and identified for the specific distribution equipment as an acceptable means of termination
(3) Incorporate, as part of the listed flexible bus system, a transition box to interface between the flexible bus system and the distribution equipment

N 371.30 Securing and Supporting. Flexible insulated bus conductors shall be supported on identified mounting means at intervals not greater than 900 mm (3 ft) for horizontal runs and 450 mm (1½ ft) for vertical runs unless otherwise permitted by the product listing. Flexible bus systems shall be secured and supported by listed associated fittings in accordance with 371.30(A) through (C).

N (A) Associated Fittings. Associated fittings shall be part of a listed flexible bus system.

N (B) Support Brackets. The support brackets for flexible bus systems shall be secured to the building structure or to other associated fittings that are secured to the building structure.

N (C) Support Tray. Flexible bus systems shall be permitted to be installed in support trays supplied as associated fittings for the listed flexible bus system. Support trays shall not be required to be continuous.

N 371.40 Short Circuit Current Rating. Flexible bus systems shall have a short circuit current rating sufficient for the available fault current.

N 371.60 Grounding. Conductive associated fitting supports for flexible bus systems shall be bonded together and grounded.

N Part III. Construction Specifications

N 371.120 Marking. Each section of flexible bus systems shall be marked with the manufacturer's name or trade designation, voltage rating, and current rating. Markings shall be located so as to be visible after installation.

N (A) System Nameplate. A system nameplate shall contain the manufacturer's name or trademark and the flexible bus system ratings. The ratings shall include the voltage, phase, current rating, short circuit current rating, and applicable environmental ratings. The nameplate shall be installed at each end of the flexible bus system. The nameplate shall be visible after installation.

N (B) Associated Fittings. Associated fittings shall be marked as suitable for flexible bus systems.

N (C) Flexible Insulated Bus. The flexible insulated bus shall be marked along the insulation with the manufacturer's name or trademark, voltage, manufacturer's part identification, and insulation temperature ratings.

ARTICLE
372 Cellular Concrete Floor Raceways

Part I. General

372.1 Scope. This article covers cellular concrete floor raceways, the hollow spaces in floors constructed of precast cellular concrete slabs, together with suitable metal fittings designed to provide access to the floor cells.

Cellular concrete floor raceways are a form of floor deck construction commonly used in high-rise office buildings. This construction method is very similar in design, application, and adaptation to cellular metal floor raceways. Basically, this wiring method consists of floor cells (that are part of the structural floor system), header ducts laid at right angles to the cells (used to carry conductors from cabinets to cells), and junction boxes.

Part II. Installations

372.12 Uses Not Permitted. Conductors shall not be installed in precast cellular concrete floor raceways as follows:

(1) Where subject to corrosive vapor
(2) In any hazardous (classified) location, except as permitted by other articles in this *Code*
(3) In commercial garages, other than for supplying ceiling outlets or extensions to the area below the floor but not above

Informational Note: See 300.8 for installation of conductors with other systems.

372.18 Cellular Concrete Floor Raceways Installation. Installation of cellular concrete floor raceways shall comply with 372.18(A) through (E).

(A) Header. The header shall be installed in a straight line at right angles to the cells. The header shall be mechanically secured to the top of the precast cellular concrete floor. The end joints shall be closed by a metal closure fitting and sealed against the entrance of concrete. The header shall be electrically continuous throughout its entire length and shall be electrically bonded to the enclosure of the distribution center.

(B) Connection to Cabinets and Other Enclosures. Connections from headers to cabinets and other enclosures shall be made by means of listed metal raceways and listed fittings.

(C) Junction Boxes. Junction boxes shall be leveled to the floor grade and sealed against the free entrance of water or concrete. Junction boxes shall be of metal and shall be mechanically and electrically continuous with the header.

(D) Inserts. Inserts shall be leveled and sealed against the entrance of concrete. Inserts shall be of metal and shall be fitted with grounded-type receptacles. An equipment grounding

conductor or bonding jumper shall connect the insert receptacles to a positive ground connection provided on the header. Where cutting through the cell wall for setting inserts or other purposes (such as providing access openings between header and cells), chips and other dirt shall not be allowed to remain in the raceway, and the tool used shall be designed so as to prevent the tool from entering the cell and damaging the conductors.

(E) Markers. A suitable number of markers shall be installed for the future location of cells.

372.20 Size of Conductors. No conductor larger than 1/0 AWG shall be installed, except by special permission.

372.22 Maximum Number of Conductors. The combined cross-sectional area of all conductors or cables shall not exceed 40 percent of the cross-sectional area of the cell or header.

372.23 Ampacity of Conductors. The ampacity adjustment factors as provided in 310.15(C) shall apply to conductors installed in cellular concrete floor raceways.

372.56 Splices and Taps. Splices and taps shall be made only in header access units or junction boxes. A continuous unbroken conductor connecting the individual outlets is not a splice or tap.

372.58 Discontinued Outlets. When an outlet is abandoned, discontinued, or removed, the sections of circuit conductors supplying the outlet shall be removed from the raceway. No splices or reinsulated conductors, such as would be the case of abandoned outlets on loop wiring, shall be allowed in raceways.

ARTICLE
374 Cellular Metal Floor Raceways

Part I. General

374.1 Scope. This article covers the use and installation requirements for cellular metal floor raceways.

Cellular metal floor raceways, as shown in Exhibit 374.1, are a listed raceway system that is a form of metal floor deck construction designed for use in steel-frame buildings and consist of sheet metal formed into shapes that are combined to form cells or raceways. The cells extend across the building and, depending on the structural strength required, can have various shapes and sizes. ANSI/UL 209, *Standard for Cellular Metal Floor Raceways and Fittings*, provides the construction, performance, and marking requirements for this equipment.

Connections to the cells are made by means of headers extending across the cells and connecting only to those cells to be used as raceways for the conductors. Two or three separate headers, connecting to different sets of cells, may be used for

EXHIBIT 374.1 Cross sections of a typical cellular metal floor raceway system. *(Courtesy of Cordeck Building Solutions)*

different systems, such as light and power, signaling, and communications systems.

374.6 Listing Requirements. Cellular metal floor raceways and associated fittings shall be listed.

Part II. Installation

374.12 Uses Not Permitted. Conductors shall not be installed in cellular metal floor raceways as follows:

(1) Where subject to corrosive vapor
(2) In any hazardous (classified) location, except as permitted by other articles in this *Code*
(3) In commercial garages, other than for supplying ceiling outlets or extensions to the area below the floor but not above

Informational Note: See 300.8 for installation of conductors with other systems.

374.18 Cellular Metal Floor Raceways Installations. Installation of cellular metal floor raceways shall comply with 374.18(A) through (D).

(A) Connection to Cabinets and Extensions from Cells. Connections between raceways and distribution centers and wall outlets shall be made by means of liquidtight flexible metal conduit, flexible metal conduit where not installed in concrete, rigid metal conduit, intermediate metal conduit, electrical metallic tubing, or approved fittings. Where there are provisions for the termination of an equipment grounding conductor, rigid polyvinyl chloride conduit, reinforced thermosetting resin conduit, electrical nonmetallic tubing, or liquidtight flexible nonmetallic conduit shall be permitted. Where installed in concrete, liquidtight flexible metal conduit and liquidtight flexible nonmetallic conduit shall be listed and marked for direct burial.

(B) Junction Boxes. Junction boxes shall be leveled to the floor grade and sealed against the free entrance of water or concrete. Junction boxes used with these raceways shall be of metal and shall be electrically continuous with the raceway.

(C) Inserts. Inserts shall be leveled to the floor grade and sealed against the entrance of concrete. Inserts shall be of metal and shall be electrically continuous with the raceway. In cutting through the cell wall and setting inserts, chips and other dirt shall not be allowed to remain in the raceway, and tools shall be used that are designed to prevent the tool from entering the cell and damaging the conductors.

(D) Markers. A suitable number of markers shall be installed for locating cells in the future.

374.20 Size of Conductors. No conductor larger than 1/0 AWG shall be installed, except by special permission.

374.22 Maximum Number of Conductors in Raceway. The combined cross-sectional area of all conductors or cables shall not exceed 40 percent of the interior cross-sectional area of the cell or header.

374.23 Ampacity of Conductors. The ampacity adjustment factors in 310.15(C) shall apply to conductors installed in cellular metal floor raceways.

374.56 Splices and Taps. Splices and taps shall be made only in header access units or junction boxes.

For the purposes of this section, so-called loop wiring (continuous unbroken conductor connecting the individual outlets) shall not be considered to be a splice or tap.

374.58 Discontinued Outlets. When an outlet is abandoned, discontinued, or removed, the sections of circuit conductors supplying the outlet shall be removed from the raceway.

No splices or reinsulated conductors, such as would be the case with abandoned outlets on loop wiring, shall be allowed in raceways.

Part III. Construction Specifications

374.100 General. Cellular metal floor raceways shall be constructed so that adequate electrical and mechanical continuity of the complete system will be secured. They shall provide a complete enclosure for the conductors. The interior surfaces shall be free from burrs and sharp edges, and surfaces over which conductors are drawn shall be smooth. Suitable bushings or fittings having smooth rounded edges shall be provided where conductors pass.

ARTICLE
376 Metal Wireways

Part I. General

376.1 Scope. This article covers the use, installation, and construction specifications for metal wireways and associated fittings.

Wireways are sheet-metal enclosures equipped with hinged or removable covers and are manufactured in 1-foot to 10-foot lengths and various widths and depths. Couplings, elbows, end plates, and accessories such as T and X fittings are available. Unlike auxiliary gutters, which are not permitted to extend more than 30 feet from the equipment they supplement, wireways may be run throughout an entire area.

Part II. Installation

376.10 Uses Permitted. The use of metal wireways shall be permitted as follows:

(1) For exposed work.
(2) In any hazardous (classified) location, as permitted by other articles in this *Code*.
(3) In wet locations where wireways are listed for the purpose.
(4) In concealed spaces as an extension that passes transversely through walls, if the length passing through the wall is unbroken. Access to the conductors shall be maintained on both sides of the wall.

376.12 Uses Not Permitted. Metal wireways shall not be used in the following:

(1) Where subject to severe physical damage
(2) Where subject to severe corrosive environments

376.20 Conductors Connected in Parallel. Where single conductor cables comprising each phase, neutral, or grounded conductor of an alternating-current circuit are connected in parallel as permitted in 310.10(G), the conductors shall be installed in groups consisting of not more than one conductor per phase, neutral, or grounded conductor.

> Informational Note: The purpose of having all parallel conductor sets within the same group is to prevent current imbalance in the paralleled conductors due to inductive reactance.

376.21 Size of Conductors. No conductor larger than that for which the wireway is designed shall be installed in any wireway.

376.22 Number of Conductors and Ampacity. The number of conductors or cables and their ampacity shall comply with 376.22(A) and (B).

(A) Cross-Sectional Areas of Wireway. The sum of the cross-sectional areas of all contained conductors and cables at any cross section of a wireway shall not exceed 20 percent of the interior cross-sectional area of the wireway.

(B) Adjustment Factors. The adjustment factors in 310.15(C)(1) shall be applied only where the number of current-carrying conductors, including neutral conductors classified as current-carrying under 310.15(E), exceeds 30 at any cross section of the wireway. Conductors for signaling circuits or controller conductors between a motor and its starter and used only for starting duty shall not be considered as current-carrying conductors.

376.23 Insulated Conductors. Insulated conductors installed in a metal wireway shall comply with 376.23(A) and (B).

(A) Deflected Insulated Conductors. Where insulated conductors are deflected within a metal wireway, either at the ends or where conduits, fittings, or other raceways or cables enter or leave the metal wireway, or where the direction of the metal wireway is deflected greater than 30 degrees, dimensions corresponding to one wire per terminal in Table 312.6(A) shall apply.

(B) Metal Wireways Used as Pull Boxes. Where insulated conductors 4 AWG or larger are pulled through a wireway, the distance between raceway and cable entries enclosing the same conductor shall not be less than that required by 314.28(A)(1) for straight pulls and 314.28(A)(2) for angle pulls. When transposing cable size into raceway size, the minimum metric designator (trade size) raceway required for the number and size of conductors in the cable shall be used.

See also

Exhibit 314.10 for an example showing calculations for splices, angle pulls, or U pulls

Exhibit 314.11 for an example showing calculations for raceways enclosing the same conductor

376.30 Securing and Supporting. Metal wireways shall be supported in accordance with 376.30(A) and (B).

(A) Horizontal Support. Wireways shall be supported where run horizontally at each end and at intervals not to exceed 1.5 m (5 ft) or for individual lengths longer than 1.5 m (5 ft) at each end or joint, unless listed for other support intervals. The distance between supports shall not exceed 3 m (10 ft).

(B) Vertical Support. Vertical runs of wireways shall be securely supported at intervals not exceeding 4.5 m (15 ft) and shall not have more than one joint between supports. Adjoining wireway sections shall be securely fastened together to provide a rigid joint.

376.56 Splices, Taps, and Power Distribution Blocks.

(A) Splices and Taps. Splices and taps shall be permitted within a wireway, provided they are accessible. The conductors, including splices and taps, shall not fill the wireway to more than 75 percent of its area at that point.

(B) Power Distribution Blocks.

(1) Installation. Power distribution blocks installed in metal wireways shall be listed. Power distribution blocks installed on the line side of the service equipment shall be marked "suitable for use on the line side of service equipment" or equivalent.

(2) Size of Enclosure. In addition to the wiring space requirement in 376.56(A), the power distribution block shall be installed in a wireway with dimensions not smaller than specified in the installation instructions of the power distribution block.

(3) Wire Bending Space. Wire bending space at the terminals of power distribution blocks shall comply with 312.6(B).

(4) Live Parts. Power distribution blocks shall not have uninsulated live parts exposed within a wireway, whether or not the wireway cover is installed.

(5) Conductors. Conductors shall be arranged so the power distribution block terminals are unobstructed following installation.

376.58 Dead Ends. Dead ends of metal wireways shall be closed.

N **376.60 Grounding.** Listed metal wireway shall be permitted as an equipment grounding conductor in accordance with 250.118(A)(13).

376.70 Extensions from Metal Wireways. Extensions from wireways shall be made with cord pendants installed in accordance with 400.14 or with any wiring method in Chapter 3 that includes a means for equipment grounding. Where a separate equipment grounding conductor is employed, connection of the equipment grounding conductors in the wiring method to the wireway shall comply with 250.8 and 250.12.

Extensions from wireways are made through knockouts provided on the wireway or are field punched. The extension wiring method must provide for equipment grounding. Cables and nonmetallic raceways as well as the wireway must include a wire-type equipment grounding equipment (EGC) to ensure effective continuation of the ground path.

Part III. Construction Specifications

376.100 Construction.

Custom-made wireways and fittings must comply with the requirements in 376.100(A) through (D).

(A) Electrical and Mechanical Continuity. Wireways shall be constructed and installed so that electrical and mechanical continuity of the complete system are assured.

(B) Substantial Construction. Wireways shall be of substantial construction and shall provide a complete enclosure for the contained conductors. All surfaces, both interior and exterior, shall be suitably protected from corrosion. Corner joints shall be made tight, and where the assembly is held together by rivets, bolts, or screws, such fasteners shall be spaced not more than 300 mm (12 in.) apart.

(C) Smooth Rounded Edges. Suitable bushings, shields, or fittings having smooth, rounded edges shall be provided where conductors pass between wireways, through partitions, around bends, between wireways and cabinets or junction boxes, and at other locations where necessary to prevent abrasion of the insulation of the conductors.

(D) Covers. Covers shall be securely fastened to the wireway.

376.120 Marking. Metal wireways shall be so marked that their manufacturer's name or trademark will be visible after installation.

ARTICLE 378   Nonmetallic Wireways

Part I. General

378.1 Scope. This article covers the use, installation, and construction specifications for nonmetallic wireways and associated fittings.

378.6 Listing Requirements. Nonmetallic wireways and associated fittings shall be listed.

Part II. Installation

378.10 Uses Permitted. The use of nonmetallic wireways shall be permitted in the following:

(1) Only for exposed work, except as permitted in 378.10(4).

(2) Where subject to corrosive environments where identified for the use.

(3) In wet locations where listed for the purpose.

Informational Note: Extreme cold may cause nonmetallic wireways to become brittle and therefore more susceptible to damage from physical contact.

(4) As extensions to pass transversely through walls if the length passing through the wall is unbroken. Access to the conductors shall be maintained on both sides of the wall.

378.12 Uses Not Permitted. Nonmetallic wireways shall not be used in the following:

(1) Where subject to physical damage
(2) In any hazardous (classified) location, except as permitted by other articles in this *Code*
(3) Where exposed to sunlight unless listed and marked as suitable for the purpose
(4) Where subject to ambient temperatures other than those for which nonmetallic wireway is listed
(5) For conductors whose insulation temperature limitations would exceed those for which the nonmetallic wireway is listed

378.20 Conductors Connected in Parallel. Where single conductor cables comprising each phase, neutral, or grounded conductor of an alternating-current circuit are connected in parallel as permitted in 310.10(G), the conductors shall be installed in groups consisting of not more than one conductor per phase, neutral, or grounded conductor to prevent current imbalance in the paralleled conductors due to inductive reactance.

378.21 Size of Conductors. No conductor larger than that for which the nonmetallic wireway is designed shall be installed in any nonmetallic wireway.

378.22 Number of Conductors. The sum of cross-sectional areas of all contained conductors or cables at any cross section of the nonmetallic wireway shall not exceed 20 percent of the interior cross-sectional area of the nonmetallic wireway. Conductors for signaling circuits or controller conductors between a motor and its starter and used only for starting duty shall not be considered as current-carrying conductors.

The adjustment factors specified in 310.15(C)(1) shall be applicable to the current-carrying conductors up to and including the 20 percent fill specified in the first paragraph of this section.

378.23 Insulated Conductors. Insulated conductors installed in a nonmetallic wireway shall comply with 378.23(A) and (B).

(A) Deflected Insulated Conductors. Where insulated conductors are deflected within a nonmetallic wireway, either at the ends or where conduits, fittings, or other raceways or cables enter or leave the nonmetallic wireway, or where the direction of the nonmetallic wireway is deflected greater than 30 degrees,

dimensions corresponding to one wire per terminal in Table 312.6(A) shall apply.

(B) Nonmetallic Wireways Used as Pull Boxes. Where insulated conductors 4 AWG or larger are pulled through a wireway, the distance between raceway and cable entries enclosing the same conductor shall not be less than that required in 314.28(A)(1) for straight pulls and in 314.28(A)(2) for angle pulls. When transposing cable size into raceway size, the minimum metric designator (trade size) raceway required for the number and size of conductors in the cable shall be used.

These requirements provide adequate space for installing and bending conductors without damaging the conductor insulation. Section 378.23(A) requires that the Table 312.6(A) column of one wire per terminal be used. Section 378.23(B) requires that the same adequate space requirements that apply to conduits entering boxes also apply to cables as they enter a wireway. Where wireways are used as pull boxes, cable entries are converted to a minimum raceway trade size to determine wireway dimensions.

378.30 Securing and Supporting. Nonmetallic wireway shall be supported in accordance with 378.30(A) and (B).

(A) Horizontal Support. Nonmetallic wireways shall be supported where run horizontally at intervals not to exceed 900 mm (3 ft), and at each end or joint, unless listed for other support intervals. In no case shall the distance between supports exceed 3 m (10 ft).

(B) Vertical Support. Vertical runs of nonmetallic wireway shall be securely supported at intervals not exceeding 1.2 m (4 ft), unless listed for other support intervals, and shall not have more than one joint between supports. Adjoining nonmetallic wireway sections shall be securely fastened together to provide a rigid joint.

378.44 Expansion Fittings. Expansion fittings for nonmetallic wireway shall be provided to compensate for thermal expansion and contraction where the length change is expected to be 6 mm (0.25 in.) or greater in a straight run.

Informational Note: See Table 352.44(A) for expansion characteristics of PVC conduit. The expansion characteristics of PVC nonmetallic wireway are identical.

378.56 Splices and Taps. Splices and taps shall be permitted within a nonmetallic wireway, provided they are accessible. The conductors, including splices and taps, shall not fill the nonmetallic wireway to more than 75 percent of its area at that point.

378.58 Dead Ends. Dead ends of nonmetallic wireway shall be closed using listed fittings.

Δ **378.60 Grounding.** Where equipment grounding is required, a separate grounding conductor shall be installed in the nonmetallic wireway. A separate equipment grounding conductor shall not be required where the grounded conductor is used to ground equipment as permitted in 250.142.

378.70 Extensions from Nonmetallic Wireways. Extensions from nonmetallic wireway shall be made with cord pendants or any wiring method of Chapter 3. A separate equipment grounding conductor shall be installed in, or an equipment grounding connection shall be made to, any of the wiring methods used for the extension.

Part III. Construction Specifications

378.120 Marking. Nonmetallic wireways shall be marked so that the manufacturer's name or trademark and interior cross-sectional area in square inches shall be visible after installation. Marking for limited smoke shall be permitted on the nonmetallic wireways that have limited smoke-producing characteristics.

ARTICLE 380 Multioutlet Assembly

Part I. General

Δ **380.1 Scope.** This article covers the use and installation requirements for multioutlet assemblies.

> Informational Note: See Article 100 for the definition of *multioutlet assembly.*

Multioutlet assemblies are metal or nonmetallic raceways that are usually surface mounted and designed to contain branch-circuit conductors and receptacles. Exhibit 380.1 provides an illustration of a multioutlet assembly. Multioutlet assemblies can be assembled at the factory or in the field with the receptacles spaced at desired intervals.

See also

220.14(H) and **Exhibit 220.2** for load calculations

Part II. Installation

380.10 Uses Permitted. The use of a multioutlet assembly shall be permitted in dry locations.

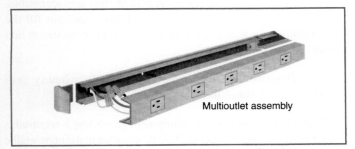

EXHIBIT 380.1 A typical multioutlet assembly shown in an exploded view.

Multioutlet assembly

380.12 Uses Not Permitted. A multioutlet assembly shall not be installed as follows:

(1) Where concealed, except that it shall be permissible to surround the back and sides of a metal multioutlet assembly by the building finish or recess a nonmetallic multioutlet assembly in a baseboard
(2) Where subject to severe physical damage
(3) Where the voltage is 300 volts or more between conductors unless the assembly is of metal having a thickness of not less than 1.02 mm (0.040 in.)
(4) Where subject to corrosive vapors
(5) In hoistways
(6) In any hazardous (classified) location, except as permitted by other articles in this *Code*
(7) Where cord and plug connected

380.23 Insulated Conductors. For field-assembled multioutlet assemblies, insulated conductors shall comply with 380.23(A) and (B), as applicable.

(A) Deflected Insulated Conductors. Where insulated conductors are deflected within a multioutlet assembly, either at the ends or where conduits, fittings, or other raceways or cables enter or leave the multioutlet assembly, or where the direction of the multioutlet assembly is deflected greater than 30 degrees, dimensions corresponding to one wire per terminal in Table 312.6(A) shall apply.

(B) Multioutlet Assemblies Used as Pull Boxes. Where insulated conductors 4 AWG or larger are pulled through a multioutlet assembly, the distance between raceway and cable entries enclosing the same conductor shall not be less than that required by 314.28(A)(1) for straight pulls and 314.28(A)(2) for angle pulls. When transposing cable size into raceway size, the minimum metric designator (trade size) raceway required for the number and size of conductors in the cable shall be used.

Safeguards to prevent overfill are provided by limiting the number of conductors that can be installed in multioutlet assemblies. For deflected insulated conductors, dimensions corresponding to the minimum width of wiring gutters must be maintained.

Where a multioutlet assembly is used as a pull box for insulated conductors of 4 AWG or larger, the distance between the raceway and the cable entries enclosing the conductor must not be less than eight times the trade size or metric designator of the raceway for straight pulls. For angle pulls, the distance must be six times the trade size or metric designator of the raceway.

380.76 Metal Multioutlet Assembly Through Dry Partitions. It shall be permissible to extend a metal multioutlet assembly through (not run within) dry partitions if arrangements are made for removing the cap or cover on all exposed portions and no outlet is located within the partitions.

Part I. General

382.1 Scope. This article covers the use, installation, and construction specifications for nonmetallic extensions.

382.6 Listing Requirements. Concealable nonmetallic extensions and associated fittings and devices shall be listed. The starting/source tap device for the extension shall contain and provide the following protection for all load-side extensions and devices:

(1) Supplementary overcurrent protection
(2) Level of protection equivalent to a Class A GFCI
(3) Level of protection equivalent to a portable GFCI
(4) Line and load-side miswire protection
(5) Provide protection from the effects of arc faults

Part II. Installation

382.10 Uses Permitted. Nonmetallic extensions shall be permitted only in accordance with 382.10(A), (B), and (C).

(A) From an Existing Outlet. The extension shall be from an existing outlet on a 15- or 20-ampere branch circuit. Where a concealable nonmetallic extension originates from a non–grounding-type receptacle, the installation shall comply with 250.130(C), 406.4(D)(2)(b), or 406.4(D)(2)(c).

(B) Exposed and in a Dry Location. The extension shall be run exposed, or concealed as permitted in 382.15, and in a dry location.

(C) Residential or Offices. For nonmetallic surface extensions mounted directly on the surface of walls or ceilings, the building shall be occupied for residential or office purposes and shall not exceed three floors above grade. Where identified for the use, concealable nonmetallic extensions shall be permitted more than three floors above grade.

> Informational Note No. 1: See 310.14(A)(3) for temperature limitation of conductors.
> Informational Note No. 2: See 362.10 for definition of *First Floor*.

382.12 Uses Not Permitted. Nonmetallic extensions shall not be used as follows:

(1) In unfinished basements, attics, or roof spaces
(2) Where the voltage between conductors exceeds 150 volts for nonmetallic surface extensions and 300 volts for aerial cable
(3) Where subject to corrosive vapors
(4) Where run through a floor or partition, or outside the room in which it originates

382.15 Exposed.

(A) Nonmetallic Extensions. One or more extensions shall be permitted to be run in any direction from an existing outlet, but not on the floor or within 50 mm (2 in.) from the floor.

(B) Concealable Nonmetallic Extensions. Where identified for the use, nonmetallic extensions shall be permitted to be concealed with paint, texture, concealing compound, plaster, wallpaper, tile, wall paneling, or other similar materials and installed in accordance with 382.15(A).

382.24 Bends.

(A) Nonmetallic Extensions. A bend that reduces the normal spacing between the conductors shall be covered with a cap to protect the assembly from physical damage.

(B) Concealable Nonmetallic Extensions. Concealable extensions shall be permitted to be folded back over themselves and flattened as required for installation.

382.30 Securing and Supporting.

(A) Nonmetallic Extensions. Nonmetallic surface extensions shall be secured in place by approved means at intervals not exceeding 200 mm (8 in.), with an allowance for 300 mm (12 in.) to the first fastening where the connection to the supplying outlet is by means of an attachment plug. There shall be at least one fastening between each two adjacent outlets supplied. An extension shall be attached to only woodwork or plaster finish and shall not be in contact with any metal work or other conductive material other than with metal plates on receptacles.

(B) Concealable Nonmetallic Extensions. All surface-mounted concealable nonmetallic extension components shall be firmly anchored to the wall or ceiling using an adhesive or mechanical anchoring system identified for this use.

382.40 Boxes and Fittings. Each run shall terminate in a fitting, connector, or box that covers the end of the assembly. All fittings, connectors, and devices shall be of a type identified for the use.

382.42 Devices.

(A) Receptacles. All receptacles, receptacle housings, and self-contained devices used with concealable nonmetallic extensions shall be identified for this use.

Δ **(B) Receptacles and Housings.** Receptacle housings and self-contained devices designed either for surface or for recessed mounting shall be permitted for use with concealable nonmetallic extensions. Receptacle housings and self-contained devices shall incorporate means for facilitating entry and termination of concealable nonmetallic extensions and for electrically connecting the housing or device. Receptacle and self-contained devices shall comply with 406.4. Power and communications outlets installed together in common housing shall be permitted in accordance with 800.133(A)(3), Exception No. 2.

382.56 Splices and Taps. Extensions shall consist of a continuous unbroken length of the assembly, without splices, and without exposed conductors between fittings, connectors, or devices. Taps shall be permitted where approved fittings completely covering the tap connections are used. Aerial cable and its tap connectors shall be provided with an approved means for polarization. Receptacle-type tap connectors shall be of the locking type.

Part III. Construction Specifications (Concealable Nonmetallic Extensions Only)

382.100 Construction. Concealable nonmetallic extensions shall be of a multilayer flat conductor design consisting of a center ungrounded conductor enclosed by a sectioned grounded conductor and an overall sectioned equipment grounding conductor.

382.104 Flat Conductors. Concealable nonmetallic extensions shall be constructed, using flat copper conductors equivalent to 14 AWG or 12 AWG conductor sizes, and constructed per 382.104(A), (B), and (C).

A multilayer flat conductor is a complete assembly of branch-circuit conductors, thinner than a business card and yet flexible enough to bend to any angle required for a customized installation. Exhibit 382.1 is one example of a concealable nonmetallic extension circuit conductor assembly of flat conductors meeting the requirements of Article 382.

A concealable extension is a five-layer design for circuit integrity and protection. Flat-sectioned equipment grounding conductor (EGC) layers fully encase the ungrounded conductor and grounded conductor layers of the cable. By design, any penetration in the flat wire cable assembly will penetrate the EGC layer first, then the grounded conductor layer, then the ungrounded conductor layer. A penetration will result in a short circuit between the three conductors described that will

EXHIBIT 382.1 One example of a flat conductor cable assembly used as a concealable nonmetallic extension. (*Courtesy of FlatWire Technologies, a division of Southwire®*)

immediately trip the overcurrent protective device (OCPD), causing automatic disconnection of the circuit.

(A) Ungrounded Conductor (Center Layer). The ungrounded conductor shall consist of one or more ungrounded flat conductor(s) enclosed in accordance with 382.104(B) and (C) and identified in accordance with 310.6(C).

(B) Grounded Conductor (Inner Sectioned Layers). The grounded conductor shall consist of two sectioned inner flat conductors that enclose the center ungrounded conductor(s). The sectioned grounded conductor shall be enclosed by the sectioned equipment grounding conductor and identified in accordance with 200.6.

(C) Equipment Grounding Conductor (Outer Sectioned Layers). The equipment grounding conductor shall consist of two overall sectioned conductors that enclose the grounded conductor and ungrounded conductor(s) and shall comply with 250.4(A)(5). The equipment grounding conductor layers shall be identified by any one of the following methods:

(1) As permitted in 250.119
(2) A clear covering
(3) One or more continuous green stripes or hash marks
(4) The term "Equipment Grounding Conductor" printed at regular intervals throughout the cable

382.112 Insulation. The ungrounded and grounded flat conductor layers shall be individually insulated and comply with 310.14(A)(3). The equipment grounding conductor shall be covered or insulated.

382.120 Marking.

(A) Cable. Concealable nonmetallic extensions shall be clearly and durably marked on both sides at intervals of not more than 610 mm (24 in.) with the information required by 310.8(A) and with the following additional information:

(1) Material of conductors
(2) Maximum temperature rating
(3) Ampacity

(B) Conductor Identification. Conductors shall be clearly and durably identified on both sides throughout their length as specified in 382.104.

ARTICLE 384 Strut-Type Channel Raceway

Part I. General

384.1 Scope. This article covers the use, installation, and construction specifications of strut-type channel raceway.

384.6 Listing Requirements. Strut-type channel raceways and accessories shall be listed and identified for such use.

Part II. Installation

384.10 Uses Permitted. The use of strut-type channel raceways shall be permitted in the following:

(1) Where exposed.

(2) In dry locations.

(3) In locations subject to corrosive vapors where protected by finishes approved for the condition.

(4) As power poles.

(5) In hazardous (classified) locations as permitted in Chapter 5.

(6) As extensions of unbroken lengths through walls, partitions, and floors where closure strips are removable from either side and the portion within the wall, partition, or floor remains covered.

(7) Ferrous channel raceways and fittings protected from corrosion solely by enamel shall be permitted only indoors.

The installation shown in Exhibit 384.1 is typical of how a strut-type channel raceway can be used.

384.12 Uses Not Permitted. Strut-type channel raceways shall not be used as follows:

(1) Where concealed.

(2) Ferrous channel raceways and fittings protected from corrosion solely by enamel shall not be permitted where subject to severe corrosive influences.

384.21 Size of Conductors. No conductor larger than that for which the raceway is listed shall be installed in strut-type channel raceways.

384.22 Number of Conductors. The number of conductors or cables permitted in strut-type channel raceways shall not exceed the percentage fill using Table 384.22 and applicable cross-sectional area of specific types and sizes of wire given in the tables in Chapter 9.

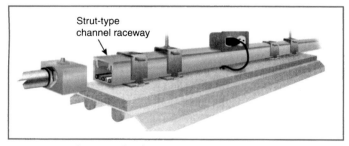

EXHIBIT 384.1 An example of a strut-type channel raceway using accessories to support and supply power to luminaires.

TABLE 384.22 *Channel Size and Inside Cross-Sectional Area*

Size Channel	Area		40% Area*		25% Area†	
	in.²	mm²	in.²	mm²	in.²	mm²
1⅝ × ¹⁵⁄₁₆	0.887	572	0.355	229	0.222	143
1⅝ × 1	1.151	743	0.460	297	0.288	186
1⅝ × 1⅜	1.677	1076	0.671	433	0.419	270
1⅝ × 1⅝	2.028	1308	0.811	523	0.507	327
1⅝ × 2⁷⁄₁₆	3.169	2045	1.267	817	0.792	511
1⅝ × 3¼	4.308	2780	1.723	1112	1.077	695
1½ × ¾	0.849	548	0.340	219	0.212	137
1½ × 1½	1.828	1179	0.731	472	0.457	295
1½ × 1⅞	2.301	1485	0.920	594	0.575	371
1½ × 3	3.854	2487	1.542	995	0.964	622

*Raceways with external joiners shall use a 40 percent wire fill calculation to determine the number of conductors permitted.
†Raceways with internal joiners shall use a 25 percent wire fill calculation to determine the number of conductors permitted.

The adjustment factors of 310.15(C)(1) shall not apply to conductors installed in strut-type channel raceways where all of the following conditions are met:

(1) The cross-sectional area of the raceway exceeds 2500 mm² (4 in.²).

(2) The current-carrying conductors do not exceed 30 in number.

(3) The sum of the cross-sectional areas of all contained conductors does not exceed 20 percent of the interior cross-sectional area of the strut-type channel raceways.

384.30 Securing and Supporting.

(A) Surface Mount. A surface mount strut-type channel raceway shall be secured to the mounting surface with retention straps external to the channel at intervals not exceeding 3 m (10 ft) and within 900 mm (3 ft) of each outlet box, cabinet, junction box, or other channel raceway termination.

(B) Suspension Mount. Strut-type channel raceways shall be permitted to be suspension mounted in air with identified methods at intervals not to exceed 3 m (10 ft) and within 900 mm (3 ft) of channel raceway terminations and ends.

384.56 Splices and Taps. Splices and taps shall be permitted in raceways that are accessible after installation by having a removable cover. The conductors, including splices and taps, shall not fill the raceway to more than 75 percent of its area at that point. All splices and taps shall be made by approved methods.

384.60 Grounding. Strut-type channel raceway enclosures providing a transition to or from other wiring methods shall have a means for connecting an equipment grounding conductor. Strut-type channel raceways shall be permitted as an equipment grounding conductor in accordance with 250.118(A)(13). Where a snap-fit metal cover for strut-type channel raceways is used to achieve electrical continuity in accordance with the listing, this cover shall not be permitted as the means for providing electrical continuity for a receptacle mounted in the cover.

Part III. Construction Specifications

384.100 Construction. Strut-type channel raceways and their accessories shall be of a construction that distinguishes them from other raceways. Raceways and their elbows, couplings, and other fittings shall be designed such that the sections can be electrically and mechanically coupled together and installed without subjecting the wires to abrasion. They shall comply with 384.100(A), (B), and (C).

(A) Material. Raceways and accessories shall be formed of steel, stainless steel, or aluminum.

(B) Corrosion Protection. Steel raceways and accessories shall be protected against corrosion by galvanizing or by an organic coating.

> Informational Note: Enamel and PVC coatings are examples of organic coatings that provide corrosion protection.

(C) Cover. Covers of strut-type channel raceways shall be either metal or nonmetallic.

384.120 Marking. Each length of strut-type channel raceway shall be clearly and durably identified as required in the first sentence of 110.21(A).

ARTICLE 386   Surface Metal Raceways

Part I. General

386.1 Scope. This article covers the use, installation, and construction specifications for surface metal raceways and associated fittings.

The installation shown in Exhibit 386.1 is an example of how a surface metal raceway can be used to extend branch-circuit wiring to new switch-controlled outlets.

386.6 Listing Requirements. Surface metal raceway and associated fittings shall be listed.

Part II. Installation

386.10 Uses Permitted. The use of surface metal raceways shall be permitted in the following:

(1) In dry locations.
(2) In Class I, Division 2 hazardous (classified) locations as permitted in 501.10(B)(3).

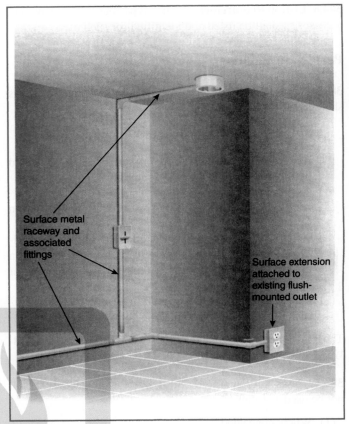

EXHIBIT 386.1 An example of a surface metal raceway extending from an existing receptacle outlet.

(3) Under raised floors, as permitted in 645.5(E)(2).
(4) Extension through walls and floors. Surface metal raceway shall be permitted to pass transversely through dry walls, dry partitions, and dry floors if the length passing through is unbroken. Access to the conductors shall be maintained on both sides of the wall, partition, or floor.

386.12 Uses Not Permitted. Surface metal raceways shall not be used in the following:

(1) Where subject to severe physical damage, unless otherwise approved
(2) Where the voltage is 300 volts or more between conductors, unless the metal has a thickness of not less than 1.02 mm (0.040 in.) nominal
(3) Where subject to corrosive vapors
(4) In hoistways
(5) Where concealed, except as permitted in 386.10

386.21 Size of Conductors. No conductor larger than that for which the raceway is designed shall be installed in surface metal raceway.

386.22 Number of Conductors or Cables. The number of conductors or cables installed in surface metal raceway shall not be greater than the number for which the raceway is designed. Cables shall be permitted to be installed where such use is not prohibited by the respective cable articles.

The adjustment factors of 310.15(C)(1) shall not apply to conductors installed in surface metal raceways where all of the following conditions are met:

(1) The cross-sectional area of the raceway exceeds 2500 mm^2 (4 in.^{2}).
(2) The current-carrying conductors do not exceed 30 in number.
(3) The sum of the cross-sectional areas of all contained conductors does not exceed 20 percent of the interior cross-sectional area of the surface metal raceway.

The number, type, and sizes of conductors permitted to be installed in a listed surface metal raceway are marked on the raceway or on the package in which it is shipped.

386.30 Securing and Supporting. Surface metal raceways and associated fittings shall be supported in accordance with the manufacturer's installation instructions.

386.56 Splices and Taps. Splices and taps shall be permitted in surface metal raceways having a removable cover that is accessible after installation. The conductors, including splices and taps, shall not fill the raceway to more than 75 percent of its area at that point. Splices and taps in surface metal raceways without removable covers shall be made only in boxes. All splices and taps shall be made by approved methods.

Taps of Type FC cable installed in surface metal raceway shall be made in accordance with 322.56(B).

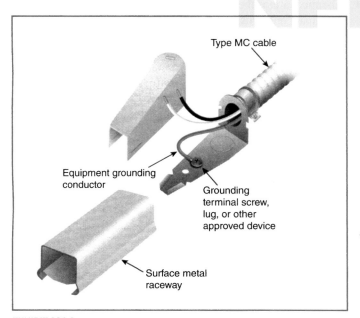

EXHIBIT 386.2 One means for terminating an EGC at a surface metal raceway.

386.60 Grounding. Surface metal raceway enclosures providing a transition from other wiring methods shall have a means for connecting an equipment grounding conductor.

As the example in Exhibit 386.2 shows, where a surface metal raceway is supplied by Type MC or NM cable, a means (e.g., grounding terminal screw or lug) for terminating the equipment grounding conductor (EGC) must be available at the surface metal raceway.

386.70 Combination Raceways. When combination surface metal raceways are used for both signaling and for lighting and power circuits, the different systems shall be run in separate compartments identified by stamping, imprinting, or color coding of the interior finish.

Part III. Construction Specifications

386.100 Construction. Surface metal raceways shall be of such construction as will distinguish them from other raceways. Surface metal raceways and their elbows, couplings, and similar fittings shall be designed so that the sections can be electrically and mechanically coupled together and installed without subjecting the wires to abrasion.

Where covers and accessories of nonmetallic materials are used on surface metal raceways, they shall be identified for such use.

386.120 Marking. Each length of surface metal raceway shall be clearly and durably identified as required in the first sentence of 110.21(A).

ARTICLE 388 Surface Nonmetallic Raceways

Part I. General

388.1 Scope. This article covers the use, installation, and construction specifications for surface nonmetallic raceways and associated fittings.

Surface nonmetallic raceways may resemble base or chair rail molding and allow for circuit conductors to be installed without the need for wall penetration. The installation shown in Exhibit 388.1 is typical of how a surface nonmetallic raceway can be used to supply power, community antenna television (CATV), and communications outlets.

388.6 Listing Requirements. Surface nonmetallic raceway and associated fittings shall be listed.

Part II. Installation

388.10 Uses Permitted. Surface nonmetallic raceways shall be permitted as follows:

EXHIBIT 388.1 An example of a surface nonmetallic raceway, resembling base molding, that supplies a receptacle outlet on the top and CATV and communications on the bottom. *(Courtesy of Legrand®)*

(1) The use of surface nonmetallic raceways shall be permitted in dry locations.
(2) Extension through walls and floors shall be permitted. Surface nonmetallic raceway shall be permitted to pass transversely through dry walls, dry partitions, and dry floors if the length passing through is unbroken. Access to the conductors shall be maintained on both sides of the wall, partition, or floor.

388.12 Uses Not Permitted. Surface nonmetallic raceways shall not be used in the following:

(1) Where concealed, except as permitted in 388.10(2)
(2) Where subject to severe physical damage
(3) Where the voltage is 300 volts or more between conductors, unless listed for higher voltage
(4) In hoistways
(5) In any hazardous (classified) location, except as permitted by other articles in this *Code*
(6) Where subject to ambient temperatures exceeding those for which the nonmetallic raceway is listed
(7) For conductors whose insulation temperature limitations would exceed those for which the nonmetallic raceway is listed

388.21 Size of Conductors. No conductor larger than that for which the raceway is designed shall be installed in surface nonmetallic raceway.

388.22 Number of Conductors or Cables. The number of conductors or cables installed in surface nonmetallic raceway shall not be greater than the number for which the raceway is designed. Cables shall be permitted to be installed where such use is not prohibited by the respective cable articles.

388.30 Securing and Supporting. Surface nonmetallic raceways and associated fittings shall be supported in accordance with the manufacturer's installation instructions.

388.56 Splices and Taps. Splices and taps shall be permitted in surface nonmetallic raceways having a cover capable of being opened in place that is accessible after installation. The conductors, including splices and taps, shall not fill the raceway to more than 75 percent of its area at that point. Splices and taps in surface nonmetallic raceways without covers capable of being opened in place shall be made only in boxes. All splices and taps shall be made by approved methods.

Δ **388.60 Grounding.** Where equipment grounding is required, a separate grounding conductor shall be installed in the raceway.

388.70 Combination Raceways. When combination surface nonmetallic raceways are used both for signaling and for lighting and power circuits, the different systems shall be run in separate compartments identified by stamping, imprinting, or color coding of the interior finish.

Part III. Construction Specifications

388.100 Construction. Surface nonmetallic raceways shall be of such construction as will distinguish them from other raceways. Surface nonmetallic raceways and their elbows, couplings, and similar fittings shall be designed so that the sections can be mechanically coupled together and installed without subjecting the wires to abrasion.

Surface nonmetallic raceways and fittings are made of suitable nonmetallic material that is resistant to moisture and chemical atmospheres. It shall also be flame retardant, resistant to impact and crushing, resistant to distortion from heat under conditions likely to be encountered in service, and resistant to low-temperature effects.

388.120 Marking. Surface nonmetallic raceways that have limited smoke-producing characteristics shall be permitted to be so identified. Each length of surface nonmetallic raceway shall be clearly and durably identified as required in the first sentence of 110.21(A).

ARTICLE
390 Underfloor Raceways

Part I. General

390.1 Scope. This article covers the use and installation requirements for underfloor raceways.

An underfloor raceway is a practical means of bringing light, power, and signal and communications systems to desks, work benches, or tables that are not located adjacent to wall space. This wiring method offers flexibility in layout when used with movable partitions and is commonly used in large retail stores and office buildings to supply power at any desired location.

Underfloor raceways are permitted beneath the surface of concrete, wood, or other flooring material. The wiring method between raceway junction boxes and cabinets or outlet boxes may be any appropriate Chapter 3 wiring method.

Article 390 for the 2020 edition was reorganized into two parts and renumbered to align with parallel numbering of similar articles in Chapter 3.

Part II. Installation

390.10 Uses Permitted. The installation of underfloor raceways shall be permitted beneath the surface of concrete or other flooring material or in office occupancies where laid flush with the concrete floor and covered with linoleum or equivalent floor covering.

390.12 Uses Not Permitted. Underfloor raceways shall not be installed (1) where subject to corrosive vapors or (2) in any hazardous (classified) locations, except as permitted by 504.20 and in Class I, Division 2 locations as permitted in 501.10(B)(3). Unless made of a material approved for the condition or unless corrosion protection approved for the condition is provided, metal underfloor raceways, junction boxes, and fittings shall not be installed in concrete or in areas subject to severe corrosive influences.

390.15 Covering. Raceway coverings shall comply with 390.15(A) through (D).

As Exhibit 390.1 illustrates, flat-top underfloor raceways over 4 inches wide and spaced less than 1 inch apart must be covered with at least 1½ inches of concrete.

Approved trench-type underfloor raceways may be installed flush with the floor surface, provided they have covers that

provide protection at least equal to those of junction box covers. Approved metal flat-top underfloor raceways, if not over 4 inches wide, may be installed flush with a concrete floor, provided they are equipped with covers that afford mechanical protection and rigidity equal to or exceeding that of junction box covers.

(A) Raceways Not over 100 mm (4 in.) Wide. Half-round and flat-top raceways not over 100 mm (4 in.) in width shall have not less than 20 mm (¾ in.) of concrete or wood above the raceway.

Exception: As permitted in 390.15(C) and (D) for flat-top raceways.

(B) Raceways over 100 mm (4 in.) Wide But Not over 200 mm (8 in.) Wide. Flat-top raceways over 100 mm (4 in.) but not over 200 mm (8 in.) wide with a minimum of 25 mm (1 in.) spacing between raceways shall be covered with concrete to a depth of not less than 25 mm (1 in.). Raceways spaced less than 25 mm (1 in.) apart shall be covered with concrete to a depth of 38 mm (1½ in.).

(C) Trench-Type Raceways Flush with Concrete. Trench-type flush raceways with removable covers shall be permitted to be laid flush with the floor surface. Such approved raceways shall be designed so that the cover plates provide adequate mechanical protection and rigidity equivalent to junction box covers.

(D) Other Raceways Flush with Concrete. In office occupancies, approved metal flat-top raceways, if not over 100 mm (4 in.) in width, shall be permitted to be laid flush with the concrete floor surface, provided they are covered with substantial linoleum that is not less than 1.6 mm (1⁄16 in.) thick or with equivalent floor covering. Where more than one and not more than three single raceways are each installed flush with the concrete, they shall be contiguous with each other and joined to form a rigid assembly.

390.20 Size of Conductors. No conductor larger than that for which the raceway is designed shall be installed in underfloor raceways.

390.22 Maximum Number of Conductors in Raceway. The combined cross-sectional area of all conductors or cables shall not exceed 40 percent of the interior cross-sectional area of the raceway.

390.23 Ampacity of Conductors. The ampacity adjustment factors in 310.15(C) shall apply to conductors installed in underfloor raceways.

390.56 Splices and Taps. Splices and taps shall be made only in junction boxes.

For the purposes of this section, so-called loop wiring (continuous, unbroken conductor connecting the individual outlets) shall not be considered to be a splice or tap.

Exception: Splices and taps shall be permitted in trench-type flush raceway having a removable cover that is accessible after

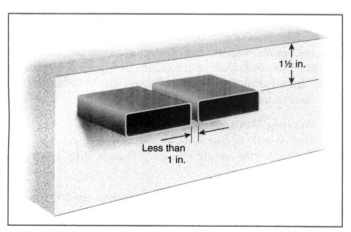

EXHIBIT 390.1 Two side-by-side underfloor raceways over 4 inches wide installed with the required covering.

installation. The conductors, including splices and taps, shall not fill more than 75 percent of the raceway area at that point.

390.57 Discontinued Outlets. When an outlet is abandoned, discontinued, or removed, the sections of circuit conductors supplying the outlet shall be removed from the raceway. No splices or reinsulated conductors, such as would be the case with abandoned outlets on loop wiring, shall be allowed in raceways.

390.70 Laid in Straight Lines. Underfloor raceways shall be laid so that a straight line from the center of one junction box to the center of the next junction box coincides with the centerline of the raceway system. Raceways shall be firmly held in place to prevent disturbing this alignment during construction.

390.71 Markers at Ends. A suitable marker shall be installed at or near each end of each straight run of raceways to locate the last insert.

390.73 Dead Ends. Dead ends of raceways shall be closed.

390.74 Junction Boxes. Junction boxes shall be leveled to the floor grade and sealed to prevent the free entrance of water or concrete. Junction boxes used with metal raceways shall be metal and shall be electrically continuous with the raceways.

390.75 Inserts. Inserts shall be leveled and sealed to prevent the entrance of concrete. Inserts used with metal raceways shall be metal and shall be electrically continuous with the raceway. Inserts set in or on fiber raceways before the floor is laid shall be mechanically secured to the raceway. Inserts set in fiber raceways after the floor is laid shall be screwed into the raceway. When cutting through the raceway wall and setting inserts, chips and other dirt shall not be allowed to remain in the raceway, and tools shall be used that are designed so as to prevent the tool from entering the raceway and damaging conductors that may be in place.

390.76 Connections to Cabinets and Wall Outlets. Connections from underfloor raceways to distribution centers and wall outlets shall be made by approved fittings or by any of the wiring methods in Chapter 3, where installed in accordance with the respective articles.

ARTICLE 392   Cable Trays

Part I. General

392.1 Scope. This article covers cable tray systems, including ladder, ventilated trough, ventilated channel, solid bottom, and other similar structures.

EXHIBIT 392.1 Type MC cable installed in a cable tray. *(Courtesy of Legrand®)*

Informational Note: See ANSI/NEMA–VE 1-2017, *Metal Cable Tray Systems*, and NECA/NEMA 105-2015, *Standard for Installing Metal Cable Tray Systems*, for further information on cable trays.

Cable trays are mechanical support systems and not raceways. See the definition of *raceway* in Article 100. Cable tray installations are typically an industrial-type wiring method. However, they are sometimes installed in commercial facilities as a wire-and-cable management system for telecommunications/data installations and for feeder and branch-circuit wiring. Exhibit 392.1 is one example of Type MC cable installed in a wire mesh-type cable tray.

Part II. Installation

392.10 Uses Permitted. Cable tray shall be permitted to be used as a support system for wiring methods containing service conductors, feeders, branch circuits, communications circuits, control circuits, and signaling circuits. Single insulated cables and single insulated conductors shall be permitted in cable tray only when installed in accordance with 392.10(B)(1). Cable tray installations shall not be limited to industrial establishments. Where exposed to direct rays of the sun, insulated conductors and jacketed cables shall be identified as being sunlight resistant. Cable trays and their associated fittings shall be identified for the intended use.

(A) Wiring Methods. The wiring methods in Table 392.10(A) shall be permitted to be installed in cable tray systems under the conditions described in their respective articles and sections.

Metal cable trays can be used in other spaces used for environmental air (plenums) to support recognized wiring methods permitted in those spaces. Metal cable trays are not the limiting factor; rather, the cable or wiring method is.

Δ **TABLE 392.10(A)** *Wiring Methods*

Wiring Method	Article
Armored cable: Type AC	320
CATV cables	800 and 820
Class 2 and Class 3 cables	722 and 725
Communications cables	800 and 805
Communications raceways	800
Electrical metallic tubing: EMT	358
Electrical nonmetallic tubing: ENT	362
Fire alarm cables	722 and 760
Flexible metal conduit: FMC	348
Flexible metallic tubing: FMT	360
Instrumentation tray cable: Type ITC	341
Intermediate metal conduit: IMC	342
Liquidtight flexible metal conduit: LFMC	350
Liquidtight flexible nonmetallic conduit: LFNC	356
Metal-clad cable: Type MC	330
Mineral-insulated, metal-sheathed cable: Type MI	332
Network-powered broadband communications cables	800 and 830
Nonmetallic-sheathed cable: Types NM, NMC, and NMS	334
Non–power-limited fire alarm cable	722 and 760
Optical fiber cables	722 and 770
Other factory-assembled, multiconductor control, signal, or power cables that are specifically approved for installation in cable trays	
Power and control tray cable: Type TC	336
Power-limited fire alarm cable	722 and 760
Power-limited tray cable	725
Rigid metal conduit: RMC	344
Rigid polyvinyl chloride conduit: PVC	352
Reinforced thermosetting resin conduit: RTRC	355
Service-entrance cable: Types SE and USE	338
Underground feeder and branch-circuit cable: Type UF	340

(B) In Industrial Establishments. The wiring methods in Table 392.10(A) shall be permitted to be used in any industrial establishment under the conditions described in their respective articles. In industrial establishments only, where conditions of maintenance and supervision ensure that only qualified persons service the installed cable tray system, any of the cables in 392.10(B)(1) and (B)(2) shall be permitted to be installed in ladder, ventilated trough, solid bottom, or ventilated channel cable trays.

Δ **(1) Single-Conductor Cables and Single Insulated Conductors.** Single-conductor cables and single insulated conductors shall be permitted to be installed in accordance with 392.10(B)(1)(a) through (B)(1)(c).

(a) Single-conductor cables and single insulated conductors shall be 1/0 AWG or larger and shall be of a type listed and marked on the surface for use in cable trays. Where 1/0 AWG through 4/0 AWG single-conductor cables and single insulated conductors are installed in ladder cable tray, the maximum allowable rung spacing for the ladder cable tray shall be 225 mm (9 in.).

(b) Welding cables shall comply with Article 630, Part IV.

See also

630.42, which requires that trays used to support welding cables be dedicated for welding cable installation

(c) Single conductors used as equipment grounding conductors shall be insulated, covered, or bare, and they shall be 4 AWG or larger.

(2) Single- and Multiconductor Medium Voltage Cables. Single- and multiconductor medium voltage cables shall be Type MV cable. Single conductors shall be installed in accordance with 392.10(B)(1).

(C) Hazardous (Classified) Locations. Cable trays in hazardous (classified) locations shall contain only the cable types and raceways permitted by other articles in this *Code*.

(D) Nonmetallic Cable Tray. In addition to the uses permitted elsewhere in 392.10, nonmetallic cable tray shall be permitted in corrosive areas and in areas requiring voltage isolation.

Fiberglass cable trays are often used to support cables in corrosive environments or in electrolytic cell rooms where voltage isolation is required.

(E) Airfield Lighting Cable Tray. In airports where maintenance and supervision conditions ensure that only qualified persons can access, install, or service the cable, airfield lighting cable used in series circuits that are rated up to 5000 volts and are powered by constant current regulators shall be permitted to be installed in cable trays.

Informational Note: Federal Aviation Administration (FAA) Advisory Circulars (ACs) provide additional practices and methods for airport lighting.

See also FAA AC 150/5345-7F, *Specification for L-824 Underground Electrical Cable for Airport Lighting Circuits*; FAA AC 150/5345-42J, *Specification for Airport Light Bases, Transformer Housings, Junction Boxes, and Accessories*; and FAA AC 150/5345-26E, *FAA Specification for L-823 Plug and Receptacle, Cable Connectors*, which are available for free download from www.faa.gov.

392.12 Uses Not Permitted. Cable tray systems shall not be used in hoistways or where subject to severe physical damage.

392.18 Cable Tray Installation.

(A) Complete System. Cable trays shall be installed as a complete system. Field bends or modifications shall be so made that the electrical continuity of the cable tray system and support for the cables is maintained. Cable tray systems shall be permitted to have mechanically discontinuous segments between cable tray runs or between cable tray runs and equipment.

Runs of cable tray are not required to be mechanically continuous from the equipment source to the equipment termination.

Breaks in the mechanical continuity of cable tray systems are permitted and often occur at tees, crossovers, elevation changes, or firestops, or for thermal contraction and expansion. Also, cable tray systems are not required to be mechanically connected to the equipment they serve.

(B) Completed Before Installation. Each run of cable tray shall be completed before the installation of cables.

(C) Covers. In portions of runs where additional protection is required, covers or enclosures providing the required protection shall be of a material that is compatible with the cable tray.

Δ **(D) Through Partitions and Walls.** Cable trays shall be permitted to extend transversely through partitions and walls or vertically through platforms and floors in wet or dry locations where the installations, complete with installed cables, are made in accordance with 300.21.

(E) Exposed and Accessible. Cable trays shall be exposed and accessible, except as permitted by 392.18(D).

(F) Adequate Access. Sufficient space shall be provided and maintained about cable trays to permit adequate access for installing and maintaining the cables.

Δ **(G) Raceways, Cables, Boxes, and Conduit Bodies Supported from Cable Tray Systems.** In industrial facilities where conditions of maintenance and supervision ensure that only qualified persons service the installation and where the cable tray systems are designed and installed to support the load, such systems shall be permitted to support raceways and cables, and boxes and conduit bodies covered in 314.1. For raceways terminating at the tray, a listed cable tray clamp or adapter shall be used to securely fasten the raceway to the cable tray system. Additional supporting and securing of the raceway shall be in accordance with the appropriate raceway article. For raceways or cables running parallel to and attached to the bottom or side of a cable tray system, fastening and supporting shall be in accordance with the appropriate raceway or cable article.

For boxes and conduit bodies attached to the bottom or side of a cable tray system, fastening and supporting shall be in accordance with 314.23.

Conduit and cable termination supports as well as outlet boxes are permitted to be supported solely by the cable tray in qualifying industrial facilities only. These items are not permitted to be supported solely by the cable tray in commercial installations.

For commercial installations (and nonqualifying industrial facilities), conduit must be supported within 3 feet of the cable tray, or within 5 feet if structural members do not permit fastening within 3 feet of the cable tray. Cables connecting to equipment outside the cable tray system must be supported according to their respective *NEC®* articles.

(H) Marking. Cable trays containing conductors operating over 600 volts shall have a permanent, legible warning notice carrying the wording "DANGER — HIGH VOLTAGE — KEEP AWAY" placed in a readily visible position on all cable trays, with the spacing of warning notices not to exceed 3 m (10 ft). The danger marking(s) or labels shall comply with 110.21(B).

Exception: Where not accessible (as applied to equipment), in industrial establishments where the conditions of maintenance and supervision ensure that only qualified persons service the installation, cable tray system warning notices shall be located where necessary for the installation to ensure safe maintenance and operation.

392.20 Cable and Conductor Installation.

(A) Multiconductor Cables Operating at 1000 Volts or Less. Multiconductor cables operating at 1000 volts or less shall be permitted to be installed in the same tray.

(B) Cables Operating at Over 1000 Volts. Cables operating at over 1000 volts and those operating at 1000 volts or less installed in the same cable tray shall comply with either of the following:

(1) The cables operating at over 1000 volts are Type MC.
(2) The cables operating at over 1000 volts are separated from the cables operating at 1000 volts or less by a solid fixed barrier of a material compatible with the cable tray.

(C) Connected in Parallel. Where single conductor cables comprising each phase, neutral, or grounded conductor of an alternating-current circuit are connected in parallel as permitted in 310.10(G), the conductors shall be installed in groups consisting of not more than one conductor per phase, neutral, or grounded conductor to prevent current imbalance in the paralleled conductors due to inductive reactance.

Single conductors shall be securely bound in circuit groups to prevent excessive movement due to fault-current magnetic forces unless single conductors are cabled together, such as triplexed assemblies.

(D) Single Conductors. Where any of the single conductors installed in ladder or ventilated trough cable trays are 1/0 through 4/0 AWG, all single conductors shall be installed in a single layer. Conductors that are bound together to comprise each circuit group shall be permitted to be installed in other than a single layer.

392.22 Number of Conductors or Cables.

(A) Number of Multiconductor Cables, Rated 2000 Volts or Less, in Cable Trays. The number of multiconductor cables, rated 2000 volts or less, permitted in a single cable tray shall not exceed the requirements of this section. The conductor sizes shall apply to both aluminum and copper conductors. Where dividers are used, fill calculations shall apply to each divided section of the cable tray.

Δ **(1) Ladder or Ventilated Trough Cable Trays Containing Any Mixture of Cables.** Where ladder or ventilated trough cable trays contain multiconductor power or lighting cables, or any mixture of multiconductor power, lighting, control, and signal cables, the maximum number of cables shall conform to 392.22(A)(1)(a) through (A)(1)(c).

(a) Where all of the cables are 4/0 AWG or larger, the sum of the diameters of all cables shall not exceed the cable tray width, and the cables shall be installed in a single layer. Where the cable ampacity is determined according to 392.80(A)(1)(c), the cable tray width shall not be less than the sum of the diameters of the cables and the sum of the required spacing widths between the cables.

(b) Where all of the cables are smaller than 4/0 AWG, the sum of the cross-sectional areas of all cables shall not exceed the maximum allowable cable fill area in Column 1 of Table 392.22(A)(1) for the appropriate cable tray width.

(c) Where 4/0 AWG or larger cables are installed in the same cable tray with cables smaller than 4/0 AWG, the sum of the cross-sectional areas of all cables smaller than 4/0 AWG shall not exceed the maximum allowable fill area resulting from the calculation in Column 2 of Table 392.22(A)(1) for the appropriate cable tray width. The 4/0 AWG and larger cables shall be installed in a single layer, and no other cables shall be placed on them.

(2) Ladder or Ventilated Trough Cable Trays Containing Multiconductor Control and/or Signal Cables Only. Where a ladder or ventilated trough cable tray having a usable inside depth of 150 mm (6 in.) or less contains multiconductor control

and/or signal cables only, the sum of the cross-sectional areas of all cables at any cross section shall not exceed 50 percent of the interior cross-sectional area of the cable tray. A depth of 150 mm (6 in.) shall be used to calculate the allowable interior cross-sectional area of any cable tray that has a usable inside depth of more than 150 mm (6 in.).

Δ **(3) Solid Bottom Cable Trays Containing Any Mixture of Cables.** Where solid bottom cable trays contain multiconductor power or lighting cables, or any mixture of multiconductor power, lighting, control, and signal cables, the maximum number of cables shall conform to 392.22(A)(3)(a) through (A)(3)(c).

(a) Where all of the cables are 4/0 AWG or larger, the sum of the diameters of all cables shall not exceed 90 percent of the cable tray width, and the cables shall be installed in a single layer.

(b) Where all of the cables are smaller than 4/0 AWG, the sum of the cross-sectional areas of all cables shall not exceed the maximum allowable cable fill area in Column 3 of Table 392.22(A)(1) for the appropriate cable tray width.

(c) Where 4/0 AWG or larger cables are installed in the same cable tray with cables smaller than 4/0 AWG, the sum of the cross-sectional areas of all cables smaller than 4/0 AWG shall not exceed the maximum allowable fill area resulting from the computation in Column 4 of Table 392.22(A)(1) for the appropriate cable tray width. The 4/0 AWG and larger cables shall be installed in a single layer, and no other cables shall be placed on them.

Δ **TABLE 392.22(A)(1)** *Allowable Cable Fill Area for Multiconductor Cables in Ladder, Ventilated Trough, or Solid Bottom Cable Trays for Cables Rated 2000 Volts or Less*

		Maximum Allowable Fill Area for Multiconductor Cables							
		Ladder or Ventilated Trough or Wire Mesh Cable Trays, 392.22(A)(1)				Solid Bottom Cable Trays, 392.22(A)(3)			
Inside Width of Cable Tray		Column 1 Applicable for 392.22(A)(1)(b) Only		Column 2[a] Applicable for 392.22(A)(1)(c) Only		Column 3 Applicable for 392.22(A)(3)(b) Only		Column 4[a] Applicable for 392.22(A)(3)(c) Only	
mm	in.	mm^2	in.^{2}	mm^2	in.^{2}	mm^2	in.^{2}	mm^2	in.^{2}
50	2.0	1,500	2.5	1,500 − (30 Sd)[b]	2.5 − (1.2 Sd)[b]	1,200	2.0	1,200 − (25 Sd)[b]	2.0 − Sd[b]
100	4.0	3,000	4.5	3,000 − (30 Sd)[b]	4.5 − (1.2 Sd)	2,300	3.5	2,300 − (25 Sd)	3.5 − Sd
150	6.0	4,500	7.0	4,500 − (30 Sd)[b]	7 − (1.2 Sd)	3,500	5.5	3,500 − (25 Sd)[b]	5.5 − Sd
200	8.0	6,000	9.5	6,000 − (30 Sd)[b]	9.5 − (1.2 Sd)	4,500	7.0	4,500 − (25 Sd)	7.0 − Sd
225	9.0	6,800	10.5	6,800 − (30 Sd)	10.5 − (1.2 Sd)	5,100	8.0	5,100 − (25 Sd)	8.0 − Sd
300	12.0	9,000	14.0	9,000 − (30 Sd)	14 − (1.2 Sd)	7,100	11.0	7,100 − (25 Sd)	11.0 − Sd
400	16.0	12,000	18.5	12,000 − (30 Sd)	18.5 − (1.2 Sd)	9,400	14.5	9,400 − (25 Sd)	14.5 − Sd
450	18.0	13,500	21.0	13,500 − (30 Sd)	21 − (1.2 Sd)	10,600	16.5	10,600 − (25 Sd)	16.5 − Sd
500	20.0	15,000	23.5	15,000 − (30 Sd)	23.5 − (1.2 Sd)	11,800	18.5	11,800 − (25 Sd)	18.5 − Sd
600	24.0	18,000	28.0	18,000 − (30 Sd)	28 − (1.2 Sd)	14,200	22.0	14,200 − (25 Sd)	22.0 − Sd
750	30.0	22,500	35.0	22,500 − (30 Sd)	35 − (1.2 Sd)	17,700	27.5	17,700 − (25 Sd)	27.5 − Sd
900	36.0	27,000	42.0	27,000 − (30 Sd)	42 − (1.2 Sd)	21,300	33.0	21,300 − (25 Sd)	33.0 − Sd

[a]The maximum allowable fill areas in Columns 2 and 4 shall be calculated. For example, the maximum allowable fill in mm^2 for a 150-mm wide cable tray in Column 2 shall be 4500 minus (30 multiplied by Sd) [the maximum allowable fill, in square inches, for a 6-in. wide cable tray in Column 2 shall be 7 minus (1.2 multiplied by Sd)].

[b]The term *Sd* in Columns 2 and 4 is equal to the sum of the diameters, in mm, of all cables 107.2 mm (in inches, of all 4/0 AWG) and larger multiconductor cables in the same cable tray with smaller cables.

(4) Solid Bottom Cable Tray Containing Multiconductor Control and/or Signal Cables Only. Where a solid bottom cable tray having a usable inside depth of 150 mm (6 in.) or less contains multiconductor control and/or signal cables only, the sum of the cross sectional areas of all cables at any cross section shall not exceed 40 percent of the interior cross-sectional area of the cable tray. A depth of 150 mm (6 in.) shall be used to calculate the allowable interior cross-sectional area of any cable tray that has a usable inside depth of more than 150 mm (6 in.).

Δ **(5) Ventilated Channel Cable Trays Containing Multiconductor Cables of Any Type.** Where ventilated channel cable trays contain multiconductor cables of any type, 392.22(A)(5)(a) and (A)(5)(b) shall apply.

(a) Where only one multiconductor cable is installed, the cross-sectional area shall not exceed the value specified in Column 1 of Table 392.22(A)(5).

(b) Where more than one multiconductor cable is installed, the sum of the cross-sectional area of all cables shall not exceed the value specified in Column 2 of Table 392.22(A)(5).

TABLE 392.22(A)(5) *Allowable Cable Fill Area for Multiconductor Cables in Ventilated Channel Cable Trays for Cables Rated 2000 Volts or Less*

Inside Width of Cable Tray		Maximum Allowable Fill Area for Multiconductor Cables			
		Column 1 One Cable		Column 2 More Than One Cable	
mm	in.	mm²	in.²	mm²	in.²
75	3	1500	2.3	850	1.3
100	4	2900	4.5	1600	2.5
150	6	4500	7.0	2450	3.8

Δ **(6) Solid Channel Cable Trays Containing Multiconductor Cables of Any Type.** Where solid channel cable trays contain multiconductor cables of any type, 392.22(A)(6)(a) and (A)(6)(b) shall apply.

(a) Where only one multiconductor cable is installed, the cross-sectional area of the cable shall not exceed the value specified in Column 1 of Table 392.22(A)(6).

(b) Where more than one multiconductor cable is installed, the sum of the cross-sectional area of all cable shall not exceed the value specified in Column 2 of Table 392.22(A)(6).

(B) Number of Single-Conductor Cables, Rated 2000 Volts or Less, in Cable Trays. The number of single conductor cables, rated 2000 volts or less, permitted in a single cable tray section shall not exceed the requirements of this section. The single conductors, or conductor assemblies, shall be evenly distributed across the cable tray. The conductor sizes shall apply to both aluminum and copper conductors.

TABLE 392.22(A)(6) *Allowable Cable Fill Area for Multiconductor Cables in Solid Channel Cable Trays for Cables Rated 2000 Volts or Less*

Inside Width of Cable Tray		Column 1 One Cable		Column 2 More Than One Cable	
mm	in.	mm²	in.²	mm²	in.²
50	2	850	1.3	500	0.8
75	3	1300	2.0	700	1.1
100	4	2400	3.7	1400	2.1
150	6	3600	5.5	2100	3.2

Δ **(1) Ladder or Ventilated Trough Cable Trays.** Where ladder or ventilated trough cable trays contain single-conductor cables, the maximum number of single conductors shall conform to 392.22(B)(1)(a) through (B)(1)(d).

(a) Where all of the cables are 1000 kcmil or larger, the sum of the diameters of all single-conductor cables shall not exceed the cable tray width, and the cables shall be installed in a single layer. Conductors that are bound together to comprise each circuit group shall be permitted to be installed in other than a single layer.

(b) Where all of the cables are from 250 kcmil through 900 kcmil, the sum of the cross-sectional areas of all single-conductor cables shall not exceed the maximum allowable cable fill area in Column 1 of Table 392.22(B)(1) for the appropriate cable tray width.

(c) Where 1000 kcmil or larger single-conductor cables are installed in the same cable tray with single-conductor cables smaller than 1000 kcmil, the sum of the cross sectional areas of all cables smaller than 1000 kcmil shall not exceed the maximum allowable fill area resulting from the computation in Column 2 of Table 392.22(B)(1) for the appropriate cable tray width.

(d) Where any of the single conductor cables are 1/0 through 4/0 AWG, the sum of the diameters of all single conductor cables shall not exceed the cable tray width.

(2) Ventilated Channel Cable Trays. Where 50 mm (2 in.), 75 mm (3 in.), 100 mm (4 in.), or 150 mm (6 in.) wide ventilated channel cable trays contain single-conductor cables, the sum of the diameters of all single conductors shall not exceed the inside width of the channel.

(C) Number of Type MV and Type MC Cables (2001 Volts or Over) in Cable Trays. The number of cables rated 2001 volts or over permitted in a single cable tray shall not exceed the requirements of this section.

The sum of the diameters of single-conductor and multiconductor cables shall not exceed the cable tray width, and the cables shall be installed in a single layer. Where single conductor cables are triplexed, quadruplexed, or bound together in circuit groups, the sum of the diameters of the single conductors shall not exceed the cable tray width, and these groups shall be installed in single layer arrangement.

TABLE 392.22(B)(1) *Allowable Cable Fill Area for Single-Conductor Cables in Ladder, Ventilated Trough, or Wire Mesh Cable Trays for Cables Rated 2000 Volts or Less*

Inside Width of Cable Tray		Maximum Allowable Fill Area for Single-Conductor Cables in Ladder, Ventilated Trough, or Wire Mesh Cable Trays			
		Column 1 Applicable for 392.22(B)(1)(b) Only		Column 2[a] Applicable for 392.22(B)(1)(c) Only	
mm	in.	mm²	in.²	mm²	in.²
50	2	1,400	2.0	1,400 − (28 Sd)[b]	2.0 − (1.1 Sd)[b]
100	4	2,800	4.5	2,800 − (28 Sd)	4.5 − (1.1 Sd)
150	6	4,200	6.5	4,200 − (28 Sd)[b]	6.5 − (1.1 Sd)[b]
200	8	5,600	8.5	5,600 − (28 Sd)	8.5 − (1.1 Sd)
225	9	6,100	9.5	6,100 − (28 Sd)	9.5 − (1.1 Sd)
300	12	8,400	13.0	8,400 − (28 Sd)	13.0 − (1.1 Sd)
400	16	11,200	17.5	11,200 − (28 Sd)	17.5 − (1.1 Sd)
450	18	12,600	19.5	12,600 − (28 Sd)	19.5 − (1.1 Sd)
500	20	14,000	21.5	14,000 − (28 Sd)	21.5 − (1.1 Sd)
600	24	16,800	26.0	16,800 − (28 Sd)	26.0 − (1.1 Sd)
750	30	21,000	32.5	21,000 − (28 Sd)	32.5 − (1.1 Sd)
900	36	25,200	39.0	25,200 − (28 Sd)	39.0 − (1.1 Sd)

[a]The maximum allowable fill areas in Column 2 shall be calculated. For example, the maximum allowable fill, in mm², for a 150-mm wide cable tray in Column 2 shall be 4200 minus (28 multiplied by Sd) [the maximum allowable fill, in square inches, for a 6-in. wide cable tray in Column 2 shall be 6.5 minus (1.1 multiplied by Sd)].

[b]The term Sd in Column 2 is equal to the sum of the diameters, in mm, of all cables 507 mm² (in inches, of all 1000 kcmil) and larger single-conductor cables in the same cable tray with small cables.

392.30 Securing and Supporting.

(A) Cable Trays. Cable trays shall be supported at intervals in accordance with the installation instructions.

(B) Cables and Conductors. Cables and conductors shall be secured to and supported by the cable tray system in accordance with the following, as applicable:

(1) In other than horizontal runs, the cables shall be fastened securely to transverse members of the cable tray.

(2) Supports shall be provided to prevent stress on cables where they enter raceways from cable tray systems.

(3) The system shall provide for the support of cables and raceway wiring methods in accordance with their corresponding articles. Where cable trays support individual conductors or multiconductor cables and where the conductors or multiconductor cables pass from one cable tray to another, or from a cable tray to raceway(s) or from a cable tray to equipment where the conductors are terminated, the distance between the cable trays or between the cable tray and the raceway(s) or the equipment shall not exceed 1.8 m (6 ft). The conductors shall be secured to the cable tray(s) at the transition, and they shall be protected, by guarding or by location, from physical damage.

(4) Cable ties shall be listed and identified for the application and for securement and support.

The 6-foot distance limit specified in 392.30(B)(3) applies to mechanically discontinuous cable tray segments for individual conductors including trays containing multiconductor cables. Cables installed within cable tray systems must meet the support requirements of the applicable *NEC* article that covers the cables. This requirement either limits the gap distance in cable tray runs and between the cable tray and the equipment enclosures or requires intermediate cable supports at the appropriate distances in place of the cable tray.

See also

336.10(7) for further information regarding multiconductor Type TC tray cable used with discontinuous cable tray

392.44 Expansion Splice Plates. Expansion splice plates for cable trays shall be provided where necessary to compensate for thermal expansion and contraction.

392.46 Bushed Conduit and Tubing. A box shall not be required where cables or conductors are installed in bushed conduit and tubing used for support or for protection against physical damage or where conductors or cables transition to a raceway wiring method from the cable tray. Conductors shall be permitted to enter equipment in accordance with 392.46(A) or (B).

(A) Through Bushed Conduit or Tubing. Individual conductors or multiconductor cables with entirely nonmetallic sheaths shall be permitted to enter enclosures where they are terminated through nonflexible bushed conduit or tubing installed for their protection provided they are secured at the point of transition from the cable tray and the conduit or tubing is sealed at the outer end using an approved means so as to prevent debris from entering the equipment through the conduit or tubing.

(B) Flanged Connections. Individual conductors or multiconductor cables with entirely nonmetallic sheaths shall be permitted to enter enclosures through openings associated with flanges from cable trays where the cable tray is attached to the flange and the flange is mounted directly to the equipment. The openings shall be made such that the conductors are protected from abrasion and the opening shall be sealed or covered to prevent debris from entering the enclosure through the opening.

> Informational Note: One method of preventing debris from entering the enclosure is to seal the outer end of the raceway or the opening with duct seal.

392.56 Cable Splices and Type MV Cable Joints. Cable splices and Type MV cable joints made and insulated by approved methods shall be permitted to be located within a cable tray, provided they are accessible. Splices and Type MV cable joints shall be permitted to project above the side rails where not subject to physical damage.

392.60 Grounding and Bonding.

(A) Metal Cable Trays. Metal cable trays shall be permitted to be used as equipment grounding conductors where continuous maintenance and supervision ensure that qualified persons service the installed cable tray system and the cable tray complies with this section. Metal cable trays that support electrical conductors shall be grounded as required for conductor enclosures in accordance with 250.96 and Part IV of Article 250. Metal cable trays containing only non-power conductors shall be electrically continuous through approved connections or the use of a bonding jumper.

> Informational Note: Examples of non-power conductors include nonconductive optical fiber cables and Class 2 and Class 3 remote-control, signaling, and power-limited circuits.

Section 392.60(A), together with 250.96, requires all cable tray systems that support electrical conductors (whether mechanically continuous or with isolated segments) to be electrically continuous and effectively bonded and grounded. This requirement applies whether the cable tray is used as an equipment grounding conductor (EGC) or is used for service conductors and connected to the grounded system conductor (or the grounding electrode conductor for ungrounded systems). Where a metal cable tray contains only non-power conductors, such as fire alarm, communications, community antenna television (CATV), or broadband conductors, the tray must be maintained electrically continuous.

(B) Steel or Aluminum Cable Tray Systems. Steel or aluminum cable tray systems shall be permitted to be used as equipment grounding conductors, provided all the following requirements are met:

(1) The cable tray sections and fittings are identified as an equipment grounding conductor.
(2) The minimum cross-sectional area of cable trays conform to the requirements in Table 392.60(B).

(3) All cable tray sections and fittings are legibly and durably marked to show the cross-sectional area of metal in channel cable trays, or cable trays of one-piece construction, and the total cross-sectional area of both side rails for ladder or trough cable trays.
(4) Cable tray sections, fittings, and connected raceways are bonded in accordance with 250.96, using bolted mechanical connectors or bonding jumpers sized and installed in accordance with 250.102.

Designers of cable tray systems, for use in establishments that qualify, can specify cables without EGCs and use the cable tray system as the required EGC, provided the cable tray system meets the requirements of 392.60(A) and (B). Exhibit 392.2 illustrates an example of the grounding and bonding of multiconductor cables in cable trays with conduit runs to power equipment. Bonding jumpers connecting discontinuous sections of cable tray used as an EGC are on the load side of the overcurrent device. The equipment bonding jumper must be sized in accordance with 250.102(D).

(C) Transitions. Where metal cable tray systems are mechanically discontinuous, as permitted in 392.18(A), a bonding jumper sized in accordance with 250.102 shall connect the two sections of the cable tray, or the cable tray and the raceway or equipment. Bonding shall be in accordance with 250.96.

The bonding of the entire cable tray system is important, especially for discontinuous cable tray segments. According to 250.96(A), properly sized and installed bonding conductors must be installed across any mechanical discontinuities in the

TABLE 392.60(B) *Metal Area Requirements for Cable Trays Used as Equipment Grounding Conductor*

Maximum Fuse Ampere Rating, Circuit Breaker Ampere Trip Setting, or Circuit Breaker Protective Relay Ampere Trip Setting for Ground-Fault Protection of Any Cable Circuit in the Cable Tray System	Minimum Cross-Sectional Area of Metal*			
	Steel Cable Trays		Aluminum Cable Trays	
	mm²	in.²	mm²	in.²
60	129	0.20	129	0.20
100	258	0.40	129	0.20
200	451.5	0.70	129	0.20
400	645	1.00	258	0.40
600	967.5	1.50†	258	0.40
1000	—	—	387	0.60
1200	—	—	645	1.00
1600	—	—	967.5	1.50
2000	—	—	1290	2.00†

*Total cross-sectional area of both side rails for ladder or trough cable trays; or the minimum cross-sectional area of metal in channel cable trays or cable trays of one-piece construction.

†Steel cable trays shall not be used as equipment grounding conductors for circuits with ground-fault protection above 600 amperes. Aluminum cable trays shall not be used as equipment grounding conductors for circuits with ground-fault protection above 2000 amperes.

EXHIBIT 392.2 An example of multiconductor cables in cable trays with conduit runs to power equipment where bonding is provided.

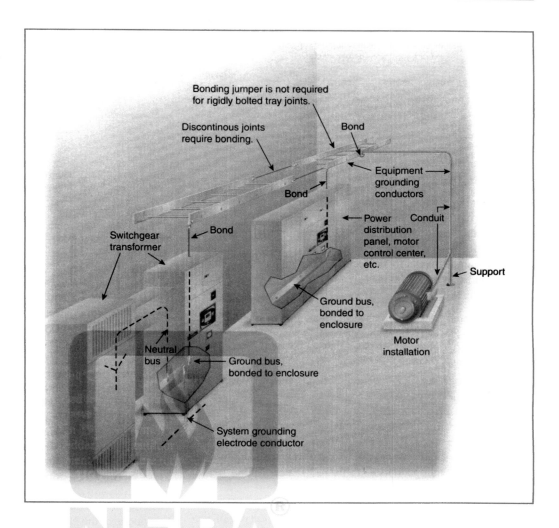

cable tray system and across any space between the cable tray and the conductor termination equipment enclosure or its equipment ground bus.

392.80 Ampacity of Conductors.

Δ **(A) Ampacity of Cables, Rated 2000 Volts or Less, in Cable Trays.**

> Informational Note: See 110.14(C) for conductor temperature limitations due to termination provisions.

(1) Multiconductor Cables. The ampacity of multiconductor cables, nominally rated 2000 volts or less, installed according to the requirements of 392.22(A) shall be as given in Table 310.16 and Table 310.18, subject to 392.80(A)(1)(a), (A)(1)(b), (A)(1)(c), and 310.14(A)(2).

(a) The adjustment factors of 310.15(C)(1) shall apply only to multiconductor cables with more than three current-carrying conductors. Adjustment factors shall be limited to the number of current-carrying conductors in the cable and not to the number of conductors in the cable tray.

(b) Where cable trays are continuously covered for more than 1.8 m (6 ft) with solid unventilated covers, not over 95 percent of the ampacities of Table 310.16 and Table 310.18 shall be permitted for multiconductor cables.

(c) Where multiconductor cables are installed in a single layer in uncovered trays, with a maintained spacing of not less than one cable diameter between cables, the ampacity shall not exceed the ambient temperature-corrected ampacities of multiconductor cables, with not more than three insulated conductors rated 0 through 2000 volts in free air, in accordance with 310.14(B).

> Informational Note: See Informative Annex B, Table B.2(3).

The cables in Exhibit 392.3, rated 2000 volts or less, are installed in a single layer in an uncovered tray, with not less than one cable diameter between cables and not more than three conductors per cable.

See also

Table B.2(3) in Informative Annex B for the ampacity of the conductors in this configuration

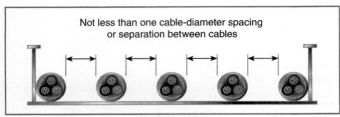

EXHIBIT 392.3 Multiconductor cables, 2000 volts or less, with not more than three conductors per cable [ampacity to be determined from Table B.2(3) in Informative Annex B].

△ **(2) Single-Conductor Cables.** The ampacity of single-conductor cables shall be as permitted by 310.14(A)(2). The adjustment factors of 310.15(C)(1) shall not apply to the ampacity of cables in cable trays. The ampacity of single-conductor cables, or single conductors cabled together (triplexed, quadruplexed, and so forth), nominally rated 2000 volts or less, shall comply with 392.80(A)(2)(a) through (A)(2)(d).

(a) Where installed according to the requirements of 392.22(B), the ampacities for 600 kcmil and larger single-conductor cables in uncovered cable trays shall not exceed 75 percent of the ampacities in Table 310.17 and Table 310.19. Where cable trays are continuously covered for more than 1.8 m (6 ft) with solid unventilated covers, the ampacities for 600 kcmil and larger cables shall not exceed 70 percent of the ampacities in Table 310.17 and Table 310.19.

(b) Where installed according to the requirements of 392.22(B), the ampacities for 1/0 AWG through 500 kcmil single-conductor cables in uncovered cable trays shall not exceed 65 percent of the ampacities in Table 310.17 and Table 310.19. Where cable trays are continuously covered for more than 1.8 m (6 ft) with solid unventilated covers, the ampacities for 1/0 AWG through 500 kcmil cables shall not exceed 60 percent of the ampacities in Table 310.17 and Table 310.19.

(c) Where single conductors are installed in a single layer in uncovered cable trays, with a maintained space of not less than one cable diameter between individual conductors, the ampacity of 1/0 AWG and larger cables shall not exceed the ampacities in Table 310.17 and Table 310.19.

Exception to (c): For solid bottom cable trays, the ampacity of single conductor cables shall be determined by 310.14(B).

(d) Where single conductors are installed in a triangular or square configuration in uncovered cable trays, with a maintained free airspace of not less than 2.15 times one conductor diameter (2.15 × O.D.) of the largest conductor contained within the configuration and adjacent conductor configurations or cables, the ampacity of 1/0 AWG and larger cables shall not exceed the ampacities of two or three single insulated conductors rated 0 through 2000 volts supported on a messenger in accordance with 310.15.

Informational Note: See Table 310.20.

The configuration of the conductors in the cable tray is the basis for determining the cable's ampacity. The installation must be

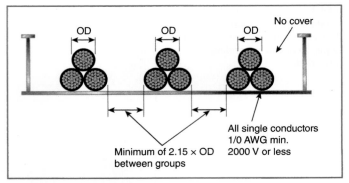

EXHIBIT 392.4 Three single conductors installed in a triangular configuration with spacing between groups of not less than 2.15 times the conductor diameter (ampacities to be determined from Table 310.20).

consistent with the specifications used in the design of the cable tray system. Section 392.80(A)(2)(d) recognizes single conductors in a triangular configuration installed in a cable tray with maintained spacing as having the same ampacity as three single insulated conductors on a messenger. The maintained spacing allows air to circulate around the cable.

Where three single conductors, nominally rated 2000 volts or less, are cabled together in a triangular configuration, with not less than 2.15 times the conductor diameter (2.15 × OD) between groups, as illustrated in Exhibit 392.4, the ampacity of the conductors is determined in accordance with Table 310.20.

Where single conductors are installed in cable trays, their ampacities are permitted to be calculated using the applicable table. Where single-conductor cables emerge from a cable tray installation and are terminated at circuit breakers, distribution switchgear, and similar electrical equipment, the temperature limitations of the electrical equipment terminals should be coordinated with the ampacity of the single-conductor cables. As stated in both the UL *Guide Information for Electrical Equipment Directory* and in 110.14(C)(1), unless the equipment is listed and marked otherwise, conductor ampacities used in determining equipment terminations must be based on Table 310.16 as modified by 310.15(B) and (C).

(3) Combinations of Multiconductor and Single-Conductor Cables. Where a cable tray contains a combination of multiconductor and single-conductor cables, the ampacities shall be as given in 392.80(A)(1) for multiconductor cables and 392.80(A)(2) for single-conductor cables, provided that the following conditions apply:

(1) The sum of the multiconductor cable fill area as a percentage of the allowable fill area for the tray calculated in accordance with 392.22(A), and the single-conductor cable fill area as a percentage of the allowable fill area for the tray calculated in accordance with 392.22(B), totals not more than 100 percent.

(2) Multiconductor cables are installed according to 392.22(A), and single-conductor cables are installed according to 392.22(B) and 392.22(C).

(B) Ampacity of Type MV and Type MC Cables (2001 Volts or Over) in Cable Trays. The ampacity of cables, rated 2001 volts, nominal, or over, installed according to 392.22(C) shall not exceed the requirements of this section.

Informational Note: See 110.40 for conductor temperature limitations due to termination provisions.

(1) Multiconductor Cables (2001 Volts or Over). The ampacity of multiconductor cables shall be as given in Table 315.60(C)(9) and Table 315.60(C)(10), subject to the following:

(1) Where cable trays are continuously covered for more than 1.8 m (6 ft) with solid unventilated covers, not more than 95 percent of the ampacities of Table 315.60(C)(9) and Table 315.60(C)(10) shall be permitted for multiconductor cables.

(2) Where multiconductor cables are installed in a single layer in uncovered cable trays, with maintained spacing of not less than one cable diameter between cables, the ampacity shall not exceed the allowable ampacities of Table 315.60(C)(5) and Table 315.60(C)(6).

(2) Single-Conductor Cables (2001 Volts or Over). The ampacity of single-conductor cables, or single conductors cabled together (triplexed, quadruplexed, and so forth), shall comply with the following:

(1) The ampacities for 1/0 AWG and larger single-conductor cables in uncovered cable trays shall not exceed 75 percent of the ampacities in Table 315.60(C)(3) and Table 315.60(C)(4). Where the cable trays are covered for more than 1.8 m (6 ft) with solid unventilated covers, the ampacities for 1/0 AWG and larger single-conductor cables shall not exceed 70 percent of the ampacities in Table 315.60(C)(3) and Table 315.60(C)(4).

(2) Where single-conductor cables are installed in a single layer in uncovered cable trays, with a maintained space of not less than one cable diameter between individual conductors, the ampacity of 1/0 AWG and larger cables shall not exceed the ampacities in Table 315.60(C)(3) and Table 315.60(C)(4).

(3) Where single conductors are installed in a triangular or square configuration in uncovered cable trays, with a maintained free air space of not less than 2.15 times the diameter (2.15 × O.D.) of the largest conductor contained within the configuration and adjacent conductor configurations or cables, the ampacity of 1/0 AWG and larger cables shall not exceed the ampacities in Table 315.60(C)(1) and Table 315.60(C)(2).

Part III. Construction Specifications

392.100 Construction.

(A) Strength and Rigidity. Cable trays shall have suitable strength and rigidity to provide adequate support for all contained wiring.

(B) Smooth Edges. Cable trays shall not have sharp edges, burrs, or projections that could damage the insulation or jackets of the wiring.

(C) Corrosion Protection. Cable tray systems shall be corrosion resistant. If made of ferrous material, the system shall be protected from corrosion as required by 300.6.

(D) Side Rails. Cable trays shall have side rails or equivalent structural members.

(E) Fittings. Cable trays shall include fittings or other suitable means for changes in direction and elevation of runs.

(F) Nonmetallic Cable Tray. Nonmetallic cable trays shall be made of flame-retardant material.

Δ **ARTICLE 393** — Low-Voltage Suspended Ceiling Power Distribution Systems

Part I. General

393.1 Scope. This article covers the installation of low-voltage suspended ceiling power distribution systems.

Low-voltage suspended ceiling power distribution systems are suspended ceiling assemblies that include a distribution bus within the supporting structure for the ceiling tiles. The bus distributes Class 2 power, which can be used for lighting and other applications. Exhibit 393.1 illustrates a typical system. Exhibit 393.2 illustrates the connection to the distribution grid bus rail.

393.6 Listing Requirements. Suspended ceiling power distribution systems and associated fittings shall be listed as in 393.6(A) or (B).

(A) Listed System. Low-voltage suspended ceiling distribution systems operating at 30 volts ac or less or 60 volts dc or less shall be listed as a complete system, with the utilization equipment, power supply, and fittings as part of the same identified system.

(B) Assembly of Listed Parts. A low-voltage suspended ceiling power distribution system assembled from the following parts, listed according to the appropriate function, shall be permitted:

(1) Listed low-voltage utilization equipment
(2) Listed Class 2 power supply
(3) Listed or identified fittings, including connectors and grid rails with bare conductors
(4) Listed low-voltage cables in accordance with 722.179, conductors in raceways, or other fixed wiring methods for the secondary circuit

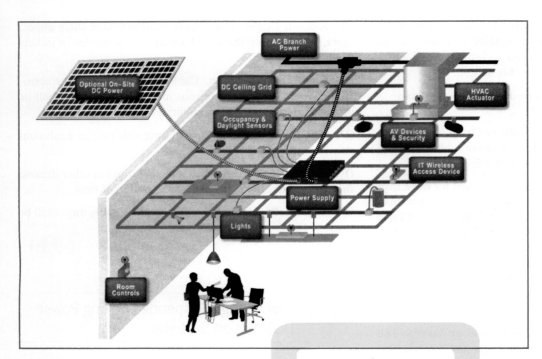

EXHIBIT 393.1 An illustration of a suspended ceiling distribution system. *(Courtesy of EMerge Alliance)*

EXHIBIT 393.2 Connectors used in a suspended ceiling distribution system. *(Courtesy of Armstrong® Ceiling Solutions)*

Part II. Installation

393.10 Uses Permitted. Low-voltage suspended ceiling power distribution systems shall be permanently connected and shall be permitted as follows:

(1) For listed utilization equipment capable of operation at a maximum of 30 volts ac (42.4 volts peak) or 60 volts dc (24.8 volts peak for dc interrupted at a rate of 10 Hz to 200 Hz) and limited to Class 2 power levels in Chapter 9, Table 11(A) and Table 11(B) for lighting, control, and signaling circuits.
(2) In indoor dry locations.
(3) For residential, commercial, and industrial installations.

(4) In other spaces used for environmental air in accordance with 300.22(C), electrical equipment having a metal enclosure, or with a nonmetallic enclosure and fittings, shall be listed for use within an air-handling space and shall have adequate fire-resistant and low-smoke-producing characteristics and associated wiring material suitable for the ambient temperature.

Informational Note: See ANSI/UL 2043-2018, *Fire Test for Heat and Visible Smoke Release for Discrete Products and Their Accessories Installed in Air-Handling Spaces*, for one method of defining adequate fire-resistant and low-smoke-producing characteristics for electrical equipment with a nonmetallic enclosure.

393.12 Uses Not Permitted. Suspended ceiling power distribution systems shall not be installed in the following:

(1) In damp or wet locations
(2) Where subject to corrosive fumes or vapors, such as storage battery rooms
(3) Where subject to physical damage
(4) In concealed locations
(5) In hazardous (classified) locations
(6) As part of a fire-rated floor-ceiling or roof-ceiling assembly, unless specifically listed as part of the assembly
(7) For lighting in general or critical patient care areas

393.14 Installation.

(A) General Requirements. Support wiring shall be installed in a neat and workmanlike manner. Cables and conductors installed exposed on the surface of ceilings and sidewalls shall be supported by the building structure in such a manner that the cable is not damaged by normal building use. Such cables shall be supported by straps, staples, hangers, cable ties listed and identified for securement and support, or similar fittings designed and installed so as not to damage the cable.

> Informational Note: Suspended ceiling low-voltage power grid distribution systems should be installed by qualified persons in accordance with the manufacturer's installation instructions.

(B) Insulated Conductors. Exposed insulated secondary circuit conductors shall be listed, of the type, and installed as described as follows:

(1) Class 2 cable supplied by a listed Class 2 power source and installed in accordance with Part I of Article 722 and Parts I and II of Article 725

(2) Wiring methods described in Chapter 3

393.21 Disconnecting Means.

(A) Location. A disconnecting means for the Class 2 supply to the power grid system shall be located so as to be accessible and within sight of the Class 2 power source for servicing or maintenance of the grid system.

(B) Multiwire Branch Circuits. Where connected to a multiwire branch circuit, the disconnecting means shall simultaneously disconnect all the supply conductors, including the grounded conductors.

393.30 Securing and Supporting.

(A) Attached to Building Structure. A suspended ceiling low-voltage power distribution system shall be secured to the mounting surface of the building structure by hanging wires, screws, or bolts in accordance with the installation and operation instructions. Mounting hardware, such as screws or bolts, shall be either packaged with the suspended ceiling low-voltage lighting power distribution system, or the installation instructions shall specify the types of mounting fasteners to be used.

(B) Attachment of Power Grid Rails. The individual power grid rails shall be mechanically secured to the overall ceiling grid assembly.

393.40 Connectors and Enclosures.

(A) Connectors. Connections to busbar grid rails, cables, and conductors shall be made with listed insulating devices, and these connections shall be accessible after installation. A soldered connection shall be made mechanically secure before being soldered. Other means of securing leads, such as push-on terminals and spade-type connectors, shall provide a secure mechanical connection. The following connectors shall be permitted to be used as connection or interconnection devices:

(1) Load connectors shall be used for power from the busbar to listed utilization equipment.

(2) A pendant connector shall be permitted to suspend low-voltage luminaires or utilization equipment below the grid

rail and to supply power from the busbar to the utilization equipment.

(3) A power feed connector shall be permitted to connect the power supply directly to a power distribution cable and to the busbar.

(4) Rail-to-rail connectors shall be permitted to interconnect busbars from one ceiling grid rail to another grid rail.

> Informational Note: See UL 310, *Standard for Electrical Quick-Connect Terminals*, for quick-connect terminals. See UL 486A-486B, *Standard for Wire Connectors*, for mechanical splicing devices.

(B) Enclosures. Where made in a wall, connections shall be installed in an enclosure in accordance with Parts I, II, and III of Article 314.

393.45 Overcurrent and Reverse Polarity (Backfeed) Protection.

(A) Overcurrent Protection. The listed Class 2 power supply or transformer primary shall be protected at not greater than 20 amperes.

(B) Interconnection of Power Sources. Listed Class 2 sources shall not have the output connections paralleled or otherwise interconnected unless listed for such interconnection.

(C) Reverse Polarity (Backfeed) Protection of Direct-Current Systems. A suspended ceiling low-voltage power distribution system shall be permitted to have reverse polarity (backfeed) protection of dc circuits by one of the following means:

(1) If the power supply is provided as part of the system, the power supply is provided with reverse polarity (backfeed) protection; or

(2) If the power supply is not provided as part of the system, reverse polarity or backfeed protection can be provided as part of the grid rail busbar or as a part of the power feed connector.

393.56 Splices.
A busbar splice shall be provided with insulation and mechanical protection equivalent to that of the grid rail busbars involved.

393.57 Connections.
Connections in busbar grid rails, cables, and conductors shall be made with listed insulating devices and be accessible after installation. Where made in a wall, connections shall be installed in an enclosure in accordance with Parts I, II, and III of Article 314, as applicable.

393.60 Equipment Grounding Conductor.
The supply side of the Class 2 power source shall be connected to an equipment grounding conductor in accordance with the applicable requirements in Part IV of Article 250.

393.61 Grounding of Load Side of Class 2 Power Source.
Class 2 load side circuits for suspended ceiling low-voltage power grid distribution systems shall be permitted to be grounded.

Part III. Construction Specifications

393.100 Sizes and Types of Conductors.

(A) Load Side Utilization Conductor Size. Current-carrying conductors for load side utilization equipment shall be copper and shall be 18 AWG minimum.

Exception: Conductors of a size smaller than 18 AWG, but not smaller than 24 AWG, shall be permitted to be used for Class 2 circuits. Where used, these conductors shall be installed using a Chapter 3 wiring method, shall be totally enclosed, shall not be subject to movement or strain, and shall comply with the ampacity requirements in Table 522.22.

(B) Power Feed Bus Rail Conductor Size. The power feed bus rail shall be 16 AWG minimum or equivalent. For a busbar with a circular cross section, the diameter shall be 1.29 mm (0.051 in.) minimum, and, for other than circular busbars, the area shall be 1.32 mm^2 (0.002 in.^{2}) minimum.

ARTICLE 394 — Concealed Knob-and-Tube Wiring

Part I. General

394.1 Scope.
This article covers the use, installation, and construction specifications of concealed knob-and-tube wiring.

Part II. Installation

394.10 Uses Permitted.
Concealed knob-and-tube wiring shall be permitted to be installed in the hollow spaces of walls and ceilings, or in unfinished attics and roof spaces as provided by 394.23, only as follows:

(1) For extensions of existing installations
(2) Elsewhere by special permission

Knob-and-tube wiring is allowed to be concealed, while open wiring on insulators (Article 398) is required to be exposed. Concealed knob-and-tube wiring is designed for use in hollow spaces of walls, ceilings, and attics and utilizes the free air in such spaces for heat dissipation.

Concealed knob-and-tube wiring is permitted to be installed only for extensions of existing installations or where special permission is granted by the AHJ.

See also

Article 100 for the definition of *special permission*

394.12 Uses Not Permitted.
Concealed knob-and-tube wiring shall not be used in the following:

(1) Commercial garages
(2) Theaters and similar locations

(3) Motion picture studios
(4) Hazardous (classified) locations
(5) Hollow spaces of walls, ceilings, and attics where such spaces are insulated by loose, rolled, or foamed-in-place insulating material that envelops the conductors

Blown-in, foamed-in, or rolled insulation prevents the dissipation of heat into the free air space, resulting in a higher conductor temperature, which could cause insulation breakdown and possible insulation ignition. This section prohibits installation of knob-and-tube wiring in hollow spaces that have been weatherized.

394.17 Through or Parallel to Framing Members.
Conductors shall comply with 398.17 where passing through holes in structural members. Where passing through wood cross members in plastered partitions, conductors shall be protected by noncombustible, nonabsorbent, insulating tubes extending not less than 75 mm (3 in.) beyond the wood member.

394.19 Clearances.

(A) General. A clearance of not less than 75 mm (3 in.) shall be maintained between conductors and a clearance of not less than 25 mm (1 in.) between the conductor and the surface over which it passes.

(B) Limited Conductor Space. Where space is too limited to provide these minimum clearances, such as at meters, panelboards, outlets, and switch points, the individual conductors shall be enclosed in flexible nonmetallic tubing, which shall be continuous in length between the last support and the enclosure or terminal point.

(C) Clearance from Piping, Exposed Conductors, and So Forth. Conductors shall comply with 398.19 for clearances from other exposed conductors, piping, and so forth.

394.23 In Accessible Attics.
Conductors in unfinished attics and roof spaces shall comply with 394.23(A) or (B).

Informational Note: See 310.14(A)(3) for temperature limitation of conductors.

(A) Accessible by Stairway or Permanent Ladder. Conductors shall be installed along the side of or through bored holes in floor joists, studs, or rafters. Where run through bored holes, conductors in the joists and in studs or rafters to a height of not less than 2.1 m (7 ft) above the floor or floor joists shall be protected by substantial running boards extending not less than 25 mm (1 in.) on each side of the conductors. Running boards shall be securely fastened in place. Running boards and guard strips shall not be required where conductors are installed along the sides of joists, studs, or rafters.

(B) Not Accessible by Stairway or Permanent Ladder. Conductors shall be installed along the sides of or through bored holes in floor joists, studs, or rafters.

Exception: In buildings completed before the wiring is installed, attic and roof spaces that are not accessible by stairway or permanent ladder and have headroom at all points less than 900 mm (3 ft), the wiring shall be permitted to be installed on the edges of rafters or joists facing the attic or roof space.

394.30 Securing and Supporting.

(A) Supporting. Conductors shall be rigidly supported on noncombustible, nonabsorbent insulating materials and shall not contact any other objects. Supports shall be installed as follows:

(1) Within 150 mm (6 in.) of each side of each tap or splice, and
(2) At intervals not exceeding 1.4 m (4½ ft).

Where it is impracticable to provide supports, conductors shall be permitted to be fished through hollow spaces in dry locations, provided each conductor is individually enclosed in flexible nonmetallic tubing that is in continuous lengths between supports, between boxes, or between a support and a box.

(B) Securing. Where solid knobs are used, conductors shall be securely tied thereto by tie wires having insulation equivalent to that of the conductor.

394.42 Devices. Switches shall comply with 404.4 and 404.10(B).

394.56 Splices and Taps. Splices shall be soldered unless approved splicing devices are used. In-line or strain splices shall not be used.

Part III. Construction Specifications

Δ **394.104 Conductors.** Conductors shall be of a type identified in Table 310.4(1).

ARTICLE 395 Outdoor Overhead Conductors over 1000 Volts

395.1 Scope. This article covers the use and installation for outdoor overhead conductors over 1000 volts, nominal.

This article is performance based. It provides the objectives for compliance, rather than prescriptive requirements. These objectives can be met through existing consensus standards for installation of overhead conductors, such as ANSI/IEEE C2, *National Electrical Safety Code (NESC®)*.

395.10 Uses Permitted. Outdoor overhead conductors over 1000 volts, nominal, shall be permitted only for systems rated over 1000 volts, nominal, as follows:

(1) Outdoors in free air
(2) For service conductors, feeders, or branch circuits

Informational Note: See IEEE C2, *National Electrical Safety Code*, and ANSI/IEEE 3001.2, *Recommended Practice for Evaluating the Electrical Service Requirements of Industrial and Commercial Power Systems*, for additional information on outdoor overhead conductors over 1000 volts.

395.30 Support.

(A) Conductors. Documentation of the engineered design by a licensed professional engineer engaged primarily in the design of such systems for the spacing between conductors shall be available upon request of the authority having jurisdiction and shall include consideration of the following:

(1) Applied voltage
(2) Conductor size
(3) Distance between support structures
(4) Type of structure
(5) Wind/ice loading
(6) Surge protection

(B) Structures. Structures of wood, metal, or concrete, or combinations of those materials, shall be provided for support of overhead conductors over 1000 volts, nominal. Documentation of the engineered design by a licensed professional engineer engaged primarily in the design of such systems and the installation of each support structure shall be available upon request of the authority having jurisdiction and shall include consideration of the following:

(1) Soil conditions
(2) Foundations and structure settings
(3) Weight of all supported conductors and equipment
(4) Weather loading and other conditions such as, but not limited to, ice, wind, temperature, and lightning
(5) Angle where change of direction occurs
(6) Spans between adjacent structures
(7) Effect of dead-end structures
(8) Strength of guy wires and guy anchors
(9) Structure size and material(s)
(10) Hardware

(C) Insulators. Insulators used to support conductors shall be rated for all of the following:

(1) Applied phase-to-phase voltage
(2) Mechanical strength required for each individual installation
(3) Impulse withstand BIL in accordance with Table 490.24(a)

Informational Note: See 395.30(A), (B), and (C), which are not all-inclusive lists.

ARTICLE 396 Messenger-Supported Wiring

Part I. General

396.1 Scope. This article covers the use, installation, and construction specifications for messenger-supported wiring.

For many years, messenger-supported wiring systems have been used in industrial installations as well as to supply services for commercial and residential installations.

See also

225.6(A) and **(B)** for references to messenger-supported wiring

Part II. Installation

396.10 Uses Permitted.

(A) Cable Types. The cable types in Table 396.10(A) shall be permitted to be installed in messenger-supported wiring under the conditions described in the article or section referenced for each.

△ **TABLE 396.10(A)** *Cable Types*

Cable Type	Section	Article
Medium-voltage cable		315
Metal-clad cable		330
Mineral-insulated, metal-sheathed cable		332
Multiconductor service-entrance cable		338
Multiconductor underground feeder and branch-circuit cable		340
Other factory-assembled, multiconductor control, signal, or power cables that are identified for the use		
Power and control tray cable		336
Power-limited tray cable	Table 722.135(B), 722.135(C), and 722.179(A)(6)	

(B) In Industrial Establishments. In industrial establishments only, where conditions of maintenance and supervision ensure that only qualified persons service the installed messenger-supported wiring, the following shall be permitted:

(1) Any of the conductor types shown in Table 310.4(1) or Table 310.4(2)
(2) MV cable

Where exposed to weather, conductors shall be listed for use in wet locations. Where exposed to direct rays of the sun, conductors or cables shall be sunlight resistant.

Some of the triplex and quadruplex cables used by utilities as service-drop cable do not use conductors recognized in Table 310.4(1) and do not meet the requirements of Article 310.

See also

310.15(B) and **Table 310.20** or two or three single-insulated conductors supported on a messenger wire

310.15(C) and **Table B.2(3)** in Informative Annex B for ampacities of conductors for other cable types

(C) Hazardous (Classified) Locations. Messenger-supported wiring shall be permitted to be used in hazardous (classified) locations where the contained cables and messenger-supported wiring are specifically permitted by other articles in this *Code*.

396.12 Uses Not Permitted. Messenger-supported wiring shall not be used in hoistways or where subject to physical damage.

396.30 Messenger.

(A) Support. The messenger shall be supported at dead ends and at intermediate locations so as to eliminate tension on the conductors. The conductors shall not be permitted to come into contact with the messenger supports or any structural members, walls, or pipes.

(B) Neutral Conductor. Where the messenger is used as a neutral conductor, it shall comply with the requirements of 225.4, 250.184(A), 250.184(B)(7), and 250.187(B).

(C) Equipment Grounding Conductor. Where the messenger is used as an equipment grounding conductor, it shall comply with the requirements of 250.32(B), 250.118, 250.184(B)(8), and 250.187(D).

396.56 Conductor Splices and Taps. Conductor splices and taps made and insulated by approved methods shall be permitted in messenger-supported wiring.

396.60 Grounding. The messenger shall be grounded as required by 250.80 and 250.86 for enclosure grounding.

ARTICLE
398 Open Wiring on Insulators

Part I. General

398.1 Scope. This article covers the use, installation, and construction specifications of open wiring on insulators.

Open wiring on insulators is an exposed wiring method that is not permitted to be concealed by the building structure or finish. It is permitted indoors or outdoors, in dry or wet locations, and where subject to corrosive vapors, provided the insulation choice from Table 310.4(1) is suitable for use in a corrosive environment.

This wiring method is not permitted for temporary lighting and power circuits on construction sites but is permitted for lighting and power circuits in agricultural buildings (see 547.20). It may also be used for services (see 230.43).

See also

Tables 310.17 and **310.19** for ampacities of conductors

Part II. Installation

398.10 Uses Permitted. Open wiring on insulators shall be permitted only for industrial or agricultural establishments on systems of 1000 volts, nominal, or less, as follows:

(1) Indoors or outdoors
(2) In wet or dry locations
(3) Where subject to corrosive vapors
(4) For services

398.12 Uses Not Permitted. Open wiring on insulators shall not be installed where concealed by the building structure.

398.15 Exposed Work.

(A) Dry Locations. In dry locations, where not exposed to physical damage, conductors shall be permitted to be separately enclosed in flexible nonmetallic tubing. The tubing shall be in continuous lengths not exceeding 4.5 m (15 ft) and secured to the surface by straps at intervals not exceeding 1.4 m (4½ ft).

(B) Entering Spaces Subject to Dampness, Wetness, or Corrosive Vapors. Conductors entering or leaving locations subject to dampness, wetness, or corrosive vapors shall have drip loops formed on them and shall then pass upward and inward from the outside of the buildings, or from the damp, wet, or corrosive location, through noncombustible, nonabsorbent insulating tubes.

> Informational Note: See 230.52 for individual conductors entering buildings or other structures.

(C) Exposed to Physical Damage. Conductors within 2.1 m (7 ft) from the floor shall be considered exposed to physical damage. Where open conductors cross ceiling joists and wall studs and are exposed to physical damage, they shall be protected by one of the following methods:

(1) Guard strips shall not be less than 25 mm (1 in.) nominal in thickness and at least as high as the insulating supports, placed on each side of and close to the wiring.
(2) A substantial running board at least 13 mm (½ in.) thick in back of the conductors with side protections. Running boards shall extend at least 25 mm (1 in.) outside the conductors, but not more than 50 mm (2 in.), and the protecting sides shall be at least 50 mm (2 in.) high and at least 25 mm (1 in.), nominal, in thickness.
(3) Boxing made in accordance with 398.15(C)(1) or (C)(2) and furnished with a cover kept at least 25 mm (1 in.) away from the conductors within. Where protecting vertical conductors on side walls, the boxing shall be closed at the top and the holes through which the conductors pass shall be bushed.
(4) Rigid metal conduit (RMC), intermediate metal conduit (IMC), rigid polyvinyl chloride conduit (PVC), reinforced thermosetting resin conduit (RTRC), or electrical metallic tubing (EMT). When installed in metal piping, the conductors shall be encased in continuous lengths of approved flexible tubing.

398.17 Through or Parallel to Framing Members. Open conductors shall be separated from contact with walls, floors, wood cross members, or partitions through which they pass by tubes or bushings of noncombustible, nonabsorbent insulating material. Where the bushing is shorter than the hole, a waterproof sleeve of noninductive material shall be inserted in the hole and an insulating bushing slipped into the sleeve at each end in such a manner as to keep the conductors absolutely out of contact with the sleeve. Each conductor shall be carried through a separate tube or sleeve.

> Informational Note: See 310.14(A)(3) for temperature limitation of conductors.

398.19 Clearances. Open conductors shall be separated at least 50 mm (2 in.) from metal raceways, piping, or other conducting material, and from any exposed lighting, power, or signaling conductor, or shall be separated therefrom by a continuous and firmly fixed nonconductor in addition to the insulation of the conductor. Where any insulating tube is used, it shall be secured at the ends. Where practicable, conductors shall pass over rather than under any piping subject to leakage or accumulations of moisture.

398.23 In Accessible Attics. Conductors in unfinished attics and roof spaces shall comply with 398.23(A) or (B).

(A) Accessible by Stairway or Permanent Ladder. Conductors shall be installed along the side of or through bored holes in floor joists, studs, or rafters. Where run through bored holes, conductors in the joists and in studs or rafters to a height of not less than 2.1 m (7 ft) above the floor or floor joists shall be protected by substantial running boards extending not less than 25 mm (1 in.) on each side of the conductors. Running boards shall be securely fastened in place. Running boards and guard strips shall not be required for conductors installed along the sides of joists, studs, or rafters.

(B) Not Accessible by Stairway or Permanent Ladder. Conductors shall be installed along the sides of or through bored holes in floor joists, studs, or rafters.

Exception: In buildings completed before the wiring is installed, in attic and roof spaces that are not accessible by stairway or permanent ladder and have headroom at all points less than 900 mm (3 ft), the wiring shall be permitted to be installed on the edges of rafters or joists facing the attic or roof space.

398.30 Securing and Supporting.

(A) Conductor Sizes Smaller Than 8 AWG. Conductors smaller than 8 AWG shall be rigidly supported on noncombustible, nonabsorbent insulating materials and shall not contact any other objects. Supports shall be installed as follows:

(1) Within 150 mm (6 in.) from a tap or splice
(2) Within 300 mm (12 in.) of a dead-end connection to a lampholder or receptacle
(3) At intervals not exceeding 1.4 m (4½ ft) and at closer intervals sufficient to provide adequate support where likely to be disturbed

(B) Conductor Sizes 8 AWG and Larger. Supports for conductors 8 AWG or larger installed across open spaces shall be permitted up to 4.5 m (15 ft) apart if noncombustible, nonabsorbent insulating spacers are used at least every 1.4 m (4½ ft) to maintain at least 65 mm (2½ in.) between conductors.

Where not likely to be disturbed in buildings of mill construction, 8 AWG and larger conductors shall be permitted to be run across open spaces if supported from each wood cross member on approved insulators maintaining 150 mm (6 in.) between conductors.

Mill construction is generally considered to be a building in which the floors and ceilings are supported by wood timbers or beams or wood cross members spaced approximately 15 feet apart. This type of construction is sometimes referred to as plank-on-timber construction. Conductors 8 AWG and larger are permitted to span the 15-foot distance where the ceilings are high and free of obstructions and the conductors are unlikely to come into contact with other objects.

(C) Industrial Establishments. In industrial establishments only, where conditions of maintenance and supervision ensure that only qualified persons service the system, conductors of sizes 250 kcmil and larger shall be permitted to be run across open spaces where supported at intervals up to 9.0 m (30 ft) apart.

The installation of open feeders on insulators mounted on the bottom of roof trusses at every bay location was once common in industrial buildings. Many bays are more than 15 feet wide.

Therefore, size 250 kcmil and larger conductors are permitted to be supported at 30-foot intervals in industrial buildings, where qualified persons must service the system.

In addition to the ease and economy of installation or alteration of open wiring, the close spacing of conductors reduces the reactance of a circuit and, hence, reduces the voltage drop.

(D) Mounting of Conductor Supports. Where nails are used to mount knobs, they shall not be smaller than tenpenny. Where screws are used to mount knobs, or where nails or screws are used to mount cleats, they shall be of a length sufficient to penetrate the wood to a depth equal to at least one-half the height of the knob and the full thickness of the cleat. Cushion washers shall be used with nails.

(E) Tie Wires. Conductors 8 AWG or larger and supported on solid knobs shall be securely tied thereto by tie wires having an insulation equivalent to that of the conductor.

398.42 Devices. Surface-type snap switches shall be mounted in accordance with 404.10(A), and boxes shall not be required. Other type switches shall be installed in accordance with 404.4.

Part III. Construction Specifications

Δ **398.104 Conductors.** Conductors shall be of a type identified in Table 310.4(1).

Equipment for General Use

ARTICLE 400

Flexible Cords and Flexible Cables

Part I. General

400.1 Scope. This article covers general requirements, applications, and construction specifications for flexible cords and flexible cables.

Flexible cords and flexible cables are not treated as a Chapter 3 wiring method. Cords and cables generally are used to connect the equipment covered in Chapter 4 to the outlet either through a cord-and-plug connection to a receptacle or, where permitted, via a direct connection to an outlet box. There are conditions in the *NEC®* in which flexible cords and cables are permitted to be used for branch circuits and feeders, such as in 555.34(A)(2) for floating and fixed piers and in 590.4 for temporary power. In general, however, the use of cords to supply electrically powered equipment is controlled by 400.10, Uses Permitted, and 400.12, Uses Not Permitted.

400.2 Other Articles. Flexible cords and flexible cables shall comply with this article and with the applicable provisions of other articles of this *Code*.

400.3 Suitability. Flexible cords and flexible cables and their associated fittings shall be suitable for the conditions of use and location.

400.4 Types. Flexible cords and flexible cables shall conform to the description in Table 400.4. The use of flexible cords and flexible cables other than those in Table 400.4 shall require permission by the authority having jurisdiction.

Most flexible cords and cables are limited to 300- or 600-volt applications. Types G, G-GC, PPE, and W portable power cables are rated 2000 volts. As new electric vehicle charging technology evolves and charging systems designed to operate above the 600-volt threshold become available, the supply cords need to be constructed and evaluated to support those higher voltage charging systems. UL 62, *Standard for Flexible Cords and Cables*, has been revised to permit Types EV, EVE, and EVT extra-hard usage electric vehicle cables to be used in systems operating at up to 1000 volts.

400.5 Ampacities for Flexible Cords and Flexible Cables.

(A) Ampacity Tables. Table 400.5(A)(1) provides the ampacities for flexible cords and flexible cables, and Table 400.5(A)(2) provides the ampacities for flexible cords and flexible cables with not more than three current-carrying conductors. These tables shall be used in conjunction with applicable end-use product standards to ensure selection of the proper size and type. Where cords and cables are used in ambient temperatures other than 30°C (86°F), the temperature correction factors from Table 310.15(B)(1)(1) that correspond to the temperature rating of the cord or cable shall be applied to the ampacity in Table 400.5(A)(1) and Table 400.5(A)(2). Cords and cables rated 105°C shall use correction factors in the 90°C column of Table 310.15(B)(1)(1) for temperature correction. Where the number of current-carrying conductors exceeds three, the ampacity of each conductor shall be reduced from the three-conductor rating as shown in Table 400.5(A)(3).

> Informational Note: See Informative Annex B, Table B.2(11), for adjustment factors for more than three current-carrying conductors in a raceway or cable with load diversity.

Where flexible power cables are used in an ambient temperature exceeding 30°C (86°F), ampacity correction factors are required to be applied to the ampacities in Table 400.5(A)(2). This requirement parallels the Article 310 requirements for ampacity correction of conductors used in elevated ambient temperatures. In fact, the ambient correction factors that are to be used for power cables are those specified in Table 310.16. The specific correction factor to be applied is based on the temperature rating of the power cable.

△ **TABLE 400.4** *Flexible Cords and Flexible Cables*

Trade Name	Type Letter	Voltage	AWG or kcmil	Number of Conductors	Insulation	AWG or kcmil	Nominal Insulation Thickness		Braid on Each Conductor	Outer Covering	Use		
							mm	mils					
Lamp cord	C	300 / 600	18–16 / 15–10	2 or more	Thermoset or thermoplastic	18–16 / 15–10	0.76 / 1.14	30 / 45	Cotton	None	Pendant or portable	Dry locations	Not hard usage
Elevator cable	E[1,2,3,4]	300 or 600	20–2	2 or more	Thermoset	20–16 / 15–12 / 12–10 / 8–2	0.51 / 0.76 / 1.14 / 1.52	20 / 30 / 45 / 60	Cotton	Three cotton; outer one flame-retardant and moisture-resistant	Elevator lighting and control	Unclassified locations	
						20–16 / 15–12 / 12–10 / 8–2	0.51 / 0.76 / 1.14 / 1.52	20 / 30 / 45 / 60	Flexible nylon jacket				
Elevator cable	EO[1,2,4]	300 or 600	20–2	2 or more	Thermoset	20–16 / 15–12 / 12–10 / 8–2	0.51 / 0.76 / 1.14 / 1.52	20 / 30 / 45 / 60	Cotton	Three cotton; outer one flame-retardant and moisture-resistant	Elevator lighting and control	Unclassified locations	
										One cotton and a neoprene jacket		Hazardous (classified) locations	
Elevator cable	ETP[2,4]	300 or 600							Rayon	Thermoplastic	Hazardous (classified) locations		
	ETT[2,4]	300 or 600							None	One cotton or equivalent and a thermoplastic jacket			
Electric vehicle cable	EV[5,6]	1000	18–500	2 or more plus equipment grounding conductor(s), plus optional hybrid data, signal communications, and optical fiber cables	Thermoset with optional nylon	18–15	0.76 (0.51)	30 (20)	Optional	Oil-resistant thermoset	Electric vehicle charging	Wet locations	Extra-hard usage
						14–10	1.14 (0.76)	45 (30)					
						8–2	1.52 (1.14)	60 (45)					
						1–4/0	2.03 (1.52)	80 (60)					
						250–500	2.41 (1.90)	95 (75)					
	EVJ[5,6]	300	18–12			18–12	0.76 (0.51)	30 (20)					Hard usage
	EVE[5,6]	1000	18–500	2 or more plus equipment grounding conductor(s), plus optional hybrid data, signal communications, and optical fiber cables	Thermoplastic elastomer with optional nylon	18–15	0.76 (0.51)	30 (20)		Oil-resistant thermoplastic elastomer			Extra-hard usage
						14–10	1.14 (0.76)	45 (30)					
						8–2	1.52 (1.14)	60 (45)					
						1–4/0	2.03 (1.52)	80 (60)					
						250–500	2.41 (1.90)	95 (75)					
	EVJE[5,6]	300	18–12			18–12	0.76 (0.51)	30 (20)					Hard usage
	EVT[5,6]	1000	18–500	2 or more plus equipment grounding conductor(s), plus optional hybrid data, signal communications, and optical fiber cables	Thermoplastic with optional nylon	18–15	0.76 (0.51)	30 (20)	Optional	Oil-resistant thermoplastic	Electric vehicle charging	Wet locations	Extra-hard usage
						14–10	1.14 (0.76)	45 (30)					
						8–2	1.52 (1.14)	60 (45)					
						1–4/0	2.03 (1.52)	80 (60)					
						250–500	2.41 (1.90)	95 (75)					
	EVJT[5,6]	300	18–12			18–12	0.76 (0.51)	30 (20)					Hard usage

Δ **TABLE 400.4** *Continued*

Trade Name	Type Letter	Voltage	AWG or kcmil	Number of Conductors	Insulation	AWG or kcmil	Nominal Insulation Thickness (mm)	(mils)	Braid on Each Conductor	Outer Covering	Use		
Portable power cable	G	2000	12–500	2–6 plus equipment grounding conductor(s)	Thermoset	12–2 1–4/0 250–500	1.52 2.03 2.41	60 80 95		Oil-resistant thermoset	Portable and extra-hard usage		
	G-GC[7]	2000	12–500	3–6 plus equipment grounding conductors and 1 ground check conductor	Thermoset	12–2 1–4/0 250–500	1.52 2.03 2.41	60 80 95		Oil-resistant thermoset			
Heater cord	HPD	300	18–12	2, 3, or 4	Thermoset	18–16 15–12	0.38 0.76	15 30	None	Cotton or rayon	Portable heaters	Dry locations	Not hard usage
Parallel heater cord	HPN[8]	300	18–12	2 or 3	Oil-resistant thermoset	18–16 15 14 12	1.14 1.52 2.41	45 60 95	None	Oil-resistant thermoset	Portable	Damp locations	Not hard usage
Thermoset jacketed heater cords	HSJ	300	18–12	2, 3, or 4	Thermoset	18–16 15–12	0.76 1.14	30 45	None	Cotton and thermoset	Portable or portable heater	Damp locations	Hard usage
	HSJW	300	18–12		Thermoset					Cotton and thermoset		Damp locations	
	HSJO	300	18–12							Cotton and oil-resistant thermoset		Damp and wet locations	
	HSJOW[9]	300	18–12									Damp locations	
	HSJOO	300	18–12		Oil-resistant thermoset								
	HSJOOW[9]	300	18–12									Damp and wet locations	
Non-integral parallel cords	NISP-1	300	20–18	2 or 3	Thermoset	20–18	0.38	15	None	Thermoset	Pendant or portable	Damp locations	Not hard usage
	NISP-2	300	18–16			18–16	0.76	30					
	NISPE-1[8]	300	20–18		Thermoplastic elastomer	20–18	0.38	15		Thermoplastic elastomer			
	NISPE-2[8]	300	18–16			18–16	0.76	30					
	NISPT-1[8]	300	20–18		Thermoplastic	20–18	0.38	15		Thermoplastic			
	NISPT-2[8]	300	18–16			18–16	0.76	30					
Twisted portable cord	PD	300 600	18–16 14–10	2 or more	Thermoset or thermoplastic	18–16 15–10	0.76 1.14	30 45	Cotton	Cotton or rayon	Pendant or portable	Dry locations	Not hard usage
Portable power cable	PPE[7]	2000	12–500	1–6 plus optional equipment grounding conductor(s)	Thermoplastic elastomer	12–2 1–4/0 250–500	1.52 2.03 2.41	60 80 95		Oil-resistant thermoplastic elastomer	Portable, extra-hard usage		
Hard service cord	S[7]	600	18–2	2 or more	Thermoset	18–15 14–10 8–2	0.76 1.14 1.52	30 45 60	None	Thermoset	Pendant or portable	Damp locations	Extra-hard usage
Flexible stage and lighting power cable	SC[7,10]	600	8–250	1 or more	Thermoset	8–2 1–4/0 250	1.52 2.03 2.41	60 80 95		Thermoset	Portable, extra-hard usage		
	SCE[7,10]	600			Thermoplastic elastomer					Thermoplastic elastomer			
	SCT[7,10]	600			Thermoplastic					Thermoplastic			

(continues)

Δ **TABLE 400.4** *Continued*

Trade Name	Type Letter	Voltage	AWG or kcmil	Number of Conductors	Insulation	AWG or kcmil	Nominal Insulation Thickness		Braid on Each Conductor	Outer Covering	Use		
							mm	mils					
Hard service cord	SE[7]	600	18–2	2 or more	Thermoplastic elastomer	18–15 14–9 8–2	0.76 1.14 1.52	30 45 60	None	Thermoplastic elastomer	Pendant or portable	Damp locations	Extra-hard usage
	SEW[7,9]	600										Damp and wet locations	
	SEO[7]	600								Oil-resistant thermoplastic elastomer		Damp locations	
	SEOW[7,9]	600										Damp and wet locations	
	SEOO[7]	600			Oil-resistant thermoplastic elastomer							Damp locations	
	SEOOW[7,9]	600										Damp and wet locations	
Junior hard service cord	SJ	300	18–10	2–6	Thermoset	18–11 10	0.76 1.14	30 45	None	Thermoset	Pendant or portable	Damp locations	Hard usage
	SJE	300			Thermoplastic elastomer					Thermoplastic elastomer		Damp locations	
	SJEW[9]	300										Damp and wet locations	
	SJEO	300								Oil-resistant thermoplastic elastomer		Damp locations	
	SJEOW[9]	300										Damp and wet locations	
	SJEOO	300			Oil-resistant thermoplastic elastomer							Damp locations	
	SJEOOW[9]	300										Damp and wet locations	
	SJO	300			Thermoset					Oil-resistant thermoset		Damp locations	
	SJOW[9]	300										Damp and wet locations	
	SJOO	300			Oil-resistant thermoset							Damp locations	
	SJOOW[9]	300										Damp and wet locations	
	SJT	300			Thermoplastic					Thermoplastic		Damp locations	
	SJTW[9]	300										Damp and wet locations	
	SJTO	300				18–12 10	0.76 1.14	30 45		Oil-resistant thermoplastic		Damp locations	
	SJTOW[9]	300										Damp and wet locations	
	SJTOO	300			Oil-resistant thermoplastic							Damp locations	
	SJTOOW[9]	300										Damp and wet locations	

Δ **TABLE 400.4** *Continued*

Trade Name	Type Letter	Voltage	AWG or kcmil	Number of Conductors	Insulation	AWG or kcmil	Nominal Insulation Thickness		Braid on Each Conductor	Outer Covering	Use		
							mm	mils					
Hard service cord	SO[7]	600	18–2	2 or more	Thermoset	18–15	0.76	30	None	Oil-resistant thermoset	Pendant or portable	Damp locations	Extra-hard usage
	SOW[7,9]	600										Damp and wet locations	
	SOO[7]	600			Oil-resistant thermoset	14–9	1.14	45				Damp locations	
						8–2	1.52	60					
	SOOW[7,9]	600										Damp and wet locations	
All thermoset parallel cord	SP-1	300	20–18	2 or 3	Thermoset	20–18	0.76	30	None	None	Pendant or portable	Damp locations	Not hard usage
	SP-2	300	18–16			18–16	1.14	45					
	SP-3	300	18–10			18–16	1.52	60			Refrigerators, room air conditioners, and as permitted in 422.16(B)		
						15–14	2.03	80					
						12	2.41	95					
						10	2.80	110					
All elastomer (thermoplastic) parallel cord	SPE-1[8]	300	20–18	2 or 3	Thermoplastic elastomer	20–18	0.76	30	Non	None	Pendant or portable	Damp locations	Not hard usage
	SPE-2[8]	300	18–16			18–16	1.14	45					
	SPE-3[8]	300	18–10			18–16	1.52	60			Refrigerators, room air conditioners, and as permitted in 422.16(B)		
						15–14	2.03	80					
						12	2.41	95					
						10	2.80	110					
All thermoplastic parallel cord	SPT-1	300	20–18	2 or 3	Thermoplastic	20–18	0.76	30	None	None	Pendant or portable	Damp locations	Not hard usage
	SPT-1W[9]	300		2								Damp and wet locations	
	SPT-2	300	18–16	2 or 3		18–16	1.14	45				Damp locations	
	SPT-2W[9]	300		2								Damp and wet locations	
	SPT-3	300	18–10	2 or 3		18–16	1.52	60			Refrigerators, room air conditioners, and as permitted in 422.16(B)	Damp locations	Not hard usage
						15–14	2.03	80					
						12	2.41	95					
						10	2.80	110					
Range, dryer cable	SRD	300	10–4	3 or 4	Thermoset	10–4	1.14	45	None	Thermoset	Portable	Damp locations	Ranges, dryers
	SRDE	300	10–4	3 or 4	Thermoplastic elastomer				None	Thermoplastic elastomer			
	SRDT	300	10–4	3 or 4	Thermoplastic				None	Thermoplastic			
Hard service cord	ST[7]	600	18–2	2 or more	Thermoplastic	18–15	0.76	30	None	Thermoplastic	Pendant or portable	Damp locations	Extra-hard usage
						14–9	1.14	45					
	STW[7,9]	600				8–2	1.52	60				Damp and wet locations	
	STO[7]	600								Oil-resistant thermoplastic		Damp locations	
	STOW[7,9]	600										Damp and wet locations	
	STOO[7]	600			Oil-resistant thermoplastic							Damp locations	
	STOOW[7]	600										Damp and wet locations	

(continues)

Δ **TABLE 400.4** *Continued*

Trade Name	Type Letter	Voltage	AWG or kcmil	Number of Conductors	Insulation	AWG or kcmil	Nominal Insulation Thickness		Braid on Each Conductor	Outer Covering	Use		
							mm	mils					
Vacuum cleaner cord	SV	300	18–16	2 or 3	Thermoset	18–16	0.38	15	None	Thermoset	Pendant or portable	Damp locations	Not hard usage
	SVE	300			Thermoplastic elastomer					Thermoplastic elastomer			
	SVEO	300								Oil-resistant thermoplastic elastomer			
	SVEOO	300			Oil-resistant thermoplastic elastomer								
	SVO	300			Thermoset					Oil-resistant thermoset			
	SVOO	300			Oil-resistant thermoset					Oil-resistant thermoset			
	SVT	300			Thermoplastic					Thermoplastic			
	SVTO	300			Thermoplastic					Oil-resistant thermoplastic			
	SVTOO	300			Oil-resistant thermoplastic								
Parallel tinsel cord	TPT[11]	300	27	2	Thermoplastic	27	0.76	30	None	Thermoplastic	Attached to an appliance	Damp locations	Not hard usage
Jacketed tinsel cord	TST[11]	300	27	2	Thermoplastic	27	0.38	15	None	Thermoplastic	Attached to an appliance	Damp locations	Not hard usage
Portable power cable	W[7]	2000	12–500 501–1000	1–6 1	Thermoset	12–2 1–4/0 250–500 501–1000	1.52 2.03 2.41 2.80	60 80 95 110		Oil-resistant thermoset	Portable, extra-hard usage		

Notes:

All types listed in Table 400.4 shall have individual conductors twisted together, except for Types HPN, SP-1, SP-2, SP-3, SPE-1, SPE-2, SPE-3, SPT-1, SPT-2, SPT-3, SPT-1W, SPT-2W, TPT, NISP-1, NISP-2, NISPT-1, NISPT-2, NISPE-1, NISPE-2, and three-conductor parallel versions of SRD, SRDE, and SRDT.

The individual conductors of all cords, except those of heat-resistant cords, shall have a thermoset or thermoplastic insulation, except that the equipment grounding conductor, where used, shall be in accordance with 400.23(B).

[1]Rubber-filled or varnished cambric tapes shall be permitted as a substitute for the inner braids.

[2]Elevator traveling cables for operating control and signal circuits shall contain nonmetallic fillers as necessary to maintain concentricity. Cables shall have steel supporting members as required for suspension by 620.41. In locations subject to excessive moisture or corrosive vapors or gases, supporting members of other materials shall be permitted. Where steel supporting members are used, they shall run straight through the center of the cable assembly and shall not be cabled with the copper strands of any conductor.

In addition to conductors used for control and signaling circuits, Types E, EO, ETP, and ETT elevator cables shall be permitted to incorporate in the construction one or more of the following: optical fibers; 24 AWG or larger telephone conductor pairs, coaxial cables, or communications cables. The 24 AWG or larger conductor pairs shall be permitted to be covered with suitable shielding for telephone, audio, data transfer, or higher frequency communications circuits; the coaxial cables shall consist of a center conductor, insulation, and a shield for use in video or other radio frequency communications circuits. The optical fiber shall be suitably covered with flame-retardant thermoplastic. The insulation of the conductors shall be rubber or thermoplastic of a thickness not less than specified for the other conductors of the particular type of cable. Metallic shields shall have their own protective covering. Where used, these components shall be permitted to be incorporated in any layer of the cable assembly but shall not run straight through the center.

[3]Insulations and outer coverings that meet the requirements as flame retardant, limited smoke, and are so listed, shall be permitted to be marked for limited smoke after the *Code* type designation.

[4]Elevator cables in sizes 20 AWG through 14 AWG are rated 300 volts, and sizes 10 AWG through 2 AWG are rated 600 volts. 12 AWG is rated 300 volts with a 0.76 mm (30 mil) insulation thickness and 600 volts with a 1.14 mm (45 mil) insulation thickness.

[5]Conductor size for Types EV, EVJ, EVE, EVJE, EVT, and EVJT cables apply to nonpower-limited circuits only. Conductors of power-limited (data, signal, or communications) circuits may extend beyond the stated AWG size range. All conductors shall be insulated for the same cable voltage rating.

[6]Insulation thickness for Types EV, EVJ, EVEJE, EVT, and EVJT cables of nylon construction is indicated in parentheses.

[7]Types G, G-GC, S, SC, SCE, SCT, SE, SEO, SEOO, SEW, SEOW, SEOOW, SO, SOO, SOW, SOOW, ST, STO, STOO, STW, STOW, STOOW, PPE, and W shall be permitted for use on theater stages, in garages, and elsewhere where flexible cords are permitted by this *Code*.

[8]The third conductor in Type HPN shall be used as an equipment grounding conductor only. The insulation of the equipment grounding conductor for Types SPE-1, SPE-2, SPE-3, SPT-1, SPT-2, SPT-3, NISPT-1, NISPT-2, NISPE-1, and NISPE-2 shall be permitted to be thermoset polymer.

[9]Cords that comply with the requirements for outdoor cords and are so listed shall be permitted to be designated as weather and water resistant with the suffix "W" after the *Code* type designation. Cords with the "W" suffix are suitable for use in wet locations and are sunlight resistant.

[10]The required outer covering on some single-conductor cables may be integral with the insulation.

[11]Types TPT and TST shall be permitted in lengths not exceeding 2.5 m (8 ft) where attached directly, or by means of a special type of plug, to a portable appliance rated at 50 watts or less and of such nature that extreme flexibility of the cord is essential.

△ **TABLE 400.5(A)(1)** *Ampacity for Flexible Cords and Flexible Cables [Based on Ambient Temperature of 30°C (86°F). See 400.13 and Table 400.4.]*

Copper Conductor Size (AWG)	Thermoplastic Types TPT, TST	Thermoset Types C, E, EO, PD, S, SJ, SJO, SJOW, SJOO, SJOOW, SO, SOW, SOO, SOOW, SP-1, SP-2, SP-3, SRD, SV, SVO, SVOO, NISP-1, NISP-2		Types HPD, HPN, HSJ, HSJO, HSJOW, HSJOO, HSJOOW
		Thermoplastic Types ETP, ETT, NISPE-1, NISPE-2, NISPT-1, NISPT-2, SE, SEW, SEO, SEOO, SEOW, SEOOW, SJE, SJEW, SJEO, SJEOO, SJEOW, SJEOOW, SJT, SJTW, SJTO, SJTOW, SJTOO, SJTOOW, SPE-1, SPE-2, SPE-3, SPT-1, SPT-1W, SPT-2, SPT-2W, SPT-3, ST, STW, SRDE, SRDT, STO, STOW, STOO, STOOW, SVE, SVEO, SVEOO, SVT, SVTO, SVTOO		
		Column A[1]	Column B[2]	
27[3]	0.5	—	—	—
20	—	5[4]	5	—
18	—	7	10	10
17	—	9	12	13
16	—	10	13	15
15	—	12	16	17
14	—	15	18	20
13	—	17	21	—
12	—	20	25	30
11	—	23	27	—
10	—	25	30	35
9	—	29	34	—
8	—	35	40	—
7	—	40	47	—
6	—	45	55	—
5	—	52	62	—
4	—	60	70	—
3	—	70	82	—
2	—	80	95	—

[1]The currents under Column A apply to three-conductor cords and other multiconductor cords connected to utilization equipment so that only three conductors are current-carrying.

[2]The currents under Column B apply to two-conductor cords and other multiconductor cords connected to utilization equipment so that only two conductors are current-carrying.

[3]Tinsel cord.

[4]Elevator cables only.

[5]7 amperes for elevator cables only; 2 amperes for other types.

A neutral conductor that carries only the unbalanced current from other conductors of the same circuit shall not be required to meet the requirements of a current-carrying conductor.

In a 3-wire circuit consisting of two phase conductors and the neutral conductor of a 4-wire, 3-phase, wye-connected system, a common conductor carries approximately the same current as the line-to-neutral currents of the other conductors and shall be considered to be a current-carrying conductor.

On a 4-wire, 3-phase, wye circuit where more than 50 percent of the load consists of nonlinear loads, there are harmonic currents present in the neutral conductor and the neutral conductor shall be considered to be a current-carrying conductor.

An equipment grounding conductor shall not be considered a current-carrying conductor.

Where a single conductor is used for both equipment grounding and to carry unbalanced current from other conductors, as provided for in 250.140 for electric ranges and electric clothes dryers, it shall not be considered as a current-carrying conductor.

TABLE 400.5(A)(2) *Ampacity of Cable Types SC, SCE, SCT, PPE, G, G-GC, and W [Based on Ambient Temperature of 30°C (86°F). See Table 400.4.]*

Copper Conductor Size (AWG or kcmil)	Temperature Rating of Cable								
	60°C (140°F)			75°C (167°F)			90°C (194°F)		
	D[1]	E[2]	F[3]	D[1]	E[2]	F[3]	D[1]	E[2]	F[3]
12	—	31	26	—	37	31	—	42	35
10	—	44	37	—	52	43	—	59	49
8	60	55	48	70	65	57	80	74	65
6	80	72	63	95	88	77	105	99	87
4	105	96	84	125	115	101	140	130	114
3	120	113	99	145	135	118	165	152	133
2	140	128	112	170	152	133	190	174	152
1	165	150	131	195	178	156	220	202	177
1/0	195	173	151	230	207	181	260	234	205
2/0	225	199	174	265	238	208	300	271	237
3/0	260	230	201	310	275	241	350	313	274
4/0	300	265	232	360	317	277	405	361	316
250	340	296	259	405	354	310	455	402	352
300	375	330	289	445	395	346	505	449	393
350	420	363	318	505	435	381	570	495	433
400	455	392	343	545	469	410	615	535	468
500	515	448	392	620	537	470	700	613	536
600	575	—	—	690	—	—	780	—	—
700	630	—	—	755	—	—	855	—	—
750	655	—	—	785	—	—	885	—	—
800	680	—	—	815	—	—	920	—	—
900	730	—	—	870	—	—	985	—	—
1000	780	—	—	935	—	—	1055	—	—

[1]The ampacities under subheading D shall be permitted for single-conductor Types SC, SCE, SCT, PPE, and W cable only where the individual conductors are not installed in raceways and are not in physical contact with each other except in lengths not to exceed 600 mm (24 in.) where passing through the wall of an enclosure.

[2]The ampacities under subheading E apply to two-conductor cables and other multiconductor cables connected to utilization equipment so that only two conductors are current-carrying.

[3]The ampacities under subheading F apply to three-conductor cables and other multiconductor cables connected to utilization equipment so that only three conductors are current-carrying.

TABLE 400.5(A)(3) *Adjustment Factors for More Than Three Current-Carrying Conductors in a Flexible Cord or Flexible Cable*

Number of Conductors	Percent of Value in Table 400.5(A)(1) and Table 400.5(A)(2)
4–6	80
7–9	70
10–20	50
21–30	45
31–40	40
41 and above	35

(B) Ultimate Insulation Temperature. In no case shall conductors be associated together in such a way with respect to the kind of circuit, the wiring method used, or the number of conductors such that the limiting temperature of the conductors is exceeded.

(C) Engineering Supervision. Under engineering supervision, conductor ampacities shall be permitted to be calculated in accordance with 310.14(B).

400.6 Markings.

(A) Standard Markings. Flexible cords and flexible cables shall be marked by means of a printed tag attached to the coil reel or carton. The tag shall contain the information required in 310.8(A). Types S, SC, SCE, SCT, SE, SEO, SEOO, SJ, SJE, SJEO, SJEOO, SJO, SJT, SJTO, SJTOO, SO, SOO, ST, STO, STOO, SEW, SEOW, SEOOW, SJEW, SJEOW, SJEOOW, SJOW, SJTW, SJTOW, SJTOOW, SOW, SOOW, STW, STOW, and STOOW flexible cords and G, G-GC, PPE, and W flexible cables shall be durably marked on the surface at intervals not exceeding 610 mm (24 in.) with the type designation, size, and number of conductors. Required markings on tags, cords, and cables shall also include the maximum operating temperature of the flexible cord or flexible cable.

(B) Optional Markings. Flexible cords and cable types listed in Table 400.4 shall be permitted to be surface marked to indicate special characteristics of the cable materials. These markings include, but are not limited to, markings for limited smoke, sunlight resistance, and so forth.

In addition to the markings identified in 400.6(A) and (B), the UL guide information, under the category Flexible Cord (ZJCZ), lists the following markings:

1. "For Mobile Home Use," "For Recreational Vehicle Use," or "For Mobile Home and Recreational Vehicle Use" followed by the current rating in amperes indicates suitability for use in mobile homes or recreational vehicles.
2. "W" indicates suitability for use outdoors and for immersion in water. The low-temperature rating for these cords is −40°C, unless otherwise marked on the cord with optional ratings of −50°C, −60°C, or −70°C. The low-temperature ratings are determined by means of a bend test (not a suppleness test) at the given temperature. The cord may be additionally marked "Water Resistant."
3. "VW-1" indicates that the cord complies with a vertical flame test. Cord that has been investigated for leakage currents between the circuit conductor and the grounding conductor and between the circuit conductor and the outer surface of the jacket may have the values marked on the cable jacket. UL 62, *Standard for Flexible Cords and Cables*, is the relevant product certification standard for these types of flexible cords.

400.10 Uses Permitted.

(A) Uses. Flexible cords and flexible cables shall be used only for the following:

(1) Pendants.
(2) Wiring of luminaires.
(3) Connection of portable luminaires, portable and mobile signs, or appliances.
(4) Elevator cables.
(5) Wiring of cranes and hoists.
(6) Connection of utilization equipment to facilitate frequent interchange.
(7) Prevention of the transmission of noise or vibration.
(8) Appliances where the fastening means and mechanical connections are specifically designed to permit ready removal for maintenance and repair, and the appliance is intended or identified for flexible cord connection.
(9) Connection of moving parts.
(10) Where specifically permitted elsewhere in this *Code*.
(11) Between an existing receptacle outlet and an inlet, where the inlet provides power to an additional single receptacle outlet. The wiring interconnecting the inlet to the single receptacle outlet shall be a Chapter 3 wiring method. The inlet, receptacle outlet, and Chapter 3 wiring method, including the flexible cord and fittings, shall be a listed assembly specific for this application.

Section 400.10(A)(11) permits a flexible cord or flexible cable to be used between an existing receptacle outlet and an inlet that provides power to a listed assembly used to supply a flat screen TV or similar equipment. The inlet is connected to a Chapter 3 wiring method that runs through the wall to a standard receptacle, into which the flat screen TV is plugged. The listed assembly is not directly connected to the wiring system of the house but is energized only when the rubber cord is plugged into an existing receptacle outlet. See Exhibit 400.1 for an example of a listed assembly.

(B) Attachment Plugs. Where used as permitted in 400.10(A) (3), (A)(6), and (A)(8), each flexible cord shall be equipped with an attachment plug and shall be energized from a receptacle outlet or cord connector body.

Exception: As permitted in 368.56.

400.12 Uses Not Permitted. Unless specifically permitted in 400.10, flexible cords, flexible cables, cord sets, and power supply cords shall not be used for the following:

(1) As a substitute for the fixed wiring of a structure
(2) Where run through holes in walls, structural ceilings, suspended ceilings, dropped ceilings, or floors
(3) Where run through doorways, windows, or similar openings
(4) Where attached to building surfaces

Exception to (4): Flexible cord and flexible cable shall be permitted to be attached to building surfaces in accordance with 368.56(B) and 590.4.

(5) Where concealed by walls, floors, or ceilings or located above suspended or dropped ceilings

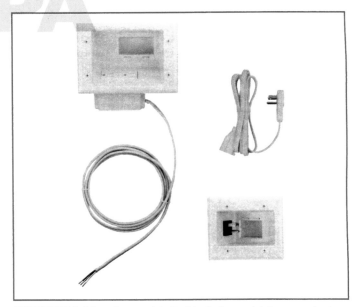

EXHIBIT 400.1 A listed assembly that includes an inlet, a receptacle outlet, a Chapter 3 wiring method, and a flexible cord. *(Courtesy of DataComm Electronics, Inc.)*

Exception to (5): Flexible cords, flexible cables, and power supply cords shall be permitted if contained within an enclosure for use in other spaces used for environmental air as permitted by 300.22(C)(3).

(6) Where installed in raceways, except as otherwise permitted in this *Code*

(7) Where subject to physical damage

> Informational Note: See UL 817, *Cord Sets and Power-Supply Cords*, and UL 62, *Flexible Cords and Cables*, for proper application.

Flexible cords and power supply cables are not limited to use with portable equipment. However, 400.12 prohibits the use of flexible cords and flexible power supply cables as a substitute for the fixed wiring of a structure or where concealed behind building walls, floors, or ceilings (including structural, suspended, or dropped-type ceilings).

The requirements of 400.10 and 400.12 need to be applied together, rather than independently. While 400.10(A)(6) permits the use of cords to supply utilization equipment that is frequently replaced or serviced, it does not specifically permit that use for an installation that is prohibited by 400.12. As an example, a listed piece of utilization equipment may be provided with a cord and attachment cap for connection to a receptacle. That is a general permitted use covered in 400.10(A)(6). However, the fact that this listed utilization equipment is equipped with a cord cannot be viewed just in the context of 400.10(A)(6). The prohibited uses in 400.12 also must be observed, and there is not a specific permitted use in 400.10 that allows listed cord-and-plug-connected equipment to be installed contrary to any of the prohibited uses in 400.12. The cord of that listed utilization equipment is not permitted to be attached to building surfaces; run through holes in walls, floors, or ceilings; or installed above a suspended ceiling unless the exceptions to 400.12(4) and (5) can be applied.

Cords run through openings in the sides, top, or bottom of cabinets, such as cords of dishwashers or combination microwave/range hood units, are not considered to be run through holes in walls, floors, or ceilings. Caution must be taken to ensure that where cords are used in this manner the opening through which the cord is passed does not present a physical damage threat to the cord.

See also

240.5, 590.4(B), and **590.4(C)** for the uses of multiconductor flexible cords for feeder and branch-circuit installations and for overcurrent protection requirements for flexible cord

410.62 for cord-connected luminaires

422.16(B)(2)(4) and **(5)** for the protection of dishwasher and trash compactor supply cords that pass through an opening in a cabinet

400.13 Splices. Flexible cord shall be used only in continuous lengths without splice or tap where initially installed in applications permitted by 400.10(A). The repair of hard-service cord and junior hard-service cord (see Trade Name column in Table 400.4) 14 AWG and larger shall be permitted if conductors are spliced in accordance with 110.14(B) and the completed splice retains the insulation, outer sheath properties, and usage characteristics of the cord being spliced.

This section permits repair of a damaged cord in such a manner that the cord will retain its original operating and use integrity. However, if the repaired cord is reused or reinstalled at a new location, the in-line repair is no longer permitted, and the cord can be used only in lengths that do not contain a splice.

The use of repaired extension and power tool cords on construction sites is covered under U.S. Occupational Health and Safety Administration (OSHA) regulations. While the repair of extension or power tool cords is not expressly prohibited, an OSHA letter of interpretation on this topic dated April 4, 2010, states that "the repairs must return the equipment to the state in which it was initially approved." More information on this topic can be found at www.osha.gov.

400.14 Pull at Joints and Terminals. Flexible cords and flexible cables shall be connected to devices and to fittings so that tension is not transmitted to joints or terminals.

Exception: Listed portable single-pole devices that are intended to accommodate such tension at their terminals shall be permitted to be used with single-conductor flexible cable.

> Informational Note: Some methods of preventing pull on a cord from being transmitted to joints or terminals include knotting the cord, winding with tape, and using support or strain-relief fittings.

400.15 In Show Windows and Showcases. Flexible cords used in show windows and showcases shall be Types S, SE, SEO, SEOO, SJ, SJE, SJEO, SJEOO, SJO, SJOO, SJT, SJTO, SJTOO, SO, SOO, ST, STO, STOO, SEW, SEOW, SEOOW, SJEW, SJEOW, SJEOOW, SJOW, SJOOW, SJTW, SJTOW, SJTOOW, SOW, SOOW, STW, STOW, or STOOW.

Exception No. 1: For the wiring of chain-supported luminaires.

Exception No. 2: As supply cords for portable luminaires and other merchandise being displayed or exhibited.

Flexible cords identified for hard usage or extra-hard usage are required in show windows and showcases, because cords in such locations are exposed to wear and tear from continual housekeeping and display changes, which can cause the worn cord to make contact with combustible materials such as fabrics or paper products, increasing the fire hazard risk.

400.16 Overcurrent Protection. Flexible cords not smaller than 18 AWG, and tinsel cords or cords having equivalent characteristics of smaller size approved for use with specific appliances, shall be considered as protected against overcurrent in accordance with 240.5.

400.17 Protection from Damage. Flexible cords and flexible cables shall be protected by bushings or fittings where passing through holes in covers, outlet boxes, or similar enclosures.

In industrial establishments where the conditions of maintenance and supervision ensure that only qualified persons service the installation, flexible cords and flexible cables shall be permitted to be installed in aboveground raceways that are no longer than 15 m (50 ft) to protect the flexible cord or flexible cable from physical damage. Where more than three current-carrying conductors are installed within the raceway, the ampacity shall be adjusted in accordance with Table 400.5(A)(3).

A variety of bushings and fittings, both insulated and noninsulated, are available for protecting flexible cords and flexible cables. Some bushings or fittings include strain-relief fittings as required in 400.14. Many insulating bushings are listed in the following product categories:

1. Conduit fittings (bushings and fittings for use on the ends of conduit in boxes and gutters)
2. Insulating devices and materials
3. Outlet bushings and fittings (for use on the ends of conduit, electrical metallic tubing, or armored cable where a change to open wiring is made)

Part II. Construction Specifications

400.20 Labels. Flexible cords shall be examined and tested at the factory and labeled before shipment.

400.21 Construction.

(A) Conductors. The individual conductors of a flexible cord or flexible cable shall have copper flexible stranding and shall not be smaller than the sizes specified in Table 400.4.

(B) Nominal Insulation Thickness. The nominal thickness of insulation for conductors of flexible cords and flexible cables shall not be less than specified in Table 400.4.

400.22 Grounded-Conductor Identification. One conductor of flexible cords that is intended to be used as a grounded circuit conductor shall have a continuous marker that readily distinguishes it from the other conductor or conductors. The identification shall consist of one of the methods indicated in 400.22(A) through (F).

(A) Colored Braid. A braid finished to show a white or gray color and the braid on the other conductor or conductors finished to show a readily distinguishable solid color or colors.

(B) Tracer in Braid. A tracer in a braid of any color contrasting with that of the braid and no tracer in the braid of the other conductor or conductors. No tracer shall be used in the braid of any conductor of a flexible cord that contains a conductor having a braid finished to show white or gray.

Exception: In the case of Types C and PD and cords having the braids on the individual conductors finished to show white or gray.

In such cords, the identifying marker shall be permitted to consist of the solid white or gray finish on one conductor, provided there is a colored tracer in the braid of each other conductor.

(C) Colored Insulation. A white or gray insulation on one conductor and insulation of a readily distinguishable color or colors on the other conductor or conductors for cords having no braids on the individual conductors.

For jacketed cords furnished with appliances, one conductor having its insulation colored light blue, with the other conductors having their insulation of a readily distinguishable color other than white or gray.

Exception: Cords that have insulation on the individual conductors integral with the jacket.

The insulation shall be permitted to be covered with an outer finish to provide the desired color.

(D) Colored Separator. A white or gray separator on one conductor and a separator of a readily distinguishable solid color on the other conductor or conductors of cords having insulation on the individual conductors integral with the jacket.

Grounded conductors in flexible cords and flexible cables are identified with a white- or gray-colored braid, a white- or gray-colored tracer in the braid, white- or gray-colored insulation, or a white- or gray-colored separator. In existing installations where a gray-colored braid, tracer, or conductor insulation is encountered, caution should be exercised because gray could also signify ungrounded conductors.

(E) Tinned Conductors. One conductor having the individual strands tinned and the other conductor or conductors having the individual strands untinned for cords having insulation on the individual conductors integral with the jacket.

(F) Surface Marking. One or more ridges, grooves, or white stripes located on the exterior of the cord so as to identify one conductor for cords having insulation on the individual conductors integral with the jacket.

One method of surface marking is to identify the grounded conductor in a cord where the conductor insulation is part of the molded jacket and is not separable. A white stripe on the cord exterior serves to identify the segment of the cord that contains the grounded conductor. An example of a type of cord with these characteristics is zip cord, which is commonly used for floor lamps and table lamps. It is important that the grounded conductor be easily identified where zip cord is used for wiring lampholders, because the *NEC* requires the grounded conductor to be connected to the device terminal that connects to the screw shell of the lampholder and to the terminal connected to the larger blade (prong) of a polarized two-wire attachment cap.

400.23 Equipment Grounding Conductor Identification. A conductor intended to be used as an equipment grounding conductor shall have a continuous identifying marker readily distinguishing it from the other conductor or conductors. Conductors having a continuous green color or a continuous green color with one or more yellow stripes shall not be used for other than equipment grounding conductors. Cords or cables consisting of integral insulation and a jacket without a nonintegral equipment grounding conductor shall be permitted to be green. The identifying marker shall consist of one of the methods in 400.23(A) or (B).

(A) Colored Braid. A braid finished to show a continuous green color or a continuous green color with one or more yellow stripes.

(B) Colored Insulation or Covering. For cords having no braids on the individual conductors, an insulation of a continuous green color or a continuous green color with one or more yellow stripes.

400.24 Attachment Plugs. Where a flexible cord is provided with an equipment grounding conductor and equipped with an attachment plug, the attachment plug shall comply with 250.138(A) and (B).

Part III. Portable Cables Over 600 Volts, up to 2000 Volts, Nominal

400.30 General. Part III applies to single and multiconductor portable cables used to connect mobile equipment and machinery.

400.31 Construction.

(A) Conductors. The conductors shall be 12 AWG copper or larger and shall employ flexible stranding.

(B) Equipment Grounding Conductor(s). An equipment grounding conductor(s) shall be provided in cables with three or more conductors. The total area shall not be less than that of the size of the equipment grounding conductor required in 250.122.

400.32 Shielding. All shields shall be connected to an equipment grounding conductor.

400.33 Equipment Grounding Conductors. Equipment grounding conductors shall be connected in accordance with Parts VI and VII of Article 250.

400.34 Minimum Bending Radii. The minimum bending radii for portable cables during installation and handling in service shall be adequate to prevent damage to the cable.

400.35 Fittings. Connectors used to connect lengths of cable in a run shall be of a type that locks or latches firmly together. Provisions shall be made to prevent opening or closing these connectors while energized. Suitable means shall be used to eliminate tension at connectors and terminations.

400.36 Splices and Terminations. Portable cables shall not contain splices unless the splices are of the permanent molded, vulcanized types in accordance with 110.14(B). Terminations on portable cables rated over 600 volts, nominal, shall be accessible only to authorized and qualified personnel.

N Part IV. Portable Power Feeder Cables Over 2000 Volts, Nominal

N 400.40 General. Part IV applies to single and multiconductor portable power feeder cables over 2000 volts nominal used to connect portable equipment and machinery.

N 400.41 Portable Power Feeder Cables. Portable power feeder cables rated greater than 2000 volts shall comply with 400.14, 400.17, and 400.23 and with the following sections.

Informational Note: See ANSI/NEMA WC 58/ ICEA S-75-381, *Portable and Power Feeder Cables for Use in Mines and Similar Applications*, for information on construction, testing, and marking of portable power feeder cables.

N 400.42 Uses Permitted. Portable power feeder cables over 2000 volts shall be used for the following:

(1) Connection of portable equipment and machinery or for wiring of cranes and hoists
(2) Temporary services and installations

N 400.43 Uses Not Permitted. Portable power feeder cables over 2000 volts shall not be used for the following:

(1) As a substitute for the fixed wiring of a structure
(2) Where run through holes in walls, ceilings, or floors
(3) Where run through doorways, windows, or similar openings
(4) Where attached to building surfaces
(5) Where concealed by walls, floors, or ceilings
(6) Where installed in raceways, except as otherwise permitted in this *Code*
(7) Where subject to physical damage

N 400.44 Construction.

N (A) Conductors. The conductors shall be 6 AWG copper or larger and shall employ flexible stranding.

N (B) Nominal Insulation Thickness. The nominal thickness of insulation for portable power feeder cables shall not be less than specified in Table 400.44(B)(1) through Table 400.44(B)(4).

N (C) Equipment Grounding Conductor(s). An equipment grounding conductor(s) shall be provided in cables with three or more conductors. The total area shall not be less than that of the size of the equipment grounding conductor required in 250.122.

N 400.45 Shielding. All shields shall be grounded at least at one end.

N **TABLE 400.44(B)(1)** *Thickness of Insulation for Three-Conductor Type G Portable Power Feeder Cables Rated 2000 Volts to 5000 Volts and Equipment Grounding Conductor Size*

Copper Conductor Size (AWG) or kcmil	Nominal Insulation Thickness of Power Conductors		Equipment Grounding Conductor
	mils	mm	Size AWG
6	110	2.79	10
4–2	110	2.79	8
1	110	2.79	7
1/0	110	2.79	6
2/0	110	2.79	5
3/0	110	2.79	4
4/0	110	2.79	3

N **TABLE 400.44(B)(2)** *Thickness of Insulation for Single Conductor Type SH Portable Power Feeder Cables Rated 2000 Volts to 25,000 Volts for 100 Percent Insulation Level*

Copper Conductor Size (AWG) or kcmil	Nominal Insulation Thickness of Power Conductors	
	mils	mm
2001 to 5000 volts		
6–4/0	110	2.79
250–500	120	3.05
5001 to 8000 volts		
4–500	150	3.81
8001 to 15,000 volts		
2–500	210	5.33
15,001 to 25,000 volts		
1–500	295	7.49

N **TABLE 400.44(B)(3)** *Thickness of Insulation for Three-Conductor Type SHD and SHD-GC Portable Power Feeder Cables Rated 2000 Volts to 25,000 Volts for 100 Percent Insulation Level and Grounding Conductor Size*

Copper Conductor Size (AWG) or kcmil	Nominal Insulation Thickness of Power Conductors		Grounding Conductor
	mils	mm	Size AWG
2000 to 5000 volts			
6–4/0	110	2.79	8
250–500	120	3.05	8
5001 to 8000 volts			
4–500	150	3.81	8
8001 to 15,000 volts			
2–4/0	210	5.33	8
15,001 to 25,000 volts			
1–4/0	295	7.49	8

N **TABLE 400.44(B)(4)** *Thickness of Insulation for Three-Conductor Type SHD-CGC Portable Power Feeder Cables Rated 2000 Volts to 5000 Volts for 100 Percent Insulation Level and Grounding Conductor Size*

Copper Conductor Size (AWG) or kcmil	Nominal Insulation Thickness of Power Conductors		Equipment Grounding Conductor
	mils	mm	Size AWG
6	110	2.79	10
4–2	110	2.79	8
1	110	2.79	7
1/0	110	2.79	6
2/0	110	2.79	5
3/0	110	2.79	4
4/0	110	2.79	3
250	120	3.05	2
300–350	120	3.05	1
500	120	3.05	2/0

N **400.46 Equipment Grounding Conductors.** Equipment grounding conductors shall be connected in accordance with Parts VI and VII of Article 250.

N **400.47 Minimum Bending Radii.** The minimum bending radii for portable power feeder cables from 2000 volts to 5000 volts shall not exceed six times the overall cable outer diameter. The minimum bending radii for portable cables from 5001 volts to 25,000 volts shall not exceed eight times the overall cable outer diameter.

N **400.48 Fittings.** The use of connectors and couplers to connect lengths of cable together in a run shall not be permitted.

N **400.49 Splices and Terminations.** Portable power feeder cables shall not contain splices. Connectors, couplers, lugs, elbows, and terminations for portable power feeder cables rated over 2000 volts, nominal, shall be accessible only to authorized and qualified personnel. Suitable means shall be used to eliminate tension at connectors, couplers, lugs, elbows, and terminations.

N **400.50 Types.** Portable power feeder cables rated greater than 2000 volts shall conform to the description in Table 400.50. Types G, SHD-PCG, and SHD-CGC shall be used only from 2000 volts to 5000 volts. Types SH, SHD, and SHD-GC shall be used from 2000 volts to 25,000 volts. Where a Type designation for portable power feeder cables over 2000 volts conflicts with a designation description in Table 400.4, the description in Table 400.50 shall apply. The use of portable power feeder cables other than those in Table 400.50 shall require permission by the authority having jurisdiction.

N **400.51 Ampacities for Portable Power Feeder Cables Rated Greater Than 2000 Volts.**

N **(A) Ampacity Tables.** Table 400.51(A)(1) provides the ampacities for single and three-conductor portable power feeder cables rated greater than 2000 volts. Where portable power feeder cables are used in ambient temperatures other than 30°C (86°F), the

N **TABLE 400.50** *Portable Power Feeder Cables*

Trade Name	Type Letter	Voltage	AWG or kcmil	Insulation	Outer Covering	Ground Check Conductor	Grounding Conductor	Shielded	Multiconductor Configuration
Portable Power Feeder Cables	G	2001–5000	8–4/0	Thermoset	Heavy duty or extra heavy duty Thermoset	No	Yes	Yes	Two parallel power conductors with a single grounding conductor
	SHD-PCG	2001–5000	3–4/0	Thermoset	Heavy duty or extra heavy duty Thermoset	No	Yes	Individually shielded power conductors	Round three power conductors that are separately covered with insulation, a tape, and a metallic shield, grounding conductor, and one or more control conductors under a unit jacket.
	SH	2001–25,000	6–500	Thermoset	Heavy duty or extra heavy duty Thermoset	No	No	Yes	N/A
	SHD	2001–25,000	6–500	Thermoset	Heavy duty or extra heavy duty Thermoset	No	Yes	Individually shielded power conductors and grounding conductors	Round three power conductors that are separately covered with insulation, a tape, and a shield, and three grounding conductors, one in each interstice.
	SHD-GC	2001–25,000	6–500	Thermoset	Heavy duty or extra heavy duty Thermoset	Yes	Yes	Individually shielded power and grounding conductors	Round three power conductors that are separately covered with insulation, a tape, and a shield, and two grounding conductors and one ground-check conductor
	SHD-CGC	2001–5000	8–500	Thermoset	Heavy duty or extra heavy duty Thermoset	Yes	Yes	Individually shielded power conductors, grounding conductors, and one ground-check conductor in center	Round three power conductors that are separately covered with insulation, a tape, and metal shield, three grounding conductors, and one center ground-check conductor.

temperature correction factors from Table 400.51(A)(2) that correspond to the differing ambient temperature shall be applied to the ampacity in Table 400.51(A)(1). Where the cable will not be completely unwound from the cord reel, the ampacity correction factor based on the number of layers remaining wound on a reel shall be applied as shown in Table 400.51(A)(3).

N **(B) Engineering Supervision.** Under engineering supervision, conductor ampacities shall be permitted to be calculated in accordance with 310.14(B).

N **400.52 Markings**

N **(A) Required Markings.** Portable power feeder cables shall be marked by means of a printed tag attached to the coil reel or carton. The tag shall contain the information required in 310.8(A). Types G, SHD-PCG, SH, SHD, SHD-GC, and SHD-CGC portable power feeder cables shall be durably marked on the surface at intervals not exceeding 610 mm (24 in.) with the following:

(1) The maximum rated voltage
(2) The proper type letter or letters for the type of portable power feeder cable as specified elsewhere in this *Code*
(3) The manufacturer's name, trademark, or other distinctive marking by which the organization responsible for the product can be readily identified
(4) The AWG size or circular mil area
(5) Maximum operating temperature

N **(B) Optional Markings.** Portable power feeder cables listed in Table 400.50 shall be permitted to be surface marked to indicate special characteristics of the cable materials. These markings shall be permitted to include, but are not limited to, markings for limited smoke, sunlight resistance, and so forth.

N **TABLE 400.51(A)(1)** *Ampacity for Portable Power Feeder Cables Over 2000 Volts [Based on Ambient Temperature of 30°C (86°F)]*

Copper Conductor Size (AWG) or kcmil	Single Conductor*			Three Conductor			Copper Conductor Size (AWG) or kcmil
	2001–8000 Volts Shielded	8001–15,000 Volts Shielded	15,001–25,000 Volts Shielded	2001–8000 Volts Shielded	8001–15,000 Volts Shielded	8001–15,000 Volts Shielded	
6	123	—	—	102	—	—	6
4	163	—	—	134	—	—	4
3	188	—	—	154	—	—	3
2	214	214	—	175	180	196	2
1	247	247	244	202	205	210	1
1/0	286	285	280	232	236	240	1/0
2/0	329	328	322	267	270	274	2/0
3/0	379	377	371	307	311	315	3/0
4/0	440	437	428	353	357	360	4/0
250	488	484	473	390	395	396	250
300	545	540	528	438	—	—	300
350	604	597	582	478	—	—	350
400	656	649	629	470	—	—	400
450	704	696	676	503	—	—	450
500	757	746	725	536	—	—	500

Note: Ampacities are based on a conductor temperature of 90°C (194°F) and an ambient air temperature of 30°C (86°F).

*Ampacities are based on single isolated cable in air operated with open-circuited shield.

N **TABLE 400.51(A)(2)** *Adjustment Factors for Different Ambient Temperatures*

Ambient Temperature, Degrees		Multiplying Correction Factor
°C	°F	
10	50	1.26
20	68	1.18
30	86	1.10
40	104	1.00
50	122	0.90

N **TABLE 400.51(A)(3)** *Adjustment Factors for Number of Layers of Cable Wound on a Reel*

Number of Layers	Multiplying Correction Factor
1	0.85
2	0.65
3	0.45
4*	0.35

*If more than four layers of cable are wound on the reel, ampacity derating should be calculated using engineering supervision.

ARTICLE 402 Fixture Wires

402.1 Scope. This article covers general requirements and construction specifications for fixture wires.

402.2 Other Articles. Fixture wires shall comply with this article and also with the applicable provisions of other articles of this *Code*.

Informational Note: See Part VI of Article 410 for application in luminaires.

402.3 Types. Fixture wires shall be of a type listed in Table 402.3, and they shall comply with all requirements of that table. The fixture wires listed in Table 402.3 are all suitable for service at 600 volts, nominal, unless otherwise specified.

Informational Note: Thermoplastic insulation may stiffen at temperatures lower than -10°C (+14°F). Thermoplastic insulation

may also be deformed at normal temperatures where subjected to pressure, such as at points of support.

402.5 Ampacities for Fixture Wires. The ampacity of fixture wire shall be as specified in Table 402.5.

No conductor shall be used under such conditions that its operating temperature exceeds the temperature specified in Table 402.3 for the type of insulation involved.

Informational Note: See 310.14(A)(3) for temperature limitation of conductors.

TABLE 402.5 *Ampacity for Fixture Wires*

Size (AWG)	Ampacity
18	6
16	8
14	17
12	23
10	28

TABLE 402.3 *Fixture Wires*

Name	Type Letter	Insulation	AWG	Thickness of Insulation		Outer Covering	Maximum Operating Temperature	Application Provisions
				mm	mils			
Heat-resistant rubber-covered fixture wire — flexible stranding	FFH-2	Heat-resistant rubber or cross-linked synthetic polymer	18–16	0.76	30	Nonmetallic covering	75°C (167°F)	Fixture wiring
	FFHH-2						90°C (194°F)	
ECTFE — solid or 7-strand	HF	Ethylene chloro-trifluoroethylene	18–14	0.38	15	None	150°C (302°F)	Fixture wiring
ECTFE — flexible stranding	HFF	Ethylene chlorotrifluoro-roethylene	18–14	0.38	15	None	150°C (302°F)	Fixture wiring
Tape insulated fixture wire — solid or 7-strand	KF-1	Aromatic polyimide tape	18–10	0.14	5.5	None	200°C (392°F)	Fixture wiring — limited to 300 volts
	KF-2	Aromatic polyimide tape	18–10	0.21	8.4	None	200°C (392°F)	Fixture wiring
Tape insulated fixture wire — flexible stranding	KFF-1	Aromatic polyimide tape	18–10	0.14	5.5	None	200°C (392°F)	Fixture wiring — limited to 300 volts
	KFF-2	Aromatic polyimide tape	18–10	0.21	8.4	None	200°C (392°F)	Fixture wiring
Perfluoro-alkoxy — solid or 7-strand (nickel or nickel-coated copper)	PAF	Perfluoro-alkoxy	18–14	0.51	20	None	250°C (482°F)	Fixture wiring (nickel or nickel-coated copper)
Perfluoro-alkoxy — flexible stranding	PAFF	Perfluoro-alkoxy	18–14	0.51	20	None	150°C (302°F)	Fixture wiring
Fluorinated ethylene propylene fixture wire — solid or 7-strand	PF	Fluorinated ethylene propylene	18–14	0.51	20	None	200°C (392°F)	Fixture wiring
Fluorinated ethylene propylene fixture wire — flexible stranding	PFF	Fluorinated ethylene propylene	18–14	0.51	20	None	150°C (302°F)	Fixture wiring
Fluorinated ethylene propylene fixture wire — solid or 7-strand	PGF	Fluorinated ethylene propylene	18–14	0.36	14	Glass braid	200°C (392°F)	Fixture wiring
Fluorinated ethylene propylene fixture wire — flexible stranding	PGFF	Fluorinated ethylene propylene	18–14	0.36	14	Glass braid	150°C (302°F)	Fixture wiring
Extruded polytetrafluoroethylene — solid or 7-strand (nickel or nickel-coated copper)	PTF	Extruded polytetrafluoro-roethylene	18–14	0.51	20	None	250°C (482°F)	Fixture wiring (nickel or nickel-coated copper)
Extruded polytetrafluoroethylene — flexible stranding 26-36 (AWG silver or nickel-coated copper)	PTFF	Extruded polytetrafluoro-roethylene	18–14	0.51	20	None	150°C (302°F)	Fixture wiring (silver or nickel-coated copper)
Heat-resistant rubber-covered fixture wire — solid or 7-strand	RFH-1	Heat-resistant rubber	18	0.38	15	Nonmetallic covering	75°C (167°F)	Fixture wiring — limited to 300 volts
	RFH-2	Heat-resistant rubber Cross-linked synthetic polymer	18–16	0.76	30	None or non-metallic covering	75°C (167°F)	Fixture wiring

TABLE 402.3 *Continued*

Name	Type Letter	Insulation	AWG	Thickness of Insulation		Outer Covering	Maximum Operating Temperature	Application Provisions
				mm	mils			
Heat-resistant cross-linked synthetic polymer-insulated fixture wire — solid or 7-strand	RFHH-2* RFHH-3*	Cross-linked synthetic polymer	18–16 18–16	0.76 1.14	30 45	None or non-metallic covering	90°C (194°F)	Fixture wiring
Silicone insulated fixture wire — solid or 7-strand	SF-1	Silicone rubber	18	0.38	15	Nonmetallic covering	200°C (392°F)	Fixture wiring — limited to 300 volts
	SF-2	Silicone rubber	18–12 10	0.76 1.14	30 45	Nonmetallic covering	200°C (392°F)	Fixture wiring
Silicone insulated fixture wire — flexible stranding	SFF-1	Silicone rubber	18	0.38	15	Nonmetallic covering	150°C (302°F)	Fixture wiring — limited to 300 volts
	SFF-2	Silicone rubber	18–12 10	0.76 1.14	30 45	Nonmetallic covering	150°C (302°F)	Fixture wiring
Thermoplastic covered fixture wire — solid or 7-strand	TF*	Thermoplastic	18–16	0.76	30	None	60°C (140°F)	Fixture wiring
Thermoplastic covered fixture wire — flexible stranding	TFF*	Thermoplastic	18–16	0.76	30	None	60°C (140°F)	Fixture wiring
Heat-resistant thermoplastic covered fixture wire — solid or 7-strand	TFN*	Thermoplastic	18–16	0.38	15	Nylon-jacketed or equivalent	90°C (194°F)	Fixture wiring
Heat-resistant thermoplastic covered fixture wire — flexible stranded	TFFN*	Thermoplastic	18–16	0.38	15	Nylon-jacketed or equivalent	90°C (194°F)	Fixture wiring
Cross-linked polyolefin insulated fixture wire — solid or 7-strand	XF*	Cross-linked polyolefin	18–14 12–10	0.76 1.14	30 45	None	150°C (302°F)	Fixture wiring — limited to 300 volts
Cross-linked polyolefin insulated fixture wire — flexible stranded	XFF*	Cross-linked polyolefin	18–14 12–10	0.76 1.14	30 45	None	150°C (302°F)	Fixture wiring — limited to 300 volts
Modified ETFE — solid or 7-strand	ZF	Modified ethylene tetrafluoro-ethylene	18–14	0.38	15	None	150°C (302°F)	Fixture wiring
Modified ETFE — flexible stranding	ZFF	Modified ethylene tetrafluoro-ethylene	18–14	0.38	15	None	150°C (302°F)	Fixture wiring
High temp. modified ETFE— solid or 7-strand	ZHF	Modified ethylene tetrafluoro-ethylene	18–14	0.38	15	None	200°C (392°F)	Fixture wiring

*Insulations and outer coverings that meet the requirements of flame retardant, limited smoke, and are so listed, shall be permitted to be marked for limited smoke after the *Code* type designation.

Fixture wire is permitted to be protected by branch-circuit overcurrent protective devices (OCPD) in accordance with 402.14, which in turn references 240.5 [specifically 240.5(B)(2)]. Fixture wire sizes 18 AWG and 16 AWG have ampacities of 6 and 8 amperes, respectively. While a 15- or 20-ampere OCPD provides a level of short-circuit and ground-fault protection for the fixture wires [limiting the length of the fixture wire per 240.5(B)(2) is necessary to limit the overall impedance under fault conditions], overload protection is not provided by the OCPD typically used for lighting branch circuits. Supplemental overload is permitted, but not required, by the *NEC*®, so protection of the fixture wires is accomplished by limiting the load of the lamps, ballasts, or drivers of a listed luminaire to not more than the ampacities specified in Table 402.5.

402.6 Minimum Size. Fixture wires shall not be smaller than 18 AWG.

402.7 Number of Conductors in Conduit or Tubing. The number of fixture wires permitted in a single conduit or tubing shall not exceed the percentage fill specified in Table 1, Chapter 9.

See also

Chapter 9, Table 4, which provides the usable area within the selected conduit or tubing

Chapter 9, Table 5, which provides the required area for each of the conductors

The following examples show how to determine the minimum size conduit where the conductors are different sizes and, where the conductors are all the same size, how to select the minimum size conduit directly from the tables in Informative Annex C.

Calculation Example 1

A remote ballast installation requires a single flexible metal conduit (FMC) to contain fourteen 16 AWG TFFN fixture wires and three 12 AWG THHN conductors. What size FMC is required?

Solution
Step 1. Using Table 1 in Chapter 9, look up the maximum percent of cross section of conduit permitted for conductors. Table 1 limits the raceway fill for more than two conductors at 40 percent of the total cross-sectional area of the raceway. Note 6 in the Notes to Tables refers to Tables 5 and 5A for conductor dimensions and Table 4 for the raceway dimensions.

Step 2. Find the individual conductor cross-sectional areas in Chapter 9, Table 5:

$$16 \text{ AWG TFFN} = 0.0072 \text{ in.}^2$$
$$12 \text{ AWG THHN} = 0.0133 \text{ in.}^2$$

Step 3. Calculate the total area occupied by the wires as follows:

$$\text{Fourteen } 16 \text{ AWG TFFN} \times 0.0072 = 0.1008 \text{ in.}^2$$
$$\text{Three } 12 \text{ AWG THHN} \times 0.0133 = \underline{0.0399 \text{ in.}^2}$$
$$\text{Total area} = 0.1407 \text{ in.}^2$$

Step 4. Using the 40-percent column of Table 4, Chapter 9, in the section entitled "Article 348, Flexible Metal Conduit (FMC)," find the appropriate FMC size based on 40-percent fill and a total conductor area fill of 0.1407 in.² Because 0.1407 in.² is greater than 0.127 and less than 0.213, select trade size ¾.

Calculation Example 2

If the conductors in the FMC are all of the same wire size (16 AWG), Informative Annex C tables can be used instead of doing the calculations. This example uses Informative Annex C tables to determine FMC size.

What size FMC is required for seventeen 16 AWG TFFN conductors?

Solution
Step 1. In Informative Annex C, Table C.3, find TFFN insulation in the first column.

Step 2. Find 16 AWG in the second column. Proceed across the table until the desired number of conductors is equal to or less than the number shown in the table for the respective conduit sizes. Trade size ½ is required.

402.8 Grounded Conductor Identification. Fixture wires that are intended to be used as grounded conductors shall be identified by one or more continuous white stripes on other than green insulation or by the means described in 400.22(A) through (E).

This requirement is similar to that required for flexible cords and cable to ensure that the grounded conductor is easily recognized. Because connection of the grounded conductor to the screw shell of lampholders is necessary to reduce exposure to an electric shock hazard when lamps are being replaced or changed, the grounded conductor must be easily recognized.

402.9 Marking.

(A) Method of Marking. Thermoplastic insulated fixture wire shall be durably marked on the surface at intervals not exceeding 610 mm (24 in.). All other fixture wire shall be marked by means of a printed tag attached to the coil, reel, or carton.

(B) Optional Marking. Fixture wire types listed in Table 402.3 shall be permitted to be surface marked to indicate special characteristics of the cable materials. These markings include, but are not limited to, markings for limited smoke, sunlight resistance, and so forth.

402.10 Uses Permitted. Fixture wires shall be permitted (1) for installation in luminaires and in similar equipment where enclosed or protected and not subject to bending or twisting in use, or (2) for connecting luminaires to the branch-circuit conductors supplying the luminaires.

Fixture wire is permitted to be used as a tap conductor to connect a luminaire(s) to the branch-circuit conductors. The transition from the branch-circuit wiring method to the fixture wire tap conductors can be accomplished via a junction box or other fitting that is allowed to contain splices.

See also

240.5(B)(2) for overcurrent protection of fixture wire tapped to branch-circuit conductors

402.12 Uses Not Permitted. Fixture wires shall not be used as branch-circuit conductors except as permitted elsewhere in this *Code*.

402.14 Overcurrent Protection. Overcurrent protection for fixture wires shall be as specified in 240.5.

ARTICLE 404 Switches

Part I. General

404.1 Scope. This article covers all switches, switching devices, and circuit breakers used as switches operating at 1000 volts and below, unless specifically referenced elsewhere in this *Code* for higher voltages.

This article does not cover wireless control equipment to which circuit conductors are not connected.

Informational Note: See 210.70 for additional information related to branch circuits that include switches or listed wall-mounted control devices.

404.2 Switch Connections.

(A) Three-Way and Four-Way Switches. Three-way and four-way switches shall be wired so that all switching is done only in the ungrounded circuit conductor. Where in metal raceways or metal-armored cables, wiring between switches and outlets shall be in accordance with 300.20(A).

Exception: Switch loops shall not require a grounded conductor.

The exception does not require a grounded conductor in a switch loop [see 300.20(A)] because the ungrounded conductor both enters and leaves the enclosure in the same cable or raceway, thus avoiding inductive heating.

(B) Grounded Conductors. Switches or circuit breakers shall not disconnect the grounded conductor of a circuit.

Exception: A switch or circuit breaker shall be permitted to disconnect a grounded circuit conductor where all circuit conductors are disconnected simultaneously, or where the device is arranged so that the grounded conductor cannot be disconnected until all the ungrounded conductors of the circuit have been disconnected.

Δ **(C) Switches Controlling Lighting Loads.** The grounded circuit conductor for the controlled lighting circuit shall be installed at the location where switches control lighting loads that are supplied by a grounded general-purpose branch circuit serving bathrooms, hallways, stairways, and habitable rooms or occupiable spaces as defined in the applicable building code. Where multiple switch locations control the same lighting load such that the entire floor area of the room or space is visible from the single or combined switch locations, the grounded circuit conductor shall only be required at one location. A grounded conductor shall not be required to be installed at lighting switch locations under any of the following conditions:

(1) Where conductors enter the box enclosing the switch through a raceway, provided that the raceway is large enough for all contained conductors, including a grounded conductor

(2) Where snap switches with integral enclosures comply with 300.15(E)

(3) Where lighting in the area is controlled by automatic means

(4) Where a switch controls a receptacle load

The grounded conductor shall be extended to any switch location as necessary and shall be connected to switching devices that require line-to-neutral voltage to operate the electronics of the switch in the standby mode and shall meet the requirements of 404.22.

Exception: The connection requirement shall not apply to replacement or retrofit switches installed in locations prior to local adoption of 404.2(C) and where the grounded conductor cannot be extended without removing finish materials. The number of electronic control switches on a branch circuit shall not exceed five, and the number connected to any feeder on the load side of a system or main bonding jumper shall not exceed 25. For the purpose of this exception, a neutral busbar, in compliance with 200.2(B) and to which a main or system bonding jumper is connected shall not be limited as to the number of electronic lighting control switches connected.

Informational Note: The provision for a grounded conductor is to complete a circuit path for electronic lighting control devices.

Many electronic lighting control devices require a standby current to maintain the ready state and detection capability of the device. This allows immediate switching of the load to the "on" condition. These devices require standby current when they are in the "off" state, that is, when there is no load current. In existing installations, many of these devices utilize the equipment grounding conductor (EGC) for the standby current flow because a grounded conductor usually was not provided in the switch box.

Section 404.2(C) does not require a grounded circuit conductor in every installation. For example, a grounded conductor is not required at initial installation if a conductor can be readily added in the future, such as in raceway installations or where the construction of the framing cavity in which the switch box is located permits access. The grounded conductor also is not required if the area served is not a bathroom, hallway, stairway, habitable room, or occupiable space.

404.3 Enclosure.

(A) General. Switches and circuit breakers shall be of the externally operable type mounted in an enclosure listed for the intended use. The minimum wire-bending space at terminals and minimum gutter space provided in switch enclosures shall be as required in 312.6.

Exception No. 1: Pendant- and surface-type snap switches and knife switches mounted on an open-face switchboard or panelboard shall be permitted without enclosures.

Exception No. 2: Switches and circuit breakers installed in accordance with 110.27(A)(1), (A)(2), (A)(3), or (A)(4) shall be permitted without enclosures.

(B) Used as a Raceway. Enclosures shall not be used as junction boxes, auxiliary gutters, or raceways for conductors feeding through or tapping off to other switches or overcurrent devices, unless the enclosure complies with 312.8.

404.4 Damp or Wet Locations.

(A) Surface-Mounted Switch or Circuit Breaker. A surface-mounted switch or circuit breaker shall be enclosed in a weatherproof enclosure or cabinet that complies with 312.2.

(B) Flush-Mounted Switch or Circuit Breaker. A flush-mounted switch or circuit breaker shall be equipped with a weatherproof cover.

(C) Switches in Tub or Shower Spaces. Switches shall not be installed within tub or shower spaces unless installed as part of a listed tub or shower assembly.

404.5 Time Switches, Flashers, and Similar Devices.
Time switches, flashers, and similar devices shall be of the enclosed type or shall be mounted in cabinets or boxes or equipment enclosures. Energized parts shall be barriered to prevent operator exposure when making manual adjustments or switching.

Exception: Devices mounted so they are accessible only to qualified persons shall be permitted without barriers, provided they are located within an enclosure such that any energized parts within 152 mm (6.0 in.) of the manual adjustment or switch are covered by suitable barriers.

404.6 Position and Connection of Switches.

(A) Single-Throw Knife Switches. Single-throw knife switches shall be placed so that gravity will not tend to close them. Single-throw knife switches, approved for use in the inverted position, shall be provided with an integral mechanical means that ensures that the blades remain in the open position when so set.

(B) Double-Throw Knife Switches. Double-throw knife switches shall be permitted to be mounted so that the throw is either vertical or horizontal. Where the throw is vertical, integral mechanical means shall be provided to hold the blades in the open position when so set.

The "integral mechanical means" does not have to be a locking device, but it must ensure that the switch blades remain disengaged regardless of their orientation when the switch is in the "off" (open) position. Many switch designs incorporate mechanical means other than a catch or a latch so that the blades cannot accidentally close from the "off" position.

(C) Connection of Switches. Single-throw knife switches and switches with butt contacts shall be connected such that their blades are de-energized when the switch is in the open position. Bolted pressure contact switches shall have barriers that prevent inadvertent contact with energized blades. Single-throw knife switches, bolted pressure contact switches, molded case switches, switches with butt contacts, and circuit breakers used as switches shall be connected so that the terminals supplying the load are de-energized when the switch is in the open position.

Exception: The blades and terminals supplying the load of a switch shall be permitted to be energized when the switch is in the open position where the switch is connected to circuits or equipment inherently capable of providing a backfeed source of power. For such installations, a permanent sign shall be installed on the switch enclosure or immediately adjacent to open switches with the following words or equivalent: WARNING — LOAD SIDE TERMINALS MAY BE ENERGIZED BY BACKFEED. The warning sign or label shall comply with 110.21(B).

Batteries, generators, photovoltaic (PV) systems, and double-ended switchboard ties are typical backfeed sources. These sources can cause the load side of the switch or circuit breaker to be energized when it is in the open position, which is a condition inherent to the circuitry.

404.7 Indicating.
General-use and motor-circuit switches, circuit breakers, and molded case switches, where mounted in an enclosure as described in 404.3, shall indicate, in a location that is visible when accessing the external operating means, whether they are in the open (off) or closed (on) position.

Where these switch or circuit breaker handles are operated vertically rather than rotationally or horizontally, the up position of the handle shall be the closed (on) position.

Exception No. 1: Vertically operated double-throw switches shall be permitted to be in the closed (on) position with the handle in either the up or down position.

Exception No. 2: On busway installations, tap switches employing a center-pivoting handle shall be permitted to be open or closed with either end of the handle in the up or down position. The switch position shall be clearly indicating and shall be visible from the floor or from the usual point of operation.

Exception No. 2 clarifies the operation of busway switches that are designed with a center pivot handle such that, at any time, one end of the handle is in the up position and the other is down, which means the handle is pulled down to turn the switch off and pulled down to turn it on. See Exhibit 404.1. The exception permits this common method of operating busway switches.

404.8 Accessibility and Grouping.

Δ **(A) Location.** All switches and circuit breakers used as switches shall be located so that they can be operated from a readily accessible place. They shall be installed such that the center of the grip of the operating handle of the switch or circuit breaker, when in its highest position, is not more than 2.0 m (6 ft 7 in.) above the floor or working platform, except as follows:

(1) On busway installations, fused switches and circuit breakers shall be permitted to be located at the same level as the

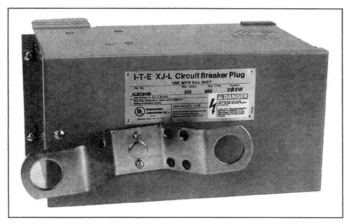

EXHIBIT 404.1 A center-pivot busway plug. *(Courtesy of Siemens Industry, Inc.)*

busway. Suitable means shall be provided to operate the handle of the device from the floor.

(2) Switches and circuit breakers installed adjacent to motors, appliances, or other equipment that they supply shall be permitted to be located higher than 2.0 m (6 ft 7 in.) and to be accessible by portable means.

(3) Hookstick operable isolating switches shall be permitted at greater heights.

(B) Voltage Between Adjacent Devices. A snap switch shall not be grouped or ganged in enclosures with other snap switches, receptacles, or similar devices, unless they are arranged so that the voltage between adjacent devices does not exceed 300 volts, or unless they are installed in enclosures equipped with identified, securely installed barriers between adjacent devices.

Barriers are required between switches that are ganged in a box and used to control 277-volt lighting on 480Y/277-volt systems where two or more phase conductors enter the box. The barriers would be required between devices fed from two different phases of this system because the voltage between the phase conductors would be 480 volts, nominal. Barriers are required even if one device space is left empty because the two remaining devices fed from different phase conductors still would be adjacent to each other.

Where switches or devices are on different systems and are in the same box, such as a switch on a 277-volt system and a receptacle on a 120-volt system, it must be determined that the voltage between adjacent device terminals does not exceed the 300-volt limitation.

Δ **(C) Multipole Snap Switches.** A multipole, general-use snap switch shall not be fed from more than a single circuit unless it is listed and marked as a two-circuit or three-circuit switch.

Informational Note: See 210.7 for disconnect requirements where more than one circuit supplies a switch.

Multiple-pole snap switches used to switch multiple circuits must be listed. Where a switch is supplied by more than one

circuit, 210.7 requires a means to simultaneously disconnect the ungrounded conductors supplying the switch.

404.9 General-Use Snap Switches, Dimmers, and Control Switches.

(A) Faceplates. Faceplates provided for snap switches, dimmers, and control switches mounted in boxes and other enclosures shall be installed so as to completely cover the opening and, where the switch is flush mounted, seat against the finished surface.

(B) Grounding. Snap switches, dimmers, and control switches shall be connected to an equipment grounding conductor and shall provide a means to connect metal faceplates to the equipment grounding conductor, whether or not a metal faceplate is installed. Metal faceplates shall be bonded to the equipment grounding conductor. Snap switches, dimmers, control switches, and metal faceplates shall be connected to an equipment grounding conductor using either of the following methods:

(1) The switch is mounted with metal screws to a metal box or metal cover that is connected to an equipment grounding conductor or to a nonmetallic box with integral means for connecting to an equipment grounding conductor.

(2) An equipment grounding conductor or equipment bonding jumper is connected to an equipment grounding termination of the snap switch.

Exception No. 1: Where no means exists within the enclosure for bonding to the equipment grounding conductor, or where the wiring method does not include or provide an equipment grounding conductor, a snap switch without a connection to an equipment grounding conductor shall be permitted for replacement purposes only. A snap switch wired under the provisions of this exception and located within 2.5 m (8 ft) vertically, or 1.5 m (5 ft) horizontally, of ground or exposed grounded metal objects shall be provided with a faceplate of nonconducting noncombustible material with nonmetallic attachment screws, unless the switch mounting strap or yoke is nonmetallic or the circuit is protected by a ground-fault circuit interrupter.

Exception No. 2: Listed kits or listed assemblies shall not be required to be bonded to an equipment grounding conductor if all of the following conditions are met:

(1) The device is provided with a nonmetallic faceplate, and the device is designed such that no metallic faceplate replaces the one provided.

(2) The device does not have mounting means to accept other configurations of faceplates.

(3) The device is equipped with a nonmetallic yoke.

(4) All parts of the device that are accessible after installation of the faceplate are manufactured of nonmetallic materials.

Exception No. 3: A snap switch with integral nonmetallic enclosure complying with 300.15(E) shall be permitted without a bonding connection to an equipment grounding conductor.

Although the non-current-carrying metal parts of switches typically are not subject to contact by personnel, metal faceplates would pose a shock hazard if they became energized. See Exhibit 404.2 for an example of the typical method by which a metal faceplate is grounded.

See also

404.9(B)(1) and **(B)(2)** for conditions under which the switch provides an effective ground fault current path to a metal faceplate

Exception No. 1 covers switch replacement where an equipment grounding conductor (EGC) is not available. This requires either a switch plate made of insulating material or ground-fault circuit-interrupter (GFCI) protection for the circuit. This exception provides additional safety to persons where older electrical installations did not provide an EGC in the wiring method.

Exception No. 2 permits listed kits or assemblies of snap switches — or other devices with nonmetallic yokes and faceplates — to be installed without an EGC connection. These assemblies are evaluated for dielectric breakdown to ensure there will not be a shock hazard. A faceplate that is not interchangeable with other types of faceplates ensures that a conductive faceplate will not be used on switches that do not provide an equipment grounding means through the yoke. Making sure that all parts accessible after installation are nonmetallic provides further protection from shock hazards.

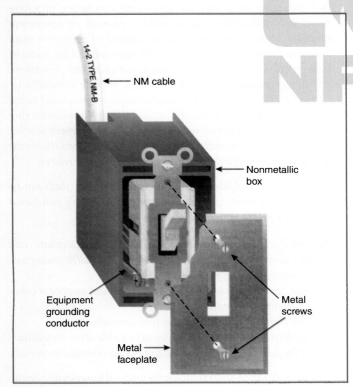

EXHIBIT 404.2 A metal faceplate grounded through attachment to the grounded yoke of a snap switch.

Exception No. 3 recognizes that listed switches of the boxless type incorporate integral nonconductive (nonmetallic) faceplates that do not permit the interchange of faceplates, which precludes the substitution of a metal faceplate. No equipment in the assembly requires grounding; therefore, an EGC is not necessary.

(C) Construction. Metal faceplates shall be of ferrous metal not less than 0.76 mm (0.030 in.) in thickness or of nonferrous metal not less than 1.02 mm (0.040 in.) in thickness. Faceplates of insulating material shall be noncombustible and not less than 2.54 mm (0.100 in.) in thickness, but they shall be permitted to be less than 2.54 mm (0.100 in.) in thickness if formed or reinforced to provide adequate mechanical strength.

404.10 Mounting of General-Use Snap Switches, Dimmers, and Control Switches.

(A) Surface Type. General-use snap switches, dimmers, and control switches used with open wiring on insulators shall be mounted on insulating material that separates the conductors at least 13 mm (½ in.) from the surface wired over.

(B) Box Mounted. Flush-type general-use snap switches, dimmers, and control switches mounted in boxes that are set back of the finished surface as permitted in 314.20 shall be installed so that the extension plaster ears are seated against the surface. Flush-type devices mounted in boxes that are flush with the finished surface or project from it shall be installed so that the mounting yoke or strap of the device is seated against the box. Screws used for the purpose of attaching a device to a box shall be of the type provided with a listed device, or shall be machine screws having 32 threads per inch or part of listed assemblies or systems, in accordance with the manufacturer's instructions.

Regardless of where a box for a flush-type switch is installed (in a wall or in a ceiling), the switch yoke or strap must be seated against the box or, where the box is set back, the yoke or strap must be against the finished surface of the wall or ceiling. Requirements for mounting screws are provided to ensure proper mounting of the device and to stop the use of drywall screws, which are not permitted.

Δ **404.11 Circuit Breakers as Switches.** A hand-operable circuit breaker equipped with a lever or handle, or a power-operated circuit breaker capable of being opened by hand in the event of a power failure, shall be permitted to serve as a switch if it has the required number of poles.

Informational Note: See 240.81 and 240.83 for requirements for circuit breakers relative to indication of state and required markings.

Circuit breakers that are capable of being hand operated must clearly indicate whether they are in the open (off) or closed (on) position.

See also

404.7 for details on handle positions

240.83(D) for switch duty (SWD) and high-intensity discharge (HID) marking for circuit breakers used as switches

404.12 Grounding of Enclosures. Metal enclosures for switches or circuit breakers shall be connected to an equipment grounding conductor as specified in Part IV of Article 250. Metal enclosures for switches or circuit breakers used as service equipment shall comply with the provisions of Part V of Article 250. Where nonmetallic enclosures are used with metal raceways or metal-armored cables, they shall comply with 314.3, Exception No. 1 or No. 2.

Except as covered in 404.9(B), Exception No. 1, nonmetallic boxes for switches shall be installed with a wiring method that provides or includes an equipment grounding conductor.

404.13 Knife Switches.

(A) Isolating Switches. Knife switches rated at over 1200 amperes at 250 volts or less, and at over 1000 amperes at 251 to 1000 volts, shall be used only as isolating switches and shall not be opened under load.

(B) To Interrupt Currents. To interrupt currents over 1200 amperes at 250 volts, nominal, or less, or over 600 amperes at 251 to 1000 volts, nominal, a circuit breaker or a switch listed for such purpose shall be used.

(C) General-Use Switches. Knife switches of ratings less than specified in 404.13(A) and (B) shall be considered general-use switches.

> Informational Note: See Article 100 for the definition of *general-use switch*.

Δ **(D) Motor-Circuit Switches.** Motor-circuit switches shall be permitted to be of the knife-switch type.

> Informational Note: See Article 100 for the definition of *motor-circuit switch*.

Δ **404.14 Rating and Use of Switches.** Switches shall be listed and marked with their ratings. Switches of the types covered in 404.14(A) through (F) shall be limited to the control of loads as specified accordingly. Switches used to control cord-and-plug-connected loads shall be limited as covered in 404.14(G).

> Informational Note No. 1: See 600.6 for switches for signs and outline lighting.
> Informational Note No. 2: See 430.83, 430.109, and 430.110 for switches controlling motors.

(A) Alternating-Current General-Use Snap Switch. This form of switch shall only be used on ac circuits and used for controlling the following:

(1) Resistive and inductive loads not exceeding the ampere rating of the switch at the voltage applied

(2) Tungsten-filament lamp loads not exceeding the ampere rating of the switch at 120 volts

(3) Electric discharge lamp loads not exceeding the marked ampere and voltage rating of the switch

(4) Motor loads not exceeding 80 percent of the ampere rating of the switch at its rated voltage

(5) Electronic ballasts, self-ballasted lamps, compact fluorescent lamps, and LED lamp loads with their associated drivers, not exceeding 20 amperes and not exceeding the ampere rating of the switch at the voltage applied

(B) Alternating-Current or Direct-Current General-Use Snap Switch. This form of switch shall be permitted on either ac or dc circuits and used only for controlling the following:

(1) Resistive loads not exceeding the ampere rating of the switch at the voltage applied.

(2) Inductive loads not exceeding 50 percent of the ampere rating of the switch at the applied voltage. Switches rated in horsepower are suitable for controlling motor loads within their rating at the voltage applied.

(3) Tungsten-filament lamp loads not exceeding the ampere rating of the switch at the applied voltage if T-rated.

(4) Electronic ballasts, self-ballasted lamps, compact fluorescent lamps, and LED lamp loads with their associated drivers, not exceeding the ampere rating of the switch at the voltage applied.

(C) CO/ALR Snap Switches. Snap switches directly connected to aluminum conductors and rated 20 amperes or less shall be marked CO/ALR.

N **(D) Snap Switch Terminations.** Snap switch terminations shall be in accordance with the following:

(1) Terminals of 15-ampere and 20-ampere snap switches not marked CO/ALR shall be used with copper and copper-clad aluminum conductors only.

(2) Terminals marked CO/ALR shall be permitted to be used with copper, aluminum, and copper-clad aluminum conductors.

(3) Snap switches connected using screwless terminals of the conductor push-in type construction (also known as conductor push-in terminals) shall be installed on not greater than 15-ampere branch circuits and shall be connected with 14 AWG solid copper wire only unless listed and marked for other types of conductors.

(E) Alternating-Current General-Use Snap Switches Rated for 347 Volts. This form of switch shall not be rated less than 15 amperes at a voltage of 347 volts ac, and they shall not be readily interchangeable in box mounting with switches covered in 404.14(A) and (B). These switches shall be used only for controlling any of the following:

(1) Noninductive loads other than tungsten-filament lamps not exceeding the ampere and voltage ratings of the switch.

(2) Inductive loads not exceeding the ampere and voltage ratings of the switch. Where particular load characteristics or

limitations are specified as a condition of the listing, those restrictions shall be observed regardless of the ampere rating of the load.

(3) Electronic ballasts, self-ballasted lamps, compact fluorescent lamps, and LED lamp loads with their associated drivers, not exceeding 20 amperes and not exceeding the ampere rating of the switch at the voltage applied.

Δ **(F) Dimmer and Electronic Control Switches.** General-use dimmer switches and electronic control switches, such as timing switches and occupancy sensors, shall be used only to control permanently connected loads, such as incandescent luminaires, unless listed for the control of other loads and installed accordingly. They shall be marked by their manufacturer with their current and voltage ratings and used for loads that do not exceed their ampere rating at the voltage applied.

General-use dimmers are not permitted to control receptacles or cord-and-plug-connected table and floor lamps. If a dimmer evaluated only for the control of incandescent luminaires is used, the potential for connecting incompatible equipment such as a cord-and-plug-connected motor-operated appliance or a portable fluorescent lamp is increased by using the dimmer to control a receptacle. Section 404.14(F) does not apply to commercial dimmers or theater dimmers that can be used for fluorescent lighting and portable lighting.

(G) Cord- and-Plug-Connected Loads. Where a snap switch or control device is used to control cord-and-plug-connected equipment on a general-purpose branch circuit, each snap switch or control device controlling receptacle outlets or cord connectors that are supplied by permanently connected cord pendants shall be rated at not less than the rating of the maximum permitted ampere rating or setting of the overcurrent device protecting the receptacles or cord connectors, as provided in 210.21(B).

> Informational Note: See 210.50(A) and 400.10(A)(1) for equivalency to a receptacle outlet of a cord connector that is supplied by a permanently connected cord pendant.

Exception: Where a snap switch or control device is used to control not more than one receptacle on a branch circuit, the switch or control device shall be permitted to be rated at not less than the rating of the receptacle.

Where the load is cord-and-plug-connected through a receptacle or cord connector on a cord pendant (i.e., interchangeable), the load varies based on what is connected. A switch supplying cord-and-plug-connected loads is required to be rated at the rating or setting of the overcurrent protective device (OCPD) supplying the branch circuit. The exception recognizes circuits that may have multiple outlets where a switch controls one receptacle on the circuit and that such a switch be rated not less than the receptacle's rating.

N **404.16 Reconditioned Equipment.**

N **(A) Lighting, Dimmer, and Electronic Control Switches.** Reconditioned lighting, dimmer, and electronic control switches shall not be permitted.

N **(B) Snap Switches.** Reconditioned snap switches of any type shall not be permitted.

N **(C) Knife Switches, Switches with Butt Contacts, and Bolted Pressure Contact Switches.** Reconditioned knife switches, switches with butt contacts, and bolted pressure contact switches shall be permitted. If equipment has been damaged by fire, products of combustion, corrosive influences, or water, it shall be specifically evaluated by its manufacturer or a qualified testing laboratory prior to being returned to service.

N **(D) Molded-Case Switches.** Reconditioned molded-case switches shall not be permitted.

Part II. Construction Specifications

404.20 Marking.

(A) Ratings. Switches shall be marked with the current, voltage, and, if horsepower rated, the maximum rating for which they are designed.

(B) Off Indication. Where in the off position, a switching device with a marked OFF position shall completely disconnect all ungrounded conductors to the load it controls.

Δ **404.22 Electronic Control Switches.** Electronic control switches shall be listed. Electronic control switches shall not introduce current on the equipment grounding conductor during normal operation.

Exception: Electronic control switches that introduce current on the equipment grounding conductor shall be permitted for applications covered by 404.2(C), Exception. Electronic control switches that introduce current on the equipment grounding conductor shall be listed and marked for use in replacement or retrofit applications only.

404.26 Knife Switches Rated 600 to 1000 Volts. Auxiliary contacts of a renewable or quick-break type or the equivalent shall be provided on all knife switches rated 600 to 1000 volts and designed for use in breaking current over 200 amperes.

404.27 Fused Switches. A fused switch shall not have fuses in parallel except as permitted in 240.8.

404.28 Wire-Bending Space. The wire-bending space required by 404.3 shall meet Table 312.6(B)(2) spacings to the enclosure wall opposite the line and load terminals.

Δ **404.30 Switch Enclosures with Doors.** Switch mechanisms mounted within enclosures with doors that, when opened, expose uninsulated live parts shall be constructed so that when the switch is in the closed position access to the switch interior is restricted. Access to the interior with the switch in the closed position shall require the use of a tool or an approved design that provides equivalent protection from access by unqualified persons.

ARTICLE 406

Receptacles, Cord Connectors, and Attachment Plugs (Caps)

406.1 Scope. This article covers the rating, type, and installation of receptacles, cord connectors, and attachment plugs (cord caps).

N 406.2 Reconditioned Equipment. Reconditioned receptacles, attachment plugs, cord connectors, and flanged surface devices shall not be permitted.

406.3 Receptacle Rating and Type.

Δ (A) Receptacles. Receptacles shall be listed and marked with the manufacturer's name or identification and voltage and ampere ratings.

(B) Rating. Receptacles and cord connectors shall be rated not less than 15 amperes, 125 volts, or 15 amperes, 250 volts, and shall be of a type not suitable for use as lampholders.

> Informational Note: See 210.21(B) for receptacle ratings where installed on branch circuits.

(C) CO/ALR Receptacles. Receptacles rated 20 amperes or less and designed for the direct connection of aluminum conductors shall be marked CO/ALR.

If the receptacle is not of the CO/ALR type, it can be connected by a copper pigtail to an aluminum branch-circuit conductor only if the wire connector is identified for such a connection and is marked with the letters AL and CU.

N (D) Receptacle Terminations. Receptacle terminations shall be in accordance with the following:

(1) Terminals of 15-ampere and 20-ampere receptacles not marked CO/ALR shall be used with copper and copper-clad aluminum conductors only.

(2) Terminals marked CO/ALR shall be permitted to be used with aluminum, copper, and copper-clad aluminum conductors.

(3) Receptacles installed using screwless terminals of the conductor push-in type construction (also known as *push-in-terminals*) shall be installed on not greater than 15-ampere branch circuits and shall be connected with 14 AWG solid copper wire only unless listed and marked for other types of conductors.

> Informational Note: See UL 498, *Attachment Plugs and Receptacles*, for information regarding screwless terminals of various type constructions employed on receptacles. Screwless terminals of the separable-terminal assembly, spring-action clamp, and insulation-displacement type constructions are not classified in UL 498 as screwless terminals of the conductor push-in type construction (also known as push-in terminals).

(E) Isolated Ground Receptacles. Receptacles incorporating an isolated equipment grounding conductor connection intended for the reduction of electromagnetic interference as permitted in 250.146(D) shall be identified by an orange triangle located on the face of the receptacle.

(1) Isolated Equipment Grounding Conductor Required. Receptacles so identified shall be used only with equipment grounding conductors that are isolated in accordance with 250.146(D).

(2) Installation in Nonmetallic Boxes. Isolated ground receptacles installed in nonmetallic boxes shall be covered with a nonmetallic faceplate.

Exception: Where an isolated ground receptacle is installed in a nonmetallic box, a metal faceplate shall be permitted if the box contains a feature or accessory that permits the connection of the faceplate to the equipment grounding conductor.

(F) Controlled Receptacle Marking. All nonlocking-type, 125-volt, 15- and 20-ampere receptacles that are controlled by an automatic control device, or that incorporate control features that remove power from the receptacle for the purpose of energy management or building automation, shall be permanently marked with the symbol shown in Figure 406.3(F) and the word "controlled."

For receptacles controlled by an automatic control device, the marking shall be located on the receptacle face and visible after installation.

In both cases where a multiple receptacle device is used, the required marking of the word "controlled" and symbol shall denote which contact device(s) are controlled.

Exception: The marking shall not be required for receptacles controlled by a wall switch that provide the required room lighting outlets as permitted by 210.70.

Controlled

FIGURE 406.3(F) *Controlled Receptacle Marking Symbol.*

Many energy efficiency codes require that a percentage of installed 125-volt, 15- and 20-ampere receptacles be controlled by automatic means, such as time clocks or occupancy devices. Controlled receptacles are required to be marked to indicate which receptacles will be automatically de-energized by the controller. This allows a different receptacle, like a computer receptacle, to be selected if the load must be supplied during overnight hours.

(G) Receptacle with USB Charger. A 125-volt 15- or 20-ampere receptacle that additionally provides Class 2 power shall be listed and constructed such that the Class 2 circuitry is integral with the receptacle.

406.4 General Installation Requirements. Receptacle outlets shall be located in branch circuits in accordance with Part III of Article 210. General installation requirements shall be in accordance with 406.4(A) through (G).

(A) Grounding Type. Except as provided in 406.4(D), receptacles installed on 15- and 20-ampere branch circuits shall be of the grounding type. Grounding-type receptacles shall be installed only on circuits of the voltage class and current for which they are rated, except as provided in 210.21(B)(1) for single receptacles or Table 210.21(B)(2) and Table 210.21(B)(3) for two or more receptacles.

(B) Connection to Equipment Grounding Conductor. Receptacles and cord connectors that have equipment grounding conductor contacts shall have those contacts connected to an equipment grounding conductor.

Exception No. 1: Receptacles mounted on portable and vehicle-mounted generator sets and generators in accordance with 250.34.

Exception No. 2: Replacement receptacles as permitted by 406.4(D).

Δ **(C) Methods of Connection to Equipment Grounding Conductor.** The equipment grounding conductor contacts of receptacles shall be connected to an equipment grounding conductor of the circuit supplying the receptacle in accordance with 250.146.

Cord connectors shall be connected to the equipment grounding conductor of the circuit supplying the cord connector.

Informational Note No. 1: See 250.118 for acceptable grounding means.
Informational Note No. 2: See 250.130 for extensions of existing branch circuits.

(D) Replacements. Replacement of receptacles shall comply with 406.4(D)(1) through (D)(8), as applicable. Arc-fault circuit-interrupter type and ground-fault circuit-interrupter type receptacles shall be installed in a readily accessible location.

(1) Grounding-Type Receptacles. Where a grounding means exists in the receptacle enclosure or an equipment grounding conductor is installed in accordance with 250.130(C), grounding-type receptacles shall be used and shall be connected to the equipment grounding conductor in accordance with 406.4(C) or 250.130(C).

(2) Non–Grounding-Type Receptacles. Where attachment to an equipment grounding conductor does not exist in the receptacle enclosure, the installation shall comply with 406.4(D)(2)(a), (D)(2)(b), or (D)(2)(c).

(a) A non–grounding-type receptacle(s) shall be permitted to be replaced with another non–grounding-type receptacle(s).

(b) A non–grounding-type receptacle(s) shall be permitted to be replaced with a ground-fault circuit interrupter-type of receptacle(s). These receptacles or their cover plates shall be marked "No Equipment Ground." An equipment grounding conductor shall not be connected from the ground-fault circuit-interrupter-type receptacle to any outlet supplied from the ground-fault circuit-interrupter receptacle.

(c) A non–grounding-type receptacle(s) shall be permitted to be replaced with a grounding-type receptacle(s) where supplied through a ground-fault circuit interrupter. Where grounding-type receptacles are supplied through the ground-fault circuit interrupter, grounding-type receptacles or their cover plates shall be marked "GFCI Protected" and "No Equipment Ground," visible after installation. An equipment grounding conductor shall not be connected between the grounding-type receptacles.

Informational Note No. 1: Some equipment or appliance manufacturers require that the branch circuit to the equipment or appliance includes an equipment grounding conductor.
Informational Note No. 2: See 250.114 for a list of a cord-and-plug-connected equipment or appliances that require an equipment grounding conductor.

(3) Ground-Fault Circuit-Interrupter Protection. Ground-fault circuit-interrupter protection for receptacles shall be provided where replacements are made at receptacle outlets that are required to be so protected elsewhere in this *Code*. Ground-fault circuit interrupters shall be listed.

Exception: Where the outlet box size will not permit the installation of the GFCI receptacle, the receptacle shall be permitted to be replaced with a new receptacle of the existing type, where GFCI protection is provided and the receptacle is marked "GFCI Protected" and "No Equipment Ground," in accordance with 406.4(D)(2)(a), (D)(2)(b), or (D)(2)(c), as applicable.

Δ **(4) Arc-Fault Circuit-Interrupter Protection.** If a receptacle located in any areas specified in 210.12(A), (B), or (C) is replaced, a replacement receptacle at this outlet shall be one of the following:

(1) A listed outlet branch-circuit type AFCI receptacle
(2) A receptacle protected by a listed outlet branch-circuit type AFCI type receptacle
(3) A receptacle protected by a listed combination type AFCI circuit breaker

Exception: Section 210.12(E), Exception, shall not apply to replacement of receptacles.

Older homes are statistically more vulnerable to electrical fires. Extra protection for older homes is provided by the gradual replacement, over time, of non-arc-fault circuit-interpreter (AFCI) protected receptacles with new AFCI-protected ones.

(5) Tamper-Resistant Receptacles. Listed tamper-resistant receptacles shall be provided where replacements are made at receptacle outlets that are required to be tamper-resistant elsewhere in this *Code*, except in one of the following cases:

(1) Where a nongrounding receptacle is replaced with another nongrounding receptacle

(2) Where aluminum branch-circuit conductors are directly terminated on a CO/ALR receptacle, installed as replacement

This requirement does not mandate receptacle replacement. It merely institutes a requirement for the receptacle if it is replaced. For example, an ordinary 15-ampere receptacle in a bedroom of a 10-year-old one-family dwelling would be required to be replaced with a tamper-resistant (TR) receptacle because tamper-resistant receptacles are required in a bedroom of a new home constructed under the current edition of the *NEC®*.

(6) Weather-Resistant Receptacles. Weather-resistant receptacles shall be provided where replacements are made at receptacle outlets that are required to be so protected elsewhere in this *Code*.

Without the requirement for weather-resistant (WR) receptacles to be installed at the time of replacement, ordinary receptacles might be installed and subject to the same failures as the receptacles they replaced.

(7) Controlled Receptacles. Automatically controlled receptacles shall be replaced with equivalently controlled receptacles. If automatic control is no longer required, the receptacle and any associated receptacles marked in accordance with 406.3(F) shall be replaced with a receptacle and faceplate not marked in accordance with 406.3(F).

N **(8) Ground-Fault Protection of Equipment (GFPE).** Receptacles shall be provided with GFPE where replacements are made at receptacle outlets that are required to be so protected elsewhere in this *Code*.

(E) Cord- and Plug-Connected Equipment. The installation of grounding-type receptacles shall not be used as a requirement that all cord-and plug-connected equipment be of the grounded type.

> Informational Note: See 250.114 for types of cord-and plug-connected equipment to be grounded.

(F) Noninterchangeable Types. Receptacles connected to circuits that have different voltages, frequencies, or types of current (ac or dc) on the same premises shall be of such design that the attachment plugs used on these circuits are not interchangeable.

N **(G) Protection of Floor Receptacles.** Protection for floor receptacles shall be in accordance with the following:

(1) Physical protection of floor receptacles shall allow floor-cleaning equipment to be operated without damage to receptacles.

(2) All 125-volt, single-phase, 15- and 20-ampere floor receptacles installed in food courts and waiting spaces of passenger transportation facilities where food or drinks are allowed shall be GFCI protected.

406.5 Receptacle Mounting. Receptacles shall be mounted in identified boxes or assemblies. The boxes or assemblies shall be securely fastened in place unless otherwise permitted elsewhere in this *Code*. Screws used for the purpose of attaching receptacles to a box shall be of the type provided with a listed receptacle, or shall be machine screws having 32 threads per inch or part of listed assemblies or systems, in accordance with the manufacturer's instructions.

Receptacles in pendant boxes are permitted if the box is supported from the flexible cord in accordance with 314.23(H)(1). A pendant box that is properly suspended is not required to be securely fastened in place.

(A) Boxes That Are Set Back. Receptacles mounted in boxes that are set back from the finished surface as permitted in 314.20 shall be installed such that the mounting yoke or strap of the receptacle is held rigidly at the finished surface.

(B) Boxes That Are Flush. Receptacles mounted in boxes that are flush with the finished surface or project therefrom shall be installed such that the mounting yoke or strap of the receptacle is held rigidly against the box or box cover.

Sections 406.5(A) through (C) allow attachment plugs to be inserted or removed without moving the receptacle. Restricting movement of the receptacle helps to maintain effective grounding continuity for contact devices or receptacle yokes. The proper installation of receptacles helps ensure that attachment plugs can be fully inserted, which provides a better contact.

See also

314.23(B) or **(C),** each of which requires that an outlet box used to enclose a receptacle be rigidly and securely supported

314.20 for more information on mounting outlet boxes with the proper setback

(C) Receptacles Mounted on Covers. Receptacles mounted to and supported by a cover shall be held rigidly against the cover by more than one screw or shall be a device assembly or box cover listed and identified for securing by a single screw.

Receptacles mounted on raised covers, such as the receptacle illustrated in Exhibit 406.1, are not permitted to be secured by a single screw unless the device is listed and identified for that method.

(D) Position of Receptacle Faces. After installation, receptacle faces shall be flush with or project from faceplates of insulating material and shall project a minimum of 0.4 mm (0.015 in.) from metal faceplates.

Requiring receptacles to project from metal faceplates prevents faults between the blades of attachment plugs and metal faceplates. Proper faceplate mounting ensures that attachment plugs can be fully inserted, which provides a better contact. The *NEC* does not specify the position (grounding blade up or down) of a common vertically mounted 15- or 20-ampere duplex receptacle. Although many drawings in this handbook show the slots for grounding blades up, the receptacle may be installed with the slots for blades down. Receptacles can also be installed horizontally as well as vertically.

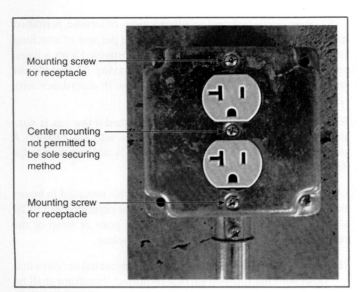

Mounting screw for receptacle

Center mounting not permitted to be sole securing method

Mounting screw for receptacle

EXHIBIT 406.1 A receptacle mounted on a raised cover secured with three screws.

Exception: Listed kits or assemblies encompassing receptacles and nonmetallic faceplates that cover the receptacle face, where the plate cannot be installed on any other receptacle, shall be permitted.

This exception allows the use of listed kits, which include the receptacle and a nonmetallic faceplate, that have been evaluated to ensure sufficient blade contact by the attachment plug when inserted in the receptacle. The kit's nonmetallic faceplate cannot fit standard-style receptacles.

(E) Receptacles in Countertops. Receptacle assemblies for installation in countertop surfaces shall be listed for countertop applications. Where receptacle assemblies for countertop applications are required to provide ground-fault circuit-interrupter protection for personnel in accordance with 210.8, such assemblies shall be permitted to be listed as GFCI receptacle assemblies for countertop applications.

(F) Receptacles in Work Surfaces. Receptacle assemblies and GFCI receptacle assemblies listed for work surface or countertop applications shall be permitted to be installed in work surfaces.

(G) Receptacle Orientation.

(1) Countertop and Work Surfaces. Receptacles shall not be installed in a face-up position in or on countertop surfaces or work surfaces unless listed for countertop or work surface applications.

(2) Under Sinks. Receptacles shall not be installed in a face-up position in the area below a sink.

(H) Receptacles in Seating Areas and Other Similar Surfaces. In seating areas or similar surfaces, receptacles shall not be installed in a face-up position unless the receptacle is any of the following:

(1) Part of an assembly listed as a furniture power distribution unit
(2) Part of an assembly listed either as household furnishings or as commercial furnishings
(3) Listed either as a receptacle assembly for countertop applications or as a GFCI receptacle assembly for countertop applications
(4) Installed in a listed floor box

(I) Exposed Terminals. Receptacles shall be enclosed so that live wiring terminals are not exposed to contact.

(J) Voltage Between Adjacent Devices. A receptacle shall not be grouped or ganged in enclosures with other receptacles, snap switches, or similar devices, unless they are arranged so that the voltage between adjacent devices does not exceed 300 volts, or unless they are installed in enclosures equipped with identified, securely installed barriers between adjacent devices.

406.6 Receptacle Faceplates (Cover Plates). Receptacle faceplates shall be installed so as to completely cover the opening and seat against the mounting surface.

Receptacle faceplates mounted inside a box having a recess-mounted receptacle shall effectively close the opening and seat against the mounting surface.

The faceplates/cover plates for recessed receptacles must fit the inside dimensions of the box. A small tolerance, typically $\frac{1}{32}$ inch or less, is required to facilitate installation of the cover plate. The recessed design and the very small opening provide the necessary protection against access to live parts. That the receptacle is required to be "effectively" closed allows some tolerance in application of the requirement.

(A) Thickness of Metal Faceplates. Metal faceplates shall be of ferrous metal not less than 0.76 mm (0.030 in.) in thickness or of nonferrous metal not less than 1.02 mm (0.040 in.) in thickness.

(B) Grounding. Metal faceplates shall be grounded.

Generally, this requirement is met by grounding the metal box. However, isolated ground receptacles installed in nonmetallic boxes are problematic because grounding the receptacle in this case does not ground the faceplate. Section 406.3(E)(2) contains two solutions concerning the receptacle faceplate. The first solution is to use only nonmetallic faceplates. The second solution, in the exception to 406.3(E)(2), allows a nonmetallic box manufacturer to add a feature or accessory to accomplish effective grounding of a metal faceplate.

(C) Faceplates of Insulating Material. Faceplates of insulating material shall be noncombustible and not less than 2.54 mm (0.10 in.) in thickness but shall be permitted to be less than 2.54 mm (0.10 in.) in thickness if formed or reinforced to provide adequate mechanical strength.

(D) Receptacle Faceplate (Cover Plates) with Integral Night Light and/or USB Charger. A flush device cover plate that additionally provides a night light and/or Class 2 output connector(s) shall be listed and constructed such that the night light and/or Class 2 circuitry is integral with the flush device cover plate.

Listed receptacle faceplates with integral night light, USB charger, or both, that rely solely on spring-tensioned contacts shall be connected to only brass or copper alloy receptacle terminal screws and shall be rated 1 watt or less.

Exception: Effective January 1, 2026, spring-tensioned contact connections to steel receptacle terminal screws shall be permitted if the receptacle faceplate is specifically listed and identified for connection to steel receptacle terminal screws.

Δ **406.7 Attachment Plugs, Cord Connectors, and Flanged Surface Devices.** All attachment plugs, cord connectors, and flanged surface devices (inlets and outlets) shall be listed and marked with the manufacturer's name or identification and voltage and ampere ratings.

(A) Construction of Attachment Plugs and Cord Connectors. Attachment plugs and cord connectors shall be constructed so that there are no exposed current-carrying parts except the prongs, blades, or pins. The cover for wire terminations shall be a part that is essential for the operation of an attachment plug or connector (dead-front construction).

(B) Connection of Attachment Plugs. Attachment plugs shall be installed so that their prongs, blades, or pins are not energized unless inserted into an energized receptacle or cord connectors. No receptacle shall be installed so as to require the insertion of an energized attachment plug as its source of supply.

An energized attachment plug cap can be dangerous. Exposed, energized blades pose a serious shock hazard to anyone handling the attachment plug. Energized blades could also make contact with metal faceplates or screws exposed on the faceplate during insertion. Attachment plug caps should never be installed in such a way that the blades can be energized without being plugged into a device.

Energized attachment plugs should not be used to supply power to a building from a portable generator when a power failure occurs. Section 406.7 prohibits the improper use of an attachment plug, where the blades are exposed and energized, to supply power to a cord body or to plug into a receptacle to backfeed it. Prongs or blades that are exposed to contact by persons must not be energized. Exhibit 406.2 illustrates a flanged inlet device that must not be energized until the cord connector is installed.

(C) Attachment Plug Ejector Mechanisms. Attachment plug ejector mechanisms shall not adversely affect engagement of the blades of the attachment plug with the contacts of the receptacle.

Attachment plug ejector mechanisms are designed for use by persons with mobility or visual impairment. They reduce the likelihood of damage to the cord when the cord is pulled to remove the plug.

EXHIBIT 406.2 Flanged inlet device. *(Courtesy of Legrand®)*

(D) Flanged Surface Inlet. A flanged surface inlet shall be installed such that the prongs, blades, or pins are not energized unless an energized cord connector is inserted into it.

406.8 Noninterchangeability. Receptacles, cord connectors, and attachment plugs shall be constructed such that receptacle or cord connectors do not accept an attachment plug with a different voltage or current rating from that for which the device is intended. However, a 20-ampere T-slot receptacle or cord connector shall be permitted to accept a 15-ampere attachment plug of the same voltage rating. Non–grounding-type receptacles and connectors shall not accept grounding-type attachment plugs.

For information on receptacle and attachment plug configurations, see NEMA WD 6, *Wiring Devices — Dimensional Specifications,* available for download at www.nema.org.

406.9 Receptacles in Damp or Wet Locations.

All 15- and 20-ampere receptacles for both damp and wet locations are required to be a listed weather-resistant (WR) type. The major differences between WR and non-WR receptacles are that the WR has additional corrosion protection, UV resistance, and cold impact resistance. As shown in Exhibit 406.3, WR receptacles are required even when the enclosures in 406.9(A) and (B) are provided.

The requirements for covers that are typically used with lower rated receptacles (15 through 60 amperes) are summarized in Exhibit 406.4.

Δ **(A) Damp Locations.** A receptacle installed outdoors in a location protected from the weather or in other damp locations shall have an enclosure for the receptacle that is weatherproof when the receptacle is covered (attachment plug cap not inserted and receptacle covers closed).

An installation suitable for wet locations shall also be considered suitable for damp locations.

A receptacle shall be considered to be in a location protected from the weather where located under roofed open porches, canopies, marquees, and the like, and will not be subjected to a beating rain or water runoff. All 125- and 250-volt nonlocking receptacles shall be a listed weather-resistant type. Hinged covers of outlet box hoods shall be able to open at least 90 degrees, or

EXHIBIT 406.3 A two-gang weatherproof cover suitable for use in wet locations. *(Courtesy of Thomas and Betts, A Member of the ABB Group)*

fully open if the cover is not designed to open 90 degrees from the closed to open position, after installation.

Informational Note: See ANSI/NEMA WD 6–2016, *Wiring Devices — Dimensional Specifications*, for the types of receptacles covered by this requirement.

(B) Wet Locations.

Δ **(1) Receptacles of 15 Amperes and 20 Amperes in a Wet Location.** Receptacles of 15 amperes and 20 amperes, 125 volts and 250 volts installed in a wet location shall have an enclosure that is weatherproof whether or not the attachment plug cap is inserted. An outlet box hood installed for this purpose shall be listed and shall be identified as extra-duty. Other listed products, enclosures, or assemblies providing weatherproof protection that do not utilize an outlet box hood need not be identified extra duty. Hinged covers of outlet box hoods shall be able to open at least 90 degrees, or fully open if the cover is not designed to open 90 degrees from the closed to open position, after installation.

Informational Note No. 1: See ANSI/UL 514D–2016, *Cover Plates for Flush-Mounted Wiring Devices*, for extra-duty outlet box hoods. Extra duty identification and requirements are not applicable to listed receptacles, faceplates, outlet boxes, enclosures, or assemblies that are identified as either being suitable for wet locations or rated as one of the outdoor enclosure–type numbers of Table 110.28 that does not utilize an outlet box hood.

EXHIBIT 406.4 Requirements for receptacle cover (enclosure) types.

Damp and Wet Receptacle Locations	Receptacle Cover (Enclosure) Type Requirements	
	Enclosure that *is not* weatherproof, with attachment plug cap inserted into receptacle 	Enclosure that *is* weatherproof, with attachment plug cap inserted into receptacle ("in-use" type) 
406.9(A): Outdoor damp locations	Minimum type required	Permitted
406.9(A): Indoor damp locations	Minimum type required	Permitted
406.9(B)(1)&(2): Wet locations	Required for receptacle types other than those rated 15 and 20 amperes, 125 and 250 volts, where the tool, appliance, or other utilization equipment plugged into the receptacle *is* attended while in use	(a) Required for receptacles rated 15 and 20 amperes, 125 and 250 volts (b) Required for receptacles other than those rated 15 and 20 amperes, 125 and 250 volts, where the tool, appliance, or other utilization equipment plugged into the receptacle *is not* attended while in use *Note: Outlet box hood must be listed and identified as "Extra Duty."*

Exception: 15- and 20-ampere, 125- through 250-volt receptacles installed in a wet location and subject to routine high-pressure spray washing shall be permitted to have an enclosure that is weatherproof when the attachment plug is removed.

All 15- and 20-ampere, 125- and 250-volt nonlocking-type receptacles shall be listed and so identified as the weather-resistant type.

Informational Note No. 2: See ANSI/NEMA WD 6–2016, *Wiring Devices — Dimensional Specifications,* for receptacle configurations. The configuration of weather-resistant receptacles covered by this requirement are identified as 5-15, 5-20, 6-15, and 6-20.

Where the cord-and-plug connection to receptacles is in a wet location, the enclosure is required to be weatherproof regardless of whether the plug is inserted. The requirement for this type of cover is not contingent on the anticipated use of the receptacle. This applies to all 15- and 20-ampere, 125- and 250-volt receptacles that are installed in wet locations at all occupancies. Exhibit 406.3 is an example of the type of receptacle enclosure required by 406.9(B)(1).

Outlet box hoods that are part of a weatherproof enclosure must be of "extra duty" durability to retain protection for the receptacles and capable of being opened at least 90 degrees after installation.

Δ **(2) Other Receptacles.** All other receptacles installed in a wet location shall be listed weather-resistant type, and installation shall comply with 406.9(B)(2)(a) or (B)(2)(b).

(a) A receptacle installed in a wet location where the product intended to be plugged into it is not attended while in use shall have an enclosure that is weatherproof with the attachment plug cap inserted or removed.

(b) A receptacle installed in a wet location where the product intended to be plugged into it will be attended while in use (e.g., portable tools) shall have an enclosure that is weatherproof when the attachment plug is removed.

Section 406.9(B)(2) does not apply to receptacles rated 15 and 20 amperes, 125 and 250 volts. Section 406.9(B)(2)(a) applies to receptacles of other ratings that supply cord-and-plug-connected equipment likely to be used outdoors or in a wet location for long periods of time. A portable pump motor is an example of such equipment. Receptacle enclosures must remain weatherproof while the portable pump is in use.

Section 406.9(B)(2)(b) applies to receptacles of other ratings that supply cord-and-plug-connected portable tools or other portable equipment likely to be used outdoors for a specific purpose and then removed.

Δ **(C) Bathtub and Shower Space.** Receptacles shall not be installed inside of the tub or shower or within a zone measured 900 mm (3 ft) horizontally from any outside edge of the bathtub or shower stall, including the space outside the bathtub or shower stall space below the zone.

The zone also includes the space measured vertically from the floor to 2.5 m (8 ft) above the top of the bathtub rim or shower stall threshold. The identified zone is all-encompassing and shall include the space directly over the bathtub or shower stall and the space below this zone, but not the space separated by a floor, wall, ceiling, room door, window, or fixed barrier.

Exception No. 1: Receptacles installed in accordance with 680.73 shall be permitted.

Exception No. 2: In bathrooms with less than the required zone, the receptacle(s) required by 210.52(D) shall be permitted to be installed opposite the bathtub rim or shower stall threshold on the farthest wall within the room.

Exception No. 3: Weight supporting ceiling receptacles (WSCR) shall be permitted to be installed for listed luminaires that employ a weight supporting attachment fitting (WSAF) in damp locations complying with 410.10(D).

Exception No. 4: In a dwelling unit, a single receptacle shall be permitted for an electronic toilet or personal hygiene device such as an electronic bidet seat. The receptacle shall be readily accessible and not located in the space between the toilet and the bathtub or shower.

Informational Note No. 1: See 210.8(A)(1) for GFCI requirements in a bathroom.

Informational Note No. 2: See 210.11(C) for bathroom branch circuits.

Informational Note No. 3: See 210.21(B)(1) for single receptacle on an individual branch.

The installation of receptacles inside bathtub or shower spaces, or within a zone measured three feet horizontally from the outer edge of the tub or shower is prohibited, even if the receptacle is installed in a weatherproof enclosure or is GFCI protected. The unprotected-line side of GFCI-protected receptacles installed in bathtub and shower spaces could become wet and therefore create a shock hazard by energizing surrounding wet surfaces. Prohibiting such installation helps minimize the use of shavers, televisions, radios, hair dryers, and so forth in these areas.

(D) Flush Mounting with Faceplate. The enclosure for a receptacle installed in an outlet box flush-mounted in a finished surface shall be made weatherproof by means of a weatherproof faceplate assembly that provides a watertight connection between the plate and the finished surface.

406.10 Grounding-Type Receptacles, Adapters, Cord Connectors, and Attachment Plugs.

(A) Grounding Poles (Connections). Grounding-type receptacles, cord connectors, and attachment plugs shall be provided with one fixed grounding pole in addition to the circuit poles. The grounding contacting pole of grounding-type plug-in ground-fault circuit interrupters shall be permitted to be of the movable, self-restoring type on circuits operating at not over 150 volts between any two conductors or any conductor and ground.

Δ **(B) Grounding-Pole (Connection) Identification.** Grounding-type receptacles, adapters, cord connections, and attachment plugs shall have a means for connection of an equipment grounding conductor to the grounding pole.

A terminal for connection to the grounding pole shall be designated by one of the following:

(1) A green-colored hexagonal-headed or -shaped terminal screw or nut, not readily removable.
(2) A green-colored pressure wire connector body (a wire barrel).
(3) A similar green-colored connection device, in the case of adapters. The grounding terminal of a grounding adapter shall be a green-colored rigid ear, lug, or similar device. The equipment grounding connection shall be so designed that it cannot make contact with current-carrying parts of the receptacle, adapter, or attachment plug. The adapter shall be polarized.
(4) If the terminal for the equipment grounding conductor is not visible, the conductor entrance hole shall be marked with the word *green* or *ground*, the letters *G* or *GR*, a grounding symbol, or otherwise identified by a distinctive green color. If the terminal for the equipment grounding conductor is readily removable, the area adjacent to the terminal shall be similarly marked.

Informational Note: See Informational Note Figure 406.10(B).

INFORMATIONAL NOTE FIGURE 406.10(B)
One Example of a Symbol Used to Identify the Termination Point for an Equipment Grounding Conductor.

Section 406.10(B)(3) requires the grounding terminal of an adapter to be a green-colored ear, lug, or similar device, thereby prohibiting use of an adapter with an attached pigtail grounding wire, which was used for many years.

(C) Grounding Terminal Use. A grounding terminal shall not be used for purposes other than connection to the equipment grounding conductor.

(D) Grounding-Pole (Connection) Requirements. Grounding-type attachment plugs and mating cord connectors and receptacles shall be designed such that the equipment grounding connection is made before the current-carrying connections. Grounding-type devices shall be so designed that grounding poles of attachment plugs cannot be brought into contact with current-carrying parts of receptacles or cord connectors.

The grounding blade of the attachment plug cap of most grounding-type combinations is longer than the circuit conductor blades and is used to ensure a "make-first, break-last" grounding connection. In some non-American National Standards Institute (ANSI) pin-and-sleeve-type configurations,

the grounding contact of the receptacle is closer to the face of the receptacle than it is to other contacts, serving the same purpose.

(E) Use. Grounding-type attachment plugs shall be used only with a cord having an equipment grounding conductor.

Informational Note: See 250.126 for identification of equipment grounding conductor terminals.

406.11 Connecting Receptacle Grounding Terminal to Box. The connection of the receptacle grounding terminal shall comply with 250.146.

Δ **406.12 Tamper-Resistant Receptacles.** All 15- and 20-ampere, 125- and 250-volt nonlocking-type receptacles in the following locations shall be listed tamper-resistant receptacles:

(1) All dwelling units, boathouses, mobile homes and manufactured homes, including their attached and detached garages, accessory buildings, and common areas
(2) Guest rooms and guest suites of hotels, motels, and their common areas
(3) Child care facilities
(4) Preschools and education facilities
(5) Within clinics, medical and dental offices, and outpatient facilities, the following spaces:

   a. Business offices accessible to the general public
   b. Lobbies, and waiting spaces
   c. Spaces of nursing homes and limited care facilities covered in 517.10(B)(2)

(6) Places of awaiting transportation, gymnasiums, skating rinks, fitness centers, and auditoriums
(7) Dormitory units
(8) Residential care/assisted living facilities, social and substance abuse rehabilitation facilities, and group homes
(9) Foster care facilities, nursing homes, and psychiatric hospitals
(10) Areas of agricultural buildings accessible to the general public and any common areas

Informational Note No. 1: See ANSI/NEMA WD 6-2016, *Wiring Devices — Dimensional Specifications*. This requirement would include receptacles identified as 5-15, 5-20, 6-15, and 6-20.

Informational Note No. 2: See NFPA 5000-2021, *Building Construction and Safety Code*, and the *International Building Code* (IBC)-2021 for more information on occupancy classifications for the types of facilities covered by this requirement.

Informational Note No. 3: Areas of agricultural building are frequently converted to hospitality areas. These areas can include petting zoos, stables, and buildings used for recreation or educational purposes where receptacles are installed.

The requirements for tamper-resistant receptacles ensure that children will be protected in all types of environments — in closely supervised areas, such as pediatric care locations and child care facilities, as well as in less structured, residential environments. Tamper-resistant construction provides the most effective and permanent means of preventing children from inserting foreign objects into receptacles. These receptacles are recognized in the U.S. General Services Administration (GSA) publication PBS 140, *Child Care Center Design Guide* (March 2003), as a critical design feature for child care areas.

Exception to (1) through (10): Receptacles in the following locations shall not be required to be tamper resistant:

(1) Receptacles located more than 1.7 m (5½ ft) above the floor

(2) Receptacles that are part of a luminaire or appliance

(3) Where the receptacle outlet is installed within the space occupied by or designated *for each appliance that, in normal use, is not easily moved from one place to another and is cord-and-plug-connected in accordance with 400.10(A) (6), (A)(7), or (A)(8) the following are permitted:*

   a. A single receptacle that is not readily accessible and supplies one appliance

   b. A duplex receptacle that is not readily accessible and supplies two appliances

(4) Nongrounding receptacles used for replacements as permitted in 406.4(D)(2)(a)

All areas specified in 210.52, as well as the pediatric areas in Article 517, require tamper-resistant receptacles. Lodging facilities require tamper-resistant receptacles to provide the same level of protection for children as they would have at home. Likewise, the receptacles in child care facilities, medical care occupancies, and some areas within assembly occupancies are required to be tamper resistant. Exhibit 406.5 shows a typical tamper-resistant receptacle.

Locking-type receptacles are not required to be tamper resistant. Only those receptacles installed at a height below 5½ feet can meet the requirements in 210.52 for wall spacing. Receptacles installed above 5½ feet are not accessible and are well out of reach of small children. Allowing the exception for a single receptacle or duplex receptacle located within dedicated space eliminates the need for tamper-resistant receptacles to be installed behind dishwashers, refrigerators, washing machines, and so forth. Non-grounding-type receptacles are exempted because there are no known 125-volt, 15- or 20-ampere

non-grounding-type receptacles listed as tamper-resistant receptacles.

406.13 Single-Pole Separable-Connector Type. Single-pole separable connectors shall be listed and labeled and shall comply with 406.13(A) through (D).

(A) Locking or Latching Type. Single-pole separable connectors shall be of either the locking or latching type and marked with the manufacturer's name or identification and voltage and ampere ratings.

(B) Identification. Connectors designated for connection to the grounded circuit conductor shall be identified by a white-colored housing; connectors designated for connection to the grounding circuit conductor shall be identified by a green-colored housing.

(C) Interchangeability. Single-pole separable connectors shall be permitted to be interchangeable for ac or dc use or for different current ratings or voltages on the same premises, provided they are listed for ac/dc use and marked in a suitable manner to identify the system to which they are intended to be connected.

(D) Connecting and Disconnecting. The use of single-pole separable connectors shall be performed by a qualified person and shall comply with at least one of the following conditions:

(1) Connection and disconnection of connectors are only possible where the supply connectors are interlocked to the source, and it is not possible to connect or disconnect connectors when the supply is energized.

(2) Line connectors are of the listed sequential-interlocking type so that load connectors are connected in the following sequence and that disconnection is in the reverse sequence:

   a. Equipment grounding conductor connection
   b. Grounded circuit conductor connection, if provided
   c. Ungrounded conductor connection

(3) A caution notice that complies with 110.21(B) is provided on the equipment employing single-pole separable

EXHIBIT 406.5 Tamper-resistant receptacle. Insertion of an object in any one side does not open the shutter *(left)*, but a two-bladed plug or grounding plug compresses the spring and simultaneously opens both shutters *(right)*. *(Courtesy of Legrand®)*

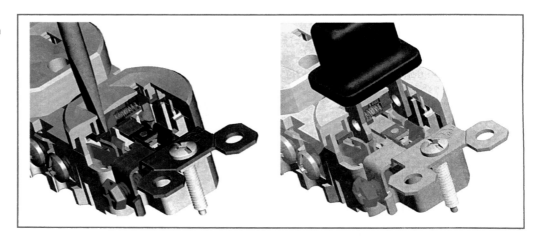

connectors, adjacent to the line connectors, indicating that connections are to be performed in the following sequence and that disconnection is in the reverse sequence:

a. Equipment grounding conductor connectors
b. Grounded circuit-conductor connectors, if provided
c. Ungrounded conductor connectors

Informational Note: See ANSI-UL 1691-2014, *Single Pole Locking-Type Separable Connectors*, for more information on single-pole locking-type separable connectors.

ARTICLE 408 Switchboards, Switchgear, and Panelboards

Part I. General

408.1 Scope. This article covers switchboards, switchgear, and panelboards. It does not apply to equipment operating at over 1000 volts, except as specifically referenced elsewhere in the *Code*.

See the definitions of *switchboard, switchgear,* and *panelboard* in Article 100. The 2014 *NEC®* modified and renamed what was formerly "metal-enclosed power switchgear" to "switchgear." This definition utilizes this generic term for both metal-clad and metal-enclosed switchgear.

See also

Article 495, which covers the general requirements for equipment operating over 1000 volts ac, 1500 volts dc, nominal

408.2 Reconditioned Equipment. The use of reconditioned equipment within the scope of this article shall be limited as described in 408.2(A) and (B). If equipment has been damaged by fire, products of combustion, corrosive influences, or water, it shall be specifically evaluated by its manufacturer or a qualified testing laboratory prior to being returned to service.

(A) Panelboards. Reconditioned panelboards shall not be permitted.

(B) Switchboards and Switchgear. Reconditioned switchboards and switchgear, or sections of switchboards or switchgear, shall be permitted.

Equipment reconditioning has long been a topic of discussion in the wake of disasters like wildfires, flooding, and hurricanes. Equipment can be new, reconditioned, refurbished, or remanufactured in response to such events. This section sets requirements for whether panelboards, switchboards, and switchgear can be reconditioned. Previously, this information was available in recommendations from manufacturers or other organizations, but it was not always readily available to those dealing with disaster recovery.

Panelboards are not permitted to be reconditioned. Switchboards and switchgear often can be reconditioned, which

typically is done in place due to the difficulties of removing and reinstalling them. The equipment must be evaluated by the manufacturer or qualified testing laboratory before being returned to operation to ensure that the original level of safety has been achieved. Additionally, reconditioning is permitted on a section-by-section basis because an event that damages a single section might not warrant condemning the entire assembly.

408.3 Support and Arrangement of Busbars and Conductors.

(A) Conductors and Busbars on a Switchboard, Switchgear, or Panelboard. Conductors and busbars on a switchboard, switchgear, or panelboard shall comply with 408.3(A)(1) and (A)(2) as applicable.

(1) Location. Conductors and busbars shall be located so as to be free from physical damage and shall be held firmly in place.

(2) Same Vertical Section. Other than the required interconnections and control wiring, only those conductors that are intended for termination in a vertical section of a switchboard or switchgear shall be located in that section.

Exception: Conductors shall be permitted to travel horizontally through vertical sections of switchboards and switchgear where such conductors are isolated from busbars by a barrier.

Conductors can be run horizontally through more than one section of a multi-section switchboard or switchgear when separated from the busbar by a service barrier. This is necessary where a raceway or cable entry is made into a section other than the one at which the conductors are terminated.

See also

230.62(C) for barrier requirements in service equipment

(B) Overheating and Inductive Effects. The arrangement of busbars and conductors shall be such as to avoid overheating due to inductive effects.

(C) Used as Service Equipment. Each switchboard, switchgear, or panelboard, if used as service equipment, shall be provided with a main bonding jumper sized in accordance with 250.28(D) or the equivalent placed within the panelboard or one of the sections of the switchboard or switchgear for connecting the grounded service conductor on its supply side to the switchboard, switchgear, or panelboard frame. All sections of a switchboard or switchgear shall be bonded together using an equipment-bonding jumper or a supply-side bonding jumper sized in accordance with 250.122 or 250.102(C)(1) as applicable.

Exception: Switchboards, switchgear, and panelboards used as service equipment on high-impedance grounded neutral systems in accordance with 250.36 shall not be required to be provided with a main bonding jumper.

(D) Terminals. In switchboards and switchgear, load terminals for field wiring shall comply with 408.18(C).

(E) Bus Arrangement.

(1) AC Phase Arrangement. Alternating-current phase arrangement on 3-phase buses shall be A, B, C from front to back, top to bottom, or left to right, as viewed from the front of the switchboard, switchgear, or panelboard. The B phase shall be that phase having the higher voltage to ground on 3-phase, 4-wire, delta-connected systems. Other busbar arrangements shall be permitted for additions to existing installations and shall be marked.

Exception: Equipment within the same single section or multisection switchboard, switchgear, or panelboard as the meter on 3-phase, 4-wire, delta-connected systems shall be permitted to have the same phase configuration as the metering equipment.

Informational Note: See 110.15 for requirements on marking the busbar or phase conductor having the higher voltage to ground where supplied from a 4-wire, delta-connected system.

The high leg is common on a 240/120-volt, 3-phase, 4-wire center-tap grounded delta system and must be designated as "B phase." Section 110.15 requires the high-leg marking to be the color orange or another effective means of identification, because the B phase voltage can be 208 volts nominal. Exhibit 408.1 is an example of a high-leg, grounded delta system.

The exception to 408.3(E)(1) permits the phase leg having the higher voltage to ground to be located at the right-hand position (C phase), making it unnecessary to transpose the panelboard, switchgear, or switchboard busbar arrangement ahead of and beyond a metering compartment. The exception recognizes the fact that metering compartments have been standardized with the high leg at the right position (C phase) rather than in the center on B phase.

> **See also**
>
> **110.15** and **230.56** for further information on identifying conductors with the higher voltage to ground

(2) DC Bus Arrangement. Direct-current ungrounded buses shall be permitted to be in any order. Arrangement of dc buses shall be field marked as to polarity, grounding system, and nominal voltage.

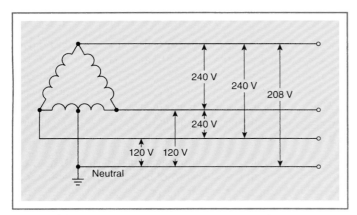

EXHIBIT 408.1 An example of a high-leg, grounded delta system.

(F) Switchboard, Switchgear, or Panelboard Identification. A caution sign(s) or a label(s) provided in accordance with 408.3(F)(1) through (F)(5) shall comply with 110.21(B).

(1) High-Leg Identification. A switchboard, switchgear, or panelboard containing a 4-wire, delta-connected system where the midpoint of one phase winding is grounded shall be legibly and permanently field marked as follows:

CAUTION _____ PHASE HAS _____ VOLTS TO GROUND

The requirement for legible marking of a switchboard, switchgear, or panelboard that contains a 3-phase, 4-wire center-tap grounded delta system resulted from injury and property damage caused by people not recognizing there is a high leg (208 volts to ground) in the switchboard, switchgear, or panelboard. This requirement helps to warn qualified individuals that a high leg exists. It is aimed at eliminating the hazards that come from accidentally connecting 120-volt equipment and devices to the high leg.

(2) Ungrounded AC Systems. A switchboard, switchgear, or panelboard containing an ungrounded ac electrical system as permitted in 250.21 shall be legibly and permanently field marked as follows:

CAUTION UNGROUNDED SYSTEM OPERATING — _____ VOLTS BETWEEN CONDUCTORS

The intent of this requirement is to delineate grounded from ungrounded electrical systems. When a ground fault occurs on a 3-phase ungrounded system, the voltage to ground on the ungrounded system equals the line-to-line voltage. The operational advantage of using an ungrounded system is continuity of operation, which in some processes might create a safer condition than would be achieved by automatic or unplanned opening of the supply circuit. Section 250.21(B) requires ungrounded systems of not less than 120 volts and not more than 1000 volts to be provided with ground detection. Ground detection will warn of the ground fault to permit an orderly shutdown of a process.

(3) High-Impedance Grounded Neutral AC System. A switchboard, switchgear, or panelboard containing a high-impedance grounded neutral ac system in accordance with 250.36 shall be legibly and permanently field marked as follows:

CAUTION: HIGH-IMPEDANCE GROUNDED NEUTRAL AC SYSTEM OPERATING — _____ VOLTS BETWEEN CONDUCTORS AND MAY OPERATE — _____ VOLTS TO GROUND FOR INDEFINITE PERIODS UNDER FAULT CONDITIONS

(4) Ungrounded DC Systems. A switchboard, switchgear, or panelboard containing an ungrounded dc electrical system in accordance with 250.169 shall be legibly and permanently field marked as follows:

CAUTION: UNGROUNDED DC SYSTEM OPERATING — _____ VOLTS BETWEEN CONDUCTORS

(5) Resistively Grounded DC Systems. A switchboard, switchgear, or panelboard containing a resistive connection between current-carrying conductors and the grounding system to stabilize voltage to ground shall be legibly and permanently field marked as follows:

CAUTION: DC SYSTEM OPERATING — _____ VOLTS BETWEEN CONDUCTORS AND MAY OPERATE — _____ VOLTS TO GROUND FOR INDEFINITE PERIODS UNDER FAULT CONDITIONS

(G) Minimum Wire-Bending Space. The minimum wire-bending space at terminals and minimum gutter space provided in switchboards, switchgear, and panelboards shall be as required in 312.6.

408.4 Descriptions Required.

(A) Circuit Directory or Circuit Description. Every circuit and circuit modification shall be provided with a legible and permanent description that complies with all of the following conditions as applicable:

(1) Located at each switch or circuit breaker in a switchboard or switchgear
(2) Included in a circuit directory that is located on the face of, inside of, or in an approved location adjacent to the panel door in the case of a panelboard
(3) Clear, evident, and specific to the purpose or use of each circuit including spare positions with an unused overcurrent device
(4) Described with a degree of detail and clarity that is unlikely to result in confusion between circuits
(5) Not dependent on transient conditions of occupancy
(6) Clear in explaining abbreviations and symbols when used

The circuit directory is an important feature for the safe operation of an electrical system. It must provide clear identification of circuit breakers and switches that service personnel or other responders may need to operate in an emergency. This requirement is specific to switchboards, switchgear, and panelboards; however, the identification requirements of 110.22 apply to all disconnecting means.

Circuits used for the same purpose must be identified by their location. For example, small-appliance branch circuits can supply outlets in the kitchen, dining room, and kitchen countertops. Identifying these circuits as small-appliance branch circuits is not acceptable; instead, they should be identified as "kitchen wall receptacles," "dining room floor receptacle," or "kitchen countertop receptacles left of sink." Circuit directories containing multiple entries with only "lights" or "outlets" do not provide the sufficient detail required by this section.

Spare devices are required to be marked to indicate that they are spares. Markings are required to indicate permanent features and not temporary conditions of occupancy. For example, for a circuit breaker supplying an office, a label with an employee's name is not useful if the employee no longer occupies that office.

(B) Source of Supply. All switchboards, switchgear, and panelboards supplied by a feeder(s) in other than one- or two-family dwellings shall be permanently marked in accordance with the following:

(1) With the identification and physical location of where the power originates
(2) With a label that is permanently affixed and of sufficient durability to withstand the environment involved
(3) Using a method that is not handwritten

Tracing a feeder circuit back to its originating switchboard, switchgear, panelboard, or other source can be a time-consuming and inaccurate process. Accurate identification of circuits promotes more efficient lockout/tagout processes, which provide a safer work environment for employees. Identification of the feeder circuit when the new feeder is being added is also more economical than the time-consuming process of tracing a circuit.

408.5 Clearance for Conductor Entering Bus Enclosures.
Where conduits or other raceways enter a switchboard, switchgear, floor-standing panelboard, or similar enclosure at the bottom, approved space shall be provided to permit installation of conductors in the enclosure. The wiring space shall not be less than shown in Table 408.5 where the conduit or raceways enter or leave the enclosure below the busbars, their supports, or other obstructions. The conduit or raceways, including their end fittings, shall not rise more than 75 mm (3 in.) above the bottom of the enclosure.

TABLE 408.5 *Clearance for Conductors Entering Bus Enclosures*

Conductor	Minimum Spacing Between Bottom of Enclosure and Busbars, Their Supports, or Other Obstructions	
	mm	in.
Insulated busbars, their supports, or other obstructions	200	8
Noninsulated busbars	250	10

408.6 Short-Circuit Current Rating.
Switchboards, switchgear, and panelboards shall have a short-circuit current rating not less than the available fault current. In other than one- and two-family dwelling units, the available fault current and the date the calculation was performed shall be field marked on the enclosure at the point of supply. The marking shall comply with 110.21(B)(3).

The available fault current must be calculated so that the short-circuit current ratings (SCCRs) of certain equipment can be determined to be adequate. Having an available fault current that is greater than the SCCR of installed equipment could lead to equipment failure during a fault. While this value is already required to be calculated and marked on service equipment in other than dwelling units, not all switchboards, switchgear, and panelboards are installed as service equipment or in other than dwelling units. The available fault current at the point where the equipment is installed must be determined to ensure that the SCCR is adequate where not installed as service equipment. It is also important to mark available fault current and when it was determined to ensure that the equipment continues to comply with 408.6 after installation modifications or expansions.

> **See also**
>
> **110.21(B)** for label durability considerations
>
> **110.24** for requirement at service equipment in other than dwellings

408.7 Unused Openings. Unused openings for circuit breakers and switches shall be closed using identified closures, or other approved means that provide protection substantially equivalent to the wall of the enclosure.

The requirement of 110.12(A) for closing unused openings such as knockouts (other than those provided for equipment mounting or drainage) applies to all electrical enclosures, including panelboard cabinets, switchgear, and switchboard enclosures. An unused opening might exist because of a renovation or an alteration of existing equipment. The requirement of this section is for closing unused openings for a circuit breaker where none is installed. Together these two requirements are necessary to restore the electrical equipment enclosure integrity to a condition that minimizes the possibility of an escaping arc, spark, or molten metal igniting surrounding combustible material and minimizes the potential for accidental contact with live parts.

N **408.9 Replacement Panelboards.** Replacement panelboards shall be permitted to be installed in existing enclosures in accordance with 408.9(A) or (B).

N **(A) Panelboards Listed for the Specific Enclosure.** If the replacement panelboard is listed for the specific enclosure identified by either catalog number or dimensional information, the panelboard shall be permitted to maintain its short-circuit current rating.

N **(B) Panelboards Not Listed for the Specific Enclosure.** If the available fault current is greater than 10,000 amperes, the completed work shall be field labeled. If the available fault current is 10,000 amperes or less, the replacement panelboard shall be identified for the application. Any previously applied listing marks on the cabinet that pertain to the panelboard shall be removed.

Part II. Switchboards and Switchgear

408.16 Switchboards and Switchgear in Damp or Wet Locations. Switchboards and switchgear in damp or wet locations shall be installed in accordance with 312.2.

408.17 Location Relative to Easily Ignitible Material. Switchboards and switchgear shall be placed so as to reduce to a minimum the probability of communicating fire to adjacent combustible materials. Where installed over a combustible floor, suitable protection thereto shall be provided.

Where flooring is combustible, one means of complying with this requirement is to form and attach a piece of sheet steel or other suitable noncombustible material to the floor under the electrical equipment.

408.18 Clearances.

(A) From Ceiling. For other than a totally enclosed switchboard or switchgear, a space not less than 900 mm (3 ft) shall be provided between the top of the switchboard or switchgear and any combustible ceiling, unless a noncombustible shield is provided between the switchboard or switchgear and the ceiling.

(B) Around Switchboards and Switchgear. Clearances around switchboards and switchgear shall comply with the provisions of 110.26.

Sufficient access and working space permit safe operation and maintenance of switchboards and switchgear. Table 110.26(A)(1) indicates minimum working clearances from 0 to 1000 volts.

> **See also**
>
> **Article 495** and **Table 110.34(A)** for switchboards and switchgear rated over 1000 volts

(C) Connections. Each section of equipment that requires rear or side access to make field connections shall be so marked by the manufacturer on the front. Section openings requiring rear or side access shall comply with 110.26. Load terminals for field wiring shall comply with 408.18(C)(1), (C)(2), or (C)(3) as applicable.

(1) Equipment Grounding Conductors. Load terminals for field wiring shall be so located that it is not necessary to reach across uninsulated ungrounded bus in order to make connections.

(2) Grounded Circuit Conductors. Where multiple branch or feeder grounded circuit conductor load terminals for field wiring are grouped together in one location, they shall be so located that it is not necessary to reach across uninsulated ungrounded bus, whether or not energized, in order to make connections.

Where only one branch or feeder set of load terminals for field wiring are grouped with its associated ungrounded load terminals, they shall be so located that it is not necessary to reach across energized uninsulated bus including other branch or feeder bus in order to make connections. Bus on the line side of service, branch, or feeder disconnects is considered energized with respect to its associated load side circuits.

(3) Ungrounded Conductors. Load terminals for ungrounded conductors shall be so located that it is not necessary to reach across energized uninsulated bus in order to make connections. Bus on the line side of service, branch, or feeder disconnects is considered energized with respect to its associated load side circuits.

408.19 Conductor Insulation. An insulated conductor used within a switchboard or switchgear shall be listed, shall be flame retardant, and shall be rated not less than the voltage applied to it and not less than the voltage applied to other conductors or busbars with which it may come into contact.

408.20 Location of Switchboards and Switchgear. Switchboards and switchgear that have any exposed live parts shall be located in permanently dry locations and then only where under competent supervision and accessible only to qualified persons. Switchboards and switchgear shall be located such that the probability of damage from equipment or processes is reduced to a minimum.

408.22 Grounding of Instruments, Relays, Meters, and Instrument Transformers on Switchboards and Switchgear. Instruments, relays, meters, and instrument transformers located on switchboards and switchgear shall be grounded as specified in 250.170 through 250.178.

408.23 Power Monitoring and Energy Management Equipment. The requirements of 312.8(B) shall apply.

Part III. Panelboards

408.30 General. All panelboards shall have a rating not less than the minimum feeder capacity required for the load calculated in accordance with Part III, IV, or V of Article 220, as applicable.

Many panelboards are suitable for use as service equipment and are so marked by the manufacturer. Listed panelboards are used with copper conductors, unless they are marked to indicate which terminals are suitable for use with aluminum conductors. Such marking must be independent of any marking on terminal connectors and must appear on a wiring diagram or other readily visible location. If all terminals are suitable for use with aluminum conductors as well as with copper conductors, the panelboard is marked "Use Copper or Aluminum Wire." A panelboard using terminals or main or branch-circuit units individually marked "AL-CU" is marked "Use Copper or Aluminum Wire" or "Use Copper Wire Only." The latter marking indicates that wiring space or other factors make the panelboard unsuitable for aluminum conductors.

Unless a panelboard is marked to indicate otherwise, the terminations are based on the use of 60°C ampacities for wire sizes 14 AWG through 1 AWG and 75°C ampacities for wire sizes 1/0 AWG and larger.

See also

110.14(C) for temperature limitations of connections and how they affect conductor ampacity

The term *lighting and appliance branch-circuit panelboards* and the term *power panelboards* are no longer used. In addition, the requirement for a maximum of 42 overcurrent devices applies only with Exception No. 1 of 408.36. All panelboards need

a single overcurrent device that protects the panelboard bus unless either of the exceptions of 408.36 applies.

408.36 Overcurrent Protection. In addition to the requirement of 408.30, a panelboard shall be protected by an overcurrent protective device having a rating not greater than that of the panelboard. This overcurrent protective device shall be located within or at any point on the supply side of the panelboard.

If a panelboard is required to have overcurrent protection, such protection can be provided by an overcurrent protective device (OCPD) in the panelboard or by an OCPD protecting the conductors that supply the panelboard. Exhibit 408.2 shows a panelboard with a main circuit breaker and provisions for inserting 60 circuit breakers. Exhibit 408.3 illustrates overcurrent protection for the panelboard feeders having a rating not greater than the rating of the panelboard. In either case, the OCPD rating is not permitted to exceed the panelboard rating. For example, a feeder protected by a 200-ampere OCPD supplies two main lugs only (MLO) panelboards, each with a 225-ampere rating. Because the panelboard is large enough to supply the calculated load and the OCPD protecting the feeder does not exceed the panelboard rating, an individual OCPD in the panelboard is not required.

Exception No. 1: Individual protection shall not be required for a panelboard protected by two main circuit breakers or two sets of fuses in other than service equipment, having a combined rating not greater than that of the panelboard. A

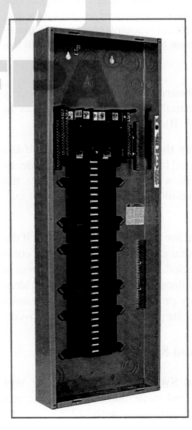

EXHIBIT 408.2 A panelboard with main circuit breaker disconnect suitable for use as service equipment. *(Courtesy of Schneider Electric)*

EXHIBIT 408.3 An arrangement of three individual panelboards with main overcurrent protection per 408.36 remote from the panelboards.

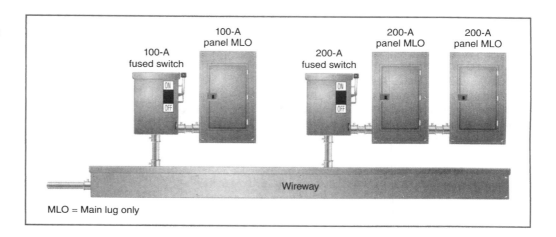

panelboard constructed or wired under this exception shall not contain more than 42 overcurrent devices. For the purposes of determining the maximum of 42 overcurrent devices, a 2-pole or a 3-pole circuit breaker shall be considered as two or three overcurrent devices, respectively.

Exception No. 2: For existing panelboards, individual protection shall not be required for a panelboard used as service equipment for an individual residential occupancy.

(A) Snap Switches Rated at 30 Amperes or Less. Panelboards equipped with snap switches rated at 30 amperes or less shall have overcurrent protection of 200 amperes or less.

(B) Supplied Through a Transformer. Where a panelboard is supplied through a transformer, the overcurrent protection required by 408.36 shall be located on the secondary side of the transformer.

Exception: A panelboard supplied by the secondary side of a transformer shall be considered as protected by the overcurrent protection provided on the primary side of the transformer where that protection is in accordance with 240.21(C)(1).

(C) Delta Breakers. A 3-phase disconnect or overcurrent device shall not be connected to the bus of any panelboard that has less than 3-phase buses. Delta breakers shall not be installed in panelboards.

(D) Back-Fed Devices. Plug-in-type overcurrent protection devices or plug-in type main lug assemblies that are backfed and used to terminate field-installed ungrounded supply conductors shall be secured in place by an additional fastener that requires other than a pull to release the device from the mounting means on the panelboard.

408.37 Panelboards in Damp or Wet Locations. Panelboards in damp or wet locations shall be installed to comply with 312.2.

408.38 Enclosure. Panelboards shall be mounted in cabinets, cutout boxes, or identified enclosures and shall be dead-front.

Where the available fault current is greater than 10,000 amperes, the panelboard and enclosure combination shall be evaluated for the application.

Exception: Panelboards other than of the dead-front, externally operable type shall be permitted where accessible only to qualified persons.

408.39 Relative Arrangement of Switches and Fuses. In panelboards, fuses of any type shall be installed on the load side of any switches.

Exception: Fuses installed as part of service equipment in accordance with the provisions of 230.94 shall be permitted on the line side of the service switch.

Sections 230.82 and 230.94 permit the service switch to be located on either the supply side or the load side of fuses such as cable limiters and other current-limiting devices. Where fuses of panelboards are accessible to other than qualified persons, such as occupants of a multifamily dwelling, 240.40 requires that disconnecting means be located on the supply side of all fuses in circuits of over 150 volts to ground and in cartridge-type fuses in circuits of any voltage. When the disconnect switch is opened, the fuses are de-energized, and danger from shock is reduced.

408.40 Grounding of Panelboards. Panelboard cabinets and panelboard frames, if of metal, shall be in physical contact with each other and shall be connected to an equipment grounding conductor. Where the panelboard is used with nonmetallic raceway or cable or where separate equipment grounding conductors are provided, a terminal bar for the equipment grounding conductors shall be secured inside the cabinet. The terminal bar shall be bonded to the cabinet and panelboard frame, if of metal; otherwise it shall be connected to the equipment grounding conductor that is run with the conductors feeding the panelboard.

A separate equipment grounding conductor (EGC) terminal bar must be installed and bonded to the panelboard for the termination of feeder and branch-circuit EGCs. Where installed within service equipment, this terminal is bonded to the neutral terminal bar. Any other connection between the equipment grounding terminal bar and the neutral bar (other than that allowed in 250.30, 250.32, and 250.142) is not permitted. If the grounding terminal bar and neutral terminal bar are connected downstream from the service equipment, current in the neutral or grounded conductor would take parallel paths through the EGCs (e.g., the raceway, the building structure, or earth) back to the service equipment. Normal load currents on the EGCs could create a shock hazard. Exposed metal parts of equipment could have a potential difference of several volts created by the load current on the grounding conductors. Another safety hazard of making this connection is arcing from loose connections at connectors and raceway fittings, which could be a potential fire hazard.

Exception: Where an isolated equipment grounding conductor for a branch circuit or a feeder is provided as permitted by 250.146(D), the insulated equipment grounding conductor that is run with the circuit conductors shall be permitted to pass through the panelboard without being connected to the panelboard's equipment grounding terminal bar.

Equipment grounding conductors shall not be connected to a terminal bar provided for grounded conductors or neutral conductors unless the bar is identified for the purpose and is located where interconnection between equipment grounding conductors and grounded circuit conductors is permitted or required by Part II and Part VII of Article 250.

The exception permits an isolated equipment grounding terminal if it is necessary for the reduction of electrical noise on the grounding circuit. The equipment grounding terminal must be grounded by an insulated EGC that is run with the circuit conductors. The isolated EGC is also permitted to pass through one or more panelboards without connection to the panelboard grounding terminal. However, it is important that the EGC terminate to the applicable separately derived system or service grounding terminal. It is not necessary to run this conductor as an isolated conductor all the way back to the service or source of the separately derived system. It should be terminated at a point in the system where the noise is mitigated. If the isolated EGC is run in a separate building, however, 250.146(D) requires the isolated EGC to terminate at a panelboard within the same building.

An isolated grounding electrode would not provide a safe path for ground-fault current. It would depend on the earth to be part of the fault return path, which might prevent a sufficient level of ground-fault current necessary to open the OCPD if a ground fault occurs.

See also

250.146(D) and **250.54** commentary for more information on isolated ground receptacles and auxiliary grounding electrodes

408.41 Grounded Conductor Terminations. Each grounded conductor shall terminate within the panelboard in an individual terminal that is not also used for another conductor.

Exception: Grounded conductors of circuits with parallel conductors shall be permitted to terminate in a single terminal if the terminal is identified for connection of more than one conductor.

In accordance with 110.14(A), conductor terminations are suitable only for a single conductor unless the terminal is marked or otherwise identified as suitable for more than one conductor. The use of a single termination point within a panelboard to connect more than one grounded conductor or to connect a grounded conductor and an EGC can be problematic if it is necessary to isolate a particular grounded conductor for testing purposes. For example, if the grounded conductors of two branch circuits were terminated at a single connection point, and it was necessary to isolate one branch circuit for the purposes of troubleshooting, the fact that the circuit not being tested remained energized could create an unsafe working condition for service personnel disconnecting the grounded conductor of the circuit being tested. In some cases, panelboard instructions are provided that permit the use of a single-conductor termination for more than one EGC.

See also

408.40 for the requirements on panelboard terminations for EGCs

408.43 Panelboard Orientation. Panelboards shall not be installed in the face-up or face-down position.

Part IV. Construction Specifications

408.50 Panels. The panels of switchboards and switchgear shall be made of moisture-resistant, noncombustible material.

408.51 Busbars. Insulated or bare busbars shall be rigidly mounted.

408.52 Protection of Instrument Circuits. Instruments, pilot lights, voltage (potential) transformers, and other switchboard or switchgear devices with potential coils shall be supplied by a circuit that is protected by standard overcurrent devices rated 15 amperes or less.

Exception No. 1: Overcurrent devices rated more than 15 amperes shall be permitted where the interruption of the circuit could create a hazard. Short-circuit protection shall be provided.

Exception No. 2: For ratings of 2 amperes or less, special types of enclosed fuses shall be permitted.

408.54 Maximum Number of Overcurrent Devices. A panelboard shall be provided with physical means to prevent the installation of more overcurrent devices than that number for which the panelboard was designed, rated, and listed.

For the purposes of this section, a 2-pole circuit breaker or fusible switch shall be considered two overcurrent devices; a

3-pole circuit breaker or fusible switch shall be considered three overcurrent devices.

408.55 Wire-Bending Space Within an Enclosure Containing a Panelboard.

(A) Top and Bottom Wire-Bending Space. The enclosure for a panelboard shall have the top and bottom wire-bending space sized in accordance with Table 312.6(B)(2) for the largest conductor entering or leaving the enclosure.

Exception No. 1: Either the top or bottom wire-bending space shall be permitted to be sized in accordance with Table 312.6(A) for a panelboard rated 225 amperes or less and designed to contain not over 42 overcurrent devices. For the purposes of this exception, a 2-pole or a 3-pole circuit breaker shall be considered as two or three overcurrent devices, respectively.

Exception No. 2: Either the top or bottom wire-bending space for any panelboard shall be permitted to be sized in accordance with Table 312.6(A) where at least one side wire-bending space is sized in accordance with Table 312.6(B)(2) for the largest conductor to be terminated in any side wire-bending space.

Exception No. 3: The top and bottom wire-bending space shall be permitted to be sized in accordance with Table 312.6(A) spacings if the panelboard is designed and constructed for wiring using only a single 90-degree bend for each conductor, including the grounded circuit conductor, and the wiring diagram shows and specifies the method of wiring that shall be used.

Exception No. 4: Either the top or the bottom wire-bending space, but not both, shall be permitted to be sized in accordance with Table 312.6(A) where there are no conductors terminated in that space.

Using Exhibit 408.4 as a reference, the general rule calls for wire-bending spaces T_1 and T_4 to be in accordance with Table 312.6(B)(2) for size M conductors (assuming those are the largest conductors entering the enclosure). Side wire-bending space T_2 must be in accordance with Table 312.6(A) for the largest wire size to be used within that side space, and T_3 must be similarly sized for the largest conductor within the enclosure's right side.

Exception No. 1 to 408.55(A) permits either T_1 or T_4 (not both) to be reduced to the space required by Table 312.6(A) for size M conductors for a panelboard rated 225 amperes or less and designed to contain not over 42 overcurrent devices.

Exception No. 2 to 408.55(A) permits either T_1 or T_4 (not both) to be reduced to the space required by Table 312.6(A) for size M conductors for any panelboard. Exception No. 2 is valid where either T_2 or T_3 (or both) is sized in accordance with Table 312.6(B)(2) for the largest conductor to be terminated in either the left- or the right-side space. Under the construction rules of 408.55, a panelboard enclosure might not be of adequate size for all manner of wiring; therefore, 312.6 must be considered when wiring is planned.

Exception No. 3 to 408.55(A) permits both the top and the bottom wire-bending space to be reduced as noted. A single 90-degree bend, meaning one and only one 90-degree bend, must be present for the ungrounded conductors. A grounded conductor is permitted to be wired straight in if spacing is provided per Table 312.6(B)(2) for the grounded conductor.

Exception No. 4 to 408.55(A) permits a reduction to the Table 312.6(A) spacing for the top or bottom space where no terminals face that space. In this case, the space is a gutter space, and measurement is on a line perpendicular to the wall of the enclosure and to the closest barrier post or side of a switch, fuse, or circuit breaker unit that is installed or may be installed in the future. Exhibit 408.4 illustrates that exception.

(B) Side Wire-Bending Space. Side wire-bending space shall be in accordance with Table 312.6(A) for the largest conductor to be terminated in that space.

(C) Back Wire-Bending Space. Where a raceway or cable entry is in the wall of the enclosure opposite a removable cover, the distance from that wall to the cover shall be permitted to comply with the distance required for one wire per terminal in Table 312.6(A). The distance between the center of the rear entry and the nearest termination for the entering conductors shall not be less than the distance given in Table 312.6(B)(2).

408.56 Minimum Spacings.
The distance between uninsulated metal parts, busbars, and other uninsulated live parts shall not be less than specified in Table 408.56.

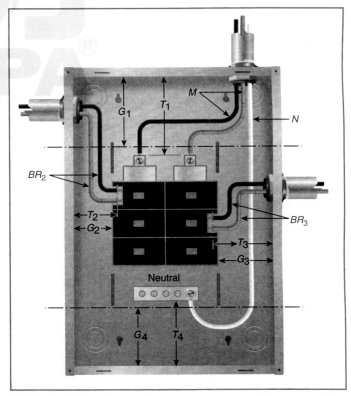

EXHIBIT 408.4 Panelboard wire-bending space per 312.6 and 408.55.

Where close proximity does not cause excessive heating, parts of the same polarity at switches, enclosed fuses, and so forth shall be permitted to be placed as close together as convenience in handling will allow.

Exception: The distance shall be permitted to be less than that specified in Table 408.56 at circuit breakers and switches and in listed components installed in switchboards, switchgear, and panelboards.

Δ **TABLE 408.56** *Minimum Spacings Between Bare Metal Parts*

AC or DC Voltage	Opposite Polarity Where Mounted on the Same Surface		Opposite Polarity Where Held Free in Air		Live Parts to Ground*	
	mm	in.	mm	in.	mm	in.
Not over 125 volts, nominal	19.1	¾	12.7	½	12.7	½
Not over 250 volts, nominal	31.8	1¼	19.1	¾	12.7	½
Not over 1000 volts, nominal	50.8	2	25.4	1	25.4	1

*For spacing between live parts and doors of cabinets, the dimensions in 312.101(A) shall apply.

408.58 Panelboard Marking. Panelboards shall be durably marked by the manufacturer with the voltage and the current rating and the number of ac phases or dc buses for which they are designed and with the manufacturer's name or trademark in such a manner so as to be visible after installation, without disturbing the interior parts or wiring.

ARTICLE 409   Industrial Control Panels

Part I.   General

Δ **409.1 Scope.** This article covers industrial control panels intended for general use and operating at 1000 volts or less.

> Informational Note: See ANSI/UL 508A, *Standard for Industrial Control Panels*, a safety standard for industrial control panels.

Field- and factory-assembled control panels are used for the control and operation of many processes, from a sewage pump station to an industrial process line. Similar in function to motor control centers in some regards, control panels also contain control, overcurrent protection, and power distribution equipment for operation of industrial heating processes, robotics, spray painting and powder coating lines, and so forth.

Δ **409.3 Other Articles.** In addition to the requirements of this article, industrial control panels that contain branch circuits

for specific loads or components, or are for control of specific types of equipment addressed in other articles of this *Code*, shall be constructed and installed in accordance with the applicable requirements from those articles.

Part II.   Installation

409.20 Conductor — Minimum Size and Ampacity. The size of the industrial control panel supply conductor shall have an ampacity not less than 125 percent of the full-load current rating of all heating loads plus 125 percent of the full-load current rating of the highest rated motor plus the sum of the full-load current ratings of all other connected motors and apparatus based on their duty cycle that may be in operation at the same time.

409.21 Overcurrent Protection.

(A) General. Industrial control panels shall be provided with overcurrent protection in accordance with Parts I and II of Article 240.

(B) Location. This protection shall be provided for each incoming supply circuit by either of the following:

(1) An overcurrent protective device located ahead of the industrial control panel.
(2) A single main overcurrent protective device located within the industrial control panel. Where overcurrent protection is provided as part of the industrial control panel, the supply conductors shall be considered as either feeders or taps as covered by 240.21.

(C) Rating. The rating or setting of the overcurrent protective device for the circuit supplying the industrial control panel shall not be greater than the sum of the largest rating or setting of the branch-circuit short-circuit and ground-fault protective device provided with the industrial control panel, plus 125 percent of the full-load current rating of all resistance heating loads, plus the sum of the full-load currents of all other motors and apparatus that could be in operation at the same time.

Exception: Where one or more instantaneous trip circuit breakers or motor short-circuit protectors are used for motor branch-circuit short-circuit and ground-fault protection as permitted by 430.52(C), the procedure specified above for determining the maximum rating of the protective device for the circuit supplying the industrial control panel shall apply with the following provision: For the purpose of the calculation, each instantaneous trip circuit breaker or motor short-circuit protector shall be assumed to have a rating not exceeding the maximum percentage of motor full-load current permitted by Table 430.52(C)(1) for the type of control panel supply circuit protective device employed.

Where no branch-circuit short-circuit and ground-fault protective device is provided with the industrial control panel for motor or combination of motor and non-motor loads, the rating or setting of the overcurrent protective device shall be based on 430.52 and 430.53, as applicable.

409.22 Short-Circuit Current Rating.

(A) Installation. An industrial control panel shall not be installed where the available fault current exceeds its short-circuit current rating as marked in accordance with 409.110(4).

(B) Documentation. If an industrial control panel is required to be marked with a short-circuit current rating in accordance with 409.110(4), the available fault current at the industrial control panel and the date the available fault current calculation was performed shall be documented and made available to those authorized to inspect, install, or maintain the installation.

409.30 Disconnecting Means. Disconnecting means that supply motor loads shall comply with Part IX of Article 430.

409.60 Bonding. Industrial control panels shall be grounded and bonded in accordance with 409.60(A) and (B).

N **(A) Grounding.** An equipment grounding conductor sized in accordance with 250.122 shall be connected to an equipment grounding bus or to an equipment grounding termination point provided in a single-section industrial control panel.

N **(B) Bonding.** Multisection industrial control panels shall be bonded together using an equipment bonding jumper sized in accordance with 250.102(D).

N **409.70 Surge Protection.** Safety circuits for personnel protection that are subject to damage from surge events shall have surge protection installed within or immediately adjacent to the control panel.

Part III. Construction Specifications

Part III provides the AHJ with a set of requirements that can be used as a benchmark for approval of a field-constructed control panel.

409.100 Enclosures. Table 110.28 shall be used as the basis for selecting industrial control panel enclosures for use in specific locations other than hazardous (classified) locations. The enclosures are not intended to protect against conditions such as condensation, icing, corrosion, or contamination that may occur within the enclosure or enter via the conduit or unsealed openings.

Δ **409.102 Busbars.** Industrial control panels utilizing busbars shall comply with 409.102(A) and (B).

(A) Support and Arrangement. Busbars shall be protected from physical damage and be held firmly in place.

(B) Phase Arrangement. The phase arrangement on 3-phase horizontal common power and vertical buses shall be A, B, C from front to back, top to bottom, or left to right, as viewed from the front of the industrial control panel. The B phase shall be that phase having the higher voltage to ground on 3-phase, 4-wire, delta-connected systems. Other busbar arrangements shall be permitted for additions to existing installations, and the phases shall be permanently marked.

409.104 Wiring Space.

(A) General. Industrial control panel enclosures shall not be used as junction boxes, auxiliary gutters, or raceways for conductors feeding through or tapping off to other switches or overcurrent devices or other equipment, unless the conductors fill less than 40 percent of the cross-sectional area of the wiring space. In addition, the conductors, splices, and taps shall not fill the wiring space at any cross section to more than 75 percent of the cross-sectional area of that space.

(B) Wire Bending Space. Wire bending space within industrial control panels for field wiring terminals shall be in accordance with the requirements in 430.10(B).

409.106 Spacings. Spacings in feeder circuits between uninsulated live parts of adjacent components, between uninsulated live parts of components and grounded or accessible non–current-carrying metal parts, between uninsulated live parts of components and the enclosure, and at field wiring terminals shall be as shown in Table 430.97(D).

Exception: Spacings shall be permitted to be less than those specified in Table 430.97(D) at circuit breakers and switches and in listed components installed in industrial control panels.

409.108 Service Equipment. Where used as service equipment, each industrial control panel shall be of the type that is suitable for use as service equipment.

Where a grounded conductor is provided, the industrial control panel shall be provided with a main bonding jumper, sized in accordance with 250.28(D), for connecting the grounded conductor, on its supply side, to the industrial control panel equipment ground bus or equipment ground terminal.

Δ **409.110 Marking.** An industrial control panel shall have permanent markings that are visible after installation. The markings in 409.110(2) and (3) shall be attached to the outside of the enclosure. The markings in 409.110(1), (4), (5), (6), and (7) shall be attached to either the inside or outside of the enclosure. The following markings shall be included:

(1) Manufacturer's name, trademark, or other descriptive marking by which the organization responsible for the product can be identified.

(2) Supply voltage, number of phases, frequency, and full-load current for each incoming supply circuit.

(3) Where the industrial control panel is supplied by more than one electrical source and where more than one disconnecting means is required to disconnect all circuits 50-volts or more within the control panel, marked to indicate that more than one disconnecting means is required to de-energize the equipment. The location of the means necessary to disconnect all circuits 50-volts or more shall be documented and available.

The person servicing the industrial control panel might not realize that more than one power source supplies the panel. Indication that multiple power sources are present and documentation that lists the location of all applicable disconnects help ensure the safety of service personnel. This requirement is similar to the requirement in Section 55.4 of UL 508A, *Standard for Industrial Control Panels,* which is the standard for listed industrial control panels.

(4) Short-circuit current rating of the industrial control panel based on one of the following:

   a. Short-circuit current rating of a listed and labeled assembly

   b. Short-circuit current rating established utilizing an approved method

Informational Note: See ANSI/UL 508A, *Standard for Industrial Control Panels,* Supplement SB, for an example of an approved method.

Exception to (4): Short-circuit current rating markings are not required for industrial control panels containing only control circuit components.

A group of components assembled in a common enclosure for the purposes of operation, control, and overcurrent protection should be able to limit and contain the effects of an internal fault (such as a short circuit or ground fault) so that the internal fault does not pose an external threat.

However, in many control panel installations, the available fault energy at the line terminals of components within the control panel is significant. In addition, there is also an interaction of the protective and control components under fault conditions that can be assessed only as part of the evaluation of the panel by a qualified testing laboratory. The panel's short-circuit current rating must be able to withstand the available short-circuit current at the panel terminals.

(5) If the industrial control panel is intended as service equipment, marked to identify it as being suitable for use as service equipment.

(6) Electrical wiring diagram, the identification number of a separate electrical wiring diagram, or a designation referenced in a separate wiring diagram.

(7) An enclosure type number.

ARTICLE 410 Luminaires, Lampholders, and Lamps

Part I. General

410.1 Scope. This article covers luminaires, portable luminaires, lampholders, pendants, incandescent filament lamps, arc lamps, electric-discharge lamps, decorative lighting products, lighting accessories for temporary seasonal and holiday use, portable flexible lighting products, and the wiring and equipment forming part of such products and lighting installations.

410.2 Reconditioned Equipment. Reconditioned luminaires, lampholders, ballasts, LED drivers, lamps, and retrofit kits shall not be permitted. If a retrofit kit is installed in a luminaire in accordance with the installation instructions, the retrofitted luminaire shall not be considered reconditioned.

410.5 Live Parts. Luminaires, portable luminaires, lampholders, and lamps shall have no live parts normally exposed to contact. Exposed accessible terminals in lampholders and switches shall not be installed in metal luminaire canopies or in open bases of portable table or floor luminaires.

Exception: Cleat-type lampholders located at least 2.5 m (8 ft) above the floor shall be permitted to have exposed terminals.

410.6 Listing Required. All luminaires, lampholders, and retrofit kits shall be listed.

410.8 Inspection. Luminaires shall be installed such that the connections between the luminaire conductors and the circuit conductors can be inspected without requiring the disconnection of any part of the wiring unless the luminaires are connected by attachment plugs and receptacles.

Part II. Luminaire Locations

410.10 Luminaires in Specific Locations.

(A) Wet and Damp Locations. Luminaires installed in wet or damp locations shall be installed such that water cannot enter or accumulate in wiring compartments, lampholders, or other electrical parts. All luminaires installed in wet locations shall be marked as suitable for wet locations. All luminaires installed in damp locations shall be marked as suitable for wet locations or suitable for damp locations.

Correct design, construction, and installation of these luminaires will prevent the entrance of rain, snow, ice, and dust. Outdoor parks and parking lots, outdoor recreational areas, car wash areas, and building exteriors are examples of wet locations.

Luminaires in locations protected from the weather and not subject to water saturation but still exposed to moisture must be marked "Suitable for Damp Locations" or "Suitable for Wet Locations." The following are examples of damp locations:

1. The underside of store or gasoline station canopies
2. Some cold-storage warehouses
3. Some agricultural buildings
4. Some basements
5. Roofed open porches and carports

Definitions of the terms *location, damp*; *location, dry*; and *location, wet* can be found in Article 100.

(B) Corrosive Locations. Luminaires installed in corrosive locations shall be of a type suitable for such locations.

(C) In Ducts or Hoods. Luminaires shall be permitted to be installed in commercial cooking hoods where all of the following conditions are met:

(1) The luminaire shall be identified for use within commercial cooking hoods and installed such that the temperature limits of the materials used are not exceeded.
(2) The luminaire shall be constructed so that all exhaust vapors, grease, oil, or cooking vapors are excluded from the lamp and wiring compartment. Diffusers shall be resistant to thermal shock.
(3) Parts of the luminaire exposed within the hood shall be corrosion resistant or protected against corrosion, and the surface shall be smooth so as not to collect deposits and to facilitate cleaning.
(4) Wiring methods and materials supplying the luminaire(s) shall not be exposed within the cooking hood.

Informational Note: See 110.11 for conductors and equipment exposed to deteriorating agents.

NFPA 96, *Standard for Ventilation Control and Fire Protection of Commercial Cooking Operations,* provides the minimum fire safety requirements related to the design, installation, operation, inspection, and maintenance of all public and private cooking operations, except in single-family residential dwellings. NFPA 96 covers residential cooking equipment where used for purposes other than residential family use — such as employee kitchens or break areas and church and meeting hall kitchens — regardless of frequency of use.

Grease can cause the deterioration of conductor insulation, resulting in short circuits or ground faults in wiring, hence the requirement prohibiting exposed wiring methods and materials (raceways, cables, lampholders) within ducts or hoods. Conventional enclosed and gasketed-type luminaires located in the path of travel of exhaust products are not permitted because a fire could result from the high temperatures on grease-coated glass bowls or globes enclosing the lamps. Recessed or surface gasketed-type luminaires intended for location within hoods must be identified as suitable for the specific purpose and should be installed with the required clearances maintained. Note that wiring systems, including rigid metal conduit, are not permitted to be run exposed within the cooking hood.

For further information, refer to UL 710, *Standard for Safety for Exhaust Hoods for Commercial Cooking Equipment.*

(D) Bathtub and Shower Areas. A luminaire installed in a bathtub or shower area shall meet all of the following requirements:

(1) No parts of cord-connected luminaires, chain-, cable-, or cord-suspended luminaires, lighting track, pendants, or ceiling-suspended (paddle) fans with luminaire (light kit) shall be located within a zone measured 900 mm (3 ft) horizontally and 2.5 m (8 ft) vertically from the top of the bathtub rim or shower stall threshold. This zone is all-encompassing and includes the space directly over the tub or shower stall.
(2) Luminaires located within the actual outside dimension of the bathtub or shower to a height of 2.5 m (8 ft) vertically from the top of the bathtub rim or shower threshold shall be marked suitable for damp locations or marked suitable for wet locations. Luminaires located where subject to shower spray shall be marked suitable for wet locations.

Where luminaires are subject to shower spray, they must be listed for a wet location. Luminaires installed in the tub or shower zone and not subject to shower spray are required to be listed for use in a damp location. GFCI protection is required only where specified in the installation instructions for the luminaire.

The intent is to keep cord-connected, chain-hanging, or pendant luminaires and suspended fans out of the reach of an individual standing on a bathtub rim. The list of prohibited items recognizes that a risk of electric shock is present for each one.

Exhibit 410.1 illustrates the restricted zone in which the specified items are prohibited. This requirement applies to

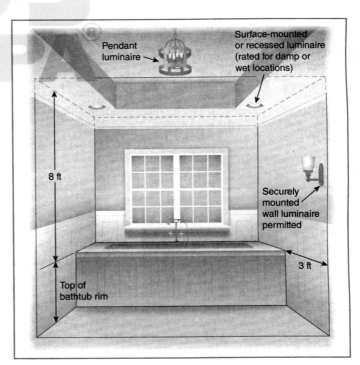

EXHIBIT 410.1 Luminaires located near a bathtub.

hydromassage bathtubs, as defined in Article 100, as well as other bathtub types and shower areas.

See also

680.43 for installation requirements for spas and hot tubs installed indoors

(E) Luminaires in Indoor Sports, Mixed-Use, and All-Purpose Facilities. Luminaires subject to physical damage, using a mercury vapor or metal halide lamp, installed in playing and spectator seating areas of indoor sports, mixed-use, or all-purpose facilities shall be of the type that protects the lamp with a glass or plastic lens. Such luminaires shall be permitted to have an additional guard.

Accidental breakage of mercury or metal halide lamp outer jackets in open luminaires has occurred in sports facilities. If the lamp is damaged, glass shards can fall on players or spectators. If the envelope is damaged, the arc tube may continue to operate even if the outer jacket is cracked or missing.

(F) Luminaires Installed in or Under Roof Decking. Luminaires installed in exposed or concealed locations under roof decking where subject to physical damage shall be installed and supported so there is not less than 38 mm (1½ in.) measured from the lowest surface of the roof decking to the top of the luminaire.

Exception: The 38 mm (1½ in.) spacing is not required where metal-corrugated sheet roof decking is covered with a minimum thickness 50 mm (2 in.) concrete slab, measured from the top of the corrugated roofing.

This requirement correlates with 300.4(E), which prohibits cables, raceways, and boxes from installation under metal-corrugated sheet roof decking. Exhibit 410.2 illustrates an installation in which the minimum clearance is not provided between the roof deck and the luminaire.

See also

300.4(E) and its commentary for more information

410.11 Luminaires Near Combustible Material. Luminaires shall be constructed, installed, or equipped with shades or guards

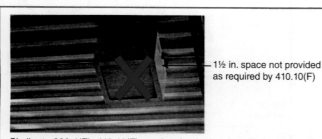

— 1½ in. space not provided as required by 410.10(F)

Similar to 300.4(E), 410.10(F) requires luminaires installed under roof decking to have a minimum 1½ in. between the top of the luminare and the decking.

EXHIBIT 410.2 A minimum 1½-inch clearance, which is necessary to prevent damage from nail penetration.

so that combustible material is not subjected to temperatures in excess of 90°C (194°F).

410.12 Luminaires over Combustible Material. Lampholders installed over highly combustible material shall be of the unswitched type. Unless an individual switch is provided for each luminaire, lampholders shall be located at least 2.5 m (8 ft) above the floor or shall be located or guarded so that the lamps cannot be readily removed or damaged.

Pendants and fixed lighting equipment may be installed above highly combustible material. If a lamp cannot be located out of reach, the requirement can be met by equipping the lamp with a suitable guard. Section 410.12 does not apply to portable lamps.

410.14 Luminaires in Show Windows. Chain-supported luminaires used in a show window shall be permitted to be externally wired. No other externally wired luminaires shall be used.

410.16 Luminaires in Clothes Closets.

N **(A) Clothes Closet Storage Space.** The clothes closet storage space shall be the volume bounded by the sides and back closet walls and planes extending from the closet floor vertically to a height of 1.8 m (6 ft) or to the highest clothes-hanging rod and parallel to the walls at a horizontal distance of 600 mm (24 in.) from the sides and back of the closet walls, respectively. The volume extends vertically to the closet ceiling parallel to the walls at a horizontal distance of 300 mm (12 in.) or the width of the shelf, whichever is greater. For a closet that permits access to both sides of a hanging rod, the clothes closet storage space includes the volume below the highest rod extending 300 mm (12 in.) on either side of the rod on a plane horizontal to the floor extending the entire length of the rod. See Figure 410.16(A).

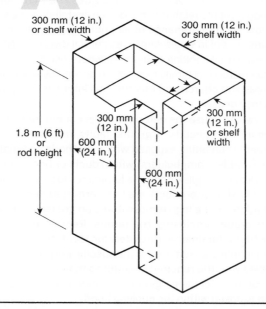

N **FIGURE 410.16(A)** *Clothes Closet Storage Space.*

The 24-inch dimension in the clothes closet storage space is intended to cover the clothes-hanging space, even if no clothes-hanging rod is installed. If such a rod is installed, the space extends from the floor to the top of the highest rod. If no clothes-hanging rod is installed, the space extends from the floor to a height of 6 feet.

In addition to the space in which clothing is hung from the closet pole or rod, this definition also establishes a 12-inch-wide shelf space to cover those installations where shelving is not in place at the time of fixture installation. If shelving is installed and the shelves are wider than 12 inches, the greater width must be applied in establishing this space.

The storage space for closets that permit access to both sides of the clothes-hanging rod is based on a horizontal plane extending 12 inches from both sides of the rod, from the rod down to the floor. This equates to the 24-inch space required for the closet rod where there is only one direction of access.

(B) Luminaire Types Permitted. Only luminaires of the following types shall be permitted in a clothes closet:

(1) Surface-mounted or recessed incandescent or LED luminaires with completely enclosed light sources
(2) Surface-mounted or recessed fluorescent luminaires
(3) Surface-mounted fluorescent or LED luminaires identified as suitable for installation within the clothes closet storage space

(C) Luminaire Types Not Permitted. Incandescent luminaires with open or partially enclosed lamps and pendant luminaires or lampholders shall not be permitted.

△ **(D) Location.** The minimum clearance between luminaires installed in clothes closets and the nearest point of a clothes closet storage space shall be as follows:

(1) 300 mm (12 in.) for surface-mounted incandescent or LED luminaires with a completely enclosed light source installed on the wall above the door or on the ceiling.
(2) 150 mm (6 in.) for surface-mounted fluorescent luminaires installed on the wall above the door or on the ceiling.
(3) 150 mm (6 in.) for recessed incandescent or LED luminaires with a completely enclosed light source installed in the wall or the ceiling.
(4) 150 mm (6 in.) for recessed fluorescent luminaires installed in the wall or the ceiling.

Exception: Surface-mounted fluorescent or LED luminaires shall be permitted to be installed within the clothes closet storage space where identified for this use.

These requirements are intended to prevent hot lamps or parts of broken lamps from coming in contact with combustibles such as boxes, cartons, and blankets stored on shelves and with clothing hung in closets. The clearance measurement for each requirement in 410.16(D) is to the luminaire, not to the lamp itself.

A luminaire in a clothes closet is not mandatory. If one is installed, however, the conditions for installation are as required by 410.16(D).

410.18 Space for Cove Lighting. Coves shall have adequate space and shall be located so that lamps and equipment can be properly installed and maintained.

Adequate space is necessary to allow easy access for relamping luminaires or replacing lampholders, ballasts, and so forth. Adequate space also improves ventilation.

△ **Part III. Luminaire Outlet Boxes, Canopies, and Pans**

410.20 Space for Conductors. Canopies and outlet boxes taken together shall provide sufficient space so that luminaire conductors and their connecting devices are capable of being installed in accordance with 314.16.

UL 1598, *Standard for Safety for Luminaires*, allows junction boxes and splice compartments that are integral to luminaires to have less free volume than required in 314.16. Section 314.16 applies to general-purpose boxes and conduit bodies where installed conductors and devices are variable, while the conductors and devices that will be contained within a luminaire junction box are known.

410.21 Temperature Limit of Conductors in Outlet Boxes. Luminaires shall be of such construction or installed so that the conductors in outlet boxes shall not be subjected to temperatures greater than that for which the conductors are rated.

Branch-circuit wiring, other than 2-wire or multiwire branch circuits supplying power to luminaires connected together, shall not be passed through an outlet box that is an integral part of a luminaire unless the luminaire is identified for through-wiring.

Informational Note: See 410.64(C) for wiring supplying power to luminaires connected together.

Branch-circuit conductors run to a lighting outlet box are not permitted to be subjected to temperatures higher than those for which they are rated. Examples of these are conductors that are rated 75°C and that supply a ceiling outlet box for the connection of a surface-mounted luminaire or are attached to the outlet box of a recessed luminaire. Listed recessed luminaires with integral boxes will specify the minimum wire gauge and insulation temperature and the maximum number of conductors to account for the heat-contributing factor of the supply conductors.

Exhibit 410.3 illustrates luminaires listed for a feed-through installation.

410.22 Outlet Boxes to Be Covered. In a completed installation, each outlet box shall be provided with a cover unless covered by means of a luminaire canopy, lampholder, receptacle that covers the box or is provided with a faceplate, or similar device.

410.23 Covering of Combustible Material at Outlet Boxes. Any combustible wall or ceiling finish exposed between

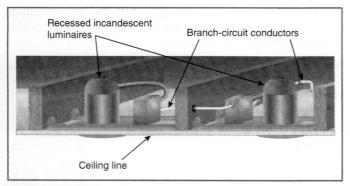

EXHIBIT 410.3 Recessed luminaires designed for feed-through branch-circuit conductors.

the edge of a luminaire canopy or pan and an outlet box having a surface area of 1160 mm² (180 in.²) or more shall be covered with noncombustible material.

Luminaires must be designed and installed not only to prevent overheating of conductors but also to prevent overheating of adjacent combustible wall or ceiling finishes. Canopy-style surface- or ceiling-mounted luminaires listed to UL 1598 are not required to have a backplate or back cover if the total surface area being covered by the canopy is less than 180 square inches. These luminaires are evaluated to ensure that temperatures on wall or ceiling surfaces on which the luminaire is mounted do not exceed 90°C. Where luminaires are not directly mounted on outlet boxes, suitable outlet box covers are required.

See also

314.20 for the requirements covering combustible finishes

410.24 Connection of Electric-Discharge and LED Luminaires.

(A) Independent of the Outlet Box. Electric-discharge and LED luminaires supported independently of the outlet box shall be connected to the branch circuit through metal raceway, non-metallic raceway, Type MC cable, Type AC cable, Type MI cable, nonmetallic sheathed cable, or by flexible cord as permitted in 410.62(B) or 410.62(C).

(B) Access to Boxes. Electric-discharge and LED luminaires surface mounted over concealed outlet, pull, or junction boxes and designed not to be supported solely by the outlet box shall be provided with suitable openings in the back of the luminaire to provide access to the wiring in the box.

Part IV. Luminaire Supports

410.30 Supports.

(A) General. Luminaires and lampholders shall be securely supported. A luminaire that weighs more than 3 kg (6 lb) or exceeds 400 mm (16 in.) in any dimension shall not be supported by the screw shell of a lampholder.

(B) Metal or Nonmetallic Poles Supporting Luminaires. Metal or nonmetallic poles shall be permitted to be used to support luminaires and as a raceway to enclose supply conductors, provided the following conditions are met:

(1) A pole shall have a handhole not less than 50 mm × 100 mm (2 in. × 4 in.) with a cover suitable for use in wet locations to provide access to the supply terminations within the pole or pole base.

Exception No. 1: No handhole shall be required in a pole 2.5 m (8 ft) or less in height abovegrade where the supply wiring method continues without splice or pull point, and where the interior of the pole and any splices are accessible by removing the luminaire.

Exception No. 2: No handhole shall be required in a pole 6.0 m (20 ft) or less in height abovegrade that is provided with a hinged base.

This exception recognizes metal light poles that do not have a handhole but instead use a hinged-base pole to permit access to splices made in the pole base. The height of the pole is limited to 20 feet. The pole and the base must be bonded in accordance with 250.96 (systems operating at 250 volts or less) or 250.97 (circuits operating at over 250 volts). Exhibit 410.4 illustrates a metal light pole with a hinged baseplate that meets the requirements of Exception No. 2.

(2) Where raceway risers or cable is not installed within the pole, a threaded fitting or nipple shall be brazed, welded,

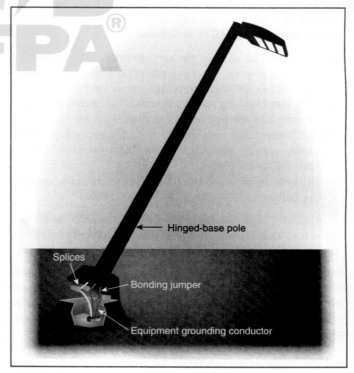

EXHIBIT 410.4 A hinged-base metal pole supporting a luminaire.

or attached to the pole opposite the handhole for the supply connection.

(3) A metal pole shall be provided with an equipment grounding terminal as follows:

    a. A pole with a handhole shall have the equipment grounding terminal accessible from the handhole.

    b. A pole with a hinged base shall have the equipment grounding terminal accessible within the base.

Exception to (3): No grounding terminal shall be required in a pole 2.5 m (8 ft) or less in height abovegrade where the supply wiring method continues without splice or pull, and where the interior of the pole and any splices are accessible by removing the luminaire.

(4) A metal pole with a hinged base shall have the hinged base and pole bonded together.

(5) Metal raceways or other equipment grounding conductors shall be bonded to the metal pole with an equipment grounding conductor recognized by 250.118 and sized in accordance with 250.122.

(6) Conductors in vertical poles used as raceway shall be supported as provided in 300.19.

Metal poles are permitted to be used as raceways, and individual single conductors can be installed to supply the luminaire supported by the pole. Where individual single conductors are installed in a luminaire pole, separation is required between the power wiring and any communications, signaling, and power-limited circuits that are also installed within the pole.

Where a light pole supports a luminaire and a security camera, the security camera signaling and power-limited wiring may be installed within the pole cavity if kept separated. Because the pole contains open circuit conductors (e.g., THW or XHHW conductors) that supply the luminaire, the separation requirement can be fulfilled by enclosing the camera conductors within a flexible raceway and installing that raceway within the pole. The use of a cable or cord assembly for the lighting circuit conductors is also an option. Section 410.30(B)(6) requires that conductors installed within poles be supported the same as in a vertical raceway in accordance with 300.19.

Section 410.30(B) is not intended to require the placement of a raceway for communications cables on the exterior of a lighting pole in every instance.

See also

725.139 and **800.133(A)(1)** as appropriate for requirements for separation from Class 2 and communications circuit conductors

410.36 Means of Support.

(A) Luminaires Supported By Outlet Boxes. Luminaires shall be permitted to be supported by outlet boxes or fittings installed as required by 314.23. The installation shall comply with the following requirements:

(1) The outlet boxes or fittings shall comply with 314.27(A)(1) and 314.27(A)(2).

(2) Luminaires shall be permitted to be supported in accordance with 314.27(E).

(3) Outlet boxes complying with 314.27(E) shall be considered lighting outlets as required by 210.70(A), (B), and (C).

(B) Suspended Ceilings. Framing members of suspended ceiling systems used to support luminaires shall be securely fastened to each other and shall be securely attached to the building structure at appropriate intervals. Luminaires shall be securely fastened to the ceiling framing member by mechanical means such as bolts, screws, or rivets. Listed clips identified for use with the type of ceiling framing member(s) and luminaire(s) shall also be permitted.

Clips that are used to support a luminaire to the framing members of a suspended ceiling must be of a type listed for the application. In addition, the ceiling framing members must be securely attached to each other and to the building structure. These requirements apply to all luminaires supported by a suspended ceiling assembly, including lay-in and surface-mounted types.

See also

300.11(B)(1) or **(2)** for the support of wiring located in the cavity of floor–ceiling assemblies

(C) Luminaire Studs. Luminaire studs that are not a part of outlet boxes, hickeys, tripods, and crowfeet shall be made of steel, malleable iron, or other material suitable for the application.

(D) Insulating Joints. Insulating joints that are not designed to be mounted with screws or bolts shall have an exterior metal casing, insulated from both screw connections.

(E) Raceway Fittings. Raceway fittings used to support a luminaire(s) shall be capable of supporting the weight of the complete fixture assembly and lamp(s).

(F) Busways. Luminaires shall be permitted to be connected to busways in accordance with 368.17(C).

(G) Trees. Outdoor luminaires and associated equipment shall be permitted to be supported by trees.

    Informational Note No. 1: See 225.26 for restrictions for support of overhead conductors.

    Informational Note No. 2: See 300.5(D) for protection of conductors.

Part V. Grounding

Δ **410.40 Equipment Grounding Conductor.** Luminaires and lighting equipment shall be connected to an equipment grounding conductor as required by Part V of this article.

Δ **410.42 Luminaire(s) with Exposed Conductive Surfaces.** Exposed conductive surfaces shall be connected to an equipment grounding conductor.

Exception: Exposed conductive surfaces that comply with any of the following shall not be required to be connected to an equipment grounding conductor:

(1) Surfaces separated from all live parts by a listed system of double insulation

(2) Surfaces on small isolated parts, such as mounting screws, clips, and decorative bands on glass spaced at least 38 mm (1½ in.) from lamp terminals

(3) Surfaces on portable luminaires with a polarized attachment plug

Δ **410.44 Connection to the Equipment Grounding Conductor.** Luminaires and equipment that require connection to an equipment grounding conductor in accordance with 410.42 shall be mechanically connected to an equipment grounding conductor.

Exception No. 1: Replacement luminaires shall be permitted to connect an equipment grounding conductor in the same manner as replacement receptacles in compliance with 250.130(C). The luminaire shall then comply with 410.42.

Exception No. 1 provides a method by which a luminaire with exposed conductive parts can be installed at an outlet where no means of grounding is provided by the existing wiring system. The means allowed by the exception is the same as is permitted for receptacles installed at outlets where no grounding means exists.

Exception No. 2: Where no equipment grounding conductor exists at the outlet, replacement luminaires that are GFCI protected or do not have exposed conductive parts shall not be required to be connected to an equipment grounding conductor.

Part VI. Wiring of Luminaires

410.50 Luminaire Wiring — General. Wiring on or within luminaires shall be neatly arranged and shall not be exposed to physical damage. Excess wiring shall be avoided. Conductors shall be arranged so that they are not subjected to temperatures above those for which they are rated.

410.51 Polarization of Luminaires. Luminaires shall be wired so that the screw shells of lampholders are connected to the same luminaire or circuit conductor or terminal. The grounded conductor, where connected to a screw shell lampholder, shall be connected to the screw shell.

Δ **410.52 Conductor Insulation.** Luminaires shall be wired with conductors having insulation suitable for the environmental conditions, current, voltage, and temperature to which the conductors will be subjected.

Informational Note: See 402.3, 402.5, and 402.6 for ampacity of fixture wire, maximum operating temperature, voltage limitations, and minimum wire size.

410.54 Pendant Conductors for Incandescent Filament Lamps.

(A) Support. Pendant lampholders with permanently attached leads, where used for other than festoon wiring, shall be hung from separate stranded rubber-covered conductors that are soldered directly to the circuit conductors but supported independently thereof.

(B) Size. Unless part of listed decorative lighting assemblies, pendant conductors shall not be smaller than 14 AWG for mogul-base or medium-base screw shell lampholders or smaller than 18 AWG for intermediate or candelabra-base lampholders.

(C) Twisted or Cabled. Pendant conductors longer than 900 mm (3 ft) shall be twisted together where not cabled in a listed assembly.

410.56 Protection of Conductors and Insulation.

(A) Properly Secured. Conductors shall be secured in a manner that does not tend to cut or abrade the insulation.

(B) Protection Through Metal. Conductor insulation shall be protected from abrasion where it passes through metal.

(C) Luminaire Stems. Splices and taps shall not be located within luminaire arms or stems.

Δ **(D) Splices and Taps.** No unnecessary splices or taps shall be made within or on a luminaire.

Informational Note: See 110.14 for approved means of making connections.

(E) Stranding. Stranded conductors shall be used for wiring on luminaire chains and on other movable or flexible parts.

(F) Tension. Conductors shall be arranged so that the weight of the luminaire or movable parts does not put tension on the conductors.

410.59 Cord-Connected Showcases. Individual showcases, other than fixed, shall be permitted to be connected by flexible cord to permanently installed receptacles, and groups of not more than six such showcases shall be permitted to be coupled together by flexible cord and separable locking-type connectors with one of the group connected by flexible cord to a permanently installed receptacle.

The installation shall comply with 410.59(A) through (E).

(A) Cord Requirements. Flexible cord shall be of the hard-service type, having conductors not smaller than the branch-circuit conductors, having ampacity at least equal to the branch-circuit overcurrent device, and having an equipment grounding conductor.

Informational Note: See Table 250.122 for size of equipment grounding conductor.

(B) Receptacles, Connectors, and Attachment Plugs. Receptacles, connectors, and attachment plugs shall be of a listed grounding type rated 15 or 20 amperes.

(C) Support. Flexible cords shall be secured to the undersides of showcases such that all of the following conditions are ensured:

(1) The wiring is not exposed to physical damage.
(2) The separation between cases is not in excess of 50 mm (2 in.), or more than 300 mm (12 in.) between the first case and the supply receptacle.
(3) The free lead at the end of a group of showcases has a female fitting not extending beyond the case.

(D) No Other Equipment. Equipment other than showcases shall not be electrically connected to showcases.

(E) Secondary Circuit(s). Where showcases are cord-connected, the secondary circuit(s) of each electric-discharge lighting ballast shall be limited to one showcase.

410.62 Cord-Connected Lampholders and Luminaires.

△ **(A) Lampholders.** Where a metal lampholder is attached to a flexible cord, the inlet shall be equipped with an insulating bushing that, if threaded, is not smaller than metric designator 12 (trade size ⅜). The cord hole shall be of a size appropriate for the cord, and all burrs and fins shall be removed in order to provide a smooth bearing surface for the cord.

Bushing having holes 7 mm (⁹⁄₃₂ in.) in diameter shall be permitted for use with plain pendant cord and holes 11 mm (¹³⁄₃₂ in.) in diameter with reinforced cord.

Metal lampholders (brass- and aluminum-shell type) used with flexible-cord pendants are required to be equipped with smooth and permanently secured insulating bushings. Nonmetallic-type lampholders do not require a bushing, because the material and the design afford equivalent protection.

(B) Adjustable Luminaires. Luminaires that require adjusting or aiming after installation shall not be required to be equipped with an attachment plug or cord connector, provided the exposed cord is suitable for hard-usage or extra-hard-usage and is not longer than that required for maximum adjustment. The cord shall not be subject to strain or physical damage.

Informational Note: See Table 400.4, "Use" column for application provisions.

(C) Electric-Discharge and LED Luminaires. Electric-discharge and LED luminaires shall comply with 410.62(C)(1), (C)(2), and (C)(3), as applicable.

(1) Cord-Connected Installation. A luminaire or a listed assembly in compliance with any of the conditions in 410.62(C)(1)(a) through (C)(1)(c) shall be permitted to be cord connected provided the luminaire is located directly below the outlet or busway, the cord is not subject to strain or physical damage, and the cord is visible over its entire length except at terminations.

Informational Note: See 400.10, Uses Permitted, and 400.12, Uses Not Permitted.

(a) A luminaire shall be permitted to be connected with a cord terminating in a grounding-type attachment plug or busway plug. If grounding is not required in accordance with 410.42, a polarized-type plug shall be permitted.

(b) A luminaire assembly equipped with a strain relief and canopy shall be permitted to use a cord connection between the luminaire assembly and the canopy. The canopy shall be permitted to include a section of raceway not over 150 mm (6 in.) in length and intended to facilitate the connection to an outlet box mounted above a suspended ceiling.

(c) Listed luminaires connected using listed assemblies that incorporate manufactured wiring system connectors in accordance with 604.100(C) shall be permitted to be cord connected.

Section 410.62(C)(1) applies to listed cord-and-plug-connected light-emitting diode (LED) and electric-discharge luminaires, such as the luminaire illustrated in Exhibit 410.5. The supply cord is not permitted to penetrate a suspended ceiling, because the cord is required to be visible along its entire length.

Section 410.62(C)(1)(c) permits a listed manufactured wiring system connector that is part of a fabricated assembly to supply the luminaires in place of a grounding-type attachment plug. Certain listed assemblies with a 6-inch maximum section of raceway are permitted to be installed above a suspended ceiling.

Supply cords cannot be used as a supporting means, and the luminaires must be suspended directly below the outlet boxes supplying each luminaire. If the luminaire is suspended below the lift-out-type ceiling, the cord is not permitted to penetrate the ceiling, unless it is part of a listed luminaire assembly as described in 410.62(C)(1)(c).

See also

400.12, which further explains the uses not permitted for cords
368.56(B), which permits luminaires to be connected to busways by cords plugged directly into the busway

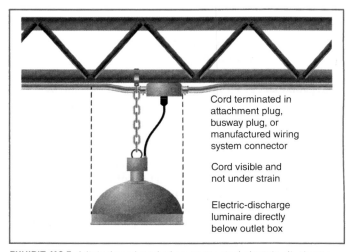

Cord terminated in attachment plug, busway plug, or manufactured wiring system connector

Cord visible and not under strain

Electric-discharge luminaire directly below outlet box

EXHIBIT 410.5 A listed cord-and-plug-connected electric-discharge luminaire.

(2) Provided with Mogul-Base, Screw Shell Lampholders. Electric-discharge luminaires provided with mogul-base, screw shell lampholders shall be permitted to be connected to branch circuits of 50 amperes or less by cords complying with 240.5. Receptacles and attachment plugs shall be permitted to be of a lower ampere rating than the branch circuit but not less than 125 percent of the luminaire full-load current.

(3) Equipped with Flanged Surface Inlet. Electric-discharge luminaires equipped with a flanged surface inlet shall be permitted to be supplied by cord pendants equipped with cord connectors. Inlets and connectors shall be permitted to be of a lower ampere rating than the branch circuit but not less than 125 percent of the luminaire load current.

410.64 Luminaires as Raceways. Luminaires shall not be used as a raceway for circuit conductors unless they comply with 410.64(A), (B), or (C).

This section does not permit luminaires to be used as raceways for circuit conductors unless specifically listed and marked for this use. According to the UL *Luminaires Marking and Application Guide,* 2016, luminaires listed for use as raceways are marked "Suitable for Use as a Raceway" and also with the maximum number, size, and type of conductor permitted in a raceway. Without those markings, a row of luminaires connected end to end cannot be used as a raceway for circuit conductors other than the 2-wire or multiwire circuit supplying the luminaires. Luminaires identified for use as a raceway have been evaluated for the heat contribution caused by additional current-carrying conductors.

(A) Listed. Luminaires listed and marked for use as a raceway shall be permitted to be used as a raceway.

(B) Through-Wiring. Luminaires identified for through-wiring, as permitted by 410.21, shall be permitted to be used as a raceway.

(C) Luminaires Connected Together. Luminaires designed for end-to-end connection to form a continuous assembly, or luminaires connected together by recognized wiring methods, shall be permitted to contain the conductors of a 2-wire branch circuit, or one multiwire branch circuit, supplying the connected luminaires and shall not be required to be listed as a raceway. One additional 2-wire branch circuit separately supplying one or more of the connected luminaires shall also be permitted.

> Informational Note: See Article 100 for the definition of *Multiwire Branch Circuit.*

Section 410.64(C) facilitates convenient switching and supply circuit arrangements for a physically continuous row of luminaires or a row that is made continuous via the wiring method. A single 2-wire or a single multiwire branch circuit supplying the luminaires is permitted to be run through the continuous row(s), and the luminaires are not required to be listed for use as a raceway. An additional 2-wire branch circuit is permitted to be run through the luminaires. This circuit can supply only luminaires in the connected row(s) and is commonly used to switch night lighting as an energy conservation method.

410.68 Feeder and Branch-Circuit Conductors and Ballasts. Feeder and branch-circuit conductors within 75 mm (3 in.) of a ballast, LED driver, power supply, or transformer shall have an insulation temperature rating not lower than 90°C (194°F), unless supplying a luminaire marked as suitable for a different insulation temperature.

Listed LED drivers (including the Class 2 output type) are limited to either 75°C or 90°C, depending on which standard was used to evaluate the device. In many ways, the installation rules established for discharge lighting ballasts over the years carried over to LED drivers. "LED driver" is a common industry term referring to the power supply for the LED.

See also

> **Table 310.4(1)** for temperature ratings, along with other insulated conductor specifications

Δ **410.69 Identification of Control Conductor Insulation.** Where control conductors are spliced, terminated, or connected in the same luminaire or enclosure as the branch-circuit conductors, the field-connected control conductor shall not be of a color reserved for the grounded branch-circuit conductor or the equipment grounding conductor.

> Informational Note: See 200.6 for identification of grounded conductor and 250.119 for identification of equipment grounding conductor.

Exception: A field-connected gray-colored control conductor shall be permitted if the insulation is permanently re-identified by marking tape, painting, or other effective means at its termination and at each location where the conductor is visible and accessible. Identification shall encircle the insulation and shall be a color other than white, gray, or green.

Conductors with the insulation colors reserved for branch-circuit conductors are not permitted to be used for luminaire control circuits if they share the same wiring compartment with the branch-circuit conductors. The exception permits field re-identification of field-connected gray control conductors. Some existing controls, drivers, and ballasts have integral lead wires that are purple and gray.

410.70 Combustible Shades and Enclosures. Air space shall be provided between lamps and shades or other enclosures of combustible material.

N **410.71 Disconnecting Means for Fluorescent or LED Luminaires that Utilize Double-Ended Lamps.**

N **(1) General.** In indoor locations other than dwellings and associated accessory structures, fluorescent or LED luminaires that utilize double-ended lamps and contain ballast(s) or LED driver(s) that can be serviced in place shall have a disconnecting means either internal or external to each luminaire. For existing installed luminaires without disconnecting means, at the time a ballast or LED driver is added or replaced a disconnecting means shall be installed. The line side terminals of the disconnecting means shall be guarded.

Exception No. 1: A disconnecting means shall not be required for luminaires installed in hazardous (classified) location(s).

Exception No. 2: A disconnecting means shall not be required for luminaires that provide emergency illumination required in 700.16.

Exception No. 3: For cord-and-plug-connected luminaires, an accessible separable connector or an accessible plug and receptacle shall be permitted to serve as the disconnecting means.

Exception No. 4: Disconnecting means shall not be required for every luminaire in a building area if all of the following conditions apply:

(1) More than one luminaire is installed in the building area
(2) The luminaires are not connected to a multiwire branch circuit
(3) The design of the installation includes disconnecting means
(4) The building area will not be left in total darkness should only one disconnect be opened

N **(2) Multiwire Branch Circuits.** When connected to multiwire branch circuits, the disconnecting means shall simultaneously break all the supply conductors to the ballast, including the grounded conductor.

N **(3) Location.** The disconnecting means shall be located so as to be accessible to qualified persons before servicing or maintaining the ballast. Where the disconnecting means is external to the luminaire, it shall be a single device, and it shall be attached to the luminaire or the luminaire shall be located within sight of the disconnecting means.

The disconnect can be either inside or outside the luminaire and must disconnect all supply conductors simultaneously, including the grounded conductor. Where the disconnecting means is external to the luminaire, it must be a single device either attached to the luminaire or within sight of the luminaire. Four exceptions to the required disconnect are provided in 410.71(1), Exception No. 4. In addition to the ungrounded conductors, the grounded or neutral conductor is also required to be disconnected. A disconnecting means is required be installed when a ballast is replaced in a luminaire that does not have an existing internal or external disconnecting means.

Part VII. Construction of Luminaires

410.80 Luminaire Rating.

(A) Marking. All luminaires shall be marked with the maximum lamp wattage or electrical rating, manufacturer's name,

trademark, or other suitable means of identification. A luminaire requiring supply wire rated higher than 60°C (140°F) shall be marked with the minimum supply wire temperature rating on the luminaire and shipping carton or equivalent.

(B) Electrical Rating. The electrical rating shall include the voltage and frequency and shall indicate the current rating of the unit, including the ballast, transformer, LED driver, power supply, or autotransformer.

410.82 Portable Luminaires. Portable luminaires shall be wired with flexible cord recognized by 400.4 and an attachment plug of the polarized or grounding type. If used with Edison-base lampholders, the grounded conductor shall be identified and attached to the screw shell and the identified blade of the attachment plug.

410.84 Cord Bushings. A bushing or the equivalent shall be provided where flexible cord enters the base or stem of a portable luminaire. The bushing shall be of insulating material unless a jacketed type of cord is used.

Part VIII. Installation of Lampholders

410.90 Screw Shell Type. Lampholders of the screw shell type shall be installed for use as lampholders only. Where supplied by a circuit having a grounded conductor, the grounded conductor shall be connected to the screw shell.

410.93 Double-Pole Switched Lampholders. Where supplied by the ungrounded conductors of a circuit, the switching device of lampholders of the switched type shall simultaneously disconnect both conductors of the circuit.

Single-pole switching may be used to interrupt the ungrounded conductor of a 2-wire circuit in which one conductor is grounded. The grounded conductor must be connected to the screw shell of the lampholder.

Where a 2-wire circuit is derived from the two ungrounded conductors of a multiwire circuit (3- or 4-wire system) or from the two ungrounded conductors of a 2-wire circuit (3-wire system) and is used with switched lampholders, the switching device is required to be double-pole and to simultaneously disconnect both ungrounded conductors of the circuit.

410.96 Lampholders in Wet or Damp Locations. Lampholders installed in wet locations shall be listed for use in wet locations. Lampholders installed in damp locations shall be listed for damp locations or shall be listed for wet locations.

410.97 Lampholders Near Combustible Material. Lampholders shall be constructed, installed, or equipped with shades or guards so that combustible material is not subjected to temperatures in excess of 90°C (194°F).

Part IX. Lamps and Auxiliary Equipment

410.100 Bases, Incandescent Lamps. An incandescent lamp for general use on lighting branch circuits shall not be equipped with a medium base if rated over 300 watts, or with a mogul base if rated over 1500 watts. Special bases or other devices shall be used for over 1500 watts.

410.104 Electric-Discharge Lamp Auxiliary Equipment.

(A) Enclosures. Auxiliary equipment for electric-discharge lamps shall be enclosed in noncombustible cases and treated as sources of heat.

The UL *Guide Information for Electrical Equipment,* which can be found at productspec.ul.com, contains information on fluorescent ballasts (FKVS) and high-intensity discharge (HID) ballasts (FLCR).

Fluorescent ballast enclosures are categorized by UL as indoor, outdoor, and weatherproof. Fluorescent ballasts can be an open type that must be installed within an enclosure, or they can be enclosed. HID ballasts are categorized the same, except there is no open-type HID ballasts.

- *Indoor ballasts* are suitable for use in an indoor, dry location only.
- *Outdoor ballasts* are designated as Type 1 or Type 2. Type 2 ballasts are provided with their own enclosure. Both types are suitable for use in outdoor equipment, wet or damp location luminaires, or outdoor signs if the ballasts are within the overall electrical enclosure.
- *Weatherproof ballasts* are suitable for use where exposed to the weather without an additional enclosure.

(B) Switching. Where supplied by the ungrounded conductors of a circuit, the switching device of auxiliary equipment shall simultaneously disconnect all conductors.

Part X. Special Provisions for Flush and Recessed Luminaires

410.110 General. Luminaires installed in recessed cavities in walls or ceilings, including suspended ceilings, shall comply with 410.115 through 410.126.

410.115 Temperature.

(A) Combustible Material. Luminaires shall be installed so that adjacent combustible material will not be subjected to temperatures in excess of 90°C (194°F).

(B) Recessed Incandescent Luminaires. Incandescent luminaires shall have thermal protection and shall be identified as thermally protected.

Exception No. 1: Thermal protection shall not be required in a recessed luminaire identified for use and installed in poured concrete.

Exception No. 2: Thermal protection shall not be required in a recessed luminaire whose design, construction, and thermal performance characteristics are equivalent to a thermally protected luminaire and are identified as inherently protected.

Because many recessed incandescent luminaires are suitable for a wide variety of lamp sizes and types and finish trims, the temperature close to the lamp can vary widely. Therefore, many manufacturers have chosen to locate thermal protectors away from the source of heat — such as in the outlet box — and to design the protector so that it detects a change in temperature resulting from the addition of thermal insulation around the luminaire. This design prevents nuisance tripping of the protector (e.g., as a result of changing lamp wattage) but still provides protection against overheating arising from thermal insulation around a recessed luminaire not designed for such use.

410.116 Clearance and Installation.

(A) Clearance from Combustible Material.

(1) Non-Type IC. A recessed luminaire that is not identified for contact with insulation shall have all recessed parts spaced not less than 13 mm (½ in.) from combustible materials. The points of support and the trim finishing off the openings in the ceiling, wall, or other finished surface shall be permitted to be in contact with combustible materials.

(2) Type IC. A recessed luminaire that is identified for contact with insulation, Type IC, shall be permitted to be in contact with combustible materials at recessed parts, points of support, and portions passing through or finishing off the opening in the building structure.

(B) Clearance from Thermal Insulation. Thermal insulation shall not be installed above a recessed luminaire or within 75 mm (3 in.) of the recessed luminaire's enclosure, wiring compartment, ballast, transformer, LED driver, or power supply unless the luminaire is identified as Type IC for insulation contact.

LED luminaires for installation in contact with thermal insulation must be identified as Type IC, which is similar to the requirements for other luminaires. Exhibit 410.6 illustrates a listed Type IC recessed luminaire installed in direct contact with thermal insulation. Thermal protection is provided to deactivate the lamp should the luminaire be mislamped so that it overheats.

(C) Installation in Fire-Resistant Construction. Luminaires marked "FOR USE IN NON-FIRE-RATED INSTALLATIONS" shall not be used in fire-rated installations. Where a luminaire is recessed in fire-resistant material in a building of fire-resistant construction, the recessed luminaire shall satisfy one of the following:

(1) The recessed luminaire shall be listed for use in a fire resistance–rated construction.
(2) The recessed luminaire shall be installed in or used with a luminaire enclosure that is listed for use in a fire resistance–rated construction.

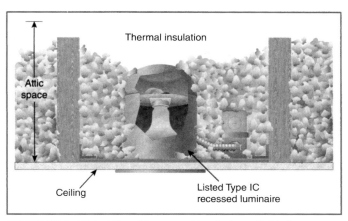

EXHIBIT 410.6 A listed Type IC recessed luminaire suitable for use in insulated ceilings and installed in direct contact with thermal insulation.

(3) The recessed luminaire shall be listed and shall be installed in accordance with a tested fire resistance–rated assembly. When a tested fire resistance–rated assembly allows the installation of a recessed fluorescent luminaire, a recessed LED luminaire of comparable construction shall be permitted.

410.117 Wiring.

(A) General. Conductors that have insulation suitable for the temperature encountered shall be used.

(B) Circuit Conductors. Branch-circuit conductors that have an insulation suitable for the temperature encountered shall be permitted to terminate in the luminaire.

(C) Tap Conductors. Tap conductors of a type suitable for the temperature encountered shall be permitted to run from the luminaire terminal connection to an outlet box placed at least 300 mm (1 ft) from the luminaire. Such tap conductors shall be in suitable raceway or Type AC or MC cable of at least 450 mm (18 in.) but not more than 1.8 m (6 ft) in length.

410.118 Access to Other Boxes.
Luminaires recessed in ceilings, floors, or walls shall not be used to access outlet, pull, or junction boxes or conduit bodies, unless the box or conduit body is an integral part of the listed luminaire.

Part XI. Construction of Flush and Recessed Luminaires

410.120 Temperature.
Luminaires shall be constructed such that adjacent combustible material is not subject to temperatures in excess of 90°C (194°F).

410.122 Lamp Wattage Marking.
Incandescent lamp luminaires shall be marked to indicate the maximum allowable wattage of lamps. The markings shall be permanently installed, in letters at least 6 mm (¼ in.) high, and shall be located where visible during relamping.

410.124 Solder Prohibited.
No solder shall be used in the construction of a luminaire recessed housing.

410.126 Lampholders.
Lampholders of the screw shell type shall be of porcelain or other suitable insulating materials.

Part XII. Special Provisions for Electric-Discharge Lighting Systems of 1000 Volts or Less

410.130 General.

(A) Open-Circuit Voltage of 1000 Volts or Less. Equipment for use with electric-discharge lighting systems and designed for an open-circuit voltage of 1000 volts or less shall be of a type identified for such service.

(B) Considered as Energized. The terminals of an electric-discharge lamp shall be considered as energized where any lamp terminal is connected to a circuit of over 300 volts.

(C) Transformers of the Oil-Filled Type. Transformers of the oil-filled type shall not be used.

(D) Additional Requirements. In addition to complying with the general requirements for luminaires, such equipment shall comply with Part XII of this article.

(E) Thermal Protection — Fluorescent Luminaires.

(1) Integral Thermal Protection. The ballast of a fluorescent luminaire installed indoors shall have integral thermal protection. Replacement ballasts shall also have thermal protection integral with the ballast.

(2) Simple Reactance Ballasts. A simple reactance ballast in a fluorescent luminaire with straight tubular lamps shall not be required to be thermally protected.

(3) Exit Luminaires. A ballast in a fluorescent exit luminaire shall not have thermal protection.

(4) Egress Luminaires. A ballast in a fluorescent luminaire that is used for egress lighting and energized only during a failure of the normal supply shall not have thermal protection.

Thermal protection that is integral with the ballast is required for fluorescent luminaires installed indoors. Thermally protected ballasts are also required as replacements for nonthermally protected ballasts in older fixtures. Thermally protected fluorescent lamp ballasts intended for use in accordance with 410.130(E) are marked "Class P." LED drivers are the LED lighting system's equivalent to ballasts. "LED driver" is a common industry term referring to the power supply for the LED.

Because different Class P ballasts have different heating characteristics, the heating characteristics should be considered when selecting replacements for nonthermally protected ballasts. This type of ballast protection is set to open the circuit at a predetermined temperature to prevent abnormal ballast heat buildup caused by a fault in one or more of the ballast components or by some lampholder or wiring fault.

Illuminated exit signs are exempt from the thermal protection requirement because overheating during high ambient conditions could cause the thermal protection to operate. This action could impair evacuation during a fire. Egress lighting is also exempt from the thermal protection requirement for the same reason that illuminated exit signs are exempt. However, this exemption applies to egress lighting that is energized only during the emergency condition.

(F) High-Intensity Discharge Luminaires.

(1) Recessed. Recessed high-intensity luminaires designed to be installed in wall or ceiling cavities shall have thermal protection and be identified as thermally protected.

(2) Inherently Protected. Thermal protection shall not be required in a recessed high-intensity luminaire whose design, construction, and thermal performance characteristics are equivalent to a thermally protected luminaire and are identified as inherently protected.

(3) Installed in Poured Concrete. Thermal protection shall not be required in a recessed high-intensity discharge luminaire identified for use and installed in poured concrete.

(4) Recessed Remote Ballasts. A recessed remote ballast for a high-intensity discharge luminaire shall have thermal protection that is integral with the ballast and shall be identified as thermally protected.

Δ **(5) Metal Halide Lamp Containment.** Luminaires that use a metal halide lamp other than a thick-glass parabolic reflector lamp (PAR) shall be provided with a containment barrier that encloses the lamp, or shall be provided with a physical means that only allows the use of a lamp that is Type O.

> Informational Note: See ANSI C78.389, *American National Standard for Electric Lamps — High Intensity Discharge, Methods of Measuring Characteristics.*

Metal halide lamps have been identified by insurers of large industrial facilities as the likely cause of ignition in fires. HID luminaires can create an ignition source when physical damage occurs to the arc tube in open luminaires. Lamp types that are suitable for use in open luminaires (those that do not require lamp enclosures) are classified by the lamp manufacturers as Type O or Type S. Type O lamps are provided with a shroud around the arc tube and are containment-tested in accordance with ANSI C78.387, *Electric Lamps — Metal-Halide Lamps — Methods of Measuring Characteristics*, and rated for use in open luminaires. The *NEC®* requires luminaires that use a metal halide lamp to have either a containment barrier that encloses the lamp or some means that allows only a Type O lamp to be installed.

410.134 Direct-Current Equipment. Luminaires installed on dc circuits shall be equipped with auxiliary equipment and resistors designed for dc operation. The luminaires shall be marked for dc operation.

410.135 Open-Circuit Voltage Exceeding 300 Volts. Equipment having an open-circuit voltage exceeding 300 volts shall not be installed in dwelling occupancies unless such equipment is designed so that there will be no exposed live parts when lamps are being inserted, are in place, or are being removed.

Luminaires intended for use in non-dwelling occupancies are so marked. Such luminaires often have maintenance features beyond the capabilities of most homeowners or operate at voltages in excess of those permitted by the *NEC* for dwelling occupancies.

See also

210.6(A) and **410.140(B)** for other references to voltage limitations within dwelling units

410.136 Luminaire Mounting.

(A) Exposed Components. Luminaires that have exposed ballasts, transformers, LED drivers, or power supplies shall be installed such that ballasts, transformers, LED drivers, or power supplies shall not be in contact with combustible material unless listed for such condition.

Δ **(B) Combustible Low-Density Cellulose Fiberboard.** Where a surface-mounted luminaire containing a ballast, transformer, LED driver, or power supply is to be installed on combustible low-density cellulose fiberboard, it shall be marked for this condition or shall be spaced not less than 38 mm (1½ in.) from the surface of the fiberboard. Where such luminaires are partially or wholly recessed, 410.110 through 410.126 shall apply.

> Informational Note: See ASTM E84-20, *Standard Test Method for Surface Burning Characteristics of Building Materials*, or ANSI/UL 723-2018, *Standard for Test for Surface Burning Characteristics of Building Materials*. Combustible low-density cellulose fiberboard includes sheets, panels, and tiles that have a density of 320 kg/m³ (20 lb/ft³) or less and that are formed of bonded plant fiber material but does not include solid or laminated wood or fiberboard that has a density in excess of 320 kg/m³ (20 lb/ft³) or is a material that has been integrally treated with fire-retarding chemicals to the degree that the flame spread index in any plane of the material will not exceed 25, determined in accordance with tests for surface burning characteristics of building materials.

410.137 Equipment Not Integral with Luminaire.

(A) Metal Cabinets. Auxiliary equipment, including reactors, capacitors, resistors, and similar equipment, where not installed as part of a luminaire assembly, shall be enclosed in accessible, permanently installed metal cabinets.

(B) Separate Mounting. Separately mounted ballasts, transformers, LED drivers, or power supplies that are listed for direct connection to a wiring system shall not be required to be additionally enclosed.

(C) Wired Luminaire Sections. Wired luminaire sections are paired, with a ballast(s) or LED driver(s) supplying a light source or light sources in both. For interconnection between paired units, it shall be permissible to use metric designator 12 (trade size ⅜) flexible metal conduit in lengths not exceeding 7.5 m (25 ft), installed in accordance with Part II of Article 348. Luminaire wire operating at line voltage, supplying only the ballast(s) or LED driver(s) of one of the paired luminaires, shall be permitted in the same raceway as the light source supply wires of the paired luminaires where the voltage rating of the light source supply wires is greater than the line voltage.

Wired luminaire sections are shipped in pairs and marked for use in pairs. Each individual unit includes lamps in odd-numbered quantities (one or three is most common), with the odd-numbered lamp in each luminaire supplied by a two-lamp ballast located in one luminaire of the pair. Two-lamp ballasts are more energy efficient than single- or three-lamp ballasts.

410.138 Autotransformers. An autotransformer that is used to raise the voltage to more than 300 volts, as part of a ballast for supplying lighting units, shall be supplied only by a grounded system.

410.139 Switches. Snap switches shall comply with 404.14.

Part XIII. Special Provisions for Electric-Discharge Lighting Systems of More Than 1000 Volts

410.140 General.

(A) Listing. Electric-discharge lighting systems with an open-circuit voltage exceeding 1000 volts shall be listed and installed in conformance with that listing.

(B) Dwelling Occupancies. Equipment that has an open-circuit voltage exceeding 1000 volts shall not be installed in or on dwelling occupancies.

(C) Live Parts. The terminal of an electric-discharge lamp shall be considered as a live part.

△ **(D) Additional Requirements.** In addition to complying with the general requirements for luminaires, such equipment shall comply with Part XIII of this article.

410.141 Control.

(A) Disconnection. Luminaires or lamp installation shall be controlled either singly or in groups by an externally operable switch or circuit breaker that opens all ungrounded primary conductors.

(B) Within Sight or Locked Type. The switch or circuit breaker shall be located within sight from the luminaires or lamps, or it shall be permitted to be located elsewhere if it is lockable open in accordance with 110.25.

410.142 Lamp Terminals and Lampholders. Parts that must be removed for lamp replacement shall be hinged or held captive. Lamps or lampholders shall be designed so that there are no exposed live parts when lamps are being inserted or removed.

410.143 Transformers.

(A) Type. Transformers shall be enclosed, identified for the use, and listed.

(B) Voltage. The secondary circuit voltage shall not exceed 15,000 volts, nominal, under any load condition. The voltage to ground of any output terminals of the secondary circuit shall not exceed 7500 volts under any load conditions.

(C) Rating. Transformers shall have a secondary short-circuit current rating of not more than 150 mA if the open-circuit voltage is over 7500 volts, and not more than 300 mA if the open-circuit voltage rating is 7500 volts or less.

(D) Secondary Connections. Secondary circuit outputs shall not be connected in parallel or in series.

410.144 Transformer Locations.

(A) Accessible. Transformers shall be accessible after installation.

(B) Secondary Conductors. Transformers shall be installed as near to the lamps as practicable to keep the secondary conductors as short as possible.

(C) Adjacent to Combustible Materials. Transformers shall be located so that adjacent combustible materials are not subjected to temperatures in excess of 90°C (194°F).

410.145 Exposure to Damage. Lamps shall not be located where normally exposed to physical damage.

410.146 Marking. Each luminaire or each secondary circuit of tubing having an open-circuit voltage of over 1000 volts shall have a clearly legible marking in letters not less than 6 mm (¼ in.) high reading "Caution _____ volts." The voltage indicated shall be the rated open-circuit voltage. The caution sign(s) or label(s) shall comply with 110.21(B).

Part XIV. Lighting Track

410.150 Installation.

(A) Lighting Track. Lighting track shall be permanently installed and permanently connected to a branch circuit. Only lighting track fittings shall be installed on lighting track. Lighting track fittings shall not be equipped with general-purpose receptacles.

A lighting track fitting differs from a *fitting* as defined in Article 100, in that it usually performs both an electrical and a mechanical function. Such assemblies are not intended to be used for locating convenience receptacles or as an alternative for required receptacle outlets such as those required in 210.62 for show windows. Lighting track can be removed and relocated and, therefore, is not a substitute for required receptacles.

(B) Connected Load. The connected load on lighting track shall not exceed the rating of the track. Lighting track shall be supplied by a branch circuit having a rating not more than that of the track. The load calculation in 220.46(B) shall not be required to limit the length of track on a single branch circuit, and it shall not be required to limit the number of luminaires on a single track.

Section 220.46(B) is intended to be used for load calculations of feeders and services. It does not limit the length of track or number of installed luminaires.

See also

220.46(B), **Calculation Example** for a load calculation method for track lighting

(C) Locations Not Permitted. Lighting track shall not be installed in the following locations:

(1) Where likely to be subjected to physical damage
(2) In wet or damp locations
(3) Where subject to corrosive vapors
(4) In storage battery rooms
(5) In hazardous (classified) locations
(6) Where concealed
(7) Where extended through walls or partitions
(8) Less than 1.5 m (5 ft) above the finished floor except where protected from physical damage or track operating at less than 30 volts rms open-circuit voltage
(9) Where prohibited by 410.10(D)

(D) Support. Fittings identified for use on lighting track shall be designed specifically for the track on which they are to be installed. They shall be securely fastened to the track, shall maintain polarization and connections to the equipment grounding conductor, and shall be designed to be suspended directly from the track.

410.153 Heavy-Duty Lighting Track. Heavy-duty lighting track is lighting track identified for use exceeding 20 amperes. Each fitting attached to a heavy-duty lighting track shall have individual overcurrent protection.

410.154 Fastening. Lighting track shall be securely mounted so that each fastening is suitable for supporting the maximum weight of luminaires that can be installed. Unless identified for supports at greater intervals, a single section 1.2 m (4 ft) or shorter in length shall have two supports, and, where installed in a continuous row, each individual section of not more than 1.2 m (4 ft) in length shall have one additional support.

410.155 Construction Requirements.

(A) Construction. The housing for the lighting track system shall be of substantial construction to maintain rigidity. The conductors shall be installed within the track housing, permitting insertion of a luminaire, and designed to prevent tampering and accidental contact with live parts. Components of lighting track systems of different voltages shall not be interchangeable. The track conductors shall be a minimum 12 AWG or equal and shall be copper. The track system ends shall be insulated and capped.

(B) Equipment Grounding Conductor. Lighting track shall be connected to the equipment grounding conductor in accordance with Part V of this article, and the track sections be securely coupled to maintain continuity of the circuitry, polarization, and grounding throughout.

Part XV. Decorative Lighting and Similar Accessories

410.160 Listing of Decorative Lighting. Decorative lighting and similar accessories used for holiday lighting and similar purposes, in accordance with 590.3(B), shall be listed.

Part XVI. Special Provisions for Horticultural Lighting Equipment

410.170 General. Luminaires complying with Parts, I. II, III, IV, V, VI, VII, IX, X, XI, and XII of this article shall be permitted to be used for horticultural lighting. Part XVI shall additionally apply to lighting equipment specifically identified for horticultural use.

Informational Note: Lighting equipment identified for horticultural use is designed to provide a spectral characteristic needed for the growth of plants and can also provide supplemental general illumination within the growing environment.

With the increasing numbers of indoor plant-growing facilities being established, requirements are necessary for the safe use of horticultural lighting. Horticultural lighting equipment provides a wavelength(s) of light that promotes plant growth. The *NEC* has requirements for agricultural buildings in Article 547. Because horticultural lighting equipment needs to be adjustable to accommodate seasonal plant diversity and growth, flexible connection to branch circuits and between luminaires is usually necessary. In many installations, horticultural lighting equipment is installed closely together.

410.172 Listing. Lighting equipment identified for horticultural use shall be listed.

410.174 Installation and Use. Lighting equipment identified for horticultural use shall be installed and used in accordance with the manufacturer's installation instructions and installation markings on the equipment as required by that listing.

410.176 Locations Not Permitted.

(A) General Lighting. Lighting equipment identified for horticultural use shall not be installed as lighting for general illumination unless such use is indicated in the manufacturer's instructions.

(B) Installed Location. Lighting equipment identified for horticultural use shall not be installed where it is likely to be subject to physical damage or where concealed.

410.178 Flexible Cord.
Flexible cord shall only be permitted when provided as part of listed lighting equipment identified for horticultural use for any of the following uses:

(1) Connecting a horticultural lighting luminaire directly to a branch circuit outlet
(2) Interconnecting horticultural lighting luminaires
(3) Connecting a horticultural lighting luminaire to a remote power source

Informational Note: Remote power sources include LED drivers, fluorescent ballasts, or HID ballasts.

410.180 Fittings and Connectors.
Fittings and connectors attached to flexible cords shall be provided as part of a listed horticultural lighting equipment device or system and installed in accordance with the instructions provided as part of that listing.

410.182 Equipment Grounding Conductor.
Lighting equipment identified for horticultural use shall be connected to the equipment grounding conductor in accordance with Part V of this article.

410.184 Ground-Fault Circuit-Interrupter (GFCI) Protection and Special Purpose Ground-Fault Circuit-Interrupter (SPGFCI) Protection.
Lighting equipment identified for horticultural use and employing flexible cord(s) with one or more separable connector(s) or attachment plug(s) shall be supplied by lighting outlets protected by a listed GFCI.

> *Exception: Circuits exceeding 150 volts to ground shall be protected by a listed SPGFCI.*

> Informational Note: See UL 943C, *Outline of Investigation for Special Purpose Ground-Fault Circuit-Interrupters*, for information on special purpose ground-fault circuit interrupters.

Δ 410.186 Support.
Fittings identified for support of horticultural lighting equipment shall be used in accordance with the installation instructions provided and shall be securely fastened.

•

N Part XVII. Special Provisions for Germicidal Irradiation Luminaires

N 410.190 General.

N 410.191 Listing.
Luminaires intended to emit germicidal irradiation shall be listed and identified as germicidal equipment.

N 410.193 Installation.
Luminaires shall be installed in accordance with the manufacturer's instructions and equipment markings.

N 410.195 Locations Not Permitted.

N (A) General Lighting. Luminaires shall not be installed as lighting for general illumination unless such use is indicated in the manufacturer's instructions.

N (B) Installed Location. Luminaires shall not be installed where likely to be subject to physical damage.

N (C) Dwellings. Luminaires shall not be installed in a dwelling unless listed and identified for use in dwellings.

N (D) Mounting Height. Luminaires installed in a building space that will be occupied during luminaire operation shall not be mounted below the minimum height specified by its listing and installation instructions.

N 410.197 Germicidal Irradiation Systems.

N (A) Listing. A germicidal irradiation system intended to provide a safeguard against UV exposure by ensuring that a building space will not be occupied during luminaire operation shall be listed and identified as a germicidal system.

N (B) System Components. All system components shall be provided by the system manufacturer or clearly specified in the installation instructions as a component that the installer is required to source separately.

N (C) Installation. A germicidal irradiation system shall be installed in accordance with the manufacturer's installation instructions and installation markings.

N (D) Dwellings. A germicidal irradiation system shall not be installed in a dwelling unless listed and identified for use in dwellings.

ARTICLE
411 Low-Voltage Lighting

Δ 411.1 Scope.
This article covers low voltage lighting systems and their associated components.

Δ 411.2 Reconditioned Equipment.
Listed low-voltage lighting systems or a lighting system assembled from listed parts shall not be reconditioned.

Δ (A) Listed System. The luminaires, power supply, and luminaire fittings (including the exposed bare conductors) of a low-voltage lighting system shall be listed for use as part of the same identified lighting system.

(B) Assembly of Listed Parts. A lighting system assembled from the following listed parts shall be permitted:

(1) Low-voltage luminaires identified for the use
(2) Power supply identified for the use
(3) Low-voltage luminaire fittings identified for the use
(4) Suitably rated cord or cable, or any Chapter 3 wiring method for the secondary circuit

A lighting system may be a complete listed system or an assembly of listed parts. Lighting systems have long been field assembled from individually listed low-voltage luminaires, a power supply, and a suitably rated cord or other fixed Chapter 3 wiring method. Installers typically verify that individually listed lighting system parts (regularly from multiple manufacturers) are intended for the use and have the needed ratings to create and assemble a low-voltage lighting system.

N 411.3 Voltage Limitations. The operating voltage of low-voltage lighting systems and their associated components shall not exceed 30 volts ac or 60 volts dc. If wet contact is likely to occur, the operating voltage of low-voltage lighting systems and their associated components shall not exceed 15 volts ac or 30 volts dc.

> Informational Note: See 680.1 for swimming pools, fountains, and similar installations.

Article 411 covers low-voltage interior and exterior (landscape) lighting systems consisting of a maximum 30-volt isolating ac or 60-volt dc power supply. In wet locations where contact with the lighting system might occur, the maximum voltage limit is 15 volts ac or 30 volts dc.

411.4 Low-Voltage Lighting Systems. Low voltage lighting systems shall consist of an isolating power supply, low-voltage luminaires, and associated equipment that are all identified for the use. The output circuits of the power supply shall be rated for 25 amperes maximum under all load conditions.

Article 411 also applies to Class 2 luminaires operating above 30 volts (42.4 volts peak voltage). This permits luminaires operating up to 60 volts dc to be installed without grounding, in accordance with 411.7(A) of the *NEC*®. Grounding is not a necessary safety measure for products operating from an isolating source, especially where voltages are within Class 2 limits. What distinguishes Article 411 from Article 410 is the voltage limitation and isolation requirement, both of which allow for a different scheme for protection against electric shock injury.

411.6 Specific Location Requirements.

(A) Walls, Floors, and Ceilings. Conductors concealed or extended through a wall, floor, or ceiling shall be in accordance with one of the following:

(1) Installed using any of the wiring methods specified in Chapter 3
(2) Installed using wiring supplied by a listed Class 2 power source and installed in accordance with 725.130

(B) Pools, Spas, Fountains, and Similar Locations. Lighting systems shall be installed not less than 3 m (10 ft) horizontally from the nearest edge of the water, unless permitted elsewhere in this *Code*.

The installation requirements of 411.6 recognize that shock and fire hazards exist even with low-voltage systems.

411.7 Secondary Circuits.

(A) Grounding. Secondary circuits shall not be grounded.

Exception: Secondary circuits supplied by a Class 2 power source listed and identified as suitable for secondary grounding shall be permitted to be grounded.

(B) Isolation. The secondary circuit shall be insulated from the branch circuit by an isolating transformer.

(C) Bare Conductors. Exposed bare conductors and current-carrying parts shall be permitted for indoor installations only. Bare conductors shall not be installed less than 2.1 m (7 ft) above the finished floor, unless specifically listed for a lower installation height.

Low-voltage bare conductor systems (i.e., trapeze lights) must be listed as a system. A system listing is necessary for these products because they often rely on special detection circuitry to address the fire risk associated with a conductive material shorting across the exposed conductors.

411.8 Branch Circuit. Lighting systems covered by this article shall be supplied from a maximum 20-ampere branch circuit.

ARTICLE
422 Appliances

Part I. General

422.1 Scope. This article covers electrical appliances used in any occupancy.

Article 422 covers appliances that are fastened in place or cord-and-plug-connected, such as air-conditioning units, dishwashers, heating appliances, water heaters, and infrared heating lamps. See Article 100 for the definition of *appliance*.

422.5 GFCI Protection.

Δ **(A) General.** Appliances identified in 422.5(A)(1) through (A)(7) 150 volts or less to ground and 60 amperes or less, single- or 3-phase, shall be provided with Class A protection for personnel. Multiple Class A protective devices shall be permitted but shall not be required.

(1) Automotive vacuum machines
(2) Drinking water coolers and bottle fill stations

(3) Cord-and-plug-connected high-pressure spray washing machines
(4) Tire inflation machines
(5) Vending machines
(6) Sump pumps
(7) Dishwashers

> Informational Note: Section 210.8 specifies requirements for GFCI protection for the branch-circuit outlet where the covered location warrants such protection.

This GFCI requirement applies to the specified appliances whether hard-wired or cord-and-plug-connected. Water coolers, such as those illustrated in Exhibit 422.1, are required to have GFCI protection. Tire inflation and automotive vacuum machines at service stations and car washes are often subject to exposure to the elements and damage from vehicles. In addition, they are used under all types of environmental conditions.

Prior to 2005, the U.S. Consumer Product Safety Commission (CPSC) investigated four separate electrocution incidents and three nonfatal shock incidents involving vending machines. Those investigations led to a requirement that vending machines be provided with GFCI protection.

(B) Type and Location. The GFCI shall be readily accessible, listed, and located in one or more of the following locations:

(1) Within the branch-circuit overcurrent device
(2) A device or outlet within the supply circuit
(3) An integral part of the attachment plug
(4) Within the supply cord not more than 300 mm (12 in.) from the attachment plug
(5) Factory installed within the appliance

Several options are available for providing GFCI protection for appliances. The requirement that GFCI protection be "readily accessible" facilitates required periodic testing as well as the resetting of a tripped device.

EXHIBIT 422.1 Two types of electrically powered water coolers required to have GFCI protection.

422.6 Listing Required. All appliances supplied by 50 volts or higher shall be listed.

Part II. Installation

422.10 Branch Circuits. Branch circuits supplying appliances shall comply with 422.10(A) or (B).

Δ **(A) Individual Branch Circuits.** Individual branch circuits supplying appliances shall comply with the following as applicable:

(1) The ampacities of branch-circuit conductors shall not be less than the marked rating of the appliance or the marked rating of an appliance having combined loads.
(2) The ampacities of branch-circuit conductors for motor-operated appliances not having a marked rating shall be in accordance with Part II of Article 430.
(3) The branch-circuit rating for an appliance that is a continuous load, other than a motor-operated appliance, shall not be less than 125 percent of the marked rating, or not less than 100 percent of the marked rating if the branch-circuit device and its assembly are listed for continuous loading at 100 percent of its rating.
(4) Branch circuits and branch-circuit conductors for household ranges and cooking appliances shall be permitted to be in accordance with Table 220.55 and shall be sized in accordance with 210.19(C).

(B) Branch Circuits Supplying Two or More Loads. For branch circuits supplying appliances and other loads, the rating shall be determined in accordance with 210.23.

422.11 Overcurrent Protection. Appliances shall be protected against overcurrent in accordance with 422.11(A) through (G) and 422.10.

(A) Branch-Circuit Overcurrent Protection. Branch circuits shall be protected in accordance with 240.4.

If a protective device rating is marked on an appliance, the branch-circuit overcurrent device rating shall not exceed the protective device rating marked on the appliance.

A listed appliance is provided with installation instructions from the manufacturer. The branch-circuit size must meet the minimum size stated in the installation instructions.

See also

110.3(B) and its associated commentary regarding the installation and use of listed or labeled equipment

(B) Household-Type Appliances with Surface Heating Elements. Household-type appliances with surface heating elements having a maximum demand of more than 60 amperes calculated in accordance with Table 220.55 shall have their power supply subdivided into two or more circuits, each of which shall be provided with overcurrent protection rated at not over 50 amperes.

(C) Infrared Lamp Commercial and Industrial Heating Appliances. Infrared lamp commercial and industrial heating appliances shall have overcurrent protection not exceeding 50 amperes.

(D) Open-Coil or Exposed Sheathed-Coil Types of Surface Heating Elements in Commercial-Type Heating Appliances. Open-coil or exposed sheathed-coil types of surface heating elements in commercial-type heating appliances shall be protected by overcurrent protective devices rated at not over 50 amperes.

(E) Single Non-Motor-Operated Appliance. If the branch circuit supplies a single non-motor-operated appliance, the rating of overcurrent protection shall comply with the following:

(1) Not exceed the overcurrent protection rating marked on the appliance.
(2) Not exceed 20 amperes if the overcurrent protection rating is not marked and the appliance is rated 13.3 amperes or less.
(3) Not exceed 150 percent of the appliance rated current if the overcurrent protection rating is not marked and the appliance is rated over 13.3 amperes. Where 150 percent of the appliance rating does not correspond to a standard overcurrent device ampere rating, the next higher standard rating shall be permitted.

(F) Electric Heating Appliances Employing Resistance-Type Heating Elements Rated More Than 48 Amperes.

(1) Electric Heating Appliances. Electric heating appliances employing resistance-type heating elements rated more than 48 amperes, other than household appliances with surface heating elements covered by 422.11(B), and commercial-type heating appliances covered by 422.11(D), shall have the heating elements subdivided. Each subdivided load shall not exceed 48 amperes, and each subdivided load shall be protected at not more than 60 amperes.

These supplementary overcurrent protective devices shall be (1) factory-installed within or on the heater enclosure or provided as a separate assembly by the heater manufacturer; (2) accessible; and (3) suitable for branch-circuit protection.

The main conductors supplying these overcurrent protective devices shall be considered branch-circuit conductors.

(2) Commercial Kitchen and Cooking Appliances. Commercial kitchen and cooking appliances using sheathed-type heating elements not covered in 422.11(D) shall be permitted to be subdivided into circuits not exceeding 120 amperes and protected at not more than 150 amperes where one of the following is met:

(1) Elements are integral with and enclosed within a cooking surface.
(2) Elements are completely contained within an enclosure identified as suitable for this use.
(3) Elements are contained within an ASME-rated and stamped vessel.

(3) Water Heaters and Steam Boilers. Resistance-type immersion electric heating elements shall be permitted to be subdivided into circuits not exceeding 120 amperes and protected at not more than 150 amperes as follows:

(1) Where contained in ASME-rated and stamped vessels
(2) Where included in listed instantaneous water heaters
(3) Where installed in low-pressure water heater tanks or open-outlet water heater vessels

Informational Note: See IEC 60335-2-21, *Household and similar electrical appliances — Safety — Particular requirements for storage water heaters*, for information on low-pressure and open-outlet heaters are atmospheric pressure water heaters

(G) Motor-Operated Appliances. Motors of motor-operated appliances shall be provided with overload protection in accordance with Part III of Article 430. Hermetic refrigerant motor-compressors in air-conditioning or refrigerating equipment shall be provided with overload protection in accordance with Part VI of Article 440. Where appliance overcurrent protective devices that are separate from the appliance are required, data for selection of these devices shall be marked on the appliance. The minimum marking shall be that specified in 430.7 and 440.4.

422.12 Central Heating Equipment. Central heating equipment other than fixed electric space-heating equipment shall be supplied by an individual branch circuit.

Exception No. 1: Auxiliary equipment, such as a pump, valve, humidifier, or electrostatic air cleaner directly associated with the heating equipment, shall be permitted to be connected to the same branch circuit.

Exception No. 2: Permanently connected air-conditioning equipment shall be permitted to be connected to the same branch circuit.

Exception No. 1 permits electric motors, ignition systems, controls, and so forth of fossil fuel-fired central heating equipment to be connected to the same individual branch circuit.

Exception No. 2 allows a permanently connected air-conditioning unit to be supplied from a branch circuit that supplies central heating equipment other than fixed electric space-heating equipment, because central heating equipment and air-conditioning equipment are considered unlikely to operate at the same time.

Δ **422.13 Storage-Type Water Heaters.** The branch-circuit overcurrent device and conductors for fixed storage-type water heaters that have a capacity of 450 L (120 gal) or less shall have an ampere rating of not less than 125 percent of the ampere rating of the water heater.

Informational Note: See 422.10 for branch-circuit rating.

422.16 Flexible Cords.

(A) General. Flexible cord shall be permitted as follows:

(1) To connect appliances to facilitate their frequent interchange or to prevent the transmission of noise or vibration.

(2) To facilitate the removal or disconnection of appliances that are fastened in place, where the fastening means and mechanical connections are specifically designed to permit ready removal for maintenance or repair and the appliance is intended or identified for flexible cord connection.

(3) All cord-and-plug-connected electrically heated appliances that produce temperatures in excess of 121°C (250°F) on surfaces with which the cord is likely to be in contact shall be provided with one of the types of heater cords listed in Table 400.4.

(B) Specific Appliances.

Δ **(1) Electrically Operated In-Sink Waste Disposers.** Electrically operated in-sink waste disposers shall be permitted to be cord-and-plug-connected with a flexible cord identified as suitable in the installation instructions of the appliance manufacturer where all of the following conditions are met:

(1) The length of the cord is not less than 450 mm (18 in.) and not exceeding 900 mm (36 in.).

(2) Receptacles are located to protect against physical damage to the flexible cord.

(3) The receptacle is accessible.

(4) The flexible cord has an equipment grounding conductor and is terminated with a grounding-type attachment plug.

Exception: A listed appliance distinctly marked to identify it as protected by a system of double insulation shall not be required to be terminated with a grounding-type attachment plug.

All cord-connected waste disposers are covered by this requirement whether they are in a sink at a kitchen, prep area at a deli counter, or a bar. The kitchen waste disposer illustrated in Exhibit 422.2 is an example of a cord-and-plug-connected appliance with mechanical connections designed to permit removal.

EXHIBIT 422.2 A cord-and-plug-connected kitchen waste disposer. *(Courtesy of the National Electrical Contractors Association)*

To facilitate control of the waste disposer from a wall location, the receptacle is permitted to be switched by a general-use snap switch as long as the load does not exceed the requirements of 404.14(A)(4). Because the waste disposer in Exhibit 422.2 is located in a kitchen, the 120-volt receptacle feeding the waste disposer requires GFCI protection based on 210.8(A)(6) requirements.

(2) Built-in Dishwashers and Trash Compactors. Built-in dishwashers and trash compactors shall be permitted to be cord-and-plug-connected with a flexible cord identified as suitable for the purpose in the installation instructions of the appliance manufacturer where all of the following conditions are met:

(1) For a trash compactor, the length of the cord is not less than 0.9 m (3 ft) and not exceeding 1.2 m (4 ft) measured from the face of the attachment plug to the plane of the rear of the appliance.

(2) For a built-in dishwasher, the length of the cord is not less than 0.9 m (3 ft) and not exceeding 2.0 m (6.5 ft) measured from the face of the attachment plug to the plane of the rear of the appliance.

(3) Receptacles are located to protect against physical damage to the flexible cord.

(4) The receptacle for a trash compactor is located in the space occupied by the appliance or adjacent thereto. If a flexible cord passes through an opening, it shall be protected against damage by a bushing, grommet, smoothed edge, or other approved means.

(5) The receptacle for a built-in dishwasher is located in the space adjacent to the space occupied by the dishwasher. If a flexible cord passes through an opening, it shall be protected against damage by a bushing, grommet, smoothed edge, or other approved means.

(6) The receptacle is accessible.

(7) The flexible cord has an equipment grounding conductor that is terminated with a grounding-type attachment plug.

Exception: A listed appliance distinctly marked to identify it as protected by a system of double insulation shall not be required to be terminated with a grounding-type attachment plug.

(3) Wall-Mounted Ovens and Counter-Mounted Cooking Units. Wall-mounted ovens and counter-mounted cooking units complete with provisions for mounting and for making electrical connections shall be permitted to be permanently connected or cord-and-plug-connected with a flexible cord identified as suitable for the purpose in the installation instructions of the appliance manufacturer.

A separable connector or a plug and receptacle combination in the supply line to an oven or cooking unit shall be identified for the temperature of the space in which it is located.

(4) Range Hoods and Microwave Oven/Range Hood Combinations. Range hoods and over-the-range microwave ovens with integral range hoods shall be permitted to be cord-and-plug-connected with a flexible cord identified as suitable for use on range hoods in the installation instructions of the appliance manufacturer, where all of the following conditions are met:

(1) The length of the cord is not less than 450 mm (18 in.) and not exceeding 1.2 m (4 ft).
(2) Receptacles are located to protect against physical damage to the flexible cord.
(3) The receptacle is supplied by an individual branch circuit.
(4) The receptacle is accessible.
(5) The flexible cord has an equipment grounding conductor and is terminated with a grounding-type attachment plug.

Exception: A listed appliance distinctly marked to identify it as protected by a system of double insulation shall not be required to be terminated with a grounding-type attachment plug.

422.17 Protection of Combustible Material. Each electrically heated appliance that is intended by size, weight, and service to be located in a fixed position shall be placed so as to provide protection between the appliance and adjacent combustible material.

Δ **422.18 Ceiling-Suspended (Paddle) Fans.**

N **(A) Support.** Ceiling-suspended (paddle) fans shall be supported independently of an outlet box or by one of the following:

(1) A listed outlet box or listed outlet box system identified for fan support installed in accordance with 314.27(C)
(2) A listed outlet box system, a listed weight-supporting ceiling receptacle, and a compatible factory-installed weight-supporting attachment fitting that is installed in accordance with 314.27(E)

The fan must be supported from the building structure if the outlet box or system is not identified to support a paddle fan. An outlet box identified for support of a paddle fan must be marked with the maximum weight it is permitted to support if the weight exceeds 35 pounds. Exhibit 422.3 shows an example of a listed outlet box used to support a paddle fan.

See also

314.27(E) and commentary, as well as commentary following the definition of the term *receptacle* in Article 100

N **(B) Location.** No metal parts of ceiling-suspended (paddle) fans in bathrooms and shower spaces shall be located within a zone measured 900 mm (3 ft) horizontally and 2.5 m (8 ft) vertically from the top of the bathtub rim or shower stall threshold. This zone is all-encompassing and shall include the space directly over the tub or shower stall.

422.19 Space for Conductors. The combined volume of the canopy of ceiling-suspended (paddle) fans and outlet box shall provide sufficient space so that conductors and their connecting devices are capable of being installed in accordance with 314.16.

EXHIBIT 422.3 Supporting a ceiling-suspended (paddle) fan (35 pounds or less) with a box identified for such use. *(Courtesy of Hubbell Incorporated)*

422.20 Outlet Boxes to Be Covered. In a completed installation, each outlet box shall be provided with a cover unless covered by means of a ceiling-suspended (paddle) fan canopy.

422.21 Covering of Combustible Material at Outlet Boxes. Any combustible ceiling finish that is exposed between the edge of a ceiling-suspended (paddle) fan canopy or pan and an outlet box and that has a surface area of 1160 mm² (180 in.²) or more shall be covered with noncombustible material.

422.22 Utilizing Separable Attachment Fittings. Appliances shall be permitted to use listed weight-supporting ceiling receptacles in combination with compatible weight-supporting attachment fittings used within their ratings and used in accordance with 314.27(E).

Part III. Disconnecting Means

422.30 General. A means shall be provided to simultaneously disconnect each appliance from all ungrounded conductors in accordance with the following sections of Part III. If an appliance is supplied by more than one branch circuit or feeder, these disconnecting means shall be grouped and identified as being the multiple disconnecting means for the appliance. Each disconnecting means shall simultaneously disconnect all ungrounded conductors that it controls.

422.31 Disconnection of Permanently Connected Appliances. For appliances that do not have a disconnecting means in accordance with 422.33 or 422.34, a disconnecting means shall be provided in accordance with 422.31(A), (B), or (C).

(A) Rated at Not over 300 Volt-Amperes or ⅛ Horsepower. For permanently connected appliances rated at not over

300 volt-amperes or ⅛ hp, the branch-circuit overcurrent device shall be permitted to serve as the disconnecting means where the switch or circuit breaker is within sight from the appliance or be capable of being locked in the open position in compliance with 110.25.

Δ **(B) Appliances Rated over 300 Volt-Amperes.** For permanently connected appliances rated over 300 volt-amperes, the branch-circuit switch or circuit breaker shall be permitted to serve as the disconnecting means where the switch or circuit breaker is within sight from the appliance or be capable of being locked in the open position in compliance with 110.25.

> Informational Note: See 422.34 for appliances employing unit switches.

Δ **(C) Motor-Operated Appliances Rated over ⅛ Horsepower.** The disconnecting means shall comply with 430.109 and 430.110. For permanently connected motor-operated appliances with motors rated over ⅛ hp, the disconnecting means shall be within sight from the appliance or be capable of being locked in the open position in compliance with 110.25.

> *Exception: If an appliance is provided with a unit switch that complies with 422.34(A), (B), or (C), the switch or circuit breaker serving as the other disconnecting means shall be permitted to be out of sight from the appliance.*

422.33 Disconnection of Cord-and-Plug-Connected or Attachment Fitting–Connected Appliances.

(A) Separable Connector or an Attachment Plug (or Attachment Fitting) and Receptacle. For cord-and-plug- (or attachment fitting–) connected appliances, an accessible separable connector or an accessible plug (or attachment fitting) and receptacle combination shall be permitted to serve as the disconnecting means. The attachment fitting shall be a factory installed part of the appliance and suitable for disconnection of the appliance. Where the separable connector or plug (or attachment fitting) and receptacle combination are not accessible, cord-and-plug-connected or attachment fitting-and-plug-connected appliances shall be provided with disconnecting means in accordance with 422.31.

Δ **(B) Connection at the Rear Base of a Range.** For cord-and-plug-connected household electric ranges, an attachment plug and receptacle connection at the rear base of a range, accessible from the front by removal of a drawer, shall be permitted.

(C) Rating. The rating of a receptacle or of a separable connector shall not be less than the rating of any appliance connected thereto.

> *Exception: Demand factors authorized elsewhere in this Code shall be permitted to be applied to the rating of a receptacle or of a separable connector.*

422.34 Unit Switch(es) as Disconnecting Means. A unit switch(es) with a marked-off position that is a part of an appliance

and disconnects all ungrounded conductors shall be permitted as the disconnecting means required by this article where other means for disconnection are provided in occupancies specified in 422.34(A) through (D).

(A) Multifamily Dwellings. In multifamily dwellings, the other disconnecting means shall be within the dwelling unit, or on the same floor as the dwelling unit in which the appliance is installed, and shall be permitted to control lamps and other appliances.

(B) Two-Family Dwellings. In two-family dwellings, the other disconnecting means shall be permitted either inside or outside of the dwelling unit in which the appliance is installed. In this case, an individual switch or circuit breaker for the dwelling unit shall be permitted and shall also be permitted to control lamps and other appliances.

(C) One-Family Dwellings. In one-family dwellings, the service disconnecting means shall be permitted to be the other disconnecting means.

(D) Other Occupancies. In other occupancies, the branch-circuit switch or circuit breaker, where readily accessible for servicing of the appliance, shall be permitted as the other disconnecting means.

422.35 Switch and Circuit Breaker to Be Indicating. Switches and circuit breakers used as disconnecting means shall be of the indicating type.

Part IV. Construction

Δ **422.40 Polarity in Cord-and-Plug-Connected Appliances.** If the appliance is provided with a manually operated, line-connected, single-pole switch for appliance on–off operation, an Edison-base lampholder, or a 15- or 20-ampere receptacle, the attachment plug shall be of the polarized or grounding type.

A 2-wire, nonpolarized attachment plug shall be permitted to be used on a listed double-insulated shaver.

> Informational Note: See 410.82 for polarity of Edison-base lampholders.

Δ **422.41 Cord-and-Plug-Connected Appliances Subject to Immersion.** Cord-and-plug-connected portable, freestanding hydromassage units and hand-held hair dryers shall be constructed to provide protection for personnel against electrocution when immersed.

Although receptacles in bathrooms of dwelling units have been required to be protected by GFCIs since the 1975 edition of the *NEC*®, many receptacles in older existing bathrooms are not so protected. Cord-and-plug-connected appliances, such as hand-held hair dryers, can fall into bathtubs and cause fatalities. Therefore, they are required to be provided with some form of protective device that is part of the appliance. Three types of protectors comply with this requirement:

1. Appliance-leakage circuit interrupters (ALCIs)
2. Immersion-detector circuit interrupters (IDCIs)
3. Ground-fault circuit interrupters (GFCIs)

ALCIs de-energize the supply to the appliance when leakage current exceeds a predetermined value. IDCIs de-energize the supply when a liquid causes a conductive path between a live part and a sensor, and GFCIs de-energize the supply when the current to ground exceeds a predetermined value.

422.42 Signals for Heated Appliances. In other than dwelling-type occupancies, each electrically heated appliance or group of appliances intended to be applied to combustible material shall be provided with a signal or an integral temperature-limiting device.

Many electrically heated appliances in commercial or industrial locations use a red light, connected to and within sight of the appliance, to indicate that the appliance is energized and operating. No signal is required for an electrically heated appliance provided with an integral high-temperature-limiting device, such as a thermostat, that limits the temperature to which the appliance can heat.

422.44 Cord-and-Plug-Connected Immersion Heaters. Electric heaters of the cord-and-plug-connected immersion type shall be constructed and installed so that current-carrying parts are effectively insulated from electrical contact with the substance in which they are immersed.

422.45 Stands for Cord-and-Plug-Connected Appliances. Each smoothing iron and other cord-and-plug-connected electrically heated appliance intended to be applied to combustible material shall be equipped with an approved stand, which shall be permitted to be a separate piece of equipment or a part of the appliance.

Δ **422.47 Water Heater Controls.** All storage or instantaneous-type water heaters shall be equipped with a temperature-limiting means in addition to its control thermostat to disconnect all ungrounded conductors. Such means shall comply with both of the following:

(1) Installed to sense maximum water temperature.
(2) Be either a trip-free, manually reset type or a type having a replacement element.

Exception No. 1: Storage water heaters that are identified as being suitable for use with a supply water temperature of 82°C (180°F) or above and a capacity of 60 kW or above.

Exception No. 2: Instantaneous-type water heaters that are identified as being suitable for such use, with a capacity of 4 L (1 gal) or less.

422.48 Infrared Lamp Industrial Heating Appliances.

(A) 300 Watts or Less. Infrared heating lamps rated at 300 watts or less shall be permitted with lampholders of the medium-base, unswitched porcelain type or other types identified as suitable for use with infrared heating lamps rated 300 watts or less.

(B) Over 300 Watts. Screw shell lampholders shall not be used with infrared lamps rated over 300 watts, unless the lampholders are identified as being suitable for use with infrared heating lamps rated over 300 watts.

Infrared (heat) radiation lamps are tungsten-filament incandescent lamps similar in appearance to lighting lamps. However, they are designed to operate at a lower temperature, thus transferring more heat radiation and less light intensity. Infrared lamps are used for a variety of heating and drying purposes in industrial locations.

Part V. Marking

422.60 Nameplate.

(A) Nameplate Marking. Each electrical appliance shall be provided with a nameplate giving the identifying name and the rating in volts and amperes, or in volts and watts. If the appliance is to be used on a specific frequency or frequencies, it shall be so marked.

Where motor overload protection external to the appliance is required, the appliance shall be so marked.

Informational Note: See 422.11 for overcurrent protection requirements.

(B) To Be Visible. Marking shall be located so as to be visible or easily accessible after installation.

422.61 Marking of Heating Elements. All heating elements that are rated over one ampere, replaceable in the field, and a part of an appliance shall be legibly marked with the ratings in volts and amperes, or in volts and watts, or with the manufacturer's part number.

422.62 Appliances Consisting of Motors and Other Loads.

(A) Nameplate Horsepower Markings. Where a motor-operated appliance nameplate includes a horsepower rating, that rating shall not be less than the horsepower rating on the motor nameplate. Where an appliance consists of multiple motors, or one or more motors and other loads, the nameplate value shall not be less than the equivalent horsepower of the combined loads, calculated in accordance with 430.110(C)(1).

(B) Additional Nameplate Markings. Appliances, other than those factory-equipped with cords and attachment plugs and with nameplates in compliance with 422.60, shall be marked in accordance with 422.62(B)(1) or (B)(2).

(1) Marking. In addition to the marking required in 422.60, the marking on an appliance consisting of a motor with other load(s) or motors with or without other load(s) shall specify the minimum supply circuit conductor ampacity and the maximum rating of the circuit overcurrent protective device. This requirement shall not apply to an appliance with a nameplate in compliance with 422.60 where both the minimum supply circuit conductor ampacity and maximum rating of the circuit overcurrent protective device are not more than 15 amperes.

(2) Alternate Marking Method. An alternate marking method shall be permitted to specify the rating of the largest motor in volts and amperes, and the additional load(s) in volts and amperes, or volts and watts in addition to the marking required in 422.60. The ampere rating of a motor ⅛ horsepower or less or a nonmotor load 1 ampere or less shall be permitted to be omitted unless such loads constitute the principal load.

ARTICLE
424 Fixed Electric Space-Heating Equipment

Part I. General

424.1 Scope. This article covers fixed electric equipment used for space heating. For the purpose of this article, heating equipment includes heating cables, unit heaters, boilers, central heating systems, or other fixed electric space-heating equipment. This article does not apply to process heating and room air conditioning.

Δ **424.3 Other Articles.** Fixed electric space-heating equipment incorporating a hermetic refrigerant motor-compressor shall additionally comply with Table 424.3 unless amended by this article.

N **TABLE 424.3** Other Articles

Equipment	Article
Air-conditioning and refrigerating equipment	440 (Parts I, II, III, IV, V, VI)

424.4 Branch Circuits.

(A) Branch-Circuit Requirements. An individual branch circuit shall be permitted to supply any volt-ampere or wattage rating of fixed electric space-heating equipment for which the branch circuit is rated.

Branch circuits supplying two or more outlets for fixed electric space-heating equipment shall be rated not over 30 amperes. In other than a dwelling unit, fixed infrared heating equipment shall be permitted to be supplied from branch circuits rated not over 50 amperes.

(B) Branch-Circuit Conductor Sizing. The branch-circuit conductor(s) ampacity shall not be less than 125 percent of the load of the fixed electric space-heating equipment and any associated motor(s).

Branch-circuit conductors supplying fixed electric space-heating equipment must be sized at 125 percent of the total load of the heaters (and motors) to protect conductors from overheating during periods of prolonged operation.

424.6 Listed Equipment. Electric baseboard heaters, heating cables, duct heaters, and radiant heating systems shall be listed and labeled.

Part II. Installation

424.10 General. Factory-installed receptacle outlets that are part of a permanently installed electric baseboard heater, or outlets provided as a separate listed assembly of an electric baseboard heater, shall be permitted in lieu of a receptacle outlet(s) that is required by 210.52. Such receptacle outlets shall not be connected to the baseboard heater circuits.

> Informational Note: Listed baseboard heaters include instructions that may not permit their installation below receptacle outlets.

This restates the permission granted in the second paragraph of 210.52 — that is, it allows factory-installed receptacle outlets in electric baseboard heaters to satisfy the spacing requirements for receptacle outlets in dwelling units according to 210.52(A).

Heating equipment and systems often have special installation instructions for spacings, types of supply wires, or special control equipment, which must be considered in determining the installation's suitability.

424.11 Supply Conductors. Fixed electric space-heating equipment requiring supply conductors with an insulation rating greater than 60°C shall be clearly and permanently marked. This marking shall be plainly visible after installation and shall be permitted to be adjacent to the field connection box.

Fixed electric space-heating equipment may require supply conductors with a temperature rating greater than 60°C, due to their proximity to the heating elements and the installation instructions provided with a listed product.

424.12 Locations.

(A) Exposed to Physical Damage. Where subject to physical damage, fixed electric space-heating equipment shall be protected in an approved manner.

Δ **(B) Damp or Wet Locations.** Heaters and related equipment installed in damp or wet locations shall be listed for such locations.

> Informational Note No. 1: See 110.11 for equipment exposed to deteriorating agents.
> Informational Note No. 2: See 680.27(C) for pool deck areas.

424.13 Spacing from Combustible Materials. Fixed electric space-heating equipment shall be installed to provide the required spacing between the equipment and adjacent combustible material, unless it is listed to be installed in direct contact with combustible material.

Part III. Control and Protection of Fixed Electric Space-Heating Equipment

424.19 Disconnecting Means. Means shall be provided to simultaneously disconnect the heater, motor controller(s), and supplementary overcurrent protective device(s) of all fixed electric space-heating equipment from all ungrounded conductors. Where heating equipment is supplied by more than one source, feeder, or branch circuit, the disconnecting means shall be grouped and identified as having multiple disconnecting means. Each disconnecting means shall simultaneously disconnect all ungrounded conductors that it controls. The disconnecting means specified in 424.19(A) and (B) shall have an ampere rating not less than 125 percent of the total load of the motors and the heaters and shall be capable of being locked in the open position in compliance with 110.25.

If there are multiple disconnecting means, they must be grouped. Each disconnecting means must simultaneously open all the ungrounded conductors that it controls to prevent the practice of disconnecting one conductor at a time at terminal blocks or similar devices. The disconnect switch must have a rating of 125 percent of the heater's total load.

A unit switch is permitted by 424.19(C) to serve as the disconnecting means, provided that it has a marked "off" position and disconnects all ungrounded conductors. In addition, other means must be provided in accordance with 424.19(C)(1) through (C)(4). Such other means are not required to be capable of being locked in the open position.

(A) Heating Equipment with Supplementary Overcurrent Protection. The disconnecting means for fixed electric space-heating equipment with supplementary overcurrent protection shall be within sight from the supplementary overcurrent protective device(s), on the supply side of these devices, if fuses, and, in addition, shall comply with either 424.19(A)(1) or (A)(2).

(1) Heater Containing No Motor Rated over ⅛ Horsepower. The disconnecting means provided shall be within sight from the motor controller(s) and the heater, or shall be lockable as specified in 424.19, or shall be a unit switch complying with 424.19(C).

(2) Heater Containing a Motor(s) Rated over ⅛ Horsepower. The disconnecting means required by 424.19 shall be permitted to serve as the required disconnecting means for both the motor controller(s) and heater under either of the following conditions:

  (1) Where the disconnecting means is in sight from the motor controller(s) and the heater and complies with Part IX of Article 430.

  (2) Where a motor(s) of more than ⅛ hp and the heater are provided with a single unit switch that complies with

422.34(A), (B), (C), or (D), the disconnecting means shall be permitted to be out of sight from the motor controller.

(B) Heating Equipment Without Supplementary Overcurrent Protection.

(1) Without Motor or with Motor Not over ⅛ Horsepower. For fixed electric space-heating equipment without a motor rated over ⅛ hp, the branch-circuit switch or circuit breaker shall be permitted to serve as the disconnecting means where the switch or circuit breaker is within sight from the heater or is capable of being locked in the open position in compliance with 110.25.

(2) Over ⅛ Horsepower. For motor-driven electric space-heating equipment with a motor rated over ⅛ hp, a disconnecting means shall be located within sight from the motor controller or shall be permitted to comply with the requirements in 424.19(A)(2).

(C) Unit Switch(es) as Disconnecting Means. A unit switch(es) with a marked "off" position that is part of a fixed heater and disconnects all ungrounded conductors shall be permitted as the disconnecting means required by this article where other means for disconnection are provided in the types of occupancies in 424.19(C)(1) through (C)(4).

(1) Multifamily Dwellings. In multifamily dwellings, the other disconnecting means shall be within the dwelling unit, or on the same floor as the dwelling unit in which the fixed heater is installed, and shall also be permitted to control general-purpose circuits and appliance circuits.

(2) Two-Family Dwellings. In two-family dwellings, the other disconnecting means shall be permitted either inside or outside of the dwelling unit in which the fixed heater is installed. In this case, an individual switch or circuit breaker for the dwelling unit shall be permitted and shall also be permitted to control general-purpose circuits and appliance circuits.

(3) One-Family Dwellings. In one-family dwellings, the service disconnecting means shall be permitted to be the other disconnecting means.

(4) Other Occupancies. In other occupancies, the branch-circuit switch or circuit breaker, where readily accessible for servicing of the fixed heater, shall be permitted as the other disconnecting means.

424.20 Thermostatically Controlled Switching Devices.

(A) Serving as Both Controllers and Disconnecting Means. Thermostatically controlled switching devices and combination thermostats and manually controlled switches shall be permitted to serve as both controllers and disconnecting means, provided they meet all of the following conditions:

  (1) Provided with a marked "off" position

  (2) Directly open all ungrounded conductors when manually placed in the "off" position

(3) Designed so that the circuit cannot be energized automatically after the device has been manually placed in the "off" position
(4) Located as specified in 424.19
(5) Located in an accessible location

(B) Thermostats That Do Not Directly Interrupt All Ungrounded Conductors. Thermostats that do not directly interrupt all ungrounded conductors and thermostats that operate remote-control circuits shall not be required to meet the requirements of 424.20(A). These devices shall not be permitted as the disconnecting means.

424.21 Switch and Circuit Breaker to Be Indicating. Switches and circuit breakers used as disconnecting means shall be of the indicating type.

424.22 Overcurrent Protection.

(A) Branch-Circuit Devices. Electric space-heating equipment, other than motor-operated equipment required to have additional overcurrent protection by Parts III and IV of Article 430 or Parts III and VI of Article 440, shall be permitted to be protected against overcurrent where supplied by one of the branch circuits in Part II of Article 210.

(B) Resistance Elements. Resistance-type heating elements in electric space-heating equipment shall be protected at not more than 60 amperes. Equipment rated more than 48 amperes and employing such elements shall have the heating elements subdivided, and each subdivided load shall not exceed 48 amperes. Where a subdivided load is less than 48 amperes, the rating of the supplementary overcurrent protective device shall comply with 424.4(B). A boiler employing resistance-type immersion heating elements contained in an ASME-rated and stamped vessel shall be permitted to comply with 424.72(A).

The reason for subdividing the overcurrent protection is to minimize the amount of damaging energy — in the form of both heat and magnetic energy — released into the heating elements during a short circuit. The damaging short-circuit energy released at the element is greatly reduced by limiting the size of the overcurrent device protecting the individual heating elements, thereby greatly reducing the risk of fire. In addition, a second benefit may be continuity of service if equipment is only partially affected.

Historically, the subdivision size of 60 amperes was selected to use the maximum fuseholder size of 60 amperes while maintaining up to a 48-ampere heating element (48 A × 125% = 60 A).

Δ **(C) Overcurrent Protective Devices.** The supplementary overcurrent protective devices for the subdivided loads specified in 424.22(B) shall meet all of the following conditions:

(1) Be factory-installed within or on the heater enclosure or supplied for use with the heater as a separate assembly by the heater manufacturer
(2) Be accessible
(3) Be suitable for branch-circuit protection

Where cartridge fuses are used to provide overcurrent protection for the subdivided loads, a single disconnecting means shall be permitted to be used as the disconnecting means for all of the subdivided loads.

Informational Note No. 1: See 240.10.
Informational Note No. 2: See 240.10 for supplementary overcurrent protection.
Informational Note No. 3: See 240.40 for disconnecting means for cartridge fuses in circuits of any voltage.

(D) Branch-Circuit Conductors. The conductors supplying the supplementary overcurrent protective devices shall be considered branch-circuit conductors.

Where the heaters are rated 50 kW or more, the conductors supplying the supplementary overcurrent protective devices specified in 424.22(C) shall be permitted to be sized at not less than 100 percent of the nameplate rating of the heater, provided all of the following conditions are met:

(1) The heater is marked with a minimum conductor size.
(2) The conductors are not smaller than the marked minimum size.
(3) A temperature-actuated device controls the cyclic operation of the equipment.

(E) Conductors for Subdivided Loads. Field-wired conductors between the heater and the supplementary overcurrent protective devices shall be sized at not less than 125 percent of the load served. The supplementary overcurrent protective devices specified in 424.22(C) shall protect these conductors in accordance with 240.4.

Where the heaters are rated 50 kW or more, the ampacity of field-wired conductors between the heater and the supplementary overcurrent protective devices shall be permitted to be not less than 100 percent of the load of their respective subdivided circuits, provided all of the following conditions are met:

(1) The heater is marked with a minimum conductor size.
(2) The conductors are not smaller than the marked minimum size.
(3) A temperature-activated device controls the cyclic operation of the equipment.

Part IV. Marking of Heating Equipment

424.28 Nameplate.

(A) Marking Required. Each unit of fixed electric space-heating equipment shall be provided with a nameplate giving the identifying name and the normal rating in volts and watts or in volts and amperes.

Electric space-heating equipment intended for use on alternating current only, direct current only, or both shall be marked to so indicate. The marking of equipment consisting of motors over ⅛ hp and other loads shall specify the rating of the motor in volts, amperes, and frequency, and the heating load in volts and watts or in volts and amperes.

(B) Location. This nameplate shall be located so as to be visible or accessible after installation.

424.29 Marking of Heating Elements. All heating elements that are replaceable in the field and are part of an electric heater shall be legibly marked with the ratings in volts and watts or in volts and amperes.

Part V. Electric Space-Heating Cables

424.34 Heating Cable Construction. Factory-assembled nonheating leads of heating cables, if any, shall be at least 2.1 m (7 ft) in length.

424.35 Marking of Heating Cables. Each unit shall be marked with the identifying name or identification symbol, catalog number, and ratings in volts and watts or in volts and amperes.

424.36 Clearances of Wiring in Ceilings. Wiring located above heated ceilings shall be spaced not less than 50 mm (2 in.) above the heated ceiling. The ampacity of conductors shall be calculated on the basis of an assumed ambient temperature of not less than 50°C (122°F), applying the correction factors in accordance with 310.15(B)(1). If this wiring is located above thermal insulation having a minimum thickness of 50 mm (2 in.), it shall be subject to the ambient correction in accordance with 310.15(B)(1).

424.38 Area Restrictions.

(A) Extending Beyond the Room or Area. Heating cables shall be permitted to extend beyond the room or area in which they originate unless prohibited by 424.38(B).

(B) Uses Not Permitted. Heating cables shall not be installed as follows:

(1) In closets, other than as noted in 424.38(C)
(2) Over the top of walls where the wall intersects the ceiling
(3) Over partitions that extend to the ceiling, unless they are isolated single runs of embedded cable
(4) Under or through walls
(5) Over cabinets whose clearance from the ceiling is less than the minimum horizontal dimension of the cabinet to the nearest cabinet edge that is open to the room or area
(6) In tub and shower walls
(7) Under cabinets or similar built-ins having no clearance to the floor

(C) In Closet Ceilings as Low-Temperature Heat Sources to Control Relative Humidity. The provisions of 424.38(B) shall not prevent the use of cable in closet ceilings as low-temperature heat sources to control relative humidity, provided they are used only in those portions of the ceiling that are unobstructed to the floor.

424.39 Clearance from Other Objects and Openings. Heating elements of cables installed in ceilings shall be separated at least 200 mm (8 in.) from the edge of outlet boxes and junction boxes that are to be used for mounting surface luminaires. A clearance of not less than 50 mm (2 in.) shall be provided from recessed luminaires and their trims, ventilating openings, and other such openings in room surfaces. No heating cable shall be covered by any ceiling surface-mounted equipment.

424.40 Splices. The length of heating cable shall only be altered using splices identified in the manufacturer's instructions.

424.41 Ceiling Installation of Heating Cables on Dry Board, in Plaster, and on Concrete.

(A) In Walls. Heating cables identified only for use in ceiling installations shall not be installed in walls unless it is necessary for an isolated single run of cable to be installed down a vertical surface to reach a dropped ceiling.

(B) Adjacent Runs. Adjacent runs of heating cable shall be installed in accordance with the manufacturer's instructions.

Δ **(C) Surfaces to Be Applied.** Heating cables shall be applied only to gypsum board, plaster lath, or other fire-resistant material. With metal lath or other electrically conductive surfaces, a coat of plaster or other means employed in accordance with the heating cable manufacturer's instructions shall be applied to completely separate the metal lath or conductive surface from the cable.

Informational Note: See 424.41(F).

(D) Splices. All heating cables, the splice between the heating cable and nonheating leads, and 75-mm (3-in.) minimum of the nonheating lead at the splice shall be embedded in plaster or dry board in the same manner as the heating cable.

(E) Ceiling Surface. The entire ceiling surface shall have a finish of thermally noninsulating sand plaster that has a nominal thickness of 13 mm (½ in.), or other noninsulating material identified as suitable for this use and applied according to specified thickness and directions.

(F) Secured. Cables shall be secured by means of approved stapling, tape, plaster, nonmetallic spreaders, or other approved means either at intervals not exceeding 400 mm (16 in.) or at intervals not exceeding 1.8 m (6 ft) for cables identified for such use. Staples or metal fasteners that straddle the cable shall not be used with metal lath or other electrically conductive surfaces.

(G) Dry Board Installations. In dry board installations, the entire ceiling below the heating cable shall be covered with gypsum board not exceeding 13 mm (½ in.) thickness. The void between the upper layer of gypsum board, plaster lath, or other fire-resistant material and the surface layer of gypsum board shall be completely filled with thermally conductive, nonshrinking plaster or other approved material or equivalent thermal conductivity.

(H) Free from Contact with Conductive Surfaces. Cables shall be kept free from contact with metal or other electrically conductive surfaces.

(I) Joists. In dry board applications, cable shall be installed parallel to the joist, leaving a clear space centered under the joist of 65 mm (2½ in.) (width) between centers of adjacent runs of cable. A surface layer of gypsum board shall be mounted so that the nails or other fasteners do not pierce the heating cable.

(J) Crossing Joists. Cables shall cross joists only at the ends of the room unless the cable is required to cross joists elsewhere in order to satisfy the manufacturer's instructions regarding clearance from ceiling penetrations and luminaires.

424.42 Finished Ceilings. Finished ceilings shall not be covered with decorative panels or beams constructed of materials that have thermal insulating properties, such as wood, fiber, or plastic. Finished ceilings shall be permitted to be covered with paint, wallpaper, or other approved surface finishes.

424.43 Installation of Nonheating Leads of Cables.

(A) Free Nonheating Leads. Free nonheating leads of cables shall be installed in accordance with Chapter 3 wiring methods, or other listed means, from the junction box to a location within the ceiling.

(B) Leads in Junction Box. Not less than 150 mm (6 in.) of free nonheating lead shall be within the junction box. The marking of the leads shall be visible in the junction box.

(C) Excess Leads. Excess leads of heating cables shall not be cut but shall be secured to the underside of the ceiling and embedded in plaster or other approved material, leaving only a length sufficient to reach the junction box with not less than 150 mm (6 in.) of free lead within the box.

424.44 Installation of Cables in Concrete or Poured Masonry Floors.

(A) Adjacent Runs. Adjacent runs of heating cable shall be installed in accordance with the manufacturer's instructions.

(B) Secured in Place. Cables shall be secured in place by nonmetallic frames or spreaders or other approved means while the concrete or other finish is applied.

(C) Leads Protected. Leads shall be protected where they leave the floor by rigid metal conduit, intermediate metal conduit, rigid polyvinyl chloride conduit, electrical metallic tubing, or by other approved means.

(D) Bushings or Approved Fittings. Bushings or approved fittings shall be used where the leads emerge within the floor slab.

(E) Ground-Fault Circuit-Interrupter Protection. In addition to the requirements in 210.8, ground-fault circuit-interrupter protection for personnel shall be provided for cables installed in

electrically heated floors of bathrooms, kitchens, and in hydromassage bathtub locations.

GFCI protection is required where cables are installed in concrete or poured masonry floors, thereby reducing shock hazards to persons walking barefoot in those areas. This requirement applies regardless of the type of floor covering over the concrete or poured masonry. Electric radiant heating cables are not permitted in spa, hot tub, or pool locations in accordance with 680.27(C)(3).

424.45 Installation of Cables Under Floor Coverings.

(A) Identification. Heating cables for installation under floor covering shall be identified as suitable for installation under floor covering.

(B) Expansion Joints. Heating cables shall not be installed where they bridge expansion joints unless provided with expansion and contraction fittings applicable to the manufacture of the cable.

(C) Connection to Conductors. Heating cables shall be connected to branch-circuit and supply wiring by wiring methods described in the installation instructions.

(D) Anchoring. Heating cables shall be positioned and secured in place under the floor covering, in accordance with the manufacturer's instructions.

(E) Ground-Fault Circuit-Interrupter Protection. In addition to the requirements in 210.8, ground-fault circuit-interrupter protection for personnel shall be provided.

(F) Grounding Braid or Sheath. Grounding means, such as copper braid, metal sheath, or other approved means, shall be provided as part of the heated length.

424.46 Inspection. Cable installations shall be made with due care to prevent damage to the cable assembly and shall be inspected and approved before cables are covered or concealed.

424.47 Label Provided by Manufacturer. The manufacturers of electric space-heating cables shall provide marking labels that indicate that the space-heating installation incorporates electric space-heating cables and instructions that the labels shall be affixed to the panelboards to identify which branch circuits supply the circuits to those space-heating installations. If the electric space-heating cable installations are visible and distinguishable after installation, the labels shall not be required to be provided and affixed to the panelboards.

N **424.48 Installation of Cables in Walls.** Unless prohibited by 424.38(B), heating cables and cable sets shall be permitted to be installed in, on, or behind walls provided all of the following are met:

(1) Heating cables and cable sets shall be identified as suitable for installation in, on, or behind walls.

(2) Heating cables and cable sets shall be GFCI protected.

(3) Grounding means, such as copper braid, metal sheath, or other approved means, shall be provided.

(4) Heating cables and cable sets shall be AFCI protected.

(5) Heating cables and cable sets shall be permitted to be installed no more than 1.2 m (4 ft) above the floor.

This requirement shall become effective January 1, 2026.

Part VI. Duct Heaters

424.57 General. Part VI shall apply to any heater mounted in the airstream of a forced-air system where the air-moving unit is not provided as an integral part of the equipment.

424.58 Identification. Heaters installed in an air duct shall be identified as suitable for the installation.

424.59 Airflow. Means shall be provided to ensure uniform airflow over the face of the heater in accordance with the manufacturer's instructions.

> Informational Note: Heaters installed within 1.2 m (4 ft) of the outlet of an air-moving device, heat pump, air conditioner, elbows, baffle plates, or other obstructions in ductwork may require turning vanes, pressure plates, or other devices on the inlet side of the duct heater to ensure an even distribution of air over the face of the heater.

424.60 Elevated Inlet Temperature. Duct heaters intended for use with elevated inlet air temperature shall be identified as suitable for use at the elevated temperatures.

424.61 Installation of Duct Heaters with Heat Pumps and Air Conditioners. Heat pumps and air conditioners having duct heaters closer than 1.2 m (4 ft) to the heat pump or air conditioner shall have both the duct heater and heat pump or air conditioner identified as suitable for such installation and so marked.

424.62 Condensation. Duct heaters used with air conditioners or other air-cooling equipment that could result in condensation of moisture shall be identified as suitable for use with air conditioners.

424.63 Fan Circuit Interlock. Means shall be provided to ensure that the fan circuit is energized when any heater circuit is energized. However, time- or temperature-controlled delay in energizing the fan motor shall be permitted.

424.64 Limit Controls. Each duct heater shall be provided with an approved, integral, automatic-reset temperature-limiting control or controllers to de-energize the circuit or circuits.

In addition, an integral independent supplementary control or controllers shall be provided in each duct heater that disconnects a sufficient number of conductors to interrupt current flow. This device shall be manually resettable or replaceable.

424.65 Location of Disconnecting Means. Duct heater controller equipment shall be accessible with the disconnecting means installed within sight from the controller or as permitted by 424.19(A).

Δ **424.66 Installation.** Duct heaters shall be installed in accordance with the manufacturer's instructions in such a manner that the operation of the duct heater does not create a hazard to persons or property. Furthermore, duct heaters shall be located with respect to building construction and other equipment so as to permit access to the heater and the heater control. Working space shall be maintained in accordance with 110.26(A)(4) and shall permit replacement of controls and heating elements and for adjusting and cleaning of controls and other parts requiring such attention.

> Informational Note: See NFPA 90A, *Standard for the Installation of Air-Conditioning and Ventilating Systems*, and NFPA 90B, *Standard for the Installation of Warm Air Heating and Air-Conditioning Systems*, for additional installation information.

Part VII. Resistance-Type Boilers

424.70 Scope. The provisions in Part VII of this article shall apply to boilers employing resistance-type heating elements. See Part VIII of this article for electrode-type boilers.

424.71 Identification. Resistance-type boilers shall be identified as suitable for the installation.

424.72 Overcurrent Protection.

(A) Boiler Employing Resistance-Type Immersion Heating Elements in an ASME-Rated and Stamped Vessel. A boiler employing resistance-type immersion heating elements contained in an ASME-rated and stamped vessel shall have the heating elements protected at not more than 150 amperes. Such a boiler rated more than 120 amperes shall have the heating elements subdivided into loads not exceeding 120 amperes.

Where a subdivided load is less than 120 amperes, the rating of the overcurrent protective device shall comply with 424.4(B).

(B) Boiler Employing Resistance-Type Heating Elements Rated More Than 48 Amperes and Not Contained in an ASME-Rated and Stamped Vessel. A boiler employing resistance-type heating elements not contained in an ASME-rated and stamped vessel shall have the heating elements protected at not more than 60 amperes. Such a boiler rated more than 48 amperes shall have the heating elements subdivided into loads not exceeding 48 amperes.

Where a subdivided load is less than 48 amperes, the rating of the overcurrent protective device shall comply with 424.4(B).

See also

424.22(B) and its commentary for an explanation of the subdivision requirement

(C) Supplementary Overcurrent Protective Devices. The supplementary overcurrent protective devices for the subdivided loads as required by 424.72(A) and (B) shall be as follows:

(1) Factory-installed within or on the boiler enclosure or provided as a separate assembly by the boiler manufacturer
(2) Accessible, but need not be readily accessible
(3) Suitable for branch-circuit protection

Where cartridge fuses are used to provide this overcurrent protection, a single disconnecting means shall be permitted for the several subdivided circuits. See 240.40.

(D) Conductors Supplying Supplementary Overcurrent Protective Devices. The conductors supplying these supplementary overcurrent protective devices shall be considered branch-circuit conductors.

Where the heaters are rated 50 kW or more, the conductors supplying the overcurrent protective device specified in 424.72(C) shall be permitted to be sized at not less than 100 percent of the nameplate rating of the heater, provided all of the following conditions are met:

(1) The heater is marked with a minimum conductor size and conductor insulation temperature rating.
(2) The conductors are not smaller than the marked minimum size.
(3) A temperature- or pressure-actuated device controls the cyclic operation of the equipment.

(E) Conductors for Subdivided Loads. Field-wired conductors between the heater and the supplementary overcurrent protective devices shall be sized at not less than 125 percent of the load served. The supplementary overcurrent protective devices specified in 424.72(C) shall protect these conductors in accordance with 240.4.

Where the heaters are rated 50 kW or more, the ampacity of field-wired conductors between the heater and the supplementary overcurrent protective devices shall be permitted to be not less than 100 percent of the load of their respective subdivided circuits, provided all of the following conditions are met:

(1) The heater is marked with a minimum conductor size.
(2) The conductors are not smaller than the marked minimum size.
(3) A temperature-actuated device controls the cyclic operation of the equipment.

424.73 Overtemperature Limit Control. Each boiler designed so that in normal operation there is no change in state of the heat transfer medium shall be equipped with a temperature-sensitive limiting means. It shall be installed to limit maximum liquid temperature and shall directly or indirectly disconnect all ungrounded conductors to the heating elements. Such means shall be in addition to a temperature-regulating system and other devices protecting the tank against excessive pressure.

424.74 Overpressure Limit Control. Each boiler designed so that in normal operation there is a change in state of the heat transfer medium from liquid to vapor shall be equipped with a pressure-sensitive limiting means. It shall be installed to limit maximum pressure and shall directly or indirectly disconnect all ungrounded conductors to the heating elements. Such means shall be in addition to a pressure-regulating system and other devices protecting the tank against excessive pressure.

Part VIII. Electrode-Type Boilers

424.80 Scope. The provisions in Part VIII of this article shall apply to boilers for operation at 600 volts, nominal, or less, in which heat is generated by the passage of current between electrodes through the liquid being heated.

Informational Note: See Part V of Article 495 for over 1000 volts.

424.81 Identification. Electrode-type boilers shall be identified as suitable for the installation.

424.82 Branch-Circuit Requirements. The size of branch-circuit conductors and overcurrent protective devices shall be calculated on the basis of 125 percent of the total load (motors not included). A contactor, relay, or other device, approved for continuous operation at 100 percent of its rating, shall be permitted to supply its full-rated load. See 210.19(A), Exception to (1). The provisions of this section shall not apply to conductors that form an integral part of an approved boiler.

Where an electrode boiler is rated 50 kW or more, the conductors supplying the boiler electrode(s) shall be permitted to be sized at not less than 100 percent of the nameplate rating of the electrode boiler, provided all the following conditions are met:

(1) The electrode boiler is marked with a minimum conductor size.
(2) The conductors are not smaller than the marked minimum size.
(3) A temperature- or pressure-actuated device controls the cyclic operation of the equipment.

424.83 Overtemperature Limit Control. Each boiler, designed so that in normal operation there is no change in state of the heat transfer medium, shall be equipped with a temperature-sensitive limiting means. It shall be installed to limit maximum liquid temperature and shall directly or indirectly interrupt all current flow through the electrodes. Such means shall be in addition to the temperature-regulating system and other devices protecting the tank against excessive pressure.

424.84 Overpressure Limit Control. Each boiler, designed so that in normal operation there is a change in state of the heat transfer medium from liquid to vapor, shall be equipped with a pressure-sensitive limiting means. It shall be installed to limit maximum pressure and shall directly or indirectly interrupt all current flow through the electrodes. Such means shall be in addition to a pressure-regulating system and other devices protecting the tank against excessive pressure.

424.85 Grounding. Boilers designed such that fault currents do not pass through the pressure vessel, and the pressure vessel is electrically isolated from the electrodes, all exposed non–current-carrying metal parts, including the pressure vessel, supply, and return connecting piping, shall be connected to an equipment grounding conductor.

For all other designs, the pressure vessel containing the electrodes shall be isolated and electrically insulated from ground.

424.86 Markings. All electrode-type boilers shall be marked to show the following:

(1) The manufacturer's name.
(2) The rating in volts, amperes, and kilowatts.
(3) The electrical supply required specifying frequency, number of phases, and number of wires.
(4) The marking "Electrode-Type Boiler."
(5) A warning marking, "All Power Supplies Shall Be Disconnected Before Servicing, Including Servicing the Pressure Vessel."

A field-applied warning marking or label shall comply with 110.21(B). The nameplate shall be located so as to be visible after installation.

Part IX. Electric Radiant Heating Panels and Heating Panel Sets

424.90 Scope. The provisions of Part IX of this article shall apply to radiant heating panels and heating panel sets.

424.92 Markings.

(A) Location. Markings shall be permanent and in a location that is visible prior to application of panel trim cover.

(B) Identified as Suitable. Each unit shall be identified as suitable for the installation.

(C) Required Markings. Each unit shall be marked with the identifying name or identification symbol, catalog number, and rating in volts and watts or in volts and amperes.

424.93 Installation.

(A) General.

(1) Manufacturer's Instructions. Heating panels and heating panel sets shall be installed in accordance with the manufacturer's instructions.

(2) Locations Not Permitted. The heating portion shall not be installed as follows:

(1) In or behind surfaces where subject to physical damage
(2) Run through or above walls, partitions, cupboards, or similar portions of structures that extend to the ceiling
(3) Run in or through thermal insulation, but shall be permitted to be in contact with the surface of thermal insulation
(4) In walls, except as permitted by 424.93(C)

(3) Separation from Outlets for Luminaires. Edges of panels and panel sets shall be separated by not less than 200 mm (8 in.) from the edges of any outlet boxes and junction boxes that are to be used for mounting surface luminaires. A clearance of not less than 50 mm (2 in.) shall be provided from recessed luminaires and their trims, ventilating openings, and other such openings in room surfaces, unless the heating panels and panel sets are listed and marked for lesser clearances, in which case they shall be permitted to be installed at the marked clearances. Sufficient area shall be provided to ensure that no heating panel or heating panel set is to be covered by any surface-mounted equipment.

(4) Surfaces Covering Heating Panels. After the heating panels or heating panel sets are installed and inspected, it shall be permitted to install a surface that has been identified by the manufacturer's instructions as being suitable for the installation. The surface shall be secured so that the nails or other fastenings do not pierce the heating panels or heating panel sets.

(5) Surface Coverings. Surfaces permitted by 424.93(A)(4) shall be permitted to be covered with paint, wallpaper, or other approved surfaces identified in the manufacturer's instructions as being suitable.

(B) Heating Panel Sets.

(1) Mounting Location. Heating panel sets shall be permitted to be secured to the lower face of joists or mounted in between joists, headers, or nailing strips.

(2) Parallel to Joists or Nailing Strips. Heating panel sets shall be installed parallel to joists or nailing strips.

(3) Installation of Nails, Staples, or Other Fasteners. Nailing or stapling of heating panel sets shall be done only through the unheated portions provided for this purpose. Heating panel sets shall not be cut through or nailed through any point closer than 6 mm (¼ in.) to the element. Nails, staples, or other fasteners shall not be used where they penetrate current-carrying parts.

(4) Installed as Complete Unit. Heating panel sets shall be installed as complete units unless identified as suitable for field cutting in an approved manner.

N (C) Installation of Heating Panels in Walls. Unless prohibited by 424.93(A)(2), heating panels shall be permitted to be installed in, on, or behind walls provided all of the following are met:

(1) Heating panels shall be identified as suitable for installation in, on, or behind walls.
(2) Heating panels shall be installed per the manufacturer's instructions and in accordance with the product listing.

(3) Heating panels shall be GFCI protected.

(4) Grounding means, such as copper braid, metal sheath, or other approved means, shall be provided.

(5) Heating panels shall be AFCI protected.

(6) Heating panels shall be permitted to be installed no more than 1.2 m (4 ft) above the floor.

Exception: Low-voltage heating panels shall not be required to be GFCI protected.

This requirement shall become effective January 1, 2026.

424.94 Clearances of Wiring in Ceilings. Wiring located above heated ceilings shall be spaced not less than 50 mm (2 in.) above the heated ceiling. The ampacity shall be calculated on the basis of an assumed ambient temperature of not less than 50°C (122°F), applying the correction factors in accordance with 310.15(B)(1). If this wiring is located above thermal insulation having a minimum thickness of 50 mm (2 in.), it shall be subject to the ambient correction in accordance with 310.15(B)(1).

424.95 Location of Branch-Circuit and Feeder Wiring in Walls.

(A) Exterior Walls. Wiring methods shall comply with 310.14(A)(3).

(B) Interior Walls. The ampacity of any wiring behind heating panels or heating panel sets located in interior walls or partitions shall be calculated on the basis of an assumed ambient temperature of 40°C (104°F), applying the correction factors in accordance with 310.15(B)(1).

424.96 Connection to Branch-Circuit Conductors.

(A) General. Heating panels or heating panel sets assembled together in the field to form a heating installation in one room or area shall be connected in accordance with the manufacturer's instructions.

(B) Heating Panels. Heating panels shall be connected to branch-circuit wiring by an approved wiring method.

(C) Heating Panel Sets.

(1) Connection to Branch-Circuit Wiring. Heating panel sets shall be connected to branch-circuit wiring by a method identified as being suitable for the purpose.

(2) Panel Sets with Terminal Junction Assembly. A heating panel set provided with terminal junction assembly shall be permitted to have the nonheating leads attached at the time of installation in accordance with the manufacturer's instructions.

424.97 Nonheating Leads. Excess nonheating leads of heating panels or heating panel sets shall be permitted to be cut to the required length as indicated in the manufacturer's installation instructions. Nonheating leads that are an integral part of a heating panel or heating panel set, either attached or provided

by the manufacturer as part of a terminal junction assembly, shall not be subjected to the ampacity requirements of 424.4(B) for branch circuits.

424.98 Installation in Concrete or Poured Masonry.

(A) Secured in Place and Identified as Suitable. Heating panels or heating panel sets shall be secured in place by means specified in the manufacturer's instructions and identified as suitable for the installation.

(B) Expansion Joints. Heating panels or heating panel sets shall not be installed where they bridge expansion joints unless provision is made for expansion and contraction.

(C) Spacings. Spacings shall be maintained between heating panels or heating panel sets and metal embedded in the floor. Grounded metal-clad heating panels shall be permitted to be in contact with metal embedded in the floor.

(D) Protection of Leads. Leads shall be protected where they leave the floor by rigid metal conduit, intermediate metal conduit, rigid polyvinyl chloride conduit, or electrical metallic tubing, or by other approved means.

(E) Bushings or Fittings Required. Bushings or approved fittings shall be used where the leads emerge within the floor slabs.

424.99 Installation Under Floor Covering.

A radiant heating mat is an example of a heating system that could be installed under a floor covering. See Exhibit 424.1.

(A) Identification. Heating panels or heating panel sets for installation under floor covering shall be identified as suitable for installation under floor covering.

(B) Installation. Listed heating panels or panel sets, if installed under floor covering, shall be installed on surfaces that are smooth and flat in accordance with the manufacturer's instructions and shall also comply with 424.99(B)(1) through (B)(6).

EXHIBIT 424.1 An example of an electric radiant heating mat for installation under flooring. *(Photo of WarmStep®, courtesy of ThermoSoft International)*

(1) Expansion Joints. Heating panels or heating panel sets shall not be installed where they bridge expansion joints unless protected from expansion and contraction.

(2) Connection to Conductors. Heating panels and heating panel sets shall be connected to branch-circuit and supply wiring by wiring methods recognized in Chapter 3.

(3) Anchoring. Heating panels and heating panel sets shall be firmly anchored to the floor using an adhesive or anchoring system identified for this use.

(4) Coverings. After heating panels or heating panel sets are installed and inspected, they shall be permitted to be covered by a floor covering that has been identified by the heater manufacturer as being suitable for the installation.

(5) GFCI Protection. In addition to the requirements in 210.8, branch circuits supplying the heating panel or heating panel sets shall have ground-fault circuit-interrupter protection for personnel.

(6) Grounding Braid or Sheath. Excluding nonheating leads, grounding means, such as copper braid, metal sheath, or other approved means, shall be provided with or as an integral part of the heating panel or heating panel set.

Part X. Low-Voltage Fixed Electric Space-Heating Equipment

424.100 Scope. Low-voltage fixed electric space-heating equipment shall consist of an isolating power supply, low-voltage heaters, and associated equipment that are all identified for use in dry locations.

424.101 Energy Source.

(A) Power Unit. The power unit shall be an isolating type with a rated output not exceeding 25 amperes, 30 volts (42.4 volts peak) ac, or 60 volts dc under all load conditions.

(B) Alternate Energy Sources. Listed low-voltage fixed electric space-heating equipment shall be permitted to be supplied directly from an alternate energy source such as solar photovoltaic (PV) or wind power. When supplied from such a source, the source and any power conversion equipment between the source and the heating equipment and its supply shall be listed and comply with the applicable section of the *NEC* for the source used. The output of the source shall meet the limits of 424.101(A).

424.102 Listed Equipment. Low-voltage fixed electric space-heating equipment shall be listed as a complete system.

424.103 Installation.

(A) General. Equipment shall be installed per the manufacturer's installation instructions.

(B) Ground. Secondary circuits shall not be grounded.

(C) Ground-Fault Protection. Ground-fault protection shall not be required.

424.104 Branch Circuit.

(A) Equipment shall be permitted to be supplied from branch circuits rated not over 30 amperes.

(B) The equipment shall be considered a continuous duty load.

ARTICLE
425 Fixed Resistance and Electrode Industrial Process Heating Equipment

Part I. General

Δ **425.1 Scope.** This article covers fixed industrial process heating employing electric resistance or electrode heating technology. For the purpose of this article, heating equipment includes boilers, electrode boilers, duct heaters, strip heaters, immersion heaters, process air heaters, or other fixed electric equipment used for industrial process heating.

Δ **425.3 Other Articles.** Fixed industrial process heating equipment incorporating a hermetic refrigerant motor-compressor shall additionally comply with Table 425.3.

N   **TABLE 425.3** Other Articles

Equipment	Article
Motors, motor circuits, and controllers	430
Air-conditioning and refrigerating equipment	440 (Parts I through IV)

425.4 Branch Circuits.

(A) Branch-Circuit Requirements. An individual branch circuit shall be permitted to supply any volt-ampere or wattage rating of fixed industrial process heating equipment for which the branch circuit is rated.

(B) Branch-Circuit Sizing. Fixed industrial process heating equipment and motors shall be considered continuous loads.

Fixed resistance and electrode industrial process heating equipment is considered a continuous load for the purpose of sizing branch circuits, feeders, service conductors, and overcurrent protective devices (OCPDs).

425.6 Listed Equipment. Fixed industrial process heating equipment shall be listed.

Part II. Installation

425.10 General.

(A) Location. Fixed industrial process heating equipment shall be accessible.

(B) Working Space. Working space about electrical enclosures for fixed industrial process heating equipment that require examination, adjustment, servicing, or maintenance while energized shall be accessible, and the work space for personnel shall comply with 110.26 and 110.34, based upon the utilization voltage to ground.

Exception: With special permission, in industrial establishments only, where conditions of maintenance and supervision ensure that only qualified persons will service the installation, working space less than that required in 110.26 or 110.34 shall be permitted.

(C) Above Grade Level, Floor, or Work Platform. Where the enclosure is located above grade, the floor, or a work platform, all of the following shall apply:

(1) The enclosure shall be accessible.
(2) The width of the working space shall be the width of the enclosure or a minimum of 762 mm (30 in.), whichever is greater.
(3) The depth of the workspace shall comply with 110.26(A) or 110.34 based upon the voltage to ground.
(4) All doors or hinged panels shall open to at least 90 degrees.

425.11 Supply Conductors.
Fixed industrial process heating equipment requiring supply conductors with over 60°C insulation shall be clearly and permanently marked. This marking shall be plainly visible after installation and shall be permitted to be adjacent to the field connection box.

Fixed resistance and electrode industrial process heating equipment may require supply conductors with a temperature rating greater than 60°C, due to their proximity to the heating elements and the installation instructions provided with a listed product.

425.12 Locations.

(A) Exposed to Physical Damage. Where subject to physical damage, fixed industrial process heating equipment shall be protected in an approved manner.

(B) Damp or Wet Locations. Fixed industrial process heating equipment installed in damp or wet locations shall be listed for such locations and shall be constructed and installed so that water or other liquids cannot enter or accumulate in or on wired sections, electrical components, or ductwork.

Informational Note: See 110.11 for equipment exposed to deteriorating agents.

425.13 Spacing from Combustible Materials.
Fixed industrial process heating equipment shall be installed to provide the required spacing between the equipment and adjacent combustible material, unless it is listed to be installed in direct contact with combustible material.

425.14 Infrared Lamp Industrial Heating Equipment.
In industrial occupancies, infrared industrial process heating equipment lampholders shall be permitted to be operated in series on circuits of over 150 volts to ground, provided the voltage rating of the lampholders is not less than the circuit voltage.

Each section, panel, or strip carrying a number of infrared lampholders, including the terminal wiring of such section, panel, or strip, shall be considered as infrared industrial heating equipment. The terminal connection block of each assembly shall be considered an individual outlet.

Part III. Control and Protection of Fixed Industrial Process Heating Equipment

425.19 Disconnecting Means.
Means shall be provided to simultaneously disconnect the heater, motor controller(s), and supplementary overcurrent protective device(s) of all fixed industrial process heating equipment from all ungrounded conductors. Where heating equipment is supplied by more than one source, feeder, or branch circuit, the disconnecting means shall be grouped and identified as having multiple disconnecting means. Each disconnecting means shall simultaneously disconnect all ungrounded conductors that it controls. The disconnecting means specified in 425.19(A) and (B) shall have an ampere rating not less than 125 percent of the total load of the motors and the heaters and shall be capable of being locked in the open position in compliance with 110.25.

If there are multiple disconnecting means, they must be grouped. Each disconnecting means must simultaneously open all the ungrounded conductors that it controls to prevent the practice of disconnecting one conductor at a time at terminal blocks or similar devices. The disconnect switch must have a rating of 125 percent of the heater's total load.

A unit switch is permitted by 425.19(C) to serve as the disconnecting means, provided that it has a marked "off" position and disconnects all ungrounded conductors.

(A) Heating Equipment with Supplementary Overcurrent Protection. The disconnecting means for fixed industrial process heating equipment with supplementary overcurrent protection shall be within sight from the supplementary overcurrent protective device(s), on the supply side of these devices, if fuses, and, in addition, shall comply with either 425.19(A)(1) or (A)(2).

(1) Heater Containing No Motor Rated over ⅛ Horsepower. The disconnecting means specified in 425.19 or unit switches complying with 425.19(C) shall be permitted to serve as the required disconnecting means for both the motor controller(s) and heater under either of the following conditions:

(1) The disconnecting means provided is also within sight from the motor controller(s) and the heater.
(2) The disconnecting means is capable of being locked in the open position in compliance with 110.25.

(2) Heater Containing a Motor(s) Rated over ⅛ Horsepower. The disconnecting means required by 425.19(A)(1) shall be permitted to serve as the required disconnecting means for both the motor controller(s) and heater under either of the following conditions:

(1) The disconnecting means is in sight from the motor controller(s) and the heater and complies with Part IX of Article 430.
(2) Motor(s) of more than ⅛ hp and the heater are provided with disconnecting means. The disconnecting means shall be permitted to be out of sight from the motor controller and shall be capable of being locked in the open position in compliance with 110.25.

(B) Heating Equipment Without Supplementary Overcurrent Protection.

(1) Without Motor or with Motor Not over ⅛ Horsepower. For fixed industrial process heating equipment without a motor rated over ⅛ hp, the branch-circuit switch or circuit breaker shall be permitted to serve as the disconnecting means where the switch or circuit breaker is within sight from the heater, shall be permitted to be out of sight from the motor controller, and shall be capable of being locked in the open position in compliance with 110.25.

(2) Over ⅛ Horsepower. For motor-driven fixed industrial process heating equipment with a motor rated over ⅛ hp, a disconnecting means shall be located within sight from the motor controller or shall be permitted to be out of sight from the motor controller and shall be capable of being locked in the open position in compliance with 110.25.

(C) Unit Switch(es) as Disconnecting Means. A unit switch(es) with a marked "off" position that is part of a fixed heater and disconnects all ungrounded conductors shall be permitted as the disconnecting means required by this article. The branch circuit switch or circuit breaker, where readily accessible for servicing of the fixed heater, shall be permitted as the other disconnecting means.

425.21 Switch and Circuit Breaker to Be Indicating. Switches and circuit breakers used as disconnecting means shall be of the indicating type.

425.22 Overcurrent Protection.

Δ **(A) Branch-Circuit Devices.** Fixed industrial process heating equipment, other than motor-operated equipment required

to have additional overcurrent protection by Parts III and IV of Article 430 or Part III of Article 440, shall be permitted to be protected against overcurrent where supplied by one of the branch circuits in Part II of Article 210.

(B) Resistance Elements. Resistance-type heating elements in fixed industrial process heating equipment shall be protected at not more than 60 amperes. Equipment rated more than 48 amperes and employing such elements shall have the heating elements subdivided, and each subdivided load shall not exceed 48 amperes.

Resistance-type heating elements in fixed industrial process heating equipment shall be permitted to be subdivided into circuits not exceeding 120 amperes and protected at not more than 150 amperes where one of the following is met:

(1) Elements are integral with and enclosed within a process heating surface.
(2) Elements are completely contained within an enclosure identified as suitable for this use.
(3) Elements are contained within an ASME-rated and stamped vessel.

Where a subdivided load is less than 48 amperes, the rating of the supplementary overcurrent protective device shall comply with 425.4(B). A boiler employing resistance-type immersion heating elements contained in an ASME-rated and stamped vessel shall be permitted to comply with 425.72(A).

The reason for subdividing the overcurrent protection is to minimize the amount of damaging energy — in the form of both heat and magnetic energy — released into the heating elements during a short circuit. The damaging short-circuit energy released at the element is greatly reduced by limiting the size of the overcurrent device protecting the individual heating elements, thereby greatly reducing the risk of fire. In addition, a second benefit may be continuity of service if equipment is only partially affected.

Historically, the subdivision size of 60 amperes was selected to use the maximum fuseholder size of 60 amperes while maintaining up to a 48-ampere heating element (48 A × 125% = 60 A).

Δ **(C) Overcurrent Protective Devices.** The supplementary overcurrent protective devices for the subdivided loads specified in 425.22(B) shall comply with the following:

(1) Be factory installed within or on the heater enclosure or supplied for use with the heater as a separate assembly by the heater manufacturer
(2) Be accessible but not be required to be readily accessible
(3) Be suitable for branch-circuit protection

Informational Note No. 1: See 240.10. Where cartridge fuses are used to provide this overcurrent protection, a single disconnecting means shall be permitted to be used for the several subdivided loads.

Informational Note No. 2: See 240.10 for supplementary overcurrent protection.

Informational Note No. 3: See 240.40 for disconnecting means for cartridge fuses in circuits of any voltage.

(D) Supplying Supplementary Overcurrent Protective Devices. The conductors supplying the supplementary overcurrent protective devices shall be considered branch-circuit conductors.

Where the heaters are rated 50 kW or more, the conductors supplying the supplementary overcurrent protective devices specified in 425.22(C) shall be permitted to be sized at not less than 100 percent of the nameplate rating of the heater, provided all of the following conditions are met:

(1) The heater is marked with a minimum conductor size.
(2) The conductors are not smaller than the marked minimum size.
(3) A temperature-actuated device controls the cyclic operation of the equipment.

(E) Conductors for Subdivided Loads. Field-wired conductors between the heater and the supplementary overcurrent protective devices for fixed industrial process heating equipment shall be sized at not less than 125 percent of the load served. The supplementary overcurrent protective devices specified in 425.22(C) shall protect these conductors in accordance with 240.4. Where the heaters are rated 50 kW or more, the ampacity of field-wired conductors between the heater and the supplementary overcurrent protective devices shall be permitted to be not less than 100 percent of the load of their respective subdivided circuits, provided all of the following conditions are met:

(1) The heater is marked with a minimum conductor size.
(2) The conductors are not smaller than the marked minimum size.
(3) A temperature-activated device controls the cyclic operation of the equipment.

Part IV. Marking of Heating Equipment

425.28 Nameplate.

(A) Marking Required. Fixed industrial process heating equipment shall be provided with a nameplate identifying the manufacturer and the rating in volts and watts or in volts and amperes.

Fixed industrial process heating equipment intended for use on alternating current only, direct current only, or both shall be marked to so indicate. The marking of equipment consisting of motors over ⅛ hp and other loads shall specify the rating of the motor in volts, amperes, and frequency and the heating load in volts and watts or in volts and amperes.

(B) Location. This nameplate shall be located so as to be permanent and shall be visible or accessible after installation.

425.29 Marking of Heating Elements.
All heating elements that are replaceable in the field and are part of industrial process heating equipment shall be legibly marked with the ratings in volts and watts or in volts and amperes.

425.45 Concealed Fixed Industrial Heating Equipment — Inspection.
Concealed fixed industrial heating equipment installations shall be made with due care to prevent damage to the heating equipment and shall be inspected and approved before heating equipment is covered or concealed.

Part V. Fixed Industrial Process Duct Heaters

425.57 General.
Part V shall apply to any heater mounted in the airstream of a forced-air system where the air-moving unit is not provided as an integral part of the equipment.

425.58 Identification.
Heaters installed in an air duct shall be identified as suitable for the installation.

425.59 Airflow.
Means shall be provided to ensure uniform airflow over the face of the heater in accordance with the manufacturer's instructions.

> Informational Note: Some heaters installed within 1.2 m (4 ft) of the outlet of an air-moving device, elbows, baffle plates, or other obstructions in ductwork use turning vanes, pressure plates, or other devices on the inlet side of the duct heater to ensure an even distribution of air over the face of the heater.

425.60 Elevated Inlet Temperature.
Duct heaters intended for use with elevated inlet air temperature shall be identified as suitable for use at the elevated temperatures.

425.63 Fan Circuit Interlock.
Means shall be provided to ensure that the fan circuit, where present, is energized when any heater circuit is energized. However, time- or temperature-controlled delay in energizing the fan motor shall be permitted.

425.64 Limit Controls.
Each duct heater shall be provided with an integral, automatic-reset temperature-limiting control or controllers to de-energize the circuit or circuits. In addition, an integral independent supplementary control or controllers shall be provided in each duct heater that disconnects a sufficient number of conductors to interrupt heating element current flow. This device shall be manually resettable or replaceable.

425.65 Location of Disconnecting Means.
Duct heater controller equipment shall be either accessible with the disconnecting means installed at or within sight from the controller or as permitted by 425.19(A).

Part VI. Fixed Industrial Process Resistance-Type Boilers

425.70 Scope.
The provisions in Part VI of this article shall apply to boilers employing resistance-type heating elements. Electrode-type boilers shall not be considered as employing resistance-type heating elements. See Part VII of this article.

425.71 Identification. Resistance-type boilers shall be identified as suitable for the installation.

425.72 Overcurrent Protection.

(A) Boiler Employing Resistance-Type Immersion Heating Elements in an ASME-Rated and Stamped Vessel. A boiler employing resistance-type immersion heating elements contained in an ASME-rated and stamped vessel shall have the heating elements protected at not more than 150 amperes. Such a boiler rated more than 120 amperes shall have the heating elements subdivided into loads not exceeding 120 amperes. Where a subdivided load is less than 120 amperes, the rating of the overcurrent protective device shall comply with 425.4(B).

(B) Boiler Employing Resistance-Type Heating Elements Rated More Than 48 Amperes and Not Contained in an ASME-Rated and Stamped Vessel. A boiler employing resistance-type heating elements not contained in an ASME-rated and stamped vessel shall have the heating elements protected at not more than 60 amperes. Such a boiler rated more than 48 amperes shall have the heating elements subdivided into loads not exceeding 48 amperes. Where a subdivided load is less than 48 amperes, the rating of the overcurrent protective device shall comply with 425.4(B).

See also

425.22(B) and its commentary for an explanation of the subdivision requirement

(C) Supplementary Overcurrent Protective Devices. The supplementary overcurrent protective devices for the subdivided loads as required by 425.72(A) and (B) shall be as follows:

(1) Factory-installed within or on the boiler enclosure or provided as a separate assembly by the boiler manufacturer.
(2) Accessible, but need not be readily accessible.

(D) Suitable for Branch-Circuit Protection. Where cartridge fuses are used to provide this overcurrent protection, a single disconnecting means shall be permitted for the several subdivided circuits. See 240.40.

(E) Conductors Supplying Supplementary Overcurrent Protective Devices. The conductors supplying these supplementary overcurrent protective devices shall be considered branch-circuit conductors. Where the heaters are rated 50 kW or more, the conductors supplying the overcurrent protective device specified in 424.72(C) shall be permitted to be sized at not less than 100 percent of the nameplate rating of the heater, provided all of the following conditions are met:

(1) The heater is marked with a minimum conductor size.
(2) The conductors are not smaller than the marked minimum size.
(3) A temperature- or pressure-actuated device controls the cyclic operation of the equipment.

(F) Conductors for Subdivided Loads. Field-wired conductors between the heater and the supplementary overcurrent protective devices shall be sized at not less than 125 percent of the load served. The supplementary overcurrent protective devices specified in 425.72(C) shall protect these conductors in accordance with 240.4. Where the heaters are rated 50 kW or more, the ampacity of field-wired conductors between the heater and the supplementary overcurrent protective devices shall be permitted to be not less than 100 percent of the load of their respective subdivided circuits, provided all of the following conditions are met:

(1) The heater is marked with a minimum conductor size.
(2) The conductors are not smaller than the marked minimum size.
(3) A temperature-activated device controls the cyclic operation of the equipment.

425.73 Overtemperature Limit Control. Each boiler designed so that in normal operation there is no change in state of the heat transfer medium shall be equipped with a temperature-sensitive limiting means. It shall be installed to limit maximum liquid temperature and shall directly or indirectly disconnect all ungrounded conductors to the heating elements. Such means shall be in addition to a temperature-regulating system and other devices protecting the tank against excessive pressure.

425.74 Overpressure Limit Control. Each boiler designed so that in normal operation there is a change in state of the heat transfer medium from liquid to vapor shall be equipped with a pressure-sensitive limiting means. It shall be installed to limit maximum pressure and shall directly or indirectly disconnect all ungrounded conductors to the heating elements. Such means shall be in addition to a pressure-regulating system and other devices protecting the tank against excessive pressure.

Part VII. Fixed Industrial Process Electrode-Type Boilers

425.80 Scope. The provisions in Part VII of this article shall apply to boilers for operation at 600 volts, nominal, or less, in which heat is generated by the passage of current between electrodes through the liquid being heated.

425.81 Identification. Electrode-type boilers shall be identified as suitable for the installation.

Δ **425.82 Branch-Circuit Requirements.** The size of branch-circuit conductors and overcurrent protective devices shall be calculated on the basis of 125 percent of the total load (motors not included). A contactor, relay, or other device, listed for continuous operation at 100 percent of its rating, shall be permitted to supply its full-rated load. See 210.19(A)(1), Exception No. 1. This section shall not apply to conductors that form an integral part of an approved boiler.

Where an electrode boiler is rated 50 kW or more, the conductors supplying the boiler electrode(s) shall be permitted to be sized at not less than 100 percent of the nameplate rating of the electrode boiler, provided all the following conditions are met:

(1) The electrode boiler is marked with a minimum conductor size.
(2) The conductors are not smaller than the marked minimum size.
(3) A temperature- or pressure-actuated switch controls the cyclic operation of the equipment.

425.83 Overtemperature Limit Control. Each boiler, designed so that in normal operation there is no change in state of the heat transfer medium, shall be equipped with a temperature-sensitive limiting means. It shall be installed to limit maximum liquid temperature and shall directly or indirectly interrupt all current flow through the electrodes. Such means shall be in addition to the temperature regulating system and other devices protecting the tank against excessive pressure.

425.84 Overpressure Limit Control. Each boiler, designed so that in normal operation there is a change in state of the heat transfer medium from liquid to vapor, shall be equipped with a pressure-sensitive limiting means. It shall be installed to limit maximum pressure and shall directly or indirectly interrupt all current flow through the electrodes. Such means shall be in addition to a pressure-regulating system and other devices protecting the tank against excessive pressure.

425.85 Grounding. Boilers designed such that fault currents do not pass through the pressure vessel, and the pressure vessel is electrically isolated from the electrodes, all exposed non–current-carrying metal parts, including the pressure vessel, supply, and return connecting piping, shall be connected to the equipment grounding conductor. For all other designs, the pressure vessel containing the electrodes shall be isolated and electrically insulated from ground.

425.86 Markings. All electrode-type boilers shall be marked to show the following:

(1) The manufacturer's name.
(2) The rating in volts, amperes, and kilowatts.
(3) The electrical supply required specifying frequency, number of phases, and number of wires.
(4) The marking "Electrode-Type Process Heating Boiler."
(5) A warning marking, "All Power Supplies Shall Be Disconnected Before Servicing, Including Servicing the Pressure Vessel."

A field-applied warning marking or label shall comply with 110.21(B). The markings shall be permanent and located so as to be visible after installation.

ARTICLE 426 — Fixed Outdoor Electric Deicing and Snow-Melting Equipment

Part I. General

426.1 Scope. This article covers fixed outdoor electric deicing and snow-melting equipment and the installation of these systems.

Article 426 includes requirements for resistance heating elements, impedance heating systems, and skin-effect heating systems used for deicing and snow melting.

(A) Embedded. Embedded in driveways, walks, steps, and other areas.

(B) Exposed. Exposed on drainage systems, bridge structures, roofs, and other structures.

Informational Note: See ANSI/IEEE 515.1-2012, *Standard for the Testing, Design, Installation and Maintenance of Electrical Resistance Trace Heating for Commercial Applications*, for further information. See IEEE 844/CSA 293 series of standards for fixed outdoor electric deicing and snow-melting equipment.

Δ **426.3 Other Articles.** Cord-and-plug-connected fixed outdoor electric deicing and snow-melting equipment shall additionally comply with Table 426.3.

N **TABLE 426.3** Other Articles

Equipment	Article
Appliances	422 (Parts I, II, III, IV, V)

426.4 Continuous Load. Fixed outdoor electric deicing and snow-melting equipment shall be considered a continuous load.

Fixed outdoor electric deicing and snow-melting equipment is considered a continuous load for the purpose of sizing branch circuits, feeders, service conductors, and overcurrent protective devices (OCPDs).

Part II. Installation

426.10 General. Equipment for outdoor electric deicing and snow melting shall be identified as suitable for the environment and installed in accordance with the manufacturer's instructions.

426.11 Use. Electric heating equipment shall be installed in such a manner as to be afforded protection from physical damage.

Underwriters Laboratories Inc. requires that manufacturers of UL-listed mat or cable deicing and snow-melting equipment provide specific installation instructions for their products. These instructions supplement the requirements contained in Article 426. For example, if the equipment can be installed only in concrete that is double poured (poured in two parts), the installation instructions are to specifically require that installation technique. Where the instructions do not specify an installation process, the use of either a single- or double-pour installation method is acceptable.

> **See also**
>
> **110.3(B)** regarding the installation and use of listed or labeled equipment

426.12 Thermal Protection. External surfaces of outdoor electric deicing and snow-melting equipment that operate at temperatures exceeding 60°C (140°F) shall be physically guarded, isolated, or thermally insulated to protect against contact by personnel in the area.

426.13 Identification. The presence of outdoor electric deicing and snow-melting equipment shall be evident by the posting of appropriate caution signs or markings where clearly visible.

Part III. Resistance Heating Elements

426.20 Embedded Deicing and Snow-Melting Equipment.

(A) Watt Density. Panels or units shall not exceed 1300 watts/m² (120 watts/ft²) of heated area.

(B) Spacing. The spacing between adjacent cable runs is dependent upon the rating of the cable and shall be not less than 25 mm (1 in.) on centers.

(C) Cover. Units, panels, or cables shall be installed as follows:

(1) On a substantial concrete, masonry, or asphalt base at least 50 mm (2 in.) thick and have at least 38 mm (1½ in.) of concrete, masonry, or asphalt applied over the units, panels, or cables; or

(2) They shall be permitted to be installed over other identified bases and embedded within 90 mm (3½ in.) of concrete, masonry, or asphalt but not less than 38 mm (1½ in.) from the top surface; or

(3) Equipment that has been listed for other forms of installation shall be installed only in the manner for which it has been identified.

(D) Secured. Cables, units, and panels shall be secured in place by frames or spreaders or other approved means while the concrete, masonry, or asphalt finish is applied.

(E) Expansion and Contraction. Cables, units, and panels shall not be installed where they bridge expansion joints unless provision is made for expansion and contraction.

426.21 Exposed Deicing and Snow-Melting Equipment.

(A) Secured. Heating element assemblies shall be secured to the surface being heated by identified means.

(B) Overtemperature. Where the heating element is not in direct contact with the surface being heated, the design of the heater assembly shall be such that its temperature limitations shall not be exceeded.

(C) Expansion and Contraction. Heating elements and assemblies shall not be installed where they bridge expansion joints unless provision is made for expansion and contraction.

(D) Flexural Capability. Where installed on flexible structures, the heating elements and assemblies shall have a flexural capability that is compatible with the structure.

426.22 Installation of Nonheating Leads for Embedded Equipment.

(A) Grounding Sheath or Braid. Except as permitted under 426.22(B), nonheating leads installed in concrete, masonry, or asphalt shall be provided with a grounding sheath or braid in accordance with 426.27 or shall be enclosed in rigid metal conduit, electrical metallic tubing, intermediate metal conduit, or other metal raceways.

(B) Splice Connections. The splice connection between the nonheating lead and heating element, within concrete, masonry, or asphalt, shall be located no less than 25 mm (1 in.) and no more than 150 mm (6 in.) from the metal raceway. The length of the nonheating lead from the metal raceway to the splice assembly shall be permitted to be provided without a grounding sheath or braid. Grounding continuity shall be maintained.

(C) Bushings. Insulating bushings shall be used in the concrete, masonry, or asphalt where the leads enter a metal raceway.

> **See also**
>
> **300.4(G)** and its commentary for more information on insulating bushings

(D) Expansion and Contraction. Leads shall be protected in expansion joints in accordance with 300.4(H) or installed in accordance with the manufacturer's instructions.

(E) Emerging from Grade. Exposed nonheating leads shall be protected by raceways or other identified means.

(F) Leads in Junction Boxes. Not less than 150 mm (6 in.) of free nonheating lead shall be within the junction box.

426.23 Installation of Nonheating Leads for Exposed Equipment.

(A) Nonheating Leads. Power supply nonheating leads (cold leads) for resistance elements shall be identified for the temperature encountered. Not less than 150 mm (6 in.) of nonheating leads shall be provided within the junction box. Preassembled factory-supplied and field-assembled nonheating leads on approved heaters shall be permitted to be shortened if the markings specified in 426.25 are retained.

(B) Protection. Nonheating power supply leads shall be enclosed in a rigid conduit, intermediate metal conduit, electrical metallic tubing, or other approved means.

426.24 Electrical Connection.

(A) Heating Element Connections. Electrical connections, other than factory connections of heating elements to nonheating elements embedded in concrete, masonry, or asphalt or on exposed surfaces, shall be made with insulated connectors identified for the use.

(B) Circuit Connections. Splices and terminations at the end of the nonheating leads, other than the heating element end, shall be installed in a box or fitting in accordance with 110.14 and 300.15.

426.25 Marking. Each factory-assembled heating unit shall be legibly marked within 75 mm (3 in.) of each end of the nonheating leads with the permanent identification symbol, catalog number, and ratings in volts and watts or in volts and amperes.

426.26 Corrosion Protection. Ferrous and nonferrous metal raceways, cable armor, cable sheaths, boxes, fittings, supports, and support hardware shall be permitted to be installed in concrete or in direct contact with the earth, or in areas subject to severe corrosive influences, where made of material suitable for the condition, or where provided with corrosion protection identified as suitable for the condition.

426.27 Grounding Braid or Sheath. Grounding means, such as copper braid, metal sheath, or other approved means, shall be provided as part of the heated section of the cable, panel, or unit.

△ **426.28 Ground-Fault Protection.** Ground-fault protection shall be provided for fixed outdoor electric deicing and snow-melting equipment. The trip level of ground-fault protection shall be as specified by the manufacturer.

Rather than protecting the entire branch circuit, the ground-fault protection requirement is focused on protecting just the equipment itself. The manufacturer and the user have an option of providing both circuit and equipment protection or just the required equipment protection. This required protection can be accomplished by the use of circuit breakers equipped with equipment ground-fault protection or an integral device supplied as part of the deicing or snow-melting equipment that is sensitive to leakage currents of 6 milliamperes to 50 milliamperes. These protection devices, if applied properly, will substantially reduce the risk of a fire being started by low-level electrical arcing.

Note that the required equipment protection is not the same as a GFCI used for personal protection that trips at 5 milliamperes (±1 milliamperes).

Part IV. Impedance Heating

426.30 Personnel Protection. Exposed elements of impedance heating systems shall be physically guarded, isolated, or thermally insulated with a weatherproof jacket to protect against contact by personnel in the area.

426.31 Isolation Transformer. An isolation transformer with a grounded shield between the primary and secondary windings shall be used to isolate the distribution system from the heating system.

426.32 Voltage Limitations. The secondary winding of the isolation transformer connected to the impedance heating elements shall not have an output voltage greater than 30 volts ac.

426.33 Induced Currents. All current-carrying components shall be installed in accordance with 300.20.

Part V. Skin-Effect Heating

426.40 Conductor Ampacity. The current through the electrically insulated conductor inside the ferromagnetic envelope shall be permitted to exceed the ampacity values shown in Table 310.16, provided it is identified as suitable for this use.

426.41 Pull Boxes. Where pull boxes are used, they shall be accessible without excavation by location in suitable vaults or above grade. Outdoor pull boxes shall be of watertight construction.

426.42 Single Conductor in Enclosure. The provisions of 300.20 shall not apply to the installation of a single conductor in a ferromagnetic envelope (metal enclosure).

426.43 Corrosion Protection. Ferromagnetic envelopes, ferrous or nonferrous metal raceways, boxes, fittings, supports, and support hardware shall be permitted to be installed in concrete or in direct contact with the earth, or in areas subjected to severe corrosive influences, where made of material suitable for the condition, or where provided with corrosion protection identified as suitable for the condition. Corrosion protection shall maintain the original wall thickness of the ferromagnetic envelope.

426.44 Equipment Grounding Conductor. The ferromagnetic envelope shall be connected to an equipment grounding conductor at both ends; and, in addition, it shall be permitted to be connected to an equipment grounding conductor at intermediate points as required by its design.

Section 250.30 shall not apply to the installation of skin-effect heating systems.

Part VI. Control and Protection

426.50 Disconnecting Means.

(A) Disconnection. All fixed outdoor deicing and snow-melting equipment shall be provided with a means for simultaneous disconnection from all ungrounded conductors. Where readily accessible to the user of the equipment, the branch-circuit switch or circuit breaker shall be permitted to serve as the disconnecting means. The disconnecting means shall be of the indicating type and be capable of being locked in the open (off) position.

(B) Cord-and-Plug-Connected Equipment. The factory-installed attachment plug of cord-and-plug-connected equipment rated 20 amperes or less and 150 volts or less to ground shall be permitted to be the disconnecting means.

426.51 Controllers.

(A) Temperature Controller with "Off" Position. Temperature-controlled switching devices that indicate an "off" position and that interrupt line current shall open all ungrounded conductors when the control device is in the "off" position. These devices shall not be permitted to serve as the disconnecting means unless they are capable of being locked in the open position in compliance with 110.25.

(B) Temperature Controller Without "Off" Position. Temperature controlled switching devices that do not have an "off" position shall not be required to open all ungrounded conductors and shall not be permitted to serve as the disconnecting means.

(C) Remote Temperature Controller. Remote controlled temperature-actuated devices shall not be required to meet the requirements of 426.51(A). These devices shall not be permitted to serve as the disconnecting means.

(D) Combined Switching Devices. Switching devices consisting of combined temperature-actuated devices and manually controlled switches that serve both as the controller and the disconnecting means shall comply with all of the following conditions:

(1) Open all ungrounded conductors when manually placed in the "off" position
(2) Be so designed that the circuit cannot be energized automatically if the device has been manually placed in the "off" position
(3) Be capable of being locked in the open position in compliance with 110.25

426.54 Cord-and-Plug-Connected Deicing and Snow-Melting Equipment.
Cord-and-plug-connected deicing and snow-melting equipment shall be listed.

According to the UL *Guide Information for Electrical Equipment*, which can be found at productspec.ul.com, category KOBQ deicing and snow-melting equipment is provided with means for permanent wiring connection, except for equipment rated 20 amperes or less and 150 volts or less to ground, which may be of cord-and-plug-connected construction.

See also

Article 100 for the definition of *listed*

Part I. General

427.1 Scope. This article covers electrically energized heating systems and the installation of these systems used with pipelines and vessels.

> Informational Note: See IEEE 515-2017, *Standard for the Testing, Design, Installation and Maintenance of Electrical Resistance Trace Heating for Industrial Applications,* for further information.

Also see applicable sections of the IEEE 844/CSA 293 series of standards for alternate technologies for fixed electric heating equipment for pipelines and vessels.

Article 427 includes requirements for impedance heating, induction heating, and skin-effect heating, in addition to resistance heating elements, where used in pipeline and vessel applications.

Δ **427.3 Other Articles.** Cord-connected pipe heating assemblies shall additionally comply with Table 427.3.

N **TABLE 427.3** Other Articles

Equipment	Article
Appliances	422 (Parts I, II, III, IV, V)

427.4 Continuous Load. Fixed electric heating equipment for pipelines and vessels shall be considered continuous load.

Fixed electric heating equipment is considered a continuous load for the purpose of sizing branch circuits, feeders, service conductors, and overcurrent protective devices (OCPDs).

Part II. Installation

427.10 General. Equipment for pipeline and vessel electric heating shall be identified as being suitable for (1) the chemical, thermal, and physical environment and (2) installation in accordance with the manufacturer's drawings and instructions.

427.11 Use. Electric heating equipment shall be installed in such a manner as to be afforded protection from physical damage.

427.12 Thermal Protection. External surfaces of pipeline and vessel heating equipment that operate at temperatures exceeding 60°C (140°F) shall be physically guarded, isolated, or thermally insulated to protect against contact by personnel in the area.

427.13 Identification. The presence of electrically heated pipelines, vessels, or both, shall be evident by the posting of appropriate caution signs or markings at intervals not exceeding 6 m (20 ft) along the pipeline or vessel and on or adjacent to equipment in the piping system that requires periodic servicing.

Part III. Resistance Heating Elements

427.14 Secured. Heating element assemblies shall be secured to the surface being heated by means other than the thermal insulation.

427.15 Not in Direct Contact. Where the heating element is not in direct contact with the pipeline or vessel being heated, means shall be provided to prevent overtemperature of the heating element unless the design of the heating assembly is such that its temperature limitations will not be exceeded.

427.16 Expansion and Contraction. Heating elements and assemblies shall not be installed where they bridge expansion joints unless provisions are made for expansion and contraction.

427.17 Flexural Capability. Where installed on flexible pipelines, the heating elements and assemblies shall have a flexural capability that is compatible with the pipeline.

427.18 Power Supply Leads.

(A) Nonheating Leads. Power supply nonheating leads (cold leads) for resistance elements shall be suitable for the temperature encountered. Not less than 150 mm (6 in.) of nonheating leads shall be provided within the junction box. Preassembled factory-supplied and field-assembled nonheating leads on approved heaters shall be permitted to be shortened if the markings specified in 427.20 are retained.

(B) Power Supply Leads Protection. Nonheating power supply leads shall be protected where they emerge from electrically heated pipeline or vessel heating units by rigid metal conduit, intermediate metal conduit, electrical metallic tubing, or other raceways identified as suitable for the application.

(C) Interconnecting Leads. Interconnecting nonheating leads connecting portions of the heating system shall be permitted to be covered by thermal insulation in the same manner as the heaters.

427.19 Electrical Connections.

(A) Nonheating Interconnections. Nonheating interconnections, where required under thermal insulation, shall be made with insulated connectors identified as suitable for this use.

(B) Circuit Connections. Splices and terminations outside the thermal insulation shall be installed in a box or fitting in accordance with 110.14 and 300.15.

427.20 Marking. Each factory-assembled heating unit shall be legibly marked within 75 mm (3 in.) of the termination end of all nonheating leads with the permanent identification symbol, catalog number, and ratings in volts and watts or in volts and amperes.

427.22 Ground-Fault Protection of Equipment. Ground-fault protection of equipment shall be provided for electric heat tracing and heating panels. This requirement shall not apply in industrial establishments where there is alarm indication of ground faults and the following conditions apply:

(1) Conditions of maintenance and supervision ensure that only qualified persons service the installed systems.
(2) Continued circuit operation is necessary for safe operation of equipment or processes.

Rather than protecting the entire branch circuit, the ground-fault protection requirement is focused on protecting just the equipment itself. Such protection affords the manufacturer and the user the option of providing both circuit and equipment protection or just the required equipment protection. Circuit breakers equipped with equipment ground-fault protection or an integral device supplied as part of the pipeline or vessel heating equipment that is sensitive to leakage currents from 6 milliamperes to 50 milliamperes will provide the required protection. These protective devices, if applied properly, substantially reduce the risk of fire being started by low-level electrical arcing.

The required equipment protection is not the same as that provided by a GFCI used for personal protection that trips at 5 milliamperes (±1 milliampere).

427.23 Grounded Conductive Covering. Electric heating equipment shall be listed and have a grounded conductive covering in accordance with 427.23(A) or (B). The conductive covering shall provide an effective ground-fault current path for operation of ground-fault protection equipment.

The grounded conductive covering is intended to provide a ground-fault current path in order to trip circuit or ground-fault protective devices, thus reducing the potential for fire and electric shock. It also provides added mechanical protection of the heating cable or panel.

(A) Heating Wires or Cables. Heating wires or cables shall have a grounded conductive covering that surrounds the heating element and bus wires, if any, and their electrical insulation.

(B) Heating Panels. Heating panels shall have a grounded conductive covering over the heating element and its electrical insulation on the side opposite the side attached to the surface to be heated.

Part IV. Impedance Heating

427.25 Personnel Protection. All accessible external surfaces of the pipeline, vessel, or both, being heated shall be physically guarded, isolated, or thermally insulated (with a weatherproof jacket for outside installations) to protect against contact by personnel in the area.

427.26 Isolation Transformer. A dual-winding transformer with a grounded shield between the primary and secondary windings shall be used to isolate the distribution system from the heating system.

427.27 Voltage Limitations. The secondary winding of the isolation transformer connected to the pipeline or vessel being heated shall not have an output voltage greater than 30 volts ac.

Exception No. 1: In industrial establishments, the isolation transformer connected to the pipeline or vessel being heated shall be permitted to have an output voltage greater than 30 but not more than 80 volts ac to ground where all of the following conditions apply:

(1) Conditions of guarding, maintenance, and supervision ensure that only qualified persons have access to the installed systems.
(2) Ground-fault protection of equipment is provided.

Exception No. 2: In industrial establishments, the isolation transformer connected to the pipeline or vessel being heated shall be permitted to have an output voltage not greater than 132 volts ac to ground where all of the following conditions apply:

(1) Conditions of guarding, maintenance, and supervision ensure that only qualified persons service the installed systems.
(2) Ground-fault protection of equipment is provided.
(3) The pipeline or vessel being heated is completely enclosed in a grounded metal enclosure.
(4) The transformer secondary connections to the pipeline or vessel being heated are completely enclosed in a grounded metal mesh or metal enclosure.

The general requirement is that the secondary winding of the isolation transformer connected to the pipeline or vessel being heated is not permitted to have an output voltage greater than 30 volts ac. However, for installations in industrial establishments, Exception No. 1 permits a maximum voltage of up to 80 volts ac to ground, provided that the two listed conditions are met. Exception No. 2 permits a voltage of not more than 132 volts ac to ground for impedance heating of the pipeline or vessel where installed in industrial establishments, provided that the four listed conditions are met.

427.28 Induced Currents. All current-carrying components shall be installed in accordance with 300.20.

427.29 Grounding. The pipeline, vessel, or both, that is being heated and operating at a voltage greater than 30 but not more than 80 shall be grounded at designated points.

427.30 Secondary Conductor Sizing. The ampacity of the conductors connected to the secondary of the transformer shall be at least 100 percent of the total load of the heater.

Part V. Induction Heating

427.36 Personnel Protection. Induction coils that operate or may operate at a voltage greater than 30 volts ac shall be enclosed in a nonmetallic or split metallic enclosure, isolated, or made inaccessible by location to protect personnel in the area.

427.37 Induced Current. Induction coils shall be prevented from inducing circulating currents in surrounding metallic equipment, supports, or structures by shielding, isolation, or insulation of the current paths. Stray current paths shall be bonded to prevent arcing.

Part VI. Skin-Effect Heating

427.45 Conductor Ampacity. The ampacity of the electrically insulated conductor inside the ferromagnetic envelope shall be permitted to exceed the values given in Table 310.16, provided it is identified as suitable for this use.

427.46 Pull Boxes. Pull boxes for pulling the electrically insulated conductor in the ferromagnetic envelope shall be permitted to be buried under the thermal insulation, provided their locations are indicated by permanent markings on the insulation jacket surface and on drawings. For outdoor installations, pull boxes shall be of watertight construction.

427.47 Single Conductor in Enclosure. The provisions of 300.20 shall not apply to the installation of a single conductor in a ferromagnetic envelope (metal enclosure).

Δ **427.48 Grounding.** The ferromagnetic envelope shall be grounded at both ends, and, in addition, it shall be permitted to be grounded at intermediate points as required by its design. The ferromagnetic envelope shall be bonded at all joints to ensure electrical continuity.

The provisions of 250.30 shall not apply to the installation of skin-effect heating systems.

Part VII. Control and Protection

427.55 Disconnecting Means.

(A) Switch or Circuit Breaker. Means shall be provided to simultaneously disconnect all fixed electric pipeline or vessel heating equipment from all ungrounded conductors. The branch-circuit switch or circuit breaker, where readily accessible to the user of the equipment, shall be permitted to serve as the disconnecting means. The disconnecting means shall be of the indicating type and shall be capable of being locked in the open (off) position. The disconnecting means shall be installed in accordance with 110.25.

(B) Cord-and-Plug-Connected Equipment. The factory-installed attachment plug of cord-and-plug-connected equipment rated 20 amperes or less and 150 volts or less to ground shall be permitted to be the disconnecting means.

427.56 Controls.

(A) Temperature Control with "Off" Position. Temperature-controlled switching devices that indicate an "off" position and that interrupt line current shall open all ungrounded conductors when the control device is in this "off" position. These devices shall not be permitted to serve as the disconnecting means unless capable of being locked in the open position.

(B) Temperature Control Without "Off" Position. Temperature controlled switching devices that do not have an "off" position shall not be required to open all ungrounded conductors and shall not be permitted to serve as the disconnecting means.

(C) Remote Temperature Controller. Remote controlled temperature-actuated devices shall not be required to meet the requirements of 427.56(A) and (B). These devices shall not be permitted to serve as the disconnecting means.

(D) Combined Switching Devices. Switching devices consisting of combined temperature-actuated devices and manually controlled switches that serve both as the controllers and the disconnecting means shall comply with all the following conditions:

(1) Open all ungrounded conductors when manually placed in the "off" position
(2) Be designed so that the circuit cannot be energized automatically if the device has been manually placed in the "off" position
(3) Be capable of being locked in the open position

427.57 Overcurrent Protection.
Heating equipment shall be considered protected against overcurrent where supplied by a branch circuit as specified in 210.20 and 210.24.

ARTICLE 430

Motors, Motor Circuits, and Controllers

Part I. General

Most electrical utilization equipment is rated in volt-ampere (VA) or watt input, but because of how motors are used to drive some form of machinery, motors typically are rated in horsepower output. Circuits supplying motors are sized according to the current input to the motor. In addition to the output of the motor, the input current is affected by the motor losses and the power factor of the motor. The losses are not the type of information found on the nameplate of a motor. The motor tables in Part XIV of Article 430 contain accurate input ampere ratings for motors based on industry standards.

Some motors are available with their output ratings expressed in watts. (One horsepower equals approximately 746 watts.) Circuits that supply motors not rated in horsepower still

must be sized according to the input of the motor, rated in amperes. Sizing circuits based solely on kilowatt output results in seriously undersized conductors (because the current requirements of the losses and the power factor are neglected) and in the improper application of overcurrent devices protecting the motor and motor circuit components.

See also

430.6 for ampacity and motor rating determination

Δ **430.1 Scope.** This article covers motors, motor branch-circuit and feeder conductors and their protection, motor overload protection, motor control circuits, motor controllers, and motor control centers.

> Informational Note No. 1: See Informational Note Figure 430.1 for the arrangement of this article.
>
> Informational Note No. 2: See 110.26(E) for installation requirements for motor control centers.

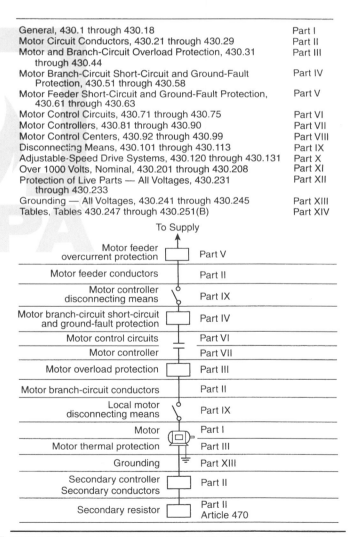

Δ **INFORMATIONAL NOTE FIGURE 430.1** *Article 430 Contents.*

Informational Note No. 3: See 440.1 for air-conditioning and refrigerating equipment.

Informational Note No. 4: See Part X for additional requirements for motors utilizing adjustable-speed drive systems.

Informational Note No. 5: See Part XI for additional requirements for motors that operate over 1000 volts, nominal.

N 430.2 Reconditioned Motors. Reconditioned motors shall be permitted if the reconditioning has been conducted in accordance with the manufacturer's instructions or, if no instructions are provided, nationally recognized standards.

Reconditioned motors identified for use in hazardous (classified) locations shall be listed as reconditioned if installed in hazardous (classified) locations.

Informational Note: See ANSI/EASA AR100-2020, *Recommended Practice for the Repair of Rotating Electrical Apparatus*, for information on the rewinding and repair of motors.

430.4 Part-Winding Motors. Where separate overload devices are used with a standard part-winding start induction motor, each half of the motor winding shall be individually protected in accordance with 430.32 and 430.37 with a trip current one-half that specified.

Each motor-winding connection shall have branch-circuit short-circuit and ground-fault protection rated at not more than one-half that specified by 430.52.

Exception: A short-circuit and ground-fault protective device shall be permitted for both windings if the device will allow the motor to start. Where time-delay (dual-element) fuses are used, they shall be permitted to have a rating not exceeding 150 percent of the motor full-load current.

430.5 Other Articles. Motors and controllers shall also comply with the applicable provisions of Table 430.5.

△ 430.6 Conductor Ampacity and Motor Rating Determination. The size of conductors supplying equipment covered by this article shall be selected from the ampacity tables in 310.15 or shall be calculated in accordance with 310.14(B). Where flexible cord is used, the size of the conductor shall be selected in accordance with 400.5. The required conductor ampacity and motor ratings shall be determined in accordance with 430.6(A), (B), (C), and (D).

(A) General Motor Applications. For general motor applications, current ratings shall be determined based on 430.6(A)(1) and (A)(2).

△ (1) Table Values. Other than for motors built for low speeds (less than 1200 RPM) or high torques, and for multispeed motors, the values given in Table 430.247, Table 430.248, Table 430.249, and Table 430.250 shall be used instead of the actual current rating marked on the motor nameplate to determine the following:

(1) Ampacity of conductors
(2) Current ratings of switches
(3) Current ratings of branch-circuit short-circuit and ground-fault protection

Where a motor is marked in amperes, but not horsepower, the horsepower rating shall be assumed to be that corresponding to the value given in Table 430.247, Table 430.248, Table 430.249, and Table 430.250, interpolated if necessary.

Exception No. 1: Multispeed motors shall be in accordance with 430.22(B) and 430.52.

Exception No. 2: For equipment that employs a shaded-pole or permanent-split capacitor-type fan or blower motor that is marked with the motor type and the marking on the equipment

TABLE 430.5 *Other Articles*

Equipment/Occupancy	Article	Section
Air-conditioning and refrigerating equipment	440	
Capacitors		460.8, 460.9
Commercial garages; aircraft hangars; motor fuel dispensing facilities; bulk storage plants; spray application, dipping, and coating processes; and inhalation anesthetizing locations	511, 513, 514, 515, 516, and 517 Part IV	
Cranes and hoists	610	
Electrically driven or controlled irrigation machines	675	
Elevators, dumbwaiters, escalators, moving walks, wheelchair lifts, and stairway chair lifts	620	
Fire pumps	695	
Hazardous (classified) locations	500–503, 505, and 506	
Industrial machinery	670	
Motion picture projectors		540.11 and 540.20
Motion picture and television studios and similar locations	530	
Resistors and reactors	470	
Theaters, audience areas of motion picture and television studios, and similar locations		520.48
Transformers and transformer vaults	450	

nameplate is not less than the current marked on the fan or blower motor nameplate, the full-load current marked on the nameplate of the appliance shall be used to determine the ampacity of branch-circuit conductors in addition to the current ratings of the following:

(1) Disconnecting means
(2) Motor controllers
(3) Short-circuit and ground-fault protective devices
(4) Separate overload protective devices

Exception No. 3: For a listed motor-operated appliance that is marked with both motor horsepower and full-load current, the motor full-load current marked on the nameplate of the appliance shall be used instead of the horsepower rating on the appliance nameplate to determine the ampacity of branch-circuit conductors in addition to the current ratings of the following:

(1) Disconnecting means
(2) Motor controllers
(3) Short-circuit and ground-fault protective devices
(4) Separate overload protective devices

(2) Nameplate Values. The motor nameplate current ratings shall be used to determine the values for the following:

(1) Separate motor overload protection
(2) For motors built for low speeds (less than 1200 RPM), high torques, canned pumps, or multispeed motors, the following:

   a. Ampacity of conductors
   b. Current ratings of switches
   c. Current ratings of branch-circuit short-circuit and ground-fault protection

(3) Large motors exceeding the values in Part XIV shall use the nameplate current rating for conductor sizing.

In 430.6(A)(1), for motors other than those built for low speeds (less than 1200 rpm), or high torque and multispeed motors, the ampacity of motor branch-circuit conductors, branch-circuit and ground-fault protection, and ampere rating of the motor disconnecting means are determined by the ampere values listed in Tables 430.247 through 430.250, rather than by the ampere values marked on the motor nameplate. Regarding 430.6(A)(2), the current rating provided on the motor nameplate is used to size the overload protective devices (i.e., thermals or heaters) that protect the motor, motor control apparatus, and motor branch-circuit conductors. If the motor is marked with a current rating rather than a horsepower rating, the horsepower rating can be found in Tables 430.247 through 430.250, by using the motor ampere rating, service factor, voltage, and single- or three-phase information from the nameplate.

(B) Torque Motors. For torque motors, the rated current shall be locked-rotor current, and this nameplate current shall be used to determine the ampacity of the branch-circuit conductors covered in 430.22 and 430.24, the current rating of the motor

overload protection, and the current rating of motor branch-circuit short-circuit and ground-fault protection in accordance with 430.52(B).

Informational Note: See 430.83(D) and 430.110 for information on motor controllers and disconnecting means.

(C) Alternating-Current Adjustable Voltage Motors. For motors used in alternating-current, adjustable voltage, variable torque drive systems, the ampacity of conductors, or current ratings of switches, branch-circuit short-circuit and ground-fault protection, and so forth, shall be based on the maximum operating current marked on the motor nameplate or the control nameplate, or both. If the maximum operating current does not appear on the nameplate, the current rating determination shall be based on 150 percent of the values given in Table 430.249 and Table 430.250.

(D) Valve Actuator Motor Assemblies. For valve actuator motor assemblies (VAMs), the rated current shall be the nameplate full-load current, and this current shall be used to determine the maximum rating or setting of the motor branch-circuit short-circuit and ground-fault protective device and the ampacity of the conductors.

430.7 Marking on Motors and Multimotor Equipment.

(A) Usual Motor Applications. A motor shall be marked with the following information:

(1) Manufacturer's name.
(2) Rated volts and full-load current. For a multispeed motor, full-load current for each speed, except shaded-pole and permanent-split capacitor motors where amperes are required only for maximum speed.
(3) Rated frequency and number of phases if an ac motor.
(4) Rated full-load speed.
(5) Rated temperature rise or the insulation system class and rated ambient temperature.
(6) Time rating. The time rating shall be 5, 15, 30, or 60 minutes, or continuous.
(7) Rated horsepower if ⅛ hp or more. For a multispeed motor rated ⅛ hp or more, rated horsepower for each speed, except shaded-pole and permanent-split capacitor motors rated ⅛ hp or more where rated horsepower is required only for maximum speed. Motors of arc welders are not required to be marked with the horsepower rating.
(8) Code letter or locked-rotor amperes if an alternating-current motor rated ½ hp or more. On polyphase wound-rotor motors, the code letter shall be omitted.

Informational Note No. 1: See 430.7(B).

(9) Design letter for design A, B, C, or D motors.

Informational Note No. 2: See ANSI/NEMA MG 1-2016, *Motors and Generators*, Part 1, Definitions, for information on motor design letter definition.

The design letters referred to in 430.7(A) indicate a motor's speed/torque characteristic curve and are not to be confused with code letters used in Table 430.7(B). For technical accuracy, code letters should be referred to as "locked-rotor indicating code letters," which are explained in 430.7(B). Design letters reflect characteristics inherent in motor design, such as locked-rotor current, slip at rated load, and locked-rotor and breakdown torque.

(10) Secondary volts and full-load current if a wound-rotor induction motor.

(11) Field current and voltage for dc excited synchronous motors.

(12) Winding — straight shunt, stabilized shunt, compound, or series, if a dc motor. Fractional horsepower dc motors 175 mm (7 in.) or less in diameter shall not be required to be marked.

(13) A motor provided with a thermal protector complying with 430.32(A)(2) or (B)(2) shall be marked "thermally protected." Thermally protected motors rated 100 watts or less and complying with 430.32(B)(2) shall be permitted to use the abbreviated marking "T.P."

(14) A motor complying with 430.32(B)(4) shall be marked "impedance protected." Impedance-protected motors rated 100 watts or less and complying with 430.32(B)(4) shall be permitted to use the abbreviated marking "Z.P."

(15) Motors equipped with electrically powered condensation prevention heaters shall be marked with the rated heater voltage, number of phases, and the rated power in watts.

(16) Motors that are electronically protected from overloads in accordance with 430.32(A)(2) and (B)(2) shall be marked "electronically protected" or "E.P."

(B) Locked-Rotor Indicating Code Letters. Code letters marked on motor nameplates to show motor input with locked rotor shall be in accordance with Table 430.7(B).

The code letter indicating motor input with locked rotor shall be in an individual block on the nameplate, properly designated.

(1) Multispeed Motors. Multispeed motors shall be marked with the code letter designating the locked-rotor kilovolt-ampere (kVA) per horsepower (hp) for the highest speed at which the motor can be started.

Exception: Constant horsepower multispeed motors shall be marked with the code letter giving the highest locked-rotor kilovolt-ampere (kVA) per horsepower (hp).

(2) Single-Speed Motors. Single-speed motors starting on wye connection and running on delta connections shall be marked with a code letter corresponding to the locked-rotor kilovolt-ampere (kVA) per horsepower (hp) for the wye connection.

TABLE 430.7(B) *Locked-Rotor Indicating Code Letters*

Code Letter	Kilovolt-Amperes per Horsepower with Locked Rotor
A	0–3.14
B	3.15–3.54
C	3.55–3.99
D	4.0–4.49
E	4.5–4.99
F	5.0–5.59
G	5.6–6.29
H	6.3–7.09
J	7.1–7.99
K	8.0–8.99
L	9.0–9.99
M	10.0–11.19
N	11.2–12.49
P	12.5–13.99
R	14.0–15.99
S	16.0–17.99
T	18.0–19.99
U	20.0–22.39
V	22.4 and up

(3) Dual-Voltage Motors. Dual-voltage motors that have a different locked-rotor kilovolt-ampere (kVA) per horsepower (hp) on the two voltages shall be marked with the code letter for the voltage giving the highest locked-rotor kilovolt-ampere (kVA) per horsepower (hp).

(4) 50/60 Hz Motors. Motors with 50- and 60-Hz ratings shall be marked with a code letter designating the locked-rotor kilovolt-ampere (kVA) per horsepower (hp) on 60 Hz.

(5) Part-Winding Motors. Part-winding start motors shall be marked with a code letter designating the locked-rotor kilovolt-ampere (kVA) per horsepower (hp) that is based on the locked-rotor current for the full winding of the motor.

The following example shows how to determine the locked-rotor current for a specific motor using Table 430.7(B).

Calculation Example

A 20-hp, 460-V, 3-phase motor has a nameplate kilovolt-ampere code letter G. Determine the maximum locked-rotor current for this motor.

Solution

Step 1. Use Table 430.7(B) to find the maximum value in the range for code letter G, which is 6.29 kVA per horsepower.

Step 2. Use the following formula to find the maximum locked-rotor current:

Locked-rotor kVA = motor hp
$\qquad$ × maximum code letter value
$\qquad$ = 20 × 6.29 = 125.8

$$\text{Locked-rotor current} = \frac{\text{locked-rotor kVA}}{\sqrt{3} \times \text{kV}}$$

For 460 V (460 V = 0.46 kV),

$$\frac{125.8}{1.73 \times 0.46} = 158 \text{ A}$$

The maximum locked-rotor current for a 20-hp, 460-V motor with code letter G is 158 A when the system voltage is 460 V.

(C) Torque Motors. Torque motors are rated for operation at standstill and shall be marked in accordance with 430.7(A), except that locked-rotor torque shall replace horsepower.

(D) Multimotor and Combination-Load Equipment.

(1) Factory-Wired. Multimotor and combination-load equipment shall be provided with a visible nameplate marked with the manufacturer's name, the rating in volts, frequency, number of phases, minimum supply circuit conductor ampacity, and the maximum ampere rating of the circuit short-circuit and ground-fault protective device. The conductor ampacity shall be calculated in accordance with 430.24 and counting all of the motors and other loads that will be operated at the same time. The short-circuit and ground-fault protective device rating shall not exceed the value calculated in accordance with 430.53. Multimotor equipment for use on two or more circuits shall be marked with the preceding information for each circuit.

Section 110.3(B) requires listed or labeled equipment to be used and installed in accordance with the manufacturer's instructions accompanying the equipment or marked on the nameplate. The nameplate marking for the maximum ampere rating of the branch-circuit short-circuit and ground-fault protective device may limit the type of protective device to a fuse by stipulating "fuse" without reference to a circuit breaker. A circuit breaker located in a panelboard, switchboard, or similar distribution equipment is permitted to supply the equipment in which the fuses are installed.

(2) Not Factory-Wired. Where the equipment is not factory-wired and the individual nameplates of motors and other loads are visible after assembly of the equipment, the individual nameplates shall be permitted to serve as the required marking.

430.8 Marking on Motor Controllers. A motor controller shall be marked with the manufacturer's name or identification, the voltage, the current or horsepower rating, the short-circuit current rating, and other necessary data to properly indicate the applications for which it is suitable.

Exception No. 1: The short-circuit current rating is not required for motor controllers applied in accordance with 430.81(A) or (B).

Exception No. 2: The short-circuit current rating is not required to be marked on the motor controller when the short-circuit current rating of the motor controller is marked elsewhere on the assembly.

Exception No. 3: The short-circuit current rating is not required to be marked on the motor controller when the assembly into which it is installed has a marked short-circuit current rating.

Exception No. 4: Short-circuit current ratings are not required for motor controllers rated less than 2 hp at 300 V or less and listed exclusively for general-purpose branch circuits.

A motor controller that includes motor overload protection suitable for group motor application shall be marked with the motor overload protection and the maximum branch-circuit short-circuit and ground-fault protection for such applications.

Combination motor controllers that employ adjustable instantaneous trip circuit breakers shall be clearly marked to indicate the ampere settings of the adjustable trip element.

Where a motor controller is built in as an integral part of a motor or of a motor-generator set, individual marking of the motor controller shall not be required if the necessary data are on the nameplate. For motor controllers that are an integral part of equipment approved as a unit, the above marking shall be permitted on the equipment nameplate.

Informational Note: See 110.10 for information on circuit impedance and other characteristics.

430.9 Terminals.

(A) Markings. Terminals of motors and controllers shall be suitably marked or colored where necessary to indicate the proper connections.

(B) Conductors. Motor controllers and terminals of control circuit devices shall be connected with copper conductors unless identified for use with a different conductor.

(C) Torque Requirements. Control circuit devices with screw-type pressure terminals used with 14 AWG or smaller copper conductors shall be torqued to a minimum of 0.8 N·m (7 lb-in.) unless identified for a different torque value.

Proper torque is essential for safe and reliable connections. A screw-type pressure terminal that has not been torqued might loosen during motor operation, resulting in overheating. Safety is enhanced by providing a minimum torque value for screw-type pressure terminals.

See also

110.14(D) and its commentary for more information on torque requirements

430.10 Wiring Space in Enclosures.

(A) General. Enclosures for motor controllers and disconnecting means shall not be used as junction boxes, auxiliary gutters, or raceways for conductors feeding through or tapping off to the other apparatus unless designs are employed that provide adequate space for this purpose.

> Informational Note: See 312.8 for switch and overcurrent-device enclosures.

(B) Wire-Bending Space in Enclosures. Minimum wire-bending space within the enclosures for motor controllers shall be in accordance with Table 430.10(B) where measured in a straight line from the end of the lug or wire connector (in the direction the wire leaves the terminal) to the wall or barrier. Where alternate wire termination means are substituted for that supplied by the manufacturer of the motor controller, they shall be of a type identified by the manufacturer for use with the motor controller and shall not reduce the minimum wire-bending space.

Δ **TABLE 430.10(B)** *Minimum Wire-Bending Space at the Terminals of Enclosed Motor Controllers*

Size of Wire (AWG or kcmil)	Wires per Terminal*			
	1		2	
	mm	in.	mm	in.
10 and smaller	Not specified		—	—
8–6	38	1½	—	—
4–3	50	2	—	—
2	65	2½	—	—
1	75	3	—	—
1/0	125	5	125	5
2/0	150	6	150	6
3/0–4/0	175	7	175	7
250	200	8	200	8
300	250	10	250	10
350–500	300	12	300	12
600–700	350	14	400	16
750–900	450	18	475	19

*Where provision for three or more wires per terminal exists, the minimum wire-bending space shall be in accordance with the requirements of 312.6(B).

Exhibit 430.1 illustrates the application of the wire-bending space requirements of either 430.10(B) or 312.6(B) within an enclosure for a motor controller.

430.11 Protection Against Liquids. Suitable guards or enclosures shall be provided to protect exposed current-carrying parts of motors and the insulation of motor leads where installed directly under equipment, or in other locations where dripping or spraying oil, water, or other liquid is capable of occurring, unless the motor is designed for the existing conditions.

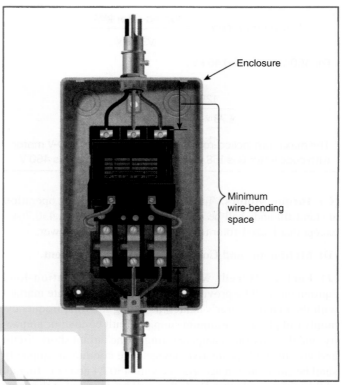

EXHIBIT 430.1 Wire-bending space in enclosures for motor controllers.

The presence of certain liquids can cause deterioration and insulation breakdown. Excess lubricants in the motor can collect dirt and clog the cooling passages of the motor, causing the motor to overheat.

430.12 Motor Terminal Housings.

(A) Material. Where motors are provided with terminal housings, the housings shall be of metal and of substantial construction.

> *Exception: In other than hazardous (classified) locations, substantial, nonmetallic, noncombustible housings shall be permitted, provided an internal grounding means between the motor frame and the equipment grounding connection is incorporated within the housing.*

(B) Dimensions and Space — Wire-to-Wire Connections. Where these terminal housings enclose wire-to-wire connections, they shall have minimum dimensions and usable volumes in accordance with Table 430.12(B).

(C) Dimensions and Space — Fixed Terminal Connections. Where these terminal housings enclose rigidly mounted motor terminals, the terminal housing shall be of sufficient size to provide minimum terminal spacings and usable volumes in accordance with Table 430.12(C)(1) and Table 430.12(C)(2).

TABLE 430.12(B) *Terminal Housings — Wire-to-Wire Connections*

Motors 275 mm (11 in.) in Diameter or Less

Horsepower	Cover Opening Minimum Dimension		Usable Volume Minimum	
	mm	in.	cm³	in.³
1 and smaller[a]	41	1⅝	170	10.5
1½, 2, and 3[b]	45	1¾	275	16.8
5 and 7½	50	2	365	22.4
10 and 15	65	2½	595	36.4

Motors Over 275 mm (11 in.) in Diameter — Alternating-Current Motors

Maximum Full Load Current for 3-Phase Motors with Maximum of 12 Leads (Amperes)	Terminal Box Cover Opening Minimum Dimension		Usable Volume Minimum		Typical Maximum Horsepower 3-Phase	
	mm	in.	cm³	in.³	230 Volt	460 Volt
45	65	2.5	595	36.4	15	30
70	84	3.3	1,265	77	25	50
110	100	4.0	2,295	140	40	75
160	125	5.0	4,135	252	60	125
250	150	6.0	7,380	450	100	200
400	175	7.0	13,775	840	150	300
600	200	8.0	25,255	1540	250	500

Direct-Current Motors

Maximum Full-Load Current for Motors with Maximum of 6 Leads (Amperes)	Terminal Box Minimum Dimensions		Usable Volume Minimum	
	mm	in.	cm³	in.³
68	65	2.5	425	26
105	84	3.3	900	55
165	100	4.0	1,640	100
240	125	5.0	2,950	180
375	150	6.0	5,410	330
600	175	7.0	9,840	600
900	200	8.0	18,040	1,100

Note: Auxiliary leads for such items as brakes, thermostats, space heaters, and exciting fields shall be permitted to be neglected if their current-carrying area does not exceed 25 percent of the current-carrying area of the machine power leads.

[a]For motors rated 1 hp and smaller, and with the terminal housing partially or wholly integral with the frame or end shield, the volume of the terminal housing shall not be less than 18.0 cm³ (1.1 in.³) per wire-to-wire connection. The minimum cover opening dimension is not specified.

[b]For motors rated 1½, 2, and 3 hp, and with the terminal housing partially or wholly integral with the frame or end shield, the volume of the terminal housing shall not be less than 23.0 cm³ (1.4 in.³) per wire-to-wire connection. The minimum cover opening dimension is not specified.

TABLE 430.12(C)(1) *Terminal Spacings — Fixed Terminals*

| Nominal Volts | Minimum Spacing | | | |
	Between Line Terminals		Between Line Terminals and Other Uninsulated Metal Parts	
	mm	in.	mm	in.
250 or less	6	¼	6	¼
Over 250 – 1000	10	⅜	10	⅜

TABLE 430.12(C)(2) *Usable Volumes — Fixed Terminals*

Power-Supply Conductor Size (AWG)	Minimum Usable Volume per Power-Supply Conductor	
	cm³	in.³
14	16	1
12 and 10	20	1¼
8 and 6	37	2¼

(D) Large Wire or Factory Connections. For motors with larger ratings, greater number of leads, or larger wire sizes, or where motors are installed as a part of factory-wired equipment, without additional connection being required at the motor terminal housing during equipment installation, the terminal housing shall be of ample size to make connections, but the foregoing provisions for the volumes of terminal housings shall not be considered applicable.

(E) Equipment Grounding Connections. A means for attachment of an equipment grounding conductor termination in accordance with 250.8 shall be provided at motor terminal housings for wire-to-wire connections or fixed terminal connections. The means for such connections shall be permitted to be located either inside or outside the motor terminal housing.

Exception: Where a motor is installed as a part of factory-wired equipment that is required to be grounded and without additional connection being required at the motor terminal housing during equipment installation, a separate means for motor grounding at the motor terminal housing shall not be required.

430.13 Bushing. Where wires pass through an opening in an enclosure, conduit box, or barrier, a bushing shall be used to protect the conductors from the edges of openings having sharp edges. The bushing shall have smooth, well-rounded surfaces where it may be in contact with the conductors. If used where oils, greases, or other contaminants may be present, the bushing shall be made of material not deleteriously affected.

Informational Note: See310.10(F) for conductors exposed to deteriorating agents.

430.14 Location of Motors.

(A) Ventilation and Maintenance. Motors shall be located so that adequate ventilation is provided and so that maintenance, such as lubrication of bearings and replacing of brushes, can be readily accomplished.

Exception: Ventilation shall not be required for submersible types of motors.

(B) Open Motors. Open motors that have commutators or collector rings shall be located or protected so that sparks cannot reach adjacent combustible material.

Exception: Installation of these motors on wooden floors or supports shall be permitted.

430.16 Exposure to Dust Accumulations.
In locations where dust or flying material collects on or in motors in such quantities as to seriously interfere with the ventilation or cooling of motors and thereby cause dangerous temperatures, suitable types of enclosed motors that do not overheat under the prevailing conditions shall be used.

> Informational Note: Especially severe conditions may require the use of enclosed pipe-ventilated motors, or enclosure in separate dusttight rooms, properly ventilated from a source of clean air.

See also

502.125 (Class II, Divisions 1 and 2) and **503.125** (Class III, Divisions 1 and 2) for requirements for motors exposed to combustible dust or readily ignitible flying material

500.5(C) (Class II locations), **500.5(D)** (Class III locations), and **506.20** (Zones 20, 21, and 22) for classification of locations

430.17 Highest Rated or Smallest Rated Motor.
In determining compliance with 430.24, 430.53(B), and 430.53(C), the highest rated or smallest rated motor shall be based on the rated full-load current as selected from Table 430.247, Table 430.248, Table 430.249, and Table 430.250.

430.18 Nominal Voltage of Rectifier Systems.
The nominal value of the ac voltage being rectified shall be used to determine the voltage of a rectifier derived system.

Exception: The nominal dc voltage of the rectifier shall be used if it exceeds the peak value of the ac voltage being rectified.

Part II. Motor Circuit Conductors

Δ **430.21 General.** Part II specifies ampacities of conductors that are capable of carrying the motor current without overheating under the conditions specified.

Part II shall not apply to motor circuits rated over 1000 volts, nominal.

> Informational Note No. 1: See Part XI for motor circuits rated over 1000 volts, nominal.

Informational Note No. 2: See 110.14(C) and 430.9(B) for equipment device terminal requirements.

430.22 Single Motor.
Conductors that supply a single motor used in a continuous duty application shall have an ampacity of not less than 125 percent of the motor full-load current rating, as determined by 430.6(A)(1), or not less than specified in 430.22(A) through (G).

The requirement that a conductor have an ampacity of at least 125 percent of the motor FLC (full-load current) rating is based on the need to provide for a sustained running current that is greater than the rated full-load current and for protection of the conductors by the motor overload protective device set above the motor full-load current rating.

The ampacity of the motor branch-circuit conductors is based on the full-load current rating values provided in Tables 430.247 through 430.250. Motor nameplate FLA (full-load amperes) generally is not to be used to size branch-circuit conductors.

Exhibit 430.2 illustrates the essential parts of a motor branch circuit. Motor feeder tap conductors are required to terminate in a branch-circuit protective device, and the conductors must be installed in accordance with 430.28.

The branch-circuit short-circuit and ground-fault protective device can be a fuse or a circuit breaker and must be capable of carrying the starting current of the motor without opening the circuit. Motor circuit conductors with an ampacity of 125 percent of the motor full-load current are reasonably protected by motor overload protective devices set to operate at nearly the same current as the ampacity of the conductors. Motor circuit conductors are permitted to be protected by Article 430 in accordance with 240.4(G).

In general, every motor must be provided with overload protective devices intended to protect the motor windings, motor-control apparatus, and motor branch-circuit conductors against excessive heating due to motor overloads and failure to start. Overload in equipment is defined as operation exceeding the normal full-load rating, which, when it persists for a sufficient length of time, causes damage or dangerous overheating. Overload in a motor includes a stalled rotor but does not include fault currents due to short circuits or ground faults.

See also

Table 430.52(C)(1) for maximum ratings of short-circuit and ground-fault protective devices

430.44 for conditions under which providing automatic opening of a motor circuit due to overload could be undesirable

(A) Direct-Current Motor–Rectifier Supplied. For dc motors operating from a rectified power supply, the conductor ampacity on the input of the rectifier shall not be less than 125 percent of the rated input current to the rectifier. For dc motors operating from a rectified single-phase power supply, the conductors between the field wiring output terminals of the rectifier and the motor shall have an ampacity of not less than the following percentages of the motor full-load current rating:

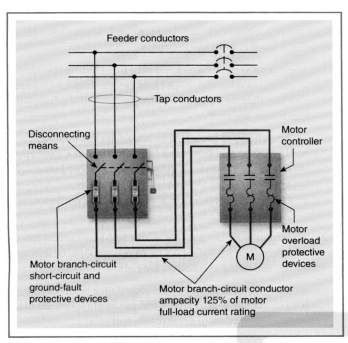

EXHIBIT 430.2 Essential parts of a motor circuit installation.

(1) Where a rectifier bridge of the single-phase, half-wave type is used, 190 percent.

(2) Where a rectifier bridge of the single-phase, full-wave type is used, 150 percent.

(B) Multispeed Motor. For a multispeed motor, the selection of branch-circuit conductors on the line side of the controller shall be based on the highest of the full-load current ratings shown on the motor nameplate. The ampacity of the branch-circuit conductors between the controller and the motor shall not be less than 125 percent of the current rating of the winding(s) that the conductors energize.

(C) Wye-Start, Delta-Run Motor. For a wye-start, delta-run connected motor, the ampacity of the branch-circuit conductors on the line side of the controller shall not be less than 125 percent of the motor full-load current as determined by 430.6(A)(1). The ampacity of the conductors between the controller and the motor shall not be less than 72 percent of the motor full-load current rating as determined by 430.6(A)(1).

Informational Note: The individual motor circuit conductors of a wye-start, delta-run connected motor carry 58 percent of the rated load current. The multiplier of 72 percent is obtained by multiplying 58 percent by 1.25.

A wye-start, delta-run winding configuration is a method of providing reduced-voltage starting for polyphase induction motor wye windings. This requires a specific type of motor controller and a delta-wired motor with all leads brought out to the terminal box. This method of starting finds wide application in certain compressors used for air conditioning and where the driven machinery is allowed to start unloaded. During starting, the

windings are arranged in a wye configuration. The wye-start configuration results in a reduced starting voltage of a mathematical ratio of $1/\sqrt{3} = 0.5774$, or 58 percent of the full line voltage, which results in approximately 58 percent starting current and about one-third of the normal starting torque. Once the motor attains speed, the windings are reconfigured to run as delta, giving full line voltage to the individual windings, which allows the motor to have full torque capability. Since the delta-connected motor load is 58 percent of the rated load current and the conductors are sized at 125 percent of the motor full-load current, the conductors would be sized at 1.25 times 58 percent, or 72 percent of the motor full-load current rating.

In Exhibit 430.3, conductors from terminals T_1, T_2, and T_3 to the motor, as well as the conductors from terminals T_4, T_5, and T_6 to the motor, are all sized at 58 percent of the full-load current used to size the conductors that supply L_1, L_2, and L_3. During START, contacts 1M and S are closed and contacts 2M are open. During RUN, contacts 1M and 2M are closed and the S contacts are open.

(D) Part-Winding Motor. For a part-winding connected motor, the ampacity of the branch-circuit conductors on the line side of the controller shall not be less than 125 percent of the motor full-load current as determined by 430.6(A)(1). The ampacity of the conductors between the controller and the motor shall not be less than 62.5 percent of the motor full-load current rating as determined by 430.6(A)(1).

Informational Note: The multiplier of 62.5 percent is obtained by multiplying 50 percent by 1.25.

A part-winding motor starter supplies power to partial sections of the primary winding of the motor. When the motor is running, load on the conductors will be approximately 50 percent of motor full-load current. The 62.5 percent of the full-load current rating is 125 percent of the running current with all windings connected.

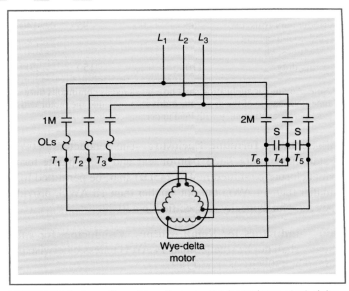

EXHIBIT 430.3 A one-line wiring diagram of a typical wye-start, delta-run motor and controller. *(Courtesy of Schneider Electric)*

(E) Other Than Continuous Duty. Conductors for a motor used in a short-time, intermittent, periodic, or varying duty application shall have an ampacity of not less than the percentage of the motor nameplate current rating shown in Table 430.22(E), unless the authority having jurisdiction grants special permission for conductors of lower ampacity.

Most motor applications are continuous duty. For motors that are not continuous duty, the motor nameplate currents and Table 430.22(E) are used to determine the branch-circuit ampacity. Branch-circuit conductors for a motor with a rated horsepower used for 5-minute short-time duty service are permitted to be sized smaller than for the same motor with a 60-minute rating, due to the cooling intervals between operating periods. For example, a 5-minute rated motor will run for 5 minutes and then be off for 55 minutes.

See also

Article 100 for definitions of the terms *duty, continuous; duty, intermittent; duty, periodic; duty, short-time;* and *duty, varying*

TABLE 430.22(E) *Duty-Cycle Service*

Classification of Service	Nameplate Current Rating Percentages			
	5-Minute Rated Motor	15-Minute Rated Motor	30- & 60-Minute Rated Motor	Continuous Rated Motor
Short-time duty operating valves, raising or lowering rolls, etc.	110	120	150	—
Intermittent duty freight and passenger elevators, tool heads, pumps, drawbridges, turntables, etc. (for arc welders, see 630.11)	85	85	90	140
Periodic duty rolls, ore- and coal-handling machines, etc.	85	90	95	140
Varying duty	110	120	150	200

Note: Any motor application shall be considered as continuous duty unless the nature of the apparatus it drives is such that the motor will not operate continuously with load under any condition of use.

(F) Separate Terminal Enclosure. The conductors between a stationary motor rated 1 hp or less and the separate terminal enclosure permitted in 430.245(B) shall be permitted to be smaller than 14 AWG but not smaller than 18 AWG, provided they have an ampacity as specified in 430.22.

(G) Conductors for Small Motors. Conductors for small motors shall not be smaller than 14 AWG unless otherwise permitted in 430.22(G)(1) or (G)(2).

(1) 18 AWG Copper. 18 AWG individual copper conductors installed in a cabinet or enclosure, copper conductors that are part of a jacketed multiconductor cable assembly, or copper conductors in a flexible cord shall be permitted, under either of the following sets of conditions:

(1) The circuit supplies a motor with a full-load current rating, as determined by 430.6(A)(1), of greater than 3.5 amperes, and less than or equal to 5 amperes, and all the following conditions are met:

    a. The circuit is protected in accordance with 430.52.
    b. The circuit is provided with maximum Class 10 or Class 10A overload protection in accordance with 430.32.
    c. Overcurrent protection is provided in accordance with 240.4(D)(1)(2).

(2) The circuit supplies a motor with a full-load current rating, as determined by 430.6(A)(1), of 3.5 amperes or less, and all the following conditions are met:

    a. The circuit is protected in accordance with 430.52.
    b. The circuit is provided with maximum Class 20 overload protection in accordance with 430.32.
    c. Overcurrent protection is provided in accordance with 240.4(D)(1)(2).

(2) 16 AWG Copper. 16 AWG individual copper conductors installed in a cabinet or enclosure, copper conductors that are part of a jacketed multiconductor cable assembly, or copper conductors in a flexible cord shall be permitted under either of the following sets of conditions:

(1) The circuit supplies a motor with a full-load current rating, as determined by 430.6(A)(1), of greater than 5.5 amperes, and less than or equal to 8 amperes, and all the following conditions are met:

    a. The circuit is protected in accordance with 430.52.
    b. The circuit is provided with maximum Class 10 or Class 10A overload protection in accordance with 430.32.
    c. Overcurrent protection is provided in accordance with 240.4(D)(2)(2).

(2) The circuit supplies a motor with a full-load current rating, as determined by 430.6(A)(1), of 5.5 amperes or less, and all the following conditions are met:

    a. The circuit is protected in accordance with 430.52.
    b. The circuit is provided with maximum Class 20 overload protection in accordance with 430.32.
    c. Overcurrent protection is provided in accordance with 240.4(D)(2)(2).

430.23 Wound-Rotor Secondary.

(A) Continuous Duty. For continuous duty, the conductors connecting the secondary of a wound-rotor ac motor to its controller shall have an ampacity not less than 125 percent of the full-load secondary current of the motor.

(B) Other Than Continuous Duty. For other than continuous duty, these conductors shall have an ampacity, in percent of full-load secondary current, not less than that specified in Table 430.22(E).

(C) Resistor Separate from Controller. Where the secondary resistor is separate from the controller, the ampacity of the conductors between controller and resistor shall not be less than that shown in Table 430.23(C).

TABLE 430.23(C) *Secondary Conductor*

Resistor Duty Classification	Ampacity of Conductor in Percent of Full-Load Secondary Current
Light starting duty	35
Heavy starting duty	45
Extra-heavy starting duty	55
Light intermittent duty	65
Medium intermittent duty	75
Heavy intermittent duty	85
Continuous duty	110

430.24 Several Motors or a Motor(s) and Other Load(s). Conductors supplying several motors, or a motor(s) and other load(s), shall have an ampacity not less than the sum of each of the following:

(1) 125 percent of the full-load current rating of the highest rated motor, as determined by 430.6(A)
(2) Sum of the full-load current ratings of all the other motors in the group, as determined by 430.6(A)
(3) 100 percent of the noncontinuous non-motor load
(4) 125 percent of the continuous non-motor load.

Informational Note: See Informative Annex D, Example No. D8.

Exception No. 1: Where one or more of the motors of the group are used for short-time, intermittent, periodic, or varying duty, the ampere rating of such motors to be used in the summation shall be determined in accordance with 430.22(E). For the highest rated motor, the greater of either the ampere rating from 430.22(E) or the largest continuous duty motor full-load current multiplied by 1.25 shall be used in the summation.

Exception No. 2: The ampacity of conductors supplying motor-operated fixed electric space-heating equipment shall comply with 424.4(B).

Exception No. 3: Where the circuitry is interlocked so as to prevent simultaneous operation of selected motors or other loads, the conductor ampacity shall be permitted to be based on the summation of the currents of the motors and other loads to be operated simultaneously that results in the highest total current.

As illustrated in Exhibit 430.4, the requirements of Article 210 and Article 430 apply where motors are connected to a 15- or

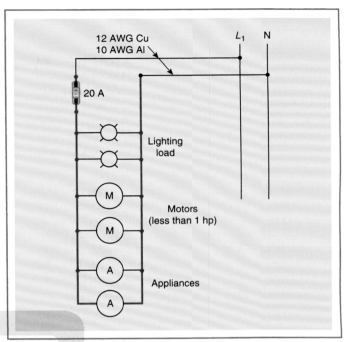

EXHIBIT 430.4 A 20-ampere branch circuit supplying lighting, small motors, and appliances.

20-ampere branch circuit that also supplies lighting or other appliance loads. Motors rated less than 1 horsepower may be connected to these circuits, and they must be provided with overload protective devices unless the motors are not permanently installed, are started manually, and are within sight of the controller location.

See also

430.32(B) and **(C)** and **430.53(A)** for additional information on the installation of motors (1 horsepower or less)

Where branch circuits or feeders serve motors and/or other electrical loads, the highest rating or setting of the branch circuit or feeder short-circuit and ground-fault protective devices for the minimum-size branch circuit or feeder conductor permitted by 430.24 is specified in 430.62.

Where two or more motors are started simultaneously, the heaviest load that a feeder will ever be required to carry occurs when the largest motor is started, and all the other motors supplied by the same feeder are running and delivering their full-rated horsepower.

This requirement and those of 430.62 for the short-circuit and ground-fault protection of the branch circuit or feeder are based on the principle that the conductors should be sized to have an ampacity equal to 125 percent of the full-load current of the largest motor plus the full-load currents of all other motors and all other loads supplied by the feeder.

Where the conductors are branch-circuit conductors to multi-motor equipment, 430.53 specifies the maximum rating of the branch-circuit short-circuit and ground-fault protective device, and 430.7(D)(1) requires the maximum ampere rating of the short-circuit and ground-fault protective device to be marked on multimotor equipment.

430.25 Multimotor and Combination-Load Equipment. The ampacity of the conductors supplying multimotor and combination-load equipment shall not be less than the minimum circuit ampacity marked on the equipment in accordance with 430.7(D). Where the equipment is not factory-wired and the individual nameplates are visible in accordance with 430.7(D)(2), the conductor ampacity shall be determined in accordance with 430.24.

Computing the load for the minimum allowable conductor size for a combination lighting (or lighting and appliance) load and motor load involves determining the lighting load in accordance with Article 220 (and other applicable articles and sections), the appliance load in accordance with Article 422, and the motor load in accordance with 430.22 (single motor) or 430.24 (two or more motors). The lighting load and the motor load are added together to determine the minimum conductor ampacity.

430.26 Feeder Demand Factor. Where reduced heating of the conductors results from motors operating on duty-cycle, intermittently, or from all motors not operating at one time, the authority having jurisdiction may grant permission for feeder conductors to have an ampacity less than specified in 430.24, provided the conductors have sufficient ampacity for the maximum load determined in accordance with the sizes and number of motors supplied and the character of their loads and duties.

> Informational Note: Demand factors determined in the design of new facilities can often be validated against actual historical experience from similar installations.

430.27 Capacitors with Motors. Where capacitors are installed in motor circuits, conductors shall comply with 460.8 and 460.9.

430.28 Feeder Taps. Feeder tap conductors shall have an ampacity not less than that required by Part II, shall terminate in a branch-circuit protective device, and, in addition, shall meet one of the following requirements:

(1) Be enclosed either by an enclosed controller or by a raceway, be not more than 3.0 m (10 ft) in length, and, for field installation, be protected by an overcurrent device on the line side of the tap conductor, the rating or setting of which shall not exceed 1000 percent of the tap conductor ampacity

(2) Have an ampacity of at least one-third that of the feeder conductors, be suitably protected from physical damage or enclosed in a raceway, and be not more than 7.5 m (25 ft) in length

(3) Have an ampacity not less than the feeder conductors

Exception: Feeder taps over 7.5 m (25 ft) long. In high-bay manufacturing buildings [over 11 m (35 ft) high at walls],

where conditions of maintenance and supervision ensure that only qualified persons service the systems, conductors tapped to a feeder shall be permitted to be not over 7.5 m (25 ft) long horizontally and not over 30.0 m (100 ft) in total length where all of the following conditions are met:

(1) The ampacity of the tap conductors is not less than one-third that of the feeder conductors.

(2) The tap conductors terminate with a single circuit breaker or a single set of fuses complying with (1) Part IV, where the load-side conductors are a branch circuit, or (2) Part V, where the load-side conductors are a feeder.

(3) The tap conductors are suitably protected from physical damage and are installed in raceways.

(4) The tap conductors are continuous from end-to-end and contain no splices.

(5) The tap conductors shall be 6 AWG copper or 4 AWG aluminum or larger.

(6) The tap conductors shall not penetrate walls, floors, or ceilings.

(7) The tap shall not be made less than 9.0 m (30 ft) from the floor.

For a single motor load, the tap conductors are sized the same as the motor branch-circuit conductors — that is, according to 430.22, which requires that motor branch-circuit conductors be sized at least 125 percent of the full-load current value for the motor given in Tables 430.248 through 430.250. The table value, rather than the nameplate value, is the full-load current used for conductor sizing according to 430.6(A).

The tap conductors must terminate in a set of fuses or a circuit breaker, thus limiting the load on the tap conductors. The reduced-size tap conductors are protected from overload by the overcurrent device where the tap conductors terminate but are protected from short circuit and ground fault only from the feeder overcurrent device.

A tap conductor installation also must meet the additional requirements associated with their tap conductor distance limits, that is, 10 feet, 25 feet, or, by exception, 100 feet. The requirements for tap conductors that supply motor loads are similar to the basic tap requirements found in 240.21.

See also

430.53(D) and associated commentary for additional information concerning a tap supplying a single motor in a group installation

Calculation Example

A 15-hp, 230-V, 3-phase, NEMA Design B, squirrel-cage induction motor with a service factor of 1.15 and a nameplate FLC of 40 A is to be supplied by a tap from a 250-kcmil feeder. Assuming three Type THWN copper conductors in a raceway and no ambient correction factor, the feeder has an ampacity of 255 A (from Table 310.16, 75°C column). The tap conductors are not over 25 ft long (see Exhibit 430.5). Determine the required branch-circuit short-circuit and ground-fault protection required by 430.28 and the overload protection required by 430.32.

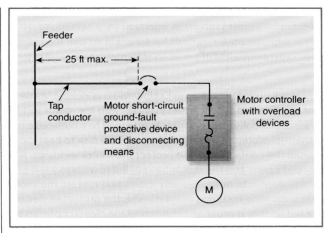

EXHIBIT 430.5 Protective devices (branch-circuit short-circuit and ground-fault) for a motor branch circuit located not more than 25 feet from the point where the conductors are tapped to a feeder.

Solution

Step 1. The tap conductors are required to have a minimum ampacity of 85 A (4 AWG) (1/3 × 255 A = 85 A).

Step 2. The FLC of the motor is 42 A based on 430.6(A) and Table 430.250. Therefore, the motor can be supplied by the 85-A (4 AWG) tap conductor.

Step 3. Because the tap conductors must terminate in a single branch-circuit protective device, the short-circuit protective device rating must be determined. According to 430.52(C)(1), the protective device for the motor cannot exceed the values given in Table 430.52 (C)(1). The maximum time-delay fuse value is 42 × 1.75 = 73.5 A. The maximum inverse time circuit breaker value is 42 × 2.50 = 105 A.

Section 430.52(C)(1)(a) allows the next higher standard size — 80 and 110 A, respectively. A higher size, based on 430.52(C)(1)(b), is allowed if the 80- or 110-A size is not adequate to start the motor.

Step 4. Because overload protection is also required for the motor, determine the rating of the overload protection. Based on 430.32, for a motor with a service factor of 1.15, the motor overload protective devices (heaters) are required to be set at a value not greater than 125 percent of the FLC marked on the motor nameplate and not at the FLC value from the table. A setting of up to 140 percent may be used according to the permissive rules in 430.32(C). With the motor overload protection set at 50 A, the circuit conductors and motor are protected from overload.

430.29 Constant Voltage Direct-Current Motors — Power Resistors.
Conductors connecting the motor controller to separately mounted power accelerating and dynamic braking resistors in the armature circuit shall have an ampacity not less

than the value calculated from Table 430.29 using motor full-load current. If an armature shunt resistor is used, the power accelerating resistor conductor ampacity shall be calculated using the total of motor full-load current and armature shunt resistor current.

Armature shunt resistor conductors shall have an ampacity of not less than that calculated from Table 430.29 using rated shunt resistor current as full-load current.

TABLE 430.29 *Conductor Rating Factors for Power Resistors*

Time in Seconds		Ampacity of Conductor in Percent of Full-Load Current
On	Off	
5	75	35
10	70	45
15	75	55
15	45	65
15	30	75
15	15	85
Continuous Duty		110

Part III. Motor and Branch-Circuit Overload Protection

Δ **430.31 General.** Part III specifies overload devices intended to protect motors, motor-control apparatus, and motor branch-circuit conductors against excessive heating due to motor overloads and failure to start.

Informational Note No. 1: See Informative Annex D, Example D8.
Informational Note No. 2: See Article 100 for the definition of *Overload*.

N **(A) Where Hazard Exists.** These provisions shall not require overload protection where a power loss would cause a hazard, such as in the case of fire pumps.

Informational Note: See 695.7 for protection of fire pump supply conductors.

N **(B) Not Over 1000 Volts.** Part III shall not apply to motor circuits rated over 1000 volts, nominal.

Informational Note: See Part XI for over 1000 volts, nominal.

Overload protection is not designed or may not be capable of breaking short-circuit current or ground-fault current.

430.32 Continuous-Duty Motors.

(A) More Than 1 Horsepower. Each motor used in a continuous duty application and rated more than 1 hp shall be protected against overload by one of the means in 430.32(A)(1) through (A)(4).

Δ **(1) Separate Overload Device.** A separate overload device that is responsive to motor current. This device shall be selected to trip or shall be rated at no more than the following percent of the motor nameplate full-load current rating:

Motors with a marked service factor 1.15 or greater   125%
Motors with a marked temperature rise 40°C or less   125%
All other motors   115%

Modification of this value shall be permitted as provided in 430.32(C). For a multispeed motor, each winding connection shall be considered separately.

Where a separate motor overload device is connected so that it does not carry the total current designated on the motor nameplate, such as for wye-delta starting, the proper percentage of nameplate current applying to the selection or setting of the overload device shall be clearly designated on the equipment, or the manufacturer's selection table shall take this into account.

> Informational Note: See 460.9 for power factor correction capacitors that are installed on the load side of the motor overload device.

To protect a motor from an overload, the motor nameplate full-load current is used to select the overload protection rather than the full-load current values from Tables 430.248 through 430.250, which are used to select the feeder and branch-circuit wiring.

A continuous-duty motor with a marked service factor of 1.15 or greater or with a marked temperature rise of 40°C or less can carry a 25-percent overload for an extended period without damage to the motor. Motors with a service factor of less than 1.15 or those with a marked temperature rise greater than 40°C might be incapable of withstanding a prolonged overload, where the motor overload protective device opens the circuit if the motor continues to draw 115 percent of its rated full-load current.

A "continuous-duty motor" is not the same as a "continuous load." The duty of a motor is determined by the application of the motor as defined in Article 100 under the five various types of *duty*.

(2) Thermal Protector or Electronically Protected. A thermal protector integral with the motor shall be approved for use with the motor it protects on the basis that it will prevent dangerous overheating of the motor due to overload and failure to start. An electronically protected motor shall be approved for use on the basis that it will prevent dangerous overheating due to the failure of the electronic control, overload, or failure to start the motor. The ultimate trip current of a thermally or electronically protected motor shall not exceed the following percentage of motor full-load current given in Table 430.248, Table 430.249, and Table 430.250:

Motor full-load current 9 amperes or less   170%
Motor full-load current from 9.1 to, and including,
   20 amperes   156%
Motor full-load current greater than 20 amperes   140%

If the motor current-interrupting device is separate from the motor and its control circuit is operated by a protective device integral with the motor, it shall be arranged so that the opening of the control circuit will result in interruption of current to the motor.

The thermal protector shown in Exhibit 430.6 is located inside the motor housing and is connected in series with the motor winding by a set of normally closed contacts attached to a bimetallic disk. The thermal protector heating coil causes the disk to heat rapidly and snap the contacts open to protect the motor windings. After the circuit opens and the motor has cooled to a normal temperature, the contacts automatically close and restart the motor. Where automatic restart is not desirable, the protective device is designed so that it must be returned to the closed position by a manually controlled reset as required by 430.43.

In lieu of electromechanical protective devices, 430.32(A)(2) also recognizes the use of electronic means to protect a motor and motor circuit conductors from overload (overheating) conditions due to excessive driven machinery loading or failure to start. Electronically protected motors must be marked in accordance with 430.7(A)(16).

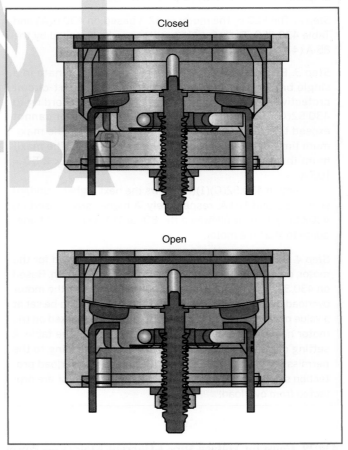

EXHIBIT 430.6 A thermal protector for a motor in which a heat-sensitive snap-action disk opens contacts and protects the motor in which it is mounted against dangerous overheating. (*Courtesy of Sensata Technologies*)

(3) Integral with Motor. A protective device integral with a motor that will protect the motor against damage due to failure to start shall be permitted if the motor is part of an approved assembly that does not normally subject the motor to overloads.

(4) Larger Than 1500 Horsepower. For motors larger than 1500 hp, a protective device having embedded temperature detectors that cause current to the motor to be interrupted when the motor attains a temperature rise greater than marked on the nameplate in an ambient temperature of 40°C.

(B) One Horsepower or Less, Automatically Started. Any motor of 1 hp or less that is started automatically shall be protected against overload by one of the following means.

(1) Separate Overload Device. By a separate overload device following the requirements of 430.32(A)(1).

For a multispeed motor, each winding connection shall be considered separately. Modification of this value shall be permitted as provided in 430.32(C).

(2) Thermal Protector or Electronically Protected. A thermal protector integral with the motor shall be approved for use with the motor that it protects on the basis that it will prevent dangerous overheating of the motor due to overload and failure to start. An electronically protected motor shall be approved for use on the basis that it will prevent dangerous overheating due to the failure of the electronic control, overload, or failure to start the motor. Where the motor current-interrupting device is separate from the motor and its control circuit is operated by a protective device integral with the motor, it shall be arranged so that the opening of the control circuit results in interruption of current to the motor.

(3) Integral with Motor. A protective device integral with a motor that protects the motor against damage due to failure to start shall be permitted (1) if the motor is part of an approved assembly that does not subject the motor to overloads, or (2) if the assembly is also equipped with other safety controls (such as the safety combustion controls on a domestic oil burner) that protect the motor against damage due to failure to start. Where the assembly has safety controls that protect the motor, it shall be so indicated on the nameplate of the assembly where it will be visible after installation.

(4) Impedance-Protected. If the impedance of the motor windings is sufficient to prevent overheating due to failure to start, the motor shall be permitted to be protected as specified in 430.32(D)(2)a. for manually started motors if the motor is part of an approved assembly in which the motor will limit itself so that it will not be dangerously overheated.

> Informational Note: Many ac motors of less than 1/20 hp, such as clock motors, series motors, and so forth, and also some larger motors such as torque motors, come within this classification. It does not include split-phase motors having automatic switches that disconnect the starting windings.

(C) Selection of Overload Device. Where the sensing element or setting or sizing of the overload device selected in accordance with 430.32(A)(1) and 430.32(B)(1) is not sufficient to start the motor or to carry the load, higher size sensing elements or incremental settings or sizing shall be permitted to be used, provided the trip current of the overload device does not exceed the following percentage of motor nameplate full-load current rating:

Motors with marked service factor 1.15 or greater	140%
Motors with a marked temperature rise 40°C or less	140%
All other motors	130%

If not shunted during the starting period of the motor as provided in 430.35, the overload device shall have sufficient time delay to permit the motor to start and accelerate its load.

> Informational Note: A Class 20 overload relay will provide a longer motor acceleration time than a Class 10 or Class 10A overload relay. A Class 30 overload relay will provide a longer motor acceleration time than a Class 20 overload relay. Use of a higher class overload relay may preclude the need for selection of a higher trip current.

(D) One Horsepower or Less, Nonautomatically Started.

(1) Permanently Installed. Overload protection shall be in accordance with 430.32(B).

(2) Not Permanently Installed.

(a) *Within Sight from Controller.* Overload protection shall be permitted to be furnished by the branch-circuit short-circuit and ground-fault protective device; such device, however, shall not be larger than that specified in Part IV of Article 430.

Exception: Any such motor shall be permitted on a nominal 120-volt branch circuit protected at not over 20 amperes.

(b) *Not Within Sight from Controller.* Overload protection shall be in accordance with 430.32(B).

(E) Wound-Rotor Secondaries. The secondary circuits of wound-rotor ac motors, including conductors, controllers, resistors, and so forth, shall be permitted to be protected against overload by the motor-overload device.

430.33 Intermittent and Similar Duty. A motor used for a condition of service that is inherently short-time, intermittent, periodic, or varying duty, as illustrated by Table 430.22(E), shall be permitted to be protected against overload by the branch-circuit short-circuit and ground-fault protective device, provided the protective device rating or setting does not exceed that specified in Table 430.52(C)(1).

Any motor application shall be considered to be for continuous duty unless the nature of the apparatus it drives is such that the motor cannot operate continuously with load under any condition of use.

Because duty-cycle service motors (short-time, intermittent, periodic, or varying) do not operate continuously, prolonged overloads are rare unless mechanical failure in the driven apparatus stalls the motor, in which case, the branch-circuit protective device would open the circuit. The omission of overload protective devices for such motors is based on the type of duty not on the time rating of the motor.

430.35 Shunting During Starting Period.

(A) Nonautomatically Started. For a nonautomatically started motor, the overload protection shall be permitted to be shunted or cut out of the circuit during the starting period of the motor if the device by which the overload protection is shunted or cut out cannot be left in the starting position and if fuses or inverse time circuit breakers rated or set at not over 400 percent of the full-load current of the motor are located in the circuit so as to be operative during the starting period of the motor.

(B) Automatically Started. The motor overload protection shall not be shunted or cut out during the starting period if the motor is automatically started.

Exception: The motor overload protection shall be permitted to be shunted or cut out during the starting period on an automatically started motor where the following apply:

(1) The motor starting period exceeds the time delay of available motor overload protective devices, and

(2) Listed means are provided to perform the following:

    a. Sense motor rotation and automatically prevent the shunting or cutout in the event that the motor fails to start, and

    b. Limit the time of overload protection shunting or cutout to less than the locked rotor time rating of the protected motor, and

    c. Provide for shutdown and manual restart if motor running condition is not reached.

430.36 Fuses — In Which Conductor.
Where fuses are used for motor overload protection, a fuse shall be inserted in each ungrounded conductor and also in the grounded conductor if the supply system is 3-wire, 3-phase ac with one conductor grounded.

430.37 Devices Other Than Fuses — In Which Conductor.
Where devices other than fuses are used for motor overload protection, Table 430.37 shall govern the minimum allowable number and location of overload units such as trip coils or relays.

All 3-phase motors, except those protected by other approved means, must be provided with three overload units, one in each phase. Examples of those motors protected by other means include specially designed or integral-type detectors, with or without supplementary external protective devices.

TABLE 430.37 *Overload Units*

Kind of Motor	Supply System	Number and Location of Overload Units, Such as Trip Coils or Relays
1-phase ac or dc	2-wire, 1-phase ac or dc ungrounded	1 in either conductor
1-phase ac or dc	2-wire, 1-phase ac or dc, one conductor grounded	1 in ungrounded conductor
1-phase ac or dc	3-wire, 1-phase ac or dc, grounded neutral conductor	1 in either ungrounded conductor
1-phase ac	Any 3-phase	1 in ungrounded conductor
2-phase ac	3-wire, 2-phase ac, ungrounded	2, one in each phase
2-phase ac	3-wire, 2-phase ac, one conductor grounded	2 in ungrounded conductors
2-phase ac	4-wire, 2-phase ac, grounded or ungrounded	2, one for each phase in ungrounded conductors
2-phase ac	Grounded neutral or 5-wire, 2-phase ac, ungrounded	2, one for each phase in any ungrounded phase wire
3-phase ac	Any 3-phase	3, one in each phase*

Exception: An overload unit in each phase shall not be required where overload protection is provided by other approved means.

430.38 Number of Conductors Opened by Overload Device.
Motor overload devices, other than fuses or thermal protectors, shall simultaneously open a sufficient number of ungrounded conductors to interrupt current flow to the motor.

430.39 Motor Controller as Overload Protection.
A motor controller shall also be permitted to serve as an overload device if the number of overload units complies with Table 430.37 and if these units are operative in both the starting and running position in the case of a dc motor, and in the running position in the case of an ac motor.

430.40 Overload Relays.
Overload relays and other devices for motor overload protection that are not capable of opening short circuits or ground faults shall be protected by fuses or circuit breakers with ratings or settings in accordance with 430.52 or by a motor short-circuit protector in accordance with 430.52.

Exception: Where approved for group installation and marked to indicate the maximum size of fuse or inverse time circuit breaker by which they must be protected, the overload devices shall be protected in accordance with this marking.

Some overload devices are marked with a maximum short-circuit and ground-fault protective device rating or setting. This rating sets the limit on the maximum rating or setting of a fuse or a circuit breaker that may be upstream from the overload device. The rating also notifies the user that coordination between the overload device and the short-circuit and ground-fault device is required, which is most often the case for group motor installation.

△ 430.42 Motors on General-Purpose Branch Circuits.
Overload protection for motors used on general-purpose branch circuits shall be provided as specified in 430.42(A), (B), (C), or (D).

(A) Not Over 1 Horsepower. One or more motors without individual overload protection shall be permitted to be connected to a general-purpose branch circuit only where the installation complies with the limiting conditions specified in 430.32(B), 430.32(D), and 430.53(A)(1) and (A)(2).

(B) Over 1 Horsepower. Motors of ratings larger than specified in 430.53(A) shall be permitted to be connected to general-purpose branch circuits only where each motor is protected by overload protection selected to protect the motor as specified in 430.32. Both the motor controller and the motor overload device shall be approved for group installation with the short-circuit and ground-fault protective device selected in accordance with 430.53.

△ (C) Cord-and-Plug-Connected. Where a motor is connected to a branch circuit by means of an attachment plug and a receptacle or cord connector, and individual overload protection is omitted in accordance with 430.42(A), the rating of the attachment plug and receptacle or cord connector shall not exceed 15 amperes at 125 volts or 250 volts. Where individual overload protection is required in accordance with 430.42(B) for a motor or motor-operated appliance that is attached to the branch circuit through an attachment plug and a receptacle or cord connector, the overload device shall be an integral part of the motor or appliance. The rating of the attachment plug and receptacle or cord connector shall determine the rating of the circuit to which the motor can be connected, in accordance with 210.21(B).

(D) Time Delay. The branch-circuit short-circuit and ground-fault protective device protecting a circuit to which a motor or motor-operated appliance is connected shall have sufficient time delay to permit the motor to start and accelerate its load.

430.43 Automatic Restarting.
A motor overload device that can restart a motor automatically after overload tripping shall not be installed unless approved for use with the motor it protects. A motor overload device that can restart a motor automatically after overload tripping shall not be installed if automatic restarting of the motor can result in injury to persons.

430.44 Orderly Shutdown.
If immediate automatic shutdown of a motor by a motor overload protective device(s) would introduce additional or increased hazard(s) to a person(s) and continued motor operation is necessary for safe shutdown of equipment or process, a motor overload sensing device(s)

complying with Part III of this article shall be permitted to be connected to a supervised alarm instead of causing immediate interruption of the motor circuit, so that corrective action or an orderly shutdown can be initiated.

Part IV. Motor Branch-Circuit Short-Circuit and Ground-Fault Protection

△ 430.51 General.
Part IV specifies devices intended to protect the motor branch-circuit conductors, the motor control apparatus, and the motors against overcurrent due to short circuits or ground faults. The devices specified in Part IV do not include the types of devices required by 210.8, 230.95, and 590.6.

> Informational Note No. 1: See Informative Annex D, Example D8, for an example of motor branch-circuit short-circuit and ground-fault protection selection.

Part IV shall not apply to motor circuits rated over 1000 volts, nominal.

> Informational Note No. 2: See Part XI for over 1000 volts, nominal.

The rules for short-circuit and ground-fault protection are specific for particular situations. A short circuit is a fault between two or more circuit conductors. A ground fault is a fault between an ungrounded conductor and ground. During a short-circuit or ground-fault condition, the extremely high current causes the protective device (typically fuses or circuit breakers in systems rated 600 volts and less) to open the circuit. Excessive current caused by an overload condition passes through the overload protective device at the motor controller, thereby causing the device to open the control-circuit or motor-circuit conductors. For small motors, a fuse or circuit breaker may be sized so that it provides all the required overcurrent (short-circuit, ground-fault, and overload) protection for the motor and motor circuit components. The single protective device must meet the applicable requirements of Parts III and IV.

430.52 Rating or Setting for Individual Motor Circuit.

(A) General. The motor branch-circuit short-circuit and ground-fault protective device shall comply with 430.52(B) and either 430.52(C) or (D), as applicable.

For certain exceptions to the maximum rating or setting of these motor branch-circuit protective devices, as specified in Table 430.52(C)(1), see 430.52, 430.53, and 430.54. In general, 430.6 requires the values given in Tables 430.247 through 430.250 (including notes) to be used instead of the actual motor nameplate current rating to determine the ampacity of conductors or ampere ratings of switches, branch-circuit overcurrent devices, and so forth. Separate motor overload protection must be based on the motor nameplate current rating.

Exhibit 430.2 illustrates a typical motor circuit in which the branch-circuit short-circuit and ground-fault protective fuse or circuit breaker rating must carry the starting current and may be sized 150 to 300 percent of the motor full-load current (depending on the type of motor).

Section 430.52 could result in a branch-circuit and ground-fault protective device rating higher than the ampacity of the motor circuit conductors.

> **See also**
>
> **430.22** and its commentary for sizing motor circuit conductors

The selected rating or setting of the branch-circuit short-circuit and ground-fault protective device should be as low as possible for maximum protection. However, if the rating or setting specified in Table 430.52(C)(1) or permitted by 430.52(C)(1)(a) is not sufficient for the starting current of the motor, a higher rating or setting is allowed per 430.52(C)(1)(b). For example, a higher rating would be allowed for a motor under severe starting conditions in which the motor and its driven machinery require an extended length of time to reach the desired speed.

(B) All Motors. The motor branch-circuit short-circuit and ground-fault protective device shall be capable of carrying the starting current of the motor.

(C) Rating or Setting.

Δ **(1) In Accordance with Table 430.52(C)(1).** A protective device that has a rating or setting not exceeding the value calculated according to the values given in Table 430.52(C)(1) shall be used unless otherwise permitted in 430.52(C)(1)(a) or (C)(1)(b).

Class CC fuses are current-limiting fuses that may also be marked as "time delay," which indicates that the fuse has been investigated in accordance with the time-delay requirements of the standard. Class CC fuses are permitted to be sized according to the requirements of non-time-delay-rated fuses because they are fast-acting. Examples of Class CC fuses are shown in Exhibit 430.7.

(a) Where the values as determined by Table 430.52(C)(1) do not correspond to the standard ampere ratings and settings provided in 240.6, the next higher standard rating or setting shall be permitted.

(b) Where the rating specified in Table 430.52(C)(1), or the rating modified by 430.52(C)(1)(a), is not sufficient for the starting current of the motor, any of the following shall apply:

(1) The rating of a nontime-delay fuse not exceeding 600 amperes or a time-delay Class CC fuse shall be permitted to be increased but shall in no case exceed 400 percent of the full-load current.
(2) The rating of a time-delay (dual-element) fuse shall be permitted to be increased but shall in no case exceed 225 percent of the full-load current.
(3) The rating of an inverse time circuit breaker shall be permitted to be increased but shall in no case exceed 400 percent for full-load currents of 100 amperes or less or 300 percent for full-load currents greater than 100 amperes.

Δ **TABLE 430.52(C)(1)** Maximum Rating or Setting of Motor Branch-Circuit Short-Circuit and Ground-Fault Protective Devices

Type of Motor	Percentage of Full-Load Current			
	Nontime Delay Fuse[1]	Dual Element (Time-Delay) Fuse[1]	Instantaneous-Trip Breaker	Inverse Time Breaker[2]
Single-phase motors	300	175	800	250
AC polyphase motors other than wound-rotor	300	175	800	250
Squirrel cage — other than Design B energy-efficient — and Design B premium efficiency	300	175	800	250
Design B energy-efficient and Design B premium efficiency	300	175	1100	250
Synchronous[3]	300	175	800	250
Wound-rotor	150	150	800	150
DC (constant voltage)	150	150	250	150

Note: See 430.54 for certain exceptions to the values specified.

[1]The values in the Nontime Delay Fuse column apply to time-delay Class CC fuses.

[2]The values given in the last column also cover the ratings of nonadjustable inverse time types of circuit breakers that can be modified as in 430.52(C)(1)(a) and (C)(1)(b).

[3]Synchronous motors of the low-torque, low-speed type (usually 450 rpm or lower), such as those used to drive reciprocating compressors, pumps, and so forth, that start unloaded, do not require a fuse rating or circuit-breaker setting in excess of 200 percent of full-load current.

EXHIBIT 430.7 Class CC fuses. *(Courtesy of Eaton, Bussmann Division)*

(4) The rating of a fuse of 601–6000 ampere classification shall be permitted to be increased but shall in no case exceed 300 percent of the full-load current.

Informational Note: See Informative Annex D, Example D8, for an example of motor branch-circuit short-circuit and ground-fault rating and setting and Informational Note Figure 430.1 for an example location.

(2) Overload Relay Table. Where maximum branch-circuit short-circuit and ground-fault protective device ratings are shown in the manufacturer's overload relay table for use with a motor controller or are otherwise marked on the equipment, they shall not be exceeded even if higher values are allowed as shown above.

Δ **(3) Instantaneous-Trip Circuit Breaker.** An instantaneous-trip circuit breaker shall be permitted if the conditions of 430.52(C)(3)(a) and (C)(3)(b) are met.

   (a) *Application.* Instantaneous-trip circuit breakers shall be adjustable and part of a listed combination motor controller having coordinated motor overload and short-circuit and ground-fault protection in each conductor.

   Informational Note No. 1: Instantaneous-trip circuit breakers are also known as motor-circuit protectors (MCPs).
   Informational Note No. 2: For the purpose of this article, instantaneous-trip circuit breakers could include a damping means to accommodate a transient motor inrush current without nuisance tripping of the circuit breaker.

   (b) *Setting.* The instantaneous-trip circuit breaker shall be adjusted to a setting in accordance with one of the following:

   (1) No greater than the value specified in Table 430.52(C)(1)
   (2) Where the value specified in Table 430.52(C)(1) is not sufficient for the starting current of the motor, one of the following settings shall be permitted:

      a. Motors other than design B energy-efficient and Design B premium efficiency motors shall be permitted to be increased but shall in no case exceed 1300 percent of the motor full-load current.
      b. Design B energy-efficient and Design B premium efficiency motors shall be permitted to be increased but shall in no case exceed 1700 percent of the motor full-load current.
      c. Where an engineering analysis determines the value is not sufficient for the starting current of the motor, it shall not be necessary to first apply the value specified in Table 430.52(C)(1).

   Informational Note No. 3: See NEMA MG 1-2016, *Motors and Generators*, Part 12.59 for additional information on the requirements for a motor to be classified "energy efficient."

   (3) Where the motor full-load current is 8 amperes or less, the setting of the instantaneous-trip circuit breaker with a continuous current rating of 15 amperes or less in a

listed combination motor controller that provides coordinated motor branch-circuit overload and short-circuit and ground-fault protection shall be permitted to be increased to the value marked on the motor controller.

(4) Multispeed Motor. For a multispeed motor, a single short-circuit and ground-fault protective device shall be permitted for two or more windings of the motor if the rating of the protective device does not exceed the above applicable percentage of the nameplate rating of the smallest winding protected.

Exception: For a multispeed motor, a single short-circuit and ground-fault protective device shall be permitted to be used and sized according to the full-load current of the highest current winding, where all of the following conditions are met:

(1) Each winding is equipped with individual overload protection sized according to its full-load current.
(2) The branch-circuit conductors supplying each winding are sized according to the full-load current of the highest full-load current winding.
(3) The motor controller for each winding has a horsepower rating not less than that required for the winding having the highest horsepower rating.

(5) Power Electronic Devices. Semiconductor fuses intended for the protection of electronic devices shall be permitted in lieu of devices listed in Table 430.52(C)(1) for power electronic devices, associated electromechanical devices (such as bypass contactors and isolation contactors), and conductors in a solid-state motor controller system if the marking for replacement fuses is provided adjacent to the fuses.

(6) Self-Protected Combination Motor Controller. A listed self-protected combination motor controller shall be permitted in lieu of the devices specified in Table 430.52(C)(1). Adjustable instantaneous-trip settings shall not exceed 1300 percent of the full-load motor current for other than Design B energy-efficient and Design B premium efficiency motors and not more than 1700 percent of the full-load motor current for Design B energy-efficient and Design B premium efficiency motors.

   Informational Note: Proper application of self-protected combination motor controllers on 3-phase systems, other than solidly grounded wye, particularly on corner grounded delta systems, considers the self-protected combination motor controllers' individual pole-interrupting capability.

A self-protected combination controller combines the functions of short-circuit protection, disconnect, controller, and overload protection into a single unit. See Exhibit 430.8 for an example of a combination controller.

Δ **(7) Motor Short-Circuit Protector.** A motor short-circuit protector shall be permitted in lieu of devices listed in Table 430.52(C)(1) if the motor short-circuit protector is part of a listed combination motor controller having coordinated motor overload protection and short-circuit and ground-fault protection in

EXHIBIT 430.8 A listed self-protected combination motor controller. *(Courtesy of Eaton Corporation)*

each conductor and it will open the circuit at currents exceeding 1300 percent of the motor full-load current for other than Design B energy-efficient and Design B premium efficiency motors and 1700 percent of the motor full-load current for Design B energy-efficient and Design B premium efficiency motors.

> Informational Note: A motor short-circuit protector, as used in this section, is a fused device and is not an instantaneous-trip circuit breaker.

(D) Torque Motors. Torque motor branch circuits shall be protected at the motor nameplate current rating in accordance with 240.4(B).

430.53 Several Motors or Loads on One Branch Circuit. Two or more motors or one or more motors and other loads shall be permitted to be connected to the same branch circuit under conditions specified in 430.53(D) and in 430.53(A), (B), or (C). The branch-circuit protective device shall be fuses or inverse time circuit breakers.

(A) Not Over 1 Horsepower. Several motors, each not exceeding 1 hp in rating, shall be permitted on a nominal 120-volt branch circuit protected at not over 20 amperes or a branch circuit of 1000 volts, nominal, or less, protected at not over 15 amperes, if all of the following conditions are met:

(1) The full-load rating of each motor does not exceed 6 amperes.
(2) The rating of the branch-circuit short-circuit and ground-fault protective device marked on any of the motor controllers is not exceeded.
(3) Individual overload protection conforms to 430.32.

Δ **(B) If Smallest Rated Motor Protected.** Two or more motors or one or more motors and other loads shall be permitted to be

connected to a branch circuit where all of the following conditions are met:

(1) The branch-circuit short-circuit and ground-fault protective device is selected not to exceed that allowed by 430.52 for the smallest rated motor supplied by the branch circuit.
(2) Each motor is provided with separate overload protection.
(3) It can be determined that the branch-circuit short-circuit and ground-fault protective device will not open under the most severe normal conditions of service that might be encountered.

Δ **(C) Other Group Installations.** Two or more motors of any size or one or more motors and other loads, with each motor having individual overload protection, shall be permitted to be connected to a branch circuit where the motor controller(s) and overload device(s) comply with 430.53(C)(1) through (C)(5).

N **(1) Types of Assemblies.** The assembly type shall meet one of the following conditions:

(1) A listed factory assembly, with the motor branch-circuit short-circuit and ground-fault protective device either provided as part of the assembly or specified by a marking on the assembly
(2) Field installation of the motor branch-circuit short-circuit and ground-fault protective device, motor controller(s), and overload device(s) as separate assemblies listed for such use and provided with manufacturers' instructions for use with each other

N **(2) Motor Overload Devices.** Each motor overload device shall meet one of the following conditions:

(1) Listed for group installation with a specified maximum rating of fuse, inverse time circuit breaker, or both
(2) Selected such that the ampere rating of the motor-branch short-circuit and ground-fault protective device does not exceed that permitted by 430.52 for that individual motor overload device and corresponding motor load

N **(3) Motor Controllers.** Each motor controller shall meet one of the following conditions:

(1) Listed for group installation with a specified maximum rating of fuse, circuit breaker, or both
(2) Selected such that the ampere rating of the motor-branch short-circuit and ground-fault protective device does not exceed that permitted by 430.52 for that individual motor controller and corresponding motor load

N **(4) Short-Circuit & Ground-Fault Protection.** The branch circuit shall be protected by fuses or listed inverse time circuit breakers having a rating not exceeding the sum of all of the following:

(1) The value specified in 430.52 for the highest rated motor connected to the branch circuit
(2) The sum of the full-load current ratings of all other motors

(3) The sum of the current ratings of other loads connected to the circuit

Where this calculation results in a rating less than the ampacity of the branch-circuit conductors, it shall be permitted to increase the maximum rating of the fuses or circuit breaker to a value not exceeding that permitted by 240.4(B). Additionally, this rating shall not be larger than allowed by 430.40 for the overload relay protecting the smallest rated motor of the group.

N **(5) Overcurrent Protection.** Loads other than motor loads shall be protected in accordance with Part I through Part VII of Article 240.

> Informational Note: See 110.10 for circuit impedance and other characteristics.

The ground-fault short-circuit protection for motors might be greater than is permitted for other loads in accordance with Article 240. Devices with the same ampere rating might have significantly different short-circuit current ratings.

Prior to the 2011 edition of the *NEC®*, all motor controllers and overload devices were required to be listed for group installation. A motor controller or overload device is not required to be marked for group motor installation if it is applied within a group installation in which the branch-circuit protection for the group is within the same size limit as what would be permitted for a single motor installation of that device.

Δ **(D) Single Motor Taps.** For group installations described in 430.53(A), (B), or (C), the conductors of any tap supplying a single motor shall not be required to have an individual branch-circuit short-circuit and ground-fault protective device if they comply with 430.53(D)(1) or (D)(2).

N **(1) Conductors to the Motor.** Conductors to the motor shall have an ampacity that is not less than the ampacity of the branch-circuit conductors unless all of the following conditions are met:

(1) The conductors from the point of the tap to the motor overload device shall not be longer than 7.5 m (25 ft).

(2) The conductor ampacity is not less than one-third the ampacity of the branch-circuit conductors. The minimum ampacity shall not be less than required in 430.22.

(3) The conductors from the point of the tap to the motor controller(s) shall be protected from physical damage by being enclosed in an approved raceway or other approved means.

N **(2) Tap Conductors Between the Branch Circuit and Listed Manual Motor Controllers.** Conductors from the point of the tap from the branch circuit to a listed manual motor controller additionally marked "Suitable for Tap Conductor Protection in Group Installations," or to a branch-circuit protective device, shall meet one of the following conditions:

(1) The length of the motor tap conductors does not exceed 3 m (10 ft) and the tap conductors comply with all of the following:

a. The ampacity of the tap conductors is not less than one-tenth of the rating or setting of the branch-circuit short-circuit ground-fault protective device.

b. The conductors from the motor controller to the motor shall have an ampacity in accordance with 430.22.

c. The conductors from the point of the tap to the motor controller(s) shall be suitably protected from physical damage and enclosed either by an enclosed motor controller or by a raceway.

Exception to (1): Physical protection of the conductors from the point of the tap to the motor controllers shall not be required if the conductors have an ampacity not less than that of the branch-circuit conductors.

(2) The length of the motor tap conductors does not exceed 7.5 m (25 ft) and the tap conductors comply with all of the following:

a. The ampacity of the tap conductors is not less than one-third of the branch-circuit conductor ampacity.

b. The conductors from the motor controller to the motor shall have an ampacity in accordance with 430.22.

c. The conductors from the point of the tap to the motor controller(s) shall be suitably protected from physical damage and enclosed either by an enclosed motor controller or by a raceway.

Exception to (2): Physical protection of the conductors from the point of the tap to the motor controllers shall not be required if the conductors have an ampacity not less than that of the branch-circuit conductors.

The conditions for applying these tap rules are like those in 430.28 covering motor supply conductors tapped to a feeder. The short-circuit ground-fault device on the line side of the tap conductors protects more than one set of conductors that supply individual motors, which eliminates the need for an individual short-circuit ground-fault device for each set of conductors that supply a motor. Additional branch-circuit protective devices (such as fuses, inverse time circuit breakers, and listed self-protected combination motor controllers) can be used in the same location in the circuit of a group installation as a manual motor controller additionally marked "Suitable for Tap Conductor Protection in Group Installations."

This approach requires that the tap conductors meet certain size, physical protection, length, and termination conditions. The tap conductors must always meet the conductor size requirements of 430.22.

Exhibit 430.9 illustrates main branch-circuit conductors supplying a motor that is part of a group installation. The tap conductors have an ampacity equal to the ampacity of the main branch-circuit conductors. Therefore, branch-circuit short-circuit and ground-fault protective devices, fuses, or circuit breakers for the conductors in the tap are not required at the point of connection of the tap conductors to the main conductors, provided that the motor controller and motor overload protective device are listed for group installation with the size of the main branch-circuit short-circuit and ground-fault protective device used.

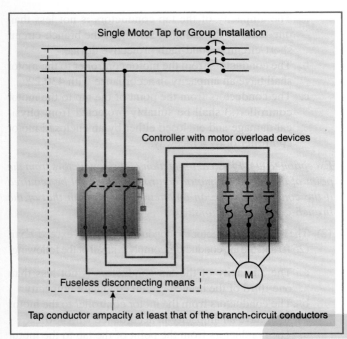

EXHIBIT 430.9 An example of the permissible omission of motor branch-circuit protective devices for tap conductors per 430.53(D)(1).

Exhibit 430.10 also illustrates main branch-circuit conductors supplying a motor that is part of a group installation. Here, the tap conductors have an ampacity at least one-third the ampacity of the main branch-circuit conductors, are not more than 25 feet in length, and are suitably protected from physical damage. The motor controller and motor overload protective device must be listed for group installation with the size of the main branch-circuit short-circuit and ground-fault protective device used.

Exhibit 430.11 illustrates the requirements of 430.53(D)(2)(1). Here the tap conductors from the point of the tap to the controller have an ampacity not less than one-tenth the rating or setting of the branch-circuit short-circuit and ground-fault protective device. These tap conductors cannot be more than 10 feet in length and must be suitably protected from physical damage. This does not apply to the conductors from the controller to the motor, which must have an ampacity in accordance with 430.22.

Exhibit 430.12 illustrates the requirements of 430.53(D)(2)(2). This requirement is similar to 430.53(D)(2)(1) but increases the maximum length to 25 feet and requires the tap conductors to be at least one-third the ampacity of the branch-circuit conductors. This requirement differs from the 25 feet tap requirement in 430.53(D)(1)(2), which requires the conductors to the motor connection to have an ampacity not less than one-third that of the branch-circuit conductors.

In these examples, the main branch-circuit fuses or circuit breakers would operate in the event of a short circuit, and the overload protective device would operate to protect the motor and tap conductors under overload conditions.

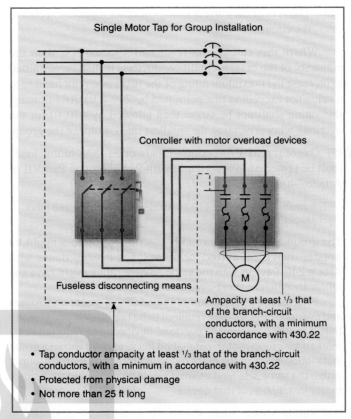

- Tap conductor ampacity at least ⅓ that of the branch-circuit conductors, with a minimum in accordance with 430.22
- Protected from physical damage
- Not more than 25 ft long

EXHIBIT 430.10 An example of the permissible omission of motor branch-circuit protective devices for tap conductors per 430.53(D)(1)(1), (2), and (3).

The tap conductors should never be of a smaller size and ampacity than the branch-circuit conductors required by 430.22. That is, a tap conductor (25 feet or less) may be one-third the ampacity of the main branch-circuit conductor to which it is connected; however, this ampacity must be equal to or larger than 125 percent of the motor's full-load current rating (see 430.22).

Calculation Example

A 25-hp, 230-V, 3-phase squirrel-cage motor is to be supplied from a branch circuit that has 2/0 AWG THW copper conductors, which have an ampacity of 175 A. See Table 310.16. Determine the correct size for the tap conductors less than 25 ft.

Solution

One-third of 175 A is 58 A. The tap conductor normally would be sized at 6 AWG THW copper (65 A). But a 6 AWG tap conductor does not meet the requirements of 430.22, that is, 125 percent of the FLC of the motor (68 A from Table 430.250), or 85 A. Therefore, the branch-circuit tap conductors are not permitted to be smaller than 4 AWG THW copper, with an ampacity of 85 A. See Table 310.16.

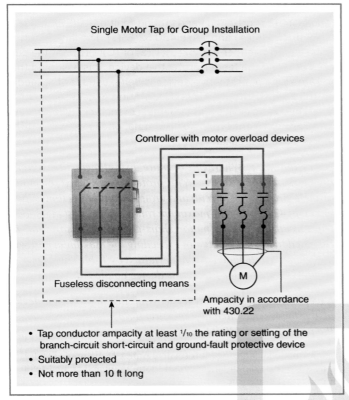

EXHIBIT 430.11 An example of the permissible omission of motor branch-circuit protective devices for tap conductors per 430.53(D)(2)(1).

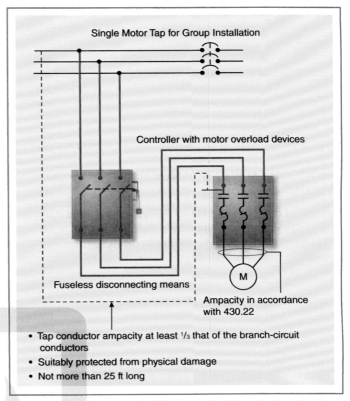

EXHIBIT 430.12 An example of the permissible omission of motor branch-circuit protective devices for tap conductors per 430.53(D)(2)(2).

430.54 Multimotor and Combination-Load Equipment. The rating of the branch-circuit short-circuit and ground-fault protective device for multimotor and combination-load equipment shall not exceed the rating marked on the equipment in accordance with 430.7(D).

430.55 Combined Overcurrent Protection. Motor branch-circuit short-circuit and ground-fault protection and motor overload protection shall be permitted to be combined in a single protective device where the rating or setting of the device provides the overload protection specified in 430.32.

Either a circuit breaker with inverse time characteristics or a dual-element (time-delay) fuse is permitted to serve both as motor overload protection and as the branch-circuit short-circuit and ground-fault protection if the requirements of 430.32 are met. These devices are not permitted to be sized as overload protection according to the values of 430.32(C). Rather, fuses are permitted to be sized as overload protection only, according to the values found in 430.32(A)(1), 430.32(B)(1), and 430.32(D)(1).

One-time, time-delay dual-element and Type S dual-element fuses and adapters are available with up to a 30-ampere rating. Type S fuses are designed to prevent oversize fusing. See 240.50 through 240.54 for more information about these fuses and adapters.

Exhibit 430.13 shows examples of time-delay, cartridge-type dual-element fuses that can withstand the normal motor starting current if sized at or near the motor full-load rating but that open when subjected to prolonged overload or open quickly during a short circuit or ground fault. The dual-element characteristics are the thermal cutout element, which permits harmless high-inrush currents to flow for short periods (but which would open the circuit during a prolonged period), and the fuse link element, which has current-limiting ability for short-circuit currents. Dual-element fuses may be used in larger sizes to provide only short-circuit and ground-fault protection.

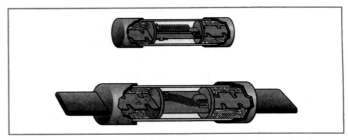

EXHIBIT 430.13 Fusetron cartridge-type fuses. *(Courtesy of Eaton, Bussman Division)*

430.56 Branch-Circuit Protective Devices — In Which Conductor. Branch-circuit protective devices shall comply with 240.15.

430.57 Size of Fuseholder. Where fuses are used for motor branch-circuit short-circuit and ground-fault protection, the fuseholders shall not be of a smaller size than required to accommodate the fuses specified by Table 430.52(C)(1).

Exception: Where fuses having time delay appropriate for the starting characteristics of the motor are used, it shall be permitted to use fuseholders sized to fit the fuses that are used.

The use of dual-element (time-delay) fuses makes it possible to use smaller fuses, thereby providing better protection because of the smaller fuses' lower ratings. They also allow for easier arrangement of equipment where space is at a premium.

430.58 Rating of Circuit Breaker. A circuit breaker for motor branch-circuit short-circuit and ground-fault protection shall have a current rating in accordance with 430.52 and 430.110.

Part V. Motor Feeder Short-Circuit and Ground-Fault Protection

430.61 General. Part V specifies protective devices intended to protect feeder conductors supplying motors against overcurrents due to short circuits or ground-faults.

> Informational Note: See Informative Annex D, Example D8, for an example of motor feeder circuit short-circuit and ground-fault protection rating and setting.

430.62 Rating or Setting — Motor Load.

Δ **(A) Specific Load.** A feeder supplying a specific fixed motor load(s) and consisting of conductor sizes in accordance with 430.24 shall be provided with a protective device having a rating or setting not greater than the largest rating or setting of the branch-circuit short-circuit and ground-fault protective device for any motor supplied by the feeder [based on the maximum permitted value for the specific type of protective device in accordance with 430.52, or 440.22(A) for hermetic refrigerant motor-compressors], plus the sum of the full-load currents of the other motors of the group.

Where the same rating or setting of the branch-circuit short-circuit and ground-fault protective device is used on two or more of the branch circuits supplied by the feeder, one of the protective devices shall be considered the largest for the above calculations.

The rating of a motor feeder short-circuit ground-fault protective device is determined by adding the rating of the largest branch-circuit short-circuit and ground-fault protective device for any motor supplied by the feeder to the sum of the full-load currents of all the other motors supplied by that feeder. The largest branch-circuit short-circuit and ground-fault protective device is based on 430.52(A) and Table 430.52(C)(1). The largest rating can be based on either 430.52(C)(1)(a) or (C)(1)(b). For the purposes of sizing the feeder protective device, it is assumed that the same type of protective device is being used for the feeder and the branch circuits. This assumption is necessary if the feeder protective device and the largest branch-circuit protective device are different types; for example, one is a fuse and the other is a circuit breaker.

Exception No. 1: Where one or more instantaneous-trip circuit breakers or motor short-circuit protectors are used for motor branch-circuit short-circuit and ground-fault protection as permitted in 430.52(C), the maximum rating of each instantaneous-trip circuit breaker or motor short-circuit protector shall be assumed to have a rating not exceeding the maximum percentage of motor full-load current permitted by Table 430.52(C)(1) for the type of feeder protective device employed.

Exception No. 2: Where the feeder overcurrent protective device also provides overcurrent protection for a motor control center, the provisions of 430.94 shall apply.

> Informational Note: See Informative Annex D, Example D8, for an example of motor feeder circuit short-circuit and ground-fault protection rating and setting.

(B) Other Installations. Where feeder conductors have an ampacity greater than required by 430.24, the rating or setting of the feeder overcurrent protective device shall be permitted to be based on the ampacity of the feeder conductors.

Exception No. 2 to 430.62(A) correlates the requirement of 430.62(B) for determining feeder short-circuit ground-fault protection with the requirements of 430.94 covering overcurrent protection for motor control centers. Where the motor feeder short-circuit ground-fault protective device is also the overcurrent protective device (OCPD) for a motor control center, its rating cannot exceed that allowed for protecting the common power bus of the motor control center.

430.63 Rating or Setting — Motor Load and Other Load(s). Where a feeder supplies a motor load and other load(s), the feeder protective device shall have a rating not less than that required for the sum of the other load(s) plus the following:

(1) For a single motor, the rating permitted by 430.52
(2) For a single hermetic refrigerant motor-compressor, the rating permitted by 440.22
(3) For two or more motors, the rating permitted by 430.62

Exception: Where the feeder overcurrent device provides the overcurrent protection for a motor control center, the provisions of 430.94 shall apply.

See the commentary following the Informational Note to 430.62(A).

Part VI. Motor Control Circuits

430.71 General. Part VI contains modifications of the general requirements and applies to the particular conditions of motor control circuits.

430.72 Overcurrent Protection.

(A) General. A motor control circuit tapped from the load side of a motor branch-circuit short-circuit and ground-fault protective device(s) and functioning to control the motor(s) connected to that branch circuit shall be protected against overcurrent in accordance with 430.72. Such a tapped control circuit shall not be considered to be a branch circuit and shall be permitted to be protected by either a supplementary or branch-circuit overcurrent protective device(s). A motor control circuit other than such a tapped control circuit shall be protected against overcurrent in accordance with 724.43 or the notes to Table 11(A) and Table 11(B) in Chapter 9, as applicable.

(B) Conductor Protection. The overcurrent protection for conductors shall be provided as specified in 430.72(B)(1) or (B)(2).

Exception No. 1: Where the opening of the control circuit would create a hazard as, for example, the control circuit of a fire pump motor, and the like, conductors of control circuits shall require only short-circuit and ground-fault protection and shall be permitted to be protected by the motor branch-circuit short-circuit and ground-fault protective device(s).

Exception No. 2: Conductors supplied by the secondary side of a single-phase transformer having only a two-wire (single-voltage) secondary shall be permitted to be protected by over-current protection provided on the primary (supply) side of the transformer, provided this protection does not exceed the value determined by multiplying the appropriate maximum rating of the overcurrent device for the secondary conductor from

Table 430.72(B)(2) by the secondary-to-primary voltage ratio. Transformer secondary conductors (other than two-wire) shall not be considered to be protected by the primary overcurrent protection.

(1) Separate Overcurrent Protection. Where the motor branch-circuit short-circuit and ground-fault protective device does not provide protection in accordance with 430.72(B)(2), separate overcurrent protection shall be provided. The overcurrent protection shall not exceed the values specified in Column A of Table 430.72(B)(2).

(2) Branch-Circuit Overcurrent Protective Device. Conductors shall be permitted to be protected by the motor branch-circuit short-circuit and ground-fault protective device and shall require only short-circuit and ground-fault protection. Where the conductors do not extend beyond the motor control equipment enclosure, the rating of the protective device(s) shall not exceed the value specified in Column B of Table 430.72(B)(2). Where the conductors extend beyond the motor control equipment enclosure, the rating of the protective device(s) shall not exceed the value specified in Column C of Table 430.72(B)(2).

(C) Control Circuit Transformer. Where a motor control circuit transformer is provided, the transformer shall be protected in accordance with 430.72(C)(1), (C)(2), (C)(3), (C)(4), or (C)(5).

Exception: Overcurrent protection shall be omitted where the opening of the control circuit would create a hazard as, for example, the control circuit of a fire pump motor and the like.

Δ **(1) Class 1 Power-Limited, Class 2, or Class 3 Circuits.** Where the transformer supplies a Class 1 power-limited circuit, the circuit shall comply with 724.30 through 724.52. Where the transformer supplies a Class 2 or Class 3 remote-control circuit, the circuit shall comply with the requirements of Part II of Article 725.

TABLE 430.72(B)(2) *Maximum Rating of Overcurrent Protective Device in Amperes*

Control Circuit Conductor Size (AWG)	Column A Separate Protection Provided		Column B Conductors Within Enclosure		Column C Conductors Extend Beyond Enclosure	
	Copper	Aluminum or Copper-Clad Aluminum	Copper	Aluminum or Copper-Clad Aluminum	Copper	Aluminum or Copper-Clad Aluminum
18	7	—	25	—	7	—
16	10	—	40	—	10	—
14	(Note 1)	—	100	—	45	—
12	(Note 1)	(Note 1)	120	100	60	45
10	(Note 1)	(Note 1)	160	140	90	75
Larger than 10	(Note 1)	(Note 1)	(Note 2)	(Note 2)	(Note 3)	(Note 3)

Notes:

1. Value specified in 310.15 as applicable.
2. 400 percent of value specified in Table 310.17 for 60°C conductors.
3. 300 percent of value specified in Table 310.16 for 60°C conductors.

(2) Transformers. Protection shall be permitted to be provided in accordance with 450.3.

(3) Less Than 50 Volt-Amperes. Control circuit transformers rated less than 50 volt-amperes (VA) and that are an integral part of the motor controller and located within the motor controller enclosure shall be permitted to be protected by primary overcurrent devices, impedance limiting means, or other inherent protective means.

(4) Primary Less Than 2 Amperes. Where the control circuit transformer rated primary current is less than 2 amperes, an overcurrent device rated or set at not more than 500 percent of the rated primary current shall be permitted in the primary circuit.

(5) Other Means. Protection shall be permitted to be provided by other approved means.

Motor control circuits can receive their power either from the load side of the motor short-circuit and ground-fault protective device or from a separate source, such as a panelboard.

Motor control circuits that receive their power from a separate source must be protected against overcurrent in accordance with 724.43 for Class 1 circuits. Conductor sizes 14 AWG and larger must be protected according to their ampacity listed in Tables 310.16 through 310.20. Conductor sizes 16 and 18 AWG must be protected at not more than 10 and 7 amperes, respectively, as specified in Table 430.72(B)(2).

If a motor control circuit is tapped from the load side of the motor branch-circuit short-circuit and ground-fault protective device, the size of the tapped conductor and the rating of the overcurrent device are based on whether the conductor stays within the motor control enclosure or leaves it. The load on a motor control circuit is similar to a motor branch-circuit load in that there is a predetermined connected load. An initial high inrush of current also occurs until the armature of the relay is seated and the current decreases to a steady state. Therefore, the overcurrent protection is similar to the short-circuit and ground-fault protection provided for a motor and is allowed to be greater than the ampacity of the control circuit conductor.

430.73 Protection of Conductors from Physical Damage. Where damage to a motor control circuit would constitute a hazard, all conductors of such a remote motor control circuit that are outside the control device itself shall be installed in a raceway or be otherwise protected from physical damage.

If damage to the control circuit conductors could result in an accidental ground fault or short circuit, causing the device to operate or rendering the device inoperative (either condition could constitute a hazard to persons or property), conductors must be installed in a raceway. Where boilers or furnaces are equipped with an automatic safety control device, damage to the conductors of the low-voltage control circuit (e.g., a thermostat) does not constitute a hazard (see Article 725, Part II).

430.74 Electrical Arrangement of Control Circuits. Where one conductor of the motor control circuit is grounded, the motor control circuit shall be arranged so that a ground fault in the control circuit remote from the motor controller will (1) not start the motor and (2) not bypass manually operated shutdown devices or automatic safety shutdown devices.

The inadvertent grounding of control circuits is a significant safety issue. Section 430.74 requires that if one side of the motor control circuit is grounded, the circuit must be arranged so that a ground fault in the remote-control device will not start the motor. For example, in the control wiring illustrated in Exhibit 430.14, the control circuit is a 120-volt, single-phase circuit derived from a 208-volt, 3-phase wye system supplying the motor, and one side of the control circuit is the grounded neutral. If the start button of the motor control circuit is connected to the grounded neutral, a ground fault on the coil side of the start button can start the motor. As shown in Exhibit 430.15, the same condition exists if the ground fault is in the wiring rather than in the control device itself. This hazardous condition can be alleviated by locating the start button in the ungrounded side of the control circuit, as shown in Exhibit 430.15.

Combinations of ground faults in motor and motor control circuits can also result in inadvertent motor starting. If the circuit

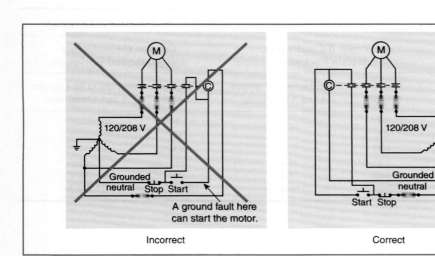

EXHIBIT 430.14 An example of control wiring in violation of 430.74 *(left)* and in compliance with 430.74 *(right)*. (For simplification, motor overload elements and disconnecting means are not shown.)

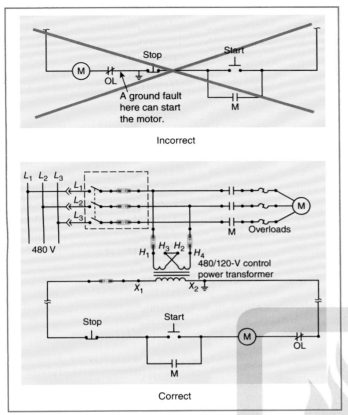

EXHIBIT 430.15 An example of control wiring using a 480/120-volt control power transformer. (The upper control circuit is not in compliance with 430.74.)

is ungrounded, the first fault could go undetected. One solution is to use double-pole control devices with one pole in each of the two control lines.

430.75 Disconnection.

(A) General. Motor control circuits shall be arranged so that they will be disconnected from all sources of supply when the disconnecting means is in the open position. The disconnecting means shall be permitted to consist of two or more separate devices, one of which disconnects the motor and the motor controller from the source(s) of power supply for the motor, and the other(s), the motor control circuit(s) from its power supply. Where separate devices are used, they shall be located immediately adjacent to each other.

Exception No. 1: Where more than 12 motor control circuit conductors are required to be disconnected, the disconnecting means shall be permitted to be located other than immediately adjacent to each other where all of the following conditions are met:

(1) Access to energized parts is limited to qualified persons in accordance with Part XII of this article.

(2) A warning sign is permanently located on the outside of each equipment enclosure door or cover permitting access

to the live parts in the motor control circuit(s), warning that motor control circuit disconnecting means are remotely located and specifying the location and identification of each disconnect. Where energized parts are not in an equipment enclosure as permitted by 430.232 and 430.233, an additional warning sign(s) shall be located where visible to persons who may be working in the area of the energized parts.

Exception No. 2: The motor control circuit disconnecting means shall be permitted to be remote from the motor controller power supply disconnecting means where the opening of one or more motor control circuit disconnecting means is capable of resulting in potentially unsafe conditions for personnel or property and the conditions of items (1) and (2) of Exception No. 1 are met.

(B) Control Transformer in Motor Controller Enclosure. Where a transformer or other device is used to obtain a reduced voltage for the motor control circuit and is located in the motor controller enclosure, such transformer or other device shall be connected to the load side of the disconnecting means for the motor control circuit.

Part VII. Motor Controllers

430.81 General. Part VII is intended to require suitable motor controllers for all motors.

(A) Stationary Motor of ⅛ Horsepower or Less. For a stationary motor rated at ⅛ hp or less that is normally left running and is constructed so that it cannot be damaged by overload or failure to start, such as clock motors and the like, the branch-circuit disconnecting means shall be permitted to serve as the motor controller.

(B) Portable Motor of ⅓ Horsepower or Less. For a portable motor rated at ⅓ hp or less, the motor controller shall be permitted to be an attachment plug and receptacle or cord connector.

430.82 Motor Controller Design.

(A) Starting and Stopping. Each motor controller shall be capable of starting and stopping the motor it controls and shall be capable of interrupting the locked-rotor current of the motor.

(B) Autotransformer. An autotransformer starter shall provide an "off" position, a running position, and at least one starting position. It shall be designed so that it cannot rest in the starting position or in any position that will render the overload device in the circuit inoperative.

(C) Rheostats. Rheostats shall be in compliance with the following:

(1) Motor-starting rheostats shall be designed so that the contact arm cannot be left on intermediate segments. The point

or plate on which the arm rests when in the starting position shall have no electrical connection with the resistor.

(2) Motor-starting rheostats for dc motors operated from a constant voltage supply shall be equipped with automatic devices that will interrupt the supply before the speed of the motor has fallen to less than one-third its normal rate.

430.83 Ratings. The motor controller shall have a rating in accordance with 430.83(A), unless otherwise permitted in 430.83(B) or (C), or in accordance with (D), under the conditions specified.

(A) General.

(1) Horsepower Ratings. Motor controllers, other than inverse time circuit breakers and molded case switches, shall have horsepower ratings at the application voltage not lower than the horsepower rating of the motor.

(2) Circuit Breaker. A branch-circuit inverse time circuit breaker rated in amperes shall be permitted as a motor controller for all motors. Where this circuit breaker is also used for overload protection, it shall conform to the appropriate provisions of this article governing overload protection.

(3) Molded Case Switch. A molded case switch rated in amperes shall be permitted as a motor controller for all motors.

A molded case switch has the same frame appearance as a molded case circuit breaker and is designed to fit in circuit-breaker enclosures. However, the device is marked with only a short-circuit current rating, which indicates that the switch does not provide overcurrent protection. Fused molded case switches that do provide overcurrent protection are marked with a short-circuit current interrupting rating. Both fused and unfused molded case switches can be used in motor circuits.

Molded case switches are permitted as motor disconnecting means per 430.109. In general, molded case switches are rated only in amperes and, where used in a motor circuit, must be sized at 115 percent of the motor full-load current rating. Disconnecting means assemblies are available that employ molded case switches marked with horsepower ratings that can be used, instead of the ampere rating of the molded case switch.

(B) Small Motors. Devices in accordance with 430.81(A) and (B) shall be permitted as a motor controller.

(C) Stationary Motors of 2 Horsepower or Less. For stationary motors rated at 2 hp or less and 300 volts or less, the motor controller shall be permitted to be either of the following:

(1) A general-use switch having an ampere rating not less than twice the full-load current rating of the motor

(2) On ac circuits, a general-use snap switch suitable only for use on ac (not general-use ac–dc snap switches) where the motor full-load current rating is not more than 80 percent of the ampere rating of the switch

(D) Torque Motors. For torque motors, the motor controller shall have a continuous-duty, full-load current rating not less than the nameplate current rating of the motor. For a motor controller rated in horsepower but not marked with the foregoing current rating, the equivalent current rating shall be determined from the horsepower rating by using Table 430.247, Table 430.248, Table 430.249, or Table 430.250.

(E) Voltage Rating. A motor controller with a straight voltage rating, for example, 240 volts or 480 volts, shall be permitted to be applied in a circuit in which the nominal voltage between any two conductors does not exceed the motor controller's voltage rating. A motor controller with a slash rating, for example, 120/240 volts or 480Y/277 volts, shall only be applied in a solidly grounded circuit in which the nominal voltage to ground from any conductor does not exceed the lower of the two values of the motor controller's voltage rating and the nominal voltage between any two conductors does not exceed the higher value of the motor controller's voltage rating.

N (F) Short-Circuit Current Rating. A motor controller shall not be installed where the available fault current exceeds the motor controller's short-circuit current rating.

Informational Note: The short-circuit current rating might be marked on the device or might be a rating for a tested combination specified in the motor controller's technical manual or instruction sheet.

430.84 Need Not Open All Conductors. The motor controller shall not be required to open all conductors to the motor.

Exception: Where the motor controller serves also as a disconnecting means, it shall open all ungrounded conductors to the motor in accordance with 430.111.

A controller that does not serve as a disconnecting means must open only as many motor circuit conductors as are necessary to stop the motor — that is, one conductor for a dc or single-phase motor circuit, two conductors for a 3-phase motor circuit, and three conductors for a 2-phase motor circuit.

430.85 In Grounded Conductors. One pole of the motor controller shall be permitted to be placed in a permanently grounded conductor if the motor controller is designed so that the pole in the grounded conductor cannot be opened without simultaneously opening all conductors of the circuit.

Generally, one conductor of a 120-volt circuit is grounded, and a single-pole device must be connected in the ungrounded conductor to serve as a controller. A 2-pole controller is permitted for such a circuit, where both conductors (grounded and ungrounded) are opened simultaneously. The same requirement can be applied to other circuits, such as 240-volt, 3-wire circuits with one conductor grounded.

430.87 Number of Motors Served by Each Motor Controller. Each motor shall be provided with an individual motor controller.

Exception No. 1: For motors rated 1000 volts or less, a single motor controller rated at not less than the equivalent horsepower, as determined in accordance with 430.110(C)(1), of all the motors in the group shall be permitted to serve the group under any of the following conditions:

(1) Where a number of motors drive several parts of a single machine or piece of apparatus, such as metal and woodworking machines, cranes, hoists, and similar apparatus

(2) Where a group of motors is under the protection of one overcurrent device in accordance with 430.53(A)

(3) Where a group of motors is located in a single room within sight from the motor controller location

Exception No. 2: A branch-circuit disconnecting means serving as the motor controller in accordance with 430.81(A) shall be permitted to serve more than one motor.

430.88 Adjustable-Speed Motors.
Adjustable-speed motors that are controlled by means of field regulation shall be equipped and connected so that they cannot be started under a weakened field.

Exception: Starting under a weakened field shall be permitted where the motor is designed for such starting.

The torque and speed of a dc motor depend on the amount of current passing through the armature. This current is a function of the shunt field strength and the RPM of the armature. A reduction of the shunt field magnetic flux causes a reduction of the counterelectromotive force in the armature, resulting in an increase in armature current, thereby increasing torque, which increases speed.

430.89 Speed Limitation.
Machines of the following types shall be provided with speed-limiting devices or other speed-limiting means:

(1) Separately excited dc motors
(2) Series motors
(3) Motor-generators and converters that can be driven at excessive speed from the dc end, as by a reversal of current or decrease in load

Exception: Separate speed-limiting devices or means shall not be required under either of the following conditions:

(1) Where the inherent characteristics of the machines, the system, or the load and the mechanical connection thereto are such as to safely limit the speed

(2) Where the machine is always under the manual control of a qualified operator

Use of dc motors is common where speed control is essential, such as electric railways and elevators, where a smooth start, controlled acceleration, and a smooth stop are necessary. If the load is removed from a series motor when it is running, the speed of the motor will increase until it is dangerously high. To produce the necessary counterelectromotive force with a weakened field,

the armature must turn correspondingly faster. Series motors are commonly used as gear-drive traction motors of electric locomotives and, thus, are continuously loaded.

Unless the exception applies, the motors, motor (compound-wound dc) generators, and (synchronous) converters must be provided with speed-limiting devices, such as a centrifugal device on the shaft of the machine or a remotely located overspeed device. This device can be set to operate a set of contacts at a predetermined speed and thereby trip a circuit breaker and de-energize the machine.

430.90 Combination Fuseholder and Switch as Motor Controller.
The rating of a combination fuseholder and switch used as a motor controller shall be such that the fuseholder will accommodate the size of the fuse specified in Part III of this article for motor overload protection.

Exception: Where fuses having time delay appropriate for the starting characteristics of the motor are used, fuseholders of smaller size than specified in Part III of this article shall be permitted.

Part VIII. Motor Control Centers

430.92 General.
Part VIII covers motor control centers installed for the control of motors, lighting, and power circuits.

Motor control centers are made up of several motor starters, controls, and disconnect switches. Motor control centers are allowed to be used as service equipment if provided with a single main disconnecting means. A second service disconnecting means, however, is permitted in the motor control center if it is provided to serve other loads.

In addition to Part VIII, installation requirements, including access and working space clearances, for motor control centers are covered in Section 110.26. The requirements of 110.26(E) specify dedicated space for a motor control center and physical protection from mechanical systems that might leak or otherwise adversely affect a motor control center.

430.94 Overcurrent Protection.
Motor control centers shall be provided with overcurrent protection in accordance with Parts I, II, and VIII of Article 240. The ampere rating or setting of the overcurrent protective device shall not exceed the rating of the common power bus. This protection shall be provided by (1) an overcurrent protective device located ahead of the motor control center or (2) a main overcurrent protective device located within the motor control center.

430.95 Service Equipment.
Where used as service equipment, each motor control center shall be provided with a single main disconnecting means to disconnect all ungrounded service conductors.

Exception No. 1: A second service disconnect shall be permitted to supply additional equipment.

TABLE 430.97(D) *Minimum Spacing Between Bare Metal Parts*

Nominal Voltage	Opposite Polarity Where Mounted on the Same Surface		Opposite Polarity Where Held Free in Air		Live Parts to Ground	
	mm	in.	mm	in.	mm	in.
Not over 125 volts, nominal	19.1	¾	12.7	½	12.7	½
Not over 250 volts, nominal	31.8	1¼	19.1	¾	12.7	½
Not over 600 volts, nominal	50.8	2	25.4	1	25.4	1

Where a grounded conductor is provided, the motor control center shall be provided with a main bonding jumper, sized in accordance with 250.28(D), within one of the sections for connecting the grounded conductor, on its supply side, to the motor control center equipment ground bus.

Exception No. 2: High-impedance grounded neutral systems shall be permitted to be connected as provided in 250.36.

430.96 Grounding. Multisection motor control centers shall be connected together with an equipment grounding conductor or an equivalent equipment grounding bus sized in accordance with Table 250.122. Equipment grounding conductors shall be connected to this equipment grounding bus or to a grounding termination point provided in a single-section motor control center.

430.97 Busbars and Conductors.

(A) Support and Arrangement. Busbars shall be protected from physical damage and be held firmly in place. Other than for required interconnections and control wiring, only those conductors that are intended for termination in a vertical section shall be located in that section.

Exception: Conductors shall be permitted to travel horizontally through vertical sections where such conductors are isolated from the busbars by a barrier.

(B) Phase Arrangement. The phase arrangement on 3-phase horizontal common power and vertical buses shall be A, B, C from front to back, top to bottom, or left to right, as viewed from the front of the motor control center. The B phase shall be that phase having the higher voltage to ground on 3-phase, 4-wire, delta-connected systems. Other busbar arrangements shall be permitted for additions to existing installations and shall be marked.

Exception: Rear-mounted units connected to a vertical bus that is common to front-mounted units shall be permitted to have a C, B, A phase arrangement where properly identified.

(C) Minimum Wire-Bending Space. The minimum wire-bending space at the motor control center terminals and minimum gutter space shall be in accordance with 312.6.

(D) Spacings. Spacings between motor control center bus terminals and other bare metal parts shall not be less than specified in Table 430.97(D).

(E) Barriers. Barriers shall be placed in all service-entrance motor control centers to isolate service busbars and terminals from the remainder of the motor control center.

430.98 Marking.

(A) Motor Control Centers. Motor control centers shall be marked according to 110.21, and the marking shall be plainly visible after installation. Marking shall also include common power bus current rating and motor control center short-circuit current rating.

(B) Motor Control Units. Motor control units in a motor control center shall comply with 430.8.

430.99 Available Fault Current. The available fault current at the motor control center and the date the available fault current calculation was performed shall be documented and made available to those authorized to inspect, install, or maintain the installation.

A motor control center is required to be marked with its short-circuit current rating per 430.98. The requirement in 430.99 requires the motor control center to be marked with the available short-circuit current. This provides a means to easily compare the equipment rating with the available fault current and to confirm compliance with 110.10.

Part IX. Disconnecting Means

430.101 General. Part IX is intended to require disconnecting means capable of disconnecting motors and motor controllers from the circuit.

430.102 Location.

Δ **(A) Motor Controller.** An individual disconnecting means shall be provided for each motor controller and shall disconnect the motor controller. The disconnecting means shall be located in sight from the motor controller location.

Exception No. 1: For motor circuits over 1000 volts, nominal, a motor controller disconnecting means lockable in accordance with 110.25 shall be permitted to be out of sight of the motor controller if the motor controller is marked with a label giving the location of the disconnecting means.

Exception No. 2: A single disconnecting means shall be permitted for a group of coordinated motor controllers that drive several parts of a single machine or piece of apparatus. The disconnecting means shall be located in sight from the motor controllers, and both the disconnecting means and the motor controllers shall be located in sight from the machine or apparatus.

Exception No. 3: The disconnecting means shall not be required to be in sight from valve actuator motor (VAM) assemblies containing the motor controller where such a location introduces additional or increased hazards to persons or property and the following conditions are met:

(1) The valve actuator motor assembly is marked with a label giving the location of the disconnecting means.

(2) The disconnecting means is lockable in accordance with 110.25.

(B) Motor. A disconnecting means shall be provided for a motor in accordance with 430.102(B)(1) or (B)(2).

(1) Separate Motor Disconnect. A disconnecting means for the motor shall be located in sight from the motor location and the driven machinery location.

Δ **(2) Motor Controller Disconnect.** The motor controller disconnecting means required in accordance with 430.102(A) shall be permitted to serve as the disconnecting means for the motor if it is in sight from the motor location and the driven machinery location.

Exception to (1) and (2): The disconnecting means for the motor shall not be required under either of the following conditions if the motor controller disconnecting means required in 430.102(A) is lockable in accordance with 110.25:

(1) Where such a location of the disconnecting means for the motor is impracticable or introduces additional or increased hazards to persons or property

Informational Note No. 1: Some examples of increased or additional hazards include, but are not limited to, motors rated in excess of 100 hp, multimotor equipment, submersible motors, motors associated with adjustable-speed drives, and motors located in hazardous (classified) locations.

(2) In industrial installations, with written safety procedures, where conditions of maintenance and supervision ensure that only qualified persons service the equipment

Informational Note No. 2: See NFPA 70E-2021, *Standard for Electrical Safety in the Workplace*, for information on lockout/tagout procedures.

The main rules of 430.102(A) and (B) require that the disconnecting means be in sight of the controller, the motor location, and the driven machinery location. The exceptions to these rules permit the disconnecting means to be out of sight under certain conditions, which include that it be capable of being locked in the open position.

A single disconnecting means may be located adjacent to a group of coordinated controllers, as illustrated in Exhibit 430.16, where the controllers are mounted on a multimotor continuous process machine.

According to 430.102(B)(2), Exception, the disconnecting means is permitted to be out of sight of the motor — as illustrated in Exhibit 430.17 — if the controller disconnecting means

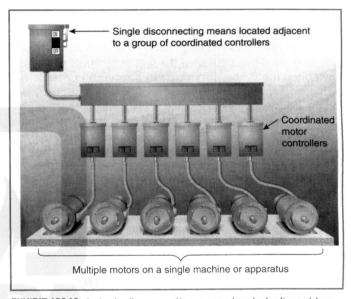

Single disconnecting means located adjacent to a group of coordinated controllers

Coordinated motor controllers

Multiple motors on a single machine or apparatus

EXHIBIT 430.16 A single disconnecting means located adjacent to a group of coordinated controllers mounted on a multimotor continuous process machine.

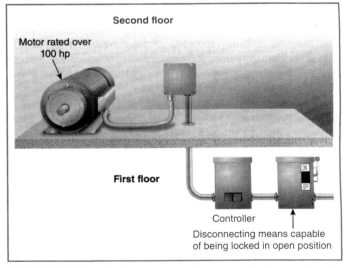

Second floor

Motor rated over 100 hp

First floor

Controller

Disconnecting means capable of being locked in open position

EXHIBIT 430.17 A controller disconnecting means that is out of sight of the motor but only for cases that meet the requirements of (1) or (2) of 430.102(B), Exception.

is individually capable of being locked in the open position and the condition of either (1) or (2) is met. Locating disconnect switches or panelboards within locked rooms or circuit breakers behind the locked door of a panelboard does not meet the requirements of 430.102.

If locating the disconnecting means close to the motor location and driven machinery is impracticable due to the type of machinery, the type of facility, lack of space for locating large equipment such as disconnecting means rated over 1000 volts, or any increased hazard to persons or property, the disconnecting means is permitted to be located remotely. Industrial facilities that comply with OSHA 29 CFR 1910.147, "The Control of Hazardous Energy (Lockout/Tagout)," are permitted to have the disconnecting means located remotely.

Section 110.2 of *NFPA 70E®*, *Standard for Electrical Safety in the Workplace®*, requires in part that all "electrical conductors and circuit parts shall not be considered to be in an electrically safe work condition until all of the requirements of Article 120 have been met." The work process specified in Article 120 includes removing the sources of energy, locking, and tagging out the disconnecting means, and verifying the absence of voltage using an approved voltage tester. Further, 120.1(A)(3) states that lockout/tagout requirements shall "apply to fixed, permanently installed equipment, temporarily installed equipment and portable equipment." The principles and procedures in *NFPA 70E* establish strict work rules requiring the locking off (out) and tagging out of disconnect switches.

430.103 Operation. The disconnecting means shall open all ungrounded supply conductors and shall be designed so that no pole can be operated independently. The disconnecting means shall be permitted in the same enclosure with the motor controller. The disconnecting means shall be designed so that it cannot be closed automatically.

> Informational Note: See 430.113 for equipment receiving energy from more than one source.

A switch, circuit breaker, or other device serves as a disconnecting means for both the controller and the motor, thereby providing safety during maintenance and inspection shutdown periods. The disconnecting means also disconnects the controller; therefore, it cannot be a part of the controller.

However, separate disconnects and controllers may be mounted on the same panel or contained in the same enclosure, such as combination fused-switch, magnetic-starter units.

Depending on the size of the motor and other conditions, the type of disconnecting means required may be a motor circuit switch, a circuit breaker, a general-use switch, an isolating switch, an attachment plug and receptacle, or a branch-circuit short-circuit and ground-fault protective device, as specified in 430.109.

If a motor stalls or is under heavy overload and the motor controller fails to properly open the circuit, the disconnecting means, which must be rated to interrupt locked-rotor current, can be used to open the circuit. In accordance with 430.109(E), for motors larger than 100 horsepower ac or 40 horsepower dc, the disconnecting means is permitted to be a general-use or an isolating switch that is plainly marked "Do not operate under load."

430.104 To Be Indicating. The disconnecting means shall plainly indicate whether it is in the open (off) or closed (on) position.

430.105 Grounded Conductors. One pole of the disconnecting means shall be permitted to disconnect a permanently grounded conductor, provided the disconnecting means is designed so that the pole in the grounded conductor cannot be opened without simultaneously disconnecting all conductors of the circuit.

430.107 Readily Accessible. At least one of the disconnecting means shall be readily accessible.

430.108 Every Disconnecting Means. Every disconnecting means in the motor circuit between the point of attachment to the feeder or branch circuit and the point of connection to the motor shall comply with the requirements of 430.109 and 430.110.

Δ **430.109 Type.** The disconnecting means shall be a type in accordance with 430.109(A), unless otherwise permitted in 430.109(B) through (G), under the conditions specified.

(A) General.

(1) Motor-Circuit Switch. A listed motor-circuit switch rated in horsepower.

(2) Molded Case Circuit Breaker. A listed molded case circuit breaker.

(3) Molded Case Switch. A listed molded case switch.

(4) Instantaneous-Trip Circuit Breaker. An instantaneous-trip circuit breaker that is part of a listed combination motor controller.

(5) Self-Protected Combination Motor Controller. Listed self-protected combination motor controller.

Δ **(6) Manual Motor Controller.** Listed manual motor controllers additionally marked "Suitable as Motor Disconnect" shall be permitted as a disconnecting means where installed between the final motor branch-circuit short-circuit protective device and the motor. Listed manual motor controllers additionally marked "Suitable as Motor Disconnect" shall be permitted as disconnecting means on the line side of the fuses in accordance with 430.52(C)(5). In this case, the fuses permitted in 430.52(C)(5) shall be considered supplementary fuses, and suitable branch-circuit short-circuit and ground-fault protective devices shall be installed on the line side of the manual motor controller additionally marked "Suitable as Motor Disconnect."

(7) System Isolation Equipment. System isolation equipment shall be listed for disconnection purposes. System isolation equipment shall be installed on the load side of the overcurrent

CLOSER LOOK: System Isolation Equipment

In large and often complex machines, repeated operation of disconnecting means for maintenance or servicing is inherent to the process, and the risk of injury to personnel is increased due to moving parts and multiple points of entry. This risk drives development of system isolation equipment (SIE). Safety procedures for personnel servicing this equipment include detailed lockout/tagout protocols for all sources of mechanical and electrical energy.

SIE helps simplify electrical lockout/tagout procedures; it can also be used to isolate energy sources such as pneumatic energy. In accordance with its definition in Article 100, *system isolation equipment* is "a redundantly monitored, remotely operated contactor-isolating system, packaged to provide the disconnection/isolation function." This type of equipment is covered in NFPA 79, *Electrical Standard for Industrial Machinery*, as a way to disconnect and isolate separately operable parts of a large industrial machine. With its inclusion in 430.109(A)(7) as a permitted type of disconnecting means, the *NEC* recognizes the use of this disconnection/isolation system in applications that do not fall within the scope of the industrial machinery standard.

In a typical configuration, the contactor is located within the system power and control panel and may control power to the entire machine or to portions of a large machine. The control equipment may be provided in several configuration options for distributing to lockout stations in single or multiplexed radial schemes according to the application.

Once an electrically safe condition is achieved (including discharge of any residual energy), verification of such condition is provided at the remote lockout station using an indicator light.

In equipment that uses lockable guarding, the same verification signal could also be used as part of the guard access system.

In contrast to a simple start/stop station and control circuit operating a magnetic contactor, the control panel for this system provides a sophisticated level of monitoring upon actuation of the remote lockout switch. If any portion of the safety system cannot be verified for proper operation, the safe condition indicator light will not illuminate at the remote lockout station. As part of the standard operating procedure, the failure to receive the safe condition signal must be considered as an indication of an unsafe condition.

Among the critical safety elements that are provided by the control panel for the isolation system are the diversity and redundancy that are integrated into the safe condition verification logic. Another element is the electrical isolation of the internal safety-related control circuits and the physical isolation of the equipment's internal components, which reduces the possibility of externally induced failure modes. The control panel modules are sealed, as are the circuits between the SIE component enclosures, to discourage tampering that could compromise the safe operation of the equipment and endanger personnel. Where the system includes multiple lockout stations, the controlled equipment cannot be re-energized until all the lockout switches are returned to the "on" position. Nominal configurations of the SIE include provisions to prevent power from unexpectedly reaching the machine upon the restoration of power from the utility source. To re-energize the machine, all lockout switches must be in the closed, or on, position while at least one lockout switch must have been in the open, or off, position or placed in the open, or off, position (and then moved to the closed, or on, position after the utility power had been restored).

---

protection and its disconnecting means. The disconnecting means shall be one of the types permitted by 430.109(A)(1) through (A)(3).

Unlike other disconnecting means recognized by 430.109(A) where the operation of the disconnecting means directly opens the supply circuit at that specific location, system isolation equipment (SIE) employs a lockable control circuit switch(es) (lockout switch) and a verification indication at the disconnecting means location (lockout station). Also, operation of the lockout switch causes power components such as a monitored magnetic contactor to open and isolate the electrical equipment associated with the machine from its power supply circuit. The SIE is classified according to its intended application with parameters that include the load characteristics, the method used to monitor the controlled load-side power circuit, the number and maximum distance to the farthest lockout station, and the available control interface functions.

(B) Stationary Motors of ⅛ Horsepower or Less. For stationary motors of ⅛ hp or less, the branch-circuit overcurrent device shall be permitted to serve as the disconnecting means.

(C) Stationary Motors of 2 Horsepower or Less. For stationary motors rated at 2 hp or less and 300 volts or less, the disconnecting means shall be permitted to be one of the following devices:

(1) A general-use switch having an ampere rating not less than twice the full-load current rating of the motor

(2) On ac circuits, a general-use snap switch suitable only for use on ac (not general-use ac–dc snap switches) where the motor full-load current rating is not more than 80 percent of the ampere rating of the switch

(3) A listed manual motor controller having a horsepower rating not less than the rating of the motor and marked "Suitable as Motor Disconnect"

(D) Autotransformer-Type Controlled Motors. For motors of over 2 hp up to and including 100 hp, the separate disconnecting means required for a motor with an autotransformer-type motor controller shall be permitted to be a general-use switch where all of the following provisions are met:

(1) The motor drives a generator that is provided with overload protection.

(2) The motor controller is capable of interrupting the locked-rotor current of the motors, is provided with a no voltage release, and is provided with running overload protection not exceeding 125 percent of the motor full-load current rating.

(3) Separate fuses or an inverse time circuit breaker rated or set at not more than 150 percent of the motor full-load current is provided in the motor branch circuit.

(E) Isolating Switches. For stationary motors rated at more than 40 hp dc or 100 hp ac, the disconnecting means shall be permitted to be a general-use or isolating switch where plainly marked "Do not operate under load."

(F) Cord-and-Plug-Connected Motors. For a cord-and-plug-connected motor, a horsepower-rated attachment plug and receptacle, flanged surface inlet and cord connector, or attachment plug and cord connector having ratings not less than the motor ratings shall be permitted to serve as the disconnecting means. Horsepower-rated attachment plugs, flanged surface inlets, receptacles, or cord connectors shall not be required for cord-and-plug-connected appliances in accordance with 422.33, room air conditioners in accordance with 440.63, or portable motors rated ⅓ hp or less.

A motor circuit switch is a horsepower-rated switch capable of interrupting the maximum overload current of a motor (see the definition of *switch, motor-circuit* in Article 100). A molded case switch (nonautomatic circuit interrupter) is a circuit breaker-like device without the overcurrent element and automatic-trip mechanism. It is rated in amperes and is suitable for use as a motor-circuit disconnect based on its ampere rating, as is a circuit breaker. The disconnecting means must be listed.

Exhibits 430.18, 430.19, 430.20, and 430.21 illustrate various methods of providing motor disconnecting means as permitted by 430.109(B), 430.109(C), 430.109(E), and 430.109(F), respectively.

Where horsepower-rated fused switches are required, marking within the enclosure usually permits a dual horsepower rating. The standard horsepower rating is based on the largest non-time-delay (non-dual-element) fuse rating that can be used in the switch and that will permit the motor to start. The maximum horsepower rating is based on the largest rated time-delay (dual-element) fuse that can be used in the switch and that will permit the motor to start. Thus, where time-delay fuses are used, smaller-size switches and fuseholders can be used (see 430.57, Exception).

(G) Torque Motors. For torque motors, the disconnecting means shall be permitted to be a general-use switch.

430.110 Current Rating and Interrupting Capacity.

(A) General. The disconnecting means for motor circuits rated 1000 volts, nominal, or less shall have a current rating not less than 115 percent of the full-load current rating of the motor.

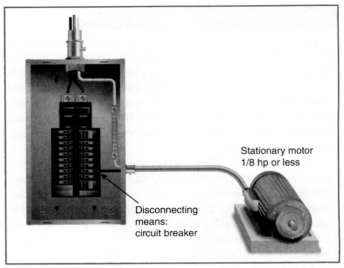

EXHIBIT 430.18 A branch-circuit overcurrent device serving as the disconnecting means for a stationary motor of 1/8 horsepower or less according to 430.109(B).

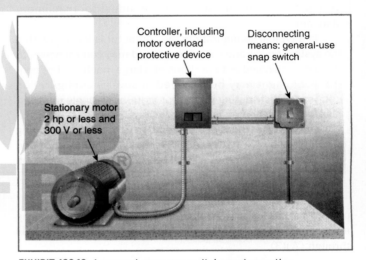

EXHIBIT 430.19 A general-use snap switch serving as the disconnecting means for a stationary motor rated at 2 horsepower or less and at 300 volts or less according to 430.109(C).

Exception: A listed unfused motor-circuit switch having a horsepower rating not less than the motor horsepower shall be permitted to have a current rating less than 115 percent of the full-load current rating of the motor.

(B) For Torque Motors. Disconnecting means for a torque motor shall have a current rating of at least 115 percent of the motor nameplate current.

(C) For Combination Loads. Where two or more motors are used together or where one or more motors are used in combination with other loads, such as resistance heaters, and where the combined load can be simultaneous on a single disconnecting means, the current and horsepower ratings of the combined

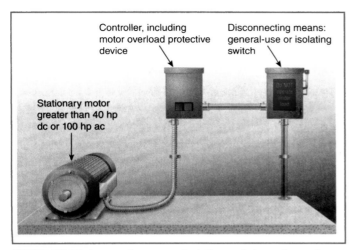

Controller, including motor overload protective device

Disconnecting means: general-use or isolating switch

Stationary motor greater than 40 hp dc or 100 hp ac

EXHIBIT 430.20 A general-use or an isolating switch serving as the disconnecting means for a stationary motor rated at more than 40 horsepower dc or 100 horsepower ac according to 430.109(E).

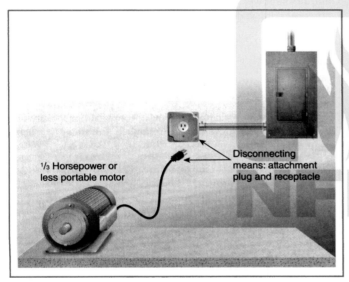

Disconnecting means: attachment plug and receptacle

⅓ Horsepower or less portable motor

EXHIBIT 430.21 An attachment plug and receptacle serving as the disconnecting means for a certain cord-and-plug-connected motor according to 430.109(F).

load shall be determined in accordance with 430.110(C)(1) through (C)(3).

Δ **(1) Horsepower Rating.** The rating of the disconnecting means shall be determined from the sum of all currents, including resistance loads, at the full-load condition and also at the locked-rotor condition. The combined full-load current and the combined locked-rotor current so obtained shall be considered as a single motor for the purpose of this requirement.

The full-load current equivalent to the horsepower rating of each motor shall be selected from Table 430.247, Table 430.248, Table 430.249, or Table 430.250. These full-load currents shall be added to the rating in amperes of other loads to obtain an equivalent full-load current for the combined load.

The locked-rotor current equivalent to the horsepower rating of each motor shall be selected from Table 430.251(A) or Table 430.251(B). The locked-rotor currents shall be added to the rating in amperes of other loads to obtain an equivalent locked-rotor current for the combined load. Where two or more motors or other loads cannot be started simultaneously, the largest sum of locked-rotor currents of a motor or group of motors that can be started simultaneously and the full-load currents of other concurrent loads shall be permitted to be used to determine the equivalent locked-rotor current for the simultaneous combined loads. In cases where different current ratings are obtained when applying these tables, the largest value obtained shall be used.

Exception No. 1: The locked-rotor current equivalent to the horsepower rating of each polyphase motor with design letter A shall be one of following:

(1) If available, the motor's marked value of locked-rotor amperes
(2) In the absence of a marked value of locked-rotor amperes for the motor, the value calculated from Equation 430.110(C)(1)a:

[430.110(C)(1)a]

$$\text{locked-rotor amperes} = \left(\frac{kVA}{hp}\right) \times \frac{(1000 \times \text{motor's marked value of rated horsepower})}{(\text{motor's marked value of rated volts}) \times (\sqrt{3})}$$

where:

kVA/hp = maximum range value of kilovolt-amperes per horsepower with locked rotor in Table 430.7(B) associated with the motor's marked locked-rotor indicating code letter

Informational Note: Equation 430.110(C)(1)a is obtained by solving for locked-rotor amperes in the formula for "kilovolt-amperes per horsepower with locked rotor," as follows:

[430.110(C)(1)b]

$$\frac{kVA}{hp} = \frac{(\sqrt{3}) \times (\text{motor's marked value of rated volts}) \times (\text{locked-rotor amperes})}{(1000 \times \text{motor's marked value of rated horsepower})}$$

The numerator of Equation 430.110(C)(1)b for kilovolt-amperes per horsepower is the apparent power input to a three-phase motor with locked rotor in units of volt-amperes. The factor of 1000 VA/kVA in the denominator converts this value to units of kilovolt-amperes and "(marked value of rated horsepower)" in the denominator converts this to kilovolt-amperes per horsepower. Note that "motor's marked value of rated volts" is a line-to-line value and "locked-rotor amperes" is a line value as opposed to a phase value.

Exception No. 2: Where part of the concurrent load is resistance load, and where the disconnecting means is a switch rated in horsepower and current, the switch used shall be permitted to have a horsepower rating not less than the combined load of the motor(s) if the current rating of the switch is not less than the locked-rotor current of the motor(s) plus the resistance load.

(2) Current Rating. The current rating of the disconnecting means shall not be less than 115 percent of the sum of all currents at the full-load condition determined in accordance with 430.110(C)(1).

Exception: A listed nonfused motor-circuit switch having a horsepower rating equal to or greater than the equivalent horsepower of the combined loads, determined in accordance with 430.110(C)(1), shall be permitted to have a current rating less than 115 percent of the sum of all currents at the full-load condition.

(3) Small Motors. For small motors not covered by Table 430.247, Table 430.248, Table 430.249, or Table 430.250, the locked-rotor current shall be assumed to be six times the full-load current.

Listed circuit breakers and molded case switches are tested under overload conditions at six times their rating, to cover motor circuit applications, and are suitable for use as a motor disconnecting means.

Calculation Example

An installation consists of one 5-hp, one 3-hp, and two ½-hp motors, plus a 10-kW heater, all rated 240 V, 3 phase. All motors are Design B motors. Determine the size of the disconnecting means required for this combination load.

Solution

Use the appropriate tables to select the full-load and locked-rotor current equivalents from the tables in Article 430 as follows:

Equivalent Full-Load and Locked-Rotor Current Rating.

Motor or Other Load	Full-Load Current Amperes from Table 430.250	Locked-Rotor Current Amperes from Table 430.251(B)
5-hp motor	15.2	92
3-hp motor	9.6	64
½-hp motor	2.2	20
½-hp motor	2.2	20
10-kW heater	24.1	24.1

$$\left[\frac{10 \times 1{,}000}{(240 \times 1.732)} = 24.1A\right]$$

	Total 53.3	Total 220.1

To determine the minimum horsepower rating for the disconnecting means, a comparison between the full-load horsepower equivalent and the locked-rotor equivalent of the combined loads must be considered.

Step 1a. Using the combined full load current 53.3 amperes and Table 430.250, the horsepower equivalent is 20 hp based on the 54-ampere value from the table.

Step 1b. Using the combined locked-rotor current of 220.1 amperes and Table 430.251 (B), the horsepower equivalent is 15 hp based on the 232-ampere value from the table.

Step 1c. Comparing the horsepower ratings determined in Steps 1a and 1b, the horsepower equivalent for the FLC (20 hp) is larger than the horsepower equivalent for the locked-rotor current (15 hp). Therefore, the disconnecting means must be rated not less than 20 hp.

Step 2. It is also necessary to determine the minimum ampere rating of the disconnecting means. Circuit breakers are rated in amperes, and many safety switches have both an ampere and horsepower rating. Using the 53.3 FLC, the minimum ampere rating is determined as follows:

$$53.3 \text{ A} \times 1.15 = 61.295 \text{ amperes}$$

Conclusion

The minimum rating of a disconnecting means would be as follows:

1. Circuit Breaker — Using standard ampere ratings for circuit breakers, a 70-ampere circuit breaker is the minimum size required.

2. Safety Switch — Using standard ratings of safety switches, a 100-ampere switch is the minimum size required. If the switch also has a horsepower rating, it will have to be not less than 20 hp. In addition, see the exception to 430.110(C)(2), which permits listed nonfused motor-circuit switches having the minimum horsepower rating to have an ampacity rating less than 115% of the FLC of the combined loads.

430.111 Switch or Circuit Breaker as Both Motor Controller and Disconnecting Means. A switch or circuit breaker shall be permitted to be used as both the motor controller and disconnecting means if it complies with 430.111(A) and is one of the types specified in 430.111(B).

(A) General. The switch or circuit breaker complies with the requirements for motor controllers specified in 430.83, opens all ungrounded conductors to the motor, and is protected by an overcurrent device in each ungrounded conductor (which shall be permitted to be the branch-circuit fuses). The overcurrent device protecting the motor controller shall be permitted to be part of the motor controller assembly or shall be permitted to be separate. An autotransformer-type motor controller shall be provided with a separate disconnecting means.

(B) Type. The device shall be one of the types specified in 430.111(B)(1), (B)(2), or (B)(3).

(1) Air-Break Switch. An air-break switch, operable directly by applying the hand to a lever or handle.

(2) Inverse Time Circuit Breaker. An inverse time circuit breaker operable directly by applying the hand to a lever or handle. The circuit breaker shall be permitted to be both power-operable and manually operable.

(3) Oil Switch. An oil switch used on a circuit whose rating does not exceed 1000 volts or 100 amperes, or by special permission on a circuit exceeding this capacity where under expert supervision. The oil switch shall be permitted to be both power-operable and manually operable.

If used as a controller, a switch or circuit breaker must meet all the requirements for controllers and be protected by branch-circuit short-circuit and ground-fault protective devices (fuses or a circuit breaker), which ensure that all ungrounded conductors will be opened.

If the controller consists of a manually operable air-break switch, an inverse time circuit breaker, or a 100-ampere maximum oil switch (higher rating by special permission), the controller is considered a satisfactory disconnecting means. The intent of 430.111 is to permit omission of an additional device to serve as a disconnecting means.

A separate disconnecting means must be provided if the controller is of the autotransformer. (This switch can be combined in the same enclosure with a motor overload protective device.)

430.112 Motors Served by Single Disconnecting Means. Each motor shall be provided with an individual disconnecting means.

Exception: A single disconnecting means shall be permitted to serve a group of motors under any one of the conditions of (1), (2), and (3). The single disconnecting means shall be rated in accordance with 430.110(C).

(1) Where a number of motors drive several parts of a single machine or piece of apparatus, such as metal- and woodworking machines, cranes, and hoists.

(2) Where a group of motors is under the protection of one set of branch-circuit protective devices as permitted by 430.53(A).

(3) Where a group of motors is in a single room within sight from the location of the disconnecting means.

A single disconnecting means must have a rating equal to the sum of the horsepower or current of each motor in the group. If the sum is over 2 horsepower, a motor circuit switch (horsepower-rated) must be used; thus, for five 2-horsepower motors, the disconnecting means should be a motor-circuit switch rated at not less than 10 horsepower.

Part (1) of the exception indicates that a single disconnecting means may be used where a number of motors drive several parts of a single machine, such as cranes (see 610.31 through 610.33), metal or woodworking machines, steel rolling mill machinery, and so forth. The single disconnecting means for multimotor machinery provides a positive means of simultaneously de-energizing all motor branch circuits, including remote-control circuits, interlocking circuits, limit-switch circuits, and operator control stations.

Part (2) of the exception refers to 430.53(A), which permits a group of motors under the protection of the same branch-circuit device, provided the device is rated not more than 20 amperes on a nominal 120-volt branch circuit, or not over 15 amperes on a branch circuit of 1000 volts, nominal or less. The motors must be rated 1 horsepower or less, and the full-load current for each motor is not permitted to exceed 6 amperes. A single disconnecting means is both practical and economical for a group of such small motors.

Part (3) of the exception covers the common situation in which a group of motors is in one room, such as a pump room, compressor room, or mixer room. It is possible to design the layout of a single disconnecting means with an unobstructed view from each motor. (See the definition of *in sight from* in Article 100.)

These conditions for an individual disconnecting means are like those specified in 430.87, which permits the use of a single controller for a group of motors.

430.113 Energy from More Than One Source. Motor and motor-operated equipment receiving electric energy from more than one source shall be provided with disconnecting means from each source of electric energy immediately adjacent to the equipment served. Each source shall be permitted to have a separate disconnecting means. Where multiple disconnecting means are provided, a permanent warning sign shall be provided on or adjacent to each disconnecting means indicating that multiple sources must be shut off to remove all power to the equipment. The sign at each disconnect shall identify the other specific circuits.

Exception No. 1: Where a motor receives electric energy from more than one source, the disconnecting means for the main power supply to the motor shall not be required to be immediately adjacent to the motor if the motor controller disconnecting means is lockable in accordance with 110.25.

Exception No. 2: A separate disconnecting means shall not be required for a Class 2 remote-control circuit complying with Parts I and II of Article 725, rated not more than 30 volts, and isolated and ungrounded.

Some motors require multiple separate sources of power to operate properly, such as a motor space heater or a speed switch. Synchronous motors commonly use dc power for excitation purposes. Section 430.113 could also apply to circuits that supply power to speed or vibration sensors mounted within or otherwise attached to the motor.

Where the individual sources have multiple disconnecting means, a permanent warning sign is required to warn the user that other power sources are present. Exception No. 2 removes the disconnect requirement only for Class 2 circuits.

Part X. Adjustable-Speed Drive Systems

430.120 General. The installation requirements for Part I through Part IX are applicable unless modified or supplemented by Part X.

Power conversion equipment used in adjustable-speed drive systems shall comply with Part X for an input or output rated 1000 volts or lower and with Part XI for an input or output rated over 1000 volts.

Adjustable-speed drives are used extensively in commercial, institutional, and industrial motor applications. Exhibit 430.22 shows 480-volt adjustable-speed drives, also known as variable-frequency drives (VFDs).

Part X consolidates requirements that are unique to these drives, which include rules regarding methods of overtemperature protection in motors (see 430.126). This is a critical area, because motors operating at reduced speed do not provide adequate air circulation over windings from a fan integral with the motor. An overload device that actuates on current exceeding full-load amperes will not operate, because the operating current at slower speeds is reduced. A thermal-sensing device integral with the motor will sense a temperature rise in the motor windings.

430.122 Conductors — Minimum Size and Ampacity.

Δ **(A) Branch/Feeder Circuit Conductors.** Circuit conductors supplying power conversion equipment included as part of an adjustable-speed drive system shall have an ampacity not less than 125 percent of the rated input current to the power conversion equipment.

> Informational Note: Power conversion equipment can have multiple power ratings and corresponding input currents.

Δ **(B) Output Conductors.** The conductors between the power conversion equipment and the motor shall have an ampacity equal to or larger than 125 percent of the motor full-load current as determined by 430.6(A) or (B).

> *Exception: If the power conversion equipment is listed and marked as "Suitable for Output Motor Conductor Protection," the conductor between the power conversion equipment and the motor shall have an ampacity equal to or greater than the larger of the following:*
>
> *(1) 125 percent of the motor full-load current as determined by 430.6(A) or (B)*
> *(2) The ampacity of the minimum conductor size marked on the power conversion equipment*

> Informational Note No. 1: See 430.130 and 430.131 for branch circuit protection requirements. The minimum ampacity required of output conductors is often different than that of the conductors supplying the power conversion equipment.

> Informational Note No. 2: Circuit conductors on the output of an adjustable-speed drive system are susceptible to breakdown under certain conditions due to the characteristics of the output waveform of the drive. Factors affecting the conductors include, but are not limited to, the output voltage, frequency, and current; the length of the conductors; the spacing between the conductors; and the dielectric strength of the conductor insulation. Methods to mitigate breakdown include consideration of one or more of these factors.

Where an adjustable-speed drive system is used in a motor circuit, two sets of branch-circuit conductors must be considered in respect to minimum size. The basis for sizing the branch-circuit conductors supplying the power conversion equipment is the input current rating of that unit. That portion of the branch circuit from the power conversion equipment to the motor (i.e., output conductors) uses the general motor branch-circuit approach, and the currents found in the Article 430 tables are the basis used for sizing. Both the input conductors and the output conductors are required to be sized by applying 125 percent to the conductor sizing parameter specified in 430.122(A) and (B).

EXHIBIT 430.22 Adjustable-speed drives for air-handling units provide significant power savings for fan appliances. *(Courtesy of International Association of Electrical Inspectors)*

Calculation Example

Given a power conversion unit to be used with a 10-hp, 480-volt, 3-phase squirrel-cage motor driving an exhaust fan, find the minimum size copper conductors to supply the input of the drive and to run from the output of the drive to the motor.

Solution

Step 1: Input Conductors. The input conductors supplying the power conversion equipment are based on the input current rating of the power conversion unit. The input current at 480 volts for a 10-hp motor is 22.2 amperes.

Per 430.122(A), 22.2 A × 1.25 = 27.75 amperes. The minimum size copper conductor based on 75°C termination rating is 10 AWG (35 amperes per Table 310.16).

Step 2: Output Conductors. The output conductors from the drive to the motor are based on the Table 430.250 full load current value for a 10 hp, 3-phase, 480-volt motor, which is 14 amperes.

Per 430.122(B), 14 A × 1.25 = 17.5 amperes. The minimum size copper conductor based on 75°C termination rating is 14 AWG (20 amperes per Table 310.16).

Note that the 14 AWG size conductor is permitted in accordance with 240.4(G), and the rating of the overcurrent device that is installed at the equipment from which the supply circuit to the power conversion equipment originates is controlled by 430.130(A) or (B).

Δ **(C) Bypass Device.** The ampacity of circuit conductors supplying power conversion equipment included as part of an adjustable-speed drive system that utilizes a bypass device shall be the larger of either of the following:

(1) 125 percent of the rated input current to the power conversion equipment

(2) 125 percent of the motor full-load current rating determined in accordance with 430.6

(D) Several Motors or a Motor and Other Loads. Conductors supplying several motors or a motor and other loads, including power conversion equipment, shall have ampacity in accordance with 430.24, using the rated input current of the power conversion equipment for purposes of calculating ampacity.

430.124 Overload Protection. Overload protection of the motor shall be provided.

(A) Included in Power Conversion Equipment. Where the power conversion equipment is marked to indicate that motor overload protection is included, additional overload protection shall not be required.

(B) Bypass Circuits. For adjustable-speed drive systems that utilize a bypass device to allow motor operation at rated full-load speed, motor overload protection as described in Article 430, Part III, shall be provided in the bypass circuit.

(C) Multiple Motor Applications. For multiple motor application, individual motor overload protection shall be provided in accordance with Article 430, Part III.

430.126 Motor Overtemperature Protection.

(A) General. Adjustable-speed drive systems shall protect against motor overtemperature conditions where the motor is not rated to operate at the nameplate rated current over the speed range required by the application. This protection shall be provided in addition to the conductor protection required in 430.32. Protection shall be provided by one of the following means:

(1) Motor thermal protector in accordance with 430.32

(2) Adjustable-speed drive system with load and speed-sensitive overload protection and thermal memory retention upon shutdown or power loss

Exception to (2): Thermal memory retention upon shutdown or power loss is not required for continuous duty loads.

(3) Overtemperature protection relay utilizing thermal sensors embedded in the motor and meeting the requirements of 430.126(A)(2)

(4) Thermal sensor embedded in the motor whose communications are received and acted upon by an adjustable-speed drive system

Informational Note: The relationship between motor current and motor temperature changes when the motor is operated by an adjustable-speed drive. In certain applications, overheating of motors can occur when operated at reduced speed, even at current levels less than a motor's rated full-load current. The overheating can be the result of reduced motor cooling when its shaft-mounted fan is operating less than rated nameplate RPM. As part of the analysis to determine whether overheating will occur, it is necessary to consider the continuous torque capability curves for the motor given the application requirements. This will assist in determining whether the motor overload protection will be able, on its own, to provide protection against overheating. These overheating protection requirements are only intended to apply to applications where an adjustable-speed drive, as defined in Article 100, is used.

For motors that utilize external forced air or liquid cooling systems, overtemperature can occur if the cooling system is not operating. Although this issue is not unique to adjustable speed applications, externally cooled motors are most often encountered with such applications. In these instances, overtemperature protection using direct temperature sensing is recommended [i.e., 430.126(A)(1), (A)(3), or (A)(4)], or additional means should be provided to ensure that the cooling system is operating (flow or pressure sensing, interlocking of adjustable-speed drive system and cooling system, etc.).

(B) Multiple Motor Applications. For multiple motor applications, individual motor overtemperature protection shall be provided as required in 430.126(A).

(C) Automatic Restarting and Orderly Shutdown. 430.43 and 430.44 shall apply to the motor overtemperature protection means.

430.128 Disconnecting Means. The disconnecting means shall be permitted to be in the incoming line to the conversion equipment and shall have a rating not less than 115 percent of the rated input current of the conversion unit.

430.130 Branch-Circuit Short-Circuit and Ground-Fault Protection for Single Motor Circuits Containing Power Conversion Equipment.

(A) Circuits Containing Power Conversion Equipment. Circuits containing power conversion equipment shall be protected by a branch-circuit short-circuit and ground-fault protective device in accordance with all of the following:

(1) The rating and type of protection shall be determined by 430.52(C)(1), (C)(3), (C)(5), or (C)(6), using the full-load current rating of the motor load as determined by 430.6(A) or (B).

Exception to (1): The rating and type of protection shall be permitted to be determined by Table 430.52(C)(1) using the power conversion equipment's rated input current where the power conversion equipment is listed and marked "Suitable for Output Motor Conductor Protection."

Informational Note No. 1: Motor conductor branch-circuit short-circuit and ground-fault protection from the power conversion equipment to the motor is provided by power conversion equipment that is listed and marked "Suitable for Output Motor Conductor Protection."

Informational Note No. 2: A motor branch circuit using power conversion equipment, including equipment listed and marked "Suitable for Output Motor Conductor Protection," includes the input circuit to the power conversion equipment.

(2) Where maximum branch-circuit short-circuit and ground-fault protective ratings are stipulated for specific device types in the manufacturer's instructions for the power conversion equipment or are otherwise marked on the equipment, they shall not be exceeded even if higher values are permitted by 430.130(A)(1).

(3) A self-protected combination motor controller shall only be permitted where specifically identified in the manufacturer's instructions for the power conversion equipment or if otherwise marked on the equipment.

Informational Note No. 3: The type of protective device, its rating, and its setting are often marked on or provided with the power conversion equipment.

(4) Where an instantaneous-trip circuit breaker or semi-conductor fuses are permitted in accordance with the drive manufacturer's instructions for use as the branch-circuit short-circuit and ground-fault protective device for listed power conversion equipment, they shall be provided as an integral part of a single listed assembly incorporating both the protective device and power conversion equipment.

(B) Bypass Circuit/Device. Branch-circuit short-circuit and ground-fault protection shall also be provided for a bypass circuit/device(s). Where a single branch-circuit short-circuit and ground-fault protective device is provided for circuits containing both power conversion equipment and a bypass circuit, the branch-circuit protective device type and its rating or setting shall be in accordance with those determined for the power conversion equipment and for the bypass circuit/device(s) equipment.

430.131 Several Motors or Loads on One Branch Circuit Including Power Conversion Equipment. For installations meeting all the requirements of 430.53 that include one or more power converters, the branch-circuit short-circuit and ground-fault protective fuses or inverse time circuit breakers shall be of a type and rating or setting permitted for use with the power conversion equipment using the full-load current rating of the connected motor load in accordance with 430.53. For the

purposes of 430.53 and 430.131, power conversion equipment shall be considered to be a motor controller.

Part XI. Over 1000 Volts, Nominal

430.201 General. Part XI recognizes the additional hazard due to the use of higher voltages. It adds to or amends the other provisions of this article.

430.202 Marking on Motor Controllers. In addition to the marking required by 430.8, a motor controller shall be marked with the control voltage.

430.203 Raceway Connection to Motors. Flexible metal conduit or liquidtight flexible metal conduit not exceeding 1.8 m (6 ft) in length shall be permitted to be employed for raceway connection to a motor terminal enclosure.

N **430.204 Wire-Bending Space in Enclosures.** Motor controllers rated over 1000 volts shall provide wire-bending space within the enclosure for conductors installed in accordance with 305.5.

430.205 Size of Conductors. The ampacities of conductors supplying equipment rated over 1000 volts, nominal, shall be determined in accordance with 315.60 or 430.205(A) and (B).

N **(A) General Motor Systems.** Conductors supplying motors shall be sized not less than the current trip setting of the motor overload protective device(s).

N **(B) Adjustable-Speed Drive Systems.** For an adjustable-speed drive system, the conductors supplying the power conversion equipment shall have an ampacity not less than 125 percent of the rated input current to the power conversion equipment.

430.206 Motor-Circuit Overcurrent Protection.

(A) General. Each motor circuit shall include coordinated protection to automatically interrupt overload and fault currents in the motor, the motor-circuit conductors, and the motor control apparatus. Adjustable-speed drive systems with input or output voltages over 1000 volts, nominal, shall comply with 430.124 and 430.126. All other motors shall comply with 430.206(B) through (C).

Exception: Where a motor is critical to an operation and the motor should operate to failure if necessary to prevent a greater hazard to persons, the sensing device(s) shall be permitted to be connected to a supervised annunciator or alarm instead of interrupting the motor circuit.

(B) Overload Protection.

(1) Type of Overload Device. Each motor shall be protected against dangerous heating due to motor overloads and failure to start by a thermal protector integral with the motor or external current-sensing devices, or both. Protective device settings for each motor circuit shall be determined under engineering supervision.

Selecting the proper overload and short-circuit protection for medium-voltage motor circuits is more complex than for low-voltage circuits. For medium-voltage motor circuits, it is critical for the overload relay to coordinate with the short-circuit protection, because some short-circuit protective devices cannot safely open below certain multiples of their rating. In these overload cases, the overload relay must open before the short-circuit protective device is asked to open. At the same time, the overload relay might not safely open beyond certain multiples of its rating, requiring the short-circuit protective device to open. The curves of both the overload relay and the short-circuit protective device must be correlated to ensure that each opens only on levels of current for which it can safely open.

(2) Wound-Rotor Alternating-Current Motors. The secondary circuits of wound-rotor ac motors, including conductors, motor controllers, and resistors rated for the application, shall be considered as protected against overcurrent by the motor overload protection means.

(3) Operation. Operation of the overload interrupting device shall simultaneously disconnect all ungrounded conductors.

(4) Automatic Reset. Overload sensing devices shall not automatically reset after trip unless resetting of the overload sensing device does not cause automatic restarting of the motor or there is no hazard to persons created by automatic restarting of the motor and its connected machinery.

(C) Fault-Current Protection.

Δ **(1) Type of Protection.** Fault-current protection shall be provided in each motor circuit as specified by either 430.206(C)(1)(a) or (C)(1)(b).

(a) A circuit breaker of suitable type and rating arranged so that it can be serviced without hazard. The circuit breaker shall simultaneously disconnect all ungrounded conductors. The circuit breaker shall be permitted to sense the fault current by means of integral or external sensing elements.

(b) Fuses of a suitable type and rating placed in each ungrounded conductor. Fuses shall be used with suitable disconnecting means, or they shall be of a type that can also serve as the disconnecting means. They shall be arranged so that they cannot be serviced while they are energized.

(2) Reclosing. Fault-current interrupting devices shall not automatically reclose the circuit.

Exception: Automatic reclosing of a circuit shall be permitted where the circuit is exposed to transient faults and where such automatic reclosing does not create a hazard to persons.

(3) Combination Protection. Overload protection and fault-current protection shall be permitted to be provided by the same device.

430.207 Rating of Motor Control Apparatus. The ultimate trip current of overcurrent (overload) relays or other motor-protective devices used shall not exceed 115 percent of the motor controller's continuous current rating. Where the motor branch-circuit disconnecting means is separate from the motor controller, the disconnecting means current rating shall not be less than the ultimate trip setting of the overcurrent relays in the circuit.

430.208 Disconnecting Means. The motor controller disconnecting means shall be a switch or circuit breaker having a voltage rating not less than that of the circuit involved, and shall be lockable in accordance with 110.25. The disconnecting means shall have a current rating of not less than 100 percent of the full-load current rating of the motor. For adjustable-speed drive systems, the disconnecting means shall have a current rating not less than 100 percent of the rated input current of the power conversion equipment.

Part XII. Protection of Live Parts — All Voltages

430.231 General. Part XII specifies that live parts shall be protected in an approved manner for the hazard involved.

430.232 Where Required. Exposed live parts of motors and motor controllers operating at 50 volts or more between terminals shall be guarded against accidental contact by enclosure or by location as follows:

(1) By installation in a room or enclosure that is accessible only to qualified persons
(2) By installation on a suitable balcony, gallery, or platform, elevated and arranged so as to exclude unqualified persons
(3) By elevation 2.5 m (8 ft) or more above the floor

Exception: Live parts of motors operating at more than 50 volts between terminals shall not require additional guarding for stationary motors that have commutators, collectors, and brush rigging located inside of motor-end brackets and not conductively connected to supply circuits operating at more than 150 volts to ground.

430.233 Guards for Attendants. Where live parts of motors or motor controllers operating at over 50 volts to ground are guarded against accidental contact only by location as specified in 430.232, and where adjustment or other attendance could be necessary during the operation of the apparatus, suitable insulating mats or platforms shall be provided so that the attendant cannot readily touch live parts unless standing on the mats or platforms.

Informational Note: See 110.26 and 110.34 for information on working space.

Part XIII. Grounding — All Voltages

430.241 General. Part XIII specifies the grounding of exposed non–current-carrying metal parts, likely to become energized, of motor and motor controller frames to limit voltage to ground in the event of accidental contact between energized parts and frames. Insulation, isolation, or guarding are suitable alternatives to grounding of motors under certain conditions.

430.242 Stationary Motors. The frames of stationary motors shall be grounded under any of the following conditions:

(1) Where supplied by metal-enclosed wiring
(2) Where in a wet location and not isolated or guarded
(3) If in a hazardous (classified) location
(4) If the motor operates with any terminal at over 150 volts to ground

Where the frame of the motor is not grounded, it shall be permanently and effectively insulated from the ground.

430.243 Portable Motors. The frames of portable motors that operate over 150 volts to ground shall be guarded or grounded.

> Informational Note No. 1: See 250.114(4) for grounding of portable appliances in other than residential occupancies.
> Informational Note No. 2: See 250.119(D) for color of equipment grounding conductor in flexible cords.

> *Exception No. 1: Listed motor-operated tools, listed motor-operated appliances, and listed motor-operated equipment shall not be required to be grounded where protected by a system of double insulation or its equivalent. Double-insulated equipment shall be distinctively marked.*

> *Exception No. 2: Listed motor-operated tools, listed motor-operated appliances, and listed motor-operated equipment connected by a cord and attachment plug other than those required to be grounded in accordance with 250.114.*

430.244 Motor Controllers. Motor controller enclosures shall be connected to the equipment grounding conductor regardless of voltage. Motor controller enclosures shall have means for attachment of an equipment grounding conductor termination in accordance with 250.8.

> *Exception: Enclosures attached to ungrounded portable equipment shall not be required to be grounded.*

430.245 Method of Grounding. Connection to the equipment grounding conductor shall be done in accordance with Part VI of Article 250.

Most motors are subject to vibration. This could require that the wiring to motors that are fixed be installed with a short section of liquidtight flexible metal conduit, liquidtight flexible nonmetallic conduit, or flexible metal conduit to the motor terminal housing to minimize the impact of the vibration. Under these conditions of use, a wire-type equipment grounding conductor (EGC) must be installed (see 250.118).

(A) Grounding Through Terminal Housings. Where the wiring to motors is metal-enclosed cable or in metal raceways, junction boxes to house motor terminals shall be provided, and the armor of the cable or the metal raceways shall be connected to them in accordance with 250.96(A) and 250.97.

Δ **(B) Separation of Junction Box from Motor.** The junction box required by 430.245(A) shall be permitted to be separated from the motor by not more than 1.8 m (6 ft) if the leads to the motor are stranded conductors within Type AC cable, interlocked metal tape Type MC cable where listed and identified in accordance with 250.118(A)(10)b., or armored cord or are stranded leads enclosed in liquidtight flexible metal conduit, flexible metal conduit, intermediate metal conduit, rigid metal conduit, or electrical metallic tubing not smaller than metric designator 12 (trade size ⅜), with the armor or raceway being connected both to the motor and to the box.

Liquidtight flexible nonmetallic conduit and rigid nonmetallic conduit shall be permitted to enclose the leads to the motor if the leads are stranded and the required equipment grounding conductor is connected to both the motor and to the box.

Where stranded leads are used, protected as specified above, each strand within the conductor shall be not larger than 10 AWG and shall comply with other requirements of this *Code* for conductors to be used in raceways.

(C) Grounding of Motor Controller-Mounted Devices. Instrument transformer secondaries and exposed non–current-carrying metal or other conductive parts or cases of instrument transformers, meters, instruments, and relays shall be grounded in accordance with 250.170 through 250.178.

Δ **Part XIV. Tables**

Tables 430.248 through 430.250 reflect the typical and most used 4- and 2-pole induction motors in use.

TABLE 430.247 *Full-Load Current in Amperes, Direct-Current Motors*

The following values of full-load currents* are for motors running at base speed.

| Horsepower | Armature Voltage Rating* | | | | | |
	90 Volts	120 Volts	180 Volts	240 Volts	500 Volts	550 Volts
¼	4.0	3.1	2.0	1.6	—	—
⅓	5.2	4.1	2.6	2.0	—	—
½	6.8	5.4	3.4	2.7	—	—
¾	9.6	7.6	4.8	3.8	—	—
1	12.2	9.5	6.1	4.7	—	—
1½	—	13.2	8.3	6.6	—	—
2	—	17	10.8	8.5	—	—
3	—	25	16	12.2	—	—
5	—	40	27	20	—	—
7½	—	58	—	29	13.6	12.2
10	—	76	—	38	18	16
15	—	—	—	55	27	24
20	—	—	—	72	34	31
25	—	—	—	89	43	38
30	—	—	—	106	51	46
40	—	—	—	140	67	61
50	—	—	—	173	83	75
60	—	—	—	206	99	90
75	—	—	—	255	123	111
100	—	—	—	341	164	148
125	—	—	—	425	205	185
150	—	—	—	506	246	222
200	—	—	—	675	330	294

*These are average dc quantities.

TABLE 430.248 *Full-Load Currents in Amperes, Single-Phase Alternating-Current Motors*

The following values of full-load currents are for motors running at usual speeds and motors with normal torque characteristics. The voltages listed are rated motor voltages. The currents listed shall be permitted for system voltage ranges of 110 to 120 and 220 to 240 volts.

Horsepower	115 Volts	200 Volts	208 Volts	230 Volts
⅙	4.4	2.5	2.4	2.2
¼	5.8	3.3	3.2	2.9
⅓	7.2	4.1	4.0	3.6
½	9.8	5.6	5.4	4.9
¾	13.8	7.9	7.6	6.9
1	16	9.2	8.8	8.0
1½	20	11.5	11.0	10
2	24	13.8	13.2	12
3	34	19.6	18.7	17
5	56	32.2	30.8	28
7½	80	46.0	44.0	40
10	100	57.5	55.0	50

Δ **TABLE 430.249** *Full-Load Current, Two-Phase Alternating-Current Motors (4-Wire)*

The following values of full-load current are for motors running at speeds usual for belted motors and motors with normal torque characteristics. Current in the common conductor of a 2-phase, 3-wire system will be 1.41 times the value given. The voltages listed are rated motor voltages. The currents listed shall be permitted for system voltage ranges of 110 to 120, 220 to 240, 440 to 480, 550 to 600, and 2300 to 2400 volts.

| Horsepower | Induction-Type Squirrel Cage and Wound Rotor (Amperes) | | | | |
	115 Volts	230 Volts	460 Volts	575 Volts	2300 Volts
½	4.0	2.0	1.0	0.8	—
¾	4.8	2.4	1.2	1.0	—
1	6.4	3.2	1.6	1.3	—
1½	9.0	4.5	2.3	1.8	—
2	11.8	5.9	3.0	2.4	—
3	—	8.3	4.2	3.3	—
5	—	13.2	6.6	5.3	—
7½	—	19	9.0	8.0	—
10	—	24	12	10	—
15	—	36	18	14	—
20	—	47	23	19	—
25	—	59	29	24	—
30	—	69	35	28	—
40	—	90	45	36	—
50	—	113	56	45	—
60	—	133	67	53	14
75	—	166	83	66	18
100	—	218	109	87	23
125	—	270	135	108	28
150	—	312	156	125	32
200	—	416	208	167	43

Δ **TABLE 430.250** *Full-Load Current, Three-Phase Alternating-Current Motors*

The following values of full-load currents are typical for motors running at speeds usual for belted motors and motors with normal torque characteristics. The voltages listed are rated motor voltages. The currents listed shall be permitted for system voltage ranges of 110 to 120, 220 to 240, 440 to 480, 550 to 600, and 2300 to 2400 volts.

	Induction-Type Squirrel Cage and Wound Rotor (Amperes)							Synchronous-Type Unity Power Factor* (Amperes)			
Horsepower	115 Volts	200 Volts	208 Volts	230 Volts	460 Volts	575 Volts	2300 Volts	230 Volts	460 Volts	575 Volts	2300 Volts
½	4.4	2.5	2.4	2.2	1.1	0.9	—	—	—	—	—
¾	6.4	3.7	3.5	3.2	1.6	1.3	—	—	—	—	—
1	8.4	4.8	4.6	4.2	2.1	1.7	—	—	—	—	—
1½	12.0	6.9	6.6	6.0	3.0	2.4	—	—	—	—	—
2	13.6	7.8	7.5	6.8	3.4	2.7	—	—	—	—	—
3	—	11.0	10.6	9.6	4.8	3.9	—	—	—	—	—
5	—	17.5	16.7	15.2	7.6	6.1	—	—	—	—	—
7½	—	25.3	24.2	22	11	9	—	—	—	—	—
10	—	32.2	30.8	28	14	11	—	—	—	—	—
15	—	48.3	46.2	42	21	17	—	—	—	—	—
20	—	62.1	59.4	54	27	22	—	—	—	—	—
25	—	78.2	74.8	68	34	27	—	53	26	21	—
30	—	92	88	80	40	32	—	63	32	26	—
40	—	120	114	104	52	41	—	83	41	33	—
50	—	150	143	130	65	52	—	104	52	42	—
60	—	177	169	154	77	62	16	123	61	49	12
75	—	221	211	192	96	77	20	155	78	62	15
100	—	285	273	248	124	99	26	202	101	81	20
125	—	359	343	312	156	125	31	253	126	101	25
150	—	414	396	360	180	144	37	302	151	121	30
200		552	528	480	240	192	49	400	201	161	40
250	—	—	—	—	302	242	60	—	—	—	—
300	—	—	—	—	361	289	72	—	—	—	—
350	—	—	—	—	414	336	83	—	—	—	—
400	—	—	—	—	477	382	95	—	—	—	—
450	—	—	—	—	515	412	103	—	—	—	—
500	—	—	—	—	590	472	118	—	—	—	—

*For 90 and 80 percent power factor, the figures shall be multiplied by 1.1 and 1.25, respectively.

Δ **TABLE 430.251(A)** *Conversion Table of Single-Phase Locked-Rotor Currents for Selection of Disconnecting Means and Controllers as Determined from Horsepower and Voltage Rating*

For use only with 430.110, 440.12, 440.41, and 455.8(C).

	Maximum Locked-Rotor Current in Amperes, Single-Phase		
Rated Horsepower	115 Volts	208 Volts	230 Volts
½	58.8	32.5	29.4
¾	82.8	45.8	41.4
1	96	53	48
1½	120	66	60
2	144	80	72
3	204	113	102
5	336	186	168
7½	480	265	240
10	1000	332	300

TABLE 430.251(B) *Conversion Table of Polyphase Design B, C, and D Maximum Locked-Rotor Currents for Selection of Disconnecting Means and Controllers as Determined from Horsepower and Voltage Rating and Design Letter*

For use only with 430.110, 440.12, 440.41, and 455.8(C).

	Maximum Motor Locked-Rotor Current in Amperes, Two- and Three-Phase, Design B, C, and D*					
	115 Volts	**200 Volts**	**208 Volts**	**230 Volts**	**460 Volts**	**575 Volts**
Rated Horsepower	**B, C, D**	**B, C, D**	**B, C, D**	**B, C, D**	**B, C, D**	**B, C, D**
½	40	23	22.1	20	10	8
¾	50	28.8	27.6	25	12.5	10
1	60	34.5	33	30	15	12
1½	80	46	44	40	20	16
2	100	57.5	55	50	25	20
3	—	73.6	71	64	32	25.6
5	—	105.8	102	92	46	36.8
7½	—	146	140	127	63.5	50.8
10	—	186.3	179	162	81	64.8
15	—	267	257	232	116	93
20	—	334	321	290	145	116
25	—	420	404	365	183	146
30	—	500	481	435	218	174
40	—	667	641	580	290	232
50	—	834	802	725	363	290
60	—	1001	962	870	435	348
75	—	1248	1200	1085	543	434
100	—	1668	1603	1450	725	580
125	—	2087	2007	1815	908	726
150	—	2496	2400	2170	1085	868
200	—	3335	3207	2900	1450	1160
250	—	—	—	—	1825	1460
300	—	—	—	—	2200	1760
350	—	—	—	—	2550	2040
400	—	—	—	—	2900	2320
450	—	—	—	—	3250	2600
500	—	—	—	—	3625	2900

*Design A motors are not limited to a maximum starting current or locked rotor current.

ARTICLE 440 Air-Conditioning and Refrigerating Equipment

Part I. General

440.1 Scope. This article applies to electric motor-driven air-conditioning and refrigerating equipment and to the branch circuits and controllers for such equipment. It provides for the special considerations necessary for circuits supplying hermetic refrigerant motor-compressors and for any air-conditioning or refrigerating equipment that is supplied from a branch circuit that supplies a hermetic refrigerant motor-compressor.

• Article 440 provides special considerations necessary for the branch circuits supplying hermetic refrigerant motor-compressors and is in addition to or amendatory of the requirements of Article 430 and other applicable articles. However, many requirements for disconnecting means, controllers, single or group

installations, and sizing of conductors, for example, are the same as or similar to those applied in Article 430. The rules for sizing conductors and overcurrent protection of feeder conductors supplying multiple air-conditioning or refrigeration units employing hermetic refrigerant motor-compressors are the same rules from Parts II and V in Article 430 that are used for sizing conductors and overcurrent protection for feeder circuits supplying multiple motors.

440.4 Marking on Hermetic Refrigerant Motor-Compressors and Equipment.

(A) Hermetic Refrigerant Motor-Compressor Nameplate. A hermetic refrigerant motor-compressor shall be provided with a nameplate that shall indicate the manufacturer's name, trademark, or symbol; identifying designation; phase; voltage; and frequency. The rated-load current in amperes of the motor-compressor shall be marked by the equipment manufacturer on either or both the motor-compressor nameplate and the nameplate of the equipment in which the motor-compressor is used. The locked-rotor current of each single-phase motor-compressor having a rated-load current of more than 9 amperes at 115 volts, or more than 4.5 amperes at 230 volts, and each polyphase motor-compressor shall be marked on the motor-compressor nameplate. Where a thermal protector complying with 440.52(A)(2) and (B)(2) is used, the motor-compressor nameplate or the equipment nameplate shall be marked with the words "thermally protected." Where a protective system complying with 440.52(A)(4) and (B)(4) is used and is furnished with the equipment, the equipment nameplate shall be marked with the words, "thermally protected system." Where a protective system complying with 440.52(A)(4) and (B)(4) is specified, the equipment nameplate shall be appropriately marked.

(B) Multimotor and Combination-Load Equipment. Multimotor and combination-load equipment shall be provided with a visible nameplate marked with the maker's name, the rating in volts, frequency and number of phases, minimum supply circuit conductor ampacity, the maximum rating of the branch-circuit short-circuit and ground-fault protective device, and the short-circuit current rating of the motor controllers or industrial control panel. The ampacity shall be calculated by using Part IV and counting all the motors and other loads that will be operated at the same time. The branch-circuit short-circuit and ground-fault protective device rating shall not exceed the value calculated by using Part III. Multimotor or combination-load equipment for use on two or more circuits shall be marked with the above information for each circuit.

Exception No. 1: Multimotor and combination-load equipment that is suitable under the provisions of this article for connection to a single 15- or 20-ampere, 120-volt, or a 15-ampere, 208- or 240-volt, single-phase branch circuit shall be permitted to be marked as a single load.

Exception No. 2: The minimum supply circuit conductor ampacity and the maximum rating of the branch-circuit short-circuit and ground-fault protective device shall not be required to be marked on a room air conditioner complying with 440.62(A).

Exception No. 3: Multimotor and combination-load equipment used in one- and two-family dwellings or cord-and-attachment-plug-connected equipment shall not be required to be marked with a short-circuit current rating.

Motor controllers or control panels associated with multimotor and combination-load air-conditioning and refrigerating equipment are required by 440.4(B) to be marked with the short-circuit current rating. Where air-conditioning equipment is installed at large commercial, institutional, and industrial complexes, the controllers and control panels of air-conditioning and refrigerating equipment are often supplied from a point on the electrical distribution system where significant short-circuit current is available.

As is the case with any electrical installation where high levels of short-circuit current are available, the short-circuit current rating marked on the air-conditioning equipment controllers and control panels provides those responsible for designing and approving the electrical installation with the necessary information to ensure compliance with the requirements of 110.9 and 110.10. Multimotor and combination-load air-conditioning equipment used in one- and two-family dwellings and cord-and-attachment-plug-connected air-conditioning equipment are not required to be marked with their short-circuit current ratings.

See also

430.8 for a similar requirement for marking the short-circuit current rating on motor controllers

(C) Branch-Circuit Selection Current. A hermetic refrigerant motor-compressor, or equipment containing such a compressor, having a protection system that is approved for use with the motor-compressor that it protects and that permits continuous current in excess of the specified percentage of nameplate rated-load current given in 440.52(B)(2) or (B)(4) shall also be marked with a branch-circuit selection current that complies with 440.52(B)(2) or (B)(4). This marking shall be provided by the equipment manufacturer and shall be on the nameplate(s) where the rated-load current(s) appears.

440.5 Marking on Controllers. A controller shall be marked with the manufacturer's name, trademark, or symbol; identifying designation; voltage; phase; full-load and locked-rotor current (or horsepower) rating; and other data as may be needed to properly indicate the motor-compressor for which it is suitable.

440.6 Ampacity and Rating. The size of conductors for equipment covered by this article shall be selected from Table 310.16 through Table 310.19 or calculated in accordance with 310.14 as applicable. The required ampacity of conductors and rating of equipment shall be determined according to 440.6(A) and 440.6(B).

(A) Hermetic Refrigerant Motor-Compressor. For a hermetic refrigerant motor-compressor, the rated-load current marked on the nameplate of the equipment in which the motor-compressor is employed shall be used in determining the rating or ampacity of the disconnecting means, the branch-circuit conductors, the controller, the branch-circuit short-circuit and ground-fault protection, and the separate motor overload protection. Where no rated-load current is shown on the equipment nameplate, the rated-load current shown on the compressor nameplate shall be used.

Exception No. 1: Where so marked, the branch-circuit selection current shall be used instead of the rated-load current to determine the rating or ampacity of the disconnecting means, the branch-circuit conductors, the controller, and the branch-circuit short-circuit and ground-fault protection.

Exception No. 2: For cord-and-plug-connected equipment, the nameplate marking shall be used in accordance with 440.22(B), Exception No. 2.

(B) Multimotor Equipment. For multimotor equipment employing a shaded-pole or permanent split-capacitor-type fan or blower motor, the full-load current for such motor marked on the nameplate of the equipment in which the fan or blower motor is employed shall be used instead of the horsepower rating to determine the ampacity or rating of the disconnecting means, the branch-circuit conductors, the controller, the branch-circuit short-circuit and ground-fault protection, and the separate overload protection. This marking on the equipment nameplate shall not be less than the current marked on the fan or blower motor nameplate.

440.7 Highest Rated (Largest) Motor. In determining compliance with this article and with 430.24, 430.53(B) and 430.53(C), and 430.62(A), the highest rated (largest) motor shall be considered to be the motor that has the highest rated-load current. Where two or more motors have the same highest rated-load current, only one of them shall be considered as the highest rated (largest) motor. For other than hermetic refrigerant motor-compressors, and fan or blower motors as covered in 440.6(B), the full-load current used to determine the highest rated motor shall be the equivalent value corresponding to the motor horsepower rating selected from Table 430.248, Table 430.249, or Table 430.250.

Exception: Where so marked, the branch-circuit selection current shall be used instead of the rated-load current in determining the highest rated (largest) motor-compressor.

440.8 Single Machine and Location. An air-conditioning or refrigerating system shall be considered to be a single machine under the provisions of 430.87, Exception No. 1, and 430.112, Exception. The motors shall be permitted to be located remotely from each other. Air-conditioning and refrigeration equipment shall not be installed within a zone measured 900 mm (3 ft) horizontally and 2.5 m (8 ft) vertically from the top of a bathtub rim or shower stall threshold. The zone shall be all-encompassing and include the space directly over the tub or shower stall.

440.9 Grounding and Bonding. Where equipment is installed outdoors on a roof, an equipment grounding conductor of the wire type shall be installed in outdoor portions of metallic raceway systems that use compression-type fittings.

Metal raceway systems with compression fittings listed for use in wet locations that are installed on rooftops to supply air-conditioning and refrigeration equipment could be subject to movement and physical damage. This is a result of rooftop activities such as snow removal or roof repair/replacement. These activities can lead to inadvertent separation of the raceway system, compromising the ground-fault return path. The installation of a wire-type equipment grounding conductor provides redundancy of the ground-fault return path to facilitate the operation of the circuit overcurrent protective device (OCPD) in the event of a line-to-ground fault at the air-conditioning or refrigeration equipment.

440.10 Short-Circuit Current Rating.

(A) Installation. Motor controllers or industrial control panels of multimotor and combination-load equipment shall not be installed where the available fault current exceeds its short-circuit current rating as marked in accordance with 440.4(B).

(B) Documentation. When motor controllers or industrial control panels of multimotor and combination-load equipment are required to be marked with a short circuit current rating, the available fault current and the date the available fault current calculation was performed shall be documented and made available to those authorized to inspect, install, or maintain the installation.

Part II. Disconnecting Means

440.11 General. Disconnecting means shall be capable of disconnecting air-conditioning and refrigerating equipment, including motor-compressors and controllers, from the circuit conductors. If the disconnecting means is readily accessible to unqualified persons, any enclosure door or hinged cover of a disconnecting means enclosure that exposes energized parts when open shall require a tool to open or be capable of being locked.

440.12 Rating and Interrupting Capacity.

(A) Hermetic Refrigerant Motor-Compressor. A disconnecting means serving a hermetic refrigerant motor-compressor shall be selected on the basis of the nameplate rated-load current or branch-circuit selection current, whichever is greater, and locked-rotor current, respectively, of the motor-compressor as follows.

(1) Ampere Rating. The ampere rating shall be at least 115 percent of the nameplate rated-load current or branch-circuit selection current, whichever is greater.

Exception: A listed unfused motor circuit switch, without fuseholders, having a horsepower rating not less than the equivalent horsepower determined in accordance with 440.12(A)(2) shall be permitted to have an ampere rating less than 115 percent of the specified current.

(2) Equivalent Horsepower. To determine the equivalent horsepower in complying with the requirements of 430.109, the horsepower rating shall be selected from Table 430.248, Table 430.249, or Table 430.250 corresponding to the rated-load current or branch-circuit selection current, whichever is greater, and also the horsepower rating from Table 430.251(A) or Table 430.251(B) corresponding to the locked-rotor current. In case the nameplate rated-load current or branch-circuit selection current and locked-rotor current do not correspond to the currents shown in Table 430.248, Table 430.249, Table 430.250, Table 430.251(A), or Table 430.251(B), the horsepower rating corresponding to the next higher value shall be selected. In case different horsepower ratings are obtained when applying these tables, a horsepower rating at least equal to the larger of the values obtained shall be selected.

(B) Combination Loads. Where the combined load of two or more hermetic refrigerant motor-compressors or one or more hermetic refrigerant motor-compressor with other motors or loads may be simultaneous on a single disconnecting means, the rating for the disconnecting means shall be determined in accordance with 440.12(B)(1) and (B)(2).

(1) Horsepower Rating. The horsepower rating of the disconnecting means shall be determined from the sum of all currents, including resistance loads, at the rated-load condition and also at the locked-rotor condition. The combined rated-load current and the combined locked-rotor current so obtained shall be considered as a single motor for the purpose of this requirement as required by 440.12(B)(1)(a) and (B)(1)(b).

(a) The full-load current equivalent to the horsepower rating of each motor, other than a hermetic refrigerant motor-compressor, and fan or blower motors as covered in 440.6(B) shall be selected from Table 430.248, Table 430.249, or Table 430.250. These full-load currents shall be added to the motor-compressor rated-load current(s) or branch-circuit selection current(s), whichever is greater, and to the rating in amperes of other loads to obtain an equivalent full-load current for the combined load.

(b) The locked-rotor current equivalent to the horsepower rating of each motor, other than a hermetic refrigerant motor-compressor, shall be selected from Table 430.251(A) or Table 430.251(B), and, for fan and blower motors of the shaded-pole or permanent split-capacitor type marked with the locked-rotor current, the marked value shall be used. The locked-rotor currents shall be added to the motor-compressor locked-rotor current(s) and to the rating in amperes of other loads to obtain an equivalent locked-rotor current for the combined load. Where two or more motors or other loads such as resistance heaters, or both, cannot be started simultaneously, appropriate combinations of locked-rotor and rated-load current or branch-circuit selection current, whichever is greater, shall be an acceptable means of determining the equivalent locked-rotor current for the simultaneous combined load.

Exception: Where part of the concurrent load is a resistance load and the disconnecting means is a switch rated in horsepower and amperes, the switch used shall be permitted to have a horsepower rating not less than the combined load to the motor-compressor(s) and other motor(s) at the locked-rotor condition, if the ampere rating of the switch is not less than this locked-rotor load plus the resistance load.

(2) Full-Load Current Equivalent. The ampere rating of the disconnecting means shall be at least 115 percent of the sum of all currents at the rated-load condition determined in accordance with 440.12(B)(1).

Exception: A listed unfused motor circuit switch, without fuseholders, having a horsepower rating not less than the equivalent horsepower determined by 440.12(B)(1) shall be permitted to have an ampere rating less than 115 percent of the sum of all currents.

(C) Small Motor-Compressors. For small motor-compressors not having the locked-rotor current marked on the nameplate, or for small motors not covered by Table 430.247, Table 430.248, Table 430.249, or Table 430.250, the locked-rotor current shall be assumed to be six times the rated-load current.

(D) Disconnecting Means. Every disconnecting means in the refrigerant motor-compressor circuit between the point of attachment to the feeder and the point of connection to the refrigerant motor-compressor shall comply with the requirements of 440.12.

(E) Disconnecting Means Rated in Excess of 100 Horsepower. Where the rated-load or locked-rotor current as determined above would indicate a disconnecting means rated in excess of 100 hp, 430.109(E) shall apply.

Δ **440.13 Cord-Connected Equipment.** For cord-connected equipment such as room air conditioners, household refrigerators and freezers, drinking water coolers, and beverage dispensers, a separable connector or an attachment plug and receptacle shall be permitted to serve as the disconnecting means.

Informational Note: See 440.63 for room air conditioners.

Δ **440.14 Location.** Disconnecting means shall be located within sight from, and readily accessible from, the air-conditioning or refrigerating equipment. The disconnecting means shall be permitted to be installed on or within the air-conditioning or refrigerating equipment. Disconnecting means shall meet the working space requirements of 110.26(A).

The disconnecting means shall not be located on panels that are designed to allow access to the air-conditioning or refrigeration equipment or where it obscures the equipment nameplate(s).

Exception No. 1: Where the disconnecting means provided in accordance with 430.102(A) is lockable in accordance with 110.25 and the refrigerating or air-conditioning equipment is essential to an industrial process in a facility with written safety procedures, and where the conditions of maintenance and supervision ensure that only qualified persons service the equipment, a disconnecting means within sight from the equipment shall not be required.

Exception No. 1 accommodates special conditions associated with process refrigerating equipment. Typically, this equipment is very large, so rated disconnects might not be available. Additionally, this equipment could be in a hazardous (classified) location, and locating the disconnecting means within sight of the motor could introduce additional hazards. The means to lock the disconnecting means in the open position must remain in place with or without the lock installed. An example of locking hardware meeting this requirement is shown in Exhibit 440.1.

Exception No. 2: Where an attachment plug and receptacle serve as the disconnecting means in accordance with 440.13, their location shall be accessible but shall not be required to be readily accessible.

EXHIBIT 440.1 An example of locking hardware (with lock installed) that remains in place with or without the lock installed. *(Courtesy of Schneider Electric)*

Informational Note: See Parts VII and IX of Article 430 for additional requirements.

The references to Parts VII and IX of Article 430 in Informational Note No. 1 are intended to call attention to the additional disconnect location requirements in 430.102, 430.107, and 430.113. The requirement of 440.14 mandates that the equipment disconnecting means be within sight from and readily accessible from the equipment, even if a remote disconnect is capable of being locked in the "open" position, in accordance with the exception to 430.102(B)(2).

This special requirement for air-conditioning and refrigerating equipment covered by Article 440 is more stringent than the requirements in Article 430. It provides protection for service personnel working on equipment located in attics, on roofs, or outside in a remote location where it is difficult to gain access to a remote lockable disconnect.

See also

440.14, Exception No. 1, for conditions where a disconnecting means within sight from the equipment is not required

Part III. Branch-Circuit Short-Circuit and Ground-Fault Protection

440.21 General. Part III specifies devices intended to protect the branch-circuit conductors, control apparatus, and motors in circuits supplying hermetic refrigerant motor-compressors against overcurrent due to short circuits and ground faults. They are in addition to or amendatory of the overcurrent protection requirements found elsewhere in this *Code*.

Where an air conditioner is listed by a qualified electrical testing laboratory with a nameplate that reads "maximum fuse size," the listing restricts the use of the unit to fuse protection only and does not cover its use with circuit breakers. If the air conditioner has been evaluated for both fuses and circuit breakers, it is marked to indicate that both types of protective devices are

acceptable. Molded-case circuit breakers, evaluated to UL 489, *Molded-Case Circuit Breakers, Molded-Case Switches and Circuit-Breaker Enclosures*, are certified for group motor protection in accordance with 430.53. The equipment covered by Article 440 quite often consists of hermetic refrigerant motors plus other types of motors, equipment, and controls. If supplemental overcurrent protection is provided within the heating, air-conditioning, and refrigeration (HACR) equipment control panel, the manufacturer will have marked the equipment with specific information on the type of branch-circuit overcurrent protective device that can be used on the line side of the HACR unit in order to provide the necessary level of short-circuit and ground-fault protection for the equipment and controls internal to the HACR unit.

Section 110.3(B) requires listed equipment to be installed and used in accordance with instructions included in the listing. In the case of air-conditioning equipment, it is important to carefully read the nameplate so that the correct type of short-circuit, ground-fault protective device is selected.

The acronym "HACR" as a prefix to circuit breaker may still be found on the nameplates of legacy HACR equipment. This abbreviation indicated that the HACR equipment manufacturer wanted to specify that if circuit breakers were used, only those evaluated for group motor applications were suitable for the branch-circuit short-circuit, and ground-fault protection. However, because UL 489 required all circuit breakers to be evaluated for protection of group installations, the HACR marking on circuit breakers was not required by the product standard. However, because legacy HACR equipment still includes nameplates specifying the specially marked circuit breakers, some circuit breaker manufacturers continue to mark their molded case circuit breakers with "HACR Type," even though it is not required by the product certification standard.

Exhibit 440.2 illustrates three supply circuit configurations in which fuses can be used to protect HACR equipment if such protection is specified on the equipment nameplate.

Current-limiting overcurrent devices, which may reduce the amount of fault current to which the equipment is subjected, can be installed in the branch circuit supplying the equipment.

See also

Article 100 for the definition of *overcurrent protective device, current-limiting (current-limiting overcurrent protective device)* and its commentary explaining short-circuit damage

110.3(B) and **110.10** (and commentary) for the installation and use of listed or labeled equipment and the selection of OCPDs (such as fuses and circuit breakers)

440.22 Application and Selection.

Δ **(A) Rating or Setting for Individual Motor-Compressor.** The motor-compressor branch-circuit short-circuit and ground-fault protective device shall be capable of carrying the starting current of the motor. A protective device having a rating or setting not exceeding 175 percent of the motor-compressor rated-load current or branch-circuit selection current, whichever is greater, shall be permitted.

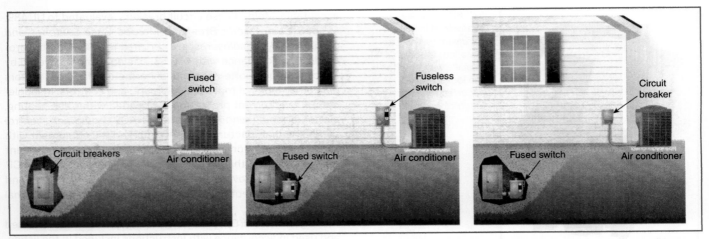

EXHIBIT 440.2 Three correct alternate wiring configurations satisfying the restriction that the equipment be protected by fuses only. (Note that the fuse rating cannot exceed the maximum fuse size specified on the air-conditioner nameplate.)

Exception No. 1: If the values for branch-circuit short-circuit and ground-fault protection in accordance with 440.22(A) do not correspond to the standard sizes or ratings of fuses, nonadjustable circuit breakers, thermal protective devices, or available settings of adjustable circuit breakers, a higher size, rating, or available setting that does not exceed the next higher standard ampere rating shall be permitted.

Exception No. 2: If the values for branch-circuit short-circuit and ground-fault protection in accordance with 440.22(A) or the rating modified by Exception No. 1 is not sufficient for the starting current of the motor, the rating or setting shall be permitted to be increased but shall not exceed 225 percent of the motor rated-load current or branch-circuit selection current, whichever is greater.

Exception No. 3: The rating of the branch-circuit short-circuit and ground-fault protective device shall not be required to be less than 15 amperes.

(B) Rating or Setting for Equipment. The equipment branch-circuit short-circuit and ground-fault protective device shall be capable of carrying the starting current of the equipment. Where the hermetic refrigerant motor-compressor is the only load on the circuit, the protection shall comply with 440.22(A). Where the equipment incorporates more than one hermetic refrigerant motor-compressor or a hermetic refrigerant motor-compressor and other motors or other loads, the equipment short-circuit and ground-fault protection shall comply with 430.53 and 440.22(B)(1) and (B)(2).

(1) Motor-Compressor Largest Load. Where a hermetic refrigerant motor-compressor is the largest load connected to the circuit, the rating or setting of the branch-circuit short-circuit and ground-fault protective device shall not exceed the value specified in 440.22(A) for the largest motor-compressor plus the sum of the rated-load current or branch-circuit selection current,

whichever is greater, of the other motor-compressor(s) and the ratings of the other loads supplied.

(2) Motor-Compressor Not Largest Load. Where a hermetic refrigerant motor-compressor is not the largest load connected to the circuit, the rating or setting of the branch-circuit short-circuit and ground-fault protective device shall not exceed a value equal to the sum of the rated-load current or branch-circuit selection current, whichever is greater, rating(s) for the motor-compressor(s) plus the value specified in 430.53(C)(4) where other motor loads are supplied, or the value specified in 240.4 where only nonmotor loads are supplied in addition to the motor-compressor(s).

Exception No. 1: Equipment that starts and operates on a 15- or 20-ampere 120-volt, or 15-ampere 208- or 240-volt single-phase branch circuit, shall be permitted to be protected by the 15- or 20-ampere overcurrent device protecting the branch circuit, but if the maximum branch-circuit short-circuit and ground-fault protective device rating marked on the equipment is less than these values, the circuit protective device shall not exceed the value marked on the equipment nameplate.

Exception No. 2: The nameplate marking of cord-and-plug-connected equipment rated not greater than 250 volts, single-phase, such as household refrigerators and freezers, drinking water coolers, and beverage dispensers, shall be used in determining the branch-circuit requirements, and each unit shall be considered as a single motor unless the nameplate is marked otherwise.

(C) Protective Device Rating Not to Exceed the Manufacturer's Values. Where maximum protective device ratings shown on a manufacturer's overload relay table for use with a motor controller are less than the rating or setting selected in accordance with 440.22(A) and (B), the protective device rating shall not exceed the manufacturer's values marked on the equipment.

Δ **Part IV. Circuit Conductors**

440.31 General. Part IV and adjustments made in accordance with Part III of Article 310 specify ampacities of conductors required to carry the motor current without overheating under the conditions specified, except as modified in 440.6(A), Exception No. 1.

These articles shall not apply to integral conductors of motors, to motor controllers and the like, or to conductors that form an integral part of approved equipment.

Δ **440.32 Single Motor-Compressor.** Branch-circuit conductors supplying a single motor-compressor shall have an ampacity not less than the greater of the following:

(1) 125 percent of the motor-compressor rated-load current
(2) 125 percent of the branch-circuit selection current

For a wye-start, delta-run connected motor-compressor, the selection of branch-circuit conductors between the motor controller and the motor-compressor shall be permitted to be based on 72 percent of either the motor-compressor rated-load current or the branch-circuit selection current, whichever is greater.

> Informational Note: The multiplier of 72 percent is obtained by multiplying 58 percent by 1.25 because the individual motor circuit conductors of wye-start, delta-run connected motor-compressors carry 58 percent of the rated-load current.

See also

430.22(C) and its commentary for more information on wye-start, delta-run motors

440.33 Motor-Compressor(s) With or Without Additional Motor Loads. Conductors supplying one or more motor-compressor(s) with or without an additional motor load(s) shall have an ampacity not less than the sum of each of the following:

(1) The sum of the rated-load or branch-circuit selection current, whichever is greater, of all motor-compressor(s)
(2) The sum of the full-load current rating of all other motors
(3) 25 percent of the highest motor-compressor or motor full load current in the group

Exception No. 1: Where the circuitry is interlocked so as to prevent the starting and running of a second motor-compressor or group of motor-compressors, the conductor size shall be determined from the largest motor-compressor or group of motor-compressors that is to be operated at a given time.

Exception No. 2: The branch-circuit conductors for room air conditioners shall be in accordance with Part VII of Article 440.

Branch circuits for listed air-conditioning and refrigerating equipment that have a nameplate marked with the branch-circuit conductor size and branch-circuit short-circuit protective device size are not required to have the branch-circuit conductors sized in accordance with 440.33. The product standard includes the 25-percent increase for the largest motor or compressor in the group plus the other nonmotor or noncompressor load; therefore, the actual nameplate full-load amperes for the complete assembly can be used to size the branch-circuit conductors.

Δ **440.34 Combination Load.** Conductors supplying a motor-compressor load(s) in addition to other load(s) shall have an ampacity sufficient for the other load(s) plus the required ampacity for the motor-compressor load(s). The motor compressor load(s) shall be determined in accordance with 440.32 or 440.33. The other load(s) shall be calculated from branch-circuit, feeder, and service load calculations.

Exception: Where the circuitry is interlocked to prevent simultaneous operation of the motor-compressor(s) and all other loads connected, the conductor size shall be determined from the largest size required for the motor-compressor(s) and other loads to be operated at a given time.

440.35 Multimotor and Combination-Load Equipment. The ampacity of the conductors supplying multimotor and combination-load equipment shall not be less than the minimum circuit ampacity marked on the equipment in accordance with 440.4(B).

Part V. Controllers for Motor-Compressors

440.41 Rating.

(A) Motor-Compressor Controller. A motor-compressor controller shall have both a continuous-duty full-load current rating and a locked-rotor current rating not less than the nameplate rated-load current or branch-circuit selection current, whichever is greater, and locked-rotor current, respectively, of the compressor. In case the motor controller is rated in horsepower but is without one or both of the foregoing current ratings, equivalent currents shall be determined from the ratings as follows. Table 430.248, Table 430.249, and Table 430.250 shall be used to determine the equivalent full-load current rating. Table 430.251(A) and Table 430.251(B) shall be used to determine the equivalent locked-rotor current ratings.

(B) Controller Serving More Than One Load. A controller serving more than one motor-compressor or a motor-compressor and other loads shall have a continuous-duty full-load current rating and a locked-rotor current rating not less than the combined load as determined in accordance with 440.12(B).

Part VI. Motor-Compressor and Branch-Circuit Overload Protection

440.51 General. Part VI specifies devices intended to protect the motor-compressor, the motor-control apparatus, and the branch-circuit conductors against excessive heating due to motor overload and failure to start.

Informational Note: See 240.4(G) for application of Parts III and VI of Article 440.

440.52 Application and Selection.

(A) Protection of Motor-Compressor. Each motor-compressor shall be protected against overload and failure to start by one of the following means:

(1) A separate overload relay that is responsive to motor-compressor current. This device shall be selected to trip at not more than 140 percent of the motor-compressor rated-load current.

(2) A thermal protector integral with the motor-compressor, approved for use with the motor-compressor that it protects on the basis that it will prevent dangerous overheating of the motor-compressor due to overload and failure to start. If the current-interrupting device is separate from the motor-compressor and its control circuit is operated by a protective device integral with the motor-compressor, it shall be arranged so that the opening of the control circuit will result in interruption of current to the motor-compressor.

(3) A fuse or inverse time circuit breaker responsive to motor current, which shall also be permitted to serve as the branch-circuit short-circuit and ground-fault protective device. This device shall be rated at not more than 125 percent of the motor-compressor rated-load current. It shall have sufficient time delay to permit the motor-compressor to start and accelerate its load. The equipment or the motor-compressor shall be marked with this maximum branch-circuit fuse or inverse time circuit breaker rating.

(4) A protective system, furnished or specified and approved for use with the motor-compressor that it protects on the basis that it will prevent dangerous overheating of the motor-compressor due to overload and failure to start. If the current-interrupting device is separate from the motor-compressor and its control circuit is operated by a protective device that is not integral with the current-interrupting device, it shall be arranged so that the opening of the control circuit will result in interruption of current to the motor-compressor.

(B) Protection of Motor-Compressor Control Apparatus and Branch-Circuit Conductors. The motor-compressor controller(s), the disconnecting means, and the branch-circuit conductors shall be protected against overcurrent due to motor overload and failure to start by one of the following means, which shall be permitted to be the same device or system protecting the motor-compressor in accordance with 440.52(A):

Exception: Overload protection of motor-compressors and equipment on 15- and 20-ampere, single-phase, branch circuits shall be permitted to be in accordance with 440.54 and 440.55.

(1) An overload relay selected in accordance with 440.52(A)(1)

(2) A thermal protector applied in accordance with 440.52(A)(2), that will not permit a continuous current in excess of 156 percent of the marked rated-load current or branch-circuit selection current

(3) A fuse or inverse time circuit breaker selected in accordance with 440.52(A)(3)

(4) A protective system, in accordance with 440.52(A)(4), that will not permit a continuous current in excess of 156 percent of the marked rated-load current or branch-circuit selection current

440.53 Overload Relays. Overload relays and other devices for motor overload protection that are not capable of opening short circuits shall be protected by fuses or inverse time circuit breakers with ratings or settings in accordance with Part III unless identified for group installation or for part-winding motors and marked to indicate the maximum size of fuse or inverse time circuit breaker by which they shall be protected.

Exception: The fuse or inverse time circuit breaker size marking shall be permitted on the nameplate of the equipment in which the overload relay or other overload device is used.

Δ **440.54 Motor-Compressors and Equipment on 15- or 20-Ampere Branch Circuits — Not Cord- and Attachment-Plug-Connected.** Overload protection for motor-compressors and equipment used on 15- or 20-ampere 120-volt, or 15-ampere 208- or 240-volt, single-phase branch circuits shall be permitted in accordance with 440.54(A) and (B).

(A) Overload Protection. The motor-compressor shall be provided with overload protection selected as specified in 440.52(A). Both the controller and motor overload protective device shall be identified for installation with the short-circuit and ground-fault protective device for the branch circuit to which the equipment is connected.

(B) Time Delay. The short-circuit and ground-fault protective device protecting the branch circuit shall have sufficient time delay to permit the motor-compressor and other motors to start and accelerate their loads.

Δ **440.55 Cord- and Attachment-Plug-Connected Motor-Compressors and Equipment on 15- or 20-Ampere Branch Circuits.** Overload protection for motor-compressors and equipment that are cord- and attachment-plug-connected and used on 15- or 20-ampere 120-volt, or 15-ampere 208- or 240-volt, single-phase branch circuits shall be permitted in accordance with 440.55(A), (B), and (C).

(A) Overload Protection. The motor-compressor shall be provided with overload protection as specified in 440.52(A). Both the controller and the motor overload protective device shall be identified for installation with the short-circuit and ground-fault protective device for the branch circuit to which the equipment is connected.

(B) Attachment Plug and Receptacle or Cord Connector Rating. The rating of the attachment plug and receptacle or cord connector shall not exceed 20 amperes at 125 volts or 15 amperes at 250 volts.

(C) Time Delay. The short-circuit and ground-fault protective device protecting the branch circuit shall have sufficient time delay to permit the motor-compressor and other motors to start and accelerate their loads.

△ Part VII. Room Air Conditioners

440.60 General. Part VII shall apply to electrically energized room air conditioners that control temperature and humidity. For the purpose of Part VII, a room air conditioner (with or without provisions for heating) shall be considered as an ac appliance of the air-cooled window, console, or in-wall type that is installed in the conditioned room and that incorporates a hermetic refrigerant motor-compressor(s). Part VII covers equipment rated not over 250 volts, single phase, and the equipment shall be permitted to be cord- and attachment-plug-connected.

A room air conditioner that is rated 3-phase or rated over 250 volts shall be directly connected to a wiring method recognized in Chapter 3, and Part VII shall not apply.

440.61 Grounding. The enclosures of room air conditioners shall be connected to the equipment grounding conductor in accordance with 250.110, 250.112, and 250.114.

440.62 Branch-Circuit Requirements.

(A) Room Air Conditioner as a Single Motor Unit. A room air conditioner shall be considered as a single motor unit in determining its branch-circuit requirements where all the following conditions are met:

(1) It is cord- and attachment-plug-connected.
(2) Its rating is not more than 40 amperes and 250 volts, single phase.
(3) Total rated-load current is shown on the room air-conditioner nameplate rather than individual motor currents.
(4) The rating of the branch-circuit short-circuit and ground-fault protective device does not exceed the ampacity of the branch-circuit conductors or the rating of the receptacle, whichever is less.

(B) Where No Other Loads Are Supplied. The total marked current rating of a cord- and attachment-plug-connected room air conditioner shall not exceed 80 percent of the current rating of a branch circuit where no other loads are supplied.

(C) Where Lighting Units or Other Appliances Are Also Supplied. The total marked current rating of a cord- and attachment-plug-connected room air conditioner shall not exceed 50 percent of the current rating of a branch circuit where lighting outlets, other appliances, or general-use receptacles are also supplied. Where the circuitry is interlocked to prevent simultaneous operation of the room air conditioner and energization of other outlets on the same branch circuit, a cord- and attachment-plug-connected room air conditioner shall not exceed 80 percent of the branch-circuit current rating.

440.63 Disconnecting Means. An attachment plug and receptacle or cord connector shall be permitted to serve as the disconnecting means for a single-phase room air conditioner rated 250 volts or less if (1) the manual controls on the room air conditioner are readily accessible and located within 1.8 m (6 ft) of the floor, or (2) an approved manually operable disconnecting means is installed in a readily accessible location within sight from the room air conditioner.

440.64 Supply Cords. Where a flexible cord is used to supply a room air conditioner, the length of such cord shall not exceed 3.0 m (10 ft) for a nominal, 120-volt rating or 1.8 m (6 ft) for a nominal, 208- or 240-volt rating.

440.65 Protection Devices. Single-phase cord- and plug-connected room air conditioners shall be provided with one of the following factory-installed devices:

(1) Leakage-current detector-interrupter (LCDI)
(2) Arc-fault circuit interrupter (AFCI)
(3) Heat detecting circuit interrupter (HDCI)

The protection device shall be an integral part of the attachment plug or be located in the power supply cord within 300 mm (12 in.) of the attachment plug.

In those areas of the country where air conditioning is not needed year round, room air conditioners are installed seasonally and stored during the heating season. Damage to the supply cords of room air conditioners can occur if care is not taken when the unit is removed and stored. The supply cords of stored units sometimes inadvertently end up under the unit with pressure from the entire weight of the unit being exerted on the cord. Damaged cords have been the cause of numerous fires around the country, thus the need to monitor the integrity of the cord.

Additionally, overheated compressors in room air-conditioning units can result in a dangerous rise in the internal temperature of the equipment, damaging the compressor and other internal components. A heat detecting circuit interrupter (HDCI) incorporates all the protection functions of a leakage-current detector-interrupter (LCDI) and also includes a thermal detecting function to provide protection against overheating of a room air-conditioner compressor.

ARTICLE
445   Generators

445.1 Scope. This article contains installation and other requirements for generators.

Article 445 covers the installation of generators. The following articles cover the use of generators in specific applications. Exhibit 445.1 shows an example of a diesel engine–driven generator.

See also

Article 695 for fire pumps

Article 700 for emergency systems

Article 701 for legally required standby systems

Article 702 for optional standby systems

Article 705 for interconnected electric power production sources

Article 708 for critical operations power systems

Δ **445.6 Listing.** Stationary generators shall be listed.

Exception: One of a kind or custom manufactured generators shall be permitted to be field labeled.

Informational Note: See UL 2200, *Standard for Stationary Engine Generator Assemblies*, for additional information.

Although model building codes contain specific listing requirements for stationary generators used as the alternate power source for emergency and legally required standby power systems, this requirement applies to all stationary generators rated 600 volts or less regardless of the type of system being powered.

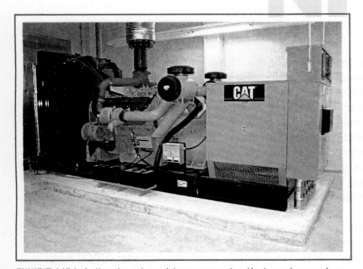

EXHIBIT 445.1 A diesel engine–driven generator that can be used as the alternate power source for an emergency system, legally required standby system, optional standby system, or critical operations standby system or as a parallel source in an interconnected power production system.

This includes stationary generators installed as the alternate power source for optional standby systems installed in residential, commercial, institutional, and industrial applications as well as stationary generators brought to a site as a temporary power source.

UL 2200, *Standard for Stationary Engine Generator Assemblies*, defines a *stationary unit (generator)* as "an engine generator that is intended to be hard-wired and/or permanently installed."

445.10 Location. Generators shall be of a type suitable for the locations in which they are installed. They shall also meet the requirements for motors in 430.14.

Informational Note: See NFPA 37-2021, *Standard for the Installation and Use of Stationary Combustion Engines and Gas Turbines*, for information on the location of generators.

445.11 Marking. Each generator shall be provided with an accessible nameplate giving the manufacturer's name, the rated frequency, the number of phases if ac, the rating in kilowatts or kilovolt-amperes, the power factor, the normal volts and amperes corresponding to the rating, and the rated ambient temperature.

Nameplates or manufacturer's instructions shall provide the following information for all stationary generators and portable generators rated more than 15 kW:

(1) Alternator subtransient, transient, synchronous, and zero sequence reactances
(2) Generator set power rating category (including but not limited to prime, standby, or continuous)
(3) Alternator temperature rise at rated load and insulation system class
(4) Indication if the generator is protected against overload by inherent design, an overcurrent protective relay, a circuit breaker, or a fuse
(5) Available fault current for inverter-based generators, in lieu of the synchronous, subtransient, and transient reactances

Marking shall be provided by the manufacturer to indicate whether or not the generator neutral is bonded to its frame. Where the bonding is modified in the field, additional marking shall be required to indicate whether the neutral is bonded to the frame.

Portable generators rated larger than 15 kilowatts and stationary generators must be marked with their reactances using the correct industry terms. The marking of the generator's maximum short-circuit current assists the inspector and the installer in verifying proper overcurrent protection in the field. Newer generators are being manufactured with inverter-based designs. Determining available fault current ratings for those generators is more complicated, and marking on the generator by the manufacturer is more appropriate and can be used instead of marking the generators' reactances.

Marking to indicate whether the generator is inherently designed to prevent overload, or whether an overcurrent protective relay is provided, assists designers and installers to size conductors from the generator output terminals for an ampacity of 100 percent, based on the exception to 445.13(A).

Marking provided by the manufacturer to indicate whether or not the generator neutral is bonded to the generator frame, plus the configuration of the transfer switch (solid neutral or switched neutral) provides installers with the necessary information to apply the requirements of Article 250 in respect to treating the generator supplied system as separately derived or non-separately derived. If the bonding of a generator is modified in the field, additional marking indicates whether the generator is configured as a separately derived or non-separately derived system.

See also

Exhibits **250.12** and **250.13** for examples of grounding and bonding connections at generators that supply premises wiring systems

445.12 Overcurrent Protection.

(A) Constant-Voltage Generators. Constant-voltage generators, except ac generator exciters, shall be protected from overload by inherent design, circuit breakers, fuses, protective relays, or other identified overcurrent protective means suitable for the conditions of use.

(B) Two-Wire Generators. Two-wire, dc generators shall be permitted to have overcurrent protection in one conductor only if the overcurrent device is actuated by the entire current generated other than the current in the shunt field. The overcurrent device shall not open the shunt field.

(C) 65 Volts or Less. Generators operating at 65 volts or less and driven by individual motors shall be considered as protected by the overcurrent device protecting the motor if these devices will operate when the generators are delivering not more than 150 percent of their full-load rated current.

(D) Balancer Sets. Two-wire, dc generators used in conjunction with balancer sets to obtain neutral points for 3-wire systems shall be equipped with overcurrent devices that disconnect the 3-wire system in case of excessive unbalancing of voltages or currents.

(E) Three-Wire, Direct-Current Generators. Three-wire, dc generators, whether compound or shunt wound, shall be equipped with overcurrent devices, one in each armature lead, and connected so as to be actuated by the entire current from the armature. Such overcurrent devices shall consist either of a double-pole, double-coil circuit breaker or of a 4-pole circuit breaker connected in the main and equalizer leads and tripped by two overcurrent devices, one in each armature lead. Such protective devices shall be interlocked so that no one pole can be opened without simultaneously disconnecting both leads of the armature from the system.

Exception to (A) through (E): Where deemed by the authority having jurisdiction that a generator is vital to the operation of an electrical system and the generator should operate to failure to prevent a greater hazard to persons, the overload sensing device(s) shall be permitted to be connected to an annunciator or alarm supervised by authorized personnel instead of interrupting the generator circuit.

445.13 Ampacity of Conductors.

(A) General. The ampacity of the conductors from the generator output terminals to the first distribution device(s) containing overcurrent protection shall not be less than 115 percent of the nameplate current rating of the generator. It shall be permitted to size the neutral conductors in accordance with 220.61. Conductors that must carry ground-fault currents shall not be smaller than required by 250.30(A). Neutral conductors of dc generators that must carry ground-fault currents shall not be smaller than the minimum required size of the largest conductor.

Exception: Where the design and operation of the generator prevent overloading, the ampacity of the conductors shall not be less than 100 percent of the nameplate current rating of the generator.

If the generator is inherently designed to prevent overload or if an overcurrent protective relay is provided, the conductors supplied from the generator output terminals are permitted to have a minimum ampacity based on 100 percent of the nameplate current rating of the generator, based upon the exception to 445.13(A).

(B) Overcurrent Protection Provided. Where the generator set is equipped with a listed overcurrent protective device or a combination of a current transformer and overcurrent relay, conductors shall be permitted to be tapped from the load side of the protected terminals in accordance with 240.21(B).

Tapped conductors shall not be permitted for portable generators rated 15 kW or less where field wiring connection terminals are not accessible.

The listed branch-circuit overcurrent protective device (OCPD), or a combination of a current transformer and overcurrent relay on the tap conductor source side terminals, provides short-circuit protection for the tap conductors. If the generator is equipped with such protection, feeder tap rules can be used with overload protection for the feeder conductors being provided by the overcurrent device that the tap conductors supply.

445.14 Protection of Live Parts. Live parts of generators operated at more than 50 volts ac or 60 volts dc to ground shall not be exposed to accidental contact where accessible to unqualified persons.

445.15 Guards for Attendants. Where necessary for the safety of attendants, the requirements of 430.233 shall apply.

445.16 Bushings. Where field-installed wiring passes through an opening in an enclosure, a conduit box, or a barrier, a bushing

shall be used to protect the conductors from the edges of an opening having sharp edges. The bushing shall have smooth, well-rounded surfaces where it may be in contact with the conductors. If used where oils, grease, or other contaminants may be present, the bushing shall be made of a material not deleteriously affected.

445.17 Generator Terminal Housings. Generator terminal housings shall comply with 430.12. Where a horsepower rating is required to determine the required minimum size of the generator terminal housing, the full-load current of the generator shall be compared with comparable motors in Table 430.247 through Table 430.250. The higher horsepower rating of Table 430.247 and Table 430.250 shall be used whenever the generator selection is between two ratings.

Exception: This section shall not apply to generators rated over 600 volts.

Δ **445.18 Disconnecting Means.**

(A) Disconnecting Means. Generators other than cord-and-plug-connected portable generators shall have one or more disconnecting means. Each disconnecting means shall simultaneously open all associated ungrounded conductors. Each disconnecting means shall be lockable open in accordance with 110.25.

The disconnecting means shall be permitted to be located within the generator behind a hinged cover, door, or enclosure panel. Where the generator disconnecting means is located within the generator, a field applied label meeting the requirements of 110.21(B) shall be provided indicating the location of the generator disconnecting means.

N **(B) Generators Installed in Parallel.** Where a generator is installed in parallel with other generators, the provisions of 445.18(A) shall be capable of isolating the generator output terminals from the paralleling system bus. The disconnecting means shall not be required to be located at the generator.

Arranging generators to operate in parallel is a common practice for large hospitals, data centers, and other large buildings requiring on-site emergency or standby power. Electrically operated circuit breakers automatically connect multiple generators to a common switchboard bus. They open and close via programmable logic controllers (PLC) commands. They close to the common bus when the generators are electrically synchronized and open when the generator shuts down or when commanded by the PLC for any preprogrammed sequence or adverse condition in the system. If the generator engine stops for any reason, the paralleling breaker in the paralleling switchboard, which is electrically interlocked with the equipment, will open. Where generators are installed in parallel, it is not necessary to provide a disconnecting means at each generator and at the paralleling equipment. Regardless of the location of the disconnecting means, compliance with 445.18(A) is required in respect to the circuit breaker or switch being capable of being locked in the open position. This protects a worker servicing a generator that

is taken out of the parallel lineup from being exposed to a back-feed at the output terminals of the generator.

N **445.19 Emergency Shutdown of Prime Mover.**

N **(A) General.** Generators shall have provisions to shut down the prime mover. The means of shutdown shall comply with all of the following:

(1) Be equipped with provisions to disable all prime mover start control circuits to render the prime mover incapable of starting

(2) Initiate a shutdown mechanism that requires a mechanical reset

The provisions to shut down the prime mover shall be permitted to satisfy the requirements of 445.18(A) where it is capable of being locked in the open position in accordance with 110.25.

The emergency engine shutdown is permitted to serve as the generator disconnecting means that isolates the generator from the equipment it supplies. Where this device is used only as the engine shutdown required by 445.19(A), it is not required to be a locking-type device. Where the device is used to shut down the engine and also serve as the disconnecting means required by 445.18(A), it is required to be capable of being locked in the "open" position. In the case of this device, the open position is that which complies with 445.19(A)(1) and (2). This is considered to be equivalent to locking a circuit breaker or safety switch in the "open," or off, position from the perspective of protecting personnel working on equipment supplied by the generator. Where two or more generators are operating in parallel, the use of the engine shutdown is not an acceptable disconnecting means because it does not meet the requirements to isolate the generator output terminals required by 445.18(B).

Exhibit 445.2 is an example of the type of device that can be used to comply with 445.19(A) and (B). It can also be used to

EXHIBIT 445.2 A lockable emergency shutdown device that can be used as a generator disconnecting means in accordance with 445.18(A) and 445.19(A). *(Courtesy of Idem Safety Switches Ltd.)*

comply with 445.18(A) because it can be locked in the position that disables the engine start circuit.

N **(B) Remote Emergency Shutdown.** For other than one- and two-family dwelling units, generators with greater than 15 kW rating shall be provided with a remote emergency stop switch to shut down the prime mover. The remote emergency stop switch shall be located outside the equipment room or generator enclosure at a readily accessible location and shall also meet the requirements of 445.19(A)(1) and (A)(2).

The remote emergency stop switch shall be permitted to be mounted on the exterior of the generator enclosure. The remote emergency stop switch shall be labeled Generator Emergency Shutdown, and the label shall meet the requirements of 110.21(B).

Requiring shutdown of the prime mover for generators greater than 15 kilowatts is necessary to provide a remote shutdown means in the event of an emergency event such as a prime mover failure. This requirement aligns with 5.6.5.8 in NFPA 110, *Standard for Emergency and Standby Power Systems*, 2022 edition. However, unlike NFPA 110, which is mandatory only for Level 1 (Article 700) and Level 2 (Article 701) emergency and standby power systems, the requirement of 445.19(B) imposes the emergency shutdown requirement on all generators having a rating greater than 15 kilowatts regardless of the type of system or installation supplied by the generator.

N **(C) Emergency Shutdown in One- and Two-Family Dwelling Units.** For other than cord-and-plug-connected portable generators, an emergency shutdown device shall be located outside the dwelling unit at a readily accessible location and shall also meet the requirements of 445.19(A)(1) and (A)(2).

An emergency shutdown device mounted on the exterior of the generator enclosure shall be permitted to satisfy the requirements of this section. The shutdown device shall be marked as the Generator Emergency Shutdown, and the label shall meet the requirements of 110.21(B)).

Where a stationary generator is used as the primary source of power or as the alternate source of power in an optional standby system installed at one- and two-family dwellings, an emergency prime mover shutdown device is required to be installed outside the dwelling unit at a readily accessible location. This rule aligns with similar emergency shutdown requirements for these types of occupancies in 230.85 for services, 480.7(B) for stationary battery systems, 706.15 for energy storage systems, and 690.12(C) for the rapid shutdown initiation device for solar photovoltaic systems. The *NEC*® requires only that these devices be installed at a readily accessible location outside the dwelling unit. Where multiple devices are required at a one- or two-family dwelling, there is no specific requirement to group them in one readily accessible location. However, because these devices are installed primarily to facilitate the shutdown of power supplies by first responders or other emergency personnel, it is advisable that the AHJ be consulted on where the devices should be installed.

Δ **445.20 Ground-Fault Circuit-Interrupter Protection for Receptacles on 15-kW or Smaller Portable Generators.** Receptacle outlets that are a part of a 15-kW or smaller portable generator shall have listed ground-fault circuit-interrupter protection (GFCI) for personnel integral to the generator or receptacle as indicated in either 445.20(A) or (B):

(A) Unbonded (Floating Neutral) Generators. Unbonded generators with both 125-volt and 125/250-volt receptacle outlets shall have listed GFCI protection for personnel integral to the generator or receptacle on all 125-volt, 15- and 20-ampere receptacle outlets.

Exception: GFCI protection shall not be required where the 125-volt receptacle outlets(s) is interlocked such that it is not available for use when any 125/250-volt receptacle(s) is in use.

(B) Bonded Neutral Generators. Bonded generators shall be provided with GFCI protection on all 125-volt, 15- and 20-ampere receptacle outlets.

Informational Note: See 590.6(A)(3) for GFCI requirements for 15-kW or smaller portable generators used for temporary electric power and lighting.

Exception to (A) and (B): If the generator was manufactured or remanufactured prior to January 1, 2015, listed cord sets or devices incorporating listed GFCI protection for personnel identified for portable use shall be permitted.

Separate requirements for different configurations of portable generators address the design differences of integral ground-fault circuit-interrupter (GFCI) protection and interlocking features of unbonded versus bonded generators. The exception to 445.20(A) addresses the problems of proper operation of a GFCI-protected 125-volt receptacle on the generator when the 125/250-volt receptacle on the generator is connected into a grounded premises wiring system via a transfer device that employs a solid neutral connection and the generator neutral is floating (not bonded to the generator frame). This connection creates multiple paths for normal circuit operation neutral current that will result in tripping of the GFCI devices protecting receptacles on the generator. If the generator is connected to a premises wiring system using a 125/250-volt receptacle on the floating neutral generator, any 125-volt receptacles not having GFCI protection have to be interlocked so they cannot export power. Where the 125/250-volt receptacle(s) is not supplying use, the 125-volt receptacles can be used to supply cord-and-plug-connected portable loads, and under the conditions of the exception, GFCI protection is not required.

The Informational Note refers the user to additional GFCI requirements for receptacles on portable generators that are used for temporary electric power and lighting at construction sites and similar locations. GFCI protection is mandatory for certain receptacle configurations on portable generators regardless of whether the generator is of the floating neutral or bonded neutral type.

ARTICLE 450 — Transformers and Transformer Vaults (Including Secondary Ties)

△ **Part I. General**

△ **450.1 Scope.** This article covers the installation of all transformers other than the following:

(1) Current transformers
(2) Dry-type transformers that constitute a component part of other apparatus and comply with the requirements for such apparatus
(3) Transformers that are an integral part of an X-ray, high-frequency, or electrostatic-coating apparatus
(4) Transformers used with Class 2 and Class 3 circuits
(5) Transformers for sign and outline lighting
(6) Transformers for electric-discharge lighting
(7) Transformers used for power-limited fire alarm circuits
(8) Transformers used for research, development, or testing, where effective arrangements are provided to safeguard persons from contacting energized parts

N **450.2 Interconnection of Transformers.** Transformers shall individually comply with the requirements of this article unless specific provisions allow for interconnection and operation as a single unit.

450.3 Overcurrent Protection. Overcurrent protection of transformers shall comply with 450.3(A), (B), or (C). As used in this section, the word *transformer* shall mean a transformer or polyphase bank of two or more single-phase transformers operating as a unit.

> Informational Note No. 1: See 240.4, 240.21, 245.26, and 245.27 for overcurrent protection of conductors.

The requirements for overcurrent protection of transformer secondaries apply only to the protection of transformers, not to the protection of conductors. The sections in Article 240 and Article 245 that are referenced in the informational note apply only to the protection of conductors, not to the protection of transformers. The overcurrent protection required by Article 450 may also satisfy the requirements in Article 240 and Article 245 for conductor protection, and vice versa, but it is also possible that they do not.

The overcurrent protection required for transformers may not provide satisfactory protection for the primary and secondary conductors. Where polyphase transformers are involved, primary and secondary conductors usually are not properly protected. The primary overcurrent device provides short-circuit protection for the primary conductors and a degree of overload protection for the transformer, and secondary overcurrent devices prevent the transformer and secondary conductors from being overloaded.

A transformer is considered the point of supply, and the conductors it supplies must be protected in accordance with their ampacity. For applications of not more than 1000 volts, nominal, 240.4(F) permits the secondary circuit conductors from a transformer to be protected by overcurrent devices in the primary circuit conductors only in two special cases: (1) a transformer with a 2-wire primary and a 2-wire secondary and (2) a 3-phase, delta-delta-connected transformer having a 3-wire, single-voltage secondary. Either case requires transformer primary protection in accordance with 450.3. Where the primary feeder to the transformer incorporates overcurrent protective devices (OCPDs) rated (or set) at a level not to exceed those prescribed herein, it is not necessary to duplicate them at the transformer.

> **See also**
>
> **240.4(F)** and **240.21(B)** for overcurrent protection requirements of transformer conductors in applications rated not more than 1000 volts, nominal
>
> **Article 240,** Part VIII, for overcurrent protection requirements for feeders and feeder taps associated with transformers in applications rated not more than 1000 volts, nominal, within supervised industrial installations
>
> **245.26 and 245.27** for overcurrent protection requirements for feeders and feeder taps associated with transformers in applications rated over 1000 volts ac, 1500 volts dc

> Informational Note No. 2: Nonlinear loads can increase heat in a transformer without operating its overcurrent protective device. See IEEE 3002.8, *Recommended Practice for Conducting Harmonic Studies and Analysis of Industrial and Commercial Power Systems*, for additional information.

The increased heating effects of nonlinear load currents must be taken into account when determining the load on a transformer. Methods for handling these heating effects include derating equipment, oversizing equipment, increasing insulation ratings, installing thermal protection systems, and using K-factor transformers. The optimum method for dealing with transformer overheating varies, depending on several technical and economic factors, and is best determined during the design phase of the electrical system.

(A) Transformers Over 1000 Volts, Nominal. Overcurrent protection shall be provided in accordance with Table 450.3(A).

Unlike the information contained in informational notes, which are explanatory in nature and not enforceable, table notes are part of the requirements of the table.

For Note 1 of Table 450.3(A), which concerns standard ratings of circuit breakers and fuses, see 240.6(A) and Table 240.6(A) for additional requirements on standard ampere ratings.

For Note 2, overcurrent protection of the secondary of a transformer is allowed to consist of not more than six sets of fuses or six circuit breakers.

For Note 3, equipment maintenance is performed by personnel who have received safety training and are familiar with proper operation of the equipment and aware of the hazards associated with it. See the definition of *qualified person* in Article 100.

For Note 4, an electronically actuated fuse responds to a signal from an electronic control rather than heat from a current. See the definition of *fuse, electronically actuated* in Article 100.

Exhibits 450.1 and 450.2 illustrate the conditions given in Note 2 of Table 450.3(A) for transformers rated over 1000 volts. The ratings or settings obtained from Table 450.3(A) are based on the type of protective device (fuse, electronic fuse, or circuit breaker), transformer-rated current and impedance, and primary and secondary voltages. The maximum ratings or settings of an OCPD for transformers rated over 1000 volts are separated into two broad categories — "any location (usually unsupervised locations)" and "supervised locations only."

The first category is not limited by location and is referred to as *any location*. The maximum ratings or settings for overcurrent devices permitted are applicable to all unsupervised locations. An "any location" transformer installation must be provided with overcurrent protection in both the primary and secondary circuits. See Exhibit 450.3 for an example of an installation using circuit breakers on the primary and the secondary for an over-1000-volt transformer with 6-percent impedance.

The second category for over-1000-volt transformers is supervised locations only. The maximum ratings or settings for overcurrent devices permitted are strictly limited to the supervised location conditions explained in Note 3. The installation shown in Exhibit 450.3 fulfills the requirements of both "any location" and "supervised locations only."

See also

450.3 and its commentary regarding the protection of transformer primary and secondary conductors

Δ **(B) Transformers 1000 Volts, Nominal, or Less.** Overcurrent protection shall be provided in accordance with Table 450.3(B)

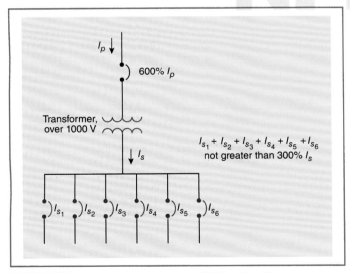

EXHIBIT 450.1 A transformer rated over 1000 volts with a secondary rated over 1000 volts, with secondary protection consisting of six circuit breakers. The sum of the ratings of the circuit breakers is not permitted to exceed 300 percent of the rated secondary current.

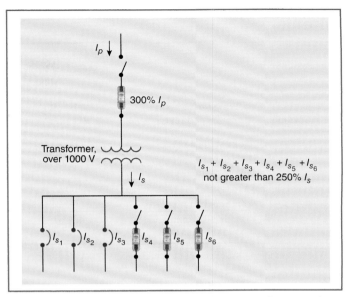

EXHIBIT 450.2 A transformer rated over 1000 volts with a secondary rated over 1000 volts, with secondary protection consisting of fuses and circuit breakers. The sum of the ratings of all the overcurrent devices is not permitted to exceed the rating permitted for fuses.

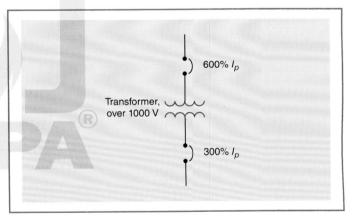

EXHIBIT 450.3 A transformer with 6-percent impedance and rated over 1000 volts using circuit-breaker protection for both the primary and the secondary. Both the primary and the secondary voltages are over 1000 volts.

unless the transformer is installed as a motor control circuit transformer in accordance with 430.72(C)(1) through (C)(5).

The ratings or settings of the OCPD obtained from Table 450.3(B) are based on the transformer-rated current and whether secondary protection is provided. According to Table 450.3(B), the maximum ratings or settings of OCPDs for transformers rated 1000 volts and less are separated into two categories — primary-only protection and primary and secondary protection.

Transformers must be protected by either of two methods. Method 1 requires primary protection only. Method 2 requires secondary side overcurrent protection at not more than 125 or

△ **TABLE 450.3(A)** *Maximum Rating or Setting of Overcurrent Protection for Transformers Over 1000 Volts (as a Percentage of Transformer-Rated Current)*

| Location Limitations | Transformer Rated Impedance | Primary Protection over 1000 Volts | | Secondary Protection[2] | | |
| | | | | Over 1000 Volts | | 1000 Volts or Less |
		Circuit Breaker[4]	Fuse Rating	Circuit Breaker[4]	Fuse Rating	Circuit Breaker or Fuse Rating
Any location	Not more than 6%	600%[1]	300%[1]	300%[1]	250%[1]	125%[1]
	More than 6% and not more than 10%	400%[1]	300%[1]	250%[1]	225%[1]	125%[1]
Supervised locations only[3]	Any	300%[1]	250%[1]	Not required	Not required	Not required
	Not more than 6%	600%	300%	300%[5]	250%[5]	250%[5]
	More than 6% and not more than 10%	400%	300%	250%[5]	225%[5]	250%[5]

[1]Where the required fuse rating or circuit breaker setting does not correspond to a standard rating or setting, a higher rating or setting that does not exceed the following shall be permitted:

(1) The next higher standard rating or setting for fuses and circuit breakers 1000 volts and below, or

(2) The next higher commercially available rating or setting for fuses and circuit breakers above 1000 volts.

[2]Where secondary overcurrent protection is required, the secondary overcurrent device shall be permitted to consist of not more than six circuit breakers or six sets of fuses grouped in one location. Where multiple overcurrent devices are utilized, the total of all the device ratings shall not exceed the allowed value of a single overcurrent device. If both circuit breakers and fuses are used as the overcurrent device, the total of the device ratings shall not exceed that allowed for fuses.

[3]A supervised location is a location where conditions of maintenance and supervision ensure that only qualified persons monitor and service the transformer installation.

[4]Electronically actuated fuses that may be set to open at a specific current shall be set in accordance with settings for circuit breakers.

[5]A transformer equipped with a coordinated thermal overload protection by the manufacturer shall be permitted to have separate secondary protection omitted.

△ **TABLE 450.3(B)** *Maximum Rating or Setting of Overcurrent Protection for Transformers 1000 Volts and Less (as a Percentage of Transformer-Rated Current)*

| Protection Method | Primary Protection | | | Secondary Protection[2] | |
	Currents of 9 Amperes or More	Currents Less Than 9 Amperes	Currents Less Than 2 Amperes	Currents of 9 Amperes or More	Currents Less Than 9 Amperes
Primary only protection	125%[1]	167%	300%	Not required	Not required
Primary and secondary protection	250%[3]	250%[3]	250%[3]	125%[1]	167%

[1]Where 125 percent of this current does not correspond to a standard rating of a fuse or nonadjustable circuit breaker, a higher rating that does not exceed the next higher standard rating shall be permitted.

[2]Where secondary overcurrent protection is required, the secondary overcurrent device shall be permitted to consist of not more than six circuit breakers or six sets of fuses grouped in one location. Where multiple overcurrent devices are utilized, the total of all the device ratings shall not exceed the allowed value of a single overcurrent device.

[3]A transformer equipped with coordinated thermal overload protection by the manufacturer and arranged to interrupt the primary current shall be permitted to have primary overcurrent protection rated or set at a current value that is not more than six times the rated current of the transformer for transformers having not more than 6 percent impedance and not more than four times the rated current of the transformer for transformers having more than 6 percent but not more than 10 percent impedance.

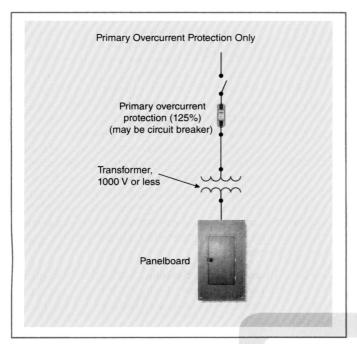

EXHIBIT 450.4 A transformer (with currents of 9 amperes or more) rated 1000 volts or less with only primary overcurrent protection.

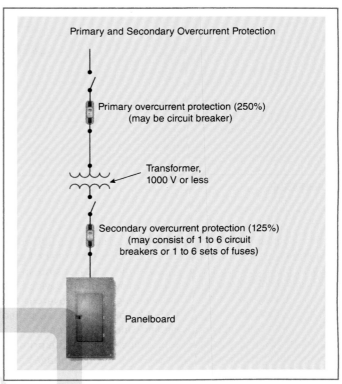

EXHIBIT 450.5 A transformer (with currents of 9 amperes or more) rated 1000 volts or less and protected by a combination of primary and secondary overcurrent protection.

167 percent (depending on secondary current rating), provided the primary side overcurrent protection is not more than 250 percent of the primary side current rating.

An example of primary-only protection is shown in Exhibit 450.4. An example of primary and secondary protection is shown in Exhibit 450.5.

> **See also**
>
> **430.72(C)** for overcurrent protection of motor control circuit transformers
>
> **450.3** and its commentary regarding the protection of transformer primary and secondary conductors

(C) Voltage (Potential) Transformers. Voltage (potential) transformers installed indoors or enclosed shall be protected with primary fuses.

> Informational Note: See 408.52 for protection of instrument circuits, including voltage transformers.

450.4 Autotransformers 1000 Volts, Nominal, or Less.

(A) Overcurrent Protection. Each autotransformer 1000 volts, nominal, or less shall be protected by an individual overcurrent device installed in series with each ungrounded input conductor. Such overcurrent device shall be rated or set at not more than 125 percent of the rated full-load input current of the autotransformer. Where this calculation does not correspond to a standard rating of a fuse or nonadjustable circuit breaker and the rated input current is 9 amperes or more, the next higher standard rating described in 240.6 shall be permitted. An overcurrent device shall not be installed in series with

the shunt winding (the winding common to both the input and the output circuits) of the autotransformer between Points A and B as shown in Figure 450.4(A).

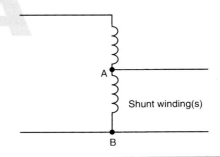

FIGURE 450.4(A) *Autotransformer.*

Exception: Where the rated input current of the autotransformer is less than 9 amperes, an overcurrent device rated or set at not more than 167 percent of the input current shall be permitted.

Because of the voltage feedback problem that may occur, an overcurrent device is not permitted between points A and B in Figure 450.4(A).

Exhibit 450.6 provides an example of overcurrent protection for an autotransformer. It shows a 2-winding, single-phase

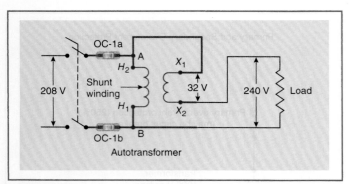

EXHIBIT 450.6 A disconnect switch with overcurrent devices properly connected to protect an autotransformer and located to meet the requirements of 450.4(A).

transformer connected to boost a 208-volt supply to 240 volts. The autotransformer is provided with a 2-pole disconnect switch with both overcurrent devices (OC-1a and OC-1b) located on the supply side of the autotransformer. If an overcurrent device were located in series with the shunt winding and the overcurrent device opened, the full 208-volt supply voltage would be applied across the 32-volt secondary winding. Under those conditions, a higher-than-normal voltage would appear across the primary winding. If the load impedance were very low, the voltage could approach $208/32 \times 208 = 1352$ V.

(B) Transformer Field-Connected as an Autotransformer. A transformer field-connected as an autotransformer shall be identified for use at elevated voltage.

Informational Note: See 210.9 and 215.11 for information on permitted uses of autotransformers.

This requirement is necessary because of the dielectric voltage withstand test requirements applied to transformers. The test is conducted at 2500 volts for windings rated 250 volts or less and at 4000 volts for higher-rated windings. A transformer intended for buck or boost operation would require that the test for the low-voltage winding be based on the sum of the primary and secondary voltage ratings.

450.5 Grounding Autotransformers.

Grounding autotransformers covered in this section are zigzag or T-connected transformers connected to 3-phase, 3-wire ungrounded systems for the purpose of creating a 3-phase, 4-wire distribution system or providing a neutral point for grounding purposes. Such transformers shall have a continuous per-phase current rating and a continuous neutral current rating. Zigzag-connected transformers shall not be installed on the load side of any system grounding connection, including those made in accordance with 250.24(C), 250.30(A)(1), or 250.32(B), Exception No. 1.

Informational Note: The phase current in a grounding autotransformer is one-third the neutral current.

The installation of grounding autotransformers on the load side of a supply system grounding connection is prohibited. This restriction applies to services, to separately derived systems, and

to feeders and branch circuits that supply separate buildings or structures. Where a zigzag transformer is used to create a neutral reference point on a circuit that is supplied from a grounded system, the current from a line-to-ground fault is shared through the supply system transformer and the zigzag transformer.

Where the rating of the circuit in which the line-to-ground fault occurs exceeds the rating of the circuit in which the zigzag transformer is used, the shared ground-fault current through the zigzag transformer has the potential to cause serious damage to the transformer. For instance, a zigzag transformer is installed on an existing 50-ampere, 3-phase, 3-wire branch circuit to create a neutral. The branch circuit is derived from a grounded wye service, from which large-capacity, 800-ampere and 1000-ampere feeders are also supplied. A line-to-ground fault in one of these feeder circuits can result in serious damage to the zigzag transformer as a result of its sharing the fault current with the system supply transformer.

(A) Three-Phase, 4-Wire System. A grounding autotransformer used to create a 3-phase, 4-wire distribution system from a 3-phase, 3-wire ungrounded system shall conform to 450.5(A)(1) through (A)(4).

(1) Connections. The transformer shall be directly connected to the ungrounded phase conductors and shall not be switched or provided with overcurrent protection that is independent of the main switch and common-trip overcurrent protection for the 3-phase, 4-wire system.

(2) Overcurrent Protection. An overcurrent sensing device shall be provided that will cause the main switch or common-trip overcurrent protection referred to in 450.5(A)(1) to open if the load on the autotransformer reaches or exceeds 125 percent of its continuous current per-phase or neutral rating. Delayed tripping for temporary overcurrents sensed at the autotransformer overcurrent device shall be permitted for the purpose of allowing proper operation of branch or feeder protective devices on the 4-wire system.

(3) Transformer Fault Sensing. A fault-sensing system that causes the opening of a main switch or common-trip overcurrent device for the 3-phase, 4-wire system shall be provided to guard against single-phasing or internal faults.

Informational Note: This can be accomplished by the use of two subtractive-connected donut-type current transformers installed to sense and signal when an unbalance occurs in the line current to the autotransformer of 50 percent or more of rated current.

(4) Rating. The autotransformer shall have a continuous neutral-current rating that is not less than the maximum possible neutral unbalanced load current of the 4-wire system.

Exhibit 450.7 illustrates the proper method of protecting a grounding autotransformer used to provide a neutral for a 3-phase system where necessary to supply a group of single-phase, line-to-neutral loads. Separate overcurrent protection is not provided for the autotransformer because there will be no control of the system line-to-neutral voltages if the

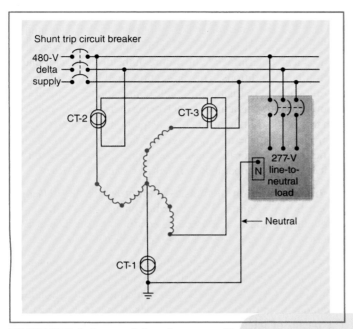

EXHIBIT 450.7 A zigzag autotransformer used to establish a neutral connection for a 480Y/277-volt, 3-phase ungrounded system to supply single-phase line-to-neutral loads.

autotransformer becomes disconnected. Consequently, simultaneous interruption of the power supply to all the line-to-neutral loads is necessary whenever the grounding autotransformer is switched off.

CT-1 is connected to an overload relay responsive to excess neutral current being supplied. See 450.5(A)(2). CT-2 and CT-3 are connected to differential-type fault-current sensing relays responsive to an unbalance of neutral current among the three phases of the grounding autotransformer (indicating an internal fault). All three relays are to be arranged to trip the circuit breaker located upstream of both the autotransformer and the line-to-neutral connected loads.

(B) Ground Reference for Fault Protection Devices. A grounding autotransformer used to make available a specified magnitude of ground-fault current for operation of a ground-responsive protective device on a 3-phase, 3-wire ungrounded system shall conform to 450.5(B)(1) and (B)(2).

(1) Rating. The autotransformer shall have a continuous neutral-current rating not less than the specified ground-fault current.

(2) Overcurrent Protection. Overcurrent protection shall comply with 450.5(B)(2)(a) and (B)(2)(b).

(a) *Operation and Interrupting Rating.* An overcurrent protective device having an interrupting rating in compliance with 110.9 and that will open simultaneously all ungrounded conductors when it operates shall be applied in the grounding autotransformer branch circuit.

(b) *Ampere Rating.* The overcurrent protection shall be rated or set at a current not exceeding 125 percent of the

autotransformer continuous per-phase current rating or 42 percent of the continuous-current rating of any series-connected devices in the autotransformer neutral connection. Delayed tripping for temporary overcurrents to permit the proper operation of ground-responsive tripping devices on the main system shall be permitted but shall not exceed values that would be more than the short-time current rating of the grounding autotransformer or any series connected devices in the neutral connection thereto.

Exception: For high-impedance grounded systems covered in 250.36, where the maximum ground-fault current is designed to be not more than 10 amperes, and where the grounding autotransformer and the grounding impedance are rated for continuous duty, an overcurrent device rated not more than 20 amperes that will simultaneously open all ungrounded conductors shall be permitted to be installed on the line side of the grounding autotransformer.

In high-impedance grounded systems, the currents are low enough that finding overcurrent devices rated at 125 percent of a typical 5-ampere system is not practical.

Exhibit 450.8 illustrates the proper method of protecting a grounding autotransformer where it is used as a ground reference for fault protective devices. The overcurrent protective device is to have a rating (or setting) not in excess of 125 percent of the rated phase current of the autotransformer (42 percent of the neutral current rating) and not more than 42 percent of the continuous current rating of the neutral grounding resistor or other current-carrying device in the neutral connection, as specified in 450.5(B)(2).

(C) Ground Reference for Damping Transitory Overvoltages. A grounding autotransformer used to limit transitory overvoltages shall be of suitable rating and connected in accordance with 450.5(A)(1).

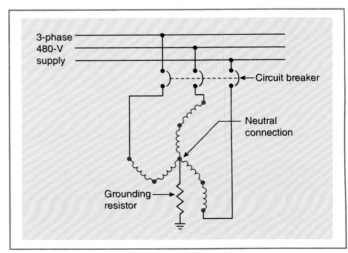

EXHIBIT 450.8 A zigzag autotransformer used to establish a reference ground-fault current for fault protective device operation or for damping transitory overvoltage surges.

For installations involving a high-resistance grounding package, the functional performance of the installation parallels that described in 450.5(B), differing only in that the magnitude of available ground-fault current would likely be a lower value. It would be appropriate to employ the connections displayed in Exhibit 450.8 and to conform to the overcurrent protection requirements prescribed in 450.5(B)(2).

With any of the grounding autotransformer applications covered by 450.5(A), (B), or (C), the use of a ganged 3-pole switching interrupter for connecting and disconnecting the autotransformer accomplishes simultaneous connection (and disconnection) of the three line terminals. If, at any time, one or two of the line connections to the autotransformer were to open, which could occur if the protective devices were single pole, the grounding autotransformer would cease to function in the desired fashion and would act as a high inductive reactance connection between the electrical system and ground. The latter connection is prone to create high-value transitory overvoltages, line-to-ground.

450.6 Secondary Ties.   As used in this article, a secondary tie is a circuit operating at 1000 volts, nominal, or less between phases that connects two power sources or power supply points, such as the secondaries of two transformers. The tie shall be permitted to consist of one or more conductors per phase or neutral. Conductors connecting the secondaries of transformers in accordance with 450.7 shall not be considered secondary ties.

As used in this section, the word *transformer* means a transformer or a bank of transformers operating as a unit.

(A) Tie Circuits.   Tie circuits shall be provided with overcurrent protection at each end as required in Parts I, II, and VIII of Article 240.

Under the conditions described in 450.6(A)(1) and 450.6(A)(2), the overcurrent protection shall be permitted to be in accordance with 450.6(A)(3).

(1) Loads at Transformer Supply Points Only.   Where all loads are connected at the transformer supply points at each end of the tie and overcurrent protection is not provided in accordance with Parts I, II, and VIII of Article 240, the ampacity of the tie shall not be less than 67 percent of the rated secondary current of the highest rated transformer supplying the secondary tie system.

(2) Loads Connected Between Transformer Supply Points.   Where load is connected to the tie at any point between transformer supply points and overcurrent protection is not provided in accordance with Parts I, II, and VIII of Article 240, the ampacity of the tie shall not be less than 100 percent of the rated secondary current of the highest rated transformer supplying the secondary tie system.

Exception: Tie circuits comprised of multiple conductors per phase shall be permitted to be sized and protected in accordance with 450.6(A)(4).

(3) Tie Circuit Protection.   Under the conditions described in 450.6(A)(1) and (A)(2), both supply ends of each ungrounded tie conductor shall be equipped with a protective device that opens at a predetermined temperature of the tie conductor under short-circuit conditions. This protection shall consist of one of the following: (1) a fusible link cable connector, terminal, or lug, commonly known as a limiter, each being of a size corresponding with that of the conductor and of construction and characteristics according to the operating voltage and the type of insulation on the tie conductors or (2) automatic circuit breakers actuated by devices having comparable time–current characteristics.

(4) Interconnection of Phase Conductors Between Transformer Supply Points.   Where the tie consists of more than one conductor per phase or neutral, the conductors of each phase or neutral shall comply with 450.6(A)(4)(a) or (A)(4)(b).

(a) *Interconnected.* The conductors shall be interconnected in order to establish a load supply point, and the protective device specified in 450.6(A)(3) shall be provided in each ungrounded tie conductor at this point on both sides of the interconnection. The means of interconnection shall have an ampacity not less than the load to be served.

(b) *Not Interconnected.* The loads shall be connected to one or more individual conductors of a paralleled conductor tie without interconnecting the conductors of each phase or neutral and without the protection specified in 450.6(A)(3) at load connection points. Where this is done, the tie conductors of each phase or neutral shall have a combined capacity ampacity of not less than 133 percent of the rated secondary current of the highest rated transformer supplying the secondary tie system, the total load of such taps shall not exceed the rated secondary current of the highest rated transformer, and the loads shall be equally divided on each phase and on the individual conductors of each phase as far as practicable.

(5) Tie Circuit Control.   Where the operating voltage exceeds 150 volts to ground, secondary ties provided with limiters shall have a switch at each end that, when open, de-energizes the associated tie conductors and limiters. The current rating of the switch shall not be less than the rated current ampacity of the conductors connected to the switch. It shall be capable of interrupting its rated current, and it shall be constructed so that it will not open under the magnetic forces resulting from short-circuit current.

(B) Overcurrent Protection for Secondary Connections.   Where secondary ties are used, an overcurrent device rated or set at not more than 250 percent of the rated secondary current of the transformers shall be provided in the secondary connections of each transformer supplying the tie system. In addition, an automatic circuit breaker actuated by a reverse-current relay set to open the circuit at not more than the rated secondary current of the transformer shall be provided in the secondary connection of each transformer.

The requirements of 450.6 apply specifically to network systems for power distribution commonly employed where the load density is high and service reliability is important. This type of distribution system introduces a variety of problems not encountered in the more common radial-type distribution system and must be designed by experienced electrical engineers. Exhibit 450.9 illustrates a typical 3-phase network system for an industrial plant fed by two primary feeders, preferably from separate substations, energized at any standard voltage up to 34,500 volts. Each of the transformers is supplied by the two primary feeders via a double-throw switch at the transformer so that the transformer can be supplied by either feeder.

Each of the network transformers is rated in the range of 300 to 1000 kilovolt-amperes and is required to be protected as illustrated in Exhibit 450.10. The primary and secondary protection is in accordance with 450.3, but an additional protective device must be provided on the secondary side. This protective device, known as a network protector, consists of a circuit breaker and a reverse-current relay. The network protector operates on reverse current to prevent power from being fed back into the transformer through the secondary ties if a fault were to occur in the transformer or a primary feeder. The reverse-current relay is set to trip the circuit breaker at a current value not more than the rated secondary current of the transformer. The relay is not designed to trip the circuit breaker in the event of an overload on the secondary of the transformer.

The secondary ties shown in Exhibit 450.9 must be protected at each end with an overcurrent device, in accordance with 450.6(A)(3). The overcurrent device most commonly provided is a special type of fuse known as a current limiter, shown in Exhibit 450.11. This high-interrupting-capacity device provides short-circuit protection for only the secondary ties by opening safely before temperatures damaging to the cable insulation are reached. The secondary ties form a closed loop equipped with

switching devices so that any part of the loop may be isolated when repairs are needed or a current limiter must be replaced.

See also

Article 100 for the definition of *overcurrent protective device, current-limiting (current-limiting overcurrent protective device)* and its commentary

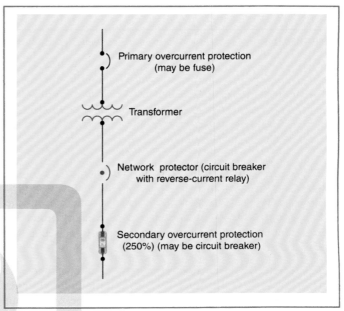

EXHIBIT 450.10 Primary and secondary overcurrent protection for a transformer in a network system, showing a network protector (an automatic circuit breaker actuated by a reverse-current relay).

EXHIBIT 450.9 A typical 3-phase network system for an industrial plant fed by two primary feeders.

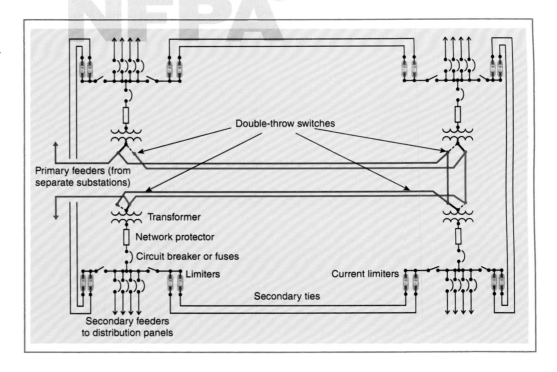

EXHIBIT 450.11 A current limiter (a special type of high-interrupting-capacity fuse). *(Courtesy of Ferraz Shawmut)*

△ **(C) Grounding.** Where the secondary tie system is grounded, each transformer secondary supplying the tie system shall be grounded in accordance with 250.30 for separately derived systems.

450.7 Parallel Operation. Transformers shall be permitted to be operated in parallel and switched as a unit, provided the overcurrent protection for each transformer meets the requirements of 450.3(A) for primary and secondary protective devices over 1000 volts, or 450.3(B) for primary and secondary protective devices 1000 volts or less.

Parallel operation of transformers that are not switched as a unit can present dangerous backfeed situations for workers performing electrical maintenance. Appropriate lockout/tagout procedures must be implemented during maintenance of electrical equipment operated or connected in parallel. See *NFPA 70E®, Standard for Electrical Safety in the Workplace®*, for safety-related work practices and appropriate lockout/tagout procedures.

450.8 Guarding. Transformers shall be guarded as specified in 450.8(A) through (D).

(A) Mechanical Protection. Appropriate provisions shall be made to minimize the possibility of damage to transformers from external causes where the transformers are exposed to physical damage.

One method of providing mechanical protection is to strategically place bollards around the transformer. This practice provides a degree of protection from vehicles.

(B) Case or Enclosure. Dry-type transformers shall be provided with a noncombustible moisture-resistant case or enclosure that provides protection against the accidental insertion of foreign objects.

(C) Exposed Energized Parts. Switches or other equipment operating at 1000 volts, nominal, or less and serving only equipment within a transformer enclosure shall be permitted to be installed in the transformer enclosure if accessible to qualified persons only. All energized parts shall be guarded in accordance with 110.27 and 110.34.

(D) Voltage Warning. The operating voltage of exposed live parts of transformer installations shall be indicated by signs or visible markings on the equipment or structures.

△ **450.9 Ventilation.** The ventilation shall dispose of the transformer full-load heat losses without creating a temperature rise that is in excess of the transformer rating.

> Informational Note No. 1: See IEEE C57.12.00-2015, *General Requirements for Liquid-Immersed Distribution, Power, and Regulating Transformers*, and IEEE C57.12.01-2020, *General Requirements for Dry-Type Distribution and Power Transformers*, for additional information.
>
> Informational Note No. 2: See IEEE C57.110-2018, *Recommended Practice for Establishing Liquid-Filled and Dry-Type Power and Distribution Transformer Capability When Supplying Nonsinusoidal Load Currents*, for more information where transformers are used with nonlinear loads that have nonsinusoidal currents that can result in additional losses and transformer heating.

Transformers with ventilating openings shall be installed so that the ventilating openings are not blocked by walls or other obstructions. The required clearances shall be clearly marked on the transformer. Transformer top surfaces that are horizontal and readily accessible shall be marked to prohibit storage.

450.10 Grounding and Bonding.

(A) Dry-Type Transformer Enclosures. Where separate equipment grounding conductors and supply-side bonding jumpers are installed, a terminal bar for all grounding and bonding conductor connections shall be secured inside the transformer enclosure. The terminal bar shall be bonded to the enclosure in accordance with 250.12 and shall not be installed on or over any vented portion of the enclosure.

Exception: Where a dry-type transformer is equipped with wire-type connections (leads), the grounding and bonding connections shall be permitted to be connected together using any of the methods in 250.8 and shall be bonded to the enclosure if of metal.

An enclosure typically is not evaluated as a grounding and bonding device. The required busbar for equipment grounding conductors (EGCs) and bonding jumpers prohibits the practice of using the transformer metal enclosure as a connection point for these conductors.

(B) Other Metal Parts. Exposed non-current-carrying metal parts of transformer installations, including fences, guards, and so forth, shall be grounded and bonded under the conditions and in the manner specified for electrical equipment and other exposed metal parts in Parts V, VI, and VII of Article 250.

450.11 Marking.

(A) General. Each transformer shall be provided with a nameplate giving the following information:

(1) Name of manufacturer
(2) Rated kilovolt-amperes
(3) Frequency
(4) Primary and secondary voltage
(5) Impedance of transformers 25 kVA and larger
(6) Required clearances for transformers with ventilating openings
(7) Amount and kind of insulating liquid where used
(8) For dry-type transformers, temperature class for the insulation system

(B) Source Marking. A transformer shall be permitted to be supplied at the marked secondary voltage, provided that the installation is in accordance with the manufacturer's instructions.

Not all transformers are designed to be backfed. Transformers are typically designed and evaluated with the supply on the primary side and the load on the secondary side. Backfeeding is permitted only when the manufacturer indicates so in the instructions.

450.12 Terminal Wiring Space. The minimum wire-bending space at fixed, 1000-volt and below terminals of transformer line and load connections shall be as required in 312.6. Wiring space for pigtail connections shall conform to Table 314.16(B)(1).

450.13 Accessibility. All transformers and transformer vaults shall be readily accessible to qualified personnel for inspection and maintenance or shall meet the requirements of 450.13(A) or 450.13(B).

Transformers are not accessible if wiring methods, or other equipment, obstruct access by a worker or prevent removal of the covers for inspection or maintenance. Practical clearance considerations required for removal and replacement of the transformer are also important.

(A) Open Installations. Dry-type transformers 1000 volts, nominal, or less, located in the open on walls, columns, or structures, shall not be required to be readily accessible.

(B) Hollow Space Installations. Dry-type transformers 1000 volts, nominal, or less and not exceeding 50 kVA shall be permitted in hollow spaces of buildings not permanently closed in by structure, provided they meet the ventilation requirements of 450.9 and separation from combustible materials requirements of 450.21(A). Transformers so installed shall not be required to be readily accessible.

Transformers are permitted by 300.22(C)(3) to be installed in hollow spaces where the space is used for environmental air, provided the transformer is in a metal enclosure (ventilated or nonventilated) and the transformer is suitable for the ambient air temperature within the hollow space.

450.14 Disconnecting Means. Transformers, other than Class 2 or Class 3 transformers, shall have a disconnecting means located either in sight of the transformer or in a remote location. Where located in a remote location, the disconnecting means shall be lockable open in accordance with 110.25, and its location shall be field marked on the transformer.

The requirement for a disconnecting means is especially important in installations utilizing the requirements of 240.21(B)(3) where several transformers in different locations could all be tapped from one feeder, and it would be impractical to de-energize the feeder to work on one of the transformers. The disconnect is required to be located within sight from the transformer but may be in a remote location if it is lockable. The location of any remote disconnect is required to be marked on the transformer.

Part II. Installation

450.21 Dry-Type Transformers Installed Indoors.

(A) Not Over 112½ kVA. Dry-type transformers installed indoors and rated 112½ kVA or less shall have a separation of at least 300 mm (12 in.) from combustible material unless separated from the combustible material by a fire-resistant, heat-insulated barrier.

Exception: This rule shall not apply to transformers rated for 1000 volts, nominal, or less that are completely enclosed, except for ventilating openings.

(B) Over 112½ kVA. Individual dry-type transformers of more than 112½ kVA rating shall be installed in a transformer room of fire-resistant construction having a minimum fire rating of 1 hour.

Exception No. 1: Transformers with Class 155 or higher insulation systems and separated from combustible material by a fire-resistant, heat-insulating barrier or by not less than 1.83 m (6 ft) horizontally and 3.7 m (12 ft) vertically shall not be required to be installed in a transformer room.

Exception No. 2: Transformers with Class 155 or higher insulation systems and completely enclosed except for ventilating openings shall not be required to be installed in a transformer room.

Informational Note: See ASTM E119-18a, *Standard Test Methods for Fire Tests of Building Construction and Materials*, for additional information on fire-resistance ratings.

Dry-type transformers with a Class 155 or higher insulation system rating are not required to be installed in transformer rooms or vaults if space separation or a fire-resistant heat-insulating barrier is provided. Although these units are designed for higher operating temperatures, the need for a transformer vault is mitigated by the fire-resistant characteristics of high-temperature insulations.

Further information on specific transformer class insulation systems may be found in UL 1561, *Dry-Type General Purpose and Power Transformers*.

(C) Over 35,000 Volts. Dry-type transformers rated over 35,000 volts shall be installed in a vault complying with Part III of this article.

Dry-type transformers depend on the surrounding air for adequate ventilation and must comply with 450.9. Where rated 112½ kilovolt-amperes or less, dry-type transformers are not required to be installed in a fire-resistant transformer room. For that reason, dry-type transformers or gas-filled or less-flammable liquid-insulated transformers, with a primary voltage of not more than 35,000 volts, are commonly used indoors.

See also

450.23 and its commentary for more information on less-flammable liquid-insulated transformers

Exhibit 450.12 shows a typical dry-type power transformer rated at 1000 kilovolt-amperes, 13,800 volts to 480 volts, 3-phase, 60 hertz. This transformer has a high-voltage flange and a low-voltage flange for connection to switchgear and a high-voltage, 2-position (double-throw), 3-pole-load air-break switch that may be attached to the case and arranged as a selector switch for connection of the transformer primary to either of two feeder sources.

Dry-type transformers rated 112½ kilovolt-amperes or less require 12 inches of separation from combustible material or separation by fire-resistant barriers. Transformers rated less than 1000 volts and completely enclosed, except for ventilating openings, are exempt from this requirement unless the manufacturer's installation instructions specify clearance distances. Transformers rated over 112½ kilovolt-amperes must be located

EXHIBIT 450.12 A dry-type transformer with a core and coil design rated at 1000 kilovolt-amperes, 13,800 volts to 480 volts, 3-phase, 60 hertz. Note cooling fans beneath each winding. *(Courtesy of Schneider Electric)*

in fire-resistant transformer rooms or vaults unless the transformers have Class 155 or higher insulation ratings.

450.22 Dry-Type Transformers Installed Outdoors. Dry-type transformers installed outdoors shall have a weatherproof enclosure.

Transformers exceeding 112½ kVA shall not be located within 300 mm (12 in.) of combustible materials of buildings unless the transformer has Class 155 insulation systems or higher and is completely enclosed except for ventilating openings.

450.23 Less-Flammable Liquid-Insulated Transformers. Transformers insulated with listed less-flammable liquids that have a fire point of not less than 300°C shall be permitted to be installed in accordance with 450.23(A) or 450.23(B).

(A) Indoor Installations. Indoor installations shall be permitted in accordance with one of the following:

(1) In Type I or Type II buildings, in areas where all of the following requirements are met:

    a. The transformer is rated 35,000 volts or less.
    b. No combustible materials are stored.
    c. A liquid confinement area is provided.
    d. The installation complies with all the restrictions provided for in the listing of the liquid.

Informational Note: Such restrictions can include, but are not limited to, maximum pressure of the tank, use of a pressure relief valve, appropriate fuse types, and proper sizing of overcurrent protection.

(2) If an automatic fire extinguishing system and a liquid confinement area is present, provided the transformer is rated 35,000 volts or less

(3) If the installation complies with 450.26

Δ **(B) Outdoor Installations.** Less-flammable liquid-filled transformers shall be permitted to be installed outdoors, attached to, adjacent to, or on the roof of buildings, if installed in accordance with either of the following:

(1) For Type I and Type II buildings, the installation shall comply with all the restrictions provided for in the listing of the liquid.

Informational Note No. 1: See NFPA 220-2021, *Standard on Types of Building Construction*, for definitions of Type I and Type II building construction.

Informational Note No. 2: Such restrictions can include, but are not limited to, maximum pressure of the tank, use of a pressure relief valve, appropriate fuse types, and proper sizing of overcurrent protection.

(2) In accordance with 450.27.

Informational Note No. 3: See 450.27 for examples of additional safeguards that can be required for installations adjacent to combustible material, fire escapes, or door and window openings.

Restrictions required by the listing of a less-flammable liquid are illustrated by the use of an FM Approvals LLC–approved liquid in a transformer tank. Pressure-relief devices must be provided. Spacing from adjacent buildings or transformers must also be provided. The spacing, as illustrated in Exhibit 450.13, is based not only on the fluid capacity of the transformer tank but also on the listing of the transformer and the building construction. In the event of a leak, the liquid confinement area prevents transformer dielectric fluid from spreading beyond the vicinity of the transformer. Further information on applications may be found in the Factory Mutual Loss Prevention Data Sheet 5-4.

The requirements in 450.23 refer to buildings of Types I and II construction. Table E.1 in Informative Annex E is a summary of the requirements for construction types. The Arabic numerals at the top of the fire resistance rating columns reflect the fire resistance ratings of the following building elements: exterior bearing walls; columns, beams, girders, trusses and arches, support bearing walls, columns, or loads from more than one floor; and the floor construction.

For example, a building of Type I, 442 construction has 4-hour fire-resistance-rated exterior bearing walls; 4-hour fire-resistance-rated columns, beams, girders, trusses, or arches; and 3-hour fire-resistance-rated floor construction. Whether a building is of Type I, Type II, or other type is determined by the requirements of the building construction code adopted by a jurisdiction.

450.24 Nonflammable Fluid-Insulated Transformers. Transformers insulated with a dielectric fluid identified as nonflammable shall be permitted to be installed indoors or outdoors. Such transformers installed indoors and rated over 35,000 volts shall be installed in a vault. Such transformers installed indoors shall be furnished with a liquid confinement area and a pressure-relief vent. The transformers shall be furnished with a means for absorbing any gases generated by arcing inside the tank, or the pressure-relief vent shall be connected to a chimney or flue that will carry such gases to an environmentally safe area.

> Informational Note: Safety may be increased if fire hazard analyses are performed for such transformer installations.

For the purposes of this section, a nonflammable dielectric fluid is one that does not have a flash point or fire point and is not flammable in air.

A liquid confinement area and a pressure-relief vent are required for nonflammable-fluid-insulated transformers. The liquid confinement area limits the extent of a spill if the tank leaks or ruptures. If a means for absorbing gases generated by arcing within the transformer is not provided, the pressure-relief vent must be connected to a chimney or flue that vents to an environmentally safe area.

The need for a gas absorption system or a chimney or flue that vents to an environmentally safe area is due to concerns about products generated during arcing. The high arc temperatures can cause the insulating medium to break down, resulting in the formation and emission of toxic or corrosive compounds.

450.25 Askarel-Insulated Transformers Installed Indoors. Askarel-insulated transformers installed indoors and rated over 25 kVA shall be furnished with a pressure-relief vent. Where installed in a poorly ventilated place, they shall be furnished with a means for absorbing any gases generated by arcing inside the case, or the pressure-relief vent shall be connected to a chimney or flue that carries such gases outside the building. Askarel-insulated transformers rated over 35,000 volts shall be installed in a vault.

Askarel-insulated transformers are no longer manufactured because askarel has been classified as a carcinogen. Very few askarel transformers are in use today. Transformers could be labeled as "PCB Transformer" without having the fire resistance properties of an askarel transformer. That is because federal regulations classify a transformer as a PCB (polychlorinated biphenyls) transformer if it contains 500 parts or more per million PCBs. In the past, transformer liquids could become contaminated with PCBs because the insulating liquid had been temporarily stored in a tank that had previously contained askarel. The information provided in the *NEC*® is for reference and for the modification of existing askarel-insulated installations. Existing askarel-insulated transformers of 35,000 volts or less are not

EXHIBIT 450.13 A transformer tank containing a less-flammable fluid listed by FM Approvals, where the spacing from adjacent combustibles to the liquid confinement area is based on the capacity of the tank.

required to be installed in vaults because askarel is considered a noncombustible fluid.

450.26 Oil-Insulated Transformers Installed Indoors. Oil-insulated transformers installed indoors shall be installed in a vault constructed as specified in Part III of this article.

Exception No. 1: Where the total capacity does not exceed 112½ kVA, the vault specified in Part III of this article shall be permitted to be constructed of reinforced concrete that is not less than 100 mm (4 in.) thick.

Exception No. 2: Where the nominal voltage does not exceed 1000, a vault shall not be required if suitable arrangements are made to prevent a transformer oil fire from igniting other materials and the total capacity in one location does not exceed 10 kVA in a section of the building classified as combustible or 75 kVA where the surrounding structure is classified as fire-resistant construction.

Exception No. 3: Electric furnace transformers that have a total rating not exceeding 75 kVA shall be permitted to be installed without a vault in a building or room of fire-resistant construction, provided suitable arrangements are made to prevent a transformer oil fire from spreading to other combustible material.

Exception No. 4: A transformer that has a total rating not exceeding 75 kVA and a supply voltage of 1000 volts or less that is an integral part of charged-particle-accelerating equipment shall be permitted to be installed without a vault in a building or room of noncombustible or fire-resistant construction, provided suitable arrangements are made to prevent a transformer oil fire from spreading to other combustible material.

Exception No. 5: Transformers shall be permitted to be installed in a detached building that does not comply with Part III of this article if neither the building nor its contents present a fire hazard to any other building or property, and if the building is used only in supplying electric service and the interior is accessible only to qualified persons.

Exception No. 6: Oil-insulated transformers shall be permitted to be used without a vault in portable and mobile surface mining equipment (such as electric excavators) if each of the following conditions is met:

(1) Provision is made for draining leaking fluid to the ground.
(2) Safe egress is provided for personnel.
(3) A minimum 6-mm (1/4-in.) steel barrier is provided for personnel protection.

450.27 Oil-Insulated Transformers Installed Outdoors. Combustible material, combustible buildings, and parts of buildings, fire escapes, and door and window openings shall be safeguarded from fires originating in oil-insulated transformers installed on roofs, attached to or adjacent to a building or combustible material.

In cases where the transformer installation presents a fire hazard, one or more of the following safeguards shall be applied according to the degree of hazard involved:

(1) Space separations
(2) Fire-resistant barriers
(3) Automatic fire suppression systems
(4) Enclosures that confine the oil of a ruptured transformer tank

Oil enclosures shall be permitted to consist of fire-resistant dikes, curbed areas or basins, or trenches filled with coarse, crushed stone. Oil enclosures shall be provided with trapped drains where the exposure and the quantity of oil involved are such that removal of oil is important.

Informational Note: See ANSI/IEEE C2-2017, *National Electrical Safety Code*, for additional information on transformers installed on poles or structures or underground.

450.28 Modification of Transformers. When modifications are made to a transformer in an existing installation that change the type of the transformer with respect to Part II of this article, such transformer shall be marked to show the type of insulating liquid installed, and the modified transformer installation shall comply with the applicable requirements for that type of transformer.

An existing askarel-insulated transformer may have the askarel replaced with either oil or a less-flammable liquid. Where such a modification takes place, the completed installation must have the same degree of safety as a new installation. For example, replacement of askarel with oil in an indoor installation without a vault may not be acceptable (see 450.26 and its exceptions). The same is true if the replacement liquid is a less-flammable liquid (see 450.23). Additional safety precautions may be necessary to compensate for the different fire characteristics of the new dielectric fluid.

Part III. Transformer Vaults

450.41 Location. Vaults shall be located where they can be ventilated to the outside air without using flues or ducts wherever such an arrangement is practicable.

450.42 Walls, Roofs, and Floors. The walls and roofs of vaults shall be constructed of materials that have approved structural strength for the conditions with a minimum fire resistance of 3 hours. The floors of vaults in contact with the earth shall be of concrete that is not less than 100 mm (4 in.) thick, but, where the vault is constructed with a vacant space or other stories below it, the floor shall have approved structural strength for the load imposed thereon and a minimum fire resistance of 3 hours. For the purposes of this section, studs and wallboard construction shall not be permitted.

Exception: Where transformers are protected with automatic sprinkler, water spray, carbon dioxide, or halon, construction of 1-hour rating shall be permitted.

Informational Note No. 1: See ASTM E119-20, *Standard Test Methods for Fire Tests of Building Construction and Materials*, for additional information.

Informational Note No. 2: A typical 3-hour construction is 150 mm (6 in.) thick reinforced concrete.

Vaults are intended primarily as passive fire protection. The need for vaults is dictated by the combustibility of the dielectric media and the size of the transformer. Transformers insulated with mineral oil have the greatest need for passive protection to prevent the spread of burning oil to other combustible materials.

Although construction of a 3-hour-rated wall may be possible using studs and wallboard, this construction method is not permitted for transformer vaults because of the concern that projectiles created in a transformer explosion be contained. A reduction in fire resistance rating from 3 hours to 1 hour is permitted for vaults equipped with an automatic fire suppression system.

See also

450.23 and its commentary, which relates to Type I and Type II building construction

450.43 Doorways. Vault doorways shall be protected in accordance with 450.43(A), (B), and (C).

(A) Type of Door. Each doorway leading into a vault from the building interior shall be provided with a tight-fitting door that has a minimum fire rating of 3 hours. The authority having jurisdiction shall be permitted to require such a door for an exterior wall opening where conditions warrant.

Exception: Where transformers are protected with automatic sprinkler, water spray, carbon dioxide, or halon, construction of 1-hour rating shall be permitted.

Informational Note: See NFPA 80-2019, *Standard for Fire Doors and Other Opening Protectives*, for additional information.

(B) Sills. A door sill or curb that is of an approved height that will confine the oil from the largest transformer within the vault shall be provided, and in no case shall the height be less than 100 mm (4 in.).

(C) Accessibility. Doors shall be equipped with locks, and doors shall be kept locked, with access being allowed only to qualified persons. Personnel doors shall be capable of opening not less than 90 degrees in the direction of egress and be equipped with listed fire exit hardware.

Section 450.43 requires transformer vault doors to open in the direction of egress and be equipped with listed fire exit hardware. An injured worker attempting to escape from a transformer vault may not be able to operate a rotating-type doorknob but would be able to escape through a door equipped with panic-type hardware.

450.45 Ventilation Openings. Where required by 450.9, openings for ventilation shall be provided in accordance with 450.45(A) through (F).

(A) Location. Ventilation openings shall be located as far as possible from doors, windows, fire escapes, and combustible material.

(B) Arrangement. A vault ventilated by natural circulation of air shall be permitted to have roughly half of the total area of openings required for ventilation in one or more openings near the floor and the remainder in one or more openings in the roof or in the sidewalls near the roof, or all of the area required for ventilation shall be permitted in one or more openings in or near the roof.

(C) Size. For a vault ventilated by natural circulation of air to an outdoor area, the combined net area of all ventilating openings, after deducting the area occupied by screens, gratings, or louvers, shall not be less than 1900 mm^2 (3 in.^{2}) per kVA of transformer capacity in service, and in no case shall the net area be less than 0.1 m^2 (1 ft^2) for any capacity under 50 kVA.

(D) Covering. Ventilation openings shall be covered with durable gratings, screens, or louvers, according to the treatment required in order to avoid unsafe conditions.

(E) Dampers. All ventilation openings to the indoors shall be provided with automatic closing fire dampers that operate in response to a vault fire. Such dampers shall possess a standard fire rating of not less than 1½ hours.

Informational Note: See ANSI/UL 555-2020, *Standard for Fire Dampers*, for additional information on fire dampers.

(F) Ducts. Ventilating ducts shall be constructed of fire-resistant material.

450.46 Drainage. Where practicable, vaults containing more than 100 kVA transformer capacity shall be provided with a drain or other means that will carry off any accumulation of oil or water in the vault unless local conditions make this impracticable. The floor shall be pitched to the drain where provided.

450.47 Water Pipes and Accessories. Any pipe or duct system foreign to the electrical installation shall not enter or pass through a transformer vault. Piping or other facilities provided for vault fire protection, or for transformer cooling, shall not be considered foreign to the electrical installation.

Automatic sprinkler protection is permitted for transformer vaults. Piping or ductwork for cooling of the transformer is also permitted to be installed in a transformer vault. No other piping or ductwork is permitted to enter or pass through a transformer vault.

450.48 Storage in Vaults. Materials shall not be stored in transformer vaults.

ARTICLE 455 Phase Converters

Part I. General

455.1 Scope. This article covers the installation and use of phase converters.

A phase converter is an electrical device that converts single-phase electrical power to 3-phase for the operation of equipment that normally operates from a 3-phase electrical supply. Phase converters are of two types: static, with no moving parts, and rotary, with an internal rotor that must be rotating before a load is applied.

See also

Article 100 for definitions of the terms *phase converter, rotary* and *phase converter, static*

Phase converters are frequently used to supply 3-phase motor loads in locations where only single-phase power is available from the local utility. Electrical installations on farms and in other remote or rural areas are examples of such locations. Although their most common loads are motors, phase converters are increasingly used to supply loads such as cellular telephone and other communication transmitter sites.

455.3 Other Articles. Phase converters shall comply with this article and with the applicable provisions of other articles of this *Code*.

455.4 Marking. Each phase converter shall be provided with a permanent nameplate indicating the following:

(1) Manufacturer's name
(2) Rated input and output voltages
(3) Frequency
(4) Rated single-phase input full-load amperes
(5) Rated minimum and maximum single load in kilovolt-amperes (kVA) or horsepower
(6) Maximum total load in kilovolt-amperes (kVA) or horsepower
(7) For a rotary-phase converter, 3-phase amperes at full load

455.5 Equipment Grounding Connection. A means for attachment of an equipment grounding conductor termination in accordance with 250.8 shall be provided.

455.6 Conductors.

(A) Ampacity. The ampacity of the single-phase supply conductors shall be determined by 455.6(A)(1) or (A)(2).

Informational Note: Single-phase conductors sized to prevent a voltage drop not exceeding 3 percent from the source of supply

to the phase converter may help ensure proper starting and operation of motor loads.

(1) Variable Loads. Where the loads to be supplied are variable, the conductor ampacity shall not be less than 125 percent of the phase converter nameplate single-phase input full-load amperes.

(2) Fixed Loads. Where the phase converter supplies specific fixed loads, and the conductor ampacity is less than 125 percent of the phase converter nameplate single-phase input full-load amperes, the conductors shall have an ampacity not less than 250 percent of the sum of the full-load, 3-phase current rating of the motors and other loads served where the input and output voltages of the phase converter are identical. Where the input and output voltages of the phase converter are different, the current as determined by this section shall be multiplied by the ratio of output to input voltage.

Calculation Example

Determine the minimum ampacity for the single-phase input conductors supplying a phase converter used to supply 3-phase receptacles for electric welders. The number of welders connected at any one time is variable. The phase converter has a 230 V, single-phase input and a 230 V, 3-phase output. The phase converter input current is 105 A.

Solution

Because the loads are variable, 455.6(A)(1) applies.

$$1.25 \times 105 \text{ A} = 131.25 \text{ A}$$

If the same phase converter is used to supply only two 3-phase, 5 hp, 230 V motors, what is the minimum ampacity of the single-phase input conductors?

Solution

Because the loads are fixed, 455.6(A)(1) applies. According to Table 430.250, the full-load current (FLC) for a 10 hp, 230 V, 3-phase motor is 15.2 A. Thus,

$$2.5 \times (15.2 + 15.2) = 76 \text{ A}$$

(B) Manufactured Phase Marking. The manufactured phase conductors shall be identified in all accessible locations with a distinctive marking. The marking shall be consistent throughout the system and premises.

The *phase, manufactured (manufactured phase)* as defined in Article 100 is the phase that is created within the rotary or static equipment and is not solidly connected to the input conductors. Identification of the manufactured phase is necessary to help installers comply with 455.9, which does not permit single-phase loads to be supplied from the manufactured phase. The method of identification is not specified by the *NEC®*; therefore, it could be by any means acceptable to the AHJ. While many phase converters are installed to supply a specific item of equipment such as a motor or heating equipment, the use of a 3-phase distribution equipment such as a panelboard supplied from the output of a phase converter is also acceptable and may be desirable as a central point from which to run branch circuits to equipment. Identification of the manufactured phase conductor at a central distribution point such as a panelboard provides the installer with visible indication of the phase conductor that cannot be used to supply any single-phase loads that have branch circuits originating from that equipment.

455.7 Overcurrent Protection. The single-phase supply conductors and phase converter shall be protected from overcurrent by 455.7(A) or (B). Where the required fuse or nonadjustable circuit breaker rating or settings of adjustable circuit breakers do not correspond to a standard rating or setting, a higher rating or setting that does not exceed the next higher standard rating shall be permitted.

(A) Variable Loads. Where the loads to be supplied are variable, overcurrent protection shall be set at not more than 125 percent of the phase converter nameplate single-phase input full-load amperes.

(B) Fixed Loads. Where the phase converter supplies specific fixed loads and the conductors are sized in accordance with 455.6(A)(2), the conductors shall be protected in accordance with their ampacity. The overcurrent protection determined from this section shall not exceed 125 percent of the phase converter nameplate single-phase input amperes.

455.8 Disconnecting Means. Means shall be provided to disconnect simultaneously all ungrounded single-phase supply conductors to the phase converter.

(A) Location. The disconnecting means shall be readily accessible and located in sight from the phase converter.

(B) Type. The disconnecting means shall be a switch rated in horsepower, a circuit breaker, or a molded-case switch. Where only nonmotor loads are served, an ampere-rated switch shall be permitted.

(C) Rating. The ampere rating of the disconnecting means shall be not less than 115 percent of the rated maximum single-phase input full-load amperes or, for specific fixed loads, shall be permitted to be selected from 455.8(C)(1) or (C)(2).

(1) Current Rated Disconnect. The disconnecting means shall be a circuit breaker or molded-case switch with an ampere rating not less than 250 percent of the sum of the following:

(1) Full-load, 3-phase current ratings of the motors
(2) Other loads served

(2) Horsepower Rated Disconnect. The disconnecting means shall be a switch with a horsepower rating. The equivalent locked rotor current of the horsepower rating of the switch shall not be less than 200 percent of the sum of the following:

(1) Nonmotor loads
(2) The 3-phase, locked-rotor current of the largest motor as determined from Table 430.251(B)
(3) The full-load current of all other 3-phase motors operating at the same time

(D) Voltage Ratios. The calculations in 455.8(C) shall apply directly where the input and output voltages of the phase converter are identical. Where the input and output voltages of the phase converter are different, the current shall be multiplied by the ratio of the output to input voltage.

455.9 Connection of Single-Phase Loads. Where single-phase loads are connected on the load side of a phase converter, they shall not be connected to the manufactured phase.

455.10 Terminal Housings. A terminal housing in accordance with the provisions of 430.12 shall be provided on a phase converter.

Part II. Specific Provisions Applicable to Different Types of Phase Converters

455.20 Disconnecting Means. The single-phase disconnecting means for the input of a static phase converter shall be permitted to serve as the disconnecting means for the phase converter and a single load if the load is within sight of the disconnecting means.

455.21 Start-Up. Power to the utilization equipment shall not be supplied until the rotary-phase converter has been started.

455.22 Power Interruption. Utilization equipment supplied by a rotary-phase converter shall be controlled in such a manner that power to the equipment will be disconnected in the event of a power interruption.

> Informational Note: Magnetic motor starters, magnetic contactors, and similar devices, with manual or time delay restarting for the load, provide restarting after power interruption.

455.23 Capacitors. Capacitors that are not an integral part of the rotary-phase conversion system but are installed for a motor load shall be connected to the line side of that motor overload protective device.

ARTICLE 460 Capacitors

N **Part I. General**

Δ **460.1 Scope.** This article covers the installation of capacitors on electrical circuits.

Surge capacitors or capacitors included as a component part of other apparatus and conforming with the requirements of such apparatus are excluded from these requirements.

460.3 Enclosing and Guarding.

(A) Containing More Than 11 L (3 gal) of Flammable Liquid. Capacitors containing more than 11 L (3 gal) of flammable liquid shall be enclosed in vaults or outdoor fenced enclosures complying with Article 110, Part III. This limit shall apply to any single unit in an installation of capacitors.

(B) Accidental Contact. Where capacitors are accessible to unauthorized and unqualified persons, they shall be enclosed, located, or guarded so that persons cannot come into accidental contact or bring conducting materials into accidental contact with exposed energized parts, terminals, or buses associated with them. However, no additional guarding is required for enclosures accessible only to authorized and qualified persons.

Part II. 1000 Volts, Nominal, or Less

460.6 Discharge of Stored Energy. Capacitors shall be provided with a means of discharging stored energy.

(A) Time of Discharge. The residual voltage of a capacitor shall be reduced to 50 volts, nominal, or less within 1 minute after the capacitor is disconnected from the source of supply.

(B) Means of Discharge. The discharge circuit shall be either permanently connected to the terminals of the capacitor or capacitor bank or provided with automatic means of connecting it to the terminals of the capacitor bank on removal of voltage from the line. Manual means of switching or connecting the discharge circuit shall not be used.

460.8 Conductors.

(A) Ampacity. The ampacity of capacitor circuit conductors shall not be less than 135 percent of the rated current of the capacitor. The ampacity of conductors that connect a capacitor to the terminals of a motor or to motor circuit conductors shall not be less than one-third the ampacity of the motor circuit conductors and in no case less than 135 percent of the rated current of the capacitor.

Capacitors are rated in reactive kilovolt-amperes (kilovars or kVAr) or kilovolt-amperes capacitive (kVAc). Both ratings are synonymous. The kVAr rating shows how many reactive kilovolt-amperes the capacitor will supply to cancel out the reactive kilovolt-amperes caused by inductance. For example, a 20-kVAr capacitor will cancel out 20 kVAr of inductive reactive kilovolt-amperes.

The capacitor circuit conductors and disconnecting means must have an ampacity not less than 135 percent of the rated current of the capacitor. Capacitors are manufactured with a tolerance of zero percent to 15 percent, so a 100-kVAr capacitor can draw a current equivalent to that of a 115-kVAr capacitor. In addition, the current draw varies directly with the line voltage, and

any variation in the line voltage from a pure sine wave form causes the capacitor to draw an increased current. Considering these factors, the increased current can amount to 135 percent of the rated current of the capacitor.

The current corresponding to the kVAr rating of a 3-phase capacitor, i_c, is computed from the following formula:

$$i_c = \frac{kVAr \times 1000}{\sqrt{3} \times V}$$

The ampacity of the conductors and the disconnecting device is then determined by multiplying i_c by 1.35.

Where harmonic-producing loads are present, adding capacitors to the electrical system can place the system in a harmonic resonance condition. The harmonic loads can excite the electrical system at the harmonic resonance frequency and cause overcurrent and overvoltage conditions. If capacitors are to be placed on electrical systems with harmonic loads, an engineering study should be conducted that evaluates the size and placement of capacitors and the reactive impedance and load of the system. Capacitors may need a reactor placed in series with them to help detune the electrical system from a harmonic resonance condition.

(B) Overcurrent Protection. An overcurrent device shall be provided in each ungrounded conductor for each capacitor bank. The rating or setting of the overcurrent device shall be as low as practicable.

Exception: A separate overcurrent device shall not be required for a capacitor connected on the load side of a motor overload protective device.

Unless the exception applies, the overcurrent device must be separate from the overcurrent device protecting any other equipment or conductor. See Exhibit 460.1, diagrams (a) and (b).

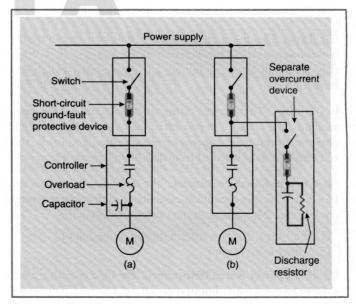

EXHIBIT 460.1 Methods of connecting capacitors in induction motor circuit for power factor correction.

(C) Disconnecting Means. A disconnecting means shall be provided in each ungrounded conductor for each capacitor bank and shall meet the following requirements:

(1) The disconnecting means shall open all ungrounded conductors simultaneously.
(2) The disconnecting means shall be permitted to disconnect the capacitor from the line as a regular operating procedure.
(3) The rating of the disconnecting means shall not be less than 135 percent of the rated current of the capacitor.

Exception: A separate disconnecting means shall not be required where a capacitor is connected on the load side of a motor controller.

460.9 Rating or Setting of Motor Overload Device. Where a motor installation includes a capacitor connected on the load side of the motor overload device, the rating or setting of the motor overload device shall be based on the improved power factor of the motor circuit.

The effect of the capacitor shall be disregarded in determining the motor circuit conductor rating in accordance with 430.22.

Where a capacitor is connected on the load side of the motor overload relays [see Exhibit 460.1, diagram (a)], the line current will be reduced due to an improved power factor, which must be taken into account when selecting the rating of a motor overload device. A value lower than that specified in 430.32 should be used for proper protection of the motor.

The most effective power factor correction is obtained where the individual capacitors are connected closest to the inductive load. Capacitor manufacturers publish tables in which the required capacitor value is obtained by referring to the speed and the horsepower of the motor. Those values improve the motor power factor to approximately 95 percent. To improve a plant power factor, capacitor manufacturers also publish tables to assist in the calculation of the total kVAr rating of capacitors required to improve the power factor to any desired value.

460.10 Grounding. Capacitor cases shall be connected to the equipment grounding conductor.

Exception: Capacitor cases shall not be connected to the equipment grounding conductor where the capacitor units are supported on a structure designed to operate at other than ground potential.

460.12 Marking. Each capacitor shall be provided with a nameplate giving the name of the manufacturer, rated voltage, frequency, kilovar or amperes, number of phases, and, if filled with a combustible liquid, the volume of liquid. Where filled with a nonflammable liquid, the nameplate shall so state. The nameplate shall also indicate whether a capacitor has a discharge device inside the case.

Part III. Over 1000 Volts, Nominal

460.24 Switching.

(A) Load Current. Switches shall be rated for switching of capacitive loads. Capacitor switch operation shall open all ungrounded conductors and the switch shall be capable of the following:

(1) Carrying continuously not less than 135 percent of the rated current of the capacitor installation
(2) Interrupting the maximum continuous load current of each capacitor, capacitor bank, or capacitor installation that will be switched as a unit
(3) Withstanding the maximum inrush current, including contributions from adjacent capacitor installations
(4) Carrying currents due to faults on capacitor side of switch

(B) Isolation.

(1) General. A means shall be installed to isolate from all sources of voltage each capacitor, capacitor bank, or capacitor installation that will be removed from service as a unit. The isolating means shall provide a visible gap in the electrical circuit adequate for the operating voltage.

Δ **(2) Isolating or Disconnecting Switches with No Interrupting Rating.** Isolating or disconnecting switches (with no interrupting rating) shall be interlocked with the load-interrupting device or be provided with prominently displayed caution signs in accordance with 495.22 to prevent switching load current.

(C) Additional Requirements for Series Capacitors. The proper switching sequence shall be ensured by use of one of the following:

(1) Mechanically sequenced isolating and bypass switches
(2) Interlocks
(3) Switching procedure prominently displayed at the switching location

460.25 Overcurrent Protection.

(A) Provided to Detect and Interrupt Fault Current. A means shall be provided to detect and interrupt fault current likely to cause dangerous pressure within an individual capacitor.

(B) Single Pole or Multipole Devices. Single-pole or multipole devices shall be permitted for this purpose.

(C) Protected Individually or in Groups. Capacitors shall be permitted to be protected individually or in groups.

(D) Protective Devices Rated or Adjusted. Protective devices for capacitors or capacitor equipment shall be rated or adjusted to operate within the limits of the safe zone for individual capacitors.

460.26 Identification.

Each capacitor shall be provided with a permanent nameplate giving the manufacturer's name, rated voltage, frequency, kilovar or amperes, number of phases, and the volume of liquid identified as flammable, if such is the case.

460.27 Grounding. Capacitor cases shall be connected to the equipment grounding conductor. If the capacitor neutral point is connected to a grounding electrode conductor, the connection shall be made in accordance with Part III of Article 250.

Exception: Capacitor cases shall not be connected to the equipment grounding conductor where the capacitor units are supported on a structure designed to operate at other than ground potential.

460.28 Means for Discharge.

(A) Means to Reduce the Residual Voltage. A means shall be provided to reduce the residual voltage of a capacitor to 50 volts or less within 5 minutes after the capacitor is disconnected from the source of supply.

(B) Connection to Terminals. A discharge circuit shall be either permanently connected to the terminals of the capacitor or provided with automatic means of connecting it to the terminals of the capacitor bank after disconnection of the capacitor from the source of supply. The windings of motors, transformers, or other equipment directly connected to capacitors without a switch or overcurrent device interposed shall meet the requirements of 460.28(A).

Means are required to drain off the stored charge in a capacitor after the supply circuit has been opened. Otherwise, a person servicing the equipment could receive a severe shock, or damage could occur to the equipment.

Exhibit 460.1, diagram (a), shows a method in which capacitors are connected in a motor circuit so that they can be switched with the motor. In that arrangement, the stored charge drains off through the windings when the circuit is opened. Diagram (b) shows another arrangement in which the capacitor is connected

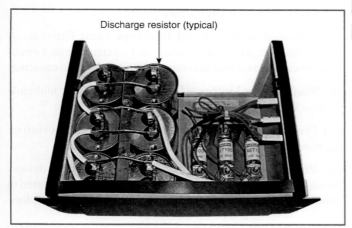

EXHIBIT 460.2 Power factor correction capacitors with internal discharge resistors (blue) and overcurrent protection. *(Courtesy of GE Energy)*

to the line side of the motor starter contacts. An automatic discharge device and a separate disconnecting means are required. As shown in Exhibit 460.2, capacitors often are equipped with built-in resistors to drain off the stored charge; however, that type of capacitor is not needed where the connection is as shown in Exhibit 460.1, diagram (a).

ARTICLE
470 Resistors and Reactors

Part I. General

Δ **470.1 Scope.** This article covers the installation of separate resistors and reactors on electrical circuits.

Exception: This article does not cover resistors and reactors that are component parts of other apparatus.

Resistors are made in many sizes and shapes and for different purposes. They can be wire, ribbon, form or edgewise wound, cast or punched steel grid, or box resistors. They can be mounted in the open or in ventilated metal boxes or cabinets, depending on their use and location. Because they give off heat, resistors must be guarded and located at safe distances from combustible materials. Where mounted on switchboards or installed in control panels, they are not required to have additional guards.

Current-limiting reactors are installed to limit the amount of current in a circuit under short circuit conditions. Reactors can be divided into two classes: those with iron cores and those with no magnetic materials in the windings. Both types can be air cooled or oil immersed.

Mechanical stresses exist between adjacent air-core reactors due to their external fields, and the manufacturer's recommendations should be followed in spacing and bracing units and fastening supporting insulators.

N **470.2 Reconditioned Equipment.**

N **(A) Resistors.** Reconditioned resistors shall not be permitted.

N **(B) Reactors.** Reconditioned reactors shall be permitted.

N ## Part II. 1000 Volts, Nominal, or Less

470.10 Location. Resistors and reactors shall not be placed where exposed to physical damage.

470.11 Space Separation. A thermal barrier shall be required if the space between the resistors or reactors and any combustible material is less than 305 mm (12 in.).

470.12 Conductor Insulation. Insulated conductors used for connections between resistance elements and controllers shall be suitable for an operating temperature of not less than 90°C (194°F).

Exception: Other conductor insulations shall be permitted for the motor starting service.

Part III. Over 1000 Volts, Nominal

470.20 General.

(A) Protected Against Physical Damage. Resistors and reactors shall be protected against physical damage.

(B) Isolated by Enclosure or Elevation. Resistors and reactors shall be isolated by enclosure or elevation to protect personnel from accidental contact with energized parts.

(C) Combustible Materials. Resistors and reactors shall not be installed in close enough proximity to combustible materials to constitute a fire hazard and shall have a clearance of not less than 305 mm (12 in.) from combustible materials.

(D) Clearances. Clearances from resistors and reactors to grounded surfaces shall be adequate for the voltage involved.

(E) Temperature Rise from Induced Circulating Currents. Metallic enclosures of reactors and adjacent metal parts shall be installed so that the temperature rise from induced circulating currents is not hazardous to personnel or does not constitute a fire hazard.

470.21 Grounding. Resistor and reactor cases or enclosures shall be connected to the equipment grounding conductor.

Exception: Resistor or reactor cases or enclosures supported on a structure designed to operate at other than ground potential shall not be connected to the equipment grounding conductor.

470.22 Oil-Filled Reactors. Installation of oil-filled reactors, in addition to the above requirements, shall comply with applicable requirements of Part II and Part III of Article 450.

ARTICLE
480 Stationary Standby Batteries

Δ **480.1 Scope.** This article applies to all installations of stationary standby batteries having a capacity greater than 3.6 MJ (1 kWh).

Informational Note No. 1: See Article 706 for installations that do not meet the definition of stationary standby batteries.

Informational Note No. 2: The following standards are frequently referenced for the installation of stationary batteries:

(1) IEEE 484, *Recommended Practice for Installation Design and Installation of Vented Lead-Acid Batteries for Stationary Applications*

(2) IEEE 485, *Recommended Practice for Sizing Vented Lead-Acid Storage Batteries for Stationary Applications*

(3) IEEE 1145, *Recommended Practice for Installation and Maintenance of Nickel-Cadmium Batteries for Photovoltaic (PV) Systems*

(4) IEEE 1187, *IEEE Recommended Practice for Installation Design, and Installation of Valve-Regulated Lead-Acid Batteries for Stationary Applications*

(5) IEEE 1375, *IEEE Guide for the Protection of Stationary Battery Systems*

(6) IEEE 1578, *Recommended Practice for Stationary Battery Electrolyte Spill Containment and Management*

(7) IEEE 1635/ASHRAE 21, *Guide for the Ventilation and Thermal Management of Batteries for Stationary Applications*

(8) UL 1973, *Standard for Batteries for Use in Stationary, Vehicle Auxiliary Power, and Light Electric Rail (LER) Applications*

(9) UL Subject 2436, *Outline of Investigation for Spill Containment for Stationary Lead Acid Battery Systems*

(10) UL 1989, *Standard for Standby Batteries*

(11) UL Subject 1974, *Standard for Evaluation of Repurposed Batteries*

(12) NFPA 855-2020, *Standard for the Installation of Stationary Energy Storage Systems*

The most common types of storage cells are the lead-acid type, the alkali (nickel-cadmium) type, and the lithium-ion type. A lead-acid cell consists of a positive plate, usually lead peroxide (a semisolid compound) mounted on a framework or grid for support, and a negative plate, made of sponge lead mounted on a grid. Grids generally are made of a lead alloy, such as lead calcium, lead antimony, or lead selenium. The electrolyte is sulfuric acid and distilled water. Lithium-ion batteries, which are used in a variety of consumer electronics products, use a variety of different chemistries. They are increasingly being used in large-scale applications because they have a very high energy density.

Lead-acid cells can be of the vented or sealed (valve-regulated) type. Under normal charging conditions, the vented type will liberate gases — hydrogen at the negative plate and oxygen at the positive plate. The valve-regulated type provides a means to recombine the gases, thus minimizing emissions from the cell.

In the alkali, or nickel-cadmium, battery, the principal active material in the positive plate is nickelous hydroxide; in the negative plate, it is cadmium hydroxide. The electrolyte is potassium hydroxide (an alkali).

In stationary installations, nickel-cadmium cells generally are of the vented type and liberate hydrogen and oxygen during normal charging. Hermetically sealed nickel-cadmium cells are sometimes used, but they require special charging equipment to prevent gas emissions.

Although some of the newer-technology batteries do not ventilate hydrogen under normal operation, they can generate hydrogen during fault conditions.

See also

Article 706 for energy storage systems

480.3 Equipment. Storage batteries and battery management equipment shall be listed. This requirement shall not apply to lead-acid batteries.

As energy demands increase, many stationary battery types will be introduced into the electrical infrastructure. Standards exist for stationary batteries to support a practical and reasonable implementation of these technologies within the electrical infrastructure. This requirement does not require listing of battery systems in Article 480 — it only addresses listing of the batteries themselves, other than the lead-acid type, and battery management equipment (such as charge controllers).

480.4 Battery and Cell Terminations.

Δ **(A) Corrosion Prevention.** Where mating dissimilar metals, antioxidant material suitable for the battery connection shall be used where recommended by the battery manufacturer's installation and instruction manual.

(B) Intercell and Intertier Conductors and Connections. The ampacity of field-assembled intercell and intertier connectors and conductors shall be of such cross-sectional area that the temperature rise under maximum load conditions and at maximum ambient temperature shall not exceed the safe operating temperature of the conductor insulation or of the material of the conductor supports.

Informational Note: Conductors sized to prevent a voltage drop exceeding 3 percent of maximum anticipated load, and where the maximum total voltage drop to the furthest point of connection does not exceed 5 percent, may not be appropriate for all battery applications. IEEE 1375-2003, *Guide for the Protection of Stationary Battery Systems,* provides guidance for overcurrent protection and associated cable sizing.

(C) Battery Terminals. Electrical connections to the battery, and the cable(s) between cells on separate levels or racks, shall not put mechanical strain on the battery terminals. Terminal plates shall be used where practicable.

Informational Note: Conductors are commonly pre-formed to eliminate stress on battery terminations. Fine stranded cables may also eliminate the stress on battery terminations. See the manufacturer's instructions for guidance.

(D) Accessibility. The terminals of all cells or multicell units shall be readily accessible for readings, inspections, and cleaning where required by the equipment design. One side of transparent battery containers shall be readily accessible for inspection of the internal components.

480.5 Wiring and Equipment Supplied from Batteries. Wiring and equipment supplied from storage batteries shall be subject to the applicable provisions of this *Code* applying to wiring and equipment operating at the same voltage, unless otherwise permitted by 480.6.

480.6 Overcurrent Protection for Prime Movers. Overcurrent protection shall not be required for conductors from a battery with a voltage of 60 volts dc or less if the battery provides power for starting, ignition, or control of prime movers. Section 300.3 shall not apply to these conductors.

The requirement to use Chapter 3 wiring methods is not applicable to battery-powered single conductors used to perform certain operational functions associated with a prime mover, such as for starting a stationary combustion engine that drives an electric generator. For example, if it is necessary to extend the conductors from the battery to the prime mover starting solenoid at a generator location, battery-powered conductors would not be required to have overcurrent protection and could be run as open, single conductors.

480.7 DC Disconnect Methods.

Δ **(A) Disconnecting Means.** A disconnecting means shall be provided for all ungrounded conductors derived from a stationary standby battery with a voltage over 60 volts dc. A disconnecting means shall be readily accessible and located within sight of the stationary standby battery.

Informational Note: See 240.21(H) for information on the location of the overcurrent device for battery conductors.

Battery systems need maintenance to remain functional. In some cases, such as in 700.3(C), the *NEC*® requires battery system maintenance. For maintenance to be performed safely on a stationary battery system, a readily accessible disconnect means located within sight of the battery system is required. It should be noted that the disconnecting means only isolates the batteries from the equipment and loads being supplied by the battery system. Similar to disconnecting means requirements in Article 690 for photovoltaic electric systems, it is recognized that the batteries, like the photovoltaic modules, cannot be shut down or de-energized. However, having a disconnecting means within sight of the power source minimizes the amount of line-side conductor that remains energized after the disconnecting means is turned off or put into the open position.

See also

240.21(H), which addresses the location of overcurrent protection devices for battery conductors

Δ **(B) Emergency Disconnect.** For one-family and two-family dwellings, a disconnecting means or its remote control for a stationary standby battery shall be located at a readily accessible location outside the building for emergency use. The disconnect shall be labeled as follows:

EMERGENCY DISCONNECT

This requirement applies only to the battery systems covered under the scope of Article 480 that are installed as a primary or standby power source at one- and two-family dwellings. If a battery-type energy storage system (ESS) is installed as a power source for a one- or two-family dwelling, the requirements of Article 706 apply to the installation.

See also

706.15 for the requirements on ESS disconnecting means at one- and two-family dwellings

(C) Disconnection of Series Battery Circuits. Battery circuits exceeding 240 volts dc nominal between conductors or to ground and subject to field servicing shall have provisions to disconnect the series-connected strings into segments not exceeding 240 volts dc nominal for maintenance by qualified persons. Non-load-break bolted or plug-in disconnects shall be permitted.

(D) Remote Actuation. Where a disconnecting means, located in accordance with 480.7(A), is provided with remote controls to activate the disconnecting means and the controls for the disconnecting means are not located within sight of the stationary standby battery, the disconnecting means shall be capable of being locked in the open position, in accordance with 110.25, and the location of the controls shall be field marked on the disconnecting means.

(E) Busway. Where a dc busway system is installed, the disconnecting means shall be permitted to be incorporated into the busway.

(F) Notification. The disconnecting means shall be legibly marked in the field. A label with the marking shall be placed in a conspicuous location near the battery if a disconnecting means is not provided. The marking shall be of sufficient durability to withstand the environment involved and shall include the following:

(1) Nominal battery voltage
(2) Available fault current derived from the stationary standby battery

Informational Note No. 1: Battery equipment suppliers can provide information about available fault current on specific battery models.

(3) An arc flash label in accordance with acceptable industry practice

Informational Note No. 2: See NFPA *70E*-2021, *Standard for Electrical Safety in the Workplace,* for assistance in determining the severity of potential exposure, planning safe work practices, arc flash labeling, and selecting personal protective equipment.

(4) Date the calculation was performed

Exception: List items (2), (3), and (4) shall not apply to one- and two-family dwellings.

(G) Identification of Power Sources. Stationary standby batteries shall be indicated by 480.7(G)(1) and (G)(2).

Δ **(1) Facilities with Utility Services and Stationary Standby Batteries.** Plaques or directories shall be installed in accordance with 705.10.

Exception: This requirement does not apply where a disconnect in 480.7(A) is not required.

(2) Facilities with Stand-Alone Systems. A permanent plaque or directory shall be installed in accordance with 710.10.

480.8 Insulation of Batteries. Batteries constructed of an electrically conductive container shall have insulating support if a voltage is present between the container and ground.

480.9 Battery Support Systems. For battery chemistries with corrosive electrolyte, the structure that supports the battery shall be resistant to deteriorating action by the electrolyte. Metallic structures shall be provided with nonconducting support members for the cells, or shall be constructed with a continuous insulating material. Paint alone shall not be considered as an insulating material.

480.10 Battery Locations. Battery locations shall conform to 480.10(A) through (G).

Batteries should be located in clean, dry rooms and be arranged to provide sufficient work space for inspection and maintenance. The voltage of the battery system, rather than the voltage of the individual batteries or cells, is to be used in determining the work space requirements in 110.26 for systems rated 1000 volts and less and 110.34 for systems rated over 1000 volts. Adequate ventilation is necessary to prevent an accumulation of an explosive mixture of the gases from batteries that generate hydrogen.

The fumes given off by some storage batteries are very corrosive; therefore, wiring and its insulation must be of a type that withstands corrosion, as required by 110.11 and 310.10(F). Special precautions are necessary to ensure that all metalwork (such as metal raceways and metal racks) is designed or treated to be corrosion resistant. The battery racks shown in Exhibit 480.1 are coated with a nonmetallic outer covering as required by 480.9. The covering insulates and provides protection against the corrosive action of fumes from batteries being charged and any

EXHIBIT 480.1 A battery room with batteries installed on corrosion-resistant racks. *(Courtesy of the International Association of Electrical Inspectors)*

electrolyte that might escape from the cells. Manufacturers sometimes suggest that aluminum or plastic conduit be used to withstand corrosive battery fumes or, if steel conduit is used, that it be zinc coated and corrosion protected with a coating of an asphaltum-type (asphalt-based) paint.

See also

300.6 for more on protection against corrosion and deterioration

Overcharging heats a battery and causes gassing and loss of water. A battery should not be allowed to reach temperatures over 110°F, because heat causes a shedding of active materials from the plates, which will eventually form a sediment buildup in the bottom of the case and short circuit the plates and the cell. Because mixtures of oxygen and hydrogen are highly explosive, flame or sparks should never be allowed near a cell, especially if the filler cap is removed.

(A) Ventilation. Provisions appropriate to the battery technology shall be made for sufficient diffusion and ventilation of gases from the battery, if present, to prevent the accumulation of an explosive mixture.

Informational Note No. 1: See NFPA 1-2021, *Fire Code*, Chapter 52, for ventilation considerations for specific battery chemistries.

Informational Note No. 2: Some battery technologies do not require ventilation.

Informational Note No. 3: See IEEE Std 1635-2012/ASHRAE Guideline 21-2012, *Guide for the Ventilation and Thermal Management of Batteries for Stationary Applications*, for additional information on the ventilation of stationary battery systems.

Ventilation is necessary to prevent classification of a battery location as a hazardous (classified) location, in accordance with Article 500.

Mechanical ventilation is not mandated. Hydrogen disperses rapidly and requires little air movement to prevent accumulation. Unrestricted natural air movement in the vicinity of the battery, together with normal air changes for occupied spaces or heat removal, normally is sufficient. If the space is confined, mechanical ventilation might be required in the vicinity of the battery.

Hydrogen is lighter than air and tends to concentrate at ceiling level, so some form of ventilation should be provided at the upper portion of the structure. Ventilation can be a fan, a roof ridge vent, or a louvered area.

Although valve-regulated batteries often are referred to as "sealed," they actually emit very small quantities of hydrogen gas under normal operation and are capable of liberating large quantities of explosive gases if overcharged. These batteries, therefore, require the same amount of ventilation as their vented counterparts. Of the primary types of battery chemistries used in storage battery systems and energy storage systems, only lithium-ion and sodium-nickel chloride batteries do not require ventilation under normal or abnormal charging conditions.

(B) Live Parts. Guarding of live parts shall comply with 110.27.

(C) Spaces About Stationary Standby Batteries. Spaces about stationary standby batteries shall comply with 110.26 and 110.34. Working space shall be measured from the edge of the battery cabinet, racks, or trays.

For battery racks, there shall be a minimum clearance of 25 mm (1 in.) between a cell container and any wall or structure on the side not requiring access for maintenance. Battery stands shall be permitted to contact adjacent walls or structures, provided that the battery shelf has a free air space for not less than 90 percent of its length.

Informational Note: Additional space is often needed to accommodate battery hoisting equipment, tray removal, or spill containment.

(D) Top Terminal Batteries. Where top terminal batteries are installed on tiered racks or on shelves of battery cabinets, working space in accordance with the battery manufacturer's instructions shall be provided between the highest point on a cell and the row, shelf, or ceiling above that point.

Informational Note: See IEEE 1187-2013, *IEEE Recommended Practice for Installation Design and Installation of Valve-Regulated Lead-Acid Batteries for Stationary Applications*, for guidance for top clearance of valve-regulated lead-acid batteries, which are commonly used in battery cabinets.

(E) Egress. Personnel doors intended for entrance to, and egress from, rooms designated as battery rooms shall open at least 90 degrees in the direction of egress and shall be equipped with listed panic or listed fire exit hardware.

(F) Piping in Battery Rooms. Gas piping shall not be permitted in dedicated battery rooms.

Δ **(G) Illumination.** Illumination shall be provided for working spaces containing stationary standby batteries. The lighting outlets shall not be controlled by automatic means only. Additional lighting outlets shall not be required where the work space is illuminated by an adjacent light source. The location of luminaires shall not result in the following:

(1) Expose personnel to energized battery components while performing maintenance on the luminaires in the battery space

(2) Create a hazard to the battery upon failure of the luminaire

480.11 Vents.

(A) Vented Cells. Each vented cell shall be equipped with a flame arrester.

Informational Note: A flame arrester prevents destruction of the cell due to ignition of gases within the cell by an external spark or flame.

(B) Sealed Cells. Where the battery is constructed such that an excessive accumulation of pressure could occur within the cell during operation, a pressure-release vent shall be provided.

Δ **480.12 Battery Interconnections.** Flexible cables, as identified in Table 400.4, in sizes 2/0 AWG and larger shall be permitted within the battery enclosure from battery terminals to a nearby junction box where they shall be connected to an approved wiring method. Flexible battery cables shall also be permitted between batteries and cells within the battery enclosure. Such cables shall be listed and identified for the environmental conditions. Flexible, fine-stranded cables shall only be used with terminals, lugs, devices, or connectors in accordance with 110.14.

480.13 Ground-Fault Detection. Battery circuits exceeding 100 volts between the conductors or to ground shall be permitted to operate with ungrounded conductors, provided a ground-fault detector and indicator is installed to monitor for ground faults.

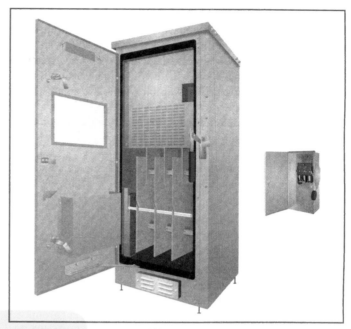

EXHIBIT 495.1 Construction differences between two classes of equipment.

N ARTICLE
495 Equipment Over 1000 Volts ac, 1500 Volts dc, Nominal

N **Part I. General**

N **495.1 Scope.** This article covers the general requirements for equipment operating at more than 1000 volts ac, 1500 volts dc, nominal.

> Informational Note No. 1: See NFPA *70E*-2021, *Standard for Electrical Safety in the Workplace*, for electrical safety requirements for employee workplaces.
> Informational Note No. 2: See ANSI Z535.4-2011, *Product Signs and Safety Labels*, for further information on hazard signs and labels.
> Informational Note No. 3: See IEEE 3001.5-2013, *Recommended Practice for the Application of Power Distribution Apparatus in Industrial and Commercial Power Systems*, for information regarding power distribution apparatus.

In developing the 2023 edition of the *NEC*®, areas of former Article 490, Equipment Over 1000 Volts, Nominal, were modified and relocated to new Article 495, Equipment Over 1000 Volts ac, 1500 Volts dc, Nominal. Some portions of former Article 490 have also been relocated into other areas of the *NEC*. For example, former 490.21, which deals with circuit-interrupter devices, has been relocated to 245.21 of new Article 245, Overcurrent Protection for Systems Rated Over 1000 Volts ac, 1500 Volts dc.

The threshold voltage for the requirements covering higher voltage equipment has been revised from 600 volts to 1000 volts beginning in the 2014 edition of the *NEC* to recognize that standard configurations commonly used in alternative energy systems operate at over 600 volts for increased efficiency and performance. One of the most impactful effects of those changes throughout the *NEC* was revision of former Article 490 to apply to equipment rated over 1000 volts rather than over 600 volts,

and with that change, product certification standards were able to be modified. One of the best examples of how that change affected products is the availability of disconnecting means (safety switches) that are rated for 1000 volts dc, rather than that class of equipment being limited to 600-volt applications because it did not meet the product safety requirements in former Article 490. Prior to the change, switches to interrupt ac or dc circuits rated over 600 volts would have had to meet the "medium voltage" equipment requirements contained in former Article 490.

Exhibit 495.1 illustrates the construction differences between two classes of equipment: a 5000-volt disconnecting means and a 1000-volt disconnecting means.

N **495.2 Reconditioned Equipment.** Except as modified within this article, reconditioned equipment shall not be permitted.

N **495.3 Other Articles.**

N **(A) Oil-Filled Equipment.** Installation of electrical equipment containing more than 38 L (10 gal) of flammable oil per unit shall meet the requirements of Parts II and III of Article 450.

N **(B) Enclosures in Damp or Wet Locations.** Enclosures in damp or wet locations shall meet the requirements of 312.2.

N **Part II. Equipment — Specific Provisions**

N **495.22 Isolating Means.** Means shall be provided to completely isolate an item of equipment from all ungrounded conductors. The use of isolating switches shall not be required where there are other ways of de-energizing the equipment for inspection and repairs, such as draw-out-type switchgear units and removable truck panels.

Isolating switches not interlocked with an approved circuit-interrupting device shall be provided with a sign warning against opening them under load. The warning sign(s) or label(s) shall comply with 110.21(B).

An identified fuseholder and fuse shall be permitted as an isolating switch.

N **495.23 Voltage Regulators.** Proper switching sequence for regulators shall be ensured by use of one of the following:

(1) Mechanically sequenced regulator bypass switch(es)
(2) Mechanical interlocks
(3) Switching procedure prominently displayed at the switching location

N **495.24 Minimum Space Separation.** In field-fabricated installations, the minimum air separation between bare live conductors and between such conductors and adjacent grounded surfaces shall not be less than the values given in Table 495.24. These values shall not apply to interior portions or exterior terminals of equipment designed, manufactured, and tested in accordance with accepted national standards.

N **495.25 Backfeed.** Installations where the possibility of backfeed exists shall comply with 495.25(A) and (B).

N **(A) Sign.** A permanent sign in accordance with 110.21(B) shall be installed on the disconnecting means enclosure or immediately adjacent to open disconnecting means with the following words or equivalent:

DANGER — CONTACTS ON EITHER SIDE OF THIS DEVICE MAY BE ENERGIZED BY BACKFEED.

N **(B) Diagram.** A permanent and legible single-line diagram of the local switching arrangement, clearly identifying each point of connection to the high-voltage section, shall be provided within sight of each point of connection.

N **Part III. Equipment — Switchgear and Industrial Control Assemblies**

N **495.30 General.** Part III covers assemblies of switchgear and industrial control equipment, including, but not limited to, switches and interrupting devices and their control, metering, protection, and regulating equipment where they are an integral part of the assembly, with associated interconnections and supporting structures.

N **TABLE 495.24** *Minimum Clearance of Live Parts*

	Impulse Withstand, Basic Impulse Level (BIL) (kV)		Minimum Clearance of Live Parts							
			Phase-to-Phase				Phase-to-Ground			
Nominal Voltage Rating (kV)			Indoors		Outdoors		Indoors		Outdoors	
	Indoors	Outdoors	mm	in.	mm	in.	mm	in.	mm	in.
2.4–4.16	60	95	115	4.5	180	7	80	3.0	155	6
7.2	75	95	140	5.5	180	7	105	4.0	155	6
13.8	95	110	195	7.5	305	12	130	5.0	180	7
14.4	110	110	230	9.0	305	12	170	6.5	180	7
23	125	150	270	10.5	385	15	190	7.5	255	10
34.5	150	150	320	12.5	385	15	245	9.5	255	10
	200	200	460	18.0	460	18	335	13.0	335	13
46	—	200	—	—	460	18	—	—	335	13
	—	250	—	—	535	21	—	—	435	17
69	—	250	—	—	535	21	—	—	435	17
	—	350	—	—	790	31	—	—	635	25
115	—	550	—	—	1350	53	—	—	1070	42
138	—	550	—	—	1350	53	—	—	1070	42
	—	650	—	—	1605	63	—	—	1270	50
161	—	650	—	—	1605	63	—	—	1270	50
	—	750	—	—	1830	72	—	—	1475	58
230	—	750	—	—	1830	72	—	—	1475	58
	—	900	—	—	2265	89	—	—	1805	71
	—	1050	—	—	2670	105	—	—	2110	83

Note: The values given are the minimum clearance for rigid parts and bare conductors under favorable service conditions. They shall be increased for conductor movement or under unfavorable service conditions or wherever space limitations permit. The selection of the associated impulse withstand voltage for a particular system voltage is determined by the characteristics of the overvoltage(surge) protective equipment.

Indicator instruments, such as voltmeters, ammeters, wattmeters, and protective relays, can be mounted on the panel doors as desired. This switchgear affords a high degree of safety because all live parts are metal-enclosed, and interlocks are provided for safe operation.

N 495.31 Arrangement of Devices in Assemblies. Arrangement of devices in assemblies shall be such that individual components can safely perform their intended function without adversely affecting the safe operation of other components in the assembly.

N 495.32 Guarding of High-Voltage Energized Parts Within a Compartment. Where access for other than visual inspection is required to a compartment that contains energized high-voltage parts, barriers shall be provided to prevent accidental contact by persons, tools, or other equipment with energized parts. Exposed live parts shall only be permitted in compartments accessible to qualified persons. Fuses and fuseholders designed to enable future replacement without de-energizing the fuseholder shall only be permitted for use by qualified persons.

An example of a high-voltage pad-mounted transformer and enclosure that may contain primary and secondary switches or circuit breakers is shown in Exhibit 495.2. The high-voltage compartment on the left of the exhibit has bayonet fusing and a load-break transformer on/off switch, which are both hot-stick operated.

N 495.33 Guarding of Energized Parts Operating at 1000 Volts, Nominal, or Less Within Compartments. Energized bare parts mounted on doors shall be guarded where the door must be opened for maintenance of equipment or removal of draw-out equipment.

EXHIBIT 495.2 Open view of a tamperproof pad-mounted transformer for loop-feed application with load-break elbow connectors. (*Courtesy of Schneider Electric*)

N 495.34 Clearance for Cable Conductors Entering Enclosure. The unobstructed space opposite terminals or opposite raceways or cables entering a switchgear or control assembly shall be approved for the type of conductor and method of termination.

N 495.35 Accessibility of Energized Parts.

N (A) High-Voltage Equipment. Doors that would provide unqualified persons access to high-voltage energized parts shall be locked. Permanent signs in accordance with 110.21(B) shall be installed on panels or doors that provide access to live parts over 1000 volts and shall read DANGER — HIGH VOLTAGE — KEEP OUT.

N (B) Control Equipment. Where operating at 1000 volts, nominal, or less, control equipment, relays, motors, and the like shall not be installed in compartments with high-voltage parts or high-voltage wiring, unless both of the following apply:

(1) The access means is interlocked with the high-voltage switch or disconnecting means to prevent the access means from being opened or removed when the high-voltage switch is in the closed position or a withdrawable disconnecting means is in the connected position.

(2) All high-voltage parts or high-voltage wiring in the compartment that remains energized when a fixed mounted high-voltage switch is in the open position or a withdrawable disconnecting means is in the isolating (fully withdrawn) position are protected by insulating or grounded metal barriers to prevent accidental contact with energized high-voltage parts or wiring.

N (C) High-Voltage Instruments or Control Transformers and Space Heaters. High-voltage instrument or control transformers and space heaters shall be permitted to be installed in the high-voltage compartment without access restrictions beyond those that apply to the high-voltage compartment generally.

N 495.37 Equipment Grounding Connections. The metal cases or frames, or both, such as those of instruments, relays, meters, and instrument and control transformers, located in or on switchgear or control assemblies, and the frames of switchgear and control assemblies shall be connected to an equipment grounding conductor or, where permitted, the grounded conductor, in accordance with 250.190.

N 495.38 Door Stops and Cover Plates. External hinged doors or covers shall be provided with stops to hold them in the open position. Cover plates intended to be removed for inspection of energized parts or wiring shall be equipped with lifting handles and shall not exceed 1.1 m² (12 ft²) in area or 27 kg (60 lb) in weight, unless they are hinged and bolted or locked.

N 495.39 Gas Discharge from Interrupting Devices. Gas discharged during operating of interrupting devices shall be directed so as not to endanger personnel.

N **495.40 Visual Inspection Windows.** Windows intended for visual inspection of disconnecting switches or other devices shall be of suitable transparent material.

N **495.41 Location of Industrial Control Equipment.** Routinely operated industrial control equipment shall meet the requirements of 495.41(A) unless infrequently operated, as covered in 495.41(B).

N **(A) Control and Instrument Transfer Switch Handles or Push Buttons.** Control and instrument transfer switch handles or push buttons shall be in a readily accessible location at an elevation of not over 2.0 m (6 ft 7 in.).

Exception: Operating handles requiring more than 23 kg (50 lb) of force shall be located no higher than 1.7 m (66 in.) in either the open or closed position.

N **(B) Infrequently Operated Devices.** Where operating handles for such devices as draw-out fuses, fused potential or control transformers and their primary disconnects, and bus transfer and isolating switches are only operated infrequently, the handles shall be permitted to be located where they are safely operable and serviceable from a portable platform.

N **495.42 Interlocks — Interrupter Switches.** Interrupter switches equipped with stored energy mechanisms shall have mechanical interlocks to prevent access to the switch compartment unless the stored energy mechanism is in the discharged or blocked position.

N **495.43 Stored Energy for Opening.** The stored energy operator shall be permitted to be left in the uncharged position after the switch has been closed if a single movement of the operating handle charges the operator and opens the switch.

N **495.44 Fused Interrupter Switches.**

N **(A) Supply Terminals.** The supply terminals of fused interrupter switches shall be installed at the top of the switch enclosure or, if the terminals are located elsewhere, the equipment shall have barriers installed to prevent persons from accidentally contacting energized parts or dropping tools or fuses into energized parts.

N **(B) Backfeed.** Where fuses can be energized by backfeed, a sign shall be placed on the enclosure door identifying this hazard.

N **(C) Switching Mechanism.** The switching mechanism shall be arranged to be operated from a location outside the enclosure where the operator is not exposed to energized parts and shall be arranged to open all ungrounded conductors of the circuit simultaneously with one operation. Switches shall be lockable open in accordance with 110.25.

N **495.45 Circuit Breakers — Interlocks.**

N **(A) Circuit Breakers.** Circuit breakers equipped with stored energy mechanisms shall be designed to prevent the release of the stored energy unless the mechanism has been fully charged.

N **(B) Mechanical Interlocks.** Mechanical interlocks shall be provided in the housing to prevent the complete withdrawal of the circuit breaker from the housing when the stored energy mechanism is in the fully charged position, unless a suitable device is provided to block the closing function of the circuit breaker before complete withdrawal.

N **495.46 Circuit Breaker Locking.** Circuit breakers shall be capable of being locked in the open position or, if they are installed in a draw-out mechanism, that mechanism shall be capable of being locked in such a position that the mechanism cannot be moved into the connected position. In either case, the provision for locking shall be lockable open in accordance with 110.25.

N **495.47 Switchgear Used as Service Equipment.** Switchgear installed as high-voltage service equipment shall include a ground bus for the connection of service cable shields and to facilitate the attachment of safety grounds for personnel protection. This bus shall be extended into the compartment where the service conductors are terminated. Where the compartment door or panel provides access to parts that can only be de-energized and visibly isolated by the serving utility, the warning sign required by 495.35(A) shall include a notice that access is limited to the serving utility or is permitted only following an authorization of the serving utility.

Switchgear must include a ground bus for the service cable shields. The bus must extend to the compartment where the service conductor terminals are located. It also provides a location for the connection of temporary protective grounding equipment for protection of personnel during servicing of the equipment.

> **See also**
>
> **120.5 of *NFPA 70E®*,** *Standard for Electrical Safety in the Workplace®*, 2021 edition, for temporary protective grounding equipment requirements, as it relates to establishing an electrically safe work condition (ESWC)

N **495.48 Substation Design, Documentation, and Required Diagram.**

N **(A) Design and Documentation.** Substations shall be designed by a qualified licensed professional engineer. Where components or the entirety of the substation is listed by a qualified electrical testing laboratory, documentation of internal design features subject to the listing investigation shall not be required. The design shall address but not be limited to the following topics, and the documentation of this design shall be made available to the authority having jurisdiction:

(1) Clearances and exits
(2) Electrical enclosures
(3) Securing and support of electrical equipment
(4) Fire protection
(5) Safety ground connection provisions
(6) Guarding live parts
(7) Transformers and voltage regulation equipment
(8) Conductor insulation, electrical and mechanical protection, isolation, and terminations

(9) Application, arrangement, and disconnection of circuit breakers, switches, and fuses
(10) Provisions for oil-filled equipment
(11) Switchgear
(12) Overvoltage (surge) protection equipment

N **(B) Diagram.** A permanent, single-line diagram of the switchgear shall be provided in a readily visible location within the same room or enclosed area with the switchgear and shall clearly identify the following:

(1) Interlocks
(2) Isolation means
(3) All possible sources of voltage to the installation under normal or emergency conditions

The marking on the switchgear shall cross-reference the diagram.

Exception: Where the equipment consists solely of a single cubicle or metal-enclosed substation containing only one high-voltage switching device, diagrams shall not be required.

N **495.49 Reconditioned Switchgear.** Reconditioned switchgear, or sections of switchgear, shall be permitted. If equipment has been damaged by fire, products of combustion, or water, it shall be specifically evaluated by its manufacturer or a qualified testing laboratory prior to being returned to service.

N **Part IV. Mobile and Portable Equipment**

N **495.61 General.**

N **(A) Covered.** The provisions of this part shall apply to installations and use of high-voltage power distribution and utilization equipment that is portable, mobile, or both, and include but not be limited to the following:

(1) Substations and switch houses mounted on skids
(2) Trailers or cars
(3) Mobile shovels
(4) Draglines
(5) Cranes
(6) Hoists
(7) Drills
(8) Dredges
(9) Compressors
(10) Pumps
(11) Conveyors
(12) Underground excavators

N **(B) Grounding and Bonding.** Grounding and bonding shall be in accordance with Part X of Article 250.

N **(C) Protection.** Approved enclosures or guarding, or both, shall be provided to protect portable and mobile equipment from physical damage.

N **(D) Disconnecting Means.** Disconnecting means shall be installed for mobile and portable high-voltage equipment according to the requirements of Part VIII of Article 230 and shall disconnect all ungrounded conductors.

N **495.62 Overcurrent Protection.** Motors driving single or multiple dc generators supplying a system operating on a cyclic load basis shall not require overload protection if the thermal rating of the ac drive motor cannot be exceeded under any operating condition. The branch-circuit protective device(s) shall provide short-circuit and locked-rotor protection and shall be permitted to be external to the equipment.

N **495.63 Enclosures.** All energized switching and control parts shall be enclosed in grounded metal cabinets or enclosures. These cabinets or enclosures shall be marked DANGER — HIGH VOLTAGE — KEEP OUT and shall be locked so that only authorized and qualified persons can enter. The danger marking(s) or label(s) shall comply with 110.21(B). Circuit breakers and protective equipment shall have the operating means projecting through the metal cabinet or enclosure so these units can be reset without opening locked doors. With doors closed, safe access for normal operation of these units shall be provided.

N **495.64 Collector Rings.** The collector ring assemblies on revolving-type machines (shovels, draglines, etc.) shall be guarded to prevent accidental contact with energized parts by personnel on or off the machine.

N **495.65 Power Cable Connections to Mobile Machines.** A metallic enclosure shall be provided on the mobile machine for enclosing the terminals of the power cable. The enclosure shall include terminal connections to the machine frame for the equipment grounding conductor. Ungrounded conductors shall be attached to insulators or be terminated in approved high-voltage cable couplers (which include equipment grounding conductor connectors) of proper voltage and ampere rating. The method of cable termination used shall prevent any strain or pull on the cable from stressing the electrical connections. The enclosure shall have provision for locking so that only authorized and qualified persons can open it and shall be marked as follows:

DANGER — HIGH VOLTAGE — KEEP OUT.

The danger marking(s) or label(s) shall comply with 110.21(B).

N **495.66 High-Voltage Portable Cable for Main Power Supply.** Flexible high-voltage cable supplying power to portable or mobile equipment shall comply with the grounding and bonding requirements in Parts V, VI, and X of Article 250 and the flexible cable requirements in Part III of Article 400.

N **Part V. Boilers**

N **495.70 General.** The provisions of Part V shall apply to boilers operating over 1000 volts, nominal, in which heat is generated by the passage of current between electrodes through the liquid being heated.

N 495.71 Electrical Supply System. Boilers shall be supplied only from a 3-phase, 4-wire solidly grounded wye system, or from isolating transformers arranged to provide such a system. Control circuit voltages shall not exceed 150 volts, shall be supplied from a grounded system, and shall have the controls in the ungrounded conductor.

N 495.72 Branch-Circuit Requirements.

N (A) Rating. Each boiler shall be supplied from an individual branch circuit rated not less than 100 percent of the total load.

N (B) Common-Trip Fault-Interrupting Device. The circuit shall be protected by a 3-phase, common-trip fault-interrupting device, which shall be permitted to automatically reclose the circuit upon removal of an overload condition but shall not reclose after a fault condition.

N (C) Phase-Fault Protection. Phase-fault protection shall be provided in each phase, consisting of a separate phase-overcurrent relay connected to a separate current transformer in the phase.

N (D) Ground Current Detection. Means shall be provided for detection of the sum of the neutral conductor and equipment grounding conductor currents and shall trip the circuit-interrupting device if the sum of those currents exceeds the greater of 5 amperes or 7½ percent of the boiler full-load current for 10 seconds or exceeds an instantaneous value of 25 percent of the boiler full-load current.

N (E) Grounded Neutral Conductor. The grounded neutral conductor shall be as follows:

(1) Connected to the pressure vessel containing the heating elements
(2) Insulated for not less than 1000 volts
(3) Have not less than the ampacity of the largest ungrounded branch-circuit conductor
(4) Installed with the ungrounded conductors in the same raceway, cable, or cable tray, or, where installed as open conductors, in close proximity to the ungrounded conductors
(5) Not used for any other circuit

N 495.73 Pressure and Temperature Limit Control. Each boiler shall be equipped with a means to limit the maximum temperature, pressure, or both, by directly or indirectly interrupting all current flow through the heating elements. Such means shall be in addition to the temperature, pressure, or both, regulating systems and pressure relief or safety valves.

N 495.74 Bonding. All exposed non-current-carrying metal parts of the boiler and associated exposed metal structures or equipment shall be bonded to the pressure vessel or to the neutral conductor to which the vessel is connected in accordance with 250.102, except the ampacity of the bonding jumper shall not be less than the ampacity of the neutral conductor.

Special
Occupancies

ARTICLE 500 Hazardous (Classified) Locations, Classes I, II, and III, Divisions 1 and 2

Articles 500 through 516 cover the requirements for electrical installations in locations classified as hazardous due to the materials handled, processed, or stored in those locations. Hazardous (classified) locations, if properly treated, are not necessarily any more dangerous to work in than other areas or locations. Hazardous locations are sometimes referred to as classified locations. As used in other NFPA codes and standards and the *NEC®*, the terms are interchangeable.

Some of the most common materials encountered in hazardous locations are flammable and combustible liquids. A flammable liquid is one that has a flash point below 100°F, while a combustible liquid has a flash point at or above 100°F. A flammable or combustible liquid must be at its flash point before an explosion can occur. For example, No. 1-D diesel fuel oil and kerosene, with flash points higher than 100°F, are combustible liquids and do not emit flammable vapors unless heated above their flash points.

Article 500 is limited to locations classified as Class I, Class II, or Class III, which are further addressed in Articles 501, 502, and 503, respectively. Articles 505 and 506 contain the requirements for using the classification by Zones. Article 504 on intrinsically safe systems is used for both methods (Division and Zone) of area classification.

Δ **500.1 Scope.**

It is outside the scope of the *NEC* to classify hazardous locations. The *NEC* provides the requirements for installing electrical equipment within locations classified as hazardous through compliance with other standards.

The *NEC* does not classify areas where explosive materials, such as ammunition, dynamite, and blasting powder, are present. Areas where such materials are present are not considered hazardous locations in accordance with Article 500. However, many organizations responsible for the safety of such areas require

equipment and wiring methods suitable for hazardous locations as part of many safety precautions, even though the equipment and wiring have not been investigated for such locations. Further information on these locations can be found in NFPA 495, *Explosive Materials Code*.

N **(A) Covered.** This article covers area classification and general requirements for electrical and electronic equipment and wiring rated at all voltages where fire or explosion hazards might exist due to flammable gases, flammable liquid–produced vapors, combustible liquid–produced vapors, combustible dusts, combustible fibers/flyings, or ignitible fibers/flyings in the following:

(1) Class I, Division 1 or Class I, Division 2 hazardous (classified) locations
(2) Class II, Division 1 or Class II, Division 2 hazardous (classified) locations
(3) Class III, Division 1 or Class III, Division 2 hazardous (classified) locations

Informational Note No. 1: See NFPA 497, *Recommended Practice for the Classification of Flammable Liquids, Gases, or Vapors and of Hazardous (Classified) Locations for Electrical Installations in Chemical Process Areas*, and NFPA 499, *Recommended Practice for the Classification of Combustible Dusts and of Hazardous (Classified) Locations for Electrical Installations in Chemical Process Areas*, for extracted information referenced in brackets. Only editorial changes were made to the extracted text to make it consistent with this *Code*.

Informational Note No. 2: See Article 100 for the definition of *restricted industrial establishment [as applied to hazardous (classified) locations]*.

N **(B) Not Covered.** This article does not cover electrical and electronic equipment and wiring rated at all voltages for the following:

(1) Zone 0, Zone 1, or Zone 2 hazardous (classified) locations
(2) Zone 20, Zone 21, or Zone 22 hazardous (classified) locations
(3) Locations subject to the unique risk and explosion hazards associated with explosives, pyrotechnics, and blasting agents

(4) Locations where pyrophoric materials are the only materials used or handled

(5) Features of equipment that involve nonelectrical potential sources of ignition (e.g., couplings, pumps, gearboxes, brakes, hydraulic and pneumatic motors, fans, engines, compressors)

Informational Note No. 1: Common nonelectrical potential sources of ignition include hot surfaces and mechanically generated sparks.

Informational Note No. 2: See ANSI/UL 80079-36, *Explosive Atmospheres — Part 36: Non-Electrical Equipment for Explosive Atmospheres — Basic Method and Requirements*, and ANSI/UL 80079-37, *Explosive Atmospheres — Part 37: Non-Electrical Equipment for Explosive Atmospheres — Non-Electrical Type of Protection Constructional Safety "c" Control of Ignition Source "b", Liquid Immersion "k"*, for additional information.

Because pyrophoric materials ignite spontaneously upon contact with air, the use of electrical equipment suitable for hazardous locations will not prevent ignition of such materials. Instead, the process containment system should be designed to prevent contact between pyrophoric material and air.

Δ **500.4 Documentation.** Areas designated as hazardous (classified) locations or determined to be unclassified shall be documented on an area classification drawing and other associated documentation. This documentation shall be available to the authority having jurisdiction (AHJ) and those authorized to design, install, inspect, maintain, or operate electrical equipment at the location.

One type of documentation consists of area classification drawings. Once the hazardous area has been classified and the hazardous area documentation has been developed, the materials and installation methods of the *NEC* are used to design and construct the electrical system for use in the indicated classified area. This approach provides the necessary information for engineers, installers, service personnel, and AHJs to ensure that the electrical equipment installed in each classified area is of the proper type. See Exhibit 500.1 for an example of an area classification drawing.

Informational Note No. 1: See the following standards for additional information on the classification of locations:

(1) NFPA 30, *Flammable and Combustible Liquids Code*
(2) NFPA 32, *Standard for Drycleaning Facilities*
(3) NFPA 33, *Standard for Spray Application Using Flammable or Combustible Materials*
(4) NFPA 34, *Standard for Dipping, Coating, and Printing Processes Using Flammable or Combustible Liquids*
(5) NFPA 35, *Standard for the Manufacture of Organic Coatings*
(6) NFPA 36, *Standard for Solvent Extraction Plants*
(7) NFPA 45, *Standard on Fire Protection for Laboratories Using Chemicals*
(8) NFPA 55, *Compressed Gases and Cryogenic Fluids Code*
(9) NFPA 58, *Liquefied Petroleum Gas Code*
(10) NFPA 59, *Utility LP-Gas Plant Code*
(11) NFPA 497, *Recommended Practice for the Classification of Flammable Liquids, Gases, or Vapors and of Hazardous (Classified) Locations for Electrical Installations in Chemical Process Areas*
(12) NFPA 499, *Recommended Practice for the Classification of Combustible Dusts and of Hazardous (Classified) Locations for Electrical Installations in Chemical Process Areas*
(13) NFPA 820, *Standard for Fire Protection in Wastewater Treatment and Collection Facilities*
(14) ANSI/API RP 500, *Recommended Practice for Classification of Locations for Electrical Installations at Petroleum Facilities Classified as Class I, Division 1 and Division 2*
(15) ISA-12.10, *Area Classification in Hazardous (Classified) Dust Locations*

Informational Note No. 2: See NFPA 77, *Recommended Practice on Static Electricity*; NFPA 780, *Standard for the*

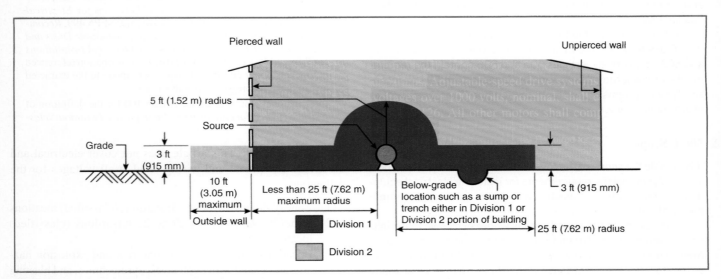

Pierced wall

Unpierced wall

5 ft (1.52 m) radius

Source

Grade

3 ft (915 mm)

10 ft (3.05 m) maximum

Outside wall

Less than 25 ft (7.62 m) maximum radius

Below-grade location such as a sump or trench either in Division 1 or Division 2 portion of building

3 ft (915 mm)

25 ft (7.62 m) radius

Division 1

Division 2

EXHIBIT 500.1 An example of an area classification drawing.

Installation of Lightning Protection Systems; and API RP 2003, *Protection Against Ignitions Arising Out of Static, Lightning, and Stray Currents*, for information on protection against static electricity and lightning hazards in hazardous (classified) locations.

Informational Note No. 3: See NFPA 30, *Flammable and Combustible Liquids Code*; and ANSI/API RP 500, *Recommended Practice for Classification of Locations for Electrical Installations at Petroleum Facilities Classified as Class I, Division 1 and Division 2*, for information on ventilation.

Informational Note No. 4: See ANSI/API RP 14F, *Recommended Practice for Design, Installation, and Maintenance of Electrical Systems for Fixed and Floating Offshore Petroleum Facilities for Unclassified and Class I, Division 1, and Division 2 Locations*, for information on electrical systems for hazardous (classified) locations on offshore oil- and gas-producing platforms, drilling rigs, and workover rigs.

Informational Note No. 5: See ANSI/UL 121203, *Portable/Personal Electronic Products Suitable for Use in Class I, Division 2, Class I, Zone 2, Class II, Division 2, Class III, Division 1, Class III, Division 2, Zone 21 and Zone 22 Hazardous (Classified) Locations*, for information on portable or transportable equipment having self-contained power supplies, such as battery-operated equipment, which could potentially become an ignition source in hazardous (classified) locations.

Informational Note No. 6: See IEC/IEEE 60079-30-2, *Explosive atmospheres — Part 30-2: Electrical resistance trace heating — Application guide for design, installation and maintenance*, for information on electrical resistance trace heating for hazardous (classified) locations.

Informational Note No. 7: See IEEE 844.2/CSA C293.2, *IEEE/CSA Standard for Skin Effect Trace Heating of Pipelines, Vessels, Equipment, and Structures — Application Guide for Design, Installation, Testing, Commissioning, and Maintenance*, for information on electric skin effect trace heating for hazardous (classified) locations.

Informational Note No. 8: See IEEE 844.4/CSA C293.4, *IEEE/CSA Standard for Impedance Heating of Pipelines and Equipment — Application Guide for Design, Installation, Testing, Commissioning, and Maintenance*, for information on electric impedance heating for hazardous (classified) locations.

The standards referenced in Articles 500 through 517 are essential for proper application of the articles. In addition to those documents and others listed in each article's Informational Notes, the following NFPA codes, standards, and recommended practices contain valuable information on hazardous locations in specific applications, occupancies, or industries:

NFPA 30A, *Code for Motor Fuel Dispensing Facilities and Repair Garages*

NFPA 51, *Standard for the Design and Installation of Oxygen–Fuel Gas Systems for Welding, Cutting, and Allied Processes*

NFPA 52, *Vehicular Natural Gas Fuel Systems Code*

NFPA 54, *National Fuel Gas Code*

NFPA 59A, *Standard for the Production, Storage, and Handling of Liquefied Natural Gas (LNG)*

NFPA 61, *Standard for the Prevention of Fires and Dust Explosions in Agricultural and Food Processing Facilities*

NFPA 85, *Boiler and Combustion Systems Hazards Code*

NFPA 99, *Health Care Facilities Code*

NFPA 407, *Standard for Aircraft Fuel Servicing*

NFPA 409, *Standard on Aircraft Hangars*

NFPA 495, *Explosive Materials Code*

NFPA 654, *Standard for the Prevention of Fire and Dust Explosions from the Manufacturing, Processing, and Handling of Combustible Particulate Solids*

NFPA 655, *Standard for Prevention of Sulfur Fires and Explosions*

500.5 Classifications of Locations.

Δ **(A) General.**

N **(1) Hazardous (Classified) Locations.** Locations shall be classified depending on the properties of the flammable gas, flammable liquid–produced vapor, combustible liquid–produced vapors, combustible dusts, or fibers/flyings that could be present, and the likelihood that a flammable or combustible concentration or quantity is present. Each room, section, or area shall be considered individually in determining its classification.

> Informational Note: Through the exercise of ingenuity in the layout of electrical installations for hazardous (classified) locations, it is frequently possible to locate much of the equipment in a reduced level of classification or in an unclassified location to reduce the amount of special equipment required.

N **(2) Refrigerant Machinery Rooms Using Ammonia.** Refrigerant machinery rooms that contain ammonia refrigeration systems and are equipped with adequate mechanical ventilation that operates continuously or is initiated by a detection system at a concentration not exceeding 150 ppm shall be permitted to be classified as "unclassified" locations.

> Informational Note: See ANSI/IIAR 2, *Standard for Design of Safe Closed-Circuit Ammonia Refrigeration Systems*, for information on classification and ventilation of areas involving closed-circuit ammonia refrigeration systems.

Determining the appropriate classification for an area requires an understanding of the following factors: (1) the type and amount of material present; (2) the potential for that presence under normal and abnormal conditions; and (3) the entire process that the material will or might undergo in given circumstances.

The *NEC* does not classify specific Class I, II, and III locations. Other standards and recommended practices of NFPA technical committees and other organizations such as American Petroleum Institute (API) with expertise in working with flammable and combustible materials can be used to classify locations.

Classification requires consideration of the specific equipment and process and depends on the materials and physical equipment and location (for example, possible leaks at flanges and machinery seals). Care must be taken when classifying an

EXHIBIT 500.2 A coal-handling operation classified as both a Class I and a Class II location. *(Courtesy of Noel Williams)*

EXHIBIT 500.3 A CNG filling station, which is a Class I location.

area that is likely to become hazardous and in determining those portions of the premises to be classified Division 1 or Division 2.

If different types of material — such as a flammable gas and a combustible dust — simultaneously exist in a process or location, the area must be classified as both a Class I and a Class II location, and protection must be provided for both hazards. Equipment that is identified for use in a Class I location may or may not be suitable for use in a Class II location and vice versa. Exhibit 500.2 shows a coal-handling operation classified as both a Class I (methane) and Class II (coal dust) location.

(B) Class I Locations. Class I locations are those in which flammable gases, flammable liquid–produced vapors, or combustible liquid–produced vapors are or may be present in the air in quantities sufficient to produce explosive or ignitible mixtures. Class I locations shall include those specified in 500.5(B)(1) and (B)(2).

For a given vapor, the vapor-in-air ratio must be within the flammable limits to be deemed a hazard. Many chemicals reach the lower limit within a few percentage points of the vapor-to-air ratio, and some are within less than 1 percentage point. The flammable range may be very narrow or very wide (such as for acetylene, which has a flammable range of 2.5 percent to nearly 100 percent).

Exhibit 500.3 shows a Class I location that dispenses compressed natural gas (CNG).

Δ **(1) Class I, Division 1.** A Class I, Division 1 location is a location:

(1) In which ignitible concentrations of flammable gases, flammable liquid–produced vapors, or combustible liquid–produced vapors can exist under normal operating conditions, or

(2) In which ignitible concentrations of such flammable gases, flammable liquid–produced vapors, or combustible

liquids above their flash points might exist frequently because of repair or maintenance operations or because of leakage, or

(3) In which breakdown or faulty operation of equipment or processes might release ignitible concentrations of flammable gases, flammable liquid–produced vapors, or combustible liquid–produced vapors and might also cause simultaneous failure of electrical equipment in such a way as to directly cause the electrical equipment to become a source of ignition

Informational Note: This classification usually includes the following locations:

(1) Where volatile flammable liquids or liquefied flammable gases are transferred from one container to another

(2) Interiors of spray booths and areas in the vicinity of spraying and painting operations where volatile flammable solvents are used

(3) Locations containing open tanks or vats of volatile flammable liquids

(4) Drying rooms or compartments for the evaporation of flammable solvents

(5) Locations containing fat- and oil-extraction equipment using volatile flammable solvents

(6) Portions of cleaning and dyeing plants where flammable liquids are used

(7) Gas generator rooms and other portions of gas manufacturing plants where flammable gas might escape

(8) Inadequately ventilated pump rooms for flammable gas or for volatile flammable liquids

(9) Interiors of refrigerators and freezers in which volatile flammable materials are stored in open, lightly stoppered, or easily ruptured containers

(10) Inside of inadequately vented enclosures containing instruments normally venting flammable gases or vapors to the interior of the enclosure

(11) Inside of vented tanks containing volatile flammable liquids

(12) Area between inner and outer roof sections of floating roof tanks containing volatile flammable fluids

(13) Inadequately ventilated areas within spraying or coating operations using volatile flammable fluids

(14) Interior of exhaust ducts used to vent ignitible concentrations of gases or vapors

(15) All other locations where ignitible concentrations of flammable vapors or gases are likely to occur during normal operations

Experience has demonstrated the prudence of avoiding the installation of instrumentation or other electrical equipment in the areas covered in list items (11) through (15). Where it cannot be avoided because it is essential to the process and other locations are not feasible, electrical equipment or instrumentation approved for the specific application or consisting of intrinsically safe systems might be considered.

Δ **(2) Class I, Division 2.** A Class I, Division 2 location is a location:

(1) In which volatile flammable gases, flammable liquid–produced vapors, or combustible liquid–produced vapors are handled, processed, or used, but in which the liquids, vapors, or gases will normally be confined within closed containers or closed systems from which they can escape only in case of accidental rupture or breakdown of such containers or systems or in case of abnormal operation of equipment, or

(2) In which ignitible concentrations of flammable gases, flammable liquid–produced vapors, or combustible liquid–produced vapors are normally prevented by positive mechanical ventilation and which might become hazardous through failure or abnormal operation of the ventilating equipment, or

(3) That is adjacent to a Class I, Division 1 location, and to which ignitible concentrations of flammable gases, flammable liquid–produced vapors, or combustible liquid–produced vapors above their flash points might occasionally be communicated unless such communication is prevented by adequate positive-pressure ventilation from a source of clean air and effective safeguards against ventilation failure are provided.

Informational Note No. 1: This classification usually includes locations where volatile flammable liquids or flammable gases or vapors are used but that, in the judgment of the authority having jurisdiction, would become hazardous only in case of an accident or of some unusual operating condition. The quantity of flammable material that might escape in case of accident, the adequacy of ventilating equipment, the total area involved, and the record of the industry or business with respect to explosions or fires are all factors that merit consideration in determining the classification and extent of each location.

Informational Note No. 2: See NFPA 30, *Flammable and Combustible Liquids Code*, and NFPA 58, *Liquefied Petroleum Gas Code*. Piping without valves, checks, meters, and similar devices would not ordinarily introduce a hazardous condition even if used for flammable liquids or gases. Depending on factors such as the quantity and size of the containers and ventilation, locations used for the storage of flammable liquids or liquefied or compressed gases in sealed containers might be considered either hazardous (classified) or unclassified locations.

(C) Class II Locations. Class II locations are those that are hazardous because of the presence of combustible dust. Class II locations shall include those specified in 500.5(C)(1) and (C)(2).

Housekeeping, settlement rates, and air velocity are all factors in determining the need for, or the extent of, a Class II classified location. A settled layer of dust could ignite at a temperature different from that of the same dust dispersed into the air as a cloud. Classification of dust layers is based on the thickness of the dust or on the amount of dust expected to settle out, usually over a set period of time.

(1) Class II, Division 1. A Class II, Division 1 location is a location:

(1) In which combustible dust is in the air under normal operating conditions in quantities sufficient to produce explosive or ignitible mixtures, or

(2) Where mechanical failure or abnormal operation of machinery or equipment might cause such explosive or ignitible mixtures to be produced, and might also provide a source of ignition through simultaneous failure of electrical equipment, through operation of protection devices, or from other causes, or

(3) In which Group E combustible dusts may be present in quantities sufficient to be hazardous in normal or abnormal operating conditions.

Informational Note: Dusts containing magnesium or aluminum are particularly hazardous, and the use of extreme precaution is necessary to avoid ignition and explosion.

Group E dusts (metal dusts, e.g., aluminum or magnesium) are particularly hazardous. For example, current through the dust can cause a sufficient temperature rise or electrical arc to trigger ignition. The classification of an area where a Group E dust is or may be present will be Division 1.

(2) Class II, Division 2. A Class II, Division 2 location is a location:

(1) In which combustible dust due to abnormal operations may be present in the air in quantities sufficient to produce explosive or ignitible mixtures; or

(2) Where combustible dust accumulations are present but are normally insufficient to interfere with the normal operation of electrical equipment or other apparatus, but could as a result of infrequent malfunctioning of handling or processing equipment become suspended in the air; or

(3) In which combustible dust accumulations on, in, or in the vicinity of the electrical equipment could be sufficient to interfere with the safe dissipation of heat from electrical equipment, or could be ignitible by abnormal operation or failure of electrical equipment.

Informational Note No. 1: The quantity of combustible dust that may be present and the adequacy of dust removal systems are factors that merit consideration in determining the classification and may result in an unclassified area.

Informational Note No. 2: Where products such as seed are handled in a manner that produces low quantities of dust, the amount of dust deposited may not warrant classification.

(D) Class III Locations. Class III locations shall be locations meeting the requirements of 500.5(D)(1) and (D)(2).

Δ **(1) Class III, Division 1.** Class III, Division 1 locations shall include those locations specified in 500.5(D)(1)(a) and (D)(1)(b).

(a) *Combustible Fibers/Flyings.* Locations where nonmetal combustible fibers/flyings are in the air under normal operating conditions in quantities sufficient to produce explosible mixtures or where mechanical failure or abnormal operation of machinery or equipment might cause combustible fibers/flyings to be produced and might also provide a source of ignition through simultaneous failure of electrical equipment, through operation of protection devices, or from other causes shall be classified as Class III, Division 1. Locations where metal combustible fibers/flyings are present shall be classified as Class II, Division 1, Group E.

Informational Note No. 1: Such locations usually include some parts of rayon, cotton, and other textile mills; associated manufacturing and processing plants; cotton gins and cotton-seed mills; flax-processing plants; clothing manufacturing plants; woodworking plants; and establishments and industries involving similar hazardous processes or conditions.

Informational Note No. 2: Combustible fibers/flyings include flat platelet-shaped particulates, such as metal flakes, and fibrous board, such as particle board.

(b) *Ignitible Fibers/Flyings.* Locations where ignitible fibers/flyings are handled, manufactured, or used shall be classified as Class III, Division 1.

Informational Note No. 1: Such locations usually include some parts of rayon, cotton, and other textile mills; associated manufacturing and processing plants; cotton gins and cotton-seed mills; flax-processing plants; clothing manufacturing plants; woodworking plants; and establishments and industries involving similar hazardous processes or conditions.

Informational Note No. 2: Ignitible fibers/flyings can include rayon, cotton (including cotton linters and cotton waste), sisal or henequen, istle, jute, hemp, tow, cocoa fiber, oakum, baled waste kapok, Spanish moss, excelsior, and other materials of similar nature.

Δ **(2) Class III, Division 2.** Class III, Division 2 locations shall include those locations specified in 500.5(D)(2)(a) and (D)(2)(b).

(a) *Combustible Fibers/Flyings.* Locations where nonmetal combustible fibers/flyings might be present in the air in quantities sufficient to produce explosible mixtures due to abnormal operations or where accumulations of nonmetal combustible fibers/flyings accumulations are present but are insufficient to interfere with the normal operation of electrical equipment or other apparatus but could, as a result of infrequent malfunctioning of handling or processing equipment, become suspended in the air shall be classified as Class III, Division 2.

(b) *Ignitible Fibers/Flyings.* Locations where ignitible fibers/flyings are stored or handled, other than in the process of manufacture, shall be classified as Class III, Division 2.

500.6 Materials.

In a hazardous location, oxygen enrichment can drastically change the explosion characteristics of materials. It lowers minimum ignition energies, increases explosion pressures, and can reduce the maximum experimental safe gap. This can render both intrinsically safe and explosionproof equipment unsafe unless the equipment has been tested for the specific conditions involved.

An oxygen-rich environment is not, in itself, a hazardous location according to Article 500; however, such an environment does pose specific design and installation challenges. Equipment found to be safe in ordinary atmospheric conditions is not necessarily safe in oxygen concentrations or pressures higher than those of the standard atmosphere. NFPA 53, *Recommended Practice on Materials, Equipment, and Systems Used in Oxygen-Enriched Atmospheres,* recommends that no electrical equipment be used in such atmospheres unless approved for use in the specific hazardous atmospheres at the maximum proposed pressure and oxygen concentration.

Δ **(A) Class I Group Classifications.** Class I groups shall be in accordance with 500.6(A)(1) through (A)(4).

Informational Note No. 1: The explosion characteristics of air mixtures of gases or vapors vary with the specific material involved. For Class I locations, Groups A, B, C, and D, the classification involves determinations of maximum explosion pressure and maximum safe clearance between parts of a clamped joint in an enclosure. It is necessary, therefore, that equipment be identified not only for class but also for the specific group of the gas or vapor that will be present.

Informational Note No. 2: Certain chemical atmospheres may have characteristics that require safeguards beyond those required for any of the Class I groups. Carbon disulfide is one of these chemicals because of its low autoignition temperature (90°C) and the small joint clearance permitted to arrest its flame.

NFPA 497, *Recommended Practice for the Classification of Flammable Liquids, Gases, or Vapors and of Hazardous (Classified) Locations for Electrical Installations in Chemical Process Areas,* contains a list of selected chemicals categorized by group. For the complete table, refer to NFPA 497, Table 4.4.2.

For dust materials, NFPA 499, *Recommended Practice for the Classification of Combustible Dusts and of Hazardous (Classified) Locations for Electrical Installations in Chemical Process Areas,* contains a list of dust materials categorized by group. For the complete table, refer to NFPA 499, Table 5.2.3.

(1) Group A. Acetylene. [**497**:3.3.5.1.1]

Δ **(2) Group B.** Flammable gas, flammable liquid–produced vapor, or combustible liquid–produced vapor mixed with air that may burn or explode, having either a maximum experimental safe gap (MESG) value less than or equal to 0.45 mm or a minimum igniting current ratio (MIC ratio) less than or equal to 0.40. [**497**:3.3.5.1.2]

Informational Note: A typical Class I, Group B material is hydrogen.

(3) Group C. Flammable gas, flammable liquid–produced vapor, or combustible liquid–produced vapor mixed with air that may burn or explode, having either a maximum experimental safe gap (MESG) value greater than 0.45 mm and less than or equal to 0.75 mm, or a minimum igniting current (MIC) ratio greater than 0.40 and less than or equal to 0.80. [**497**:3.3.5.1.3]

Informational Note: A typical Class I, Group C material is ethylene.

Δ **(4) Group D.** Flammable gas, flammable liquid–produced vapor, or combustible liquid–produced vapor mixed with air that may burn or explode, having either a maximum experimental safe gap (MESG) value greater than 0.75 mm or a minimum igniting current (MIC) ratio greater than 0.80. [**497**:3.3.5.1.4]

Informational Note No. 1: A typical Class I, Group D material is propane. [**497**:3.3.5.1.4]

Informational Note No. 2: See ANSI/ASHRAE 15, *Safety Standard for Refrigeration Systems*, for information on the classification of areas involving ammonia atmospheres.

Flammable gases, flammable liquid–produced vapors, and combustible liquid–produced vapors are separated into four Class I groups — A, B, C, and D (or three Class I Zone 0 or Zone 1 groups — IIC, IIB, and IIA; see 505.6), depending on their properties. By grouping explosive mixtures that have similar igniting current ratios and maximum safe clearances between parts of a joint in an enclosure, equipment can be designed for the entire group rather than an individual chemical. This method makes it easier to properly select equipment designed for use in the particular group involved.

Selected combustible materials have been evaluated for the purpose of designating the appropriate gas group — A, B, C, or D (or IIC, IIB, or IIA) — and this information is used to properly select electrical equipment for use in Class I locations. These materials, with their group classification and relevant physical properties, are listed in NFPA 497, *Recommended Practice for the Classification of Flammable Liquids, Gases, or Vapors and of Hazardous (Classified) Locations for Electrical Installations in Chemical Process Areas.* For the complete table, refer to NFPA 497, Table 4.4.2.

(B) Class II Combustible Dust Group Classifications. Combustible dust shall be grouped in accordance with 500.6(B)(1) through (B)(3).

(1) Group E. Atmospheres containing combustible metal dusts, including aluminum, magnesium, and their commercial alloys, or other combustible dusts whose particle size, abrasiveness, and conductivity present similar hazards in the use of electrical equipment. [**499**:3.3.8.1.1]

Informational Note: Certain metal dusts may have characteristics that require safeguards beyond those required for atmospheres containing the dusts of aluminum, magnesium, and their commercial alloys. For example, zirconium, thorium, and uranium dusts have extremely low ignition temperatures [as low as 20°C (68°F)] and minimum ignition energies lower than any material classified in any of the Class I or Class II groups.

Δ **(2) Group F.** Atmospheres containing combustible carbonaceous dusts that have more than 8 percent total entrapped volatiles (see ASTM D3175-2017, *Standard Test Method for Volatile Matter in the Analysis Sample of Coal and Coke*, for coal and coke dusts) or that have been sensitized by other materials so that they present an explosion hazard. [**499**:3.3.8.1.2] Coal, carbon black, charcoal, and coke dusts are examples of carbonaceous dusts. [**499**:A.3.3.8.1.2]

Informational Note: Testing of specific dust samples, following established ASTM testing procedures, is a method used to identify the combustibility of a specific dust and the need to classify those locations containing that material as Group F.

Δ **(3) Group G.** Atmospheres containing combustible dusts not included in Group E or Group F, including flour, grain, wood, plastic, and chemicals. [**499**:3.3.8.1.3]

NFPA 664, *Standard for the Prevention of Fires and Explosions in Wood Processing and Woodworking Facilities*, establishes minimum requirements for industrial, commercial, or institutional facilities that process wood or that manufacture wood products, creating wood chips, particles, or dust.

Informational Note No. 1: See NFPA 499, *Recommended Practice for the Classification of Combustible Dusts and of Hazardous (Classified) Locations for Electrical Installations in Chemical Process Areas*, for information on group classification of Class II materials.

Informational Note No. 2: The explosion characteristics of air mixtures of dust vary with the materials involved. For Class II locations, Groups E, F, and G, the classification involves the tightness of the joints of assembly and shaft openings to prevent the entrance of dust in the dust-ignitionproof enclosure, the blanketing effect of layers of dust on the equipment that may cause overheating, and the ignition temperature of the dust. It is necessary, therefore, that equipment be identified not only for the class but also for the specific group of dust that will be present.

Informational Note No. 3: See ANSI/IEEE C2, *National Electrical Safety Code*, Section 127A, Coal Handling Areas. Certain dusts might require additional precautions due to chemical phenomena that can result in the generation of ignitible gases.

Section 500.6(B) separates combustible dusts into three Class II groups — E, F, and G — depending on their properties.

Selected combustible dusts have been evaluated for the purpose of designating the appropriate dust group — E, F, or G — and this information is used to select electrical equipment for use in Class II locations. The selected materials, with their group classification and relevant physical properties, are listed in NFPA 499, *Recommended Practice for the Classification of Combustible Dusts and of Hazardous (Classified) Locations for Electrical Installations in Chemical Process Areas.* For the complete table, refer to NFPA 499, Table 5.2.3.

(C) Class III Combustible Fibers/Flyings. Combustible fibers/flyings shall not be further grouped.

(D) Class III Ignitible Fibers/Flyings. Ignitible fibers/flyings shall not be further grouped.

500.7 Protection Techniques. Electrical and electronic equipment in hazardous (classified) locations shall be protected by one or more of the techniques in 500.7(A) through (P). Suitability of the protection techniques for specific hazardous locations is shown in Chapter 9, Table 13.

(A) Explosionproof Equipment. This protection technique shall be permitted for equipment in Class I, Division 1 or 2 locations.

(B) Dust Ignitionproof. This protection technique shall be permitted for equipment in Class II, Division 1 or 2 locations.

(C) Dusttight. This protection technique shall be permitted for equipment in Class II, Division 2 or Class III, Division 1 or 2 locations.

(D) Purged and Pressurized. This protection technique shall be permitted for equipment in any hazardous (classified) location for which it is identified.

Purging is the process of supplying an enclosure with a protective gas at a sufficient flow and positive pressure to reduce the initial concentration of any flammable gases, flammable liquid–produced vapors, or combustible liquid–produced vapors to an acceptable level. Pressurization is the process of supplying an enclosure with a protective gas, with or without continuous flow, at sufficient pressure to prevent the entrance of a material.

NFPA 496, *Standard for Purged and Pressurized Enclosures for Electrical Equipment*, provides requirements for these enclosures in Class I and Class II hazardous locations. In Class I locations, purged and pressurized enclosures are used to eliminate or reduce, within the enclosure, the Class I hazardous location classification. In Class II locations, pressurized enclosures prevent the entrance of dusts into an enclosure. Purged and pressurized enclosures make it possible for equipment that is not otherwise acceptable for Class I and Class II locations to be used in those locations.

The types of pressurizing are as follows:

1. Type X reduces the classification within a protected enclosure from Division 1 or Zone 1 to unclassified.
2. Type Y reduces the classification within a protected enclosure from Division 1 to Division 2 or from Zone 1 to Zone 2.
3. Type Z reduces the classification within a protected enclosure from Division 2 or Zone 2 to unclassified.

Exhibit 500.4 shows a panel that uses the Type X purge and pressurization technique.

Δ **(E) Intrinsic Safety.** This protection technique shall be permitted for equipment in Class I, Division 1 or Division 2; Class II, Division 1 or Division 2; or Class III, Division 1 or Division 2 locations.

EXHIBIT 500.4 A panel using Type X pressurization. *(Courtesy of Pepperl+Fuchs, Inc.)*

(F) Nonincendive Circuit. This protection technique shall be permitted for equipment in Class I, Division 2; Class II, Division 2; or Class III, Division 1 or 2 locations.

(G) Nonincendive Equipment. This protection technique shall be permitted for equipment in Class I, Division 2; Class II, Division 2; or Class III, Division 1 or 2 locations.

(H) Nonincendive Component. This protection technique shall be permitted for equipment in Class I, Division 2; Class II, Division 2; or Class III, Division 1 or 2 locations.

(I) Oil Immersion. This protection technique shall be permitted for current-interrupting contacts in Class I, Division 2 locations as described in 501.115(B)(1)(2).

(J) Hermetically Sealed. This protection technique shall be permitted for equipment in Class I, Division 2; Class II, Division 2; or Class III, Division 1 or 2 locations.

Δ **(K) Detection System for Flammable Gases.** A detection system for flammable gases shall be permitted as a means of protection in restricted industrial establishments.

Δ **(1) General.** Any gas detection system used as a protection technique shall meet all of the requirements in 500.7(K)(1)(a) through (K)(1)(e).

(a) The gas detection equipment used shall be listed for Class I, Division 1 and listed for the detection of the specific gas or vapor to be encountered.

(b) The gas detection system shall not use portable or transportable equipment or temporary wiring methods.

(c) The gas detection system shall only use point-type sensors. The system shall be permitted to be augmented with open-path (line-of-sight)–type sensors, but open-path–type sensors shall not be the basis for this protection technique.

(d) The type of detection equipment and its listing, installation location(s), alarm and shutdown criteria, and calibration frequency shall be documented where gas detectors are used as a protection technique.

(e) The applications for the use of gas detection systems as a protection technique shall be limited to 500.7(K)(2), (K)(3), or (K)(4).

The gas detection system must be suitable for the original division classification of the area, even though the remainder of installed equipment is permitted to be suitable for the next lower division.

> Informational Note No. 1: See ANSI/UL 121303, *Guide for Use of Detectors for Flammable Gases*, or ANSI/FM 121303, *Guide for Use of Detectors for Flammable Gases*, for additional information.
>
> Informational Note No. 2: See ANSI/UL 60079-29-1, *Explosive Atmospheres — Part 29-1: Gas Detectors — Performance Requirements of Detectors for Flammable Gases*, or ANSI/FM 60079-29-1, *Explosive Atmospheres — Part 29-1: Gas Detectors — Performance Requirements of Detectors for Flammable Gases*, for additional information.
>
> Informational Note No. 3: See ANSI/API RP 500, *Recommended Practice for Classification of Locations for Electrical Installations at Petroleum Facilities Classified as Class I, Division 1 and Division 2*, for additional information.
>
> Informational Note No. 4: See ANSI/UL 60079-29-2, *Explosive Atmospheres — Part 29-2: Gas Detectors — Selection, Installation, Use and Maintenance of Detectors for Flammable Gases and Oxygen*, or ANSI/FM-60079-29-2, *Explosive atmospheres — Part 29-2: Gas Detectors — Selection, Installation, Use and Maintenance of Detectors for Flammable Gases and Oxygen*, for additional information.

Δ **(2) Inadequate Ventilation.** A location, enclosed space, or building that is classified as a Class I, Division 1 location due to inadequate ventilation and is provided with a detection system for flammable gases shall be permitted to use electrical equipment, installation methods, and wiring practices suitable for Class I, Division 2 installations. Sensing a gas concentration of not more than 40 percent of the lower flammable limit or a gas detector system malfunction shall activate an alarm (audible or visual, or both, as most appropriate for the area).

Section 17.11 of NFPA 30, *Flammable and Combustible Liquids Code*, and 3.3.1 of NFPA 497, *Recommended Practice for the Classification of Flammable Liquids, Gases, or Vapors and of Hazardous (Classified) Locations for Electrical Installations in Chemical Process Areas*, provide information on what is considered adequate ventilation.

Δ **(3) Interior of a Building or Enclosed Space.** Any building or enclosed space that does not contain a source of flammable gases or vapors that is located in, or has an opening into, a Class I, Division 2 hazardous (classified) location and is provided with

a detection system for flammable gases shall be permitted to use electrical equipment, installation methods, and wiring practices suitable for unclassified installations under all of the following conditions:

(1) An alarm (audible or visual, or both) shall be sounded at not more than 20 percent of the lower flammable limit.

(2) Sensing a gas concentration of not more than 40 percent of the lower flammable limit or a gas detector system malfunction shall activate an alarm (audible or visual, or both, as most appropriate for the area) and initiate automatic disconnection of power from all electrical devices in the area that are not suitable for Class I, Division 2.

(3) The power disconnecting device(s) shall be suitable for Class I, Division 1 if located inside the building or enclosed space. If the disconnecting device(s) is located outside the building or enclosed space, it shall be suitable for the location in which it is installed.

Redundant or duplicate equipment (such as sensors) shall be permitted to be installed to avoid disconnecting electrical power when equipment malfunctions are indicated.

When automatic shutdown could introduce additional or increased hazard, this technique shall not be permitted.

Δ **(4) Interior of a Control Panel.** Inside the interior of a control panel containing instrumentation or other equipment using or measuring flammable liquids, gases, or vapors which is provided with a detection system for flammable gases shall be permitted to use electrical equipment, installation methods, and wiring practices suitable for Class I, Division 2 installations.

An alarm (audible or visual, or both) shall be sounded at not more than 40 percent of the lower flammable limit.

(L) Inherently Safe Optical Radiation "op is." This protection technique shall be permitted for equipment in Class I or II, Division 1 or 2 locations for which the equipment is identified.

> Informational Note: The identified class and division depends on the intended explosive atmosphere and the number of faults applied as part of the protection technique evaluation.

(M) Protected Optical Radiation "op pr." This protection technique shall be permitted for equipment in Class I or II, Division 2 locations for which the equipment is identified.

> Informational Note: The identified class and division depends on the intended explosive atmosphere as part of the protection technique evaluation.

(N) Optical System With Interlock "op sh." This protection technique shall be permitted for equipment in Class I or II, Division 1 or 2 locations for which the equipment is identified.

> Informational Note: The identified class and division depends on the intended explosive atmosphere and the number of faults applied as part of the protection technique evaluation.

(O) Protection by Skin Effect Trace Heating "IEEE 844.1". This protection technique shall be permitted for skin effect trace heating equipment in Class I, Division 2; Class II, Division 2; or Class III, Division 2 locations for which it is listed.

N **(P) Protection by Electrical Resistance Trace Heating "60079-30-1".** This protection technique shall be permitted for electrical resistance trace heating equipment in Class I, Division 1; Class I, Division 2; Class II, Division 1; Class II, Division 2; Class III, Division 1; or Class III, Division 2 locations for which it is listed.

N **(Q) Protection by Impedance Heating "IEEE 844.3".** This protection technique shall be permitted for impedance heating equipment in Class I, Division 2; Class II, Division 2; or Class III, Division 2 locations for which it is listed.

N **(R) Enclosed-Break.** This protection technique shall be permitted for equipment in Class I, Division 2 locations.

N **(S) Nonsparking.** This protection technique shall be permitted for equipment in Class I, Division 2 locations.

N **(T) Sealed.** This protection technique shall be permitted for equipment in Class I, Division 2; Class II, Division 2; Class III, Division 1; or Class III, Division 2 locations.

Δ **(U) Special Protection Techniques.** Protection techniques not specified in 500.7(A) through (T) shall be permitted for use in equipment listed for use in hazardous (classified) locations.

Informational Note: See ANSI/UL 60079-33, *Explosive Atmospheres — Part 33: Equipment Protection by Special Protection "s"*, for additional information.

Δ **500.8 Equipment.** Explosionproof or dust-ignitionproof equipment shall not be permitted for use at temperatures lower than -25°C (-13°F) unless they are identified for low-temperature service.

Informational Note: At low ambient temperatures, flammable concentrations of vapors might not exist in a location classified as Class I, Division 1 at normal ambient temperature.

Δ **(A) Suitability.** Suitability of identified equipment shall be determined by one of the following:

(1) Equipment listing or labeling
(2) Evidence of equipment evaluation from a qualified testing laboratory or inspection agency concerned with product evaluation
(3) Evidence acceptable to the authority having jurisdiction such as a manufacturer's self-evaluation or an owner's engineering judgment

Informational Note: Additional documentation for equipment might include certificates demonstrating compliance with applicable equipment standards, indicating special conditions of use, and providing other pertinent information.

Several testing and product evaluation agencies list electrical equipment that is suitable for use in hazardous locations. Testing laboratories outside the United States provide listings of equipment for use in hazardous locations, but they might not be

testing and investigating the equipment for use in hazardous locations as defined in Article 500. These foreign laboratories certify equipment for installation according to an International Electrotechnical Commission (IEC) classification scheme similar to that in Articles 505 and 506. However, equipment certified to a product standard used in another country might not comply with the *NEC*.

(B) Approval for Class and Properties.

Δ **(1) Equipment Identification.** Equipment shall be identified not only for the class of location but also for the explosive, combustible, or ignitible properties of the specific gas, vapor, dust, or fibers/flyings that will be present. In addition, Class I equipment shall not have any exposed surface that operates at a temperature in excess of the autoignition temperature of the specific gas or vapor. Class II equipment shall not have an external temperature higher than that specified in 500.8(D)(2). Class III equipment shall not exceed the maximum surface temperatures specified in 503.5.

Exception No. 1: Group D equipment shall be permitted to be used for atmospheres containing butadiene if all conduit runs into explosionproof equipment are provided with explosionproof seals installed within 450 mm (18 in.) of the enclosure.

Exception No. 2: Group C equipment shall be permitted to be used for atmospheres containing allyl glycidyl ether, n-butyl glycidyl ether, ethylene oxide, propylene oxide, and acrolein if all conduit runs into explosionproof equipment are provided with explosionproof seals installed within 450 mm (18 in.) of the enclosure.

Informational Note: See 500.8(C)(6)(a) regarding general-purpose equipment. Luminaires and other heat-producing apparatus, switches, circuit breakers, and plugs and receptacles are potential sources of ignition and are investigated for suitability in classified locations. Such types of equipment, as well as cable terminations for entry into explosionproof enclosures, are available as listed for Class I, Division 2 locations. Fixed wiring, however, might use wiring methods that are not evaluated with respect to classified locations. Therefore, wiring products such as cable, raceways, boxes, and fittings are not marked as being suitable for Class I, Division 2 locations.

Where installed in a Class I or Class II location, equipment must be suitable for the specific group indicated on the classification document (see 500.4). An explosionproof enclosure suitable only for Group D, for example, is not acceptable for Group B. Enclosures are often identified for more than one class or group.

Some portable devices — cell phones, multimeters, and flashlights — have the capacity to cause ignition of a hazardous atmosphere. Although such electrical equipment is outside the scope of the *NEC*, all equipment used should be suitable for the specific hazardous location.

Δ **(2) Equipment Application.** Equipment identified for a Division 1 location shall be permitted in a Division 2 location of the same class, group, and temperature class and shall comply with the requirements of 500.8(B)(2)(a) or (B)(2)(b) as applicable.

(a) Intrinsically safe apparatus having a control drawing requiring the installation of associated apparatus for a Division 1 installation shall be permitted to be installed in a Division 2 location if the same associated apparatus is used for the Division 2 installation.

(b) Equipment required to be explosionproof shall incorporate seals in accordance with 501.15(A) or (D) when the wiring methods of 501.10(B) are employed.

(3) General-Purpose Equipment. Where specifically permitted in Part III of Articles 501, 502, and 503, general-purpose equipment or equipment in general-purpose enclosures shall be permitted to be installed in Division 2 locations if the equipment does not constitute a source of ignition under normal operating conditions.

(4) Process Seals. Equipment that depends on a single compression seal, diaphragm, or tube to prevent flammable or combustible fluids from entering the equipment shall be identified for a Class I, Division 2 location even if installed in an unclassified location. Equipment installed in a Class I, Division 1 location shall be identified for the Class I, Division 1 location.

Informational Note: Equipment used for flow measurement is an example of equipment having a single compression seal, diaphragm, or tube.

(5) Motors. Unless otherwise specified, normal operating conditions for motors shall be assumed to be rated full-load steady conditions.

Locked-rotor or other abnormal motor conditions, such as single phasing, are not considered when evaluating motor-operating temperatures (internal and external) in Class I, Division 2 locations. However, such abnormal load conditions must be considered when the external temperatures of explosionproof motors for Class I, Division 1 locations and dust-ignitionproof motors for Class II, Division 1 locations are evaluated. Awareness of increased temperatures in motors controlled by variable-speed drives is important when they are operated at lower speeds and are dependent on the fan for cooling.

Δ **(6) Simultaneous Classifications.** Where flammable gases, flammable liquid–produced vapors, or combustible liquid–produced vapors and combustible dusts are or might be present at the same time, the simultaneous presence of the specific materials shall be considered when determining the safe operating temperature of the electrical equipment.

Examples of where flammable liquid and dust might be simultaneously present are at a coal-handling facility, where there is methane gas and coal dust, and in an automotive paint spray shop, where flammable paint and powdered metal flecks are sprayed. In the presence of such a combination of simultaneous hazards, less energy may be needed, and the accumulation of gas need not be in the flammable range for an ignition to occur.

(C) Marking. Equipment shall be marked to show the environment for which it has been evaluated. Unless otherwise specified

or allowed in 500.8(C)(6), the marking shall include the information specified in 500.8(C)(1) through (C)(5).

(1) Class. The marking shall specify the class(es) for which the equipment is suitable.

Δ **(2) Division.** The marking shall specify the division if the equipment is suitable for Division 2 only. Equipment suitable for Division 1 shall be permitted to omit the division marking.

Informational Note: See 500.8(B)(2). Equipment not marked to indicate a division, or marked "Division 1" or "Div. 1," is suitable for both Division 1 and Division 2 locations. Equipment marked "Division 2" or "Div. 2" is suitable for Division 2 locations only.

(3) Material Classification Group. The marking shall specify the applicable material classification group(s) or specific gas, vapor, dust, or fiber/flying in accordance with 500.6.

Exception: Fixed luminaires marked for use only in Class I, Division 2 or Class II, Division 2 locations shall not be required to indicate the group.

Informational Note: A specific gas, vapor, dust, or fiber/flying is typically identified by the generic name, chemical formula, CAS number, or combination thereof.

(4) Equipment Temperature. The marking shall specify the temperature class or operating temperature at a 40°C ambient temperature, or at the higher ambient temperature if the equipment is rated and marked for an ambient temperature of greater than 40°C. For equipment installed in a Class II, Division 1 location, the temperature class or operating temperature shall be based on operation of the equipment when blanketed with the maximum amount of dust that can accumulate on the equipment. The temperature class, if provided, shall be indicated using the temperature class (T codes) shown in Table 500.8(C)(4). Equipment for Class I and Class II shall be marked with the maximum

TABLE 500.8(C)(4) *Classification of Maximum Surface Temperature*

Maximum Temperature		Temperature Class (T Code)
°C	°F	
450	842	T1
300	572	T2
280	536	T2A
260	500	T2B
230	446	T2C
215	419	T2D
200	392	T3
180	356	T3A
165	329	T3B
160	320	T3C
135	275	T4
120	248	T4A
100	212	T5
85	185	T6

safe operating temperature, as determined by simultaneous exposure to the combinations of Class I and Class II conditions.

Exception: Equipment of the non–heat-producing type, such as junction boxes, conduit, and fittings, and equipment of the heat-producing type having a maximum temperature not more than 100°C shall not be required to have a marked operating temperature or temperature class.

Informational Note: More than one marked temperature class or operating temperature, for gases and vapors, dusts, and different ambient temperatures, may appear.

(5) Ambient Temperature Range. Electrical equipment designed for use in the ambient temperature range between –25°C to +40°C shall require no ambient temperature marking. For equipment rated for a temperature range other than –25°C to +40°C, the marking shall specify the special range of ambient temperatures in degrees Celsius. The marking shall include either the symbol "Ta" or "Tamb."

Informational Note: As an example, such a marking might be "–30°C ≤ Ta ≤ +40°C."

(6) Special Allowances.

(a) *General-Purpose Equipment.* Fixed general-purpose equipment in Class I locations, other than fixed luminaires, that is acceptable for use in Class I, Division 2 locations shall not be required to be marked with the class, division, group, temperature class, or ambient temperature range.

See also

Article 501, Part III, for which equipment is permitted to be installed in a general-purpose enclosure for Class I, Division 2 locations

(b) *Dusttight Equipment.* Fixed dusttight equipment, other than fixed luminaires, that is acceptable for use in Class II, Division 2 and Class III locations shall not be required to be marked with the class, division, group, temperature class, or ambient temperature range.

(c) *Associated Apparatus.* Associated intrinsically safe apparatus and associated nonincendive field wiring apparatus that are not protected by an alternative type of protection shall not be marked with the class, division, group, or temperature class. Associated intrinsically safe apparatus and associated nonincendive field wiring apparatus shall be marked with the class, division, and group of the apparatus to which it is to be connected.

(d) *Simple Apparatus.* "Simple apparatus" as defined in Article 100 Part III, shall not be required to be marked with class, division, group, temperature class, or ambient temperature range.

(D) Temperature.

△ **(1) Class I Temperature.** The temperature marking specified in 500.8(C) shall not exceed the autoignition temperature of the specific gas or vapor to be encountered.

Informational Note: See NFPA 497, *Recommended Practice for the Classification of Flammable Liquids, Gases, or Vapors and of Hazardous (Classified) Locations for Electrical Installations in Chemical Process Areas,* for information on autoignition temperatures of gases and vapors.

Electrical equipment intended for installation in a hazardous location is evaluated for maximum temperatures regardless of the type of protection afforded the equipment. Gases and vapors are qualified with an autoignition temperature, and dusts are qualified with a layer or cloud ignition temperature, each of which is the temperature at which the material ignites. The equipment temperature is compared to these material temperatures to determine if a potential for a thermal ignition exists.

The autoignition temperature of a solid, liquid, or gaseous substance is the minimum temperature required to initiate or cause self-sustained combustion independent of the heating or heated element. The flash point is the minimum temperature at which a liquid gives off enough vapor to form an ignitible mixture with air. The ignition temperature and the flash point are unrelated properties, except that the flash point is always lower than the ignition temperature.

△ **(2) Class II Temperature.** The temperature marking specified in 500.8(C) shall be less than the ignition temperature of the specific dust or metal fiber/flying to be encountered. For organic dusts that might dehydrate or carbonize, the temperature marking shall not exceed the lower of either the ignition temperature or 165°C (329°F).

Informational Note: See NFPA 499, *Recommended Practice for the Classification of Combustible Dusts and of Hazardous (Classified) Locations for Electrical Installations in Chemical Process Areas,* for minimum ignition temperatures of specific dusts.

△ **(3) Class III Temperature.** The temperature marking specified in 500.8(C) shall be less than the ignition temperature of the specific fiber/flying to be encountered, except as specified in 500.8(D)(3)(a) or (D)(3)(b).

(a) For nonmetal combustible fibers/flyings that might dehydrate or carbonize, the temperature marking shall not exceed the lower of either the ignition temperature or 165°C (329°F).

(b) When ignitible fibers/flyings are present, the maximum surface temperatures under operating conditions shall not exceed 165°C (329°F) for equipment that is not subject to overloading, and 120°C (248°F) for equipment (such as motors or power transformers) that might be overloaded.

(E) Threading. The supply connection entry thread form shall be NPT or metric. Conduit and fittings shall be made wrenchtight to prevent sparking when fault current flows through the conduit system, and to ensure the explosionproof integrity of the conduit system where applicable. Equipment provided with threaded entries for field wiring connections shall be installed in accordance with 500.8(E)(1) or (E)(2) and with (E)(3).

To ensure the integrity of the ground-fault current path of the conduit system, all conduit joints must be made up wrenchtight to prevent ignition-capable arcing between the conduit and the coupling, fitting, or enclosure under ground-fault conditions. The use of a bonding jumper in lieu of a wrenchtight connection is not permitted.

Δ **(1) Equipment Provided with Threaded Entries for NPT-Threaded Conduit or Fittings.** For equipment provided with threaded entries for NPT-threaded conduit or fittings, listed conduit, listed conduit fittings, or listed cable fittings shall be used. All NPT-threaded conduit and fittings shall be threaded with a National (American) Standard Pipe Taper (NPT) thread.

NPT-threaded entries into explosionproof equipment shall be made up with at least five threads fully engaged.

Exception: For listed explosionproof equipment, joints with factory-threaded NPT entries shall be made up with at least four and one-half threads fully engaged.

Informational Note No. 1: See ASME B1.20.1, *Pipe Threads, General Purpose (Inch)*, for thread specifications for male NPT threads.

Informational Note No. 2: See ASME B1.20.1, *Pipe Threads, General Purpose (Inch)*, and ANSI/UL 1203, *Explosion-Proof and Dust-Ignition-Proof Electrical Equipment for Use in Hazardous (Classified) Locations*, for information on female NPT-threaded entries using modified National Standard Pipe Taper (NPT) threads.

Δ **(2) Equipment Provided with Threaded Entries for Metric-Threaded Fittings.** For equipment with metric-threaded entries, listed conduit fittings or listed cable fittings shall be used. Such entries shall be identified as being metric, or listed adapters to permit connection to conduit or NPT-threaded fittings shall be provided with the equipment and shall be used for connection to conduit or NPT-threaded fittings.

Metric-threaded fittings installed into explosionproof equipment shall have a class of fit of at least 6g/6H and shall be made up with at least five threads fully engaged.

Informational Note: See ISO 965-1, *ISO general purpose metric screw threads — Tolerances — Part 1: Principles and basic data*, and ISO 965-3, *ISO general purpose metric screw threads — Tolerances — Part 3: Deviations for constructional screw threads*, for threading specifications for metric-threaded entries.

(3) Unused Openings. All unused openings shall be closed with blanking elements or close-up plugs that are listed for the location. The thread engagement shall comply with the requirements of 500.8(E)(1) or (E)(2).

Δ **(F) Optical Fiber Cables.** An optical fiber cable, with or without current-carrying conductors (hybrid optical fiber cable), shall be installed to address the associated fire hazard and sealed to address the associated explosion hazard in accordance with Part II of Articles 501, 502, or 503, as applicable.

Δ **(G) Equipment Involving Optical Radiation.** The risk of ignition from optical radiation shall be evaluated for laser equipment, optical fiber equipment, and any other convergent light sources or beams where light is focused in one single point within a hazardous area with a wavelength range of 380 nm to 10 μm. This requirement shall include optical equipment that is located outside the explosive atmosphere, but whose emitted optical radiation enters such atmospheres.

Informational Note: See ANSI/UL 60079-28, *Explosive Atmospheres — Part 28: Protection of Equipment and Transmission Systems Using Optical Radiation*, for information on types of protection that can be applied to minimize the risk of ignition in explosive atmospheres from optical radiation.

ARTICLE 501 Class I Locations

Part I. General

Δ **501.1 Scope.** This article covers the requirements for electrical and electronic equipment and wiring for all voltages in Class I, Division 1 and Division 2 locations where flammable gases, flammable liquid–produced vapors, or combustible liquid–produced vapors are or might be present in the air in quantities sufficient to produce explosive or ignitible mixtures.

Where ignitible concentrations (concentrations within flammable or explosive limits) are present, the atmosphere can be ignited by an arc, spark, or high temperature. NFPA 497, *Recommended Practice for the Classification of Flammable Liquids, Gases, or Vapors and of Hazardous (Classified) Locations for Electrical Installations in Chemical Process Areas*, includes information on the flammable limits of liquids and gases.

Electrical equipment that can cause ignition-capable arcs or sparks should be kept out of Class I locations, or the equipment must be identified for the appropriate hazardous location. It is practically impossible to make threaded joints gastight. The conduit system and apparatus enclosure "breathe" due to temperature changes, and any flammable gases or vapors in the room may slowly enter the conduit or enclosure, creating an explosive mixture. If an arc occurs, an explosion could also occur. When an explosion occurs within the enclosure or conduit system, the burning mixture or hot gases must be sufficiently confined within the system to prevent ignition of any explosive mixture outside the system.

501.5 Zone Equipment. Equipment listed and marked in accordance with 505.9(C)(2) for use in Zone 0, 1, or 2 locations shall be permitted in Class I, Division 2 locations for the same gas and with a suitable temperature class. Equipment listed and marked in accordance with 505.9(C)(2) for use in Zone 0 locations shall be permitted in Class I, Division 1 or Division 2 locations for the same gas and with a suitable temperature class.

Part II. Wiring

501.10 Wiring Methods. Wiring methods shall comply with 501.10(A) or (B).

(A) Class I, Division 1.

Δ **(1) General.** In Class I, Division 1 locations, the following wiring methods shall be permitted:

> Informational Note No. 1: See Article 100 for the definition of *restricted industrial establishment [as applied to hazardous (classified) locations].*

(1) Threaded rigid metal conduit (RMC) or threaded intermediate metal conduit (IMC), including RMC or IMC conduit systems with supplemental corrosion protection coatings.

(2) PVC conduit, RTRC conduit, or HDPE conduit, where encased in a concrete envelope a minimum of 50 mm (2 in.) thick and provided with not less than 600 mm (24 in.) of cover measured from the top of the conduit to grade. The concrete encasement shall be permitted to be omitted where it is in accordance with 514.8(C) or 515.8(A). RMC or IMC conduit shall be used for the last 600 mm (24 in.) of the underground run to emergence or to the point of connection to the aboveground raceway. An equipment grounding conductor shall be included to provide for electrical continuity of the raceway system and for grounding of non–current-carrying metal parts.

(3) Type MI cable terminated with fittings listed for the location. Type MI cable shall be installed and supported to avoid tensile stress at the termination fittings.

A termination fitting used in a Division 1 location with Type MI cable must be specifically listed for use in Class I, Division 1 hazardous locations. Exhibit 501.1 shows an example of this type of cable and fitting in which the screw-on pot contains field-installed sealing compound to seal the end of the cable. The threaded gland has threads for connection to explosion-proof enclosures.

(4) In restricted industrial establishments, Type MC-HL cable listed for use in Class I, Zone 1 or Division 1 locations, with a gas/vaportight continuous corrugated metallic sheath, an overall jacket of suitable polymeric material, and a separate equipment grounding conductor(s) in accordance with 250.122, and terminated with fittings listed for the application. If installed in a ladder, ventilated trough, or ventilated channel cable tray, the cable shall be installed in accordance with 392.22. Type MC-HL cable shall be installed in accordance with Part II of Article 330.

(5) In restricted industrial establishments, Type ITC-HL cable listed for use in Class I, Division 1 or Zone 1 locations, with a gas/vaportight continuous corrugated metallic sheath and an overall jacket of suitable polymeric material, terminated with fittings listed for the application, and installed in accordance with 335.4.

(6) Optical fiber cable Type OFNP, Type OFCP, Type OFNR, Type OFCR, Type OFNG, Type OFCG, Type OFN, or Type OFC installed in raceways in accordance with 501.10(A). These optical fiber cables shall be sealed in accordance with 501.15.

(7) In restricted industrial establishments for applications limited to 600 volts nominal or less, and where the cable is not subject to physical damage and is terminated with fittings listed for the location, Type TC-ER-HL cable. If installed in a ladder, ventilated trough, or ventilated channel cable tray, the cable shall be installed in accordance with 392.22. Type TC-ER-HL cable shall be listed for use in Class I, Division 1 or Zone 1 locations and shall be installed in accordance with 336.10.

> Informational Note No. 2: See ANSI/UL 2225, *Cables and Cable-Fittings for Use in Hazardous (Classified) Locations*, for information on construction, testing, and marking of cables and cable fittings.

(8) In restricted industrial establishments, listed Type P cable with metal braid armor and an overall jacket, terminated with fittings listed for the location, and installed in accordance with Part II of Article 337. If installed in a ladder, ventilated trough, or ventilated channel cable tray, the cable shall be installed in accordance with 392.22.

> Informational Note No. 3: See UL 1309A, *Outline of Investigation for Cable for Use in Mobile Installations*, for information on construction, testing, and marking of Type P cable.
>
> Informational Note No. 4: See ANSI/UL 2225, *Cables and Cable-Fittings for Use in Hazardous (Classified) Locations*, for information on construction, testing, and marking of cable.

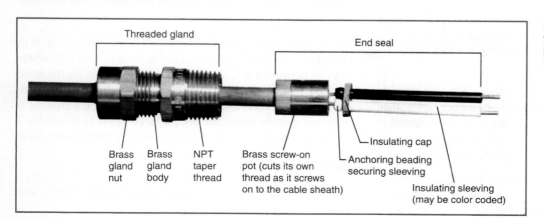

EXHIBIT 501.1 Type MI cable and fitting listed for use in hazardous locations. *(Courtesy of Tyco Thermal Controls)*

Δ (2) **Flexible Connections.** If flexibility is necessary to minimize the transmission of vibration from equipment during operation or to allow for movement after installation during maintenance, one of the following shall be permitted:

(1) Flexible fittings listed for the location.
(2) Flexible cord in accordance with 501.140, terminated with cord connectors listed for the location.
(3) In restricted industrial establishments, for applications limited to 600 volts nominal or less where the cable is not subject to physical damage and is terminated with fittings listed for the location, Type TC-ER-HL cable. The cable shall be listed for use in Class I, Division 1 or Zone 1 locations and shall be installed in accordance with 336.10.

> Informational Note No. 1: See ANSI/UL 2225, *Cables and Cable-Fittings for Use in Hazardous (Classified) Locations*, for information on construction, testing, and marking of cables and cable fittings.

(4) In restricted industrial establishments, listed Type P cable with metal braid armorand an overall jacket where the cable is terminated with fittings listed for the location and installed in accordance with Part II of Article 337.

> Informational Note No. 2: See UL 1309A, *Outline of Investigation for Cable for Use in Mobile Installations*, for information on construction, testing, and marking of Type P cable fittings.
> Informational Note No. 3: See ANSI/UL 2225, *Cables and Cable-Fittings for Use in Hazardous (Classified) Locations*, for information on construction, testing, and marking of cable fittings.

This section refers to a connection to equipment that requires flexibility, not to a flexible wiring method. For wiring methods where a flexible connection is no longer necessary, refer to 501.10(A)(1). Four options are presented if a length of flexible connection is necessary. Listed flexible fittings commonly used at motor connections can withstand continuous vibration for long periods and provide maximum protection to enclosed conductors. Limited use of flexible cord is permitted in accordance with 501.140 for specific applications where flexibility is made necessary by the type of equipment being supplied. Type TC-ER-HL or Type P is permitted in industrial establishments where a flexible connection is necessary. The requirement in 501.15(A)(1) to provide a seal within 18 inches of an explosionproof enclosure applies where flexible connections are used.

(3) Boxes and Fittings. All boxes and fittings shall be identified for Class I, Division 1.

> Informational Note No. 1: See ANSI/UL 2225, *Cables and Cable-Fittings for Use in Hazardous (Classified) Locations*, for information on construction, testing, and marking of cables, explosionproof cable fittings, and explosionproof cord connectors for entry into enclosures required to be explosionproof.
> Informational Note No. 2: See ANSI/UL 1203, *Explosion-Proof and Dust-Ignition-Proof Electrical Equipment for Use in Hazardous (Classified) Locations*, for information on construction, testing, and marking of explosionproof conduit fittings for entry into enclosures required to be explosionproof.

EXHIBIT 501.2 An explosionproof junction box with a screw-type cover. *(Courtesy of O-Z/Gedney, Emerson Electric Co.)*

Exhibit 501.2 shows an explosionproof junction box with two hubs and a threaded opening for the screw-type cover. Unused openings must be effectively closed by inserting threaded metal plugs that engage at least five full threads [4½ permitted in accordance with the exception to 500.8(E)] and afford protection equivalent to that of the box wall.

(B) Class I, Division 2.

Δ (1) **General.** In Class I, Division 2 locations, all wiring methods in accordance with 501.10(A) and the following wiring methods shall be permitted:

> Informational Note No. 1: See Article 100 for the definition of *restricted industrial establishment [as applied to hazardous (classified) locations].*

(1) Rigid metal conduit (RMC) or intermediate metal conduit (IMC) with listed threaded or threadless fittings, including RMC or IMC conduit systems with supplemental corrosion protection coatings.
(2) Enclosed gasketed busways and enclosed gasketed wireways.
(3) Type PLTC cable or Type PLTC-ER cable used for Class 2 and Class 3 circuits, including installation in cable tray systems. The cable shall be terminated with listed fittings. Type PLTC-ER cable shall include an equipment grounding conductor in addition to a drain wire that might be present.
(4) Type ITC cable or Type ITC-ER cable as permitted in 335.4 and terminated with listed fittings. Type ITC-ER cable shall include an equipment grounding conductor in addition to a drain wire.
(5) Type MC, Type MV, Type TC, or Type TC-ER cable, including installation in cable tray systems. Type TC-ER cable shall include an equipment grounding conductor in addition to a drain wire that might be present. All cable types shall be terminated with listed fittings.
(6) Where metal conduit will not provide the corrosion resistance needed for the installation environment, any of the following shall be permitted:

a. Listed reinforced thermosetting resin conduit (RTRC), factory elbows, and associated fittings, all marked with the suffix -XW

b. PVC-coated RMC, factory elbows, and associated fittings

c. PVC-coated IMC, factory elbows, and associated fittings

d. In restricted industrial establishments, Schedule 80 PVC conduit, factory elbows, and associated fittings

(7) Optical fiber cable Type OFNP, Type OFCP, Type OFNR, Type OFCR, Type OFNG, Type OFCG, Type OFN, or Type OFC installed in cable trays or any other raceway in accordance with 501.10(B). Optical fiber cables shall be sealed in accordance with 501.15.

(8) Cablebus.

(9) In restricted industrial establishments, listed Type P cable with or without metal braid armor, with an overall jacket, and terminated with fittings listed for the location when entering explosionproof, flameproof, or pressurized equipment. The cable shall be installed in accordance with Part II of Article 337.

> Informational Note No. 2: See ANSI/UL 1309A, *Outline of Investigation for Cable for Use in Mobile Installations*, for information on construction, testing, and marking of Type P cable.

> Informational Note No. 3: See ANSI/UL 2225, *Cables and Cable-Fittings for Use in Hazardous (Classified) Locations*, for information on construction, testing, and marking of cable fittings.

Δ **(2) Flexible Connections.** If flexibility is necessary to minimize the transmission of vibration from equipment during operation or to allow for movement after installation during maintenance, one or more of the following shall be permitted:

(1) Listed flexible metal fittings

(2) Flexible metal conduit with listed fittings and bonded in accordance with 501.30(B)

(3) Interlocked armor Type MC cable with listed fittings

(4) Liquidtight flexible metal conduit with listed fittings and bonded in accordance with 501.30(B)

(5) Liquidtight flexible nonmetallic conduit with listed fittings

(6) Flexible cord listed for extra-hard usage and terminated with listed fittings, with a conductor for use as an equipment grounding conductor

(7) For elevator use, an identified elevator cable of Type EO, Type ETP, or Type ETT, shown under the "use" column in Table 400.4 for "hazardous (classified) locations" and terminated with listed fittings

(8) In restricted industrial establishments, listed Type P cable with or without metal braid armor, with an overall jacket, terminated with listed fittings and installed in accordance with Part II of Article 337

Where necessary, limited flexibility is provided through use of wiring methods with listed fittings such as flexible metal conduit (FMC), liquidtight flexible metal conduit (LFMC), and extra-hard-usage flexible cord. However, the fittings are not required to be specifically identified for Class I locations.

See also

501.10(A)(2) and its commentary for more information on flexible connections

(3) Nonincendive Field Wiring. Nonincendive field wiring shall be permitted using any of the wiring methods permitted for unclassified locations. Nonincendive field wiring systems shall be installed in accordance with the control drawing(s). Simple apparatus, not shown on the control drawing, shall be permitted in a nonincendive field wiring circuit if the simple apparatus does not interconnect the nonincendive field wiring circuit to any other circuit.

> Informational Note: See Article 100 for the definition of *simple apparatus*.

Separate nonincendive field wiring circuits shall be installed in accordance with one of the following:

(1) In separate cables

(2) In multiconductor cables where the conductors of each circuit are within a grounded metal shield

(3) In multiconductor cables or in raceways, where the conductors of each circuit have insulation with a minimum thickness of 0.25 mm (0.01 in.)

Although many low-voltage, low-energy circuits, including some communications circuits and thermocouple circuits, are of the nonincendive type, a Class 2 or Class 3 circuit, as defined in Article 725, is not necessarily nonincendive.

See also

Article 100 for the definitions of the terms *nonincendive circuit* and *nonincendive field wiring*

(4) Boxes and Fittings. Boxes and fittings shall be explosionproof if required by 501.105(B)(2), 501.115(B)(1), or 501.150(B)(1).

> Informational Note No. 1: See ANSI/UL 2225, *Cables and Cable-Fittings for Use in Hazardous (Classified) Locations*, for information on construction, testing, and marking of cable for entry into enclosures required to be explosionproof.

> Informational Note No. 2: See ANSI/UL 1203, *Explosion-Proof and Dust-Ignition-Proof Electrical Equipment for Use in Hazardous (Classified) Locations*, for information on construction, testing, and marking of explosionproof conduit fittings for entry into enclosures required to be explosionproof.

In Class I, Division 2 locations, boxes, fittings, and joints are not required to be explosionproof if they contain no arcing devices. They also are not required to be explosionproof for lighting outlets or enclosures containing nonarcing devices (such as solid-state relays, solenoids, and control transformers), if the maximum operating temperature of any exposed surface does not exceed 80 percent of the ignition temperature.

501.15 Sealing and Drainage. Seals in conduit and cable systems shall comply with 501.15(A) through (F). Sealing compound shall be used in Type MI cable termination fittings to exclude moisture and other fluids from the cable insulation.

Commentary Table 501.1 summarizes the sealing requirements of 501.15(A) through (F).

COMMENTARY TABLE 501.1 Conduit and Cable Sealing Requirements

Classification	Application	Location of Seal
Conduit Seals Class I, Division 1	Switch enclosure Circuit breaker enclosure Fuse enclosure Relay enclosure Resistor enclosure Arcing or sparking apparatus High-temperature apparatus	In conduit run within 18 in. of enclosure.
	Explosionproof enclosure containing arcing or sparking contacts that are hermetically sealed against gas or vapor entry Explosionproof enclosure containing arcing or sparking contacts that are immersed in oil, in accordance with 501.115(B)(1)(2)	In conduit runs smaller than trade size 2, no seal is required. If conduit is trade size 2 or larger, in conduit run within 18 in. of enclosure.
	Enclosure containing terminals, splices, or taps; fitting containing terminals, splices, or taps	In conduit runs smaller than trade size 2, no seal is required. If conduit is trade size 2 or larger, in conduit run within 18 in. of enclosure.
	Two explosionproof enclosures with a conduit run between them of 36 in. or less	In conduit run within 18 in. of each enclosure. Permitted to use a single seal in each run as long as the seal is within 18 in. of each enclosure.
	Two explosionproof enclosures with a conduit run between them greater than 36 in.	In conduit run within 18 in. of each enclosure.
	Conduit run leaving Division 1 location	Within 10 ft of either side of boundary. No unions, couplings, boxes, or fittings (other than explosionproof reducers) permitted between the seal fitting and the point where the conduit leaves the Division 1 location.
	Metal conduit containing no unions, couplings, boxes, or fittings that passes completely through a Class I, Division 1 location, with no fittings less than 12 in. beyond each boundary	Not required to be sealed if the termination points of the unbroken conduit are in unclassified locations.
Class I, Division 2	Enclosure required to be explosionproof	Seal as required for similar equipment in Division 1 location.
	Conduit run leaving Division 2 location	Within 10 ft of either side of boundary. No unions, couplings, boxes, or fittings (other than explosionproof reducers) permitted between the seal fitting and the point where the conduit leaves the Division 2 location. Not required to be explosionproof seal.
	Metal conduit containing no unions, couplings, boxes, or fittings that passes completely through a Division 2 location with no fittings less than 12 in. beyond each boundary	Not required to be sealed if the termination points of the unbroken conduit are in unclassified locations.
	Conduit systems terminating at an unclassified location where a wiring method transition is made to cable tray, cablebus, ventilated busway, Type MI cable, or open wiring	Not required to be sealed if passing from the Class I, Division 2 location into an outdoor unclassified location or an indoor location if the conduit system is all in one room. The conduits do not terminate at an enclosure containing an ignition source in normal operation.
	Conduit leaving a purged enclosure or room that is unclassified due to pressurization and entering a Division 2 location	Not required to be sealed at the boundary.
Cable Seals Class I, Division 1	Enclosure with integral seal	Conduit seal fitting not required.
	Multiconductor Type MC-HL cables with a gastight/vaportight continuous corrugated metallic sheath and an overall jacket of suitable polymeric material	Seal at all terminations with a listed fitting after removing the jacket and any other covering, so that the sealing compound surrounds each individual insulated conductor.
	Cables in conduit with a gastight/vaportight continuous sheath capable of transmitting gases or vapors through the cable core	Seal in the Division 1 location after removing the jacket and any other coverings, so that the sealing compound surrounds each individual insulated conductor and the outer jacket.

(continues)

COMMENTARY TABLE 501.1 Continued

Classification	Application	Location of Seal
	Multiconductor cables with a gastight/vaportight continuous sheath capable of transmitting gases or vapors through the cable core	Permitted to be considered a single conductor by sealing the cable in the conduit within 18 in. of the enclosure and the cable end within the enclosure by an approved means, to minimize the entrance of gases or vapors and prevent the propagation of flame into the cable core, or by other approved methods.
	For shielded cables and twisted pair cables	Removal of the shielding material or separation of the twisted pair is not required. Sealing the cable in the conduit and the cable end within the enclosure by an approved means to minimize the entrance of gases or vapors and prevent the propagation of flame into the cable core, or by other approved methods.
	Each multiconductor cable in conduit if the cable is incapable of transmitting gases or vapors through the cable core	Considered a single conductor. These cables are sealed in accordance with 501.15(A).
Class I, Division 2	Cables entering enclosures that are required to be explosionproof for Class I locations	Sealed at the point of entrance. These cables are sealed in accordance with the requirements of Division 1 locations.
	Multiconductor cables with a gastight/vaportight continuous sheath capable of transmitting gases or vapors through the cable core	Sealed in a listed fitting in the Division 2 location after removing the jacket and any other coverings, so that the sealing compound surrounds each individual insulated conductor.
	Multiconductor cables in conduit	Sealed in accordance with the requirements for Division 1 locations.
	Cables with a gastight/vaportight continuous sheath that will not transmit gases or vapors through the cable core in excess of the quantity permitted for seal fittings. The minimum length of such cable run is not less than that length that limits gas or vapor flow through the cable core to the rate permitted for seal fittings (0.007 ft^3 per hour of air at a pressure of 6 in. of water).	Not required to be sealed unless entering an enclosure that is required to be explosionproof.
	Cables with a gastight/vaportight continuous sheath capable of transmitting gases or vapors through the cable core	Not required to be sealed unless entering an enclosure that is required to be explosionproof or unless the cable is attached to process equipment or devices that may cause a pressure in excess of 6 in. of water to be exerted at a cable end, in which case a seal, barrier, or other means is provided to prevent migration of flammables into an unclassified area.
	Cables with an unbroken gastight/vaportight continuous sheath that pass through a Class I, Division 2 location	No seal required.
	Cables that do not have a gastight/vaportight continuous sheath	Sealed at the boundary of the Division 2 and unclassified location in such a manner as to minimize the passage of gases or vapors into an unclassified location.

Informational Note No. 1: Seals are provided in conduit and cable systems to minimize the passage of gases and vapors and prevent the passage of flames from one portion of the electrical installation to another through the conduit. Such communication through Type MI cable is inherently prevented by construction of the cable. Unless specifically designed and tested for the purpose, conduit and cable seals are not intended to prevent the passage of liquids, gases, or vapors at a continuous pressure differential across the seal. Even at differences in pressure across the seal equivalent to a few inches of water, there may be a slow passage of gas or vapor through a seal and through conductors passing through the seal. Temperature extremes and highly corrosive liquids and vapors can affect the ability of seals to perform their intended function.

Informational Note No. 2: Gas or vapor leakage and propagation of flames may occur through the interstices between the strands

of standard stranded conductors larger than 2 AWG. Special conductor constructions, such as compacted strands or sealing of the individual strands, are means of reducing leakage and preventing the propagation of flames.

Because the sealing compound used in conduit seal fittings is typically somewhat porous, gases, particularly those under slight pressure and those with small molecules such as hydrogen, could pass slowly through the compound. Because the seal is around the insulation on the conductor, gases can be transmitted slowly through the air spaces (the interstices) between strands of stranded conductors. Under normal conditions for smaller conductors with only normal atmospheric pressure differentials across the seal, the passage of gas through a seal is not sufficient to result in a hazard. For larger conductors, gas or vapor leakage and flame propagation can occur through the interstices. Special conductor constructions, such as compacted strands or sealing individual strands, may reduce leakage and prevent flame propagation.

Different sealing compounds have different rates of expansion and contraction that can affect their performance within a given fitting. Tapes or compounds on conduit threads can weaken the seal fitting and interrupt the equipment grounding path.

(A) Conduit Seals, Class I, Division 1. In Class I, Division 1 locations, conduit seals shall be located in accordance with 501.15(A)(1) through (A)(4).

(1) Entering Enclosures. Each conduit entry into an explosionproof enclosure shall have a conduit seal where either of the following conditions apply:

(1) The enclosure contains apparatus, such as switches, circuit breakers, fuses, relays, or resistors that may produce arcs, sparks, or temperatures that exceed 80 percent of the autoignition temperature, in degrees Celsius, of the gas or vapor involved in normal operation.

Exception: Seals shall not be required for conduit entering an enclosure under any one of the following conditions:

(1) The switch, circuit breaker, fuse, relay, or resistor is enclosed within a chamber hermetically sealed against the entrance of gases or vapors.
(2) The switch, circuit breaker, fuse, relay, or resistor is immersed in oil in accordance with 501.115(B)(1)(2).
(3) The switch, circuit breaker, fuse, relay, or resistor is enclosed within an enclosure, identified for the location, and marked "Leads Factory Sealed," or "Factory Sealed," "Seal not Required," or equivalent.
(4) The switch, circuit breaker, fuse, relay, or resistor is part of a nonincendive circuit.

(2) The entry is metric designator 53 (trade size 2) or larger, and the enclosure contains terminals, splices, or taps.

A seal fitting must be placed within 18 inches of the entrance of trade size 2 or larger conduit into any explosionproof enclosure, regardless of whether the enclosure contains arcing or sparking equipment or only splices, taps, or terminals.

An enclosure, identified for the location, and marked "Leads Factory Sealed", or "Factory Sealed," or "Seal not Required," or equivalent shall not be considered to serve as a seal for another adjacent enclosure that is required to have a conduit seal.

Conduit seals shall be installed within 450 mm (18 in.) from the enclosure or as required by the enclosure marking. Only threaded couplings, or explosionproof fittings such as unions, reducers, elbows, and capped elbows that are not larger than the trade size of the conduit, shall be permitted between the sealing fitting and the explosionproof enclosure.

Conduit seals are to prevent an explosion from traveling through the conduit to another enclosure and to minimize the passage of gases or vapors from hazardous locations to non-hazardous locations.

△ **(2) Pressurized Enclosures.** Conduit seals shall be installed within 450 mm (18 in.) of the enclosure in each conduit entry into a pressurized enclosure where the conduit is not pressurized as part of the protection system.

> Informational Note No. 1: Installing the seal as close as possible to the enclosure will reduce problems with purging the dead airspace in the pressurized conduit.
> Informational Note No. 2: See NFPA 496, *Standard for Purged and Pressurized Enclosures for Electrical Equipment*, for information regarding pressurized enclosures.

(3) Two or More Explosionproof Enclosures. Where two or more explosionproof enclosures that require conduit seals are connected by nipples or runs of conduit not more than 900 mm (36 in.) long, a single conduit seal in each such nipple connection or run of conduit shall be considered sufficient if the seal is located not more than 450 mm (18 in.) from either enclosure.

If two enclosures are spaced within 36 inches of each other, as illustrated by Enclosures No. 1 and No. 2, in Exhibit 501.3, a single seal may be placed between two connecting nipples if the seal is located not more than 18 inches from either enclosure and if no additional conduit fittings exist between the enclosures, such as a "T" fitting. If such a "T", or other conduit body, fitting exists,

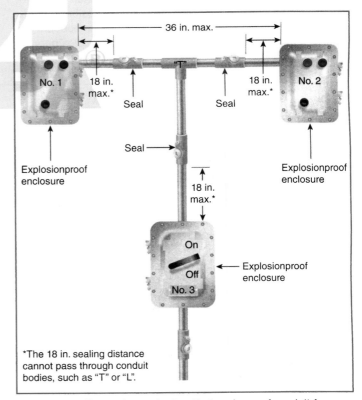

EXHIBIT 501.3 Two seals required so that each run of conduit from Enclosure No. 1 is sealed.

then a seal between each enclosure and the "T", or conduit body fitting, is necessary.

(4) Class I, Division 1 Boundary. A conduit seal shall be required in each conduit run leaving a Division 1 location. The sealing fitting shall be permitted to be installed on either side of the boundary within 3.05 m (10 ft) of the boundary, and it shall be designed and installed to minimize the amount of gas or vapor within the portion of the conduit installed in the Division 1 location that can be communicated beyond the seal. The conduit run between the conduit seal and the point at which the conduit leaves the Division 1 location shall contain no union, coupling, box, or other fitting except for a listed explosionproof reducer installed at the conduit seal.

Where the seal is located on the Division 2 side of the boundary, the Division 1 wiring method shall extend into the Division 2 area to the seal.

A seal fitting is required at the boundary where the conduit leaves a Division 1 location or passes from a Division 2 location to an unclassified location per 501.15(B)(2). The seal is permitted on either side of the boundary, and no union, coupling, box, or similar fitting is permitted between the seal and the boundary. However, listed explosionproof reducers are permitted to be installed at the conduit seal.

The seal is best located on the nonhazardous side of the boundary, where it serves two purposes: completion of the explosionproof wiring method and completion of the explosionproof enclosure system. For example, even though a conduit seal is not required for ½-inch conduit connected to an explosionproof box that contains only splices, the required seal at the boundary of the Division 1 location serves to complete the explosionproof

system. The seal at the boundary also prevents the conduit system from serving as a pipe to transmit flammable mixtures from either a Division 1 or a Division 2 location to an unclassified location.

Exception No. 1: Metal conduit that contains no unions, couplings, boxes, or fittings, that passes completely through a Division 1 location with no fittings installed within 300 mm (12 in.) of either side of the boundary, shall not require a conduit seal if the termination points of the unbroken conduit are located in unclassified locations.

Exception No. 2: For underground conduit installed in accordance with 300.5 where the boundary is below grade, the sealing fitting shall be permitted to be installed after the conduit emerges from below grade, but there shall be no union, coupling, box, or fitting, other than listed explosionproof reducers at the sealing fitting, in the conduit between the sealing fitting and the point at which the conduit emerges from below grade.

Exhibit 501.4 illustrates a Class I, Division 1 location where the enclosures for the disconnecting means and motor controller for the motor (right portion of the drawing) are placed on the other side of the wall in a nonhazardous location and are not required to be explosionproof.

Each of the three conduits shown is sealed on the nonhazardous side before it enters the hazardous location, in accordance with 501.15(A)(4). The pigtail leads of both motors are factory sealed at the motor-terminal housing, and, unless the size of the flexible fitting entering the motor-terminal housing is trade size 2 or larger, no other seals are needed at that point.

EXHIBIT 501.4 A Class I, Division 1 location where threaded metal conduits, seal fittings, explosionproof fittings, and equipment for power and lighting are used.

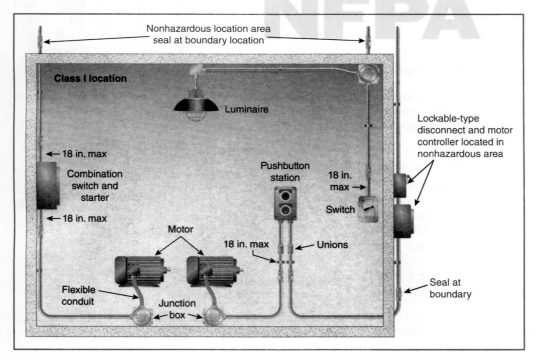

Because the pushbutton control station and the combination switch and starter are considered arc-producing devices, conduits are sealed within 18 inches of the entrance to those enclosures. Seals are required even if the contacts are immersed in oil if the conduit is trade size 2 or larger.

Additionally, Exhibit 501.4 shows a seal provided within 18 inches of the switch controlling the lighting. Because the explosionproof luminaire has an explosionproof chamber for the wiring that is separated or sealed from the lamp compartment, a separate seal is not required adjacent to this luminaire.

(B) Conduit Seals, Class I, Division 2. In Class I, Division 2 locations, conduit seals shall be located in accordance with 501.15(B)(1) and (B)(2).

(1) Entering Enclosures. For connections to enclosures that are required to be explosionproof, a conduit seal shall be provided in accordance with 501.15(A)(1)(1) and (A)(3). All portions of the conduit run or nipple between the seal and enclosure shall comply with 501.10(A).

An enclosure will not be explosionproof if the necessary conduit seals are not provided. Seals complete the explosionproof enclosure.

Δ **(2) Class I, Division 2 Boundary.** A conduit seal shall be required in each conduit run leaving a Class I, Division 2 location. The sealing fitting shall be permitted to be installed on either side of the boundary within 3.05 m (10 ft) of the boundary and it shall be designed and installed to minimize the amount of gas or vapor within the portion of the conduit installed in the Division 2 location that can be communicated beyond the seal. Wiring methods permitted in 501.10(B)(1)(1) or (B)(1)(6) shall be used between the sealing fitting and the point at which the conduit leaves the Division 2 location, and a threaded connection shall be used at the sealing fitting. The conduit run between the conduit seal and the point at which the conduit leaves the Division 2 location shall contain no union, coupling, box, or other fitting except for a listed explosionproof reducer installed at the conduit seal. Such seals shall not be required to be explosionproof but shall be identified for the purpose of minimizing the passage of gases permitted under normal operating conditions and shall be accessible.

> Informational Note No. 1: See ANSI/UL 514B, *Conduit, Tubing, and Cable Fittings*, for additional information.

Exception No. 1: Metal conduit that contains no unions, couplings, boxes, or fittings and that passes completely through a Division 2 location with no fittings installed within 300 mm (12 in.) of either side of the boundary shall not require a seal if the termination points of the unbroken conduit are located in unclassified locations.

Exception No. 2: Conduit terminating in an unclassified location where the metal conduit transitions to cable tray, cablebus, ventilated busway, or Type MI cable, or to cable not installed in any cable tray or raceway system, shall not require a seal

where passing from the Division 2 location into the unclassified location under the following conditions:

(1) The unclassified location is outdoors, or the unclassified location is indoors and the conduit system is entirely in one room.

(2) The conduits do not terminate at an enclosure containing an ignition source in normal operation.

Exception No. 3: Conduit passing from an enclosure or a room permitted to use general-purpose equipment as a result of pressurization into a Division 2 location shall not require a seal at the boundary.

> Informational Note No. 2: See NFPA 496, *Standard for Purged and Pressurized Enclosures for Electrical Equipment*, for further information.

Exception No. 4: Aboveground conduit shall not require a seal where passing from a Division 2 location into an unclassified location if all of the following conditions are met:

(1) No part of the conduit passes through a Division 1 location where the conduit contains unions, couplings, boxes, or fittings that are located within 300 mm (12 in.) of the Division 1 location.

(2) The conduit is located entirely outdoors.

(3) The conduit is not directly connected to canned pumps, process or service connections for flow, pressure, or analysis measurement, and so forth, that depend on a single compression seal, diaphragm, or tube to prevent flammable or combustible fluids from entering the conduit system.

(4) The conduit contains only threaded metal conduit, unions, couplings, conduit bodies, and fittings in the unclassified location.

(5) The conduit is sealed at its entry to each enclosure or fitting located in the Division 2 location that contains terminals, splices, or taps.

(C) Class I, Divisions 1 and 2. Seals installed in Class I, Division 1 and Division 2 locations shall comply with 501.15(C)(1) through (C)(6).

Exception: Seals that are not required to be explosionproof by 501.15(B)(2) or 504.70 shall not be required to comply with 501.15(C).

(1) Fittings. Enclosures that contain connections or equipment shall be provided with an integral sealing means, or sealing fittings listed for the location shall be used. Sealing fittings shall be listed for use with one or more specific compounds and shall be accessible.

Fittings should be sealed with only the sealing compound or compounds specified by the manufacturer's instructions furnished with the fitting.

(2) Compound. The compound shall provide a seal to minimize the passage of gas and/or vapors through the sealing fitting and shall not be affected by the surrounding atmosphere or liquids. The melting point of the compound shall not be less than 93°C (200°F).

(3) Thickness of Compounds. The thickness of the sealing compound installed in completed seals, other than listed cable sealing fittings, shall not be less than the metric designator (trade size) of the sealing fitting expressed in the units of measurement employed; however, in no case shall the thickness of the compound be less than 16 mm (⅝ in.).

Exhibit 501.5 is a cutaway of a seal fitting containing sealing compound and conductors. All conductors must be separated to permit the compound to run between them. The compound must have a minimum thickness of not less than the trade size of the conduit and never less than ⅝ inch. To eliminate the time-consuming task of field-poured seals, a factory-sealed device with the seal designed into the device is permissible.

(4) Splices and Taps. Splices and taps shall not be made in fittings intended only for sealing with compound; nor shall other fittings in which splices or taps are made be filled with compound.

(5) Assemblies. An entire assembly shall be identified for the location where the equipment that may produce arcs, sparks, or high temperatures is located in a compartment that is separate from the compartment containing splices or taps, and an integral seal is provided where conductors pass from one compartment to the other. In Division 1 locations, seals shall be provided in conduit connecting to the compartment containing splices or taps where required by 501.15(A)(1)(2).

(6) Conductor or Optical Fiber Fill. The cross-sectional area of the conductors or optical fiber tubes (metallic or nonmetallic) permitted in a seal shall not exceed 25 percent of the

cross-sectional area of a rigid metal conduit of the same trade size unless the seal is specifically identified for a higher percentage of fill.

The maximum permitted fill for most conduit seals is 25 percent, which is less than permitted for most conduit applications. If the conduit fill exceeds 25 percent of the cross-sectional area of the seal fitting, a larger trade size or expanded seal might be required. Reducers are allowed for connection of a larger trade size seal fitting to the conduit per 501.15(A)(1).

(D) Cable Seals, Class I, Division 1. In Division 1 locations, cable seals shall be located according to 501.15(D)(2) through (D)(3).

Δ **(1) At Terminations.** Cables shall be sealed at all terminations with sealing fittings. The seals at all terminations shall be in accordance with 501.15(C) and shall be installed within 450 mm (18 in.) of the enclosure or as required by the enclosure marking. Only threaded couplings or explosionproof fittings such as unions, reducers, elbows, and capped elbows not larger than the trade size of the conduit shall be permitted between the sealing fitting and the enclosure.

Type MC-HL cable with a gas/vaportight continuous corrugated metallic sheath and an overall jacket of suitable polymeric material, Type TC-ER-HL cable, and Type P cable shall be sealed with a listed fitting after the jacket and any other covering have been removed so that the sealing compound can surround each individual insulated conductor to minimize the passage of gases and vapors.

Shielded cables and twisted pair cables that have their conductors sealed in accordance with the instructions provided with their listed fitting to minimize the entrance of gases or vapors and prevent propagation of flame into the cable core shall not be required to have the shielding material removed or the twisted pairs separated.

The sealing requirements for cables installed in Class I, Division 1 locations differ from the requirements for sealing conduits. Unlike 501.15(A) for conduits, 501.15(D)(1) requires cables to be sealed at all terminations regardless of the type of equipment contained in the enclosure or the cable diameter.

Type MC-HL cable is specifically listed for use as a wiring method in Class I, Division 1 locations. Exhibit 501.6 shows an example of a cable seal fitting for Type MC-HL cable.

Δ **(2) Cables Capable of Transmitting Gases or Vapors.** Cables with a gas/vaportight continuous sheath capable of transmitting gases or vapors through the cable core, installed in conduit, shall be sealed in the Class I, Division 1 location after the jacket and any other coverings have been removed so that the sealing compound can surround each individual insulated conductor or optical fiber tube and the outer jacket.

Exception: Multiconductor cables with a gas/vaportight continuous sheath capable of transmitting gases or vapors through the cable core shall be permitted to be considered a single

EXHIBIT 501.5 A seal fitting to minimize the passage of gases from one portion of the electrical installation to another. *(Courtesy of Appleton™, Emerson Electric Co.)*

EXHIBIT 501.6 An explosionproof seal fitting for Type MC-HL cable. *(Courtesy of Eaton, Crouse-Hinds Division)*

conductor *if the cable is sealed in the conduit within 450 mm (18 in.) of the enclosure and the cable end is sealed within the enclosure by an approved means to minimize the entrance of gases or vapors and prevent the propagation of flame into the cable core, or by other approved methods. If both requirements are met, the shielding material shall not be required to be removed and the twisted pairs of shielded cables and twisted pair cables shall not be required to be separated.*

In addition to the conduit seal, the cable within the conduit also must be sealed to prevent gases from passing through the cable. A single conduit seal can serve both purposes, sealing the conduit and sealing the cable.

(3) Cables Incapable of Transmitting Gases or Vapors. Each multiconductor cable installed in conduit shall be considered as a single conductor if the cable is incapable of transmitting gases or vapors through the cable core. These cables shall be sealed in accordance with 501.15(A).

(E) Cable Seals, Class I, Division 2. In Division 2 locations, cable seals shall be located in accordance with 501.15(E)(1) through (E)(4).

Exception: Cables with an unbroken gas/vaportight continuous sheath shall be permitted to pass through a Division 2 location without seals.

Δ **(1) Terminations.** Cables entering enclosures that are required to be explosionproof shall be sealed at the point of entrance into the enclosure. The sealing fitting shall comply with 501.15(B)(1) or be explosionproof. Multiconductor or optical multifiber cables with a gas/vaportight continuous sheath capable of transmitting gases or vapors through the cable core that are installed in a Division 2 location shall be sealed with a listed fitting after the jacket and any other coverings have been removed such that the sealing compound surrounds each individual insulated conductor or optical fiber tube to minimize the passage of gases and vapors. Multiconductor or optical multifiber cables installed in conduit shall be sealed in accordance with 501.15(D).

Exception No. 1: Cables leaving an enclosure or room that is permitted to use general-purpose equipment as a result of Type Z pressurization and entering a Division 2 location shall not require a seal at the boundary.

Exception No. 2: Removal of shielding material from shielded cables and separation of twisted pair cables shall not be required if the conductors are sealed in accordance with instructions provided with the listed fitting to minimize the entrance of gases or vapors and prevent propagation of flame into the cable core.

(2) Cables That Do Not Transmit Gases or Vapors. Cables that have a gas/vaportight continuous sheath and do not transmit gases or vapors through the cable core in excess of the quantity permitted for seal fittings shall not be required to be sealed except as required in 501.15(E)(1). The minimum length of such a cable run shall be not less than the length needed to limit gas or vapor flow through the cable core, excluding the interstices of the conductor strands, to the rate permitted for seal fittings [200 cm^3/hr (0.007 ft^3/hr) of air at a pressure of 1500 pascals (6 in. of water)].

The ability of a cable to transmit gases or vapors through the core (primarily between insulated conductors) depends not only on how tightly packed the conductors are within the outer sheaths and the location and composition of fillers but also on how the cable has been handled and the geometry of the cable run. If any concern that the cable run is capable of transmitting gases or vapors through the core exists, a seal fitting should be installed.

See also

501.15, Informational Note No. 2, and its commentary for more on seal fittings

(3) Cables Capable of Transmitting Gases or Vapors. Cables with a gas/vaportight continuous sheath capable of transmitting gases or vapors through the cable core shall not be required to be sealed except as required in 501.15(E)(1), unless the cable is attached to process equipment or devices that may cause a pressure in excess of 1500 pascals (6 in. of water) to be exerted at a cable end, in which case a seal, a barrier, or other means shall be provided to prevent migration of flammables into an unclassified location.

(4) Cables Without Gas/Vaportight Sheath. Cables that do not have a gas/vaportight continuous sheath shall be sealed at the boundary of the Division 2 and unclassified location in such a manner as to minimize the passage of gases or vapors into an unclassified location.

(F) Drainage.

Exhibit 501.7 shows a seal designed for use in a vertical run of conduit to provide drainage for any condensation trapped above the enclosure by the seal. Any accumulation of water runs down over the surface of the sealing compound, flowing through an explosionproof drain.

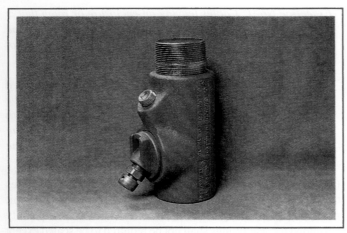

EXHIBIT 501.7 A seal with an automatic drain plug. *(Courtesy of Appleton™, Emerson Electric Co.)*

Exhibit 501.8 shows a combination drain and breather fitting. This fitting is specifically designed to permit the escape of accumulated water through its drain and to allow the continuous circulation of air through the breather, preventing condensation of moisture that could be present while still providing explosion-proof protection.

(1) Control Equipment. Where there is a probability that liquid or other condensed vapor may be trapped within enclosures for control equipment or at any point in the raceway system, approved means shall be provided to prevent accumulation or to permit periodic draining of such liquid or condensed vapor.

(2) Motors and Generators. Where liquid or condensed vapor may accumulate within motors or generators, joints and conduit systems shall be arranged to minimize the entrance of liquid. If means to prevent accumulation or to permit periodic draining are necessary, such means shall be provided at the time of manufacture and shall be considered an integral part of the machine.

Δ **501.17 Process Sealing.** Process-connected equipment, including, but not limited to, canned pumps, submersible pumps, and flow, pressure, temperature, or analysis measurement instruments, shall be sealed with process seals. A process seal shall be a device that prevents the migration of process fluids from the designed containment into the external electrical system. Process-connected electrical equipment that incorporates a single

EXHIBIT 501.8 A combination breather-drainage fitting. *(Courtesy of Appleton™, Emerson Electric Co.)*

process seal, such as a single compression seal, diaphragm, or tube to prevent flammable or combustible fluids from entering a conduit or cable system capable of transmitting fluids, shall be provided with an additional means to mitigate a single process seal failure. The additional means might include, but are not limited to, the following:

(1) A suitable barrier meeting the process temperature and pressure conditions that the barrier will be subjected to upon failure of the single process seal. There shall be a vent or drain between the single process seal and the suitable barrier. Indication of the single process seal failure shall be provided by visible leakage, an audible whistle, or other means of monitoring.

(2) A listed Type MI cable assembly, rated at not less than 125 percent of the process pressure and not less than 125 percent of the maximum process temperature (in degrees Celsius), installed between the cable or conduit and the single process seal.

(3) A drain or vent located between the single process seal and a conduit or cable seal. The drain or vent shall be sufficiently sized to prevent overpressuring the conduit or cable seal above 6 in. water column (1493 Pa). Indication of the single process seal failure shall be provided by visible leakage, an audible whistle, or other means of monitoring.

(4) An add-on secondary seal marked "secondary seal" and rated for the pressure and temperature conditions to which it will be subjected upon failure of the single process seal.

Process-connected electrical equipment that does not rely on a single process seal or is listed and marked "single seal", "dual seal", or "dual seal without annunciation" shall not be required to be provided with an additional means of sealing.

Process-connected electrical equipment marked "single seal — install conduit or cable seal" shall be sealed in accordance with 501.15.

Informational Note: See ANSI/UL 122701, *Requirements for Process Sealing Between Electrical Systems and Flammable or Combustible Process Fluids*, for construction and testing requirements for process sealing of listed and marked single seal, dual seal, or secondary seal equipment.

In addition to the primary seal provided with canned pumps and other process equipment that operate above atmospheric pressure, 501.17 requires a second means to prevent fluid from entering the electrical conduit or cable system if the primary seal fails. The method employed is not limited to the four common techniques described.

501.20 Conductor Insulation, Class I, Divisions 1 and 2. Where condensed vapors or liquids may collect on, or come in contact with, the insulation on conductors, such insulation shall be of a type identified for use under such conditions; or the insulation shall be protected by a sheath of lead or by other approved means.

Type THWN conductors are commonly used in areas where they could be exposed to gasoline because of their ease of handling. Not all Type THWN conductors are suitable where they could be exposed to gasoline. THWN wire suitable for exposure to liquid gasoline and gasoline vapors at ordinary ambient temperature is marked ''Gasoline and Oil Resistant I'' (or "GR1") or ''Gasoline and Oil Resistant II'' (or "GR2"). See the UL *Guide Information for Electrical Equipment,* which can be found at productspec.ul.com, for further information on these cables.

501.25 Uninsulated Exposed Parts, Class I, Divisions 1 and 2. There shall be no uninsulated exposed parts, such as electrical conductors, buses, terminals, or components, that operate at more than 30 volts (15 volts in wet locations). These parts shall additionally be protected by a protection technique according to 500.7(E), (F), or (G) that is suitable for the location.

The intrinsically safe or nonincendive techniques listed in 500.7 limit the circuit's energy to a level incapable of causing ignition in the hazardous area.

Δ **501.30 Grounding and Bonding.** Regardless of the voltage of the electrical system, wiring systems and equipment shall comply with 501.30(A) and (B).

N **(A) Grounding.** Wiring systems and equipment shall be grounded in accordance with Part I and Part VI of Article 250, as applicable.

Δ **(B) Bonding.** Bonding shall comply with Part I and Part V of Article 250, as applicable, and 501.30(B)(1) and (B)(2).

N **(1) Specific Bonding Means.** Bonding shall comply with 501.30(B)(1)(a) and (B)(1)(b).

(a) The locknut-bushing and double-locknut types of contacts shall not be depended on for bonding purposes, but bonding jumpers with identified fittings or other approved means of bonding shall be used. These bonding means shall apply to all metal raceways, fittings, boxes, cable trays, and enclosures, and other parts of raceway systems between Class I locations and the point of grounding for service equipment or point of grounding for a separately derived system. Metal struts, angles, or channels provided for support and mechanical or physical protection as permitted in 335.4(5), 336.10(7)(c), or 722.135(C) shall be bonded in accordance with 250.102.

(b) Where the branch-circuit overcurrent protection is located on the load side of the disconnecting means, the specific bonding means shall be permitted to end at the nearest point where the grounded circuit conductor and the grounding electrode conductor are connected together on the line side of the building or structure disconnecting means as specified in 250.32(B).

The specific bonding methods mentioned in this section are intended to provide a mechanical/electrical connection that is low impedance and free from accidental arcing due to loose connections; they apply to raceways and raceway-to-enclosure connections both inside and outside the hazardous location. Section 250.100 specifies this enhanced level of bonding for all raceways and enclosures, and the requirement is not contingent on the circuit voltage. This includes metal raceways and enclosures containing signaling, communications, or other power-limited circuits.

Section 250.100 clarifies that the installation of a wire-type equipment grounding conductor (EGC) in a metal raceway does not negate the special raceway and enclosure bonding requirements. The electrical continuity of raceways and raceway-to-enclosure connections must always be ensured through compliance with 250.100 and 501.30(A), regardless of whether a wire-type EGC has been installed in the raceway.

Section 501.30(B)(1)(b) covers the grounding and bonding requirements that are specific to hazardous locations where the installation occurs at a multibuilding or multistructure setting. If the service equipment and the electrical equipment supplying the hazardous location are not located in the same building or structure, applying the bonding requirement of 501.30(A) from the hazardous location back to the service equipment is not required. It is necessary only to apply the bonding requirement from the hazardous location back to the grounding electrode connection to the grounded conductor on the line side of the building or structure disconnecting means.

N **(2) Flexible Metal Conduit and Liquidtight Flexible Metal Conduit.** Flexible metal conduit and liquidtight flexible metal conduit shall comply with 501.30(B)(2)(a) and (B)(2)(b).

(a) Flexible metal conduit and liquidtight flexible metal conduit shall include an equipment bonding jumper of the wire type in accordance with 250.102.

(b) In Class I, Division 2 locations, the bonding jumper shall not be required where all of the following conditions are met:

(1) Listed liquidtight flexible metal conduit 1.8 m (6 ft) or less in length, with fittings listed for grounding, is used.
(2) Overcurrent protection in the circuit is limited to 10 amperes or less.
(3) The load is part of a meter, instrument, or relay circuit.

501.35 Surge Protection.

(A) Class I, Division 1. Surge arresters, surge-protective devices, and capacitors shall be installed in enclosures identified for Class I, Division 1 locations. Surge-protective capacitors shall be of a type designed for specific duty.

(B) Class I, Division 2. Surge arresters and surge-protective devices shall be nonarcing, such as metal-oxide varistor (MOV) sealed type, and surge-protective capacitors shall be of a type designed for specific duty. Enclosures shall be permitted to be of the general-purpose type. Surge protection of types other than described in this paragraph shall be installed in enclosures identified for Class I, Division 1 locations.

In Class I, Division 2 locations, only spark-producing types of surge arresters require installation in an enclosure identified for the location. Nonarcing, sealed, and solid state–type surge-protective devices (SPDs) are permitted where installed in a general-purpose-type enclosure. Surge arresters can also be installed in oil-filled enclosures or have the arcing or sparking contacts enclosed in hermetically sealed chambers.

Part III. Equipment

501.100 Transformers and Capacitors.

(A) Class I, Division 1. In Class I, Division 1 locations, transformers and capacitors shall comply with 501.100(A)(1) and (A)(2).

(1) Containing Liquid That Will Burn. Transformers and capacitors containing a liquid that will burn shall be installed only in vaults that comply with 450.41 through 450.48 and with (1) through (4) as follows:

(1) There shall be no door or other communicating opening between the vault and the Division 1 location.
(2) Ample ventilation shall be provided for the continuous removal of flammable gases or vapors.
(3) Vent openings or ducts shall lead to a safe location outside of buildings.
(4) Vent ducts and openings shall be of sufficient area to relieve explosion pressures within the vault, and all portions of vent ducts within the buildings shall be of reinforced concrete construction.

(2) Not Containing Liquid That Will Burn. Transformers and capacitors that do not contain a liquid that will burn shall be installed in vaults complying with 501.100(A)(1) or be identified for Class I locations.

(B) Class I, Division 2. In Class I, Division 2 locations, transformers shall comply with 450.21 through 450.27, and capacitors shall comply with 460.3 through 460.28.

501.105 Meters, Instruments, and Relays.

△ **(A) Class I, Division 1.** In Class I, Division 1 locations, meters, instruments, and relays, including kilowatt-hour meters, instrument transformers, resistors, rectifiers, and thermionic tubes, shall be provided with enclosures identified for Class I, Division 1 locations. Enclosures for Class I, Division 1 locations include explosionproof enclosures and purged and pressurized enclosures.

Informational Note: See NFPA 496, *Standard for Purged and Pressurized Enclosures for Electrical Equipment.*

(B) Class I, Division 2. In Class I, Division 2 locations, meters, instruments, and relays shall comply with 501.105(B)(2) through (B)(6).

(1) General-Purpose Assemblies. Where an assembly is made up of components for which general-purpose enclosures are acceptable as provided in 501.105(B)(1), (B)(2), and (B)(3),

a single general-purpose enclosure shall be acceptable for the assembly. Where such an assembly includes any of the equipment described in 501.105(B)(1), 501.105(B)(2), and 501.105(B)(3), the maximum obtainable surface temperature of any component of the assembly that exceeds 100°C shall be clearly and permanently indicated on the outside of the enclosure. Alternatively, equipment shall be permitted to be marked to indicate the temperature class for which it is suitable, using the temperature class (T Code) of Table 500.8(C)(4).

(2) Contacts. Switches, circuit breakers, and make-and-break contacts of pushbuttons, relays, alarm bells, and horns shall have enclosures identified for Class I, Division 1 locations in accordance with 501.105(A).

Exception: General-purpose enclosures shall be permitted if current-interrupting contacts comply with one of the following:

(1) Are immersed in oil
(2) Are enclosed within a chamber that is hermetically sealed against the entrance of gases or vapors
(3) Are in nonincendive circuits
(4) Are listed for Division 2 locations

This exception identifies the conditions under which general-purpose enclosures are permitted for certain equipment in Class 1, Division 2 locations. Although gasketed seals can be very effective, depending on the gasket material used, gasket materials can be damaged and deteriorate rapidly if exposed to atmospheres that contain solvent vapors. In accordance with the definition in Article 100 of *hermetically sealed*, such enclosures cannot be used to satisfy those requirements in which hermetic sealing is recognized as a protection technique.

This does not mean that the entire enclosure or circuit is required to qualify as being protected by one of these techniques, only that the contacts need such protection.

(3) Resistors and Similar Equipment. Resistors, resistance devices, thermionic tubes, rectifiers, and similar equipment that are used in or in connection with meters, instruments, and relays shall comply with 501.105(A).

Exception: General-purpose-type enclosures shall be permitted if such equipment is without make-and-break or sliding contacts [other than as provided in 501.105(B)(2)] and if the marked maximum operating temperature of any exposed surface will not exceed 80 percent of the autoignition temperature in degrees Celsius of the gas or vapor involved or has been tested and found incapable of igniting the gas or vapor. This exception shall not apply to thermionic tubes.

(4) Without Make-or-Break Contacts. Transformer windings, impedance coils, solenoids, and other windings that do not incorporate sliding or make-or-break contacts shall be provided with enclosures. General-purpose-type enclosures shall be permitted.

(5) Fuses. Where general-purpose enclosures are permitted in 501.105(B)(2) through (B)(4), fuses for overcurrent protection of instrument circuits not subject to overloading in normal use shall be permitted to be mounted in general-purpose enclosures if each such fuse is preceded by a switch complying with 501.105(B)(2).

(6) Connections. To facilitate replacements, process control instruments shall be permitted to be connected through flexible cord and attachment plug and receptacle if all of the following conditions apply:

(1) The attachment plug and receptacle are listed for use in Class I, Division 2 locations and listed for use with flexible cords.

Exception No. 1 to (1): A Class I, Division 2 listing shall not be required if the circuit involves only nonincendive field wiring.

Exception No. 2 to (1): In restricted industrial establishments, the Class I, Division 2 listing shall not be required if the requirements of 501.105(B)(6)(2), (B)(6)(3), and (B)(6)(4) are satisfied and the receptacle carries a label warning against plugging or unplugging when energized.

(2) The flexible cord does not exceed 900 mm (3 ft), is of a type listed for extra-hard usage, or is listed for hard usage and protected by location.

(3) Only necessary receptacles are provided.

(4) Unless the attachment plug and receptacle are interlocked mechanically or electrically, or otherwise designed so that they cannot be separated when the contacts are energized and the contacts cannot be energized when the plug and socket outlet are separated, a switch complying with 501.105(B)(2) is provided so that the attachment plug or receptacle is not necessary to interrupt current.

Exception to (4): The switch shall not be required if the circuit is nonincendive field wiring.

501.115 Switches, Circuit Breakers, Motor Controllers, and Fuses.

(A) Class I, Division 1. In Class I, Division 1 locations, switches, circuit breakers, motor controllers, and fuses, including pushbuttons, relays, and similar devices, shall be provided with enclosures, and the enclosure in each case, together with the enclosed apparatus, shall be identified as a complete assembly for use in Class I locations.

Exhibit 501.9 shows an explosionproof panelboard that consists of an assembly of branch-circuit devices enclosed in a cast metal explosionproof housing. Explosionproof panelboards are provided with bolted access covers and threaded conduit-entry hubs designed to withstand the force of an internal explosion.

Exhibit 501.10 shows a cylindrical-type (spin-top) combination motor controller, motor control starter, and circuit breaker in an explosionproof enclosure. The top and bottom covers are threaded for quick removal for installation and servicing.

Exhibit 501.11 illustrates a standard toggle switch in an explosionproof enclosure.

EXHIBIT 501.9 An explosionproof panelboard. *(Courtesy of Appleton™, Emerson Electric Co.)*

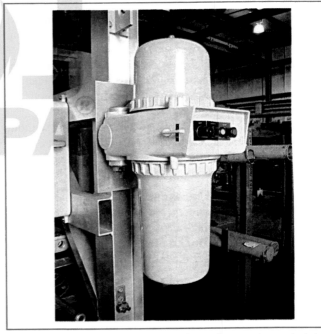

EXHIBIT 501.10 An explosionproof enclosure for a motor control starter and circuit breaker. *(Courtesy of Appleton™, Emerson Electric Co.)*

(B) Class I, Division 2. Switches, circuit breakers, motor controllers, and fuses in Class I, Division 2 locations shall comply with 501.115(B)(1) through (B)(4).

(1) Type Required. Circuit breakers, motor controllers, and switches intended to interrupt current in the normal performance

EXHIBIT 501.11 A standard toggle switch in an explosionproof enclosure. *(Courtesy of Appleton™, Emerson Electric Co.)*

of the function for which they are installed shall be provided with enclosures identified for Class I, Division 1 locations in accordance with 501.105(A), unless general-purpose enclosures are provided and any of the following apply:

(1) The interruption of current occurs within a chamber hermetically sealed against the entrance of gases and vapors.
(2) The current make-and-break contacts are oil-immersed and of the general-purpose type having a 50-mm (2-in.) minimum immersion for power contacts and a 25-mm (1-in.) minimum immersion for control contacts.
(3) The interruption of current occurs within an enclosure, identified for the location, and marked "Leads Factory Sealed", or "Factory Sealed", or "Seal not Required", or equivalent.
(4) The device is a solid state, switching control without contacts, where the surface temperature does not exceed 80 percent of the autoignition temperature in degrees Celsius of the gas or vapor involved.

(2) Isolating Switches. Fused or unfused disconnect and isolating switches for transformers or capacitor banks that are not intended to interrupt current in the normal performance of the function for which they are installed shall be permitted to be installed in general-purpose enclosures.

(3) Fuses. For the protection of motors, appliances, and lamps, other than as provided in 501.115(B)(4), standard plug or cartridge fuses shall be permitted, provided they are placed within enclosures identified for the location; or fuses shall be permitted if they are within general-purpose enclosures, and if they are of a type in which the operating element is immersed in oil or other approved liquid, or the operating element is enclosed within a chamber hermetically sealed against the entrance of gases and vapors, or the fuse is a nonindicating, filled, current-limiting type.

(4) Fuses Internal to Luminaires. Listed cartridge fuses shall be permitted as supplementary protection within luminaires.

501.120 Control Transformers and Resistors. Transformers, impedance coils, and resistors used as, or in conjunction with, control equipment for motors, generators, and appliances shall comply with 501.120(A) and (B).

(A) Class I, Division 1. In Class I, Division 1 locations, transformers, impedance coils, and resistors, together with any switching mechanism associated with them, shall be provided with enclosures identified for Class I, Division 1 locations in accordance with 501.105(A).

(B) Class I, Division 2. In Class I, Division 2 locations, control transformers and resistors shall comply with 501.120(B)(1) through (B)(3).

(1) Switching Mechanisms. Switching mechanisms used in conjunction with transformers, impedance coils, and resistors shall comply with 501.115(B).

(2) Coils and Windings. Enclosures for windings of transformers, solenoids, or impedance coils shall be permitted to be of the general-purpose type.

(3) Resistors. Resistors shall be provided with enclosures; and the assembly shall be identified for Class I locations, unless resistance is nonvariable and maximum operating temperature, in degrees Celsius, will not exceed 80 percent of the autoignition temperature of the gas or vapor involved or the resistor has been tested and found incapable of igniting the gas or vapor.

501.125 Motors and Generators.

(A) Class I, Division 1. In Class I, Division 1 locations, motors, generators, and other rotating electrical machinery shall be one of the following:

(1) Identified for Class I, Division 1 locations
(2) Of the totally enclosed type supplied with positive-pressure ventilation from a source of clean air with discharge to a safe area, so arranged to prevent energizing of the machine until ventilation has been established and the enclosure has been purged with at least 10 volumes of air, and so arranged to automatically de-energize the equipment when the air supply fails
(3) Of the totally enclosed inert gas–filled type supplied with a suitable reliable source of inert gas for pressurizing the enclosure, with devices provided to ensure a positive pressure in the enclosure and arranged to automatically de-energize the equipment when the gas supply fails
(4) For machines that are for use only in restricted industrial establishments, the machine is permitted to be of a type designed to be submerged in a liquid that is flammable only when vaporized and mixed with air, or in a gas or vapor at a pressure greater than atmospheric and that is flammable only when mixed with air; and the machine is so arranged to prevent energizing it until it has been purged with the liquid or gas to exclude air, and so arranged to automatically de-energize the equipment when the supply

of liquid or gas or vapor fails or the pressure is reduced to atmospheric

Totally enclosed motors of the types specified in 501.125(A)(2) or (A)(3) shall have no external surface with an operating temperature in degrees Celsius in excess of 80 percent of the autoignition temperature of the gas or vapor involved. Appropriate devices shall be provided to detect and automatically de-energize the motor or provide an adequate alarm if there is any increase in temperature of the motor beyond designed limits. Auxiliary equipment shall be of a type identified for the location in which it is installed.

Δ **(B) Class I, Division 2.** In Class I, Division 2 locations, motors, generators, and other rotating electrical machinery shall comply with (1), (2), or (3). They shall also comply with (4) and (5), if applicable.

Motors with arcing devices, such as commutators, are required to be provided with an enclosure identified for the location, such as an explosionproof enclosure. Other motor types without arcing devices, such as a squirrel-cage induction motor, are permitted without special enclosures. Additionally, indication of the maximum temperature present in a motor is critical to proper installation. Many motor heaters are de-energized automatically when the motor is running. However, the heater ratings usually are low compared with the normal heat generated during motor operation. Unless otherwise indicated on the motor wiring diagram or in instructions provided with the motor, there is no need to de-energize the heater except to save energy. The heater temperature must be marked on the motor, or the heater must be identified for the location.

(1) Be identified for Class I, Division 2 locations, or

(2) Be identified for Class I, Division 1 locations where sliding contacts, centrifugal or other types of switching mechanism (including motor overcurrent, overloading, and over-temperature devices), or integral resistance devices, either while starting or while running, are employed, or

(3) Be open or nonexplosionproof enclosed motors, such as squirrel-cage induction motors without brushes, switching mechanisms, or similar arc-producing devices that are not identified for use in a Class I, Division 2 location.

(4) The exposed surface of space heaters used to prevent condensation of moisture during shutdown periods shall not exceed 80 percent of the autoignition temperature in degrees Celsius of the gas or vapor involved when operated at rated voltage, and the maximum space heater surface temperature [based on a 40°C or higher marked ambient] shall be permanently marked on a visible nameplate mounted on the motor. Otherwise, space heaters shall be identified for Class I, Division 2 locations.

(5) A sliding contact shaft bonding device used for the purpose of maintaining the rotor at ground potential, shall be permitted where the potential discharge energy is determined to be nonincendive for the application. The shaft bonding device shall be permitted to be installed on the inside or the outside of the motor.

Informational Note No. 1: It is important to consider the temperature of internal and external surfaces that might be exposed to the flammable atmosphere.

Informational Note No. 2: It is important to consider the risk of ignition due to currents arcing across discontinuities and over-heating of parts in multisection enclosures of large motors and generators. Such motors and generators might need equipotential bonding jumpers across joints in the enclosure and from enclosure to ground. Where the presence of ignitible gases or vapors is suspected, clean-air purging might be needed immediately prior to and during start-up periods.

Informational Note No. 3: See IEEE 1349, *IEEE Guide for the Application of Electric Machines in Zone 2 and Class I, Division 2 Hazardous (Classified) Locations*, for information on the application of rotating electric machines including shaft bonding devices and potential discharge energy calculations.

Informational Note No. 4: See ANSI/UL 122001, *General Requirements for Electrical Ignition Systems for Internal Combustion Engines in Class I, Division 2 or Zone 2, Hazardous (Classified) Locations*, for reciprocating engine–driven generators, compressors, and other equipment installed in Class I, Division 2 locations. Reciprocating engine–driven generators, compressors, and other equipment installed in Class I, Division 2 locations might present a risk of ignition of flammable materials associated with fuel, starting, and compression due to inadvertent release or equipment malfunction by the engine ignition system and controls.

Motor types used where flammable gases or vapors with very low ignition temperatures might be present should be carefully selected. Modern motors with high-temperature insulation systems, such as Class H [180°C (356°F)], could operate close to or above the ignition temperature of the flammable mixture.

Informational Note No. 5: See UL 1836, *Outline of Investigation for Electric Motors and Generators for Use in Class I, Division 2, Class I, Zone 2, Class II, Division 2 and Zone 22 Hazardous (Classified) Locations*, for details of the evaluation process to determine incendivity. Refer to Annex A and Figure A1.

501.130 Luminaires. Luminaires shall comply with 501.130(A) or (B).

(A) Class I, Division 1. In Class I, Division 1 locations, luminaires shall comply with 501.130(A)(1) through (A)(4).

(1) Luminaires. Each luminaire shall be identified as a complete assembly for the Class I, Division 1 location and shall be clearly marked to indicate the maximum wattage of lamps for which it is identified. Luminaires intended for portable use shall be specifically listed as a complete assembly for that use.

(2) Physical Damage. Each luminaire shall be protected against physical damage by a suitable guard or by location.

(3) Pendant Luminaires. Pendant luminaires shall be suspended by and supplied through threaded rigid metal conduit stems or threaded steel intermediate conduit stems, and threaded joints shall be provided with set-screws or other effective means to prevent loosening. For stems longer than 300 mm (12 in.), permanent and effective bracing against lateral displacement shall be provided at a level not more than 300 mm (12 in.) above the lower end of the stem, or flexibility in the form of a fitting or flexible connector identified for the Class I, Division 1 location shall be provided not more than 300 mm (12 in.) from the point of attachment to the supporting box or fitting.

(4) Supports. Boxes, box assemblies, or fittings used for the support of luminaires shall be identified for Class I locations.

(B) Class I, Division 2. In Class I, Division 2 locations, luminaires shall comply with 501.130(B)(1) through (B)(6).

(1) Luminaires. Where lamps are of a size or type that may, under normal operating conditions, reach surface temperatures exceeding 80 percent of the autoignition temperature in degrees Celsius of the gas or vapor involved, luminaires shall comply with 501.130(A)(1) or shall be of a type that has been tested in order to determine the marked operating temperature or temperature class (T code).

(2) Physical Damage. Luminaires shall be protected from physical damage by suitable guards or by location. Where there is danger that falling sparks or hot metal from lamps or luminaires might ignite localized concentrations of flammable vapors or gases, suitable enclosures or other effective protective means shall be provided.

(3) Pendant Luminaires. Pendant luminaires shall be suspended by threaded rigid metal conduit stems, threaded steel intermediate metal conduit stems, or other approved means. For rigid stems longer than 300 mm (12 in.), permanent and effective bracing against lateral displacement shall be provided at a level not more than 300 mm (12 in.) above the lower end of the stem, or flexibility in the form of an identified fitting or flexible connector shall be provided not more than 300 mm (12 in.) from the point of attachment to the supporting box or fitting.

Δ **(4) Portable Lighting Equipment.** Portable lighting equipment shall comply with 501.130(B)(4)(a) or (B)(4)(b).

(a) Portable lighting equipment shall comply with 501.130(B)(1).

(b) Portable lighting equipment mounted on movable stands and connected by flexible cords in accordance with 501.140 shall be permitted to comply with 501.130(B)(1), where mounted in any position, if it is protected from physical damage in accordance with 501.130(B)(2).

(5) Switches. Switches that are a part of a luminaire or of an individual lampholder shall comply with 501.115(B)(1).

(6) Starting Equipment. Starting and control equipment for electric-discharge lamps shall comply with 501.120(B).

Exception: A thermal protector potted into a thermally protected fluorescent lamp ballast if the luminaire is identified for the location.

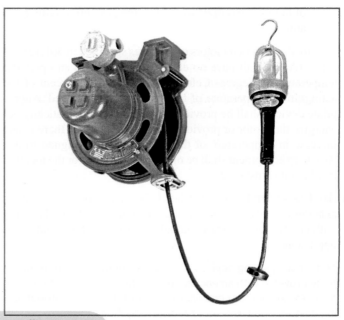

EXHIBIT 501.12 An explosionproof hand lamp for use in Class I locations. *(Courtesy of Appleton™, Emerson Electric Co.)*

Operating temperatures must be considered if the area is a Class I location. Luminaires must be identified for this location and properly marked. Generally, enclosed and gasketed luminaires — without guards, if breakage is unlikely — or luminaires identified for Class I, Division 2 locations are required in Division 2 locations.

Portable luminaires are required to be specifically listed as a complete assembly for use in Class 1, Division 1 or 2 locations. Exhibit 501.12 shows an explosionproof hand lamp. Lamp compartments must be sealed from the terminal compartment. Provisions must be made for the connection of 3-conductor (one must be a grounding conductor) flexible, extra-hard-usage cord in accordance with 501.140(A)(1).

501.135 Utilization Equipment.

(A) Class I, Division 1. In Class I, Division 1 locations, all utilization equipment shall be identified for Class I, Division 1 locations.

(B) Class I, Division 2. In Class I, Division 2 locations, all utilization equipment shall comply with 501.135(B)(1) through (B)(4).

Δ **(1) Heaters.** Electrically heated utilization equipment shall meet one of the following requirements:

(1) The heater shall not exceed 80 percent of the autoignition temperature in degrees Celsius of the gas or vapor involved on any surface that is exposed to the gas or vapor when continuously energized at the maximum rated ambient temperature. If a temperature controller is not provided, these conditions shall apply when the heater is operated at 120 percent of rated voltage.

Exception No. 1 to (1): For motor-mounted anticondensation space heaters, see 501.125.

Exception No. 2 to (1): Where a current-limiting device is applied to the circuit serving the heater to limit the current in the heater to a value less than that required to raise the heater surface temperature to 80 percent of the autoignition temperature.

(2) The heater shall be identified for Class I, Division 1 locations.

Exception to (2): Electrical resistance and skin effect heat tracing identified for Class I, Division 2 locations shall be permitted.

> Informational Note No. 1: See ANSI/UL 60079-30-1, *Explosive Atmospheres — Part 30-1: Electrical Resistance Trace Heating — General and Testing Requirements*, for information on electric resistance heat tracing.
>
> Informational Note No. 2: See IEEE 844.1/CSA C22.2 No. 293.1, *IEEE/CSA Standard for Skin Effect Trace Heating of Pipelines, Vessels, Equipment, and Structures — General, Testing, Marking, and Documentation Requirements*, for information on electric skin effect heat tracing.
>
> Informational Note No. 3: See IEEE 844.3/CSA C22.2 No. 293.3, IEEE/CSA *Standard for Impedance Heating of Pipelines and Equipment — General, Testing, Marking, and Documentation Requirements*, for information on electric impedance heating.

(2) Motors. Motors of motor-driven utilization equipment shall comply with 501.125(B).

(3) Switches, Circuit Breakers, and Fuses. Switches, circuit breakers, and fuses shall comply with 501.115(B).

N **(4) Luminaires.** Luminaires shall comply with 501.130(B).

501.140 Flexible Cords, Class I, Divisions 1 and 2.

(A) Permitted Uses. Flexible cord shall be permitted as follows:

(1) For connection between portable lighting equipment or other portable utilization equipment and the fixed portion of their supply circuit. The flexible cord shall be attached to the utilization equipment with a cord connector listed for the protection technique of the equipment wiring compartment. An attachment plug in accordance with 501.140(B)(4) shall be employed.

(2) For that portion of the circuit where the fixed wiring methods of 501.10(A) cannot provide the necessary degree of movement for fixed and mobile electrical utilization equipment, and the flexible cord is protected by location or by a suitable guard from damage and only in a restricted industrial establishment.

(3) For electric submersible pumps with means for removal without entering the wet-pit. The extension of the flexible cord within a suitable raceway between the wet-pit and the power source shall be permitted.

EXHIBIT 501.13 A temporary portable assembly. (*Courtesy of Killark, a division of Hubbell Incorporated*)

(4) For electric mixers intended for travel into and out of open-type mixing tanks or vats.

(5) For temporary portable assemblies consisting of receptacles, switches, and other devices that are not considered portable utilization equipment but are individually listed for the location.

An example of a portable assembly that is not considered utilization equipment is a power cart (see Exhibit 501.13) that provides power during servicing or maintenance. The flexible cord is required to be continuous from the power source to the assembly and from the assembly to the utilization equipment.

(B) Installation. Where flexible cords are used, the cords shall comply with all of the following:

(1) Be of a type listed for extra-hard usage

(2) Contain, in addition to the conductors of the circuit, an equipment grounding conductor complying with 400.23

(3) Be supported by clamps or by other suitable means in such a manner that there is no tension on the terminal connections

(4) In Division 1 locations or in Division 2 locations where the boxes, fittings, or enclosures are required to be explosion-proof, the cord shall be terminated with a cord connector or attachment plug listed for the location or a listed cord connector installed with a seal listed for the location. In Division 2 locations where explosionproof equipment is not required, the cord shall be terminated with a listed cord connector or listed attachment plug.

(5) Be of continuous length. Where 501.140(A)(5) is applied, cords shall be of continuous length from the power source to the temporary portable assembly and from the temporary portable assembly to the utilization equipment.

> Informational Note: See 501.20 for flexible cords exposed to liquids having a deleterious effect on the conductor insulation.

N **501.141 Flexible Cables, Class I, Division 2.** Flexible cables installed in Class I, Division 2 locations shall comply with 501.141(A) and (B).

N **(A) Permitted Uses.** Flexible cables shall be permitted to be installed in accordance with 501.141(A)(1) and (A)(2).

N **(1) Other Than Nonincendive Field Wiring Applications.** Flexible cables in other than nonincendive field wiring applications shall be permitted in accordance with the following:

    (1) Flexible cables shall be permitted to connect two pieces of electrical equipment by means of a cable assembly installed in accordance with 501.141(B)(2)(a) or (B)(2)(b).

    (2) Flexible cables shall be permitted to connect a piece of electrical equipment to the premises wiring by means of a cable assembly installed in accordance with 501.141(B)(2)(c).

N **(2) Nonincendive Field Wiring Applications.** Flexible cables in nonincendive field wiring applications shall be permitted to be used in accordance with 501.10(B)(3).

N **(B) Installation.** If flexible cables are used as permitted in 501.141(A), the associated cable assemblies shall comply with 501.141(B)(1) through (B)(3).

N **(1) Cable Types.** Listed Type P cables shall comply with 501.141(A)(1) and shall be installed as required in Part II of Article 337. The associated cable assemblies shall comply with the requirements of 501.141(B)(2).

N **(2) Termination Means.** Terminations shall comply with 501.141(B)(2)(a), (B)(2)(b), or (B)(2)(c).

    (a) *Connecting Two Devices or Pieces of Electrical Utilization Equipment Together.* The cable connectors on each end of the cable shall be listed for use in Class I, Division 2 locations and listed for the type of cable being used.

    (b) *Connecting Two Devices or Pieces of Electrical Utilization Equipment Together.* A cable connector listed for Class I, Division 2 and listed for the type of cable being used shall be used on one end and a fitting listed for the type of protection and the type of cable being used shall be used on the other end.

    (c) *Connecting an Electrical Device or Utilization Equipment to Premises Wiring.* The cable connectors used on both ends shall be listed for Class I, Division 2 locations and for the type of cable being used. On one end of the cable, the cable connector shall also be listed for the type of protection.

N **(3) Disconnection.** Flexible cable shall be installed in accordance with 501.141(B)(3)(a) through (B)(3)(c) to protect against the disconnection of the cable connectors when energized.

    (a) *Switch.* A switch complying with the requirements of 501.105(B)(2) shall be provided to disconnect power so that cable connectors are not depended on as a disconnecting means.

    (b) *Cable Connectors Mechanically or Electrically Interlocked.* Switches shall not be required where the cable connectors are interlocked mechanically or electrically, or are otherwise designed to ensure the cable connectors cannot be separated when energized and cannot be energized when separated.

    (c) *Warning Label.* The fixed equipment and the cable assembly shall both carry a label warning against plugging or unplugging when energized, with both labels as close to the cable connector termination as possible.

501.145 Receptacles and Attachment Plugs, Class I, Division 1 and Division 2. Receptacles and attachment plugs shall be listed for the location, except as permitted by 501.105(B)(6).

(A) Receptacles. Receptacles shall be part of the premises wiring, except as permitted by 501.140(A).

Δ **(B) Attachment Plugs.** Attachment plugs shall be of the type that provides connection to the equipment grounding conductor of a permitted flexible cord.

Exhibit 501.14 shows an explosionproof receptacle and attachment plug with an interlocking switch. The design of this device is such that when the switch is in the "on" position, the plug cannot be removed. Also, the switch cannot be placed in the "on" position when the plug has been removed. The plug is to be used with Type S or equivalent extra-hard-service flexible cord having an EGC.

Exhibit 501.15 shows a 30-ampere, 4-pole receptacle and attachment plug assembly that is suitable for use without a switch. The design is such that the mating parts of the receptacle and plug are enclosed in a chamber that seals the arc and, by delayed-action construction, prevents complete removal of the plug until the arc or hot metal has cooled.

501.150 Signaling, Alarm, Remote-Control, and Communications Systems.

(A) Class I, Division 1. In Class I, Division 1 locations, all apparatus and equipment of signaling, alarm, remote-control, and communications systems, regardless of voltage, shall be

EXHIBIT 501.14 A receptacle and attachment plug of the explosionproof type with an interlocking switch. (*Courtesy of Appleton™, Emerson Electric Co.*)

EXHIBIT 501.15 A 4-pole (delayed action) explosionproof receptacle and attachment plug suitable for use without a switch. *(Courtesy of Appleton™, Emerson Electric Co.)*

EXHIBIT 501.16 An audible signaling device for use in hazardous locations. *(Courtesy of Eaton, Crouse-Hinds Division)*

identified for Class I, Division 1 locations, and all wiring shall comply with 501.10(A), 501.15(A), and 501.15(C).

(B) Class I, Division 2. In Class I, Division 2 locations, signaling, alarm, remote-control, and communications systems shall comply with 501.150(B)(1) through (B)(4).

(1) Contacts. Switches, circuit breakers, and make-and-break contacts of pushbuttons, relays, alarm bells, and horns shall have enclosures identified for Class I, Division 1 locations in accordance with 501.105(A).

Exception: General-purpose enclosures shall be permitted if current-interrupting contacts are one of the following:

(1) Immersed in oil
(2) Enclosed within a chamber hermetically sealed against the entrance of gases or vapors
(3) In nonincendive circuits
(4) Part of a listed nonincendive component

See also

501.105(B)(2) and its commentary regarding arcing contacts and the use of general-purpose enclosures in Class I, Division 2 locations

(2) Resistors and Similar Equipment. Resistors, resistance devices, thermionic tubes, rectifiers, and similar equipment shall comply with 501.105(B)(3).

(3) Protectors. Enclosures shall be provided for lightning protective devices and for fuses. Such enclosures shall be permitted to be of the general-purpose type.

(4) Wiring and Sealing. All wiring shall comply with 501.10(B), 501.15(B), and 501.15(C).

Some audible signaling devices may contain make-and-break contacts that are capable of producing a spark of sufficient

energy to cause ignition of a hazardous atmospheric mixture. If used in Class I locations, this type of equipment must be contained in explosionproof or purged and pressurized enclosures, wiring methods must comply with 501.10, and seal fittings must be provided in accordance with 501.15. (See Exhibit 501.16.) Electronic signal devices without make-and-break contacts usually do not require explosionproof enclosures in Division 2 locations.

ARTICLE
502 Class II Locations

Part I. General

502.1 Scope. This article covers the requirements for electrical and electronic equipment and wiring for all voltages in Class II, Division 1 and 2 locations where fire or explosion hazards may exist due to combustible dust.

Class II, Division 1 and 2 locations are defined in 500.5(C) as "hazardous because of the presence of combustible dust." Two different types of dust environments typically warrant a Class II, Division 1 area classification. The first is where a cloud of combustible dust is likely to be present continuously or intermittently under normal operating conditions or because of repair or maintenance operations or leakage. The other environment is one in which a dust layer is likely to accumulate to a depth greater than $1/8$ inch on major horizontal surfaces over a defined period of time, usually 24 hours. A Class II, Division 2 location is typically one where these conditions exist infrequently or under abnormal conditions.

NFPA 484, *Standard for Combustible Metals*, provides methods to minimize the occurrence of, and resulting damage from, fire or explosion in areas where combustible metals or metal dusts are produced, processed, finished, handled, stored, and used.

502.5 Explosionproof Equipment. Explosionproof equipment and wiring shall not be required and shall not be acceptable in Class II locations unless also identified for such locations.

The electrical equipment required in Class II locations is different from that required for Class I locations. Class II equipment is designed to prevent the ignition of layers of dust, which can also cause an increase in equipment operating temperature, while Class I equipment does not address this concern. To protect against explosions in hazardous locations, all electrical equipment exposed to the hazardous atmosphere must be suitable for the location. Equipment suitable for one class and group is not necessarily suitable for any other class and group.

Class I equipment is not necessarily suitable for a Class II location because the hazard contemplated in the equipment design is different. Grain dust, for example, ignites at a temperature lower than that of most flammable vapors. Motors listed for use in Class I locations might not have dust shields on the bearings to prevent entrance of dust into the bearing race, thereby causing overheating of the bearing and resulting in ignition of dust on the motor. Class I equipment is not designed for dust layering. Dust-ignitionproof enclosures are not required to be explosionproof. Explosionproof enclosures are not necessarily dust-ignitionproof.

502.6 Zone Equipment. Equipment listed and marked in accordance with 506.9(C)(2) for Zone 20 locations shall be permitted in Class II, Division 1 locations for the same dust atmosphere; and with a suitable temperature class.

Equipment listed and marked in accordance with 506.9(C)(2) for Zone 20, 21, or 22 locations shall be permitted in Class II, Division 2 locations for the same dust atmosphere and with a suitable temperature class.

Part II. Wiring

502.10 Wiring Methods. Wiring methods shall comply with 502.10(A) or (B).

Informational Note: See Article 100 for the definition of *restricted industrial establishment [as applied to hazardous (classified) locations]*.

(A) Class II, Division 1.

Δ **(1) General.** In Class II, Division 1 locations, the following wiring methods shall be permitted:

  (1) Threaded rigid metal conduit (RMC) or threaded intermediate metal conduit (IMC), including conduit systems with supplemental corrosion protection coatings.

  (2) Type MI cable with termination fittings listed for the location. Type MI cable shall be installed and supported in a manner to avoid tensile stress at the termination fittings.

  (3) In restricted industrial establishments, Type MC-HL cable, listed for use in Class II, Division 1 locations, with a gas/vaportight continuous corrugated metallic sheath, an overall jacket of suitable polymeric material, a separate equipment grounding conductor(s) in accordance with 250.122, and provided with termination fittings listed for the location, shall be permitted.

  (4) Optical fiber cable Type OFNP, Type OFCP, Type OFNR, Type OFCR, Type OFNG, Type OFCG, Type OFN, or Type OFC shall be permitted to be installed in raceways in accordance with 502.10(A). Optical fiber cables shall be sealed in accordance with 502.15.

  (5) In restricted industrial establishments, listed Type ITC-HL cable with a gas/vaportight continuous corrugated metallic sheath and an overall jacket of suitable polymeric material, and terminated with fittings listed for the application, and installed in accordance with 335.4.

  (6) In restricted industrial establishments, for applications limited to 600 volts nominal or less, where the cable is not subject to physical damage and is terminated with fittings listed for the location, listed Type TC-ER-HL cable. When installed in ladder, ventilated trough, or ventilated channel cable trays, cables shall be installed in a single layer, with a space not less than the larger cable diameter between the two adjacent cables unless otherwise protected against dust buildup resulting in increased heat, Type TC-ER-HL cable shall be installed in accordance with 336.10.

Informational Note No. 1: See ANSI/UL 2225, *Cables and Cable-Fittings for Use in Hazardous (Classified) Locations*, for construction, testing, and marking of cables and cable fittings.

  (7) In restricted industrial establishments, listed Type P cable with metal braid armor, with an overall jacket, that is terminated with fittings listed for the location and installed in accordance with 337.10. When installed in ladder, ventilated trough, or ventilated channel cable trays, cables shall be installed in a single layer, with a space not less than the larger cable diameter between the two adjacent cables, unless otherwise protected against dust buildup resulting in increased heat.

Informational Note No. 2: See ANSI/UL 1309, *Marine Shipboard Cable*, for information on construction, testing, and marking of Type P cable.

Informational Note No. 3: See ANSI/UL 2225, *Cables and Cable-Fittings for Use in Hazardous (Classified) Locations*, for information on construction, testing, and marking of cable fittings.

Δ **(2) Flexible Connections.** Where flexible connections are necessary, one or more of the following shall also be permitted:

  (1) Dusttight flexible connectors.

  (2) Liquidtight flexible metal conduit (LFMC) with listed fittings and bonded in accordance with 502.30(B).

  (3) Liquidtight flexible nonmetallic conduit (LFNC) with listed fittings.

  (4) Interlocked armor Type MC cable having an overall jacket of suitable polymeric material and provided with termination fittings listed for Class II, Division 1 locations.

  (5) Flexible cord listed for extra-hard usage and terminated with listed dusttight cord connectors. Where used, flexible cords shall comply with 502.140.

  (6) For elevator use, an identified elevator cable of Type EO, Type ETP, or Type ETT, shown under the "use" column

in Table 400.4 for "hazardous (classified) locations" and terminated with listed dusttight fittings.

(7) In restricted industrial establishments, for applications limited to 600 volts nominal or less, and where the cable is not subject to physical damage and is terminated with fittings listed for the location, listed Type TC-ER-HL cable. Type TC-ER-HL cable shall be installed in accordance with 336.10.

> Informational Note No. 1: See ANSI/UL 2225, *Cables and Cable-Fittings for Use in Hazardous (Classified) Locations*, for construction, testing, and marking of cables and cable fittings.

(8) In restricted industrial establishments, listed Type P cable with metal braid armor, with an overall jacket, terminated with fittings listed for the location, and installed in accordance with 337.10.

> Informational Note No. 2: See UL 1309A, *Outline of Investigation for Cable for Use in Mobile Installations*, for information on construction, testing, and marking of Type P cable.
>
> Informational Note No. 3: See ANSI/UL 2225, *Cables and Cable-Fittings for Use in Hazardous (Classified) Locations*, for information on construction, testing, and marking of cable fittings.

See also

501.10(A)(2) and its commentary for information on the use of a flexible fitting

(3) Boxes and Fittings. Boxes and fittings shall be provided with threaded bosses for connection to conduit or cable terminations and shall be dusttight. Boxes and fittings in which taps, joints, or terminal connections are made, or that are used in Group E locations, shall be identified for Class II locations.

> Informational Note: See ANSI/UL 2225, *Cables and Cable-Fittings for Use in Hazardous (Classified) Locations*, for information on construction, testing, and marking of cables, dust-ignitionproof cable fittings, and dust-ignitionproof cord connectors for entry into enclosures required to be dust-ignitionproof.

(B) Class II, Division 2. Wiring methods installed in Class II, Division 2 locations shall be in accordance with 502.10(B) (1) through (B)(4).

Δ **(1) General.** In Class II, Division 2 locations, the following wiring methods shall be permitted:

(1) All wiring methods permitted in 502.10(A).
(2) Rigid metal conduit (RMC) or intermediate metal conduit (IMC) with listed threaded or threadless fittings, including conduit systems with supplemental corrosion protection coatings.
(3) Dusttight wireways or electrical metallic tubing (EMT) with listed compression-type connectors or listed compression-type couplings.
(4) Type MC, Type MV, Type TC, or Type TC-ER cable, including installation in cable tray systems. Type TC-ER cable shall include an equipment grounding conductor in addition to a drain wire that might be present. The cable shall be terminated with listed fittings.

(5) Type PLTC cable or Type PLTC-ER cable used in Class 2 or Class 3 circuits, including installation in cable tray systems. The cable shall be terminated with listed fittings. Type PLTC-ER cable shall include an equipment grounding conductor in addition to a drain wire that might be present.
(6) Type ITC cable or Type ITC-ER cable as permitted in 335.4 and terminated with listed fittings. Type ITC-ER cable shall include an equipment grounding conductor in addition to a drain wire.
(7) In restricted industrial establishments where wiring methods in 502.10(B)(1)(1)(2) will not provide the corrosion resistance required for the installation environment, either of the following:

a. Listed reinforced thermosetting resin conduit (RTRC), factory elbows, and associated fittings, all marked with suffix -XW
b. Schedule 80 PVC conduit, factory elbows, and associated fittings

(8) Optical fiber cable Type OFNP, Type OFCP, Type OFNR, Type OFCR, Type OFNG, Type OFCG, Type OFN, or Type OFC, installed in cable trays or any other raceway in accordance with 502.10(B). Optical fiber cables shall be sealed in accordance with 502.15.
(9) Cablebus.
(10) In restricted industrial establishments, listed Type P cable with or without metal braid armor, with an overall jacket, that is terminated with listed fittings and installed in accordance with 337.10.

> Informational Note: See UL 1309A, *Outline of Investigation for Cable for Use in Mobile Installations*, for information on construction, testing, and marking of Type P cable.

(2) Flexible Connections. If flexibility is necessary, 502.10(A) (2) shall apply.

(3) Nonincendive Field Wiring. Nonincendive field wiring shall be permitted using any of the wiring methods permitted for unclassified locations. Nonincendive field wiring systems shall be installed in accordance with the control drawing(s). Simple apparatus, not shown on the control drawing, shall be permitted in a nonincendive field wiring circuit if the simple apparatus does not interconnect the nonincendive field wiring circuit to any other circuit.

> Informational Note: See Article 100 for the definition of *simple apparatus*.

Separate nonincendive field wiring circuits shall be installed in accordance with one of the following:

(1) In separate cables
(2) In multiconductor cables where the conductors of each circuit are within a grounded metal shield
(3) In multiconductor cables or in raceways where the conductors of each circuit have insulation with a minimum thickness of 0.25 mm (0.01 in.)

EXHIBIT 502.1 Junction box with threaded hubs, suitable for use in Class II, Group E hazardous atmospheres. *(Courtesy of Appleton™, Emerson Electric Co.)*

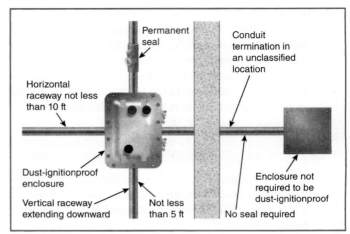

EXHIBIT 502.2 Three methods for preventing dust from entering a dust-ignitionproof enclosure through the raceway.

(4) Boxes and Fittings. All boxes and fittings shall be dusttight.

Boxes and fittings in a Class II, Division 2 location need only be dusttight. In Division 1 locations, however, boxes containing taps, joints, or terminal connections, in addition to being dusttight, must be provided with threaded hubs and must be identified for use in Class II locations. Exhibit 502.1 shows a dusttight cover.

Δ **502.15 Sealing, Class II, Divisions 1 and 2.** If a raceway provides communication between an enclosure that is required to be dust-ignitionproof and one that is not, suitable means shall be provided to prevent the entrance of dust into the dust-ignitionproof enclosure through the raceway. One of the following means shall be permitted:

(1) A permanent and effective seal
(2) A horizontal raceway not less than 3.05 m (10 ft) long
(3) A vertical raceway not less than 1.5 m (5 ft) long and extending downward from the dust-ignitionproof enclosure
(4) A raceway installed in a manner equivalent to 502.15(2) or (3) that extends only horizontally and downward from the dust-ignition proof enclosures
(5) Electrical sealing putty

If a raceway provides communication between an enclosure that is required to be dust-ignitionproof and an enclosure in an unclassified location, seals shall not be required.

Sealing fittings shall be accessible and shall not be required to be explosionproof.

Four suitable ways are provided in 502.15 to prevent dust from entering dust-ignitionproof enclosures through the raceway. Three of these methods for Class II locations are shown in Exhibit 502.2.

The requirement to provide a seal applies if a raceway connects two enclosures in a hazardous location — from one enclosure that is required to be dust-ignitionproof to one that is not required to be dust-ignitionproof. Dust could enter the system through the other enclosure that is not dust-ignitionproof. If a

raceway extends from a dust-ignitionproof enclosure to an enclosure in an unclassified location, a seal in that raceway is not required since dust will not enter through the conduit system.

Seal fittings designed for use in Class I locations are acceptable for Class II locations. However, because the Class I location pressure-piling considerations are not inherent in Class II locations, conduit seals are not required to be explosionproof. Conduit seals are expected only to prevent the migration of dust into dust-ignitionproof enclosures.

502.25 Uninsulated Exposed Parts, Class II, Divisions 1 and 2. There shall be no uninsulated exposed parts, such as electrical conductors, buses, terminals, or components, that operate at more than 30 volts (15 volts in wet locations). These parts shall additionally be protected by a protection technique according to 500.7(E), (F), or (G) that is suitable for the location.

Δ **502.30 Grounding and Bonding.** Regardless of the voltage of the electrical system, wiring systems and equipment shall comply with 502.30(A) and (B).

N **(A) Grounding.** Wiring systems and equipment shall be grounded in accordance with Part I and Part VI of Article 250, as applicable.

Δ **(B) Bonding.** Bonding shall comply with Part I and Part V of Article 250, as applicable, and 502.30(B)(1) and (B)(2).

N **(1) Specific Bonding Means.** Bonding shall comply with 502.30(B)(1)(a) and (B)(1)(b).

(a) The locknut-bushing and double-locknut types of contacts shall not be depended on for bonding purposes, but bonding jumpers with identified fittings or other approved means of bonding shall be used. These bonding means shall apply to all metal raceways, fittings, boxes, cable trays, and enclosures, and other parts of raceway systems between hazardous (classified) locations and the point of grounding for service equipment or point of grounding for a separately derived system. Metal struts, angles, or channels provided for support and mechanical or physical protection as permitted in 335.4(5), 336.10(7)(c), or 722.135(C) shall be bonded in accordance with 250.102.

(b) Where the branch-circuit overcurrent protection is located on the load side of the disconnecting means, the specific bonding means shall be permitted to end at the nearest point where the grounded circuit conductor and the grounding electrode conductor are connected together on the line side of the building or structure disconnecting means as specified in 250.32(B).

The requirements for enhanced bonding in Class II locations are the same as those given in 501.30(A) for Class I locations.

See also

501.30(B)(1) and its commentary for more information on specific bonding methods

250.100 for additional requirements applying to bonding in hazardous locations

N **(2) Liquidtight Flexible Metal Conduit.** Liquidtight flexible metal conduit shall comply with 502.30(B)(2)(a) and (B)(2)(b).

(a) Liquidtight flexible metal conduit shall include an equipment bonding jumper of the wire type in accordance with 250.102.

(b) In Class II, Division 2 locations, the bonding jumper shall not be required where all of the following conditions are met:

(1) Listed liquidtight flexible metal conduit 1.8 m (6 ft) or less in length, with fittings listed for grounding, is used.
(2) Overcurrent protection in the circuit is limited to 10 amperes or less.
(3) The load is part of a meter, instrument, or relay circuit.

502.35 Surge Protection — Class II, Divisions 1 and 2. Surge arresters and surge-protective devices installed in a Class II, Division 1 location shall be in suitable enclosures. Surge-protective capacitors shall be of a type designed for specific duty.

Part III. Equipment

When designing an electrical installation for hazardous locations, the preferred location of service equipment, switchboards, panelboards, and much of the electrical equipment is in less hazardous areas, usually in a separate room. The use of pressurized rooms, as described in NFPA 496, *Standard for Purged and Pressurized Enclosures for Electrical Equipment*, is a common method of protecting panelboards and switchboards in grain elevators and similar locations.

502.100 Transformers and Capacitors.

(A) Class II, Division 1. In Class II, Division 1 locations, transformers and capacitors shall comply with 502.100(A)(1) through (A)(3).

(1) Containing Liquid That Will Burn. Transformers and capacitors containing a liquid that will burn shall be installed only in vaults complying with 450.41 through 450.48, and, in addition, (1), (2), and (3) shall apply.

(1) Doors or other openings communicating with the Division 1 location shall have self-closing fire doors on both sides of the wall, and the doors shall be carefully fitted and provided with suitable seals (such as weather stripping) to minimize the entrance of dust into the vault.
(2) Vent openings and ducts shall communicate only with the outside air.
(3) Suitable pressure-relief openings communicating with the outside air shall be provided.

(2) Not Containing Liquid That Will Burn. Transformers and capacitors that do not contain a liquid that will burn shall be installed in vaults complying with 450.41 through 450.48 or be identified as a complete assembly, including terminal connections.

(3) Group E. No transformer or capacitor shall be installed in a Class II, Division 1, Group E location.

(B) Class II, Division 2. In Class II, Division 2 locations, transformers and capacitors shall comply with 502.100(B)(1) through (B)(3).

(1) Containing Liquid That Will Burn. Transformers and capacitors containing a liquid that will burn shall be installed in vaults that comply with 450.41 through 450.48.

(2) Containing Askarel. Transformers containing askarel and rated in excess of 25 kVA shall be as follows:

(1) Provided with pressure-relief vents
(2) Provided with a means for absorbing any gases generated by arcing inside the case, or the pressure-relief vents shall be connected to a chimney or flue that will carry such gases outside the building
(3) Have an airspace of not less than 150 mm (6 in.) between the transformer cases and any adjacent combustible material

(3) Dry-Type Transformers. Dry-type transformers shall be installed in vaults or shall have their windings and terminal connections enclosed in tight metal housings without ventilating or other openings and shall operate at not over 600 volts, nominal.

502.115 Switches, Circuit Breakers, Motor Controllers, and Fuses.

(A) Class II, Division 1. In Class II, Division 1 locations, switches, circuit breakers, motor controllers, fuses, push-buttons, relays, and similar devices shall be provided with enclosures identified for the location.

(B) Class II, Division 2. In Class II, Division 2 locations, enclosures for fuses, switches, circuit breakers, and motor controllers, including push buttons, relays, and similar devices, shall be dust-tight or otherwise identified for the location.

Exhibits 502.3 and 502.4 show dust-ignitionproof equipment that is suitable for use in Class II, Division 1 locations. Dust-ignitionproof equipment enclosures can be used in Class II, Division 2 locations, but because of the reduced level of hazard associated with Division 2, dusttight equipment enclosures also are permitted. In addition to being suitable for Class II and the specific division, the equipment also must be suitable for the dust group(s) (i.e., Groups E, F, and G) present in a specific hazardous location.

EXHIBIT 502.3 A dust-ignitionproof pushbutton control station suitable for use in Class II, Group E, F, and G locations. *(Courtesy of Appleton™, Emerson Electric Co.)*

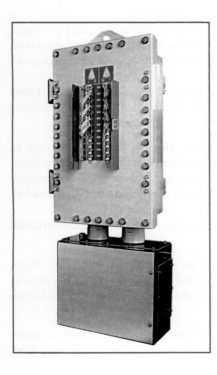

EXHIBIT 502.4 A dust-ignitionproof panelboard for use in Class II, Division 2, Group F and G locations. *(Courtesy of Eaton, Crouse-Hinds Division)*

502.120 Control Transformers and Resistors.

(A) Class II, Division 1. In Class II, Division 1 locations, control transformers, solenoids, impedance coils, resistors, and any overcurrent devices or switching mechanisms associated with them shall be provided with enclosures identified for the location.

(B) Class II, Division 2. In Class II, Division 2 locations, transformers and resistors shall comply with 502.120(B)(1) through (B)(3).

(1) Switching Mechanisms. Switching mechanisms (including overcurrent devices) associated with control transformers, solenoids, impedance coils, and resistors shall be provided with enclosures that are dusttight or otherwise identified for the location.

(2) Coils and Windings. Where not located in the same enclosure with switching mechanisms, control transformers, solenoids, and impedance coils shall be provided with enclosures that are dusttight or otherwise identified for the location.

(3) Resistors. Resistors and resistance devices shall have dust-ignitionproof enclosures that are dusttight or otherwise identified for the location.

502.125 Motors and Generators.

(A) Class II, Division 1. In Class II, Division 1 locations, motors, generators, and other rotating electrical machinery shall be in conformance with either of the following:

(1) Identified for the location
(2) Totally enclosed pipe-ventilated

In a pipe-ventilated system, the supply and exhaust of air from/to outside the Class II area is through a closed pipe system. This prevents accumulation of dust inside the totally enclosed equipment.

(B) Class II, Division 2. In Class II, Division 2 locations, motors, generators, and other rotating electrical equipment shall be totally enclosed nonventilated, totally enclosed pipe-ventilated, totally enclosed water-air-cooled, totally enclosed fan-cooled, or dust-ignitionproof, for which maximum full-load external temperature shall be in accordance with 500.8(D)(2) for normal operation when operating in free air (not dust blanketed) and shall have no external openings.

Exception: If the authority having jurisdiction believes accumulations of nonconductive, nonabrasive dust will be moderate and if machines can be easily reached for routine cleaning and maintenance, the following shall be permitted to be installed:

(1) Standard open-type machines without sliding contacts, centrifugal or other types of switching mechanism (including motor overcurrent, overloading, and overtemperature devices), or integral resistance devices

(2) Standard open-type machines with such contacts, switching mechanisms, or resistance devices enclosed within dusttight housings without ventilating or other openings

(3) Self-cleaning textile motors of the squirrel-cage type

(4) Machines with sealed bearings, bearing isolators, and seals

Totally enclosed motors are permitted in Class II, Division 2 locations if the external surface temperatures, without a dust blanket, do not exceed the temperatures indicated under the maximum full-load (normal operation) conditions in 500.8(D)(2). Totally enclosed fan-cooled (TEFC) motors should be examined carefully to verify that there are no external openings, even though the motor may be marked TEFC.

502.128 Ventilating Piping. Ventilating pipes for motors, generators, or other rotating electrical machinery, or for enclosures for electrical equipment, shall be of metal not less than 0.53 mm (0.021 in.) in thickness or of equally substantial noncombustible material and shall comply with all of the following:

(1) Lead directly to a source of clean air outside of buildings
(2) Be screened at the outer ends to prevent the entrance of small animals or birds
(3) Be protected against physical damage and against rusting or other corrosive influences

Ventilating pipes shall also comply with 502.128(A) and (B).

(A) Class II, Division 1. In Class II, Division 1 locations, ventilating pipes, including their connections to motors or to the dust-ignitionproof enclosures for other equipment, shall be dusttight throughout their length. For metal pipes, seams and joints shall comply with one of the following:

(1) Be riveted and soldered
(2) Be bolted and soldered
(3) Be welded
(4) Be rendered dusttight by some other equally effective means

(B) Class II, Division 2. In Class II, Division 2 locations, ventilating pipes and their connections shall be sufficiently tight to prevent the entrance of appreciable quantities of dust into the ventilated equipment or enclosure and to prevent the escape of sparks, flame, or burning material that might ignite dust accumulations or combustible material in the vicinity. For metal pipes, lock seams and riveted or welded joints shall be permitted; and tight-fitting slip joints shall be permitted where some flexibility is necessary, as at connections to motors.

502.130 Luminaires.

(A) Class II, Division 1. In Class II, Division 1 locations, luminaires for fixed and portable lighting shall comply with 502.130(A)(1) through (A)(4).

(1) Marking. Each luminaire shall be identified for the location and shall be clearly marked to indicate the type and maximum wattage of the lamp for which it is designed.

(2) Physical Damage. Each luminaire shall be protected against physical damage by a suitable guard or by location.

(3) Pendant Luminaires. Pendant luminaires shall be suspended by threaded rigid metal conduit stems, by threaded steel intermediate metal conduit stems, by chains with approved fittings, or by other approved means. For rigid stems longer than 300 mm (12 in.), permanent and effective bracing against lateral displacement shall be provided at a level not more than 300 mm (12 in.) above the lower end of the stem, or flexibility in the form of a fitting or a flexible connector listed for the location shall be provided not more than 300 mm (12 in.) from the point of attachment to the supporting box or fitting. Threaded joints shall be provided with set screws or other effective means to prevent loosening. Where wiring between an outlet box or fitting and a pendant luminaire is not enclosed in conduit, flexible cord listed for hard usage shall be permitted to be used in accordance with 502.10(A)(2)(5). Flexible cord shall not serve as the supporting means for a luminaire.

(4) Supports. Boxes, box assemblies, or fittings used for the support of luminaires shall be identified for Class II locations.

(B) Class II, Division 2. In Class II, Division 2 locations, luminaires shall comply with 502.130(B)(1) through (B)(5).

(1) Portable Lighting Equipment. Portable lighting equipment shall be identified for the location. They shall be clearly marked to indicate the maximum wattage of lamps for which they are designed.

(2) Fixed Lighting. Luminaires for fixed lighting shall be provided with enclosures that are dusttight or otherwise identified for the location. Each luminaire shall be clearly marked to indicate the maximum wattage of the lamp that shall be permitted without exceeding an exposed surface temperature in accordance with 500.8(D)(2) under normal conditions of use.

(3) Physical Damage. Luminaires for fixed lighting shall be protected from physical damage by suitable guards or by location.

(4) Pendant Luminaires. Pendant luminaires shall be suspended by threaded rigid metal conduit stems, by threaded steel intermediate metal conduit stems, by chains with approved fittings, or by other approved means. For rigid stems longer than 300 mm (12 in.), permanent and effective bracing against lateral displacement shall be provided at a level not more than 300 mm (12 in.) above the lower end of the stem, or flexibility in the form of an identified fitting or a flexible connector shall be provided not more than 300 mm (12 in.) from the point of attachment to the supporting box or fitting. Where wiring between an outlet box or fitting and a pendant luminaire is not enclosed in conduit, flexible cord listed for hard usage shall be permitted if terminated with a listed cord connector that maintains the protection technique. Flexible cord shall not serve as the supporting means for a luminaire.

(5) Electric-Discharge Lamps. Starting and control equipment for electric-discharge lamps shall comply with the requirements of 502.120(B).

All lighting equipment in Class II, Division 1 locations and all portable lighting equipment in a Class II, Division 2 location must be identified for such use. If not identified for use in a Class II location, all fixed lighting equipment in a Class II, Division 2 location must be dusttight.

Section 502.130 requires luminaires in Class II locations to be marked to indicate maximum lamp wattage. In addition, lamps must be guarded or protected by their location to prevent damage that could allow the escape of sparks or burning material.

Exhibit 502.5 shows a listed luminaire suitable for use in Class II, Group E, F, and G locations. The portable hand lamp shown in Exhibit 501.12 is listed as a complete assembly for use in Class I locations and in any Class II, Group F or G location.

502.135 Utilization Equipment.

(A) Class II, Division 1. In Class II, Division 1 locations, all utilization equipment shall be identified for the location.

(B) Class II, Division 2. In Class II, Division 2 locations, all utilization equipment shall comply with 502.135(B)(1) through (B)(4).

(1) Heaters. Electrically heated utilization equipment shall be identified for the location.

Exception: Metal-enclosed radiant heating panel equipment shall be permitted to be dusttight and marked in accordance with 500.8(C).

EXHIBIT 502.5 A typical luminaire for use in Class II, Division 1 locations. *(Courtesy of Eaton, Crouse-Hinds Division)*

(2) Motors. Motors of motor-driven utilization equipment shall comply with 502.125(B).

(3) Switches, Circuit Breakers, and Fuses. Enclosures for switches, circuit breakers, and fuses shall comply with 502.115(B).

(4) Transformers, Solenoids, Impedance Coils, and Resistors. Transformers, solenoids, impedance coils, and resistors shall comply with 502.120(B).

502.140 Flexible Cords — Class II, Divisions 1 and 2.

(A) Permitted Uses. Flexible cords used in Class II locations shall be permitted as follows:

(1) For connection between portable lighting equipment or other portable utilization equipment and the fixed portion of its supply circuit. The flexible cord shall be attached to the utilization equipment with a cord connector listed for the protection technique of the equipment wiring compartment. An attachment plug in accordance with 502.145 shall be employed.

(2) Where flexible cord is permitted by 502.10(A)(2) for fixed and mobile electrical utilization equipment; where the flexible cord is protected by location or by a suitable guard from damage; and only in a restricted industrial establishment.

(3) For electric submersible pumps with means for removal without entering the wet-pit. The extension of the flexible cord within a suitable raceway between the wet-pit and the power source shall be permitted.

(4) For electric mixers intended for travel into and out of open-type mixing tanks or vats.

(5) For temporary portable assemblies consisting of receptacles, switches, and other devices that are not considered portable utilization equipment but are individually listed for the location.

(B) Installation. Where flexible cords are used, the cords shall comply with all of the following:

(1) Be of a type listed for extra-hard usage.

Exception to (1): Flexible cord listed for hard usage as permitted by 502.130(A)(3) and (B)(4).

(2) Contain, in addition to the conductors of the circuit, an equipment grounding conductor complying with 400.23.

(3) Be supported by clamps or by other suitable means in such a manner that there will be no tension on the terminal connections.

(4) In Division 1 locations, the cord shall be terminated with a cord connector listed for the location or a listed cord connector installed with a seal listed for the location. In

Division 2 locations, the cord shall be terminated with a listed dusttight cord connector.

(5) Be of continuous length. Where 502.140(A)(5) is applied, cords shall be of continuous length from the power source to the temporary portable assembly and from the temporary portable assembly to the utilization equipment.

502.145 Receptacles and Attachment Plugs. Receptacles and attachment plugs shall be identified for the location.

(A) Class II, Division 1.

(1) Receptacles. In Class II, Division 1 locations, receptacles shall be part of the premises wiring.

(2) Attachment Plugs. Attachment plugs shall be of the type that provides for connection to the equipment grounding conductor of the flexible cord.

(B) Class II, Division 2.

(1) Receptacles. In Class II, Division 2 locations, receptacles shall be part of the premises wiring.

(2) Attachment Plugs. Attachment plugs shall be of the type that provides for connection to the equipment grounding conductor of the flexible cord.

Δ **502.150 Signaling, Alarm, Remote-Control, and Communications Systems; and Meters, Instruments, and Relays.**

(A) Class II, Division 1. In Class II, Division 1 locations, signaling, alarm, remote-control, and communications systems; and meters, instruments, and relays shall comply with 502.150(A)(1) through (A)(3).

Δ **(1) Contacts.** Enclosures containing contacts shall comply with the requirements of 502.150(A)(1)(a) or (A)(1)(b).

(a) Switches, circuit breakers, relays, contactors, fuses, and current-breaking contacts for bells, horns, howlers, sirens, and other devices in which sparks or arcs might be produced shall be provided with enclosures identified for the location.

(b) Where current-breaking contacts are immersed in oil or where the interruption of current occurs within a chamber sealed against the entrance of dust, enclosures shall be permitted to be of the general-purpose type.

Δ **(2) Resistors and Similar Equipment.** Enclosures containing resistors shall comply with the requirements of 502.150(A)(2)(a) or (A)(2)(b).

(a) Resistors, transformers, choke coils, rectifiers, thermionic tubes, and other heat-generating equipment shall be provided with enclosures identified for the location.

(b) Where resistors or similar equipment are immersed in oil or enclosed in a chamber sealed against the entrance of dust, enclosures shall be permitted to be of the general-purpose type.

(3) Rotating Machinery. Motors, generators, and other rotating electrical machinery shall comply with 502.125(A).

(B) Class II, Division 2. In Class II, Division 2 locations, signaling, alarm, remote-control, and communications systems; and meters, instruments, and relays shall comply with 502.150(B)(1) through (B)(4).

Δ **(1) Contacts.** Enclosures for contacts shall comply with the requirements of 502.150(B)(1)(a) or (B)(1)(b).

(a) Contacts shall comply with the requirements of 502.150(A)(1) or be installed in enclosures that are dusttight or otherwise identified for the location.

(b) Enclosures in nonincendive circuits shall be permitted to be of the general-purpose type.

(2) Transformers and Similar Equipment. The windings and terminal connections of transformers, choke coils, and similar equipment shall comply with 502.120(B)(2).

(3) Resistors and Similar Equipment. Resistors, resistance devices, thermionic tubes, rectifiers, and similar equipment shall comply with 502.120(B)(3).

(4) Rotating Machinery. Motors, generators, and other rotating electrical machinery shall comply with 502.125(B).

Δ **(5) Connections.** To facilitate replacements, process control instruments shall be permitted to be connected through flexible cord, attachment plug, and receptacle if all of the following conditions apply:

(1) Attachment plug and receptacle are listed for use in Class II, Division 2 locations, and listed for use with flexible cords.

Exception No. 1 to (1): A Class II, Division 2 listing shall not be required if the circuit involves only nonincendive field wiring.

Exception No. 2 to (1): In restricted industrial establishments, the Class II, Division 2 listing shall not be required when the requirements of 502.150(B)(5)(2), (B)(5)(3), and (B)(5)(4) are satisfied and the receptacle carries a label warning against plugging or unplugging when energized.

(2) The flexible cord does not exceed 900 mm (3 ft), is of a type listed for extra-hard usage, or, if listed for hard usage, is protected by location.

(3) Only necessary receptacles are provided.

(4) Unless the attachment plug and receptacle are interlocked mechanically or electrically, or otherwise designed so that they cannot be separated when the contacts are energized, and the contacts cannot be energized when the plug and socket outlet are separated, a switch complying with 502.115(B) is provided so that the attachment plug or receptacle is not depended on to interrupt current.

Exception to (4): The switch shall not be required if the circuit is nonincendive field wiring.

ARTICLE
503 Class III Locations

Part I. General

503.1 Scope. This article covers the requirements for electrical and electronic equipment and wiring for all voltages in Class III, Division 1 and Division 2 locations where fire or explosion hazards might exist due to nonmetal combustible fibers/flyings or ignitible fibers/flyings.

Class III locations usually include textile mills that process cotton, rayon, and other fabrics, where easily ignitible fibers/flyings are present in the manufacturing process. Sawmills and other woodworking plants, where sawdust, wood shavings, and combustible fibers/flyings are present, also can become hazardous locations. However, if wood flour (dust) is present, the location is a Class II, Group G location and not a Class III location.

Fibers/flyings are hazardous not only because they are easily ignited, but also because flames quickly spread through them. Such fires travel with a rapidity approaching an explosion and are commonly called flash fires.

Class III, Division 1 applies to locations where material is handled, manufactured, or used. Division 2 applies to locations where material is stored or handled but where no manufacturing processes are performed. Unlike Class I and Class II locations, Class III locations do not have material group designations.

Δ **503.5 General.** Equipment installed in Class III locations shall be able to function at full rating without developing surface temperatures high enough to cause excessive dehydration or gradual carbonization of accumulated fibers/flyings.

> Informational Note No. 1: See NFPA 505, *Fire Safety Standard for Powered Industrial Trucks Including Type Designations, Areas of Use, Conversions, Maintenance, and Operations*, for information on electric trucks.
> Informational Note No. 2: Organic material that is carbonized or excessively dry is highly susceptible to spontaneous ignition.

503.6 Zone Equipment. Equipment listed and marked in accordance with 506.9(C)(2) for Zone 20 locations and with a temperature marking in accordance with 500.8(D)(3) shall be permitted in Class III, Division 1 locations.

Equipment listed and marked in accordance with 506.9(C)(2) for Zone 20, Zone 21, or Zone 22 locations and with a temperature marking in accordance with 500.8(D)(3) shall be permitted in Class III, Division 2 locations.

Part II. Wiring

503.10 Wiring Methods. Wiring methods shall comply with 503.10(A) or (B).

> Informational Note: See Article 100 for the definition of *restricted industrial establishment [as applied to hazardous (classified) locations]*.

(A) Class III, Division 1.

Δ **(1) General.** In Class III, Division 1 locations, the following wiring methods shall be permitted:

(1) Rigid metal conduit (RMC), PVC conduit, RTRC conduit, intermediate metal conduit (IMC), electrical metallic tubing (EMT), dusttight wireways, or Type MC or Type MI cable with listed termination fittings.

(2) Type PLTC cable or Type PLTC-ER cable used in Class 2 and Class 3 circuits, including installation in cable tray systems. The cable shall be terminated with listed fittings. Type PLTC-ER cable shall include an equipment grounding conductor in addition to a drain wire that might be present.

(3) Type ITC cable or Type ITC-ER cable as permitted in 335.4 and terminated with listed fittings. Type ITC-ER cable shall include an equipment grounding conductor in addition to a drain wire.

(4) Type MV, Type TC, or Type TC-ER cable, including installation in cable tray systems. Type TC-ER cable shall include an equipment grounding conductor in addition to a drain wire that might be present. The cable shall be terminated with listed fittings.

(5) Cablebus.

(6) In restricted industrial establishments, listed Type P cable with metal braid armor, with an overall jacket, that is terminated with fittings listed for the location, and installed in accordance with Part II of Article 337. If installed in ladder, ventilated trough, or ventilated channel cable trays, cables shall be installed in a single layer, with a space not less than the larger cable diameter between the two adjacent cables unless otherwise protected against dust buildup resulting in increased heat.

> Informational Note No. 1: See UL 1309A, *Outline of Investigation for Cable for Use in Mobile Installations*, for information on construction, testing, and marking of Type P cable.
> Informational Note No. 2: See ANSI/UL 2225, *Cables and Cable-Fittings for Use in Hazardous (Classified) Locations*, for information on construction, testing, and marking of cable fittings.

(2) Boxes and Fittings. All boxes and fittings shall be dusttight.

Δ **(3) Flexible Connections.** Where flexible connections are necessary, one or more of the following shall be permitted:

(1) Dusttight flexible connectors
(2) Liquidtight flexible metal conduit (LFMC) with listed fittings
(3) Liquidtight flexible nonmetallic conduit (LFNC) with listed fittings and bonded in accordance with 503.30(B)
(4) Interlocked armor Type MC cable having an overall jacket of suitable polymeric material and installed with listed dusttight termination fittings
(5) Flexible cord in accordance with 503.140
(6) For elevator use, an identified elevator cable of Type EO, Type ETP, or Type ETT, shown under the "use" column

in Table 400.4 for "hazardous (classified) locations" and terminated with listed dusttight fittings

(7) In restricted industrial establishments, listed Type P cable with metal braid armor, with an overall jacket, that is terminated with fittings listed for the location and installed in accordance with Part II of Article 337

Informational Note: See UL 1309A, *Outline of Investigation for Cable for Use in Mobile Installations*, for information on construction, testing, and marking of Type P cable.

(4) Nonincendive Field Wiring. Nonincendive field wiring shall be permitted using any of the wiring methods permitted for unclassified locations. Nonincendive field wiring systems shall be installed in accordance with the control drawing(s). Simple apparatus, not shown on the control drawing, shall be permitted in a nonincendive field wiring circuit if the simple apparatus does not interconnect the nonincendive field wiring circuit to any other circuit.

Informational Note: See Article 100 for the definition of *simple apparatus*.

Separate nonincendive field wiring circuits shall be installed in accordance with one of the following:

(1) In separate cables
(2) In multiconductor cables where the conductors of each circuit are within a grounded metal shield
(3) In multiconductor cables where the conductors of each circuit have insulation with a minimum thickness of 0.25 mm (0.01 in.)

Δ **(B) Class III, Division 2.** Wiring methods in Class III, Division 2 locations shall be in accordance with the following:

(1) The wiring shall comply with 503.10(A).
(2) In sections, compartments, or areas that do not contain machinery and are used solely for storage, open wiring on insulators shall be permitted where installed in accordance with Part II of Article 398, including the condition required by 398.15(C) that protection be provided where conductors are not run in roof spaces and are well out of reach of sources of physical damage.

503.25 Uninsulated Exposed Parts, Class III, Divisions 1 and 2. There shall be no uninsulated exposed parts, such as electrical conductors, buses, terminals, or components, that operate at more than 30 volts (15 volts in wet locations). These parts shall additionally be protected by a protection technique according to 500.7(E), (F), or (G) that is suitable for the location.

Exception: As provided in 503.155.

Exposed live parts are permitted in Class III, Division 1 and 2 locations provided the voltage does not exceed 30 volts in dry locations or 15 volts in wet locations. Protection techniques permitted for these parts are intrinsically safe or nonincendive. These techniques limit the circuit's energy to a level incapable of causing ignition of the hazardous area.

Δ **503.30 Grounding and Bonding.** Regardless of the voltage of the electrical system, wiring systems and equipment shall comply with 503.30(A) and (B).

N **(A) Grounding.** Wiring systems and equipment shall be grounded in accordance with Part I and Part VI of Article 250, as applicable.

Δ **(B) Bonding.** Bonding shall comply with Part I and Part V of Article 250, as applicable, and 503.30(B)(1) and (B)(2).

N **(1) Specific Bonding Means.** Bonding shall comply with 503.30(B)(1)(a) and (B)(2)(b).

(a) The locknut-bushing and double-locknut types of contacts shall not be depended on for bonding purposes, but bonding jumpers with identified fittings or other approved means of bonding shall be used. These bonding means shall apply to all metal raceways, fittings, boxes, cable trays, and enclosures, and other parts of raceway systems between hazardous (classified) locations and the point of grounding for service equipment or point of grounding for a separately derived system. Metal struts, angles, or channels provided for support and mechanical or physical protection as permitted in 335.4(5), 336.10(7)(c), or 722.135(C) shall be bonded in accordance with 250.102.

(b) Where the branch-circuit overcurrent protection is located on the load side of the disconnecting means, the specific bonding means shall be permitted to end at the nearest point where the grounded circuit conductor and the grounding electrode conductor are connected together on the line side of the building or structure disconnecting means as specified in 250.32(B).

The requirements for enhanced bonding in Class III locations are the same as those for Class I and Class II locations.

See also

501.30(B)(1) and its commentary for more information on grounding and bonding requirements

250.100 for additional requirements applying to bonding in hazardous locations

N **(2) Liquidtight Flexible Metal Conduit.** Liquidtight flexible metal conduit shall comply with 503.30(B)(2)(a) and (B)(2)(b).

(a) Liquidtight flexible metal conduit shall include an equipment bonding jumper of the wire type in accordance with 250.102.

(b) In Class III locations, the bonding jumper shall not be required where all of the following conditions are met:

(1) Listed liquidtight flexible metal conduit 1.8 m (6 ft) or less in length, with fittings listed for grounding, is used.
(2) Overcurrent protection in the circuit is limited to 10 amperes or less.
(3) The load is part of a meter, instrument, or relay circuit.

Part III. Equipment

503.100 Transformers and Capacitors — Class III, Divisions 1 and 2. Transformers and capacitors shall comply with 502.100(B).

503.115 Switches, Circuit Breakers, Motor Controllers, and Fuses — Class III, Divisions 1 and 2. Switches, circuit breakers, motor controllers, and fuses, including pushbuttons, relays, and similar devices, shall be provided with dusttight enclosures.

503.120 Control Transformers and Resistors — Class III, Divisions 1 and 2. Transformers, impedance coils, and resistors used as, or in conjunction with, control equipment for motors, generators, and appliances shall be provided with dusttight enclosures complying with the temperature limitations in 503.5.

503.125 Motors and Generators — Class III, Division 1 and Division 2. In Class III, Division 1 and Division 2 locations, motors, generators, and other rotating machinery shall be totally enclosed nonventilated, totally enclosed pipe ventilated, or totally enclosed fan cooled.

Exception: In locations where, in the judgment of the authority having jurisdiction, only moderate accumulations of ignitible fibers/flyings are likely to collect on, in, or in the vicinity of a rotating electrical machine and where such machine is readily accessible for routine cleaning and maintenance, one of the following shall be permitted:

(1) Self-cleaning textile motors of the squirrel-cage type
(2) Standard open-type machines without sliding contacts or centrifugal or other types of switching mechanisms, including motor overload devices
(3) Standard open-type machines having such contacts, switching mechanisms, or resistance devices enclosed within tight housings without ventilating or other openings

503.128 Ventilating Piping — Class III, Divisions 1 and 2. Ventilating pipes for motors, generators, or other rotating electrical machinery, or for enclosures for electric equipment, shall be of metal not less than 0.53 mm (0.021 in.) in thickness, or of equally substantial noncombustible material, and shall comply with the following:

(1) Lead directly to a source of clean air outside of buildings
(2) Be screened at the outer ends to prevent the entrance of small animals or birds
(3) Be protected against physical damage and against rusting or other corrosive influences

Ventilating pipes shall be sufficiently tight, including their connections, to prevent the entrance of appreciable quantities of fibers/flyings into the ventilated equipment or enclosure and to prevent the escape of sparks, flame, or burning material that might ignite accumulations of fibers/flyings or combustible material in the vicinity. For metal pipes, lock seams and riveted or welded joints shall be permitted; and tight-fitting slip joints shall be permitted where some flexibility is necessary, as at connections to motors.

503.130 Luminaires — Class III, Divisions 1 and 2.

(A) Fixed Lighting. Luminaires for fixed lighting shall provide enclosures for lamps and lampholders that are designed to minimize entrance of fibers/flyings and to prevent the escape of sparks, burning material, or hot metal. Each luminaire shall be clearly marked to show the maximum wattage of the lamps that shall be permitted without exceeding an exposed surface temperature of 165°C (329°F) under normal conditions of use.

(B) Physical Damage. A luminaire that may be exposed to physical damage shall be protected by a suitable guard.

(C) Pendant Luminaires. Pendant luminaires shall be suspended by stems of threaded rigid metal conduit, threaded intermediate metal conduit, threaded metal tubing of equivalent thickness, or by chains with approved fittings. For stems longer than 300 mm (12 in.), permanent and effective bracing against lateral displacement shall be provided at a level not more than 300 mm (12 in.) above the lower end of the stem, or flexibility in the form of an identified fitting or a flexible connector shall be provided not more than 300 mm (12 in.) from the point of attachment to the supporting box or fitting.

(D) Portable Lighting Equipment. Portable lighting equipment shall be equipped with handles and protected with substantial guards. Lampholders shall be of the unswitched type with no provision for receiving attachment plugs. There shall be no exposed current-carrying metal parts, and all exposed non–current-carrying metal parts shall be grounded. In all other respects, portable lighting equipment shall comply with 503.130(A).

503.135 Utilization Equipment — Class III, Divisions 1 and 2.

(A) Heaters. Electrically heated utilization equipment shall be identified for Class III locations.

(B) Motors. Motors of motor-driven utilization equipment shall comply with 503.125.

(C) Switches, Circuit Breakers, Motor Controllers, and Fuses. Switches, circuit breakers, motor controllers, and fuses shall comply with 503.115.

503.140 Flexible Cords — Class III, Divisions 1 and 2. Flexible cords shall comply with the following:

(1) Be of a type listed for extra-hard usage
(2) Contain, in addition to the conductors of the circuit, an equipment grounding conductor complying with 400.23
(3) Be supported by clamps or other suitable means in such a manner that there will be no tension on the terminal connections
(4) Be terminated with a listed dusttight cord connector

503.145 Receptacles and Attachment Plugs — Class III, Division 1 and Division 2. Receptacles and attachment plugs shall be of the grounding type, shall be designed to minimize the accumulation or the entry of fibers/flyings, and shall prevent the escape of sparks or molten particles.

Exception: In locations where, in the judgment of the authority having jurisdiction, only moderate accumulations of ignitible fibers/flyings are likely to collect in the vicinity of a receptacle, and where such receptacle is readily accessible for routine cleaning and mounted to minimize the entry of fibers/flyings, general-purpose grounding-type receptacles shall be permitted.

503.150 Signaling, Alarm, Remote-Control, and Local Loudspeaker Intercommunications Systems — Class III, Division 1 and Division 2.

Signaling, alarm, remote-control, and local loudspeaker intercommunications systems shall comply with the requirements of this article regarding wiring methods, switches, transformers, resistors, motors, luminaires, and related components.

503.155 Electric Cranes, Hoists, and Similar Equipment — Class III, Divisions 1 and 2.

Where installed for operation over combustible fibers or accumulations of flyings, traveling cranes and hoists for material handling, traveling cleaners for textile machinery, and similar equipment shall comply with 503.155(A) through (D).

In Class III locations, two hazards can be introduced by cranes, equipped with rolling or sliding collectors that contact with bare conductors, that are installed over accumulations of fibers/flyings.

The first hazard results from arcing between a conductor and a collector rail igniting combustible fibers or lint that has accumulated on or near the bare conductor. This hazard can be prevented by maintaining the proper alignment of the bare conductor, by using a collector designed so that proper contact is always maintained, and by using guards or shields to confine hot metal particles that result from arcing.

The second hazard occurs if enough moisture is present and fibers/flyings accumulating on the insulating supports of the bare conductors form a conductive path between the conductors or from one conductor to ground, permitting enough current to flow to ignite the fibers. If the system is ungrounded, a current flow to ground is unlikely to start a fire.

A suitable recording ground detector sounds an alarm and automatically de-energizes contact conductors when the insulation resistance is lowered by an accumulation of fibers on the insulators or in case of a fault to ground. A ground-fault indicator that maintains an alarm until the system is de-energized or the ground fault is cleared is permitted.

(A) Power Supply. The power supply to contact conductors shall be electrically isolated from all other systems, ungrounded, and shall be equipped with an acceptable ground detector that gives an alarm and automatically de-energizes the contact conductors in case of a fault to ground or gives a visual and audible alarm as long as power is supplied to the contact conductors and the ground fault remains.

(B) Contact Conductors. Contact conductors shall be located or guarded so as to be inaccessible to other than authorized persons and shall be protected against accidental contact with foreign objects.

(C) Current Collectors. Current collectors shall be arranged or guarded so as to confine normal sparking and prevent escape of sparks or hot particles. To reduce sparking, two or more separate surfaces of contact shall be provided for each contact conductor. Reliable means shall be provided to keep contact conductors and current collectors free of accumulations of lint or flyings.

(D) Control Equipment. Control equipment shall comply with 503.115 and 503.120.

503.160 Storage Battery Charging Equipment — Class III, Divisions 1 and 2.

Storage battery charging equipment shall be located in separate rooms built or lined with substantial noncombustible materials. The rooms shall be constructed to prevent the entrance of ignitible amounts of flyings or lint and shall be well ventilated.

ARTICLE 504 Intrinsically Safe Systems

Δ **504.1 Scope.** This article covers the installation of intrinsically safe (I.S.) apparatus, wiring, and systems for hazardous (classified) locations.

Informational Note: See ANSI/ISA RP 12.06.01, *Recommended Practice for Wiring Methods for Hazardous (Classified) Locations Instrumentation — Part 1: Intrinsic Safety,* for additional information.

There are two standards used in the United States for construction and performance requirements for intrinsically safe (IS) systems: ANSI/UL 913, *Standard for Intrinsically Safe Apparatus and Associated Apparatus for Use in Class I, II, and III, Division 1, Hazardous (Classified) Locations,* and ANSI/UL 60079-11, *Electrical Apparatus for Explosive Gas Atmospheres — Part 11: Intrinsic Safety "I,"* which is based on the IEC 60079-11 standard. The NEC® offers the choice of designating hazardous locations as two divisions (1 and 2) or three zones (0, 1, and 2). Equipment certified by a testing laboratory for Zone 1 would not necessarily meet UL 913 requirements for Division 1.

Due to its physical and electrical characteristics, an IS circuit does not develop sufficient electrical energy (millijoules) in an arc or spark to cause ignition or sufficient thermal energy resulting from an overload condition to cause the temperature of the installed circuit to exceed the ignition temperature of a specified gas or vapor under normal or abnormal operating conditions.

An abnormal condition may occur due to damage, failure of electrical components, excessive voltage, or improper adjustment or maintenance of the equipment. Abnormal conditions are mitigated by associated apparatus such as the IS barrier shown in Exhibit 504.1.

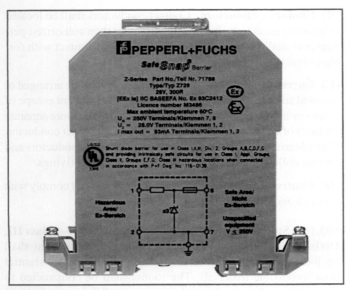

EXHIBIT 504.1 A typical IS barrier that limits the energy available to the hazardous location. *(Courtesy of Pepperl+Fuchs, Inc.)*

504.3 Application of Other Articles. Except as modified by this article, all applicable articles of this *Code* shall apply.

Because IS wiring must be low energy, the wiring itself is most likely to be a Class 2 or a power-limited fire-protective signaling circuit. See Article 725 or 760, as appropriate, for the requirements for such wiring. The installation may also fall under the scope of Article 800. The associated apparatus, on the other hand, may be supplied by ordinary power circuits, in which case other *NEC* requirements may apply.

The associated apparatus is not normally suitable for a hazardous location. Therefore, another protection technique, such as installing the associated apparatus in an explosionproof enclosure, is commonly used if it must be installed in a hazardous location. IS systems are not exempt from the grounding and bonding requirements of 501.30, 502.30, 503.30, and 505.30.

504.4 Equipment. All intrinsically safe apparatus and associated apparatus shall be listed.

Exception: Simple apparatus, as described on the control drawing, shall not be required to be listed.

504.10 Equipment Installation.

(A) Control Drawing. Intrinsically safe apparatus, associated apparatus, and other equipment shall be installed in accordance with the control drawing(s).

A simple apparatus, whether or not shown on the control drawing(s), shall be permitted to be installed provided the simple apparatus does not interconnect intrinsically safe circuits.

An example of the control drawing required to be followed to correctly install an IS system is shown in Exhibit 504.2. This drawing is normally provided by the associated equipment manufacturer. A similar drawing is provided by the IS equipment manufacturer. Compliance with the requirements of both drawings is required to properly install an IS system.

Informational Note No. 1: The control drawing identification is marked on the apparatus.

Informational Note No. 2: Associated apparatus with a marked Um of less than 250 V may require additional overvoltage

EXHIBIT 504.2 A sample zener carrier control drawing.

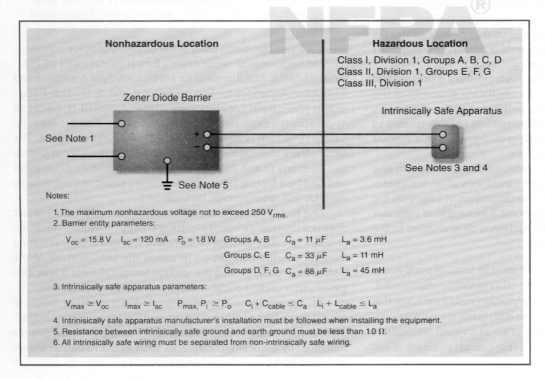

Notes:

1. The maximum nonhazardous voltage not to exceed 250 V$_{rms}$.
2. Barrier entity parameters:

$V_{oc} = 15.8$ V $I_{sc} = 120$ mA $P_o = 1.8$ W   Groups A, B $C_a = 11$ μF $L_a = 3.6$ mH

                                                  Groups C, E $C_a = 33$ μF $L_a = 11$ mH

                                                  Groups D, F, G $C_a = 88$ μF $L_a = 45$ mH

3. Intrinsically safe apparatus parameters:

$V_{max} \geq V_{oc}$ $I_{max} \geq I_{sc}$ $P_{max}, P_i \geq P_o$ $C_i + C_{cable} \leq C_a$ $L_i + L_{cable} \leq L_a$

4. Intrinsically safe apparatus manufacturer's installation must be followed when installing the equipment.
5. Resistance between intrinsically safe ground and earth ground must be less than 1.0 Ω.
6. All intrinsically safe wiring must be separated from non-intrinsically safe wiring.

protection at the inputs to limit any possible fault voltages to less than the Um marked on the product.

An IS system is required to be installed according to the control drawings, which may put limitations on cables and on the separation of circuits in the system. Control drawings also illustrate what is permitted to be connected in the system. Compliance with all the conditions in the control drawings is essential if intrinsic safety is to be maintained. The investigation of the equipment by third-party testing laboratories is based on installation in accordance with the control drawing. See Exhibit 504.2 for an example of a control drawing.

(B) Location. Intrinsically safe apparatus shall be permitted to be installed in any hazardous (classified) location for which it has been identified.

Associated apparatus shall be permitted to be installed in any hazardous (classified) location for which it has been identified.

Simple apparatus shall be permitted to be installed in any hazardous (classified) location in accordance with 504.10(D).

(C) Enclosures. General-purpose enclosures shall be permitted for intrinsically safe apparatus and associated apparatus unless otherwise specified in the manufacturer's documentation.

(D) Simple Apparatus. Simple apparatus shall be permitted to be installed in any hazardous (classified) location in which the maximum surface temperature of the simple apparatus does not exceed the ignition temperature of the flammable gases or vapors, flammable liquids, combustible dusts, or ignitible fibers/flyings present. The maximum surface temperature can be determined from the values of the output power from the associated apparatus or apparatus to which it is connected to obtain the temperature class. The temperature class can be determined by:

(1) Reference to Table 504.10(D)
(2) Calculation using the following equation:

$$T = P_O R_{th} \times T_{amb} \qquad \text{[504.10(D)]}$$

where:

T = surface temperature

P_O = output power marked on the associated apparatus or intrinsically safe apparatus

R_{th} = thermal resistance of the simple apparatus

T_{amb} = ambient temperature (normally 40°C) and reference Table 500.8(C)(4)

TABLE 504.10(D) *Assessment for T4 Classification According to Component Size and Temperature*

Total Surface Area Excluding Lead Wires	Requirement for T4 Classification
<20 mm²	Surface temperature ≤275°C
≥20 mm² ≤10 cm²	Surface temperature ≤200°C
≥20 mm²	Power not exceeding 1.3 W*

*Based on 40°C ambient temperature. Reduce to 1.2 W with an ambient of 60°C or 1.0 W with 80°C ambient temperature.

In addition, components with a surface area smaller than 10 cm² (excluding lead wires) may be classified as T5 if their surface temperature does not exceed 150°C.

Simple apparatus stores little or no energy. Simple apparatus is permitted to be used without requiring the apparatus to be listed or to be specifically mentioned on the control drawing. See the first informational note following the definition of *simple apparatus* in Article 100 for examples of simple apparatus.

504.20 Wiring Methods. Any of the wiring methods suitable for unclassified locations, including those covered by Chapter 7 and Chapter 8, shall be permitted for installing intrinsically safe apparatus. Sealing shall be as provided in 504.70, and separation shall be as provided in 504.30.

An IS system evaluation also includes wiring faults and cable parameters (e.g., short circuits and cable capacitance). Any of the wiring methods for unclassified locations may be used for IS systems, as long as the conditions specified in the control drawings are followed.

See also

504.3 and its commentary for more information on the types of circuits typically used in an IS system

504.30 Separation of Intrinsically Safe Conductors.

It is essential that non-IS circuits and IS circuits be physically and electrically separated to prevent unsafe energy from being introduced into the IS system by a wiring fault. Other low-voltage, low-energy circuits, such as Class 2 and communications circuits, are not IS circuits and must not be installed in the same raceways or cables as IS circuits in either a hazardous or a nonhazardous location.

(A) From Nonintrinsically Safe Circuit Conductors.

Δ **(1) In Raceways, Cable Trays, and Cables.** Conductors of intrinsically safe circuits shall not be placed in any raceway, cable tray, or cable with conductors of any nonintrinsically safe circuit, unless they meet the requirements of one of the following methods:

(1) Separated from conductors of nonintrinsically safe circuits in accordance with one of the following:

   a. By a distance of at least 50 mm (2 in.) and secured
   b. By a grounded metal partition that is 0.91 mm (0.0359 in.) or thicker
   c. An approved insulating partition

(2) All of the intrinsically safe circuit conductors or nonintrinsically safe circuit conductors are in Type MC cable, Type MI cable, or other approved grounded metal-sheathed or metal-clad cables where the sheathing or cladding is capable of carrying fault current to ground

(3) In a Division 2 or Zone 2 location, installed in a raceway, cable tray, or cable along with nonincendive field wiring circuits when installed in accordance with 504.30(B)

(4) Where passing through a Division 2 or Zone 2 location to supply apparatus that is located in a Division 1, Zone 0 or Zone 1 location, installed in a raceway, cable tray, or cable along with nonincendive field wiring circuits when installed in accordance with 504.30(B)

Items No. 3 and No. 4 permit IS circuits in a Division 2 or Zone 2 location to be installed with nonincendive field wiring circuits if the circuits are installed as required for two separate IS circuits.

Δ **(2) Within Enclosures.** Conductors of intrinsically safe circuits shall be secured so that any conductor that might come loose from a terminal is unlikely to come into contact with another terminal. The conductors shall be separated from conductors of nonintrinsically safe circuits by one of the following methods:

(1) Separation by at least 50 mm (2 in.) from conductors of any nonintrinsically safe circuits, and secured

(2) Separation from conductors of nonintrinsically safe circuits by use of a grounded metal partition 0.91 mm (0.0359 in.) or thicker or approved restricted access wiring ducts separated from other wiring ducts by a minimum of 19 mm (¾ in.)

(3) Separation from conductors of nonintrinsically safe circuits by use of rigid insulating partition 0.91 mm (0.0359 in,) or thicker that extends to within 1.5 mm (0.0625 in.) of the enclosure walls

(4) Use of separate wiring compartments for intrinsically safe and nonintrinsically safe terminals

(5) Either all intrinsically safe circuit conductors or all nonintrinsically safe circuit conductors are installed in grounded metal-sheathed or metal-clad cables, where the sheathing or cladding is capable of carrying fault current to ground

Δ **(3) Other (Not in Raceway or Cable Tray Systems).** Conductors and cables of intrinsically safe circuits run in other than raceway or cable tray systems shall be separated by at least 50 mm (2 in.) and secured from conductors and cables of any nonintrinsically safe circuits unless one of the following applies:

(1) All of the intrinsically safe circuit conductors are in Type MI or MC cables.

(2) All of the nonintrinsically safe circuit conductors are in raceways or Type MI or Type MC cables where the sheathing or cladding is capable of carrying fault current to ground.

Even where not installed in an enclosure, raceway, or cable tray, IS circuit conductors are required to be securely separated from other conductors.

(B) From Different Intrinsically Safe Circuit Conductors. The clearance between two terminals for connection of field wiring of different intrinsically safe circuits shall be at least 6 mm (0.25 in.), unless this clearance is permitted to be reduced by the control drawing. Different intrinsically safe circuits shall be separated from each other by one of the following means:

(1) The conductors of each circuit are within a grounded metal shield.

(2) The conductors of each circuit have insulation with a minimum thickness of 0.25 mm (0.01 in.).

Exception: Unless otherwise identified.

The minimum required clearance provides a safeguard against an inadvertent connection between adjacent terminals (with different IS circuits) that could occur during maintenance or connection of a new circuit to an existing terminal block for IS circuits.

(C) From Grounded Metal. The clearance between the uninsulated parts of field wiring conductors connected to terminals and grounded metal or other conducting parts shall be at least 3 mm (0.125 in.).

504.50 Grounding.

Δ **(A) Intrinsically Safe Apparatus, Enclosures, and Raceways.** Intrinsically safe apparatus, enclosures, and raceways, if of metal, shall be connected to the equipment grounding conductor.

Informational Note: See ANSI/ISA RP 12.06.01, *Recommended Practice for Wiring Methods for Hazardous (Classified) Locations Instrumentation — Part 1: Intrinsic Safety.* In addition to an equipment grounding conductor connection, a connection to a grounding electrode might be needed for some associated apparatus, such as zener diode barriers, if specified in the control drawing.

Δ **(B) Associated Apparatus and Cable Shields.** Associated apparatus and cable shields shall be grounded in accordance with the required control drawing. See 504.10(A).

Informational Note: See ANSI/ISA RP 12.06.01, *Recommended Practice for Wiring Methods for Hazardous (Classified) Locations Instrumentation — Part 1: Intrinsic Safety.* In addition to an equipment grounding conductor connection, a connection to a grounding electrode might be needed for some associated apparatus, such as zener diode barriers, if specified in the control drawing.

Exhibit 504.3 illustrates a common type of zener diode barrier, which is also called a shunt diode barrier. (See Exhibit 504.1.) Maintaining a low-impedance path to ground for zener diode barrier systems is important because such systems shunt fault currents to ground.

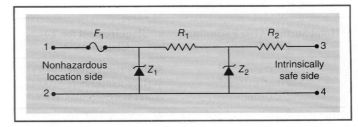

EXHIBIT 504.3 Zener, or shunt, diode barrier.

(C) Connection to Grounding Electrodes. Where connection to a grounding electrode is required, the grounding electrode shall be as specified in 250.52(A)(1), (A)(2), (A)(3), and (A)(4) and shall comply with 250.30(A)(4). Sections 250.52(A)(5), (A)(7), and (A)(8) shall not be used if any of the electrodes specified in 250.52(A)(1), (A)(2), (A)(3), or (A)(4) are present.

Where a grounding electrode is necessary for associated apparatus, the electrodes specified in 250.52(A)(1) through (A)(4) (metal underground water pipes, metal in-ground support structures, concrete-encased electrodes, and ground rings) are required to be used if present. These electrodes usually provide lower resistance grounds than ground rods and plate electrodes, which are covered in 250.52(A)(5) and (A)(7).

See also

504.50(B), Informational Note, for where this connection might be required

504.60 Bonding.

(A) Intrinsically Safe Apparatus. Intrinsically safe apparatus, if of metal, shall be bonded in the hazardous (classified) location in accordance with 501.30(B), 502.30(B), 503.30(B), 505.30(B), or 506.30(B), as applicable.

(B) Metal Raceways. Where metal raceways are used for intrinsically safe system wiring, bonding at all ends of the raceway, regardless of the location, shall be in accordance with 501.30(B), 502.30(B), 503.30(B), 505.30(B), or 506.30(B), as applicable.

504.70 Sealing. Conduits and cables that are required to be sealed by 501.15, 502.15, 505.16, and 506.16 shall be sealed to minimize the passage of gases, vapors, or dusts. Such seals shall not be required to be explosionproof or flameproof but shall be identified for the purpose of minimizing passage of gases, vapors, or dusts under normal operating conditions and shall be accessible.

Exception: Seals shall not be required for enclosures that contain only intrinsically safe apparatus, except as required by 501.17.

The use of an IS system does not remove the need to seal interconnecting cables. Any cable capable of transmitting material to another location must be sealed. These seals are not required to be explosionproof or flameproof, but they must be identified to minimize the passage of gases or dust and must be accessible.

504.80 Identification. Labels required by this section shall be suitable for the environment where they are installed, with consideration given to exposure to chemicals and sunlight.

(A) Terminals. Intrinsically safe circuits shall be identified at terminal and junction locations in a manner that is intended to prevent unintentional interference with the circuits during testing and servicing.

(B) Wiring. Raceways, cable trays, and other wiring methods for intrinsically safe system wiring shall be identified with permanently affixed labels with the wording "Intrinsic Safety Wiring" or equivalent. The labels shall be located so as to be visible after installation and placed so that they may be readily traced through the entire length of the installation. Intrinsic safety circuit labels shall appear in every section of the wiring system that is separated by enclosures, walls, partitions, or floors. Spacing between labels shall not be more than 7.5 m (25 ft).

Exception: Circuits run underground shall be permitted to be identified where they become accessible after emergence from the ground.

Informational Note No. 1: Wiring methods permitted in unclassified locations may be used for intrinsically safe systems in hazardous (classified) locations. Without labels to identify the application of the wiring, enforcement authorities cannot determine that an installation is in compliance with this *Code*.

Informational Note No. 2: In unclassified locations, identification is necessary to ensure that nonintrinsically safe wire will not be inadvertently added to existing raceways at a later date.

(C) Color Coding. Color coding shall be permitted to identify intrinsically safe conductors where they are colored light blue and where no other conductors colored light blue are used. Likewise, color coding shall be permitted to identify raceways, cable trays, and junction boxes where they are colored light blue and contain only intrinsically safe wiring.

ARTICLE
505 Zone 0, 1, and 2 Locations

Δ **505.1 Scope.**

N **(A) Covered.** This article covers the requirements for the zone classification system as an alternative to the division classification system covered in 500.1 for electrical and electronic equipment and wiring for all voltages where fire or explosion hazards might exist due to flammable gases, vapors, or liquids for the following:

(1) Zone 0 hazardous (classified) locations
(2) Zone 1 hazardous (classified) locations
(3) Zone 2 hazardous (classified) locations

Informational Note No. 1: The term "Class I" was originally included as a prefix to Zone 0, Zone 1, and Zone 2 locations and references as an identifier for flammable gases, vapors, or liquids to differentiate from Class II and Class III locations. Zone 0, Zone 1, and Zone 2 only apply to flammable gases, vapors, or liquids, so the "Class I" prefix is redundant and has been deleted. However, the marking of "Class I" is left as an optional marking within this Article.

Informational Note No. 2: See NFPA 497-2021, *Recommended Practice for the Classification of Flammable Liquids, Gases, or Vapors and of Hazardous (Classified) Locations for Electrical Installations in Chemical Process Areas*, for extracted text that is followed by a reference in brackets. Only editorial changes were made to the extracted text to make it consistent with this *Code*.

Informational Note No. 3: See Article 100 for the definition of *restricted industrial establishment [as applied to hazardous (classified) locations]*.

The requirements in this article parallel those in Articles 500 and 501. The zone classification concept, based on the standards developed by the International Electrotechnical Commission (IEC), offers an alternative method of classifying Class I hazardous locations. The IEC classification scheme includes underground mines, whereas in the United States, underground mines are under the jurisdiction of the Mine Safety and Health Administration (MSHA) and are outside the scope of the *NEC®*.

N (B) Not Covered. This article does not cover electrical and electronic equipment and wiring in the following:

(1) Class I, Class II, or Class III, Division 1 or Division 2 hazardous (classified) locations

(2) Zone 20, Zone 21, or Zone 22 hazardous (classified) locations

(3) Locations subject to the unique risk and explosion hazards associated with explosives, pyrotechnics, and blasting agents

(4) Locations where pyrophoric materials are the only materials used or handled

(5) Features of equipment that involve nonelectrical potential sources of ignition (e.g., couplings, pumps, gearboxes, brakes, hydraulic and pneumatic motors, fans, engines, compressors)

Informational Note No. 1: Common nonelectrical potential sources of ignition include hot surfaces and mechanically generated sparks.

Informational Note No. 2: See ANSI/UL 80079-36, *Explosive Atmospheres — Part 36: Non-Electrical Equipment for Explosive Atmospheres — Basic Method and Requirements*, and ANSI/UL 80079-37, *Explosive Atmospheres — Part 37: Non-Electrical Equipment for Explosive Atmospheres — Non-Electrical Type of Protection Constructional Safety "c" Control of Ignition Source "b", Liquid Immersion "k"*, for additional information.

△ 505.4 Documentation. Areas designated as hazardous (classified) locations or as unclassified shall be documented on an area classification drawing and other associated documentation. This documentation shall be made available to the AHJ and those authorized to design, install, inspect, maintain, or operate electrical equipment at the location.

Informational Note No. 1: See ANSI/API RP 505, *Recommended Practice for Classification of Locations for Electrical Installations at Petroleum Facilities Classified as Class I, Zone 0, Zone 1, and Zone 2*; ANSI/ISA 60079-10-1 (12.24.01), *Explosive Atmospheres — Part 10-1: Classification of areas — Explosive gas atmospheres*; and EI 15, *Model Code of Safe Practice, Part 15: Area Classification for Installations Handling Flammable Fluids*, for examples of area classification drawings.

Informational Note No. 2: See 505.8(I)(2), (I)(3), or (I)(4) for information on where gas detection equipment is used as a means

of protection. The documentation typically includes the type of detection equipment, its listing, the installation location(s), the alarm and shutdown criteria, and the calibration frequency.

Informational Note No. 3: See NFPA 497, *Recommended Practice for the Classification of Flammable Liquids, Gases, or Vapors and of Hazardous (Classified) Locations for Electrical Installations in Chemical Process Areas*; ANSI/API RP 505, *Recommended Practice for Classification of Locations for Electrical Installations at Petroleum Facilities Classified as Class I, Zone 0, Zone 1, and Zone 2*; ANSI/ISA 60079-10-1 (12.24.01), *Explosive Atmospheres — Part 10-1: Classification of areas — Explosive gas atmospheres*; and EI 15, *Model Code of Safe Practice, Part 15: Area Classification for Installations Handling Flammable Fluids*, for information on the classification of locations.

Informational Note No. 4: See NFPA 77, *Recommended Practice on Static Electricity*; NFPA 780, *Standard for the Installation of Lightning Protection Systems*, and API RP 2003, *Protection Against Ignitions Arising Out of Static, Lightning, and Stray Currents*, for information on protection against static electricity and lightning hazards in hazardous (classified) locations.

Informational Note No. 5: See NFPA 30, *Flammable and Combustible Liquids Code*, and ANSI/API RP 505, *Recommended Practice for Classification of Locations for Electrical Installations at Petroleum Facilities Classified as Class I, Zone 0, Zone 1, and Zone 2*, for information on ventilation.

Informational Note No. 6: See ANSI/API RP 14FZ, *Recommended Practice for Design, Installation, and Maintenance of Electrical Systems for Fixed and Floating Offshore Petroleum Facilities for Unclassified and Class I, Zone 0, Zone 1, and Zone 2 Locations*, for information on electrical systems for hazardous (classified) locations on offshore oil and gas producing platforms, drilling rigs, and workover rigs.

Informational Note No. 7: See UL 120101, *Definitions and Information Pertaining to Electrical Apparatus in Hazardous Locations*, and ANSI/UL 60079-0, *Explosive Atmospheres — Part 0: Equipment — General Requirements*, for further information on the general application of electrical equipment in hazardous (classified) locations.

Informational Note No. 8: See ANSI/UL 121203, *Portable/Personal Electronic Products Suitable for Use in Class I, Division 2, Class I, Zone 2, Class II, Division 2, Class III, Division 1, Class III, Division 2, Zone 21 and Zone 22 Hazardous (Classified) Locations*, for information on whether portable or transportable equipment having self-contained power supplies, such as battery-operated equipment, could potentially become an ignition source in hazardous (classified) locations.

Informational Note No. 9: See ANSI/UL 60079-28, *Explosive Atmospheres — Part 28: Protection of Equipment and Transmission Systems Using Optical Radiation*, for information concerning the installation of equipment utilizing optical emissions technology (such as laser equipment) that could potentially become an ignition source in hazardous (classified) locations.

Informational Note No. 10: See IEC/IEEE 60079-30-2, *Explosive Atmospheres — Part 30-2: Electrical Resistance Trace Heating — Application Guide for Design, Installation and Maintenance*, for information on electrical resistance trace heating for hazardous (classified) locations.

Informational Note No. 11: See IEEE 844.2/CSA C293.2, *IEEE/CSA Standard for Skin Effect Trace Heating of Pipelines, Vessels, Equipment, and Structures — Application Guide for Design, Installation, Testing, Commissioning, and Maintenance*, for information on electric skin effect trace heating for hazardous (classified) locations.

Informational Note No. 12: See IEEE 844.4/CSA C293.4, *IEEE/CSA Standard for Impedance Heating of Pipelines and Equipment — Application Guide for Design, Installation, Testing, Commissioning, and Maintenance,* for information on electric impedance heating for hazardous (classified) locations.

505.5 Classifications of Locations.

Δ **(A) General.**

N **(1) Hazardous (Classified) Locations.** Locations shall be classified depending on the properties of the flammable gases, flammable liquid–produced vapors, combustible liquid–produced vapors, combustible dusts, or fibers/flyings that could be present and the likelihood that a flammable or combustible concentration or quantity is present. Each room, section, or area shall be considered individually in determining its classification.

Informational Note No. 1: See 505.7 for restrictions on area classification.

Informational Note No. 2: Through the exercise of ingenuity in the layout of electrical installations for hazardous (classified) locations, it is frequently possible to locate much of the equipment in a reduced level of classification or in an unclassified location to reduce the amount of special equipment required.

N **(2) Refrigerant Machinery Rooms Using Ammonia.** Refrigerant machinery rooms that contain ammonia refrigeration systems and are equipped with adequate mechanical ventilation that operates continuously or is initiated by a detection system at a concentration not exceeding 150 ppm shall be permitted to be classified as "unclassified" locations.

Informational Note: See ANSI/IIAR 2, *Standard for Safe Design of Closed-Circuit Ammonia Refrigeration Systems,* for information on the classification and ventilation of areas involving closed-circuit ammonia refrigeration systems.

(B) Zone 0, 1, and 2 Locations. Zone 0, 1, and 2 locations are those in which flammable gases or vapors are or may be present in the air in quantities sufficient to produce explosive or ignitible mixtures. Zone 0, 1, and 2 locations shall include those specified in 505.5(B)(1), (B)(2), and (B)(3).

Δ **(1) Zone 0.** A Zone 0 location is a location in which one of the following conditions exists:

(1) Ignitible concentrations of flammable gases or vapors are present continuously
(2) Ignitible concentrations of flammable gases or vapors are present for long periods of time

Informational Note No. 1: See ANSI/API RP 505, *Recommended Practice for Classification of Locations for Electrical Installations at Petroleum Facilities Classified as Class I, Zone 0, Zone 1, and Zone 2,* and ANSI/ISA 60079-10-1 (12.24.01), *Explosive Atmospheres — Part 10-1: Classification of Areas — explosive gas atmospheres,* for information for when flammable gases or vapors are present continuously or for long periods of time.

Informational Note No. 2: This classification includes the following locations:

(1) Inside vented tanks or vessels that contain volatile flammable liquids
(2) Inside inadequately vented spraying or coating enclosures where volatile flammable solvents are used
(3) Between the inner and outer roof sections of a floating roof tank containing volatile flammable liquids
(4) Inside open vessels, tanks, and pits containing volatile flammable liquids
(5) Interior of an exhaust duct used to vent ignitible concentrations of gases or vapors
(6) Inside inadequately ventilated enclosures that contain normally venting instruments using or analyzing flammable fluids and venting to the inside of the enclosures

(2) Zone 1. A Zone 1 location is a location

(1) In which ignitible concentrations of flammable gases or vapors are likely to exist under normal operating conditions; or
(2) In which ignitible concentrations of flammable gases or vapors may exist frequently because of repair or maintenance operations or because of leakage; or
(3) In which equipment is operated or processes are carried on, of such a nature that equipment breakdown or faulty operations could result in the release of ignitible concentrations of flammable gases or vapors and also cause simultaneous failure of electrical equipment in a mode to cause the electrical equipment to become a source of ignition; or
(4) That is adjacent to a Zone 0 location from which ignitible concentrations of vapors could be communicated, unless communication is prevented by adequate positive pressure ventilation from a source of clean air and effective safeguards against ventilation failure are provided.

Informational Note No. 1: Normal operation is considered the situation when plant equipment is operating within its design parameters. Minor releases of flammable material may be part of normal operations. Minor releases include the releases from mechanical packings on pumps. Failures that involve repair or shutdown (such as the breakdown of pump seals and flange gaskets, and spillage caused by accidents) are not considered normal operation.

Informational Note No. 2: This classification usually includes the following locations:

(1) Where volatile flammable liquids or liquefied flammable gases are transferred from one container to another
(2) Areas in the vicinity of spraying and painting operations where flammable solvents are used
(3) Adequately ventilated drying rooms or compartments for evaporation of flammable solvents
(4) Adequately ventilated locations containing fat and oil extraction equipment using volatile flammable solvents
(5) Portions of cleaning and dyeing plants where volatile flammable liquids are used
(6) Adequately ventilated gas generator rooms and other portions of gas manufacturing plants where flammable gas might escape

(7) Inadequately ventilated pump rooms for flammable gas or for volatile flammable liquids

(8) Interiors of refrigerators and freezers in which volatile flammable materials are stored in the open, lightly stoppered, or in easily ruptured containers

(9) Other locations where ignitible concentrations of flammable vapors or gases are likely to occur in the course of normal operation but are not classified Zone 0

(3) Zone 2. A Zone 2 location is a location

(1) In which ignitible concentrations of flammable gases or vapors are not likely to occur in normal operation and, if they do occur, will exist only for a short period; or

(2) In which volatile flammable liquids, flammable gases, or flammable vapors are handled, processed, or used but in which the liquids, gases, or vapors normally are confined within closed containers of closed systems from which they can escape, only as a result of accidental rupture or breakdown of the containers or system, or as a result of the abnormal operation of the equipment with which the liquids or gases are handled, processed, or used; or

(3) In which ignitible concentrations of flammable gases or vapors normally are prevented by positive mechanical ventilation but which may become hazardous as a result of failure or abnormal operation of the ventilation equipment; or

(4) That is adjacent to a Zone 1 location, from which ignitible concentrations of flammable gases or vapors could be communicated, unless such communication is prevented by adequate positive-pressure ventilation from a source of clean air and effective safeguards against ventilation failure are provided.

Informational Note: The Zone 2 classification usually includes locations where volatile flammable liquids or flammable gases or vapors are used but which would become hazardous only in case of an accident or of some unusual operating condition.

Δ **505.6 Material Groups.** For purposes of testing, approval, and area classification, various air mixtures (not oxygen enriched) shall be grouped as required in 505.6(A), (B), and (C).

Informational Note No. 1: See 90.2(D). This Code does not apply to installations underground in mines. Group I is intended for use in describing atmospheres that contain firedamp: a mixture of gases, composed mostly of methane, found underground, usually in mines.

Informational Note No. 2: See ANSI/UL 60079-11, *Explosive Atmospheres — Part 11: Equipment Protection by Intrinsic Safety "i"*. The gas and vapor subdivision is based on the maximum experimental safe gap (MESG), minimum igniting current (MIC), or both.

Informational Note No. 3: See ISO/IEC 80079-20-1, *Explosive atmospheres — Part 20-1: Material characteristics for gas and vapor classification — Test methods and data*, for information on the test equipment for determining MIC in the classification of gases or vapors according to their maximum experimental safe gaps and minimum igniting currents.

Informational Note No. 4: Group II is currently subdivided into Group IIA, Group IIB, and Group IIC. Prior marking requirements permitted some types of protection to be marked without a subdivision, showing only Group II.

Informational Note No. 5: It is necessary that the meanings of the different equipment markings and Group II classifications be carefully observed to avoid confusion with Class I, Division 1 and Division 2, Groups A, B, C, and D.

Zone 0, 1, and 2, groups shall be as follows:

Δ **(A) Group IIC.** Atmospheres containing acetylene, hydrogen, or flammable gas, flammable liquid–produced vapor, or combustible liquid–produced vapor mixed with air that may burn or explode, having either a maximum experimental safe gap (MESG) value less than or equal to 0.50 mm or minimum igniting current (MIC) ratio less than or equal to 0.45. [**497**:3.3.5.2.3]

Informational Note: See 500.6(A)(1) and (A)(2) for Class I, Group A and Class I, Group B classifications. Group IIC is equivalent to a combination of Class I, Group A and Class I, Group B.

(B) Group IIB. Atmospheres containing acetaldehyde, ethylene, or flammable gas, flammable liquid–produced vapor, or combustible liquid–produced vapor mixed with air that may burn or explode, having either maximum experimental safe gap (MESG) values greater than 0.50 mm and less than or equal to 0.90 mm or minimum igniting current ratio (MIC ratio) greater than 0.45 and less than or equal to 0.80. [**497**:3.3.5.2.2]

Informational Note No. 1: See 500.6(A)(3). Class I, Group C is equivalent to Group IIB.

Informational Note No. 2: Class I, Group B is equivalent to Group IIB + H_2.

(C) Group IIA. Atmospheres containing acetone, ammonia, ethyl alcohol, gasoline, methane, propane, or flammable gas, flammable liquid–produced vapor, or combustible liquid–produced vapor mixed with air that may burn or explode, having either a maximum experimental safe gap (MESG) value greater than 0.90 mm or minimum igniting current (MIC) ratio greater than 0.80. [**497**:3.3.5.2.1]

Informational Note: See 500.6(A)(4). Class I, Group D is equivalent to Group IIA.

The zone classification system for gases is different from the division system used in Articles 500 and 501. Commentary Table 505.1 contrasts the classification of the two systems. Note that Group I is used for the classification of gases normally

COMMENTARY TABLE 505.1 Comparison of Groups Defined Under the Zone and Division Classification Systems

Zone Classification	Division Classification
Group IIC	Groups A and B
Group IIB	Group C
Group IIA	Group D
Group I	Group D

encountered in mining applications. The group consists primarily of methane, which is known in some countries as firedamp.

Δ **505.7 Special Precaution.** This article requires equipment construction and installation that ensures safe performance under conditions of proper use and maintenance.

> Informational Note No. 1: It is important that inspection authorities and users exercise more than ordinary care regarding the installation and maintenance of electrical equipment in hazardous (classified) locations.
>
> Informational Note No. 2: Electrical equipment that is dependent on the protection technique permitted by 505.8(A) might not be suitable for use at temperatures lower than -20°C (-4°F) unless they are identified for use at lower temperatures. Low ambient conditions require special consideration. At low ambient temperatures, flammable concentrations of vapors might not exist in a location classified at normal ambient temperature.

(A) Implementation of Zone Classification System. Classification of areas, engineering and design, selection of equipment and wiring methods, installation, and inspection shall be performed by qualified persons.

(B) Dual Classification. In instances of areas within the same facility classified separately, Zone 2 locations shall be permitted to abut, but not overlap, Class I, Division 2 locations. Zone 0 or Zone 1 locations shall not abut Class I, Division 1 or Division 2 locations.

An installation is permitted to be designed using either the classification scheme of Article 500 or the classification scheme of Article 505. Both schemes cannot be used for classifying the same area.

(C) Reclassification Permitted. A Class I, Division 1 or Division 2 location shall be permitted to be reclassified as a Zone 0, Zone 1, or Zone 2 location, provided all of the space that is classified because of a single flammable gas or vapor source is reclassified under the requirements of this article.

(D) Solid Obstacles. Flameproof equipment with flanged joints shall not be installed such that the flange openings are closer than the distances shown in Table 505.7(D) to any solid obstacle that is not a part of the equipment (such as steelworks, walls, weather guards, mounting brackets, pipes, or other electrical equipment) unless the equipment is listed for a smaller distance of separation.

TABLE 505.7(D) *Minimum Distance of Obstructions from Flameproof "d" Flange Openings*

Gas Group	Minimum Distance	
	mm	in.
IIC	40	$1^{37}/_{64}$
IIB	30	$1^{3}/_{16}$
IIA	10	$^{25}/_{64}$

(E) Simultaneous Presence of Flammable Gases and Combustible Dusts or Fibers/Flyings. Where flammable gases, combustible dusts, or fibers/flyings are or may be present at the same time, the simultaneous presence shall be considered during the selection and installation of the electrical equipment and the wiring methods, including the determination of the safe operating temperature of the electrical equipment.

(F) Available Fault Current for Type of Protection "e". Unless listed and marked for connection to circuits with higher available fault current, the available fault current for electrical equipment using type of protection "e" for the field wiring connections in Zone 1 locations shall be limited to 10,000 rms symmetrical amperes to reduce the likelihood of ignition of a flammable atmosphere by an arc during a short-circuit event.

> Informational Note: Limitation of the available fault current to this level may require the application of current-limiting fuses or current-limiting circuit breakers.

The limit on available short-circuit current is due to the rating of terminals and terminal blocks of equipment evaluated under UL 508A, *Standard for Industrial Control Panels,* using the Type "e" protection technique.

505.8 Protection Techniques. Acceptable protection techniques for electrical and electronic equipment in hazardous (classified) locations shall be as described in 505.8(A) through (Q).

> Informational Note No. 1: See ANSI/UL 120101, *Definitions and Information Pertaining to Electrical Equipment in Hazardous Locations,* and ANSI/UL 60079-0, *Explosive Atmospheres — Part 0: Equipment — General Requirements,* for additional information.
>
> Informational Note No. 2: See Chapter 9, Table 13 for descriptions of subdivisions of protection techniques.

Where the area is classified in accordance with the zone method, 505.8(A) through (N) identify the methods for protecting electrical and electronic equipment. Many of the protection methods are different from those in Article 501, and only those specified are suitable for zone installations. Of the many protection techniques, intrinsic safety, flameproof, and increased safety are the most common for Zone 1 locations.

See also

505.9(C) for marking requirements

(A) Flameproof Enclosure "d". This protection technique shall be permitted for equipment in Zone 1 or Zone 2 locations.

Equipment identified as flameproof is similar to explosionproof equipment. Flameproof protection is commonly combined with increased safety protection. For example, motor control and other switching contacts are commonly protected by flameproof enclosures, with the field wiring terminals protected in a separate but attached enclosure by increased safety. The conductors between the enclosures are protected by flameproof feed-through insulators. The equipment shown in Exhibit 505.1 employs this combination of protection techniques.

EXHIBIT 505.1 Typical control stations with the combination of flameproof and increased safety types of protection suitable for use in Zone 1 areas. *(Courtesy of Eaton, Crouse-Hinds Division)*

(B) Pressurized Enclosure "p". This protection technique shall be permitted for equipment in those Zone 1 or Zone 2 locations for which it is identified.

(C) Intrinsic Safety "i". This protection technique shall be permitted for apparatus and associated apparatus for Zone 0, Zone 1, or Zone 2 locations for which it is listed.

The identifying letter for intrinsic safety is "i" followed by either "a," "b," or "c," indicating whether the equipment is suitable for Zone 0 (ia), Zone 1 (ib), or Zone 2 (ic). The associated apparatus is identified by the same letters in brackets, that is, [ia], [ib], or [ic].

The "ic" classification extends the intrinsic safety concept to Zone 2 locations. The concept is similar to the nonincendive technique in that faults to the circuit or system are not applied.

(D) Type of Protection "n". This protection technique shall be permitted for equipment in Zone 2 locations. Type of protection "n" is further subdivided into nA, nC, and nR.

(E) Liquid Immersion "o". This protection technique shall be permitted for equipment in Zone 1 or Zone 2 locations.

(F) Increased Safety "e". This protection technique shall be permitted for equipment in Zone 1 or Zone 2 locations.

The increased safety protection technique is commonly used for terminal boxes, fluorescent luminaires, motors, and generators (see 505.22).

(G) Encapsulation "m". This protection technique shall be permitted for equipment in Zone 0, Zone 1, or Zone 2 locations for which it is identified.

(H) Powder Filling "q". This protection technique shall be permitted for equipment in Zone 1 or Zone 2 locations.

Δ **(I) Detection Systems for Flammable Gases.** A detection system for flammable gases shall be permitted as a means of protection in restricted industrial establishments.

Δ **(1) General.** Any gas detection system used as a protection technique shall meet all of the requirements in 505.8(I)(1)(a) through (I)(1)(e).

(a) The gas detection equipment used shall be listed for Zone 1 and listed for the detection of the specific gas or vapor to be encountered.

(b) The gas detection system shall not use portable or transportable equipment, or temporary wiring methods.

(c) The gas detection system shall only use point-type sensors. The system shall be permitted to be augmented with open-path (line-of-sight)–type sensors, but open-path–type sensors shall not be the basis for this protection technique.

(d) The type of detection equipment, its listing, the installation location(s), the alarm and shutdown criteria, and the calibration frequency shall be documented where gas detectors are used as a protection technique.

(e) The applications for the use of gas detection systems as a protection technique shall be limited to 505.8(I)(2), (I)(3), or (I)(4).

Informational Note No. 1: See ANSI/UL 121303, *Guide for Use of Detectors for Flammable Gases*, or ANSI/FM 121303, *Guide for Use of Detectors for Flammable Gases*, for additional information.

Informational Note No. 2: See ANSI/UL 60079-29-1, *Explosive Atmospheres — Part 29-1: Gas Detectors — Performance Requirements of Detectors for Flammable Gases*, or ANSI/FM 60079-29-1, *Explosive Atmospheres — Part 29-1: Gas Detectors — Performance Requirements of Detectors for Flammable Gases*, for additional information.

Informational Note No. 3: See ANSI/API RP 505, *Recommended Practice for Classification of Locations for Electrical Installations at Petroleum Facilities Classified as Class I, Zone 0, Zone 1, and Zone 2*, for additional information.

Informational Note No. 4: See ANSI/UL 60079-29-2, *Explosive Atmospheres –- Part 29-2: Gas Detectors — Selection, Installation, Use and Maintenance of Detectors for Flammable Gases and Oxygen*, or ANSI/FM 60079-29-2, *Explosive Atmospheres — Part 29-2: Gas Detectors — Selection, Installation, Use and Maintenance of Detectors for Flammable Gases and Oxygen*, for additional information.

Δ **(2) Inadequate Ventilation.** A location, enclosed space, or building that is classified as a Zone 1 location due to inadequate ventilation and that is provided with a detection system for flammable gases shall be permitted to use electrical equipment, installation methods, and wiring practices suitable for Zone 2 installations. Sensing a gas concentration of not more than 40 percent of the lower flammable limit or a gas detector system malfunction shall activate an alarm (audible or visual, or both, as most appropriate for the area).

Section 17.11 of NFPA 30, *Flammable and Combustible Liquids Code*, and 3.3.1 of NFPA 497, *Recommended Practice for the Classification of Flammable Liquids, Gases, or Vapors and of Hazardous (Classified) Locations for Electrical Installations in Chemical Process Areas*, provide information on what is considered adequate ventilation.

Δ **(3) Interior of a Building or Enclosed Space.** Any building or enclosed space that does not contain a source of flammable gas or vapors that is located in, or with an opening into, a Zone 2 hazardous (classified) location that is provided with a detection system for flammable gases shall be permitted to use electrical equipment, installation methods, and wiring practices suitable for unclassified installations under all of the following conditions:

(1) An alarm (audible or visual, or both) shall be sounded at not more than 20 percent of the lower flammable limit.

(2) Sensing a gas concentration of not more than 40 percent of the lower flammable limit or a gas detector system malfunction shall activate an alarm (audible or visual, or both, as most appropriate for the area) and initiate automatic disconnection of power from all electrical devices in the area that are not suitable for Zone 2.

(3) The power disconnecting device(s) shall be suitable for Zone 1 if located inside the building or enclosed space. If the disconnecting device(s) is located outside the building or enclosed space, it shall be suitable for the location in which it is installed.

Redundant or duplicate equipment (such as sensors) shall be permitted to be installed to avoid disconnecting electrical power when equipment malfunctions are indicated.

When automatic shutdown could introduce additional or increased hazard, this technique shall not be permitted.

(4) Interior of a Control Panel. Inside the interior of a control panel containing instrumentation or other equipment using or measuring flammable liquids, gases, or vapors and that is provided with a detection system for flammable gases equipment shall be allowed to use electrical equipment, installation methods, and wiring practices suitable for Zone 2 installations.

An alarm (audible or visual, or both) shall be sounded at not more than 40 percent of the lower flammable limit.

The gas detection system must be suitable for the original zone classification of the area even though the remainder of installed equipment is permitted to be suitable for one zone lower.

(J) Protection by Electrical Resistance Trace Heating "60079-30-1". This protection technique shall be permitted for electrical resistance trace heating equipment in Zone 1 or Zone 2 for which it is listed.

(K) Inherently Safe Optical Radiation "op is". This protection technique shall be permitted for equipment in Zone 0, Zone 1, or Zone 2 locations for which the equipment is identified.

(L) Protected Optical Radiation "op pr". This protection technique shall be permitted for equipment in Zone 1 or Zone 2 locations for which the equipment is identified.

(M) Optical System With Interlock "op sh". This protection technique shall be permitted for equipment in Zone 0, Zone 1, or Zone 2 locations for which the equipment is identified.

(N) Protection by Skin Effect Trace Heating "IEEE 844.1". This protection technique shall be permitted for skin effect trace heating equipment in Zone 1 or Zone 2 locations for which it is listed.

N **(O) Protection by Impedance Heating "IEEE 844.3".** This protection technique shall be permitted for impedance heating of pipelines, and equipment in Zone 2 locations for which it is listed.

N **(P) Pressurized Room "p".** This protection technique shall be permitted for equipment in Zone 1 or Zone 2 locations for which it is identified.

N **(Q) Special Protection "s".** This protection technique shall be permitted for equipment in Zone 0, Zone 1, or Zone 2 locations for which it is listed.

505.9 Equipment.

(A) Suitability. Suitability of identified equipment shall be determined by one of the following:

(1) Equipment listing or labeling

(2) Evidence of equipment evaluation from a qualified testing laboratory or inspection agency concerned with product evaluation

(3) Evidence acceptable to the authority having jurisdiction such as a manufacturer's self-evaluation or an owner's engineering judgment

Informational Note: Additional documentation for equipment may include certificates demonstrating compliance with applicable equipment standards, indicating special conditions of use, and other pertinent information.

(B) Listing.

(1) Equipment that is listed for a Zone 0 location shall be permitted in a Zone 1 or Zone 2 location of the same gas or vapor, provided that it is installed in accordance with the requirements for the marked type of protection. Equipment that is listed for a Zone 1 location shall be permitted in a Zone 2 location of the same gas or vapor, provided that it is installed in accordance with the requirements for the marked type of protection.

(2) Equipment shall be permitted to be listed for a specific gas or vapor, specific mixtures of gases or vapors, or any specific combination of gases or vapors.

Informational Note: One common example is equipment marked for "IIB. + H2."

(C) Marking. Equipment shall be marked in accordance with 505.9(C)(1) or (C)(2).

Δ **(1) Division Equipment.** Equipment identified for Class I, Division 1 or Class I, Division 2 shall, in addition to being marked in accordance with 500.8(C), be permitted to be marked with all of the following:

(1) Class I, Zone 1 or Zone 1; Class I, Zone 2 or Zone 2 (as applicable)
(2) Applicable gas classification group(s) in accordance with Table 505.9(C)(1)
(3) Temperature classification in accordance with 505.9(D)

TABLE 505.9(C)(1) *Material Groups*

Material Group	Comment
IIC	See 505.6(A)
IIB	See 505.6(B)
IIA	See 505.6(C)

Δ **(2) Zone Equipment.** Equipment meeting one or more of the protection techniques described in 505.8 shall be marked with all of the following in the order shown:

(1) Class I shall be an optional marking. If it is included in the equipment marking, the Class I marking shall precede the zone marking.
(2) Zone in accordance with Chapter 9, Table 13.
(3) Symbol "AEx".
(4) Protection technique(s) in accordance with Chapter 9, Table 13.
(5) Applicable material group in accordance with Table 505.9(C)(1) or a specific gas or vapor.
(6) Temperature classification in accordance with 505.9(D).
(7) Equipment protection level (EPL).

The symbol AEx identifies the equipment as meeting American national standards. In European Union countries, the symbol EEx was used historically but no longer appears in current European standards. Only equipment marked AEx has been evaluated for use in electrical systems and hazardous locations covered under the *NEC*. The equipment protection level (EPL) Ga, Gb, or Gc is required as part of the equipment markings. The EPL indicates the level of protection provided by the equipment and is correlated to the Zone in which the equipment will be installed and operated.

Exception No. 1: Associated apparatus NOT suitable for installation in a hazardous (classified) location shall be required to be marked only with 505.9(C)(2)(3), (C)(2)(4), and (C)(2)(5), but BOTH the symbol AEx (3) and the symbol for the type of protection (4) shall be enclosed within the same square brackets, for example, [AEx ia Ga] IIC.

Exception No. 2: Simple apparatus as defined in Article 100 shall not be required to have a marked operating temperature or temperature class.

Exception No. 3: Fittings for the termination of cables shall not be required to have a marked operating temperature or temperature class.

Informational Note No. 1: See Informational Note Figure 505.9(C)(2), for an explanation of the marking that is required. An example of the required marking for intrinsically safe apparatus

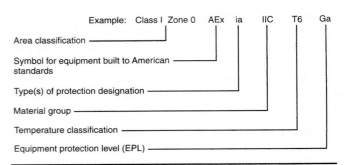

INFORMATIONAL NOTE FIGURE 505.9(C)(2) *Zone Equipment Marking.*

for installation in Zone 0 is "Class I, Zone 0, AEx ia IIC T6 Ga" or "Zone 0, AEx, ia, IIC T6 Gb."

Informational Note No. 2: An example of the required marking for intrinsically safe associated apparatus mounted in a flameproof enclosure for installation in Zone 1 is "Class I, Zone 1 AEx db[ia] IIC T4 Gb" or "Zone 1, AEx, db[ia Ga] IIC T4 Gb."

Informational Note No. 3: An example of the required marking for intrinsically safe associated apparatus NOT for installation in a hazardous (classified) location is "[AEx ia Ga] IIC."

Informational Note No. 4: EPLs are designated as G for gas, or D for dust, and are then followed by a letter (a, b, or c) to give the user a better understanding as to whether the equipment provides (a) a "very high," (b) a "high," or (c) an "enhanced" level of protection against ignition of an explosive atmosphere. For example, a Zone 1, AEx db IIC T4 Gb motor (which is suitable by protection concept for application in Zone 1) is marked with an EPL of "Gb" to indicate that it was provided with a high level of protection.

Informational Note No. 5: See ANSI/UL 60079-26, *Explosive Atmospheres — Part 26: Equipment with Equipment Protection Level (EPL) Ga*, for additional information. Equipment installed outside a Zone 0 location and electrically connected to equipment located inside a Zone 0 location might be marked Zone 0/1. The "/" indicates that equipment contains a separation element and can be installed at the boundary between a Zone 0 and a Zone 1 location.

Δ **(D) Temperature Classification Marking.** The temperature marking shall not exceed the autoignition temperature of the specific gas or vapor to be encountered.

Informational Note No. 1: See ANSI/UL 60079-26, *Explosive Atmospheres — Part 26: Equipment with Equipment Protection Level (EPL) Ga*, for more information. Equipment installed outside a Zone 0 location and electrically connected to equipment located inside a Zone 0 location might be marked Zone 0/1. The "/" indicates that equipment contains a separation element and can be installed at the boundary between a Zone 0 and a Zone 1 location.

Equipment shall be marked to show the operating temperature or temperature class referenced to a 40°C ambient, or at the higher ambient temperature if the equipment is rated and marked for an ambient temperature of greater than 40°C. The temperature class, if provided, shall be indicated using the temperature class (T code) shown in Table 505.9(D).

N TABLE 505.9(D) *Temperature Classification Marking of Maximum Surface Temperature for Group II Electrical Equipment*

Temperature Class (T Code)	Maximum Surface Temperature (°C)
T1	≤450
T2	≤300
T3	≤200
T4	≤135
T5	≤100
T6	≤85

Electrical equipment designed for use in the ambient temperature range between −20°C and +40°C shall require no ambient temperature marking.

Electrical equipment that is designed for use in a range of ambient temperatures other than −20°C to +40°C is considered to be special; and the ambient temperature range shall then be marked on the equipment, including either the symbol "Ta" or "Tamb" together with the special range of ambient temperatures, in degrees Celsius.

> Informational Note No. 2: For example, such a marking might be "−30°C to +40°C."

Exception No. 1: Equipment of the non–heat-producing type, such as conduit fittings, and equipment of the heat-producing type having a maximum temperature of not more than 100°C (212°F) shall not be required to have a marked operating temperature or temperature class.

Exception No. 2: Equipment identified for Class I, Division 1 or Division 2 locations as permitted by 505.20(A), (B), and (C) shall be permitted to be marked in accordance with 505.8(C) and Table 500.8(C)(4).

(E) Threading. The supply connection entry thread form shall be NPT or metric. Conduit and fittings shall be made wrenchtight to prevent sparking when fault current flows through the conduit system, and to ensure the explosionproof or flameproof integrity of the conduit system where applicable. Equipment provided with threaded entries for field wiring connections shall be installed in accordance with 505.9(E)(1) or (E)(2) and with (E)(3).

Δ **(1) Equipment Provided with Threaded Entries for NPT Threaded Conduit or Fittings.** For equipment provided with threaded entries for NPT threaded conduit or fittings, listed conduit, listed conduit fittings, or listed cable fittings shall be used.

All NPT threaded conduit and fittings shall be threaded with a National (American) Standard Pipe Taper (NPT) thread.

NPT threaded entries into explosionproof or flameproof equipment shall be made up with at least five threads fully engaged.

Exception: For listed explosionproof or flameproof equipment, factory-threaded NPT entries shall be made up with at least 4½ threads fully engaged.

> Informational Note No. 1: See ASME B1.20.1, *Pipe Threads, General Purpose (Inch)*, for thread specifications for male NPT threads.
> Informational Note No. 2: See ANSI/UL 60079-1, *Explosive Atmospheres — Part 1: Equipment Protection by Flameproof Enclosures "d"*, and ASME B1.20.1, *Pipe Threads, General Purpose (Inch)*, for information on female NPT threaded entries using modified National Standard Pipe Taper (NPT) thread.

Δ **(2) Equipment Provided with Threaded Entries for Metric Threaded Conduit or Fittings.** For equipment with metric threaded entries, listed conduit fittings or listed cable fittings shall be used. Such entries shall be identified as being metric, or listed adapters to permit connection to conduit or NPT threaded fittings shall be provided with the equipment and shall be used for connection to conduit or NPT threaded fittings.

Metric threaded fittings installed into explosionproof or flameproof equipment entries shall have a class of fit of at least 6g/6H and be made up with at least five threads fully engaged.

> Informational Note: See ISO 965-1, *ISO general purpose metric screw threads — Tolerances — Part 1: Principles and basic data*, and ISO 965-3, *ISO general purpose metric screw threads — Tolerances — Part 3: Deviations for constructional screw threads*, for threading specifications for metric threaded entries.

Listed fittings must be used with metric threaded entries to ensure the integrity of the conduit system. In addition, either the entries must be identified as being metric or adapters must be provided for making connection to NPT (National Pipe Taper) fittings. Exhibit 505.2 is an example of an adapter that provides a means of connecting conduit or fitting with NPT threads to a Type "e" (increased safety) enclosure that has metric threads.

(3) Unused Openings. All unused openings shall be closed with blanking elements or close-up plugs that are listed for the location and will maintain the type of protection. Thread engagement shall comply with 505.9(E)(1) or (E)(2).

EXHIBIT 505.2 A typical hub providing an NPT-threaded entry for conduit or cable into an increased safety enclosure. *(Courtesy of Eaton, Crouse-Hinds Division)*

Δ **(F) Optical Fiber Cables.** An optical fiber cable, with or without current-carrying conductors (hybrid optical fiber cable), shall be installed to address the associated fire hazard and sealed to address the associated explosion hazard in accordance with 505.15 and 505.16.

Δ **(G) Equipment Involving Optical Radiation.** For equipment involving sources of optical radiation (such as laser or LED sources) in the wavelength range from 380 nm to 10 μm, the risk of ignition from optical radiation shall be considered for all electrical parts and circuits that may be exposed to the radiation, both inside and outside the optical equipment. This includes optical equipment, which itself is located outside the explosive atmosphere, but its emitted optical radiation enters such atmospheres.

> Informational Note: See ANSI/UL 60079-28, *Explosive Atmospheres — Part 28: Protection of Equipment and Transmission Systems Using Optical Radiation,* for information on types of protection to minimize the risk of ignition in explosive atmospheres from optical radiation.

Exception: All luminaires (fixed, portable, or transportable) and hand lights, intended to be supplied by mains (with or without galvanic isolation) or powered by batteries, with any continuous divergent light source, including LEDs, shall be excluded from this requirement.

505.15 Wiring Methods. Wiring methods shall maintain the integrity of protection techniques and shall comply with 505.15(A) through (C).

Δ **(A) Zone 0.** In Zone 0 locations, equipment protected by intrinsic safety "ia" and equipment protected by encapsulation "ma" shall be connected using intrinsically safe "ia" circuits with wiring methods in accordance with 504.20.

This requirement is one of the most significant differences between the zone and division area classification requirements. The degree of hazard within a Zone 0 area is considered so severe that all wiring in this area must be intrinsically safe (technique "ia"). In general, only instrumentation and signaling circuits installed in accordance with Article 504 can be used in a Zone 0 area.

(B) Zone 1.

Δ **(1) General.** In Zone 1 locations, the following wiring methods shall be permitted:

> Informational Note No. 1: See Article 100 for the definition of *restricted industrial establishment [as applied to hazardous (classified) locations].*

(1) All wiring methods permitted by 505.15(A).

(2) In restricted industrial establishments where the cable is not subject to physical damage, Type MC-HL cable listed for use in Zone 1 or Class I, Division 1 locations, with a gas/vaportight continuous corrugated metallic sheath, an overall jacket of suitable polymeric material, and a separate equipment grounding conductor(s) in accordance

with 250.122. Type MC-HL cable shall be terminated with fittings listed for the application and installed in accordance with Part II of Article 330.

(3) In restricted industrial establishments where the cable is not subject to physical damage, Type ITC-HL cable listed for use in Zone 1 or Class I, Division 1 locations, with a gas/vaportight continuous corrugated metallic sheath and an overall jacket of suitable polymeric material. Type ITC-HL cable shall be terminated with fittings listed for the application and installed in accordance with 335.4

(4) Type MI cable terminated with fittings listed for Zone 1 or Class I, Division 1 locations. Type MI cable shall be installed and supported in a manner to avoid tensile stress at the termination fittings.

(5) Threaded rigid metal conduit (RMC) or threaded intermediate metal conduit (IMC), including RMC or IMC conduit systems with supplemental corrosion protection coatings.

(6) Where encased in a concrete envelope a minimum of 50 mm (2 in.) thick and provided with not less than 600 mm (24 in.) of cover measured from the top of the conduit to grade, PVC or RTRC conduit. RMC or IMC conduit shall be used for the last 600 mm (24 in.) of the underground run to emergence or to the point of connection to the aboveground raceway. An equipment grounding conductor shall be included to provide for electrical continuity of the raceway system and for grounding of non–current-carrying metal parts.

(7) Intrinsic safety type of protection "ib" using the wiring methods in accordance with 504.20.

(8) Optical fiber cable Type OFNP, Type OFCP, Type OFNR, Type OFCR, Type OFNG, Type OFCG, Type OFN, or Type OFC installed in raceways in accordance with 505.15(B). Optical fiber cable shall be sealed in accordance with 505.16.

(9) In restricted industrial establishments for applications limited to 600 volts nominal or less where the cable is not subject to physical damage, Type TC-ER-HL shall be terminated with fittings listed for the location and installed in accordance with 336.10.

> Informational Note No. 2: See ANSI/UL 2225, *Cables and Cable-Fittings for Use in Hazardous (Classified) Locations,* for information on construction, testing, and marking of cables and cable fittings.

(10) In restricted industrial establishments, listed Type P cable with metal braid armor and an overall jacket. Type P cable shall be terminated with fittings listed for the location and installed in accordance with Part II of Article 337.

> Informational Note No. 3: See UL 1309A, *Outline of Investigation for Cable for use in Mobile Installations,* for information on construction, testing, and marking of Type P cable.
>
> Informational Note No. 4: See ANSI/UL 2225, *Cables and Cable-Fittings for Use in Hazardous (Classified) Locations,* for information on construction, testing, and marking of cable fittings.

Δ **(2) Flexible Connections.** If flexibility is necessary to minimize the transmission of vibration from equipment during operation or to allow for movement after installation during maintenance, one of the following shall be permitted:

(1) Flexible fittings listed for the location.

(2) Flexible cord in accordance with 505.17(A), terminated with cord connectors listed for the location.

(3) In restricted industrial establishments for applications limited to 600 volts nominal or less, where the cable is not subject to physical damage and is terminated with fittings listed for the location, Type TC-ER-HL cable. Type TC-ER-HL cable shall be listed for use in Class I, Division 1 or Zone 1 locations and shall be installed in accordance with 336.10.

> Informational Note No. 1: See ANSI/UL 2225, *Cables and Cable-Fittings for Use in Hazardous (Classified) Locations*, for information on construction, testing, and marking of cables and cable fittings.

(4) In restricted industrial establishments listed Type P cable with metal braid armor and an overall jacket. Type P cable shall be terminated with fittings listed for the location and installed in accordance with Part II of Article 337.

> Informational Note No. 2: See UL 1309A, *Outline of Investigation for Cable for Use in Mobile Installations*, for information on construction, testing, and marking of Type P cable.

> Informational Note No. 3: See ANSI/UL 2225, *Cables and Cable-Fittings for Use in Hazardous (Classified) Locations*, for information on construction, testing, and marking of cable fittings.

"Flexible connections" refers only to the fittings. The intent of this section is not to permit a flexible wiring method. The fitting should be no longer than is needed. Limited use of flexible cord is permitted in accordance with 505.17 for specific applications where flexibility of the wiring method is made necessary by the type of equipment being supplied.

The requirement in 505.16(B)(1) to provide a seal within 2 inches of an enclosure and the requirement in 505.16(B)(2) to provide a seal within 18 inches of an enclosure apply where flexible connections are used.

(C) Zone 2.

Δ **(1) General.** In Zone 2 locations, the following wiring methods shall be permitted:

> Informational Note No. 1: See Article 100 for the definition of *restricted industrial establishment [as applied to hazardous (classified) locations]*.

(1) All wiring methods permitted by 505.15(B).

(2) Type MC, Type MV, Type TC, or Type TC-ER cable, including installation in cable tray systems. Type TC-ER shall include a separate equipment grounding conductor in addition to a drain wire that might be present. The cable shall be terminated with listed fittings. Single conductor Type MV cables shall be shielded or metallic-armored.

(3) Type ITC cable or Type ITC-ER cable as permitted in 335.4 and terminated with listed fittings. Type ITC-ER shall include a separate insulated equipment grounding conductor in addition to a drain wire.

(4) Type PLTC cable or Type PLTC-ER cable used for Class 2 or Class 3 circuits, including installation in cable tray systems. The cable shall be terminated with listed fittings. Type PLTC-ER shall include a separate insulated equipment grounding conductor in addition to a drain wire that might be present.

(5) Enclosed gasketed busways or enclosed gasketed wireways.

(6) In restricted industrial establishments and where metal conduit does not provide the corrosion resistance needed for the environment where it is installed, listed reinforced thermosetting resin conduit (RTRC), factory elbows, and associated fittings, all marked with the suffix -XW, and Schedule 80 PVC conduit, factory elbows, and associated fittings. Where seals are required for boundary conditions as defined in 505.16(C)(1)(b), the Zone 1 wiring method shall extend into the Zone 2 area to the seal, which shall be located on the Zone 2 side of the Zone 1/Zone 2 boundary.

(7) Intrinsic safety type of protection "ic" using any of the wiring methods permitted for unclassified locations. Intrinsic safety type of protection "ic" systems shall be installed in accordance with the control drawing(s). Simple apparatus, not shown on the control drawing, shall be permitted in an intrinsic safety type of protection "ic" circuit if the simple apparatus does not interconnect the intrinsic safety type of protection "ic" systems to any other circuit. Separate intrinsic safety type of protection "ic" systems shall be installed in accordance with one of the following:

   a. In separate cables
   b. In multiconductor cables where the conductors of each circuit are within a grounded metal shield
   c. In multiconductor cables where the conductors of each circuit have insulation with a minimum thickness of 0.25 mm (0.01 in.)

> Informational Note No. 2: See Article 100 for the definition of *simple apparatus*.

(8) Optical fiber cable of Type OFNP, Type OFCP, Type OFNR, Type OFCR, Type OFNG, Type OFCG, Type OFN, or Type OFC installed in cable trays or any other raceway in accordance with 505.15(C). Optical fiber cable shall be sealed in accordance with 505.16.

(9) Cablebus.

(10) In restricted industrial establishments, listed Type P cable with or without metal braid armor and an overall jacket. Type P cable shall be terminated with fittings listed for the location and installed in accordance with Part II of Article 337.

Informational Note No. 3: See UL 1309A, *Outline of Investigation for Cable for Use in Mobile Installations,* for information on construction, testing, and marking of Type P cable.

Informational Note No. 4: See ANSI/UL 2225, *Cables and Cable-Fittings for Use in Hazardous (Classified) Locations,* for information on construction, testing, and marking of cable fittings.

Δ **(2) Flexible Connections.** Where flexibility is necessary to minimize the transmission of vibration from equipment during operation or to allow for movement after installation during maintenance, one or more of the following wiring methods shall be permitted:

(1) Listed flexible metal fittings
(2) Flexible metal conduit with listed fittings
(3) Interlocked armor Type MC cable with listed fittings
(4) Type P cable
(5) Type TC-ER or Type TC-ER-HL cable
(6) Type ITC-ER or Type ITC-HL cable
(7) Type PLTC-ER cable
(8) Liquidtight flexible metal conduit with listed fittings
(9) Liquidtight flexible nonmetallic conduit with listed fittings
(10) Flexible cord in accordance with 505.17, terminated with a listed cord connector that maintains the type of protection of the terminal compartment
(11) For elevator use, an identified elevator cable of Type EO, Type ETP, or Type ETT, shown under the "use" column in Table 400.4 for "hazardous (classified) locations" and terminated with listed fittings

If flexible conduit is used, it shall be grounded in accordance with 505.30(A) and bonded in accordance with 505.30(B).

Δ **505.16 Sealing and Drainage.** Seals in conduit and cable systems shall comply with 505.16(A) through (E). Sealing compound shall be used in Type MI cable termination fittings to exclude moisture and other fluids from the cable insulation.

See also

501.15 commentary for further information regarding conduit and cable sealing for hazardous locations

Informational Note No. 1: See 505.16(C)(2)(c). Seals are provided in conduit and cable systems to minimize the passage of gases and vapors and prevent the passage of flames from one portion of the electrical installation to another through the conduit. Such communication through Type MI cable is inherently prevented by construction of the cable. Unless specifically designed and tested for the purpose, conduit and cable seals are not intended to prevent the passage of liquids, gases, or vapors at a continuous pressure differential across the seal. Even at differences in pressure across the seal equivalent to a few inches of water, there might be a slow passage of gas or vapor through a seal and through conductors passing through the seal.

Informational Note No. 2: See 505.16(D)(2). Temperature extremes and highly corrosive liquids and vapors can affect the ability of seals to perform their intended function.

Informational Note No. 3: Gas or vapor leakage and propagation of flames might occur through the interstices between the strands of standard stranded conductors larger than 2 AWG. Special conductor constructions, for example, compacted strands or sealing of the individual strands, are means of reducing leakage and preventing the propagation of flames.

(A) Zone 0. In Zone 0 locations, seals shall be located according to 505.16(A)(1), (A)(2), and (A)(3).

(1) Conduit Seals. Seals shall be provided within 3.05 m (10 ft) of where a conduit leaves a Zone 0 location. There shall be no unions, couplings, boxes, or fittings, except listed reducers at the seal, in the conduit run between the seal and the point at which the conduit leaves the location.

Exception: A rigid unbroken conduit that passes completely through the Zone 0 location with no fittings less than 300 mm (12 in.) beyond each boundary shall not be required to be sealed if the termination points of the unbroken conduit are in unclassified locations.

(2) Cable Seals. Seals shall be provided on cables at the first point of termination after entry into the Zone 0 location.

(3) Not Required to Be Explosionproof or Flameproof. Seals shall not be required to be explosionproof or flameproof.

(B) Zone 1. In Zone 1 locations, seals shall be located in accordance with 505.16(B)(1) through (B)(8).

(1) Type of Protection "d", "db", "e", or "eb" Enclosures. Conduit seals shall be provided within 50 mm (2 in.) for each conduit entering enclosures having type of protection "d", "db", "e", or "eb".

Exception No. 1: Where the enclosure having type of protection "d" or "db" is marked to indicate that a seal is not required.

Exception No. 2: For type of protection "e" or "eb", conduit and fittings employing only NPT to NPT raceway joints or fittings listed for type of protection "e" or "eb" shall be permitted between the enclosure and the seal, and the seal shall not be required to be within 50 mm (2 in.) of the entry.

Informational Note: Examples of fittings employing other than NPT threads include conduit couplings, capped elbows, unions, and breather drains.

Exception No. 3: For conduit installed between type of protection "e" or "eb" enclosures employing only NPT to NPT raceway joints or conduit fittings listed for type of protection "e" or "eb", a seal shall not be required.

(2) Explosionproof Equipment. Conduit seals shall be provided for each conduit entering explosionproof equipment according to 505.16(B)(2)(a), (B)(2)(b), and (B)(2)(c).

(a) In each conduit entry into an explosionproof enclosure where either of the following conditions apply:

(1) The enclosure contains apparatus, such as switches, circuit breakers, fuses, relays, or resistors that may produce arcs,

sparks, or high temperatures that are considered to be an ignition source in normal operation. For the purposes of this section, high temperatures shall be considered to be any temperatures exceeding 80 percent of the autoignition temperature in degrees Celsius of the gas or vapor involved.

Exception: Seals shall not be required for conduit entering an enclosure where such switches, circuit breakers, fuses, relays, or resistors comply with one of the following:

(1) Are enclosed within a chamber hermetically sealed against the entrance of gases or vapors.
(2) Are immersed in oil.
(3) Are enclosed within an enclosure, identified for the location, and marked "Leads Factory Sealed," "Factory Sealed," "Seal not Required," or equivalent.

(2) The entry is metric designator 53 (trade size 2) or larger and the enclosure contains terminals, splices, or taps.

An enclosure, identified for the location, and marked "Leads Factory Sealed," or "Factory Sealed," "Seal not Required," or equivalent shall not be considered to serve as a seal for another adjacent explosionproof enclosure that is required to have a conduit seal.

(b) Conduit seals shall be installed within 450 mm (18 in.) from the enclosure. Only threaded couplings, or explosionproof fittings such as unions, reducers, elbows, and capped elbows that are not larger than the trade size of the conduit, shall be permitted between the sealing fitting and the explosionproof enclosure.

(c) Where two or more explosionproof enclosures for which conduit seals are required under 505.16(B)(2) are connected by nipples or by runs of conduit not more than 900 mm (36 in.) long, a single conduit seal in each such nipple connection or run of conduit shall be considered sufficient if located not more than 450 mm (18 in.) from either enclosure.

Δ **(3) Pressurized Enclosures and Pressurized Rooms.** Conduit seals shall be provided in each conduit entry into a pressurized enclosure or pressurized room where the conduit is not pressurized as part of the protection system. Conduit seals shall be installed within 450 mm (18 in.) from the pressurized enclosure or pressurized room.

> Informational Note No. 1: Installing the seal as close as possible to the enclosure reduces problems with purging the dead airspace in the pressurized conduit.
> Informational Note No. 2: See NFPA 496, *Standard for Purged and Pressurized Enclosures for Electrical Equipment*, for information on pressurized equipment.
> Informational Note No. 3: See UL 60079-13, *Explosive Atmospheres — Part 13: Equipment Protection by Pressurized Room "p" and Artificially Ventilated Room "v"*, for additional information.

(4) Zone 1 Boundary. Conduit seals shall be provided in each conduit run leaving a Zone 1 location. The sealing fitting shall be permitted on either side of the boundary of such location within 3.05 m (10 ft) of the boundary and shall be designed and installed so as to minimize the amount of gas or vapor within the Zone 1 portion of the conduit from being communicated to the conduit beyond the seal. Except for listed explosionproof reducers at the conduit seal, there shall be no union, coupling, box, or fitting between the conduit seal and the point at which the conduit leaves the Zone 1 location.

Exception: Metal conduit containing no unions, couplings, boxes, or fittings and passing completely through a Zone 1 location with no fittings less than 300 mm (12 in.) beyond each boundary shall not require a conduit seal if the termination points of the unbroken conduit are in unclassified locations.

(5) Cables Capable of Transmitting Gases or Vapors. Conduits containing cables with a gas/vaportight continuous sheath capable of transmitting gases or vapors through the cable core shall be sealed in the Zone 1 location after removing the jacket and any other coverings so that the sealing compound surrounds each individual insulated conductor or optical fiber tube and the outer jacket.

Exception: Multiconductor cables with a gas/vaportight continuous sheath capable of transmitting gases or vapors through the cable core shall be permitted to be considered as a single conductor by sealing the cable in the conduit within 450 mm (18 in.) of the enclosure and the cable end within the enclosure by an approved means to minimize the entrance of gases or vapors and prevent the propagation of flame into the cable core, or by other approved methods. For shielded cables and twisted pair cables, it shall not be required to remove the shielding material or separate the twisted pair.

In addition to the conduit seal, the cable within the conduit also must be sealed to prevent gases from passing through the cable. A single conduit seal can serve both purposes — sealing the conduit and sealing the cable.

(6) Cables Incapable of Transmitting Gases or Vapors. Each multiconductor or optical multifiber cable in conduit shall be considered as a single conductor or single optical fiber tube if the cable is incapable of transmitting gases or vapors through the cable core. These cables shall be sealed in accordance with 505.16(D).

(7) Cables Entering Enclosures. Cable seals shall be provided for each cable entering flameproof or explosionproof enclosures. The seal shall comply with 505.16(D).

(8) Zone 1 Boundary. Cables shall be sealed at the point at which they leave the Zone 1 location.

Exception: Where cable is sealed at the termination point.

(C) Zone 2. In Zone 2 locations, seals shall be installed in accordance with 505.16(C)(1) and (C)(2).

Δ **(1) Conduit Seals.** Conduit seals shall be provided in accordance with 505.16(C)(1)(a) through (C)(1)(f).

(a) *Flameproof and Explosionproof Enclosures.* Conduit seals shall be required for connections to enclosures that are required to be flameproof or explosionproof, in accordance with 505.16(B)(1) and (B)(2). All portions of the conduit run or nipple between the seal and enclosure shall comply with 505.16(B).

(b) *Conduits Between Zone 2 and Unclassified Locations.* Conduit seals shall be required in each conduit run passing from a Zone 2 location into an unclassified location. The conduit seals and conduit run shall comply with all the following:

(1) The sealing fitting shall be permitted on either side of the boundary of the location within 3.05 m (10 ft) of the boundary.

(2) The sealing fitting shall be designed and installed to minimize the amount of gas or vapor within the Zone 2 portion of the conduit from being communicated to the conduit beyond the seal.

(3) Threaded rigid metal conduit (RMC) or threaded intermediate metal conduit (IMC) shall be used between the sealing fitting and the point at which the conduit leaves the Zone 2 location, and a threaded connection shall be used at the sealing fitting.

(4) There shall be no union, coupling, box, or fitting between the conduit seal and the point at which the conduit leaves the Zone 2 location except for listed explosionproof reducers at the conduit seal.

(5) Conduits shall be sealed to minimize the amount of gas or vapor within the Zone 2 portion of the conduit from being communicated to the conduit beyond the seal.

(6) Such seals shall not be required to be flameproof or explosionproof but shall be identified for the purpose of minimizing passage of gases under normal operating conditions and shall be accessible.

(c) *Conduits Passing Through a Zone 2 Location.* Metal conduit containing no unions, couplings, boxes, or fittings and passing completely through a Zone 2 location with no fittings less than 300 mm (12 in.) beyond each boundary shall not be required to be sealed if the termination points of the unbroken conduit are in unclassified locations.

(d) *Conduit Systems Ending in an Unclassified Location.* Conduit systems terminating in an unclassified location where a wiring method transition is made to cable tray, cablebus, ventilated busway, Type MI cable, or cable that is not installed in a raceway or cable tray system shall not be required to be sealed where passing from the Zone 2 location into the unclassified location. The unclassified location shall be outdoors or, if the conduit system is all in one room, it shall be permitted to be indoors. The conduits shall not terminate at an enclosure containing an ignition source in normal operation.

(e) *Pressurized Enclosures or Pressurized Rooms.* Conduit systems passing from enclosures or rooms that permit general-purpose equipment as a result of pressurization into a Zone 2 location shall not require a seal at the boundary.

Informational Note: See NFPA 496, *Standard for Purged and Pressurized Enclosures for Electrical Equipment*, for information on purged and pressurized equipment.

(f) *Outdoor Conduit System Segments.* Segments of above-ground conduit systems shall not be required to be sealed where passing from a Zone 2 location into an unclassified location if all the following conditions are met:

(1) The conduit system segment shall not pass through a Zone 0 or Zone 1 location where the conduit contains unions, couplings, boxes, or fittings within 300 mm (12 in.) of the Zone 0 or Zone 1 location.

(2) The conduit system segment shall be located entirely in an outdoor location.

(3) The conduit system segment shall not be directly connected to canned pumps, process or service connections for flow, pressure, or analysis measurement, and so forth, that depend on a single compression seal, diaphragm, or tube to prevent flammable or combustible fluids from entering the conduit system.

(4) The conduit system segment shall only have threaded rigid metal conduit (RMC) and threaded intermediate metal conduit (IMC) with threaded unions, couplings, conduit bodies, and fittings in the unclassified location.

(5) The conduit system segment shall be sealed at its entry to each enclosure or fitting housing terminals, splices, or taps in Zone 2 locations.

Δ **(2) Cable Seals.** Cable seals shall be installed in accordance with 505.16(C)(2)(a) through (C)(2)(c).

(a) *Explosionproof and Flameproof Enclosures.* Cables entering enclosures required to be flameproof or explosionproof shall be sealed at the point of entrance. The seal shall comply with 505.16(D). Multiconductor or optical multifiber cables with a gas/vaportight continuous sheath capable of transmitting gases or vapors through the cable core shall be sealed in the Zone 2 location after removing the jacket and any other coverings so that the sealing compound surrounds each individual insulated conductor or optical fiber tube to minimize the passage of gases and vapors. Multiconductor or optical multifiber cables in conduit shall be sealed as described in 505.16(B)(4).

Exception No. 1: Cables passing from an enclosure or room that is unclassified as a result of Type Z pressurization into a Zone 2 location shall not require a seal at the boundary.

Exception No. 2: Shielded cables and twisted pair cables shall not require removal of the shielding material or separation of the twisted pairs if the termination is by an approved means to minimize the entrance of gases or vapors and prevent propagation of flame into the cable core.

(b) *Restricted Breathing Enclosures "nR".* Cables entering restricted breathing enclosures required to be restricted breathing shall be sealed at the point of entrance into the enclosure. These seals shall be installed in accordance with 505.16(D).

Multiconductor cables or multifiber optical fiber cables with a gas/vaportight continuous sheath capable of transmitting gases or vapors through the cable core shall be sealed in the Zone 2 location. The jacket and any other coverings shall be removed to allow the sealing compound to surround each individual insulated conductor or optical fiber tube to minimize the passage of gases and vapors. Multiconductor cables or optical fiber cables in conduit shall be sealed as described in 505.16(C)(1)(b).

Exception No. 1: Cables passing from an enclosure or room that is unclassified as a result of Type Z pressurization into a Zone 2 location shall not require a seal at the boundary.

Exception No. 2: Shielded cables and twisted pair cables terminated with fittings listed for the location shall not require removal of the shielding material or separation of the twisted pairs.

(c) *Cables That Will Not Transmit Gases or Vapors.* Cables with a gas/vaportight continuous sheath that will not transmit gases or vapors through the cable core in excess of the quantity permitted for seal fittings shall not be required to be sealed except as required in 505.16(C)(2)(b). The minimum length of such cable run shall not be less than the length that limits gas or vapor flow through the cable core to the rate permitted for seal fittings [200 cm³/hr (0.007 ft³/hr) of air at a pressure of 1500 pascals (6 in. of water)].

Informational Note No. 1: See ANSI/UL 2225, *Cables and Cable-Fittings for Use in Hazardous (Classified) Locations*, for further information on construction, testing, and marking of cables, cable fittings, and cord connectors.

Informational Note No. 2: The cable core does not include the interstices of the conductor strands.

(d) *Cables Capable of Transmitting Gases or Vapors.* Cables with a gas/vaportight continuous sheath capable of transmitting gases or vapors through the cable core shall not be required to be sealed except as required in 505.16(C)(2)(b), unless the cable is attached to process equipment or devices that might cause a pressure in excess of 1500 pascals (6 in. of water) to be exerted at a cable end, in which case a seal, barrier, or other means shall be provided to prevent migration of flammables into an unclassified area.

Exception: Cables with an unbroken gas/vaportight continuous sheath shall be permitted to pass through a Zone 2 location without seals.

(e) *Cables Without a Gas/Vaportight Continuous Sheath.* Cables that do not have a gas/vaportight continuous sheath shall be sealed at the boundary of the Zone 2 and unclassified location to minimize the passage of gases or vapors into an unclassified location.

Informational Note: The cable sheath can be either metal or a nonmetallic material.

(D) Zones 0, 1, and 2. Where required, seals in Zones 0, 1, and 2 locations shall comply with 505.16(D)(1) through (D)(5).

(1) Fittings. Enclosures for connections or equipment shall be provided with an integral means for sealing, or sealing fittings listed for the location shall be used. Sealing fittings shall be listed for use with one or more specific compounds and shall be accessible.

(2) Compound. The compound shall provide a seal against passage of gas or vapors through the seal fitting, shall not be affected by the surrounding atmosphere or liquids, and shall not have a melting point less than 93°C (200°F).

(3) Thickness of Compounds. In a completed seal, the minimum thickness of the sealing compound shall not be less than the trade size of the sealing fitting and, in no case, less than 16 mm (⅝ in.).

Exception: Listed cable sealing fittings shall not be required to have a minimum thickness equal to the trade size of the fitting.

(4) Splices and Taps. Splices and taps shall not be made in fittings intended only for sealing with compound, nor shall other fittings in which splices or taps are made be filled with compound.

(5) Conductor or Optical Fiber Fill. The cross-sectional area of the conductors or optical fiber tubes (metallic or nonmetallic) permitted in a seal shall not exceed 25 percent of the cross-sectional area of a rigid metal conduit of the same trade size unless it is specifically listed for a higher percentage of fill.

(E) Drainage.

(1) Control Equipment. Where there is a probability that liquid or other condensed vapor may be trapped within enclosures for control equipment or at any point in the raceway system, approved means shall be provided to prevent accumulation or to permit periodic draining of such liquid or condensed vapor.

(2) Motors and Generators. Where liquid or condensed vapor may accumulate within motors or generators, joints and conduit systems shall be arranged to minimize entrance of liquid. If means to prevent accumulation or to permit periodic draining are necessary, such means shall be provided at the time of manufacture and shall be considered an integral part of the machine.

505.17 Flexible Cables, Cords and Connections.

Δ **(A) Flexible Cords, Zone 1 and Zone 2.** A flexible cord shall be permitted for connection between portable lighting equipment or other portable utilization equipment and the fixed portion of their supply circuit. Flexible cord shall also be permitted in restricted industrial establishments for any segment of the circuit where installation of one of the fixed wiring methods of 505.15(B) and (C) will not provide the flexibility needed to minimize the transmission of vibration from equipment during operation or to allow for movement after installation during maintenance operations. The flexible cord shall be protected against physical damage and be continuous for the entire length of the cord from equipment connection to equipment connection.

Where flexible cords are used, the cords shall comply with the following:

(1) Be of a type listed for extra-hard usage
(2) Contain, in addition to the conductors of the circuit, an equipment grounding conductor complying with 400.23
(3) Be connected to terminals or to supply conductors in an approved manner
(4) Be supported by clamps or by other suitable means in such a manner that there will be no tension on the terminal connections
(5) Where the flexible cord enters boxes, fittings, or enclosures that are required to be explosionproof or flameproof, be terminated with a listed cord connector that maintains the type of protection
(6) Where entering an increased safety "e" enclosure, be terminated with a listed increased safety "e" cord connector

Informational Note No. 1: See 400.10 for permitted uses of flexible cords.

Informational Note No. 2: See Article 100 for the definition of *restricted industrial establishment [as applied to hazardous (classified) locations]*.

Electric submersible pumps with means for removal without entering the wet-pit shall be considered portable utilization equipment. The extension of the flexible cord within a suitable raceway between the wet-pit and the power source shall be permitted.

Electric mixers intended for travel into and out of open-type mixing tanks or vats shall be considered portable utilization equipment.

Informational Note No. 3: See 505.18 for flexible cords exposed to liquids having a deleterious effect on the conductor insulation.

(B) Instrumentation Connections for Zone 2. To facilitate replacements, process control instruments shall be permitted to be connected through flexible cords, attachment plugs, and receptacles, provided that all of the following conditions apply:

(1) A switch listed for Zone 2 is provided so that the attachment plug is not depended on to interrupt current, unless the circuit is type "ia," "ib," or "ic" protection, in which case the switch is not required.
(2) The current does not exceed 3 amperes at 120 volts, nominal.
(3) The power-supply cord does not exceed 900 mm (3 ft), is of a type listed for extra-hard usage or for hard usage if protected by location, and is supplied through an attachment plug and receptacle of the locking and grounding type.
(4) Only necessary receptacles are provided.
(5) The receptacle carries a label warning against unplugging under load.

505.18 Conductors and Conductor Insulation.

(A) Conductors. For type of protection "e," field wiring conductors shall be copper. Every conductor (including spares) that enters Type "e" equipment shall be terminated at a Type "e" terminal.

(B) Conductor Insulation. Where condensed vapors or liquids may collect on, or come in contact with, the insulation on conductors, such insulation shall be of a type identified for use under such conditions, or the insulation shall be protected by a sheath of lead or by other approved means.

505.19 Uninsulated Exposed Parts. There shall be no uninsulated exposed parts, such as electrical conductors, buses, terminals, or components, that operate at more than 30 volts (15 volts in wet locations). These parts shall additionally be protected by type of protection "ia", "ib", or "ic" that is suitable for the location.

505.20 Equipment Requirements.

(A) Zone 0. In Zone 0 locations, only equipment specifically listed and marked as suitable for the location shall be permitted.

Exception: Intrinsically safe apparatus listed for use in Class I, Division 1 locations for the same gas, or as permitted by 505.9(B)(2), and with a suitable temperature class shall be permitted.

(B) Zone 1. In Zone 1 locations, only equipment specifically listed and marked as suitable for the location shall be permitted.

Exception No. 1: Equipment identified for use in Class I, Division 1 or listed for use in Zone 0 locations for the same gas, or as permitted by 505.9(B)(2), and with a suitable temperature class shall be permitted.

Exception No. 2: Equipment identified for Zone 1 or Zone 2 type of protection "p" shall be permitted.

Δ **(C) Zone 2.** In Zone 2 locations, only equipment specifically listed and marked as suitable for the location shall be permitted.

Exception No. 1: Equipment listed for use in Zone 0 or Zone 1 locations for the same gas, or as permitted by 505.9(B)(2), and with a suitable temperature class shall be permitted.

Exception No. 2: Equipment identified for Zone 1 or Zone 2 type of protection "p" shall be permitted.

Exception No. 3: Equipment identified for use in Class I, Division 1 or Division 2 locations for the same gas, or as permitted by 505.9(B)(2), and with a suitable temperature class shall be permitted.

Exception No. 4: In Zone 2 locations, the installation of open or nonexplosionproof or nonflameproof enclosed motors such as squirrel-cage induction motors without brushes, switching mechanisms, or similar arc-producing devices that are not identified for use in a Zone 2 location shall be permitted.

Exception No. 5: The exposed surface of space heaters used to reduce condensation of moisture during shutdown periods shall not exceed 80 percent of the autoignition temperature in degrees Celsius of the gas or vapor involved when operated at rated voltage, and the maximum space heater surface temperature [based on a 40°C or higher marked ambient] shall be permanently marked on a visible nameplate mounted on the motor. Otherwise, space heaters shall be identified for Class I, Division 2 or Zone 2 locations.

Exception No. 6: A sliding contact shaft bonding device used for the purpose of maintaining the rotor at ground potential shall be permitted where the potential discharge energy is determined to be nonincendive for the application. The shaft bonding device shall be permitted to be installed on the inside or the outside of the motor.

Informational Note No. 1: It is important to consider the temperature of internal and external surfaces that might be exposed to the flammable atmosphere.

Informational Note No. 2: It is important to consider the risk of ignition due to currents arcing across discontinuities and overheating of parts in multisection enclosures of large motors and generators. Such motors and generators might need equipotential bonding jumpers across joints in the enclosure and from enclosure to ground. Where the presence of ignitible gases or vapors is suspected, clean air purging might be needed immediately prior to and during start-up periods.

Informational Note No. 3: See IEEE STD 1349, *IEEE Guide for the Application of Electric Machines in Zone 2 and Class I, Division 2 Hazardous (Classified) Locations,* for information on the application of rotating electric machines including shaft bonding devices and potential discharge energy calculations.

Informational Notes 1 and 2 identify unique issues that apply to motors in Zone 2 locations. High-inertia loads can cause increased rotor heating during starting. In addition, sparking can occur between motor housing assemblies during starting.

(D) Materials. Equipment marked Group IIC shall be permitted for applications requiring Group IIA or Group IIB equipment. Similarly, equipment marked Group IIB shall be permitted for applications requiring Group IIA equipment.

Equipment marked for a specific gas or vapor shall be permitted for applications where the specific gas or vapor may be encountered.

Informational Note: One common example combines these markings with equipment marked IIB +H2. This equipment is suitable for applications requiring Group IIA equipment, Group IIB equipment, or equipment for hydrogen atmospheres.

(E) Manufacturer's Instructions. Electrical equipment installed in hazardous (classified) locations shall be installed in accordance with the instructions (if any) provided by the manufacturer.

Δ **505.22 Increased Safety "e" Motors and Generators.** In Zone 1 locations, increased safety "e" motors and generators of all voltage ratings shall be listed for Zone 1 locations, and shall comply with all of the following:

(1) Motors shall be marked with the current ratio, I_A/I_N, and time, t_E.

(2) Motors shall have controllers marked with the model or identification number, output rating (horsepower or kilowatt), full-load amperes, starting current ratio (I_A/I_N), and time (t_E) of the motors that they are intended to protect; the controller marking shall also include the specific overload protection type (and setting, if applicable) that is listed with the motor or generator.

(3) Connections shall be made with the specific terminals listed with the motor or generator.

(4) Terminal housings shall be permitted to be of substantial, nonmetallic, nonburning material, provided an internal grounding means between the motor frame and the equipment grounding connection is incorporated within the housing.

(5) The provisions of Part III of Article 430 shall apply regardless of the voltage rating of the motor.

(6) The motors shall be protected against overload by a separate overload device that is responsive to motor current. This device shall be selected to trip or shall be rated in accordance with the listing of the motor and its overload protection.

(7) Sections 430.32(C) and 430.44 shall not apply to such motors.

(8) The motor overload protection shall not be shunted or cut out during the starting period.

Informational Note: See ANSI/UL 122001, *General Requirements for Electrical Ignition Systems for Internal Combustion Engines in Class I, Division 2 or Zone 2, Hazardous (Classified) Locations,* for information on ignition systems for reciprocating engines installed in Zone 2 hazardous (classified) locations. Reciprocating engine–driven generators, compressors, and other equipment installed in Zone 2 locations might present a risk of ignition of flammable materials associated with fuel, starting, and compression due to inadvertent release or equipment malfunction by the engine ignition system and controls.

Δ **505.26 Process Sealing.** Process-connected equipment including, but not limited to, canned pumps, submersible pumps, and flow, pressure, temperature, or analysis measurement instruments shall be sealed with a process seal to prevent the migration of process fluids from the designed containment into the external electrical system. Process-connected electrical equipment that incorporates a single process seal, such as a single compression seal, diaphragm, or tube to prevent flammable or combustible fluids from entering a conduit or cable system capable of transmitting fluids, shall be provided with an additional means to mitigate a single process seal failure. The additional means might include, but is not limited to, the following:

(1) A suitable barrier meeting the process temperature and pressure conditions to which the barrier is subjected upon failure of the single process seal. There shall be a vent or drain between the single process seal and the suitable

barrier. Indication of the single process seal failure shall be provided by visible leakage, an audible whistle, or other means of monitoring.

(2) A listed Type MI cable assembly, rated at not less than 125 percent of the process pressure and not less than 125 percent of the maximum process temperature (in degrees Celsius), installed between the cable or conduit and the single process seal.

(3) A drain or vent located between the single process seal and a conduit or cable seal. The drain or vent shall be sufficiently sized to prevent overpressuring the conduit or cable seal above 6 in. water column (1493 Pa). Indication of the single process seal failure shall be provided by visible leakage, an audible whistle, or other means of monitoring.

(4) An add-on secondary seal marked "secondary seal" and rated for the pressure and temperature conditions to which it will be subjected upon failure of the single process seal.

Process-connected electrical equipment that does not rely on a single process seal or is listed and marked "single seal", "dual seal", or "dual seal without annunciation" shall not be required to be provided with an additional means of sealing.

Process-connected electrical equipment marked "single seal — install conduit or cable seal" shall be sealed in accordance with 505.16.

> Informational Note: See ANSI/UL 122701, *Requirements for Process Sealing Between Electrical Systems and Flammable or Combustible Process Fluids*, for construction and testing requirements for process sealing of listed and marked single seal, dual seal, or secondary seal equipment.

The requirements for process sealing clarify the sealing, venting, and primary seal failure indication methods for the additional process seal.

See also

501.17 commentary for further information on process sealing

Δ **505.30 Grounding and Bonding.** Regardless of the voltage of the electrical system, wiring systems and equipment shall comply with 505.30(A) and (B).

N **(A) Grounding.** Wiring systems and equipment shall be grounded in accordance with Part I and Part VI of Article 250, as applicable.

Δ **(B) Bonding.** Wiring systems and equipment shall be bonded in accordance with Part I and Part V of Article 250, as applicable, and 505.30(B)(1) and (B)(2).

N **(1) Specific Bonding Means.** Bonding shall comply with 505.30(B)(1)(a) and (B)(1)(b).

(a) The locknut-bushing and double-locknut types of contacts shall not be depended on for bonding purposes, but bonding jumpers with identified fittings or other approved means

of bonding shall be used. These bonding means shall apply to all metal raceways, fittings, boxes, cable trays, and enclosures, and other parts of raceway systems between hazardous (classified) locations and the point of grounding for service equipment or point of grounding for a separately derived system. Metal struts, angles, or channels provided for support and mechanical or physical protection as permitted in 335.4(5), 336.10(7)(c), or 722.135(C) shall be bonded in accordance with 250.102.

(b) Where the branch-circuit overcurrent protection is located on the load side of the disconnecting means, the specific bonding means shall be permitted to end at the nearest point where the grounded circuit conductor and the grounding electrode conductor are connected together on the line side of the building or structure disconnecting means as specified in 250.32(B).

The grounding and bonding requirements for the zone classification system are identical to those for the division classification system.

See also

501.30 and its commentary for information on grounding and bonding requirements in Class I locations

N **(2) Flexible Metal Conduit and Liquidtight Flexible Metal Conduit.** Flexible metal conduit and liquidtight flexible metal conduit shall comply with 505.30(B)(2)(a) and (B)(2)(b).

(a) Flexible metal conduit and liquidtight flexible metal conduit shall include an equipment bonding jumper of the wire type in accordance with 250.102.

(b) In Zone 2 locations, the bonding jumper shall not be required where all of the following conditions are met:

(1) Listed liquidtight flexible metal conduit 1.8 m (6 ft) or less in length, with fittings listed for grounding, is used.
(2) Overcurrent protection in the circuit is limited to 10 amperes or less.
(3) The load is part of a meter, instrument, or relay circuit.

ARTICLE
506
Zone 20, 21, and 22 Locations for Combustible Dusts or Ignitible Fibers/Flyings

Δ **506.1 Scope.**

Zones 20, 21, and 22 are analogous to both Class II and III, Division 1 and 2 hazardous locations in Articles 502 and 503 in order to address combustible dusts and ignitible fibers/flyings. Hazardous locations containing combustible dusts or ignitible fibers/flyings are not subdivided into classes of material under the zone method. However, similar to the division classification scheme, combustible dusts and ignitible fibers are placed into groups (IIIC, IIIB, and IIIA).

N **(A) Covered.** This article covers the requirements for the zone classification system for electrical and electronic equipment and wiring for all voltages where fire and explosion hazards might exist due to combustible dusts, combustible fibers/flyings, or ignitible fibers/flyings for the following:

(1) Zone 20 hazardous (classified) locations
(2) Zone 21 hazardous (classified) locations
(3) Zone 22 hazardous (classified) locations

> Informational Note No. 1: See 505.20 or 505.22 for Zone 0, Zone 1, or Zone 2 hazardous (classified) locations where fire or explosion hazards might exist due to flammable gases, flammable vapors, or flammable liquids.
>
> Informational Note No. 2: Zone 20, Zone 21, and Zone 22 area classifications are based on the modified IEC area classification system as defined in ANSI/ISA 60079-10-2 (12.10.05), *Explosive Atmospheres — Part 10-2: Classification of Areas — Combustible Dust Atmospheres*.
>
> Informational Note No. 3: See NFPA 499, *Recommended Practice for the Classification of Combustible Dusts and of Hazardous (Classified) Locations for Electrical Installations in Chemical Process Areas*, for information regarding classification of hazardous (classified) locations using Zone methodology.

N **(B) Not Covered.** This article does not cover electrical and electronic equipment and wiring of all voltages in the following:

(1) Class I, Class II, or Class III, Division 1 or Division 2 hazardous (classified) locations.
(2) Zone 0, Zone 1, or Zone 2 hazardous (classified) locations.
(3) Locations subject to the unique risk and explosion hazards associated with explosives, pyrotechnics, or blasting agents.
(4) Locations where pyrophoric materials are the only materials used or handled.
(5) Features of equipment that involve nonelectrical potential sources of ignition (e.g., couplings, pumps, gearboxes, brakes, hydraulic and pneumatic motors, fan, engine, compressor).

> Informational Note No. 1: Common nonelectrical potential sources of ignition include hot surfaces and mechanically generated sparks.
>
> Informational Note No. 2: See ANSI/UL 80079-36, *Explosive Atmospheres — Part 36: Non-Electrical Equipment for Explosive Atmospheres — Basic Method and Requirements,* and ANSI/UL 80079-37, *Explosive Atmospheres — Part 37: Non-Electrical Equipment for Explosive Atmospheres — Non-Electrical Type of Protection Constructional Safety "c" Control of Ignition Source "b", Liquid Immersion "k",* for additional information.

Δ **506.4 Documentation.** Areas designated as hazardous (classified) or unclassified locations shall be documented on an area classification drawing and other associated documentation. This documentation shall be made available to the AHJ and to those authorized to design, install, inspect, maintain, or operate electrical equipment.

> Informational Note No. 1: See ANSI/UL 60079-28, *Explosive Atmospheres — Part 28: Protection of equipment and*

transmission systems using optical radiation, for information concerning the installation of equipment using optical emissions technology (such as laser equipment) that could potentially become an ignition source in hazardous (classified) locations.

> Informational Note No. 2: See IEC/IEEE 60079-30-2, *Explosive atmospheres — Part 30-2: Electrical Resistance Trace Heating — Application Guide for Design, Installation and Maintenance*, for information on electrical resistance trace heating for hazardous (classified) locations.
>
> Informational Note No. 3: See IEEE 844.2/CSA C293.2, *IEEE/CSA Standard for Skin Effect Trace Heating of Pipelines, Vessels, Equipment, and Structures — Application Guide for Design, Installation, Testing, Commissioning, and Maintenance*, for information on electric skin effect trace heating for hazardous (classified) locations.
>
> Informational Note No. 4: See IEEE 844.4/CSA C293.4, *IEEE/CSA Standard for Impedance Heating of Pipelines and Equipment — Application Guide for Design, Installation, Testing, Commissioning, and Maintenance*, for information on electric impedance heating for hazardous (classified) locations.

506.5 Classification of Locations.

Δ **(A) Classifications of Locations.** Locations shall be classified on the basis of the properties of the combustible dust, combustible fibers/flyings, or ignitible fibers/flyings that might be present, and the likelihood that a combustible or ignitible concentration or quantity is present. Each room, section, or area shall be considered individually in determining its classification.

Δ **(B) Zone 20, Zone 21, and Zone 22 Locations.** Zone 20, Zone 21, and Zone 22 locations are those in which combustible dust, combustible fibers/flyings, or ignitible fibers/flyings are or might be present in the air or in layers, in quantities sufficient to produce explosible or ignitible mixtures. Zone 20, Zone 21, and Zone 22 locations shall include those specified in 506.5(B)(1), (B)(2), and (B)(3).

> Informational Note: Through the exercise of ingenuity in the layout of electrical installations for hazardous (classified) locations, it is frequently possible to locate much of the equipment in a reduced level of classification to reduce the amount of special equipment required.

(1) Zone 20. A Zone 20 location is a location where one of the following apply:

(1) Ignitible concentrations of combustible dust, combustible fibers/flyings, or ignitible fibers/flyings are present continuously or for long periods of time.
(2) Group IIIC combustible dusts are present in hazardous quantities continuously or for long periods of time.

Δ **(2) Zone 21.** A Zone 21 location is a location where one of the following apply:

(1) Ignitible concentrations of combustible dust, combustible fibers/flyings, or ignitible fibers/flyings are likely to exist occasionally under normal operating conditions.
(2) Ignitible concentrations of combustible dust, combustible fibers/flyings, or ignitible fibers/flyings might exist

frequently because of repair or maintenance operations or because of leakage.

(3) Equipment is operated or processes are carried on of such a nature that equipment breakdown or faulty operations could result in the release of ignitible concentrations of combustible dust, combustible fibers/flyings, or ignitible fibers/flyings and also cause simultaneous failure of electrical equipment in a mode to cause the electrical equipment to become a source of ignition.

(4) The location is adjacent to a Zone 20 location from which ignitible concentrations of combustible dust, combustible fibers/flyings, or ignitible fibers/flyings could be communicated.

Exception: When communication from an adjacent Zone 20 location is minimized by adequate positive pressure ventilation from a source of clean air, and effective safeguards against ventilation failure are provided.

(5) Group IIIC combustible dusts are present in hazardous quantities occasionally, under normal or abnormal operating conditions, or frequently because of repair or maintenance operations or because of leakage.

Informational Note No. 1: See ANSI/ISA 60079-10-2 (12.10.05), *Explosive Atmospheres — Part 10-2: Classification of Areas — Combustible Dust Atmospheres*, regarding the classification of Zone 21 locations.

Informational Note No. 2: This classification usually includes the following:

(1) Locations outside dust containment and in the immediate vicinity of access doors subject to frequent removal or opening for operation purposes when internal combustible mixtures are present

(2) Locations outside dust containment in the proximity of filling and emptying points, feed belts, sampling points, truck dump stations, belt dump over points, and so on, where no measures are employed to prevent the formation of combustible mixtures

(3) Locations outside dust containment where dust accumulates and where, due to process operations, the dust layer is likely to be disturbed and form combustible mixtures

(4) Locations inside dust containment where explosible dust clouds are likely to occur (but neither continuously, nor for long periods, nor frequently), for example, silos (if filled and/or emptied only occasionally) and the dirty side of filters if large self-cleaning intervals are occurring

Δ **(3) Zone 22.** A Zone 22 location is a location where one of the following apply:

(1) Ignitible concentrations of combustible dust, combustible fibers/flyings, or ignitible fibers/flyings are not likely to occur in normal operation and, if they do occur, will only persist for a short period.

(2) Combustible dust, combustible fibers/flyings, or ignitible fibers/flyings are handled, processed, or used, but the

dust or fibers/flyings are normally confined within closed containers of closed systems from which they can escape only as a result of the abnormal operation of the equipment with which the dust or fibers/flyings are handled, processed, or used.

(3) The location is adjacent to a Zone 21 location, from which ignitible concentrations of combustible dust, combustible fibers/flyings, or ignitible fibers/flyings could be communicated.

Exception No. 1: When communication from an adjacent Zone 21 location is minimized by adequate positive pressure ventilation from a source of clean air, and effective safeguards against ventilation failure are provided.

Exception No. 2: For Group IIIC combustible dusts or metal combustible fibers/flyings, there shall only be Zone 20 or 21 locations.

Informational Note No. 1: See ANSI/ISA 60079-10-2 (12.10.05), *Explosive Atmospheres — Part 10-2: Classification of Areas — Combustible Dust Atmospheres*, regarding the classification of Zone 22 locations.

Informational Note No. 2: Zone 22 locations usually include the following:

(1) Outlets from bag filter vents (in the event of a malfunction, there can be emission of combustible mixtures)

(2) Locations near equipment that has to be opened at infrequent intervals or equipment that from experience can easily form leaks where, due to pressure above atmospheric, dust will blow out

(3) Pneumatic equipment or flexible connections that can become damaged

(4) Storage locations for bags containing dusty product (failure of bags can occur during handling, causing dust leakage)

(5) Locations where controllable dust layers are formed that are likely to be raised into explosible dust–air mixtures

Only if the layer is removed by cleaning before hazardous dust–air mixtures can be formed is the area designated unclassified.

Informational Note No. 3: Protective measures to reduce the formation of explosible dust–air mixtures can often result in a Zone 21 location being classified as a Zone 22 location, or possibly unclassified. Such measures include local exhaust ventilation.

Δ **506.6 Material Groups.** For the purposes of testing, approval, and area classification, various air mixtures (not oxygen enriched) shall be grouped as follows:

(1) Group IIIC: Combustible metal dust, including combustible metal fibers/flyings. [**499**:3.3.8.2.1]

(2) Group IIIB: Combustible dust other than combustible metal dust. [**499**:3.3.8.2.2]

(3) Group IIIA: Combustible fibers/flyings or ignitible fibers/flyings other than metal. [**499**:3.3.8.2.3]

Informational Note No. 1: Group IIIA materials are larger particle-size Group IIIB materials and do not include metal dust or metal fibers/flyings. [**499**:A.3.3.8.2.3]

Informational Note No. 2: Examples of ignitible fibers/flyings include rayon, cotton (including cotton linters and cotton waste), sisal, jute, hemp, cocoa fiber, oakum, and baled waste kapok.

Informational Note No. 3: Combustible fibers/flyings include flat platelet-shaped particulates, such as metal flakes, and fibrous board, such as particle board.

The zone classification system includes both dusts and combustible fibers and flyings. The grouping of material by hazard also differs from the system used in Articles 500, 502, and 503. Commentary Table 506.1 contrasts the classification in the two systems.

COMMENTARY TABLE 506.1 Comparison of Zone and Division Classification Systems

Hazard	Zone Classification	Division Classification
Combustible metal dusts, fibers, and flyings	Group IIIC	Class II, Group E
Coal, coke, and other carbonaceous dusts	Group IIIB	Class II, Group F
Combustible dusts other than combustible metal dust	Group IIIB	Class II, Group G
Combustible fibers and flyings other than metals	Group IIIA	Class III

506.7 Special Precaution. This article shall require equipment construction and installation that ensures safe performance under conditions of proper use and maintenance.

(A) Implementation of Zone Classification System. Classification of areas, engineering and design, selection of equipment and wiring methods, installation, and inspection shall be performed by qualified persons.

(B) Dual Classification. In instances of areas within the same facility classified separately, Zone 22 locations shall be permitted to abut, but not overlap, Class II or Class III, Division 2 locations. Zone 20 or Zone 21 locations shall not abut Class II or Class III, Division 1 or Division 2 locations.

(C) Reclassification Permitted. A Class II or Class III, Division 1 or Division 2 location shall be permitted to be reclassified as a Zone 20, Zone 21, or Zone 22 location if all of the space that is classified because of a single combustible dust, combustible fiber/flying, or ignitible fiber/flying source is reclassified under the requirements of this article.

(D) Simultaneous Presence of Flammable Gases and Combustible Dusts or Fibers/Flyings. Where flammable gases, combustible dusts, combustible fibers/flyings, or ignitible fibers/flyings are or might be present at the same time, the simultaneous presence shall be considered during the selection and installation of the electrical equipment and the wiring methods, including the determination of the safe operating temperature of the electrical equipment.

506.8 Protection Techniques. Acceptable protection techniques for electrical and electronic equipment in hazardous (classified) locations shall be as described in 506.8(A) through (P).

Informational Note No. 1: See ANSI/UL 120101, *Definitions and Information Pertaining to Electrical Equipment in Hazardous Locations*; and ANSI/UL 60079–0, *Explosive Atmospheres — Part 0: Equipment — General Requirements*, for additional information.

Informational Note No. 2: See Chapter 9, Table 13 for descriptions of subdivisions of protection techniques.

(A) Dust Ignitionproof. This protection technique shall be permitted for equipment in Zone 20, Zone 21, and Zone 22 locations for which it is identified.

(B) Pressurized. This protection technique shall be permitted for equipment in Zone 21 and Zone 22 locations for which it is identified.

(C) Intrinsic Safety. This protection technique shall be permitted for equipment in Zone 20, Zone 21, and Zone 22 locations for which it is identified.

(D) Dusttight. This protection technique shall be permitted for equipment in Zone 22 locations for which it is identified.

(E) Protection by Encapsulation "m". This protection technique shall be permitted for equipment in Zone 20, Zone 21, and Zone 22 locations for which it is identified.

(F) Nonincendive Equipment. This protection technique shall be permitted for equipment in Zone 22 locations for which it is identified.

(G) Protection by Enclosure "t". This protection technique shall be permitted for equipment in Zone 20, Zone 21, and Zone 22 locations for which it is identified.

(H) Protection by Pressurized Enclosure "p". This protection technique shall be permitted for equipment in Zone 21 and Zone 22 locations for which it is identified.

(I) Protection by Intrinsic Safety "i". This protection technique shall be permitted for equipment in Zone 20, Zone 21, and Zone 22 locations for which it is listed.

(J) Protection by Electrical Resistance Trace Heating "60079-30-1". This protection technique shall be permitted for electrical resistance trace heating equipment in Zone 21 or Zone 22 for which it is listed.

(K) Inherently Safe Optical Radiation "op is". This protection technique shall be permitted for equipment in Zone 20, 21, or 22 locations for which the equipment is identified.

(L) Protected Optical Radiation "op pr". This protection technique shall be permitted for equipment in Zone 21 or 22 locations for which the equipment is identified.

(M) Optical System with Interlock "op sh". This protection technique shall be permitted for equipment in Zone 20, 21, or 22 locations for which the equipment is identified.

(N) Protection by Skin Effect Trace Heating "IEEE 844.1". This protection technique shall be permitted for skin effect trace heating equipment in Zone 21 or Zone 22 for which it is listed.

N **(O) Pressurized Room "p".** This protection technique shall be permitted in Zone 21 and Zone 22 locations for which it is identified.

N **(P) Special Protection "s".** This protection technique shall be permitted for equipment in Zone 20, Zone 21, or Zone 22 locations for which they are listed.

506.9 Equipment Requirements.

(A) Suitability. Suitability of identified equipment shall be determined by one of the following:

(1) Equipment listing or labeling
(2) Evidence of equipment evaluation from a qualified testing laboratory or inspection agency concerned with product evaluation
(3) Evidence acceptable to the authority having jurisdiction such as a manufacturer's self-evaluation or an owner's engineering judgment

Informational Note: Additional documentation for equipment might include certificates demonstrating compliance with applicable equipment standards, indicating special conditions of use, and other pertinent information.

(B) Listing. Equipment that is listed for Zone 20 shall be permitted in a Zone 21 or Zone 22 location of the same combustible dust, combustible fiber/flyings, or ignitible fiber/flyings. Equipment that is listed for Zone 21 shall be permitted in a Zone 22 location of the same combustible dust, combustible fiber/flyings, or ignitible fiber/flyings.

(C) Marking.

(1) Division Equipment. Equipment identified for Class II, Division 1, Class II, Division 2, Class III, Division 1, or Class III, Division 2 shall, in addition to being marked in accordance with 500.8(C), be permitted to be marked with all of the following:

(1) Zone 20, 21, or 22 (as applicable)
(2) Material group in accordance with 506.6
(3) Maximum surface temperature in accordance with 506.9(D), marked as a temperature value in degrees C, preceded by "T" and followed by the symbol "°C"

Δ **(2) Zone Equipment.** Equipment meeting one or more of the protection techniques described in 506.8 shall be marked with the following in the order shown:

(1) Zone in accordance with Chapter 9, Table 13
(2) Symbol "AEx"

The symbol AEx identifies the equipment as meeting American national standards. In European Union countries, the symbol EEx was used historically but no longer appears in current European standards. Only equipment marked AEx has been evaluated for use in electrical systems and hazardous locations covered by the *NEC®*.

(3) Protection technique(s) in accordance with Chapter 9, Table 13
(4) Material group in accordance with 506.6
(5) Maximum surface temperature in accordance with 506.9(D), marked as a temperature value in degrees Celsius, preceded by "T" and followed by the symbol "°C"
(6) Ambient temperature marking in accordance with 506.9(D)
(7) Equipment protection level (EPL)

Informational Note: EPLs are designated as G for gas, or D for dust, and are then followed by a letter (a, b, or c) to give the user a better understanding as to whether the equipment provides (a) a "very high," (b) a "high," or (c) an "enhanced" level of protection against ignition of an explosive atmosphere. For example, a Zone 21 AEx pb IIIB T165°C Db motor is marked with an EPL of "Db".

Exception: Associated apparatus NOT suitable for installation in a hazardous (classified) location shall be required to be marked only with 506.9(C)(2)(2) and (C)(2)(3), and where applicable (C)(2)(4), but BOTH the symbol AEx in 506.9(C)(2)(2) and the symbol for the type of protection in 506.9(C)(2)(3) shall be enclosed within the same square brackets; for example, [AEx ia] IIIC.

(D) Temperature Classifications. Equipment shall be marked to show the maximum surface temperature referenced to a 40°C ambient, or at the higher marked ambient temperature if the equipment is rated and marked for an ambient temperature of greater than 40°C. For equipment installed in a Zone 20 or Zone 21 location, the operating temperature shall be based on operation of the equipment when blanketed with the maximum amount of dust (or with dust-simulating fibers/flyings) that can accumulate on the equipment. Electrical equipment designed for use in the ambient temperature range between −20°C and +40°C shall require no additional ambient temperature marking. Electrical equipment that is designed for use in a range of ambient temperatures other than −20°C and +40°C is considered to be special, and the ambient temperature range shall then be marked on the equipment, including either the symbol "Ta" or "Tamb" together with the special range of ambient temperatures.

Informational Note: As an example, such a marking might be "−30°C ≤ Ta ≤ +40°C."

Exception No. 1: Equipment of the non–heat-producing type, such as conduit fittings, shall not be required to have a marked operating temperature.

Exception No. 2: Equipment identified for Class II, Division 1 or Class II, Division 2 locations as permitted by 506.20(B) and (C) shall be permitted to be marked in accordance with 500.8(C) and Table 500.8(C)(4).

(E) Threading. The supply connection entry thread form shall be NPT or metric. Conduit and fittings shall be made wrenchtight to prevent sparking when the fault current flows through the conduit system and to ensure the integrity of the conduit system. Equipment provided with threaded entries for field wiring connections shall be installed in accordance with 506.9(E)(1) or (E)(2) and with (E)(3).

(1) Equipment Provided with Threaded Entries for NPT-Threaded Conduit or Fittings. For equipment provided with threaded entries for NPT-threaded conduit or fittings, listed conduit fittings or listed cable fittings shall be used. All NPT-threaded conduit and fittings shall be threaded with a National (American) Standard Pipe Taper (NPT) thread.

> Informational Note: See ASME B1.20.1, *Pipe Threads, General Purpose (Inch),* for thread specifications for NPT threads.

(2) Equipment Provided with Threaded Entries for Metric-Threaded Fittings. For equipment with metric-threaded entries, listed conduit fittings or listed cable fittings shall be used. Such entries shall be identified as being metric, or listed adapters to permit connection to conduit or NPT-threaded fittings shall be provided with the equipment and shall be used for connection to conduit or NPT-threaded fittings. Metric-threaded fittings installed into equipment entries shall be made up with at least five threads fully engaged.

(3) Unused Openings. All unused openings shall be closed with blanking elements or close-up plugs that are listed for the location and will maintain the type of protection. Thread engagement shall comply with the requirements of 506.9(E)(1) or (E)(2).

Δ **(F) Optical Fiber Cables.** An optical fiber cable, with or without current-carrying conductors (hybrid optical fiber cable), shall be installed to address the associated fire hazard and sealed to address the associated explosion hazard in accordance with 506.15 and 506.16.

The requirements for fiber optic cables with conductors capable of carrying current (including those that are grounded) are required to follow the general wiring and sealing methods for Zone 20, 21, or 22 locations.

Δ **(G) Equipment Involving Optical Radiation.** For equipment involving sources of optical radiation (such as laser or LED sources) in the wavelength range from 380 nm to 10 μm, the risk of ignition from optical radiation shall be considered for all electrical parts and circuits that might be exposed to the radiation, both inside and outside the optical equipment. This includes optical equipment, which itself is located outside the explosive atmosphere but its emitted optical radiation enters such atmospheres.

> Informational Note: See ANSI/UL 60079-28, *Explosive Atmospheres — Part 28: Protection of Equipment and Transmission Systems Using Optical Radiation,* for information on types of

protection that can be applied to minimize the risk of ignition in explosive atmospheres from optical radiation.

Exception: All luminaires (fixed, portable, or transportable) and hand lights intended to be supplied by mains (with or without galvanic isolation) or powered by batteries, with any continuous divergent light source, including LEDs, shall be excluded from this requirement.

506.15 Wiring Methods. Wiring methods shall maintain the integrity of the protection techniques and shall comply with 506.15(A), (B), or (C).

> Informational Note: See Article 100 for the definition of *restricted industrial establishment [as applied to hazardous (classified) locations].*

Δ **(A) Zone 20.** In Zone 20 locations, the following wiring methods shall be permitted:

(1) Threaded rigid metal conduit (RMC) or threaded intermediate metal conduit (IMC).

(2) Type MI cable terminated with fittings listed for the location. Type MI cable shall be installed and supported in a manner to avoid tensile stress at the termination fittings.

Exception No. 1: Type MI cable and fittings listed for Class II, Division 1 locations shall be permitted to be used.

Exception No. 2: Equipment identified as intrinsically safe "ia" shall be permitted to be connected using the wiring methods identified in 504.20.

(3) In restricted industrial establishments, Type MC-HL cable listed for use in Zone 20 locations, with a continuous corrugated metallic sheath, an overall jacket of suitable polymeric material, and a separate equipment grounding conductor(s) in accordance with 250.122, and terminated with fittings listed for the application. Type MC-HL cable shall be installed in accordance with Part II of Article 330.

Exception: Type MC-HL cable and fittings listed for Class II, Division 1 locations shall be permitted to be used.

(4) In restricted industrial establishments, and where the cable is not subject to physical damage, Type ITC-HL cable listed for use in Zone 1 or Class I, Division 1 locations, with a gas/vaportight continuous corrugated metallic sheath and an overall jacket of suitable polymeric material, and terminated with fittings listed for the application. Type ITC-HL cable shall be installed in accordance with 335.4.

(5) Fittings and boxes shall be identified for use in Zone 20 locations.

Exception: Boxes and fittings listed for Class II, Division 1 locations shall be permitted to be used.

(6) If flexible connections are necessary, liquidtight flexible metal conduit (LFMC) with listed fittings, liquidtight

flexible nonmetallic conduit (LFNC) with listed fittings, or flexible cord listed for extra-hard usage and provided with listed fittings. Where flexible cords are used, they shall also comply with 506.17 and be terminated with a listed cord connector that maintains the type of protection of the terminal compartment. If flexible connections are subject to oil or other corrosive conditions, the insulation of the conductors shall be of a type listed for the condition or be protected by means of a suitable sheath.

Exception No. 1: Liquidtight flexible conduit (LFMC or LFNC), flexible conduit fittings, and cord fittings listed for Class II, Division 1 locations shall be permitted.

Exception No. 2: For elevator use, an identified elevator cable of Type EO, Type ETP, or Type ETT, shown under the "use" column in Table 400.4 for "hazardous (classified) locations," and terminated with listed connectors that maintain the type of protection of the terminal compartment shall be permitted.

Informational Note No. 1: See 506.30 for grounding requirements where flexible conduit is used.

Informational Note No. 2: See ANSI/UL 2225, *Cables and Cable-Fittings for Use in Hazardous (Classified) Locations*, for information on construction, testing, and marking of cables, cable fittings, and cord connectors.

(7) Optical fiber cable Type OFNP, Type OFCP, Type OFNR, Type OFCR, Type OFNG, Type OFCG, Type OFN, or Type OFC installed in raceways in accordance with 506.15(A). Optical fiber cables shall be sealed in accordance with 506.16.

Δ **(B) Zone 21.** In Zone 21 locations, the following wiring methods shall be permitted:

(1) All wiring methods permitted in 506.15(A)
(2) Fittings and boxes that are dusttight, that are provided with threaded bosses for connection to conduit, and in which taps, joints, or terminal connections are not made and are not used in locations where metal dust is present

Informational Note: See ANSI/UL 2225, *Cables and Cable-Fittings for Use in Hazardous (Classified) Locations*, for information on construction, testing, and marking of cables, cable fittings, and cord connectors.

Exception: Equipment identified as intrinsically safe "ib" shall be permitted to be connected using the wiring methods identified in 504.20.

Δ **(C) Zone 22.** In Zone 22 locations, the following wiring methods shall be permitted:

(1) All wiring methods permitted in 506.15(B).
(2) Rigid metal conduit (RMC) or intermediate metal conduit (IMC) with listed threaded or threadless fittings.
(3) Electrical metallic tubing (EMT) or dusttight wireways.
(4) Type MC or Type MI cable with listed termination fittings.

(5) Type PLTC cable or Type PLTC-ER cable used in Class 2 or Class 3 circuits, including installation in cable tray systems. The cable shall be terminated with listed fittings. Type PLTC-ER cable shall include an equipment grounding conductor in addition to a drain wire that might be present.
(6) Type ITC cable or Type ITC-ER cable as permitted in 335.4 and terminated with listed fittings. Type ITC-ER cable shall include an equipment grounding conductor in addition to a drain wire.
(7) Type MV, Type TC, or Type TC-ER cable, including installation in cable tray systems. Type TC-ER cable shall include an equipment grounding conductor in addition to a drain wire that might be present. The cable shall be terminated with listed fittings.
(8) Intrinsic safety type of protection "ic" using any of the wiring methods permitted for unclassified locations. Intrinsic safety type of protection "ic" systems shall be installed in accordance with the control drawing(s). Simple apparatus, not shown on the control drawing, shall be permitted in a circuit of intrinsic safety type of protection "ic", provided that the simple apparatus does not interconnect the intrinsic safety type of protection "ic" circuit to any other circuit. Separation of circuits of intrinsic safety type of protection "ic" shall be in accordance with one of the following:

a. Be in separate cables
b. Be in multiconductor cables where the conductors of each circuit are within a grounded metal shield
c. Be in multiconductor cables where the conductors have insulation with a minimum thickness of 0.25 mm (0.01 in.)

Informational Note: See Article 100 for the definition of *simple apparatus*.

(9) Boxes and fittings shall be dusttight.
(10) Optical fiber cable Type OFNP, Type OFCP, Type OFNR, Type OFCR, Type OFNG, Type OFCG, Type OFN, or Type OFC installed in cable trays or any raceway in accordance with 506.15(C). Optical fiber cables shall be sealed in accordance with 506.16.
(11) Cablebus.

506.16 Sealing. Where necessary to protect against the ingress of combustible dust, combustible fibers/flyings, or ignitible fibers/flyings, or to maintain the type of protection, seals shall be provided. The seal shall be identified as capable of preventing the ingress of combustible dust, combustible fiber/flying, or ignitible fiber/flying and maintaining the type of protection but need not be explosionproof or flameproof.

Δ **506.17 Flexible Cords.** Flexible cords used in Zone 20, Zone 21, and Zone 22 locations shall comply with all of the following:

(1) Be of a type listed for extra-hard usage

(2) Contain, in addition to the conductors of the circuit, an equipment grounding conductor complying with 400.23

(3) Be connected to terminals or to supply conductors in an approved manner

(4) Be supported by clamps or by other suitable means in such a manner to minimize tension on the terminal connections

(5) Be terminated with a listed cord connector that maintains the protection technique of the terminal compartment

Informational Note: See ANSI/UL 2225, *Cables and Cable-Fittings for Use in Hazardous (Classified) Locations*, for information on construction, testing, and marking of cables, cable fittings, and cord connectors.

506.20 Equipment Installation.

Δ **(A) Zone 20.** In Zone 20 locations, only equipment listed and marked as suitable for the location shall be permitted.

Exception No. 1: Equipment listed for use in Class II, Division 1 locations with a suitable temperature class shall be permitted.

Exception No. 2: For locations involving Group IIIA materials, equipment listed for use in Class III, Division 1 locations with a suitable temperature in accordance with 500.8(D)(3) shall be permitted.

Δ **(B) Zone 21.** In Zone 21 locations, only equipment listed and marked as suitable for the location shall be permitted.

Exception No. 1: Apparatus listed for use in Class II, Division 1 locations with a suitable temperature class shall be permitted.

Exception No. 2: Pressurized equipment identified for Class II, Division 1 shall be permitted.

Exception No. 3: For locations involving Group IIIA materials, equipment listed for use in Class III, Division 1 locations with a suitable temperature in accordance with 500.8(D)(3) shall be permitted.

Δ **(C) Zone 22.** In Zone 22 locations, only equipment listed and marked as suitable for the location shall be permitted.

Exception No. 1: Apparatus listed for use in Class II, Division 1 or Class II, Division 2 locations with a suitable temperature class shall be permitted.

Exception No. 2: Pressurized equipment identified for Class II, Division 1 or Division 2 shall be permitted.

Exception No. 3: For Group IIIA materials, equipment listed for use in Class III, Division 1 or Class III, Division 2 locations with a suitable temperature in accordance with 500.8(D)(3) shall be permitted.

(D) Material Group. Equipment marked Group IIIC shall be permitted for applications requiring Group IIIA or Group IIIB equipment. Similarly, equipment marked Group IIIB shall be permitted for applications requiring Group IIIA equipment.

(E) Manufacturer's Instructions. Electrical equipment installed in hazardous (classified) locations shall be installed in accordance with the manufacturer's instructions, if provided.

Δ **(F) Temperature.** The temperature marking specified in 506.9(C)(2)(5) shall comply with 506.20(F)(1) or (F)(2):

(1) Combustible dusts or combustible fibers/flyings shall be less than the lower of either the layer or cloud ignition temperature of the specific combustible dust or combustible fiber/flying. For nonmetal dusts or nonmetal combustible fibers/flyings that might dehydrate or carbonize, the temperature marking shall not exceed the lower of either the ignition temperature or 165°C (329°F).

(2) For ignitible fibers/flyings, less than 165°C (329°F) for equipment that is not subject to overloading, or 120°C (248°F) for equipment (such as motors or power transformers) that may be overloaded.

Informational Note: See NFPA 499, *Recommended Practice for the Classification of Combustible Dusts and of Hazardous (Classified) Locations for Electrical Installations in Chemical Process Areas*, for minimum ignition temperatures of specific dusts.

Δ **506.30 Grounding and Bonding.** Regardless of the voltage of the electrical system, wiring systems and equipment shall comply with 506.30(A) and (B).

N **(A) Grounding.** Wiring systems and equipment shall be grounded in accordance with Part I and Part VI of Article 250, as applicable.

Δ **(B) Bonding.** Bonding shall comply with Part I and Part V of Article 250, as applicable, and 506.30(B)(1) and (B)(2).

N **(1) Specific Bonding Means.** Bonding shall comply with 506.30(B)(1)(a) and (B)(1)(b).

(a) The locknut-bushing and double-locknut types of contacts shall not be depended on for bonding purposes, but bonding jumpers with identified fittings or other approved means of bonding shall be used. These bonding means shall apply to all metal raceways, fittings, boxes, cable trays, and enclosures, and other parts of raceway systems between hazardous (classified) locations and the point of grounding for service equipment or point of grounding for a separately derived system. Metal struts, angles, or channels provided for support and mechanical or physical protection as permitted in 335.4(5), 336.10(7)(c), or 722.135(C) shall be bonded in accordance with 250.102.

(b) Where the branch-circuit overcurrent protection is located on the load side of the disconnecting means, the specific bonding means shall be permitted to end at the nearest point where the grounded circuit conductor and the grounding electrode conductor are connected together on the line side of the building or structure disconnecting means as specified in 250.32(B).

N (2) Liquidtight Flexible Metal Conduit. Liquidtight flexible metal conduit shall comply with 506.30(B)(2)(a) and (B)(2)(b).

(a) Liquidtight flexible metal conduit shall include an equipment bonding jumper of the wire type in accordance with 250.102.

(b) In Zone 22 locations, the bonding jumper shall not be required where all of the following conditions are met:

(1) Listed liquidtight flexible metal conduit 1.8 m (6 ft) or less in length, with fittings listed for grounding, is used.

(2) Overcurrent protection in the circuit is limited to 10 amperes or less.

(3) The load is part of a meter, instrument, or relay circuit.

ARTICLE 511 Commercial Garages, Repair and Storage

△ 511.1 Scope. These occupancies shall include locations used for service and repair operations in connection with self-propelled vehicles (including, but not limited to, passenger automobiles, buses, trucks, and tractors) in which volatile flammable liquids or flammable gases are used for fuel or power.

> Informational Note: See NFPA 30A-2021, *Code for Motor Fuel Dispensing Facilities and Repair Garages,* for extracted text that is followed by a reference in brackets. Only editorial changes were made to the extracted text to make it consistent with this *Code.*

Article 100 defines *garage* as "a building or portion of a building in which one or more self-propelled vehicles can be kept for use, sale, storage, rental, repair, exhibition, or demonstration purposes." Article 511 applies to commercial garages in which the primary operation is the service and repair of self-propelled vehicles that use flammable gases or liquids for fuel. Garages are further defined in Article 100 as either a major repair garage or minor repair garage. Service operations in which minor repairs, such as oil changes, occur are covered under the requirements of this article.

The requirements of Article 511 mitigate the potential for an ignition-capable arc or spark from electrical wiring or equipment used in or above hazardous locations.

See also

511.3 and its associated commentary for area classification of garages

555.12, which requires that the repair facilities for boats and other marine craft comply with the requirements of Article 511

N 511.2 Other Articles. In addition to the requirements of this article, these occupancies shall comply with Table 511.2, as applicable, except as modified by this article.

N TABLE 511.2 *Other Articles*

Requirement	Division Classified Locations	Zone Classified Locations
Area classification	500.5, 500.6	505.5, 505.6, 505.7
Equipment	Part III of 501, 500.7, 500.8, 501.5	505.8, 505.9, 505.20, 505.22
Wiring	Part II of 501	505.15, 505.16, 505.17, 505.18, 505.19, 505.26, 505.30

511.3 Area Classification, General. Where Class I liquids or gaseous fuels are stored, handled, or transferred, electrical wiring and electrical utilization equipment shall be designed in accordance with the requirements for Class I, Division 1 or 2 hazardous (classified) locations as classified in accordance with 500.5 and 500.6, and this article. A Class I location shall not extend beyond an unpierced wall, roof, or other solid partition that has no openings. [**30A:**8.3.1, 8.3.3]

Where the term "Class I" is used with respect to Zone classifications within this article of the *Code,* it shall apply to Zone 0, Zone 1, and Zone 2 designations.

> Informational Note: The term "Class I" was originally included as a prefix to Zone 0, Zone 1, and Zone 2 locations and references as an identifier for flammable gases, vapors, or liquids to differentiate from Class II and Class III locations. Zone 0, Zone 1, and Zone 2 only apply to flammable gases, vapors, or liquids so the "Class I" prefix is redundant and has been deleted, except for text that is extracted from other documents or to remain consistent throughout this article.

The classification of areas in a garage is dependent upon the level of repair (major or minor) being conducted and the handling of flammable liquids or gaseous fuels other than for dispensing purposes. As stated in 511.3(B), the classification of dispensing areas in a repair garage is addressed in Article 514.

The term *transferred* is used in determining the requirements for classifying locations where a significant quantity of flammable or gaseous liquids is exposed to the atmosphere by a motor vehicle repair operation (e.g., major engine overhauls or repairs that require draining of the motor vehicle fuel tank). Minor repair garages, by definition, are not permitted to conduct such types of repair operations involving the transfer of flammable or gaseous liquids.

The term *Class I liquids* refers to *flammable liquids* as defined in NFPA 30, *Flammable and Combustible Liquids Code,* which are liquids with a flash point below 100°F (38°C). Gasoline is a common Class I liquid, whereas diesel fuel, with a flash point above 100°F, is classified as a Class II combustible liquid. The use of Class I, Class II, and Class III in NFPA 30 for the classification of liquids has no direct correlation to the use of Class I, Class II, and Class III in the *NEC*® to designate hazardous locations.

The need to establish hazardous locations can be mitigated through the use of mechanical ventilation that meets the specified air exchange parameters. The *NEC* indicates different classification areas dependent on whether or not ventilation is provided. Where it is determined that hazardous locations exist within a commercial garage, all applicable requirements of Article 501 for installing wiring and equipment in Class I, Division 1 and 2 locations must be followed.

Many steps are required to properly classify a hazardous location. Although the *NEC* provides general area classifications, it does not classify specific locations. The *NEC* classifications have been extracted from other NFPA documents. The classifications from those documents are based on the premise that all applicable requirements of the document have been met. Deviations in on-site conditions, such as process conditions, area ventilation, and room construction, from those assumed by the document may alter the general classification. Those responsible for the specific area classification must consider the basis for the general classifications to determine the applicability to a specific location.

See also

NFPA 30A, *Code for Motor Fuel Dispensing Facilities and Repair Garages,* for the specific construction and installation requirements used to develop the area classifications in Article 511

Δ **(A) Parking Garages.** Parking garages used for parking or storage shall be permitted to be unclassified.

> Informational Note: See NFPA 88A, *Standard for Parking Structures*, and NFPA 30A, *Code for Motor Fuel Dispensing Facilities and Repair Garages*, for additional information.

In accordance with NFPA 88A, *Standard for Parking Structures*, a mechanical ventilating system that is capable of continuously providing a ventilation rate of 1 cubic foot per minute for each square foot of floor area is required for all enclosed, basement, and underground parking garages.

(B) Repair Garages, with Dispensing. Major and minor repair garages that dispense motor fuels into the fuel tanks of vehicles, including flammable liquids having a flash point below 38°C (100°F) such as gasoline, or gaseous fuels such as natural gas, hydrogen, or LPG, shall have the dispensing functions and components classified in accordance with Table 514.3(B)(1) in addition to any classification required by this section. Where Class I liquids, other than fuels, are dispensed, the area within 900 mm (3 ft) of any fill or dispensing point, extending in all directions, shall be a Class I, Division 2 location.

Δ **(C) Repair Garages, Major and Minor.** Where vehicles using Class I liquids or heavier-than-air gaseous fuels (such as LPG) are repaired, hazardous area classification shall be in accordance with Table 511.3(C).

> Informational Note: See NFPA 30A, *Code for Motor Fuel Dispensing Facilities and Repair Garages*, Table 8.3.2, for additional information.

The Class I, Division 2 location above grade within a major repair garage in which Class I liquids or heavier-than-air gaseous fuels are transferred extends 18 inches above floor level, unless proper mechanical ventilation is provided. The same 18 inches applies down from ceiling areas where lighter-than-air-fuel vehicles are maintained, unless ventilation similar to the floor area of the ceiling area is provided. Areas suitably cut off and areas adjacent to unclassified, ventilated garages are not classified as hazardous.

The Class I, Division 1 location below grade extends from the floor of the pit or depression to the garage floor level. A pit or depression can be classified as Class I, Division 2. However, ventilation providing at least 1 cubic foot per minute for each square foot of floor space must exhaust air from within 12 inches of the floor level of the pit. See Exhibit 511.1 for an illustration of classified and unclassified locations in a major repair garage.

Some minor repair garages primarily offer oil and filter change and lubrication-type services but do not transfer fuel. If the lower-level work area is provided with exhaust ventilation at the rate specified, the lower level is not classified as a hazardous location.

Δ **(D) Repair Garages, Major.** Where vehicles using lighter-than-air gaseous fuels (such as hydrogen and natural gas) are repaired or stored, hazardous area classification shall be in accordance with Table 511.3(D).

> Informational Note: See NFPA 30A, *Code for Motor Fuel Dispensing Facilities and Repair Garages*, Table 8.3.2, for additional information.

(E) Modifications to Classification.

Δ **(1) Specific Areas Adjacent to Classified Locations.** Areas adjacent to classified locations in which flammable vapors are not likely to be released, such as stock rooms, switchboard rooms, and other similar locations, shall be unclassified where mechanically ventilated at a rate of four or more air changes per hour, designed with positive air pressure, or separated by an unpierced wall, roof, or other solid partition.

(2) Alcohol-Based Windshield Washer Fluid. The area used for storage, handling, or dispensing into motor vehicles of alcohol-based windshield washer fluid in repair garages shall be unclassified unless otherwise classified by a provision of 511.3. [30A:8.3.1, Exception]

511.4 Wiring and Equipment in Class I Locations.

(A) Wiring Located in Class I Locations. Wiring located within Class I locations as classified in 511.3 shall conform with the requirements of Part II of Article 501 or 504.20, as applicable.

(B) Equipment Located in Class I Locations. Within Class I locations as defined in 511.3, equipment shall conform with the requirements of Part III of Article 501 or 504.10, as applicable.

Most battery-operated and portable equipment is capable of igniting the atmosphere in a hazardous location. Therefore, hand-held, portable, and mobile equipment brought into hazardous locations should be suitable for the location.

Δ **(1) Fuel-Dispensing Units.** Where fuel-dispensing units (other than liquid petroleum gas, which is prohibited) are located within buildings, 514.1 shall apply.

Where mechanical ventilation is provided in the dispensing area, the control shall be interlocked so that the dispenser cannot operate without ventilation, in accordance with 500.5(B)(2).

See also

Figure 514.3 and **Exhibit 514.1** for more information on classified areas in the vicinity of dispensing units

△ **TABLE 511.3(C)** *Extent of Classified Locations for Major and Minor Repair Garages with Heavier-Than-Air Fuel*

Location	Class I Division (Group D)	Class I Zone (Group IIA)	Extent of Classified Location
Repair garage, major (where Class I liquids or gaseous fuels are transferred or dispensed*)	1	1	Entire space within any pit, belowgrade work area, or subfloor work area that is not ventilated
	2	2	Entire space within any pit, belowgrade work area, or subfloor work area that is provided with ventilation of at least 0.3 m³/min/m² (1 ft³/min/ft²) of floor area, with suction taken from a point within 300 mm (12 in.) of floor level
	2	2	Up to 450 mm (18 in.) above floor level of the room, except as noted below, for entire floor area
	Unclassified	Unclassified	Up to 450 mm (18 in.) above floor level of the room where room is provided with ventilation of at least 0.3 m³/min/m² (1 ft³/min/ft²) of floor area, with suction taken from a point within 300 mm (12 in.) of floor level
	2	2	Within 0.9 m (3 ft) of any fill or dispensing point, extending in all directions
Specific areas adjacent to classified locations	Unclassified	Unclassified	Areas adjacent to classified locations where flammable vapors are not likely to be released, such as stock rooms, switchboard rooms, and other similar locations, where mechanically ventilated at a rate of four or more air changes per hour or designed with positive air pressure or where effectively separated by walls or partitions
Repair garage, minor (where Class I liquids or gaseous fuels are not transferred or dispensed*)	2	2	Entire space within any pit, belowgrade work area, or subfloor work area that is not ventilated
	2	2	Up to 450 mm (18 in.) above floor level, extending 0.9 m (3 ft) horizontally in all directions from opening to any pit, belowgrade work area, or subfloor work area that is not ventilated
	Unclassified	Unclassified	Entire space within any pit, belowgrade work area, or subfloor work area that is provided with ventilation of at least 0.3 m³/min/m² (1 ft³/min/ft²) of floor area, with suction taken from a point within 300 mm (12 in.) of floor level
Specific areas adjacent to classified locations	Unclassified	Unclassified	Areas adjacent to classified locations where flammable vapors are not likely to be released, such as stock rooms, switchboard rooms, and other similar locations, where mechanically ventilated at a rate of four or more air changes per hour or designed with positive air pressure or where effectively separated by walls or partitions

*Includes draining of Class I liquids from vehicles.

△ **TABLE 511.3(D)** *Extent of Classified Locations for Major Repair Garages with Lighter-than-Air Fuel*

Location	Class I Division[2]	Class I Zone[3]	Extent of Classified Location
Repair garage, major (where lighter-than-air gaseous fueled[1] vehicles are repaired or stored)	2	2	Within 450 mm (18 in.) of ceiling, except as noted below
	Unclassified	Unclassified	Within 450 mm (18 in.) of ceiling where ventilation of at least 0.3 m³/min/m² (1 ft³/min/ft²) of floor area is provided, with suction taken from a point within 450 mm (18 in.) of the highest point in the ceiling
Specific areas adjacent to classified locations	Unclassified	Unclassified	Areas adjacent to classified locations where flammable vapors are not likely to be released, such as stock rooms, switchboard rooms, and other similar locations, where mechanically ventilated at a rate of four or more air changes per hour, designed with positive air pressure, or effectively separated by walls or partitions

[1]Includes fuels such as hydrogen and natural gas, but not LPG.
[2]For hydrogen (lighter than air) Group B, or natural gas Group D.
[3]For hydrogen (lighter than air) Group IIC or IIB+H2, or natural gas Group IIA.

EXHIBIT 511.1 Classification of locations in commercial garages.

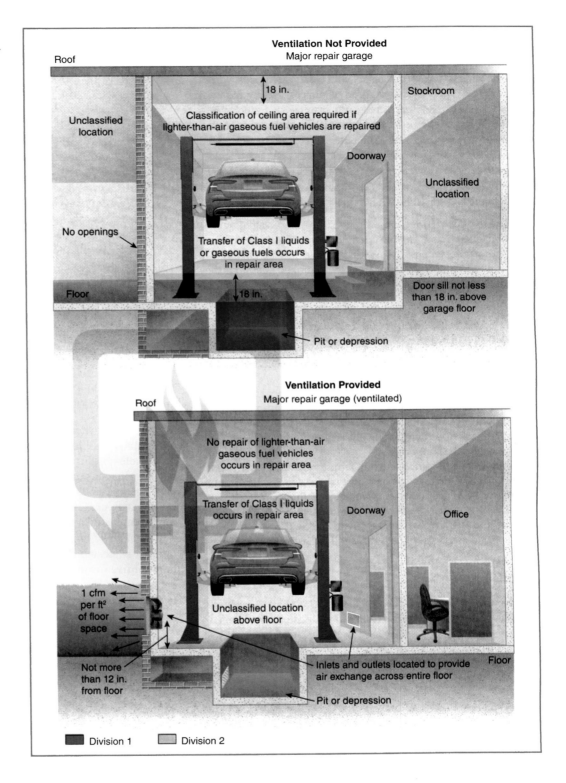

Ventilation Not Provided
Major repair garage

Roof

18 in.

Unclassified location

Classification of ceiling area required if lighter-than-air gaseous fuel vehicles are repaired

Stockroom

Doorway

Unclassified location

No openings

Transfer of Class I liquids or gaseous fuels occurs in repair area

Floor

18 in.

Door sill not less than 18 in. above garage floor

Pit or depression

Ventilation Provided
Major repair garage (ventilated)

Roof

No repair of lighter-than-air gaseous fuel vehicles occurs in repair area

Transfer of Class I liquids occurs in repair area

Doorway

Office

1 cfm per ft² of floor space

Unclassified location above floor

Not more than 12 in. from floor

Inlets and outlets located to provide air exchange across entire floor

Floor

Pit or depression

Division 1   Division 2

(2) Portable Lighting Equipment. Portable lighting equipment shall be equipped with a handle, lampholder, hook, and substantial guard attached to the lampholder or handle. All exterior surfaces that might come in contact with battery terminals, wiring terminals, or other objects shall be of nonconducting material or shall be effectively protected with insulation. Lampholders shall be of an unswitched type and shall not provide means for plug-in of attachment plugs. The outer shell shall be of molded composition or other suitable material. Unless the lamp and its cord are supported or arranged in such a manner that they cannot be used in the locations classified in 511.3, they shall be of a type identified for Class I, Division 1 locations.

Portable luminaires often are used for supplemental lighting during vehicle servicing. Unless specifically identified for use in a Class I, Division 1 location, reel-type portable hand luminaires are permitted only if the cord and lamp are arranged such that they cannot be used inside the boundary of the hazardous locations within the garage.

511.7 Wiring and Equipment Installed Above Hazardous (Classified) Locations.

The wiring system in the area above the hazardous location must not produce an ignition-capable arc, or if sparks can be produced, they must not reach the hazardous location. Wiring installed above unclassified areas can be selected from the methods covered in Chapter 3, provided the article covering the wiring method does not contain any restrictions that would limit its use in commercial garages.

Operations that involve open flames or electric arcs, including fusion gas welding and electric welding, as well as the requirements for heat-producing appliances, are found in NFPA 30A, *Code for Motor Fuel Dispensing Facilities and Repair Garages*. Repair work that involves an open flame or electric arcs must be restricted to areas specifically provided for such purposes.

(A) Wiring in Spaces Above Hazardous (Classified) Locations.

Δ **(1) Fixed Wiring Above Hazardous (Classified) Locations.** Fixed wiring above hazardous (classified) locations shall be permitted to be one or more of the following:

(1) Rigid metal conduit (RMC) or intermediate metal conduit (IMC) with listed threaded or threadless fittings, or electrical metallic conduit (EMT) with listed fittings.
(2) Rigid polyvinyl chloride conduit (PVC), reinforced thermosetting resin conduit (RTRC), or electrical nonmetallic tubing (ENT).
(3) Flexible metal conduit (FMC), liquidtight flexible metal conduit (LFMC), or liquidtight flexible nonmetallic conduit (LFNC), with listed fittings.
(4) Type MC cable, Type AC cable, Type TC cable, or Type TC-ER cable, including installation in cable trays. Type TC-ER cable shall include an equipment grounding conductor (EGC) in addition to any drain wire. All cable types shall have listed fittings.

(5) Type MI cable terminated with listed fittings and supported in a manner to avoid tensile stress.
(6) Manufactured wiring systems.
(7) Type PLTC cable or Type PLTC-ER cable in Class I, Class 2, or Class 3 circuits. Type PLTC-ER cable shall include an equipment grounding conductor (EGC) in addition to any drain wire.
(8) Type ITC cable or Type ITC-ER cable in accordance with 335.4 and 335.5, terminated with listed fittings. Type ITC-ER cable shall include an equipment grounding conductor (EGC) in addition to any drain wire.
(9) Cellular metal floor raceways or cellular concrete floor raceways only for supplying ceiling outlets or extensions to the area below the floor. Such raceways shall have no connections leading into or through any Class I location above the floor.

(2) Pendant. For pendants, flexible cord suitable for the type of service and listed for hard usage shall be used.

(B) Electrical Equipment Installed Above Hazardous (Classified) Locations.

Δ **(1) Fixed Electrical Equipment.** Electrical equipment in a fixed position shall be located above the level of any defined hazardous (classified) location or shall be identified for the location.

(a) *Arcing Equipment.* Equipment that is less than 3.7 m (12 ft) above the floor level and that might produce arcs, sparks, or particles of hot metal, such as cutouts, switches, charging panels, generators, motors, or other equipment (excluding receptacles, lamps, and lampholders) having make-and-break or sliding contacts, shall be of the totally enclosed type or constructed to prevent the escape of sparks or hot metal particles.

(b) *Fixed Lighting.* Lamps and lampholders for fixed lighting that is located over lanes through which vehicles are commonly driven or that might otherwise be exposed to physical damage shall be located not less than 3.7 m (12 ft) above floor level, unless of the totally enclosed type or constructed to prevent escape of sparks or hot metal particles.

Δ **511.8 Underground Wiring Below Hazardous (Classified) Locations.** Underground wiring shall be installed in accordance with one of the following wiring methods:

(1) Threaded rigid metal conduit (RMC) or threaded intermediate metal conduit (IMC) with listed threaded fittings.
(2) Rigid polyvinyl chloride conduit (PVC), reinforced thermosetting resin conduit (RTRC), or high-density polyethylene conduit (HDPE) where buried under not less than 600 mm (2 ft) of cover. Where PVC conduit, RTRC conduit, or HDPE conduit is used, threaded rigid metal conduit or threaded intermediate metal conduit shall be used for the last 600 mm (2 ft) of the underground run to emergence or to the point of connection to the aboveground raceway, and an equipment grounding conductor shall be included

to provide electrical continuity of the raceway system and for grounding of non–current-carrying metal parts.

511.9 Sealing. Seals complying with the requirements of 501.15 and 501.15(B)(2) shall be provided and shall apply to horizontal as well as vertical boundaries of the defined Class I locations.

Note that the general rules of 501.15(A)(4) and (B)(2) on providing seals at hazardous location boundaries apply where raceway installations in a commercial garage pass from classified to unclassified locations. In accordance with 501.15(A)(4), Exception No. 2, where a raceway runs from a Class I, Division 1 location into an underground unclassified location and then emerges from below ground into an unclassified location, the boundary seal is permitted to be located more than 10 feet from the actual boundary, provided the seal is the first fitting(s) at the point the conduit emerges from below ground into the unclassified location.

Exhibit 511.2 depicts two receptacle outlet enclosures that are located at least 12 inches above an area that has been classified as Class I. The rigid metal conduit passes unbroken from the outlet boxes through the Class I location into the unclassified underground location beneath the floor. The conduit coupling is located 12 inches or more from the penetration into the hazardous location. No seals are required for this installation in accordance with 501.15(A)(4), Exception No. 1.

511.10 Special Equipment.

(A) Battery Charging Equipment. Battery chargers and their control equipment, and batteries being charged, shall not be located within locations classified in 511.3.

(B) Electric Vehicle Charging Equipment.

(1) General. All electrical equipment and wiring shall be installed in accordance with Part III of Article 625, except as

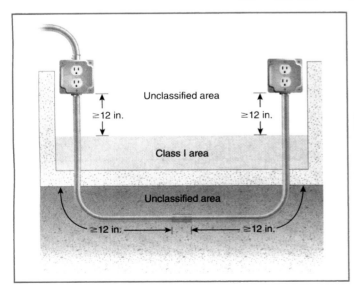

EXHIBIT 511.2 Seals not required for conduits that pass unbroken through the Class I location.

required by 511.10(B)(2) and (B)(3). Flexible cords shall be of a type identified for extra-hard usage.

(2) Connector Location. No connector shall be located within a Class I location as defined in 511.3.

(3) Plug Connections to Vehicles. Where the cord is suspended from overhead, it shall be arranged so that the lowest point of sag is at least 150 mm (6 in.) above the floor. Where an automatic arrangement is provided to pull both cord and plug beyond the range of physical damage, no additional connector shall be required in the cable or at the outlet.

511.12 Ground-Fault Circuit-Interrupter Protection for Personnel. Ground-fault circuit-interrupter protection for personnel shall be provided as required in 210.8(B).

511.16 Grounding and Bonding Requirements.

(A) General Grounding Requirements. All metal raceways, the metal armor or metallic sheath on cables, and all non–current-carrying metal parts of fixed or portable electrical equipment, regardless of voltage, shall be grounded.

(B) Supplying Circuits with Grounded and Grounding Conductors in Class I Locations. Grounding in Class I locations shall comply with 501.30.

Δ **(1) Circuits Supplying Portable Equipment or Pendants.** Where a circuit supplies portables or pendants and includes a grounded conductor in accordance with 200.3, receptacles, attachment plugs, connectors, and similar devices shall be of the grounding type and the grounded conductor of the flexible cord shall be connected to the screw shell of any lampholder or to the grounded terminal of any utilization equipment supplied.

(2) Approved Means. Approved means shall be provided for maintaining continuity of the equipment grounding conductor between the fixed wiring system and the non–current-carrying metal portions of pendant luminaires, portable luminaires, and portable utilization equipment.

N **ARTICLE 512** Cannabis Oil Equipment and Cannabis Oil Systems Using Flammable Materials

N **Part I. General**

N **512.1 Scope.** This article covers cannabis oil preparatory equipment, extraction equipment, booths, post-processing equipment, and systems using flammable materials (flammable gas, flammable liquid–produced vapor, combustible liquid–produced vapor) in commercial and industrial facilities.

Informational Note No. 1: See ANSI/UL 1389, *Plant Oil Extraction Equipment for Installation and Use in Ordinary*

(Unclassified) Locations and Hazardous (Classified) Locations, for information on cannabis oil equipment and systems for hazardous (classified) locations.

Informational Note No. 2: See NFPA 1, *Fire Code*; NFPA 55, *Compressed Gases and Cryogenic Fluids Code*; NFPA 58, *Liquefied Petroleum Gas Code*; and ICC IFC, *International Fire Code*, together with the manufacturer's installation instructions, for information on the installation of cannabis oil equipment and systems.

The extraction of cannabis oil can pose many safety issues, including the risk of fire and explosion. In many cases, the extraction process itself creates a hazardous (classified) location due to the use of flammable solvents such as butane, pentane, hexane, propane, and ethanol, which can be released during the processing and extraction of plant oils. Article 512 has been added to the *NEC®* to assist in the safe installation of equipment and systems utilized for cannabis oil extraction within commercial and industrial facilities.

N 512.2 Other Articles. In addition to the requirements of this article, cannabis oil equipment and cannabis oil systems using flammable materials shall comply with Table 512.2, as applicable, except as modified by this article.

N TABLE 512.2 *Other Articles*

Requirement	Division Classified Locations	Zone Classified Locations
Area classification	500.5, 500.6	505.5, 505.6, 505.7
Equipment	Part III of 501, 500.7, 500.8, 501.5	505.8, 505.9, 505.20, 505.22
Wiring	Part II of 501	505.15, 505.16, 505.17, 505.18, 505.19, 505.26, 505.30

N 512.3 Classified Locations. Cannabis oil equipment and systems that can release flammable materials during operation shall be classified in accordance with 512.3(A) and (B).

Informational Note No. 1: Some cannabis oil applications can result in the release of heavier-than-air flammable gases or vapors into the surrounding atmosphere as a normal part of the overall extraction process (e.g., during disconnecting or opening of vessels containing flammable solvents, or during off-gassing of spent material or extracted plant oil). Cannabis oil equipment and systems can also include the connection of external containers, or other external sources, of flammable solvent.

Informational Note No. 2: See NFPA 30, *Flammable and Combustible Liquids Code*; NFPA 33, *Standard for Spray Application Using Flammable or Combustible Materials*; and NFPA 497, *Recommended Practice for the Classification of Flammable Liquids, Gases, or Vapors and of Hazardous (Classified) Locations for Electrical Installations in Chemical Process Areas*, for information on area classification.

Informational Note No. 3: See NFPA 36, *Standard for Solvent Extraction Plants*, for information on area classification in commercial-scale extraction processes.

N (A) Cannabis Oil Equipment and Systems Other Than Booths.

N (1) Where Flammable Gases or Vapors Are Released. For sources of gases or vapors from a flammable material, the location shall be classified in accordance with the following and as shown in Figure 512.3(A)(1):

(1) The space within 915 mm (3 ft) in all directions from any such equipment or container and extending to the floor or grade level shall be classified as Class I, Division 1 or Zone 1, whichever is applicable.

(2) The space extending 610 mm (2 ft) beyond the Class I, Division 1 or Zone 1 location shall be classified as Class I, Division 2 or Zone 2, whichever is applicable.

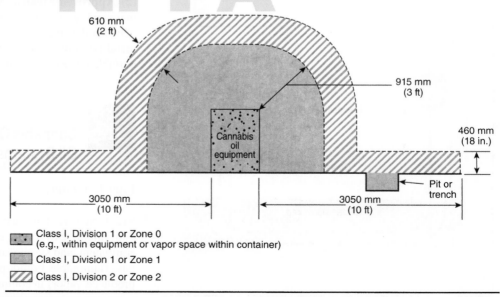

N FIGURE 512.3(A)(1) *Area Classification for Equipment and Systems Other than Booths, Where Flammable Gases or Vapors Are Released.*

(3) The space extending 1525 mm (5 ft) horizontally beyond the space described in 512.3(A)(1)(2) up to a height of 460 mm (18 in.) above the floor or grade level shall be classified as Class I, Division 2 or Zone 2, whichever is applicable.

(4) The space inside of a tank or container and the inside of equipment that contains a flammable material shall be classified as Class I, Division 1 or Zone 0, whichever is applicable.

(5) Sumps, pits, or belowgrade channels within 3.05 m (10 ft) horizontally of a vapor source shall be classified as Class I, Division 1 or Zone 1. If the sump, pit, or channel extends beyond 3.05 m (10 ft) horizontally from the vapor source, it shall be provided with a vapor stop or classified as Class I, Division 1 or Zone 1 for its entire length.

N **(2) Where Flammable Gases or Vapors Are Not Released, Except During Disconnection or Opening.** Where listed equipment is marked to indicate that the level of release during disconnection or opening is maintained below 25 percent LFL without ventilation, flammable solvents shall not be released during the extraction process except during disconnecting or opening of vessels containing flammable solvents, or during off-gassing of spent material or extracted plant oil.

For sources of gas or vapor from a flammable material, the location shall be classified in accordance with the following and as shown in Figure 512.3(A)(2):

(1) The space within 915 mm (3 ft) in all directions from any such equipment or container and extending to the floor or grade level shall be classified as Class I, Division 2 or Zone 2, whichever is applicable.

(2) The space extending beyond the Division 2 or Zone 2 area shall be unclassified.

(3) The space inside of a tank or container and the inside of equipment that contains a flammable material shall be classified as Class I, Division 1 or Zone 0, whichever is applicable.

(4) The space extending 2134 mm (7 ft) horizontally beyond the space described in 512.3(A)(2)(1) up to a height of 460 mm (18 in.) above the floor or grade level shall be classified as Class I, Division 2 or Zone 2, whichever is applicable.

(5) Sumps, pits, or belowgrade channels within 3.05 m (10 ft) horizontally of a vapor source shall be classified as Class I, Division 1 or Zone 1. If the sump, pit, or channel extends beyond 3.05 m (10 ft) horizontally from the vapor source, it shall be provided with a vapor stop or it shall be classified as Class I, Division 1 or Zone 1 for its entire length.

N **(B) Cannabis Oil Booths.** Air exhausted from the booths shall not be recirculated or exhausted from the booths into the room in which the booths are installed. Ventilation other than exhaust ventilation can be provided to the booth, but cannot be recirculated or exhausted from the booth into the room in which the booth is installed.

N **(1) Where Flammable Gases or Vapors Are Released.** For sources of gas or vapor from a flammable material, the location shall be classified in accordance with the following and as shown in Figure 512.3(B)(1):

(1) The space within the booth shall be classified as Class I, Division 1 or Zone 1, whichever is applicable.

(2) The space within 915 mm (3 ft) of any opening shall be classified as Class I, Division 2 or Zone 2, whichever is applicable.

(3) The interior of fresh air supply ducts and fresh air supply plenums shall be unclassified.

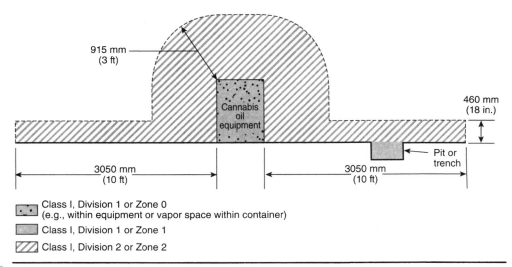

915 mm (3 ft)

460 mm (18 in.)

Cannabis oil equipment

Pit or trench

3050 mm (10 ft)

3050 mm (10 ft)

⬚ Class I, Division 1 or Zone 0 (e.g., within equipment or vapor space within container)

⬛ Class I, Division 1 or Zone 1

▨ Class I, Division 2 or Zone 2

N **FIGURE 512.3(A)(2)** *Area Classification for Equipment and Systems Other than Booths, Where Flammable Gases or Vapors Are Not Released Except During Disconnection or Opening.*

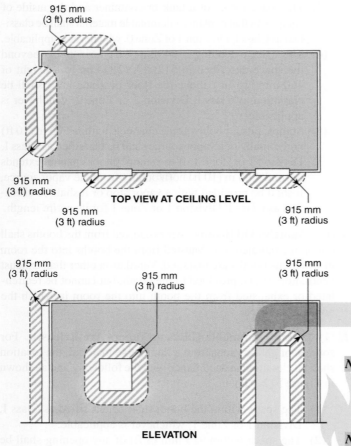

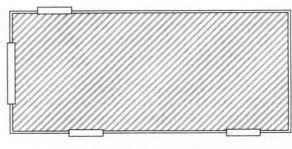

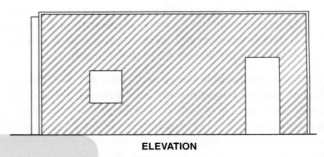

N **FIGURE 512.3(B)(2)** *Area Classification for Booths Where Flammable Gases or Vapors Are Not Released Except During Disconnection or Opening.*

N **Part II. Wiring**

N **512.10 Wiring Installation and Operation.** All wiring installed or operated within any of the hazardous (classified) locations defined in 512.3 shall comply with the requirements of Part II of Article 501 or 505.15, as applicable, for the division or zone location in which it is installed.

N **512.13 Wiring Installed Above Hazardous (Classified) Locations.** Other than above cannabis oil booths, all fixed wiring installed at an elevation above hazardous (classified) locations *[see Figure 512.3(A)(1) and Figure 512.3(A)(2)]* shall be in metal raceways, PVC conduit, RTRC conduit, or ENT conduit. Where used, cables shall be Type MI, Type TC, or Type MC.

N **Part III. Equipment**

N **512.20 Equipment and Systems.** Installation of cannabis oil equipment and systems shall be in a distinct room or area located at commercial or industrial facilities. Where all electrical equipment within cannabis oil booths is listed for Class I, Division 1 or Zone 1 locations, gas detection shall not be required to be provided within the booth. Where gas detection in accordance with 500.7(K) is provided within booths listed

Class I, Division 1 or Zone 1

Class I, Division 2 or Zone 2

N **FIGURE 512.3(B)(1)** *Area Classification for Booths Where Flammable Gases or Vapors Are Released.*

N **(2) Where Flammable Gases or Vapors Are Not Released, Except During Disconnection or Opening.** Where listed equipment is marked to indicate that the level of release during disconnection or opening is maintained below 25 percent LFL without ventilation, flammable solvents shall not be released during the extraction process except during disconnecting or opening of vessels containing flammable solvents, or during off-gassing of spent material or extracted plant oil.

For sources of gas or vapor from a flammable material, the location shall be classified in accordance with the following and as shown in Figure 512.3(B)(2):

(1) The space within the booth shall be classified as Class I, Division 2 or Zone 2, whichever is applicable.
(2) The space outside any opening shall be unclassified.
(3) The interior of fresh air supply ducts and fresh air supply plenums shall be unclassified.

for Class I, Division 2 or Zone 2 locations, electrical equipment shall be permitted.

N **(A) Cannabis Oil Preparatory Equipment.** Equipment that is used to prepare the plant material for subsequent extraction of the plant oil (e.g., trimming, deseeding, drying/curing) shall be listed for the location.

N **(B) Cannabis Oil Extraction Equipment.** Equipment that uses flammable materials (solvents) in the process of extracting the plant oil from the plant material shall be listed for the location.

> Informational Note: Extraction equipment can use flammable materials as solvents to extract the plant oil from the plant material by saturating the plant material in a vented container, sealed container, or pressure vessel. Typical flammable materials used in the extraction process include butane, ethanol, hexane, pentane, propane, and LPG.

N **(C) Cannabis Oil Booths.** Enclosed areas used to house cannabis oil equipment and systems shall be listed for the location.

> Informational Note: Cannabis oil booths can be designed to house a single piece or multiple pieces of cannabis oil equipment. Booths range in size and can be large enough to permit entrance of personnel to perform the processing tasks.

N **(D) Cannabis Oil Post-Processing Equipment.** Equipment that is used in the final processing stages of the extracted plant oil (e.g., vacuum ovens, rotary evaporators, solvent recovery pumps) shall be listed for the location.

N **(E) Cannabis Oil Systems.** Any combination of cannabis oil equipment needed for the overall extraction process (e.g., cannabis oil preparatory equipment, cannabis oil extraction equipment, cannabis oil booths, cannabis oil post-processing equipment) shall be listed for the location.

> Informational Note: See NFPA 70B, *Recommended Practice for Electrical Equipment Maintenance*, for information related to general electrical equipment maintenance and developing an effective electrical preventive maintenance (EPM) program.

N **512.22 Equipment Installed in Hazardous (Classified) Locations.** All equipment installed or operated within any of the classified locations defined in 512.3 shall comply with the requirements of Part III of Article 501 or 505.9, as applicable, for the division or zone area in which they are used.

N **512.30 Equipment Installed Above Hazardous (Classified) Locations.** Equipment that could produce arcs, sparks, or hot metal particles, such as lamps and lampholders for fixed lighting, cutouts, switches, receptacles, motors, or other equipment having make-and-break or sliding contacts, where installed above a classified location other than cannabis oil booths, shall be of the totally enclosed type or be constructed to prevent the escape of sparks or hot metal particles.

N **512.32 Marking.** Cannabis oil preparatory equipment, extraction equipment, booths, and post-processing equipment

shall be listed and marked to show the hazardous (classified) location for which it is permitted to be installed.

N **(A) Division Equipment.** Equipment for Class I, Division 1 or Class I, Division 2 shall be marked in accordance with 500.8(C).

N **(B) Zone Equipment.** Equipment for Zone 1 or Zone 2 shall be marked in accordance with 500.8(C)(2).

ARTICLE 513 Aircraft Hangars

Δ **513.1 Scope.** This article shall apply to buildings or structures in any part of which aircraft containing Class I (flammable) liquids or Class II (combustible) liquids whose temperatures are above their flash points are housed or stored and in which aircraft might undergo service, repairs, or alterations. It shall not apply to locations used exclusively for aircraft that have never contained fuel or unfueled aircraft.

> Informational Note No. 1: See NFPA 409, *Standard on Aircraft Hangars*, for definitions of *aircraft hangar* and *unfueled aircraft*.
> Informational Note No. 2: See NFPA 30, *Flammable and Combustible Liquids Code*, for information on fuel classification.

Article 513 does not apply to areas in which the only fuel contained in the aircraft is a Class II combustible liquid, unless the fuel will be used or stored above its flash point within the hangar. A Class II liquid has a closed-cup flash point at or above 100°F. Some aviation fuel, such as Jet-A, is a Class II combustible liquid. An aircraft manufacturing plant in which the aircraft under construction have never contained fuel is an example of a facility not covered by the requirements of Article 513.

Many steps are required to properly classify a hazardous location. Although the *NEC*® provides general area classifications in Articles 501, 502, 503, 505, and 506, it does not classify specific locations. The *NEC* classifications for specific occupancies or processes have been extracted from other NFPA documents. The classifications from those documents are based on the premise that all applicable requirements of the document have been met. Deviations in on-site conditions, such as process conditions, area ventilation, and room construction, from those assumed by the document may alter the general classification. Those responsible for the specific area classification must consider the basis for the general classifications to determine the applicability to a specific location.

N **513.2 Other Articles.** In addition to the requirements of this article, aircraft hangars shall comply with Table 513.2, as applicable, except as modified by this article.

513.3 Classification of Locations. Where the term "Class I" is used with respect to Zone classifications within this article of the Code, it shall apply to Zone 0, Zone 1, and Zone 2 designations.

N TABLE 513.2 *Other Articles*

Requirement	Division Classified Locations	Zone Classified Locations
Area classification	500.5, 500.6	505.5, 505.6, 505.7
Equipment	Part III of 501, 500.7, 500.8, 501.5	505.8, 505.9, 505.20, 505.22
Wiring	Part II of 501	505.15, 505.16, 505.17, 505.18, 505.19, 505.26, 505.30

Informational Note: The term "Class I" was originally included as a prefix to Zone 0, Zone 1, and Zone 2 locations and references as an identifier for flammable gases, vapors, or liquids to differentiate from Class II and Class III locations. Zone 0, Zone 1, and Zone 2 only apply to flammable gases, vapors, or liquids so the "Class I" prefix is redundant and has been deleted, except for text that is extracted from other documents or to remain consistent throughout this article.

(A) Below Floor Level. Any pit or depression below the level of the hangar floor shall be classified as a Class I, Division 1 or Zone 1 location that shall extend up to said floor level.

(B) Areas Not Separated or Ventilated. The entire area of the hangar, including any adjacent and communicating areas not suitably separated from the hangar, shall be classified as a Class I, Division 2 or Zone 2 location up to a level 450 mm (18 in.) above the floor.

(C) Vicinity of Aircraft.

Δ **(1) Aircraft Maintenance and Storage Hangars.** The area within 1.5 m (5 ft) horizontally from aircraft power plants or aircraft fuel tanks shall be classified as a Class I, Division 2 or Zone 2 location that shall extend upward from the floor to a level 1.5 m (5 ft) above the upper surface of wings and engine enclosures.

Information on aircraft parking patterns, the types of aircraft, and the operations to be performed in the hangar are needed to properly classify hangar areas. Much of the hangar area classification is dependent on the dimensional outline of the aircraft; therefore, consideration of future aircraft and parking patterns is appropriate to avoid costs associated with changes in the area classification. Exhibit 513.1 illustrates area classifications in aircraft hangars. In aircraft painting hangars, the classified areas expand to include the entire surface area of the aircraft per 513.3(C)(2), not only the engines and fuel tanks.

Δ **(2) Aircraft Painting Hangars.** The area within 3 m (10 ft) horizontally from aircraft surfaces from the floor to 3 m (10 ft) above the aircraft shall be classified as Class I, Division 1 or Zone 1. The area horizontally from aircraft surfaces between 3.0 m (10 ft) and 9.0 m (30 ft) from the floor to 9.0 m (30 ft) above the aircraft surface shall be classified as Class I, Division 2 or Zone 2.

Informational Note: See NFPA 33, *Standard for Spray Application Using Flammable or Combustible Materials*, for information

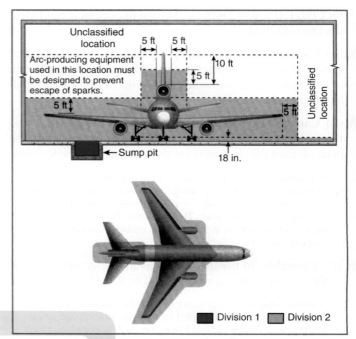

EXHIBIT 513.1 Area classification in aircraft hangars.

on ventilation and grounding for static protection in spray painting areas.

Δ **(D) Areas Suitably Separated and Ventilated.** Adjacent areas in which flammable liquids or vapors are not likely to be released, such as stock rooms, electrical control rooms, and other similar locations, shall be unclassified where mechanically ventilated at a rate of four or more air changes per hour, designed with positive air pressure, or effectively separated from the hangar itself by walls or partitions.

513.4 Wiring and Equipment in Class I Locations.

Δ **(A) General.** All wiring and equipment that is installed or operated within any of the hazardous (classified) locations defined in 513.3 shall comply with the applicable requirements of the hazardous (classified) locations.

Attachment plugs and receptacles shall be identified for the location or shall be designed so they cannot be energized while the connections are being made or broken.

Δ **(B) Stanchions, Rostrums, and Docks.** Electrical wiring, outlets, and equipment (including lamps) on or attached to stanchions, rostrums, or docks that are located in a hazardous (classified) location as defined in 513.3(C) shall comply with the applicable requirements of Parts II and III of Article 501 or 505.17 through 505.30, as applicable.

513.7 Wiring and Equipment Not Installed in Class I Locations.

(A) Fixed Wiring. All fixed wiring in a hangar but not installed in a Class I location as classified in 513.3 shall be installed in metal raceways or shall be Type MI, TC, or MC cable.

Exception: Wiring in unclassified locations, as described in 513.3(D), shall be permitted to be any suitable type wiring method recognized in Chapter 3.

(B) Pendants. For pendants, flexible cord suitable for the type of service and identified for hard usage or extra-hard usage shall be used. Each such cord shall include a separate equipment grounding conductor.

(C) Arcing Equipment. In locations above those described in 513.3, equipment that is less than 3.0 m (10 ft) above wings and engine enclosures of aircraft and that may produce arcs, sparks, or particles of hot metal, such as lamps and lampholders for fixed lighting, cutouts, switches, receptacles, charging panels, generators, motors, or other equipment having make-and-break or sliding contacts, shall be of the totally enclosed type or constructed so as to prevent the escape of sparks or hot metal particles.

Exception: Equipment in areas described in 513.3(D) shall be permitted to be of the general-purpose type.

(D) Lampholders. Lampholders of metal-shell, fiber-lined types shall not be used for fixed incandescent lighting.

(E) Stanchions, Rostrums, or Docks. Where stanchions, rostrums, or docks are not located or likely to be located in a Class I location, as defined in 513.3(C), wiring and equipment shall comply with 513.7, except that such wiring and equipment not more than 457 mm (18 in.) above the floor in any position shall comply with 513.4(B). Receptacles and attachment plugs shall be of a locking type that will not readily disconnect.

Δ **(F) Mobile Stanchions.** Mobile stanchions with electrical equipment complying with 513.7(E) shall carry at least one permanently affixed warning sign with the following words or equivalent:

WARNING
KEEP 5 FT CLEAR OF AIRCRAFT
ENGINES AND FUEL TANK AREAS

or

WARNING
KEEP 1.5 FT CLEAR OF AIRCRAFT
ENGINES AND FUEL TANK AREAS

513.8 Underground Wiring.

(A) Wiring and Equipment Embedded, Under Slab, or Underground. All wiring installed in or under the hangar floor shall comply with the requirements for Class I, Division 1 locations. Where such wiring is located in vaults, pits, or ducts, adequate drainage shall be provided.

(B) Uninterrupted Raceways, Embedded, Under Slab, or Underground. Uninterrupted raceways that are embedded in a

EXHIBIT 513.2 Explosionproof conduit seal fittings that can be used to meet the requirements for boundary seals in runs of conduit that transition from classified to unclassified locations. *(Courtesy of Appleton™, Emerson Electric Co.)*

hangar floor or buried beneath the hangar floor shall be considered to be within the Class I location above the floor, regardless of the point at which the raceway descends below or rises above the floor.

Wiring and equipment embedded in or buried below the hangar floor is considered to be in a Class I, Division 1 location, whereas uninterrupted raceways that are embedded in or buried below the hangar floor are considered to be in the same hazardous location that exists above the floor. Raceways that rise out of the floor in unclassified locations must be provided with boundary seals in accordance with 501.15. Exhibit 513.2 shows two examples of seal fittings that can be used to comply with the requirements to provide boundary seals for raceways installed in or under hangar floors.

513.9 Sealing. Seals shall be provided in accordance with 501.15 or 505.16, as applicable. Sealing requirements specified shall apply to horizontal as well as to vertical boundaries of the defined Class I locations.

513.10 Special Equipment.

(A) Aircraft Electrical Systems.

(1) De-energizing Aircraft Electrical Systems. Aircraft electrical systems shall be de-energized when the aircraft is stored in a hangar and, whenever possible, while the aircraft is undergoing maintenance.

(2) Aircraft Batteries. Aircraft batteries shall not be charged where installed in an aircraft located inside or partially inside a hangar.

Δ **(B) Aircraft Battery Charging and Equipment.** Battery chargers and their control equipment shall not be located or operated within any of the Class I locations defined in 513.3 and shall be located in a separate building or in an area defined in 513.3(D). Mobile chargers shall carry at least one permanently affixed warning sign with the following words or equivalent:

**WARNING
KEEP 5 FT CLEAR OF AIRCRAFT ENGINES
AND FUEL TANK AREAS**

or

**WARNING
KEEP 1.5 METERS CLEAR OF AIRCRAFT ENGINES
AND FUEL TANK AREAS
WARNING
KEEP 1.5 METERS CLEAR OF AIRCRAFT ENGINES
AND FUEL TANK AREAS**

Tables, racks, trays, and wiring shall not be located within a Class I location and shall comply with the requirements of 480.9 and 480.10.

(C) External Power Sources for Energizing Aircraft.

(1) Not Less Than 450 mm (18 in.) Above Floor. Aircraft energizers shall be designed and mounted such that all electrical equipment and fixed wiring will be at least 450 mm (18 in.) above floor level and shall not be operated in a Class I location as defined in 513.3(C).

Δ **(2) Marking for Mobile Units.** Mobile energizers shall carry at least one permanently affixed warning sign with the following words or equivalent:

**WARNING
KEEP 5 FT CLEAR OF AIRCRAFT ENGINES
AND FUEL TANK AREAS**

or

**WARNING
KEEP 1.5 METERS CLEAR OF AIRCRAFT ENGINES
AND FUEL TANK AREAS**

(3) Cords. Flexible cords for aircraft energizers and ground support equipment shall be identified for the type of service and extra-hard usage and shall include an equipment grounding conductor.

(D) Mobile Servicing Equipment with Electrical Components.

Δ **(1) General.** Mobile servicing equipment (such as vacuum cleaners, air compressors, air movers) having electrical wiring and equipment not suitable for Class I, Division 2 or Zone 2 locations shall be so designed and mounted that all such fixed wiring and equipment will be at least 450 mm (18 in.) above the floor. Such mobile equipment shall not be operated within the Class I location defined in 513.3(C) and shall carry at least one permanently affixed warning sign with the following words or equivalent:

**WARNING
KEEP 5 FT CLEAR OF AIRCRAFT ENGINES
AND FUEL TANK AREAS**

or

**WARNING
KEEP 1.5 METERS CLEAR OF AIRCRAFT ENGINES
AND FUEL TANK AREAS**

(2) Cords and Connectors. Flexible cords for mobile equipment shall be suitable for the type of service and identified for extra-hard usage and shall include an equipment grounding conductor. Attachment plugs and receptacles shall be identified for the location in which they are installed and shall provide for connection of the equipment grounding conductor.

(3) Restricted Use. Equipment that is not identified as suitable for Class I, Division 2 locations shall not be operated in locations where maintenance operations likely to release flammable liquids or vapors are in progress.

(E) Portable Equipment.

(1) Portable Lighting Equipment. Portable lighting equipment that is used within a hangar shall be identified for the location in which they are used. For portable luminaires, flexible cord suitable for the type of service and identified for extra-hard usage shall be used. Each such cord shall include a separate equipment grounding conductor.

(2) Portable Utilization Equipment. Portable utilization equipment that is or may be used within a hangar shall be of a type suitable for use in Class I, Division 2 or Zone 2 locations. For portable utilization equipment, flexible cord suitable for the type of service and approved for extra-hard usage shall be used. Each such cord shall include a separate equipment grounding conductor.

513.12 Ground-Fault Circuit-Interrupter Protection for Personnel. Ground-fault circuit-interrupter protection for personnel shall be provided as required in 210.8(B).

513.16 Grounding and Bonding Requirements.

(A) General Grounding Requirements. All metal raceways, the metal armor or metallic sheath on cables, and all non–current-carrying metal parts of fixed or portable electrical equipment, regardless of voltage, shall be grounded. Grounding in Class I locations shall comply with 501.30 for Class I, Division 1 and 2 locations and 505.30 for Zone 0, 1, and 2 locations.

(B) Supplying Circuits with Grounded and Equipment Grounding Conductors in Class I Locations.

(1) Circuits Supplying Portable Equipment or Pendants. Where a circuit supplies portables or pendants and includes a grounded conductor, receptacles, attachment plugs, connectors, and similar devices shall be of the grounding type, and the grounded conductor of the flexible cord shall be connected to the screw shell of any lampholder or to the grounded terminal of any utilization equipment supplied.

(2) Approved Means. Approved means shall be provided for maintaining continuity of the equipment grounding conductor between the fixed wiring system and the non–current-carrying metal portions of pendant luminaires, portable luminaires, and portable utilization equipment.

ARTICLE 514 Motor Fuel Dispensing Facilities

Δ **514.1 Scope.** This article shall apply to motor fuel dispensing facilities, marine/motor fuel dispensing facilities, motor fuel dispensing facilities located inside buildings, and fleet vehicle motor fuel dispensing facilities.

> Informational Note: See NFPA 30A-2021, *Code for Motor Fuel Dispensing Facilities and Repair Garages*, for information regarding safeguards for motor fuel dispensing facilities and for extracted text that is followed by a reference in brackets. Only editorial changes were made to the extracted text to make it consistent with this *Code*.

Article 514 encompasses all locations where volatile flammable liquids or gases are dispensed into the fuel tanks of self-propelled vehicles or other approved fuel tanks. It also includes dispensing at marine facilities such as marinas and boatyards. The phrase "approved containers" in the definition of *motor fuel dispensing facility* in Article 100 covers portable gasoline containers and also applies to dispensing locations for liquefied petroleum gas (LPG), including those locations that do not serve self-propelled vehicles.

N **514.2 Other Articles.** In addition to the requirements of this article, motor fuel dispensing facilities shall comply with Table 514.2, as applicable, except as modified by this article.

N **TABLE 514.2** *Other Articles*

Requirement	Division Classified Locations	Zone Classified Locations
Area classification	500.5, 500.6	505.5, 505.6, 505.7
Equipment	Part III of 501, 500.7, 500.8, 501.5	505.8, 505.9, 505.20, 505.22
Wiring	Part II of 501	505.15, 505.16, 505.17, 505.18, 505.19, 505.26, 505.30

514.3 Classification of Locations. Where the term "Class I" is used with respect to Zone classifications within this article of the *Code*, it shall apply to Zone 0, Zone 1, and Zone 2 designations.

> Informational Note: The term "Class I" was originally included as a prefix to Zone 0, Zone 1, and Zone 2 locations and references

as an identifier for flammable gases, vapors, or liquids to differentiate from Class II and Class III locations. Zone 0, Zone 1, and Zone 2 only apply to flammable gases, vapors, or liquids so the "Class I" prefix is redundant and has been deleted, except for text that is extracted from other documents or to remain consistent throughout this article.

[See Figure 514.3.]

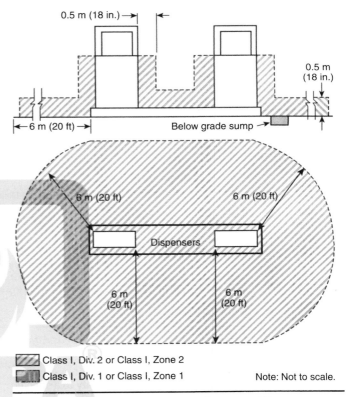

| | Class I, Div. 2 or Class I, Zone 2 |
| | Class I, Div. 1 or Class I, Zone 1 |

Note: Not to scale.

Δ **FIGURE 514.3** *Classified Areas Adjacent to Dispensers. [30A:Figure 8.3.3(a)]*

(A) Unclassified Locations. Where the authority having jurisdiction can satisfactorily determine that flammable liquids having a flash point below 38°C (100°F), such as gasoline, will not be handled, such location shall not be required to be classified.

(B) Classified Locations. [See Figure 514.3(B).]

Many steps are required to properly classify a hazardous location. Although the *NEC*® provides general area classifications in Articles 501, 502, 503, 505, and 506, it does not classify specific locations. The *NEC* classifications for specific occupancies or processes have been extracted from other NFPA documents. The classifications from those documents are based on the premise that all applicable requirements of the document have been met. Deviations in on-site conditions, such as process conditions, area ventilation, and room construction, from those assumed by the document may alter the general classification. Those responsible for the specific area classification must consider the basis for the

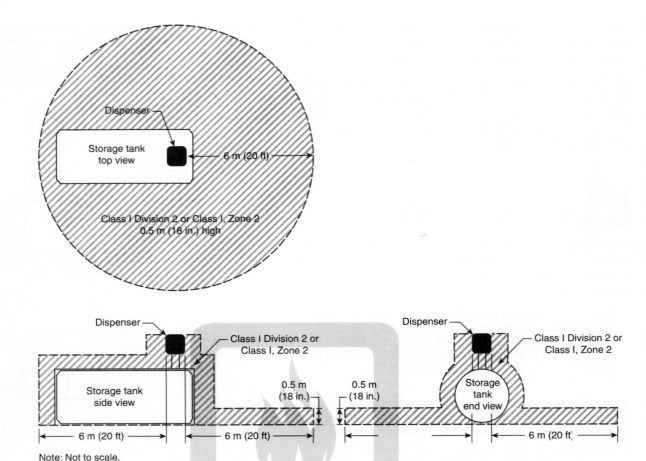

Note: Not to scale.

△ **FIGURE 514.3(B)** *Classified Areas Adjacent to Dispenser Mounted on Aboveground Storage Tank.* [*30A:Figure 8.3.3(b)*]

general classifications to determine the applicability to a specific location.

Tables 514.3(B)(1) and (B)(2) are extracted from NFPA 30A, *Code for Motor Fuel Dispensing Facilities and Repair Garages.* Also see Exhibit 514.1 for an illustration of the Class I location around overhead motor fuel dispensing units.

Aboveground tanks with dispensing equipment are frequently used at fleet motor fuel dispensing facilities. Hazardous area classification for aboveground tank installations also is determined in accordance with Table 514.3(B)(1).

See also

NFPA 30A, which contains the specific construction and installation requirements used to develop the area classifications in Article 514

(1) Class I Locations. Table 514.3(B)(1) shall be applied where Class I liquids are stored, handled, or dispensed and shall be used to delineate and classify motor fuel dispensing facilities and commercial garages as defined in Article 100. Table 515.3 shall be used for the purpose of delineating and classifying aboveground tanks. A Class I location shall not extend beyond an unpierced wall, roof, or other solid partition. [**30A:**8.1, 8.2, 8.3]

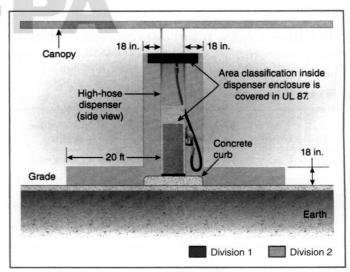

EXHIBIT 514.1 Extent of Class I location around overhead motor fuel dispensing units in accordance with Table 514.3(B)(1).

△ **(2) Compressed Natural Gas, Liquefied Natural Gas, and Liquefied Petroleum Gas Areas.** Table 514.3(B)(2) shall be

TABLE 514.3(B)(1) *Class I Locations — Motor Fuel Dispensing Facilities*

Location	Division (Group D)	Zone (Group IIA)	Extent of Classified Location[1]
Dispensing Device (except Overhead Type) [2,3]			
Under dispenser containment	1	1	Entire space within and under dispenser pit or containment
Dispenser	2	2	Within 450 mm (18 in.) of dispenser enclosure or that portion of dispenser enclosure containing liquid-handling components, extending horizontally in all directions and down to grade level
Outdoor	2	2	Up to 450 mm (18 in.) above grade level, extending 6 m (20 ft) horizontally in all directions from dispenser enclosure
Indoor			
– with mechanical ventilation	2	2	Up to 450 mm (18 in.) above floor level, extending 6 m (20 ft) horizontally in all directions from dispenser enclosure
– with gravity ventilation	2	2	Up to 450 mm (18 in.) above floor level, extending 7.5 m (25 ft) horizontally in all directions from dispenser enclosure
Dispensing Device — Overhead Type[4]	1	1	Space within dispenser enclosure and all electrical equipment integral with dispensing hose or nozzle
	2	2	Within 450 mm (18 in.) of dispenser enclosure, extending horizontally in all directions and down to grade level
	2	2	Up to 450 mm (18 in.) above grade level, extending 6 m (20 ft) horizontally in all directions from a point vertically below edge of dispenser enclosure
Remote Pump —			
Outdoor	1	1	Entire space within any pit or box below grade level, any part of which is within 3 m (10 ft) horizontally from any edge of pump
	2	2	Within 900 mm (3 ft) of any edge of pump, extending horizontally in all directions
	2	2	Up to 450 mm (18 in.) above grade level, extending 3 m (10 ft) horizontally in all directions from any edge of pump
Indoor	1	1	Entire space within any pit
	2	2	Within 1.5 m (5 ft) of any edge of pump, extending in all directions
	2	2	Up to 900 mm (3 ft) above floor level, extending 7.5 m (25 ft) horizontally in all directions from any edge of pump
Sales, Storage, Rest Rooms including structures (such as the attendant's kiosk) on or adjacent to dispensers	unclassified	unclassified	Except as noted below
	1	1	Entire volume, if there is any opening to room within the extent of a Division 1 or Zone 1 location
	2	2	Entire volume, if there is any opening to room within the extent of a Division 2 or Zone 2 location
Tank, Aboveground			
Inside tank	1	0	Entire inside volume
Shell, ends, roof, dike area	1	1	Entire space within dike, where dike height exceeds distance from tank shell to inside of dike wall for more than 50 percent of tank circumference
	2	2	Entire space within dike, where dike height does not exceed distance from tank shell to inside of dike wall for more than 50 percent of tank circumference
Vent	2	2	Within 3 m (10 ft) of shell, ends, or roof of tank
	1	1	Within 1.5 m (5 ft) of open end of vent, extending in all directions
	2	2	Between 1.5 m and 3 m (5 ft and 10 ft) from open end of vent, extending in all directions
Tank, Underground			
Inside tank	1	0	Entire inside volume
Fill opening	1	1	Entire space within any pit or box below grade level, any part of which is within a Division 1 or Division 2 classified location or within a Zone 1 or Zone 2 classified location
	2	2	Up to 450 mm (18 in.) above grade level, extending 1.5 m (5 ft) horizontally in all directions from any tight-fill connection and extending 3 m (10 ft) horizontally in all directions from any loose-fill connection

(continues)

TABLE 514.3(B)(1) *Continued*

Location	Division (Group D)	Zone (Group IIA)	Extent of Classified Location[1]
Vent	1	1	Within 1.5 m (5 ft) of open end of vent, extending in all directions
	2	2	Between 1.5 m and 3 m (5 ft and 10 ft) from open end of vent, extending in all directions
Vapor Processing System			
Pits	1	1	Entire space within any pit or box below grade level, any part of which: (1) is within a Division 1 or Division 2 classified location; (2) is within a Zone 1 or Zone 2 classified location; (3) houses any equipment used to transfer or process vapors
Equipment in protective enclosures	2	2	Entire space within enclosure
Equipment *not* within protective enclosure	2	2	Within 450 mm (18 in.) of equipment containing flammable vapors or liquid, extending horizontally in all directions and down to grade level
	2	2	Up to 450 mm (18 in.) above grade level within 3 m (10 ft) horizontally of the vapor processing equipment
– Equipment enclosure	1	1	Entire space within enclosure, if flammable vapor or liquid is present under normal operating conditions
	2	2	Entire space within enclosure, if flammable vapor or liquid is not present under normal operating conditions
– Vacuum assist blower	2	2	Within 450 mm (18 in.) of blower, extending horizontally in all directions and down to grade level
	2	2	Up to 450 mm (18 in.) above grade level, extending 3 m (10 ft) horizontally in all directions
Vault	1	1	Entire interior space, if Class I liquids are stored within

[1]For marine application, *grade level* means the surface of a pier, extending down to water level.
[2]Refer to Figure 514.3 and Figure 514.3(B) for an illustration of classified location around dispensing devices.
[3]Area classification inside the dispenser enclosure is covered in UL 87, *Standard for Power-Operated Dispensing Devices for Petroleum Products.*
[4]Ceiling-mounted hose reel. [**30A:**Table 8.3.1]

TABLE 514.3(B)(2) *Electrical Equipment Classified Areas for Dispensing Devices*

Dispensing Device	Extent of Classified Area	
	Class I, Division 1	**Class I, Division 2**
Compressed natural gas (CNG)	Entire space within the dispenser enclosure	1.5 m (5 ft) in all directions from dispenser enclosure
Liquefied natural gas (LNG)	Entire space within the dispenser enclosure	3 m (10 ft) in all directions from the dispenser enclosure
Liquefied petroleum gas (LP-Gas)	Entire space within the dispenser enclosure; 450 mm (18 in.) from the exterior surface of the dispenser enclosure to an elevation of 1.22 m (4 ft) above the base of the dispenser; the entire pit or open space beneath the dispenser and within 6 m (20 ft) horizontally from any edge of the dispenser when the pit or trench is not mechanically ventilated	Up to 450 mm (18 in.) above ground and within 6 m (20 ft) horizontally from any edge of the dispenser enclosure, including pits or trenches within this area when provided with adequate mechanical ventilation

[**30A:**Table 12.6.2]

used to delineate and classify areas where CNG, LNG, compressed or liquefied hydrogen, LP-Gas, or combinations of these, are dispensed as motor vehicle fuels along with Class I or Class II liquids that are also dispensed as motor vehicle fuels. [**30A:**12.1]

Where CNG or LNG dispensers are installed beneath a canopy or enclosure, either the canopy or enclosure shall be designed to prevent accumulation or entrapment of ignitible vapors or all electrical equipment installed beneath the canopy or enclosure shall be suitable for Class I, Division 2 hazardous (classified) locations. [**30A:**12.4]

Dispensing devices for LP-Gas shall be located as follows:

(1) At least 3 m (10 ft) from any dispensing device for Class I liquids

(2) At least 1.5 m (5 ft) from any dispensing device for Class I liquids where the following conditions exist:

a. The LP-Gas deliver nozzle and filler valve release no more than 4 cm³ (0.1 oz) of liquid upon disconnection.

b. The fixed maximum liquid level gauge remains closed during the entire refueling process.

[**30A:**12.5.2]

Informational Note No. 1: See NFPA 58, *Liquefied Petroleum Gas Code*, for requirements on dispensing devices for LP-Gas.

Informational Note No. 2: See NFPA 58, *Liquefied Petroleum Gas Code*, and NFPA 59, *Utility LP-Gas Plant Code*, for information on classified areas pertaining to LP-Gas systems other than residential or commercial.

Informational Note No. 3: See 514.3(C) for motor fuel dispensing stations in marinas and boatyards.

Δ **(3) Fuel Storage.**

(a) Aboveground tanks storing CNG or LNG shall be separated from any adjacent property line that is or can be built upon, any public way, and the nearest important building on the same property. [**30A:**12.3.1]

Informational Note: See NFPA 52, *Vehicular Natural Gas Fuel Systems Code*, Section 8.4, for the relevant distances for CNG and LNG.

(b) Aboveground tanks storing hydrogen shall be separated from any adjacent property line that is or can be built upon, any public way, and the nearest important building on the same property. [**30A:**12.3.2]

Informational Note: See NFPA 2, *Hydrogen Technologies Code*, for the relevant distances for hydrogen.

(c) Aboveground tanks storing LP-Gas shall be separated from any adjacent property line that is or can be built upon, any public way, and the nearest important building on the same property. [**30A:**12.3.3]

Informational Note: See NFPA 58, *Liquefied Petroleum Gas Code*, Section 6.3, for the relevant distances for LP-Gas.

(d) Aboveground tanks storing CNG, LNG, or LP-Gas shall be separated from each other by at least 6 m (20 ft) and from dispensing devices that dispense liquid or gaseous motor vehicle fuels by at least 6 m (20 ft). [**30A:**12.3.3]

Exception No. 1: The required separation shall not apply to tanks or dispensers storing or handling fuels of the same chemical composition.

Exception No. 2: The required separation shall not apply when both the gaseous fuel storage and dispensing equipment are at least 15 m (50 ft) from any other aboveground motor fuel storage or dispensing equipment.

Informational Note: See NFPA 52, *Vehicular Natural Gas Fuel Systems Code*, or NFPA 58, *Liquefied Petroleum Gas Code*, for additional information.

(e) *Dispenser Installations Beneath Canopies.* Where CNG or LNG dispensers are installed beneath a canopy or enclosure, either the canopy or enclosure shall be designed to prevent accumulation or entrapment of ignitible vapors or all electrical equipment installed beneath the canopy or enclosure shall be suitable for Class I, Division 2 hazardous (classified) locations. [**30A:**12.4]

(f) *Specific Requirements for LP-Gas Dispensing Devices.* [**30A:**12.5] Dispensing devices for LP-Gas shall be located as follows:

(1) At least 3 m (10 ft) from any dispensing device for Class I liquids

(2) At least 1.5 m (5 ft) from any dispensing device for Class I liquids where the following conditions exist:

a. The LP-Gas deliver nozzle and filler valve release no more than 4 cm³ (0.1 oz) of liquid upon disconnection.

b. The fixed maximum liquid level gauge remains closed during the entire refueling process. [**30A:**12.5.2]

Table 514.3(B)(2) shall be used to delineate and classify areas for the purpose of installation of electrical wiring and electrical utilization equipment.

Δ **(C) Motor Fuel Dispensing Stations in Boatyards and Marinas.**

Informational Note: See NFPA 303, *Fire Protection Standard for Marinas and Boatyards*, and NFPA 30A, *Code for Motor Fuel Dispensing Facilities and Repair Garages*, for additional information.

(1) General. Electrical wiring and equipment located at or serving motor fuel dispensing locations shall be installed on the side of the wharf, pier, or dock opposite from the liquid piping system.

Δ **(2) Classification of Class I, Division 1 and 2 Areas.** The criteria provided in 514.3(C)(2)(a) and (C)(2)(b) shall be used for the purposes of applying Table 514.3(B)(1) and Table 514.3(B)(2) to motor fuel dispensing equipment on floating or fixed piers, wharfs, or docks.

(a) *Closed Construction.* Where the construction of floating docks, piers, or wharfs is closed so that there is no space between the bottom of the dock, pier, or wharf and the water, as in the case of concrete-enclosed expanded foam or similar construction, and the construction includes integral service boxes with supply chases, the following shall apply:

(1) The space above the surface of the floating dock, pier, or wharf shall be a Class I, Division 2 location with distances in accordance with Table 514.3(B)(1) for dispenser and outdoor locations.

(2) Spaces below the surface of the floating dock, pier, or wharf that have areas or enclosures, such as tubs, voids, pits, vaults, boxes, depressions, fuel piping chases, or similar spaces, where flammable liquid or vapor can accumulate shall be a Class I, Division 1 location.

Exception No. 1: Dock, pier, or wharf sections that do not support fuel dispensers and abut, but are located 6.0 m (20 ft) or more from, dock sections that support a fuel dispenser(s) shall be permitted to be Class I, Division 2 locations where documented air space is provided between dock sections to allow flammable liquids or vapors to dissipate without traveling to such dock sections. The documentation shall comply with the requirements of 500.4.

Exception No. 2: Dock, pier, or wharf sections that do not support fuel dispensers and do not directly abut sections that support fuel dispensers shall be permitted to be unclassified where documented air space is provided and where flammable liquids or vapors cannot travel to such dock sections. The documentation shall comply with the requirements of 500.4.

(b) *Open Construction.* Where the construction of piers, wharfs, or docks is open, as in the case of decks built on stringers supported by pilings, floats, pontoons, or similar construction, the following shall apply:

(1) The area 450 mm (18 in.) above the surface of the dock, pier, or wharf and extending 6.0 m (20 ft) horizontally in all directions from the outside edge of the dispenser and down to the water level shall be a Class 1, Division 2 location.

(2) Enclosures such as tubs, voids, pits, vaults, boxes, depressions, piping chases, or similar spaces where flammable liquids or vapors can accumulate within 6.0 m (20 ft) of the dispenser shall be a Class I, Division 1 location.

• Δ **514.4 Wiring and Equipment Installed in Hazardous (Classified) Locations.** All electrical equipment and wiring installed in the hazardous (classified) locations specified in 514.3 shall comply with Parts II and III of Article 501. Conductor insulation in these locations shall comply with 501.20.

The applicable wiring methods and equipment in Article 501 must be used within the Class I areas at a motor fuel dispensing facility. An explosionproof junction box of the type frequently used in gasoline dispensing units is shown in Exhibit 514.2. The

EXHIBIT 514.2 A typical explosionproof junction box used in a dispenser application. *(Courtesy of Eaton, Crouse-Hinds Division)*

branch-circuit conductors for dispenser power, lighting, or both connect to the internal wiring of the dispenser in the explosionproof junction box.

514.7 Wiring and Equipment Above Hazardous (Classified) Locations. Fixed wiring and equipment above hazardous (classified) locations shall be installed in accordance with 514.3 and shall be one or more of the following:

(1) Rigid metal conduit (RMC) or intermediate metal conduit (IMC) with listed threaded or threadless fittings, or electrical metallic tubing (EMT) with listed fittings.

(2) Rigid polyvinyl chloride conduit (PVC), reinforced thermosetting resin conduit (RTRC), or electrical nonmetallic tubing (ENT).

(3) Flexible metal conduit (FMC), liquidtight flexible metal conduit (LFMC), or liquidtight flexible nonmetallic conduit (LFNC), with listed fittings.

(4) Type MC cable, Type AC cable, Type TC cable, or Type TC-ER cable, including installation in cable trays, with listed fittings. Type TC-ER cable shall include an equipment grounding conductor (EGC) in addition to any drain wire.

(5) Type MI cable terminated with listed fittings and supported to avoid tensile stress.

(6) Manufactured wiring systems.

(7) Type PLTC cable or Type PLTC-ER cable used in Class 2 or Class 3 circuits. Type PLTC-ER cable shall include an equipment grounding conductor (EGC) in addition to any drain wire.

(8) Type ITC cable or ITC-ER cable in accordance with 335.4 and 335.5 and terminated with listed fittings. Type ITC-ER cable shall include an equipment grounding conductor (EGC) in addition to any drain wire.

(9) Cellular metal floor raceways or cellular concrete floor raceways only for supplying ceiling outlets or extensions to the area below the floor. Such raceways shall have no connections leading into or through any Class I location above the floor.

Δ **514.8 Underground Wiring.** All underground wiring shall comply with 514.8(A), (B), or (C).

If fuel spilled in the vicinity of gasoline dispensers seeps into the ground, it could migrate into underground electrical conduits. Therefore, all conduits installed below the hazardous locations of a motor fuel dispensing facility are required to be sealed within 10 feet of the point of emergence from below grade. This boundary seal minimizes the passage of gasoline or other fuel vapors into unclassified locations where the electrical equipment is not explosionproof or otherwise protected. Tables 514.3(B)(1) and (B)(2) define the extent of the aboveground Class I, Divisions 1 and 2 locations.

N **(A) Metal Conduit.** Threaded rigid metal conduit (RMC) or threaded intermediate metal conduit (IMC) with listed threaded fittings shall be permitted. Any portion of electrical wiring that is below the surface of a Class I, Division 1 or Division 2 location [as classified in Table 514.3(B)(1) and Table 514.3(B)(2)] shall be sealed within 3.05 m (10 ft) of the point of emergence above grade. The conduit shall not contain any unions, couplings, boxes, or fittings between the conduit seal and the point of emergence above grade.

N **(B) Type MI Cable.** Type MI cable shall be permitted where it is installed in accordance with Part II of Article 332.

N **(C) Nonmetallic Conduit.** Rigid polyvinyl chloride conduit (PVC), reinforced thermosetting resin conduit (RTRC), or high-density polyethylene conduit (HDPE) shall be permitted where buried under not less than 600 mm (2 ft) of cover. Where PVC conduit, RTRC conduit, or HDPE conduit is used, threaded rigid metal conduit (RMC) or threaded intermediate metal conduit (IMC) shall be used for the last 600 mm (2 ft) of the underground run to emergence or to the point of connection to the aboveground raceway. An equipment grounding conductor (EGC) shall be included to provide electrical continuity of the raceway system and for grounding of non–current-carrying metal parts.

Where polyvinyl chloride conduit (PVC), reinforced thermosetting resin conduit (RTRC), or high-density polyethylene (HDPE) conduit is used for underground wiring, threaded rigid metal conduit or threaded steel intermediate metal conduit must be used for the last 2 feet of the underground run to the point of emergence or to the point of connection to the aboveground raceway. These rigid nonmetallic conduits, including any nonmetallic conduit elbows and fittings, must be located not less than 2 feet below grade, as shown in Exhibit 514.3.

If rigid nonmetallic conduit is used, an equipment grounding conductor (EGC) must be included and must be bonded to the explosionproof raceway system that is installed inside the

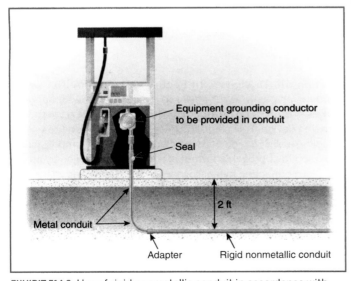

EXHIBIT 514.3 Use of rigid nonmetallic conduit in accordance with 514.8(C).

dispenser. Installation is accomplished by terminating the EGC on the ground screw (or other means) provided in the dispenser junction box.

514.9 Sealing.

(A) At Dispenser. A listed seal shall be provided in each conduit run entering or leaving a dispenser or any cavities or enclosures in direct communication therewith. The sealing fitting or listed explosionproof reducer at the seal shall be the first fitting after the conduit emerges from the earth or concrete.

(B) At Boundary. Additional seals shall be provided in accordance with 501.15. Sections 501.15(A)(4) and (B)(2) shall apply to horizontal as well as to vertical boundaries of the defined Class I locations.

Seal fittings are required in all conduits entering or leaving a dispenser and leaving a Class I location. Even though a conduit runs from dispenser to dispenser and does not leave the hazardous location, seals are necessary where the conduit enters each dispenser. Conduits passing under the boundaries of the hazardous locations (20-foot radius from dispenser, 10-foot radius from a loose-fill tank connection, and 5-foot radius from a tight-fill tank connection) are considered to be in a Class I location, and the seal is to be the first fitting at the point of emergence.

Panelboards generally are located in a room or area that is an unclassified location; however, any conduit coming from the dispenser or passing under the hazardous location boundaries is required to be sealed at the panelboard location to minimize the likelihood of gas migration into the remote location. If the panelboard is located in the lube or repair room, all conduits emerging into the 18-inch hazardous location are required to be sealed. See Exhibits 514.4 and 514.5.

514.11 Circuit Disconnects.

Δ **(A) Emergency Electrical Disconnects.** Fuel dispensing systems shall be provided with one or more clearly identified emergency shutoff devices or electrical disconnects. Such devices or disconnects shall be installed in approved locations but not less than 6 m (20 ft) or more than 30 m (100 ft) from the fuel dispensing devices that they serve. Emergency shutoff devices or electrical disconnects shall disconnect power to all dispensing devices; to all remote pumps serving the dispensing devices; to all associated power, control, and signal circuits; and to all other electrical equipment in the hazardous (classified) locations surrounding the fuel dispensing devices. When more than one emergency shutoff device or electrical disconnect is provided, all devices shall be interconnected. Resetting from an emergency shutoff condition shall require manual intervention and the manner of resetting shall be approved by the authority having jurisdiction. [**30A:**6.7] The emergency shutoff device shall disconnect simultaneously from the source of supply, all conductors of the circuits, including the grounded conductor, if any. Equipment grounding conductors shall remain connected.

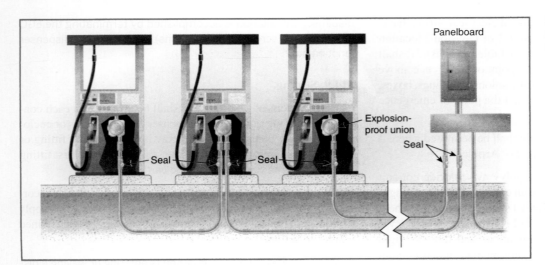

EXHIBIT 514.4 A gasoline dispenser installation indicating locations for seal fittings.

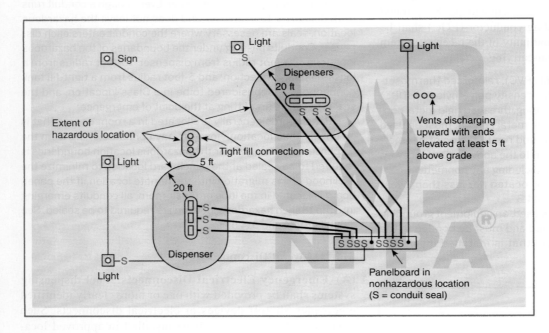

EXHIBIT 514.5 Required seals at points marked "S." Seals are not required at the sign and two of the lights because conduit runs do not pass through a hazardous location.

Exception: Intrinsically safe systems shall not be required to meet this requirement.

(B) Attended Self-Service Motor Fuel Dispensing Facilities. At attended motor fuel dispensing facilities, the devices or disconnects shall be readily accessible to the attendant. [**30A:**6.7.1]

(C) Unattended Self-Service Motor Fuel Dispensing Facilities. At unattended motor fuel dispensing facilities, the devices or disconnects shall be readily accessible to patrons and at least one additional device or disconnect shall be readily accessible to each group of dispensing devices on an individual island. [**30A:**6.7.2]

Section 514.11 aligns with Section 6.7 of NFPA 30A, which requires an easily accessible and clearly identified emergency power shutoff to be provided at a location remote from the dispensing device. This shutoff is necessary because a fire or large gasoline spill at the dispensing island may make it impossible for a person to approach and shut off the flow of gasoline by operating a disconnecting means located at the dispensing island.

The distance requirements in 514.11(A), extracted from NFPA 30A, address both attended and unattended self-service dispensing facilities. The term *clearly identified* means that a sign must be posted indicating where the shutoff switch is located. This emergency power shutoff must be readily accessible and not blocked by such things as tires, cases of lubricating oil, or display merchandise. All dispensing facility operators as well as

EXHIBIT 514.6 A self-service dispensing island *(left)* and associated emergency disconnecting means *(right)*.

responding fire fighters should know the location of the emergency power shutoff.

Exhibit 514.6 shows a typical dispensing island at a self-service gas station (left) and the required clearly identified emergency disconnect (right). During the evening hours, this location is an unattended self-service dispensing facility. The disconnecting means must be located where acceptable to the AHJ and is required to be located between 20 feet and 100 feet from the dispensers.

514.13 Provisions for Maintenance and Service of Dispensing Equipment. Each dispensing device shall be provided with a means to remove all external voltage sources, including power, communications, data, and video circuits and including feedback, during periods of maintenance and service of the dispensing equipment. The location of this means shall be permitted to be other than inside or adjacent to the dispensing device. The means shall be capable of being locked in the open position in accordance with 110.25.

Because sophisticated control circuitry is integrated into dispensing equipment, simply shutting off the main power source to the dispenser or remote pump does not necessarily ensure that the equipment has been isolated from all sources of voltage. For the safety of personnel servicing dispensing equipment, a means to disconnect all external voltage sources from each dispensing device, including sources that may backfeed into the dispenser, must be installed so that the equipment is completely isolated from all voltage sources. Exhibit 514.7 shows the interior of an electronic dispensing unit. Power is required to be removed from printers, displays, keypads, and card readers.

514.16 Grounding and Bonding. All metal raceways, the metal armor or metallic sheath on cables, and all non–current-carrying metal parts of fixed and portable electrical equipment, regardless of voltage, shall be grounded and bonded. Grounding and bonding in Class I locations shall comply with 501.30.

EXHIBIT 514.7 An example of the complexity of the interior of electronic dispensing units.

ARTICLE 515 Bulk Storage Plants

Δ **515.1 Scope.** This article covers a property or portion of a property where flammable liquids are received by tank vessel, pipelines, tank car, or tank vehicle and are stored or blended in bulk for the purpose of distributing such liquids by tank vessel, pipeline, tank car, tank vehicle, portable tank, or container.

> Informational Note: See NFPA 30-2021, *Flammable and Combustible Liquids Code*, for extracted text that is followed by a reference in brackets. Only editorial changes were made to the extracted text to make it consistent with this *Code*.

Article 515 covers facilities that store (in bulk) and distribute flammable liquids, as opposed to dispensing liquids into fuel tanks of vehicles. Flammable liquid dispensing locations, including those within the bulk storage facility, are covered under Article 514. Bulk storage tanks may be located inside buildings or outside, either above ground or underground. This article addresses the hazardous locations in the vicinity of the storage tank, as well as the tank vehicle, pier, or wharf from which the liquids are loaded and off-loaded. This article also covers the classification of the areas around drum storage containers. Area classification information for facilities at which liquified natural gas is received or stored in bulk is found in NFPA 59A, *Standard for the Production, Storage and Handling of Liquefied Natural Gas (LNG)*.

N **515.2 Other Articles.** In addition to the requirements of this article, bulk storage plants shall comply with Table 515.2, as applicable, except as modified by this article.

N **TABLE 515.2** *Other Articles*

Requirement	Division Classified Locations	Zone Classified Locations
Area classification	500.5, 500.6	505.5, 505.6, 505.7
Equipment	Part III of 501, 500.7, 500.8, 501.5	505.8, 505.9, 505.20, 505.22
Wiring	Part II of 501	505.15, 505.16, 505.17, 505.18, 505.19, 505.26, 505.30

Δ **515.3 Classified Locations.** Where the term "Class I" is used with respect to Zone classifications within this article of the *Code*, it shall apply to Zone 0, Zone 1, and Zone 2 designations.

> Informational Note No. 1: The term "Class I" was originally included as a prefix to Zone 0, Zone 1, and Zone 2 locations and references as an identifier for flammable gases, vapors, or liquids to differentiate from Class II and Class III locations. Zone 0, Zone 1, and Zone 2 only apply to flammable gases, vapors, or liquids so the "Class I" prefix is redundant and has been deleted, except for text that is extracted from other documents or to remain consistent throughout this article.

Table 515.3 shall be applied where Class I flammable liquids are stored, handled, or dispensed and shall be used to delineate and classify bulk storage plants. The classified location shall not extend beyond a floor, wall, roof, or other solid partition that has no communicating openings.

Where the installation does not meet the requirements found in Table 515.3, the authority having jurisdiction shall have the authority to classify the extent of the classified space.

> Informational Note No. 2: See NFPA 30, *Flammable and Combustible Liquids Code*, Chapter 5, for the area classifications listed in Table 515.3 that are based on the premise that the installation meets all the applicable requirements.
> Informational Note No. 3: See 514.3(C) through (E) for gasoline dispensing stations in marinas and boatyards.
> Informational Note No. 4: See NFPA 30, *Flammable and Combustible Liquids Code*, Section 7.3, for additional information.

Many steps are required to properly classify a hazardous location. Although the *NEC®* provides general area classifications in Articles 501, 502, 503, 505, and 506, it does not classify specific locations. The *NEC* classifications for specific occupancies or processes have been extracted from other NFPA documents. The classifications from those documents are based on the premise that all applicable requirements of the document have been met. Deviations in on-site conditions, such as process conditions, area ventilation, and room construction, from those assumed by the document may alter the general classification. Those responsible for the specific area classification must consider the basis for the general classifications to determine the applicability to a specific location.

NFPA 30, *Flammable and Combustible Liquids Code*, contains the specific construction and installation requirements used to develop the area classifications in Article 515. Table 515.3 is extracted from NFPA 30, Table 7.3.3. The area classifications listed in the table are based on the premise that all applicable requirements of NFPA 30 have been met as conveyed by Informational Note No. 1 to 515.3.

Exhibits 515.1 through 515.5 illustrate the hazardous locations associated with several types of flammable liquid containers and operations.

Δ **TABLE 515.3** *Electrical Area Classifications*

Location	Division	Zone	Extent of Classified Area
Indoor equipment installed where flammable vapor–air mixtures can exist under normal operation (see Informational Note)	1	0	The entire area associated with such equipment where flammable gases or vapors are present continuously or for long periods of time
	1	1	Area within 1.5 m (5 ft) of any edge of such equipment, extending in all directions
	2	2	Area between 1.5 m and 2.5 m (5 ft and 8 ft) of any edge of such equipment, extending in all directions; also, space up to 900 mm (3 ft) above floor or grade level within 1.5 m to 7.5 m (5 ft to 25 ft) horizontally from any edge of such equipment[1]
Outdoor equipment installed where flammable vapor–air mixtures can exist under normal operation	1	0	The entire area associated with such equipment where flammable gases or vapors are present continuously or for long periods of time
	1	1	Area within 900 mm (3 ft) of any edge of such equipment, extending in all directions
	2	2	Area between 900 mm (3 ft) and 2.5 m (8 ft) of any edge of such equipment, extending in all directions; also, space up to 900 mm (3 ft) above floor or grade level within 900 mm to 3.0 m (3 ft to 10 ft) horizontally from any edge of such equipment
Tank storage installations inside buildings	1	1	All equipment located below grade level
	2	2	Any equipment located at or above grade level
Tank — aboveground, fixed roof	1	0	Inside fixed roof tank
	1	1	Area inside dike where dike height is greater than the distance from the tank to the dike for more than 50 percent of the tank circumference
	2	2	Within 3.0 m (10 ft) from shell, ends, or roof of tank; also, area inside dike to level of top of dike wall
	1	0	Area inside of vent piping or opening
	1	1	Within 1.5 m (5 ft) of open end of vent, extending in all directions
	2	2	Area between 1.5 m and 3.0 m (5 ft and 10 ft) from open end of vent, extending in all directions
Tank — aboveground, floating roof			
With fixed outer roof	1	0	Area between the floating and fixed roof sections and within the shell
With no fixed outer roof	1	1	Area above the floating roof and within the shell
Tank vault — interior	1	1	Entire interior volume, if Class I liquids are stored within
Underground tank fill opening	1	1	Any pit, box, or space below grade level, if any part is within a Division 1 or 2, or Zone 1 or 2 classified location
	2	2	Up to 450 mm (18 in.) above grade level within a horizontal radius of 3.0 m (10 ft) from a loose fill connection, and within a horizontal radius of 1.5 m (5 ft) from a tight fill connection
Vent — discharging upward	1	0	Area inside of vent piping or opening
	1	1	Within 900 mm (3 ft) of open end of vent, extending in all directions
	2	2	Area between 900 mm and 1.5 m (3 ft and 5 ft) of open end of vent, extending in all directions
Drum and container filling — outdoors or indoors	1	0	Area inside the drum or container
	1	1	Within 900 mm (3 ft) of vent and fill openings, extending in all directions
	2	2	Area between 900 mm and 1.5 m (3 ft and 5 ft) from vent or fill opening, extending in all directions; also, up to 450 mm (18 in.) above floor or grade level within a horizontal radius of 3.0 m (10 ft) from vent or fill opening
Pumps, bleeders, withdrawal fittings			
Indoor	2	2	Within 1.5 m (5 ft) of any edge of such devices, extending in all directions; also, up to 900 mm (3 ft) above floor or grade level within 7.5 m (25 ft) horizontally from any edge of such devices
Outdoor	2	2	Within 900 mm (3 ft) of any edge of such devices, extending in all directions. Also, up to 450 mm (18 in.) above grade level within 3.0 m (10 ft) horizontally from any edge of such devices

(continues)

Δ   **TABLE 515.3**   *Continued*

Location	Division	Zone	Extent of Classified Area
Pits and sumps			
Without mechanical ventilation	1	1	Entire area within a pit or sump if any part is within a Division 1 or 2 or Zone 1 or 2 classified location
With adequate mechanical ventilation	2	2	Entire area within a pit or sump if any part is within a Division 1 or 2 or Zone 1 or 2 classified location
Containing valves, fittings, or piping, and not within a Division 1 or 2 or Zone 1 or 2 classified location	2	2	Entire pit or sump
Drainage ditches, separators, impounding basins			
Outdoor	2	2	Area up to 450 mm (18 in.) above ditch, separator, or basin; also, area up to 450 mm (18 in.) above grade within 4.5 m (15 ft) horizontally from any edge
Indoor			Same as pits and sumps
Tank vehicle and tank car[2]			
Loading through open dome	1	0	Area inside of the tank
	1	1	Within 900 mm (3 ft) of edge of dome, extending in all directions
	2	2	Area between 900 mm and 4.5 m (3 ft and 15 ft) from edge of dome, extending in all directions
Loading through bottom connections with atmospheric venting	1	0	Area inside of the tank
	1	1	Within 900 mm (3 ft) of point of venting to atmosphere, extending in all directions
	2	2	Area between 900 mm and 4.5 m (3 ft and 15 ft) from point of venting to atmosphere, extending in all directions; also, up to 450 mm (18 in.) above grade within a horizontal radius of 3.0 m (10 ft) from point of loading connection
Loading through closed dome with atmospheric venting	1	1	Within 900 mm (3 ft) of open end of vent, extending in all directions
	2	2	Area between 900 mm and 4.5 m (3 ft and 15 ft) from open end of vent, extending in all directions; also, within 900 mm (3 ft) of edge of dome, extending in all directions
Loading through closed dome with vapor control	2	2	Within 900 mm (3 ft) of point of connection of both fill and vapor lines extending in all directions
Bottom loading with vapor control or any bottom unloading	2	2	Within 900 mm (3 ft) of point of connections, extending in all directions; also, up to 450 mm (18 in.) above grade within a horizontal radius of 3.0 m (10 ft) from point of connections
Storage and repair garage for tank vehicles	1	1	All pits or spaces below floor level
	2	2	Area up to 450 mm (18 in.) above floor or grade level for entire storage or repair garage
Garages for other than tank vehicles	Unclassified		If there is any opening to these rooms within the extent of an outdoor classified location, the entire room shall be classified the same as the area classification at the point of the opening.
Outdoor drum storage	Unclassified		
Inside rooms or storage lockers used for the storage of Class I liquids	2	2	Entire room or locker
Indoor warehousing where there is no flammable liquid transfer	Unclassified		If there is any opening to these rooms within the extent of an indoor classified location, the classified location shall extend through the opening to the same extent as if the wall, curb, or partition did not exist.
Office and rest rooms	Unclassified		If there is any opening to these rooms within the extent of an indoor classified location, the room shall be classified the same as if the wall, curb, or partition did not exist.
Piers and wharves			See Figure 515.3.

[1]The release of Class I liquids can generate vapors to the extent that the entire building, and possibly an area surrounding it, should be considered a Class I, Division 2 or Zone 2 location.

[2]When classifying extent of area, consideration shall be given to the fact that tank cars or tank vehicles can be spotted at varying points. Therefore, the extremities of the loading or unloading positions shall be used. [**30:**Table 7.3.3]

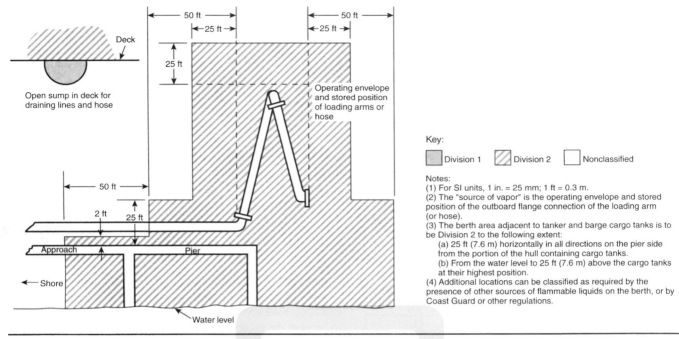

FIGURE 515.3 *Area Classification for a Marine Terminal Handling Flammable Liquids.* [*30:Figure 29.3.22*]

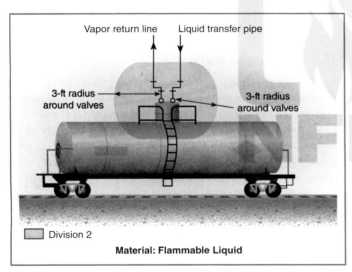

EXHIBIT 515.1 Tank car/tank truck loading and unloading via closed system with transfer through dome only.

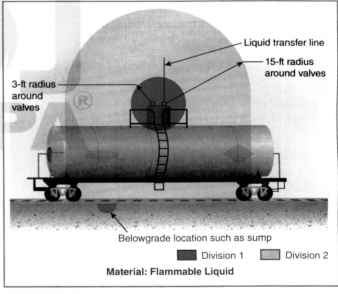

EXHIBIT 515.2 Open system with top or bottom product transfer.

Exhibits 515.1 and 515.2 depict the classification difference between using a closed and an open transfer system on a tank car.

Δ **515.4 Wiring and Equipment Located in Hazardous (Classified) Locations.** All electrical wiring and equipment within the hazardous (classified) locations specified in 515.3 shall comply with the applicable requirements of Table 515.2.

515.7 Wiring and Equipment Above Hazardous (Classified) Locations.

Δ **(A) Fixed Wiring.** All fixed wiring above hazardous (classified) locations shall comply with 501.10(B) or 505.15(C), as applicable.

Δ **(B) Fixed Equipment.** Fixed equipment that might produce arcs, sparks, or particles of hot metal, such as lamps and

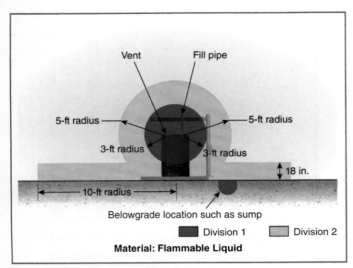

EXHIBIT 515.3 Drum filling station, outdoors or indoors, with adequate ventilation.

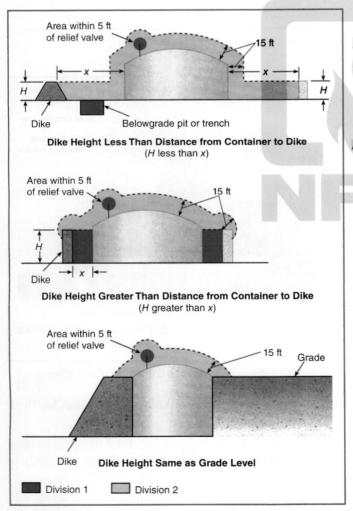

EXHIBIT 515.4 Storage tanks for cryogenic liquids. [Source: Adapted from NFPA 59A-2019, Figures 11.9.2(a) through 11.9.2(c)]

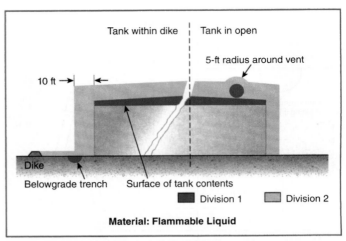

EXHIBIT 515.5 Fixed roof storage tank, outdoors at grade.

lampholders for fixed lighting, cutouts, switches, receptacles, motors, or other equipment having make-and-break or sliding contacts, shall be of the totally enclosed type or be constructed to prevent the escape of sparks or hot metal particles.

△ **(C) Portable Luminaires or Other Utilization Equipment.** Portable luminaires or other utilization equipment and their flexible cords shall comply with Part III of Article 501 or 505.17 for the class of location above which they are connected or used.

515.8 Underground Wiring.

△ **(A) Wiring Method.** Underground wiring shall be installed in threaded rigid metal conduit or threaded steel intermediate metal conduit or, where buried under not less than 600 mm (2 ft) of cover, shall be permitted in PVC conduit, RTRC conduit, or a listed cable. Where PVC conduit or RTRC conduit is used, threaded rigid metal conduit or threaded steel intermediate metal conduit shall be used for not less than the last 600 mm (2 ft) of the conduit run to the conduit point of emergence from the underground location or to the point of connection to an aboveground raceway. Where cable is used, it shall be enclosed in threaded rigid metal conduit or threaded steel intermediate metal conduit from the point of lowest buried cable level to the point of connection to the aboveground raceway.

See also

514.8 and its commentary regarding underground wiring. Note that polyvinyl chloride conduit (PVC) and reinforced thermosetting resin conduit (RTRC) are the only nonmetallic raceways permitted for occupancies covered in Article 515.

(B) Insulation. Conductor insulation shall comply with 501.20.

△ **(C) Nonmetallic Wiring.** Where PVC conduit, RTRC conduit, or cable with a nonmetallic sheath is used, an equipment grounding conductor shall be included to provide for electrical continuity of the raceway system and for grounding of non–current-carrying metal parts.

515.9 Sealing. Sealing requirements shall apply to horizontal as well as to vertical boundaries of the defined Class I locations. Buried raceways and cables under defined Class I locations shall be considered to be within a Class I, Division 1 or Zone 1 location.

515.10 Special Equipment — Motor Fuel Dispensers. In addition to the requirements of this article, dispensers for gasoline or other volatile flammable liquids or liquified flammable gases shall comply with the requirements for motor fuel dispensing facilities, as applicable, except as modified by this article.

Δ **515.16 Grounding and Bonding.** All metal raceways, the metal armor or metallic sheath on cables, and all non–current-carrying metal parts of fixed or portable electrical equipment, regardless of voltage, shall be grounded and bonded.

Grounding and bonding in Class I locations shall comply with 501.30 for Class I, Division 1 and 2 locations and 505.30 for Zone 0, 1, and 2 locations.

> Informational Note: See NFPA 30, *Flammable and Combustible Liquids Code*, 6.5.4, for information on grounding for static protection.

ARTICLE
516
Spray Application, Dipping, Coating, and Printing Processes Using Flammable or Combustible Materials

Part I. General

Δ **516.1 Scope.** This article covers the regular or frequent application of flammable liquids, combustible liquids, and combustible powders by spray operations and the application of flammable liquids or combustible liquids at temperatures above their flashpoint by spraying, dipping, coating, printing, or other means.

> Informational Note No. 1: See NFPA 33-2021, *Standard for Spray Application Using Flammable or Combustible Materials*, or NFPA 34-2021, *Standard for Dipping, Coating, and Printing Processes Using Flammable or Combustible Liquids*, for extracted text that is followed by a reference in brackets. Only editorial changes were made to the extracted text to make it consistent with this *Code*.

> Informational Note No. 2: See NFPA 91, *Standard for Exhaust Systems for Air Conveying of Vapors, Gases, Mists, and Particulate Solids*, for information regarding ventilation.

Industry experience has shown that the largest fire losses and frequency of fires occur where the proper codes and standards have not been used or applied. NFPA 33, *Standard for Spray Application Using Flammable or Combustible Materials*, covers the spray application of flammable or combustible materials by means of compressed air atomization, airless or hydraulic atomization, electrostatic application methods, or any other means in continuous or intermittent processes. NFPA 33 also covers the application of combustible powders applied by powder spray guns, electrostatic powder spray guns, and the fluidized bed application method or electrostatic fluidized bed application method. NFPA 33 contains requirements for the maintenance of safe conditions as well as personal safety.

NFPA 34, *Standard for Dipping, Coating, and Printing Processes Using Flammable or Combustible Liquids*, covers the fire and explosion hazards of dipping and coating processes that use flammable or combustible liquids or that use waterborne, water-based, and water-reducible materials that contain flammable or combustible liquids or that produce combustible deposits or residues.

Flammable or combustible liquids or powders and their vapors or mists and highly combustible residues or powders are the principal hazards of these spray application locations. Careful consideration of the location and installation of extinguishing equipment can reduce the possibility of fire spreading. NFPA 33 and NFPA 34 address these additional nonelectrical hazards.

N **516.2 Other Articles.** In addition to the requirements of this article, spray application, dipping, coating, and printing processes using flammable or combustible materials shall comply with Table 516.2, as applicable, except as modified by this article.

N **TABLE 516.2** *Other Articles*

Requirement	Division Classified Locations	Zone Classified Locations
Area classification	500.5, 500.6	505.5, 505.6, 505.7, 506.5, 506.6, 506.7
Equipment	500.7, 500.8, 501.5, 502.5, 502.6, Part III of 501, Part III of 502	505.8, 505.9, 505.20, 505.22, 506.8, 506.9
Wiring	Part II of 501, Part II of 502	505.15, 505.16, 505.17, 505.18, 505.19, 505.26, 505.30, 506.15, 506.17, 506.20, 506.30

516.3 Class I Locations. Where the term *Class I* is used with respect to Zone classifications within this article of the *Code*, it shall apply to Zone 0, Zone 1, and Zone 2 designations.

> Informational Note: The term *Class I* was originally included as a prefix to Zone 0, Zone 1, and Zone 2 locations and references as an identifier for flammable gases, vapors, or liquids to differentiate from Class II and Class III locations. Zone 0, Zone 1, and Zone 2 only apply to flammable gases, vapors, or liquids so the *Class I* prefix is redundant and has been deleted, except for text that is extracted from other documents or to remain consistent throughout this article.

Part II. Open Containers

Δ **516.4 Area Classification.** Area classification for open containers, supply containers, waste containers, spray gun cleaners, and solvent distillation units that contain Class I liquids and are located in ventilated areas shall be in accordance with the following:

(1) The area within 915 mm (3 ft) in all directions from any such container or equipment and extending to the floor or grade level shall be classified as Class I, Division 1 or Zone 1, whichever is applicable. [**33**:6.5.5.1(1)]

(2) The area extending 610 mm (2 ft) beyond the Division 1 or Zone 1 location shall be classified as Class I, Division 2 or Zone 2, whichever is applicable. [**33**:6.5.5.1(2)]

(3) The area extending 1525 mm (5 ft) horizontally beyond the area described in 516.4(2) up to a height of 460 mm (18 in.) above the floor or grade level shall be classified as Class I, Division 2 or Zone 2, whichever is applicable. [**33**:6.5.5.1(3)]

(4) The area inside any tank or container shall be classified as Class I, Division 1 or Zone 0, whichever is applicable. [**33**:6.5.5.1(4)]

(5) Sumps, pits, or belowgrade channels within 3.05 m (10 ft) horizontally of a vapor source shall be classified as Class I, Division 1 or Zone 1. If the sump, pit, or channel extends beyond 3.05 m (10 ft) from the vapor source, it shall be provided with a vapor stop or be classified as Class I, Division 1 or Zone 1 for its entire length.

For the purposes of electrical area classification, the Division system and the Zone system shall not be intermixed for any given source of release. [**33**:6.2.3]

Electrical wiring and utilization equipment installed in these areas shall be suitable for the location, as shown in Figure 516.4. [**33**:6.5.5.2]

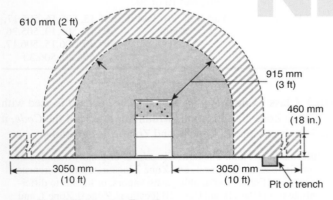

Class I, Division 1 or Zone 0
(e.g., vapor space in container)

Class I, Division 1 or Zone 1

Class I, Division 2 or Zone 2

FIGURE 516.4 *Electrical Area Classification for Class I Liquid Operations Around Open Containers, Supply Containers, Waste Containers, Spray Gun Cleaners, and Solvent Distillation Units.* [**33**:Figure 6.5.5.2]

Part III. Spray Application Processes

516.5 Area Classification. For spray application processes, the area classification is based on quantities of flammable vapors, combustible mists, residues, dusts, or deposits that are present or might be present in quantities sufficient to produce ignitable or explosive mixtures with air.

Many steps are required to properly classify a hazardous location. Although the *NEC®* provides general area classifications in Articles 501, 502, 503, 505, and 506, it does not classify specific locations. The *NEC* classifications for specific occupancies or processes have been extracted from other NFPA documents. The classifications from those documents are based on the premise that all applicable requirements of the document have been met. Deviations in on-site conditions, such as process conditions, area ventilation, and room construction, from those assumed by the document may alter the general classification. Those responsible for the specific area classification must consider the basis for the general classifications to determine the applicability to a specific location.

The determination of the extent of hazardous areas involved in spray application requires an understanding of the multiple hazards of flammable vapors, mists, powders, and highly combustible deposits applied at each location.

See also

NFPA **33** and NFPA **34,** which contain the specific construction and installation requirements used to develop the area classifications in Article 516

(A) Zone Classification of Locations.

(1) Classification of Locations. The Zone system of electrical area classification shall be applied as follows:

(1) The inside of closed containers or vessels shall be considered a Class I, Zone 0 location.

(2) A Class I, Division 1 location shall be permitted to be alternatively classified as a Class I, Zone 1 location.

(3) A Class I, Division 2 location shall be permitted to be alternatively classified as a Class I, Zone 2 location.

(4) A Class II, Division 1 location shall be permitted to be alternatively classified as a Zone 21 location.

(5) A Class II, Division 2 location shall be permitted to be alternatively classified as a Zone 22 location. [**33**:6.2.2]

(2) Classification Systems. For the purposes of electrical area classification, the Division system and the Zone system shall not be intermixed for any given source of release. [**33**:6.2.3]

In instances of areas within the same facility classified separately, Class I, Zone 2 locations shall be permitted to abut, but not overlap, Class I, Division 2 locations. Class I, Zone 0 or Zone 1 locations shall not abut Class I, Division 1 or Division 2 locations. [**33**:6.2.4]

(3) Equipment. Open flames, spark-producing equipment or processes, and equipment whose exposed surfaces exceed the autoignition temperature of the material being sprayed shall not be located in a spray area or in any surrounding area that is classified as Division 2, Zone 2, or Zone 22. [**33**:6.2.5]

Exception: This requirement shall not apply to drying, curing, or fusing apparatus.

Any utilization equipment or apparatus that is capable of producing sparks or particles of hot metal and that is located above or adjacent to either the spray area or the surrounding Division 2, Zone 2, or Zone 22 areas shall be of the totally enclosed type or shall be constructed to prevent the escape of sparks or particles of hot metal. [**33**:6.2.6]

Δ **(B) Class I, Division 1 or Class I, Zone 0 Locations.** The interior of any open or closed container or vessel of a flammable liquid shall be considered Class I, Division 1, or Class I, Zone 0, as applicable.

> Informational Note: See NFPA 33, *Standard for Spray Application Using Flammable or Combustible Materials*, Chapter 6, for additional information.

(C) Class I, Division 1; Class I, Zone 1; Class II, Division 1; or Zone 21 Locations. The following spaces shall be considered Class I, Division 1; Class I, Zone 1; Class II, Division 1; or Zone 21 locations, as applicable:

(1) The interior of spray booths and rooms except as specifically provided in 516.5(D).
(2) The interior of exhaust ducts.
(3) Any area in the direct path of spray operations.
(4) Sumps, pits, or below grade channels within 7620 mm (25 ft) horizontally of a vapor source. If the sump, pit, or channel extends beyond 7620 mm (25 ft) from the vapor source, it shall be provided with a vapor stop or it shall be classified as Class I, Division 1 for its entire length. [**34**:6.4.1]
(5) All space in all directions outside of but within 900 mm (3 ft) of open containers, supply containers, spray gun cleaners, and solvent distillation units containing flammable liquids.
(6) For limited finishing workstations, the area inside the curtains or partitions. *[See Figure 516.5(D)(5).]*

(D) Class I, Division 2; Class I, Zone 2; Class II, Division 2; or Zone 22 Locations. The spaces listed in 516.5(D)(1) through (D)(5) shall be considered Class I, Division 2; Class I, Zone 2; Class II, Division 2; or Zone 22 as applicable.

(1) Unenclosed Spray Processes. Electrical wiring and utilization equipment located outside but within 6100 mm (20 ft) horizontally and 3050 mm (10 ft) vertically of an enclosed spray area and not separated from the spray area by partitions extending to the boundaries of the area designated as Division 2, Zone 2 or Zone 22 in Figure 516.5(D)(1) shall be suitable for Class I, Division 2; Class I, Zone 2; Class II, Division 2; or Zone 22 locations, whichever is applicable. [**33**:6.5.1] *[See Figure 516.5(D)(1).]*

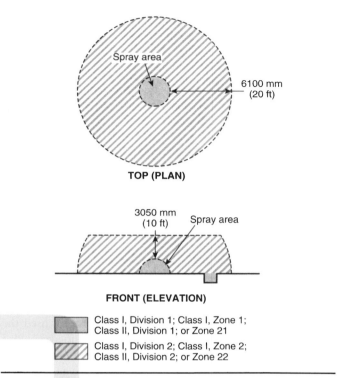

FRONT (ELEVATION)

| | Class I, Division 1; Class I, Zone 1; Class II, Division 1; or Zone 21 |
| Class I, Division 2; Class I, Zone 2; Class II, Division 2; or Zone 22 |

FIGURE 516.5(D)(1) *Electrical Area Classification for Unenclosed Spray Areas. [**33**:Figure 6.5.1]*

(2) Closed-Top, Open-Face, and Open-Front Spray Booths and Spray Rooms. If spray application operations are conducted within a closed-top, open-face, or open-front booth or room, as shown in Figure 516.5(D)(2), any electrical wiring or utilization equipment located outside of the booth or room but within 915 mm (3 ft) of any opening shall be suitable for Class I,

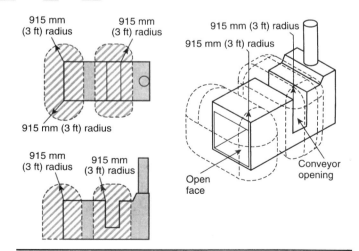

FIGURE 516.5(D)(2) *Class I, Division 2; Class I, Zone 2; Class II, Division 2; or Zone 22 Locations Adjacent to a Closed Top, Open Face, or Open Front Spray Booth or Room. [**33**:Figure 6.5.2(a)]*

Division 2; Class I, Zone 2; Class II, Division 2; or Zone 22 locations, whichever is applicable. The Class I, Division 2; Class I, Zone 2; Class II, Division 2; or Zone 22 locations shown in Figure 516.5(D)(2) shall extend from the edges of the open face or open front of the booth or room.

(3) Open-Top Spray Booths. For spraying operations conducted within an open top spray booth, the space 915 mm (3 ft) vertically above the booth and within 915 mm (3 ft) of other booth openings shall be considered Class I, Division 2; Class I, Zone 2; Class II, Division 2; or Zone 22 whichever is applicable. [**33**:6.5.3]

(4) Enclosed Spray Booths and Spray Rooms. For spray application operations confined to an enclosed spray booth or room, electrical area classification shall be as follows:

(1) The area within 915 mm (3 ft) of any opening shall be classified as Class I, Division 2; Class I, Zone 2; Class II, Division 2; or Zone 22 locations, whichever is applicable, as shown in Figure 516.5(D)(4).

(2) Where automated spray application equipment is used, the area outside the access doors shall be unclassified provided the door interlock prevents the spray application operations when the door is open.

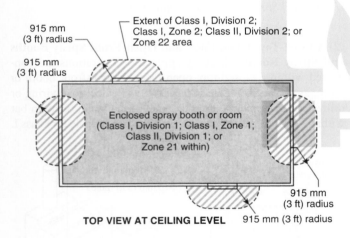

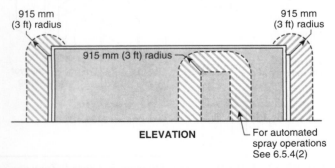

FIGURE 516.5(D)(4) *Class I, Division 2; Class I, Zone 2; Class II, Division 2; or Zone 22 Locations Adjacent to an Enclosed Spray Booth or Spray Room.* [*33:Figure 6.5.4*]

EXHIBIT 516.1 An enclosed paint spray room.

(3) Where exhaust air is permitted to be recirculated, both of the following shall apply:

  a. The interior of any recirculation path from the secondary particulate filters up to and including the air supply plenum shall be classified as Class I, Division 2; Class I, Zone 2; Class II, Division 2; or Zone 22 locations, whichever is applicable.

  b. The interior of fresh air supply ducts shall be unclassified.

(4) Where exhaust air is not recirculated, the interior of fresh air supply ducts and fresh air supply plenums shall be unclassified.

[**33**:6.5.4]

Spray booths with adequate mechanical ventilation must discharge the vapor or powder to a safe location to reduce the possibility of an explosion and to control the accumulation of overspray residues, many of which are not only highly combustible but also subject to spontaneous ignition.

Exhibit 516.1 shows an enclosed spray room. Note the illumination that meets the requirements of 516.6(C) and the absence of other electrical equipment during spray operations in accordance with 516.6(D).

Δ **(5) Limited Finishing Workstations.**

(a) For limited finishing workstations, the area inside the 915 mm (3 ft) space horizontally and vertically beyond the volume enclosed by the outside surface of the curtains or partitions shall be classified as Class I, Division 2; Class I, Zone 2; Class II, Division 2; or Zone 22, as shown in Figure 516.5(D)(5).

(b) A limited finishing workstation shall be designed and constructed to have all of the following:

(1) A dedicated make-up air supply

(2) Curtains or partitions that are noncombustible or limited combustible

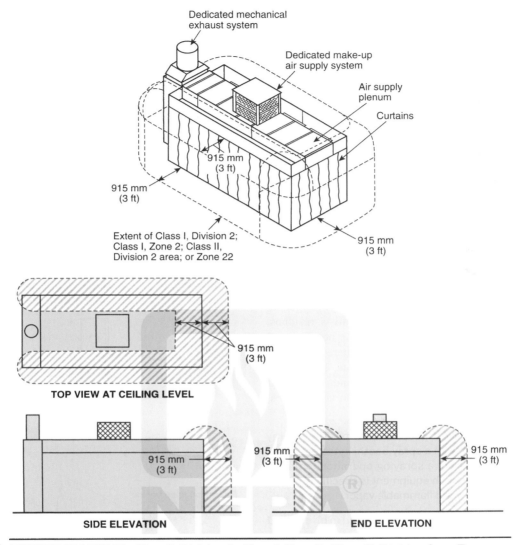

△ **FIGURE 516.5(D)(5)** *Class I, Division 2; Class I, Zone 2; Class II, Division 2; or Zone 22 Locations Adjacent to a Limited Finishing Workstation.* [*33:Figure 14.3.6.1*]

(3) A dedicated mechanical exhaust and filtration system

(4) An approved automatic extinguishing system

[**33**:14.3.1]

> Informational Note: See NFPA 701, *Standard Methods of Fire Tests for Flame Propagation of Textiles and Films*, for information on limited combustible curtains or partitions.

(c) The amount of material sprayed in a limited finishing workstation shall not exceed 3.8 L (1 gal) in any 8-hour period. [**33**:14.3.2]

(d) Curtains or partitions shall be fully closed during any spray operations. [**33**:14.3.4]

(e) The equipment within the limited finishing workstation shall be interlocked such that the spray application equipment cannot be operated unless the exhaust ventilation system is operating and functioning properly and spray application is automatically stopped if the exhaust ventilation system fails.

(f) Any limited finishing workstation used for spray application operations shall not be used for any operation that is capable of producing sparks or particles of hot metal or for operations that involve open flames or electrical utilization equipment capable of producing sparks or particles of hot metal. [**33**:14.3.6]

(g) Where industrial air heaters are used to elevate the air temperature for drying, curing, or fusing operations, a high limit switch shall be provided to automatically shut off the drying apparatus if the air temperature in the limited finishing workstation exceeds the maximum discharge-air temperature allowed by the standard that the heater is listed to or 93°C (200°F), whichever is less. [**33**:14.3.7.1]

(h) A means shall be provided to show that the limited finishing workstation is in the drying or curing mode of operation and that the limited finishing workstation is to be unoccupied. [**33**:14.3.7.2]

(i) Any containers of flammable or combustible liquids shall be removed from the limited finishing workstation before the drying apparatus is energized. [**33**:14.3.7.3]

(j) Portable spot-drying, curing, or fusion apparatus shall be permitted to be used in a limited finishing workstation, provided that it is not located within the hazardous (classified) location defined in 14.3.5 of NFPA 33 when spray application operations are being conducted. [**33**:14.3.8]

(k) Recirculation of exhaust air shall be permitted when the provisions of 516.5(D)(4)(3) are both met. [**33**:14.3.9]

516.6 Wiring and Equipment in Class I Locations.

(A) Wiring and Equipment — Vapors. All electrical wiring and equipment within the hazardous (classified) locations (containing vapor only — not residues) defined in 516.5 shall comply with the requirements of Part II and Part III of Article 501 or with 505.17 through 505.30, as applicable.

(B) Wiring and Equipment — Vapors and Residues. Unless specifically listed for locations containing deposits of dangerous quantities of flammable or combustible vapors, mists, residues, dusts, or deposits (as applicable), there shall be no electrical equipment in any spray area as herein defined whereon deposits of combustible residue could readily accumulate, except wiring in rigid metal conduit, intermediate metal conduit, Type MI cable, or in metal boxes or fittings containing no taps, splices, or terminal connections. [**33**:6.4.2]

Electrical equipment generally is not permitted inside any spray booth, in the exhaust duct from a spray booth, in the entrained air of an exhaust system from a spraying operation, or in the direct path of spray, unless such equipment is specifically listed for readily ignitible deposits and flammable vapor.

(C) Illumination. Luminaires shall be permitted to be installed as follows:

(1) Luminaires, like that shown in Figure 516.6(C)(1), that are attached to the walls or ceiling of a spray area but that are outside any classified area and are separated from the spray area by glass panels shall be suitable for use in unclassified locations. Such fixtures shall be serviced from outside the spray area. [**33**:6.6.1]

(2) Luminaires, like that shown in Figure 516.6(C)(1), that are attached to the walls or ceiling of a spray area; that are separated from the spray area by glass panels and that are located within a Class I, Division 2; a Class I, Zone 2; a Class II, Division 2; or a Zone 22 location shall be suitable for such location. Such fixtures shall be serviced from outside the spray area. [**33**:6.6.2]

(3) Luminaires, like that shown in Figure 516.6(C)(2), that are an integral part of the walls or ceiling of a spray area shall be permitted to be separated from the spray area by glass panels that are an integral part of the fixture. Such fixtures shall be listed for use in Class I, Division 2; Class I, Zone 2; Class II, Division 2; or Zone 22 locations, whichever is applicable,

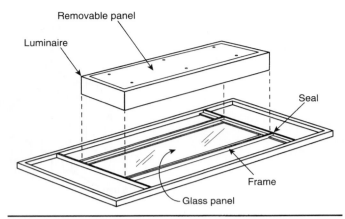

Δ **FIGURE 516.6(C)(1)** *Example of a Luminaire Mounted Outside the Spray Area and Serviced from Outside the Spray Area.* [**33**:*Figure 6.6.1.1*]

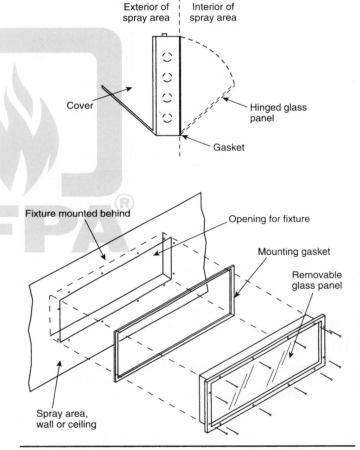

Δ **FIGURE 516.6(C)(2)** *Example of Luminaires That Are Integral Parts of the Spray Area and That Are Serviced from Inside the Spray Area.* [**33**:*Figure 6.6.3.1*]

and also shall be listed for accumulations of deposits of combustible residues. Such fixtures shall be permitted to be serviced from inside the spray area. [**33**:6.6.3]

(4) Glass panels used to separate luminaires from the spray area or that are an integral part of the luminaire shall meet the following requirements:

a. Panels for luminaires or for observation shall be of heat-treated glass, laminated glass, wired glass, or hammered-wired glass and shall be sealed to confine vapors, mists, residues, dusts, and deposits to the spray area. [**33**:5.5.1]

Exception to a.: Listed spray booth assemblies that have vision panels constructed of other materials shall be permitted.

b. Panels for luminaires shall be separated from the luminaire to prevent the surface temperature of the panel from exceeding 93°C (200°F). [**33**:5.5.2]

c. The panel frame and method of attachment shall be designed to not fail under fire exposure before the vision panel fails. [**33**:5.5.3]

Sufficient permanent illumination should be provided at the time the equipment is installed to avoid the use of temporary luminaires connected to ordinary extension cords in any spray area. See 516.6(D). A practical method of lighting is the use of wired or tempered glass panels in the top or sides of the spray booth, with electrical luminaires located outside the booth, avoiding the direct path of the spray.

Even where a panel of glass or translucent material separates a lighting unit from the readily ignitible location, the unit may be in a hazardous area and must be suitable for use in the specific location. Lighting units that are integral to the paint spray booth must be listed both for the location and for deposits of readily combustible paint residues on the side of the luminaire that forms part of the interior surface of the spray booth.

(D) Portable Equipment. Portable electric luminaires or other utilization equipment shall not be used in a spray area during spray operations.

Exception No. 1: Where portable electric luminaires are required for operations in spaces not readily illuminated by fixed lighting within the spraying area, they shall be of the type identified for Class I, Division 1 or Class 1, Zone 1 locations where readily ignitible residues could be present. [33:6.9 Exception]

Exception No. 2: Where portable electric drying apparatus is used in spray booths and the following requirements are met:

(1) The apparatus and its electrical connections are not located within the spray enclosure during spray operations.
(2) Electrical equipment within 450 mm (18 in.) of the floor is identified for Class I, Division 2 or Class I, Zone 2 locations.
(3) All metallic parts of the drying apparatus are electrically bonded and grounded.
(4) Interlocks are provided to prevent the operation of spray equipment while drying apparatus is within the spray enclosure, to allow for a 3-minute purge of the enclosure before energizing the drying apparatus and to shut off drying apparatus on failure of ventilation system.

Δ **(E) Electrostatic Equipment.** Electrostatic spraying or detearing equipment shall be installed and used only as provided in 516.10.

> Informational Note: See NFPA 33, *Standard for Spray Application Using Flammable or Combustible Materials,* for additional information.

(F) Static Electric Discharges. All persons and all electrically conductive objects, including any metal parts of the process equipment or apparatus, containers of material, exhaust ducts, and piping systems that convey flammable or combustible liquids, shall be electrically grounded. [**34**:6.8.1]

516.7 Wiring and Equipment Not Within Hazardous (Classified) Locations.

(A) Wiring. All fixed wiring above hazardous (classified) locations shall be permitted to be one or more of the following:

(1) Rigid metal conduit (RMC) or intermediate metal conduit (IMC) with listed threaded or threadless fittings, or electrical metallic tubing (EMT) with listed fittings.
(2) Rigid polyvinyl chloride conduit (PVC), reinforced thermosetting resin conduit (RTRC), or electrical nonmetallic tubing (ENT).
(3) Type MC cable, Type TC cable, or Type TC-ER cable, including installation in cable trays, terminated with listed fittings. Type TC-ER cable shall include an equipment grounding conductor (EGC) in addition to any drain wire.
(4) Type MI cable terminated with listed fittings and installed and supported to avoid tensile stress.
(5) Type PLTC cable or Type PLTC-ER cable used in Class 2 or Class 3 circuits. Type PLTC-ER cable shall include an equipment grounding conductor (EGC) in addition to any drain wire.
(6) Type ITC cable or Type ITC-ER cable in accordance with 335.4 and 335.5 and terminated with listed fittings. Type ITC-ER cable shall include an EGC in addition to any drain wire.
(7) Cellular metal raceways supplying ceiling outlets or as extensions to the area below the floor of a hazardous (classified) location. If cellular metal raceways are used, they shall not have connections leading into or passing through the hazardous (classified) location unless suitable seals are provided.

Δ **(B) Equipment.** Equipment that could produce arcs, sparks, or particles of hot metal, such as lamps and lampholders for fixed lighting, cutouts, switches, receptacles, motors, or other equipment having make-and-break or sliding contacts, where installed above a classified location or above a location where freshly finished goods are handled, shall be of the totally enclosed type or be constructed to prevent the escape of sparks or hot metal particles.

Even though areas adjacent to spray booths (particularly where coating-material stocks are located) are assumed to have ventilation sufficient to prevent the presence of flammable vapors or deposits, luminaires should be totally enclosed to prevent hot particles from falling in any area that may have freshly painted stock, accidentally spilled flammable or combustible materials, readily ignitible refuse, or flammable or combustible liquid containers that have been left open accidentally. Where luminaires are in areas subject to atmospheres of flammable vapor, lamps should be replaced while electricity is off; otherwise, sparking might occur.

516.10 Special Equipment.

(A) Fixed Electrostatic Equipment. This section shall apply to any equipment using electrostatically charged elements for the atomization, charging, and/or precipitation of hazardous materials for coatings on articles or for other similar purposes in which the charging or atomizing device is attached to a mechanical support or manipulator, including robotic devices. This section shall not apply to devices that are held or manipulated by hand. Where robot or programming procedures involve manual manipulation of the robot arm while spraying with the high voltage on, the requirements of 516.10(B) shall apply. The installation of electrostatic spraying equipment shall comply with the requirements of 516.10(A)(1) through (A)(10). Spray equipment shall be listed. All automatic electrostatic equipment systems shall comply with the requirements of 516.6(B) through (D) and 516.6(F).

(1) Power and Control Equipment. Transformers, high-voltage supplies, control apparatus, and all other electrical portions of the equipment shall be installed outside of the Class I location or be of a type identified for the location.

Exception: High-voltage grids, electrodes, electrostatic atomizing heads, and their connections shall be permitted within the Class I location.

(2) Electrostatic Equipment. Electrodes and electrostatic atomizing heads shall be adequately supported in permanent locations and shall be effectively insulated from ground. Electrodes and electrostatic atomizing heads that are permanently attached to their bases, supports, reciprocators, or robots shall be deemed to comply with this section.

(3) High-Voltage Leads. High-voltage leads shall be properly insulated and protected from mechanical damage or exposure to destructive chemicals. Any exposed element at high voltage shall be effectively and permanently supported on suitable insulators and shall be effectively guarded against accidental contact or grounding.

(4) Support of Goods. Goods being coated using this process shall be supported on conveyors or hangers. The conveyors or hangers shall be arranged (1) to ensure that the parts being coated are electrically connected to ground with a resistance of 1 megohm or less and (2) to prevent parts from swinging.

(5) Automatic Controls. Electrostatic apparatus shall be equipped with automatic means that will rapidly de-energize the high-voltage elements under any of the following conditions:

(1) Stoppage of ventilating fans or failure of ventilating equipment from any cause
(2) Stoppage of the conveyor carrying goods through the high-voltage field unless stoppage is required by the spray process
(3) Occurrence of excessive current leakage at any point in the high-voltage system
(4) De-energizing the primary voltage input to the power supply

Δ **(6) Grounding.** All electrically conductive objects in the spray area, except those objects required by the process to be at high voltage, shall be adequately grounded. This requirement shall apply to paint containers, wash cans, guards, hose connectors, brackets, and any other electrically conductive objects or devices in the area.

> Informational Note: See NFPA 33, *Standard for Spray Application Using Flammable or Combustible Materials*; NFPA 34, *Standard for Dipping, Coating, and Printing Processes Using Flammable or Combustible Liquids*; and NFPA 77, *Recommended Practice on Static Electricity*, for information on grounding and bonding for static electricity purposes.

All electrically conductive objects, including metal parts of spray booths, exhaust ducts, piping systems conveying flammable or combustible liquids or paint, solvent tanks, and canisters should be properly grounded to prevent sparks from the accumulation of static electricity.

The same informational note appears in 516.10(B)(4) and (C)(4) to increase awareness of three standards that deal with spray applications, dipping and coating processes, and static electricity for grounding and bonding. The requirements of Article 250 do not address grounding and bonding for the purpose of mitigating static electricity hazards in areas where easily ignitable materials are used or stored.

(7) Isolation. Safeguards such as adequate booths, fencing, railings, interlocks, or other means shall be placed about the equipment or incorporated therein so that they, either by their location, character, or both, ensure that a safe separation of the process is maintained.

(8) Signs. Signs shall be conspicuously posted to convey the following:

(1) Designate the process zone as dangerous with regard to fire and accident
(2) Identify the grounding requirements for all electrically conductive objects in the spray area
(3) Restrict access to qualified personnel only

(9) Insulators. All insulators shall be kept clean and dry.

(10) Other Than Nonincendive Equipment. Spray equipment that cannot be classified as nonincendive shall comply with 516.10(A)(10)(a) and (A)(10)(b).

(a) Conveyors, hangers, and application equipment shall be arranged so that a minimum separation of at least twice the sparking distance is maintained between the workpiece or material being sprayed and electrodes, electrostatic atomizing heads, or charged conductors. Warnings defining this safe distance shall be posted. [**33**:11.4.1]

(b) The equipment shall provide an automatic means of rapidly de-energizing the high-voltage elements in the event the distance between the goods being painted and the electrodes or electrostatic atomizing heads falls below that specified in 516.10(A)(10)(a). [**33**:11.3.8]

(B) Hand-Spraying Electrostatic Equipment. This section shall apply to any equipment using electrostatically charged elements for the atomization, charging, or precipitation of flammable and combustible materials for coatings on articles, or for other similar purposes in which the charging or atomizing device is hand-held and manipulated during the spraying operation. Electrostatic hand-spraying equipment and devices used in connection with paint-spraying operations shall be of listed types and shall comply with 516.10(B)(1) through (B)(5).

(1) General. The high-voltage circuits shall be designed so as not to produce a spark of sufficient intensity to ignite the most readily ignitible of those vapor–air mixtures likely to be encountered or result in appreciable shock hazard upon coming in contact with a grounded object under all normal operating conditions. The electrostatically charged exposed elements of the handgun shall be capable of being energized only by an actuator that also controls the coating material supply.

(2) Power Equipment. Transformers, power packs, control apparatus, and all other electrical portions of the equipment shall be located outside of the Class I location or be identified for the location.

Exception: The handgun itself and its connections to the power supply shall be permitted within the Class I location.

(3) Handle. The handle of the spraying gun shall be electrically connected to ground by a conductive material and be constructed so that the operator in normal operating position is in electrical contact with the grounded handle with a resistance of not more than 1 megohm to prevent buildup of a static charge on the operator's body. Signs indicating the necessity for grounding other persons entering the spray area shall be conspicuously posted.

Δ **(4) Electrostatic Equipment.** All electrically conductive objects in the spraying area, except those objects required by the process to be at high voltage shall be electrically connected to ground with a resistance of not more than 1 megohm. This requirement shall apply to paint containers, wash cans, and any other electrical conductive objects or devices in the area. The equipment shall carry a prominent, permanently installed warning regarding the necessity for this grounding feature.

Informational Note: See NFPA 33, *Standard for Spray Application Using Flammable or Combustible Materials*; NFPA 34, *Standard for Dipping, Coating, and Printing Processes Using Flammable or Combustible Liquids*; and NFPA 77, *Recommended Practice on Static Electricity*, for information on grounding and bonding for static electricity purposes.

(5) Support of Objects. Objects being painted shall be maintained in electrical contact with the conveyor or other grounded support. Hooks shall be regularly cleaned to ensure adequate grounding of 1 megohm or less. Areas of contact shall be sharp points or knife edges where possible. Points of support of the object shall be concealed from random spray where feasible, and, where the objects being sprayed are supported from a conveyor, the point of attachment to the conveyor shall be located so as to not collect spray material during normal operation.

(C) Powder Coating. This section shall apply to processes in which combustible dry powders are applied. The hazards associated with combustible dusts are present in such a process to a degree, depending on the chemical composition of the material, particle size, shape, and distribution.

Δ **(1) Electrical Equipment and Sources of Ignition.** Electrical equipment and other sources of ignition shall comply with the requirements of Part III of Article 502 or 506.20, as applicable. Portable electric luminaires and other utilization equipment shall not be used within a Class II location during operation of the finishing processes. Such luminaires or utilization equipment used during cleaning or repairing operations shall be of a type identified for Class II, Division 1 locations and all exposed metal parts shall be connected to an equipment grounding conductor.

Exception: Portable electric luminaires shall be of the type listed for Class II, Division 1 locations where required for operations in spaces not readily illuminated by fixed lighting within the spraying area and where readily ignitible residues might be present.

(2) Fixed Electrostatic Spraying Equipment. The provisions of 516.10(A) and 516.10(C)(1) shall apply to fixed electrostatic spraying equipment.

(3) Electrostatic Hand-Spraying Equipment. The provisions of 516.10(B) and 516.10(C)(1) shall apply to electrostatic hand-spraying equipment.

Δ **(4) Electrostatic Fluidized Beds.** Electrostatic fluidized beds and associated equipment shall be of identified types. The high-voltage circuits shall be designed such that any discharge produced when the charging electrodes of the bed are approached or contacted by a grounded object shall not be of sufficient intensity to ignite any powder–air mixture likely to be encountered or to result in an appreciable shock hazard.

(a) Transformers, power packs, control apparatus, and all other electrical portions of the equipment shall be located outside the powder-coating area or shall otherwise comply with the requirements of 516.10(C)(1).

Exception: The charging electrodes and their connections to the power supply shall be permitted within the powder-coating area.

(b) All electrically conductive objects within the powder-coating area shall be adequately grounded. The powder-coating equipment shall carry a prominent, permanently installed warning regarding the necessity for grounding these objects.

Informational Note: See NFPA 33, *Standard for Spray Application Using Flammable or Combustible Materials*; NFPA 34, *Standard for Dipping, Coating, and Printing Processes Using Flammable or Combustible Liquids*; and NFPA 77, *Recommended Practice on Static Electricity*, for information on grounding and bonding for static electricity purposes.

(c) Objects being coated shall be maintained in electrical contact (less than 1 megohm) with the conveyor or other support in order to ensure proper grounding. Hangers shall be regularly cleaned to ensure effective electrical contact. Areas of electrical contact shall be sharp points or knife edges where possible.

(d) The electrical equipment and compressed air supplies shall be interlocked with a ventilation system so that the equipment cannot be operated unless the ventilating fans are in operation. [**33**:Chapter 15]

516.16 Grounding. All metal raceways, the metal armors or metallic sheath on cables, and all non–current-carrying metal parts of fixed or portable electrical equipment, regardless of voltage, shall be grounded and bonded. Grounding and bonding shall comply with 501.30, 502.30, or 505.30, as applicable.

Part IV. Spray Application Operations in Membrane Enclosures

Δ **516.18 Area Classification for Temporary Membrane Enclosures.** Electrical area classification shall be as follows:

(1) The area within the membrane enclosure shall be considered a Class I, Division 1 area, as shown in Figure 516.18.

(2) A 1.5 m (5 ft) zone outside of the membrane enclosure shall be considered Class I, Division 2, as shown in Figure 516.18.

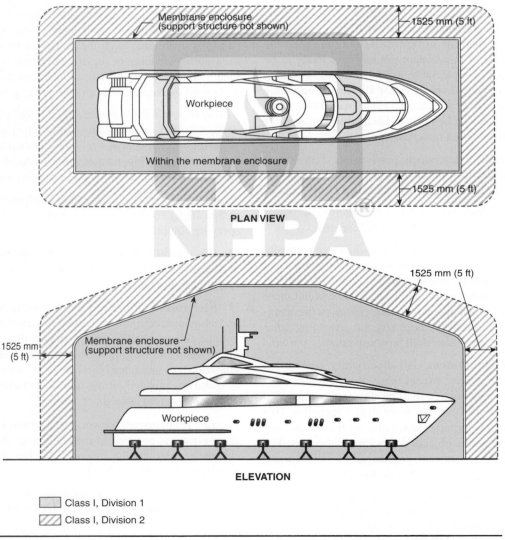

PLAN VIEW

ELEVATION

☐ Class I, Division 1
▨ Class I, Division 2

Δ **FIGURE 516.18** *Electrical Classifications for Outdoor Membrane Enclosures.* [**33**:*Figure 18.6.1.2*]

Informational Note No. 1: See NFPA 33, *Standard for Spray Application Using Flammable or Combustible Materials*, for information on occupancy, ventilation, fire protection, and permitting for spray application operations in membrane enclosures. This document limits spray application operations within both outdoor and indoor temporary membrane enclosures, as well as use and time constraints. The risks to people and property are unique when spray painting within the confined spaces of temporary membrane enclosures.

Informational Note No. 2: See NFPA 33, *Standard for Spray Application Using Flammable or Combustible Materials*, Section 18.6, for the limits of material used in a vertical plane for membrane enclosures.

Informational Note No. 3: See NFPA 701, *Standard Methods of Fire Tests for Flame Propagation of Textiles and Films*, Test Method 2, for construction information.

Informational Note No. 4: See NFPA 33, *Standard for Spray Application Using Flammable or Combustible Materials*, 18.3.2.1.1, for membrane installation beneath sprinklers.

Informational Note No. 5: See NFPA 13, *Standard for the Installation of Sprinkler Systems*, 8.15.15, for information on the protection of membrane structures.

516.23 Electrical and Other Sources of Ignition. Electrical wiring and utilization equipment used within the classified areas inside and outside of membrane enclosures during spray painting shall be suitable for the location and shall comply with all of the following:

(1) All power to the workpiece shall be removed during spray painting.
(2) Workpieces shall be grounded.
(3) Spray paint equipment shall be grounded.
(4) Scaffolding shall be bonded to the workpiece and grounded by an approved method.

Part V. Printing, Dipping, and Coating Processes

Δ **516.29 Classification of Locations.** Classification is based on quantities of flammable vapors, combustible mists, residues, dusts, or deposits that are present or might be present in quantities sufficient to produce ignitable or explosive mixtures with air. Electrical wiring and electrical utilization equipment located adjacent to open processes shall comply with the requirements as follows. Examples of these requirements are illustrated in Figure 516.29(1), Figure 516.29(2), Figure 516.29(3), and Figure 516.29(4).

Informational Note: See NFPA 33, *Standard for Spray Application Using Flammable or Combustible Materials*, Chapter 6, and NFPA 34, *Standard for Dipping, Coating, and Printing Processes Using Flammable or Combustible Liquids*, Chapter 6, for additional information.

(1) Electrical wiring and electrical utilization equipment located in any sump, pit, or below grade channel that is within 7620 mm (25 ft) horizontally of a vapor source, as defined by this standard, shall be suitable for Class I, Division 1 or Class I, Zone 1 locations. If the sump, pit, or channel extends beyond 7620 mm (25 ft) of the vapor source, it shall be provided with a vapor stop, or it shall be classified as Class I, Division 1 or Class I, Zone 1 for its entire length. [**34**:6.4.1]
(2) Electrical wiring and electrical utilization equipment located within 1525 mm (5 ft) of a vapor source shall be suitable for Class I, Division 1 or Class I, Zone 1 locations. The space inside a dip tank, ink fountain, ink reservoir, or ink tank shall be classified as Class I, Division 1 or Class I, Zone 0, whichever is applicable.

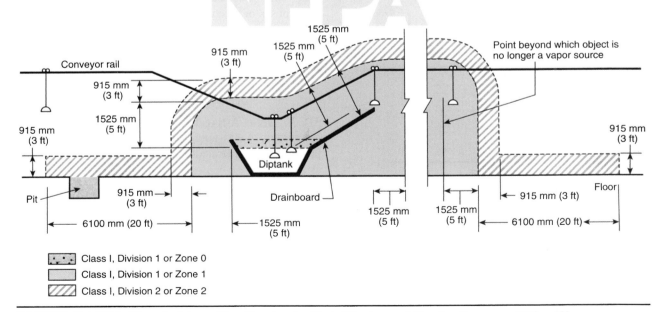

FIGURE 516.29(1) *Electrical Area Classification for Open Dipping and Coating Processes Without Vapor Containment or Ventilation.* [*34:Figure 6.4(a)*]

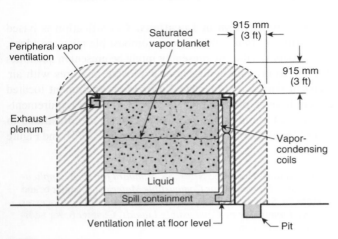

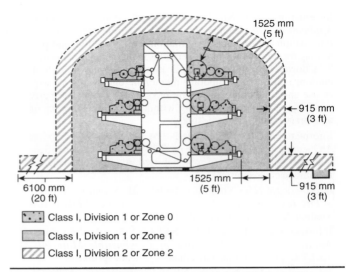

FIGURE 516.29(4) *Electrical Area Classification for a Typical Printing Process. [34:Figure 6.4(d)]*

Class I, Division 1 or Zone 0

Class I, Division 1 or Zone 1

Class I, Division 2 or Zone 2

Definitions

Freeboard: The distance from the maximum solvent or coating material level to the top of the tank

Freeboard ratio: The freeboard height divided by the smaller of the interior length or interior width of the tank

FIGURE 516.29(2) *Electrical Area Classification for Open Dipping and Coating Processes with Peripheral Vapor Containment and Ventilation — Vapors Confined to Process Equipment. [34:Figure 6.4(b)]*

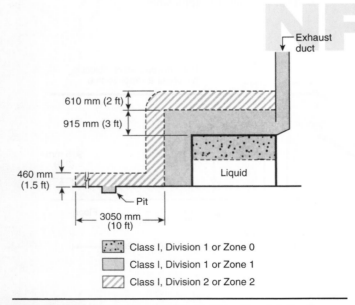

Class I, Division 1 or Zone 0

Class I, Division 1 or Zone 1

Class I, Division 2 or Zone 2

FIGURE 516.29(3) *Electrical Area Classification for Open Dipping and Coating Processes with Partial Peripheral Vapor Containment and Ventilation — Vapors NOT Confined to Process Equipment. [34:Figure 6.4(c)]*

(3) Electrical wiring and electrical utilization equipment located within 915 mm (3 ft) of the Class I, Division 1 or Class I, Zone 1 location shall be suitable for Class I, Division 2 or Class I, Zone 2 locations, whichever is applicable.

(4) The space 915 mm (3 ft) above the floor and extending 6100 mm (20 ft) horizontally in all directions from the Class I, Division 1 or Class I, Zone 1 location shall be classified as Class I, Division 2 or Class I, Zone 2, and electrical wiring and electrical utilization equipment located within this space shall be suitable for Class I, Division 2 or Class I, Zone 2 locations, whichever is applicable.

(5) This space shall be permitted to be nonclassified for purposes of electrical installations if the surface area of the vapor source does not exceed 0.5 m² (5 ft²), the contents of the dip tank, ink fountain, ink reservoir, or ink tank do not exceed 19 L (5 gal), and the vapor concentration during operating and shutdown periods does not exceed 25 percent of the lower flammable limit.

516.35 Areas Adjacent to Enclosed Dipping and Coating Processes. Areas adjacent to enclosed dipping and coating processes are illustrated by Figure 516.35 and shall be classified as follows:

(1) The interior of any enclosed dipping or coating process or apparatus shall be a Class I, Division 1 or Class I, Zone 1 location, and electrical wiring and electrical utilization equipment located within this space shall be suitable for Class I, Division 1 or Class I, Zone 1 locations, whichever is applicable. The area inside the dip tank shall be classified as Class I, Division 1 or Class I, Zone 0, whichever is applicable.

(2) The space within 915 mm (3 ft) in all directions from any opening in the enclosure and extending to the floor

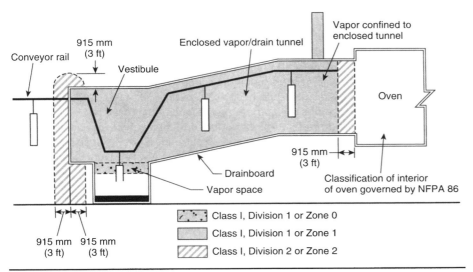

FIGURE 516.35 Electrical Area Classification Around Enclosed Dipping and Coating Processes. [34:Figure 6.5]

or grade level shall be classified as Class I, Division 2 or Class I, Zone 2, and electrical wiring and electrical utilization equipment located within this space shall be suitable for Class I, Division 2 locations or Class I, Zone 2 locations, whichever is applicable.

(3) All other spaces adjacent to an enclosed dipping or coating process or apparatus shall be classified as nonhazardous for purposes of electrical installations.

516.36 Equipment and Containers in Ventilated Areas. Open containers, supply containers, waste containers, and solvent distillation units that contain Class I liquids shall be located in areas ventilated in accordance with 516.4.

516.37 Luminaires. For printing, coating, and dipping equipment where the process area is enclosed by glass panels that are sealed to confine vapors and mists to the inside of the enclosure, luminaires that are attached to the walls or ceilings of a process enclosure and that are located outside of any classified area shall be permitted to be of general purpose construction. Such luminaires shall be serviced from outside the enclosure.

Luminaires that are attached to the walls or ceilings of a process enclosure, are located within the Class I, Division 2 or Class I, Zone 2 location, and are separated from the process area by glass panels that are sealed to confine vapors and mists shall be suitable for use in that location. Such fixtures shall be serviced from outside the enclosure.

516.38 Wiring and Equipment Not Within Hazardous (Classified) Locations.

(A) Wiring. Fixed wiring above hazardous (classified) locations shall be permitted to be one or more of the following:

(1) Rigid metal conduit (RMC) or intermediate metal conduit (IMC) with listed threaded or threadless fittings, or electrical metallic tubing (EMT) or electrical nonmetallic tubing (Type ENT) with listed fittings.

(2) Rigid polyvinyl chloride conduit (PVC) or reinforced thermosetting resin conduit (RTRC).

(3) Type MC cable or Type TC cable with listed fittings.

(4) Type MI cable terminated with listed fittings and installed and supported to avoid tensile stress.

(5) Cellular metal floor raceways only to supply ceiling outlets or as extensions to the area below the floor of a hazardous (classified) location. If cellular metal raceways are used, they shall not have connections leading into or passing through the hazardous (classified) location unless suitable seals are provided.

Δ **(B) Equipment.** Equipment that is capable of producing arcs, sparks, or particles of hot metal, such as lamps and lampholders for fixed lighting, cutouts, switches, receptacles, motors, or other equipment having make-and-break or sliding contacts, where installed above a classified location or above a location where freshly finished goods are handled, shall be of the totally enclosed type or be constructed to prevent the escape of sparks or hot metal particles.

Δ **516.40 Static Electric Discharges.** All persons and all electrically conductive objects, including any metal parts of the process equipment or apparatus, containers of material, exhaust ducts, and piping systems that convey flammable or combustible liquids, shall be electrically grounded.

Provision shall be made to dissipate static electric charges from all nonconductive substrates in printing processes.

Informational Note: See NFPA 77, *Recommended Practice on Static Electricity,* for information on reducing the risk of ignition from electrostatic discharges.

ARTICLE 517 Health Care Facilities

Part I. General

517.1 Scope. This article applies to electrical construction and installation criteria in health care facilities that provide services to human beings.

The requirements of this article shall specify the installation criteria and wiring methods that minimize electrical hazards by the maintenance of adequately low potential differences only between exposed conductive surfaces that are likely to become energized and could be contacted by a patient.

> Informational Note No. 1: In a health care facility, it is difficult to prevent the occurrence of a conductive or capacitive path from the patient's body to some grounded object, because that path might be established accidentally or through instrumentation directly connected to the patient. Other electrically conductive surfaces that might make an additional contact with the patient, or instruments that might be connected to the patient, then become possible sources of electric currents that can traverse the patient's body. The hazard is increased as more apparatus is associated with the patient, therefore more intensive precautions are needed. Control of electric shock hazard requires the limitation of electric current that might flow in an electrical circuit involving the patient's body by raising the resistance of the conductive circuit that includes the patient, or by insulating exposed conductive surfaces that might become energized, in addition to reducing the potential difference that can appear between exposed conductive surfaces in the patient care vicinity, or by combinations of these methods. A special problem is presented by the patient with an externalized direct conductive path to the heart muscle. The patient could be electrocuted at current levels so low that additional protection in the design of appliances, insulation of the catheter, and control of medical practice is required.

The requirements in Parts II and III not only apply to single-function buildings but are also intended to be individually applied to their respective forms of occupancy within a multifunction building [e.g., a doctor's examining room located within a limited care facility would be required to meet 517.10(A)].

> Informational Note No. 2: For information concerning performance, maintenance, and testing criteria, refer to the appropriate health care facilities documents.
>
> Informational Note No. 3: Text that is followed by a reference in brackets has been extracted from NFPA 99-2021, *Health Care Facilities Code*, or NFPA 101-2021, *Life Safety Code*. Only editorial changes were made to the extracted text to make it consistent with this *Code*.

The requirements of Article 517 apply to all types of health care facilities. Some areas of health care facilities are not directly used for the treatment of patients. For example, in a suite of doctors' offices within an office building, a doctor's business office is treated as a business occupancy and the electrical installation is required to comply with the applicable requirements of Chapters 1 through 4. The wiring and electrical equipment in the examining rooms and other patient care areas within the office suite are subject to the requirements of Article 517.

The scope of Article 517 also includes mobile health care facilities, but it does not include veterinary offices or animal hospitals. In-home health care is governed by the requirements of Chapters 1 through 4. As the use of in-home medical electrical equipment (including life support medical equipment that can be used outside of a health care setting) increases, there are considerations for safe use and reliability that have been compiled in Informative Annex K, Use of Medical Electrical Equipment in Dwellings and Residential Board-and-Care Occupancies.

Other standards referenced in Article 517 include NFPA 99, *Health Care Facilities Code*; NFPA 101®, *Life Safety Code*®; and NFPA 20, *Standard for the Installation of Stationary Pumps for Fire Protection*.

See also

Article 100 for general definitions and for definitions of terms used exclusively in Article 517

N **517.6 Patient Care–Related Electrical Equipment.** The reconditioning requirements of this *Code* shall not apply to patient care–related electrical equipment.

> Informational Note No. 1: Patient care–related electrical equipment is differentiated from electrical equipment as described in 110.21(A)(2).
>
> Informational Note No. 2: If patient care–related electrical equipment is relocated, it is expected to be recommissioned or recertified in accordance with the U.S. *Federal Food, Drug, and Cosmetic Act (FDCA)*.

Because of its specialized use and construction, relocated or reconditioned patient care–related electrical equipment such as magnetic resonance imaging (MRI), fluoroscopy, computerized tomography (CT), and X-ray units are regulated under federal requirements and are not subject to the general reconditioned electrical equipment requirements in Article 110. However, electrical equipment that is part of the electrical infrastructure of a health care facility is covered by the reconditioned electrical equipment requirements in Chapters 1 through 4 of the *NEC*®.

Part II. Wiring and Protection

517.10 Applicability.

(A) Applicability. Part II shall apply to patient care space of all health care facilities.

The designation of the types of patient care spaces is the responsibility of the governing body of the health care facility. Both the design and the inspection of a patient care space are based on the level of patient care anticipated in the space.

Δ **(B) Not Covered.** Part II shall not apply to the following:

(1) Business offices, corridors, waiting rooms, and the like in clinics, medical and dental offices, and outpatient facilities

(2) Spaces of nursing homes and limited care facilities wired in accordance with Chapters 1 through 4 of this *Code* where these spaces are used exclusively as patient sleeping rooms, as determined by the health care facility's governing body

Informational Note No. 1: See 406.12(5) for receptacles located in health care facility business offices, corridors, and waiting rooms that are required to be tamper resistant.

Informational Note No. 2: See 210.12(D) for branch circuits supplying outlets and receptacles located in patient sleeping rooms in nursing homes and limited care facilities that are connected to arc-fault circuit-interrupter circuits.

(3) Areas used exclusively for any of the following purposes:
   a. Intramuscular injections (immunizations)
   b. Psychiatry and psychotherapy
   c. Alternative medicine
   d. Optometry
   e. Pharmacy services not contiguous to health care facilities

Informational Note No. 3: See NFPA *101*-2021, *Life Safety Code.*

Areas in nursing homes that are designated as patient sleeping rooms are not considered to be patient care spaces, even if residents require assistance to attend to their personal needs and safety. However, areas of nursing homes, including patient bedrooms, in which residents are intended to be examined or treated can be considered examining rooms and thus be classified in the broader category as general care space.

Article 517 is about mitigating electrical hazards in health care settings. As today's health care environment shifts from the traditional medical office or hospital settings to outpatient clinics and business occupancies where a person can receive certain minor levels of treatment or preventative procedures, it is necessary for the *NEC* to address this new direction to provide the necessary level of electrical safety for patients. The specific areas in 517.10(B)(3) are locations where patients are not being treated or examined in a manner that exposes them to an electrical hazard. For example, flu shots or other tests or immunizations given at a retail pharmacy do not result in the pharmacy being classed as a health care facility. Likewise, alternative health care practices such as acupuncture or homeopathy do not present an electrical hazard to patients. Optometry does not involve invasive procedures using electromedical equipment and is also excluded from the special wiring requirements in Part II of Article 517.

See also

Article 100 for the definition of the term *patient care space category*

517.12 Wiring Methods. Except as modified in this article, wiring methods shall comply with Chapters 1 through 4 of this *Code.*

517.13 Equipment Grounding Conductor for Receptacles and Fixed Electrical Equipment in Patient Care Spaces. Wiring serving patient care spaces shall comply with the requirements of 517.13(A) and (B).

Exception: Luminaires more than 2.3 m (7½ ft) above the floor and switches located outside of the patient care vicinity shall be permitted to be connected to an equipment grounding return path complying with the requirements of 517.13(A) or (B).

Luminaires mounted 7½ feet above the floor and switches located outside the patient care vicinity are not required to have a separate insulated equipment grounding conductor (EGC); it is unlikely that a patient would contact these items or that an attendant would contact one of these items and a patient at the same time. The patient care vicinity consists of an area 6 feet horizontally in all directions from the bed and vertically to 7½ feet above the floor.

(A) Wiring Methods. All branch circuits serving patient care spaces shall be provided with an effective ground-fault current path by installation in a metal raceway system or a cable having a metallic armor or sheath assembly. The metal raceway system, metallic cable armor, or sheath assembly shall itself qualify as an equipment grounding conductor in accordance with 250.118.

These requirements apply to the branch circuits in spaces used for patient care and are not limited to patient rooms. Examining rooms, therapy areas, recreational areas, solaria, and certain patient corridors are also covered by these requirements. The branch-circuit wiring method used in these areas is one component of a two-part ground-fault current return path arrangement that is unique to patient care spaces. The metal raceway, metal cable armor, or metal cable sheath must qualify as an EGC in accordance with 250.118, independent of the second component of this equipment grounding method covered by 517.13(B).

Metal-sheathed cable assemblies are not permitted as a general wiring method for life safety and critical branch circuits, because 517.31(C)(3) requires such wiring to be protected by installation in metal raceways. However, listed flexible raceway and metal-sheathed cable assemblies are allowed for limited applications as described in 517.31(C)(3)(3) a. through f.

See also

517.31(C)(3) for additional requirements covering the mechanical protection of essential electrical system conductors

(B) Insulated Equipment Grounding Conductors and Insulated Equipment Bonding Jumpers.

Δ **(1) General.** An insulated copper equipment grounding conductor that is clearly identified along its entire length by green insulation and installed with the branch circuit conductors within the wiring method in accordance with 517.13(A) shall be connected to the following:

   (1) Grounding terminals of all receptacles other than isolated ground receptacles
   (2) Metal outlet boxes, metal device boxes, or metal enclosures
   (3) Non-current-carrying conductive surfaces of fixed electrical equipment likely to become energized that are subject to personal contact, operating at over 100 volts

Exception No. 1: For other than isolated ground receptacles, an insulated equipment bonding jumper that directly connects to the equipment grounding conductor shall be permitted to connect the box and receptacle(s) to the equipment grounding conductor. Isolated ground receptacles shall be connected in accordance with 517.16.

Exception No. 2: Metal faceplates shall be connected to an effective ground-fault current path by means of a metal mounting screw(s) securing the faceplate to a metal yoke or strap of a receptacle or to a metal outlet box.

These requirements cover the second component of the equipment grounding method. An insulated copper EGC, either solid or stranded, sized in accordance with 250.122, must be installed with the branch-circuit conductors in a wiring method that meets the requirements of 517.13(A). The requirement to install an insulated EGC applies only to branch circuits. Feeders supplying the panelboards from which the patient care space branch circuits originate are not subject to the requirements of 517.13.

(2) Sizing. Equipment grounding conductors and equipment bonding jumpers shall be sized in accordance with 250.122.

517.14 Panelboard Bonding. The equipment grounding terminal buses of the normal and essential branch-circuit panelboards serving the same individual patient care vicinity shall be connected together with an insulated continuous copper conductor not smaller than 10 AWG. Where two or more panelboards serving the same individual patient care vicinity are served from separate transfer switches on the essential electrical system, the equipment grounding terminal buses of those panelboards shall be connected together with an insulated continuous copper conductor not smaller than 10 AWG. This conductor shall be permitted to be broken in order to terminate on the equipment grounding terminal bus in each panelboard.

Exception: The insulated continuous copper conductor not smaller than 10 AWG shall be permitted to be terminated on listed connections to aluminum or copper busbars not smaller than 6 mm thick × 50 mm wide (¼ in. thick × 2 in. wide) and of sufficient length to accommodate the number of terminations necessary for the bonding of the panelboards. The busbar shall be securely fastened and installed in an accessible location.

This requirement applies to multiple panelboards that might be supplied from the same system or from different systems serving the same individual patient care vicinity. The exception emulates requirements from Article 250 permitting the use of busbar as a common connection point for grounding and/or bonding conductors. The use of busbar installed in an accessible location facilitates inspection of these important bonding connections that are used to eliminate differences of potential on exposed surfaces of equipment supplied from different sources.

517.16 Use of Isolated Ground Receptacles. An isolated ground receptacle, if used, shall not defeat the purposes of the safety features of the grounding systems detailed in 517.13. [**99:**6.3.2.2.5(A)]

The EGC of isolated ground receptacles does not provide the functional benefit of the multiple equipment grounding paths specified in 517.13. For that reason, isolated-ground-type receptacles (Type IG) are not permitted to be installed in the patient care vicinity. They are permitted to be installed in patient care spaces outside the patient care vicinity. If isolated ground receptacles are installed in these spaces, the installation has to comply with 517.13(A) and (B), and an additional insulated EGC is required to be installed and connected to the receptacle equipment grounding terminal. The additional EGC is required to be identified by green insulation with one or more yellow stripes. The insulated EGC required by 517.13(B) is required to be identified by green insulation having no yellow stripes.

Type IG receptacles are identified by an orange triangle located on the face of the receptacle. Because they are not permitted in the patient care vicinity, it is not necessary for them to be additionally listed as hospital grade.

(A) Inside of a Patient Care Vicinity. An isolated ground receptacle shall not be installed within a patient care vicinity. [**99:**6.3.2.2.5(B)]

(B) Outside of a Patient Care Vicinity. Isolated ground receptacle(s) installed in patient care spaces outside of a patient care vicinity(s) shall comply with 517.16(B)(1) and (B)(2).

(1) The equipment grounding terminals of isolated ground receptacles installed in branch circuits for patient care spaces shall be connected to an insulated equipment grounding conductor in accordance with 250.146(D) installed in a wiring method described in 517.13(A).

The equipment grounding conductor connected to the equipment grounding terminals of isolated ground receptacles in patient care spaces shall be clearly identified along the equipment grounding conductor's entire length by green insulation with one or more yellow stripes.

(2) The insulated equipment grounding conductor required in 517.13(B)(1) shall be clearly identified along its entire length by green insulation, with no yellow stripes, and shall not be connected to the grounding terminals of isolated equipment ground receptacles but shall be connected to the box or enclosure indicated in 517.13(B)(1)(2) and to non–current-carrying conductive surfaces of fixed electrical equipment indicated in 517.13(B)(1)(3).

Informational Note No. 1: This type of installation is typically used where a reduction of electrical noise (electromagnetic interference) is necessary, and parallel grounding paths are to be avoided.

Informational Note No. 2: Care should be taken in specifying a system containing isolated ground receptacles, because the impedance of the effective ground-fault current path is dependent upon the equipment grounding conductor(s) and does not benefit from any conduit or building structure in parallel with the equipment grounding conductor.

517.17 Ground-Fault Protection of Equipment.

If ground-fault protection of equipment (GFPE) is applied to the service providing power to a health care facility, an additional level of ground-fault protection is required downstream. Section 517.17(B) requires GFPE for every feeder.

Δ **(A) Applicability.** The requirements of 517.17 shall apply to buildings or portions of buildings containing health care facilities with Category 1 spaces or utilizing electrical life-support equipment, and buildings that provide the required essential utilities or services for the operation of Category 1 spaces or electrical life-support equipment.

Buildings used exclusively as health care facilities and mixed- or multiple-use buildings in which portions are used as health care facilities are subject to the requirements of 517.17 if the level of care provided is classified as Category 1 (critical care). Because most hospitals have areas that provide this level of care, the requirement applies, and it would also apply to an outpatient surgical center that was an occupancy within a mixed-use building.

Based on this criterion, most general medical and dental practices located in multiple occupancy buildings are not required to have GFPE. In addition, if the service is not provided with GFPE (either because it is not required by the *NEC* or, if optional, has not been incorporated as part of the design), the second level of GFPE is unnecessary.

The selectivity required by 517.17(C) is accomplished through the installation of GFPE for all feeder disconnecting means in the first level of distribution downstream of the GFPE-protected service equipment or building disconnecting means specified in 215.10 or 230.95. Therefore, in a mixed-use building containing health care (with Category 1 patient care spaces) and other types of occupancies, any feeder supplied from this level of distribution is required to have GFPE even if it is supplying an occupancy that is not a health care facility.

In the case of existing multiple occupancy buildings that have GFPE for the service equipment, a tenant buildout or a renovation for a new health care occupancy could result in the need to also provide second-level GFPE for all other occupancies. Careful analysis of the impact of this requirement on the existing service equipment might warrant an alternative approach such as installation of another service if permitted by 230.2(A) through (D).

Section 517.45 requires clinics or ambulatory care facilities that use life support equipment or that have areas designated as critical care to be provided with an essential electrical system that includes an alternate power source. It is not intended that ground-fault protection be installed between the on-site generator and the transfer switch or on the load side of the essential electrical system transfer switch.

See also

230.95(C), Informational Note No. 3, which calls attention to problems that could arise when ground-fault-protected systems are transferred to another supply system

(B) Feeders. Where ground-fault protection of equipment is provided for operation of the service disconnecting means or feeder disconnecting means as specified by 230.95 or 215.10, an additional step of ground-fault protection shall be provided in all next level feeder disconnecting means downstream toward the load. Such protection shall consist of overcurrent devices and current transformers or other protective equipment that shall cause the feeder disconnecting means to open.

The additional levels of ground-fault protection of equipment shall not be installed on the load side of an essential electrical system transfer switch.

This requirement is unlike the requirements of 210.13, 215.10, and 230.95, in which mandatory GFPE is based on the rating of the disconnecting means (1000 amperes or more). The second level of GFPE is based on the need to provide selectivity between the feeder protective devices and the service or building supply protective devices. With proper ground-fault coordination per 517.17(C), this additional level of protection is intended to limit a ground fault to a single feeder and thereby prevent a total power outage of the entire health care facility. Coordination includes consideration of the trip setting, the time setting, and the time required for operation (opening time) of each level of the ground-fault protection system.

(C) Selectivity. Ground-fault protection of equipment for operation of the service and feeder disconnecting means shall be fully selective such that the feeder device, but not the service device, shall open on ground faults on the load side of the feeder device. Separation of ground-fault protection time-current characteristics shall conform to manufacturer's recommendations and shall consider all required tolerances and disconnect operating time to achieve 100 percent selectivity.

Informational Note: See 230.95, Informational Note, for transfer of alternate source where ground-fault protection is applied.

(D) Testing. When ground-fault protection of equipment is first installed, each level shall be performance tested to ensure compliance with 517.17(C). This testing shall be conducted by a qualified person(s) using a test process in accordance with the instruction provided with the equipment. A written record of this testing shall be made and shall be available to the authority having jurisdiction.

See also

230.95(C) for more information on GFPE performance testing

Δ **517.18 Category 2 Spaces.**

Δ **(A) Patient Bed Location.** Each patient bed location shall be supplied by at least two branch circuits, one from the critical branch and one from the normal system. All branch circuits from the normal system shall originate in the same panelboard. The electrical receptacles or the cover plate for the electrical receptacles supplied from the critical branch shall have a distinctive color or marking so as to be readily identifiable and shall also indicate the panelboard and branch-circuit number supplying them.

Branch circuits serving patient bed locations shall not be part of a multiwire branch circuit.

Exception No. 1: Branch circuits serving only special-purpose outlets or receptacles, such as portable X-ray outlets, shall not be required to be served from the same distribution panel or panels.

Exception No. 2: The requirements of 517.18(A) shall not apply to patient bed locations in clinics, medical and dental offices, and outpatient facilities; psychiatric, substance abuse, and rehabilitation hospitals; sleeping rooms of nursing homes; and limited care facilities meeting the requirements of 517.10(B)(2).

Exception No. 3: A Category 2 patient bed location served from two separate transfer switches on the critical branch shall not be required to have circuits from the normal system.

Exception No. 4: Circuits served by Type 2 essential electrical systems shall be permitted to be fed by the equipment branch of the essential electrical system.

Patient bed locations in general care spaces are prohibited from deriving all their branch circuits from the essential electrical system. At least one branch circuit for each location must originate in a normal system panelboard. This supply circuit arrangement correlates with the limitations imposed by 517.34 regarding the types of loads permitted to be supplied by the critical branch of the essential electrical system.

Exception No. 3 allows both of the required branch circuits for a general care patient bed location to be supplied by the critical branch, provided they are supplied by two separate transfer switches. Two critical branch circuits have a higher reliability than one normal and one critical branch circuit.

Exception No. 4 recognizes that Category 2 spaces are permitted to be served by a Type 1 or Type 2 essential electrical system (EES). A Type 2 EES consists of the Life Safety and Equipment Branch. Section 517.44(A)(1) permits these types of configurations.

Multiwire branch circuits are required by 210.4(B) to be simultaneously disconnected from all ungrounded conductors, which could result in unintended and potentially dangerous interruption of power to lighting or receptacle loads at a patient bed location. Because of that concern, multiwire branch circuits are not permitted to be used to meet the requirements of 517.18(A).

(B) Patient Bed Location Receptacles.

(1) Minimum Number and Supply. Each patient bed location shall be provided with a minimum of eight receptacles.

(2) Receptacle Requirements. The receptacles required in 517.18(B)(1) shall be permitted to be of the single, duplex, or quadruplex type or any combination of the three. All receptacles shall be listed "hospital grade" and shall be so identified. The grounding terminal of each receptacle shall be connected to an insulated copper equipment grounding conductor sized in accordance with Table 250.122.

Exception No. 1: The requirements of 517.18(B)(1) and (B)(2) shall not apply to psychiatric, substance abuse, and rehabilitation hospitals meeting the requirements of 517.10(B)(2).

Exception No. 2: Psychiatric security rooms shall not be required to have receptacle outlets installed in the room.

Informational Note: It is not intended that there be a total, immediate replacement of existing non–hospital grade receptacles. It is intended, however, that non–hospital grade receptacles be replaced with hospital grade receptacles upon modification of use, renovation, or as existing receptacles need replacement.

Δ **(C) Designated Category 2 Pediatric Locations.** Receptacles that are located within patient rooms, bathrooms, playrooms, and activity rooms of pediatric units or spaces with similar risk as determined by the health care facility's governing body by conducting a risk assessment, other than infant nurseries, shall be listed and identified as "tamper resistant" or shall employ a listed tamper-resistant cover. [**99:**6.3.2.2.1(D)]

Unlike the requirement in 406.12, this requirement applies to receptacles of any rating installed in specified rooms of pediatric locations. Safeguarding can be achieved through the use of either listed tamper-resistant receptacles or listed tamper-resistant covers. The use of locking covers over ordinary receptacles does not meet this requirement. Only 125-volt, 15- and 20-ampere tamper-resistant receptacles are available; therefore, other receptacle types are likely to require a listed tamper-resistant cover.

Exhibit 517.1 shows a listed hospital-grade tamper-resistant receptacle that can be used to comply with 517.18(C). The receptacle is identified by a green dot on its face and the letters "TR" located at the top right of the receptacle. Exhibit 517.2 shows a listed tamper-resistant, hospital-grade, GFCI receptacle installation in an examination room.

EXHIBIT 517.1 A tamper-resistant, hospital-grade receptacle. *(Courtesy of Legrand®)*

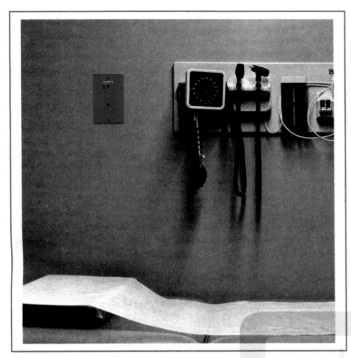

EXHIBIT 517.2 Hospital-grade, tamper-resistant, GFCI in a hospital environment. *(Courtesy of Leviton Manufacturing Co., Inc.)*

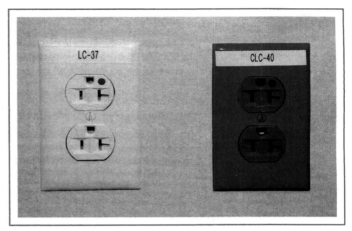

EXHIBIT 517.3 A receptacle from the normal system *(left)* and the critical branch *(right).* Note the label that indicates the panelboard and the circuit number. *(Courtesy of the International Association of Electrical Inspectors)*

△ 517.19 Category 1 Spaces.

△ **(A) Patient Bed Location Branch Circuits.** Each patient bed location shall be supplied by at least two branch circuits, one or more from the critical branch and one or more from the normal system. At least one branch circuit from the critical branch shall supply an outlet(s) only at that bed location.

The electrical receptacles or the cover plates for the electrical receptacles supplied from the life safety and critical branches shall have a distinctive color or marking so as to be readily identifiable. [**99:**6.7.2.2.5(B)]

All branch circuits from the normal system shall be from a single panelboard. Critical branch receptacles shall be identified and shall also indicate the panelboard and circuit number supplying them.

Branch circuits serving patient bed locations shall not be part of a multiwire branch circuit.

Exception No. 1: Branch circuits serving only special-purpose receptacles or equipment in Category 1 spaces shall be permitted to be served by other panelboards.

Exception No. 2: Category 1 spaces served from two separate critical branch transfer switches shall not be required to have circuits from the normal system.

Each patient bed location must be served by receptacles supplied from at least two circuits. The circuits must originate from the normal system and the critical branch, or both circuits can originate from at least two separate transfer switches of the critical branch. The critical branch receptacles are required to be labeled with the panelboard and the circuit supplying them. They are also required to be identified as being supplied from the critical branch, which is often accomplished through the use of a color code for receptacles established for that facility (the color red is used in many health care facilities to identify such receptacles), as illustrated in Exhibit 517.3.

Multiwire branch circuits are required by 210.4(B) to be simultaneously disconnected from all ungrounded conductors. This could result in unintended and potentially dangerous interruption of power to lighting or receptacle loads at a patient bed location. Because of this concern, these circuits are not permitted to be used to meet the requirements of 517.19(A). Exception No. 2 covers the special case in which two separate transfer switches supply a single patient care area. Critical branch circuits supplied from two separate transfer switches provide an equivalent level of redundancy to that specified by the main requirement.

(B) Patient Bed Location Receptacles.

△ **(1) Minimum Number and Supply.** Each patient bed location shall be provided with a minimum of 14 receptacles, with at least one connected to either of the following:

(1) The normal system branch circuit required in 517.19(A)
(2) A critical branch circuit supplied by a different transfer switch than the other receptacles at the same patient bed location

(2) Receptacle Requirements. The receptacles required in 517.19(B)(1) shall be permitted to be of the single, duplex, or quadruplex type or any combination of the three. All receptacles shall be listed "hospital grade" and shall be so identified. The grounding terminal of each receptacle shall be connected to the reference grounding point by means of an insulated copper equipment grounding conductor.

Each patient bed location must be provided with at least 14 receptacles that may be of the single, duplex, or quadruplex type, provided they are listed hospital-grade type and are identified with a green dot (shown in Exhibit 517.1). A duplex receptacle is two receptacles on one mounting strap or yoke. A quadruplex receptacle is considered four receptacles.

Each patient bed location in critical care spaces must be supplied by at least two branch circuits, one from the normal system and one from the critical branch, as shown in Exhibit 517.4. The normal circuits must be supplied from the same panel (L-1). The critical branch circuits are permitted to be supplied from different panels (EES-1 and EES-2). However, the critical branch circuit to patient bed location A cannot supply receptacles for patient bed location B.

Patient bed location receptacles can also be supplied by two different critical branch circuits, instead of one critical branch and one normal, provided the critical branch circuits are supplied from two different transfer switches. The requirements for the number and type of branch circuits in critical care spaces are intended to ensure that equipment for critical care patients will not be without electrical power.

(C) Operating Room Receptacles.

(1) Minimum Number and Supply. Each operating room shall be provided with a minimum of 36 receptacles divided between at least two branch circuits. At least 12 receptacles, but no more than 24, shall be connected to either of the following:

(1) The normal system branch circuit required in 517.19(A)
(2) A critical branch circuit supplied by a different transfer switch than the other receptacles at the same location

(2) Receptacle Requirements. The receptacles shall be permitted to be of the locking or nonlocking type and of the single, duplex, or quadruplex types or any combination of the three.

All nonlocking-type receptacles shall be listed hospital grade and so identified. The grounding terminal of each receptacle shall be connected to the reference grounding point by means of an insulated copper equipment grounding conductor.

(D) Patient Care Vicinity Grounding and Bonding (Optional). A patient care vicinity shall be permitted to have a patient equipment grounding point. The patient equipment grounding point, where supplied, shall be permitted to contain one or more listed grounding and bonding jacks. An equipment bonding jumper not smaller than 10 AWG shall be used to connect the grounding terminal of all grounding-type receptacles to the patient equipment grounding point. The bonding conductor shall be permitted to be arranged centrically or looped as convenient.

Informational Note: Where there is no patient equipment grounding point, it is important that the distance between the reference grounding point and the patient care vicinity be as short as possible to minimize any potential differences.

A patient care vicinity is permitted to have a patient equipment grounding point with multiple grounding or bonding jacks. See Exhibit 517.5.

(E) Equipment Grounding and Bonding. Where a grounded electrical distribution system is used and metal feeder raceway or Type MC or MI cable that qualifies as an equipment grounding

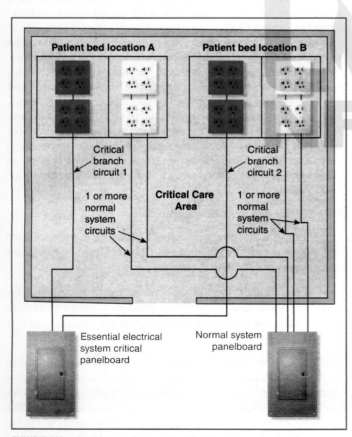

EXHIBIT 517.4 Examples of normal and critical branch circuits supplying patient bed locations in a critical care space.

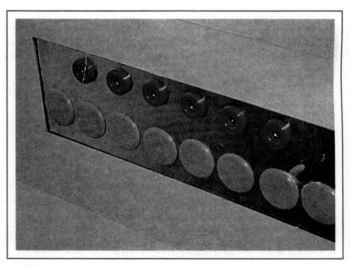

EXHIBIT 517.5 A patient equipment grounding point. *(Courtesy of the International Association of Electrical Inspectors)*

conductor in accordance with 250.118 is installed, grounding of enclosures and equipment, such as panelboards, switchboards, and switchgear, shall be ensured by one of the following bonding means at each termination or junction point of the metal raceway or Type MC or MI cable:

(1) A grounding bushing and a continuous copper bonding jumper, sized in accordance with 250.122, with the bonding jumper connected to the junction enclosure or the ground bus of the panel

(2) Connection of feeder raceways or Type MC or MI cable to threaded hubs or bosses on terminating enclosures

(3) Other approved devices such as bonding-type locknuts or bushings. Standard locknuts shall not be used for bonding.

Δ **(F) Additional Protective Techniques in Category 1 Spaces (Optional).** Isolated power systems shall be permitted to be used for Category 1 spaces, and, if used, the isolated power system equipment shall be listed as isolated power equipment. The isolated power system shall be designed and installed in accordance with 517.160.

> *Exception: The audible and visual indicators of the line isolation monitor shall be permitted to be located at the nursing station for the area being served.*

(G) Isolated Power System Equipment Grounding. Where an isolated ungrounded power source is used and limits the first-fault current to a low magnitude, the equipment grounding conductor associated with the secondary circuit shall be permitted to be run outside of the enclosure of the power conductors in the same circuit.

> Informational Note: Although it is permitted to run the equipment grounding conductor outside of the conduit, it is safer to run it with the power conductors to provide better protection in case of a second ground fault.

Installing the EGC inside the raceway with the conductors delivering the fault current reduces the impedance of the grounding path.

(H) Special-Purpose Receptacle Grounding. The equipment grounding conductor for special-purpose receptacles, such as the operation of mobile X-ray equipment, shall be extended to the reference grounding points of branch circuits for all locations likely to be served from such receptacles. Where such a circuit is served from an isolated ungrounded system, the equipment grounding conductor shall not be required to be run with the power conductors; however, the equipment grounding terminal of the special-purpose receptacle shall be connected to the reference grounding point.

517.20 Wet Procedure Locations.

Δ **(A) Receptacles and Fixed Equipment.** Wet procedure locations shall be provided with special protection against electric shock. [**99**:6.3.2.3.1]

This special protection shall be provided by one of the following:

(1) Isolated power systems that remain in operation in the event of a single line-to-ground fault condition that inherently limits the possible ground-fault current due to a first fault to a low value, without interrupting the power supply

> Informational Note No. 1: Isolated power systems can eliminate the danger of electric shock to patients who might be more susceptible to leakage current and unable to move in their beds.

(2) Power distribution system in which the power supply is interrupted if the ground-fault current does, in fact, exceed the trip value of a Class A GFCI

> Informational Note No. 2: See Annex E of ANSI/UL 943-2018, *Ground-Fault Circuit-Interrupters*, and 110.3(B) for the manufacturers' installation instructions of listed ground-fault circuit interrupters for information on the supply connection of life-support equipment to circuits providing ground-fault circuit-interrupter (GFCI) protection of personnel at outlets.

[**99**:6.3.2.3.2]

See also

Article 100 for the definition of *location, wet procedure* and its associated commentary

Exception: Branch circuits supplying only listed, fixed, therapeutic, and diagnostic equipment shall be permitted to be supplied from a grounded service, single- or 3-phase system if the following conditions are met:

(1) Wiring for grounded and isolated circuits does not occupy the same raceway.

(2) All conductive surfaces of the equipment are connected to an insulated copper equipment grounding conductor.

Δ **(B) Isolated Power Systems.** Where an isolated power system is utilized, the isolated power equipment shall be listed as isolated power equipment, and the isolated power system shall be designed and installed in accordance with 517.160.

> Informational Note: See Part IV of Article 680 for requirements on the installation of therapeutic pools and tubs.

Δ **517.21 Ground-Fault Circuit-Interrupter Protection for Personnel in Category 2 and Category 1 Spaces.** Receptacles shall not be required in bathrooms or toilet rooms. [**99**:6.3.2.2.2(D)]

Receptacles located in patient bathrooms and toilet rooms in Category 2 spaces shall have ground-fault circuit-interrupter protection in accordance with 210.8(B)(1).

Ground-fault circuit-interrupter protection for personnel shall not be required for receptacles installed in those Category 2 and Category 1 spaces where a basin, sink, or other similar plumbing fixture is installed in the patient bed location.

> Informational Note: See ANSI/UL 943-2018, *Ground-Fault Circuit-Interrupters*, Annex E, and, in accordance with 110.3(B), the manufacturers' installation instructions of listed ground-fault

circuit interrupters for information on the supply connection of life-support equipment to circuits providing ground-fault circuit-interrupter (GFCI) protection of personnel at outlets.

A basin and a toilet may be located within the patient room or as part of the bed assembly. Although the presence of a basin and a toilet meets the definition of a bathroom, the receptacles serving the critical care space are exempt from the GFCI requirement because of the specialized use of a critical care space that includes the need for uninterrupted power for cord-and-plug-connected life-support equipment. A separate bathroom within the patient room is required to have GFCI protection for receptacles. Receptacles in other bathrooms for patients, staff, or the public are required to be GFCI protected in accordance with 210.8(B).

N 517.22 Demand Factors. Demand factors for receptacle loads supplied by branch circuits not exceeding 150 volts to ground and installed in Category 1, Category 2, Category 3, and Category 4 patient care spaces shall be in accordance with 220.110.

> Informational Note: See Article 100 for the definitions of patient care space categories.

Part III. Essential Electrical System (EES)

517.25 Essential Electrical Systems for Health Care Facilities. Type 1 and Type 2 essential electrical systems (EES) for health care facilities shall comprise separate branches capable of supplying a limited amount of lighting and power service, which is considered essential for life safety and orderly cessation of procedures during the time normal electrical service is interrupted for any reason.

> Informational Note: See NFPA 99-2021, *Health Care Facilities Code*, for information on essential electrical systems.

Δ 517.26 Application of Other Articles. The life safety branch of the essential electrical system shall meet the requirements of Article 700, except as amended as follows:

(1) Section 700.4 shall not apply.
(2) Section 700.10(D) shall not apply.
(3) Section 700.17 shall be replaced with the following: Branch circuits that supply emergency lighting shall be installed to provide service from a source in accordance with 700.12 when normal supply for lighting is interrupted or where single circuits supply luminaires containing secondary batteries.
(4) Section 700.32 shall not apply.

> Informational Note No. 1: See NFPA 110-2019, *Standard for Emergency and Standby Power Systems*, for additional information.
>
> Informational Note No. 2: See 517.29 and NFPA 99-2021, *Health Care Facilities Code*, for additional information.

Where a requirement in Article 517 differs from a requirement in Article 700, Article 517 takes precedence. For example, 517.31(B)

differs from 700.5 in smaller facilities with a maximum demand on the essential electrical system of 150 kilovolt-amperes or less. In those cases, a single transfer switch is permitted to supply the entire essential electrical system, and separation of the branches of the essential electrical system is not required. Additionally, the requirements for selective coordination in 700.32 are modified by Article 517 for all branches of the essential electrical system, and coordination of overcurrent protective devices (OCPDs) is covered in 517.31(G).

See also

517.31(C)(1) for the physical separation requirements for circuits supplied by essential electrical systems in hospitals

517.31(G) for the requirement covering coordination of OCPDs in the essential and life safety branches of hospital essential electrical systems

517.42(D) for the physical separation requirements for circuits supplied by essential electrical systems in nursing homes and limited care facilities

517.29 Type 1 Essential Electrical Systems.

> Informational Note: Type 1 essential electrical systems are comprised of three separate branches capable of supplying a limited amount of lighting and power service that is considered essential for life safety and effective facility operation during the time the normal electrical service is interrupted for any reason. These three separate branches are the life safety, critical, and equipment branches. [**99**:A.6.7.2.3]

Δ (A) Applicability. The requirements of 517.29 through 517.35 shall apply to Type 1 essential electrical systems. Type 1 systems shall be required for Category 1 spaces. Type 1 systems shall be permitted to serve Category 2, Category 3, and Category 4 spaces.

> Informational Note No. 1: See NFPA 99-2021, *Health Care Facilities Code*, for performance, maintenance, and testing requirements of essential electrical systems in hospitals. See NFPA 20-2019, *Standard for the Installation of Stationary Pumps for Fire Protection*, for installation of centrifugal fire pumps.
>
> Informational Note No. 2: See NFPA 99-2021, *Health Care Facilities Code*, 6.7.5 and 6.7.6, for additional information on Type 1 and Type 2 essential electrical systems.

Δ (B) Type 1 Essential Electrical Systems. Category 1 spaces shall be served by a Type 1 essential electrical system. [**99**:6.4.1]

Category 1 spaces shall not be served by a Type 2 EES. [**99**:6.4.2]

517.30 Sources of Power.

(A) Two Independent Power Sources. Essential electrical systems (EES) shall have two or more independent sources (or sets of sources). One on-site source (or sets of sources) shall be sized to supply the entire EES. The other independent source (or sets of sources) shall be sized to supply the entire EES and shall be permitted to be located on-site or off-site. Additional sources other than the first two independent sources shall be permitted to be sized to supply the intended load.

Informational Note: An example of a set of sources may be several generators that combined serve the entire EES.

Δ **(B) Power Sources for the EES.** Power sources for the EES shall be permitted to be any of those specified in 517.30(B)(1) through (B)(5).

N **(1) Utility Supply Power.** Where utility power is used as the normal source, utility power shall not be used as the alternate source unless permitted elsewhere in this article.

> Informational Note: See 517.35 and 517.45 for essential system loads that can be supplied from dual sources of utility supply power.

Δ **(2) Generating Units.**

Δ **(3) Fuel Cell Systems.** Fuel cell systems shall be permitted to serve as the alternate power source for all or part of an EES. [**99:**6.7.1.5.1]

(a) Installation of fuel cells shall comply with the requirements in Parts I through VII of Article 692 for 1000 volts or less and Part VIII for over 1000 volts.

(b) N + 1 units shall be provided where N units have sufficient capacity to supply the demand load of the portion of the system served.

(c) Systems shall be able to assume loads within 10 seconds of loss of normal power source.

(d) Systems shall have a continuing source of fuel supply, together with sufficient on-site fuel storage for the essential system type.

(e) Where life safety and critical portions of the distribution system are present, a connection shall be provided for a portable diesel generator.

> Informational Note: See NFPA 853-2020, *Standard for the Installation of Stationary Fuel Cell Power Systems*, for information on installation of stationary fuel cells.

N **(4) Energy Storage Systems.** Energy storage systems shall be permitted to serve as the alternate source for all or part of an EES.

> Informational Note: See NFPA 111-2022, *Standard on Stored Electrical Energy Emergency and Standby Power Systems*, for information on the installation of energy storage systems.

N **(5) Health Care Microgrid.** EES shall be permitted to be supplied by a health care microgrid that also supplies nonessential loads. The health care microgrid shall be permitted to share distributed resources with the normal system. Health care microgrid systems shall be designed with sufficient reliability to provide effective facility operation consistent with the facility emergency operations plan. Health care microgrid system components shall not be compromised by failure of the normal source.

> Informational Note: See NFPA 99-2021, *Health Care Facilities Code, for information on health care microgrids.*

(C) Location of EES Components. EES components shall be located to minimize interruptions caused by natural forces common to the area (e.g., storms, floods, earthquakes, or hazards created by adjoining structures or activities). [**99:**6.2.4.1]

Δ **(1) Services.** Installation of electrical service distribution equipment shall be located to reduce possible interruption of normal electrical services resulting from natural or manmade causes as well as internal wiring and equipment failures.

(2) Feeders. Feeders shall be located to provide physical separation of the feeders of the alternate source and from the feeders of the normal electrical source to prevent possible simultaneous interruption. [**99:**6.2.4.3]

> Informational Note: Facilities in which the normal source of power is supplied by two or more separate central station-fed services experience greater than normal electrical service reliability than those with only a single feed. Such a dual source of normal power consists of two or more electrical services fed from separate generator sets or a utility distribution network that has multiple power input sources and is arranged to provide mechanical and electrical separation so that a fault between the facility and the generating sources is not likely to cause an interruption of more than one of the facility service feeders.

517.31 Requirements for the Essential Electrical System.

(A) Separate Branches. Type 1 essential electrical systems shall be comprised of three separate branches capable of supplying a limited amount of lighting and power service that is considered essential for life safety and effective hospital operation during the time the normal electrical service is interrupted for any reason. The three branches are life safety, critical, and equipment.

The division between the branches shall occur at transfer switches where more than one transfer switch is required. [**99:**6.7.2.3.1]

(B) Transfer Switches. Transfer switches shall be in accordance with one of the following:

(1) The number of transfer switches to be used shall be based on reliability and design. Each branch of the essential electrical system shall have one or more transfer switches.

(2) One transfer switch shall be permitted to serve one or more branches in a facility with a continuous load on the switch of 150 kVA (120 kW) or less. [**99:**6.7.6.2.1.4]

> Informational Note No. 1: See NFPA 99-2021, *Health Care Facilities Code*, 6.7.3.1, 6.7.2.2.5, 6.7.2.2.5.15, and 6.7.2.2.7, for more information on transfer switches.
>
> Informational Note No. 2: See Informational Note Figure 517.31(B)(1).
>
> Informational Note No. 3: See Informational Note Figure 517.31(B)(2).

In larger health care facilities, 517.31(B) requires one or more transfer switches to supply each branch of the essential electrical system. In a small health care facility with an essential load not exceeding 150 kilovolt-amperes, the essential electrical system can be served by a single transfer switch that can handle all loads. This is based on the assumption that the alternate source of power is sufficiently large to handle the simultaneous transfer of all systems in the event of a normal power loss.

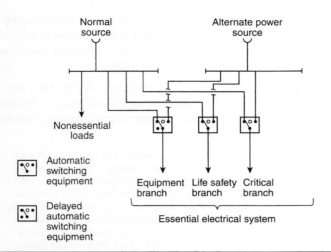

△ **INFORMATIONAL NOTE FIGURE 517.31(B)(1)** *Type 1 Essential Electrical System — Minimum Requirement (Greater Than 150 kVA) for Transfer Switch Arrangement.*

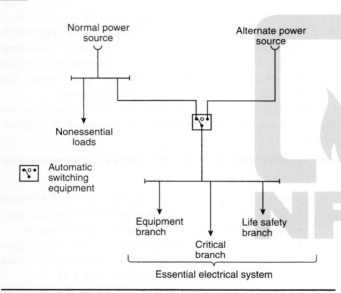

△ **INFORMATIONAL NOTE FIGURE 517.31(B)(2)** *Type 1 Essential Electrical System — Minimum Requirement (150 kVA or Less) for Transfer Switch Arrangement.*

(1) Optional Loads. Loads served by the generating equipment not specifically named in this article shall be served by their own transfer switches such that the following conditions apply:

(1) These loads shall not be transferred if the transfer will overload the generating equipment.
(2) These loads shall be automatically shed upon generating equipment overloading.

(2) Contiguous Facilities. Hospital power sources and alternate power sources shall be permitted to serve the essential electrical systems of contiguous or same-site facilities.

(C) Wiring Requirements.

△ **(1) Separation from Other Circuits.** The life safety branch and critical branch [of the essential electrical system] shall be kept independent of all other wiring and equipment. [**99:**6.7.5.2.1]

(a) Raceways, cables, or enclosures of the life safety and critical branch shall be readily identified as components of the essential electrical system (EES). Boxes and enclosures (including transfer switches, generators, and power panels) shall be field- or factory-marked and identified as components of the EES. Raceways and cables shall be field- or factory-marked as components of the EES at intervals not to exceed 7.6 m (25 ft).

(b) Conductors of the life safety branch or critical branch shall not enter the same raceways, boxes, or cabinets with each other or any other wiring system. Branch conductors shall be permitted to occupy common equipment, raceways, boxes, or cabinets of other circuits not part of the life safety branch and critical branch where such wiring complies with one of the following:

(1) Is in transfer equipment enclosures
(2) Is in exit or emergency luminaires supplied from two sources
(3) Is in a common junction box attached to exit or emergency luminaires supplied from two sources
(4) Is for two or more circuits supplied from the same branch and same transfer switch

(c) The wiring of the equipment branch shall be permitted to occupy the same raceways, boxes, or cabinets of other circuits that are not part of the essential electrical system.

(d) Where Category 2 locations are served from two separate transfer switches on the essential electrical system in accordance with 517.18(A), Exception No. 3, the Category 2 circuits from the two separate systems shall be kept independent of each other.

(e) Where Category 1 locations are served from two separate transfer switches on the essential electrical system in accordance with 517.19(A), Exception No. 2, the critical care circuits from the two separate systems shall be kept independent of each other.

The life safety branch and the critical branch of the essential electrical system are not permitted to occupy the same raceways, boxes, or cabinets with each other or with other wiring. However, circuits on the life safety or critical branch are allowed to be in the same raceway or cable with other circuits of the same branch.

Where Category 2 or Category 1 spaces are supplied from two transfer switches on an essential electrical system, the separate feeder and branch circuits are to be kept independent of each other. The issue of a failure in one circuit affecting the other is the same whether there is a normal system circuit and an essential system circuit or two essential system circuits.

(2) Isolated Power Systems. Where isolated power systems are installed in any of the areas in 517.34(A)(1) and (A)(2), each system shall be supplied by an individual circuit serving no other load.

Δ **(3) Mechanical Protection of the Essential Electrical System.** The wiring of the life safety and critical branches shall be mechanically protected by raceways. Where installed as branch circuits in patient care spaces, the installation shall comply with the requirements of 517.13(A) and (B) and 250.118. Only the following wiring methods shall be permitted:

To increase reliability of power delivery to life safety and patient care equipment, the feeder and branch-circuit wiring of the life safety and critical branch requires additional protection against mechanical damage that is not normally mandated for other occupancies. Although 517.31(C)(3) permits nonmetallic wiring methods, they are permitted only for branch circuits that do not supply patient care spaces and for feeders. The wiring methods for branch circuits in patient care spaces must comply with this section and in addition are required to comply with 517.13(A) and (B). This means metal-jacketed or metal-sheathed cables that do comply with 517.13(A) and (B) can be used for branch circuits supplied from the normal power system, but only metal raceways can be used for the branch circuits supplied by the critical and life safety branches of the essential electrical system. Section 517.31(C)(3)(3) does provide for limited use of metal-jacketed or metal-sheathed cables supplied from the life safety or critical branches.

(1) Nonflexible metal raceways, Type MI cable, RTRC marked with the suffix –XW, or Schedule 80 PVC conduit. Nonmetallic raceways shall not be used for branch circuits that supply patient care spaces.

(2) Where encased in not less than 50 mm (2 in.) of concrete, Schedule 40 PVC conduit, flexible nonmetallic or jacketed metallic raceways, or jacketed metallic cable assemblies listed for installation in concrete. Nonmetallic raceways shall not be used for branch circuits that supply patient care spaces.

(3) Listed flexible metal raceways and listed metal sheathed cable assemblies, as follows:

   a. Where used in listed prefabricated medical headwalls
   b. In listed office furnishings
   c. Where fished into existing walls or ceilings, not otherwise accessible and not subject to physical damage
   d. Where necessary for flexible connection to equipment
   e. For equipment that requires a flexible connection due to movement, vibration, or operation
   f. Luminaires installed in ceiling structures

(4) Flexible power cords of appliances or other utilization equipment connected to the essential electrical system.

(5) Cables for Class 2 or Class 3 systems permitted in Part VI of this article, with or without raceways.

Informational Note: See 517.13 for additional grounding requirements in patient care areas.

Section 517.31(C)(3)(3)(c) permits fishing flexible metal raceways and metal-sheathed cables in existing installations. This facilitates installations in renovated areas where the existing walls or ceilings remain intact. The secondary conductors of limited energy systems, such as nurse call, telephone, and alarm circuits, are exempt from being run in raceways, provided they comply with their applicable articles. Although this requirement allows substantial latitude in the wiring method, the restrictions of 300.22 (ducts, plenums, and other air-handling spaces) apply, unless cables specifically listed for use in those environments are used.

Δ **(D) Capacity of Systems.** The essential electrical system shall have the capacity and rating to meet the maximum actual demand likely to be produced by the connected load.

Feeders shall be sized in accordance with 215.2 and Part III of Article 220. The alternate power source(s) required in 517.30 shall have the capacity and rating to meet the demand produced by the load at any given time.

Demand calculations for sizing of the alternate power source(s) shall be based on any of the following:

(1) Prudent demand factors and historical data
(2) Connected load
(3) Feeder calculations
(4) Any combination of the above

The sizing requirements in 700.4 and 701.4 shall not apply to alternate sources.

(E) Receptacle Identification. The electrical receptacles or the cover plates for the electrical receptacles supplied from the life safety and critical branches shall have a distinctive color or marking so as to be readily identifiable. [**99:**6.7.2.5(B)]

(F) Feeders from Alternate Power Source. A single feeder supplied by a local or remote alternate power source shall be permitted to supply the essential electrical system to the point at which the life safety, critical, and equipment branches are separated. Installation of the transfer equipment shall be permitted at other than the location of the alternate power source.

Δ **(G) Coordination.** Overcurrent protective devices serving the essential electrical system shall be coordinated for the period of time that a fault's duration extends beyond 0.1 second.

Exception No. 1: Coordination shall not be required between transformer primary and secondary overcurrent protective devices where only one overcurrent protective device or set of overcurrent protective devices exists on the transformer secondary.

Exception No. 2: Coordination shall not be required between overcurrent protective devices of the same size (ampere rating) in series.

Informational Note No. 1: The terms *coordination* and *coordinated* as used in this section do not cover the full range of overcurrent conditions.

Informational Note No. 2: See 517.17(C) for information on requirements for the coordination of ground-fault protection.

517.32 Branches Requiring Automatic Connection.

(A) Life Safety and Critical Branch Used in a Type 1 EES. Those functions of patient care depending on lighting or appliances that are connected to the essential electrical system shall be divided into the life safety branch and the critical branch, as described in 517.33 and 517.34.

(B) Life Safety and Critical Branch Used in a Type 2 EES. The life safety and critical branches shall be installed and connected to the alternate power source specified in 517.41(A) and (B) so that all functions specified herein for the life safety and critical branches are automatically restored to operation within 10 seconds after interruption of the normal source. [**99:**6.7.5.3.1]

517.33 Life Safety Branch. The life safety branch shall be limited to circuits essential to life safety. [**99:**6.7.5.1.2.3]

No functions other than those listed in 517.33(A) through (H) shall be connected to the life safety branch. The life safety branch shall supply power as follows:

(A) Illumination of Means of Egress. Illumination of means of egress such as lighting required for corridors, passageways, stairways, and landings at exit doors, and all necessary ways of approach to exits. Switching arrangements to transfer patient corridor lighting in hospitals from general illumination circuits to night illumination circuits shall be permitted, if only one of two circuits can be selected and both circuits cannot be extinguished at the same time.

Informational Note: See NFPA *101*-2021, *Life Safety Code*, Sections 7.8 and 7.9.

(B) Exit Signs. Exit signs and exit directional signs.

Informational Note: See NFPA *101*-2021, *Life Safety Code*, Section 7.10.

(C) Alarm and Alerting Systems. Alarm and alerting systems including the following:

(1) Fire alarm systems
(2) Alarm and alerting systems (other than fire alarm systems) shall be connected to the life safety branch or critical branch. [**99:**6.7.5.1.2.5]
(3) Alarms for systems used for the piping of nonflammable medical gases
(4) Mechanical, control, and other accessories required for effective life safety systems operation shall be permitted to be connected to the life safety branch.

Heating, ventilation, and air conditioning (HVAC) controls are permitted to be on the life safety branch because the operation of an HVAC system and associated dampers can affect smoke control and life safety.

(D) Communications Systems. Hospital communications systems, where used for issuing instructions during emergency conditions. [**99:**6.7.5.1.2.4(3)]

(E) Generator Set Locations. Generator set locations as follows:

(1) Task illumination
(2) Battery charger for emergency battery-powered lighting unit(s)
(3) Select receptacles at the generator set location and essential electrical system transfer switch locations

[**99:**6.7.5.1.2.4(4)]

(F) Generator Set Accessories. Loads dedicated to a specific generator, including the fuel transfer pump(s), ventilation fans, electrically operated louvers, controls, cooling system, and other generator accessories essential for generator operation, shall be connected to the life safety branch or to the output terminals of the generator with overcurrent protective devices. [**99:**6.7.5.1.2.6]

Proper operation of a generator used as the alternate source for an essential electrical system is highly dependent on accessory equipment, such as fuel pumps, cooling and ventilation fans and louvers, and water pumps. This equipment must have an alternate source of power either through supply from the life safety branch or taps (with overcurrent protection) from the generator output terminals.

(G) Elevators. Elevator cab lighting, control, communications, and signal systems. [**99:**6.7.5.1.2.4(5)]

(H) Automatic Doors. Electrically powered doors used for building egress. [**99:**6.7.5.1.2.4(6)]

517.34 Critical Branch.

Δ **(A) Task Illumination, Fixed Equipment, and Select Receptacles.** The critical branch shall supply power for task illumination, fixed equipment, select receptacles, and select power circuits serving the following spaces and functions related to patient care:

(1) Category 1 spaces where deep sedation or general anesthesia is administered, task illumination, select receptacles, and fixed equipment
(2) Task illumination and select receptacles in the following:

   a. Patient care spaces, including infant nurseries, selected acute nursing areas, psychiatric bed areas (omit receptacles), and ward treatment rooms
   b. Medication preparation spaces
   c. Pharmacy dispensing spaces
   d. Nurses' stations — unless adequately lighted by corridor luminaires

(3) Additional specialized patient care task illumination and receptacles, where needed
(4) Nurse call systems
(5) Blood, bone, and tissue banks
(6) Telecommunications entrance facility, telecommunications equipment rooms, and telecommunication rooms and equipment in these rooms

(7) Task illumination, select receptacles, and select power circuits for the following areas:

a. Category 1 or 2 spaces with at least one duplex receptacle per patient bed location, and task illumination as required by the governing body of the health care facility

b. Angiographic labs

c. Cardiac catheterization labs

d. Coronary care units

e. Hemodialysis rooms or areas

f. Emergency room treatment areas (select)

g. Human physiology labs

h. Intensive care units

i. Postoperative recovery rooms (select)

(8) Clinical IT-network equipment

(9) Wireless phone and paging equipment for clinical staff communications

(10) Additional task illumination, receptacles, and select power circuits needed for effective facility operation, including single-phase fractional horsepower motors, which are permitted to be connected to the critical branch

[**99**:6.7.5.1.3.2]

The critical branch is intended to serve a limited number of receptacles and locations in order to reduce the load and minimize the chances of a fault condition that causes the circuit overcurrent protective device to open. Receptacles in general patient care area corridors are permitted on the critical branch as that area could be used for triage under extraordinary circumstances such as a mass casualty event, but they must be identified in some manner (color-coded or labeled) as part of the essential electrical system, in accordance with 517.31(E).

(B) Switching. It shall be permitted to control task illumination on the critical branch.

Δ **(C) Subdivision of the Critical Branch.** The critical branch shall be permitted to be subdivided into two or more branches. [**99**:6.7.5.1.3.1]

Informational Note: It is important to analyze the consequences of supplying an area with only critical branch power when failure occurs between the area and the transfer switch. Some proportion of normal and critical power or critical power from separate transfer switches might be appropriate.

517.35 Equipment Branch Connection to Alternate Power Source. The equipment branch shall be installed and connected to the alternate power source such that the equipment described in 517.35(A) is automatically restored to operation at appropriate time-lag intervals following the energizing of the life safety and critical branches. [**99**:6.7.5.1.4.2(A)]

The arrangement of the connection to the alternate power source shall also provide for the subsequent connection of equipment described in 517.35(B). [**99**:6.7.5.1.4.2(B)]

Exception: For essential electrical systems under 150 kVA, deletion of the time-lag intervals feature for delayed automatic connection to the equipment system shall be permitted.

(A) Equipment for Delayed Automatic Connection. The following equipment shall be permitted to be arranged for delayed automatic connection to the alternate power source:

(1) Central suction systems serving medical and surgical functions, including controls, with such suction systems permitted to be placed on the critical branch

(2) Sump pumps and other equipment required to operate for the safety of major apparatus, including associated control systems and alarms

(3) Compressed air systems serving medical and surgical functions, including controls with such air systems permitted to be placed on the critical branch

(4) Smoke control and stair pressurization systems

(5) Kitchen hood supply or exhaust systems, or both, if required to operate during a fire in or under the hood

(6) Supply, return, and exhaust ventilating systems for the following:

a. Airborne infectious/isolation rooms

b. Protective environment rooms

c. Exhaust fans for laboratory fume hoods

d. Nuclear medicine areas where radioactive material is used

e. Ethylene oxide evacuation

f. Anesthetic evacuation

[**99**:6.7.5.1.4.3(A)]

Where delayed automatic connection is not appropriate, the ventilation systems specified in 517.35(A)(6) shall be permitted to be placed on the critical branch. [**99**:6.7.5.1.4.3(B)]

(7) Supply, return, and exhaust ventilating systems for operating and delivery rooms

(8) Supply, return, exhaust ventilating systems and/or air-conditioning systems serving telephone equipment rooms and closets and data equipment rooms and closets

Exception: Sequential delayed automatic connection to the alternate power source to prevent overloading the generator shall be permitted where engineering studies indicate it is necessary.

Δ **(B) Equipment for Delayed Automatic or Manual Connection.** The following equipment shall be permitted to be arranged for either delayed automatic or manual connection to the alternate power source:

(1) Heating equipment to provide heating for operating, delivery, labor, recovery, intensive care, coronary care, nurseries, infection/isolation rooms, emergency treatment spaces, and general patient rooms and pressure maintenance (jockey or make-up) pump(s) for water-based fire protection systems

Exception: Heating of general patient rooms and infection/ isolation rooms during disruption of the normal source shall not be required under any of the following conditions:

(1) The outside design temperature is higher than −6.7°C (20°F).

(2) The outside design temperature is lower than −6.7°C (20°F), and where a selected room(s) is provided for the needs of all confined patients, only such room(s) need be heated.

(3) The facility is served by a dual source of normal power.

Informational Note No. 1: The design temperature is based on the 97.5 percent design value as shown in Chapter 24 of the ASHRAE *Handbook of Fundamentals* (2013).

Informational Note No. 2: See 517.30(C) for a description of a dual source of normal power.

(2) An elevator(s) selected to provide service to patient, surgical, obstetrical, and ground floors during interruption of normal power. In instances where interruption of normal power would result in other elevators stopping between floors, throw-over facilities shall be provided to allow the temporary operation of any elevator for the release of patients or other persons who may be confined between floors.

(3) Hyperbaric facilities.

(4) Hypobaric facilities.

(5) Automatically operated doors.

(6) Minimal electrically heated autoclaving equipment shall be permitted to be arranged for either automatic or manual connection to the alternate source.

(7) Controls for equipment listed in 517.35.

(8) Other selected equipment shall be permitted to be served by the equipment system. [**99**:6.7.5.1.4.4]

Δ **517.40 Type 2 Essential Electrical Systems.**

> Informational Note No. 1: Nursing homes and other limited care facilities can contain Category 1 and/or Category 2 patient care spaces, depending on the design and type of care administered in the facility. For Category 1 spaces, see 517.29 through 517.35. For Category 2 spaces not served by Type 1 essential electrical systems, see 517.40 through 517.44.
>
> Informational Note No. 2: Type 2 essential electrical systems are comprised of two separate branches capable of supplying a limited amount of lighting and power service that is considered essential for the protection of life and safety and effective operation of the institution during the time normal electrical service is interrupted for any reason. These two separate branches are the life safety and equipment branches. The number of transfer switches to be used should be based upon reliability, design, and load considerations. Each branch of the essential electrical system should have one or more transfer switches. One transfer switch should be permitted to serve one or more branches in a facility with a maximum demand on the essential electrical system of 150 kVA (120 kW). [**99**:A.6.7.6.2.1]

For the smaller, less complex facility, only a minimum alternate lighting and alarm service needs to be furnished. At nursing homes or limited care facilities where patients are sustained by electrical life-support equipment or inpatient hospital care is provided, 517.40(B) requires that the facility be provided with an essential electrical system that meets the requirements in 517.29 through 517.35 for hospitals. Because the level of care is comparable to that provided in a hospital, a "hospital-grade" essential electrical system is required for this type of nursing home.

Δ **(A) Applicability.** The requirements of 517.40(C) through 517.44 shall apply to Category 2 spaces.

Exception: The requirements of 517.40(C) through 517.44 shall not apply to freestanding buildings used as nursing homes and limited care facilities if the following apply:

(1) Admitting and discharge policies are maintained that preclude the provision of care for any patient or resident who might need to be sustained by electrical life-support equipment.

(2) No surgical treatment requiring general anesthesia is offered.

(3) An automatic battery-operated system(s) or equipment shall be effective for at least 1½ hours and is otherwise in accordance with 700.12 and that shall be capable of supplying lighting for exit lights, exit corridors, stairways, nursing stations, medical preparation areas, boiler rooms, and communications areas. This system shall also supply power to operate all alarm systems.

Informational Note: See NFPA *101*-2021, *Life Safety Code.*

Δ **(B) Category 1 Spaces in Inpatient Hospital Care Facilities.** For those nursing homes and limited care facilities that admit patients who need to be sustained by electrical life-support equipment, the essential electrical system from the source to the portion of the facility where such patients are treated shall comply with the requirements of 517.29 through 517.35.

Δ **(C) Facilities Contiguous or Located on the Same Site with Hospitals.** Nursing homes and limited care facilities that are contiguous or located on the same site with a hospital shall be permitted to have their essential electrical systems supplied by the hospital.

A single alternate power supply is permitted to serve a single building or campus with multiple types of health care occupancies. This is the same allowance specified in 517.31(B)(2) for a campus having only hospital occupancies. The use of multiple alternate sources or a health care microgrid is permitted and may be desirable to ensure reliability.

See also

Article 100 for the definition of the term *microgrid, health care*

517.41 Required Power Sources.

Δ **(A) Independent Power Sources.** Essential electrical systems (EES) shall have two or more independent sources (or sets of sources). One on-site source (or sets of sources) shall be sized to supply the entire EES. The other independent source (or sets of sources) shall be sized to supply the entire EES and shall be permitted to be located on-site or off-site. Additional sources other than the first two independent sources shall be permitted to be sized to supply the intended load.

> Informational Note: An example of a set of sources may be several generators that combined serve the entire EES.

(B) Location of EES Components. EES components shall be located to minimize interruptions caused by natural forces common to the area (e.g., storms, floods, earthquakes, or hazards created by adjoining structures or activities). [**99**:6.2.4.1]

Installations of electrical services shall be located to reduce possible interruption of normal electrical services resulting from similar causes as well as possible disruption of normal electrical service due to internal wiring and equipment failures. [**99**:6.2.4.2]

Feeders shall be located to provide physical separation of the feeders of the alternate source and from the feeders of the normal electrical source to prevent possible simultaneous interruption. [**99**:6.2.4.3]

517.42 Essential Electrical Systems for Nursing Homes and Limited Care Facilities.

(A) General. The [Type 2] essential electrical system shall be divided into the following two branches:

(1) Life safety branch
(2) Equipment branch

[**99**:6.7.6.2.1.2]

The division between the branches shall occur at transfer switches where more than one transfer switch is required. [**99**:6.7.2.2.1]

> Informational Note No. 1: Type 2 essential electrical systems are comprised of two separate branches capable of supplying a limited amount of lighting and power service that is considered essential for the protection of life and safety and effective operation of the institution during the time normal electrical service is interrupted for any reason. These two separate branches are the life safety and equipment branches. [**99**:A.6.7.6.2.1]
>
> Informational Note No. 2: The number of transfer switches to be used should be based upon reliability, design, and load considerations. Each branch of the essential electrical system should have one or more transfer switches. One transfer switch should be permitted to serve one or more branches in a facility with a maximum demand on the essential electrical system of 150 kVA (120 kW). [**99**:A.6.7.6.2.1]
>
> Informational Note No. 3: See NFPA 99-2021, *Health Care Facilities Code*, 6.7.2.2, for more information.

Δ **(B) Transfer Switches.** The number of transfer switches to be used shall be based upon reliability, design, and load considerations. [**99**:6.7.2.2.3]

Transfer switches shall be in accordance with one of the following:

(1) Each branch of the essential electrical system shall have one or more transfer switches. [**99**:6.7.2.2.3.1]
(2) One transfer switch shall be permitted to serve one or more branches in a facility with a continuous load on the switch of 150 kVA (120 kW) or less. [**99**:6.7.2.2.3.2]

Informational Note No. 1: See NFPA 99-2021, *Health Care Facilities Code*, 6.7.2.2.4, 6.7.2.2.5, 6.7.2.2.5.15, and 6.7.2.2.7 for more information on transfer switches.

Informational Note No. 2: See Informational Note Figure 517.42(B)(1).

Informational Note No. 3: See Informational Note Figure 517.42(B)(2).

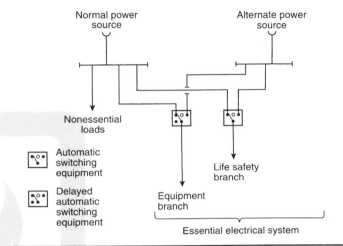

Δ **INFORMATIONAL NOTE FIGURE 517.42(B)(1)** *Type 2 Essential Electrical Systems (Nursing Home and Limited Health Care Facilities) — Minimum Requirement (Greater Than 150 kVA) for Transfer Switch Arrangement.*

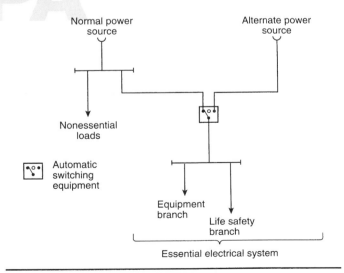

Δ **INFORMATIONAL NOTE FIGURE 517.42(B)(2)** *Type 2 Essential Electrical Systems (Nursing Home and Limited Health Care Facilities) — Minimum Requirement (150 kVA or Less) for Transfer Switch Arrangement.*

(C) Capacity of System. The essential electrical system shall have capacity to meet the demand for the operation of all functions and equipment to be served by each branch at one time.

(D) Separation from Other Circuits. The life safety branch and equipment branch shall be kept entirely independent of all other wiring and equipment. [**99**:6.7.6.3.1]

These circuits shall not enter the same raceways, boxes, or cabinets with other wiring except as follows:

(1) In transfer switches
(2) In exit or emergency luminaires supplied from two sources
(3) In a common junction box attached to exit or emergency luminaires supplied from two sources

(E) Receptacle Identification. The electrical receptacles or the cover plates for the electrical receptacles supplied from the life safety or equipment branches shall have a distinctive color or marking so as to be readily identifiable. [**99**:6.7.6.3.2]

Informational Note: If color is used to identify these receptacles, the same color should be used throughout the facility. [**99**:A.6.7.6.3.2]

517.43 Automatic Connection to Life Safety and Equipment Branch.

The life safety and equipment branches shall be installed and connected to the alternate source of power specified in 517.41 so that all functions specified herein for the life safety and equipment branches are automatically restored to operation within 10 seconds after interruption of the normal source. [**99**:6.7.6.4.1]

No functions other than those listed in 517.43(A) through (G) shall be connected to the life safety branch. [**99**:6.7.6.2.1.5(D)]

The life safety branch shall supply power as follows:

(A) Illumination of Means of Egress. Illumination of means of egress as is necessary for corridors, passageways, stairways, landings, and exit doors and all ways of approach to exits. Switching arrangement to transfer patient corridor lighting from general illumination circuits shall be permitted if only one of two circuits can be selected and both circuits cannot be extinguished at the same time.

Informational Note: See NFPA *101-2021*, *Life Safety Code*, Sections 7.8 and 7.9.

(B) Exit Signs. Exit signs and exit directional signs.

Informational Note: See NFPA *101-2021*, *Life Safety Code*, Section 7.10.

(C) Alarm and Alerting Systems. Alarm and alerting systems, including the following:

(1) Fire alarms

Informational Note No. 1: See NFPA *101-2021*, *Life Safety Code*, Sections 9.6 and 18.3.4.

(2) Alarms required for systems used for the piping of nonflammable medical gases

Informational Note No. 2: See NFPA 99-2021, *Health Care Facilities Code*, 6.7.5.1.2.5.

(D) Communications Systems. Communications systems, where used for issuing instructions during emergency conditions. [**99**:6.7.5.1.2.4(3)]

(E) Generator Set Location. Task illumination and select receptacles at the generator set location and essential electrical system transfer switch locations.

(F) Elevators. Elevator cab lighting, control, communications, and signal systems. [**99**:6.7.5.1.2.4(5)]

(G) AC Equipment for Nondelayed Automatic Connection. Generator accessories, including, but not limited to, the transfer fuel pump, electrically operated louvers, and other generator accessories essential for generator operation shall be arranged for automatic connection to the alternate power source. [**99**:6.7.6.2.1.6(C)]

517.44 Connection to Equipment Branch.
The equipment branch shall be installed and connected to the alternate power source such that equipment described in 517.35(A)(6) is automatically restored to operation at appropriate time-lag intervals following the energizing of the life safety and critical branches. [**99**:6.7.5.1.4.2(A)]

The equipment branch arrangement shall also provide for the additional connection of equipment listed in 517.44(B).

Exception: For essential electrical systems under 150 kVA, deletion of the time-lag intervals feature for delayed automatic connection to the equipment branch shall be permitted.

(A) Delayed Automatic Connections to Equipment Branch. The following equipment shall be permitted to be connected to the equipment branch and shall be arranged for delayed automatic connection to the alternate power source:

(1) Task illumination and select receptacles in the following: [**99**:6.7.6.2.1.6(D)(1)]

    a. Patient care spaces [**99**:6.7.6.2.1.6(D)(1)(a)]
    b. Medication preparation spaces [**99**:6.7.6.2.1.6(D)(1)(b)]
    c. Pharmacy dispensing space [**99**:6.7.6.2.1.6(D)(1)(c)]
    d. Nurses' stations — unless adequately lighted by corridor luminaires [**99**:6.7.6.2.1.6(D)(1)(d)]

(2) Supply, return, and exhaust ventilating systems for airborne infectious isolation rooms [**99**:6.7.6.2.1.6(D)(2)]
(3) Sump pumps and other equipment required to operate for the safety of major apparatus and associated control systems and alarms [**99**:6.7.6.2.1.6(D)(3)]
(4) Smoke control and stair pressurization systems [**99**:6.7.6.2.1.6(D)(4)]
(5) Kitchen hood supply or exhaust systems, or both, if required to operate during a fire in or under the hood [**99**:6.7.6.2.1.6(D)(5)]
(6) Nurse call systems [**99**:6.7.6.2.1.6(D)(6)]

Δ **(B) Delayed-Automatic or Manual Connection to the Equipment Branch.** The equipment specified in 517.44(B)(1) through (B)(4) shall be permitted to be connected to the equipment branch and shall be arranged for either delayed-automatic or manual connection to the alternate power source.

N **(1) Heating Equipment to Provide Heating for General Patient Rooms.** Heating of general patient rooms during disruption of the normal source shall not be required under any of the following conditions:

(1) The outside design temperature is higher than −6.7°C (20°F).

(2) The outside design temperature is lower than −6.7°C (20°F) and, where a selected room(s) is provided for the needs of all confined patients, then only such room(s) need be heated.

(3) The facility is served by a dual source of normal power as described in 517.30(C), Informational Note.

Informational Note: The outside design temperature is based on the 97.5 percent design values, as shown in Chapter 24 of the ASHRAE *Handbook of Fundamentals* (2013).

N **(2) Elevator Service.** In instances where interruptions of power would result in elevators stopping between floors, throw-over facilities shall be provided to allow the temporary operation of any elevator for the release of passengers.

N **(3) Optional Connections to the Equipment Branch.** Additional illumination, receptacles, and equipment shall be permitted to be connected only to the equipment branch.

N **(4) Multiple Systems.** Where one switch serves multiple systems as permitted in 517.43, transfer for all loads shall be nondelayed automatic.

[**99:**6.7.6.2.1.6(E)]

Informational Note: See 517.43(G) for elevator cab lighting, control, and signal system requirements. [**99:**A.6.7.6.2.1.6(E)(2)]

517.45 Essential Electrical Systems for Other Health Care Facilities.

Δ **(A) Essential Electrical Distribution.** If required by the governing body, the essential electrical distribution system for Category 3 patient care spaces shall be comprised of an alternate power system capable of supplying a limited amount of lighting and power service for the orderly cessation of procedures during a time normal electrical service is interrupted.

Informational Note: See NFPA 99-2021, *Health Care Facilities Code*.

(B) Electrical Life Support Equipment. Where electrical life support equipment is required, the essential electrical distribution system shall be as described in 517.29 through 517.30.

Δ **(C) Category 1 Patient Care Spaces.** Where Category 1 patient care spaces are present, the essential electrical distribution system shall be in accordance with 517.29 through 517.30.

Δ **(D) Category 2 Patient Care Spaces.** Where Category 2 patient care spaces are present, the essential electrical distribution system shall be in accordance with 517.40 through 517.45.

(E) Power Systems. If required, alternate power sources acceptable to the governing body shall comply with the requirements of NFPA 99-2021, *Health Care Facilities Code*.

Depending on the type and level of patient care, medical and dental offices and ambulatory health care facilities may require an alternate source of power similar to that provided at a hospital. The level of patient care may be essentially the same as a hospital, even though the facility is not called or licensed as a hospital.

Part IV. Inhalation Anesthetizing Locations

Informational Note: See NFPA 99-2021, *Health Care Facilities Code*, for further information regarding safeguards for anesthetizing locations.

Δ **517.60 Anesthetizing Location Classification.**

Informational Note: See 517.20 if either of the anesthetizing locations in 517.60(A) or 517.60(B) is designated a wet procedure location.

(A) Hazardous (Classified) Location.

(1) Use Location. In a location where flammable anesthetics are employed, the entire area shall be considered to be a Class I, Division 1 location that extends upward to a level 1.52 m (5 ft) above the floor. The remaining volume up to the structural ceiling is considered to be above a hazardous (classified) location.

(2) Storage Location. Any room or location in which flammable anesthetics or volatile flammable disinfecting agents are stored shall be considered to be a Class I, Division 1 location from floor to ceiling.

Some countries still use flammable anesthetics and rely on these safety measures. There are no known medical schools in the United States still teaching the use of flammable anesthetics or health care facilities in the United States using flammable anesthetics. Use of these precautions would be necessary should flammable anesthetics be re-instituted.

Section 517.60 designates anesthetizing locations either as hazardous locations, where flammable or nonflammable anesthetics may be interchangeably employed [517.60(A)], or as other-than-hazardous locations, where only nonflammable anesthetics are used [517.60(B)].

(B) Unclassified Location. Any inhalation anesthetizing location designated for the exclusive use of nonflammable anesthetizing agents shall be considered to be an unclassified location.

517.61 Wiring and Equipment.

(A) Within Hazardous (Classified) Anesthetizing Locations.

(1) Isolation. Except as permitted in 517.160, each power circuit within, or partially within, a flammable anesthetizing location as referred to in 517.60 shall be isolated from any distribution system by the use of an isolated power system.

(2) Design and Installation. Where an isolated power system is utilized, the isolated power equipment shall be listed as isolated power equipment, and the isolated power system shall be designed and installed in accordance with 517.160.

(3) Equipment Operating at More Than 10 Volts. In hazardous (classified) locations referred to in 517.60, all fixed wiring and equipment and all portable equipment, including lamps and other utilization equipment, operating at more than 10 volts between conductors shall comply with the requirements of 501.1 through 501.25, and 501.100 through 501.150, and 501.30(A) and (B) for Class I, Division 1 locations. All such equipment shall be specifically approved for the hazardous atmospheres involved.

(4) Extent of Location. Where a box, fitting, or enclosure is partially, but not entirely, within a hazardous (classified) location(s), the hazardous (classified) location(s) shall be considered to be extended to include the entire box, fitting, or enclosure.

(5) Receptacles and Attachment Plugs. Receptacles and attachment plugs in a hazardous (classified) location(s) shall be listed for use in Class I, Group C hazardous (classified) locations and shall have provision for the connection of an equipment grounding conductor.

See also

517.19(B)(2) commentary regarding receptacles listed for hospital use

(6) Flexible Cord Type. Flexible cords used in hazardous (classified) locations for connection to portable utilization equipment, including lamps operating at more than 8 volts between conductors, shall be of a type approved for extra-hard usage in accordance with Table 400.4 and shall include an additional equipment grounding conductor.

(7) Flexible Cord Storage. A storage device for the flexible cord shall be provided and shall not subject the cord to bending at a radius of less than 75 mm (3 in.).

(B) Above Hazardous (Classified) Anesthetizing Locations.

(1) Wiring Methods. Wiring above a hazardous (classified) location referred to in 517.60 shall be installed in rigid metal conduit, electrical metallic tubing, intermediate metal conduit, Type MI cable, or Type MC cable that employs a continuous, gas/vaportight metal sheath.

(2) Equipment Enclosure. Installed equipment that may produce arcs, sparks, or particles of hot metal, such as lamps and lampholders for fixed lighting, cutouts, switches, generators, motors, or other equipment having make-and-break or sliding contacts, shall be of the totally enclosed type or be constructed so as to prevent escape of sparks or hot metal particles.

Exception: Wall-mounted receptacles installed above the hazardous (classified) location in flammable anesthetizing locations shall not be required to be totally enclosed or have openings guarded or screened to prevent dispersion of particles.

(3) Luminaires. Surgical and other luminaires shall conform to 501.130(B).

Exception No. 1: The surface temperature limitations set forth in 501.130(B)(1) shall not apply.

Exception No. 2: Integral or pendant switches that are located above and cannot be lowered into the hazardous (classified) location(s) shall not be required to be explosionproof.

(4) Seals. Listed seals shall be provided in conformance with 501.15, and 501.15(A)(4) shall apply to horizontal as well as to vertical boundaries of the defined hazardous (classified) locations.

(5) Receptacles and Attachment Plugs. Receptacles and attachment plugs located above hazardous (classified) anesthetizing locations shall be listed for hospital use for services of prescribed voltage, frequency, rating, and number of conductors with provision for the connection of the equipment grounding conductor. This requirement shall apply to attachment plugs and receptacles of the 2-pole, 3-wire grounding type for single-phase, 120-volt, nominal, ac service.

(6) 250-Volt Receptacles and Attachment Plugs Rated 50 and 60 Amperes. Receptacles and attachment plugs rated 250 volts, for connection of 50-ampere and 60-ampere ac medical equipment for use above hazardous (classified) locations, shall be arranged so that the 60-ampere receptacle will accept either the 50-ampere or the 60-ampere plug. Fifty-ampere receptacles shall be designed so as not to accept the 60-ampere attachment plug. The attachment plugs shall be of the 2-pole, 3-wire design with a third contact connecting to the insulated (green or green with yellow stripe) equipment grounding conductor of the electrical system.

(C) Unclassified Anesthetizing Locations.

(1) Wiring Methods. Wiring serving unclassified locations, as defined in 517.60, shall be installed in a metal raceway system or cable assembly. The metal raceway system or cable armor or sheath assembly shall qualify as an equipment grounding conductor in accordance with 250.118. Type MC and Type MI cable shall have an outer metal armor, sheath, or sheath assembly that is identified as an equipment grounding conductor.

Exception: Pendant receptacle installations that employ listed Type SJO or equivalent hard usage or extra-hard usage, flexible cords suspended not less than 1.8 m (6 ft) from the floor shall not be required to be installed in a metal raceway or cable assembly.

(2) Receptacles and Attachment Plugs. Receptacles and attachment plugs installed and used in unclassified locations shall be listed "hospital grade" for services of prescribed voltage, frequency, rating, and number of conductors with provision for connection of the equipment grounding conductor. This requirement shall apply to 2-pole, 3-wire grounding type for single-phase, 120-, 208-, or 240-volt, nominal, ac service.

(3) 250-Volt Receptacles and Attachment Plugs Rated 50 Amperes and 60 Amperes. Receptacles and attachment plugs rated 250 volts, for connection of 50-ampere and 60-ampere ac medical equipment for use in unclassified locations, shall be arranged so that the 60-ampere receptacle will accept either the 50-ampere or the 60-ampere plug. Fifty-ampere receptacles shall be designed so as not to accept the 60-ampere attachment plug. The attachment plugs shall be of the 2-pole, 3-wire design with a third contact connecting to the insulated (green or green with yellow stripe) equipment grounding conductor of the electrical system.

517.62 Grounding. In any anesthetizing area, all metal raceways and metal-sheathed cables and all normally non–current-carrying conductive portions of fixed electrical equipment shall be connected to an equipment grounding conductor. Grounding and bonding in Class I locations shall comply with 501.30.

Exception: Equipment operating at not more than 10 volts between conductors shall not be required to be connected to an equipment grounding conductor.

The grounding requirements for anesthetizing locations apply to metal raceways, metal-sheathed cables, and electrical equipment. Carts, tables, and other nonelectrical items are not required to be grounded.

517.63 Grounded Power Systems in Anesthetizing Locations.

(A) Battery-Powered Lighting Units. One or more battery-powered lighting units shall be provided and shall be permitted to be wired to the critical lighting circuit in the area and connected ahead of any local switches.

Failure of the emergency circuit feeder that supplies an operating room will ordinarily plunge the room into darkness. Unless an uninterruptible power supply is installed, a delay in the restoration of illumination could occur until the alternate source of the essential electrical system comes on-line. Even though such a delay is limited to 10 seconds (per 517.32), loss of illumination at a critical point in a surgical procedure could result in danger to the patient or operating room personnel. To safeguard against being thrust into complete darkness upon interruption of normal power, at least one battery-operated emergency lighting unit is required to be installed. This type of unit provides immediate illumination upon loss of power, helping to mitigate the impact of sudden interruption of the normal illumination. It is permitted to connect the lighting unit to the critical branch circuit.

(B) Branch-Circuit Wiring. Branch circuits supplying only listed, fixed, therapeutic and diagnostic equipment, permanently installed above the hazardous (classified) location and in unclassified locations, shall be permitted to be supplied from a normal grounded service, single- or three-phase system, provided the following apply:

(1) Wiring for grounded and isolated circuits does not occupy the same raceway or cable.
(2) All conductive surfaces of the equipment are connected to an equipment grounding conductor.
(3) Equipment (except enclosed X-ray tubes and the leads to the tubes) is located at least 2.5 m (8 ft) above the floor or outside the anesthetizing location.
(4) Switches for the grounded branch circuit are located outside the hazardous (classified) location.

Exception: Sections 517.63(B)(3) and (B)(4) shall not apply in unclassified locations.

(C) Fixed Lighting Branch Circuits. Branch circuits supplying only fixed lighting shall be permitted to be supplied by a normal grounded service, provided the following apply:

(1) Such luminaires are located at least 2.5 m (8 ft) above the floor.
(2) All conductive surfaces of luminaires are connected to an equipment grounding conductor.
(3) Wiring for circuits supplying power to luminaires does not occupy the same raceway or cable for circuits supplying isolated power.
(4) Switches are wall-mounted and located above hazardous (classified) locations.

Exception: Sections 517.63(C)(1) and (C)(4) shall not apply in unclassified locations.

(D) Remote-Control Stations. Wall-mounted remote-control stations for remote-control switches operating at 24 volts or less shall be permitted to be installed in any anesthetizing location.

(E) Location of Isolated Power Systems. Where an isolated power system is utilized, the isolated power equipment shall be listed as isolated power equipment. Isolated power system equipment and its supply circuit shall be permitted to be located in an anesthetizing location, provided it is installed above a hazardous (classified) location or in an unclassified location.

(F) Circuits in Anesthetizing Locations. Except as permitted above, each power circuit within, or partially within, a flammable anesthetizing location as referred to in 517.60 shall be isolated from any distribution system supplying other-than-anesthetizing locations.

517.64 Low-Voltage Equipment and Instruments.

(A) Equipment Requirements. Low-voltage equipment that is frequently in contact with the bodies of persons or has exposed current-carrying elements shall comply with one of the following:

(1) Operate on an electrical potential of 10 volts or less
(2) Be approved as intrinsically safe or double-insulated equipment
(3) Be moisture resistant

(B) Power Supplies. Power shall be supplied to low-voltage equipment from one of the following:

(1) An individual portable isolating transformer (autotransformers shall not be used) connected to an isolated power circuit receptacle by means of an appropriate cord and attachment plug

(2) A common low-voltage isolating transformer installed in an unclassified location

(3) Individual dry-cell batteries

(4) Common batteries made up of storage cells located in an unclassified location

(C) Isolated Circuits. Isolating-type transformers for supplying low-voltage circuits shall have both of the following:

(1) Approved means for insulating the secondary circuit from the primary circuit

(2) The core and case connected to an equipment grounding conductor

(D) Controls. Resistance or impedance devices shall be permitted to control low-voltage equipment but shall not be used to limit the maximum available voltage to the equipment.

(E) Battery-Powered Appliances. Battery-powered appliances shall not be capable of being charged while in operation unless their charging circuitry incorporates an integral isolating-type transformer.

(F) Receptacles or Attachment Plugs. Any receptacle or attachment plug used on low-voltage circuits shall be of a type that does not permit interchangeable connection with circuits of higher voltage.

> Informational Note: Any interruption of the circuit, even circuits as low as 10 volts, either by any switch or loose or defective connections anywhere in the circuit, may produce a spark that is sufficient to ignite flammable anesthetic agents.

Part V. Diagnostic Imaging and Treatment Equipment

517.70 Applicability. Nothing in this part shall be construed as specifying safeguards against possible radiation or magnetic fields.

> Informational Note No. 1: Radiation safety and performance requirements of several classes of X-ray equipment are regulated under Public Law 90-602 and are enforced by the Department of Health and Human Services.
>
> Informational Note No. 2: Information on radiation protection by the National Council on Radiation Protection and Measurements is published as *Reports of the National Council on Radiation Protection and Measurement*. These reports are obtainable from NCRP Publications, P.O. Box 30175, Washington, DC 20014.
>
> Informational Note No. 3: Examples of diagnostic imaging equipment can include, but are not limited to, the following:

(1) General radiographic (X-ray) equipment (mobile and fixed)
(2) General fluoroscopic equipment (mobile and fixed)
(3) Interventional equipment (mobile and fixed)
(4) Bone mineral density equipment
(5) Dental equipment
(6) Computerized tomography (CT) equipment
(7) Positron emission tomography (PET) equipment
(8) Nuclear medicine equipment
(9) Mammography equipment
(10) Magnetic resonance (MR) equipment
(11) Diagnostic ultrasound equipment
(12) Electrocardiogram equipment

> Informational Note No. 4: Examples of treatment equipment can include, but are not limited to, the following:

(1) Linear accelerators
(2) Gamma knife
(3) Cyber knife
(4) Proton therapy
(5) Tomotherapy

517.71 Connection to Supply Circuit.

(A) Fixed and Stationary Diagnostic Imaging and Treatment Equipment. Fixed and stationary diagnostic imaging and treatment equipment shall be connected to the power supply by means of a wiring method complying with applicable requirements of Chapters 1 through 4 of this *Code*, as modified by this article.

Exception: Equipment properly supplied by a branch circuit rated at not over 30 amperes shall be permitted to be supplied through a suitable attachment plug and hard-service cable or cord.

(B) Portable, Mobile, and Transportable Diagnostic Imaging and Treatment Equipment. Individual branch circuits shall not be required for portable, mobile, and transportable medical diagnostic imaging and treatment equipment requiring a capacity of not over 60 amperes.

(C) Over 1000-Volt Supply. Circuits and equipment operated on a supply circuit of over 1000 volts shall comply with Parts I through IV of Article 490.

517.72 Disconnecting Means.

(A) Capacity. A disconnecting means rated for at least 50 percent of the input required for the momentary rating or 100 percent of the input required for the long-time rating of the diagnostic imaging and treatment equipment, whichever is greater, shall be provided in the supply circuit.

(B) Location. The disconnecting means shall be operable from a location readily accessible from the control location.

(C) Portable, Mobile, and Transportable Diagnostic Imaging and Treatment Equipment. For equipment connected to a 120-volt branch circuit of 30 amperes or less, a grounding-type attachment plug and receptacle of proper rating shall be permitted to serve as a disconnecting means.

517.73 Rating of Supply Conductors and Overcurrent Protection.

(A) Branch Circuits. The ampacity of supply branch-circuit conductors and the current rating of overcurrent protective devices shall not be less than 50 percent of the momentary rating or 100 percent of the long-time rating, whichever is greater.

△ (B) Feeders. The ampacity of supply feeders and the current rating of overcurrent protective devices supplying two or more branch circuits supplying diagnostic imaging and treatment equipment shall not be less than 50 percent of the momentary demand rating of the largest unit, plus 25 percent of the momentary demand rating of the next largest unit, plus 10 percent of the momentary demand rating of each additional unit.

> Informational Note No. 1: The minimum conductor size for branch and feeder circuits is also governed by voltage regulation requirements. For a specific installation, the manufacturer usually specifies minimum distribution transformer and conductor sizes, rating of disconnecting means, and overcurrent protection.
>
> Informational Note No. 2: The ampacity of the branch-circuit conductors and the ratings of disconnecting means and overcurrent protection for diagnostic imaging and treatment equipment are usually designated by the manufacturer for the specific installation.

517.74 Control Circuit Conductors.

(A) Number of Conductors in Raceway. The number of control circuit conductors installed in a raceway shall be determined in accordance with 300.17.

(B) Minimum Size of Conductors. Size 18 AWG or 16 AWG fixture wires in accordance with 724.49 and flexible cords shall be permitted for the control and operating circuits of diagnostic imaging and treatment equipment and auxiliary equipment where protected by not larger than 20-ampere overcurrent devices.

517.76 Transformers and Capacitors.
Transformers and capacitors that are part of diagnostic imaging and treatment equipment shall not be required to comply with Parts I and II of Articles 450 and 460.

Capacitors shall be mounted within enclosures of insulating material or grounded metal.

△ 517.77 Installation of Cables with Grounded Shields. Cables with grounded shields shall be permitted to be installed in cable trays or cable troughs along with control and power supply conductors without the need for barriers to separate the wiring.

517.78 Guarding and Grounding.

△ (A) High-Voltage Parts. All high-voltage parts shall be mounted within grounded enclosures. The connection from the high-voltage equipment to other high-voltage components shall be made with high-voltage shielded cables.

(B) Low-Voltage Cables. Low-voltage cables connecting to oil-filled units that are not completely sealed, such as transformers, condensers, oil coolers, and high-voltage switches, shall have insulation of the oil-resistant type.

△ (C) Non–Current-Carrying Metal Parts. Non–current-carrying metal parts of diagnostic imaging and treatment equipment (e.g., controls, tables, transformer tanks, shielded cables) shall be connected to an equipment grounding conductor in accordance with Part VII of Article 250, as modified by 517.13(A) and (B).

Part VI. Communications, Signaling Systems, Data Systems, Fire Alarm Systems, and Systems Less Than 120 Volts, Nominal

517.80 Patient Care Spaces. Equivalent insulation and isolation to that required for the electrical distribution systems in patient care areas shall be provided for communications, signaling systems, data system circuits, fire alarm systems, and systems less than 120 volts, nominal.

Class 2 and Class 3 signaling and communications systems, Class 2 circuits that transmit power and data to a powered device, and power-limited fire alarm systems shall not be required to comply with the grounding requirements of 517.13, to comply with the mechanical protection requirements of 517.31(C)(3)(5), or to be enclosed in raceways, unless otherwise specified by Chapters 7 or 8.

Secondary circuits of transformer-powered communications or signaling systems shall not be required to be enclosed in raceways unless otherwise specified by Chapters 7 or 8. [**99**:6.7.2.2.7]

> Informational Note: See ANSI/NEMA C137.3-2017, *American National Standard for Lighting Systems — Minimum Requirements for Installation of Energy Efficient Power over Ethernet (PoE) Lighting Systems*, for information on installation of cables for PoE lighting systems.

A primary objective of Article 517 is to minimize patients' exposure to any level of current that could injure them. The equivalent insulation and isolation of these circuits from contact by the patient is for protection from any level of shock hazard that could result from inadvertent contact with energized circuit conductors or parts associated with the limited energy systems. Nurse call, intercom, speaker, cable television, fire alarm systems, and equipment powered and receiving data from Class 2 circuits are examples of the types of circuits covered by this requirement.

See also

517.1, Informational Note No. 1, for more information on sensitivity to shock hazards

Class 2 and Class 3 remote control and signaling circuits, Class 2 power circuits that provide power and data to powered equipment, as well as power-limited fire alarm circuits that are installed in patient care spaces are not required to comply with the same grounding or mechanical protection requirements as branch circuits for power and lighting. This does not remove applicable installation requirements, such as the use of a raceway or other protection method, specified in the applicable requirements of Chapters 7 or 8.

517.81 Other-Than-Patient-Care Spaces. In other-than-patient-care spaces, installations shall be in accordance with other parts of this *Code*.

517.82 Signal Transmission Between Appliances.

(A) General. Permanently installed signal cabling from an appliance in a patient location to remote appliances shall employ a signal transmission system that prevents hazardous grounding interconnection of the appliances.

> Informational Note: See 517.13(A) for additional grounding requirements in patient care spaces.

(B) Common Signal Grounding Wire. Common signal grounding wires (i.e., the chassis ground for single-ended transmission) shall be permitted to be used between appliances all located within the patient care vicinity, provided the appliances are served from the same reference grounding point.

Part VII. Isolated Power Systems

517.160 Isolated Power Systems.

(A) Installations.

(1) Isolated Power Circuits. Each isolated power circuit shall be controlled by a switch or circuit breaker that has a disconnecting pole in each isolated circuit conductor to simultaneously disconnect all power. Such isolation shall be accomplished by means of one or more isolation transformers, by means of generator sets, or by means of electrically isolated batteries. Conductors of isolated power circuits shall not be installed in cables, raceways, or other enclosures containing conductors of another system.

(2) Circuit Characteristics. Circuits supplying primaries of isolating transformers shall operate at not more than 600 volts between conductors and shall be provided with proper overcurrent protection. The secondary voltage of such transformers shall not exceed 600 volts between conductors of each circuit. All circuits supplied from such secondaries shall be ungrounded and shall have an approved overcurrent device of proper ratings in each conductor. Circuits supplied directly from batteries or from motor generator sets shall be ungrounded and shall be protected against overcurrent in the same manner as transformer-fed secondary circuits. If an electrostatic shield is present, it shall be connected to the reference grounding point.

(3) Equipment Location. The isolating transformers, motor generator sets, batteries and battery chargers, and associated primary or secondary overcurrent devices shall not be installed in hazardous (classified) locations. The isolated secondary circuit wiring extending into a hazardous anesthetizing location shall be installed in accordance with 501.10.

Δ **(4) Isolation Transformers.** An isolation transformer shall not serve more than one operating room except as covered in 517.160(A)(4)(a) and (A)(4)(b).

For purposes of this section, anesthetic induction rooms are considered part of the operating room or rooms served by the induction rooms.

(a) *Induction Rooms.* Where an induction room serves more than one operating room, the isolated circuits of the induction room shall be permitted to be supplied from the isolation transformer of any one of the operating rooms served by that induction room.

(b) *Higher Voltages.* Isolation transformers shall be permitted to serve single receptacles in several patient areas where the following apply:

(1) The receptacles are reserved for supplying power to equipment requiring 150 volts or higher, such as portable X-ray units.
(2) The receptacles and mating plugs are not interchangeable with the receptacles on the local isolated power system.

(5) Conductor Identification. The isolated circuit conductors shall be identified as follows:

(1) Isolated Conductor No. 1 — Orange with at least one distinctive colored stripe other than white, green, or gray along the entire length of the conductor
(2) Isolated Conductor No. 2 — Brown with at least one distinctive colored stripe other than white, green, or gray along the entire length of the conductor

For 3-phase systems, the third conductor shall be identified as yellow with at least one distinctive colored stripe other than white, green, or gray along the entire length of the conductor. Where isolated circuit conductors supply 125-volt, single-phase, 15- and 20-ampere receptacles, the striped orange conductor(s) shall be connected to the terminal(s) on the receptacles that are identified in accordance with 200.10(B) for connection to the grounded circuit conductor.

(6) Wire-Pulling Compounds. Wire-pulling compounds that increase the dielectric constant shall not be used on the secondary conductors of the isolated power supply.

> Informational Note No. 1: It is desirable to limit the size of the isolation transformer to 10 kVA or less and to use conductor insulation with low leakage to meet impedance requirements.
>
> Informational Note No. 2: Minimizing the length of branch-circuit conductors and using conductor insulations with a dielectric constant less than 3.5 and insulation resistance constant greater than 6100 megohm-meters (20,000 megohm-feet) at 16°C (60°F) reduces leakage from line to ground, reducing the hazard current.

(B) Line Isolation Monitor.

(1) Characteristics. In addition to the usual control and overcurrent protective devices, each isolated power system shall be provided with a listed continually operating line isolation monitor that indicates total hazard current. The monitor shall be designed such that a green signal lamp, conspicuously visible to persons in each area served by the isolated power system, remains lighted when the system is adequately isolated from ground. An adjacent red signal lamp and an audible warning signal (remote if desired) shall be energized when the total hazard current (consisting of possible resistive and capacitive leakage currents) from either isolated conductor to ground reaches a threshold value of 5 mA under nominal line voltage conditions. The line monitor shall not alarm for a fault hazard of less than 3.7 mA or for a total hazard current of less than 5 mA.

Exception: A system shall be permitted to be designed to operate at a lower threshold value of total hazard current. A line isolation monitor for such a system shall be permitted to be approved, with the provision that the fault hazard current shall be permitted to be reduced but not to less than 35 percent of the corresponding threshold value of the total hazard current, and the monitor hazard current is to be correspondingly reduced to not more than 50 percent of the alarm threshold value of the total hazard current.

See Exhibit 517.6 for an example of a hospital isolated power system panel.

(2) Impedance. The line isolation monitor shall be designed to have sufficient internal impedance such that, when properly connected to the isolated system, the maximum internal current that can flow through the line isolation monitor, when any point of the isolated system is grounded, shall be 1 mA.

EXHIBIT 517.6 An example of a hospital isolated power system panel with built-in isolation transformer, line isolation monitor, load center, and grounded busbar. *(Courtesy of Schneider Electric)*

Exception: The line isolation monitor shall be permitted to be of the low-impedance type such that the current through the line isolation monitor, when any point of the isolated system is grounded, will not exceed twice the alarm threshold value for a period not exceeding 5 milliseconds.

Informational Note: Reduction of the monitor hazard current, provided this reduction results in an increased "not alarm" threshold value for the fault hazard current, will increase circuit capacity.

(3) Ammeter. An ammeter calibrated in the total hazard current of the system (contribution of the fault hazard current plus monitor hazard current) shall be mounted in a plainly visible place on the line isolation monitor with the "alarm on" zone at approximately the center of the scale.

Exception: The line isolation monitor shall be permitted to be a composite unit, with a sensing section cabled to a separate display panel section on which the alarm or test functions are located.

Informational Note: It is desirable to locate the ammeter so that it is conspicuously visible to persons in the anesthetizing location.

ARTICLE 518 Assembly Occupancies

518.1 Scope. Except for the assembly occupancies explicitly covered by 520.1, this article covers all buildings or portions of buildings or structures designed or intended for the gathering together of 100 or more persons for such purposes as deliberation, worship, entertainment, eating, drinking, amusement, awaiting transportation, or similar purposes.

Informational Note: See NFPA *101*-2021, *Life Safety Code*, or the local building code for methods of determining population capacity.

Article 518 applies to assembly occupancies designed or intended for 100 or more persons with the population capacity determined by methods utilized in *NFPA 101®*, *Life Safety Code®*. Article 518 would apply, for example, to a church chapel or an auditorium for occupancy of 100 or more persons but not to a supermarket. Even though a supermarket might contain 100 or more persons at any given time, it is not specifically designed or intended for the assembly of persons. Article 518 does not apply to office buildings or schools, even though such buildings, as a rule, are designed for occupancy by 100 or more persons. The article does, however, apply to assembly halls, restaurants, and assembly areas such as large meeting or conference rooms, cafeterias, gymnasiums, and auditoriums within an office or school building if these parts of the building are designed or intended for the assembly of 100 or more persons. Note that the production and audience areas of theaters within a building are subject to the requirements of Article 520.

The following information for determining new assembly occupancy capacity is extracted from *NFPA 101*:

12.1.7 Occupant Load.

12.1.7.1 General. The occupant load, in number of persons for whom means of egress and other provisions are required, shall be determined on the basis of the occupant load factors of Table 7.3.1.2 [Commentary Table 518.1] that are characteristic of the use of the space or shall be determined as the maximum probable population of the space under consideration, whichever is greater.

7.3.1.1.2 For other than existing means of egress, where more than one means of egress is required, the means of egress shall be of such width and capacity that the loss of any one means of egress leaves available not less than 50 percent of the required capacity.

7.3.1.2 Occupant Load Factor. The occupant load in any building or portion thereof shall be not less than the number of persons determined by dividing the floor area assigned to that use by the occupant load factor for that use as specified in Table 7.3.1.2 [Commentary Table 518.1]. Where both gross and net area figures are given for the same occupancy, calculations shall be made by applying the gross area figure to the gross area of the portion of the building devoted to the use for which the gross area figure is specified and by applying the net area figure to the net area of the portion of the building devoted to the use for which the net area figure is specified.

518.2 General Classification.

(A) Examples. Assembly occupancies shall include, but not be limited to, the following:

(1) Armories
(2) Assembly halls
(3) Auditoriums
(4) Bowling lanes
(5) Casinos and gaming facilities
(6) Club rooms
(7) Conference rooms
(8) Courtrooms
(9) Dance halls
(10) Dining and drinking facilities
(11) Exhibition halls
(12) Gymnasiums
(13) Mortuary chapels
(14) Multipurpose rooms
(15) Museums
(16) Places of awaiting transportation
(17) Places of religious worship
(18) Pool rooms
(19) Restaurants
(20) Skating rinks

(B) Multiple Occupancies. Where an assembly occupancy forms a portion of a building containing other occupancies, Article 518 applies only to that portion of the building considered an assembly occupancy. Occupancy of any room or space

for assembly purposes by less than 100 persons in a building of other occupancy, and incidental to such other occupancy, shall be classified as part of the other occupancy.

(C) Theatrical Areas. Where any such building or structure, or portion of a building or structure, contains a projection booth or stage platform or area for the presentation of theatrical or musical productions, either fixed or portable, the wiring for that area, including associated audience seating areas, and all equipment that is used in the referenced area, and portable equipment and wiring for use in the production that will not be connected to permanently installed wiring, shall comply with Article 520.

> Informational Note: See NFPA *101*-2021, *Life Safety Code*, or the local building code for methods of determining population capacity.

The requirements in Article 520 apply to theatrical areas within assembly occupancies. See the commentary following 520.1.

518.3 Temporary Wiring. In exhibition halls used for display booths, as in trade shows, the temporary wiring shall be permitted to be installed in accordance with Article 590. Flexible cables and cords approved for hard or extra-hard usage shall be permitted to be laid on floors where protected from contact by the general public. The ground-fault circuit-interrupter requirements of 590.6 shall not apply. All other ground-fault circuit-interrupter requirements of this *Code* shall apply.

Where ground-fault circuit-interrupter protection for personnel is cord-and-plug-connected to the branch circuit or to the feeder, the ground-fault circuit-interrupter protection shall be listed as portable ground-fault circuit-interrupter protection or provide a level of protection equivalent to a portable ground-fault circuit interrupter, whether assembled in the field or at the factory.

> *Exception: Where conditions of supervision and maintenance ensure that only qualified persons will service the installation, flexible cords or cables identified in Table 400.4 for hard usage or extra-hard usage shall be permitted in cable trays used only for temporary wiring. All cords or cables shall be installed in a single layer. A permanent sign shall be attached to the cable tray at intervals not to exceed 7.5 m (25 ft) and read as follows:*

CABLE TRAY FOR TEMPORARY WIRING ONLY

GFCI requirements for temporary wiring installations are covered in 590.6, but those requirements do not apply to temporary installations under Article 518. Temporary installations in exhibition halls must meet all other applicable GFCI protection requirements such as those specified in 210.8. Although trade show booths are connected to temporary wiring, GFCI protection is required for water features at garden shows, receptacles near sinks for food vendors, and other, similar installations.

Permanent GFCI protection differs from portable GFCI protection. Product standards for GFCI equipment require portable

COMMENTARY TABLE 518.1 Occupant Load Factor

Use	(ft²/person)ᵃ	(m²/person)ᵃ
Assembly Use		
Concentrated use, without fixed seating	7 net	0.65 net
Less concentrated use, without fixed seating	15 net	1.4 net
Bench-type seating	1 person/18 linear in.	1 person/455 linear mm
Fixed seating	Use number of fixed seats	Use number of fixed seats
Waiting spaces	See 12.1.7.2 and 13.1.7.2 [of *NFPA 101*].	See 12.1.7.2 and 13.1.7.2 [of *NFPA 101*].
Kitchens	100	9.3
Library stack areas	100	9.3
Library reading rooms	50 net	4.6 net
Swimming pools	50 (water surface)	4.6 (water surface)
Swimming pool decks	30	2.8
Exercise rooms with equipment	50	4.6
Exercise rooms without equipment	15	1.4
Stages	15 net	1.4 net
Lighting and access catwalks, galleries, gridirons	100 net	9.3 net
Casinos and similar gaming areas	11	1
Skating rinks	50	4.6
Business Use (other than below)	150	14
Concentrated business useᶠ	50	4.6
Airport traffic control tower observation levels	40	3.7
Collaboration rooms/spaces ≤450 ft² (41.8 m²) in areaᶠ	30	2.8
Collaboration rooms/spaces >450 ft² (41.8 m²) in areaᶠ	15	1.4
Day-Care Use	35 net	3.3 net
Detention and Correctional Use	120	11.1
Educational Use		
Classrooms	20 net	1.9 net
Shops, laboratories, vocational rooms	50 net	4.6 net
Health Care Use		
Inpatient treatment departments	240	22.3
Sleeping departments	120	11.1
Ambulatory health care	150	14
Industrial Use		
General and high hazard industrial	100	9.3
Special-purpose industrial	NA	NA
Mercantile Use		
Sales area on street floorᵇ·ᶜ	30	2.8
Sales area on two or more street floorsᶜ	40	3.7
Sales area on floor below street floorᶜ	30	2.8
Sales area on floors above street floorᶜ	60	5.6
Floors or portions of floors used only for offices	See business use.	See business use.
Floors or portions of floors used only for storage, receiving, and shipping, and not open to general public	300	27.9
Mall structuresᵈ	Per factors applicable to use of spaceᵉ	
Residential Use		
Hotels and dormitories	200	18.6
Apartment buildings	200	18.6
Board and care, large	200	18.6
Storage Use		
In storage occupancies	NA	NA
In mercantile occupancies	300	27.9
In other than storage and mercantile occupancies	500	46.5

Note: NA = not applicable. The occupant load is the maximum probable number of occupants present at any time.

ᵃAll factors are expressed in gross area unless marked "net."

Notes b through f contain specific load or egress considerations, such as when no direct egress to a street is available, for the occupancies referenced. Refer to *NFPA 101* for further information.

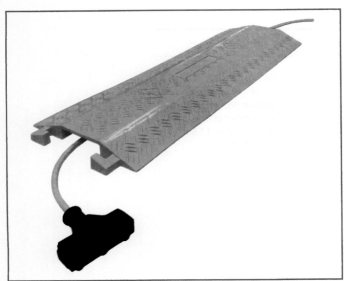

EXHIBIT 518.1 Protective cover for flexible cords and cables laid on floors or on the ground. (*Courtesy of Electriduct*)

devices to de-energize the contacts when the grounded conductor is open, a grounded conductor is transposed (miswired), and/or an ungrounded conductor is open. For that reason, GFCI protection for temporary wiring must be listed as portable or provide equivalent protection to a portable GFCI device.

Exhibit 518.1 is an example of a protective cover used where cords are subject to physical damage, such as pedestrian walkways or light vehicle pathways.

518.4 Wiring Methods.

Δ **(A) General.** The wiring method shall qualify as an equipment grounding conductor in accordance with 250.118 or shall contain an equipment grounding conductor sized in accordance with Table 250.122, and shall be any of the following:

(1) Metal raceways
(2) Flexible metal raceways
(3) Nonmetallic raceways encased in not less than 50 mm (2 in.) of concrete
(4) Type MI, Type MC, or Type AC cable

N **(B) Communications, Signaling Systems, Data Systems, Fire Alarm Systems, and Systems Less Than 120 Volts, Nominal.** Fixed wiring methods for specific installations shall be as follows:

(1) Audio signal processing, amplification, and reproduction equipment — 640.9
(2) Communications systems — Part IV of Article 805 and Part VI of Article 840
(3) Class 2 and Class 3 remote control and signaling circuits — Article 725, Part III
(4) Class 2 circuits that transmit power, data, or both to a powered device

Informational Note: See ANSI/NEMA C137.3-2017, *American National Standard for Lighting Systems — Minimum Requirements for Installation of Energy Efficient Power over Ethernet (PoE) Lighting Systems*, for information on installation of cables for PoE lighting systems. See Part III of Article 760 for information on fire alarm circuits.

(C) Nonrated Construction. In addition to the wiring methods permitted by 518.4(A), nonmetallic-sheathed cable, electrical nonmetallic tubing, and rigid nonmetallic conduit shall be permitted to be installed in those buildings or portions of those buildings that are not required to be of fire-rated construction by the applicable building code.

Informational Note: Fire-rated construction is the fire-resistive classification used in building codes.

Δ **(D) Spaces with Finish Rating.** Electrical nonmetallic tubing and rigid nonmetallic conduit shall be permitted to be installed in club rooms, conference and meeting rooms in hotels or motels, courtrooms, dining facilities, restaurants, mortuary chapels, museums, libraries, and places of religious worship where the following apply:

(1) The tubing or conduit is installed concealed within walls, floors, and ceilings where the walls, floors, and ceilings provide a thermal barrier of material that has at least a 15-minute finish rating as identified in listings of fire-rated assemblies.
(2) The tubing or conduit is installed above suspended ceilings where the suspended ceilings provide a thermal barrier of material that has at least a 15-minute finish rating as identified in listings of fire-rated assemblies.

Electrical nonmetallic tubing and rigid nonmetallic conduit are not recognized for use in other space used for environmental air in accordance with 300.22(C).

Informational Note: A finish rating is established for assemblies containing combustible (wood) supports. The finish rating is defined as the time at which the wood stud or wood joist reaches an average temperature rise of 121°C (250°F) or an individual temperature rise of 163°C (325°F) as measured on the plane of the wood nearest the fire. A finish rating is not intended to represent a rating for a membrane ceiling.

The wiring methods identified in 518.4(A) and (B) apply to any wall, floor, or ceiling within an assembly occupancy, as classified in 518.2. The requirements of 518.4(C) apply to those portions of the building and those assembly occupancies not required to be fire rated. Determination of whether a building or portions of a building are required to be constructed using fire-rated construction methods is a function of the building code in effect in a jurisdiction. Under certain height, area, and occupancy type parameters, a building intended for the assembly of 100 or more persons could be constructed using methods that are not fire-rated. The use of electrical nonmetallic tubing and rigid nonmetallic conduit as permitted in 518.4(D) applies only to the specific occupancies described, provided the wiring methods are installed concealed behind a surface that has a 15-minute finish rating.

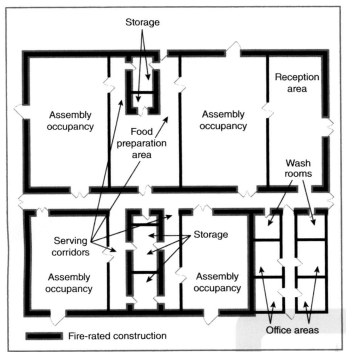

EXHIBIT 518.2 Floor plan of a single-story facility. The walls required by the local building code to be of fire-rated construction are represented by the red highlighted lines; the thin black lines represent walls not required by the local building code to be of fire-rated construction.

Exhibit 518.2 illustrates a single-story facility in which the washrooms and office area are not assembly occupancies, as defined in 518.2, and therefore require no special wiring methods. Ordinary wiring methods may be used on the inside surface of the storage area walls and on or in the partitions between storage areas, because those areas are not assembly occupancies. However, the main requirements of 518.4 apply inside any hollow spaces of the fire-rated storage area walls, because the serving corridors are part of the assembly occupancies as a result of the particular building design. If the hollow spaces of fire-rated walls or ceiling also provide a 15-minute finish rating and are not "other spaces used for environmental air" as described in 300.22(C), electrical nonmetallic tubing as well as rigid nonmetallic conduit is permitted for specifically described occupancies, as specified in 362.10. Also, wiring in ceilings or floors that are required to be of fire-rated construction in the assembly occupancy must also comply with 518.4.

Assembly occupancies frequently require emergency wiring, particularly for emergency illumination and exit lighting.

See also

700.10(D) for special fire protection requirements for emergency circuits in assembly occupancies with an occupant capacity of 1000 or more

Δ **518.5 Supply.** Portable switchboards, portable power distribution equipment, and commercial appliance outlet centers shall be installed in accordance with 518.5(A) through (C).

N **(A) Power Outlets and Commercial Appliance Outlet Centers.**

N **(1) Overcurrent Protection.** Power outlets and commercial appliance outlet centers shall provide overcurrent protection or shall be protected by overcurrent devices.

N **(2) Accessibility.** Overcurrent devices, power outlets, and commercial appliance outlet centers shall not be accessible to the general public.

N **(3) Equipment Grounding Conductor Connections.** Connecting means for an equipment grounding conductor shall be provided.

N **(4) Markings.** Power outlets and commercial appliance outlet centers shall be marked as follows:

FOR USE BY QUALIFIED PERSONS ONLY.
RISK OF ELECTRIC SHOCK.
Disconnect all power before servicing.
Disconnecting means location:

N **(5) Panelboard Orientation.** A panelboard installed in a listed commercial appliance outlet center designed for in-floor mounting shall be permitted to be orientated in the face-up position, if such orientation is part of the listing, and 408.43 shall not apply.

N **(B) Portable Switchboards and Portable Power Distribution Equipment.** Portable switchboards and portable power distribution equipment shall be supplied only from listed power outlets or listed commercial appliance outlet centers, each having sufficient voltage and ampere ratings.

N **(C) Neutral Conductor of Feeders Supplying Solid State Dimmer Systems.** The neutral conductor of feeders supplying solid-state phase control, 3-phase, 4-wire dimmer systems shall be considered a current-carrying conductor for purposes of ampacity adjustment.

The neutral conductor of feeders supplying solid-state sine wave, 3-phase, 4-wire dimming systems shall not be considered a current-carrying conductor for purposes of ampacity adjustment.

Exception: The neutral conductor of feeders supplying systems that use or are capable of using both phase-control and sine-wave dimmers shall be considered as current-carrying for purposes of ampacity adjustment.

Informational Note: See Article 100 for definitions of solid-state dimmer types.

Portable switchboards and portable power distribution equipment must be supplied from listed power outlets, such as the one shown in Exhibit 518.3, or from listed commercial appliance outlet centers. This equipment must be rated for the voltage and current for which they are used. Power outlets, commercial appliance outlet centers, and their associated overcurrent protective devices must not be accessible to the general public.

Some professional performance lighting systems use solid-state, sine wave, 3-phase, 4-wire dimming systems. These dimmers

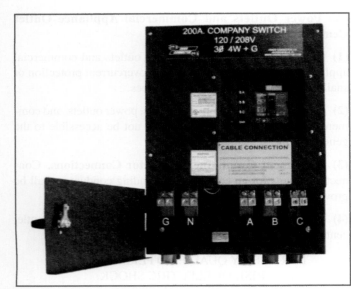

EXHIBIT 518.3 A listed power outlet for connection of portable switchboards in an assembly occupancy. *(Courtesy of Union Connector Co., Inc.)*

vary with the amplitude of the applied voltage waveform, without any of the nonlinear switching found in phase-control solid-state dimmers. Because solid-state sine wave dimmers are linear loads, they do not require the neutral to be considered a current-carrying conductor. This characteristic impacts the number of current-carrying conductors in a raceway or cable and the need to apply ampacity adjustment factors.

518.6 Illumination. Illumination shall be provided for all working spaces about fixed service equipment, switchboards, switchgear, panelboards, or motor control centers installed outdoors that serve assembly occupancies. Control by automatic means only shall not be permitted. Additional lighting outlets shall not be required where the workspace is illuminated by an adjacent light source.

In some assembly occupancies, fixed service equipment is installed outdoors. Illumination of indoor service equipment is required by 110.26(D). Section 518.6 requires illumination of outdoor service equipment supplying assembly occupancies.

ARTICLE

520   Theaters, Audience Areas of Motion Picture and Television Studios, Performance Areas, and Similar Locations

Part I. General

520.1 Scope. This article covers all buildings or that part of a building or structure, indoor or outdoor, designed or used for

presentation, dramatic, musical, motion picture projection, or similar purposes and to specific audience seating areas within motion picture or television studios.

The special requirements of Article 520 apply only to that part of a building used as a theater or for a similar purpose and do not necessarily apply to the entire building. In a school building, for example, the requirements of Article 520 apply to an auditorium used for dramatic or other performances. The special requirements of this article apply to the stage, auditorium, dressing rooms, and main corridors leading to the auditorium but not to other parts of the building not involved in the use of the auditorium for performances or entertainment. Audience areas of motion picture and television studios, as defined and covered in Article 530, and stage areas of concert and festival venues are also covered by the requirements of Article 520.

520.5 Wiring Methods.

Δ **(A) General.** The fixed wiring method shall be any of the following:

(1) Metal raceways
(2) Nonmetallic raceways encased in at least 50 mm (2 in.) of concrete
(3) Type MI cable, Type MC cable, or Type AC cable containing an insulated equipment grounding conductor sized in accordance with Table 250.122

N **(B) Communications, Signaling Systems, Data Systems, Fire Alarm Systems, and Systems Less Than 120 Volts, Nominal.** Fixed wiring methods for specific installations shall be as follows:

(1) Audio signal processing, amplification, and reproduction equipment — 640.9
(2) Communications systems — Parts I and IV of Article 800, Part IV of Article 805, and Part VI of Article 840
(3) Class 2 and Class 3 remote control and signaling circuits — Part III of Article 725
(4) Class 2 circuits that transmit power, data, or both to a powered device

Informational Note: See ANSI/NEMA C137.3-2017, *American National Standard for Lighting Systems — Minimum Requirements for Installation of Energy Efficient Power over Ethernet (PoE) Lighting Systems*, for information on installation of cables for PoE lighting systems. See Part III of Article 760 for information on fire alarm circuits.

(C) Portable Equipment. The wiring for portable switchboards, stage set lighting, stage effects, and other wiring not fixed as to location shall be permitted with approved flexible cords and cables as provided elsewhere in Article 520. Fastening such cables and cords by uninsulated staples or nailing shall not be permitted.

(D) Nonrated Construction. Nonmetallic-sheathed cable, Type AC cable, electrical nonmetallic tubing, and rigid nonmetallic conduit shall be permitted to be installed in those buildings or portions of buildings that are not required to be of fire-rated construction by the applicable building code.

Because theaters and similar buildings are usually required to be of fire-rated construction as determined by applicable building codes, the fixed wiring methods for power and lighting circuits are limited to those specified in 520.5(A).

See also

518.4 and its commentary for wiring methods where fire-rated construction is required

Section 520.5(B) permits the installation of communications circuits, Class 2 and Class 3 remote-control and signaling circuits, sound-reproduction wiring, and fire alarm circuits using wiring methods, including power-limited circuit cables covered in Article 722 and communications cables covered in Chapter 8. Where portability, flexibility, and adjustments are necessary, suitable cords and cables are permitted. Section 520.5(D) permits Type AC cable as one of the wiring methods in buildings or portions of buildings that are not required to be of fire-rated construction. Under this condition, Type AC cable is not required to contain an insulated equipment grounding conductor (EGC).

520.6 Number of Conductors in Raceway. The number of conductors permitted in any metal conduit, rigid nonmetallic conduit as permitted in this article, or electrical metallic tubing for circuits or for remote-control conductors shall not exceed the percentage fill shown in Table 1 of Chapter 9. Where contained within an auxiliary gutter or a wireway, the sum of the cross-sectional areas of all contained conductors at any cross section shall not exceed 20 percent of the interior cross-sectional area of the auxiliary gutter or wireway. The 30-conductor limitation of 366.22 and 376.22 shall not apply.

520.9 Branch Circuits. A branch circuit of any size supplying one or more receptacles shall be permitted to supply stage set lighting. The voltage rating of the receptacles shall be not less than the circuit voltage. Receptacle ampere ratings and branch-circuit conductor ampacity shall be not less than the branch-circuit overcurrent device ampere rating. Table 210.21(B)(2) and 210.23 shall not apply. The requirements in 210.8(B), other than 210.8(B)(6), shall apply.

The occupancies referenced in Article 520 are excluded from all the general requirements relating to connector rating and branch circuit loading found elsewhere in the *NEC®*, such as in Table 210.21(B)(2). Connectors must be rated sufficiently for the parameters involved, thus permitting connectors with voltage and current ratings higher than the branch-circuit rating to be used. However, per 406.4(F), the same connector cannot be used for a different voltage or current on the same premises.

The stage set lighting and associated equipment, such as stage effects, both fixed and portable, must be as flexible as possible. Connectors often are used for different purposes and are therefore marked on a show-by-show basis to designate the voltage, current, and type of current actually employed. Stage set lighting is usually planned in advance, and the loads on each receptacle are known. Loads are not casually connected as they might be at a typical general-use wall

receptacle. Care is taken to ensure that circuits are not overloaded, thereby avoiding unwanted tripping during a performance. Outdoor circuits are often dimmed and are exempt from the GFCI requirement of 210.8(B)(6) because dimmer-rated GFCI devices are not readily available.

520.10 Portable Equipment Used Outdoors. Portable stage and studio lighting equipment and portable power distribution equipment not identified for outdoor use shall be permitted for temporary use outdoors if the equipment is supervised by qualified personnel while energized and barriered from the general public.

Informational Note: See ANSI/ESTA E1.58-2017, *Electrical Safety Standard for Portable Stage and Studio Equipment Used Outdoors*, for information on the use of portable stage and studio lighting equipment outdoors.

Portable indoor stage or studio equipment that is not marked as suitable for wet or damp locations is permitted to be used temporarily in outdoor locations. If rain occurs, this equipment is typically de-energized, and a protective cover is installed before it is re-energized. At the end of the day, the equipment is either de-energized and protected or dismantled and stored.

Part II. Fixed Stage Switchboards

520.21 General. Fixed stage switchboards shall comply with the following:

(1) Fixed stage switchboards shall be listed.
(2) Fixed stage switchboards shall be readily accessible but shall not be required to be located on or adjacent to the stage. Multiple fixed stage switchboards shall be permitted at different locations.
(3) A fixed stage switchboard shall contain overcurrent protective devices for all branch circuits supplied by that switchboard.
(4) A fixed stage switchboard shall be permitted to supply both stage and nonstage equipment.
(5) Fixed stage switchboards shall comply with the marking and working space requirements in 408.18(C) but shall not be required to comply with the load terminal location requirements in 408.18(C)(1), (C)(2), and (C)(3).

520.25 Dimmers. Dimmers shall comply with 520.25(A) through (C).

A high-density digital dimmer rack typically contains one dimmer (usually of 20-, 50-, or 100-ampere capacity) for each branch circuit connected to it. The rack is usually supplied by a 3-phase, 4-wire-plus-ground feeder, which is distributed via buses to all dimmers in the rack. Typical dimmer racks contain between 12 and 96 dimmers and can have total power capacities of up to 288 kilowatts. In large theatrical systems, many racks may be bused together. A central control electronics module drives multiple dimmers in the rack. A digital data link can connect the dimmer rack to a remotely located computer control console.

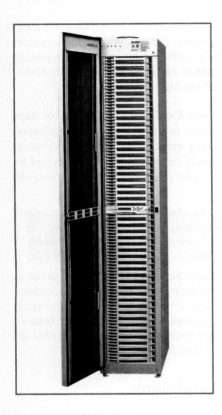

EXHIBIT 520.1 A typical high-density digital SCR dimmer switchboard. (*Courtesy of Electronic Theatre Controls, Inc.*)

EXHIBIT 520.2 An electronic computer lighting control console for remotely controlling solid-state-type dimmers. (*Courtesy of Electronic Theatre Controls, Inc.*)

Exhibit 520.1 shows a high-density digital silicon-controlled rectifier (SCR) dimmer switchboard. Dimmers for individual circuits are contained in dual plug-in dimmer modules, which also contain circuit breakers for overcurrent protection and filter chokes to minimize acoustic noise from the lamp filaments. The digital control electronics in this exhibit are contained in a plug-in module with front-panel controls for configuration and testing.

(A) Disconnection and Overcurrent Protection. If dimmers are installed in ungrounded conductors, each dimmer shall have overcurrent protection not greater than 125 percent of the dimmer rating and shall be disconnected from all ungrounded conductors where the master or individual switch or circuit breaker supplying such dimmer is in the open position.

(B) Autotransformer-Type Dimmers. The circuit supplying an autotransformer-type dimmer shall not exceed 150 volts between conductors. The grounded conductor shall be common to the input and output circuits.

> Informational Note: See 210.9 for circuits derived from autotransformers.

Any desired voltage may be applied to the lamps, from full-line voltage to voltage so low that the lamps provide no illumination, by means of a movable contact tap. This type of dimmer produces very little heat and operates at high efficiency. Its dimming effect, within its maximum rating, is independent of the wattage of the load.

(C) Solid-State-Type Dimmers. The circuit supplying a solid-state dimmer shall not exceed 150 volts between

conductors unless the dimmer is listed specifically for higher voltage operation. Where a grounded conductor supplies a dimmer, it shall be common to the input and output circuits. Dimmer chassis shall be connected to the equipment grounding conductor.

Solid-state stage dimmers are often used since stage switchboards are usually remote controlled. The switchboard or dimmer rack normally is located offstage in a dimmer room, where proper climate control can be furnished and noise from the rack cooling fans does not interfere with the performance onstage. Branch circuits usually are connected to the dimmer rack on a dimmer-per-circuit basis. One or more control cables connect the dimmer rack to a remote lighting control console, such as the computer-style one shown in Exhibit 520.2, which can be located onstage or in the auditorium in view of the stage.

520.26 Type of Switchboard. A stage switchboard shall be either one or a combination of the types specified in 520.26(A), (B), (C), and (D).

(A) Manual. Dimmers and switches are operated by handles mechanically linked to the control devices.

(B) Remotely Controlled. Devices are operated electrically from a pilot-type control console or panel. Pilot control panels either shall be part of the switchboard or shall be permitted to be at another location.

(C) Intermediate. A stage switchboard with circuit interconnections is a secondary switchboard (patch panel) or panelboard remote to the primary stage switchboard. It shall contain overcurrent protection. Where the required branch-circuit overcurrent protection is provided in the dimmer panel, it shall be permitted to be omitted from the intermediate switchboard.

The intermediate-stage switchboard located between the dimmer switchboard and the branch circuits is usually called a patch panel. Its purpose is to break down larger dimmer circuits to smaller branch circuits, to select the branch circuits to be controlled by a dimmer, or both.

(D) Constant Power. A stage switchboard containing only overcurrent protective devices and no control elements.

520.27 Stage Switchboard Feeders.

(A) Type of Feeder. Feeders supplying stage switchboards shall be one of the types in 520.27(A)(1) through (A)(3).

(1) Single Feeder. A single feeder disconnected by a single disconnect device.

(2) Multiple Feeders to Intermediate Stage Switchboard (Patch Panel). Multiple feeders of unlimited quantity shall be permitted, provided that all multiple feeders are part of a single system. Where combined, neutral conductors in a given raceway shall be of sufficient ampacity to carry the maximum unbalanced current supplied by multiple feeder conductors in the same raceway, but they need not be greater than the ampacity of the neutral conductor supplying the primary stage switchboard. Parallel neutral conductors shall comply with 310.10(G).

Stage switchboard feeders are often many dimmer-controlled circuits at 100 amperes or less, single phase, so they can be distributed to different combinations of the same size or smaller branch circuits. This type of installation usually requires a common neutral, and because of the quantity of circuits, many installations require several parallel neutrals running in several raceways. Generally, these parallel neutrals are sized as follows: (1) size the common neutral to the feeder of the primary switchboard, then (2) split the neutral into multiple parallel conductors, one per raceway. In no case are the ungrounded conductors permitted to be installed in one raceway and the common neutral installed in another.

(3) Separate Feeders to Single Primary Stage Switchboard (Dimmer Bank). Installations with separate feeders to a single primary stage switchboard shall have a disconnecting means for each feeder. The primary stage switchboard shall have a permanent and obvious label stating the number and location of disconnecting means. If the disconnecting means are located in more than one distribution switchboard, the primary stage switchboard shall be provided with barriers to correspond with these multiple locations.

Large primary stage switchboards usually consist of several sections, often called dimmer racks, that form a dimmer bank. The dimmer racks can be fed separately or bused together to accept one or more feeder circuits. In older theaters where an intermediate stage switchboard is connected to a primary stage switchboard, a single large feeder usually supplies the primary stage switchboard, because the intermediate stage switchboard patches only the ungrounded conductors and requires a common neutral.

(B) Neutral Conductor. For the purpose of ampacity adjustment, the following shall apply:

(1) The neutral conductor of feeders supplying solid-state, phase-control 3-phase, 4-wire dimming systems shall be considered a current-carrying conductor.
(2) The neutral conductor of feeders supplying solid-state, sine wave 3-phase, 4-wire dimming systems shall not be considered a current-carrying conductor.
(3) The neutral conductor of feeders supplying systems that use or are capable of using both phase-control and sine wave dimmers shall be considered as current-carrying.

The neutral is not always considered a current-carrying conductor with the use of solid-state dimmers. If the sine wave–type dimmer is the only type in use, the neutral of the circuits supplying it need not be considered a current-carrying conductor. However, if phase-control dimmers are used, or if combinations of phase control– and sine wave–type dimmers are connected to the same feeder or branch circuit, the neutral conductor must be considered a current-carrying conductor. The neutral of feeders supplying solid-state, 3-phase, 4-wire dimming systems carries triplen-harmonic currents that are present even under balanced load conditions.

Δ **(C) Supply Capacity.** For the purposes of calculating supply capacity to switchboards, considering the maximum load that the switchboard is intended to control in a given installation shall be permitted if the following apply:

(1) All feeders supplying the switchboard shall be protected by an overcurrent device with a rating not greater than the ampacity of the feeder.
(2) The opening of the overcurrent device shall not affect the proper operation of the egress or emergency lighting systems.

Informational Note: See 220.40 for calculation of stage switchboard feeder loads.

Part III. Fixed Stage Equipment Other Than Switchboards

520.40 Stage Lighting Hoists.
Where a stage lighting hoist is listed as a complete assembly and contains an integral cable-handling system and cable to connect a moving wiring device to a fixed junction box for connection to permanent wiring, the extra-hard usage requirement of 520.44(C)(1) shall not apply.

A listed stage lighting hoist, such as the one shown in Exhibit 520.3, is part of the rigging system. It provides an automated system — as opposed to a manual counterweight system — for lifting and supporting heavy integrated lighting equipment over a stage.

520.41 Circuit Loads.

(A) Circuits Rated 20 Amperes or Less. Footlights, border lights, and proscenium sidelights shall be arranged so that no

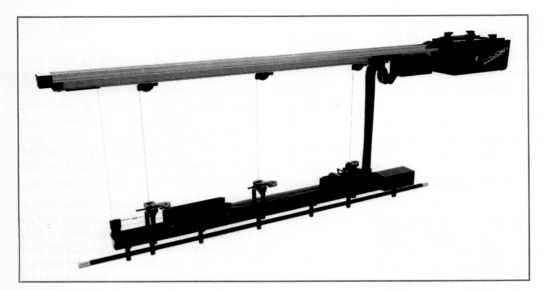

EXHIBIT 520.3 A stage lighting hoist. *(Courtesy of Electronic Theatre Controls, Inc.)*

branch circuit supplying such equipment carries a load exceeding 20 amperes.

(B) Circuits Rated Greater Than 20 Amperes. Where only heavy-duty lampholders are used, such circuits shall be permitted to comply with Article 210 for circuits supplying heavy-duty lampholders.

In accordance with 210.23(C) and (D), 30-, 40-, or 50-ampere branch circuits are permitted if heavy-duty lampholders, such as medium- or mogul-base Edison screw shell types, are used for fixed lighting.

520.42 Conductor Insulation. Foot, border, proscenium, or portable strip lights and connector strips shall be wired with conductors that have insulation suitable for the temperature at which the conductors are operated, but not less than 125°C (257°F). The ampacity of the 125°C (257°F) conductors shall be that of 60°C (140°F) conductors. All drops from connector strips shall be 90°C (194°F) wire sized to the ampacity of 60°C (140°F) cords and cables with no more than 150 mm (6 in.) of conductor extending into the connector strip. Section 310.15(C)(1) shall not apply.

Informational Note: See Table 310.4(1) for conductor types.

The 125°C minimum temperature rating is due to the heat from the lamps raising the ambient temperature where the wiring is located. Drops from connector strips usually are flexible cord. The derating factor for more than three current-carrying conductors may not be necessary if the conductors are (1) not all energized at one time or are not often energized at full intensity (dimmed) and (2) not energized continuously.

520.43 Footlights.

(A) Metal Trough Construction. Where metal trough construction is employed for footlights, the trough containing the

circuit conductors shall be made of sheet metal not lighter than 0.81 mm (0.032 in.) and treated to prevent oxidation. Lampholder terminals shall be kept at least 13 mm (½ in.) from the metal of the trough. The circuit conductors shall be soldered to the lampholder terminals.

(B) Other-Than-Metal Trough Construction. Where the metal trough construction specified in 520.43(A) is not used, footlights shall consist of individual outlets with lampholders wired with rigid metal conduit, intermediate metal conduit, or flexible metal conduit, Type MC cable, or mineral-insulated, metal-sheathed cable. The circuit conductors shall be soldered to the lampholder terminals.

(C) Disappearing Footlights. Disappearing footlights shall be arranged so that the current supply is automatically disconnected when the footlights are replaced in the storage recesses designed for them.

Historically, the footlights described in 520.43(A) and (B) were built in the field. Modern footlights are compartmentalized, factory-wired assemblies for field installation and may be permanently exposed or be of the disappearing type. Disappearing footlights must automatically disconnect the current supply when the footlights are in the closed position, thereby preventing heat entrapment that could cause a fire. Disconnection is accomplished by switches in the terminal compartment.

520.44 Borders, Proscenium Sidelights, Drop Boxes, and Connector Strips.

(A) General. Borders and proscenium sidelights shall be as follows:

(1) Constructed as specified in 520.43
(2) Suitably stayed and supported

(3) Designed so that the flanges of the reflectors or other guards protect the lamps from mechanical damage and from accidental contact with scenery or other combustible material

These types of stage lighting instruments must be suitably supported and protected from mechanical damage. Commonly, lampholders are wired alternately on three or four circuits. A splice box is provided on top of the housing for enclosing connections between the cable supplying the border light and the border light's internal wiring, which consists of wiring from the splice box to the lamp sockets in a trough extending the length of the border.

(B) Connector Strips and Drop Boxes. Connector strips and drop boxes shall be as follows:

(1) Suitably stayed and supported
(2) Listed as stage and studio wiring devices

(C) Cords and Cables for Border Lights, Drop Boxes, and Connector Strips.

(1) General. Cords and cables for supply to border lights, drop boxes, and connector strips shall be listed for extra-hard usage. The cords and cables shall be suitably supported. Such cords and cables shall be employed only where flexible conductors are necessary. Ampacity of the conductors shall be as provided in 400.5.

Border lights typically are supported by steel cables to facilitate height adjustment for cleaning and lamp replacement, and the circuit conductors supplying the border lights are carried to each border light in flexible cable. Each flexible cable usually contains

many circuits; however, its overall length is limited by its ability to travel up and down without getting tangled. See Exhibit 520.4.

(2) Cords and Cables Not in Contact with Heat-Producing Equipment. Listed multiconductor extra-hard usage type cords and cables not in direct contact with equipment containing heat-producing elements shall be permitted to have their ampacity determined by Table 520.44(C)(2)(1). Maximum load current in any conductor with an ampacity determined by Table 520.44(C)(2)(1) shall not exceed the values in Table 520.44(C)(2)(1).

Extra-hard-usage cords and cables not in direct contact with heat-producing equipment are permitted to have their ampacity determined by Table 520.44(C)(2)(1) instead of 400.5(A).

Table 520.44(C)(2)(1) is based on a minimum 50-percent diversity factor. It makes allowance for the fact that not all circuits are on at the same time, not all circuits are at full intensity (dimmed), and not all circuits are on for a long period of time. If the load diversity does not follow this pattern, such as border lights that are all left on at full intensity to light the stage for rehearsal, lecture, or classroom purposes, Table 520.44(C)(2)(1) must not be used.

Flexible cords and cables are permitted to be used only where necessary, such as for border lights requiring height adjustment, otherwise Chapter 3 wiring methods are required for fixed connections. These fixed conductors are required to follow the ampacity calculations in Article 310, but the adjustment factors in 310.15(C)(1) do not take into account load diversity. Informational Note No. 1 to 310.15(C)(1) refers to Annex B [see Table B.2(11)] for adjustment factors with at least a 50-percent load diversity. This annex table for Chapter 3 wiring methods correlates with the flexible cord adjustment factors in Table 520.44(C)(2)(1) for a load diversity of 50 percent.

EXHIBIT 520.4 A suspended connector strip with spotlights attached.

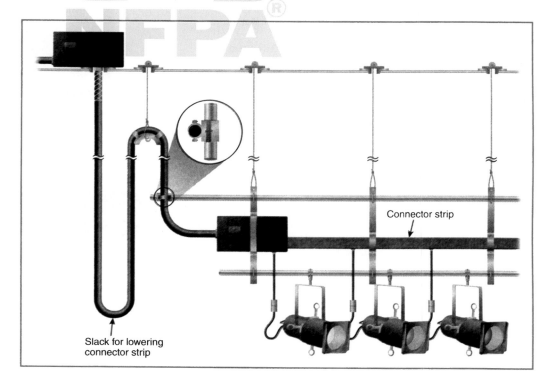

Connector strip

Slack for lowering connector strip

Δ **TABLE 520.44(C)(2)(1)** *Ampacity of Listed Extra-Hard Usage Cords and Cables with Temperature Ratings of 75°C (167°F) and 90°C (194°F) [Based on Ambient Temperature of 30°C (86°F)]*

Size (AWG)	Temperature Rating of Cords and Cables		Maximum Rating of Overcurrent Device
	75°C (167°F)	90°C (194°F)	
14	24	28	15
12	32	35	20
10	41	47	25
8	57	65	35
6	77	87	45
4	101	114	60
2	133	152	80

Note: Ampacity shown is the ampacity for multiconductor cords and cables where only three copper conductors are current-carrying in accordance with 400.5. If the number of current-carrying conductors in a cord or cable exceeds three and the load diversity is 50 percent or less, the ampacity of each conductor shall be reduced as shown in Table 520.44(C)(2)(2):

N **TABLE 520.44(C)(2)(2)** *Ampacity Adjustment Factors for More Than Three Current-Carrying Conductors in a Cord or Cable Where Load Diversity Is 50 Percent or Less*

Number of Conductors	Percent of Ampacity Value in Table 520.44(C)(2)(1)
4–6	80
7–24	70
25–42	60
43 and above	50

Note: Ultimate insulation temperature. In no case shall conductors be associated together in such a way with respect to the kind of circuit, the wiring method used, or the number of conductors such that the temperature limit of the conductors is exceeded.

A neutral conductor that carries only the unbalanced current from other conductors of the same circuit need not be considered as a current-carrying conductor.

In a 3-wire circuit consisting of two-phase conductors and the neutral conductor of a 4-wire, 3-phase, wye-connected system, the neutral conductor carries approximately the same current as the line-to-neutral currents of the other conductors and shall be considered to be a current-carrying conductor.

On a 4-wire, 3-phase wye circuit where the major portion of the load consists of nonlinear loads, there are harmonic currents in the neutral conductor. Therefore, the neutral conductor shall be considered to be a current-carrying conductor.

Informational Note: For the purposes of Table 520.44(C)(2)(1), load diversity is the percentage of the total current of all simultaneously energized circuits fed by the cable to the sum of the ampacities of all pairs of circuit conductors in that cable.

(3) Identification of Conductors in Multiconductor Extra-Hard-Usage Cords and Cables. Neutral conductors shall be white without stripe or shall be identified by a distinctive white marking at their terminations. Equipment grounding conductors shall be green with or without yellow stripe or shall be identified by a distinctive green marking at their terminations.

520.45 Receptacles. Receptacles for electrical equipment on stages shall be rated in amperes. Conductors supplying receptacles shall be in accordance with Articles 310 and 400.

520.46 Connector Strips, Drop Boxes, Floor Pockets, and Other Outlet Enclosures. Receptacles for the connection of portable stage-lighting equipment shall be pendant or mounted in pockets or enclosures and shall comply with 520.45. Supply cables for connector strips and drop boxes shall be as specified in 520.44(C).

Exhibit 520.4 shows a hanging connector strip with its associated hardware and flexible cable allowing for height adjustment. Exhibits 520.5 and 520.6 illustrate two types of connections for portable stage lighting equipment.

520.47 Backstage Lamps (Bare Bulbs). Lamps (bare bulbs) installed in backstage and ancillary areas where they can come in contact with scenery shall be located and guarded so as to be free from physical damage and shall provide an air space of not less than 50 mm (2 in.) between such lamps and any combustible material.

Exception: Decorative lamps installed in scenery shall not be considered to be backstage lamps for the purpose of this section.

520.48 Curtain Machines. Curtain machines shall be listed.

EXHIBIT 520.5 A 4-gang, 4-receptacle pin-plug outlet box designed for flush mounting. *(Courtesy of Electronic Theatre Controls, Inc.)*

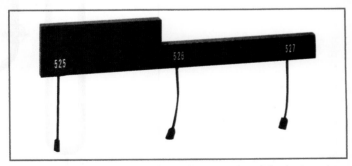

EXHIBIT 520.6 A typical three-circuit connector strip designed for wall or pipe mounting. *(Courtesy of Electronic Theatre Controls, Inc.)*

520.49 Smoke Ventilator Control. Where stage smoke ventilators are released by an electrical device, the circuit operating the device shall be normally closed and shall be controlled by at least two externally operable switches, one switch being placed at a readily accessible location on stage and the other where designated by the authority having jurisdiction. The device shall be designed for the full voltage of the circuit to which it is connected, no resistance being inserted. The device shall be enclosed in a metal box having a door that shall remain closed except during service to the equipment.

In addition to the two externally operable switches at different locations, the design of a normally closed circuit ensures that smoke ventilators operate when the circuit opens for any reason, such as a circuit breaker tripping or a fuse blowing.

Part IV. Portable Switchboards on Stage

520.50 Road Show Connection Panel (A Type of Patch Panel). A panel designed to allow for road show connection of portable stage switchboards to fixed lighting outlets by means of permanently installed supplementary circuits. The panel, supplementary circuits, and outlets shall comply with 520.50(A) through (D).

Also known as a road show interconnect or intercept panel, this panel is designed to connect the load side of a portable switchboard to the fixed building branch circuits and associated outlets. It may also provide for the fixed branch circuits to be connected to a fixed switchboard when the portable switchboard is not installed.

(A) Load Circuits. Circuits shall originate from grounding-type polarized inlets of current and voltage rating that match the fixed-load receptacle.

The required grounding-type polarized inlets may be flush or pendant. The fixed-load receptacle is where the portable switchboard connects to the house circuits to control the theater lights.

(B) Circuit Transfer. Circuits that are transferred between fixed and portable switchboards shall have all circuit conductors transferred simultaneously.

(C) Overcurrent Protection. The supply devices of these supplementary circuits shall be protected by branch-circuit overcurrent protective devices. Each supplementary circuit, within the road show connection panel and theater, shall be protected by branch-circuit overcurrent protective devices installed within the road show connection panel.

Because inlets feeding permanent circuits are cord-and-plug-connected to a portable switchboard, there is no guarantee that the branch circuit overcurrent protective devices (OCPDs) in the switchboard will be rated correctly for the ampacity of the permanent circuit conductors fed by the panel. That is why supplemental overcurrent protection is provided for every inlet in a road show connection panel.

(D) Enclosure. Panel construction shall be in accordance with Article 408.

520.51 Supply. Portable switchboards shall be supplied only from power outlets of sufficient voltage and ampere rating. Such power outlets shall include only externally operable, enclosed fused switches or circuit breakers mounted on stage or at the permanent switchboard in locations readily accessible from the stage floor. Provisions for connection of an equipment grounding conductor shall be provided. For the purposes of ampacity adjustment, the requirements of 520.27(B) shall apply.

Power outlets, known in the entertainment industry as company switches or bull switches, are the point in the wiring system where portable feeder cables connect to the fixed building wiring. They can be as simple as an overcurrent-protected multipole receptacle designed to accept the supply cable described in 520.54(K), Exception, or they can be multiple sets of parallel single-conductor feeder cables. These single-conductor feeder cables, as described in 520.54(C), can be terminated via single-pole separable connectors, as described in 520.53(C), or directly to busbars, fused disconnect switches, or circuit breakers with wire connectors (lugs).

520.52 Overcurrent Protection for Branch Circuits. Portable switchboards shall contain overcurrent protection for branch circuits. The requirements of 210.23 shall not apply.

520.53 Construction. Portable stage switchboards shall be listed and shall comply with 520.53(A) through (E). The load

EXHIBIT 520.7 A large, portable SCR dimmer switchboard (rolling rack). *(Courtesy of Electronic Theatre Controls, Inc.)*

terminal location requirements in 408.18(C)(1), (C)(2), and (C)(3) shall not apply to portable stage switchboards.

Exhibit 520.7 illustrates a portable switchboard known as a rolling rack.

(A) Pilot Light. A pilot light shall be provided for each ungrounded conductor feeding the switchboard. The pilot light(s) shall be connected to the incoming feeder so that operation of the main overcurrent protective device or master switch shall not affect the operation of the pilot light(s).

This requirement applies to switchboards with or without a main disconnect provided on the switchboard. The pilot light serves as a warning at the switchboard to indicate the presence of power, whether or not there is a main disconnect in the portable switchboard, and before an integral main disconnect is activated.

(B) Neutral Terminal. In portable switchboard equipment designed for use with 3-phase, 4-wire with ground supply, the current rating of the supply neutral terminal, and the ampacity of its associated busbar or wiring, or both, shall have an ampacity equal to at least twice the ampacity of the largest ungrounded supply terminal.

Exception: Where portable switchboard equipment is specifically constructed and identified to be internally converted in the field, in an approved manner, from use with a balanced 3-phase, 4-wire with ground supply to a balanced single-phase, 3-wire with ground supply, the supply neutral terminal and its associated busbar, wiring, or both, shall have an ampacity equal to at least that of the largest ungrounded single-phase supply terminal.

If a 3-phase, 4-wire portable switchboard is brought into a space that has only single-phase, 3-wire service, the switchboard most likely will be connected with one leg feeding two phases and the other leg feeding the third phase of the switchboard. This type of connection increases the current flowing through the neutral, so the neutral terminal and busbar must be rated for double size to allow for that possibility. The exception to 520.53(B) provides for a neutral sized for the single-phase feed where a switchboard contains devices that can divide the B-phase load equally between the A-phase and C-phase buses for single-phase operation.

Δ **(C) Single-Pole Separable Connectors.** Single-pole separable connectors shall comply with 406.13. Sections 400.14, 406.7, and 406.8 shall not apply to listed single-pole separable connectors and single-conductor cable assemblies utilizing listed single-pole separable connectors. Where paralleled sets of current-carrying, single-pole separable connectors are provided as input devices, they shall be prominently labeled with a warning indicating the presence of internal parallel connections.

A listed, special type of connection device suitable for connecting single-conductor feeder cables must be of the locking type to reduce the likelihood of its separating while under load. The connectors must be used in sets because they are only single-pole

types. The connector sets must be arranged to reduce the likelihood that connections are made in the incorrect order, in accordance with one of the following methods:

1. The supply disconnect cannot be energized until all conductor connectors are connected. This includes equipment grounding conductors (EGCs), grounded conductors (if used), and ungrounded conductors.
2. The connectors are precluded from being connected in any order other than the proper one (first make/last break of the grounding conductor and connect next-to-first and disconnect next-to-last for the grounded conductor).
3. The individual connectors, free of any special electromechanical intervention, are marked with instructions to the user regarding proper connection.

Single-pole separable connectors are quick-connect feeder splicing and terminating devices, not attachment plugs or receptacles. They are designed to be sized, terminated, and inspected by a qualified person before being energized and are to be guarded from accidental disconnection before being de-energized.

See also

Article 100 for the definition of *single-pole separable connector*

406.13 for the general requirements covering the use of single-pole separable connectors

(D) Supply Feed-Through. Where a portable stage switchboard contains a feed-through outlet of the same rating as its supply inlet, the feed-through outlet shall not require overcurrent protection in the switchboard.

(E) Interior Conductors. All conductors other than busbars within the switchboard enclosure shall be stranded.

520.54 Supply Conductors.

(A) General. The supply to a portable stage switchboard shall be by means of listed extra-hard usage cords or cables. The supply cords or cables shall terminate within the switchboard enclosure in an externally operable fused master switch or circuit breaker or in an identified connector assembly. The supply cords or cable (and connector assembly) shall have current ratings not less than the total load connected to the switchboard and shall be protected by overcurrent devices.

As with the supply end described in 520.51, the termination connection required in 520.54(A) could be as simple as a permanently terminated multiconductor supply cord or multipole connector assembly (inlet) or as complex as a set of parallel single-conductor feeder cables. These cables can be field-connected to an assembly of single-pole connectors (inlet) or directly connected, with wire connectors, to busbars or a fused switch or breaker.

Road shows are permitted to size the feeder to the actual connected load rather than sizing it based on the overcurrent protection rating. Section 220.40 provides the requirements for sizing the feeder or service. Sections 520.54(D) and (E) provide rules under which conductors are permitted to be reduced in size.

(B) Conductor Sizing. The power supply conductors for portable stage switchboards utilizing solid-state phase control dimmers shall be sized considering the neutral conductor as a current-carrying conductor for ampacity adjustment purposes. The power supply conductors for portable stage switchboards utilizing only solid-state sine wave dimmers shall be sized considering the neutral conductor as a non–current carrying conductor for ampacity adjustment purposes.

(C) Single-Conductor Cables. Single-conductor portable supply cable sets shall be not smaller than 2 AWG conductors. The equipment grounding conductor shall not be smaller than 6 AWG conductor. Single-conductor grounded neutral cables for a supply shall be sized in accordance with 520.54(J). Where single conductors are paralleled for increased ampacity, the paralleled conductors shall be of the same length and size. Single-conductor supply cables shall be grouped together but not bundled. The equipment grounding conductor shall be permitted to be of a different type if it meets the other requirements of this section, and it shall be permitted to be reduced in size in accordance with 250.122. Grounded (neutral) and equipment grounding conductors shall be identified in accordance with 200.6, 250.119, and 310.6. Grounded conductors shall be permitted to be identified by marking at least the first 150 mm (6 in.) from both ends of each length of conductor with white or gray. Equipment grounding conductors shall be permitted to be identified by marking at least the first 150 mm (6 in.) from both ends of each length of conductor with green or green with yellow stripes. Where more than one nominal voltage exists within the same premises, each ungrounded conductor shall be identified by system.

(D) Supply Conductors Not Over 3 m (10 ft) Long. Where supply conductors do not exceed 3 m (10 ft) in length between supply and switchboard or supply and a subsequent overcurrent device, the supply conductors shall be permitted to be reduced in size where all of the following conditions are met:

(1) The ampacity of the supply conductors shall be at least one-quarter of the current rating of the supply overcurrent protective device.

(2) The supply conductors shall terminate in a single overcurrent protective device that will limit the load to the ampacity of the supply conductors. This single overcurrent device shall be permitted to supply additional overcurrent devices on its load side.

(3) The supply conductors shall not penetrate walls, floors, or ceilings or be run through doors or traffic areas. The supply conductors shall be protected from physical damage.

(4) The supply conductors shall be suitably terminated in an approved manner.

(5) Conductors shall be continuous without splices or connectors.

(6) Conductors shall not be bundled.

(7) Conductors shall be supported above the floor in an approved manner.

(E) Supply Conductors Not Over 6 m (20 ft) Long. Where supply conductors do not exceed 6 m (20 ft) in length between supply and switchboard or supply and a subsequent overcurrent protection device, the supply conductors shall be permitted to be reduced in size where all of the following conditions are met:

(1) The ampacity of the supply conductors shall be at least one-half of the current rating of the supply overcurrent protective device.

(2) The supply conductors shall terminate in a single overcurrent protective device that limits the load to the ampacity of the supply conductors. This single overcurrent device shall be permitted to supply additional overcurrent devices on its load side.

(3) The supply conductors shall not penetrate walls, floors, or ceilings or be run through doors or traffic areas. The supply conductors shall be adequately protected from physical damage.

(4) The supply conductors shall be suitably terminated in an approved manner.

(5) The supply conductors shall be supported in an approved manner at least 2.1 m (7 ft) above the floor except at terminations.

(6) The supply conductors shall not be bundled.

(7) Tap conductors shall be in unbroken lengths.

Similar to the requirements for feeder taps in 240.21(B), 520.54(D) and (E) permit supply conductors for portable switchboards to be sized according to their overcurrent protection, not by the total connected load. Loads of 144 kilovolt-amperes and greater are not uncommon, even on portable switchboard equipment. Installations in the field include lighting for theatrical-type productions with large numbers of stage lighting fixtures. However, only a fraction of the many fixtures installed are used at any one time.

These rules are intended to allow one or more switchboards with smaller feeders to be connected to larger supplies (company switches). If the requirements cannot be met, properly sized fixed or portable OCPDs must be provided for each of the smaller switchboards.

Column D of Table 400.5(A)(2) can be used if the conductors are not bundled. If the conductors are bundled, column F and all applicable derating factors apply. Some devices used in the theater to terminate single-conductor cables are rated for use at 90°C. However, if single-conductor cables are terminated directly to a circuit breaker or fused switch, the temperature limitations of the terminations might be lower than 90°C.

(F) Supply Conductors Not Reduced in Size. Supply conductors not reduced in size under 520.54(D) or (E) shall be permitted to pass through holes in walls specifically designed for the purpose. If penetration is through the fire-resistant–rated wall, it shall be in accordance with 300.21.

(G) Protection of Supply Conductors and Connectors. All supply conductors and connectors shall be protected against physical damage by an approved means. This protection shall not be required to be raceways.

(H) Number of Supply Interconnections. Where connectors are used in a supply conductor, there shall be a maximum number of three interconnections (mated connector pairs) where the total length from supply to switchboard does not exceed 30 m (100 ft). In cases where the total length from supply to switchboard exceeds 30 m (100 ft), one additional interconnection shall be permitted for each additional 30 m (100 ft) of supply conductor.

Excessive numbers of interconnections could jeopardize the mechanical and electrical integrity of the supply conductors.

(I) Single-Pole Separable Connectors. Where single-pole portable cable connectors are used, they shall be listed and of the locking type. Sections 406.7 and 406.8 shall not apply to listed single-pole separable connectors and single-conductor cable assemblies utilizing listed single-pole separable connectors.

(J) Supply Neutral Conductor. Supply neutral conductors shall comply with 520.54(J)(1) and (J)(2).

(1) Marking. Grounded neutral conductors shall be permitted to be identified by marking at least the first 150 mm (6 in.) from both ends of each length of conductor with white or gray.

(2) Conductor Sizing. Where single-conductor feeder cables not installed in raceways are used on multiphase circuits feeding portable stage switchboards containing solid-state phase-control dimmers, the grounded neutral conductor shall have an ampacity of at least 130 percent of the ungrounded circuit conductors feeding the portable stage switchboard. Where such feeders are supplying only solid-state sine wave dimmers, the grounded neutral conductor shall have an ampacity of at least 100 percent of the ungrounded circuit conductors feeding the portable stage switchboard.

Three-phase, 4-wire switchboards that contain solid-state phase-control dimming devices must, when connected to a 3-phase, 4-wire supply, be connected to that supply with a multiconductor cable sized by counting the neutral as a current-carrying conductor or with a set of single-conductor cables where the neutral is sized at 130 percent of the ungrounded conductors.

Solid-state sine-wave dimmers are linear devices that do not add nonlinear loads to the neutral conductor. Where feeders supply solid-state sine-wave dimmers, the neutral conductor is sized by considering it a non−current-carrying conductor. However, it must have an ampacity of at least 100 percent of the ampacity of the phase conductors.

(K) Qualified Persons. The routing of portable supply conductors, the making and breaking of supply connectors and other supply connections, and the energization and de-energization of supply services shall be performed by qualified persons, and portable switchboards shall be so marked, indicating this requirement in a permanent and conspicuous manner.

Exception: A portable switchboard shall be permitted to be connected to a permanently installed supply receptacle by other than qualified persons provided that the supply receptacle is protected for its current rating by an overcurrent device of not greater than 150 amperes, and where the receptacle, interconnection, and switchboard comply with all of the following:

(1) They employ listed multipole connectors for every supply interconnection.

(2) They prevent access to all supply connections by the general public.

(3) They employ listed extra-hard usage multiconductor cords or cables with an ampacity not less than the load and not less than the ampere rating of the connectors.

This requirement divides the acceptable practices and requirements for professional and professional-grade educational venues (qualified) from those for amateur or amateur-grade educational venues (other than qualified). The requirements allow for such things as single-conductor feeder systems, feeders sized for the current-connected load, tap rules, and so forth, and require the services of a qualified person. The exception provides for a conventional feeder system suitable for use by an untrained person.

Part V. Portable Stage Equipment Other Than Switchboards

520.61 Arc Lamps. Arc lamps, including enclosed arc lamps and associated ballasts, shall be listed. Interconnecting cord sets and interconnecting cords and cables shall be extra-hard usage type and listed.

520.62 Portable Power Distribution Units. Portable power distribution units shall comply with the requirements of 520.62(A) and (B).

N **(A) Listing.** Portable power distribution units shall be listed.

(B) Single-Conductor Feeder Systems. Portable power distribution equipment fed by single-conductor feeder systems shall comply with the requirements of 520.53(C) and (D) and 520.54.

520.63 Bracket Fixture Wiring.

(A) Bracket Wiring. Brackets for use on scenery shall be wired internally, and the fixture stem shall be carried through to the back of the scenery where a bushing shall be placed on the end of the stem. Externally wired brackets or other fixtures shall be permitted where wired with cords designed for hard usage that extend through scenery and without joint or splice in canopy of fixture back and terminate in an approved-type stage connector located, where practical, within 450 mm (18 in.) of the fixture.

(B) Mounting. Fixtures shall be securely fastened in place.

520.64 Portable Strips. Portable strips shall be constructed in accordance with the requirements for border lights and proscenium sidelights in 520.44(A). The supply cable shall be protected by bushings where it passes through metal and shall be arranged so that tension on the cable will not be transmitted to the connections.

> Informational Note No. 1: See 520.42 for wiring of portable strips.
> Informational Note No. 2: See 520.68(A)(4) for insulation types required on single conductors.

520.65 Festoons. Joints in festoon wiring shall be staggered. Where such lampholders have terminals of a type that puncture the insulation and make contact with the conductors, they shall be attached only to conductors of the stranded type. Lamps enclosed in lanterns or similar devices of combustible material shall be equipped with guards.

Staggering joints in festoon wiring ensures that connections are not opposite one another. Joints that are not staggered could cause sparking due to improper insulation or unraveling of insulation, which, in turn, could ignite lanterns or other combustible material enclosing lamps. Where lampholders have terminals that puncture the conductor insulation to make contact with the conductors, stranded conductors must be used.

See also

Article 100 for the definition of *festoon lighting*

520.66 Special Effects. Electrical devices used for simulating lightning, waterfalls, and the like shall be constructed and located so that flames, sparks, or hot particles cannot come in contact with combustible material.

520.67 Multipole Branch-Circuit Cable Connectors. Multipole branch-circuit cable connectors, male and female, for flexible conductors shall be constructed so that tension on the cord or cable is not transmitted to the connections. The female half shall be attached to the load end of the power supply cord or cable. The connector shall be rated in amperes and designed so that differently rated devices cannot be connected together; however, a 20-ampere T-slot receptacle shall be permitted to accept a 15-ampere attachment plug of the same voltage rating. Alternating-current multipole connectors shall be polarized and comply with 406.7 and 406.10.

> Informational Note: See 400.14 for pull at terminals.

520.68 Conductors for Portables.

(A) Conductor Type.

(1) General. Flexible conductors, including cable extensions, used to supply portable stage equipment shall be listed extra-hard usage cords or cables.

(2) Protected Applications. Listed, hard usage (junior hard service) cord or cable shall be permitted where all of the following conditions are met:

(1) The cord or cable is protected from physical damage by attachment over its entire length to a pipe, tower, truss, scaffold, or other substantial support structure, or installed in a location that inherently prevents physical damage to the cord.
(2) The cord or cable is connected to a branch circuit protected by an overcurrent protective device rated at not over 20 amperes.
(3) The cord or cable does not exceed 30 m (100 ft) in length.

(3) Stand Lamps. Listed, hard usage cord shall be permitted to supply stand lamps where the cord is not subject to physical damage and is protected by an overcurrent device rated at not over 20 amperes.

One of the most common stand lamps is a ghostlight, as shown in Exhibit 520.8. Ghostlights typically are placed on a stage outside of performance and rehearsal times to keep the theater from being in total darkness.

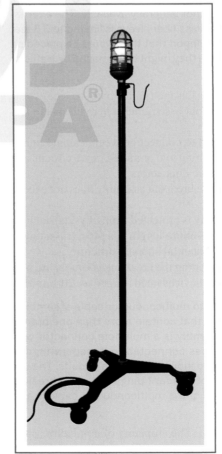

EXHIBIT 520.8 A stand lamp known as a ghostlight. *(Courtesy of Behind the Scenes Foundation)*

(4) Luminaire Supply Cords. Listed hard usage supply cords shall be permitted to supply luminaires if all of the following conditions are met:

(1) The supply cord is not longer than 2.0 m (6.6 ft).
(2) The supply cord is attached at one end to the luminaire or a luminaire-specific listed connector that mates with a panel-mounted inlet on the body of the luminaire.
(3) The supply cord is protected by an overcurrent protective device of not more than 20 amperes.
(4) The luminaire is listed.
(5) The supply cord is not subject to physical damage.

(5) High-Temperature Applications. A special assembly of conductors in sleeving not longer than 1.0 m (3.3 ft) shall be permitted to be employed in lieu of flexible cord if the individual wires are stranded and rated not less than 125°C (257°F) and the outer sleeve is glass fiber with a wall thickness of at least 0.635 mm (0.025 in.).

Portable stage equipment requiring flexible supply conductors with a higher temperature rating where one end is permanently attached to the equipment shall be permitted to employ alternate conductors as determined by a qualified testing laboratory and recognized test standards.

Stage equipment, such as stage lighting fixtures, often operate at elevated temperatures. High-temperature (150°C to 250°C), extra-hard-usage cords are not generally available. The alternative use of conductors in a glass fiber sleeve is limited to 3.3 feet in length to reduce the likelihood that they would be placed on the floor or other area where they might be damaged by traffic or moving scenery.

(6) Breakouts. Listed, hard usage (junior hard service) cords shall be permitted in breakout assemblies where all of the following conditions are met:

(1) The cords are utilized to connect between a single multi-pole connector containing two or more branch circuits and multiple 2-pole, 3-wire connectors.
(2) The longest cord in the breakout assembly does not exceed 6.0 m (20 ft).
(3) The breakout assembly is protected from physical damage by attachment over its entire length to a pipe, truss, tower, scaffold, or other substantial support structure.
(4) All branch circuits feeding the breakout assembly are protected by overcurrent devices rated at not over 20 amperes.

These requirements apply to multiconductor cable assemblies with multipole connectors that contain more than one branch circuit. The breakout assembly is a multipole connector with several pendant receptacles connected to it, separating the multiple branch circuits into individual branch circuits. The use of a similar arrangement of pendant plugs to form a break-in assembly on the other end of the multiconductor cable is also possible.

(B) Conductor Ampacity. The ampacity of conductors shall be as given in 400.5, except multiconductor, listed, extra-hard

usage portable cords that are not in direct contact with equipment containing heat-producing elements shall be permitted to have their ampacity determined by Table 520.44(C)(2)(1). Maximum load current in any conductor with an ampacity determined by Table 520.44(C)(2)(1) shall not exceed the values in Table 520.44(C)(2)(1). Where the ampacity adjustment factors of Table 520.44(C)(2)(2) are applied for more than three current-carrying conductors in a portable cord, the load diversity shall be 50 percent or less.

Exception: Where alternate conductors are allowed in 520.68(A)(5), their ampacity shall be as given in the appropriate table in this Code for the types of conductors employed.

Listed portable, multiconductor cable is permitted to be sized in accordance with Table 520.44(C)(2)(1) and Table 520.44(C)(2)(2), similar to the method used for border light cable. A cable, used in lieu of a connector strip, directly above heat-producing equipment should be spaced sufficiently above that equipment to avoid the elevated temperatures or should be sized in accordance with 400.5.

(C) Overcurrent Protection. Overcurrent protection of conductors for portables shall comply with 240.5.

N (D) Special-Purpose Multicircuit Cable Systems. Special-purpose multicircuit cable systems shall comply with the following requirements:

(1) Branch circuits shall be rated at not more than 20 amperes and not more than 150 volts to ground.
(2) Trunk cable types shall be extra-hard usage (hard service) or hard usage (junior hard service).
(3) The ampacity of trunk cables shall be determined in accordance with Table 520.44(C)(2)(1).
(4) Trunk cables, breakout assemblies, and multicircuit enclosures shall be listed.
(5) Section 406.4(F) shall not apply to multicircuit, multipole plugs or receptacles that are part of a special-purpose multicircuit cable system.
(6) All multicircuit, multipole connectors shall be clearly marked with the voltage of the branch circuits serviced by the connector.
(7) Installation and operation shall be performed by qualified persons.

520.69 Adapters. Adapters, two-fers, and other single- and multiple-circuit outlet devices shall comply with 520.69(A), (B), and (C).

(A) No Reduction in Current Rating. Each receptacle and its corresponding cable shall have the same current and voltage rating as the plug supplying it. It shall not be utilized in a stage circuit with a greater current rating.

(B) Connectors. All connectors shall be wired in accordance with 520.67.

Plugs and receptacles must be of the same rating even though available adapters allow connector bodies to be connected to a plug of a larger rating. For example, a 12 AWG conductor with an ampacity of 20 amperes could be connected to a 100-ampere circuit via an adapter. An overload could result in a fire because the circuit breaker or fuse would not provide adequate protection.

(C) Conductor Type. Conductors for adapters and two-fers shall be listed extra-hard usage or listed hard usage (junior hard service) cord. Hard usage (junior hard service) cord shall be restricted in overall length to 2.0 m (6.6 ft).

Part VI. Dressing Rooms, Dressing Areas, and Makeup Areas.

520.71 Pendant Lampholders. Pendant lampholders shall not be installed in dressing or makeup rooms.

520.72 Lamp Guards. All exposed lamps in dressing or makeup areas including rooms where they are less than 2.5 m (8 ft) from the floor shall be equipped with open-end guards riveted to the outlet box cover or otherwise sealed or locked in place. Recessed lamps shall not be required to be equipped with guards.

Lamps in dressing rooms are required to be provided with suitable open-end guards that permit relamping and that are not easily removed. Guards make it difficult to circumvent their purpose of preventing contact between the lamps and flammable materials.

520.73 Switches Required. All luminaires, lampholders, and any receptacles adjacent to the mirror(s) and above the dressing or makeup counter(s) installed in dressing or makeup rooms shall be controlled by wall switches installed in the dressing or makeup room(s). Other outlets installed in the dressing or makeup rooms shall not be required to be switched.

520.74 Pilot Lights Required. Each switch required in 520.73 shall be provided with a pilot light located outside of and adjacent to the door of the room being controlled to indicate when the circuit is energized. Each pilot light shall be permanently identified indicating a description of the circuit controlled. Pilot lights shall be neon, LED, or other extended-life lamp. Pilot lights shall be recessed or provided with a mechanical guard.

Part VII. Equipment Grounding Conductor

520.81 Equipment Grounding Conductor. All metal raceways and metal-sheathed cables shall be connected to an equipment grounding conductor. The metal frames and enclosures of all equipment, including border lights and portable luminaires, shall be connected to an equipment grounding conductor.

ARTICLE 522 Control Systems for Permanent Amusement Attractions

Part I. General

522.1 Scope. This article covers the installation of control circuit power sources and control circuit conductors for electrical equipment, including associated control wiring in or on all structures, that are an integral part of a permanent amusement attraction.

Article 522 provides requirements for permanent amusement attractions and theme parks. Article 525 applies to temporary attractions, such as carnivals, circuses, and fairs, where most of the attractions consist of portable modules that are moved from place to place. In contrast, theme parks are permanent facilities that have entertainment features fixed in place so that they are not readily portable. In the United States, approximately 475 amusement and theme parks operate a wide variety of permanent entertainment features.

Article 522 addresses the unique applications and installations utilized in the theme park and amusement industry and covers the wiring requirements for the control circuit power source and control circuit conductors, allowing for installation methods that are not recognized in the requirements of Articles 724 and 725. The control voltage used is a maximum of 150 volts ac to ground or 300 volts dc to ground.

522.5 Voltage Limitations. Control voltage shall be a maximum of 150 volts, nominal, ac to ground or 300 volts dc to ground.

522.7 Maintenance. The conditions of maintenance and supervision shall ensure that only qualified persons service the permanent amusement attraction.

Part II. Control Circuits

522.10 Power Sources for Control Circuits.

(A) Power-Limited Control Circuits. Power-limited control circuits shall be supplied from a source that has a rated output of not more than 30 volts and 1000 volt-amperes.

(1) Control Transformers. Transformers used to supply power-limited control circuits shall comply with the applicable sections within Parts I and II of Article 450.

(2) Other Power-Limited Control Power Sources. Power-limited control power sources, other than transformers, shall be protected by overcurrent devices rated at not more than 167 percent of the volt-ampere rating of the source divided by the rated voltage. The fusible overcurrent devices shall not be interchangeable with fusible overcurrent devices of higher ratings. The overcurrent device shall be permitted to be an integral part of the power source.

To comply with the 1000 volt-ampere limitation of 522.10(A), the maximum output of power sources, other than transformers, shall be limited to 2500 volt-amperes, and the product of the maximum current and maximum voltage shall not exceed 10,000 volt-amperes. These ratings shall be determined with any overcurrent-protective device bypassed.

(B) Non–Power-Limited Control Circuits. Non–power-limited control circuits shall not exceed 300 volts. The power output of the source shall not be required to be limited.

(1) Control Transformers. Transformers used to supply non–power-limited control circuits shall comply with the applicable sections within Parts I and II of Article 450.

(2) Other Non–Power-Limited Control Power Sources. Non–power-limited control power sources, other than transformers, shall be protected by overcurrent devices rated at not more than 125 percent of the volt-ampere rating of the source divided by the rated voltage. The fusible overcurrent devices shall not be interchangeable with fusible overcurrent devices of higher ratings. The overcurrent device shall be permitted to be an integral part of the power source.

Part III. Control Circuit Wiring Methods

522.20 Conductors, Busbars, and Slip Rings. Insulated control circuit conductors shall be copper and shall be permitted to be stranded or solid. Listed multiconductor cable assemblies shall be permitted.

Exception No. 1: Busbars and slip rings shall be permitted to be materials other than copper.

Exception No. 2: Conductors used as specific-purpose devices, such as thermocouples and resistive thermal devices, shall be permitted to be materials other than copper.

522.21 Conductor Sizing.

(A) Conductors Within a Listed Component or Assembly. Conductors of size 30 AWG or larger shall be permitted within a listed component or as part of the wiring of a listed assembly.

(B) Conductors Within an Enclosure or Operator Station. Conductors of size 30 AWG or larger shall be permitted in a listed and jacketed multiconductor cable within an enclosure or operator station. Conductors in a non-jacketed multiconductor cable, such as ribbon cable, shall not be smaller than 26 AWG. Single conductors shall not be smaller than 24 AWG.

Exception: Single conductors 30 AWG or larger shall be permitted for jumpers and special wiring applications.

(C) Conductors Outside of an Enclosure or Operator Station. The size of conductors in a listed and jacketed,

multiconductor cable shall not be smaller than 26 AWG. Single conductors shall not be smaller than 18 AWG and shall be installed only where part of a recognized wiring method of Chapter 3.

522.22 Conductor Ampacity. Ampacities for conductors sized 16 AWG and smaller shall be as specified in Table 522.22.

Δ **TABLE 522.22** *Conductor Ampacity Based on Copper Conductors with 60°C and 75°C Insulation in an Ambient Temperature of 30°C*

Conductor Size (AWG)	Ampacity	
	60°C	75°C
30	–	0.5
28	–	0.8
26	–	1
24	2	2
22	3	3
20	5	5
18	7	7
16	10	10

Notes:
1. For ambient temperatures other than 30°C, temperature correction factors provided in 310.15(B)(1) shall be used.
2. Ampacity for conductors with 90°C or greater insulation shall be based on ampacities in the 75°C column.

522.23 Overcurrent Protection for Conductors. Conductors 30 AWG through 16 AWG shall have overcurrent protection in accordance with the appropriate conductor ampacity in Table 522.22. Conductors larger than 16 AWG shall have overcurrent protection in accordance with the appropriate conductor ampacity in Table 310.16.

522.24 Conductors of Different Circuits in the Same Cable, Cable Tray, Enclosure, or Raceway. Control circuits shall be permitted to be installed with other circuits as specified in 522.24(A) and (B).

(A) Two or More Control Circuits. Control circuits shall be permitted to occupy the same cable, cable tray, enclosure, or raceway without regard to whether the individual circuits are alternating current or direct current, provided all conductors are insulated for the maximum voltage of any conductor in the cable, cable tray, enclosure, or raceway.

(B) Control Circuits with Power Circuits. Control circuits shall be permitted to be installed with power conductors as specified in 522.24(B)(1) through (B)(3).

(1) In a Cable, Enclosure, or Raceway. Control circuits and power circuits shall be permitted to occupy the same cable, enclosure, or raceway only where the equipment powered is functionally associated.

(2) In Factory- or Field-Assembled Control Centers. Control circuits and power circuits shall be permitted to be installed in factory- or field-assembled control centers.

(3) In a Manhole. Control circuits and power circuits shall be permitted to be installed as underground conductors in a manhole in accordance with one of the following:

(1) The power or control circuit conductors are in a metal-enclosed cable or Type UF cable
(2) The conductors are permanently separated from the power conductors by a continuous fixed nonconductor, such as flexible tubing, in addition to the insulation on the wire
(3) The conductors are permanently and effectively separated from the power conductors and securely fastened to racks, insulators, or other approved supports
(4) In cable trays, where the control circuit conductors and power conductors not functionally associated with them are separated by a solid fixed barrier of a material compatible with the cable tray, or where the power or control circuit conductors are in a metal-enclosed cable

522.25 Ungrounded Control Circuits. Separately derived ac circuits and systems 50 volts or greater and 2-wire dc circuits and systems 60 volts or greater shall be permitted to be ungrounded, provided that all the following conditions are met:

(1) Continuity of control power is required for orderly shutdown.
(2) Ground detectors are installed on the control system.

522.28 Control Circuits in Wet Locations. Where wet contact is likely to occur, ungrounded 2-wire direct-current control circuits shall be limited to 30 volts maximum for continuous dc or 12.4 volts peak for direct current that is interrupted at a rate of 10 to 200 Hz.

ARTICLE 525 Carnivals, Circuses, Fairs, and Similar Events

Δ **Part I. General**

525.1 Scope. This article covers the installation of portable wiring and equipment for carnivals, circuses, fairs, and similar functions, including wiring in or on all structures.

Article 525 addresses the installation of portable wiring and equipment for temporary attractions, such as carnivals, circuses, and fairs. Article 525 is intended to apply to all wiring in or on portable structures, whereas Articles 518, 520, and 522 apply to permanent structures.

525.3 Other Articles.

(A) Portable Wiring and Equipment. Wherever the requirements of other articles of this *Code* and Article 525 differ, the requirements of Article 525 shall apply to the portable wiring and equipment.

(B) Attractions Utilizing Pools, Fountains, and Similar Installations with Contained Volumes of Water. This equipment shall be installed to comply with the applicable requirements of Parts I, II, III, and V of Article 680.

525.5 Overhead Conductor Clearances.

(A) Vertical Clearances. Conductors shall have a vertical clearance to ground in accordance with 225.18. These clearances shall apply only to wiring installed outside of tents and concessions.

(B) Clearance to Portable Structures.

(1) 600 Volts (or Less). Portable structures shall be maintained not less than 4.5 m (15 ft) in any direction from overhead conductors operating at 600 volts or less, except for the conductors supplying the portable structure. Portable structures included in 525.3(B) shall comply with Table 680.9(A).

(2) Over 600 Volts. Portable structures shall not be located under or within a space that is located 4.5 m (15 ft) horizontally and extending vertically to grade of conductors operating in excess of 600 volts.

Portable structures, which include rides, attractions, and vendor booths, are not permitted in the area that extends 15 feet horizontally from the overhead conductors and down to grade level. Exhibit 525.1 depicts the restricted area.

EXHIBIT 525.1 The separation distance required around overhead conductors.

525.6 Protection of Electrical Equipment. Electrical equipment and wiring methods in or on portable structures shall be provided with mechanical protection where such equipment or wiring methods are subject to physical damage.

Part II. Power Sources

A power source can be a service or a separately derived system, such as a generator or transformer, or a combination of multiple sources. In addition to the requirements in 525.10(A) and (B), the requirements for services in Article 230 are applicable.

525.10 Services. Services shall comply with 525.10(A) and (B).

Service equipment must be installed in accordance with Article 230 and must be lockable where accessible to unqualified persons. Fairs, carnivals, and similar events generate significant pedestrian traffic throughout the sites, including those areas where electrical equipment is located. This requirement helps safeguard the general public from accidentally coming in contact with energized service equipment.

(A) Guarding. Service equipment shall not be installed in a location that is accessible to unqualified persons, unless the equipment is lockable.

(B) Mounting and Location. Service equipment shall be securely fastened to a solid backing and be installed so as to be protected from the weather, unless of weatherproof construction.

525.11 Multiple Sources of Supply. Where multiple services or separately derived systems, or both, supply portable structures, the equipment grounding conductors of all the sources of supply that serve such structures separated by less than 3.7 m (12 ft) shall be bonded together at the portable structures. The bonding conductor shall be copper and sized in accordance with Table 250.122 based on the largest overcurrent device supplying the portable structures, but not smaller than 6 AWG.

To maintain an equal potential between exposed, non–current-carrying metal parts of portable structures that have a physical separation less than 12 feet, they must be bonded to each other using a copper conductor sized per Table 250.122, but not smaller than 6 AWG.

Part III. Wiring Methods

525.20 Wiring Methods.

(A) Type. Where flexible cords or cables are used, they shall be listed for extra-hard usage. Where flexible cords or cables are used and are not subject to physical damage, they shall be permitted to be listed for hard usage. Where used outdoors, flexible cords and cables shall also be listed for wet locations and shall be sunlight resistant. Extra-hard usage flexible cords or cables shall be permitted for use as permanent wiring on portable amusement rides and attractions where not subject to physical damage.

(B) Single-Conductor. Single-conductor cable shall be permitted only in sizes 2 AWG or larger.

(C) Open Conductors. Open conductors shall be prohibited except as part of a listed assembly or festoon lighting installed in accordance with Article 225.

(D) Splices. Flexible cords or cables shall be continuous without splice or tap between boxes or fittings.

(E) Cord Connectors. Cord connectors shall not be laid on the ground unless listed for wet locations. Connectors and cable connections shall not be placed in audience traffic paths or within areas accessible to the public unless guarded.

(F) Support. Wiring for an amusement ride, attraction, tent, or similar structure shall not be supported by any other ride or structure unless specifically designed for the purpose.

(G) Protection. Flexible cords or cables accessible to the public shall be arranged to minimize the tripping hazard and shall be permitted to be covered with nonconductive matting secured to the walkway surface or protected with another approved cable protection method, provided that the matting or other protection method does not constitute a greater tripping hazard than the uncovered cables. Burying cables shall be permitted. The requirements of 300.5 shall not apply.

(H) Boxes and Fittings. A box or fitting shall be installed at each connection point, outlet, switchpoint, or junction point.

525.21 Rides, Tents, and Concessions.

(A) Disconnecting Means. A means to disconnect each portable structure from all ungrounded conductors shall be provided. The disconnecting means shall be located within sight of and within 1.8 m (6 ft) of the operator's station. The disconnecting means shall be readily accessible to the operator, including when the ride is in operation. If accessible to unqualified persons, the disconnecting means shall be lockable. A shunt trip device that opens the fused disconnect or circuit breaker if a switch located in the ride operator's console is closed shall be a permissible method of opening the circuit.

(B) Portable Wiring Inside Tents and Concessions. Electrical wiring for lighting, where installed inside of tents and concessions, shall be securely installed and, where subject to physical damage, shall be provided with mechanical protection. All lamps for general illumination shall be protected from accidental breakage by a luminaire or lampholder with a guard.

525.22 Portable Distribution or Termination Boxes. Portable distribution or termination boxes shall comply with 525.22(A) through (D).

(A) Construction. Boxes shall be designed so that no live parts are exposed except where necessary for examination, adjustment, servicing, or maintenance by qualified persons. If installed outdoors, the box shall be of weatherproof construction and mounted so that the bottom of the enclosure is not less than 150 mm (6 in.) above the ground.

Requiring equipment to be mounted so that the bottom of the enclosure is at least 6 inches above the ground prevents excessive moisture from entering the equipment and allows for proper radius of bend on conductors entering and exiting the equipment from below.

(B) Busbars and Terminals. Busbars shall have an ampere rating not less than the overcurrent device supplying the feeder supplying the box. Where conductors terminate directly on busbars, busbar connectors shall be provided.

(C) Receptacles and Overcurrent Protection. Receptacles shall have overcurrent protection installed within the box. The overcurrent protection shall not exceed the ampere rating of the receptacle, except as permitted in Article 430 for motor loads.

(D) Single-Pole Connectors. Where single-pole connectors are used, they shall comply with 530.10.

525.23 Ground-Fault Circuit-Interrupter (GFCI) Protection.

Section 525.23 provides three categories: where GFCIs are required, where GFCIs are not required, and where GFCIs are not permitted to be installed. The application where GFCI protection is not required is very specific. The receptacles must be locking, must be quick disconnect/reconnect, and must not be accessible from grade. GFCI protection is not allowed on circuits that supply means-of-egress illumination. GFCI receptacles that are supplied by branch circuits that utilize flexible cord are required to be listed for portable use. This requirement ensures that the GFCI devices will have open neutral protection.

(A) Where GFCI Protection Is Required. In addition to the requirements of 210.8(B), GFCI protection for personnel shall be provided for the following:

(1) All 125-volt, single-phase, 15- and 20-ampere non-locking-type receptacles used for disassembly and reassembly or readily accessible to the general public
(2) Equipment that is readily accessible to the general public and supplied from a 125-volt, single-phase, 15- or 20-ampere branch circuit

The GFCI shall be permitted to be an integral part of the attachment plug or located in the power-supply cord within 300 mm (12 in.) of the attachment plug. Listed cord sets incorporating GFCI for personnel shall be permitted.

(B) Where GFCI Protection Is Not Required. Receptacles that are not accessible from grade level and that only facilitate quick disconnecting and reconnecting of electrical equipment shall not be required to be provided with GFCI protection. These receptacles shall be of the locking type.

(C) Where GFCI Protection Is Not Permitted. Egress lighting shall not be protected by a GFCI.

(D) Receptacles Supplied by Portable Cords. Where GFCI protection is provided through the use of GFCI receptacles, and the branch circuits supplying receptacles use flexible cord, the GFCI protection shall be listed, labeled, and identified for portable use.

Part IV. Equipment Grounding and Bonding

525.30 Equipment Bonding. The following equipment connected to the same source shall be bonded:

(1) Metal raceways and metal-sheathed cable
(2) Metal enclosures of electrical equipment
(3) Metal frames and metal parts of portable structures, trailers, trucks, or other equipment that contain or support electrical equipment

Where the metal frames or parts of the equipment in 525.30(1), (2), or (3) are likely to become energized in the event of a fault, the equipment grounding conductor of the supply circuit shall be permitted to serve as the bonding means.

Δ **525.31 Equipment Grounding.** The equipment grounding conductor shall be connected to the system grounded conductor at the service disconnecting means or, in the case of a separately derived system such as a generator, at the generator or first disconnecting means supplied by the generator.

525.32 Equipment Grounding Conductor Continuity Assurance. The continuity of the equipment grounding conductors shall be verified each time that portable electrical equipment is connected.

The transient nature of the events covered under Article 525 and, in some cases, the entire associated electrical distribution system increases the possibility that continuity of the equipment grounding conductor (EGC) system could be interrupted. Verification of the grounding system continuity each time equipment is reconnected helps ensure the safety of workers and members of the general public who may come in contact with electrical equipment.

ARTICLE
530 Motion Picture and Television Studios and Remote Locations

Part I. General

Δ **530.1 Scope.** The requirements of this article shall apply to motion picture and television studios in facilities and locations staffed by qualified persons, except as provided in 520.1. Such occupancies shall include those using either electronic or film cameras for image capture.

Informational Note: See NFPA 40-2019, *Standard for the Storage and Handling of Cellulose Nitrate Film*, for methods of protecting against cellulose nitrate film hazards.

The requirements for motion picture studios and those for television studios are virtually the same and are intended to apply only to those locations presenting special hazards, for example, where film is handled or for temporary structures constructed of wood or other combustible material. It is important to note that Article 530 considers these occupancies specialized workplaces, not assembly occupancies, unless an audience is present. If an audience is present, the requirements of Article 520 apply to audience seating areas. Equipment listed for use under Article 520 is often listed for use under Article 530.

See also

NFPA 140, *Standard on Motion Picture and Television Production Studio Soundstages, Approved Production Facilities, and Production Locations,* which addresses additional aspects of these facilities

N 530.3 Restricted Public Access.

N (A) Studios. The electrical equipment covered in this article shall be used in motion picture or television studios with restricted public access.

N (B) Remote Locations. Where the equipment is deployed on remote locations, restricted public access shall be provided in the form of physical barriers or other access control measures.

N 530.4 Supervision by Qualified Personnel. Portable electrical equipment, including distribution systems, generators, battery systems, and other power sources, shall be deployed, energized, and, while energized, operated and continuously supervised by trained, qualified, and employer-authorized personnel.

Δ 530.5 Wiring Methods. Wiring methods for permanent installations shall be in accordance with 530.5(A) or (B).

N (A) General. The permanent wiring shall be permitted to be any of the following:
(1) Metal raceways
(2) Nonmetallic raceways encased in not less than 50 mm (2 in.) of concrete
(3) Type MI cable, Type MC cable, or Type AC cable containing an insulated equipment grounding conductor sized in accordance with Table 250.122

N (B) Communications, Signaling Systems, Data Systems, Fire Alarm Systems, and Systems Less than 120 Volts, Nominal. Permanent wiring methods for communications, signaling, data, fire alarm systems, and systems operating at less than 120 volts, nominal, shall be in accordance with the following:
(1) Audio signal processing, amplification, and reproduction equipment — 640.9
(2) Communications systems — Parts I and IV of Article 800; Part IV of Article 805; and Part IV of Article 840

(3) Class 2 and Class 3 remote control and signaling circuits — Part III of Article 725
(4) Class 2 circuits that transmit power, data or both to a powered device

Informational Note: See ANSI/NEMA C137.3-2017, *American National Standard for Lighting Systems — Minimum Requirements for Installation of Energy Efficient Power over Ethernet (PoE) Lighting Systems,* for information on installation of cables for PoE lighting systems. See Part III of Article 760 for information on fire alarm circuits.

530.7 Sizing of Feeder Conductors for Motion Picture and/ or Television Studio Sets. Applying the demand factors listed in Table 530.7 to the portion of the maximum possible connected load for studio or stage set lighting for all permanently installed feeders between substations and stages and to all permanently installed feeders between the main stage switchboard and stage distribution centers shall be permitted.

TABLE 530.7 Demand Factors for Stage Set Lighting

Portion of Stage Set Lighting Load to Which Demand Factor Applied (volt-amperes)	Feeder Demand Factor (percent)
First 50,000 or less at	100
From 50,001 to 100,000 at	75
From 100,001 to 200,000 at	60
Remaining over 200,000 at	50

Δ 530.8 Equipment Grounding Conductor. Permanent wiring systems shall include an equipment grounding conductor of a type in accordance with 250.118 installed with the supply conductors and connected to the building or structure disconnecting means and the grounding electrode(s). The grounding electrode(s) shall also be connected to the building or structure disconnecting means. Equipment grounding conductors shall be sized in accordance with 250.122. This shall not apply to pendant and portable stage lighting, stage sound equipment, or other special stage equipment operating at not over 150 volts dc to ground.

530.9 Plugs and Receptacles. Plugs and receptacles shall be in accordance with 530.9(A), (B), and (C).

(A) Rating. Plugs and receptacles, including cord connectors and flanged surface devices, shall be rated in amperes. The voltage rating of the plugs and receptacles shall not be less than the nominal circuit voltage. Plug and receptacle ampere ratings for ac circuits shall not be less than the feeder or branch-circuit overcurrent device ampere rating. Table 210.21(B)(2) shall not apply.

The occupancies referenced in Article 530 are excluded from all the general requirements relating to connector rating and branch-circuit loading found elsewhere in the *NEC*®, such as in Table 210.21(B)(2). Connectors must be rated for the parameters involved, thus permitting connectors with voltage and current ratings higher than the branch-circuit rating to be used.

N **(B) Construction.** Plugs and receptacles shall be constructed so that differently rated devices cannot be connected together. Alternating-current multipole connectors shall be polarized and comply with the requirements of 406.7 and 406.10.

Exception: 125-volt, 20-ampere, nonlocking (T-slot) receptacles shall be permitted to accept a 15-ampere attachment plug of the same voltage rating.

Δ **(C) Interchangeability.** Plugs and receptacles used in portable professional motion picture and television equipment shall be permitted to be interchangeable for ac or dc use if they are on the same premises, listed for ac and dc use, and clearly marked to identify the system to which they are connected.

The studio set lighting and associated equipment, both fixed and portable, must be as flexible as possible. Connectors are often used for different purposes and are therefore marked on a show-by-show basis to designate the voltage, current, and type of current actually employed. The receptacle loads on studio set lighting are known and are not casually connected as they might be at a typical general-use wall receptacle. Care is taken to ensure that circuits are not overloaded, thereby avoiding nuisance tripping.

530.10 Single-Pole Separable Connectors.

Δ **(A) General.** Single-pole separable connectors shall comply with the requirements of 406.13. Sections 400.14, 406.7, and 406.8 shall not apply to listed single-pole separable connectors.

N **(B) Paralleled Input Devices.** Where paralleled sets of current-carrying single-pole separable connectors are provided as input devices, they shall be prominently labeled with a warning indicating the presence of internal parallel connections. All paralleled input devices other than primary input devices shall be guarded against accidental contact.

N **(C) Supply Feed-Through Outlets.** Where portable equipment contains a feed-through outlet of the same rating as its supply inlet, the feed-through outlet shall not require overcurrent protection in the equipment.

530.11 Branch Circuits. A branch circuit of any size supplying one or more receptacles shall be permitted to supply stage set lighting loads.

The GFCI requirements of 210.8(B), excluding 210.8(B)(6), shall apply.

Branch circuits supplying egress lighting, life-critical stunts, life-critical special effects, or any other condition where a nonorderly shutdown might introduce additional or increased hazards shall not be protected by GFCIs.

Outdoor circuits are often dimmed and are exempt from the GFCI requirement of 210.8(B)(6) because dimmer-rated GFCI devices are not readily available. Due to the unique requirements of motion picture production, the use of GFCIs for outdoor circuits

must be evaluated on a case-by-case basis. For example, one popular way to perform a high fall stunt from a building is for the stunt performer to land onto an air bag. The air bag must be constantly inflated by a high power fan which is supplied by a dedicated generator; that circuit is thus a life-critical circuit and should not be protected by a GFCI, even if a rain effect is part of the stunt.

530.12 Enclosing and Guarding Live Parts.

(A) Live Parts. Parts of electrical equipment that are live or are likely to become energized shall be enclosed, guarded, or located so persons cannot accidentally come in contact with them or bring conductive material into contact with them.

(B) Switches. All switches shall be of the externally operable type.

Part II. Portable Equipment In Production Areas of Studios and Remote Locations

Δ **530.21 Portable Equipment.**

N **(A) Listing.** Portable stage and studio electrical equipment shall be listed or approved. Field-assembled extension cords and multiconductor cable assemblies consisting of listed connectors and cable shall be permitted in production areas.

N **(B) Outdoor Use.** Portable stage and studio equipment and portable power distribution equipment not identified for outdoor use shall be permitted for temporary use if the equipment is supervised by qualified personnel while energized and barriered from the general public.

Informational Note No. 1: See ANSI/ESTA E1.58, *Electrical Safety Standard for Portable Stage and Studio Equipment Used Outdoors*, for requirements covering temporary outdoor use of equipment not identified for outdoor use.

Informational Note No. 2: See ANSI/ESTA E1.19-2015, *Recommended Practice for the use of Class A Ground-Fault Circuit Interrupters (GFCIs) intended for personnel protection in the Entertainment Industry*, for guidance on the use of GFCIs in wet locations.

Portable indoor stage or studio equipment is permitted to be *temporarily* used outdoors, provided it is supervised by a qualified person and not accessible to the general public. If it rains, the equipment is typically de-energized and covered. At the end of the day, the equipment is either de-energized and protected or dismantled and stored.

530.22 Portable Wiring.

(A) Stage Set Wiring. The wiring for stage set lighting and other supply wiring not fixed in place shall use listed hard usage flexible cords and cables. Where subject to physical damage, such wiring shall use listed extra-hard usage flexible cords and cables. Splices and taps in cables shall be permitted if the total connected load does not exceed the maximum ampacity of the cable.

Δ **(B) Stage or Special Effects and Electrical Equipment Used as Stage Properties.** The wiring for stage effects and electrical equipment used as stage properties shall be permitted to be wired with single- or multiconductor listed flexible cords or cables if the conductors are protected from physical damage and secured to the scenery by approved cable ties or insulated staples. Splices or taps shall be permitted where such are made with listed devices and the circuit is protected at not more than 20 amperes.

(C) Other Electrical Equipment. Cords and cables other than extra-hard usage, where supplied as a part of a listed assembly, shall be permitted.

N **(D) Portable Feeder Cable Penetration of Walls, Floors, or Ceilings.** Portable feeder cables shall be permitted to temporarily penetrate fire-rated walls, floors, or ceilings where all of the following apply:

(1) The opening is of noncombustible material.
(2) When in use, the penetration is sealed with a temporary seal of a listed firestop material.
(3) When not in use, the opening shall be capped with a material of equivalent fire rating.

Δ **(E) Cable Protection.** Cables shall be protected by bushings where they pass through enclosures and shall be arranged so that tension on the cable is not transmitted to the connections. Where power conductors pass through metal, the requirements of 300.20 shall apply.

N **(F) Special-Purpose Multicircuit Cable Systems.** Special-purpose multicircuit cable systems shall comply with 520.68(D).

Δ **530.23 Overcurrent Protection.** Overcurrent protective devices in production areas shall comply with 530.23(A) through (D).

Δ **(A) Portable Stage Cables.** Overcurrent protection for portable stage cables shall comply with the requirements of 240.5.

N **(B) Portable Single Conductor Feeder Cables Using Single-Pole Separable Connectors.** Portable feeder cables in production areas shall be protected by means of overcurrent devices set at not more than 400 percent of the ampacity of the cable listed in Table 400.5(A)(2). The maximum load on a single conductor portable feeder cable shall not exceed the cable ampacity in Table 400.5(A)(2).

Δ **(C) DC Plugging Boxes.** Cables and cords supplied through plugging boxes shall include only copper conductors. Cables and cords smaller than 8 AWG shall be attached to a plugging box by means of a plug containing two cartridge fuses or a 2-pole circuit breaker. Plugging boxes shall not be permitted on ac systems. Receptacles in dc plugging boxes shall be rated at not less than 30 amperes.

(D) Alternating-Current Power Distribution Boxes. Alternating-current power distribution boxes used in production areas shall contain receptacles of a polarized, grounding type. A feed-through outlet of the same rating as its supply inlet shall not require overcurrent protection in the equipment. Alternating-current power distribution boxes shall be listed.

> Informational Note: See ANSI/UL 1640-2016, *Standard for Portable Power-Distribution Equipment*, for information on alternating-current power distribution boxes.

N **530.24 Purpose-Built Luminaires, Lighting, and Effects Equipment.** Purpose-built luminaires, lighting, and effects equipment shall not be required to be listed but shall be required to be approved.

Δ **530.26 Portable Luminaires.**

N **(A) Listing.** Portable luminaires shall be listed.

Exception: Portable luminaires used as properties on a motion picture set or television stage set, on a studio stage or lot, or on location shall not be considered portable luminaires for the purpose of this section.

(B) Portable Enclosed-Arc Luminaires. Portable enclosed-arc lamps, and associated ballasts shall be listed. Interconnecting cord sets and interconnecting cords and cables for enclosed-arc luminaires shall be of extra-hard usage type and listed.

> Informational Note: See ANSI/ESTA E1.16-2007, *Entertainment Technology — Configuration Standard for Metal-Halide Power Cables*, for information on enclosed-arc luminaire interconnecting cord sets and cables.

N **Part III. Portable Equipment in Support Areas**

N **530.41 Restricted Public Access.** The electrical equipment used in non-production areas shall be restricted from access by the general public.

N **530.42 Overcurrent Protection for Portable Cable.** Overcurrent protection of conductors for portable cables shall comply with the requirements of 240.5(A).

N **530.43 Portable Generators.**

N **(A) Location.** Portable and vehicle-mounted generators shall be located away from flammable materials, and exhaust shall be vented away from structures and other areas where people might congregate.

N **(B) Ventilation.** Portable and vehicle-mounted generators shall not be operated in areas without natural or provided ventilation to prevent the buildup of exhaust.

N **530.44 Ground-Fault Circuit-Interrupter (GFCI) Protection.**

N **(A) Where GFCI Protection is Required.** In addition to the requirements of 210.8(B), GFCI protection for personnel shall be provided for the following:

(1) All 125-volt, single-phase, 15- and 20-ampere receptacles that are readily accessible to unqualified personnel and that are used for other than motion picture and television production equipment

(2) Equipment, other than motion picture and television production equipment, that is readily accessible to unqualified personnel and supplied from a 125-volt, single-phase, 15- or 20-ampere branch circuit

Listed GFCI protection for personnel that is identified for portable use shall be permitted to be an integral part of the attachment plug or be located in the power-supply cord within 300 mm (12 in.) of the attachment plug. Listed cord sets incorporating GFCI for personnel shall be permitted.

N (B) Where GFCI Protection is Not Permitted. Egress lighting shall not be protected by a GFCI.

N (C) Receptacles Supplied by Portable Cords. GFCI protection shall be listed, labeled, and identified for portable use where it is provided using GFCI receptacles and the branch circuits supplying receptacles use flexible cord.

N 530.45 Production Vehicles and Trailers. Where the wiring of production vehicles and trailers are supplied by a grounded ac service or by a grounded separately derived ac source, they shall comply with 530.45(A) through (F) of this section.

N (A) Internal Panelboards (Where Used). A listed and appropriately rated panelboard or other equipment specifically listed for this purpose shall be used. The grounded conductor termination bar shall be insulated from the enclosure.

N (B) Grounding. The panelboard shall have an equipment grounding bus with terminals for all equipment grounding conductors or other approved equipment grounding means.

N (C) Power-Supply Grounding. The equipment grounding conductor in the supply cord or feeder shall be connected to the equipment grounding bus or other approved equipment ground means in the panelboard.

N (D) Insulated Grounded Conductor (Neutral Conductor). The grounded circuit conductor (neutral conductor) shall be insulated from the equipment grounding conductors and from equipment enclosures and other grounded parts.

N (E) Required Bonding. All exposed non–current-carrying metal parts that are likely to become energized shall be effectively bonded to the grounding terminal or enclosure of the panelboard. A bonding conductor shall be connected between any panelboard and an accessible terminal on the chassis of the portable trailer or vehicle.

N (F) Production Vehicles and Trailers with Onboard Generators. Production vehicles and trailers with onboard generators shall comply with the requirements of 551.30.

N 530.46 Protection. Flexible cords and cables accessible to personnel shall be arranged to minimize tripping hazard potential,

and shall be permitted to be covered with a nonconductive matting secured to the walkway surface or protected with another approved cable protection method if the matting or other protection method does not constitute a greater tripping hazard than the uncovered cables.

Part IV. Dressing Rooms

Δ **530.61 Fixed Wiring in Dressing Rooms.** Fixed wiring in dressing rooms shall be installed in accordance with the wiring methods covered in Chapter 3.

Part V. Portable Substations

530.71 General. Wiring and equipment in portable substations rated 50 to 1000 volts, nominal, shall conform to the requirements of the sections applying to installations in permanently fixed substations. Where limited space is available, when approved, working spaces shall be permitted to be reduced where the following conditions apply:

(1) The equipment is arranged so that the qualified operator can work safely.

(2) The equipment is guarded so that other persons in the vicinity cannot accidentally come into contact with current-carrying parts or bring conducting objects into contact with them while they are energized.

530.72 Over 1000 Volts, Nominal. Wiring and equipment of portable substations rated over 1000 volts, nominal, shall comply with the requirements of Part IV of Article 490.

ARTICLE 540 — Motion Picture Projection Rooms

Part I. General

540.1 Scope. This article applies to motion picture projection rooms, motion picture projectors, and associated equipment of the professional and nonprofessional types using incandescent, carbon arc, xenon, or other light source equipment that develops hazardous gases, dust, or radiation.

> Informational Note: See NFPA 40-2019, *Standard for the Storage and Handling of Cellulose Nitrate Film*, for further information.

Motion picture projection rooms are not hazardous locations as classified in Article 500. Some older types of film, such as cellulose nitrate film, are highly flammable and rarely used today. The more commonly used cellulose acetate film is not volatile at ordinary temperatures and does not emit flammable gases. Therefore, wiring methods for projection rooms are not required to be suitable for hazardous locations.

Part II. Equipment and Projectors of the Professional Type

△ **540.10 Motion Picture Projection Room Required.** Every professional-type projector shall be located within a projection room. Every projection room shall be of permanent construction and approved for the type of building in which it is located. All projection ports, spotlight ports, viewing ports, and similar openings shall be provided with glass or other approved material to completely close the opening. Such rooms shall not be considered hazardous (classified) locations as defined in Article 500.

> Informational Note: See NFPA *101*-2021, *Life Safety Code*, for further information on protecting openings in projection rooms handling cellulose nitrate motion picture film.

The audience area of a motion picture theater is covered by Article 520. Article 540 addresses projection rooms for theaters. Every professional projector is required to be located within a permanent projection room that is closed from the theater by glass or other material over all projection, spotlight, viewing, or other ports. Modern projectors contain rectifiers as an integral part of their equipment, thereby eliminating generators and other associated equipment described in 540.11. In accordance with 540.31, nonprofessional projectors that utilize cellulose acetate (safety) film are not required to be installed within a projection room.

540.11 Location of Associated Electrical Equipment.

(A) Motor Generator Sets, Transformers, Rectifiers, Rheostats, and Similar Equipment. Motor-generator sets, transformers, rectifiers, rheostats, and similar equipment for the supply or control of current to projection or spotlight equipment shall, where nitrate film is used, be located in a separate room. Where placed in the projection room, they shall be located or guarded so that arcs or sparks cannot come in contact with film, and the commutator end or ends of motor-generator sets shall comply with one of the conditions in 540.11(A)(1) through (A)(6).

(1) Types. Be of the totally enclosed, enclosed fan-cooled, or enclosed pipe-ventilated type.

(2) Separate Rooms or Housings. Be enclosed in separate rooms or housings built of noncombustible material constructed so as to exclude flyings or lint with approved ventilation from a source of clean air.

(3) Solid Metal Covers. Have the brush or sliding-contact end of motor-generator enclosed by solid metal covers.

(4) Tight Metal Housings. Have brushes or sliding contacts enclosed in tight metal housings.

(5) Upper and Lower Half Enclosures. Have the upper half of the brush or sliding-contact end of the motor-generator enclosed by a wire screen or perforated metal and the lower half enclosed by solid metal covers.

(6) Wire Screens or Perforated Metal. Have wire screens or perforated metal placed at the commutator of brush ends. No dimension of any opening in the wire screen or perforated metal shall exceed 1.27 mm (0.05 in.), regardless of the shape of the opening and of the material used.

(B) Switches, Overcurrent Devices, or Other Equipment. Switches, overcurrent devices, or other equipment not normally required or used for projectors, sound reproduction, flood or other special effect lamps, or other equipment shall not be installed in projection rooms.

Exception No. 1: In projection rooms approved for use only with cellulose acetate (safety) film, the installation of appurtenant electrical equipment used in conjunction with the operation of the projection equipment and the control of lights, curtains, and audio equipment, and so forth, shall be permitted. In such projection rooms, a sign reading "Safety Film Only Permitted in This Room" shall be posted on the outside of each projection room door and within the projection room itself in a conspicuous location.

Exception No. 2: Remote-control switches for the control of auditorium lights or switches for the control of motors operating curtains and masking of the motion picture screen shall be permitted to be installed in projection rooms.

(C) Emergency Systems. Control of emergency systems shall comply with Article 700.

540.12 Work Space. Each motion picture projector, floodlight, spotlight, or similar equipment shall have clear working space not less than 750 mm (30 in.) wide on each side and at the rear of the equipment.

Exception: One such space shall be permitted between adjacent pieces of equipment.

540.13 Conductor Size. Conductors supplying outlets for arc and xenon projectors of the professional type shall not be smaller than 8 AWG and shall have an ampacity not less than the projector current rating. Conductors for incandescent-type projectors shall conform to normal wiring standards as provided in 210.24.

540.14 Conductors on Lamps and Hot Equipment. Insulated conductors having a rated operating temperature of not less than 200°C (392°F) shall be used on all lamps or other equipment where the ambient temperature at the conductors as installed will exceed 50°C (122°F).

540.15 Flexible Cords. Cords approved for hard usage, as provided in Table 400.4, shall be used on portable equipment.

540.20 Listing Requirements. Projectors and enclosures for arc, xenon, and incandescent lamps and rectifiers, transformers, rheostats, and similar equipment shall be listed.

540.21 Marking. Projectors and other equipment shall be marked with the manufacturer's name or trademark and with the voltage and current for which they are designed in accordance with 110.21.

Part III. Nonprofessional Projectors

540.31 Motion Picture Projection Room Not Required. Projectors of the nonprofessional or miniature type, where employing cellulose acetate (safety) film, shall be permitted to be operated without a projection room.

540.32 Listing Requirements. Projection equipment shall be listed.

ARTICLE
545 Manufactured Buildings and Relocatable Structures

Part I. General

545.1 Scope. This article covers requirements for manufactured buildings, building components, relocatable structures, and the conductors that connect relocatable structures to a supply of electricity.

The term *building, manufactured (manufactured building)* is defined in Article 100. The distinction between manufactured buildings covered in Article 545 and manufactured homes covered Article 550 and defined in Article 100 is important. The most distinguishing feature between the two types of structures is how they are placed on the building site. Manufactured homes are built on a chassis and installed on site with or without a permanent foundation. Manufactured buildings are generally constructed within a factory or assembly plant and then transported to the building site. They are not built on a chassis and are designed to be installed on a permanent foundation. Mobile homes that are used as other than dwelling units have been incorporated into Article 545.

In addition, the organizations responsible for construction standards for these units differ. In the case of manufactured homes, the U.S. Department of Housing and Urban Development, 24 CFR 3280, "Manufactured Home Construction and Safety Standards," contains construction requirements for manufactured homes. Manufactured homes bear a nameplate documenting that the unit was constructed in accordance with the federal standard in force at the time of manufacture. In accordance with federal law, this nameplate is an identifying mark that is universally recognized throughout the United States.

Manufactured building construction standards generally are promulgated through state or local units of government. Manufactured building construction can be affected by differences in building construction regulations among the jurisdictions where the buildings will be delivered. The building typically will have an information sheet (often inside the cabinet below the kitchen sink or on a closet wall) indicating the applicable building, electrical, plumbing, and mechanical codes to which the building was constructed.

545.4 Wiring Methods.

(A) Methods Permitted. All raceway and cable wiring methods included in this *Code* and other wiring systems specifically intended and listed for use in manufactured buildings shall be permitted with listed fittings and with fittings listed and identified for manufactured buildings.

(B) Securing Cables. In closed construction, cables shall be permitted to be secured only at cabinets, boxes, or fittings where 10 AWG or smaller conductors are used and protection against physical damage is provided.

545.5 Supply Conductors. Provisions shall be made to route the service-entrance conductors, underground service conductors, service-lateral, feeder, or branch-circuit supply to the service or building disconnecting means conductors.

545.6 Installation of Service-Entrance Conductors. Service-entrance conductors shall be installed after erection at the building site.

Exception: The service-entrance conductors shall be permitted to be installed prior to the erection at the building site where the point of attachment is known prior to manufacture.

545.7 Service Equipment. Service equipment shall be installed in accordance with 230.70.

545.8 Protection of Conductors and Equipment. Protection shall be provided for exposed conductors and equipment during processes of manufacturing, packaging, in transit, and erection at the building site.

545.9 Boxes.

(A) Other Dimensions. Boxes of dimensions other than those required in Table 314.16(A) shall be permitted to be installed where tested, identified, and listed to applicable standards.

(B) Not Over 1650 cm³ (100 in.³). Any box not over 1650 cm³ (100 in.³) in size, intended for mounting in closed construction, shall be affixed with anchors or clamps so as to provide a rigid and secure installation.

545.10 Receptacle or Switch with Integral Enclosure. A receptacle or switch with integral enclosure and mounting means, where tested, identified, and listed to applicable standards, shall be permitted to be installed.

See also

300.15(E) and its commentary for more information about wiring devices with integral enclosures

545.11 Bonding and Grounding. Prewired panels and building components shall provide for the bonding, or bonding and grounding, of all exposed metals likely to become energized, in accordance with Article 250, Parts V, VI, and VII.

545.12 Grounding Electrode Conductor. Provisions shall be made to route a grounding electrode conductor from the service, feeder, or branch-circuit supply to the point of attachment to the grounding electrode.

545.13 Component Interconnections. Fittings and connectors that are intended to be concealed at the time of on-site assembly, where tested, identified, and listed to applicable standards, shall be permitted for on-site interconnection of modules or other building components. Such fittings and connectors shall be equal to the wiring method employed in insulation, temperature rise, and fault-current withstand and shall be capable of enduring the vibration and minor relative motions occurring in the components of manufactured buildings.

The structural components or modules usually are constructed in manufacturing facilities and then transported over the road to a building site for complete assembly of the structure, such as a dwelling unit, motel, or office building. Each module may be prewired at the factory and supplied with fittings and connectors. At the on-site location, the connectors are used to interconnect two or more modules.

Part II. Relocatable Structures

545.20 Application Provisions. Relocatable structures shall comply with Part II of this article and the applicable sections of Part I.

545.22 Power Supply.

Δ **(A) Feeder.** A relocatable structure shall be supplied by a feeder. The feeder shall include insulated color-coded conductors, one of which shall be an equipment grounding conductor. The equipment grounding conductor shall be permitted to be uninsulated if part of a listed cable assembly.

> Informational Note: See 590.4(B) for temporary installation of feeder conductors.

(B) Number of Supplies. Where two or more relocatable structures are structurally connected to form a single unit and there is a factory-installed panelboard in each relocatable structure, each panelboard shall be permitted to be supplied by a separate feeder.

(C) Identification. The identification requirements in 225.37 shall not apply to relocatable structures structurally connected provided the following conditions are met:

(1) The relocatable structures are located on an industrial or commercial establishment where the conditions of

maintenance and supervision ensure qualified individuals will service the installation.

(2) The individual panelboard enclosures or covers have been marked to indicate to location of their supply disconnecting means. The marking shall be visible without removing the cover and shall be of sufficient durability to withstand the environment involved.

Δ **(D) Grounding.**

N **(1) Feeders.** The feeder(s) shall be grounded in accordance with Parts I, II, and III of Article 250.

N **(2) Two or More Relocatable Structures.** Where two or more relocatable structures are structurally connected to form a single unit, and a common grounding electrode conductor and tap arrangement as specified in 250.64(D)(1) is utilized, it shall be permitted to use the chassis bonding conductor specified in 545.26 as the tap conductor.

545.24 Disconnecting Means and Branch-Circuit Overcurrent Protection.

(A) Disconnecting Means. A single disconnecting means consisting of a circuit breaker, or a switch and fuses and its accessories, shall be provided in a readily accessible location for each relocatable structure.

(B) Branch-Circuit Protective Equipment and Panelboards. Branch-circuit distribution equipment shall be installed in each relocatable structure and shall include overcurrent protection for each branch circuit consisting of either circuit breakers or fuses.

> Panelboards shall be installed in a readily accessible location.

545.26 Bonding of Exposed Non–Current-Carrying Metal Parts. All exposed non–current-carrying metal parts that are likely to become energized shall be effectively bonded to the grounding terminal or enclosure of the panelboard. A bonding conductor shall be connected between the panelboard and an accessible terminal on the chassis.

545.27 Intersystem Bonding. Where two or more relocatable structures are structurally connected to form a single unit, it shall be permissible to install one communication system bonding termination in accordance with 250.94 provided all the following conditions are met:

(1) There is a factory-installed panelboard in each relocatable structure.
(2) There is a bonding conductor between the grounding terminal in each panelboard and chassis in accordance with 545.26.
(3) There is a minimum 6 AWG copper conductor that extends from the communication system bonding termination that is connected to each chassis bonding conductor required by 545.26.

The communication bonding termination shall be permitted to be located in the same area as the primary protector or common communications equipment supplying the relocatable structures.

545.28 Ground-Fault Circuit-Interrupters (GFCI). In addition to the requirements of 210.8(B), all receptacle outlets installed in compartments accessible from outside the relocatable structure shall have GFCI protection for personnel.

ARTICLE 547 Agricultural Buildings

N Part I. General

547.1 Scope. This article applies to the following agricultural buildings or that part of a building or adjacent areas of similar or like nature as specified in 547.1(A) or (B).

Article 547 applies not only to agricultural buildings but also to adjacent areas similar in nature. The requirements address the severe environmental conditions that regularly exist on agricultural premises. Damp and wet conditions, dust from feed and litter, and corrosive agents from livestock excrement are all present in these settings as part of normal operating conditions.

(A) Excessive Dust and Dust with Water. Agricultural buildings where excessive dust and dust with water may accumulate, including all areas of poultry, livestock, and fish confinement systems, where litter dust or feed dust, including mineral feed particles, may accumulate.

(B) Corrosive Atmosphere. Agricultural buildings where a corrosive atmosphere exists. Such buildings include areas where the following conditions exist:

(1) Poultry and animal excrement may cause corrosive vapors.
(2) Corrosive particles may combine with water.
(3) The area is damp and wet by reason of periodic washing for cleaning and sanitizing with water and cleansing agents.
(4) Similar conditions exist.

547.3 Other Articles. For buildings and structures not having conditions as specified in 547.1, the electrical installations shall be made in accordance with the applicable articles in this *Code*.

547.4 Surface Temperatures. Electrical equipment or devices installed in accordance with this article shall be installed in a manner such that they will function at full rating without developing surface temperatures in excess of the specified normal safe operating range of the equipment or device.

Informational Note: See Part III of Article 502 for use of equipment in Class 2 locations.

N Part II. Installations

547.20 Wiring Methods. Wiring methods shall be limited to the following:

(1) Type UF
(2) Type NMC
(3) Type SE cable-copper
(4) Jacketed Type MC cable
(5) Raceways identified for the locations specified in 547.1(A) and (B)

The wiring methods of Article 502, Part II, shall be permitted for areas described in 547.1.

Informational Note: See 300.7, 352.44, and 355.44 for installation of raceway systems exposed to widely different temperatures.

547.21 Mounting. All cables shall be secured within 200 mm (8 in.) of each cabinet, box, or fitting. Nonmetallic boxes, fittings, conduit, and cables shall be permitted to be mounted directly to any building surface covered by this article without maintaining the 6 mm (¼ in.) airspace in accordance with 300.6(D).

The 8-inch support distance is less than that required for cables in other types of occupancies. Locating the nonmetallic equipment and wiring methods directly on the interior surface of the building allows a sealant to be placed along the wiring method to facilitate cleaning. Decreasing the support spacing along with eliminating the ¼-inch airspace also reduces the potential for physical damage to cable-type wiring methods in agricultural buildings.

See also

300.6(D), Exception, regarding installation without airspace on certain surfaces

547.22 Equipment Enclosures, Boxes, Conduit Bodies, and Fittings. Equipment enclosures, boxes, conduit bodies, and fittings installed in areas of buildings where excessive dust could be present shall be designed to minimize the entrance of dust and shall have no openings through which dust could enter the enclosure.

547.23 Damp or Wet Locations. In damp or wet locations, equipment enclosures, boxes, conduit bodies, and fittings shall be placed or equipped so as to prevent moisture from entering or accumulating within the enclosure, box, conduit body, or fitting. In wet locations, including normally dry or damp locations where surfaces are periodically washed or sprayed with water, boxes, conduit bodies, and fittings shall be listed for use in wet locations, and equipment enclosures shall be weatherproof.

547.24 Corrosive Atmosphere. Where wet dust, excessive moisture, corrosive gases or vapors, or other corrosive conditions could be present, equipment enclosures, boxes, conduit bodies, and fittings shall have corrosion resistance properties suitable for the conditions.

Informational Note No. 1: See Table 110.28 for appropriate enclosure type designations.

Informational Note No. 2: Aluminum and magnetic ferrous materials can corrode in agricultural environments.

547.25 Flexible Connections. Where necessary to employ flexible connections, one or more of the following shall be permitted:

(1) Dusttight flexible connectors
(2) Liquidtight flexible metal conduit (LFMC) with listed fittings
(3) Liquidtight flexible nonmetallic conduit (LFNC) with listed fittings
(4) Flexible cord listed and identified for hard usage and terminated with listed dusttight cord connectors

547.26 Physical Protection. All electrical wiring and equipment subject to physical damage shall be protected.

Nonmetallic cables shall not be permitted to be concealed within walls and above ceilings of buildings (i.e., offices, lunchrooms, ancillary areas, etc.) or portions thereof, which are contiguous with or physically adjoined to livestock confinement areas.

Informational Note: Rodents and other pests are common around such installations and will damage nonmetallic cable by chewing the cable jacket and conductor insulation concealed within walls and ceilings of livestock containment areas of agricultural buildings.

Δ **547.27 Separate Equipment Grounding Conductor.** Where a separate equipment grounding conductor, not part of a listed cable assembly, is installed underground within a location falling under the scope in 547.1, it shall be insulated.

Informational Note: See 250.120(B) for further information on aluminum and copper-clad aluminum conductors.

This requirement improves the longevity of equipment grounding conductors (EGCs) installed underground in the highly corrosive locations that are typical of many farm buildings.

547.28 Ground-Fault Circuit-Interrupter Protection. Ground-fault circuit-interrupter (GFCI) protection shall be provided as required in 210.8(B) for areas of agricultural buildings not included in the scope of this article. GFCI protection shall not be required for other than 125-volt, 15- and 20-ampere receptacles installed within the following areas:

(1) Areas requiring an equipotential plane
(2) Outdoors
(3) Damp or wet locations
(4) Dirt confinement areas for livestock

547.29 Switches, Receptacles, Circuit Breakers, Controllers, and Fuses. Switches, including pushbuttons, relays, and similar devices, receptacles, circuit breakers, controllers, and fuses, shall be provided with enclosures as specified in 547.22.

547.30 Motors. Motors and other rotating electrical machinery shall be totally enclosed or designed so as to minimize the entrance of dust, moisture, or corrosive particles.

547.31 Luminaires. Luminaires shall comply with 547.31(A) through (C).

(A) Minimize the Entrance of Dust. Luminaires shall be installed to minimize the entrance and accumulation of dust, foreign matter, moisture, and corrosive material.

(B) Exposed to Physical Damage. Luminaires exposed to physical damage shall be protected by a suitable guard.

(C) Exposed to Water. Luminaires exposed to water from condensation, building cleansing water, or solution shall be listed for use in wet locations.

N **Part III. Distribution**

Δ **547.40 Electrical Supply to Building(s) or Structure(s) from a Distribution Point.** Any agricultural building or structure for livestock located on the same premises shall be supplied from a distribution point. More than one distribution point on the same premises shall be permitted.

Any existing agricultural building or structure for other than livestock not under the scope of Article 547 shall be permitted to be supplied in accordance with 250.32(B)(1), Exception No. 1.

The requirements in 547.40 cover the installation of conductors that originate from an electrical distribution point and supply agricultural buildings. A distribution point, sometimes referred to as the center yard pole, or meter pole is often used as a means of centrally locating the origin of the electrical distribution system.

See also

Article 100 for the definition of the term *distribution point*

Many agricultural sites consist of multiple buildings. A distribution point often supplies multiple buildings via an overhead distribution system. A disconnecting means, referred to as the site-isolating device, disconnects all ungrounded conductors run to the buildings and structures. This device provides a single location for the disconnection of all power to the buildings. It is not considered the service disconnecting means.

The site-isolating device is required to be pole-mounted to the height required for the conductors it supplies, thereby rendering the device inaccessible. The remote operating handle for the device must be readily accessible to personnel. If the supply system includes a grounded conductor, it must be connected to a grounding electrode system at the site-isolating device.

The site-isolating device is not required to provide overcurrent protection. Based on the requirements in 547.41(B) and 547.42, the location of the service disconnecting means and overcurrent protection is on the load side of the site-isolating device. Where the site is supplied by more than one distribution point, a plaque or directory that provides information about the location of each distribution point and the buildings served by each distribution point is required at each site-isolating device.

In Exhibit 547.1, the pole-mounted site-isolating device is located at the distribution point. A set of overhead service conductors is run to each of the three structures, and a service disconnecting means and overcurrent protection are installed at each building. The supply conductors are considered service conductors because there is no overcurrent protection at the site-isolating device. A grounding electrode system is required at the distribution point, and a grounding electrode conductor (GEC) connection to the supply system grounded conductor must be made at the site-isolating device. A grounding electrode system is also required at each of the buildings.

Any portion of an EGC installed underground must be insulated in accordance with 547.27. This is to protect the EGC from the corrosive influences inherent to agricultural premises and to reduce leakage current in those areas where livestock are kept. Prevention of stray voltage at agricultural premises is extremely important.

N **547.41 Overhead Service.** Overhead service installations shall comply with 547.41(A)(1) through (A)(9), 547.41(B)(1) through (B)(3), or 547.42.

(A) Site-Isolating Device. Site-isolating devices shall comply with 547.41(A)(1) through (A)(9).

(1) Where Required. A site-isolating device shall be installed at the distribution point where two or more buildings or structures are supplied from the distribution point.

(2) Location. The site-isolating device shall be pole-mounted and be not less than the height above grade required by 230.24 for the conductors it supplies.

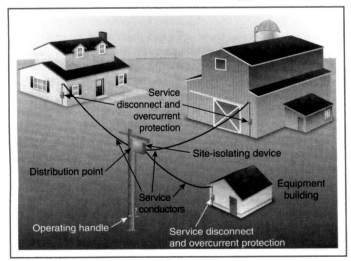

EXHIBIT 547.1 Site-isolating device located at the distribution point with service conductors run to each building. A service disconnecting means and overcurrent protection are installed at each building.

(3) Operation. The site-isolating device shall simultaneously disconnect all ungrounded service conductors from the premises wiring.

(4) Grounding and Bonding. At the site-isolating device, the grounding electrode conductor shall be connected to the system grounded conductor and the enclosure.

(5) Rating. The site-isolating device shall be rated for the calculated load as determined by Part V of Article 220.

(6) Overcurrent Protection. The site-isolating device shall not be required to provide overcurrent protection.

(7) Accessibility. The site-isolating device shall be capable of being remotely operated by an operating handle installed at a readily accessible location. The operating handle of the site-isolating device, when in its highest position, shall not be more than 2.0 m (6 ft 7 in.) above grade or a working platform.

(8) Series Devices. An additional site-isolating device for the premises wiring system shall not be required where a site-isolating device meeting all applicable requirements of this section is provided by the serving utility as part of their service requirements.

(9) Marking. A site-isolating device shall be permanently marked to identify it as a site-isolating device. This marking shall be located on the operating handle or immediately adjacent thereto.

(B) Service Disconnecting Means and Overcurrent Protection at the Building(s) or Structure(s). Where the service disconnecting means and overcurrent protection are located at the building(s) or structure(s), the requirements of 547.41(B)(1) through (B)(3) shall apply.

(1) Conductor Sizing. The supply conductors shall be sized in accordance with Part V of Article 220.

Δ **(2) Conductor Installation.** The supply conductors shall be installed in accordance with Part II of Article 225.

Δ **(3) Grounding and Bonding.** For each building or structure, grounding and bonding of the supply conductors shall be in accordance with 250.32, and the following conditions shall be met:

(1) The equipment grounding conductor is not smaller than the largest supply conductor if of the same material or is adjusted in size in accordance with the equivalent size columns of Table 250.122 if of different materials.
(2) The equipment grounding conductor is connected to the grounded circuit conductor and the site-isolating device enclosure at the distribution point.

547.42 Service Disconnecting Means and Overcurrent Protection at the Distribution Point. The service disconnecting means and overcurrent protection for each set of feeders or branch circuits shall be located at the distribution point. The service disconnecting means shall be installed in accordance with Part VI of Article 230. The feeders or branch circuits supplied to buildings or structures shall comply with 250.32 and Article 225, Parts I and II.

> Informational Note: Methods to reduce neutral-to-earth voltages in livestock facilities include supplying buildings or structures with 4-wire single-phase services, sizing 3-wire single-phase service and feeder conductors to limit voltage drop to 2 percent, and connecting loads line-to-line, will provide reasonable efficiency of operation.

547.43 Identification. Where a site is supplied by more than one distribution point, a permanent plaque or directory shall be installed at each of these distribution points denoting the location of each of the other distribution points and the buildings or structures served by each.

547.44 Equipotential Planes and Bonding of Equipotential Planes. The installation and bonding of equipotential planes shall comply with 547.44(A) and (B). For the purposes of this section, the term *livestock* shall not include poultry.

Grounding and bonding requirements unique to agricultural settings are necessary due to the sensitivity of livestock to slight differences in potential between surfaces with which they are in direct contact. The wet or damp concrete common to animal confinement areas enhances that sensitivity.

Δ **(A) Where Required.** Equipotential planes shall be required in the following areas:

(1) Indoors. Equipotential planes shall be installed in confinement areas with concrete floors where metallic equipment is located that may become energized and is accessible to livestock.

(2) Outdoors. Equipotential planes shall be installed in concrete slabs where metallic equipment is located that may become energized and is accessible to livestock.

The equipotential plane shall encompass the area where the livestock stands while accessing metallic equipment that may become energized.

(B) Bonding. Equipotential planes shall be bonded to the grounding electrode system or an equipment grounding terminal in any panelboard of the electrical grounding system associated with the equipotential plane. The bonding conductor shall be solid copper, insulated, covered or bare, and not smaller than 8 AWG. The means of bonding to wire mesh or conductive elements shall be by pressure connectors or clamps of brass, copper, copper alloy, or other approved means. Slatted floors that are supported by structures that are a part of an equipotential plane shall not require bonding.

> Informational Note No. 1: See ASEA/ASABE EP473.2-2001 (R2015), *Equipotential Planes in Animal Containment Areas*, for methods to establish equipotential planes.

> Informational Note No. 2: See ASEA/ASABE EP342.3-2010 (R2015), *Safety for Electrically Heated Livestock Waterers*, for methods for safe installation of livestock waterers.

> Informational Note No. 3: Low grounding electrode system resistances may reduce voltage differences in livestock facilities.

Electrically heated livestock watering troughs could pose an electric shock hazard for livestock and personnel. The referenced document provides information on the proper installation of such equipment.

ARTICLE 550 Mobile Homes, Manufactured Homes, and Mobile Home Parks

Part I. General

550.1 Scope. This article covers the electrical conductors and equipment installed within or on mobile and manufactured homes, the conductors that connect mobile and manufactured homes to a supply of electricity, and the installation of electrical wiring, luminaires, equipment, and appurtenances related to electrical installations within a mobile home park up to the mobile home service-entrance conductors or, if none, the mobile home service equipment.

> Informational Note: See NFPA 501-2017, *Standard on Manufactured Housing*, and Part 3280, *Manufactured Home Construction and Safety Standards*, of the Federal Department of Housing and Urban Development for additional information on manufactured housing.

The *Manufactured Home Construction and Safety Standards*, issued by the U.S. Housing and Urban Development Administration (HUD), incorporates many of the requirements of Article 550 of the *NEC®*. The federal standard contains the requirements for electrical systems, conductors, and equipment installed within or on mobile homes and the conductors that connect mobile homes to a supply of electricity. Mobile homes are defined as manufactured homes in the HUD regulations. For the purposes of the *NEC*, and unless otherwise indicated, the term *mobile home* includes manufactured homes.

The regulations pertaining to electrical systems are in 24 CFR 3280.801–3280.816. They require that new manufactured homes comply with the federal standard. In some cases, HUD has delegated the enforcement of this standard to state and private inspection agencies and qualified testing laboratories. The service equipment and feeders installed at the mobile or manufactured home site are covered by the requirements in Part III of Article 550.

See also

Article 545 for requirements for mobiles homes that are not intended as a dwelling unit

545.1 commentary for information on the distinction between manufactured homes and manufactured buildings

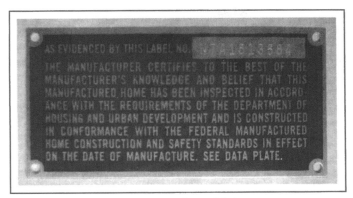

EXHIBIT 550.1 A HUD label required for mobile homes.

Manufactured homes and mobile homes are required to meet federal standards. Manufacturers attach a label certifying that the construction is in accordance with those standards. The certification facilitates the transport of these homes nationwide by allowing AHJ approval without having to dismantle the home to verify the construction and electrical installation. Although the federal process supersedes local codes for the home construction, local codes are applicable for the on-site work. Refer to Exhibit 550.1 for an example of the HUD label.

550.4 General Requirements.

(A) In Other Than Mobile Home Parks. Mobile homes installed in other than mobile home parks shall comply with the provisions of this article.

(B) Connection to Wiring System. This article shall apply to mobile homes intended for connection to a wiring system rated 120/240 volts, nominal, 3-wire ac, with a grounded neutral conductor.

(C) Listed and Labeled. All electrical materials, devices, appliances, fittings, and other equipment shall be listed and labeled by a qualified testing agency and shall be connected in an approved manner when installed.

Part II. Mobile and Manufactured Homes

550.10 Power Supply.

(A) Feeder. The power supply to the mobile home shall be a feeder assembly consisting of not more than one listed 50-ampere mobile home power-supply cord or a permanently installed feeder.

Exception No. 1: A mobile home that is factory equipped with gas or oil-fired central heating equipment and cooking appliances shall be permitted to be provided with a listed mobile home power-supply cord rated 40 amperes.

Exception No. 2: A feeder assembly shall not be required for manufactured homes constructed in accordance with 550.32(B).

(B) Power-Supply Cord. If the mobile home has a power-supply cord, it shall be permanently attached to the panelboard, or to a junction box permanently connected to the panelboard, with the free end terminating in an attachment plug cap.

Cords with adapters and pigtail ends, extension cords, and similar items shall not be attached to, or shipped with, a mobile home.

A suitable clamp or the equivalent shall be provided at the panelboard knockout to afford strain relief for the cord to prevent strain from being transmitted to the terminals when the power-supply cord is handled in its intended manner.

The cord shall be a listed type with four conductors, one of which shall be identified by a continuous green color or a continuous green color with one or more yellow stripes for use as the equipment grounding conductor.

(C) Attachment Plug Cap. The attachment plug cap shall be a 3-pole, 4-wire, grounding type, rated 50 amperes, 125/250 volts with a configuration as shown in Figure 550.10(C) and intended for use with the 50-ampere, 125/250-volt receptacle configuration shown in Figure 550.10(C). It shall be listed, by itself or as part of a power-supply cord assembly, for the purpose and shall be molded to or installed on the flexible cord so that it is secured tightly to the cord at the point where the cord enters the attachment plug cap. If a right-angle cap is used, the configuration shall be oriented so that the grounding member is farthest from the cord.

Informational Note: See ANSI/NEMA WD 6-2016, *Wiring Devices — Dimensional Specifications*, Figure 14-50, for complete details of the 50-ampere plug and receptacle configuration.

125/250-V, 50-A, 3-pole, 4-wire, grounding type

FIGURE 550.10(C) *50-Ampere, 125/250-Volt Receptacle and Attachment Plug Cap Configurations, 3-Pole, 4-Wire, Grounding-Types, Used for Mobile Home Supply Cords and Mobile Home Parks.*

(D) Overall Length of a Power-Supply Cord. The overall length of a power-supply cord, measured from the end of the cord, including bared leads, to the face of the attachment plug cap shall not be less than 6.4 m (21 ft) and shall not exceed 11 m (36½ ft). The length of the cord from the face of the attachment plug cap to the point where the cord enters the mobile home shall not be less than 6.0 m (20 ft).

(E) Marking. The power-supply cord shall bear the following marking:

FOR USE WITH MOBILE HOMES — 40 AMPERES

or

FOR USE WITH MOBILE HOMES — 50 AMPERES

(F) Point of Entrance. The point of entrance of the feeder assembly to the mobile home shall be in the exterior wall, floor, or roof.

(G) Protected. Where the cord passes through walls or floors, it shall be protected by means of conduits and bushings or equivalent. The cord shall be permitted to be installed within the mobile home walls, provided a continuous raceway having a maximum size of 32 mm (1¼ in.) is installed from the branch-circuit panelboard to the underside of the mobile home floor.

(H) Protection Against Corrosion and Mechanical Damage. Permanent provisions shall be made for the protection of the attachment plug cap of the power-supply cord and any connector cord assembly or receptacle against corrosion and mechanical damage if such devices are in an exterior location while the mobile home is in transit.

(I) Mast Weatherhead or Raceway. Where the calculated load exceeds 50 amperes or where a permanent feeder is used, the supply shall be by means of either of the following:

(1) One mast weatherhead installation, installed in accordance with Part II of Article 230, containing four continuous, insulated, color-coded feeder conductors, one of which shall be an equipment grounding conductor

(2) A rigid metal conduit, intermediate metal conduit, rigid polyvinyl chloride conduit, or other raceways identified for the location, from the disconnecting means in the mobile home to the underside of the mobile home, with provisions for the attachment to a suitable junction box or fitting to the raceway on the underside of the mobile home [with or without conductors as in 550.10(I)(1)]. The manufacturer shall provide written installation instructions stating the proper feeder conductor sizes for the raceway and the size of the junction box to be used.

Cord-and-plug connection is permitted for loads that do not exceed 50 amperes, but many larger mobile and manufactured homes often contain a considerable amount of electrical equipment and often exceed this rating. A permanently connected feeder is required if the calculated load exceeds 50 amperes.

A raceway is required from the distribution panelboard in the mobile home to the underside of the mobile home. Typically, the feeder conductors in the raceway are installed when the mobile home is located at its site. The raceway provides a means to install the feeder conductors to the mobile home panelboard without damaging the interior finish. The feeder assembly must comprise four continuous, insulated, color-coded conductors, as indicated in 550.10(I)(1) and 550.33(B).

550.11 Disconnecting Means and Branch-Circuit Protective Equipment. The branch-circuit equipment shall be permitted to be combined with the disconnecting means as a single assembly. Such a combination shall be permitted to be designated as a panelboard. If a fused panelboard is used, the maximum fuse size for the mains shall be plainly marked with lettering at least 6 mm (¼ in.) high and visible when fuses are changed.

Where plug fuses and fuseholders are used, they shall be tamper-resistant Type S, enclosed in dead-front fuse panelboards. Electrical panelboards containing circuit breakers shall also be dead-front type.

Informational Note: See 110.22 concerning identification of each disconnecting means and each service, feeder, or branch circuit at the point where it originated and the type marking needed.

(A) Disconnecting Means. A single disconnecting means shall be provided in each mobile home consisting of a circuit breaker, or a switch and fuses and its accessories installed in a readily accessible location near the point of entrance of the supply cord or conductors into the mobile home. The main circuit breakers or fuses shall be plainly marked "Main." This equipment shall contain a solderless type of grounding connector or bar for the purposes of grounding, with sufficient terminals for all grounding conductors. The terminations of the grounded circuit conductors shall be insulated in accordance with 550.16(A). The disconnecting equipment shall have a rating not less than the calculated load. The distribution equipment, either circuit breaker or fused type, shall be located a minimum of 600 mm (24 in.) from the bottom of such equipment to the floor level of the mobile home.

Informational Note: See 550.20(B) for information on disconnecting means for branch circuits designed to energize heating or air-conditioning equipment, or both, located outside the mobile home, other than room air conditioners.

A panelboard shall be rated not less than 50 amperes and employ a 2-pole circuit breaker rated 40 amperes for a 40-ampere supply cord, or 50 amperes for a 50-ampere supply cord. A panelboard employing a disconnect switch and fuses shall be rated 60 amperes and shall employ a single 2-pole, 60-ampere fuseholder with 40- or 50-ampere main fuses for 40- or 50-ampere supply cords, respectively. The outside of the panelboard shall be plainly marked with the fuse size.

The panelboard shall be located in an accessible location but shall not be located in a bathroom or a clothes closet. A clear working space at least 750 mm (30 in.) wide and 750 mm (30 in.) in front of the panelboard shall be provided. This space shall extend from the floor to the top of the panelboard.

(B) Branch-Circuit Protective Equipment. Branch-circuit distribution equipment shall be installed in each mobile home and shall include overcurrent protection for each branch circuit consisting of either circuit breakers or fuses.

The branch-circuit overcurrent devices shall be rated as follows:

(1) Not more than the circuit conductors; and

(2) Not more than 150 percent of the rating of a single appliance rated 13.3 amperes or more that is supplied by an individual branch circuit; but

(3) Not more than the overcurrent protection size and of the type marked on the air conditioner or other motor-operated appliance.

(C) Two-Pole Circuit Breakers. Where circuit breakers are provided for branch-circuit protection, 240-volt circuits shall be protected by a 2-pole common or companion trip, or by circuit breakers with identified handle ties.

(D) Electrical Nameplates. A metal nameplate on the outside adjacent to the feeder assembly entrance shall read as follows:

THIS CONNECTION FOR 120/240-VOLT,
3-POLE, 4-WIRE, 60-HERTZ,
_____ AMPERE SUPPLY

The correct ampere rating shall be marked in the blank space.

Exception: For manufactured homes, the manufacturer shall provide in its written installation instructions or in the data plate the minimum ampere rating of the feeder assembly or, where provided, the service-entrance conductors intended for connection to the manufactured home. The rating provided shall not be less than the minimum load calculated in accordance with 550.18.

550.12 Branch Circuits. The number of branch circuits required shall be determined in accordance with 550.12(A) through (E).

(A) Lighting. The number of branch circuits shall be based on 33 volt-amperes/m² (3 VA/ft²) times outside dimensions of the mobile home (coupler excluded) divided by 120 volts to determine the number of 15- or 20-ampere lighting area circuits, for example,

$$\frac{3 \times \text{length} \times \text{width}}{120 \times 15 \text{ (or 20)}} \qquad \textbf{[550.12(A)]}$$

= No. of 15- (or 20-) ampere circuits

(B) Small Appliances. In kitchens, pantries, dining rooms, and breakfast rooms, two or more 20-ampere small-appliance circuits, in addition to the number of circuits required elsewhere in this section, shall be provided for all receptacle outlets required by 550.13(D) in these rooms. Such circuits shall have no other outlets.

Exception No. 1: Receptacle outlets installed solely for the electrical supply and support of an electric clock in any the rooms specified in 550.12(B) shall be permitted.

Exception No. 2: Receptacle outlets installed to provide power for supplemental equipment and lighting on gas-fired ranges, ovens, or counter-mounted cooking units shall be permitted.

Exception No. 3: A single receptacle for refrigeration equipment shall be permitted to be supplied from an individual branch circuit rated 15 amperes or greater.

Countertop receptacle outlets installed in the kitchen shall be supplied by not less than two small-appliance circuit branch circuits, either or both of which shall be permitted to supply receptacle outlets in the kitchen and other locations specified in 550.12(B).

(C) Laundry Area. Where a laundry area is provided, a 20-ampere branch circuit shall be provided to supply the laundry receptacle outlet(s). This circuit shall have no other outlets.

(D) General Appliances. (Including furnace, water heater, range, and central or room air conditioner, etc.). There shall be one or more circuits of adequate rating in accordance with the following:

Informational Note: See Article 440 for central air conditioning.

(1) The ampere rating of fixed appliances shall be not over 50 percent of the circuit rating if lighting outlets (receptacles, other than kitchen, dining area, and laundry, considered as lighting outlets) are on the same circuit.
(2) For fixed appliances on a circuit without lighting outlets, the sum of rated amperes shall not exceed the branch-circuit rating. Motor loads or continuous loads shall not exceed 80 percent of the branch-circuit rating.
(3) The rating of a single cord-and-plug-connected appliance on a circuit having no other outlets shall not exceed 80 percent of the circuit rating.
(4) The rating of a range branch circuit shall be based on the range demand as specified for ranges in 550.18(B)(5).

(E) Bathrooms. Bathroom receptacle outlets shall be supplied by at least one 20-ampere branch circuit. Such circuits shall have no outlets other than as provided for in 550.13(E)(2).

550.13 Receptacle Outlets.

(A) Grounding-Type Receptacle Outlets. All receptacle outlets shall comply with the following:

(1) Be of grounding type
(2) Be installed according to 406.4
(3) Except where supplying specific appliances, be 15- or 20-ampere, 125-volt, either single or multiple type, and accept parallel-blade attachment plugs

Δ **(B) Ground-Fault Circuit Interrupters (GFCI).** Ground-fault circuit-interrupter protection shall be provided as required in 210.8(A). In addition, in the following areas within a mobile or manufactured home, GFCI protection is limited to 125-volt, 15- and 20-ampere receptacles or outlets:

(1) Compartments accessible from outside the unit
(2) Bathrooms, including receptacles in luminaires
(3) Kitchens, where receptacles are installed to serve countertop surfaces
(4) Sinks, where receptacles are installed within 1.8 m (6 ft) from the top inside edge of the sink
(5) Dishwashers

Informational Note: See 422.5(A) for information on protection of dishwashers.

The locations where GFCI protection is required for receptacles installed in a mobile or manufactured home parallel those specified in 210.8(A) for site-built dwelling units.

(C) Cord-Connected Fixed Appliance. A grounding-type receptacle outlet shall be provided for each cord-connected fixed appliance installed.

(D) Receptacle Outlets Required. Except in the bath, closet, and hallway areas, receptacle outlets shall be installed at wall spaces 600 mm (2 ft) wide or more so that no point along the floor line is more than 1.8 m (6 ft) measured horizontally from an outlet in that space. In addition, a receptacle outlet shall be installed in the following locations:

(1) Over or adjacent to countertops in the kitchen [at least one on each side of the sink if countertops are on each side and are 300 mm (12 in.) or over in width].
(2) Adjacent to the refrigerator and freestanding gas-range space. A multiple-type receptacle shall be permitted to serve as the outlet for a countertop and a refrigerator.
(3) At countertop spaces for built-in vanities.
(4) At countertop spaces under wall-mounted cabinets.
(5) In the wall at the nearest point to where a bar-type counter attaches to the wall.
(6) In the wall at the nearest point to where a fixed room divider attaches to the wall.
(7) In laundry areas within 1.8 m (6 ft) of the intended location of the laundry appliance(s).
(8) At least one receptacle outlet located outdoors and accessible at grade level and not more than 2.0 m (6½ ft) above grade. A receptacle outlet located in a compartment accessible from the outside of the unit shall be considered an outdoor receptacle.
(9) At least one receptacle outlet shall be installed in bathrooms within 900 mm (36 in.) of the outside edge of each basin. The receptacle outlet shall be located above or adjacent to the basin location. This receptacle shall be in addition to any receptacle that is a part of a luminaire or appliance. The receptacle shall not be enclosed within a bathroom cabinet or vanity.

(E) Pipe Heating Cable(s) Outlet. For the connection of pipe heating cable(s), a receptacle outlet shall be located on the underside of the unit as follows:

(1) Within 600 mm (2 ft) of the cold water inlet.
(2) Connected to an interior branch circuit, other than a small-appliance branch circuit. It shall be permitted to use a bathroom receptacle circuit for this purpose.
(3) On a circuit where all of the outlets are on the load side of the ground-fault circuit interrupter.
(4) This outlet shall not be considered as the receptacle required by 550.13(D)(8).

A receptacle outlet (sometimes referred to as a heat tape outlet) is required on the underside of mobile homes to supply cord-and-plug-connected pipe-heating cables. The receptacle must be GFCI protected and connected to a branch circuit that serves the interior of the mobile home. Requiring all outlets on this branch circuit to be on the load (downstream) side of the GFCI is to allow supervision of the power supply and GFCI for this outlet from inside the mobile home. If the overcurrent protective device (OCPD) or GFCI device opens, the occupants of the home are more likely to notice it than if the heating outlet was supplied by a dedicated circuit.

(F) Receptacle Outlets Not Permitted. Receptacle outlets shall not be permitted in the following locations:

(1) Receptacle outlets shall not be installed within or directly over a bathtub or shower space.
(2) A receptacle shall not be installed in a face-up position in any countertop.
(3) Receptacle outlets shall not be installed above electric baseboard heaters, unless provided for in the listing or manufacturer's instructions.

(G) Receptacle Outlets Not Required. Receptacle outlets shall not be required in the following locations:

(1) In the wall space occupied by built-in kitchen or wardrobe cabinets
(2) In the wall space behind doors that can be opened fully against a wall surface
(3) In room dividers of the lattice type that are less than 2.5 m (8 ft) long, not solid, and within 150 mm (6 in.) of the floor
(4) In the wall space afforded by bar-type counters

550.14 Luminaires and Appliances.

(A) Fasten Appliances in Transit. Means shall be provided to securely fasten appliances when the mobile home is in transit. (See 550.16 for provisions on grounding.)

(B) Accessibility. Every appliance shall be accessible for inspection, service, repair, or replacement without removal of permanent construction.

(C) Pendants. Listed pendant-type luminaires or pendant cords shall be permitted.

(D) Bathtub and Shower Luminaires. Where a luminaire is installed over a bathtub or in a shower stall, it shall be of the enclosed and gasketed type listed for wet locations.

550.15 Wiring Methods and Materials.
Except as specifically limited in this section, the wiring methods and materials included in this *Code* shall be used in mobile homes. Where conductors are terminated, they shall be used with equipment listed and identified for the conductor materials.

(A) Nonmetallic Boxes. Nonmetallic boxes shall be permitted only with nonmetallic cable or nonmetallic raceways.

(B) Nonmetallic Cable Protection. Nonmetallic cable located 380 mm (15 in.) or less above the floor, if exposed, shall be protected from physical damage by covering boards, guard strips, or raceways. Cable likely to be damaged by stowage shall be so protected in all cases.

(C) Metal-Covered and Nonmetallic Cable Protection. Metal-covered and nonmetallic cables shall be permitted to pass through the centers of the wide side of 2 by 4 studs. However, they shall be protected where they pass through 2 by 2 studs or at other studs or frames where the cable or armor would be less than 32 mm (1¼ in.) from the inside or outside surface of the studs where the wall covering materials are in contact with the studs. Steel plates on each side of the cable, or a tube, with not less than 1.35 mm (0.053 in.) wall thickness shall be required to protect the cable. These plates or tubes shall be securely held in place.

(D) Metal Faceplates. Where metal faceplates are used, the installation shall comply with 404.9(B) and 406.6(B).

(E) Installation Requirements. Where a range, clothes dryer, or other appliance is connected by metal-covered cable or flexible metal conduit, a length of not less than 900 mm (3 ft) of unsupported cable or conduit shall be provided to service the appliance. The cable or flexible metal conduit shall be secured to the wall. Type NM or Type SE cable shall not be used to connect a range or dryer. This shall not prohibit the use of Type NM or Type SE cable between the branch-circuit overcurrent protective device and a junction box or range or dryer receptacle.

(F) Raceways. Where rigid metal conduit or intermediate metal conduit is terminated at an enclosure with a locknut and bushing connection, two locknuts shall be provided, one inside and one outside of the enclosure. Rigid nonmetallic conduit, electrical nonmetallic tubing, or surface raceway shall be permitted. All cut ends of conduit and tubing shall be reamed or otherwise finished to remove rough edges.

(G) Switches. Switches shall be rated as follows:

(1) For lighting circuits, switches shall be rated not less than 10 amperes, 120 to 125 volts, and in no case less than the connected load.
(2) Switches for motor or other loads shall comply with 404.14.

(H) Under-Chassis Wiring (Exposed to Weather).

(1) Where outdoor or under-chassis line-voltage (120 volts, nominal, or higher) wiring is exposed, it shall be protected by a conduit or raceway identified for use in wet locations. The conductors shall be listed for use in wet locations.

(2) Where wiring is exposed to physical damage, it shall be protected by a raceway, conduit, or other means.

The under-chassis wiring method is not restricted to rigid or intermediate conduit as long as the raceway is suitable for installation in wet locations and, if necessary, for locations where it is subject to physical damage. This allows for the standard practice in manufactured home construction of installing polyvinyl chloride (PVC) conduit or reinforced thermosetting resin conduit (RTRC) under the chassis.

(I) Boxes, Fittings, and Cabinets. Boxes, fittings, and cabinets shall be securely fastened in place and shall be supported from a structural member of the home, either directly or by using a substantial brace.

Exception: Snap-in-type boxes. Boxes provided with special wall or ceiling brackets and wiring devices with integral enclosures that securely fasten to walls or ceilings and are identified for the use shall be permitted without support from a structural member or brace. The testing and approval shall include the wall and ceiling construction systems for which the boxes and devices are intended to be used.

(J) Appliance Terminal Connections. Appliances having branch-circuit terminal connections that operate at temperatures higher than 60°C (140°F) shall have circuit conductors as described in the following:

(1) Branch-circuit conductors having an insulation suitable for the temperature encountered shall be permitted to be run directly to the appliance.
(2) Conductors having an insulation suitable for the temperature encountered shall be run from the appliance terminal connection to a readily accessible outlet box placed at least 300 mm (1 ft) from the appliance. These conductors shall be in a suitable raceway or Type AC or MC cable of at least 450 mm (18 in.) but not more than 1.8 m (6 ft) in length.

(K) Component Interconnections. Fittings and connectors that are intended to be concealed at the time of assembly shall be listed and identified for the interconnection of building components. Such fittings and connectors shall be equal to the wiring method employed in insulation, temperature rise, and fault-current withstanding and shall be capable of enduring the vibration and shock occurring in mobile home transportation.

Informational Note: See 550.19 for interconnection of multiple section units.

550.16 Grounding. Grounding of both electrical and nonelectrical metal parts in a mobile home shall be through connection to a grounding bus in the mobile home panelboard and shall be connected through the green-colored insulated conductor in the supply cord or the feeder wiring to the grounding bus in the service-entrance equipment located adjacent to the mobile home location. Neither the frame of the mobile home nor the frame of any appliance shall be connected to the grounded circuit conductor in the mobile home. Where the panelboard is the service equipment as permitted by 550.32(B), the neutral conductors and the equipment grounding bus shall be connected.

(A) Grounded Conductor.

(1) Insulated. The grounded circuit conductor shall be insulated from the equipment grounding conductors and from equipment enclosures and other grounded parts. The grounded circuit conductor terminals in the panelboard and in ranges, clothes dryers, counter-mounted cooking units, and wall-mounted ovens shall be insulated from the equipment enclosure. Bonding screws, straps, or buses in the panelboard or in appliances shall be removed and discarded. Where the panelboard is the service equipment as permitted by 550.32(B), the neutral conductors and the equipment grounding bus shall be connected.

The feeder assembly must consist of a listed cord or four color-coded insulated conductors, one of which is the grounded conductor (white) and one of which is used for grounding purposes (green). Thus, the grounded and grounding conductors are kept independent of each other and are connected only at the service equipment [at the point of connection of the grounding electrode conductor (GEC)]. Grounding of metal parts, including the frame of the mobile home or the frame of any appliance, is accomplished by connection to the equipment grounding bus and never to the grounded conductor (neutral bus). This prevents incidental contact between the grounded conductor and non–current-carrying metal parts of electrical equipment. Without the separation of the grounded and grounding conductors, contact could result in the metal structure or metal sheathing of the mobile home becoming a parallel path for neutral current.

Many ranges and clothes dryers have a factory-installed bonding jumper. Removing this jumper does not compromise or void the listing of the product, because isolation of the metal appliance frame from the grounded circuit conductor is required by the *NEC*.

(2) Connections of Ranges and Clothes Dryers. Connections of ranges and clothes dryers with 120/240-volt, 3-wire ratings shall be made with 4-conductor cord and 3-pole, 4-wire, grounding-type plugs or by Type AC cable, Type MC cable, or conductors enclosed in flexible metal conduit.

(B) Equipment Grounding Means.

(1) Supply Cord or Permanent Feeder. The green-colored insulated grounding wire in the supply cord or permanent feeder wiring shall be connected to the grounding bus in the panelboard or disconnecting means.

(2) Electrical System. In the electrical system, all exposed metal parts, enclosures, frames, luminaire canopies, and so forth, shall be effectively bonded to the grounding terminal or enclosure of the panelboard.

(3) Cord-Connected Appliances. Cord-connected appliances, such as washing machines, clothes dryers, and refrigerators, and the electrical system of gas ranges and so forth, shall be grounded by means of a cord with an equipment grounding conductor and grounding-type attachment plug.

(C) Bonding of Non-Current-Carrying Metal Parts.

(1) Exposed Non-Current-Carrying Metal Parts. All exposed non-current-carrying metal parts that are likely to become energized shall be effectively bonded to the grounding terminal or enclosure of the panelboard. A bonding conductor shall be connected between the panelboard and an accessible terminal on the chassis. Chassis of multiple mobile home sections shall be bonded together with a solid copper, 8 AWG minimum, insulated or bare, bonding conductor with terminations in accordance with 250.8 and 250.12.

(2) Grounding Terminals. Grounding terminals shall be of the solderless type and listed as pressure-terminal connectors recognized for the wire size used. The bonding conductor shall be solid or stranded, insulated or bare, and shall be 8 AWG copper minimum, or equivalent. The bonding conductor shall be routed so as not to be exposed to physical damage.

(3) Metallic Piping and Ducts. Metallic gas, water, and waste pipes and metallic air-circulating ducts shall be considered bonded if they are connected to the terminal on the chassis [see 550.16(C)(1)] by clamps, solderless connectors, or by suitable grounding-type straps.

(4) Metallic Roof and Exterior Coverings. Any metallic roof and exterior covering shall be considered bonded if the following conditions are met:

(1) The metal panels overlap one another and are securely attached to the wood or metal frame parts by metallic fasteners.
(2) The lower panel of the metallic exterior covering is secured by metallic fasteners at a cross member of the chassis by two metal straps per mobile home unit or section at opposite ends.

The bonding strap material shall be a minimum of 100 mm (4 in.) in width of material equivalent to the skin or a material of equal or better electrical conductivity. The straps shall be fastened with paint-penetrating fittings such as screws and starwashers or equivalent.

550.17 Testing.

(A) Dielectric Strength Test. The wiring of each mobile home shall be subjected to a 1-minute, 900-volt, dielectric strength test (with all switches closed) between live parts (including neutral conductor) and the mobile home ground. Alternatively, the test shall be permitted to be performed at 1080 volts for 1 second. This test shall be performed after branch circuits are complete and after luminaires or appliances are installed.

Exception: Listed luminaires or appliances shall not be required to withstand the dielectric strength test.

(B) Continuity and Operational Tests and Polarity Checks. Each mobile home shall be subjected to all of the following:

(1) An electrical continuity test to ensure that all exposed electrically conductive parts are properly bonded

(2) An electrical operational test to demonstrate that all equipment, except water heaters and electric furnaces, is connected and in working order

(3) Electrical polarity checks of permanently wired equipment and receptacle outlets to determine that connections have been properly made

550.18 Calculations. The following method shall be employed in calculating the supply-cord and distribution-panelboard load for each feeder assembly for each mobile home in lieu of the procedure shown in Article 220 and shall be based on a 3-wire, 120/240-volt supply with 120-volt loads balanced between the two ungrounded conductors of the 3-wire system.

(A) Lighting, Small-Appliance, and Laundry Load.

(1) Lighting Volt-Amperes. Length times width of mobile home floor (outside dimensions) times 33 volt-amperes/m² (3 VA/ft²)— for example, length × width × 3 = lighting volt-amperes.

Δ **(2) Small-Appliance Volt-Amperes.** Number of circuits times 1500 volt-amperes for each 20-ampere appliance receptacle circuit — for example, number of circuits × 1500 = small-appliance volt-amperes.

(3) Laundry Area Circuit Volt-Amperes. 1500 volt-amperes.

(4) Total Volt-Amperes. Lighting volt-amperes plus small-appliance volt-amperes plus laundry area volt-amperes equals total volt-amperes.

(5) Net Volt-Amperes. First 3000 total volt-amperes at 100 percent plus remainder at 35 percent equals volt-amperes to be divided by 240 volts to obtain current (amperes) per leg.

Δ **(B) Total Load for Determining Power Supply.** Total load for determining power supply is the sum of the following:

(1) Lighting and small-appliance load as calculated in 550.18(A)(5).
(2) Nameplate amperes for motors and heater loads (exhaust fans, air conditioners, electric, gas, or oil heating). Omit smaller of the heating and cooling loads, except include blower motor if used as air-conditioner evaporator motor. Where an air conditioner is not installed and a 40-ampere power-supply cord is provided, allow 15 amperes per leg for air conditioning.
(3) Twenty-five percent of current of largest motor in Table 550.18(B).
(4) Total of nameplate amperes for waste disposer, dishwasher, water heater, clothes dryer, wall-mounted oven, cooking units. Where the number of these appliances exceeds three, use 75 percent of total.
(5) Derive amperes for freestanding range (as distinguished from separate ovens and cooking units) by dividing the following values by 240 volts as shown in Table 550.18(B).
(6) If outlets or circuits are provided for other than factory-installed appliances, include the anticipated load.

TABLE 550.18(B) *Freestanding Range Load*

Nameplate Rating (watts)	Use (volt-amperes)
0–10,000	80 percent of rating
Over 10,000–12,500	8,000
Over 12,500–13,500	8,400
Over 13,500–14,500	8,800
Over 14,500–15,500	9,200
Over 15,500–16,500	9,600
Over 16,500–17,500	10,000

Informational Note: See Informative Annex D, Example D11, for an illustration of the application of this calculation.

(C) Optional Method of Calculation for Lighting and Appliance Load. The optional method for calculating lighting and appliance load shown in 220.82 shall be permitted.

550.19 Interconnection of Multiple-Section Mobile or Manufactured Home Units.

(A) Wiring Methods. Approved and listed fixed-type wiring methods shall be used to join portions of a circuit that must be electrically joined and are located in adjacent sections after the home is installed on its support foundation. The circuit's junction shall be accessible for disassembly when the home is prepared for relocation.

Informational Note: See 550.15(K) for component interconnections.

(B) Disconnecting Means. Expandable or multiunit manufactured homes, not having permanently installed feeders, that are to be moved from one location to another shall be permitted to have disconnecting means with branch-circuit protective equipment in each unit when so located that after assembly or joining together of units, the requirements of 550.10 will be met.

550.20 Outdoor Outlets, Luminaires, Air-Cooling Equipment, and So Forth.

(A) Listed for Outdoor Use. Outdoor luminaires and equipment shall be listed for wet locations or outdoor use. Outdoor receptacles shall comply with 406.9. Where located on the underside of the home or located under roof extensions or similarly protected locations, outdoor luminaires and equipment shall be listed for use in damp locations.

(B) Outside Heating Equipment, Air-Conditioning Equipment, or Both. A mobile home provided with a branch circuit designed to energize outside heating equipment, air-conditioning equipment, or both, located outside the mobile home, other than room air conditioners, shall have such branch-circuit conductors terminate in a listed outlet box, or disconnecting means, located on the outside of the mobile home. A label shall be permanently affixed adjacent to the outlet box and shall contain the following information:

THIS CONNECTION IS FOR HEATING
AND/OR AIR-CONDITIONING EQUIPMENT.
THE BRANCH CIRCUIT IS RATED
AT NOT MORE THAN
_____ AMPERES, AT _____ VOLTS, 60 HERTZ,
_____ CONDUCTOR AMPACITY.
A DISCONNECTING MEANS SHALL BE LOCATED
WITHIN SIGHT OF THE EQUIPMENT.

The correct voltage and ampere rating shall be given. The tag shall be not less than 0.51 mm (0.020 in.) thick etched brass, stainless steel, anodized or alclad aluminum, or equivalent. The tag shall not be less than 75 mm by 45 mm (3 in. by 1¾ in.) minimum size.

550.25 Arc-Fault Circuit-Interrupter Protection. All 120-volt, single-phase, 15- and 20-ampere branch circuits supplying outlets or devices installed in mobile homes and manufactured homes shall comply with 210.12.

Arc-fault circuit-interrupter (AFCI) protection is required in accordance with 210.12. The branch circuits covered by 550.25 include those that fall within the voltage and current ratings specified and those that supply lighting outlets, receptacle outlets, smoke alarm outlets, and other power outlets. This requirement does not supersede the current HUD 24 CFR 3280 requirements for factory-installed wiring in manufactured homes.

Part III. Services and Feeders

550.30 Distribution System. The mobile home park secondary electrical distribution system to mobile home lots shall be single-phase, 120/240 volts, nominal.

The distribution systems at mobile home parks must supply 120/240 volts to the mobile home lot. Because appliances and other equipment are usually installed during the manufacturing process of mobile homes and are nominally rated 120/240 volts, the home is therefore intended to connect to a 120/240-volt, 3-wire ac, grounded neutral system.

550.31 Allowable Demand Factors. Park electrical wiring systems shall be calculated (at 120/240 volts) on the larger of the following:

(1) 16,000 volt-amperes for each mobile home lot
(2) The load calculated in accordance with 550.18 for the largest typical mobile home that each lot will accept

It shall be permissible to calculate the feeder or service load in accordance with Table 550.31. No demand factor shall be allowed for any other load, except as provided in this *Code*.

Mobile home park electrical wiring systems must be calculated based on the larger of (1) not less than 16,000 volt-amperes (at

TABLE 550.31 *Demand Factors for Services and Feeders*

Number of Mobile Homes	Demand Factor (%)
1	100
2	55
3	44
4	39
5	33
6	29
7–9	28
10–12	27
13–15	26
16–21	25
22–40	24
41–60	23
61 and over	22

120/240 volts) for each mobile home lot or (2) the calculated load of the largest typical mobile home the lot accommodates. However, the ampacity of the feeder-circuit conductors to each mobile home lot cannot be less than 100 amperes (at 120/240 volts), per 550.32(C).

550.32 Service Equipment.

Δ **(A) Mobile Home Service Equipment.** The mobile home service equipment shall not be mounted in or on the mobile home. The service equipment shall be rated not less than that required in accordance with 550.32(C), mounted in a readily accessible outdoor location, and within sight from the mobile home it serves. The mobile home service disconnect shall be permitted to be used as the emergency disconnect in accordance with 230.85.

Mobile home service equipment must be within sight from the mobile home it serves. The exterior mobile home service disconnect is also allowed to serve as the emergency disconnect required in 230.85.

(B) Manufactured Home Service Equipment. The manufactured home service equipment shall be permitted to be installed in or on a manufactured home, provided that all of the following conditions are met:

(1) The manufacturer shall include in its written installation instructions information indicating that the home shall be secured in place by an anchoring system or installed on and secured to a permanent foundation.
(2) The installation of the service shall comply with Part I through Part VII of Article 230.
(3) Means shall be provided for the connection of a grounding electrode conductor to the service equipment and routing it outside the structure.
(4) Bonding and grounding of the service shall be in accordance with Part I through Part V of Article 250.
(5) The manufacturer shall include in its written installation instructions one method of grounding the service

equipment at the installation site. The instructions shall clearly state that other methods of grounding are found in Article 250.

(6) The minimum size grounding electrode conductor shall be specified in the instructions.

(7) A warning label shall be mounted on or adjacent to the service equipment. The label shall meet the requirements in 110.21(B) and shall state the following:

> WARNING
> DO NOT PROVIDE ELECTRICAL POWER
> UNTIL THE GROUNDING ELECTRODE(S)
> IS INSTALLED AND CONNECTED
> (SEE INSTALLATION INSTRUCTIONS).

Where the service equipment is not installed in or on the unit, the installation shall comply with the other requirements of this section.

This section specifies the conditions required for installing the service equipment in or on a manufactured home. The concern over the unit being moved off site without the ability to disconnect the electrical supply is addressed in condition (1). A manufactured home with a service in or on the unit must be anchored in place or secured to a permanent foundation.

The other specified conditions cover the need to provide proper grounding and bonding conductors, systems, and connections and the need to install the service equipment in accordance with the applicable requirements in Article 230. These requirements apply only to *manufactured homes* as defined in Article 100.

(C) Rating. Mobile home service equipment shall be rated at not less than 100 amperes at 120/240 volts, and provisions shall be made for connecting a mobile home feeder assembly by a permanent wiring method. Power outlets used as mobile home service equipment shall also be permitted to contain receptacles rated up to 50 amperes with appropriate overcurrent protection. Fifty-ampere receptacles shall conform to the configuration shown in Figure 550.10(C).

> Informational Note: See ANSI/NEMA WD 6-2016, *Wiring Devices — Dimensional Specifications*, Figure 14-50, for complete details of the 50-ampere plug and receptacle configuration.

(D) Additional Outside Electrical Equipment. Means for connecting a mobile home accessory building or structure or additional electrical equipment located outside a mobile home by a fixed wiring method shall be provided in either the mobile home service equipment or the local external disconnecting means permitted in 550.32(A).

(E) Additional Receptacles. Receptacles located outside a mobile or manufactured home shall be provided with ground-fault circuit-interrupter protection as specified by 210.8(A). Where receptacles provide power to a mobile or manufactured home in accordance with 550.10, ground-fault circuit-interrupter protection shall not be required.

(F) Mounting Height. Outdoor mobile home disconnecting means shall be installed so the bottom of the enclosure containing the disconnecting means is not less than 600 mm (2 ft) above finished grade or working platform. The disconnecting means shall be installed so that the center of the grip of the operating handle, when in the highest position, is not more than 2.0 m (6 ft 7 in.) above the finished grade or working platform.

(G) Marking. Where a 125/250-volt receptacle is used in mobile home service equipment, the service equipment shall be marked as follows:

> TURN DISCONNECTING SWITCH OR
> CIRCUIT BREAKER OFF BEFORE INSERTING
> OR REMOVING PLUG. PLUG MUST BE FULLY
> INSERTED OR REMOVED.

The marking shall be located on the service equipment adjacent to the receptacle outlet.

550.33 Feeder.

N **(A) Feeder Equipment.** The feeder assembly, including the disconnecting means, shall not be mounted in or on the mobile home. A manufactured home feeder disconnecting means shall be permitted to be installed in or on the manufactured home in accordance with the requirements of 550.32(B). The feeder equipment shall be rated not less than that required in 550.32(C), mounted in a readily accessible outdoor location, and within sight from the mobile home or manufactured home it serves. Grounding of the disconnecting means shall be in accordance with 250.32.

(B) Feeder Conductors. Feeder conductors shall comply with the following:

(1) Feeder conductors shall consist of either a listed cord, factory installed in accordance with 550.10(B), or a permanently installed feeder consisting of four insulated, color-coded conductors that shall be identified by the factory or field marking of the conductors in compliance with 310.6. Equipment grounding conductors shall not be identified by stripping the insulation.

(2) Feeder conductors shall be installed in compliance with 250.32(B).

Exception: An existing feeder installed without an equipment grounding conductor shall be permitted to comply with 250.32(B)(1) Exception No. 1.

(C) Feeder Capacity. Mobile home and manufactured home feeder circuit conductors shall have a capacity not less than the loads supplied, shall have an ampacity of not less than 100 amperes, and shall be permitted to be sized in accordance with 310.12.

ARTICLE 551 Recreational Vehicles and Recreational Vehicle Parks

Part I. General

551.1 Scope. This article covers the electrical conductors and equipment other than low-voltage and automotive vehicle circuits or extensions thereof, installed within or on recreational vehicles, the conductors that connect recreational vehicles to a supply of electricity, and the installation of equipment and devices related to electrical installations within a recreational vehicle park.

> Informational Note: See NFPA 1192-2021, *Standard on Recreational Vehicles,* and ANSI/RVIA LV-2020, *Standard for Low Voltage Systems in Conversion and Recreational Vehicles*, for information on low-voltage systems.

Laws in many states require a factory inspection of recreational vehicles (RVs) by either a governmental or a private inspection agency. NFPA 1192, *Standard on Recreational Vehicles*, is widely accepted by the RV industry and AHJs who are responsible for ensuring that RVs are built to a recognized safety standard. Section 4.4 of NFPA 1192 requires compliance with Parts I through V of the *NEC®* and with ANSI/RVIA LV, *Low Voltage Systems in Conversion and Recreational Vehicles*, for the RV electrical systems rated 24 volts, nominal, or less.

N 551.3 Electrical Datum Plane Distances. The electrical datum plane distance(s) is determined by the normal high water level and encompasses the areas subject to tidal movement and areas in which the water level is affected by the conditions such as climate (rain or snowfall) or by human intervention (the opening and closing of dams or floodgates). The distance does not consider extremes due to natural or manmade disasters.

N (A) Areas Subject to Tidal Fluctuations. In land areas subject to tidal fluctuation, the electrical datum plane shall be a horizontal plane that is 606 mm (2 ft) above the highest high tide level for the area occurring under normal circumstances.

N (B) Areas Not Subject to Tidal Fluctuations. In land areas not subject to tidal fluctuation, the electrical datum plane shall be a horizontal plane that is 606 mm (2 ft) above the normal high water level for the area occurring under normal circumstances.

551.4 General Requirements.

(A) Not Covered. A recreational vehicle not used for the purposes as defined in 551.2 shall not be required to meet the requirements of Part IV pertaining to the number or capacity of circuits required. It shall, however, meet all other applicable requirements of this article if the recreational vehicle is provided with an electrical installation intended to be energized from a 120-volt, 208Y/120-volt, or 120/240-volt, nominal, ac power-supply system.

(B) Systems. This article covers combination electrical systems, generator installations, and 120-volt, 208Y/120-volt, or 120/240-volt, nominal, systems.

> Informational Note: See NFPA 1192-2021, *Standard on Recreational Vehicles*, and ANSI/RVIA 12V-2020, *Standard for Low Voltage Systems in Conversion and Recreational Vehicles*, for information on low-voltage systems.

Δ (C) Labels. Labels required by Article 551 shall be made of etched, metal-stamped, or embossed brass; stainless steel; plastic laminates not less than 0.13 mm (0.005 in.) thick; or anodized or alclad aluminum not less than 0.5 mm (0.020 in.) thick or the equivalent.

> Informational Note: See ANSI Z535.4-2011, *Product Safety Signs and Labels*, for guidance on other label criteria used in the recreational vehicle industry.

Part II. Combination Electrical Systems

551.20 Combination Electrical Systems.

(A) General. Vehicle wiring suitable for connection to a battery or dc supply source shall be permitted to be connected to a 120-volt source, provided the entire wiring system and equipment are rated and installed in full conformity with Parts I, II, III, IV, and V requirements of this article covering 120-volt electrical systems. Circuits fed from ac transformers shall not supply dc appliances.

(B) Voltage Converters (120-Volt Alternating Current to Low-Voltage Direct Current). The 120-volt ac side of the voltage converter shall be wired in full conformity with the requirements of Parts I, II, and IV of this article for 120-volt electrical systems.

Exception: Converters supplied as an integral part of a listed appliance shall not be subject to 551.20(B).

All converters and transformers shall be listed for use in recreational vehicles and designed or equipped to provide over-temperature protection. To determine the converter rating, the following percentages shall be applied to the total connected load, including average battery-charging rate, of all 12-volt equipment:

The first 20 amperes of load at 100 percent plus
The second 20 amperes of load at 50 percent plus
All load above 40 amperes at 25 percent

Exception: A low-voltage appliance that is controlled by a momentary switch (normally open) that has no means for holding in the closed position or refrigerators with a 120-volt function shall not be considered as a connected load when determining the required converter rating. Momentarily energized appliances shall be limited to those used to prepare the vehicle for occupancy or travel.

(C) Bonding Voltage Converter Enclosures. The non–current-carrying metal enclosure of the voltage converter shall be connected to the frame of the vehicle with a minimum 8 AWG copper conductor. The voltage converter shall be provided with a separate chassis bonding conductor that shall not be used as a current-carrying conductor.

This requirement reduces the possibility of damage to the power-supply cord by large dc fault currents that may find their way back to the vehicle frame or battery through the ac grounding conductor of the converter. Metal enclosures of listed converters are provided with an external pressure terminal connector for this purpose.

(D) Dual-Voltage Fixtures, Including Luminaires or Appliances. Fixtures, including luminaires, or appliances having both 120-volt and low-voltage connections shall be listed for dual voltage.

In the dual-voltage fixtures, barriers are used to separate the 120-volt and the 12-volt wiring connections.

(E) Autotransformers. Autotransformers shall not be used.

(F) Receptacles and Plug Caps. Where a recreational vehicle is equipped with an ac system, a low-voltage system, or both, receptacles and plug caps of the low-voltage system shall differ in configuration from those of the ac system. Where a vehicle equipped with a battery or other low-voltage system has an external connection for low-voltage power, the connector shall have a configuration that will not accept ac power.

Part III. Other Power Sources

551.30 Generator Installations.

(A) Mounting. Generators shall be mounted in such a manner as to be effectively bonded to the recreational vehicle chassis. The connection of the electrical system produced by the generator shall provide an effective ground-fault return path when operational.

(B) Generator Protection. Equipment shall be installed to ensure that the current-carrying conductors from the engine generator and from an outside source are not connected to a vehicle circuit at the same time. Automatic transfer switches in such applications shall be listed for use in one of the following:

(1) Emergency systems
(2) Optional standby systems

Receptacles used as disconnecting means shall be accessible (as applied to wiring methods) and capable of interrupting their rated current without hazard to the operator.

(C) Installation of Storage Batteries and Generators. Storage batteries and internal-combustion-driven generator units (subject to the provisions of this *Code*) shall be secured in place to avoid displacement from vibration and road shock.

(D) Ventilation of Generator Compartments. Compartments accommodating internal-combustion-driven generator units shall be provided with ventilation in accordance with instructions provided by the manufacturer of the generator unit.

> Informational Note: See NFPA 1192-2021, *Standard on Recreational Vehicles*, for generator compartment construction requirements.

(E) Supply Conductors. The supply conductors from the engine generator to the first termination on the vehicle shall be of the stranded type and be installed in listed flexible conduit or listed liquidtight flexible conduit. The point of first termination shall be in one of the following:

(1) Panelboard
(2) Junction box with a blank cover
(3) Junction box with a receptacle
(4) Enclosed transfer switch
(5) Receptacle assembly listed in conjunction with the generator

The panelboard, enclosed transfer switch, or junction box with a receptacle shall be installed within 450 mm (18 in.) of the point of entry of the supply conductors into the vehicle. A junction box with a blank cover shall be mounted on the compartment wall inside or outside the compartment; to any part of the generator-supporting structure (but not to the generator); to the vehicle floor on the outside of the vehicle; or within 450 mm (18 in.) of the point of entry of the supply conductors into the vehicle. A receptacle assembly listed in conjunction with the generator shall be mounted in accordance with its listing.

551.31 Multiple Supply Source.

(A) Multiple Supply Sources. Where a multiple supply system consisting of an alternate power source and a power-supply cord is installed, the feeder from the alternate power source shall be protected by an overcurrent protective device. Installation shall be in accordance with 551.30(A), 551.30(B), and 551.40.

(B) Multiple Supply Sources Capacity. The multiple supply sources shall not be required to be of the same capacity.

(C) Alternate Power Sources Exceeding 30 Amperes. If an alternate power source exceeds 30 amperes, 120 volts, nominal, it shall be permissible to wire it as a 120-volt, nominal, system, a 208Y/120-volt, nominal, system, or a 120/240-volt, nominal, system, provided an overcurrent protective device of the proper rating is installed in the feeder.

(D) Feeder Assembly Not Less Than 30 Amperes. The external feeder assembly shall be permitted to be less than the calculated load but not less than 30 amperes and shall have overcurrent protection not greater than the capacity of the external feeder assembly.

551.32 Other Sources.
Other sources of ac power, such as inverters, motor generators, or engine generators, shall be listed for use in recreational vehicles and shall be installed in accordance with the terms of the listing. Other sources of ac power shall be

wired in full conformity with the requirements in Parts I, II, III, IV, and V of this article covering 120-volt electrical systems.

551.33 Alternate Source Restrictions. Transfer equipment, if not integral with the listed power source, shall be installed to ensure that the current-carrying conductors from other sources of ac power and from an outside source are not connected to the vehicle circuit at the same time. Automatic transfer switches in such applications shall be listed for use in one of the following:

(1) Emergency systems
(2) Optional standby systems

Automatic transfer switches with relays might simultaneously connect multiple sources in the event of a relay failure. Therefore, the transfer switch must be listed for use in emergency or standby systems.

Part IV. Nominal 120-Volt or 120/240-Volt Systems

551.40 120-Volt or 120/240-Volt, Nominal, Systems.

Δ **(A) General Requirements.** The electrical equipment and material of recreational vehicles indicated for connection to a wiring system rated 120 volts, nominal, 2-wire with equipment grounding conductor, or a wiring system rated 120/240 volts, nominal, 3-wire with equipment grounding conductor, shall be listed and installed in accordance with Parts I, II, III, IV, and V of this article. Electrical equipment connected line-to-line shall have a nominal voltage rating of 208–230 volts.

(B) Materials and Equipment. Electrical materials, devices, appliances, fittings, and other equipment installed in, intended for use in, or attached to the recreational vehicle shall be listed. All products shall be used only in the manner in which they have been tested and found suitable for the intended use.

(C) Ground-Fault Circuit-Interrupter Protection. The internal wiring of a recreational vehicle having only one 15- or 20-ampere branch circuit as permitted in 551.42(A) and (B) shall have ground-fault circuit-interrupter protection for personnel. The ground-fault circuit interrupter shall be installed at the point where the power supply assembly terminates within the recreational vehicle. Where a separable cord set is not employed, the ground-fault circuit interrupter shall be permitted to be an integral part of the attachment plug of the feeder assembly. The ground-fault circuit interrupter shall provide protection also under the conditions of an open grounded circuit conductor, interchanged circuit conductors, or both.

N **(D) Loss of Ground Device.** Each recreational vehicle shall have a listed grounding monitor interrupter permanently installed between the feeder assembly connection to the vehicle and before either a transfer switch if installed or the panelboard. This requirement shall become effective January 1, 2026.

551.41 Receptacle Outlets Required.

(A) Spacing. Receptacle outlets shall be installed at wall spaces 600 mm (2 ft) wide or more so that no point along the floor line is more than 1.8 m (6 ft), measured horizontally, from an outlet in that space.

Exception No. 1: Bath and hallway areas shall not be required to comply with 551.41(A).

Exception No. 2: Wall spaces occupied by kitchen cabinets, wardrobe cabinets, built-in furniture, behind doors that can open fully against a wall surface, or similar facilities shall not be required to comply with 551.41(A).

Exception No. 3: Wall spaces in the special transportation area of a recreational vehicle designed to transport internal combustion vehicles shall not be required to comply with 551.41(A).

(B) Location. Receptacle outlets shall be installed as follows:

(1) Adjacent to countertops in the kitchen [at least one on each side of the sink if countertops are on each side and are 300 mm (12 in.) or over in width and depth].
(2) Adjacent to the refrigerator and gas range space, except where a gas-fired refrigerator or cooking appliance, requiring no external electrical connection, is factory installed.
(3) Adjacent to countertop spaces of 300 mm (12 in.) or more in width and depth that cannot be reached from a receptacle required in 551.41(B)(1) by a cord of 1.8 m (6 ft) without crossing a traffic area, cooking appliance, or sink.
(4) Rooftop decks that are accessible from inside the recreational vehicle shall have at least one receptacle installed within the perimeter of the rooftop deck. The receptacle shall not be located more than 1.2 m (4 ft) above the balcony, deck, or porch surface. The receptacle shall comply with the requirements of 406.9(B) for wet locations.
(5) A special transportation area of recreational vehicles designed to transport internal combustion engine vehicles shall have at least one receptacle installed.

(C) Ground-Fault Circuit-Interrupter Protection. Where provided, each 125-volt, single-phase, 15- or 20-ampere receptacle outlet shall have ground-fault circuit-interrupter protection for personnel in the following locations:

(1) Adjacent to a bathroom lavatory

The walls of an RV often do not provide the necessary depth for the installation of a GFCI receptacle. This requirement does not prohibit a bathroom receptacle from being mounted in the side of a lavatory cabinet.

(2) Where the receptacles are installed to serve the countertop surfaces and are within 1.8 m (6 ft) of any lavatory or sink

Exception No. 1: Receptacles installed for appliances in dedicated spaces, such as for disposals, refrigerators, and freezers shall not require ground-fault circuit-interrupter protection.

Exception No. 2: Single receptacles for interior connections of expandable room sections shall not require ground-fault circuit-interrupter protection.

Exception No. 3: De-energized receptacles that are within 1.8 m (6 ft) of any sink or lavatory due to the retraction of the expandable room section shall not require ground-fault circuit-interrupter protection.

(3) In the area occupied by a toilet, shower, tub, or any combination thereof
(4) On the exterior of the vehicle

Exception: Receptacles that are located inside of an access panel that is installed on the exterior of the vehicle to supply power for an installed appliance shall not be required to have ground-fault circuit-interrupter protection.

(5) In the special transportation area of a recreational vehicle that is separated from the living area by a wall

The receptacle outlet shall be permitted in a listed luminaire. A receptacle outlet shall not be installed in a tub or combination tub–shower compartment.

(D) Face-Up Position. A receptacle shall not be installed in a face-up position in any countertop or similar horizontal surface.

551.42 Branch Circuits Required. Each recreational vehicle containing an ac electrical system shall contain one of the circuit arrangements in 551.42(A) through (D).

(A) One 15-Ampere Circuit. One 15-ampere circuit to supply lights, receptacle outlets, and fixed appliances. Such recreational vehicles shall be equipped with one 15-ampere switch and fuse or one 15-ampere circuit breaker.

(B) One 20-Ampere Circuit. One 20-ampere circuit to supply lights, receptacle outlets, and fixed appliances. Such recreational vehicles shall be equipped with one 20-ampere switch and fuse or one 20-ampere circuit breaker.

(C) Two to Five 15- or 20-Ampere Circuits. Two to five 15- or 20-ampere circuits to supply lights, receptacle outlets, and fixed appliances shall be permitted. Such recreational vehicles shall be permitted to be equipped with panelboards rated 120 volts maximum or 120/240 volts maximum and listed for 30-ampere application supplied by the appropriate feeder assemblies. Not more than two 120-volt thermostatically controlled appliances shall be installed in such systems unless appliance isolation switching, energy management systems, or similar methods are used.

Exception No. 1: Additional 15- or 20-ampere circuits shall be permitted where a listed energy management system rated at 30-amperes maximum is employed within the system.

Exception No. 2: Six 15- or 20-ampere circuits shall be permitted without employing an energy management system, provided that the added sixth circuit serves only the power converter, and the combined load of all six circuits does not exceed the allowable load that was designed for use by the original five circuits.

> Informational Note: See 210.23(B) for permissible loads. See 551.45(C) for main disconnect and overcurrent protection requirements.

(D) More Than Five Circuits Without a Listed Energy Management System. A 50-ampere, 120/208-240-volt feeder assembly and a minimum 50-ampere-rated panelboard shall be used where six or more circuits are employed. The load distribution shall ensure a reasonable current balance between phases.

551.43 Branch-Circuit Protection.

(A) Rating. The branch-circuit overcurrent devices shall be rated as follows:

(1) Not more than the circuit conductors, and
(2) Not more than 150 percent of the rating of a single appliance rated 13.3 amperes or more and supplied by an individual branch circuit, but
(3) Not more than the overcurrent protection size marked on an air conditioner or other motor-operated appliances

(B) Protection for Smaller Conductors. A 20-ampere fuse or circuit breaker shall be permitted for protection for fixtures, including luminaires, leads, cords, or small appliances, and 14 AWG tap conductors, not over 1.8 m (6 ft) long for recessed luminaires.

(C) Fifteen-Ampere Receptacles Considered Protected by 20 Amperes. If more than one receptacle or load is on a branch circuit, 15-ampere receptacles shall be permitted to be protected by a 20-ampere fuse or circuit breaker.

551.44 Feeder Assembly. Each recreational vehicle shall have only one of the main feeder assemblies covered in 551.44(A) through (D).

(A) Fifteen-Ampere Main Feeder Assembly. Recreational vehicles wired in accordance with 551.42(A) shall use a listed 15-ampere or larger main feeder assembly.

(B) Twenty-Ampere Main Feeder Assembly. Recreational vehicles wired in accordance with 551.42(B) shall use a listed 20-ampere or larger main feeder assembly.

(C) Thirty-Ampere Main Feeder Assembly. Recreational vehicles wired in accordance with 551.42(C) shall use a listed 30-ampere or larger main feeder assembly.

(D) Fifty-Ampere Feeder Assembly. Recreational vehicles wired in accordance with 551.42(D) shall use a listed 50-ampere, 120/208-240-volt main feeder assembly.

551.45 Panelboard.

(A) Listed and Appropriately Rated. A listed and appropriately rated panelboard or other equipment specifically listed for this purpose shall be used. The grounded conductor termination bar shall be insulated from the enclosure as provided in 551.54(C). An equipment grounding terminal bar shall be attached inside the enclosure of the panelboard.

(B) Location. The panelboard shall be installed in a readily accessible location with the RV in the setup mode. Working clearance for the panelboard with the RV in the setup mode shall be not less than 600 mm (24 in.) wide and 750 mm (30 in.) deep.

Exception No. 1: Where the panelboard cover is exposed to the inside aisle space, one of the working clearance dimensions shall be permitted to be reduced to a minimum of 550 mm (22 in.). A panelboard is considered exposed where the panelboard cover is within 50 mm (2 in.) of the aisle's finished surface or not more than 25 mm (1 in.) from the backside of doors that enclose the space.

Exception No. 2: Compartment doors used for access to a generator shall be permitted to be equipped with a locking system.

Some RVs have expandable room sections (or slide-out rooms) that are in a stowed position when the vehicle travels from place to place. The specified working clearance in front of the panelboard located in an expanded section must be maintained only when the section is extended.

(C) Dead-Front Type. The panelboard shall be of the dead-front type and shall consist of one or more circuit breakers or Type S fuseholders. A main disconnecting means shall be provided where fuses are used or where more than two circuit breakers are employed. A main overcurrent protective device not exceeding the feeder assembly rating shall be provided where more than two branch circuits are employed.

551.46 Means for Connecting to Power Supply.

(A) Assembly. The feeder assembly or assemblies shall be factory supplied or factory installed and be of one of the types specified herein.

(1) Separable. Where a separable feeder assembly consisting of a cord with a female connector and molded attachment plug cap is provided, the vehicle shall be equipped with a permanently mounted, flanged surface inlet (male, recessed-type motor-base attachment plug) wired directly to the panelboard by an approved wiring method. The attachment plug cap shall be of a listed type.

(2) Permanently Connected. Each feeder assembly shall be connected directly to the terminals of the panelboard or conductors within a junction box and provided with means to prevent strain from being transmitted to the terminals. The ampacity of the conductors between each junction box and the terminals of each panelboard shall be at least equal to the ampacity of the power-supply cord. The supply end of the assembly shall be equipped with an attachment plug of the type described in 551.46(C). Where the cord passes through the walls or floors, it shall be protected by means of conduit and bushings or

equivalent. The cord assembly shall have permanent provisions for protection against corrosion and mechanical damage while the vehicle is in transit and while the cord assembly is being stored or removed for use.

(B) Cord. The cord exposed usable length shall be measured from the point of entrance to the recreational vehicle or the face of the flanged surface inlet (motor-base attachment plug) to the face of the attachment plug at the supply end.

The cord exposed usable length, measured to the point of entry on the vehicle exterior, shall be a minimum of 7.5 m (25 ft) where the point of entrance is at the side of the vehicle or shall be a minimum 9.0 m (30 ft) where the point of entrance is at the rear of the vehicle.

Where the cord entrance into the vehicle is more than 900 mm (3 ft) above the ground, the minimum cord lengths above shall be increased by the vertical distance of the cord entrance heights above 900 mm (3 ft).

Informational Note: See 551.46(E) for location of point of entrance of a power-supply assembly on the recreational vehicle exterior.

(C) Attachment Plugs.

(1) Units with One 15-Ampere Branch Circuit. Recreational vehicles having only one 15-ampere branch circuit as permitted by 551.42(A) shall have an attachment plug that shall be 2-pole, 3-wire grounding type, rated 15 amperes, 125 volts, conforming to the configuration shown in Figure 551.46(C)(1).

Informational Note: See ANSI/NEMA WD 6-2016, *Wiring Devices — Dimensional Specifications*, Figure 5.15, for complete details of this configuration.

Receptacles   **Caps**

125-V, 20-A, 2-pole, 3-wire, grounding type

125-V, 15-A, 2-pole, 3-wire, grounding type

20-A, 125-V, 2-pole, 3-wire, grounding type

30-A, 125-V, 2-pole, 3-wire, grounding type

50-A, 125/250-V, 3-pole, 4-wire, grounding type

FIGURE 551.46(C)(1) Configurations for Grounding-Type Receptacles and Attachment Plug Caps Used for Recreational Vehicle Supply Cords and Recreational Vehicle Lots.

(2) Units with One 20-Ampere Branch Circuit. Recreational vehicles having only one 20-ampere branch circuit as permitted in 551.42(B) shall have an attachment plug that shall be 2-pole, 3-wire grounding type, rated 20 amperes, 125 volts, conforming to the configuration shown in Figure 551.46(C)(1).

> Informational Note: See ANSI/NEMA WD 6-2016, *Wiring Devices — Dimensional Specifications*, Figure 5.20, for complete details of this configuration.

(3) Units with Two to Five 15- or 20-Ampere Branch Circuits. Recreational vehicles wired in accordance with 551.42(C) shall have an attachment plug that shall be 2-pole, 3-wire grounding type, rated 30 amperes, 125 volts, conforming to the configuration shown in Figure 551.46(C)(1) intended for use with units rated at 30 amperes, 125 volts.

> Informational Note: See ANSI/NEMA WD 6-2016, *Wiring Devices — Dimensional Specifications*, Figure TT, for complete details of this configuration.

The 30-ampere plug and receptacle configuration in Figure 551.46(C)(1) is not a standard 5-30P plug and 5-30R receptacle. The configuration is unique to RVs.

(4) Units with 50-Ampere Power-Supply Assembly. Recreational vehicles having a power-supply assembly rated 50 amperes as permitted by 551.42(D) shall have a 3-pole, 4-wire grounding-type attachment plug rated 50 amperes, 125/250 volts, conforming to the configuration shown in Figure 551.46(C)(1).

> Informational Note: See ANSI/NEMA WD 6-2016, *Wiring Devices — Dimensional Specifications*, Figure 14.50, for complete details of this configuration.

(D) Labeling at Electrical Entrance. Each recreational vehicle shall have a safety label with the signal word WARNING in minimum 6-mm (¼-in.) high letters and body text in minimum 3-mm (⅛-in.) high letters on a contrasting background. The safety label shall be affixed to the exterior skin, at or near the point of entrance of the power-supply cord(s), and shall read, using one of the following warnings, as appropriate:

WARNING
THIS CONNECTION IS FOR 110–125-VOLT AC,
60 HZ, _____ AMPERE SUPPLY.
DO NOT EXCEED CIRCUIT RATING.
EXCEEDING THE CIRCUIT RATING CAN CAUSE A
FIRE AND RESULT IN DEATH OR SERIOUS INJURY.

or

WARNING
THIS CONNECTION IS FOR 208Y/120-VOLT or
120/240-VOLT AC, 3-POLE, 4-WIRE, 60 HZ, _____
AMPERE SUPPLY.
DO NOT EXCEED CIRCUIT RATING.
EXCEEDING THE CIRCUIT RATING CAN CAUSE A
FIRE AND RESULT IN DEATH OR SERIOUS INJURY.

The correct ampere rating shall be marked in the blank space.

(E) Location. The point of entrance of a power-supply assembly shall be located within 4.5 m (15 ft) of the rear, on the left (road) side or at the rear, left of the longitudinal center of the vehicle, within 450 mm (18 in.) of the outside wall.

Exception No. 1: A recreational vehicle equipped with only a listed flexible drain system or a side-vent drain system shall be permitted to have the electrical point of entrance located on either side, provided the drain(s) for the plumbing system is (are) located on the same side.

Exception No. 2: A recreational vehicle shall be permitted to have the electrical point of entrance located more than 4.5 m (15 ft) from the rear. Where this occurs, the distance beyond the 4.5-m (15-ft) dimension shall be added to the cord's minimum length as specified in 551.46(B).

Exception No. 3: Recreational vehicles designed for transporting livestock shall be permitted to have the electrical point of entrance located on either side or the front.

551.47 Wiring Methods.

(A) Wiring Systems. Cables and raceways installed in accordance with Articles 320, 322, 330 through 340, 342 through 362, 386, and 388 shall be permitted in accordance with their applicable article, except as otherwise specified in this article. An equipment grounding means shall be provided in accordance with 250.118.

> **See also**
>
> **348.60** and its commentary for information regarding the use of flexible metal conduit as an equipment grounding conductor (EGC)

(B) Conduit and Tubing. Where rigid metal conduit or intermediate metal conduit is terminated at an enclosure with a locknut and bushing connection, two locknuts shall be provided, one inside and one outside of the enclosure. All cut ends of conduit and tubing shall be reamed or otherwise finished to remove rough edges.

> **See also**
>
> **344.28, 358.28(A), 300.4(G),** and associated commentary for the protection of conductor insulation against abrasion at conduit and tubing terminations

(C) Nonmetallic Boxes. Nonmetallic boxes shall be acceptable only with nonmetallic-sheathed cable or nonmetallic raceways.

(D) Boxes. In walls and ceilings constructed of wood or other combustible material, boxes and fittings shall be flush with the finished surface or project therefrom.

Δ **(E) Mounting.** Wall and ceiling boxes shall be mounted in accordance with 314.23.

Exception No. 1: Snap-in-type boxes or boxes provided with special wall or ceiling brackets that securely fasten boxes in walls or ceilings shall be permitted.

Exception No. 2: A wooden plate providing a 38-mm (1½-in.) minimum width backing around the box and of a thickness of 13 mm (½ in.) or greater (actual) attached directly to the wall panel shall be considered as approved means for mounting outlet boxes.

Exception No. 2 permits the mounting of outlet boxes by screws to a wooden plate that is secured directly to the back of the wall panel. The wooden plate must extend at least 1½ inches around the box. The exception recognizes the special construction of RV walls, which often makes it difficult or impossible to attach an outlet box to a structural member, as required by 314.23(B).

(F) Raceway and Cable Continuity. Raceways and cable sheaths shall be continuous between boxes and other enclosures.

(G) Protected. Metal-clad, Type AC, or nonmetallic-sheathed cables and electrical nonmetallic tubing shall be permitted to pass through the centers of the wide side of 2 by 4 wood studs. However, they shall be protected where they pass through 2 by 2 wood studs or at other wood studs or frames where the cable or tubing would be less than 32 mm (1¼ in.) from the inside or outside surface. Steel plates on each side of the cable or tubing or a steel tube, with not less than 1.35 mm (0.053 in.) wall thickness, shall be installed to protect the cable or tubing. These plates or tubes shall be securely held in place. Where nonmetallic-sheathed cables pass through punched, cut, or drilled slots or holes in metal members, the cable shall be protected by bushings or grommets securely fastened in the opening prior to installation of the cable.

(H) Bends. No bend shall have a radius of less than five times the cable diameter.

(I) Cable Supports. Where connected with cable connectors or clamps, cables shall be secured and supported within 300 mm (12 in.) of outlet boxes, panelboards, and splice boxes on appliances. Supports and securing shall be provided at intervals not exceeding 1.4 m (4½ ft) at other places.

(J) Nonmetallic Box Without Cable Clamps. Nonmetallic-sheathed cables shall be secured and supported within 200 mm (8 in.) of a nonmetallic outlet box without cable clamps. Where wiring devices with integral enclosures are employed with a loop of extra cable to permit future replacement of the device, the cable loop shall be considered as an integral portion of the device.

(K) Physical Damage. Where subject to physical damage, exposed nonmetallic cable shall be protected by covering boards, guard strips, raceways, or other means.

(L) Receptacle Faceplates. Metal faceplates shall comply with 406.6(A). Nonmetallic faceplates shall comply with 406.6(C).

(M) Metal Faceplates Grounded. Metal faceplates shall be installed in compliance with 404.9(B) and 404.6(B).

(N) Moisture or Physical Damage. Wiring shall be protected in accordance with the following:

(1) Where outdoor or under-chassis line-voltage (120 volts, nominal, or higher) wiring is exposed, it shall be protected by a conduit or raceway identified for use in wet locations. The conductors shall be listed for use in wet locations.
(2) Where wiring is exposed to physical damage, it shall be protected by a raceway.

(O) Component Interconnections. Fittings and connectors that are intended to be concealed at the time of assembly shall be listed and identified for the interconnection of building components. Such fittings and connectors shall be equal to the wiring method employed in insulation, temperature rise, and fault-current withstanding and shall be capable of enduring the vibration and shock occurring in recreational vehicles.

(P) Method of Connecting Expandable Units. The method of connecting expandable units to the main body of the vehicle shall comply with 551.47(P)(1) or (P)(2).

(1) Cord-and-Plug-Connected. Cord-and-plug connections shall comply with 551.47(P)(1)(a) through (P)(1)(d).

(a) That portion of a branch circuit that is installed in an expandable unit shall be permitted to be connected to the portion of the branch circuit in the main body of the vehicle by means of an attachment plug and cord listed for hard usage. The cord and its connections shall comply with Part I and Part II, as applicable, of Article 400 and shall be considered as a permitted use under 400.10. Where the attachment plug and cord are located within the vehicle's interior, use of plastic thermoset or elastomer parallel cord Type SPT-3, SP-3, or SPE shall be permitted.

(b) Where the receptacle provided for connection of the cord to the main circuit is located on the outside of the vehicle, it shall be protected with a ground-fault circuit interrupter for personnel and be listed for wet locations. A cord located on the outside of a vehicle shall be identified for outdoor use.

(c) Unless removable or stored within the vehicle interior, the cord assembly shall have permanent provisions for protection against corrosion and mechanical damage while the vehicle is in transit.

(d) The attachment plug and cord shall be installed so as not to permit exposed live attachment plug pins.

(2) Direct Wired. That portion of a branch circuit that is installed in an expandable unit shall be permitted to be connected to the portion of the branch circuit in the main body of the vehicle by means of flexible cord installed in accordance with 551.47(P)(2)(a) through (P)(2)(e) or other approved wiring method.

(a) The flexible cord shall be listed for hard usage and for use in wet locations.

(b) The flexible cord shall be permitted to be exposed on the underside of the vehicle.

(c) The flexible cord shall be permitted to pass through the interior of a wall or floor assembly or both a maximum concealed length of 600 mm (24 in.) before terminating at an outlet or junction box.

(d) Where concealed, the flexible cord shall be installed in nonflexible conduit or tubing that is continuous from the outlet or junction box inside the recreational vehicle to a weatherproof outlet box, junction box, or strain relief fitting listed for use in wet locations that is located on the underside of the recreational vehicle. The outer jacket of the flexible cord shall be continuous into the outlet or junction box.

(e) Where the flexible cord passes through the floor to an exposed area inside of the recreational vehicle, it shall be protected by means of conduit and bushings or equivalent.

Where subject to physical damage, the flexible cord shall be protected with RMC, IMC, Schedule 80 PVC, reinforced thermosetting resin conduit (RTRC) listed for exposure to physical damage, or other approved means and shall extend at least 150 mm (6 in.) above the floor. A means shall be provided to secure the flexible cord where it enters the recreational vehicle.

Section 551.47(P) covers the interconnection between the main body of a vehicle and an expandable unit. Two methods of interconnection are permitted—one by means of cord-and-plug connections with the cord listed for hard usage and the other by means of a direct-wired connection, using flexible cord with the outer jacket intact, installed in nonflexible conduit or tubing. Where the cord could be subject to physical damage, rigid metal conduit (RMC), intermediate metal conduit (IMC), Schedule 80 polyvinyl chloride (PVC), reinforced thermosetting resin conduit (RTRC), or other approved means must be used to protect the cord.

(Q) Prewiring for Air-Conditioning Installation. Prewiring installed for the purpose of facilitating future air-conditioning installation shall comply with the applicable portions of this article and the following:

(1) An overcurrent protective device with a rating compatible with the circuit conductors shall be installed in the panelboard and wiring connections completed.

(2) The load end of the circuit shall terminate in a junction box with a blank cover or other listed enclosure. Where a junction box with a blank cover is used, the free ends of the conductors shall be adequately capped or taped.

(3) A safety label with the signal word WARNING in minimum 6-mm (¼-in.) high letters and body text in minimum 3-mm (⅛-in.) high letters on a contrasting background shall be affixed on or adjacent to the junction box and shall read as follows:

WARNING
AIR-CONDITIONING CIRCUIT.
THIS CONNECTION IS FOR AIR CONDITIONERS
RATED 110–125-VOLT AC, 60 HZ,
___ AMPERES MAXIMUM.
DO NOT EXCEED CIRCUIT RATING.
EXCEEDING THE CIRCUIT RATING MAY
CAUSE A FIRE AND RESULT IN DEATH
OR SERIOUS INJURY.

An ampere rating, not to exceed 80 percent of the circuit rating, shall be legibly marked in the blank space.

(4) The circuit shall serve no other purpose.

Δ **(R) Prewiring for Generator Installation.** Prewiring installed for the purpose of facilitating future generator installation shall comply with the other applicable portions of this article and the following:

(1) Circuit conductors shall be appropriately sized in relation to the anticipated load as stated on the label required in 551.47(R)(4).

(2) Where junction boxes are utilized at either of the circuit originating or terminus points, free ends of the conductors shall be adequately capped or taped.

(3) Where devices such as receptacle outlet, transfer switch, and so forth, are installed, the installation shall be complete, including circuit conductor connections.

(4) A safety label with the signal word WARNING in minimum 6 mm (¼ in.) high letters and body text in minimum 3 mm (⅛ in.) high letters on a contrasting background shall be affixed on the cover of each junction box or transfer switch containing incomplete circuitry and shall read, using one of the following warnings, as appropriate:

WARNING
GENERATOR
ONLY INSTALL A GENERATOR LISTED
SPECIFICALLY FOR RV USE
HAVING OVERCURRENT PROTECTION
RATED 110–125-VOLT AC,
60 HZ, _____ AMPERES MAXIMUM.

or

WARNING
GENERATOR
ONLY INSTALL A GENERATOR LISTED
SPECIFICALLY FOR RV USE
HAVING OVERCURRENT PROTECTION
RATED 120–240-VOLT AC,
60 HZ, _____ AMPERES MAXIMUM.

The correct ampere rating shall be legibly marked in the blank space.

Δ **(S) Prewiring for Other Circuits.** Prewiring installed for the purpose of installing other appliances or devices shall comply with the applicable portions of this article and the following:

(1) An overcurrent protection device with a rating compatible with the circuit conductors shall be installed in the panelboard with wiring connections completed.

(2) The load end of the circuit shall terminate in a junction box with a blank cover or a device listed for the purpose. Where a junction box with blank cover is used, the free ends of the conductors shall be adequately capped or taped.

(3) A safety label with the signal word WARNING in minimum 6-mm (¼-in.) high letters and body text in minimum 3-mm (⅛-in.) high letters on a contrasting background shall be affixed on or adjacent to the junction box or device listed for the purpose and shall read as follows:

WARNING
THIS CONNECTION IS FOR _____ RATED _____ VOLT AC, 60 HZ, _____ AMPERES MAXIMUM. DO NOT EXCEED CIRCUIT RATING. EXCEEDING THE CIRCUIT RATING MAY CAUSE A FIRE AND RESULT IN DEATH OR SERIOUS INJURY.

An ampere rating not to exceed 80 percent of the circuit rating shall be legibly marked in the blank space.

551.48 Conductors and Boxes. The maximum number of conductors permitted in boxes shall be in accordance with 314.16.

551.49 Grounded Conductors. The identification of grounded conductors shall be in accordance with 200.6.

551.50 Connection of Terminals and Splices. Conductor splices and connections at terminals shall be in accordance with 110.14.

551.51 Switches.

(A) Rating. Switches shall be rated in accordance with 551.51(A)(1) and (A)(2).

(1) Lighting Circuits. For lighting circuits, switches shall be rated not less than 10 amperes, 120–125 volts and in no case less than the connected load.

(2) Motors or Other Loads. Switches for motors or other loads shall comply with 404.14.

(B) Location. Switches shall not be installed within wet locations in tub or shower spaces unless installed as part of a listed tub or shower assembly.

551.52 Receptacles. All receptacle outlets shall be of the grounding type and installed in accordance with 406.4 and 210.21.

551.53 Luminaires and Other Equipment.

(A) General. Any combustible wall or ceiling finish exposed between the edge of a canopy or pan of a luminaire or ceiling-suspended (paddle) fan and the outlet box shall be covered with noncombustible material.

(B) Shower Luminaires. If a luminaire is provided over a bathtub or in a shower stall, it shall be of the enclosed and gasketed type and listed for the type of installation, and it shall be ground-fault circuit-interrupter protected.

Due to the low ceilings in RVs, luminaires installed above a tub or shower enclosure could be easily reached by most persons standing in the enclosure. Accordingly, only luminaires that are listed for wet locations and have GFCI protection are permitted to be installed above a tub or shower enclosure.

(C) Outdoor Outlets, Luminaires, Air-Cooling Equipment, and So On. Outdoor luminaires and other equipment shall be listed for outdoor use.

551.54 Grounding. (See also 551.56 on bonding of non–current-carrying metal parts.)

(A) Power-Supply Grounding. The equipment grounding conductor in the supply cord or feeder shall be connected to the equipment grounding bus or other approved equipment grounding means in the panelboard.

(B) Panelboard. The panelboard shall have an equipment grounding bus with terminals for all equipment grounding conductors or other approved equipment grounding means.

Δ **(C) Insulated Neutral Conductor.** The neutral conductor shall be insulated from the equipment grounding conductors and from equipment enclosures and other grounded parts. The neutral conductor terminals in the panelboard and in ranges, clothes dryers, counter-mounted cooking units, and wall-mounted ovens shall be insulated from the equipment enclosure. Bonding screws, straps, or buses in the panelboard or in appliances shall be removed and discarded. Connection of electric ranges and electric clothes dryers utilizing a grounded conductor, if cord-connected, shall be made with 4-conductor cord and 3-pole, 4-wire grounding-type plug caps and receptacles.

551.55 Interior Equipment Grounding.

(A) Exposed Metal Parts. In the electrical system, all exposed metal parts, enclosures, frames, luminaire canopies, and so forth, shall be effectively bonded to the grounding terminals or enclosure of the panelboard.

(B) Equipment Grounding and Bonding Conductors. Bare wires, insulated wire with an outer finish that is green or green with one or more yellow stripes, shall be used for equipment grounding or bonding conductors only.

(C) Grounding of Electrical Equipment. Grounding of electrical equipment shall be accomplished by one or more of the following methods:

(1) Connection of metal raceway, the sheath of Type MC and Type MI cable where the sheath is identified for grounding, or the armor of Type AC cable to metal enclosures.

(2) A connection between the one or more equipment grounding conductors and a metal enclosure by means of a grounding screw, which shall be used for no other purpose, or a listed grounding device.

(3) The equipment grounding conductor in nonmetallic-sheathed cable shall be permitted to be secured under a

screw threaded into the luminaire canopy other than a mounting screw or cover screw, or attached to a listed grounding means (plate) in a nonmetallic outlet box for luminaire mounting. [Grounding means shall also be permitted for luminaire attachment screws.]

(D) Grounding Connection in Nonmetallic Box. A connection between the one or more equipment grounding conductors brought into a nonmetallic outlet box shall be so arranged that a connection of the equipment grounding conductor can be made to any fitting or device in that box that requires grounding.

(E) Grounding Continuity. Where more than one equipment grounding or bonding conductor of a branch circuit enters a box, all such conductors shall be connected together using a method specified in 250.8, and the arrangement shall be such that the disconnection or removal of a receptacle, luminaire, or other device fed from the box will not interfere with or interrupt the grounding continuity.

(F) Cord-Connected Appliances. Cord-connected appliances, such as washing machines, clothes dryers, refrigerators, and the electrical system of gas ranges, and so forth, shall be grounded by means of an approved cord with equipment grounding conductor and grounding-type attachment plug.

551.56 Bonding of Non–Current-Carrying Metal Parts.

(A) Required Bonding. All exposed non–current-carrying metal parts that are likely to become energized shall be effectively bonded to the grounding terminal or enclosure of the panelboard.

(B) Bonding Chassis. A bonding conductor shall be connected between any panelboard and an accessible terminal on the chassis. Bonding terminations shall be suitable for the environment in which the conductors and terminations are installed.

Exception: Any recreational vehicle that employs a unitized metal chassis-frame construction to which the panelboard is securely fastened with a bolt(s) and nut(s) or by welding or riveting shall be considered to be bonded.

(C) Bonding Conductor Requirements. Grounding terminals shall be of the solderless type and listed as pressure terminal connectors recognized for the wire size used. The bonding conductor shall be solid or stranded, insulated or bare, and shall be 8 AWG copper minimum, or equal.

(D) Metallic Roof and Exterior Bonding. The metal roof and exterior covering shall be considered bonded where both of the following conditions apply:

(1) The metal panels overlap one another and are securely attached to the wood or metal frame parts by metal fasteners.
(2) The lower panel of the metal exterior covering is secured by metal fasteners at each cross member of the chassis, or the lower panel is connected to the chassis by a metal strap.

(E) Gas, Water, and Waste Pipe Bonding. The gas, water, and waste pipes shall be considered grounded if they are bonded to the chassis.

(F) Furnace and Metal Air Duct Bonding. Furnace and metal circulating air ducts shall be bonded.

551.57 Appliance Accessibility and Fastening. Every appliance shall be accessible for inspection, service, repair, and replacement without removal of permanent construction. Means shall be provided to securely fasten appliances in place when the recreational vehicle is in transit.

Part V. Factory Tests

551.60 Factory Tests (Electrical). Each recreational vehicle designed with a 120-volt or a 120/240-volt electrical system shall withstand the applied voltage without electrical breakdown of a 1-minute, 900-volt ac or 1280-volt dc dielectric strength test, or a 1-second, 1080-volt ac or 1530-volt dc dielectric strength test, with all switches closed, between ungrounded and grounded conductors and the recreational vehicle ground. During the test, all switches and other controls shall be in the "on" position. Fixtures, including luminaires and permanently installed appliances, shall not be required to withstand this test. The test shall be performed after branch circuits are complete prior to energizing the system and again after all outer coverings and cabinetry have been secured. The dielectric test shall be performed in accordance with the test equipment manufacturer's written instructions.

Each recreational vehicle shall be subjected to all of the following:

(1) A continuity test to ensure that all metal parts are properly bonded
(2) Operational tests to demonstrate that all equipment is properly connected and in working order
(3) Polarity checks to determine that connections have been properly made
(4) GFCI test to demonstrate that the ground fault protection device(s) installed on the recreational vehicle are operating properly

Part VI. Recreational Vehicle Parks

551.71 Type Receptacles Provided.

(A) 20-Ampere. Every recreational vehicle site with electrical supply shall be equipped with recreational vehicle site supply equipment with at least one 20-ampere, 125-volt weather-resistant receptacle. This receptacle, when used in recreational vehicle site electrical equipment, shall not be required to be tamper-resistant in accordance with 406.12.

(B) 30-Ampere. A minimum of 70 percent of all recreational vehicle sites with electrical supply shall each be equipped with a 30-ampere, 125-volt weather-resistant receptacle conforming to Figure 551.46(C)(1). This supply shall be permitted to include additional receptacle configurations conforming to 551.81. The remainder of all recreational vehicle sites with electrical supply shall be equipped with one or more of the receptacle configurations conforming to 551.81.

Δ **(C) 50-Ampere.** A minimum of 20 percent of existing and 40 percent of all new recreational vehicle sites with electrical supply, shall each be equipped with a 50-ampere, 125/250-volt weather-resistant receptacle conforming to the configuration as identified in Figure 551.46(C)(1). Every recreational vehicle site equipped with a 50-ampere receptacle shall also be equipped with a 30-ampere, 125-volt receptacle conforming to Figure 551.46(C)(1). These electrical supplies shall be permitted to include additional receptacles that have configurations in accordance with 551.81. The weather-resistant requirement for 50-ampere, 125/250-volt receptacles shall become effective January 1, 2026.

> Informational Note: The percentage of 50 ampere sites required by 551.71 could be inadequate for seasonal recreational vehicle sites serving a higher percentage of recreational vehicles with 50-ampere electrical systems. In that type of recreational vehicle park, the percentage of 50-ampere sites could approach 100 percent.

At least one 20-ampere, 125-volt receptacle must be installed at each RV campsite. Many RVs require a 30-ampere connection, and 70 percent of sites must also provide a 30-ampere receptacle.

Some RVs have a 50-ampere, 120/240-volt supply installed, and 20 percent of existing and 40 percent of new RV sites must be provided with a 50-ampere receptacle to accommodate the larger electrical system. This receptacle is in addition to the 20- and 30-ampere receptacles required for the site. This requirement increases the load capacity for RV park services and feeders.

See also

Figure 551.46(C)(1), which shows receptacle configurations
551.81, which provides receptacle ratings

(D) Tent Sites. Dedicated tent sites with a 15- or 20-ampere electrical supply shall be permitted to be excluded when determining the percentage of recreational vehicle sites with 30- or 50-ampere receptacles.

(E) Additional Receptacles. Additional receptacles shall be permitted for the connection of electrical equipment outside the recreational vehicle within the recreational vehicle park.

Δ **(F) GFCI Protection.**

N **(1) Receptacles Installed in Other Than Recreational Vehicle Site Equipment.** Ground-fault circuit-interrupter protection shall be provided as required in 210.8(B).

N **(2) Receptacles Installed in Recreational Vehicle Site Equipment.** Ground-fault circuit-interrupter protection shall

only be required for 125-volt, single-phase, 15- and 20-ampere receptacles.

> Informational Note No. 1: Appliances used within the recreational vehicle can create leakage current levels at the supply receptacle(s) that could exceed the limits of a Class A GFCI device.
> Informational Note No. 2: The definition of *Feeder Assembly* clarifies that the power supply cord to a recreational vehicle is considered a feeder.

551.72 Distribution System.

Δ **(A) Systems.** Distribution systems shall provide the voltage and have a capacity for the receptacles provided in the recreational vehicle (RV) site supply equipment as calculated according to 551.73 and shall have an ampacity not less than 30 amperes. Systems permitted include single-phase 120 volts, single-phase 120/240 volts, or single-phase 120/208 volts — two ungrounded and one neutral conductor taken from a 208Y/120-volt system.

(B) Three-Phase Systems. Feeders from 208Y/120-volt, 3-phase systems shall be permitted to include two ungrounded conductors and shall include one grounded conductor and one equipment grounding conductor. So far as practicable, the loads shall be equally distributed on the 3-phase system.

Δ **(C) Receptacles.** Receptacles rated at 50 amperes shall be supplied from a circuit of the voltage class and rating of the receptacle. Other recreational vehicle sites with 125-volt, 20- and 30-ampere receptacles shall be permitted to be derived from any grounded distribution system that supplies 120-volt, single-phase power.

Δ **(D) Neutral Conductors.** Neutral conductors shall be permitted to be reduced in size below the minimum required size of the ungrounded conductors for 240-volt, line-to-line, permanently connected loads only. The neutral conductors shall not be reduced in size below the size of the ungrounded conductors for the site distribution.

> Informational Note: Due to the long circuit lengths typical in most recreational vehicle parks, feeder conductor sizes found in the ampacity tables of Article 310 could be inadequate to maintain the voltage regulation suggested in 215.2(A), Informational Note No. 2. Total circuit voltage drop is a sum of the voltage drops of each serial circuit segment, where the load for each segment is calculated using the load that segment sees and the demand factors shown in Table 551.73(A).

Δ **(E) Connected Devices.** The use of listed surge protective devices shall be permitted.

> Informational Note: Use of multiple autotransformers on the load side of RV pedestals, supplied by a single feeder, can result in increased current on the RV park or campground distribution system.

(F) Connection to Recreational Vehicle Site Equipment. Each recreational vehicle shall be powered by only one 30-ampere or one 50-ampere external power supply cord.

Informational Note: The requirement in 551.72(F) does not preclude the use of the 15- or 20-ampere receptacle convenience outlet on the recreational vehicle supply equipment.

551.73 Calculated Load.

Δ **(A) Basis of Calculations.** Electrical services and feeders shall be calculated on the basis of not less than all of the following:

(1) 12,000 volt-amperes per site equipped with 50-ampere, 208Y/120-volt or 120/240-volt supply facilities
(2) 3600 volt-amperes per site equipped with both 20-ampere and 30-ampere supply facilities
(3) 2400 volt-amperes per site equipped with only 20-ampere supply facilities
(4) 600 volt-amperes per site equipped with only 20-ampere supply facilities that are dedicated to tent sites

The demand factors set forth in Table 551.73(A) shall be the minimum allowable demand factors that shall be permitted in calculating load for service and feeders. Where the electrical supply for a recreational vehicle site has more than one receptacle.

Where the electrical supply is in a location that serves two recreational vehicles, the equipment for both sites shall comply with 551.77, and the calculated load shall only be calculated for the two receptacles with the highest rating.

TABLE 551.73(A) *Demand Factors for Site Feeders and Service-Entrance Conductors for Park Sites*

Number of Recreational Vehicle Sites	Demand Factor (%)
1	100
2	90
3	80
4	75
5	65
6	60
7–9	55
10–12	50
13–15	48
16–18	47
19–21	45
22–24	43
25–35	42
36 plus	41

(B) Demand Factors. The demand factor for a given number of sites shall apply to all sites indicated. For example, 20 sites calculated at 45 percent of 3600 volt-amperes results in a permissible demand of 1620 volt-amperes per site or a total of 32,400 volt-amperes for 20 sites.

Informational Note: These demand factors may be inadequate in areas of extreme hot or cold temperature with loaded circuits for heating or air conditioning.

Loads for other amenities such as, but not limited to, service buildings, recreational buildings, and swimming pools shall be calculated separately and then be added to the value calculated for the recreational vehicle sites where they are all supplied by a common service.

551.74 Overcurrent Protection.
Overcurrent protection shall be provided in accordance with Article 240.

551.76 Grounding — Recreational Vehicle Site Supply Equipment.

(A) Grounding Electrode. Recreational vehicle site supply equipment, other than those used as service equipment, shall not be required to have a grounding electrode. An auxiliary grounding electrode(s) in accordance with 250.54 shall be permitted to be installed.

(B) Exposed Non–Current-Carrying Metal Parts. Exposed non–current-carrying metal parts of fixed equipment, metal boxes, cabinets, and fittings that are not electrically connected to grounded equipment shall be grounded by an equipment grounding conductor run with the circuit conductors from the service equipment or from the transformer of a secondary distribution system. Equipment grounding conductors shall be sized in accordance with 250.122 and shall be permitted to be spliced by listed means.

The arrangement of equipment grounding connections shall be such that the disconnection or removal of a receptacle or other device will not interfere with, or interrupt, the grounding continuity.

(C) Secondary Distribution System. Each secondary distribution system shall be grounded at the transformer.

(D) Grounded Conductor Not to Be Used as an Equipment Ground. The grounded conductor shall not be used as an equipment grounding conductor for recreational vehicles or equipment within the recreational vehicle park.

(E) No Connection on the Load Side. No connection to a grounding electrode shall be made to the grounded conductor on the load side of the service disconnecting means except as covered in 250.30(A) for separately derived systems, and 250.32(B), Exception No. 1 for separate buildings.

551.77 Recreational Vehicle Site Supply Equipment.
Recreational vehicle site supply equipment shall be listed for use as recreational vehicle site supply equipment and shall comply with 551.77(A) through (F).

(A) Location. Where provided on back-in sites, the recreational vehicle site electrical supply equipment shall be located on the left (road) side of the parked vehicle, on a line that is 1.5 m to 2.1 m (5 ft to 7 ft) from the left edge (driver's side of the parked RV) of the recreational vehicle stand and shall be located at any point on this line from the rear of the recreational vehicle stand to 4.5 m (15 ft) forward of the rear of the recreational vehicle stand.

For pull-through sites, the electrical supply equipment shall be permitted to be located at any point along the line that is 1.5 m to 2.1 m (5 ft to 7 ft) from the left edge (driver's side of the parked RV) from 4.9 m (16 ft) forward of the rear of the recreational vehicle stand to the center point between the two roads that gives access to and egress from the pull-through sites.

The left edge (driver's side of the parked RV) of the recreational vehicle stand shall be marked.

(B) Disconnecting Means. A disconnecting switch or circuit breaker shall be provided in the site supply equipment for disconnecting the power supply to the recreational vehicle.

(C) Access. All site supply equipment shall be accessible by an unobstructed entrance or passageway not less than 600 mm (2 ft) wide and 2.0 m (6 ft 6 in.) high.

(D) Mounting Height. Site supply equipment shall be located not less than 600 mm (2 ft) above the electrical datum plane for that RV site and no more than 2.0 m (6 ft 6 in.) above the electrical datum plane unless platform provisions are made to reach the circuit protection devices that are no more than 2.0 m (6 ft 6 in.) above that platform.

(E) Working Space. Sufficient space shall be provided and maintained about all electrical equipment to permit ready and safe operation, in accordance with 110.26.

(F) Marking. Where the site supply equipment contains a 125/250-volt receptacle, the equipment shall be marked as follows: "Turn disconnecting switch or circuit breaker off before inserting or removing plug. Plug must be fully inserted or removed." The marking shall be located on the equipment adjacent to the receptacle outlet.

The marking is to reduce the possibility of a partially engaged attachment plug, which could result in intermittent neutral (grounded conductor) contact. Loss of the neutral could momentarily apply the line-to-line voltage (240 volts) across 125-volt equipment, causing damage to equipment and wiring within the vehicle.

551.78 Protection of Outdoor Equipment.

(A) Wet Locations. All switches, circuit breakers, receptacles, control equipment, and metering devices located in wet locations shall be weatherproof.

(B) Meters. If secondary meters are installed, meter sockets without meters installed shall be blanked off with an approved blanking plate.

△ **551.79 Clearance for Overhead Conductors.** Open conductors of not over 1000 volts, nominal, shall have a vertical clearance of not less than 5.5 m (18 ft) and a horizontal clearance of not less than 900 mm (3 ft) in all areas subject to recreational vehicle movement. In all other areas, clearances shall conform to 235.360 and 235.361.

Informational Note: See 235.360 and 235.361, for clearances of conductors over 600 volts, nominal.

551.80 Underground Service, Feeder, Branch-Circuit, and Recreational Vehicle Site Feeder-Circuit Conductors.

(A) General. All direct-burial conductors, including the equipment grounding conductor if of aluminum, shall be insulated and identified for the use. All conductors shall be continuous from equipment to equipment. All splices and taps shall be made in approved junction boxes or by use of listed material.

(B) Protection Against Physical Damage. Direct-buried conductors and cables entering or leaving a trench shall be protected by rigid metal conduit, intermediate metal conduit, electrical metallic tubing with supplementary corrosion protection, rigid polyvinyl chloride conduit (PVC), nonmetallic underground conduit with conductors (NUCC), high density polyethylene conduit (HDPE), reinforced thermosetting resin conduit (RTRC), liquidtight flexible nonmetallic conduit, liquidtight flexible metal conduit, or other approved raceways or enclosures. Where subject to physical damage, the conductors or cables shall be protected by rigid metal conduit, intermediate metal conduit, Schedule 80 PVC conduit, or RTRC listed for exposure to physical damage. All such protection shall extend at least 450 mm (18 in.) into the trench from finished grade.

Informational Note: See 300.5 and Article 340 for conductors or Type UF cable used underground or in direct burial in earth.

551.81 Receptacles. A receptacle to supply electric power to a recreational vehicle shall be one of the configurations shown in Figure 551.46(C)(1) in the following ratings:

(1) 50-ampere — 125/250-volt, 50-ampere, 3-pole, 4-wire grounding type for 120/240-volt systems
(2) 30-ampere — 125-volt, 30-ampere, 2-pole, 3-wire grounding type for 120-volt systems
(3) 20-ampere — 125-volt, 20-ampere, 2-pole, 3-wire grounding type for 120-volt systems

Informational Note: See ANSI/NEMA WD 6-2016, *Wiring Devices — Dimensional Specifications*, Figures 14-50, TT, and 5-20, for complete details of these configurations.

ARTICLE
552 Park Trailers

Part I. General

552.1 Scope. The provisions of this article cover the electrical conductors and equipment installed within or on park trailers not covered fully under Articles 550 and 551.

This article covers park trailers that have a single chassis and wheels, that do not exceed 400 ft² (set up), and that are not used as permanent residences. Article 552 does not apply to units that meet the definition of the term *park trailer* (see Article 100) but are used for commercial purposes (see 552.4).

Park trailers equipped with electrical loads similar to those used in mobile homes are not uncommon. It also is not uncommon for a park trailer to be located in the same park trailer community for several years without relocation.

Park trailers are somewhat similar to mobile homes and recreational vehicles (RVs), and many requirements in Article 552 parallel those contained in Articles 550 and 551. Article 552, therefore, is similar in structure to Articles 550 and 551.

552.4 General Requirements. A park trailer is intended for seasonal use. It is not intended as a permanent dwelling unit or for commercial uses such as banks, clinics, offices, or similar. Units designed for such purposes are classified as relocatable structures and are covered in Part II of Article 545.

552.5 Labels. Labels required by Article 552 shall be made of etched, metal-stamped, or embossed brass or stainless steel; plastic laminates not less than 0.13 mm (0.005 in.) thick; or anodized or alclad aluminum not less than 0.5 mm (0.020 in.) thick or the equivalent.

> Informational Note: See ANSI Z535.4-2011, *Product Safety Signs and Labels,* for guidance on other label criteria used in the park trailer industry.

Part II. Low-Voltage Systems

In some park trailers, 12-volt systems are used for interior lighting and other small loads. The 12-volt system often is supplied from an on-board battery or through a transfer switch from a 120/12-volt transformer in conjunction with a full-wave rectifier.

552.10 Low-Voltage Systems.

(A) Low-Voltage Circuits. Low-voltage circuits furnished and installed by the park trailer manufacturer, other than those related to braking, shall be subject to this *Code.* Circuits supplying lights subject to federal or state regulations shall comply with applicable government regulations and this *Code.*

The requirements of Part II apply to the low-voltage wiring within the park trailer that would be used in place of 120-volt ac supplies. These requirements do not apply to the trailer braking circuits.

(B) Low-Voltage Wiring.

(1) Material. Copper conductors shall be used for low-voltage circuits.

Exception: A metal chassis or frame shall be permitted as the return path to the source of supply.

The sidewalls or the roof of a park trailer are not permitted to serve as the ground return path. See the definition of the term *frame* in Article 100.

(2) Conductor Types. Conductors shall conform to the requirements for Type GXL, HDT, SGT, SGR, or Type SXL or shall have insulation in accordance with Table 310.4(1) or the equivalent. Conductor sizes 6 AWG through 18 AWG or SAE shall be listed. Single-wire, low-voltage conductors shall be of the stranded type.

> Informational Note: See SAE J1128-2015, *Low Voltage Primary Cable,* for Types GXL, HDT, and SXL, and SAE J1127-2015, *Low Voltage Battery Cable,* for Types SGT and SGR.

(3) Marking. All insulated low-voltage conductors shall be surface marked at intervals not greater than 1.2 m (4 ft) as follows:

(1) Listed conductors shall be marked as required by the listing agency.
(2) SAE conductors shall be marked with the name or logo of the manufacturer, specification designation, and wire gauge.
(3) Other conductors shall be marked with the name or logo of the manufacturer, temperature rating, wire gauge, conductor material, and insulation thickness.

(C) Low-Voltage Wiring Methods.

(1) Physical Protection. Conductors shall be protected against physical damage and shall be secured. Where insulated conductors are clamped to the structure, the conductor insulation shall be supplemented by an additional wrap or layer of equivalent material, except that jacketed cables shall not be required to be so protected. Wiring shall be routed away from sharp edges, moving parts, or heat sources.

(2) Splices. Conductors shall be spliced or joined with splicing devices that provide a secure connection or by brazing, welding, or soldering with a fusible metal or alloy. Soldered splices shall first be spliced or joined to be mechanically and electrically secure without solder, and then soldered. All splices, joints, and free ends of conductors shall be covered with an insulation equivalent to that on the conductors.

(3) Separation. Battery and other low-voltage circuits shall be physically separated by at least a 13-mm (½-in.) gap or other approved means from circuits of a different power source. Acceptable methods shall be by clamping, routing, or equivalent means that ensure permanent total separation. Where circuits of different power sources cross, the external jacket of the nonmetallic-sheathed cables shall be deemed adequate separation.

(4) Ground Connections. Ground connections to the chassis or frame shall be made in an accessible location and shall be mechanically secure. Ground connections shall be by means of copper conductors and copper or copper-alloy terminals of the solderless type identified for the size of wire used. The surface on which ground terminals make contact shall be cleaned and be free from oxide or paint or shall be electrically connected through the use of a cadmium, tin, or zinc-plated internal/external-toothed lockwasher or locking terminals. Ground terminal attaching screws, rivets or bolts, nuts, and lockwashers shall be cadmium, tin, or zinc-plated except rivets shall be permitted to be unanodized aluminum where attaching to aluminum structures.

The chassis-grounding terminal of the battery shall be connected to the unit chassis with a minimum 8 AWG copper conductor. In the event the unbonded lead from the battery exceeds 8 AWG, the bonding conductor size shall be not less than that of the unbonded lead.

In a combination ac/dc appliance, this ground connection arrangement minimizes the possibility of low-voltage circuit-fault currents passing through the ac panelboard bonding conductor and the equipment grounding conductor (EGC) and subsequently passing through the negative dc conductor feeding the appliance; that conductor may also be bonded to the external metal cover of the appliance. The ac EGC of the appliance might not have sufficient ampacity to safely conduct the dc fault current, which would necessitate installation of the battery bonding conductor. Some RVs already have one side of the battery circuit bonded to the frame by an 8 AWG or larger copper conductor.

(D) Battery Installations. Storage batteries subject to this *Code* shall be securely attached to the unit and installed in an area vaportight to the interior and ventilated directly to the exterior of the unit. Where batteries are installed in a compartment, the compartment shall be ventilated with openings having a minimum area of 1100 mm² (1.7 in.²) at both the top and at the bottom. Where compartment doors are equipped for ventilation, the openings shall be within 50 mm (2 in.) of the top and bottom. Batteries shall not be installed in a compartment containing spark- or flame-producing equipment.

(E) Overcurrent Protection.

(1) Rating. Low-voltage circuit wiring shall be protected by overcurrent protective devices rated not in excess of the ampacity of copper conductors, in accordance with Table 552.10(E)(1).

TABLE 552.10(E)(1)   *Low-Voltage Overcurrent Protection*

Wire Size (AWG)	Ampacity	Wire Type
18	6	Stranded only
16	8	Stranded only
14	15	Stranded or solid
12	20	Stranded or solid
10	30	Stranded or solid

(2) Type. Circuit breakers or fuses shall be of an approved type, including automotive types. Fuseholders shall be clearly marked with maximum fuse size and shall be protected against shorting and physical damage by a cover or equivalent means.

Informational Note: See ANSI/SAE J554-1987, *Standard for Electric Fuses (Cartridge Type)*; SAE J1284-1988, *Standard for Blade Type Electric Fuses*; and UL 275-2005, *Standard for Automotive Glass Tube Fuses*, for further information.

(3) Appliances. Appliances such as pumps, compressors, heater blowers, and similar motor-driven appliances shall be installed in accordance with the manufacturer's instructions.

Motors that are controlled by automatic switching or by latching-type manual switches shall be protected in accordance with 430.32(B).

(4) Location. The overcurrent protective device shall be installed in an accessible location on the unit within 450 mm (18 in.) of the point where the power supply connects to the unit circuits. If located outside the park trailer, the device shall be protected against weather and physical damage.

Exception: External low-voltage supply shall be permitted to have the overcurrent protective device within 450 mm (18 in.) after entering the unit or after leaving a metal raceway.

(F) Switches. Switches shall have a dc rating not less than the connected load.

(G) Luminaires. All low-voltage interior luminaires rated more than 4 watts, employing lamps rated more than 1.2 watts, shall be listed.

Part III. Combination Electrical Systems

552.20 Combination Electrical Systems.

(A) General. Unit wiring suitable for connection to a battery or other low-voltage supply source shall be permitted to be connected to a 120-volt source, provided that the entire wiring system and equipment are rated and installed in full conformity with Parts I, III, IV, and V requirements of this article covering 120-volt electrical systems. Circuits fed from ac transformers shall not supply dc appliances.

(B) Voltage Converters (120-Volt Alternating Current to Low-Voltage Direct Current). The 120-volt ac side of the voltage converter shall be wired in full conformity with the requirements of Parts I and IV of this article for 120-volt electrical systems.

Exception: Converters supplied as an integral part of a listed appliance shall not be subject to 552.20(B).

All converters and transformers shall be listed for use in recreation units and designed or equipped to provide over-temperature protection. To determine the converter rating, the following percentages shall be applied to the total connected load, including average battery-charging rate, of all 12-volt equipment:

    The first 20 amperes of load at 100 percent plus
    The second 20 amperes of load at 50 percent plus
    All load above 40 amperes at 25 percent

Exception: A low-voltage appliance that is controlled by a momentary switch (normally open) that has no means for holding in the closed position shall not be considered as a connected load when determining the required converter rating. Momentarily energized appliances shall be limited to those used to prepare the unit for occupancy or travel.

(C) Bonding Voltage Converter Enclosures. The non–current-carrying metal enclosure of the voltage converter shall be connected to the frame of the unit with an 8 AWG copper conductor minimum. The equipment grounding conductor for the battery and the metal enclosure shall be permitted to be the same conductor.

(D) Dual-Voltage Fixtures Including Luminaires or Appliances. Fixtures, including luminaires, or appliances having both 120-volt and low-voltage connections shall be listed for dual voltage.

In the dual-voltage fixtures, barriers are used to separate the 120-volt and the 12-volt wiring connections.

(E) Autotransformers. Autotransformers shall not be used.

(F) Receptacles and Plug Caps. Where a park trailer is equipped with a 120-volt or 120/240-volt ac system, a low-voltage system, or both, receptacles and plug caps of the low-voltage system shall differ in configuration from those of the 120-volt or 120/240-volt system. Where a unit equipped with a battery or dc system has an external connection for low-voltage power, the connector shall have a configuration that will not accept 120-volt power.

Part IV. Nominal 120-Volt or 120/240-Volt Systems

552.40 120-Volt or 120/240-Volt, Nominal, Systems.

Δ **(A) General Requirements.** The electrical equipment and material of park trailers indicated for connection to a wiring system rated 120 volts, nominal, 2-wire with an equipment grounding conductor, or a wiring system rated 120/240 volts, nominal, 3-wire with an equipment grounding conductor, shall be listed and installed in accordance with Parts I, III, IV, and V of this article.

(B) Materials and Equipment. Electrical materials, devices, appliances, fittings, and other equipment installed, intended for use in, or attached to the park trailer shall be listed. All products shall be used only in the manner in which they have been tested and found suitable for the intended use.

552.41 Receptacle Outlets Required.

(A) Spacing. Receptacle outlets shall be installed at wall spaces 600 mm (2 ft) wide or more so that no point along the floor line is more than 1.8 m (6 ft), measured horizontally, from an outlet in that space.

Exception No. 1: Bath and hallway areas shall not be required to comply with 552.41(A).

Exception No. 2: Wall spaces occupied by kitchen cabinets, wardrobe cabinets, built-in furniture; behind doors that could open fully against a wall surface; or similar facilities.

(B) Location. Receptacle outlets shall be installed as follows:

(1) Adjacent to countertops in the kitchen [at least one on each side of the sink if countertops are on each side and are 300 mm (12 in.) or over in width and depth]

(2) Adjacent to the refrigerator and gas range space, except where a gas-fired refrigerator or cooking appliance, requiring no external electrical connection, is factory-installed

(3) Adjacent to countertop spaces of 300 mm (12 in.) or more in width and depth that cannot be reached from a receptacle required in 552.41(B)(1) by a cord of 1.8 m (6 ft) without crossing a traffic area, cooking appliance, or sink

(C) Ground-Fault Circuit-Interrupter Protection. Each 125-volt, single-phase, 15- or 20-ampere receptacle shall have ground-fault circuit-interrupter protection for personnel in the following locations:

(1) Where the receptacles are installed to serve kitchen countertop surfaces

(2) Within 1.8 m (6 ft) of any lavatory or sink

Exception: Receptacles installed for appliances in dedicated spaces, such as for dishwashers, disposals, refrigerators, freezers, and laundry equipment.

(3) In the area occupied by a toilet, shower, tub, or any combination thereof

(4) On the exterior of the unit

Exception: Receptacles that are located inside of an access panel that is installed on the exterior of the unit to supply power for an installed appliance shall not be required to have ground-fault circuit-interrupter protection.

The receptacle outlet shall be permitted in a listed luminaire. A receptacle outlet shall not be installed in a tub or combination tub–shower compartment.

The walls of a park trailer often do not provide the necessary depth for the installation of a GFCI receptacle. This requirement does not prohibit a bathroom receptacle from being mounted in the side of a lavatory cabinet. Receptacles of any type are not permitted in a tub or tub–shower compartment.

(D) Pipe Heating Cable Outlet. Where a pipe heating cable outlet is installed, the outlet shall be as follows:

(1) Located within 600 mm (2 ft) of the cold water inlet

(2) Connected to an interior branch circuit, other than a small-appliance branch circuit

(3) On a circuit where all of the outlets are on the load side of the ground-fault circuit-interrupter protection for personnel

(4) Mounted on the underside of the park trailer and shall not be considered to be the outdoor receptacle outlet required in 552.41(E)

(E) Outdoor Receptacle Outlets. At least one receptacle outlet shall be installed outdoors. A receptacle outlet located in a compartment accessible from the outside of the park trailer shall be considered an outdoor receptacle. Outdoor receptacle outlets shall be protected as required in 552.41(C)(4).

(F) Receptacle Outlets Not Permitted.

(1) Shower or Bathtub Space. Receptacle outlets shall not be installed in or within reach [750 mm (30 in.)] of a shower or bathtub space.

(2) Face-Up Position. A receptacle shall not be installed in a face-up position in any countertop or other similar horizontal surface.

552.42 Branch-Circuit Protection.

(A) Rating. The branch-circuit overcurrent devices shall be rated as follows:

(1) Not more than the circuit conductors
(2) Not more than 150 percent of the rating of a single appliance rated 13.3 amperes or more and supplied by an individual branch circuit
(3) Not more than the overcurrent protection size marked on an air conditioner or other motor-operated appliances.

(B) Protection for Smaller Conductors. A 20-ampere fuse or circuit breaker shall be permitted for protection for fixtures, including luminaires, leads, cords, or small appliances, and 14 AWG tap conductors, not over 1.8 m (6 ft) long for recessed luminaires.

(C) Fifteen-Ampere Receptacle Considered Protected by 20 Amperes. If more than one receptacle or load is on a branch circuit, 15-ampere receptacles shall be permitted to be protected by a 20-ampere fuse or circuit breaker.

552.43 Power Supply.

(A) Feeder. The power supply to the park trailer shall be a feeder assembly consisting of not more than one listed 30-ampere or 50-ampere park trailer power-supply cord, with an integrally molded or securely attached cap, or a permanently installed feeder.

(B) Power-Supply Cord. If the park trailer has a power-supply cord, it shall be permanently attached to the panelboard, or to a junction box permanently connected to the panelboard, with the free end terminating in a molded-on attachment plug cap.

Cords with adapters and pigtail ends, extension cords, and similar items shall not be attached to, or shipped with, a park trailer.

A suitable clamp or the equivalent shall be provided at the panelboard knockout to afford strain relief for the cord to prevent strain from being transmitted to the terminals when the power-supply cord is handled in its intended manner.

The cord shall be a listed type with 3-wire, 120-volt or 4-wire, 120/240-volt conductors, one of which shall be identified by a continuous green color or a continuous green color with one or more yellow stripes for use as the equipment grounding conductor.

(C) Mast Weatherhead or Raceway. Where the calculated load exceeds 50 amperes or where a permanent feeder is used, the supply shall be by means of one of the following:

(1) One mast weatherhead installation, installed in accordance with Article 230, containing four continuous, insulated, color-coded feeder conductors, one of which shall be an equipment grounding conductor
(2) A rigid metal conduit, intermediate metal conduit, rigid polyvinyl chloride conduit, or other raceways identified for the location from the disconnecting means in the park trailer to the underside of the park trailer

552.44 Cord.

(A) Permanently Connected. Each feeder assembly shall be factory supplied or factory installed and connected directly to the terminals of the panelboard or conductors within a junction box and provided with means to prevent strain from being transmitted to the terminals. The ampacity of the conductors between each junction box and the terminals of each panelboard shall be at least equal to the ampacity of the feeder cord. The supply end of the assembly shall be equipped with an attachment plug of the type described in 552.44(C). Where the cord passes through the walls or floors, it shall be protected by means of conduit and bushings or equivalent. The cord assembly shall have permanent provisions for protection against corrosion and mechanical damage while the unit is in transit.

(B) Cord Length. The cord-exposed usable length shall be measured from the point of entrance to the park trailer or the face of the flanged surface inlet (motor-base attachment plug) to the face of the attachment plug at the supply end.

The cord-exposed usable length, measured to the point of entry on the unit exterior, shall be a minimum of 7.0 m (23 ft) where the point of entrance is at the side of the unit, or shall be a minimum 8.5 m (28 ft) where the point of entrance is at the rear of the unit. The maximum length shall not exceed 11 m (36½ ft).

Where the cord entrance into the unit is more than 900 mm (3 ft) above the ground, the minimum cord lengths above shall be increased by the vertical distance of the cord entrance heights above 900 mm (3 ft).

(C) Attachment Plugs.

(1) Units with Two to Five 15- or 20-Ampere Branch Circuits. Park trailers wired in accordance with 552.46(A) shall have an attachment plug that shall be 2-pole, 3-wire grounding type, rated 30 amperes, 125 volts, conforming to the configuration shown in Figure 552.44(C)(1) intended for use with units rated at 30 amperes, 125 volts.

Informational Note: See ANSI/NEMA WD 6-2016, *Wiring Devices — Dimensional Specifications*, Figure TT, for complete details of this configuration.

(2) Units with 50-Ampere Feeder Assembly. Park trailers having a feeder assembly rated 50 amperes as permitted by 552.43(B) shall have a 3-pole, 4-wire grounding-type attachment plug rated 50 amperes, 125/250 volts, conforming to the configuration shown in Figure 552.44(C)(1).

30-A, 125-V, 2-pole, 3-wire, grounding type

50-A, 125/250-V, 3-pole, 4-wire, grounding type

FIGURE 552.44(C)(1) Attachment Cap and Receptacle Configurations.

Informational Note: See ANSI/NEMA WD 6-2016, *Wiring Devices — Dimensional Specifications*, Figure 14-50, for complete details of this configuration.

The 30-ampere plug and receptacle configuration in Figure 552.44(C)(1) is unique to RVs. They are not a standard 5-30P plug or 5-30R receptacle.

(D) Labeling at Electrical Entrance. Each park trailer shall have a safety label with the signal word WARNING in minimum 6 mm (¼ in.) high letters and body text in minimum 3 mm (⅛ in.) high letters on a contrasting background. The safety label shall be affixed to the exterior skin, at or near the point of entrance of the feeder assembly and shall read, as appropriate:

WARNING:
THIS CONNECTION IS FOR 110–125-VOLT AC,
60 HZ, 30-AMPERE SUPPLY

or

WARNING:
THIS CONNECTION IS FOR 208Y/120-VOLT OR 120/240-
VOLT AC, 3-POLE, 4-WIRE, 60 HZ, _____-AMPERE
SUPPLY.

followed by

DO NOT EXCEED THE CIRCUIT RATING.
EXCEEDING THE CIRCUIT RATING CAN CAUSE A
FIRE AND RESULT IN DEATH OR SERIOUS INJURY.

The correct ampere rating shall be marked in the blank space and the label shall meet the requirements in 110.21(B).

(E) Location. The point of entrance of a feeder assembly shall be located on either side or the rear, within 450 mm (18 in.), of an outside wall.

552.45 Panelboard.

(A) Listed and Appropriately Rated. A listed and appropriately rated panelboard shall be used. The grounded conductor termination bar shall be insulated from the enclosure as provided in 552.55(C). An equipment grounding terminal bar shall be attached inside the metal enclosure of the panelboard.

(B) Location. The panelboard shall be installed in a readily accessible location. Working clearance for the panelboard shall be not less than 600 mm (24 in.) wide and 750 mm (30 in.) deep.

Exception: Where the panelboard cover is exposed to the inside aisle space, one of the working clearance dimensions shall be permitted to be reduced to a minimum of 550 mm (22 in.). A panelboard shall be considered exposed where the panelboard cover is within 50 mm (2 in.) of the aisle's finished surface or not more than 25 mm (1 in.) from the backside of doors that enclose the space.

(C) Dead-Front Type. The panelboard shall be of the dead-front type. A main disconnecting means shall be provided where fuses are used or where more than two circuit breakers are employed. A main overcurrent protective device not exceeding the feeder assembly rating shall be provided where more than two branch circuits are employed.

552.46 Branch Circuits. Branch circuits shall be determined in accordance with 552.46(A) and (B).

(A) Two to Five 15- or 20-Ampere Circuits. A maximum of five 15- or 20-ampere circuits to supply lights, receptacle outlets, and fixed appliances shall be permitted. Such park trailers shall be permitted to be equipped with panelboards rated at 120 volts maximum or 120/240 volts maximum and listed for a 30-ampere-rated feeder assembly. Not more than two 120-volt thermostatically controlled appliances shall be installed in such systems unless appliance isolation switching, energy management systems, or similar methods are used.

Exception No. 1: Additional 15- or 20-ampere circuits shall be permitted where a listed energy management system rated at 30 amperes maximum is employed within the system.

Exception No. 2: Six 15- or 20-ampere circuits shall be permitted without employing an energy management system, provided that the added sixth circuit serves only the power converter, and the combined load of all six circuits does not exceed the allowable load that was designed for use by the original five circuits.

Informational Note: See 210.23(B) for permissible loads. See 552.45(C) for main disconnect and overcurrent protection requirements.

(B) More Than Five Circuits. Where more than five circuits are needed, they shall be determined in accordance with 552.46(B)(1), (B)(2), and (B)(3).

(1) Lighting. Based on 33 volt-amperes/m² (3 VA/ft²) multiplied by the outside dimensions of the park trailer (coupler excluded) divided by 120 volts to determine the number of 15- or 20-ampere lighting area circuits, for example,

$$\frac{3 \times length \times width}{120 \times 15 \ (or \ 20)} \quad \textbf{[552.46(B)(1)]}$$

= No. of 15- (or 20-) ampere circuits

The lighting circuits shall be permitted to serve listed cord-connected kitchen waste disposers and to provide power for supplemental equipment and lighting on gas-fired ranges, ovens, or counter-mounted cooking units.

(2) Small Appliances. Small-appliance branch circuits shall be installed in accordance with 210.11(C)(1).

Δ **(3) General Appliances.** (including furnace, water heater, space heater, range, and central or room air conditioner, etc.) An individual branch circuit shall be permitted to supply any load for which it is rated. There shall be one or more circuits of adequate rating in accordance with 552.46(B)(3)(a) through (B)(3)(d).

Informational Note No. 1: See 210.11(C)(2) for laundry branch circuit.

Informational Note No. 2: See Article 440 for central air conditioning.

(a) The total rating of fixed appliances shall not exceed 50 percent of the circuit rating if lighting outlets, general-use receptacles, or both are also supplied.

(b) For fixed appliances with a motor(s) larger than ⅛ horsepower, the total calculated load shall be based on 125 percent of the largest motor plus the sum of the other loads. Where a branch circuit supplies continuous load(s) or any combination of continuous and noncontinuous loads, the branch-circuit conductor size shall be in accordance with 210.19(A).

(c) The rating of a single cord-and-plug-connected appliance supplied by other than an individual branch circuit shall not exceed 80 percent of the circuit rating.

(d) The rating of a range branch circuit shall be based on the range demand as specified for ranges in 552.47(B)(5).

552.47 Calculations. The following method shall be employed in computing the supply-cord and distribution-panelboard load for each feeder assembly for each park trailer in lieu of the procedure shown in Article 220 and shall be based on a 3-wire, 208Y/120-volt or 120/240-volt supply with 120-volt loads balanced between the two phases of the 3-wire system.

(A) Lighting and Small-Appliance Load. Lighting Volt-Amperes: Length times width of park trailer floor (outside dimensions) times 33 volt-amperes/m² (3 VA/ft²). For example,

Length × width × 3 = lighting volt-amperes

Small-Appliance Volt-Amperes: Number of circuits times 1500 volt-amperes for each 20-ampere appliance receptacle circuit (see definition of *Appliance, Portable* with fine print note) including 1500 volt-amperes for laundry circuit. For example,

No. of circuits × 1500 = small-appliance volt-amperes

Total: Lighting volt-amperes plus small-appliance volt-amperes = total volt-amperes

First 3000 total volt-amperes at 100 percent plus remainder at 35 percent = volt-amperes to be divided by 240 volts to obtain current (amperes) per leg.

Δ **(B) Total Load for Determining Power Supply.** Total load for determining power supply is the sum of the following:

(1) Lighting and small-appliance load as calculated in 552.47(A).
(2) Nameplate amperes for motors and heater loads (exhaust fans, air conditioners, electric, gas, or oil heating). Omit smaller of the heating and cooling loads, except include blower motor if used as air-conditioner evaporator motor. Where an air conditioner is not installed and a 50-ampere power-supply cord is provided, allow 15 amperes per phase for air conditioning.
(3) Twenty-five percent of current of largest motor in 552.47(B)(2).
(4) Total of nameplate amperes for disposal, dishwasher, water heater, clothes dryer, wall-mounted oven, cooking units. Where the number of these appliances exceeds three, use 75 percent of total.
(5) Derive amperes for freestanding range (as distinguished from separate ovens and cooking units) by dividing the following values by 240 volts as shown in Table 552.47(B).
(6) If outlets or circuits are provided for other than factory-installed appliances, include the anticipated load.

Informational Note: See Informative Annex D, Example D12, for an illustration of the application of this calculation.

TABLE 552.47(B) *Minimum Loads for Freestanding Electric Ranges*

Nameplate Rating (watts)	Use (volt-amperes)
0–10,000	80 percent of rating
Over 10,000–12,500	8,000
Over 12,500–13,500	8,400
Over 13,500–14,500	8,800
Over 14,500–15,500	9,200
Over 15,500–16,500	9,600
Over 16,500–17,500	10,000

(C) Optional Method of Calculation for Lighting and Appliance Load. For park trailers, the optional method for calculating lighting and appliance load shown in 220.82 shall be permitted.

552.48 Wiring Methods.

(A) Wiring Systems. Cables and raceways installed in accordance with Articles 320, 322, 330 through 340, 342 through 362, 386, and 388 shall be permitted in accordance with their applicable article, except as otherwise specified in this article. An equipment grounding means shall be provided in accordance with 250.118.

> **See also**
>
> **348.60** for information regarding the use of flexible metal conduit (FMC) as an EGC

(B) Conduit and Tubing. Where rigid metal conduit or intermediate metal conduit is terminated at an enclosure with a locknut and bushing connection, two locknuts shall be provided, one inside and one outside of the enclosure. All cut ends of conduit and tubing shall be reamed or otherwise finished to remove rough edges.

> **See also**
>
> **344.28, 358.28(A), 300.4(G),** and associated commentary for more information on the protection of conductor insulation against abrasion at conduit and tubing terminations

(C) Nonmetallic Boxes. Nonmetallic boxes shall be acceptable only with nonmetallic-sheathed cable or nonmetallic raceways.

(D) Boxes. In walls and ceilings constructed of wood or other combustible material, boxes and fittings shall be flush with the finished surface or project therefrom.

Δ **(E) Mounting.** Wall and ceiling boxes shall be mounted in accordance with 314.23.

Exception No. 1: Snap-in-type boxes or boxes provided with special wall or ceiling brackets that securely fasten boxes in walls or ceilings shall be permitted.

Exception No. 2: A wooden plate providing a 38-mm (1½-in.) minimum width backing around the box and of a thickness of 13 mm (½ in.) or greater (actual) attached directly to the wall panel shall be considered as approved means for mounting outlet boxes.

Exception No. 2 permits the mounting of outlet boxes by screws to a wooden plate that is secured directly to the back of a wall panel. The wooden plate must extend at least 1½ inches around the box. This requirement recognizes the special construction of RV walls, which often makes it difficult or impossible to attach an outlet box to a structural member, as required by 314.23(B).

(F) Cable Sheath. The sheath of nonmetallic-sheathed cable, and the armor of metal-clad cable and Type AC cable, shall be continuous between outlet boxes and other enclosures.

(G) Protected. Metal-clad, Type AC, or nonmetallic-sheathed cables and electrical nonmetallic tubing shall be permitted to pass through the centers of the wide side of 2 by 4 wood studs. However, they shall be protected where they pass through 2 by 2 wood studs or at other wood studs or frames where the cable

or tubing would be less than 32 mm (1¼ in.) from the inside or outside surface. Steel plates on each side of the cable or tubing, or a steel tube, with not less than 1.35 mm (0.053 in.) wall thickness, shall be installed to protect the cable or tubing. These plates or tubes shall be securely held in place. Where nonmetallic-sheathed cables pass through punched, cut, or drilled slots or holes in metal members, the cable shall be protected by bushings or grommets securely fastened in the opening prior to installation of the cable.

(H) Cable Supports. Where connected with cable connectors or clamps, cables shall be secured and supported within 300 mm (12 in.) of outlet boxes, panelboards, and splice boxes on appliances. Supports and securing shall be provided at intervals not exceeding 1.4 m (4½ ft) at other places.

(I) Nonmetallic Box Without Cable Clamps. Nonmetallic-sheathed cables shall be secured and supported within 200 mm (8 in.) of a nonmetallic outlet box without cable clamps. Where wiring devices with integral enclosures are employed with a loop of extra cable to permit future replacement of the device, the cable loop shall be considered as an integral portion of the device.

(J) Physical Damage. Where subject to physical damage, exposed nonmetallic cable shall be protected by covering boards, guard strips, raceways, or other means.

(K) Receptacle Faceplates. Metal faceplates shall comply with 406.6(A). Nonmetallic faceplates shall comply with 406.6(C).

(L) Metal Faceplates Grounded. Where metal faceplates are used, they shall be grounded.

(M) Moisture or Physical Damage. Where outdoor or under-chassis wiring is 120 volts, nominal, or over and is exposed to moisture or physical damage, the wiring shall be protected by rigid metal conduit, by intermediate metal conduit, by electrical metallic tubing, by rigid polyvinyl chloride conduit, by other raceways identified for the location, or by Type MI cable that is closely routed against frames and equipment enclosures or other raceway or cable identified for the application.

(N) Component Interconnections. Fittings and connectors that are intended to be concealed at the time of assembly shall be listed and identified for the interconnection of building components. Such fittings and connectors shall be equal to the wiring method employed in insulation, temperature rise, and fault-current withstanding, and shall be capable of enduring the vibration and shock occurring in park trailers.

(O) Method of Connecting Expandable Units. The method of connecting expandable units to the main body of the park trailer shall comply with 552.48(O)(1) and 552.48(O)(2) as applicable.

(1) Cord-and-Plug Connected. Cord-and-plug connections shall comply with 552.48(O)(1)(a) through (O)(1)(d).

(a) The portion of a branch circuit that is installed in an expandable unit shall be permitted to be connected to the portion of the branch circuit in the main body of the vehicle by means of an attachment plug and cord listed for hard usage. The cord and its connections shall comply with Parts I and II of Article 400 and shall be considered as a permitted use under 400.10. Where the attachment plug and cord are located within the park trailer's interior, use of plastic thermoset or elastomer parallel cord Type SPT-3, SP-3, or SPE shall be permitted.

(b) Where the receptacle provided for connection of the cord to the main circuit is located on the outside of the park trailer, it shall be protected with a ground-fault circuit interrupter for personnel and be listed for wet locations. A cord located on the outside of a park trailer shall be identified for outdoor use.

(c) Unless removable or stored within the park trailer interior, the cord assembly shall have permanent provisions for protection against corrosion and mechanical damage while the park trailer is in transit.

(d) The attachment plug and cord shall be installed so as not to permit exposed live attachment plug pins.

(2) Direct Wires Connected. That portion of a branch circuit that is installed in an expandable unit shall be permitted to be connected to the portion of the branch circuit in the main body of the park trailer by means of flexible cord installed in accordance with 552.48(O)(2)(a) through (O)(2)(f) or other approved wiring method.

(a) The flexible cord shall be listed for hard usage and for use in wet locations.

(b) The flexible cord shall be permitted to be exposed on the underside of the vehicle.

(c) The flexible cord shall be permitted to pass through the interior of a wall or floor assembly or both a maximum concealed length of 600 mm (24 in.) before terminating at an outlet or junction box.

(d) Where concealed, the flexible cord shall be installed in nonflexible conduit or tubing that is continuous from the outlet or junction box inside the park trailer to a weatherproof outlet box, junction box, or strain relief fitting listed for use in wet locations that is located on the underside of the park trailer. The outer jacket of flexible cord shall be continuous into the outlet or junction box.

(e) Where the flexible cord passes through the floor to an exposed area inside of the park trailer, it shall be protected by means of conduit and bushings or equivalent.

(f) Where subject to physical damage, the flexible cord shall be protected with RMC, IMC, Schedule 80 PVC, reinforced thermosetting resin conduit (RTRC) listed for exposure to physical damage, or other approved means and shall extend at least 150 mm (6 in.) above the floor. A means shall be provided to secure the flexible cord where it enters the park trailer.

(P) Prewiring for Air-Conditioning Installation. Prewiring installed for the purpose of facilitating future air-conditioning installation shall comply with the applicable portions of this article and the following:

(1) An overcurrent protective device with a rating compatible with the circuit conductors shall be installed in the panelboard and wiring connections completed.

(2) The load end of the circuit shall terminate in a junction box with a blank cover or other listed enclosure. Where a junction box with a blank cover is used, the free ends of the conductors shall be adequately capped or taped.

(3) A safety label with the word WARNING in minimum 6 mm (¼ in.) high letters and body text in minimum 3 mm (⅛ in.) high letters on a contrasting background shall be affixed on or adjacent to the junction box and shall read as follows:

WARNING
AIR-CONDITIONING CIRCUIT.
THIS CONNECTION IS FOR AIR CONDITIONERS
RATED 110–125-VOLT AC, 60 HZ,
____ AMPERES MAXIMUM.
DO NOT EXCEED CIRCUIT RATING.

EXCEEDING THE CIRCUIT RATING MAY
CAUSE A FIRE AND RESULT IN
DEATH OR SERIOUS INJURY

An ampere rating not to exceed 80 percent of the circuit rating shall be legibly marked in the blank space.

(4) The circuit shall serve no other purpose.

(Q) Prewiring for Other Circuits. Prewiring installed for the purpose of installing other appliances or devices shall comply with the applicable portions of this article and the following:

(1) An overcurrent protection device with a rating compatible with the circuit conductors shall be installed in the panelboard with wiring connections completed.

(2) The load end of the circuit shall terminate in a junction box with a blank cover or a device listed for the purpose. Where a junction box with blank cover is used, the free ends of the conductors shall be adequately capped or taped.

(3) A safety label with the signal word WARNING in minimum 6 mm (¼ in.) high letters and body text in minimum 3 mm (⅛ in.) high letters on a contrasting background shall be affixed on or adjacent to the junction box or device listed for the purpose and shall read as follows:

WARNING
THIS CONNECTION IS FOR ____ RATED ____ VOLT
AC, 60 HZ, ____ AMPERES MAXIMUM. DO NOT
EXCEED CIRCUIT RATING. EXCEEDING THE CIRCUIT
RATING MAY CAUSE A FIRE AND RESULT IN DEATH
OR SERIOUS INJURY.

An ampere rating not to exceed 80 percent of the circuit rating shall be legibly marked in the blank space.

552.49 Maximum Number of Conductors in Boxes. The maximum number of conductors permitted in boxes shall be in accordance with 314.16.

552.50 Grounded Conductors. The identification of grounded conductors shall be in accordance with 200.6.

552.51 Connection of Terminals and Splices. Conductor splices and connections at terminals shall be in accordance with 110.14.

552.52 Switches. Switches shall be rated as required by 552.52(A) and (B).

(A) Lighting Circuits. For lighting circuits, switches shall be rated not less than 10 amperes, 120/125 volts, and in no case less than the connected load.

(B) Motors or Other Loads. For motors or other loads, switches shall have ampere or horsepower ratings, or both, adequate for loads controlled. (An ac general-use snap switch shall be permitted to control a motor 2 hp or less with full-load current not over 80 percent of the switch ampere rating.)

(C) Location. Switches shall not be installed within wet locations in tub or shower spaces unless installed as part of a listed tub or shower assembly.

552.53 Receptacles. All receptacle outlets shall be of the grounding type and installed in accordance with 210.21 and 406.4.

552.54 Luminaires.

(A) General. Any combustible wall or ceiling finish exposed between the edge of a canopy or pan of a luminaire or ceiling suspended (paddle) fan and the outlet box shall be covered with noncombustible material or a material identified for the purpose.

(B) Shower Luminaires. If a luminaire is provided over a bathtub or in a shower stall, it shall be of the enclosed and gasketed type and listed for the type of installation, and it shall be ground-fault circuit-interrupter protected.

Due to the low ceilings in park trailers, luminaires installed above a tub or shower enclosure could be easily reached by most persons standing in the enclosure. Accordingly, only luminaires that are listed for wet locations and have GFCI protection are permitted to be installed above a tub or shower enclosure.

(C) Outdoor Outlets, Luminaires, Air-Cooling Equipment, and So On. Outdoor luminaires and other equipment shall be listed for outdoor use or wet locations.

552.55 Grounding. (See also 552.57 on bonding of non–current-carrying metal parts.)

(A) Power-Supply Grounding. The equipment grounding conductor in the supply cord or feeder shall be connected to the equipment grounding bus or other approved equipment grounding means in the panelboard.

(B) Panelboard. The panelboard shall have an equipment grounding bus with sufficient terminals for all equipment grounding conductors or other approved grounding means.

(C) Insulated Grounded Conductor. The grounded circuit conductor shall be insulated from the equipment grounding conductors and from equipment enclosures and other grounded parts. The grounded circuit conductor terminals in the panelboard and in ranges, clothes dryers, counter-mounted cooking units, and wall-mounted ovens shall be insulated from the equipment enclosure. Bonding screws, straps, or buses in the panelboard or in appliances shall be removed and discarded. Connection of electric ranges and electric clothes dryers utilizing a grounded conductor, if cord-connected, shall be made with 4-conductor cord and 3-pole, 4-wire, grounding-type plug caps and receptacles.

552.56 Interior Equipment Grounding.

(A) Exposed Metal Parts. In the electrical system, all exposed metal parts, enclosures, frames, luminaire canopies, and so forth, shall be effectively bonded to the grounding terminals or enclosure of the panelboard.

(B) Equipment Grounding Conductors. Bare conductors or conductors with insulation or individual covering that is green or green with one or more yellow stripes shall be used for equipment grounding conductors only.

(C) Grounding of Electrical Equipment. Where grounding of electrical equipment is specified, it shall be permitted as follows:

(1) Connection of metal raceway (conduit or electrical metallic tubing), the sheath of Type MC and Type MI cable where the sheath is identified for grounding, or the armor of Type AC cable to metal enclosures.
(2) A connection between the one or more equipment grounding conductors and a metal box by means of a grounding screw, which shall be used for no other purpose, or a listed grounding device.
(3) The equipment grounding conductor in nonmetallic-sheathed cable shall be permitted to be secured under a screw threaded into the luminaire canopy other than a mounting screw or cover screw or attached to a listed grounding means (plate) in a nonmetallic outlet box for luminaire mounting (grounding means shall also be permitted for luminaire attachment screws).

(D) Grounding Connection in Nonmetallic Box. A connection between the one or more grounding conductors brought into a nonmetallic outlet box shall be arranged so that a connection can be made to any fitting or device in that box that requires grounding.

(E) Grounding Continuity. Where more than one equipment grounding conductor of a branch circuit enters a box, all such conductors shall be in good electrical contact with each other, and the arrangement shall be such that the disconnection or removal of a receptacle, fixture, including a luminaire, or other device fed from the box will not interfere with or interrupt the grounding continuity.

(F) Cord-Connected Appliances. Cord-connected appliances, such as washing machines, clothes dryers, refrigerators, and the electrical system of gas ranges, and so on, shall be grounded by means of an approved cord with equipment grounding conductor and grounding-type attachment plug.

552.57 Bonding of Non–Current-Carrying Metal Parts.

(A) Required Bonding. All exposed non–current-carrying metal parts that are likely to become energized shall be effectively bonded to the grounding terminal or enclosure of the panelboard.

(B) Bonding Chassis. A bonding conductor shall be connected between any panelboard and an accessible terminal on the chassis. Bonding terminations shall be suitable for the environment in which the conductors and terminations are installed.

Exception: Any park trailer that employs a unitized metal chassis-frame construction to which the panelboard is securely fastened with a bolt(s) and nut(s) or by welding or riveting shall be considered to be bonded.

(C) Bonding Conductor Requirements. Grounding terminals shall be of the solderless type and listed as pressure terminal connectors recognized for the wire size used. The bonding conductor shall be solid or stranded, insulated or bare, and shall be 8 AWG copper minimum or equivalent.

(D) Metallic Roof and Exterior Bonding. The metal roof and exterior covering shall be considered bonded where both of the following conditions apply:

(1) The metal panels overlap one another and are securely attached to the wood or metal frame parts by metal fasteners.
(2) The lower panel of the metal exterior covering is secured by metal fasteners at each cross member of the chassis, or the lower panel is connected to the chassis by a metal strap.

(E) Gas, Water, and Waste Pipe Bonding. The gas, water, and waste pipes shall be considered grounded if they are bonded to the chassis.

(F) Furnace and Metal Air Duct Bonding. Furnace and metal circulating air ducts shall be bonded.

552.58 Appliance Accessibility and Fastening.
Every appliance shall be accessible for inspection, service, repair, and replacement without removal of permanent construction. Means shall be provided to securely fasten appliances in place when the park trailer is in transit.

552.59 Outdoor Outlets, Fixtures, Including Luminaires, Air-Cooling Equipment, and So On.

(A) Listed for Outdoor Use. Outdoor fixtures, including luminaires, and equipment shall be listed for outdoor use. Outdoor receptacle outlets shall be in accordance with 406.9(A) and (B). Switches and circuit breakers installed outdoors shall comply with 404.4.

(B) Outside Heating Equipment, Air-Conditioning Equipment, or Both. A park trailer provided with a branch circuit designed to energize outside heating equipment or air-conditioning equipment, or both, located outside the park trailer, other than room air conditioners, shall have such branch-circuit conductors terminate in a listed outlet box or disconnecting means located on the outside of the park trailer. A safety label with the word WARNING in minimum 6 mm (¼ in.) high letters and body text in minimum 3 mm (⅛ in.) high letters on a contrasting background shall be affixed within 150 mm (6 in.) from the listed box or disconnecting means and shall read as follows:

> WARNING
> THIS CONNECTION IS FOR HEATING
> AND/OR AIR-CONDITIONING EQUIPMENT.
> THE BRANCH CIRCUIT IS RATED AT NOT MORE
> THAN _____ AMPERES, AT _____ VOLTS, 60 HZ,
> _____ CONDUCTOR AMPACITY.
> A DISCONNECTING MEANS SHALL BE
> LOCATED WITHIN SIGHT OF THE EQUIPMENT.
> EXCEEDING THE CIRCUIT RATING MAY CAUSE A
> FIRE AND RESULT IN DEATH OR SERIOUS INJURY.

The correct voltage and ampere rating shall be given.

Part V. Factory Tests

552.60 Factory Tests (Electrical).
Each park trailer shall be subjected to the tests required by 552.60(A) and (B).

(A) Circuits of 120 Volts or 120/240 Volts. Each park trailer designed with a 120-volt or a 120/240-volt electrical system shall withstand the applied voltage without electrical breakdown of a 1 minute, 900-volt dielectric strength test, or a 1 second, 1080-volt dielectric strength test, with all switches closed, between ungrounded and grounded conductors and the park trailer ground. During the test, all switches and other controls shall be in the "on" position. Fixtures, including luminaires, and permanently installed appliances shall not be required to withstand this test.

Each park trailer shall be subjected to the following:

(1) A continuity test to ensure that all metal parts are properly bonded
(2) Operational tests to demonstrate that all equipment is properly connected and in working order
(3) Polarity checks to determine that connections have been properly made
(4) Receptacles requiring GFCI protection shall be tested for correct function by the use of a GFCI testing device

(B) Low-Voltage Circuits. An operational test of low-voltage circuits shall be conducted to demonstrate that all equipment is connected and in electrical working order. This test shall be performed in the final stages of production after all outer coverings and cabinetry have been secured.

Marinas, Boatyards, Floating Buildings, and Commercial and Noncommercial Docking Facilities

Part I. General

555.1 Scope. This article covers the installation of wiring and equipment in the areas comprising fixed or floating piers, wharves, docks, floating buildings, and other areas in marinas, boatyards, boat basins, boathouses, yacht clubs, boat condominiums, docking facilities associated with one-family dwellings, two-family dwellings, multifamily dwellings, and residential condominiums; any multiple docking facility or similar occupancies; and facilities that are used, or intended for use, for the purpose of repair, berthing, launching, storage, or fueling of small craft and the moorage of floating buildings.

> Informational Note No. 1: See NFPA 303-2016, *Fire Protection Standard for Marinas and Boatyards,* for additional information.
> Informational Note No. 2: Where boats, floating buildings, docks, and similar structures are connected to an electrical source or a supply of electricity, hazardous voltages and currents may create serious safety concerns.
> Informational Note No. 3: Text that is followed by a reference in brackets has been extracted from NFPA 303-2016, *Fire Protection Standard for Marinas and Boatyards*, and NFPA 307-2016, *Standard for the Construction and Fire Protection of Marine Terminals, Piers, and Wharves.* Only editorial changes were made to the extracted text to make it consistent with this *Code.*

The requirements of Article 555 apply to public and private docking, storage, repair, and fueling facilities for small craft. It also applies to floating buildings. The term *small craft* is not defined in the *NEC®*. Based on the scope of NFPA 303, *Fire Protection Standard for Marinas and Boatyards*, the term *small craft* includes recreational and commercial boats, yachts, and other craft that do not exceed 300 gross tons. For facilities that serve larger craft and ships, see NFPA 307, *Standard for the Construction and Fire Protection of Marine Terminals, Piers, and Wharves.* Requirements for floating buildings, including floating dwelling units, which previously were contained in Article 553 until the 2020 NEC, are covered in Article 555.

This article also applies to docking facilities associated with one-family dwellings, two-family dwellings, and multifamily dwellings.

555.3 Electrical Datum Plane Distances.

(A) Floating Piers. The electrical datum plane for floating piers and boat landing stages that is (1) installed to permit rise and fall response to water level and without lateral movement, and (2) that are so equipped that piers and landing stages can rise to the datum plane established for 555.3(B) or (C), shall be a horizontal plane 762 mm (30 in.) above the water level at the floating pier or boat landing stage and a minimum of 305 mm (12 in.) above the level of the deck.

(B) Areas Subject to Tidal Fluctuations. In land areas subject to tidal fluctuation, the electrical datum plane shall be a horizontal plane that is 606 mm (2 ft) above the highest tide level for the area occurring under normal circumstances, based on the highest high tide.

(C) Areas Not Subject to Tidal Fluctuations. In land areas not subject to tidal fluctuation, the electrical datum plane shall be a horizontal plane that is 606 mm (2 ft) above the highest water level for the area occurring under normal circumstances.

Throughout Article 555, the physical location of electrical equipment is referenced to the electrical datum plane, which is used as a horizontal benchmark on land and on floating piers. The definition of the term *electrical datum plane* encompasses areas subject to tidal movement and areas in which the water level is affected only by conditions such as climate (rain or snowfall) or by human intervention (the opening or closing of dams and floodgates). In either case, the term covers the normal highest water level, such as astronomical high tides. The term does not cover extremes due to natural or manmade disasters.

See also

Article 100 for the definition of *electrical datum plane*

555.4 Location of Service Equipment. The service equipment for a floating building, dock, or marina shall be located on land no closer than 1.5 m (5 ft) horizontally from and adjacent to the structure served, but not on or in the structure itself or any other floating structure. Service equipment shall be elevated a minimum of 300 mm (12 in.) above the electrical datum plane.

This requirement ensures that supply conductors to a floating building, dock, or pier can be disconnected in an emergency, such as during a storm, when the floating structure has to be moved quickly. Service equipment is not permitted to be installed on the floating building and any other floating structure, such as a pier, and is now subject to horizontal clearances from these structures. Additionally, minimum elevation requirements above the electrical datum plane provide clearances above normally experienced high-water levels, as covered by the definition of *electrical datum plane* in Article 100. This requirement applies only to equipment that is covered by the definition of *service equipment*. A minimum elevation requirement above docks and piers for other electrical equipment, not meeting the definition of *service equipment*, is covered by the requirements of 555.30(A).

555.5 Maximum Voltage. Pier power distribution systems shall not exceed 250 volts phase to phase. Pier power distribution systems, where qualified personnel service the equipment under engineering supervision, shall be permitted to exceed 250 volts but these systems shall not exceed 600 volts.

Δ **555.6 Load Calculations for Service and Feeder Conductors.** General lighting and other loads shall be calculated in accordance with Part III of Article 220, and, in addition, the demand factors set forth in 220.120 shall be permitted for each service and/or feeder circuit supplying receptacles that provide shore power for boats.

All load calculations for the types of facilities covered by Article 555 are located in Article 220. Part VII of Article 220 contains demand factors that are permitted to be applied to receptacles supplying shore power to watercraft.

555.7 Transformers.

(A) General. Transformers and enclosures shall be identified for wet locations. The bottom of transformer enclosures shall not be located below the electrical datum plane.

(B) Replacements. Transformers and enclosures shall be identified for wet locations where replacements are made.

555.8 Marine Hoists, Railways, Cranes, and Monorails.
Motors and controls for marine hoists, railways, cranes, and monorails shall not be located below the electrical datum plane. Where it is necessary to provide electric power to a mobile crane or hoist in the yard and a trailing cable is utilized, it shall be a listed portable power cable rated for the conditions of use and be provided with an outer jacket of distinctive color for safety.

555.10 Signage.
Permanent safety signs shall be installed to give notice of electrical shock hazard risks to persons using or swimming near a docking facility, boatyard, or marina and shall comply with all of the following:

(1) The signage shall comply with 110.21(B)(1) and be of sufficient durability to withstand the environment.
(2) The signs shall be clearly visible from all approaches to a marina, docking facility, or boatyard facility.
(3) The signs shall state "WARNING — POTENTIAL SHOCK HAZARD — ELECTRICAL CURRENTS MAY BE PRESENT IN THE WATER."

Electrical shock drowning is only one of many hazards that exist in the water around marinas and boatyards. Part of an effective plan to reduce the number of incidents is a no swimming policy. Prohibiting recreational swimming in the immediate vicinity of boats and docks using ac electrical power will protect the public against the dangers associated with using electrical power in marinas and boatyards. The warnings provided by signage (see Exhibit 555.1), along with enforcement by marina and boatyard operators, can help promote changes in the behavior of those using facilities covered by Article 555 with the intended result being to save lives and prevent injuries that have occurred too frequently in bodies of water associated with public and private marinas and docking facilities.

555.11 Motor Fuel Dispensing Stations — Hazardous (Classified) Locations.
Electrical wiring and equipment located at or serving motor fuel dispensing locations shall comply with Article 514 in addition to the requirements of this article.

EXHIBIT 555.1 An example of signage that includes the messaging required by 555.10.

See also

NFPA 303, *Fire Protection Standard for Marinas and Boatyards*

NFPA 30A, *Code for Motor Fuel Dispensing Facilities and Repair Garages*

555.12 Repair Facilities — Hazardous (Classified) Locations.
Electrical wiring and equipment located at facilities for the repair of marine craft containing flammable or combustible liquids or gases shall comply with Article 511 in addition to the requirements of this article.

555.13 Bonding of Non-Current-Carrying Metal Parts.
All metal parts in contact with the water, all metal piping, and all non-current-carrying metal parts that are likely to become energized and that are not connected to a branch circuit or feeder equipment grounding conductor, shall be connected to the grounding bus in the panelboard using solid copper conductors; insulated, covered, or bare; not smaller than 8 AWG. Connections to bonded parts shall be made in accordance with 250.8.

N 555.14 Equipotential Planes and Bonding of Equipotential Planes.
An equipotential plane shall be installed where required in this section to mitigate step and touch voltages at electrical equipment. The parts specified in this section shall be bonded together and to the electrical grounding system. The bonding conductor shall be solid copper conductors; insulated, covered, or bare; not smaller than 8 AWG.

N (A) Areas Requiring Equipotential Planes.
Equipotential planes shall be installed adjacent to all outdoor service equipment or disconnecting means that control equipment in or on water where the following conditions exist:

(1) Where the system voltage exceeds 250 volts to ground
(2) Where the equipment is located within 3 m (10 ft) of the body of water

The equipotential plane shall include all metallic enclosures and controls that are likely to become energized and are accessible to personnel. The equipotential plane shall encompass the area around the equipment and shall extend from the area directly below the equipment out not less than 900 mm (36 in.) in all directions from which a person would be able to stand and come in contact with the equipment.

N **(B) Areas Not Requiring Equipotential Planes.** Equipotential planes shall not be required for the controlled utilization equipment on the docking facility or floating building supplied by the service equipment or disconnecting means.

Equipotential bonding of electrical equipment to mitigate the hazard of potential differences is a proven safety feature used in many settings where a shock hazard exists due to touch and step potentials between items of electrical equipment or between electrical equipment and ground. The increased likelihood of step and touch potentials due to wet or damp conditions resulting from the close proximity of electrical equipment to water is the hazard being addressed by this requirement. If outdoor service equipment or other disconnecting means is used to supply equipment in or on the water and the supply system voltage is rated over 250 volts to ground (i.e., a 277/480-volt wye system or a 240-volt delta system), all electrical equipment enclosures within 36 inches horizontally of the service equipment or disconnecting means must be connected together with an 8 AWG conductor. This connection equalizes the touch potential of all exposed conductive surfaces that are *likely to become energized*. Creating the equipotential plane is only required if the service equipment or disconnecting means is located within 10 feet of the body of water.

> **See also**
>
> **Article 100** for the definition of *energized, likely to become*
>
> **682.33** for a parallel requirement covering electrical equipment installed in proximity to "natural and artificially made bodies of water"

N **555.15 Replacement of Equipment.** When modifications or replacements of electrical enclosures, devices, or wiring methods are necessary on a docking facility, they shall be required to comply with the requirements of this *Code*, and the installation shall require an inspection of the circuit. Existing equipment that has been damaged shall be identified, documented, and repaired by a qualified person to the minimum requirements of the edition of this *Code* to which it was originally installed.

> Informational Note: NFPA 303-2021, *Fire Protection Standard for Marinas and Boatyards*, is a resource for guiding the electrical inspection of a marina.

Newly installed electrical equipment that complies with the requirements of Article 555 is considered to be safe. Once the installation has been approved by the AHJ, the responsibility for ongoing safe operation of the electrical system is transferred to the owner/operator of the docking facility. An unsafe presumption by many is that if the equipment is functioning, it must be safe. Unfortunately, that is not always true. Electrical

equipment installed at docking facilities is subject to extreme environmental exposure and also to physical damage that is unintended in most cases but is inherent to the use of these types of facilities. While it may still be functioning, damaged electrical equipment can manifest into an unsafe condition, such as enclosures and pedestals that no longer prevent the entrance of moisture. Internal damage to electrical equipment can go unseen until it fails or an accident occurs.

Section 555.15 has the sole purpose of preventing electrical accidents due to the use of electrical equipment that no longer provides the requisite level of electrical safety. The NFPA standard cited in the Informational Note provides for yearly inspection of docking facilities, including a list of observations to make on docking facility electrical systems. Electricity and water are normally not a good mix, but *NEC* requirements such as those contained in Articles 555 and 680 provide for safe interface with electrical equipment while near or in water. However, this safe interface is only as good as how well the original level of safety provided by a compliant installation is maintained.

> **See also**
>
> **90.2(B),** in which the necessity of maintenance of electrical systems is discussed

Part II. Marinas, Boatyards, and Docking Facilities

555.30 Electrical Equipment and Connections.

Δ **(A) General.** All electrical components within electrical equipment (excluding wiring methods) and connections not intended for operation while submerged shall be located at least 305 mm (12 in.) above the deck of a fixed or floating structure, but not below the electrical datum plane. Conductor splices, within junction boxes identified for wet locations, utilizing sealed wire connector systems listed and identified for submersion shall be required for floating structures where located above the waterline but below the electrical datum plane.

Δ **(B) Replacements.** Replacement electrical connections shall be located at least 305 mm (12 in.) above the deck of a floating or fixed structure. Conductor splices, within junction boxes identified for wet locations, utilizing sealed wire connector systems listed and identified for submersion shall be required where located above the waterline but below the electrical datum plane.

Not all listed sealed wire-connector systems provide the same degree of protection from moisture ingress. Some sealed wire-connector systems are marked "Watertight" or "Submersible," as applicable.

555.31 Electrical Equipment Enclosures.

(A) Securing and Supporting. Electrical equipment enclosures installed on piers above deck level shall be securely and substantially supported by structural members, independent of any conduit connected to them. If enclosures are not attached to mounting surfaces by means of external ears or lugs, the internal screw heads shall be sealed to prevent seepage of water through mounting holes.

(B) Location. Electrical equipment enclosures on piers shall be located so as not to interfere with mooring lines.

555.32 Circuit Breakers, Switches, Panelboards, and Marina Power Outlets. Circuit breakers and switches installed in gasketed enclosures shall be arranged to permit required manual operation without exposing the interior of the enclosure. All such enclosures shall be arranged with a weep hole to discharge condensation.

555.33 Receptacles. Receptacles shall be mounted not less than 305 mm (12 in.) above the deck surface of the pier and not below the electrical datum plane on a fixed pier.

The location of enclosures for receptacles on fixed and floating piers is based on the *electrical datum plane* as defined in Article 100. For floating piers, the datum plane is 12 inches above the deck of the pier. The purpose of this requirement is to prevent submersion of receptacle enclosures.

The requirements for enclosures in 555.33(A)(1) address their exposure to the severe weather (wind-driven rain) and environmental conditions (splashing from breaking waves or wakes) frequently encountered at marine locations.

(A) Shore Power Receptacles.

(1) Enclosures. Receptacles intended to supply shore power to boats shall be enclosed in listed marina power outlets, enclosures listed for wet locations, or shall be installed in listed enclosures protected from the weather. The integrity of the assembly shall not be affected when the receptacles are in use with any type of booted or nonbooted attachment plug/cap inserted.

(2) Strain Relief. Means shall be provided where necessary to reduce the strain on the plug and receptacle caused by the weight and catenary angle of the shore power cord.

(3) Branch Circuits. Each single receptacle that supplies shore power to boats shall be supplied from a marina power outlet or panelboard by an individual branch circuit of the voltage class and rating corresponding to the rating of the receptacle.

> Informational Note: Supplying receptacles at voltages other than the voltages marked on the receptacle may cause overheating or malfunctioning of connected equipment, for example, supplying single-phase, 120/240-volt, 3-wire loads from a 208Y/120-volt, 3-wire source.

The requirement that each single receptacle that supplies shore power to boats be supplied from an individual branch circuit can be met through the use of multiwire branch circuits derived from single-phase, 3-wire systems or from 3-phase, 4-wire systems. Although the ungrounded conductors of a multiwire branch circuit share the same grounded (neutral) conductor, this configuration can be considered multiple branch circuits in accordance

with 210.4(A). If multiwire branch circuits are used, they must be provided with a disconnecting means that simultaneously opens all ungrounded circuit conductors at the point the circuit originates in accordance with 210.4(B). This requirement enhances the safety of those who have to work on a particular receptacle, but it could be inconvenient for boaters due to power to more receptacles being interrupted when the multiwire branch-circuit disconnecting means is opened.

See also

300.13(B) and its commentary regarding device removal for multiwire branch circuits

(4) Ratings. Shore power for boats shall be provided by single receptacles rated not less than 30 amperes.

> Informational Note: See NFPA 303-2016, *Fire Protection Standard for Marinas and Boatyards*, for locking- and grounding-type receptacles for auxiliary power to boats.

(a) Receptacles rated 30 amperes and 50 amperes shall be of the locking and grounding type.

> Informational Note: See ANSI/NEMA WD 6-2016, *Wiring Devices — Dimensional Specifications*, for various configurations and ratings of locking- and grounding-type receptacles and caps.

(b) Receptacles rated 60 amperes or higher shall be of the pin and sleeve type.

> Informational Note: See ANSI/UL 1686, *UL Standard for Safety Pin and Sleeve Configurations*, for various configurations and ratings of pin and sleeve receptacles.

Single locking- and grounding-type receptacles and attachment caps are required for providing shore power to boats. This facilitates proper connections and prevents unintentional disconnection of on-board equipment, such as bilge pumps, refrigerators, and so forth. Exhibit 555.2 illustrates a chart of grounding-type locking plug and receptacle configurations. Exhibit 555.3 shows pin-and-sleeve-type receptacle configurations.

(B) Other Than Shore Power. Receptacles other than those supplying shore power to boats shall be permitted to be enclosed in marina power outlets with the receptacles that provide shore power to boats, provided the receptacles are marked to clearly indicate that the receptacles are not to be used to supply power to boats.

All receptacle configurations and locations covered by 210.8, other than those supplying shore power to boats, are required to be GFCI protected.

(C) Replacement Receptacles. The requirements in 555.33 shall apply to the replacement of marina receptacles.

555.34 Wiring Methods and Installation.

(A) Wiring Methods.

(1) General. Wiring methods of Chapter 3 shall be permitted where identified for use in wet locations and shall contain a wire-type insulated equipment grounding conductor.

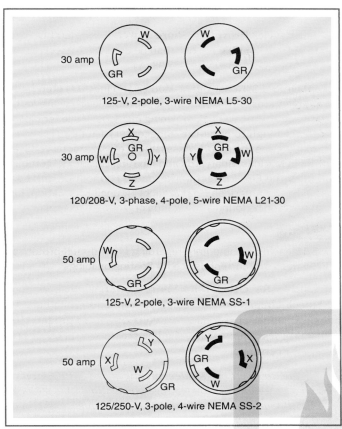

EXHIBIT 555.2 Typical configurations from 30 amperes to 50 amperes for single locking- and grounding-type receptacles and attachment plug caps used to provide shore power for boats in marinas and boatyards.

(2) Portable Power Cables. Extra-hard usage cord and extra-hard usage portable power cables rated not less than 75°C (167°F) and 600 volts, listed for use in the environment within which it is installed, shall be permitted as follows:

(1) As permanent wiring on the underside of piers (floating or fixed)
(2) Where flexibility is necessary as on piers composed of floating sections

The cable construction requirements are necessary due to the cables' exposure to extremes in weather conditions and to operational hazards such as oil and gasoline spills. Not all portable power cables are suitable for exposure to gasoline. A cable evaluated for oil and gasoline resistance at 75°C is marked "GASOLINE AND OIL RESISTANT II," or "GR2."

(B) Installation.

(1) Overhead Wiring. Overhead wiring shall be installed to avoid possible contact with masts and other parts of boats being moved in the yard.

Conductors and cables shall be routed to avoid wiring closer than 6.0 m (20 ft) from the outer edge or any portion of the yard

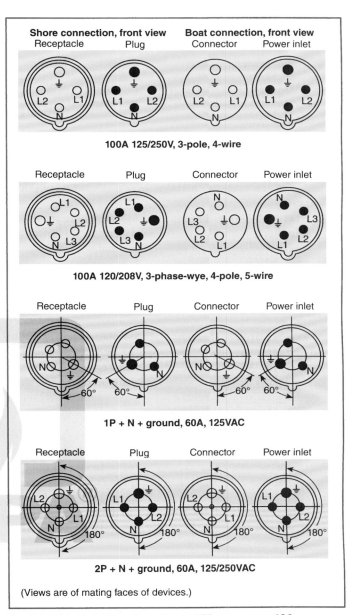

EXHIBIT 555.3 Typical configurations of 60 amperes or 100 amperes for safety pin-and-sleeve-type receptacles, plugs, connectors, and power inlets used to provide shore power for boats in marinas and boatyards.

that can be used for moving vessels or stepping or unstepping masts.

(2) Outdoor Branch Circuits and Feeders. Multiple feeders and branch circuits shall be permitted and clearances for overhead branch-circuit and feeder wiring in locations of the boatyard other than those described in 555.34(B)(1) shall be located not less than 5.49 m (18 ft) above grade. Only Part I of Article 225 shall apply to marina installations.

Approval for wiring over and under navigable water by the AHJ may include federal and local agencies, such as the Army Corps of Engineers, the Coast Guard, or local harbormasters, that have specific authority over the waterways.

(3) Portable Power Cables.

(a) Where portable power cables are permitted by 555.34(A)(2), the installation shall comply with the following:

(1) Cables shall be properly supported.
(2) Cables shall be located on the underside of the pier.
(3) Cables shall be securely fastened by nonmetallic clips to structural members other than the deck planking.
(4) Cables shall not be installed where subject to physical damage.
(5) Where cables pass through structural members, they shall be protected against chafing by a permanently installed oversized sleeve of nonmetallic material.

(b) Where portable power cables are used as permitted in 555.34(A)(2)(2), there shall be a junction box of corrosion-resistant construction with permanently installed terminal blocks on each pier section to which the feeders and feeder extensions are to be connected. A listed marina power outlet employing terminal blocks/bars shall be permitted in lieu of a junction box. Metal junction boxes and covers, and metal screws and parts that are exposed externally to the boxes, shall be of corrosion-resistant materials or protected by material resistant to corrosion.

Δ **(4) Protection.** Rigid metal conduit, intermediate metal conduit, reinforced thermosetting resin conduit (RTRC) listed for aboveground use, or rigid polyvinyl chloride (PVC) conduit suitable for the location shall be used to protect wiring to a point at least 2.5 m (8 ft) above docks, decks of piers, and landing stages. The conduit shall be connected to the enclosure by full standard threads or fittings listed for use in damp or wet locations, as applicable.

555.35 Ground-Fault Protection of Equipment (GFPE) and Ground-Fault Circuit Interrupter. For other than floating buildings, ground-fault protection for docking facilities shall be provided in accordance with 555.35(A) through (D).

N **(A) Feeder.** Listed GFPE, rated not more than 100 milliamperes, shall be provided for feeders installed on docking facilities. Coordination with downstream GFPE shall be permitted at the feeder overcurrent protective device.

Exception: Transformer secondary conductors of a separately derived system that do not exceed 3 m (10 ft) and are installed in a raceway shall be permitted to be installed without ground-fault protection.

This exception shall also apply to the supply terminals of the equipment supplied by the transformer secondary conductors.

N **(B) Branch-Circuits**

N **(1) Receptacles Providing Shore Power.** Listed GFPE, rated not more than 30 milliamperes, shall be provided for receptacles installed in accordance with 555.33(A).

According to U.S. Coast Guard studies and industry standards, 30 milliamperes represents an acceptable level to prevent a majority of electrical shock drowning incidents while remaining practical enough to minimize unnecessary tripping. For more information, see the report "Assessment of Hazardous Voltage/Current in Marinas, Boatyards and Floating Buildings," commissioned by the Fire Protection Research Foundation and conducted by the American Boat & Yacht Council Foundation, Inc. The report can be found at www.nfpa.org/news-and-research.

See also

555.53 and its commentary regarding ground-fault protection of the main overcurrent device for floating buildings

N **(2) Outlets for Other than Shore Power.** Outlets supplied by branch circuits not exceeding 150 volts to ground and 60 amperes, single phase, and 150 volts or less to ground, 100 amperes or less, three phase, shall be provided with GFCI protection for personnel.

Exception to (B): Low-voltage circuits not requiring grounding, not exceeding the low-voltage contact limit and supplied by listed transformers or power supplies that comply with 680.23(A)(2) shall be permitted to be installed without ground-fault protection.

The characteristics of the branch circuits covered by this requirement are similar to the parameters of 210.8(B) for GFCI protection of receptacles installed in or at other than dwelling units. An important distinction between this requirement and that of 210.8(B) is the use of the term *outlet*. An *outlet* includes a point in a wiring system at which a receptacle is installed and also includes locations in the wiring system where power is taken to supply utilization equipment using other than a cord-and-plug connection, such "hard-wired" or "direct-wired" connections.

N **(C) Boat Hoists.** GFCI protection for personnel shall be provided for outlets not exceeding 240 volts that supply a boat hoist installed at docking facilities. GFCI protected receptacles for other than shore power shall be permitted to supply boat hoists.

Compliance with this GFCI requirement can be accomplished through the use of a device installed at the outlet or a device installed in the branch circuit supplying the outlet. Specifying GFCI protection "for outlets" that supply a boat hoist extends this requirement to direct- or hard-wired models as well as cord-and-plug-connected models.

(D) Leakage Current Measurement Device. Where more than three receptacles supply shore power to boats, a listed leakage current measurement device for use in marina applications shall be available and be used to determine leakage current from each boat that will utilize shore power. The listing requirement for the leakage current measurement device for use in marina applications shall become effective January 1, 2026.

> Informational Note No. 1: Leakage current measurement will provide the capability to determine when an individual boat has defective wiring or other problems contributing to hazardous voltage and current. The use of a test device will allow the facility operator to identify a boat that is creating problems. In some cases a single boat could cause an upstream GFPE device protecting a feeder to operate even though multiple boats are supplied from the same feeder. The use of a test device will help the facility operator prevent a particular boat from contributing to hazardous voltage and current in the marina area.

> Informational Note No. 2: An annual test of each boat with the leakage current measurement device is a prudent step toward determining if a boat has defective wiring that could be contributing hazardous voltage and current. Where the leakage current measurement device reveals that a boat is contributing hazardous voltage and current, repairs should be made to the boat before it is permitted to utilize shore power.

Exception: Where the shore power equipment includes a leakage indicator and leakage alarm, a separate leakage test device shall not be required.

555.36 Disconnecting Means for Shore Power Connection(s). Disconnecting means shall be provided to isolate each boat from its supply connection(s).

(A) Type. The disconnecting means shall consist of a circuit breaker, switch, or both, and shall be properly identified as to which receptacle it controls.

Δ **(B) Location.** The disconnecting means shall be readily accessible, located not more than 762 mm (30 in.) from the receptacle it controls, and located in the supply circuit ahead of the receptacle. Circuit breakers or switches located in marina power outlets complying with this section shall be permitted as the disconnecting means.

N **(C) Emergency Electrical Disconnect.** Each marina power outlet or enclosure that provides shore power to boats shall be provided with a listed emergency shutoff device or electrical disconnect that is clearly marked "Emergency Shutoff" in accordance with 110.22(A). The emergency shutoff device or electrical disconnect shall be within sight of the marina power outlet or other enclosure that provides shore power to boats, readily accessible, externally operable, manually resettable, and listed for use in wet locations. The emergency shutoff device or electrical disconnect shall de-energize the power supply to all circuits supplied by the marina power outlet(s) or enclosure(s) that provide shore power to boats. A circuit breaker handle shall not be used for this purpose.

Requiring a readily accessible means to disconnect shore power connections to watercraft enables marina and boatyard users to respond to an emergency condition without increasing the danger to themselves. The natural reaction of many is to enter the water if a disabled or distressed swimmer is observed and there is no marked or readily accessible means to shut the power off. Unfortunately, this only puts this person at peril of also becoming a victim of the current that exists in the water. A marked, local, and externally operable means of interrupting the power at the marina outlet is a far safer method of response to an incident involving a person fully or partially submerged in the water becoming disabled due to swimming or wading through a voltage gradient that results in electrical current through his or her body or by being in water and touching a surface that is exposed and conductive and at a different potential than the water.

555.37 Equipment Grounding Conductor.

(A) Equipment to Be Connected to Equipment Grounding Conductor. The following items shall be connected to an equipment grounding conductor run with the circuit conductors in the same raceway, cable, or trench:

(1) Metal boxes, metal cabinets, and all other metal enclosures
(2) Metal frames of utilization equipment
(3) Grounding terminals of grounding-type receptacles

Δ **(B) Type of Equipment Grounding Conductor.** An equipment grounding conductor shall be of the wire-type, insulated, and sized in accordance with 250.122 but not smaller than 12 AWG.

The use of an insulated aluminum or copper wire–type equipment grounding conductor (EGC) ensures a high-integrity path for ground-fault current. Because of the corrosive conditions in marinas and boatyards, metal raceways are not permitted to serve as the sole EGC.

Δ **(C) Feeder Equipment Grounding Conductor.** Where a feeder supplies a remote panelboard or other distribution equipment, an insulated equipment grounding conductor shall extend from a grounding terminal in the service to a grounding terminal and busbar in the remote panelboard or other distribution equipment.

(D) Branch-Circuit Equipment Grounding Conductor. The insulated equipment grounding conductor for branch circuits shall terminate at a grounding terminal in a remote panelboard, in other distribution equipment, or in the main service equipment.

N **(E) Cord-and-Plug-Connected Appliances.** Unless double-insulated, cord-and-plug-connected appliances shall be grounded by means of an equipment grounding conductor in the cord and a grounding-type attachment plug.

Exception: An equipment grounding conductor shall be permitted to be uninsulated if a part of a listed cable assembly identified for the environment and not subject to atmospheres or environments such as, but not limited to, storm water basins, sewage treatment ponds, and natural bodies of water containing salt water.

N **555.38 Luminaires.**

N **(A) General.** All luminaires and retrofit kits shall be listed and identified for use in their intended environment. Luminaires and their supply connections shall be secured to structural elements of the marina to limit damage from watercraft and prevent entanglement of and interaction with sea life.

N **(B) Underwater Luminaires.** Luminaires installed below the highest high tide level or electrical datum plane and likely to be periodically submersed shall be limited to those luminaires that comply with the following:

(1) Identified as submersible
(2) Operate below the low-voltage contact limit defined in Article 100
(3) Supplied by an isolating transformer or power supply in accordance with 680.23(A)(2)

Limiting sources of electricity in the water due to equipment failure resulting from not being suitable for the harsh environment at marinas is the objective of this requirement for listed or certified luminaires and retrofit kits. Durability and resistance of the equipment to the harsh effects of outdoor conditions are considerations in evaluating products for the marina environment. Additionally, limiting voltage levels and using isolating transformers or power supplies — as is done with underwater luminaires for swimming pools and other bodies of water covered by Article 680 — is another protection technique for reducing electrical hazards associated with luminaires within marinas and boatyards that are intentionally or likely to be periodically submerged.

Part III. Floating Buildings

Although the *NEC* does not cover electrical installations on ships or watercraft, it does cover installations for floating buildings. Floating buildings are permanently moored in one location, for example, restaurants, aquariums, and dwelling units. All other applicable articles apply to these floating buildings.

See also

90.2(C)(1) for the scope statement on floating buildings covered by the *NEC*

90.2(D)(1) for the scope statement on ships and watercraft (other than floating buildings) not covered by the *NEC*

555.50 Service Conductors. One set of service conductors shall be permitted to serve more than one set of service equipment.

555.51 Feeder Conductors. Each floating building shall be supplied by a single set of feeder conductors from its service equipment.

Exception: Where the floating building has multiple occupancy, each occupant shall be permitted to be supplied by a single set of feeder conductors extended from the occupant's service equipment to the occupant's panelboard.

555.52 Installation of Services and Feeders.

(A) Flexibility. Flexibility of the wiring system shall be maintained between floating buildings and the supply conductors. All wiring shall be installed so that motion of the water surface and changes in the water level will not result in unsafe conditions.

(B) Wiring Methods. Liquidtight flexible metal conduit or liquidtight flexible nonmetallic conduit with approved fittings shall be permitted for feeders and where flexible connections are required for services. Extra-hard usage portable power cable listed for both wet locations and sunlight resistance shall be permitted for a feeder to a floating building where flexibility is required. Other raceways suitable for the location shall be permitted to be installed where flexibility is not required.

If a portable power cable is used as a feeder to a floating building, it is important to understand the differences between cable types. A cable designation with a "W" as the last letter is suitable for use in wet locations and is sunlight resistant. If "W" is the only letter, this indicates that the cable is a portable power cable. Where Type W cables from Table 400.4 are used to supply a floating building, they must be listed for use in wet locations and for sunlight resistance. Not all Type W cables are listed for wet location applications.

555.53 Ground-Fault Protection. The main overcurrent protective device that feeds the floating building shall have ground-fault protection not exceeding 100 mA. Ground-fault protection of each individual branch or feeder circuit shall be permitted as a suitable alternative. Outdoor outlets, shore power outlets, and boat hoists located at floating buildings shall comply with 555.35(B) and (C).

Overcurrent protection for supply conductors is provided by the service equipment. Since these conductors can develop leakage, ground-fault protection is required at the main device or, alternatively, for each feeder or branch circuit. Factors such as corrosion and equipment degradation or lack of maintenance can cause ground faults to the metal surfaces of floating buildings or shore-powered vessels. Persons in contact with those metal surfaces, in proximity to the water surrounding the metal surface, or attempting to exit the water via a metal swim platform or ladder could be subjected to an electric shock.

While branch-circuit GFCI devices, which may trip at as low as 4 milliamperes, are permitted to be used, this is not practical for all floating buildings due to normal leakage currents of appliances and other equipment in or on the floating building. Therefore, the ground-fault current level of the device is not permitted to exceed 100 milliamperes. Devices operating at current levels higher than those specified for a Class A GFCI in ANSI/UL 943, *Ground-Fault Circuit Interrupters*, do not provide GFCI protection of personnel, but they will open the circuit at much lower levels of ground-fault current than will a standard overcurrent protection device (OCPD). See the Informational Note for the term *ground-fault circuit interrupter* in Article 100 for details on the operational current levels of GFCIs.

Outlets for other than shore power and outlets rated 240 volts or less for boat hoists are required to be provided with GFCI protection, as defined in Article 100. Ground-fault protection equipment (GFPE) is not a permitted protection method for these outlets per 555.35(C) and (D).

555.54 Grounding. Grounding at floating buildings shall comply with 555.54(A) through (D).

(A) Grounding of Electrical and Nonelectrical Parts. Grounding of both electrical and nonelectrical parts in a floating building shall be through connection to a grounding bus in the building panelboard.

(B) Installation and Connection of Equipment Grounding Conductor. The equipment grounding conductor shall be installed with the feeder conductors and connected to a grounding terminal in the service equipment.

(C) Identification of Equipment Grounding Conductor. The equipment grounding conductor shall be an insulated copper conductor with a continuous outer finish that is either green or green with one or more yellow stripes. For conductors larger than 6 AWG, or where multiconductor cables are used, re-identification of conductors allowed in 250.119(B)(2)b. and (B)(2)c. shall be permitted.

(D) Grounding Electrode Conductor Connection. The grounding terminal in the service equipment shall be grounded by connection through an insulated grounding electrode conductor to a grounding electrode on shore.

555.55 Insulated Neutral. The grounded circuit conductor (neutral) shall be an insulated conductor identified in compliance with 200.6. The neutral conductor shall be connected to the equipment grounding terminal in the service equipment, and, except for that connection, it shall be insulated from the equipment grounding conductors, equipment enclosures, and all other grounded parts. The neutral conductor terminals in the panelboard and in ranges, clothes dryers, counter-mounted cooking units, and the like shall be insulated from the enclosures.

555.56 Equipment Grounding.

(A) Electrical Systems. All enclosures and exposed metal parts of electrical systems shall be connected to the grounding bus.

(B) Cord-Connected Appliances. Where required to be grounded, cord-connected appliances shall be grounded by means of an equipment grounding conductor in the cord and a grounding-type attachment plug.

Temporary Installations

590.1 Scope. The provisions of this article apply to temporary electric power and lighting installations.

Temporary installations are temporary as approved by the AHJ. Article 590 applies to any temporary installation whether it is at a transient or a permanent location. The installation could be at a construction site, a store parking lot, or a local craft fair in a field.

590.2 All Wiring Installations.

(A) Other Articles. Except as specifically modified in this article, all other requirements of this *Code* for permanent wiring shall apply to temporary wiring installations.

Temporary installations of electrical equipment must be installed in accordance with all applicable permanent installation requirements except as modified by the rules in this article. For example, the requirements of 300.15 specify that a box or other enclosure must be used where splices are made. This rule is amended by 590.4(G), which, for construction sites, permits splices to be made in multiconductor cords and cables without the use of a box.

Dismissing the need to comply with the requirements of the *NEC®* because the installation "is only temporary" reduces the level of safety for users of the temporary installation. For instance, contrary to what is believed by some to be acceptable, there is no permission in Article 590 allowing temporary services to be grounded any differently than a permanently installed service. Where used, rod-type electrodes must comply with all of the requirements of 250.53, including the need to install a supplemental electrode, unless the 25-ohm earth resistance condition can be met. Only under that condition can one ground rod be used.

Electrical accidents do not discriminate and can occur in any installation, permanent or temporary, if the requirements of the *NEC* are not followed. Due to the nature of work occurring at construction sites and the higher probability of wiring systems being damaged and compromised, following the requirements of Article 590 is essential to electrical safety. Bypassing GFCI protection because the device is "nuisance tripping" is another compromise of the safety system that occurs on construction sites. In the vast majority of these cases, the GFCI is doing what it is designed and intended to do — prevent electrical injuries and deaths.

(B) Approval. Temporary wiring methods shall be acceptable only if approved based on the conditions of use and any special requirements of the temporary installation.

Temporary wiring methods are approved based on criteria such as length of time in service, severity of physical abuse, exposure to weather, and other special requirements. Special requirements can range from tunnel construction projects to tent cities constructed after a natural disaster to flammable hazardous material reclamation projects.

590.3 Time Constraints.

(A) During the Period of Construction. Temporary electric power and lighting installations shall be permitted during the period of construction, remodeling, maintenance, repair, or demolition of buildings, structures, equipment, or similar activities.

(B) 90 Days. Temporary electric power and lighting installations shall be permitted for a period not to exceed 90 days for holiday decorative lighting and similar purposes.

The 90-day time limit applies only to temporary electrical installations associated with holiday displays. Other installations are not bound by this time limit, and, in fact, some construction projects extend over multiple years in which temporary power is necessary for the duration of the project.

(C) Emergencies and Tests. Temporary electric power and lighting installations shall be permitted during emergencies and for tests, experiments, and developmental work.

(D) Removal. Temporary wiring shall be removed immediately upon completion of construction or purpose for which the wiring was installed.

Because temporary wiring installations might not meet all the requirements for a permanent installation due to the modifications permitted by Article 590, all temporary wiring must not only be disconnected but must also be removed from the building, structure, or other location of installation.

See also

590.4(B) and (C) for the permission to use nonmetallic jacketed cables and flexible cords for temporary branch circuits and feeders

590.4 General.

(A) Services. Services shall be installed in conformance with Parts I through VIII of Article 230, as applicable.

(B) Feeders. Overcurrent protection shall be provided in accordance with 240.4, 240.5, 245.26, 445.12, and 445.13. Conductors shall be permitted within cable assemblies or within multiconductor cords or cables of a type identified in Table 400.4 for hard usage or extra-hard usage. For the purpose of this section, the following wiring methods shall be permitted:

(1) Type NM, Type NMC, and Type SE cables shall be permitted to be used in any dwelling, building, or structure without any height limitation or limitation by building construction type and without concealment within walls, floors, or ceilings.
(2) Type SE cable shall be permitted to be installed in a raceway in an underground installation.

Exception: Single insulated conductors shall be permitted where installed for the purpose(s) specified in 590.3(C) and accessible only to qualified persons.

(C) Branch Circuits. All branch circuits shall originate in an approved power outlet, switchgear, switchboard or panelboard, motor control center, or fused switch enclosure. Conductors shall be permitted within cable assemblies or within multiconductor cord or cable of a type identified in Table 400.4 for hard usage or extra-hard usage. Conductors shall be protected from overcurrent as provided in 240.4, 240.5, and 245.26. For the purposes of this section, the following wiring methods shall be permitted:

(1) Type NM, Type NMC, and Type SE cables shall be permitted to be used in any dwelling, building, or structure without any height limitation or limitation by building construction type and without concealment within walls, floors, or ceilings.
(2) Type SE cable shall be permitted to be installed in a raceway in an underground installation.

Types NM, NMC, and SE cable may be used in any building or structure regardless of building height and construction type.

Temporary feeders and branch circuits are permitted to be cable assemblies, multiconductor cords, or single-conductor cords. Cords must be identified for hard or extra-hard usage according to Table 400.4. Individual conductors, as described in Table 310.4(1), are not permitted as open conductors but can be part of a cable assembly or used in a raceway system. Open or individual conductor feeders are permitted only during emergencies or tests by the exception to 590.4(B).

The basic requirement is that temporary wiring be located and installed so that it will not be physically damaged. Note that hard-usage or extra-hard-usage extension cords are permitted to be laid on the floor.

Exception: Branch circuits installed for the purposes specified in 590.3(B) or 590.3(C) shall be permitted to be run as single insulated conductors. Where the wiring is installed in accordance with 590.3(B), the voltage to ground shall not exceed 150 volts, the wiring shall not be subject to physical damage, and the conductors shall be supported on insulators at intervals of not more than 3.0 m (10 ft); or, for festoon lighting, the conductors shall be so arranged that excessive strain is not transmitted to the lampholders.

(D) Receptacles.

(1) All Receptacles. All receptacles shall be of the grounding type. Unless installed in a continuous metal raceway that qualifies as an equipment grounding conductor in accordance with 250.118 or a continuous metal-covered cable that qualifies as an equipment grounding conductor in accordance with 250.118, all branch circuits shall include a separate equipment grounding conductor, and all receptacles shall be electrically connected to the equipment grounding conductor(s). Receptacles on construction sites shall not be installed on any branch circuit that supplies temporary lighting.

Conductors for lighting and receptacle loads are required to be separate so that the activation of an overcurrent device or GFCI does not de-energize the lighting circuit. Metal cables or raceways must be continuous and qualify as an equipment grounding conductor (EGC). Temporary wiring is subject to frequent handling and relocation, and maintaining the continuity of EGCs needs to be a focus of those responsible for operation of a safe temporary electrical system.

(2) Receptacles in Wet Locations. All 15- and 20-ampere, 125- and 250-volt receptacles installed in a wet location shall comply with 406.9(B)(1).

(E) Disconnecting Means. Suitable disconnecting switches or plug connectors shall be installed to permit the disconnection of all ungrounded conductors of each temporary circuit. Multiwire branch circuits shall be provided with a means to disconnect simultaneously all ungrounded conductors at the power outlet or panelboard where the branch circuit originated. Identified handle ties shall be permitted.

(F) Lamp Protection. All lamps for general illumination shall be protected from accidental contact or breakage by a suitable luminaire or lampholder with a guard.

Metal guarded sockets shall not be used unless the metal guard is connected to the circuit equipment grounding conductor.

(G) Splices. A box, conduit body, or other enclosure, with a cover installed, shall be required for all splices.

Exception No. 1: On construction sites, a box, conduit body, or other enclosure shall not be required for either of the following conditions:

(1) The circuit conductors being spliced are all from nonmetallic multiconductor cord or cable assemblies, provided that the equipment grounding continuity is maintained with or without the box.

(2) The circuit conductors being spliced are all from metalsheathed cable assemblies terminated in listed fittings that mechanically secure the cable sheath to maintain effective electrical continuity.

Exception No. 2: On construction sites, branch circuits that are permanently installed in framed walls and ceilings and are used to supply temporary power or lighting, and that are GFCI protected, the following shall be permitted:

(1) A box cover shall not be required for splices installed completely inside of junction boxes with plaster rings.

(2) Listed pigtail-type lampholders shall be permitted to be installed in ceiling-mounted junction boxes with plaster rings.

(3) Finger safe devices shall be permitted for supplying and connection of devices.

(H) Protection from Accidental Damage. Flexible cords and cables shall be protected from accidental damage. Sharp corners and projections shall be avoided. Where passing through doorways or other pinch points, protection shall be provided to avoid damage.

One of the modifications to a Chapter 4 requirement [i.e., 400.12(3)] is to permit flexible cords and cables to pass through doorways due to the need to provide power in areas that do not have permanent installed receptacles and other power outlets.

(I) Termination(s) at Devices. Flexible cords and cables entering enclosures containing devices requiring termination shall be secured to the box with fittings listed for connecting flexible cords and cables to boxes designed for the purpose.

(J) Support. Cable assemblies and flexible cords and cables shall be supported in place at intervals that ensure that they will be protected from physical damage. Support shall be in the form of staples, cable ties, straps, or similar type fittings installed so as not to cause damage. Cable assemblies and flexible cords and cables installed as branch circuits or feeders shall not be installed on the floor or on the ground. Extension cords shall not be required to comply with 590.4(J). Vegetation shall not be used for support of overhead spans of branch circuits or feeders.

Exception: For holiday lighting in accordance with 590.3(B), where the conductors or cables are arranged with strain relief devices, tension take-up devices, or other approved means to avoid damage from the movement of the live vegetation, trees shall be permitted to be used for support of overhead spans of branch-circuit conductors or cables.

Temporary wiring methods do not have to be supported in accordance with the permanent installation requirements (from Chapter 3) for the particular wiring method. Adequate support is needed only to minimize the possibility of damage to the wiring method during its temporary period of use. The use of vegetation as a support structure for overhead spans of branch-circuit and feeder conductors is not permitted.

The exception allows branch-circuit and feeder cables supplying holiday lighting to be installed and supported by trees for a period of not more than 90 days, provided the wiring is arranged with proper strain relief devices, tension take-up devices, or other means to prevent damage to the conductors from the tree swaying. All temporary wiring must be removed at the end of the temporary period or project.

590.5 Listing of Decorative Lighting. Decorative lighting used for holiday lighting and similar purposes, in accordance with 590.3(B), shall be listed and shall be labeled on the product.

590.6 Ground-Fault Protection for Personnel. Ground-fault protection for personnel for all temporary wiring installations shall be provided to comply with 590.6(A) and (B). This section shall apply only to temporary wiring installations used to supply temporary power to equipment used by personnel during construction, remodeling, maintenance, repair, or demolition of buildings, structures, equipment, or similar activities. This section shall apply to power derived from an electric utility company or from an on-site-generated power source.

(A) Receptacle Outlets. Temporary receptacle installations used to supply temporary power to equipment used by personnel during construction, remodeling, maintenance, repair, or demolition of buildings, structures, equipment, or similar activities shall comply with the requirements of 590.6(A)(1) through (A)(3), as applicable.

Exception: In industrial establishments only, where conditions of maintenance and supervision ensure that only qualified personnel are involved, an assured equipment grounding conductor program as specified in 590.6(B)(2) shall be permitted for only those receptacle outlets used to supply equipment that would create a greater hazard if power were interrupted or having a design that is not compatible with GFCI protection.

A cord-and-plug-connected ventilation fan for personnel working in toxic environments is an example of where the loss of power poses a greater hazard to personnel. Some electrically operated testing equipment has proven to be incompatible with GFCI protection.

(1) Receptacle Outlets Not Part of Permanent Wiring. All 125-volt, single-phase, 15-, 20-, and 30-ampere receptacle outlets that are not a part of the permanent wiring of the building or structure and that are in use by personnel shall have ground-fault circuit-interrupter protection for personnel. In addition to this required ground-fault circuit-interrupter protection for personnel, listed cord sets or devices incorporating listed ground-fault circuit-interrupter protection for personnel identified for portable use shall be permitted.

(2) Receptacle Outlets Existing or Installed as Permanent Wiring. Ground-fault circuit-interrupter protection for personnel shall be provided for all 125-volt, single-phase, 15-, 20-, and 30-ampere receptacle outlets installed or existing as part of the permanent wiring of the building or structure and used for temporary electric power. Listed cord sets or devices incorporating listed ground-fault circuit-interrupter protection for personnel identified for portable use shall be permitted.

(3) Receptacles on 15-kW or less Portable Generators. All 125-volt and 125/250-volt, single-phase, 15-, 20-, and 30-ampere receptacle outlets that are a part of a 15-kW or smaller portable generator shall have listed ground-fault circuit-interrupter protection for personnel. All 15- and 20-ampere, 125- and 250-volt receptacles, including those that are part of a portable generator, used in a damp or wet location shall comply with 406.9(A) and (B). Listed cord sets or devices incorporating listed ground-fault circuit-interrupter protection for personnel identified for portable use shall be permitted for use with 15-kW or less portable generators manufactured or remanufactured prior to January 1, 2015.

Requiring GFCI protection of all temporarily installed, 125-volt, single-phase, 15-, 20-, and 30-ampere receptacles is intended to protect personnel using these receptacles from shock hazards that could be encountered during construction and maintenance activities. Receptacles on a construction site provide power via the temporary wiring system or the permanent premises wiring system of the structure. The latter method can be used when the premises wiring is available prior to project completion. This requirement applies even where the final occupancy would not require GFCI protection for the receptacle being utilized.

Section 590.6(A)(3) specifically addresses GFCI protection for small generators that are common at construction sites. Generators manufactured prior to January 1, 2011, were not required to provide this protection. Therefore, listed cord sets or other devices are permitted to provide GFCI protection. Exhibits 590.1 and 590.2 show examples of ways to implement the GFCI requirements for temporary installations.

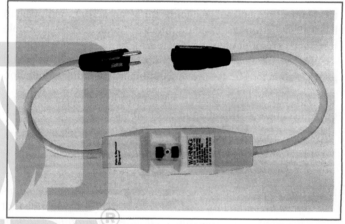

EXHIBIT 590.1 A raintight, portable GFCI with open neutral protection designed for use on the line end of a flexible cord. (*Courtesy of Legrand®*)

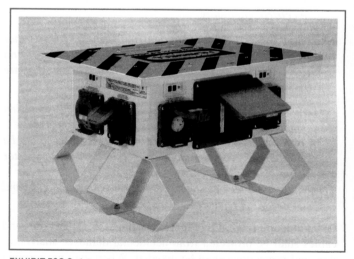

EXHIBIT 590.2 A temporary power outlet unit commonly used on construction sites with a variety of configurations, including GFCI protection. (*Courtesy of Hubbell Wiring Device—Kellems*)

(B) Other Receptacle Outlets. For temporary wiring installations, receptacles, other than those covered by 590.6(A)(1) through (A)(3) used to supply temporary power to equipment used by personnel during construction, remodeling, maintenance, repair, or demolition of buildings, structures, or equipment, or similar activities, shall have protection in accordance with 590.6(B)(1) or the assured equipment grounding conductor program in accordance with 590.6(B)(2).

(1) GFCI Protection. Ground-fault circuit-interrupter protection for personnel.

(2) Assured Equipment Grounding Conductor Program. A written assured equipment grounding conductor program continuously enforced at the site by one or more designated persons to ensure that equipment grounding conductors for all cord sets, receptacles that are not a part of the permanent wiring of the building or structure, and equipment connected by cord and plug are installed and maintained in accordance with the applicable requirements of 250.114, 250.138, 406.4(C), and 590.4(D).

(a) The following tests shall be performed on all cord sets, receptacles that are not part of the permanent wiring of the building or structure, and cord-and-plug-connected equipment required to be connected to an equipment grounding conductor:

(1) All equipment grounding conductors shall be tested for continuity and shall be electrically continuous.
(2) Each receptacle and attachment plug shall be tested for correct attachment of the equipment grounding conductor. The equipment grounding conductor shall be connected to its proper terminal.
(3) All required tests shall be performed as follows:

    a. Before first use on site
    b. When there is evidence of damage
    c. Before equipment is returned to service following any repairs
    d. At intervals not exceeding 3 months

(b) The tests required in 590.6(B)(2)(a) shall be recorded and made available to the authority having jurisdiction.

The assured equipment grounding conductor program shall be documented and made available to the authority having jurisdiction.

> Informational Note: See OSHA 29 CFR 1910 and 1926 for requirements for assured equipment grounding conductor programs. See NFPA *70E*-2018, *Standard for Electrical Safety in the Workplace*, for additional information.

The environmental conditions encountered during construction or demolition could subject personnel to an increased exposure to electrical shock hazards. Receptacle configurations other than 125 volts, single phase, 15, 20, and 30 amperes supplying temporary power must also be GFCI protected or be installed and maintained in accordance with the documented assured EGC program specified in 590.6(B)(2).

The Occupational Safety and Health Administration (OSHA) test requirements are very similar to the *NEC* requirements for an assured grounding program. According to OSHA 29 CFR 1926.404(b)(1)(iii), those test requirements are as follows:

> The employer shall establish and implement an assured equipment grounding conductor program on construction sites covering all cord sets, receptacles which are not a part of the building or structure, and equipment connected by cord and plug which are available for use or used by employees. This program shall comply with the following minimum requirements:
>
> (A) A written description of the program, including the specific procedures adopted by the employer, shall be available at the jobsite for inspection and copying by the Assistant Secretary and any affected employee.
>
> (B) The employer shall designate one or more competent persons to implement the program.

590.7 Guarding. For wiring over 600 volts, nominal, suitable fencing, barriers, or other effective means shall be provided to limit access only to authorized and qualified personnel.

590.8 Overcurrent Protective Devices.

Δ **(A) Where Reused.** Overcurrent protective devices that have been previously used and are installed in a temporary installation shall be examined to ensure they have been properly installed and properly maintained, and there is no evidence of impending failure.

> Informational Note: See the following standards for further information for properly maintained equipment:
>
> (1) NEMA AB 4, *Guidelines for Inspection and Preventive Maintenance of Molded-Case Circuit Breakers Used in Commercial and Industrial Applications*
> (2) NFPA 70B, *Recommended Practice for Electrical Equipment Maintenance*
> (3) NEMA GD 1, *Evaluating Water-Damaged Electrical Equipment*
> (4) IEEE 1458, *IEEE Recommended Practice for the Selection, Field Testing, and Life Expectancy of Molded-Case Circuit Breakers for Industrial Applications*

Δ **(B) Service Overcurrent Protective Devices.** Overcurrent protective devices for solidly grounded wye electrical services of more than 150 volts to ground but not exceeding 1000 volts phase-to-phase, available fault current greater than 10,000 amperes, shall be current limiting.

It is not uncommon for temporary electrical equipment to be used more than once. Temporary equipment is often subjected to conditions that permanently installed equipment is not. Weather, transportation, storage, and use can play a role in creating an unsafe condition within the temporary equipment. For that reason, equipment must be examined before being put back in use.

For equipment that poses a greater risk of causing an injury or death to workers, the overcurrent protective device (OCPD) at the temporary service must be current limiting. This reduces the amount of incident energy available in a temporary system, thereby potentially reducing the risk to employees. This requirement applies only to service overcurrent protective devices rated to interrupt fault currents exceeding 10,000 amperes, which is not uncommon in larger capacity temporary installations such as commercial and industrial construction projects and temporary services for large events such as a music festival or similar event.

Special Equipment

ARTICLE 600 — Electric Signs and Outline Lighting

Part I. General

600.1 Scope. This article covers the installation of conductors, equipment, and field wiring for electric signs, retrofit kits, and outline lighting, regardless of voltage. All installations and equipment using neon tubing, such as signs, decorative elements, skeleton tubing, or art forms, are covered by this article.

> Informational Note: Sign and outline lighting illumination systems include, but are not limited to, cold cathode neon tubing, high-intensity discharge lamps (HID), fluorescent or incandescent lamps, light-emitting diodes (LEDs), and electroluminescent and inductance lighting.

Covered under the requirements of this article are signs of the fixed, stationary, and portable self-contained types. Electric signs and outline lighting frequently include sources of illumination identical to those of luminaires; however, the structure and electrical operation of many electric signs are far more complex than simply a set of fluorescent lamps within an enclosure. The terms *electric sign* and *outline lighting* as defined in Article 100 distinguish the function and use of equipment covered by these requirements from the equipment covered by the requirements of Article 410.

Neon tubing is used extensively in the sign industry, and its uses go far beyond the typical electric sign or outline lighting applications. Neon tubing is used in decorative and artistic applications to enhance the indoor and outdoor appearance of buildings and structures. Neon art forms are mounted on enclosures, sign bodies, and other support structures, or they are field-installed skeleton tubing. Depending on how neon art forms are constructed and installed, they are subject to the requirements of either Part I or Parts I and II of Article 600.

600.3 Listing. Fixed, mobile, or portable electric signs, section signs, outline lighting, photovoltaic (PV) powered signs, and retrofit kits, regardless of voltage, shall be listed and labeled, provided with installation instructions, and installed in conformance with that listing, unless otherwise approved by special permission.

Electric signs and outline lighting are occasionally designed for a very specific purpose and may be a one-time construction for which the manufacturer might not obtain listing. Listing or approval by special permission helps ensure that electrical equipment does not pose a shock or fire hazard.

(A) Field-Installed Skeleton Tubing. Field-installed skeleton tubing shall not be required to be listed where installed in conformance with this *Code*.

(B) Outline Lighting. Outline lighting shall not be required to be listed as a system when it consists of listed luminaires wired in accordance with Chapter 3.

600.4 Markings.

(A) Signs and Outline Lighting Systems. Signs and outline lighting systems shall be listed and labeled; marked with the manufacturer's name, trademark, or other means of identification; and input voltage and current rating.

(B) Signs with a Retrofitted Illumination System. Signs with a retrofitted illumination system shall contain the following:

(1) The sign shall be marked that the illumination system has been replaced.
(2) The marking shall include the kit providers and installer's name, logo, or unique identifier.
(3) Signs equipped with tubular light-emitting diode lamps powered by the existing sign sockets shall include a label alerting the service personnel that the sign has been modified. The label shall meet the requirements of 110.21(B). The label shall also include a warning not to install fluorescent lamps and shall also be visible during relamping.

Retrofit kits for electric signs must be listed in accordance with 600.3. As part of the listing, installation instructions are required to be provided. The installation instructions specify the manufacturer and the type of sign in which the kit is intended to be installed. The marking required by this section will assist the AHJ in verifying that the kit is installed in accordance with the manufacturer's installation instructions.

The label required by item (3) addresses installations in which tubular light-emitting diode (LED) lamps are used with the existing sign sockets. The label is important because installing a fluorescent lamp into retrofitted sockets with unregulated line voltage can be a potential hazard for service personnel.

(C) Signs with Lampholders for Incandescent Lamps. Signs and outline lighting systems with lampholders for incandescent lamps shall be marked to indicate the maximum allowable lamp wattage per lampholder. The markings shall be permanently installed, in letters at least 6 mm (¼ in.) high, and shall be located where visible during relamping.

The required markings are only required to be visible during relamping of the sign. The markings can be placed within the interior of a sign body or sign equipment enclosure.

(D) Visibility. The markings required in 600.4(A) and listing labels shall be visible after installation and shall be permanently applied in a location visible prior to servicing. The marking shall be permitted to be installed in a location not viewed by the public.

(E) Installation Instructions. All signs, outline lighting, skeleton tubing systems, and retrofit kits shall be marked to indicate that field wiring and installation instructions are required.

Exception: Portable, cord-connected signs are not required to be marked.

600.5 Branch Circuits.

Δ **(A) Required Branch Circuit.** Each commercial building and each commercial occupancy accessible to pedestrians shall be provided with at least one outlet in an accessible location at each entrance to each tenant space for sign or outline lighting system use. The outlet(s) shall be supplied by a branch circuit rated at least 20 amperes that supplies no other load.

> *Exception No. 1: A sign or outline lighting outlet shall not be required at entrances for deliveries, service corridors, or service hallways that are intended to be used only by service personnel or employees.*

> *Exception No. 2: The required branch circuit shall be permitted to supply loads directly related to the control of the sign such as electronic or electromechanical controllers.*

This requirement is not contingent on whether an electric sign will be installed at the time an occupant moves in, since it is common to install an electric sign after the space is occupied or when a new occupant moves into an existing space.

(B) Marking. A disconnecting means for a sign, outline lighting system, or controller shall be marked to identify the sign, outline lighting system, or controller it controls.

Exception: An external disconnecting means that is mounted on the sign body, sign enclosure, sign pole, or controller shall not be required to identify the sign or outline lighting system it controls.

(C) Rating. Branch circuits that supply signs shall be rated in accordance with 600.5(C)(1) or (C)(2) and shall be considered to be continuous loads for the purposes of calculations.

(1) Neon Signs. Branch circuits that supply neon tubing installations shall not be rated in excess of 30 amperes.

(2) All Other Signs. Branch circuits that supply all other signs and outline lighting systems shall be rated not to exceed 20 amperes.

Large signs often have load requirements that exceed the maximum rating specified by 600.5(C). These signs typically are supplied by a feeder that in turn supplies branch circuits, which must be rated as specified by this requirement. In some cases, particularly for signs installed along highways or large free-standing signs, a utility service dedicated to the sign is provided. The rating of the feeder or service is not limited by this requirement. Because sign loads are continuous, the conductors and overcurrent protective devices (OCPDs) for circuits supplying sign loads have to be sized in accordance with the requirements for continuous loads contained in Articles 210, 215, and 230.

(D) Wiring Methods. Wiring methods used to supply signs shall comply with 600.5(D)(1), (D)(2), and (D)(3).

(1) Supply. The wiring method used to supply signs and outline lighting systems shall terminate within a sign, an outline lighting system enclosure, a suitable box, a conduit body, or panelboard.

Δ **(2) Enclosures as Pull Boxes.**

(a) Listed and labeled electrical enclosures integral to the sign shall be permitted to be used for voltages up to 600 volts as pull or junction boxes for conductors supplying the following:

(1) Other adjacent signs
(2) Outline lighting systems
(3) Floodlights that are part of a sign

(b) The enclosures in 600.5(D)(2)(a) shall be permitted to contain both branch and secondary circuit conductors.

(c) Listed and labeled neon transformer boxes shall be permitted to contain multiple voltages over 1000 volts. A disconnecting means shall be provided to de-energize all ungrounded conductors in the enclosures.

Δ **(3) Metal or Nonmetallic Poles.** Metal or nonmetallic poles used to support signs shall be permitted to enclose supply conductors.

Δ **600.6 Disconnects.** Each sign and outline lighting system, feeder conductors, or branch circuits supplying a sign, outline lighting system, or skeleton tubing shall be controlled by an externally operable switch or circuit breaker that opens all ungrounded conductors and controls no other load. Signs and outline lighting systems located within fountains shall have the disconnect located in accordance with 680.13.

Exception No. 1: A disconnecting means shall not be required for an exit directional sign located within a building.

Exception No. 2: A disconnecting means shall not be required for cord-connected signs with an attachment plug.

Informational Note: The location of the disconnect is intended to allow service or maintenance personnel and first responders complete and local control of the disconnecting means.

Δ **(A) Location.** The disconnecting means shall be accessible and located in accordance with 600.6(A)(1), 600.6(A)(2), or 600.6(A)(3). If the disconnecting means is remote from the sign it controls, it shall comply with 600.6(A)(4).

Δ **(1) At Point of Entry to a Sign.** The disconnect shall be located at the point the feeder circuit or branch circuits supplying a sign or outline lighting system enters a sign enclosure, a sign body, or a pole in accordance with 600.5(D)(3). The disconnect shall open all ungrounded conductors where it enters the enclosure of the sign or pole.

Exception No. 1: A disconnect shall not be required for branch circuits or feeder conductors passing through the sign where not accessible and enclosed in a Chapter 3 listed raceway or metal-jacketed cable identified for the location.

Exception No. 2: A disconnect shall not be required at the point of entry to a sign enclosure or sign body for branch circuits or feeder conductors that supply an internal panelboards in a sign enclosure or sign body. The conductors shall be enclosed where not accessible in a Chapter 3 listed raceway or metal-jacketed cable identified for the location. A field-applied permanent hazard label that is visible during servicing shall be applied to the raceway at or near the point of entry into the sign enclosure or sign body. The danger label shall state the following: "Danger. This raceway contains energized conductors." The marking shall include the location of the disconnecting means for the energized conductors. The disconnecting means shall be capable of being locked in the open position.

Δ **(2) Within Sight of the Sign.** The disconnecting means shall be within sight of the sign or outline lighting system that it controls. Where the disconnecting means is out of the line of sight from any section that is able to be energized, the disconnecting means shall be lockable in accordance with 110.25. A permanent field-applied marking identifying the location of the disconnecting means shall be applied to the sign in a location visible during servicing.

Δ **(3) Within Sight of the Controller.** The following shall apply for signs or outline lighting systems operated by electronic or electromechanical controllers located external to the sign or outline lighting system:

(1) The disconnecting means shall be located within sight of the controller or in the same enclosure with the controller.
(2) The disconnecting means shall disconnect the sign or outline lighting system and the controller from all ungrounded supply conductors.
(3) The disconnecting means shall be designed such that no pole can be operated independently and shall be lockable in accordance with 110.25.

Exception: Where the disconnecting means is not located within sight of the controller, a permanent field-applied marking identifying the location of the disconnecting means shall be applied to the controller in a location visible during servicing.

Δ **(4) Remote Location.** The disconnecting means, if located remote from the sign, sign body, or pole, shall be mounted at an accessible location available to first responders and service personnel. The location of the disconnect shall be marked with a label at the sign location and marked as the disconnect for the sign or outline lighting system.

A disconnecting means that is not within sight of the sign that it controls is required to be installed where it can be operated by maintenance personnel and by emergency first responders. In accordance with 600.6(A)(2), a remote disconnecting means must be constructed or equipped to be locked in the open position in accordance with 110.25. Additionally, the sign controlled by the remote disconnecting means must have a permanent and durable label that denotes the location of the disconnecting means. The disconnecting means is required to be marked in accordance with 110.22(A).

(B) Control Switch Rating. Switches, flashers, and similar devices controlling transformers and electronic power supplies shall be rated for controlling inductive loads or have a current rating not less than twice the current rating of the transformer or the electronic power supply.

A switching device that controls the primary circuit of a transformer supplying a luminous gas tube is subject to a highly inductive load that causes severe arcing of its contacts. Therefore, the switch or flasher is required to be rated for the inductive load, or it must have a current rating that is at least twice the current rating of the transformer it controls.

600.7 Grounding and Bonding.

(A) Grounding.

(1) Equipment Grounding Conductor. Metal equipment of signs, outline lighting, and skeleton tubing systems shall be grounded by connection to the equipment grounding conductor of the supply branch circuit(s) or feeder using the types of equipment grounding conductors specified in 250.118.

Exception: Portable cord-connected signs shall not be required to be connected to the equipment grounding conductor where protected by a system of double insulation or its equivalent. Double insulated equipment shall be distinctively marked.

(2) Size of Equipment Grounding Conductor. The equipment grounding conductor size shall be in accordance with 250.122.

(3) Connections of Equipment Grounding Conductor. Equipment grounding conductor connections shall be made in accordance with 250.130 and in a method specified in 250.8.

(4) Auxiliary Grounding Electrode. Auxiliary grounding electrode(s) shall be permitted for electric signs and outline lighting systems covered by this article and shall meet the requirements of 250.54.

(5) Metal Building Parts. Metal parts of a building shall not be permitted as a secondary return conductor or an equipment grounding conductor.

(B) Bonding.

(1) Bonding of Metal Parts. Metal parts and equipment of signs and outline lighting systems shall be bonded together and to the associated transformer or power-supply equipment grounding conductor of the branch circuit or feeder supplying the sign or outline lighting system and shall meet the requirements of 250.90.

Exception: Remote metal parts of a section sign or outline lighting system only supplied by a remote Class 2 power supply shall not be required to be bonded to an equipment grounding conductor.

(2) Bonding Connections. Bonding connections shall be made in accordance with 250.8.

Δ **(3) Metal Building Parts.** Metal parts of a building shall not be used as a means for bonding metal parts and equipment of signs or outline lighting systems together or to the transformer or power-supply equipment grounding conductor of the supply circuit.

(4) Flexible Metal Conduit Length. Listed flexible metal conduit or listed liquidtight flexible metal conduit that encloses the secondary circuit conductor from a transformer or power supply for use with neon tubing shall be permitted as a bonding means if the total accumulative length of the conduit in the secondary circuit does not exceed 30 m (100 ft).

Listed flexible metal conduit (FMC) and listed liquidtight flexible metal conduit (LFMC) are suitable as a bonding means in lengths up to 100 feet because the purpose of the bonding specified by 600.7(B)(1) is not to operate an OCPD on the primary side of the transformer or power supply but to minimize differences of potential between the metal parts of signs or outline lighting systems.

(5) Small Metal Parts. Small metal parts not exceeding 50 mm (2 in.) in any dimension, not likely to be energized, and spaced at least 19 mm (¾ in.) from neon tubing shall not require bonding.

(6) Nonmetallic Conduit. Where listed nonmetallic conduit is used to enclose the secondary circuit conductor from a transformer or power supply and a bonding conductor is required, the bonding conductor shall be installed separate and remote from the nonmetallic conduit and be spaced at least 38 mm (1½ in.) from the conduit when the circuit is operated at 100 Hz or less or 45 mm (1¾ in.) when the circuit is operated at over 100 Hz.

Secondary circuit raceways normally contain only one conductor, which is connected to one side of the neon tube. Where nonmetallic conduit is used and any sign parts are required to be bonded, the bonding conductor(s) must be run outside of and be separated from the nonmetallic conduit. Installing bonding conductors inside the nonmetallic conduit with secondary power-supply conductors is prohibited because doing so could increase the chance of failure of the conductor or nonmetallic tubing.

Δ **(7) Bonding Conductors.** Bonding conductors installed outside of a sign or raceway shall be protected from physical damage. Bonding conductors shall comply with 250.120 and 250.122. Bonding conductor size shall also comply with one of the following:

(1) Bonding conductors shall be copper and not smaller than 14 AWG.
(2) Bonding conductors shall be copper-clad aluminum and not smaller than 12 AWG.

Δ **(8) Signs in Fountains.** Signs or outline lighting installed inside a fountain shall have all metal parts bonded to the equipment grounding conductor of the branch circuit for the fountain recirculating system. The bonding connection shall be as near as practicable to the fountain and shall be permitted to be made to metal piping systems that are bonded in accordance with 680.54(B).

Informational Note: See 600.32(J) for restrictions on length of high-voltage secondary conductors.

600.8 Enclosures. Live parts, other than lamps, and neon tubing shall be enclosed. Transformers and power supplies provided with an integral enclosure, including a primary and secondary circuit splice enclosure, shall not require an additional enclosure.

(A) Strength. Enclosures shall have ample structural strength and rigidity.

(B) Material. Sign and outline lighting system enclosures shall be constructed of metal or shall be listed.

(C) Minimum Thickness of Enclosure Metal. Sheet copper or aluminum shall be at least 0.51 mm (0.020 in.) thick. Sheet steel shall be at least 0.41 mm (0.016 in.) thick.

(D) Protection of Metal. Metal parts of equipment shall be protected from corrosion.

600.9 Location.

(A) Vehicles. Sign or outline lighting system equipment shall be at least 4.3 m (14 ft) above areas accessible to vehicles unless protected from physical damage.

(B) Pedestrians. Neon tubing, other than listed, dry-location, portable signs, readily accessible to pedestrians shall be protected from physical damage.

Informational Note: See 600.41(D) for additional requirements.

(C) Adjacent to Combustible Materials. Signs and outline lighting systems shall be installed so that adjacent combustible materials are not subjected to temperatures in excess of 90°C (194°F).

The spacing between wood or other combustible materials and an incandescent or HID lamp or lampholder shall not be less than 50 mm (2 in.).

(D) Wet Location. Signs and outline lighting system equipment for wet location use, other than listed watertight type, shall be weatherproof and have drain holes, as necessary, in accordance with the following:

(1) Drain holes shall not be larger than 13 mm (½ in.) or smaller than 6 mm (¼ in.).
(2) Every low point or isolated section of the equipment shall have at least one drain hole.
(3) Drain holes shall be positioned such that there will be no external obstructions.

600.10 Portable or Mobile Signs.

(A) Support. Portable or mobile signs shall be adequately supported and readily movable without the use of tools.

(B) Attachment Plug. An attachment plug shall be provided for each portable or mobile sign.

(C) Wet or Damp Location. Portable or mobile signs in wet or damp locations shall comply with 600.10(C)(1) and (C)(2).

(1) Cords. All cords shall be junior hard-service or hard-service types as designated in Table 400.4 and have an equipment grounding conductor.

(2) Ground-Fault Circuit Interrupter. In addition to the requirements in 210.8, the manufacturer of portable or mobile signs shall provide listed ground-fault circuit-interrupter protection for personnel. The ground-fault circuit interrupter shall be an integral part of the attachment plug or shall be located in the power-supply cord within 300 mm (12 in.) of the attachment plug.

The type of GFCI required for portable electric signs provides integral open-neutral protection in accordance with UL 48, *Standard for Electric Signs*. An interruption of the neutral conductor on the supply side of the GFCI disables the protection circuitry. Open-neutral protection ensures that if damage to the supply cord causes a break in the grounded conductor, both conductors on the load-side circuit to the portable sign will be opened and no voltage will be present at the sign. These protective devices are required to be original equipment installed by the manufacturer as an integrated GFCI device in the attachment plug or as part of the cord as shown in Exhibit 600.1.

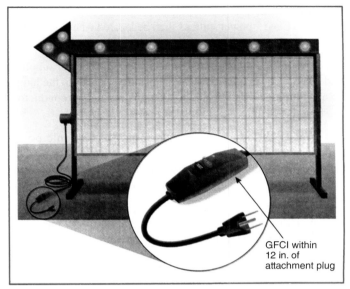

GFCI within 12 in. of attachment plug

EXHIBIT 600.1 A factory-installed GFCI device located in the power-supply cord within 12 inches of the attachment plug.

(D) Dry Location. Portable or mobile signs in dry locations shall meet the following:

(1) Cords shall be SP-2, SPE-2, SPT-2, or heavier, as designated in Table 400.4.
(2) The cord shall not exceed 4.5 m (15 ft) in length.

600.12 Field-Installed Secondary Wiring. Field-installed secondary circuit wiring for electric signs, retrofit kits, outline lighting systems, skeleton tubing, and photovoltaic (PV) powered sign systems shall be in accordance with their installation instructions and 600.12(A), (B), or (C).

(A) 1000 Volts or Less. Neon and secondary circuit wiring of 1000 volts or less shall comply with 600.31.

(B) Over 1000 Volts. Neon secondary circuit wiring of over 1000 volts shall comply with 600.32.

(C) Class 2. Where the installation complies with 600.33 and the power source provides a Class 2 output that complies with 600.24, either of the following wiring methods shall be permitted as determined by the installation instructions and conditions.

(1) Wiring methods identified in Chapter 3
(2) Class 2 cables complying with Table 600.33(A)(1) and Table 600.33(A)(2)

600.21 Ballasts, Transformers, Electronic Power Supplies, and Class 2 Power Sources. Ballasts, transformers, electronic power supplies, and Class 2 power sources shall be of the self-contained type or be enclosed by placement in a listed sign body or listed separate enclosure.

(A) Accessibility. Ballasts, transformers, electronic power supplies, and Class 2 power sources shall be located where accessible and shall be securely fastened in place.

(B) Location. Ballasts, transformers, electronic power supplies, and Class 2 power sources shall be installed as near to the lamps or neon tubing as practicable to keep the secondary conductors as short as possible.

(C) Wet Location. Ballasts, transformers, electronic power supplies, and Class 2 power sources used in wet locations shall be of the weatherproof type or be of the outdoor type and protected from the weather by placement in a sign body or separate enclosure.

(D) Working Space. A working space at least 900 mm (3 ft) high × 900 mm (3 ft) wide × 900 mm (3 ft) deep shall be provided at each ballast, transformer, electronic power supply, and Class 2 power source or at its enclosure where not installed in a sign.

(E) Attic and Soffit Locations. Ballasts, transformers, electronic power supplies, and Class 2 power sources shall be permitted to be located in attics and soffits, provided there is an access door at least 900 mm × 562.5 mm (36 in. × 22½ in.) and a passageway of at least 900 mm (3 ft) high × 600 mm (2 ft) wide with a suitable permanent walkway at least 300 mm (12 in.) wide extending from the point of entry to each component. At least one lighting outlet containing a switch or controlled by a wall switch shall be installed in such spaces. At least one point of control shall be at the usual point of entry to these spaces. The lighting outlet shall be provided at or near the equipment requiring servicing.

(F) Suspended Ceilings. Ballasts, transformers, electronic power supplies, and Class 2 power sources shall be permitted to be located above suspended ceilings, provided that their enclosures are securely fastened in place and not dependent on the suspended-ceiling grid for support. Ballasts, transformers, and electronic power supplies installed in suspended ceilings shall not be connected to the branch circuit by flexible cord.

600.22 Ballasts.

(A) Type. Ballasts shall be identified for the use and shall be listed.

(B) Thermal Protection. Ballasts shall be thermally protected.

600.23 Transformers and Electronic Power Supplies.

(A) Type. Transformers and electronic power supplies shall be identified for the use and shall be listed.

(B) Secondary-Circuit Ground-Fault Protection. Transformers and electronic power supplies other than the following shall have secondary-circuit ground-fault protection:

(1) Transformers with isolated ungrounded secondaries and with a maximum open circuit voltage of 7500 volts or less

(2) Transformers with integral porcelain or glass secondary housing for the neon tubing and requiring no field wiring of the secondary circuit

(C) Voltage. Secondary-circuit voltage shall not exceed 15,000 volts, nominal, under any load condition. The voltage to ground of any output terminals of the secondary circuit shall not exceed 7500 volts, under any load condition.

(D) Rating. Transformers and electronic power supplies shall have a secondary-circuit current rating of not more than 300 mA.

(E) Secondary Connections. Secondary circuit outputs shall not be connected in parallel or in series.

(F) Marking. Transformers and electronic power supplies that are equipped with secondary-circuit ground-fault protection shall be so marked.

600.24 Class 2 Power Sources.
Class 2 transformers, power supplies, and power sources shall comply with the requirements of Class 2 circuits and 600.24(A), (B), (C), and (D).

(A) Listing. Class 2 power supplies and power sources shall be listed for use with electric signs and outline lighting systems or shall be a component in a listed electric sign.

(B) Equipment Grounding Conductor. Metal parts of Class 2 power supplies and power sources shall be connected to the equipment grounding conductor.

(C) Wiring Methods on the Supply Side of the Class 2 Power Supply. Conductors and equipment on the supply side of the power source shall be installed in accordance with the appropriate requirements of Chapter 3.

(D) Secondary Wiring. Secondary wiring on the load side of a Class 2 power source shall comply with 600.12(C) and 600.33.

Part II. Field-Installed Skeleton Tubing, Outline Lighting, and Secondary Wiring

600.30 Applicability.
Part II of this article shall apply to all of the following:

(1) Field-installed skeleton tubing
(2) Field-installed secondary circuits
(3) Outline lighting
(4) Field-installed retrofit kits

These requirements shall be in addition to the requirements of Part I.

600.31 Neon Secondary-Circuit Wiring, 1000 Volts or Less, Nominal.

(A) Wiring Method. Conductors shall be installed using any wiring method included in Chapter 3 suitable for the conditions.

(B) Insulation and Size. Conductors shall be listed, insulated, and not smaller than 18 AWG.

(C) Number of Conductors in Raceway. The number of conductors in a raceway shall be in accordance with Table 1 of Chapter 9.

(D) Installation. Conductors shall be installed so they are not subject to physical damage.

(E) Protection of Leads. Bushings shall be used to protect wires passing through an opening in metal.

600.32 Neon Secondary-Circuit Wiring, over 1000 Volts, Nominal.

(A) Wiring Methods.

(1) Installation. Conductors shall be installed in rigid metal conduit, intermediate metal conduit, liquidtight flexible nonmetallic conduit, flexible metal conduit, liquidtight flexible metal conduit, electrical metallic tubing, metal enclosures; on insulators in metal raceways; or in other equipment listed for use with neon secondary circuits over 1000 volts.

(2) Number of Conductors. Conduit or tubing shall contain only one conductor.

(3) Size. Conduit or tubing shall be a minimum of metric designator 16 (trade size ½).

Δ **(4) Spacing from Grounded Parts.** Other than at the location of connection to a metal enclosure or sign body, nonmetallic conduit or flexible nonmetallic conduit shall comply with the following:

(1) Be spaced not less than 38 mm (1½ in.) from grounded or bonded parts when the conduit contains a conductor operating at 100 Hz or less, or
(2) Be spaced not less than 45 mm (1¾ in.) from grounded or bonded parts when the conduit contains a conductor operating at more than 100 Hz

Where installed in nonmetallic conduit, GTO (gas tube and oil burner ignition) cable located in close proximity to a grounded surface could result in damaging stress to the cable insulation due to capacitive coupling and the resulting production of ozone.

(5) Metal Building Parts. Metal parts of a building shall not be permitted as a secondary return conductor or an equipment grounding conductor.

(B) Insulation and Size. Conductors shall be insulated, listed as gas tube sign and ignition cable type GTO, rated for 5, 10, or 15 kV, not smaller than 18 AWG, and have a minimum temperature rating of 105°C (221°F).

(C) Installation. Conductors shall be so installed that they are not subject to physical damage.

(D) Bends in Conductors. Sharp bends in insulated conductors shall be avoided.

(E) Spacing. Secondary conductors shall be separated from each other and from all objects other than insulators or neon tubing by a spacing of not less than 38 mm (1½ in.). GTO cable installed in metal conduit or tubing shall not require spacing between the cable insulation and the conduit or tubing.

(F) Insulators and Bushings. Insulators and bushings for conductors shall be listed for use with neon secondary circuits over 1000 volts.

(G) Conductors in Raceways. The insulation on all conductors shall extend not less than 65 mm (2½ in.) beyond the metal conduit or tubing.

(H) Between Neon Tubing and Midpoint Return. Conductors shall be permitted to run between the ends of neon tubing or to the secondary circuit midpoint return of listed transformers or listed electronic power supplies and provided with terminals or leads at the midpoint.

(I) Dwelling Occupancies. Equipment having an open circuit voltage exceeding 1000 volts shall not be installed in or on dwelling occupancies.

(J) Length of Secondary Circuit Conductors.

(1) Secondary Conductor to the First Electrode. The length of secondary circuit conductors from a high-voltage terminal or lead of a transformer or electronic power supply to the first neon tube electrode shall not exceed the following:

(1) 6 m (20 ft) where installed in metal conduit or tubing
(2) 15 m (50 ft) where installed in nonmetallic conduit

(2) Other Secondary Circuit Conductors. All other sections of secondary circuit conductor in a neon tube circuit shall be as short as practicable.

(K) Splices. Splices in high-voltage secondary circuit conductors shall be made in listed enclosures rated over 1000 volts. Splice enclosures shall be accessible after installation and listed for the location where they are installed.

600.33 Class 2 Sign Illumination Systems, Secondary Wiring. The wiring methods and materials used shall be in accordance with the sign manufacturer's installation instructions using any applicable wiring methods from Chapter 3, Wiring Methods, or the requirements for Class 2 circuits contained in 600.12(C), 600.24, and 600.33(A), (B), (C), and (D).

(A) Insulation and Sizing of Class 2 Conductors. Class 2 cable listed for the application that complies with Table 600.33(A)(1) or Table 600.33(A)(2) for substitutions shall be installed on the load side of the Class 2 power source. The conductors shall have an ampacity not less than the load to be supplied and shall not be sized smaller than 18 AWG.

TABLE 600.33(A)(1) *Applications of Power Limited Cable in Signs and Outline Lighting*

Location	CL2	CL3	CL2R	CL3R	CL2P	CL3P	PLTC
Nonconcealed spaces inside buildings	Y	Y	Y	Y	Y	Y	Y
Concealed spaces inside buildings that are not used as plenums or risers	Y	Y	Y	Y	Y	Y	Y
Environmental air spaces plenums	N	N	N	N	Y	Y	N
Environmental air spaces risers	N	N	Y	Y	Y	Y	N
Wet locations	N	N	N	N	N	N	Y

Y = Permitted. N = Not Permitted.

TABLE 600.33(A)(2) *Class 2 Cable Substitutions*

Cable Type	Permitted Substitutions
CL3P	CMP
CL2P	CMP, CL3P
CL3R	CMP, CL3P, CMR
CL2R	CMP, CL3P, CL2P, CMR, CL3R
CL3	CMP, CL3P, CMR, CL3R, CMG, CM, PLTC
CL2	CMP, CL3P, CL2P, CMR, CL3R, CL2R, CMG, CM, PLTC, CL3
CL3X	CMP, CL3P, CMR, CL3R, CMG, CM, PLTC, CL3, CMX
CL2X	CMP, CL3P, CL2P, CMR, CL3R, CL2R, CMG, CM, PLTC, CL3, CL2, CMX, CL3X
PLTC	None

(1) General Use. CL2 or CL3, PLTC, or any listed applicable cable for general use shall be installed within and on buildings or structures.

(2) Other Building Locations. In other locations, any listed applicable cable permitted in 600.33(A)(1), (A)(2), (A)(3), and (A)(4) and Table 600.33(A)(1) and Table 600.33(A)(2) shall be permitted to be used as follows:

(1) CL2P or CL3P — Ducts, plenums, or other spaces used for environmental air
(2) CL2R or CL3R — Vertical shafts and risers
(3) Substitutions from Table 600.33(A)(2)

(3) Wet Locations. Class 2 cable used in a wet location shall be listed and marked suitable for use in a wet location.

(4) Other Locations. Class 2 cable exposed to sunlight shall be listed and marked "sunlight resistant — suitable for outdoor use."

Exception: Listed PLTC not marked as sunlight resistant shall be permitted.

Informational Note: PLTC is tested for exposure to sunlight but might not be so marked.

(B) Installation. Secondary wiring shall be installed in accordance with 600.33(B)(1) and (B)(2).

(1) Wiring shall be installed and supported in a neat and workmanlike manner. Cables and conductors installed exposed on the surface of ceilings and sidewalls shall be supported by the building structure in such a manner that the cable is not damaged by normal building use. The cable shall be supported and secured at intervals not exceeding 1.8 m (6 ft). Such cables shall be supported by straps, staples, hangers, cable ties, or similar fittings designed and installed so as not to damage the cable. The installation shall also comply with 300.4(D).

(2) Connections in cable and conductors shall be made with listed insulating devices and be accessible after installation. Where made in a wall, connections shall be enclosed in a listed box.

(C) Protection Against Physical Damage. If subject to physical damage, the conductors shall be protected and installed in accordance with 300.4. All through-wall penetrations shall be protected by a listed bushing or raceway.

(D) Grounding and Bonding. Grounding and bonding shall be in accordance with 600.7.

600.34 Photovoltaic (PV) Powered Sign. All field wiring of components and subassemblies for an off-grid stand-alone, on-grid interactive, or non-grid interactive PV installation shall be installed in accordance with Article 690, as applicable, 600.34, and the PV powered sign installation instructions.

(A) Equipment. Inverters, motor generators, PV modules, PV panels, ac PV modules, dc combiners, dc-ac converters, and charge controllers intended for use in PV powered sign systems shall be listed for PV application.

(B) Wiring. Wiring from a photovoltaic panel or wiring external to the PV sign body shall be:

(1) Listed, labeled, and suitable for photovoltaic applications
(2) Routed to closely follow the sign body or enclosure
(3) As short as possible and secured at intervals not exceeding 0.91 m (3 ft)
(4) Protected where subject to physical damage

(C) Flexible Cords and Cables. Flexible cords and cables shall comply with Article 400 and be identified as extra hard usage, rated for outdoor use, and water and sunlight resistant.

(D) Grounding. Grounding a PV powered sign shall comply with Article 690, Part V and 600.7.

(E) Disconnecting Means. The disconnecting means for a PV powered sign shall comply with Article 690, Part III and 600.6.

(F) Battery Compartments. Battery compartments shall require a tool to open.

600.35 Retrofit Kits.

(A) General. A general-use or sign-specific retrofit kit for a sign or outline lighting system shall include installation instructions and requirements for field conversion of a host sign. The retrofit kit shall be listed and labeled.

See also

Article 100 for definitions of *host sign; retrofit kit, general use (general use retrofit kit);* and *retrofit kit, sign specific (sign specific retrofit kit)*

(B) Damaged Parts. All parts that are not replaced by a retrofit kit shall be inspected for damage. Any part found to be damaged or damaged during conversion of the sign shall be replaced or repaired to maintain the sign or outline lighting system's dry, damp, or wet location rating.

(C) Marking. The retrofitted sign shall be marked in accordance with 600.4(B).

600.41 Neon Tubing.

(A) Design. The length and design of the tubing shall not cause a continuous overcurrent beyond the design loading of the transformer or electronic power supply.

(B) Support. Tubing shall be supported by listed tube supports. The neon tubing shall be supported within 150 mm (6 in.) from the electrode connection.

(C) Spacing. A spacing of not less than 6 mm (¼ in.) shall be maintained between the tubing and the nearest surface, other than its support.

(D) Protection. Field-installed skeleton tubing shall not be subject to physical damage. Where the tubing is readily accessible to other than qualified persons, field-installed skeleton tubing

shall be provided with suitable guards or protected by other approved means.

600.42 Electrode Connections.

(A) Points of Transition. Where the high-voltage secondary circuit conductors emerge from the wiring methods specified in 600.32(A), they shall be enclosed in a listed assembly.

(B) Accessibility. Terminals of the electrode shall not be accessible to unqualified persons.

(C) Electrode Connections. Connections shall be made by use of a connection device, twisting of the wires together, or use of an electrode receptacle. Connections shall be electrically and mechanically secure and shall be in an enclosure listed for the purpose.

(D) Support. Neon secondary conductor(s) shall be supported not more than 150 mm (6 in.) from the electrode connection to the tubing.

(E) Receptacles. Electrode receptacles shall be listed.

(F) Bushings. Where electrodes penetrate an enclosure, bushings listed for the purpose shall be used unless receptacles are provided.

(G) Wet Locations. A listed cap shall be used to close the opening between neon tubing and a receptacle where the receptacle penetrates a building. Where a bushing or neon tubing penetrates a building, the opening between neon tubing and the bushing shall be sealed.

(H) Electrode Enclosures. Electrode enclosures shall be listed.

(1) Dry Locations. Electrode enclosures that are listed, labeled, and identified for use in dry, damp, or wet locations shall be permitted to be installed and used in such locations.

(2) Damp and Wet Locations. Electrode enclosures installed in damp and wet locations shall be specifically listed, labeled, and identified for use in such locations.

Informational Note: See 110.3(B) covering installation and use of electrical equipment.

ARTICLE 604 Manufactured Wiring Systems

604.1 Scope. This article applies to field-installed wiring using off-site manufactured subassemblies for branch circuits, remote-control circuits, signaling circuits, and communications circuits in accessible areas.

604.6 Listing Requirements. Manufactured wiring systems and associated components shall be listed.

Informational Note: See ANSI/UL 183, *Standard for Manufacturing Wiring Systems*, the safety standard for manufactured wiring systems.

604.7 Installation. Manufactured wiring systems shall be secured and supported in accordance with the applicable cable or conduit article for the cable or conduit type employed.

The securing and supporting requirements for manufactured wiring systems are taken from the specific article in Chapter 3 that covers the wiring method employed in the system construction. Typically, manufactured wiring systems are constructed of armored cable (Type AC), covered by Article 320, or of metal-clad cable (Type MC), covered by Article 330.

604.10 Uses Permitted. Manufactured wiring systems shall be permitted in accessible and dry locations and in ducts, plenums, and other air-handling spaces where listed for this application and installed in accordance with 300.22.

Exception No. 1: In concealed spaces, one end of tapped cable shall be permitted to extend into hollow walls for direct termination at switch and outlet points.

Exception No. 2: Manufactured wiring system assemblies installed outdoors shall be listed for use in outdoor locations.

Manufactured wiring systems are constructed of Type AC or Type MC cable and are provided with factory connectors and receptacles. The connection devices used with these systems facilitate ease of initial installation and future relocation of equipment. These systems are used extensively for the installation of branch-circuit and tap conductors supplying luminaires in accessible locations, including open and suspended-ceiling construction. Manufactured wiring systems employing flexible conduits, flexible cords, busways, and surface-mounted raceways also are permitted by this article.

604.12 Uses Not Permitted. Manufactured wiring system types shall not be permitted where limited by the applicable article in Chapter 3 for the wiring method used in its construction.

604.100 Construction.

(A) Cable, Conduit, and Tubing Types.

(1) Cables. Cable shall be listed Type AC cable or listed Type MC cable containing nominal 600-volt, 8 AWG to 12 AWG insulated copper-clad aluminum or copper conductors.

Other cables as listed in 722.135, 800.113, and 830.179 shall be permitted in manufactured wiring systems for wiring of equipment within the scope of their respective articles.

(2) Conduits and Tubing. Conduit shall be listed flexible metal conduit (FMC), listed liquidtight flexible metal conduit (LFMC), liquidtight flexible nonmetallic conduit (LFNC), or electrical metallic tubing (EMT) containing nominal 600-volt, 8 AWG to 12 AWG insulated copper-clad aluminum or copper conductors with a bare or insulated copper-clad aluminum or

copper equipment grounding conductor equivalent in size to the ungrounded conductor.

Exception No. 1 to (1) and (2): A luminaire tap, no longer than 1.8 m (6 ft) and intended for connection to a single luminaire, shall be permitted to contain conductors smaller than 12 AWG but not smaller than 18 AWG.

Exception No. 2 to (1) and (2): Listed manufactured wiring assemblies containing conductors smaller than 12 AWG shall be permitted for remote-control, signaling, or communications circuits.

Exception No. 3 to (2): Listed manufactured wiring systems containing unlisted flexible metal conduit of noncircular cross section or trade sizes smaller than permitted by 348.20(A), or both, shall be permitted where the wiring systems are supplied with fittings and conductors at the time of manufacture.

(3) Flexible Cord. Flexible cord suitable for hard usage, with minimum 12 AWG conductors, shall be permitted as part of a listed factory-made assembly not exceeding 1.8 m (6 ft) in length when making a transition between components of a manufactured wiring system and utilization equipment not permanently secured to the building structure. The cord shall be visible for the entire length, shall not be subject to physical damage, and shall be provided with identified strain relief.

Flexible cord facilitates a transition between manufactured wiring systems and utilization equipment found in display cases, merchandise racks, temporary workstations, and the like. This transition is limited, however, to hard-usage cord not over 6 feet in length to minimize damage, as illustrated in Exhibit 604.1. Examples of polarized receptacles and connectors are shown in Exhibit 604.2.

Exception: Listed electric-discharge luminaires that comply with 410.62(C) shall be permitted with conductors smaller than 12 AWG.

This exception and the requirements in 410.62(C)(1) permit the use of flexible cord equipped with a manufactured wiring system connector as a means to supply listed electric-discharge luminaires such as fluorescent or high-intensity discharge types. In this application, the cord-equipped luminaires are supplied from branch-circuit conductors installed using a manufactured wiring system. This method of supplying luminaires is permitted only where the cord is visible for its entire length, from its attachment to the luminaire to its interface with the branch-circuit conductors of the manufactured wiring system. Where used for connection of listed electric-discharge luminaires, listed manufactured wiring system cord assemblies not longer than 6 feet and containing conductors smaller than 12 AWG copper are permitted.

(4) Busways. Busways shall be listed continuous plug-in type containing factory-mounted, bare or insulated conductors, which shall be copper or aluminum bars, rods, or tubes. The busway shall be provided with an equipment ground. The busway shall

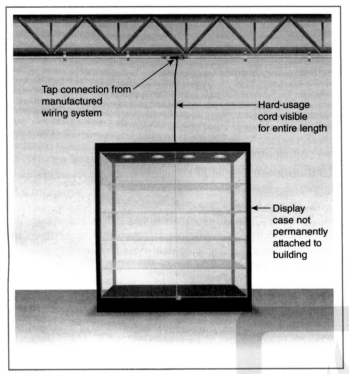

EXHIBIT 604.1 Transition wiring between a manufactured wiring system and utilization equipment.

Tap connection from manufactured wiring system

Hard-usage cord visible for entire length

Display case not permanently attached to building

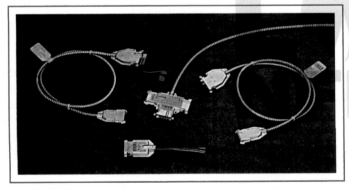

EXHIBIT 604.2 Polarized receptacles and connectors of a manufactured wiring system. *(Courtesy of RELOC® Wiring Solutions, an Acuity Brands Company)*

be rated nominal 600 volts, 20, 30, or 40 amperes. Busways shall be installed in accordance with 368.12, 368.17(D), and 368.30.

(5) Raceway. Prewired, modular, surface-mounted raceways shall be listed for the use, rated nominal 600 volts, 20 amperes, and installed in accordance with 386.12, 386.30, 386.60, and 386.100.

Metal and nonmetallic surface-mounted raceways prewired as a manufactured wiring system are required to be listed specifically for the application. ANSI/UL 183, *Standard for Manufactured Wiring Systems,* covers the construction of these systems. Article 380 contains the requirements for the installation of surface,

flush, or freestanding raceways that contain conductors and receptacles. Although multioutlet assemblies are permitted to be assembled in the field, listed multioutlet assemblies are covered by ANSI/UL 5, *Surface Metal Raceways and Fittings,* and ANSI/UL 5A, *Nonmetallic Surface Raceways and Fittings.*

(B) Marking. Each section shall be marked to identify the type of cable, flexible cord, or conduit.

(C) Receptacles and Connectors. Receptacles and connectors shall be of the locking type, uniquely polarized and identified for the purpose, and shall be part of a listed assembly for the appropriate system. All connector openings shall be designed to prevent inadvertent contact with live parts or capped to effectively close the connector openings.

(D) Other Component Parts. Other component parts shall be listed for the appropriate system.

ARTICLE 605 Office Furnishings

605.1 Scope.

(A) Covered. This article covers electrical equipment, lighting accessories, and wiring systems used to connect, contained within, or installed on office furnishings.

(B) Not Covered. This article does not apply to individual office furnishings not connected to a system, such as chairs, freestanding desks, tables, storage units, and shelving units.

This article covers electrical equipment and conductors installed in office furnishings. Office furnishings may be partitions that are freestanding or fixed but are not as permanent as a conventional stud-and-wallboard type of partition. They may also be storage units, desks, and workstations that are interconnected much in the same way traditional office partitions are connected. This article does not cover office furnishings that are not connected to a powered office furnishing system. The electrical equipment and devices shown in Exhibit 605.1 are components of a modular office furnishing system.

605.3 General. Wiring systems shall be identified as suitable for providing power for lighting accessories and utilization equipment used within office furnishings. A wired partition shall not extend from floor to ceiling.

Exception: Where permitted by the authority having jurisdiction, these relocatable wired partitions shall be permitted to extend to, but shall not penetrate, the ceiling.

605.4 Wireways. All conductors and connections shall be contained within wiring channels of metal or other material identified as suitable for the conditions of use. Wiring channels shall be free of projections or other conditions that might damage conductor insulation.

EXHIBIT 605.1 Example of a fixed-type office partition where the branch-circuit wiring is run inside the partition.

A wiring channel that is separate from the channel containing the branch circuits for light and power may be provided within the system components for the routing of communications, signaling, and optical fiber cables.

605.5 Office Furnishing Interconnections. The electrical connection between office furnishings shall be a flexible assembly identified for use with office furnishings or shall be permitted to be installed using flexible cord, provided that all the following conditions are met:

(1) The cord is extra-hard usage type with 12 AWG or larger conductors, with an insulated equipment grounding conductor.
(2) The office furnishings are mechanically contiguous.
(3) The cord is not longer than necessary for maximum positioning of the office furnishing but is in no case to exceed 600 mm (2 ft).
(4) The cord is terminated at an attachment plug-and-cord connector with strain relief.

605.6 Lighting Accessories. Lighting equipment shall be listed, labeled, and identified for use with office furnishings and shall comply with 605.6(A), (B), and (C).

(A) Support. A means for secure attachment or support shall be provided.

(B) Connection. Where cord and plug connection is provided, it shall comply with all of the following:

(1) The cord length shall be suitable for the intended application but shall not exceed 2.7 m (9 ft) in length.
(2) The cord shall not be smaller than 18 AWG.
(3) The cord shall contain an equipment grounding conductor, except as specified in 605.6(B)(4).

(4) Cords on the load side of a listed Class 2 power source shall not be required to contain an equipment grounding conductor.
(5) The cord shall be of the hard usage type, except as specified in 605.6(B)(6).
(6) A cord provided on a listed Class 2 power source shall be of the type provided with the listed luminaire assembly or of the type specified in 725.130 and 725.127.
(7) Connection by other means shall be identified as suitable for the conditions of use.

(C) Receptacle Outlet. Receptacles shall not be permitted in lighting accessories.

605.7 Fixed-Type Office Furnishings. Office furnishings that are fixed (secured to building surfaces) shall be permanently connected to the building electrical system by one of the wiring methods of Chapter 3.

605.8 Freestanding-Type Office Furnishings. Office furnishings of the freestanding type (not fixed) shall be permitted to be connected to the building electrical system by one of the wiring methods of Chapter 3.

Office furnishings that are attached to the building are required by 605.7 to be connected to the premises wiring with a permanent wiring method. Freestanding furnishings are permitted by 605.8 to connect with a permanent wiring method or by 605.9 to connect with a flexible cord. Although multiwire branch circuits are not permitted in cord-connected furnishings by 605.9(D), they may be supplied by multiple branch circuits.

605.9 Freestanding-Type Office Furnishings, Cord-and Plug-Connected. Individual office furnishings of the freestanding type, or groups of individual office furnishings that are electrically connected, are mechanically contiguous, and do not exceed 9.0 m (30 ft) when assembled, shall be permitted to be connected to the building electrical system by a single flexible cord and plug, provided that all of the conditions of 605.9(A) through (D) are met.

(A) Flexible Power-Supply Cord. The flexible power supply cord shall be extra-hard usage type with 12 AWG or larger conductors, with an insulated equipment grounding conductor, and shall not exceed 600 mm (2 ft) in length.

(B) Receptacle Supplying Power. The receptacle(s) supplying power shall be on a separate circuit serving only the office furnishing and no other loads and shall be located not more than 300 mm (12 in.) from the office furnishing that is connected to it.

(C) Receptacle, Maximum. An individual office furnishing or groups of interconnected individual office furnishings shall not contain more than 13 15-ampere, 125-volt receptacles. For purposes of this requirement, a receptacle is considered (1) up to two (simplex) receptacles provided within a single enclosure and that are within 0.3 m (1 ft) of each other or (2) one duplex receptacle.

(D) Multiwire Circuits, Not Permitted. An individual office furnishing or groups of interconnected office furnishings shall not contain multiwire circuits.

Informational Note: See 210.4 for circuits supplying office furnishings in 605.7 and 605.8.

ARTICLE 610 Cranes and Hoists

Part I. General

610.1 Scope. This article covers the installation of electrical equipment and wiring used in connection with cranes, monorail hoists, hoists, and all runways.

Informational Note: See ASME B30, *Safety Standards for Cableways, Cranes, Derricks, Hoists, Hooks, Jacks, and Slings*, for further information.

The requirements of Article 610 should be closely followed when providing power to electric cranes to help ensure overall electrical safety. Electric cranes, such as the ones shown in Exhibit 610.1, present unique challenges to ensuring that electrical safety is maintained. Constant movement of the crane requires flexibility of power and control wiring or the use of contact conductors installed along the crane runway or bridge. The duty cycle of crane motors is addressed in Table 610.14(A), which covers conductor ampacities for short-time rated crane and hoist motors.

610.3 Special Requirements for Particular Locations.

(A) Hazardous (Classified) Locations. All equipment that operates in a hazardous (classified) location shall conform to Article 500.

EXHIBIT 610.1 Example of crane installations and the electrical equipment and wiring used. *(Getty Images)*

(1) Class I Locations. Equipment used in locations that are hazardous because of the presence of flammable gases or vapors shall conform to Article 501.

(2) Class II Locations. Equipment used in locations that are hazardous because of combustible dust shall conform to Article 502.

(3) Class III Locations. Equipment used in locations that are hazardous because of the presence of easily ignitible fibers or flyings shall conform to Article 503.

See also

503.155 commentary for more details on cranes and hoists in Class III, Divisions 1 and 2, locations

(B) Combustible Materials. Where a crane, hoist, or monorail hoist operates over readily combustible material, the resistors shall be located as permitted in the following:

(1) A well ventilated cabinet composed of noncombustible material constructed so that it does not emit flames or molten metal
(2) A cage or cab constructed of noncombustible material that encloses the sides of the cage or cab from the floor to a point at least 150 mm (6 in.) above the top of the resistors

(C) Electrolytic Cell Lines. See 668.32.

Special precautions are necessary on electrolytic cell lines to prevent the introduction of exposed grounded parts. Conductive surfaces of cranes in the cell line work zone are to be insulated from ground as described in 668.32.

Part II. Wiring

610.11 Wiring Method. Conductors shall be enclosed in raceways or be Type AC cable with insulated equipment grounding conductor, Type MC cable, or Type MI cable unless otherwise permitted or required in 610.11(A) through (E).

For Type AC cable, an insulated wire-type equipment grounding conductor (EGC) terminated on the grounding terminals of crane- and hoist-associated equipment is required to ensure the continuity of the grounding and bonding connection to equipment that is frequently subject to vibration.

(A) Contact Conductor. Contact conductors shall not be required to be enclosed in raceways.

(B) Exposed Conductors. Short lengths of exposed conductors at resistors, collectors, and other equipment shall not be required to be enclosed in raceways.

Short runs of open conductors facilitate connection to resistors, collectors, and similar equipment. Each conductor is required by 610.12 to be provided with separately bushed holes in boxes as well as in cable and raceway fittings used where the transition to open wiring is made.

(C) Flexible Connections to Motors and Similar Equipment. Where flexible connections are necessary, flexible stranded conductors shall be used. Conductors shall be in flexible metal conduit, liquidtight flexible metal conduit, liquidtight flexible nonmetallic conduit, multiconductor cable, or an approved nonmetallic flexible raceway.

(D) Pushbutton Station Multiconductor Cable. Where multiconductor cable is used with a suspended pushbutton station, the station shall be supported in some satisfactory manner that protects the electrical conductors against strain.

Exhibit 610.2 shows an example of suitable strain relief for a cord that supports a control pushbutton station for an overhead crane.

(E) Flexibility to Moving Parts. Where flexibility is required for power or control to moving parts, listed festoon cable or a cord suitable for the purpose shall be permitted, provided the following apply:

(1) Suitable strain relief and protection from physical damage is provided.
(2) In Class I, Division 2 locations, the cord is approved for extra-hard usage.

610.12 Raceway or Cable Terminal Fittings. Conductors leaving raceways or cables shall comply with either 610.12(A) or (B).

(A) Separately Bushed Hole. A box or terminal fitting that has a separately bushed hole for each conductor shall be used wherever a change is made from a raceway or cable to exposed wiring. A fitting used for this purpose shall not contain taps or splices and shall not be used at luminaire outlets.

(B) Bushing in Lieu of a Box. A bushing shall be permitted to be used in lieu of a box at the end of a rigid metal conduit, intermediate metal conduit, or electrical metallic tubing where the raceway terminates at unenclosed controls or similar equipment, including contact conductors, collectors, resistors, brakes, power-circuit limit switches, and dc split-frame motors.

610.13 Types of Conductors. Conductors shall comply with Table 310.4(1) unless otherwise permitted in 610.13(A) through (C).

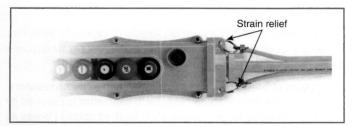

EXHIBIT 610.2 A suitable strain relief grip for a cord-suspended pushbutton station. *(Courtesy of Magnetek)*

(A) Exposed to External Heat or Connected to Resistors. A conductor(s) exposed to external heat or connected to resistors shall have a flame-resistant outer covering or be covered with flame-resistant tape individually or as a group.

(B) Contact Conductors. Contact conductors along runways, crane bridges, and monorails shall be permitted to be bare and shall be copper, aluminum, steel, or other alloys or combinations thereof in the form of hard-drawn wire, tees, angles, tee rails, or other stiff shapes.

(C) Flexibility. Where flexibility is required, listed flexible cord or cable, or listed festoon cable, shall be permitted to be used and, where necessary, cable reels or take-up devices shall be used.

610.14 Rating and Size of Conductors.

(A) Ampacity. The ampacities of conductors shall be as shown in Table 610.14(A).

Informational Note: See 430.23 for the ampacities of conductors between controllers and resistors.

(B) Secondary Resistor Conductors. Where the secondary resistor is separate from the controller, the minimum size of the conductors between controller and resistor shall be calculated by multiplying the motor secondary current by the appropriate factor from Table 610.14(B) and selecting a wire from Table 610.14(A).

(C) Minimum Size. Conductors external to motors and controls shall be not smaller than 16 AWG unless otherwise permitted in either of the following:

(1) 18 AWG wire in multiconductor cord shall be permitted for control circuits not exceeding 7 amperes.
(2) Wires not smaller than 20 AWG shall be permitted for electronic circuits.

(D) Contact Conductors. Contact wires shall have an ampacity not less than that required by Table 610.14(A) for 75°C (167°F) wire, and in no case shall they be smaller than as shown in Table 610.14(D).

(E) Calculation of Motor Load.

(1) Single Motor. For one motor, 100 percent of motor nameplate full-load ampere rating shall be used.

(2) Multiple Motors on Single Crane or Hoist. For multiple motors on a single crane or hoist, the minimum ampacity of the power supply conductors shall be the nameplate full-load ampere rating of the largest motor or group of motors for any single crane motion, plus 50 percent of the nameplate full-load ampere rating of the next largest motor or group of motors, using that column of Table 610.14(A) that applies to the longest time-rated motor.

Δ **TABLE 610.14(A)** *Ampacities of Insulated Copper Conductors Used with Short-Time Rated Crane and Hoist Motors. Based on Ambient Temperature of 30°C (86°F)*

Maximum Operating Temperature	Up to Four Simultaneously Energized Conductors in Raceway or Cable[1]				Up to Three ac[2] or Four dc[1] Simultaneously Energized Conductors in Raceway or Cable		Maximum Operating Temperature
	75°C (167°F)		90°C (194°F)		125°C (257°F)		
Size (AWG or kcmil)	Types MTW, RHW, THW, THWN, XHHW, USE, ZW		Types TA, TBS, SA, SIS, PFA, FEP, FEPB, RHH, THHN, XHHW, Z, ZW		Types FEP, FEPB, PFA, PFAH, SA, TFE, Z, ZW		Size (AWG or kcmil)
	60 Min	30 Min	60 Min	30 Min	60 Min	30 Min	
16	10	12	—	—	—	—	16
14	25	26	31	32	38	40	14
12	30	33	36	40	45	50	12
10	40	43	49	52	60	65	10
8	55	60	63	69	73	80	8
6	76	86	83	94	101	119	6
5	85	95	95	106	115	134	5
4	100	117	111	130	133	157	4
3	120	141	131	153	153	183	3
2	137	160	148	173	178	214	2
1	143	175	158	192	210	253	1
1/0	190	233	211	259	253	304	1/0
2/0	222	267	245	294	303	369	2/0
3/0	280	341	305	372	370	452	3/0
4/0	300	369	319	399	451	555	4/0
250	364	420	400	461	510	635	250
300	455	582	497	636	587	737	300
350	486	646	542	716	663	837	350
400	538	688	593	760	742	941	400
450	600	765	660	836	818	1042	450
500	660	847	726	914	896	1143	500

Δ
AMPACITY CORRECTION FACTORS

Ambient Temperature (°C)	For ambient temperatures other than 30°C (86°F), multiply the ampacities shown above by the appropriate factor shown below.						Ambient Temperature (°F)
21–25	1.05	1.05	1.04	1.04	1.02	1.02	70–77
26–30	1.00	1.00	1.00	1.00	1.00	1.00	79–86
31–35	0.94	0.94	0.96	0.96	0.97	0.97	88–95
36–40	0.88	0.88	0.91	0.91	0.95	0.95	97–104
41–45	0.82	0.82	0.87	0.87	0.92	0.92	106–113
46–50	0.75	0.75	0.82	0.82	0.89	0.89	115–122
51–55	0.67	0.67	0.76	0.76	0.86	0.86	124–131
56–60	0.58	0.58	0.71	0.71	0.83	0.83	133–140
61–70	0.33	0.33	0.58	0.58	0.76	0.76	142–158
71–80	—	—	0.41	0.41	0.69	0.69	160–176
81–90	—	—	—	—	0.61	0.61	177–194
91–100	—	—	—	—	0.51	0.51	195–212
101–120	—	—	—	—	0.40	0.40	213–248

Note: Other insulations shown in Table 310.4(1) and approved for the temperature and location shall be permitted to be substituted for those shown in Table 610.14(A). The allowable ampacities of conductors used with 15-minute motors shall be the 30-minute ratings increased by 12 percent.

[1]For 5 to 8 simultaneously energized power conductors in raceway or cable, the ampacity of each power conductor shall be reduced to a value of 80 percent of that shown in this table.

[2]For 4 to 6 simultaneously energized 125°C (257°F) ac power conductors in raceway or cable, the ampacity of each power conductor shall be reduced to a value of 80 percent of that shown in this table.

TABLE 610.14(B) *Secondary Conductor Rating Factors*

Time in Seconds		Ampacity of Wire in Percent of Full-Load Secondary Current
On	**Off**	
5	75	35
10	70	45
15	75	55
15	45	65
15	30	75
15	15	85
Continuous Duty		110

TABLE 610.14(D) *Minimum Contact Conductor Size Based on Distance Between Supports*

Minimum Size of Wire (AWG)	Maximum Distance Between End Strain Insulators or Clamp-Type Intermediate Supports
6	9.0 m (30 ft) or less
4	18 m (60 ft) or less
2	Over 18 m (60 ft)

(3) Multiple Cranes or Hoists on a Common Conductor System. For multiple cranes, hoists, or both, supplied by a common conductor system, calculate the motor minimum ampacity shall be calculated for each crane as defined in 610.14(E), added them together, and the sum multiplied by the appropriate demand factor from Table 610.14(E)(3).

TABLE 610.14(E)(3) *Demand Factors*

Number of Cranes or Hoists	Demand Factor
2	0.95
3	0.91
4	0.87
5	0.84
6	0.81
7	0.78

(F) Other Loads. Additional loads, such as heating, lighting, and air conditioning, shall be provided for by application of the appropriate sections of this *Code*.

(G) Nameplate. Each crane, monorail, or hoist shall be provided with a visible nameplate marked with the manufacturer's name, rating in volts, frequency, number of phases, and circuit amperes as calculated in 610.14(E) and (F).

610.15 Common Return. Where a crane or hoist is operated by more than one motor, a common-return conductor of proper ampacity shall be permitted.

Part III. Contact Conductors

610.21 Installation of Contact Conductors. Contact conductors shall comply with 610.21(A) through (H).

(A) Locating or Guarding Contact Conductors. Runway contact conductors shall be guarded, and bridge contact conductors shall be located or guarded in such a manner that persons cannot inadvertently touch energized current-carrying parts.

(B) Contact Wires. Wires that are used as contact conductors shall be secured at the ends by means of approved strain insulators and shall be mounted on approved insulators so that the extreme limit of displacement of the wire does not bring the latter within less than 38 mm (1½ in.) from the surface wired over.

(C) Supports Along Runways. Main contact conductors carried along runways shall be supported on insulating supports placed at intervals not exceeding 6.0 m (20 ft) unless otherwise permitted in 610.21(F).

Such conductors shall be separated at not less than 150 mm (6 in.), other than for monorail hoists where a spacing of not less than 75 mm (3 in.) shall be permitted. Where necessary, intervals between insulating supports shall be permitted to be increased up to 12 m (40 ft), the separation between conductors being increased proportionally.

(D) Supports on Bridges. Bridge wire contact conductors shall be kept at least 65 mm (2½ in.) apart, and, where the span exceeds 25 m (80 ft), insulating saddles shall be placed at intervals not exceeding 15 m (50 ft).

(E) Supports for Rigid Conductors. Conductors along runways and crane bridges, that are of the rigid type specified in 610.13(B) and not contained within an approved enclosed assembly, shall be carried on insulating supports spaced at intervals of not more than 80 times the vertical dimension of the conductor, but in no case greater than 4.5 m (15 ft), and spaced apart sufficiently to give a clear electrical separation of conductors or adjacent collectors of not less than 25 mm (1 in.).

(F) Track as Circuit Conductor. Monorail, tram rail, or crane runway tracks shall be permitted as a conductor of current for one phase of a 3-phase, ac system furnishing power to the carrier, crane, or trolley, provided all of the following conditions are met:

(1) The conductors supplying the other two phases of the power supply are insulated.
(2) The power for all phases is obtained from an insulating transformer.
(3) The voltage does not exceed 300 volts.
(4) The rail serving as a conductor shall be bonded to the equipment grounding conductor at the transformer and also shall be permitted to be grounded by the fittings used for the suspension or attachment of the rail to a building or structure.

Crane runway tracks are permitted as a current-carrying conductor where part of a 3-phase system is furnishing power to the crane. The track is also permitted to be grounded through the metal supporting means attached to the building's metal frame.

(G) Electrical Continuity of Contact Conductors. All sections of contact conductors shall be mechanically joined to provide a continuous electrical connection.

(H) Not to Supply Other Equipment. Contact conductors shall not be used as feeders for any equipment other than the crane(s) or hoist(s) that they are primarily designed to serve.

610.22 Collectors. Collectors shall be designed so as to reduce to a minimum sparking between them and the contact conductor; and, where operated in rooms used for the storage of easily ignitible combustible fibers and materials, they shall comply with 503.155.

Part IV. Disconnecting Means

610.31 Runway Conductor Disconnecting Means. A disconnecting means that has a continuous ampere rating not less than that calculated in 610.14(E) and (F) shall be provided between the runway contact conductors and the power supply. The disconnecting means shall comply with 430.109. This disconnecting means shall be as follows:

(1) Readily accessible and operable from the ground or floor level
(2) Lockable open in accordance with 110.25
(3) Open all ungrounded conductors simultaneously
(4) Placed within view of the runway contact conductors

Exception: The runway conductor disconnecting means for electrolytic cell lines shall be permitted to be placed out of view of the runway contact conductors where either of the following conditions are met:

(1) Where a location in view of the contact conductors is impracticable or introduces additional or increased hazards to persons or property
(2) In industrial installations, with written safety procedures, where conditions of maintenance and supervision ensure that only qualified persons service the equipment

610.32 Disconnecting Means for Cranes and Monorail Hoists. A disconnecting means in compliance with 430.109 shall be provided in the leads from the runway contact conductors or other power supply on all cranes and monorail hoists. The disconnecting means shall be lockable open in accordance with 110.25.

Where a monorail hoist or hand-propelled crane bridge installation meets all of the following, the disconnecting means shall be permitted to be omitted:

(1) The unit is controlled from the ground or floor level.
(2) The unit is within view of the power supply disconnecting means.

(3) No fixed work platform has been provided for servicing the unit.

Means shall be provided at the operating station to open the power circuit to all motors of the crane or monorail hoist.

Many crane installations are not arranged so that the unit is within view of the power-supply disconnecting means. When one crane is being serviced, another unit on the same system could remain energized and could be run into the person performing maintenance on the crane. Therefore, a disconnecting means (lock-open type) must be provided in the contact conductors to disconnect all power to the system.

610.33 Rating of Disconnecting Means. The continuous ampere rating of the switch or circuit breaker required by 610.32 shall not be less than 50 percent of the combined short-time ampere rating of the motors or less than 75 percent of the sum of the short-time ampere rating of the motors required for any single motion.

Part V. Overcurrent Protection

610.41 Feeders, Runway Conductors.

(A) Single Feeder. The runway supply conductors and main contact conductors of a crane or monorail shall be protected by an overcurrent device(s) that shall not be greater than the largest rating or setting of any branch-circuit protective device plus the sum of the nameplate ratings of all the other loads with application of the demand factors from Table 610.14(E)(3).

(B) More Than One Feeder Circuit. Where more than one feeder circuit is installed to supply runway conductors, each feeder circuit shall be sized and protected in compliance with 610.41(A).

Multiple feeders are sometimes used to supply long runway conductors to minimize voltage drops on the runway conductors.

610.42 Branch-Circuit Short-Circuit and Ground-Fault Protection. Branch circuits shall be protected in accordance with 610.42(A). Branch-circuit taps, where made, shall comply with 610.42(B).

(A) Fuse or Circuit Breaker Rating. Crane, hoist, and monorail hoist motor branch circuits shall be protected by fuses or inverse-time circuit breakers that have a rating in accordance with Table 430.52(C)(1). Where two or more motors operate a single motion, the sum of their nameplate current ratings shall be considered as that of a single motor.

(B) Taps.

(1) Multiple Motors. Where two or more motors are connected to the same branch circuit, each tap conductor to an individual motor shall have an ampacity not less than one-third that of the branch circuit. Each motor shall be protected from overload according to 610.43.

(2) Control Circuits. Where taps to control circuits originate on the load side of a branch-circuit protective device, each tap and piece of equipment shall be protected in accordance with 430.72.

610.43 Overload Protection.

(A) Motor and Branch-Circuit Overload Protection. Each motor, motor controller, and branch-circuit conductor shall be protected from overload by one of the following means:

(1) A single motor shall be considered as protected where the branch-circuit overcurrent device meets the rating requirements of 610.42.
(2) Overload relay elements in each ungrounded circuit conductor, with all relay elements protected from short circuit by the branch-circuit protection.
(3) Thermal sensing devices, sensitive to motor temperature or to temperature and current, that are thermally in contact with the motor winding(s). Hoist functions shall be considered to be protected if the sensing device limits the hoist to lowering only during an overload condition. Traverse functions shall be considered to be protected if the sensing device limits the travel in both directions for the affected function during an overload condition of either motor.

(B) Manually Controlled Motor. If the motor is manually controlled, with spring return controls, the overload protective device shall not be required to protect the motor against stalled rotor conditions.

(C) Multimotor. Where two or more motors drive a single trolley, truck, or bridge and are controlled as a unit and protected by a single set of overload devices with a rating equal to the sum of their rated full-load currents, a hoist or trolley shall be considered to be protected if the sensing device is connected in the hoist's upper limit switch circuit so as to prevent further hoisting during an overtemperature condition of either motor.

(D) Hoists and Monorail Hoists. Hoists and monorail hoists and their trolleys that are not used as part of an overhead traveling crane shall not require individual motor overload protection, provided the largest motor does not exceed 7½ hp and all motors are under manual control of the operator.

Part VI. Control

610.51 Separate Controllers. Each motor shall be provided with an individual controller unless otherwise permitted in 610.51(A) or (B).

(A) Motions with More Than One Motor. Where two or more motors drive a single hoist, carriage, truck, or bridge, they shall be permitted to be controlled by a single controller.

(B) Multiple Motion Controller. One controller shall be permitted to be switched between motors, under the following conditions:

(1) The controller has a horsepower rating that is not lower than the horsepower rating of the largest motor.
(2) Only one motor is operated at one time.

610.53 Overcurrent Protection. Conductors of control circuits shall be protected against overcurrent. Control circuits shall be considered as protected by overcurrent devices that are rated or set at not more than 300 percent of the ampacity of the control conductors, unless otherwise permitted in 610.53(A) or (B).

(A) Taps to Control Transformers. Taps to control transformers shall be considered as protected where the secondary circuit is protected by a device rated or set at not more than 200 percent of the rated secondary current of the transformer and not more than 200 percent of the ampacity of the control circuit conductors.

(B) Continuity of Power. Where the opening of the control circuit would create a hazard, as for example, the control circuit of a hot metal crane, the control circuit conductors shall be considered as being properly protected by the branch-circuit overcurrent devices.

610.57 Clearance. The dimension of the working space in the direction of access to live parts that are likely to require examination, adjustment, servicing, or maintenance while energized shall be a minimum of 750 mm (2½ ft). Where controls are enclosed in cabinets, the door(s) shall either open at least 90 degrees or be removable.

Part VII. Grounding and Bonding

610.61 Grounding and Bonding. All exposed non–current-carrying metal parts of cranes, monorail hoists, hoists, and accessories, including pendant controls, shall be bonded either by mechanical connections or bonding jumpers, where applicable, so that the entire crane or hoist is an effective ground-fault current path by connection to the equipment grounding conductor of the branch circuit or feeder as required or permitted by Article 250, Parts I, V, VI, and VII.

Moving parts, other than removable accessories, or attachments that have metal-to-metal bearing surfaces, shall be considered to be electrically bonded to each other through bearing surfaces for the purpose of establishing an effective ground-fault current path. The trolley frame and bridge frame shall not be considered as electrically bonded through the bridge and trolley wheels and its respective tracks. A separate bonding conductor shall be provided.

These requirements are not intended to allow the trolley frame or bridge frame to serve as the EGC for electrical equipment on a crane. However, the frame and all other non–current-carrying metal parts of the crane must be bonded together so as to form an effective ground-fault current path in the event of a fault. The EGCs that are run with the circuit conductors are required to be one of the types described in 250.118. Metal-to-metal bearing surfaces of moving parts are considered to be a suitable grounding and bonding connection. However, the bridge and trolley wheel contact with their tracks is not permitted to be used as a reliable grounding and bonding connection. Because dirt or other foreign surfaces could impede the effectiveness of the wheel-to-track contact as a reliable grounding and bonding connection, the bridge and trolley frames of an electric crane are required to be bonded through the use of a separate conductor.

ARTICLE 620 Elevators, Dumbwaiters, Escalators, Moving Walks, Platform Lifts, and Stairway Chairlifts

Part I. General

Δ **620.1 Scope.** This article covers the installation of electrical equipment and wiring used in connection with elevators, dumbwaiters, escalators, moving walks, platform lifts, and stairway chairlifts.

> Informational Note No. 1: See ASME A17.1/CSA B44, *Safety Code for Elevators and Escalators*, for information on the installation of elevators and escalators.

Informational Note No. 2: See CSA B44.1/ASME A17.5, *Elevator and escalator electrical equipment*, for information on elevator and escalator electrical equipment.

Informational Note No. 3: See ASME A18.1, *Safety Standard for Platform Lifts and Stairway Chairlifts*, for information on installation of platform lifts and stairway chairlifts. The term *wheelchair lift* has been changed to platform lift.

Informational Note No. 4: The motor controller, motion controller, and operation controller are located in a single enclosure or a combination of enclosures.

Informational Note No. 5: See Informational Note Figure 620.1 for information only.

620.3 Voltage Limitations. The supply voltage shall not exceed 300 volts between conductors unless otherwise permitted in 620.3(A) through (C).

(A) Power Circuits. Branch circuits to door operator controllers and door motors and branch circuits and feeders to motor controllers, driving machine motors, machine brakes, and motor-generator sets shall not have a circuit voltage in excess of 1000 volts. Internal voltages of power conversion equipment and functionally associated equipment, and the operating voltages of wiring interconnecting the equipment, shall be permitted to be higher, provided that all such equipment and wiring shall be listed for the higher voltages. Where the voltage exceeds 600 volts, warning labels or signs that read "DANGER — HIGH VOLTAGE" shall be attached to the equipment and shall be plainly visible. The danger sign(s) or label(s) shall comply with 110.21(B).

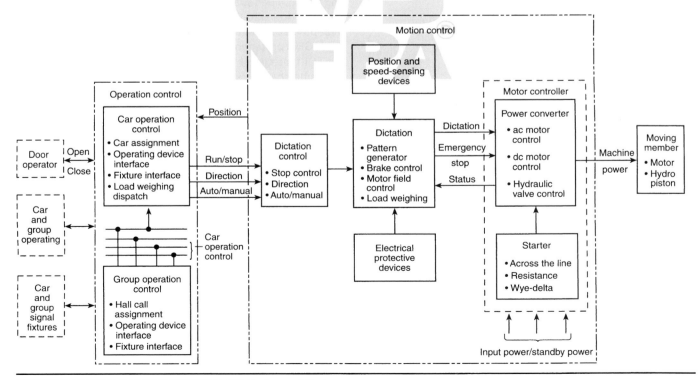

INFORMATIONAL NOTE FIGURE 620.1 *Control System.*

(B) Lighting Circuits. Lighting circuits shall comply with the requirements of Article 410.

(C) Heating and Air-Conditioning Circuits. Branch circuits for heating and air-conditioning equipment located on the elevator car shall not have a circuit voltage in excess of 1000 volts.

620.4 Live Parts Enclosed. All live parts of electrical apparatus in the hoistways, at the landings, in or on the cars of elevators and dumbwaiters, in the wellways or the landings of escalators or moving walks, or in the runways and machinery spaces of platform lifts and stairway chairlifts shall be enclosed to protect against accidental contact.

> Informational Note: See 110.27 for guarding of live parts (1000 volts, nominal, or less).

620.5 Working Clearances. Working space shall be provided about controllers, disconnecting means, and other electrical equipment in accordance with 110.26(A).

Where conditions of maintenance and supervision ensure that only qualified persons examine, adjust, service, and maintain the equipment, the clearance requirements of 110.26(A) shall not be required where any of the conditions in 620.5(A) through (D) are met.

(A) Flexible Connections to Equipment. Electrical equipment in the following is provided with flexible leads to all external connections so that it can be repositioned to meet the clear working space requirements of 110.26:

(1) Controllers and disconnecting means for dumbwaiters, escalators, moving walks, platform lifts, and stairway chairlifts installed in the same space with the driving machine

(2) Controllers and disconnecting means for elevators installed in the hoistway or on the car

(3) Controllers for door operators

(4) Other electrical equipment installed in the hoistway or on the car

Due to the physical constraints of the locations where this equipment typically is installed and the necessity of performing diagnostic work on it while it is energized, 620.5(A) permits flexible leads on equipment so it can be moved to a location that meets the working clearance requirements of 110.26(A).

(B) Guards. Live parts of the electrical equipment are suitably guarded, isolated, or insulated to reduce the likelihood of inadvertent contact with live parts operating at voltages greater than 30 volts ac rms, 42 volts ac peak, or 60 volts dc, and the equipment can be examined, adjusted, serviced, or maintained while energized without removal of this protection.

(C) Examination, Adjusting, and Servicing. Electrical equipment is not required to be examined, adjusted, serviced, or maintained while energized.

(D) Low Voltage. Uninsulated parts are at a voltage not greater than 30 volts rms, 42 volts peak, or 60 volts dc.

Δ **620.6 Ground-Fault Circuit-Interrupter Protection for Personnel.**

N **(A) Pits, Hoistways, and on Cars.** Each 125-volt, single-phase, 15- and 20-ampere receptacle installed in pits, in hoistways, on the cars of elevators and dumbwaiters associated with wind turbine tower elevators, on the platforms or in the runways and machinery spaces of platform lifts and stairway chairlifts, and in escalator and moving walk wellways shall be a listed Class A ground-fault circuit-interrupter type.

N **(B) Machine Rooms, Control Spaces, Machinery Spaces, Control Rooms, and Truss Interiors.** All 125-volt, single-phase, 15- and 20-ampere receptacles installed in machine rooms, control spaces, machinery spaces, control rooms, and truss interiors shall have listed Class A ground-fault circuit-interrupter protection for personnel.

N **(C) Sump Pumps.** A permanently installed sump pump shall be permanently wired or shall be supplied by a receptacle that is protected by a listed Class A ground-fault circuit-interrupter.

These GFCI requirements are intended to reduce the shock hazard to maintenance personnel who service elevator equipment using portable hand tools and temporary lighting.

Section 620.6(A) requires GFCI-type receptacles based on the premise that the reset pushbutton for a tripped GFCI receptacle should be within easy reach of an elevator mechanic working in a confined space.

Section 620.6(B) requires that all 15- and 20-ampere receptacles installed in machine rooms and machinery spaces have GFCI protection. This protection can be afforded by either a GFCI-type circuit breaker or a GFCI-type receptacle because machine spaces usually do not cause access hazards for service personnel.

Section 620.6(C) requires that sump pumps for equipment installed under the scope of Article 620 have GFCI protection, which aligns with other areas of the *NEC®* that require GFCI protection of sump pumps, such as 210.8(D) and 422.5(A).

Part II. Conductors

Δ **620.11 Insulation of Conductors.** The insulation of conductors shall comply with 620.11(A) through (D).

> Informational Note: See UL 2556-2015, *Wire and Cable Test Methods,* for one method of determining that the insulation of conductors is flame retardant by testing the conductors or cables to the FV-2/VW-1 Test.

(A) Hoistway Door Interlock Wiring. The conductors to the hoistway door interlocks from the hoistway riser shall be one of the following:

(1) Flame retardant and suitable for a temperature of not less than 200°C (392°F). Conductors shall be Type SF or equivalent.

(2) Physically protected using an approved method, such that the conductor assembly is flame retardant and suitable for a temperature of not less than 200°C (392°F).

(B) Traveling Cables. Traveling cables used as flexible connections between the elevator or dumbwaiter car or counterweight and the raceway shall be of the types of elevator cable listed in Table 400.4 or other approved types.

(C) Other Wiring. All conductors in raceways shall have flame-retardant insulation.

Conductors shall be Type MTW, TF, TFF, TFN, TFFN, THHN, THW, THWN, TW, XHHW, hoistway cable, or any other conductor with insulation designated as flame retardant. Shielded conductors shall be permitted if such conductors are insulated for the maximum nominal circuit voltage applied to any conductor within the cable or raceway system.

(D) Insulation. All conductors shall have an insulation voltage rating equal to at least the maximum nominal circuit voltage applied to any conductor within the enclosure, cable, or raceway. Insulations and outer coverings that are marked for limited smoke and are so listed shall be permitted.

620.12 Minimum Size of Conductors. The minimum size of conductors, other than conductors that form an integral part of control equipment, shall be in accordance with 620.12(A) and (B).

(A) Traveling Cables.

(1) Lighting Circuits. For lighting circuits, 14 AWG copper, 20 AWG copper or larger conductors shall be permitted in parallel, provided the ampacity is equivalent to at least that of 14 AWG copper.

Section 310.10(G) provides the conditions under which conductors can be installed in parallel for power and lighting circuits. One of those conditions stipulates that the minimum size for parallel conductors is 1/0 AWG. In high-rise structures, the length of the elevator traveling cables makes it hard to maintain an acceptable level of voltage drop for equipment on or within the car. To require compliance with the 310.10(G) requirements for parallel conductors would result in exceptionally large traveling cables.

Section 620.12(A)(1) amends these general requirements for parallel conductors and permits 20 AWG and larger conductors to be installed in parallel for lighting circuits, provided that the combined ampacity of the paralleled conductors is not less than that of a 14 AWG copper conductor (e.g., 15 amperes for 60°C). This requirement is unique to Article 620 and is an example of the structure of the *NEC* as set forth in 90.3.

N **(2) Class 2 and Communications Circuits.** Communications cables used for Class 2 or communications circuits shall have a current limit equal to or greater than the current required to power the powered Class 2 or communications device. Communications cables shall comply with 800.179. The minimum conductor size for communications circuits shall be 24 AWG.

(3) Other Circuits. For other circuits, the minimum size conductor shall be 20 AWG copper.

N **(4) Paralleled Conductors.** Where ampacity requirements or voltage drop conditions in a traveling cable circuit prevent the use of a single conductor of AWG 14 or smaller, conductors shall be permitted in parallel in compliance with all the following:

(1) Each conductor shall be no smaller than 20 AWG copper.
(2) The paralleled conductors shall be the same type and have the same ampacity rating.
(3) No more than 3 conductors shall be paralleled.
(4) The overcurrent protection shall be such that the ampacity of each individual conductor will not be exceeded if one of the parallel conductors becomes inadvertently disconnected.

(B) Other Wiring. 24 AWG copper. Smaller size listed conductors shall be permitted.

With the extensive use of electronics with lower currents, conductors smaller than 24 AWG are permitted by 620.12(B), provided that they are listed and have the necessary strength and durability for the conditions to which they will be exposed. One application is the shielded cables interconnecting various microprocessors in an elevator distributed control system.

Δ **620.13 Feeder and Branch-Circuit Conductors.** Conductors shall have an ampacity in accordance with 620.13(A) through (D). With generator field control, the conductor ampacity shall be based on the nameplate current rating of the driving motor of the motor-generator set that supplies power to the elevator motor.

Informational Note No. 1: The heating of conductors depends on root-mean-square current values, which, with generator field control, are reflected by the nameplate current rating of the motor-generator driving motor rather than by the rating of the elevator motor, which represents actual but short-time and intermittent full-load current values.

Informational Note No. 2: See Informational Note Figure 620.13.

(A) Conductors Supplying Single Motor. Conductors supplying a single motor shall have an ampacity not less than the percentage of motor nameplate current determined from 430.22(A) and (E).

Informational Note: Some elevator motor currents, or those motor currents of similar function, exceed the motor nameplate value. Heating of the motor and conductors is dependent on the root-mean square (rms) current value and the length of operation time. Because this motor application is inherently intermittent duty, conductors are sized for duty cycle service as shown in Table 430.22(E).

(B) Conductors Supplying a Single Motor Controller. Conductors supplying a single motor controller shall have an ampacity not less than the motor controller nameplate current rating, plus all other connected loads. Motor controller nameplate current ratings shall be permitted to be derived based on the rms value of the motor current using an intermittent duty cycle and other control system loads, if present.

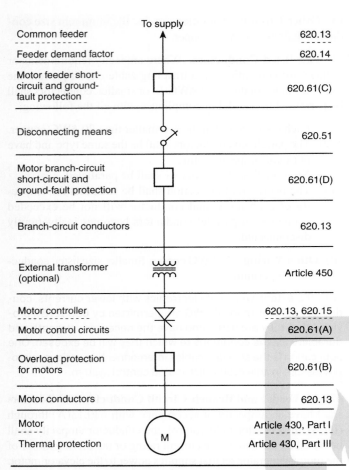

Common feeder	620.13
Feeder demand factor	620.14
Motor feeder short-circuit and ground-fault protection	620.61(C)
Disconnecting means	620.51
Motor branch-circuit short-circuit and ground-fault protection	620.61(D)
Branch-circuit conductors	620.13
External transformer (optional)	Article 450
Motor controller	620.13, 620.15
Motor control circuits	620.61(A)
Overload protection for motors	620.61(B)
Motor conductors	620.13
Motor	Article 430, Part I
Thermal protection	Article 430, Part III

INFORMATIONAL NOTE FIGURE 620.13 Single-Line Diagram.

(C) Conductors Supplying a Single Power Transformer. Conductors supplying a single power transformer shall have an ampacity not less than the nameplate current rating of the power transformer plus all other connected loads.

> Informational Note No. 1: The nameplate current rating of a power transformer supplying a motor controller reflects the nameplate current rating of the motor controller at line voltage (transformer primary).
>
> Informational Note No. 2: See Informative Annex D, Example No. D10.

(D) Conductors Supplying More Than One Motor, Motor Controller, or Power Transformer. Conductors supplying more than one motor, motor controller, or power transformer shall have an ampacity not less than the sum of the nameplate current ratings of the equipment plus all other connected loads. The ampere ratings of motors to be used in the summation shall be determined from Table 430.22(E), 430.24, and 430.24, Exception No. 1.

> Informational Note: See Informative Annex D, Example Nos. D9 and D10.

620.14 Feeder Demand Factor. Feeder conductors of less ampacity than required by 620.13 shall be permitted, subject to the requirements of Table 620.14.

TABLE 620.14 *Feeder Demand Factors for Elevators*

Number of Elevators on a Single Feeder	Demand Factor*
1	1.00
2	0.95
3	0.90
4	0.85
5	0.82
6	0.79
7	0.77
8	0.75
9	0.73
10 or more	0.72

* Demand factors are based on 50 percent duty cycle (i.e., half time on and half time off).

620.15 Motor Controller Rating. The motor controller rating shall comply with 430.83. The rating shall be permitted to be less than the nominal rating of the elevator motor, when the controller inherently limits the available power to the motor and is marked as power limited.

> Informational Note: See 430.8 for controller markings.

The inherent power-limiting ability of certain adjustable-speed drive controllers is the basis for permitting the controller to have a lower current or horsepower rating than that of the motor. For a controller to be used in this manner, the manufacturer's marking must indicate that it is power limiting.

620.16 Short-Circuit Current Rating.

(A) Marking. Where an elevator control panel is installed, it shall be marked with its short-circuit current rating, based on one of the following:

(1) Short-circuit current rating of a listed assembly
(2) Short-circuit current rating established utilizing an approved method

> Informational Note: UL 508A-2013, *Standard for Industrial Control Panels*, Supplement SB, is an example of an approved method.

(B) Installation. The elevator control panel shall not be installed where the available fault current exceeds its short-circuit current rating, as marked in accordance with 620.16(A).

Part III. Wiring

620.21 Wiring Methods. Conductors, cables, and optical fiber cables located in hoistways, escalator and moving walk wellways, platform lifts, stairway chairlift runways, machinery spaces, control spaces, in or on cars, machine rooms, and control rooms, not including the traveling cables connecting the car or counterweight and hoistway wiring, shall be installed in rigid metal conduit, intermediate metal conduit, electrical metallic tubing, rigid nonmetallic conduit, or wireways, or shall be Type MC, MI, or AC cable unless otherwise permitted in 620.21(A) through (C). Unused conductors in an enclosure shall be insulated or protected from accidental contact with exposed live parts.

Exception: Cords and cables of listed cord-and-plug-connected equipment shall not be required to be installed in a raceway.

Informational Note: When an elevator is classified as a fire service access elevator or occupant evacuation operation elevator, some building codes require additional protection for conductors that are located outside of the elevator hoistway and machine room.

(A) Elevators.

(1) Hoistways and Pits.

(a) Types CL2P, CL2R, and CL2 cables shall be permitted, provided the cables are supported and protected from physical damage. Substitute cables for Class 2 cables installed in accordance with 722.135(E) shall be permitted.

(b) Flexible cords and cables that are components of listed equipment and used in circuits operating at 30 volts rms or less or 42 volts dc or less shall be permitted, provided the cords and cables are supported and protected from physical damage and are of a jacketed and flame-retardant type.

(c) The following wiring methods shall be permitted in the hoistway in lengths not to exceed 1.8 m (6 ft):

(1) Flexible metal conduit.
(2) Liquidtight flexible metal conduit.
(3) Liquidtight flexible nonmetallic conduit.
(4) Flexible cords and cables, or conductors grouped together and taped or corded, shall be permitted to be installed without a raceway. They shall be located to be protected from physical damage, shall be of a flame-retardant type, and shall be part of one of the following:

  a. Listed equipment
  b. Driving machine
  c. Driving machine brake

Exception to 620.21(A)(1)(c)(1), (A)(1)(c)(2), and (A)(1)(c)(3): The conduit length shall not be required to be limited between risers and limit switches, interlocks, operating buttons, and similar devices.

(d) A sump pump or oil recovery pump located in the pit shall be permitted to be cord connected. The cord shall be a hard usage oil-resistant type, of a length not to exceed 1.8 m (6 ft), and shall be located to be protected from physical damage.

(e) Hard-service cords and junior hard-service cords that conform to the requirements of Article 400 (Table 400.4) shall

be permitted as flexible connections between the fixed wiring in the hoistway and hoistway access switches when located in the hoistway door sight guard.

Informational Note: See ASME A17.1-2019/CSA B44-19, *Safety Code for Elevators and Escalators.*

Δ **(2) Cars.**

(a) Flexible metal conduit, liquidtight flexible metal conduit, or liquidtight flexible nonmetallic conduit of metric designator 12 (trade size ⅜), or larger, not exceeding 1.8 m (6 ft) in length, shall be permitted on cars where so located as to be free from oil and if securely fastened in place.

Exception: Liquidtight flexible nonmetallic conduit (LFNC-B) of metric designator 12 (trade size ⅜) or larger shall be permitted in lengths in excess of 1.8 m (6 ft).

(b) Hard-service cords and junior hard-service cords that conform to the requirements of Article 400 (Table 400.4) shall be permitted as flexible connections between the fixed wiring on the car and devices on the car doors or gates. Hard-service cords only shall be permitted as flexible connections for the top-of-car operating device or the car-top work light. Devices or luminaires shall be grounded by means of an equipment grounding conductor run with the circuit conductors. Cables with smaller conductors and other types and thicknesses of insulation and jackets shall be permitted as flexible connections between the fixed wiring on the car and devices on the car doors or gates, if listed for this use.

(c) Flexible cords and cables that are components of listed equipment and used in circuits operating at 30 volts rms or less or 42 volts dc or less shall be permitted, provided the cords and cables are supported and protected from physical damage and are of a jacketed and flame-retardant type.

(d) The following wiring methods shall be permitted on the car assembly in lengths not to exceed 1.8 m (6 ft):

(1) Flexible metal conduit
(2) Liquidtight flexible metal conduit
(3) Liquidtight flexible nonmetallic conduit
(4) Flexible cords and cables, or conductors grouped together and taped or corded, shall be permitted to be installed without a raceway. They shall be located to be protected from physical damage and shall be of a flame-retardant type and shall be part of one of the following:

  a. Listed equipment
  b. A driving machine
  c. A driving machine brake

Δ **(3) Within Machine Rooms, Control Rooms, and Machinery Spaces and Control Spaces.**

(a) Flexible metal conduit, liquidtight flexible metal conduit, or liquidtight flexible nonmetallic conduit of metric designator 12 (trade size ⅜), or larger, not exceeding 1.8 m (6 ft) in length, shall be permitted between control panels and machine motors, machine brakes, motor-generator sets, disconnecting means, and pumping unit motors and valves.

Exception: Liquidtight flexible nonmetallic conduit (LFNC-B) metric designator 12 (trade size ⅜) or larger shall be permitted to be installed in lengths in excess of 1.8 m (6 ft).

(b) Where motor-generators, machine motors, or pumping unit motors and valves are located adjacent to or underneath control equipment and are provided with extra-length terminal leads not exceeding 1.8 m (6 ft) in length, such leads shall be permitted to be extended to connect directly to controller terminal studs without regard to the carrying-capacity requirements of Articles 430 and 445. Auxiliary gutters shall be permitted in machine and control rooms between controllers, starters, and similar apparatus.

(c) Flexible cords and cables that are components of listed equipment and used in circuits operating at 30 volts rms or less or 42 volts dc or less shall be permitted, provided the cords and cables are supported and protected from physical damage and are of a jacketed and flame-retardant type.

(d) On existing or listed equipment, conductors shall also be permitted to be grouped together and taped or corded without being installed in a raceway. Such cable groups shall be supported at intervals not over 900 mm (3 ft) and located so as to be protected from physical damage.

(e) Flexible cords and cables in lengths not to exceed 1.8 m (6 ft) that are of a flame-retardant type and located to be protected from physical damage shall be permitted in these rooms and spaces without being installed in a raceway. They shall be part of one of the following:

(1) Listed equipment
(2) A driving machine
(3) A driving machine brake

(4) Counterweight. The following wiring methods shall be permitted on the counterweight assembly in lengths not to exceed 1.8 m (6 ft):

(1) Flexible metal conduit
(2) Liquidtight flexible metal conduit
(3) Liquidtight flexible nonmetallic conduit
(4) Flexible cords and cables, or conductors grouped together and taped or corded, shall be permitted to be installed without a raceway. They shall be located to be protected from physical damage, shall be of a flame-retardant type, and shall be part of one of the following:

   a. Listed equipment
   b. A driving machine
   c. A driving machine brake

(B) Escalators.

Δ **(1) Wiring Methods.** Flexible metal conduit, liquidtight flexible metal conduit, or liquidtight flexible nonmetallic conduit shall be permitted in escalator and moving walk wellways.

Flexible metal conduit or liquidtight flexible conduit of metric designator 12 (trade size ⅜) shall be permitted in lengths not in excess of 1.8 m (6 ft).

Exception: Metric designator 12 (trade size ⅜), nominal or larger liquidtight flexible nonmetallic conduit (LFNC-B) shall be permitted to be installed in lengths in excess of 1.8 m (6 ft).

(2) Class 2 Circuit Cables. Types CL2P, CL2R, and CL2 cables shall be permitted to be installed within escalators and moving walkways, provided the cables are supported and protected from physical damage. Substitute cables for Class 2 cables installed in accordance with 722.135(E) shall be permitted.

(3) Flexible Cords. Hard-service cords that conform to the requirements of Article 400 (Table 400.4) shall be permitted as flexible connections on escalators and moving walk control panels and disconnecting means where the entire control panel and disconnecting means are arranged for removal from machine spaces as permitted in 620.5.

(C) Platform Lifts and Stairway Chairlift Raceways.

Δ **(1) Wiring Methods.** Flexible metal conduit or liquidtight flexible metal conduit shall be permitted in platform lifts and stairway chairlift runways and machinery spaces. Flexible metal conduit or liquidtight flexible conduit of metric designator 12 (trade size ⅜) shall be permitted in lengths not in excess of 1.8 m (6 ft).

Exception: Metric designator 12 (trade size ⅜) or larger liquidtight flexible nonmetallic conduit (LFNC-B) shall be permitted to be installed in lengths in excess of 1.8 m (6 ft).

(2) Class 2 Circuit Cables. Types CL2P, CL2R, and CL2 cables shall be permitted to be installed within platform lifts and stairway chairlift runways and machinery spaces, provided the cables are supported and protected from physical damage. Substitute cables for Class 2 cables installed in accordance with 722.135(E) shall be permitted.

(3) Flexible Cords and Cables. Flexible cords and cables that are components of listed equipment and used in circuits operating at 30 volts rms or less or 42 volts dc or less shall be permitted in lengths not to exceed 1.8 m (6 ft), provided the cords and cables are supported and protected from physical damage and are of a jacketed and flame-retardant type.

620.22 Branch Circuits for Car Lighting, Receptacle(s), Ventilation, Heating, and Air-Conditioning.

Δ **(A) Car Light Receptacles, Auxiliary Lighting, and Ventilation.** A separate branch circuit shall supply the car lights. The car lights branch circuit shall be permitted to supply receptacles (alarm devices, emergency responder radio coverage (ERRC), car ventilation purification systems, monitoring devices not part of the control system), auxiliary lighting power source, car emergency signaling, communications devices (including their associated charging circuits), and ventilation on each elevator car or inside the operation controller. The overcurrent device protecting the branch circuit shall be located in the elevator machine room, control room, machinery space, or control space. Where there is no machine room, control room, machinery space, or control space outside the hoistway, the overcurrent device shall be located outside the hoistway and accessible to qualified persons only.

Required lighting shall not be connected to the load side of a ground-fault circuit interrupter.

A service receptacle installed on an elevator car top is required to be a GFCI-type device in accordance with 620.6. Because the car lights and receptacle are supplied from the same branch circuit, the line-side connection ensures that GFCI operation will not also interrupt power to the car lighting. The same requirement is found in 620.23(A) and 620.24(A) for machine room lighting and hoistway pit lighting.

(B) Air-Conditioning and Heating Source. A separate branch circuit shall supply the air-conditioning and heating units on each elevator car. The overcurrent device protecting the branch circuit shall be located in the elevator machine room, control room, machinery space, or control space. Where there is no machine room, control room, machinery space, or control space outside the hoistway, the overcurrent device shall be located outside the hoistway and accessible only to qualified persons.

620.23 Branch Circuits for Machine Room, Control Room/Machinery Space, Control Space, or Truss Interior Lighting and Receptacle(s).

(A) Separate Branch Circuits. The branch circuits supplying the lighting for machine rooms, control rooms, machinery spaces, control spaces, or truss interiors, where required, shall be separate from the branch circuits supplying the receptacles in those places. These circuits shall supply no other loads.

Required lighting shall not be connected to the load side of a ground-fault circuit interrupter.

(B) Lighting Switch. The machine room, control room/machinery space, or control space lighting switch shall be located at the point of entry.

Δ **(C) Duplex Receptacle.** At least one 125-volt, single-phase, 15- or 20-ampere duplex receptacle shall be provided in each machine room, control room and machinery space, control space, and in truss interiors where required.

> Informational Note: See ASME A17.1/CSA B44, *Safety Code for Elevators and Escalators*, for illumination levels and receptacle requirements.

The receptacles required by 620.23 and 620.24 are required by 620.6 to be provided with GFCI protection. Luminaires are not permitted to be connected to the load side of GFCI devices. This placement prevents power interruption to the machine room

lighting if the GFCI operates. ASME A17.1, *Safety Code for Elevators and Escalators,* requires a minimum of 5 foot-candles (54 lux) at the pit floor and requires luminaires to be externally guarded to prevent accidental breakage. Luminaires in pits should be mounted so that the car or counterweight does not strike them when on fully compressed buffers.

620.24 Branch Circuit for Hoistway Pit Lighting and Receptacles.

(A) Separate Branch Circuits. Separate branch circuits shall supply the hoistway pit lighting and receptacles.

Required lighting shall not be connected to the load side of a ground-fault circuit interrupter.

(B) Lighting Switch. The lighting switch shall be so located as to be readily accessible from the pit access door.

(C) Duplex Receptacle. At least one 125-volt, single-phase, 15- or 20-ampere duplex receptacle shall be provided in the hoistway pit.

> Informational Note No. 1: See ASME A17.1-2016/CSA B44-16, *Safety Code for Elevators and Escalators*, for illumination levels.
> Informational Note No. 2: See 620.6 for ground-fault circuit-interrupter requirements.

620.25 Branch Circuits for Other Utilization Equipment.

(A) Additional Branch Circuits. Additional branch circuit(s) shall supply utilization equipment not identified in 620.22, 620.23, and 620.24. Other utilization equipment shall be restricted to that equipment identified in 620.1.

(B) Overcurrent Devices. The overcurrent devices protecting the branch circuit(s) shall be located in the elevator machine room, control room, machinery space, or control space. Where there is no machine room, control room, machinery space, or control space outside the hoistway, or for escalator and moving walk applications, the overcurrent device shall be located outside the hoistway and accessible only to qualified persons.

Part IV. Installation of Conductors

620.32 Metal Wireways and Nonmetallic Wireways. The sum of the cross-sectional area of the individual conductors in a wireway shall not be more than 50 percent of the interior cross-sectional area of the wireway.

Vertical runs of wireways shall be securely supported at intervals not exceeding 4.5 m (15 ft) and shall have not more than one joint between supports. Adjoining wireway sections shall be securely fastened together to provide a rigid joint.

620.33 Number of Conductors in Raceways. The sum of the cross-sectional area of the individual conductors in raceways shall not exceed 40 percent of the interior cross-sectional area of the raceway, except as permitted in 620.32 for wireways.

620.34 Supports. Supports for cables or raceways in a hoistway or in an escalator or moving walk wellway or platform lift and stairway chairlift runway shall be securely fastened to the guide rail; escalator or moving walk truss; or to the hoistway, wellway, or runway construction.

620.35 Auxiliary Gutters. Auxiliary gutters shall not be subject to the restrictions of 366.10(C) covering length or of 366.22 covering number of conductors.

620.36 Different Systems in One Raceway or Traveling Cable. Optical fiber cables and conductors for operating devices, operation and motion control, power, signaling, fire alarm, lighting, heating, and air-conditioning circuits of 1000 volts or less shall be permitted to be run in the same traveling cable or raceway system if all conductors are insulated for the maximum voltage applied to any conductor within the cables or raceway system and if all live parts of the equipment are insulated from ground for this maximum voltage. Traveling cable or raceway shall also be permitted to include shielded pairs, coaxial cables, and other communications circuits. Type CMP-LP or CMR-LP cables complying with 800.179 shall be permitted in raceways.

The use of greater numbers of individual cables and of much longer cables in tall buildings increases the likelihood of the multiple cable loops becoming twisted. To prevent the practice of tying other cables to the traveling cable, one elevator cable or raceway is permitted to enclose optical fiber cables and all the conductors for power, control, lighting, video, fire alarm, and communications circuits.

620.37 Wiring in Hoistways, Machine Rooms, Control Rooms, Machinery Spaces, and Control Spaces.

Δ **(A) Uses Permitted.** Electrical wiring, raceways, and cables used directly in connection with the elevator or dumbwaiter shall be permitted inside the hoistway, machine rooms, control rooms, machinery spaces, and control spaces, including wiring for the following:

(1) Signals
(2) Communications with the car
(3) Fire detection systems
(4) Pit sump pumps
(5) Branch circuits in 620.24
(6) Heating, lighting, and ventilating the hoistway
(7) Heating, air conditioning, lighting, and ventilating the elevator car

Δ **(B) Lightning Protection.** Bonding of elevator rails (car and/or counterweight) to a lightning protection system down conductor(s) shall be permitted. The lightning protection system down conductor(s) shall not be located within the hoistway. Elevator rails or other hoistway equipment shall not be used as the down conductor for lightning protection systems.

Informational Note No. 1: See 250.106 for bonding requirements.
Informational Note No. 2: See NFPA 780-2020, *Standard for the Installation of Lightning Protection Systems*, for further information.

Where a lightning protection system is provided with the system grounding "down" conductor(s) located outside the hoistway within a critical horizontal distance of the elevator rails, bonding of the rails to the lightning protection system grounding down conductor(s) is required by NFPA 780, *Standard for the Installation of Lightning Protection Systems*, to provide potential equalization. The requirements of 620.37(B) provide the necessary correlation for this bonding of the elevator rails to occur. Bonding prevents a dangerous side flash between the lightning protection system grounding down conductor(s) and the elevator rails. A lightning strike on the building air terminal will be conducted through the lightning protection system grounding down conductor(s), and, if the elevator rails are not at the same potential as the lightning protection system grounding down conductor(s), a side flash may occur. Generally, down conductors are installed vertically near the structure's perimeter.

Δ **(C) Feeders.** Feeders for supplying power to elevators and dumbwaiters shall be installed outside the hoistway unless as follows:

(1) By special permission, feeders for elevators shall be permitted within an existing hoistway if no conductors are spliced within the hoistway.
(2) Feeders shall be permitted inside the hoistway for elevators with driving machine motors located in the hoistway or on the car or counterweight.

Δ **620.38 Electrical Equipment in Garages and Similar Occupancies.** Electrical equipment and wiring used for elevators, dumbwaiters, escalators, moving walks, and platform lifts and stairway chairlifts in garages shall comply with the requirements of 511.3(A).

Informational Note: Garages used for parking or storage and where no repair work is done in accordance with 511.3(A) are not classified.

Part V. Traveling Cables

620.41 Suspension of Traveling Cables. Traveling cables shall be suspended at the car and hoistways' ends, or counterweight end where applicable, so as to reduce the strain on the individual copper conductors to a minimum.

Traveling cables shall be supported, utilizing listed components, by one of the following methods:

(1) By their steel supporting member(s)
(2) By looping the cables around supports for unsupported lengths less than 30 m (100 ft)
(3) By suspending from the supports by a means that automatically tightens around the cable when tension is increased for unsupported lengths up to 60 m (200 ft)

Unsupported length for the hoistway suspension means shall be that length of cable measured from the point of suspension in the hoistway to the bottom of the loop, with the elevator car located at the bottom landing. Unsupported length for the car suspension means shall be that length of cable measured from the point of suspension on the car to the bottom of the loop, with the elevator car located at the top landing.

620.42 Hazardous (Classified) Locations. In hazardous (classified) locations, traveling cables shall be of a type approved for hazardous (classified) locations as permitted in 501.10(B)(2)(7), 502.10(B)(2)(6), 503.10(A)(3)(6), 505.15(C)(2), and 506.15(A)(6).

620.43 Location of and Protection for Cables. Traveling cable supports shall be located so as to reduce to a minimum the possibility of damage due to the cables coming in contact with the hoistway construction or equipment in the hoistway. Where necessary, suitable guards shall be provided to protect the cables against damage.

620.44 Installation of Traveling Cables. Traveling cables that are suitably supported and protected from physical damage shall be permitted to be run without the use of a raceway in either or both of the following:

(1) When used inside the hoistway, on the elevator car, hoistway wall, counterweight, or controllers and machinery that are located inside the hoistway, provided the cables are in the original sheath.
(2) From inside the hoistway, to elevator controller enclosures and to elevator car and machine room, control room, machinery space, and control space connections that are located outside the hoistway for a distance not exceeding 1.8 m (6 ft) in length as measured from the first point of support on the elevator car or hoistway wall, or counterweight where applicable, provided the conductors are grouped together and taped or corded, or in the original sheath. These traveling cables shall be permitted to be continued to this equipment.

Traveling cables between fixed suspension points are not required to be installed in a raceway. If the fixed suspension point is on top of the car, the cables on the side of the car might be exposed. If the suspension point is under the car, the cables might be run up the side of the car to the car's top junction box. Suitable guards might be necessary to protect the cables from damage. In order to connect to equipment, the traveling cable is permitted to continue out of the hoistway, up to a length of 6 feet from the first support, without the use of a raceway. See Exhibit 620.1.

Part VI. Disconnecting Means and Control

620.51 Disconnecting Means. A single means for disconnecting all ungrounded main power supply conductors

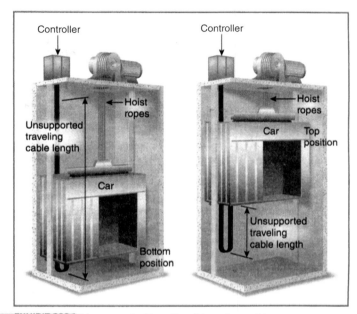

EXHIBIT 620.1 Unsupported lengths of traveling cable.

for each elevator, dumbwaiter, escalator, moving walk, platform lift, or stairway chairlift shall be provided and be designed so that no pole can be operated independently. Where multiple driving machines are connected to a single elevator, escalator, moving walk, or pumping unit, there shall be one disconnecting means to disconnect the motor(s) and control valve operating magnets.

The disconnecting means for the main power supply conductors shall not disconnect the branch circuits required in 620.22, 620.23, and 620.24.

The branch circuits that supply elevator car lighting, receptacles, ventilation, air conditioning, and heating are required to be independent of the control portion of the elevator. In addition, the branch circuits supplying hoistway pit lighting and receptacles and machine room or control room lights and receptacles are not permitted to be disconnected by the main elevator power disconnect. This requirement provides for passenger safety and comfort and for the safety of elevator maintenance personnel during an inadvertent or emergency shutdown of the main power circuit to the elevator. The disconnecting means must always be located outside the hoistway.

(A) Type. The disconnecting means shall be an enclosed externally operable fused motor circuit switch or circuit breaker that is lockable only in the open position in accordance with 110.25.

The disconnecting means shall be a listed device.

Informational Note No. 1: See ASME A17.1-2019/CSA B44-19, *Safety Code for Elevators and Escalators*, for additional information.
Informational Note No. 2: See ASME A18.1-2017, *Safety Standard for Platform Lifts and Stairway Chairlifts*, for additional information.

Exception No. 1: Where an individual branch circuit supplies a platform lift, the disconnecting means required by 620.51(C)(4) shall be permitted to comply with 430.109(C). This disconnecting means shall be listed and shall be lockable open in accordance with 110.25.

Exception No. 2: Where an individual branch circuit supplies a stairway chairlift or where a stairway chairlift is supplied by batteries as the primary source, the stairway chairlift shall be permitted to be cord-and-plug-connected, provided it complies with 422.16(A) and the cord does not exceed 1.8 m (6 ft) in length.

(B) Operation. No provision shall be made to open or close this disconnecting means from any other part of the premises. If sprinklers are installed in hoistways, machine rooms, control rooms, machinery spaces, or control spaces, the disconnecting means shall be permitted to automatically open the power supply to the affected elevator(s) prior to the application of water. No provision shall be made to automatically close this disconnecting means. Power shall only be restored by manual means.

> Informational Note: To reduce hazards associated with water on live elevator electrical equipment.

Where sprinklers are installed in hoistways, machine rooms, or machinery spaces, a means must be provided to automatically disconnect the main line power supply to the affected elevator(s) upon or prior to the application of water in accordance with Section 2.8.3.3.2 of ASME A17.1, *Safety Code for Elevators and Escalators*. Water on elevator electrical equipment can result in hazards such as uncontrolled car movement (wet machine brakes), movement of an elevator with open doors (water on safety circuits bypassing car and/or hoistway door interlocks), and shock hazards.

Automatic disconnection of the main line power supply is not required by ASME A17.1 where hoistways and machine rooms are not sprinklered. NFPA 13, *Standard for the Installation of Sprinkler Systems*, provides requirements for sprinkler installation in machine rooms, hoistways, and pits.

Elevator shutdown generally is accomplished through the use of heat detectors located near sprinkler heads. The heat detectors are designed to actuate and generate an alarm signal prior to water discharge from the sprinkler heads. An output control relay powered by the fire alarm system then provides a monitored output to the main line disconnecting means control circuit, which activates the shunt trip. This ensures that all components have secondary power and are monitored for integrity as required by *NFPA 72®, National Fire Alarm and Signaling Code®*.

Elevator shutdown can occur even if the car is not at a landing. In order to avoid trapping occupants in elevator car(s) when the main power to the elevator driving machine is interrupted due to sprinkler activation, Section 2.27.3.2 of ASME A17.1 requires installation of a fire alarm initiating device(s) in sprinklered hoistways for the purposes of initiating elevator

car(s) recall before the main line power is disconnected. This operation is referred to as "Phase I Emergency Recall" in the ASME code.

See also

Section 21.4 of **NFPA 72** for additional requirements relating to the fire alarm system and elevator shutdown

(C) Location. The disconnecting means shall be located where it is readily accessible to qualified persons.

(1) On Elevators Without Generator Field Control. On elevators without generator field control, the disconnecting means shall be located within sight of the motor controller. Where the motor controller is located in the elevator hoistway, the disconnecting means required by 620.51(A) shall be located outside the hoistway and accessible to qualified persons only. An additional fused or non-fused, enclosed, externally operable motor-circuit switch that is lockable open in accordance with 110.25 to disconnect all ungrounded main power-supply conductors shall be located within sight of the motor controller. The additional switch shall be a listed device and shall comply with 620.91(C).

Driving machines or motion and operation controllers not within sight of the disconnecting means shall be provided with a manually operated switch installed in the control circuit to prevent starting. The manually operated switch(es) shall be installed adjacent to this equipment.

Where the driving machine of an electric elevator or the hydraulic machine of a hydraulic elevator is located in a remote machine room or remote machinery space, a single means for disconnecting all ungrounded main power-supply conductors shall be provided and be lockable open in accordance with 110.25.

A common installation is a machine room, containing the driving machine, motor controller, motion controller, and operation controller, located outside of the hoistway. A disconnecting means must be within sight of the motor controller. Any driving machine or motion and operation controller not within sight of this disconnecting means must be provided with a manual switch in its control circuit to prevent it from starting. Exhibit 620.2 illustrates the requirement on disconnecting means for driving machines or motion and operation controllers not within sight of the main line disconnecting means.

(2) On Elevators with Generator Field Control. On elevators with generator field control, the disconnecting means shall be located within sight of the motor controller for the driving motor of the motor-generator set. Driving machines, motor-generator sets, or motion and operation controllers not within sight of the disconnecting means shall be provided with a manually operated switch installed in the control circuit to prevent starting. The manually operated switch(es) shall be installed adjacent to this equipment.

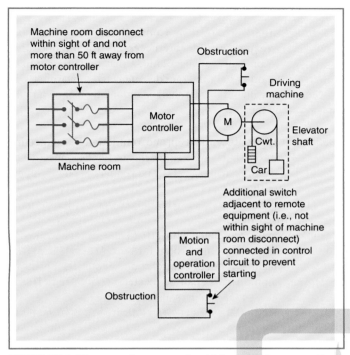

EXHIBIT 620.2 Disconnecting means for driving machines or motion and operation controllers not within sight of the main line disconnecting means. (*Courtesy of ASME*)

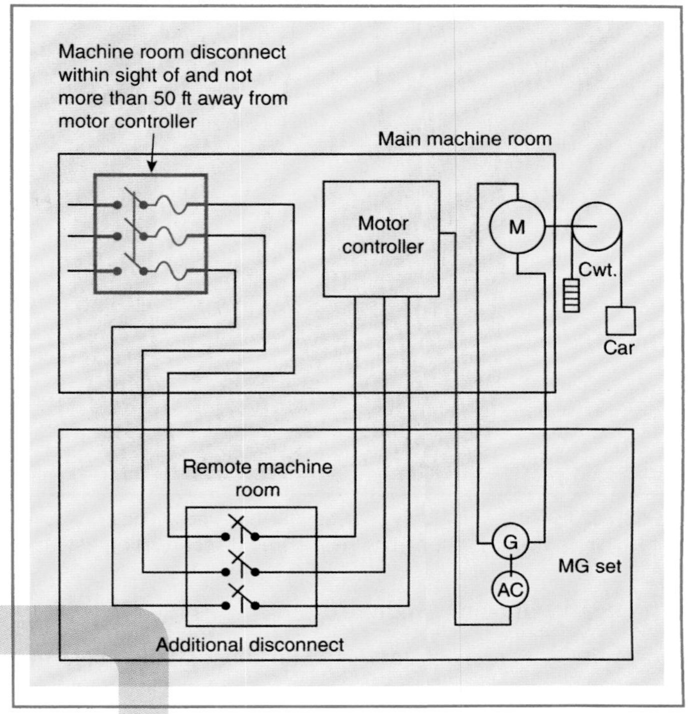

EXHIBIT 620.3 Disconnecting means for a motor-generator set in a remote location. (*Courtesy of ASME*)

Where the driving machine or the motor-generator set is located in a remote machine room or remote machinery space, a single means for disconnecting all ungrounded main power-supply conductors shall be provided and be lockable open in accordance with 110.25.

Where the driving machine is located in a remote machine room away from the control room, a disconnecting means must be within sight of the motor controller. Additionally, a means for disconnecting all ungrounded main power-supply conductors must be provided in the remote machine room. See Exhibits 620.3 and 620.4 for examples of disconnecting means for a motor-generator set and for driving machines in remote locations.

(3) On Escalators and Moving Walks. On escalators and moving walks, the disconnecting means shall be installed in the space where the controller is located.

The local emergency stop control at the escalator location shown in Exhibit 620.5, which is required by Section 6.1.6.3.1 of ASME A.17.1, *Safety Code for Elevators and Escalators,* for passenger safety, cannot be used as the disconnecting means required by this section.

(4) On Platform Lifts and Stairway Chairlifts. On platform lifts and stairway chairlifts, the disconnecting means shall be located within sight of the motor controller.

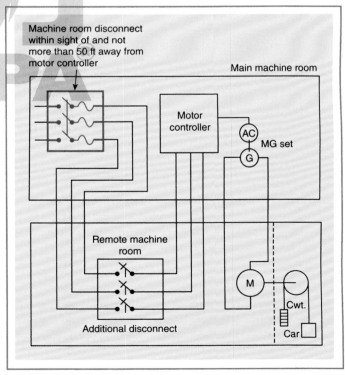

EXHIBIT 620.4 Disconnecting means for driving machines in a remote location. (*Courtesy of ASME*)

EXHIBIT 620.5 The emergency stop button installed at the escalator location is not considered to be the required disconnecting means.

(D) Identification and Signs.

(1) Available Fault Current Field Marking. The disconnecting means shall be legibly marked in the field with the available fault current at its line terminals. The field marking(s) shall include the date the available fault current calculation was performed and be of sufficient durability to withstand the environment involved.

When modifications to the electrical installation occur that affect the available fault current at the disconnecting means, the available fault current shall be verified or recalculated as necessary to ensure the elevator equipment's short-circuit current rating is sufficient for the available fault current at the line terminals of the equipment. The required field marking(s) shall be adjusted to reflect the new level of available fault current.

(E) Surge Protection. Where any of the disconnecting means in 620.51 has been designated as supplying an emergency system load, a legally required system load, or a critical operation power system load, a listed SPD shall be installed.

620.52 Power from More Than One Source.

(A) Single-Car and Multicar Installations. On single-car and multicar installations, equipment receiving electrical power from more than one source shall be provided with a disconnecting means for each source of electrical power. The disconnecting means shall be within sight of the equipment served.

(B) Warning Sign for Multiple Disconnecting Means. Where multiple disconnecting means are used and parts of the controllers remain energized from a source other than the one disconnected, a warning sign shall be mounted on or next to the disconnecting means. The sign shall be clearly legible and shall read as follows:

WARNING
PARTS OF THE CONTROLLER ARE NOT
DE-ENERGIZED BY THIS SWITCH.

The warning sign(s) or label(s) shall comply with 110.21(B).

(C) Interconnection Multicar Controllers. Where interconnections between controllers are necessary for the operation of the system on multicar installations that remain energized from a source other than the one disconnected, a warning sign in accordance with 620.52(B) shall be mounted on or next to the disconnecting means.

620.53 Car Light, Receptacle(s), and Ventilation Disconnecting Means. Elevators shall have a single means for disconnecting all ungrounded car light, receptacle(s), and ventilation power-supply conductors for that elevator car.

The disconnecting means shall be an enclosed, externally operable, fused motor-circuit switch or circuit breaker that is lockable open in accordance with 110.25 and shall be located in the machine room or control room for that elevator car. Where there is no machine room or control room outside the hoistway, the disconnecting means shall be located outside the hoistway and accessible to qualified persons only.

Disconnecting means shall be numbered to correspond to the identifying number of the elevator car whose light source they control.

The disconnecting means shall be provided with a sign to identify the location of the supply side overcurrent protective device.

Exception: Where a separate branch circuit supplies car lighting, a receptacle(s), and a ventilation motor not exceeding 2 hp, the disconnecting means required by 620.53 shall be permitted to comply with 430.109(C). This disconnecting means shall be listed and shall be lockable open in accordance with 110.25.

Section 430.109(C) permits the use of general-use snap switches and listed manual motor controllers as disconnecting means for motors rated 2 horsepower or less and 300 volts or less. General-use snap switches used as a disconnecting means for these motors are not required to have a horsepower rating, but their ampere rating cannot be less than twice the full-load current rating of the motor. All the methods permitted by this section are required to be capable of being locked in the open position.

620.54 Heating and Air-Conditioning Disconnecting Means. Elevators shall have a single means for disconnecting all ungrounded car heating and air-conditioning power-supply conductors for that elevator car.

The disconnecting means shall be an enclosed, externally operable, fused motor-circuit switch or circuit breaker that is lockable open in accordance with 110.25 and shall be located in the machine room or control room for that elevator car. Where there is no machine room or control room outside the hoistway, the disconnecting means shall be located outside the hoistway and accessible to qualified persons only.

Where there is equipment for more than one elevator car in the machine room, the disconnecting means shall be numbered to correspond to the identifying number of the elevator car whose heating and air-conditioning source they control.

The disconnecting means shall be provided with a sign to identify the location of the supply side overcurrent protective device.

620.55 Utilization Equipment Disconnecting Means. Each branch circuit for other utilization equipment shall have a single means for disconnecting all ungrounded conductors. The disconnecting means shall be lockable open in accordance with 110.25.

Where there is more than one branch circuit for other utilization equipment, the disconnecting means shall be numbered to correspond to the identifying number of the equipment served. The disconnecting means shall be provided with a sign to identify the location of the supply side overcurrent protective device.

Part VII. Overcurrent Protection

620.61 Overcurrent Protection. Overcurrent protection shall be provided in accordance with 620.61(A) through (D).

Δ **(A) Operating Devices and Control and Signaling Circuits.** Operating devices and control and signaling circuits shall be protected against overcurrent in accordance with 724.43 and 724.45.

Class 2 power-limited circuits shall be protected against overcurrent in accordance with Chapter 9, Notes to Tables 11(A) and 11(B).

Δ **(B) Overload Protection for Motors.** Motor and branch-circuit overload protection shall conform to Article 430, Part III, and 620.61(B)(1) through (B)(4).

(1) Duty Rating on Elevator, Dumbwaiter, and Motor-Generator Sets Driving Motors. Duty on elevator and dumbwaiter driving machine motors and driving motors of motor-generators used with generator field control shall be rated as intermittent. Such motors shall be permitted to be protected against overload in accordance with 430.33.

(2) Duty Rating on Escalator Motors. Duty on escalator and moving walk driving machine motors shall be rated as continuous. Such motors shall be protected against overload in accordance with 430.32.

(3) Overload Protection. Escalator and moving walk driving machine motors and driving motors of motor-generator sets shall be protected against running overload as provided in Table 430.37.

(4) Duty Rating and Overload Protection on Platform Lift and Stairway Chairlift Motors. Duty on platform lift and stairway chairlift driving machine motors shall be rated as intermittent. Such motors shall be permitted to be protected against overload in accordance with 430.33.

Informational Note: See 430.44 for further information for orderly shutdown.

(C) Motor Feeder Short-Circuit and Ground-Fault Protection. Motor feeder short-circuit and ground-fault protection shall be as required in Article 430, Part V.

(D) Motor Branch-Circuit Short-Circuit and Ground-Fault Protection. Motor branch-circuit short-circuit and ground-fault protection shall be as required in Article 430, Part IV.

Δ **620.62 Selective Coordination.** Where more than one driving machine disconnecting means is supplied by the same source, the overcurrent protective devices in each disconnecting means shall be selectively coordinated with any other supply-side overcurrent protective devices.

Selective coordination shall be selected by a licensed professional engineer or other qualified person engaged primarily in the design, installation, or maintenance of electrical systems. The selection and device settings shall be documented and made available to those authorized to design, install, inspect, maintain, and operate the system.

Exception: Selective coordination shall not be required between two overcurrent devices located in series if no loads are connected in parallel with the downstream device.

Coordination of the overcurrent protection devices (OCPDs) is important to ensure continuity of power where more than one elevator is supplied by a single feeder. For example, if a building contains three elevators and a fault occurs in the circuit conductors to one of the elevators, only the overcurrent device ahead of that faulted circuit should open. Coordination leaves the remaining two elevators in operation. This arrangement is especially important because elevators are commonly used to carry fire fighters and equipment closer to the fire during fire-fighting operations. Where the overcurrent devices in the elevator room do not have proper coordination with the upstream feeder overcurrent device, the potential for interruption of power to all three elevators is increased.

The one-line diagram on the right side of Exhibit 620.6 illustrates the overcurrent protection arrangement required by 620.62, while the left side shows the potential for unnecessary power interruption to all the elevators where the overcurrent protection is not selectively coordinated. Feeder overcurrent devices in the main distribution panel should not open, so two elevators can remain in use.

See also

620.51 for the requirement regarding power-supply disconnecting means

620.65 Signage. Equipment enclosures containing selectively coordinated overcurrent devices shall be legibly marked in the field to indicate that the overcurrent devices are selectively coordinated. The marking shall meet the requirements of 110.21(B), shall be readily visible, and shall state the following:

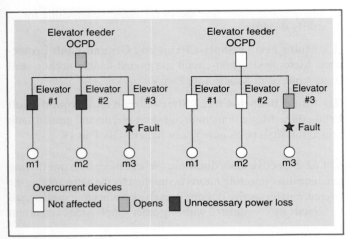

EXHIBIT 620.6 Examples of a system of OCPDs that are not selectively coordinated (*left*) and a system where selectively coordinated overcurrent protection limits the power outage to only the elevator circuit in which the fault has occurred (*right*).

CAUTION:
OVERCURRENT DEVICES IN THIS ENCLOSURE ARE SELECTIVELY COORDINATED. EQUIVALENT REPLACEMENTS AND TRIP SETTINGS ARE REQUIRED.

Part VIII. Machine Rooms, Control Rooms, Machinery Spaces, and Control Spaces

620.71 Guarding Equipment. Elevator, dumbwaiter, escalator, and moving walk driving machines; motor-generator sets; motor controllers; and disconnecting means shall be installed in a room or space set aside for that purpose unless otherwise permitted in 620.71(A) or (B). The room or space shall be secured against unauthorized access.

(A) Motor Controllers. Motor controllers shall be permitted outside the spaces herein specified, provided they are in enclosures with doors or removable panels that are capable of being locked in the closed position and the disconnecting means is located adjacent to or is an integral part of the motor controller. Motor controller enclosures for escalator or moving walks shall be permitted in the balustrade on the side located away from the moving steps or moving treadway. If the disconnecting means is an integral part of the motor controller, it shall be operable without opening the enclosure.

(B) Driving Machines. Elevators with driving machines located on the car, on the counterweight, or in the hoistway, and driving machines for dumbwaiters, platform lifts, and stairway lifts, shall be permitted outside the spaces herein specified.

Part IX. Grounding and Bonding

620.81 Metal Raceways Attached to Cars. Metal raceways, Type MC cable, Type MI cable, or Type AC cable attached to elevator cars shall be bonded to metal parts of the car that are bonded to the equipment grounding conductor.

620.82 Electric Elevators. For electric elevators, the frames of all motors, elevator machines, controllers, and the metal enclosures for all electrical equipment in or on the car or in the hoistway shall be bonded in accordance with Article 250, Parts V and VII.

620.83 Nonelectric Elevators. For elevators other than electric having any electrical conductors attached to the car, the metal frame of the car, where normally accessible to persons, shall be bonded in accordance with Article 250, Parts V and VII.

620.84 Escalators, Moving Walks, Platform Lifts, and Stairway Chairlifts. Escalators, moving walks, platform lifts, and stairway chairlifts shall comply with Article 250.

Part X. Emergency and Standby Power Systems

620.91 Emergency and Standby Power Systems. Elevators shall be permitted to be powered by an emergency or standby power system.

> Informational Note No. 1: See ASME A17.1-2016/CSA B44-16, *Safety Code for Elevators and Escalators*, 2.27.2, for additional information.
>
> Informational Note No. 2: When an elevator is classified as a fire service access elevator or occupant evacuation operation elevator, some building codes require the elevator equipment, elevator hoistway lighting, ventilation and cooling equipment for elevator machine rooms, control rooms, machine spaces, and control spaces as well as elevator car lighting to be supplied by standby power systems in compliance with Article 701.

(A) Regenerative Power. For elevator systems that regenerate power back into the power source that is unable to absorb the regenerative power under overhauling elevator load conditions, a means shall be provided to absorb this power.

(B) Other Building Loads. Other building loads, such as power and lighting, shall be permitted as the energy absorption means required in 620.91(A), provided that such loads are automatically connected to the emergency or standby power system operating the elevators and are large enough to absorb the elevator regenerative power.

(C) Disconnecting Means. The disconnecting means required by 620.51 shall disconnect the elevator from both the emergency or standby power system and the normal power system.

Where an additional power source is connected to the load side of the disconnecting means, which allows automatic movement of the car to permit evacuation of passengers, the disconnecting means required in 620.51 shall be provided with an auxiliary contact that is positively opened mechanically, and the opening shall not be solely dependent on springs. This contact shall cause the additional power source to be disconnected from its load when the disconnecting means is in the open position.

ARTICLE
625 Electric Vehicle Power Transfer System

Part I. General

The automotive industry has provided consumers with a large variety of electric vehicles (EVs) and combination electric–fossil fuel, or hybrid, vehicles that have overcome the "range anxiety" concerns associated with many of the early EVs. EV charging occurs in all occupancies, including residential, commercial, retail, and public sites. The National Institute of Standards and Technology (NIST) projects that many homes and businesses will soon have EV charging infrastructure included as a standard part of the premises wiring system. Legislation in several U.S. states requires new homes to be EV ready, meaning that infrastructure needs to be provided to support the installation of electric vehicle supply equipment (EVSE). Additionally, laws in some states require new public parking areas to be equipped with a certain number of spaces that have EVSE.

Article 625 covers installation safety requirements for charging equipment and for connecting the charging (supply) equipment to the EV through either a conductive, inductive, or wireless power transfer. In addition, ventilation requirements aimed at preventing an ignitible air-hydrogen mixture are included in the requirements. The fundamental purpose of the *NEC®* to minimize fire and shock hazards is conveyed through the Article 625 requirements covering the charger-vehicle interface and the environment that is created by the charging of some types of batteries.

Use of the EV as a power source for standby power or as an interconnected source is also covered in Article 625. This bi-directional connection is attractive to consumers who would otherwise be using an engine generator or other alternate source in the event of an interruption of the normal utility supply.

See also

90.2(C)(6) for the *NEC* scope statement on connecting electric vehicles to the premises wiring system

Δ **625.1 Scope.** This article covers the electrical conductors and equipment connecting an electric vehicle to premises wiring for the purposes of charging, power export, or bidirectional current flow.

> Informational Note No. 1: See NFPA 505-2018, *Fire Safety Standard for Powered Industrial Trucks Including Type Designations, Areas of Use, Conversions, Maintenance, and Operations*, for information on fire protection of industrial trucks.
>
> Informational Note No. 2: See UL 2594-2016, *Electric Vehicle Supply Equipment*, for information on conductive electric vehicle supply equipment.
>
> Informational Note No. 3: See UL 2202-2009, *Electric Vehicle Charging System Equipment*, for information on conductive electric vehicle charging equipment.
>
> Informational Note No. 4: See UL 2750-2020, *Outline of Investigation for Wireless Power Transfer Equipment for Electric Vehicles*, for information on wireless power transfer equipment for transferring power to an electric vehicle.

> Informational Note No. 5: See NECA 413-2019, *Installing and Maintaining Electric Vehicle Supply Equipment (EVSE)*, for information on the procedures for installing and maintaining AC Level 1, AC Level 2, and fast-charging dc electric vehicle supply equipment (EVSE).

Article 625 covers all electrical wiring and equipment installed between the service point and the automotive-type EV. Automotive-type EVs are emphasized because they are much different from other commonly used electric vehicles, such as industrial fork-lifts, hoists, lifts, transports, golf carts, and airport personnel trams. The charging systems and other exterior electrical connections for these off-road vehicles are usually serviced and maintained by trained mechanics or technicians.

625.4 Voltages. Unless other voltages are specified, the nominal ac system voltages of 120, 120/240, 208Y/120, 240, 480Y/277, 480, 600Y/347, 600, or 1000 volts or dc system input voltages of up to 1000 volts shall be used to supply equipment covered by this article. Output voltages to the electric vehicle are not specified.

625.6 Listed. Electric vehicle power transfer system equipment for the purposes of charging, power export, or bidirectional current flow shall be listed.

Part II. Equipment Construction

625.17 Cords and Cables.

(A) Power-Supply Cord. The cable for cord-connected electric vehicle supply equipment (EVSE) shall comply with all of the following:

(1) Be any of the types specified in 625.17(B)(1) or hard service cord, junior hard service cord, or portable power cable types in accordance with Table 400.4. Hard service cord, junior hard service cord, or portable power cable types shall be listed, as applicable, for exposure to oil and damp and wet locations.

(2) Have an ampacity as specified in Table 400.5(A)(1) or, for 8 AWG and larger, in the 60°C (140°F) columns of Table 400.5(A)(2).

(3) Have an overall length as specified in either of the following:

    a. When the interrupting device of the personnel protection system specified in 625.22 is located within the enclosure of the supply equipment or charging system, the power-supply cord shall be not more than the length indicated in (i) or (ii):

        (i) For portable equipment in accordance with 625.44(A), the power-supply cord shall be not more than 300 mm (12 in.) long.

        (ii) For fastened-in-place equipment in accordance with 625.44(B), the power-supply cord shall be not more than 1.8 m (6 ft) long and the equipment shall be installed at a height that prevents the power-supply cord from contacting the floor when it is connected to the proper receptacle.

b. When the interrupting device of the personnel protection system specified in 625.22 is located at the attachment plug, or within the first 300 mm (12 in.) of the power-supply cord, the overall cord length shall be not greater than 4.6 m (15 ft).

Δ **(B) Output Cable to Electric Vehicles.** The output cable to electric vehicles shall be one of the following:

(1) Listed Type EV, EVJ, EVE, EVJE, EVT, or EVJT flexible cable as specified in Table 400.4
(2) An integral part of listed electric vehicle supply equipment

> Informational Note No. 1: See UL 2594-2016, *Standard for Electric Vehicle Supply Equipment*, for information on conductive electric vehicle supply equipment.
> Informational Note No. 2: See UL 2202-2009, *Standard for Electric Vehicle (EV) Charging System Equipment*, for information on conductive electric vehicle charging equipment.

(C) Overall Cord and Cable Length. The overall usable length shall not exceed 7.5 m (25 ft) unless equipped with a cable management system that is part of the listed electric vehicle supply equipment.

(1) Portable Equipment. For portable EVSE, the cord-exposed usable length shall be measured from the face of the attachment plug to the face of the electric vehicle connector.

(2) Fastened-in-Place. Where the EVSE is fastened-in-place, the usable length of the output cable to the electric vehicle shall be measured from the cable exit of the electric vehicle supply equipment to the face of the electric vehicle connector.

> Where the wireless power transfer equipment (WPTE) is fastened-in-place, the output cable to the primary pad shall be measured from the cable exit of the control box to the cable inlet at the primary pad.

The maximum 25-foot cable length takes into account both the power supply cable and the output cable to the EV. For example, 625.17(A)(3)(b) allows a maximum 15-foot-long power cord, which limits the length of the output cable to 10 feet. Cable lengths in excess of 25 feet are permitted where provided with a listed cable management system. This provides commercial parking areas with flexibility in the number of charging spaces they can provide. As illustrated in Exhibit 625.1, wireless power transfer equipment (WPTE) still requires a maximum 25-foot cable length from where the cable exits the control box to where it enters the primary wireless charging pad.

(D) Interconnecting Cabling Systems. Other cabling systems that are integral parts of listed EVSE and are intended to interconnect pieces of equipment within an EVSE system using approved installation methods shall be permitted.

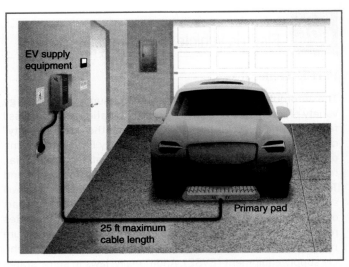

EXHIBIT 625.1 Wireless power transfer equipment (WPTE) allowing the transmission of energy between the premises wiring system and the electric vehicle without physical electrical contact, such as plugging in a supply cord connector.

625.22 Personnel Protection System. EVSE shall have a listed system of protection against electric shock of personnel. Where cord-and-plug-connected equipment is used, the interrupting device of a listed personnel protection system shall be provided according to 625.17(A). A personnel protection system shall not be required for EVSE that supplies less than 60 volts dc.

The listed personnel protection system may consist of one or more components that provide protection against electric shock for different portions of the EV supply equipment circuitry, which might be operating at frequencies other than 50/60 hertz, at direct-current potentials, and/or voltages above 150 volts to ground. Standard GFCI devices do not provide the range of protection needed for the various types of charging systems available. For systems operating above 150 volts to ground, the protective system may include monitoring systems to ensure that proper grounding is provided and maintained during charging.

Part III. Installation

Δ **625.40 Electric Vehicle Branch Circuit.** Each outlet installed for the purpose of supplying EVSE greater than 16 amperes or 120 volts shall be supplied by an individual branch circuit.

> *Exception: Branch circuits shall be permitted to feed multiple EVSEs as permitted by 625.42(A) or (B).*

Δ **625.41 Overcurrent Protection.** Overcurrent protection for feeders and branch circuits supplying EVSE and WPTE, including bidirectional EVSE and WPTE, shall be sized for continuous duty and shall have a current rating of not less than 125 percent of the maximum load of the equipment. Where noncontinuous loads are supplied from the same feeder, the overcurrent device shall have a current rating of not less than the sum of the noncontinuous loads plus 125 percent of the continuous loads.

Δ **625.42 Rating.** The EVSE shall have sufficient rating to supply the load served. Electric vehicle charging loads shall be considered to be continuous loads for the purposes of this article. Service and feeder shall be sized in accordance with the product ratings, unless the overall rating of the installation can be limited through controls as permitted by 625.42(A) or (B).

N **(A) Energy Management System (EMS).** Where an EMS in accordance with 750.30 provides load management of EVSE, the maximum equipment load on a service and feeder shall be the maximum load permitted by the EMS. The EMS shall be permitted to be integral to one piece of equipment or integral to a listed system consisting of more than one piece of equipment. When one or more pieces of equipment are provided with an integral load management control, the system shall be marked to indicate this control is provided.

N **(B) EVSE with Adjustable Settings.** EVSE with restricted access to an ampere adjusting means complying with 750.30(C) shall be permitted. If adjustments have an impact on the rating label, those changes shall be in accordance with manufacturer's instructions, and the adjusted rating shall appear on the rating label with sufficient durability to withstand the environment involved. EVSE as referenced shall be permitted to have ampere ratings that are equal to the adjusted current setting.

Three methods for EV power transfer — referred to as Level 1, Level 2, and DC fast charging, sometimes called Level 3 — cover the range of power levels anticipated for charging EVs. EV charging loads are considered to be continuous loads for the purposes of applying 625.42.

Level 1 allows broad access to charge an EV, often by plugging into a grounded 120-volt electrical receptacle (NEMA 5-15R or 5-20R). Level 1 chargers are commonly a portable style, likely provided by an EV manufacturer with the vehicle, which means they would need to be connected to the premises wiring system by one of the means listed in 625.44(A). Although it is most common for portable chargers to be plugged into a 120-volt receptacle outlet, the *NEC* permits receptacle outlets as large as 50 amperes and 250 volts, single phase.

Level 2 is the primary method of EV power transfer at both private and public facilities. It requires special equipment and connection to an electric power supply dedicated to EV power transfer. Fastened-in-place and fixed EVSE have ratings of 120/240 volts, 208Y/120 volts, and even 480Y/277 volts in some cases. Equipment can be manufactured in both single- or three-phase configurations. See 625.41 for sizing overcurrent protective devices (OCPDs), which are required to have a rating of not less than 125 percent of the maximum load of the equipment.

DC Fast Charging (DCFC) is the EV equivalent of a commercial gasoline dispensing station. Initially, DCFC were called Level 3 because they exceeded UL 2594, *Standard for Electric Vehicle*

Supply Equipment, thresholds for Level 2. Although sometimes still referenced as Level 3, manufacturers have moved toward referencing this level of EV power transfer as DCFC. This high-speed, high-power method charges an EV in about the same time it takes to refuel a conventional vehicle. Because of individual supply requirements and available source voltages, exact voltage and load specifications for DCFC have not been defined in the same way that Level 1 and Level 2 have. These power requirements are specified by the equipment manufacturer, but at present the maximum current is specified as 400 amperes with 350 kilowatts of continuous power supplied, at a rate that continues to grow.

Connection of electric vehicle supply equipment (EVSE) to an energy management system (EMS) can preclude the need for a service or feeder upgrade to an existing electrical installation. Where an EMS is utilized for EVs, it is required to be installed in accordance with 625.42, which provides requirements around the load management of EMS. An EVSE equipped with restricted access to a means that adjusts the output can be utilized, provided that the requirements listed in 625.42(B) and 750.30(C) are met. Where utilized, special attention should be paid to the adjusted ampacity marking requirements, as listed in 625.42(B). Where the EVSE is determined to be a continuous load, the adjusted rating should be calculated as such, and the rating of the circuit conductors and the OCPD must be sized accordingly.

625.43 Disconnecting Means. For EVSE and WPTE rated more than 60 amperes or more than 150 volts to ground, the disconnecting means shall be provided and installed in a readily accessible location. If the disconnecting means is installed remote from the equipment, a plaque shall be installed on the equipment denoting the location of the disconnecting means. The disconnecting means shall be lockable open in accordance with 110.25.

625.44 Equipment Connection. EVSE and WPTE shall be connected to the premises wiring system in accordance with one of the methods in 625.44(A) through (C).

Some manufacturers produce 125-volt, single-phase, 15- or 20-ampere portable charging units for convenience charging. These charging units can be stored in the vehicle. Fastened EVSE and WPTE rated up to 250 volts and 60 amperes may be cord-and-plug-connected under the specified conditions. All other EVSE and WPTE must be mounted as fixed-in-place, which requires using fasteners that require a tool for removal, and permanently (direct) wired.

(A) Portable Equipment. Portable equipment shall be connected to the premises wiring system by one or more of the following methods:

(1) A nonlocking, 2-pole, 3-wire grounding-type receptacle outlet rated at 125 volts, single phase, 15 or 20 amperes
(2) A nonlocking, 2-pole, 3-wire grounding-type receptacle outlet rated at 250 volts, single phase, 15 or 20 amperes

(3) A nonlocking, 2-pole, 3-wire or 3-pole, 4-wire grounding-type receptacle outlet rated at 250 volts, single phase, 30 or 50 amperes, or 125/250 volts, single-phase, 30, 50, or 60 amperes

(4) A nonlocking, 2-pole, 3-wire grounding-type receptacle outlet rated at 60 volts dc maximum, 15 or 20 amperes

(B) Fastened-in-Place Equipment. Equipment that is fastened-in-place shall be connected to the premises wiring system by one of the following methods:

(1) A nonlocking, 2-pole, 3-wire grounding-type receptacle outlet rated 125 volts or 250 volts, single phase, up to 50 amperes

(2) A nonlocking, 3-pole, 4-wire grounding-type receptacle outlet rated 250 volts, three phase, up to 50 amperes

(3) A nonlocking, 3-pole, 4-wire grounding-type receptacle outlet rated 125/250 volts, single phase, 30, 50, or 60 amperes

(4) A nonlocking, 2-pole, 3-wire grounding-type receptacle outlet rated 60 volts dc maximum, 15 or 20 amperes

(C) Fixed-in-Place Equipment. All other EVSE and WPTE shall be permanently wired and fixed-in-place to the supporting surface.

625.46 Loss of Primary Source. Means shall be provided such that, upon loss of voltage from the utility or other electrical system(s), energy cannot be back fed through the electric vehicle and the supply equipment to the premises wiring system unless permitted by 625.48.

625.47 Multiple Feeder or Branch Circuits. Where equipment is identified for the application, more than one feeder or branch circuit shall be permitted to supply equipment.

> **See also**
>
> **225.30(A)(7)** for correlating permission regarding electric vehicle power transfer systems that are listed, labeled, and identified as being able to be supplied by more than one branch circuit or feeder

Δ **625.48 Interactive Equipment.** EVSE or WPTE that incorporates a power export function and that is part of an interactive system that serves as an optional standby system, an electric power production source, or a bidirectional power feed shall be listed and marked as suitable for that purpose. When used as an optional standby system, the requirements of Parts I and II of Article 702 shall apply; when used as an electric power production source, the requirements of Parts I and II of Article 705 shall apply. EVPE that provides a receptacle outlet as its point of power export shall be in accordance with 625.60.

> Informational Note No. 1: See UL 1741, *Inverters, Converters, Controllers and Interconnection System Equipment for Use with Distributed Energy Resources,* for further information on supply equipment.

> Informational Note No. 2: See UL 9741, *Bidirectional Electric Vehicle (EV) Charging System Equipment,* for vehicle interactive systems.

> Informational Note No. 3: See SAE J3072, *Standard for Interconnection Requirements for Onboard, Utility-Interactive Inverter Systems,* for further information.

N **625.49 Island Mode.** EVPE and bidirectional EVSE that incorporate a power export function shall be permitted to be a part of an interconnected power system operating in island mode.

625.50 Location. The EVSE shall be located for direct electrical coupling of the EV connector (conductive or inductive) to the electric vehicle. Unless specifically listed and marked for the location, the coupling means of the EVSE shall be stored or located at a height of not less than 450 mm (18 in.) above the floor level for indoor locations or 600 mm (24 in.) above the grade level for outdoor locations. This requirement does not apply to portable EVSE constructed in accordance with 625.44(A).

625.52 Ventilation. The ventilation requirement for charging an electric vehicle in an indoor enclosed space shall be determined by 625.52(A) or (B).

Where the EV charging operation is conducted in outdoor or open locations, the resulting off-gassing of hydrogen does not pose the same risk of creating an ignitible environment as in indoor locations. The lighter-than-air hydrogen readily diffuses into the atmosphere. In addition to driveways and parking lots, structures with adequate natural ventilation — such as carports and open parking structures — do not require mechanical ventilation. NFPA 88A, *Standard for Parking Structures,* provides a quantifiable definition of the term *open parking structure.*

(A) Ventilation Not Required. Where electric vehicle storage batteries are used or where the equipment is listed for charging electric vehicles indoors without ventilation, mechanical ventilation shall not be required.

Most batteries used in EVs do not emit hydrogen gas in quantities that could cause an explosion. Preventive measures such as mechanical or passive ventilation are not required, because the EV batteries and charging systems are designed to prevent or limit the emission of hydrogen during charging. The Society of Automotive Engineers (SAE) recommended practice SAE J1718, *Measurement of Hydrogen Gas Emission from Battery-Powered Passenger Cars and Light Trucks During Battery Charging,* can be used to assess suitability for indoor charging. This recommended practice includes procedures for tests during normal charging operations and potential equipment failure modes.

(B) Ventilation Required. Where the equipment is listed for charging electric vehicles that require ventilation for indoor charging, mechanical ventilation, such as a fan, shall be provided. The ventilation shall include both supply and exhaust equipment and shall be permanently installed and located to intake from, and vent directly to, the outdoors. Positive-pressure ventilation systems shall be permitted only in vehicle charging buildings or areas that have been specifically designed and approved for that application. Mechanical ventilation requirements shall be determined by one of the methods specified in 625.52(B)(1) through (B)(4).

(1) Table Values. For supply voltages and currents specified in Table 625.52(B)(1)(1) or Table 625.52(B)(1)(2), the minimum ventilation requirements shall be as specified in Table 625.52(B)(1)(1) or Table 625.52(B)(1)(2) for each of the total number of electric vehicles that can be charged at one time.

(2) Other Values. For supply voltages and currents other than specified in Table 625.52(B)(1)(1) or Table 625.52(B)(1)(2), the

TABLE 625.52(B)(1)(1) *Minimum Ventilation Required in Cubic Meters per Minute (m³/min) for Each of the Total Number of Electric Vehicles That Can Be Charged at One Time*

Branch-Circuit Ampere Rating	Branch-Circuit Voltage							
		Single Phase			3 Phase			
	DC ≥ 50 V	120 V	208 V	240 V or 120/240 V	208 V or 208Y/120 V	240 V	480 V or 480Y/277 V	600 V or 600Y/347 V
15	0.5	1.1	1.8	2.1	—	—	—	—
20	0.6	1.4	2.4	2.8	4.2	4.8	9.7	12
30	0.9	2.1	3.6	4.2	6.3	7.2	15	18
40	1.2	2.8	4.8	5.6	8.4	9.7	19	24
50	1.5	3.5	6.1	7.0	10	12	24	30
60	1.8	4.2	7.3	8.4	13	15	29	36
100	2.9	7.0	12	14	21	24	48	60
150	—	—	—	—	31	36	73	91
200	—	—	—	—	42	48	97	120
250	—	—	—	—	52	60	120	150
300	—	—	—	—	63	73	145	180
350	—	—	—	—	73	85	170	210
400	—	—	—	—	84	97	195	240

TABLE 625.52(B)(1)(2) *Minimum Ventilation Required in Cubic Feet per Minute (cfm) for Each of the Total Number of Electric Vehicles That Can Be Charged at One Time*

Branch-Circuit Ampere Rating	Branch-Circuit Voltage							
		Single Phase			3 Phase			
	DC ≥ 50 V	120 V	208 V	240 V or 120/240 V	208 V or 208Y/120 V	240 V	480 V or 480Y/277 V	600 V or 600Y/347 V
15	15.4	37	64	74	—	—	—	—
20	20.4	49	85	99	148	171	342	427
30	30.8	74	128	148	222	256	512	641
40	41.3	99	171	197	296	342	683	854
50	51.3	123	214	246	370	427	854	1066
60	61.7	148	256	296	444	512	1025	1281
100	102.5	246	427	493	740	854	1708	2135
150	—	—	—	—	1110	1281	2562	3203
200	—	—	—	—	1480	1708	3416	4270
250	—	—	—	—	1850	2135	4270	5338
300	—	—	—	—	2221	2562	5125	6406
350	—	—	—	—	2591	2989	5979	7473
400	—	—	—	—	2961	3416	6832	8541

minimum ventilation requirements shall be calculated by means of the following general formulas, as applicable:

(1) Single-phase ac or dc:

Ventilation$_{\text{single-phase ac or dc}}$ in cubic meters per minute (m^3/min) =

$$\frac{(\text{volts})(\text{amperes})}{1718} \quad \textbf{[625.52(B)(2)a]}$$

Ventilation$_{\text{single-phase ac or dc}}$ in cubic feet per minute (cfm) =

$$\frac{(\text{volts})(\text{amperes})}{48.7} \quad \textbf{[625.52(B)(2)b]}$$

(2) Three-phase ac:

Ventilation$_{\text{3-phase}}$ in cubic meters per minute (m^3/min) =

$$\frac{1.732(\text{volts})(\text{amperes})}{1718} \quad \textbf{[625.52(B)(2)c]}$$

Ventilation$_{\text{3-phase}}$ in cubic feet per minute (cfm) =

$$\frac{1.732(\text{volts})(\text{amperes})}{48.7} \quad \textbf{[625.52(B)(2)d]}$$

(3) Engineered Systems. For an equipment ventilation system designed by a person qualified to perform such calculations as an integral part of a building's total ventilation system, the minimum ventilation requirements shall be permitted to be determined in accordance with calculations specified in the engineering study.

(4) Supply Circuits. The supply circuit to the mechanical ventilation equipment shall be electrically interlocked with the equipment and shall remain energized during the entire electric vehicle charging cycle. Equipment receptacles rated at 125 volts, single phase, 15 and 20 amperes shall be switched and the mechanical ventilation system shall be electrically interlocked through the switch supply power to the receptacle. Equipment supplied from less than 50 volts dc shall be switched and the mechanical ventilation system shall be electrically interlocked through the switch supply power to the equipment.

The sufficient diffusion and dilution of hydrogen gas from gas-emitting batteries prevent a hazardous condition. During the charging process, certain batteries used in some EVs emit hydrogen gas, which is colorless, odorless, tasteless, nontoxic, and flammable. At atmospheric pressure, the flammable range for hydrogen is 4 to 75 percent by volume in air.

NFPA 69, *Standard on Explosion Prevention Systems*, establishes requirements to ensure safety with flammable mixtures. Section 8.3, Design and Operating Requirements, of NFPA 69-2019 specifies that combustible gas concentrations be restricted to 25 percent of the lower flammable limit to provide a safety margin for personnel. Safety is accomplished by keeping the concentration of hydrogen below 25 percent of the lower flammability limit. That is 1 percent (25 percent × 4 percent = 1 percent) hydrogen by volume in air, or below 10,000 parts per million (ppm) hydrogen.

A ventilation system for a typical residential-type garage includes both supply and mechanical exhaust equipment and is permanently installed. The system brings outdoor air into the space, circulates the air through the space, and exhausts the air directly to the outdoors. Typically, the equipment includes a passive vent for intake on one side of the enclosed space and an exhaust fan vented to the outside on the other side.

In enclosed commercial garages and other structures, additional ventilation is not required if the exhaust, as required by the building code for carbon monoxide or other purposes, is greater than the quantity listed in the table. Other engineered EV ventilation systems are allowed as part of the building ventilation system.

The ventilation system and the charging system must be interlocked to prevent charging if the ventilation is not operating. This charging arrangement can be used with EVs equipped with a self-contained charging system in which activation of the charging system does not depend on a signal from the EV. A manually operated switch controls the receptacle used to supply the vehicle charging system, and it is also interlocked with the power supply to the ventilation fan. This arrangement ensures that the ventilation fan is operating whenever the vehicle charging receptacle is energized.

Δ **625.54 Ground-Fault Circuit-Interrupter Protection for Personnel.** All receptacles installed for the connection of electric vehicle charging shall have ground-fault circuit-interrupter protection for personnel.

Portable and fastened-in-place EVSE and WPTE that are permitted to be cord-and plug-connected must be supplied through a GFCI-protected receptacle. This includes all the single- and three-phase receptacle configurations specified in 625.44(A) and (B). The outlet supplying direct-connected EVSE is not required to be GFCI protected, unless specified in the manufacturer's instructions.

625.56 Receptacle Enclosures. All receptacles installed in a wet location for electric vehicle charging shall have an enclosure that is weatherproof with the attachment plug cap inserted or removed. An outlet box hood installed for this purpose shall be listed and shall be identified as extra duty. Other listed products, enclosures, or assemblies providing weatherproof protection that do not utilize an outlet box hood shall not be required to be marked extra duty.

625.60 AC Receptacle Outlets Used for EVPE. AC receptacles installed in electric vehicles and intended to allow for connection of off-board utilization equipment shall comply with 625.60(A) through (D).

EVs equipped with a receptacle(s) for the purposes of exporting power to supply utilization equipment must follow similar safety requirements as required by the *NEC* for other sources, such as portable generators. Product certification, receptacle configuration limitations, overcurrent protection, and ground-fault circuit-interrupter protection for personnel are features required for receptacles that will be used to supply portable utilization equipment or to supply power to a premises as a standby (alternate) or stand-alone power source. The requirements for power export receptacles, like similar safety features required by Article 445 for portable generators, provide a safe interface for users of these electric vehicle power export equipment (EVPE) receptacle outlets.

(A) Type. The receptacle outlet shall be listed.

(B) Rating. The receptacle outlet shall be rated 250 volts maximum, single phase 50 amperes maximum.

(C) Overcurrent Protection. Electric vehicles provided with receptacle outlets for power export shall be provided with overcurrent protection integral to the power export system. The overcurrent protection shall have a nominal rating sufficient for the receptacle it protects. The overcurrent protection shall also be sufficiently rated for the maximum available fault current at the receptacle and shall be included in the interactive equipment evaluation. See 625.48.

(D) GFCI Protection for Personnel. Ground-fault circuit-interrupter protection for personnel shall be provided for all receptacles. The ground-fault circuit-interrupter indication and reset shall be installed in a readily accessible location.

Informational Note: There are various methods available to achieve ground-fault circuit-interrupter protection.

Part IV. Wireless Power Transfer Equipment

625.101 Grounding. The primary pad base plate shall be of a nonferrous metal and shall be connected to the circuit equipment grounding conductor unless the listed WPTE employs a double-insulation system. The base plate shall be sized to match the size of the primary pad enclosure.

625.102 Installation.

(A) General. The control pad, if included in the WPTE configuration, shall comply with 625.102(B). The primary pad shall comply with 625.102(C).

Δ **(B) Control Box.** The control box enclosure shall be suitable for the environment and shall be mounted at a height not less than 450 mm (18 in.) above the floor level for indoor locations or 600 mm (24 in.) above grade level for outdoor locations. The control box shall be mounted in one of the following forms:

(1) Pedestal
(2) Wall or pole
(3) Building or structure
(4) Raised concrete pad

Δ **(C) Primary Pad.** The primary pad shall be installed secured to the surface or embedded in the surface of the floor with its top flush with the surface or below the surface, all per manufacturer's instructions and the following:

(1) If the primary pad is located in an area requiring snow removal, it shall not be located on or above the surface.

Exception: Where installed on private property where snow removal is done manually, the primary pad shall be permitted to be installed on or above the surface.

(2) The primary pad enclosure shall be suitable for the environment. If the primary pad is located in an area subject to severe climatic conditions (e.g., flooding), the enclosure shall be suitably rated for those conditions.

Δ **(D) Protection of Cords and Cables to the Primary Pad.** The output cable to the primary pad shall be secured in place over its entire length for the purpose of restricting its movement and to prevent strain at the connection points. If installed in conditions where drive-over could occur, the cable shall be provided with supplemental protection.

Where there is no control box, the cord or cable supplying power to the primary pad shall be secured in place in order to restrict movement and to prevent strain at the connection points. Where subject to vehicular traffic, supplemental protection shall be provided.

(E) Other Wiring Systems. Other wiring systems and fittings specifically listed for use on the WPTE shall be permitted.

ARTICLE
626 Electrified Truck Parking Spaces

Part I. General

626.1 Scope. This article covers the electrical conductors and equipment external to the truck or transport refrigerated unit that connects nonpropulsion electrical elements of trucks or transport refrigerated units to a supply of electricity, and the installation of equipment and devices related to electrical installations within an electrified truck parking space.

Stringent federal and state mandates to reduce diesel engine emissions have led to using electric power for operation of transport truck heating and refrigeration equipment necessary to maintain their loads, as well as convenience power to the cab for the driver, while the truck is parked and not running. Exhibit 626.1 shows an electrified truck parking space that is providing non-vehicular power to the vehicle. Because much of the transport industry is interstate commerce, this article provides for standardization of truck parking space equipment so that driver interface with electrical connection devices can be safely accomplished from coast to coast.

626.3 Other Articles. Wherever the requirements of other articles of this *Code* and Article 626 differ, the requirements of Article 626 shall apply. Unless electrified truck parking space wiring systems are supported or arranged in such a manner that they cannot be used in or above locations classified in 511.3

EXHIBIT 626.1 An example of an electrified truck parking space providing power to a truck while it is parked and not running. *(Courtesy of SRP)*

or 514.3, or both, they shall comply with 626.3(A) and (B) in addition to the requirements of this article.

(A) Vehicle Repair and Storage Facilities. Electrified truck parking space electrical wiring systems located at facilities for the repair or storage of self-propelled vehicles that use volatile flammable liquids or flammable gases for fuel or power shall comply with Article 511.

(B) Motor Fuel Dispensing Stations. Electrified truck parking space electrical wiring systems located at or serving motor fuel dispensing stations shall comply with Article 514.

> Informational Note: See NFPA 88A-2019, *Standard for Parking Structures*, and NFPA 30A-2021, *Code for Motor Fuel Dispensing Facilities and Repair Garages*, for additional information.

626.4 General Requirements.

(A) Not Covered. This article shall not apply to that portion of other equipment in residential, commercial, or industrial facilities that requires electric power used to load and unload cargo, operate conveyors, and for other equipment used on the site or truck.

(B) Distribution System Voltages. Unless other voltages are specified, the nominal ac system voltages of 120, 120/240, 208Y/120, 240, or 480Y/277 shall be used to supply equipment covered by this article.

(C) Connection to Wiring System. This article shall apply to the electrified truck parking space supply equipment intended for connection to a wiring system as defined in 626.4(B).

Part II. Electrified Truck Parking Space Electrical Wiring Systems

626.10 Branch Circuits. Electrified truck parking space single-phase branch circuits shall be derived from a 208Y/120-volt, 3-phase, 4-wire system or a 120/240-volt, single-phase, 3-wire system.

Exception: A 120-volt distribution system shall be permitted to supply existing electrified truck parking spaces.

626.11 Feeder and Service Load Calculations.

(A) Parking Space Load. The calculated load of a feeder or service shall be not less than the sum of the loads on the branch circuits. Electrical service and feeders shall be calculated on the basis of not less than 11 kVA per electrified truck parking space.

(B) Demand Factors. Electrified truck parking space electrical wiring system demand factors shall be based upon the climatic temperature zone in which the equipment is installed. The demand factors set forth in Table 626.11(B) shall be the minimum allowable demand factors that shall be permitted for calculating load for service and feeders. No demand factor shall be allowed for any other load, except as provided in this article.

> Informational Note: The U.S. Department of Agriculture (USDA) has developed a commonly used "Plant Hardiness Zone" map that

TABLE 626.11(B) *Demand Factors for Services and Feeders*

Climatic Temperature Zone (USDA Hardiness Zone) (See Note)	Demand Factor (%)
1	70
2a	67
2b	62
3a	59
3b	57
4a	55
4b	51
5a	47
5b	43
6a	39
6b	34
7a	29
7b	24
8a	21
8b	20
9a	20
9b	20
10a	21
10b	23
11	24

Note: The climatic temperature zones shown in Table 626.11(B) correlate with those found on the "USDA Plant Hardiness Zone Map," and the climatic temperature zone selected for use with the table shall be determined through the use of this map based on the installation location.

is publicly available. The map provides guidance for determining the Climatic Temperature Zone. Data indicate that the HVAC has the highest power requirement in cold climates, with the heating demand representing the greatest load, which in turn is dependent on outside temperature. In very warm climates, where no heating load is necessary, the cooling load increases as the outdoor temperature rises. These demand factors do not apply to the portion of electrical wiring systems that supply the transport refrigerated units (TRUs).

(C) Two or More Electrified Truck Parking Spaces. Where the electrified truck parking space wiring system is in a location that serves two or more electrified truck parking spaces, the equipment for each space shall comply with 626.11(A), and the calculated load shall be calculated on the basis of each parking space.

(D) Conductor Rating. Truck space branch-circuit supplied loads shall be considered to be continuous.

Part III. Electrified Truck Parking Space Supply Equipment

626.22 Wiring Methods and Materials.

(A) Electrified Truck Parking Space Supply Equipment Type. The electrified truck parking space supply equipment shall be provided in one of the following forms:

(1) Pedestal
(2) Overhead gantry
(3) Raised concrete pad

(B) Mounting Height. Post, pedestal, and raised concrete pad types of electrified truck parking space supply equipment shall be not less than 600 mm (2 ft) aboveground or above the point identified as the prevailing highest water level mark or an equivalent benchmark based on seasonal or storm-driven flooding from the authority having jurisdiction.

(C) Access to Working Space. All electrified truck parking space supply equipment shall be accessible by an unobstructed entrance or passageway not less than 600 mm (2 ft) wide and not less than 2.0 m (6 ft 6 in.) high.

(D) Disconnecting Means. A disconnecting switch or circuit breaker shall be provided to disconnect one or more electrified truck parking space supply equipment sites from a remote location. The disconnecting means shall be provided and installed in a readily accessible location and shall be lockable open in accordance with 110.25.

626.23 Overhead Gantry or Cable Management System.

(A) Cable Management. Electrified truck parking space equipment provided from either overhead gantry or cable management systems shall utilize a permanently attached power supply cable in electrified truck parking space supply equipment. Other cable types and assemblies listed as being suitable for the purpose, including optional hybrid communications, signal, and composite optical fiber cables, shall be permitted.

(B) Strain Relief. Means to prevent strain from being transmitted to the wiring terminals shall be provided. Permanently attached power supply cable(s) shall be provided with a means to de-energize the cable conductors and power service delivery device upon exposure to strain that could result in either cable damage or separation from the power service delivery device and exposure of live parts.

626.24 Electrified Truck Parking Space Supply Equipment Connection Means.

(A) General. Each truck shall be supplied from electrified truck parking space supply equipment through suitable extra-hard service cables or cords. Each connection to the equipment shall be by a single separable power supply cable assembly.

Δ **(B) Receptacle.** All receptacles shall be listed and of the grounding type. Every truck parking space with electrical supply shall be equipped with the following:

(1) A maximum of three receptacles, each 2-pole, 3-wire grounding type and rated 20 amperes, 125 volts, and two of the three connected to two separate branch circuits.

Informational Note No. 1: See ANSI/NEMA WD 6-2016, *Wiring Devices — Dimensional Specifications*, Figure 5-20, for the non-locking-type and grounding-type 20-ampere receptacle configuration.

(2) One single receptacle, 3-pole, 4-wire grounding type, single phase rated either 30 amperes 208Y/120 volts or 125/250 volts. The 125/250-volt receptacle shall be permitted to be used on a 208Y/120-volt, single-phase circuit.

Informational Note No. 2: See UL1686–2012, *Pin and Sleeve Configurations*, Figure C2.9 or Part C3, for various configurations of 30-ampere pin and sleeve receptacles.

Exception: Where electrified truck parking space supply equipment provides the heating, air-conditioning, and comfort-cooling function without requiring a direct electrical connection at the truck, only two receptacles identified in 626.24(B)(1) shall be required.

(C) Disconnecting Means, Parking Space. The electrified truck parking space supply equipment shall be provided with a switch or circuit breaker for disconnecting the power supply to the electrified truck parking space. A disconnecting means shall be provided and installed in a readily accessible location and shall be lockable open in accordance with 110.25.

(D) Ground-Fault Circuit-Interrupter Protection for Personnel. In addition to the requirements in 210.8, the electrified truck parking space equipment shall be designed and constructed such that all receptacle outlets in 626.24 are provided with ground-fault circuit-interrupter protection for personnel.

626.25 Separable Power-Supply Cable Assembly. A separable power-supply cable assembly, consisting of a power-supply cord, a cord connector, and an attachment plug intended for connection with a truck flanged surface inlet, shall be of a listed type. The power-supply cable assembly or assemblies shall be identified and be one of the types and ratings specified in 626.25(A) and (B). Cords with adapters and pigtail ends, extension cords, and similar items shall not be used.

(A) Rating(s).

(1) Twenty-Ampere Power-Supply Cable Assembly. Equipment with a 20-ampere, 125-volt receptacle, in accordance with 626.24(B)(1), shall use a listed 20-ampere power-supply cable assembly.

Exception: It shall be permitted to use a listed separable power-supply cable assembly, either hard service or extra-hard service and rated 15 amperes, 125 volts, for connection to an engine block heater for legacy vehicles.

(2) Thirty-Ampere Power-Supply Cable Assembly. Equipment with a 30-ampere, 208Y/120-volt or 125/250-volt receptacle, in accordance with 626.24(B)(2), shall use a listed 30-ampere main power-supply cable assembly.

(B) Power-Supply Cord.

(1) Conductors. The cord shall be a listed type with three or four conductors, for single-phase connection, one conductor of which shall be identified in accordance with 400.23.

Exception: It shall be permitted to use a separate listed three-conductor separable power-supply cable assembly, one conductor of which shall be identified in accordance with 400.23 and rated 15 amperes, 125 volts for connection to an engine block heater for existing vehicles.

(2) Cord. Extra-hard usage flexible cords and cables rated not less than 90°C (194°F), 600 volts; listed for both wet locations and sunlight resistance; and having an outer jacket rated to be resistant to temperature extremes, oil, gasoline, ozone, abrasion, acids, and chemicals shall be permitted where flexibility is necessary between the electrified truck parking space supply equipment, the panel board, and flanged surface inlet(s) on the truck.

Exception: Cords for the separable power supply cable assembly for 15- and 20-ampere connections shall be permitted to be a hard service type.

(3) Cord Overall Length. The exposed cord length shall be measured from the face of the attachment plug to the point of entrance to the truck or the face of the flanged surface inlet or to the point where the cord enters the truck. The overall length of the cable shall not exceed 7.5 m (25 ft) unless equipped with a cable management system that is listed as suitable for the purpose.

Δ **(4) Attachment Plug.** The attachment plug(s) shall be listed, by itself or as part of a cord set, for the purpose and shall be molded

to or installed on the flexible cord so that it is secured tightly to the cord at the point where the cord enters the attachment plug. If a right-angle cap is used, the configuration shall be oriented so that the grounding member is farthest from the cord. Where a flexible cord is provided, the attachment plug shall comply with 250.138(A).

(a) *Connection to 20-Ampere Receptacle.* A separable power-supply cable assembly for connection to a truck flanged surface inlet, rated at 20 amperes, shall have a non-locking-type attachment plug that shall be 2-pole, 3-wire grounding type rated 20 amperes, 125 volts and intended for use with the 20-ampere, 125-volt receptacle.

Exception: A separable power-supply cable assembly, provided for the connection of only an engine block heater, shall have an attachment plug of the 2-pole, 3-wire grounding type, rated 15 amperes, 125 volts.

Informational Note: See ANSI/NEMA WD 6-2016, *Wiring Devices — Dimensional Specifications*, Figure 5-15 or Figure 5-20, for non-locking- and grounding-type 15- or 20-ampere plug and receptacle configurations.

(b) *Connection to 30-Ampere Receptacle.* A separable power-supply cable assembly for connection to a truck flanged surface inlet, rated at 30 amperes, shall have an attachment plug that shall be 3-pole, 4-wire grounding type rated 30 amperes, 208Y/120 volts or 125/250 volts, and intended for use with the receptacle in accordance with 626.24(B)(2). The 125/250-volt attachment plug shall be permitted to be used on a 208Y/120-volt, single-phase circuit.

Informational Note: See UL 1686-2012, *Pin and Sleeve Configurations*, Figure C2.10 or Part C3, for various configurations of 30-ampere pin and sleeve plugs.

Δ **(5) Cord Connector.** The cord connector for a separable power-supply cable assembly, as specified in 626.25(A)(1), shall be a 2-pole, 3-wire grounding type rated 20 amperes, 125 volts. The cord connector for a separable power-supply cable assembly, as specified in 626.25(A)(2), shall be a 3-pole, 4-wire grounding type rated 30 amperes, 208Y/120 volts or 125/250 volts. The 125/250-volt cord connector shall be permitted to be used on a 208Y/120-volt, single-phase circuit.

Exception: The cord connector for a separable power supply cable assembly, rated 15 amperes, provided for the connection of an engine block heater for existing vehicles, shall have an attachment plug that shall be 2-pole, 3-wire grounding type rated 15 amperes, 125 volts.

Informational Note: See UL 1686-2012, *Pin and Sleeve Configurations*, Figure C2.9 or Part C3, for various configurations of 30-ampere cord connectors.

626.26 Loss of Primary Power. Means shall be provided such that, upon loss of voltage from the utility or other electric supply system(s), energy cannot be back-fed through the truck and the truck supply equipment to the electrified truck parking space wiring system unless permitted in 626.27.

626.27 Interactive Systems. Electrified truck parking space supply equipment and other parts of a system, either on-board or off-board the vehicle, that are identified for and intended to be interconnected to a vehicle and also serve as an optional standby system or an electric power production source or provide for bidirectional power feed shall be listed as suitable for that purpose. When used as an optional standby system, the requirements of Article 702 shall apply, and when used as an electric power production source, the requirements of Article 705 shall apply.

Part IV. Transport Refrigerated Units (TRUs)

626.30 Transport Refrigerated Units. Electrified truck parking spaces intended to supply transport refrigerated units (TRUs) shall include an individual branch circuit and receptacle for operation of the refrigeration/heating units. The receptacle associated with the TRUs shall be provided in addition to the receptacles required in 626.24(B).

(A) Branch Circuits. TRU spaces shall be supplied from 208-volt, 3-phase, 240-volt, 3-phase, or 480-volt, 3-phase branch circuits and with an equipment grounding conductor.

(B) Electrified Truck Parking Space Supply Equipment. The electrified truck parking space supply equipment, or portion thereof, providing electric power for the operation of TRUs shall be independent of the loads in Part III of Article 626.

626.31 Disconnecting Means and Receptacles.

(A) Disconnecting Means. Disconnecting means shall be provided to isolate each refrigerated unit from its supply connection. A disconnecting means shall be provided and installed in a readily accessible location and shall be lockable open in accordance with 110.25.

(B) Location. The disconnecting means shall be readily accessible, located not more than 750 mm (30 in.) from the receptacle it controls, and located in the supply circuit ahead of the receptacle. Circuit breakers or switches located in power outlets complying with this section shall be permitted as the disconnecting means.

Δ **(C) Receptacles.** All receptacles shall be listed and of the grounding type. Every electrified truck parking space intended to provide an electrical supply for TRUs shall be equipped with one or more of the following:

(1) A 30-ampere, 480-volt, 3-phase, 3-pole, 4-wire receptacle
(2) A 60-ampere, 208-volt, 3-phase, 3-pole, 4-wire receptacle
(3) A 20-ampere, 1000-volt, 3-phase, 3-pole, 4-wire receptacle, pin and sleeve type
(4) A 60-ampere, 250-volt, 3-phase, 3-pole, 4-wire receptacle
(5) A 60-ampere, 480-volt, 3-phase, 3-pole, 4-wire receptacle

Informational Note: See UL 1686-2012, *Pin and Sleeve Configurations*, Figure C2.11, for complete details of the 30-ampere pin and sleeve receptacle configuration for refrigerated containers (TRUs) and for various configurations of 60-ampere pin and sleeve receptacles.

626.32 Separable Power Supply Cable Assembly. A separable power supply cable assembly, consisting of a cord with an attachment plug and cord connector, shall be one of the types and ratings specified in 626.32(A), (B), and (C). Cords with adapters and pigtail ends, extension cords, and similar items shall not be used.

(A) Rating(s). The power supply cable assembly shall be listed and rated in accordance with one of the following:

(1) A 30-ampere, 480-volt, 3-phase assembly
(2) A 60-ampere, 208-volt, 3-phase assembly
(3) A 20-ampere, 1000-volt, 3-phase assembly
(4) A 60-ampere, 480-volt, 3-phase assembly
(5) A 60-ampere, 250-volt, 3-phase assembly

(B) Cord Assemblies. The cord shall be a listed type with four conductors, for 3-phase connection, one of which shall be identified in accordance with 400.23 for use as the equipment grounding conductor. Extra-hard usage cables rated not less than 90°C (194°F), 600 volts, listed for both wet locations and sunlight resistance, and having an outer jacket rated to be resistant to temperature extremes, oil, gasoline, ozone, abrasion, acids, and chemicals, shall be permitted where flexibility is necessary between the electrified truck parking space supply equipment and the inlet(s) on the TRU.

Δ **(C) Attachment Plug(s) and Cord Connector(s).** Where a flexible cord is provided with an attachment plug and cord connector, they shall comply with 250.138(A). The attachment plug(s) and cord connector(s) shall be listed, by itself or as part of the power-supply cable assembly, for the purpose and shall be molded to or installed on the flexible cord so that it is secured tightly to the cord at the point where the cord enters the attachment plug or cord connector. If a right-angle cap is used, the configuration shall be oriented so that the grounding member is farthest from the cord. An attachment plug and cord connector for the connection of a truck or trailer shall be rated in accordance with one of the following:

(1) 30-ampere, 480-volt, 3-phase, 3-pole, 4-wire and intended for use with 30-ampere, 480-volt, 3-phase, 3-pole, 4-wire receptacles and inlets, respectively
(2) 60-ampere, 208-volt, 3-phase, 3-pole, 4-wire and intended for use with 60-ampere, 208-volt, 3-phase, 3-pole, 4-wire receptacles and inlets, respectively
(3) 20-ampere, 1000-volt, 3-phase, 3-pole, 4-wire and intended for use with 20-ampere, 1000-volt, 3-phase, 3-pole, 4-wire receptacles and inlets, respectively
(4) 60-ampere, 480-volt, 3-phase, 3-pole, 4-wire and intended for use with 60-ampere, 480-volt, 3-phase, 3-pole, 4-wire receptacles and inlets, respectively

(5) 60-ampere, 250-volt, 3-phase, 3-pole, 4-wire and intended for use with 60-ampere, 250-volt, 3-phase, 3-pole, 4-wire receptacles and inlets, respectively

Informational Note: See UL 1686-2012, *Pin and Sleeve Configurations*, Figures C2.12 and C2.11, for complete details of the 30-ampere pin and sleeve receptacle configuration for refrigerated containers (TRUs) and for various configurations of 60-ampere pin and sleeve receptacles.

Transport refrigeration units (TRUs) are permitted to be supplied by 208-, 240-, or 480-volt, three-phase branch circuits. The current rating for these branch circuits cannot exceed 60 amperes. The connection to the TRU is accomplished using a flexible cord with a pin-and-sleeve-type cord connector that is connected to a flanged surface inlet mounted on the refrigerated box truck or refrigerated trailer. The flexible cords must be rated for wet locations, sunlight resistance, and extra-hard usage. The cords must have three insulated conductors to supply the three-phase power and one insulated and identified equipment grounding conductor. Section 626.31(C) specifies the type of receptacle(s) required at each truck parking space that is intended to accommodate refrigerated trucks or tractor-trailers.

ARTICLE 630 Electric Welders

Part I. General

630.1 Scope. This article covers apparatus for electric arc welding, resistance welding, plasma cutting, and other similar welding and cutting process equipment that is connected to an electrical supply system.

The two general types of electric welding are resistance welding and arc welding. Resistance welding, or "spot" welding, is the process of electrically fusing two or more metal sheets or parts. The metal parts are placed between two electrodes or welding points, and a high current at a low voltage is passed through the electrodes. The resistance of the metal parts to the flow of current heats them to a molten state, and a weld is made.

Arc welding is the butting of two metal parts, then striking an arc at the joint with a metal electrode (a flux-coated wire rod). The electrode itself is melted and supplies the extra metal necessary for joining the metal parts.

Article 630 also covers electrically supplied equipment associated with plasma cutting operations. This electrically powered equipment controls the flammable gas or gases used for cutting.

630.6 Listing. All welding and cutting power equipment under the scope of this article shall be listed.

N **630.8 Ground-Fault Circuit-Interrupter Protection for Personnel.** All 125-volt, 15- and 20-ampere receptacles for electrical hand tools or portable lighting equipment, supplied by single-phase branch circuits rated 150 volts or less to ground, installed in work areas where welders are operated shall have ground-fault circuit-interrupter protection for personnel.

Part II. Arc Welders

630.11 Ampacity of Supply Conductors. The ampacity of conductors for arc welders shall be in accordance with 630.11(A) and (B).

(A) Individual Welders. The ampacity of the supply conductors shall be not less than the I_{1eff} value on the rating plate. Alternatively, if the I_{1eff} is not given, the ampacity of the supply conductors shall not be less than the current value determined by multiplying the rated primary current in amperes given on the welder rating plate by the factor shown in Table 630.11(A) based on the duty cycle of the welder.

TABLE 630.11(A) *Duty Cycle Multiplication Factors for Arc Welders*

| | Multiplier for Arc Welders | |
Duty Cycle	Nonmotor Generator	Motor Generator
100	1.00	1.00
90	0.95	0.96
80	0.89	0.91
70	0.84	0.86
60	0.78	0.81
50	0.71	0.75
40	0.63	0.69
30	0.55	0.62
20 or less	0.45	0.55

(B) Group of Welders. Minimum conductor ampacity shall be based on the individual currents determined in 630.11(A) as the sum of 100 percent of the two largest welders, plus 85 percent of the third largest welder, plus 70 percent of the fourth largest welder, plus 60 percent of all remaining welders.

Exception: Percentage values lower than those given in 630.11(B) shall be permitted in cases where the work is such that a high-operating duty cycle for individual welders is impossible.

Informational Note: Duty cycle considers welder loading based on the use to be made of each welder and the number of welders supplied by the conductors that will be in use at the same time. The load value used for each welder considers both the magnitude and the duration of the load while the welder is in use.

Even under high-production conditions, the loads on transformer arc welders are considered intermittent. Therefore, the minimum ampacity of feeder conductors supplying several transformers (three or more) is permitted to be determined by applying the percentage values specified in 630.11(B). See also 630.31(B). The ampacity of the conductors is based on the I_{1eff} rating on the welder rating plate. If the I_{1eff} rating is not available, a calculation is done by selecting the appropriate factor from Table 630.11(A) based on the type of welder and the duty cycle of the welder. The selected factor is then multiplied by the primary current rating from the welder rating plate to determine the minimum ampacity of the supply conductors.

630.12 Overcurrent Protection. Overcurrent protection for arc welders shall be as provided in 630.12(A) and (B). Where the values as determined by this section do not correspond to the standard ampere ratings provided in 240.6 or where the rating or setting specified results in unnecessary opening of the overcurrent device, the next higher standard rating or setting shall be permitted.

(A) For Welders. Each welder shall have overcurrent protection rated or set at not more than 200 percent of I_{1max}. Alternatively, if the I_{1max} is not given, the overcurrent protection shall be rated or set at not more than 200 percent of the rated primary current of the welder.

An overcurrent device shall not be required for a welder that has supply conductors protected by an overcurrent device rated or set at not more than 200 percent of I_{1max} or at the rated primary current of the welder.

If the supply conductors for a welder are protected by an overcurrent device rated or set at not more than 200 percent of I_{1max} or at the rated primary current of the welder, a separate overcurrent device shall not be required.

(B) For Conductors. Conductors that supply one or more welders shall be protected by an overcurrent device rated or set at not more than 200 percent of the conductor ampacity.

> Informational Note: I_{1max} is the maximum value of the rated supply current at maximum rated output. I_{1eff} is the maximum value of the effective supply current, calculated from the rated supply current (I_1), the corresponding duty cycle (duty factor) (X), and the supply current at no-load (I_0) by the following equation:

$$I_{1eff} = \sqrt{I_1^2 X + I_0^2(1 - X)} \qquad \textbf{[630.12(B)]}$$

Some arc welding machines have a welding range involving an excess secondary-current output capacity beyond that indicated by the secondary rating marked on the machines. This excess capacity (generally not more than 150 percent of the marked output capacity) is usually supplied by means of secondary taps in addition to the tap(s) intended for normal output current; the higher currents thus available are intended to provide for heavier welding work, including the use of larger-sized electrodes. This excess capacity is somewhat analogous to the inherent overload capacity of motors and transformers. Because the use of this excess current capacity and the overloading of welding machines, except for relatively short periods of time, could be hazardous, it should be undertaken with caution.

Calculation Example

A motor-generator-type electric arc welder has a nameplate primary current rating of 95 amperes and a duty cycle of 80 percent. Determine the minimum ampacity of the branch-circuit conductors and the maximum rating or setting for the branch-circuit overcurrent protective device (OCPD).

Solution

Step 1. Determine the minimum ampacity for the supply circuit conductors [630.11(A)]:

$$95 \text{ A} \times 0.91 \text{ (duty cycle factor)} = 86.45 \text{ A}$$

Copper THWN conductor selected from the 75°C column of Table 310.16: 3 AWG (100-A allowable ampacity)

Step 2. Determine the maximum rating of setting for the OCPD for the welder and the branch-circuit conductors [630.12(A) and (B)]:

$$95 \text{ A} \times 200\% = 190 \text{ A}$$

Next standard size OCPD: 200 A

Conclusion. The minimum conductor ampacity is 86.45 amperes, and the maximum rating or setting for the branch-circuit OCPD is 200 amperes. This rating or setting is the maximum permitted for a circuit supplying a single welder. However, the *NEC*® does not prohibit the use of a smaller-size OCPD, because that is a performance consideration related to the intended use of the welder.

Δ **630.13 Disconnecting Means.** A disconnecting means shall be provided in the supply circuit for each arc welder that is not equipped with a disconnect mounted as an integral part of the welder.

The disconnecting means shall be a switch, circuit breaker, or listed cord-and-plug connector, and its rating shall be not less than that necessary to accommodate overcurrent protection as specified in 630.12.

630.14 Marking. A rating plate shall be provided for arc welders giving the following information:
(1) Name of manufacturer
(2) Frequency
(3) Number of phases
(4) Primary voltage
(5) I_{1max} and I_{1eff}, or rated primary current
(6) Maximum open-circuit voltage
(7) Rated secondary current
(8) Basis of rating, such as the duty cycle

630.15 Grounding of Welder Secondary Circuit. The secondary circuit conductors of an arc welder, consisting of the electrode conductor and the work conductor, shall not be considered as premises wiring for the purpose of applying Article 250.

Informational Note: Connecting welder secondary circuits to grounded objects can create parallel paths and can cause objectionable current over equipment grounding conductors.

In theory and in accordance with the *NEC* definition in Article 100, the secondary circuit of an arc welder could be viewed as a *separately derived system*. However, the function of a welder is to create a high-current circuit between the electrode and the work surface. In the normal operation of an ac power distribution system, such an event would be considered a fault, and the operation of an overcurrent device to open the circuit and clear the fault is a fundamental concept of Articles 240 and 250. In the case of an arc welder, the opening of an overcurrent device is not intended unless the welding operation significantly exceeds the operating parameters of the welder. Grounding of a welder secondary terminal has the potential to cause excessive and potentially degrading parallel currents on power system equipment grounding conductors (EGCs).

This requirement clarifies that for the purposes of Article 250 — specifically, the requirements covering grounding of separately derived systems — the secondary circuit of a welder is not treated as premises wiring and is not required to be grounded as such. This removes any potential conflict where grounding in the welder secondary circuit occurs at the work object.

Part III. Resistance Welders

630.31 Ampacity of Supply Conductors. The ampacity of the supply conductors for resistance welders shall be in accordance with 630.31(A) and (B).

Informational Note: The ampacity of the supply conductors for resistance welders necessary to limit the voltage drop to a value permissible for the satisfactory performance of the welder is usually greater than that required to prevent overheating.

(A) Individual Welders. The ampacity of conductors for individual welders shall comply with the following:

(1) The ampacity of the supply conductors for a welder that can be operated at different times at different values of primary current or duty cycle shall not be less than 70 percent of the rated primary current for seam and automatically fed welders, and 50 percent of the rated primary current for manually operated nonautomatic welders.

(2) The ampacity of the supply conductors for a welder wired for a specific operation for which the actual primary current and duty cycle are known and remain unchanged shall not be less than the product of the actual primary current and the multiplier specified in Table 630.31(A) for the duty cycle at which the welder will be operated.

(B) Groups of Welders. The ampacity of conductors that supply two or more welders shall not be less than the sum of the value obtained in accordance with 630.31(A) for the largest welder supplied and 60 percent of the values obtained for all the other welders supplied.

Informational Note: Explanation of Terms

(1) The *rated primary current* is the rated kilovolt-amperes (kVA) multiplied by 1000 and divided by the rated primary voltage, using values given on the nameplate.

(2) The *actual primary current* is the current drawn from the supply circuit during each welder operation at the particular heat tap and control setting used.

(3) The *duty cycle* is the percentage of the time during which the welder is loaded. For instance, a spot welder supplied by a 60-Hz system (216,000 cycles per hour) and making 400 15-cycle welds per hour would have a duty cycle of 2.8 percent (400 multiplied by 15, divided by 216,000, multiplied by 100). A seam welder operating 2 cycles "on" and 2 cycles "off" would have a duty cycle of 50 percent.

TABLE 630.31(A) *Duty Cycle Multiplication Factors for Resistance Welders*

Duty Cycle (%)	Multiplier
50	0.71
40	0.63
30	0.55
25	0.50
20	0.45
15	0.39
10	0.32
7.5	0.27
5 or less	0.22

The ampacity of supply conductors for a welder that is not wired for a specific function (i.e., one operated at varying intervals for different applications, such as dissimilar metals or thicknesses) is permitted by 630.31(A)(1) to be 70 percent of the rated primary current for automatically fed welders and 50 percent of the rated primary current for manually operated welders. The rated primary current can be determined using the following equation with the values given on the welder nameplate:

$$\text{Related primary voltage} = \frac{\text{welder kVA} \times 1000}{\text{rated primary voltage}}$$

Where the actual primary current and the duty cycle are known, such as for a welder wired for a specific operation, the ampacity of the supply conductors is not permitted to be less than the product of the actual primary current (current drawn during weld operation) and the multiplier, as provided in Table 630.31(A), for the duty cycle at which the welder will be operated.

Calculation Example

A seam welder is set to draw current for 3 cycles and to be off for 4 cycles during every 7-cycle period. The welder's duty cycle is calculated as follows:

$$\frac{3}{7} \times 100\% = 42.9\% \text{ (duty cycle)}$$

The duty cycle is set for a specific operation by adjusting the controller for the welder. An instrument capable of measuring current impulses for 3 cycles (1/20 second) is required to measure the actual primary current as required by 630.31(A)(2) in order to size the conductors. For the sizing of supply conductors, voltage drop should be limited to a value permissible for the satisfactory performance of the welder.

630.32 Overcurrent Protection. Overcurrent protection for resistance welders shall be as provided in 630.32(A) and (B). Where the values as determined by this section do not correspond with the standard ampere ratings provided in 240.6 or where the rating or setting specified results in unnecessary opening of the overcurrent device, a higher rating or setting that does not exceed the next higher standard ampere rating shall be permitted.

(A) For Welders. Each welder shall have an overcurrent device rated or set at not more than 300 percent of the rated primary current of the welder. If the supply conductors for a welder are protected by an overcurrent device rated or set at not more than 200 percent of the rated primary current of the welder, a separate overcurrent device shall not be required.

(B) For Conductors. Conductors that supply one or more welders shall be protected by an overcurrent device rated or set at not more than 300 percent of the conductor ampacity.

630.33 Disconnecting Means. A switch or circuit breaker shall be provided by which each resistance welder and its control equipment can be disconnected from the supply circuit. The ampere rating of this disconnecting means shall not be less than the supply conductor ampacity determined in accordance with 630.31. The supply circuit switch shall be permitted as the welder disconnecting means where the circuit supplies only one welder.

630.34 Marking. A nameplate shall be provided for each resistance welder, giving the following information:

(1) Name of manufacturer
(2) Frequency
(3) Primary voltage
(4) Rated kilovolt-amperes (kVA) at 50 percent duty cycle
(5) Maximum and minimum open-circuit secondary voltage
(6) Short-circuit secondary current at maximum secondary voltage
(7) Specified throat and gap setting

Part IV. Welding Cable

630.41 Conductors. Insulation of conductors intended for use in the secondary circuit of electric welders shall be flame retardant.

Listed welding cable is intended to be used for the secondary circuits of electric welders and cannot be used as "building wire" for circuits operating at 1000 volts or less unless the cable is also

one of the types covered in Table 310.4(1). The fine stranding allows for the flexibility necessary in manual and automatic welding operations. Terminals used with this type of cable must be suitable for use with the fine stranding used in this type of cable construction. See 110.14 for more information regarding terminations used with conductors having other than Class B or C stranding.

630.42 Installation. Cables shall be permitted to be installed in a dedicated cable tray as provided in 630.42(A), (B), and (C).

(A) Cable Support. The cable tray shall provide support at not greater than 150-mm (6-in.) intervals.

(B) Spread of Fire and Products of Combustion. The installation shall comply with 300.21.

(C) Signs. A permanent sign shall be attached to the cable tray at intervals not greater than 6.0 m (20 ft). The sign shall read as follows:

CABLE TRAY FOR WELDING CABLES ONLY

ARTICLE 640 Audio Signal Processing, Amplification, and Reproduction Equipment

Part I. General

640.1 Scope.

(A) Covered. This article covers equipment and wiring for audio signal generation, recording, processing, amplification, and reproduction; distribution of sound; public address; speech input systems; temporary audio system installations; and electronic organs or other electronic musical instruments. This also includes audio systems subject to Article 517, Part VI, and Articles 518, 520, 525, and 530.

> Informational Note: Examples of permanently installed distributed audio system locations include, but are not limited to, restaurant, hotel, business office, commercial and retail sales environments, churches, and schools. Both portable and permanently installed equipment locations include, but are not limited to, residences, auditoriums, theaters, stadiums, and movie and television studios. Temporary installations include, but are not limited to, auditoriums, theaters, stadiums (which use both temporary and permanently installed systems), and outdoor events such as fairs, festivals, circuses, public events, and concerts.

Equipment covered by Article 640 includes amplifiers; public address (PA) systems and centralized sound systems used in schools, factories, businesses, stadiums, and similar locations; intercommunications devices and systems; and devices used for recording or reproducing voice or music. The scope is limited to equipment whose main function is the processing, distribution, amplification, and reproduction of audio frequency bandwidth signals. This limitation does not preclude equipment that uses radio frequency or other forms of transmission between equipment components, such as wireless microphone systems.

Electronic organs are synthesizers, and synthesizers also generate audio signals. For the sake of clarity, electronic organs are uniquely cited in the scope, and electronic musical instruments are included to cover all other forms of electronic tone generation. Electronic musical instruments create an electronic signal as their sole or primary output and require amplification and reproduction equipment to be audible.

(B) Not Covered. This article does not cover the installation and wiring of fire and burglary alarm signaling devices.

640.3 Locations and Other Articles. Circuits and equipment shall comply with 640.3(A) through (N), as applicable.

(A) Spread of Fire or Products of Combustion. Section 300.21 shall apply.

Δ **(B) Ducts, Plenums, and Other Air-Handling Spaces.** Section 300.22(B) shall apply to circuits and equipment installed in ducts specifically fabricated for environmental air. Section 300.22(C) shall apply to circuits and equipment installed in other spaces used for environmental air (plenums).

Exception No. 1: Class 2 and Class 3 cables installed in accordance with 722.135(B) shall be permitted to be installed in ducts specifically fabricated for environmental air.

Exception No. 2: Class 2 and Class 3 cables installed in accordance with 722.135(B) shall be permitted to be installed in other spaces used for environmental air (plenums).

Informational Note: See NFPA 90A, *Standard for the Installation of Air-Conditioning and Ventilating Systems*, 4.3.11.2.6.5, which permits loudspeakers, loudspeaker assemblies, and their accessories listed in accordance with UL 2043-2013, *Fire Test for Heat and Visible Smoke Release for Discrete Products and Their Accessories Installed in Air-Handling Spaces*, to be installed in other spaces used for environmental air (ceiling cavity plenums).

N **(C) Communications Cables.** Types CMP, CMR, CMG, and CM communications cables shall be permitted to substitute for Class 2 and Class 3 cables in accordance with 722.135(E).

Δ **(D) Cable Trays.** Cable trays and cable tray systems shall be installed in accordance with Part II of Article 392. The installation of Class 2, Class 3, and Type PLTC cables in cable trays shall be in accordance with 722.135(B).

(E) Hazardous (Classified) Locations. Equipment used in hazardous (classified) locations shall comply with the applicable requirements of Chapter 5.

The examples of assembly occupancies described in 518.2(A) are some of the most common locations for the installation of both distributed audio systems (e.g., background music) and centralized systems (permanently installed sound reinforcement systems for meeting rooms, auditoriums, gymnasiums, and so forth).

(F) Assembly Occupancies. Equipment used in assembly occupancies shall comply with Article 518.

(G) Theaters, Audience Areas of Motion Picture and Television Studios, and Similar Locations. Equipment used in theaters, audience areas of motion picture and television studios, and similar locations shall comply with Article 520.

(H) Carnivals, Circuses, Fairs, and Similar Events. Equipment used in carnivals, circuses, fairs, and similar events shall comply with Article 525.

(I) Motion Picture and Television Studios. Equipment used in motion picture and television studios shall comply with Article 530.

(J) Swimming Pools, Fountains, and Similar Locations. Audio equipment used in or near swimming pools, fountains, and similar locations shall comply with Article 680.

The underwater installation of audio equipment in swimming pools is covered in 680.27(A).

> **See also**
>
> **640.10** for the acceptable placement, wiring, and use of equipment used near (rather than immersed in) bodies of water, both natural and artificial

Δ **(K) Combination Systems.** Where the authority having jurisdiction permits audio systems for paging or music, or both, to be combined with fire alarm systems, the wiring shall comply with Article 760.

Informational Note: See *NFPA 72, National Fire Alarm and Signaling Code*, and NFPA *101, Life Safety Code*, for installation requirements for such combination systems.

In addition to alarm tones, fire alarm systems frequently use loudspeakers for verbal announcements. All such systems must comply with Article 760 and with *NFPA 72®, National Fire Alarm and Signaling Code®*. Audio systems that use a paging or background music system are permitted to be used as part of a fire alarm warning system, but they must comply with Article 760. The installation of fire alarm systems is governed by *NFPA 72*. Refer to *NFPA 101®, Life Safety Code®*, for multipurpose systems.

(L) Antennas. Equipment used in audio systems that contain an audio or video tuner and an antenna input shall comply with Article 810. Wiring other than antenna wiring that connects such equipment to other audio equipment shall comply with this article.

The term *receiver* is commonly used in the consumer market to mean an amplifier combined with a radio tuner (typically AM/FM) and other signal processing and/or switching functions. Except for the tuner function and the antenna input, the signal processing functions are the same as those provided by equipment covered in Article 640.

See also

810, Part II, which covers the antenna installation for such equipment

810.3, which references Parts I and II of Article 640 as appropriate for wiring requirements (other than for the antenna)

(M) Generators. Generators shall be installed in accordance with 445.10 through 445.12, 445.14 through 445.16, and 445.18. Grounding of portable and vehicle-mounted generators shall be in accordance with 250.34.

(N) Organ Pipes. Additions of pipe organ pipes to an electronic organ shall be in accordance with 650.4 through 650.9.

640.4 Protection of Electrical Equipment. Amplifiers, loudspeakers, and other equipment shall be so located or protected as to guard against environmental exposure or physical damage, such as might result in fire, shock, or personal hazard.

640.5 Access to Electrical Equipment Behind Panels Designed to Allow Access. Access to equipment shall not be denied by an accumulation of wires and cables that prevents removal of panels, including suspended ceiling panels.

640.6 Mechanical Execution of Work.

(A) Installation of Audio Distribution Cables. Cables installed exposed on the surface of ceilings and sidewalls shall be supported in such a manner that the audio distribution cables will not be damaged by normal building use. Such cables shall be secured by straps, staples, cable ties, hangers, or similar fittings designed and installed so as not to damage the cable. The installation shall conform to 300.4 and 300.11(A).

If installed in the hollow space above a suspended, dropped, or similar ceiling, cables of audio systems covered in Article 640 are required to be supported in accordance with 300.11. Without specific instructions permitting the use of the ceiling system support wires as a means to support wiring methods, an independent support system for the cables must be installed. Additional ceiling wires installed to support the audio system wiring are required to be secured in place. Attachment of the additional support wires to the ceiling system and to the building structure above the ceiling provides the secure support required by 300.11(B)(1) and (B)(2). The use of securing hardware such as straps, staples, cable ties, hangers, or other approved means is required, and this hardware must be installed so as not to damage the audio cable. The *NEC*® does not specify the distance between securing points.

(B) Abandoned Audio Distribution Cables. The accessible portion of abandoned audio distribution cables shall be removed.

(C) Installed Audio Distribution Cable Identified for Future Use.

(1) Cable Identification Means. Cables identified for future use shall be marked with a tag of sufficient durability to withstand the environment involved.

(2) Cable Tag Criteria. Cable tags shall have the following information:

(1) Date cable was identified for future use
(2) Date of intended use
(3) Information related to the intended future use of cable

640.7 Grounding.

(A) General. Wireways and auxiliary gutters shall be connected to an equipment grounding conductor(s), to an equipment bonding jumper, or to the grounded conductor where permitted or required by 250.92(B)(1) or 250.142. Where the wireway or auxiliary gutter does not contain power-supply wires, the equipment grounding conductor shall not be required to be larger than 14 AWG copper or its equivalent. Where the wireway or auxiliary gutter contains power-supply wires, the equipment grounding conductor shall not be smaller than specified in 250.122.

(B) Separately Derived Systems with 60 Volts to Ground. Grounding of separately derived systems with 60 volts to ground shall be in accordance with 647.6.

(C) Isolated Ground Receptacles. Isolated grounding-type receptacles shall be permitted as described in 250.146(D), and for the implementation of other technical power systems in compliance with Article 250. For separately derived systems with 60 volts to ground, the branch-circuit equipment grounding conductor shall be terminated as required in 647.6(B).

Informational Note: See 406.3(E) for grounding-type receptacles and required identification.

The reference to 647.6 provides requirements for grounding separately derived systems operating at 60 volts to ground. These separately derived systems are used for the reduction of electromagnetic noise in audio and video systems.

Section 640.7(B) addresses the proper use of isolated ground receptacles when used with technical power systems of the type that are separately derived systems with 60 volts to ground.

640.8 Grouping of Conductors. Insulated conductors of different systems grouped or bundled so as to be in close physical contact with each other in the same raceway or other enclosure, or in portable cords or cables, shall comply with 300.3(C)(1).

640.9 Wiring Methods.

(A) Wiring to and Between Audio Equipment.

(1) Power Wiring. Wiring and equipment from source of power to and between devices connected to the premises wiring systems shall comply with the requirements of Chapters 1 through 4, except as modified by this article.

(2) Separately Derived Power Systems. Separately derived systems shall comply with the applicable articles of this *Code*, except as modified by this article. Separately derived systems with 60 volts to ground shall be permitted for use in audio system installations as specified in Article 647.

(3) Other Wiring. All wiring not connected to the premises wiring system or to a wiring system separately derived from the premises wiring system shall comply with Part II of Article 725.

(B) Auxiliary Power Supply Wiring. Equipment that has a separate input for an auxiliary power supply shall be wired in compliance with Article 725. Battery installation shall be in accordance with Article 480. This section shall not apply to the use of uninterruptible power supply (UPS) equipment, or other sources of supply, that are intended to act as a direct replacement for the primary circuit power source and are connected to the primary circuit input.

> Informational Note: See *NFPA 72-2019, National Fire Alarm and Signaling Code*, where equipment is used for a fire alarm system.

Audio equipment with a separate input for an auxiliary power supply is typically used for emergency paging or fire alarm systems. These auxiliary power supply inputs typically range from 12 to 48 volts dc. Article 480 covers installation and overcurrent protection of battery circuits of this type. The term *auxiliary* is used to indicate that the equipment is also capable of being powered by the premises wiring system through an independent input connector, cord, or cable.

A replacement source for the premises wiring system such as an uninterruptible power supply (UPS) or a standby generator is not covered by the requirements of this section unless it is directly connected to the auxiliary power supply input and supplies the audio equipment with a dc voltage.

Δ **(C) Output Wiring and Listing of Amplifiers.** Amplifiers with output circuits carrying audio program signals shall be permitted to employ Class 1, Class 2, or Class 3 wiring where the amplifier is listed and marked for use with the specific class of wiring method. Such listing shall ensure the energy output is equivalent to the shock and fire risk of the same class as stated in Articles 724 and 725. Overcurrent protection shall be provided and shall be permitted to be inherent in the amplifier.

Audio amplifier output circuits wired using Class 1 wiring methods shall be considered equivalent to Class 1 circuits and shall be installed in accordance with 724.46, where applicable.

Audio amplifier output circuits wired using Class 2 or Class 3 wiring methods shall be considered equivalent to Class 2 or Class 3 circuits, respectively. They shall use conductors insulated at not less than the requirements of 722.179 and shall be installed in accordance with 722.135 and 725.136 through 725.144.

> Informational Note No. 1: See UL 1711-2016, *Amplifiers for Fire Protective Signaling Systems*, which contains requirements for the listing of amplifiers used for fire alarm systems in compliance with *NFPA 72-2019, National Fire Alarm and Signaling Code*.
> Informational Note No. 2: See UL 813-1996, *Commercial Audio Equipment*; UL 1419-2016, *Professional Video and Audio Equipment*; ANSI/UL 1492-1996, *Audio-Video Products and Accessories*; UL 6500-1999, *Audio/Video and Musical Instrument Apparatus for Household, Commercial, and Similar Use*; and

UL 62368-1-2014, *Audio/Video, Information and Communication Technology Equipment — Part 1: Safety Requirements*, for examples of requirements for listing amplifiers used in residential, commercial, and professional use.

(D) Use of Audio Transformers and Autotransformers. Audio transformers and autotransformers shall be used only for audio signals in a manner so as not to exceed the manufacturer's stated input or output voltage, impedance, or power limitations. The input or output wires of an audio transformer or autotransformer shall be allowed to connect directly to the amplifier or loudspeaker terminals. No electrical terminal or lead shall be required to be grounded or bonded.

Audio transformers and autotransformers are commonly used between the amplifier output and the loudspeaker input for the following reasons:

1. At the output of the amplifier to change the amplifier's operating voltage to match the design impedance of the loudspeaker
2. At the loudspeaker, where the inherently low voice coil impedance is raised, to match the output voltage of the amplifier (or autotransformer)
3. Between the amplifier output and the loudspeaker input as an attenuating device (volume control)

Audio autotransformers are similar in concept to autotransformers used for light and power. Audio transformers are commonly used to provide electrical isolation of the speakers from the signal source. Either type of audio transformer (two windings or an autotransformer) is referred to as an "impedance matching transformer."

The last sentence of 640.9(D) specifically addresses the fact that electrical terminals are not to be treated in the same manner as transformers used for light and power might be (e.g., grounding the common terminal of an autotransformer). Some amplifier outputs are deliberately isolated from equipment ground, in which case such a connection could damage the amplifier and violate the manufacturer's recommended use. The frame of the transformer may or may not require bonding, depending on the manufacturer's installation instructions.

640.10 Audio Systems Near Bodies of Water. Audio systems near bodies of water, either natural or artificial, shall be subject to the restrictions specified in 640.10(A) and (B).

Exception: This section does not include audio systems intended for use on boats, yachts, or other forms of land or water transportation used near bodies of water, whether or not supplied by branch-circuit power.

> Informational Note: See 680.27(A) for installation of underwater audio equipment.

(A) Equipment Supplied by Branch-Circuit Power. Audio system equipment supplied by branch-circuit power shall not be placed horizontally within 1.5 m (5 ft) of the inside wall of a pool, spa, hot tub, or fountain, or within 1.5 m (5 ft) of the prevailing or tidal high water mark. In addition to the requirements in 210.8(B), the equipment shall be provided with branch-circuit power protected by a ground-fault circuit interrupter where required by other articles.

The particular application of underwater loudspeakers is unique in construction and wiring to pools and is addressed in Article 680. Other locations where audio equipment might be used "near bodies of water" are not addressed in Article 680. Locations where audio equipment is used near bodies of water are covered by 640.10.

The term *prevailing or tidal high water mark* recognizes that the edges of natural bodies of water can advance or recede. Such water level changes must be anticipated.

Unless required by other sections of the *NEC*, the requirement for GFCI protection does not apply where the equipment (specifically an amplifier or a receiver) is not installed near a body of water.

(B) Equipment Not Supplied by Branch-Circuit Power. Audio system equipment powered by a listed Class 2 power supply or by the output of an amplifier listed as permitting the use of Class 2 wiring shall be restricted in placement only by the manufacturer's recommendations.

> Informational Note: See 640.10(A) for placement of the power supply or amplifier if supplied by branch-circuit power.

Part II. Permanent Audio System Installations

640.21 Use of Flexible Cords and Cables.

(A) Between Equipment and Branch-Circuit Power. Power supply cords for audio equipment shall be suitable for the use and shall be permitted to be used where the interchange, maintenance, or repair of such equipment is facilitated through the use of a power-supply cord.

(B) Between Loudspeakers and Amplifiers or Between Loudspeakers. Cables used to connect loudspeakers to each other or to an amplifier shall comply with Article 722. Other listed cable types and assemblies, including optional hybrid communications, signal, and hybrid optical fiber cables, shall be permitted.

Some loudspeakers are identified as being for outdoor use, as shown in Exhibit 640.1. The conductors supplying outdoor speakers must be identified for the environment.

See also

110.11, which specifies electrical equipment and conductors be identified for use in the operating environment and applies to audio equipment and its conductors

(C) Between Equipment. Cables used for the distribution of audio signals between equipment shall comply with Article 722. Other listed cable types and assemblies, including optional hybrid communications, signal, and hybrid optical fiber cables, shall be permitted. Other cable types and assemblies specified by the equipment manufacturer as acceptable for the use shall be permitted in accordance with 110.3(B).

EXHIBIT 640.1 Loudspeakers for outdoor use above ground or partially in ground. *(Courtesy of Bose)*

> Informational Note: See 770.3 for the classification of composite optical fiber cables.

Δ **(D) Between Equipment and Power Supplies Other Than Branch-Circuit Power.** The following power supplies, other than branch-circuit power supplies, shall be installed and wired between equipment in accordance with this *Code* for the voltage and power delivered:

(1) Storage batteries
(2) Transformers
(3) Transformer rectifiers
(4) Other ac or dc power supplies

> Informational Note: For some equipment, these sources such as in items (1) and (2) serve as the only source of power. These could, in turn, be supplied with intermittent or continuous branch-circuit power.

(E) Between Equipment Racks and Premises Wiring System. Flexible cords and cables shall be permitted for the electrical connection of permanently installed equipment racks to the premises wiring system to facilitate access to equipment or for the purpose of isolating the technical power system of the rack from the premises ground. Connection shall be made either by using approved plugs and receptacles or by direct connection within an approved enclosure. Flexible cords and cables shall not be subjected to physical manipulation or abuse while the rack is in use.

640.22 Wiring of Equipment Racks and Enclosures. Metal equipment racks and enclosures shall be bonded and grounded. Bonding shall not be required if the rack is connected to a technical power ground.

Wires, cables, structural components, or other equipment shall not be placed in such a manner as to prevent reasonable access to equipment power switches and resettable or replaceable circuit overcurrent protection devices.

Supply cords or cables, if used, shall terminate within the equipment rack enclosure in an identified connector assembly. The supply cords or cable (and connector assembly if used) shall have sufficient ampacity to carry the total load connected to the equipment rack and shall be protected by overcurrent devices.

640.23 Conduit or Tubing.

(A) Number of Conductors. The number of conductors permitted in a single conduit or tubing shall not exceed the percentage fill specified in Table 1, Chapter 9.

(B) Nonmetallic Conduit or Tubing and Insulating Bushings. The use of nonmetallic conduit or tubing and insulating bushings shall be permitted where a technical power system is employed and shall comply with applicable articles.

640.24 Wireways, Gutters, and Auxiliary Gutters. The use of metallic and nonmetallic wireways, gutters, and auxiliary gutters shall be permitted for use with audio signal conductors and shall comply with applicable articles with respect to permitted locations, construction, and fill.

640.25 Loudspeaker Installation in Fire Resistance–Rated Partitions, Walls, and Ceilings. Loudspeakers installed in a fire resistance–rated partition, wall, or ceiling shall be listed and labeled, or identified as speaker assemblies for fire resistance, or installed in an enclosure or recess that maintains the fire resistance rating.

> Informational Note: Fire-rated construction is the fire-resistive classification used in building codes.

The enclosure must maintain the fire resistance rating of the wall or ceiling in which a flush-mounted loudspeaker is installed. Listed enclosures are available for this purpose. Site-built enclosures may be installed with the approval of the AHJ and have been used as a method to maintain the fire resistance rating of the wall or ceiling.

Part III. Portable and Temporary Audio System Installations

While the equipment used for portable and temporary audio systems might not differ fundamentally from that used in permanent installations, the enclosures that serve as portable equipment racks must provide both transit protection and mechanical protection while the equipment is in use. Such enclosures can be constructed of metal, wood, plastic, or reinforced plastic construction. Nonmetallic construction enclosures frequently do not comply with EIA/ECA 310-D, *Cabinets, Racks, Panels, and Associated Equipment*.

640.41 Multipole Branch-Circuit Cable Connectors. Multipole branch-circuit cable connectors, male and female, for power-supply cords and cables shall be so constructed that tension on the cord or cable is not transmitted to the connections. The female half shall be attached to the load end of the power supply cord or cable. The connector shall be rated in amperes and designed so that differently rated devices cannot be connected together. Alternating-current multipole connectors shall be polarized and comply with 406.7(A) and (B) and 406.10. Alternating-current or direct-current multipole connectors utilized for connection between loudspeakers and amplifiers, or between loudspeakers, shall not be compatible with nonlocking 15- or 20-ampere rated connectors intended for branch-circuit power or with connectors rated 250 volts or greater and of either the locking or nonlocking type. Signal cabling not intended for such loudspeaker and amplifier interconnection shall not be permitted to be compatible with multipole branch-circuit cable connectors of any accepted configuration.

> Informational Note: See 400.14 for pull at terminals.

640.42 Use of Flexible Cords and Cables.

(A) Between Equipment and Branch-Circuit Power. Power supply cords for audio equipment shall be listed and shall be permitted to be used where the interchange, maintenance, or repair of such equipment is facilitated through the use of a power-supply cord.

(B) Between Loudspeakers and Amplifiers, or Between Loudspeakers. Installation of flexible cords and cables used to connect loudspeakers to each other or to an amplifier shall comply with Part I of Article 400 and Parts I, II, and III of Article 725, respectively. Cords and cables listed for portable use, either hard or extra-hard usage as defined by Article 400, shall also be permitted. Other listed cable types and assemblies, including optional hybrid communications, signal, and hybrid optical fiber cables, shall be permitted.

(C) Between Equipment and/or Between Equipment Racks. Installation of flexible cords and cables used for the distribution of audio signals between equipment shall comply with Parts I and II of Article 400 and Parts I, II, and III of Article 725, respectively. Cords and cables listed for portable use, either hard or extra-hard service as defined by Article 400, shall also be permitted. Other listed cable types and assemblies, including optional hybrid communications, signal, and hybrid optical fiber cables, shall be permitted.

Δ **(D) Between Equipment, Equipment Racks, and Power Supplies Other Than Branch-Circuit Power.** Wiring between the following power supplies, other than branch-circuit power supplies, shall be installed, connected, or wired in accordance with this *Code* for the voltage and power required:

(1) Storage batteries
(2) Transformers
(3) Transformer rectifiers
(4) Other ac or dc power supplies

(E) Between Equipment Racks and Branch-Circuit Power. The supply to a portable equipment rack shall be by means of listed extra-hard usage cords or cables, as defined in Table 400.4. For outdoor portable or temporary use, the cords or cables shall be further listed as being suitable for wet locations and sunlight resistant. Sections 520.5, 520.10, and 525.3 shall apply as appropriate when the following conditions exist:

(1) Where equipment racks include audio and lighting and/or power equipment
(2) When using or constructing cable extensions, adapters, and breakout assemblies

640.43 Wiring of Equipment Racks. Equipment racks fabricated of metal shall be bonded and grounded. Nonmetallic racks with covers (if provided) removed shall not allow access to Class 1, Class 3, or primary circuit power without the removal of covers over terminals or the use of tools.

Wires, cables, structural components, or other equipment shall not be placed in such a manner as to prevent reasonable access to equipment power switches and resettable or replaceable circuit overcurrent protection devices.

Wiring that exits the equipment rack for connection to other equipment or to a power supply shall be relieved of strain or otherwise suitably terminated such that a pull on the flexible cord or cable will not increase the risk of damage to the cable or connected equipment such as to cause an unreasonable risk of fire or electric shock.

640.44 Environmental Protection of Equipment. Portable equipment not listed for outdoor use shall be permitted only where appropriate protection of such equipment from adverse weather conditions is provided to prevent risk of fire or electric shock. Where the system is intended to remain operable during adverse weather, arrangements shall be made for maintaining operation and ventilation of heat-dissipating equipment.

640.45 Protection of Wiring. Where accessible to the public, flexible cords and cables laid or run on the ground or on the floor shall be covered with approved nonconductive mats. Cables and mats shall be arranged so as not to present a tripping hazard. The cover requirements of 300.5 shall not apply to wiring protected by burial.

640.46 Equipment Access. Equipment likely to present a risk of fire, electric shock, or physical injury to the public shall be protected by barriers or supervised by qualified personnel so as to prevent public access.

Δ **ARTICLE**
645 Information Technology Equipment

Δ **645.1 Scope.** This article covers equipment, power-supply wiring, equipment interconnecting wiring, and grounding of information technology equipment and systems in an information technology equipment room.

> Informational Note: See NFPA 75, *Standard for the Fire Protection of Information Technology Equipment*, which covers the

requirements for the protection of information technology equipment and information technology equipment areas.

The term *information technology equipment* (ITE) is also used by UL 60950-1, *Information Technology Equipment — Safety — Part 1: General Requirements*, as well as by international standards, as a more inclusive term for the equipment addressed by Article 645.

Article 645 applies only to equipment and systems, including the associated wiring, located within the ITE room. An ITE room is an enclosed area that contains computer-based business and industrial equipment. It is designed to comply with the special construction and fire protection provisions of NFPA 75 as well as by 645.4.

Small terminals, such as remote telephone terminal units, remote data terminals, personal computers, and cash registers in stores and supermarkets, are not covered by Article 645.

645.3 Other Articles. Circuits and equipment shall comply with 645.3(A) through (I), as applicable.

(A) Spread of Fire or Products of Combustion. Sections 300.21, 770.26, and 800.26 shall apply to penetrations of the fire-resistant room boundary.

Δ **(B) Wiring and Cabling in Other Spaces Used for Environmental Air (Plenums).** The following sections and tables shall apply to wiring and cabling in other spaces used for environmental air (plenums) above an information technology equipment room:

(1) Wiring methods: 300.22(C)(1)
(2) Class 2, Class 3, and PLTC cables: 722.135(B)
(3) Fire alarm systems: 760.53(B)(2) and Table 760.154
(4) Optical fiber cables: 770.113(C) and Table 770.154(a)
(5) Communications circuits: 800.133(C) and Table 800.154(a)
(6) CATV and radio distribution systems: 800.113(C) and Table 800.154(a)

(C) Bonding and Grounding. The non-current-carrying conductive members of optical fiber cables in an information technology equipment room shall be bonded and grounded in accordance with 770.114.

Δ **(D) Electrical Classification of Data Circuits.** Section 725.60(A)(4) shall apply to the electrical classification of listed information technology equipment signaling circuits.

(E) Fire Alarm Cables and Equipment. Parts I, II, and III of Article 760 shall apply to fire alarm systems cables and equipment installed in an information technology equipment room. Only fire alarm cables listed in accordance with Part IV of Article 760 and listed fire alarm equipment shall be permitted to be installed in an information technology equipment room.

Δ **(F) Cable Routing Assemblies, Communications Wires, Cables, Raceways, and Equipment.** Sections 800.110, 800.113, and 800.154 shall apply to cable routing assemblies and communications raceways. Parts I, II, III, IV, and V of

Articles 800 and 805 shall apply to communications wires, cables, and equipment installed in an information technology equipment room. Only communications wires and cables listed in accordance with 800.179, cable routing assemblies, and communications raceways listed in accordance with 800.182, and communications equipment listed in accordance with 800.171 shall be permitted to be installed in an information technology equipment room. Article 645 shall apply to the powering of communications equipment in an information technology equipment room.

> Informational Note: See Article 100, Definitions, for a definition of *communications equipment.*

(G) Community Antenna Television and Radio Distribution Systems Cables and Equipment. Parts I, II, III, IV, and V of Articles 800 and 820 shall apply to community antenna television and radio distribution systems cables and equipment installed in an information technology equipment room. Only community antenna television and radio distribution cables listed in accordance with 800.179 and listed CATV equipment shall be permitted to be installed in an information technology equipment room. Article 645 shall apply to the powering of community antenna television and radio distribution systems equipment installed in an information technology equipment room.

(H) Optical Fiber Cables. Only optical fiber cables listed in accordance with 770.179 shall be permitted to be installed in an information technology equipment room.

(I) Cables Not in Information Technology Equipment Room. Cables extending beyond the information technology equipment room shall be subject to the applicable requirements of this *Code.*

Δ **645.4 Special Requirements for Information Technology Equipment Room.** The alternative wiring methods to Chapter 3 and Parts I and II of Article 725 for signaling wiring and Parts I and V of Article 770 for optical fiber cabling shall be permitted where all of the following conditions are met:

(1) Disconnecting means complying with 645.10 are provided.
(2) A heating/ventilating/air-conditioning (HVAC) system is provided in one of the methods identified in the following:

     a. A separate HVAC system that is dedicated for information technology equipment use and is separated from other areas of occupancy
     b. An HVAC system that serves other occupancies and meets all of the following:

         (i) Also serves the information technology equipment room
         (ii) Provides fire/smoke dampers at the point of penetration of the room boundary
         (iii) Activates the damper operation upon initiation by smoke detector alarms, by operation of the disconnecting means required by 645.10, or by both

Informational Note No. 1: See NFPA 75-2020, *Standard for the Fire Protection of Information Technology Equipment,* Chapter 11, Section 11.1, 11.1.1, 11.1.2, and 11.1.3, for further information.

(3) All information technology and communications equipment installed in the room is listed.
(4) The room is occupied by, and accessible to, only those personnel needed for the maintenance and functional operation of the installed information technology equipment.
(5) The room is separated from other occupancies by fire-resistant-rated walls, floors, and ceilings with protected openings.

Informational Note No. 2: See NFPA 75-2020, *Standard for the Fire Protection of Information Technology Equipment,* Chapter 6, for further information on room construction requirements.

(6) Only electrical equipment and wiring associated with the operation of the information technology room is installed in the room.

Informational Note No. 3: HVAC systems, communications systems, and monitoring systems such as telephone, fire alarm systems, security systems, water detection systems, and other related protective equipment are examples of equipment associated with the operation of the information technology room.

Use of the wiring methods permitted by Article 645 is based on the construction of the ITE room meeting the requirements in NFPA 75. For such ITE rooms, Article 645 contains wiring method installation requirements — for example, requirements for wiring methods in the space beneath the raised floor used for environmental air of an ITE room — that are less stringent than those in Chapter 3 for the same type of space.

Application of these modified requirements is contingent on the ITE room construction and equipment meeting all six conditions specified in 645.4. If any one of the six conditions is not met, wiring methods installed in the ITE room must follow the applicable requirements of Chapter 3, Parts I and II of Article 725 for signal wiring, and Parts I and V of Article 770 for optical fiber cabling.

645.5 Supply Circuits and Interconnecting Cables.

(A) Branch-Circuit Conductors. The branch-circuit conductors supplying one or more units of information technology equipment shall have an ampacity not less than 125 percent of the total connected load.

Δ **(B) Power-Supply Cords.** Information technology equipment shall be permitted to be connected to branch circuits by power-supply cords that comply with the following:

(1) Power-supply cords shall not exceed 4.5 m (15 ft).
(2) Power-supply cords shall be listed and a type permitted for use on listed information technology equipment or shall be constructed of listed flexible cord and listed attachment plugs and cord connectors of a type permitted for information technology equipment.

(3) Plugs and receptacles used to connect the power-supply cords shall be listed and identified for the system voltage and current applied.

> Informational Note No. 1: See UL 60950-1, *Safety of Information Technology Equipment — Safety — Part 1: General Requirements*; or UL 62368-1, *Audio/Video, Information and Communication Technology Equipment — Part 1: Safety Requirements*, for one method of determining if cords are of a permitted type.
>
> Informational Note No. 2: See ANSI/NEMA WD-6, *Wiring Devices — Dimensional Specifications*, which identifies plug and receptacle configurations L25-30P and L25-30R for 240 Vac and L26-30P and L26-30R for 240/415 Vac.

(C) Interconnecting Cables. Separate information technology equipment units shall be permitted to be interconnected by means of listed cables and cable assemblies. The 4.5 m (15 ft) limitation in 645.5(B)(1) shall not apply to interconnecting cables.

(D) Physical Protection. Where exposed to physical damage, power-supply cords, branch-circuit supply conductors, and interconnecting cables shall be protected.

(E) Under Raised Floors. Where the area under the floor is accessible and openings minimize the entrance of debris beneath the floor, power-supply cords, communications cables, connecting cables, interconnecting cables, cord-and-plug connections, and receptacles associated with the information technology equipment shall be permitted under a raised floor of approved construction. The installation requirement shall comply with 645.5(E)(1) through (E)(3).

Δ **(1) Installation Requirements for Branch-Circuit Supply Conductors Under a Raised Floor.**

(a) The supply conductors shall be installed in accordance with 300.11.

(b) In addition to the wiring methods of 300.22(C), the following wiring methods shall also be permitted:

(1) Rigid metal conduit
(2) Rigid nonmetallic conduit
(3) Intermediate metal conduit
(4) Electrical metallic tubing
(5) Electrical nonmetallic tubing
(6) Metal wireway
(7) Nonmetallic wireway
(8) Surface metal raceway with metal cover
(9) Surface nonmetallic raceway
(10) Flexible metal conduit
(11) Liquidtight flexible metal conduit
(12) Liquidtight flexible nonmetallic conduit
(13) Type MI cable
(14) Type MC cable
(15) Type AC cable
(16) Associated metallic and nonmetallic boxes or enclosures
(17) Type TC power and control tray cable

Branch-circuit conductors installed under the raised floor of an ITE room using any of the wiring methods listed are required to conform to the specific article for the wiring method used.

See also

300.11(A), which requires raceways, cables, and boxes to be securely fastened in place, even though they are installed below a raised floor

Δ **(2) Installation Requirements for Power-Supply Cords, Data Cables, Interconnecting Cables, and Grounding Conductors Under a Raised Floor.** The following cords, cables, and conductors shall be permitted to be installed under a raised floor:

(1) Power-supply cords of listed information technology equipment in accordance with 645.5(B).
(2) Interconnecting cables enclosed in a raceway.
(3) Equipment grounding conductors.
(4) Where the air space under a raised floor is protected by an automatic fire suppression system, in addition to wiring installed in compliance with 722.135(B), Types CL2R, CL3R, CL2, and CL3 and substitute cables, including CMP, CMR, CM, and CMG, installed in accordance with 722.135(E) shall be permitted under raised floors.
(5) Where the air space under a raised floor is not protected by an automatic fire suppression system, in addition to wiring installed in compliance with 722.135(B), substitute cable Type CMP installed in accordance with 722.135(E) shall be permitted under raised floors.
(6) Listed Type DP cable having adequate fire-resistant characteristics suitable for use under raised floors of an information technology equipment room.

> Informational Note: See CSA "Vertical Flame Test-Cables in Cable Trays" as described in CSA C22.2 No. 0.3, *Test Methods for Electrical Wires and Cables*, for one method of defining resistance to the spread of fire where the damage (char length) of the cable does not exceed 1.5 m (4 ft 11 in.) or "UL Flame Exposure, Vertical Flame Tray Test" in UL 1685, *Standard for Safety for Vertical-Tray Fire-Propagation and Smoke-Release Test for Electrical and Optical-Fiber Cables*. The smoke measurements in the test method are not applicable.

Supply cords of ITE are permitted to be run through holes in a raised floor to connect to receptacles located below the raised floor. Openings in a raised floor through which cords and cables are run must be made so the cords and cables are not subject to abrasion. Allowing cords through openings in a raised floor is an amendment to the general prohibition of this practice found in 400.12.

Other than branch-circuit conductors and power supply cords, interconnecting cables used under raised floors are required to be enclosed in a raceway, be listed as Type DP (data processing) cables, or be of the appropriate cable type permitted by 645.5(E)(2)(4).

(3) Installation Requirements for Optical Fiber Cables Under a Raised Floor. The installation of optical fiber cables shall comply with either of the following:

(1) Where the air space under a raised floor is protected by an automatic fire suppression system, optical fiber cables installed in accordance with 770.113(C), Types OFNR,

OFCR, OFNG, OFCG, OFN, and OFC shall be permitted under raised floors.

(2) Where the air space under a raised floor is not protected by an automatic fire suppression system, only optical fiber cables installed in accordance with 770.113(C) shall be permitted under raised floors.

Δ **(F) Securing in Place.** Power-supply cords; communications cables, connecting cables, interconnecting cables, and associated boxes, connectors, plugs, and receptacles that are listed as part of, or for, information technology equipment shall not be required to be secured in place where installed under raised floors.

> Informational Note: See 300.11 for securement requirements for raceways and cables not listed as part of, or for, information technology equipment.

(G) Abandoned Supply Circuits and Interconnecting Cables. The accessible portion of abandoned supply circuits and interconnecting cables shall be removed unless contained in a raceway.

(H) Installed Supply Circuits and Interconnecting Cables Identified for Future Use.

(1) Cable Identification Means. Supply circuits and interconnecting cables identified for future use shall be marked with a tag of sufficient durability to withstand the environment involved.

(2) Cable Tag Criteria. Supply circuit tags and interconnecting cable tags shall have the following information:

(1) Date identified for future use
(2) Date of intended use
(3) Information relating to the intended future use

645.10 Disconnecting Means. An approved means shall be provided to disconnect power to all electronic equipment in the information technology equipment room or in designated zones within the room. There shall also be a similar approved means to disconnect the power to all dedicated HVAC systems serving the room or designated zones and to cause all required fire/smoke dampers to close. The disconnecting means shall comply with either 645.10(A) or (B).

> *Exception: These requirements shall not apply to installations complying with Article 685.*

(A) Remote Disconnect Controls.

(1) Emergency Access. Remote disconnect controls shall be located at approved locations readily accessible in case of fire to authorized personnel and emergency responders.

(2) Disconnect Identification. The remote disconnect means for the control of electronic equipment power and HVAC systems shall be grouped and identified. A single means to control both systems shall be permitted.

(3) Fire/Smoke Zone Isolation. Where multiple zones are created, each zone shall have an approved means to confine fire or products of combustion to within the zone.

Δ **(4) System Operation Continuity.** Additional means to prevent unintentional operation of remote disconnect controls shall be permitted.

> Informational Note: See NFPA 75, *Standard for the Fire Protection of Information Technology Equipment*, for further information.

Typically, the circuit supplying the ITE and the circuit supplying the HVAC system will be controlled through separate disconnecting means. However, operation of these disconnecting means can be accomplished through the use of a single remote control, such as one pushbutton. The disconnecting means is required to disconnect the conductors of each circuit from their supply source and close all required fire/smoke dampers. (See the definition of the term *disconnecting means* in Article 100.) The disconnecting means is permitted to be remote-controlled switching devices, such as relays, with pushbutton stations at the principal exit doors.

Δ **(B) Critical Operations Data Systems.** Remote disconnecting controls shall not be required for critical operations data systems when all of the following conditions are met:

(1) An approved procedure has been established and maintained for removing power and air movement within the room or zone.
(2) Qualified personnel are continuously available to advise emergency responders and to instruct them of disconnecting methods.
(3) A smoke-sensing fire detection system is in place.

> Informational Note: See *NFPA 72, National Fire Alarm and Signaling Code*, for further information.

(4) An approved fire suppression system suitable for the application is in place.
(5) Cables installed under a raised floor, other than branch-circuit wiring, and power cords are installed in compliance with 645.5(E)(2) or (E)(3), or in compliance with Table 645.10(B).

Only those data systems designated as critical in function based on the definition of *critical operations data system* found in Article 100, are permitted to implement the provision for not installing the remote disconnect control covered in 645.10(A).

Δ **645.11 Uninterruptible Power Supply (UPS).** UPS systems installed within the information technology equipment room and their supply and output circuits shall comply with 645.10, except for the following installations and constructions:

Δ **TABLE 645.10(B)** Cables Installed Under Raised Floors

Cable Type	Applicable Sections
Branch circuits under raised floors	645.5(E)(1)
Supply cords of listed information technology equipment	645.5(E)(2)(1), 300.22(C)
Class 2 and Class 3 remote control and PLTC cables in other spaces used for environmental air (plenums)	722.135(B)
Optical fiber cable in other spaces used for environmental air (plenums)	770.113(C) and Table 770.154(a)
Communications wires and cables, cable routing assemblies, and communications raceways in other spaces used for environmental air (plenums)	800.113(C) and Tables 800.154(a), (b), and (c)
Coaxial CATV and radio distribution cables in other spaces used for environmental air (plenums)	800.113(C) and Table 800.154(a)

(1) Installations complying with Parts I and II of Article 685
(2) Power sources limited to 750 volt-amperes or less derived either from UPS equipment or from battery circuits integral to electronic equipment

The disconnecting means shall also disconnect the battery from its load.

> Informational Note: See UL 1778, *Uninterruptible Power Systems*, and UL 62368-1, *Audio/Video, Information and Communication Technology Equipment — Part 1: Safety Requirements*, for information on product listings for electronic equipment disconnecting means and backup battery power sources.

645.14 System Grounding. Separately derived power systems shall be installed in accordance with Parts I and II of Article 250. Power systems derived within listed information technology equipment that supply information technology systems through receptacles or cable assemblies supplied as part of this equipment shall not be considered separately derived for the purpose of applying 250.30.

Δ **645.15 Equipment Grounding and Bonding.** All exposed non-current-carrying metal parts of an information technology system shall be bonded to the equipment grounding conductor in accordance with Parts I, V, VI, VII, and VIII of Article 250 or shall be double insulated. Where signal reference structures are installed, they shall be bonded to the equipment grounding conductor provided for the information technology equipment. Any auxiliary grounding electrode(s) installed for information technology equipment shall be installed in accordance with 250.54.

> Informational Note: See 250.146(D) and 406.3(E) for information on isolated grounding-type receptacles.

645.16 Marking. Each unit of an information technology system supplied by a branch circuit shall be provided with a manufacturer's nameplate, which shall also include the input power requirements for voltage, frequency, and maximum rated load in amperes.

645.17 Power Distribution Units. Power distribution units that are used for information technology equipment shall be permitted to have multiple panelboards within a single cabinet if the power distribution unit is utilization equipment listed for information technology application.

Power distribution units (PDUs) are specialized electrical distribution equipment used to supply multiple bays of rack-mounted modules installed in an ITE room. Due to the large number of overcurrent protective devices (OCPDs) used in this type of application, PDUs are built with multiple panelboards installed in a single cabinet.

645.18 Surge Protection for Critical Operations Data Systems. A listed surge-protective device (SPD) shall be installed for critical operations data systems in accordance with Part II of Article 242.

645.25 Engineering Supervision. As an alternative to the feeder and service load calculations required by Parts III and IV of Article 220, feeder and service load calculations for new or existing loads shall be permitted to be used if provided by qualified persons under engineering supervision.

An engineered alternative to the load calculations in Parts III and IV of Article 220 recognizes that the loads associated with computer hardware vary according to the operating system and software being used. Therefore, identical pieces of equipment installed in a facility will each have a load based on how that equipment is being used and applied.

The engineered alternative provides for a customized load calculation based on how a facility or a specific industry applies its computer hardware.

645.27 Selective Coordination. Critical operations data system(s) overcurrent protective devices shall be selectively coordinated with all supply-side overcurrent protective devices.

Selective coordination shall be selected by a licensed professional engineer or other qualified persons engaged primarily in the design, installation, or maintenance of electrical systems. The selection shall be documented and made available to those authorized to design, install, inspect, maintain, and operate the system.

ARTICLE 646   Modular Data Centers

Part I. General

△ **646.1 Scope.** This article covers modular data centers.

> Informational Note No. 1: Modular data centers include the installed information technology equipment (ITE) and support equipment, electrical supply and distribution, wiring and protection, working space, grounding, HVAC, and the like, that are located in an equipment enclosure.

> Informational Note No. 2: See NFPA 75, *Standard for the Fire Protection of Information Technology Equipment*, which covers the requirements for the protection of information technology equipment and systems in an information technology equipment room.

> Informational Note No. 3: See UL 60950-1, *Information Technology Equipment — Safety — Part 1: General Requirements*, and UL 62368-1, *Audio/Video, Information and Communication Technology Equipment — Part 1: Safety Requirements*, for information on listing requirements for both information technology equipment and communications equipment contained within a modular data center.

> Informational Note No. 4: *Modular data centers* are sometimes referred to as containerized data centers.

> Informational Note No. 5: Equipment enclosures housing only support equipment (e.g., HVAC or power distribution equipment) that are not part of a specific modular data center are not considered a modular data center.

A modular data center (MDC) is similar to an information technology equipment (ITE) room, which is covered under Article 645. The distinction between the two is that the MDC is a preassembled and packaged enclosure delivered with the ITE installed, while the ITE room is built and equipped on site. The MDC can be supplied through the premises wiring system or through a separate MDC enclosure. Data, fire alarm, communications, control, audio, and visual circuits from the MDC typically are brought outside the MDC and into the facility. An MDC is large enough for personnel to enter; therefore, access, working space, and emergency lighting are specifically addressed.

646.3 Other Articles. Circuits and equipment shall comply with 646.3(A) through (M) as applicable. Wherever the requirements of other articles of this *Code* and Article 646 differ, the requirements of Article 646 shall apply.

(A) Spread of Fire or Products of Combustion. Sections 300.21, 770.26, and 800.26 shall apply to penetrations of a fire-resistant room boundary, if provided.

△ **(B) Wiring and Cabling in Other Spaces Used for Environmental Air (Plenums).** The following sections and tables shall apply to wiring and cabling in other spaces used for environmental air (plenums) within a modular data center space:

(1) Wiring methods: 300.22(C)(1)
(2) Class 2, Class 3, and PLTC cables: 722.135(B)

(3) Fire alarm systems: 760.53(B)(2) and Table 760.154
(4) Optical fiber cables: 770.113(C) and Table 770.154(a)
(5) Communications circuits: 800.113(C) and Table 800.154(a)
(6) CATV and radio distribution systems: 800.113(C) and Table 800.154(a)

> Informational Note: Environmentally controlled working spaces, aisles, and equipment areas in an MDC are not considered a plenum.

△ **(C) Grounding and Bonding.** The non-current-carrying conductive members of optical fiber cables in an MDC shall be grounded in accordance with 770.114. Grounding and bonding of communications protectors, cable shields, and non-current-carrying metallic members of cable shall comply with Part IV of Article 805.

△ **(D) Electrical Classification of Data Circuits.** Section 725.60(A)(4) shall apply to the electrical classification of listed information technology equipment signaling circuits.

(E) Fire Alarm Equipment. Parts I, II, and III of Article 760 shall apply to fire alarm systems, cables, and equipment installed in an MDC, where provided. Only fire alarm cables listed in accordance with Part IV of Article 760 and listed fire alarm equipment shall be permitted to be installed in an MDC.

△ **(F) Cable Routing Assemblies and Communications Wires, Cables, Raceways, and Equipment.** Sections 800.110, 800.113, and 800.154 shall apply to cable routing assemblies and communications raceways. Parts I, II, III, IV, and V of Articles 800 and 805 shall apply to communications wires, cables, and equipment installed in an MDC. Only communications wires and cables listed in accordance with 800.179, cable routing assemblies and communications raceways listed in accordance with 800.182, and communications equipment listed in accordance with 800.171 shall be permitted to be installed in an MDC.

> Informational Note: See Article 100 for a definition of *communications equipment*.

(G) Community Antenna Television and Radio Distribution Systems Cables and Equipment. Parts I, II, III, IV, and V of Articles 800 and 820 shall apply to community antenna television and radio distribution systems equipment installed in an MDC. Only community antenna television and radio distribution cables listed in accordance with 800.179 and listed CATV equipment shall be permitted to be installed in an MDC.

(H) Surge-Protective Devices (SPDs). Where provided, surge-protective devices shall be listed and labeled and installed in accordance with Part II of Article 242.

(I) Lighting. Lighting shall be installed in accordance with Parts I through XIV of Article 410.

(J) Power Distribution Wiring and Wiring Protection. Power distribution wiring and wiring protection within an MDC shall comply with Parts I, II, and III of Article 210 for branch circuits.

(K) Wiring Methods and Materials. Wiring methods and materials shall comply with the following:

(1) Unless modified elsewhere in this article, wiring methods and materials for power distribution shall comply with Chapter 3. Wiring shall be suitable for its use and installation and shall be listed and labeled.

Exception: This requirement shall not apply to wiring that is part of listed and labeled equipment.

(2) The following wiring methods shall not be permitted:

    a. Integrated gas spacer cable: Type IGS (Article 326)
    b. Concealed knob-and-tube wiring (Article 394)
    c. Messenger-supported wiring (Article 396)
    d. Open wiring on insulators (Article 398)
    e. Outdoor overhead conductors over 600 volts (Article 395)

(3) Wiring in areas under a raised floor that are constructed and used for ventilation as described in 645.5(E) shall be permitted to use the wiring methods described in 645.5(E) if the conditions of 645.4 are met.

(4) Installation of wiring for remote-control, signaling, and power-limited circuits shall comply with Part II of Article 725.

(5) Installation of optical fiber cables shall comply with Part V of Article 770.

(6) Alternate wiring methods as permitted by Article 645 shall be permitted for MDCs, provided that all of the conditions of 645.4 are met.

(L) Service Equipment. For an MDC that is designed such that it can be powered from a separate electrical service, the service equipment for control and protection of services and their installation shall comply with Parts I, V, VI, and VII of Article 230. The service equipment and their arrangement and installation shall permit the installation of the service-entrance conductors in accordance with Parts I and IV of Article 230. Service equipment shall be listed and labeled and marked as being suitable for use as service equipment.

(M) Disconnecting Means. An approved means shall be provided to disconnect power to all electronic equipment in the MDC in accordance with 645.10. There shall also be a similar approved means to disconnect the power to all dedicated HVAC systems serving the MDC that shall cause all required fire/smoke dampers to close.

Δ **646.4 Applicable Requirements.** All MDCs shall be listed and labeled and comply with 646.3(M) and 646.5 through 646.9 or comply with this article.

    Informational Note: See UL Subject 2755, *Outline of Investigation for Modular Data Centers,* for information on listing requirements for MDCs.

An MDC must be listed and labeled. An evaluation of the equipment, installed wiring, lighting, and work space is conducted as part of the listing. Any field-installed wiring — including supply circuits and data circuits — is required to comply with the appropriate *NEC®* article.

Δ **646.5 Nameplate Data.** A permanent nameplate shall be attached to each equipment enclosure of an MDC and shall be plainly visible after installation. The nameplate shall include the following information, as applicable:

(1) Supply voltage, number of phases, frequency, and full-load current. The full-load current shown on the nameplate shall not be less than the sum of the full-load currents required for all motors and other equipment that may be in operation at the same time under normal conditions of use. Where unusual type loads, duty cycles, and so forth, require oversized conductors or permit reduced-size conductors, the required capacity shall be included in the marked full-load current. Where more than one incoming supply circuit is to be provided, the nameplate shall state the preceding information for each circuit. For listed equipment, the full-load current shown on the nameplate shall be permitted to be the maximum, measured, 15-minute, average full-load current.

    Informational Note No. 1: See 430.22(E) and 430.26 for duty cycle requirements.

(2) For MDCs powered by a separate service, the short-circuit current rating of the service equipment provided as part of the MDC.

    Informational Note No. 2: This rating may be part of the service equipment marking.

(3) For MDCs powered by a separate service, if the required service as determined by Parts III and IV of Article 220 is less than the rating of the service panel used, the required service shall be included on the nameplate. As an alternative to the feeder and service load calculations required by Parts III and IV of Article 220, feeder and service load calculations for new, future, or existing loads shall be permitted to be used if performed by qualified persons under engineering supervision.

    Informational Note No. 3: Branch circuits supplying ITE loads are assumed to be loaded not less than 80 percent of the branch-circuit rating with a 100 percent duty cycle.

(4) Electrical diagram number(s) or the number of the index to the electrical drawings.

(5) For MDC equipment enclosures that are not powered by a separate service, feeder, or branch circuit, a reference to the powering equipment.

(6) Manufacturer's name or trademark.

646.6 Supply Conductors and Overcurrent Protection.

A permanent nameplate is required on each enclosure to indicate the required supply and the full-load current. The full-load current is not necessarily determined with all equipment operating simultaneously as indicated in 646.5(1). The MDC is considered a continuous load; therefore, supply conductors must be sized for 125 percent of the marked full-load current.

(A) Size. The size of the supply conductor shall be such as to have an ampacity not less than 125 percent of the full-load current rating.

> Informational Note No. 1: See the 0–2000-volt ampacity tables of Article 310 for ampacity of conductors rated 600 V and below.
> Informational Note No. 2: See 430.22(E) and 430.26 for duty cycle requirements.

(B) Overcurrent Protection. Where overcurrent protection for supply conductors is furnished as part of the MDC, overcurrent protection for each supply circuit shall comply with 646.6(B)(1) through (B)(2).

(1) Service Equipment — Overcurrent Protection. Service conductors shall be provided with overcurrent protection in accordance with 230.90 through 230.95.

(2) Taps and Feeders. Where overcurrent protection for supply conductors is furnished as part of the MDC as permitted by 240.21, the overcurrent protection shall comply with the following:

(1) The overcurrent protection shall consist of a single circuit breaker or set of fuses.
(2) The MDC shall be marked "OVERCURRENT PROTECTION PROVIDED AT MDC SUPPLY TERMINALS."
(3) The supply conductors shall be considered either as feeders or as taps and be provided with overcurrent protection complying with 240.21.

646.7 Short-Circuit Current Rating.

(A) Service Equipment. The service equipment of an MDC that connects directly to a service shall have a short-circuit current rating not less than the available fault current of the service.

(B) MDCs Connected to Branch Circuits and Feeders. Modular data centers that connect to a branch circuit or a feeder circuit shall have a short-circuit current rating not less than the available fault current of the branch circuit or feeder. The short-circuit current rating of the MDC shall be based on the short-circuit current rating of a listed and labeled MDC or the short-circuit current rating established using an approved method.

Exception: This requirement shall not apply to listed and labeled equipment connected to branch circuits located inside of the MDC equipment enclosure.

> Informational Note: See UL 508A-2018, *Standard for Industrial Control Panels, Supplement SB*, for an example of an approved method.

(C) MDCs Powered from Separate MDC System Enclosures. Modular data center equipment enclosures, powered from a separate MDC system enclosure that is part of the specific MDC system, shall have a short-circuit current rating coordinated with the powering module in accordance with 110.10.

> Informational Note: See UL 508A-2018, *Standard for Industrial Control Panels, Supplement SB*, for an example of an approved method for determining short-circuit current ratings.

646.8 Field-Wiring Compartments. A field-wiring compartment in which service or feeder connections are to be made shall be readily accessible and comply with the following:

(1) Permit the connection of the supply wires after the MDC is installed
(2) Permit the connection to be introduced and readily connected
(3) Be located so that the connections may be readily inspected after the MDC is installed

646.9 Flexible Power Cords and Cables for Connecting Equipment Enclosures of an MDC System.

(A) Uses Permitted. Flexible power cords and cables shall be permitted to be used for connections between equipment enclosures of an MDC system where not subject to physical damage.

> Informational Note: One example of flexible power cord usage for connections between equipment enclosures of an MDC system is between an MDC enclosure containing only servers and one containing power distribution equipment.

(B) Uses Not Permitted. Flexible power cords and cables shall not be used for connection to external sources of power.

> Informational Note: Examples of external sources of power are electrical services, feeders, and premises branch circuits.

(C) Listing. Where flexible power cords or cables are used, they shall be listed as suitable for extra-hard usage. Where used outdoors, flexible power cords and cables shall also be listed as suitable for wet locations and shall be sunlight resistant.

(D) Single-Conductor Cable. Single-conductor power cable shall be permitted to be used only in sizes 2 AWG or larger.

Part II. Equipment

646.10 Electrical Supply and Distribution. Equipment used for electrical supply and distribution in an MDC, including fittings, devices, luminaires, apparatus, machinery, and the like, shall comply with Parts I and II of Article 110.

646.11 Distribution Transformers.

(A) Utility-Owned Transformers. Utility-owned distribution transformers shall not be permitted in an MDC.

(B) Non-Utility-Owned Premises Transformers. Non-utility-owned premises distribution transformers installed in the vicinity of an MDC shall be of the dry type or the type filled with a noncombustible dielectric medium. Such transformers shall be installed in accordance with Parts I and II of Article 450. Non-utility-owned premises distribution transformers shall not be permitted in an MDC.

(C) Power Transformers. Power transformers that supply power only to the MDC shall be permitted to be installed in the MDC equipment enclosure. Only dry-type transformers shall be permitted to be installed in the MDC equipment enclosure. Such transformers shall be installed in accordance with Parts I, II, and III of Article 450.

646.12 Receptacles. At least one 125-volt ac, 15- or 20-ampere-rated duplex convenience outlet shall be provided in each work area of the MDC to facilitate the powering of test and measurement equipment that may be required during routine maintenance and servicing, without having to route flexible power cords through or across doorways or around line-ups of equipment, or the like.

646.13 Other Electrical Equipment. Electrical equipment that is an integral part of the MDC, including information technology equipment, lighting, control, power, HVAC (heating, ventilation, and air-conditioning), emergency lighting, alarm circuits, and so forth, shall comply with the requirements for its use and installation and shall be listed and labeled.

646.14 Installation and Use. Listed and labeled equipment shall be installed and used in accordance with any instructions or limitations included in the listing.

Part III. Lighting

646.15 General Illumination. Illumination shall be provided for all workspaces and areas that are used for exit access and exit discharge. The illumination shall be arranged so that the failure of any single lighting unit does not result in a complete loss of illumination.

> Informational Note: See NFPA *101*-2018, *Life Safety Code*, Section 7.8, for information on illumination of means of egress.

646.16 Emergency Lighting. Areas that are used for exit access and exit discharge shall be provided with emergency lighting. Emergency lighting systems shall be listed and labeled equipment installed in accordance with the manufacturer's instructions.

> Informational Note: See NFPA *101*-2018, *Life Safety Code*, Section 7.9, for information on emergency lighting.

646.17 Emergency Lighting Circuits. No appliances or lamps, other than those specified as required for emergency use, shall be supplied by emergency lighting circuits. Branch circuits supplying emergency lighting shall be installed to provide service from storage batteries, generator sets, UPS, separate service, fuel cells, or unit equipment. No other equipment shall be connected to these circuits unless the emergency lighting system includes a backup system where only the lighting is supplied by battery circuits under power failure conditions. All boxes and enclosures (including transfer switches, generators, and power panels) for emergency circuits shall be marked to identify them as components of an emergency circuit or system.

Part IV. Workspace

646.18 General. Space about electrical equipment shall comply with 110.26.

646.19 Entrance to and Egress from Working Space. For equipment over 1.8 m (6 ft) wide or deep, there shall be one entrance to and egress from the required working space not less than 610 mm (24 in.) wide and 2.0 m (6½ ft) high at each end of the working space. Doors shall open to the full extent of their designed egress opening and be equipped with listed panic hardware or listed fire exit hardware. A single entrance to and egress from the required working space shall be permitted where either of the conditions in 646.19(A) or (B) is met.

(A) Unobstructed Egress. Where the location permits a continuous and unobstructed way of egress travel, a single entrance to the working space shall be permitted.

(B) Extra Working Space. Where the depth of the working space is twice that required by 110.26(A)(1), a single entrance shall be permitted. It shall be located such that the distance from the equipment to the nearest edge of the entrance is not less than the minimum clear distance specified in Table 110.26(A)(1) for equipment operating at that voltage and in that condition.

646.20 Working Space for ITE.

(A) Low-Voltage Circuits. The working space about ITE where any live parts that may be exposed during routine servicing operate at not greater than 30 volts rms, 42 volts peak, or 60 volts dc shall not be required to comply with the workspace requirements of 646.19.

(B) Other Circuits. Any areas of ITE that require servicing of parts that are greater than 30 volts rms, 42 volts peak, or 60 volts dc shall comply with the workspace requirements of 646.19.

> Informational Note No. 1: For example, field-wiring compartments for ac mains connections, power distribution units, and so forth.
> Informational Note No. 2: It is assumed that ITE operates at voltages not exceeding 1000 volts.

646.21 Work Areas and Working Space About Batteries. Working space about a battery system shall comply with 110.26. Working space shall be measured from the edges of the battery racks, cabinets, or trays.

646.22 Workspace for Routine Service and Maintenance. Workspace shall be provided to facilitate routine servicing and maintenance (those tasks involving operations that can be accomplished by employees and where extensive disassembly of equipment is not required). Routine servicing and maintenance shall be able to be performed without exposing the worker to a risk of electric shock or personal injury.

> Informational Note: An example of such routine maintenance is cleaning or replacing an air filter.

ARTICLE 647   Sensitive Electronic Equipment

647.1 Scope. This article covers the installation and wiring of separately derived systems operating at 120 volts line-to-line and 60 volts to ground for sensitive electronic equipment.

This type of supply system is employed as a means to reduce objectionable noise and its adverse effect on the performance of electronic audio and video equipment. Article 647 permits the use of this type of supply system for all commercial and industrial applications where sensitive audio/video or similar electronic equipment is used. Such systems can be used only in areas that are under the close supervision of qualified individuals.

647.3 General. Use of a separately derived 120-volt single-phase 3-wire system with 60 volts on each of two ungrounded conductors to an equipment grounding conductor shall be permitted for the purpose of reducing objectionable noise in sensitive electronic equipment locations, provided the following conditions apply:

(1) The system is installed only in commercial or industrial occupancies.
(2) The system's use is restricted to areas under close supervision by qualified personnel.
(3) All of the requirements in 647.4 through 647.8 are met.

647.4 Wiring Methods.

(A) Panelboards and Overcurrent Protection. Use of standard single-phase panelboards and distribution equipment with a higher voltage rating shall be permitted. The system shall be clearly marked on the face of the panel or on the inside of the panel doors. Common trip two-pole circuit breakers or a combination two-pole fused disconnecting means that are identified for use at the system voltage shall be provided for both ungrounded conductors in all feeders and branch circuits. Branch circuits and feeders shall be provided with a means to simultaneously disconnect all ungrounded conductors.

Circuit breakers and fuses are acceptable means of providing overcurrent protection for technical power circuits. Additionally, all technical power feeder and branch circuits are required to be provided with a disconnecting means that simultaneously opens all ungrounded conductors of the circuit.

(B) Junction Boxes. All junction box covers shall be clearly marked to indicate the distribution panel and the system voltage.

(C) Conductor Identification. All feeders and branch-circuit conductors installed under this section shall be identified as to system at all splices and terminations by color, marking, tagging, or equally effective means. The means of identification shall be posted at each branch-circuit panelboard and at the disconnecting means for the building.

(D) Voltage Drop. The voltage drop on any branch circuit shall not exceed 1.5 percent. The combined voltage drop of feeder and branch-circuit conductors shall not exceed 2.5 percent.

Unlike electrical distribution systems that supply lighting and appliance branch circuits, the supply systems covered by Article 647 are subject to mandatory voltage-drop requirements. The voltage-drop requirements are needed to ensure the operation of overcurrent devices in order to protect conductors and equipment supplied by these systems. Because the use of standard overcurrent devices and distribution equipment with higher voltage ratings is permitted, the impedance in circuits supplied by these systems under fault conditions is a primary concern.

(1) Fixed Equipment. The voltage drop on branch circuits supplying equipment connected using wiring methods in Chapter 3 shall not exceed 1.5 percent. The combined voltage drop of feeder and branch-circuit conductors shall not exceed 2.5 percent.

(2) Cord-Connected Equipment. The voltage drop on branch circuits supplying receptacles shall not exceed 1 percent. For the purposes of making this calculation, the load connected to the receptacle outlet shall be considered to be 50 percent of the branch-circuit rating. The combined voltage drop of feeder and branch-circuit conductors shall not exceed 2.0 percent.

> Informational Note: The purpose of this provision is to limit voltage drop to 1.5 percent where portable cords may be used as a means of connecting equipment.

647.5 Three-Phase Systems. Where 3-phase power is supplied, a separately derived 6-phase "wye" system with 60 volts to ground installed under this article shall be configured as three separately derived 120-volt single-phase systems having a combined total of no more than six disconnects.

647.6 Grounding.

(A) General. The transformer secondary center tap of the 60/120-volt, 3-wire system shall be grounded as provided in 250.30.

A technical power system has two ungrounded conductors with 120 volts between them and a grounded reference conductor at 60 volts with respect to the ungrounded conductors.

(B) Equipment Grounding Conductors Required. Permanently wired utilization equipment and receptacles shall be grounded by means of an equipment grounding conductor run with the circuit conductors and connected to an equipment grounding bus prominently marked "Technical Equipment Ground" in the branch-circuit panelboard. The equipment grounding bus shall be connected to the grounded conductor on the line side of disconnecting means supplied by the separately derived system. The equipment grounding conductor shall not be smaller than that specified in Table 250.122 and run with the feeder conductors. The technical equipment grounding bus shall not be required to be bonded to the panelboard enclosure. Other equipment grounding methods authorized elsewhere in this *Code* shall be permitted where the impedance of the equipment grounding return path does not exceed the impedance of equipment grounding conductors sized and installed in accordance with this article.

> Informational Note No. 1: See 250.122 for equipment grounding conductor sizing requirements where circuit conductors are adjusted in size to compensate for voltage drop.
>
> Informational Note No. 2: These requirements limit the impedance of the ground fault return path where only 60 volts apply to a fault condition instead of the usual 120 volts.

647.7 Receptacles.

Δ **(A) General.** Where receptacles are used as a means of connecting equipment, the following conditions shall be met:

(1) All 15- and 20-ampere receptacles shall be GFCI protected.
(2) All receptacle outlet strips, adapters, receptacle covers, and faceplates shall be marked with the following words or equivalent:

WARNING — TECHNICAL POWER
Do not connect to lighting equipment.
For electronic equipment use only.
60/120 V. 1Φac
GFCI protected

The warning sign(s) or label(s) shall comply with 110.21(B).

(3) A 125-volt, single-phase, 15- or 20-ampere-rated receptacle having one of its current-carrying poles connected to a grounded circuit conductor shall be located within 1.8 m (6 ft) of all permanently installed 15- or 20-ampere-rated 60/120-volt technical power-system receptacles.
(4) All 125-volt receptacles used for 60/120-volt technical power shall have a unique configuration and be identified for use with this class of system.

Exception: Receptacles and attachment plugs rated 125-volt, single-phase, 15- or 20-amperes, and that are identified for use with grounded circuit conductors, shall be permitted in machine rooms, control rooms, equipment rooms, equipment racks, and other similar locations that are restricted to use by qualified personnel.

(B) Isolated Ground Receptacles. Isolated ground receptacles shall be permitted as described in 250.146(D); however, the branch-circuit equipment grounding conductor shall be terminated as required in 647.6(B).

647.8 Lighting Equipment. Lighting equipment installed under this article for the purpose of reducing electrical noise originating from lighting equipment shall meet the conditions of 647.8(A) through (C).

(A) Disconnecting Means. All luminaires connected to separately derived systems operating at 60 volts to ground, and associated control equipment if provided, shall have a disconnecting means that simultaneously opens all ungrounded conductors. The disconnecting means shall be located within sight of the luminaire or be lockable open in accordance with 110.25.

(B) Luminaires. All luminaires shall be permanently installed and listed for connection to a separately derived system at 120 volts line-to-line and 60 volts to ground.

(C) Screw Shell. Luminaires installed under this section shall not have an exposed lamp screw shell.

ARTICLE 650 Pipe Organs

650.1 Scope. This article covers those electrical circuits and parts of electrically operated pipe organs that are employed for the control of the keyboards and of the pipe organ sounding apparatus, typically organ pipes.

> Informational Note: The typical pipe organ is a very large musical instrument that is built as part of a building or structure.

Δ **650.3 Other Articles.** Installations of circuits and equipment shall comply with 650.3(A) and (B) as applicable. Wherever the requirements of other articles in Chapters 1 through 7 of this *Code* and Article 650 differ, the requirements of Article 650 shall apply.

(A) Electronic Organ Equipment. Installations of digital/analog–sampled sound production technology and associated audio signal processing, amplification, reproduction equipment, and wiring installed as part of a pipe organ shall be in accordance with Article 640.

Some pipe organ installations incorporate digital/analog–sampled sound technology. The requirements in Article 640 are necessary for electronic sound production, amplification, signal processing, and other sound reproduction circuits and equipment installed as part of a pipe organ.

(B) Optical Fiber Cable. Installations of optical fiber cables shall be in accordance with Parts I and V of Article 770.

650.4 Source of Energy. DC power shall be supplied by a listed dc power supply with a maximum output of 30 volts.

> Informational Note: Class 1 power-limited power supplies are often utilized in pipe organ applications.

650.5 Grounding or Double Insulation of the DC Power Supply. The installation of the dc power supply shall comply with either of the following:

(1) The dc power supply shall be double insulated.
(2) The metallic case of the dc power supply shall be bonded to the input equipment grounding conductor.

650.6 Conductors. Conductors shall comply with 650.6(A) through (D).

(A) Size. The minimum conductor size shall be not less than 28 AWG for electronic signal circuits and not less than 26 AWG for electromagnetic valve supply and the like. The minimum conductor size of a main common-return conductor in the electromagnetic supply shall not be less than 14 AWG.

(B) Insulation. Conductors shall have thermoplastic or thermosetting insulation.

(C) Conductors to Be Cabled. Except for the common-return conductor and conductors inside the organ proper, the organ sections and the organ console conductors shall be cabled. The common-return conductors shall be permitted under an additional covering enclosing both cable and return conductor, or they shall be permitted as a separate conductor and shall be permitted to be in contact with the cable.

Δ **(D) Cable Covering.** Each cable shall be provided with an outer covering, either overall or on each of any subassemblies of grouped conductors. Tape shall be permitted in place of a covering. Where not installed in metal raceway, the covering shall be resistant to flame spread, or the cable or each cable subassembly shall be covered with a closely wound listed fireproof tape.

> Informational Note: See UL 2556-2015, *Wire, Cables and Cable Test Methods*, for one method of determining that cable is resistant to flame spread by testing the cable to the FV-2/VW-1 Test.

650.7 Installation of Conductors. Cables shall be securely fastened in place and shall be permitted to be attached directly to the organ structure without insulating supports. Splices shall not be required to be enclosed in boxes or other enclosures. Control equipment and busbars connecting common-return conductors shall be permitted to be attached directly to the organ structure without insulation supports. Abandoned cables that are not terminated at equipment shall be identified with a tag of sufficient durability to withstand the environment involved.

650.8 Overcurrent Protection. Circuits shall be so arranged that 20 AWG through 28 AWG conductors shall be protected by an overcurrent device rated at not more than 6 amperes. Other conductor sizes shall be protected in accordance with their ampacity. A common return conductor shall not require overcurrent protection.

650.9 Protection from Accidental Contact. The wiring of the pipe organ sounding apparatus shall be within the lockable enclosure (organ chamber) where the exterior pipes shall be permitted to form part of the enclosure.

> Informational Note: Access to the pipe organ sounding apparatus and the associated circuitry is restricted by an enclosure. In most pipe organ installations, exterior pipes form part of the enclosure. In other installations, the pipes are covered by millwork that permits the passage of sound.

Part I. General

660.1 Scope. This article covers all X-ray equipment operating at any frequency or voltage for industrial or other nonmedical or nondental use.

> Informational Note: See Article 517, Part V, for X-ray installations in health care facilities.

Nothing in this article shall be construed as specifying safeguards against the useful beam or stray X-ray radiation.

> Informational Note No. 1: Radiation safety and performance requirements of several classes of X-ray equipment are regulated under Public Law 90-602 and are enforced by the Department of Health and Human Services.
>
> Informational Note No. 2: In addition, information on radiation protection by the National Council on Radiation Protection and Measurements is published as *Reports of the National Council on Radiation Protection and Measurement*. These reports can be obtained from NCRP Publications, 7910 Woodmont Ave., Suite 1016, Bethesda, MD 20814.

Article 660 covers X-ray equipment in industrial facilities or similar locations, where it is commonly used for inspecting a process or product. This permits nondestructive testing without dismantling or applying stress to detect cracks, flaws, or structural defects. Welded joints frequently are inspected with X-ray equipment to detect hidden defects that can cause failure under stress.

The most common industrial application of X-rays is radiography, in which shadow pictures of the object are produced. The type and thickness of the material involved govern the voltage to be employed, which can range from a few thousand volts (kV) to millions of volts (MV). Metal objects that are as much as 20 inches thick can be X-rayed.

Fluoroscopy is another X-ray technique used for industrial and commercial applications. Fluoroscopy is similar to radiography, but it operates at less than 250 kilovolts. Most of these systems project a shadow picture on a screen, similar to those used for security checks of luggage at airport terminals. Fluoroscopy is capable of detecting minute flaws or defects.

660.3 Hazardous (Classified) Locations. Unless identified for the location, X-ray and related equipment shall not be installed or operated in hazardous (classified) locations.

Informational Note: See Article 517, Part IV, for additional information.

660.4 Connection to Supply Circuit.

(A) Fixed and Stationary Equipment. Fixed and stationary X-ray equipment shall be connected to the power supply by means of a wiring method meeting the general requirements of this *Code*. Equipment properly supplied by a branch circuit rated at not over 30 amperes shall be permitted to be supplied through a suitable attachment plug cap and hard-service cable or power-supply cord.

(B) Portable, Mobile, and Transportable Equipment. Individual branch circuits shall not be required for portable, mobile, and transportable X-ray equipment requiring a capacity of not over 60 amperes. Portable and mobile types of X-ray equipment of any capacity shall be supplied through a suitable hard-service cable or power-supply cord. Transportable X-ray equipment of any capacity shall be permitted to be connected to its power supply by suitable connections and hard-service cable or power-supply cord.

(C) Over 1000 Volts, Nominal. Circuits and equipment operated at more than 1000 volts, nominal, shall comply with Article 490.

660.5 Disconnecting Means.
A disconnecting means of adequate capacity for at least 50 percent of the input required for the momentary rating, or 100 percent of the input required for the long-time rating, of the X-ray equipment, whichever is greater, shall be provided in the supply circuit. The disconnecting means shall be located within sight from the X-ray control and readily accessible.

Exception: The disconnecting means for the X-ray equipment shall not be required under either of the following conditions, provided that the controller disconnecting means is lockable open in accordance with 110.25:

(1) Where such a location of the disconnecting means for the X-ray equipment is impracticable or introduces additional or increased hazards to persons or property

(2) In industrial installations, with written safety procedures, where conditions of maintenance and supervision ensure that only qualified persons service the equipment

660.6 Rating of Supply Conductors and Overcurrent Protection.

(A) Branch-Circuit Conductors. The ampacity of supply branch-circuit conductors and the overcurrent protective devices shall not be less than 50 percent of the momentary rating or 100 percent of the long-time rating, whichever is greater.

(B) Feeder Conductors. The ampacity of conductors and the rating of overcurrent devices of a feeder for two or more branch circuits supplying X-ray units shall not be less than 100 percent of the momentary demand rating [as determined by 660.6(A)] of the two largest X-ray apparatus plus 20 percent of the momentary ratings of other X-ray apparatus.

Informational Note: The minimum conductor size for branch and feeder circuits is also governed by voltage regulation requirements. For a specific installation, the manufacturer usually specifies minimum distribution transformer and conductor sizes, rating of disconnect means, and overcurrent protection.

660.7 Wiring Terminals.
X-ray equipment not provided with a permanently attached power-supply cord shall be provided with suitable wiring terminals or leads for the connection of power-supply conductors of the size required by the rating of the branch circuit for the equipment.

660.9 Minimum Size of Conductors.
Size 18 AWG or 16 AWG fixture wires, as specified in 724.49, and flexible cords shall be permitted for the control and operating circuits of X-ray and auxiliary equipment where protected by not larger than 20-ampere overcurrent devices.

660.10 Equipment Installations.
All equipment for new X-ray installations and all used or reconditioned X-ray equipment moved to and reinstalled at a new location shall be of an approved type.

Part II. Control

660.20 Fixed and Stationary Equipment.

(A) Separate Control Device. A separate control device, in addition to the disconnecting means, shall be incorporated in the X-ray control supply or in the primary circuit to the high-voltage transformer. This device shall be a part of the X-ray equipment but shall be permitted in a separate enclosure immediately adjacent to the X-ray control unit.

A control device provides means for initiating and terminating X-ray exposures and automatically times their duration.

(B) Protective Device. A protective device, which shall be permitted to be incorporated into the separate control device, shall be provided to control the load resulting from failures in the high-voltage circuit.

660.21 Portable and Mobile Equipment.
Portable and mobile equipment shall comply with 660.20, but the manually controlled device shall be located in or on the equipment.

660.23 Industrial and Commercial Laboratory Equipment.

(A) Radiographic and Fluoroscopic Types. All radiographic- and fluoroscopic-type equipment shall be effectively enclosed or shall have interlocks that de-energize the equipment automatically to prevent ready access to live current-carrying parts.

(B) Diffraction and Irradiation Types. Diffraction- and irradiation-type equipment or installations not effectively enclosed or not provided with interlocks to prevent access to uninsulated live parts during operation shall be provided with a positive means to indicate when they are energized. The indicator shall be a pilot light, readable meter deflection, or equivalent means.

660.24 Independent Control. Where more than one piece of equipment is operated from the same high-voltage circuit, each piece or each group of equipment as a unit shall be provided with a high-voltage switch or equivalent disconnecting means. This disconnecting means shall be constructed, enclosed, or located so as to avoid contact by persons with its live parts.

Part III. Transformers and Capacitors

660.35 General. Transformers and capacitors that are part of an X-ray equipment shall not be required to comply with Articles 450 and 460.

High-ratio step-up transformers that are an integral part of X-ray equipment are not required to comply with Article 450 and are generally used to provide the high voltage necessary for X-ray tubes. Because the fire hazard is lower due to the low primary voltage, X-ray transformers are not required to be installed in fire-resistant vaults.

660.36 Capacitors. Capacitors shall be mounted within enclosures of insulating material or grounded metal.

Part IV. Guarding and Grounding

660.47 General.

(A) High-Voltage Parts. All high-voltage parts, including X-ray tubes, shall be mounted within grounded enclosures. Air, oil, gas, or other suitable insulating media shall be used to insulate the high voltage from the grounded enclosure. The connection from the high-voltage equipment to X-ray tubes and other high-voltage components shall be made with high-voltage shielded cables.

(B) Low-Voltage Cables. Low-voltage cables connecting to oil-filled units that are not completely sealed, such as transformers, condensers, oil coolers, and high-voltage switches, shall have insulation of the oil-resistant type.

Δ **660.48 Grounding.** Battery-operated X-ray equipment shall not be required to comply with the grounding requirements of this *Code*.

According to 90.3, the general requirements of Chapter 2 apply to all electrical installations. Therefore, Part I of Article 250, specifically 250.4(A)(2) and (3), would apply to the grounding of X-ray equipment.

ARTICLE 665 Induction and Dielectric Heating Equipment

Part I. General

665.1 Scope. This article covers the construction and installation of dielectric heating, induction heating, induction melting, and induction welding equipment and accessories for industrial and scientific applications. Medical or dental applications, appliances, or line frequency pipeline and vessel heating are not covered in this article.

Informational Note: See Article 427, Part V, for line frequency induction heating of pipelines and vessels.

To prevent spurious radiation caused by induction and dielectric heating equipment, the Federal Communications Commission (FCC) has established rules that govern the use of this type of industrial heating equipment operating above 10 kilohertz (FCC, 47 CFR 18).

See also

NFPA 86, *Standard for Ovens and Furnaces*, for more information on electric heating systems using an induction heater or a dielectric heater in ovens and furnaces

665.5 Output Circuit. The output circuit shall include all output components external to the converting device, including contactors, switches, busbars, and other conductors. The current flow from the output circuit to ground under operating and ground-fault conditions shall be limited to a value that does not cause 50 volts or more to ground to appear on any accessible part of the heating equipment and its load. The output circuit shall be permitted to be isolated from ground.

If the load (object being heated) accidentally contacts the output coil, a voltage to ground will appear on the load, depending on the various impedances to ground of the coil and the load. If the voltage on the load is limited to less than 50 volts, guarding per 110.27(A) is not required. If the coil is isolated from ground and the load is grounded through an impedance that is low (less than 1 percent) relative to the coil impedance to ground, the voltage of the load to ground will be low no matter where the load contacts the coil.

In induction melting furnaces, an additional reason for isolating the coil from ground is to limit the fault current when a coil does go to ground. Limiting the fault current prevents severe damage to the water-cooled coil, resulting in a water leak and the potential for a water-molten metal explosion. If water is trapped under molten metal, the rapid transfer of heat to the water causes the water to turn almost instantly into steam. The resulting 1600-to-1 expansion of the steam results in the ejection of molten metal from the furnace.

CLOSER LOOK: Solid-State Converter Power Circuit

A solid-state power converter consists of three sections: the rectifier section, the inverter section, and the output section, which includes the load coil and is usually located outside the power supply. Exhibit 665.1 shows an example of an enclosed power supply for an induction heating process.

The rectifier section converts 3-phase, line frequency voltage to direct current. The output of the rectifier section supplies energy for the inverter section. The inverter section converts the energy to a variable frequency for the output circuit. The variable output frequency controls the power delivered to the load.

The output section consists of a capacitor in parallel (current fed) or in series (voltage fed) with a coil. Capacitance and inductance operate at a resonant frequency, and as the output frequency approaches the resonant frequency, the power to the load approaches its maximum. The output power is very low at minimum frequency.

Induction Heating

Induction heating occurs when an electrically conductive material (load) is placed in a varying magnetic field generated by a coil (inductor) around or adjacent to the workpiece to be heated. The varying magnetic field induces current in the load. Heat is generated by the resulting I^2R losses in the load. Induction heating can be further subdivided into heating, melting, and welding.

Induction heating raises the temperature of the load to a temperature below its melting point, usually for the purposes of hardening, tempering, annealing, forging, extruding, or rolling. Frequencies used for heating range from about 50 hertz to 500 kilohertz. Power levels range from 5 kilowatts to 42 megawatts.

EXHIBIT 665.1 *Enclosed power supply for an induction heating process. (Courtesy of Ajax Tocco Magnethermic, Park Ohio Industries)*

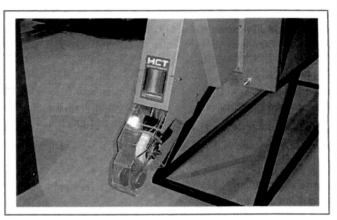

EXHIBIT 665.2 *A solid-state induction welding machine. (Courtesy of Thermatool Corp.)*

Induction melting raises the temperature of the load to a temperature above its melting point, so the molten material can be alloyed, homogenized, and/or poured. Frequencies used for melting range from about 50 hertz to 10 kilohertz. Power levels range from 5 kilowatts to 16.5 megawatts.

Induction welding is primarily used in the manufacture of welded pipe and tubing. In this process, a high-frequency current is passed through an induction coil in the proximity of the conducting metal surfaces to be joined. Selected portions are heated nearly instantaneously to the forging temperature, then are joined under pressure to produce a forge weld. Frequencies used for welding range from about 100 to 800 kilohertz. Power levels range from 20 kilowatts to 1 megawatt. Exhibit 665.2 shows an induction welding machine used in the manufacture of pipe and tubing.

Dielectric Heating

Dielectric heating equipment is similar to induction heating equipment, except that it is used to heat nonmetallic materials as opposed to metals. Typical applications include the drying of textiles after dyeing, drying of water-based coatings on paper, preheating of wood fibers for the medium-density fiberboard (MDF) industry, welding of plastic materials, and food processing.

At radio frequencies, the material to be heated forms a dielectric when placed between metal capacitor plates connected across the output of the generator. A high-frequency alternating electric field is created between the electrode plates. The molecules vibrate in the dielectric field, causing dissipation of energy through the material and frictional heating of the dielectric material. At higher (microwave) frequencies, a similar process occurs, but the generator is coupled to a resonant cavity into which the dielectric material is placed.

The frequency of operation of dielectric heating equipment is considerably higher than for induction heating. These machines operate at the assigned radio frequencies of 13.56, 27.12, and 40.68 megahertz or at microwave frequencies of 915 and 2450 megahertz.

The power for these machines ranges from 0.5 kilowatt to 1 megawatt.

665.7 Remote Control.

(A) Multiple Control Points. Where multiple control points are used for applicator energization, a means shall be provided and interlocked so that the applicator can be energized from only one control point at a time. A means for de-energizing the applicator shall be provided at each control point.

(B) Foot Switches. Switches operated by foot pressure shall be provided with a shield over the contact button to avoid accidental closing of a foot switch.

665.10 Ampacity of Supply Conductors. The ampacity of supply conductors shall be determined by 665.10(A) or (B).

(A) Nameplate Rating. The ampacity of conductors supplying one or more pieces of equipment shall be not less than the sum of the nameplate ratings for the largest group of machines capable of simultaneous operation, plus 100 percent of the standby currents of the remaining machines. Where standby currents are not given on the nameplate, the nameplate rating shall be used as the standby current.

(B) Motor-Generator Equipment. The ampacity of supply conductors for motor-generator equipment shall be determined in accordance with Article 430, Part II.

665.12 Disconnecting Means. A readily accessible disconnecting means shall be provided to disconnect each heating equipment from its supply circuit. The disconnecting means shall be located within sight from the controller or be lockable open in accordance with 110.25.

The rating of this disconnecting means shall not be less than the nameplate rating of the heating equipment. Motor-generator equipment shall comply with Article 430, Part IX. The supply circuit disconnecting means shall be permitted to serve as the heating equipment disconnecting means where only one heating equipment is supplied.

Part II. Guarding, Grounding, and Labeling

665.19 Component Interconnection. The interconnection components required for a complete heating equipment installation shall be guarded.

665.20 Enclosures. The converting device (excluding the component interconnections) shall be completely contained within an enclosure(s) of noncombustible material.

665.21 Control Panels. All control panels shall be of dead-front construction.

665.22 Access to Internal Equipment. Access doors or detachable access panels shall be employed for internal access to heating equipment. Access doors to internal compartments containing equipment employing voltages from 150 volts to 1000 volts ac or dc shall be capable of being locked closed or shall be interlocked to prevent the supply circuit from being energized while the door(s) is open. The provision for locking or adding a lock to the access doors shall be installed on or at the access door and shall remain in place with or without the lock installed.

Access doors to internal compartments containing equipment employing voltages exceeding 1000 volts ac or dc shall be provided with a disconnecting means equipped with mechanical lockouts to prevent access while the heating equipment is energized, or the access doors shall be capable of being locked closed and interlocked to prevent the supply circuit from being energized while the door(s) is open. Detachable panels not normally used for access to such parts shall be fastened in a manner that makes them inconvenient to remove.

665.23 Hazard Labels or Signs. Labels or signs that read "DANGER — HIGH VOLTAGE — KEEP OUT" shall be attached to the equipment and shall be plainly visible where persons might come in contact with energized parts when doors are open or closed or when panels are removed from compartments containing over 150 volts ac or dc. Hazard signs or labels shall comply with 110.21(B).

665.24 Capacitors. The time and means of discharge shall be in accordance with 460.6 for capacitors rated 600 volts, nominal, and under. The time and means of discharge shall be in accordance with 460.28 for capacitors rated over 600 volts, nominal. Capacitor internal pressure switches connected to a circuit-interrupter device shall be permitted for capacitor overcurrent protection.

Enhanced protection against rupture of capacitor cases is needed when capacitors are operated at the higher frequencies used for induction and dielectric heating. A high-resistance fault condition can cause case pressure to build up inside the capacitor over a very short time. Capacitor internal pressure switches are the preferred method to detect this type of failure.

Consider a 5000-kVAr (kilovolt-ampere reactive), 2500-volt, 300-hertz capacitor. Nominal current is 2000 amperes. A "high-resistance" fault of 10 ohms results in 250 amperes of resistive current, or a total capacitor current of 2016 amperes root-mean square (rms). This small increase in rms current will not result in the opening of an overcurrent device even though 625 kilowatts of thermal energy are being generated inside the capacitor, which is designed to dissipate about 1.5 kilowatts of losses.

665.25 Dielectric Heating Applicator Shielding. Protective cages or adequate shielding shall be used to guard dielectric heating applicators. Interlock switches shall be used on all hinged access doors, sliding panels, or other easy means of access to the applicator. All interlock switches shall be connected in such a manner as to remove all power from the applicator when any one of the access doors or panels is open.

665.26 Grounding and Bonding. Bonding to the equipment grounding conductor or inter-unit bonding, or both, shall be used wherever required for circuit operation, and for limiting to a safe value radio frequency voltages between all exposed non–current-carrying parts of the equipment and earth ground, between all equipment parts and surrounding objects, and between such objects and earth ground. Such connection to the equipment grounding conductor and bonding shall be installed in accordance with Article 250, Parts II and V.

> Informational Note: Under certain conditions, contact between the object being heated and the applicator results in an unsafe condition, such as eruption of heated materials. Grounding of the object being heated and ground detection can be used to prevent this unsafe condition.

Because of stray currents between units of equipment or between equipment and the ground, bonding presents special problems at radio frequencies. Special bonding requirements are particularly needed at dielectric heating frequencies (100 to 200 megahertz) because of the differences in radio frequency potential that can exist between the equipment and surrounding metal units or other units of the installation. Bonding has been accomplished by placing all units of the equipment on a flooring or base consisting of a copper or aluminum sheet, then thoroughly bonding by soldering, welding, or bolting. Such special bonding holds the radio frequency resistance and reactance between units to a minimum, and any stray circulating currents flowing through the bonding will not cause a dangerous voltage drop.

The operator can be protected from high radio frequency potentials by shielding at dielectric heating frequencies. Interference with radio communications systems at such high frequencies can be eliminated by totally enclosing all components in a shielding of copper or aluminum.

665.27 Marking. Each heating equipment shall be provided with a nameplate giving the manufacturer's name and model identification and the following input data: line volts, frequency, number of phases, maximum current, full-load kilovolt-amperes (kVA), and full-load power factor. Additional data shall be permitted.

ARTICLE
668 Electrolytic Cells

668.1 Scope. This article applies to the installation of the electrical components and accessory equipment of electrolytic cells, electrolytic cell lines, and process power supply for the production of aluminum, cadmium, chlorine, copper, fluorine, hydrogen peroxide, magnesium, sodium, sodium chlorate, and zinc.

Not covered by this article are cells used as a source of electric energy and for electroplating processes and cells used for the production of hydrogen.

> Informational Note No. 1: In general, any cell line or group of cell lines operated as a unit for the production of a particular metal,

gas, or chemical compound may differ from any other cell lines producing the same product because of variations in the particular raw materials used, output capacity, use of proprietary methods or process practices, or other modifying factors to the extent that detailed *Code* requirements become overly restrictive and do not accomplish the stated purpose of this *Code*.

> Informational Note No. 2: See IEEE 463-2013, *Standard for Electrical Safety Practices in Electrolytic Cell Line Working Zones*, for further information.

Within a cell line working zone, both an electrolytic cell line and its direct current (dc) process power-supply circuit are treated as an individual machine supplied from a single source, even though they might cover acres of space, have a load current in excess of 400,000 amperes dc, or have a circuit voltage in excess of 1000 volts dc. The cell line process current passes through each cell in a series connection, and the load current cannot be subdivided the way it can in the heating circuit of a resistance-type electric furnace. Because a cell line is supplied by its individual dc rectifier system, the rectifier or the entire cell line circuit is de-energized by removing its primary power source.

In some electrolytic cell systems, the terminal voltage of the process supply can be significant. The voltage to ground of exposed live parts from one end of a cell line to the other is variable between the limits of the terminal voltage. Hence, operating and maintenance personnel and their tools are required to be insulated from ground.

668.3 Other Articles.

(A) Lighting, Ventilating, Material Handling. Chapters 1 through 4 shall apply to services, feeders, branch circuits, and apparatus for supplying lighting, ventilating, material handling, and the like that are outside the electrolytic cell line working zone.

(B) Systems Not Electrically Connected. Those elements of a cell line power-supply system that are not electrically connected to the cell supply system, such as the primary winding of a two-winding transformer, the motor of a motor-generator set, feeders, branch circuits, disconnecting means, motor controllers, and overload protective equipment, shall be required to comply with all applicable sections of this *Code*.

(C) Electrolytic Cell Lines. Electrolytic cell lines shall comply with the provisions of Chapters 1 through 4 except as amended in 668.3(C)(1) through (C)(4).

(1) Conductors. The electrolytic cell line conductors shall not be required to comply with Articles 110, 210, 215, 220, and 225. See 668.12.

(2) Overcurrent Protection. Overcurrent protection of electrolytic cell dc process power circuits shall not be required to comply with the requirements of Article 240.

(3) Grounding. Except as required by this article, equipment located or used within the electrolytic cell line working zone or associated with the cell line dc power circuits shall not be required to comply with Article 250.

(4) Working Zone. The electrolytic cells, cell line attachments, and the wiring of auxiliary equipment and devices within the cell line working zone shall not be required to comply with Articles 110, 210, 215, 220, and 225. See 668.30.

> Informational Note: See 668.15 for equipment, apparatus, and structural component grounding.

668.10 Cell Line Working Zone.

(A) Area Covered. The space envelope of the cell line working zone shall encompass spaces that meet any of the following conditions:

(1) Is within 2.5 m (96 in.) above energized surfaces of electrolytic cell lines or their energized attachments
(2) Is below energized surfaces of electrolytic cell lines or their energized attachments, provided the headroom in the space beneath is less than 2.5 m (96 in.)
(3) Is within 1.0 m (42 in.) horizontally from energized surfaces of electrolytic cell lines or their energized attachments or from the space envelope described in 668.10(A)(1) or (A)(2)

(B) Area Not Covered. The cell line working zone shall not be required to extend through or beyond walls, floors, roofs, partitions, barriers, or the like.

668.11 Direct-Current Cell Line Process Power Supply.

(A) Not Grounded. The direct-current cell line process power-supply conductors shall not be required to be grounded.

(B) Metal Enclosures Grounded. All metal enclosures of power-supply apparatus for the direct-current cell line process operating with a power supply over 50 volts shall be grounded by either of the following means:

(1) Through protective relaying equipment
(2) By a minimum 2/0 AWG copper grounding electrode conductor or a conductor of equal or greater conductance

(C) Grounding Requirements. The grounding electrode connections required by 668.11(B) shall be installed in accordance with 250.8, 250.10, 250.12, 250.68, and 250.70.

668.12 Cell Line Conductors.

(A) Insulation and Material. Cell line conductors shall be either bare, covered, or insulated and of copper, aluminum, copper-clad aluminum, steel, or other suitable material.

(B) Size. Cell line conductors shall be of such cross-sectional area that the temperature rise under maximum load conditions and at maximum ambient shall not exceed the safe operating temperature of the conductor insulation or the material of the conductor supports.

(C) Connections. Cell line conductors shall be joined by bolted, welded, clamped, or compression connectors.

668.13 Disconnecting Means.

(A) More Than One Process Power Supply. Where more than one direct-current cell line process power supply serves the same cell line, a disconnecting means shall be provided on the cell line circuit side of each power supply to disconnect it from the cell line circuit.

(B) Removable Links or Conductors. Removable links or removable conductors shall be permitted to be used as the disconnecting means.

668.14 Shunting Means.

(A) Partial or Total Shunting. Partial or total shunting of cell line circuit current around one or more cells shall be permitted.

(B) Shunting One or More Cells. The conductors, switches, or combination of conductors and switches used for shunting one or more cells shall comply with the applicable requirements of 668.12.

668.15 Grounding.
For equipment, apparatus, and structural components that are required to be grounded in accordance with Article 668, Article 250, Part III, for a local grounding electrode system shall apply, except a water pipe electrode shall not be required to be used. Any electrode or combination of electrodes described in 250.52 shall be permitted.

668.20 Portable Electrical Equipment.

(A) Portable Electrical Equipment Not to Be Grounded. The frames and enclosures of portable electrical equipment used within the cell line working zone shall not be grounded.

Exception No. 1: Where the cell line voltage does not exceed 200 volts dc, these frames and enclosures shall be permitted to be grounded.

Exception No. 2: These frames and enclosures shall be permitted to be grounded where guarded.

(B) Isolating Transformers. Electrically powered, hand-held, cord-connected portable equipment with ungrounded frames or enclosures used within the cell line working zone shall be connected to receptacle circuits that have only ungrounded conductors such as a branch circuit supplied by an isolating transformer with an ungrounded secondary.

(C) Marking. Ungrounded portable electrical equipment shall be distinctively marked and shall employ plugs and receptacles of a configuration that prevents connection of this equipment to grounding receptacles and that prevents inadvertent interchange of ungrounded and grounded portable electrical equipment.

668.21 Power-Supply Circuits and Receptacles for Portable Electrical Equipment.

(A) Isolated Circuits. Circuits supplying power to ungrounded receptacles for hand-held, cord-connected equipment shall be electrically isolated from any distribution system supplying areas other than the cell line working zone and shall be ungrounded. Power for these circuits shall be supplied through isolating transformers. Primaries of such transformers shall operate at not more than 1000 volts between conductors and shall be provided with proper overcurrent protection. The secondary voltage of such transformers shall not exceed 300 volts between conductors, and all circuits supplied from such secondaries shall be ungrounded and shall have an approved overcurrent device of proper rating in each conductor.

(B) Noninterchangeability. Receptacles and their mating plugs for ungrounded equipment shall not have provision for an equipment grounding conductor and shall be of a configuration that prevents their use for equipment required to be grounded.

(C) Marking. Receptacles on circuits supplied by an isolating transformer with an ungrounded secondary shall be a distinctive configuration, shall be distinctively marked, and shall not be used in any other location in the plant.

668.30 Fixed and Portable Electrical Equipment.

(A) Electrical Equipment Not Required to Be Grounded. Alternating-current systems supplying fixed and portable electrical equipment within the cell line working zone shall not be required to be grounded.

(B) Exposed Conductive Surfaces Not Required to Be Grounded. Exposed conductive surfaces, such as electrical equipment housings, cabinets, boxes, motors, raceways, and the like, that are within the cell line working zone shall not be required to be grounded.

Δ **(C) Wiring Methods.** Auxiliary electrical equipment such as motors, transducers, sensors, control devices, and alarms, mounted on an electrolytic cell or other energized surface, shall be connected to premises wiring systems using any of the following:

(1) Multiconductor hard usage cord.
(2) Wire or cable in suitable raceways or metal or nonmetallic cable trays. If metal conduit, cable tray, armored cable, or similar metallic systems are used, they shall be installed with insulating breaks such that they do not cause a potentially hazardous electrical condition.

(D) Circuit Protection. Circuit protection shall not be required for control and instrumentation that are totally within the cell line working zone.

(E) Bonding. Bonding of fixed electrical equipment to the energized conductive surfaces of the cell line, its attachments, or auxiliaries shall be permitted. Where fixed electrical equipment is mounted on an energized conductive surface, it shall be bonded to that surface.

668.31 Auxiliary Nonelectrical Connections. Auxiliary nonelectrical connections, such as air hoses, water hoses, and the like, to an electrolytic cell, its attachments, or auxiliary equipment shall not have continuous conductive reinforcing wire, armor, braids, and the like. Hoses shall be of a nonconductive material.

668.32 Cranes and Hoists.

(A) Conductive Surfaces to Be Insulated from Ground. The conductive surfaces of cranes and hoists that enter the cell line working zone shall not be required to be grounded. The portion of an overhead crane or hoist that contacts an energized electrolytic cell or energized attachments shall be insulated from ground.

(B) Hazardous Electrical Conditions. Remote crane or hoist controls that could introduce hazardous electrical conditions into the cell line working zone shall employ one or more of the following systems:

(1) Isolated and ungrounded control circuit in accordance with 668.21(A)
(2) Nonconductive rope operator
(3) Pendant pushbutton with nonconductive supporting means and having nonconductive surfaces or ungrounded exposed conductive surfaces
(4) Radio

668.40 Enclosures. General-purpose electrical equipment enclosures shall be permitted where a natural draft ventilation system prevents the accumulation of gases.

ARTICLE 669 Electroplating

669.1 Scope. This article applies to the installation of the electrical components and accessory equipment that supply the power and controls for electroplating, anodizing, electropolishing, and electrostripping. For purposes of this article, the term *electroplating* shall be used to identify any or all of these processes.

Because of the extremely high currents and low voltages normally involved, conventional wiring methods cannot be used in electroplating, anodizing, electropolishing, and electrostripping processes. Section 669.6 permits the use of bare conductors supported from insulators in systems exceeding 50 volts direct current (dc). Some systems in the aluminum anodizing process have potentials up to 240 volts. Warning signs in accordance with 669.7 are required to be posted to indicate the presence of bare conductors.

669.3 General. Equipment for use in electroplating processes shall be identified for such service.

669.5 Branch-Circuit Conductors. Branch-circuit conductors supplying one or more units of equipment shall have an ampacity of not less than 125 percent of the total connected load. The ampacities for busbars shall be in accordance with 366.23.

669.6 Wiring Methods. Conductors connecting the electrolyte tank equipment to the conversion equipment shall be in accordance with 669.6(A) and (B).

(A) Systems Not Exceeding 60 Volts Direct Current. Insulated conductors shall be permitted to be run without insulated support, provided they are protected from physical damage. Bare copper or aluminum conductors shall be permitted where supported on insulators.

(B) Systems Exceeding 60 Volts Direct Current. Insulated conductors shall be permitted to be run on insulated supports, provided they are protected from physical damage. Bare copper or aluminum conductors shall be permitted where supported on insulators and guarded against accidental contact up to the point of termination in accordance with 110.27.

669.7 Warning Signs. Warning signs shall be posted to indicate the presence of bare conductors. The warning sign(s) or label(s) shall comply with 110.21(B).

669.8 Disconnecting Means.

(A) More Than One Power Supply. Where more than one power supply serves the same dc system, a disconnecting means shall be provided on the dc side of each power supply.

(B) Removable Links or Conductors. Removable links or removable conductors shall be permitted to be used as the disconnecting means.

669.9 Overcurrent Protection. Direct-current conductors shall be protected from overcurrent by one or more of the following:

(1) Fuses or circuit breakers
(2) A current-sensing device that operates a disconnecting means
(3) Other approved means

ARTICLE 670 Industrial Machinery

Δ **670.1 Scope.** This article covers the nameplate data for, overvoltage protection for, and the size and overcurrent protection of supply conductors to industrial machinery.

> Informational Note No. 1: See NFPA 79, *Electrical Standard for Industrial Machinery*, for further information.

The equipment and wiring of industrial machinery, for which different component parts may be purchased and assembled at the location of use, must be installed in accordance with the applicable articles in the *NEC*®. Machinery assembled by the manufacturer, in accordance with NFPA 79, *Electrical Standard for Industrial Machinery*, then disassembled for shipping and reassembled at its place of use, comes under only Article 670 and any *NEC* sections referenced herein. In this case, the machinery is treated as a package unit.

The information to be included on the nameplate allows for proper conductor sizing, overcurrent protection of the feeder or branch circuit supplying the industrial machine, and integration of the machine into the facility electrical system.

> Informational Note No. 2: See 110.26 for information on the workspace requirements for equipment containing supply conductor terminals.
> Informational Note No. 3: See NFPA 79, *Electrical Standard for Industrial Machinery*, for information on the workspace requirements for machine power and control equipment.

Working clearances around control equipment enclosures and compartments containing equipment operating at 1000 volts or less and that are an integral part of an industrial machine are contained in Section 11.5 in NFPA 79. The requirements of NFPA 79 closely parallel those found in 110.26(A), but some requirements in NFPA 79 allow smaller clearances under very specific conditions of operation and equipment construction.

670.3 Machine Nameplate Data.

> Informational Note: See 430.22(E) and 430.26 for duty cycle requirements.

Δ **(A) Permanent Nameplate.** A permanent nameplate shall be attached to the outside of the control equipment enclosure or on the machine immediately adjacent to the main control equipment enclosure that is visible after installation. The nameplate shall include the following information:

(1) Supply voltage, number of phases, frequency, and full-load current
(2) Maximum ampere rating of the short-circuit and ground-fault protective device
(3) Ampere rating of largest motor, from the motor nameplate, or load
(4) Short-circuit current rating of the machine industrial control panel based on one of the following:

   a. Short-circuit current rating of a listed and labeled machine control enclosure or assembly
   b. Short-circuit current rating established using an approved method

> Informational Note: See UL 508A-2017, *Industrial Control Panels, Supplement SB*, for an example of an approved method.

(5) Electrical diagram number(s) or the number of the index to the electrical drawings

The full-load current shown on the nameplate shall not be less than the sum of the full-load currents required for all motors and other equipment that can be in operation at the same time under normal conditions of use. Where unusual type loads, duty cycles, and so forth require oversized conductors or permit reduced-size conductors, the required capacity shall be included in the marked "full-load current." Where more than one incoming supply circuit is to be provided, the nameplate shall state the preceding information for each circuit.

An industrial machine's nameplate must provide the short-circuit current rating of the machine's industrial control panel. That rating is established either as part of the listing of the control enclosure or assembly or, for assemblies that are not listed, by an approved method of determining the short-circuit current rating.

In the absence of product listing, Supplement SB to UL 508A, *Standard for Industrial Control Panels*, is referred to as one example of a method for determining the short-circuit current rating of a control panel or assembly that could be used as a basis for equipment approval.

The second paragraph of 670.3(A) recognizes that the operating characteristics of an industrial machine may permit the use of over-sized or reduced-sized feeder conductors. An example of this is an industrial machine containing motors sized for high torque but, in normal operation, run at close to no-load current values. In this case, it may be appropriate to reduce the full-load current marking on the machine nameplate.

See also

430.26, which covers feeder demand factors

(B) Overcurrent Protection. Where overcurrent protection is provided in accordance with 670.4(C), the machine shall be marked "overcurrent protection provided at machine supply terminals."

670.4 Supply Conductors and Overcurrent Protection.

(A) Size. The size of the supply conductor shall be such as to have an ampacity not less than 125 percent of the full-load current rating of all resistance heating loads plus 125 percent of the full-load current rating of the highest rated motor plus the sum of the full-load current ratings of all other connected motors and apparatus, based on their duty cycle, that may be in operation at the same time.

> Informational Note No. 1: See Table 310.16 through Table 310.20 for ampacity of conductors rated 2000 volts and below.
>
> Informational Note No. 2: See 430.22(E) and 430.26 for duty cycle requirements.

The duty cycle of motors and apparatus must be considered when determining the minimum ampacity of a supply circuit conductor for an industrial machine. Depending on the operating characteristics of the motor, the duty cycle of the apparatus might not always result in reduction of the supply conductor ampacity. Where motors are used in other than a continuous-duty mode of operation, Table 430.22(E) provides percentages by which the full-load current of a given motor is increased or decreased for the purpose of sizing motor circuit conductors. A motor that is loaded continuously under any conditions of use is an example of a continuous-duty application.

(B) Disconnecting Means. A machine shall be considered as an individual unit and therefore shall be provided with disconnecting means. The disconnecting means shall be permitted to be supplied by branch circuits protected by either fuses or circuit breakers. The disconnecting means shall not be required to incorporate overcurrent protection.

> Informational Note: See NFPA *70E, Standard for Electrical Safety in the Workplace*, which provides guidance for creating an electrically safe work condition for performing maintenance or other work on the machine.

Regarding the machine disconnecting means, the 2021 edition of NFPA 79, *Electrical Standard for Industrial Machinery*, states, in part: "The center of the grip of the operating handle of a supply circuit disconnecting means, when in its highest position, shall be not more than 2.0 m (6 ft 7 in.) above the servicing level. A permanent operating platform, readily accessible by means of a permanent stair(s) or ladder, shall be considered the servicing level."

The disconnecting means shall "be provided with a permanent means permitting it to be locked in the off (open) position only (e.g., by padlocks) independent of the enclosure door or enclosure cover position. When so locked, remote as well as local closing into the on position shall be prevented."

(C) Overcurrent Protection. Where furnished as part of the machine, overcurrent protection for each supply circuit shall consist of a single circuit breaker or set of fuses, the machine shall bear the marking required in 670.3, and the supply conductors shall be considered either as feeders or as taps as covered by 240.21.

The rating or setting of the overcurrent protective device for the circuit supplying the machine shall not be greater than the sum of the largest rating or setting of the branch-circuit short-circuit and ground-fault protective device provided with the machine, plus 125 percent of the full-load current rating of all resistance heating loads, plus the sum of the full-load currents of all other motors and apparatus that could be in operation at the same time.

Exception: Where one or more instantaneous trip circuit breakers or motor short-circuit protectors are used for motor branch-circuit short-circuit and ground-fault protection as permitted by 430.52(C), the procedure specified in 670.4(C) for determining the maximum rating of the protective device for the circuit supplying the machine shall apply with the following provision: For the purpose of the calculation, each instantaneous trip circuit breaker or motor short-circuit protector shall be assumed to have a rating not exceeding the maximum percentage of motor full-load current permitted by Table 430.52(C)(1) for the type of machine supply circuit protective device employed.

Where no branch-circuit short-circuit and ground-fault protective device is provided with the machine, the rating or setting of the overcurrent protective device shall be based on 430.52 and 430.53, as applicable.

The nameplate provides the necessary information to size the branch-circuit or feeder conductors, the machine disconnecting means, and overcurrent protection. The computation of motor and nonmotor loads is reflected on the nameplate as full-load amperes, and no further calculation is necessary. Sizing of circuit conductors and overcurrent protection beyond the machine disconnecting means is under the scope of NFPA 79, *Electrical Standard for Industrial Machinery*.

670.5 Short-Circuit Current Rating.

(A) Installation. Industrial machinery shall not be installed where the available fault current exceeds its short-circuit current rating as marked in accordance with 670.3(A)(4).

(B) Available Short-Circuit Current Field Marking. Industrial machinery shall be legibly marked in the field with the available fault current. The field marking(s) shall include the date the available fault current calculation was performed and be of sufficient durability to withstand the environment involved.

670.6 Overvoltage Protection. Industrial machinery with safety circuits shall have overvoltage protection.

A study commissioned by the Fire Protection Research Foundation, "Data Assessment for Electrical Surge Protection Devices," provides results of a 2013–2014 survey of facility managers concerning surge damage. It shows that 26 percent had damage to safety interlocking systems on industrial machines due to surges. Safety interlocking systems are in place to protect workers and maintenance personnel from contact with exposed live parts and electric shock. The report can be found at www.nfpa.org/news-and-research.

EXHIBIT 675.1 An electrically driven irrigation machine. *(Courtesy of Valmont Industries, Inc.)*

EXHIBIT 675.2 Example of control equipment for an irrigation machine. *(Courtesy of Valmont Industries, Inc.)*

ARTICLE 675 Electrically Driven or Controlled Irrigation Machines

Part I. General

675.1 Scope. This article applies to electrically driven or controlled irrigation machines, and to the branch circuits and controllers for such equipment.

Electric pump motors used to supply water to irrigation machines are covered by the general requirements of the *NEC®* and not by Article 675. Exhibit 675.1 shows an electrically driven irrigation machine, which is covered by the requirements of Article 675. An example of control equipment used for an irrigation machine is shown in Exhibit 675.2.

675.4 Irrigation Cable.

(A) Construction. The cable used to interconnect enclosures on the structure of an irrigation machine shall be an assembly of stranded, insulated conductors with nonhygroscopic and nonwicking filler in a core of moisture- and flame-resistant nonmetallic material overlaid with a metallic covering and jacketed with a moisture-, corrosion-, and sunlight-resistant nonmetallic material.

The conductor insulation shall be of a type listed in Table 310.4(1) for an operating temperature of 75°C (167°F) or higher and for use in wet locations. The core insulating material thickness shall not be less than 0.76 mm (30 mils), and the metallic overlay thickness shall be not less than 0.20 mm (8 mils). The jacketing material thickness shall be not less than 1.27 mm (50 mils).

A composite of power, control, and grounding conductors in the cable shall be permitted.

(B) Alternate Wiring Methods. Installation of other listed cables complying with the construction requirements of 675.4(A) shall be permitted.

(C) Supports. Irrigation cable shall be secured by straps, hangers, or similar fittings identified for the purpose and so installed as not to damage the cable. Cable shall be supported at intervals not exceeding 1.2 m (4 ft).

(D) Fittings. Fittings shall be used at all points where irrigation cable terminates. The fittings shall be designed for use with the cable and shall be suitable for the conditions of service.

675.5 More Than Three Conductors in a Raceway or Cable. The signal and control conductors of a raceway or cable shall not be counted for the purpose of ampacity adjustment as required in 310.15(C)(1).

675.6 Marking on Main Control Panel. The main control panel shall be provided with a nameplate giving the following information:

(1) The manufacturer's name, the rated voltage, the phase, and the frequency
(2) The current rating of the machine
(3) The rating of the main disconnecting means and size of overcurrent protection required

675.7 Equivalent Current Ratings. Where intermittent duty is not involved, Article 430 shall be used for determining ratings for controllers, disconnecting means, conductors, and the like. Where irrigation machines have inherent intermittent duty, the determinations of equivalent current ratings in 675.7(A) and (B) shall be used.

(A) Continuous-Current Rating. The equivalent continuous-current rating for the selection of branch-circuit conductors and overcurrent protection shall be equal to 125 percent of the motor nameplate full-load current rating of the largest motor, plus a quantity equal to the sum of each of the motor nameplate full-load current ratings of all remaining motors on the circuit, multiplied by the maximum percent duty cycle at which they can continuously operate.

(B) Locked-Rotor Current. The equivalent locked-rotor current rating shall be equal to the numerical sum of the locked-rotor current of the two largest motors plus 100 percent of the sum of the motor nameplate full-load current ratings of all the remaining motors on the circuit.

675.8 Disconnecting Means.

(A) Main Controller. A controller that is used to start and stop the complete machine shall meet all of the following requirements:

(1) An equivalent continuous current rating not less than specified in 675.7(A) or 675.22(A)
(2) A horsepower rating not less than the value from Table 430.251(A) and Table 430.251(B), based on the equivalent locked-rotor current specified in 675.7(B) or 675.22(B)

Exception: A listed molded case switch shall not require a horsepower rating.

A listed molded case switch used as a motor controller is not required to have a horsepower rating, but it is required to have a continuous-current (ampere) rating not less than that specified by 675.7(A) or 675.22(A).

(B) Main Disconnecting Means. The main disconnecting means for the machine shall provide overcurrent protection, shall be at the point of connection of electric power to the machine, or shall be in sight from the machine, and it shall be readily accessible and lockable open in accordance with 110.25. This disconnecting means shall have a horsepower and current rating not less than required for the main controller.

Exception No. 1: Circuit breakers without marked horsepower ratings shall be permitted in accordance with 430.109.

Exception No. 2: A listed molded case switch without marked horsepower ratings shall be permitted.

The main disconnecting means is permitted to be up to 50 feet from the machine but must be in sight, readily accessible, and capable of being locked in the open position. This eliminates one set of overcurrent protective devices (OCPDs) and one disconnecting means where the circuit originates at the motor control panel for the irrigation pump and the panel is located within 50 feet of the center pivot machine. It also alleviates some potential problems with machines designed to be towed to a second site.

(C) Disconnecting Means for Individual Motors and Controllers. A disconnecting means shall be provided to simultaneously disconnect all ungrounded conductors for each motor and controller and shall be located as required by Article 430, Part IX. The disconnecting means shall not be required to be readily accessible.

675.9 Branch-Circuit Conductors. The branch-circuit conductors shall have an ampacity not less than specified in 675.7(A) or 675.22(A).

675.10 Several Motors on One Branch Circuit.

(A) Protection Required. Several motors, each not exceeding 2 hp rating, shall be permitted to be used on an irrigation machine circuit protected at not more than 30 amperes at 1000 volts, nominal, or less, provided all of the following conditions are met:

(1) The full-load rating of any motor in the circuit shall not exceed 6 amperes.
(2) Each motor in the circuit shall have individual overload protection in accordance with 430.32.
(3) Taps to individual motors shall not be smaller than 14 AWG copper and not more than 7.5 m (25 ft) in length.

(B) Individual Protection Not Required. Individual branch-circuit short-circuit protection for motors and motor controllers shall not be required where the requirements of 675.10(A) are met.

675.11 Collector Rings.

(A) Transmitting Current for Power Purposes. Collector rings shall have a current rating not less than 125 percent of the full-load current of the largest device served plus the full-load current of all other devices served, or as determined from 675.7(A) or 675.22(A).

(B) Control and Signal Purposes. Collector rings for control and signal purposes shall have a current rating not less than 125 percent of the full-load current of the largest device served plus the full-load current of all other devices served.

(C) Grounding. The collector ring used for grounding shall have a current rating not less than that sized in accordance with 675.11(A).

(D) Protection. Collector rings shall be protected from the expected environment and from accidental contact by means of a suitable enclosure.

△ **675.12 Grounding.** The following equipment shall be grounded:

(1) All electrical equipment on the irrigation machine
(2) All electrical equipment associated with the irrigation machine
(3) Metal junction boxes and enclosures
(4) Control panels or control equipment that supplies or controls electrical equipment to the irrigation machine

Exception: Grounding shall not be required on machines where all of the following provisions are met:

(1) The machine is electrically controlled but not electrically driven.
(2) The control voltage is 30 volts or less.
(3) The control or signal circuits are current limited as specified in Chapter 9, Tables 11(A) and 11(B).

675.13 Methods of Grounding. Machines that require grounding shall have a non–current-carrying equipment grounding conductor provided as an integral part of each cord, cable, or raceway. This equipment grounding conductor shall be sized not less than the largest supply conductor in each cord, cable, or raceway. Feeder circuits supplying power to irrigation machines shall have an equipment grounding conductor sized according to Table 250.122.

675.14 Bonding. Where electrical grounding is required on an irrigation machine, the metallic structure of the machine, metallic conduit, or metallic sheath of cable shall be connected to the equipment grounding conductor. Metal-to-metal contact with a part that is connected to the equipment grounding conductor and the non–current-carrying parts of the machine shall be considered as an acceptable bonding path.

675.15 Lightning Protection. If an irrigation machine has a stationary point, a grounding electrode system in accordance with Article 250, Part III, shall be connected to the machine at the stationary point for lightning protection.

If the electrical power supply to irrigation machine equipment is a service, the requirements of Article 250 for grounding the system and equipment are applicable. Due to the physical location of irrigation equipment, the most likely grounding electrode is a driven ground rod or ground plate. Where lightning protection is installed, NFPA 780, *Standard for the Installation of Lightning Protection Systems,* requires an electrode for that system. In accordance with 250.60, a common electrode is not permitted to serve the dual function of grounding the electric service and grounding the lightning protection system. However, the separate electrode systems are required to be bonded together but separated from each other in accordance with NFPA 780.

675.16 Energy from More Than One Source. Equipment within an enclosure receiving electric energy from more than one source shall not be required to have a disconnecting means for the additional source if its voltage is 30 volts or less and it meets the requirements of Part II of Article 725.

675.17 Connectors. External plugs and connectors on the equipment shall be of the weatherproof type.

Unless provided solely for the connection of circuits meeting the requirements of Part II of Article 725, external plugs and connectors shall be constructed as specified in 250.124(A).

Part II. Center Pivot Irrigation Machines

675.21 General. Part II covers additional special requirements that are peculiar to center pivot irrigation machines. Article 100 for the definition of *Center Pivot Irrigation Machine.*

675.22 Equivalent Current Ratings. To establish ratings of controllers, disconnecting means, conductors, and the like, for the inherent intermittent duty of center pivot irrigation machines, the determinations in 675.22(A) and (B) shall be used.

(A) Continuous-Current Rating. The equivalent continuous-current rating for the selection of branch-circuit conductors and branch-circuit devices shall be equal to 125 percent of the motor nameplate full-load current rating of the largest motor plus 60 percent of the sum of the motor nameplate full-load current ratings of all remaining motors on the circuit.

(B) Locked-Rotor Current. The equivalent locked-rotor current rating shall be equal to the numerical sum of two times the locked-rotor current of the largest motor plus 80 percent of the sum of the motor nameplate full-load current ratings of all the remaining motors on the circuit.

ARTICLE 680 Swimming Pools, Fountains, and Similar Installations

Part I. General

680.1 Scope. The provisions of this article apply to the construction and installation of electrical wiring for, and equipment in or adjacent to, all swimming, wading, therapeutic, and decorative pools; fountains; hot tubs; spas; and hydromassage bathtubs, whether permanently installed or storable, and to metallic auxiliary equipment, such as pumps, filters, and similar equipment. The term *body of water* used throughout Part I applies to all bodies of water covered in this scope unless otherwise amended.

The installations covered by this article can be indoors or outdoors, permanent or storable. This article also applies to pools used in religious services such as baptisms. Pools used for such purposes are defined as *immersion pools*. Requirements for natural and artificially made bodies of water not covered by Article 680 are contained in Article 682.

Studies indicate that a person in a swimming pool can receive a severe electric shock by reaching out and touching the energized enclosure of an appliance, such as a radio, because the immersed person's body, which has a lower resistance to electric current, establishes a conductive path through the water to earth. Also, a person not in contact with an appliance or any grounded object can receive an electric shock and be rendered immobile by a potential gradient in the water itself. The level of electrical current necessary to cause immobilization might not cause electrocution, but it could lead to drowning, which is known as electric shock drowning (ESD). Shock hazards in and around a swimming pool can result from faulty electrical equipment directly associated with the pool or from faulty electrical equipment not associated with but in close proximity to the pool.

Accordingly, the requirements of Article 680 covering effective bonding and grounding, installation of receptacles and luminaires, use of GFCIs, modified wiring methods, and so forth, apply not only to the installation of the pool but also to installations and equipment adjacent to or associated with the pool.

To ensure the safety of persons using the bodies of water covered by Article 680, the risk of electric shock is minimized by the use of one or more of the following means:

1. GFCI protection and low-voltage equipment
2. Double-insulated equipment
3. Insulation and isolation
4. Equipotential bonding
5. Physical separation and restricted locations
6. Robust physical protection requirements for circuit conductors

680.4 Inspections After Installation. The authority having jurisdiction shall be permitted to require periodic inspection and testing.

New swimming pools that have been installed in accordance with the requirements of Article 680 provide protection against electrical shock in a potentially dangerous environment. New installations performed by competent contractors and inspected by qualified inspection personnel provide the necessary level of safety. However, once the project has been completed, there is rarely any follow-up inspection (some states require ongoing safety inspections of commercial and institutional swimming pools) to ensure that the safety features of the electrical system, such as bonding connections and GFCI protection, remain fully functional. This new requirement provides the AHJ(s) with the necessary *NEC®* text to support an ongoing follow-up compliance program if they choose to provide that service.

Δ **680.5 Ground-Fault Circuit-Interrupter (GFCI) and Special Purpose Ground-Fault Circuit-Interrupter (SPGFCI) Protection.**

N **(A) General.** The GFCI and SPGFCI requirements in this article, unless otherwise noted, are in addition to the requirements in 210.8.

N **(B) 150 Volts or Less to Ground.** Where required in this article, ground-fault protection of receptacles and outlets on branch circuits rated 150 volts or less to ground and 60 amperes or less, single- or 3-phase, shall be provided with a Class A GFCI.

Exception: Receptacles and outlets that are part of listed equipment with ratings not exceeding the low-voltage contact limit that are supplied by listed transformers or power supplies that comply with 680.23(A)(2) shall not be required to be provided with ground-fault protection.

Informational Note: The high leg of a 120/240-volt 4-wire delta-connected system, and the two ungrounded phases of a corner-grounded delta system have a voltage to ground greater than 150 volts, exceeding the limit for a Class A GFCI.

N **(C) Above 150 Volts to Ground.** Where required in this article, ground-fault protection of receptacles and outlets on branch circuits operating at voltages above 150 volts to ground, not exceeding 480 volts phase-to-phase, single- or 3-phase, shall be provided with SPGFCI protection not to exceed 20-mA ground-fault trip current.

Informational Note: See UL 943C, *Outline of Investigation for Special Purpose Ground-Fault Circuit Interrupters*, for information on Classes C, D, and E ground-fault circuit interrupters.

Δ **680.6 Listing Requirements.** All electrical equipment covered by this article shall be listed.

Δ **680.7 Grounding and Bonding.**

N **(A) Feeders and Branch Circuits.** Feeders and branch circuits installed in a corrosive environment or wet location shall contain an EGC that is an insulated copper conductor sized in accordance with Table 250.122, but not smaller than 12 AWG.

The equipment grounding conductor (EGC) installed with the feeder can be an aluminum conductor if the location in which it is installed is not a corrosive environment. Where a panelboard is supplied by a separately derived system, such as a transformer, the rules covering the EGC apply only to the feeder between the separately derived system (in the case of a transformer, the secondary conductors) and the panelboard, and not to any feeder conductors on the supply side of the separately derived source. The feeder is also required to be installed in a raceway where installed in corrosive environments in accordance with 680.14.

See Exhibit 680.1 for an illustration of applying the wiring method and grounding requirements to a feeder-supplied panelboard that provides branch circuits for swimming pool–related equipment. Also note the branch-circuit wiring methods shown in the illustration.

N **(B) Cord-and-Plug Connections.** The flexible cord shall contain an EGC that is an insulated copper conductor sized in accordance with Table 250.122, but not smaller than 12 AWG. The flexible cord shall terminate in a grounding-type attachment plug having a fixed grounding contact member.

N **(C) Terminals.** Terminals used for bonding and equipment grounding shall be identified for use in wet locations. Field-installed terminals in damp or wet locations or corrosive environments shall be composed of copper, copper alloy, or stainless steel and shall be listed for direct burial use.

680.8 Cord-and-Plug-Connected Equipment. Fixed or stationary equipment, other than underwater luminaires, for a permanently installed pool shall be permitted to be connected with a flexible cord and plug to facilitate the removal or disconnection for maintenance or repair.

(A) Length. For other than storable pools, the flexible cord shall not exceed 900 mm (3 ft) in length.

(B) Equipment Grounding. The flexible cord shall have a copper equipment grounding conductor sized in accordance with 250.122 but not smaller than 12 AWG. The cord shall terminate in a grounding-type attachment plug.

(C) Construction. The equipment grounding conductors shall be connected to a fixed metal part of the assembly. The removable part shall be mounted on or bonded to the fixed metal part.

In some climates, disconnecting and removing a permanent pool's filter pump for cold-weather months is necessary. A 3-foot cord is permitted to facilitate the removal of fixed or stationary equipment for maintenance and storage. Listed filter pumps for use with storable pools (covered in Part III) are considered portable and are permitted to be equipped with cords longer than 3 feet.

680.9 Overhead Conductor Clearances. Overhead conductors shall meet the clearance requirements in this section. Where a minimum clearance from the water level is given, the measurement shall be taken from the maximum water level of the specified body of water.

Δ **(A) Power.** Overhead conductors and open overhead wiring not in a raceway shall comply with the minimum clearances given in Table 680.9(A) and illustrated in Figure 680.9(A).

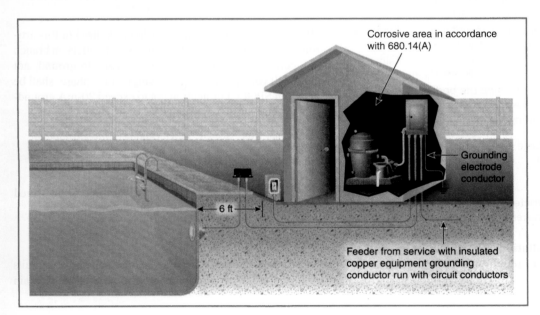

EXHIBIT 680.1 Wiring method and grounding requirements for a feeder panelboard supplying swimming pool equipment.

Corrosive area in accordance with 680.14(A)

Grounding electrode conductor

6 ft

Feeder from service with insulated copper equipment grounding conductor run with circuit conductors

TABLE 680.9(A) *Overhead Conductor Clearances*

Clearance Parameters	Insulated Cables, 0–750 Volts to Ground, Supported on and Cabled Together with a Solidly Grounded Bare Messenger or Solidly Grounded Neutral Conductor		All Other Conductors Voltage to Ground			
			0 through 15 kV		Over 15 through 50 kV	
	m	ft	m	ft	m	ft
A. Clearance in any direction to the water level, edge of water surface, base of diving platform, or permanently anchored raft	6.9	22.5	7.5	25	8.0	27
B. Clearance in any direction to the observation stand, tower, or diving platform	4.4	14.5	5.2	17	5.5	18
C. Horizontal limit of clearance measured from inside wall of the pool	This limit shall extend to the outer edge of the structures listed in A and B of this table but not less than 3 m (10 ft).					

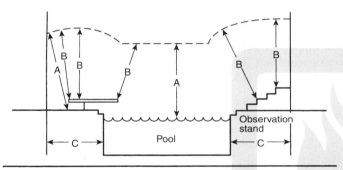

FIGURE 680.9(A) *Clearances from Pool Structures.*

Δ **(B) Communications Systems.** Communications, radio, and television coaxial cables within the scope of Chapter 8 shall be permitted at a height of not less than 3.0 m (10 ft) above the maximum water level of swimming and wading pools, and diving structures, observation stands, towers, or platforms.

(C) Network-Powered Broadband Communications Systems. The minimum clearances for overhead network-powered broadband communications systems conductors from pools or fountains shall comply with the provisions in Table 680.9(A) for conductors operating at 0 to 750 volts to ground.

These clearances consider factors such as the use of skimmers with aluminum handles. They are intended to provide sufficient separation between the conductors and the pool. The maximum water level (see definition in Article 100) of the body of water (i.e., pool, spa, hot tub, or other) is used to determine compliance with 680.9.

Δ **680.10 Electric Pool Water Heaters Incorporating Resistive Heating Elements and Electrically Powered Swimming Pool Heat Pumps and Chillers.**

N **(A) Electric Pool Water Heaters.** All electric pool water heaters incorporating resistive heating elements shall have the heating elements subdivided into loads not exceeding 48 amperes and

protected at not over 60 amperes. The ampacity of the branch-circuit conductors and the rating or setting of overcurrent protective devices shall be 125 percent of the total nameplate-rated load or greater.

N **(B) Electrically Powered Swimming Pool Heat Pumps and Chillers.** Electrically powered swimming pool heat pumps and chillers using the circulating water system and providing heating, cooling, or both, shall be listed and rated for their intended use. The ampacity of the branch-circuit conductors and the rating or setting of overcurrent protective devices shall be sized to comply with the nameplate.

680.11 Underground Wiring. Underground wiring shall comply with 680.11(A) and (B).

(A) Underground Wiring. Underground wiring within 1.5 m (5 ft) horizontally from the inside wall of the pool shall be permitted. The following wiring methods shall be considered suitable for the conditions in these locations provided they are installed complete between outlets, junctions, or splicing points:

(1) Rigid metal conduit
(2) Intermediate metal conduit
(3) Rigid polyvinyl chloride conduit
(4) Reinforced thermosetting resin conduit
(5) Jacketed Type MC cable that is listed for burial use
(6) Liquidtight flexible nonmetallic conduit listed for direct burial use
(7) Liquidtight flexible metal conduit listed for direct burial use

(B) Wiring Under Pools. Underground wiring shall not be permitted under the pool unless this wiring is necessary to supply pool equipment permitted by this article.

Δ **680.12 Equipment Rooms, Vaults, and Pits.**

N **(A) Drainage.** Electrical equipment shall not be installed in rooms, vaults, or pits that do not have drainage that prevents water accumulation during normal operation or maintenance unless the equipment is rated and identified for submersion.

> Informational Note: Chemicals such as chlorine cause severe corrosive and deteriorating effects on electrical connections, equipment, and enclosures when stored and kept in the same vicinity. Adequate ventilation of indoor spaces such as equipment and storage rooms is addressed by ANSI/APSP-11, *Standard for Water Quality in Public Pools and Spas*, and can reduce the likelihood of the accumulation of corrosive vapors.

N **(B) Receptacles.** At least one GFCI-protected 125-volt, 15- or 20-ampere receptacle supplied from a general purpose branch circuit shall be located within an equipment room. All other receptacles supplied by branch circuits rated 150 volts or less to ground within an equipment room and any receptacles supplied by a branch circuit rated 150 volts or less to ground in a vault or pit shall be GFCI protected.

680.13 Maintenance Disconnecting Means. One or more means to simultaneously disconnect all ungrounded conductors shall be provided for all utilization equipment other than lighting. Each means shall be readily accessible and within sight from its equipment and shall be located not less than 1.5 m (5 ft) horizontally from the inside walls of a pool, spa, fountain, or hot tub unless separated from the open water by a permanently installed barrier that provides a 1.5 m (5 ft) reach path or greater. This horizontal distance shall be measured from the water's edge along the shortest path required to reach the disconnect.

A readily accessible disconnecting means is required to be located within sight of pool, spa, fountain, and hot tub equipment. This provides service personnel with the ability to safely disconnect power for servicing equipment such as motors, heaters, and control panels. The intent of the 5-foot distance from the inside walls of the pool or separation by a permanent barrier is that the disconnecting means not be in reach of someone in the pool.

△ 680.14 Corrosive Environments.

N **(A) Wiring Methods.** Wiring methods shall be suitable for use in corrosive environments. Rigid metal conduit, intermediate metal conduit, rigid polyvinyl chloride conduit, reinforced thermosetting resin conduit, and liquidtight flexible nonmetallic conduit shall be considered suitable for use. Aluminum conduit and tubing shall not be permitted.

N **(B) Other Equipment.** Other equipment shall be suitable for use in corrosive environments or be installed in identified corrosion-resistant enclosures. Equipment listed for pool and spa use shall be considered suitable for use.

Part II. Permanently Installed Pools

680.20 General. Electrical installations at permanently installed pools shall comply with the provisions of Part I and Part II of this article.

680.21 Motors.

△ (A) Wiring Methods. The wiring to a pool motor shall comply with 680.21(A)(1) or (A)(2).

(1) Flexible Connections. Where necessary to employ flexible connections at or adjacent to the motor, liquidtight flexible metal, liquidtight flexible nonmetallic conduit, or MC cable suitable for the use shall be permitted.

△ (2) Cord-and-Plug Connections. Pool-associated motors shall be permitted to employ cord-and-plug connections. The flexible cord shall not exceed 900 mm (3 ft) in length.

(B) Double-Insulated Pool Pumps. A listed cord-and-plug-connected pool pump incorporating an approved system of double insulation that provides a means for grounding only the internal and nonaccessible, non-current-carrying metal parts of the pump shall be connected to any wiring method recognized in Chapter 3 that is suitable for the location. Where the equipment grounding conductor of the motor circuit is connected to the equipotential bonding means in accordance with the second sentence of 680.26(B)(6)(a), the branch-circuit wiring shall comply with 680.21(A).

The internal metal parts of a swimming pool pump incorporating a system of double insulation are grounded; however, they are not required to be incorporated into the bonding system required by 680.26(B), because bonding defeats the double insulation system.

(C) Ground-Fault Protection. Outlets serving pool motors shall have ground-fault protection complying with 680.5(B) or (C), as applicable.

Exception: Listed low-voltage motors not requiring grounding, with ratings not exceeding the low-voltage contact limit that are supplied by listed transformers or power supplies that comply with 680.23(A)(2), shall be permitted to be installed without ground-fault protection.

(D) Pool Pump Motor Replacement. Where a pool pump motor in 680.21(C) is replaced or repaired, the replacement or repaired pump motor shall be provided with ground-fault protection complying with 680.5(B) or (C), as applicable.

Even though the original installation might not have provided GFCI protection for the swimming pool pump motor, a new or rebuilt pump motor installed to replace a failed motor is required to have GFCI protection. This applies to all direct-connected (hard-wired) and cord-and-plug-connected swimming pool pump motors required to be provided with GFCI protection by 680.21(C).

680.22 Lighting, Receptacles, and Equipment.

(A) Receptacles.

The requirements of 680.22(A) apply to receptacles located near a permanently installed pool or near a fountain that has water

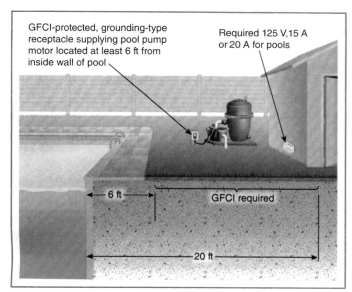

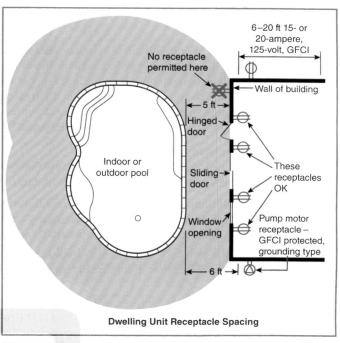

Dwelling Unit Receptacle Spacing

EXHIBIT 680.2 An example of a receptacle installed according to 680.22(A).

EXHIBIT 680.3 Acceptable receptacle locations within 20 feet of a permanently installed swimming pool.

common to a pool. They do not apply to direct-connected equipment.

> **See also**
>
> **680.21(C),** which covers direct-connected pool pump motors
>
> **680.50,** which covers pools with associated fountains

As required by 680.22(A)(1), a permanently installed pool is required to have at least one receptacle on a general-purpose branch circuit that is located at least 6 feet from the pool and not more than 20 feet from the pool. This allows ordinary appliances to be safely plugged in and used near the pool but avoids the need for extension cords in the pool vicinity. The 6-foot minimum dimension reduces the likelihood that an appliance with a 6-foot cord could be accidentally knocked into the pool.

Section 680.22(A)(4) applies to pools located outdoors or indoors, permanently installed, and for residential or commercial use. Because people within 20 feet of a pool are normally subjected to dampness and moisture, the GFCI requirement within the 20-foot space is warranted.

Examples of receptacles meeting the requirements of 680.22(A) are shown in Exhibits 680.2 and 680.3. Receptacles located in a structure are permitted to be less than 6 feet from the pool, because the structure is considered to provide a permanent barrier separating the receptacle from the pool. The one required receptacle precludes having to run the cord of an appliance used on the pool deck through a doorway.

(1) Required Receptacle, Location. Where a permanently installed pool is installed, no fewer than one 125-volt, 15- or 20-ampere receptacle on a general-purpose branch circuit shall be located not less than 1.83 m (6 ft) from, and not more than 6.0 m (20 ft) from, the inside wall of the pool. This receptacle shall be located not more than 2.0 m (6 ft 6 in.) above the floor, platform, or grade level serving the pool.

(2) Circulation and Sanitation System, Location. Receptacles that provide power for water-pump motors or for other loads directly related to the circulation and sanitation system shall be located at least 1.83 m (6 ft) from the inside walls of the pool. These receptacles shall have GFCI protection and be of the grounding type.

(3) Other Receptacles, Location. Other receptacles shall be not less than 1.83 m (6 ft) from the inside walls of a pool.

(4) Ground-Fault Circuit-Interrupter (GFCI) and Special Purpose Ground-Fault Circuit-Interrupter (SPGFCI) Protection. All receptacles rated 125 volts through 250 volts, 60 amperes or less, located within 6.0 m (20 ft) of the inside walls of a pool shall have GFCI protection complying with 680.5(B) or SPGFCI protection complying with 680.5(C), as applicable.

(5) Measurements. In determining the dimensions in this section addressing receptacle spacings, the distance to be measured shall be the shortest path the supply cord of an appliance connected to the receptacle would follow without piercing a floor, wall, ceiling, doorway with hinged or sliding door, window opening, or other effective permanent barrier.

(B) Luminaires, Lighting Outlets, and Ceiling-Suspended (Paddle) Fans.

(1) New Outdoor Installation Clearances. In outdoor pool areas, luminaires, lighting outlets, and ceiling-suspended (paddle) fans installed above the pool or the area extending 1.5 m (5 ft) horizontally from the inside walls of the pool shall be installed at a height not less than 3.7 m (12 ft) above the maximum water level of the pool.

(2) Indoor Clearances. For installations in indoor pool areas, the clearances shall be the same as for outdoor areas unless modified as provided in this paragraph. If the branch circuit supplying the equipment is protected by a ground-fault circuit interrupter, the following equipment shall be permitted at a height not less than 2.3 m (7 ft 6 in.) above the maximum pool water level:

(1) Totally enclosed luminaires
(2) Ceiling-suspended (paddle) fans identified for use beneath ceiling structures such as provided on porches or patios

(3) Existing Installations. Existing luminaires and lighting outlets located less than 1.5 m (5 ft) measured horizontally from the inside walls of a pool shall be not less than 1.5 m (5 ft) above the surface of the maximum water level, shall be rigidly attached to the existing structure, and shall be protected by a ground-fault circuit interrupter.

(4) Ground-Fault Circuit-Interrupter (GFCI) and Special Purpose Ground-Fault Circuit-Interrupter (SPGFCI) Protection in Adjacent Areas. Luminaires, lighting outlets, and ceiling-suspended (paddle) fans installed in the area extending between 1.5 m (5 ft) and 3.0 m (10 ft) horizontally from the inside walls of a pool shall have GFCI protection complying with 680.5(B) or SPGFCI protection complying with 680.5(C), as applicable, unless installed not less than 1.5 m (5 ft) above the maximum water level and rigidly attached to the structure adjacent to or enclosing the pool.

(5) Cord-and-Plug-Connected Luminaires. Cord-and-plug-connected luminaires shall comply with the requirements of 680.8 where installed within 4.9 m (16 ft) of any point on the water surface, measured radially.

Exhibit 680.4 clarifies the limitations of 680.22(B) for the areas surrounding outdoor and indoor pools.

(6) Low-Voltage Luminaires. Listed low-voltage luminaires not requiring grounding, not exceeding the low-voltage contact limit, and supplied by listed transformers or power supplies that comply with 680.23(A)(2) shall be permitted to be located less than 1.5 m (5 ft) from the inside walls of the pool.

(7) Low-Voltage Gas-Fired Luminaires, Decorative Fireplaces, Fire Pits, and Similar Equipment. Listed low-voltage gas-fired-luminaires, decorative fireplaces, fire pits, and similar equipment using low-voltage ignitors that do not require grounding, and are supplied by listed transformers or power supplies that comply with 680.23(A)(2) with outputs that do not exceed the low-voltage contact limit shall be permitted to be located less than 1.5 m (5 ft) from the inside walls of the pool. Metallic equipment shall be bonded in accordance with the requirements in 680.26(B). Transformers or power supplies supplying this type of equipment shall be installed in accordance with the requirements in 680.24. Metallic gas piping shall be bonded in accordance with the requirements in 250.104(B) and 680.26(B)(7).

(8) Measurements. In determining the dimensions in this section addressing luminaires, the distance to be measured shall be the shortest path an imaginary cord connected to the luminaire would follow without piercing a floor, wall, ceiling, doorway with hinged or sliding door, window opening, or other effective permanent barrier.

(C) Switching Devices. Switching devices shall be located at least 1.5 m (5 ft) horizontally from the inside walls of a pool unless separated from the pool by a solid fence, wall, or other permanent barrier that provides at least a 1.5 m (5 ft) reach distance. Alternatively, a switch that is listed as being acceptable for use within 1.5 m (5 ft) shall be permitted.

(D) Other Outlets. Other outlets shall be not less than 3.0 m (10 ft) from the inside walls of the pool. Measurements shall be determined in accordance with 680.22(A)(5).

> Informational Note: Other outlets may include, but are not limited to, remote-control, signaling, fire alarm, and communications circuits.

(E) Other Equipment. Other equipment with ratings exceeding the low-voltage contact limit shall be located at least 1.5 m (5 ft) horizontally from the inside walls of a pool unless separated from the pool by a solid fence, wall, or other permanent barrier.

680.23 Underwater Luminaires. This section covers all luminaires installed below the maximum water level of the pool.

(A) General.

(1) Luminaire Design, Normal Operation. The design of an underwater luminaire supplied from a branch circuit either directly or by way of a transformer or power supply meeting the requirements of this section shall be such that, where the luminaire is properly installed without a GFCI, there is no shock hazard with any likely combination of fault conditions during normal use (not relamping).

(2) Transformers and Power Supplies. Transformers and power supplies used for the supply of underwater luminaires, together with the transformer or power supply enclosure, shall be listed, labeled, and identified for swimming pool and spa use. The transformer or power supply shall incorporate either a transformer of the isolated winding type, with an ungrounded secondary that has a grounded metal barrier between the primary and secondary windings, or one that incorporates an approved system of double insulation between the primary and secondary windings.

See also

110.3(B) for requirements on the installation and use of listed electrical equipment

(3) GFCI Protection, Lamping, Relamping, and Servicing. GFCI protection shall be installed in the branch circuit supplying luminaires operating at voltages greater than the low-voltage contact limit.

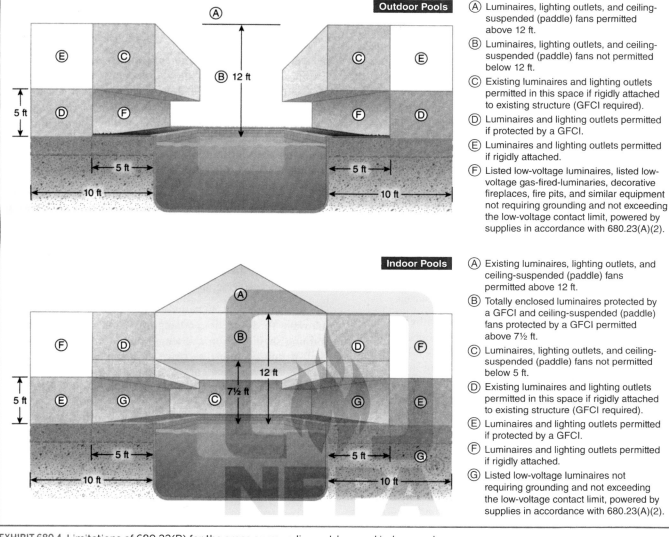

Outdoor Pools

Ⓐ Luminaires, lighting outlets, and ceiling-suspended (paddle) fans permitted above 12 ft.

Ⓑ Luminaires, lighting outlets, and ceiling-suspended (paddle) fans not permitted below 12 ft.

Ⓒ Existing luminaires and lighting outlets permitted in this space if rigidly attached to existing structure (GFCI required).

Ⓓ Luminaires and lighting outlets permitted if protected by a GFCI.

Ⓔ Luminaires and lighting outlets permitted if rigidly attached.

Ⓕ Listed low-voltage luminaires, listed low-voltage gas-fired-luminaires, decorative fireplaces, fire pits, and similar equipment not requiring grounding and not exceeding the low-voltage contact limit, powered by supplies in accordance with 680.23(A)(2).

Indoor Pools

Ⓐ Existing luminaires, lighting outlets, and ceiling-suspended (paddle) fans permitted above 12 ft.

Ⓑ Totally enclosed luminaires protected by a GFCI and ceiling-suspended (paddle) fans protected by a GFCI permitted above 7½ ft.

Ⓒ Luminaires, lighting outlets, and ceiling-suspended (paddle) fans not permitted below 5 ft.

Ⓓ Existing luminaires and lighting outlets permitted in this space if rigidly attached to existing structure (GFCI required).

Ⓔ Luminaires and lighting outlets permitted if protected by a GFCI.

Ⓕ Luminaires and lighting outlets permitted if rigidly attached.

Ⓖ Listed low-voltage luminaires not requiring grounding and not exceeding the low-voltage contact limit, powered by supplies in accordance with 680.23(A)(2).

EXHIBIT 680.4 Limitations of 680.22(B) for the areas surrounding outdoor and indoor pools.

Dry-niche, no-niche, or wet-niche underwater luminaires operating at a voltage exceeding the low-voltage contact limit require GFCI protection.

See also

Article 100 for the definition of the term *low-voltage contact limit* and its associated commentary

(4) Voltage Limitation. No luminaires shall be installed for operation on supply circuits over 150 volts between conductors.

(5) Location, Wall-Mounted Luminaires. Luminaires mounted in walls shall be installed with the top of the luminaire lens not less than 450 mm (18 in.) below the normal water level of the pool, unless the luminaire is listed and identified for use at lesser depths. No luminaire shall be installed less than 100 mm (4 in.) below the normal water level of the pool.

(6) Bottom-Mounted Luminaires. A luminaire facing upward shall comply with either (1) or (2):

(1) Have the lens guarded to prevent contact by any person
(2) Be listed for use without a guard

(7) Dependence on Submersion. Luminaires that depend on submersion for safe operation shall be inherently protected against the hazards of overheating when not submerged.

(8) Compliance. Compliance with these requirements shall be obtained by the use of a listed underwater luminaire and by installation of a listed GFCI in the branch circuit or a listed transformer or power supply for luminaires operating at not more than the low-voltage contact limit.

(B) Wet-Niche Luminaires.

(1) Forming Shells. Forming shells shall be installed for the mounting of all wet-niche underwater luminaires and shall be equipped with provisions for conduit entries. Metal parts of the luminaire and forming shell in contact with the pool water shall be of brass or other approved corrosion-resistant metal. All forming shells used with nonmetallic conduit systems, other than those that are part of a listed low-voltage lighting system not requiring grounding, shall include provisions for terminating an 8 AWG copper conductor.

(2) Wiring Extending Directly to the Forming Shell. Conduit shall be installed from the forming shell to a junction box or other enclosure conforming to the requirements in 680.24. Conduit shall be rigid metal, intermediate metal, liquidtight flexible nonmetallic, or rigid polyvinyl chloride conduit.

(a) *Metal Conduit.* Metal conduit shall be listed and shall be red brass or stainless steel.

Informational Note: See UL 6A, *Electrical Rigid Metal Conduit - Aluminum, Red Brass, and Stainless Steel*, for information on the listing criteria for red brass and stainless steel conduit.

(b) *Nonmetallic Conduit.* Where a nonmetallic conduit is used, an 8 AWG insulated solid or stranded copper bonding jumper shall be installed in this conduit unless a listed low-voltage lighting system not requiring grounding is used. The bonding jumper shall be terminated in the forming shell, junction box or transformer enclosure, or ground-fault circuit-interrupter enclosure. The termination of the 8 AWG bonding jumper in the forming shell shall be covered with, or encapsulated in, a listed potting compound to protect the connection from the possible deteriorating effect of pool water.

An 8 AWG insulated copper bonding jumper is required to be installed in the conduit to provide electrical continuity between the forming shell and the junction box or other enclosure. This bonding conductor is in addition to the EGC required by 680.23(F)(2).

The function of this conductor is twofold: (1) It permanently bonds all non–current-carrying metal surfaces of the forming shell to any non–current-carrying parts of the deck box and to the EGC of the circuit that supplies the wet-niche luminaire, and (2) it serves as the path for ground-fault current in the event of a ground fault when the wet-niche luminaire is removed from the forming shell, which is typically done during relamping. Damage to the wet-niche luminaire supply cord could result in such a ground-fault scenario.

Low-voltage lighting systems that are listed for installation without an EGC or a bonding conductor are exempt from this requirement.

(3) Equipment Grounding Provisions for Cords. Other than listed low-voltage lighting systems not requiring grounding, wet-niche luminaires that are supplied by a flexible cord or cable shall have all exposed non–current-carrying metal parts connected to an insulated copper equipment grounding conductor that is an integral part of the cord or cable. This equipment grounding conductor shall be connected to a grounding terminal in the supply junction box, transformer enclosure, or other enclosure. The equipment grounding conductor shall not be smaller than the supply conductors and not smaller than 16 AWG.

(4) Luminaire Grounding Terminations. The end of the flexible-cord jacket and the flexible-cord conductor terminations within a luminaire shall be covered with, or encapsulated in, a suitable potting compound to prevent the entry of water into the luminaire through the cord or its conductors. If present, the connection of the equipment grounding conductor within a luminaire shall be similarly treated to protect such connection from the deteriorating effect of pool water in the event of water entry into the luminaire.

(5) Luminaire Bonding. The luminaire shall be bonded to, and secured to, the forming shell by a positive locking device that ensures a low-resistance contact and requires a tool to remove the luminaire from the forming shell. Bonding shall not be required for luminaires that are listed for the application and have no non–current-carrying metal parts.

(6) Servicing. Wet-niche luminaires shall be removable from the water for inspection, relamping, or other maintenance. The forming shell location and length of cord in the forming shell shall permit personnel to place the removed luminaire on the deck or other dry location for such maintenance. The luminaire maintenance location shall be accessible without entering or going in the pool water.

In spa locations where wet-niche luminaires are installed low in the foot well of the spa, the luminaire shall only be required to reach the bench location, where the spa can be drained to make the bench location dry.

(C) Dry-Niche Luminaires.

(1) Construction. A dry-niche luminaire shall have provision for drainage of water. Other than listed low-voltage luminaires not requiring grounding, a dry-niche luminaire shall have means for accommodating one equipment grounding conductor for each conduit entry.

(2) Junction Box. A junction box shall not be required but, if used, shall not be required to be elevated or located as specified in 680.24(A)(2) if the luminaire is specifically identified for the purpose.

(D) No-Niche Luminaires. A no-niche luminaire shall meet the construction requirements of 680.23(B)(3) and be installed in accordance with 680.23(B). Where connection to a forming shell is specified, the connection shall be to the mounting bracket.

(E) Through-Wall Lighting Assembly. A through-wall lighting assembly shall be equipped with a threaded entry or hub, or a nonmetallic hub, for the purpose of accommodating the termination of the supply conduit. A through-wall lighting assembly shall meet the construction requirements of 680.23(B)(3) and be installed in accordance with 680.23. Where connection to a forming shell is specified, the connection shall be to the conduit termination point.

(F) Branch-Circuit Wiring.

Δ **(1) Wiring Methods.** Where branch-circuit wiring, on the supply side of enclosures and junction boxes connected to conduits run to underwater luminaires, is installed in corrosive environments, the wiring method of that portion of the branch circuit shall be in accordance with 680.14.

Exception: Where connecting to transformers or power supplies for pool lights, liquidtight flexible metal conduit shall be permitted. The length shall not exceed 1.8 m (6 ft) for any one length or exceed 3.0 m (10 ft) in total length used.

Δ **(2) Equipment Grounding.** Other than listed low-voltage luminaires not requiring grounding, all through-wall lighting assemblies, wet-niche, dry-niche, or no-niche luminaires shall be connected to an insulated copper equipment grounding conductor installed with the circuit conductors. The equipment grounding conductor shall be installed without joint or splice except as permitted in 680.23(F)(2)(a) and (F)(2)(b). The equipment grounding conductor shall be sized in accordance with 250.122 but shall not be smaller than 12 AWG.

Exception: An equipment grounding conductor between the wiring chamber of the secondary winding of a transformer and a junction box shall be sized in accordance with the transformer secondary overcurrent protection provided.

(a) If more than one underwater luminaire is supplied by the same branch circuit, the equipment grounding conductor, installed between the junction boxes, transformer enclosures, or other enclosures in the supply circuit to wet-niche luminaires, or between the field-wiring compartments of dry-niche luminaires, shall be permitted to be terminated on grounding terminals.

(b) If the underwater luminaire is supplied from a transformer, ground-fault circuit interrupter, clock-operated switch, or a manual snap switch that is located between the panelboard and a junction box connected to the conduit that extends directly to the underwater luminaire, the equipment grounding conductor shall be permitted to terminate on grounding terminals on the transformer, ground-fault circuit interrupter, clock-operated switch enclosure, or an outlet box used to enclose a snap switch.

In addition to the bonding jumper and EGC of the cord or cable contained in the nonmetallic conduit between the forming shell and the deck box, the wiring method from the deck box to the power source is required to contain a separate EGC regardless of the type of conduit installed. This EGC must be insulated, copper, and not smaller than 12 AWG. The grounding terminals within the deck (junction) box are used to terminate and bond together all the equipment grounding and bonding conductors associated with the circuit supplying the underwater luminaire(s).

Exhibit 680.5 illustrates an installation of a forming shell for a wet-niche luminaire and a flush junction (deck) box installed according to 680.24(A)(2). (A surface deck box is also shown in Exhibit 680.5.)

See also

680.24(D) for commentary on listed pool junction boxes

(3) Conductors. Conductors on the load side of a GFCI or of a transformer, used to comply with the provisions of 680.23(A)(8), shall not occupy raceways, boxes, or enclosures containing other conductors unless one of the following conditions applies:

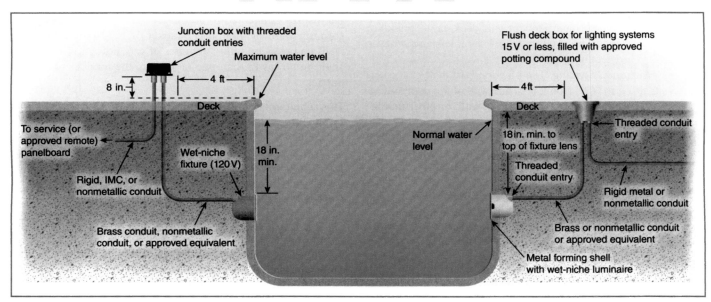

EXHIBIT 680.5 A flush junction (deck) box and a forming shell for a wet-niche luminaire.

(1) The other conductors are GFCI protected.
(2) The other conductors are equipment grounding conductors and bonding jumpers as required per 680.23(B)(2)(b).
(3) The other conductors are supply conductors to a feed-through-type GFCI.
(4) GFCIs shall be permitted in a panelboard that contains circuits protected by other than ground-fault circuit interrupters.

680.24 Junction Boxes and Electrical Enclosures for Transformers or Ground-Fault Circuit Interrupters.

(A) Junction Boxes. A junction box connected to a conduit that extends directly to a forming shell or mounting bracket of a no-niche luminaire shall meet the requirements of this section.

(1) Construction. The junction box shall be listed, labeled, and identified as a swimming pool junction box and shall comply with the following conditions:

(1) Be equipped with threaded entries or hubs or a nonmetallic hub
(2) Be comprised of copper, brass, suitable plastic, or other approved corrosion-resistant material
(3) Be provided with electrical continuity between every connected metal conduit and the grounding terminals by means of copper, brass, or other approved corrosion-resistant metal that is integral with the box

Δ **(2) Installation.** Where the luminaire operates over the low-voltage contact limit, the junction box location shall comply with 680.24(A)(2)(a) and (A)(2)(b). Where the luminaire operates at the low-voltage contact limit or less, the junction box location shall be permitted to comply with 680.24(A)(2)(c).

(a) *Vertical Spacing.* The junction box shall be located not less than 100 mm (4 in.), measured from the inside of the bottom of the box, above the ground level, or pool deck, or not less than 200 mm (8 in.) above the maximum pool water level, whichever provides the greater elevation.

(b) *Horizontal Spacing.* The junction box shall be located not less than 1.2 m (4 ft) from the inside wall of the pool, unless separated from the pool by a solid fence, wall, or other permanent barrier.

(c) *Flush Deck Box.* If used on a lighting system operating at the low-voltage contact limit or less, a flush deck box shall be permitted if both of the following conditions are met:

(1) Potting compound is used to fill the box to prevent the entrance of moisture.
(2) The flush deck box is located not less than 1.2 m (4 ft) from the inside wall of the pool.

(B) Other Enclosures. An enclosure for a transformer, ground-fault circuit interrupter, or a similar device connected to a conduit that extends directly to a forming shell or mounting bracket of a no-niche luminaire shall meet the requirements of this section.

(1) Construction. The enclosure shall be listed and labeled for the purpose and meet the following requirements:

(1) Equipped with threaded entries or hubs or a nonmetallic hub
(2) Comprised of copper, brass, suitable plastic, or other approved corrosion-resistant material
(3) Provided with sealing compound identified for use with cable insulation, conductor insulation, bare conductor, shield, or other components, sat the conduit connection, that prevents circulation of air between the conduit and the enclosures
(4) Provided with electrical continuity between every connected metal conduit and the grounding terminals by means of copper, brass, or other approved corrosion-resistant metal that is integral with the box

(2) Installation.

(a) *Vertical Spacing.* The enclosure shall be located not less than 100 mm (4 in.), measured from the inside of the bottom of the box, above the ground level, or pool deck, or not less than 200 mm (8 in.) above the maximum pool water level, whichever provides the greater elevation.

(b) *Horizontal Spacing.* The enclosure shall be located not less than 1.2 m (4 ft) from the inside wall of the pool, unless separated from the pool by a solid fence, wall, or other permanent barrier.

(C) Protection. Junction boxes and enclosures mounted above the grade of the finished walkway around the pool shall not be located in the walkway unless afforded additional protection, such as by location under diving boards, adjacent to fixed structures, and the like.

(D) Grounding Terminals. Junction boxes, transformer and power-supply enclosures, and ground-fault circuit-interrupter enclosures connected to a conduit that extends directly to a forming shell or mounting bracket of a no-niche luminaire shall be provided with a number of grounding terminals that shall be no fewer than one more than the number of conduit entries.

This requirement ensures the availability of integral grounding terminals necessary for the grounding and bonding of underwater luminaires. A box that is not specifically listed for use with swimming pools might not provide the correct number of integral grounding and bonding terminals. The number of grounding terminals in a box or enclosure is required to be one more than the number of conduit entries for which the box is designed.

(E) Strain Relief. The termination of a flexible cord of an underwater luminaire within a junction box, transformer or power-supply enclosure, ground-fault circuit interrupter, or other enclosure shall be provided with a strain relief.

(F) Grounding. The grounding terminals of a junction box, transformer enclosure, or other enclosure in the supply circuit to a wet-niche or no-niche luminaire and the field-wiring chamber of a dry-niche luminaire shall be connected to the equipment grounding terminal of the panelboard. This terminal shall be directly connected to the panelboard enclosure.

680.26 Equipotential Bonding.

Δ **(A) Performance.** The equipotential bonding required by 680.26(B) and (C) to reduce voltage gradients in the pool area shall be installed for pools with or without associated electrical equipment related to the pool.

The function of equipotential bonding differs from the primary function of bonding to meet the requirements of Article 250. Providing a path for ground-fault current is not the function of the equipotential bonding grid and associated bonding conductors. The only function of the 8 AWG solid copper conductor required by 680.26(B) is equipotential bonding to eliminate the voltage gradient in the pool area. The bonding conductor is not required to extend or connect to any parts or equipment other than those covered in 680.26(B)(1) through (B)(7) and to a pool water bonding element covered in 680.26(C).

Creating an electrically safe environment in and around permanently installed swimming pools requires the installation of a bonding system to establish equal electrical potential (voltage) in the vicinity of the swimming pool. A person who is immersed in a pool or who is lying on or walking on a conductive perimeter surface is susceptible to differences in electrical potential that may be present in the pool area. Bonding reduces possible injurious or disabling shock hazards created by stray currents in the ground or piping connected to the swimming pool. See Exhibit 680.6.

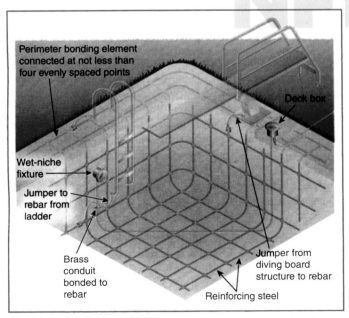

Perimeter bonding element connected at not less than four evenly spaced points

Deck box

Wet-niche fixture

Jumper to rebar from ladder

Brass conduit bonded to rebar

Jumper from diving board structure to rebar

Reinforcing steel

EXHIBIT 680.6 Bonding of conductive metal equipment and parts associated with a swimming pool.

(B) Bonded Parts. The parts specified in 680.26(B)(1) through (B)(7) shall be bonded together using solid copper conductors, insulated, covered, or bare, not smaller than 8 AWG or with rigid metal conduit of brass or other identified corrosion-resistant metal. Connections to bonded parts shall be made in accordance with 250.8. An 8 AWG or larger solid copper bonding conductor provided to reduce voltage gradients in the pool area shall not be required to be extended or attached to remote panelboards, service equipment, or electrodes.

(1) Conductive Pool Shells. Bonding to conductive pool shells shall be provided as specified in 680.26(B)(1)(a) or (B)(1)(b). Cast-in-place concrete, pneumatically applied or sprayed concrete, and concrete block with painted or plastered coatings shall all be considered conductive materials due to water permeability and porosity. Reconstructed pool shells shall also meet the requirements of this section. Vinyl liners and fiberglass composite shells shall be considered to be nonconductive materials and not subject to these requirements.

(a) *Structural Reinforcing Steel.* Unencapsulated structural reinforcing steel shall be bonded together by steel tie wires or the equivalent. Where structural reinforcing steel is encapsulated in a nonconductive compound, a copper conductor grid shall be installed in accordance with 680.26(B)(1)(b).

Encapsulated reinforcing steel does not provide the conductivity necessary to establish the required common bonding grid around the contour of a conductive pool shell. Therefore, a bonding connection to the encapsulated reinforcing steel, such as epoxy-coated rebar, is not required. However, a copper bonding grid around the contour of a conductive pool shell must be provided and constructed as prescribed in 680.26(B)(1)(b).

In Exhibits 680.6 and 680.7, structural reinforcing steel serves as a common point to which all non–current-carrying metal parts in the pool area are connected. This connection method is one way of satisfying the requirement to bond all metal parts together. Individual pieces of hardware such as the hooks used to attach safety or lane ropes that are less than 4 inches in any dimension and do not penetrate into the pool structure more than 1 inch are not required to be bonded per 680.26(B)(5).

(b) *Copper Conductor Grid.* A copper conductor grid shall be provided and shall comply with the following:

(1) Be constructed of minimum 8 AWG bare solid copper conductors bonded to each other at all points of crossing in accordance with 250.8 or other approved means
(2) Conform to the contour of the pool
(3) Be arranged in a 300 mm (12 in.) by 300 mm (12 in.) network of conductors in a uniformly spaced perpendicular grid pattern with a tolerance of 100 mm (4 in.)
(4) Be secured within or under the pool no more than 150 mm (6 in.) from the outer contour of the pool shell

Δ **(2) Perimeter Surfaces.** The perimeter surface to be bonded shall be considered to extend for 1 m (3 ft) horizontally beyond

EXHIBIT 680.7 Structural reinforcing steel for a poured-concrete pool that serves as the pool shell bonding grid.

the inside walls of the pool and shall include unpaved surfaces and other types of paving. Perimeter surfaces separated from the pool by a permanent wall or building 1.5 m (5 ft) in height or more shall require equipotential bonding only on the pool side of the permanent wall or building. Bonding to perimeter surfaces shall be provided as specified in 680.26(B)(2)(a), (B)(2)(b), or (B)(2)(c) and shall be attached to the pool reinforcing steel or copper conductor grid at a minimum of four points uniformly spaced around the perimeter of the pool. For nonconductive pool shells, bonding at four points shall not be required.

The requirement for bonding perimeter surfaces applies to paved and unpaved surfaces, such as a lawn surrounding a permanently installed aboveground swimming pool. Where the paved portion of the perimeter surface extends less than 3 feet horizontally from the inside walls of the pool, the perimeter bonding grid must be continued under the adjacent unpaved perimeter surface. If walls or other physical barriers prevent the perimeter from extending 3 feet beyond the inside walls of the pool, the bonding grid is required only to extend under the available perimeter area.

The perimeter bonding grid can comprise structural reinforcing metal (rebar or welded wire mesh) that is conductive to the perimeter surface and installed in or under the perimeter surface. Where structural reinforcing steel is not available, a single, bare, solid 8 AWG or larger copper conductor can be installed around the pool's perimeter in an area measuring between 18 inches and 24 inches from the inside pool walls. The 8 AWG bonding conductor can be installed in the paving material (i.e., in the concrete), or it can be buried in the material (subgrade) below the paving material. Where buried, the bonding conductor must not be less than 4 inches and not

more than 6 inches below the surface level of the subgrade material.

If structural reinforcing steel is not available, another choice for bonding the perimeter surfaces is through the use of a grid comprising bare 8 AWG solid copper conductors installed to follow the contour of the perimeter surface. The copper conductor grid pattern is arranged using the same space and tolerances as required for bonding grids installed for conductive pool shells in accordance with 680.26(B)(1)(b).

This requirement does not apply to decks constructed of nonconductive materials such as wood, plastic, or fiberglass. However, the perimeter surface under an elevated deck around a pool installed on or partially in the ground requires bonding.

The required perimeter surface bond must be connected at four evenly spaced points around the pool perimeter to the conductive pool shell. Connection between the perimeter surface bond and nonconductive pool shells is not required.

(a) *Structural Reinforcing Steel.* Structural reinforcing steel shall be bonded in accordance with 680.26(B)(1)(a).

(b) *Copper Ring.* Where structural reinforcing steel is not available or is encapsulated in a nonconductive compound, a copper conductor(s) shall be used where the following requirements are met:

(1) At least one minimum 8 AWG bare solid copper conductor shall be provided.
(2) The conductors shall follow the contour of the perimeter surface.
(3) Only listed splicing devices or exothermic welding shall be permitted.
(4) The required conductor shall be 450 mm to 600 mm (18 in. to 24 in.) from the inside walls of the pool.

(5) The required conductor shall be secured within or under the perimeter surface 100 mm to 150 mm (4 in. to 6 in.) below the subgrade.

(c) *Copper Grid*. Where structural reinforcing steel is not available or is encapsulated in a nonconductive compound, copper grid shall be used where the following requirements are met:

(1) The copper grid shall be constructed of 8 AWG solid bare copper and be arranged in accordance with 680.26(B)(1)(b)(3).

(2) The copper grid shall follow the contour of the perimeter surface extending 1 m (3 ft) horizontally beyond the inside walls of the pool.

(3) Only listed splicing devices or exothermic welding shall be permitted.

(4) The copper grid shall be secured within or under the deck or unpaved surfaces between 100 mm to 150 mm (4 in. to 6 in.) below the subgrade.

(3) Metallic Components. All metallic parts of the pool structure, including reinforcing metal not addressed in 680.26(B)(1)(a), shall be bonded. Where reinforcing steel is encapsulated with a nonconductive compound, the reinforcing steel shall not be required to be bonded.

(4) Underwater Lighting. All metal forming shells and mounting brackets of no-niche luminaires shall be bonded.

Exception: Listed low-voltage lighting systems with nonmetallic forming shells shall not require bonding.

Δ **(5) Metal Fittings.** All metal fittings within or attached to the pool structure shall be bonded.

Exception: The following shall not be required to be bonded:

(1) Isolated parts that are not over 100 mm (4 in.) in any dimension and do not penetrate into the pool structure more than 25 mm (1 in.)

(2) Metallic pool cover anchors intended for insertion in a concrete or masonry deck surface, 25 mm (1 in.) or less in any dimension and 51 mm (2 in.) or less in length

(3) Metallic pool cover anchors intended for insertion in a wood or composite deck surface, 51 mm (2 in.) or less in any flange dimension and 51 mm (2 in.) or less in length

(6) Electrical Equipment. Metal parts of the following electrical equipment shall be bonded:

(1) Electrically powered pool cover(s)

(2) Pool water circulation, treatment, heating, cooling, or dehumidification equipment

(3) Unless separated from the pool by a permanent barrier that prevents contact by a person, any other electrical equipment within 1.5 m (5 ft) measured horizontally from the inside wall of the pool, or 3.7 m (12 ft) measured vertically above the maximum water level of the pool, or as measured vertically above any observation stands, towers, or platforms, or any diving structures

Exception: Metal parts of listed equipment incorporating an approved system of double insulation shall not be bonded.

(a) *Double-Insulated Water Pump Motors.* Where a double-insulated water pump motor is installed under the provisions of this rule, a solid 8 AWG copper conductor of sufficient length to make a bonding connection to a replacement motor shall be extended from the swimming pool equipotential bonding means to an accessible point in the vicinity of the pool pump motor. Where there is no connection between the swimming pool equipotential bonding means and the equipment grounding system for the premises, this bonding conductor shall be connected to the equipment grounding conductor of the motor circuit.

(b) *Pool Water Heaters.* For pool water heaters rated at more than 50 amperes and having specific instructions regarding bonding and grounding, only those parts designated to be bonded shall be bonded and only those parts designated to be grounded shall be grounded.

Δ **(7) Fixed Metal Parts.** All fixed metal parts, including, but not limited to, metal-sheathed cables and raceways, metal piping, metal awnings, metal fences, and metal door and window frames, shall be bonded where located no greater than either of the following:

(1) 1.5 m (5 ft) horizontally from the inside walls of the pool

(2) 3.7 m (12 ft) vertically above the maximum water level of the pool, observation stands, towers, or platforms, or any diving structures

Exception: Those separated from the pool by a permanent barrier that prevents contact by a person shall not be required to be bonded.

The metal parts required to be bonded include all metal parts of electrical equipment associated with the water-circulating system of the pool, all metal parts of the pool structure, and all fixed metal parts within 5 feet of the inside walls of the pool and not separated by a permanent barrier. The bonding of these parts can be accomplished by one or more of the following methods using a solid 8 AWG or larger copper conductor that is insulated, covered, or bare:

1. Connecting the parts directly to each other in series or parallel configurations

2. Connecting the parts to the unencapsulated structural metal forming the shell of a conductive pool or connecting the parts to a copper conductor grid system used around the contour of a conductive pool shell

3. Connecting the parts using the pool shell constructed of bolted or welded steel as a common connection point

4. Connecting the parts to the perimeter bonding grid consisting of either structural reinforcing steel (rebar or welded wire mesh) or a solid 8 AWG bare copper conductor encircling the pool's perimeter

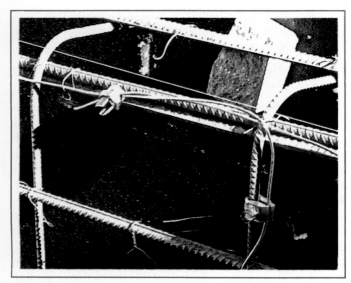

EXHIBIT 680.8 Use of listed connectors for making bonding connections at a swimming pool.

Brass or other corrosion-resistant rigid metal conduit (RMC) can also be used as a bonding conductor for connecting metal parts together. See Exhibit 680.6 for an example of using brass RMC as the method of bonding two electrical enclosures and as a point to connect bonding jumpers run to the pool reinforcing steel and stainless steel ladder.

As specified in 250.8(A), exothermic welding, listed pressure connectors and clamps, and other listed means are permitted as the method of connecting bonding conductors to swimming pool equipment. Connections in pool areas must be suitable for wet conditions and exposure to pool chemicals. Pool chemicals in swimming pool water can make the vicinity of the swimming pool area a corrosive environment. The integrity of the bonding connections should be periodically inspected, particularly those bonding connections between the solid 8 AWG copper conductor and an aluminum (or other dissimilar metal) ladder. See Exhibit 680.8 for an illustration of two acceptable methods of making swimming pool bonding connections.

(C) Pool Water. Where none of the bonded parts as specified in 680.26(B)(1) through (B)(7) are in direct connection with the pool water, the pool water shall be in direct contact with an approved corrosion-resistant conductive surface that exposes not less than 5800 mm^2 (9 in.^{2}) of surface area to the pool water at all times. The conductive surface shall be located where it is not exposed to physical damage or dislodgement during usual pool activities, and it shall be bonded in accordance with 680.26(B).

Where bonded items such as ladders, rails, or underwater luminaires are in direct contact with the pool water and provide the required surface area, it is not necessary to provide another conductive element. A conductive pool shell in contact with the water also satisfies this requirement. However, where the pool does not include any of those items, it is necessary to install a conductive element. Devices have been specifically listed as a means to provide this contact with the pool water.

680.27 Specialized Pool Equipment.

(A) Underwater Audio Equipment. All underwater audio equipment shall be identified as underwater audio equipment.

(1) Speakers. Each speaker shall be mounted in an approved metal forming shell, the front of which is enclosed by a captive metal screen, or equivalent, that is bonded to, and secured to, the forming shell by a positive locking device that ensures a low-resistance contact and requires a tool to open for installation or servicing of the speaker. The forming shell shall be installed in a recess in the wall or floor of the pool.

(2) Wiring Methods. Rigid metal conduit of brass or other identified corrosion-resistant metal, liquidtight flexible nonmetallic conduit (LFNC), rigid polyvinyl chloride conduit, or reinforced thermosetting resin conduit shall extend from the forming shell to a listed junction box or other enclosure as provided in 680.24. Where rigid polyvinyl chloride conduit, reinforced thermosetting resin conduit, or liquidtight flexible nonmetallic conduit is used, an 8 AWG insulated solid or stranded copper bonding jumper shall be installed in this conduit. The bonding jumper shall be terminated in the forming shell and the junction box. The termination of the 8 AWG bonding jumper in the forming shell shall be covered with, or encapsulated in, a listed potting compound to protect such connection from the possible deteriorating effect of pool water.

(3) Forming Shell and Metal Screen. The forming shell and metal screen shall be of brass or other approved corrosion-resistant metal. All forming shells shall include provisions for terminating an 8 AWG copper conductor.

(B) Electrically Operated Pool Covers.

(1) Motors and Controllers. The electric motors, controllers, and wiring shall be located not less than 1.5 m (5 ft) from the inside wall of the pool unless separated from the pool by a wall, cover, or other permanent barrier. Electric motors installed below grade level shall be of the totally enclosed type. The device that controls the operation of the motor for an electrically operated pool cover shall be located such that the device operator has full view of the pool.

Exception: Motors that are part of listed systems with ratings not exceeding the low-voltage contact limit that are supplied by listed transformers or power supplies that comply with 680.23(A)(2) shall be permitted to be located less than 1.5 m (5 ft) from the inside walls of the pool.

(2) Protection. The electric motor and controller shall be connected to a GFCI-protected branch circuit.

Exception: Motors that are part of listed systems with ratings not exceeding the low-voltage contact limit that are supplied by listed transformers or power supplies that comply with 680.23(A)(2) shall not be required to have GFCI protection.

(C) Deck Area Heating. The provisions of this section shall apply to all pool deck areas, including a covered pool, where electrically operated comfort heating units are installed within 6.0 m (20 ft) of the inside wall of the pool.

(1) Unit Heaters. Unit heaters shall be rigidly mounted to the structure and shall be of the totally enclosed or guarded type. Unit heaters shall not be mounted over the pool or within the area extending 1.5 m (5 ft) horizontally from the inside walls of a pool.

(2) Permanently Wired Radiant Heaters. Radiant electric heaters shall be suitably guarded and securely fastened to their mounting device(s). Heaters shall not be installed over a pool or within the area extending 1.5 m (5 ft) horizontally from the inside walls of the pool and shall be mounted at least 3.7 m (12 ft) vertically above the pool deck unless otherwise approved.

(3) Radiant Heating Cables Not Permitted. Radiant heating cables embedded in or below the deck shall not be permitted.

680.28 Gas-Fired Water Heater. Circuits serving gas-fired swimming pool and spa water heaters operating at voltages above the low-voltage contact limit shall be provided with GFCI protection.

Part III. Storable Pools, Storable Spas, Storable Hot Tubs, and Storable Immersion Pools

680.30 General. Electrical installations at storable pools, storable spas, storable hot tubs, or storable immersion pools shall comply with the provisions of Part I and Part III of this article.

Pools, spas, and hot tubs of any dimension with inflatable or molded polymeric walls are considered storable. Other storable units are those that can be readily disassembled and are limited to a maximum water depth of 42 inches. See the defined terms *pool, storable (storable immersion pool)* or *spa or hot tub, storable (storable spa or hot tub)* in Article 100. A storable pool, its associated equipment, and perimeter do not require equipotential bonding as specified in 680.26. However, the filter pump must be listed (see 680.6), be double insulated, and have an EGC that is an integral part of the flexible cord. The 3-foot-length limitation for flexible cords in 680.8 does not apply to the power cord of equipment listed for use with a storable swimming pool. According to the product standard for storable pool pump/filter units, the cord must be at least 25 feet long to discourage the use of extension cords. All electrical equipment used with a storable pool is required to have GFCI protection for personnel. Exhibit 680.9 illustrates an inflatable swimming pool that is considered to be a storable swimming pool regardless of the water depth.

EXHIBIT 680.9 Example of an inflatable swimming pool subject to the requirements in Parts I and III of Article 680.

The UL marking requirement for these listed units includes the wording "Do Not Use with Permanently Installed Pools."

Δ **680.31 Pumps.** A cord-connected pool filter pump shall incorporate an approved system of double insulation or its equivalent and shall be provided with means for the termination of an equipment grounding conductor for the connection to the internal and nonaccessible non–current-carrying metal parts of the pump.

Cord-connected pool filter pumps shall be provided with a GFCI that is an integral part of the attachment plug or located in the power-supply cord within 300 mm (12 in.) of the attachment plug.

Δ **680.32 Ground-Fault Circuit-Interrupter (GFCI) and Special Purpose Ground-Fault Circuit-Interrupter (SPGFCI) Protection.** All electrical equipment, including power-supply cords, used with storable pools shall have GFCI protection complying with 680.5(B) or SPGFCI protection complying with 680.5(C), as applicable.

All receptacles rated 125 volts through 250 volts, 60 amperes or less, located within 6.0 m (20 ft) of the inside walls of a storable pool, storable spa, or storable hot tub shall have GFCI protection complying with 680.5(B) or SPGFCI protection complying with 680.5(C), as applicable. In determining these dimensions, the distance to be measured shall be the shortest path the supply cord of an appliance connected to the receptacle would follow without piercing a floor, wall, ceiling, doorway with hinged or sliding door, window opening, or other effective permanent barrier.

680.33 Luminaires. An underwater luminaire, if installed, shall be installed in or on the wall of the storable pool, storable spa, or storable hot tub. It shall comply with either 680.33(A) or (B).

(A) Within the Low-Voltage Contact Limit. A luminaire shall be part of a cord-and-plug-connected lighting assembly. This assembly shall be listed as an assembly for the purpose and have the following construction features:

(1) No exposed metal parts
(2) A luminaire lamp that is suitable for use at the supplied voltage
(3) An impact-resistant polymeric lens, luminaire body, and transformer enclosure
(4) A transformer or power supply meeting the requirements of 680.23(A)(2) with a primary rating not over 150 V

(B) Over the Low-Voltage Contact Limit But Not over 150 Volts. A lighting assembly without a transformer or power supply and with the luminaire lamp(s) operating at not over 150 volts shall be permitted to be cord-and-plug-connected where the assembly is listed as an assembly for the purpose. The installation shall comply with 680.23(A)(5), and the assembly shall have the following construction features:

(1) No exposed metal parts
(2) An impact-resistant polymeric lens and luminaire body
(3) A GFCI with open neutral conductor protection as an integral part of the assembly
(4) The luminaire lamp permanently connected to the GFCI with open-neutral protection
(5) Compliance with the requirements of 680.23(A)

680.34 Receptacle Locations. Receptacles shall not be located less than 1.83 m (6 ft) from the inside walls of a storable pool, storable spa, or storable hot tub. In determining these dimensions, the distance to be measured shall be the shortest path the supply cord of an appliance connected to the receptacle would follow without piercing a floor, wall, ceiling, doorway with hinged or sliding door, window opening, or other effective permanent barrier.

680.35 Storable and Portable Immersion Pools. Storable and portable immersion pools shall additionally comply with 680.35(A) through (G).

(A) Cord Connection for Self-Contained Storable and Portable Immersion Pools. Self-contained storable and portable packaged immersion pools with identified integral permanently attached switches and/or controls, pumps, and/or heaters, including circulation heaters, rated 120 volts and 20 amperes or less, single phase, shall be permitted to be cord-and-plug-connected with a cord not shorter than 1.83 m (6 ft) and not longer than 4.6 m (15 ft), and shall be GFCI protected. The cord shall be minimum hard usage. If the GFCI is provided as an integral part of the cord assembly, it shall be located at the attachment plug or in the power supply cord within 300 mm (12 in.) of the attachment plug.

(B) Storable and Portable Pumps. A cord-connected storable or portable pump utilized with, but not built-in or permanently attached as an integral part of, a storable or portable immersion pool shall be listed, labeled, and identified for swimming pool and spa use, and shall meet the requirements of 680.31.

(C) Storable and Portable Heaters. A storable or portable heater, including circulation heaters, used with, but not built-in or permanently attached as an integral part of, a storable or portable immersion pool shall be rated 120 volts and 20 amperes or less or 250 volts and 30 amperes or less, single phase; shall be identified for swimming pool and spa use; shall be permitted to be cord-and-plug-connected with a cord not shorter than 1.83 m (6 ft) and not longer than 4.6 m (15 ft); and heaters supplied by branch circuits rated 150 volts or less to ground shall be provided with GFCI protection. The cord shall be minimum hard usage. If the GFCI is provided as an integral part of the cord assembly, it shall be located at the attachment plug or in the power supply cord within 300 mm (12 in.) of the attachment plug.

(D) Audio Equipment. Audio equipment shall not be installed in or on a self-contained storable or portable immersion pool. All audio equipment operating at greater than the low-voltage contact limit and located within 1.83 m (6 ft) from the inside walls of a storable or portable immersion pool shall be grounded and shall be GFCI protected.

(E) Location Proximate to Luminaires, Lighting Outlets, and Ceiling-Suspended (Paddle) Fans. The storable or portable immersion pool shall be installed at least 3 m (10 ft), measured diagonally from the nearest point on each luminaire, lighting outlet, and ceiling-suspended (paddle) fan operating at greater than the low-voltage contact limit, to the nearest point on the top rim of the immersion pool. Luminaires shall not be installed in or on a storable or portable immersion pool.

(F) Location Proximate to Switches. The storable or portable immersion pool shall be installed at least 1.5 m (5 ft), measured horizontally, from all switches operating above the low-voltage contact limit. Switches that are packaged with the immersion pool as an integral part of its internal electrical system, or that are integral to cord-and-plug-connected electrical equipment utilized in the pool's water circulation or drain system, including storable and portable pumps, heaters, and circulation heaters, shall be permitted to be located less than 1.5 m (5 ft) from the inside walls of the pool.

(G) Receptacles. All receptacles rated 250 volts, 50 amperes or less, located within 6.0 m (20 ft) of the inside walls of the installed storable or portable immersion pool and utilized for supplying power to heaters or other electrical equipment serving the immersion pool, shall meet the requirements of 680.32 and 680.34.

Part IV. Permanently Installed and Self-Contained Spas and Hot Tubs and Permanently Installed Immersion Pools

680.40 General. Electrical installations at spas and hot tubs shall comply with the provisions of Part I and Part IV of this article.

Δ 680.41 Location of Other Equipment.

N (A) Emergency Switch for Spas and Hot Tubs. For other than one-family dwellings, a clearly labeled emergency shutoff or control switch for the purpose of stopping the motor(s) that provides power to the recirculation system and jet system shall be installed at a point readily accessible to the users and not less than 1.5 m (5 ft) away, adjacent to, and within sight of the spa or hot tub.

A local disconnecting device that is capable of being used in an emergency is required for spas and hot tubs to allow rapid disconnection if someone becomes entrapped by the water recirculation system intake. For more information regarding this issue, visit the U.S. Consumer Product Safety Commission online at www.poolsafely.gov.

See Exhibit 680.10 for an illustration of the emergency shutoff switch location. The shutoff switch can be either a line-operated device or a remote-control circuit that causes the pump circuit to open.

N (B) Equipment Exceeding the Low-Voltage Contact Limit. Except for self-contained spas and hot tubs, equipment with ratings exceeding the low-voltage contact limit shall be located at least 1.5 m (5 ft) horizontally from the inside walls of a spa or hot tub, unless separated from the spa or hot tub by a solid fence, wall, or other permanent barrier.

680.42 Outdoor Installations. A spa or hot tub installed outdoors shall comply with the provisions of Parts I and II of this article, except as permitted in 680.42(A) and (B), that would otherwise apply to pools installed outdoors.

(A) Flexible Connections. Listed packaged spa or hot tub equipment assemblies or self-contained spas or hot tubs utilizing a factory-installed or assembled control panel or panelboard shall be permitted to use flexible connections as covered in 680.42(A)(1) and (A)(2).

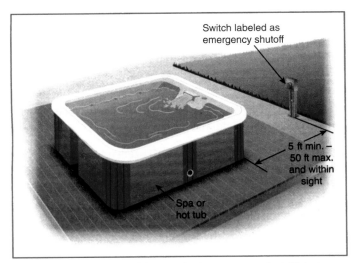

EXHIBIT 680.10 Location of the required emergency shutoff device.

(1) Flexible Conduit. Liquidtight flexible metal conduit or liquidtight flexible nonmetallic conduit shall be permitted.

(2) Cord-and-Plug Connections. Cord-and-plug connections with a cord not longer than 4.6 m (15 ft) shall be permitted if GFCI protected.

Δ (B) Bonding. Bonding by metal-to-metal mounting on a common frame or base shall be permitted. The metal bands or hoops used to secure wooden staves shall not be required to be bonded as required in 680.26.

Equipotential bonding of perimeter surfaces in accordance with 680.26(B)(2) shall not be required to be provided for spas and hot tubs where all of the following conditions apply:

(1) The spa or hot tub shall be listed, labeled, and identified as a self-contained spa for aboveground use.
(2) The spa or hot tub shall not be identified as suitable only for indoor use.
(3) The installation shall be in accordance with the manufacturer's instructions and shall be located on or above grade.
(4) The top rim of the spa or hot tub shall be at least 710 mm (28 in.) above all perimeter surfaces that are within 760 mm (30 in.), measured horizontally from the spa or hot tub. The height of nonconductive external steps for entry to or exit from the self-contained spa shall not be used to reduce or increase this rim height measurement.

Informational Note: See ANSI/UL 1563, *Standard for Electric Spas, Equipment Assemblies, and Associated Equipment*, for information regarding listing requirements for self-contained spas and hot tubs.

Δ (C) Underwater Luminaires. Wiring to an underwater luminaire shall comply with 680.23 or 680.33.

680.43 Indoor Installations. A spa or hot tub installed indoors shall comply with the provisions of Parts I and II of this article except as modified by this section and shall be connected by the wiring methods of Chapter 3.

Exception No. 1: Listed spa and hot tub packaged units rated 20 amperes or less shall be permitted to be cord-and-plug-connected to facilitate the removal or disconnection of the unit for maintenance and repair.

Exception No. 2: The equipotential bonding requirements for perimeter surfaces in 680.26(B)(2) shall not apply to a listed self-contained spa or hot tub installed above a finished floor.

Exception No. 3: For a dwelling unit(s) only, where a listed spa or hot tub is installed indoors, the wiring method requirements of 680.42(C) shall also apply.

(A) Receptacles. At least one 125-volt, 15- or 20-ampere receptacle on a general-purpose branch circuit shall be located not less than 1.83 m (6 ft) from, and not exceeding 3.0 m (10 ft) from, the inside wall of the spa or hot tub.

(1) Location. All receptacles shall be located at least 1.83 m (6 ft) measured horizontally from the inside walls of the spa or hot tub.

(2) Ground-Fault Circuit-Interrupter (GFCI) and Special Purpose Ground-Fault Circuit-Interrupter (SPGFCI) Protection for Receptacles, General. All receptacles rated 125 volts through 250 volts, 60 amperes or less, located within 3.0 m (10 ft) of the inside walls of a spa or hot tub shall have GFCI protection complying with 680.5(B) or SPGFCI protection complying with 680.5(C), as applicable.

(3) Protection, Spa or Hot Tub Supply Receptacle. Receptacles that provide power for a spa or hot tub shall not exceed 150 volts to ground and shall be GFCI protected.

(4) Measurements. In determining the dimensions in this section addressing receptacle spacings, the distance to be measured shall be the shortest path the supply cord of an appliance connected to the receptacle would follow without piercing a floor, wall, ceiling, doorway with hinged or sliding door, window opening, or other effective permanent barrier.

(B) Installation of Luminaires, Lighting Outlets, and Ceiling-Suspended (Paddle) Fans.

Δ **(1) Elevation.** Luminaires, except as covered in 680.43(B)(2), lighting outlets, and ceiling-suspended (paddle) fans located over the spa or hot tub or within 1.5 m (5 ft) from the inside walls of the spa or hot tub shall comply with the clearances specified in 680.43(B)(1)(a), (B)(1)(b), and (B)(1)(c) above the maximum water level.

(a) *Without GFCI.* Where no GFCI protection is provided, the mounting height shall be not less than 3.7 m (12 ft).

(b) *With GFCI.* Where GFCI protection is provided, the mounting height shall be permitted to be not less than 2.3 m (7 ft 6 in.).

(c) *Below 2.3 m (7 ft 6 in.).* Luminaires meeting the requirements of item (1) or (2) and protected by a ground-fault circuit interrupter shall be permitted to be installed less than 2.3 m (7 ft 6 in.) over a spa or hot tub:

(1) Recessed luminaires with a glass or plastic lens, nonmetallic or electrically isolated metal trim, and suitable for use in damp locations

(2) Surface-mounted luminaires with a glass or plastic globe, a nonmetallic body, or a metallic body isolated from contact, and suitable for use in damp locations

(2) Underwater Applications. Underwater luminaires shall comply with the provisions of 680.23 or 680.33.

(C) Switches. Switches shall be located at least 1.5 m (5 ft), measured horizontally, from the inside walls of the spa or hot tub.

(D) Bonding. The following parts shall be bonded together:

(1) All metal fittings within or attached to the spa or hot tub structure

(2) Metal parts of electrical equipment associated with the spa or hot tub water circulating system, including pump motors, unless part of a listed, labeled, and identified self-contained spa or hot tub

(3) Metal raceway and metal piping that are within 1.5 m (5 ft) of the inside walls of the spa or hot tub and that are not separated from the spa or hot tub by a permanent barrier

(4) All metal surfaces that are within 1.5 m (5 ft) of the inside walls of the spa or hot tub and that are not separated from the spa or hot tub area by a permanent barrier

Exception: Small conductive surfaces not likely to become energized, such as air and water jets and drain fittings, where not connected to metallic piping, towel bars, mirror frames, and similar nonelectrical equipment, shall not be required to be bonded.

(5) Non-current-carrying metal parts of electrical devices and controls that are not associated with the spas or hot tubs and that are located within 1.5 m (5 ft) of such units

(E) Methods of Bonding. All metal parts associated with the spa or hot tub shall be bonded by any of the following methods:

(1) The interconnection of threaded metal piping and fittings

(2) Metal-to-metal mounting on a common frame or base

(3) The provisions of a solid copper bonding jumper, insulated, covered, or bare, not smaller than 8 AWG

Bonding requirements for spas and hot tubs are similar to those in Part II of Article 680, except that equipment mounted on a common frame or base with metal-to-metal contact is an acceptable bonding method.

(F) Grounding. The following equipment shall be connected to the equipment grounding conductor:

(1) All electrical equipment located within 1.5 m (5 ft) of the inside wall of the spa or hot tub

(2) All electrical equipment associated with the circulating system of the spa or hot tub

Exception to (1) and (2): Electrical equipment listed for operation at the low-voltage contact limit or less and supplied by a transformer or power supply that complies with 680.23(A)(2) shall not be required to be connected to the equipment grounding conductor.

(G) Underwater Audio Equipment. Underwater audio equipment shall comply with the provisions of Part II of this article.

Δ **680.44 Ground-Fault Circuit-Interrupter (GFCI) and Special Purpose Ground-Fault Circuit-Interrupter (SPGFCI) Protection.**

N **(A) General.** Except as otherwise provided in this section, the outlet(s) that supplies a self-contained spa or hot tub, a packaged spa or hot tub equipment assembly, or a field-assembled spa or hot tub shall have GFCI protection complying with 680.5(B) or SPGFCI protection complying with 680.5(C), as applicable.

△ **(B) Listed Units.** If so marked, a listed self-contained unit or a listed packaged equipment assembly that includes integral GFCI protection for all electrical parts within the unit or assembly (pumps, air blowers, heaters, lights, controls, sanitizer generators, wiring, and so forth) shall be permitted without additional ground-fault protection.

N **(C) Gas-Fired Water Heaters.** Circuits serving gas-fired spa and hot tub water heaters operating separately from the spa or hot tub they serve, and operating at voltages above the low-voltage contact limit, shall be GFCI protected.

680.45 Permanently Installed Immersion Pools. Electrical installations at permanently installed immersion pools, whether installed indoors or outdoors, shall comply with the provisions of Part I, Part II, and Part IV of this article except as modified by this section and shall be connected by the wiring methods of Chapter 3. With regard to provisions in Part IV of this article, an immersion pool shall be considered to be a spa or hot tub.

Exception No. 1: The equipotential bonding requirements in 680.26(B) shall not apply to immersion pools that incorporate no permanently installed or permanently connected electrical equipment, and that are installed with all portions located on or above a finished floor.

Exception No. 2: The equipotential bonding requirements for perimeter surfaces in 680.26(B)(2) shall not apply to nonconductive perimeter surfaces, such as steps, treads, and walking surfaces made of fiberglass composite.

(A) Cord-and-Plug Connections. To facilitate the removal or disconnection of the unit(s) for maintenance, storage, and repair, self-contained portable packaged immersion pools with integral pumps and/or heaters, including circulation heaters, rated 120 volts and 20 amperes or less shall be permitted to be cord-and-plug-connected with a cord not shorter than 1.83 m (6 ft) and not longer than 4.6 m (15 ft) and shall be GFCI protected. The cord shall ground all non-current-carrying metal parts of the electrical equipment. If the GFCI is provided as an integral part of the cord assembly, it shall be located at the attachment plug or in the power-supply cord within 300 mm (12 in.) of the attachment plug.

△ **(B) Storable and Portable Pumps.** A cord-connected storable or portable pump used with, but not built-in or permanently attached as an integral part of, a permanently installed immersion pool shall be identified for swimming pool and spa use. It shall incorporate an approved system of double insulation or its equivalent and shall be provided with means for grounding only the internal and nonaccessible non-current-carrying metal parts of the appliance. Cord-connected pool filter pumps shall be provided with a GFCI that is an integral part of the attachment plug or located in the power-supply cord within 300 mm (12 in.) of the attachment plug.

(C) Heaters. Heaters used with permanently installed immersion pools shall comply with either 680.45(C)(1) or (C)(2).

(1) Permanently Installed Heaters. A permanently installed heater, including immersion heaters, circulation heaters, and combination pump-heater units, built-in or permanently attached as an integral part of a permanently installed immersion pool, rated 120 volts or 250 volts, shall be identified for swimming pool and spa use; shall be grounded and bonded; and heaters supplied by branch circuits rated 150 volts or less to ground shall be provided with GFCI protection. Permanently installed immersion heaters, rated 120 volts and 20 amperes or less or 250 volts and 30 amperes or less, single phase, are permitted to be cord-and-plug-connected with a cord not shorter than 1.83 m (6 ft) and not longer than 4.6 m (15 ft), shall be GFCI protected, and shall be provided with means for grounding all non-current-carrying metal parts of the appliance. If the GFCI is provided as an integral part of the cord assembly, it shall be located at the attachment plug or in the power-supply cord within 300 mm (12 in.) of the attachment plug.

△ **(2) Storable and Portable Heaters.** A cord-connected storable or portable heater, including immersion heaters, circulation heaters, and combination pump-heater units, used with, but not permanently installed or attached as an integral part of a permanently installed immersion pool, rated 120 volts and 20 amperes or less or 250 volts and 30 amperes or less, single phase, shall be identified for swimming pool and spa use; shall be cord-and-plug-connected with a cord not shorter than 1.83 m (6 ft) and not longer than 4.6 m (15 ft), heaters supplied by branch circuits rated 150 volts or less to ground shall be provided with Class A ground-fault circuit-interrupter protection, and shall be provided with means for grounding all non-current-carrying metal parts of the appliance. If the ground-fault circuit interrupter is provided as an integral part of the cord assembly, it shall be located at the attachment plug or in the power-supply cord within 300 mm (12 in.) of the attachment plug.

(D) Audio Equipment. Audio equipment shall not be installed in or on a permanently installed immersion pool. All audio equipment operating at greater than the low-voltage contact limit and located within 1.83 m (6 ft) from the inside walls of a permanently installed immersion pool shall be connected to the equipment grounding conductor and GFCI protected.

Part V. Fountains

Part V applies to permanently installed decorative fountains and reflecting pools in the ground, partially in the ground, or in a building. These units are primarily for aesthetic value and are not intended for swimming or wading.

See also

Article 682 for installations in natural lakes, rivers, or ponds, which are not covered in Part V

△ **680.50 General.** Part I and Part V of this article shall apply to all permanently installed fountains.

N **(A) Additional Requirements.**

(1) Fountains that have water common to a pool shall also comply with Part II of Article 680.

(2) Fountains intended for recreational use by pedestrians, including splash pads, shall also comply with the requirements in 680.26.

(3) Part V does not apply to self-contained, portable fountains, which shall comply with Parts II and III of Article 422.

N **(B) Location of Equipment Exceeding the Low-Voltage Contact Limit.** Equipment with ratings exceeding the low-voltage contact limit shall be located at least 1.5 m (5 ft) horizontally from the inside walls of a fountain, unless separated from the fountain by a solid fence, wall, or other permanent barrier.

680.51 Luminaires, Submersible Pumps, and Other Submersible Equipment.

(A) Ground-Fault Circuit Interrupter. Luminaires, submersible pumps, and other submersible equipment, unless listed for operation at low-voltage contact limit or less and supplied by a transformer or power supply that complies with 680.23(A)(2), shall be GFCI protected.

(B) Operating Voltage. No luminaires shall be installed for operation on supply circuits over 150 volts between conductors. Submersible pumps and other submersible equipment shall operate at 300 volts or less between conductors.

(C) Luminaire Lenses. Luminaires shall be installed with the top of the luminaire lens below the normal water level of the fountain unless listed for above-water locations. A luminaire facing upward shall comply with either (1) or (2):

(1) Have the lens guarded to prevent contact by any person

(2) Be listed for use without a guard

(D) Overheating Protection. Electrical equipment that depends on submersion for safe operation shall be protected against overheating by a low-water cutoff or other approved means when not submerged.

(E) Wiring. Equipment shall be equipped with provisions for threaded conduit entries or be provided with a suitable flexible cord. The maximum length of each exposed cord in the fountain shall be limited to 3.0 m (10 ft). Cords extending beyond the fountain perimeter shall be enclosed in approved wiring enclosures. Metal parts of equipment in contact with water shall be of brass or other approved corrosion-resistant metal.

(F) Servicing. All equipment shall be removable from the water for relamping or normal maintenance. Luminaires shall not be permanently embedded into the fountain structure such that the water level must be reduced or the fountain drained for relamping, maintenance, or inspection.

(G) Stability. Equipment shall be inherently stable or be securely fastened in place.

680.52 Junction Boxes and Other Enclosures.

(A) General. Junction boxes and other enclosures used for other than underwater installation shall comply with 680.24.

(B) Underwater Junction Boxes and Other Underwater Enclosures. Junction boxes and other underwater enclosures shall meet the requirements of 680.52(B)(1) and (B)(2).

(1) Construction.

(a) Underwater enclosures shall be equipped with provisions for threaded conduit entries or compression glands or seals for cord entry.

(b) Underwater enclosures shall be listed and rated for prolonged submersion and made of copper, brass, or other corrosion-resistant material.

Δ **(2) Installation.** Underwater enclosure installations shall comply with 680.52(B)(2)(a) and (B)(2)(b).

(a) Underwater enclosures shall be filled with a listed potting compound to prevent the entry of moisture.

(b) Underwater enclosures shall be firmly attached to the supports or directly to the fountain surface and bonded as required. Where the junction box is supported only by conduits in accordance with 314.23(E) and (F), the conduits shall be of copper, brass, stainless steel, or other corrosion-resistant metal. Where the box is fed by nonmetallic conduit, it shall have additional supports and fasteners of copper, brass, or other corrosion-resistant material.

680.54 Grounding and Bonding.

Δ **(A) Grounding.** The following equipment shall be connected to an equipment grounding conductor:

(1) All electrical equipment located within the fountain or within 1.5 m (5 ft) of the inside wall of the fountain, other than listed low-voltage luminaires not requiring grounding

(2) All electrical equipment associated with the recirculating system of the fountain

(3) Panelboards that are not part of the service equipment and that supply any electrical equipment associated with the fountain

(B) Bonding. If a conductor is used for bonding, it shall be a minimum 8 AWG solid copper conductor. The following parts shall be bonded together and connected to an equipment grounding conductor for a branch circuit supplying fountain equipment:

(1) All metal piping systems associated with the fountain

(2) All metal fittings within or attached to the fountain

(3) Metal parts of electrical equipment associated with the fountain water-circulating system, including pump motors

(4) Metal raceways that are within 1.5 m (5 ft) of the inside wall or perimeter of the fountain and that are not separated from the fountain by a permanent barrier

(5) All metal surfaces that are within 1.5 m (5 ft) of the inside wall or perimeter of the fountain and that are not separated from the fountain by a permanent barrier

(6) Electrical equipment located less than 1.5 m (5 ft) of the inside wall or perimeter of the fountain

N (C) Equipotential Bonding of Splash Pads. For the purpose of equipotential bonding, the shell of a splash pad shall comprise the area traversed by pedestrians bounded by the extent of the footing of the splash pad and rising to its exposed surface(s) and its collection basin area. The boundary of this area shall be considered to be the inside wall for the purpose of perimeter bonding.

680.55 Methods of Grounding.

(A) Applied Provisions. The provisions of 680.7(A), 680.21(A), 680.23(B)(3), 680.23(F)(1) and (F)(2), and 680.24(F) shall apply.

(B) Supplied by a Flexible Cord. Electrical equipment that is supplied by a flexible cord shall have all exposed non–current-carrying metal parts grounded by an insulated copper equipment grounding conductor that is an integral part of this cord. The equipment grounding conductor shall be connected to an equipment grounding terminal in the supply junction box, transformer enclosure, power supply enclosure, or other enclosure.

680.56 Cord-and-Plug-Connected Equipment.

(A) GFCI Protection. All electrical equipment, including power-supply cords, shall be GFCI protected.

(B) Cord Type. Flexible cord immersed in or exposed to water shall be of a type for extra-hard usage, as designated in Table 400.4, and shall be a listed type with a "W" suffix.

(C) Sealing. The end of the flexible cord jacket and the flexible cord conductor termination within equipment shall be covered with, or encapsulated in, a suitable potting compound to prevent the entry of water into the equipment through the cord or its conductors. In addition, the ground connection within equipment shall be similarly treated to protect such connections from the deteriorating effect of water that may enter into the equipment.

(D) Terminations. Connections with flexible cord shall be permanent, except that grounding-type attachment plugs and receptacles shall be permitted to facilitate removal or disconnection for maintenance, repair, or storage of fixed or stationary equipment not located in any water-containing part of a fountain.

680.57 Signs.

(A) General. This section covers electric signs installed within a fountain or within 3.0 m (10 ft) of the fountain edge.

(B) GFCI Protection. Branch circuits or feeders supplying the sign shall have GFCI protection.

(C) Location.

(1) Fixed or Stationary. A fixed or stationary electric sign installed within a fountain shall be not less than 1.5 m (5 ft) inside the fountain measured from the outside edges of the fountain.

(2) Portable. A portable electric sign shall not be placed within a pool or fountain or within 1.5 m (5 ft) measured horizontally from the inside walls of the fountain.

(D) Disconnect. A sign shall have a local disconnecting means in accordance with 600.6 and 680.13.

(E) Bonding and Grounding. A sign shall be grounded and bonded in accordance with 600.7.

Electric signs in fountains are required to have GFCI protection. This protection is required in the feeder or branch circuit supplying the sign. To prevent contact by persons around the fountain, fixed or stationary signs must be at least 5 feet from the edge of the fountain (see Exhibit 680.11). This also applies to signs located within 10 feet of the outside edge of the fountain.

Δ 680.58 Ground-Fault Circuit-Interrupter (GFCI) and Special Purpose Ground-Fault Circuit-Interrupter (SPGFCI) Protection for Adjacent Receptacle Outlets. All receptacles rated 125 volts through 250 volts, 60 amperes or less, located within 6.0 m (20 ft) of a fountain edge shall have GFCI protection complying with 680.5(B) or SPGFCI protection complying with 680.5(C), as applicable.

680.59 Ground-Fault Circuit-Interrupter (GFCI) and Special Purpose Ground-Fault Circuit-Interrupter (SPGFCI) Protection for Permanently Installed Nonsubmersible Pumps. Outlets supplying all permanently installed nonsubmersible pump motors shall have GFCI protection

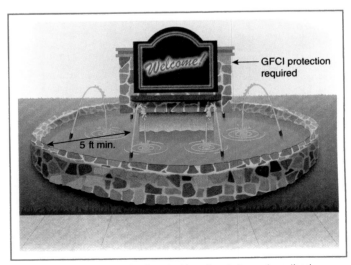

EXHIBIT 680.11 Electric sign located in a fountain as described in 680.57.

complying with 680.5(B) or SPGFCI protection complying with 680.5(C), as applicable.

Exception: Listed low-voltage motors not requiring grounding, with ratings not exceeding the low-voltage contact limit that are supplied by listed transformers or power supplies that comply with 680.23(A)(2), shall be permitted to be installed without GFCI or SPGFCI protection.

Part VI. Pools and Tubs for Therapeutic Use

Part VI recognizes therapeutic equipment in other locations, such as athletic training rooms and health care facilities. Exhibit 680.12 is an example of a therapeutic tub that is equipped with an electrically powered pump and jet system for hydrotherapy to treat and prevent injuries.

△ **680.60 General.** The provisions of Part I and Part VI of this article shall apply to pools and tubs for therapeutic use in health care facilities, gymnasiums, athletic training rooms, and similar areas. Portable therapeutic appliances shall comply with Parts II and III of Article 422.

Portable therapeutic appliances are required by 422.41 to provide protection from shock or electrocution while in the "on" or "off" position. The device used is an immersion detection circuit interrupter (IDCI).

680.61 Permanently Installed Therapeutic Pools. Therapeutic pools that are constructed in the ground, on the

EXHIBIT 680.12 A whirlpool tub of the type common to athletic training and sports injury treatment facilities. *(Courtesy of Whitehall Manufacturing, a Division of Morris Group International)*

ground, or in a building in such a manner that the pool cannot be readily disassembled shall comply with Parts I and II of this article.

Exception: The limitations of 680.22(B)(1) through (B)(4) shall not apply where all luminaires are of the totally enclosed type.

680.62 Therapeutic Tubs (Hydrotherapeutic Tanks). Therapeutic tubs, used for the submersion and treatment of patients, that are not easily moved from one place to another in normal use or that are fastened or otherwise secured at a specific location, including associated piping systems, shall comply with Part VI.

(A) Protection. Except as otherwise provided in this section, the outlet(s) that supplies a self-contained therapeutic tub or hydrotherapeutic tank, a packaged therapeutic tub or hydrotherapeutic tank, or a field-assembled therapeutic tub or hydrotherapeutic tank shall be GFCI protected.

(1) Listed Units. If so marked, a listed, labeled, and identified self-contained unit or a listed, labeled, and identified packaged equipment assembly that includes integral ground-fault circuit-interrupter protection for all electrical parts within the unit or assembly (pumps, air blowers, heaters, lights, controls, sanitizer generators, wiring, and so forth) shall be permitted without additional GFCI protection.

(2) Other Units. A therapeutic tub or hydrotherapeutic tank rated 3 phase or rated over 250 volts or with a heater load of more than 50 amperes shall not require the supply to be protected by a ground-fault circuit interrupter.

(B) Bonding. The following parts shall be bonded together:

(1) All metal fittings within or attached to the tub structure
(2) Metal parts of electrical equipment associated with the tub water circulating system, including pump motors
(3) Metal-sheathed cables and raceways and metal piping that are within 1.5 m (5 ft) of the inside walls of the tub and not separated from the tub by a permanent barrier
(4) All metal surfaces that are within 1.5 m (5 ft) of the inside walls of the tub and not separated from the tub area by a permanent barrier
(5) Electrical devices and controls that are not associated with the therapeutic tubs and located within 1.5 m (5 ft) from such units.

Exception: Small conductive surfaces not likely to become energized, such as air and water jets and drain fittings not connected to metallic piping, and towel bars, mirror frames, and similar nonelectrical equipment not connected to metal framing, shall not be required to be bonded.

(C) Methods of Bonding. All metal parts required to be bonded by this section shall be bonded by any of the following methods:

(1) The interconnection of threaded metal piping and fittings
(2) Metal-to-metal mounting on a common frame or base
(3) Connections by suitable metal clamps
(4) By the provisions of a solid copper bonding jumper, insulated, covered, or bare, not smaller than 8 AWG

(D) Grounding. The following fixed or stationary equipment shall be connected to the equipment grounding conductor:

(1) All electrical equipment located within 1.5 m (5 ft) of the inside wall of the tub
(2) All electrical equipment associated with the circulating system of the tub

(E) Receptacles. All receptacles within 1.83 m (6 ft) of a therapeutic tub shall have ground-fault protection complying with 680.5(B) or (C), as applicable.

(F) Luminaires. All luminaires used in therapeutic tub areas shall be of the totally enclosed type.

Part VII. Hydromassage Bathtubs

680.70 General. Hydromassage bathtubs as defined in Article 100 shall comply with Part VII of this article. They shall not be required to comply with other parts of this article.

680.71 Protection. Hydromassage bathtubs and their associated electrical components shall be on an individual branch circuit(s) and protected by a readily accessible GFCI. All 125-volt, single-phase receptacles not exceeding 30 amperes and located within 1.83 m (6 ft) measured horizontally of the inside walls of a hydromassage tub shall be GFCI protected.

Hydromassage bathtubs (see definition in Article 100) are treated the same as ordinary bathtubs in regard to the installation and location of luminaires, switches, and other electrical equipment. Exhibit 680.13 is an example of a hydromassage bathtub.

See also

410.10(D) for special requirements relating to cord-connected luminaires, hanging luminaires, and pendants near bathtubs

210.8(A) and **(B)** for requirements for GFCI protection of bathroom receptacles

Article 100 for the definition of *accessible, readily*

Δ **680.72 Other Electrical Equipment.** Luminaires, switches, receptacles, and other electrical equipment located in the same room, and not directly associated with a hydromassage bathtub, shall be installed in accordance with Chapters 1 through 4 in this *Code* covering the installation of that equipment in bathrooms.

680.73 Accessibility. Hydromassage bathtub electrical equipment shall be accessible without damaging the building structure or building finish. Where the hydromassage bathtub is cord-and-plug-connected with the supply receptacle accessible only through a service access opening, the receptacle shall be

EXHIBIT 680.13 A hydromassage bathtub.

installed so that its face is within direct view and not more than 300 mm (1 ft) of the opening.

Where a GFCI-type receptacle is installed under a hydromassage tub, the receptacle is required by 680.71 to be readily accessible (see definition of *accessible, readily* in Article 100). Access can be either an integral part of the tub or an access panel door that is provided in the finish that encloses the tub that allows for monthly testing of the GFCI receptacle. If the access panel requires a tool to open, such as a screwdriver, the installation may not be deemed readily accessible. Where utilization equipment associated with a hydromassage bathtub is cord-and-plug-connected, the supply receptacle must be installed within 1 foot of the access opening and be positioned so that the face of the receptacle is visible. This requirement facilitates connection and disconnection of the cord-and-plug-connected equipment.

680.74 Bonding.

(A) General. The following parts shall be bonded together:

(1) All metal fittings within or attached to the tub structure that are in contact with the circulating water
(2) Metal parts of electrical equipment associated with the tub water circulating system, including pump and blower motors
(3) Metal-sheathed cables, metal raceways, and metal piping within 1.5 m (5 ft) of the inside walls of the tub and not separated from the tub by a permanent barrier
(4) All exposed metal surfaces that are within 1.5 m (5 ft) of the inside walls of the tub and not separated from the tub area by a permanent barrier
(5) Non-current-carrying metal parts of electrical devices and controls that are not associated with the hydromassage tubs within 1.5 m (5 ft) from such units

Exception No. 1: Small conductive surfaces not likely to become energized, such as air and water jets, supply valve assemblies, and drain fittings not connected to metallic piping, and towel bars, mirror frames, and similar nonelectrical equipment not connected to metal framing shall not be required to be bonded.

Exception No. 2: Double-insulated motors and blowers shall not be bonded.

Exception No. 3: Small conductive surfaces of electrical equipment not likely to become energized, such as the mounting strap or yoke of a listed light switch or receptacle that is grounded, shall not be required to be bonded.

(B) Bonding Conductor. All metal parts required to be bonded by this section shall be bonded together using a solid copper bonding jumper, insulated, covered, or bare, not smaller than 8 AWG. The bonding jumper(s) shall be required for equipotential bonding in the area of the hydromassage bathtub and shall not be required to be extended or attached to any remote panelboard, service equipment, or any electrode. In all installations a bonding jumper long enough to terminate on a replacement non-double-insulated pump or blower motor shall be provided and shall be terminated to the equipment grounding conductor of the branch circuit of the motor when a double-insulated circulating pump or blower motor is used.

Part VIII. Electrically Powered Pool Lifts

680.80 General. Electrically powered pool lifts as defined in Article 100 shall comply with Part VIII of this article. Part VIII shall not be subject to the requirements of other parts of this article except where the requirements are specifically referenced.

680.81 Equipment Approval. Lifts shall be listed, labeled, and identified for swimming pool and spa use.

Exception No. 1: Lifts where the battery is removed for charging at another location and the battery is rated less than or equal to the low-voltage contact limit shall not be required to be listed or labeled.

Exception No. 2: Solar-operated or solar-recharged lifts where the solar panel is attached to the lift and the battery is rated less than or equal to 24 volts shall not be required to be listed or labeled.

Exception No. 3: Lifts that are supplied from a source not exceeding the low-voltage contact limit and supplied by listed transformers or power supplies that comply with 680.23(A)(2) shall not be required to be listed or labeled.

680.82 Protection. Pool lifts connected to premises wiring and operated above the low-voltage contact limit shall be provided with GFCI protection and comply with 680.5.

680.83 Equipotential Bonding. Lifts shall be bonded in accordance with 680.26(B)(5) and (B)(7) using solid copper

conductors, insulated, covered, or bare, not smaller than 8 AWG. Connections to bonded parts shall be made in accordance with 250.8. An 8 AWG or larger solid copper bonding conductor provided to reduce voltage gradients in the pool lift area shall not be required to be extended or attached to remote panelboards, service equipment, or electrodes.

680.84 Switching Devices and Receptacles. Switches and switching devices that are operated above the low-voltage contact limit shall comply with 680.22(C). Receptacles for electrically powered pool lifts that are operated above the low-voltage contact limit shall comply with 680.22(A)(3) and (A)(4).

680.85 Nameplate Marking. Electrically powered pool lifts shall be provided with a nameplate giving the identifying name and model and rating in volts and amperes, or in volts and watts. If the lift is to be used on a specific frequency or frequencies, it shall be so marked. Battery-powered pool lifts shall indicate the type reference of the battery or battery pack to be used. Batteries and battery packs shall be provided with a battery type reference and voltage rating.

Exception: Nameplate ratings for battery-powered pool lifts shall only need to provide a rating in volts in addition to the identifying name and model.

ARTICLE 682 Natural and Artificially Made Bodies of Water

Part I. General

682.1 Scope. This article applies to the installation of electrical wiring for, and equipment in and adjacent to, natural or artificially made bodies of water not covered by other articles in this *Code*, such as, but not limited to, aeration ponds, fish farm ponds, storm retention basins, treatment ponds, and irrigation (channels) facilities.

Electrical equipment such as pumps, luminaires, and their associated supply wiring are frequently installed in lakes, ponds, aeration and treatment basins, and similar bodies of water. The requirements in Article 682 are designed to minimize the shock hazards inherent in those wet and damp locations. Artificially made bodies of water do not include the pools, fountains, and spas that are specifically covered under the scope of Article 680.

682.3 Other Articles. If the water is subject to boat traffic, the wiring shall comply with 555.34(B).

682.4 Industrial Application. This article shall not apply in industrial applications where there is alarm indication of equipment faults and the following conditions are in place:

(1) Conditions of maintenance and supervision ensure that only qualified persons service and operate the installed systems.

(2) Continued circuit operation is necessary for safe operation of equipment or processes.

682.5 Electrical Datum Plane Distances. The electrical datum plane shall consist of one of the following:

(1) In land areas subject to tidal fluctuation, the electrical datum plane shall be a horizontal plane 600 mm (2 ft) above the highest tide level for the area occurring under normal circumstances, that is, highest high tide.

(2) In land areas not subject to tidal fluctuation, the electrical datum plane shall be a horizontal plane 600 mm (2 ft) above the highest water level for the area occurring under normal circumstances.

(3) In land areas subject to flooding, the electrical datum plane based on (1) or (2) above shall be a horizontal plane 600 mm (2 ft) above the point identified as the prevailing high water mark or an equivalent benchmark based on seasonal or storm-driven flooding from the authority having jurisdiction.

The electrical datum plane for floating structures and landing stages that are (a) installed to permit rise and fall response to water level, without lateral movement, and (b) that are so equipped that they can rise to the datum plane established for (1) or (2) above, shall be a horizontal plane 750 mm (30 in.) above the water level at the floating structure or landing stage and a minimum of 300 mm (12 in.) above the level of the deck.

> **See also**
>
> **Article 100** for the definition of *electrical datum plane* and associated commentary

Part II. Installation

682.10 Electrical Equipment and Transformers. Electrical equipment and transformers, including their enclosures, shall be specifically approved for the intended location. No portion of an enclosure for electrical equipment not identified for operation while submerged shall be located below the electrical datum plane.

> **See also**
>
> **640.10** for requirements covering the installation of audio system equipment near a body of water

682.11 Location of Electrical Distribution Equipment. On land, the equipment serving feeders shall comply with one of the following:

(1) Be located no closer than 1.5 m (5 ft) horizontally from the shoreline, and live parts of the equipment are elevated a minimum of 300 mm (12 in.) above the electrical datum plane

(2) Be located no closer than the shoreline, and live parts of the equipment are located a minimum of 3 m (10 ft) above the electrical datum plane

682.12 Electrical Connections. Electrical connections shall comply with 682.12(A) or (B).

N **(A) General.** All electrical connections not intended for operation while submerged shall be located at least 300 mm (12 in.) above the deck of a floating or fixed structure but not below the electrical datum plane on fixed structures. Conductor splices, within junction boxes identified for wet locations, using sealed wire connector systems listed and identified for submersion shall be required for floating structures where located above the waterline but below the electrical datum plane.

N **(B) Replacements.** Replacement electrical connections located below the electrical datum plane shall be identified for submersion.

682.13 Wiring Methods and Installation. Wiring methods and installation shall comply with 682.13(A) through (C).

N **(A) Wiring Methods.** Wiring methods suitable for use in wet locations that include an insulated equipment grounding conductor shall be permitted.

N **(B) Portable Power Cables.** Extra-hard usage portable power cables that contain an insulated equipment grounding conductor rated not less than 75°C (167°F) and 600 volts shall be permitted as follows:

(1) As permanent wiring on the underside of piers (floating or fixed)

(2) Where flexibility is necessary as on piers composed of floating sections

N **(C) Submersible or Floating Equipment Power Connection(s).** Submersible or floating equipment shall be cord-and-plug-connected, using extra-hard usage cord, as designated in Table 400.4 and listed with a "W" suffix.

Δ **682.14 Disconnecting Means.** A disconnecting means shall be provided to isolate each submersible or floating electrical equipment from its supply connection(s) without requiring the plug to be removed from the receptacle. Plug and receptacle combinations shall be arranged to be suitable for the location while in use.

Exception: Equipment listed for direct connection and equipment anchored in place and incapable of routine movement caused by water currents or wind shall be permitted to be connected using wiring methods covered in 682.13.

Δ **(A) Type and Marking.** The disconnecting means shall consist of a circuit breaker, a switch, or both, and shall be specifically marked to designate which receptacle or other outlet it controls.

(B) Location. The disconnecting means shall be readily accessible on land, located not more than 750 mm (30 in.) from the receptacle it controls, and shall be located in the supply circuit ahead of the receptacle. The disconnecting means shall be located within sight of but not closer than 1.5 m (5 ft) from the shoreline and shall be elevated not less than 300 mm (12 in.) above the datum plane.

There is no restriction on the length of the flexible cord required by this section. In the case of tidal waters, the amount of fluctuation between low and high tides has a direct impact on the datum plane and shoreline benchmarks and, consequently, on the length of the flexible cord from the receptacle to the floating or submersible equipment.

See also

Article 100 for the definition of the term *shoreline*

682.15 Ground-Fault Protection. The GFCI requirements in this article, unless otherwise noted, shall be in addition to the requirements in 210.8. Ground-fault protective devices shall be listed and provided in accordance with 682.15(A) and (B). The protective device shall be located not less than 300 mm (12 in.) above the established electrical datum plane.

(A) Outlets. Outlets supplied by branch circuits not exceeding 150 volts to ground and 60 amperes, single-phase, shall be provided with ground-fault circuit-interrupter protection for personnel.

Based on the Article 100 definition of *outlet*, GFCI protection is required for all receptacle, lighting, and other power outlets that are associated with a natural or artificial body of water. For example, a water pump rated 240 volts and 50 amperes that is supplied either through a cord-and-plug or direct-wired connection would have GFCI protection because it is supplied via a receptacle outlet or an outlet intended for direct (hard-wired) connection of utilization equipment.

△ **(B) Feeder and Branch Circuits on Piers.** Feeder and branch-circuit conductors that are installed on piers shall be provided with ground-fault protection not exceeding 30 mA. Coordination with downstream ground-fault protection shall be permitted at the feeder overcurrent protective device.

Exception No. 1: Transformer secondary conductors of a separately derived ac system, operating at voltages exceeding 15 volts ac, that do not exceed 3 m (10 ft) and are installed in a raceway shall be permitted to be installed without ground-fault protection. This exception shall also apply to the supply terminals of the equipment supplied by the transformer secondary conductors.

Exception No. 2: Low-voltage circuits not requiring grounding, not exceeding the low-voltage contact limit as defined in Article

100, and supplied by listed transformers or power supplies that comply with 680.23(A)(2) shall be permitted to be installed without ground-fault protection.

Part III. Grounding and Bonding

△ **682.30 Grounding.** Wiring and equipment within the scope of this article shall be grounded as specified in 555.37, 555.54, and the requirements in Part III of this article.

682.31 Equipment Grounding Conductors.

N **(A) Equipment to Be Connected to Equipment Grounding Conductor.** The following shall be connected to an equipment grounding conductor run with the circuit conductors in the same raceway, cable, or trench:

(1) Metal boxes, metal cabinets, and all other metal enclosures
(2) Metal frames of utilization equipment
(3) Grounding terminals of grounding-type receptacles

(B) Type. Equipment grounding conductors shall be insulated copper conductors sized in accordance with 250.122 but not smaller than 12 AWG.

(C) Feeders. Where a feeder supplies a remote panelboard or other distribution equipment, an insulated equipment grounding conductor shall extend from a grounding terminal in the service to a grounding terminal and busbar in the remote panelboard or other distribution equipment.

(D) Branch Circuits. The insulated equipment grounding conductor for branch circuits shall terminate at a grounding terminal in a remote panelboard or other distribution equipment or the grounding terminal in the main service equipment.

(E) Cord-and-Plug-Connected Appliances. Unless double insulated, cord-and-plug-connected appliances shall be grounded by means of an equipment grounding conductor in the cord and a grounding-type attachment plug.

Exception: An equipment grounding conductor shall be permitted to be uninsulated if the EGC is part of a listed cable assembly identified for the environment and not subject to environments such as, but not limited to, storm water basins, sewage treatment ponds, and natural bodies of water containing salt.

682.32 Bonding of Non–Current-Carrying Metal Parts. All metal parts in contact with the water, all metal piping, tanks, and all non–current-carrying metal parts that are likely to become energized shall be bonded to the grounding terminal in the distribution equipment.

682.33 Equipotential Planes and Bonding of Equipotential Planes. An equipotential plane shall be installed where required in this section to mitigate step and touch voltages at electrical equipment.

(A) Areas Requiring Equipotential Planes. Equipotential planes shall be installed adjacent to all outdoor service equipment or disconnecting means that control equipment in or on water, that have a metallic enclosure and controls accessible to personnel, and that are likely to become energized. The equipotential plane shall encompass the area around the equipment and shall extend from the area directly below the equipment out not less than 900 mm (36 in.) in all directions from which a person would be able to stand and come in contact with the equipment.

(B) Areas Not Requiring Equipotential Planes. Equipotential planes shall not be required for the controlled equipment supplied by the service equipment or disconnecting means.

(C) Bonding.

(1) Bonded Parts. The parts specified in 682.33(C)(1) through (C)(3) shall be bonded together and to the electrical grounding system. Bonding conductors shall be solid copper, insulated, covered or bare, and not smaller than 8 AWG. Connections shall be made by exothermic welding or by listed pressure connectors or clamps that are labeled as being suitable for the purpose and are of stainless steel, brass, copper, or copper alloy.

(2) Outdoor Service Equipment and Disconnects. Outdoor service equipment or disconnecting means that control equipment in or on water, that have a metallic enclosure and controls accessible to personnel, and that are likely to become energized shall be bonded to the equipotential plane.

(3) Walking Surfaces. Surfaces directly below the equipment specified in 682.33(C)(2) but not less than 900 mm (36 in.) in all directions from the equipment from which a person would be able to stand and come in contact with the equipment shall be bonded to the equipotential plane. Bonding to this surface shall be wire mesh or other conductive elements on, embedded in, or placed under the walk surface within 75 mm (3 in.).

ARTICLE 685 Integrated Electrical Systems

Part I. General

685.1 Scope. This article covers integrated electrical systems, other than unit equipment, in which orderly shutdown is necessary to ensure safe operation. An *integrated electrical system* as used in this article is a unitized segment of an industrial wiring system where all of the following conditions are met:

(1) An orderly shutdown is required to minimize personnel hazard and equipment damage.
(2) The conditions of maintenance and supervision ensure that qualified persons service the system. The name(s) of the qualified person(s) shall be kept in a permanent record at the office of the establishment in charge of the completed installation.

A person designated as a qualified person shall possess the skills and knowledge related to the construction and operation of the electrical equipment and installation and shall have received documented safety training on the hazards involved. Documentation of their qualifications shall be on file with the office of the establishment in charge of the completed installation.

(3) Effective safeguards approved by the authority having jurisdiction are established and maintained.

The integrated electrical systems commonly used in large and complex industrial processes are designed, installed, and operated under stringent on-site engineering supervision. The control equipment, including overcurrent devices, is located so that it is accessible to qualified personnel, but that location might not meet — and is not required to meet — the conditions described in the Article 100 definition of the term *accessible, readily (readily accessible)*. Locating overcurrent devices and their associated disconnecting means so that they are not readily accessible to unqualified personnel is one of the preventive measures used to help maintain continuity of operation.

For some industrial processes, the sudden loss of electric power to vital equipment is an unacceptable level of risk, and an orderly shutdown procedure is necessary to prevent severe equipment damage, injury to personnel, or — in some extreme cases — catastrophic failure. Orderly shutdown is commonly employed in nuclear power–generating facilities, paper mills, and other areas with hazardous processes.

685.3 Application of Other Articles. The articles/sections in Table 685.3 apply to particular cases of installation of conductors and equipment, where there are orderly shutdown requirements that are in addition to those of this article or are modifications of them.

TABLE 685.3 *Application of Other Articles*

Conductor/Equipment	Section
More than one building or other structure	225, Part II
Ground-fault protection of equipment	230.95, Exception
Protection of conductors	240.4
Electrical system coordination	240.12
Ground-fault protection of equipment	240.13(1)
Grounding ac systems of 50 volts to less than 1000 volts	250.21
Equipment protection	427.22
Orderly shutdown	430.44
Disconnection	430.75, Exception Nos. 1 and 2
Disconnecting means in sight from controller	430.102(A), Exception No. 2
Energy from more than one source	430.113, Exception Nos. 1 and 2
Disconnecting means	645.10, Exception
Uninterruptible power supplies (UPS)	645.11(1)
Point of connection	705.12

Part II. Orderly Shutdown

685.10 Location of Overcurrent Devices in or on Premises. Location of overcurrent devices that are critical to integrated electrical systems shall be permitted to be accessible, with mounting heights permitted to ensure security from operation by unqualified personnel.

685.12 Direct-Current System Grounding. Two-wire dc circuits shall be permitted to be ungrounded.

685.14 Ungrounded Control Circuits. Where operational continuity is required, control circuits of 150 volts or less from separately derived systems shall be permitted to be ungrounded.

ARTICLE 690 Solar Photovoltaic (PV) Systems

Part I. General

Δ **690.1 Scope.** This article applies to solar PV systems, other than those covered by Article 691, including the array circuit(s), inverter(s), and controller(s) for such systems. The systems covered by this article include those interactive with other electric power production sources or stand-alone, or both. These PV systems may have ac or dc output for utilization.

> Informational Note No. 1: See Informational Note Figure 690.1.
> Informational Note No. 2: Article 691 covers the installation of large-scale PV electric supply stations.

The use of photovoltaic (PV) systems as interactive or stand-alone power-supply systems has steadily increased as the technology of PV equipment has evolved and its availability has improved. Exhibit 690.1 shows a typical installation of a PV array in a field. The requirements of Article 690 cover the use of stand-alone and interactive PV systems. The requirements of Article 691 cover large-scale PV systems with a capacity of 5000

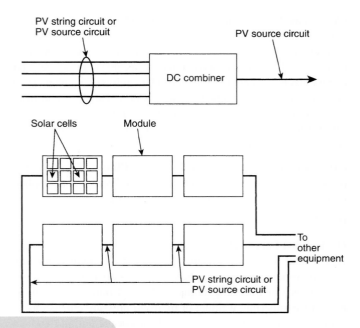

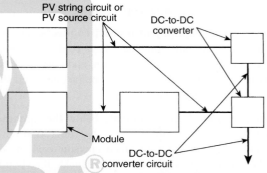

Δ **INFORMATIONAL NOTE FIGURE 690.1** *Illustration of PV System DC Circuits and PV System Components in a Typical PV Installation.*

kilowatts (kW) or more. Interactive photovoltaic systems are also subject to the requirements for interconnected electric power production sources contained in Article 705.

690.4 General Requirements.

(A) PV Systems. PV systems shall be permitted to supply a building or other structure in addition to any other electrical supply system(s).

(B) Equipment. Electronic power converters, motor generators, PV modules, ac modules and ac module systems, dc combiners, PV rapid shutdown equipment (PVRSE), PV hazard control equipment (PVHCE), PV hazard control systems (PVHCS), dc circuit controllers, and charge controllers intended for use in PV systems shall be listed or be evaluated for the application and have a field label applied.

The simplified circuit schematic in Exhibit 690.2 illustrates components in a PV system. Specific requirements for overcurrent protection, disconnecting means, and grounding are covered in

EXHIBIT 690.1 A PV array. (*Getty Images*)

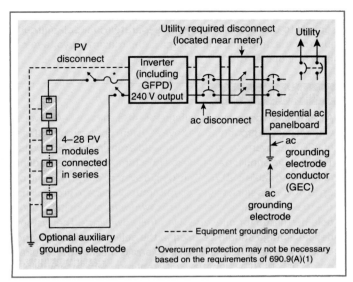

EXHIBIT 690.2 Simplified circuit schematic of a rooftop grid-connected system.

other sections of Article 690 and should not be assumed based on this drawing. Instructions for or labels on the PV module might require additional overcurrent devices that are not shown.

Equipment listed for marine, mobile, telecommunications, or other applications might not be suitable for installation in permanent PV power systems.

See also

NFPA 790, *Standard for Competency of Third-Party Field Evaluation Bodies*, and **NFPA 791**, *Recommended Practice and Procedures for Unlabeled Electrical Equipment Evaluation*, for information on field evaluation of equipment

Δ **(C) Qualified Personnel.** The installation of equipment, associated wiring, and interconnections shall be performed only by qualified persons.

(D) Multiple PV Systems. Multiple PV systems shall be permitted to be installed in or on a single building or structure. Where the PV systems are remotely located from each other, a directory in accordance with 705.10 shall be provided at each PV system disconnecting means.

(E) Locations Not Permitted. PV system equipment and disconnecting means shall not be installed in bathrooms.

Δ **(F) Electronic Power Converters Mounted in Not Readily Accessible Locations.** Electronic power converters and their associated devices shall be permitted to be mounted on roofs or other areas that are not readily accessible. Disconnecting means shall be installed in accordance with 690.15.

N **(G) PV Equipment Floating on Bodies of Water.** PV equipment floating on or attached to structures floating on bodies of water shall be identified as being suitable for the purpose and shall utilize wiring methods that allow for any expected movement of the equipment.

Informational Note: PV equipment in these installations are often subject to increased levels of humidity, corrosion, and mechanical and structural stresses. Expected movement of floating PV arrays is often included in the structural design.

690.6 Alternating-Current (ac) Modules and Systems.

(A) Photovoltaic Source Circuits. The requirements of Article 690 pertaining to PV source circuits shall not apply to ac modules or ac module systems. The PV source circuit, conductors, and inverters shall be considered as internal components of an ac module or ac module system.

(B) Output Circuit. The output of an ac module or ac module system shall be considered an inverter output circuit.

Pre-engineered alternating current (ac) module systems have components specified in their instructions that will interconnect multiple ac modules into a system with a single output circuit. The output circuit from such a system can be treated like an inverter output circuit just as a single ac module's output is. All direct current (dc) and ac wiring up to the system termination point, as identified in its instructions, can be considered internal to the system. Considerations for ampacity, disconnection, and other requirements that generally apply to the interconnection of multiple ac modules are addressed in the pre-engineering of the system.

Part II. Circuit Requirements

Δ **690.7 Maximum Voltage.** The maximum voltage shall be used to determine the voltage and voltage to ground of circuits in the application of this *Code*. Maximum voltage shall be used for conductors, cables, equipment, working space, and other applications where voltage limits and ratings are used. The maximum voltage of PV system dc circuits shall be the highest voltage between any two conductors of a circuit or any conductor and ground and shall comply with the following:

(1) PV system dc circuits shall not exceed 1000 volts within or originating from arrays located on or attached to buildings and PV system dc circuits inside buildings.
(2) PV system dc circuits shall not exceed 600 volts on or in one- and two-family dwellings.
(3) PV system dc circuits exceeding 1000 volts shall comply with 690.31(G).

Δ **(A) Photovoltaic Source Circuits.** The maximum dc voltage for a PV source circuit shall be calculated in accordance with one of the following methods:

(1) The sum of the PV module–rated open-circuit voltage of the series-connected modules in the PV string circuit corrected for the lowest expected ambient temperature using the open-circuit voltage temperature coefficients in accordance with the instructions included in the listing or labeling of the module
(2) For crystalline and multicrystalline silicon modules, the sum of the PV module–rated open-circuit voltage of the series-connected modules in the PV string circuit corrected

TABLE 690.7(A) *Voltage Correction Factors for Crystalline and Multicrystalline Silicon Modules*

Correction Factors for Ambient Temperatures Below 25°C (77°F). (Multiply the rated open-circuit voltage by the appropriate correction factor shown below.)

Ambient Temperature (°C)	Factor	Ambient Temperature (°F)
24 to 20	1.02	76 to 68
19 to 15	1.04	67 to 59
14 to 10	1.06	58 to 50
9 to 5	1.08	49 to 41
4 to 0	1.10	40 to 32
−1 to −5	1.12	31 to 23
−6 to −10	1.14	22 to 14
−11 to −15	1.16	13 to 5
−16 to −20	1.18	4 to −4
−21 to −25	1.20	−5 to −13
−26 to −30	1.21	−14 to −22
−31 to −35	1.23	−23 to −31
−36 to −40	1.25	−32 to −40

for the lowest expected ambient temperature using the correction factors provided in Table 690.7(A)

(3) For PV systems with an inverter generating capacity of 100 kW or greater, a documented and stamped PV system design, using an industry standard method maximum voltage calculation provided by a licensed professional electrical engineer

Informational Note No. 1: One source for lowest-expected, ambient temperature design data for various locations the chapter titled "Extreme Annual Mean Minimum Design Dry Bulb Temperature" found in the *ASHRAE Handbook — Fundamentals, 2017*. These temperature data can be used to calculate maximum voltage.

Informational Note No. 2: See SAND 2004-3535, *Photovoltaic Array Performance Model*, for one industry standard method for calculating maximum voltage of a PV system.

A PV source is not a constant-voltage source, and the difference between the rated operating voltage determined under controlled laboratory conditions and the open-circuit voltage, adjusted for lowest expected ambient temperature, under field-installed conditions can be significant. Consequently, the higher-rated open-circuit voltage must be used to select circuit components with proper voltage ratings.

The voltage (both open circuit and operating) of a PV power source increases as the temperature decreases. The installer should note the temperature conditions for which the PV device was rated. If the anticipated lowest temperature at the installation site is lower than the rating condition (25°C), Table 690.7(A) must be used to adjust the maximum open-circuit voltage of crystalline systems before conductors, overcurrent devices, and switchgear are selected. For other than crystalline systems, see the manufacturer's instructions.

Where a listed PV module includes open-circuit voltage temperature coefficients in the installation instructions, the temperature coefficients provide a more accurate maximum system voltage than those from Table 690.7(A) and are required to be used instead of applying the table.

Bipolar PV systems (with positive and negative voltages) are required to be separated into two separate monopolar circuits; thus, the maximum circuit voltage is the maximum voltage of a single monopole.

Application Example

A system with open-circuit voltages of +480 volts and −480 volts with respect to ground would have a system open-circuit voltage of 480 volts even though the pole-to-pole maximum is 960 volts. This voltage should be multiplied by a temperature-dependent factor from Table 690.7(A), yielding a system design voltage of up to 600 volts since this is the maximum voltage to ground and the maximum voltage between the two wires of a single monopole permitted for one- and two-family dwellings. The system design voltage should be used in the selection of cables and other equipment. Certain bipolar PV arrays meeting the requirements of 690.7(C) might have different requirements for calculating the maximum system voltage.

Δ **(B) DC-to-DC Converter Circuits.** In PV dc-to-dc converter circuits, the maximum voltage shall be calculated in accordance with 690.7(B)(1) or (B)(2).

(1) Single DC-to-DC Converter. For circuits connected to the output of a single dc-to-dc converter, the maximum voltage shall be determined in accordance with the instructions included in the listing or labeling of the dc-to-dc converter. If the instructions do not provide a method to determine the maximum voltage, the maximum voltage shall be the maximum rated voltage output of the dc-to-dc converter.

(2) Two or More Series-Connected DC-to-DC Converters. For circuits connected to the output of two or more series-connected dc-to-dc converters, the maximum voltage shall be determined in accordance with the instructions included in the listing or labeling of the dc-to-dc converter. If the instructions do not provide a method to determine the maximum voltage, the maximum voltage shall be the sum of the maximum rated voltage output of the dc-to-dc converters in series.

Δ **(C) Bipolar PV Source Circuits.** For monopole subarrays in bipolar systems, the maximum voltage shall be the highest voltage between the monopole circuit conductors where one conductor of the monopole circuit is connected to the functionally grounded reference. To prevent overvoltage in the event of a ground fault or arc fault, the monopole circuits shall be isolated from ground.

(D) Marking DC PV Circuits. A permanent readily visible label indicating the highest maximum dc voltage in a PV system, calculated in accordance with 690.7, shall be provided by the installer at one of the following locations:

(1) DC PV system disconnecting means
(2) PV system electronic power conversion equipment
(3) Distribution equipment associated with the PV system

690.8 Circuit Sizing and Current.

(A) Calculation of Maximum Circuit Current. The maximum current for the specific circuit shall be calculated in accordance with one of the methods in 690.8(A)(1) or (A)(2).

Δ **(1) PV System Circuits.** The maximum current shall be calculated in accordance with 690.8(A)(1)(a) through (A)(1)(c).

(a) *Photovoltaic Source Circuit Currents* The maximum current shall be as calculated in either of the following:

(1) The maximum current shall be the sum of the short-circuit current ratings of the PV modules connected in parallel multiplied by 125 percent.

(2) For PV systems with an inverter generating capacity of 100 kW or greater, a documented and stamped PV system design, using an industry standard method maximum current calculation provided by a licensed professional electrical engineer, shall be permitted. The calculated maximum current value shall be based on the highest 3-hour current average resulting from the simulated local irradiance on the PV array accounting for elevation and orientation. The current value used by this method shall not be less than 70 percent of the value calculated using 690.8(A)(1)(a)(1).

Informational Note: See SAND 2004-3535, *Photovoltaic Array Performance Model*, for one industry standard method for calculating maximum current of a PV system. This model is used by the System Advisor Model simulation program provided by the National Renewable Energy Laboratory.

Calculating PV source maximum circuit currents, as described in 690.8(A)(1)(a)(1), uses the array short-circuit current, which allows for proper sizing of conductors to handle the current generated during periods of operation under a short-circuit condition. The 125-percent factor is required because PV modules can deliver output currents higher than the rated short-circuit currents for more than 3 hours near solar noon, which means they should be considered as a continuous load.

The other option for calculating PV source maximum circuit currents, as described in 690.8(A)(1)(a)(2), permits a licensed professional electrical engineer to perform simulations to establish the highest 3-hour current for a specific system.

(b) *PV DC-to-DC Converter Circuit Current.* The maximum current shall be the sum of parallel connected dc-to-dc converter continuous output current ratings.

(c) *Inverter Output Circuit Current.* The maximum current shall be the inverter continuous output current rating.

Informational Note: Modules that can produce electricity when exposed to light on multiple surfaces are labeled with applicable short-circuit currents. Additional guidance is provided in the instructions included with the listing.

Both stand-alone and interactive inverters are power-limited devices. Output circuits connected to these devices are sized on the continuous-rated outputs of the inverter. Exhibit 690.3 shows an inverter label displaying the maximum output circuit current along with other necessary ratings.

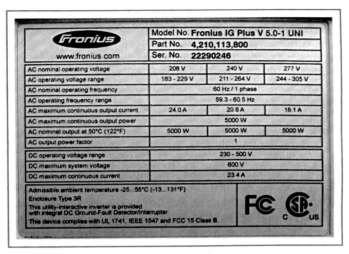

Model No. Fronius IG Plus V 5.0-1 UNI		
Part No. 4,210,113,800		
Ser. No. 22290246		
AC nominal operating voltage	208 V	240 V	277 V
AC operating voltage range	183 - 229 V	211 - 264 V	244 - 305 V
AC nominal operating frequency	60 Hz / 1 phase		
AC operating frequency range	59.3 - 60.5 Hz		
AC maximum continuous output current	24.0 A	20.8 A	18.1 A
AC maximum continuous output power	5000 W		
AC nominal output at 50°C (122°F)	5000 W	5000 W	5000 W
AC output power factor	1		
DC operating voltage range	230 - 500 V		
DC maximum system voltage	600 V		
DC maximum continuous current	23.4 A		

Admissible ambient temperature -25...55°C (-13...131°F)
Enclosure Type 3R
This utility-interactive inverter is provided with integral DC Ground-Fault Detector/Interrupter
This device complies with UL 1741, IEEE 1547 and FCC 15 Class B.

EXHIBIT 690.3 An interactive inverter label.

(2) Circuits Connected to the Input of Electronic Power Converters. Where a circuit is protected with an overcurrent device not exceeding the conductor ampacity, the maximum current shall be permitted to be the rated input current of the electronic power converter input to which it is connected.

(B) Conductor Ampacity. Circuit conductors shall have an ampacity not less than the larger of 690.8(B)(1) or (B)(2).

(1) Without Adjustment and Correction Factors. The minimum conductor size with an ampacity not less than the maximum currents calculated in 690.8(A) multiplied by 125 percent.

Exception: Circuits containing an assembly, together with its overcurrent device(s), that is listed for continuous operation at 100 percent of its rating shall be permitted to be used at 100 percent of its rating.

(2) With Adjustment and Correction Factors. The maximum currents calculated in 690.8(A) with adjustment and correction factors.

(C) Systems with Multiple Direct-Current Voltages. For a PV power source that has multiple output circuit voltages and employs a common-return conductor, the ampacity of the common-return conductor shall not be less than the sum of the ampere ratings of the overcurrent devices of the individual output circuits.

Δ **(D) Multiple PV String Circuits.** Where an overcurrent device is used to protect more than one set of parallel-connected PV string circuits, the ampacity of each conductor protected by the device shall not be less than the sum of the following:

(1) The rating of the overcurrent device
(2) The sum of the maximum currents as calculated in 690.8(A)(1)(a) for the other parallel-connected PV string circuits protected by the overcurrent device

Normally, labels or module instructions require reverse overcurrent protection for each module or string of modules. In some cases, modules with low-rated short-circuit currents and high values of the required series protective fuse could allow the use of one overcurrent device to provide reverse-current protection for multiple modules or strings of modules and overcurrent protection for the conductors. The PV module manufacturer should be contacted for specific information regarding allowable source circuit configurations.

Δ **690.9 Overcurrent Protection.**

(A) Circuits and Equipment. PV system dc circuit and inverter output conductors and equipment shall be protected against overcurrent. Circuits sized in accordance with 690.8(A)(2) are required to be protected against overcurrent with overcurrent protective devices. Each circuit shall be protected from overcurrent in accordance with 690.9(A)(1), (A)(2), or (A)(3).

(1) Circuits Where Overcurrent Protection Not Required. Overcurrent protective devices shall not be required where both of the following conditions are met:

(1) The conductors have sufficient ampacity for the maximum circuit current.
(2) The currents from all sources do not exceed the maximum overcurrent protective device rating specified for the PV module or electronic power converter.

(2) Circuits Where Overcurrent Protection is Required on One End. A circuit conductor connected at one end to a current-limited supply, where the conductor is rated for the maximum circuit current from that supply, and also connected to sources having an available maximum circuit current greater than the ampacity of the conductor, shall be protected from overcurrent at the point of connection to the higher current source.

> Informational Note: Photovoltaic system dc circuits and electronic power converter outputs powered by these circuits are current-limited and in some cases do not need overcurrent protection. Where these circuits are connected to higher current sources, such as parallel-connected PV system dc circuits, energy storage systems, or a utility service, the overcurrent device is often installed at the higher current source end of the circuit conductor.

It may be possible for other PV source circuits, other supply sources through the inverter, and energy storage system (ESS) circuits to supply current to source circuits in the event of a fault. An overcurrent device is required for each conductor at each connection point to limit the fault current on that conductor, unless the conductors are sized for the maximum available current. Where more than two strings of PV modules are connected in parallel, overcurrent devices might be required in the dc PV source or output circuits.

(3) Other Circuits. Circuits that do not comply with 690.9(A)(1) or (A)(2) shall be protected with one of the following methods:

(1) Conductors not greater than 3 m (10 ft) in length and not in buildings, protected from overcurrent on one end

(2) Conductors not greater than 3 m (10 ft) in length and in buildings, protected from overcurrent on one end and in a raceway or metal clad cable
(3) Conductors protected from overcurrent on both ends
(4) Conductors not installed on or in buildings are permitted to be protected from overcurrent on one end of the circuit where the circuit complies with all of the following conditions:

  a. The conductors are installed in metal raceways or metal-clad cables, or installed in enclosed metal cable trays, or underground, or where directly entering pad-mounted enclosures.
  b. The conductors for each circuit terminate on one end at a single circuit breaker or a single set of fuses that limit the current to the ampacity of the conductors.
  c. The overcurrent device for the conductors is an integral part of a disconnecting means or shall be located within 3 m (10 ft) of conductor length of the disconnecting means.
  d. The disconnecting means for the conductors is installed outside of a building, or at a readily accessible location nearest the point of entrance of the conductors inside of a building, including installations complying with 230.6.

Section 690.9(A)(3) addresses circuits with sources of overcurrent on both ends with four protection options.

Section 690.9(A)(3)(1) permits short conductor lengths to be protected from overcurrent on one end. Short conductors are common where combiner boxes are installed next to inverters, and the language in this requirement will reduce the need for fuses on both ends of a short wire. The "tap rule" in 240.21(B) is an example of where the 10-foot length is used for feeder taps with remote overcurrent protection.

Section 690.9(A)(3)(2) permits conductors of 10 feet or less in buildings if they are in raceways or metal-clad cables, which is similar to the requirement in 240.21.

Section 690.9(A)(3)(3) covers conductors in general with sources of overcurrent on both ends since they would require overcurrent protection on both ends as required by Article 240.

Section 690.9(A)(3)(4) covers longer runs with overcurrent on one end where the conductors are located outside of the building. Since conductors in PV systems have very limited short-circuit current on at least one end of the conductor, overcurrent protection is typically located on one end of the circuit. With the language related to inputs to electronic conversion devices, many dc circuits will be required to have overcurrent protection on both ends of the circuit where on buildings and where greater than 10 feet in length, in accordance with 690.9(A)(3)(1).

(B) Device Ratings. Overcurrent devices used in PV source circuits shall be listed for use in PV systems. Electronic devices that are listed to prevent backfeed current in PV system dc circuits shall be permitted to prevent overcurrent of conductors on the PV array side of the device. Overcurrent devices, where required, shall be rated in accordance with one of the following and permitted to be rounded up to the next higher standard size in accordance with 240.4(B):

(1) Overcurrent devices shall be rated not less than 125 percent of the maximum currents calculated in 690.8(A).
(2) An assembly, together with its overcurrent device(s), that is listed for continuous operation at 100 percent of its rating shall be permitted to be used at 100 percent of its rating.

Informational Note: Some electronic devices prevent backfeed current, which in some cases is the only source of overcurrent in PV system dc circuits.

Because these circuits are subject to environmental stresses, the overcurrent devices for the dc circuits are required to be specifically listed for use in PV systems. The overcurrent devices can be either supplemental or branch-circuit devices. Direct-current fault currents are considerably harder to interrupt than ac faults. Overcurrent devices marked or listed only for ac use should not be used in dc circuits. Automotive- and marine-type fuses, although used in dc systems, might not have the proper ratings for use in PV systems.

Δ **(C) PV System DC Circuits.** A single overcurrent protective device, where required, shall be permitted to protect the PV modules, dc-to-dc converters, and conductors of each circuit. Where single overcurrent protection devices are used to protect circuits, all overcurrent devices shall be placed in the same polarity for all circuits within a PV system. The overcurrent devices shall be accessible but shall not be required to be readily accessible.

Informational Note: Due to improved ground-fault protection required in PV systems by 690.41(B), a single overcurrent protective device in either the positive or negative conductors of a PV system in combination with this ground-fault protection provides adequate overcurrent protection.

Δ **(D) Transformers.** Overcurrent protection for power transformers shall be installed in accordance with 705.30(F).

Exception: A power transformer with a current rating on the side connected toward the interactive inverter output, not less than the rated continuous output current of the inverter, shall be permitted without overcurrent protection from the inverter.

Δ **690.11 Arc-Fault Circuit Protection (dc).** Photovoltaic systems with PV system dc circuits operating at 80 volts dc or greater between any two conductors shall be protected by a listed PV arc-fault circuit interrupter or other system components listed to provide equivalent protection. The system shall detect and interrupt arcing faults resulting from a failure in the intended continuity of a conductor, connection, module, or other system component in the PV system dc circuits.

Exception: PV system dc circuits that utilize metal-clad cables, are installed in metal raceways or enclosed metal cable trays, or are underground shall be permitted without arc-fault circuit protection if the installation complies with at least one of the following:

(1) The PV system dc circuits are not installed in or on buildings.
(2) The PV system dc circuits are located in or on detached structures whose sole purpose is to support or contain PV system equipment.

The arc-fault protective device used to meet this requirement must be listed for dc use and listed for use in PV systems. Listed components that provide protection equivalent to arc-fault protection also are permitted by this requirement. The exception allows PV output circuits on ground-mounted PV systems meeting the requirements and those installed in metal raceway and metal clad cables to be installed without arc-fault protection.

690.12 Rapid Shutdown of PV Systems on Buildings. PV system circuits installed on or in buildings shall include a rapid shutdown function to reduce shock hazard for firefighters in accordance with 690.12(A) through (D).

Exception No. 1: Ground-mounted PV system circuits that enter buildings, of which the sole purpose is to house PV system equipment, shall not be required to comply with 690.12.

Exception No. 2: PV equipment and circuits installed on nonenclosed detached structures including but not limited to parking shade structures, carports, solar trellises, and similar structures shall not be required to comply with 690.12.

Informational Note: Exceptions for rapid shutdown are intended to be consistent with building and fire codes that have limitations as to the types of buildings on which firefighters typically perform rooftop operations.

Fire fighters must contend with elements of a PV system that remain energized after the service disconnect is opened. This rapid shutdown requirement reduces the potential for shock within 30 seconds of activation of shutdown. Methods and designs for achieving proper rapid shutdown are not addressed by the *NEC®* but instead are addressed in the product standards for this type of equipment.

(A) Controlled Conductors. Requirements for controlled conductors shall apply to the following:

(1) PV system dc circuits
(2) Inverter output circuits originating from inverters located within the array boundary

Informational Note: The rapid shutdown function reduces the risk of electrical shock that dc circuits in a PV system could pose for firefighters. The ac output conductors from PV systems that include inverters will either be de-energized after shutdown initiation or will remain energized by other sources such as a utility service. To prevent PV arrays with attached inverters from having energized ac conductors within the PV array(s), those circuits are also specifically controlled after shutdown initiation.

Exception: PV system circuits originating within or from arrays not attached to buildings that terminate on the exterior of buildings and PV system circuits installed in accordance with 230.6 shall not be considered controlled conductors for the purposes of 690.12.

(B) Controlled Limits. The use of the term *array boundary* in this section is defined as 305 mm (1 ft) from the array in all directions. Controlled conductors outside the array boundary shall comply with 690.12(B)(1) and inside the array boundary shall comply with 690.12(B)(2). Equipment and systems shall be permitted to meet the requirements of both inside and outside the array as defined by the manufacturer's instructions included with the listing.

(1) Outside the Array Boundary. Controlled conductors located outside the boundary or more than 1 m (3 ft) from the point of entry inside a building shall be limited to not more than 30 volts within 30 seconds of rapid shutdown initiation. Voltage shall be measured between any two conductors and between any conductor and ground.

Δ **(2) Inside the Array Boundary.** The PV system shall comply with one of the following:

(1) The PV system shall provide shock hazard control for firefighters through the use of a PVHCS installed in accordance with the instructions included with the listing or field labeling. Where a PVHCS requires initiation to transition to a controlled state, the rapid shutdown initiation device required in 690.12(C) shall perform this initiation.

Informational Note No. 1: A listed or field-labeled PVHCS is comprised of either an individual piece of equipment that fulfills the necessary functions or multiple pieces of equipment coordinated to perform the functions as described in the installation instructions to reduce the risk of electric shock hazard within a damaged PV array for firefighters. See UL 3741, *Photovoltaic Hazard Control*.

(2) The PV system shall provide shock hazard control for firefighters by limiting the highest voltage inside equipment or between any two conductors of a circuit or any conductor and ground inside array boundary to not more than 80 volts within 30 seconds of rapid shutdown initiation.

Informational Note No. 2: Common methods include the use of PV equipment with a limited maximum voltage of 80 volts as determined by 690.7, PVRSE, PVHCE, or any combination of these.

Exhibit 690.4 illustrates the array boundary and the controlled conductors and limits.

Δ **(C) Initiation Device.** Where circuits identified in 690.12(A) are required to meet the requirements in 690.12(B), an initiation device(s) shall be provided and shall initiate the rapid shutdown function. The device's "off" position shall indicate that the rapid shutdown function has been initiated for all PV systems connected to that device. For one- and two-family dwellings, an initiation device(s), where required, shall be located at a readily accessible outdoor location.

For a single PV system, the rapid shutdown initiation shall occur by the operation of any single initiation device. Devices shall consist of at least one or more of the following:

(1) Service disconnecting means
(2) PV system disconnecting means
(3) Readily accessible switch that plainly indicates whether it is in the "off" or "on" position

Where multiple PV systems are installed with rapid shutdown functions on a single service, the initiation device(s) shall consist of not more than six switches or six sets of circuit breakers, or a combination of not more than six switches and sets of circuit breakers, mounted in a single enclosure, or in a group of

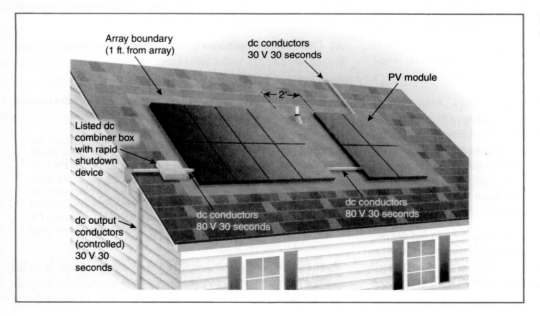

EXHIBIT 690.4 Controlled conductors and limits.

Array boundary (1 ft. from array)
dc conductors 30 V 30 seconds
PV module
2'
Listed dc combiner box with rapid shutdown device
dc conductors 80 V 30 seconds
dc conductors 80 V 30 seconds
dc output conductors (controlled) 30 V 30 seconds

separate enclosures. These initiation device(s) shall initiate the rapid shutdown of all PV systems with rapid shutdown functions on that service.

Multiple rapid shutdown initiating devices are permitted, as long as each initiating device is capable of initiating shutdown.

(D) Buildings with Rapid Shutdown. Buildings with PV systems shall have a permanent label located at each service equipment location to which the PV systems are connected or at an approved readily visible location and shall indicate the location of rapid shutdown initiation devices. The label shall include a simple diagram of a building with a roof and shall include the following words:

> SOLAR PV SYSTEM IS EQUIPPED
> WITH RAPID SHUTDOWN.
> TURN RAPID SHUTDOWN SWITCH
> TO THE "OFF" POSITION TO
> SHUT DOWN PV SYSTEM AND
> REDUCE SHOCK HAZARD IN ARRAY.

The title "SOLAR PV SYSTEM IS EQUIPPED WITH RAPID SHUTDOWN" shall have these letters capitalized and having a minimum height of 9.5 mm (⅜ in.). All text shall be legible and contrast the background.

Informational Note: See Informational Note Figure 690.12(D).

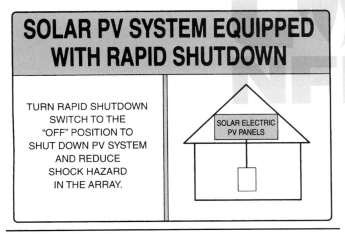

INFORMATIONAL NOTE FIGURE 690.12(D) Label for Roof-Mounted PV Systems with Rapid Shutdown.

Exhibit 690.5 shows a typical rapid shutdown device.

(1) Buildings with More Than One Rapid Shutdown Type. For buildings that have PV systems with more than one rapid shutdown type or PV systems with no rapid shutdown, a detailed plan view diagram of the roof shall be provided showing each different PV system with a dotted line around areas that remain energized after rapid shutdown is initiated.

EXHIBIT 690.5 Rapid shutdown device.

(2) Rapid Shutdown Switch. A rapid shutdown switch shall have a label that includes the following wording located on or no more than 1 m (3 ft) from the switch:

> RAPID SHUTDOWN SWITCH FOR
> SOLAR PV SYSTEM

The label shall be reflective, with all letters capitalized and having a minimum height of 9.5 mm (⅜ in.) in white on red background.

Part III. Disconnecting Means

690.13 Photovoltaic System Disconnecting Means. Means shall be provided to disconnect the PV system from all wiring systems including power systems, energy storage systems, and utilization equipment and its associated premises wiring.

(A) Location.

(1) Readily Accessible. The PV system disconnecting means shall be installed at a readily accessible location.

(2) Enclosure Doors and Covers. Where a disconnecting means for circuits operating above 30 volts is readily accessible to unqualified persons, an enclosure door or hinged cover that exposes energized parts when open shall have its door or cover locked or require a tool to be opened.

(B) Marking. Each PV system disconnecting means shall plainly indicate whether in the open (off) or closed (on) position and be permanently marked "PV SYSTEM DISCONNECT" or equivalent. Additional markings shall be permitted based upon the specific system configuration. For PV system disconnecting means where the line and load terminals may be energized in the open position, the device shall be marked with the following words or equivalent:

WARNING
ELECTRIC SHOCK HAZARD
TERMINALS ON THE LINE AND LOAD SIDES MAY BE
ENERGIZED IN THE OPEN POSITION

The warning sign(s) or label(s) shall comply with 110.21(B).

△ (C) Maximum Number of Disconnects. Each PV system disconnecting means shall consist of not more than six switches or six sets of circuit breakers, or a combination of not more than six switches and sets of circuit breakers, mounted in a single enclosure, or in a group of separate enclosures. A single PV system disconnecting means shall be permitted for the combined ac output of one or more inverters or ac modules.

> Informational Note: This requirement does not limit the number of PV systems connected to a service as permitted in 690.4(D). This requirement allows up to six disconnecting means to disconnect a single PV system. For PV systems where all power is converted through interactive inverters, a dedicated circuit breaker, in 705.12(B)(1), is an example of a single PV system disconnecting means.

If a building has multiple sources of power, such as the utility, a PV system, a backup generator, and a wind system, no more than six disconnects for each source of power to the building are permitted. However, this does not require the disconnects for all the sources to be grouped together.

Interactive ac PV modules are designed to produce power only if they are connected to an external power source at the correct voltage and frequency. A single disconnecting means removes the external source and turns off the output of all ac PV modules connected to that disconnecting device.

(D) Ratings. The PV system disconnecting means shall have ratings sufficient for the maximum circuit current, available fault current, and voltage that is available at the terminals of the PV system disconnect.

(E) Type of Disconnect. The PV system disconnecting means shall simultaneously disconnect the PV system conductors that are not solidly grounded from all conductors of other wiring systems. The PV system disconnecting means or its remote operating device or the enclosure providing access to the disconnecting means shall be capable of being locked in accordance with 110.25. The PV system disconnecting means shall be one of the following:

(1) A manually operable switch or circuit breaker
(2) A connector meeting the requirements of 690.33(D)(1) or (D)(3)
(3) A pull-out switch with the required interrupting rating
(4) A remote-controlled switch or circuit breaker that is operable locally and opens automatically when control power is interrupted
(5) A device listed or approved for the intended application

> Informational Note: Circuit breakers marked "line" and "load" may not be suitable for backfeed or reverse current.

Exhibit 690.6 shows an ac PV system disconnect.

EXHIBIT 690.6 PV system disconnect.

690.15 Disconnecting Means for Isolating Photovoltaic Equipment. Disconnecting means of the type required in 690.15(A) shall be provided to disconnect ac PV modules, fuses, dc-to-dc converters, inverters, and charge controllers from all conductors that are not solidly grounded.

△ (A) Type of Disconnecting Means. Where a disconnect is required to isolate equipment, the disconnecting means shall be one of the following:

(1) An equipment disconnecting means in accordance with 690.15(C)
(2) An isolating device as part of listed equipment where an interlock or similar means prevents the opening of the isolating device under load
(3) For circuits with a maximum circuit current of 30 amperes or less, an isolating device in accordance with 690.15(B)

(B) Isolating Device. An isolating device shall not be required to have an interrupting rating. Where an isolating device is not rated for interrupting the circuit current, it shall be marked "Do Not Disconnect Under Load" or "Not for Current Interrupting." An isolating device shall not be required to simultaneously disconnect all current-carrying conductors of a circuit. The isolating device shall be one of the following:

(1) A mating connector meeting the requirements of 690.33 and listed and identified for use with specific equipment
(2) A finger-safe fuse holder
(3) An isolating device that requires a tool to place the device in the open (off) position
(4) An isolating device listed for the intended application

△ (C) Equipment Disconnecting Means. Equipment disconnecting means shall comply with the following:

(1) Have ratings sufficient for the maximum circuit current, available fault current, and voltage that is available at the terminals.

(2) Simultaneously disconnect all current-carrying conductors that are not solidly grounded to the circuit to which it is connected.

(3) Be externally operable without exposing the operator to contact with energized parts and shall indicate whether in the open (off) or closed (on) position. Where not within sight or not within 3 m (10 ft) of the equipment, the disconnecting means or its remote operating device or the enclosure providing access to the disconnecting means shall be capable of being locked in accordance with 110.25.

(4) Be one of the types in 690.13(E)(1) through (E)(5).

Equipment disconnecting means, other than those complying with 690.33, shall be marked in accordance with the warning in 690.13(B) if the line and load terminals can be energized in the open position.

Informational Note: A common installation practice is to terminate PV source-side dc conductors in the same manner that utility source-side ac conductors are generally connected on the line side of a disconnecting means. This practice is more likely to de-energize load-side terminals, blades, and fuses when the disconnect is in the open position and no energized sources are connected to the load side of the disconnect.

(D) Location and Control. Isolating devices or equipment disconnecting means shall comply with one or more of the following:

(1) Located within the equipment
(2) Located in sight from and readily accessible from the equipment for those to whom access is required
(3) Lockable in accordance with 110.25
(4) Provided with remote controls to activate the disconnecting means where the remote controls comply with one of the following:

    a. The disconnecting means and their controls are located within the same equipment.
    b. The disconnecting means is lockable in accordance with 110.25, and the location of the controls are marked on the disconnecting means.

Part IV. Wiring Methods and Materials

690.31 Wiring Methods.

Δ **(A) Wiring Systems.**

N **(1) Serviceability.** Where wiring devices with integral enclosures are used, sufficient length of cable shall be provided to facilitate replacement.

N **(2) Where Readily Accessible.** Where not guarded, PV system dc circuit conductors operating at voltages greater than 30 volts that are readily accessible to unqualified persons shall be installed in Type MC cable, in multiconductor jacketed cable, or in raceway.

N **(3) Conductor Ampacity.** The ampacity of 105°C (221°F) and 125°C (257°F) conductors shall be permitted to be determined by Table 690.31(A)(3)(1). For ambient temperatures greater than 30°C (86°F), the ampacities of these conductors shall be corrected in accordance with Table 690.31(A)(3)(2).

N **(4) Special Equipment.** In addition to wiring methods included elsewhere in this *Code*, other wiring systems specifically listed for use in PV systems shall be permitted.

TABLE 690.31(A)(3)(1) *Ampacities of Insulated Conductors Rated Up To and Including 2000 Volts, 105°C Through 125°C (221°F Through 257°F), Not More Than Three Current-Carrying Conductors in Raceway, Cable, or Earth (Directly Buried), Based on Ambient Temperature of 30°C (86°F)*

	Types	
AWG	PVC, CPE, XLPE 105°C	XLPE, EPDM 125°C
18	15	16
16	19	20
14	29	31
12	36	39
10	46	50
8	64	69
6	81	87
4	109	118
3	129	139
2	143	154
1	168	181
1/0	193	208
2/0	229	247
3/0	263	284
4/0	301	325

TABLE 690.31(A)(3)(2) *Correction Factors*

Ambient Temperature (°C)	Temperature Rating of Conductor		Ambient Temperature (°F)
	105°C (221°F)	125°C (257°F)	
30	1	1	86
31–35	0.97	0.97	87–95
36–40	0.93	0.95	96–104
41–45	0.89	0.92	105–113
46–50	0.86	0.89	114–122
51–55	0.82	0.86	123–131
56–60	0.77	0.83	132–140
61–65	0.73	0.79	141–149
66–70	0.68	0.76	150–158
71–75	0.63	0.73	159–167
76–80	0.58	0.69	168–176
81–85	0.52	0.65	177–185
86–90	0.45	0.61	186–194
91–95	0.37	0.56	195–203
96–100	0.26	0.51	204–212
101–105	—	0.46	213–221
106–110	—	0.4	222–230
111–115	—	0.32	231–239
116–120	—	0.23	240–248

Informational Note: See 110.14(C) for conductor temperature limitations due to termination provisions.

All cables and conductors installed outdoors and exposed to direct sunlight and wet conditions must be suitable for such conditions. Conductors inside raceways installed in wet locations are required to be identified or listed as suitable for wet locations.

See also

310.10(C) for the requirements on conductors installed in wet locations

Open, single conductors are permitted where listed and identified as "Photovoltaic Wire," "Photovoltaic Cable," "PV Wire," or "PV Cable." These conductors are evaluated for use where exposed to direct sunlight and wet conditions.

Most PV modules do not have means for attaching raceways. These circuits might have to be made "not readily accessible" by use of physical barriers such as wire plastic or metal guards.

Δ **(B) Identification and Grouping.**

N **(1) Conductors of Different Systems.** Where not otherwise allowed in an equipment's listing, PV system dc circuits shall not occupy the same equipment wiring enclosure, cable, or raceway as other non-PV systems or inverter output circuits unless separated from other circuits by a barrier or partition.

Exception: Where all conductors or cables have an insulation rating equal to at least the maximum circuit voltage applied to any conductor within the same wiring method, the following shall be permitted:

(1) Multiconductor jacketed cables for remote control, signaling, or power-limited circuits shall be permitted within the same wiring enclosure, cable, or raceway as PV system dc circuits where all circuits serve the PV system.

(2) Inverter output circuits shall be permitted to occupy the same junction box, pull box, or wireway with PV system dc circuits that are identified and grouped as required by 690.31(B)(2) and (B)(3).

(3) PV system dc circuits utilizing multiconductor jacketed cable or metal-clad cable assemblies or listed wiring harnesses identified for the application shall be permitted to occupy the same wiring method as inverter output circuits and other non-PV systems.

AC branch-circuit conductors that supply an exterior luminaire installed near a roof-mounted PV array are examples of conductors that must not share the same raceway or cable with PV source or output circuit conductors. Conductors of different systems are permitted in the same cable tray where separated from the PV source or output circuit conductors by a barrier or a partition of a material compatible with the cable tray. *Barrier* is a more generic term than *partition* and would allow approved methods to keep these circuits from direct contact with each other. The exception permits dc multiconductor jacketed cable or metal-clad cable assemblies, listed

for the application, to be installed with inverter output circuits and other non-PV system conductors.

DC conductors directly related to a specific PV system are permitted in the same raceway as PV source and output conductors, provided they are grouped and identified and meet the separation requirements of 300.3(C).

Δ **(2) Identification.** PV system dc circuit conductors shall be identified at all termination, connection, and splice points by color coding, marking tape, tagging, or other approved means in accordance with 690.31(B)(2)(a) and (B)(2)(b).

Exception: Where the identification of the conductors is evident by spacing or arrangement, further identification shall not be required.

(a) Conductors that rely on other than color coding for polarity identification shall be identified by an approved permanent marking means such as labeling, sleeving, or shrink-tubing that is suitable for the conductor size.

(b) The permanent marking means for nonsolidly grounded positive conductors shall include imprinted plus signs (+) or the word POSITIVE or POS durably marked on insulation of a color other than green, white, or gray. The permanent marking means for nonsolidly grounded negative conductors shall include imprinted negative signs (−) or the word NEGATIVE or NEG durably marked on insulation of a color other than green, white, gray, or red. Only solidly grounded PV system dc circuit conductors shall be marked in accordance with 200.6.

Δ **(3) Grouping.** Where ac and dc conductors of PV systems occupy the same junction box, pull box, or wireway, the ac and dc circuit conductors shall be grouped separately by cable ties or similar means at least once and at intervals not to exceed 1.8 m (6 ft).

Exception: The requirement for grouping shall not apply if the circuit enters from a cable or raceway unique to the circuit that makes the grouping obvious.

(C) Cables. Type PV wire or cable and Type distributed generation (DG) cable shall be listed.

Informational Note: See UL 4703, *Standard for Photovoltaic Wire*, for PV wire and UL 3003, *Distributed Generation Cables*, for DG cable. PV wire and cable and DG cable have a nonstandard outer diameter.

Δ **(1) Single-Conductor Cable.** Single-conductor cables shall comply with 690.31(C)(1)(a) through (C)(1)(c).

(a) Single-conductor cable in exposed outdoor locations in PV system dc circuits within the PV array shall be permitted to be one of the following:

(1) PV wire or cable
(2) Single-conductor cable marked sunlight resistant and Type USE-2 and Type RHW-2

(b) Exposed cables sized 8 AWG or smaller shall be supported and secured at intervals not to exceed 600 mm (24 in.)

by cable ties, straps, hangers, or similar fittings listed and identified for securement and support in outdoor locations. PV wire or cable shall be permitted in all locations where RHW-2 is permitted.

PV systems meeting the requirements of 691.4 shall be permitted to have support and securement intervals as defined in the engineered design.

(c) Exposed cables sized larger than 8 AWG shall be supported and secured at intervals not to exceed 1400 mm (54 in.) by cable ties, straps, hangers, or similar fittings listed and identified for securement and support in outdoor locations.

Most PV modules are designed for a direct series connection by using factory-installed leads and connectors. To accommodate such a connection, use of a single-conductor PV wire or cable or Type USE-2 cable and single-conductor cable listed for PV applications is permitted in PV source circuits. Type USE-2 cables must be marked RHW-2 and sunlight resistant. Cables connected to PV modules are exposed to sunlight, which can result in ultraviolet degradation of the insulation.

Extremely long runs of separated conductors (with loop inductance and distributed capacitance) and the resulting long-time constants in dc circuits can result in improper operation of overcurrent devices. Running both positive and negative conductors of each circuit and the equipment grounding conductor (EGC) as close together as possible also decreases induced currents from nearby lightning strikes. Because PV modules might operate at high temperatures and are installed in outdoor, exposed locations, the use of high-temperature conductors rated for wet locations, such as USE-2 or RHW-2, is often necessary. See 310.15(B)(2) for requirements on the ampacities of conductors in raceways or cables installed on rooftops exposed to sunlight. Single-conductor cables listed and labeled for use in PV applications will be identified as "PV Wire," "PV Cable," "Photovoltaic Wire," or "Photovoltaic Cable."

Δ **(2) Cable Tray.** Single-conductor PV wire or cable of all sizes or distributed generation (DG) cable of all sizes, with or without a cable tray rating, shall be permitted in cable trays installed in outdoor locations, provided that the cables are supported at intervals not to exceed 300 mm (12 in.) and secured at intervals not to exceed 1400 mm (54 in.).

Where installed in uncovered cable trays, ampacity of single-conductor PV wire smaller than 1/0 AWG, the adjustment factors for 1/0 AWG single conductor cable in 392.80(A)(2) shall be permitted to be used.

Where single-conductor PV wire smaller than 1/0 AWG is installed in ladder ventilated trough cable trays, the following shall apply:

(1) All single conductors shall be installed in a single layer.
(2) Conductors that are bound together to comprise each circuit pair shall be permitted to be installed in other than a single layer.
(3) The sum of diameters of all single conductor cables shall not exceed the cable tray width.

(3) Multiconductor Jacketed Cables. Where part of a listed PV assembly, multiconductor jacketed cables shall be installed in accordance with the included instructions. Where not part of a listed assembly, or where not otherwise covered in this *Code*, multiconductor jacketed cables, including DG cable, shall be installed in accordance with the product listing and shall be permitted in PV systems. These cables shall be installed in accordance with the following:

(1) In raceways, where on or in buildings other than rooftops
(2) Where not in raceways, in accordance with the following:

  a. Marked sunlight resistant in exposed outdoor locations
  b. Protected or guarded, where subject to physical damage
  c. Closely follow the surface of support structures
  d. Secured at intervals not exceeding 1.8 m (6 ft)
  e. Secured within 600 mm (24 in.) of mating connectors or entering enclosures
  f. Marked direct burial, where buried in the earth

(4) Flexible Cords and Cables Connected to Tracking PV Arrays. Flexible cords and flexible cables, where connected to moving parts of tracking PV arrays, shall comply with Article 400 and shall be of a type identified as a hard service cord or portable power cable; they shall be suitable for extra-hard usage, listed for outdoor use, water resistant, and sunlight resistant. Allowable ampacities shall be in accordance with 400.5. Stranded copper PV wire shall be permitted to be connected to moving parts of tracking PV arrays in accordance with the minimum number of strands specified in Table 690.31(C)(4).

TABLE 690.31(C)(4) *Minimum PV Wire Strands*

PV Wire AWG	Minimum Strands
18	17
16–10	19
8–4	49
2	130
1 AWG–1000 MCM	259

(5) Flexible, Fine-Stranded Cables. Flexible, fine-stranded cables shall be terminated only with terminals, lugs, devices, or connectors in accordance with 110.14.

Section 110.14 requires connectors and terminals for conductors more finely stranded than Class B and Class C stranding, as shown in Chapter 9, Table 10, to be identified for the specific conductor class or classes.

(6) Small-Conductor Cables. Single-conductor cables listed for outdoor use that are sunlight resistant and moisture resistant in sizes 16 AWG and 18 AWG shall be permitted for module interconnections where such cables meet the ampacity requirements of 400.5. Section 310.14 shall be used to determine the cable ampacity adjustment and correction factors.

Because the smaller cables might not be marked with standard code-recognized markings (such as USE-2), the PV module manufacturer or installer should verify that these cables are listed and labeled for PV use, which would indicate that they have the necessary sunlight and moisture resistance and are suitable for exposed, outdoor use.

In accordance with 200.6(A)(5), grounded conductors that are smaller than 6 AWG and used in PV source circuits are permitted to be marked at the time of installation with a white marking at all terminations. Only solidly grounded PV systems are allowed to have white markings.

△ **(D) Direct-Current Circuits on or in Buildings.** Wiring methods on or in buildings shall comply with the installation requirements in 690.31(D)(1) and (D)(2).

The use of metal raceways, Type MC cable, or metal enclosures inside a building provides additional physical protection for these circuits. Metal raceways also provide additional fire resistance should faults develop in the cable, and they provide an additional ground-fault detection path for the dc ground-fault detector-interrupter (GFDI) protection required by 690.41(B).

N (1) Metal Raceways and Enclosures. Where inside buildings, PV system dc circuits that exceed 30 volts or 8 amperes shall be contained in metal raceways, in Type MC metal-clad cable that complies with 250.118(A)(10)(b) or (A)(10)(c), or in metal enclosures.

Exception: PVHCS installed in accordance with 690.12(B)(2)(1) shall be permitted to be provided with or listed for use with nonmetallic enclosure(s), nonmetallic raceway(s), and cables other than Type MC metal-clad cable(s), at the point of penetration of the surface of the building.

△ **(2) Marking and Labeling.** Unless located and arranged so the purpose is evident, the following wiring methods and enclosures that contain PV system dc circuit conductors shall be marked with the wording PHOTOVOLTAIC POWER SOURCE or SOLAR PV DC CIRCUIT by means of permanently affixed labels or other approved permanent marking:

(1) Exposed raceways, cable trays, and other wiring methods
(2) Covers or enclosures of pull boxes and junction boxes
(3) Conduit bodies in which any of the available conduit openings are unused

The labels or markings shall be visible after installation. All letters shall be capitalized and shall be a minimum height of 9.5 mm (⅜ in.) in white on a red background. Labels shall appear on every section of the wiring system that is separated by enclosures, walls, partitions, ceilings, or floors. Spacing between labels or markings, or between a label and a marking, shall not be more than 3 m (10 ft). Labels required by this section shall be suitable for the environment where they are installed.

The objective of the requirements contained in 690.31(D)(2) is to protect persons from inadvertently damaging PV source and output circuit conductors. Where the location of the PV circuit

conductors is not obvious, fire fighters, other first responders, and maintenance personnel could be exposed to shock hazards. Ventilating roofs containing PV source or output circuits by cutting the membrane with saws could expose personnel to shock hazards and the building to further damage resulting from the ignition of combustible members due to arcing from damaged conductors.

(E) Bipolar Photovoltaic Systems. Where the sum, without consideration of polarity, of the voltages of the two monopole circuits exceeds the rating of the conductors and connected equipment, monopole circuits in a bipolar PV system shall be physically separated, and the electrical output circuits from each monopole circuit shall be installed in separate raceways until connected to the inverter. The disconnecting means and overcurrent protective devices for each monopole circuit output shall be in separate enclosures. All conductors from each separate monopole circuit shall be routed in the same raceway. Solidly grounded bipolar PV systems shall be clearly marked with a permanent, legible warning notice indicating that the disconnection of the grounded conductor(s) may result in overvoltage on the equipment.

Exception: Listed switchgear rated for the maximum voltage between circuits and containing a physical barrier separating the disconnecting means for each monopole circuit shall be permitted to be used instead of disconnecting means in separate enclosures.

(F) Wiring Methods and Mounting Systems. Roof-mounted PV array mounting systems shall be permitted to be held in place with an approved means other than those required by 110.13 and shall utilize wiring methods that allow any expected movement of the array.

Informational Note: Expected movement of unattached PV arrays is often included in structural calculations.

N (G) Over 1000 Volts DC. Equipment and wiring methods containing PV system dc circuits with a maximum voltage greater than 1000 volts shall comply with the following:

(1) Shall not be permitted on or in one- and two-family dwellings.
(2) Shall not be permitted within buildings containing habitable rooms.
(3) Where installed on the exterior of buildings shall be located less than 3 m (10 ft) above grade. Wiring methods containing PV system dc circuits connected to this equipment shall not be permitted to attach to the building greater than 10 m (33 ft) along the building surface from the equipment.

690.32 Component Interconnections. Fittings and connectors that are intended to be concealed at the time of on-site assembly, where listed for such use, shall be permitted for on-site interconnection of modules or other array components. Such fittings and connectors shall be equal to the wiring method employed in insulation, temperature rise, and short-circuit current rating, and shall be capable of resisting the effects of the environment in which they are used.

690.33 Mating Connectors. Mating connectors, other than connectors covered by 690.32, shall comply with 690.33(A) through (D).

(A) Configuration. The mating connectors shall be polarized and shall have a configuration that is noninterchangeable with receptacles in other electrical systems on the premises.

(B) Guarding. The mating connectors shall be constructed and installed so as to guard against inadvertent contact with live parts by persons.

(C) Type. The mating connectors shall be of the latching or locking type. Mating connectors that are readily accessible and that are used in circuits operating at over 30 volts dc or 15 volts ac shall require a tool for opening. Where mating connectors are not of the identical type and brand, they shall be listed and identified for intermatability, as described in the manufacturer's instructions.

The requirements for intermatability refer to the connection of two different brands or types of mating connectors that can be installed together only where they have been listed and identified for this use. The term *intermatability* is used in UL 6703, *Connectors for Use in Photovoltaic Systems.*

(D) Interruption of Circuit. Mating connectors shall be one of the following:

(1) Rated for interrupting current without hazard to the operator
(2) A type that requires the use of a tool to open and marked "Do Not Disconnect Under Load" or "Not for Current Interrupting"
(3) Supplied as part of listed equipment and used in accordance with instructions provided with the listed connected equipment

Informational Note: Some listed equipment, such as microinverters, are evaluated to make use of mating connectors as disconnect devices even though the mating connectors are marked as "Do Not Disconnect Under Load" or "Not for Current Interrupting."

The three options for connectors in this requirement provide for safe disconnection of circuit connectors either by allowing them to be opened under load or by requiring a warning indicating that load disconnection is necessary prior to opening the connector. Connectors that are not rated for disconnection under load cannot be opened or disconnected without the use of a tool. Connectors that are included as part of the listing of listed equipment also are permitted.

690.34 Access to Boxes. Junction, pull, and outlet boxes located behind modules or panels shall be so installed that the wiring contained in them can be rendered accessible directly or by displacement of a module(s) or panel(s) secured by removable fasteners and connected by a flexible wiring system.

Part V. Grounding and Bonding

690.41 PV System DC Circuit Grounding and Protection.

Δ **(A) PV System DC Circuit Grounding Configurations.** One or more of the following system configurations shall be employed for PV system dc circuits:

(1) 2-wire circuits with one functionally grounded conductor
(2) Bipolar circuits according to 690.7(C) with a functional ground reference (center tap)
(3) Circuits not isolated from the grounded inverter output circuit
(4) Ungrounded circuits
(5) Solidly grounded circuits as permitted in 690.41(B)
(6) Circuits protected by equipment listed and identified for the use

(B) DC Ground-Fault Detector-Interrupter (GFDI) Protection. PV system dc circuits that exceed 30 volts or 8 amperes shall be provided with GFDI protection meeting the requirements of 690.41(B)(1) and (B)(2) to reduce fire hazards.

Solidly grounded PV source circuits with not more than two modules in parallel and not on or in buildings shall be permitted without GFDI protection.

Informational Note: Not all inverters, charge controllers, or dc-to-dc converters include dc GFDI protection. Equipment that does not have GFDI protection often includes the following statement in the manual: "Warning: This unit is not provided with a GFDI device."

(1) Ground-Fault Detection. The GFDI device or system shall detect ground fault(s) in the PV system dc circuits, including any functionally grounded conductors, and be listed for providing GFDI protection. For dc-to-dc converters not listed as providing GFDI protection, where required, listed GFDI protection equipment identified for the combination of the dc-to-dc converter and the GFDI device shall be installed to protect the circuit.

Informational Note: Some dc-to-dc converters without integral GFDI protection on their input (source) side can prevent other GFDI protection equipment from properly functioning on portions of PV system dc circuits.

GFDI protection for the dc portions of PV systems should not be confused with the requirements for ac circuit GFCI protection. [See the definitions for *ground-fault circuit interrupter (GFCI)* and *ground-fault detector-interrupter, dc (GFDI)* in Article 100.] A GFCI is intended for the protection of personnel in single-phase and some three-phase ac systems. The ac GFCI functions to open the ungrounded conductor when fault current, typically 5-milliamps or more, is detected. In contrast, GFDIs meeting this requirement are intended to prevent fires in dc PV circuits due to ground faults.

(2) Faulted Circuits. The faulted circuits shall be controlled by one of the following methods:

(1) The current-carrying conductors of the faulted circuit shall be automatically disconnected.
(2) The device providing GFDI protection fed by the faulted circuit shall automatically cease to supply power to output circuits and interrupt the faulted PV system dc circuits from the ground reference in a functionally grounded system.

(3) Indication of Faults. The GFDI protection equipment shall provide indication of ground faults at a readily accessible location.

> Informational Note: Examples of indication include, but are not limited to, the following: remote indicator light, display, monitor, signal to a monitored alarm system, or receipt of notification by web-based services.

Δ **690.42 Point of PV System DC Circuit Grounding Connection.**

N **(A) Circuits with GFDI Protection.** Circuits protected by GFDI equipment in accordance with 690.41(B) shall have any circuit-to-ground connection made by the GFDI equipment.

N **(B) Solidly Grounded Circuits.** For solidly grounded PV system dc circuits, the grounding connection shall be made from any single point on the PV dc system to a point in the grounding electrode system in 690.47(A).

690.43 Equipment Grounding and Bonding. Exposed non-current-carrying metal parts of PV module frames, electrical equipment, and conductor enclosures of PV systems shall be connected to an equipment grounding conductor in accordance with 250.134 or 250.136, regardless of voltage. Equipment grounding conductors and devices shall comply with 690.43(A) through (D).

Δ **(A) Photovoltaic Module Mounting Systems and Devices.** Devices and systems used for mounting PV modules that are also used for bonding module frames shall be listed, labeled, and identified for bonding PV modules.

> Informational Note: See UL 2703, *Standard for Mounting Systems, Mounting Devices, Clamping/Retention Devices, and Ground Lugs for Use with Flat-Plate Photovoltaic Modules and Panels for PV Module Clamps*, and UL 3703, *Standard for Solar Trackers*.

(B) Equipment Secured to Grounded Metal Supports. Devices listed, labeled, and identified for bonding and grounding the metal parts of PV systems shall be permitted to bond the equipment to grounded metal supports. Metallic support structures shall have identified bonding jumpers connected between separate metallic sections or shall be identified for equipment bonding and shall be connected to the equipment grounding conductor.

(C) Location. Equipment grounding conductors shall be permitted to be run separately from the PV system conductors within the PV array. Where PV system circuit conductors leave the vicinity of the PV array, equipment grounding conductors shall comply with 250.134.

(D) Bonding for Over 250 Volts. The bonding requirements contained in 250.97 shall apply only to solidly grounded PV system circuits operating over 250 volts to ground.

690.45 Size of Equipment Grounding Conductors. Equipment grounding conductors for PV system circuits shall be sized in accordance with 250.122. Where no overcurrent protective device is used in the circuit, an assumed overcurrent device rated in accordance with 690.9(B) shall be used when applying Table 250.122.

Increases in equipment grounding conductor size to address voltage drop considerations shall not be required.

Δ **690.47 Grounding Electrode System.**

Δ **(A) Buildings or Structures Supporting a PV System.** A building or structure(s) supporting a PV system shall utilize a grounding electrode system installed in accordance with 690.47(B).

PV array equipment grounding conductors shall be connected to a grounding electrode system in accordance with Part VII of Article 250. This connection shall be in addition to any other equipment grounding conductor requirements in 690.43(C). The PV array equipment grounding conductors shall be sized in accordance with 690.45. For specific PV system grounding configurations permitted in 690.41(A), one of the following conditions shall apply:

(1) For PV systems that are not solidly grounded, the equipment grounding conductor for the output of the PV system, where connected to associated distribution equipment connected to a grounding electrode system, shall be permitted to be the only connection to ground for the system.
(2) For solidly grounded PV systems, as permitted in 690.41(A)(5), the grounded conductor shall be connected to a grounding electrode system by means of a grounding electrode conductor sized in accordance with 250.166.

> Informational Note: Most PV systems are functionally grounded systems rather than solidly grounded systems as defined in this *Code*. For functionally grounded PV systems with an interactive inverter output, the ac equipment grounding conductor is connected to associated grounded ac distribution equipment. This connection is most often the connection to ground for ground-fault protection and equipment grounding of the PV array.

(B) Grounding Electrodes and Grounding Electrode Conductors. Additional grounding electrodes shall be permitted to be installed in accordance with 250.52 and 250.54. Grounding electrodes shall be permitted to be connected directly to the PV module frame(s) or support structure. A grounding electrode conductor shall be sized according to 250.66. A support structure for a ground-mounted PV array shall be permitted to be considered a grounding electrode if it meets the requirements of 250.52. PV arrays mounted to buildings shall be permitted to use the metal structural frame of the building if the requirements of 250.68(C)(2) are met.

Part VI. Source Connections

690.56 Identification of Power Sources. Plaques or directories shall be installed in accordance with 705.10.

Δ **690.59 Connection to Other Sources.** PV systems connected to other sources shall be installed in accordance with Parts I and II of Article 705.

The requirements for inverters in 705.40 prevent energizing of otherwise de-energized system conductors or output conductors of other off-site sources (such as an electrical utility) and are intended to prevent electric shock. The ability to automatically de-energize output upon loss of voltage is normally a feature of the interactive inverter.

690.72 Self-Regulated PV Charge Control. The PV source circuit shall be considered to comply with the requirements for charge control of a battery without the use of separate charge control equipment if the circuit meets both of the following:

(1) The PV source circuit is matched to the voltage rating and charge current requirements of the interconnected battery cells.
(2) The maximum charging current multiplied by 1 hour is less than 3 percent of the rated battery capacity expressed in ampere-hours or as recommended by the battery manufacturer.

ARTICLE 691
Large-Scale Photovoltaic (PV) Electric Supply Stations

Δ **691.1 Scope.** This article covers the installation of large-scale PV electric supply stations not under exclusive utility control.

> Informational Note No. 1: Facilities covered by this article have specific design and safety features unique to large-scale PV facilities outlined in 691.4 and are operated for the sole purpose of providing electric supply to a system operated by a regulated utility for the transfer of electric energy.
>
> Informational Note No. 2: See 90.2(B)(5) for additional information about utility-owned properties not covered under this *Code*. See ANSI/IEEE C2-2017, *National Electrical Safety Code*, for additional information on electric supply stations.
>
> Informational Note No. 3: See Informational Note Figure 691.1.

The systems covered by this article are a minimum size of 5000 kilowatts (kW), or 5 megawatts (MW). Based on their immense size, these large-scale photovoltaic (PV) installations are often referenced as "solar farms." As of 2021, the largest solar farm in the United States, and 13th largest in the world, was the Copper Mountain Solar Facility in Boulder City, Nevada, sized at 802 megawatts. In the same year, the largest solar farm in the world, Bhadla Solar Park in India, nearly tripled the size of Copper Mountain, having a capacity of roughly 2245 megawatts. Exhibit 691.1 shows the Nyngan Solar Farm in Australia, a large-scale PV installation generating 102 megawatts of power through roughly 1.35 million solar PV modules. Because installations of this magnitude are much too large to be installed on buildings, the installation requirements for Article 691 differ from those of Article 690, which is primarily aimed at installation on buildings.

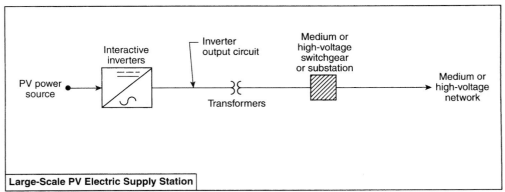

Notes:
(1) The diagram is for informational purposes only and is not representative of all potential configurations.
(2) Custom designs occur in each configuration, and some components are optional.

Δ **INFORMATIONAL NOTE FIGURE 691.1** *Identification of Large-Scale PV Electric Supply Station Components.*

EXHIBIT 691.1 The Nyngan Solar Farm, Australia. (*Getty Images*)

691.4 Special Requirements for Large-Scale PV Electric Supply Stations. Large-scale PV electric supply stations shall be accessible only to authorized personnel and comply with the following:

(1) Electrical circuits and equipment shall be maintained and operated only by qualified persons.

> Informational Note No. 1: See NFPA *70E-2021, Standard for Electrical Safety in the Workplace*, for electrical safety requirements.

(2) Access to PV electric supply stations shall be restricted in accordance with 110.31. Field-applied hazard markings shall be applied in accordance with 110.21(B).

(3) The connection between the PV electric supply station and the system operated by a utility for the transfer of electrical energy shall be through medium- or high-voltage switch gear, substation, switch yard, or similar methods whose sole purpose shall be to interconnect the two systems.

(4) The electrical loads within the PV electric supply station shall only be used to power auxiliary equipment for the generation of the PV power.

(5) Large-scale PV electric supply stations shall not be installed on buildings.

(6) The station shall be monitored from a central command center.

(7) The station shall have an inverter generating capacity of at least 5000 kW.

> Informational Note No. 2: Some individual sites with capacities less than 5000 kW are operated as part of a group of facilities with a total generating capacity exceeding 5000 kW.

Large-scale PV systems are required to be accessible to qualified personnel only. Section 691.9 requires an engineered design installation to have documented procedures and means of isolation of equipment.

691.5 Equipment. All electrical equipment shall be approved for installation by one of the following:

(1) Listing and labeling

(2) Be evaluated for the application and have a field label applied

(3) Where products complying with 691.5(1) or (2) are not available, by engineering review validating that the electrical equipment is evaluated and tested to relevant standards or industry practice

691.6 Engineered Design. Documentation of the electrical portion of the engineered design of the electric supply station shall be stamped and provided upon request of the AHJ. Additional stamped independent engineering reports detailing compliance of the design with applicable electrical standards and industry practice shall be provided upon request of the AHJ. The independent engineer shall be a licensed professional electrical engineer retained by the system owner or installer. This documentation shall include details of conformance of the design with Article 690, and any alternative methods to Article 690, or other articles of this *Code*.

691.7 Conformance of Construction to Engineered Design. Documentation that the construction of the electric supply station conforms to the electrical engineered design shall be provided upon request of the AHJ. Additional stamped independent engineering reports detailing the construction conforms with this *Code*, applicable standards and industry practice shall be provided upon request of the AHJ. The independent engineer shall be a licensed professional electrical engineer retained by the system owner or installer. This documentation, where requested, shall be available prior to commercial operation of the station.

691.8 Direct Current Operating Voltage. For large-scale PV electric supply stations, calculations shall be included in the documentation required in 691.6.

691.9 Disconnecting Means for Isolating Photovoltaic Equipment. Disconnecting means for equipment shall not be required within sight of equipment and shall be permitted to be located remotely from equipment. The engineered design required by 691.6 shall document disconnection procedures and means of isolating equipment.

> Informational Note: See NFPA 70B-2019, *Recommended Practice for Electrical Equipment Maintenance*, for information on electrical system maintenance. See NFPA 70E-2021, *Standard for Electrical Safety in the Workplace*, for information on written procedures and conditions of maintenance, including lockout/tagout procedures.

Buildings whose sole purpose is to house and protect supply station equipment shall not be required to comply with 690.12. Written standard operating procedures shall be available at the site detailing necessary shutdown procedures in the event of an emergency.

691.10 Fire Mitigation. PV systems that do not comply with the requirements of 690.11 shall include details of fire mitigation plans to address dc arc-faults in the documentation required in 691.6.

> Informational Note: Fire mitigation plans are typically reviewed by the local fire agency and include topics such as access roads within the facility.

691.11 Fence Bonding and Grounding. Fence grounding requirements and details shall be included in the documentation required in 691.6.

> Informational Note: See 250.194 for fence bonding and grounding requirements enclosing substation portions of an electric supply station. Grounding requirements for other portions of electric supply station fencing are assessed based on the presence of overhead conductors, proximity to generation and distribution equipment, and associated step and touch potential.

ARTICLE 692 Fuel Cell Systems

Part I. General

692.1 Scope. This article applies to the installation of fuel cell systems.

> Informational Note: Some fuel cell systems can be interactive with other electrical power production sources, are stand-alone, or both. Some fuel cell systems are connected to electric energy storage systems such as batteries. Fuel cell systems can have ac output(s), dc output(s), or both for utilization.

The rising demand for electric power has led to the development of power sources that are viable alternatives to, or can be interconnected with, electric utility distribution systems. Article 692 covers the installation of on-premises electrical supply systems in which the power is derived from an electrochemical system that consumes fuel to generate an electric current.

The principle of operation is that dc is generated through a chemical reaction in which hydrogen-rich fuel such as natural gas, liquefied petroleum gas (LP-Gas), or hydrogen is consumed. The consumption of the fuel gas is via an electrochemical process, as opposed to internal combustion prime movers, which consume fuel using a combustion process. A power inverter converts the dc to ac. The installation requirements of Article 692 allow power derived from fuel cells to be safely delivered into residential and light commercial occupancies as the sole source of electric power or as an integrated source with a utility or other power source.

692.4 Installation.

(A) Fuel Cell System. A fuel cell system shall be permitted to supply a building or other structure in addition to any service(s) of another electricity supply system(s).

(B) Identification of Power Sources. Fuel cell systems shall be marked with a plaque or directory installed in accordance with 705.10.

(C) System Installation. The construction and operation of equipment, associated wiring, and interconnections shall be performed only by qualified persons.

> Informational Note: See Article 100 for the definition of *qualified person.*

692.6 Listing Requirement. The fuel cell system shall be approved for the application in accordance with one of the following:

(1) Be listed for the application
(2) Be evaluated for the application and have a field label applied

Part II. Circuit Requirements

692.8 Circuit Sizing and Current.

(A) Nameplate Rated Circuit Current. The nameplate(s) rated circuit current shall be the rated current indicated on the fuel cell nameplate(s).

(B) Conductor Ampacity and Overcurrent Device Ratings. The ampacity of the feeder circuit conductors from the fuel cell system(s) to the premises wiring system shall not be less than the greater of (1) nameplate(s) rated circuit current or (2) the rating of the fuel cell system(s) overcurrent protective device(s).

(C) Ampacity of Grounded or Neutral Conductor. If an interactive single-phase, 2-wire fuel cell output(s) is connected to the grounded or neutral conductor and a single ungrounded conductor of a 3-wire system or of a 3-phase, 4-wire, wye-connected system, the maximum unbalanced neutral load current plus the fuel cell system(s) output rating shall not exceed the ampacity of the grounded or neutral conductor.

692.9 Overcurrent Protection.

(A) Circuits and Equipment. If the fuel cell system is provided with overcurrent protection sufficient to protect the circuit conductors that supply the load, additional circuit overcurrent devices shall not be required. Equipment and conductors connected to more than one electrical source shall be protected.

(B) Accessibility. Overcurrent devices shall be readily accessible.

Part III. Disconnecting Means

692.13 All Conductors. Means shall be provided to disconnect all current-carrying conductors of a fuel cell system power source from all other conductors in a building or other structure.

△ **692.17 Switch or Circuit Breaker.** The disconnecting means for ungrounded conductors shall consist of readily accessible, manually operable switch(es) or circuit breaker(s).

Where all terminals of the disconnecting means may be energized in the open position, a warning sign shall be mounted on or adjacent to the disconnecting means. The sign shall be clearly legible and shall have the following words or equivalent:

> DANGER
> ELECTRIC SHOCK HAZARD.
> DO NOT TOUCH TERMINALS.
> TERMINALS ON BOTH THE LINE AND
> LOAD SIDES MAY BE ENERGIZED
> IN THE OPEN POSITION.

The danger sign(s) or label(s) shall comply with 110.21(B).

Part IV. Wiring Methods

△ **692.31 Wiring Systems.** In addition to wiring methods included in Chapter 3 of this *Code*, wiring methods and fittings specifically listed and identified for use with fuel cell systems shall be permitted.

Part V. Marking

692.50 Fuel Cell Power Sources. A marking specifying the fuel cell system, output voltage, output power rating, and continuous output current rating shall be provided at the disconnecting means for the fuel cell power source at an accessible location on the site.

692.51 Fuel Shut-Off. The location of the manual fuel shut-off valve shall be marked at the location of the primary disconnecting means of the building or circuits supplied.

△ **692.52 Stored Energy.** A fuel cell system that stores electrical energy shall require the following warning sign, or equivalent, at the location of the service disconnecting means of the premises:

> WARNING
> FUEL CELL POWER SYSTEM CONTAINS
> ELECTRICAL ENERGY STORAGE DEVICES.

The warning sign(s) or label(s) shall comply with 110.21(B).

Part VI. Connection to Other Circuits

N **692.60 Connection to Other Systems.** Fuel cell systems connected to other sources shall be installed in accordance with Parts I and II of Article 705.

692.61 Transfer Switch. A transfer switch shall be required in non-grid-interactive systems that use utility grid backup.

The transfer switch shall maintain isolation between the electrical production and distribution network and the fuel cell system. The transfer switch shall be permitted to be located externally or internally to the fuel cell system unit. Where the utility service conductors of the structure are connected to the transfer switch, the switch shall comply with Article 230, Part V.

ARTICLE
694 Wind Electric Systems

Part I. General

△ **694.1 Scope.** This article applies to wind (turbine) electric systems that consist of one or more wind electric generators and their related alternators, generators, inverters, controllers, and associated equipment.

> Informational Note: Some wind electric systems are interactive with other electric power sources (*see Informational Note Figure 694.1*). Some systems have ac output and some have dc output. Some systems contain electrical energy storage, such as batteries.

Like photovoltaic and fuel cell systems, wind-driven turbines as a stand-alone, or an interconnected power production source, are available for use as part of the premises wiring system. Due to an increased desire for renewable energy, these systems have seen a significant increase in use. According to the U.S. Department of Energy, although the United States reached 10 gigawatts (GW) of wind power capacity in 25 years, it took only 4 years to add an additional 40 gigawatts (2008–2012). As of 2020, the United States' total cumulative wind power capacity of 122 gigawatts was second in the world, as compared to the largest worldwide generator of wind power, China, at approximately 288 gigawatts. Wind power installations often are land based but can also be installed offshore. In 2020 alone, the United States increased its offshore wind power generation capacity over 24%, to roughly 35 gigawatts.

Wind turbine farms are becoming more common. Some are utility owned, while others are owned by private investors. Most wind electric systems consist of a single wind turbine, such as the one shown in Exhibit 694.1.

Many of the requirements in Article 694 are similar to those contained in Articles 690 and 692. The requirements apply to all wind turbines within the scope of the *NEC®*, regardless of the kilowatt rating.

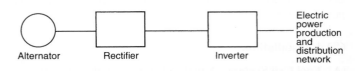

INFORMATIONAL NOTE FIGURE 694.1 *Identification of Wind Electric System Components — Interactive System.*

EXHIBIT 694.1 A wind electric system consisting of a single wind turbine.

△ **694.7 Construction and Maintenance.** The construction and maintenance, associated wiring, and interconnections shall be performed only by qualified persons.

> Informational Note: See Article 100 for the definition of *qualified person*.

Wind-powered systems present some unique hazards, including the danger of working in elevated, confined spaces. Therefore, personnel working on these systems have to be trained to recognize and avoid all hazards associated with the installation and servicing of this type of power generation system.

(A) Wind Electric Systems. A wind electric system(s) shall be permitted to supply a building or other structure in addition to other sources of supply. These requirements apply to both onshore and offshore installations.

△ **(B) Equipment.** Wind electric systems shall comply with one of the following:

(1) Be listed
(2) Be evaluated for the application and have a field label applied

Wind electric systems undergoing evaluation for type certification and listing shall be permitted to be operated in a controlled location with access limited to qualified personnel.

> Informational Note: See UL 6141, *Standard for Wind Turbines Permitting Entry of Personnel*, and UL 6142, *Standard for Small Wind Turbine Systems*, for further information on wind turbine equipment. Ratings for wind turbines could include limitations on installation locations such as onshore or offshore. Testing is typically performed under supervision of a qualified electrical testing organization.

Three documents published by Underwriters Laboratories — Subject 6140, *Outline of Investigation for Wind Turbine Generating Systems*; Subject 6141, *Outline of Investigation for Wind Turbine Converters and Interconnection Systems Equipment*; and UL 6142, *Small Wind Turbine Systems* — provide the basis for

certifying (product listing or classification) the overall wind turbine generator and its associated components or individual components, such as inverters and interconnection hardware, that are associated with a wind electric system. Some small wind turbines come listed as a unit, while others are assembled from individual listed components and require a field evaluation and label to assist the AHJ in approval of the system.

(C) Diversion Load Controllers. A wind electric system employing a diversion load controller as the primary means of regulating the speed of a wind turbine rotor shall be equipped with an additional, independent, reliable means to prevent overspeed operation. An interconnected utility service shall not be considered to be a reliable diversion load.

(D) Overvoltage Protection. A listed surge protective device shall be installed between a wind electric system and any loads served by the premises electrical system. The SPD shall be permitted to be a Type 3 SPD on the circuit serving a wind electric system or a Type 2 SPD located anywhere on the load side of the service disconnect. SPDs shall be installed in accordance with Part II of Article 242.

Because the towers associated with wind electric systems will generally be the tallest structures in the vicinity, the use of surge-protective devices covered in Article 242, Overvoltage Protection, is mandatory to help protect the premises wiring systems against the effects of lightning.

> **See also**
>
> **694.40(B)(3)** and **(4),** which cover grounding of towers and guy wires
>
> **Chapter 9** of **NFPA 780,** *Standard for the Installation of Lightning Protection Systems,* for the protection of wind turbines from lightning strikes

(E) Receptacles. A receptacle shall be permitted to be supplied by a wind electric system branch or feeder circuit for maintenance or data acquisition use. Receptacles shall be protected with an overcurrent device with a rating not to exceed the current rating of the receptacle. In addition to the requirements in 210.8, all 125-volt, single-phase, 15- and 20-ampere receptacles installed for maintenance of the wind turbine shall have ground-fault circuit-interrupter protection for personnel.

Receptacles installed for the maintenance of a wind turbine often are installed in areas similar to those required to be GFCI protected in other areas of the *NEC*. The shock hazard to personnel servicing wind turbines is similar, and GFCI protection for personnel is required for 125-volt, single-phase, 15- and 20-ampere receptacles. For example, a receptacle installed in a shed for system maintenance requires GFCI protection similar to that for a receptacle in a dwelling unit garage.

(F) Poles or Towers Supporting Wind Turbines Used as a Raceway. A pole or tower shall be permitted to be used as a raceway if approved in accordance with one of the following:

(1) Be evaluated as part of the listing for the wind turbine

(2) Be listed for the application

(3) Be evaluated for the application and have a field label applied

(G) Working Clearances. Working space shall be provided for electrical cabinets and other electrical equipment in accordance with 110.26(A).

For large wind turbines where service personnel enter the equipment, where conditions of maintenance and supervision ensure that only qualified persons perform the work, working clearances shall be permitted to comply with Table 694.7(G) for systems up to 1000 volts nominal.

TABLE 694.7(G) *Working Spaces*

Nominal Voltage to Ground	Condition 1	Condition 2	Condition 3
0–150	900 mm (3 ft)	900 mm (3 ft)	900 mm (3 ft)
151–1000	900 mm (3 ft)	1.0 m (3 ft 6 in.)	1.2 m (4 ft)

Part II. Circuit Requirements

694.10 Maximum Voltage.

(A) Wind Turbine Output Circuits. Wind turbine output circuits on or in one- and two-family dwellings shall be permitted to have a maximum voltage up to 600 volts.

(B) Direct-Current Utilization Circuits. The voltage of dc utilization circuits shall comply with 210.6.

(C) Circuits over 150 Volts to Ground. In one- and two-family dwellings, live parts in circuits over 150 volts to ground shall not be accessible to other than qualified persons while energized.

Informational Note: See 110.27 for guarding of live parts and 210.6 for branch circuit voltage limitations.

694.12 Circuit Sizing and Current.

(A) Calculation of Maximum Circuit Current. The maximum current for a circuit shall be calculated in accordance with 694.12(A)(1) through (A)(3).

(1) Turbine Output Circuit Currents. The maximum current shall be based on the circuit current of the wind turbine operating at maximum output power.

(2) Inverter Output Circuit Current. The maximum output current shall be the inverter continuous output current rating.

(3) Stand-Alone Inverter Input Circuit Current. The maximum input current shall be the stand-alone continuous inverter input current rating of the inverter producing rated power at the lowest input voltage.

(B) Ampacity and Overcurrent Device Ratings.

(1) Continuous Current. Wind turbine electric system currents shall be considered to be continuous.

(2) Sizing of Conductors and Overcurrent Devices. Circuit conductors and overcurrent devices shall be sized to carry not less than 125 percent of the maximum current as calculated in 694.12(A). The rating or setting of overcurrent devices shall be permitted in accordance with 240.4(B) and (C).

Exception: Circuits containing an assembly, together with its overcurrent devices, listed for continuous operation at 100 percent of its rating shall be permitted to be used at 100 percent of its rating.

694.15 Overcurrent Protection.

Δ **(A) Circuits and Equipment.** Turbine output circuits, inverter output circuits, and storage battery circuit conductors and equipment shall be protected in accordance with 240.4 and 240.5. Circuits connected to more than one electrical source shall have overcurrent devices located so as to provide overcurrent protection from all sources.

Exception: An overcurrent device shall not be required for circuit conductors sized in accordance with 694.12(B) where the maximum current from all sources does not exceed the ampacity of the conductors.

Informational Note: Possible backfeed of current from any source of supply, including a supply through an inverter to the wind turbine output circuit, is a consideration in determining whether overcurrent protection from all sources is provided. Some wind electric systems rely on the turbine output circuit to regulate turbine speed. Inverters may also operate in reverse for turbine startup or speed control.

(B) Power Transformers. Overcurrent protection for a transformer with sources on each side shall be provided in accordance with 450.3 by considering first one side of the transformer, then the other side of the transformer, as the primary.

Exception: A power transformer with a current rating on the side connected to the inverter output, which is not less than the rated continuous output current rating of the inverter, shall not be required to have overcurrent protection at the inverter.

(C) Direct-Current Rating. Overcurrent devices, either fuses or circuit breakers, used in any dc portion of a wind electric system shall be listed for use in dc circuits and shall have appropriate voltage, current, and interrupting ratings.

Part III. Disconnecting Means

694.20 All Conductors. Means shall be provided to disconnect all current-carrying conductors of a wind electric power source from all other conductors in a building or other structure. A switch, circuit breaker, or other device, either ac or dc, shall not be installed in a grounded conductor if operation of that switch, circuit breaker, or other device leaves the marked, grounded conductor in an ungrounded and energized state.

Exception: A wind turbine that uses the turbine output circuit for regulating turbine speed shall not require a turbine output circuit disconnecting means.

694.22 Additional Provisions. Disconnecting means shall comply with 694.22(A) through (D).

Δ **(A) Disconnecting Means.** The disconnecting means shall not be required to be suitable for use as service equipment. The disconnecting means for ungrounded conductors shall consist of manually operable switches or circuit breakers complying with all of the following requirements:

(1) They shall be located where readily accessible.
(2) They shall be externally operable without exposing the operator to contact with live parts.
(3) They shall plainly indicate whether in the open or closed position.
(4) They shall have an interrupting rating sufficient for the nominal circuit voltage and the current that is available at the line terminals of the equipment.

Where all terminals of the disconnecting means are capable of being energized in the open position, a warning sign shall be mounted on or adjacent to the disconnecting means. The sign shall be clearly legible and shall have the following words or equivalent:

> WARNING.
> ELECTRIC SHOCK HAZARD.
> DO NOT TOUCH TERMINALS.
> TERMINALS ON BOTH THE LINE
> AND LOAD SIDES MAY BE
> ENERGIZED IN THE OPEN POSITION.

The warning sign(s) or label(s) shall comply with 110.21(B).

(B) Equipment. Equipment such as rectifiers, controllers, output circuit isolating and shorting switches, and overcurrent devices shall be permitted on the wind turbine side of the disconnecting means.

(C) Requirements for Disconnecting Means.

(1) Location. The wind electric system disconnecting means shall be installed at a readily accessible location either on or adjacent to the turbine tower, on the outside of a building or structure, or inside at the point of entrance of the wind system conductors.

Exception: Installations that comply with 694.30(C) shall be permitted to have the disconnecting means located remotely from the point of entry of the wind system conductors.

A wind turbine disconnecting means shall not be required to be located at the nacelle or tower.

The disconnecting means shall not be installed in bathrooms.

For one-family and two-family dwellings, a disconnecting means or manual shutdown button or switch shall be located at a readily accessible location outside the building.

The general requirement for locating the wind electric system disconnecting means is similar to 230.70(A) for services. For one- and two-family dwelling units, a means must be provided on the outside of the building to disconnect or shut down the wind generation system. This requirement brings wind generation systems into alignment with methods of protecting emergency personnel, such as the requirements for an emergency disconnect for services or rapid shutdown for solar photovoltaic systems. The exception permits the system disconnecting means to be located at any readily accessible location within a building or structure, except a bathroom, provided the wiring method between the conductor point of entry and the disconnecting means is a metal raceway or the conductors are protected by a metal enclosure. Supply conductors are permitted to be installed without a disconnecting means until they reach the building or structure.

(2) Marking. Each turbine system disconnecting means shall be permanently marked to identify it as a wind electric system disconnect.

(3) Suitable for Use. Turbine system disconnecting means shall be suitable for the prevailing conditions.

(4) Maximum Number of Disconnects. The turbine disconnecting means shall consist of not more than six switches or six circuit breakers mounted in a single enclosure, in a group of separate enclosures, or in or on a switchgear.

(D) Equipment That Is Not Readily Accessible. Rectifiers, controllers, and inverters shall be permitted to be mounted in nacelles or other exterior areas that are not readily accessible.

694.23 Turbine Shutdown.

(A) Manual Shutdown. Wind turbines shall be required to have a readily accessible manual shutdown button or switch. Operation of the button or switch shall result in a parked turbine state that shall either stop the turbine rotor or allow limited rotor speed combined with a means to de-energize the turbine output circuit.

Exception: Turbines with a swept area of less than 50 m² (538 ft²) shall not be required to have a manual shutdown button or switch.

(B) Shutdown Procedure. The shutdown procedure for a wind turbine shall be defined and permanently posted at the location of a shutdown means and at the location of the turbine controller or disconnect, if the location is different.

Although the exception to 694.23(A) permits a smaller turbine without a manual shutdown switch, a shutdown procedure is required for all turbines.

Δ **694.24 Disconnection of Wind Electric System Equipment.** Means shall be provided to disconnect equipment, such as inverters, batteries, and charge controllers, from all ungrounded conductors of all sources. If the equipment is energized from more than one source, the disconnecting means shall be grouped and identified.

A single disconnecting means in accordance with 694.22 shall be permitted for the combined ac output of one or more inverters.

A shorting switch or plug shall be permitted to be used as an alternative to a disconnect in systems that regulate turbine speed using the turbine output circuit.

Exception: Equipment housed in a turbine nacelle shall not be required to have a disconnecting means.

694.26 Fuses. Means shall be provided to disconnect a fuse from all sources of supply where the fuse is energized from both directions and is accessible to other than qualified persons. Switches, pullouts, or similar devices that are rated for the application shall be permitted to serve as a means to disconnect fuses from all sources of supply.

694.28 Installation and Service of a Wind Turbine. Open circuiting, short circuiting, or mechanical brakes shall be used to disable a turbine for installation and service.

> Informational Note: Some wind turbines rely on the connection from the alternator to a remote controller for speed regulation. Opening turbine output circuit conductors may cause mechanical damage to a turbine and create excessive voltages that could damage equipment or expose persons to electric shock.

Part IV. Wiring Methods

694.30 Permitted Methods.

Δ **(A) Wiring Systems.** In addition to wiring methods included in Chapter 3 of this *Code*, wiring methods and fittings specifically intended listed and identified for use on wind turbines shall be permitted. In readily accessible locations, turbine output circuits that operate at voltages greater than 30 volts shall be installed in raceways.

(B) Flexible Cords and Cables. Flexible cords and cables, where used to connect the moving parts of turbines or where used for ready removal for maintenance and repair, shall comply with Article 400 and shall be of a type identified as hard service cord or portable power cable, shall be suitable for extra-hard usage, shall be listed for outdoor use, and shall be water resistant. Cables exposed to sunlight shall be sunlight resistant. Flexible, fine-stranded cables shall be terminated only with terminals, lugs, devices, or connectors in accordance with 110.14(A).

To provide a greater degree of flexibility, the conductors used in flexible cords and flexible cables are more finely stranded than conductors with Class B or C stranding. Terminals, connectors, and devices used with classes of stranding other than Class B or C are required to be identified for the classes of stranding for which they are suitable.

See also

110.14 and its commentary for more information on conductor terminations

(C) Direct-Current Turbine Output Circuits Inside a Building. Direct-current turbine output circuits installed inside a building or structure shall be enclosed in metal raceways or installed in metal enclosures, or run in Type MC metal-clad cable that complies with 250.118(A)(10), from the point of penetration of the surface of the building or structure to the first readily accessible disconnecting means.

Part V. Grounding and Bonding

694.40 Equipment Grounding and Bonding.

(A) General. Exposed non–current-carrying metal parts of towers, turbine nacelles, other equipment, and conductor enclosures shall be grounded and bonded to the premises grounding and bonding system. Attached metal parts, such as turbine blades and tails that are not likely to become energized, shall not be required to be grounded or bonded.

(B) Tower Grounding and Bonding.

(1) Grounding Electrodes and Grounding Electrode Conductors. A wind turbine tower shall be connected to a grounding electrode system. Where installed in close proximity to galvanized foundation or tower anchor components, galvanized grounding electrodes shall be used.

> Informational Note: Copper and copper-clad grounding electrodes, where used in highly conductive soils, can cause electrolytic corrosion of galvanized foundation and tower anchor components.

(2) Bonding Conductor. Equipment grounding conductors or supply-side bonding jumpers, as applicable, shall be required between turbines, towers, and the premises grounding system.

(3) Tower Connections. Equipment grounding, bonding, and grounding electrode conductors, where used, shall be connected to metallic towers using listed means. All mechanical elements used to terminate these conductors shall be accessible.

(4) Guy Wires. Guy wires used to support turbine towers shall not be required to be connected to an equipment grounding conductor or to comply with the requirements of 250.110.

> Informational Note: Guy wires supporting grounded towers are unlikely to become energized under normal conditions, but partial lightning currents could flow through guy wires when exposed to a lightning environment. Grounding of metallic guy wires may be required by lightning standards. See NFPA 780-2017, *Standard for the Installation of Lightning Protection Systems,* for information on lightning protection systems.

Part VI. Marking

694.52 Power Systems Employing Energy Storage. Wind electric systems employing energy storage shall be marked with the maximum operating voltage, any equalization voltage, and the polarity of the grounded circuit conductor.

694.54 Identification of Power Sources. Wind turbine systems shall be marked with a plaque or directory installed in accordance with 705.10.

694.56 Instructions for Disabling Turbine. A plaque shall be installed at or adjacent to the turbine location providing basic instructions for disabling the turbine.

Part VII. Connection to Other Sources

694.60 Identified Interactive Equipment. Only inverters that are listed, labeled, and identified as interactive shall be permitted in interactive systems.

Δ **694.62 Installation.** Wind electric systems connected to other sources shall be installed in accordance with Parts I and II of Article 705.

694.66 Operating Voltage Range. Wind electric systems connected to dedicated branch or feeder circuits shall be permitted to exceed normal voltage operating ranges on these circuits, provided that the voltage at any distribution equipment supplying other loads remains within normal ranges.

> Informational Note: Wind turbines might use the electric grid to dump energy from short-term wind gusts. See ANSI C84.1-2006, *Voltage Ratings for Electric Power Systems and Equipment (60 Hz)*, for information on normal operating voltages.

ARTICLE 695 Fire Pumps

Δ **695.1 Scope.**

(A) Covered. This article covers the installation of the following:

(1) Electric power sources and interconnecting circuits
(2) Switching and control equipment dedicated to fire pump drivers

> Informational Note: Text that is followed by a reference in brackets has been extracted from NFPA 20-2019, *Standard for the Installation of Stationary Pumps for Fire Protection*. Only editorial changes were made to the extracted text to make it consistent with this *Code*.

Δ **(B) Not Covered.** This article does not cover the following:

(1) The performance, maintenance, and acceptance testing of the fire pump system and the internal wiring of the components of the system

(2) The installation of pressure maintenance (jockey or makeup) pumps

> Informational Note No. 1: See Article 430 for the installation of pressure maintenance (jockey or makeup) pumps supplied by the fire pump circuit or another source.

(3) Transfer equipment upstream of the fire pump transfer switch(es)

> Informational Note No. 2: See NFPA 20-2019, *Standard for the Installation of Stationary Pumps for Fire Protection*, for further information.

(4) Water pumps installed in one- and two-family dwellings and used for fire suppression

> Informational Note No. 3: See NFPA 13D-2019, *Standard for the Installation of Sprinkler Systems in One- and Two-Family Dwellings and Manufactured Homes*, for further information.

The requirements covering reliable power supplies for electric fire pump motors correlate with those in NFPA 20, *Standard for the Installation of Stationary Pumps for Fire Protection*. However, the *NEC®* and NFPA 20 have a distinct division of responsibility for fire pump requirements. Performance issues, including the determination of power supply reliability, are under the jurisdiction of the NFPA Technical Committee on Fire Pumps, while electrical installation requirements are within the purview of the National Electrical Code Committee.

An electric motor–driven fire pump such as the one shown in Exhibit 695.1 is covered by the requirements of Article 695. If a pump is installed in accordance with NFPA 13D, *Standard for the Installation of Sprinkler Systems in One- and Two-Family Dwellings and Manufactured Homes*, the requirements of Article 695 and NFPA 20 do not apply to the electrical installation.

Although the installation requirements for pressure maintenance (jockey) pumps are not covered by Article 695, these pumps are permitted by NFPA 20 to be supplied by a fire pump service or feeder or by a separate supply.

In respect to the power supply for a fire pump, Article 695 permits a single source if it is deemed to be reliable by the AHJ. This differs from the approach taken in Article 700 for emergency systems, where in all cases there has to be some form of alternate power in the event of an interruption of the normal power source. There are a number of similarities between Articles 695 and 700 in respect to minimizing potential failure points in the circuit supplying equipment that performs the critical building and life safety functions covered in each article. Requirements such as protecting feeder circuit conductors with a 2-hour enclosure, a 2-hour fire-resistive cable system, or encasement in 2 inches of concrete are contained in both Articles 695 and 700.

N **695.2 Reconditioned Equipment.** Reconditioned fire pump controllers and transfer switches shall not be permitted.

Δ **695.3 Power Source(s) for Electric Motor-Driven Fire Pumps.** Electric motor-driven fire pumps shall have a reliable source of power.

EXHIBIT 695.1 An electric motor specifically listed for fire pump service. *(Courtesy of Liberty Mutual Insurance)*

Informational Note: See NFPA 20-2019, *Standard for the Installation of Stationary Pumps for Fire Protection*, 9.3.2 and A.9.3.2, for guidance on the determination of power source reliability.

(A) Individual Sources. Where reliable, and where capable of carrying indefinitely the sum of the locked-rotor current of the fire pump motor(s) and the pressure maintenance pump motor(s) and the full-load current of the associated fire pump accessory equipment when connected to this power supply, the power source for an electric motor driven fire pump shall be one or more of the following.

These two main requirements, reliability and locked-rotor capacity, ensure that the fire pump operates in the event of a fire without interruption of power and that the fire pump continues to operate until the fire is extinguished, the fire pump is purposely shut down, or the pump itself is destroyed.

The determination of whether the serving electric utility is a reliable source of power is an issue for the AHJ. The following excerpt from A.9.3.2 in Annex A of NFPA 20 elaborates on several key characteristics of a reliable power supply:

A reliable power source possesses the following characteristics:

(1) The source power plant has not experienced any shutdowns longer than 10 continuous hours in the year prior

to plan submittal. NFPA 25 requires special undertakings (i.e., fire watches) when a water-based fire protection system is taken out of service for longer than 10 hours. If the normal source power plant has been intentionally shut down for longer than 10 hours in the past, it is reasonable to require a backup source of power.

(2) Power outages have not routinely been experienced in the area of the protected facility caused by failures in generation or transmission. This standard is not intended to require that the normal source of power be infallible to deem the power reliable. NFPA 20 does not intend to require a backup source of power for every installation using an electric motor–driven fire pump.

(3) The normal source of power is not supplied by overhead conductors outside the protected facility. Fire departments responding to an incident at the protected facility will not operate aerial apparatus near live overhead power lines, without exception. A backup source of power is required in case this scenario occurs and the normal source of power must be shut off. Additionally, many utility providers will remove power to the protected facility by physically cutting the overhead conductors. If the normal source of power is provided by overhead conductors, which will not be identified, the utility provider could mistakenly cut the overhead conductor supplying the fire pump.

(4) Only the disconnect switches and overcurrent protection devices permitted by 9.2.3 [of NFPA 20] are installed in the normal source of power. Power disconnection and activated overcurrent protection should occur only in the fire pump controller. The provisions of 9.2.2 [of NFPA 20] for the disconnect switch and overcurrent protection essentially require disconnection and overcurrent protection to occur in the fire pump controller. If unanticipated disconnect switches or overcurrent protection devices are installed in the normal source of power that do not meet the requirements of 9.2.2 [of NFPA 20], the normal source of power must be considered not reliable and a backup source of power is necessary.

Performance requirements for the alternate source of electric power can be found in NFPA 110, *Standard for Emergency and Standby Power Systems*.

(1) Electric Utility Service Connection. A fire pump shall be permitted to be supplied by a separate service, or from a connection located ahead of and not within the same cabinet, enclosure, vertical switchgear section, or vertical switchboard section as the service disconnecting means. The connection shall be located and arranged so as to minimize the possibility of damage by fire from within the premises and from exposing hazards. A tap ahead of the service disconnecting means shall comply with 230.82(5). The service equipment shall comply with the labeling requirements in 230.2 and the location requirements in 230.72(B). [**20**:9.2.2(1)]

EXHIBIT 695.2 Two permitted configurations for connecting to an electric utility–supplied service.

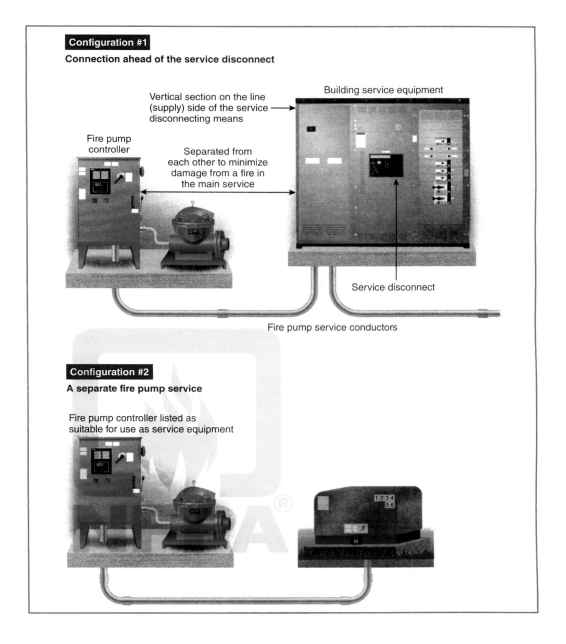

Configuration #1

Connection ahead of the service disconnect

Vertical section on the line (supply) side of the service disconnecting means

Building service equipment

Fire pump controller

Separated from each other to minimize damage from a fire in the main service

Service disconnect

Fire pump service conductors

Configuration #2

A separate fire pump service

Fire pump controller listed as suitable for use as service equipment

The top section of Exhibit 695.2 shows a single service in which a dedicated set of service-entrance conductors for the fire pump is tapped to the incoming service conductors ahead of and separate from the service disconnecting means. This tap is permitted under the conditions specified in 230.40, Exception No. 5, which references systems covered by 230.82(5). The bottom section of Exhibit 695.2 shows a dedicated service supplying the fire pump as permitted by 230.2(A)(1). Both configurations shown in Exhibit 695.2 minimize the potential for accidental disconnection during response to a fire.

(2) On-Site Power Production Facility. A fire pump shall be permitted to be supplied by an on-site power production facility. The source facility shall be located and protected to minimize the possibility of damage by fire. [**20:**9.2.2(3)]

On-site power production facilities are permitted to be used as the sole power source for an electrically driven fire pump motor. See Article 100 for the definition of the term *facility, on-site power production*.

(3) Dedicated Feeder. A dedicated feeder shall be permitted where it is derived from a service connection as described in 695.3(A)(1). [**20:**9.2.2(3)]

> Informational Note: See NFPA 20-2019, *Standard for the Installation of Stationary Pumps for Fire Protection*, 9.2.2, for more information on normal power sources. Subsection 9.2.2(3) permits a "dedicated feeder" to be derived from a "dedicated service" disconnecting means. Subsection 9.2.2(5) permits a "dedicated transformer connection" that is supplied directly from a "dedicated service disconnecting means" where the service is not at utilization voltage.

(B) Multiple Sources. If reliable power cannot be obtained from a source described in 695.3(A), power shall be supplied by one of the following: [**20:**9.3.2]

(1) Individual Sources. An approved combination of two or more of the sources from 695.3(A).

(2) Individual Source and On-site Standby Generator. An approved combination of one or more of the sources in 695.3(A) and an on-site standby generator complying with 695.3(D). [**20:**9.3.4]

Exception to 695.3(B)(1) and (B)(2): An alternate source of power shall not be required where a back-up engine-driven fire pump, back-up steam turbine-driven fire pump, or back-up electric motor-driven fire pump with an independent power source in accordance with 695.3(A) or (C) is installed.

If none of the power supply sources specified in 695.3(A)(1) through (A)(3) can individually provide reliable power with adequate capacity, 695.3(B) permits an approved combination (two or more) of those sources or a combination of one or more of those sources with an on-site standby generator.

In lieu of installing an on-site standby generator, an engine- or steam turbine–driven fire pump can be provided as backup for an electric fire pump. In that instance, the electric fire pump is permitted to be supplied by only a single power source. This allowance provides some design options for augmenting an electric fire pump that is supplied by an unreliable source.

(C) Multibuilding Campus-Style Complexes. If the sources in 695.3(A) are not practicable and the installation is part of a multibuilding campus-style complex, feeder sources shall be permitted if approved by the authority having jurisdiction and installed in accordance with either 695.3(C)(1) and (C)(3) or (C)(2) and (C)(3).

(1) Feeder Sources. Two or more feeders shall be permitted as more than one power source if such feeders are connected to, or derived from, separate utility services. The connection(s), overcurrent protective device(s), and disconnecting means for such feeders shall meet the requirements of 695.4(B)(1)(b).

Δ **(2) Feeder and Alternate Source.** A feeder shall be permitted as a normal power source if an alternate power source independent from the feeder is provided. The connection(s), overcurrent protective device(s), and disconnecting means for such feeders shall meet the requirements of 695.4(B)(1)(b).

Section 695.3(C) allows fire pumps to be supplied by feeder circuits that are part of a medium- or high-voltage premises wiring system. This distribution arrangement is common in industrial and institutional campus settings. The conductors supplied by the higher voltage level distribution systems are not service conductors, because the service point and the service-disconnecting means generally are located at a campus distribution switchyard

or distribution building. Also, all the distribution conductors on the load side of the service equipment, even though they resemble electric utility–type distribution, are considered to be feeders or — in some cases where the circuit supplies a single piece of utilization equipment — branch circuits.

A fire pump supplied by a radial loop type of distribution system (commonly used for medium- and high-voltage distribution) where the two feeders originate from a single substation has to be augmented by an on-site standby generator. In the system shown in Exhibit 695.3, the two feeders originate from different utility substations, a distribution arrangement that allows the two feeders, without an on-site standby generator, to be multiple sources for the electric fire pump as permitted by 695.3(C)(1).

(3) Selective Coordination. Overcurrent protective device(s) shall be selectively coordinated with all supply-side overcurrent protective device(s).

Selective coordination shall be selected by a licensed professional engineer or other qualified persons engaged primarily in the design, installation, or maintenance of electrical systems. The selection shall be documented and made available to those authorized to design, install, maintain, and operate the system.

Exception: Selective coordination shall not be required between two overcurrent devices located in series if no loads are connected in parallel with the downstream device.

The term *coordination, selective (selective coordination)*, as defined in Article 100, indicates that a selectively coordinated system is one in which the operation of the overcurrent protective scheme localizes an overcurrent condition to the circuit conductors or equipment in which an overload or fault (short circuit or ground fault) has occurred. Because a fire suppression system is intended to provide occupants time to exit the building and to protect property, a selectively coordinated overcurrent protection scheme that localizes and minimizes the extent of an interruption of power due to the opening of a protective device is a critical safety element.

Design and verification of electrical system coordination can be achieved only through a coordination study performed by an engineer or other person with experience in designing coordinated electrical systems. A coordination study entails detailed analysis of electrical supply system fault-current characteristics. The design must integrate overcurrent protective devices (OCPDs) that interact by localizing the overcurrent condition and isolating that part of the emergency system. Modifications to the electrical system after the initial design and installation can affect the original implementation of the coordinated system.

(D) On-Site Standby Generator as Alternate Source. An on-site standby generator(s) used as an alternate source of power shall comply with 695.3(D)(1) through (D)(3). [**20:**9.6.2.1]

(1) Capacity. The generator shall have sufficient capacity to allow normal starting and running of the motor(s) driving the fire pump(s) while supplying all other simultaneously operated load(s). [**20:**9.6.1.1]

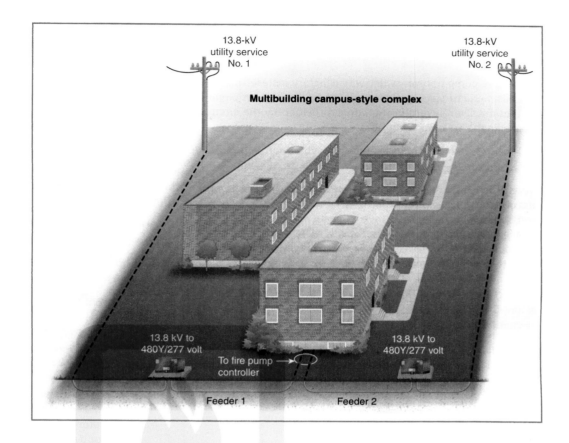

EXHIBIT 695.3 Multiple feeder sources for campus-style application.

13.8-kV utility service No. 1

13.8-kV utility service No. 2

Multibuilding campus-style complex

13.8 kV to 480Y/277 volt

13.8 kV to 480Y/277 volt

To fire pump controller

Feeder 1   Feeder 2

Automatic shedding of one or more optional standby loads in order to comply with this capacity requirement shall be permitted.

Only the sources specified in 695.3(A)(1) through (A)(3) are required to be capable of indefinitely carrying the locked-rotor current of the fire pump motor. On-site standby generators are required only to be capable of carrying the starting and running current of the fire pump motor. The generator disconnecting means and the OCPD(s) for the electric-driven fire pump are not required to be sized for locked-rotor current of the fire pump motor(s).

(2) Connection. A tap ahead of the generator disconnecting means shall not be required. [20:9.6.1.2]

(3) Adjacent Disconnects. The requirements of 430.113 shall not apply.

(E) Arrangement. All power supplies shall be located and arranged to protect against damage by fire from within the premises and exposing hazards. [20:9.1.4]

Multiple power sources shall be arranged so that a fire at one source does not cause an interruption at the other source.

A risk assessment should be performed to determine potential fire hazard exposures to the installation. Determining compliance of the installation with this section requires review of individual building or structure characteristics. The type of construction, type of content, proximity of the building to other hazard exposures, and the location of the primary and alternate power sources for the fire pump should be considered.

(F) Transfer of Power. Transfer of power to the fire pump controller between the individual source and one alternate source shall take place within the pump room. [20:9.6.4]

(G) Power Source Selection. Selection of power source shall be performed by a transfer switch listed for fire pump service. [20:10.8.1.3.1]

Transfer switches specifically listed for fire pump service must meet additional requirements in the product standards, such as the following:

- They must be electrically operated and mechanically held. This eliminates contactor-type transfer switches or schemes that do not include a mechanical latching mechanism.
- They must be horsepower or ampere rated. Where rated in horsepower, the transfer switch must have a horsepower rating at least equal to the motor horsepower. The listing requirements for short-circuit current rating of transfer switches is a function of the OCPD installed ahead of the switch. Because the OCPD must be sized to at least 600 percent of the full-load current of the fire pump motor(s), the switch must be sized to coordinate with the OCPD rating.

(H) Overcurrent Device Selection. An instantaneous trip circuit breaker shall be permitted in lieu of the overcurrent devices specified in 695.4(B)(2)(a)(1), provided that it is part of a transfer switch assembly listed for fire pump service that complies with 695.4(B)(2)(a)(2).

A listed fire pump transfer switch with a factory-installed instantaneous circuit breaker provides ground-fault and short-circuit protection. Overcurrent protection is provided by the circuit breaker in the fire pump controller.

(I) Phase Converters. Phase converters shall not be used to supply power to a fire pump. [**20**:9.1.7]

A phase converter used in a fire pump circuit would be in continuous operation because the controller has to be constantly powered. Voltage imbalance between phases under unloaded or lightly loaded conditions could adversely affect electronics integral to the controller. This is the reason phase converters are not permitted for fire pump service.

695.4 Continuity of Power. Circuits that supply electric motor–driven fire pumps shall be supervised from inadvertent disconnection as covered in 695.4(A) or (B).

(A) Direct Connection. The supply conductors shall directly connect the power source to a listed fire pump controller, a listed combination fire pump controller and power transfer switch, or a listed fire pump power transfer switch.

(B) Connection Through Disconnecting Means and Overcurrent Device.

Section 695.4(B) permits, but does not require, the installation of a disconnecting means and associated overcurrent protection between a power source and the fire pump control devices described in 695.4(B)(1). Other *NEC* requirements — such as 230.70 and 225.31 — could necessitate the installation of the disconnecting means and overcurrent protection covered in the requirements of 695.4(B). While not always possible, the best method to provide continuity of power is the direct connection of the source to the fire pump control equipment in accordance with 695.4(A). The permitted additional disconnect facilitates the creation of an electrically safe work condition as required by *NFPA 70E®, Standard of Electrical Safety in the Workplace®*, while maintenance of the fire pump controller is being performed.

(1) Number of Disconnecting Means.

(a) *General.* A single disconnecting means and associated overcurrent protective device(s) shall be permitted to be installed between the fire pump power source(s) and one of the following: [**20**:9.2.3]

(1) A listed fire pump controller
(2) A listed fire pump power transfer switch
(3) A listed combination fire pump controller and power transfer switch

(b) *Feeder Sources.* For systems installed under the provisions of 695.3(C) only, additional disconnecting means and the associated overcurrent protective device(s) shall be permitted.

(c) *On-Site Standby Generator.* Where an on-site standby generator is used to supply a fire pump, an additional disconnecting means and an associated overcurrent protective device(s) shall be permitted.

An on-site standby generator equipped with an integral disconnecting means and overcurrent protection is allowed in addition to a disconnecting means and overcurrent protection installed elsewhere in the alternate supply circuit to the fire pump. The second disconnecting means and overcurrent device could be located in distribution equipment and is required to comply with the requirements of 695.4(B)(2)(b) and 695.4(B)(3)(b) through (3)(e).

(2) Overcurrent Device Selection. Overcurrent devices shall comply with 695.4(B)(2)(a) or (B)(2)(b).

(a) *Individual Sources.* Overcurrent protection for individual sources shall comply with the following:

(1) Overcurrent protective device(s) shall be rated to carry indefinitely the sum of the locked-rotor current of the largest fire pump motor and the full-load current of all of the other pump motors and accessory equipment. [**20**:9.2.3.4] Where the locked-rotor current value does not correspond to a standard overcurrent device size, the next standard overcurrent device size shall be used in accordance with 240.6. The requirement to carry the locked-rotor currents indefinitely shall not apply to conductors or devices other than overcurrent devices in the fire pump motor circuit(s).

Exception: The requirement to carry the locked-rotor currents indefinitely shall not apply to feeder overcurrent protective devices installed in accordance with 695.3(C).

A key factor in the reliable power source equation is sizing the overcurrent protection in a supervised fire pump disconnecting means so it is able to carry locked-rotor current (LRC) indefinitely. Opening of the circuit by an overcurrent device installed in a fire pump circuit cannot be tolerated, except under short circuits or ground faults. The circuit has to perform as if a direct connection exists to the power source. Sizing for LRC applies only to OCPDs and not to the conductors or other devices in the fire pump motor circuit. Similar requirements are contained in 695.5(B) and 695.5(C)(2). Alternately, a listed fire pump assembly complying with 695.4(B)(2) is permitted.

It is unlikely that all fire pumps on a circuit will be under simultaneous locked-rotor conditions. Therefore, only the LRC of the largest fire pump motor is required. Full-load current is used for all other pump motors and accessory equipment.

(2) Overcurrent protection shall be provided by an assembly listed for fire pump service and complying with the following:

a. The overcurrent protective device shall not open within 2 minutes at 600 percent of the full-load current of the fire pump motor(s).

b. The overcurrent protective device shall not open with a re-start transient of 24 times the full-load current of the fire pump motor(s).

c. The overcurrent protective device shall not open within 10 minutes at 300 percent of the full-load current of the fire pump motor(s).

d. The trip point for circuit breakers shall not be field adjustable. [**20**:9.2.3.4.1]

(b) *On-Site Standby Generators.* Overcurrent protective devices between an on-site standby generator and a fire pump controller shall be selected and sized to allow for instantaneous pickup of the full pump room load, but shall not be larger than the value selected to comply with 430.62 to provide short-circuit protection only. [**20**:9.6.1.1]

This requirement correlates with 695.3(D)(1) covering the required capacity of an on-site standby generator. OCPDs supplied by an on-site standby generator are not required to be sized to carry the locked-rotor current (LRC) of the fire pump(s) indefinitely. The on-site standby generator is not limited to supplying only the fire pump. Where other pump room loads such as lights or fans are supplied, the generator must have sufficient capacity to instantaneously carry the entire load supplied. The OCPDs are not required to provide overload protection and are required to be sized per 430.62.

(3) Disconnecting Means. All disconnecting devices that are unique to the fire pump loads shall comply with items 695.4(B)(3)(a) through (B)(3)(e).

(a) *Features and Location — Normal Power Source.* The disconnecting means for the normal power source shall comply with all of the following: [**20**:9.2.3.1]

(1) Be identified as suitable for use as service equipment.

(2) Be lockable in the closed position. The provision for locking or adding a lock to the disconnecting means shall be installed on or at the switch or circuit breaker used as the disconnecting means and shall remain in place with or without the lock installed.

(3) Not be located within the same enclosure, panelboard, switchboard, switchgear, or motor control center, with or without common bus, that supplies loads other than the fire pump.

(4) Be located sufficiently remote from other building or other fire pump source disconnecting means such that inadvertent operation at the same time would be unlikely.

Exception to 695.4(B)(3)(a): For a multibuilding campus-style complex(s) installed under the provisions of 695.3(C), only the requirements in 695.4(B)(3)(a)(2) shall apply for normal power source disconnects.

A disconnecting means supplied by one of the individual sources specified in 695.3(A) cannot be installed in distribution equipment that supplies other than fire pump loads. Although Article 700 permits separate switchboard sections for other standby loads, 695.4(B)(3)(a)(3) does not permit it for fire pump source disconnecting means.

(b) *Features and Location — On-Site Standby Generator.* The disconnecting means for an on-site standby generator(s) used as the alternate power source shall be installed in accordance with 700.10(B)(6) for emergency circuits and shall be lockable in the closed position. The provision for locking or adding a lock to the disconnecting means shall be installed on or at the switch or circuit breaker used as the disconnecting means and shall remain in place with or without the lock installed.

A disconnecting means supplied by an on-site standby generator is permitted to be installed in equipment that supplies other loads. However, compliance with 700.10(B)(6) is required. The effect of this requirement is that a fire pump feeder cannot be supplied from equipment in which the fire pump conductors are installed in the same enclosure or vertical switchboard section with conductors supplying loads that are designated or classed as legally required standby (Article 701) or optional standby (Article 702) loads. To help minimize inadvertent opening of the fire pump circuit, the disconnecting means is required to be capable of being locked in the closed (on) position.

(c) *Disconnect Marking.* The disconnecting means shall be marked "Fire Pump Disconnecting Means." The letters shall be at least 25 mm (1 in.) in height, and they shall be visible without opening enclosure doors or covers. [**20**:9.2.3.1(5)]

(d) *Controller Marking.* A placard shall be placed adjacent to the fire pump controller, stating the location of this disconnecting means and the location of the key (if the disconnecting means is locked). [**20**:9.2.3.2]

(e) *Supervision.* The disconnecting means shall be supervised in the closed position by one of the following methods:

(1) Central station, proprietary, or remote station signal device

(2) Local signaling service that causes the sounding of an audible signal at a constantly attended point

(3) Locking the disconnecting means in the closed position

(f) Sealing of disconnecting means and approved weekly recorded inspections when the disconnecting means are located within fenced enclosures or in buildings under the control of the owner [**20**:9.2.3.3]

Supervision of the disconnecting means is required to ensure an uninterrupted power supply to the fire pump. Ideally, power supply conductors are run directly to the listed fire pump control and/or transfer equipment without the need for an additional service disconnecting means and overcurrent protection. However, this arrangement is not always possible; therefore, the single disconnecting means in 695.4(B) is permitted, provided it is locked in the closed position or monitored to ensure that it remains in the closed position.

Supervision (monitoring) of the disconnecting means by a local (protected premises) fire alarm system, central station, proprietary supervising station, or remote supervising station requires a connection to the premises fire alarm system. A fire alarm system initiating device circuit will generate a supervisory signal at the fire alarm control unit on loss of voltage to the fire pump controller. For more information on this interface with the fire alarm system, see *NFPA 72®, National Fire Alarm and Signaling Code®*.

Calculation Example

A fusible service disconnect switch supplies power to a 100-hp, 460-V, 3-phase fire pump and to a 1½-hp, 460-V, 3-phase jockey pump. Determine the sizes of the disconnecting means and the OCPD for the system. Also determine the minimum ampacity of the feeder conductors.

Solution

Step 1. Determine the minimum ratings of the disconnecting means and the OCPD.

According to the motor nameplates, the locked-rotor current (LRC) is 725 A for the 100-hp motor and 20 A for the 1½-hp motor. If the locked-rotor amperes are not on the nameplates, the LRCs found in Table 430.251(B) must be used. Calculate the size by summing the LRC of both motors and then going to the next larger standard-size OCPD, as follows:

$$100\text{-hp, 3-phase LRC} = 725 \text{ A}$$
$$1\tfrac{1}{2}\text{-hp, 3-phase LRC} = \underline{20 \text{ A}}$$
$$\text{Total LRC} = 745 \text{ A}$$

The next larger standard-size disconnect switch and overcurrent device is 800 A. An adjustable-trip circuit breaker of 750 A also is permitted, because it, too, will carry the LRC indefinitely.

Step 2. Determine the minimum ampacity for the fire pump feeder conductor.

Even though the disconnect switch and the overcurrent device are sized according to LRCs, the feeder conductors to the fire pump and associated equipment are required to have an ampacity not less than 125 percent of the full-load current (FLC) rating of the fire pump motor(s) and pressure maintenance pump motor(s), plus 100 percent of associated accessory equipment. Calculate the size of the feeder to the fire pump controller using 430.6(A)(1) and Table 430.250 for the FLC of the motors:

100-hp, 3-phase FLC

$$124 \text{ A} \times 1.25 = 155.0 \text{ A}$$

1½-hp, 3-phase FLC

$$3 \text{ A} \times 1.25 = \underline{3.75 \text{ A}}$$
$$\text{Total FLC} = 158.75 \text{ A or } 159 \text{ A}$$

Thus, the minimum ampacity for the feeder conductors is 159 A. Using the 75°C column, per 110.14(C)(1)(b), from Table 310.16, a 2/0 copper conductor is the minimum size required.

695.5 Transformers. Where the service or system voltage is different from the utilization voltage of the fire pump motor, transformer(s) protected by disconnecting means and overcurrent protective devices shall be permitted to be installed between the system supply and the fire pump controller in accordance with 695.5(A) and (B), or with (C). Only transformers covered in 695.5(C) shall be permitted to supply loads not directly associated with the fire pump system.

(A) Size. Where a transformer supplies an electric motor driven fire pump, it shall be rated at a minimum of 125 percent of the sum of the fire pump motor(s) and pressure maintenance pump(s) motor loads, and 100 percent of the associated fire pump accessory equipment supplied by the transformer.

(B) Overcurrent Protection. The primary overcurrent protective device(s) shall be selected or set to carry indefinitely the sum of the locked-rotor current of the fire pump motor(s) and the pressure maintenance pump motor(s) and the full-load current of the associated fire pump accessory equipment when connected to this power supply. Secondary overcurrent protection shall not be permitted. The requirement to carry the locked-rotor currents indefinitely shall not apply to conductors or devices other than overcurrent devices in the fire pump motor circuit(s).

The sizing of dedicated transformers and overcurrent protection can be broken down into three basic requirements. Generally stated, they are as follows:

1. The transformer must be sized to at least 125 percent of the sum of the loads.
2. The transformer primary overcurrent device must be the sum of locked-rotor current (LRC) of the fire pump motor(s) and the pressure maintenance pump motor(s) and the full-load current (FLC) of the associated fire pump accessory equipment. This differs from the general requirements of sizing overcurrent protection for fire pump motors.
3. The transformer secondary must not contain any overcurrent devices.

See Exhibit 695.4 for a one-line diagram on applying the dedicated fire pump transformer overcurrent protection requirements. Alternately, the jockey pump could be supplied by a separate panelboard.

Calculation Example

A 4160/480-V, 3-phase, dedicated transformer supplies power to a 100-hp, 460-V, 3-phase, code letter G fire pump and to a 1½-hp, 460-V, 3-phase, code letter H jockey pump. Determine the sizes of the dedicated transformer and its primary overcurrent protection.

Solution

Step 1. Determine the minimum standard-size transformer. First, to determine the minimum current value for use in the 3-phase power calculation, add the FLCs of the fire pump motor(s) and jockey pump motor(s). The FLCs of the two motors, using the FLC values from Table 430.250, are as follows:

100-hp, 3-phase FLC = 124 A
1½-hp, 3-phase FLC = 3 A
Total FLC = 127 A

Now, increase the sum of the fire pump motor and the jockey pump motor to 125 percent:

$$127 \text{ A} \times 1.25 = 158.75 \text{ A}$$

Then, size the transformer as follows:

$$\text{Transformer kVA} = \frac{\text{volts} \times \text{amperes} \times \sqrt{3}}{1000}$$

$$= \frac{480 \times 158.75 \times \sqrt{3}}{1000}$$

$$= 131.98 \text{ kVA}$$

The minimum-size transformer permitted is 131.98 kVA. The next larger standard-size transformer available is 150 kVA, but any larger size is permitted.

Step 2. Calculate the minimum-size primary OCPD permitted for this transformer. According to 695.5(B), the minimum primary OCPD must allow the transformer secondary to supply the LRC to the fire pump and, in this case, the jockey pump. The LRC of each motor must be individually calculated if it is not available on the motor nameplate. In this example, however, only the kVA code letters are assumed to be available. According to 430.7(B) and using the maximum values for the individual code letters per Table 430.7(B), calculate the maximum LRCs, as follows:

For the 100-hp motor, code letter G:

$$\text{LRC} = \text{motor hp} \times \text{max. code letter value}$$

$$\times \frac{1000}{\text{rated motor voltage} \times \text{3-phase factor}}$$

$$= 100 \text{ hp} \times \frac{6.29 \text{ kVA}}{\text{hp}} \times \frac{1000}{460 \times \sqrt{3}} = 789.46 \text{ A}$$

For the 1½-hp motor, code letter H (using the same formula):

$$\text{LCR} = 1\tfrac{1}{2} \text{ hp} \times \frac{7.09 \text{ kVA}}{\text{hp}} \times \frac{1000}{460 \times \sqrt{3}} = 13.35 \text{ A}$$

For the total LRC:

100-hp LRC = 789.46 A
1½-hp LRC = 13.35 A
Total LRC = 802.81 A or 803 A

Now, calculate the equivalent LRC on the primary side of the transformer, based on the calculated LRC of the secondary of the transformer, as follows:

$$\text{LRC}_{\text{primary}} = \frac{\text{secondary voltage}}{\text{primary voltage}} \times \text{LRC}_{\text{secondary}}$$

$$= \frac{480 \text{ V}}{4160 \text{ V}} \times 803 \text{ A}$$

$$= 92.65 \text{ A or } 93 \text{ A}$$

This 93 A represents the secondary LRC reflected to the primary side of the transformer. Because this value is the absolute smallest OCPD permitted, the next larger standard size, according to 240.6, is 100 A.

Conclusions.
1. The smallest standard-size transformer that is permitted is 150 kVA.
2. The smallest standard-size OCPD permitted on the primary of the transformer is 100 A.
3. A secondary OCPD is not permitted.

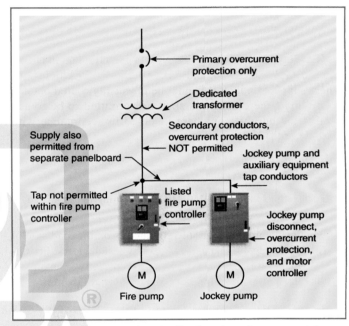

EXHIBIT 695.4 Overcurrent protection for a transformer supplying a fire pump and associated equipment. The transformer primary OCPD must be capable of carrying LRCs of the fire pump motor and the jockey pump motor indefinitely.

(C) Feeder Source. Where a feeder source is provided in accordance with 695.3(C), transformers supplying the fire pump system shall be permitted to supply other loads. All other loads shall be calculated in accordance with Article 220, including demand factors as applicable.

(1) Size. Transformers shall be rated at a minimum of 125 percent of the sum of the fire pump motor(s) and pressure maintenance pump(s) motor loads, and 100 percent of the remaining load supplied by the transformer.

(2) Overcurrent Protection. The transformer size, the feeder size, and the overcurrent protective device(s) shall be coordinated such that overcurrent protection is provided for the transformer in accordance with 450.3 and for the feeder in accordance with 215.3, and such that the overcurrent protective device(s) is selected or set to carry indefinitely the sum of the locked-rotor current of the fire pump motor(s), the pressure maintenance pump motor(s), the full-load current of the associated fire pump accessory equipment, and 100 percent of the remaining loads supplied by the transformer. The requirement to carry the locked-rotor currents indefinitely shall not apply to conductors or devices other than overcurrent devices in the fire pump motor circuit(s).

695.6 Power Wiring. Power circuits and wiring methods shall comply with the requirements in 695.6(A) through (J), and as permitted in 230.90(A), Exception No. 4; 230.94, Exception No. 4; 240.13; 230.208; 240.4(A); and 430.31.

(A) Supply Conductors.

(1) Services and On-Site Power Production Facilities. Service conductors and conductors supplied by on-site power production facilities shall be physically routed outside a building(s) and shall be installed as service-entrance conductors in accordance with 230.6, 230.9, and Parts III and IV of Article 230. Where supply conductors cannot be physically routed outside of buildings, the conductors shall be permitted to be routed through the building(s) where installed in accordance with 230.6(1) or (2).

> *Exception: The supply conductors within the fire pump room shall not be required to meet 230.6(1) or (2).*

> Informational Note: See 250.24(C) for routing the grounded conductor to the service equipment.

Δ **(2) Feeders.** Fire pump supply conductors on the load side of the final disconnecting means and overcurrent device(s) permitted by 695.4(B) or conductors that connect directly to an on-site standby generator shall comply with all of the following:

(1) *Independent Routing.* The conductors shall be kept entirely independent of all other wiring.
(2) *Associated Fire Pump Loads.* The conductors shall supply only loads that are directly associated with the fire pump system.
(3) *Protection from Potential Damage.* The conductors shall be protected from potential damage by fire, structural failure, or operational accident.
(4) *Inside of a Building.* Where routed through a building, the conductors shall be protected from fire for 2 hours using one of the following methods:

    a. The cable or raceway is encased in a minimum 50 mm (2 in.) of concrete.
    b. The cable or raceway is part of a listed fire-resistive cable system.

> Informational Note No. 1: See UL 2196, *Fire Test for Circuit Integrity of Fire-Resistive Power, Instrumentation, Control and Data Cables,* for one method of defining a fire-resistive cable system.
> Informational Note No. 2: See UL *Guide Information for Electrical Circuit Integrity Systems* (FHIT) for identifying the system and its installation limitations to maintain a minimum 2-hour fire-resistive rating.

> Informational Note No. 3: The listing organization provides information for fire-resistive cable systems on proper installation requirements to maintain the fire rating.

    c. The cable or raceway is protected by a listed electrical circuit protective system.

> Informational Note No. 4: See UL 1724, *Fire Tests for Electrical Circuit Protective Systems,* for one method of defining an electrical circuit protective system.
> Informational Note No. 5: See UL *Guide Information for Electrical Circuit Integrity Systems* (FHIT) for identifying the system and its installation limitations to maintain a minimum 2-hour fire-resistive rating.
> Informational Note No. 6: The listing organization provides information for electrical circuit protective systems on proper installation requirements to maintain the fire rating.

> *Exception to 695.6(A)(2)(4): The supply conductors located in the electrical equipment room where they originate and in the fire pump room shall not be required to have the minimum 2-hour fire separation or fire-resistance rating unless otherwise required by 700.10(D) of this Code.*

Feeder conductors are those that are installed from the load side of the supervised disconnecting means permitted by 695.4(B)(1)(a), (b), or (c). In addition, conductors directly connected to the output of an on-site standby generator are also covered by 695.6(A)(2). Unlike service conductors, feeder conductors are not required to be installed on the outside of a building or structure. However, if the feeder conductors are run through a building, they are required to be protected from damage by fire to ensure that power to the fire pump is not interrupted.

The difference between a 2-hour fire rating of an electrical circuit, such as a conduit with wires, and a 2-hour fire resistance rating of a structural member, such as a wall, is that at the end of a 2-hour fire test on an electrical conduit with wires, the circuit must function electrically (no short circuits, grounds, or opens are permitted) and its insulation must be intact. A wall subjected to a 2-hour fire resistance test must only prevent a fire from passing through or past the wall, without regard to damage to the wall. All fire ratings and fire resistance ratings are based on the assumption that the structural supports for the assembly are not impaired by the effects of the fire.

The UL *Fire Resistance Directory* describes three categories of products that can be used in the fire protection of electrical circuits for fire pumps: electrical circuit integrity systems (FHIT), electrical circuit protective materials (FHIY), and fire-resistive cables (FHJR). (The four-letter codes in parentheses are the UL product category guide designations.) For information on electrical circuit protective systems, see UL 1724, *Outline of Investigation for Fire Tests for Electrical Circuit Protective Systems.*

(B) Conductor Size.

Δ **(1) Fire Pump Motors and Other Equipment.** Conductors supplying a fire pump motor(s), pressure maintenance pumps, and associated fire pump accessory equipment shall have an ampacity of not less than the sum of the following:

(1) 125 percent of the sum of the fire pump motor(s) and pressure maintenance motor(s) full-load current(s), as determined by 430.6(A)

(2) 100 percent of the associated fire pump accessory equipment full-load current(s)

(2) Fire Pump Motors Only. Conductors supplying only a fire pump motor shall have a minimum ampacity in accordance with 430.22 and shall comply with the voltage drop requirements in 695.7.

Calculation Example

Listed fire pump controller and pump combinations are available in a wye-start, delta-run configuration as well as variable speed drive configurations. In the wye-delta configuration, six circuit conductors are run from the controller to the motor; when the motor is in the run mode, the conductors that supply each winding are connected in parallel.

See also

430.22(C) and its commentary for wye-start, delta-run operation

Determine the minimum size for the line- and load-side conductors of a controller with a fire pump with a 50-hp, 3-phase, 460-V motor. The pump motor and controller are configured for a wye-start, delta-run operation.

- Table 430.250 specifies full-load current (FLC) for a 50-hp motor as 65 A.
- Section 430.22(C) requires a controller line-side minimum conductor ampacity based on 125 percent of motor FLC.
- Section 430.22(C) requires a controller load-side minimum conductor ampacity based on 72 percent of motor FLC.

Solution

Step 1. Determine minimum conductor ampacity.

(a) Load side: 65 A × 0.72 = 47 A
(b) Line side: 65 A × 1.25 = 81 A

Step 2. Determine Type THWN copper conductor minimum size using Table 310.16 and assuming 75°C terminations in the controller.

(a) Load side: 50 A requires 8 AWG conductors. The combined ampacity of the two 8 AWG circuit conductors connected in parallel to each winding in the run mode is 100 A.
(b) Line side: 81 A requires 4 AWG conductors. The minimum size for the conductors may have to be increased to comply with the mandatory voltage-drop performance requirements in 695.7.

(C) Overload Protection. Power circuits shall not have automatic protection against overloads. Except for protection of transformer primaries provided in 695.5(C)(2), branch-circuit and feeder conductors shall be protected against short circuit only. Where a tap is made to supply a fire pump, the wiring shall be treated as service conductors in accordance with 230.6. The applicable distance and size restrictions in 240.21 shall not apply.

Exception No. 1: Conductors between storage batteries and the engine shall not require overcurrent protection or disconnecting means.

Exception No. 2: For an on-site standby generator(s) rated to produce continuous current in excess of 225 percent of the full-load amperes of the fire pump motor, the conductors between the on-site generator(s) and the combination fire pump transfer switch controller or separately mounted transfer switch shall be installed in accordance with 695.6(A)(2).

The protection provided shall be in accordance with the short-circuit current rating of the combination fire pump transfer switch controller or separately mounted transfer switch.

Δ **(D) Pump Wiring.**

N **(1) Wiring Methods.** Wiring from the controller(s) to the pump motor shall be in rigid metal conduit, intermediate metal conduit, electrical metallic tubing, liquidtight flexible metal conduit, or liquidtight flexible nonmetallic conduit LFNC-B, listed Type MC cable with an impervious covering, or Type MI cable. [**20:**9.4.4.1]

N **(2) Fittings.** Fittings shall be listed for use in wet locations.

N **(3) Connections.** Electrical connections at motor terminal boxes shall be made with a listed means of connection. Twist-on, insulation-piercing–type, and soldered wire connectors shall not be used for this purpose. [**20:**9.4.4.2, 9.4.4.3]

(E) Loads Supplied by Controllers and Transfer Switches. A fire pump controller and fire pump power transfer switch, if provided, shall not serve any load other than the fire pump for which it is intended.

(F) Mechanical Protection. All wiring from engine controllers and batteries shall be protected against physical damage and shall be installed in accordance with the controller and engine manufacturer's instructions.

(G) Ground-Fault Protection of Equipment. Ground-fault protection of equipment shall not be installed in any fire pump power circuit. [**20:**9.1.8.1]

The continued operation of the fire pumps until the fire is extinguished is essential. Ground-fault protection of equipment is not permitted to be used to protect components of a fire pump installation. Ground-fault detection that provides only an alarm is not prohibited by this requirement.

(H) Listed Electrical Circuit Protective System to Controller Wiring. Electrical circuit protective system installation shall comply with any restrictions provided in the listing of the electrical circuit protective system used, and the following also shall apply:

(1) A junction box shall be installed ahead of the fire pump controller a minimum of 300 mm (12 in.) beyond the fire-rated ceiling, wall, or floor bounding the fire zone.

The junction box mounted inside the fire pump room allows for a transition between solid conductors that are used in some electrical circuit protective systems (e.g., Type MI cable) and stranded conductors that are required at the supply terminals of the controller by the controller manufacturer and its listing. Splices are not permitted in the controller enclosure, in accordance with NFPA 20, *Standard for the Installation of Stationary Pumps for Fire Protection*. In addition, where an electrical circuit protective system employs single conductor cables, such as Type MI cable, the necessity to modify enclosures to prevent inductive heating can result in a compromise of the controller enclosure's resistance to water infiltration.

(2) Where required by the manufacturer of a listed electrical circuit protective system or by the listing, or as required elsewhere in this *Code*, the raceway between a junction box and the fire pump controller shall be sealed at the junction box end as required and in accordance with the instructions of the manufacturer. [**20**:9.8.2]

When fire-resistive cables pass through an area where there is a fire, they will likely produce flammable gases and smoke. These cables are not allowed to enter the fire pump controller, to prevent damage to the controller, and to prevent an explosion when gases are ignited by the motor contactor(s) or other sparking or arcing components, such as control relays or circuit breakers.

(3) Standard wiring between the junction box and the controller shall be permitted. [**20**:9.8.3]

(I) Junction Boxes. Where fire pump wiring to or from a fire pump controller is routed through a junction box, the following requirements shall be met:

(1) The junction box shall be securely mounted. [**20**:9.7(1)]
(2) Mounting and installing of a junction box shall not violate the enclosure type rating of the fire pump controller(s). [**20**:9.7(2)]
(3) Mounting and installing of a junction box shall not violate the integrity of the fire pump controller(s) and shall not affect the short-circuit current rating of the controller(s).
(4) As a minimum, a Type 2, drip-proof enclosure (junction box) shall be used where installed in the fire pump room. The enclosure shall be listed to match the fire pump controller enclosure type rating. [**20**:9.7(4)]

These requirements maintain the controller enclosure's environmental rating. Use of conduit hubs having the same environmental rating as the controller enclosure minimizes the entry of water or other liquids into the enclosure.

See also

110.28, Table 110.28, and associated commentary for information on environmental ratings of electrical equipment enclosures

(5) Terminals, junction blocks, wire connectors, and splices, where used, shall be listed. [**20**:9.7(5)]
(6) A fire pump controller or fire pump power transfer switch, where provided, shall not be used as a junction box to supply other equipment, including a pressure maintenance (jockey) pump(s).

(J) Terminations. Where raceways or cable are terminated at a fire pump controller, the following requirements shall be met:

(1) Raceway or cable fittings listed and identified for use in wet locations shall be used.
(2) The type rating of the raceway or cable fittings shall be at least equal to that of the fire pump controller.
(3) The installation instructions of the manufacturer of the fire pump controller shall be followed.
(4) Alterations to the fire pump controller, other than raceway or cable terminations as allowed elsewhere in this *Code*, shall be approved by the authority having jurisdiction.

695.7 Voltage Drop.

Δ **(A) Motor Starting.** Unless the requirements of 695.7(B) or (C) are met, the voltage at the fire pump controller line terminals shall not drop more than 15 percent below normal (controller-rated voltage) under motor starting conditions. [**20**:9.4.1]

N **(B) Emergency Run.** The requirements of 695.7(A) shall not apply to emergency-run mechanical starting, provided a successful start can be demonstrated on the standby generator system. [**20**:9.4.2]

N **(C) Bypass Mode.** The requirements of 695.7(A) shall not apply to the bypass mode of a variable speed pressure limiting control, provided a successful start can be demonstrated on the standby gen-set. [**20**:9.4.3]

Δ **(D) Motor Running.** The voltage at the contactor load terminals to which the motor is connected shall not drop more than 5 percent below the voltage rating of the motor when the motor is operating at 115 percent of the full-load current rating of the motor. [**20**:9.4.4]

Δ **695.10 Listed Equipment.** Diesel engine fire pump controllers, electric fire pump controllers, electric motors, fire pump power transfer switches, foam pump controllers, and limited service controllers shall be listed for fire pump service. [**20**:9.5.1.1, 10.1.2.1, 12.1.3.1]

Prior to being shipped to the installation site, listed fire pump controllers should be matched with the listed electric motor(s) they will control, to ensure compatibility of the individually listed components. The combination fire pump controller and transfer switch shown in Exhibit 695.5 are examples of listed equipment.

EXHIBIT 695.5 Listed combination fire pump controller and power transfer switch.

695.12 Equipment Location.

(A) Controllers and Transfer Switches. Electric motor-driven fire pump controllers and power transfer switches shall be located as close as practicable to, and within sight of, the motors that they control.

(B) Engine-Drive Controllers. Engine-drive fire pump controllers shall be located as close as is practical to, and within sight of, the engines that they control.

(C) Storage Batteries. Storage batteries for fire pump engine drives shall be supported above the floor, secured against displacement, and located where they are not subject to physical damage, flooding with water, excessive temperature, or excessive vibration.

(D) Energized Equipment. All energized equipment parts shall be located at least 300 mm (12 in.) above the floor level.

(E) Protection Against Pump Water. Fire pump controller and power transfer switches shall be located or protected so that they are not damaged by water escaping from pumps or pump connections.

(F) Mounting. All fire pump control equipment shall be mounted in a substantial manner on noncombustible supporting structures.

NFPA 20, *Standard for the Installation of Stationary Pumps for Fire Protection*, specifies a suitable space for fire pump equipment. This space must be free from hazards that could impair operation of the fire pump. Neither the *NEC* nor NFPA 20 mandates a dedicated room for the fire pump.

Even though 695.12(A) requires fire pump controllers and transfer switches to be "as close as practicable" to their associated fire pump motor, the minimum working space required by 110.26 must be maintained.

Fire pump controllers are housed in enclosures suitable to protect the contents against limited amounts of falling water and dirt. In addition, all energized parts in the enclosure must be mounted at least 12 inches above the floor. Typically, the floor space for this area is equipped with a floor drain.

695.14 Control Wiring.

(A) Control Circuit Failures. External control circuits that extend outside the fire pump room shall be arranged so that failure of any external circuit (open or short circuit) shall not prevent the operation of a pump(s) from all other internal or external means. Breakage, disconnecting, shorting of the wires, or loss of power to these circuits could cause continuous running of the fire pump but shall not prevent the controller(s) from starting the fire pump(s) due to causes other than these external control circuits. All control conductors within the fire pump room that are not fault tolerant shall be protected against physical damage. [**20**:10.5.2.6, 12.5.2.5]

NFPA 20 permits the installation of up to three interconnected fire pumps in different locations in a building, such as a high-rise. NFPA 20 requires the control wiring between those locations to be protected from fire and physical damage in the same manner as power conductors.

(B) Sensor Functioning. No undervoltage, phase-loss, frequency-sensitive, or other sensor(s) shall be installed that automatically or manually prohibits actuation of the motor contactor. [**20**:10.4.5.6]

Exception: A phase-loss sensor(s) shall be permitted only as a part of a listed fire pump controller.

(C) Remote Device(s). No remote device(s) shall be installed that will prevent automatic operation of the transfer switch. [**20**:10.8.1.3]

(D) Engine-Drive Control Wiring. All wiring between the controller and the diesel engine shall be stranded and sized to continuously carry the charging or control currents as required by the controller manufacturer. Such wiring shall be protected against physical damage. Controller manufacturer's specifications for distance and wire size shall be followed. [**20**:12.3.5.1]

(E) Electric Fire Pump Control Wiring Methods. All electric motor–driven fire pump control wiring shall be in rigid metal conduit, intermediate metal conduit, liquidtight flexible metal conduit, electrical metallic tubing, liquidtight flexible nonmetallic conduit, listed Type MC cable with an impervious covering, or Type MI cable.

Δ **(F) Generator Control Wiring Methods.** Control conductors installed between the fire pump power transfer switch and the standby generator supplying the fire pump during normal power loss shall be kept entirely independent of all other wiring. The integrity of the generator remote start circuit shall be monitored for broken, disconnected, or shorted wires. Loss of integrity shall start the generator(s).

Informational Note: See NFPA 20-2019, *Standard for the Installation of Stationary Pumps for Fire Protection*, 3.3.7.2, for more information on fault-tolerant external control circuits.

The control conductors shall be protected to resist potential damage by fire or structural failure. Where routed through a building, the conductors shall be protected from fire for 2 hours using one of the following methods:

(1) The cable or raceway is encased in a minimum 50 mm (2 in.) of concrete.
(2) The cable or raceway is part of a listed fire-resistive cable system.

Informational Note No. 1: See UL 2196-2017, *Standard for Fire Test for Circuit Integrity of Fire-Resistive Power, Instrumentation, Control and Data Cables*, for testing requirements for fire-resistive cables.
Informational Note No. 2: The listing organization provides information for fire-resistive cable systems on proper installation requirements to maintain the fire rating.

(3) The cable or raceway is protected by a listed electrical circuit protective system.

Informational Note No. 3: See UL 1724, *Fire Tests for Electrical Circuit Protection Systems*, for testing requirements for circuit protective systems.

Informational Note No. 4: Electrical circuit protective systems could include, but are not limited to, thermal barriers or a protective shaft.

Informational Note No. 5: The listing organization provides information for electrical circuit protective systems on proper installation requirements to maintain the fire rating.

Having the power wiring protected against fire damage is only one reliability consideration. For the generator to provide power, it has to receive the necessary signal to start. It is also critical to protect the control circuit wiring between the fire pump transfer switch/controller and the on-site standby generator.

695.15 Surge Protection. A listed surge protective device (SPD) shall be installed in or on the fire pump controller.

Informational Note: See UL 1449-2021, *Standard for Surge Protective Devices*, for proper application of SPD types.

Exception: Surge-protective devices shall not be required in or on a fire pump controller for diesel fire pumps.

Electronic components are used for control functions within fire pump controllers. These components are sensitive to overvoltages, and the proper operation of the electrically driven fire pump could be compromised due to component failure. As cited in the Fire Protection Research Foundation (FPRF) report *Data Assessment for Electrical Surge Protective Devices*, a survey of facility managers conducted by the National Electrical Manufacturers Association (NEMA) for the years 2013 and 2014 documented that 12 percent of the respondents indicated they had experienced surge damage to fire pump associated equipment. Considering that most electric motor-driven fire pumps and associated controllers are supplied directly by services, the exposure of such equipment to line surges due to lightning or utility switching is extremely high. The type of surge protective device (SPD) is largely driven by where the fire pump and controller are located within the electrical distribution system. The SPD for the fire pump controller can be field or factory installed.

The FPRF report can be viewed at: www.nfpa.org/News-and-Research/Data-research-and-tools/Electrical/Data-Assessment-for-Electrical-Surge-Protection-Devices.

See also

Article 242 for the requirements covering types and installation of SPDs

CHAPTER 7

Special Conditions

ARTICLE
700 | Emergency Systems

Part I. General

Δ **700.1 Scope.** This article applies to the electrical safety of the installation, operation, and maintenance of emergency systems consisting of circuits and equipment intended to supply, distribute, and control electricity for illumination, power, or both, to required facilities when the normal electrical supply or system is interrupted.

> Informational Note No. 1: Emergency systems are generally installed in places of assembly where artificial illumination is required for safe exiting and for panic control in buildings subject to occupancy by large numbers of persons, such as hotels, theaters, sports arenas, health care facilities, and similar institutions. Emergency systems may also provide power for such functions as ventilation where essential to maintain life, fire detection and alarm systems, elevators, fire pumps, public safety communications systems, industrial processes where current interruption would produce serious life safety or health hazards, and similar functions.
>
> Informational Note No. 2: See Article 517, Health Care Facilities, for further information regarding wiring and installation of emergency systems in health care facilities.
>
> Informational Note No. 3: See NFPA 99-2018, *Health Care Facilities Code,* for further information regarding performance and maintenance of emergency systems in health care facilities.
>
> Informational Note No. 4: See NFPA *101*-2018, *Life Safety Code,* for specification of locations where emergency lighting is considered essential to life safety.
>
> Informational Note No. 5: See NFPA 110-2019, *Standard for Emergency and Standby Power Systems,* and NFPA 111-2019, *Standard on Stored Electrical Energy Emergency and Standby Power Systems,* for further information regarding performance of emergency and standby power systems. Emergency systems are considered Level 1 systems when applying NFPA 110.

Emergency systems are designed and installed to maintain a specific level of illumination for means of egress and to provide power for fire alarm systems, fire pumps, automatic doors, and similar equipment if the normal power supply fails.

Article 700 applies to the installation of emergency systems that are essential for safety to human life and are legally required by municipal, state, federal, or other codes or by a governmental agency having jurisdiction. Article 700 does not mandate whether emergency systems are required or where emergency or exit lights should be located. These determinations rely on NFPA 101®, *Life Safety Code*®, or the adopted building code. The systems covered by Article 700 are generally viewed as what is necessary for an orderly and safe evacuation from a building or other occupancy type that is required by the applicable codes to be provided with emergency power.

The approach of Article 708 differs in that it provides requirements for power facilities that must be kept continuously operational throughout the duration of an emergency. Critical operations power systems (COPS) are generally installed in vital infrastructure facilities — those that, if destroyed or incapacitated, would disrupt national security, the economy, public health, or safety — and in areas where enhanced electrical infrastructure for continuity of operation has been deemed necessary by governmental authority.

N **700.2 Reconditioned Equipment.** Reconditioned transfer switches shall not be permitted.

700.3 Tests and Maintenance.

(A) Commissioning Witness Test. The authority having jurisdiction shall conduct or witness the commissioning of the complete system upon installation and periodically afterward.

> Informational Note: See NECA 90, *Standard for Commissioning Building Electrical Systems.*

(B) Tested Periodically. Systems shall be tested periodically on a schedule approved by the authority having jurisdiction to ensure the systems are maintained in proper operating condition.

(C) Maintenance. Emergency system equipment shall be maintained in accordance with manufacturer instructions and industry standards.

(D) Written Record. A written record shall be kept of such tests and maintenance.

(E) Testing Under Load. Means for testing all emergency lighting and power systems during maximum anticipated load conditions shall be provided.

> Informational Note: See NFPA 110-2019, *Standard for Emergency and Standby Power Systems*, for information on testing and maintenance of emergency power supply systems (EPSSs).

Emergency system testing can be divided into two general categories — acceptance testing and operational testing. Section 700.3 requires both types of testing as well as written records of testing and maintenance.

Acceptance testing is performed after the emergency system has been installed but before the system is placed into service. Acceptance testing ensures that the emergency system meets the original installation specification.

Operational testing ensures that the emergency system remains functional. Generally, actual emergency system loads are smaller than the design capacity of the emergency generator system.

Visual inspections are an important aspect of routine maintenance. For lead-acid batteries, it is important that electrolyte levels be maintained. Transparent cases for lead-acid batteries allow easy viewing of electrolyte levels. Terminals should be inspected for signs of corrosion, even on batteries that are labeled maintenance free.

Further information on tests and maintenance can be found in NFPA 70B, *Recommended Practice for Electrical Equipment Maintenance*; NFPA 99, *Health Care Facilities Code*; NFPA 101, *Life Safety Code*, NFPA 110, *Standard for Emergency and Standby Power Systems*; and NFPA 111, *Standard on Stored Electrical Energy Emergency and Standby Power Systems*.

Δ **(F) Temporary Source of Power for Maintenance or Repair of the Alternate Source of Power.** If the emergency system relies on a single alternate source of power, which will be disabled for maintenance or repair, the emergency system shall include permanent switching means to connect a portable or temporary alternate source of power that shall be available for the duration of the maintenance or repair. The permanent switching means to connect a portable or temporary alternate source of power shall comply with the following:

(1) Connection to the portable or temporary alternate source of power shall not require modification of the permanent system wiring.

(2) Transfer of power between the normal power source and the emergency power source shall be in accordance with 700.12.

(3) The connection point for the portable or temporary alternate source shall be marked with the phase rotation and system bonding requirements.

(4) The switching means, including the interlocks, shall be listed and provided with mechanical or mechanical and

electrical interlocking to prevent inadvertent interconnection of power sources.

(5) The switching means shall include a contact point that shall annunciate at a location remote from the generator or at another facility monitoring system to indicate that the permanent emergency source is disconnected from the emergency system.

(6) The permanent connection point for the temporary generator shall be located outdoors and shall not have cables from the connection point to the temporary generator routed through exterior windows, doors, or similar openings.

(7) A permanent label shall be field applied at the permanent connection point to identify the system voltage, maximum amperage, short-circuit current rating of the load side of equipment supplied, and ungrounded conductor identification in accordance with 210.5.

It shall be permissible to use manual switching to switch from the permanent source of power to the portable or temporary alternate source of power and to use the switching means for connection of a load bank.

> Informational Note: See Informational Note Figure 700.3(F) for one example of many possible methods to achieve the requirements of 700.3(F).

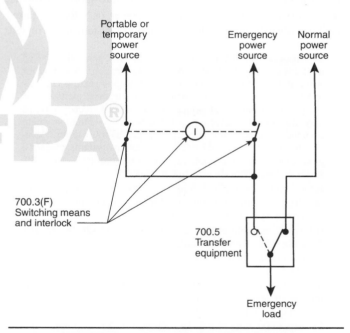

INFORMATIONAL NOTE FIGURE 700.3(F)

Exception: The permanent switching means to connect a portable or temporary alternate source of power for the duration of the maintenance or repair shall not be required where any of the following conditions exists:

(1) All processes that rely on the emergency system source are capable of being disabled during maintenance or repair of the emergency source of power.

(2) The building or structure is unoccupied and fire protection systems are fully functional and do not require an alternate power source.

(3) Other temporary means can be substituted for the emergency system.

(4) A permanent alternate emergency source, such as, but not limited to, a second on-site standby generator or separate electric utility service connection, capable of supporting the emergency system, exists.

700.4 Capacity and Rating.

(A) Capacity. An emergency system shall have adequate capacity in accordance with Parts I through IV of Article 220 or by another approved method. The system capacity shall be sufficient for the rapid load changes and transient power and energy requirements associated with any expected loads.

The emergency system must be designed with adequate capacity and rating to safely carry, at one time, the entire load connected to it. The system must be capable of restarting emergency loads that have been interrupted, such as motors that may have stopped, and it must be suitable for the available fault current.

See also

517.31(D) for the requirement covering acceptable methods to calculate the electrical demand of a hospital's essential electrical system

Δ **(B) Selective Load Management.** The alternate power source shall be permitted to supply emergency, legally required standby, and optional standby system loads where the source has adequate capacity or where load management (that includes automatic selective load pickup and load shedding) is provided as needed to ensure adequate power to the following in order of priority:

(1) Emergency circuits
(2) Legally required standby circuits
(3) Optional standby circuits

If a generator is used for peak load shaving, supplying backup power, and other uses, priority loads must be properly and reliably served. Selective load pickup and load shedding are not required where the generator has the capacity to supply all loads served.

If a generator is used for peak load shaving or in a cogeneration system, the increase in wear and tear will likely result in an increase in downtime for maintenance. Also, using the emergency generator on a regular basis for nonemergency loads provides assurance that the emergency generator will supply emergency power when needed.

N **(C) Parallel Operation.** Parallel operation of the emergency source(s) shall consist of the sources specified in 700.4(C)(1) and (C)(2).

N **(1) Normal Source.** The emergency source shall be permitted to operate in parallel with the normal source in compliance with Part I or Part II of Article 705 where the capacity required to supply the emergency load is maintained at all times. Any operating condition that results in less than the required emergency source capacity shall initiate a system malfunction signal in accordance with 700.6(A).

Parallel operation shall be permitted for satisfying the test requirements of 700.3(B), provided all other conditions of 700.3 are met.

Informational Note: Peak load shaving is one application for parallel source operation.

N **(2) Emergency Source.** Emergency sources shall be permitted to operate in parallel where the necessary equipment to establish and maintain a synchronous condition is provided.

700.5 Transfer Equipment.

Double-throw automatic transfer switches (ATS) typically are used for emergency and standby power generation systems rated 1000 volts or less. These transfer switches do not normally incorporate overcurrent protection. ATS are available in ratings up to 38 kilovolts. For reliability, those used for emergency and legally required standby systems must be electrically operated and mechanically held. System grounding is determined by the type of transfer switch employed.

See also

250.30 and associated commentary regarding separately derived systems

Transfer switches should be located close to the load. It may be advantageous to use multiple transfer switches of lower current rating located near the loads rather than one large transfer switch at the point of incoming service. For information on time-delay devices for ATS, see the accompanying Closer Look feature.

CLOSER LOOK: Time-Delay Devices on Automatic Transfer Switches

The normal power source is usually a service, and the emergency power source is an automatically started engine generator set that starts when the normal source fails. Time-delay controls are essential to the operation of the ATS.

To avoid unnecessary starting and transfer to the alternate supply, a time delay can override momentary interruptions and temporary reductions in normal power source voltage but still allow starting and transfer if the reduction or outage is sustained. However, the time delay should be set fast enough to effectively operate the transfer switch.

The delay generally is set at 1 second but can be set higher if reclosers or circuit breakers on the utility power lines take longer to operate or if momentary power dips exceed 1 second. If longer delay settings are used, care must be taken to ensure that sufficient time remains to meet 10-second power restoration requirements. The AHJ might determine that an outage is

---

not a longer-term power failure until the utility automatic protective devices fail to restore power to the facility. For example, the 10-second power restoration requirements would become effective after the 2-second recloser cycle.

Once the load is transferred to the alternate source, another timer delays transfer back to the normal power source until that source has time to stabilize. Another important function of this timer required by 700.12(D)(1) is to allow an engine generator to operate under load to ensure continued good performance of the set and its starting system. This delay should be automatically nullified if the alternate source fails, and the normal source is available.

Engine generator manufacturers often recommend a cool-down period for their sets that allows them to run unloaded after the load is transferred back to the normal source. A third time delay, usually 5 minutes, is provided for that purpose. Running an unloaded engine longer is usually not recommended, because it can cause deterioration in engine performance.

If more than one ATS is connected to the same engine generator, it is sometimes recommended that transfer of the loads be sequenced to the alternate source. Using a sequencing scheme can reduce starting capacity requirements of the generator. A fourth timer, adjustable from 0 to 5 minutes, will delay transfer to the emergency supply source for this and other similar requirements.

Δ **(A) General.** Transfer equipment shall be automatic, listed, and marked for emergency use. Transfer equipment shall be designed and installed to prevent the inadvertent interconnection of normal and emergency sources of supply in any operation of the transfer equipment. Transfer equipment and electric power production systems installed to permit operation in parallel with the normal source shall meet the requirements of Article 705. Meter-mounted transfer switches shall not be permitted for emergency system use.

Most ATS are not designed to permit parallel operation of generation equipment and the normal source. Compliance with Article 705 is necessary only where the transfer allows parallel operation with the normal supply. Certain ATS configurations are intentionally designed to briefly (for a few cycles) parallel the generation equipment with the normal source upon load transfer from generator to normal source. That load transfer method may result in minimal disturbance or effect on the load. If continuous parallel operation of generation equipment and the source is desired, paralleling switchgear or paralleling equipment with appropriate protection is required.

(B) Bypass Isolation Transfer Switches. Means shall be permitted to bypass and isolate the transfer equipment. Where bypass isolation transfer switches are used, inadvertent parallel operation shall be prevented.

Δ **(C) Automatic Transfer Switches.** Automatic transfer switches shall be electrically operated and mechanically held.

N **(D) Redundant Transfer Equipment.** If emergency loads are supplied by a single feeder, the emergency power system shall include redundant transfer equipment or a bypass isolation transfer switch to facilitate maintenance as required in 700.3(C) without jeopardizing continuity of power. If the redundant transfer equipment or bypass isolation transfer switch is manual (or nonautomatic), then it shall be actively supervised by a qualified person when the primary (automatic) transfer equipment is disabled for maintenance or repair.

Exception: The requirement for redundancy with the transfer equipment shall not apply where any of the following conditions exist:

(1) All processes that rely on the emergency system source are capable of being disabled during maintenance or repair activities without jeopardizing the safety to human life.

(2) The building or structure is unoccupied and fire protection systems are fully functional and do not require an alternate power source.

(3) Other temporary means shall be permitted to be substituted for the emergency system.

(4) A written emergency plan that includes mitigation actions and responsibilities for qualified persons to address the recognized site hazards for the duration of the maintenance or repair activities shall be developed and implemented. The emergency plan shall be made available to the authority having jurisdiction.

(E) Use. Transfer equipment shall supply only emergency loads.

Informational Note: Transfer equipment that supplies emergency loads provides separation of this load type from any others and is independent of any equipment used to combine or parallel sources.

The alternate power source can supply emergency loads as well as other loads. However, the emergency system transfer switch is limited to supplying emergency loads. Legally required standby loads and optional standby loads (covered by Articles 701 and 702) require separate transfer switches. A typical emergency system transfer switch is shown in Exhibit 700.1.

(F) Documentation. The short-circuit current rating of the transfer equipment, based on the specific overcurrent protective device type and settings protecting the transfer equipment, shall be field marked on the exterior of the transfer equipment.

In this requirement, the term *short-circuit current rating* includes all the various options by which the product standard evaluates transfer switches for fault currents, such as short-circuit withstand and closing rating, short-time current rating, and the common industry term withstand/close-on rating.

Product standards require transfer equipment to be marked with the short-circuit withstand/closing or short-time current rating (short-circuit current rating). Some transfer switches are marked by the manufacturer with several options, each resulting in a different short-circuit current rating. Short-circuit current rating values can vary based upon the overcurrent protective

EXHIBIT 700.1 An emergency system transfer switch. *(Courtesy of the International Association of Electrical Inspectors)*

device (OCPD) type, ampere rating, and setting. The field marking required by this section documents the specifics of the protection scheme and verifies compliance with 110.3(B) and 110.10.

700.6 Signals. Audible, visual, and facility or network remote annunciation devices shall be provided, where applicable, for the purpose described in 700.6(A) through (D).

(A) Malfunction. Malfunction signals indicate a malfunction of the emergency source.

(B) Carrying Load. Load carrying signals indicate that the emergency source is carrying load.

(C) Storage Battery Charging Malfunction. Storage battery charging malfunction signals indicate a charging malfunction on a battery required for source readiness, including starting the prime mover, is not functioning.

Inadequate testing and maintenance can result in emergency equipment failure. Battery-operated unit equipment generally has a test switch that simulates failure of the normal system and an indicating light that glows brightly while the system is charging and that dims when the system is ready. Installing signal devices that annunciate trouble where personnel familiar with the operation of the emergency equipment can see or hear them allows action to be taken to maintain system function.

(D) Ground Fault. Ground-fault signals indicate a ground fault in solidly grounded wye emergency systems of more than 150 volts to ground and circuit-protective devices rated 1000 amperes or more. The sensor for the ground-fault signal devices

shall be located at, or ahead of, the main system disconnecting means for the emergency source, and the maximum setting of the signal devices shall be for a ground-fault current of 1200 amperes. Instructions on the course of action to be taken in the event of indicated ground fault shall be located at or near the sensor location.

For systems with multiple emergency sources connected to a paralleling bus, the ground fault sensor and the system bonding jumper shall be permitted to be at an alternative location.

Ground-fault protection is not required on emergency systems because it could interrupt the system when it is needed. However, ground faults must be detected and indicated so that the ground fault can be cleared as soon as practical.

See also

700.31 for more on ground-fault protection of equipment

700.7 Signs.

(A) Emergency Sources. A sign shall be placed at the service-entrance equipment, indicating type and location of each on-site emergency power source.

Exception: A sign shall not be required for individual unit equipment as specified in 700.12(H).

(B) Grounding. Where removal of a grounding or bonding connection in normal power source equipment interrupts the grounding electrode conductor connection to the alternate power source(s) grounded conductor, a warning sign shall be installed at the normal power source equipment stating:

WARNING
SHOCK HAZARD EXISTS IF GROUNDING
ELECTRODE CONDUCTOR OR BONDING JUMPER
CONNECTION IN THIS EQUIPMENT IS REMOVED
WHILE ALTERNATE SOURCE(S) IS ENERGIZED.

The warning sign(s) or label(s) shall comply with 110.21(B).

Emergency and standby systems that have a solid (unswitched) neutral in the transfer equipment (nonseparately derived system) rely on the grounding and bonding connections in the normal source supply equipment to ensure that the ground-fault current path is completed from a ground fault to the alternate source. If a main or system bonding jumper is removed [e.g., to perform testing on ground-fault protection of equipment (GFPE) systems], service personnel could inadvertently become part of the current path if a ground fault occurs.

700.8 Surge Protection. A listed SPD shall be installed in or on all emergency system switchgear, switchboards, and panelboards.

See also

695.15 commentary for a discussion on the use of surge protective devices (SPDs) to protect electronic components that are integral to the operation of equipment performing a critical safety function

Part II. Circuit Wiring

700.10 Wiring, Emergency System.

(A) Identification. Emergency circuits shall be permanently marked so they will be readily identified as a component of an emergency circuit or system by the following methods:

(1) All boxes and enclosures (including transfer switches, generators, and power panels) for emergency circuits shall be permanently marked as a component of an emergency circuit or system.

(2) Where boxes or enclosures are not encountered, exposed cable or raceway systems shall be permanently marked to be identified as a component of an emergency circuit or system, at intervals not to exceed 7.6 m (25 ft).

Receptacles supplied from the emergency system shall have a distinctive color or marking on the receptacle cover plates or the receptacles.

The required marking can be by color code, the words "emergency system," or any other method acceptable to the AHJ that identifies the box or enclosure as a component of the emergency system.

(B) Wiring. Wiring from an emergency source or emergency source distribution overcurrent protection to emergency loads shall be kept entirely independent of all other wiring and equipment unless otherwise permitted in the following:

(1) Wiring from the normal power source located in transfer equipment enclosures

(2) Wiring supplied from two sources in exit or emergency luminaires

(3) Wiring from two sources in a listed load control relay supplying exit or emergency luminaires, or in a common junction box, attached to exit or emergency luminaires

(4) Wiring within a common junction box attached to unit equipment, containing only the branch circuit supplying the unit equipment and the emergency circuit supplied by the unit equipment

Except as noted in this section, wiring for the emergency circuits must be completely independent of all other wiring and equipment. This practice ensures that a fault in any other system wiring will not affect the performance of the emergency wiring or equipment.

(5) Wiring within a traveling cable to an elevator

(6) Wiring from an emergency source to supply emergency and other (nonemergency) loads in accordance with the following:

a. Separate vertical switchgear sections or separate vertical switchboard sections, with or without a common bus, or individual disconnects mounted in separate enclosures shall be used to separate emergency loads from all other loads.

b. The common bus of separate sections of the switchgear, separate sections of the switchboard, or the individual enclosures shall be either of the following:

(i) Supplied by single or multiple feeders without overcurrent protection at the source

(ii) Supplied by single or multiple feeders with overcurrent protection, provided that the overcurrent protection that is common to an emergency system and any nonemergency system(s) is selectively coordinated with the next downstream overcurrent protective device in the nonemergency system(s)

Informational Note: See Informational Note Figure 700.10(B)(1) and Informational Note Figure 700.10(B)(2) for further information.

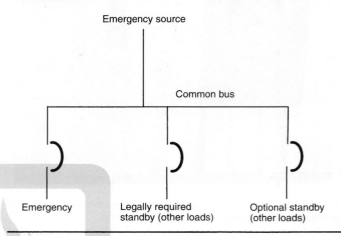

INFORMATIONAL NOTE FIGURE 700.10(B)(1) *Single or Multiple Feeders Without Overcurrent Protection.*

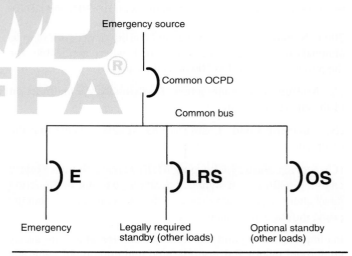

INFORMATIONAL NOTE FIGURE 700.10(B)(2) *Single or Multiple Feeders with Overcurrent Protection.*

c. Emergency circuits shall not originate from the same vertical switchgear section, vertical switchboard section, panelboard enclosure, or individual disconnect enclosure as other circuits.

d. It shall be permissible to use single or multiple feeders to supply distribution equipment between an emergency source and the point where the emergency loads are separated from all other loads.

e. At the emergency power source, such as a generator, multiple integral overcurrent protective devices shall each be permitted to supply a designated emergency or a designated nonemergency load, provided that there is complete separation between emergency and nonemergency loads beginning immediately after the overcurrent protective device line-side connections.

Wiring of two or more emergency circuits supplied from the same source shall be permitted in the same raceway, cable, box, or cabinet.

A switchboard may further distribute and provide power for the emergency, legally required, and optional standby systems, provided separate vertical switchboard sections are used.

Separate vertical switchboard sections provide the required physical separation between the wiring of emergency and standby systems that are supplied from a common power source. This physical separation cannot occur within a panelboard enclosure because of its open design. The supply tap box on generators equipped with disconnects with or without overcurrent protection is not generally designed or manufactured for the installation of multiple transfer switches to serve separate circuits for emergency systems (which can include fire pump loads), legally required standby systems, and optional standby systems. In addition, large systems might employ multiple generators. Any combination of those systems can be supplied from a single feeder or multiple feeders, or from separate vertical sections of a switchboard that are either supplied by a common bus or supplied individually.

Exhibits 700.2, 700.3, and 700.4 show illustrations of feeder configurations.

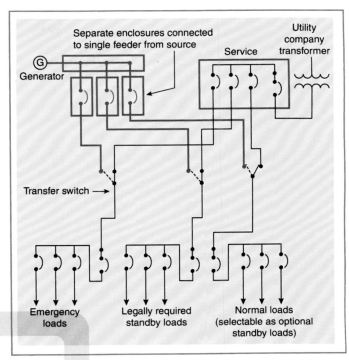

EXHIBIT 700.3 Illustration of a single feeder that supplies multiple transfer switches.

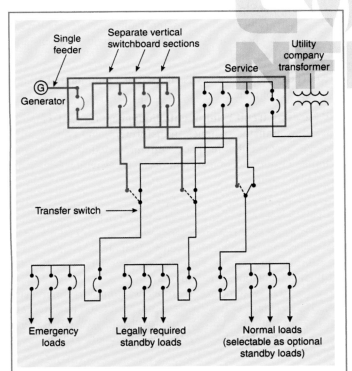

EXHIBIT 700.2 Illustration of a single feeder that supplies separate vertical sections of the switchboard.

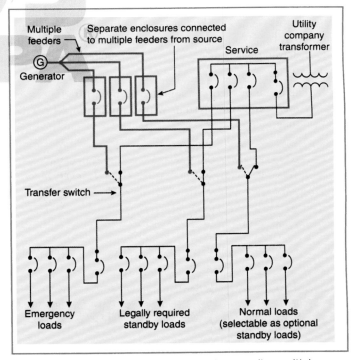

EXHIBIT 700.4 Illustration of a generator that supplies multiple feeders at its terminals.

A single common feeder is permitted to be installed between the alternate source and the point in the distribution system at which the physical separation of the emergency, legally required standby, and optional standby system conductors occurs, such as at a switchboard or other distribution equipment. Many large campus facilities with multiple buildings, such as medical centers, colleges or universities, prisons, and shopping malls, rely on central generation of emergency power.

(C) Wiring Design and Location. Emergency wiring circuits shall be designed and located so as to minimize the hazards that might cause failure due to flooding, fire, icing, vandalism, and other adverse conditions.

(D) Fire Protection.

(1) Occupancies. Emergency systems shall meet the additional requirements in 700.10(D)(2) through (D)(4) in the following occupancies:

(1) Assembly occupancies for not less than 1000 persons
(2) Buildings above 23 m (75 ft) in height
(3) Educational occupancies with more than 300 occupants

Δ **(2) Feeder-Circuit Wiring.** Feeder-circuit wiring shall meet one of the following conditions:

(1) The cable or raceway is installed in spaces or areas that are fully protected by an approved automatic fire protection system.

If emergency system feeders are installed above a suspended ceiling, sprinklers must be provided above the suspended ceiling for the system to be considered as fully protected by a fire suppression system. Ceiling-installed sprinklers that protect the space below the ceiling do not provide the required protection for the emergency system feeder(s).

(2) The cable or raceway is protected by a listed electrical circuit protective system with a minimum 2-hour fire rating.

> Informational Note No. 1: See UL 1724, *Fire Tests for Electrical Circuit Protection Systems,* for one method of defining an electrical circuit protective system. The UL *Guide Information for Electrical Circuit Integrity Systems* (FHIT) contains information to identify the system and its installation limitations to maintain a minimum 2-hour fire-resistive rating and is available from the certification body.

(3) The cable or raceway is a listed fire-resistive cable system with a minimum 2-hour fire rating.

> Informational Note No. 2: See UL 2196-2017, *Standard for Fire Test for Circuit Integrity of Fire-Resistive Power, Instrumentation, Control and Data Cables,* for one method of defining a fire-resistive cable system.

(4) The cable or raceway is protected by a listed fire-rated assembly that has a minimum fire rating of 2 hours and contains only emergency circuits.

If feeders are not located in building spaces that are fully protected by a fire suppression system, other fire protection techniques that can be used include the following:

- **Listed electrical circuit protective systems,** which are described in the UL *Guide Information for Electrical Equipment* and can be accessed at https://iq.ulprospector.com/en. The four-letter codes (shown in parentheses) are the UL product category guide designations. Examples of these systems include electrical circuit protective systems (FHIT), electrical circuit protective materials (FHIY), and fire-resistive cables (FHJR). Circuit integrity cable is covered under category FHJR.
- **Listed thermal barrier systems (XCLF),** which also can be found at https://iq.ulprospector.com/en. An example of the thermal barrier protection technique is batts and blankets (XCLR) wrapped over the wiring method to achieve a predetermined fire rating.
- **Fire-rated assemblies,** which also are found at https://iq.ulprospector.com/en. All fire ratings and fire resistance ratings assume that the structural supports for the assembly are not impaired by the fire. There is a difference between a 2-hour fire rating of an electrical circuit, such as a conduit with wires, and a 2-hour fire resistance rating of a structural member, such as a wall. At the end of a 2-hour fire test on an electrical conduit with wires, its insulation must be intact, and the circuit must function electrically; no short circuits, grounds, or opens are permitted. A wall subjected to a 2-hour fire resistance test must only prevent a fire from passing through or past the wall, regardless of damage to the wall.
- **Encasement in concrete,** which has been successful for many years in protecting premises from faults in service conductors per 230.6. Encasement in 2 inches of concrete is possible after original construction.

(5) The cable or raceway is encased in a minimum of 50 mm (2 in.) of concrete.

(3) Feeder-Circuit Equipment. Equipment for feeder circuits (including transfer switches, transformers, and panelboards) shall be located either in spaces fully protected by an approved automatic fire protection system or in spaces with a 2-hour fire resistance rating.

Fire protection requirements for both the emergency system feeder circuits and the equipment ensure the integrity as well as the performance of the emergency electrical system. If feeders and equipment are in building spaces that are fully protected by an approved fire suppression system, generally no further fire protection techniques are required.

Sprinkler systems are the most common fire suppression systems. Building spaces that are fully protected by automatic sprinkler systems meet the requirements of 700.10(D). Requirements for fire suppression systems are included in the following standards:

NFPA 12, *Standard on Carbon Dioxide Extinguishing Systems*
NFPA 12A, *Standard on Halon 1301 Fire Extinguishing Systems*

NFPA 13, *Standard for the Installation of Sprinkler Systems*

NFPA 15, *Standard for Water Spray Fixed Systems for Fire Protection*

NFPA 17, *Standard for Dry Chemical Extinguishing Systems*

NFPA 2001, *Standard on Clean Agent Fire Extinguishing Systems*

If feeder-circuit equipment is not located in a space that is fully protected by a fire suppression system, the space must have a 2-hour fire resistance rating.

See also

700.10(D)(2)(4) commentary regarding fire-rated assemblies

(4) **Source Control Wiring.** Control conductors installed between the emergency power supply system/stored-energy power supply system (EPSS/SEPSS) and transfer equipment or control systems that initiate the operation of emergency sources or initiate the automatic connection to emergency loads shall be kept entirely independent of all other wiring and shall meet the conditions of 700.10(D)(2). The integrity of source control wiring shall be monitored for broken, disconnected, or shorted wires. Loss of integrity shall result in the following actions:

(1) *Generators.* Shall start the generator(s).
(2) *All other sources.* Shall be considered a system malfunction and initiate the designated signal(s) in 700.6(A).

N **700.11 Wiring, Class-2-Powered Emergency Lighting Systems.**

N **(A) General.** Line voltage supply wiring and installation of Class 2 emergency lighting control devices shall comply with 700.10. Class 2 emergency circuits shall comply with 700.11(B) through (D).

N **(B) Identification.** Emergency circuits shall be permanently marked so they will be readily identified as a component of an emergency circuit or system by the following methods:

(1) All boxes and enclosures for Class 2 emergency circuits shall be permanently marked as a component of an emergency circuit or system.
(2) Exposed cable, cable tray, or raceway systems shall be permanently marked to be identified as a component of an emergency circuit or system, within 900 mm (3 ft) of each connector and at intervals not to exceed 7.6 m (25 ft).

N **(C) Separation of Circuits.** Class 2 emergency circuits shall be wired in a listed, jacketed cable or with one of the wiring methods of Chapter 3. If installed alongside nonemergency Class 2 circuits that are bundled, Class 2 emergency circuits shall be bundled separately. If installed alongside nonemergency Class 2 circuits that are not bundled, Class 2 emergency circuits shall be separated by a nonconductive sleeve or nonconductive barrier from all other Class 2 circuits. Separation from other circuits shall comply with 725.136.

N **(D) Protection.** Wiring shall comply with the requirements of 300.4 and be installed in a raceway, armored or metal-clad cable, or cable tray.

Exception No. 1: Section 700.11(D) shall not apply to wiring that does not exceed 1.83 m (6 ft) in length and that terminates at an emergency luminaire or an emergency lighting control device.

Exception No. 2: Section 700.11(D) shall not apply to locked rooms or locked enclosures that are accessible only to qualified persons.

Informational Note: Locked rooms accessible only to qualified persons include locked telecommunications rooms, locked electrical equipment rooms, or other access-controlled areas.

Part III. Sources of Power

700.12 General Requirements. Current supply shall be such that, in the event of failure of the normal supply to, or within, the building or group of buildings concerned, emergency lighting, emergency power, or both shall be available within the time required for the application but not to exceed 10 seconds. The supply system for emergency purposes, in addition to the normal services to the building and meeting the general requirements of this section, shall be one or more of the types of systems described in 700.12(C) through (H). Unit equipment in accordance with 700.12(H) shall satisfy the applicable requirements of this article.

At least two sources of power must be provided — one normal supply and one or more of the emergency systems described in 700.12. The sources (see Exhibits 700.5 and 700.6) can be one of the following:

1. Two services — one normal supply and one emergency supply (preferably from separate utility stations)
2. One normal service and a storage battery (or unit equipment) system
3. One normal service and a generator set
4. One normal service and an uninterruptible power supply (UPS)
5. One normal service and a fuel cell system
6. Direct-current microgrid and a utility service, an engine generator, a storage battery system, a UPS system, or a fuel cell

A means must be provided to transfer the emergency loads to the alternate supply within 10 seconds of normal supply interruption. If a separate service is used, both may operate normally, but equipment for emergency lighting and power must be arranged to be energized from either service.

If the alternate or emergency source of supply is a storage battery or generator set, the single emergency system usually is operated on the normal service, and the battery (or batteries) or generator operates only if the normal service fails. However, a generator can be used for peak load shaving and other standby systems in accordance with 700.4.

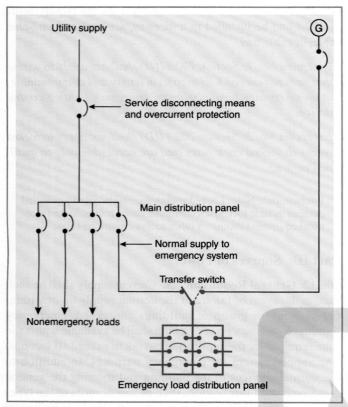

EXHIBIT 700.5 Emergency load arranged to be supplied from a generator, as permitted by 700.12(D).

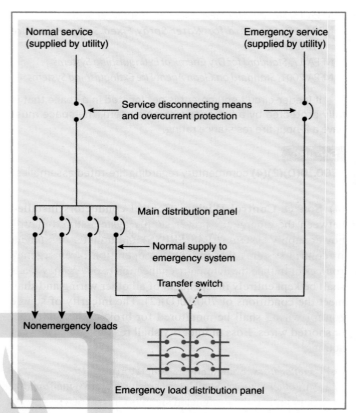

EXHIBIT 700.6 Emergency load arranged to be supplied from two widely separated services, as permitted by 700.12(F).

Disconnecting means and overcurrent protection (see Exhibits 700.5 and 700.6) must be provided for emergency systems as required by Articles 225 and 230.

(A) Power Source Considerations. In selecting an emergency source of power, consideration shall be given to the occupancy and the type of service to be rendered, whether of minimum duration, as for evacuation of a theater, or longer duration, as for supplying emergency power and lighting due to an indefinite period of current failure from trouble either inside or outside the building.

Δ **(B) Equipment Design and Location.** Equipment shall be designed and located so as to minimize the hazards that might cause complete failure due to flooding, fires, icing, and vandalism.

The design and selection of a location must consider all hazards, including geographical, that could impair reliability. See Exhibit 700.7.

Equipment for sources of power as described in 700.12(C) through (H) shall be installed either in spaces fully protected by approved automatic fire protection systems or in spaces with a 2-hour fire rating where located within the following:

(1) Assembly occupancies for more than 1000 persons
(2) Buildings above 23 m (75 ft) in height
(3) Educational occupancies with more than 300 occupants

EXHIBIT 700.7 Alternate source of power in a heated and secure enclosure that is in an area not subject to flooding. *(Courtesy of the International Association of Electrical Inspectors)*

Informational Note No. 1: See NFPA *101*-2021, *Life Safety Code*, Section 6.1, for information on occupancy classifications.
Informational Note No. 2: See IEEE 3006.5-2014, *Recommended Practice for the Use of Probability Methods for Conducting a Reliability Analysis of Industrial and Commercial Power Systems*, for information regarding power system reliability.

Δ **(C) Supply Duration.** The emergency power source shall be of suitable rating and capacity to supply and maintain the total load for the duration determined by the system design. In no case shall the duration be less than 2 hours of system operation unless used for emergency illumination in 700.12(C)(4) or unit equipment in 700.12(H). Additionally, the power source shall comply with 700.12(C)(1) through (C)(5) as applicable.

> Informational Note: See NFPA 110-2022, *Standard for Emergency and Standby Power Systems*, for information on classification of emergency power supply systems (EPSS).

N **(1) On-Site Fuel Supply.** An on-site fuel supply shall be provided, sufficient for not less than 2 hours operation of the system.

N **(2) Fuel Transfer Pumps.** Where power is needed for the operation of the fuel transfer pumps to deliver fuel to the source, these pumps shall be connected to the emergency power system.

Engine-driven generators that use an electric fuel transfer pump might not start or might not continue operating if the fuel pump is not operating. Such pumps, which transfer fuel into a day tank, must be supplied by the emergency system.

N **(3) Public Gas System, Municipal Water Supply.** Sources shall not be solely dependent on a public utility gas system for their fuel supply or municipal water supply for their cooling systems.

> *Exception: Where approved by the authority having jurisdiction, the use of other than on-site fuels shall be permitted where there is a low probability of a simultaneous failure of both the off-site fuel delivery system and power from the outside electrical utility company. Where the public gas system is approved, the requirements of 700.12(C)(1) shall not apply.*

Off-site fuel supplies such as natural gas or piped steam can be used where experience has demonstrated their reliability. Off-site fuel supplies also can be used where they provide greater reliability than gasoline or diesel engines. Natural gas delivered by a gas utility often is used in isolated areas where refueling could be a problem.

N **(4) Storage Batteries and UPS.** Storage batteries and UPS used to supply emergency illumination shall be of suitable rating and capacity to supply and maintain the total load for a minimum period of 1½ hours, without the voltage applied to the load falling below 87½ percent of nominal voltage. Automotive-type batteries shall not be used. An automatic battery charging means shall be provided.

Battery storage systems supplied by photovoltaic (PV) and wind-powered systems might not provide the reliable battery charging needed for an emergency system. Deep-cycle lead acid batteries are more suitable for use with emergency systems because of their ability to supply sustained loads for longer periods (at least 1.5 hours at 87.5 percent) and because they can be deeply discharged and recharged on a regular basis. Standard automotive batteries are designed for high-current, short duration (such as engine starting) cycles rather than sustained loading conditions.

See also

Article 480 for more information on storage batteries

NFPA 72®, *National Fire Alarm and Signaling Code®* (Sections 10.6.7.2.1, 10.6.7.2.2, and 10.6.10) for more information on storage batteries used as the secondary power supply for fire alarm systems

N **(5) Automatic Fuel Transfer.** Where dual fuel sources are used, means shall be provided for automatically transferring from one fuel source to another.

(D) Generator Set.

(1) Prime Mover-Driven. For a generator set driven by a prime mover approved by the authority having jurisdiction and sized in accordance with 700.4, means shall be provided for automatically starting the prime mover on failure of the normal power source and for automatic transfer and operation of all required electrical circuits. A time-delay feature shall be provided to avoid retransfer in case of short-time reestablishment of the normal source.

Prime movers can be internal-combustion engines, steam or gas turbines, or other approved types of mechanical drivers. A storage battery used to start the prime mover must be provided with an automatic battery-charging means. An on-site fuel supply that meets 700.12(C)(1) must also be available.

Some types of drivers, particularly large ones, can take longer than 10 seconds to accelerate and develop generator voltage. Gas and steam turbines and large internal-combustion engines can have prolonged starting times. Depending on the specific loads, short-time supply could be provided by a UPS, a generator shared with other loads, or a generator with limited emergency supply, such as an expander, a steam turbine, or a waste heat system.

(2) Battery Power and Dampers. Where a storage battery is used for control or signal power or as the means of starting the prime mover, it shall be suitable for the purpose and shall be equipped with an automatic charging means independent of the generator set. Where the battery charger is required for the operation of the generator set, it shall be connected to the emergency system. Where power is required for the operation of dampers used to ventilate the generator set, the dampers shall be connected to the emergency system.

(3) Auxiliary Power Supply. Generator sets that require more than 10 seconds to develop power shall be permitted if an auxiliary power supply energizes the emergency system until the generator can pick up the load.

(4) Outdoor Generator Sets. Where an outdoor-housed generator set is equipped with a readily accessible disconnecting means in accordance with 445.18, and the disconnecting means is located within sight of the building or structure supplied, an additional disconnecting means shall not be required where ungrounded conductors serve or pass through the building or structure. Where the generator supply conductors terminate at a disconnecting means in or on a building or structure, the disconnecting means shall meet the requirements of 225.36.

The disconnecting means on the generator can be used as the disconnecting means required in 225.31, provided the disconnecting means is readily accessible and within sight of the building. [See the definitions of the terms *accessible, readily (readily accessible)* and *in sight from (within sight from) (within sight)* in Article 100.]

See also

445.18 for requirements covering generator disconnecting means

Exception: For installations under single management, where conditions of maintenance and supervision ensure that only qualified persons will monitor and service the installation and where documented safe switching procedures are established and maintained for disconnection, the generator set disconnecting means shall not be required to be located within sight of the building or structure served.

The circuit between the generator and the building or structure is a feeder. Therefore, the requirements for outdoor feeders contained in Article 225 must be followed, including those covering disconnecting means for outdoor branch circuits and feeders. Section 700.12(D)(4) modifies the requirement in 225.31(B) for the location of disconnecting means. The feeder disconnecting means is permitted to be located at the generator location provided the disconnecting means, and not just the generator, is within sight and readily accessible from the building being supplied.

Δ **(E) Stored-Energy Power Supply Systems (SEPSS).** Stored energy power supply systems shall comply with 700.12(E)(1) and (E)(2).

N **(1) Types.** Systems shall consist of one or more of the following system types:

(1) Uninterruptible power supply (UPS)

Informational Note: See UL 1778, *Uninterruptible Power Systems,* for further information.

UPS generally include a rectifier, a storage battery, and an inverter to convert dc to ac. Uninterruptible power supplies can be very complex systems with redundant components and high-speed solid-state switching. A common practice is to include an automatic bypass for UPS malfunction to permit maintenance. UPS systems often are used to provide a "ride-through" power source between when the normal source is lost and when the engine generator is started and brought on line.

(2) Fuel cell system
(3) Energy storage system (ESS)
(4) Storage battery
(5) Other approved equivalent stored energy sources that comply with 700.12

N **(2) Fire Protection, Suppression, Ventilation, and Separation.** The systems in 700.12(E)(1) shall be installed with the fire protection, suppression, ventilation, and separation requirements specified in the manufacturer's instructions or equipment listing.

Informational Note: See NFPA 853-2020, *Standard for the Installation of Stationary Fuel Cell Power Systems,* and NFPA 855-2020, *Standard for the Installation of Stationary Energy Storage Systems,* for additional information on fire protection installation requirements.

(F) Separate Service. Where approved by the authority having jurisdiction as suitable for use as an emergency source of power, an additional service shall be permitted. This service shall be in accordance with the applicable provisions of Article 230 and the following additional requirements:

(1) Separate overhead service conductors, service drops, underground service conductors, or service laterals shall be installed.
(2) The service conductors for the separate service shall be installed sufficiently remote electrically and physically from any other service conductors to minimize the possibility of simultaneous interruption of supply.

The use of a separate service requires a judgment by the AHJ. Such judgment should be based on the nature of the emergency loads and the expected reliability of the other available sources.

Δ **(G) Microgrid Systems.** On-site sources, designated as emergency sources, shall be permitted to be connected to a microgrid system.

The system shall isolate the emergency system from all nonemergency loads when the normal electric supply is interrupted or shall meet the requirements of 700.4(B). Interruption or partial or complete failure of the normal or nonemergency source(s) shall not impact the availability, capacity, and duration provided by the designated emergency sources.

The designated stored-energy electrical emergency power source(s) of a microgrid system shall be permitted to remain interconnected to any available power production source during operation of the emergency source(s) where the lack of, or failure of, the interconnected power production source(s), or related controls, does not impact system operation. Interconnected power production sources, other than the designated stored emergency power source(s), shall not be required to meet the requirements of this article.

(H) Battery-Equipped Emergency Luminaires.

Δ **(1) Listing.** All battery-equipped emergency luminaires shall be listed.

Informational Note No. 1: See ANSI/UL 924, *Emergency Lighting and Power Equipment,* for the requirements covering battery-equipped emergency luminaires and emergency battery packs. A listed emergency battery pack installed in a listed luminaire will provide similar functionality as a listed battery-equipped emergency luminaire.

Informational Note No. 2: Unit equipment is a type of battery-equipped emergency luminaire.

Battery-equipped emergency luminaires or, as some may remember, unit equipment must be permanently fixed in place, usually by mounting screws that are accessible only from within the unit. One or more lamps may be mounted on or remote from the unit. The unit should be located where it can be readily checked or tested for proper performance. See Exhibit 700.8.

Unit equipment is intended to provide illumination for the area where it is installed. For instance, if an emergency luminaire is in a corridor, connecting it to the branch circuit supplying the normal corridor lights (on the line side of any switching arrangements) provides the most reliable emergency lighting arrangement. If the normal lighting circuit power is interrupted or if there is a larger-scale power outage, the unit automatically energizes the unit lamps, restoring emergency illumination to the corridor. A separate circuit is permitted for unit equipment as noted in 700.12(H)(2)(3)(c). It should be noted that under this condition, failure of the normal corridor circuit would not result in operation of the unit equipment and the corridor would remain dark unless there are multiple normal lighting circuits supplying the corridor. The separate branch circuit feeding the unit must be identified at the panelboard (the same panelboard that supplies the normal lighting) and be provided with an accessory that allows the circuit breaker or branch-circuit switch to be locked in the on or closed position.

Δ **(2) Installation.** Battery-equipped emergency luminaires shall be installed in accordance with the following:

(1) Battery-equipped emergency luminaires shall be permanently fixed in place (i.e., not portable).

(2) Wiring to each luminaire shall be installed in accordance with the requirements of any of the wiring methods in Chapter 3 unless otherwise specified in Part II, IV, or V of this article. Flexible cord-and-plug connection shall be permitted for unit equipment, provided that the cord does not exceed 900 mm (3 ft) in length. Flexible cord, with or without a plug, shall also be permitted for battery-equipped emergency luminaires installed in accordance with 410.62(C)(1).

(3) The branch circuit feeding the battery-equipped emergency luminaire shall be one of the following:

    a. The same branch circuit as that serving the normal lighting in the area and connected ahead of any local switches.

    b. The same or a different branch circuit as that serving the normal lighting in the area if that circuit is equipped with means to monitor the status of that area's normal lighting branch circuit ahead of any local switches.

    c. A separate branch circuit originating from the same panelboard as one or more normal lighting circuits. This separate branch circuit disconnecting means shall be provided with a lock-on feature.

Unlike the accessories specified in 110.25 that are used to lock a circuit breaker in the off position for worker safety, the lock-on feature is an accessory that is intended to preclude unintentional opening or turning off a circuit breaker. If the normal power to the battery unit is interrupted, the lamps will be supplied by the battery until normal power is restored. If this should go undetected, it could render the battery unit inoperable or in a diminished capacity if the normal lighting fails due to the battery being deeply discharged. These devices help ensure that the battery unit will be functional upon interruption of the normal lighting in the area served by the battery unit.

Exhibit 700.9 shows an example of the type of accessory required by 700.12(H)(2)(3)(c). These devices do not prevent automatic operation of the circuit breaker's internal mechanism that opens the circuit in the event of an overcurrent condition.

EXHIBIT 700.8 Self-contained, fully automatic unit equipment for operating emergency lighting located on the unit or for remotely located exit signs or lighting heads. (*Courtesy of The Exit Light Co., Inc.*)

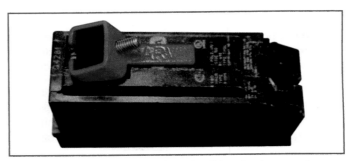

EXHIBIT 700.9 A lock-on device for a circuit breaker to prevent inadvertent turning off of the normal power circuit supplying battery-operated unit equipment. (*Courtesy of Garvin, A Southwire Business*)

(4) The branch circuit that feeds battery-equipped emergency luminaires shall be clearly identified at the distribution panel.

(5) Emergency luminaires that obtain power from a battery-equipped emergency luminaire shall be wired to the battery-equipped emergency luminaires as required in Part II, IV, or V of this article.

(6) Remote luminaires providing lighting for the exterior of an exit door shall be permitted to be supplied by the battery-equipped emergency luminaire serving the area immediately inside the exit door.

Unit equipment serving the area immediately inside the exit door is allowed to supply remote emergency luminaires installed outside the exit door. The normal lighting branch circuit for the area inside the exit door can be used to supply unit equipment that in turn supplies emergency luminaires installed inside and outside the exit door. If the power to the normal lighting branch circuit for the area is interrupted, the indoor and outdoor emergency luminaires will activate, even if the normal branch circuit for exterior lighting remains energized.

Part IV. Emergency System Circuits for Lighting and Power

700.15 Loads on Emergency Branch Circuits. No appliances and no lamps, other than those specified as required for emergency use, shall be supplied by emergency lighting circuits.

700.16 Emergency Illumination.

(A) General. Emergency illumination shall include means of egress lighting, illuminated exit signs, and all other luminaires specified as necessary to provide required illumination.

Δ **(B) System Reliability.** Emergency lighting systems shall be designed and installed so that the failure of any illumination source cannot leave in total darkness any space that requires emergency illumination. Emergency lighting control devices in the emergency lighting system shall be listed for use in emergency systems. Listed unit equipment in accordance with 700.12(H) shall be considered as meeting the provisions of this section.

> Informational Note: See 700.23 through 700.26 for applications of emergency system control devices.

For unit equipment, a second lamp ensures that the area is not left in total darkness. This section does not require redundant batteries, control circuitry, or light-emitting diode (LED) drivers.

(C) Discharge Lighting. Where high-intensity discharge lighting such as high- and low-pressure sodium, mercury vapor, and metal halide is used as the sole source of normal illumination, the emergency lighting system shall be required to operate until normal illumination has been restored.

High-intensity discharge (HID) luminaires take some time to fully illuminate once they are energized. Therefore, if HID luminaires are the sole source of normal illumination in an area, the *NEC®*

requires that the emergency lighting system operate not only until the normal system is returned to service but also until the HID luminaires provide illumination. This requirement does not apply if another type of luminaire, such as an incandescent, also normally illuminates the area.

(D) Disconnecting Means. Where an emergency system is installed, emergency illumination shall be provided in the area of the disconnecting means required by 225.31 and 230.70, as applicable, where the disconnecting means are installed indoors.

Exception: Alternative means that ensure that the emergency lighting illumination level is maintained shall be permitted.

Δ **700.17 Branch Circuits for Emergency Lighting.** Branch circuits that supply emergency lighting shall be installed to provide service from a source complying with 700.12 when the normal supply for lighting is interrupted. Such installations shall provide either of the following:

(1) An emergency lighting supply, independent of the normal lighting supply, with provisions for automatically transferring the emergency lights upon the event of failure of the normal lighting supply.

(2) Two or more branch circuits supplied from separate and complete systems with independent power sources. One of the two power sources and systems shall be part of the emergency system, and the other shall be permitted to be part of the normal power source and system. Each system shall provide sufficient power for emergency lighting purposes.

Unless both systems are used for regular lighting purposes and both are kept lighted, means shall be provided for automatically energizing either system upon failure of the other. Either system or both systems shall be permitted to be a part of the general lighting of the protected occupancy if circuits supplying lights for emergency illumination are installed in accordance with other sections of this article.

The emergency lighting supply must be independent of the normal lighting supply and must automatically operate when there is a failure of the branch circuit(s) supplying the normal lighting.

Section 700.17(2) requires emergency lighting to be supplied by a minimum of two branch circuits from separate systems with different power sources. Where a failure of the normal lighting branch circuit activates the emergency lighting supply, an area supplied by only one lighting branch circuit would be in total darkness if that branch circuit were to fail. For example, if a single branch circuit, supplied by an emergency circuit panelboard, supplies the lighting in a stairwell (means of egress), a failure of that branch circuit leaves the stairwell in total darkness. If two branch circuits from separate power sources are run to the stairwell, it is unlikely that both circuits to the stairway would fail simultaneously; therefore, the risk to occupants created by total darkness is minimized.

700.18 Circuits for Emergency Power. For branch circuits that supply equipment classed as emergency, there shall be an emergency system supply source to which the load will be transferred automatically upon the failure of the normal supply.

700.19 Multiwire Branch Circuits. The branch circuit serving emergency lighting and power circuits shall not be part of a multiwire branch circuit.

Part V. Control — Emergency Lighting Circuits

700.20 Switch Requirements. The switch or switches installed in emergency lighting circuits shall be arranged so that only authorized persons have control of emergency lighting.

Exception No. 1: Where two or more single-throw switches are connected in parallel to control a single circuit, at least one of these switches shall be accessible only to authorized persons.

Exception No. 2: Additional switches that act only to put emergency lights into operation but not disconnect them shall be permissible.

Switches connected in series or 3- and 4-way switches shall not be used.

700.21 Switch Location. All manual switches for controlling emergency circuits shall be in locations convenient to authorized persons responsible for their actuation. In facilities covered by Articles 518 and 520, a switch for controlling emergency lighting systems shall be located in the lobby or at a place conveniently accessible thereto.

In no case shall a control switch for emergency lighting be placed in a motion-picture projection booth or on a stage or platform.

Exception: Where multiple switches are provided, one such switch shall be permitted in such locations where arranged so that it can only energize the circuit but cannot de-energize the circuit.

700.22 Exterior Lights. Those lights on the exterior of a building that are not required for illumination when there is sufficient daylight shall be permitted to be controlled by an automatic light-actuated device.

700.23 Dimmer and Relay Systems. A dimmer or relay system containing more than one dimmer or relay and listed for use in emergency systems shall be permitted to be used as a control device for energizing emergency lighting circuits. Upon failure of normal power, the dimmer or relay system shall be permitted to selectively energize only those branch circuits required to provide minimum emergency illumination using a control bypass function. Where the dimmer or relay system is fed by a normal/emergency power source from an upstream transfer switch, normal power sensing for this function shall be permitted to be from a normal-only power source upstream of the transfer switch. All branch circuits supplied by the dimmer or relay system cabinet shall comply with the wiring methods of Part II of Article 700.

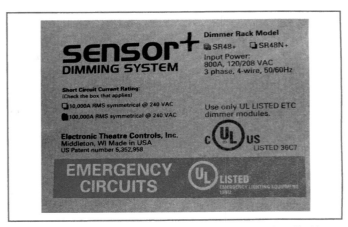

EXHIBIT 700.10 An example of a label for a dimmer system that is listed for emergency system use. *(Courtesy of Electronic Theatre Controls, Inc.)*

Dimmer systems that are listed for emergency system use include a method to sense failure of normal power and selectively energize branch circuits fed from the dimmer cabinet, regardless of the setting of control switches or panels normally used to control the dimmer system. Dimmer systems usually are supplied by a feeder that is transferred from the normal system to the emergency system by a transfer switch. See Exhibit 700.10.

700.24 Directly Controlled Emergency Luminaires. Where emergency illumination is provided by one or more directly controlled emergency luminaires that, upon loss of normal power, respond to an external control input to establish the required emergency illumination level, such directly controlled emergency luminaries shall be listed for use in emergency systems. Luminaires that are energized to the required emergency illumination level by disconnection of their control input by a listed emergency lighting control device shall not be required to be listed for use in emergency systems.

Directly controlled emergency luminaires are those containing any type of control input that is used in setting the luminaire to the required emergency illumination level upon failure of normal power and emergency power is present. This control input can be as simple as built-in sensing of normal power. It can also be an analog control input such as 0–10 volt or a more complex digital input. Some types of directly controlled emergency luminaires force their output to full brightness by disconnection of the control input from the normal control system using an automatic load control relay or other listed bypass device. Other types rely on an active signal from an emergency control system to set the required illumination level. Section 700.24 recognizes that input or loss of input qualifies as direct control of the emergency luminaires. Required listing for use in emergency systems of directly controlled emergency luminaires and controlling devices provides evaluation for predictable operation and reliability under the wide variety of conditions that might be present at failure of normal power.

See also

700.16(B) for the requirement covering the listing of control devices installed in emergency lighting systems

700.25 Branch Circuit Emergency Lighting Transfer Switch. Emergency lighting loads supplied by branch circuits rated at not greater than 20 amperes shall be permitted to be transferred from the normal branch circuit to an emergency branch circuit using a listed branch circuit emergency lighting transfer switch. The mechanically held requirement of 700.5(C) shall not apply to listed branch circuit emergency lighting transfer switches.

700.26 Automatic Load Control Relay. If an emergency lighting load is automatically energized upon loss of the normal supply, a listed automatic load control relay shall be permitted to energize the load. The load control relay shall not be used as transfer equipment.

Automatic load control relays traditionally were part of emergency unit equipment, but stand-alone devices are now listed under ANSI/UL 924, *Standard for Emergency Lighting and Power Equipment*. Load control relays listed under UL 924 are not to be used to transfer a load between two nonsynchronous power sources. Load control relays do not have mechanisms required by UL 1008, *Transfer Switch Equipment*, to prevent inadvertent connection of the normal and emergency sources, and they do not undergo the fault-current evaluation that is required of UL 1008 for transfer switches.

The UL Product IQ *Guide Information* differentiates automatic transfer switches (product category WPWR) from automatic load control relays (product category FTBR).

N **700.27 Class 2 Powered Emergency Lighting Systems.** Devices that combine control signals with Class 2 emergency power on a single circuit shall be listed as emergency lighting control devices.

> Informational Note: An example of a device combining control signals with Class 2 emergency power sources is a Power over Ethernet (PoE) switch.

Part VI. Overcurrent Protection

700.30 Accessibility. The branch-circuit overcurrent devices in emergency circuits shall be accessible to authorized persons only.

700.31 Ground-Fault Protection of Equipment. The alternate source for emergency systems shall not be required to provide ground-fault protection of equipment with automatic disconnecting means. Ground-fault indication at the emergency source shall be provided in accordance with 700.6(D) if ground-fault protection of equipment with automatic disconnecting means is not provided.

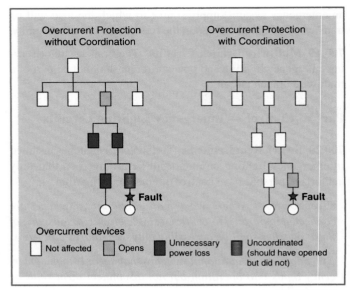

EXHIBIT 700.11 Overcurrent protection schemes without system coordination and with system coordination.

Δ **700.32 Selective Coordination.**

Coordination, selective (selective coordination), as defined in Article 100, is a protection method in which the operation of the overcurrent protective scheme localizes an overcurrent condition to the circuit conductors or equipment in which an overload or fault (short circuit or ground fault) has occurred. This selective operation prevents power loss to unaffected loads.

Continuity of operation of lighting and life-safety equipment is necessary for safe occupant evacuation. This requirement minimizes the possibility that an overload, short circuit, or ground fault in a 20-ampere branch circuit would cause the feeder protective device supplying the branch-circuit panelboard to open. Coordination must be carried through each level of distribution that supplies power to the emergency system.

Design and verification of electrical system coordination can be achieved only through a coordination study. A coordination study entails detailed analysis of electrical supply system fault-current characteristics. Modifications to the electrical system after the initial design and installation can affect the original implementation of the coordinated system.

Examples of overcurrent protection with and without coordinated protection are illustrated in Exhibit 700.11.

N **(A) General.** Emergency system(s) overcurrent protective devices (OCPDs) shall be selectively coordinated with all supply-side and load-side OCPDs.

Selective coordination shall be selected by a licensed professional engineer or other qualified persons engaged primarily in the design, installation, or maintenance of electrical systems. The selection shall be documented and made available to those authorized to design, install, inspect, maintain, and operate the system.

N **(B) Replacements.** Where emergency system(s) OCPDs are replaced, they shall be reevaluated to ensure selective coordination is maintained with all supply-side and load-side OCPDs.

N **(C) Modifications.** If modifications, additions, or deletions to the emergency system(s) occur, selective coordination of the emergency system(s) OCPDs with all supply-side and load-side OCPDs shall be reevaluated.

Exception: Selective coordination shall not be required between two overcurrent devices located in series if no loads are connected in parallel with the downstream device.

The exception to 700.32(C) recognizes a series-rated system in which the device immediately upstream is designed to open before the downstream device under short-circuit conditions.

> Informational Note: See Informational Note Figure 700.32(C) for an example of how emergency system OCPDs selectively coordinate with all supply-side OCPDs.
> OCPD D selectively coordinates with OCPDs C, F, E, B, and A.
> OCPD C selectively coordinates with OCPDs F, E, B, and A.
> OCPD F selectively coordinates with OCPD E.
> OCPD B is not required to selectively coordinate with OCPD A because OCPD B is not an emergency system OCPD.

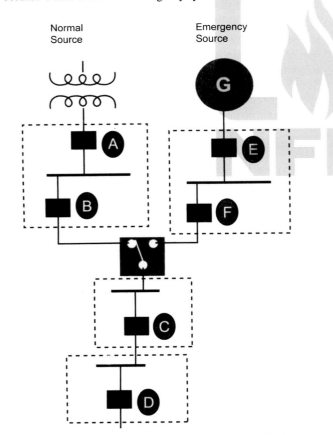

N **INFORMATIONAL NOTE FIGURE 700.32(C)** *Emergency System Selective Coordination.*

ARTICLE 701 Legally Required Standby Systems

Part I. General

701.1 Scope. This article applies to the electrical safety of the installation, operation, and maintenance of legally required standby systems consisting of circuits and equipment intended to supply, distribute, and control electricity to required facilities for illumination or power, or both, when the normal electrical supply or system is interrupted.

The systems covered by this article consist only of those that are permanently installed in their entirety, including the power source.

> Informational Note No. 1: See NFPA 99-2018, *Health Care Facilities Code*, for further information.
> Informational Note No. 2: See NFPA 110-2019, *Standard for Emergency and Standby Power Systems*, for further information regarding performance of emergency and standby power systems.
> Informational Note No. 3: See ANSI/IEEE 446-1995, *Recommended Practice for Emergency and Standby Power Systems for Industrial and Commercial Applications*, for further information.
> Informational Note No. 4: Legally required standby systems are typically installed to serve loads, such as heating and refrigeration systems, communications systems, ventilation and smoke removal systems, sewage disposal systems, lighting systems, and industrial processes that, when stopped during any interruption of the normal electrical supply, could create hazards or hamper rescue or firefighting operations.
> Informational Note No. 5: Legally required standby systems are considered level one systems when failure to perform could result in loss of human life or serious injuries and level two systems when failure of legally required standby systems to perform is less critical to human life and safety when applying NFPA 110-2019, *Standard for Emergency Standby Power Systems*.

Legally required standby systems provide electric power to aid in firefighting, rescue operations, control of health hazards, and similar operations that are code or AHJ mandated. In comparison, emergency systems (see Article 700) are those systems essential for safety to life. Optional standby systems (see Article 702) are those in which failure can cause, for example, physical discomfort, interruption of an industrial process, damage to process equipment, or disruption of business.

The requirements for legally required standby systems are similar to those for emergency systems, including the need for a transfer switch listed for emergency use. However, there are a few differences. When normal power is lost, legally required systems must be able to supply standby power in 60 seconds or less, instead of the 10 seconds or less required of emergency systems. Wiring for legally required standby systems may occupy the same raceways, cables, boxes, and cabinets as other general wiring, whereas wiring for emergency systems must be kept entirely independent of other wiring.

Legally required standby systems take second priority to emergency systems if they are involved in sharing an alternate supply and/or load shedding or peak shaving schemes.

N 701.2 Reconditioned Equipment. Reconditioned transfer switches shall not be permitted.

701.3 Commissioning and Maintenance.

(A) Commissioning Witness Test. The authority having jurisdiction shall conduct or witness the commissioning of the complete system upon installation.

(B) Tested Periodically. Systems shall be tested periodically on a schedule and in a manner approved by the authority having jurisdiction to ensure the systems are maintained in proper operating condition.

(C) Maintenance. Legally required standby system equipment shall be maintained in accordance with manufacturer instructions and industry standards.

(D) Written Record. A written record shall be kept on such tests and maintenance.

(E) Testing Under Load. Means for testing legally required standby systems under load shall be provided.

Informational Note: See NFPA 110-2019, *Standard for Emergency and Standby Power Systems*, for information on testing and maintenance of emergency power supply systems (EPSSs).

701.4 Capacity and Rating.

(A) Rating. Legally required standby system equipment shall be suitable for the available fault current at its terminals.

(B) Capacity. A legally required standby system shall have adequate capacity in accordance with Parts I through IV of Article 220 or by another approved method. The system capacity shall be sufficient for the rapid load changes and transient power and energy requirements associated with any expected loads.

(C) Load Management. The alternate power source shall be permitted to supply legally required standby and optional standby system loads where the alternate source has adequate capacity or where load management (that includes automatic selective load pickup and load shedding) is provided that will ensure adequate power to the legally required standby circuits.

N (D) Parallel Operation. Parallel operation shall comply with Part I or Part II of Article 705 where the legally required source capacity required to supply the legally required load is maintained at all times. Parallel operation of the legally required source(s) shall consist of the sources specified in 701.4(D)(1) and (D)(2).

N (1) Normal Source. The alternate power source shall be permitted to operate in parallel with the normal source in compliance with Part I or Part II of Article 705 where the capacity

required to supply the legally required standby load is maintained at all times. Any operating condition that results in less than the required source capacity shall initiate a legally required standby source malfunction signal in 701.6(A).

Parallel operation shall be permitted for satisfying the test requirements of 701.3(B), provided all other conditions of 701.3 are met.

Informational Note: Peak load shaving is one application for parallel source operation.

N (2) Alternate Source. Legally required standby sources shall be permitted to operate in parallel where the necessary equipment to establish and maintain a synchronous condition is provided.

701.5 Transfer Equipment.

Δ (A) General. Transfer equipment shall be automatic, listed, and marked for emergency system or legally required standby use. Transfer equipment shall be designed and installed to prevent the inadvertent interconnection of normal and alternate sources of supply in any operation of the transfer equipment. Transfer equipment and electric power production systems installed to permit operation in parallel with the normal source shall meet the requirements of Article 705. Meter-mounted transfer switches shall not be permitted for legally required system use.

Not all automatic transfer switches (ATS) permit parallel operation of generation equipment and the normal source; therefore, those transfer switches do not need to comply with Article 705. Some ATS equipment is designed to briefly allow (for a few cycles) parallel operation of the generation equipment with the normal source upon load transfer. This load transfer can occur with minimal disturbance or effect on the load. Transfer switches that employ paralleling must comply with Article 705.

(B) Bypass Isolation Switches. Means to bypass and isolate the transfer switch equipment shall be permitted. Where bypass isolation switches are used, inadvertent parallel operation shall be avoided.

Δ (C) Automatic Transfer Switches. Automatic transfer switches shall be electrically operated and mechanically held.

This requirement correlates with NFPA 110, *Standard for Emergency and Standby Power Systems,* and requires the relay contacts to be mechanically held in the event of coil failure.

When standby systems are tested, both the normal system and the standby system are energized. If the two sources are not synchronized, as much as twice the rated voltage could exist across the transfer switch contacts. Some listed transfer switches are designed and tested to be suitable for switching between out-of-phase power sources. Other protection methods can be employed, such as a mechanical interlock that prevents inadvertent interconnection or an electronic method that prevents both systems from being interconnected.

(D) Documentation. The short-circuit current rating of the transfer equipment, based on the specific overcurrent protective device type and settings protecting the transfer equipment, shall be field marked on the exterior of the transfer equipment.

In this requirement, the term *short-circuit current rating* (see Article 100) includes all the various options by which the product standard evaluates transfer switches for fault currents, such as short-circuit withstand and closing rating, short-time current rating, and the common industry term withstand/close-on rating.

Product standards require transfer equipment to be marked with the short-circuit withstand/closing or short-time current rating (short-circuit current rating). Some transfer switches are marked by the manufacturer with several options, each resulting in a different short-circuit current rating. Short-circuit current rating values can vary based upon the overcurrent protective device (OCPD) type, ampere rating, and setting. The field marking required by this section documents the specifics of the protection scheme and verifies compliance with 110.3(B) and 110.10.

701.6 Signals. Audible and visual signal devices shall be provided, where practicable, for the purposes described in 701.6(A), (B), (C), and (D).

(A) Malfunction. Malfunction signals indicate a malfunction of the standby source.

(B) Carrying Load. Load carrying signals indicate that the standby source is carrying load.

(C) Battery Charging Malfunction. Battery charging malfunction signals indicate charging malfunction on a battery required for source readiness, including the prime mover starting battery.

Informational Note: See NFPA 110-2019, *Standard for Emergency and Standby Power Systems*, for signals for generator sets.

(D) Ground Fault. Ground-fault signals indicate a ground fault in solidly grounded wye, legally required standby systems of more than 150 volts to ground and circuit-protective devices rated 1000 amperes or more. The sensor for the ground-fault signal devices shall be located at, or ahead of, the main system disconnecting means for the legally required standby source, and the maximum setting of the signal devices shall be for a ground-fault current of 1200 amperes. Instructions on the course of action to be taken in the event of an indicated ground fault shall be located at or near the sensor location.

For systems with multiple emergency sources connected to a paralleling bus, the ground-fault sensor shall be permitted at an alternate location.

Informational Note: See NFPA 110-2019, *Standard for Emergency and Standby Power Systems*, for signals for generator sets.

Although 701.31 specifies that automatic ground-fault protection of equipment is not required to be provided on the alternate source, ground faults can occur on such systems, and they can result in equipment burndown. Because of the importance of legally required systems, automatic disconnection of power in the event of a ground fault is inappropriate. Detection of such a fault, however, is required so that the condition can be corrected.

If multiple sources are connected to a paralleling bus, the ground-fault sensor can be located on feeders connected to the bus, rather than to each of the sources. With this arrangement, ground-fault protection can easily discriminate between more critical and less critical loads. The system would not be disabled for a downstream fault and critical loads such as fire pumps and emergency circuits can be alarmed, while less critical circuits can be tripped. This results in overall better protection for less critical systems and better reliability for critical circuits.

701.7 Signs.

(A) Mandated Standby. A sign shall be placed at the service entrance indicating type and location of each on-site legally required standby power source.

Exception: A sign shall not be required for individual unit equipment as specified in 701.12(I).

Δ **(B) Grounding.** Where removal of a grounding or bonding connection in normal power source equipment interrupts the grounding electrode conductor connection to the alternate power source(s) grounded conductor, a warning sign shall be installed at the normal power source equipment stating:

WARNING
SHOCK HAZARD EXISTS IF GROUNDING
ELECTRODE CONDUCTOR OR BONDING JUMPER
CONNECTION IN THIS EQUIPMENT IS REMOVED
WHILE ALTERNATE SOURCE(S) IS ENERGIZED.

The warning sign(s) or label(s) shall comply with 110.21(B).

Legally required standby systems that have a solid (unswitched) neutral in the transfer equipment (nonseparately derived system) rely on the grounding and bonding connections in the normal source supply equipment to ensure that the ground-fault current path is completed from a ground fault to the alternate source. If a main or system bonding jumper is removed [e.g., to perform testing on ground-fault protection of equipment (GFPE) systems], service personnel could inadvertently become part of the current path if a ground fault occurs.

Part II. Circuit Wiring

Δ **701.10 Wiring Legally Required Standby Systems.**

N **(A) General.** The legally required standby system wiring shall be permitted to occupy the same raceways, cables, boxes, and cabinets with other general wiring.

Unlike emergency system wiring, which must be kept entirely independent of other wiring to ensure reliability, the wiring for legally required standby systems may be installed with other general (normal) wiring because legally required standby system loads are not essential for life safety.

N **(B) Wiring.** Wiring from a legally required source to supply legally required and other (nonlegally required) loads shall be in accordance with the following:

(1) The common bus of switchgear, sections of a switchboard, or individual enclosures shall be either of the following:

    a. Supplied by single or multiple feeders without overcurrent protection at the source

    b. Supplied by single or multiple feeders with overcurrent protection, provided that the overcurrent protection that is common to a legally required system and any nonlegally required system(s) is selectively coordinated with the next downstream overcurrent protective device in the nonlegally required system(s)

Informational Note: See Informational Note Figure 701.10(B)(1) and Informational Note Figure 701.10(B)(2) for further information.

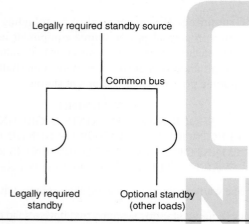

N **INFORMATIONAL NOTE FIGURE 701.10(B)(1)** *Single or Multiple Feeders Without Overcurrent Protection.*

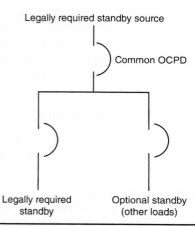

N **INFORMATIONAL NOTE FIGURE 701.10(B)(2)** *Single or Multiple Feeders with Overcurrent Protection.*

Part III. Sources of Power

701.12 General Requirements. Current supply shall be such that, in the event of failure of the normal supply to, or within, the building or group of buildings concerned, legally required standby power will be available within the time required for the application but not to exceed 60 seconds. The supply system for legally required standby purposes, in addition to the normal services to the building, shall be permitted to comprise one or more of the types of systems described in 701.12(A) through (I). Unit equipment in accordance with 701.12(I) shall satisfy the applicable requirements of this article.

(A) Power Source Considerations. In selecting a legally required standby source of power, consideration shall be given to the type of service to be rendered, whether of short-time duration or long duration.

(B) Equipment Design and Location. Consideration shall be given to the location or design, or both, of all equipment to minimize the hazards that might cause complete failure due to floods, fires, icing, and vandalism.

Informational Note: See ANSI/IEEE 493-2007, *Recommended Practice for the Design of Reliable Industrial and Commercial Power Systems*, for further information.

Δ **(C) Supply Duration.** The alternate power source shall be of suitable rating and capacity to supply and maintain the total load for the duration determined by the system design. In no case shall the duration be less than 2 hours of system operation. Additionally, the power source shall comply with 701.12(C)(1) through (C)(5) as applicable.

Informational Note: See NFPA 110-2022, *Standard for Emergency and Standby Power Systems*, for information on classification of emergency power supply systems (EPSS).

N **(1) On-Site Fuel Supply.** An on-site fuel supply shall be provided, sufficient for not less than 2 hours operation of the system.

Δ **(2) Fuel Transfer Pumps.** Where power is needed for the operation of the fuel transfer pumps to deliver fuel to the source, these pumps shall be connected to the legally required standby power system.

Δ **(3) Public Gas System, Municipal Water Supply.** Sources shall not be solely dependent on a public utility gas system for their fuel supply or on a municipal water supply for their cooling systems.

Exception: Where approved by the authority having jurisdiction, the use of other than on-site fuels shall be permitted where there is a low probability of a simultaneous failure of both the off-site fuel delivery system and power from the outside electrical utility company. Where a public gas system is approved, the requirements of 701.12(C)(1) shall not apply.

N **(4) Storage Batteries and UPS.** Storage batteries and UPS used to supply standby illumination shall be of suitable rating and capacity to supply and maintain the total load for a minimum period of 1½ hours, without the voltage applied to the load falling below 87½ percent of nominal voltage. Automotive-type batteries shall not be used. An automatic battery charging means shall be provided.

N (5) Automatic Fuel Source Transfer. Where dual fuel sources are used, means shall be provided for automatically transferring from one fuel source to another.

(D) Generator Set.

(1) Prime Mover-Driven. For a generator set driven by a prime mover approved by the authority having jurisdiction and sized in accordance with 701.4, means shall be provided for automatically starting the prime mover upon failure of the normal power source and for automatic transfer and operation of all required electrical circuits. A time-delay feature permitting a 15-minute setting shall be provided to avoid retransfer in case of short-time reestablishment of the normal source.

(2) Battery Power. Where a storage battery is used for control or signal power or as the means of starting the prime mover, it shall be suitable for the purpose and shall be equipped with an automatic charging means independent of the generator set.

(3) Outdoor Generator Sets. If an outdoor-housed generator set is equipped with a readily accessible disconnecting means in accordance with 445.18, and the disconnecting means is located within sight of the building or structure supplied, an additional disconnecting means shall not be required where ungrounded conductors serve or pass through the building or structure. Where the generator supply conductors terminate at a disconnecting means in or on a building or structure, the disconnecting means shall meet the requirements of 225.36.

The disconnecting means on an outdoor generator set can be used as the disconnecting means required in 225.31, provided the disconnecting means, and not just the generator, is readily accessible and is within sight of the building. [See the definitions of the terms *accessible, readily (readily accessible)* and *in sight from (within sight from) (within sight)* in Article 100.]

(E) Stored-Energy Power Supply Systems (SEPSS). Stored energy power supply systems shall comply with 701.12(E)(1) and (E)(2).

N (1) Types. Systems shall consist of one or more of the following system types:

  a. Uninterruptible power supply (UPS)

  Informational Note: See UL 1778, *Uninterruptable Power Systems*, and UL 924, *Emergency Lighting and Power Equipment*, for further information.

  b. Fuel cell system
  c. Energy storage system (ESS)
  d. Storage battery
  e. Other approved equivalent stored energy sources that comply with 701.12

N (2) Fire Protection, Suppression, Ventilation, and Separation. The systems in 701.12(E)(1) shall be installed with the fire protection, suppression, ventilation, and separation requirements specified in the manufacturer's instructions or equipment listing.

  Informational Note: See NFPA 853-2020, *Standard for the Installation of Stationary Fuel Cell Power Systems,* and NFPA 855-2020, *Standard for the Installation of Stationary Energy Storage Systems*, for additional information on fire protection installation requirements.

(F) Separate Service. Where approved, by the authority having jurisdiction as suitable for use as a legally required source of power, an additional service shall be permitted. This service shall be in accordance with Article 230 and the following additional requirements:

  (1) Separate overhead service conductors, service drops, underground service conductors, or service laterals shall be installed.
  (2) The service conductors for the separate service shall be installed sufficiently remote electrically and physically from any other service conductors to minimize the possibility of simultaneous interruption of supply.

(G) Connection Ahead of Service Disconnecting Means. Where approved by the authority having jurisdiction, connections located ahead of and not within the same cabinet, enclosure, vertical switchgear section, or vertical switchboard section as the service disconnecting means shall be permitted. The legally required standby service shall be sufficiently separated from the normal main service disconnecting means to minimize simultaneous interruption of supply through an occurrence within the building or groups of buildings served.

  Informational Note: See 230.82 for equipment permitted on the supply side of a service disconnecting means.

If a legally required standby system is supplied by a connection on the line side of the normal service disconnecting means, 230.82 requires that the tapped conductors be installed in accordance with all the requirements for service-entrance conductors and that the conductors terminate in equipment suitable for use as service equipment. Those requirements help ensure that the legally required standby system disconnecting means is rated for the fault current available from the utility.

Δ (H) Microgrid Systems. On-site sources, designated as legally required standby sources, shall be permitted to be connected to a microgrid system.

  The system shall isolate the legally required standby system from all nonlegally required loads when the normal electric supply is interrupted or shall meet the requirements of 701.4(C). Interruption or partial or complete failure of the normal source(s) shall not impact the availability, capacity, and duration provided by the designated legally required standby sources.

  The designated stored-energy legally required standby power source(s) of a microgrid system shall be permitted to remain interconnected to any available power production source during operation of the legally required standby source(s) where the lack of, or failure of, the interconnected power production source(s), or related controls, does not impact system operation. Interconnected power production sources, other than the designated SEPSS, shall not be required to meet the requirements of this article.

Δ **(I) Battery-Equipped Emergency Luminaires, Used for Legally Required Standby Systems.** Battery-equipped emergency luminaires used for legally required standby systems shall comply with 701.12(H).

Part IV. Overcurrent Protection

701.30 Accessibility. The branch-circuit overcurrent devices in legally required standby circuits shall be accessible to authorized persons only.

701.31 Ground-Fault Protection of Equipment. The alternate source for legally required standby systems shall not be required to provide ground-fault protection of equipment with automatic disconnecting means. Ground-fault indication at the legally required standby source shall be provided in accordance with 701.6(D) if ground-fault protection of equipment with automatic disconnecting means is not provided.

Δ **701.32 Selective Coordination.**

Coordination, selective (selective coordination), as defined in Article 100, is a protection method in which the operation of the overcurrent protective scheme localizes an overcurrent condition to the circuit conductors or equipment in which an overload or fault (short circuit or ground fault) has occurred. This selective operation prevents power loss to unaffected loads.

Continuity of operation of lighting and life-safety equipment is necessary for safe occupant evacuation. This requirement minimizes the possibility that an overload, short circuit, or ground fault in a 20-ampere branch circuit would cause the feeder protective device supplying the branch-circuit panelboard to open. Coordination must be carried through each level of distribution that supplies power to the emergency system.

Design and verification of electrical system coordination can be achieved only through a coordination study. A coordination study entails detailed analysis of electrical supply system fault-current characteristics. Modifications to the electrical system after the initial design and installation can affect the original implementation of the coordinated system. For examples of overcurrent protection with and without coordinated protection, see Exhibit 700.11 within Article 700.

N **(A) General.** Legally required standby system(s) overcurrent protective devices (OCPDs) shall be selectively coordinated with all supply-side and load-side OCPDs.

Selective coordination shall be selected by a licensed professional engineer or other qualified persons engaged primarily in the design, installation, or maintenance of electrical systems. The selection shall be documented and made available to those authorized to design, install, inspect, maintain, and operate the system.

N **(B) Replacements.** Where legally required standby OCPDs are replaced, they shall be reevaluated to ensure selective coordination is maintained with all supply-side and load-side OCPDs.

N **(C) Modifications.** If modifications, additions, or deletions to the legally required standby system(s) occur, selective coordination of the legally required system(s) OCPDs with all supply-side and load-side OCPDs shall be reevaluated.

Exception: Selective coordination shall not be required between two overcurrent devices located in series if no loads are connected in parallel with the downstream device.

Informational Note: See Informational Note Figure 701.32(C) for an example of how legally required standby system OCPDs selectively coordinate with all supply-side OCPDs.
OCPD D selectively coordinates with OCPDs C, F, E, B, and A.
OCPD C selectively coordinates with OCPDs F, E, B, and A.
OCPD F selectively coordinates with OCPD E.
OCPD B is not required to selectively coordinate with OCPD A because OCPD B is not a legally required standby system OCPD.

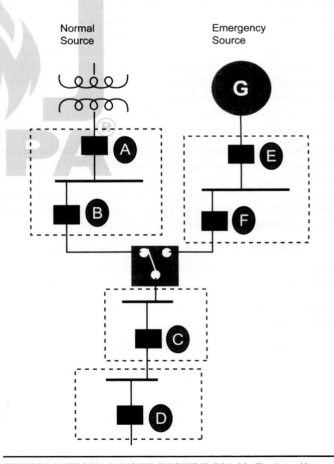

INFORMATIONAL NOTE FIGURE 701.32(C) *Legally Required Standby System Selective Coordination.*

ARTICLE 702 Optional Standby Systems

Part I. General

702.1 Scope. This article applies to the installation and operation of optional standby systems.

The systems covered by this article consist of those that are permanently installed in their entirety, including prime movers, and those that are arranged for a connection to a premises wiring system from a portable alternate power supply.

> Informational Note: Optional standby systems are typically installed to provide an alternate source of electric power for such facilities as industrial and commercial buildings, farms, and residences and to serve loads such as heating and refrigeration systems, data processing and communications systems, and industrial processes that, when stopped during any power outage, could cause discomfort, serious interruption of the process, damage to the product or process, or the like.

Article 702 applies not only to permanently installed generators and prime movers but also to portable alternate power supplies that can be connected to an optional standby system. For example, upon failure of an optional standby generator at a frozen food processing plant, a vehicle-mounted generator can be brought in and connected to the plant's optional standby system, which has the capability for such a connection. See Exhibit 702.1.

Generators used in residential-type occupancies are installed as optional standby systems because there is no code or AHJ mandate to provide an alternate source of power. Some of the largest and most sophisticated standby systems installed are optional standby systems that are installed to support uninterrupted operation of data systems associated with businesses such as financial institutions, mercantile operations, and Internet service providers.

See also

Article 700 for emergency systems

Article 701 for legally required standby systems

N 702.2 Reconditioned Equipment. Reconditioned transfer switches shall not be permitted.

702.4 Capacity and Rating.

(A) System Capacity.

(1) Manual and Nonautomatic Load Connection. If the connection of load is manual or nonautomatic, an optional standby system shall have adequate capacity and rating for the supply of all equipment intended to be operated at one time. The user of the optional standby system shall be permitted to select the load connected to the system.

> Informational Note: Manual and nonautomatic transfer equipment require human intervention.

EXHIBIT 702.1 A trailer- (vehicle-) mounted portable generator. *(Courtesy of the International Association of Electrical Inspectors)*

The use of manual transfer equipment requires the user to manually switch to the standby source. The user can then select the necessary loads. The standby source must have adequate capacity to supply the user-selected loads that are intended to be operated at the same time.

Δ (2) Automatic Load Connection. If the connection of load is automatic, an optional standby system shall comply with 702.4(A)(2)(a) or (A)(2)(b) in accordance with Parts I through IV of Article 220 or by another approved method.

For standby systems employing automatic transfer switches (ATS), the source must have the capacity to supply all the loads connected to it, unless an automatic load management system (sometimes referred to as load shedding) is used to ensure that the transferred load does not overload the source.

(a) *Full Load.* The standby source shall be capable of supplying the full load that is automatically connected.

(b) *Energy Management System (EMS).* Where a system is employed in accordance with 750.30 that will automatically manage the connected load, the standby source shall have a capacity sufficient to supply the maximum load that will be connected by the EMS.

702.5 Interconnection or Transfer Equipment.

(A) General. Interconnection or transfer equipment shall be required for all standby systems subject to the requirements of this article. Equipment shall be suitable for the intended use and shall be listed, designed, and installed so as to prevent the inadvertent interconnection of all sources of supply in any operation of the equipment.

Traditional automatic transfer switches (ATS) are not designed to permit parallel operation of generation equipment and the normal source and need not comply with Article 705. However, certain ATS configurations are intentionally designed to briefly allow (for a few cycles) parallel operation of the generation equipment with the normal source upon load transfer from generator to normal source. This load transfer can occur with minimal disturbance or effect on the load. Transfer switches that employ this type of paralleling must comply with Article 705.

Exception: Temporary connection of a portable generator without transfer equipment shall be permitted where conditions of maintenance and supervision ensure that only qualified persons service the installation and where the normal supply is physically isolated by a lockable disconnecting means or by disconnection of the normal supply conductors.

The exception provides requirements for the connection of loads to a generator without the use of a transfer switch. Supervision by qualified personnel is critical to ensuring that power produced by the generator is not fed back onto the utility distribution lines through the loads, potentially creating a dangerous electrical shock hazard to line workers.

Δ **(B) Meter-Mounted Transfer Switches.** Transfer switches installed between the utility meter and the meter enclosure shall be listed meter-mounted transfer switches and shall be approved.

> Informational Note No. 1: See UL 1008M, *Transfer Switch Equipment, Meter Mounted*, for more information.
>
> Informational Note No. 2: Manual and nonautomatic transfer equipment use human intervention.

Because most meters and their associated enclosures are under the control of the serving electric utility, their approval is necessary in order to install a meter-mounted transfer switch. In addition to this requirement in Article 702, the permission to install a meter-mounted transfer switch, such as the type shown in Exhibit 702.2, ahead of service equipment is specified in 230.82(11). The meter-mounted transfer switch is required to be marked indicating that it is not service equipment. However, the fact that it is not service equipment does not preclude making a grounding electrode conductor connection to the grounded (neutral) service conductor in the meter enclosure as long as it is accessible, as required by 250.68(A). Additionally, the meter-mounted transfer switch is not suitable for use as the emergency disconnect required by 230.85 for one- and two-family dwellings.

(C) Documentation. In other than dwelling units, the short-circuit current rating of the transfer equipment, based on the specific overcurrent protective device type and settings protecting the transfer equipment, shall be field marked on the exterior of the transfer equipment.

Product standards require transfer equipment to be marked with the short-circuit withstand/closing or short-time current rating (short-circuit current rating). Typically, a transfer switch is marked by the manufacturer with several options, resulting in many short-circuit current rating values. Those values can vary based upon the overcurrent protective device (OCPD) type, ampere rating, and setting. For a specific installation, the short-circuit current rating of the transfer switch is based on the overcurrent protection provided. The field marking required by this section documents the specifics of the protection scheme and verifies compliance with 110.3(B) and 110.10.

(D) Parallel Installation. Systems installed to permit operation in parallel with the normal source shall also meet Part I or Part II of Article 705.

702.6 Signals. Audible and visual signal devices shall be provided, where practicable, for the following purposes specified in 702.6(A) and (B).

(A) Malfunction. To indicate malfunction of the optional standby source.

(B) Carrying Load. To indicate that the optional standby source is carrying load.

> *Exception: Signals shall not be required for portable standby power sources.*

702.7 Signs.

Δ **(A) Standby.** A sign shall be placed at the service equipment for other than one- and two-family dwellings that indicates the type and location of each on-site optional standby power source. For one- and two-family dwelling units, a sign shall be placed at the disconnecting means required in 230.85 that indicates the location of each permanently installed on-site optional standby power source disconnect or means to shut down the prime mover as required in 445.19(C).

Δ **(B) Grounding.** Where removal of a grounding or bonding connection in normal power source equipment interrupts the grounding electrode conductor connection to the alternate power

EXHIBIT 702.2 A meter-mounted transfer switch suitable for use in optional standby systems. *(Courtesy of Global Power Products, Inc.)*

source(s) grounded conductor, a warning sign shall be installed at the normal power source equipment stating:

WARNING:
SHOCK HAZARD EXISTS IF GROUNDING ELECTRODE CONDUCTOR OR BONDING JUMPER CONNECTION IN THIS EQUIPMENT IS REMOVED WHILE ALTERNATE SOURCE(S) IS ENERGIZED.

The warning sign(s) or label(s) shall comply with 110.21(B).

Optional standby systems that have a solid (unswitched) neutral in the transfer equipment (nonseparately derived system) rely on the grounding and bonding connections in the normal source supply equipment to ensure that the ground-fault current path is completed from a ground fault to the alternate source. If a main or system bonding jumper is removed [e.g., to perform testing on ground-fault protection of equipment (GFPE) systems], service personnel could inadvertently become part of the current path if a ground fault occurs.

Δ **(C) Power Inlet.** Where a power inlet is used for a temporary connection to a portable generator, a warning sign shall be placed near the inlet to indicate the type of derived system that the system is capable of based on the wiring of the transfer equipment. The sign shall display one of the following warnings:

WARNING:
FOR CONNECTION OF A SEPARATELY DERIVED (BONDED NEUTRAL) SYSTEM ONLY
or
WARNING:
FOR CONNECTION OF A NONSEPARATELY DERIVED (FLOATING NEUTRAL) SYSTEM ONLY

Part II. Wiring

702.10 Wiring Optional Standby Systems. The optional standby system wiring shall be permitted to occupy the same raceways, cables, boxes, and cabinets with other general wiring.

702.11 Portable Generator Grounding.

(A) Separately Derived System. Where a portable optional standby source is used as a separately derived system, it shall be grounded to a grounding electrode in accordance with 250.30.

(B) Nonseparately Derived System. Where a portable optional standby source is used as a nonseparately derived system, the equipment grounding conductor shall be bonded to the system grounding electrode.

702.12 Outdoor Generator Sets.

(A) Portable Generators Greater Than 15 kW and Permanently Installed Generators. Where an outdoor housed generator set is equipped with a readily accessible disconnecting

means in accordance with 445.18, and the disconnecting means is located within sight of the building or structure supplied, an additional disconnecting means shall not be required where ungrounded conductors serve or pass through the building or structure. Where the generator supply conductors terminate at a disconnecting means in or on a building or structure, the disconnecting means shall meet the requirements of 225.36.

The disconnecting means on an outdoor generator set can be used as the disconnecting means required in 225.31, provided the disconnecting means, and not just the generator, is readily accessible and is within sight of the building.

(B) Portable Generators 15 kW or Less. Where a portable generator, rated 15 kW or less, is installed using a flanged inlet or other cord-and-plug-type connection, a disconnecting means shall not be required where ungrounded conductors serve or pass through a building or structure. The flanged inlet or other cord-and-plug-type connection shall be located outside of a building or structure.

Δ **(C) Power Inlets Rated at 100 Amperes or Greater, for Portable Generators.** Equipment containing power inlets for the connection of a generator source shall be listed for the intended use. Systems with power inlets not rated as a disconnecting means shall be equipped with an interlocked disconnecting means.

Exception: Supervised industrial installations where permanent space is identified for the portable generator located within line of sight of the power inlets shall not be required to have interlocked disconnecting means nor inlets rated as disconnects.

This requirement ensures that a portable generator can be safely disconnected from a power inlet rated 100 amperes or more. A generator that is disconnected (unplugged) under load can present a safety hazard if the inlet is not rated for load break. The power inlet is required to be equipped with an interlocked disconnecting means to ensure that the disconnecting means is opened prior to disengaging the inlet. The exceptions recognize power inlets that are load break rated and those installed in supervised industrial installations where certain conditions exist.

ARTICLE 705 Interconnected Electric Power Production Sources

Part I. General

705.1 Scope. This article covers installation of one or more electric power production sources operating in parallel with a primary source(s) of electricity.

Informational Note No. 1: Examples of the types of primary sources include a utility supply or an on-site electric power source(s).
Informational Note No. 2: See Informational Note Figure 705.1.

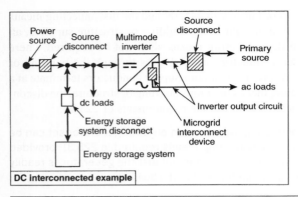

DC interconnected example

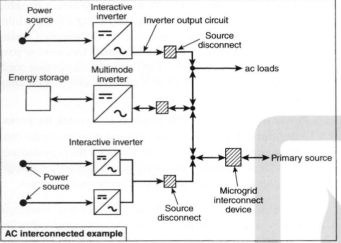

AC interconnected example

Notes:
(1) These diagrams are intended to be a means of identification for power source components, circuits, and connections.
(2) The power source disconnect in these diagrams separates the power source from other systems.
(3) Equipment disconnecting means not shown.
(4) System grounding and equipment grounding are not shown.
(5) Custom designs occur in each configuration, and some components are optional.

N **INFORMATIONAL NOTE FIGURE 705.1** *Identification of Power Source Components in Common Configurations*

Article 705 contains requirements for interconnecting power production sources that operate in parallel as distributed generation. It does not cover sources that are connected by a transfer switch that permits operation of a single source. The primary source is not required to be a utility source. The requirements of Article 705 are not dependent on the type of generating source employed. These installations are subject to the entire *NEC®*, as detailed in 90.3.

Δ **705.5 Parallel Operation.**

N **(A) Output Compatibility.** Power production sources operating in parallel with a primary source of electricity or other power production sources shall have compatible voltage, wave shape, and frequency ratings.

N **(B) Synchronous Generators.** Synchronous generators operating in parallel with a primary power source shall be installed with the required synchronizing equipment.

> Informational Note: See IEEE 1547, *Standard for Interconnection and Interoperability of Distributed Energy Resources with Associated Electric Power Systems Interfaces,* and UL 1741, *Standard for Inverters, Converters, Controllers and Interconnection System Equipment for Use with Distributed Energy Resources,* for utility interconnection.

Control of the power production source should include real power, reactive power, and harmonic content of the output. The output characteristics of a rotating generator are significantly different from those of a solid-state power source.

The parallel operation of generators is a complex balance of several variables that are design parameters and therefore beyond the scope of the *NEC*.

Some inverters, uninterruptible power supplies (UPS), or solid-state variable-speed drives may produce harmonic currents. (See Exhibit 705.1.) The multiples of the basic supply frequency (usually 60 hertz) can cause additional heating, which could require derating of generators, transformers, cables, and motors. Special generator voltage control systems are required to avoid erratic operation or destruction of control devices. Circuit breakers could require derating if the higher harmonics become significant.

In Exhibit 705.1 (middle), motors and transformers will be driven by harmonic-rich voltage and might require derating. In Exhibit 705.1 (bottom), the source generator could require derating, and special voltage control might be needed.

Δ **705.6 Equipment Approval.** Interconnection and interactive equipment intended to connect to or operate in parallel with power production sources shall be listed for the required interactive function or be evaluated for the interactive function and have a field label applied, or both.

> Informational Note No. 1: See UL 1741, *Standard for Inverters, Converters, Controllers and Interconnection System Equipment for Use with Distributed Energy Resources,* for evaluating interconnected equipment. Sources identified as stand-alone, interactive, or multimode are specifically identified and certified to operate in these operational modes. Stand-alone sources operate in island mode, interactive sources operate in interactive mode, and multimode sources operate in either island mode or interactive mode. Stand-alone sources are not evaluated for interactive capabilities.

> Informational Note No. 2: An interactive function is common in equipment such as microgrid interconnect devices, power control systems, interactive inverters, synchronous engine generators, ac energy storage systems, and ac wind turbines.

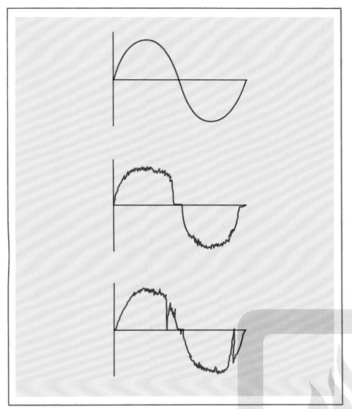

EXHIBIT 705.1 Typical output wave shapes: *(top)* with rotating generator and system wave shape normally encountered with motor, lighting, and heating loads; *(middle)* with inverter source; and *(bottom)* with variable speed drive, rectifier, and uninterruptible power to supply loads.

705.8 System Installation. Installation of one or more electrical power production sources operating in parallel with a primary source(s) of electricity shall be performed only by qualified persons.

> Informational Note: See Article 100 for the definition of *Qualified Person.*

Interconnected power production sources introduce hazards unique to systems operating in parallel. Installation and maintenance personnel must be qualified in parallel operation of electrical systems. The *NEC* defines a qualified person but does not detail the additional training necessary to be deemed a qualified person (see *NFPA 70E®, Standard for Electrical Safety in the Workplace®*). Special training for persons working on interconnected systems is key to ensuring that personnel can work safely on these systems.

Δ **705.10 Identification of Power Sources.** Permanent plaques, labels, or directories shall be installed at each service equipment location, or at an approved readily visible location in accordance with the following:

(1) Denote the location of each power source disconnecting means for the building or structure.

Exception: Installations with multiple colocated power production sources shall be permitted to be identified as a group(s). The plaque, label, or directory shall not be required to identify each power source individually.

(2) Indicate the emergency telephone numbers of any off-site entities servicing the power source systems.

> Informational Note: See NFPA 1-2021, *Fire Code,* 11.12.2.1.5 for installer information.

(3) Be marked with the wording "CAUTION: MULTIPLE SOURCES OF POWER." The marking shall comply with 110.21(B).

Δ **705.11 Source Connections to a Service.**

Electric power production sources are permitted to be connected on the supply side of the service disconnecting means, or they can be connected on the load side. See Exhibit 705.2.

N **(A) Service Connections.** An electric power production source shall be permitted to be connected to a service by one of the following methods:

(1) To a new service in accordance with 230.2(A)
(2) To the supply side of the service disconnecting means in accordance with 230.82(6)
(3) To an additional set of service entrance conductors in accordance with 230.40, Exception No. 5

These connections shall comply with 705.11(B) through (F).

Δ **(B) Conductors.** Service conductors connected to power production sources shall comply with the following:

(1) The ampacity of the service conductors connected to the power production source service disconnecting means shall not be less than the sum of the power production source maximum circuit current in 705.28(A).
(2) The service conductors connected to the power production source service disconnecting means shall be sized in accordance with 705.28 and not be smaller than 6 AWG copper or 4 AWG aluminum or copper-clad aluminum.
(3) The ampacity of any other service conductors to which the power production sources are connected shall not be less than that required in 705.11(B).

(C) Connections. Connections to service conductors or equipment shall comply with 705.11(C)(1) through (C)(3).

N **(1) Splices or Taps.** Service conductorsplices and taps shall be made in accordance with 230.33 or 230.46 and comply with all applicable enclosure fill requirements.

N **(2) Existing Equipment.** Any modifications to existing equipment shall be made in accordance with the manufacturer's instructions, or the modification must be field evaluated for the application and be field labeled.

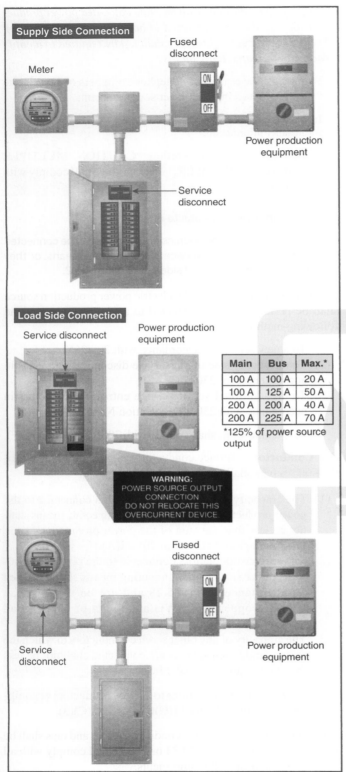

EXHIBIT 705.2 The points of interconnection permitted by 705.11 and 705.12.

N **(3) Utility-Controlled Equipment.** For meter socket enclosures or other equipment under the exclusive control of the electric utility, only connections approved by the electric utility shall be permitted.

N **(D) Service Disconnecting Means.** A disconnecting means in accordance with Parts VI through VII of Article 230 shall be provided to disconnect all ungrounded conductors of a power production source from the conductors of other systems.

N **(E) Bonding and Grounding.** All metal enclosures, metal wiring methods, and metal parts associated with the service connected to a power production source shall be bonded in accordance with Parts II through V and VIII of Article 250.

Δ **(F) Overcurrent Protection.** The power production source service conductors shall be protected from overcurrent in accordance with Part VII of Article 230. The rating of the overcurrent protection device of the power production source service disconnecting means shall be used to determine if ground-fault protection of equipment is required in accordance with 230.95.

Δ **705.12 Load-Side Source Connections.** The output of an interconnected electric power source shall be permitted to be connected to the load side of the service disconnecting means of the other source(s) at any distribution equipment on the premises. Where distribution equipment or feeders are fed simultaneously by a primary source of electricity and one or more other power source(s), the feeders or distribution equipment shall comply with relevant sections of 705.12(A) and (B). Currents from power source connections to feeders or busbars shall be based on the maximum circuit currents calculated in 705.28(A). The ampacity of feeders and taps shall comply with 705.12(A), and the ampere ratings of busbars shall comply with 705.12(B).

(A) Feeders and Feeder Taps. Where the power source output connection is made to a feeder, the following shall apply:

(1) The feeder ampacity is greater than or equal to 125 percent of the power-source output circuit current.

(2) Where the power-source output connection is made at a location other than the opposite end of the feeder from the primary source overcurrent device, that portion of the feeder on the load side of the power source output connection shall be protected by one of the following:

   a. The feeder ampacity shall be not less than the sum of the rating of the primary source overcurrent device and 125 percent of the power-source output circuit current.

   b. An overcurrent device at the load side of the power source connection point shall be rated not greater than the ampacity of the feeder.

(3) For taps sized in accordance with 240.21(B)(2) or (B)(4), the ampacity of taps conductors shall not be less than one-third of the sum of the rating of the overcurrent device protecting the feeder plus the ratings of any power source overcurrent devices connected to the feeder.

(B) Busbars. For power source connections to distribution equipment with no specific listing and instructions for combining multiple sources, one of the following methods shall be used to determine the required ampere ratings of busbars:

(1) The sum of 125 percent of the power source(s) output circuit current and the rating of the overcurrent device protecting the busbar shall not exceed the busbar ampere rating.

Informational Note: This general rule assumes no limitation in the number of the loads or sources applied to busbars or their locations.

(2) Where two sources, one a primary power source and the other another power source, are located at opposite ends of a busbar that contains loads, the sum of 125 percent of the power-source(s) output circuit current and the rating of the overcurrent device protecting the busbar shall not exceed 120 percent of the busbar ampere rating. The busbar shall be sized for the loads connected in accordance with Article 220. A permanent warning label shall be applied to the distribution equipment adjacent to the back-fed breaker from the power source that displays the following or equivalent wording:

WARNING:
POWER SOURCE OUTPUT
DO NOT RELOCATE THIS OVERCURRENT DEVICE.

The warning sign(s) or label(s) shall comply with 110.21(B).

(3) The sum of the ampere ratings of all overcurrent devices on panelboards, both load and supply devices, excluding the rating of the overcurrent device protecting the busbar, shall not exceed the ampacity of the busbar. The rating of the overcurrent device protecting the busbar shall not exceed the rating of the busbar. Permanent warning labels shall be applied to distribution equipment displaying the following or equivalent wording:

WARNING:
EQUIPMENT FED BY MULTIPLE SOURCES.
TOTAL RATING OF ALL OVERCURRENT DEVICES
EXCLUDING MAIN SUPPLY OVERCURRENT DEVICE
SHALL NOT EXCEED AMPACITY OF BUSBAR.

The warning sign(s) or label(s) shall comply with 110.21(B).

(4) A connection at either end of a center-fed panelboard in dwellings shall be permitted where the sum of 125 percent of the power-source(s) output circuit current and the rating of the overcurrent device protecting the busbar does not exceed 120 percent of the busbar ampere rating.

(5) Connections shall be permitted on busbars of panelboards that supply lugs connected to feed-through conductors or are supplied by feed-through conductors. The feed-through conductors shall be sized in accordance with 705.12(A). Where an overcurrent device is installed at either end of the feed-through conductors, panelboard busbars on either side of the feed-through conductors shall be permitted to be sized in accordance with 705.12(B)(1) through (B)(3).

(6) Connections shall be permitted on switchgear, switchboards, and panelboards in configurations other than those permitted in 705.12(B)(1) through (B)(5) where designed under engineering supervision that includes available fault-current and busbar load calculations.

Informational Note: Specifically designed equipment exists, listed to UL 1741, *Inverters, Converters, Controllers and Interconnection System Equipment for Use With Distributed Energy Resources*, for the combination and distribution of sources to supply loads. The options provided in 705.12(B) are for equipment with no specific listing for combining sources.

The required ampacity of a feeder or bus connected to the interactive inverter is based on the inverter output circuit current, rather than on the overcurrent protective device (OCPD) in the inverter. Except where 705.12(A)(2)(b) and (B)(3) apply, the required conductor ampacity is determined by adding the ampacity of the primary OCPD protecting a busbar or feeder and 125 percent of inverter output current.

Where a tap is made to a feeder supplied by the inverter and the normal source, the calculated sum is used as the rating of the overcurrent device to determine the ampacity of the tap conductors in 240.21(B).

Unlike in service equipment where the number or rating of overcurrent devices is not limited, 705.12(B)(3) places a limit on panelboard overcurrent devices. The sum of the ratings of all overcurrent devices (excluding the main overcurrent device) supplying and/or being supplied by the panelboard is limited to the busbar rating. In addition, the main overcurrent device also must be limited to the ampacity of the busbar.

Δ **705.13 Energy Management Systems (EMS).** An EMS in accordance with 750.30 shall be permitted to limit current and loading on the busbars and conductors supplied by the output of one or more interconnected electric power production or energy storage sources.

Informational Note: A listed power control system (PCS) is a type of EMS that is capable of monitoring multiple power sources and controlling the current on busbars and conductors to prevent overloading. See UL 1741, *Inverters, Converters, Controllers and Interconnection System Equipment for Use with Distributed Energy Resources*, and UL 916, *Energy Management Equipment*, for information on PCS and EMS.

The use of a power control system (PCS) (see Exhibit 705.3) that involves monitoring and control of all the individual sources can prevent the feeder or busbar from being overloaded while making efficient use of the variable resources.

Δ **705.20 Source Disconnecting Means.** Means shall be provided to disconnect power source output conductors of electric power production equipment from conductors of other systems. A single disconnecting means shall be permitted to disconnect multiple power sources from conductors of other systems.

Informational Note: See 480.7, Part II of Article 445, Part III of Article 690, Part III of Article 692, Part III of Article 694, and Part II of Article 706 for specific source disconnecting means requirements.

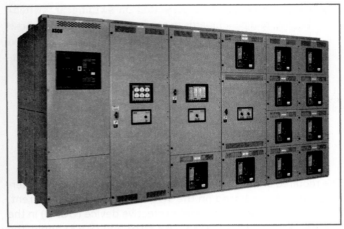

EXHIBIT 705.3 PCS for monitoring and controlling individual power sources. *(Courtesy of ASCO Power Technologies)*

The disconnecting means shall comply with the following:

(1) Be one of the following types:

    a. A manually operable switch or circuit breaker

    b. A load-break-rated pull-out switch

    c. A power-operated or remote-controlled switch or circuit breaker that is manually operable locally and opens automatically when control power is interrupted

    d. A device listed or approved for the intended application

(2) Simultaneously disconnect all ungrounded conductors of the circuit

(3) Located where readily accessible

(4) Externally operable without exposed live parts

(5) Plainly indicate whether in the open (off) or closed (on) position

(6) Have ratings sufficient for the maximum circuit current, available fault current, and voltage that is available at the terminals

(7) Where the line and load terminals are capable of being energized in the open position, be marked with the following words or equivalent:

WARNING
ELECTRIC SHOCK HAZARD TERMINALS ON THE
LINE AND LOAD SIDES
MAY BE ENERGIZED IN THE OPEN POSITION.

Informational Note: With interconnected power sources, some equipment, including switches and fuses, is capable of being energized from both directions.

705.25 Wiring Methods. Power source output conductors shall comply with 705.25(A) through (C).

Δ **(A) General.** Wiring methods and fittings listed for use with power production systems shall be permitted in addition to general wiring methods and fittings permitted elsewhere in this *Code*.

Δ **(B) Flexible Cords and Cables.** Flexible cords and cables, where used to connect the moving parts of power production equipment, or where used for ready removal for maintenance and repair, shall be listed and identified as DG cable, or other cable suitable for extra hard use, and shall be water resistant. Cables exposed to sunlight shall be sunlight resistant. Flexible, fine-stranded cables shall be terminated only with terminals, lugs, devices, or connectors in accordance with 110.14(A).

(C) Multiconductor Cable Assemblies. Multiconductor cable assemblies used in accordance with their listings shall be permitted.

Informational Note: See UL 3003, *Distributed Generation Cables*, and UL 9703, *Outline of Investigation for Distributed Generation Wiring Harnesses*, for additional information on DG cable (distributed generation cable) and harnesses. An ac module harness is one example of a multiconductor cable assembly.

705.28 Circuit Sizing and Current.

(A) Power Source Output Maximum Current. Where not elsewhere required or permitted in this *Code*, the maximum current for power sources shall be calculated using one of the following methods:

(1) The sum of the continuous output current ratings of the power production equipment at the circuit nominal system voltage

(2) For power production equipment controlled by an EMS, the current setpoint of the EMS

(3) Where sources controlled by an EMS are combined with other sources on the same power source output circuit, the sum of 705.28(A)(1) and (A)(2)

(B) Conductor Ampacity. Where not elsewhere required or permitted in this *Code*, the power source output conductors shall have an ampacity not less than the larger of the following and comply with 110.14(C):

(1) The maximum currents in 705.28(A) multiplied by 125 percent without adjustment or correction factors

Exception No. 1: If the assembly, including the overcurrent devices protecting the circuit, is listed for operation at 100 percent of its rating, the ampacity of the conductors shall be permitted to be not less than the calculated maximum current of 705.28(A).

Exception No. 2: Where a portion of a circuit is connected at both its supply and load ends to separately installed pressure connections as covered in 110.14(C)(2), it shall be permitted to have an ampacity not less than the calculated maximum current of 705.28(A). No portion of the circuit installed under this exception shall extend into an enclosure containing either the circuit supply or the circuit load terminations, as covered in 110.14(C)(1).

Exception No. 3: Grounded conductors that are not connected to an overcurrent device shall be permitted to be sized at 100 percent of the calculated maximum current of 705.28(A).

(2) The maximum currents in 705.28(A) after the application of adjustment and correction factors in accordance with 310.14

(3) Where connected to feeders, if smaller than the feeder conductors, the ampacity as calculated in 240.21(B) based on the over-current device protecting the feeder

(C) Neutral Conductors. Neutral conductors shall be permitted to be sized in accordance with either 705.28(C)(1) or (C)(2).

(1) Single-Phase Line-to-Neutral Power Sources. Where not elsewhere required or permitted in this *Code*, the ampacity of a neutral conductor to which a single-phase line-to-neutral power source is connected shall not be smaller than the ampacity in 705.28(B).

(2) Neutral Conductor Used Solely for Instrumentation, Voltage, Detection, or Phase Detection. A power production equipment neutral conductor used solely for instrumentation, voltage detection, or phase detection shall be permitted to be sized in accordance with 250.102.

705.30 Overcurrent Protection.

Power production sources and conductors are required to follow the applicable article for determining overcurrent protection. These installations are subject to the entire *NEC*, as detailed in 90.3.

Δ **(A) Circuit and Equipment.** Power source output conductors and equipment shall be provided with overcurrent protection. Circuits connected to more than one electrical source shall have overcurrent devices located to provide overcurrent protection from all sources.

(B) Overcurrent Device Ratings. The overcurrent devices in other than generator systems shall be sized to carry not less than 125 percent of the maximum currents as calculated in 705.28(A). The rating or setting of overcurrent devices shall be permitted in accordance with 240.4(B) and (C).

Exception: Circuits containing an assembly together with its overcurrent device(s) that is listed for continuous operation at 100 percent of its rating shall be permitted to be utilized at 100 percent of its rating.

N **(C) Marking.** Equipment containing overcurrent devices supplied from interconnected power sources shall be marked to indicate the presence of all sources.

N **(D) Suitable for Backfeed.** Fused disconnects, unless otherwise marked, shall be considered suitable for backfeed. Circuit breakers not marked "line" and "load" shall be considered suitable for backfeed. Circuit breakers marked "line" and "load" shall be considered suitable for backfeed or reverse current if specifically rated.

N **(E) Fastening.** Listed plug-in-type circuit breakers backfed from electric power sources that are listed and identified as interactive shall be permitted to omit the additional fastener normally required by 408.36(D) for such applications.

Δ **(F) Transformers.** The following apply to the installation of transformers:

(1) For the purpose of overcurrent protection, the primary side of transformers with sources on each side shall be the side connected to the largest source of available fault current.

(2) Transformer secondary conductors shall be protected in accordance with 240.21(C).

705.32 Ground-Fault Protection. Where ground-fault protection of equipment is installed in ac circuits as required elsewhere in this *Code*, the output of interconnected power production equipment shall be connected to the supply side of the ground-fault protection equipment.

Exception: Connection of power production equipment shall be permitted to be made to the load side of ground-fault protection equipment where installed in accordance with 705.11 or where there is ground-fault protection for equipment from all ground-fault current sources.

705.40 Loss of Primary Source. The output of interactive electric power production equipment shall be automatically disconnected from all ungrounded conductors of the primary source when one or more of the phases of the primary source to which it is connected opens. The interactive electric power production equipment shall not be reconnected to the primary source until all the phases of the primary source to which it is connected are restored. This requirement shall not be applicable to electric power production equipment providing power to an emergency or legally required standby system.

Exception: A listed interactive inverter shall trip or shall be permitted to automatically cease exporting power when one or more of the phases of the interconnected primary source opens and shall not be required to automatically disconnect all ungrounded conductors from the primary source. A listed interactive inverter shall be permitted to automatically or manually resume exporting power to the interconnected system once all phases of the source to which it is connected are restored.

Informational Note No. 1: Risks to personnel and equipment associated with the primary source could occur if an interactive electric power production source can operate as an intentional island. Special detection methods are required to determine that a primary source supply system outage has occurred and whether there should be automatic disconnection. When the primary source supply system is restored, special detection methods are typically required to limit exposure of power production sources to out-of-phase reconnection.

Informational Note No. 2: Induction-generating equipment connected on systems with significant capacitance can become self-excited upon loss of the primary source and experience severe overvoltage as a result.

Interactive power production equipment shall be permitted to operate in island mode to supply loads that have been disconnected from the electric utility or other electric power production and distribution network.

When two interconnected power systems separate, they can drift out of synchronization. Damage to the system(s) can occur if restoration of one system occurs out of phase with the other system(s). If the reconnection is out of phase, violent electromechanical stresses can destroy mechanical components such as gears, couplings, and shafts and can displace coils.

See Part II of Article 705 for microgrid system requirements, including the necessary interconnect device if a microgrid system is operating in stand-alone mode (islanded).

705.45 Unbalanced Interconnections.

(A) Single Phase. Single-phase power sources in interactive systems shall be connected to 3-phase power systems in order to limit unbalanced voltages at the point of interconnection to not more than 3 percent.

> Informational Note: For interactive power sources, unbalanced voltages can be minimized by the same methods that are used for single-phase loads on a 3-phase power system. See ANSI/C84.1-2016, *Electric Power Systems and Equipment — Voltage Ratings (60 Hertz)*.

(B) Three Phase. Three-phase power sources in interactive systems shall have all phases automatically de-energized upon loss of, or unbalanced, voltage in one or more phases unless the interconnected system is designed so that significant unbalanced voltages will not result.

Part II. Microgrid Systems

The requirements in Article 705 apply to power production systems that operate in parallel with a primary supply. Section 705.40 requires that the ungrounded conductors of the power production source be automatically disconnected from the ungrounded conductors of the primary source. Section 705.40 permits inverters to provide the disconnection of the load from the primary source and the power production source to operate as a stand-alone system.

705.50 System Operation. Interconnected microgrid systems shall be capable of operating in interactive mode with a primary source of power, or electric utility, or other electric power production and distribution network. Microgrid systems shall be permitted to disconnect from other sources and operate in island mode.

> Informational Note No. 1: Microgrid systems often include a single source or a compatible interconnection of multiple sources such as engine generators, solar PV, wind, or ESS.
> Informational Note No. 2: See Article 517 for health care facilities incorporating microgrids.

705.60 Primary Power Source Connection. Connections to primary power sources that are external to the microgrid system shall comply with the requirements of 705.11, 705.12, or 705.13. Power source conductors connecting to a microgrid system, including conductors supplying distribution equipment, shall be considered as power source output conductors.

705.65 Reconnection to Primary Power Source. Microgrid systems that reconnect to primary power sources shall be provided with the necessary equipment to establish a synchronous transition.

Δ **705.70 Microgrid Interconnect Devices (MID).** Microgrid interconnect devices shall comply with the following:

(1) Be required for any connection between a microgrid system and a primary power source
(2) Be evaluated for the application and have a field label applied or be listed for the application
(3) Have overcurrent devices located to provide overcurrent protection from all sources

> Informational Note: MID functionality is often incorporated in an interactive or multimode inverter, energy storage system, or similar device identified for interactive operation.

N **705.76 Microgrid Control System (MCS).** Microgrid control systems shall comply with the following:

(1) Coordinate interaction between multiple power sources of similar or different types, manufacturers, and technologies (including energy storage)
(2) Be evaluated for the application and have a field label applied, or be listed, or be designed under engineering supervision
(3) Monitor and control microgrid power production and power quality
(4) Monitor and control transitions with a primary source external to the microgrid

> Informational Note: MID functionality is often incorporated in an interactive or multimode inverter, energy storage system, or similar device identified for interactive operation.

N ## Part III. Interconnected Systems Operating in Island Mode

N **705.80 Power Source Capacity.** For interconnected power production sources that operate in island mode, capacity shall be calculated using the sum of all power source output maximum currents for the connected power production source.

N **705.81 Voltage and Frequency Control.** Power sources operating in island mode shall be controlled so that voltage and frequency are supplied within limits compatible with the connected loads.

N **705.82 Single 120-Volt Supply.** Systems operating in island mode shall be permitted to supply 120 volts to single-phase, 3-wire, 120/240-volt distribution equipment where there are no 240-volt outlets and where there are no multiwire branch circuits. In all installations, the sum of the ratings of the power sources shall be less than the rating of the neutral bus in the distribution equipment. This equipment shall be marked with the following words or equivalent:

> WARNING: SINGLE 120-VOLT SUPPLY
> DO NOT CONNECT MULTIWIRE
> BRANCH CIRCUITS

The warning sign(s) or label(s) shall comply with 110.21(B).

ARTICLE 706 Energy Storage Systems

Part I. General

△ **706.1 Scope.** This article applies to all energy storage systems (ESS) having a capacity greater than 3.6 MJ (1 kWh) that may be stand-alone or interactive with other electric power production sources. These systems are primarily intended to store and provide energy during normal operating conditions.

Energy storage systems (ESS) store energy for later use. Wind power and photovoltaic (PV) systems generate power when the resource is available, not necessarily when the energy is needed. Energy storage optimizes the use of stand-alone alternative energy by storing energy during peak production times so that the energy can be used at a time when wind or sunlight is not available. Increasingly, energy storage is being used to take advantage of utility generating capacity during overnight hours when demand for energy is low and the time-of-use energy rates are customer friendly. Storing energy might reduce the need to build additional generating stations. Energy storage can be installed at the generating facility, or it can be geographically distributed throughout the service area.

The overall scope of Article 706 covers the complete assembly for storing and exporting electrical energy. In accordance with 706.5, this assembly of components is required to be listed as an ESS. UL 9540, *Energy Storage Systems and Equipment*, contains product safety requirements for electrochemical, chemical, mechanical, and thermal ESS. A battery-based ESS certified in accordance with UL 9540 is subject to the installation requirements of Article 706, rather than with Article 480. A group of separate components that includes storage batteries; that is provided with support systems (racks), charge controller(s), and inverters; and that does not have an overall listing as an ESS is a storage battery system and as such is subject to the requirements of Article 480.

> Informational Note No. 1: See Article 480 for installations that meet the definition of *stationary standby batteries*.
> Informational Note No. 2: For batteries rated in ampere hours, kWh is equal to the nominal rated voltage times ampere-hour rating divided by 1000.
> Informational Note No. 3: The following standards are frequently referenced for the installation of ESSs:
>
> (1) NFPA 1-2021, *Fire Code*
> (2) NFPA 111-2019, *Standard on Stored Electrical Energy Emergency and Standby Power Systems*
> (3) NECA 416-2016, *Recommended Practice for Installing Energy Storage Systems (ESS)*
> (4) UL 810A, *Electrochemical Capacitors*
> (5) NFPA 855-2020, *Standard for the Installation of Stationary Energy Storage Systems*
> (6) UL 1973, *Standard for Batteries for Use in Stationary, Vehicle Auxiliary Power, and Light Electric Rail (LER) Applications*
> (7) UL 1989, *Standard for Standby Batteries*
> (8) UL 9540, *Standard for Safety Energy Storage Systems and Equipment*
> (9) UL Subject 2436, *Spill Containment For Stationary Lead Acid Battery Systems*

706.3 Qualified Personnel. The installation and maintenance of ESS equipment and all associated wiring and interconnections shall be performed only by qualified persons.

> Informational Note: See Article 100 for the definition of *qualified person*.

706.4 System Requirements. Each ESS shall be provided with a nameplate plainly visible after installation and marked with the following:

(1) Manufacturer's name, trademark, or other descriptive marking by which the organization responsible for supplying the ESS can be identified
(2) Rated frequency
(3) Number of phases, if ac
(4) Rating (kW or kVA)
(5) Available fault current derived by the ESS at the output terminals
(6) Maximum output and input current of the ESS at the output terminals
(7) Maximum output and input voltage of the ESS at the output terminals
(8) Utility-interactive capability, if applicable

706.5 Listing. Energy storage systems shall be listed.

706.6 Multiple Systems. Multiple ESSs shall be permitted to be installed on the same premises.

△ **706.7 Commissioning and Maintenance.**

N **(A) Commissioning.** ESSs shall be commissioned upon installation. This shall not apply in one- and two-family dwellings.

> Informational Note: See NFPA 855-2020, *Standard for the Installation of Stationary Energy Storage Systems*, for information related to the commissioning of ESSs.

N **(B) Maintenance.** ESSs shall be maintained in proper and safe operating condition. The required maintenance shall be in accordance with the manufacturer's requirements and industry standards. A written record of the system maintenance shall be kept and shall include records of repairs and replacements necessary to maintain the system in proper and safe operating condition. This shall not apply in one- and two-family dwellings.

> Informational Note: See NFPA 70B-2019, *Recommended Practice for Electrical Equipment Maintenance*, or ANSI/NETA ATS-2017, *Standard for Acceptance Testing Specifications for Electrical Power Equipment and Systems*, for information related to general electrical equipment maintenance and developing an effective electrical preventive maintenance (EPM) program.

706.9 Maximum Voltage. The maximum voltage of an ESS shall be the rated ESS input and output voltage(s) indicated on the ESS nameplate(s) or system listing.

Part II. Disconnecting Means

706.15 Disconnecting Means.

Δ **(A) ESS Disconnecting Means.** Means shall be provided to disconnect the ESS from all wiring systems, including other power systems, utilization equipment, and its associated premises wiring.

N **(B) Location and Control.** The disconnecting means shall be readily accessible and shall comply with one or more of the following:

(1) Located within the ESS
(2) Located within sight and within 3 m (10 ft) from the ESS
(3) Where not located within sight of the ESS, the disconnecting means, or the enclosure providing access to the disconnecting means, shall be capable of being locked in accordance with 110.25

Where controls to activate the disconnecting means of an ESS are used and are not located within sight of the ESS, the disconnecting means shall be lockable in accordance with 110.25, and the location of the controls shall be marked on the disconnecting means.

For one- and two-family dwellings, an ESS shall include an emergency shutdown function to cease the export of power from the ESS to premises wiring of other systems. An initiation device(s) shall be located at a readily accessible location outside the building and shall plainly indicate whether in the "off" or "on" position. The "off" position of the device(s) shall perform the ESS emergency shutdown function.

See also

Section 230.85 and associated commentary for information on emergency disconnects for services supplying one- and two-family dwellings

Δ **(C) Notification and Marking.** Each ESS disconnecting means shall plainly indicate whether it is in the open (off) or closed (on) position and be permanently marked as follows:

"ENERGY STORAGE SYSTEM DISCONNECT"

The disconnecting means shall be legibly marked in the field to indicate the following:

(1) Nominal ESS output voltage
(2) Available fault current derived from the ESS
(3) An arc-flash label applied in accordance with acceptable industry practice
(4) Date the calculation was performed

Exception: List items (2), (3), and (4) shall not apply to one- and two-family dwellings.

Informational Note No. 1: See NFPA *70E*-2018, *Standard for Electrical Safety in the Workplace*, for industry practices for equipment labeling. This standard provides specific criteria for developing arc-flash labels for equipment that provides nominal system voltage, incident energy levels, arc-flash boundaries, minimum required levels of personal protective equipment, and so forth.

Informational Note No. 2: ESS electronics could include inverters or other types of power conversion equipment.

For ESS disconnecting means where the line and load terminals could be energized in the open position, the device shall be marked with the following words or equivalent:

WARNING
ELECTRIC SHOCK HAZARD
TERMINALS ON THE LINE AND LOAD
SIDES MAY BE ENERGIZED IN THE OPEN POSITION

The notification(s) and marking(s) shall comply with 110.21(B).

(D) Partitions Between Components. Where circuits from the input or output terminals of energy storage components in an ESS pass through a wall, floor, or ceiling, a readily accessible disconnecting means shall be provided within sight of the energy storage component. Fused disconnecting means or circuit breakers shall be permitted to be used.

N **(E) Disconnecting Means for Batteries.** In cases where the battery is separate from the ESS electronics and is subject to field servicing, 706.15(E)(1) through (E)(4) shall apply.

Informational Note: Batteries could include an enclosure, battery monitoring and controls, or other related battery components.

N **(1) Disconnecting Means.** A disconnecting means shall be provided for all ungrounded conductors. A disconnecting means shall be readily accessible and located within sight of the battery.

Informational Note: See 240.21(H) for information on the location of the overcurrent device for battery conductors.

N **(2) Disconnection of Series Battery Circuits.** Battery circuits exceeding 240 volts dc nominal between conductors or to ground shall have provisions to disconnect the series-connected strings into segments not exceeding 240 volts dc nominal for maintenance by qualified persons. Non-load-break bolted or plug-in disconnects shall be permitted.

N **(3) Remote Activation.** Where a disconnecting means is provided with remote controls to activate the disconnecting means and the controls for the disconnecting means are not located within sight of the battery, the disconnecting means shall be capable of being locked in the open position, in accordance with 110.25, and the location of the controls shall be field marked on the disconnecting means.

N **(4) Notification.** The disconnecting means shall be legibly marked in the field. The marking shall be of sufficient durability to withstand the environment involved and shall include the following:

(1) Nominal battery voltage
(2) Available fault current derived from the stationary standby battery system

Informational Note No. 1: Battery equipment suppliers can provide information about available fault current on any particular battery model.

(3) An arc-flash label in accordance with acceptable industry practice

Informational Note No. 2: See NFPA *70E*-2021, *Standard for Electrical Safety in the Workplace*, for assistance in determining the severity of potential exposure, planning safe work practices, determining arc-flash labeling, and selecting personal protective equipment.

(4) Date the calculation was performed

706.16 Connection to Energy Sources. The connection of an ESS to sources of energy shall comply with 706.16(A) through (F).

(A) Source Disconnect. A disconnect that has multiple sources of power shall disconnect all energy sources when in the off position.

(B) Identified Interactive Equipment. ESS that operate in parallel with other ac sources shall use inverters that are listed and identified as interactive.

(C) Loss of Interactive System Power. Upon loss of a primary source of power, an ESS with a utility-interactive inverter shall comply with the requirements of 705.40.

(D) Unbalanced Interconnections. Unbalanced ac connections between an ESS and other ac electric power production sources shall be in accordance with 705.45

Δ **(E) Other Energy Sources.** The connection of an ESS to other energy sources shall be in accordance with 705.12.

(F) Stand-Alone Operation. Where the output of an ESS is capable of operating in stand-alone mode, the requirements of 710.15 shall apply.

> **See also**
>
> **Section 710.1** and associated commentary for more information on power production sources operating in stand-alone mode being used as the alternate source for an optional standby system

Part III. Installation Requirements

706.20 General.

Δ **(A) Ventilation.** Provisions appropriate to the energy storage technology shall be made for sufficient diffusion and ventilation of any possible gases from the storage device, if present, to prevent the accumulation of an explosive mixture. Ventilation of an ESS shall be permitted to be provided in accordance with the manufacturer's recommendations and listing for the system.

Informational Note No. 1: See NFPA 855-2020, *Standard for the Installation of Stationary Energy Storage Systems*, for technology-specific guidance. Not all ESS technologies require ventilation. Informational Note No. 2: See IEEE 1635-2018/ASHRAE Guideline 21-2018, *Guide for the Ventilation and Thermal Management of Batteries for Stationary Applications*, as a source for design of ventilation of batteries.

(B) Dwelling Units. An ESS for one- and two-family dwelling units shall not exceed 100 volts dc between conductors or to ground.

Exception: Where live parts are not accessible during routine ESS maintenance, a maximum ESS voltage of 600 volts dc shall be permitted.

(C) Spaces About ESS Components.

(1) General. Working spaces for ESS shall comply with 110.26 and 110.34.

(2) Space Between Components. ESSs shall be permitted to have space between components in accordance with the manufacturer's instructions and listing.

Informational Note: Additional space may be needed to accommodate ESS hoisting equipment, tray removal, or spill containment.

706.21 Directory (Identification of Power Sources). ESS shall be indicated by markings or labels that shall be in accordance with 110.21(B).

Δ **(A) Facilities with Utility Services and ESS.** Plaques or directories shall be installed in accordance with 705.10.

(B) Facilities with Stand-Alone Systems. Plaques or directories shall be installed in accordance with 710.10.

Part IV. Circuit Requirements

706.30 Circuit Sizing and Current.

(A) Maximum Rated Current for a Specific Circuit. The maximum current for the specific circuit shall be calculated in accordance with 706.30(A)(1) through (A)(5).

(1) Nameplate-Rated Circuit Current. Circuit current shall be the rated current indicated on the ESS nameplate(s) or system listing. Where the ESS has separate input (charge) and output (discharge) circuits or ratings, these shall be considered individually. Where the same terminals on the ESS are used for charging and discharging, the rated current shall be the greater of the two.

(2) Inverter Output Circuit Current. The maximum current shall be the inverter continuous output current rating.

(3) Inverter Input Circuit Current. The maximum current shall be the continuous inverter input current rating when the inverter is producing rated power at the lowest input voltage.

(4) Inverter Utilization Output Circuit Current. The maximum current shall be the continuous ac output current rating of the inverter when the inverter is producing rated power.

(5) DC to DC Converter Output Current. The maximum current shall be the dc-to-dc converter continuous output current rating.

Δ **(B) Conductor Ampacity.** The ampacity of the output circuit conductors of the ESS(s) connected to the wiring system serving the loads to be serviced by the system shall not be less than the greater of the nameplate(s)-rated circuit current as determined in accordance with 706.30(A)(1) or the rating of the ESS(s) overcurrent protective device(s).

(C) Ampacity of Grounded or Neutral Conductor. If the output of a single-phase, 2-wire ESS output(s) is connected to the grounded or neutral conductor and a single ungrounded conductor of a 3-wire system or of a 3-phase, 4-wire, wye-connected system, the maximum unbalanced neutral load current plus the ESS(s) output rating shall not exceed the ampacity of the grounded or neutral conductor.

706.31 Overcurrent Protection.

Δ **(A) Circuits and Equipment.** Protection devices for ESS circuits shall be in accordance with 706.31(B) through (F). Circuits shall be protected at the source from overcurrent. A circuit conductor connected at one end to a supply with integral fault protection, where the conductor is rated for the maximum circuit current from that supply, and also connected to sources having an available maximum circuit current greater than the ampacity of the conductor, shall be protected from overcurrent at the point of connection to the higher current source.

> Informational Note: Listed electronic power converter circuits powered by an ESS have integral fault protection. Where these circuits are connected to higher current sources such as a utility service, the overcurrent device is more appropriately installed at the higher current source end of the circuit conductor.

Δ **(B) Overcurrent Device Ampere Ratings.** Overcurrent protective devices, where required, shall be not less than 125 percent of the maximum currents calculated in 706.30(A).

Exception: Where the assembly, including the overcurrent protective devices, is listed for operation at 100 percent of its rating, the ampere rating of the overcurrent devices shall be permitted to be not less than the maximum currents calculated in 706.30(B).

(C) Direct Current Rating. Overcurrent protective devices, either fuses or circuit breakers, used in any dc portion of an ESS shall be listed for dc and shall have the appropriate voltage, current, and interrupting ratings for the application.

(D) Current Limiting. A listed and labeled current-limiting overcurrent protective device shall be installed adjacent to the ESS for each dc output circuit.

Exception: Where current-limiting overcurrent protection is provided for the dc output circuits of a listed ESS, additional current-limiting overcurrent devices shall not be required.

(E) Fuses. Means shall be provided to disconnect any fuses associated with ESS equipment and components when the fuse is energized from both directions and is accessible to other than qualified persons. Switches, pullouts, or similar devices that are rated for the application shall be permitted to serve as a means to disconnect fuses from all sources of supply.

(F) Location. Where circuits from the input or output terminals of energy storage components in an ESS pass through a wall, floor, or ceiling, overcurrent protection shall be provided at the energy storage component end of the circuit.

706.33 Charge Control.

(A) General. Provisions shall be provided to control the charging process of the ESS. All adjustable means for control of the charging process shall be accessible only to qualified persons.

(B) Diversion Charge Controller.

(1) Sole Means of Regulating Charging. An ESS employing a diversion charge controller as the sole means of regulating charging shall be equipped with a second independent means to prevent overcharging of the storage device.

(2) Circuits with Diversion Charge Controller and Diversion Load. Circuits containing a diversion charge controller and a diversion load shall comply with the following:

    (1) The current rating of the diversion load shall be less than or equal to the current rating of the diversion load charge controller. The voltage rating of the diversion load shall be greater than the maximum ESS voltage. The power rating of the diversion load shall be at least 150 percent of the power rating of the charging source.

    (2) The conductor ampacity and the rating of the overcurrent device for this circuit shall be at least 150 percent of the maximum current rating of the diversion charge controller.

(3) ESS Using Interactive Inverters. Systems using interactive inverters to control energy storage state-of-charge by diverting excess power into an alternate electric power production and distribution system, such as utility, shall comply with 706.33(B)(3)(a) and (B)(3)(b).

    (a) These systems shall not be required to comply with 706.33(B)(2).

    (b) These systems shall have a second, independent means of controlling the ESS charging process for use when the alternate system is not available or when the primary charge controller fails or is disabled.

(C) Charge Controllers and DC-to-DC Converters. Where charge controllers and other DC-to-DC power converters that increase or decrease the output current or output voltage with respect to the input current or input voltage are installed, all of the following shall apply:

    (1) The ampacity of the conductors in output circuits shall be based on the maximum rated continuous output current of the charge controller or converter for the selected output voltage range.

    (2) The voltage rating of the output circuits shall be based on the maximum voltage output of the charge controller or converter for the selected output voltage range.

Part V. Flow Battery ESSs

Part V applies to ESSs composed of or containing flow batteries.

> Informational Note: Due to the unique design features and difference in operating characteristics of flow batteries as compared with that of storage batteries such as lead acid or lithium ion batteries, the requirements for flow batteries have been included herein (Article 706, Part V).

Δ **706.40 General.** The system and system components shall also meet Parts I, II, and III of this article.

> Informational Note: See NFPA 855-2020, *Standard for the Installation of Stationary Energy Storage Systems*, for installation requirements for ESS, including requirements for flow batteries.

706.41 Electrolyte Classification. The electrolyte(s) that are acceptable for use in the batteries associated with the ESS shall be identified by name and chemical composition. Such identification shall be provided by readily discernable signage adjacent to every location in the system where the electrolyte can be put into or taken out of the system.

706.42 Electrolyte Containment. Flow battery systems shall be provided with a means for electrolyte containment to prevent spills of electrolyte from the system. An alarm system shall be provided to signal an electrolyte leak from the system. Electrical wiring and connections shall be located and routed in a manner that mitigates the potential for exposure to electrolytes.

706.43 Flow Controls. Controls shall be provided to safely shut down the system in the event of electrolyte blockage.

706.44 Pumps and Other Fluid Handling Equipment. Pumps and other fluid handling equipment are to be rated/specified suitable for exposure to the electrolytes.

Part VI. Other Energy Storage Technologies

Part VI applies to ESSs using other technologies intended to store energy and when there is a demand for electrical power to use the stored energy to generate the needed power.

706.50 General. All electrical connections to and from the system and system components shall be in accordance with the applicable provisions of this *Code*. The systems shall comply with Parts I, II, III, and IV of this article.

N **706.51 Flywheel ESS (FESS).** Flywheel ESS (FESS) using flywheels as the storage mechanism shall also comply with all of the following:

(1) FESS shall not be used for one- or two-family dwelling units.

> Informational Note No. 1: FESS are intended for high-power shorter term applications. They contain parts that rotate under

high speed with hazardous kinetic energy and include parts such as magnetic bearings that require ongoing monitoring and maintenance and, therefore, are not suitable for residential-type applications.

(2) FESS shall be provided with bearing monitoring and controls that can identify bearing wear or damage to avoid catastrophic failure.

> Informational Note No. 2: The bearing monitoring controls should be evaluated as part of the listing evaluation.

(3) FESS shall be provided with a containment means to contain moving parts that could break from the system upon catastrophic failure.

> Informational Note No. 3: The containment means should be evaluated as part of the listing evaluation.

(4) The spin-down time of the FESS shall be provided in the maintenance documentation.

ARTICLE 708 Critical Operations Power Systems (COPS)

Part I. General

Δ **708.1 Scope.** This article applies to the installation, operation, monitoring, control, and maintenance of the portions of the premises wiring system intended to supply, distribute, and control electricity to designated critical operations areas (DCOA) in the event of disruption to elements of the normal system.

Critical operations power systems are those systems so classed by municipal, state, federal, or other codes by any governmental agency having jurisdiction or by facility engineering documentation establishing the necessity for such a system. These systems include but are not limited to power systems, HVAC, fire alarm, security, communications, and signaling for designated critical operations areas.

> Informational Note No. 1: Critical operations power systems are generally installed in vital infrastructure facilities that, if destroyed or incapacitated, would disrupt national security, the economy, public health or safety; and where enhanced electrical infrastructure for continuity of operation has been deemed necessary by governmental authority.
> Informational Note No. 2: See *NFPA 1600*-2019, *Standard on Continuity, Emergency, and Crisis Management*, for further information on disaster and emergency management.
> Informational Note No. 3: See NFPA 110-2019, *Standard for Emergency and Standby Power Systems*, for further information regarding performance of emergency and standby power systems.
> Informational Note No. 4: See NFPA *101*-2021, *Life Safety Code*, or the applicable building code, for specification of locations where emergency lighting is considered essential to life safety.

Informational Note No. 5: See NFPA 730-2020, *Guide for Premises Security*, and ANSI/TIA-5017-2016, *Telecommunications Physical Network Security Standard*, for further information regarding physical security.

Informational Note No. 6: See *NFPA 1600*-2019, *Standard on Continuity, Emergency, and Crisis Management*, A.5.3.2. Threats to facilities that may require transfer of operation to the critical systems include both naturally occurring hazards and human-caused events.

Informational Note No. 7: See Informative Annex F, Availability and Reliability for Critical Operations Power Systems; and Development and Implementation of Functional Performance Tests (FPTs) for Critical Operations Power Systems.

Informational Note No. 8: See Informative Annex G, Supervisory Control and Data Acquisition (SCADA).

Informational Note No. 9: Text that is followed by a reference in brackets has been extracted from *NFPA 1600*-2019, *Standard on Continuity, Emergency, and Crisis Management*. Only editorial changes were made to the extracted text to make it consistent with this *Code*.

Article 708 addresses homeland security issues for facilities that are mission critical. These requirements go beyond those of Article 700, in that these electrical systems must continue to operate during the full duration of an emergency and beyond. See Exhibit 708.1. Examples of facilities that would use a critical operations power system (COPS) include police stations and fire stations. Not all such facilities within an area would be designated as COPS. Because power must operate continuously with a robust power supply, only those facilities that are designated as critical would be included.

EXHIBIT 708.1 New Hampshire Department of Safety Emergency Management Center, a typical COPS facility: *(top)* facility and *(bottom)* facility signage.

N 708.2 Reconditioned Equipment. Reconditioned transfer switches shall not be permitted.

708.4 Risk Assessment. Risk assessment for critical operations power systems shall be documented and shall be conducted in accordance with 708.4(A) through (C).

> Informational Note: See *NFPA 1600*-2019, *Standard on Continuity, Emergency, and Crisis Management*, Chapter 5, which provides additional guidance concerning risk assessment and hazard analysis.

(A) Conducting Risk Assessment. In critical operations power systems, risk assessment shall be performed to identify hazards, the likelihood of their occurrence, and the vulnerability of the electrical system to those hazards.

Δ (B) Identification of Hazards. Hazards to be considered at a minimum shall include, but shall not be limited to, the following:

(1) Naturally occurring hazards (geological, meteorological, and biological)

(2) Human-caused events (accidental and intentional)

(C) Developing Mitigation Strategy. Based on the results of the risk assessment, a strategy shall be developed and implemented to mitigate the hazards that have not been sufficiently mitigated by the prescriptive requirements of this *Code*.

708.5 Physical Security. Physical security shall be provided for critical operations power systems in accordance with 708.5(A) and (B).

(A) Risk Assessment. Based on the results of the risk assessment, a strategy for providing physical security for critical operations power systems shall be developed, documented, and implemented.

(B) Restricted Access. Electrical circuits and equipment for critical operations power systems shall be accessible to qualified personnel only.

708.6 Testing and Maintenance.

(A) Conduct or Witness Test. The authority having jurisdiction shall conduct or witness a test of the complete system upon installation and periodically afterward.

(B) Tested Periodically. Systems shall be tested periodically on a schedule approved by the authority having jurisdiction to ensure the systems are maintained in proper operating condition.

(C) Maintenance. The authority having jurisdiction shall require a documented preventive maintenance program for critical operations power systems.

> Informational Note: See NFPA 70B-2019, *Recommended Practice for Electrical Equipment Maintenance*, for information concerning maintenance.

(D) Written Record. A written record shall be kept of such tests and maintenance.

Δ **(E) Testing Under Load.** Means for testing all critical power systems during maximum anticipated load conditions shall be provided.

Informational Note: See NFPA 110-2019, *Standard for Emergency and Standby Power Systems*, for information concerning testing and maintenance of emergency power supply systems (EPSSs) that are also applicable to COPS.

N **708.7 Cybersecurity.** COPS that are connected to a communication network and have the capability to permit control of any portion of the premises COPS shall comply with either of the following:

(1) The ability to control the system is limited to a direct connection through a local nonnetworked interface.
(2) It is connected through a networked interface complying with one of the following methods:

   a. The system and associated software are identified as being evaluated for cybersecurity.
   b. A cybersecurity assessment is conducted on the connected system to determine vulnerabilities to cyberattacks.

The cybersecurity assessment shall be conducted when the system configuration changes and at not more than 5-year intervals.

Documentation of the evaluation, assessment, and certification shall be made available to those authorized to inspect, operate, and maintain the system.

Informational Note No. 1: See ANSI/ISA 62443, Cybersecurity Standards series; UL 2900, Cybersecurity Standards series; or the NIST Framework for Improving Critical infrastructure Cybersecurity, Version 1.1, for assessment requirements.

Informational Note No. 2: Examples of the commissioning certification used to demonstrate the system has been investigated for cybersecurity vulnerabilities could be one of the following:

(1) The ISA Security Compliance Institute (ISCI) conformity assessment program
(2) Certification of compliance by a nationally recognized test laboratory
(3) Manufacturer certification for the specific type and brand of system provided

708.8 Commissioning.

(A) Commissioning Plan. A commissioning plan shall be developed and documented.

Informational Note No. 1: See NFPA 70B-2019, *Recommended Practice for Electrical Equipment Maintenance*, for further information on developing a commissioning program.

Informational Note No. 2: See 708.7 for cybersecurity assessments.

(B) Component and System Tests. The installation of the equipment shall undergo component and system tests to ensure that, when energized, the system will function properly.

(C) Baseline Test Results. A set of baseline test results shall be documented for comparison with future periodic maintenance testing to identify equipment deterioration.

(D) Functional Performance Tests. A functional performance test program shall be established, documented, and executed upon complete installation of the critical system in order to establish a baseline reference for future performance requirements.

Informational Note: See Informative Annex F, Availability and Reliability for Critical Operations Power Systems; and Development and Implementation of Functional Performance Tests (FPTs) for Critical Operations Power Systems, for more information on developing and implementing a functional performance test program.

Part II. Circuit Wiring and Equipment

708.10 Feeder and Branch Circuit Wiring.

(A) Identification.

(1) Boxes and Enclosures. In a building or at a structure where a critical operations power system and any other type of power system are present, all boxes and enclosures (including transfer switches, generators, and power panels) for critical operations power system circuits shall be permanently marked so they will be readily identified as a component of the critical operations power system.

(2) Receptacle Identification. In a building in which COPS are present with other types of power systems described in other sections in this article, the cover plates for the receptacles or the receptacles themselves supplied from the COPS shall have a distinctive color or marking so as to be readily identifiable. Nonlocking-type, 125-volt, 15- and 20-ampere receptacles supplied from the COPS shall have an illuminated face or an indicator light to indicate that there is power to the receptacle.

Exception: If the COPS supplies power to a DCOA that is a stand-alone building, receptacle cover plates or the receptacles themselves shall not be required to have distinctive marking.

Δ **(B) Wiring.** Wiring of two or more COPS circuits supplied from the same source shall be permitted in the same raceway, cable, box, or cabinet. In other than transfer equipment enclosures, wiring from a COPS source or COPS source distribution overcurrent protection to critical loads shall be kept entirely independent of all other wiring and equipment.

(C) COPS Feeder Wiring Requirements. COPS feeders shall comply with 708.10(C)(1) through (C)(3).

Δ **(1) Protection Against Physical Damage.** The wiring of the COPS system shall be protected against physical damage. Only the following wiring methods shall be permitted:

(1) Rigid metal conduit, intermediate metal conduit, or Type MI cable.

(2) Where encased in not less than 50 mm (2 in.) of concrete, any of the following wiring methods shall be permitted:

    a. Schedule 40 or Schedule 80 rigid polyvinyl chloride conduit (PVC)
    b. Reinforced thermosetting resin conduit (RTRC)
    c. Electrical metallic tubing (EMT)
    d. Flexible nonmetallic or jacketed metallic raceways
    e. Jacketed metallic cable assemblies listed for installation in concrete

(3) Where provisions must be made for flexibility at equipment connection, one or more of the following shall also be permitted:

    a. Flexible metal fittings
    b. Flexible metal conduit with listed fittings
    c. Liquidtight flexible metal conduit with listed fittings

Δ **(2) Fire Protection for Feeders.** Feeders shall meet one of the following conditions:

(1) The cable or raceway is protected by a listed electrical circuit protective system with a minimum 2-hour fire rating.

> Informational Note No. 1: See UL 1724, *Fire Tests for Electrical Circuit Protective Systems*, for one method of defining an electrical circuit protective system, by establishing a rating when tested. UL *Guide Information for Electrical Circuit Integrity Systems* (FHIT) contains information to identify the system and its installation limitations to maintain a minimum 2-hour fire resistive rating.

(2) The cable or raceway is a listed fire-resistive cable system with a minimum 2-hour fire rating.

> Informational Note No. 2: See UL 2196-2017, *Standard for Fire Test for Circuit Integrity of Fire-Resistive Power, Instrumentation, Control and Data Cables*, for testing requirements for fire-resistive cables.

> Informational Note No. 3: The listing organization provides information for fire-resistive cable systems on proper installation requirements to maintain the fire rating.

(3) The cable or raceway is protected by a listed fire-rated assembly that has a minimum fire rating of 2 hours.

Unlike the emergency system feeders covered in Article 700, COPS feeders are required to employ another fire protection technique even where located in building spaces that are fully protected by a fire suppression system.

The feeder-circuit wiring requires a minimum 2-hour fire rating provided by a listed electrical circuit protective system or a listed fire-rated assembly unless encased in 2 inches of concrete.

It is important to understand the difference between a 2-hour fire rating of an electrical circuit, such as a conduit with wires, and a 2-hour fire resistance rating of a structural member, such as a wall. At the end of a 2-hour fire test on an electrical conduit with wires, its insulation must be intact and the circuit must function electrically; no short circuits, grounds, or opens are permitted. A wall subjected to a 2-hour fire resistance test must only prevent a fire from passing through or past the wall,

without regard to damage to the wall. All fire ratings and fire resistance ratings are based on the assumption that the structural supports for the assembly are not impaired by the effects of the fire.

Listed electrical circuit protective systems are described in the UL *Guide Information for Electrical Equipment*. The four-letter codes (shown in parentheses) are the UL product category guide designations. Examples of these systems include electrical circuit protective systems (FHIT), electrical circuit protective materials (FHIY), and fire-resistive cables (FHJR). Circuit integrity cable is covered under category FHJR.

(4) The cable or raceway is encased in a minimum of 50 mm (2 in.) of concrete.

(3) Floodplain Protection. Where COPS feeders are installed below the level of the 100-year floodplain, the insulated circuit conductors shall be listed for use in a wet location and be installed in a wiring method that is permitted for use in wet locations.

(D) COPS Branch Circuit Wiring.

(1) *Outside the DCOA.* COPS branch circuits installed outside the DCOA shall comply with the physical and fire protection requirements of 708.10(C)(1) through (C)(3).

(2) *Within the DCOA.* Any of the wiring methods recognized in Chapter 3 of this *Code* shall be permitted within the DCOA.

708.11 Branch Circuit and Feeder Distribution Equipment.

(A) Branch Circuit Distribution Equipment. COPS branch circuit distribution equipment shall be located within the same DCOA as the branch circuits it supplies.

(B) Feeder Distribution Equipment. Equipment for COPS feeder circuits (including transfer equipment, transformers, and panelboards) shall comply with the following:

(1) Be located in spaces with a 2-hour fire resistance rating

(2) Be located above the 100-year floodplain

708.12 Feeders and Branch Circuits Supplied by COPS. Feeders and branch circuits supplied by the COPS shall supply only equipment specified as required for critical operations use.

708.14 Wiring of HVAC, Fire Alarm, Security, Emergency Communications, and Signaling Systems. All conductors or cables shall be installed using any of the metal wiring methods permitted by 708.10(C)(1) and, in addition, shall comply with the following, as applicable:

(1) All cables for fire alarm, security, signaling systems, and emergency communications shall be shielded twisted pair cables or installed to comply with the performance requirements of the system.

(2) Shields of cables for fire alarm, security, signaling systems, and emergency communications shall be arranged in accordance with the manufacturer's published installation instructions.

(3) Optical fiber cables shall be used for connections between two or more buildings on the property and under single management.

(4) A listed primary protector shall be provided on all communications circuits. Listed secondary protectors shall be provided at the terminals of the communications circuits.

(5) Conductors for all control circuits rated above 50 volts shall be rated not less than 600 volts.

(6) Communications, fire alarm, and signaling circuits shall use relays with contact ratings that exceed circuit voltage and current ratings in the controlled circuit.

(7) All cables for fire alarm, security, emergency communications, and signaling systems shall be riser-rated and shall be part of a listed 2-hour fire-resistive cable system or protected by a listed 2-hour electrical circuit protective system.

(8) Control, monitoring, and power wiring to HVAC systems shall be part of a listed 2-hour fire-resistive cable system or protected by a listed 2-hour electrical circuit protective system.

Part III. Power Sources and Connection

708.20 Sources of Power.

(A) General Requirements. Current supply shall be such that, in the event of failure of the normal supply to the DCOA, critical operations power shall be available within the time required for the application. The supply system for critical operations power, in addition to the normal services to the building and meeting the general requirements of this section, shall be one or more of the types of systems described in 708.20(E) through (H).

> Informational Note No. 1: Assignment of degree of reliability of the recognized critical operations power system depends on the careful evaluation in accordance with the risk assessment.
>
> Informational Note No. 2: See IEEE 3006.5–2014, *Recommended Practice for the Use of Probability Methods for Conducting a Reliability Analysis of Industrial and Commercial Power Systems,* for guidance about determining degree of reliability.

(B) Fire Protection. Where located within a building, equipment for sources of power as described in 708.20(E) through (H) shall be installed either in spaces fully protected by an approved automatic fire protection system or in spaces with a 2-hour fire rating.

(C) Grounding. All sources of power shall be grounded as a separately derived source in accordance with 250.30.

> *Exception: Where installed in accordance with 708.10(C) and 708.11(B), equipment containing the main bonding jumper or system bonding jumper for the normal source and the feeder*

wiring to the transfer equipment shall not be required to be grounded as a separately derived source.

(D) Surge Protection Devices. Surge protection devices shall be provided at all facility distribution voltage levels.

(E) Storage Battery. An automatic battery charging means shall be provided. Batteries shall be compatible with the charger for that particular installation. Automotive-type batteries shall not be used.

(F) Generator Set.

(1) Prime Mover-Driven. Generator sets driven by a prime mover shall be provided with means for automatically starting the prime mover on failure of the normal power source. A time-delay feature permitting a minimum 15-minute setting shall be provided to avoid retransfer in case of short-time reestablishment of the normal source.

(2) Power for fuel transfer pumps. Where power is needed for the operation of the fuel transfer pumps to deliver fuel to a generator set day tank, this pump shall be connected to the COPS.

(3) Dual Supplies. Prime movers shall not be solely dependent on a public utility gas system for their fuel supply or municipal water supply for their cooling systems. Means shall be provided for automatically transferring from one fuel supply to another where dual fuel supplies are used.

(4) Battery Power and Dampers. Where a storage battery is used for control or signal power or as the means of starting the prime mover, it shall be suitable for the purpose and shall be equipped with an automatic charging means independent of the generator set. Where the battery charger is required for the operation of the generator set, it shall be connected to the COPS. Where power is required for the operation of dampers used to ventilate the generator set, the dampers shall be connected to the COPS.

(5) Outdoor Generator Sets.

(a) *Permanently Installed Generators and Portable Generators Greater Than 15 kW.* Where an outdoor housed generator set is equipped with a readily accessible disconnecting means in accordance with 445.18, and the disconnecting means is located within sight of the building or structure supplied, an additional disconnecting means shall not be required where ungrounded conductors serve or pass through the building or structure. Where the generator supply conductors terminate at a disconnecting means in or on a building or structure, the disconnecting means shall meet the requirements of 225.36.

(b) *Portable Generators 15 kW or Less.* Where a portable generator, rated 15 kW or less, is installed using a flanged inlet or other cord-and-plug-type connection, a disconnecting means shall not be required where ungrounded conductors serve or pass through a building or structure.

(6) Means for Connecting Portable or Vehicle-Mounted Generator. Where the COPS is supplied by a single generator, a means to connect a portable or vehicle-mounted generator shall be provided.

(7) On-Site Fuel Supply. Where internal combustion engines are used as the prime mover, an on-site fuel supply shall be provided. The on-site fuel supply shall be secured and protected in accordance with the risk assessment.

(G) Uninterruptible Power Supplies. Uninterruptible power supplies used as the sole source of power for COPS shall comply with 708.20(E) and (F).

(H) Fuel Cell System. Installation of a fuel cell system shall meet the requirements of Parts II through VIII of Article 692.

708.21 Ventilation. Adequate ventilation shall be provided for the alternate power source for continued operation under maximum anticipated ambient temperatures.

> Informational Note: See NFPA 110-2019, *Standard for Emergency and Standby Power Systems*, and NFPA 111-2019, *Standard on Stored Electrical Energy Emergency and Standby Power Systems*, for additional information on ventilation air for combustion and cooling.

Air-cooled radiators, air intake for combustion engines, and discharge of generated heat are some factors affecting adequate ventilation. In addition, internal combustion engines require proper exhaust ventilation to remove carbon monoxide.

708.22 Capacity of Power Sources.

(A) Capacity and Rating. A COPS shall have capacity and rating for all loads to be operated simultaneously for continuous operation with variable load for an unlimited number of hours, except for required maintenance of the power source. A portable, temporary, or redundant alternate power source shall be available for use whenever the COPS power source is out of service for maintenance or repair.

(B) Selective Load Management. The alternate power source shall be permitted to supply COPS emergency, legally required standby, and optional loads where the source has adequate capacity or where load management (that includes automatic selective load pickup and load shedding) is provided as needed to ensure adequate power to (1) the COPS and emergency circuits, (2) the legally required standby circuits, and (3) the optional standby circuits, in that order of priority. The alternate power source shall be permitted to be used for peak load shaving, provided these conditions are met.

Peak load-shaving operation shall be permitted for satisfying the test requirement of 708.6(B), provided all other conditions of 708.6 are met.

(C) Duration of COPS Operation. The alternate power source shall be capable of operating the COPS for a minimum of 72 hours at full load of DCOA with a steady-state voltage within ±10 percent of nominal utilization voltage.

708.24 Transfer Equipment.

Δ **(A) General.** Transfer equipment, including automatic transfer switches, shall be automatic, listed, and identified for emergency use. Transfer equipment shall be designed and installed to prevent the inadvertent interconnection of normal and critical operations sources of supply in any operation of the transfer equipment. Transfer equipment and electric power production systems installed to permit operation in parallel with the normal source shall meet the requirements of Parts I and II of Article 705.

(B) Bypass Isolation Transfer Switches. Means shall be permitted to bypass and isolate the transfer equipment. If bypass isolation transfer switches are used, inadvertent parallel operation shall be avoided.

(C) Automatic Transfer Switches. If used with sources that are not inherently synchronized, automatic transfer switches shall comply with the following:

(1) Automatic transfer switches shall be listed for emergency use.
(2) Automatic transfer switches shall be electrically operated and mechanically held.

(D) Redundant Transfer Equipment. If COPS loads are supplied by a single feeder, the COPS shall include redundant transfer equipment or a bypass isolation transfer switch to facilitate maintenance as required in 708.6(C) without jeopardizing continuity of power. If the redundant transfer equipment or bypass isolation transfer switch is manual (or nonautomatic), then it shall be actively supervised by a qualified person when the primary (automatic) transfer equipment is disabled for maintenance or repair.

(E) Use. Transfer equipment shall supply only COPS loads.

(F) Documentation. The short-circuit current rating of the transfer equipment, based on the specific overcurrent protective device type and settings protecting the transfer equipment, shall be field marked on the exterior of the transfer equipment.

708.30 Branch Circuits Supplied by COPS. Branch circuits supplied by the COPS shall only supply equipment specified as required for critical operations use.

Part IV. Overcurrent Protection

708.50 Accessibility. The feeder- and branch-circuit overcurrent devices shall be accessible to authorized persons only.

708.52 Ground-Fault Protection of Equipment.

(A) Applicability. The requirements of 708.52 shall apply to critical operations (including multiple occupancy buildings) with critical operation areas.

(B) Feeders. Where ground-fault protection is provided for operation of the service disconnecting means or feeder disconnecting means as specified by 230.95 or 215.10, an additional step of ground-fault protection shall be provided in all next level feeder disconnecting means downstream toward the load. Such protection shall consist of overcurrent devices and current transformers or other equivalent protective equipment that causes the feeder disconnecting means to open.

(C) Testing. When equipment ground-fault protection is first installed, each level shall be tested to ensure that ground-fault protection is operational.

> Informational Note: Testing is intended to verify the ground-fault function is operational. The performance test is not intended to verify selectivity in 708.52(D), as this is often coordinated similarly to circuit breakers by reviewing time and current curves and properly setting the equipment. (Selectivity of fuses and circuit breakers is not performance tested for overload and short circuit.)

(D) Selectivity. Ground-fault protection for operation of the service and feeder disconnecting means shall be fully selective such that the feeder device, but not the service device, shall open on ground faults on the load side of the feeder device. Separation of ground-fault protection time-current characteristics shall conform to the manufacturer's recommendations and shall consider all required tolerances and disconnect operating time to achieve 100 percent selectivity.

> Informational Note: See 230.95, Informational Note No. 4, for transfer of alternate source where ground-fault protection is applied.

Δ **708.54 Selective Coordination.**

N **(A) General.** Critical operations power system(s) overcurrent protective devices (OCPDs) shall be selectively coordinated with all supply-side and load-side OCPDs.

Selective coordination shall be selected by a licensed professional engineer or other qualified persons engaged primarily in the design, installation, or maintenance of electrical systems. The selection shall be documented and made available to those authorized to design, install, inspect, maintain, and operate the system.

N **(B) Replacements.** Where critical operations power system(s) OCPDs are replaced, they shall be reevaluated to ensure selective coordination is maintained with all supply-side and load-side OCPDs.

N **(C) Modifications.** If modifications, additions, or deletions to the critical operations power system(s) occur, selective coordination of the critical operations power system(s) OCPDs with all supply-side and load-side OCPDs shall be reevaluated.

> *Exception: Selective coordination shall not be required between two overcurrent devices located in series if no loads are connected in parallel with the downstream device.*

> Informational Note: See Informational Note Figure 708.54(C) for an example of how critical operations power system OCPDs selectively coordinate with all supply-side OCPDs.
> OCPD D selectively coordinates with OCPDs C, F, E, B, and A.
> OCPD C selectively coordinates with OCPDs F, E, B, and A.
> OCPD F selectively coordinates with OCPD E.

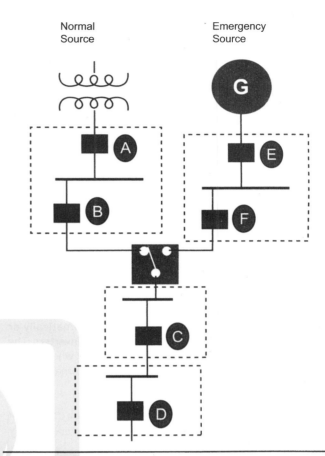

Normal Source *Emergency Source*

INFORMATIONAL NOTE FIGURE 708.54(C) *Critical Operations Power System Selective Coordination.*

OCPD B is not required to selectively coordinate with OCPD A because OCPD B is not a critical operations power system OCPD.

Part V. System Performance and Analysis

Δ **708.64 Emergency Operations Plan.** A facility with a COPS shall have a documented emergency operations plan. The plan shall consider emergency operations and response, recovery, and continuity of operations.

> Informational Note: See *NFPA 1600*-2019, *Standard on Continuity, Emergency, and Crisis Management*, Section 5.7, which provides guidance for the development and implementation of emergency plans.

ARTICLE
710 Stand-Alone Systems

Δ **710.1 Scope.** This article covers electric power production systems that operate in island mode not connected to an electric utility or other electric power production and distribution network.

Informational Note: These systems operate independently from an electric utility and include isolated microgrid systems. Stand-alone systems often include a single or a compatible interconnection of sources such as engine generators, solar PV, wind, ESS, or batteries.

This article addresses the operating parameters for electric power production sources in stand-alone mode. If a stand-alone system is interconnected with the alternating-current (ac) system, the requirements of Article 705 apply. Power production sources such as a generator, photovoltaic (PV) system, fuel cell, and wind electric system supplying a sign, lights, irrigation system, or remote facilities such as a cabin are a few examples of stand-alone systems. These systems can include energy storage or backup power supplies.

Article 710, which covers stand-alone systems, and Article 702, which covers optional standby systems, are not mutually exclusive. In respect to how the two articles interact, a stand-alone system operating in island mode (such as a generator or a solar PV system) serves as the alternate electric power production source for the optional standby system loads covered by Article 702. The connection of the stand-alone system to the optional standby system loads can be done automatically or manually using transfer equipment or multimode inverters. The capacity of the stand-alone source is affected by whether the optional standby system loads are connected automatically or manually as specified in 702.4(A)(1) and (2).

Δ **710.6 Equipment Approval.** All power production equipment or systems shall be approved for use in island mode and comply with one of the following:

(1) Be listed
(2) Be evaluated for the application and have a field label applied

Δ **710.10 Identification of Power Sources.** A permanent plaque, label, or directory shall be installed at a building supplied by a stand-alone system at the power source disconnecting means location, or at an approved readily visible location. The plaque, label, or directory shall denote the location of each power source disconnecting means for the building or be grouped with other plaques or directories for other on-site sources. Where multiple sources supply the building, markings shall comply with 705.10.

710.12 Stand-Alone Inverter Input Circuit Current. The maximum current shall be the stand-alone continuous inverter input current rating when the inverter is producing rated power at the lowest input voltage.

710.15 General. Premises wiring systems shall be adequate to meet the requirements of this *Code* for similar installations supplied by a feeder or service. The wiring on the supply side of the building or structure disconnecting means shall comply with the requirements of this *Code*, except as modified by 710.15(A) through (G).

(A) Supply Output. Power supply to premises wiring systems fed by stand-alone or isolated microgrid power sources shall be

permitted to have less capacity than the calculated load. The capacity of the sum of all sources of the stand-alone supply shall be equal to or greater than the load posed by the largest single utilization equipment connected to the system. Calculated general lighting loads shall not be considered as a single load.

Informational Note: For general-use loads the system capacity can be calculated using the sum of the capacity of the firm sources, such as generators and ESS inverters. For specialty loads intended to be powered directly from a variable source, the capacity can be calculated using the sum of the variable sources, such as PV or wind inverters, or the combined capacity of both firm and variable sources.

Even though a stand-alone installation may have disconnecting means rated at 100 or 200 amperes at 120/240 volts, the PV source is not required to provide either the full current rating or the dual voltages of the service equipment. A PV installation usually is designed so that the actual ac demands on the system are sized to the output rating of the PV system. The inverter output is required to have sufficient capacity to power the largest single piece of utilization equipment to be supplied by the PV system, but the inverter output does not have to be rated for potential multiple loads to be simultaneously connected to it.

(B) Sizing and Protection. The circuit conductors between a stand-alone source and a building or structure disconnecting means shall be sized based on the sum of the output ratings of the stand-alone source(s). For three-phase interconnections, the phase loads shall be controlled or balanced to be compatible with specifications of the sum of the power supply capacities.

Δ **(C) Single 120-Volt Supply.** Stand-alone and isolated microgrid systems shall be permitted to supply 120 volts to single-phase, 3-wire, 120/240-volt service equipment or distribution panels where there are no 240-volt outlets and where there are no multiwire branch circuits. In all installations, the sum of the ratings of the power sources shall be less than the rating of the neutral bus in the service equipment. This equipment shall be marked with the following words or equivalent:

WARNING:
SINGLE 120-VOLT SUPPLY.
DO NOT CONNECT MULTIWIRE
BRANCH CIRCUITS!

The warning sign(s) or label(s) shall comply with 110.21(B).

If multiwire branch circuits are connected to a normal 120/240-volt ac service, the currents in the neutral conductors subtract or are at most no larger than the rating of the branch-circuit overcurrent device. If the electrical system consists of a single 120-volt electrical system supplying the two buses in the panelboard, the currents in the grounded conductor for each multiwire branch circuit add rather than subtract. Because the two buses are in phase, there is no neutral conductor. The currents in these conductors can be as high as twice the rating of the branch-circuit overcurrent device, and overloading is possible.

(D) Three-phase Supply. Stand-alone and microgrid systems shall be permitted to supply three-phase, 3-wire or 4-wire systems.

(E) Energy Storage or Backup Power System Requirements. Energy storage or backup power supplies shall not be required.

(F) Voltage and Frequency Control. The stand-alone power sources shall be controlled during operation so that voltage and frequency are supplied within limits compatible with the connected loads.

ARTICLE 722 Cables for Power-Limited Circuits and Fault-Managed Power Circuits

N Part I. General

N 722.1 Scope. This article covers the general requirements for the installation of single- and multiple-conductor cables used in Class 2 and Class 3 power-limited circuits, power-limited fire alarm (PLFA) circuits, and Class 4 fault-managed power circuits.

To remove redundancy throughout most of Chapter 7, Code-Making Panel No. 3 (CMP-3) consolidated all general cable requirements into a new Article 722. The intent of this consolidation is to help with clarity and usability of the *NEC®* while making room for future technologies as they evolve.

N 722.3 Other Articles. In addition to the requirements of this article, installation of cables shall comply with the articles or sections listed in 722.3(A) through (O). Only those sections of Article 300 referenced in this article shall apply.

N (A) Installation of Cables and Conductors in Raceway. The number and size of conductors and cables, as well as raceway sizing, shall comply with 300.17.

N (B) Spread of Fire or Products of Combustion. Installation of power-limited circuits shall comply with 300.21.

N (C) Ducts, Plenums, and Other Air-Handling Spaces. Power-limited circuits installed in ducts, plenums, or other space used for environmental air shall comply with 300.22.

Exception No. 1: Cables selected in accordance with Table 722.135(B) and installed in accordance with 300.22(B), Exception shall be permitted to be installed in ducts specifically fabricated for environmental air.

Exception No. 2: Cables selected in accordance with Table 722.135(B) shall be permitted to be installed in other spaces used for environmental air (plenums).

N (D) Cables in Ducts for Dust, Loose Stock, or Vapor Removal. Section 300.22(A) for wiring systems shall apply.

Exception: Nonconductive optical fiber cables shall be permitted in ducts used for dust, loose stock, or vapor removal.

N (E) Cable Trays. Cable tray installations shall comply with Parts I and II of Article 392.

N (F) Instrumentation Tray Cable. Circuits wired using instrumentation tray cable shall comply with 335.1 and 335.4 through 335.9.

N (G) Raceways or Sleeves Exposed to Different Temperatures. Section 300.7(A) shall apply.

Condensation often forms in raceways that run between spaces of differing temperatures, such as a walk-in cooler or freezer. Section 722.3(G) brings the requirements of 300.7(A) into Article 722.

> **See also**
> **300.7(A)** and its associated commentary

N (H) Vertical Support for Fire-Resistive Cables and Conductors. Vertical installations of circuit integrity (CI) cables and conductors installed in a raceway or conductors and cables of electrical circuit protective systems and fire resistive-cable systems shall be installed in accordance with 300.19.

N (I) Installation of Cables with Other Systems. Section 300.8 shall apply.

N (J) Corrosive, Damp, or Wet Locations. The installation of power-limited cables shall comply with the applicable requirements in 110.11, 300.5(B), 300.6, 300.9, and 310.10(F) when installed in corrosive, damp, or wet locations.

N (K) Cable Routing Assemblies. Cables installed in cable routing assemblies shall be selected in accordance with Table 800.154(c), listed in accordance with 800.182, and installed in accordance with 800.110(C)(1)), 800.110(C)(2), and 800.113.

N (L) Communications Raceways. Cables communications raceways shall be selected in accordance with Table 800.154(b), listed in accordance with 800.182, and installed in accordance with 800.113 and 362.24 through 362.56, where the requirements applicable to electrical nonmetallic tubing (ENT) apply.

N (M) Temperature Limitation of Cables. The requirements of 310.14(A)(3) on the temperature limitation of conductors shall apply to power-limited circuit cables and fault-managed power cables.

N (N) Identification of Equipment Grounding Conductors. Equipment grounding conductors shall be identified in accordance with 250.119.

Exception: Cables that do not contain an equipment grounding conductor shall be permitted to use a conductor with green insulation, or green insulation with one or more yellow stripes, for other than equipment grounding purposes.

N (O) Specific Requirements. As appropriate, the installation of wires and cables shall also comply with the following:

(1) Class 2 and Class 3 cables — Part II of Article 725

(2) Class 4 cables — Part IV of Article 726

(3) Fire alarm cables — Part III of Article 760

(4) Optical fiber cables — Part V of Article 770

N **722.10 Hazardous (Classified) Locations.** Class 4 cables shall be permitted to be used in hazardous (classified) locations where specifically permitted by other articles of this *Code*.

N **722.12 Uses Not Permitted.** Class 4 cables shall not be permitted for any applications that are not part of a Class 4 system.

Exception: Use of Class 4 cable for other applications shall be permitted if the cable has been listed as suitable for the other applications.

N **722.21 Access to Electrical Equipment Behind Panels Designed to Allow Access.** Access to electrical equipment shall not be denied by an accumulation of cables that prevents removal of panels, including suspended ceiling panels.

Excess accumulation of wires and cables, often due to improper installation, can limit access to electrical equipment by preventing the removal of access panels and ceiling tiles. To safely service, rearrange, or install electrical equipment, the worker must have an accessible work space. See Exhibit 722.1.

See also

300.11(B), which permits the use of support wires and approved fittings that are independent of the suspended ceiling support wires

N **722.24 Mechanical Execution of Work.**

N **(A) General.** Cables shall be installed in a neat and workmanlike manner. Cables installed exposed on the surface of ceilings and sidewalls shall be supported by the building structure in such a manner that the cable will not be damaged by normal building use. Such cables shall be secured by hardware, including straps, staples, hangers, listed cable ties identified for securement and support, or similar fittings, designed and installed so as not to damage the cable. The installation shall conform to 300.4 and 300.11.

A bushing shall be installed where cables emerge from raceway used for mechanical support or protection in accordance with 300.15(C).

Nonmetallic cable ties and other nonmetallic cable accessories used to secure and support cables in other spaces used for environmental air (plenums) shall be listed as having low smoke and heat release properties in accordance with 300.22(C).

Informational Note No. 1: See NFPA 90A-2021, *Standard for the Installation of Air-Conditioning and Ventilating Systems*, for discrete combustible components.

Informational Note No. 2: Paint, plaster, cleaners, abrasives, corrosive residues, or other contaminants could result in an undetermined alteration of cable properties.

Cable must be attached to or supported by the building structure by cable ties, straps, clamps, hangers, and so forth.

The installation method must not damage the cable. In addition, the location of the cable should be carefully evaluated to ensure that activities and processes within the building do not cause damage to the cable. See Exhibit 722.1.

Section 300.4(D) requires protection of cables that are installed on or in framing members. Such cables are required to be installed in a manner that protects them from nail or screw penetration. This section permits attachment to baseboards and non-load-bearing walls, which are not structural components.

N **(B) Support of Cables.** Cables shall not be strapped, taped, or attached by any means to the exterior of any conduit or other raceway as a means of support.

Exception No. 1: Class 2 circuit conductors or cables shall be permitted to be installed as permitted by 300.11(C)(2).

Exception No. 2: Overhead (aerial) spans of optical fiber cables shall be permitted to be attached to the exterior of a raceway-type mast intended for the attachment and support of such cables.

N **(C) Circuit Integrity (CI) Cable.** Circuit integrity (CI) cable shall be supported at a distance not exceeding 610 mm (24 in.). Cable shall be secured to the noncombustible surface of the building structure. Cable supports and fasteners shall be steel.

N **722.25 Abandoned Cables.** The accessible portion of abandoned cables shall be removed. Where cables are identified

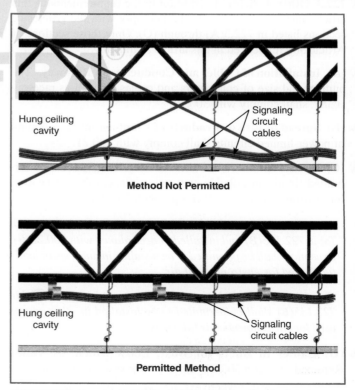

EXHIBIT 722.1 Incorrect cable installation (*upper diagram*) and correct method (*lower diagram*).

for future use with a tag, the tag shall be of sufficient durability to withstand the environment involved.

N **722.31 Safety-Control Equipment.** Where damage to power-limited circuits can result in a failure of safety-control equipment that would introduce a direct fire or life hazard, the power limited circuits shall be installed using Class 1 circuit wiring methods in accordance with 724.46. All conductors of such circuits shall be installed in rigid metal conduit, intermediate metal conduit, rigid nonmetallic conduit, electrical metallic tubing, Type MI cable, or Type MC cable, or be otherwise suitably protected from physical damage.

N **722.135 Installation of Cables.** The installation of cables shall comply with 722.135(A) through (I), as applicable.

N **(A) Listing.** Cables installed in buildings shall be listed.

N **(B) Cables in Buildings.** The installation of cables shall comply with Table 722.135(B).

> Informational Note No. 1: See NFPA 90A-2021, *Standard for the Installation of Air-Conditioning and Ventilating Systems*, 4.3.4 and 4.3.11.3.3, for information on fire protection of wiring installed in ducts specifically fabricated for environmental air and other spaces used for environmental air (plenums).
>
> Informational Note No. 2: See 300.21 for firestop requirements for floor penetrations.
>
> Informational Note No. 3: See Chapter 3 for the installation requirements for PLTC cables installed outdoors in cable trays.
>
> Informational Note No. 4: See UL 2024, *Cable Routing Assemblies and Communications Raceways*, for applicable requirements for plenum, riser, and general-purpose cable routing assemblies and raceways.

N **(C) Industrial Establishments.** In industrial establishments where the conditions of maintenance and supervision ensure that only qualified persons service the installation, Type PLTC cable shall be permitted in accordance with either of the following:

(1) Where the cable is not subject to physical damage, Type PLTC cable that complies with the crush and impact requirements of Type MC cable and is identified as Type PLTC-ER for such use shall be permitted to be exposed between the cable tray and the utilization equipment or device. The cable shall be continuously supported and protected against physical damage using mechanical protection such as dedicated struts, angles, or channels. The cable shall be supported and secured at intervals not exceeding 1.8 m (6 ft). Where not subject to physical damage, Type PLTC-ER cable shall be permitted to transition between cable trays and between cable trays and utilization equipment or devices for a distance not to exceed 1.8 m (6 ft) without continuous support. The cable shall be mechanically supported where exiting the cable tray to ensure that the minimum bending radius is not exceeded.

(2) Type PLTC cable, with a metallic sheath or armor in accordance with 722.179(A)(6), shall be permitted to be installed exposed. The cable shall be continuously supported and protected against physical damage using mechanical protection such as dedicated struts, angles, or channels. The cable shall be secured at intervals not exceeding 1.8 m (6 ft).

N **(D) In Hoistways.** In hoistways, cables shall be installed in rigid metal conduit, rigid nonmetallic conduit, intermediate metal conduit, liquidtight flexible nonmetallic conduit, or electrical metallic tubing. For elevators or similar equipment, these conductors shall be permitted to be installed as provided in 620.21.

N **(E) Cable Substitutions.** The substitutions for cables listed in Table 722.135(E) shall be permitted. Where substitute cables are installed, the installation requirements of the articles described in 722.3(O) shall also apply. CI cables shall be permitted to be installed to provide 2-hour circuit integrity. See 722.135(F).

> Informational Note: See 800.179 for information on Types CMP, CMR, CM, and CMX.

N **(F) Circuit Integrity (CI) Cable, Fire-Resistive Cable System, or Electrical Circuit Protective System.** CI cable, a fire-resistive cable system, or a listed electrical circuit protective system shall be permitted for use in systems that supply critical circuits to ensure survivability for continued circuit operation for a specified time under fire conditions.

N **(G) Thermocouple Circuits.** Conductors in Type PLTC cables used for Class 2 thermocouple circuits shall be permitted to be any of the materials used for thermocouple extension wire.

N **(H) Bundling of 4-Pair Cables Transmitting Power and Data.** Where 4-pair cables are used to transmit power and data to a powered device, 725.144 shall apply.

N **(I) Installation of Circuit Conductors Extending Beyond One Building.** Circuit conductors that extend beyond one building and are run such that they are subject to accidental contact with electric light or power conductors operating over 300 volts to ground, or are exposed to lightning on interbuilding circuits on the same premises, shall comply with the following:

(1) For other than coaxial conductors, 800.44, 800.53, 800.100, 805.50, 805.93, 805.170(A), and 805.170(B)

(2) For coaxial conductors, 800.44, 820.93, and 820.100

(3) The installation requirements of Part I of Article 300

N **Part II. Listing Requirements**

N **722.179 Listing and Marking of Cables.** Cables installed in buildings shall be listed in accordance with 722.179(A) and marked in accordance with 722.179(B), and they shall be permitted to be marked in accordance with 722.179(C).

N **TABLE 722.135(B)** *Installation of Listed Cables in Buildings*

Applications		Cable Type[1]					
		Plenum	Riser	General-Purpose	Limited-Use	Under Carpet	PLTC
In ducts specifically fabricated for environmental air as described in 300.22(B)[2]	Cables in lengths as short as practicable to perform the required function	Y	N	N	N	N	N
	In metal raceway that complies with 300.22(B)	Y	Y	Y	Y	N	Y
In other spaces used for environmental air (plenums) as described in 300.22(C)	Cables in other spaces used for environmental air	Y	N	N	N	N	N
	Cables in metal raceway that complies with 300.22(C)	Y	Y	Y	Y	N	Y
	Cables in plenum communications raceways	Y	N	N	N	N	N
	Cables in plenum cable routing assemblies	Y	N	N	N	N	N
	Cables supported by open metal cable trays	Y	N	N	N	N	N
	Cables or cables installed in raceways or cable routing assemblies supported by solid bottom metal cable trays with solid metal covers	Y	Y	Y	Y	N	Y
In risers and vertical runs	Cables in vertical runs penetrating one or more floors and in vertical runs in a shaft	Y	Y	N	N	N	N
	Cables in metal raceways	Y	Y	Y	Y	N	Y
	Cables in fireproof shafts	Y	Y	Y	N	N	Y
	Cables in plenum communications raceways	Y	Y	N	N	N	N
	Cables in plenum cable routing assemblies	Y	Y	N	N	N	N
	Cables in riser communications raceways	Y	Y	N	N	N	N
	Cables in riser cable routing assemblies	Y	Y	N	N	N	N
	Cables in one- and two-family dwellings	Y	Y	Y	Y[3]	N	Y
Cables and innerducts installed in metal raceways in a riser having firestops at each floor[2]	Cables	Y	Y	Y	Y	N	Y
	Cables in plenum communications raceways (innerduct)	Y	Y	Y	Y	N	Y
	Cables in riser communications raceways (innerduct)	Y	Y	Y	Y	N	Y
	Cables in general-purpose communications raceways (innerduct)	Y	Y	Y	Y	N	Y
In fireproof riser shafts having firestops at each floor[2]	Cables	Y	Y	Y	N	N	Y
	Cables in plenum communications raceways or plenum cable routing assemblies	Y	Y	Y	N	N	Y
	Cables in riser communications raceways or riser cable routing assemblies	Y	Y	Y	N	N	Y
	Cables in general-purpose communications raceways or general-purpose cable routing assemblies	Y	Y	Y	N	N	Y
In cable trays	Outdoors	N	N	N	N	N	Y
	Cables, or cables in plenum, riser, or general-purpose communications raceways, installed indoors	Y	Y	Y	N	N	Y
In cross-connect arrays	Cables, and cables in plenum, riser, or general-purpose communications raceways or cable routing assemblies	Y	Y	Y	N	N	Y
In one-, two-, and multifamily dwellings, and in building locations other than the locations covered above	Cables	Y	Y	Y	Y[3]	N	Y
	Cables in plenum, riser, or general-purpose communications raceways or cable routing assemblies, or raceways recognized in Chapter 3	Y	Y	Y	Y	N	Y
	Cables in nonconcealed spaces	Y	Y	Y	Y[4]	Y	Y
	Under carpet, floor covering, modular flooring, and planks	N	N	N	N	Y	N

[1]"N" indicates that the cable type shall not be installed in the application. "Y" indicates that the cable type shall be permitted to be installed in the application, subject to any limitations described in this article or the articles described in 722.3(O).

[2]In 300.22(B), cables shall be permitted in ducts specifically fabricated for environmental air only if directly associated with the air distribution system.

[3]Limited-use cable shall be permitted to be installed only in one-, two-, and multifamily dwellings and only if the cable is smaller in diameter than 6.35 mm (0.25 in.).

[4]The exposed length of cable shall not exceed 3.05 m (10 ft).

N **TABLE 722.135(E)** *Cable Substitutions*

Cable Type	Permitted Substitutions
CL3P	CMP
CL2P	CMP, CL3P
CL3R	CMP, CL3P, CMR
CL2R	CMP, CL3P, CL2P, CMR, CL3R
PLTC	None
CL3	CMP, CL3P, CMR, CL3R, CMG, CM, PLTC
CL2	CMP, CL3P, CL2P, CMR, CL3R, CL2R, CMG, CM, PLTC, CL3
CL3X	CMP, CL3P, CMR, CL3R, CMG, CM, PLTC, CL3, CMX
CL2X	CMP, CL3P, CL2P, CMR, CL3R, CL2R, CMG, CM, PLTC, CL3, CL2, CMX, CL3X
FPLP	CMP
FPLR	CMP, FPLP, CMR
FPL	CMP, FPLP, CMR, FPLR, CMG, CM
OFNP	None
OFCP	OFNP
OFNR	OFNP
OFCR	OFNP, OFCP, OFNR
OFNG, OFN	OFNP, OFNR
OFCG, OFC	OFNP, OFCP, OFNR, OFCR, OFNG, OFN
CMUC	None

Exception: Optical fiber cables that are installed in compliance with 770.48 shall not be required to be listed.

N **(A) Listing of Cables.** Cables installed as wiring methods within buildings shall be listed as resistant to the spread of fire and other criteria in accordance with 722.179(A)(1) through (A)(16).

Informational Note No. 1: See UL 13, *Standard for Power-Limited Circuit Cables*, for applicable requirements for listing of Class 2 and Class 3 cable and power-limited tray cable (PLTC).

Informational Note No. 2: See UL 1424, *Cables for Power-Limited Fire-Alarm Circuits*, for applicable requirements for listing of power-limited fire alarm cable.

Informational Note No. 3: See UL 1651, *Optical Fiber Cable*, for applicable requirements for listing of optical fiber cable.

Informational Note No. 4: See UL 1400-2, *Outline for Fault-Managed Power Systems — Part 2: Requirements for Class 4 Cables*, for applicable requirements for listing of Class 4 cable.

N **(1) Plenum Cable.** Plenum cable shall be listed as suitable for use in ducts, plenums, and other space for environmental air and shall be listed as having adequate fire-resistant and low-smoke producing characteristics. Refer to Table 722.179(B) for plenum cable types.

Informational Note: See NFPA 262-2019, *Standard Method of Test for Flame Travel and Smoke of Wires and Cables for Use in Air-Handling Spaces*, for the test method used to determine that a cable is low-smoke producing and fire resistant, exhibiting a maximum peak optical density of 0.50 or less, an average optical density of 0.15 or less, and a maximum flame spread distance of 1.52 m (5 ft) or less.

NFPA 262, *Standard Method of Test for Flame Travel and Smoke of Wires and Cables for Use in Air-Handling Spaces*, is a test method for electrical wires and cables that are to be installed without raceways in plenums and other spaces used for environmental air. NFPA 262 was originally developed as an adaptation of the Steiner Tunnel test (ASTM E84/UL 723, *Standard Test Method for Surface Burning Characteristics of Building Materials*).

NFPA 262 does not list pass/fail criteria. The criteria for acceptance of a given application are given in the appropriate sections of the *NEC*. In addition to the informational note in this section, see the informational notes that follow 722.179(A)(1), 770.179(A), and 800.182(A).

A Class 2 or Class 3 cable that has passed the requirements of this test can be used in ducts, plenums, or other air-handling spaces. In addition, such cable can be used anywhere in a building where Class 2 or Class 3 cable is permitted. [See Table 722.135(E).]

N **(2) Riser Cable.** Riser cable shall be listed as suitable for use in a vertical run in a shaft or from floor to floor and shall be listed as having fire-resistant characteristics capable of preventing the carrying of fire from floor to floor.

Informational Note: See ANSI/UL 1666-2012, *Test for Flame Propagation Height of Electrical and Optical-Fiber Cable Installed Vertically in Shafts*, for the cable requirements defining fire-resistant characteristics capable of preventing the carrying of fire from floor to floor.

In the fire test covered in UL 1666, *Test for Flame Propagation Height of Electrical and Optical-Fiber Cables Installed Vertically in Shafts*, cables are arranged in a simulated vertical shaft and subjected to an ignition source. The shaft is a 19-foot-high concrete shaft divided into two compartments by a 1-foot by 2-foot opening at the 12-foot level. To pass, cables must not propagate flame to the top of the 12-foot-high compartment during the 30-minute test.

N **(3) General-Purpose Cable.** General-purpose cable shall be listed as resistant to the spread of fire and as suitable for general-purpose use, except for use in risers, ducts, plenums, and other space used for environmental air.

Informational Note: See UL 2556, *Wire and Cable Test Methods*, for defining resistant to the spread of fire. One method is to demonstrate that the cables do not spread fire to the top of the tray in the UL Flame Exposure, Vertical Tray Flame Test. The smoke measurements in the test method are not applicable.

A method of defining resistant to the spread of fire is for the damage (char length) not to exceed 1.5 m (4 ft 11 in.) when performing the FT4 Vertical Flame Test.

N **(4) Alternative General-Purpose Cable.** Alternative general-purpose optical fiber cable shall be listed as suitable for general-purpose use, with the exception of risers and plenums, and shall also be resistant to the spread of fire.

Informational Note: See CSA C22.2 No. 0.3-M-2001, *Test Methods for Electrical Wires and Cables*, for the CSA vertical flame test — cables in cable trays, that can also be used to define resistance to the spread of fire when the damage (char length) does not exceed 1.5 m (4 ft 11 in.).

N **(5) Limited-Use Cable.** Limited-use cable shall be listed as suitable for use in dwellings and raceways and shall be listed as resistant to flame spread.

> Informational Note: See ANSI/UL 2556, *Standard for Wire and Cable Test Methods,* for one method of determining that cable is resistant to flame spread by testing the cable to the FV-2/VW-1 test.

N **(6) Type PLTC.** Type PLTC nonmetallic-sheathed, power-limited tray cable shall be listed as being suitable for cable trays, resistant to the spread of fire, and sunlight- and moisture-resistant. Type PLTC cable used in a wet location shall be listed for use in wet locations and marked "wet" or "wet location."

> Informational Note: See ANSI/UL 1685-2010, *Standard for Safety for Vertical-Tray Fire-Propagation and Smoke-Release Test for Electrical and Optical-Fiber Cables,* for the UL flame exposure, vertical tray flame test that is used to determine resistance to the spread of fire when cables do not spread fire to the top of the tray. The smoke measurements in the test method are not applicable.
>
> See CSA C22.2 No. 0.3-M-2001, *Test Methods for Electrical Wires and Cables,* for the CSA vertical flame test — cables in cable trays that can also be used to define resistance to the spread of fire when the damage (char length) does not exceed 1.5 m (4 ft 11 in.).

N **(7) Circuit Integrity (CI) Cable, Fire-Resistive Cable System, or Electrical Circuit Protective System.** Cables that are used for survivability of critical circuits under fire conditions shall comply with either 722.179(A)(7)(a), (A)(7)(b), or (A)(7)(c).

Section 722.179(A)(7) permits the use of circuit integrity (CI) cable for applications where continuity of the operations of critical circuits is needed during a fire. Such circuits could be essential to fire-fighting operations or could be circuits whose interruption could cause a more dangerous condition to occur. A smoke removal system is an example of where it could be necessary to use CI cables for control circuits to ensure that the dampers operate during a fire.

> Informational Note: See *NFPA 72, National Fire Alarm and Signaling Code,* 12.4.3 and 12.4.4, for additional information on fire alarm CI cable, fire-resistive cable systems, or electrical circuit protective systems used for fire alarm circuits to comply with the survivability requirements to maintain the circuit's electrical function during fire conditions for a defined period of time.

(a) *CI Cables.* CI cables of the types specified in 722.179(A)(1), (A)(2), (A)(3), (A)(4), and (A)(6) and used for survivability of critical circuits shall be marked with the additional classification using the suffix "CI." To maintain its listed fire-resistive rating, CI cable shall only be installed in free air in accordance with 722.24(C). CI cables shall only be permitted to be installed in a raceway where specifically listed and marked as part of a fire-resistive cable system as covered in 722.179(A)(7)(b).

> Informational Note: See UL 2196, *Fire Test for Circuit Integrity of Fire-Resistive Power, Instrumentation, Control and Data Cables,* and UL 1425, *Cables for Non–Power-Limited Fire-Alarm Circuits,* for information on establishing a rating for CI cable. The *UL Guide Information for Nonpower-limited Fire Alarm Circuits* (HNHT) contains information to identify the cable and its installation limitations to maintain the fire-resistive rating.

(b) *Fire-Resistive Cables.* Fire-resistive cables of the types specified in 722.179(A)(1), (A)(2), (A)(3), (A)(4), (A)(6), and (A)(7)(a) that are part of a fire-resistive cable system shall be identified with the system identifier and hourly rating marked on the protectant or the smallest unit container and installed in accordance with the listing of the system.

> Informational Note: See UL 2196, *Fire Test for Circuit Integrity of Fire-Resistive Power, Instrumentation, Control and Data Cables,* for information on establishing a rating for a fire-resistive cable system. The *UL Guide Information for Electrical Circuit Integrity Systems* (FHIT) contains information to identify the system and its installation limitations to maintain a minimum fire-resistive rating.

(c) *Electrical Circuit Protective System.* Protectants for cables of the types specified in 722.179(A)(1), (A)(2), (A)(3), (A)(4), and (A)(6) that are part of an electrical circuit protective system shall be identified with the protective system identifier and hourly rating marked on the protectant or the smallest unit container and installed in accordance with the listing of the protective system.

> Informational Note: See UL 1724, *Fire Tests for Electrical Circuit Protective Systems,* for information on establishing a rating for an electrical circuit protective system. The *UL Guide Information for Electrical Circuit Integrity Systems* (FHIT) contains information to identify the system and its installation limitations to maintain the fire-resistive rating.

N **(8) Class 3 Single Conductors.** Class 3 single conductors used as other wiring within buildings shall be listed Type CL3 and shall not be smaller than 18 AWG.

> Informational Note: See ANSI/UL 1685-2010, *Standard for Safety for Vertical-Tray Fire-Propagation and Smoke-Release Test for Electrical and Optical-Fiber Cables,* for the UL flame exposure, vertical tray flame test that is used to determine resistance to the spread of fire when cables do not spread fire to the top of the tray. The smoke measurements in the test method are not applicable.
>
> See CSA C22.2 No. 0.3-M-2001, *Test Methods for Electrical Wires and Cables,* for the CSA vertical flame test — cables in cable trays that can also be used to define resistance to the spread of fire when the damage (char length) does not exceed 1.5 m (4 ft 11 in.).

N **(9) Limited Power (LP) Cable.** Class 2 and Class 3 LP cables shall be listed as suitable for carrying power and data up to a specified current limit for each conductor without exceeding the temperature rating of the cable. The cables shall be marked with the suffix "-LP (XXA)" where XXA designates the current limit in amperes per conductor.

> Informational Note: An example of the marking on 23 AWG, 4-pair, Class 2 cable rated 75°C with an LP current rating of 0.6 amperes per conductor is "CL2-LP (0.6A) 75°C 23 AWG 4-pair."

N (10) Undercarpet Cables. Undercarpet cable shall be listed as suitable for use under carpet, floor covering, modular tiles, and planks.

> Informational Note: See UL 444, *Standard for Safety for Communications Cables*, for the compressive loading test used to determine the suitability of cable for undercarpet use.

N (11) Wet Locations. Cable used in a wet location shall be listed for use in wet locations and be marked "wet" or "wet location" or have a moisture-impervious metal sheath.

N (12) Field-Assembled Optical Fiber Cables. Field-assembled optical fiber cable shall comply with 722.179(A)(12)(a) through (d).

    (a) The specific combination of jacket and optical fibers intended to be installed as a field-assembled optical fiber cable shall be one of the types in 722.179(A)(1), (A)(2), or (A)(3) and shall be marked in accordance with Table 179(B).

    (b) The jacket of a field-assembled optical fiber cable shall have a surface marking indicating the specific optical fibers with which it is identified for use.

    (c) The optical fibers shall have a permanent marking, such as a marker tape, indicating the jacket with which they are identified for use.

    (d) The jacket without fibers shall meet the listing requirements for communications raceways in 800.182(A), (B), or (C) in accordance with the cable marking.

N (13) Cables Containing Optical Fibers. Composite optical fiber cables shall be listed as electrical cables based on the type of electrical conductors.

N (14) Class 2 and Class 3 Cable Voltage and Temperature Ratings. Class 2 cables shall have a voltage rating of not less than 150 volts. Class 3 cables shall have a voltage rating of not less than 300 volts. Class 2 and Class 3 cables shall have a temperature rating of not less than 60°C (140°F).

N (15) Power-Limited Fire Alarm (PLFA) Cables. PFLA cables shall comply with 722.179(A)(15)(a) through (A)(15)(d).

    (a) Conductors for cables, other than coaxial cables, shall be solid or stranded copper. Coaxial cables shall be permitted to use 30 percent conductivity copper-covered steel center conductor wire.

    (b) The size of conductors in a multiconductor cable shall not be smaller than 26 AWG. Single conductors shall not be smaller than 18 AWG. Conductors of 26 AWG shall be permitted only where spliced with a connector listed as suitable for 26 AWG to 24 AWG or larger conductors that are terminated on equipment or where the 26 AWG conductors are terminated on equipment listed as suitable for 26 AWG conductors.

    (c) Cables shall have a voltage rating of not less than 300 volts.

    (d) Cables shall have a temperature rating of not less than 60°C (140°F).

N (16) Class 4 Cable Construction.

N (1) Sizes. Conductors of sizes not smaller than 24 AWG shall be permitted to be used.

N (2) Insulation. Insulation on conductors shall be rated not less than 450 volts dc.

N (3) Voltage Rating. Cables shall have a voltage rating of not less than 450 volts dc. Voltage ratings shall not be marked on the cables.

N (4) Temperature Rating. Cables shall have a temperature rating of not less than 60°C (140°F).

N (5) Cabling. Cables shall comply with any requirements provided in the listing of the system.

> Informational Note: See UL 1400-1, *Outline for Fault-Managed Power Distribution Technologies — Part 1: General Requirements*, for information on determining applicable requirements for the listing of Class 4 power systems. Excessive cable lengths can result in higher capacitance which could affect the safety of the circuit.

N (B) Marking. Cables shall be durably marked on the surface in accordance with the following:

    (1) The AWG size or circular mil area shall be repeated at intervals not exceeding 610 mm (24 in.).
    (2) All other markings shall be repeated at intervals not exceeding 1.0 m (40 in.).
    (3) The proper type designation for the type of cable shall be marked in accordance with Table 722.179(B).
    (4) The manufacturer's name, trademark, or other distinctive marking by which the organization responsible for the product can be readily identified shall be marked.
    (5) The AWG size or circular mil area shall be marked.

> Informational Note No. 1: See Chapter 9, Table 8, for conductor area expressed in SI units for conductor sizes specified in AWG or circular mil area.

    (6) The temperature rating for a temperature rating exceeding 60°C (140°F) shall be marked.

> Informational Note No. 2: A minimum temperature rating of 60°C is assumed for cables not marked with a temperature rating.

    (7) Voltage ratings shall not be marked on the cables.

N   *TABLE 722.179(B)*   *Cable Type Markings*

Cable Type	Cable Marking
Class 4 plenum cable	CL4P
Class 3 plenum cable	CL3P
Class 2 plenum cable	CL2P
Power-limited fire alarm plenum cable	FPLP
Nonconductive optical fiber plenum cable	OFNP
Conductive optical fiber plenum cable	OFCP
Class 4 riser cable	CL4R
Class 3 riser cable	CL3R
Class 2 riser cable	CL2R
Power-limited fire alarm riser cable	FPLR
Nonconductive optical fiber riser cable	OFNR
Conductive optical fiber riser cable	OFCR
Class 4 general-purpose cable	CL4
Class 3 general-purpose cable	CL3
Class 2 general-purpose cable	CL2
Power-limited fire alarm cable	FPL
Nonconductive general-purpose optical fiber cable	OFN
Conductive general-purpose optical fiber cable	OFC
Alternative nonconductive general-purpose optical fiber cable	OFNG
Alternative conductive general-purpose optical fiber cable	OFCG
Class 3 cable — limited use	CL3X
Class 2 cable — limited use	CL2X
Undercarpet cable	CMUC

Note: All types of CL2, CL3, and FPL cables containing optical fibers are provided with the suffix "-OF."

Exception: Voltage markings shall be permitted where the cable has multiple listings and a voltage marking is required for one or more of the listings.

> Informational Note No. 3: Voltage markings on cables could be misinterpreted to suggest that the cables may be suitable for Class 1 electric light and power applications.
>
> Informational Note No. 4: Cable types are listed in descending order of fire resistance rating.

N   **(C) Optional Markings.** Cables shall be permitted to be surface marked to indicate special characteristics of the cable materials.

> Informational Note No. 1: Examples of these characteristics include, but are not limited to, limited smoke, halogen free, low smoke and halogen free, and sunlight resistant.
>
> Informational Note No. 2: Some examples of optional markings are ST1 to indicate limited smoke characteristics. See UL 2556, *Wire and Cable Test Methods*; HF to indicate halogen free. See in UL 2885, *Outline of Investigation for Acid Gas, Acidity and Conductivity of Combusted Materials*; and LSHF to indicate halogen free and low-smoke characteristics. See IEC 61034-2, *Measurement of smoke density of cables burning under defined conditions — Part 2: Test procedure and requirements*.

ARTICLE 724

Class 1 Power-Limited Circuits and Class 1 Power-Limited Remote-Control and Signaling Circuits

N   **724.1 Scope.** This article covers Class 1 circuits, including power-limited Class 1 remote-control and signaling circuits, that are not an integral part of a device or utilization equipment.

> Informational Note: See 300.26 for classifications of remote-control and signaling circuits.

Article 724 may include systems such as security system circuits, access control circuits, nurse call circuits, some computer network systems, some control circuits for lighting dimmer systems, and some low-voltage control circuits that originate from listed appliances or from listed computer equipment.

The installation requirements for the wiring of information technology equipment (electronic data processing and computer equipment) located within the confines of a room that is constructed according to the requirements of NFPA 75, *Standard for the Fire Protection of Information Technology Equipment*, are not covered by Article 724. The wiring within those specially constructed rooms is covered by Article 645.

In addition, if listed computer equipment is interconnected and all the interconnected equipment is in close proximity, the wiring is considered an integral part of the equipment and, therefore, not subject to the requirements of Article 724. If the wiring leaves the group of equipment to connect to other devices in the same room or elsewhere in the building, the wiring is considered "wiring within buildings" and is subject to the requirements of Article 724.

According to 90.3, the general wiring methods found in Chapters 1 through 4 of the *NEC®* apply to remote-control, signaling, and power-limited circuits, except as amended by Article 724 for specified conditions.

A remote-control, signaling, or power-limited circuit is the portion of the wiring system between the load side of the overcurrent device or the power-limited supply and all connected equipment.

Class 1 circuits are not permitted to exceed 30 volts and 1000 volt-amperes. In many cases, Class 1 circuits are extensions of power systems and are subject to the requirements of the power systems, except under the following conditions:

1. Conductors size 16 AWG and 18 AWG may be used. (See 724.43.)
2. Where damage to the circuit would introduce a hazard, the circuit must be mechanically protected. (See 724.31.)
3. The adjustment factors of 310.15(C) apply only if such conductors carry a continuous load exceeding 10 percent of the ampacity of the conductor. (See 724.51.)

Class 1 remote-control circuits are commonly used to operate motor controllers in conjunction with moving equipment or mechanical processes, elevators, conveyors, and other such equipment. Class 1 remote-control circuits also can be used as shunt-trip circuits for circuit breakers.

N 724.3 Other Articles. In addition to the requirements of this article, circuits and equipment shall comply with 724.3(A) through (J).

N (A) Number and Size of Conductors in Raceway. The number and size of conductors shall comply with 300.17.

N (B) Spread of Fire or Products of Combustion. Installation of Class 1 circuits shall comply with 300.21.

N (C) Ducts, Plenums, and Other Air-Handling Spaces. Class 1 circuits installed in ducts, plenums, and other spaces used for environmental air shall comply with 300.22.

N (D) Hazardous (Classified) Locations. Class 1 circuits shall not be installed in any hazardous (classified) locations except as permitted by other articles of this *Code*.

N (E) Cable Trays. Cable tray installations shall comply with Parts I and II of Article 392.

N (F) Raceways Exposed to Different Temperatures. Installation of raceways shall comply with 300.7(A).

Condensation often forms in conduits that are exposed to different temperatures, such as a walk-in refrigerator or an air-conditioned area to a space that is warmer. Section 724.3(F) brings the requirements of 300.7(A) into Article 724.

> **See also**
>
> **300.7(A)** and its associated commentary

N (G) Vertical Support for Fire-Rated Cables and Conductors. Vertical installations of circuit integrity (CI) cables and conductors installed in a raceway or conductors and cables of electrical circuit protective systems shall comply with 300.19.

N (H) Bushings. Bushings shall be installed where cables emerge from raceways used for mechanical support or protection in accordance with 300.15(C).

N (I) Installation of Conductors With Other Systems. Installation of conductors with other systems shall comply with 300.8.

N (J) Identification of Equipment Grounding Conductors. Equipment grounding conductors shall be identified in accordance with 250.119.

N 724.21 Access to Electrical Equipment Behind Panels Designed to Allow Access. Access to electrical equipment shall not be denied by an accumulation of wires and cables preventing the removal of panels, including suspended ceiling panels.

An excess accumulation of wires and cables can limit access to electrical equipment by preventing the removal of access panels or ceiling panels. To safely service, rearrange, or install electrical equipment, the worker must have accessible workspace. See Exhibit 724.1.

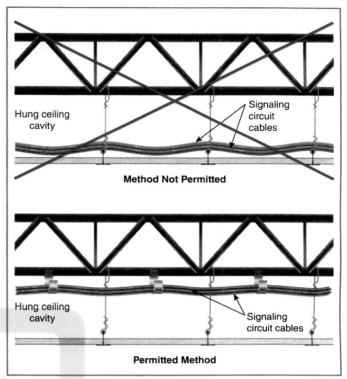

EXHIBIT 724.1 Incorrect cable installation *(upper diagram)* and correct method *(lower diagram)*.

> **See also**
>
> **300.11(B),** which permits the use of support wires and approved fittings that are independent of the suspended ceiling support wires

N 724.24 Mechanical Execution of Work. Class 1 circuits shall be installed in a neat and workmanlike manner. Cables and conductors installed exposed on the surfaces of ceilings and sidewalls shall be supported by the building structure such that the cable will not be damaged by normal building use. Such cables shall be supported by straps, staples, hangers, cable ties, or similar fittings that are designed and installed to not damage the cable. The installation shall also comply with the requirements of 300.4 and 300.11.

> Informational Note: Paint, plaster, cleaners, abrasives, corrosive residues, or other contaminants can result in an undetermined alteration of Class 1 cable properties.

Cable must be attached to or supported by the building structure by cable ties, straps, clamps, hangers, and so forth. The installation method must not damage the cable. In addition, the location of the cable should be carefully evaluated to ensure that activities and processes within the building do not cause damage to the cable. (See Exhibit 724.1.)

Section 300.4(D) requires protection of cables that are installed on framing members. Such cables are required to be installed in a manner that protects them from nail or screw penetration. This section permits attachment to baseboards and non-load-bearing walls, which are not structural components.

N **724.30 Class 1 Circuit Identification.** Class 1 circuits shall be identified at terminal and junction locations in a manner that prevents unintentional interference with other circuits during testing and servicing.

N **724.31 Safety-Control Equipment.** If controlling safety-control equipment, Class 1 circuits shall be provided with physical protection if the failure of such equipment to operate introduces a direct fire or life hazard. All conductors of such circuits shall be installed in rigid metal conduit, intermediate metal conduit, rigid nonmetallic conduit, electrical metallic tubing, Type MI cable, or Type MC cable, or be otherwise suitably protected from physical damage.

The remote-control circuits to safety-control devices are required to be classified as Class 1 if failure of the safety-control circuit could cause a direct fire or life hazard. One example of the direct link between a failure and the initiation of a fire hazard is a boiler explosion caused by failure of the low-water cutoff circuit. See Exhibit 724.2.

Generally, signaling systems such as a nurse call system do not fit this category. Those systems do not have a direct link to the initiation of fire or the initiation of a life hazard but, rather, serve as the reporting or warning link of a hazard initiated by some other (indirect) cause.

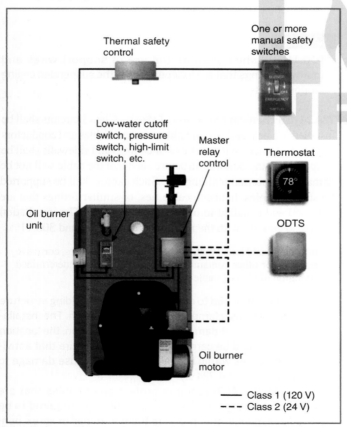

EXHIBIT 724.2 Typical installation of an automatic oil burner unit for a boiler employing a safety shutdown circuit required to be Class 1.

N **724.40 Class 1 Circuits.** Class 1 circuits shall be supplied from a source with a rated output of not more than 30 volts and 1000 volt-amperes.

N **(A) Class 1 Transformers.** Transformers shall be permitted to supply Class 1 circuits.

> Informational Note: See Parts I and II of Article 450 for information on transformers used to supply a Class 1 circuit.

N **(B) Other Class 1 Power Sources.** Power sources other than transformers shall be protected by overcurrent devices rated at not more than 167 percent of the volt-ampere rating of the source divided by the rated voltage. The overcurrent devices shall not be interchangeable with overcurrent devices of higher ratings. The overcurrent device shall be permitted to be an integral part of the power supply.

To comply with the 1000 volt-ampere limitation of 724.40, the maximum output (VA_{max}) of power sources other than transformers shall be limited to 2500 volt-amperes, and the product of the maximum current (I_{max}) and maximum voltage (V_{max}) shall not exceed 10,000 volt-amperes. These ratings shall be determined with any overcurrent-protective device bypassed.

VA_{max} is the maximum volt-ampere output after one minute of operation regardless of load and with overcurrent protection bypassed, if used. Current-limiting impedance shall not be bypassed when determining VA_{max}.

I_{max} is the maximum output current under any noncapacitive load, including short circuit, and with overcurrent protection bypassed, if used. Current-limiting impedance should not be bypassed when determining I_{max}. Where a current-limiting impedance listed for the purpose or as part of a listed product is used in combination with a stored energy source, such as a storage battery, to limit the output current, I_{max} limits apply after 5 seconds.

V_{max} is the maximum output voltage regardless of load with rated input applied.

N **724.43 Class 1 Circuit Overcurrent Protection.** Overcurrent protection for conductors 14 AWG and larger shall be provided in accordance with the conductor ampacity, without applying the ampacity adjustment and correction factors specified in 310.15 to the ampacity calculation. Overcurrent protection shall not exceed 7 amperes for 18 AWG conductors and 10 amperes for 16 AWG.

Exception: Where other articles of this Code permit or require other overcurrent protection.

N **724.45 Class 1 Circuit Overcurrent Device Location.** Overcurrent devices shall be located as specified in 724.45(A) through (E).

N **(A) Point of Supply.** Overcurrent devices shall be located at the point where the conductor to be protected receives its supply.

N **(B) Feeder Taps.** Class 1 circuit conductors shall be permitted to be tapped, without overcurrent protection at the tap, where the overcurrent device protecting the circuit conductor is sized to protect the tap conductor.

N **(C) Branch-Circuit Taps** Class 1 circuit conductors 14 AWG and larger that are tapped from the load side of the overcurrent protective device(s) of a controlled light and power circuit shall require only short-circuit and ground-fault protection and shall be permitted to be protected by the branch-circuit overcurrent protective device(s) where the rating of the protective device(s) is not more than 300 percent of the ampacity of the Class 1 circuit conductor.

N **(D) Primary Side of Transformer.** Class 1 circuit conductors supplied by the secondary of a single-phase transformer having only a 2-wire (single-voltage) secondary shall be permitted to be protected by overcurrent protection provided on the primary side of the transformer if the protection is in accordance with 450.3 and does not exceed the value determined by multiplying the secondary conductor ampacity by the secondary-to-primary transformer voltage ratio. Transformer secondary conductors other than 2-wire shall not be considered to be protected by the primary overcurrent protection.

N **(E) Input Side of Electronic Power Source.** Class 1 circuit conductors supplied by the output of a single-phase, listed electronic power source other than a transformer having only a 2-wire (single-voltage) output for connection to Class 1 circuits shall be permitted to be protected by overcurrent protection provided on the input side of the electronic power source if the protection does not exceed the value determined by multiplying the Class 1 circuit conductor ampacity by the output-to-input voltage ratio. Electronic power source outputs other than 2-wire (single voltage) shall not be considered to be protected by the primary overcurrent protection.

N **724.46 Class 1 Circuit Wiring Methods.** Class 1 circuits shall be installed in accordance with 300.2 through 300.26.

> *Exception No. 1: The requirements of 724.48 through 724.51 shall be permitted to apply in installations of Class 1 circuits.*

> *Exception No. 2: Methods permitted or required by other articles of this Code shall apply to installations of Class 1 circuits.*

N **724.48 Conductors of Different Circuits in the Same Cable, Cable Tray, Enclosure, or Raceway.** Class 1 circuits shall be permitted to be installed with other circuits as specified in 724.48(A) and (B).

N **(A) Two or More Class 1 Circuits.** Class 1 circuits shall be permitted to occupy the same cable, cable tray, enclosure, or raceway regardless of whether the individual circuits are alternating current or direct current if all conductors are insulated for the maximum voltage of any conductor in the cable, cable tray, enclosure, or raceway.

N **(B) Class 1 Circuits with Power-Supply Circuits.** Class 1 circuits shall be permitted to be installed with power-supply conductors as specified in 724.48(B)(1) through (B)(4).

N **(1) In Cables, Enclosures, or Raceways.** Class 1 circuits and power-supply circuits shall be permitted to occupy the same cable, enclosure, or raceway without a barrier only where the equipment powered is functionally associated. Class 1 circuits shall be permitted to be installed together with the conductors of electric light, power, non-power-limited fire alarm systems, and medium-power network-powered broadband communications circuits where separated by a barrier.

N **(2) In Factory- or Field-Assembled Control Centers.** Class 1 circuits and power-supply circuits shall be permitted to be installed in factory- or field-assembled control centers.

N **(3) In Manholes.** Class 1 circuits and power-supply circuits shall be permitted to be installed as underground conductors in manholes in accordance with one of the following:

(1) The power-supply or Class 1 circuit conductors are in metal-enclosed cable or Type UF cable.
(2) The conductors are permanently separated from power-supply conductors by continuous firmly fixed nonconductors, such as flexible tubing, in addition to insulation on the wire.
(3) The conductors are permanently and effectively separated from power-supply conductors and securely fastened to racks, insulators, or other approved supports.

Class 1 power-limited circuit conductors are permitted to be installed in manholes with wiring of power-supply conductors where one of the permanent separation requirements that follow are used:

1. Power-supply or Class 1 circuits are in metal-enclosed cable or Type UF cable.
2. Conductors are permanently separated by continuous firmly fixed nonconductors, such as flexible tubing.
3. Conductors are permanently separated by securely fastening to racks, insulators, or other approved supports.

Section 724.48(B)(1) permits Class 1 conductors to occupy the same cable, enclosure, or raceway with power supply conductors if they are functionally associated.

N **(4) In Cable Trays.** Installations in cable trays shall comply with the requirements of one of the following:

(1) Class 1 circuit conductors and power-supply conductors not functionally associated with the Class 1 circuit conductors shall be separated by a solid fixed barrier of a material compatible with the cable tray.
(2) Class 1 circuit conductors and power-supply conductors not functionally associated with the Class 1 circuit conductors shall be permitted to be installed in a cable tray

without barriers where all of the conductors are installed with separate multiconductor Type AC, Type MC, Type MI, or Type TC cables and all the conductors in the cables are insulated at 600 volts or greater.

N 724.49 Class 1 Circuit Conductors.

N (A) Sizes and Use. Conductors that are 18 AWG and 16 AWG shall be permitted to be used if they supply loads that do not exceed the ampacities specified in 402.5 and are installed in a raceway, an approved enclosure, or a listed cable. Conductors larger than 16 AWG shall not supply loads greater than the ampacities specified in 310.14. Flexible cords shall comply with the requirements of Article 400.

N (B) Insulation. Insulation on conductors shall be rated for the system voltage and not less than 600 volts. Conductors larger than 16 AWG shall comply with the requirements of Article 310. Conductors that are 18 AWG and 16 AWG shall be Type FFH-2, Type KF-2, Type KFF-2, Type PAF, Type PAFF, Type PF, Type PFF, Type PGF, Type PGFF, Type PTF, Type PTFF, Type RFH-2, Type RFHH-2, Type RFHH-3, Type SF-2, SFF-2, Type TF, Type TFF, Type TFFN, Type TFN, Type ZF, or Type ZFF. Conductors with other types and thicknesses of insulation shall be permitted if listed for Class 1 circuit use.

Class 1 circuit conductor insulation is required to be rated not less than 600 volts, which effectively requires Class 1 circuits to be wired using the wiring methods found in Chapter 3 or to use conductors specifically listed for Class 1 circuits.

N 724.51 Number of Conductors in Cable Trays and Raceways, and Ampacity Adjustment.

N (A) Class 1 Circuit Conductors. Where only Class 1 circuit conductors are in a raceway, the number of conductors shall be determined in accordance with 300.17. The ampacity adjustment factors specified in 310.15(C)(1) shall apply only if such conductors carry continuous loads in excess of 10 percent of the ampacity of each conductor.

N (B) Power-Supply Conductors and Class 1 Circuit Conductors. Where power-supply conductors and Class 1 circuit conductors are permitted in a raceway in accordance with 724.48, the number of conductors shall be determined in accordance with 300.17. The ampacity adjustment factors specified in 310.15(C)(1) shall apply as follows:

(1) To all conductors where the Class 1 circuit conductors carry continuous loads in excess of 10 percent of the ampacity of each conductor and where the total number of conductors is more than three

(2) To the power-supply conductors only, where the Class 1 circuit conductors do not carry continuous loads in excess of 10 percent of the ampacity of each conductor and where the number of power-supply conductors is more than three

N (C) Class 1 Circuit Conductors in Cable Trays. Where Class 1 circuit conductors are installed in cable trays, they shall comply with the requirements of 392.22 and 392.80(A).

N 724.52 Circuits Extending Beyond One Building. Class 1 circuits that extend aerially beyond one building shall also meet the requirements of Part I of Article 225.

Δ | ARTICLE
725 Class 2 and Class 3 Power-Limited Circuits

Part I. General

Δ **725.1 Scope.** This article covers power-limited circuits, including power-limited remote-control and signaling circuits, that are not an integral part of a device or of utilization equipment.

Informational Note No. 1: The circuits described herein are characterized by usage and electrical power limitations that differentiate them from electric light and power circuits; therefore, alternative requirements are given regarding minimum wire sizes, ampacity adjustment and correction factors, overcurrent protection, insulation requirements, and wiring methods and materials.

Informational Note No. 2: See 300.26 for classifications of remote-control and signaling circuits.

Article 725 can include systems such as security system circuits, access control circuits, nurse call circuits, some computer network systems, some control circuits for lighting dimmer systems, and some low-voltage control circuits that originate from listed appliances or from listed computer equipment.

The installation requirements for the wiring of information technology equipment (electronic data processing and computer equipment) located within the confines of a room that is constructed according to the requirements of NFPA 75, *Standard for the Fire Protection of Information Technology Equipment*, are not covered by Article 725. The wiring within those specially constructed rooms is covered by Article 645.

In addition, if listed computer equipment is interconnected and all the interconnected equipment is in close proximity, the wiring is considered an integral part of the equipment and, therefore, not subject to the requirements of Article 725. If the wiring leaves the group of equipment to connect to other devices in the same room or elsewhere in the building, the wiring is considered "wiring within buildings" and is subject to the requirements of Article 725.

According to 90.3, the general wiring methods found in Chapters 1 through 4 of the *NEC®* apply to Class 2 and Class 3 power-limited circuits and power-limited remote-control and signaling circuits, except as amended by Article 725 for specified conditions.

Conductors and equipment on the supply side of overcurrent protection, transformers, or current-limiting devices of

Class 2 and Class 3 circuits must be installed according to the applicable requirements of Chapter 3. Load-side conductors and equipment must comply with Article 725. Class 2 and Class 3 conductors are required to be separated from and not occupy the same raceways, cable trays, cables, or enclosures as electric light, power, and Class 1 conductors, except as noted in 725.136.

Many batteries can be considered Class 2 power supplies if the voltage is 30 volts or less and listed and identified as Class 2. See 725.60(A)(5).

725.3 Other Articles. In addition to the requirements of this article, circuits and equipment shall comply with the articles or sections listed in 725.3(A) through (E). Only those sections of Article 300 referenced in this article shall apply to Class 2 and Class 3 circuits.

(A) Spread of Fire or Products of Combustion. Installation of Class 2 and Class 3 circuits shall comply with 300.21.

Δ **(B) Ducts, Plenums, and Other Air-Handling Spaces.** Class 2 and Class 3 circuits installed in ducts, plenums, or other space used for environmental air shall comply with 300.22.

(C) Motor Control Circuits. Motor control circuits tapped from the load side of the motor branch-circuit protective device(s) as specified in 430.72(A) shall comply with Part IV of Article 430.

Δ **(D) Identification of Equipment Grounding Conductors.** Equipment grounding conductors shall be identified in accordance with 250.119.

N **(E) Cables for Class 2 and Class 3 Circuits.** The listing and installation of cables for Class 2 and Class 3 circuits shall comply with Part I and Part II of Article 722.

N **725.10 Hazardous (Classified) Locations.** Cables and equipment shall be permitted to be used in hazardous (classified) locations where specifically permitted by other articles in this *Code*.

725.21 Access to Electrical Equipment Behind Panels Designed to Allow Access. Access to electrical equipment shall not be denied by an accumulation of wires and cables that prevents removal of panels, including suspended ceiling panels.

An excess accumulation of wires and cables can limit access to electrical equipment by preventing the removal of access panels. To safely service, rearrange, or install electrical equipment, the worker must have an accessible work space. See Exhibit 725.1.

See also

300.11(B), which permits the use of support wires and approved fittings that are independent of the suspended ceiling support wires

Δ **725.24 Mechanical Execution of Work.** Class 2 and Class 3 equipment shall be installed in a neat and workmanlike manner. The installation shall also comply with 300.4 and 300.11.

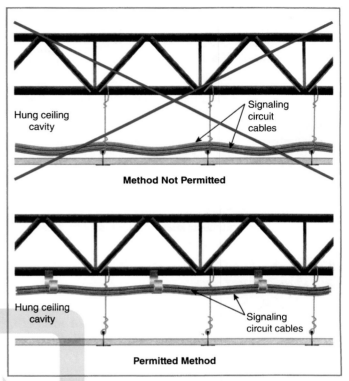

EXHIBIT 725.1 Incorrect cable installation *(upper diagram)* and correct method *(lower diagram)*.

Cable must be attached to or supported by the building structure by cable ties, straps, clamps, hangers, and so forth. See Exhibit 725.1. The installation method must not damage the cable. In addition, the location of the cable should be carefully evaluated to ensure that activities and processes within the building do not cause damage to the cable. See 300.11(C).

Section 300.4(D) requires protection of cables that are installed on framing members. Such cables are required to be installed in a manner that protects them from nail or screw penetration. This section permits attachment to baseboards and non-load-bearing walls, which are not structural components.

725.30 Class 2 and Class 3 Circuit Identification. Class 2 and Class 3 circuits shall be identified at terminal and junction locations in a manner that prevents unintentional interference with other circuits during testing and servicing.

725.31 Safety-Control Equipment. Where damage to power-limited circuits can result in a failure of safety-control equipment that would introduce a direct fire or life hazard, the power-limited circuits shall be installed in accordance with 724.31. Room thermostats, water temperature regulating devices, and similar controls used in conjunction with electrically controlled household heating and air conditioning shall not be considered safety-control equipment.

Part II. Class 2 and Class 3 Circuits

725.60 Power Sources for Class 2 and Class 3 Circuits.

Δ **(A) Power Source.** The power source for a Class 2 or a Class 3 circuit shall be as follows:

> Informational Note No. 1: Informational Note Figure 725.60 illustrates the relationships between Class 2 or Class 3 power sources, their supply, and the Class 2 or Class 3 circuits.
>
> Informational Note No. 2: See Chapter 9, Table 11(A) and Table 11(B), for requirements for listed Class 2 and Class 3 power sources.

(1) A listed Class 2 or Class 3 transformer
(2) A listed Class 2 or Class 3 power supply
(3) Other listed equipment marked to identify the Class 2 or Class 3 power source

Exception No. 1 to (3): Thermocouples shall not require listing as a Class 2 power source.

Exception No. 2 to (3): Limited power circuits of listed equipment where these circuits have energy levels rated at or below the limits established in Chapter 9, Table 11(A) and Table 11(B).

> Informational Note No. 3: Examples of other listed equipment are as follows:
>
> (1) A circuit card listed for use as a Class 2 or Class 3 power source where used as part of a listed assembly
> (2) A current-limiting impedance, listed for the purpose, or part of a listed product, used in conjunction with a non-power-limited transformer or a stored energy source, for example, storage battery, to limit the output current
> (3) A thermocouple
> (4) Limited voltage/current or limited impedance secondary communications circuits of listed industrial control equipment

(4) Listed audio/video, information technology (computer), communications, and industrial equipment limited-power circuits

> Informational Note No. 4: One way to determine applicable requirements for listing of information technology (computer) equipment is to refer to UL 60950-1-2011, *Standard for Safety of Information Technology Equipment*. Another way to determine applicable requirements for listing of audio/video, information technology, and communications equipment is to refer to UL 62368-1-2014, *Safety of audio/video, information and communication technology equipment*. Typically such circuits are used to interconnect data circuits for the purpose of exchanging information data. One way to determine applicable requirements for listing of industrial equipment is to refer to UL 61010-2-201, *Safety requirements for electrical equipment for measurement, control, and laboratory use — Part 2-201: Particular requirements for control equipment*, and/or UL 61800-5-1, *Adjustable speed electrical power drive systems — Part 5-1: Safety requirements — Electrical, thermal and energy*.

(5) A battery source or battery source system that is listed and identified as Class 2

(B) Interconnection of Power Sources. Class 2 or Class 3 power sources shall not have the output connections paralleled or otherwise interconnected unless listed for such interconnection.

Δ **(C) Marking.** The equipment supplying the circuits shall be durably marked where plainly visible to indicate each circuit that is a Class 2 or Class 3 circuit. The power sources for limited power circuits in 725.60(A)(3), limited power circuits for listed audio/video equipment, listed information technology equipment, listed communications equipment, and listed industrial equipment in 725.60(A)(4) shall have a label indicating the maximum voltage and rated current output per conductor for each connection point on the power source. Where multiple connection points have the same rating, a single label shall be permitted to be used.

> Informational Note No. 1: Rated current for power sources covered in 725.144 is the output current per conductor the power source is designed to deliver to an operational load at normal operating conditions, as declared by the manufacturer.
>
> Informational Note No. 2: An example of a label is "52V @ 0.433A, 57V MAX" for an IEEE 802.3 compliant Class 8 power source.

Δ **725.127 Wiring Methods on Supply Side of the Class 2 or Class 3 Power Source.** Conductors and equipment on the supply side of the power source shall be installed in accordance with the appropriate requirements of Chapters 1 through 4.

Exception: The input leads of a transformer or other power source supplying Class 2 and Class 3 circuits shall be permitted to be smaller than 14 AWG but not smaller than 18 AWG if they are protected by an overcurrent device rated not over 20 amperes, are not over 305 mm (12 in.) long, and have insulation that complies with 724.49(B).

Δ **725.130 Wiring Methods and Materials on Load Side of the Class 2 or Class 3 Power Source.** Class 2 and Class 3 circuits on the load side of the power source shall be permitted to be installed using wiring methods and materials in accordance with

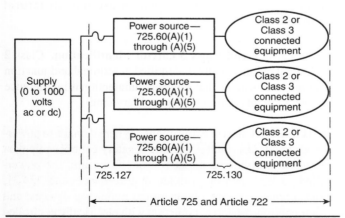

Δ *INFORMATIONAL NOTE FIGURE 725.60 Class 2 and Class 3 Circuits.*

725.130(A), (B), or a combination of both. Parts I and II of Article 722 shall apply.

Δ **(A) Class 1 Wiring Methods and Materials.** Use of Class 1 wiring methods for Class 2 and Class 3 circuits shall be permitted. Separation from electric light, power, Class 1, non-power-limited fire alarm circuit conductors, and medium-power network-powered broadband communications cables shall comply with 725.136.

> *Exception: The ampacity adjustment factors given in 310.15(C)(1) shall not apply.*

Δ **(B) Class 2 and Class 3 Wiring Methods and Materials.** Conductors on the load side of the power source shall be insulated in accordance with 722.179 and be installed in accordance with 722.135 and 725.136 through 725.144.

> *Exception No. 1: As provided for in 620.21 for elevators and similar equipment.*
>
> *Exception No. 2: Other wiring methods and materials installed in accordance with 725.3 shall be permitted to extend or replace the conductors and cables described in 722.179(A) and permitted by 725.130(B).*
>
> *Exception No. 3: Bare Class 2 conductors shall be permitted as part of a listed intrusion protection system where installed in accordance with the listing instructions for the system.*

N **725.136 Separation from Electric Light, Power, Class 1, Non-Power-Limited Fire Alarm Circuit Conductors, and Medium-Power Network-Powered Broadband Communications Cables.**

N **(A) General.** Cables and conductors of Class 2 and Class 3 circuits shall not be placed in any cable, cable tray, compartment, enclosure, manhole, outlet box, device box, raceway, or similar fitting with conductors of electric light, power, Class 1, non-power-limited fire alarm circuits, and medium-power network-powered broadband communications circuits unless permitted by 725.136(B) through (I).

> Generally, the lower voltage ratings of listed Class 2 and Class 3 cables do not allow them to be installed with electric light, power, Class 1, non–power-limited fire alarm circuits, and medium-power network-powered broadband communications cables. Failure of the cable insulation due to a fault could lead to hazardous voltages being imposed on the Class 2 or Class 3 circuit conductors. However, see 725.136(I) for methods to provide separation.

N **(B) Separated by Barriers.** Class 2 and Class 3 circuits shall be permitted to be installed together with the conductors of electric light, power, Class 1, non-power-limited fire alarm, and medium-power network-powered broadband communications circuits where they are separated by a barrier.

N **(C) Raceways Within Enclosures.** In enclosures, Class 2 and Class 3 circuits shall be permitted to be installed in a raceway to separate them from Class 1, non-power-limited fire alarm, and medium-power network-powered broadband communications circuits.

N **(D) Associated Systems Within Enclosures.** Class 2 and Class 3 circuit conductors in compartments, enclosures, device boxes, outlet boxes, or similar fittings shall be permitted to be installed with electric light, power, Class 1, non-power-limited fire alarm, and medium-power network-powered broadband communications circuits where they are introduced solely to connect the equipment connected to Class 2 and Class 3 circuits, and where one of the following applies:

(1) The electric light, power, Class 1, non-power-limited fire alarm, and medium-power network-powered broadband communications circuit conductors are routed to maintain a minimum of 6 mm (0.25 in.) separation from the conductors and cables of Class 2 and Class 3 circuits.

(2) The circuit conductors operate at 150 volts or less to ground and comply with one of the following:

a. The Class 2 and Class 3 circuits are installed using Type CL3, Type CL3R, or Type CL3P or permitted substitute cables if these Class 3 cable conductors extending beyond the jacket are separated by a minimum of 6 mm (0.25 in.) or by a nonconductive sleeve or nonconductive barrier from all other conductors.

b. The Class 2 and Class 3 circuit conductors are installed as a Class 1 circuit in accordance with 724.40.

> An example of associated systems is a Class 2 circuit source that is the secondary of a control transformer installed in the same motor-starter enclosure. In such an installation, the Class 2 conductor insulation is not required to have the same voltage rating as the insulation on the power conductors in the same enclosure, provided the installation complies with 725.136(D)(1) or (2).

N **(E) Enclosures with Single Opening.** Class 2 and Class 3 circuit conductors entering compartments, enclosures, device boxes, outlet boxes, or similar fittings shall be permitted to be installed with Class 1, non-power-limited fire alarm, and medium-power network-powered broadband communications circuits where they are introduced solely to connect the equipment connected to Class 2 and Class 3 circuits. Where Class 2 and Class 3 circuit conductors must enter an enclosure that is provided with a single opening, they shall be permitted to enter through a single fitting (such as a tee) if the conductors are separated from the conductors of the other circuits by a continuous and firmly fixed nonconductor, such as flexible tubing.

N **(F) Manholes.** Underground Class 2 and Class 3 circuit conductors in a manhole shall be permitted to be installed with Class 1, non-power-limited fire alarm, and medium-power network-powered broadband communications circuits where one of the following conditions is met:

(1) The electric light, power, Class 1, non-power-limited fire alarm, and medium-power network-powered broadband communications circuit conductors are in a metal-enclosed cable or Type UF cable.

(2) The Class 2 and Class 3 circuit conductors are permanently and effectively separated from the conductors of other circuits by a continuous and firmly fixed nonconductor, such as flexible tubing, in addition to the insulation or covering on the wire.

(3) The Class 2 and Class 3 circuit conductors are permanently and effectively separated from conductors of the other circuits and securely fastened to racks, insulators, or other approved supports.

N (G) Cable Trays. Class 2 and Class 3 circuit conductors shall be permitted to be installed in cable trays where the conductors of the electric light, Class 1, and non-power-limited fire alarm circuits are separated by a solid fixed barrier of a material compatible with the cable tray or where the Class 2 or Class 3 circuits are installed in Type MC cable.

N (H) Where Protected. Class 2 and Class 3 circuits shall be permitted to be installed together with the conductors of electric light, power, Class 1, non-power-limited fire alarm, and medium-power network-powered broadband communications circuits where they are installed using Class 1 wiring methods in accordance with 724.46 and where they are protected by an approved raceway.

N (I) Other Applications. For other applications, conductors of Class 2 and Class 3 circuits shall be separated by at least 50 mm (2 in.) from conductors of any electric light, power, Class 1, non-power-limited fire alarm, or medium-power network-powered broadband communications circuits unless one of the following conditions is met:

(1) Either all of the electric light, power, Class 1, non-power-limited fire alarm, and medium-power network-powered broadband communications circuit conductors or all of the Class 2 and Class 3 circuit conductors are in a raceway or in metal-sheathed, metal-clad, nonmetallic-sheathed, Type TC, or Type UF cables.

(2) All of the electric light, power, Class 1, non-power-limited fire alarm, and medium-power network-powered broadband communications circuit conductors are permanently separated from all of the Class 2 and Class 3 circuit conductors by a continuous and firmly fixed nonconductor, such as porcelain tubes or flexible tubing, in addition to the insulation on the conductors.

N 725.139 Installation of Conductors of Different Circuits in the Same Cable, Enclosure, Cable Tray, Raceway, or Cable Routing Assembly.

N (A) Two or More Class 2 Circuits. Conductors of two or more Class 2 circuits shall be permitted within the same cable, enclosure, raceway, or cable routing assembly.

N (B) Two or More Class 3 Circuits. Conductors of two or more Class 3 circuits shall be permitted within the same cable, enclosure, raceway, or cable routing assembly.

N (C) Class 2 Circuits with Class 3 Circuits. Conductors of one or more Class 2 circuits shall be permitted within the same cable, enclosure, raceway, or cable routing assembly with conductors of Class 3 circuits if the insulation of the Class 2 circuit conductors in the cable, enclosure, raceway, or cable routing assembly is at least that required for Class 3 circuits.

N (D) Class 2 and Class 3 Circuits with Communications Circuits.

N (1) Communications Cables. Conductors of one or more Class 2 or Class 3 circuits shall be permitted in the same cable with conductors of communications circuits if the cable is a listed communications cable installed in accordance with Part V of Article 800. The cables shall be listed as communications cables.

N (2) Composite Cables. Cables constructed of individually listed Class 2, Class 3, and communications cables under a common jacket shall be permitted to be classified as communications cables. The fire resistance rating of the composite cable shall be determined by the performance of the composite cable.

N (E) Class 2 or Class 3 Cables with Other Circuit Cables. Jacketed cables of Class 2 or Class 3 circuits shall be permitted in the same enclosure, cable tray, raceway, or cable routing assembly with jacketed cables of any of the following:

(1) Power-limited fire alarm systems in compliance with Parts I and III of Article 760

(2) Nonconductive and conductive optical fiber cables in compliance with Parts I and IV of Article 770

(3) Communications circuits in compliance with Parts I and IV of Article 805

(4) Community antenna television and radio distribution systems in compliance with Parts I and IV of Article 820

(5) Low-power, network-powered broadband communications in compliance with Parts I and IV of Article 830

N (F) Class 2 or Class 3 Conductors or Cables and Audio System Circuits. Audio system circuits described in 640.9(C) and installed using Class 2 or Class 3 wiring methods in compliance with 722.135 shall not be installed in the same cable, raceway, or cable routing assembly with Class 2 or Class 3 conductors or cables.

N 725.144 Bundling of Cables Transmitting Power and Data. Sections 725.144(A) and (B) shall apply to Class 2 and Class 3 circuits that transmit power and data to a powered device over listed cabling. Section 300.11 and Parts I and III of Article

725 shall apply to Class 2 and Class 3 circuits that transmit power and data. The conductors that carry power for the data circuits shall be copper. The current in the power circuit shall not exceed the current limitation of the connectors.

Informational Note No. 1: One example of the use of cables that transmit power and data is the connection of closed-circuit TV cameras (CCTV).

Informational Note No. 2: The 8P8C connector is in widespread use with powered communications systems. IEC 60603-7-2008, *Connectors for electronic equipment — Part 7-1: Detail specification for 8-way, unshielded, free and fixed connectors,* specifies these connectors to have a current-carrying capacity per contact of 1.0 amperes maximum at 60°C (149°F). See IEC 60603-7 for more information on current-carrying capacity at higher and lower temperatures.

Informational Note No. 3: The requirements of Table 725.144 were derived for carrying power and data over 4-pair copper balanced twisted pair cabling. This type of cabling is described in ANSI/TIA 568-C.2-2009, *Commercial Building Telecommunications Cabling Standard — Part 2: Balanced Twisted-Pair Telecommunications Cabling and Components.*

Informational Note No. 4: See TIA-TSB-184-A-2017, *Guidelines for Supporting Power Delivery Over Balanced Twisted-Pair Cabling,* for information on installation and management of balanced twisted pair cabling supporting power delivery.

Informational Note No. 5: See ANSI/NEMA C137.3-2017, *American National Standard for Lighting Systems — Minimum Requirements for Installation of Energy Efficient Power over Ethernet (PoE) Lighting Systems,* for information on installation of cables for PoE lighting systems.

Informational Note No. 6: Rated current for power sources covered in 725.144 is the output current per conductor the power source is designed to deliver to an operational load at normal operating conditions, as declared by the manufacturer. In the design of these systems, the actual current in a given conductor might vary from the rated current per conductor by as much as 20 percent. An increase in current in one conductor is offset by a corresponding decrease in current in one or more conductors of the same cable.

This section provides requirements for cables that are used for transmission of data and power. This is commonly referred to as Power over Ethernet (PoE). Common applications include telephones, wireless access points, and security cameras powered from a Class 2 power supply that also uses conductors in the same cable for data transmission. These devices may be identified as complying to IEEE 802.3 networking standards, including IEEE 802.3af, *Data Terminal Equipment (DTE) Power via Media Dependent Interface (MDI)*; IEEE 802.3at, *Data Terminal Equipment (DTE) Power via Media Dependent Interface (MDI) Enhancements*; or IEEE 802.3bt, *Physical Layer and Management Parameters for Power over Ethernet over 4 pairs.* In all but the highest power classes of the IEEE 802.3 standards, the rated current output per conductor is less than 0.3 amperes.

Section 725.144 protects against situations in which current flow in a large number of bundled cables could cause an increased temperature in a conductor or cable that can degrade the insulation. The ampacities in Table 725.144 are based on a fact-finding report by Underwriters Laboratories and were updated for the 2020 edition of the *NEC* to provide greater precision.

By far, the most common installations are designed in accordance with IEEE 802.3 networking standards, for which power sources are rated as less than 0.433 amperes per conductor, usually 0.3 amperes per conductor or less. The 4-pair local area network (LAN) cabling typically is 24 AWG or larger.

N **TABLE 725.144** *Ampacities of Each Conductor in Amperes in 4-Pair Class 2 or Class 3 Balanced Twisted-Pair Cables Based on Copper Conductors at an Ambient Temperature of 30°C (86°F) with All Conductors in All Cables Carrying Current, 60°C (140°F), 75°C (167°F), and 90°C (194°F) Rated Cables*

	Number of 4-Pair Cables in a Bundle																	
	1–7			8–19			20–37			38–61			62–91			92–192		
	Temperature Rating			Temperature Rating			Temperature Rating			Temperature Rating			Temperature Rating			Temperature Rating		
AWG	60°C	75°C	90°C	60°C	75°C	90°C	60°C	75°C	90°C	60°C	75°C	90°C	60°C	75°C	90°C	60°C	75°C	90°C
26	1.00	1.23	1.42	0.71	0.87	1.02	0.55	0.68	0.78	0.46	0.57	0.67	0.45	0.55	0.64	NA	NA	NA
24	1.19	1.46	1.69	0.81	1.01	1.17	0.63	0.78	0.91	0.55	0.67	0.78	0.46	0.56	0.65	0.40	0.48	0.55
23	1.24	1.53	1.78	0.89	1.11	1.28	0.77	0.95	1.10	0.66	0.80	0.93	0.58	0.71	0.82	0.45	0.55	0.63
22	1.50	1.86	2.16	1.04	1.28	1.49	0.77	0.95	1.11	0.66	0.82	0.96	0.62	0.77	0.89	0.53	0.63	0.72

Notes:

1. For bundle sizes over 192 cables, or for conductor sizes smaller than 26 AWG, ampacities shall be permitted to be determined by qualified personnel under engineering supervision.

2. Where only half of the conductors in each cable are carrying current, the values in the table shall be permitted to be increased by a factor of 1.4.

Informational Note No. 1: Elevated cable temperatures can reduce a cable's data transmission performance. For information on practices for 4-pair balanced twisted pair cabling, see TIA-TSB-184-A and 6.4.7, 6.6.3, and Annex G of ANSI/TIA-568-C.2, which provide guidance on adjustments for operating temperatures between 20°C and 60°C.

Informational Note No. 2: The per-contact current rating of connectors can limit the maximum allowable current below the ampacity shown in Table 725.144.

COMMENTARY TABLE 725.1 Performance Characteristics for Each IEEE Class of PoE

IEEE Class	PSE Power (W)	PD Power (W)	Maximum Cable Current (mA)	Maximum Rated Current per Conductor (mA)	
				2-Pair	4-Pair
1	4	3.84	91	45	20
2	7	6.49	159	80	35
3	15.4	13	350	175	77
4	30	25.5	600	300	150
5	45	40	900	N/A	225
6	60	51	1200	N/A	300
7	75	62	1442	N/A	361
8	90	71.3	1731	N/A	433

Table Notes:
PSE = power sourcing equipment (the power supply).
PD = powered device (the load).
Maximum Cable Current = combined one-way current through all conductors.
Maximum Rated Current per Conductor = single conductor current for use with Table 725.144.

PoE installations designed to IEEE 802.3 are comprised of power sourcing equipment (PSE) and a powered device (PD). The PSE is an *NEC* Class 2 power supply. The PD is the load that is being supplied the power. The PSEs and PDs are connected via an eight-conductor category cable. The PD can use four or eight of the conductors to draw current from and return current to the PSE. PSEs and PDs are defined by one of eight IEEE 802.3 classes, ranging from 3.84 watts to 90 watts, detailed in Commentary Table 725.1. Above a certain IEEE class (above Class 4), the PD must use all eight conductors. The category cable between the PSE and the PD can be 328 feet long. As a result, the PSE power and the PD power are not equivalent, to account for cable losses. This can be seen as the difference between the PSE power and the PD power in Commentary Table 725.1. The maximum cable current (i.e., the combined one-way current through all the conductors) can be calculated from the PSE power and voltage. The calculated worst-case cable current is in the fourth column of the commentary table.

Table 725.144 provides the rated current per conductor for a PSE, which is the maximum one-way cable current draw divided either between two conductors or among four conductors. The divisor is chosen depending on whether the PD is drawing and then returning current over a total of only four conductors or all eight conductors. Commentary Table 725.1 shows the worst-case rated current for each IEEE class, one for four conductors (the 2-Pair column) and one for eight conductors (the 4-Pair column). The currents in the last column, Maximum Rated Current per Conductor (mA), are the currents to be used in reference to Table 725.144.

For the common IEEE standard PoE described above, the following items provide a simple path to qualify an installation according to 725.144:

1. When cabling is 24 AWG or larger and the rated current is less than 0.3 ampere per conductor, the installation meets 725.144.
2. If the plans call for higher current (above 0.3 amperes per conductor but below 0.433 amperes per conductor), the installation meets 725.144 if the answer to any of the following questions is YES:

   a. Is the cabling rated as limited power (LP) 0.6 or greater?
   b. Are fewer than 37 ports of PoE provided?
   c. Are the cables bundled in groups of 37 or fewer?

   If neither item 1 nor 2 applies, then the bundle size and temperature rating of the cable will have to be checked.

3. Is 75°C cabling of 23 AWG or larger being used in bundles of 192 or fewer?

   Note: TIA-568.2-D Category 6A cabling ("Cat 6a") is typically 23 AWG; other categories of cabling may be in 23 AWG variants.

4. Is 60°C cabling of 23 AWG or larger being used in bundles of 61 or fewer?
5. Is 75°C cabling of 24 AWG or larger being used in bundles of 91 or fewer?

   Note: TIA-568.2-D Category 5E cabling ("Cat 5e") is typically 24 AWG.

The values in Table 725.144 are specified at an ambient temperature of 30°C. The cable bundles are referred to as "horizontal cables" in the TIA/EIA 568 standards, which specify an ambient

COMMENTARY TABLE 725.2 *NEC* Table 725.144 Adjusted for Ambient Temperature of 45°C (113°F)

AWG	Number of 4-Pair Cables in a Bundle																	
	1–7			8–19			20–37			38–61			62–91			92–192		
	Temperature Rating			Temperature Rating			Temperature Rating			Temperature Rating			Temperature Rating			Temperature Rating		
	60°C	75°C	90°C	60°C	75°C	90°C	60°C	75°C	90°C	60°C	75°C	90°C	60°C	75°C	90°C	60°C	75°C	90°C
26	0.71	1.00	1.23	0.50	0.71	0.88	0.39	0.56	0.68	0.33	0.47	0.58	0.32	0.45	0.55	NA	NA	NA
24	0.84	1.19	1.46	0.57	0.82	1.01	0.45	0.64	0.79	0.39	0.55	0.68	0.33	0.46	0.56	0.28	0.39	0.48
23	0.88	1.25	1.54	0.63	0.91	1.11	0.54	0.78	0.95	0.47	0.65	0.81	0.41	0.58	0.71	0.32	0.45	0.55
22	1.06	1.52	1.87	0.74	1.05	1.29	0.54	0.78	0.96	0.47	0.67	0.83	0.44	0.63	0.77	0.37	0.51	0.62

temperature of 45°C. Commentary Table 725.2 is Table 725.144 adjusted to an ambient temperature of 45°C using equation 310.15(B).

(A) Use of 4-Pair Class 2 or Class 3 Cables to Transmit Power and Data. Where Type CL3P, Type CL2P, Type CL3R, Type CL2R, Type CL3, or Type CL2 4-pair cables transmit power and data, the rated current per conductor of the power source shall not exceed the ampacities in Table 725.144 at an ambient temperature of 30°C (86°F). For ambient temperatures above 30°C (86°F), the correction factors in Table 310.15(B)(1)(1) or in Equation 310.15(B) shall apply.

Exception: Compliance with Table 725.144 shall not be required for installations where conductors are 24 AWG or larger and the rated current per conductor of the power source does not exceed 0.3 amperes.

Informational Note: One example of the use of Class 2 cables is a network of closed-circuit TV cameras using 24 AWG, 60°C rated, Type CL2R, Category 5e balanced twisted-pair cabling.

Δ **(B) Use of Class 2-LP or Class 3-LP Cables to Transmit Power and Data.** Type CL3P-LP, Type CL2P-LP, Type CL3R-LP, Type CL2R-LP, Type CL3-LP, or Type CL2-LP cables shall be permitted to supply power to equipment from a power source with a rated current per conductor up to the marked current limit located immediately following the suffix "-LP" and shall be permitted to transmit data to the equipment. Where the number of bundled LP cables is 192 or less and the selected ampacity of the cables in accordance with Table 725.144 exceeds the marked current limit of the cable, the ampacity determined from the table shall be permitted to be used. For ambient temperatures above 30°C (86°F), the correction factors of Table 310.15(B)(1)(1) or Equation 310.15(B) shall apply. The Class 2-LP and Class 3-LP cables shall comply with the following, as applicable:

(1) Cables with the suffix "-LP" shall be permitted to be installed in bundles, raceways, cable trays, communications raceways, and cable routing assemblies.
(2) Cables with the suffix "-LP" and a marked current limit shall follow the substitution hierarchy of 722.135(E) for the cable type without the suffix "-LP" and without the marked current limit.
(3) System design shall be permitted by qualified persons under engineering supervision.

Informational Note: An example of the marking on a 23 AWG, 4-pair, Class 2 cable rated 75°C with an LP current rating of 0.6 amperes per conductor is "CL2-LP(0.6A) 75°C 23 AWG 4-pair". See 722.179(A)(9).

Part III. Listing Requirements

725.160 Listing and Marking of Equipment for Power and Data Transmission. The listed power source for circuits intended to provide power and data over Class 2 cables to remote equipment shall be as specified in 725.60(A)(1), (A)(2), (A)(3), or (A)(4). In accordance with 725.60(B), the power sources shall not have the output connections paralleled or otherwise interconnected, unless listed for such interconnection. Powered devices connected to a circuit supplying data and power shall be listed. Marking of equipment output connections shall be in accordance with 725.60(C).

N **ARTICLE 726 Class 4 Fault-Managed Power Systems**

N **Part I. General**

N **726.1 Scope.** This article covers the installation of wiring systems and equipment, including utilization equipment, of Class 4 fault-managed power (FMP) systems.

Informational Note No. 1: Class 4 fault-managed power systems consist of a Class 4 power transmitter and a Class 4 power receiver connected by a Class 4 cabling system. These systems are characterized by monitoring the circuit for faults and controlling the source current to ensure the energy delivered into any fault is limited. Class 4 systems differ from Class 1, Class 2, and Class 3 systems in that they are not limited for power delivered to an appropriate load. They are current limited for faults between the Class 4 transmitter and Class 4 receiver.

Informational Note No. 2: The circuits described in this article are characterized by monitoring and control systems that differentiate them from electric light and power circuits; therefore, alternative requirements to those of Chapters 1 through 4 are given.

Class 4, or fault-managed power (FMP), systems are not Power over Ethernet (PoE) systems, which are Class 2 or Class 3 systems. Class 4 systems are designed to enable high-voltage power, not over 450-volts peak ac or dc, to be delivered with enhanced safety, as both line-to-line and line-to-ground faults are mitigated within milliseconds of occurrence resulting in no harm to people or property. Class 4 system transmitters (TX) and receivers (RX) must be listed and from the same manufacturer to function properly. The TX and RX are interconnected with listed Class 4 cables. See UL 1400-2, *Outline of Investigation for Fault-Managed Power Systems — Part 2 Requirements for Cables*, at https://iq.ulprospector.com/en/. Class 4 circuits must be durably marked where plainly visible with the maximum voltage and current output.

N **726.3 Other Articles.** The listing and installation of cables for Class 4 circuits shall comply with Article 722. Only those sections of Article 300 referenced in Article 722 shall apply to Class 4 circuits.

N **726.10 Hazardous (Classified) Locations.** Class 4 power systems shall be permitted to be used in hazardous (classified) locations where specifically permitted by other articles in this *Code*.

N **726.12 Uses Not Permitted.** Class 4 power systems shall not be permitted in dwelling units.

N **726.24 Mechanical Execution of Work.** Class 4 equipment shall be installed in a neat and workmanlike manner. The installation shall also comply with 300.4 and 300.11.

N ## Part II. Class 4 Circuits

N **726.121 Power Sources for Class 4 Circuits.** The power source shall be a listed Class 4 power transmitter or a listed Class 4 power transmitter as part of a transmitter/receiver system and shall provide the protections in accordance with 726.121(A). Class 4 circuits shall be supplied from a power source (transmitter) that has a voltage output of not more than 450 volts peak or dc.

> Informational Note No. 1: Informational Note Figure 726.121 illustrates the relationships between Class 4 power transmitters (power sources), Class 4 circuits, Class 4 power receivers, and utilization equipment.
>
> Informational Note No. 2: See UL 1400-1, *Outline for Fault-Managed Power Systems — Part 1: General Requirements*, for information on determining applicable requirements for the listing of Class 4 power systems.

N **(A) Fault Management.** For listing purposes, a transmitter shall interrupt an energized circuit when any of the following conditions occur on the circuit between the transmitter and receiver:

(1) A short circuit
(2) A line-to-line fault condition that presents an unacceptable risk of fire or electric shock
(3) A ground-fault condition that presents an unacceptable risk of fire or electric shock

(4) An overcurrent condition
(5) A malfunction of the monitoring or control system that presents an unacceptable risk of fire or electric shock
(6) Any other condition that presents an unacceptable risk of fire or electric shock

> Informational Note: See UL 1400-1, *Outline for Fault-Managed Power Systems — Part 1: General Requirements*, for information on determining applicable requirements for the listing of Class 4 power systems, including safe operation and limiting the risk of fire and electric shock.

N **726.122 Class 4 Loads.** Outputs of a Class 4 receiver and power outputs of Class 4 utilization equipment shall be considered a separately derived system if the outputs are used as a supply for a feeder or branch circuit.

> Informational Note: Class 4 utilization equipment that does not provide power outputs is not subject to these requirements.

Exception: A Class 4 receiver with limited-power circuit outputs shall be permitted to meet the requirements of Part II of Article 725.

N **726.124 Class 4 Marking.**

N **(A) Class 4 Transmitter Marking.** The equipment supplying the Class 4 circuits shall be durably marked where plainly visible to indicate each circuit that is a Class 4 circuit. The marking shall also include the maximum voltage and current output for each connection point. Where multiple connection points have the same rating, a single label shall be permitted to be used.

> Informational Note: An example of marking is "Class 4: +/–190V, 5A" for a Class 4 transmitter capable of delivering 1.9 kW from 380 volts line to line.

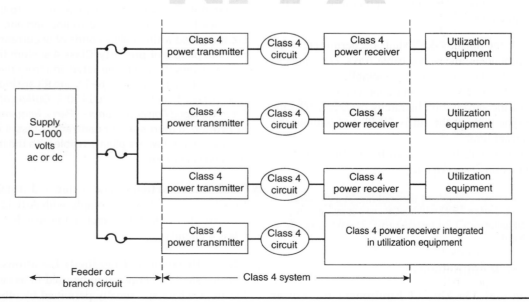

N **INFORMATIONAL NOTE FIGURE 726.121** *Class 4 Circuits.*

N **(B) Class 4 Receiver Marking.**

N **(1) Class 4 Circuits.** A Class 4 receiver or Class 4 utilization equipment shall be durably marked where plainly visible to indicate each circuit that is a Class 4 circuit. The marking shall include the maximum input voltage and current for each connection point.

N **(2) Output Terminals and Socket Outlets.** Where the Class 4 receiver or Class 4 utilization equipment has outputs, terminals, or socket outlets for providing power to other equipment, each output shall be durably marked where plainly visible. The marking shall include the maximum output voltage and current for each connection point. Where multiple connection points have the same rating, a single label shall be permitted to be used. Class 1, Class 2, and Class 3 circuits shall be identified in accordance with 724.30 or Part II of Article 725.

N **726.130 Terminals and Connectors.**

N **(A) Listing.** Connecting hardware used on Class 4 distribution systems shall be listed.

N **(B) Noninterchangeability.** Connectors for Class 4 circuits shall be designed such that they are not interchangeable with non-power-limited sources located on the same premises.

N **(C) Guarding.** Any junctions and mating connectors shall be constructed and installed to guard against inadvertent contact with live parts by persons.

N **726.136 Separation from Electric Light, Power, Class 1, Non–Power-Limited Fire Alarm Circuit, and Medium-Power Network-Powered Broadband Communications Cables.**

N **(A) General.** Cables and conductors of Class 4 circuits shall not be placed in any cable, cable tray, compartment, enclosure, manhole, outlet box, device box, raceway, or similar fitting with conductors of electric light, power, Class 1, non–power-limited fire alarm, and medium-power network-powered broadband communications circuits unless permitted by 726.136(B) through (H).

In general, because of the potential higher voltage and strict monitoring of Class 4 circuits, the cables are not allowed to be installed with electric light, power, Class 1, non–power-limited fire alarm circuits, and medium-power network-powered broadband communications cables. Class 4 circuit power and current are continually monitored by the transmitters (TX) and receivers (RX) to clear any line-to-line or line-to-ground faults. However, 726.136(B) through (H) does provide for approved methods of separation.

N **(B) Separated by Barriers.** Class 4 circuits shall be permitted to be installed together with the conductors of electric light, power, Class 1, non–power-limited fire alarm, and medium-power network-powered broadband communications circuits where they are separated by a barrier.

N **(C) Raceways Within Enclosures.** In enclosures, Class 4 circuits shall be permitted to be installed in a raceway to separate them from Class 1, non–power-limited fire alarm, and medium-power network-powered broadband communications circuits.

N **(D) Associated Systems Within Enclosures.** Class 4 circuit conductors in compartments, enclosures, device boxes, outlet boxes, or similar fittings shall be permitted to be installed with electric light, power, Class 1, non–power-limited fire alarm, and medium-power network-powered broadband communications circuits where they are introduced solely to connect the equipment connected to Class 4 circuits, and where either of the following applies:

(1) The electric light, power, Class 1, non–power-limited fire alarm, and medium-power network-powered broadband communications circuit conductors are routed to maintain a minimum of 6 mm (0.25 in.) separation from the conductors and cables of Class 4 circuits.

(2) The non–Class 4 circuit conductors operate at 150 volts or less to ground and the Class 4 circuits are installed using Type CL4, Type CL4R, or Type CL4P cables if any CL4 cable conductors extending beyond the jacket are separated by a minimum of 6 mm (0.25 in.) or by a nonconductive sleeve or nonconductive barrier from all other conductors.

An example of associated systems is a programmable circuit breaker or alarm system. In such an installation, the Class 4 conductor insulation is not required to have the same voltage rating as the insulation on the power conductors in the same enclosure. They do require a 6 millimeter or ¼ inch separation at all points.

N **(E) Enclosures with Single Openings.** Class 4 circuit conductors entering compartments, enclosures, device boxes, outlet boxes, or similar fittings shall be permitted to be installed with Class 1, non–power-limited fire alarm, and medium-power network-powered broadband communications circuits where they are introduced solely to connect the equipment connected to Class 4 circuits. Where Class 4 circuit conductors must enter an enclosure that is provided with a single opening, they shall be permitted to enter through a single fitting (such as a tee) if the conductors are separated from the conductors of the other circuits by a continuous and firmly fixed nonconductor, such as flexible tubing.

N **(F) Manholes.** Underground Class 4 circuit conductors in a manhole shall be permitted to be installed with Class 1, non–power-limited fire alarm, and medium-power network-powered broadband communications circuits where one of the following conditions is met:

(1) The electric light, power, Class 1, non–power-limited fire alarm, and medium-power network-powered broadband communications circuit conductors are in a metal-enclosed cable or Type UF cable.

(2) The Class 4 circuit conductors are permanently and effectively separated from the conductors of other circuits by a continuous and firmly fixed nonconductor, such as flexible tubing, in addition to the insulation or covering on the wire.

(3) The Class 4 circuit conductors are permanently and effectively separated from conductors of the other circuits and securely fastened to racks, insulators, or other approved supports.

N **(G) Cable Trays.** Class 4 circuit conductors shall be permitted to be installed in cable trays where the conductors of the electric light, Class 1, and non–power-limited fire alarm circuits are separated by a solid fixed barrier of a material compatible with the cable tray or where the Class 4 circuits are installed in Type MC cable.

N **(H) Other Applications.** For other applications, conductors of Class 4 circuits shall be separated by at least 50 mm (2 in.) from conductors of any electric light, power, Class 1, non–power-limited fire alarm, or medium-power network-powered broadband communications circuits unless one of the following conditions is met:

(1) Either all of the electric light, power, Class 1, non–power-limited fire alarm, and medium-power network-powered broadband communications circuit conductors or all of the Class 4 circuit conductors are in a raceway or in metal-sheathed, metal-clad, non–metallic-sheathed, Type TC, or Type UF cables

(2) All of the electric light, power, Class 1, non–power-limited fire alarm, and medium-power network-powered broadband communications circuit conductors are permanently separated from all of the Class 4 circuit conductors by a continuous and firmly fixed nonconductor, such as porcelain tubes or flexible tubing, in addition to the insulation on the conductors

N **726.139 Installation of Conductors of Different Circuits in the Same Cable, Enclosure, Cable Tray, Raceway, or Cable Routing Assembly.**

N **(A) Two or More Class 4 Circuits.** Conductors of two or more Class 4 circuits shall be permitted within the same cable, enclosure, raceway, or cable routing assembly.

N **(B) Class 4 Circuits With Class 2, Class 3, or Communications Circuits.** Conductors of one or more Class 4 circuits shall be permitted within the same cable assembly as conductors of Class 2, Class 3, or communications circuits if the insulation of the Class 2, Class 3, or communications circuit conductors in the cable is at least that required for Class 4 circuits. Class 4 cables shall be permitted within the same enclosure, raceway, or cable routing assembly as Class 2, Class 3, or communications circuits.

N **(C) Class 4 Cables With Other Circuit Cables.** Jacketed cables of Class 4 circuits shall be permitted in the same enclosure, cable tray, raceway, or cable routing assembly with jacketed cables of any of the following:

(1) Power-limited fire alarm systems in compliance with Parts I and III of Article 760

(2) Nonconductive and conductive optical fiber cables in compliance with Parts I and IV of Article 770

(3) Communications circuits in compliance with Parts I and IV of Article 805

(4) Community antenna television and radio distribution systems in compliance with Parts I and IV of Article 820

(5) Low-power, network-powered broadband communications in compliance with Parts I and IV of Article 830

N **726.144 Ampacity.** The ampacity of Class 4 cables shall comply with 300.15 based on the temperature rating of the Class 4 cable for conductors sized 16 AWG to 6 AWG. For conductors sized 24 AWG to 17 AWG, the Class 4 cable shall be rated for the intended ampacity as evidenced by the marking FMP-XXA, where XX is the maximum allowable ampacity permitted.

> Informational Note No. 1: See 722.179(A)(16) for additional Class 4 cable requirements.
>
> Informational Note No. 2: See UL 1400-1, *Outline of Investigation for Fault-Managed Power Systems — Part 1: General Requirements,* and UL 1400-2, *Outline of Fault-Managed Power Systems — Part 2: Requirements for Class 4 Cables,* for information on determining maximum allowable ampacities.

N **Part III. Listing Requirements**

N **726.170 Listing of Equipment for Class 4 Systems.** The active components of a Class 4 system shall be listed as a Class 4 device. The listing information shall include compatible devices if a listed Class 4 device depends on specific system devices for interoperability, monitoring, or control.

> Informational Note No. 1: See UL 1400-1, *Outline for Fault-Managed Power Systems — Part I: General Requirements,* for information on determining applicable requirements for the listing of Class 4 power systems.
>
> Informational Note No. 2: An example of a dependent active device in a Class 4 system is a transmitter that relies on a particular receiver or receivers as part of the monitoring and control system.

ARTICLE
728 **Fire-Resistive Cable Systems**

728.1 Scope. This article covers the installation of fire-resistive cables, fire-resistive conductors, and other system components used for survivability of critical circuits to ensure continued operation during a specified time under fire conditions as required in this *Code.*

728.3 Other Articles. Wherever the requirements of other articles of this *Code* and Article 728 differ, the requirements of Article 728 shall apply.

This article does not require the use of fire-resistive cable systems. It provides requirements if such systems are required to be installed. Some of the common applications of fire-resistive cable systems are fire pump installations and critical operations power systems (COPS). Often, the use of fire-resistive cable systems will be a design consideration to ensure circuit integrity.

△ 728.4 General. Fire-resistive cables and conductors and their components shall be tested and listed as a complete system, shall be designated for use in a specific system, and shall not be interchangeable between systems.

> Informational Note: One method of defining the fire rating is by testing the system in accordance with UL 2196, *Fire Test for Circuit Integrity of Fire-Resistive Power, Instrumentation, Control and Data Cables.*

728.5 Installations. Fire-resistive cable systems installed outside the fire-rated rooms that they serve, such as the electrical room or the fire pump room, shall comply with the requirements of 728.5(A) through (H) and all other installation instructions provided in the listing.

(A) Mounting. The fire-resistive cable system shall be secured to the building structure in accordance with the listing and the manufacturer's installation instructions.

(B) Supports. The fire-resistive cable system shall be supported in accordance with the listing and the manufacturer's installation instructions.

> Informational Note: The supports are critical for survivability of the system. Each system has its specific support requirements.

△ (C) Raceways and Couplings. Where fire-resistive cable is listed to be installed in a raceway, the raceway enclosing the cable, any couplings, and any connectors shall be listed as part of the fire-resistive cable system.

The raceway fill for each system shall comply with the listing requirements for the system and shall not be greater than the fill permitted in Chapter 9, Table 1.

> Informational Note: Raceway fill might not be the same for all listed fire-resistive cable systems.

(D) Cable Trays. Cable trays used as part of a fire-resistive cable system shall be listed as part of the fire-resistive cable system.

(E) Boxes. Boxes or enclosures used as part of a fire-resistive cable system shall be listed as part of the fire-resistive cable system and shall be secured to the building structure independently of the raceways or cables listed in the system.

△ (F) Pulling Lubricants. Fire-resistive cable installed in a raceway shall only use pulling lubricants listed as part of the fire-resistive cable system.

(G) Vertical Supports. Cables and conductors installed in vertical raceways shall be supported in accordance with the listing of the fire-resistive cable system and in accordance with 300.19.

(H) Splices. Only splices that are part of the listing for the fire-resistive cable system shall be used. Splices shall have manufacturer's installation instructions.

△ 728.60 Equipment Grounding Conductor. Fire-resistive cables installed in a raceway requiring an equipment grounding conductor shall use the same fire-resistive cable described in the system unless alternative equipment grounding conductors are listed with the system. Any alternative equipment grounding conductors shall be marked with the system number. The system shall specify a permissible equipment grounding conductor. If not specified, the equipment grounding conductor shall be the same as the fire-resistive cable described in the system.

728.120 Marking. In addition to the marking required in 310.8, system cables and conductors shall be surface marked with the suffix "FRR" (fire-resistive rating), along with the circuit integrity duration in hours, and with the system identifier.

ARTICLE 750 Energy Management Systems

750.1 Scope. This article applies to the installation and operation of energy management systems.

> Informational Note: Performance provisions in other codes establish prescriptive requirements that may further restrict the requirements contained in this article.

This article addresses the installation and operation of energy management systems. Energy management systems have become integral elements of the electrical infrastructure through the control of utilization equipment, energy storage, and power protection. Energy management has two basic aspects: monitoring the system and controlling some part of the system. These two elements must be separated in order to allow the system to monitor and possibly restrict those areas of control that would adversely affect the electrical system. The requirements ensure that an energy management system does not overload a branch circuit, feeder, or service or override a load-shedding system for an alternate power source for fire pumps and other emergency systems.

N 750.6 Listing. Energy management systems shall be one of the following:

(1) Listed as a complete energy management system
(2) Listed as a kit for field installation in switch or overcurrent device enclosures
(3) Listed individual components assembled as a system

750.20 Alternate Power Sources. An energy management system shall not override any control necessary to ensure continuity of an alternate power source for the following:

(1) Fire pumps
(2) Health care facilities

(3) Emergency systems
(4) Legally required standby systems
(5) Critical operations power systems

750.30 Load Management. Energy management systems shall be permitted to monitor and control electrical loads and sources in accordance with 750.30(A) through (C).

Systems necessary for life safety, fire protection, and critical operations must continue to operate even with the loss of primary power. Load shedding is often employed for the alternate power source of a backup system to give priority to needed equipment. Disconnecting some equipment — such as a ventilation system that prevents an explosive concentration from being reached — could introduce safety hazards and must be avoided. An energy management system must not disconnect or override the control of those systems.

(A) Load Shedding Controls. An energy management system shall not override the load shedding controls put in place to ensure the minimum electrical capacity for the following:

(1) Fire pumps
(2) Emergency systems
(3) Legally required standby systems
(4) Critical operations power systems

Δ **(B) Disconnection of Power.** An energy management system shall not cause disconnection of power to the following:

(1) Elevators, escalators, moving walks, or stairway lift chairs
(2) Positive mechanical ventilation for hazardous (classified) locations
(3) Ventilation used to exhaust hazardous gas or reclassify an area
(4) Circuits supplying emergency lighting
(5) Essential electrical systems in health care facilities

(C) Capacity of Branch Circuit, Feeder, or Service. An energy management system shall not cause a branch circuit, feeder, or service to be overloaded. If an EMS is used to limit the current on a conductor, 750.30(C)(1) through (C)(4) shall apply:

N **(1) Current Setpoint.** A single value equal to the maximum ampere setpoint of the EMS shall be permitted for one or more of the following:

(1) For calculating the connected load per 220.70
(2) For the maximum source current permitted by EMS control

N **(2) System Malfunction.** The EMS shall use monitoring and controls to automatically cease current flow upon malfunction of the EMS.

N **(3) Settings.** Adjustable settings shall be permitted if access to the settings is accomplished by at least one of the following:

(1) Located behind removable and sealable covers over the adjustment means

(2) Located behind a cover or door that requires the use of a tool to open
(3) Located behind locked doors accessible only to qualified personnel
(4) Password protected with password accessible only to qualified personnel
(5) Software that has password protected access to the adjusting means accessible to qualified personnel only

N **(4) Marking.** The equipment that supplies the branch circuit, feeder, or service shall be field marked with the following information:

(1) Maximum current setting
(2) Date of calculation and setting
(3) Identification of loads and sources associated with the current limiting feature
(4) The following or equivalent wording: "The setting for the EMS current limiting feature shall not be bypassed"

The markings shall meet the requirements in 110.21(B) and shall be located such that they are clearly visible to qualified persons before examination, adjustment, servicing, or maintenance of the equipment.

Δ **750.50 Directory.** Where an energy management system is employed to control electrical power through the use of a remote means, a directory identifying the controlled device(s) and circuit(s) shall be posted on the enclosure of the controller, disconnect, or branch-circuit overcurrent device.

ARTICLE 760   Fire Alarm Systems

Part I. General

Δ **760.1 Scope.** This article covers the installation of wiring and equipment of fire alarm systems, including all circuits controlled and powered by the fire alarm system.

> Informational Note No. 1: Fire alarm systems include fire detection and alarm notification, guard's tour, sprinkler waterflow, and sprinkler supervisory systems. Circuits controlled and powered by the fire alarm system include circuits for the control of building systems safety functions, elevator capture, elevator shutdown, door release, smoke doors and damper control, fire doors and damper control, and fan shutdown, but only where these circuits are powered by and controlled by the fire alarm system.
>
> Informational Note No. 2: See *NFPA 72, National Fire Alarm and Signaling Code,* for further information on the installation and monitoring for integrity requirements for fire alarm systems.

Article 760 covers only circuits that are powered and controlled by the fire alarm system, including fire safety features such as smoke door control, damper control, fan shutdown, and elevator recall. Circuits powered and controlled by other building systems such as heating, ventilating, and air conditioning (HVAC); security; lighting controls; and time recording are covered by Article 725.

NFPA 72®, *National Fire Alarm and Signaling Code®*, requires that all wiring, cable, and equipment be in accordance with the *NEC®* and specifically with Article 760. *NFPA 72* provides the requirements for the listing, selection, installation, performance, use, testing, and maintenance of fire alarm system components. To determine if a specific occupancy is required to have a fire alarm system, see *NFPA 101®, Life Safety Code®*, or other local codes.

Examples of fire alarm equipment and devices are shown in Exhibits 760.1 and 760.2. Single- and multiple-station smoke alarms, such as those commonly installed in dwelling units, are supplied through 120-volt branch circuits rather than through a fire alarm signaling circuit that is powered and controlled by a fire alarm control panel. Branch circuits supplying power to single- and multiple-station smoke alarms are not subject to the requirements of Article 760.

760.3 Other Articles. Circuits and equipment shall comply with 760.3(A) through (O). Only those sections of Article 300 referenced in this article shall apply to fire alarm systems.

EXHIBIT 760.1 Typical fire alarm control unit. *(Courtesy of the International Association of Electrical Inspectors)*

EXHIBIT 760.2 Typical spot-type smoke detector.

(A) Spread of Fire or Products of Combustion. Installation of fire alarm circuits shall comply with 300.21.

(B) Ducts, Plenums, and Other Air-Handling Spaces. Power-limited and non-power-limited fire alarm cables installed in ducts, plenums, or other spaces used for environmental air shall comply with 300.22.

Exception No. 1: Power-limited fire alarm cables selected in accordance with Table 760.154 and installed in accordance with 722.135 and 300.22(B), Exception, shall be permitted to be installed in ducts specifically fabricated for environmental air.

Exception No. 2: Power-limited fire alarm cables selected in accordance with Table 760.154 and installed in accordance with 722.135 shall be permitted to be installed in other spaces used for environmental air (plenums).

See also

300.22(B), 300.22(C), and associated commentary for more information on wiring installed in ducts, plenums, or other air-handling spaces

Δ **(C) Corrosive, Damp, or Wet Locations.** Fire alarm circuits and equipment installed in corrosive, damp, or wet locations shall comply with 110.11, 300.5(B), 300.6, 300.9, and 310.10(F).

Cables and equipment used in wet or damp locations, high ambient temperature areas, or corrosive locations must be identified as suitable for the environment. Underground installations are considered wet locations.

Δ **(D) Building Control Circuits.** Building control systems (e.g., elevator capture, fan shutdown) associated with the fire alarm system shall comply with Article 725.

(E) Optical Fiber Cables. Where optical fiber cables are utilized for fire alarm circuits, the cables shall be installed in accordance with Article 770.

(F) Installation of Conductors with Other Systems. Installations shall comply with 300.8.

(G) Raceways or Sleeves Exposed to Different Temperatures. Installations shall comply with 300.7(A).

Condensation often forms in conduit that runs between spaces having different temperatures, such as from an air-conditioned space to a space that is not air conditioned or into a freezer. Section 300.7(A) requires blocking of the circulation of warm air into the colder section of the raceway.

(H) Vertical Support for Fire-Resistive Cables and Conductors. Vertical installations of circuit integrity (CI) cables and conductors installed in a raceway or conductors and cables of fire-resistive cable systems shall be installed in accordance with 300.19.

Support requirements for fire-rated cable are located in 300.19(B). The strength of cables and conductors decreases with heat, and they could break if not properly supported. Support of vertical runs of fire-rated cables helps ensure continued performance if the cable is exposed directly to a fire or to high temperatures caused by a fire.

(I) Installation of Cables and Conductors in Raceway. The number and size of cables and conductors shall comply with 300.17.

Compliance with the raceway fill requirement protects conductors from abrasion or other physical damage resulting from too many conductors being pulled through a raceway. In addition, the ability to install and withdraw conductors simplifies maintenance.

(J) Bushing. A bushing shall be installed where cables emerge from raceway used for mechanical support or protection in accordance with 300.15(C).

(K) Cable Routing Assemblies. Power-limited fire alarm cables shall be permitted to be installed in plenum cable routing assemblies, riser cable routing assemblies, and general-purpose cable routing assemblies selected in accordance with Table 800.154(c), listed in accordance with 800.182, and installed in accordance with 800.110(C) and 800.113.

(L) Communications Raceways. Power-limited fire alarm cables shall be permitted to be installed in plenum communications raceways, riser communications raceways, and general-purpose communications raceways selected in accordance with Table 800.154(b), listed in accordance with 800.182, and installed in accordance with 800.113 and 362.24 through 362.56, where the requirements applicable to electrical nonmetallic tubing apply.

(M) Temperature Limitations of Power-Limited and Non–Power-Limited Fire Alarm Cables. The requirements of 310.14(A)(3) on the temperature limitation of conductors shall apply to power-limited fire alarm cables and non–power-limited fire alarms cables.

(N) Identification of Equipment Grounding Conductors. Equipment grounding conductors shall be identified in accordance with 250.119.

Exception: Conductors with green insulation shall be permitted to be used as ungrounded signal conductors for Types FPLP, FPLR, FPL, and substitute cables installed in accordance with 760.154(A).

N **(O) Cables for Power-Limited Fire Alarm (PLFA) Circuits.** The listing and installation of cables for power-limited fire alarm circuits shall comply with Part III of this article and Parts I and II of Article 722.

N **760.10 Hazardous (Classified) Locations.** Cables and equipment shall be permitted to be used in hazardous (classified) locations where specifically permitted by other articles in this *Code*.

760.21 Access to Electrical Equipment Behind Panels Designed to Allow Access. Access to electrical equipment shall not be denied by an accumulation of conductors and cables that prevents removal of panels, including suspended ceiling panels.

An excess accumulation of wires and cables can limit access to equipment by preventing the removal of access panels. See Exhibit 760.3.

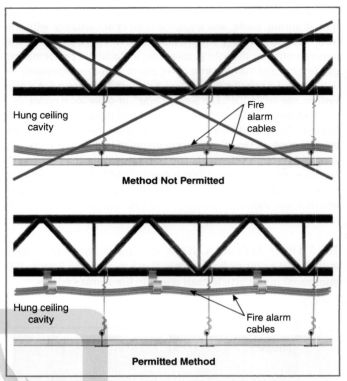

EXHIBIT 760.3 Incorrect cable installation *(upper diagram)* and correct method *(lower diagram)*.

760.24 Mechanical Execution of Work.

(A) General. Fire alarm circuits shall be installed in a neat and workmanlike manner. Cables and conductors installed exposed on the surface of ceilings and sidewalls shall be supported by the building structure in such a manner that the cable will not be damaged by normal building use. Such cables shall be supported by hardware, including straps, staples, hangers, listed cable ties identified for securement and support, or similar fittings designed and installed so as not to damage the cable. The installation shall also comply with 300.4 and 300.11.

Informational Note: Paint, plaster, cleaners, abrasives, corrosive residues, or other contaminants might result in an undetermined alteration of PLFA and NPLFA cable properties.

The location of fire alarm cables should be carefully evaluated to ensure that activities and processes within the building do not cause damage to the cable.

The reference to 300.4 calls attention to the hazards to which cables are exposed where they are installed on framing members. Such cables are required to be installed in a manner that protects them from nail or screw penetration. This section permits attachment to baseboards and non-load-bearing walls, which are not structural components.

(B) Circuit Integrity (CI) Cable. Circuit integrity (CI) cables shall be supported at a distance not exceeding 610 mm (24 in.). Where located within 2.1 m (7 ft) of the floor in accordance with 760.53(A)(1) and 760.130(B)(1), as applicable, the cable shall be fastened in an approved manner at intervals of not more than 450 mm (18 in.). Cable supports and fasteners shall be steel.

760.25 Abandoned Cables. The accessible portion of abandoned fire alarm cables shall be removed. Where cables are identified for future use with a tag, the tag shall be of sufficient durability to withstand the environment involved.

Abandoned cables increase fire loading unnecessarily and, if installed in plenums, can affect airflow.

See also

Article 100 for the definition of the term *cable, abandoned* *(abandoned cable)*

760.30 Fire Alarm Circuit Identification. Fire alarm circuits shall be identified at terminal and junction locations in a manner that helps to prevent unintentional signals on fire alarm system circuit(s) during testing and servicing of other systems.

One way to facilitate circuit identification is to use a terminal cabinet with permanently mounted and labeled terminals, such as the one shown in Exhibit 760.4. Another common method is to paint fire alarm system circuit junction box covers red and/or label them with the words "FIRE ALARM." Some jurisdictions require that all conduits carrying fire alarm system circuits be red. Other jurisdictions require a red stripe every 10 feet or red fittings where specific lengths of conduit are joined for fire alarm system circuit conduits.

Δ **760.32 Fire Alarm Circuits Extending Beyond One Building.** Non–power-limited fire alarm circuits and power-limited fire alarm circuits that extend beyond one building and run outdoors shall meet the installation requirements of Parts II, III, and IV of Article 805 and shall meet the installation requirements of Part I of Article 300.

N **760.33 Supply-Side Overvoltage Protection.** A listed surge-protective device (SPD) shall be installed on the supply side of a fire alarm control panel in accordance with Part II of Article 242.

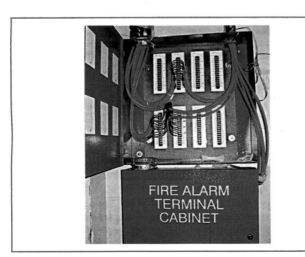

EXHIBIT 760.4 Fire alarm terminal cabinet. *(Courtesy of JENSEN HUGHES, Warwick, RI)*

760.35 Fire Alarm Circuit Requirements. Fire alarm circuits shall comply with 760.35(A) and (B).

(A) Non–Power-Limited Fire Alarm (NPLFA) Circuits. See Parts I and II.

(B) Power-Limited Fire Alarm (PLFA) Circuits. See Parts I and III.

Power source limitations for power-limited fire alarm circuits used by testing laboratories are found in Chapter 9, Tables 12(A) and 12(B). Table 12(A) covers alternating-current (ac) source limitations, and Table 12(B) covers direct-current (dc) source limitations.

Part II. Non–Power-Limited Fire Alarm (NPLFA) Circuits

760.41 NPLFA Circuit Power Source Requirements.

(A) Power Source. The power source of non–power-limited fire alarm circuits shall comply with Chapters 1 through 4, and the output voltage shall be not more than 600 volts, nominal. The fire alarm circuit disconnect shall be permitted to be secured in the "on" position.

This section correlates with *NFPA 72, National Fire Alarm and Signaling Code,* which requires the circuit disconnecting means to be accessible only to authorized personnel. Limiting access decreases the chance that the power to the fire alarm system is turned off.

Δ **(B) Branch Circuit.** The branch circuit supplying the fire alarm equipment(s) shall supply no other loads. The location of the branch-circuit overcurrent protective device shall be permanently identified at the fire alarm control unit. The circuit disconnecting means shall have red identification, shall be accessible only to qualified personnel, and shall be identified as "FIRE ALARM CIRCUIT." The red identification shall not damage the overcurrent protective devices or obscure the manufacturer's markings. This branch circuit shall not be supplied through ground-fault circuit interrupters or arc-fault circuit-interrupters.

NFPA 72, National Fire Alarm and Signaling Code, requires that the power to the fire alarm system be supplied from a branch circuit dedicated to the fire alarm system. The dedicated circuit can be used to power other equipment that is part of the system, but this power circuit cannot be used to power other equipment, such as telephone switches, computer stations, and other equipment that is not directly associated with the fire alarm system functions. Further, *NFPA 72* requires the location of the branch-circuit disconnecting means to be permanently identified at the control unit.

To minimize interruption of the normal ac power, non–power-limited fire alarm equipment is not permitted to be supplied by a branch circuit protected with a GFCI device or with an AFCI device. The AFCI protection requirements apply to outlets supplying single- or multiple-station smoke alarms where the smoke alarm outlet is located in an area covered by the requirements of 210.12(A). These smoke alarms are supplied by a branch circuit covered by the requirements of Article 210, not by the fire alarm control panel. In new construction, single- and multiple-station smoke alarms are required by *NFPA 72* to have a backup battery that will supply power in the event that the branch-circuit power is interrupted due to the operation of an AFCI device.

760.43 NPLFA Circuit Overcurrent Protection. Overcurrent protection for conductors 14 AWG and larger shall be provided in accordance with the conductor ampacity without applying the ampacity adjustment and correction factors of 310.14 to the ampacity calculation. Overcurrent protection shall not exceed 7 amperes for 18 AWG conductors and 10 amperes for 16 AWG conductors.

Exception: Where other articles of this Code permit or require other overcurrent protection.

760.45 NPLFA Circuit Overcurrent Device Location. Overcurrent devices shall be located at the point where the conductor to be protected receives its supply.

Exception No. 1: Where the overcurrent device protecting the larger conductor also protects the smaller conductor.

Exception No. 2: Transformer secondary conductors. Non–power-limited fire alarm circuit conductors supplied by the secondary of a single-phase transformer that has only a 2-wire (single-voltage) secondary shall be permitted to be protected by overcurrent protection provided by the primary (supply) side of the transformer, provided the protection is in accordance with 450.3 and does not exceed the value determined by multiplying the secondary conductor ampacity by the secondary-to-primary transformer voltage ratio. Transformer secondary conductors other than 2-wire shall not be considered to be protected by the primary overcurrent protection.

Exception No. 3: Electronic power source output conductors. Non–power-limited circuit conductors supplied by the output of a single-phase, listed electronic power source, other than a transformer, having only a 2-wire (single-voltage) output for connection to non–power-limited circuits shall be permitted to be protected by overcurrent protection provided on the input side of the electronic power source, provided this protection does not exceed the value determined by multiplying the non–power-limited circuit conductor ampacity by the output-to-input voltage ratio. Electronic power source outputs, other than 2-wire (single voltage), connected to non–power-limited circuits shall not be considered to be protected by overcurrent protection on the input of the electronic power source.

Informational Note: A single-phase, listed electronic power supply whose output supplies a 2-wire (single-voltage) circuit is an example of a non–power-limited power source that meets the requirements of 760.41.

760.46 NPLFA Circuit Wiring. Installation of non–power-limited fire alarm circuits shall be in accordance with 110.3(B), 300.7, 300.11, 300.15, 300.17, 300.19(B), and other appropriate articles of Chapter 3.

Exception No. 1: As provided in 760.48 through 760.53.

EXHIBIT 760.5 Typical manual fire alarm pull station box. *(Courtesy of the Protectowire Fire Systems)*

Exception No. 2: Where other articles of this Code require other methods.

Section 300.15 requires non–power-limited circuit terminations to be made in a box or conduit body. However, 300.15(E) permits devices with integral terminal enclosures and mounting brackets to be used without a box. Devices must be mounted on a box or conduit body where the instructions or the listing requires the use of a box. Fire alarm system components such as manual fire alarm boxes are tested frequently. Therefore, secure mounting is necessary to ensure that the manual fire alarm device will remain in place. (See Exhibit 760.5.)

760.48 Conductors of Different Circuits in Same Cable, Enclosure, or Raceway.

(A) Class 1 with NPLFA Circuits. Class 1 and non–power-limited fire alarm circuits shall be permitted to occupy the same cable, enclosure, or raceway without regard to whether the individual circuits are alternating current or direct current, provided all conductors are insulated for the maximum voltage of any conductor in the enclosure or raceway.

(B) Fire Alarm with Power-Supply Circuits. Power-supply and fire alarm circuit conductors shall be permitted in the same cable, enclosure, or raceway only where connected to the same equipment.

760.49 NPLFA Circuit Conductors.

(A) Sizes and Use. Only copper conductors shall be permitted to be used for fire alarm systems. Size 18 AWG and 16 AWG conductors shall be permitted to be used, provided they supply loads that do not exceed the ampacities given in Table 402.5 and are installed in a raceway, an approved enclosure, or a listed cable. Conductors larger than 16 AWG shall not supply loads greater than the ampacities given in 310.14, as applicable.

NFPA 72, National Fire Alarm and Signaling Code, requires fire alarm device and appliance voltages to be between 85 and

110 percent of nominal rated voltage. Calculations should be made to ensure that all devices or appliances will be operating within those limits at full circuit load. If future circuit extensions are anticipated, larger conductors should be considered. Some manufacturers specify maximum circuit loop resistances. The equipment specifications should be consulted to ensure that maximum allowable loop resistances are not exceeded.

(B) Insulation. Insulation on conductors shall be rated for the system voltage and not less than 600 volts. Conductors larger than 16 AWG shall comply with Article 310. Conductors 18 AWG and 16 AWG shall be Type KF-2, KFF-2, PAFF, PTFF, PF, PFF, PGF, PGFF, RFH-2, RFHH-2, RFHH-3, SF-2, SFF-2, TF, TFF, TFN, TFFN, ZF, or ZFF. Conductors with other types and thickness of insulation shall be permitted if listed for non–power-limited fire alarm circuit use.

Informational Note: See Table 402.3 for application provisions.

(C) Conductor Materials. Conductors shall be solid or stranded copper.

Exception to (B) and (C): Wire Types PAF and PTF shall be permitted only for high-temperature applications between 90°C (194°F) and 250°C (482°F).

760.51 Number of Conductors in Cable Trays and Raceways, and Ampacity Adjustment Factors.

(A) NPLFA Circuits and Class 1 Circuits. Where only non-power-limited fire alarm circuit and Class 1 circuit conductors are in a raceway, the number of conductors shall be determined in accordance with 300.17. The ampacity adjustment factors given in 310.15(C)(1) shall apply if such conductors carry continuous load in excess of 10 percent of the ampacity of each conductor.

(B) Power-Supply Conductors and NPLFA Circuit Conductors. Where power-supply conductors and non-power-limited fire alarm circuit conductors are permitted in a raceway in accordance with 760.48, the number of conductors shall be determined in accordance with 300.17. The ampacity adjustment factors given in 310.15(C)(1) shall apply as follows:

(1) To all conductors where the fire alarm circuit conductors carry continuous loads in excess of 10 percent of the ampacity of each conductor and where the total number of conductors is more than three

(2) To the power-supply conductors only, where the fire alarm circuit conductors do not carry continuous loads in excess of 10 percent of the ampacity of each conductor and where the number of power-supply conductors is more than three

(C) Cable Trays. Where fire alarm circuit conductors are installed in cable trays, they shall comply with 392.22 and 392.80(A).

760.53 Multiconductor NPLFA Cables. Multiconductor non–power-limited fire alarm cables that meet the requirements of 760.176 shall be permitted to be used on fire alarm circuits

operating at 150 volts or less and shall be installed in accordance with 760.53(A) and (B).

(A) NPLFA Wiring Method. Multiconductor non-power-limited fire alarm circuit cables shall be installed in accordance with 760.53(A)(1), (A)(2), and (A)(3).

(1) In Raceways, Exposed on Ceilings or Sidewalls, or Fished in Concealed Spaces. Cable splices or terminations shall be made in listed fittings, boxes, enclosures, fire alarm devices, or utilization equipment. Where installed exposed, cables shall be adequately supported and installed in such a way that maximum protection against physical damage is afforded by building construction such as baseboards, door frames, ledges, and so forth. Where located within 2.1 m (7 ft) of the floor, cables shall be securely fastened in an approved manner at intervals of not more than 450 mm (18 in.).

(2) Passing Through a Floor or Wall. Cables shall be installed in metal raceway or rigid nonmetallic conduit where passing through a floor or wall to a height of 2.1 m (7 ft) above the floor, unless adequate protection can be afforded by building construction such as detailed in 760.53(A)(1), or unless an equivalent solid guard is provided.

(3) In Hoistways. Cables shall be installed in rigid metal conduit, rigid nonmetallic conduit, intermediate metal conduit, liquidtight flexible nonmetallic conduit, or electrical metallic tubing where installed in hoistways.

Exception: As provided for in 620.21 for elevators and similar equipment.

(B) Applications of Listed NPLFA Cables. The use of non–power-limited fire alarm circuit cables shall comply with 760.53(B)(1) through (B)(4).

(1) Ducts Specifically Fabricated for Environmental Air. Multiconductor non–power-limited fire alarm circuit cables, Types NPLFP, NPLFR, and NPLF, shall not be installed exposed in ducts specifically fabricated for environmental air.

Informational Note: See 300.22(B).

Cables marked NPLFP are not permitted to be installed in ducts, but they are permitted in other spaces for environmental air, as specified in 760.53(B)(2). The higher possible voltages and currents of non–power-limited fire alarm circuits preclude the use of the listed cables inside ducts.

(2) Other Spaces Used for Environmental Air (Plenums). Cables installed in other spaces used for environmental air shall be Type NPLFP.

Exception No. 1: Types NPLFR and NPLF cables installed in compliance with 300.22(C).

Exception No. 2: Other wiring methods in accordance with 300.22(C) and conductors in compliance with 760.49(C).

Spaces over suspended ceilings used as an environmental air-handling return are considered by the *NEC* as "other spaces used for environmental air." Non–power-limited cables in other spaces used for environmental air must be marked NPLFP, as specified in 760.176(C).

Exception No. 3: Type NPLFP-CI cable shall be permitted to be installed to provide a 2-hour circuit integrity rated cable.

(3) Riser. Cables installed in vertical runs and penetrating one or more floors, or cables installed in vertical runs in a shaft, shall be Type NPLFR. Floor penetrations requiring Type NPLFR shall contain only cables suitable for riser or plenum use.

Exception No. 1: Type NPLF or other cables that are specified in Chapter 3 and are in compliance with 760.49(C) and encased in metal raceway.

Exception No. 2: Type NPLF cables located in a fireproof shaft having firestops at each floor.

Informational Note: See 300.21 for firestop requirements for floor penetrations.

Exception No. 3: Type NPLF-CI cable shall be permitted to be installed to provide a 2-hour circuit integrity rated cable.

(4) Other Wiring Within Buildings. Cables installed in building locations other than the locations covered in 760.53(B)(1), (B)(2), and (B)(3) shall be Type NPLF.

Exception No. 1: Chapter 3 wiring methods with conductors in compliance with 760.49(C).

Exception No. 2: Type NPLFP or Type NPLFR cables shall be permitted.

Exception No. 3: Type NPLFR-CI cable shall be permitted to be installed to provide a 2-hour circuit integrity rated cable.

Part III. Power-Limited Fire Alarm (PLFA) Circuits

760.121 Power Sources for PLFA Circuits.

Δ **(A) Power Source.** The power source for a power-limited fire alarm circuit shall be as specified in the following:

Informational Note No. 1: See Chapter 9, Tables 12(A) and 12(B), for the listing requirements for power-limited fire alarm circuit sources.

(1) A listed PLFA or Class 3 transformer
(2) A listed PLFA or Class 3 power supply
(3) Listed equipment marked to identify the PLFA power source

Informational Note No. 2: Examples of listed equipment are a fire alarm control panel with integral power source; a circuit card listed for use as a PLFA source, where used as part of a listed assembly; a current-limiting impedance, listed for the purpose or part of a listed product, used in conjunction with a non-power-limited transformer or a stored energy source, for example, storage battery, to limit the output current.

(B) Branch Circuit. The branch circuit supplying the fire alarm equipment(s) shall comply with the following requirements:

(1) The branch circuit shall supply no other loads.

(2) The branch circuit shall not be supplied through ground-fault circuit interrupters or arc-fault circuit interrupters.
(3) The location of the branch-circuit overcurrent protective device shall be permanently identified at the fire alarm control unit.
(4) The circuit disconnecting means shall have red identification, shall be accessible only to qualified personnel, and shall be identified with the following words: "FIRE ALARM CIRCUIT." The red identification shall not damage the overcurrent protective devices or obscure the manufacturer's markings.
(5) The fire alarm branch-circuit disconnecting means shall be permitted to be secured in the "on" position.

Informational Note: See 210.8(A)(5), Exception, for requirements on receptacles in dwelling-unit unfinished basements that supply power for fire alarm systems.

To help ensure that any branch circuit that supplies power to fire alarm equipment does not get turned off, 760.121(B)(5) permits the use of a locking device to secure the disconnecting means in the "on" position. In Exhibit 760.6, a circuit breaker is being used as the branch-circuit disconnecting means for fire alarm equipment with a circuit breaker lock in place to keep it locked in the "on" position. The circuit breaker lock is also red in color and clearly identified as "FIRE ALARM," which further assists in maintaining that the branch-circuit power to the fire alarm equipment remains on and is supplying the necessary power.

See also

760.41(B) for more information on branch-circuit requirements for fire alarm systems

Δ **760.124 Circuit Marking.** The equipment supplying PLFA circuits shall be durably marked where plainly visible to indicate each circuit that is a power-limited fire alarm circuit.

760.127 Wiring Methods on Supply Side of the PLFA Power Source. Conductors and equipment on the supply side of the

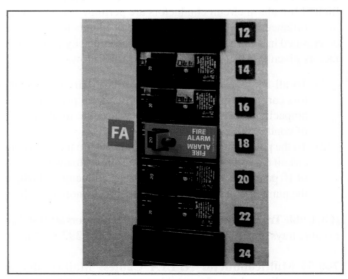

EXHIBIT 760.6 Circuit breaker lock. *(Courtesy of Space Age Electronics, Inc., Sterling, MA)*

power source shall be installed in accordance with the appropriate requirements of Part II and Chapters 1 through 4. Transformers or other devices supplied from power-supply conductors shall be protected by an overcurrent device rated not over 20 amperes.

Exception: The input leads of a transformer or other power source supplying power-limited fire alarm circuits shall be permitted to be smaller than 14 AWG, but not smaller than 18 AWG, if they are not over 300 mm (12 in.) long and if they have insulation that complies with 760.49(B).

760.130 Wiring Methods and Materials on Load Side of the PLFA Power Source. Fire alarm circuits on the load side of the power source shall be permitted to be installed using wiring methods and materials in accordance with 760.130(A), (B), or a combination of both. Parts I and II of Article 722 shall apply.

Δ **(A) NPLFA Wiring Methods and Materials.** NPLFA wiring methods shall be permitted when used in accordance with 760.46, 760.49, or 760.53 for PLFA circuits. Conductors shall be solid or stranded copper. Separation from electric light, power, Class 1, non-power-limited fire alarm circuit conductors, and medium-power network-powered broadband communications cables shall comply with 760.136.

Exception: The ampacity adjustment factors specified in 310.15(C)(1) shall not apply.

If power-limited circuits are installed as non–power-limited circuits, the power-limited marking must be removed from equipment, overcurrent protection must be provided in accordance with 760.43, and reclassified circuits must maintain separation from power-limited circuits in accordance with 760.48 and 760.133.

Δ **(B) PLFA Wiring Methods and Materials.** Power-limited fire alarm conductors and cables described in 722.179 shall be installed as detailed in 722.135 and 760.130(B)(1) through (B)(4). Devices shall be installed in accordance with 110.3(B), 300.11(A), and 300.15.

Δ **(1) In Raceways, Exposed on Ceilings or Sidewalls, or Fished in Concealed Spaces.** Cable splices or terminations shall be made in listed fittings, boxes, enclosures, fire alarm devices, or utilization equipment. Where installed exposed, cables shall be adequately supported and installed such that maximum protection against physical damage is afforded by building construction such as baseboards, door frames, ledges, and so forth. Where located within 2.1 m (7 ft) of the floor, cables shall be securely fastened in an approved manner at intervals of not more than 450 mm (18 in.).

Δ **(2) Passing Through a Floor or Wall.** Cables shall be installed in metal raceways or rigid nonmetallic conduit where passing through a floor or wall to a height of 2.1 m (7 ft) above the floor, unless adequate protection can be afforded by building construction such as detailed in 760.130(B)(1) or unless an equivalent solid guard is provided.

Δ **(3) Nonconcealed Spaces.** Cables specified in Chapter 3 and meeting the requirements of 722.179(A)(15)(a) and (A)(15)(b) shall be permitted to be installed in nonconcealed spaces where the exposed length of cable does not exceed 3 m (10 ft).

N **(4) Portable Fire Alarm Systems.** A portable fire alarm system provided to protect a stage or set when not in use shall be permitted to use wiring methods in accordance with 530.12.

760.133 Installation of Conductors and Equipment in Cables, Compartments, Cable Trays, Enclosures, Manholes, Outlet Boxes, Device Boxes, Raceways, and Cable Routing Assemblies for Power-Limited Fire Alarm Circuits. Conductors and equipment for power-limited fire alarm circuits shall be installed in accordance with Parts I and II of Article 722 and 760.136 through 760.143.

760.136 Separation from Electric Light, Power, Class 1, NPLFA, and Medium-Power Network-Powered Broadband Communications Circuit Conductors.

(A) General. Power-limited fire alarm circuit cables and conductors shall not be placed in any cable, cable tray, compartment, enclosure, manhole, outlet box, device box, raceway, or similar fitting with conductors of electric light, power, Class 1, non–power-limited fire alarm circuits, and medium-power network-powered broadband communications circuits unless permitted by 760.136(B) through (G).

Failure of the cable insulation due to a fault could lead to hazardous voltages being imposed on the power-limited fire alarm circuit conductors.

(B) Separated by Barriers. Power-limited fire alarm circuit cables shall be permitted to be installed together with Class 1, non–power-limited fire alarm, and medium-power network-powered broadband communications circuits where they are separated by a barrier.

(C) Raceways Within Enclosures. In enclosures, power-limited fire alarm circuits shall be permitted to be installed in a raceway within the enclosure to separate them from Class 1, non–power-limited fire alarm, and medium-power network-powered broadband communications circuits.

Δ **(D) Associated Systems Within Enclosures.** Power-limited fire alarm conductors in compartments, enclosures, device boxes, outlet boxes, or similar fittings shall be permitted to be installed with electric light, power, Class 1, non–power-limited fire alarm, and medium-power network-powered broadband communications circuits where they are introduced solely to connect the equipment connected to power-limited fire alarm circuits, and shall comply with either of the following conditions:

(1) The electric light, power, Class 1, non-power-limited fire alarm, and medium-power network-powered broadband communications circuit conductors are routed to maintain a minimum of 6 mm ($\frac{1}{4}$ in.) separation from the conductors and cables of power-limited fire alarm circuits.

(2) The circuit conductors operate at 150 volts or less to ground and also comply with one of the following conditions:

    a. The fire alarm power-limited circuits are installed using Type FPL, Type FPLR, Type FPLP, or permitted substitute cables if these power-limited cable conductors extending beyond the jacket are separated by a minimum of 6 mm (¼ in.) or by a nonconductive sleeve or nonconductive barrier from all other conductors.

    b. The power-limited fire alarm circuit conductors are installed as non-power-limited circuits in accordance with 760.46.

(E) Enclosures with Single Opening. Power-limited fire alarm circuit conductors entering compartments, enclosures, device boxes, outlet boxes, or similar fittings shall be permitted to be installed with electric light, power, Class 1, non–power-limited fire alarm, and medium-power network-powered broadband communications circuits where they are introduced solely to connect the equipment connected to power-limited fire alarm circuits or to other circuits controlled by the fire alarm system to which the other conductors in the enclosure are connected. Where power-limited fire alarm circuit conductors must enter an enclosure that is provided with a single opening, they shall be permitted to enter through a single fitting (such as a tee), provided the conductors are separated from the conductors of the other circuits by a continuous and firmly fixed nonconductor, such as flexible tubing.

(F) In Hoistways. In hoistways, power-limited fire alarm circuit conductors shall be installed in rigid metal conduit, rigid nonmetallic conduit, intermediate metal conduit, liquidtight flexible nonmetallic conduit, or electrical metallic tubing. For elevators or similar equipment, these conductors shall be permitted to be installed as provided in 620.21.

N **(G) Where Protected.** PLFA circuits shall be permitted to be installed together with the conductors of electric light, power, Class 1, non–power-limited fire alarm, and medium-power network-powered broadband communications circuits where they are installed using NPFLA wiring methods and materials in accordance with Part II of Article 760 and are protected by an approved method.

(H) Other Applications. For other applications, power-limited fire alarm circuit conductors shall be separated by at least 50 mm (2 in.) from conductors of any electric light, power, Class 1, non–power-limited fire alarm, or medium-power network-powered broadband communications circuits unless one of the following conditions is met:

    (1) Either (a) all of the electric light, power, Class 1, non–power-limited fire alarm, and medium-power network-powered broadband communications circuit conductors or (b) all of the power-limited fire alarm circuit conductors are in a raceway or in metal-sheathed, metal-clad, nonmetallic-sheathed, or Type UF cables.

    (2) All of the electric light, power, Class 1, non–power-limited fire alarm, and medium-power network-powered

broadband communications circuit conductors are permanently separated from all of the power-limited fire alarm circuit conductors by a continuous and firmly fixed nonconductor, such as porcelain tubes or flexible tubing, in addition to the insulation on the conductors.

760.139 Installation of Conductors of Different PLFA Circuits, Class 2, Class 3, and Communications Circuits in the Same Cable, Enclosure, Cable Tray, Raceway, or Cable Routing Assembly.

Δ **(A) Two or More PLFA Circuits.** Cable and conductors of two or more power-limited fire alarm circuits shall be permitted within the same cable, enclosure, cable tray, raceway, or cable routing assembly.

(B) Class 2 Circuits with PLFA Circuits. Conductors of one or more Class 2 circuits shall be permitted within the same cable, enclosure, cable tray, raceway, or cable routing assembly with conductors of power-limited fire alarm circuits if the insulation of the Class 2 circuit conductors in the cable, enclosure, raceway, or cable routing assembly is at least that required by the power-limited fire alarm circuits.

N **(C) Class 3 and Communications Circuits with PLFA Circuits.** Cable and conductors of Class 3 and communications circuits shall be permitted within the same cable, enclosure, cable tray, raceway, or cable routing assembly with cables and conductors of power-limited fire alarm circuits.

(D) Low-Power Network-Powered Broadband Communications Cables and PLFA Cables. Low-power network-powered broadband communications circuits shall be permitted in the same enclosure, cable tray, raceway, or cable routing assembly with PLFA cables.

Δ **(E) Audio System Circuits and PLFA Circuits.** Audio system circuits described in 640.9(C) and installed using Class 2 or Class 3 wiring methods in compliance with 722.135 shall not be installed in the same cable, cable tray, raceway, or cable routing assembly with power-limited conductors or cables.

Audio circuits that are installed as Class 2 or Class 3 circuits are prohibited from being installed in the same cable or raceway with power-limited fire alarm wiring. A fault between audio amplifier circuits and power-limited fire alarm circuits has the potential to impair the fire alarm system.

760.142 Conductor Size. Conductors of 26 AWG shall be permitted only where spliced with a connector listed as suitable for 26 AWG to 24 AWG or larger conductors that are terminated on equipment or where the 26 AWG conductors are terminated on equipment listed as suitable for 26 AWG conductors. Single conductors shall not be smaller than 18 AWG.

Circuit conductors as small as 26 AWG are permitted to be used in digitally addressable fire alarm systems according to the listing or installation instructions of the fire alarm equipment.

760.143 Support of Conductors. Power-limited fire alarm circuit conductors shall not be strapped, taped, or attached by any means to the exterior of any conduit or other raceway as a means of support.

See also

760.24(A) for more information on the support of conductors

760.145 Current-Carrying Continuous Line-Type Fire Detectors.

(A) Application. Listed continuous line-type fire detectors, including insulated copper tubing of pneumatically operated detectors, employed for both detection and carrying signaling currents shall be permitted to be used in power-limited circuits.

(B) Installation. Continuous line-type fire detectors shall be installed in accordance with 760.124 through 760.130 and 760.133.

760.154 Applications of Listed PLFA Cables. PLFA cables shall comply with the requirements described in Table 760.154

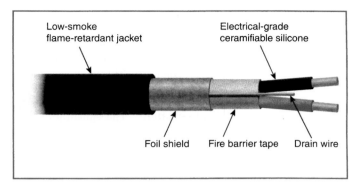

EXHIBIT 760.7 Type CI cable.

or where cable substitutions are made as shown in 760.154(A). Where substitute cables are installed, the wiring requirements of Article 760, Parts I and III, shall apply. Types FPLP-CI, FPLR-CI, and FPL-CI cables shall be permitted to be installed to provide 2-hour circuit integrity rated cables.

Δ **TABLE 760.154** *Applications of Listed PLFA Cables in Buildings*

Applications		Cable Type		
		FPLP & FPLP-CI	FPLR & FPLR-CI	FPL & FPL-CI
In fabricated ducts as described in 300.22(B)	In fabricated ducts	Y*	N	N
	In metal raceway that complies with 300.22(B)	Y*	Y*	Y*
In other spaces used for environmental air as described in 300.22(C)	In other spaces used for environmental air	Y*	N	N
	In metal raceway that complies with 300.22(C)	Y*	Y*	Y*
	In plenum communications raceways	Y*	N	N
	In plenum cable routing assemblies	Y*	N	N
	Supported by open metal cable trays	Y*	N	N
	Supported by solid bottom metal cable trays with solid metal covers	Y*	Y*	Y*
In risers	In vertical runs	Y*	Y*	N
	In metal raceways	Y*	Y*	Y*
	In fireproof shafts	Y*	Y*	Y*
	In plenum communications raceways	Y*	Y*	N
	In plenum cable routing assemblies	Y*	Y*	N
	In riser communications raceways	Y*	Y*	N
	In riser cable routing assemblies	Y*	Y*	N
	In one- and two-family dwellings	Y*	Y*	Y*
Within buildings in other than air-handling spaces and risers	General	Y*	Y*	Y*
	Supported by cable trays	Y*	Y*	Y*
	In any raceway recognized in Chapter 3	Y*	Y*	Y*
	In plenum communications raceway	Y*	Y*	Y*
	In plenum cable routing assemblies	Y*	Y*	Y*
	In riser communications raceways	Y*	Y*	Y*
	In riser cable routing assemblies	Y*	Y*	Y*
	In general-purpose communications raceways	Y*	Y*	Y*
	In general-purpose cable routing assemblies	Y*	Y*	Y*

Note:

"N" indicates that the cable type shall not be permitted to be installed in the application.

"Y*" indicates that the cable type shall be permitted to be installed in the application subject to the limitations described in 760.130 through 760.145.

Type CI cable is permitted for applications where survivability of fire alarm circuits is needed during a fire. Such circuits could be essential to communicating evacuation or relocation instructions to building occupants under fire or other emergency conditions. The construction of typical CI cable is illustrated in Exhibit 760.7.

Δ **(A) Fire Alarm Cable Substitutions.** The substitutions for fire alarm cables listed in Table 760.154(A) and illustrated in Figure 760.154(A) shall be permitted. Where substitute cables are installed, the wiring requirements of Article 760, Parts I and III, shall apply.

> Informational Note: See 800.179 for information on communications cables (CMP, CMR, CMG, CM).

TABLE 760.154(A) *Cable Substitutions*

Cable Type	Permitted Substitutions
FPLP	CMP
FPLR	CMP, FPLP, CMR
FPL	CMP, FPLP, CMR, FPLR, CMG, CM

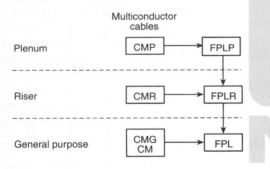

Type CM—Communications wires and cables
Type FPL—Power-limited fire alarm cables

A → B   Cable A shall be permitted to be used in place of cable B, 26 AWG minimum

FIGURE 760.154(A) *Cable Substitution Hierarchy.*

Part IV. Listing Requirements

Δ **760.176 Listing and Marking of NPLFA Cables.** Non-power-limited fire alarm cables installed as wiring within buildings shall be listed in accordance with 760.176(A) and (B), be listed as resistant to the spread of fire in accordance with 760.176(C) through (F), and be marked in accordance with 760.176(G). Cable used in a wet location shall be listed for use in wet locations or have a moisture-impervious metal sheath. Non-power-limited fire alarm cables shall have a temperature rating of not less than 60°C (140°F). Non-power-limited fire alarm cables shall be permitted to contain optical fibers.

> Informational Note: See UL 1425, *Standard for Cables for Non-Power-Limited Fire-Alarm Circuits*, for information on non-power-limited fire alarm cables.

(A) NPLFA Conductor Materials. Conductors shall be 18 AWG or larger solid or stranded copper.

(B) Insulated Conductors. Insulation on conductors shall be rated for the system voltage and not less than 600 V. Insulated conductors 14 AWG and larger shall be one of the types listed in Table 310.4(1) or one that is identified for such use. Insulated conductors 18 AWG and 16 AWG shall be in accordance with 760.49.

Δ **(C) Type NPLFP.** Type NPLFP non-power-limited fire alarm cable for use in other space used for environmental air shall be listed as being suitable for use in other space used for environmental air as described in 300.22(C) and shall also be listed as having adequate fire-resistant and low smoke–producing characteristics.

> Informational Note: See NFPA 262, *Standard Method of Test for Flame Travel and Smoke of Wires and Cables for Use in Air-Handling Spaces*, for one method of defining a cable that is low-smoke producing and fire-resistant if the cable exhibits a maximum peak optical density of 0.50 or less, an average optical density of 0.15 or less, and a maximum flame spread distance of 1.52 m (5 ft) or less when tested.

See also

760.53(B)(2) and its commentary, which discusses other spaces used for environmental air

Δ **(D) Type NPLFR.** Type NPLFR non–power-limited fire alarm riser cable shall be listed as being suitable for use in a vertical run in a shaft or from floor to floor and shall also be listed as having fire-resistant characteristics capable of preventing the carrying of fire from floor to floor.

> Informational Note: See UL 1666, *Test for Flame Propagation Height of Electrical and Optical-Fiber Cables Installed Vertically in Shafts*, for one method of defining fire-resistant characteristics capable of preventing the carrying of fire from floor to floor.

Δ **(E) Type NPLF.** Type NPLF non–power-limited fire alarm cable shall be listed as being suitable for general-purpose fire alarm use, with the exception of use in risers, ducts, plenums, and other space used for environmental air, and shall also be listed as being resistant to the spread of fire.

> Informational Note: See UL 2556, *Wire and Cable Test Methods*, for one method of defining *resistant to the spread of fire*. One method is to demonstrate that the cables do not spread fire to the top of the tray in the "UL Flame Exposure, Vertical Tray Flame Test." The smoke measurements in the test method are not applicable.
> Another method of defining *resistant to the spread of fire* is for the damage (char length) not to exceed 1.5 m (4 ft 11 in.) when performing the FT4 "Vertical Flame Test."

Δ **(F) Circuit Integrity (CI) Cable, Fire-Resistive Cable System, or Electrical Circuit Protective System.** Cables that are used for survivability of critical circuits under fire conditions shall be listed and meet the requirements of 760.176(F)(1), (F)(2), or (F)(3).

> Informational Note: See *NFPA 72, National Fire Alarm and Signaling Code*, 12.4.3 and 12.4.4, for additional information on circuit integrity (CI) cable, fire-resistive cable systems, or electrical circuit protective systems used for fire alarm circuits to comply with the survivability requirements to maintain the circuit's electrical function during fire conditions for a defined period of time.

Type CI cable is designed to retain vital electrical performance during and immediately after fire exposure.

Δ **(1) Circuit Integrity (CI) Cables.** Circuit integrity (CI) cables specified in 760.176(C), (D), and (E) and used for survivability of critical circuits shall be marked for an additional classification using the suffix "-CI." In order to maintain its listed fire-resistive rating, CI cables shall only be installed in free air in accordance with 760.24(B). CI cables shall only be permitted to be installed in a raceway where specifically listed and marked as part of an electrical circuit protective fire-resistive cable system as covered in 760.176(F)(2). CI cables shall only be permitted to be installed in a raceway where specifically listed and marked as part of an electrical circuit protective system as covered in 760.176(F)(2).

> Informational Note: See UL 2196, *Fire Test for Circuit Integrity of Fire-Resistive Power, Instrumentation, Control and Data Cables*, and UL 1425, *Cables for Non-Power-Limited Fire-Alarm Circuits*, for information on establishing a rating for CI cable. The *UL Guide Information for Nonpower-limited Fire Alarm Circuits* (HNHT) contains information for identifying the cable and its installation limitations to maintain the fire-resistive rating.

Δ **(2) Fire-Resistive Cable Systems.** Cables specified in 760.176(C), (D), (E), and (F)(1) that are part of a fire-resistive cable system shall be identified with the system identifier and hourly rating marked on the protectant or the smallest unit container and installed in accordance with the listing of the system.

> Informational Note: See UL 2196, *Fire Test for Circuit Integrity of Fire-Resistive Power, Instrumentation, Control and Data Cables*, for information on establishing a rating for a fire-resistive cable system. The *UL Guide Information for Electrical Circuit Integrity Systems* (FHIT) contains information for identifying the system and its installation limitations to maintain a minimum fire-resistive rating.

N **(3) Electrical Circuit Protective System.** Protectants for cables specified in 760.176(C), (D), and (E) that are part of an electrical circuit protective system shall be identified with the protective system identifier and hourly rating marked on the protectant or the smallest unit container and installed in accordance with the listing of the protective system.

> Informational Note: See UL 1724, *Fire Tests for Electrical Circuit Protective Systems*, for information on establishing a rating for an electrical circuit protective system. The *UL Guide Information for Electrical Circuit Integrity Systems* (FHIT) contains information

for identifying the system and its installation limitations to maintain the fire-resistive rating.

(G) NPLFA Cable Markings. Multiconductor non–power-limited fire alarm cables shall be marked in accordance with Table 760.176(G). Non–power-limited fire alarm circuit cables shall be permitted to be marked with a maximum usage voltage rating of 150 volts. Cables that are listed for circuit integrity shall be identified with the suffix "-CI" as defined in 760.176(F). The temperature rating shall be marked on the jacket of NPLFA cables that have a temperature rating exceeding 60°C (140°F). The jacket of NPLFA cables shall be marked with the conductor size.

> Informational Note: Cable types are listed in descending order of fire performance.

Δ *TABLE 760.176(G) NPLFA Cable Markings*

Cable Marking	Type	Reference
NPLFP	Non–power-limited fire alarm circuit cable for use in other space used for environmental air	760.176(C) and (G)
NPLFR	Non–power-limited fire alarm circuit riser cable	760.176(D) and (G)
NPLF	Non–power-limited fire alarm circuit cable	760.176(E) and (G)

Notes:
1. Cables identified in 760.176(C), (D), and (E) and meeting the requirements for circuit integrity shall have the additional classification using the suffix "-CI" (for example, NPLFP-CI, NPLFR-CI, and NPLF-CI).
2. Cables containing optical fibers shall be provided with the suffix "-OF".

Δ **760.179 Listing and Marking of Insulated Continuous Line-Type Fire Detectors.** Insulated continuous line-type fire detectors shall be listed in accordance with 760.179(A) through (D). Cable used in a wet location shall be listed for use in wet locations or have a moisture-impervious metal sheath.

N **(A) Listing.** The cable shall be listed as being resistant to the spread of fire in accordance with 722.179(A)(1), (A)(2), and (A)(3).

N **(B) Voltage and Temperature Rating.** The cable shall have a voltage rating of not less than 300 volts. The cable shall have a temperature rating of not less than 60°C (140°F).

N **(C) Markings.** The cable shall be marked as fire resistance Type FPLP, Type FPLR, or Type FPL in accordance with 722.179(B). The voltage rating shall not be marked on the cable. The temperature rating shall be marked on the jacket of cables that have a temperature rating exceeding 60°C (140°F). The jacket of PLFA cables shall be marked with the conductor size.

> Informational Note: Voltage ratings on cables might be misinterpreted to suggest that the cables could be suitable for Class 1, electric light, and power applications.

Exception: Voltage markings shall be permitted where the cable has multiple listings and voltage marking is required for one or more of the listings.

N (D) Cable Jacket Compound. The cable jacket compound shall have a high degree of abrasion resistance.

ARTICLE
770   Optical Fiber Cables

Part I. General

770.1 Scope. This article covers the installation of optical fiber cables. This article does not cover the construction of optical fiber cables.

Article 770 permits the use of optical fiber technology in conjunction with electrical conductors for communications, signaling, and control circuits in lieu of metallic conductors. Optical fiber cables may be nonconductive, or they may be composite, containing electrical conductors. See Exhibits 770.1 and 770.2. The most common optical fiber cable used in buildings is nonconductive. Because they are not affected by electrical noise, optical fiber cables to transmit data or other communications can be desirable in circumstances where electrical noise is a problem.

770.3 Other Articles. Installations of optical fiber cables shall comply with 770.3(A) through (D). Only those sections of Chapter 2 and Article 300 referenced in this article shall apply to optical fiber cables.

Δ (A) Hazardous (Classified) Locations. Listed optical fiber cables shall be permitted to be installed in hazardous (classified) locations. The cables shall be sealed in accordance with 501.15, 502.15, 505.16, or 506.16, as applicable.

(B) Cables in Ducts for Dust, Loose Stock, or Vapor Removal. The requirements of 300.22(A) for wiring systems shall apply to conductive optical fiber cables.

EXHIBIT 770.1 An example of a nonconductive optical fiber cable.

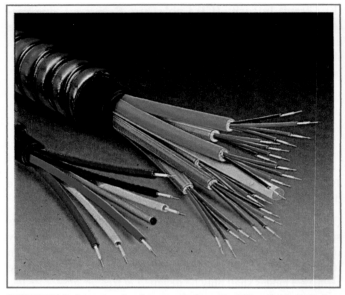

EXHIBIT 770.2 An example of a composite optical fiber cable that also meets the requirements of Article 330 and is referred to as Type MC cable. *(Courtesy of AFC Cable Systems, a Part of Atkore International)*

(C) Hybrid Cables. Hybrid optical fiber cables shall be classified as electrical cables in accordance with the type of electrical conductors. They shall be constructed, listed, and marked in accordance with the appropriate article for each type of electrical cable.

N (D) Vertical Support for Fire-Resistive Cables. Vertical installations of circuit integrity (CI) cables installed in a raceway or cables of fire-resistive cable systems shall be installed in accordance with their listing.

770.21 Access to Electrical Equipment Behind Panels Designed to Allow Access. Access to electrical equipment shall not be denied by an accumulation of optical fiber cables that prevents removal of panels, including suspended ceiling panels.

An excess accumulation of wires and cables can limit access to equipment by preventing the removal of access panels. Exhibit 770.3 shows both incorrect and correct installation methods for the installation of communication cables. The correct installation method has the cables supported well above the acoustical ceiling panels, allowing for proper access when needed.

Δ 770.24 Mechanical Execution of Work.

N (A) General. Optical fiber cables shall be installed in a neat and workmanlike manner. Cables installed exposed on the surface of ceilings and sidewalls shall be supported by the building structure in such a manner that the cable will not be damaged by normal building use. Such cables shall be secured by hardware, including straps; staples; cable ties listed and identified for securement and support; and hangers, or similar fittings, designed and installed so as not to damage the cable.

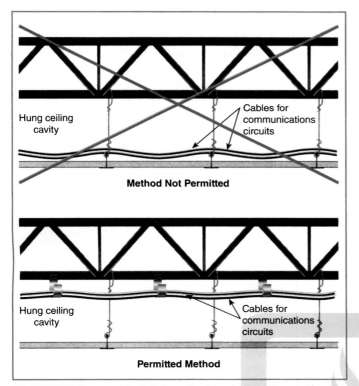

EXHIBIT 770.3 Incorrect installation of cables *(upper diagram)* and correct method *(lower diagram)*.

The installation shall also conform to 300.4 and 300.11. Plenum cable ties and other nonmetallic cable accessories used to secure and support cables in other spaces used for environmental air (plenums) shall be listed as having low smoke and heat release properties in accordance with 800.170.

> Informational Note No. 1: See ANSI/NECA/FOA 301-2016, *Standard for Installing and Testing Fiber Optic Cables*, ANSI/TIA-568.0-D-2015, *Generic Telecommunications Cabling for Customer Premises*, and ANSI/TIA 568.3-D-2016, *Optical Fiber Cabling and Components Standard*, for accepted industry practices.
> Informational Note No. 2: See NFPA 90A-2021, *Standard for the Installation of Air-Conditioning and Ventilating Systems*, for discrete combustible components installed in accordance with 300.22(C).

Informational Note No. 2 references two requirements in NFPA 90A, *Standard for the Installation of Air-Conditioning and Ventilating Systems*, that have an influence on installations covered in the *NEC®*. This is not intended to require products covered by this section to be listed for other than their smoke and heat properties.

> Informational Note No. 3: Paint, plaster, cleaners, abrasives, corrosive residues, or other contaminants may result in an undetermined alteration of optical fiber cable properties.

N **(B) Circuit Integrity (CI) Cable.** Circuit integrity (CI) cable shall be supported at a distance not exceeding 610 mm (24 in.).

Cable shall be secured to the noncombustible surface of the building structure. Cable supports and fasteners shall be steel.

770.25 Abandoned Cables. The accessible portion of abandoned optical fiber cables shall be removed. Where cables are identified for future use with a tag, the tag shall be of sufficient durability to withstand the environment involved.

See Article 100 for the definition of *accessible (as applied to wiring methods)*. Abandoned cable unnecessarily increases fire loading, and, where installed in plenums, it can affect airflow.

See also

> **Article 100** for definition of the term cable, *abandoned (abandoned cable)*

770.26 Spread of Fire or Products of Combustion. Installations of optical fiber cables and communications raceways in hollow spaces, vertical shafts, and ventilation or air-handling ducts shall be made so that the possible spread of fire or products of combustion will not be substantially increased. Openings around penetrations of optical fiber cables and communications raceways through fire-resistant–rated walls, partitions, floors, or ceilings shall be firestopped using approved methods to maintain the fire resistance rating.

Informational Note: Directories of electrical construction materials published by qualified testing laboratories contain many listing installation restrictions necessary to maintain the fire-resistive rating of assemblies where penetrations or openings are made. Building codes also contain restrictions on membrane penetrations on opposite sides of a fire resistance–rated wall assembly. An example is the 600-mm (24-in.) minimum horizontal separation that usually applies between boxes installed on opposite sides of the wall. Assistance in complying with 770.26 can be found in building codes, fire resistance directories, and product listings.

N **770.27 Temperature Limitation of Optical Fiber Cables.** Optical fiber cable shall not be used in such a manner that its operating temperature exceeds that of its rating.

Part II. Cables Outside and Entering Buildings

770.44 Overhead (Aerial) Optical Fiber Cables. Overhead optical fiber cables containing a non–current-carrying metallic member entering buildings shall comply with 800.44(A) and (B).

(A) On Poles and In-Span. Where outside plant optical fiber cables and electric light or power conductors are supported by the same pole or are run parallel to each other in-span, the conditions described in 770.44(A)(1) through (A)(4) shall be met.

(1) Relative Location. Where practicable, the outside plant optical fiber cables shall be located below the electric light or power conductors.

(2) Attachment to Cross-Arms. Attachment of outside plant optical fiber cables to a cross-arm that carries electric light or power conductors shall not be permitted.

(3) Climbing Space. The climbing space through outside plant optical fiber cables shall comply with the requirements of 225.14(B).

(4) Clearance. Supply service drops and sets of overhead service conductors of 0 to 750 volts running above and parallel to optical fiber cable service drops shall have a minimum separation of 300 mm (12 in.) at any point in the span, including the point of their attachment to the building. Clearance of not less than 1.0 m (40 in.) shall be maintained between the two services at the pole.

(B) Above Roofs. Outside plant optical fiber cables shall have a vertical clearance of not less than 2.5 m (8 ft) from all points of roofs above which they pass.

Exception No. 1: The requirement of 770.44(B) shall not apply to auxiliary buildings such as garages and the like.

Exception No. 2: A reduction in clearance above only the overhanging portion of the roof to not less than 450 mm (18 in.) shall be permitted if (1) not more than 1.2 m (4 ft) of optical fiber cable service drop cable passes above the roof overhang, and (2) the cable is terminated at a through- or above-the-roof raceway or approved support.

Exception No. 3: Where the roof has a slope of not less than 100 mm in 300 mm (4 in. in 12 in.), a reduction in clearance to not less than 900 mm (3 ft) shall be permitted.

Informational Note: See ANSI/IEEE C2-2017, *National Electric Safety Code, Part 2, Safety Rules for Overhead Lines*, for additional information regarding overhead wires and cables.

770.47 Underground Optical Fiber Cables Entering Buildings. Underground optical fiber cables entering buildings shall comply with 770.47(A) and (B).

(A) Underground Systems with Electric Light, Power, Class 1, or Non–Power-Limited Fire Alarm Circuit Conductors. Underground conductive optical fiber cables entering buildings with electric light, power, Class 1, or non–power-limited fire alarm circuit conductors in a raceway, handhole enclosure, or manhole shall be located in a section separated from such conductors by means of brick, concrete, or tile partitions or by means of a suitable barrier.

Δ **(B) Direct-Buried Cables and Raceways.** Direct-buried conductive optical fiber cables shall be separated by at least 300 mm (12 in.) from conductors of any electric light, power, non–power-limited fire alarm circuit conductors, or Class 1 circuit.

Exception No. 1: Separation shall not be required where the electric service conductors are installed in raceways or have metal cable armor.

Exception No. 2: Separation shall not be required where electric light or power branch-circuit or feeder conductors, non–power-limited fire alarm circuit conductors, or Class 1 circuit conductors are installed in a raceway or in metal-sheathed, metal-clad, or Type UF or Type USE cables.

770.48 Unlisted Cables Entering Buildings.

(A) Conductive and Nonconductive Cables. Unlisted conductive and nonconductive outside plant optical fiber cables shall be permitted to be installed in building spaces, other than risers, ducts used for environmental air, plenums used for environmental air, and other spaces used for environmental air, where the length of the cable within the building, measured from its point of entrance, does not exceed 15 m (50 ft) and the cable enters the building from the outside and is terminated in an enclosure.

The point of entrance shall be permitted to be extended from the penetration of the external wall, roof, or floor slab by continuously enclosing the entrance optical fiber cables in rigid metal conduit (RMC) or intermediate metal conduit (IMC) to the point of emergence.

Informational Note: Splice cases or terminal boxes, both metallic and plastic types, typically are used as enclosures for splicing or terminating optical fiber cables.

Unlisted optical fiber cables are permitted to be installed within a building provided they originate outside the building. They are limited to 50 feet of cable measured from the point at which they enter the building. The point from which this measurement is taken can be extended within the building by enclosing the cables within RMC or IMC.

Δ **(B) Nonconductive Cables in Raceway.** Unlisted nonconductive outside plant optical fiber cables shall be permitted to enter the building from the outside and shall be permitted to be installed in any of the following raceways:

(1) Intermediate metal conduit (IMC)
(2) Rigid metal conduit (RMC)
(3) Rigid polyvinyl chloride conduit (PVC)
(4) Electrical metallic tubing (EMT)

Unlisted nonconductive outside plant cables installed in rigid polyvinyl chloride conduit (PVC) or electrical metallic tubing (EMT) shall not be installed in risers, ducts used for environmental air, plenums used for environmental air, and other spaces used for environmental air.

770.49 Metal Entrance Conduit Grounding. Metal conduit containing optical fiber entrance cable shall be connected by a bonding conductor or grounding electrode conductor to a grounding electrode or, where present, the building grounding electrode system in accordance with 770.100(B).

Part III. Protection

770.93 Grounding, Bonding, or Interruption of Non–Current-Carrying Metallic Members of Optical Fiber Cables. Optical fiber cables entering the building or terminating on the outside of the building shall comply with 770.93(A) or (B).

(A) Entering Buildings. In installations where an optical fiber cable is exposed to contact with electric light or power conductors and the cable enters the building, the non–current-carrying metallic members shall be either grounded or bonded as specified in 770.100 or interrupted by an insulating joint or equivalent device. The grounding or interruption shall be as close as practicable to the point of entrance.

(B) Terminating on the Outside of Buildings. In installations where an optical fiber cable is exposed to contact with electric light or power conductors and the cable is terminated on the outside of the building, the non–current-carrying metallic members shall be either grounded or bonded as specified in 770.100 or interrupted by an insulating joint or equivalent device. The grounding, bonding, or interruption shall be as close as practicable to the point of termination of the cable.

Part IV. Grounding Methods

770.100 Entrance Cable Bonding and Grounding. If required, the non-current-carrying metallic members of optical fiber cables entering buildings shall be bonded or grounded as specified in 770.100(A) through (D).

(A) Bonding Conductor or Grounding Electrode Conductor.

(1) Insulation. The bonding conductor or grounding electrode conductor shall be listed and shall be permitted to be insulated, covered, or bare.

(2) Material. The bonding conductor or grounding electrode conductor shall be copper or other corrosion-resistant conductive material, stranded or solid.

(3) Size. The bonding conductor or grounding electrode conductor shall not be smaller than 14 AWG. It shall have a current-carrying capacity not less than that of the grounded metallic member(s). The bonding conductor or grounding electrode conductor shall not be required to exceed 6 AWG.

(4) Length. The bonding conductor or grounding electrode conductor shall be as short as practicable. In one- and two-family dwellings, the bonding conductor or grounding electrode conductor shall be as short as practicable not to exceed 6.0 m (20 ft) in length.

> Informational Note: Similar bonding conductor or grounding electrode conductor length limitations applied at apartment buildings and commercial buildings help to reduce voltages that may develop between the building's power and communications systems during lightning events.

Exception: In one- and two-family dwellings if it is not practicable to achieve an overall maximum bonding conductor or grounding electrode conductor length of 6.0 m (20 ft), a separate ground rod meeting the minimum dimensional criteria of 770.100(B)(3)(2) shall be driven, the grounding electrode conductor shall be connected to the separate ground rod in accordance with 770.100(C), and the separate ground rod shall

be bonded to the power grounding electrode system in accordance with 770.100(D).

(5) Run in Straight Line. The bonding conductor or grounding electrode conductor shall be run in as straight a line as practicable.

(6) Physical Protection. Bonding conductors and grounding electrode conductors shall be protected where exposed to physical damage. Where the bonding conductor or grounding electrode conductor is installed in a metal raceway, both ends of the raceway shall be bonded to the contained conductor or to the same terminal or electrode to which the bonding conductor or grounding electrode conductor is connected.

(B) Electrode. The bonding conductor and grounding electrode conductor shall be connected in accordance with 770.100(B)(1), (B)(2), or (B)(3).

(1) In Buildings or Structures with an Intersystem Bonding Termination. If the building or structure served has an intersystem bonding termination as required by 250.94, the bonding conductor shall be connected to the intersystem bonding termination.

> Informational Note: See Informational Note Figure 800.100(B)(1) for an illustration of the application of the bonding conductor in buildings or structures equipped with an intersystem bonding termination.

Δ **(2) In Buildings or Structures with Grounding Means.** If an intersystem bonding termination is established, 250.94(A) shall apply.

If the building or structure served has no intersystem bonding termination, the bonding conductor or grounding electrode conductor shall be connected to the nearest accessible location on one of the following:

(1) The building or structure grounding electrode system as covered in 250.50
(2) The power service accessible means external to enclosures using the options identified in 250.94(A), Exception
(3) The nonflexible metal power service raceway
(4) The service equipment enclosure
(5) The grounding electrode conductor or the grounding electrode conductor metal enclosure of the power service
(6) The grounding electrode conductor or the grounding electrode of a building or structure disconnecting means that is connected to a grounding electrode as covered in 250.32
(7) The grounded interior metal water piping system, within 1.5 m (5 ft) from its point of entrance to the building, as covered in 250.52

> Informational Note: See Informational Note Figure 800.100(B)(2) for an illustration of the application of the bonding conductor in buildings or structures not equipped with an intersystem bonding termination or terminal block providing access to the building grounding electrode system.

Δ **(3) In Buildings or Structures Without Intersystem Bonding Termination or Grounding Means.** If the building or structure served has no intersystem bonding termination or grounding means, as described in 770.100(B)(2), the grounding electrode conductor shall be connected to either of the following:

(1) To any one of the individual grounding electrodes described in 250.52(A)(1), (A)(2), (A)(3), or (A)(4).

(2) If the building or structure served has no grounding means, as described in 770.100(B)(2) or (B)(3)(1), to any one of the individual grounding electrodes described in 250.52(A)(7) and (A)(8) or to a ground rod or pipe not less than 1.5 m (5 ft) in length and 12.7 mm (½ in.) in diameter, driven, where practicable, into permanently damp earth and separated from lightning protection system conductors as covered in 800.53 and at least 1.8 m (6 ft) from electrodes of other systems. Steam, hot water pipes, or lightning protection system conductors shall not be employed as electrodes for non–current-carrying metallic members.

(C) Electrode Connection. Connections to grounding electrodes shall comply with 250.70.

(D) Bonding of Electrodes. A bonding jumper not smaller than 6 AWG copper or equivalent shall be connected between the grounding electrode and power grounding electrode system at the building or structure served where separate electrodes are used.

Exception: At mobile homes as covered in 770.106.

Informational Note No. 1: See 250.60 for connection to a lightning protection system.

Informational Note No. 2: Bonding together of all separate electrodes limits potential differences between them and between their associated wiring systems.

770.106 Grounding and Bonding of Entrance Cables at Mobile Homes.

Δ **(A) Grounding.** Grounding shall comply with 770.106(A)(1) and (A)(2).

N **(1) Installations Without Mobile Home Service Equipment.** If there is no mobile home service equipment located within 9.0 m (30 ft) of the exterior wall of the mobile home it serves, the non-current-carrying metallic members of optical fiber cables entering the mobile home shall be grounded in accordance with 770.100(B)(3).

N **(2) Installations Without Mobile Home Disconnecting Means.** If there is no mobile home disconnecting means grounded in accordance with 250.32 and located within 9.0 m (30 ft) of the exterior wall of the mobile home it serves, the non-current-carrying metallic members of optical fiber cables entering the mobile home shall be grounded in accordance with 770.100(B)(3).

(B) Bonding. The grounding electrode shall be bonded to the metal frame or available grounding terminal of the mobile home with a copper conductor or other equivalent corrosion-resistant material not smaller than 12 AWG under either of the following conditions:

(1) If there is no mobile home service equipment or disconnecting means as in 770.106(A)

(2) If the mobile home is supplied by cord and plug

Part V. Installation Methods Within Buildings

770.110 Raceways, Cable Routing Assemblies, and Cable Trays for Optical Fiber Cables.

(A) Types of Raceways. Optical fiber cables shall be permitted to be installed in any raceway that complies with either 770.110(A)(1) or (A)(2).

(1) Raceways Recognized in Chapter 3. Optical fiber cables shall be permitted to be installed in any raceway included in Chapter 3. The raceways shall be installed in accordance with Chapter 3.

(2) Communications Raceways. Optical fiber cables shall be permitted to be installed in listed communications raceways selected in accordance with Table 800.154(b).

(B) Raceway Fill for Optical Fiber Cables. Raceway fill for optical fiber cables shall comply with either 770.110(B)(1) or (B)(2).

(1) Without Electric Light or Power Conductors. Where optical fiber cables are installed in raceway without electric light or power conductors, the raceway fill requirements of Chapters 3 and 9 shall not apply.

(2) Nonconductive Optical Fiber Cables with Electric Light or Power Conductors. Where nonconductive optical fiber cables are installed with electric light or power conductors in a raceway, the raceway fill requirements of Chapters 3 and 9 shall apply.

(C) Cable Routing Assemblies. Optical fiber cables shall be permitted to be installed in listed cable routing assemblies selected in accordance with Table 800.154(c).

(D) Cable Trays. Optical fiber cables shall be permitted to be installed in metal or listed nonmetallic cable tray systems.

N **770.111 Innerduct for Optical Fiber Cables.** Listed plenum communications raceways, listed riser communications raceways, and listed general-purpose communications raceways selected in accordance with Table 800.154(b) shall be permitted to be installed as innerduct in any type of listed raceway permitted in Chapter 3.

770.113 Installation of Optical Fiber Cables. Installation of optical fiber cables shall comply with 770.113(A) through

(J). Installation of raceways and cable routing assemblies shall comply with 770.110.

(A) Listing. Optical fiber cables installed in buildings shall be listed in accordance with 770.179 and installed in accordance with the limitations of the listing.

Exception: Optical fiber cables that are installed in compliance with 770.48 shall not be required to be listed.

Δ **(B) Ducts Specifically Fabricated for Environmental Air.** Installations of optical fiber cables in ducts specifically fabricated for environmental air shall be in accordance with 770.113(B)(1) and (B)(2).

N **(1) Uses Permitted.** The following cables shall be permitted in ducts specifically fabricated for environmental air as described in 300.22(B) if they are directly associated with the air distribution system:

(1) Up to 1.22 m (4 ft) of Types OFNP and OFCP
(2) Types OFNP, OFCP, OFNR, OFCR, OFNG, OFCG, OFN, and OFC installed in raceways that are installed in compliance with 300.22(B)

Informational Note: For information on fire protection of wiring installed in fabricated ducts, see NFPA 90A-2018, *Standard for the Installation of Air-Conditioning and Ventilating Systems.*

N **(2) Uses Not Permitted.** Types OFNR, OFCR, OFNG, OFCG, OFN, and OFC shall not be permitted to be installed in ducts specifically fabricated for environmental air as described in 300.22(B).

Informational Note: See NFPA 90A-2021, *Standard for the Installation of Air-Conditioning and Ventilating Systems,* for information on fire protection of wiring installed in fabricated ducts.

Δ **(C) Other Spaces Used for Environmental Air (Plenums).** Installations of optical fiber cables in other spaces used for environmental air shall be in accordance with 770.13(C)(1) and (C)(2).

N **(1) Uses Permitted.** The following cables shall be permitted in other spaces used for environmental air as described in 300.22(C):

(1) Types OFNP and OFCP
(2) Types OFNP and OFCP installed in plenum communications raceways
(3) Types OFNP and OFCP installed in plenum cable routing assemblies
(4) Types OFNP and OFCP supported by open metal cable tray systems
(5) Types OFNP, OFCP, OFNR, OFCR, OFNG, OFCG, OFN, and OFC installed in raceways that are installed in compliance with 300.22(C)
(6) Types OFNP, OFCP, OFNR, OFCR, OFNG, OFCG, OFN, and OFC supported by solid bottom metal cable trays with

solid metal covers in other spaces used for environmental air (plenums), as described in 300.22(C)
(7) Types OFNP, OFCP, OFNR, OFCR, OFNG, OFCG, OFN, and OFC installed in plenum riser and general-purpose communications raceways supported by solid bottom metal cable trays with solid metal covers in other spaces used for environmental air (plenums), as described in 300.22(C)

N **(2) Uses Not Permitted.** Types OFNR, OFCR, OFNG, OFCG, OFN, and OFC shall not be permitted to be installed in other spaces used for environmental air (plenums).

Informational Note: See NFPA 90A-2018, *Standard for the Installation of Air-Conditioning and Ventilating Systems,* for information on fire protection of wiring installed in other spaces used for environmental air.

Δ **(D) Risers — Cables in Vertical Runs.** Installations of optical fiber cables in vertical runs shall be in accordance with 770.113(D)(1) and (D)(2).

N **(1) Uses Permitted.** The following cables shall be permitted in vertical runs penetrating one or more floors and in vertical runs in a shaft:

(1) Types OFNP, OFCP, OFNR, and OFCR
(2) Types OFNP, OFCP, OFNR, and OFCR installed in the following:
   a. Plenum communications raceways
   b. Plenum cable routing assemblies
   c. Riser communications raceways
   d. Riser cable routing assemblies

N **(2) Uses Not Permitted.** Types OFNG, OFCG, OFN, and OFC shall not be permitted to be installed in vertical runs.

Informational Note: See 770.26 for firestop requirements for floor penetrations.

(E) Risers — Cables Permitted in Metal Raceways. The following cables shall be permitted in metal raceways in a riser having firestops at each floor:

(1) Types OFNP, OFCP, OFNR, OFCR, OFNG, OFCG, OFN, and OFC
(2) Types OFNP, OFCP, OFNR, OFCR, OFNG, OFCG, OFN, and OFC installed in the following:
   a. Plenum communications raceways (innerduct)
   b. Riser communications raceways (innerduct)
   c. General-purpose communications raceways (innerduct)

Informational Note: See 770.26 for firestop requirements for floor penetrations.

(F) Risers — Cables Permitted in Fireproof Shafts. The following cables shall be permitted to be installed in fireproof riser shafts having firestops at each floor:

(1) Types OFNP, OFCP, OFNR, OFCR, OFNG, OFCG, OFN, and OFC

(2) Types OFNP, OFCP, OFNR, OFCR, OFNG, OFCG, OFN, and OFC installed in the following:

  a. Plenum communications raceways
  b. Plenum cable routing assemblies
  c. Riser communications raceways
  d. Riser cable routing assemblies
  e. General-purpose communications raceways
  f. General-purpose cable routing assemblies

Informational Note: See 770.26 for firestop requirements for floor penetrations.

(G) Risers — Cables Permitted in One- and Two-Family Dwellings. The following cables shall be permitted in one- and two-family dwellings:

(1) Types OFNP, OFCP, OFNR, OFCR, OFNG, OFCG, OFN, and OFC

(2) Types OFNP, OFCP, OFNR, OFCR, OFNG, OFCG, OFN, and OFC installed in the following:

  a. Plenum communications raceways
  b. Plenum cable routing assemblies
  c. Riser communications raceways
  d. Riser cable routing assemblies
  e. General-purpose communications raceways
  f. General-purpose cable routing assemblies

(H) Cable Trays — Cables Permitted. The following cables shall be permitted to be supported by cable trays:

(1) Types OFNP, OFCP, OFNR, OFCR, OFNG, OFCG, OFN, and OFC

(2) Types OFNP, OFCP, OFNR, OFCR, OFNG, OFCG, OFN, and OFC installed in the following:

  a. Plenum communications raceways
  b. Riser communications raceways
  c. General-purpose communications raceways

(I) Distributing Frames and Cross-Connect Arrays — Cables Permitted. The following cables shall be permitted to be installed in distributing frames and cross-connect arrays:

(1) Types OFNP, OFCP, OFNR, OFCR, OFNG, OFCG, OFN, and OFC

(2) Types OFNP, OFCP, OFNR, OFCR, OFNG, OFCG, OFN, and OFC installed in the following:

  a. Plenum communications raceways
  b. Plenum cable routing assemblies
  c. Riser communications raceways
  d. Riser cable routing assemblies
  e. General-purpose communications raceways
  f. General-purpose cable routing assemblies

(J) Other Building Locations — Cables Permitted. The following cables shall be permitted to be installed in building locations other than the locations covered in 770.113(B) through (I):

(1) Types OFNP, OFCP, OFNR, OFCR, OFNG, OFCG, OFN, and OFC

(2) Types OFNP, OFCP, OFNR, OFCR, OFNG, OFCG, OFN, and OFC installed in:

  a. Plenum communications raceways
  b. Plenum cable routing assemblies
  c. Riser communications raceways
  d. Riser cable routing assemblies
  e. General-purpose communications raceways
  f. General-purpose cable routing assemblies

(3) Types OFNP, OFCP, OFNR, OFCR, OFNG, OFCG, OFN, and OFC installed in a raceway of a type recognized in Chapter 3

Δ **770.114 Grounding.** Non-current-carrying conductive members of optical fiber cables shall be bonded to a grounded equipment rack or enclosure, or grounded in accordance with the grounding methods specified by 770.100(B) using a conductor specified in 770.100(A).

770.133 Installation of Optical Fibers and Electrical Conductors.

(A) In Cable Trays and Raceways. Conductive optical fiber cables contained in an armored or metal-clad-type sheath and nonconductive optical fiber cables shall be permitted to occupy the same cable tray or raceway with conductors for electric light, power, Class 1, non-power-limited fire alarm, Type ITC, or medium-power network-powered broadband communications circuits operating at 1000 volts or less. Conductive optical fiber cables without an armored or metal-clad-type sheath shall not be permitted to occupy the same cable tray or raceway with conductors for electric light, power, Class 1, non-power-limited fire alarm, Type ITC, or medium-power network-powered broadband communications circuits, unless all of the conductors of electric light, power, Class 1, non-power-limited fire alarm, and medium-power network-powered broadband communications circuits are separated from all of the optical fiber cables by a permanent barrier or listed divider.

(B) In Cabinets, Outlet Boxes, and Similar Enclosures. Nonconductive optical fiber cables shall not be permitted to occupy the same cabinet, outlet box, panel, or similar enclosure housing the electrical terminations of an electric light, power, Class 1, non-power-limited fire alarm, or medium-power network-powered broadband communications circuit unless one or more of the following conditions exist:

(1) The nonconductive optical fiber cables are functionally associated with the electric light, power, Class 1, non-power-limited fire alarm, or medium-power network-powered broadband communications circuit.

(2) The conductors for electric light, power, Class 1, non-power-limited fire alarm, Type ITC, or medium-power

network-powered broadband communications circuits operate at 1000 volts or less.

(3) The nonconductive optical fiber cables and the electrical terminations of electric light, power, Class 1, non-power-limited fire alarm, or medium-power network-powered broadband communications circuit are installed in factory- or field-assembled control centers.

(4) The nonconductive optical fiber cables are installed in an industrial establishment where conditions of maintenance and supervision ensure that only qualified persons service the installation.

When optical fibers are within the same hybrid cable for electric light, power, Class 1, non-power-limited fire alarm, or medium-power network-powered broadband communications circuits operating at 1000 volts or less, they shall be permitted to be installed only where the functions of the optical fibers and the electrical conductors are associated.

Optical fibers in hybrid optical fiber cables containing only current-carrying conductors for electric light, power, or Class 1 circuits rated 1000 volts or less shall be permitted to occupy the same cabinet, cable tray, outlet box, panel, raceway, or other termination enclosure with conductors for electric light, power, or Class 1 circuits operating at 1000 volts or less.

Optical fibers in hybrid optical fiber cables containing current-carrying conductors for electric light, power, or Class 1 circuits rated over 1000 volts shall be permitted to occupy the same cabinet, cable tray, outlet box, panel, raceway, or other termination enclosure with conductors for electric light, power, or Class 1 circuits in industrial establishments, where conditions of maintenance and supervision ensure that only qualified persons service the installation.

Δ **(C) With Other Circuits.** Conductive and nonconductive optical fiber cables shall be permitted in the same raceway, cable tray, box, enclosure, or cable routing assembly, with conductors of any of the following:

(1) Class 2 and Class 3 remote-control, signaling, and power-limited circuits in compliance with 645.5(E)(2) or Parts I and II of Article 725

(2) Power-limited fire alarm systems in compliance with Parts I and III of Article 760

(3) Communications circuits in compliance with Parts I and V of Article 805

(4) Community antenna television and radio distribution systems in compliance with Parts I and V of Article 820

(5) Low-power network-powered broadband communications circuits in compliance with Parts I and V of Article 830

(D) Support of Optical Fiber Cables. Raceways shall be used for their intended purpose. Optical fiber cables shall not be strapped, taped, or attached by any means to the exterior of any conduit or raceway as a means of support.

Exception: Overhead (aerial) spans of optical fiber cables shall be permitted to be attached to the exterior of a raceway-type mast intended for the attachment and support of such cables.

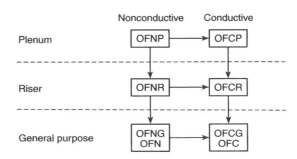

A ─► B Cable A shall be permitted to be used in place of cable B.

N **FIGURE 770.154** *Cable Substitution Hierarchy.*

N **770.154 Applications of Listed Optical Fiber Cables.** Permitted and nonpermitted applications of listed optical fiber cables shall be as indicated in Table 770.154(a). The permitted applications shall be subject to the installation requirements of 770.110 and 770.113. The substitutions for optical fiber cables in Table 770.154(b) and illustrated in Figure 770.154 shall be permitted.

Part VI. Listing Requirements

770.179 Optical Fiber Cables. Optical fiber cables shall be listed and identified in accordance with 770.179(A) through (G) and shall be marked in accordance with Table 770.179. Optical fiber cables shall have a temperature rating of not less than 60°C (140°F). The temperature rating shall be marked on the jacket of optical fiber cables that have a temperature rating exceeding 60°C (140°F).

> Informational Note: See UL 1651-2015, *Standard for Optical Fiber Cable*, for information on optical fiber cables.

Optical fiber cables must have a temperature rating of not less than 60°C (140°F) to correlate with requirements for wires and cables that are addressed in 800.179.

Δ **(A) Types OFNP and OFCP.** Types OFNP and OFCP nonconductive and conductive optical fiber plenum cables shall be suitable for use in ducts, plenums, and other space used for environmental air and shall also have adequate fire-resistant and low-smoke-producing characteristics.

> Informational Note: See NFPA 262-2019, *Standard Method of Test for Flame Travel and Smoke of Wires and Cables for Use in Air-Handling Spaces*, for one method of defining that a cable has adequate fire-resistant and low-smoke-producing characteristics where the cable exhibits a maximum peak optical density of 0.50 or less, an average optical density of 0.15 or less, and a maximum flame spread distance of 1.52 m (5 ft) or less.

(B) Types OFNR and OFCR. Types OFNR and OFCR nonconductive and conductive optical fiber riser cables shall be suitable for use in a vertical run in a shaft or from floor to floor and shall also have the fire-resistant characteristics capable of preventing the carrying of fire from floor to floor.

N   **TABLE 770.154(a)** *Applications of Listed Optical Fiber Cables in Buildings*

Applications		OFNP, OFCP	OFNR, OFCR	OFNG, OFCG, OFN, OFC
		Listed Optical Fiber Cable Type		
In ducts specifically fabricated for environmental air as described in 300.22(B)	In fabricated ducts	Y*	N	N
	In metal raceway that complies with 300.22(B)	Y*	Y*	Y*
In other spaces used for environmental air (plenums) as described in 300.22(C)	In other spaces used for environmental air	Y*	N	N
	In metal raceway that complies with 300.22(C)	Y*	Y*	Y*
	In plenum communications raceways	Y*	N	N
	In plenum cable routing assemblies	Y*	N	N
	Supported by open metal cable trays	Y*	N	N
	Supported by solid bottom metal cable trays with solid metal covers	Y*	Y*	Y*
In risers	In vertical runs	Y*	Y*	N
	In metal raceways	Y*	Y*	Y*
	In fireproof shafts	Y*	Y*	Y*
	In plenum communications raceways	Y*	Y*	N
	In plenum cable routing assemblies	Y*	Y*	N
	In riser communications raceways	Y*	Y*	N
	In riser cable routing assemblies	Y*	Y*	N
	In one- and two-family dwellings	Y*	Y*	Y*
Within buildings in other than air-handling spaces and risers	General	Y*	Y*	Y*
	Supported by cable trays	Y*	Y*	Y*
	In distributing frames and cross-connect arrays	Y*	Y*	Y*
	In any raceway recognized in Chapter 3	Y*	Y*	Y*
	In plenum communications raceway	Y*	Y*	Y*
	In plenum cable routing assemblies	Y*	Y*	Y*
	In riser communications raceways	Y*	Y*	Y*
	In riser cable routing assemblies	Y*	Y*	Y*
	In general-purpose communications raceways	Y*	Y*	Y*
	In general-purpose cable routing assemblies	Y*	Y*	Y*

Note: "N" indicates that the cable type shall not be permitted to be installed in the application. "Y*" indicates that the cable type shall be permitted to be installed in the application subject to the limitations described in 770.110 and 770.113.

Informational Note No. 1: Part V of Article 770 covers installation methods within buildings. This table covers the applications of listed optical fiber cables in buildings. The definition of *Point of Entrance* is in 770.2.

Informational Note No. 2: For information on the restrictions to the installation of optical fiber cables in ducts specifically fabricated for environmental air, see 770.113(B).

N   **TABLE 770.154(b)** *Cable Substitutions*

Cable Type	Permitted Substitutions
OFNP	None
OFCP	OFNP
OFNR	OFNP
OFCR	OFNP, OFCP, OFNR
OFNG, OFN	OFNP, OFNR
OFCG, OFC	OFNP, OFCP, OFNR, OFCR, OFNG, OFN

Informational Note: See ANSI/UL 1666-2017, *Standard Test for Flame Propagation Height of Electrical and Optical-Fiber Cable Installed Vertically in Shafts*, for one method of defining fire-resistant characteristics capable of preventing the carrying of fire from floor to floor.

Δ **(C) Types OFNG and OFCG.** Types OFNG and OFCG nonconductive and conductive general-purpose optical fiber cables shall be suitable for general-purpose use, with the exception of risers and plenums, and shall also be resistant to the spread of fire.

TABLE 770.179 *Cable Markings*

Cable Marking	Type
OFNP	Nonconductive optical fiber plenum cable
OFCP	Conductive optical fiber plenum cable
OFNR	Nonconductive optical fiber riser cable
OFCR	Conductive optical fiber riser cable
OFNG	Nonconductive optical fiber general-purpose cable
OFCG	Conductive optical fiber general-purpose cable
OFN	Nonconductive optical fiber general-purpose cable
OFC	Conductive optical fiber general-purpose cable

Informational Note No 1: See CSA Vertical Flame Test — Cables in Cable Trays, as described in CSA C22.2 No. 0.3-2009 (R2019), *Test Methods for Electrical Wires and Cables*, for one method of defining *resistant to the spread of fire* for the damage (char length) not to exceed 1.5 m (4 ft 11 in.) when performing the test.

Informational Note No. 2: See ANSI/UL 1685-2015, *Standard for Safety for Vertical-Tray Fire Propagation and Smoke-Release Test for Electrical and Optical-Fiber Cables*, for another method of defining *resistant to the spread of fire* where the cables do not spread fire to the top of the tray in the UL flame exposure, vertical tray flame test. The smoke measurements in the test method are not applicable.

Δ **(D) Types OFN and OFC.** Types OFN and OFC nonconductive and conductive optical fiber cables shall be suitable for general-purpose use, with the exception of risers, plenums, and other spaces used for environmental air, and shall also be resistant to the spread of fire.

Informational Note No. 1: See ANSI/UL 1685-2015, *Standard for Safety for Vertical-Tray Fire-Propagation and Smoke-Release Test for Electrical and Optical-Fiber Cables*, for one method of defining *resistant to the spread of fire* where the cables do not spread fire to the top of the tray in the UL flame exposure, vertical tray flame test. The smoke measurements in the test method are not applicable.

Informational Note No. 2: See CSA Vertical Flame Test — Cables in Cables Trays, as described in CSA C22.2 No. 0.3-2009 (R2019), *Test Methods for Electrical Wires and Cables*, for another method of defining *resistant to the spread of fire* where the damage (char length) does not exceed 1.5 m (4 ft 11 in.).

Informational Note No. 3: Cable types are listed in descending order of fire resistance rating. Within each fire resistance rating, nonconductive cable is listed first because it is often substituted for conductive cable.

Δ **(E) Circuit Integrity (CI), Fire-Resistive Cable System, or Electrical Circuit Protective System.** Cables that are used for survivability of critical circuits under fire conditions shall meet either 770.179(E)(1), (E)(2), or (E)(3).

Δ **(1) Circuit Integrity (CI) Cables.** Cables specified in 770.179(A) through (D), and used for survivability of critical circuits, shall be marked with the additional classification using the suffix "CI."

In order to maintain its listed fire rating, CI cable shall only be installed in free air in accordance with 770.24. CI cables shall only be permitted to be installed in a raceway where specifically listed and marked as part of a fire-resistive cable system as covered in 770.179(E)(2).

Informational Note: See UL 2196, *Standard for Fire Test for Circuit Integrity of Fire-Resistive Power, Instrumentation, Control and Data Cables*, for one method of defining CI cable for establishing a minimum 2-hour fire resistance rating for the cable as specified in UL 1651, *Optical Fiber Cable*. UL *Guide Information for Optical Cable Fiber (QAYK)* contains information to identify the cable and its installation limitations to maintain the fire-resistive rating.

Δ **(2) Fire-Resistive Cables.** Cables specified in 770.179(A) through (D) and 770.179(E)(1) that are part of an electrical circuit protective system shall be fire-resistive cable and identified with the protective system number on the product or on the smallest unit container in which the product is packaged and installed in accordance with the listing of the protective system.

Informational Note: See UL 2196, *Standard for Fire Test for Circuit Integrity of Fire-Resistive Power, Instrumentation, Control and Data Cables*, for one method of defining an electrical circuit protective system for establishing a rating for the system. UL *Guide Information for Electrical Circuit Integrity Systems (FHIT)* contains information to identify the system and its installation limitations to maintain a minimum fire-resistive rating.

(F) Field-Assembled Optical Fiber Cables. Field-assembled optical fiber cable shall comply with the following:

(1) The specific combination of jacket and optical fibers intended to be installed as a field-assembled optical fiber cable shall be one of the types in 770.179(A), (B), or (D) and shall be marked in accordance with Table 770.179.

(2) The jacket of a field-assembled optical fiber cable shall have a surface marking indicating the specific optical fibers with which it is identified for use.

(3) The optical fibers shall have a permanent marking, such as a marker tape, indicating the jacket with which they are identified for use.

(4) The jacket without fibers shall meet the listing requirements for communications raceways in 800.182(A), (B), or (C) in accordance with the cable marking.

Δ **(G) Optional Markings.** Cables shall be permitted to be surface marked to indicate special characteristics of the cable materials.

Informational Note: These markings can include, but are not limited to, markings for limited-smoke halogen-free, low-smoke halogen-free, and sunlight resistance.

770.180 Grounding Devices. Where bonding or grounding is required, devices used to connect a shield, a sheath, or non–current-carrying metallic members of a cable to a bonding conductor or grounding electrode conductor shall be listed or be part of listed equipment.

Communications Systems

CHAPTER 8

ARTICLE 800 — General Requirements for Communications Systems

Part I. General

Δ **800.1 Scope.** This article covers general requirements for communications systems. These general requirements apply to communications circuits, community antenna television and radio distribution systems, network-powered broadband communications systems, and premises-powered broadband communications systems, unless modified by Articles 805, 820, 830, or 840.

> Informational Note No. 1: See 90.2(D)(4) for installations of circuits and equipment that are not covered.
>
> Informational Note No. 2: See Part II of Article 725 for information on the installation of Class 2 and Class 3 circuits and 722.135(E) for the substitution of communications cables for Class 2 and Class 3 cables.
>
> Informational Note No. 3: See Part II of Article 760 for information on the installation of power-limited fire alarm circuits, including the substitution of communications cables for power-limited fire alarm cables.

Section 90.3 specifies that Chapter 8 covers communications systems and is not subject to the requirements of Chapters 1 through 7, other than where Chapter 8 specifies a requirement. Chapter 8 was reorganized for the 2020 edition of the *NEC*®. With the intent of minimizing redundant requirements across Chapter 8, a new Article 800 was created to contain the "general" requirements for communications systems. What was Article 800 in previous editions of the *NEC* is now Article 805, while the rest of Chapter 8 remains the same with respect to numbers and titles.

800.2 Reconditioned Equipment. The requirements of 110.21(A)(2) shall apply.

800.3 Other Articles. Only those sections of Chapters 1 through 7 referenced in Chapter 8 shall apply to Chapter 8. The definitions from Article 100 apply to Chapter 8. Installations of circuits and equipment shall comply with 800.3(A) through (I).

N **(A) Output Circuits.** As appropriate for the services provided, the output circuits derived from a network-powered broadband communications system's network interface unit (NIU) or from a premises-powered broadband communications system's network terminal shall comply with the requirements of the following:

(1) Installations of Class 2 and Class 3 circuits — Part II of Article 725 and Parts I and II of Article 722

(2) Installations of power-limited fire alarm circuits — Part III of Article 760

(3) Installations of optical fiber cables — Part V of Article 770

(4) Installations of communications circuits — Part IV of Article 805

> Informational Note: The communications circuits covered by Article 805 are commonly referred to as POTS (plain old telephone service) circuits.

(5) Installations of premises (within buildings) community antenna television and radio distribution circuits — Part V of Article 820

(B) Hazardous (Classified) Locations. Circuits and equipment installed in a location that is classified in accordance with 500.5 and 505.5 shall comply with the applicable requirements of Chapter 5.

(C) Wiring in Ducts for Dust, Loose Stock, or Vapor Removal. The requirements of 300.22(A) shall apply.

(D) Equipment in Other Space Used for Environmental Air. The requirements of 300.22(C)(3) shall apply.

(E) Installation and Use. The requirements of 110.3(B) shall apply.

(F) Optical Fiber Cable. Where optical fiber cable is used to provide a communications circuit within a building, Article 770 shall apply.

N (G) Vertical Support for Fire-Resistive Cables and Conductors. Vertical installations of circuit integrity (CI) cables and conductors installed in a raceway or conductors and cables of fire-resistive cable systems shall be installed in accordance with 300.19.

N (H) Bonding and Grounding of Cable Shields. The requirements of 250.4(A)(5) shall apply to the metal shields of cables used for communications.

800.21 Access to Electrical Equipment Behind Panels Designed to Allow Access. Access to electrical equipment shall not be denied by an accumulation of wires and cables that prevents removal of panels, including suspended ceiling panels.

An excess accumulation of wires and cables can limit access to equipment by preventing the removal of access panels or ceiling tiles. Exhibit 800.1 shows examples of proper and improper installation of communications cables above ceiling tiles that are designed to allow access to the space above.

Δ 800.24 Mechanical Execution of Work.

N (A) General. Circuits and equipment shall be installed in a neat and workmanlike manner. Cables installed exposed on the surface of ceilings and sidewalls shall be supported by the building structure in such a manner that the cable will not be damaged by normal building use. Such cables shall be secured by hardware, including straps; staples: cable ties listed and identified for securement and support; and hangers, or similar fittings, designed and installed so as not to damage the cable. The installation shall also conform to 300.4 and 300.11. Plenum cable ties and other nonmetallic cable accessories used to secure and support cables in other spaces used for environmental air (plenums) shall be listed as having low smoke and heat release properties in accordance with 800.170.

> Informational Note No. 1: See ANSI/BICSI N1-2019, *Installation Practices for Telecommunications and IC Cabling and Related Cabling Infrastructure*; ANSI/TIA-568.1- E-2020, *Commercial Building Telecommunications Infrastructure Standard*; ANSI/TIA-569-E-2019, *Telecommunications Pathways and Spaces*; ANSI/TIA-570-C-2012, *Residential Telecommunications Infrastructure Standard*; ANSI/TIA-1005-A-2012, *Telecommunications Infrastructure Standard for Industrial Premises*; ANSI/TIA-1179-A-2017, *Healthcare Facility Telecommunications Infrastructure Standard*; ANSI/TIA-4966-2014, *Telecommunications Infrastructure Standard for Educational Facilities*; and other ANSI-approved installation standards for accepted industry practices.
>
> Informational Note No. 2: See NFPA 90A-2021, *Standard for the Installation of Air-Conditioning and Ventilating Systems*, for discrete combustible components installed in accordance with 300.22(C).
>
> Informational Note No. 3: Paint, plaster, cleaners, abrasives, corrosive residues, or other contaminants may result in an undetermined alteration of wire and cable properties.

This requirement does not contain specific supporting and securing intervals. It does reference 300.4 and 300.11 as a general rule for securing equipment and cables, as well as for protection of cables from physical damage. Securing and supporting intervals are typically found in the manufacturer's installation instructions.

See also

800.110(C) for support of cable routing assemblies

N (B) Circuit Integrity (CI) Cable. CI cable shall be supported at a distance not exceeding 610 mm (24 in.). Cable shall be secured to the noncombustible surface of the building structure. Cable supports and fasteners shall be steel.

800.25 Abandoned Cables. The accessible portion of abandoned cables shall be removed. Where cables are identified for future use with a tag, the tag shall be of sufficient durability to withstand the environment involved.

See Article 100 for the definition of *accessible (as applied to wiring methods)*. Abandoned cable unnecessarily increases fire loading and can affect airflow where installed in plenums.

See also

Article 100 for the definition of the term *cable, abandoned (abandoned cable)*

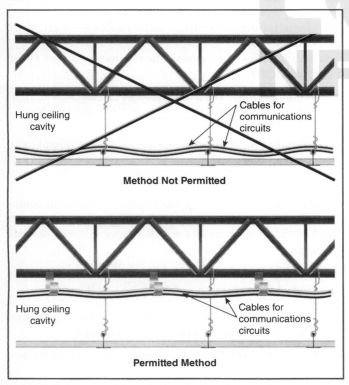

EXHIBIT 800.1 Incorrect installation of cables (*upper diagram*) and correct method (*lower diagram*).

Within the image:
Hung ceiling cavity
Cables for communications circuits
Method Not Permitted
Hung ceiling cavity
Cables for communications circuits
Permitted Method

800.26 Spread of Fire or Products of Combustion. Installations of cables, communications raceways, cable routing assemblies in hollow spaces, vertical shafts, and ventilation or air-handling ducts shall be made so that the possible spread of fire or products of combustion will not be substantially increased. Openings around penetrations of cables, communications raceways, and cable routing assemblies through fire-resistant-rated walls, partitions, floors, or ceilings shall be firestopped using approved methods to maintain the fire resistance rating.

> Informational Note: Directories of electrical construction materials published by qualified testing laboratories contain many listing installation restrictions necessary to maintain the fire-resistive rating of assemblies where penetrations or openings are made. Building codes also contain restrictions on membrane penetrations on opposite sides of a fire resistance–rated wall assembly. An example is the 600 mm (24 in.) minimum horizontal separation that usually applies between boxes installed on opposite sides of the wall. Assistance in complying with 800.26 can be found in building codes, fire resistance directories, and product listings.

800.27 Temperature Limitation of Wires and Cables. No wire or cable shall be used in such a manner that its operating temperature exceeds that of its rating.

Part II. Wires and Cables Outside and Entering Buildings

800.44 Overhead (Aerial) Wires and Cables. Overhead (aerial) communications wires and cables and CATV-type coaxial cables entering buildings shall comply with 800.44(A) through (D).

> Informational Note: See ANSI C2-2017, *National Electrical Safety Code, Part 2 Safety Rules for Overhead Lines*, for additional information regarding overhead (aerial) wires and cables.

(A) On Poles, In-Span, Above Roofs, on Masts, or Between Buildings. If communications wires and cables or CATV-type coaxial cables and electric light or power conductors are supported by the same pole or are run parallel to each other in-span, the conditions described in 800.44(A)(1) through (A)(4) shall be met.

(1) Relative Location. If practicable, the communications wires and cables and CATV- type coaxial cables shall be located below the electric light or power conductors.

(2) Attachment to Cross-Arms. Communications wires and cables and CATV-type coaxial cables shall not be attached to a cross-arm that carries electric light or power conductors.

(3) Climbing Space. The climbing space through wires and cables shall comply with the requirements of 225.14(B).

Δ **(4) Clearance.** Supply service drops and sets of overhead service conductors of 0 volts to 750 volts running above and parallel to communications wires and cables and CATV-type coaxial service drops shall have a minimum separation of 300 mm (12 in.)

at any point in the span, including the point of their attachment to the building, provided that the ungrounded conductors are insulated and that a clearance of not less than 1.0 m (40 in.) is maintained between the two services at the pole.

(B) Above Roofs. Communications wires and cables and CATV-type coaxial cables shall have a vertical clearance of not less than 2.5 m (8 ft) from all points of roofs above which they pass.

> *Exception No. 1: Communications wires and cables and CATV-type coaxial cables shall not be required to have a vertical clearance of not less than 2.5 m (8 ft) above auxiliary buildings, such as garages and the like.*

> *Exception No. 2: A reduction in clearance above only the overhanging portion of the roof to not less than 450 mm (18 in.) shall be permitted if (1) not more than 1.2 m (4 ft) of communications and CATV-type service-drop conductors pass above the roof overhang and (2) they are terminated at a through- or above-the-roof raceway or approved support.*

> *Exception No. 3: Where the roof has a slope of not less than 100 mm in 300 mm (4 in. in 12 in.), a reduction in clearance to not less than 900 mm (3 ft) shall be permitted.*

> Informational Note: See ANSI/IEEE C2-2017, *National Electrical Safety Code, Part 2, Safety Rules for Overhead Lines*, for additional information regarding overhead (aerial) wire and cables.

(C) On Masts. Overhead (aerial) communications wires and cables and CATV-type coaxial cables shall be permitted to be attached to an above-the-roof raceway mast that does not enclose or support conductors of electric light or power circuits.

(D) Between Buildings. Communications and CATV-type coaxial cables extending between buildings or structures, and also the supports or attachment fixtures, shall be identified and shall have sufficient strength to withstand the loads to which they might be subjected.

> *Exception: If a communications cable or a CATV-type coaxial cable does not have sufficient strength to be self-supporting, it shall be attached to a supporting messenger cable that, together with the attachment fixtures or supports, shall be acceptable for the purpose and shall have sufficient strength to withstand the loads to which they may be subjected.*

In addition to the weight of the cable itself, wind and ice loads should be considered because they can damage cables and attachment points.

N **(E) On Buildings.** Where attached to buildings, communications wires and cables and CATV-type coaxial cables shall be securely fastened in such a manner that they will be separated from other conductors in accordance with 800.44(E)(1) and (E)(2).

N **(1) Electric Light or Power.** The communications wires and cables and CATV-type coaxial cables shall have a separation of at least 100 mm (4 in.) from electric light, power, Class 1, or non-power-limited fire alarm circuit conductors not in raceway or cable, or shall be permanently separated from conductors of the other system by a continuous and firmly fixed nonconductive barrier in addition to the insulation on the wires.

N (2) Other Communications Systems. Communications wires and cables and CATV-type coaxial cables shall be installed so that there will be no unnecessary interference in the maintenance of the separate systems. In no case shall the wires, cables, messenger strand, or equipment of one system cause abrasion to the wires, cables, messenger strand, or equipment of any other system.

N 800.47 Underground Systems Entering Buildings. Underground communications wires and cables, CATV-type coaxial cables, and network-powered broadband communications cables entering buildings shall comply with 800.47(A) and (B). The requirements of 310.10(C) shall not apply to communications wires and cables and CATV-type coaxial cables.

N (A) Underground Systems with Electric Light, Power, Class 1, or Non-Power-Limited Fire Alarm Circuit Conductors. Underground communications wires and cables, CATV-type coaxial cables, and network-powered broadband communications cables in a raceway, pedestal, handhole enclosure, or manhole containing electric light, power, Class 1, or non-power-limited fire alarm circuit conductors shall be in a section separated from such conductors by means of brick, concrete, or tile partitions or by means of a suitable barrier.

N (B) Direct-Buried Cables and Raceways. Direct-buried communications wires and cables, CATV-type coaxial cables, and network-powered broadband communications cables shall be separated at least 300 mm (12 in.) from conductors of any light or power, non-power-limited fire alarm circuit conductors, or Class 1 circuit.

Exception No. 1: Separation shall not be required if electric service conductors or all the direct-buried communications wires and cables, CATV-type coaxial cables, and network-powered broadband communications cables are installed in raceways or have metal cable armor.

Exception No. 2: Separation shall not be required under one of the following conditions:

(1) If the electric light or power branch-circuit or feeder conductors or Class 1 circuit conductors are installed in a raceway or in metal-sheathed, metal-clad, or Type UF or Type USE cables

(2) If all the direct-buried communications wires cables, CATV-type coaxial cables, and network-powered broadband communications cables have metal cable armor or are installed in raceway

N 800.48 Unlisted Cables Entering Buildings. Unlisted outside plant communications cables and unlisted outside plant CATV-type

coaxial cables shall be permitted to be installed in building spaces other than risers, ducts used for environmental air, plenums used for environmental air, and other spaces used for environmental air if all of the following applies:

(1) The length of the cable within the building, measured from its point of entrance, does not exceed 15 m (50 ft).
(2) The cable enters the building from the outside.
(3) The unlisted outside plant communications cable is terminated in an enclosure or on a listed primary protector, or the unlisted outside plant CATV type coaxial cable is terminated at a grounding block.

The point of entrance shall be permitted to be extended from the penetration of the external wall, roof, or floor slab by continuously enclosing the entrance cables in rigid metal conduit (RMC) or intermediate metal conduit (IMC) to the point of emergence.

Informational Note No. 1: Splice cases or terminal boxes, both metallic and plastic types, are typically used as enclosures for splicing or terminating communications cables.

Informational Note No. 2: This section limits the length of unlisted outside plant cable to 15 m (50 ft) from the point of entrance, while 805.90(B) requires that the primary protector be located as close as practicable to the point of entrance of the cable. Therefore, in installations requiring a primary protector, the outside plant cable may not extend 15 m (50 ft) into the building if it is practicable to place the primary protector closer to the point of entrance.

In general, because they are a part of the building fuel load, communications cables within buildings are required to be listed to ensure that they have met certain testing parameters. Unlisted cables, such as outside plant cables, have not been tested to the same parameters and, therefore, are limited in length within a building unless installed in rigid metal conduit (RMC) or intermediate metal conduit (IMC) where it emerges from the slab.

800.49 Metal Entrance Conduit Grounding. Metal conduit containing entrance wire or cable shall be connected by a bonding conductor or grounding electrode conductor to a grounding electrode or, where present, the building grounding electrode system in accordance with 800.100(B).

Δ 800.53 Separation from Lightning Conductors. Where practicable on buildings, a separation of at least 1.8 m (6 ft) shall be maintained between lightning protection conductors and all communications wires and cables and CATV-type coaxial cables.

Informational Note No. 1: See ANSI C2-2017 *National Electrical Safety Code, Part 2, Safety Rules for Overhead Lines,* for additional information regarding overhead (aerial) wires and cables.

Informational Note No. 2: See NFPA 780-2020, *Standard for the Installation of Lightning Protection Systems,* for information on calculation of separation distances using the sideflash equation.

Part III. Grounding Methods

800.100 Cable and Primary Protector Bonding and Grounding.

(A) Bonding Conductor or Grounding Electrode Conductor.

(1) Insulation. The bonding conductor or grounding electrode conductor shall be listed and shall be permitted to be insulated, covered, or bare.

(2) Material. The bonding conductor or grounding electrode conductor shall be copper or other corrosion-resistant conductive material, stranded or solid.

(3) Size. The bonding conductor or grounding electrode conductor shall not be smaller than 14 AWG. The bonding conductor or grounding electrode conductor shall have a current-carrying capacity not less than the aggregate of the grounded metal cable sheath member, the metal strength member(s), and the protected conductor(s) of the communications cable, or the outer sheath of the coaxial cable, as applicable. The bonding conductor or grounding electrode conductor shall not be required to exceed 6 AWG.

(4) Length. The bonding conductor or grounding electrode conductor shall be as short as practicable. In one- and two-family dwellings, the bonding conductor or grounding electrode conductor shall be as short as practicable, not to exceed 6.0 m (20 ft) in length.

Informational Note: Similar bonding conductor or grounding electrode conductor length limitations applied at apartment buildings and commercial buildings help to reduce voltages that may be developed between the building's power and communications systems during lightning events. See ANSI/TIA-607-D-2019, *Generic Telecommunications Bonding and Grounding (Earthing) for Customer Premises*, which includes useful information to reduce such voltages.

Exception: In one- and two-family dwellings if it is not practicable to achieve an overall maximum bonding conductor or grounding electrode conductor length of 6.0 m (20 ft), a separate ground rod meeting the minimum dimensional criteria of 800.100(B)(3)(2) or (B)(3)(3) shall be driven, the bonding conductor or grounding electrode conductor shall be connected to the ground rod in accordance with 800.100(C), and the ground rod shall be connected to the power grounding electrode system in accordance with 800.100(D).

Limiting the bonding conductor length reduces the impedance of the conductor, resulting in a low, if any, potential difference between conductors and equipment of the communications system and the electrical conductors and equipment in the building. The low-impedance bonding connection reduces the fire and shock hazard in the event that electric utility power lines come in contact with communications conductors.

The informational note to 800.100(A)(4) provides guidance for the treatment of the cable and primary protector grounding conductor length at apartment and commercial buildings that is consistent with the 20-foot rule for one- and two-family dwellings. However, a specific length is not specified in the *NEC* because such a limitation might not be practical in some installations.

(5) Run in Straight Line. The bonding conductor or grounding electrode conductor shall be run in as straight a line as practicable.

(6) Physical Protection. Bonding conductors and grounding electrode conductors shall be protected where exposed to physical damage. If the bonding conductor or grounding electrode conductor is installed in a metal raceway, both ends of the raceway shall be bonded to the contained conductor or to the same terminal or electrode to which the bonding conductor or grounding electrode conductor is connected.

(B) Electrode. The bonding conductor or grounding electrode conductor shall be connected in accordance with 800.100(B)(1), (B)(2), or (B)(3).

(1) In Buildings or Structures with an Intersystem Bonding Termination. If the building or structure served has an intersystem bonding termination as required by 250.94, the bonding conductor shall be connected to the intersystem bonding termination.

Informational Note: Informational Note Figure 800.100(B)(1) illustrates the connection of the bonding conductor in buildings or structures equipped with an intersystem bonding termination or a terminal block providing access to the building grounding means.

Δ **(2) In Buildings or Structures with Grounding Means.** If an intersystem bonding termination is established, 250.94(A) shall apply. If the building or structure served has no intersystem bonding termination, the bonding conductor or grounding electrode

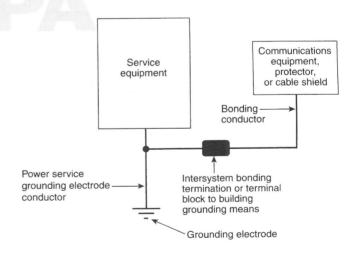

Δ *INFORMATIONAL NOTE FIGURE 800.100(B)(1)*
Illustration of a Bonding Conductor in a Communications Installation Equipped With an Intersystem Bonding Termination or Terminal Block Providing Access To the Building Grounding Means.

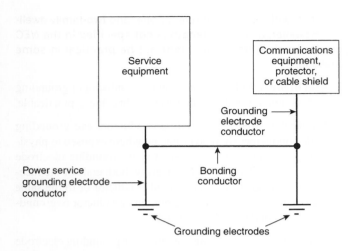

conductor shall be connected to the nearest accessible location on one of the following:

(1) The building or structure grounding electrode system as covered in 250.50
(2) The power service accessible means external to enclosures using the options identified in 250.94(A), Exception
(3) The nonflexible metal power service raceway
(4) The service equipment enclosure
(5) The grounding electrode conductor or the grounding electrode conductor metal enclosure of the power service
(6) The grounding electrode conductor or the grounding electrode of a building or structure disconnecting means that is connected to a grounding electrode as covered in 250.32
(7) The grounded interior metal water piping system, within 1.5 m (5 ft) from its point of entrance to the building, as covered in 250.52

See also

250.52(A)(1) for more information on the use of a metal water piping system as a grounding electrode

250.68(C)(1) and its commentary for more information on connecting to interior metal water piping

A bonding device intended to provide a termination point for the bonding conductor (intersystem bonding) shall not interfere with the opening of an equipment enclosure. A bonding device shall be mounted on nonremovable parts. A bonding device shall not be mounted on a door or cover even if the door or cover is nonremovable.

For purposes of this section, the mobile home service equipment or the mobile home disconnecting means located within 9.0 m (30 ft) of the exterior wall of the mobile home it serves, or at a mobile home disconnecting means connected to an electrode by a grounding electrode conductor in accordance with 250.32 and located within 9.0 m (30 ft) of the exterior wall of the mobile home it serves, shall be considered to meet the requirements of this section.

Informational Note: See Informational Note Figure 800.100(B)(2) for an illustration of a grounding electrode conductor and a bonding conductor in a communications installation not equipped with an intersystem bonding termination or terminal block.

(3) In Buildings or Structures Without an Intersystem Bonding Termination or Grounding Means. If the building or structure served has no intersystem bonding termination or grounding means, as described in 800.100(B)(2), the grounding electrode conductor shall be connected to one of the following:

(1) To any one of the individual grounding electrodes described in 250.52(A)(1), (A)(2), (A)(3), or (A)(4)
(2) If the building or structure served has no intersystem bonding termination or grounding means, as described in 800.100(B)(2) or (B)(3)(1), to any one of the individual grounding electrodes described in 250.52(A)(5), (A)(7), and (A)(8)
(3) For communications circuits covered in Article 805 or network-powered broadband communications systems covered in Article 830, to a ground rod or pipe not less than 1.5 m (5 ft) in length and 12.7 mm (0.5 in.) in diameter, driven, where practicable, into permanently damp earth and separated from lightning protection system conductors, as covered in 800.53, and at least 1.8 m (6 ft) from electrodes of other systems

Steam pipes, hot water pipes, or lightning protection system conductors shall not be employed as grounding electrodes or as a bonding or grounding electrode conductor for protectors and grounded metal members.

(C) Electrode Connection. Connections to grounding electrodes shall comply with 250.70.

(D) Bonding of Electrodes. A bonding jumper not smaller than 6 AWG copper or equivalent shall be connected between the grounding electrode and power grounding electrode system at the building or structure served if separate electrodes are used.

Exception: Bonding of electrodes at mobile homes shall be in accordance with 800.106.

Informational Note No. 1: See 250.60 for connection to a lightning protection system.

Informational Note No. 2: Bonding together of all separate electrodes limits potential differences between them and between their associated wiring systems.

Bonding of communications system and power system grounding electrodes is required where they are installed at the same building or structure. A common error is often made when grounding community antenna television (CATV) systems. Many times, the coaxial cable sheath is connected to a rod-type grounding electrode driven by the CATV installer at a convenient location near the point of cable entry to the building, instead of bonding it to the electrical service grounding electrode system. This creates two grounding electrode systems at the same structure and potentially creates a difference in potential between the electrodes. This can result in a difference in potential between the grounded components connected to the sheath of the coaxial cable and any other electrical equipment that is grounded to the electrode for the other building systems.

Section 250.94 requires that the intersystem bonding termination (IBT) device have three or more termination points that are accessible and external to the service equipment for making the bonding and grounding connection for other systems. For existing installations where an intersystem bonding termination is not available, alternative bonding means are described in 800.100(B)(2). A separate grounding electrode is permitted by 800.100(B)(3) only if the building or structure has neither an intersystem bonding termination nor a grounding means, which is rare. The earth cannot be used as the bonding conductor, as specified in 250.54, because it does not provide the required low-impedance path.

Both communications systems and power systems are subject to current surges. If the grounded conductors and parts of the two systems are not bonded by a low-impedance path, such line surges can raise the potential difference between the two systems to many thousands of volts. This can result in arcing between the two systems — for example, wherever the coaxial cable jacket contacts a grounded part, such as a metal water pipe or metal structural member — inside the building.

If a person is the interface between two systems that are not properly bonded, a high-voltage surge could result in electric shock. More common, however, is burnout of a television tuner, a part that is almost always an interface between the two systems. The tuner is connected to the power system ground through the grounded neutral of the power supply, even if the television set itself is not provided with an equipment grounding conductor (EGC).

See also

250.92(B) and its commentary for more information on bonding for services

800.106 Primary Protector Grounding and Bonding at Mobile Homes.

(A) Grounding. Grounding shall comply with 800.106(A)(1) and (A)(2).

Δ **(1) Mobile Home Service Equipment.** Where there is no mobile home service equipment located within 9.0 m (30 ft) of the exterior wall of the mobile home it serves, grounding shall comply with one of the following:

(1) The following components (if present) shall be connected to a grounding electrode in accordance with 800.100(B)(3):

  a. Primary protector grounding terminal
  b. Network interface unit
  c. Coaxial cable shield ground
  d. Surge arrester grounding terminal
  e. Network-powered broadband communications cable shield
  f. Network-powered broadband communications cable metal members not used for communications or powering

(2) The non-current-carrying metal members of optical fiber cables shall be connected to a grounding electrode in accordance with 770.106(A)(1). The network terminal, if required to be grounded, shall be connected to a grounding electrode in accordance with 800.106(A)(1)(1). The grounding electrode shall be bonded in accordance with 770.106(B).

(2) Mobile Home Feeder Disconnecting Means. Where there is no mobile home disconnecting means grounded in accordance with 250.32 and located within 9.0 m (30 ft) of the exterior wall of the mobile home it serves, grounding shall comply with one of the following:

(1) The following components (if present) shall be connected to a grounding electrode in accordance with 800.100(B)(3):

  a. Primary protector grounding terminal
  b. Network interface unit
  c. Network-powered broadband communications shield
  d. Network-powered broadband communications cable metal members not used for communications or powering

(2) The non-current-carrying metal members of optical fiber cables shall be connected to a grounding electrode in accordance with 770.106(A)(2). The network terminal, if required to be grounded, shall be connected to a grounding electrode in accordance with 800.106(A)(2). The grounding electrode shall be bonded in accordance with 770.106(B).

(B) Bonding. The primary protector grounding terminal or grounding electrode, network-powered broadband communications cable grounding terminal, or network interface unit grounding terminal shall be bonded together and connected to the metal frame or available grounding terminal of the mobile home with a copper conductor not smaller than 12 AWG under either of the following conditions:

(1) If there is no mobile home service equipment or disconnecting means as in 800.106(A)
(2) If the mobile home is supplied by cord and plug

Part IV. Installation Methods Within Buildings

800.110 Raceways, Cable Routing Assemblies, and Cable Trays.

Δ **(A) Types of Raceways.** Wires and cables shall be permitted to be installed in raceways that comply with 800.110(A)(1), 800.110(A)(2), or 800.110(A)(3). Medium-power networked-powered broadband communications cables shall not be installed in raceways that comply with 800.110(A)(2).

(1) Raceways Recognized in Chapter 3. Wires and cables shall be permitted to be installed in any raceway included in Chapter 3. The raceways shall be installed in accordance with Chapter 3.

(2) Communications Raceways. Wires and cables shall be permitted to be installed in plenum communications raceways, riser communications raceways, and general-purpose communications raceways selected in accordance with Table 800.154(b), listed in accordance with 800.182, and installed in accordance with 800.113 and 362.24 through 362.56, where the requirements applicable to electrical nonmetallic tubing (ENT) apply.

(3) Innerduct for Communications Wires and Cables, Coaxial Cables, or Network-Powered Broadband Communications Cables. Listed plenum communications raceways, listed riser communications raceways, and listed general-purpose communications raceways selected in accordance with Table 800.154(b) shall be permitted to be installed as innerduct in any type of listed raceway permitted in Chapter 3.

(B) Raceway Fill. The raceway fill requirements of Chapters 3 and 9 shall apply to medium-power network-powered broadband communications cables.

(C) Cable Routing Assemblies. Cables shall be permitted to be installed in plenum cable routing assemblies, riser cable routing assemblies, and general-purpose cable routing assemblies selected in accordance with Table 800.154(c), listed in accordance with 800.182, and installed in accordance with 800.110(C)(1) and (C)(2) and 800.113.

(1) Horizontal Support. Cable routing assemblies shall be supported where run horizontally at intervals not to exceed 900 mm (3 ft) and at each end or joint, unless listed for other support intervals. In no case shall the distance between supports exceed 3 m (10 ft).

(2) Vertical Support. Vertical runs of cable routing assemblies shall be supported at intervals not exceeding 1.2 m (4 ft), unless listed for other support intervals, and shall not have more than one joint between supports.

(D) Cable Trays. Wires and cables and communications raceways shall be permitted to be installed in metal or listed nonmetallic cable tray systems. Ladder cable trays shall be permitted to support cable routing assemblies.

800.113 Installation of Cables Used for Communications Circuits, Communications Wires, Cable Routing Assemblies, and Communications Raceways.

Installation of wires, cables, cable routing assemblies, and communications raceways shall comply with 800.113(A) through (L). Installation of cable routing assemblies and communications raceways shall comply also with 800.110. Types of cables used by this section are identified in Table 800.113.

N **TABLE 800.113** *Cables Used for Communications Circuits*

	Listed Cable Types
Plenum cables	CMP, CATVP, BLP, OFNP, OFCP
Riser cables	CMR, CATVR, BMR, BLR, OFNR, OFCR
General-purpose cables	CMG, CM, CATV, BM, BL, OFNG, OFN, OFCG, OFC
Limited-use cables	CMX, CATVX, BLX
Undercarpet	CMUC
Underground	BMU, BLU

Δ **(A) Listing.** Cables used for communications circuits, communications wires, cable routing assemblies, and communications raceways installed in buildings shall be listed and installed in accordance with the limitations of the listing.

Exception: Cables installed in compliance with 800.48 shall not be required to be listed.

Δ **(B) Ducts Specifically Fabricated for Environmental Air.** Installations of cables used for communications circuits, communications wires, cable routing assemblies, and communications raceways in ducts specifically fabricated for environmental air shall be in accordance with 800.113(B)(1) and (B)(2).

Δ **(1) Uses Permitted.** The following cables shall be permitted in ducts specifically fabricated for environmental air as described in 300.22(B) if they are directly associated with the air distribution system:

(1) Plenum cables up to 1.22 m (4 ft) in length
(2) Plenum, rise, general-purpose, and limited-use cables installed in raceways that are installed in compliance with 300.22(B)

N **(2) Uses Not Permitted.** The following cables, wires, cable routing assemblies, and communications raceways shall not be permitted in ducts specifically fabricated for environmental air as described in 300.22(B):

(1) Plenum, riser, and general-purpose communications raceways
(2) Plenum, riser, and general-purpose cable routing assemblies
(3) Riser, general-purpose, and limited-use cables
(4) Type CMUC cables and wires
(5) Types BMU and BLU cables
(6) Communications wires
(7) Hybrid power and communications cables

Informational Note: See NFPA 90A-2021, *Standard for the Installation of Air-Conditioning and Ventilating Systems*, for information on fire protection of wiring installed in fabricated ducts.

Δ **(C) Other Spaces Used for Environmental Air (Plenums).** Installations of cables used for communications circuits, communications wires, cable routing assemblies, and communications raceways in other spaces used for environmental air (plenums) shall be in accordance with 800.113(C)(1) and (C)(2).

Δ **(1) Uses Permitted.** The following cables, wires, cable routing assemblies, and communications raceways shall be permitted in other spaces used for environmental air as described in 300.22(C):

(1) Plenum cables
(2) Plenum communications raceways
(3) Plenum cable routing assemblies
(4) Plenum cables installed in plenum communications raceways
(5) Plenum cables installed in plenum cable routing assemblies
(6) Plenum cables and plenum communications raceways supported by open metal cable tray systems
(7) Plenum, riser, general-purpose, and limited-use cables, and communications wires installed in raceways that are installed in compliance with 300.22(C)
(8) Plenum, rise, general-purpose, limited-use cables and plenum, riser, and general-purpose communications raceways supported by solid bottom metal cable trays with solid metal covers in other spaces used for environmental air (plenums) as described in 300.22(C)
(9) Plenum, riser, general-purpose, and limited-use cables installed in plenum, riser, and general-purpose communications raceways supported by solid bottom metal cable trays with solid metal covers in other spaces used for environmental air (plenums) as described in 300.22(C)

N **(2) Uses Not Permitted.** The following cables, wires, cable routing assemblies, and communications raceways shall not be permitted in other spaces used for environmental air as described in 300.22(C):

(1) Riser, general-purpose, and limited-use cables
(2) Riser and general-purpose communications raceways
(3) Riser and general-purpose cable routing assemblies
(4) Type CMUC cables and wires
(5) Types BMR, BM, BMU, and BLU cables
(6) Communications wires
(7) Hybrid power and communications cables

Informational Note: See NFPA 90A-2021, *Standard for the Installation of Air-Conditioning and Ventilating Systems*, for information on fire protection of wiring installed in other spaces used for environmental air.

Δ **(D) Risers — Cables, Cable Routing Assemblies, and Communications Raceways in Vertical Runs.** Installations of cables used for communications circuits, communications wires, cable routing assemblies, and communications raceways in risers shall be in accordance with 800.113(D)(1) and (D)(2).

N **(1) Uses Permitted.** The following cables, cable routing assemblies, and communications raceways shall be permitted in vertical runs penetrating one or more floors and in vertical runs in a shaft:

(1) Plenum and riser cables
(2) Plenum and riser communications raceways
(3) Plenum and riser cable routing assemblies
(4) Plenum and riser cables installed in the following:

   a. Plenum communications raceways
   b. Riser communications raceways
   c. Plenum cable routing assemblies
   d. Riser cable routing assemblies

N **(2) Uses Not Permitted.** The following cables, wires, cable routing assemblies, and communications raceways shall not be permitted in risers:

(1) General-purpose and limited-use cables
(2) General-purpose communications raceways
(3) General-purpose cable routing assemblies
(4) Type CMUC cables and wires
(5) Types BMR, BM, BMU, and BLU cables
(6) Communications wires
(7) Hybrid power and communications cables

Informational Note: See 800.26 for firestop requirements for floor penetrations.

Δ **(E) Risers — Cables and Innerducts in Metal Raceways.** Installations of cables used for communications circuits, communications wires, cable routing assemblies, and communications raceways in metal raceways in a riser having firestops at each floor shall be in accordance with 800.113(E)(1) and (E)(2).

Δ **(1) Uses Permitted.** The following cables and innerducts shall be permitted in metal raceways in a riser having firestops at each floor:

(1) Plenum, riser, general-purpose, and limited-use cables
(2) Plenum, riser, and general-purpose communications raceways (innerduct)
(3) Plenum, riser, general-purpose, and limited-use cables installed in the following:

   a. Plenum communications raceways (innerduct)
   b. Riser communications raceways (innerduct)
   c. General-purpose communications raceways (innerduct)

N **(2) Uses Not Permitted.** The following cables, wires, cable routing assemblies, and communications raceways shall not be permitted in metal raceways in a riser having firestops at each floor:

(1) Plenum, riser, and general-purpose cable routing assemblies
(2) Type CMUC cables and wires

(3) Types BMR, BM, BMU, and BLU cables
(4) Communications wires
(5) Hybrid power and communications cables

Informational Note: See 800.26 for firestop requirements for floor penetrations.

Δ **(F) Risers — Cables, Cable Routing Assemblies, and Communications Raceways in Fireproof Shafts.** Installations of cables used for communications circuits, communications wires, cable routing assemblies, and communications raceways in fireproof riser shafts having firestops at each floor shall be in accordance with 800.113(F)(1) and (F)(2).

Δ **(1) Uses Permitted.** The following cables, cable routing assemblies, and communications raceways shall be permitted to be installed in fireproof riser shafts having firestops at each floor:

(1) Plenum, riser, general-purpose, and limited-use cables
(2) Plenum, riser, and general-purpose communications raceways
(3) Plenum, riser, and general-purpose cable routing assemblies
(4) Plenum, riser, general-purpose, and limited-use cables installed in the following:

   a. Plenum communications raceways
   b. Riser communications raceways
   c. General-purpose communications raceways
   d. Plenum cable routing assemblies
   e. Riser cable routing assemblies
   f. General-purpose cable routing assemblies

N **(2) Uses Not Permitted.** The following cables, wires, cable routing assemblies, and communications raceways shall not be permitted in metal raceways in fireproof riser shafts having firestops at each floor:

(1) Type CMUC cables and wires
(2) Type BMU and BLU cables
(3) Communications wires
(4) Hybrid power and communications cables

Informational Note: See 800.26 for firestop requirements for floor penetrations.

Δ **(G) Risers — One- and Two-Family Dwellings.** Installations of cables used for communications circuits, communications wires, cable routing assemblies, and communications raceways in risers in one- and two-family dwellings shafts shall be in accordance with 800.113(G)(1) and (G)(2).

Δ **(1) Uses Permitted.** The following cables, cable routing assemblies, and communications raceways shall be permitted in one- and two-family dwellings:

(1) Plenum, riser, and general-purpose cables
(2) Limited-use cables less than 6 mm (0.25 in.) in diameter
(3) Plenum, riser, and general-purpose communications raceways
(4) Plenum, riser, and general-purpose cable routing assemblies

(5) Plenum, riser, and general-purpose cables installed in the following:

   a. Plenum communications raceways
   b. Riser communications raceways
   c. General-purpose communications raceways
   d. Plenum cable routing assemblies
   e. Riser cable routing assemblies
   f. General-purpose cable routing assemblies

N **(2) Uses Not Permitted.** The following cables and wires shall not be permitted in risers in one- and two-family dwellings:

(1) Type CMUC cables and wires
(2) Type BMU and BLU cables
(3) Communications wires
(4) Hybrid power and communications cables

Δ **(H) Cable Trays.** Installations of cables used for communications circuits, communications wires, cable routing assemblies, and communications raceways supported by cable trays shall be in accordance with 800.113(H)(1) and (H)(2).

Δ **(1) Uses Permitted.** The following wires, cables, and communications raceways shall be permitted to be supported by cable trays:

(1) Plenum, riser, and general-purpose cables
(2) Plenum, riser, and general-purpose communications raceways
(3) Communications wires, plenum, riser, and general-purpose cables installed in the following:

   a. Plenum communications raceways
   b. Riser communications raceways
   c. General-purpose communications raceways

N **(2) Uses Not Permitted.** The following cables and wires shall not be supported by cable trays:

(1) Limited-use cables
(2) Type CMUC cables and wires
(3) Type BMU and BLU cables
(4) Communications wires
(5) Hybrid power and communications cables

Δ **(I) Distributing Frames and Cross-Connect Arrays.** Installations of cables used for communications circuits, communications wires, cable routing assemblies, and communications raceways in distributing frames and cross-connect arrays shall be in accordance with 800.113(I)(1) and (I)(2).

Δ **(1) Uses Permitted.** The following wires, cables, cable routing assemblies, and communications raceways shall be permitted to be installed in distributing frames and cross-connect arrays:

(1) Plenum, riser, and general-purpose cables and communications wires
(2) Plenum, riser, and general-purpose communications raceways

(3) Plenum, riser, and general-purpose cable routing assemblies

(4) Communications wires, plenum, riser, and general-purpose cables installed in the following:

   a. Plenum communications raceways

   b. Riser communications raceways

   c. General-purpose communications raceways

   d. Plenum cable routing assemblies

   e. Riser cable routing assemblies

   f. General-purpose cable routing assemblies

N **(2) Uses Not Permitted.** The following cables and wires shall not be installed in distributing frames and cross-connect arrays:

(1) Types BMR, BM, BMU, and BLU cables

(2) Limited-use cables

(3) Type CMUC cables and wires

(4) Hybrid power and communications cables

Δ **(J) Other Building Locations.** Installations of cables used for communications circuits, cable communications wires, routing assemblies, and communications raceways in building locations other than those covered in 800.113(B) through (I) shall be in accordance with 800.113(J)(1) and (J)(2).

Δ **(1) Uses Permitted.** The following wires, cables, cable routing assemblies, and communications raceways shall be permitted to be installed in building locations other than the locations covered in 800.113(B) through (I):

(1) Plenum, riser, and general-purpose cables

(2) Limited-use cables with a maximum of 3 m (10 ft) of exposed length in nonconcealed spaces

(3) Plenum, riser, and general-purpose communications raceways

(4) Plenum, riser, and general-purpose cable routing assemblies

(5) Communications wires, plenum, riser, and general-purpose cables installed in the following:

   a. Plenum communications raceways

   b. Riser communications raceways

   c. General-purpose communications raceways

(6) Plenum, riser, and general-purpose cables installed in the following:

   a. Plenum cable routing assemblies

   b. Riser cable routing assemblies

   c. General-purpose cable routing assemblies

(7) Communications wires and plenum, riser, general-purpose, and limited-use cables installed in raceways recognized in Chapter 3

(8) Type CMUC undercarpet communications wires and cables installed under carpet, modular flooring, and planks

N **(2) Uses Not Permitted.** The following cables, wires, cable routing assemblies, and communications raceways shall not be installed in building locations other than the locations covered in 800.113(B) through (I):

(1) Types BMU and BLU cables

(2) Communications wires

(3) Hybrid power and communications cables

Δ **(K) Multifamily Dwellings.** Installations of cables used for communications circuits, communications wires, cable routing assemblies, and communications raceways in multifamily dwellings shall be in accordance with 800.113(K)(1) and (K)(2).

Δ **(1) Uses Permitted.** The following cables, cable routing assemblies, and communications raceways shall be permitted to be installed in multifamily dwellings in locations other than the locations covered in 800.113(B) through (G):

(1) Plenum, riser, and general-purpose cables

(2) Limited-use cables less than 6 mm (0.25 in.) in diameter in nonconcealed spaces

(3) Plenum, riser, and general-purpose communications raceways

(4) Plenum, riser, and general-purpose cable routing assemblies

(5) Communications wires and plenum, riser, and general-purpose cables installed in the following:

   a. Plenum communications raceways

   b. Riser communications raceways

   c. General-purpose communications raceways

(6) Plenum, riser, and general-purpose cables installed in the following:

   a. Plenum cable routing assemblies

   b. Riser cable routing assemblies

   c. General-purpose cable routing assemblies

(7) Communications wires and plenum, riser, general-purpose, and limited-use cables installed in raceways recognized in Chapter 3

(8) Type CMUC under-carpet communications wires and cables installed under carpet, modular flooring, and planks

N **(2) Uses Not Permitted.** The following cables, cable routing assemblies, and communications raceways shall not be installed in multifamily dwellings in locations other than the locations covered in 800.113(B) through (G):

(1) Types BMU and BLU cables

(2) Communications wires

(3) Hybrid power and communications cables

Δ **(L) One- and Two-Family Dwellings.** Installations of cables used for communications circuits, communications wires, cable routing assemblies, and communications raceways in one- and two-family dwellings in locations other than those covered in 800.113(B) through (F) shall be in accordance with 800.113(L)(1) and (L)(2).

Δ **(1) Uses Permitted.** The following wires, cables, cable routing assemblies, and communications raceways shall be permitted to be installed in one- and two-family dwellings in locations other than the locations covered in 800.113(B) through (F):

(1) Plenum, riser, and general-purpose cables
(2) Limited-use cables less than 6 mm (0.25 in.) in diameter
(3) Plenum, riser, and general-purpose communications raceways
(4) Plenum, riser, and general-purpose cable routing assemblies
(5) Communications wires, plenum, riser, and general-purpose cables installed in the following:

    a. Plenum communications raceways
    b. Riser communications raceways
    c. General-purpose communications raceways

(6) Plenum, riser, and general-purpose cables installed in the following:

    a. Plenum cable routing assemblies
    b. Riser cable routing assemblies
    c. General-purpose cable routing assemblies

(7) Communications wires and plenum, riser, general-purpose, and limited-use cables installed in raceways recognized in Chapter 3
(8) Type CMUC under-carpet communications wires and cables installed under carpet, modular flooring, and planks
(9) Hybrid power and communications cable listed in accordance with 800.179

N **(2) Uses Not Permitted.** The following cables, wires, cable routing assemblies, and communications raceways shall not be installed in one- and two-family dwellings in locations other than those covered in 800.113(B) through (F):

(1) Types BMU and BLU cables
(2) Communications wires

N **800.133 Installation of Communications Wires and Cables and CATV-Type Coaxial Cables.** Installation of communications wires and cables, from the protector to the equipment, or where no protector is required, communications wires and cables attached to the outside or inside of the building, shall comply with 800.133(A) and 800.133(B). Installation of CATV-type coaxial cables, beyond the point of grounding as defined in 820.93, shall comply with 800.133(A) through (C).

Jackets of coaxial cable do not have sufficient construction specifications to permit them to be installed with electric light, power, Class 1, non–power-limited fire alarm circuits, and medium- and high-power network-powered broadband communications cable. Failure of the cable insulation due to a fault could lead to hazardous voltages being imposed on the Class 2 or Class 3 circuit conductors.

N **(A) In Raceways, Cable Trays, Boxes, Cables, Enclosures, and Cable Routing Assemblies.**

N **(1) Other Circuits.** Communications cables and CATV-type coaxial cables shall be permitted in the same raceway, cable tray, box, enclosure, or cable routing assembly together and with jacketed cables of any of the following:

(1) Class 2 and Class 3 remote-control, signaling, and power-limited circuits in compliance with 645.5(E)(2) or Parts I and II of Article 725
(2) Power-limited fire alarm systems in compliance with Parts I and III of Article 760
(3) Nonconductive and conductive optical fiber cables in compliance with Parts I and V of Article 770
(4) Communications circuits in compliance with Parts I and IV of Articles 800 and 805
(5) Community antenna television and radio distribution systems in compliance with Parts I and V of Articles 800 and 820
(6) Low-power network-powered broadband communications circuits in compliance with Parts I and V of Articles 800 and 830

N **(2) Class 2 and Class 3 Circuits.** Class 1 circuits shall not be run in the same cable with communications circuits. Class 2 and Class 3 circuit conductors shall be permitted in the same listed communications cable with communications circuits.

N **(3) Electric Light, Power, Class 1, Non-Power-Limited Fire Alarm, and Medium-Power Network-Powered Broadband Communications Circuits in Raceways, Compartments, and Boxes.** Communications wires and cables and CATV-type coaxial cables shall not be placed in any raceway, compartment, outlet box, junction box, or similar fitting with conductors of electric light, power, Class 1, non-power-limited fire alarm, or medium-power network-powered broadband communications circuits.

Exception No. 1: Communications wires and cables and CATV-type coaxial cables shall be permitted to be placed in any raceway, compartment, outlet box, junction box, or other enclosures with conductors of electric light, power, Class 1, non-power-limited fire alarm, or medium-power network-powered broadband communications circuits where all of the conductors of electric light, power, Class 1, non-power-limited fire alarm, and medium-power network-powered broadband communications circuits are separated from all of the communications wires and cables and CATV-type coaxial cables by a permanent barrier or listed divider.

Exception No. 2: Communications wires and cables and CATV-type coaxial cables shall be permitted to be placed in outlet boxes, junction boxes, or similar fittings or compartments with power conductors where such conductors are introduced solely for power supply to the communications and coaxial cable system distribution equipment. The power circuit conductors shall be routed within the enclosure to maintain a minimum 6 mm (¼ in.) separation from the communications wires and cables and the CATV- type coaxial cables.

Exception No. 3: Separation of circuits shall not be required in elevator traveling cables constructed in accordance with 620.36.

N **(B) Other Applications.** Communications wires and cables and CATV-type coaxial cables shall be separated at least 50 mm (2 in.) from conductors of any electric light, power, Class 1, non-power-limited fire alarm, or medium-power network-powered broadband communications circuits.

Exception No. 1: Separation shall not be required where either (1) all of the conductors of electric light, power, Class 1, non-power-limited fire alarm, and medium-power network-powered broadband communications circuits are in a raceway or in metal-sheathed, metal-clad, nonmetallic-sheathed, Type AC or Type UF cables, or (2) all of the communications wires and cables and all of the CATV-type coaxial cables are encased in raceway.

Exception No. 2: Separation shall not be required where the communications wires and cables and CATV-type coaxial cables are permanently separated from the conductors of electric light, power, Class 1, non-power-limited fire alarm, and medium-power network-powered broadband communications circuits by a continuous and firmly fixed nonconductor, such as porcelain tubes or flexible tubing, in addition to the insulation on the wire.

N **(C) Support of Communications Wires and Cables and CATV-Type Coaxial Cables.** Raceways shall be used for their intended purpose. Communications wires and cables and CATV-type coaxial cables shall not be strapped, taped, or attached by any means to the exterior of any raceway as a means of support.

Exception: Overhead (aerial) spans of communications drop wires, communications cables, and CATV-type coaxial cables shall be permitted to be attached to the exterior of a raceway-type mast intended for the attachment and support of such wires and cables.

In some instances, the only way to achieve the proper clearance above roadways, driveways, or structures is by use of a mast. The exception permits overhead spans of communications cable to be attached to the exterior of a raceway-type mast only if the mast is installed to support communications cable. The attachment of communications cable to a service mast or to a feeder and/or branch-circuit mast is prohibited by 230.28 and 225.17, respectively.

800.154 Applications of Listed Communications Wires, Cables, and Raceways, and Listed Cable Routing Assemblies. Permitted and nonpermitted applications of listed communications wires, cables, coaxial cables, network-powered broadband communications system cables and raceways, and listed cable routing assemblies, shall be in accordance with one of the following:

(1) Listed communications wires and cables as indicated in Table 800.154(a)
(2) Listed communications raceways as indicated in Table 800.154(b)
(3) Listed cable routing assemblies as indicated in Table 800.154(c)

The permitted applications shall be subject to the installation requirements of 800.110 and 800.113.

Communications system wires, cables, raceways, and routing assemblies generally are required to be listed where installed within a building. Unlisted cables are permitted to enter the building, but their length is limited. The length of unlisted cable permitted in a building depends on how and where it is terminated.

Exhibit 800.2 illustrates applications of listed communications cables.

See also

800.48 for unlisted outside plant communications cables entering buildings

805.90(B) for location of the primary protector

Part V. Listing Requirements

N **800.170 Plenum Cable Ties.** Cable ties intended for use in other space used for environmental air (plenums) shall be listed as having low smoke and heat release properties.

Informational Note: See NFPA 90A-2018, *Standard for the Installation of Air-Conditioning and Ventilating Systems*, and ANSI/UL 2043-2013, *Standard for Safety Fire Test for Heat and Visible Smoke Release for Discrete Products and Their Accessories Installed in Air-Handling Spaces*, for information on listing discrete products as having low smoke and heat release properties.

N **800.171 Communications Equipment.** Communications equipment shall be listed as being suitable for electrical connection to a communications network.

Informational Note No. 1: See ANSI/UL 60950-1-2014, *Standard for Safety of Information Technology Equipment*, ANSI/UL 1863-2012, *Standard for Safety Communications Circuit Accessories*, or ANSI/UL 62368-1-2014 or ANSI/UL 62368-1-2018, *Audio/Video, Information and Communication Technology Equipment — Part 1: Safety Requirements*.

Informational Note No. 2: See ANSI/ATIS 0600337-2016, *Requirements for Maximum Voltage, Current, and Power Levels Used in Communications Circuits*, for additional information regarding voltages, currents, and power allowed on communications circuits.

UL 1863, *Standard for Communications Circuit Accessories*, and UL 60950, *Standard for Safety of Information Technology Equipment — Safety — Part 1: General Requirements*, are two safety standards that contain requirements for determining whether equipment connected to a telecommunications network is suitable for the intended purpose. Listed equipment that is connected to the telecommunications network and evaluated according to other U.S. safety standards is also subject to telecommunications requirements appropriate for the equipment. Examples include information technology equipment, audio-video equipment, and signaling equipment connected to a central station. The appropriate requirements contained within the applicable safety standard are extracted from UL 1863, UL 60950, or both.

EXHIBIT 800.2 Applications of listed communications cables at various locations in a building.

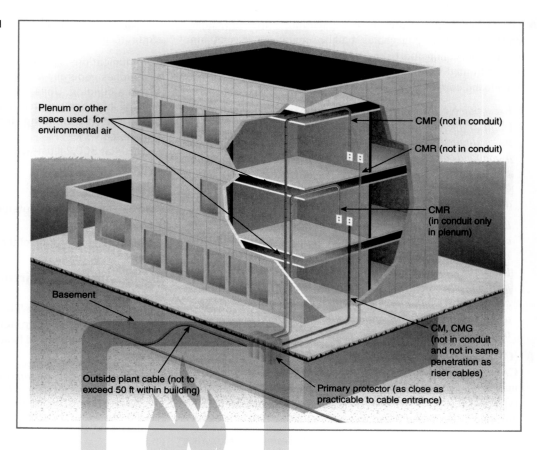

TABLE 800.154(a) *Applications of Listed Communications Wires, Cables, and Network-Powered Broadband Communications System Cables in Buildings*

Applications		Wire and Cable Type									
		Plenum	Riser	BMR	General-Purpose	BM	Limited-Use	Under-carpet	BMU, BLU	Hybrid Power and Communications Cables	Communications Wires
In ducts specifically fabricated for environmental air as described in 300.22(B)	In fabricated ducts	Y	N	N	N	N	N	N	N	N	N
	In metal raceway that complies with 300.22(B)	Y	Y	Y	Y	Y	Y	N	N	N	Y
In other spaces used for environmental air (plenums) as described in 300.22(C)	In other spaces used for environmental air	Y	N	N	N	N	N	N	N	N	N
	In metal raceway that complies with 300.22(C)	Y	Y	Y	Y	Y	Y	N	N	N	Y
	In plenum communications raceways	Y	N	N	N	N	N	N	N	N	N
	In plenum cable routing assemblies	Y	N	N	N	N	N	N	N	N	N
	Supported by open metal cable trays	Y	N	N	N	N	N	N	N	N	N

△ **TABLE 800.154(a)** *Continued*

Applications		Wire and Cable Type									
		Plenum	Riser	BMR	General-Purpose	BM	Limited-Use	Under-carpet	BMU, BLU	Hybrid Power and Communications Cables	Communications Wires
	Supported by solid bottom metal cable trays with solid metal covers	Y	Y	Y	Y	Y	Y	N	N	N	N
In risers	In vertical runs	Y	Y	Y	N	N	N	N	N	N	N
	In metal raceways	Y	Y	Y	Y	Y	Y	N	N	N	N
	In fireproof shafts	Y	Y	Y	Y	Y	Y	N	N	N	N
	In plenum communications raceways	Y	Y	N	N	N	N	N	N	N	N
	In plenum cable routing assemblies	Y	Y	N	N	N	N	N	N	N	N
	In riser communications raceways	Y	Y	N	N	N	N	N	N	N	N
	In riser cable routing assemblies	Y	Y	N	N	N	N	N	N	N	N
	In one- and two-family dwellings	Y	Y	Y	Y	Y	Y	N	N	Y	N
Within buildings in other than air-handling spaces and risers	General	Y	Y	Y	Y	Y	Y	N	N	N	N
	In one- and two-family dwellings	Y	Y	Y	Y	Y	Y	Y	N	Y	N
	In multifamily dwellings	Y	Y	Y	Y	Y	Y	Y	N	N	N
	In nonconcealed spaces	Y	Y	Y	Y	Y	Y	Y	N	N	N
	Supported by cable trays	Y	Y	Y	Y	Y	N	N	N	N	N
	Under carpet, modular flooring, and planks	N	N	N	N	N	N	Y	N	N	N
	In distributing frames and cross-connect arrays	Y	Y	N	Y	N	N	N	N	N	Y
	In rigid metal conduit (RMC) and intermediate metal conduit (IMC)	Y	Y	Y	Y	Y	Y	Y	Y	Y	Y
	In any raceway recognized in Chapter 3	Y	Y	Y	Y	Y	Y	N	N	N	Y
	In plenum communications raceways	Y	Y	N	Y	N	N	N	N	N	Y
	In plenum cable routing assemblies	Y	Y	N	Y	N	N	N	N	N	Y
	In riser communications raceways	Y	Y	N	Y	N	N	N	N	N	Y
	In riser cable routing assemblies	Y	Y	N	Y	N	N	N	N	N	Y
	In general-purpose communications raceways	Y	Y	N	Y	N	N	N	N	N	Y
	In general-purpose cable routing assemblies	Y	Y	N	Y	N	N	N	N	N	Y

Note: An "N" in the table indicates that the cable type shall not be installed in the application. A "Y" indicates that the cable type shall be permitted to be installed in the application subject to the limitations described in 800.113. The Riser column includes all riser cables except BMR, and the General-Purpose column includes all general-purpose cables except BM.

Informational Note No. 1: Part IV of Article 800 covers installation methods within buildings. This table covers the applications of listed communications wires, cables, and raceways in buildings.

Informational Note No. 2: For information on the restrictions to the installation of communications cables in fabricated ducts, see 800.113(B).

Δ **TABLE 800.154(b)** *Applications of Listed Communications Raceways in Buildings*

Applications		Listed Communications Raceway Type		
		Plenum	Riser	General-Purpose
In ducts specifically fabricated for environmental air as described in 300.22(B)	In fabricated ducts	N	N	N
	In metal raceway that complies with 300.22(B)	N	N	N
In other spaces used for environmental air (plenums) as described in 300.22(C)	In other spaces used for environmental air	Y	N	N
	In metal raceway that complies with 300.22(C)	Y	Y	Y
	In plenum cable routing assemblies	N	N	N
	Supported by open metal cable trays	Y	N	N
	Supported by solid bottom metal cable trays with solid metal covers	Y	Y	Y
In risers	In vertical runs	Y	Y	N
	In metal raceways	Y	Y	Y
	In fireproof shafts	Y	Y	Y
	In plenum cable routing assemblies	N	N	N
	In riser cable routing assemblies	N	N	N
	In one- and two-family dwellings	Y	Y	Y
Within buildings in other than air-handling spaces and risers	General	Y	Y	Y
	In one- and two-family dwellings	Y	Y	Y
	In multifamily dwellings	Y	Y	Y
	In nonconcealed spaces	Y	Y	Y
	Supported by cable trays	Y	Y	Y
	Under carpet, modular flooring, and planks	N	N	N
	In distributing frames and cross-connect arrays	Y	Y	Y
	In any raceway recognized in Chapter 3	Y	Y	Y
	In plenum cable routing assemblies	N	N	N
	In riser cable routing assemblies	N	N	N
	In general-purpose cable routing assemblies	N	N	N

Note: An "N" in the table indicates that the communications raceway type shall not be installed in the application. A "Y" indicates that the communications raceway type shall be permitted to be installed in the application, subject to the limitations described in 800.110 and 800.113.

Δ **TABLE 800.154(c)** *Applications of Listed Cable Routing Assemblies in Buildings*

Applications		Listed Cable Routing Assembly Type		
		Plenum	Riser	General-Purpose
In ducts specifically fabricated for environmental air as described in 300.22(B)	In fabricated ducts	N	N	N
	In metal raceway that complies with 300.22(B)	N	N	N
In other spaces used for environmental air (plenums) as described in 300.22(C)	In other spaces used for environmental air	Y	N	N
	In metal raceway that complies with 300.22(C)	N	N	N
	In plenum communications raceways	N	N	N
	Supported by open metal cable trays	Y	N	N
	Supported by solid bottom metal cable trays with solid metal covers	N	N	N
In risers	In vertical runs	Y	Y	N
	In metal raceways	N	N	N
	In fireproof shafts	Y	Y	Y
	In plenum communications raceways	N	N	N
	In riser communications raceways	N	N	N
	In one- and two-family dwellings	Y	Y	Y

Δ **TABLE 800.154(c)** *Continued*

Applications		Listed Cable Routing Assembly Type		
		Plenum	Riser	General-Purpose
Within buildings in other than air-handling spaces and risers	General	Y	Y	Y
	In one- and two-family dwellings	Y	Y	Y
	In multifamily dwellings	Y	Y	Y
	In nonconcealed spaces	Y	Y	Y
	Supported by cable trays	Y	Y	Y
	Under carpet, modular flooring, and planks	N	N	N
	In distributing frames and cross-connect arrays	Y	Y	Y
	In any raceway recognized in Chapter 3	N	N	N
	In plenum communications raceways	N	N	N
	In riser communications raceways	N	N	N
	In general-purpose communications raceways	N	N	N

Note: An "N" in the table indicates that the cable routing assembly type shall not be installed in the application. A "Y" indicates that the cable routing assembly type shall be permitted to be installed in the application subject to the limitations described in 800.113.

800.179 Wires and Cables. Communications wires and cables, community antenna television cables, and network-powered broadband communications cables shall be listed in accordance with 800.179(A) through (L) and shall have a temperature rating of not less than 60°C (140°F). The temperature rating shall be marked on the jacket of cables that have a temperature rating exceeding 60°C (140°F). Conductors in communications cables, other than in a coaxial cable, shall be copper. Cables shall be permitted to contain optical fibers. Cables containing optical fibers shall be marked with the suffix "-OF."

Communications wires and cables and network-powered communications cables shall have a voltage rating of not less than 300 volts; the insulation for the individual conductors, other than the outer conductor of a coaxial cable, shall be rated for 300 volts minimum. The cable voltage rating shall not be marked on the cable or on the under-carpet communications wire.

Exception: Voltage markings shall be permitted where the cable has multiple listings and voltage marking is required for one or more of the listings.

Informational Note No. 1: Voltage markings on cables could be misinterpreted to suggest that the cables may be suitable for Class 1, electric light, and power applications.
Informational Note No. 2: See UL 444-2017, *Standard for Communications Cables*, for information on communications cables.
Informational Note No. 3: See UL1655-2009, *Standard for Community-Antenna Television Cables*, for information on community-antenna television cables.

Conductor insulation rating of at least 300 volts is required for the following reasons:

1. To coordinate with protector installation requirements (i.e., protectors are not required within a block unless the cable is exposed to over 300 volts)
2. To recognize the fact that primary protectors are designed to allow voltages below 300 volts to pass
3. To accommodate the voltages ordinarily found on a telephone line [48 volts dc plus ringing voltage up to 130 volts root-mean square (rms)]
4. To permit communications cable to substitute for 300-volt power-limited fire-protective signaling cable

Δ **(A) Plenum Cables.** Type CMP communications plenum cables, Type CATVP community antenna television plenum coaxial cables, and Type BLP network-powered broadband communication low-power plenum cables shall be listed as being suitable for use in ducts, plenums, and other spaces used for environmental air and shall also be listed as having adequate fire-resistant and low-smoke-producing characteristics.

Informational Note: See NFPA 262-2019, *Standard Method of Test for Flame Travel and Smoke of Wires and Cables for Use in Air-Handling Spaces*, for one method of defining a cable that is low-smoke-producing cable and fire-resistant cable so that the cable exhibits a maximum peak optical density of 0.50 or less, an average optical density of 0.15 or less, and a maximum flame spread distance of 1.52 m (5 ft).

Δ **(B) Riser Cables.** Type CMR communications riser cables, Type CATVR community antenna television riser coaxial cables, Type BMR network-powered broadband communications medium-power riser cables, and Type BLR network-powered broadband communications low-power riser cables shall be listed as being suitable for use in a vertical run in a shaft or from floor to floor and shall also be listed as having fire-resistant characteristics capable of preventing the carrying of fire from floor to floor.

Informational Note: See ANSI/UL 1666-2017, *Standard Test for Flame Propagation Height of Electrical and Optical-Fiber Cable Installed Vertically in Shafts*, for one method of defining fire-resistant characteristics of the cable capable of preventing the carrying of fire from floor to floor.

Δ **(C) General-Purpose Cables.**

N **(1) Type CMG.** Type CMG communications general-purpose cables shall be listed as being suitable for general-purpose use, with the exception of risers and plenums, and shall also be listed as being resistant to the spread of fire.

Informational Note: See CSA Vertical Flame Test — Cables in Cable Trays as described in CSA C22.2 No. 0.3-09 (R2019), *Test Methods for Electrical Wires and Cables*, for one method of defining resistance to the spread of fire where the damage (char length) of the cable does not exceed 1.5 m (4 ft 11 in.) or FT4 Flame Test in ANSI/UL 1685-2015, *Standard for Safety for Vertical-Tray Fire-Propagation and Smoke-Release Test for Electrical and Optical-Fiber Cables*. The smoke measurements in the test methods are not applicable.

N **(2) Types CM, CATV, BM, and BL.** Type CM communications general-purpose cables, Type CATV community antenna television coaxial general-purpose cables, Type BM network-powered broadband communications medium-power general-purpose cables, and Type BL network-powered broadband communications low-power general-purpose cables shall be listed as being suitable for general-purpose use, with the exception of risers and plenums, and shall also be listed as being resistant to the spread of fire.

Informational Note: See UL Flame Exposure in ANSI/UL 1685-2015, *Standard for Safety for Vertical-Tray Fire-Propagation and Smoke-Release Test for Electrical and Optical-Fiber Cables,* for one method of defining *resistance to the spread of fire* where the damage (char length) of the cable does not exceed 244 cm (8 ft 0 in.). The smoke measurements in the test method are not applicable.

(D) Limited-Use Cables. Type CMX limited-use communications cables, Type CATVX limited-use community antenna television coaxial cables, and Type BLX limited-use network-powered broadband low-power cables shall be listed as being suitable for use in dwellings and for use in raceway and shall also be listed as being resistant to flame spread.

Informational Note: See ANSI/UL 2556, *Standard for Wire and Cable Test Method*, for one method of determining that cable is resistant to flame spread is by testing the cable to the FV-2/VW-1 flame test.

N **(E) Circuit Integrity (CI) Cable, Fire-Resistive Cable System, or Electrical Circuit Protective System.** Cables that are used for survivability of critical circuits under fire conditions shall be listed and meet either 800.179(E)(1), (E)(2), or (E)(3).

N **(1) CI Cables.** Cables specified in 800.179(A) through (C) and used for survivability of critical circuits shall be marked with the additional classification using the suffix "CI." In order to maintain its listed fire rating, CI cable shall only be installed in free air in accordance with 800.24. CI cables shall only be permitted to be installed in a raceway where specifically listed and marked as part of a fire-resistive cable system as covered in 800.179(E)(2).

Informational Note: See UL 2196, *Standard for Fire Test for Circuit Integrity of Fire-Resistant Power, Instrumentation, Control, and Data Cables*, for one method of defining CI cable by establishing a minimum 2-hour fire resistance rating for the cable as specified in UL 444, *Standard for Safety Communications Cables*.

N **(2) Fire-Resistive Cable Systems.** Cables specified in 800.179(A) through (C) and 800.179(E)(1) that are part of an electrical circuit protective system shall be fire-resistive cable identified with the protective system number on the product, or on the smallest unit container in which the product is packaged, and shall be installed in accordance with the listing of the protective system.

Informational Note No. 1: See UL 2196, *Fire Test for Circuit Integrity of Fire-Resistive Power, Instrumentation, Control and Data Cables*, for one method of defining an electrical circuit protective system rating for the system. UL *Guide Information for Electrical Circuit Integrity Systems (FHIT)* contains information to identify the system and its installation limitations to maintain a minimum fire-resistive rating.
Informational Note No. 2: The listing organization provides information for electrical circuit protective systems (FHIT), including installation requirements for maintaining the fire rating.

N **(3) Electrical Circuit Protective System.** Protectants for cables specified in 800.179(A) through (E), which are part of an electrical circuit protective system, shall be identified with the protective system identifier and hourly rating marked on the protectant or the smallest unit container and installed in accordance with the listing of the system.

Informational Note: See UL 1724, *Fire Tests for Electrical Circuit Protective Systems*, for one method of defining an electrical circuit protective system. UL *Guide Information for Electrical Circuit Integrity Systems (FHIT)* contains information to identify the system and its installation limitations to maintain the fire-resistive rating.

N **(F) Types CMP-LP, CMR-LP, CMG-LP, and CM-LP Limited Power (LP) Cables.** Types CMP-LP, CMR-LP, CMG-LP, and CM-LP communications limited power cables shall be listed as suitable for carrying power and data up to a specified current limit for each conductor without exceeding the temperature rating of the cable where the cable is installed in cable bundles in free air or installed within a raceway, cable tray, or cable routing assembly. The cables shall be marked with the suffix "-LP(XXA)," where XX designates the current limit in amperes per conductor.

Informational Note: An example of the marking on a communications cable with an LP rating is "CMP-LP (0.6A)(75°C) 23 AWG 4 pair," which indicates that it is a 4-pair plenum cable with 23 AWG conductors, a temperature rating of 75°C, and a current limit of 0.6 amperes.

N **(G) Type CMUC Undercarpet Wires and Cables.** Type CMUC undercarpet communications wires and cables shall be listed as being suitable for undercarpet use and shall also be listed as being resistant to flame spread.

> Informational Note: See ANSI/UL 2556, *Standard for Wire and Cable Test Methods*, for one method of determining that cable is resistant to flame spread in accordance with the FV-2/VW-1 flame test.

N **(H) Communications Wires.** Communications wires, such as distributing frame wire and jumper wire, shall be listed as being resistant to the spread of fire.

> Informational Note No. 1: See UL Flame Exposure, Vertical Flame Tray Test in ANSI/UL 1685-2015, *Standard for Safety for Vertical-Tray Fire-Propagation and Smoke-Release Test for Electrical and Optical-Fiber Cables*, for one method of defining *cable flame resistance to the spread of fire* where the cables do not spread fire to the top of the tray. The smoke measurements in the test method are not applicable.
>
> Informational Note No. 2: See CSA Vertical Flame Test — Cables in Cable Trays, as described in CSA C22.2 No. 0.3-09 (R2019), *Test Methods for Electrical Wires and Cables*, for another method of defining *resistance to the spread of fire* is for the damage (char length) of the cable to not exceed 1.5 m (4 ft 11 in.).

N **(I) Optional Markings.** Cables shall be permitted to be surface marked to indicate special characteristics of the cable materials.

> Informational Note: These markings can include, but are not limited to, markings for limited-smoke, halogen-free, low-smoke halogen-free, and sunlight resistance.

800.180 Grounding Devices. Where bonding or grounding is required, devices used to connect a shield, a sheath, or non-current-carrying metallic members of a cable to a bonding conductor or grounding electrode conductor shall be listed or be part of listed equipment.

Δ **800.182 Cable Routing Assemblies and Communications Raceways.** Cable routing assemblies and communications raceways shall be listed in accordance with 800.182(A) through (C). Cable routing assemblies shall be marked in accordance with Table 800.182(a). Communications raceways shall be marked in accordance with Table 800.182(b).

> Informational Note: See ANSI/UL 2024-5-2015, *Cable Routing Assemblies and Communications Raceways*, for information on listing requirements for both communications raceways and cable routing assemblies.

TABLE 800.182(a) *Cable Routing Assembly Markings*

Type	Marking
Plenum Cable Routing Assembly	Plenum Cable Routing Assembly
Riser Cable Routing Assembly	Riser Cable Routing Assembly
General-Purpose Cable Routing Assembly	General-Purpose Cable Routing Assembly

TABLE 800.182(b) *Communications Raceway Markings*

Type	Marking
Plenum Communications Raceway	Plenum Communications Raceway
Riser Communications Raceway	Riser Communications Raceway
General-Purpose Communications Raceway	General-Purpose Communications Raceway

Δ **(A) Plenum Cable Routing Assemblies and Plenum Communications Raceways.** Plenum cable routing assemblies and plenum communications raceways shall be listed as having adequate fire-resistant and low-smoke-producing characteristics.

> Informational Note No. 1: See ASTM E84-19B, *Standard Test Method for Surface Burning Characteristics of Building Materials*, or ANSI/UL 723-2018, *Standard Test Method for Surface Burning Characteristics of Building Materials*, for one method of defining cable routing assemblies and communications raceways that have adequate fire-resistant and low-smoke-producing characteristics and exhibit a maximum flame spread index of 25 and a maximum smoke developed index of 50.
>
> Informational Note No. 2: See NFPA 262-2019, *Standard Method of Test for Flame Travel and Smoke of Wires and Cables for Use in Air-Handling Spaces*, for another method of defining communications raceways that have adequate fire-resistant and low-smoke-producing characteristics and exhibit a maximum peak optical density of 0.50 or less, an average optical density of 0.15 or less, and a maximum flame spread distance of 1.52 m (5 ft) or less.
>
> Informational Note No. 3: See 4.3.11.2.6 or 4.3.11.5.5 of NFPA 90A-2021, *Standard for the Installation of Air-Conditioning and Ventilating Systems*, for information on materials exposed to the airflow in ceiling cavity and raised floor plenums.

Δ **(B) Riser Cable Routing Assemblies and Riser Communications Raceways.** Riser cable routing assemblies and riser communications raceways shall be listed as having adequate fire-resistant characteristics capable of preventing the carrying of fire from floor to floor.

> Informational Note: See ANSI/UL 1666-2017, *Standard Test for Flame Propagation Height of Electrical and Optical-Fiber Cable Installed Vertically in Shafts*, for one method of defining fire-resistant characteristics capable of preventing the carrying of fire from floor to floor of the cable routing assemblies and communications raceways.

Δ **(C) General-Purpose Cable Routing Assemblies and General-Purpose Communications Raceways.** General-purpose cable routing assemblies and general-purpose communications raceways shall be listed as being resistant to the spread of fire.

> Informational Note: See ANSI/UL 1685-2015, *Standard for Safety for Vertical-Tray Fire-Propagation and Smoke-Release Test for Electrical and Optical-Fiber Cables*, for one method of defining resistance to the spread of fire where the cable routing assemblies and communications raceways do not spread fire to the top of the tray.

The application of listed communications raceways and cable routing assemblies is summarized in Tables 800.154(b) and (c). The installation location dictates the type of cable permitted within the raceway or assembly as summarized in Table 800.154(a).

A raceway marked "plenum" on its surface or on a marker tape is suitable for use in ducts, plenums, or other spaces used for environmental air in accordance with 800.154. A "plenum" raceway is also suitable for installation in risers, for general-purpose use, and for dwellings.

A raceway or routing assembly marked "riser" on its surface or on a marker tape is suitable for installation in risers in accordance with 800.154. A "riser" raceway or routing assembly is also suitable for general-purpose use and for dwellings.

A raceway or routing assembly marked as "general purpose" is suitable for installation in general-purpose areas in accordance with 800.154 and for dwellings.

ARTICLE 805 Communications Circuits

Part I. General

805.1 Scope. This article covers communications circuits and equipment.

Article 805 contains the requirements that are unique within Chapter 8 for communications circuits. In previous editions of the *NEC®*, these requirements could be found in Article 800, which now contains requirements that apply generally across all of Chapter 8.

Although information technology equipment systems often are used for or with communications systems, Article 805 does not cover wiring of that equipment.

See also

Article 645, which provides requirements for wiring contained solely within an information technology equipment room

Article 722, which provides general wiring requirements that extend beyond a computer room and for wiring of local area networks within buildings

Article 760, which covers wiring requirements for fire alarm systems

805.18 Installation of Equipment. Equipment electrically connected to a communications network shall be listed in accordance with 800.171.

Exception: This listing requirement shall not apply to test equipment that is intended for temporary connection to a telecommunications network by qualified persons during the course of installation, maintenance, or repair of telecommunications equipment or systems.

Part II. Wires and Cables Outside and Entering Buildings

805.47 Underground Communications Wires and Cables Entering Buildings — Underground Block Distribution. Where

the entire street circuit is run underground and the circuit within the block is placed so as to be free from the likelihood of accidental contact with electric light or power circuits of over 300 volts to ground, the insulation requirements of 805.50(A) and 805.50(C) shall not apply, insulating supports shall not be required for the conductors, and bushings shall not be required where the conductors enter the building.

805.50 Circuits Requiring Primary Protectors. Circuits that require primary protectors as provided in 805.90 shall comply with 805.50(A), 805.50(B), and 805.50(C).

(A) Insulation, Wires, and Cables. Communications wires and cables without a metal shield, running from the last outdoor support to the primary protector, shall be listed in accordance with 805.173.

(B) On Buildings. Communications wires and cables in accordance with 805.50(A) shall be separated at least 100 mm (4 in.) from electric light or power conductors not in a raceway or cable or be permanently separated from conductors of the other systems by a continuous and firmly fixed nonconductor in addition to the insulation on the wires, such as porcelain tubes or flexible tubing. Communications wires and cables in accordance with 805.50(A) exposed to accidental contact with electric light and power conductors operating at over 300 volts to ground and attached to buildings shall be separated from woodwork by being supported on glass, porcelain, or other insulating material.

Exception: Separation from woodwork shall not be required where fuses are omitted as provided for in 805.90(A)(1), or where conductors are used to extend circuits to a building from a cable having a grounded metal sheath.

Δ **(C) Entering Buildings.**

N **(1) Installed Inside Buildings.** If a primary protector is installed inside the building, the communications wires and cables shall enter the building either through a noncombustible, nonabsorbent insulating bushing or through a metal raceway.

Exception: The insulating bushing shall not be required if the entering communications wires and cables meet one or more of the following conditions:

(1) Is a metal-sheathed cable
(2) Pass through masonry
(3) Meet the requirements of 805.50(A) and fuses are omitted in accordance with 805.90(A)(1)
(4) Meet the requirements of 805.50(A) and are used to extend circuits to a building from a cable having a grounded metal sheath

N **(2) Orientation of Raceways or Bushings.** Raceways or bushings shall slope upward from the outside, or, where this cannot be done, drip loops shall be formed in the communications wires and cables immediately before they enter the building.

N **(3) Service Head.** Raceways shall be equipped with an approved service head. More than one communications wire and cable shall be permitted to enter through a single raceway or bushing. Conduits or other metal raceways located ahead of the primary protector shall be grounded.

Part III. Protection

805.90 Protective Devices.

(A) Application. A listed primary protector shall be provided on each circuit run partly or entirely in aerial wire or aerial cable not confined within a block. Also, a listed primary protector shall be provided on each circuit, aerial or underground, located within the block containing the building served so as to be exposed to accidental contact with electric light or power conductors operating at over 300 volts to ground. In addition, where there exists a lightning exposure, each interbuilding circuit on a premises shall be protected by a listed primary protector at each end of the interbuilding circuit. Installation of primary protectors shall also comply with 110.3(B).

> Informational Note No. 1: On a circuit not exposed to accidental contact with power conductors, providing a listed primary protector in accordance with this article helps protect against other hazards, such as lightning and above-normal voltages induced by fault currents on power circuits in proximity to the communications circuit.
>
> Informational Note No. 2: Interbuilding circuits are considered to have a lightning exposure unless one or more of the following conditions exist:
>
> (1) Circuits in large metropolitan areas where buildings are close together and sufficiently high to intercept lightning.
> (2) Interbuilding cable runs of 42 m (140 ft) or less, directly buried or in underground conduit, where a continuous metallic cable shield or a continuous metal conduit containing the cable is connected to each building grounding electrode system.
> (3) Areas having an average of five or fewer thunderstorm days per year and earth resistivity of less than 100 ohm-meters. Such areas are found along the Pacific coast.

Telephone utility companies ordinarily provide primary protectors if telephone lines are exposed to lightning. Installers of private networks that include interbuilding cable should also install primary protectors where cables are exposed to lightning. A primary protector is required at each end of an interbuilding communications circuit where lightning exposure exists. See Exhibit 805.1 for an example of a primary protector unit typically installed in commercial buildings.

> Informational Note No. 3: See NFPA 780-2020, *Standard for the Installation of Lightning Protection Systems*, for information on lightning protection systems.

Δ **(1) Fuseless Primary Protectors.** Fuseless-type primary protectors shall be permitted under any of the following conditions:

(1) Where conductors enter a building through a cable with grounded metallic sheath member(s) and where the

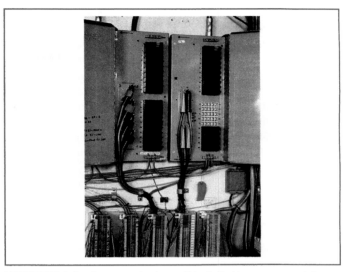

EXHIBIT 805.1 A primary protector unit installed in a commercial building that is the interface to the outside plant cable.

conductors in the cable safely fuse on all currents greater than the current-carrying capacity of the primary protector and of the primary protector bonding conductor or grounding electrode conductor

(2) Where insulated conductors in accordance with 805.50(A) are used to extend circuits to a building from a cable with an effectively grounded metallic sheath member(s) and where the conductors in the cable or cable stub, or the connections between the insulated conductors and the plant exposed to accidental contact with electric light or power conductors operating at greater than 300 volts to ground, safely fuse on all currents greater than the current-carrying capacity of the primary protector, or the associated insulated conductors and of the primary protector bonding conductor or grounding electrode conductor

(3) Where insulated conductors in accordance with 805.50(A) or (B) are used to extend circuits to a building from other than a cable with metallic sheath member(s), where (a) the primary protector is listed as being suitable for this purpose for application with circuits extending from other than a cable with metallic sheath members and (b) the connections of the insulated conductors to the plant exposed to accidental contact with electric light or power conductors operating at greater than 300 volts to ground or the conductors of the plant exposed to accidental contact with electric light or power conductors operating at greater than 300 volts to ground safely fuse on all currents greater than the current-carrying capacity of the primary protector, or associated insulated conductors and of the primary protector bonding conductor or grounding electrode conductor

(4) Where insulated conductors in accordance with 805.50(A) are used to extend circuits aerially to a building from a buried or underground circuit that is unexposed to accidental

contact with electric light or power conductors operating at greater than 300 volts to ground

(5) Where insulated conductors in accordance with 805.50(A) are used to extend circuits to a building from cable with an effectively grounded metallic sheath member(s), and where (a) the combination of the primary protector and insulated conductors is listed as being suitable for this purpose for application with circuits extending from a cable with an effectively grounded metallic sheath member(s) and (b) the insulated conductors safely fuse on all currents greater than the current-carrying capacity of the primary protector and of the primary protector bonding conductor or grounding electrode conductor

Informational Note: See ANSI/IEEE C2-2017, *National Electrical Safety Code*, Section 9, for examples of methods of protective grounding that can achieve effective grounding of communications cable sheaths for cables from which communications circuits are extended.

(2) Fused Primary Protectors. Where the requirements listed under 805.90(A)(1)(a) through (A)(1)(e) are not met, fused-type primary protectors shall be used. Fused-type primary protectors shall consist of an arrester connected between each line conductor and ground, a fuse in series with each line conductor, and an appropriate mounting arrangement. Primary protector terminals shall be marked to indicate line, instrument, and ground, as applicable.

(B) Location. The primary protector shall be located in, on, or immediately adjacent to the structure or building served and as close as practicable to the point of entrance.

For purposes of this section, primary protectors located at mobile home service equipment within 9.0 m (30 ft) of the exterior wall of the mobile home it serves, or at a mobile home disconnecting means connected to an electrode by a grounding electrode conductor in accordance with 250.32 and located within 9.0 m (30 ft) of the exterior wall of the mobile home it serves, shall be considered to meet the requirements of this section.

Informational Note: Selecting a primary protector location to achieve the shortest practicable primary protector bonding conductor or grounding electrode conductor helps limit potential differences between communications circuits and other metallic systems.

(C) Hazardous (Classified) Locations. The primary protector shall not be located in any hazardous (classified) locations, as defined in 500.5 and 505.5, or in the vicinity of easily ignitible material.

Exception: As permitted in 501.150, 502.150, and 503.150.

(D) Secondary Protectors. Where a secondary protector is installed in series with the indoor communications wire and cable between the primary protector and the equipment, it shall be listed for the purpose in accordance with 805.170(B).

Informational Note: Secondary protectors on circuits exposed to accidental contact with electric light or power conductors operating at greater than 300 volts to ground are not intended for use without primary protectors.

805.93 Grounding, Bonding, or Interruption of Non–Current-Carrying Metallic Sheath Members of Communications Cables. Communications cables entering the building or terminating on the outside of the building shall comply with 805.93(A) or (B).

(A) Entering Buildings. In installations where the communications cable enters a building, the metallic sheath members of the cable shall be grounded or bonded as specified in 800.100 or interrupted by an insulating joint or equivalent device. The grounding, bonding, or interruption shall be as close as practicable to the point of entrance.

(B) Terminating on the Outside of Buildings. In installations where the communications cable is terminated on the outside of the building, the metallic sheath members of the cable shall be grounded or bonded as specified in 800.100 or interrupted by an insulating joint or equivalent device. The grounding, bonding, or interruption shall be as close as practicable to the point of termination of the cable.

Part IV. Installation Methods Within Buildings

805.154 Substitutions for Listed Communications Cables. The substitutions for communications cables listed in Table 805.154 and illustrated in Figure 805.154 shall be permitted.

TABLE 805.154 *Cable Substitutions*

Cable Type	Permitted Substitutions
CMR	CMP
CMG, CM	CMP, CMR
CMX	CMP, CMR, CMG, CM

805.156 Dwelling Unit Communications Outlet. For new construction, a minimum of one communications outlet shall be installed within the dwelling in a readily accessible area and cabled to the service provider demarcation point.

The location of the communications outlet within a dwelling is not specified, but cable must be installed between the outlet location and the point at which the communications services provider installs its equipment or the point at which it connects to owner-supplied equipment. Although an increasing number of dwelling unit owners or occupants use cellular or personal communications services (PCS) telephones exclusively, having at least one wired communications outlet in every dwelling facilitates connection of dial-up devices used in home fire detection and security systems.

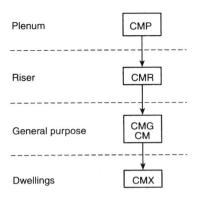

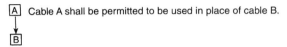

Type CM—Communications cables

A Cable A shall be permitted to be used in place of cable B.

B

FIGURE 805.154 *Cable Substitution Hierarchy.*

Part V. Listing Requirements

Δ **805.170 Protectors.** Protectors shall be listed in accordance with 805.170(A) or 805.170(B).

(A) Primary Protectors. The primary protector shall be listed and consist of an arrester connected between each line conductor and ground in an appropriate mounting. Primary protector terminals shall be marked to indicate line and ground as applicable.

> Informational Note: See ANSI/UL 497-2017, *Standard for Protectors for Paired Conductor Communications Circuits,* to determine applicable requirements for a listed primary protector.

(B) Secondary Protectors. The secondary protector shall be listed as suitable to provide means to safely limit currents to less than the current-carrying capacity of listed indoor communications wire and cable, listed telephone set line cords, and listed communications terminal equipment having ports for external wire line communications circuits. Any overvoltage protection, arresters, or grounding connection shall be connected on the equipment terminals side of the secondary protector current-limiting means.

> Informational Note: See ANSI/UL 497A-2019, *Standard for Secondary Protectors for Communications Circuits,* to determine applicable requirements for a listed secondary protector.

805.173 Drop Wire and Cable. Communications wires and cables without a metallic shield, running from the last outdoor support to the primary protector, shall be listed as being suitable for the purpose and shall have current-carrying capacity as specified in 805.90(A)(1)(b) or (A)(1)(c).

ARTICLE 810 Antenna Systems

Part I. General

810.1 Scope. This article covers antenna systems for radio and television receiving equipment, amateur and citizen band radio transmitting and receiving equipment, and certain features of transmitter safety. This article covers antennas such as wire-strung type, multi-element, vertical rod, flat, or parabolic and also covers the wiring and cabling that connect them to equipment. This article does not cover equipment and antennas used for coupling carrier current to power line conductors.

Article 810 covers wiring requirements for television and radio receiving equipment, specifically including digital satellite receiving equipment for television signals and wiring for amateur radio equipment and citizens band (CB) radio equipment. Chapters 1 through 4 cover wiring to the power supply.

Δ **810.3 Other Articles.** Wiring from the source of power to and between devices connected to the interior wiring system shall comply with the following:

(1) Chapters 1 through 4 other than as modified by Parts I and II of Article 640.
(2) Coaxial cables that connect antennas to equipment shall comply with the appropriate article of Chapter 8.
(3) Wiring and equipment installed in hazardous (classified) locations shall comply with the appropriate requirements in Chapter 5.

810.4 Community Television Antenna. The installation of the antenna shall comply with this article. The installation of the distribution system shall comply with the appropriate article of Chapter 8.

810.5 Radio Noise Suppression. Radio interference eliminators, interference capacitors, or noise suppressors connected to power-supply leads shall be of a listed type. They shall not be exposed to physical damage.

810.6 Antenna Lead-In Protectors. If an antenna lead-in surge protector is installed, it shall be listed as being suitable for limiting surges on the cable that connects the antenna to the receiver/transmitter electronics and shall be connected between the conductors and the grounded shield or other ground connection. The antenna lead-in protector shall be grounded using a bonding conductor or grounding electrode conductor installed in accordance with 810.21(F).

> Informational Note: See UL 497E, *Outline of Investigation for Protectors for Antenna Lead-In Conductors,* for information concerning protectors for antenna lead-in conductors.

810.7 Grounding Devices. If bonding or grounding is required, devices used to connect a shield, a sheath, non-current-carrying metal members of a cable, or metal parts of equipment or antennas to a bonding conductor or grounding electrode conductor shall be listed or be part of listed equipment.

Part II. Receiving Equipment — Antenna Systems

810.11 Material. Antennas and lead-in conductors shall be of hard-drawn copper, bronze, aluminum alloy, copper-clad steel, or other high-strength, corrosion-resistant material.

Exception: Soft-drawn or medium-drawn copper shall be permitted for lead-in conductors if the maximum span between points of support is less than 11 m (35 ft).

810.12 Supports. Outdoor antennas and lead-in conductors shall be securely supported. The antennas or lead-in conductors shall not be attached to the electric service mast. They shall not be attached to poles or similar structures carrying open electric light or power wires or trolley wires of over 250 volts between conductors. Insulators supporting the antenna conductors shall have sufficient mechanical strength to safely support the conductors. Lead-in conductors shall be securely attached to the antennas.

810.13 Avoidance of Contacts with Conductors of Other Systems. Outdoor antennas and lead-in conductors from an antenna to a building shall not cross over open conductors of electric light or power circuits and shall be kept well away from all such circuits so as to avoid the possibility of accidental contact. Where proximity to open electric light or power service conductors of less than 250 volts between conductors cannot be avoided, the installation shall be such as to provide a clearance of at least 600 mm (2 ft).

Where practicable, antenna conductors shall be installed so as not to cross under open electric light or power conductors.

One of the leading causes of electric shock and electrocution, according to statistical reports, is the accidental contact of antennas, ladders, and other equipment with overhead light or power conductors. Extreme caution should be exercised during installation. Antennas should be visually inspected periodically to ensure that they continue to be safe from exposure to overhead power conductors.

810.14 Splices. Splices and joints in antenna spans shall be made mechanically secure with approved splicing devices or by such other means and be suitable for the conditions of use and location in compliance with 110.14(A) and (B).

Δ **810.15 Grounding or Bonding.** Masts and metal structures supporting antennas shall be grounded or bonded in accordance

with 810.21, unless the antenna and its related supporting mast or structure are within a zone of protection defined by a 46 m (150 ft) radius rolling sphere.

> Informational Note: See NFPA 780-2020, *Standard for the Installation of Lightning Protection Systems*, 4.7.3.1, for the application of the term *rolling sphere*.

810.16 Size of Wire-Strung Antenna — Receiving Station.

(A) Size of Antenna Conductors. Outdoor antenna conductors for receiving stations shall be of a size not less than given in Table 810.16(A).

TABLE 810.16(A) *Size of Receiving Station Outdoor Antenna Conductors*

Material	Minimum Size of Conductors (AWG) Where Maximum Open Span Length Is		
	Less Than 11 m (35 ft)	11 m to 45 m (35 ft to 150 ft)	Over 45 m (150 ft)
Aluminum alloy, hard-drawn copper	19	14	12
Copper-clad steel, bronze, or other high-strength material	20	17	14

(B) Self-Supporting Antennas. Outdoor antennas, such as vertical rods and flat, parabolic, or dipole structures, shall be of corrosion-resistant materials and of strength suitable to withstand ice and wind loading conditions and shall be located well away from overhead conductors of electric light and power circuits of over 150 volts to ground, so as to avoid the possibility of the antenna or structure falling into or making accidental contact with such circuits.

810.17 Size of Lead-in — Receiving Station. Lead-in conductors from outside antennas for receiving stations shall, for various maximum open span lengths, be of such size as to have a tensile strength at least as great as that of the conductors for antennas as specified in 810.16. If the lead-in consists of two or more conductors that are twisted together, are enclosed in the same covering, or are concentric, the conductor size shall, for various maximum open span lengths, be such that the tensile strength of the combination is at least as great as that of the conductors for antennas as specified in 810.16.

810.18 Clearances — Receiving Stations.

Δ **(A) Outside of Buildings.** Lead-in conductors attached to buildings shall be installed so that they cannot swing closer than 600 mm (2 ft) to the conductors of circuits of 250 volts or less between conductors, or 3.0 m (10 ft) to the conductors of circuits of over 250 volts between conductors, except that in the case of circuits not over 150 volts between conductors, if all conductors involved are supported so as to ensure permanent separation, the clearance shall be permitted to be reduced but shall not be less than 100 mm (4 in.). The clearance between lead-in conductors and any conductor forming a part of a lightning protection system shall not be less than 1.8 m (6 ft). Underground conductors shall be separated at least 300 mm (12 in.) from conductors of any light or power circuits or Class 1 circuits.

Exception: The separation and clearance requirements shall not apply if the electric light or power conductors, Class 1 conductors, or lead-in conductors are installed in raceways or metal cable armor.

Informational Note No. 1: See 250.60 for grounding associated with lightning protection components — strike termination devices. See NFPA 780-2020, *Standard for the Installation of Lightning Protection Systems,* for detailed information on grounding, bonding, and spacing from lightning protection systems, and the calculation of specific separation distances using the sideflash equation in Section 4.6.

Informational Note No. 2: See NFPA 780-2020, *Standard for the Installation of Lighting Protection Systems,* for information on bonding or separation of metal raceways, enclosures, frames, and other non-current-carrying metal parts of electrical equipment installed on a building equipped with a lightning protection system. Separation from lightning protection conductors is typically 1.8 m (6 ft) through air or 900 mm (3 ft) through dense materials such as concrete, brick, or wood.

Δ **(B) Antennas and Lead-ins — Indoors.** Indoor antennas and indoor lead-ins shall not be run nearer than 50 mm (2 in.) to conductors of other wiring systems in the premises unless one of the following conditions applies:

(1) The other conductors are in metal raceways or cable armor.
(2) The indoor antennas and indoor lead-ins are permanently separated from such other conductors by a continuous firmly fixed nonconductor.

(C) In Boxes or Other Enclosures. Indoor antennas and indoor lead-ins shall be permitted to occupy the same box or enclosure with conductors of other wiring systems if separated from such other conductors by an effective permanently installed barrier.

810.19 Electrical Supply Circuits Used in Lieu of Antenna — Receiving Stations. If an electrical supply circuit is used in lieu of an antenna, the device by which the radio receiving set is connected to the supply circuit shall be listed.

The connecting device is usually a small, fixed capacitor connecting the antenna terminal of the receiver and one wire of the supply circuit. As is the case with most receivers, the capacitor should be designed for operation at not less than 300 volts. This voltage rating ensures a high degree of safety and minimizes the possibility of a breakdown in the capacitor, thereby avoiding a short circuit to ground through the antenna coil of the set.

810.20 Antenna Discharge Units — Receiving Stations.

Δ **(A) General Requirement.** Each lead-in conductor from an outdoor antenna shall be provided with a listed antenna discharge unit.

Exception: A separate antenna discharge unit is not required if the lead-in conductors are enclosed in a continuous metal shield that complies with one of the following:

(1) *Is grounded or bonded with a conductor in accordance with 810.21*
(2) *Is protected by an antenna discharge unit*

An antenna discharge unit (lightning arrester) is not required if the lead-in conductors are enclosed in a continuous metal shield, such as rigid metal conduit (RMC) or intermediate metal conduit (IMC), electrical metallic tubing, or any metal raceway or metal-shielded cable that is effectively grounded. A lightning discharge will take the path of lower impedance and jump from the lead-in conductors to the metal raceway or shield rather than take the path through the antenna coil of the receiver.

(B) Location. Antenna discharge units shall be located outside the building or inside the building between the point of entrance of the lead-in and the radio set or transformers and as near as practicable to the entrance of the conductors to the building. The antenna discharge unit shall not be located near combustible material or in a hazardous (classified) location as defined in accordance with 500.5 and 505.5.

(C) Grounding or Bonding. The antenna discharge unit shall be grounded or bonded in accordance with 810.21.

810.21 Bonding Conductors and Grounding Electrode Conductors — Receiving Stations. Bonding conductors and grounding electrode conductors shall comply with 810.21(A) through 810.21(K).

(A) Material. The bonding conductor or grounding electrode conductor shall be of copper, aluminum, copper-clad steel, copper-clad aluminum, bronze, or similar corrosion-resistant material. Aluminum or copper-clad aluminum bonding conductors or grounding electrode conductors shall not be used if subject to corrosive conditions or in direct contact with masonry or the earth or where subject to corrosive conditions. If used outside, aluminum or copper-clad aluminum conductors shall not be installed within 450 mm (18 in.) of the earth.

(B) Insulation. Insulation on bonding conductors or grounding electrode conductors shall not be required.

(C) Supports. The bonding conductor or grounding electrode conductor shall be securely fastened in place and shall be permitted to be directly attached to the surface wired over without the use of insulating supports.

Exception: Where proper support cannot be provided, the size of the bonding conductors or grounding electrode conductors shall be increased proportionately.

(D) Physical Protection. Bonding conductors and grounding electrode conductors shall be protected where exposed to physical damage. Where the bonding conductor or grounding electrode conductor is installed in a metal raceway, both ends of the raceway shall be bonded to the contained conductor or to the same terminal or electrode to which the bonding conductor or grounding electrode conductor is connected.

Where metal raceways are used to enclose the grounding electrode conductor (GEC), a connection between the GEC and the metal conduit must be provided at both ends of the conduit to ensure an adequate low-impedance current path to ground.

See also

250.64(E), which covers raceways, cable armor, and enclosures for GECs

(E) Run in Straight Line. The bonding conductor or grounding electrode conductor for an antenna mast or antenna discharge unit shall be run in as straight a line as practicable.

(F) Electrode. The bonding conductor or grounding electrode conductor shall be connected as required in 810.21(F)(1) through 810.21(F)(3).

(1) In Buildings or Structures with an Intersystem Bonding Termination. If the building or structure served has an intersystem bonding termination as required by 250.94, the bonding conductor shall be connected to the intersystem bonding termination.

△ **(2) In Buildings or Structures with Grounding Means.** If the building or structure served has no intersystem bonding termination, the bonding conductor or grounding electrode conductor shall be connected to the nearest accessible location on one of the following:

(1) The building or structure grounding electrode system as covered in 250.50
(2) The power service accessible means external to the building, as covered in 250.94
(3) The nonflexible metal power service raceway
(4) The service equipment enclosure
(5) The grounding electrode conductor or the grounding electrode conductor metal enclosures of the power service
(6) The grounded interior metal water piping systems, within 1.52 m (5 ft) from its point of entrance to the building, as covered in 250.52

See also

250.52(A)(1) for more information on the use of a metal water piping system

250.68(C) for more information on connections to GEC connections

A bonding device intended to provide a termination point for the bonding conductor (intersystem bonding) shall not interfere with the opening of an equipment enclosure. A bonding device shall be mounted on nonremovable parts. A bonding device shall

not be mounted on a door or cover even if the door or cover is nonremovable.

(3) In Buildings or Structures Without an Intersystem Bonding Termination or Grounding Means. If the building or structure served has no intersystem bonding termination or grounding means as described in 810.21(F)(2), the grounding electrode conductor shall be connected to a grounding electrode as described in 250.52.

(G) Inside or Outside Building. The bonding conductor or grounding electrode conductor shall be permitted to be run either inside or outside the building.

(H) Size. The bonding conductor or grounding electrode conductor shall not be smaller than 10 AWG copper, 8 AWG aluminum, or 17 AWG copper-clad steel or bronze.

(I) Common Ground. A single bonding conductor or grounding electrode conductor shall be permitted for both protective and operating purposes.

(J) Bonding of Electrodes. A bonding jumper not smaller than 6 AWG copper or equivalent shall be connected between the radio and television equipment grounding electrode and the power grounding electrode system at the building or structure served if separate electrodes are used.

Antenna masts must be bonded to the same grounding electrode used for the building's electrical system to ensure that all exposed, non-current-carrying metal parts are at the same potential. In many cases, masts are connected incorrectly to conveniently located vent pipes, metal gutters, or downspouts. Such a connection could create potential differences between lead-in conductors and various metal parts located in or on buildings, resulting in possible shock and fire hazards. An underground gas piping system is not permitted to be used as a grounding electrode.

The use of separate radio/television grounding electrodes is not required. However, where they are provided, 810.21(J) requires the radio/television system grounding electrode to be connected, via a bonding jumper, to the grounding electrode of the electrical distribution system of the building or structure.

(K) Electrode Connection. Connections to grounding electrodes shall comply with 250.70.

Part III. Amateur and Citizen Band Transmitting and Receiving Stations — Antenna Systems

810.51 Other Sections. In addition to complying with Part III, antenna systems for amateur and citizen band transmitting and receiving stations shall also comply with 810.11 through 810.15.

810.52 Size of Conductors. Antenna conductors for transmitting and receiving stations shall be of a size not less than given in Table 810.52.

△ **TABLE 810.52** *Size of Outdoor Antenna Conductors*

	Minimum Size of Conductors (AWG) If Maximum Open Span Length Is	
Material	Less Than 45 m (150 ft)	Over 45 m (150 ft)
Hard-drawn copper	14	10
Copper-clad steel, bronze, or other high-strength material	14	12

810.53 Size of Lead-in Conductors. Lead-in conductors for transmitting stations shall, for various maximum span lengths, be of a size at least as great as that of conductors for antennas as specified in 810.52.

810.54 Clearance on Building. Antenna conductors for transmitting stations, attached to buildings, shall be firmly mounted at least 75 mm (3 in.) clear of the surface of the building on nonabsorbent insulating supports, such as treated pins or brackets equipped with insulators having not less than 75-mm (3-in.) creepage and airgap distances. Lead-in conductors attached to buildings shall also comply with these requirements.

Exception: If the lead-in conductors are enclosed in a continuous metal shield that is grounded with a conductor in accordance with 810.58, they shall not be required to comply with these requirements. If grounded, the metal shield shall also be permitted to be used as a conductor.

The term *creepage distance*, as defined in NFPA 791, *Recommended Practice and Procedures for Unlabeled Electrical Equipment Evaluation*, is the shortest distance along the surface of the insulating material between two conductive parts. Creepage distance is measured from the conductor across the face of the supporting insulator to the building surface. Airgap distance is measured from the conductor (at its closest point) across the air space (not necessarily in a straight line) to the surface of the building.

810.55 Entrance to Building. Except where protected with a continuous metallic shield that is grounded with a conductor in accordance with 810.58, lead-in conductors for transmitting stations shall enter buildings by one of the following methods:

(1) Through a rigid, noncombustible, nonabsorbent insulating tube or bushing
(2) Through an opening provided for the purpose in which the entrance conductors are firmly secured so as to provide a clearance of at least 50 mm (2 in.)
(3) Through a drilled window pane

810.56 Protection Against Accidental Contact. Lead-in conductors to radio transmitters shall be located or installed so as to make accidental contact with them difficult.

△ **810.57 Antenna Discharge Units — Transmitting Stations.** Each lead-in conductor for outdoor antennas shall be provided with an antenna discharge unit or other suitable means that drain static charges from the antenna system.

Exception No. 1: If the lead-in conductor is protected by a continuous metal shield that is grounded with a conductor in accordance with 810.58, an antenna discharge unit or other suitable means shall not be required for the lead-in conductor.

Exception No. 2: If the antenna is grounded or bonded with a conductor in accordance with 810.58, an antenna discharge unit or other suitable means shall not be required.

If an antenna discharge unit is not installed at a transmitting station, protection against lightning can be provided by a switch that connects the lead-in conductors to ground during the times the station is not in operation.

810.58 Bonding Conductors and Grounding Electrode Conductors — Amateur and Citizen Band Transmitting and Receiving Stations. Bonding conductors and grounding electrode conductors shall comply with 810.58(A) through 810.58(C).

(A) Other Sections. All bonding conductors and grounding electrode conductors for amateur and citizen band transmitting and receiving stations shall comply with 810.21(A) through 810.21(C).

(B) Size of Protective Bonding Conductor or Grounding Electrode Conductor. The protective bonding conductor or grounding electrode conductor for transmitting stations shall be as large as the lead-in but not smaller than 10 AWG copper, bronze, or copper-clad steel.

(C) Size of Operating Bonding Conductor or Grounding Electrode Conductor. The operating bonding conductor or grounding electrode conductor for transmitting stations shall not be less than 14 AWG copper or its equivalent.

Part IV. Interior Installation — Transmitting Stations

△ **810.70 Separation from Other Conductors.** All conductors inside the building shall be separated at least 100 mm (4 in.) from the conductors of any electric light, power, or signaling circuit unless one of the following conditions applies:

(1) The conductors of a permanent audio system are installed in compliance with Parts I and II of Article 640.
(2) The conductors of portable and temporary audio systems are installed in compliance with Parts I and III of Article 640.
(3) The conductors are separated from such other conductors by a continuous and firmly fixed nonconductor.

810.71 General. Transmitters shall comply with 810.71(A) through (C).

(A) Enclosing. The transmitter shall be enclosed in a metal frame or grille or separated from the operating space by a barrier or other equivalent means, all metallic parts of which are effectively connected to a bonding conductor or grounding electrode conductor.

(B) Grounding of Controls. All external metal handles and controls accessible to the operating personnel shall be effectively connected to an equipment grounding conductor if the transmitter is powered by the premises wiring system or grounded with a conductor in accordance with 810.21.

(C) Interlocks on Doors. All access doors shall be provided with interlocks that disconnect all voltages of over 350 volts between conductors when any access door is opened.

ARTICLE 820 Community Antenna Television and Radio Distribution Systems

Part I. General

820.1 Scope. This article covers coaxial cable distribution of radio frequency signals typically employed in community antenna television (CATV) systems.

Article 820 covers the installation of coaxial cable for closed-circuit television, cable television, and security television cameras. This article also covers coaxial cable for radio and television receiving equipment. Article 830 covers network-powered broadband system installations. Many of the requirements that were similar or redundant to other Chapter 8 articles have been consolidated within the new Article 800.

820.3 Other Articles. The wiring methods of Article 830 shall be permitted to substitute for the wiring methods of Article 820.

Informational Note: Use of Article 830 wiring methods will facilitate the upgrading of Article 820 installations to network-powered broadband applications.

820.15 Power Limitations. Coaxial cable shall be permitted to deliver power to equipment that is directly associated with the radio frequency distribution system if the voltage is not over 60 volts and if the current is supplied by a transformer or other device that has power-limiting characteristics.

Power shall be blocked from premises devices on the network that are not intended to be powered via the coaxial cable.

Part III. Protection

820.93 Grounding of the Outer Conductive Shield of Coaxial Cables. Coaxial cables entering buildings or attached

to buildings shall comply with 820.93(A) or (B). Where the outer conductive shield of a coaxial cable is grounded, no other protective devices shall be required. For purposes of this section, grounding located at mobile home service equipment located within 9.0 m (30 ft) of the exterior wall of the mobile home it serves, or at a mobile home disconnecting means grounded in accordance with 250.32 and located within 9.0 m (30 ft) of the exterior wall of the mobile home it serves, shall be considered to meet the requirements of this section.

Informational Note: Selecting a grounding block location to achieve the shortest practicable bonding conductor or grounding electrode conductor helps limit potential differences between CATV and other metallic systems.

Proper bonding of the community antenna television (CATV) system coaxial cable sheath to the electrical power grounding electrode is needed to prevent potential fire and shock hazards. Failure to bond the two systems together can lead to a difference in potential between normally non–current-carrying parts.

See also

250.94 for more information regarding bonding of communications systems

(A) Entering Buildings. In installations where the coaxial cable enters the building, the outer conductive shield shall be grounded in accordance with 820.100. The grounding shall be as close as practicable to the point of entrance.

(B) Terminating Outside of the Building. In installations where the coaxial cable is terminated outside of the building, the outer conductive shield shall be grounded in accordance with 820.100. The grounding shall be as close as practicable to the point of attachment or termination.

(C) Location. Where installed, a listed primary protector shall be applied on each community antenna and radio distribution (CATV) cable external to the premises. The listed primary protector shall be located as close as practicable to the entrance point of the cable on either side or integral to the ground block.

(D) Hazardous (Classified) Locations. If a primary protector or equipment providing the primary protection function is used, it shall not be located in any hazardous (classified) location as defined in 500.5 and 505.5 or in the vicinity of easily ignitible material.

Exception: Primary protection equipment shall be used only if permitted by 501.150, 502.150, and 503.150.

Part IV. Grounding Methods

Δ **820.100 Cable Bonding and Grounding.** The shield of the coaxial cable shall be bonded or grounded as specified in 820.100(A) and (B).

Exception: For communications systems using coaxial cable completely contained within the building (i.e., they do not

exit the building) or the exterior zone of protection defined by a 46 m (150 ft) radius rolling sphere and isolated from outside cable plant, the shield shall be permitted to be grounded by a connection to an equipment grounding conductor as described in 250.118. Connecting to an equipment grounding conductor through a grounded receptacle using a dedicated bonding jumper and a permanently connected listed device shall be permitted. Use of a cord and plug for the connection to an equipment grounding conductor shall not be permitted.

Informational Note: See NFPA 780-2020, *Standard for the Installation of Lightning Protection Systems,* 4.7.3.1, for the application of the term *rolling sphere.*

(A) General Requirements. The installation shall be in accordance with 800.100.

(B) Shield Protection Devices. Grounding of a coaxial drop cable shield by means of a protective device that does not interrupt the grounding system within the premises shall be permitted.

The electric utility supply, the CATV system, and the premises wiring are all grounded. When a ground fault occurs, the current tries to return to its source. Such ground faults can cause current on the CATV shield, the primary function of which is to prevent radio frequency (RF) leakage out of the cable. The fault current can cause the cable shield to burn open and also damage the cable insulation. A device that can safely conduct current at 60 hertz and block current at the higher frequencies can be connected between the cable shield and ground, thereby maintaining grounding integrity. An ordinary fuse, for example, would not be suitable.

820.103 Equipment Grounding. Unpowered equipment and enclosures or equipment powered by the coaxial cable shall be considered grounded where connected to the metallic cable shield.

Part V. Installation Methods Within Buildings

820.154 Substitutions of Listed CATV Cables. The substitutions for coaxial cables in Table 820.154 and illustrated in Figure 820.154 shall be permitted.

Informational Note: The substitute cables in Table 820.154 and Figure 820.154 are only coaxial-type cables.

The application of coaxial cables, communications raceways, and cable routing assemblies is summarized in Table 800.154(a). All communication wires, cables and raceways, and cable routing assemblies in Table 800.154(a) must be listed. The installation location dictates the type of coaxial cable permitted within the raceway or assembly and is subject to the installation requirements of 800.110 and 800.113.

TABLE 820.154 *Coaxial Cable Uses and Permitted Substitutions*

Cable Type	Permitted Substitutions
CATVP	CMP, BLP
CATVR	CATVP, CMP, CMR, BMR, BLP, BLR
CATV	CATVP, CMP, CATVR, CMR, CMG, CM, BMR, BM, BLP, BLR, BL
CATVX	CATVP, CMP, CATVR, CMR, CATV, CMG, CM, BMR, BM, BLP, BLR, BL, BLX

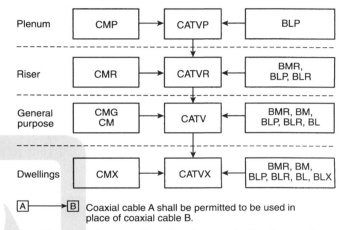

Type BL—Network-powered broadband communications low-power cables
Type BM—Network-powered broadband communications medium-power cables
Type CATV—Community antenna television cables
Type CM—Communications cables

FIGURE 820.154 *Coaxial Cable Substitution Hierarchy.*

<table>
<tr><td>ARTICLE
830</td><td>Network-Powered Broadband Communications Systems</td></tr>
</table>

Part I. General

830.1 Scope. This article covers network-powered broadband communications systems that provide any combination of voice, audio, video, data, and interactive services through a network interface unit.

Informational Note: A typical basic system configuration includes a cable supplying power and broadband signal to a network interface unit that converts the broadband signal to the component signals. Typical cables are coaxial cable with both broadband signal and power on the center conductor, composite metallic cable with a coaxial member(s) or twisted pair members for the broadband signal and twisted pair members for power, and hybrid optical fiber cable with a pair of conductors for power. Larger systems may also include network components such as amplifiers that require network power.

Network-powered broadband communications circuits provide a wide array of subscriber services, including voice, data (such as Internet access), interactive services, and television signals.

Article 830 contains requirements for wiring both the inside and the outside of buildings. Other articles cover the wiring derived from the network interface unit (NIU) into the premises. The major difference between Articles 820 and 830 is the voltage present on the circuit conductors. Article 820 systems are limited to 60 volts, but Article 830 systems are permitted to have ratings as high as 150 volts. Higher voltages allow systems to power more sophisticated electronics and to provide a wider variety of services.

△ 830.15 Power Limitations. Network-powered broadband communications systems shall be classified as having low- or medium-power sources as specified in the following:

(1) Sources shall be classified as defined in Table 830.15.
(2) Direct-current power sources exceeding 150 volts to ground, but no more than 200 volts to ground, with the current to ground limited to 10 mA dc, that meet the current and power limitation for medium-power sources in Table 830.15 shall be classified as medium-power sources.

Informational Note: See UL 60950-21-2007, *Standard for Safety for Information Technology Equipment — Safety — Part 21: Remote Power Feeding*, for listing information on equipment that complies with 830.15(2).

△ TABLE 830.15 *Limitations for Network-Powered Broadband Communications Systems*

Network Power Source	Low	Medium
Circuit voltage, V_{max} (volts)[1]	0–100	0–150
Power limitation, VA_{max} (volt-amperes)[1]	250	250
Current limitation, I_{max} (amperes)[1]	$1000/V_{max}$	$1000/V_{max}$
Maximum power rating (volt-amperes)	100	100
Maximum voltage rating (volts)	100	150
Maximum overcurrent protection (amperes)[2]	$100/V_{max}$	NA

[1]V_{max}, I_{max}, and VA_{max} are determined with the current-limiting impedance in the circuit (not bypassed) as follows:

V_{max} — Maximum system voltage regardless of load with rated input applied

I_{max} — Maximum system current under any noncapacitive load, including short circuit, and with overcurrent protection bypassed if used. I_{max} limits apply after 1 minute of operation

VA_{max} — Maximum volt-ampere output after 1 minute of operation regardless of load and overcurrent protection bypassed if used

[2]Overcurrent protection is not required if the current-limiting device provides equivalent current limitation and the current-limiting device does not reset until power or the load is removed.

Only network-powered broadband systems that operate within the voltage, current, and power parameters specified in Table 830.15 or a direct current (dc) system operating at not more than 200 volts and 10 milliamperes to ground are permitted. The dc systems must meet the current and power limitations for medium-power systems specified in Table 830.15.

Part II. Cables Outside and Entering Buildings

830.40 Entrance Cables. Network-powered broadband communications cables located outside and entering buildings shall comply with 830.40(A) and (B).

(A) Medium-Power Circuits. Medium-power network-powered broadband communications circuits located outside and entering buildings shall be installed using Type BMU, Type BM, or Type BMR network-powered broadband communications medium-power cables.

△ (B) Low-Power Circuits. Low-power network-powered broadband communications circuits located outside and entering buildings shall be installed using Type BLU or Type BLX low-power network-powered broadband communications cables. Cables shown in Table 830.154 shall be permitted to substitute.

830.44 Overhead (Aerial) Cables. Overhead (aerial) network-powered broadband communications cables shall comply with 830.44(A) through (F).

Network-powered broadband communications systems can contain sufficient energy to pose an electric shock hazard. For that reason, they are subject to requirements similar to those for overhead power conductors.

Overhead (aerial) spans of network-powered broadband communications cables must be of sufficient size and strength to maintain clearances and must avoid possible contact with light or power conductors. Splices and joints must be made with approved connectors or other means that provide sufficient mechanical strength so that conductors are not weakened, which could cause them to break and come into contact with higher-voltage conductors.

See also

800.44 for general requirements for overhead (aerial) communications wires and cables

(A) On Poles and In-Span or Above Roofs. Where network-powered broadband communications cables are installed on poles and in-span or above roofs, they shall comply with 800.44.

△ (B) Clearance from Ground. Overhead (aerial) spans of network-powered broadband communications cables shall conform to not less than the following:

(1) 2.9 m (9½ ft) — above finished grade, sidewalks, or from any platform or projection from which they might be reached and accessible to pedestrians only

(2) 3.5 m (11½ ft) — over residential property and driveways, and those commercial areas not subject to truck traffic

(3) 4.7 m (15½ ft) — over public streets, alleys, roads, parking areas subject to truck traffic, driveways on other than residential property, and other land traversed by vehicles such as cultivated, grazing, forest, and orchard

Informational Note: See ANSI/IEEE C2-2017, *National Electrical Safety Code*, Table 232-1, which provides for clearances of wires, conductors, and cables above ground and roadways, rather than using the clearances referenced in 225.18.

(C) Over Pools. Clearance of network-powered broadband communications cable in any direction from the water level, edge of pool, base of diving platform, or anchored raft shall comply with those clearances in 680.9.

(D) Final Spans. Final spans of network-powered broadband communications cables without an outer jacket shall be permitted to be attached to the building, but they shall be kept not less than 900 mm (3 ft) from windows that are designed to be opened, doors, porches, balconies, ladders, stairs, fire escapes, or similar locations.

Exception: Conductors run above the top level of a window shall be permitted to be less than the 900-mm (3-ft) requirement above.

Overhead (aerial) network-powered broadband communications cables shall not be installed beneath openings through which materials might be moved, such as openings in farm and commercial buildings, and shall not be installed where they obstruct entrance to these building openings.

(E) Between Buildings. Network-powered broadband communications cables extending between buildings or structures, and also the supports or attachment fixtures, shall be identified as suitable for outdoor aerial applications and shall have sufficient strength to withstand the loads to which they may be subjected.

Exception: Where a network-powered broadband communications cable does not have sufficient strength to be self-supporting, it shall be attached to a supporting messenger cable that, together with the attachment fixtures or supports, shall be acceptable for the purpose and shall have sufficient strength to withstand the loads to which they may be subjected.

In addition to the weight of the cable itself, wind and ice loads also must be considered because they can damage cables and attachment points.

(F) On Buildings. Where attached to buildings, network-powered broadband communications cables shall be securely fastened in such a manner that they are separated from other conductors in accordance with 830.44(F)(1) through (F)(3).

(1) Electric Light or Power. The network-powered broadband communications cable shall have a separation of at least 100 mm (4 in.) from electric light, power, Class 1, or non-power-limited fire alarm circuit conductors not in raceway or cable, or

be permanently separated from conductors of the other system by a continuous and firmly fixed nonconductor in addition to the insulation on the wires.

(2) Other Communications Systems. Network-powered broadband communications cables shall be installed so that there will be no unnecessary interference in the maintenance of the separate systems. In no case shall the conductors, cables, messenger strand, or equipment of one system cause abrasion to the conductors, cables, messenger strand, or equipment of any other system.

(3) Protection from Damage. Network-powered broadband communications cables attached to buildings or structures and located within 2.5 m (8 ft) of finished grade shall be protected by enclosures, raceways, or other approved means.

Exception: A low-power network-powered broadband communications circuit that is equipped with a listed fault protection device, appropriate to the network-powered broadband communications cable used, and located on the network side of the network-powered broadband communications cable shall not be required to be additionally protected by enclosures, raceways, or other approved means.

830.47 Underground Network-Powered Broadband Communications Cables Entering Buildings. Underground network-powered broadband communications cables entering buildings shall comply with 830.47(A) and (B).

Δ **(A) Protection from Physical Damage.** Direct-buried cable, conduit, or other raceways shall be installed to meet the minimum cover requirements of Table 830.47(A). In addition, direct-buried cables emerging from the ground shall be protected by enclosures, raceways, or other approved means extending from the minimum cover distance required by Table 830.47(A) below grade to a point at least 2.5 m (8 ft) above finished grade. In no case shall the protection be required to exceed 450 mm (18 in.) below finished grade. Types BMU and BLU direct-buried cables emerging from the ground shall be installed in rigid metal conduit (RMC), intermediate metal conduit (IMC), rigid nonmetallic conduit, or other approved means extending from the minimum cover distance required by Table 830.47(A) below grade to the point of entrance.

Exception: Protection from physical damage shall not be required if a low-power network-powered broadband communications circuit is equipped with a listed fault protection device that is located on the network side of the network-powered broadband cable being protected and the device is appropriate to the network-powered broadband communications cable used.

(B) Pools. Cables located under the pool or within the area extending 1.5 m (5 ft) horizontally from the inside wall of the pool shall meet those clearances and requirements specified in 680.11.

Δ **TABLE 830.47(A)** *Network-Powered Broadband Communications Systems Minimum Cover Requirements*

Location of Wiring Method or Circuit	Direct Burial Cables		Rigid Metal Conduit (RMC) or Intermediate Metal Conduit (IMC)		Nonmetallic Raceways Listed for Direct Burial; Without Concrete Encasement or Other Approved Raceways	
	mm	in.	mm	in.	mm	in.
All locations not specified below	450	18	150	6	300	12
In trench below 50 mm (2 in.) thick concrete or equivalent	300	12	150	6	150	6
Under a building (in raceway only)	0	0	0	0	0	0
Under minimum of 100 mm (4 in.) thick concrete exterior slab with no vehicular traffic and the slab extending not less than 150 mm (6 in.) beyond the underground installation	300	12	100	4	100	4
One- and two-family dwelling driveways and outdoor parking areas and used only for dwelling-related purposes	300	12	300	12	300	12

Notes:
1. Cover is the shortest distance measured between a point on the top surface of any direct-buried cable, conduit, or other raceway and the top surface of finished grade, concrete, or similar cover.
2. Raceways approved for burial only where concrete encased shall require a concrete envelope not less than 50 mm (2 in.) thick.
3. Lesser depths shall be permitted where cables rise for terminations or splices or where access is otherwise required.
4. Where solid rock is encountered, all wiring shall be installed in metal or nonmetallic raceway permitted for direct burial. The raceways shall be covered by a minimum of 50 mm (2 in.) of concrete extending down to rock.

Part III. Protection

830.90 Primary Electrical Protection.

(A) Application. Primary electrical protection shall be provided on all network-powered broadband communications conductors that are neither grounded nor interrupted and are run partly or entirely in aerial cable not confined within a block. Also, primary electrical protection shall be provided on all aerial or underground network-powered broadband communications conductors that are neither grounded nor interrupted and are located within the block containing the building served so as to be exposed to lightning or accidental contact with electric light or power conductors operating at over 300 volts to ground.

Exception: Primary electrical protection shall not be required on the network-powered broadband communications conductors where electrical protection is provided on the derived circuit(s) (output side of the NIU) in accordance with 830.90(B)(3).

Informational Note No. 1: On network-powered broadband communications conductors not exposed to lightning or accidental contact with power conductors, providing primary electrical protection in accordance with this article helps protect against other hazards, such as ground potential rise caused by power fault currents, and above-normal voltages induced by fault currents on power circuits in proximity to the network-powered broadband communications conductors.

Informational Note No. 2: Network-powered broadband communications circuits are considered to have a lightning exposure unless one or more of the following conditions exist:

(1) Circuits in large metropolitan areas where buildings are close together and sufficiently high to intercept lightning.

(2) Areas having an average of five or fewer thunderstorm days each year and earth resistivity of less than 100 ohm-meters. Such areas are found along the Pacific coast.

Utility companies may provide primary protectors if conductors are exposed to lightning. Typically, cables are not considered to be exposed to lightning if one or both of the conditions in Informational Note No. 2 exist. A primary protector is required at each end of a communications circuit where lightning exposure exists, unless protection is provided on the output side of the NIU.

Informational Note No. 3: See NFPA 780-2020, *Standard for the Installation of Lightning Protection Systems*, for information on lightning protection systems.

(1) Fuseless Primary Protectors. Fuseless-type primary protectors shall be permitted where power fault currents on all protected conductors in the cable are safely limited to a value no greater than the current-carrying capacity of the primary protector and of the primary protector bonding conductor or grounding electrode conductor.

(2) Fused Primary Protectors. Where the requirements listed in 830.90(A)(1) are not met, fused-type primary protectors shall be used. Fused-type primary protectors shall consist of an arrester connected between each conductor to be protected and ground, a fuse in series with each conductor to be protected, and an appropriate mounting arrangement. Fused primary protector terminals shall be marked to indicate line, instrument, and ground, as applicable.

(B) Location. The location of the primary protector, where required, shall comply with the following:

(1) A listed primary protector shall be applied on each network-powered broadband communications cable external to and on the network side of the network interface unit.

(2) The primary protector function shall be an integral part of and contained in the network interface unit. The network interface unit shall be listed as being suitable for application with network-powered broadband communications systems and shall have an external marking indicating that it contains primary electrical protection.

(3) The primary protector(s) shall be provided on the derived circuit(s) (output side of the NIU), and the combination of the NIU and the protector(s) shall be listed as being suitable for application with network-powered broadband communications systems.

A primary protector, whether provided integrally or external to the network interface unit, shall be located as close as practicable to the point of entrance.

For purposes of this section, a network interface unit and any externally provided primary protectors located at mobile home service equipment located in sight from and not more than 9.0 m (30 ft) from the exterior wall of the mobile home it serves, or at a mobile home disconnecting means grounded in accordance with 250.32 and located in sight from and not more than 9.0 m (30 ft) from the exterior wall of the mobile home it serves, shall be considered to meet the requirements of this section.

> Informational Note: Selecting a network interface unit and primary protector location to achieve the shortest practicable primary protector bonding conductor or grounding electrode conductor helps limit potential differences between communications circuits and other metallic systems.

(C) Hazardous (Classified) Locations. The primary protector or equipment providing the primary protection function shall not be located in any hazardous (classified) location as defined in 500.5 and 505.5 or in the vicinity of easily ignitible material.

Exception: As permitted in 501.150, 502.150, and 503.150.

Part IV. Grounding Methods

830.93 Grounding or Interruption of Metallic Members of Network-Powered Broadband Communications Cables. Network-powered communications cables entering buildings or attaching to buildings shall comply with 830.93(A) or (B).

For purposes of this section, grounding located at mobile home service equipment located within 9.0 m (30 ft) of the exterior wall of the mobile home it serves, or at a mobile home disconnecting means grounded in accordance with 250.32 and located within 9.0 m (30 ft) of the exterior wall of the mobile

home it serves, shall be considered to meet the requirements of this section.

> Informational Note: Selecting a grounding location to achieve the shortest practicable bonding conductor or grounding electrode conductor helps limit potential differences between the network-powered broadband communications circuits and other metallic systems.

Proper bonding of the network-powered broadband communications system cable sheath to the electrical power grounding electrode is needed to prevent potential fire and shock hazards. Failure to bond the two systems together can lead to a difference in potential between normally non–current-carrying parts.

See also

250.94 for more information regarding bonding of communications systems

(A) Entering Buildings. In installations where the network-powered communications cable enters the building, the shield shall be grounded in accordance with 800.100, and metallic members of the cable not used for communications or powering shall be grounded in accordance with 800.100 or interrupted by an insulating joint or equivalent device. The grounding or interruption shall be as close as practicable to the point of entrance.

(B) Terminating Outside of the Building. In installations where the network-powered communications cable is terminated outside of the building, the shield shall be grounded in accordance with 800.100, and metallic members of the cable not used for communications or powering shall be grounded in accordance with 800.100 or interrupted by an insulating joint or equivalent device. The grounding or interruption shall be as close as practicable to the point of attachment of the NIU.

Part V. Installation Methods Within Buildings

830.133 Installation of Network-Powered Broadband Communications Cables and Equipment. Cable and equipment installations within buildings shall comply with 830.133(A) through (C), as applicable.

(A) Separation of Conductors.

(1) In Raceways, Cable Trays, Boxes, Enclosures, and Cable Routing Assemblies.

(a) *Low- and Medium-Power Network-Powered Broadband Communications Circuit Cables.* Low- and medium-power network-powered broadband communications cables shall be permitted in the same raceway, cable tray, box, enclosure, or cable routing assembly.

(b) *Low-Power Network-Powered Broadband Communications Circuit Cables with Other Circuits.* Low-power network-powered broadband communications cables shall be permitted in the same raceway, cable tray, box, enclosure, or cable routing assembly with jacketed cables of any of the following circuits:

(1) Class 2 and Class 3 remote-control, signaling, and power-limited circuits in compliance with Parts I and II of Article 725

(2) Power-limited fire alarm systems in compliance with Parts I and III of Article 760

(3) Communications circuits in compliance with Parts I and IV of Article 805

(4) Nonconductive and conductive optical fiber cables in compliance with Parts I and V of Article 770

(5) Community antenna television and radio distribution systems in compliance with Parts I and V of Article 820

(c) *Medium-Power Network-Powered Broadband Communications Circuit Cables with Optical Fiber Cables and Other Communications Cables.* Medium-power network-powered broadband communications cables shall not be permitted in the same raceway, cable tray, box, enclosure, or cable routing assembly with conductors of any of the following circuits:

(1) Communications circuits in compliance with Parts I and IV of Article 805

(2) Conductive optical fiber cables in compliance with Parts I and V of Article 770

(3) Community antenna television and radio distribution systems in compliance with Parts I and V of Article 820

(d) *Medium-Power Network-Powered Broadband Communications Circuit Cables with Other Circuits.* Medium-power network-powered broadband communications cables shall not be permitted in the same raceway, cable tray, box, enclosure, or cable routing assembly with conductors of any of the following circuits:

(1) Class 2 and Class 3 remote-control, signaling, and power-limited circuits in compliance with Parts I and II of Article 725

(2) Power-limited fire alarm systems in compliance with Parts I and III of Article 760

(e) *Electric Light, Power, Class 1, Nonpowered Broadband Communications Circuit Cables.* Network-powered broadband communications cable shall not be placed in any raceway, cable tray, compartment, outlet box, junction box, or similar fittings with conductors of electric light, power, Class 1, or non-power-limited fire alarm circuit cables.

Exception No. 1: Network-powered broadband communications cable shall be permitted to be placed in a raceway, cable tray, compartment, outlet box, junction box, or similar fittings with conductors of electric light, power, Class 1, or non-power-limited fire alarm circuit cables where all of the conductors of electric light, power, Class 1, non-power-limited fire alarm circuits are separated from all of the network-powered

broadband communications cables by a permanent barrier or listed divider.

Exception No. 2: Where power circuit conductors in outlet boxes, junction boxes, or similar fittings or compartments where such conductors are introduced solely for power supply to the network-powered broadband communications system distribution equipment, the power circuit conductors shall be routed within the enclosure to maintain a minimum 6 mm (¼ in.) separation from network-powered broadband communications cables.

Δ **(2) Other Applications.** Network-powered broadband communications cable shall be separated at least 50 mm (2 in.) from conductors of any electric light, power, Class 1, and non-power-limited fire alarm circuits.

Exception No. 1: Separation shall not be required where: (1) all of the conductors of electric light, power, Class 1, and non-power-limited fire alarm circuits are in a raceway or in metal-sheathed, metal-clad, nonmetallic-sheathed, Type AC, or Type UF cables, or (2) all of the network-powered broadband communications cables are encased in a raceway.

Exception No. 2: Separation shall not be required where the network-powered broadband communications cables are permanently separated from the conductors of electric light, power, Class 1, and non-power-limited fire alarm circuits by a continuous and firmly fixed nonconductor, such as porcelain tubes or flexible tubing, in addition to the insulation on the wire.

(B) Support of Network-Powered Broadband Communications Cables. Raceways shall be used for their intended purpose. Network-powered broadband communications cables shall not be strapped, taped, or attached by any means to the exterior of any conduit or raceway as a means of support.

N **(C) Splicing of Medium-Powered Network-Powered Communications Cables.** Where a medium-powered network-powered broadband communications cable is spliced or extended, a listed junction box or listed patch panel shall be used.

830.154 Substitutions of Network-Powered Broadband Communications System Cables. The substitutions for network-powered broadband system cables listed in Table 830.154 shall be permitted.

The applications for the cable, communications raceways, and cable routing assemblies are summarized in Table 800.154(a). The installation location dictates the type of cable permitted within the raceway or assembly and is subject to the installation requirements of 800.110, 800.113, and 830.40. Table 830.154 reflects permitted interchangeability between cable types that may be utilized within the applications listed in Table 800.154(a).

TABLE 830.154 *Cable Substitutions*

Cable Type	Permitted Cable Substitutions
BM	BMR
BLP	CMP, CL3P
BLR	CMP, CL3P, CMR, CL3R, BLP, BMR
BL	CMP, CMR, CM, CMG, CL3P, CL3R, CL3, BMR, BM, BLP, BLR
BLX	CMP, CMR, CM, CMG, CMX, CL3P, CL3R, CL3, CL3X, BMR, BM, BLP, BRP, BL

830.160 Bends. Bends in network broadband cable shall be made so as not to damage the cable. The radius of the curve of the inner edge of any bend shall not be less than 10 times the diameter of the cable.

> Informational Note: See ANSI/TIA-568.0-E *Generic Telecommunications Cabling for Customer Premises*, for information on bend radii of network broadband cable during different types of installation conditions.

Part VI. Listing Requirements

830.179 Network-Powered Broadband Communications Equipment and Cables. Network-powered broadband communications equipment and cables shall be listed and marked in accordance with 830.179(A) through (C).

Exception No. 1: This listing requirement shall not apply to community antenna television and radio distribution system coaxial cables that were installed prior to January 1, 2000, in accordance with Article 820 and are used for low-power network-powered broadband communications circuits.

Exception No. 2: Substitute cables for network-powered broadband communications cables shall be permitted as shown in Table 830.154.

(A) General Requirements. The general requirements in 800.179 shall apply.

(B) Network-Powered Broadband Communications Medium-Power Cables. Network-powered broadband communications medium-power cables shall be factory-assembled cables consisting of a jacketed coaxial cable, a jacketed combination of coaxial cable and multiple individual conductors, or a jacketed combination of an optical fiber cable and multiple individual conductors. The insulation for the individual conductors shall be rated for 300 volts minimum. Cables intended for outdoor use shall be listed as suitable for the application. Cables shall be marked in accordance with 310.8. Type BMU cables shall be jacketed and listed as being suitable for outdoor underground use.

An insulation rating of 300 volts is necessary for the following reasons:

1. This rating coordinates with protector installation requirements.
2. Primary protectors are designed to allow voltages below 300 to pass.

3. Network-powered broadband communications circuits typically operate in a voltage range up to 150 volts root-mean square (rms).

(C) Network-Powered Broadband Communication Low-Power Cables. Network-powered broadband communications low-power cables shall be factory-assembled cables consisting of a jacketed coaxial cable, a jacketed combination of coaxial cable and multiple individual conductors, or a jacketed combination of an optical fiber cable and multiple individual conductors. The insulation for the individual conductors shall be rated for 300 volts minimum. Cables intended for outdoor use shall be listed as suitable for the application. Cables shall be marked in accordance with 310.8. Type BLU cables shall be jacketed and listed as being suitable for outdoor underground use.

ARTICLE 840 Premises-Powered Broadband Communications Systems

Part I. General

△ **840.1 Scope.** This article covers premises-powered broadband communications systems.

> Informational Note: A typical basic system configuration consists of an optical fiber, twisted pair, or coaxial cable to the premises supplying a broadband signal to a network terminal that converts the broadband signal into component signals, such as traditional telephone, video, high-speed Internet, and interactive services. Powering for the network terminal and network devices is typically accomplished through a premises power supply that might be built into the network terminal or provided as a separate unit. In order to provide communications in the event of a power interruption, a battery backup unit or an uninterruptible power supply (UPS) is typically part of the powering system.

Although similar to Article 830, which addresses network-powered broadband communications systems, Article 840 covers premises-powered broadband communications systems.

Premises-powered broadband communications systems provide a wide array of subscriber services, including voice, video, data (such as Internet access), and interactive services, through an optical network terminal (ONT).

Article 840 contains requirements for wiring both the inside and the outside of buildings. Other articles cover the wiring derived from the ONT into the premises.

See also

Article 725, which covers wiring of Class 2 and Class 3 circuits

Article 760, which covers the wiring of fire alarm systems

Article 770, which covers the installation of optical fiber cables

Article 800, which covers general requirements for all communications systems

Article 820, which covers coaxial cable installations for television signals

Part II. Cables Outside and Entering Buildings

840.47 Underground Wires and Cables Entering Buildings. Direct-buried cables shall be installed to have a minimum cover of 150 mm (6 in.).

Part III. Protection

840.90 Protective Devices. The requirements of 805.90 shall apply.

840.93 Grounding or Interruption. Non-current-carrying metallic members of optical fiber cables, communications cables, or coaxial cables entering buildings or attaching to buildings shall comply with 840.93(A), (B), or (C), respectively.

(A) Non-Current-Carrying Metallic Members of Optical Fiber Cables. Non-current-carrying metallic members of optical fiber cables entering a building or terminating on the outside of a building shall comply with 770.93(A) or (B).

(B) Communications Cables. The grounding or interruption of the metallic sheath of communications cable shall comply with 805.93.

(C) Coaxial Cables. Where the network terminal is installed inside or outside of the building, with coaxial cables terminating at the network terminal, and is either entering, exiting, or attached to the outside of the building, 820.93 shall apply.

840.94 Premises Circuits Leaving the Building. Where circuits leave the building to power equipment remote to the building or outside the exterior zone of protection defined by a 46 m (150 ft) radius rolling sphere, 805.90 and 805.93 shall apply.

> Informational Note: See NFPA 780-2020, *Standard for the Installation of Lightning Protection Systems*, for the theory of the term *rolling sphere*.

Part IV. Grounding Methods

840.101 Premises Circuits Not Leaving the Building. If the network terminal is served by a nonconductive optical fiber cable, or where any non-current-carrying metal member of a conductive optical fiber cable is interrupted by an insulating joint or equivalent device, and circuits that terminate at the network terminal are completely contained within the building (i.e., they do not exit the building), 840.101(A), (B), or (C) shall apply, as applicable.

(A) Coaxial Cable Shield Grounding. The shield of coaxial cable shall be grounded by one of the following:

(1) Any of the methods described in 820.100 or 800.106
(2) A fixed connection to an equipment grounding conductor as described in 250.118
(3) Connection to the network terminal grounding terminal provided that the terminal is connected to ground by one

of the methods described in 820.100 or 800.106, or to an equipment grounding conductor through a listed grounding device that will retain the ground connection if the network terminal is unplugged

The coaxial shield is permitted to be grounded through the optical network terminal (ONT) if the ONT grounding connection is permanent or the connection is to an equipment grounding conductor (EGC) through a listed grounding device that will retain the grounding connection if the ONT is unplugged.

(B) Communications Circuit Grounding. Communications circuits shall not be required to be grounded.

(C) Network Terminal Grounding. The network terminal shall not be required to be grounded unless required by its listing. If the coaxial cable shield is separately grounded as described in 840.101(A)(1) or 840.101(A)(2), the use of a cord and plug for the connection to the network terminal grounding connection shall be permitted.

> Informational Note: If required to be grounded, a listed device that extends the equipment grounding conductor from the receptacle to the network terminal equipment grounding terminal is permitted. Sizing of the extended equipment grounding conductor is covered in Table 250.122.

840.102 Premises Circuits Leaving the Building. If circuits leave the building to power equipment remote to the building or outside the exterior zone of protection defined by a 46 m (150 ft) radius rolling sphere, the installation of communications wires and cables shall comply with 800.100 and 800.106, and the installation of coaxial cables shall comply with 820.100 and 800.106.

> Informational Note: See NFPA 780-2020, *Standard for the Installation of Lightning Protection Systems*, for the application of the term *rolling sphere*.

Part VI. Premises Powering of Communications Equipment over Communications Cables

Δ **840.160 Powering Circuits.** Listed communications cables, in addition to carrying the communications circuit, shall also be permitted to carry circuits for powering listed communications equipment. The power source shall be listed in accordance with 840.170(C). Installation of the listed 4-pair communications cables for a communications circuit or installation where 4-pair communications cables are substituted for Class 2 and Class 3 cables in accordance with 722.135(E) shall comply with 725.144.

Exception: Installing communications cables in compliance with 725.144 shall not be required for listed 4-pair communications cables where the rated current of the power source does not exceed 0.3 amperes in any conductor 24 AWG or larger.

> Informational Note No. 1: A typical communications cable for this application is a 4-pair cable sometimes referred to as Category 5e (or higher) LAN cable or balanced twisted pair cable.

These types of cables are often used to provide Ethernet- and Power over Ethernet (PoE)-type services.

Informational Note No. 2: See 725.144 for requirements to manage the temperature rise of bundles of cables that provide power.

Power over Ethernet (PoE) systems are allowed to provide power to communications equipment via the communications cable. As more power is delivered to the equipment, overheating the cables becomes more of a concern. Therefore, how the cables are bundled together must be monitored and accounted for. Cables are required to comply with 725.144 for communications cables used as Class 2 and Class 3 cables. Table 725.144 contains ampacity values for various conductor sizes with respect to the number of cables within a given bundle. This requirement does not apply when the power source limits the current imposed on a conductor to less than 0.3 ampere when the conductor is 24 AWG or larger.

Part VII. Listing Requirements

840.170 Equipment and Cables. Premises-powered broadband communications systems equipment and cables shall comply with 840.170(A) through (D).

(A) Network Terminal. The network terminal and applicable grounding means shall be listed for application with premises-powered broadband communications systems.

Informational Note No. 1: See ANSI/UL 60950-1-2014, *Standard for Safety of Information Technology Equipment*; ANSI/UL 498A-2015, *Current Taps and Adapters*; ANSI/UL 467-2013, *Grounding and Bonding Equipment;* or ANSI/UL 62368-1-2014, *Audio/Video, Information and Communication Technology Equipment – Part 1: Safety Requirements.*

Informational Note No. 2: There are no requirements on the network terminal and its grounding methodologies except for those covered by the listing of the product.

(B) Premises Communications Wires and Cables. Communications wires and cables shall be listed and marked in accordance with 800.179.

(C) Power Source. The power source for circuits intended to provide power over communications cables to remote equipment shall be limited in accordance with Table 11(B) in Chapter 9 for voltage sources up to 60 volts dc and be listed as specified in either of the following:

(1) A power source shall be listed as specified in 725.60(A)(1), (A)(2), (A)(3), or (A)(4). The power sources shall not have the output connections paralleled or otherwise interconnected unless listed for such interconnection.

(2) A power source shall be listed as communications equipment for limited-power circuits.

Informational Note: See ANSI/UL 60950-1-2014, *Standard for Safety of Information Technology Equipment-Safety — Part 1*, or ANSI/UL 62368-1-2014, *Audio/Video, Information and Communication Technology Equipment — Part 1: Safety Requirements*. Typically, such circuits are used to interconnect equipment for the purpose of exchanging information (data).

(D) Accessory Equipment. Communications accessory equipment and/or assemblies shall be listed for application with premises-powered communications systems.

Informational Note: See ANSI/UL 1863-2004, *Communications-Circuit Accessories.*

Tables

The tables in Chapter 9 are part of the mandatory requirements of the *NEC®*. Tables 1 through 10 deal with conductors and raceways. The following four tables — Tables 11(A), 11(B), 12(A), and 12(B) — provide parameters for power limitations for Class 2 and Class 3 power-limited circuits and for power-limited fire alarm circuits. Table 13 was added for 2023 in order to assist in clarifying protection techniques and types of protection for hazardous (classified) locations.

As the adoption and use of the *NEC* has increased in areas of the world where the metric system is the standard, providing a means to allow for the use of electrical products with metric measurements or designations has become necessary for assimilation of *NEC* requirements in the metric world. For every requirement that specifies the size of a conduit or tubing with a cylindrical cross-section size, two size designations, the metric designator and the trade size, are given per Table 300.1(C). This designation affects all metal, nonmetallic, rigid, and flexible conduit and tubing types that have a cylindrical cross section. For example, ½-inch conduit and tubing are referred to as metric designator 16 or trade size ½, ¾ inch is metric designator 21 or trade size ¾, 4 inches is metric designator 103 or trade size 4, and so forth.

See also

90.9 and **300.1(C)** commentary for more information on metrication and the revised way of providing conduit and tubing sizes in the *NEC*

Because conduits and tubing from different manufacturers have different internal diameters for the same trade size, Table 4 provides the diameter and the actual area of different conduit and tubing types at fill percentages of 100, 60, 53 (one wire), 31 (two wires), and 40 (over two wires). The 60-percent fill is provided in Table 4 to correlate Note 4 (found in the Notes to Tables section of this chapter) with the conduit and tubing fill tables, which permits conduit or tubing nipples 24 inches or less in length to have a conductor fill of up to 60 percent. Separate sections in Table 4 cover metal, nonmetallic, rigid, and flexible conduit and tubing types. Examples of how to use the conduit and tubing conductor fill tables are included in the commentary both here and in Informative Annex C.

Informative Annex C contains conductor fill tables for each of 12 types of conduit and tubing. The Informative Annex C tables — which are based on the dimensions given in Tables 1 and 4 of Chapter 9 for conduit and tubing fill and on the dimensions for conductors in Table 5 of Chapter 9 — provide conductor fill information based on the specific conduit or tubing and on the conductor insulation type, size, and stranding characteristics. Examples of how to use these tables are included in the commentary both here and in Informative Annex C.

TABLE 1 *Percent of Cross Section of Conduit and Tubing for Conductors and Cables*

Number of Conductors and/or Cables	Cross-Sectional Area (%)
1	53
2	31
Over 2	40

Informational Note No. 1: Table 1 is based on common conditions of proper cabling and alignment of conductors where the length of the pull and the number of bends are within reasonable limits. It should be recognized that, for certain conditions, a larger size conduit or a lesser conduit fill should be considered.

Table 1 establishes the maximum fill permitted for the circular conduit and tubing types. It is the basis for Table 4 and for the information on conduit and tubing fill provided in the Informative Annex C tables. Informational Note No. 1 advises that factors such as the length of the run or the number and total radius of bends can increase the difficulty of pulling conductors into the raceway and in extreme cases could result in damage to conductor insulation. To mitigate such adverse effects and to facilitate the ease of installing the conductors in the conduit or tubing, it is recommended, where a difficult installation is anticipated, that the maximum number of conductors permitted not be installed or that the size of the conduit or tubing be increased by at least one trade size larger than the minimum required by the *NEC*.

Informational Note No. 2: When pulling three conductors or cables into a raceway, if the ratio of the raceway (inside diameter) to the conductor or cable (outside diameter) is between 2.8 and

3.2, jamming can occur. While jamming can occur when pulling four or more conductors or cables into a raceway, the probability is very low.

Informational Note No. 2 warns of another potential pitfall associated with pulling conductors into conduit or tubing. Conductor jamming can occur during the installation (pulling) of conductors into a conduit even if fill allowances of 40 percent are observed. During the installation of three conductors or cables into the raceway, one conductor could slip between the other two conductors. This is more likely to take place at bends, where the raceway might be slightly oval.

As an example, Table C.1 in Informative Annex C permits three 8 AWG conductors in trade size ½ electrical metallic tubing (EMT). An 8 AWG conductor has an outside diameter (OD) of 0.216 inch (from Table 5), and a ½ inch EMT has an internal diameter (ID) of 0.622 inch (from Table 4).

The EMT in a straight run has an ID of 0.622 inch, but because it may not be round at a bend, one conductor could slip between the other two and cause a jam as the conductors exit the bend. In a straight run, assuming no variation in the EMT's ID or in a conductor's OD, one conductor usually cannot slip between the other two, because the total of the ODs of the conductors (3×0.216 inch $= 0.648$ inch) is greater than the EMT's ID of 0.622 inch. At a bend, however, the major ID of the raceway could increase due to bending, particularly in tubing, to a diameter slightly larger than 0.648 inch, permitting the middle conductor to be pulled between the outer two conductors. As the conductors exit the bend and the raceway returns to its normal shape with an ID of 0.622 inch, the conductors could jam. This can also occur in straight runs where the ratio of the raceway's ID to the conductor's OD approaches 3. The jam ratio is calculated as follows:

$$\text{Jam ratio} = \frac{\text{ID of raceway}}{\text{OD of conductor}} = \frac{0.622}{0.216} = 2.88$$

To avoid difficult conductor installations and potential conductor insulation damage due to jamming within the conduit or tubing, a jam ratio between 2.8 and 3.2 should be avoided.

Notes to Tables

(1) See Informative Annex C for the maximum number of conductors and fixture wires, all of the same size (total cross-sectional area including insulation) permitted in trade sizes of the applicable conduit or tubing.

(2) Table 1 applies only to complete conduit or tubing systems and is not intended to apply to sections of conduit or tubing used to protect exposed wiring and cable from physical damage.

The maximum fill requirements do not apply to short sections of conduit or tubing used for the physical protection of conductors and cables. Cables commonly are protected from physical damage by conduit or tubing sleeves sized to enable the cable to be passed through with relative ease without injuring or abrading the protective jacket of the cable. The requirement of 300.5(D)(1) regarding physical protection of direct-buried cables and conductors as they emerge from below grade is an example of

conduit or tubing being used as a protective sleeve and not as a continuous raceway system per 300.12. However, a fitting is required on the end(s) of the conduit or tubing to protect the conductors or cables from abrasion, as specified in 300.15(C).

(3) Equipment grounding or bonding conductors, where installed, shall be included when calculating conduit or tubing fill. The actual dimensions of the equipment grounding or bonding conductor (insulated or bare) shall be used in the calculation.

All insulated, covered, and bare conductors occupy space within a raceway. Therefore, all installed conductors must be included in the raceway fill calculation, including non–current-carrying conductors such as equipment grounding conductors, bonding conductors, and bonding jumpers. The only exception to this rule is the addition of an equipment grounding conductor permitted in trade size ⅜ flexible metal conduit (see the note to Table 348.22). The dimensions of bare conductors are given in Table 8.

(4) Where conduit or tubing nipples, not including connectors, having a maximum length not to exceed 600 mm (24 in.) are installed between boxes, cabinets, and similar enclosures, the nipples shall be permitted to be filled to 60 percent of their total cross-sectional area, and 310.15(C)(1) adjustment factors need not apply to this condition.

(5) For conductors not included in Chapter 9, such as multiconductor cables and optical fiber cables, the actual dimensions shall be used.

For conductors not included in Chapter 9, such as high-voltage types, the cross-sectional area can be calculated in the following manner, using the actual dimensions of each conductor:

$$\text{cross-sectional area} = d^2 \text{ cmil}$$

where:

d = outside diameter of a conductor (including insulation) [1 inch = 1000 mil (1 mil = 0.001 inch)]

cmil = circular mil, a unit measure of area equal to π/4 (3.1416/4 = 0.7854) square mil. In other words, 1 cmil = 0.7854 square mil.

Calculation Example

Three 15-kV single conductors are to be installed in rigid metal conduit (RMC). The outside diameter of each conductor measures 1⅝ in., or 1.625 in. What size RMC will accommodate the three conductors?

Solution

Step 1. Find the cross-sectional area within the conduit to be displaced by the three conductors:

$$(1.625 \text{ in.})^2 \times 0.7854 \times 3 = 6.2218 \text{ in.}^2 \text{ or } 6.222 \text{ in.}^2$$

Step 2. Determine the correct conduit size to accommodate the three conductors. Table 1 allows 40-percent conduit fill for three or more conductors, and Table 4 indicates that 40 percent of trade size 5 RMC is 8.085 in.². Thus, trade size 5 RMC will accommodate three 15-kV single conductors.

(6) For combinations of conductors of different sizes, use actual dimensions or Table 5 and Table 5A for dimensions of conductors and Table 4 for the applicable conduit or tubing dimensions.

The following two examples demonstrate how to calculate the minimum trade size conduit or tubing required for conductors of different sizes.

Calculation Example 1

A 200-A feeder is routed in various wiring methods [EMT; polyvinyl chloride conduit (PVC) (Schedule 40); and RMC] from the main switchboard in one building to a distribution panelboard in another building. The circuit consists of four 4/0 AWG XHHW copper conductors and one 6 AWG XHHW copper conductor. Select the proper trade size for the various types of conduit and tubing to be used for the feeder.

Solution
All the raceways for this example require conduit fill to be calculated according to Table 1 in Chapter 9, which permits conduit fill to a maximum of 40 percent where more than two conductors are installed. (See 344.22 for RMC, 352.22 for PVC, and 358.22 for EMT.) Note 6 refers to Table 5 for the area required for each insulated conductor. Note 6 also refers to Table 4 for selection of the appropriate trade size conduit or tubing. Table 4 contains the allowable cross-sectional area for conduit and tubing based on conductor-occupied space (40 percent maximum in this example).

Step 1. Calculate the total area occupied by the conductors, using the approximate areas listed in Table 5:

Four 4/0 AWG XHHW:

$$4 \times 0.3197 \text{ in.}^2 = 1.2788 \text{ in.}^2$$

One 6 AWG XHHW:

$$1 \times 0.0590 \text{ in.}^2 = 0.0590 \text{ in.}^2$$

Total area = 1.3378 in.^{2} or 1.338 in.^{2}

Step 2. Determine the proper trade size EMT, RMC, and PVC (Schedule 40) from Table 4. The portion of this feeder installed in EMT requires a minimum trade size 2, which has 1.342 in.^{2} of available space for 40-percent fill. RMC also requires a minimum trade size 2, because trade size 2 RMC has 1.363 in.^{2} of available space for the conductors. PVC (Schedule 40), however, requires a minimum trade size 2½. Trade size 2 PVC has 1.316 in.^{2} allowable space, which is less than the 1.338 in.^{2} required for this combination of conductors. Therefore, it is necessary to increase the PVC size to 2½ trade size, the next standard size increment.

Calculation Example 2

Determine the minimum size RMC allowed for the 10 mixed conductor sizes and types described as follows:

Quantity	Wire Size and Type	Cross-Sectional Area of Each Wire (from Table 5)	Cross-Sectional Area
4	12 AWG THWN	0.0133	0.0532
3	8 AWG TW	0.0437	0.1311
3	6 AWG THW	0.0726	0.2178
		Total	0.4021

Solution
The "Over 2 Wires" column in Table 4 indicates that 40 percent of a trade size 1¼ RMC is 0.610 in.^{2} Therefore, trade size 1¼ is the minimum size RMC allowed for this combination of 10 conductors.

(7) When calculating the maximum number of conductors or cables permitted in a conduit or tubing, all of the same size (total cross-sectional area including insulation), the next higher whole number shall be used to determine the maximum number of conductors permitted when the calculation results in a decimal greater than or equal to 0.8. When calculating the size for conduit or tubing permitted for a single conductor, one conductor shall be permitted when the calculation results in a decimal greater than or equal to 0.8.

Calculation Example

Determine how many 10 AWG THHN conductors are permitted in a trade size 1¼ RMC.

Solution
Table 1 permits 40-percent fill for over two conductors. From Table 4, 40-percent fill for trade size 1¼ RMC is 0.610 in., and from Table 5, the cross-sectional area of a 10 AWG THHN conductor is 0.0211 in.^{2} The number of conductors permitted is calculated as follows:

$$\frac{0.610 \text{ in.}^2}{0.0211 \text{ in.}^2 \text{ per conductor}} = 28.910 \text{ conductors}$$

Based on the maximum allowable fill, the number of 10 AWG THHN conductors in trade size 1¼ RMC cannot exceed 28. However, in accordance with Note 7, an increase to the next whole number of 29 conductors is permitted in this case, because 0.910 is greater than 0.8, which is the benchmark for determining whether an increase to the next whole number for the maximum number of conductors is permitted. Although increasing the total to 29 conductors results in the raceway fill exceeding 40 percent, the amount by which it is exceeded is a fraction of 1 percent and will not adversely affect the installation of the conductors. This number of conductors does have to be addressed from a mutual heating effect in accordance with 310.15(C)(1). Bear in mind that this is the maximum number of conductors permitted, and, in accordance with Informational Note No. 1 to Table 1 in Chapter 9, an installation with fewer than the maximum number of conductors allowed may be prudent.

Verification of this solution can be found in Informative Annex C, Table C.8, which lists twenty-nine 10 AWG THWN conductors as the maximum number permitted in trade size 1¼ RMC. Several examples of the application of Note 7 are found in the Informative Annex C tables. Where the calculation results in a decimal value less than 0.8, the maximum number of conductors permitted is based on the next lower whole number.

(8) Where bare conductors are permitted by other sections of this *Code*, the dimensions for bare conductors in Table 8 shall be permitted.
(9) A multiconductor cable, optical fiber cable, or flexible cord of two or more conductors shall be treated as a single conductor for calculating percentage conduit or tubing fill area. For cables that have elliptical cross sections, the cross-sectional area calculation shall be based on using the major diameter of the ellipse as a circle diameter. Assemblies of single insulated conductors without an overall covering shall not be considered a cable when determining conduit or tubing fill area. The conduit or tubing fill for the assemblies shall be calculated based upon the individual conductors.

Note 9 to the table recognizes that some triplexed assemblies of conductors are available without an outer sheath, which reduces the assemblies' diameter compared to the same size conductors with an overall outer covering.

(10) The values for approximate conductor diameter and area shown in Table 5 are based on worst-case scenario and indicate round concentric-lay-stranded conductors. Solid and round concentric-lay-stranded conductor values are grouped together for the purpose of Table 5. Round compact-stranded conductor values are shown in Table 5A. If the actual values of the conductor diameter and area are known, they shall be permitted to be used.

Prior to the 2005 edition of the *NEC*, Table 2 was included in Article 344 as Table 344.24. The requirements on minimum bending radius (depending on the type of bending equipment employed) apply to all the rigid, flexible, metallic, and nonmetallic conduit and tubing types. Because of its widespread application for all the circular conduit and tubing types (refer to Section __.24 of the applicable article), this table is more appropriately located in Chapter 9 and referred to in the respective conduit and tubing articles.

TABLE 2 *Radius of Conduit and Tubing Bends*

Conduit or Tubing Size		One Shot and Full Shoe Benders		Other Bends	
Metric Designator	Trade Size	mm	in.	mm	in.
16	½	101.6	4	101.6	4
21	¾	114.3	4½	127	5
27	1	146.05	5¾	152.4	6
35	1¼	184.15	7¼	203.2	8
41	1½	209.55	8¼	254	10
53	2	241.3	9½	304.8	12
63	2½	266.7	10½	381	15
78	3	330.2	13	457.2	18
91	3½	381	15	533.4	21
103	4	406.4	16	609.6	24
129	5	609.6	24	762	30
155	6	762	30	914.4	36

TABLE 4 *Dimensions and Percent Area of Conduit and Tubing (Areas of Conduit or Tubing for the Combinations of Wires Permitted in Table 1, Chapter 9)*

Article 358 — Electrical Metallic Tubing (EMT)

Metric Designator	Trade Size	Over 2 Wires 40%		60%		1 Wire 53%		2 Wires 31%		Nominal Internal Diameter		Total Area 100%	
		mm²	in.²	mm²	in.²	mm²	in.²	mm²	in.²	mm	in.	mm²	in.²
16	½	78	0.122	118	0.182	104	0.161	61	0.094	15.8	0.622	196	0.304
21	¾	137	0.213	206	0.320	182	0.283	106	0.165	20.9	0.824	343	0.533
27	1	222	0.346	333	0.519	295	0.458	172	0.268	26.6	1.049	556	0.864
35	1¼	387	0.598	581	0.897	513	0.793	300	0.464	35.1	1.380	968	1.496
41	1½	526	0.814	788	1.221	696	1.079	407	0.631	40.9	1.610	1314	2.036

Δ **TABLE 4** *Continued*

Article 358 — Electrical Metallic Tubing (EMT)

Metric Designator	Trade Size	Over 2 Wires 40%		60%		1 Wire 53%		2 Wires 31%		Nominal Internal Diameter		Total Area 100%	
		mm²	in.²	mm²	in.²	mm²	in.²	mm²	in.²	mm	in.	mm²	in.²
53	2	866	1.342	1299	2.013	1147	1.778	671	1.040	52.5	2.067	2165	3.356
63	2½	1513	2.343	2270	3.515	2005	3.105	1173	1.816	69.4	2.731	3783	5.858
78	3	2280	3.538	3421	5.307	3022	4.688	1767	2.742	85.2	3.356	5701	8.846
91	3½	2980	4.618	4471	6.927	3949	6.119	2310	3.579	97.4	3.834	7451	11.545
103	4	3808	5.901	5712	8.852	5046	7.819	2951	4.573	110.1	4.334	9521	14.753
129	5	5220	8.085	7830	12.127	6916	10.713	4045	6.266	128.9	5.073	13050	20.212
155	6	7528	11.663	11292	17.495	9975	15.454	5834	9.039	154.8	6.093	18821	29.158

Article 362 — Electrical Nonmetallic Tubing (ENT)

Metric Designator	Trade Size	Over 2 Wires 40%		60%		1 Wire 53%		2 Wires 31%		Nominal Internal Diameter		Total Area 100%	
		mm²	in.²	mm²	in.²	mm²	in.²	mm²	in.²	mm	in.	mm²	in.²
16	½	73	0.114	110	0.171	97	0.151	57	0.088	15.3	0.602	184	0.285
21	¾	131	0.203	197	0.305	174	0.269	102	0.157	20.4	0.804	328	0.508
27	1	215	0.333	322	0.499	284	0.441	166	0.258	26.1	1.029	537	0.832
35	1¼	375	0.581	562	0.872	497	0.770	291	0.450	34.5	1.36	937	1.453
41	1½	512	0.794	769	1.191	679	1.052	397	0.616	40.4	1.59	1281	1.986
53	2	849	1.316	1274	1.975	1125	1.744	658	1.020	52	2.047	2123	3.291
63	2½	—	—	—	—	—	—	—	—	—	—	—	—
78	3	—	—	—	—	—	—	—	—	—	—	—	—
91	3½	—	—	—	—	—	—	—	—	—	—	—	—

Article 348 — Flexible Metal Conduit (FMC)

Metric Designator	Trade Size	Over 2 Wires 40%		60%		1 Wire 53%		2 Wires 31%		Nominal Internal Diameter		Total Area 100%	
		mm²	in.²	mm²	in.²	mm²	in.²	mm²	in.²	mm	in.	mm²	in.²
12	⅜	30	0.046	44	0.069	39	0.061	23	0.036	9.7	0.384	74	0.116
16	½	81	0.127	122	0.190	108	0.168	63	0.098	16.1	0.635	204	0.317
21	¾	137	0.213	206	0.320	182	0.283	106	0.165	20.9	0.824	343	0.533
27	1	211	0.327	316	0.490	279	0.433	163	0.253	25.9	1.020	527	0.817
35	1¼	330	0.511	495	0.766	437	0.677	256	0.396	32.4	1.275	824	1.277
41	1½	480	0.743	720	1.115	636	0.985	372	0.576	39.1	1.538	1201	1.858
53	2	843	1.307	1264	1.961	1117	1.732	653	1.013	51.8	2.040	2107	3.269
63	2½	1267	1.963	1900	2.945	1678	2.602	982	1.522	63.5	2.500	3167	4.909
78	3	1824	2.827	2736	4.241	2417	3.746	1414	2.191	76.2	3.000	4560	7.069
91	3½	2483	3.848	3724	5.773	3290	5.099	1924	2.983	88.9	3.500	6207	9.621
103	4	3243	5.027	4864	7.540	4297	6.660	2513	3.896	101.6	4.000	8107	12.566

(continues)

Δ **TABLE 4** *Continued*

Δ

Article 342 — Intermediate Metal Conduit (IMC)

Metric Designator	Trade Size	Over 2 Wires 40%		60%		1 Wire 53%		2 Wires 31%		Nominal Internal Diameter		Total Area 100%	
		mm²	in.²	mm²	in.²	mm²	in.²	mm²	in.²	mm	in.	mm²	in.²
12	⅜	—	—	—	—	—	—	—	—	—	—	—	—
16	½	89	0.137	133	0.205	117	0.181	69	0.106	16.8	0.660	222	0.342
21	¾	151	0.235	226	0.352	200	0.311	117	0.182	21.9	0.864	377	0.586
27	1	248	0.384	372	0.575	329	0.508	192	0.297	28.1	1.105	620	0.959
35	1¼	425	0.659	638	0.988	564	0.873	330	0.510	36.8	1.448	1064	1.647
41	1½	573	0.890	859	1.335	759	1.179	444	0.690	42.7	1.683	1432	2.225
53	2	937	1.452	1405	2.178	1241	1.924	726	1.125	54.6	2.150	2341	3.630
63	2½	1323	2.054	1985	3.081	1753	2.722	1026	1.592	64.9	2.557	3308	5.135
78	3	2046	3.169	3069	4.753	2711	4.199	1586	2.456	80.7	3.176	5115	7.922
91	3½	2729	4.234	4093	6.351	3616	5.610	2115	3.281	93.2	3.671	6822	10.584
103	4	3490	5.452	5235	8.179	4624	7.224	2705	4.226	105.4	4.166	8725	13.631
129	5	5455	8.528	8183	12.792	7229	11.30	4228	6.610	131.78	5.210	13639	21.32
155	6	7878	12.304	11817	18.456	10439	16.302	6106	9.536	158.36	6.258	19696	30.76

Article 356 — Liquidtight Flexible Nonmetallic Conduit (LFNC-A*)

Metric Designator	Trade Size	Over 2 Wires 40%		60%		1 Wire 53%		2 Wires 31%		Nominal Internal Diameter		Total Area 100%	
		mm²	in.²	mm²	in.²	mm²	in.²	mm²	in.²	mm	in.	mm²	in.²
12	⅜	50	0.077	75	0.115	66	0.102	39	0.060	12.6	0.495	125	0.192
16	½	80	0.125	121	0.187	107	0.165	62	0.097	16.0	0.630	201	0.312
21	¾	139	0.214	208	0.321	184	0.283	107	0.166	21.0	0.825	346	0.535
27	1	221	0.342	331	0.513	292	0.453	171	0.265	26.5	1.043	552	0.854
35	1¼	387	0.601	581	0.901	513	0.796	300	0.466	35.1	1.383	968	1.502
41	1½	520	0.807	781	1.211	690	1.070	403	0.626	40.7	1.603	1301	2.018
53	2	863	1.337	1294	2.006	1143	1.772	669	1.036	52.4	2.063	2157	3.343

*Corresponds to 356.2(1).

Article 356 — Liquidtight Flexible Nonmetallic Conduit (LFNC-B*)

Metric Designator	Trade Size	Over 2 Wires 40%		60%		1 Wire 53%		2 Wires 31%		Nominal Internal Diameter		Total Area 100%	
		mm²	in.²	mm²	in.²	mm²	in.²	mm²	in.²	mm	in.	mm²	in.²
12	⅜	49	0.077	74	0.115	65	0.102	38	0.059	12.5	0.494	123	0.192
16	½	81	0.125	122	0.188	108	0.166	63	0.097	16.1	0.632	204	0.314
21	¾	140	0.216	210	0.325	185	0.287	108	0.168	21.1	0.830	350	0.541
27	1	226	0.349	338	0.524	299	0.462	175	0.270	26.8	1.054	564	0.873
35	1¼	394	0.611	591	0.917	522	0.810	305	0.474	35.4	1.395	984	1.528
41	1½	510	0.792	765	1.188	676	1.050	395	0.614	40.3	1.588	1276	1.981
53	2	836	1.298	1255	1.948	1108	1.720	648	1.006	51.6	2.033	2091	3.246

*Corresponds to 356.2(2).

Δ **TABLE 4** *Continued*

Article 356 — Liquidtight Flexible Nonmetallic Conduit (LFNC-C*)

Metric Designator	Trade Size	Over 2 Wires 40%		60%		1 Wire 53%		2 Wires 31%		Nominal Internal Diameter		Total Area 100%	
		mm²	in.²	mm²	in.²	mm²	in.²	mm²	in.²	mm	in.	mm²	in.²
12	⅜	47.7	0.074	71.5	0.111	63.2	0.098	36.9	0.057	12.3	0.485	119.19	0.185
16	½	77.9	0.121	116.9	0.181	103.2	0.160	60.4	0.094	15.7	0.620	194.778	0.302
21	¾	134.6	0.209	201.9	0.313	178.4	0.276	104.3	0.162	20.7	0.815	336.568	0.522
27	1	215.0	0.333	322.5	0.500	284.9	0.442	166.6	0.258	26.2	1.030	537.566	0.833
35	1¼	380.4	0.590	570.6	0.884	504.1	0.781	294.8	0.457	34.8	1.370	951.039	1.474
41	1½	509.2	0.789	763.8	1.184	674.7	1.046	394.6	0.612	40.3	1.585	1272.963	1.973
53	2	847.6	1.314	1271.4	1.971	1123.1	1.741	656.9	1.018	51.9	2.045	2119.063	3.285

*Corresponds to 356.2(3).

Article 350 — Liquidtight Flexible Metal Conduit (LFMC)

Metric Designator	Trade Size	Over 2 Wires 40%		60%		1 Wire 53%		2 Wires 31%		Nominal Internal Diameter		Total Area 100%	
		mm²	in.²	mm²	in.²	mm²	in.²	mm²	in.²	mm	in.	mm²	in.²
12	⅜	49	0.077	74	0.115	65	0.102	38	0.059	12.5	0.494	123	0.192
16	½	81	0.125	122	0.188	108	0.166	63	0.097	16.1	0.632	204	0.314
21	¾	140	0.216	210	0.325	185	0.287	108	0.168	21.1	0.830	350	0.541
27	1	226	0.349	338	0.524	299	0.462	175	0.270	26.8	1.054	564	0.873
35	1¼	394	0.611	591	0.917	522	0.810	305	0.474	35.4	1.395	984	1.528
41	1½	510	0.792	765	1.188	676	1.050	395	0.614	40.3	1.588	1276	1.981
53	2	836	1.298	1255	1.948	1108	1.720	648	1.006	51.6	2.033	2091	3.246
63	2½	1259	1.953	1888	2.929	1668	2.587	976	1.513	63.3	2.493	3147	4.881
78	3	1931	2.990	2896	4.485	2559	3.962	1497	2.317	78.4	3.085	4827	7.475
91	3½	2511	3.893	3766	5.839	3327	5.158	1946	3.017	89.4	3.520	6277	9.731
103	4	3275	5.077	4912	7.615	4339	6.727	2538	3.935	102.1	4.020	8187	12.692
129	5	—	—	—	—	—	—	—	—	—	—	—	—
155	6	—	—	—	—	—	—	—	—	—	—	—	—

Article 344 — Rigid Metal Conduit (RMC)

Metric Designator	Trade Size	Over 2 Wires 40%		60%		1 Wire 53%		2 Wires 31%		Nominal Internal Diameter		Total Area 100%	
		mm²	in.²	mm²	in.²	mm²	in.²	mm²	in.²	mm	in.	mm²	in.²
12	⅜	—	—	—	—	—	—	—	—	—	—	—	—
16	½	81	0.125	122	0.188	108	0.166	63	0.097	16.1	0.632	204	0.314
21	¾	141	0.220	212	0.329	187	0.291	109	0.170	21.2	0.836	353	0.549
27	1	229	0.355	344	0.532	303	0.470	177	0.275	27.0	1.063	573	0.887
35	1¼	394	0.610	591	0.916	522	0.809	305	0.473	35.4	1.394	984	1.526
41	1½	533	0.829	800	1.243	707	1.098	413	0.642	41.2	1.624	1333	2.071
53	2	879	1.363	1319	2.045	1165	1.806	681	1.056	52.9	2.083	2198	3.408
63	2½	1255	1.946	1882	2.919	1663	2.579	972	1.508	63.2	2.489	3137	4.866

(continues)

Δ **TABLE 4** *Continued*

Article 344 — Rigid Metal Conduit (RMC)

Metric Designator	Trade Size	Over 2 Wires 40%		60%		1 Wire 53%		2 Wires 31%		Nominal Internal Diameter		Total Area 100%	
		mm²	in.²	mm²	in.²	mm²	in.²	mm²	in.²	mm	in.	mm²	in.²
78	3	1936	3.000	2904	4.499	2565	3.974	1500	2.325	78.5	3.090	4840	7.499
91	3½	2584	4.004	3877	6.006	3424	5.305	2003	3.103	90.7	3.570	6461	10.010
103	4	3326	5.153	4990	7.729	4408	6.828	2578	3.994	102.9	4.050	8316	12.882
129	5	5220	8.085	7830	12.127	6916	10.713	4045	6.266	128.9	5.073	13050	20.212
155	6	7528	11.663	11292	17.495	9975	15.454	5834	9.039	154.8	6.093	18821	29.158

Article 352 — Rigid PVC Conduit (PVC), Schedule 80

Metric Designator	Trade Size	Over 2 Wires 40%		60%		1 Wire 53%		2 Wires 31%		Nominal Internal Diameter		Total Area 100%	
		mm²	in.²	mm²	in.²	mm²	in.²	mm²	in.²	mm	in.	mm²	in.²
12	⅜	—	—	—	—	—	—	—	—	—	—	—	—
16	½	56	0.087	85	0.130	75	0.115	44	0.067	13.4	0.526	141	0.217
21	¾	105	0.164	158	0.246	139	0.217	82	0.127	18.3	0.722	263	0.409
27	1	178	0.275	267	0.413	236	0.365	138	0.213	23.8	0.936	445	0.688
35	1¼	320	0.495	480	0.742	424	0.656	248	0.383	31.9	1.255	799	1.237
41	1½	442	0.684	663	1.027	585	0.907	342	0.530	37.5	1.476	1104	1.711
53	2	742	1.150	1113	1.725	983	1.523	575	0.891	48.6	1.913	1855	2.874
63	2½	1064	1.647	1596	2.471	1410	2.183	825	1.277	58.2	2.290	2660	4.119
78	3	1660	2.577	2491	3.865	2200	3.414	1287	1.997	72.7	2.864	4151	6.442
91	3½	2243	3.475	3365	5.213	2972	4.605	1738	2.693	84.5	3.326	5608	8.688
103	4	2907	4.503	4361	6.755	3852	5.967	2253	3.490	96.2	3.786	7268	11.258
129	5	4607	7.142	6911	10.713	6105	9.463	3571	5.535	121.1	4.768	11518	17.855
155	6	6605	10.239	9908	15.359	8752	13.567	5119	7.935	145.0	5.709	16513	25.598

Articles 352 and 353 — Rigid PVC Conduit (PVC), Schedule 40, and HDPE Conduit (HDPE)

Metric Designator	Trade Size	Over 2 Wires 40%		60%		1 Wire 53%		2 Wires 31%		Nominal Internal Diameter		Total Area 100%	
		mm²	in.²	mm²	in.²	mm²	in.²	mm²	in.²	mm	in.	mm²	in.²
12	⅜	—	—	—	—	—	—	—	—	—	—	—	—
16	½	74	0.114	110	0.171	97	0.151	57	0.088	15.3	0.602	184	0.285
21	¾	131	0.203	196	0.305	173	0.269	101	0.157	20.4	0.804	327	0.508
27	1	214	0.333	321	0.499	284	0.441	166	0.258	26.1	1.029	535	0.832
35	1¼	374	0.581	561	0.872	495	0.770	290	0.450	34.5	1.360	935	1.453
41	1½	513	0.794	769	1.191	679	1.052	397	0.616	40.4	1.590	1282	1.986
53	2	849	1.316	1274	1.975	1126	1.744	658	1.020	52.0	2.047	2124	3.291
63	2½	1212	1.878	1817	2.817	1605	2.488	939	1.455	62.1	2.445	3029	4.695
78	3	1877	2.907	2816	4.361	2487	3.852	1455	2.253	77.3	3.042	4693	7.268
91	3½	2511	3.895	3766	5.842	3327	5.161	1946	3.018	89.4	3.521	6277	9.737
103	4	3237	5.022	4855	7.532	4288	6.654	2508	3.892	101.5	3.998	8091	12.554
129	5	5099	7.904	7649	11.856	6756	10.473	3952	6.126	127.4	5.016	12748	19.761
155	6	7373	11.427	11060	17.140	9770	15.141	5714	8.856	153.2	6.031	18433	28.567

Δ **TABLE 4** *Continued*

Article 352 — Type A, Rigid PVC Conduit (PVC)

Metric Designator	Trade Size	Over 2 Wires 40%		60%		1 Wire 53%		2 Wires 31%		Nominal Internal Diameter		Total Area 100%	
		mm²	in.²	mm²	in.²	mm²	in.²	mm²	in.²	mm	in.	mm²	in.²
16	½	100	0.154	149	0.231	132	0.204	77	0.119	17.8	0.700	249	0.385
21	¾	168	0.260	251	0.390	222	0.345	130	0.202	23.1	0.910	419	0.650
27	1	279	0.434	418	0.651	370	0.575	216	0.336	29.8	1.175	697	1.084
35	1¼	456	0.707	684	1.060	604	0.937	353	0.548	38.1	1.500	1140	1.767
41	1½	600	0.929	900	1.394	795	1.231	465	0.720	43.7	1.720	1500	2.324
53	2	940	1.459	1410	2.188	1245	1.933	728	1.131	54.7	2.155	2350	3.647
63	2½	1406	2.181	2109	3.272	1863	2.890	1090	1.690	66.9	2.635	3515	5.453
78	3	2112	3.278	3169	4.916	2799	4.343	1637	2.540	82.0	3.230	5281	8.194
91	3½	2758	4.278	4137	6.416	3655	5.668	2138	3.315	93.7	3.690	6896	10.694
103	4	3543	5.489	5315	8.234	4695	7.273	2746	4.254	106.2	4.180	8858	13.723
129	5	—	—	—	—	—	—	—	—	—	—	—	—
155	6	—	—	—	—	—	—	—	—	—	—	—	—

Article 352 — Type EB, Rigid PVC Conduit (PVC)

Metric Designator	Trade Size	Over 2 Wires 40%		60%		1 Wire 53%		2 Wires 31%		Nominal Internal Diameter		Total Area 100%	
		mm²	in.²	mm²	in.²	mm²	in.²	mm²	in.²	mm	in.	mm²	in.²
16	½	—	—	—	—	—	—	—	—	—	—	—	—
21	¾	—	—	—	—	—	—	—	—	—	—	—	—
27	1	—	—	—	—	—	—	—	—	—	—	—	—
35	1¼	—	—	—	—	—	—	—	—	—	—	—	—
41	1½	—	—	—	—	—	—	—	—	—	—	—	—
53	2	999	1.550	1499	2.325	1324	2.053	774	1.201	56.4	2.221	2498	3.874
63	2½	—	—	—	—	—	—	—	—	—	—	—	—
78	3	2248	3.484	3373	5.226	2979	4.616	1743	2.700	84.6	3.330	5621	8.709
91	3½	2932	4.546	4397	6.819	3884	6.023	2272	3.523	96.6	3.804	7329	11.365
103	4	3726	5.779	5589	8.669	4937	7.657	2887	4.479	108.9	4.289	9314	14.448
129	5	5726	8.878	8588	13.317	7586	11.763	4437	6.881	135.0	5.316	14314	22.195
155	6	8133	12.612	12200	18.918	10776	16.711	6303	9.774	160.9	6.336	20333	31.530

TABLE 5 *Dimensions of Insulated Conductors and Fixture Wires*

Type	Size (AWG or kcmil)	Approximate Area		Approximate Diameter	
		mm²	in.²	mm	in.
Type: FFH-2, RFH-1, RFH-2, RFHH-2, RHH*, RHW*, RHW-2*, RHH, RHW, RHW-2, SF-1, SF-2, SFF-1, SFF-2, TF, TFF, THHW, THW, THW-2, TW, XF, XFF					
RFH-2, FFH-2, RFHH-2	18	9.355	0.0145	3.454	0.136
	16	11.10	0.0172	3.759	0.148
RHH, RHW, RHW-2	14	18.90	0.0293	4.902	0.193
	12	22.77	0.0353	5.385	0.212
	10	28.19	0.0437	5.994	0.236
	8	53.87	0.0835	8.280	0.326
	6	67.16	0.1041	9.246	0.364
	4	86.00	0.1333	10.46	0.412
	3	98.13	0.1521	11.18	0.440
	2	112.9	0.1750	11.99	0.472
	1	171.6	0.2660	14.78	0.582
	1/0	196.1	0.3039	15.80	0.622
	2/0	226.1	0.3505	16.97	0.668
	3/0	262.7	0.4072	18.29	0.720
	4/0	306.7	0.4754	19.76	0.778
	250	405.9	0.6291	22.73	0.895
	300	457.3	0.7088	24.13	0.950
	350	507.7	0.7870	25.43	1.001
	400	556.5	0.8626	26.62	1.048
	500	650.5	1.0082	28.78	1.133
	600	782.9	1.2135	31.57	1.243
	700	874.9	1.3561	33.38	1.314
	750	920.8	1.4272	34.24	1.348
	800	965.0	1.4957	35.05	1.380
	900	1057	1.6377	36.68	1.444
	1000	1143	1.7719	38.15	1.502
	1250	1515	2.3479	43.92	1.729
	1500	1738	2.6938	47.04	1.852
	1750	1959	3.0357	49.94	1.966
	2000	2175	3.3719	52.63	2.072
SF-2, SFF-2	18	7.419	0.0115	3.073	0.121
	16	8.968	0.0139	3.378	0.133
	14	11.10	0.0172	3.759	0.148
SF-1, SFF-1	18	4.194	0.0065	2.311	0.091
RFH-1, TF, TFF, XF, XFF	18	5.161	0.0088	2.692	0.106
TF, TFF, XF, XFF	16	7.032	0.0109	2.997	0.118
TW, XF, XFF, THHW, THW, THW-2	14	8.968	0.0139	3.378	0.133
TW, THHW, THW, THW-2	12	11.68	0.0181	3.861	0.152
	10	15.68	0.0243	4.470	0.176
	8	28.19	0.0437	5.994	0.236
RHH*, RHW*, RHW-2*	14	13.48	0.0209	4.140	0.163
RHH*, RHW*, RHW-2*, XF, XFF	12	16.77	0.0260	4.623	0.182

TABLE 5 *Continued*

Type	Size (AWG or kcmil)	Approximate Area		Approximate Diameter	
		mm²	in.²	mm	in.
Type: RHH*, RHW*, RHW-2*, THHN, THHW, THW, THW-2, TFN, TFFN, THWN, THWN-2, XF, XFF					
RHH,* RHW,* RHW-2,* XF, XFF	10	21.48	0.0333	5.232	0.206
RHH*, RHW*, RHW-2*	8	35.87	0.0556	6.756	0.266
TW, THW, THHW, THW-2, RHH*, RHW*, RHW-2*	6	46.84	0.0726	7.722	0.304
	4	62.77	0.0973	8.941	0.352
	3	73.16	0.1134	9.652	0.380
	2	86.00	0.1333	10.46	0.412
	1	122.6	0.1901	12.50	0.492
	1/0	143.4	0.2223	13.51	0.532
	2/0	169.3	0.2624	14.68	0.578
	3/0	201.1	0.3117	16.00	0.630
	4/0	239.9	0.3718	17.48	0.688
	250	296.5	0.4596	19.43	0.765
	300	340.7	0.5281	20.83	0.820
	350	384.4	0.5958	22.12	0.871
	400	427.0	0.6619	23.32	0.918
	500	509.7	0.7901	25.48	1.003
	600	627.7	0.9729	28.27	1.113
	700	710.3	1.1010	30.07	1.184
	750	751.7	1.1652	30.94	1.218
	800	791.7	1.2272	31.75	1.250
	900	874.9	1.3561	33.38	1.314
	1000	953.8	1.4784	34.85	1.372
	1250	1200	1.8602	39.09	1.539
	1500	1400	2.1695	42.21	1.662
	1750	1598	2.4773	45.11	1.776
	2000	1795	2.7818	47.80	1.882
TFN, TFFN	18	3.548	0.0055	2.134	0.084
	16	4.645	0.0072	2.438	0.096
THHN, THWN, THWN-2	14	6.258	0.0097	2.819	0.111
	12	8.581	0.0133	3.302	0.130
	10	13.61	0.0211	4.166	0.164
	8	23.61	0.0366	5.486	0.216
	6	32.71	0.0507	6.452	0.254
	4	53.16	0.0824	8.230	0.324
	3	62.77	0.0973	8.941	0.352
	2	74.71	0.1158	9.754	0.384
	1	100.8	0.1562	11.33	0.446
	1/0	119.7	0.1855	12.34	0.486
	2/0	143.4	0.2223	13.51	0.532
	3/0	172.8	0.2679	14.83	0.584
	4/0	208.8	0.3237	16.31	0.642
	250	256.1	0.3970	18.06	0.711
	300	297.3	0.4608	19.46	0.766

(continues)

TABLE 5 *Continued*

Type	Size (AWG or kcmil)	Approximate Area		Approximate Diameter	
		mm²	in.²	mm	in.
Type: FEP, FEPB, PAF, PAFF, PF, PFA, PFAH, PFF, PGF, PGFF, PTF, PTFF, TFE, THHN, THWN, THWN-2, Z, ZF, ZFF, ZHF					
THHN, THWN, THWN-2	350	338.2	0.5242	20.75	0.817
	400	378.3	0.5863	21.95	0.864
	500	456.3	0.7073	24.10	0.949
	600	559.7	0.8676	26.70	1.051
	700	637.9	0.9887	28.50	1.122
	750	677.2	1.0496	29.36	1.156
	800	715.2	1.1085	30.18	1.188
	900	794.3	1.2311	31.80	1.252
	1000	869.5	1.3478	33.27	1.310
PF, PGFF, PGF, PFF, PTF, PAF, PTFF, PAFF	18	3.742	0.0058	2.184	0.086
	16	4.839	0.0075	2.489	0.098
PF, PGFF, PGF, PFF, PTF, PAF, PTFF, PAFF, TFE, FEP, PFA, FEPB, PFAH	14	6.452	0.0100	2.870	0.113
TFE, FEP, PFA, FEPB, PFAH	12	8.839	0.0137	3.353	0.132
	10	12.32	0.0191	3.962	0.156
	8	21.48	0.0333	5.232	0.206
	6	30.19	0.0468	6.198	0.244
	4	43.23	0.0670	7.417	0.292
	3	51.87	0.0804	8.128	0.320
	2	62.77	0.0973	8.941	0.352
TFE, PFAH, PFA	1	90.26	0.1399	10.72	0.422
TFE, PFA, PFAH, Z	1/0	108.1	0.1676	11.73	0.462
	2/0	130.8	0.2027	12.90	0.508
	3/0	158.9	0.2463	14.22	0.560
	4/0	193.5	0.3000	15.70	0.618
ZF, ZFF, ZHF	18	2.903	0.0045	1.930	0.076
	16	3.935	0.0061	2.235	0.088
Z, ZF, ZFF, ZHF	14	5.355	0.0083	2.616	0.103
Z	12	7.548	0.0117	3.099	0.122
	10	12.32	0.0191	3.962	0.156
	8	19.48	0.0302	4.978	0.196
	6	27.74	0.0430	5.944	0.234
	4	40.32	0.0625	7.163	0.282
	3	55.16	0.0855	8.382	0.330
	2	66.39	0.1029	9.195	0.362
	1	81.87	0.1269	10.21	0.402
Type: KF-1, KF-2, KFF-1, KFF-2, XHH, XHHW, XHHW-2, ZW					
XHHW, ZW, XHHW-2, XHH	14	8.968	0.0139	3.378	0.133
	12	11.68	0.0181	3.861	0.152
	10	15.68	0.0243	4.470	0.176
	8	28.19	0.0437	5.994	0.236
	6	38.06	0.0590	6.960	0.274
	4	52.52	0.0814	8.179	0.322
	3	62.06	0.0962	8.890	0.350
	2	73.94	0.1146	9.703	0.382

TABLE 5 *Continued*

Type	Size (AWG or kcmil)	Approximate Area		Approximate Diameter	
		mm²	in.²	mm	in.
Type: KF-1, KF-2, KFF-1, KFF-2, XHH, XHHW, XHHW-2, ZW					
XHHW, XHHW-2, XHH	1	98.97	0.1534	11.23	0.442
	1/0	117.7	0.1825	12.24	0.482
	2/0	141.3	0.2190	13.41	0.528
	3/0	170.5	0.2642	14.73	0.58
	4/0	206.3	0.3197	16.21	0.638
	250	251.9	0.3904	17.91	0.705
	300	292.6	0.4536	19.30	0.76
	350	333.3	0.5166	20.60	0.811
	400	373.0	0.5782	21.79	0.858
	500	450.6	0.6984	23.95	0.943
	600	561.9	0.8709	26.75	1.053
	700	640.2	0.9923	28.55	1.124
	750	679.5	1.0532	29.41	1.158
	800	717.5	1.1122	30.23	1.190
	900	796.8	1.2351	31.85	1.254
	1000	872.2	1.3519	33.32	1.312
	1250	1108	1.7180	37.57	1.479
	1500	1300	2.0156	40.69	1.602
	1750	1492	2.3127	43.59	1.716
	2000	1682	2.6073	46.28	1.822
KF-2, KFF-2	18	2.000	0.003	1.575	0.062
	16	2.839	0.0043	1.88	0.074
	14	4.129	0.0064	2.286	0.090
	12	6.000	0.0092	2.743	0.108
	10	8.968	0.0139	3.378	0.133
KF-1, KFF-1	18	1.677	0.0026	1.448	0.057
	16	2.387	0.0037	1.753	0.069
	14	3.548	0.0055	2.134	0.084
	12	5.355	0.0083	2.616	0.103
	10	8.194	0.0127	3.226	0.127

*Types RHH, RHW, and RHW-2 without outer covering.

TABLE 5A *Compact Copper and Aluminum Building Wire Nominal Dimensions** and Areas*

Size (AWG or kcmil)	Bare Conductor Diameter		Type RHH* Approximate Diameter		Type RHH* Approximate Area		Types THW and THHW Approximate Diameter		Types THW and THHW Approximate Area		Type THHN Approximate Diameter		Type THHN Approximate Area		Type XHHW Approximate Diameter		Type XHHW Approximate Area		Size (AWG or kcmil)
	mm	in.	mm	in.	mm²	in.²	mm	in.	mm²	in.²	mm	in.	mm²	in.²	mm	in.	mm²	in.²	
8	3.404	0.134	6.604	0.260	34.25	0.0531	6.477	0.255	32.90	0.0510	—	—	—	—	5.690	0.224	25.42	0.0394	8
6	4.293	0.169	7.493	0.295	44.10	0.0683	7.366	0.290	42.58	0.0660	6.096	0.240	29.16	0.0452	6.604	0.260	34.19	0.0530	6
4	5.410	0.213	8.509	0.335	56.84	0.0881	8.509	0.335	56.84	0.0881	7.747	0.305	47.10	0.0730	7.747	0.305	47.10	0.0730	4
2	6.807	0.268	9.906	0.390	77.03	0.1194	9.906	0.390	77.03	0.1194	9.144	0.360	65.61	0.1017	9.144	0.360	65.61	0.1017	2
1	7.595	0.299	11.81	0.465	109.5	0.1698	11.81	0.465	109.5	0.1698	10.54	0.415	87.23	0.1352	10.54	0.415	87.23	0.1352	1
1/0	8.534	0.336	12.70	0.500	126.6	0.1963	12.70	0.500	126.6	0.1963	11.43	0.450	102.6	0.1590	11.43	0.450	102.6	0.1590	1/0
2/0	9.550	0.376	13.72	0.540	147.8	0.2290	13.84	0.545	150.5	0.2332	12.57	0.495	124.1	0.1924	12.45	0.490	121.6	0.1885	2/0
3/0	10.74	0.423	14.99	0.590	176.3	0.2733	14.99	0.590	176.3	0.2733	13.72	0.540	147.7	0.2290	13.72	0.540	147.7	0.2290	3/0
4/0	12.07	0.475	16.26	0.640	207.6	0.3217	16.38	0.645	210.8	0.3267	15.11	0.595	179.4	0.2780	14.99	0.590	176.3	0.2733	4/0
250	13.21	0.520	18.16	0.715	259.0	0.4015	18.42	0.725	266.3	0.4128	17.02	0.670	227.4	0.3525	16.76	0.660	220.7	0.3421	250
300	14.48	0.570	19.43	0.765	296.5	0.4596	19.69	0.775	304.3	0.4717	18.29	0.720	262.6	0.4071	18.16	0.715	259.0	0.4015	300
350	15.65	0.616	20.57	0.810	332.3	0.5153	20.83	0.820	340.7	0.5281	19.56	0.770	300.4	0.4656	19.30	0.760	292.6	0.4536	350
400	16.74	0.659	21.72	0.855	370.5	0.5741	21.97	0.865	379.1	0.5876	20.70	0.815	336.5	0.5216	20.32	0.800	324.3	0.5026	400
500	18.69	0.736	23.62	0.930	438.2	0.6793	23.88	0.940	447.7	0.6939	22.48	0.885	396.8	0.6151	22.35	0.880	392.4	0.6082	500
600	20.65	0.813	26.29	1.035	542.8	0.8413	26.67	1.050	558.6	0.8659	25.02	0.985	491.6	0.7620	24.89	0.980	486.6	0.7542	600
700	22.28	0.877	27.94	1.100	613.1	0.9503	28.19	1.110	624.3	0.9676	26.67	1.050	558.6	0.8659	26.67	1.050	558.6	0.8659	700
750	23.06	0.908	28.83	1.135	652.8	1.0118	29.21	1.150	670.1	1.0386	27.31	1.075	585.5	0.9076	27.69	1.090	602.0	0.9331	750
900	25.37	0.999	31.50	1.240	779.3	1.2076	31.09	1.224	759.1	1.1766	30.33	1.194	722.5	1.1196	29.69	1.169	692.3	1.0733	900
1000	26.92	1.060	32.64	1.285	836.6	1.2968	32.64	1.285	836.6	1.2968	31.88	1.255	798.1	1.2370	31.24	1.230	766.6	1.1882	1000

*Types RHH and RHW without outer coverings.

**Dimensions are from industry sources.

Most aluminum building wire in Types THW, THHW, THWN/THHN, and XHHW conductors is compact stranded. Table 5A provides appropriate dimensions for those types of wire.

Δ **TABLE 8** *Conductor Properties*

Size (AWG or kcmil)	Area mm²	Area Circular mils	Stranding Quantity	Stranding Diameter mm	Stranding Diameter in.	Overall Diameter mm	Overall Diameter in.	Overall Area mm²	Overall Area in.²	Uncoated ohm/km	Uncoated ohm/kFT	Coated ohm/km	Coated ohm/kFT	Aluminum ohm/km	Aluminum ohm/kFT
18	0.823	1620	1	—	—	1.02	0.040	0.823	0.001	25.5	7.77	26.5	8.08	42.0	12.8
18	0.823	1620	7	0.39	0.015	1.16	0.046	1.06	0.002	26.1	7.95	27.7	8.45	42.8	13.1
16	1.31	2580	1	—	—	1.29	0.051	1.31	0.002	16.0	4.89	16.7	5.08	26.4	8.05
16	1.31	2580	7	0.49	0.019	1.46	0.058	1.68	0.003	16.4	4.99	17.3	5.29	26.9	8.21
14	2.08	4110	1	—	—	1.63	0.064	2.08	0.003	10.1	3.07	10.4	3.19	16.6	5.06
14	2.08	4110	7	0.62	0.024	1.85	0.073	2.68	0.004	10.3	3.14	10.7	3.26	16.9	5.17
12	3.31	6530	1	—	—	2.05	0.081	3.31	0.005	6.34	1.93	6.57	2.01	10.45	3.18
12	3.31	6530	7	0.78	0.030	2.32	0.092	4.25	0.006	6.50	1.98	6.73	2.05	10.69	3.25
10	5.261	10380	1	—	—	2.588	0.102	5.26	0.008	3.984	1.21	4.148	1.26	6.561	2.00
10	5.261	10380	7	0.98	0.038	2.95	0.116	6.76	0.011	4.070	1.24	4.226	1.29	6.679	2.04
8	8.367	16510	1	—	—	3.264	0.128	8.37	0.013	2.506	0.764	2.579	0.786	4.125	1.26
8	8.367	16510	7	1.23	0.049	3.71	0.146	10.76	0.017	2.551	0.778	2.653	0.809	4.204	1.28
6	13.30	26240	7	1.56	0.061	4.67	0.184	17.09	0.027	1.608	0.491	1.671	0.510	2.652	0.808
4	21.15	41740	7	1.96	0.077	5.89	0.232	27.19	0.042	1.010	0.308	1.053	0.321	1.666	0.508
3	26.67	52620	7	2.20	0.087	6.60	0.260	34.28	0.053	0.802	0.245	0.833	0.254	1.320	0.403
2	33.62	66360	7	2.47	0.097	7.42	0.292	43.23	0.067	0.634	0.194	0.661	0.201	1.045	0.319
1	42.41	83690	19	1.69	0.066	8.43	0.332	55.80	0.087	0.505	0.154	0.524	0.160	0.829	0.253
1/0	53.49	105600	19	1.89	0.074	9.45	0.372	70.41	0.109	0.399	0.122	0.415	0.127	0.660	0.201
2/0	67.43	133100	19	2.13	0.084	10.62	0.418	88.74	0.137	0.3170	0.0967	0.329	0.101	0.523	0.159
3/0	85.01	167800	19	2.39	0.094	11.94	0.470	111.9	0.173	0.2512	0.0766	0.2610	0.0797	0.413	0.126
4/0	107.2	211600	19	2.68	0.106	13.41	0.528	141.1	0.219	0.1996	0.0608	0.2050	0.0626	0.328	0.100
250	127	—	37	2.09	0.082	14.61	0.575	168	0.260	0.1687	0.0515	0.1753	0.0535	0.2778	0.0847
300	152	—	37	2.29	0.090	16.00	0.630	201	0.312	0.1409	0.0429	0.1463	0.0446	0.2318	0.0707
350	177	—	37	2.47	0.097	17.30	0.681	235	0.364	0.1205	0.0367	0.1252	0.0382	0.1984	0.0605
400	203	—	37	2.64	0.104	18.49	0.728	268	0.416	0.1053	0.0321	0.1084	0.0331	0.1737	0.0529
500	253	—	37	2.95	0.116	20.65	0.813	336	0.519	0.0845	0.0258	0.0869	0.0265	0.1391	0.0424
600	304	—	61	2.52	0.099	22.68	0.893	404	0.626	0.0704	0.0214	0.0732	0.0223	0.1159	0.0353
700	355	—	61	2.72	0.107	24.49	0.964	471	0.730	0.0603	0.0184	0.0622	0.0189	0.0994	0.0303
750	380	—	61	2.82	0.111	25.35	0.998	505	0.782	0.0563	0.0171	0.0579	0.0176	0.0927	0.0282
800	405	—	61	2.91	0.114	26.16	1.030	538	0.834	0.0528	0.0161	0.0544	0.0166	0.0868	0.0265
900	456	—	61	3.09	0.122	27.79	1.094	606	0.940	0.0470	0.0143	0.0481	0.0147	0.0770	0.0235
1000	507	—	61	3.25	0.128	29.26	1.152	673	1.042	0.0423	0.0129	0.0434	0.0132	0.0695	0.0212
1250	633	—	91	2.98	0.117	32.74	1.289	842	1.305	0.0338	0.0103	0.0347	0.0106	0.0554	0.0169
1500	760	—	91	3.26	0.128	35.86	1.412	1011	1.566	0.02814	0.00858	0.02814	0.00883	0.0464	0.0141
1750	887	—	127	2.98	0.117	38.76	1.526	1180	1.829	0.02410	0.00735	0.02410	0.00756	0.0397	0.0121
2000	1013	—	127	3.19	0.126	41.45	1.632	1349	2.092	0.02109	0.00643	0.02109	0.00662	0.0348	0.0106

Column group headers: **Conductors** — Stranding (Diameter), Overall (Diameter, Area); **Direct-Current Resistance at 75°C (167°F)** — Copper (Uncoated, Coated), Aluminum.

Notes:

1. These resistance values are valid **only** for the parameters as given. Using conductors having coated strands, different stranding type, and, especially, other temperatures changes the resistance.

2. Equation for temperature change: $R_2 = R_1 [1 + a (T_2 - 75)]$, where $\alpha_{cu} = 0.00323$, $\alpha_{AL} = 0.00330$ at 75°C.

3. Conductors with compact and compressed stranding have smaller bare conductor diameters than those shown. See Table 5A for actual compact cable dimensions.

4. The IACS conductivities used: bare copper = 100%, aluminum = 61%.

5. Class B stranding is listed as well as solid for some sizes. Its overall diameter and area are those of its circumscribing circle.

Informational Note: NEMA WC/70-2009, *Power Cables Rated 2000 Volts or Less for the Distribution of Electrical Energy*, or ANSI/UL 1581-2017, *Reference Standard for Electrical Wires, Cables, and Flexible Cords*, is the source for the construction information.

National Bureau of Standards Handbook 100, dated 1966, and *Handbook 109*, dated 1972, is the reference where the resistance is calculated.

In addition to traditional wire sizes expressed as American Wire Gage (AWG), circular mil (cmil) area, or thousands of circular mil (kcmil) area, wire is available with its cross-sectional area expressed in square millimeters (mm²).

The *NEC* requires that insulated conductors be marked with their sizes and that the sizes be expressed in either AWG or circular mil area. [See 110.6 and 310.8(A)(4).] The *NEC* allows no exceptions to either of those two requirements. Because Article 310 does not specifically prohibit optional marking on insulated conductors, the *NEC* permits square millimeter (mm²) markings on conductors, but only if they are in addition to the required traditional markings of AWG or circular mil area.

According to IEEE/ASTM SI 10-2016, *American National Standard for Metric Practice*, conversion from circular mils to square meters is done by multiplying circular mils by 5.067075 × 10⁻¹⁰. However, because square millimeters, rather than square meters, is the standard marking for wire size and because the reciprocal is more appropriate for this conversion, a simpler conversion factor to convert from square millimeters to circular mils (approximately) follows:

$$k = 1973.53 \ \frac{\text{circular mils}}{\text{mm}^2}$$

The following example provides a comparison of the square millimeter wire gauge to traditional wire sizes.

Calculation Example

What traditional wire size does the size 125 mm² represent (approximately)?

Solution

Circular mil area = wire size (mm²) × conversion factor

$$= 125 \text{ mm}^2 \times 1973.53 \ \frac{\text{circular mils}}{\text{mm}^2}$$

$$= 246{,}691 \text{ circular mils or } 246.691 \text{ kcmil}$$

Therefore, the 125-mm² wire is larger than 4/0 AWG (211.6 kcmil) but smaller than a 250-kcmil conductor.

If a 125-mm² wire is determined to be the minimum or recommended size conductor, it is important to understand that size 250 kcmil would be the only Table 8 conductor with equivalent cross-sectional area because 4/0 AWG is simply not enough metal. It is important, however, to note that the 125-mm² conductor ampacity could not be used for a 250-kcmil conductor, because the metric conductor size is smaller. The ampacity of a 4/0 AWG can be used, or the ampacity can be calculated under engineering supervision.

TABLE 9 *Alternating-Current Resistance and Reactance for 600-Volt Cables, 3-Phase, 60 Hz, 75°C (167°F) — Three Single Conductors in Conduit*

			Alternating-Current Resistance for Uncoated Copper Wires			Alternating-Current Resistance for Aluminum Wires			Effective Z at 0.85 *PF* for Uncoated Copper Wires			Effective Z at 0.85 *PF* for Aluminum Wires			
Size (AWG or kcmil)	X_L (Reactance) for All Wires		PVC Conduit	Aluminum Conduit	Steel Conduit	PVC Conduit	Aluminum Conduit	Steel Conduit	PVC Conduit	Aluminum Conduit	Steel Conduit	PVC Conduit	Aluminum Conduit	Steel Conduit	Size (AWG or kcmil)
	PVC, Aluminum Conduits	Steel Conduit													
14	0.190	0.240	10.2	10.2	10.2	—	—	—	8.9	8.9	8.9	—	—	—	14
	0.058	0.073	3.1	3.1	3.1	—	—	—	2.7	2.7	2.7	—	—	—	
12	0.177	0.223	6.6	6.6	6.6	10.5	10.5	10.5	5.6	5.6	5.6	9.2	9.2	9.2	12
	0.054	0.068	2.0	2.0	2.0	3.2	3.2	3.2	1.7	1.7	1.7	2.8	2.8	2.8	
10	0.164	0.207	3.9	3.9	3.9	6.6	6.6	6.6	3.6	3.6	3.6	5.9	5.9	5.9	10
	0.050	0.063	1.2	1.2	1.2	2.0	2.0	2.0	1.1	1.1	1.1	1.8	1.8	1.8	
8	0.171	0.213	2.56	2.56	2.56	4.3	4.3	4.3	2.26	2.26	2.30	3.6	3.6	3.6	8
	0.052	0.065	0.78	0.78	0.78	1.3	1.3	1.3	0.69	0.69	0.70	1.1	1.1	1.1	
6	0.167	0.210	1.61	1.61	1.61	2.66	2.66	2.66	1.44	1.48	1.48	2.33	2.36	2.36	6
	0.051	0.064	0.49	0.49	0.49	0.81	0.81	0.81	0.44	0.45	0.45	0.71	0.72	0.72	
4	0.157	0.197	1.02	1.02	1.02	1.67	1.67	1.67	0.95	0.95	0.98	1.51	1.51	1.51	4
	0.048	0.060	0.31	0.31	0.31	0.51	0.51	0.51	0.29	0.29	0.30	0.46	0.46	0.46	
3	0.154	0.194	0.82	0.82	0.82	1.31	1.35	1.31	0.75	0.79	0.79	1.21	1.21	1.21	3
	0.047	0.059	0.25	0.25	0.25	0.40	0.41	0.40	0.23	0.24	0.24	0.37	0.37	0.37	
2	0.148	0.187	0.62	0.66	0.66	1.05	1.05	1.05	0.62	0.62	0.66	0.98	0.98	0.98	2
	0.045	0.057	0.19	0.20	0.20	0.32	0.32	0.32	0.19	0.19	0.20	0.30	0.30	0.30	

TABLE 9 *Continued*

| | | | Ohms to Neutral per Kilometer | | | | | | | | | | | | | |
| | | | Ohms to Neutral per 1000 Feet | | | | | | | | | | | | | |

Size (AWG or kcmil)	X_L (Reactance) for All Wires		Alternating-Current Resistance for Uncoated Copper Wires			Alternating-Current Resistance for Aluminum Wires			Effective Z at 0.85 PF for Uncoated Copper Wires			Effective Z at 0.85 PF for Aluminum Wires			Size (AWG or kcmil)
	PVC, Aluminum Conduits	Steel Conduit	PVC Conduit	Aluminum Conduit	Steel Conduit	PVC Conduit	Aluminum Conduit	Steel Conduit	PVC Conduit	Aluminum Conduit	Steel Conduit	PVC Conduit	Aluminum Conduit	Steel Conduit	
1	0.151	0.187	0.49	0.52	0.52	0.82	0.85	0.82	0.52	0.52	0.52	0.79	0.79	0.82	1
	0.046	0.057	0.15	0.16	0.16	0.25	0.26	0.25	0.16	0.16	0.16	0.24	0.24	0.25	
1/0	0.144	0.180	0.39	0.43	0.39	0.66	0.69	0.66	0.43	0.43	0.43	0.62	0.66	0.66	1/0
	0.044	0.055	0.12	0.13	0.12	0.20	0.21	0.20	0.13	0.13	0.13	0.19	0.20	0.20	
2/0	0.141	0.177	0.33	0.33	0.33	0.52	0.52	0.52	0.36	0.36	0.36	0.52	0.52	0.52	2/0
	0.043	0.054	0.10	0.10	0.10	0.16	0.16	0.16	0.11	0.11	0.11	0.16	0.16	0.16	
3/0	0.138	0.171	0.253	0.269	0.259	0.43	0.43	0.43	0.289	0.302	0.308	0.43	0.43	0.46	3/0
	0.042	0.052	0.077	0.082	0.079	0.13	0.13	0.13	0.088	0.092	0.094	0.13	0.13	0.14	
4/0	0.135	0.167	0.203	0.220	0.207	0.33	0.36	0.33	0.243	0.256	0.262	0.36	0.36	0.36	4/0
	0.041	0.051	0.062	0.067	0.063	0.10	0.11	0.10	0.074	0.078	0.080	0.11	0.11	0.11	
250	0.135	0.171	0.171	0.187	0.177	0.279	0.295	0.282	0.217	0.230	0.240	0.308	0.322	0.33	250
	0.041	0.052	0.052	0.057	0.054	0.085	0.090	0.086	0.066	0.070	0.073	0.094	0.098	0.10	
300	0.135	0.167	0.144	0.161	0.148	0.233	0.249	0.236	0.194	0.207	0.213	0.269	0.282	0.289	300
	0.041	0.051	0.044	0.049	0.045	0.071	0.076	0.072	0.059	0.063	0.065	0.082	0.086	0.088	
350	0.131	0.164	0.125	0.141	0.128	0.200	0.217	0.207	0.174	0.190	0.197	0.240	0.253	0.262	350
	0.040	0.050	0.038	0.043	0.039	0.061	0.066	0.063	0.053	0.058	0.060	0.073	0.077	0.080	
400	0.131	0.161	0.108	0.125	0.115	0.177	0.194	0.180	0.161	0.174	0.184	0.217	0.233	0.240	400
	0.040	0.049	0.033	0.038	0.035	0.054	0.059	0.055	0.049	0.053	0.056	0.066	0.071	0.073	
500	0.128	0.157	0.089	0.105	0.095	0.141	0.157	0.148	0.141	0.157	0.164	0.187	0.200	0.210	500
	0.039	0.048	0.027	0.032	0.029	0.043	0.048	0.045	0.043	0.048	0.050	0.057	0.061	0.064	
600	0.128	0.157	0.075	0.092	0.082	0.118	0.135	0.125	0.131	0.144	0.154	0.167	0.180	0.190	600
	0.039	0.048	0.023	0.028	0.025	0.036	0.041	0.038	0.040	0.044	0.047	0.051	0.055	0.058	
750	0.125	0.157	0.062	0.079	0.069	0.095	0.112	0.102	0.118	0.131	0.141	0.148	0.161	0.171	750
	0.038	0.048	0.019	0.024	0.021	0.029	0.034	0.031	0.036	0.040	0.043	0.045	0.049	0.052	
1000	0.121	0.151	0.049	0.062	0.059	0.075	0.089	0.082	0.105	0.118	0.131	0.128	0.138	0.151	1000
	0.037	0.046	0.015	0.019	0.018	0.023	0.027	0.025	0.032	0.036	0.040	0.039	0.042	0.046	

Notes:

1. These values are based on the following constants: UL-Type RHH wires with Class B stranding, in cradled configuration. Wire conductivities are 100 percent IACS copper and 61 percent IACS aluminum, and aluminum conduit is 45 percent IACS. Capacitive reactance is ignored, since it is negligible at these voltages. These resistance values are valid only at 75°C (167°F) and for the parameters as given, but are representative for 600-volt wire types operating at 60 Hz.

2. *Effective Z* is defined as $R \cos(\theta) + X \sin(\theta)$, where θ is the power factor angle of the circuit. Multiplying current by effective impedance gives a good approximation for line-to-neutral voltage drop. Effective impedance values shown in this table are valid only at 0.85 power factor. For another circuit power factor (PF), effective impedance (Ze) can be calculated from R and X_L values given in this table as follows: $Ze = R \times PF + X_L \sin[\arccos(PF)]$.

Voltage-drop calculations using the dc resistance formula are not always accurate for ac circuits, especially for those with a less-than-unity power factor or for those that use conductors larger than 2 AWG. Table 9, which allows *NEC* users to perform simple ac voltage-drop calculations, was compiled using the Neher–McGrath ac resistance calculation method, and the values presented are both reliable and conservative. This table contains completed calculations of effective impedance (*Z*) for the average ac circuit with an 85-percent power factor (see Calculation Example 1). If calculations with a different power factor are necessary, Table 9 also contains the appropriate values of inductive reactance and ac resistance (see Calculation Example 2). The basic assumptions and the limitations of Table 9 are as follows:

1. Capacitive reactance is ignored.
2. Three conductors are in a raceway.
3. The calculated voltage-drop values are approximate.
4. For circuits with other parameters, the Neher–McGrath ac resistance calculation method is used.

Calculation Example 1

A feeder has a 100-A continuous load. The system source is 240 V, 3 phase, and the supplying circuit breaker is 125 A. The feeder is in a trade size 1¼ aluminum conduit with three 1 AWG THHN copper conductors operating at their maximum temperature rating of 75°C. The circuit length is 150 ft, and the power factor is 85 percent. Using Table 9, determine the approximate voltage drop of this circuit.

Solution

Step 1. Find the approximate line-to-neutral voltage drop. Using the Table 9 column "Effective *Z* at 0.85 *PF* for Uncoated Copper Wires," select aluminum conduit and size 1 AWG copper wire. Use the given value of 0.16 ohm per 1000 ft in the following formula:

$$\text{Voltage drop}_{(line\text{-}to\text{-}neutral)} = \text{table value} \times \frac{\text{circuit length}}{1000 \text{ ft}} \times \text{circuit load}$$

$$= 0.16 \text{ ohm} \times \frac{150 \text{ ft}}{1000 \text{ ft}} \times 100 \text{ A}$$

$$= 2.40 \text{ V}$$

Step 2. Find the line-to-line voltage drop:

$$\text{Voltage drop}_{(line\text{-}to\text{-}line)} = \text{voltage drop}_{(line\text{-}to\text{-}neutral)} \times \sqrt{3}$$

$$= 2.40 \text{ V} \times 1.732$$

$$= 4.157 \text{ V}$$

Step 3. Find the voltage present at the load end of the circuit:

$$240 \text{ V} - 4.157 \text{ V} = 235.84 \text{ V}$$

Calculation Example 2

A 270-A continuous load is present on a feeder. The circuit consists of a single 4-in. PVC conduit with three 600-kcmil XHHW/USE aluminum conductors fed from a 480-V, 3-phase, 3-wire source. The conductors are operating at their maximum rated temperature of 75°C. If the power factor is 0.7 and the circuit length is 250 ft, is the voltage drop excessive?

Solution

Step 1. Using the Table 9 column "X_L (Reactance) for All Wires," select PVC conduit and the row for size 600 kcmil. A value of 0.039 ohm per 1000 ft is given as the X_L. Next, using the column "Alternating-Current Resistance for Aluminum Wires," select PVC conduit and the row for size 600 kcmil. A value of 0.036 ohm per 1000 ft is given as the *R*.

Step 2. Find the angle representing a power factor of 0.7. Using a calculator with trigonometric functions or a trigonometric function table, find the arccosine ($\cos^{-1}$) θ of 0.7, which is 45.57 degrees. For this example, call this angle θ.

Step 3. Find the impedance (Z) corrected to 0.7 power factor (Z_c):

$$Z_c = (R \times \cos \theta) + (X_L \times \sin \theta)$$

$$= (0.036 \times 0.7) + (0.039 \times 0.7141)$$

$$= 0.0252 + 0.0279$$

$$= 0.0531 \text{ ohm to neutral}$$

Step 4. As in Calculation Example 1, find the approximate line-to-neutral voltage drop:

$$\text{Voltage drop}_{(line\text{-}to\text{-}neutral)} = Z_c \times \frac{\text{circuit length}}{1000 \text{ ft}} \times \text{circuit load}$$

$$= 0.0531 \times \frac{250 \text{ ft}}{1000 \text{ ft}} \times 270 \text{ A}$$

$$= 3.584 \text{ V}$$

Step 5. Find the approximate line-to-line voltage drop:

$$\text{Voltage drop}_{(line\text{-}to\text{-}line)} = \text{voltage drop}_{(line\text{-}to\text{-}line)} \times \sqrt{3}$$

$$= 3.584 \text{ V} \times 1.732$$

$$= 6.208 \text{ V}$$

Step 6. Find the approximate voltage drop expressed as a percentage of the circuit voltage:

$$\text{Percentage voltage drop}_{(line\text{-}to\text{-}line)} = \frac{6.208 \text{ V}}{480 \text{ V}} \times 100$$

$$= 1.29\% \text{ voltage drop}$$

Step 7. Find the voltage present at the load end of the circuit:

$$480 \text{ V} - 6.208 \text{ V} = 473.8 \text{ V}$$

According to 210.19, Informational Note, this voltage drop does not appear to be excessive.

TABLE 10 *Conductor Stranding*

Conductor Size		Number of Strands		
		Copper		Aluminum
AWG or kcmil	mm²	Class B[a]	Class C	Class B[a]
24–30	0.20–0.05	[b]	—	—
22	0.32	7	—	—
20	0.52	10	—	—
18	0.82	16	—	—
16	1.3	26	—	—
14–2	2.1–33.6	7	19	7[c]
1–4/0	42.4–107	19	37	19
250–500	127–253	37	61	37
600–1000	304–508	61	91	61
1250–1500	635–759	91	127	91
1750–2000	886–1016	127	271	127

[a]Conductors with a lesser number of strands shall be permitted based on an evaluation for connectability and bending.

[b]Number of strands vary.

[c]Aluminum 14 AWG (2.1 mm²) is not available.

With the permission of Underwriters Laboratories, Inc., material is reproduced from UL Standard 486A-486B, *Wire Connectors*, which is copyrighted by Underwriters Laboratories, Inc., Northbrook, Illinois. While use of this material has been authorized, UL shall not be responsible for the manner in which the information is presented, nor for any interpretations thereof. For more information on UL or to purchase standards, please visit our Standards website at www.comm-2000.com or call 1-888-853-3503.

Table 11(A) and Table 11(B)

For listing purposes, Table 11(A) and Table 11(B) provide the required power source limitations for Class 2 and Class 3 power sources. Table 11(A) applies for alternating-current sources, and Table 11(B) applies for direct-current sources.

The power for Class 2 and Class 3 circuits shall be either (1) inherently limited, requiring no overcurrent protection, or (2) not inherently limited, requiring a combination of power source and overcurrent protection. Power sources designed for interconnection shall be listed for the purpose.

As part of the listing, the Class 2 or Class 3 power source shall be durably marked where plainly visible to indicate the class of supply and its electrical rating. A Class 2 power source not suitable for wet location use shall be so marked.

Exception: Limited power circuits used by listed information technology equipment.

Overcurrent devices, where required, shall be located at the point where the conductor to be protected receives its supply and shall not be interchangeable with devices of higher ratings. The overcurrent device shall be permitted as an integral part of the power source.

Because Class 2 and Class 3 power supplies have a listing requirement in 725.60(A), Tables 11(A) and 11(B) are not directly referenced by the typical installer. The information has been retained in Chapter 9 to provide direction for organizations properly equipped and qualified to evaluate and list these products.

TABLE 11(A) *Class 2 and Class 3 Alternating-Current Power Source Limitations*

Power Source		Inherently Limited Power Source (Overcurrent Protection Not Required)				Not Inherently Limited Power Source (Overcurrent Protection Required)			
		Class 2			Class 3	Class 2		Class 3	
Source voltage V_{max} (volts) (see Note 1)		0 through 20*	Over 20 and through 30*	Over 30 and through 150	Over 30 and through 100	0 through 20*	Over 20 and through 30*	Over 30 and through 100	Over 100 and through 150
Power limitations VA_{max} (volt-amperes) (see Note 1)		—	—	—	—	250 (see Note 3)	250	250	N.A.
Current limitations I_{max} (amperes) (see Note 1)		8.0	8.0	0.005	$150/V_{max}$	$1000/V_{max}$	$1000/V_{max}$	$1000/V_{max}$	1.0
Maximum overcurrent protection (amperes)		—	—	—	—	5.0	$100/V_{max}$	$100/V_{max}$	1.0
Power source maximum nameplate rating	VA (volt-amperes)	$5.0 \times V_{max}$	100	$0.005 \times V_{max}$	100	$5.0 \times V_{max}$	100	100	100
	Current (amperes)	5.0	$100/V_{max}$	0.005	$100/V_{max}$	5.0	$100/V_{max}$	$100/V_{max}$	$100/V_{max}$

Note: Notes for this table can be found following Table 11(B).

*Voltage ranges shown are for sinusoidal ac in indoor locations or where wet contact is not likely to occur. For nonsinusoidal or wet contact conditions, see Note 2.

TABLE 11(B) *Class 2 and Class 3 Direct-Current Power Source Limitations*

Power Source		Inherently Limited Power Source (Overcurrent Protection Not Required)					Not Inherently Limited Power Source (Overcurrent Protection Required)			
		Class 2				Class 3	Class 2		Class 3	
Source voltage V_{max} (volts) (see Note 1)		0 through 20*	Over 20 and through 30*	Over 30 and through 60*	Over 60 and through 150	Over 60 and through 100	0 through 20*	Over 20 and through 60*	Over 60 and through 100	Over 100 and through 150
Power limitations VA_{max} (volt-amperes) (see Note 1)		—	—	—	—	—	250 (see Note 3)	250	250	N.A.
Current limitations I_{max} (amperes) (see Note 1)		8.0	8.0	$150/V_{max}$	0.005	$150/V_{max}$	$1000/V_{max}$	$1000/V_{max}$	$1000/V_{max}$	1.0
Maximum overcurrent protection (amperes)		—	—	—	—	—	5.0	$100/V_{max}$	$100/V_{max}$	1.0
Power source maximum nameplate rating	VA (volt-amperes)	$5.0 \times V_{max}$	100	100	$0.005 \times V_{max}$	100	$5.0 \times V_{max}$	100	100	100
	Current (amperes)	5.0	$100/V_{max}$	$100/V_{max}$	0.005	$100/V_{max}$	5.0	$100/V_{max}$	$100/V_{max}$	$100/V_{max}$

*Voltage ranges shown are for continuous dc in indoor locations or where wet contact is not likely to occur.

For interrupted dc or wet contact conditions, see Note 4.

Notes for Table 11(A) and Table 11(B)

1. V_{max}, I_{max}, and VA_{max} are determined with the current-limiting impedance in the circuit (not bypassed) as follows:

V_{max}: Maximum output voltage regardless of load with rated input applied.

I_{max}: Maximum output current under any noncapacitive load, including short circuit, and with overcurrent protection bypassed if used. Where a transformer limits the output current, I_{max} limits apply after 1 minute of operation. Where a current-limiting impedance, listed for the purpose, or as part of a listed product, is used in combination with a non–power-limited transformer or a stored energy source, e.g., storage battery, to limit the output current, I_{max} limits apply after 5 seconds.

VA_{max}: Maximum volt-ampere output after 1 minute of operation regardless of load and overcurrent protection bypassed if used.

2. For nonsinusoidal ac, V_{max} shall not be greater than 42.4 volts peak. Where wet contact (immersion not included) is likely to occur, Class 3 wiring methods shall be used or V_{max} shall not be greater than 15 volts for sinusoidal ac and 21.2 volts peak for nonsinusoidal ac.

3. If the power source is a transformer, VA_{max} is 350 or less when V_{max} is 15 or less.

4. For dc interrupted at a rate of 10 to 200 Hz, V_{max} shall not be greater than 24.8 volts peak. Where wet contact (immersion not included) is likely to occur, Class 3 wiring methods shall be used, or V_{max} shall not be greater than 30 volts for continuous dc; 12.4 volts peak for dc that is interrupted at a rate of 10 to 200 Hz.

Table 12(A) and Table 12(B)

For listing purposes, Table 12(A) and Table 12(B) provide the required power source limitations for power-limited fire alarm sources. Table 12(A) applies for alternating-current sources, and Table 12(B) applies for direct-current sources. The power for power-limited fire alarm circuits shall be either (1) inherently limited, requiring no overcurrent protection, or (2) not inherently limited, requiring the power to be limited by a combination of power source and overcurrent protection.

As part of the listing, the PLFA power source shall be durably marked where plainly visible to indicate that it is a power-limited fire alarm power source. The overcurrent device, where required, shall be located at the point where the conductor to be protected receives its supply and shall not be interchangeable with devices of higher ratings. The overcurrent device shall be permitted as an integral part of the power source.

Because power-limited fire alarm (PLFA) power supplies have a listing requirement in 760.121(A), this information is not directly referenced by the typical installer. Tables 12(A) and 12(B) provide direction for organizations properly equipped and qualified to evaluate and list these products.

TABLE 12(A) *PLFA Alternating-Current Power Source Limitations*

Power Source		Inherently Limited Power Source (Overcurrent Protection Not Required)			Not Inherently Limited Power Source (Overcurrent Protection Required)		
Circuit voltage V_{max} (volts) (see Note 1)		0 through 20	Over 20 and through 30	Over 30 and through 100	0 through 20	Over 20 and through 100	Over 100 and through 150
Power limitations VA_{max} (volt-amperes) (see Note 1)		—	—	—	250 (see Note 2)	250	N.A.
Current limitations I_{max} (amperes) (see Note 1)		8.0	8.0	$150/V_{max}$	$1000/V_{max}$	$1000/V_{max}$	1.0
Maximum overcurrent protection (amperes)		—	—	—	5.0	$100/V_{max}$	1.0
Power source maximum nameplate ratings	VA (volt-amperes)	$5.0 \times V_{max}$	100	100	$5.0 \times V_{max}$	100	100
	Current (amperes)	5.0	$100/V_{max}$	$100/V_{max}$	5.0	$100/V_{max}$	$100/V_{max}$

Note: Notes for this table can be found following Table 12(B).

TABLE 12(B) *PLFA Direct-Current Power Source Limitations*

Power Source		Inherently Limited Power Source (Overcurrent Protection Not Required)			Not Inherently Limited Power Source (Overcurrent Protection Required)		
Circuit voltage V_{max} (volts) (see Note 1)		0 through 20	Over 20 and through 30	Over 30 and through 100	0 through 20	Over 20 and through 100	Over 100 and through 150
Power limitations VA_{max} (volt-amperes) (see Note 1)		—	—	—	250 (see Note 2)	250	N.A.
Current limitations I_{max} (amperes) (see Note 1)		8.0	8.0	$150/V_{max}$	$1000/V_{max}$	$1000/V_{max}$	1.0
Maximum overcurrent protection (amperes)		—	—	—	5.0	$100/V_{max}$	1.0
Power source maximum nameplate ratings	VA (volt-amperes)	$5.0 \times V_{max}$	100	100	$5.0 \times V_{max}$	100	100
	Current (amperes)	5.0	$100/V_{max}$	$100/V_{max}$	5.0	$100/V_{max}$	$100/V_{max}$

Notes for Table 12(A) and Table 12(B)

1. V_{max}, I_{max}, and VA_{max} are determined as follows:

V_{max}: Maximum output voltage regardless of load with rated input applied.

I_{max}: Maximum output current under any noncapacitive load, including short circuit, and with overcurrent protection bypassed if used. Where a transformer limits the output current, I_{max} limits apply after 1 minute of operation. Where a current-limiting impedance, listed for the purpose, is used in combination with a nonpower-limited transformer or a stored energy source, e.g., storage battery, to limit the output current, I_{max} limits apply after 5 seconds.

VA_{max}: Maximum volt-ampere output after 1 minute of operation regardless of load and overcurrent protection bypassed if used. Current limiting impedance shall not be bypassed when determining I_{max} and VA_{max}.

2. If the power source is a transformer, VA_{max} is 350 or less when V_{max} is 15 or less.

N **TABLE 13** *Equipment Suitable for Hazardous (Classified) Locations*

Area Classification	Type (Level) of Protection	
Zone 0	Intrinsically safe	Intrinsically safe for Class I, Division 1
	Intrinsic safety (Group II)	ia
	Encapsulation (Group II)	ma
	Flameproof (Group II)	da[1]
	Inherently safe optical radiation	op is, with EPL Ga[2]
	Optical system with interlock	op sh, with EPL Ga[2]
	Special protection (Group II)	sa
	EPL[3]	Ga
Zone 1	Equipment suitable for use in Zone 0	
	Equipment suitable for use in Class I, Division 1	
	Flameproof (Group II)	d, db
	Intrinsic safety (Group II)	ib
	Increased safety (Group II)	e, eb
	Pressurized enclosure (Group II)	p, px, pxb, py, pyb
	Encapsulation (Group II)	m, mb
	Pressurized room (Group II)	pb
	Powder filling (Group II)	q, qb
	Liquid immersion (Group II)	o, ob
	Electrical resistance trace heating	60079-30-1, with EPL Gb[2]
	Skin effect trace heating	IEEE 844.1, with EPL Gb[2]
	Inherently safe optical radiation	op is, with EPL Gb[2]
	Optical system with interlock	op sh, with EPL Gb[2]
	Protected optical radiation	op pr, with EPL Gb[2]
	Special protection (Group II)	sb
	EPL[3]	Gb
Zone 2	Equipment suitable for use in Zone 0	
	Equipment suitable for use in Zone 1	
	Equipment suitable for use in Class I, Division 1	
	Equipment suitable for use in Class I, Division 2	
	Type of protection "n" (Group II)	nA, nC, nL, nR
	Pressurized enclosure (Group II)	pz, pzc
	Intrinsic safety (Group II)	ic
	Flameproof (Group II)	dc
	Increased safety (Group II)	ec
	Liquid immersion (Group II)	oc
	Encapsulation (Group II)	mc
	Pressurized room (Group II)	pc
	Electrical resistance trace heating	60079-30-1, with EPL Gc[2]
	Skin effect trace heating	IEEE 844.1, with EPL Gc[2]
	Impedance heating	IEEE 844.3, with EPL Gc[2]
	Inherently safe optical radiation	op is, with EPL Gc[2]
	Optical system with interlock	op sh, with EPL Gc[2]
	Protected optical radiation	op pr, with EPL Gc[2]
	Special protection (Group II)	sc
	EPL[3]	Gc
	Other electrical apparatus[4]	

N **TABLE 13** *Continued*

Area Classification	Type (Level) of Protection	
Unclassified	Associated apparatus for Zone 0 (Group II)	[ia]
	Associated apparatus for Zone 1 (Group II)	[ib]
	Associated apparatus for Zone 2 (Group II)	[ic]
	Associated pressurization equipment (Group II)	[p]
	Associated optical radiation equipment (Group II)	[op is]
	Associated optical radiation equipment (Group II)	[op sh]
Zone 20	Equipment suitable for use in Class II, Division 1	
	Intrinsic safety (Group III)	ia
	Intrinsically safe	Intrinsically safe for Class II Division 1
	Protection by enclosure (Group III)	ta
	Encapsulation (Group III)	ma
	Inherently safe optical radiation	op is, with EPL Da[2]
	Optical system with interlock	op sh, with EPL Da[2]
	Special protection (Group III)	sa
	EPL[3]	Da
Group IIIA Only	Equipment suitable for use in Class III, Division 1	
Zone 21	Equipment suitable for use in Zone 20	
	Equipment suitable for use in Class II, Division 1	
	Intrinsic safety (Group III)	ib
	Protection by enclosure (Group III)	tb
	Pressurized enclosure (Group III)	p, px, pxb, py, pyb
	Encapsulation (Group III)	mb
	Pressurized room (Group III)	pb
	Electrical resistance trace heating	60079-30-1, with EPL Db[2]
	Skin effect trace heating	IEEE 844.1, with EPL Db[2]
	Impedance heating	IEEE 844.3, with EPL Db[2]
	Inherently safe optical radiation	op is, with EPL Db[2]
	Optical system with interlock	op sh, with EPL Db[2]
	Protected optical radiation	op pr, with EPL Db[2]
	Special protection (Group III)	sb
	EPL[3]	Db
Group IIIA Only	Equipment suitable for use in Class III, Division 1	
Zone 22	Equipment suitable for use in Zone 20	
	Equipment suitable for use in Zone 21	
	Equipment suitable for use in Class II, Division 1	
	Equipment suitable for use in Class II, Division 2	
	Intrinsic safety (Group III)	ic
	Protection by enclosure (Group III)	tc
	Pressurized enclosure (Group III)	pz, pzc
	Encapsulation (Group III)	mc
	Pressurized room (Group III)	pc
	Electrical resistance trace heating	60079-30-1, with EPL Dc[2]
	Skin effect trace heating	IEEE 844.1, with EPL Dc[2]
	Impedance heating	IEEE 844.3, with EPL Dc[2]
	Inherently safe optical radiation	op is, with EPL Dc[2]
	Optical system with interlock	op sh, with EPL Dc[2]
	Protected optical radiation	op pr, with EPL Dc[2]
	Type 22 vacuum cleaners and dust collectors	62784
	Special protection (Group III)	sc
	EPL[3]	Dc
	Other electrical apparatus[4]	

(continues)

N **TABLE 13** *Continued*

Area Classification	Type (Level) of Protection	
Group IIIA Only	Equipment suitable for use in Class III, Division 2	
Unclassified	Associated apparatus for Zone 20 (Group III)	[ia]
	Associated apparatus for Zone 21 (Group III)	[ib]
	Associated apparatus for Zone 22 (Group III)	[ic]
	Associated pressurization equipment (Group III)	[p]
	Associated optical radiation equipment (Group III)	[op is]
	Associated optical radiation equipment (Group III)	[op sh]
Class I, Division 1	Equipment marked for use in Class I, Division 1[5]	
	Intrinsically safe	Intrinsically safe for Class I
	Pressurized enclosure	Type X, for Class I
	Pressurized enclosure	Type Y, for Class I
	Equipment suitable for use in Zone 0	
	Intrinsic safety (Group II)	ia
	Encapsulation (Group II)	ma
	Inherently safe optical radiation	op is, with EPL Ga[2]
	Optical system with interlock	op sh, with EPL Ga[2]
	Special protection (Group II)	sa
Class I, Division 2	Equipment suitable for use in Class I, Division 1	
	Equipment marked for use in Class I, Division 2[5]	
	Pressurized enclosure	Type Z, for Class I
	Equipment suitable for use in Zone 0, Zone 1 or Zone 2	
	Type of protection "n" (Group II)	nA, nC, nL, nR
	Pressurized enclosure (Group II)	px, pxb, py, pyb, pz, pzc
	Intrinsic safety (Group II)	ia, ib, ic
	Flameproof (Group II)	da, db, dc
	Increased safety (Group II)	eb, ec
	Liquid immersion (Group II)	ob, oc
	Encapsulation (Group II)	ma, mb, mc
	Pressurized room (Group II)	pb
	Pressurized room (Group II)	pc
	Electrical resistance trace heating	60079-30-1, with EPL Gb or Gc[2]
	Skin effect trace heating	IEEE 844.1, with EPL Gb or Gc[2]
	Impedance heating	IEEE 844.3, with EPL Gb or Gc[2]
	Inherently safe optical radiation	op is, with EPL Ga, Gb or Gc[2]
	Optical system with interlock	op sh, with EPL Ga, Gb or Gc[2]
	Protected optical radiation	op pr, with EPL Gb or Gc[2]
	Special protection (Group II)	sa, sb, sc
	Other electrical apparatus[4]	
Unclassified	Associated apparatus for Class I, Division 1	
Class II, Division 1[6]	Equipment marked for use in Class II, Division 1[5]	
	Intrinsically safe	Intrinsically safe for Class II
	Pressurized enclosure	Type X, for Class II
	Pressurized enclosure	Type Y, for Class II
	Equipment suitable for use in Zone 20	
	Intrinsic safety (Group III)	ia
	Protection by enclosure (Group III)	ta
	Encapsulation (Group III)	ma
	Inherently safe optical radiation	op is, with EPL Da2[2]
	Optical system with interlock	op sh, with EPL Da2[2]
	Special protection (Group III)	sa

N **TABLE 13** *Continued*

Area Classification	Type (Level) of Protection	
Class II, Division 2[6]	Equipment suitable for use in Class II, Division 1	
	Equipment marked for use in Class II, Division 2[5]	
	Pressurized enclosure	Type Z, for Class II
	Equipment suitable for use in Zone 20, Zone 21 or Zone 22	
	Intrinsic safety (Group III)	ia, ib, ic
	Protection by enclosure (Group III)	ta, tb, tc
	Pressurized enclosure (Group III)	px, pxb, py, pyb, pz, pzc
	Encapsulation (Group III)	ma, mb, mc
	Pressurized room (Group III)	pb
	Pressurized room (Group III)	pc
	Electrical resistance trace heating	60079-30-1, with EPL Db or Dc[2]
	Skin effect trace heating	IEEE 844.1, with EPL Db or Dc[2]
	Impedance heating	IEEE 844.3, with EPL Db or Dc[2]
	Inherently safe optical radiation	op is, with EPL Da, Db or Dc[2]
	Optical system with interlock	op sh, with EPL Da, Db or Dc[2]
	Protected optical radiation	op pr, with EPL Db or Dc[2]
	Special protection (Group III)	sa, sb, sc
	Other electrical apparatus[4]	
Unclassified	Associated apparatus for Class II, Division 1	
Class III, Division 1	Equipment suitable for use in Class II, Division 1	
	Equipment marked for use in Class III, Division 1[5,7]	
	Intrinsically safe	Intrinsically safe for Class ll or Class III
	Equipment suitable for use in Zone 20 or Zone 21	
	Intrinsic safety (Group III)	ia, ib
	Protection by enclosure (Group III)	ta, tb
	Encapsulation (Group III)	ma, mb
	Inherently safe optical radiation	op is, with EPL Da or Db[2]
	Optical system with interlock	op sh, with EPL Da or Db[2]
	Special protection (Group III)	sa
Class III, Division 2	Equipment suitable for use in Class II, Division 1	
	Equipment suitable for use in Class II, Division 2	
	Equipment suitable for use in Class III, Division 1	
	Equipment marked for use in Class III, Division 2[5]	
	Intrinsically safe	Intrinsically safe for Class ll or Class III
	Equipment suitable for use in Zone 20, Zone 21 or Zone 22	
	Intrinsic safety (Group III)	ia, ib, ic
	Protection by enclosure (Group III)	ta, tb, tc
	Pressurized enclosure (Group III)	px, pxb, py, pyb, pz, pzc
	Encapsulation (Group III)	ma, mb, mc
	Electrical resistance trace heating	60079-30-1, with EPL Db or Dc[2]
	Skin effect trace heating	IEEE 844.1, with EPL Db or Dc[2]
	Impedance heating	IEEE 844.3, with EPL Db or Dc[2]
	Inherently safe optical radiation	op is, with EPL Da, Db or Dc[2]
	Optical system with interlock	op sh, with EPL Da, Db or Dc[2]
	Protected optical radiation	op pr, with EPL Db or Dc[2]
	Special protection (Group III)	sa, sb, sc
	Other electrical apparatus[4]	

(continues)

N **TABLE 13** *Continued*

Area Classification	Type (Level) of Protection
Unclassified	Associated apparatus for Class III, Division 1

Note: This table is structured to show the area classification on the left side and the permitted equipment on the right side. Zone equipment is suitable for use in some class/division locations and vice versa. This is indicated by the phrase "Equipment suitable for use in…", for example, in Class I, Division 1 locations, "Equipment suitable for use in Zone 0" means all equipment listed under Zone 0 can be used with an appropriate equipment group and temperature class.

[1]"da" is limited to sensors of portable combustible gas detectors.

[2]Equipment marked with these types of protection is available in multiple levels of protection not specifically identified within the Ex marking.

[3]The EPL takes precedence over the type of protection, for example "AEx ia __ __ Gb" is suitable for Zone 1, not Zone 0; "AEx op is __ __ Db" is suitable for Zone 21, not Zone 20; and "AEx 60079-30-1 __ __ Gc" is suitable for Zone 2, not Zone 1. Selection according to the mark EPL is critical to the safe application of this equipment.

[4]"Other electrical apparatus" indicates electrical apparatus complying with the requirements of a recognized standard for industrial electrical apparatus that does not in normal service (a) have ignition-capable hot surfaces, or (b) produce incendive arcs or sparks. *[See 501.105(B), 501.115(B), 501.125(B), 502.100(B), 502.115(B), 502.125(B), or 502.130(B).]* "Other electrical apparatus" also includes equipment or systems currently acceptable as alternative means of protection *(see 500.7)*.

[5]With the exception of intrinsically safe or purged/pressurized equipment, equipment for use in a Class XX, Division XX, location is not required to be marked with a type of protection, only the location where that equipment is permitted to be installed.

[6]For use in Class II and Class III locations, such (zone-acceptable) equipment is subject to the requirements of 502.6 or 503.6, respectively. Group IIIA equipment is not suitable for use in Class II locations.

[7]In Class III, Division 1 locations, switches, controllers, circuit breakers, fuses, control transformers, resistors, utilization equipment (fixed and portable), electric cranes, hoists, and similar equipment can be housed in dusttight enclosures.

Product Safety Standards

Informative Annex A is not a part of the requirements of this NFPA document but is included for informational purposes only.

This informative annex provides a list of product safety standards used for product listing where that listing is required by this *Code*. It is recognized that this list is current at the time of publication but that new standards or modifications to existing standards can occur at any time while this edition of the *Code* is in effect.

This informative annex does not form a mandatory part of the requirements of this *Code* but is intended to identify for the *Code* users the standards upon which *Code* requirements have been based.

A key element of a safe and *NEC*®-compliant electrical installation is adherence to product installation requirements imposed by product-testing organizations as part of their evaluation of an electrical product. Section 110.3(B) requires compliance with the installation and use instructions included with listed and labeled products. Numerous requirements in the *NEC* specify the use of listed products. For those requirements where product listing is mandatory, Informative Annex A is a compilation of applicable product safety standards. The product safety standards included in Informative Annex A are only those for which the *NEC* has a mandatory listing requirement. Many other safety standards are associated with products that do not have a mandatory listing or labeling requirement in the *NEC*. For more information on

product standards, consult the product directories available from testing organizations. Many testing organizations offer online tools that provide a list of vendors that offer listed products, along with information about the product listing standard and any special requirements that are part of the listing.

Product safety standards, installation codes such as the *NEC*, and qualified electrical inspection are separate but not mutually exclusive components of the North American electrical safety system. The effectiveness of this system strongly depends on a close working relationship among the organizations responsible for the development of product standards and installation codes and the electrical inspection community. In addition, proper electrical installation by *qualified persons* (see the definition of this term in Article 100) trained on the installation codes and application of listed products is a key component of this interdependent system that places public safety as its primary objective. All these components must be in place for the electrical safety system to be effective.

The function of Informative Annex A is to provide users of the *NEC* with the name, number, and developing organization for all product standards related to mandatory requirements for the use of listed products.

See also

110.3(B) for more information on listing and labeling

N **TABLE A.1(a)** *Product Safety Standards for Conductors and Equipment That Have an Associated Listing Requirement*

Article	Standard Number	Standard Title
110	UL 10C	Positive Pressure Fire Tests of Door Assemblies
	UL 305	Panic Hardware
	UL 486D	Sealed Wire Connector Systems
	UL 2043	Fire Test for Heat and Visible Smoke Release for Discrete Products and Their Accessories Installed in Air-Handling Spaces
	UL 62275	Cable Management Systems — Cable Ties for Electrical Installation
210	UL 498	Attachment Plugs and Receptacles
	UL 935	Fluorescent-Lamp Ballasts
	UL 943	Ground Fault Circuit Interrupters
	UL 1029	High-Intensity-Discharge Lamp Ballast
	UL 1699	Arc-Fault Circuit-Interrupters
	UL 1699A	Outlet Branch Circuit AFCIs
225	UL 6	Electrical Rigid Metal Conduit — Steel
	UL 6A	Electrical Rigid Metal Conduit — Aluminum, Red Brass and Stainless Steel
	UL 360	Liquid-Tight Flexible Metal Conduit
	UL 651	Schedule 40, 80, Type EB and A Rigid PVC Conduit and Fittings
	UL 1242	Electrical Intermediate Metal Conduit — Steel
	UL 1660	Liquid-Tight Flexible Nonmetallic Conduit
	UL 2515	Aboveground Reinforced Thermosetting Resin Conduit (RTRC) and Fittings
230	UL 6	Electrical Rigid Metal Conduit — Steel
	UL 6A	Electrical Rigid Metal Conduit — Aluminum, Red Brass and Stainless Steel
	UL 67	Panelboards
	UL 98	Enclosed and Dead-Front Switches
	UL 218	Fire Pump Controllers
	UL 231	Power Outlets
	UL 347	Medium-Voltage AC Contactors, Controllers, and Control Centers
	UL 360	Liquid-Tight Flexible Metal Conduit
	UL 414	Meter Sockets
	UL 486A-486B	Wire Connectors
	UL 486C	Splicing Wire Connectors
	UL 489	Molded-Case Circuit Breakers, Molded-Case Switches and Circuit-Breaker Enclosures
	UL 508	Industrial Control Equipment
	UL 508A	Industrial Control Panels
	UL 514B	Conduit, Tubing and Cable Fittings
	UL 651	Schedule 40, 80, Type EB and A Rigid PVC Conduit and Fittings
	UL 845	Motor Control Centers
	UL 857	Busways
	UL 869A	Reference Standard for Service Equipment
	UL 891	Switchboards
	UL 977	Fused Power-Circuit Devices
	UL 1008	Transfer Switch Equipment
	UL 1008A	Transfer Switch Equipment, Over 1000 Volts
	UL 1008M	Meter-Mounted Transfer Switches
	UL 1008S	Solid-State Transfer Switches
	UL 1053	Ground-Fault Sensing and Relaying Equipment
	UL 1062	Unit Substations
	UL 1066	Low-Voltage AC and DC Power Circuit Breakers Used in Enclosures
	UL 1242	Electrical Intermediate Metal Conduit — Steel
	UL 1429	Pullout Switches
	UL 1449	Surge Protective Devices
	UL 1558	Metal-Enclosed Low-Voltage Power Circuit Breaker Switchgear
	UL 1660	Liquid-Tight Flexible Nonmetallic Conduit
	UL 1740	Robots and Robotic Equipment
	UL 1953	Power Distribution Blocks
	UL 2011	Machinery
	UL 2200	Stationary Engine Generator Assemblies
	UL 2416	Audio/Video, Information and Communication Technology Equipment Cabinet, Enclosure and Rack Systems
	UL 2446	Unitary Boiler Room Systems

N *TABLE A.1(a)* *Continued*

Article	Standard Number	Standard Title
	UL 2565	Industrial Metalworking and Woodworking Machine Tools
	UL 2735	Electric Utility Meters
	UL 2745	Meter Socket Adapters for Communications Equipment
	UL 2876	Remote Racking Devices for Switchgear and Controlgear
	UL 4248-1	Fuseholders — Part 1: General Requirements
	UL 60947-1	Low-Voltage Switchgear and Controlgear — Part 1: General Rules
	UL 61800-5-1	Adjustable Speed Electrical Power Drive Systems — Part 5-1: Safety Requirements — Electrical, Thermal and Energy
240	UL 248-1	Low-Voltage Fuses — Part 1: General Requirements
	UL 248-2	Low-Voltage Fuses — Part 2: Class C Fuses
	UL 248-3	Low-Voltage Fuses — Part 2: Class CA and CB Fuses
	UL 248-4	Low-Voltage Fuses — Part 4: Class CC Fuses
	UL 248-5	Low-Voltage Fuses — Part 5: Class G Fuses
	UL 248-6	Low-Voltage Fuses — Part 6: Class H Non-Renewable Fuses
	UL 248-8	Low-Voltage Fuses — Part 8: Class J Fuses
	UL 248-9	Low-Voltage Fuses — Part 9: Class K Fuses
	UL 248-10	Low-Voltage Fuses — Part 10: Class L Fuses
	UL 248-11	Low-Voltage Fuses — Part 11: Plug Fuses
	UL 248-12	Low-Voltage Fuses — Part 12: Class R Fuses
	UL 248-15	Low-Voltage Fuses — Part 15: Class T Fuses
	UL 248-17	Low-Voltage Fuses — Part 17: Class CF Fuses
	UL 248-18	Low-Voltage Fuses — Part 18: Class CD Fuses
	UL 489	Molded-Case Circuit Breakers, Molded-Case Switches, and Circuit-Breaker Enclosures
	UL 489I	Solid State Molded-Case Circuit Breakers
	UL 943	Ground-Fault Circuit-Interrupters
	UL 1053	Ground-Fault Sensing and Relaying Equipment
	UL 1066	Low-Voltage AC and DC Power Circuit Breakers Used in Enclosures
	UL 4248-1	Fuseholders — Part 1: General Requirements
242	UL 1449	Surge Protective Devices
250	UL 1	Flexible Metal Conduit
	UL 4	Armored Cable
	UL 5	Surface Metal Raceways and Fittings
	UL 6	Electrical Rigid Metal Conduit — Steel
	UL 6A	Electrical Rigid Metal Conduit — Aluminum, Red Brass and Stainless Steel
	UL 360	Liquid-Tight Flexible Metal Conduit
	UL 467	Grounding and Bonding Equipment
	UL 486A-486B	Wire Connectors
	UL 486C	Splicing Wire Connectors
	UL 486D	Sealed Wire Connector Systems
	UL 498	Attachment Plugs and Receptacles
	UL 504	Mineral-Insulated, Metal-Sheathed Cable
	UL 514A	Metallic Outlet Boxes
	UL 514B	Conduit, Tubing, and Cable Fittings
	UL 797	Electrical Metallic Tubing — Steel
	UL 797A	Electrical Metallic Tubing — Aluminum
	UL 1242	Electrical Intermediate Metal Conduit — Steel
	UL 1569	Metal-Clad Cables
	UL 1652	Flexible Metallic Tubing
300	UL 4	Armored Cable
	UL 44	Thermoset-Insulated Wires and Cables
	UL 83	Thermoplastic-Insulated Wires and Cables
	UL 83A	Fluoropolymer Insulated Wire
	UL 263	Fire Tests of Building Construction and Materials
	UL 504	Mineral-Insulated, Metal-Sheathed Cable
	UL 746C	Polymeric Materials — Use in Electrical Equipment Evaluations
	UL 1569	Metal-Clad Cable
	UL 1581	Reference Standard for Electrical Wires, Cables, and Flexible Cords

(continues)

Article	Standard Number	Standard Title
	UL 2239	Hardware for Support of Conduit, Tubing and Cable
	UL 2556	Wire and Cable Test Methods
	UL 62275	Cable Management Systems — Cable Ties for Electrical Installation
310	UL 44	Thermoset-Insulated Wires and Cables
	UL 83	Thermoplastic-Insulated Wires and Cables
	UL 83A	Fluoropolymer Insulated Wire
	UL 224	Extruded Insulating Tubing
	UL 1063	Machine-Tool Wires and Cables
	UL 1441	Coated Electrical Sleeving
315	ANSI C119.4	Electric Connectors — Connectors for Use between Aluminum-to-Aluminum and Aluminum-to-Copper Conductors Designed for Normal Operation at or Below 93°C and Copper-to-Copper Conductors Designed for Normal Operation at or Below 100°C
	IEEE 48	IEEE Standard for Test Procedures and Requirements for Alternating-Current Cable Terminations Used on Shielded Cables Having Laminated Insulation Rated 2.5 kV through 765 kV or Extruded Insulation Rated 2.5 kV through 500 kV
	IEEE 386	IEEE Standard for Separable Insulated Connector Systems for Power Distribution Systems Rated 2.5 kV through 35 kV
	IEEE 404	IEEE Standard for Extruded and Laminated Dielectric Shielded Cable Joints Rated 2.5 kV to 500 kV
	UL 4	Armored Cable
	UL 504	Mineral-Insulated, Metal-Sheathed Cable
	UL 1072	Medium Voltage Power Cables
	UL 1569	Metal-Clad Cable
312	UL 50	Enclosures for Electrical Equipment
	UL 50E	Enclosures for Electrical Equipment, Environmental Considerations
	UL 514C	Nonmetallic Outlet Boxes, Flush-Device Boxes, and Covers
	UL 916	Energy Management Equipment
	UL 2808	Energy Monitoring Equipment
	UL 61010-1 and UL 61010-2-030	Electrical Equipment for Measurement, Control, and Laboratory Use — Part 2-030: Particular Requirements for Testing and Measuring Circuits
314	UL 50	Enclosures for Electrical Equipment
	UL 50E	Enclosures for Electrical Equipment, Environmental Considerations
	UL 486D	Sealed Wire Connector Systems
	UL 498	Attachment Plugs and Receptacles
	UL 498B	Receptacles with Integral Switching Means
	UL 498D	Attachment Plugs, Cord Connectors and Receptacles with Arcuate (Locking Type) Contacts
	UL 498E	Attachment Plugs, Cord Connectors and Receptacles — Enclosure Types for Environmental Protection
	UL 514A	Metallic Outlet Boxes
	UL 514B	Conduit, Tubing, and Cable Fittings
	UL 514C	Nonmetallic Outlet Boxes, Flush-Device Boxes, and Covers
	UL 514D	Cover Plates for Flush-Mounted Wiring Devices
	UL 1953	Power Distribution Blocks
320	UL 4	Armored Cable
	UL 44	Thermoset-Insulated Wires and Cables
	UL 83	Thermoplastic-Insulated Wires and Cables
	UL 83A	Fluoropolymer Insulated Wire
	UL 514B	Conduit, Tubing, and Cable Fittings
	UL 514C	Nonmetallic Outlet Boxes, Flush-Device Boxes, and Covers
	UL 1063	Machine-Tool Wires and Cables
	UL 1565	Positioning Devices
	UL 2239	Hardware for the Support of Conduit, Tubing, and Cable
322	UL 486A-486B	Wire Connectors
	UL 498	Attachment Plugs and Receptacles
	UL 514A	Metallic Outlet Boxes
324	UL 486A-486B	Wire Connectors
	UL 498	Attachment Plugs and Receptacles
330	UL 44	Thermoset-Insulated Wires and Cables
	UL 66	Fixture Wire
	UL 83	Thermoplastic-Insulated Wires and Cables
	UL 83A	Fluoropolymer Insulated Wire

N *TABLE A.1(a)* *Continued*

Article	Standard Number	Standard Title
	UL 514B	Conduit, Tubing, and Cable Fittings
	UL 1063	Machine-Tool Wires and Cables
	UL 1565	Positioning Devices
	UL 1569	Metal-Clad Cables
	UL 2225	Cables and Cable-Fittings For Use In Hazardous (Classified) Locations
	UL 2239	Hardware for the Support of Conduit, Tubing, and Cable
332	UL 504	Mineral-Insulated, Metal-Sheathed Cable
	UL 514B	Conduit, Tubing and Cable Fittings
334	UL 719	Nonmetallic-Sheathed Cables
	UL 2256	Nonmetallic Sheathed Cable Interconnects
	UL 62275	Cable Management Systems — Cable Ties for Electrical Installations
335	UL 2250	Instrumentation Tray Cable
336	UL 514B	Conduit, Tubing, and Cable Fittings
	UL 1277	Electrical Power and Control Tray Cables with Optional Optical-Fiber Members
	UL 2225	Cables and Cable-Fittings For Use In Hazardous (Classified) Locations
337	UL 1309A	Cable for Use in Mobile Installations
338	UL 514B	Conduit, Tubing, and Cable Fittings
	UL 854	Service-Entrance Cables
340	UL 514B	Conduit, Tubing, and Cable Fittings
340	UL 493	Thermoplastic-Insulated Underground Feeder and Branch-Circuit Cables
342	UL 514B	Conduit, Tubing, and Cable Fittings
	UL 1242	Electrical Intermediate Metal Conduit — Steel
344	UL 6	Electrical Rigid Metal Conduit — Steel
	UL 6A	Electrical Rigid Metal Conduit — Aluminum, Red Brass and Stainless Steel
	UL 514B	Conduit, Tubing, and Cable Fittings
348	UL 1	Flexible Metal Conduit
	UL 62275	Cable Management Systems — Cable Ties for Electrical Installation
350	UL 360	Liquid-Tight Flexible Steel Conduit
	UL 514B	Conduit, Tubing, and Cable Fittings
	UL 62275	Cable Management Systems — Cable Ties for Electrical Installation
352	UL 651	Schedule 40, 80, Type EB and A Rigid PVC Conduit and Fittings
353	UL 651A	Schedule 40 and 80 High Density Polyethylene (HDPE) Conduit
354	UL 1990	Nonmetallic Underground Conduit with Conductors
355	UL 2420	Belowground Reinforced Thermosetting Resin Conduit (RTRC) and Fittings
	UL 2515	Aboveground Reinforced Thermosetting Resin Conduit (RTRC) and Fittings
	UL 2515A	Supplemental Requirements for Extra-Heavy Wall Reinforced Thermosetting Resin Conduit (RTRC) and Fittings
356	UL 1660	Liquid-Tight Flexible Nonmetallic Conduit
	UL 62275	Cable Management Systems — Cable Ties for Electrical Installation
358	UL 514B	Conduit, Tubing, and Cable Fittings
	UL 797	Electrical Metallic Tubing — Steel
	UL 797A	Electrical Metallic Tubing — Aluminum and Stainless Steel
360	UL 514B	Conduit, Tubing, and Cable Fittings
	UL 1652	Flexible Metallic Tubing
362	UL 1653	Electrical Nonmetallic Tubing
	UL 62275	Cable Management Systems — Cable Ties for Electrical Installation
366	UL 870	Wireways, Auxiliary Gutters, and Associated Fittings
368	UL 509	Bus Drop Cable
370	ANSI/CSA C22.2 No. 273	Cablebus
374	UL 209	Cellular Metal Floor Raceways and Fittings
	UL 360	Liquid-Tight Flexible Metal Conduit
	UL 1660	Liquid-Tight Flexible Nonmetallic Conduit
376	UL 870	Wireways, Auxiliary Gutters and Associated Fittings
	UL 1953	Power Distribution Blocks
378	UL 870	Wireways, Auxiliary Gutters, and Associated Fittings

(continues)

N **TABLE A.1(a)** *Continued*

Article	Standard Number	Standard Title
382	UL 5A	Nonmetallic Surface Raceways and Fittings
	UL183	Manufactured Wiring Systems
	UL 467	Grounding and Bonding Equipment
	UL 498	Attachment Plugs and Receptacles
	UL 498D	Attachment Plugs, Cord Connectors and Receptacles with Arcuate (Locking Type) Contacts
	UL 498E	Attachment Plugs, Cord Connectors and Receptacles — Enclosure Types for Environmental Protection
	UL 498F	Plugs, Socket-Outlets and Couplers with Arcuate (Locking Type) Contacts
	UL 498M	Marine Shore Power Inlets
	UL 514D	Cover Plates for Flush-Mounted Wiring Devices
	UL 746C	Polymeric Materials — Use in Electrical Equipment Evaluations
	UL 943	Ground-Fault Circuit-Interrupters
	UL 991	Tests for Safety-Related Controls Employing Solid-State Devices
	UL 1077	Supplementary Protectors for Use in Electrical Equipment
	UL 1699	Arc-Fault Circuit-Interrupters
	UL 1998	Software in Programmable Components
384	UL 5B	Strut-Type Channel Raceways and Fittings
386	UL 5	Surface Metal Raceways and Fittings
388	UL 5A	Nonmetallic Surface Raceways and Fittings
392	UL 62275	Cable Management Systems — Cable Ties for Electrical Installation
393	UL 13	Power-Limited Circuit Cables
	UL 50	Enclosures for Electrical Equipment, Non-Environmental Considerations
	UL 50E	Enclosures for Electrical Equipment, Environmental Considerations
	UL 514C	Nonmetallic Outlet Boxes, Flush-Device Boxes, and Covers
	UL 1310	Class 2 Power Units
	UL 2043	Fire Test for Heat and Visible Smoke Release for Discrete Products and Their Accessories Installed in Air-Handling Spaces
	UL 2577	Suspended Ceiling Power Grid Systems and Equipment
	UL 62368-1	Audio/Video, Information and Communication Technology Equipment — Part 1: Safety Requirements
396	UL 1072	Medium-Voltage Power Cables
404	UL 20	General-Use Snap Switches
	UL 98	Enclosed and Dead-Front Switches
	UL 98A	Open-Type Switches
	UL 363	Knife Switches
	UL 489	Automatic Electrical Controls for Household and Similar Use; Part 2: Particular Requirements for Timers and Time Switches
	UL 773	Plug-In Locking Type Photocontrols for Use with Area Lighting
	UL 773A	Nonindustrial Photoelectric Switches for Lighting Control
	UL 917	Clock-Operated Switches
	UL 977	Fused Power-Circuit Devices
	UL 1066	Low-Voltage AC and DC Power Circuit Breakers Used in Enclosures
	UL 1472	Solid-State Dimming Controls
	UL 1429	Pullout Switches
	UL 60730-1	Automatic Electrical Controls — Part 1: General Requirements
	UL 60730-2	Automatic Electrical Controls for Household and Similar Use; Part 2: Particular Requirements for Timers and Time Switches
	UL 60730-2-7	Automatic Electrical Controls for Household and Similar Use; Part 2: Particular Requirements for Timers and Time Switches
	ANSI/NEMA WD 6–2016	Wiring Devices — Dimensional Specifications
406	UL 498	Attachment Plugs and Receptacles
	UL 498B	Receptacles with Integral Switching Means
	UL 498D	Attachment Plugs, Cord Connectors and Receptacles with Arcuate (Locking Type) Contacts
	UL 498E	Attachment Plugs, Cord Connectors and Receptacles — Enclosure Types for Environmental Protection
	UL 498F	Plugs, Socket-Outlets and Couplers with Arcuate (Locking Type) Contacts
	UL 498M	Marine Shore Power Inlets
	UL 514A	Metallic Outlet Boxes
	UL 514C	Nonmetallic Outlet Boxes, Flush-Device Boxes, and Covers
	UL 514D	Cover Plates for Flush-Mounted Wiring Devices
	UL 943	Ground-Fault Circuit-Interrupters

N **TABLE A.1(a)** *Continued*

Article	Standard Number	Standard Title
	UL 943B	Appliance Leakage-Current Interrupters
	UL 943C	Special Purpose Ground-Fault Circuit-Interrupters
	UL 970	Retail Fixtures and Merchandising Displays
	UL 1286	Office Furnishings Systems
	UL 1310	Class 2 Power Units
	UL 1682	Plugs, Receptacles, and Cable Connectors, of the Pin and Sleeve Type
	UL 1691	Single Pole Locking-Type Separable Connectors
	UL 1699	Arc-Fault Circuit-Interrupters
	UL 2999	Individual Commercial Office Furnishings
408	UL 44	Thermoset-Insulated Wires and Cables
	UL 67	Panelboards
	UL 891	Switchboards
	UL 1558	Metal-Enclosed Low-Voltage Power Circuit Breaker Switchgear
409	UL 508	Industrial Control Equipment
	UL 508A	Industrial Control Panels
410	ANSI/CSA-C22.2 No. 184.2	Solid-State Controls for Lighting Systems (SSCLS)
	UL 153	Portable Electric Luminaires
	UL 496	Lampholders
	UL 498	Attachment Plugs and Receptacles
	UL 498B	Receptacles with Integral Switching Means
	UL 498D	Attachment Plugs, Cord Connectors and Receptacles with Arcuate (Locking Type) Contacts
	UL 498E	Attachment Plugs, Cord Connectors and Receptacles — Enclosure Types for Environmental Protection
	UL 498F	Plugs, Socket-Outlets and Couplers with Arcuate (Locking Type) Contacts
	UL 542	Fluorescent Lamp Starters
	UL 588	Seasonal and Holiday Decorative Products
	UL 935	Fluorescent-Lamp Ballasts
	UL 943	Ground-Fault Circuit-Interrupters
	UL 970	Retail Fixtures and Merchandising Displays
	UL 1029	High-Intensity-Discharge Lamp Ballasts
	UL 1029A	Ignitors and Related Auxiliaries for HID Lamp Ballasts
	UL 1574	Track Lighting Systems
	UL 1598	Luminaires
	UL 1598B	Luminaire Reflector Kits for Installation on Previously Installed Fluorescent Luminaires, Supplemental Requirements
	UL 1598C	Light-Emitting Diode (LED) Retrofit Luminaire Conversion Kits
	UL 1993	Self-Ballasted Lamps and Lamp Adapters
	UL 2388	Flexible Lighting Products
	UL 8750	Light Emitting Diode (LED) Equipment for Use in Lighting Products
	UL 8752	Organic Light Emitting Diode (OLED) Panels
	UL 8753	Field-Replaceable Light Emitting Diode (LED) Light Engines
	UL 8754	Holders, Bases and Connectors for Solid-State (LED) Light Engines and Arrays
	UL 8800	Horticultural Lighting Equipment and Systems
411	UL 1310	Class 2 Power Units
	UL 1838	Low-Voltage Landscape Lighting Systems
	UL 2108	Low-Voltage Lighting Systems
	UL 5085-3	Low Voltage Transformers — Part 3: Class 2 and Class 3 Transformers
422	ANSI/CSA-C22.2 No. 339	Hand-held motor-operated electric tools — Safety — Particular requirements for chain beam saws
	UL 22	Amusement and Gaming Machines
	UL 73	Motor-Operated Appliances
	UL 82	Electric Gardening Appliances
	UL 122	Photographic Equipment
	UL 141	Garment Finishing Appliances
	UL 174	Household Electric Storage Tank Water Heaters
	UL 197	Commercial Electric Cooking Appliances
	UL 283	Air Fresheners and Deodorizers

(continues)

N **TABLE A.1(a)** *Continued*

Article	Standard Number	Standard Title
	UL 399	Drinking Water Coolers
	UL 430	Waste Disposers
	UL 498	Attachment Plugs and Receptacles
	UL 498D	Attachment Plugs, Cord Connectors and Receptacles with Arcuate (Locking Type) Contacts
	UL 498E	Attachment Plugs, Cord Connectors and Receptacles — Enclosure Types for Environmental Protection
	UL 498F	Plugs, Socket-Outlets and Couplers with Arcuate (Locking Type) Contacts
	UL 499	Electric Heating Appliances
	UL 507	Electric Fans
	UL 514A	Metallic Outlet Boxes
	UL 515	Electric Resistance Trace Heating for Commercial Applications
	UL 561	Floor Finishing Machines
	UL 574	Electric Oil Heaters
	UL 621	Ice Cream Makers
	UL 705	Power Ventilators
	UL 710B	Recirculating Systems
	UL 749	Household Dishwashers
	UL 751	Vending Machines
	UL 763	Motor-Operated Commercial Food Preparing Machines
	UL 778	Motor-Operated Water Pumps
	UL 834	Heating, Water Supply, and Power Boilers — Electric
	UL 858	Household Electric Ranges
	UL 875	Electric Dry-Bath Heaters
	UL 921	Commercial Dishwashers
	UL 923	Microwave Cooking Appliances
	UL 943	Ground-Fault Circuit-Interrupters
	UL 962	Household and Commercial Furnishings
	UL 962A	Furniture Power Distribution Units
	UL 979	Water Treatment Appliances
	UL 982	Motor-Operated Household Food Preparing Machines
	UL 987	Stationary and Fixed Electric Tools
	UL 1017	Vacuum Cleaners, Blower Cleaners, and Household Floor Finishing Machines
	UL 1026	Household Electric Cooking and Food Serving Appliances
	UL 1086	Household Trash Compactors
	UL 1090	Electric Snow Movers
	UL 1206	Electric Commercial Clothes-Washing Equipment
	UL 1240	Electric Commercial Clothes-Drying Equipment
	UL 1278	Movable and Wall- or Ceiling-Hung Electric Room Heaters
	UL 1447	Electric Lawn Mowers
	UL 1450	Motor-Operated Air Compressors, Vacuum Pumps, and Painting Equipment
	UL 1453	Electric Booster and Commercial Storage Tank Water Heaters
	UL 1576	Flashlights and Lanterns
	UL 1594	Sewing and Cutting Machines
	UL 1647	Motor-Operated Massage and Exercise Machines
	UL 1727	Commercial Electric Personal Grooming Appliances
	UL 1776	High-Pressure Cleaning Machines
	UL 2157	Electric Clothes Washing Machines and Extractors
	UL 2158	Electric Clothes Dryers
	UL 2565	Industrial Metalworking and Woodworking Machine Tools
	UL 60335-2-3	Household and Similar Electrical Appliances, Part 2: Particular Requirements for Electric Irons
	UL 60335-2-8	Household and Similar Electrical Appliances, Part 2: Particular Requirements for Shavers, Hair Clippers, and Similar Appliances
	UL 60335-2-24	Household and Similar Electrical Appliances, Part 2: Particular Requirements for Refrigerating Appliances, Ice-Cream Appliances, and Ice-Makers
	UL 60335-2-40	Household and Similar Electrical Appliances, Part 2: Particular Requirements for Electrical Heat Pumps, Air-Conditioners and Dehumidifiers
	UL 60335-2-67	Household and Similar Electrical Appliances — Safety — Part 2-67: Particular Requirements for Floor Treatment Machines, For Commercial Use
	UL 60335-2-68	Household and Similar Electrical Appliances — Safety — Part 2-68: Particular Requirements for Spray Extraction Machines, for Commercial Use

N **TABLE A.1(a)** *Continued*

Article	Standard Number	Standard Title
	UL 60335-2-79	Household and Similar Electrical Appliances — Safety — Part 2-79: Particular Requirements for High Pressure Cleaners and Steam Cleaners
	UL 60730-2-9	Automatic Electrical Controls; Part 2: Particular Requirements for Temperature Sensing Controls
	UL 60745-1	Hand-Held Motor-Operated Electric Tools — Safety — Part 1: General Requirements
	UL 60745-2-1	Hand-Held Motor-Operated Electric Tools — Safety — Part 2-1: Particular Requirements for Drills and Impact Drills
	UL 60745-2-2	Hand-Held Motor-Operated Electric Tools — Safety — Part 2-2: Particular Requirements for Screwdrivers and Impact Wrenches
	UL 60745-2-3	Hand-Held Motor-Operated Electric Tools — Safety — Part 2-3: Particular Requirements for Grinders, Polishers, and Disk-Type Sanders
	UL 60745-2-4	Hand-Held Motor-Operated Electric Tools — Safety — Part 2-4: Particular Requirements for Sanders and Polishers Other Than Disk Type
	UL 60745-2-5	Hand-Held Motor-Operated Electric Tools — Safety — Part 2-5: Particular Requirements for Circular Saws
	UL 60745-2-6	Hand-Held Motor-Operated Electric Tools — Safety — Part 2-6: Particular Requirements for Hammers
	UL 60745-2-8	Hand-Held Motor-Operated Electric Tools — Safety — Part 2-8: Particular Requirements for Shears and Nibblers
	UL 60745-2-9	Hand-Held Motor-Operated Electric Tools — Safety — Part 2-9: Particular Requirements for Tappers
	UL 60745-2-11	Hand-Held Motor-Operated Electric Tools — Safety — Part 2-11: Particular Requirements for Reciprocating Saws
	UL 60745-2-12	Hand-Held Motor-Operated Electric Tools — Safety — Part 2-12: Particular Requirements For Concrete Vibrators
	UL 60745-2-13	Hand-Held Motor-Operated Electric Tools — Safety — Part 2-13: Particular Requirements For Chain Saws
	UL 60745-2-14	Hand-Held Motor-Operated Electric Tools — Safety — Part 2-14: Particular Requirements for Planers
	UL 60745-2-15	Hand-Held Motor-Operated Electric Tools — Safety — Part 2-15: Particular Requirements for Hedge Trimmers
	UL 60745-2-16	Hand-Held Motor-Operated Electric Tools — Safety — Part 2-16: Particular Requirements for Tackers
	UL 60745-2-17	Hand-Held Motor-Operated Electric Tools — Safety — Part 2-17: Particular Requirements for Routers and Trimmers
	UL 60745-2-18	Hand-Held Motor-Operated Electric Tools — Safety — Part 2-18: Particular Requirements For Strapping Tools
	UL 60745-2-19	Hand-Held Motor-Operated Electric Tools — Safety — Part 2-19: Particular Requirements for Jointers
	UL 60745-2-20	Hand-Held Motor-Operated Electric Tools — Safety — Part 2-20: Particular Requirements for Band Saws
	UL 60745-2-21	Hand-Held Motor-Operated Electric Tools — Safety — Part 2-21: Particular Requirements For Drain Cleaners
	UL 60745-2-22	Hand-Held Motor-Operated electric Tools — Safety — Part 2-22: Particular Requirements for Cut-Off Machines
	UL 60745-2-23	Hand-Held Motor-Operated electric Tools — Safety — Part 2-23: Particular Requirements for Die Grinders and Small Rotary Tools
	UL 62841-1	Electric Motor-Operated Hand-Held Tools, Transportable Tools And Lawn And Garden Machinery — Safety — Part 1: General Requirements
	UL 62841-2-1	Electric Motor-Operated Hand-Held Tools, Transportable Tools And Lawn And Garden Machinery — Safety — Part 2-1: Particular Requirements For Hand-Held Drills and Impact Drills
	UL 62841-2-2	Electric Motor-Operated Hand-Held Tools, Transportable Tools And Lawn And Garden Machinery — Safety — Part 2-2: Particular Requirements For Screwdrivers And Impact Wrenches
	UL 62841-2-3	Electric Motor-Operated Hand-Held Tools, Transportable Tools And Lawn And Garden Machinery — Safety — Part 2-3: Particular Requirements For Hand-Held Grinders, Polishers, and Disk-Type Sanders
	UL 62841-2-4	Electric Motor-Operated Hand-Held Tools, Transportable Tools And Lawn And Garden Machinery — Safety — Part 2-4: Particular Requirements For Hand-Held Sanders And Polishers Other Than Disc Type
	UL 62841-2-5	Electric Motor-Operated Hand-Held Tools, Transportable Tools And Lawn And Garden Machinery — Safety — Part 2-5: Particular Requirements For Hand-Held Circular Saws
	UL 62841-2-8	Electric Motor-Operated Hand-Held Tools, Transportable Tools And Lawn And Garden Machinery — Safety — Part 2-8: Particular Requirements For Hand-Held Shears and Nibblers

(continues)

N **TABLE A.1(a)** *Continued*

Article	Standard Number	Standard Title
	UL 62841-2-9	Electric Motor-Operated Hand-Held Tools, Transportable Tools And Lawn And Garden Machinery — Safety — Part 2-9: Particular Requirements For Hand-Held Tappers And Threaders
	UL 62841-2-10	Electric Motor-Operated Hand-Held Tools, Transportable Tools And Lawn And Garden Machinery — Safety — Part 2-10: Particular Requirements For Hand-Held Mixers
	UL 62841-2-11	Electric Motor-Operated Hand-Held Tools, Transportable Tools And Lawn And Garden Machinery — Safety — Part 2-11: Particular Requirements for Hand-Held Reciprocating Saws
	UL 62841-2-14	Electric Motor-Operated Hand-Held Tools, Transportable Tools And Lawn And Garden Machinery — Safety — Part 2-14: Particular Requirements For Hand-Held Planers
	UL 62841-2-17	Electric Motor-Operated Hand-Held Tools, Transportable Tools And Lawn And Garden Machinery — Safety — Part 2-17: Particular Requirements For Hand-Held Routers
	UL 62841-2-21	Electric Motor-Operated Hand-Held Tools, Transportable Tools And Lawn And Garden Machinery — Safety — Part 2-21: Particular Requirements For Hand-Held Drain Cleaners
	UL 62841-3-1	Electric Motor-Operated Hand-Held Tools, Transportable Tools And Lawn And Garden Machinery — Safety — Part 3-1: Particular Requirements For Transportable Table Saws
	UL 62841-3-4	Electric Motor-Operated Hand-Held Tools, Transportable Tools And Lawn And Garden Machinery — Safety — Part 3-4: Particular Requirements for Transportable Bench Grinders
	UL 62841-3-6	Electric Motor-Operated Hand-Held Tools, Transportable Tools And Lawn And Garden Machinery — Safety — Part 3-6: Particular Requirements For Transportable Diamond Drills with Liquid System
	UL 62841-3-9	Electric Motor-Operated Hand-Held Tools, Transportable Tools And Lawn And Garden Machinery — Safety — Part 3-9: Particular Requirements For Transportable Mitre Saws
	UL 62841-3-10	Electric Motor-Operated Hand-Held Tools, Transportable Tools And Lawn And Garden Machinery — Safety — Part 3-10: Particular Requirements for Transportable Cut-Off Machines
	UL 62841-3-12	Electric Motor-Operated Hand-Held Tools, Transportable Tools And Lawn And Garden Machinery — Safety — Part 3-12: Particular Requirements for Transportable Threading Machines
	UL 62841-3-13	Electric Motor-Operated Hand-Held Tools, Transportable Tools And Lawn And Garden Machinery — Safety — Part 3-13: Particular Requirements For Transportable Drills
	UL 62841-3-14	Electric Motor-Operated Hand-Held Tools, Transportable Tools And Lawn And Garden Machinery — Safety — Part 3-14: Particular Requirements for Transportable Drain Cleaners
	UL 62841-3-1000	Electric Motor-Operated Hand-Held Tools, Transportable Tools And Lawn And Garden Machinery — Safety — Part 3-1000: Particular Requirements for Transportable Laser Engravers
	UL 62841-4-1	Electric Motor-Operated Hand-Held Tools, Transportable Tools And Lawn And Garden Machinery — Safety — Part 4-1: Particular Requirements for Chain Saws
	UL 62841-4-2	Electric Motor-Operated Hand-Held Tools, Transportable Tools And Lawn And Garden Machinery — Safety — Part 4-2: Particular Requirements for Hedge Trimmers
	UL 62841-4-1000	Electric Motor-Operated Hand-Held Tools, Transportable Tools And Lawn And Garden Machinery — Safety — Part 4-1000: Particular Requirements For Utility Machines
424	UL 499	Electric Heating Appliances
	UL 1042	Electric Baseboard Heating Equipment
	UL 1673	Electric Space Heating Cables
	UL 1693	Electric Radiant Heating Panels and Heating Panel Sets
	UL 1995	Heating and Cooling Equipment
	UL 1996	Electric Duct Heaters
	UL 2021	Fixed and Location-Dedicated Electric Room Heaters
	UL 2683	Electric Heating Products for Floor and Ceiling Installation
425	UL 508	Industrial Control Equipment
	UL 2021	Fixed and Location-Dedicated Electric Room Heaters
426	IEEE 515	Testing, Design, Installation and Maintenance of Electrical Resistance Trace Heating for Industrial Applications
	UL 1588	Roof and Gutter De-Icing Cable Units
	UL 2049	Residential Pipe Heating Cable
427	IEEE 515	Testing, Design, Installation and Maintenance of Electrical Resistance Trace Heating for Industrial Applications
	UL 515	Electrical Resistance Heat Tracing for Commercial Applications
	UL 2049	Residential Pipe Heating Cable
430	UL 4	Armored Cable
	UL 98	Enclosed and Dead-Front Switches
	UL 347	Medium-Voltage AC Contactors, Controllers, and Control Centers
	UL 347A	Medium Voltage Power Conversion Equipment
	UL 489	Molded-Case Circuit Breakers, Molded-Case Switches and Circuit-Breaker Enclosures

N **TABLE A.1(a)** *Continued*

Article	Standard Number	Standard Title
	UL 508	Industrial Control Equipment
	UL 705	Power Ventilators
	UL 745-1	Portable Electric Tools
	UL 845	Motor Control Centers
	UL 987	Stationary and Fixed Electric Tools
	UL 1004-1	Rotating Electrical Machines — General Requirements
	UL 1004-2	Impedance Protected Motors
	UL 1004-3	Thermally Protected Motors
	UL 1004-6	Servo and Stepper Motors
	UL 1004-7	Electronically Protected Motors
	UL 1004-8	Inverter Duty Motors
	UL 1004-9	Form Wound and Medium Voltage Rotating Electrical Machines
	UL 1066	Low-Voltage AC and DC Power Circuit Breakers Used in Enclosures
	UL 1569	Metal Clad Cables
	UL 1812	Ducted Heat Recovery Ventilators
	UL 1815	Nonducted Heat Recovery Ventilators
	UL 2565	Industrial Metalworking and Woodworking Machine Tools
	UL 60034-1	Rotating Electrical Machines — Part 1: Rating and Performance
	UL 60335-2-40	Household and Similar Electrical Appliances — Part 2: Particular Requirements for Electrical Heat Pumps, Air-Conditioners and Dehumidifiers
	UL 60730-2-22	Automatic Electrical Controls — Part 2: Particular Requirements for Thermal Motor Protectors
	UL 60745-1	Electric Motor-Operated Hand-Held Tools, Transportable Tools And Lawn And Garden Machinery — Safety — Part 1: General Requirements
	UL 60745-2-1	Electric Motor-Operated Hand-Held Tools, Transportable Tools And Lawn And Garden Machinery — Safety — Part 2-1: Particular Requirements For Hand-Held Drills and Impact Drills
	UL 60745-2-2	Electric Motor-Operated Hand-Held Tools, Transportable Tools And Lawn And Garden Machinery — Safety — Part 2-2: Particular Requirements For Screwdrivers And Impact Wrenches
	UL 60745-2-3	Electric Motor-Operated Hand-Held Tools, Transportable Tools And Lawn And Garden Machinery — Safety — Part 2-3: Particular Requirements For Hand-Held Grinders, Polishers, and Disk-Type Sanders
	UL 60745-2-4	Electric Motor-Operated Hand-Held Tools, Transportable Tools And Lawn And Garden Machinery — Safety — Part 2-4: Particular Requirements For Hand-Held Sanders And Polishers Other Than Disc Type
	UL 60745-2-5	Electric Motor-Operated Hand-Held Tools, Transportable Tools And Lawn And Garden Machinery — Safety — Part 2-5: Particular Requirements For Hand-Held Circular Saws
	UL 60745-2-8	Electric Motor-Operated Hand-Held Tools, Transportable Tools And Lawn And Garden Machinery — Safety — Part 2-8: Particular Requirements For Hand-Held Shears and Nibblers
	UL 60947-1	Low-Voltage Switchgear and Controlgear — Part 1: General Rules
	UL 60947-4-1	Low-Voltage Switchgear and Controlgear — Part 4-1: Contactors and Motor-Starters — Electromechanical Contactors and Motor-Starters
	UL 60947-4-2	Low-Voltage Switchgear and Controlgear — Part 4-2: Contactors and Motor-Starters — AC Semiconductor Motor Controllers and Starters
	UL 60947-5-1	Low-Voltage Switchgear and Controlgear — Part 5-1: Control Circuit Devices and Switching Elements — Electromechanical Control Circuit Devices
	UL 60947-5-2	Low-Voltage Switchgear and Controlgear — Part 5-2: Control Circuit Devices and Switching Elements — Proximity Switches
	UL 61800-5-1	Adjustable Speed Electrical Power Drive Systems — Part 5-1: Safety Requirements — Electrical, Thermal and Energy
	UL 62841-2-9	Electric Motor-Operated Hand-Held Tools, Transportable Tools And Lawn And Garden Machinery — Safety — Part 2-9: Particular Requirements For Hand-Held Tappers And Threaders
	UL 62841-2-10	Electric Motor-Operated Hand-Held Tools, Transportable Tools And Lawn And Garden Machinery — Safety — Part 2-10: Particular Requirements For Hand-Held Mixers
	UL 62841-2-11	Electric Motor-Operated Hand-Held Tools, Transportable Tools And Lawn And Garden Machinery — Safety — Part 2-11: Particular Requirements for Hand-Held Reciprocating Saws
	UL 62841-2-14	Electric Motor-Operated Hand-Held Tools, Transportable Tools And Lawn And Garden Machinery — Safety — Part 2-14: Particular Requirements For Hand-Held Planers
	UL 62841-2-17	Electric Motor-Operated Hand-Held Tools, Transportable Tools And Lawn And Garden Machinery — Safety — Part 2-17: Particular Requirements For Hand-Held Routers

(continues)

Article	Standard Number	Standard Title
	UL 62841-2-21	Electric Motor-Operated Hand-Held Tools, Transportable Tools And Lawn And Garden Machinery — Safety — Part 2-21: Particular Requirements For Hand-Held Drain Cleaners
	UL 62841-3-1	Electric Motor-Operated Hand-Held Tools, Transportable Tools And Lawn And Garden Machinery — Safety — Part 3-1: Particular Requirements For Transportable Table Saws
	UL 62841-3-4	Electric Motor-Operated Hand-Held Tools, Transportable Tools And Lawn And Garden Machinery — Safety — Part 3-4: Particular Requirements for Transportable Bench Grinders
	UL 62841-3-6	Electric Motor-Operated Hand-Held Tools, Transportable Tools And Lawn And Garden Machinery — Safety — Part 3-6: Particular Requirements For Transportable Diamond Drills with Liquid System
	UL 62841-3-9	Electric Motor-Operated Hand-Held Tools, Transportable Tools And Lawn And Garden Machinery — Safety — Part 3-9: Particular Requirements For Transportable Mitre Saws
	UL 62841-3-10	Electric Motor-Operated Hand-Held Tools, Transportable Tools And Lawn And Garden Machinery — Safety — Part 3-10: Particular requirements for Transportable Cut-Off Machines
	UL 62841-3-12	Electric Motor-Operated Hand-Held Tools, Transportable Tools And Lawn And Garden Machinery — Safety — Part 3-12: Particular requirements for Transportable Threading Machines
	UL 62841-3-13	Electric Motor-Operated Hand-Held Tools, Transportable Tools And Lawn And Garden Machinery — Safety — Part 3-13: Particular Requirements For Transportable Drills
	UL 62841-3-14	Electric Motor-Operated Hand-Held Tools, Transportable Tools And Lawn And Garden Machinery — Safety — Part 3-14: Particular requirements for Transportable Drain Cleaners
	UL 62841-3-1000	Electric Motor-Operated Hand-Held Tools, Transportable Tools And Lawn And Garden Machinery — Safety — Part 3-1000: Particular Requirements for Transportable Laser Engravers
	UL 62841-4-1	Electric Motor-Operated Hand-Held Tools, Transportable Tools And Lawn And Garden Machinery — Safety — Part 4-1: Particular Requirements for Chain Saws
	UL 62841-4-2	Electric Motor-Operated Hand-Held Tools, Transportable Tools And Lawn And Garden Machinery — Safety — Part 4-2: Particular Requirements for Hedge Trimmers
	UL 62841-4-1000	Electric Motor-Operated Hand-Held Tools, Transportable Tools And Lawn And Garden Machinery — Safety — Part 4-1000: Particular Requirements For Utility Machines
440	UL 98	Enclosed and Dead-Front Switches
	UL 416	Refrigerated Medical Equipment
	UL 484	Room Air Conditioners
	UL 489	Molded-Case Circuit Breakers, Molded-Case Switches and Circuit-Breaker Enclosures
	UL 508	Industrial Control Equipment
	UL 541	Refrigerated Vending Machines
	UL 563	Ice Makers
	UL 1429	Pullout Switches
	UL 1995	Heating and Cooling Equipment
	UL 60335-2-24	Household and Similar Electrical Appliances, Part 2: Particular Requirements for Refrigerating Appliances, Ice-Cream Appliances and Ice-Makers
	UL 60335-2-40	Household and Similar Electrical Appliances, Part 2: Particular Requirements for Electrical Heat Pumps, Air-Conditioners and Dehumidifiers
	UL 60335-2-89	Household and Similar Electrical Appliances — Safety — Part 2-89: Particular Requirements for Commercial Refrigerating Appliances with an Incorporated or Remote Refrigerant Unit or Compressor
	UL 60947-4-1	Low-Voltage Switchgear and Controlgear — Part 4-1: Contactors and Motor-Starters — Electromechanical Contactors and Motor-Starters
	UL 60947-4-2	Low-Voltage Switchgear and Controlgear — Part 4-2: Contactors and Motor-Starters — AC Semiconductor Motor Controllers and Starters
	UL 61800-5-1	Adjustable Speed Electrical Power Drive Systems — Part 5-2: Safety Requirements — Functional
445	UL 508	Industrial Control Equipment
	UL 943	Ground-Fault Circuit-Interrupters
	UL 943C	Special Purpose Ground-Fault Circuit-Interrupters
	UL 1004-4	Electric Generators
	UL 2200	Stationary Engine Generator Assemblies
450	UL 10C	Positive Pressure Fire Tests of Door Assemblies
	UL 305	Panic Hardware
	UL 340	Tests for Comparative Flammability of Liquids
	UL 60730-2-14	Automatic Electrical Controls; Part 2: Particular Requirements for Electric Actuators
480	UL 10C	Positive Pressure Fire Tests of Door Assemblies
	UL 305	Panic Hardware
	UL 1642	Lithium Batteries

N **TABLE A.1(a)** *Continued*

Article	Standard Number	Standard Title
	UL 1973	Batteries for Use in Stationary, Vehicle Auxiliary Power and Light Electric Rail (LER) Applications
	UL 1989	Standby Batteries
	UL 2054	Household and Commercial Batteries
	UL 4127	Low Voltage Battery Cable
	UL 4128	Intercell and Intertier Connectors for use in Electrochemical Battery System Applications
490	UL 347	Medium-Voltage AC Contactors, Controllers, and Control Centers
	UL 347A	Medium Voltage Power Conversion Equipment
	UL 347B	Medium Voltage Motor Controllers, Up to 15kV
	UL 347C	Medium Voltage Solid State Resistive Load Controllers, Up to 15kV
	UL 1008A	Transfer Switch Equipment, Over 1000 Volts
500	FM 121303	Guide for Use of Detectors for Flammable Gases
	IEEE 844.1	Skin Effect Trace Heating of Pipelines, Vessels, Equipment, and Structures — General, Testing, Marking, and Documentation Requirements
	IEEE 1349	Guide for the Application of Electric Machines in Zone 2 and Class I, Division 2 Hazardous (Classified) Locations
	NFPA 33	Standard for Spray Application Using Flammable or Combustible Materials
	NFPA 34	Standard for Dipping, Coating, and Printing Processes Using Flammable or Combustible Liquids
	NFPA 496	Standard for Purged and Pressurized Enclosures for Electrical Equipment
	UL 674	Electric Motors and Generators for Use in Hazardous (Classified) Locations
	UL 698A	Industrial Control Panels Relating to Hazardous (Classified) Locations
	UL 783	Electric Flashlights and Lanterns for Use in Hazardous (Classified) Locations
	UL 823	Electric Heaters For Use in Hazardous (Classified) Locations
	UL 844	Electric Heaters For Use in Hazardous (Classified) Locations
	UL 913	Intrinsically Safe Apparatus and Associated Apparatus for Use in Class I, II, and III, Division 1, Hazardous (Classified) Locations
	UL 1203	Explosionproof and Dust-Ignition-Proof Electrical Equipment for Use in Hazardous (Classified) Locations
	UL 1389	Plant Oil Extraction Equipment for Installation and Use in Ordinary (Unclassified) Locations and Hazardous (Classified) Locations
	UL 1836	Electric Motors and Generators for Use in Class I, Division 2, Class I, Zone 2, Class II, Division 2 and Zone 22 Hazardous (Classified) Locations
	UL 2225	Cable and Cable Fittings for Use in Hazardous (Classified) Locations
	UL 60079-28	Explosive Atmospheres — Part 28: Protection of Equipment and Transmission Systems Using Optical Radiation.
	UL 60079-29-1	Explosive Atmospheres — Part 29-1: Gas Detectors — Performance Requirements of Detectors for Flammable Gases
	UL 60079-29-4	Explosive Atmospheres — Part 29-4: Gas Detectors — Performance Requirements of Open Path Detectors for Flammable Gases
	UL 60079-30-1	Explosive Atmospheres — Electrical Resistance Trace Heating — General and Testing Requirements
	UL 60079-33	Explosive Atmospheres — Part 33: Equipment Protection by Special Protection "s"
	UL 121201	Nonincendive Electrical Equipment for Use in Class I and II, Division 2 and Class III, Divisions 1 and 2 Hazardous (Classified) Locations
	UL 121303	Guide for Use of Detectors for Flammable Gases
	UL 122001	General Requirements for Electrical Ignition Systems for Internal Combustion Engines in Class I, Division 2 or Zone 2, Hazardous (Classified) Locations
	UL 122701	Requirements for Process Sealing Between Electrical Systems and Potentially Flammable or Combustible Process Fluids
501	IEEE 844.1	Skin Effect Trace Heating of Pipelines, Vessels, Equipment, and Structures — General, Testing, Marking, and Documentation Requirements
	IEEE 1349	Guide for the Application of Electric Machines in Zone 2 and Class I, Division 2 Hazardous (Classified) Locations
	NFPA 496	Standard for Purged and Pressurized Enclosures for Electrical Equipment
	UL 674	Electric Motors and Generators for Use in Hazardous (Classified) Locations
	UL 783	Electric Flashlights and Lanterns for Use in Hazardous (Classified) Locations
	UL 823	Standard for Electric Heaters For Use in Hazardous (Classified) Locations
	UL 844	Luminaires for Use in Hazardous (Classified) Locations
	UL 1072	Medium-Voltage Power Cables

(continues)

N *TABLE A.1(a)* *Continued*

Article	Standard Number	Standard Title
	UL 1203	Explosionproof and Dust-Ignition-Proof Electrical Equipment for Use in Hazardous (Classified) Locations
	UL 1277	Electrical Power and Control Tray Cables with Optional Optical-Fiber Members
	UL 1309A	Cable for Use in Mobile Applications
	UL 1836	Electric Motors and Generators for Use in Class I, Division 2, Class I, Zone 2, Class II, Division 2 and Zone 22 Hazardous (Classified) Locations
	UL 2225	Cable and Cable Fittings for Use in Hazardous (Classified) Locations
	UL 60079-28	Explosive Atmospheres — Part 28: Protection of Equipment and Transmission Systems Using Optical Radiation
	UL 60079-29-1	Explosive Atmospheres — Part 29-1: Gas Detectors — Performance Requirements of Detectors for Flammable Gases
	UL 60079-29-4	Explosive Atmospheres — Part 29-4: Gas Detectors — Performance Requirements of Open Path Detectors for Flammable Gases
	UL 60079-30-1	Part 30-1: Electrical Resistance Trace Heating — General and Testing Requirements
	UL 60079-33	Explosive Atmospheres — Part 33: Equipment Protection by Special Protection "s"
	UL 121201	Nonincendive Electrical Equipment for Use in Class I and II, Division 2 and Class III, Divisions 1 and 2 Hazardous (Classified) Locations
	UL 122001	General Requirements for Electrical Ignition Systems for Internal Combustion Engines in Class I, Division 2 or Zone 2, Hazardous (Classified) Locations
	UL 122701	Requirements for Process Sealing Between Electrical Systems and Potentially Flammable or Combustible Process Fluids
502	IEEE 844.1	Skin Effect Trace Heating of Pipelines, Vessels, Equipment, and Structures — General, Testing, Marking, and Documentation Requirements
	NFPA 496	Standard for Purged and Pressurized Enclosures for Electrical Equipment
	UL 674	Electric Motors and Generators for Use in Hazardous (Classified) Locations
	UL 783	Electric Flashlights and Lanterns for Use in Hazardous (Classified) Locations
	UL 823	Electric Heaters For Use in Hazardous (Classified) Locations
	UL 844	Luminaires for Use in Hazardous (Classified) Locations
	UL 1203	Explosionproof and Dust-Ignition-Proof Electrical Equipment for Use in Hazardous (Classified) Locations
	UL 1309A	Cable for Mobile Installations
	UL 1836	Outline of Investigation for Electric Motors and Generators for Use in Class I, Division 2, Class I, Zone 2, Class II, Division 2 and Zone 22 Hazardous (Classified) Locations
	UL 2225	Cable and Cable Fittings for Use in Hazardous (Classified) Locations
	UL 60079-28	Part 30-1: Part 28: Protection of Equipment and Transmission Systems Using Optical Radiation
	UL 60079-30-1	Explosive Atmospheres — Electrical Resistance Trace Heating — General and Testing Requirements
	UL 60079-33	Explosive Atmospheres — Part 33: Equipment Protection by Special Protection "s"
	UL 121201	Nonincendive Electrical Equipment for Use in Class I and II, Division 2 and Class III, Divisions 1 and 2 Hazardous (Classified) Locations
503	IEEE 844.1	Skin Effect Trace Heating of Pipelines, Vessels, Equipment, and Structures — General, Testing, Marking, and Documentation Requirements
	UL 823	Standard for Electric Heaters For Use in Hazardous (Classified) Locations
	UL 844	Luminaires for Use in Hazardous (Classified) Locations
	UL 1836	Electric Motors and Generators for Use in Class I, Division 2, Class I, Zone 2, Class II, Division 2 and Zone 22 Hazardous (Classified) Locations
	UL 60079-30-1	Explosive Atmospheres — Electrical Resistance Trace Heating — General and Testing Requirements
	UL 121201	Nonincendive Electrical Equipment for Use in Class I and II, Division 2 and Class III, Divisions 1 and 2 Hazardous (Classified) Locations
504	UL 698A	Standard for Industrial Control Panels Relating to Hazardous (Classified) Locations
	UL 913	Intrinsically Safe Apparatus and Associated Apparatus for Use in Class I, II, and III, Division 1, Hazardous (Classified) Locations
	UL 120202	Recommendations for the Preparation, Content, and Organization of Intrinsic Safety Control Drawings
505	FM 121303	Guide for Use of Detectors for Flammable Gases
	IEEE 844.1	Skin Effect Trace Heating of Pipelines, Vessels, Equipment, and Structures — General, Testing, Marking, and Documentation Requirements
	IEEE 1349	Guide for the Application of Electric Machines in Zone 2 and Class I, Division 2 Hazardous (Classified) Locations
	UL 1309A	Cable for Mobile Installations
	UL 2225	Cable and Cable Fittings for Use in Hazardous (Classified) Locations

N **TABLE A.1(a)** *Continued*

Article	Standard Number	Standard Title
	UL 60079-0	Explosive Atmospheres — Part 0: Equipment — General Requirements
	UL 60079-1	Explosive Atmospheres — Part 1: Equipment Protection by Flameproof Enclosures "d"
	UL 60079-2	Explosive Atmospheres — Part 2: Equipment protection by pressurized enclosure "p"
	UL 60079-5	Explosive Gas Atmospheres — Part 5: Type of Protection — Powder Filling "q"
	UL 60079-6	Explosive Atmospheres — Part 6: Equipment Protection by Liquid Immersion "o"
	UL 60079-7	Explosive Atmospheres — Part 7: Equipment Protection by Increased Safety "e"
	UL 60079-10-1	Explosive Atmospheres — Part 10-1: Classification of Areas — Explosive Gas Atmospheres
	UL 60079-11	Explosive Atmospheres — Part 11: Equipment Protection by Intrinsic Safety "i"
	UL 60079-13	Explosive Atmospheres — Part 13: Equipment Protection by Pressurized Room "p" and Artificially Ventilated Room "v"
	UL 60079-15	Explosive Atmospheres — Part 15: Equipment Protection by Type of Protection "n"
	UL 60079-18	Explosive Atmospheres — Part 18: Equipment Protection by Encapsulation "m"
	UL 60079-25	Explosive Atmospheres — Part 25: Intrinsically Safe Electrical Systems
	UL 60079-26	Explosive Atmospheres — Part 26: Equipment with Equipment Protection Level (EPL) Ga
	UL 60079-28	Explosive Atmospheres — Part 28: Protection of Equipment and Transmission Systems Using Optical Radiation
	UL 60079-29-1	Explosive Atmospheres — Part 29-1: Gas Detectors — Performance Requirements of Detectors for Flammable Gases
	UL 60079-29-4	Explosive Atmospheres — Part 29-4: Gas Detectors — Performance Requirements of Open Path Detectors for Flammable Gases
	UL 60079-30-1	Explosive Atmospheres — Part 30-1: Electrical Resistance Trace Heating — General and Testing Requirements
	UL 60079-33	Explosive Atmospheres — Part 33: Equipment Protection by Special Protection "s"
	UL 80079-36	Explosive Atmospheres — Part 36: Non-Electrical Equipment for Explosive Atmospheres — Basic Method and Requirements
	UL 80079-37	Explosive Atmospheres — Part 37: Non-Electrical Equipment for Explosive Atmospheres — Non Electrical Type of Protection Constructional Safety "c", Control of Ignition Source "b", Liquid Immersion "k"
	UL 121303	Guide for Use of Detectors for Flammable Gases
	UL 122701	Requirements for Process Sealing Between Electrical Systems and Flammable or Combustible Process Fluids
506	IEEE 844.1	Skin Effect Trace Heating of Pipelines, Vessels, Equipment, and Structures — General, Testing, Marking, and Documentation Requirements
	UL 698A	Industrial Control Panels Relating to Hazardous (Classified) Locations
	UL 2225	Cable and Cable Fittings for Use in Hazardous (Classified) Locations
	UL 60079-0	Explosive Atmospheres — Part 0: Equipment — General Requirements
	UL 60079-2	Explosive atmospheres — Part 2: Equipment protection by pressurized enclosure "p"
	UL 60079-11	Explosive Atmospheres — Part 11: Equipment Protection by Intrinsic Safety "i"
	UL 60079-18	Explosive Atmospheres — Part 18: Equipment Protection by Encapsulation "m"
	UL 60079-25	Explosive Atmospheres — Part 25: Intrinsically Safe Electrical Systems
	UL 60079-28	Explosive Atmospheres — Part 28: Protection of Equipment and Transmission Systems Using Optical Radiation
	UL 60079-30-1	Part 30-1: Electrical Resistance Trace Heating — General and Testing Requirements
	UL 60079-31	Explosive Atmospheres — Part 31: Equipment Dust Ignition Protection by Enclosure "t"
	UL 60079-33	Explosive Atmospheres — Part 33: Equipment Protection by Special Protection "s"
	UL 62784	Vacuum Cleaners and Dust Extractors Providing Equipment Protection Level Dc for the Collection of Combustible Dusts — Particular Requirements
	UL 80079-36	Explosive Atmospheres — Part 36: Non-Electrical Equipment for Explosive Atmospheres — Basic Method and Requirements
	UL 80079-37	Explosive Atmospheres — Part 37: Non-Electrical Equipment for Explosive Atmospheres — Non Electrical Type of Protection Constructional Safety "c", Control of Ignition Source "b", Liquid Immersion "k"
512	UL 1389	Plant Oil Extraction Equipment for Installation and Use in Ordinary (Unclassified) Locations and Hazardous (Classified) Locations
516	NFPA 33	Standard for Spray Application Using Flammable or Combustible Materials
	NFPA 34	Standard for Dipping, Coating, and Printing Processes Using Flammable or Combustible Liquids
	UL 844	Luminaires for Use in Hazardous (Classified) Locations

(continues)

N **TABLE A.1(a)** *Continued*

Article	Standard Number	Standard Title
517	AAMI ES 60601-1	Medical electrical equipment — Part 1: General requirements for basic safety and essential performance
	UL 5	Surface Metal Raceways and Fittings
	UL 5A	Nonmetallic Surface Raceways and Fittings
	UL 467	Grounding and Bonding Equipment
	UL 498	Attachment Plugs and Receptacles
	UL 498D	Attachment Plugs, Cord Connectors and Receptacles with Arcuate (Locking Type) Contacts
	UL 498E	Attachment Plugs, Cord Connectors and Receptacles — Enclosure Types for Environmental Protection
	UL 498F	Plugs, Socket-Outlets and Couplers with Arcuate (Locking Type) Contacts
	UL 651	Schedule 40, 80, Type EB and A Rigid PVC Conduit and Fittings
	UL 1022	Line Isolation Monitors
	UL 1047	Isolated Power Systems Equipment
	UL 1286	Office Furnishing Systems
	UL 2930	Cord-and-Plug-connected Health Care Facility Outlet Assemblies
	UL 60601-1	Medical Electrical Equipment — Part 1: General Requirements for Safety
	UL 122701	Requirements for Process Sealing Between Electrical Systems and Flammable or Combustible Process Fluids
518	UL 498	Attachment Plugs and Receptacles
	UL 498D	Attachment Plugs, Cord Connectors and Receptacles with Arcuate (Locking Type) Contacts
	UL 498E	Attachment Plugs, Cord Connectors and Receptacles — Enclosure Types for Environmental Protection
	UL 498F	Plugs, Socket-Outlets and Couplers with Arcuate (Locking Type) Contacts
	UL 943	Ground-Fault Circuit-Interrupters
	UL 943C	Special Purpose Ground-Fault Circuit-Interrupters
	UL 2305	Exhibition Display Units, Fabrication and Installation
	UL 2305A	Convention Center Cord Sets
520	UL 62	Flexible Cords and Cables
	UL 334	Theater Lighting Distribution and Control Equipment
	UL 1573	Stage and Studio Luminaires and Connector Strips
	UL 1640	Portable Power-Distribution Equipment
	UL 1691	Single Pole Locking-Type Separable Connectors
522	UL 13	Power Limited Circuit Cables
	UL 1063	Machine-Tool Wires and Cables
	UL 2250	Instrumentation Tray Cable
525	UL 62	Flexible Cords and Cables
	UL 817	Cord Sets and Power-Supply Cords
	UL 943	Ground-Fault Circuit-Interrupters
	UL 943C	Special Purpose Ground-Fault Circuit-Interrupters
	UL 1691	Single Pole Locking-Type Separable Connectors
530	UL 62	Flexible Cords and Cables
	UL 1479	Fire Tests of Penetration Firestops
	UL 1573	Stage and Studio Luminaires and Connector Strips
	UL 1680	Stage and Lighting Cables
	UL 1691	Single Pole Locking-Type Separable Connectors
	UL 1836	Electric Motors and Generators for Use in Class I, Division 2, Class I, Zone 2, Class II, Division 2 and Zone 22 Hazardous (Classified) Locations
	UL 62368-1	Audio/Video, Information and Communication Technology Equipment — Part 1: Safety Requirements
540	UL 62368-1	Audio/Video, Information and Communication Technology Equipment — Part 1: Safety Requirements
545	UL 5	Surface Metal Raceways and Fittings
	UL 5A	Nonmetallic Surface Raceways and Fittings
	UL 5B	Strut-Type Channel Raceways and Fittings
	UL 5C	Surface Raceways and Fittings for Use with Data, Signal, and Control Circuits
	UL 20	General Use Snap Switches
	UL 209	Cellular Metal Floor Raceways and Fittings
	UL 498	Attachment Plugs and Receptacles
	UL 498D	Attachment Plugs, Cord Connectors and Receptacles with Arcuate (Locking Type) Contacts
	UL 498E	Attachment Plugs, Cord Connectors and Receptacles — Enclosure Types for Environmental Protection
	UL 498F	Plugs, Socket-Outlets and Couplers with Arcuate (Locking Type) Contacts
	UL 514A	Metallic Outlet Boxes
	UL 514C	Nonmetallic Outlet Boxes, Flush-Device Boxes, and Covers
	UL 2024	Cable Routing Assemblies and Communications Raceways

N **TABLE A.1(a)** *Continued*

Article	Standard Number	Standard Title
547	UL 50	Enclosures for Electrical Equipment, Non-Environmental Considerations
	UL 50E	Enclosures for Electrical Equipment, Environmental Considerations
	UL 62	Flexible Cords and Cables
	UL 514A	Metallic Outlet Boxes
	UL 514B	Conduit, Tubing, and Cable Fittings
	UL 514C	Nonmetallic Outlet Boxes, Flush-Device Boxes, and Covers
	UL 1598	Luminaires
	UL 2225	Cable and Cable Fittings for Use in Hazardous (Classified) Locations
550	UL 6	Electrical Rigid Metal Conduit — Steel
	UL 6A	Electrical Rigid Metal Conduit — Aluminum, Red Brass and Stainless Steel
	UL 83	Thermoplastic-Insulated Wires and Cables
	UL 307A	Liquid Fuel-Burning Heating Appliances for Manufactured Homes and Recreational Vehicles
	UL 307B	Gas-Burning Heating Appliances for Manufactured Homes and Recreational Vehicles
	UL 360	Liquid-Tight Flexible Metal Conduit
	UL 467	Grounding and Bonding Equipment
	UL 498	Attachment Plugs and Receptacles
	UL 498D	Attachment Plugs, Cord Connectors and Receptacles with Arcuate (Locking Type) Contacts
	UL 498E	Attachment Plugs, Cord Connectors and Receptacles — Enclosure Types for Environmental Protection
	UL 498F	Plugs, Socket-Outlets and Couplers with Arcuate (Locking Type) Contacts
	UL 651	Schedule 40, 80, Type EB and A Rigid PVC Conduit and Fittings
	UL 817	Cord Sets and Power-Supply Cords
	UL 1242	Electrical Intermediate Metal Conduit — Steel
	UL 1462	Mobile Home Pipe Heating Cable
	UL 1598	Luminaires
	UL 1660	Liquid-Tight Flexible Nonmetallic Conduit
	UL 2108	Low-Voltage Lighting Systems
	UL 2515	Aboveground Reinforced Thermosetting Resin Conduit (RTRC) and Fittings
551	UL 6	Electrical Rigid Metal Conduit — Steel
	UL 6A	Electrical Rigid Metal Conduit — Aluminum, Red Brass and Stainless Steel
	UL 62	Flexible Cords and Cables
	UL 231	Power Outlets
	UL 234	Low Voltage Lighting Fixtures for use in Recreational Vehicles
	UL 360	Liquid-Tight Flexible Metal Conduit
	UL 467	Grounding and Bonding Equipment
	UL 486C	Splicing Wire Connectors
	UL 498	Attachment Plugs and Receptacles
	UL 498D	Attachment Plugs, Cord Connectors and Receptacles with Arcuate (Locking Type) Contacts
	UL 498E	Attachment Plugs, Cord Connectors and Receptacles — Enclosure Types for Environmental Protection
	UL 498F	Plugs, Socket-Outlets and Couplers with Arcuate (Locking Type) Contacts
	UL 514A	Metallic Outlet Boxes
	UL 514C	Nonmetallic Outlet Boxes, Flush-Device Boxes, and Covers
	UL 514D	Cover Plates for Flush-Mounted Wiring Devices
	UL 651	Schedule 40, 80, Type EB and A Rigid PVC Conduit and Fittings
	UL 817	Cord Sets and Power-Supply Cords
	UL 943	Ground-Fault Circuit-Interrupters
	UL 1004-4	Electric Generators
	UL 1008	Transfer Switch Equipment
	UL 1008M	Transfer Switch Equipment, Meter Mounted
	UL 1008S	Solid-State Transfer Switches
	UL 1242	Electrical Intermediate Metal Conduit — Steel
	UL 1449	Surge Protective Devices
	UL 1598	Luminaires
	UL 1660	Liquid-Tight Flexible Nonmetallic Conduit
	UL 2200	Stationary Engine Generator Assemblies
	UL 2515	Aboveground Reinforced Thermosetting Resin Conduit (RTRC) and Fittings
	UL 60730-1	Automatic Electrical Controls; Part 1: General Requirements
	UL 60730-2-9	Automatic Electrical Controls; Part 2: Particular Requirements for Temperature Sensing Controls

(continues)

TABLE A.1(a) *Continued*

Article	Standard Number	Standard Title
552	SAE J1128-2015	Low Voltage Primary Cable, for Types GXL, HDT, and SXL
	SAE J1127-2015	Low Voltage Battery Cable, for Types SGT and SGR
	UL 6	Electrical Rigid Metal Conduit — Steel
	UL 6A	Electrical Rigid Metal Conduit — Aluminum, Red Brass and Stainless Steel
	UL 50	Enclosures for Electrical Equipment, Non-Environmental Considerations
	UL 50E	Enclosures for Electrical Equipment, Environmental Considerations
	UL 62	Flexible Cords and Cables
	UL 67	Panelboards
	UL 231	Power Outlets
	UL 234	Low Voltage Lighting Fixtures for Use in Recreational Vehicles
	UL 360	Liquid-Tight Flexible Metal Conduit
	UL 430	Waste Disposers
	UL 467	Grounding and Bonding Equipment
	UL 514A	Metallic Outlet Boxes
	UL 514B	Conduit, Tubing, and Cable Fittings
	UL 514C	Nonmetallic Outlet Boxes, Flush-Device Boxes, and Covers
	UL 651	Schedule 40, 80, Type EB and A Rigid PVC Conduit and Fittings
	UL 817	Cord Sets and Power-Supply Cords
	UL 916	Energy Management Equipment
	UL 943	Ground-Fault Circuit-Interrupters
	UL 1004-4	Electric Generators
	UL 1242	Electrical Intermediate Metal Conduit — Steel
	UL 1563	Electric Spas, Equipment Assemblies, and Associated Equipment
	UL 1598	Luminaires
	UL 1660	Liquid-Tight Flexible Nonmetallic Conduit
	UL 2108	Low Voltage Lighting Systems
	UL 2200	Stationary Engine Generator Assemblies
	UL 2515	Aboveground Reinforced Thermosetting Resin Conduit (RTRC) and Fittings
555	UL 6	Electrical Rigid Metal Conduit — Steel
	UL 6A	Electrical Rigid Metal Conduit — Aluminum, Red Brass and Stainless Steel
	UL 231	Power Outlets
	UL 486D	Sealed Wire Connector Systems
	UL 651	Schedule 40, 80, Type EB and A Rigid PVC Conduit and Fittings
	UL 676	Underwater Luminaires and Submersible Junction Boxes
	UL 943	Ground-Fault Circuit-Interrupters
	UL 1053	Ground-Fault Sensing and Relaying Equipment
	UL 1650	Portable Power Cable
	UL 2515	Aboveground Reinforced Thermosetting Resin Conduit (RTRC) and Fittings
590	UL 496	Lampholders
	UL 514B	Conduit, Tubing, and Cable Fittings
	UL 588	Seasonal and Holiday Decorative Products
	UL 817	Cord Sets
	UL 943	Ground-Fault Circuit-Interrupters
	UL 1088	Temporary Lighting Strings
	UL 1377	Wire used in Low Voltage Seasonal Lighting Products In Circuits With a Maximum Available Power of 15W
600	UL 1	Flexible Metal Conduit
	UL 5	Surface Metal Raceways and Fittings
	UL 5A	Nonmetallic Surface Raceways and Fittings
	UL 13	Power-Limited Circuit Cables
	UL 48	Electric Signs
	UL 50	Enclosures for Electrical Equipment, Non-Environmental Considerations
	UL 50E	Enclosures for Electrical Equipment, Environmental Considerations
	UL 98B	Enclosed and Dead-Front Switches for Use in Photovoltaic Systems
	UL 248-19	Low-Voltage Fuses — Part 19: Photovoltaic Fuses
	UL 360	Liquid-Tight Flexible Metal Conduit
	UL 489B	Molded-Case Circuit Breakers, Molded-Case Switches, and Circuit-Breaker Enclosures For Use With Photovoltaic (PV) Systems
	UL 508I	Disconnect Switches Intended for Use in Photovoltaic Systems

N **TABLE A.1(a)** *Continued*

Article	Standard Number	Standard Title
	UL 814	Gas-Tube-Sign Cable
	UL 879	Electric Sign Components
	UL 879A	LED Sign and Sign Retrofit Kits
	UL 879B	Polymeric Enclosure Systems for the Splice Between Neon Tubing Electrode Leads and GTO Cable, and the GTO Cable Leading to the Splice
	UL 943	Ground-Fault Circuit-Interrupters
	UL 1310	Class 2 Power Units
	UL 1660	Liquid-Tight Flexible Nonmetallic Conduit
	UL 1699B	Photovoltaic (PV) DC Arc-Fault Circuit Protection
	Ul 1741	Inverters, Converters, Controllers and Interconnection System Equipment for Use With Distributed Energy Resources
	UL 2161	Neon Transformers and Power Supplies
	UL 2703	Mounting Systems, Mounting Devices, Clamping/Retention Devices, and Ground Lugs for Use with Flat-Plate Photovoltaic Modules and Panels
	UL 3001	Distributed Energy Generation and Storage Systems
	UL 3003	Distributed Generation Cables
	UL 3703	Solar Trackers
	UL 4703	Photovoltaic Wire
	UL 6703	Connectors for Use in Photovoltaic Systems
	UL 7103	Investigation for Building-Integrated Photovoltaic Roof Coverings
	UL 8703	Concentrator Photovoltaic Modules and Assemblies
	UL 9703	Distributed Generation Wiring Harnesses
	UL 61730-1	Photovoltaic (PV) Module Safety Qualification — Part 1: Requirements For Construction
	UL 61730-2	Photovoltaic (PV) Module Safety Qualification — Part 2: Requirements For Testing
	UL 62109	Power Converters for Use in Photovoltaic Power Systems — Part 1: General Requirements
	UL 62368-1	Audio/Video, Information and Communication Technology Equipment — Part 1: Safety Requirements
604	UL 1	Flexible Metal Conduit
	UL 4	Armored Cable
	UL 5	Surface Metal Raceways and Fittings
	UL 5A	Nonmetallic Surface Raceways and Fittings
	UL 5B	Strut-Type Channel Raceways and Fittings
	UL 5C	Surface Raceways and Fittings for Use with Data, Signal, and Control Circuits
	UL 62	Flexible Cords and Cables
	UL 183	Manufactured Wiring Systems
	UL 209	Cellular Metal Floor Raceways and Fittings
	UL 360	Liquid-Tight Flexible Metal Conduit
	UL 797	Electrical Metallic Tubing — Steel
	UL 797A	Electrical Metallic Tubing — Aluminum and Stainless Steel
	UL 857	Busways
	UL 1569	Metal-Clad Cables
	UL 2024	Cable Routing Assemblies and Communications Raceways
605	UL 962	Household and Commercial Furnishings
	UL 1286	Office Furnishings Systems
	UL 1310	Class 2 Power Units
	UL 2999	Individual Commercial Office Furnishings
	UL 5085-3	Low Voltage Transformers — Part 3: Class 2 and Class 3 Transformers
	UL 62368-1	Audio/Video, Information and Communication Technology Equipment — Part 1: Safety Requirements
610	UL 62	Flexible Cords and Cables
	UL 2273	Festoon Cable
620	UL 62	Flexible Cords and Cables
	UL 83	Thermoplastic-Insulated Wires and Cables
	UL 98	Enclosed and Dead-Front Switches
	UL 104	Elevator Door Locking Devices and Contacts
	UL 489	Molded-Case Circuit Breakers, Molded-Case Switches and Circuit-Breaker Enclosures
	UL 508	Industrial Control Equipment
	UL 508A	Industrial Control Panels
	UL 1066	Low-Voltage AC and DC Power Circuit Breakers Used in Enclosures

(continues)

TABLE A.1(a) *Continued*

Article	Standard Number	Standard Title
	UL 1310	Class 2 Power Units
	UL 1449	Surge Protective Devices
	UL 1685	Vertical-Tray Fire-Propagation and Smoke-Release Test for Electrical and Optical-Fiber Cables
	UL 2556	Wire and Cable Test Methods
	UL 62368-1	Audio/Video, Information and Communication Technology Equipment — Part 1: Safety Requirements
625	UL 62	Flexible Cords And Cables
	UL 1650	Portable Power Cable
	UL 2202	Electric Vehicle (EV) Charging System Equipment
	UL 2231-1	Personnel Protection Systems for Electric Vehicle (EV) Supply Circuits — Part 1: General Requirements
	UL 2231-2	Personnel Protection Systems for Electric Vehicle (EV) Supply Circuits — Part 2: Particular Requirements for Protection Devices for Use in Charging Systems
	UL 2251	Plugs, Receptacles and Couplers for Electrical Vehicles
	UL 2580	Batteries for Use in Electric Vehicles
	UL 2594	Electric Vehicle Supply Equipment
	UL 9741	Electric Vehicle Power Export Equipment (EVPE)
626	UL 62	Flexible Cords and Cables
	UL 231	Power Outlets
	UL 498	Attachment Plugs and Receptacles
	UL 498D	Attachment Plugs, Cord Connectors and Receptacles with Arcuate (Locking Type) Contacts
	UL 498E	Attachment Plugs, Cord Connectors and Receptacles — Enclosure Types for Environmental Protection
	UL 498F	Plugs, Socket-Outlets and Couplers with Arcuate (Locking Type) Contacts
	UL 817	Cord Sets and Power-Supply Cords
	UL 1651	Optical Fiber Cable
	UL 1686	Pin and Sleeve Configurations
630	UL 551	Transformer-Type Arc-Welding Machines
640	UL 13	Power Limited Circuit Cables
	UL 62	Flexible Cords and Cables
	UL 813	Commercial Audio Equipment
	UL 1310	Class 2 Power Units
	UL 1419	Professional Video and Audio Equipment
	UL 1492	Audio-Video Products and Accessories
	UL 1711	Amplifiers for Fire Protective Signaling Systems
	UL 2269	Optical Fiber/Communications/Signaling/Coaxial Cable Outlet Boxes
	UL 6500	Audio/Video and Musical Instrument Apparatus for Household, Commercial, and Similar General Use
	UL 60065	Audio, Video and Similar Electronic Apparatus — Safety Requirements
	UL 62368-1	Audio/Video, Information and Communication Technology Equipment — Part 1: Safety Requirements
645	UL 38	Manual Signaling Boxes for Fire Alarm Systems
	UL 268	Smoke Detectors for Fire Alarm Systems
	UL 444	Communications Cables
	UL 464	Audible Signaling Devices for Fire Alarm and Signaling Systems, Including Accessories
	UL 497B	Protectors for Data Communications and Fire Alarm Circuits
	UL 833	Control Units and Accessories for Fire Alarm Systems
	UL 864	Control Units and Accessories for Fire Alarm Systems
	UL 1424	Cables for Power-Limited Fire-Alarm Circuits
	UL 1425	Cables for Non-Power-Limited Fire-Alarm Circuits
	UL 1449	Surge Protective Devices
	UL 1480	Speakers for Fire Alarm and Signaling Systems, Including Accessories
	UL 1638	Visible Signaling Devices for Fire Alarm and Signaling Systems, Including Accessories
	UL 1651	Optical Fiber Cable
	UL 1685	Vertical-Tray Fire-Propagation and Smoke-Release Test for Electrical and Optical-Fiber Cables
	UL 1690	Data-Processing Cable
	UL 1778	Uninterruptible Power Systems
	UL 2024	Cable Routing Assemblies and Communications Raceways
	UL 60950-1	Information Technology Equipment Safety — Part 1: General Requirements
	UL 60950-21	Information Technology Equipment Safety — Part 21: Remote Power Feeding
	UL 60950-22	Information Technology Equipment Safety — Part 22: Equipment to be Installed Outdoors
	UL 60950-23	Information Technology Equipment Safety — Part 23: Large Data Storage Equipment
	UL 62368-1	Audio/Video, Information and Communication Technology Equipment — Part 1: Safety Requirements

N **TABLE A.1(a)** *Continued*

Article	Standard Number	Standard Title
646	UL 10C	Positive Pressure Fire Tests of Door Assemblies
	UL 62	Flexible Cords and Cables
	UL 67	Panelboards
	UL 98	Enclosed and Dead-Front Switches
	UL 305	Panic Hardware
	UL 347	Medium-Voltage AC Contactors, Controllers, and Control Centers
	UL 489	Molded-Case Circuit Breakers, Molded-Case Switches and Circuit-Breaker Enclosures
	UL 508	Industrial Control Equipment
	UL 508A	Industrial Control Panels
	UL 845	Motor Control Centers
	UL 869A	Reference Standard for Service Equipment
	UL 891	Switchboards
	UL 924	Emergency Lighting and Power Equipment
	UL 977	Fused Power-Circuit Devices
	UL 1008	Transfer Switch Equipment
	UL 1008A	Transfer Switch Equipment, Over 1000 Volts
	UL 1008M	Meter-Mounted Transfer Switches
	UL 1008S	Solid-State Transfer Switches
	UL 1062	Unit Substations
	UL 1066	Low-Voltage AC and DC Power Circuit Breakers Used in Enclosures
	UL 1429	Pullout Switches
	UL 1449	Surge Protective Devices
	UL 1655	Community-Antenna Television Cables
	UL 1989	Standby Batteries
	UL 2755	Modular Data Centers
	UL 62368-1	Audio/Video, Information and Communication Technology Equipment — Part 1: Safety Requirements
647	UL 1598	Luminaires
650	UL 1310	Class 2 Power Units
	UL 1581	Reference Standard for Electrical Wires, Cables, and Flexible Cords
	UL 62368-1	Audio/Video, Information and Communication Technology Equipment — Part 1: Safety Requirements
670	ANSI/CSA-C22.2 No. 19085-1	Woodworking machines — Safety — Part 1: Common requirements
	UL 508	Industrial Control Equipment
	UL 61800-5-1	Adjustable Speed Electrical Power Drive Systems — Part 5-1: Safety Requirements — Electrical, Thermal and Energy
675	UL 493	Thermoplastic-Insulated Underground Feeder and Branch-Circuit Cables
	UL 1581	Reference Standard for Electrical Wires, Cables, and Flexible Cords
680	UL 6	Electrical Rigid Metal Conduit — Steel
	UL 6A	Electrical Rigid Metal Conduit — Aluminum, Red Brass and Stainless Steel
	UL 20	General Use Snap-Switches
	UL 62	Flexible Cords and Cables
	UL 360	Liquid-Tight Flexible Metal Conduit
	UL 379	Power Units for Fountain, Swimming Pool, and Spa Luminaires
	UL 467	Grounding and Bonding Equipment
	UL 486D	Sealed Wire Connector Systems
	UL 489	Molded-Case Circuit Breakers, Molded-Case Switches and Circuit-Breaker Enclosures
	UL 651	Schedule 40, 80, Type EB and A Rigid PVC Conduit and Fittings
	UL 676	Underwater Luminaires and Submersible Junction Boxes
	UL 676A	Potting Compounds for Swimming Pool, Fountain, and Spa Equipment
	UL 943	Ground-Fault Circuit-Interrupters
	UL 943C	Special Purpose Ground-Fault Circuit-Interrupters
	UL 1004-10	Pool Pump Motors
	UL 1081	Swimming Pool Pumps, Filters, and Chlorinators
	UL 1241	Junction Boxes for Swimming Pool Luminaires
	UL 1242	Electrical Intermediate Metal Conduit — Steel
	UL 1261	Electric Water Heaters for Pools and Tubs
	UL 1563	Electric Spas, Equipment Assemblies, and Associated Equipment

(continues)

TABLE A.1(a) *Continued*

Article	Standard Number	Standard Title
	UL 1569	Metal-Clad Cables
	UL 1660	Liquid-Tight Flexible Nonmetallic Conduit
	UL 1795	Hydromassage Bathtubs
	UL 2420	Belowground Reinforced Thermosetting Resin Conduit (RTRC) and Fittings
	UL 2452	Electric Swimming Pool and Spa Cover Operators
	UL 2515	Aboveground Reinforced Thermosetting Resin Conduit (RTRC) and Fittings
	UL 2515A	Supplemental Requirements for Extra Heavy Wall Reinforced Thermosetting Resin Conduit (RTRC) and Fittings
	UL 2995	Lifts for Swimming Pools and Spas
	UL 60335-2-1000	Household and Similar Electrical Appliances: Particular Requirements for Electrically Powered Pool Lifts
682	UL 486D	Sealed Wire Connector Systems
	UL 1650	Portable Power Cable
	UL 1838	Low Voltage Landscape Lighting Systems
690	UL 98B	Enclosed and Dead-Front Switches for Use in Photovoltaic Systems
	UL 248-19	Low-Voltage Fuses — Part 19: Photovoltaic Fuses
	UL 467	Grounding and Bonding Equipment
	UL 489B	Molded-Case Circuit Breakers, Molded-Case Switches, and Circuit-Breaker Enclosures For Use With Photovoltaic (PV) Systems
	UL 508I	Disconnect Switches Intended for Use in Photovoltaic Systems
	UL 1569	Metal-Clad Cables
	UL 1699B	Photovoltaic (PV) DC Arc-Fault Circuit Protection
	UL 1703	Flat-Plate Photovoltaic Modules and Panels
	UL 1741	Inverters, Converters, Controllers and Interconnection System Equipment for Use with Distributed Energy Resources
	UL 2703	Mounting Systems, Mounting Devices, Clamping/Retention Devices, and Ground Lugs for Use with Flat-Plate Photovoltaic Modules and Panels
	UL 3001	Distributed Energy Generation and Storage Systems
	UL 3003	Distributed Generation Cables
	UL 3005	Distributed Energy Resource Management Systems
	UL 3703	Solar Trackers
	UL 3730	Photovoltaic Junction Boxes
	UL 3741	Photovoltaic Hazard Control
	UL 4703	Photovoltaic Wire
	UL 6703	Connectors for Use in Photovoltaic Systems
	UL 7103	Investigation for Building-Integrated Photovoltaic Roof Coverings
	UL 8703	Concentrator Photovoltaic Modules and Assemblies
	UL 8801	Photovoltaic Luminaire Systems
	UL 9703	Distributed Generation Wiring Harnesses
	UL 61730-1	Photovoltaic (PV) Module Safety Qualification — Part 1: Requirements for Construction
	UL 61730-2	Photovoltaic (PV) Module Safety Qualification — Part 2: Requirements for Testing
	UL 62109-1	Power Converters for Use in Photovoltaic Power Systems — Part 1: General Requirements
	UL 62275	Cable Management Systems — Cable Ties for Electrical Installation
692	UL 2262	Fuel Cell Modules for Use in Portable and Stationary Equipment
	UL 2262A	Borohydride Fuel Cartridges with Integral Fuel Processing for Use with Portable Fuel Cell Power Systems or Similar Equipment
	UL 2265	Fuel Cell Power Units and Fuel Storage Containers for Portable Devices
	UL 2265A	Hand-held or Hand-Transportable Fuel Cell Power Units with Disposable Methanol Fuel Cartridges for use in Original Equipment Manufacturer's Information Technology Equipment
	UL 2265C	Hand-Held or Hand-Transportable Alkaline (Direct Borohydride) Fuel Cell Power Units and Borohydride Fuel Cartridges For Use With Consumer Electronics or Information Technology Equipment
	UL 2266	Electromagnetic Compatibility, Electrical Safety, and Physical Protection of Stationary and Portable Fuel Cell Power Systems for Use with Commercial Network Telecommunications Equipment
	UL 2267	Fuel Cell Power Systems for Installation in Industrial Electric Trucks
694	UL 467	Grounding and Bonding Equipment
	UL 489C	Molded-Case Circuit Breakers and Molded-Case Switches for Use with Wind Turbines
	UL 1741	Inverters, Converters, Controllers and Interconnection System Equipment for Use With Distributed Energy Resources

N *TABLE A.1(a)* *Continued*

Article	Standard Number	Standard Title
	UL 2227	Flexible Motor Supply Cable and Wind Turbine Tray Cable
	UL 2736	Single Pole Separable Interconnecting Cable Connectors for Use with Wind Turbine Generating Systems
	UL 4143	Wind Turbine Generator — Life Time Extension (LTE)
	UL 6141	Wind Turbines Permitting Entry of Personnel
	UL 6142	Wind Turbine Generating Systems — Small
695	UL 6	Electrical Rigid Metal Conduit — Steel
	UL 6A	Electrical Rigid Metal Conduit — Aluminum, Red Brass and Stainless Steel
	UL 218	Fire Pump Controllers
	UL 448	Centrifugal Stationary Pumps for Fire-Protection Service
	UL 448B	Residential Fire Pumps Intended for One- and Two-Family Dwellings and Manufactured Homes
	UL 448C	Stationary, Rotary-Type, Positive-Displacement Pumps for Fire Protection Service
	UL 651	Schedule 40, 80, Type EB and A Rigid PVC Conduit and Fittings
	UL 1004-5	Fire Pump Motors
	UL 1242	Electrical Intermediate Metal Conduit — Steel
	UL 1569	Metal-Clad Cables
	UL 1724	Fire Tests for Electrical Circuit Protective Systems
	UL 2196	Fire Test for Circuit Integrity of Fire-Resistive Power, Instrumentation, Control and Data Cables
	UL 2515	Aboveground Reinforced Thermosetting Resin Conduit (RTRC) and Fittings
700	UL 924	Emergency Lighting and Power Equipment
	UL 1008	Transfer Switch Equipment
	UL 1008A	Transfer Switch Equipment, Over 1000 Volts
	UL 1449	Surge Protective Devices
	UL 1724	Fire Tests for Electrical Circuit Protective Systems
	UL 2196	Fire Test for Circuit Integrity of Fire-Resistive Power, Instrumentation, Control and Data Cables
	UL 2200	Stationary Engine Generator Assemblies
701	UL 924	Emergency Lighting and Power Equipment
	UL 1008	Transfer Switch Equipment
	UL 1008A	Transfer Switch Equipment, Over 1000 Volts
702	UL 98	Enclosed and Dead-Front Switches
	UL 1008	Transfer Switch Equipment
	UL 1008A	Transfer Switch Equipment, Over 1000 Volts
	UL 1008M	Meter-Mounted Transfer Switches
	UL 1008S	Solid-State Transfer Switches
705	UL 62	Flexible Cords and Cables
	UL 98	Enclosed and Dead-Front Switches
	UL 486D	Sealed Wire Connector Systems
	UL 489	Molded-Case Circuit Breakers, Molded-Case Switches and Circuit-Breaker Enclosures
	UL 1066	Low-Voltage AC and DC Power Circuit Breakers Used in Enclosures
	UL 1429	Pullout Switches
	UL 1741	Inverters, Converters, Controllers and Interconnection System Equipment for Use With Distributed Energy Resources
	UL 2200	Stationary Engine Generator Assemblies
	UL 3003	Distributed Generation Cables
	UL 6141	Wind Turbines Permitting Entry of Personnel
	UL 6142	Small Wind Turbine Systems
	UL 9540	Energy Storage Systems and Equipment
	UL 62109-2	Power Converters for Use in Photovoltaic Power Systems — Part 2: Particular Requirements for Inverters
706	UL 248-2	Low-Voltage Fuses — Part 2: Class C Fuses
	UL 248-3	Low-Voltage Fuses — Part 3: Class CA and CB Fuses
	UL 248-4	Low-Voltage Fuses — Part 4: Class CC Fuses
	UL 248-5	Low-Voltage Fuses — Part 5: Class G Fuses
	UL 248-6	Low-Voltage Fuses — Part 6: Class H Non-Renewable Fuses
	UL 248-8	Low-Voltage Fuses — Part 8: Class J Fuses
	UL 248-9	Low-Voltage Fuses — Part 9: Class K Fuses
	UL 248-10	Low-Voltage Fuses — Part 10: Class L Fuses

(continues)

N *TABLE A.1(a)* *Continued*

Article	Standard Number	Standard Title
	UL 248-12	Low-Voltage Fuses — Part 12: Class R Fuses
	UL 248-15	Low-Voltage Fuses — Part 15: Class T Fuses
	UL 248-17	Low-Voltage Fuses — Part 17: Class CF Fuses
	UL 248-18	Low-Voltage Fuses — Part 18: Class CD Fuses
	UL 489	Molded-Case Circuit Breakers, Molded-Case Switches and Circuit-Breaker Enclosures
	UL 489H	Molded-Case Circuit Breakers, Molded-Case Switches, and Circuit-Breaker Enclosures, for Use with Direct Current (DC) Microgrids
	UL 1066	Low-Voltage AC and DC Power Circuit Breakers Used in Enclosures
	UL 1741	Inverters, Converters, Controllers and Interconnection System Equipment for Use With Distributed Energy Resources
	UL 9540	Energy Storage Systems and Equipment
708	UL 1	Flexible Metal Conduit
	UL 4	Armored Cable
	UL 83	Thermoplastic-Insulated Wires and Cables
	UL 360	Liquid-Tight Flexible Metal Conduit
	UL 493	Thermoplastic-Insulated Underground Feeder and Branch-Circuit Cables
	UL 497A	Secondary Protectors for Communications Circuits
	UL 1008	Transfer Switch Equipment
	UL 1008A	Transfer Switch Equipment, Over 1000 Volts
	UL 1008M	Meter-Mounted Transfer Switches
	UL 1008S	Solid-State Transfer Switches
	UL 1569	Metal-Clad Cables
	UL 2196	Fire Test for Circuit Integrity of Fire-Resistive Power, Instrumentation, Control and Data Cables
710	UL 1741	Inverters, Converters, Controllers and Interconnection System Equipment for Use With Distributed Energy Resources
	UL 2200	Stationary Engine Generator Assemblies
	UL 8801	Photovoltaic Luminaire Systems
	UL 9540	Energy Storage Systems and Equipment
	UL 62109-1	Power Converters for use in Photovoltaic Power Systems — Part 1: General Requirements
	UL 62109-2	Power Converters for Use in Photovoltaic Power Systems — Part 2: Particular Requirements for Inverters
722	UL 13	Standard for Power-Limited Circuit Cables
	UL 444	Standard for Safety for Communications Cables
	UL 1424	Cables for Power-Limited Fire-Alarm Circuits
	UL 1651	Optical Fiber Cable
	UL 1666	Test for Flame Propagation Height of Electrical and Optical-Fiber Cable Installed Vertically in Shafts
	UL 1685	Standard for Safety for Vertical-Tray Fire-Propagation and Smoke- Release Test for Electrical and Optical-Fiber Cables
	UL 1724	Fire Tests for Electrical Circuit Protective Systems
	UL 2024	Standard for Safety for Communications Cables
	UL 2196	Fire Test for Circuit Integrity of Fire-Resistive Power, Instrumentation, Control and Data Cables
	UL 2556	Standard for Wire and Cable Test Methods
725	UL 1310	Class 2 Power Units
	UL 5085-3	Low Voltage Transformers — Part 3: Class 2 and Class 3 Transformers
	UL 9990	Information and Communication Technology (ICT) Power Cables
	UL 61010-2-201	Safety Requirements for Electrical Equipment for Measurement, Control, and Laboratory Use — Part 2-201: Particular Requirements for Control Equipment
	UL 61800-5-1	Adjustable Speed Electrical Power Drive Systems — Part 5-1: Safety Requirements — Electrical, Thermal and Energy
	UL 62368-1	Audio/Video, Information and Communication Technology Equipment — Part 1: Safety Requirements
726	UL 1400-1	Fault-Managed Power Systems — Part 1 General Requirements
	UL 1400-2	Fault-Managed Power Systems — Part 2 Requirements for Cables
	UL 1666	Test for Flame Propagation Height of Electrical and Optical-Fiber Cables Installed Vertically in Shafts
	UL 1685	Vertical-Tray Fire-Propagation and Smoke-Release Test for Electrical and Optical-Fiber Cables
	UL 2556	Wire and Cable Test Methods
728	UL 5	Surface Metal Raceways and Fittings
	UL 5A	Nonmetallic Surface Raceways and Fittings
	UL 5B	Strut-Type Channel Raceways and Fittings
	UL 5C	Surface Raceways and Fittings for Use with Data, Signal, and Control Circuits

N **TABLE A.1(a)** *Continued*

Article	Standard Number	Standard Title
	UL 209	Cellular Metal Floor Raceways and Fittings
	UL 467	Grounding and Bonding Equipment
	UL 514A	Metallic Outlet Boxes
	UL 514C	Nonmetallic Outlet Boxes, Flush-Device Boxes, and Covers
	UL 568	Nonmetallic Cable Tray Systems
	UL 884	Underfloor Raceways and Fittings
	UL 1724	Fire Tests for Electrical Circuit Protective Systems
	UL 2024	Cable Routing Assemblies and Communications Raceways
	UL 2196	Fire Test for Circuit Integrity of Fire-Resistive Power, Instrumentation, Control and Data Cables
760	UL 268	Smoke Detectors for Fire Alarm Signaling Systems
	UL 268A	Smoke Detectors for Duct Application
	UL 486C	Splicing Wire Connectors
	UL 497B	Protectors for Data Communication and Fire Alarm Circuits
	UL 1424	Cables for Power-Limited Fire-Alarm Circuits
	UL 1425	Cables for Non–Power-Limited Fire-Alarm Circuits
	UL 1480	Speakers for Fire Alarm and Signaling Systems, Including Accessories
	UL 1666	Test for Flame Propagation Height of Electrical and Optical-Fiber Cables Installed Vertically in Shafts
	UL 1685	Vertical-Tray Fire-Propagation and Smoke-Release Test for Electrical and Optical-Fiber Cables
	UL 2196	Fire Test for Circuit Integrity of Fire-Resistive Power, Instrumentation, Control and Data Cables
	UL 60730-2-14	Automatic Electrical Controls; Part 2: Particular Requirements for Electric Actuators
770	UL 467	Grounding and Bonding Equipment
	UL 568	Nonmetallic Cable Tray Systems
	UL 1651	Optical Fiber Cable
	UL 2024	Optical Fiber and Communication Cable Raceway
	UL 2196	Fire Test for Circuit Integrity of Fire-Resistive Power, Instrumentation, Control and Data Cables
	UL 62275	Cable Management Systems — Cable Ties for Electrical Installation
800	UL 444	Communications Cables
	UL 467	Grounding and Bonding Equipment
	UL 489A	Circuit Breakers for Use in Communication Equipment
	UL 497	Protectors for Paired-Conductor Communications Circuits
	UL 497A	Secondary Protectors for Communications Circuits
	UL 497C	Protectors for Coaxial Communications Circuits
	UL 497E	Protectors for Antenna Lead-In Conductors
	UL 523	Telephone Service Drop Wire
	UL 568	Nonmetallic Cable Tray Systems
	UL 723	Test for Surface Burning Characteristics of Building Materials
	UL 1581	Reference Standard for Electrical Wires, Cables, and Flexible Cords
	UL 1666	Test for Flame Propagation Height of Electrical and Optical-Fiber Cables Installed Vertically in Shafts
	UL 1685	Vertical-Tray Fire-Propagation and Smoke-Release Test for Electrical and Optical-Fiber Cables
	UL 1863	Communication Circuit Accessories
	UL 2024	Cable Routing Assemblies and Communications Raceways
	UL 62275	Cable Management Systems — Cable Ties for Electrical Installation
805	UL 444	Communications Cables
	UL 497	Protectors for Paired-Conductor Communications Circuits
	UL 497A	Secondary Protectors for Communications Circuits
	UL 497C	Protectors for Coaxial Communications Circuits
	UL 497E	Protectors for Antenna Lead-In Conductors
	UL 523	Telephone Service Drop Wire
	UL 719	Nonmetallic-Sheathed Cables
	UL 1310	Class 2 Power Units
	UL 1581	Reference Standard for Electrical Wires, Cables, and Flexible Cords
	UL 1685	Vertical-Tray Fire-Propagation and Smoke-Release Test for Electrical and Optical-Fiber Cables
	UL 1863	Communication Circuit Accessories
	UL 2043	Fire Test for Heat and Visible Smoke Release for Discrete Products and Their Accessories Installed in Air-Handling Spaces
	UL 62275	Cable Management Systems — Cable Ties for Electrical Installation
	UL 62368-1	Audio/Video, Information and Communication Technology Equipment — Part 1: Safety Requirements

(continues)

N *TABLE A.1(a)* *Continued*

Article	Standard Number	Standard Title
810	UL 150	Antenna Rotators
	UL 452	Antenna-Discharge Units
	UL 467	Grounding and Bonding Equipment
	UL 497E	Protectors for Antenna Lead-In Conductors
820	UL 444	Communications Cables
	UL 497E	Protectors for Antenna Lead-In Conductors
	UL 1655	Community-Antenna Television Cables
830	UL 444	Communications Cables
	UL 497A	Secondary Protectors for Communications Circuits
	UL 497C	Protectors for Coaxial Communications Circuits
	UL 497E	Protectors for Antenna Lead-In Conductors
	UL 62368-1	Audio/Video, Information and Communication Technology Equipment — Part 1: Safety Requirements
840	UL 444	Communications Cables
	UL 467	Grounding and Bonding Equipment
	UL 498A	Current Taps and Adapters
	UL 1310	Class 2 Power Units
	UL 1651	Optical Fiber Cable
	UL 1863	Communication Circuit Accessories
	UL 2024	Cable Routing Assemblies and Communications Raceways
	UL 62368-1	Audio/Video, Information and Communication Technology Equipment — Part 1: Safety Requirements
Tables 11(A)	UL 1310	Class 2 Power Units
and 11(B)	UL 1434	Thermistor-Type Devices
	UL 5085-3	Low Voltage Transformers — Part 3: Class 2 and Class 3 Transformers
	UL 62368-1	Audio/Video, Information and Communication Technology Equipment — Part 1: Safety Requirements
Tables 12(A)	UL 1310	Class 2 Power Units
and 12(B)	UL 1434	Thermistor-Type Devices
	UL 5085-3	Low Voltage Transformers — Part 3: Class 2 and Class 3 Transformers
	UL 62368-1	Audio/Video, Information and Communication Technology Equipment — Part 1: Safety Requirements

N *TABLE A.1(b)* *Product Safety Standards for Conductors and Equipment That Do Not Have an Associated Listing Requirement*

Article	Standard Number	Standard Title
110	UL 969	Marking and Labeling Systems
	UL 9691	Recommended Practice for Nameplates for Use in Electrical Installations
300	UL 635	Insulating Bushings
314	UL 514C	Conduit, Tubing, and Cable Fittings
	UL 2239	Hardware for the Support of Conduit, Tubing and Cable
320	UL 514A	Metallic Outlet Boxes
	UL 2239	Hardware for the Support of Conduit, Tubing and Cable
322	UL 5	Surface Metal Raceways and Fittings
	UL 2239	Hardware for the Support of Conduit, Tubing and Cable
324	UL 5	Surface Metal Raceways and Fittings
	UL 2239	Hardware for the Support of Conduit, Tubing and Cable
330	UL 2239	Hardware for the Support of Conduit, Tubing and Cable
332	UL 1565	Positioning Devices
	UL 2239	Hardware for the Support of Conduit, Tubing and Cable
334	UL 6	Electrical Rigid Metal Conduit — Steel
	UL 6A	Electrical Rigid Metal Conduit — Aluminum, Red Brass and Stainless Steel
	UL 514B	Conduit, Tubing, and Cable Fittings
	UL 651	Schedule 40 and 80 Rigid PVC Conduit
	UL 797	Electrical Metallic Tubing — Steel
	UL 797A	Electrical Metallic Tubing — Aluminum and Stainless Steel
	UL 1242	Electrical Intermediate Metal Conduit — Steel

N **TABLE A.1(b)** *Continued*

Article	Standard Number	Standard Title
	UL 1565	Positioning Devices
	UL 2239	Hardware for the Support of Conduit, Tubing and Cable
	UL 2420	Belowground Reinforced Thermosetting Resin Conduit (RTRC) and Fittings
	UL 2515	Aboveground Reinforced Thermosetting Resin Conduit (RTRC) and Fittings
	UL 2515A	Supplemental Requirements for Extra Heavy Wall Reinforced Thermosetting Resin Conduit (RTRC) and Fittings.
335	UL 2250	Instrumentation Tray Cable
337	UL 1565	Positioning Devices
	UL 2239	Hardware for the Support of Conduit, Tubing and Cable
340	UL 493	Thermoplastic-Insulated Underground Feeder and Branch-Circuit Cables
342	UL 635	Insulating Bushings
	UL 2239	Hardware for the Support of Conduit, Tubing and Cable
344	UL 635	Insulating Bushings
	UL 2239	Hardware for the Support of Conduit, Tubing and Cable
348	UL 2239	Hardware for the Support of Conduit, Tubing and Cable
350	UL 2239	Hardware for the Support of Conduit, Tubing and Cable
352	UL 635	Insulating Bushings
	UL 2239	Hardware for the Support of Conduit, Tubing and Cable
353	UL 635	Insulating Bushings
355	UL 635	Insulating Bushings
	UL 2239	Hardware for the Support of Conduit, Tubing and Cable
356	UL 2239	Hardware for the Support of Conduit, Tubing and Cable
358	UL 2239	Hardware for the Support of Conduit, Tubing and Cable
362	UL 2239	Hardware for the Support of Conduit, Tubing and Cable
368	UL 857	Busways
392	UL 568	Nonmetallic Cable Tray Systems
400	UL 62	Flexible Cords and Cables
	UL 498	Attachment Plugs and Receptacles
	UL 498B	Receptacles with Integral Switching Means
	UL 498D	Attachment Plugs, Cord Connectors and Receptacles with Arcuate (Locking Type) Contacts
	UL 498E	Attachment Plugs, Cord Connectors and Receptacles — Enclosure Types for Environmental Protection
	UL 514B	Conduit, Tubing, and Cable Fittings
	UL 817	Cord Sets and Power-Supply Cords
	UL 1650	Portable Power Cable
	UL 1680	Stage and Lighting Cables
402	UL 66	Fixture Wire
408	UL 50	Enclosures for Electrical Equipment, Non-Environmental Considerations
	UL 50E	Enclosures for Electrical Equipment, Environmental Considerations
424	UL 834	Heating, Water Supply, and Power Boilers — Electric
	UL 1693	Electric Radiant Heating Panels and Heating Panel Sets
	UL 1995	Heating and Cooling Equipment
	UL 1996	Electric Duct Heaters
	UL 60335-1	Safety of Household and Similar Electrical Appliances, Part 1: General Requirements
	UL 60335-2-40	Household and Similar Electrical Appliances, Part 2–40
425	UL 834	Heating, Water Supply, and Power Boilers — Electric
426	UL 1588	Roof and Gutter De-Icing Cable Units
427	UL 515	Electrical Resistance Trace Heating for Commercial Applications
	UL 1462	Mobile Home Pipe Heating Cable
	UL 2049	Residential Pipe Heating Cable
430	UL 248-13	Low Voltage Fuses — Part 13: Semiconductor Fuses
445	UL 3001	Distributed Energy Generation and Storage Systems
	UL 3010	Single Site Energy Systems
450	UL 50	Enclosures for Electrical Equipment, Non-Environmental Considerations
	UL 50E	Enclosures for Electrical Equipment, Environmental Considerations
	UL 248-1	Low-Voltage Fuses — Part 1: General Requirements
	UL 248-2	Low-Voltage Fuses — Part 2: Class C Fuses
	UL 248-3	Low-Voltage Fuses — Part 3: Class CA and CB Fuses

(continues)

N **TABLE A.1(b)** *Continued*

Article	Standard Number	Standard Title
	UL 248-4	Low-Voltage Fuses — Part 4: Class CC Fuses
	UL 248-5	Low-Voltage Fuses — Part 5: Class G Fuses
	UL 248-8	Low-Voltage Fuses — Part 8: Class J Fuses
	UL 248-9	Low-Voltage Fuses — Part 9: Class K Fuses
	UL 489	Molded-Case Circuit Breakers, Molded-Case Switches and Circuit-Breaker Enclosures
	UL 1561	Dry-Type General Purpose and Power Transformers
	UL 5085-2	Low Voltage Transformers — Part 2: General Purpose Transformers
460	UL 810	Capacitors
	UL 1283	Electromagnetic Interference Filters
	UL 60384-14	Fixed Capacitors for Use in Electronic Equipment — Part 14: Sectional Specification: Fixed Capacitors for Electromagnetic Interference Suppression and Connection to the Supply Mains
470	UL 508	Industrial Control Equipment
	UL 1283	Electromagnetic Interference Filters
500	ANSI/IEEE C2	National Electrical Safety Code, Section 127A, Coal Handling Areas
	API RP 14F	Recommended Practice for Design and Installation of Electrical Systems for Fixed and Floating Offshore Petroleum Facilities for Unclassified and Class I, Division 1 and Division 2 Locations
	API RP 500	Recommended Practice for Classification of Locations of Electrical Installations at Petroleum Facilities Classified as Class I, Division 1 and Division 2
	API RP 2003	Protection Against Ignitions Arising Out of Static Lightning and Stray Currents.
	ASHRAE 15	Safety Standard for Refrigeration Systems.
	ASME B1.20.1	Pipe Threads, General Purpose (Inch)
	IEEE 844.2	Standard for Skin Effect Trace Heating of Pipelines, Vessels, Equipment, and Structures — Application Guide for Design, Installation, Testing, Commissioning, and Maintenance
	IEEE 60079-30-2	IEEE/IEC International Standard for Explosive atmospheres — Part 30-2: Electrical resistance trace heating — Application guide for design, installation, and maintenance
	IIAR 2	Standard for Safe Design of Closed-Circuit Ammonia Refrigeration Systems
	ISA-12.10	Area Classification in Hazardous (Classified) Dust Locations
	ISO 965-1	ISO general purpose metric screw threads — Tolerances — Part 1: Principles and basic data
	ISO 965-3	ISO general purpose metric screw threads — Tolerances — Part 3: Deviations for constructional screw threads
	NFPA 30	Flammable and Combustible Liquids Code
	NFPA 32	Standard for Drycleaning Facilities
	NFPA 33	Standard for Spray Application Using Flammable or Combustible Materials
	NFPA 34	Standard for Dipping, Coating and Printing Processes Using Flammable or Combustible Liquids
	NFPA 35	Standard for the Manufacture of Organic Coatings
	NFPA 36	Standard for Solvent Extraction Plants
	NFPA 45	Standard on Fire Protection for Laboratories Using Chemicals
	NFPA 55	Compressed Gases and Cryogenic Fluids Code
	NFPA 58	Liquefied Petroleum Gas Code
	NFPA 59	Utility LP-Gas Plant Code
	NFPA 77	Recommended Practice on Static Electricity
	NFPA 497	Recommended Practice for the Classification of Flammable Liquids, Gases, or Vapors and of Hazardous (Classified) Locations for Electrical Installations in Chemical Process Areas
	NFPA 499	Recommended Practice for the Classification of Combustible Dusts and of Hazardous (Classified) Locations for Electrical Installation in Chemical Process Areas
	NFPA 780	Standard for the Installation of Lightning Protection Systems
	NFPA 820	Standard for Fire Protection in Wastewater Treatment and Collection Facilities
	UL 60079-29-2	Explosive Atmospheres — Part 29-2: Gas detectors — Selection, installation, use and maintenance of detectors for flammable gases and oxygen
	UL 120002	Certificate Standard for AEx Equipment for Hazardous (Classified) Locations
	UL 120101	Definitions and Information Pertaining to Electrical Equipment in Hazardous (Classified) Locations
	UL 121303	Guide for Combustible Gas Detection as a Method of Protection
	UL RP 121203	Recommended Practice for Portable/Personal Electronic Products Suitable for Use in Class I, Division 2, Class I, Zone 2, Class II, Division 2, Class III, Division 1, Class III, Division 2, Zone 21 and 22 Hazardous (Classified) Locations
501	UL 62	Flexible Cord and Cable
	UL 504	Mineral-Insulated, Metal-Sheathed Cable
502	UL RP 121203	Recommended Practice for Portable/Personal Electronic Products Suitable for Use in Class I, Division 2, Class I, Zone 2, Class II, Division 2, Class III, Division 1, Class III, Division 2, Zone 21 and Zone 22 Hazardous (Classified) Locations

N *TABLE A.1(b)* *Continued*

Article	Standard Number	Standard Title
503	NFPA 505	Fire Safety Standard for Powered Industrial Trucks Including Type Designations, Areas of Use, Conversions, Maintenance, and Operations
	UL RP 121203	Recommended Practice for Portable/Personal Electronic Products Suitable for Use in Class I, Division 2, Class I, Zone 2, Class II, Division 2, Class III, Division 1, Class III, Division 2, Zone 21 and Zone 22 Hazardous (Classified) Locations
504	ISA-RP 12.06.01	Recommended Practice for Wiring Methods for Hazardous (Classified) Locations Instrumentation — Part 1: Intrinsic Safety
505	ANSI/API RP 14FZ	Recommended Practice for Design and Installation of Electrical Systems for Fixed and Floating Offshore Petroleum Facilities for Unclassified and Class I, Zone 0, Zone 1, and Zone 2 Locations
	API RP 505	Recommended Practice for Classification of Locations for Electrical Installations at Petroleum Facilities Classified as Class I, Zone 0, Zone 1, and Zone 2
	API RP 2003	Protection Against Ignitions Arising Out of Static Lightning and Stray Currents.
	ASME B1.20.1	Pipe Threads, General Purpose (Inch)
	EI 15	Model Code of Safe Practice, Part 15: Area Classification Code for Installations Handling Flammable Fluids
	IEEE 844.2	Skin Effect Trace Heating of Pipelines, Vessels, Equipment, and Structures — Application Guide for Design, Installation, Testing, Commissioning, and Maintenance
	IEEE 60079-30-2	Explosive Atmospheres — Part 30-2: Electrical resistance trace heating — Application guide for design, installation and maintenance
	IIAR 2	Standard for Safe Design of Closed-Circuit Ammonia Refrigeration Systems
	ISA-60079-10-1 (12.24.01)	Explosive Atmospheres — Part 10-1: Classification of Areas — Explosive gas atmospheres
	ISA-60079-29-2	Explosive Atmospheres — Part 29-2: Gas detectors — Selection, installation, use and maintenance of detectors for flammable gases and oxygen
	ISO 965-1	ISO general purpose metric screw threads — Tolerances — Part 1: Principles and basic data
	ISO 965-3	ISO general purpose metric screw threads — Tolerances — Part 3: Deviations for constructional screw threads
	NFPA 30	Flammable and Combustible Liquids Code
	NFPA 77	Recommended Practice on Static Electricity
	NFPA 497	Recommended Practice for the Classification of Flammable Liquids, Gases, or Vapors and of Hazardous (Classified) Locations for Electrical Installations in Chemical Process Areas
	NFPA 780	Standard for the Installation of Lightning Protection Systems
	UL 80079-20-1	Explosive Atmospheres — Part 20-1: Material Characteristics for Gas and Vapour Classification — Test Methods and Data
	UL 120101	Definitions and Information Pertaining to Electrical Equipment in Hazardous (Classified) Locations
	UL 121303	Guide for Use of Detectors for Flammable Gases
	UL RP 121203	Recommended Practice for Portable/Personal Electronic Products Suitable for Use in Class I, Division 2, Class I, Zone 2, Class II, Division 2, Class III, Division 1, Class III, Division 2, Zone 21 and Zone 22 Hazardous (Classified) Locations
506	ASME B1.20.1	Pipe Threads, General Purpose (Inch)
	IEEE 844.2	Skin Effect Trace Heating of Pipelines, Vessels, Equipment, and Structures — Application Guide for Design, Installation, Testing, Commissioning, and Maintenance
	IEEE 60079-30-2	Explosive Atmospheres — Part 30-2: Electrical resistance trace heating — Application guide for design, installation and maintenance
	ISA-60079-10-2 (12.10.05)	Explosive Atmospheres — Part 10-2: Classification of Areas — Combustible Dust Atmospheres
	NFPA 499	Recommended Practice for the Classification of Combustible Dusts and of Hazardous (Classified) Locations for Electrical Installation in Chemical Process Areas
	UL RP 121203	Recommended Practice for Portable/Personal Electronic Products Suitable for Use in Class I, Division 2, Class I, Zone 2, Class II, Division 2, Class III, Division 1, Class III, Division 2, Zone 21 and Zone 22 Hazardous (Classified) Locations
511	NFPA 30A	Code for Motor Fuel Dispensing Facilities and Repair Garages
	NFPA 88A	Standard for Parking Structures
512	ICC IFC	International Fire Code
	NFPA 1	Fire Code
	NFPA 30	Flammable and Combustible Liquids Code
	NFPA 33	Standard for Spray Application Using Flammable or Combustible Materials
	NFPA 36	Standard for Solvent Extraction Plants

(continues)

N *TABLE A.1(b)* *Continued*

Article	Standard Number	Standard Title
	NFPA 58	Liquefied Petroleum Gas Code
	NFPA 70B	Recommended Practice for Electrical Equipment Maintenance
	NFPA 497	Recommended Practice for the Classification of Flammable Liquids, Gases, or Vapors and of Hazardous (Classified) Locations for Electrical Installations in Chemical Process Areas
513	NFPA 30	Flammable and Combustible Liquids Code
	NFPA 33	Standard for Spray Application Using Flammable or Combustible Materials
	NFPA 409	Standard on Aircraft Hangars
514	NFPA 2	Hydrogen Technologies Code
	NFPA 30A	Code for Motor Fuel Dispensing Facilities and Repair Garages
	NFPA 52	Vehicular Natural Gas Fuel Systems Code
	NFPA 58	Liquefied Petroleum Gas Code
	NFPA 59	Utility LP-Gas Plant Code
	NFPA 303	Fire Protection Standard for Marinas and Boatyards
515	NFPA 30	Flammable and Combustible Liquids Code
516	NFPA 13	Standard for the Installation of Sprinkler Systems
	NFPA 33	Standard for Spray Application Using Flammable or Combustible Materials
	NFPA 34	Standard for Dipping, Coating and Printing Processes Using Flammable or Combustible Liquids
	NFPA 77	Recommended Practice on Static Electricity
	NFPA 91	Standard for Exhaust Systems for Air Conveying of Vapors, Gases, Mists, and Particulate Solids
	NFPA 701	Standard Methods of Fire Tests for Flame Propagation of Textiles and Films
620	UL 4	Armored Cable
	UL 44	Thermoset-Insulated Wires and Cables
	UL 66	Fixture Wire
	UL 504	Mineral Insulated Wire
	UL 1063	Machine-Tool Wires and Cables
	UL 1569	Metal Clad Cable
625	UL 3001	Distributed Energy Generation and Storage Systems
	UL 3010	Single Site Energy Systems
630	UL 1276	Welding Cable
650	UL 1651	Optical Fiber Cable
660	UL 62	Flexible Cords and Cables
	UL 817	Cord Sets and Power Supply Cords
668	UL 4	Armored Cable
	UL 62	Flexible Cords and Cables
670	UL 2011	Machinery
675	UL 44	Thermoset-Insulated Wires and Cables
	UL 83	Thermoplastic-Insulated Wires and Cables
	UL 83A	Fluoropolymer Insulated Wire
	UL 1063	Machine-Tool Wires and Cables
	UL 1263	Irrigation Cable
690	UL 3001	Distributed Energy Generation and Storage Systems
	UL 3010	Single Site Energy Systems
691	UL 3001	Distributed Energy Generation and Storage Systems
	UL 3010	Single Site Energy Systems
692	UL 44	Thermoset-Insulated Wires and Cables
	UL 83	Thermoplastic-Insulated Wires and Cables
	UL 83A	Fluoropolymer Insulated Wire
	UL 1063	Machine-Tool Wires and Cables
	UL 3001	Distributed Energy Generation and Storage Systems
	UL 3010	Single Site Energy Systems
694	UL 44	Thermoset-Insulated Wires and Cables
	UL 62	Flexible Cords and Cables
	UL 83	Thermoplastic-Insulated Wires and Cables
	UL 83A	Fluoropolymer Insulated Wire
	UL 1063	Machine-Tool Wires and Cables
	UL 3001	Distributed Energy Generation and Storage Systems
	UL 3010	Single Site Energy Systems

N ***TABLE A.1(b)*** *Continued*

Article	Standard Number	Standard Title
700	UL 3001	Distributed Energy Generation and Storage Systems
701	UL 3001	Distributed Energy Generation and Storage Systems
702	UL 3001	Distributed Energy Generation and Storage Systems
705	UL 3001	Distributed Energy Generation and Storage Systems
	UL 3010	Single Site Energy Systems
710	UL 3001	Distributed Energy Generation and Storage Systems
	UL 3010	Single Site Energy Systems

Application Information for Ampacity Calculation

This informative annex is not a part of the requirements of this NFPA document but is included for informational purposes only.

Informative Annex B is based on calculations using the Neher–McGrath conductor ampacity formula. The use of this method to calculate conductor ampacity is permitted under engineering supervision per 310.14(A)(1) and 310.14(B). The formula is found in 310.15(B).

Although conductor ampacities calculated using this formula might exceed those found in a table of ampacities, such as Table 310.16, the limitations for connecting to equipment terminals specified in 110.14(C) must be followed. For equipment 1000 volts and under, the ampacities used with equipment terminals are based on Table 310.16.

B.1 Equation Application Information. This informative annex provides application information for ampacities calculated under engineering supervision.

B.2 Typical Applications Covered by Tables. Typical ampacities for conductors rated 0 through 2000 volts are shown in Table B.2(1) through Table B.2(10). Table B.2(11) provides the adjustment factors for more than three current-carrying conductors in a raceway or cable with load diversity. Underground electrical duct bank configurations, as detailed in Figure B.2(2), Figure B.2(3), and Figure B.2(4), are utilized for conductors rated 0 through 5000 volts. In Figure B.2(1) through Figure B.2(4), where adjacent duct banks are used, a separation of 1.5 m (5 ft) between the centerlines of the closest ducts in each bank or 1.2 m (4 ft) between the extremities of the concrete envelopes is sufficient to prevent derating of the conductors due to mutual heating. These ampacities were calculated as detailed in the basic ampacity paper, AIEE Paper 57-660, *The Calculation of the Temperature Rise and Load Capability of Cable Systems,* by J. H. Neher and M. H. McGrath. For additional information concerning the application of these ampacities, see IEEE STD 835, *Standard Power Cable Ampacity Tables.*

Typical values of thermal resistivity (Rho) are as follows:

Average soil (90 percent of USA) = 90
Concrete = 55
Damp soil (coastal areas, high water table) = 60
Paper insulation = 550
Polyethylene (PE) = 450
Polyvinyl chloride (PVC) = 650
Rubber and rubber-like = 500
Very dry soil (rocky or sandy) = 120

Thermal resistivity, as used in this informative annex, refers to the heat transfer capability through a substance by conduction. It is the reciprocal of thermal conductivity and is normally expressed in the units°C-cm/watt. For additional information on determining soil thermal resistivity (Rho), see IEEE STD 442, *Guide for Soil Thermal Resistivity Measurements*.

If other factors remain the same, a soil resistivity higher than 90 reduces ampacities to values below those listed in Tables B.2(5) through B.2(10). Conversely, a load factor less than 100 percent increases ampacities if other factors remain the same. See Table B.2(7) for allowable adjustments if the load factor is less than 100 percent. Reduced load factors are used in Informational Note Figures B.2(2) through B.2(4).

B.3 Criteria Modifications. Where values of load factor and Rho are known for a particular electrical duct bank installation and they are different from those shown in a specific table or figure, the ampacities shown in the table or figure can be modified by the application of factors derived from the use of Figure B.3.

Where two different ampacities apply to adjacent portions of a circuit, the higher ampacity can be used beyond the point of transition, a distance equal to 3 m (10 ft) or 10 percent of the circuit length calculated at the higher ampacity, whichever is less.

Where the burial depth of direct burial or electrical duct bank circuits are modified from the values shown in a figure or table, ampacities can be modified as shown in (a) and (b) as follows.

(a) Where burial depths are increased in part(s) of an electrical duct run to avoid underground obstructions, no decrease in ampacity of the conductors is needed, provided the total length of parts of the duct run increased in depth to avoid obstructions is less than 25 percent of the total run length.

(b) Where burial depths are deeper than shown in a specific underground ampacity table or figure, an ampacity derating factor of 6 percent per increased 300 mm (foot) of depth for all values of Rho can be utilized. No rating change is needed where the burial depth is decreased.

The information provided in B.3 on how to address different conductor ampacities that apply to adjacent portions of a circuit is also specified in the exception to 310.14(A)(2). See the Calculation Example following that exception.

Exhibit B.1 illustrates an installation in which a change in the burial depth results in an analysis of whether a modification to the conductor ampacity is necessary. If that portion deeper than 30 inches does not exceed 25 percent of the total run length, no decrease in ampacity is required, even if part of the run is more than 30 inches deep, which is the maximum depth assumed in Figure B.2(1), Note 1, to maintain the accuracy of the tables.

For example, in accordance with Table B.2(7) and Figure B.2(1), the ampacity of six parallel runs of 500 kcmil, Type XHHW copper conductors [as shown in Detail 3 in Figure B.2(1)], where Rho = 90, is 6 × 273 A = 1638 A at a depth measuring no more than 30 inches to the top duct in the bank. If the burial depth is 6 feet, the ampacity is calculated as follows: 1638 A − (3.5 × 0.06 × 1638 A) = 1294 A.

B.4 Electrical Ducts. The term *electrical duct(s)* is defined in Article 100.

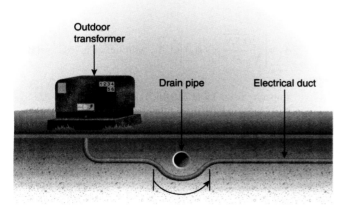

EXHIBIT B.1 The presence of a drain pipe results in a change of burial depth for that portion of the electrical installation under the drain pipe.

B.5 Table B.2(6) and Table B.2(7).

(a) To obtain the ampacity of cables installed in two electrical ducts in one horizontal row with 190-mm (7.5-in.) center-to-center spacing between electrical ducts, similar to Figure B.2(1), Detail 1, multiply the ampacity shown for one duct in Table B.2(6) and Table B.2(7) by 0.88.

(b) To obtain the ampacity of cables installed in four electrical ducts in one horizontal row with 190-mm (7.5-in.) center-to-center spacing between electrical ducts, similar to Figure B.2(1), Detail 2, multiply the ampacity shown for three electrical ducts in Table B.2(6) and Table B.2(7) by 0.94.

The underground ampacity tables [Tables B.2(5) through B.2(10)] are based on the 7.5-inch center-to-center spacing illustrated in Figure B.2(1). Although moving directly buried cables or electrical ducts farther apart will increase ampacities, the effect is surprisingly small. One calculation indicates that two side-by-side electrical ducts buried 30 inches below grade would have to be spaced about 5 feet apart before they could be considered single electrical ducts as shown in Detail 1, Figure B.2(1). Each of these factors — decreasing the burial depth, decreasing the thermal resistivity of the earth or other surrounding medium, and decreasing the load factor — has a much greater effect in increasing ampacity than does increasing the horizontal spacing.

B.6 Electrical Ducts Used in Figure B.2(1). If spacing between electrical ducts, as shown in Figure B.2(1), is less than as specified where electrical ducts enter equipment enclosures from underground, the ampacity of conductors contained within such electrical ducts need not be reduced.

B.7 Examples Showing Use of Figure B.3 for Electrical Duct Bank Ampacity Modifications. Figure B.3 is used for interpolation or extrapolation for values of Rho and load factor for cables installed in electrical ducts. The upper family of curves shows the variation in ampacity and Rho at unity load factor in terms of I_1, the ampacity for Rho = 60, and 50 percent load factor. Each curve is designated for a particular ratio I_2/I_1, where I_2 is the ampacity at Rho = 120 and 100 percent load factor.

The lower family of curves shows the relationship between Rho and load factor that will give substantially the same ampacity as the indicated value of Rho at 100 percent load factor.

As an example, to find the ampacity of a 500-kcmil copper cable circuit for six electrical ducts as shown in Table B.2(5): At the Rho = 60, LF = 50, I_1 = 583; for Rho = 120 and LF = 100, I_2 = 400. The ratio I_2/I_1 = 0.686. Locate Rho = 90 at the bottom of the chart and follow the 90 Rho line to the intersection with 100 percent load factor where the equivalent Rho = 90. Then follow the 90 Rho line to I_2/I_1 ratio of 0.686 where F = 0.74. The desired ampacity = 0.74 × 583 = 431, which agrees with the table for Rho = 90, LF = 100.

To determine the ampacity for the same circuit where Rho = 80 and LF = 75, using Figure B.3, the equivalent Rho = 43, F = 0.855, and the desired ampacity = 0.855 × 583 = 498 amperes.

Δ **TABLE B.2(1)** *Ampacities of Two or Three Insulated Conductors, Rated 0 Through 2000 Volts, Within an Overall Covering (Multiconductor Cable), in Raceway in Free Air Based on Ambient Air Temperature of 30°C (86°F)**

Size (AWG or kcmil)	Temperature Rating of Conductor. [See Table 310.4(1).]						Size (AWG or kcmil)
	60°C (140°F)	75°C (167°F)	90°C (194°F)	60°C (140°F)	75°C (167°F)	90°C (194°F)	
	Types TW, UF	Types RHW, THHW, THW, THWN, XHHW, ZW	Types THHN, THHW, THW-2, THWN-2, RHH, RWH-2, USE-2, XHHW, XHHW-2, ZW-2	Type TW	Types RHW, THHW, THW, THWN, XHHW	Types THHN, THHW, THW-2, THWN-2, RHH, RWH-2, USE-2, XHHW, XHHW-2, ZW-2	
	COPPER			ALUMINUM OR COPPER-CLAD ALUMINUM			
14**	16	18	21	—	—	—	14
12**	20	24	27	16	18	21	12
10**	27	33	36	21	25	28	10
8	36	43	48	28	33	37	8
6	48	58	65	38	45	51	6
4	66	79	89	51	61	69	4
3	76	90	102	59	70	79	3
2	88	105	119	69	83	93	2
1	102	121	137	80	95	106	1
1/0	121	145	163	94	113	127	1/0
2/0	138	166	186	108	129	146	2/0
3/0	158	189	214	124	147	167	3/0
4/0	187	223	253	147	176	197	4/0
250	205	245	276	160	192	217	250
300	234	281	317	185	221	250	300
350	255	305	345	202	242	273	350
400	274	328	371	218	261	295	400
500	315	378	427	254	303	342	500
600	343	413	468	279	335	378	600
700	376	452	514	310	371	420	700
750	387	466	529	321	384	435	750
800	397	479	543	331	397	450	800
900	415	500	570	350	421	477	900
1000	448	542	617	382	460	521	1000

*Refer to 310.15 for the ampacity correction factors where the ambient temperature is other than 30°C (86°F).

**Refer to 240.4(D) for conductor overcurrent protection limitations.

Values for using Figure B.3 are found in the electrical duct bank ampacity tables of this informative annex.

Where the load factor is less than 100 percent and can be verified by measurement or calculation, the ampacity of electrical duct bank installations can be modified as shown. Different values of Rho can be accommodated in the same manner.

Informational Note: The ampacity limit for 10 through 85 current-carrying conductors is based on the following equation. For more than 85 conductors, special calculations are required that are beyond the scope of this table.

$$A_2 = \left\lfloor \sqrt{\frac{0.5N}{E}} \times (A_1) \right\rfloor \text{ or } A_1, \text{ whichever is less} \quad \textbf{[B.7a]}$$

where:

A_1 = ampacity from Table 310.16, Table 310.18, Table B.2(1), Table B.2(6), or Table B.2(7) multiplied by the appropriate adjustment factor from Table B.2(11).

N = total number of conductors used to select adjustment factor from Table B.2(11)

E = number of conductors carrying current simultaneously in the raceway or cable

A_2 = ampacity limit for the current-carrying conductors in the raceway or cable

Δ **TABLE B.2(3)** *Ampacities of Multiconductor Cables with Not More Than Three Insulated Conductors, Rated 0 Through 2000 Volts, in Free Air Based on Ambient Air Temperature of 40°C (104°F) (for Types TC, MC, MI, UF, and USE Cables)**

Size (AWG or kcmil)	Temperature Rating of Conductor. [See Table 310.4(1).]								Size (AWG or kcmil)
	60°C (140°F)	75°C (167°F)	85°C (185°F)	90°C (194°F)	60°C (140°F)	75°C (167°F)	85°C (185°F)	90°C (194°F)	
	COPPER				ALUMINUM OR COPPER-CLAD ALUMINUM				
18	—	—	—	11	—	—	—	—	18
16	—	—	—	16	—	—	—	—	16
14**	18	21	24	25	—	—	—	—	14
12**	21	28	30	32	18	21	24	25	12
10**	28	36	41	43	21	28	30	32	10
8	39	50	56	59	30	39	44	46	8
6	52	68	75	79	41	53	59	61	6
4	69	89	100	104	54	70	78	81	4
3	81	104	116	121	63	81	91	95	3
2	92	118	132	138	72	92	103	108	2
1	107	138	154	161	84	108	120	126	1
1/0	124	160	178	186	97	125	139	145	1/0
2/0	143	184	206	215	111	144	160	168	2/0
3/0	165	213	238	249	129	166	185	194	3/0
4/0	190	245	274	287	149	192	214	224	4/0
250	212	274	305	320	166	214	239	250	250
300	237	306	341	357	186	240	268	280	300
350	261	337	377	394	205	265	296	309	350
400	281	363	406	425	222	287	317	334	400
500	321	416	465	487	255	330	368	385	500
600	354	459	513	538	284	368	410	429	600
700	387	502	562	589	306	405	462	473	700
750	404	523	586	615	328	424	473	495	750
800	415	539	604	633	339	439	490	513	800
900	438	570	639	670	362	469	514	548	900
1000	461	601	674	707	385	499	558	584	1000

*Refer to 310.15 for the ampacity correction factors where the ambient temperature is other than 40°C (104°F).

**Refer to 240.4(D) for conductor overcurrent protection limitations.

Example 1

Calculate the ampacity limit for twelve 14 AWG THWN current-carrying conductors (75°C) in a raceway that contains 24 conductors that may, at different times, be current-carrying.

$$A_2 = \sqrt{\frac{(0.5)(24)}{12}} \times 20(0.7) \qquad \textbf{[B.7b]}$$

$$= 14 \text{ amperes (i.e., 50 percent diversity)}$$

Example 2

Calculate the ampacity limit for eighteen 14 AWG THWN current-carrying conductors (75°C) in a raceway that contains 24 conductors that may, at different times, be current-carrying.

$$A_2 = \sqrt{\frac{(0.5)(24)}{18}} \times 20(0.7) = 11.5 \text{ amperes} \qquad \textbf{[B.7c]}$$

Δ **TABLE B.2(5)** *Ampacities of Single Insulated Conductors, Rated 0 Through 2000 Volts, in Nonmagnetic Underground Electrical Ducts (One Conductor per Electrical Duct), Based on Ambient Earth Temperature of 20°C (68°F), Electrical Duct Arrangement in Accordance with Figure B.2(1), Conductor Temperature 75°C (167°F)*

Size (kcmil)	3 Electrical Ducts [Fig. B.2(1), Detail 2] Types RHW, THHW, THW, THWN, XHHW, USE			6 Electrical Ducts [Fig. B.2(1), Detail 3] Types RHW, THHW, THW, THWN, XHHW, USE			9 Electrical Ducts [Fig. B.2(1), Detail 4] Types RHW, THHW, THW, THWN, XHHW, USE			3 Electrical Ducts [Fig. B.2(1), Detail 2] Types RHW, THHW, THW, THWN, XHHW, USE			6 Electrical Ducts Fig. B.2(1), Detail 3] Types RHW, THHW, THW, THWN, XHHW, USE			9 Electrical Ducts [Fig. B.2(1), Detail 4] Types RHW, THHW, THW, THWN, XHHW, USE			Size (kcmil)
	COPPER									ALUMINUM OR COPPER-CLAD ALUMINUM									
	RHO 60 LF 50	RHO 90 LF 100	RHO 120 LF 100	RHO 60 LF 50	RHO 90 LF 100	RHO 120 LF 100	RHO 60 LF 50	RHO 90 LF 100	RHO 120 LF 100	RHO 60 LF 50	RHO 90 LF 100	RHO 120 LF 100	RHO 60 LF 50	RHO 90 LF 100	RHO 120 LF 100	RHO 60 LF 50	RHO 90 LF 100	RHO 120 LF 100	
250	410	344	327	386	295	275	369	270	252	320	269	256	302	230	214	288	211	197	250
350	503	418	396	472	355	330	446	322	299	393	327	310	369	277	258	350	252	235	350
500	624	511	484	583	431	400	545	387	360	489	401	379	457	337	313	430	305	284	500
750	794	640	603	736	534	494	674	469	434	626	505	475	581	421	389	538	375	347	750
1000	936	745	700	864	617	570	776	533	493	744	593	557	687	491	453	629	432	399	1000
1250	1055	832	781	970	686	632	854	581	536	848	668	627	779	551	508	703	478	441	1250
1500	1160	907	849	1063	744	685	918	619	571	941	736	689	863	604	556	767	517	477	1500
1750	1250	970	907	1142	793	729	975	651	599	1026	796	745	937	651	598	823	550	507	1750
2000	1332	1027	959	1213	836	768	1030	683	628	1103	850	794	1005	693	636	877	581	535	2000

Ambient Temp. (°C)	Correction Factors						Ambient Temp. (°F)
6–10	1.09	1.09	1.09	1.09	1.09	1.09	43–50
11–15	1.04	1.04	1.04	1.04	1.04	1.04	52–59
16–20	1.00	1.00	1.00	1.00	1.00	1.00	61–68
21–25	0.95	0.95	0.95	0.95	0.95	0.95	70–77
26–30	0.90	0.90	0.90	0.90	0.90	0.90	79–86

Δ **TABLE B.2(6)** *Ampacities of Three Insulated Conductors, Rated 0 Through 2000 Volts, Within an Overall Covering (Three-Conductor Cable) in Underground Electrical Ducts (One Cable per Electrical Duct) Based on Ambient Earth Temperature of 20°C (68°F), Electrical Duct Arrangement in Accordance with Figure B.2(1), Conductor Temperature 75°C (167°F)*

Size (AWG or kcmil)	1 Electrical Duct [Fig. B.2(1), Detail 1] Types RHW, THHW, THW, THWN, XHHW, USE			3 Electrical Ducts [Fig. B.2(1), Detail 2] Types RHW, THHW, THW, THWN, XHHW, USE			6 Electrical Ducts [Fig. B.2(1), Detail 3] Types RHW, THHW, THW, THWN, XHHW, USE			1 Electrical Duct [Fig. B.2(1), Detail 1] Types RHW, THHW, THW, THWN, XHHW, USE			3 Electrical Ducts [Fig. B.2(1), Detail 2] Types RHW, THHW, THW, THWN, XHHW, USE			6 Electrical Ducts [Fig. B.2(1), Detail 3] Types RHW, THHW, THW, THWN, XHHW, USE			Size (AWG or kcmil)
	COPPER									ALUMINUM OR COPPER-CLAD ALUMINUM									
	RHO 60 LF 50	RHO 90 LF 100	RHO 120 LF 100	RHO 60 LF 50	RHO 90 LF 100	RHO 120 LF 100	RHO 60 LF 50	RHO 90 LF 100	RHO 120 LF 100	RHO 60 LF 50	RHO 90 LF 100	RHO 120 LF 100	RHO 60 LF 50	RHO 90 LF 100	RHO 120 LF 100	RHO 60 LF 50	RHO 90 LF 100	RHO 120 LF 100	
8	58	54	53	56	48	46	53	42	39	45	42	41	43	37	36	41	32	30	8
6	77	71	69	74	63	60	70	54	51	60	55	54	57	49	47	54	42	39	6
4	101	93	91	96	81	77	91	69	65	78	72	71	75	63	60	71	54	51	4
2	132	121	118	126	105	100	119	89	83	103	94	92	98	82	78	92	70	65	2
1	154	140	136	146	121	114	137	102	95	120	109	106	114	94	89	107	79	74	1
1/0	177	160	156	168	137	130	157	116	107	138	125	122	131	107	101	122	90	84	1/0
2/0	203	183	178	192	156	147	179	131	121	158	143	139	150	122	115	140	102	95	2/0
3/0	233	210	204	221	178	158	205	148	137	182	164	159	172	139	131	160	116	107	3/0
4/0	268	240	232	253	202	190	234	168	155	209	187	182	198	158	149	183	131	121	4/0
250	297	265	256	280	222	209	258	184	169	233	207	201	219	174	163	202	144	132	250
350	363	321	310	340	267	250	312	219	202	285	252	244	267	209	196	245	172	158	350
500	444	389	375	414	320	299	377	261	240	352	308	297	328	254	237	299	207	190	500
750	552	478	459	511	388	362	462	314	288	446	386	372	413	314	293	374	254	233	750
1000	628	539	518	579	435	405	522	351	321	521	447	430	480	361	336	433	291	266	1000

Ambient Temp. (°C)	Correction Factors						Ambient Temp. (°F)
6–10	1.09	1.09	1.09	1.09	1.09	1.09	43–50
11–15	1.04	1.04	1.04	1.04	1.04	1.04	52–59
16–20	1.00	1.00	1.00	1.00	1.00	1.00	61–68
21–25	0.95	0.95	0.95	0.95	0.95	0.95	70–77
26–30	0.90	0.90	0.90	0.90	0.90	0.90	79–86

Δ **TABLE B.2(7)** *Ampacities of Three Single Insulated Conductors, Rated 0 Through 2000 Volts, in Underground Electrical Ducts (Three Conductors per Electrical Duct) Based on Ambient Earth Temperature of 20°C (68°F), Electrical Duct Arrangement in Accordance with Figure B.2(1), Conductor Temperature 75°C (167°F)*

Size (AWG or kcmil)	1 Electrical Duct [Fig. B.2(1), Detail 1] Types RHW, THHW, THW, THWN, XHHW, USE			3 Electrical Ducts [Fig. B.2(1), Detail 2] Types RHW, THHW, THW, THWN, XHHW, USE			6 Electrical Ducts [Fig. B.2(1), Detail 3] Types RHW, THHW, THW, THWN, XHHW, USE			1 Electrical Duct [Fig. B.2(1), Detail 1] Types RHW, THHW, THW, THWN, XHHW, USE			3 Electrical Ducts [Fig. B.2(1), Detail 2] Types RHW, THHW, THW, THWN, XHHW, USE			6 Electrical Ducts [Fig. B.2(1), Detail 3] Types RHW, THHW, THW, THWN, XHHW, USE			Size (AWG or kcmil)
	COPPER									ALUMINUM OR COPPER-CLAD ALUMINUM									
	RHO 60 LF 50	RHO 90 LF 100	RHO 120 LF 100	RHO 60 LF 50	RHO 90 LF 100	RHO 120 LF 100	RHO 60 LF 50	RHO 90 LF 100	RHO 120 LF 100	RHO 60 LF 50	RHO 90 LF 100	RHO 120 LF 100	RHO 60 LF 50	RHO 90 LF 100	RHO 120 LF 100	RHO 60 LF 50	RHO 90 LF 100	RHO 120 LF 100	
8	63	58	57	61	51	49	57	44	41	49	45	44	47	40	38	45	34	32	8
6	84	77	75	80	67	63	75	56	53	66	60	58	63	52	49	59	44	41	6
4	111	100	98	105	86	81	98	73	67	86	78	76	79	67	63	77	57	52	4
3	129	116	113	122	99	94	113	83	77	101	91	89	83	77	73	84	65	60	3
2	147	132	128	139	112	106	129	93	86	115	103	100	108	87	82	101	73	67	2
1	171	153	148	161	128	121	149	106	98	133	119	115	126	100	94	116	83	77	1
1/0	197	175	169	185	146	137	170	121	111	153	136	132	144	114	107	133	94	87	1/0
2/0	226	200	193	212	166	156	194	136	126	176	156	151	165	130	121	151	106	98	2/0
3/0	260	228	220	243	189	177	222	154	142	203	178	172	189	147	138	173	121	111	3/0
4/0	301	263	253	280	215	201	255	175	161	235	205	198	219	168	157	199	137	126	4/0
250	334	290	279	310	236	220	281	192	176	261	227	218	242	185	172	220	150	137	250
300	373	321	308	344	260	242	310	210	192	293	252	242	272	204	190	245	165	151	300
350	409	351	337	377	283	264	340	228	209	321	276	265	296	222	207	266	179	164	350
400	442	376	361	394	302	280	368	243	223	349	297	284	321	238	220	288	191	174	400
500	503	427	409	460	341	316	412	273	249	397	338	323	364	270	250	326	216	197	500
600	552	468	447	511	371	343	457	296	270	446	373	356	408	296	274	365	236	215	600
700	602	509	486	553	402	371	492	319	291	488	408	389	443	321	297	394	255	232	700
750	632	529	505	574	417	385	509	330	301	508	425	405	461	334	309	409	265	241	750
800	654	544	520	597	428	395	527	338	308	530	439	418	481	344	318	427	273	247	800
900	692	575	549	628	450	415	554	355	323	563	466	444	510	365	337	450	288	261	900
1000	730	605	576	659	472	435	581	372	338	597	494	471	538	385	355	475	304	276	1000

Ambient Temp. (°C)	Correction Factors						Ambient Temp. (°F)
6–10	1.09	1.09	1.09	1.09	1.09	1.09	43–50
11–15	1.04	1.04	1.04	1.04	1.04	1.04	52–59
16–20	1.00	1.00	1.00	1.00	1.00	1.00	61–68
21–25	0.95	0.95	0.95	0.95	0.95	0.95	70–77
26–30	0.90	0.90	0.90	0.90	0.90	0.90	79–86

TABLE B.2(8) *Ampacities of Two or Three Insulated Conductors, Rated 0 Through 2000 Volts, Cabled Within an Overall (Two- or Three-Conductor) Covering, Directly Buried in Earth, Based on Ambient Earth Temperature of 20°C (68°F), Electrical Duct Arrangement in Accordance with Figure B.2(1), 100 Percent Load Factor, Thermal Resistance (Rho) of 90*

Size (AWG or kcmil)	1 Cable [Fig. B.2(1), Detail 5]		2 Cables [Fig. B.2(1), Detail 6]		1 Cable [Fig. B.2(1), Detail 5]		2 Cables [Fig. B.2(1), Detail 6]		Size (AWG or kcmil)
	60°C (140°F)	75°C (167°F)	60°C (140°F)	75°C (167°F)	60°C (140°F)	75°C (167°F)	60°C (140°F)	75°C (167°F)	
	TYPES				TYPES				
	UF	RHW, THHW, THW, THWN, XHHW, USE	UF	RHW, THHW, THW, THWN, XHHW, USE	UF	RHW, THHW, THW, THWN, XHHW, USE	UF	RHW, THHW, THW, THWN, XHHW, USE	
	COPPER				ALUMINUM OR COPPER-CLAD ALUMINUM				
8	64	75	60	70	51	59	47	55	8
6	85	100	81	95	68	75	60	70	6
4	107	125	100	117	83	97	78	91	4
2	137	161	128	150	107	126	110	117	2
1	155	182	145	170	121	142	113	132	1
1/0	177	208	165	193	138	162	129	151	1/0
2/0	201	236	188	220	157	184	146	171	2/0
3/0	229	269	213	250	179	210	166	195	3/0
4/0	259	304	241	282	203	238	188	220	4/0
250	—	333	—	308	—	261	—	241	250
350	—	401	—	370	—	315	—	290	350
500	—	481	—	442	—	381	—	350	500
750	—	585	—	535	—	473	—	433	750
1000	—	657	—	600	—	545	—	497	1000
Ambient Temp. (°C)	Correction Factors								**Ambient Temp. (°F)**
6–10	1.12	1.09	1.12	1.09	1.12	1.09	1.12	1.09	43–50
11–15	1.06	1.04	1.06	1.04	1.06	1.04	1.06	1.04	52–59
16–20	1.00	1.00	1.00	1.00	1.00	1.00	1.00	1.00	61–68
21–25	0.94	0.95	0.94	0.95	0.94	0.95	0.94	0.95	70–77
26–30	0.87	0.90	0.87	0.90	0.87	0.90	0.87	0.90	79–86

Note: For ampacities of Type UF cable in underground electrical ducts, multiply the ampacities shown in the table by 0.74.

△ **TABLE B.2(9)** *Ampacities of Three Triplexed Single Insulated Conductors, Rated 0 Through 2000 Volts, Directly Buried in Earth Based on Ambient Earth Temperature of 20°C (68°F), Electrical Duct Arrangement in Accordance with Figure B.2(1), 100 Percent Load Factor, Thermal Resistance (Rho) of 90*

Size (AWG or kcmil)	See Fig. B.2(1), Detail 7		See Fig. B.2(1), Detail 8		See Fig. B.2(1), Detail 7		See Fig. B.2(1), Detail 8		Size (AWG or kcmil)
	60°C (140°F)	75°C (167°F)	60°C (140°F)	75°C (167°F)	60°C (140°F)	75°C (167°F)	60°C (140°F)	75°C (167°F)	
	TYPES				TYPES				
	UF	USE	UF	USE	UF	USE	UF	USE	
	COPPER				ALUMINUM OR COPPER-CLAD ALUMINUM				
8	72	84	66	77	55	65	51	60	8
6	91	107	84	99	72	84	66	77	6
4	119	139	109	128	92	108	85	100	4
2	153	179	140	164	119	139	109	128	2
1	173	203	159	186	135	158	124	145	1
1/0	197	231	181	212	154	180	141	165	1/0
2/0	223	262	205	240	175	205	159	187	2/0
3/0	254	298	232	272	199	233	181	212	3/0
4/0	289	339	263	308	226	265	206	241	4/0
250	—	370	—	336	—	289	—	263	250
350	—	445	—	403	—	349	—	316	350
500	—	536	—	483	—	424	—	382	500
750	—	654	—	587	—	525	—	471	750
1000	—	744	—	665	—	608	—	544	1000
Ambient Temp. (°C)	**Correction Factors**								**Ambient Temp. (°F)**
6–10	1.12	1.09	1.12	1.09	1.12	1.09	1.12	1.09	43–50
11–15	1.06	1.04	1.06	1.04	1.06	1.04	1.06	1.04	52–59
16–20	1.00	1.00	1.00	1.00	1.00	1.00	1.00	1.00	61–68
21–25	0.94	0.95	0.94	0.95	0.94	0.95	0.94	0.95	70–77
26–30	0.87	0.90	0.87	0.90	0.87	0.90	0.87	0.90	79–86

Δ **TABLE B.2(10)** *Ampacities of Three Single Insulated Conductors, Rated 0 Through 2000 Volts, Directly Buried in Earth Based on Ambient Earth Temperature of 20°C (68°F), Electrical Duct Arrangement in Accordance with Figure B.2(1), 100 Percent Load Factor, Thermal Resistance (Rho) of 90*

Size (AWG or kcmil)	See Fig. B.2(1), Detail 9		See Fig. B.2(1), Detail 10		See Fig. B.2(1), Detail 9		See Fig. B.2(1), Detail 10		Size (AWG or kcmil)
	60°C (140°F)	75°C (167°F)	60°C (140°F)	75°C (167°F)	60°C (140°F)	75°C (167°F)	60°C (140°F)	75°C (167°F)	
	TYPES				TYPES				
	UF	USE	UF	USE	UF	USE	UF	USE	
	COPPER				ALUMINUM OR COPPER-CLAD ALUMINUM				
8	84	98	78	92	66	77	61	72	8
6	107	126	101	118	84	98	78	92	6
4	139	163	130	152	108	127	101	118	4
2	178	209	165	194	139	163	129	151	2
1	201	236	187	219	157	184	146	171	1
1/0	230	270	212	249	179	210	165	194	1/0
2/0	261	306	241	283	204	239	188	220	2/0
3/0	297	348	274	321	232	272	213	250	3/0
4/0	336	394	309	362	262	307	241	283	4/0
250	—	429	—	394	—	335	—	308	250
350	—	516	—	474	—	403	—	370	350
500	—	626	—	572	—	490	—	448	500
750	—	767	—	700	—	605	—	552	750
1000	—	887	—	808	—	706	—	642	1000
1250	—	979	—	891	—	787	—	716	1250
1500	—	1063	—	965	—	862	—	783	1500
1750	—	1133	—	1027	—	930	—	843	1750
2000	—	1195	—	1082	—	990	—	897	2000
Ambient Temp. (°C)	Correction Factors								Ambient Temp. (°F)
6–10	1.12	1.09	1.12	1.09	1.12	1.09	1.12	1.09	43–50
11–15	1.06	1.04	1.06	1.04	1.06	1.04	1.06	1.04	52–59
16–20	1.00	1.00	1.00	1.00	1.00	1.00	1.00	1.00	61–68
21–25	0.94	0.95	0.94	0.95	0.94	0.95	0.94	0.95	70–77
26–30	0.87	0.90	0.87	0.90	0.87	0.90	0.87	0.90	79–86

Δ **TABLE B.2(11)** *Adjustment Factors for More Than Three Current-Carrying Conductors in a Raceway or Cable with Load Diversity*

Number of Conductors*	Percent of Values in Tables as Adjusted for Ambient Temperature if Necessary
4–6	80
7–9	70
10–24	70**
25–42	60**
43–85	50**

*Number of conductors is the total number of conductors in the raceway or cable adjusted in accordance with 310.15(E) and (F).

**These factors include the effects of a load diversity of 50 percent.

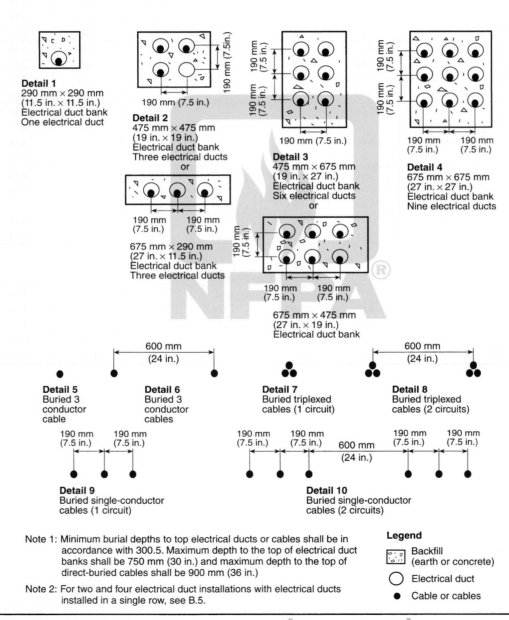

Note 1: Minimum burial depths to top electrical ducts or cables shall be in accordance with 300.5. Maximum depth to the top of electrical duct banks shall be 750 mm (30 in.) and maximum depth to the top of direct-buried cables shall be 900 mm (36 in.).

Note 2: For two and four electrical duct installations with electrical ducts installed in a single row, see B.5.

Legend

Backfill (earth or concrete)

Electrical duct

Cable or cables

Δ **FIGURE B.2(1)** *Cable Installation Dimensions for Use with Table B.2(5) Through Table B.2(10).*

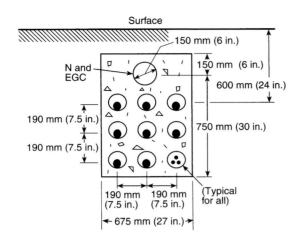

Design Criteria
Neutral and Equipment
  Grounding conductor (EGC)
  Duct = 150 mm (6 in.)
Phase Ducts = 75 to 125 mm (3 to 5 in.)
Conductor Material = Copper
Number of Cables per Duct = 3

Number of Cables per Phase = 9
Rho Concrete = Rho Earth – 5
Rho PVC Duct = 650
Rho Cable Insulation = 500
Rho Cable Jacket = 650

Notes:
1. Neutral configuration per 300.5(I), Exception No. 2, for isolated phase installations in nonmagnetic ducts.
2. Phasing is A, B, C in rows or columns. Where magnetic electrical ducts are used, conductors are installed A, B, C per electrical duct with the neutral and all equipment grounding conductors in the same electrical duct. In this case, the 6-in. trade size neutral duct is eliminated.
3. Maximum harmonic loading on the neutral conductor cannot exceed 50 percent of the phase current for the ampacities shown in the table below.
4. Metallic shields of Type MV-90 cable shall be grounded at one point only where using A, B, C phasing in rows or columns.

Size kcmil	TYPES RHW, THHW, THW, THWN, XHHW, USE, OR MV-90*			Size kcmil
	Total per Phase Ampere Rating			
	RHO EARTH 60 LF 50	RHO EARTH 90 LF 100	RHO EARTH 120 LF 100	
250	2340 (260A/Cable)	1530 (170A/Cable)	1395 (155A/Cable)	250
350	2790 (310A/Cable)	1800 (200A/Cable)	1665 (185A/Cable)	350
500	3375 (375A/Cable)	2160 (240A/Cable)	1980 (220A/Cable)	500

Ambient Temp. (°C)	For ambient temperatures other than 20°C (68°F), multiply the ampacities shown above by the appropriate factor shown below.				Ambient Temp. (°F)
6–10	1.09	1.09	1.09	1.09	43–50
11–15	1.04	1.04	1.04	1.04	52–59
16–20	1.00	1.00	1.00	1.00	61–68
21–25	0.95	0.95	0.95	0.95	70–77
26–30	0.90	0.90	0.90	0.90	79–86

*Limited to 75°C conductor temperature.

FIGURE B.2(2) *Ampacities of Single Insulated Conductors Rated 0 Through 5000 Volts in Underground Electrical Ducts (Three Conductors per Electrical Duct), Nine Single-Conductor Cables per Phase Based on Ambient Earth Temperature of 20°C (68°F), Conductor Temperature 75°C (167°F).*

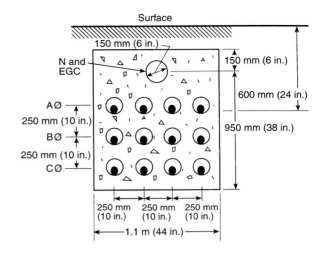

Design Criteria
Neutral and Equipment
  Grounding conductor (EGC)
  Duct = 150 mm (6 in.)
Phase Ducts = 75 mm (3 in.)
Conductor Material = Copper
Number of Cables per Duct = 1

Number of Cables per Phase = 4
Rho Concrete = Rho Earth – 5
Rho PVC Duct = 650
Rho Cable Insulation = 500
Rho Cable Jacket = 650

Notes:
1. Neutral configuration per 300.5(I), Exception No 2.
2. Maximum harmonic loading on the neutral conductor cannot exceed 50 percent of the phase current for the ampacities shown in the table below.
3. Metallic shields of Type MV-90 cable shall be grounded at one point only.

Size kcmil	TYPES RHW, THHW, THW, THWN, XHHW, USE, OR MV-90*			Size kcmil
	Total per Phase Ampere Rating			
	RHO EARTH 60 LF 50	RHO EARTH 90 LF 100	RHO EARTH 120 LF 100	
750	2820 (705A/Cable)	1860 (465A/Cable)	1680 (420A/Cable)	750
1000	3300 (825A/Cable)	2140 (535A/Cable)	1920 (480A/Cable)	1000
1250	3700 (925A/Cable)	2380 (595A/Cable)	2120 (530A/Cable)	1250
1500	4060 (1015A/Cable)	2580 (645A/Cable)	2300 (575A/Cable)	1500
1750	4360 (1090A/Cable)	2740 (685A/Cable)	2460 (615A/Cable)	1750

Ambient Temp. (°C)	For ambient temperatures other than 20°C (68°F), multiply the ampacities shown above by the appropriate factor shown below.				Ambient Temp. (°F)
6–10	1.09	1.09	1.09	1.09	43–50
11–15	1.04	1.04	1.04	1.04	52–59
16–20	1.00	1.00	1.00	1.00	61–68
21–25	0.95	0.95	0.95	0.95	70–77
26–30	0.90	0.90	0.90	0.90	79–86

*Limited to 75°C conductor temperature.

FIGURE B.2(3) *Ampacities of Single Insulated Conductors Rated 0 Through 5000 Volts in Nonmagnetic Underground Electrical Ducts (One Conductor per Electrical Duct), Four Single-Conductor Cables per Phase Based on Ambient Earth Temperature of 20°C (68°F), Conductor Temperature 75°C (167°F).*

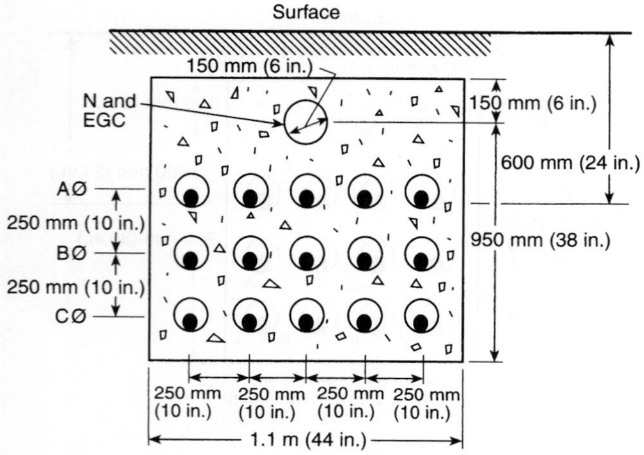

Design Criteria
Neutral and Equipment
Grounding conductor (EGC)
Duct = 150 mm (6 in.)
Phase Ducts = 75 mm (3 in.)
Conductor Material = Copper
Number of Cables per Duct = 1

Number of Cables per Phase = 5
Rho Concrete = Rho Earth – 5
Rho PVC Duct = 650
Rho Cable Insulation = 500
Rho Cable Jacket = 650

Notes:
1. Neutral configuration per 300.5(I), Exception No. 2.
2. Maximum harmonic loading on the neutral conductor cannot exceed 50 percent of the phase current for the ampacities shown in the table below.
3. Metallic shields of Type MV-90 cable shall be grounded at one point only.

Size kcmil	TYPES RHW, THHW, THW, THWN, XHHW, USE, OR MV-90* Total per Phase Ampere Rating			Size kcmil
	RHO EARTH 60 LF 50	RHO EARTH 90 LF 100	RHO EARTH 120 LF 100	
2000	5575 (1115A/Cable)	3375 (675A/Cable)	3000 (600A/Cable)	2000

Ambient Temp. (°C)	For ambient temperatures other than 20°C (68°F), multiply the ampacities shown above by the appropriate factor shown below.					Ambient Temp. (°F)
6–10	1.09	1.09	1.09	1.09	1.09	43–50
11–15	1.04	1.04	1.04	1.04	1.04	52–59
16–20	1.00	1.00	1.00	1.00	1.00	61–68
21–25	0.95	0.95	0.95	0.95	0.95	70–77
26–30	0.90	0.90	0.90	0.90	0.90	79–86

*Limited to 75°C conductor temperature.

FIGURE B.2(4) *Ampacities of Single Insulated Conductors Rated 0 Through 5000 Volts in Nonmagnetic Underground Electrical Ducts (One Conductor per Electrical Duct), Five Single-Conductor Cables per Phase Based on Ambient Earth Temperature of 20°C (68°F), Conductor Temperature 75°C (167°F).*

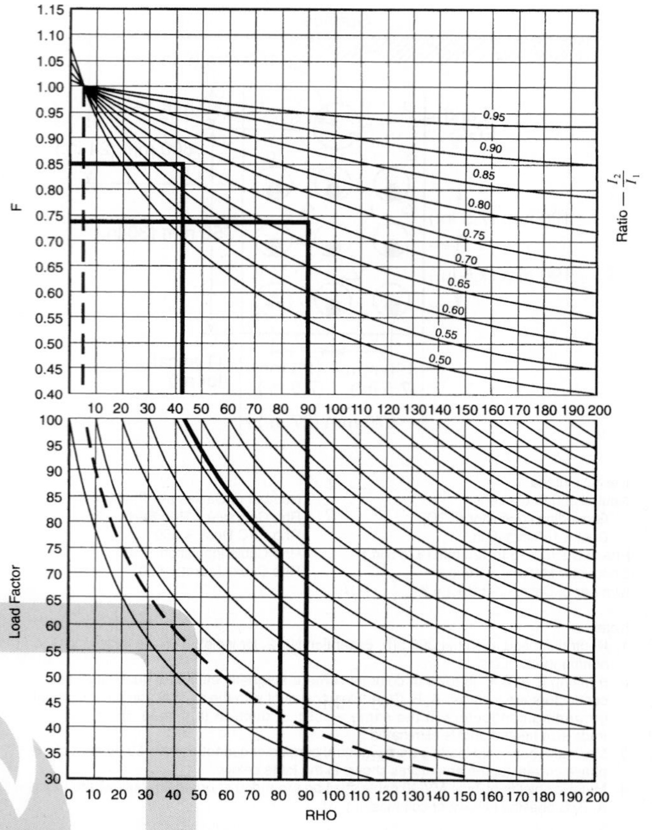

FIGURE B.3 *Interpolation Chart for Cables in a Duct Bank.* I_1 *= ampacity for Rho = 60, 50 LF;* I_2 *= ampacity for Rho = 120, 100 LF (load factor); desired ampacity =* $F \times I_1$.

Conduit, Tubing, and Cable Tray Fill Tables for Conductors and Fixture Wires of the Same Size

INFORMATIVE
ANNEX

C

This informative annex is not a part of the requirements of this NFPA document but is included for informational purposes only.

The Informative Annex C conduit and tubing conductor fill tables are provided only as an informational tool and are not part of the mandatory requirements of the *NEC*®. These conductor fill values have been calculated based on the conductor fill requirements, conduit and tubing dimensions, and conductor dimensions from Chapter 9, Tables 1, 4, 5, and 5A. Where all the conductors in conduit or tubing are the same physical size and have the same insulation characteristics, the use of the Informative Annex C tables to determine the maximum number of conductors ensures compliance with the requirements for raceway fill found in 300.17 and the respective conduit and tubing articles. Although the values for maximum conduit or tubing fill do not exceed those permitted by Chapter 9, Table 1, the advice provided in Informational Note No. 1 to that table on considering a lesser conductor fill or a larger size conduit to that table should always be considered where the wire-pulling conditions are not optimum.

Chapter 9, Table 1, sets forth the percentage fill required, and Tables 4, 5, and 5A list the accurate conduit, tubing, and wire dimensions. Users can calculate the percent fill or use the tables in Informative Annex C, all of which were generated using Chapter 9, Tables 1, 4, 5, and 5A.

The 13 sets of tables in Informative Annex C correspond to the 13 conduit and tubing wiring methods in Chapter 3 of the *NEC*. Each set of tables is subdivided into three conductor categories: (1) conductors for general wiring (Article 310), (2) fixture wires (Article 402), and (3) compact stranded conductors. In Informative Annex C, tables that use compact stranding are listed as "A" tables.

To select the correct metric designator or trade size conduit or tubing, proceed as follows:

Step 1. Select the appropriate table using the lists of metal and nonmetallic wiring methods.

Step 2. Choose the appropriate conductors (general wiring conductors, fixture wires, or compact conductors).

Step 3. Choose the appropriate insulation.

Step 4. Select the correct trade size conduit or tubing for the given quantity and size of conductors required.

The following examples detail how to determine the correct trade size conduit or tubing.

Application Example 1

An installation requires ten 10 AWG THWN-2 copper conductors in an underground conduit across a parking lot for exterior lighting. What size polyvinyl chloride (PVC) conduit will be required?

Solution

Commentary Table C.1 lists minimum sizes of PVC conduit for this application.

COMMENTARY TABLE C.1 Types of PVC Conduit

Wiring Method	Table	Minimum Trade Size PVC Conduit
Rigid PVC conduit, Schedule 40 and HDPE	Table C.11	1
Type A, rigid PVC conduit	Table C.12	¾
Rigid PVC conduit, Schedule 80	Table C.10	1
Type EB, PVC conduit	Table C.13	Available only in trade sizes 2 and larger

Application Example 2

An underground service lateral requires four 600-kcmil XHHW compact-stranded aluminum conductors. What trade size conduit will be required for RMC, IMC, PVC Schedule 40, PVC Schedule 80, and PVC Type EB?

Solution

See Commentary Table C.2 for the minimum trade size for the respective conduit types needed for the four 600-kcmil aluminum conductors.

COMMENTARY TABLE C.2 Minimum Trade Size Conduit

Wiring Method	Table	Minimum Trade Size Conduit
RMC	Table C.9(A)	3
IMC	Table C.4(A)	3
Rigid PVC conduit, Schedule 40	Table C.11(A)	3
Rigid PVC conduit, Schedule 80	Table C.10(A)	3½
Type EB, PVC conduit	Table C.13(A)	3

Most aluminum building wire in Types THW, THWN/THHN, and XHHW is compact stranded.

Application Example 3

A motor will be supplied by three 4 AWG THW conductors. What size metal conduit or tubing will be required? What size metal flex will be required at the motor termination?

Solution

See Commentary Table C.3 for the minimum size metal conduit or metal tubing for this application.

COMMENTARY TABLE C.3 Types of Metal Conduit and Metal Flex

Wiring Method	Table	Trade Size Conduit or Tubing
EMT	Table C.1	1
IMC	Table C.4	1
RMC	Table C.9	1
FMC	Table C.3	1
Liquidtight FMC	Table C.8	1

If a wire-type equipment grounding conductor (EGC) (likely to be smaller than 4 AWG based on rating or size of motor circuit short-circuit, ground-fault protective device, and Table 250.122) is installed in any one of these raceway types, there is a mixture of conductor sizes. For that reason, the minimum size conduit or tubing has to be calculated based on Chapter 9, Tables 4, 5, and 8 if a wire-type EGC is installed in the conduit or tubing. Of course, based on the Informative Annex C tables, the conduit size could be increased to trade size 1¼. Because six 4 AWG THW conductors can be installed in trade size 1¼, it would be of sufficient size for the three circuit conductors and the wire-type EGC. According to 250.122(A), a wire-type EGC is not required to be larger than the circuit conductors. This approach might not yield the minimum conduit or tubing size, but it will result in a compliant installation and could facilitate an easier installation if the conduit run is particularly long or has a substantial number of bends. See the commentary following Informational Note No. 1 to Chapter 9, Table 1.

Application Example 4

A fire alarm system installation requires the riser to contain twenty-one 16 AWG TFF conductors. If the riser conductors are installed in electrical metallic tubing, what minimum size tubing is required?

Solution

According to Table C.1, trade size 1 EMT is required.

Δ **TABLE C.1** *Maximum Number of Conductors or Fixture Wires in Electrical Metallic Tubing (EMT) (Based on Chapter 9: Table 1, Table 4, and Table 5)*

Type	Conductor Size (AWG/ kcmil)	Trade Size (Metric Designator)												
		⅜ (12)	½ (16)	¾ (21)	1 (27)	1¼ (35)	1½ (41)	2 (53)	2½ (63)	3 (78)	3½ (91)	4 (103)	5 (129)	6 (155)
						CONDUCTORS								
RHH, RHW, RHW-2	14	—	4	7	11	20	27	46	80	120	157	201	302	427
	12	—	3	6	9	17	23	38	66	100	131	167	251	354
	10	—	2	5	8	13	18	30	53	81	105	135	203	286
	8	—	1	2	4	7	9	16	28	42	55	70	106	150
	6	—	1	1	3	5	8	13	22	34	44	56	85	120
	4	—	1	1	2	4	6	10	17	26	34	44	66	94
	3	—	1	1	1	4	5	9	15	23	30	38	58	82
	2	—	1	1	1	3	4	7	13	20	26	33	50	71
	1	—	0	1	1	1	3	5	9	13	17	22	33	47
	1/0	—	0	1	1	1	2	4	7	11	15	19	29	41
	2/0	—	0	1	1	1	2	4	6	10	13	17	25	35
	3/0	—	0	0	1	1	1	3	5	8	11	14	21	30
	4/0	—	0	0	1	1	1	3	5	7	9	12	18	26
	250	—	0	0	0	1	1	1	3	5	7	9	14	20
	300	—	0	0	0	1	1	1	3	5	6	8	12	17
	350	—	0	0	0	1	1	1	3	4	6	7	11	16
	400	—	0	0	0	1	1	1	2	4	5	7	10	14
	500	—	0	0	0	0	1	1	2	3	4	6	8	12
	600	—	0	0	0	0	1	1	1	3	4	5	7	10
	700	—	0	0	0	0	0	1	1	2	3	4	6	9
	750	—	0	0	0	0	0	1	1	2	3	4	6	8
	800	—	0	0	0	0	0	1	1	2	3	4	6	8
	900	—	0	0	0	0	0	1	1	1	3	3	5	7
	1000	—	0	0	0	0	0	1	1	1	2	3	5	7
	1250	—	0	0	0	0	0	0	1	1	1	2	3	5
	1500	—	0	0	0	0	0	0	1	1	1	1	3	4
	1750	—	0	0	0	0	0	0	1	1	1	1	3	4
	2000	—	0	0	0	0	0	0	1	1	1	1	2	3
TW, THHW, THW, THW-2	14	—	8	15	25	43	58	96	168	254	332	424	638	900
	12	—	6	11	19	33	45	74	129	195	255	326	490	691
	10	—	5	8	14	24	33	55	96	145	190	243	365	515
	8	—	2	5	8	13	18	30	53	81	105	135	203	286
RHH*, RHW*, RHW-2*	14	—	6	10	16	28	39	64	112	169	221	282	424	599
	12	—	4	8	13	23	31	51	90	136	177	227	341	481
	10	—	3	6	10	18	24	40	70	106	138	177	266	376
	8	—	1	4	6	10	14	24	42	63	83	106	159	225
TW, THW, THHW, THW-2, RHH*, RHW*, RHW-2*	6	—	1	3	4	8	11	18	32	48	63	81	122	172
	4	—	1	1	3	6	8	13	24	36	47	60	91	128
	3	—	1	1	3	5	7	12	20	31	40	52	78	110
	2	—	1	1	2	4	6	10	17	26	34	44	66	94
	1	—	1	1	1	3	4	7	12	18	24	31	46	66
	1/0	—	0	1	1	2	3	6	10	16	20	26	40	56
	2/0	—	0	1	1	1	3	5	9	13	17	22	33	47
	3/0	—	0	1	1	1	2	4	7	11	15	19	28	40
	4/0	—	0	0	1	1	1	3	6	9	12	16	24	33

(continues)

Δ **TABLE C.1** *Continued*

Type	Conductor Size (AWG/kcmil)	⅜ (12)	½ (16)	¾ (21)	1 (27)	1¼ (35)	1½ (41)	2 (53)	2½ (63)	3 (78)	3½ (91)	4 (103)	5 (129)	6 (155)
								Trade Size (Metric Designator)						
	250	—	0	0	1	1	1	3	5	7	10	13	19	27
	300	—	0	0	1	1	1	2	4	6	8	11	16	23
	350	—	0	0	0	1	1	1	4	6	7	10	15	21
	400	—	0	0	0	1	1	1	3	5	7	9	13	19
	500	—	0	0	0	1	1	1	3	4	6	7	11	16
	600	—	0	0	0	1	1	1	2	3	4	6	9	13
	700	—	0	0	0	0	1	1	1	3	4	5	8	11
	750	—	0	0	0	0	1	1	1	3	4	5	7	10
	800	—	0	0	0	0	1	1	1	3	3	5	7	10
	900	—	0	0	0	0	0	1	1	2	3	4	6	9
	1000	—	0	0	0	0	0	1	1	2	3	4	6	8
	1250	—	0	0	0	0	0	1	1	1	2	3	4	6
	1500	—	0	0	0	0	0	1	1	1	1	2	4	5
	1750	—	0	0	0	0	0	0	1	1	1	2	3	5
	2000	—	0	0	0	0	0	0	1	1	1	1	3	4
THHN, THWN, THWN-2	14	—	12	22	35	61	84	138	241	364	476	608	914	1290
	12	—	9	16	26	45	61	101	176	266	347	443	666	941
	10	—	5	10	16	28	38	63	111	167	219	279	420	593
	8	—	3	6	9	16	22	36	64	96	126	161	242	342
	6	—	2	4	7	12	16	26	46	69	91	116	175	247
	4	—	1	2	4	7	10	16	28	43	56	71	107	152
	3	—	1	1	3	6	8	13	24	36	47	60	91	128
	2	—	1	1	3	5	7	11	20	30	40	51	76	108
	1	—	1	1	1	4	5	8	15	22	29	37	56	80
	1/0	—	1	1	1	3	4	7	12	19	25	32	47	67
	2/0	—	0	1	1	2	3	6	10	16	20	26	40	56
	3/0	—	0	1	1	1	3	5	8	13	17	22	33	46
	4/0	—	0	1	1	1	2	4	7	11	14	18	27	38
	250	—	0	0	1	1	1	3	6	9	11	15	22	31
	300	—	0	0	1	1	1	3	5	7	10	13	19	27
	350	—	0	0	1	1	1	2	4	6	9	11	17	24
	400	—	0	0	0	1	1	1	4	6	8	10	15	21
	500	—	0	0	0	1	1	1	3	5	6	8	12	17
	600	—	0	0	0	1	1	1	2	4	5	7	10	14
	700	—	0	0	0	1	1	1	2	3	4	6	9	12
	750	—	0	0	0	0	1	1	1	3	4	5	8	12
	800	—	0	0	0	0	1	1	1	3	4	5	8	11
	900	—	0	0	0	0	1	1	1	3	3	4	7	10
	1000	—	0	0	0	0	1	1	1	2	3	4	6	9
FEP, FEPB, PFA, PFAH, TFE	14	—	12	21	34	60	81	134	234	354	462	590	886	1252
	12	—	9	15	25	43	59	98	171	258	337	430	647	913
	10	—	6	11	18	31	42	70	122	185	241	309	464	655
	8	—	3	6	10	18	24	40	70	106	138	177	266	376
	6	—	2	4	7	12	17	28	50	75	98	126	189	267
	4	—	1	3	5	9	12	20	35	53	69	88	132	187
	3	—	1	2	4	7	10	16	29	44	57	73	110	155
	2	—	1	1	3	6	8	13	24	36	47	60	91	128
PFA, PFAH, TFE	1	—	1	1	2	4	6	9	16	25	33	42	63	89
PFA, PFAH, TFE, Z	1/0	—	1	1	1	3	5	8	14	21	27	35	53	74
	2/0	—	0	1	1	3	4	6	11	17	22	29	43	61
	3/0	—	0	1	1	2	3	5	9	14	18	24	36	51
	4/0	—	0	1	1	1	2	4	8	11	15	19	29	41

Δ **TABLE C.1** *Continued*

Type	Conductor Size (AWG/ kcmil)	3/8 (12)	1/2 (16)	3/4 (21)	1 (27)	1¼ (35)	1½ (41)	2 (53)	2½ (63)	3 (78)	3½ (91)	4 (103)	5 (129)	6 (155)
Z	14	—	14	25	41	72	98	161	282	426	556	711	1068	1508
	12	—	10	18	29	51	69	114	200	302	394	504	758	1070
	10	—	6	11	18	31	42	70	122	185	241	309	464	655
	8	—	4	7	11	20	27	44	77	117	153	195	293	414
	6	—	3	5	8	14	19	31	54	82	107	137	206	291
	4	—	1	3	5	9	13	21	37	56	74	94	142	200
	3	—	1	2	4	7	9	15	27	41	54	69	103	146
	2	—	1	1	3	6	8	13	22	34	45	57	86	121
	1	—	1	1	2	4	6	10	18	28	36	46	70	98
XHHW, ZW, XHHW-2, XHH	14	—	8	15	25	43	58	96	168	254	332	424	638	900
	12	—	6	11	19	33	45	74	129	195	255	326	490	691
	10	—	5	8	14	24	33	55	96	145	190	243	365	515
	8	—	2	5	8	13	18	30	53	81	105	135	203	286
	6	—	1	3	6	10	14	22	39	60	78	100	150	212
	4	—	1	2	4	7	10	16	28	43	56	72	109	153
	3	—	1	1	3	6	8	14	24	36	48	61	92	130
	2	—	1	1	3	5	7	11	20	31	40	51	77	109
XHHW, XHHW-2, XHH	1	—	1	1	1	4	5	8	15	23	30	38	57	81
	1/0	—	1	1	1	3	4	7	13	19	25	32	48	68
	2/0	—	0	1	1	2	3	6	10	16	21	27	40	57
	3/0	—	0	1	1	1	3	5	9	13	17	22	33	47
	4/0	—	0	1	1	1	2	4	7	11	14	18	27	39
	250	—	0	0	1	1	1	3	6	9	12	15	22	32
	300	—	0	0	1	1	1	3	5	8	10	13	19	27
	350	—	0	0	1	1	1	2	4	7	9	11	17	24
	400	—	0	0	0	1	1	1	4	6	8	10	15	21
	500	—	0	0	0	1	1	1	3	5	6	8	12	18
	600	—	0	0	0	1	1	1	2	4	5	6	10	14
	700	—	0	0	0	0	1	1	2	3	4	6	9	12
	750	—	0	0	0	0	1	1	1	3	4	5	8	12
	800	—	0	0	0	0	1	1	1	3	4	5	8	11
	900	—	0	0	0	0	1	1	1	3	3	4	7	10
	1000	—	0	0	0	0	0	1	1	2	3	4	6	9
	1250	—	0	0	0	0	0	1	1	1	2	3	5	7
	1500	—	0	0	0	0	0	1	1	1	1	3	4	6
	1750	—	0	0	0	0	0	0	1	1	1	2	4	5
	2000	—	0	0	0	0	0	0	1	1	1	1	3	5

FIXTURE WIRES

Type	Conductor Size (AWG/ kcmil)	3/8 (12)	1/2 (16)	3/4 (21)	1 (27)	1¼ (35)	1½ (41)	2 (53)	2½ (63)	3 (78)	3½ (91)	4 (103)	5 (129)	6 (155)
RFH-2, FFH-2, RFHH-2	18	—	8	14	24	41	56	92	161	244	318	407	611	863
	16	—	7	12	20	34	47	78	136	205	268	343	515	728
SF-2, SFF-2	18	—	10	18	30	52	71	116	203	307	401	513	771	1088
	16	—	8	15	25	43	58	96	168	254	332	424	638	900
	14	—	7	12	20	34	47	78	136	205	268	343	515	728
SF-1, SFF-1	18	—	18	33	53	92	125	206	360	544	710	908	1364	1926
RFH-1, TF, TFF, XF, XFF	18	—	14	24	39	68	92	152	266	402	524	670	1007	1422
	16	—	11	19	31	55	74	123	215	324	423	541	813	1148
XF, XFF	14	—	8	15	25	43	58	96	168	254	332	424	638	900
TFN, TFFN	18	—	22	38	63	109	148	244	426	643	839	1073	1612	2276
	16	—	17	29	48	83	113	186	325	491	641	819	1231	1738

(continues)

Δ **TABLE C.1** *Continued*

Type	Conductor Size (AWG/kcmil)	⅜ (12)	½ (16)	¾ (21)	1 (27)	1¼ (35)	1½ (41)	2 (53)	2½ (63)	3 (78)	3½ (91)	4 (103)	5 (129)	6 (155)
PF, PFF, PGF, PGFF, PAF, PTF, PTFF, PAFF	18	—	21	36	59	103	140	231	404	610	796	1017	1528	2158
	16	—	16	28	46	79	108	179	312	471	615	787	1182	1669
	14	—	12	21	34	60	81	134	234	354	462	590	886	1252
ZF, ZFF, ZHF	18	—	27	47	77	133	181	298	520	786	1026	1311	1970	2782
	16	—	20	35	56	98	133	220	384	580	757	967	1453	2052
	14	—	14	25	41	72	98	161	282	426	556	711	1068	1508
KF-2, KFF-2	18	—	40	71	115	199	271	447	781	1179	1539	1967	2955	4173
	16	—	28	49	80	139	189	312	545	823	1074	1372	2062	2911
	14	—	19	33	54	93	127	209	366	553	721	922	1385	1956
	12	—	13	23	37	65	88	146	254	384	502	641	963	1360
	10	—	8	15	25	43	58	96	168	254	332	424	638	900
KF-1, KFF-1	18	—	46	82	133	230	313	516	901	1361	1776	2269	3410	4815
	16	—	33	57	93	161	220	363	633	956	1248	1595	2396	3383
	14	—	22	38	63	109	148	244	426	643	839	1073	1612	2276
	12	—	14	25	41	72	98	161	282	426	556	711	1068	1508
	10	—	9	16	27	47	64	105	184	278	363	464	698	985
XF, XFF	12	—	4	8	13	23	31	51	90	136	177	227	341	481
	10	—	3	6	10	18	24	40	70	106	138	177	266	376

Notes:

1. This table is for concentric stranded conductors only. For compact stranded conductors, Table C.1(A) should be used.

2. Two-hour fire-rated RHH cable has ceramifiable insulation, which has much larger diameters than other RHH wires. Consult manufacturer's conduit fill tables.

*Types RHH, RHW, and RHW-2 without outer covering.

Δ **TABLE C.1(A)** *Maximum Number of Conductors or Fixture Wires in Electrical Metallic Tubing (EMT)*
(Based on Chapter 9: Table 1, Table 4, and Table 5A)

Type	Conductor Size (AWG/ kcmil)	Trade Size (Metric Designator)												
		⅜ (12)	½ (16)	¾ (21)	1 (27)	1¼ (35)	1½ (41)	2 (53)	2½ (63)	3 (78)	3½ (91)	4 (103)	5 (129)	6 (155)
COMPACT CONDUCTORS														
THW, THW-2, THHW	8	—	2	4	6	11	16	26	46	69	90	115	174	245
	6	—	1	3	5	9	12	20	35	53	70	89	134	189
	4	—	1	2	4	6	9	15	26	40	52	67	100	142
	2	—	1	1	3	5	7	11	19	29	38	49	74	105
	1	—	1	1	1	3	4	8	13	21	27	34	52	73
	1/0	—	1	1	1	3	4	7	12	18	23	30	45	63
	2/0	—	0	1	1	2	3	5	10	15	20	25	38	53
	3/0	—	0	1	1	1	3	5	8	13	17	21	32	46
	4/0	—	0	1	1	1	2	4	7	11	14	18	27	38
	250	—	0	0	1	1	1	3	5	8	11	14	21	30
	300	—	0	0	1	1	1	3	5	7	9	12	18	26
	350	—	0	0	1	1	1	2	4	6	8	11	16	23
	400	—	0	0	0	1	1	1	4	6	8	10	15	21
	500	—	0	0	0	1	1	1	3	5	6	8	12	18
	600	—	0	0	0	1	1	1	2	4	5	7	10	14
	700	—	0	0	0	1	1	1	2	3	4	6	9	13
	750	—	0	0	0	0	1	1	1	3	4	5	8	12
	900	—	0	0	0	0	1	1	1	3	4	5	7	10
	1000	—	0	0	0	0	1	1	1	2	3	4	7	9
THHN, THWN, THWN-2	8	—	—	—	—	—	—	—	—	—	—	—	—	—
	6	—	2	4	7	13	18	29	52	78	102	130	196	277
	4	—	1	3	4	8	11	18	32	48	63	81	121	171
	2	—	1	1	3	6	8	13	23	34	45	58	87	123
	1	—	1	1	2	4	6	10	17	26	34	43	65	92
	1/0	—	1	1	1	3	5	8	14	22	29	37	55	78
	2/0	—	1	1	1	3	4	7	12	18	24	30	46	65
	3/0	—	0	1	1	2	3	6	10	15	20	25	38	54
	4/0	—	0	1	1	1	3	5	8	12	16	21	32	45
	250	—	0	1	1	1	1	4	6	10	13	16	25	35
	300	—	0	0	1	1	1	3	5	8	11	14	21	30
	350	—	0	0	1	1	1	3	5	7	10	12	19	27
	400	—	0	0	1	1	1	2	4	6	9	11	17	24
	500	—	0	0	0	1	1	1	4	5	7	9	14	20
	600	—	0	0	0	1	1	1	3	4	6	7	11	16
	700	—	0	0	0	1	1	1	2	4	5	7	10	14
	750	—	0	0	0	1	1	1	2	4	5	6	9	13
	900	—	0	0	0	0	1	1	1	3	4	5	8	11
	1000	—	0	0	0	0	1	1	1	3	3	4	7	10

(continues)

Δ **TABLE C.1(A)** *Continued*

Type	Conductor Size (AWG/ kcmil)	Trade Size (Metric Designator)												
		⅜ (12)	½ (16)	¾ (21)	1 (27)	1¼ (35)	1½ (41)	2 (53)	2½ (63)	3 (78)	3½ (91)	4 (103)	5 (129)	6 (155)
XHHW, XHHW-2	8	—	3	5	8	15	20	34	59	90	117	149	225	317
	6	—	1	4	6	11	15	25	44	66	87	111	167	236
	4	—	1	3	4	8	11	18	32	48	63	81	121	171
	2	—	1	1	3	6	8	13	23	34	45	58	87	123
	1	—	1	1	2	4	6	10	17	26	34	43	65	92
	1/0	—	1	1	1	3	5	8	14	22	29	37	55	78
	2/0	—	1	1	1	3	4	7	12	18	24	31	47	66
	3/0	—	0	1	1	2	3	6	10	15	20	25	38	54
	4/0	—	0	1	1	1	3	5	8	13	17	21	32	46
	250	—	0	1	1	1	2	4	7	10	13	17	26	36
	300	—	0	0	1	1	1	3	6	9	11	14	22	31
	350	—	0	0	1	1	1	3	5	8	10	13	19	27
	400	—	0	0	1	1	1	2	4	7	9	11	17	25
	500	—	0	0	0	1	1	1	4	6	7	9	14	20
	600	—	0	0	0	1	1	1	3	4	6	8	11	16
	700	—	0	0	0	1	1	1	2	4	5	7	10	14
	750	—	0	0	0	1	1	1	2	3	5	6	9	13
	900	—	0	0	0	0	1	1	1	3	4	5	8	11
	1000	—	0	0	0	0	1	1	1	3	4	5	7	10

Definition: *Compact stranding* is the result of a manufacturing process where the stranded conductor is compressed to the extent that the interstices (voids between strand wires) are virtually eliminated.

Δ **TABLE C.2** *Maximum Number of Conductors or Fixture Wires in Electrical Nonmetallic Tubing (ENT)*
(Based on Chapter 9: Table 1, Table 4, and Table 5)

Type	Conductor Size (AWG/ kcmil)	Trade Size (Metric Designator)												
		⅜ (12)	½ (16)	¾ (21)	1 (27)	1¼ (35)	1½ (41)	2 (53)	2½ (63)	3 (78)	3½ (91)	4 (103)	5 (129)	6 (155)
CONDUCTORS														
RHH, RHW, RHW-2	14	—	4	7	11	20	27	45	—	—	—	—	—	—
	12	—	3	5	9	16	22	37	—	—	—	—	—	—
	10	—	2	4	7	13	18	30	—	—	—	—	—	—
	8	—	1	2	4	7	9	15	—	—	—	—	—	—
	6	—	1	1	3	5	7	12	—	—	—	—	—	—
	4	—	1	1	2	4	6	10	—	—	—	—	—	—
	3	—	1	1	1	4	5	8	—	—	—	—	—	—
	2	—	1	1	1	3	4	7	—	—	—	—	—	—
	1	—	0	1	1	1	3	5	—	—	—	—	—	—
	1/0	—	0	1	1	1	2	4	—	—	—	—	—	—
	2/0	—	0	0	1	1	1	3	—	—	—	—	—	—
	3/0	—	0	0	1	1	1	3	—	—	—	—	—	—
	4/0	—	0	0	1	1	1	2	—	—	—	—	—	—
	250	—	0	0	0	1	1	1	—	—	—	—	—	—
	300	—	0	0	0	1	1	1	—	—	—	—	—	—
	350	—	0	0	0	1	1	1	—	—	—	—	—	—
	400	—	0	0	0	1	1	1	—	—	—	—	—	—
	500	—	0	0	0	0	1	1	—	—	—	—	—	—
	600	—	0	0	0	0	1	1	—	—	—	—	—	—
	700	—	0	0	0	0	0	1	—	—	—	—	—	—
	750	—	0	0	0	0	0	1	—	—	—	—	—	—
	800	—	0	0	0	0	0	1	—	—	—	—	—	—
	900	—	0	0	0	0	0	1	—	—	—	—	—	—
	1000	—	0	0	0	0	0	1	—	—	—	—	—	—
	1250	—	0	0	0	0	0	0	—	—	—	—	—	—
	1500	—	0	0	0	0	0	0	—	—	—	—	—	—
	1750	—	0	0	0	0	0	0	—	—	—	—	—	—
	2000	—	0	0	0	0	0	0	—	—	—	—	—	—
TW, THHW, THW, THW-2	14	—	8	14	24	42	57	94	—	—	—	—	—	—
	12	—	6	11	18	32	44	72	—	—	—	—	—	—
	10	—	4	8	13	24	32	54	—	—	—	—	—	—
	8	—	2	4	7	13	18	30	—	—	—	—	—	—
RHH*, RHW*, RHW-2*	14	—	5	9	16	28	38	63	—	—	—	—	—	—
	12	—	4	8	13	22	30	50	—	—	—	—	—	—
	10	—	3	6	10	17	24	39	—	—	—	—	—	—
	8	—	1	3	6	10	14	23	—	—	—	—	—	—
TW, THW, THHW, THW-2, RHH*, RHW*, RHW-2*	6	—	1	2	4	8	11	18	—	—	—	—	—	—
	4	—	1	1	3	6	8	13	—	—	—	—	—	—
	3	—	1	1	3	5	7	11	—	—	—	—	—	—
	2	—	1	1	2	4	6	10	—	—	—	—	—	—
	1	—	0	1	1	3	4	7	—	—	—	—	—	—
	1/0	—	0	1	1	2	3	6	—	—	—	—	—	—
	2/0	—	0	1	1	1	3	5	—	—	—	—	—	—
	3/0	—	0	1	1	1	2	4	—	—	—	—	—	—
	4/0	—	0	0	1	1	1	3	—	—	—	—	—	—

(continues)

Δ **TABLE C.2** *Continued*

Type	Conductor Size (AWG/ kcmil)	⅜ (12)	½ (16)	¾ (21)	1 (27)	1¼ (35)	1½ (41)	2 (53)	2½ (63)	3 (78)	3½ (91)	4 (103)	5 (129)	6 (155)
								Trade Size (Metric Designator)						
	250	—	0	0	1	1	1	3	—	—	—	—	—	—
	300	—	0	0	1	1	1	2	—	—	—	—	—	—
	350	—	0	0	0	1	1	1	—	—	—	—	—	—
	400	—	0	0	0	1	1	1	—	—	—	—	—	—
	500	—	0	0	0	1	1	1	—	—	—	—	—	—
	600	—	0	0	0	0	1	1	—	—	—	—	—	—
	700	—	0	0	0	0	1	1	—	—	—	—	—	—
	750	—	0	0	0	0	1	1	—	—	—	—	—	—
	800	—	0	0	0	0	1	1	—	—	—	—	—	—
	900	—	0	0	0	0	0	1	—	—	—	—	—	—
	1000	—	0	0	0	0	0	1	—	—	—	—	—	—
	1250	—	0	0	0	0	0	1	—	—	—	—	—	—
	1500	—	0	0	0	0	0	1	—	—	—	—	—	—
	1750	—	0	0	0	0	0	0	—	—	—	—	—	—
	2000	—	0	0	0	0	0	0	—	—	—	—	—	—
THHN, THWN, THWN-2	14	—	11	21	34	60	82	135	—	—	—	—	—	—
	12	—	8	15	25	43	59	99	—	—	—	—	—	—
	10	—	5	9	15	27	37	62	—	—	—	—	—	—
	8	—	3	5	9	16	21	36	—	—	—	—	—	—
	6	—	1	4	6	11	15	26	—	—	—	—	—	—
	4	—	1	2	4	7	9	16	—	—	—	—	—	—
	3	—	1	1	3	6	8	13	—	—	—	—	—	—
	2	—	1	1	3	5	7	11	—	—	—	—	—	—
	1	—	1	1	1	3	5	8	—	—	—	—	—	—
	1/0	—	1	1	1	3	4	7	—	—	—	—	—	—
	2/0	—	0	1	1	2	3	6	—	—	—	—	—	—
	3/0	—	0	1	1	1	3	5	—	—	—	—	—	—
	4/0	—	0	1	1	1	2	4	—	—	—	—	—	—
	250	—	0	0	1	1	1	3	—	—	—	—	—	—
	300	—	0	0	1	1	1	3	—	—	—	—	—	—
	350	—	0	0	1	1	1	2	—	—	—	—	—	—
	400	—	0	0	0	1	1	1	—	—	—	—	—	—
	500	—	0	0	0	1	1	1	—	—	—	—	—	—
	600	—	0	0	0	1	1	1	—	—	—	—	—	—
	700	—	0	0	0	0	1	1	—	—	—	—	—	—
	750	—	0	0	0	0	1	1	—	—	—	—	—	—
	800	—	0	0	0	0	1	1	—	—	—	—	—	—
	900	—	0	0	0	0	1	1	—	—	—	—	—	—
	1000	—	0	0	0	0	0	1	—	—	—	—	—	—
FEP, FEPB, PFA, PFAH, TFE	14	—	11	20	33	58	79	131	—	—	—	—	—	—
	12	—	8	15	24	42	58	96	—	—	—	—	—	—
	10	—	6	10	17	30	41	69	—	—	—	—	—	—
	8	—	3	6	10	17	24	39	—	—	—	—	—	—
	6	—	2	4	7	12	17	28	—	—	—	—	—	—
	4	—	1	3	5	8	12	19	—	—	—	—	—	—
	3	—	1	2	4	7	10	16	—	—	—	—	—	—
	2	—	1	1	3	6	8	13	—	—	—	—	—	—
PFA, PFAH, TFE	1	—	1	1	2	4	5	9	—	—	—	—	—	—
PFA, PFAH, TFE, Z	1/0	—	1	1	1	3	4	8	—	—	—	—	—	—
	2/0	—	0	1	1	3	4	6	—	—	—	—	—	—
	3/0	—	0	1	1	2	3	5	—	—	—	—	—	—
	4/0	—	0	1	1	1	2	4	—	—	—	—	—	—

Δ **TABLE C.2** *Continued*

Type	Conductor Size (AWG/ kcmil)	⅜ (12)	½ (16)	¾ (21)	1 (27)	1¼ (35)	1½ (41)	2 (53)	2½ (63)	3 (78)	3½ (91)	4 (103)	5 (129)	6 (155)
Z	14	—	13	24	40	70	95	158	—	—	—	—	—	—
	12	—	9	17	28	49	68	112	—	—	—	—	—	—
	10	—	6	10	17	30	41	69	—	—	—	—	—	—
	8	—	3	6	11	19	26	43	—	—	—	—	—	—
	6	—	2	4	7	13	18	30	—	—	—	—	—	—
	4	—	1	3	5	9	12	21	—	—	—	—	—	—
	3	—	1	2	4	6	9	15	—	—	—	—	—	—
	2	—	1	1	3	5	7	12	—	—	—	—	—	—
	1	—	1	1	2	4	6	10	—	—	—	—	—	—
XHHW, ZW, XHHW-2, XHH	14	—	8	14	24	42	57	94	—	—	—	—	—	—
	12	—	6	11	18	32	44	72	—	—	—	—	—	—
	10	—	4	8	13	24	32	54	—	—	—	—	—	—
	8	—	2	4	7	13	18	30	—	—	—	—	—	—
	6	—	1	3	5	10	13	22	—	—	—	—	—	—
	4	—	1	2	4	7	9	16	—	—	—	—	—	—
	3	—	1	1	3	6	8	13	—	—	—	—	—	—
	2	—	1	1	3	5	7	11	—	—	—	—	—	—
XHHW, XHHW-2, XHH	1	—	1	1	1	3	5	8	—	—	—	—	—	—
	1/0	—	0	1	1	3	4	7	—	—	—	—	—	—
	2/0	—	0	1	1	2	3	6	—	—	—	—	—	—
	3/0	—	0	1	1	1	3	5	—	—	—	—	—	—
	4/0	—	0	1	1	1	2	4	—	—	—	—	—	—
	250	—	0	0	1	1	1	3	—	—	—	—	—	—
	300	—	0	0	1	1	1	3	—	—	—	—	—	—
	350	—	0	0	1	1	1	2	—	—	—	—	—	—
	400	—	0	0	0	1	1	1	—	—	—	—	—	—
	500	—	0	0	0	1	1	1	—	—	—	—	—	—
	600	—	0	0	0	1	1	1	—	—	—	—	—	—
	700	—	0	0	0	0	1	1	—	—	—	—	—	—
	750	—	0	0	0	0	1	1	—	—	—	—	—	—
	800	—	0	0	0	0	1	1	—	—	—	—	—	—
	900	—	0	0	0	0	1	1	—	—	—	—	—	—
	1000	—	0	0	0	0	0	1	—	—	—	—	—	—
	1250	—	0	0	0	0	0	1	—	—	—	—	—	—
	1500	—	0	0	0	0	0	1	—	—	—	—	—	—
	1750	—	0	0	0	0	0	0	—	—	—	—	—	—
	2000	—	0	0	0	0	0	0	—	—	—	—	—	—
FIXTURE WIRES														
RFH-2, FFH-2, RFHH-2	18	—	8	14	23	40	54	90	—	—	—	—	—	—
	16	—	6	12	19	33	46	76	—	—	—	—	—	—
SF-2, SFF-2	18	—	10	17	29	50	69	114	—	—	—	—	—	—
	16	—	8	14	24	42	57	94	—	—	—	—	—	—
	14	—	6	12	19	33	46	76	—	—	—	—	—	—
SF-1, SFF-1	18	—	17	31	51	89	122	202	—	—	—	—	—	—
RFH-1, TF, TFF, XF, XFF	18	—	13	23	38	66	90	149	—	—	—	—	—	—
	16	—	10	18	30	53	73	120	—	—	—	—	—	—
XF, XFF	14	—	8	14	24	42	57	94	—	—	—	—	—	—
TFN, TFFN	18	—	20	37	60	105	144	239	—	—	—	—	—	—
	16	—	16	28	46	80	110	183	—	—	—	—	—	—

(continues)

△ **TABLE C.2** *Continued*

Type	Conductor Size (AWG/ kcmil)	Trade Size (Metric Designator)												
		⅜ (12)	½ (16)	¾ (21)	1 (27)	1¼ (35)	1½ (41)	2 (53)	2½ (63)	3 (78)	3½ (91)	4 (103)	5 (129)	6 (155)
PF, PFF, PGF, PGFF, PAF, PTF, PTFF, PAFF	18	—	19	35	57	100	137	227	—	—	—	—	—	—
	16	—	15	27	44	77	106	175	—	—	—	—	—	—
	14	—	11	20	33	58	79	131	—	—	—	—	—	—
ZF, ZFF, ZHF	18	—	25	45	74	129	176	292	—	—	—	—	—	—
	16	—	18	33	54	95	130	216	—	—	—	—	—	—
	14	—	13	24	40	70	95	158	—	—	—	—	—	—
KF-2, KFF-2	18	—	38	67	111	193	265	439	—	—	—	—	—	—
	16	—	26	47	77	135	184	306	—	—	—	—	—	—
	14	—	18	31	52	91	124	205	—	—	—	—	—	—
	12	—	12	22	36	63	86	143	—	—	—	—	—	—
	10	—	8	14	24	42	57	94	—	—	—	—	—	—
KF-1, KFF-1	18	—	44	78	128	223	305	506	—	—	—	—	—	—
	16	—	31	55	90	157	214	355	—	—	—	—	—	—
	14	—	20	37	60	105	144	239	—	—	—	—	—	—
	12	—	13	24	40	70	95	158	—	—	—	—	—	—
	10	—	9	16	26	45	62	103	—	—	—	—	—	—
XF, XFF	12	—	4	8	13	22	30	50	—	—	—	—	—	—
	10	—	3	6	10	17	24	39	—	—	—	—	—	—

Notes:

1. This table is for concentric stranded conductors only. For compact stranded conductors, Table C.2(A) should be used.

2. Two-hour fire-rated RHH cable has ceramifiable insulation, which has much larger diameters than other RHH wires. Consult manufacturer's conduit fill tables.

*Types RHH, RHW, and RHW-2 without outer covering.

TABLE C.2(A) *Maximum Number of Conductors or Fixture Wires in Electrical Nonmetallic Tubing (ENT)*
(Based on Chapter 9: Table 1, Table 4, and Table 5A)

Type	Conductor Size (AWG/kcmil)	Trade Size (Metric Designator)												
		⅜ (12)	½ (16)	¾ (21)	1 (27)	1¼ (35)	1½ (41)	2 (53)	2½ (63)	3 (78)	3½ (91)	4 (103)	5 (129)	6 (155)
COMPACT CONDUCTORS														
THW, THW-2, THHW	8	—	1	4	6	11	15	26	—	—	—	—	—	—
	6	—	1	3	5	9	12	20	—	—	—	—	—	—
	4	—	1	1	3	6	9	15	—	—	—	—	—	—
	2	—	1	1	2	5	6	11	—	—	—	—	—	—
	1	—	1	1	1	3	4	7	—	—	—	—	—	—
	1/0	—	0	1	1	3	4	6	—	—	—	—	—	—
	2/0	—	0	1	1	2	3	5	—	—	—	—	—	—
	3/0	—	0	1	1	1	3	5	—	—	—	—	—	—
	4/0	—	0	1	1	1	2	4	—	—	—	—	—	—
	250	—	0	0	1	1	1	3	—	—	—	—	—	—
	300	—	0	0	1	1	1	2	—	—	—	—	—	—
	350	—	0	0	1	1	1	2	—	—	—	—	—	—
	400	—	0	0	0	1	1	1	—	—	—	—	—	—
	500	—	0	0	0	1	1	1	—	—	—	—	—	—
	600	—	0	0	0	1	1	1	—	—	—	—	—	—
	700	—	0	0	0	0	1	1	—	—	—	—	—	—
	750	—	0	0	0	0	1	1	—	—	—	—	—	—
	900	—	0	0	0	0	1	1	—	—	—	—	—	—
	1000	—	0	0	0	0	1	1	—	—	—	—	—	—
THHN, THWN, THWN-2	8	—	—	—	—	—	—	—	—	—	—	—	—	—
	6	—	2	4	7	13	17	29	—	—	—	—	—	—
	4	—	1	2	4	8	11	18	—	—	—	—	—	—
	2	—	1	1	3	5	8	13	—	—	—	—	—	—
	1	—	1	1	2	4	6	9	—	—	—	—	—	—
	1/0	—	1	1	1	3	5	8	—	—	—	—	—	—
	2/0	—	0	1	1	3	4	7	—	—	—	—	—	—
	3/0	—	0	1	1	2	3	5	—	—	—	—	—	—
	4/0	—	0	1	1	1	3	4	—	—	—	—	—	—
	250	—	0	0	1	1	1	3	—	—	—	—	—	—
	300	—	0	0	1	1	1	3	—	—	—	—	—	—
	350	—	0	0	1	1	1	3	—	—	—	—	—	—
	400	—	0	0	1	1	1	2	—	—	—	—	—	—
	500	—	0	0	0	1	1	1	—	—	—	—	—	—
	600	—	0	0	0	1	1	1	—	—	—	—	—	—
	700	—	0	0	0	1	1	1	—	—	—	—	—	—
	750	—	0	0	0	1	1	1	—	—	—	—	—	—
	900	—	0	0	0	0	1	1	—	—	—	—	—	—
	1000	—	0	0	0	0	1	1	—	—	—	—	—	—

(continues)

TABLE C.2(A) *Continued*

Type	Conductor Size (AWG/ kcmil)	⅜ (12)	½ (16)	¾ (21)	1 (27)	1¼ (35)	1½ (41)	2 (53)	2½ (63)	3 (78)	3½ (91)	4 (103)	5 (129)	6 (155)
							Trade Size (Metric Designator)							
XHHW, XHHW-2	8	—	3	5	8	14	20	33	—	—	—	—	—	—
	6	—	1	4	6	11	15	25	—	—	—	—	—	—
	4	—	1	2	4	8	11	18	—	—	—	—	—	—
	2	—	1	1	3	5	8	13	—	—	—	—	—	—
	1	—	1	1	2	4	6	9	—	—	—	—	—	—
	1/0	—	1	1	1	3	5	8	—	—	—	—	—	—
	2/0	—	1	1	1	3	4	7	—	—	—	—	—	—
	3/0	—	0	1	1	2	3	5	—	—	—	—	—	—
	4/0	—	0	1	1	1	3	5	—	—	—	—	—	—
	250	—	0	0	1	1	1	4	—	—	—	—	—	—
	300	—	0	0	1	1	1	3	—	—	—	—	—	—
	350	—	0	0	1	1	1	3	—	—	—	—	—	—
	400	—	0	0	1	1	1	2	—	—	—	—	—	—
	500	—	0	0	0	1	1	1	—	—	—	—	—	—
	600	—	0	0	0	1	1	1	—	—	—	—	—	—
	700	—	0	0	0	1	1	1	—	—	—	—	—	—
	750	—	0	0	0	1	1	1	—	—	—	—	—	—
	900	—	0	0	0	0	1	1	—	—	—	—	—	—
	1000	—	0	0	0	0	1	1	—	—	—	—	—	—

Definition: *Compact stranding* is the result of a manufacturing process where the stranded conductor is compressed to the extent that the interstices (voids between strand wires) are virtually eliminated.

Δ **TABLE C.3** *Maximum Number of Conductors or Fixture Wires in Flexible Metal Conduit (FMC)*
(Based on Chapter 9: Table 1, Table 4, and Table 5)

Type	Conductor Size (AWG/ kcmil)	⅜ (12)	½ (16)	¾ (21)	1 (27)	1¼ (35)	1½ (41)	2 (53)	2½ (63)	3 (78)	3½ (91)	4 (103)	5 (129)	6 (155)
						CONDUCTORS								
RHH, RHW, RHW-2	14	1	4	7	11	17	25	44	67	96	131	171	—	—
	12	1	3	6	9	14	21	37	55	80	109	142	—	—
	10	1	3	5	7	11	17	30	45	64	88	115	—	—
	8	0	1	2	4	6	9	15	23	34	46	60	—	—
	6	0	1	1	3	5	7	12	19	27	37	48	—	—
	4	0	1	1	2	4	5	10	14	21	29	37	—	—
	3	0	1	1	1	3	5	8	13	18	25	33	—	—
	2	0	1	1	1	3	4	7	11	16	22	28	—	—
	1	0	0	1	1	1	2	5	7	10	14	19	—	—
	1/0	0	0	1	1	1	2	4	6	9	12	16	—	—
	2/0	0	0	1	1	1	1	3	5	8	11	14	—	—
	3/0	0	0	0	1	1	1	3	5	7	9	12	—	—
	4/0	0	0	0	1	1	1	2	4	6	8	10	—	—
	250	0	0	0	0	1	1	1	3	4	6	8	—	—
	300	0	0	0	0	1	1	1	2	4	5	7	—	—
	350	0	0	0	0	1	1	1	2	3	5	6	—	—
	400	0	0	0	0	0	1	1	1	3	4	6	—	—
	500	0	0	0	0	0	1	1	1	3	4	5	—	—
	600	0	0	0	0	0	1	1	1	2	3	4	—	—
	700	0	0	0	0	0	0	1	1	1	3	3	—	—
	750	0	0	0	0	0	0	1	1	1	2	3	—	—
	800	0	0	0	0	0	0	1	1	1	2	3	—	—
	900	0	0	0	0	0	0	1	1	1	2	3	—	—
	1000	0	0	0	0	0	0	1	1	1	1	3	—	—
	1250	0	0	0	0	0	0	0	1	1	1	1	—	—
	1500	0	0	0	0	0	0	0	1	1	1	1	—	—
	1750	0	0	0	0	0	0	0	1	1	1	1	—	—
	2000	0	0	0	0	0	0	0	0	1	1	1	—	—
TW, THHW, THW, THW-2	14	3	9	15	23	36	53	94	141	203	277	361	—	—
	12	2	7	11	18	28	41	72	108	156	212	277	—	—
	10	1	5	8	13	21	30	54	81	116	158	207	—	—
	8	1	3	5	7	11	17	30	45	64	88	115	—	—
RHH*, RHW*, RHW-2*	14	1	6	10	15	24	35	62	94	135	184	240	—	—
	12	1	5	8	12	19	28	50	75	108	148	193	—	—
	10	1	4	6	10	15	22	39	59	85	115	151	—	—
	8	1	1	4	6	9	13	23	35	51	69	90	—	—
TW, THW, THHW, THW-2, RHH*, RHW*, RHW-2*	6	1	1	3	4	7	10	18	27	39	53	69	—	—
	4	0	1	1	3	5	7	13	20	29	39	51	—	—
	3	0	1	1	3	4	6	11	17	25	34	44	—	—
	2	0	1	1	2	4	5	10	14	21	29	37	—	—
	1	0	1	1	1	2	4	7	10	15	20	26	—	—
	1/0	0	0	1	1	1	3	6	9	12	17	22	—	—
	2/0	0	0	1	1	1	3	5	7	10	14	19	—	—
	3/0	0	0	1	1	1	2	4	6	9	12	16	—	—
	4/0	0	0	0	1	1	1	3	5	7	10	13	—	—

(continues)

△ TABLE C.3 *Continued*

Type	Conductor Size (AWG/ kcmil)	⅜ (12)	½ (16)	¾ (21)	1 (27)	1¼ (35)	1½ (41)	2 (53)	2½ (63)	3 (78)	3½ (91)	4 (103)	5 (129)	6 (155)
								Trade Size (Metric Designator)						
	250	0	0	0	1	1	1	3	4	6	8	11	—	—
	300	0	0	0	1	1	1	2	3	5	7	9	—	—
	350	0	0	0	0	1	1	1	3	4	6	8	—	—
	400	0	0	0	0	1	1	1	3	4	6	7	—	—
	500	0	0	0	0	1	1	1	2	3	5	6	—	—
	600	0	0	0	0	0	1	1	1	3	4	5	—	—
	700	0	0	0	0	0	1	1	1	2	3	4	—	—
	750	0	0	0	0	0	1	1	1	2	3	4	—	—
	800	0	0	0	0	0	1	1	1	1	3	4	—	—
	900	0	0	0	0	0	0	1	1	1	3	3	—	—
	1000	0	0	0	0	0	0	1	1	1	2	3	—	—
	1250	0	0	0	0	0	0	1	1	1	1	2	—	—
	1500	0	0	0	0	0	0	0	1	1	1	1	—	—
	1750	0	0	0	0	0	0	0	1	1	1	1	—	—
	2000	0	0	0	0	0	0	0	1	1	1	1	—	—
THHN, THWN, THWN-2	14	4	13	22	33	52	76	135	202	291	396	518	—	—
	12	3	9	16	24	38	56	98	147	212	289	378	—	—
	10	1	6	10	15	24	35	62	93	134	182	238	—	—
	8	1	3	6	9	14	20	35	53	77	105	137	—	—
	6	1	2	4	6	10	14	25	38	55	76	99	—	—
	4	0	1	2	4	6	9	16	24	34	46	61	—	—
	3	0	1	1	3	5	7	13	20	29	39	51	—	—
	2	0	1	1	3	4	6	11	17	24	33	43	—	—
	1	0	1	1	1	3	4	8	12	18	24	32	—	—
	1/0	0	1	1	1	2	4	7	10	15	20	27	—	—
	2/0	0	0	1	1	1	3	6	9	12	17	22	—	—
	3/0	0	0	1	1	1	2	5	7	10	14	18	—	—
	4/0	0	0	1	1	1	1	4	6	8	12	15	—	—
	250	0	0	0	1	1	1	3	5	7	9	12	—	—
	300	0	0	0	1	1	1	3	4	6	8	11	—	—
	350	0	0	0	1	1	1	2	3	5	7	9	—	—
	400	0	0	0	0	1	1	1	3	5	6	8	—	—
	500	0	0	0	0	1	1	1	2	4	5	7	—	—
	600	0	0	0	0	0	1	1	1	3	4	5	—	—
	700	0	0	0	0	0	1	1	1	3	4	5	—	—
	750	0	0	0	0	0	1	1	1	2	3	4	—	—
	800	0	0	0	0	0	1	1	1	2	3	4	—	—
	900	0	0	0	0	0	0	1	1	1	3	4	—	—
	1000	0	0	0	0	0	0	1	1	1	3	3	—	—
FEP, FEPB, PFA, PFAH, TFE	14	4	12	21	32	51	74	130	196	282	385	502	—	—
	12	3	9	15	24	37	54	95	143	206	281	367	—	—
	10	2	6	11	17	26	39	68	103	148	201	263	—	—
	8	1	4	6	10	15	22	39	59	85	115	151	—	—
	6	1	2	4	7	11	16	28	42	60	82	107	—	—
	4	1	1	3	5	7	11	19	29	42	57	75	—	—
	3	0	1	2	4	6	9	16	24	35	48	62	—	—
	2	0	1	1	3	5	7	13	20	29	39	51	—	—
PFA, PFAH, TFE	1	0	1	1	2	3	5	9	14	20	27	36	—	—
PFA, PFAH, TFE, Z	1/0	0	1	1	1	3	4	8	11	17	23	30	—	—
	2/0	0	1	1	1	2	3	6	9	14	19	24	—	—
	3/0	0	0	1	1	1	3	5	8	11	15	20	—	—
	4/0	0	0	1	1	1	2	4	6	9	13	16	—	—

Δ **TABLE C.3** *Continued*

Type	Conductor Size (AWG/kcmil)	⅜ (12)	½ (16)	¾ (21)	1 (27)	1¼ (35)	1½ (41)	2 (53)	2½ (63)	3 (78)	3½ (91)	4 (103)	5 (129)	6 (155)
Z	14	5	15	25	39	61	89	157	236	340	463	605	—	—
	12	4	11	18	28	43	63	111	168	241	329	429	—	—
	10	2	6	11	17	26	39	68	103	148	201	263	—	—
	8	1	4	7	11	17	24	43	65	93	127	166	—	—
	6	1	3	5	7	12	17	30	45	65	89	117	—	—
	4	1	1	3	5	8	12	21	31	45	61	80	—	—
	3	0	1	2	4	6	8	15	23	33	45	58	—	—
	2	0	1	1	3	5	7	12	19	27	37	49	—	—
	1	0	1	1	2	4	6	10	15	22	30	39	—	—
XHHW, ZW, XHHW-2, XHH	14	3	9	15	23	36	53	94	141	203	277	361	—	—
	12	2	7	11	18	28	41	72	108	156	212	277	—	—
	10	1	5	8	13	21	30	54	81	116	158	207	—	—
	8	1	3	5	7	11	17	30	45	64	88	115	—	—
	6	1	1	3	5	8	12	22	33	48	65	85	—	—
	4	0	1	2	4	6	9	16	24	34	47	61	—	—
	3	0	1	1	3	5	7	13	20	29	40	52	—	—
	2	0	1	1	3	4	6	11	17	24	33	44	—	—
XHH, XHHW, XHHW-2	1	0	1	1	1	3	5	8	13	18	25	32	—	—
	1/0	0	1	1	1	2	4	7	10	15	21	27	—	—
	2/0	0	0	1	1	2	3	6	9	13	17	23	—	—
	3/0	0	0	1	1	1	3	5	7	10	14	19	—	—
	4/0	0	0	1	1	1	2	4	6	9	12	15	—	—
	250	0	0	0	1	1	1	3	5	7	10	13	—	—
	300	0	0	0	1	1	1	3	4	6	8	11	—	—
	350	0	0	0	1	1	1	2	4	5	7	9	—	—
	400	0	0	0	0	1	1	1	3	5	6	8	—	—
	500	0	0	0	0	1	1	1	3	4	5	7	—	—
	600	0	0	0	0	0	1	1	1	3	4	5	—	—
	700	0	0	0	0	0	1	1	1	3	4	5	—	—
	750	0	0	0	0	0	1	1	1	2	3	4	—	—
	800	0	0	0	0	0	1	1	1	2	3	4	—	—
	900	0	0	0	0	0	0	1	1	1	3	4	—	—
	1000	0	0	0	0	0	0	1	1	1	3	3	—	—
	1250	0	0	0	0	0	0	1	1	1	1	3	—	—
	1500	0	0	0	0	0	0	1	1	1	1	2	—	—
	1750	0	0	0	0	0	0	0	1	1	1	1	—	—
	2000	0	0	0	0	0	0	0	1	1	1	1	—	—
FIXTURE WIRES														
RFH-2, FFH-2, RFHH-2	18	3	8	14	22	35	51	90	135	195	265	346	—	—
	16	2	7	12	19	29	43	76	114	164	223	292	—	—
SF-2, SFF-2	18	4	11	18	28	44	64	113	170	246	334	437	—	—
	16	3	9	15	23	36	53	94	141	203	277	361	—	—
	14	2	7	12	19	29	43	76	114	164	223	292	—	—
SF-1, SFF-1	18	7	19	33	50	78	114	201	302	435	592	773	—	—
RFH-1, TF, TFF, XF, XFF	18	5	14	24	37	58	84	148	223	321	437	571	—	—
	16	4	11	19	30	47	68	120	180	259	353	461	—	—

(continues)

△ **TABLE C.3** *Continued*

Type	Conductor Size (AWG/ kcmil)	⅜ (12)	½ (16)	¾ (21)	1 (27)	1¼ (35)	1½ (41)	2 (53)	2½ (63)	3 (78)	3½ (91)	4 (103)	5 (129)	6 (155)
						Trade Size (Metric Designator)								
XF, XFF	14	3	9	15	23	36	53	94	141	203	277	361	—	—
TFN, TFFN	18	8	23	38	59	93	135	237	357	514	699	914	—	—
	16	6	17	29	45	71	103	181	272	392	534	698	—	—
PF, PFF, PGF, PGFF, PAF, PTF, PTFF, PAFF	18	8	22	36	56	88	128	225	338	487	663	866	—	—
	16	6	17	28	43	68	99	174	262	377	513	670	—	—
	14	4	12	21	32	51	74	130	196	282	385	502	—	—
ZF, ZFF, ZHF	18	10	28	47	72	113	165	290	436	628	855	1117	—	—
	16	7	20	35	53	83	122	214	322	463	631	824	—	—
	14	5	15	25	39	61	89	157	236	340	463	605	—	—
KF-2, KFF-2	18	15	42	71	109	170	247	436	654	942	1282	1675	—	—
	16	10	29	49	76	118	173	304	456	657	895	1169	—	—
	14	7	20	33	51	80	116	204	307	442	601	785	—	—
	12	5	13	23	35	55	80	142	213	307	418	546	—	—
	10	3	9	15	23	36	53	94	141	203	277	361	—	—
KF-1, KFF-1	18	18	48	82	125	196	286	503	755	1087	1480	1933	—	—
	16	12	34	57	88	138	201	353	530	764	1040	1358	—	—
	14	8	23	38	59	93	135	237	357	514	699	914	—	—
	12	5	15	25	39	61	89	157	236	340	463	605	—	—
	10	3	10	16	25	40	58	103	154	222	303	395	—	—
XF, XFF	12	1	5	8	12	19	28	50	75	108	148	193	—	—
	10	1	4	6	10	15	22	39	59	85	115	151	—	—

Notes:

1. This table is for concentric stranded conductors only. For compact stranded conductors, Table C.3(A) should be used.

2. Two-hour fire-rated RHH cable has ceramifiable insulation, which has much larger diameters than other RHH wires. Consult manufacturer's conduit fill tables.

*Types RHH, RHW, and RHW-2 without outer covering.

TABLE C.3(A) *Maximum Number of Conductors or Fixture Wires in Flexible Metal Conduit (FMC)*
(Based on Chapter 9: Table 1, Table 4, and Table 5A)

Type	Conductor Size (AWG/ kcmil)	Trade Size (Metric Designator)												
		⅜ (12)	½ (16)	¾ (21)	1 (27)	1¼ (35)	1½ (41)	2 (53)	2½ (63)	3 (78)	3½ (91)	4 (103)	5 (129)	6 (155)
COMPACT CONDUCTORS														
THW, THW-2, THHW	8	1	2	4	6	10	14	25	38	55	75	98	—	—
	6	1	1	3	5	7	11	20	29	43	58	76	—	—
	4	0	1	2	3	5	8	15	22	32	43	57	—	—
	2	0	1	1	2	4	6	11	16	23	32	42	—	—
	1	0	1	1	1	3	4	7	11	16	22	29	—	—
	1/0	0	1	1	1	2	3	6	10	14	19	25	—	—
	2/0	0	0	1	1	1	3	5	8	12	16	21	—	—
	3/0	0	0	1	1	1	2	4	7	10	14	18	—	—
	4/0	0	0	1	1	1	1	4	6	8	11	15	—	—
	250	0	0	0	1	1	1	3	4	7	9	12	—	—
	300	0	0	0	1	1	1	2	4	6	8	10	—	—
	350	0	0	0	1	1	1	2	3	5	7	9	—	—
	400	0	0	0	0	1	1	1	3	5	6	8	—	—
	500	0	0	0	0	1	1	1	3	4	5	7	—	—
	600	0	0	0	0	0	1	1	1	3	4	6	—	—
	700	0	0	0	0	0	1	1	1	3	4	5	—	—
	750	0	0	0	0	0	1	1	1	2	3	5	—	—
	900	0	0	0	0	0	1	1	1	2	3	4	—	—
	1000	0	0	0	0	0	0	1	1	1	3	4	—	—
THHN, THWN, THWN-2	8	—	—	—	—	—	—	—	—	—	—	—	—	—
	6	1	3	4	7	11	16	29	43	62	85	111	—	—
	4	1	1	3	4	7	10	18	27	38	52	69	—	—
	2	0	1	1	3	5	7	13	19	28	38	49	—	—
	1	0	1	1	2	3	5	9	14	21	28	37	—	—
	1/0	0	1	1	1	3	4	8	12	17	24	31	—	—
	2/0	0	1	1	1	2	4	6	10	14	20	26	—	—
	3/0	0	0	1	1	1	3	5	8	12	17	22	—	—
	4/0	0	0	1	1	1	2	4	7	10	14	18	—	—
	250	0	0	1	1	1	1	3	5	8	11	14	—	—
	300	0	0	0	1	1	1	3	5	7	9	12	—	—
	350	0	0	0	1	1	1	3	4	6	8	10	—	—
	400	0	0	0	1	1	1	2	3	5	7	9	—	—
	500	0	0	0	0	1	1	1	3	4	6	8	—	—
	600	0	0	0	0	1	1	1	2	3	5	6	—	—
	700	0	0	0	0	0	1	1	1	3	4	6	—	—
	750	0	0	0	0	0	1	1	1	3	4	5	—	—
	900	0	0	0	0	0	1	1	1	2	3	4	—	—
	1000	0	0	0	0	0	0	1	1	1	3	4	—	—

(continues)

TABLE C.3(A) *Continued*

Type	Conductor Size (AWG/ kcmil)	⅜ (12)	½ (16)	¾ (21)	1 (27)	1¼ (35)	1½ (41)	2 (53)	2½ (63)	3 (78)	3½ (91)	4 (103)	5 (129)	6 (155)
											Trade Size (Metric Designator)			
XHHW, XHHW-2	8	1	3	5	8	13	19	33	50	71	97	127	—	—
	6	1	2	4	6	9	14	24	37	53	72	95	—	—
	4	1	1	3	4	7	10	18	27	38	52	69	—	—
	2	0	1	1	3	5	7	13	19	28	38	49	—	—
	1	0	1	1	2	3	5	9	14	21	28	37	—	—
	1/0	0	1	1	1	3	4	8	12	17	24	31	—	—
	2/0	0	1	1	1	2	4	7	10	15	20	26	—	—
	3/0	0	0	1	1	1	3	5	8	12	17	22	—	—
	4/0	0	0	1	1	1	2	4	7	10	14	18	—	—
	250	0	0	1	1	1	1	4	5	8	11	14	—	—
	300	0	0	0	1	1	1	3	5	7	9	12	—	—
	350	0	0	0	1	1	1	3	4	6	8	11	—	—
	400	0	0	0	1	1	1	2	4	5	7	10	—	—
	500	0	0	0	0	1	1	1	3	4	6	8	—	—
	600	0	0	0	0	1	1	1	2	3	5	6	—	—
	700	0	0	0	0	0	1	1	1	3	4	6	—	—
	750	0	0	0	0	0	1	1	1	3	4	5	—	—
	900	0	0	0	0	0	1	1	1	2	3	4	—	—
	1000	0	0	0	0	0	1	1	1	2	3	4	—	—

Definition: *Compact stranding* is the result of a manufacturing process where the stranded conductor is compressed to the extent that the interstices (voids between strand wires) are virtually eliminated.

Δ **TABLE C.4** *Maximum Number of Conductors or Fixture Wires in Intermediate Metal Conduit (IMC)*
(Based on Chapter 9: Table 1, Table 4, and Table 5)

Type	Conductor Size (AWG/ kcmil)	Trade Size (Metric Designator)												
		⅜ (12)	½ (16)	¾ (21)	1 (27)	1¼ (35)	1½ (41)	2 (53)	2½ (63)	3 (78)	3½ (91)	4 (103)	5 (129)	6 (155)
CONDUCTORS														
RHH, RHW, RHW-2	14	—	4	8	13	22	30	49	70	108	144	186	291	419
	12	—	4	6	11	18	25	41	58	89	120	154	241	348
	10	—	3	5	8	15	20	33	47	72	97	124	195	281
	8	—	1	3	4	8	10	17	24	38	50	65	102	147
	6	—	1	1	3	6	8	14	19	30	40	52	81	118
	4	—	1	1	3	5	6	11	15	23	31	41	63	92
	3	—	1	1	2	4	6	9	13	21	28	36	56	80
	2	—	1	1	1	3	5	8	11	18	24	31	48	70
	1	—	0	1	1	2	3	5	7	12	16	20	32	46
	1/0	—	0	1	1	1	3	4	6	10	14	18	28	40
	2/0	—	0	1	1	1	2	4	6	9	12	15	24	35
	3/0	—	0	0	1	1	1	3	5	7	10	13	20	30
	4/0	—	0	0	1	1	1	3	4	6	9	11	17	25
	250	—	0	0	1	1	1	1	3	5	6	8	13	19
	300	—	0	0	0	1	1	1	3	4	6	7	12	17
	350	—	0	0	0	1	1	1	2	4	5	7	10	15
	400	—	0	0	0	1	1	1	2	3	5	6	9	14
	500	—	0	0	0	1	1	1	1	3	4	5	8	12
	600	—	0	0	0	0	1	1	1	2	3	4	7	10
	700	—	0	0	0	0	1	1	1	2	3	4	6	9
	750	—	0	0	0	0	1	1	1	1	3	4	5	8
	800	—	0	0	0	0	0	1	1	1	3	3	5	8
	900	—	0	0	0	0	0	1	1	1	2	3	5	7
	1000	—	0	0	0	0	0	1	1	1	2	3	4	6
	1250	—	0	0	0	0	0	1	1	1	1	1	3	5
	1500	—	0	0	0	0	0	0	1	1	1	1	3	4
	1750	—	0	0	0	0	0	0	1	1	1	1	2	4
	2000	—	0	0	0	0	0	0	1	1	1	1	2	3
TW, THHW, THW, THW-2	14	—	10	17	27	47	64	104	147	228	304	392	613	885
	12	—	7	13	21	36	49	80	113	175	234	301	471	679
	10	—	5	9	15	27	36	59	84	130	174	224	350	506
	8	—	3	5	8	15	20	33	47	72	97	124	195	281
RHH*, RHW*, RHW-2*	14	—	6	11	18	31	42	69	98	151	202	261	408	588
	12	—	5	9	14	25	34	56	79	122	163	209	328	473
	10	—	4	7	11	19	26	43	61	95	127	163	256	369
	8	—	2	4	7	12	16	26	37	57	76	98	153	221
TW, THW, THHW, THW-2, RHH*, RHW*, RHW-2*	6	—	1	3	5	9	12	20	28	43	58	75	117	169
	4	—	1	2	4	6	9	15	21	32	43	56	87	126
	3	—	1	1	3	6	8	13	18	28	37	48	75	108
	2	—	1	1	3	5	6	11	15	23	31	41	63	92
	1	—	1	1	1	3	4	7	11	16	22	28	44	64
	1/0	—	1	1	1	3	4	6	9	14	19	24	38	55
	2/0	—	0	1	1	2	3	5	8	12	16	20	32	46
	3/0	—	0	1	1	1	3	4	6	10	13	17	27	39
	4/0	—	0	1	1	1	2	4	5	8	11	14	22	33

(continues)

Δ **TABLE C.4** *Continued*

Type	Conductor Size (AWG/ kcmil)	⅜ (12)	½ (16)	¾ (21)	1 (27)	1¼ (35)	1½ (41)	2 (53)	2½ (63)	3 (78)	3½ (91)	4 (103)	5 (129)	6 (155)
								Trade Size (Metric Designator)						
	250	—	0	0	1	1	1	3	4	7	9	12	18	26
	300	—	0	0	1	1	1	2	4	6	8	10	16	23
	350	—	0	0	1	1	1	2	3	5	7	9	14	20
	400	—	0	0	0	1	1	1	3	4	6	8	12	18
	500	—	0	0	0	1	1	1	2	4	5	7	10	15
	600	—	0	0	0	1	1	1	1	3	4	5	8	12
	700	—	0	0	0	0	1	1	1	3	4	5	7	11
	750	—	0	0	0	0	1	1	1	2	3	4	7	10
	800	—	0	0	0	0	1	1	1	2	3	4	6	10
	900	—	0	0	0	0	1	1	1	2	3	4	6	9
	1000	—	0	0	0	0	0	1	1	1	3	3	5	8
	1250	—	0	0	0	0	0	1	1	1	1	3	4	6
	1500	—	0	0	0	0	0	1	1	1	1	2	3	5
	1750	—	0	0	0	0	0	0	1	1	1	1	3	4
	2000	—	0	0	0	0	0	0	1	1	1	1	3	4
THHN, THWN, THWN-2	14	—	14	24	39	68	91	149	211	326	436	562	879	1268
	12	—	10	17	29	49	67	109	154	238	318	410	641	925
	10	—	6	11	18	31	42	69	97	150	200	258	404	583
	8	—	3	6	10	18	24	39	56	86	115	149	233	336
	6	—	2	4	7	13	17	28	40	62	83	107	168	242
	4	—	1	3	4	8	11	17	25	38	51	66	103	149
	3	—	1	2	4	6	9	15	21	32	43	56	87	126
	2	—	1	1	3	5	7	12	17	27	36	47	73	106
	1	—	1	1	2	4	5	9	13	20	27	35	54	78
	1/0	—	1	1	1	3	4	8	11	17	23	29	45	66
	2/0	—	1	1	1	3	4	6	9	14	19	24	38	55
	3/0	—	0	1	1	2	3	5	7	12	16	20	31	45
	4/0	—	0	1	1	1	2	4	6	9	13	17	26	38
	250	—	0	0	1	1	1	3	5	8	10	13	21	30
	300	—	0	0	1	1	1	3	4	7	9	12	18	26
	350	—	0	0	1	1	1	2	4	6	8	10	16	23
	400	—	0	0	1	1	1	2	3	5	7	9	14	20
	500	—	0	0	0	1	1	1	3	4	6	7	12	17
	600	—	0	0	0	1	1	1	2	3	5	6	9	14
	700	—	0	0	0	1	1	1	1	3	4	5	8	12
	750	—	0	0	0	1	1	1	1	3	4	5	8	11
	800	—	0	0	0	0	1	1	1	3	4	5	7	11
	900	—	0	0	0	0	1	1	1	2	3	4	6	9
	1000	—	0	0	0	0	1	1	1	2	3	4	6	9
FEP, FEPB, PFA, PFAH, TFE	14	—	13	23	38	66	89	145	205	317	423	545	852	1230
	12	—	10	17	28	48	65	106	150	231	309	398	622	898
	10	—	7	12	20	34	46	76	107	166	221	285	446	644
	8	—	4	7	11	19	26	43	61	95	127	163	256	369
	6	—	3	5	8	14	19	31	44	67	90	116	182	262
	4	—	1	3	5	10	13	21	30	47	63	81	127	183
	3	—	1	3	4	8	11	18	25	39	52	68	106	153
	2	—	1	2	4	6	9	15	21	32	43	56	87	126
PFA, PFAH, TFE	1	—	1	1	2	4	6	10	14	22	30	39	60	87

△ **TABLE C.4** *Continued*

Type	Conductor Size (AWG/kcmil)	⅜ (12)	½ (16)	¾ (21)	1 (27)	1¼ (35)	1½ (41)	2 (53)	2½ (63)	3 (78)	3½ (91)	4 (103)	5 (129)	6 (155)
PFA, PFAH, TFE, Z	1/0	—	1	1	1	4	5	8	12	19	25	32	50	73
	2/0	—	1	1	1	3	4	7	10	15	21	27	42	60
	3/0	—	0	1	1	2	3	6	8	13	17	22	34	49
	4/0	—	0	1	1	1	3	5	7	10	14	18	28	41
Z	14	—	16	28	46	79	107	175	247	381	510	657	1027	1482
	12	—	11	20	32	56	76	124	175	271	362	466	728	1051
	10	—	7	12	20	34	46	76	107	166	221	285	446	644
	8	—	4	7	12	22	29	48	68	105	140	180	282	407
	6	—	3	5	9	15	20	33	47	73	98	127	198	286
	4	—	1	3	6	10	14	23	33	50	67	87	136	196
	3	—	1	2	4	7	10	17	24	37	49	63	99	143
	2	—	1	1	3	6	8	14	20	30	41	53	82	119
	1	—	1	1	3	5	7	11	16	25	33	43	67	96
XHHW, ZW, XHHW-2, XHH	14	—	10	17	27	47	64	104	147	228	304	392	613	885
	12	—	7	13	21	36	49	80	113	175	234	301	471	679
	10	—	5	9	15	27	36	59	84	130	174	224	350	506
	8	—	3	5	8	15	20	33	47	72	97	124	195	281
	6	—	1	4	6	11	15	24	35	53	71	92	144	208
	4	—	1	3	4	8	11	18	25	39	52	67	104	151
	3	—	1	2	4	7	9	15	21	33	44	56	88	127
	2	—	1	1	3	5	7	12	18	27	37	47	74	107
XHHW, XHHW-2, XHH	1	—	1	1	2	4	6	9	13	20	27	35	55	80
	1/0	—	1	1	1	3	5	8	11	17	23	30	46	67
	2/0	—	1	1	1	3	4	6	9	14	19	25	38	56
	3/0	—	0	1	1	2	3	5	7	12	16	20	32	46
	4/0	—	0	1	1	1	2	4	6	10	13	17	26	38
	250	—	0	0	1	1	1	3	5	8	11	14	21	31
	300	—	0	0	1	1	1	3	4	7	9	12	18	27
	350	—	0	0	1	1	1	3	4	6	8	10	16	23
	400	—	0	0	1	1	1	2	3	5	7	9	14	21
	500	—	0	0	0	1	1	1	3	4	6	8	12	17
	600	—	0	0	0	1	1	1	2	3	5	6	9	14
	700	—	0	0	0	1	1	1	1	3	4	5	8	12
	750	—	0	0	0	1	1	1	1	3	4	5	8	11
	800	—	0	0	0	0	1	1	1	3	4	5	7	11
	900	—	0	0	0	0	1	1	1	2	3	4	6	9
	1000	—	0	0	0	0	1	1	1	2	3	4	6	9
	1250	—	0	0	0	0	0	1	1	1	2	3	4	7
	1500	—	0	0	0	0	0	1	1	1	1	2	4	6
	1750	—	0	0	0	0	0	1	1	1	1	2	3	5
	2000	—	0	0	0	0	0	0	1	1	1	1	3	4
FIXTURE WIRES														
RFH-2, FFH-2, RFHH-2	18	—	9	16	26	45	61	100	141	218	292	376	588	848
	16	—	8	13	22	38	51	84	119	184	246	317	495	715
SF-2, SFF-2	18	—	12	20	33	57	77	126	178	275	368	474	741	1069
	16	—	10	17	27	47	64	104	147	228	304	392	613	885
	14	—	8	13	22	38	51	84	119	184	246	317	495	715

(continues)

△ **TABLE C.4** *Continued*

Type	Conductor Size (AWG/ kcmil)	⅜ (12)	½ (16)	¾ (21)	1 (27)	1¼ (35)	1½ (41)	2 (53)	2½ (63)	3 (78)	3½ (91)	4 (103)	5 (129)	6 (155)
								Trade Size (Metric Designator)						
SF-1, SFF-1	18	—	21	36	59	101	137	223	316	487	651	839	1312	1892
RFH-1, TF, TFF, XF, XFF	18	—	15	26	43	75	101	165	233	360	481	619	969	1398
	16	—	12	21	35	60	81	133	188	290	388	500	782	1128
XF, XFF	14	—	10	17	27	47	64	104	147	228	304	392	613	885
TFN, TFFN	18	—	25	42	69	119	162	264	373	576	769	991	1550	2237
	16	—	19	32	53	91	123	201	285	440	588	757	1184	1708
PF, PFF, PGF, PGFF, PAF, PTF, PTFF, PAFF	18	—	23	40	66	113	153	250	354	546	730	940	1470	2121
	16	—	18	31	51	88	118	193	274	422	564	727	1137	1640
	14	—	13	23	38	66	89	145	205	317	423	545	852	1230
ZF, ZFF, ZHF	18	—	30	52	85	146	197	322	456	704	941	1211	1895	2734
	16	—	22	38	63	108	146	238	336	519	694	894	1398	2017
	14	—	16	28	46	79	107	175	247	381	510	657	1027	1482
KF-2, KFF-2	18	—	45	78	128	219	296	484	684	1056	1411	1817	2842	4101
	16	—	32	54	89	153	207	337	477	737	984	1268	1983	2861
	14	—	21	36	60	103	139	227	321	495	661	852	1332	1922
	12	—	15	25	41	71	96	158	223	344	460	592	926	1337
	10	—	10	17	27	47	64	104	147	228	304	392	613	885
KF-1, KFF-1	18	—	52	90	147	253	342	558	790	1218	1628	2097	3280	4732
	16	—	37	63	103	178	240	392	555	856	1144	1473	2304	3325
	14	—	25	42	69	119	162	264	373	576	769	991	1550	2237
	12	—	16	28	46	79	107	175	247	381	510	657	1027	1482
	10	—	10	18	30	52	70	114	161	249	333	429	671	968
XF, XFF	12	—	5	9	14	25	34	56	79	122	163	209	—	—
	10	—	4	7	11	19	26	43	61	95	127	163	—	—

Notes:

1. This table is for concentric stranded conductors only. For compact stranded conductors, Table C.4(A) should be used.

2. Two-hour fire-rated RHH cable has ceramifiable insulation, which has much larger diameters than other RHH wires. Consult manufacturer's conduit fill tables.

*Types RHH, RHW, and RHW-2 without outer covering.

TABLE C.4(A) *Maximum Number of Conductors or Fixture Wires in Intermediate Metal Conduit (IMC)*
(Based on Chapter 9: Table 1, Table 4, and Table 5A)

Type	Conductor Size (AWG/ kcmil)	⅜ (12)	½ (16)	¾ (21)	1 (27)	1¼ (35)	1½ (41)	2 (53)	2½ (63)	3 (78)	3½ (91)	4 (103)	5 (129)	6 (155)
						Trade Size (Metric Designator)								
					COMPACT CONDUCTORS									
THW, THW-2, THHW	8	—	2	4	7	13	17	28	40	62	83	107	—	—
	6	—	1	3	6	10	13	22	31	48	64	82	—	—
	4	—	1	2	4	7	10	16	23	36	48	62	—	—
	2	—	1	1	3	5	7	12	17	26	35	45	—	—
	1	—	1	1	1	4	5	8	12	18	25	32	—	—
	1/0	—	1	1	1	3	4	7	10	16	21	27	—	—
	2/0	—	0	1	1	3	4	6	9	13	18	23	—	—
	3/0	—	0	1	1	2	3	5	7	11	15	20	—	—
	4/0	—	0	1	1	1	2	4	6	9	13	16	—	—
	250	—	0	0	1	1	1	3	5	7	10	13	—	—
	300	—	0	0	1	1	1	3	4	6	9	11	—	—
	350	—	0	0	1	1	1	2	4	6	8	10	—	—
	400	—	0	0	1	1	1	2	3	5	7	9	—	—
	500	—	0	0	0	1	1	1	3	4	6	8	—	—
	600	—	0	0	0	1	1	1	2	3	5	6	—	—
	700	—	0	0	0	1	1	1	1	3	4	5	—	—
	750	—	0	0	0	1	1	1	1	3	4	5	—	—
	900	—	0	0	0	0	1	1	1	2	3	4	—	—
	1000	—	0	0	0	0	1	1	1	2	3	4	—	—
THHN, THWN, THWN-2	8	—	—	—	—	—	—	—	—	—	—	—	—	—
	6	—	3	5	8	14	19	32	45	70	93	120	—	—
	4	—	1	3	5	9	12	20	28	43	58	74	—	—
	2	—	1	1	3	6	8	14	20	31	41	53	—	—
	1	—	1	1	3	5	6	10	15	23	31	40	—	—
	1/0	—	1	1	2	4	5	9	13	20	26	34	—	—
	2/0	—	1	1	1	3	4	7	10	16	22	28	—	—
	3/0	—	0	1	1	3	4	6	9	14	18	24	—	—
	4/0	—	0	1	1	2	3	5	7	11	15	19	—	—
	250	—	0	1	1	1	2	4	6	9	12	15	—	—
	300	—	0	0	1	1	1	3	5	7	10	13	—	—
	350	—	0	0	1	1	1	3	4	7	9	11	—	—
	400	—	0	0	1	1	1	2	4	6	8	10	—	—
	500	—	0	0	1	1	1	2	3	5	7	9	—	—
	600	—	0	0	0	1	1	1	2	4	5	7	—	—
	700	—	0	0	0	1	1	1	2	3	5	6	—	—
	750	—	0	0	0	1	1	1	1	3	4	6	—	—
	900	—	0	0	0	0	1	1	1	3	3	5	—	—
	1000	—	0	0	0	0	1	1	1	2	3	4	—	—

(continues)

TABLE C.4(A) *Continued*

Type	Conductor Size (AWG/ kcmil)	⅜ (12)	½ (16)	¾ (21)	1 (27)	1¼ (35)	1½ (41)	2 (53)	2½ (63)	3 (78)	3½ (91)	4 (103)	5 (129)	6 (155)
XHHW, XHHW-2	8	—	3	6	9	16	22	37	52	80	107	138	—	—
	6	—	2	4	7	12	16	27	38	59	80	103	—	—
	4	—	1	3	5	9	12	20	28	43	58	74	—	—
	2	—	1	1	3	6	8	14	20	31	41	53	—	—
	1	—	1	1	3	5	6	10	15	23	31	40	—	—
	1/0	—	1	1	2	4	5	9	13	20	26	34	—	—
	2/0	—	1	1	1	3	4	7	11	17	22	29	—	—
	3/0	—	0	1	1	3	4	6	9	14	18	24	—	—
	4/0	—	0	1	1	2	3	5	7	11	15	20	—	—
	250	—	0	1	1	1	2	4	6	9	12	16	—	—
	300	—	0	0	1	1	1	3	5	8	10	13	—	—
	350	—	0	0	1	1	1	3	4	7	9	12	—	—
	400	—	0	0	1	1	1	3	4	6	8	11	—	—
	500	—	0	0	1	1	1	2	3	5	7	9	—	—
	600	—	0	0	0	1	1	1	2	4	5	7	—	—
	700	—	0	0	0	1	1	1	2	3	5	6	—	—
	750	—	0	0	0	1	1	1	1	3	4	6	—	—
	900	—	0	0	0	1	1	1	1	3	4	5	—	—
	1000	—	0	0	0	0	1	1	1	2	3	4	—	—

Definition: *Compact stranding* is the result of a manufacturing process where the stranded conductor is compressed to the extent that interstices (voids between strand wires) are virtually eliminated.

Δ **TABLE C.5** *Maximum Number of Conductors or Fixture Wires in Liquidtight Flexible Nonmetallic Conduit (LFNC-A)*
(Based on Chapter 9: Table 1, Table 4, and Table 5)

Type	Conductor Size (AWG/ kcmil)	Trade Size (Metric Designator)												
		⅜ (12)	½ (16)	¾ (21)	1 (27)	1¼ (35)	1½ (41)	2 (53)	2½ (63)	3 (78)	3½ (91)	4 (103)	5 (129)	6 (155)
CONDUCTORS														
RHH, RHW, RHW-2	14	2	4	7	11	20	27	45	—	—	—	—	—	—
	12	1	3	6	9	17	23	38	—	—	—	—	—	—
	10	1	3	5	8	13	18	30	—	—	—	—	—	—
	8	1	1	2	4	7	9	16	—	—	—	—	—	—
	6	1	1	1	3	5	7	13	—	—	—	—	—	—
	4	0	1	1	2	4	6	10	—	—	—	—	—	—
	3	0	1	1	1	4	5	8	—	—	—	—	—	—
	2	0	1	1	1	3	4	7	—	—	—	—	—	—
	1	0	0	1	1	1	3	5	—	—	—	—	—	—
	1/0	0	0	1	1	1	2	4	—	—	—	—	—	—
	2/0	0	0	1	1	1	1	4	—	—	—	—	—	—
	3/0	0	0	0	1	1	1	3	—	—	—	—	—	—
	4/0	0	0	0	1	1	1	3	—	—	—	—	—	—
	250	0	0	0	0	1	1	1	—	—	—	—	—	—
	300	0	0	0	0	1	1	1	—	—	—	—	—	—
	350	0	0	0	0	1	1	1	—	—	—	—	—	—
	400	0	0	0	0	1	1	1	—	—	—	—	—	—
	500	0	0	0	0	0	1	1	—	—	—	—	—	—
	600	0	0	0	0	0	1	1	—	—	—	—	—	—
	700	0	0	0	0	0	0	1	—	—	—	—	—	—
	750	0	0	0	0	0	0	1	—	—	—	—	—	—
	800	0	0	0	0	0	0	1	—	—	—	—	—	—
	900	0	0	0	0	0	0	1	—	—	—	—	—	—
	1000	0	0	0	0	0	0	1	—	—	—	—	—	—
	1250	0	0	0	0	0	0	0	—	—	—	—	—	—
	1500	0	0	0	0	0	0	0	—	—	—	—	—	—
	1750	0	0	0	0	0	0	0	—	—	—	—	—	—
	2000	0	0	0	0	0	0	0	—	—	—	—	—	—
TW, THHW, THW, THW-2	14	5	9	15	24	43	58	96	—	—	—	—	—	—
	12	4	7	12	19	33	44	74	—	—	—	—	—	—
	10	3	5	9	14	24	33	55	—	—	—	—	—	—
	8	1	3	5	8	13	18	30	—	—	—	—	—	—
RHH*, RHW*, RHW-2*	14	3	6	10	16	28	38	64	—	—	—	—	—	—
	12	3	5	8	13	23	31	51	—	—	—	—	—	—
	10	1	3	6	10	18	24	40	—	—	—	—	—	—
	8	1	1	4	6	11	14	24	—	—	—	—	—	—
TW, THW, THHW, THW-2, RHH*, RHW*, RHW-2*	6	1	1	3	4	8	11	18	—	—	—	—	—	—
	4	1	1	1	3	6	8	13	—	—	—	—	—	—
	3	1	1	1	3	5	7	11	—	—	—	—	—	—
	2	0	1	1	2	4	6	10	—	—	—	—	—	—
	1	0	1	1	1	3	4	7	—	—	—	—	—	—
	1/0	0	0	1	1	2	3	6	—	—	—	—	—	—
	2/0	0	0	1	1	1	3	5	—	—	—	—	—	—
	3/0	0	0	1	1	1	2	4	—	—	—	—	—	—
	4/0	0	0	0	1	1	1	3	—	—	—	—	—	—

(continues)

Δ **TABLE C.5** *Continued*

Type	Conductor Size (AWG/ kcmil)	⅜ (12)	½ (16)	¾ (21)	1 (27)	1¼ (35)	1½ (41)	2 (53)	2½ (63)	3 (78)	3½ (91)	4 (103)	5 (129)	6 (155)
														Trade Size (Metric Designator)
	250	0	0	0	1	1	1	3	—	—	—	—	—	—
	300	0	0	0	1	1	1	2	—	—	—	—	—	—
	350	0	0	0	0	1	1	1	—	—	—	—	—	—
	400	0	0	0	0	1	1	1	—	—	—	—	—	—
	500	0	0	0	0	1	1	1	—	—	—	—	—	—
	600	0	0	0	0	1	1	1	—	—	—	—	—	—
	700	0	0	0	0	0	1	1	—	—	—	—	—	—
	750	0	0	0	0	0	1	1	—	—	—	—	—	—
	800	0	0	0	0	0	1	1	—	—	—	—	—	—
	900	0	0	0	0	0	0	1	—	—	—	—	—	—
	1000	0	0	0	0	0	0	1	—	—	—	—	—	—
	1250	0	0	0	0	0	0	1	—	—	—	—	—	—
	1500	0	0	0	0	0	0	1	—	—	—	—	—	—
	1750	0	0	0	0	0	0	0	—	—	—	—	—	—
	2000	0	0	0	0	0	0	0	—	—	—	—	—	—
THHN, THWN, THWN-2	14	8	13	22	35	62	83	138	—	—	—	—	—	—
	12	5	9	16	25	45	60	100	—	—	—	—	—	—
	10	3	6	10	16	28	38	63	—	—	—	—	—	—
	8	1	3	6	9	16	22	36	—	—	—	—	—	—
	6	1	2	4	6	12	16	26	—	—	—	—	—	—
	4	1	1	2	4	7	9	16	—	—	—	—	—	—
	3	1	1	1	3	6	8	13	—	—	—	—	—	—
	2	1	1	1	3	5	7	11	—	—	—	—	—	—
	1	0	1	1	1	4	5	8	—	—	—	—	—	—
	1/0	0	1	1	1	3	4	7	—	—	—	—	—	—
	2/0	0	0	1	1	2	3	6	—	—	—	—	—	—
	3/0	0	0	1	1	1	3	5	—	—	—	—	—	—
	4/0	0	0	1	1	1	2	4	—	—	—	—	—	—
	250	0	0	0	1	1	1	3	—	—	—	—	—	—
	300	0	0	0	1	1	1	3	—	—	—	—	—	—
	350	0	0	0	1	1	1	2	—	—	—	—	—	—
	400	0	0	0	0	1	1	1	—	—	—	—	—	—
	500	0	0	0	0	1	1	1	—	—	—	—	—	—
	600	0	0	0	0	1	1	1	—	—	—	—	—	—
	700	0	0	0	0	1	1	1	—	—	—	—	—	—
	750	0	0	0	0	0	1	1	—	—	—	—	—	—
	800	0	0	0	0	0	1	1	—	—	—	—	—	—
	900	0	0	0	0	0	1	1	—	—	—	—	—	—
	1000	0	0	0	0	0	0	1	—	—	—	—	—	—
FEP, FEPB, PFA, PFAH, TFE	14	7	12	21	34	60	80	133	—	—	—	—	—	—
	12	5	9	15	25	44	59	97	—	—	—	—	—	—
	10	4	6	11	18	31	42	70	—	—	—	—	—	—
	8	1	3	6	10	18	24	40	—	—	—	—	—	—
	6	1	2	4	7	13	17	28	—	—	—	—	—	—
	4	1	1	3	5	9	12	20	—	—	—	—	—	—
	3	1	1	2	4	7	10	16	—	—	—	—	—	—
	2	1	1	1	3	6	8	13	—	—	—	—	—	—
PFA, PFAH, TFE	1	0	1	1	2	4	5	9	—	—	—	—	—	—

Δ **TABLE C.5** *Continued*

Type	Conductor Size (AWG/ kcmil)	⅜ (12)	½ (16)	¾ (21)	1 (27)	1¼ (35)	1½ (41)	2 (53)	2½ (63)	3 (78)	3½ (91)	4 (103)	5 (129)	6 (155)
PFA, PFAH, TFE, Z	1/0	0	1	1	1	3	5	8	—	—	—	—	—	—
	2/0	0	1	1	1	3	4	6	—	—	—	—	—	—
	3/0	0	0	1	1	2	3	5	—	—	—	—	—	—
	4/0	0	0	1	1	1	2	4	—	—	—	—	—	—
Z	14	9	15	25	41	72	97	161	—	—	—	—	—	—
	12	6	10	18	29	51	69	114	—	—	—	—	—	—
	10	4	6	11	18	31	42	70	—	—	—	—	—	—
	8	2	4	7	11	20	26	44	—	—	—	—	—	—
	6	1	3	5	8	14	18	31	—	—	—	—	—	—
	4	1	1	3	5	9	13	21	—	—	—	—	—	—
	3	1	1	2	4	7	9	15	—	—	—	—	—	—
	2	1	1	1	3	6	8	13	—	—	—	—	—	—
	1	1	1	1	2	4	6	10	—	—	—	—	—	—
XHHW, ZW, XHHW-2, XHH	14	5	9	15	24	43	58	96	—	—	—	—	—	—
	12	4	7	12	19	33	44	74	—	—	—	—	—	—
	10	3	5	9	14	24	33	55	—	—	—	—	—	—
	8	1	3	5	8	13	18	30	—	—	—	—	—	—
	6	1	1	3	5	10	13	22	—	—	—	—	—	—
	4	1	1	2	4	7	10	16	—	—	—	—	—	—
	3	1	1	1	3	6	8	14	—	—	—	—	—	—
	2	1	1	1	3	5	7	11	—	—	—	—	—	—
XHHW, XHHW-2, XHH	1	0	1	1	1	4	5	8	—	—	—	—	—	—
	1/0	0	1	1	1	3	4	7	—	—	—	—	—	—
	2/0	0	0	1	1	2	3	6	—	—	—	—	—	—
	3/0	0	0	1	1	1	3	5	—	—	—	—	—	—
	4/0	0	0	1	1	1	2	4	—	—	—	—	—	—
	250	0	0	0	1	1	1	3	—	—	—	—	—	—
	300	0	0	0	1	1	1	3	—	—	—	—	—	—
	350	0	0	0	1	1	1	2	—	—	—	—	—	—
	400	0	0	0	0	1	1	1	—	—	—	—	—	—
	500	0	0	0	0	1	1	1	—	—	—	—	—	—
	600	0	0	0	0	1	1	1	—	—	—	—	—	—
	700	0	0	0	0	1	1	1	—	—	—	—	—	—
	750	0	0	0	0	0	1	1	—	—	—	—	—	—
	800	0	0	0	0	0	1	1	—	—	—	—	—	—
	900	0	0	0	0	0	1	1	—	—	—	—	—	—
	1000	0	0	0	0	0	0	1	—	—	—	—	—	—
	1250	0	0	0	0	0	0	1	—	—	—	—	—	—
	1500	0	0	0	0	0	0	1	—	—	—	—	—	—
	1750	0	0	0	0	0	0	0	—	—	—	—	—	—
	2000	0	0	0	0	0	0	0	—	—	—	—	—	—
FIXTURE WIRES														
RFH-2, FFH-2, RFHH-2	18	5	8	14	23	41	55	92	—	—	—	—	—	—
	16	4	7	12	20	35	47	77	—	—	—	—	—	—
SF-2, SFF-2	18	6	11	18	29	52	70	116	—	—	—	—	—	—
	16	5	9	15	24	43	58	96	—	—	—	—	—	—
	14	4	7	12	20	35	47	77	—	—	—	—	—	—
SF-1, SFF-1	18	12	19	33	52	92	124	205	—	—	—	—	—	—

(continues)

Δ   **TABLE C.5** *Continued*

	Conductor Size (AWG/ kcmil)	Trade Size (Metric Designator)												
Type		⅜ (12)	½ (16)	¾ (21)	1 (27)	1¼ (35)	1½ (41)	2 (53)	2½ (63)	3 (78)	3½ (91)	4 (103)	5 (129)	6 (155)
RFH-1, TF, TFF,	18	8	14	24	39	68	91	152	—	—	—	—	—	—
XF, XFF	16	7	11	19	31	55	74	122	—	—	—	—	—	—
XF, XFF	14	5	9	15	24	43	58	96	—	—	—	—	—	—
TFN, TFFN	18	14	22	39	62	109	146	243	—	—	—	—	—	—
	16	10	17	29	47	83	112	185	—	—	—	—	—	—
PF, PFF, PGF,	18	13	21	37	59	103	139	230	—	—	—	—	—	—
PGFF, PAF,	16	10	16	28	45	80	107	178	—	—	—	—	—	—
PTF, PTFF, PAFF	14	7	12	21	34	60	80	133	—	—	—	—	—	—
ZF, ZFF, ZHF	18	17	27	47	76	133	179	297	—	—	—	—	—	—
	16	12	20	35	56	98	132	219	—	—	—	—	—	—
	14	9	15	25	41	72	97	161	—	—	—	—	—	—
KF-2, KFF-2	18	25	41	71	114	200	269	445	—	—	—	—	—	—
	16	18	29	49	79	139	187	311	—	—	—	—	—	—
	14	12	19	33	53	94	126	209	—	—	—	—	—	—
	12	8	13	23	37	65	87	145	—	—	—	—	—	—
	10	5	9	15	24	43	58	96	—	—	—	—	—	—
KF-1, KFF-1	18	29	48	82	131	231	310	514	—	—	—	—	—	—
	16	20	33	58	92	162	218	361	—	—	—	—	—	—
	14	14	22	39	62	109	146	243	—	—	—	—	—	—
	12	9	15	25	41	72	97	161	—	—	—	—	—	—
	10	6	10	17	27	47	63	105	—	—	—	—	—	—
XF, XFF	12	3	5	8	13	23	31	51	—	—	—	—	—	—
	10	1	3	6	10	18	24	40	—	—	—	—	—	—

Notes:

1. This table is for concentric stranded conductors only. For compact stranded conductors, Table C.6(A) should be used.

2. Two-hour fire-rated RHH cable has ceramifiable insulation, which has much larger diameters than other RHH wires. Consult manufacturer's conduit fill tables.

*Types RHH, RHW, and RHW-2 without outer covering.

Δ **TABLE C.5(A)** *Maximum Number of Conductors or Fixture Wires in Liquidtight Flexible Nonmetallic Conduit (LFNC-A) (Based on Chapter 9: Table 1, Table 4, and Table 5A)*

Type	Conductor Size (AWG/ kcmil)	⅜ (12)	½ (16)	¾ (21)	1 (27)	1¼ (35)	1½ (41)	2 (53)	2½ (63)	3 (78)	3½ (91)	4 (103)	5 (129)	6 (155)
							Trade Size (Metric Designator)							
COMPACT CONDUCTORS														
THW, THW-2, THHW	8	1	2	4	6	11	16	26	—	—	—	—	—	—
	6	1	1	3	5	9	12	20	—	—	—	—	—	—
	4	1	1	2	4	7	9	15	—	—	—	—	—	—
	2	1	1	1	3	5	6	11	—	—	—	—	—	—
	1	0	1	1	1	3	4	8	—	—	—	—	—	—
	1/0	0	1	1	1	3	4	7	—	—	—	—	—	—
	2/0	0	0	1	1	2	3	5	—	—	—	—	—	—
	3/0	0	0	1	1	1	3	5	—	—	—	—	—	—
	4/0	0	0	1	1	1	2	4	—	—	—	—	—	—
	250	0	0	0	1	1	1	3	—	—	—	—	—	—
	300	0	0	0	1	1	1	3	—	—	—	—	—	—
	350	0	0	0	1	1	1	2	—	—	—	—	—	—
	400	0	0	0	0	1	1	1	—	—	—	—	—	—
	500	0	0	0	0	1	1	1	—	—	—	—	—	—
	600	0	0	0	0	1	1	1	—	—	—	—	—	—
	700	0	0	0	0	1	1	1	—	—	—	—	—	—
	750	0	0	0	0	0	1	1	—	—	—	—	—	—
	900	0	0	0	0	0	1	1	—	—	—	—	—	—
	1000	0	0	0	0	0	1	1	—	—	—	—	—	—
THHN, THWN, THWN-2	8	—	—	—	—	—	—	—	—	—	—	—	—	—
	6	1	2	4	7	13	18	29	—	—	—	—	—	—
	4	1	1	3	4	8	11	18	—	—	—	—	—	—
	2	1	1	1	3	6	8	13	—	—	—	—	—	—
	1	0	1	1	2	4	6	10	—	—	—	—	—	—
	1/0	0	1	1	1	3	5	8	—	—	—	—	—	—
	2/0	0	1	1	1	3	4	7	—	—	—	—	—	—
	3/0	0	0	1	1	2	3	6	—	—	—	—	—	—
	4/0	0	0	1	1	1	3	5	—	—	—	—	—	—
	250	0	0	1	1	1	1	3	—	—	—	—	—	—
	300	0	0	0	1	1	1	3	—	—	—	—	—	—
	350	0	0	0	1	1	1	3	—	—	—	—	—	—
	400	0	0	0	1	1	1	2	—	—	—	—	—	—
	500	0	0	0	0	1	1	1	—	—	—	—	—	—
	600	0	0	0	0	1	1	1	—	—	—	—	—	—
	700	0	0	0	0	1	1	1	—	—	—	—	—	—
	750	0	0	0	0	1	1	1	—	—	—	—	—	—
	900	0	0	0	0	0	1	1	—	—	—	—	—	—
	1000	0	0	0	0	0	1	1	—	—	—	—	—	—

(continues)

Δ   **TABLE C.5(A)** *Continued*

Type	Conductor Size (AWG/ kcmil)	⅜ (12)	½ (16)	¾ (21)	1 (27)	1¼ (35)	1½ (41)	2 (53)	2½ (63)	3 (78)	3½ (91)	4 (103)	5 (129)	6 (155)
XHHW, XHHW-2	8	1	3	5	8	15	20	34	—	—	—	—	—	—
	6	1	2	4	6	11	15	25	—	—	—	—	—	—
	4	1	1	3	4	8	11	18	—	—	—	—	—	—
	2	1	1	1	3	6	8	13	—	—	—	—	—	—
	1	0	1	1	2	4	6	10	—	—	—	—	—	—
	1/0	0	1	1	1	3	5	8	—	—	—	—	—	—
	2/0	0	1	1	1	3	4	7	—	—	—	—	—	—
	3/0	0	0	1	1	2	3	6	—	—	—	—	—	—
	4/0	0	0	1	1	1	3	5	—	—	—	—	—	—
	250	0	0	1	1	1	2	4	—	—	—	—	—	—
	300	0	0	0	1	1	1	3	—	—	—	—	—	—
	350	0	0	0	1	1	1	3	—	—	—	—	—	—
	400	0	0	0	1	1	1	2	—	—	—	—	—	—
	500	0	0	0	0	1	1	1	—	—	—	—	—	—
	600	0	0	0	0	1	1	1	—	—	—	—	—	—
	700	0	0	0	0	1	1	1	—	—	—	—	—	—
	750	0	0	0	0	1	1	1	—	—	—	—	—	—
	900	0	0	0	0	0	1	1	—	—	—	—	—	—
	1000	0	0	0	0	0	1	1	—	—	—	—	—	—

Definition: *Compact stranding* is the result of a manufacturing process where the stranded conductor is compressed to the extent that the interstices (voids between strand wires) are virtually eliminated.

Δ **TABLE C.6** *Maximum Number of Conductors or Fixture Wires in Liquidtight Flexible Nonmetallic Conduit (LFNC-B*)*
(Based on Chapter 9: Table 1, Table 4, and Table 5)

Type	Conductor Size (AWG/ kcmil)	Trade Size (Metric Designator)												
		⅜ (12)	½ (16)	¾ (21)	1 (27)	1¼ (35)	1½ (41)	2 (53)	2½ (63)	3 (78)	3½ (91)	4 (103)	5 (129)	6 (155)
CONDUCTORS														
RHH, RHW, RHW-2	14	2	4	7	12	21	27	44	—	—	—	—	—	—
	12	1	3	6	10	17	22	36	—	—	—	—	—	—
	10	1	3	5	8	14	18	29	—	—	—	—	—	—
	8	1	1	2	4	7	9	15	—	—	—	—	—	—
	6	1	1	1	3	6	7	12	—	—	—	—	—	—
	4	0	1	1	2	4	6	9	—	—	—	—	—	—
	3	0	1	1	1	4	5	8	—	—	—	—	—	—
	2	0	1	1	1	3	4	7	—	—	—	—	—	—
	1	0	0	1	1	1	3	5	—	—	—	—	—	—
	1/0	0	0	1	1	1	2	4	—	—	—	—	—	—
	2/0	0	0	1	1	1	1	3	—	—	—	—	—	—
	3/0	0	0	0	1	1	1	3	—	—	—	—	—	—
	4/0	0	0	0	1	1	1	2	—	—	—	—	—	—
	250	0	0	0	0	1	1	1	—	—	—	—	—	—
	300	0	0	0	0	1	1	1	—	—	—	—	—	—
	350	0	0	0	0	1	1	1	—	—	—	—	—	—
	400	0	0	0	0	1	1	1	—	—	—	—	—	—
	500	0	0	0	0	1	1	1	—	—	—	—	—	—
	600	0	0	0	0	0	1	1	—	—	—	—	—	—
	700	0	0	0	0	0	0	1	—	—	—	—	—	—
	750	0	0	0	0	0	0	1	—	—	—	—	—	—
	800	0	0	0	0	0	0	1	—	—	—	—	—	—
	900	0	0	0	0	0	0	1	—	—	—	—	—	—
	1000	0	0	0	0	0	0	1	—	—	—	—	—	—
	1250	0	0	0	0	0	0	0	—	—	—	—	—	—
	1500	0	0	0	0	0	0	0	—	—	—	—	—	—
	1750	0	0	0	0	0	0	0	—	—	—	—	—	—
	2000	0	0	0	0	0	0	0	—	—	—	—	—	—
TW, THHW, THW, THW-2	14	5	9	15	25	44	57	93	—	—	—	—	—	—
	12	4	7	12	19	33	43	71	—	—	—	—	—	—
	10	3	5	9	14	25	32	53	—	—	—	—	—	—
	8	1	3	5	8	14	18	29	—	—	—	—	—	—
RHH*, RHW*, RHW-2*	14	3	6	10	16	29	38	62	—	—	—	—	—	—
	12	3	5	8	13	23	30	50	—	—	—	—	—	—
	10	1	3	6	10	18	23	39	—	—	—	—	—	—
	8	1	1	4	6	11	14	23	—	—	—	—	—	—
TW, THW, THHW, THW-2, RHH*, RHW*, RHW-2*	6	1	1	3	5	8	11	18	—	—	—	—	—	—
	4	1	1	1	3	6	8	13	—	—	—	—	—	—
	3	1	1	1	3	5	7	11	—	—	—	—	—	—
	2	0	1	1	2	4	6	9	—	—	—	—	—	—
	1	0	1	1	1	3	4	7	—	—	—	—	—	—
	1/0	0	0	1	1	2	3	6	—	—	—	—	—	—
	2/0	0	0	1	1	2	3	5	—	—	—	—	—	—
	3/0	0	0	1	1	1	2	4	—	—	—	—	—	—
	4/0	0	0	0	1	1	1	3	—	—	—	—	—	—

(continues)

△ **TABLE C.6** *Continued*

Type	Conductor Size (AWG/ kcmil)	Trade Size (Metric Designator)												
		⅜ (12)	½ (16)	¾ (21)	1 (27)	1¼ (35)	1½ (41)	2 (53)	2½ (63)	3 (78)	3½ (91)	4 (103)	5 (129)	6 (155)
	250	0	0	0	1	1	1	3	—	—	—	—	—	—
	300	0	0	0	1	1	1	2	—	—	—	—	—	—
	350	0	0	0	0	1	1	1	—	—	—	—	—	—
	400	0	0	0	0	1	1	1	—	—	—	—	—	—
	500	0	0	0	0	1	1	1	—	—	—	—	—	—
	600	0	0	0	0	1	1	1	—	—	—	—	—	—
	700	0	0	0	0	0	1	1	—	—	—	—	—	—
	750	0	0	0	0	0	1	1	—	—	—	—	—	—
	800	0	0	0	0	0	1	1	—	—	—	—	—	—
	900	0	0	0	0	0	0	1	—	—	—	—	—	—
	1000	0	0	0	0	0	0	1	—	—	—	—	—	—
	1250	0	0	0	0	0	0	1	—	—	—	—	—	—
	1500	0	0	0	0	0	0	0	—	—	—	—	—	—
	1750	0	0	0	0	0	0	0	—	—	—	—	—	—
	2000	0	0	0	0	0	0	0	—	—	—	—	—	—
THHN, THWN, THWN-2	14	8	13	22	36	63	81	134	—	—	—	—	—	—
	12	5	9	16	26	46	59	97	—	—	—	—	—	—
	10	3	6	10	16	29	37	61	—	—	—	—	—	—
	8	1	3	6	9	16	21	35	—	—	—	—	—	—
	6	1	2	4	7	12	15	25	—	—	—	—	—	—
	4	1	1	2	4	7	9	15	—	—	—	—	—	—
	3	1	1	1	3	6	8	13	—	—	—	—	—	—
	2	1	1	1	3	5	7	11	—	—	—	—	—	—
	1	0	1	1	1	4	5	8	—	—	—	—	—	—
	1/0	0	1	1	1	3	4	7	—	—	—	—	—	—
	2/0	0	0	1	1	2	3	6	—	—	—	—	—	—
	3/0	0	0	1	1	1	3	5	—	—	—	—	—	—
	4/0	0	0	1	1	1	2	4	—	—	—	—	—	—
	250	0	0	0	1	1	1	3	—	—	—	—	—	—
	300	0	0	0	1	1	1	3	—	—	—	—	—	—
	350	0	0	0	1	1	1	2	—	—	—	—	—	—
	400	0	0	0	0	1	1	1	—	—	—	—	—	—
	500	0	0	0	0	1	1	1	—	—	—	—	—	—
	600	0	0	0	0	1	1	1	—	—	—	—	—	—
	700	0	0	0	0	1	1	1	—	—	—	—	—	—
	750	0	0	0	0	0	1	1	—	—	—	—	—	—
	800	0	0	0	0	0	1	1	—	—	—	—	—	—
	900	0	0	0	0	0	1	1	—	—	—	—	—	—
	1000	0	0	0	0	0	0	1	—	—	—	—	—	—
FEP, FEPB, PFA, PFAH, TFE	14	7	12	21	35	61	79	130	—	—	—	—	—	—
	12	5	9	15	25	44	58	94	—	—	—	—	—	—
	10	4	6	11	18	32	41	68	—	—	—	—	—	—
	8	1	3	6	10	18	23	39	—	—	—	—	—	—
	6	1	2	4	7	13	17	27	—	—	—	—	—	—
	4	1	1	3	5	9	12	19	—	—	—	—	—	—
	3	1	1	2	4	7	10	16	—	—	—	—	—	—
	2	1	1	1	3	6	8	13	—	—	—	—	—	—

△ **TABLE C.6** *Continued*

Type	Conductor Size (AWG/ kcmil)	⅜ (12)	½ (16)	¾ (21)	1 (27)	1¼ (35)	1½ (41)	2 (53)	2½ (63)	3 (78)	3½ (91)	4 (103)	5 (129)	6 (155)
PFA, PFAH, TFE	1	0	1	1	2	4	5	9	—	—	—	—	—	—
PFA, PFAH, TFE, Z	1/0	0	1	1	1	3	4	7	—	—	—	—	—	—
	2/0	0	1	1	1	3	4	6	—	—	—	—	—	—
	3/0	0	0	1	1	2	3	5	—	—	—	—	—	—
	4/0	0	0	1	1	1	2	4	—	—	—	—	—	—
Z	14	9	15	26	42	73	95	156	—	—	—	—	—	—
	12	6	10	18	30	52	67	111	—	—	—	—	—	—
	10	4	6	11	18	32	41	68	—	—	—	—	—	—
	8	2	4	7	11	20	26	43	—	—	—	—	—	—
	6	1	3	5	8	14	18	30	—	—	—	—	—	—
	4	1	1	3	5	9	12	20	—	—	—	—	—	—
	3	1	1	2	4	7	9	15	—	—	—	—	—	—
	2	1	1	1	3	6	7	12	—	—	—	—	—	—
	1	1	1	1	2	5	6	10	—	—	—	—	—	—
XHHW, ZW, XHHW-2, XHH	14	5	9	15	25	44	57	93	—	—	—	—	—	—
	12	4	7	12	19	33	43	71	—	—	—	—	—	—
	10	3	5	9	14	25	32	53	—	—	—	—	—	—
	8	1	3	5	8	14	18	29	—	—	—	—	—	—
	6	1	1	3	6	10	13	22	—	—	—	—	—	—
	4	1	1	2	4	7	9	16	—	—	—	—	—	—
	3	1	1	1	3	6	8	13	—	—	—	—	—	—
	2	1	1	1	3	5	7	11	—	—	—	—	—	—
XHHW, XHHW-2, XHH	1	0	1	1	1	4	5	8	—	—	—	—	—	—
	1/0	0	1	1	1	3	4	7	—	—	—	—	—	—
	2/0	0	0	1	1	2	3	6	—	—	—	—	—	—
	3/0	0	0	1	1	1	3	5	—	—	—	—	—	—
	4/0	0	0	1	1	1	2	4	—	—	—	—	—	—
	250	0	0	0	1	1	1	3	—	—	—	—	—	—
	300	0	0	0	1	1	1	3	—	—	—	—	—	—
	350	0	0	0	1	1	1	2	—	—	—	—	—	—
	400	0	0	0	1	1	1	1	—	—	—	—	—	—
	500	0	0	0	0	1	1	1	—	—	—	—	—	—
	600	0	0	0	0	1	1	1	—	—	—	—	—	—
	700	0	0	0	0	1	1	1	—	—	—	—	—	—
	750	0	0	0	0	0	1	1	—	—	—	—	—	—
	800	0	0	0	0	0	1	1	—	—	—	—	—	—
	900	0	0	0	0	0	1	1	—	—	—	—	—	—
	1000	0	0	0	0	0	0	1	—	—	—	—	—	—
	1250	0	0	0	0	0	0	1	—	—	—	—	—	—
	1500	0	0	0	0	0	0	1	—	—	—	—	—	—
	1750	0	0	0	0	0	0	0	—	—	—	—	—	—
	2000	0	0	0	0	0	0	0	—	—	—	—	—	—
FIXTURE WIRES														
RFH-2, FFH-2, RFHH-2	18	5	8	15	24	42	54	89	—	—	—	—	—	—
	16	4	7	12	20	35	46	75	—	—	—	—	—	—
SF-2, SFF-2	18	6	11	19	30	53	69	113	—	—	—	—	—	—
	16	5	9	15	25	44	57	93	—	—	—	—	—	—
	14	4	7	12	20	35	46	75	—	—	—	—	—	—
SF-1, SFF-1	18	12	19	33	53	94	122	199	—	—	—	—	—	—

(continues)

Δ **TABLE C.6** *Continued*

Type	Conductor Size (AWG/ kcmil)	⅜ (12)	½ (16)	¾ (21)	1 (27)	1¼ (35)	1½ (41)	2 (53)	2½ (63)	3 (78)	3½ (91)	4 (103)	5 (129)	6 (155)
						Trade Size (Metric Designator)								
RFH-1, TF, TFF, XF, XFF	18	8	14	24	39	69	90	147	—	—	—	—	—	—
	16	7	11	20	32	56	72	119	—	—	—	—	—	—
XF, XFF	14	5	9	15	25	44	57	93	—	—	—	—	—	—
TFN, TFFN	18	14	23	39	63	111	144	236	—	—	—	—	—	—
	16	10	17	30	48	85	110	180	—	—	—	—	—	—
PF, PFF, PGF, PGFF, PAF, PTF, PTFF, PAFF	18	13	21	37	60	105	136	224	—	—	—	—	—	—
	16	10	16	29	46	81	105	173	—	—	—	—	—	—
	14	7	12	21	35	61	79	130	—	—	—	—	—	—
ZF, ZFF, ZHF	18	17	28	48	77	136	176	288	—	—	—	—	—	—
	16	12	20	35	57	100	130	213	—	—	—	—	—	—
	14	9	15	26	42	73	95	156	—	—	—	—	—	—
KF-2, KFF-2	18	25	42	72	116	203	264	433	—	—	—	—	—	—
	16	18	29	50	81	142	184	302	—	—	—	—	—	—
	14	12	19	34	54	95	124	203	—	—	—	—	—	—
	12	8	13	23	38	66	86	141	—	—	—	—	—	—
	10	5	9	15	25	44	57	93	—	—	—	—	—	—
KF-1, KFF-1	18	29	48	83	134	235	304	499	—	—	—	—	—	—
	16	20	34	58	94	165	214	351	—	—	—	—	—	—
	14	14	23	39	63	111	144	236	—	—	—	—	—	—
	12	9	15	26	42	73	95	156	—	—	—	—	—	—
	10	6	10	17	27	48	62	102	—	—	—	—	—	—
XF, XFF	12	3	5	8	13	23	30	50	—	—	—	—	—	—
	10	1	3	6	10	18	23	39	—	—	—	—	—	—

Notes:

1. This table is for concentric stranded conductors only. For compact stranded conductors, Table C.5(A) should be used.

2. Two-hour fire-rated RHH cable has ceramifiable insulation, which has much larger diameters than other RHH wires. Consult manufacturer's conduit fill tables.

*Types RHH, RHW, and RHW-2 without outer covering.

TABLE C.6(A) *Maximum Number of Conductors or Fixture Wires in Liquidtight Flexible Nonmetallic Conduit (LFNC-B)* *(Based on Chapter 9: Table 1, Table 4, and Table 5A)*

Type	Conductor Size (AWG/ kcmil)	Trade Size (Metric Designator)												
		⅜ (12)	½ (16)	¾ (21)	1 (27)	1¼ (35)	1½ (41)	2 (53)	2½ (63)	3 (78)	3½ (91)	4 (103)	5 (129)	6 (155)
		COMPACT CONDUCTORS												
THW, THW-2, THHW	8	1	2	4	7	12	15	25	—	—	—	—	—	—
	6	1	1	3	5	9	12	19	—	—	—	—	—	—
	4	1	1	2	4	7	9	14	—	—	—	—	—	—
	2	1	1	1	3	5	6	11	—	—	—	—	—	—
	1	0	1	1	1	3	4	7	—	—	—	—	—	—
	1/0	0	1	1	1	3	4	6	—	—	—	—	—	—
	2/0	0	0	1	1	2	3	5	—	—	—	—	—	—
	3/0	0	0	1	1	1	3	4	—	—	—	—	—	—
	4/0	0	0	1	1	1	2	4	—	—	—	—	—	—
	250	0	0	0	1	1	1	3	—	—	—	—	—	—
	300	0	0	0	1	1	1	2	—	—	—	—	—	—
	350	0	0	0	1	1	1	2	—	—	—	—	—	—
	400	0	0	0	0	1	1	1	—	—	—	—	—	—
	500	0	0	0	0	1	1	1	—	—	—	—	—	—
	600	0	0	0	0	1	1	1	—	—	—	—	—	—
	700	0	0	0	0	1	1	1	—	—	—	—	—	—
	750	0	0	0	0	0	1	1	—	—	—	—	—	—
	900	0	0	0	0	0	1	1	—	—	—	—	—	—
	1000	0	0	0	0	0	1	1	—	—	—	—	—	—
THHN, THWN, THWN-2	8	—	—	—	—	—	—	—	—	—	—	—	—	—
	6	1	2	4	7	13	17	28	—	—	—	—	—	—
	4	1	1	3	4	8	11	17	—	—	—	—	—	—
	2	1	1	1	3	6	7	12	—	—	—	—	—	—
	1	0	1	1	2	4	6	9	—	—	—	—	—	—
	1/0	0	1	1	1	4	5	8	—	—	—	—	—	—
	2/0	0	1	1	1	3	4	6	—	—	—	—	—	—
	3/0	0	0	1	1	2	3	5	—	—	—	—	—	—
	4/0	0	0	1	1	1	3	4	—	—	—	—	—	—
	250	0	0	1	1	1	1	3	—	—	—	—	—	—
	300	0	0	0	1	1	1	3	—	—	—	—	—	—
	350	0	0	0	1	1	1	2	—	—	—	—	—	—
	400	0	0	0	1	1	1	2	—	—	—	—	—	—
	500	0	0	0	0	1	1	1	—	—	—	—	—	—
	600	0	0	0	0	1	1	1	—	—	—	—	—	—
	700	0	0	0	0	1	1	1	—	—	—	—	—	—
	750	0	0	0	0	1	1	1	—	—	—	—	—	—
	900	0	0	0	0	0	1	1	—	—	—	—	—	—
	1000	0	0	0	0	0	1	1	—	—	—	—	—	—

(continues)

Δ TABLE C.6(A) *Continued*

Type	Conductor Size (AWG/ kcmil)	Trade Size (Metric Designator)												
		⅜ (12)	½ (16)	¾ (21)	1 (27)	1¼ (35)	1½ (41)	2 (53)	2½ (63)	3 (78)	3½ (91)	4 (103)	5 (129)	6 (155)
XHHW, XHHW-2	8	1	3	5	9	15	20	33	—	—	—	—	—	—
	6	1	2	4	6	11	15	24	—	—	—	—	—	—
	4	1	1	3	4	8	11	17	—	—	—	—	—	—
	2	1	1	1	3	6	7	12	—	—	—	—	—	—
	1	0	1	1	2	4	6	9	—	—	—	—	—	—
	1/0	0	1	1	1	4	5	8	—	—	—	—	—	—
	2/0	0	1	1	1	3	4	7	—	—	—	—	—	—
	3/0	0	0	1	1	2	3	5	—	—	—	—	—	—
	4/0	0	0	1	1	1	3	4	—	—	—	—	—	—
	250	0	0	1	1	1	1	3	—	—	—	—	—	—
	300	0	0	0	1	1	1	3	—	—	—	—	—	—
	350	0	0	0	1	1	1	3	—	—	—	—	—	—
	400	0	0	0	1	1	1	2	—	—	—	—	—	—
	500	0	0	0	0	1	1	1	—	—	—	—	—	—
	600	0	0	0	0	1	1	1	—	—	—	—	—	—
	700	0	0	0	0	1	1	1	—	—	—	—	—	—
	750	0	0	0	0	1	1	1	—	—	—	—	—	—
	900	0	0	0	0	0	1	1	—	—	—	—	—	—
	1000	0	0	0	0	0	1	1	—	—	—	—	—	—

Definition: *Compact stranding* is the result of a manufacturing process where the stranded conductor is compressed to the extent that the interstices (voids between strand wires) are virtually eliminated.

Δ **TABLE C.7** *Maximum Number of Conductors of Fixture Wires in Liquidtight Flexible Nonmetallic Conduit (LFNC-C)*
(Based on Chapter 9: Table 1, Table 4, and Table 5)

Type	Conductor Size (AWG/ kcmil)	Trade Size (Metric Designator)												
		⅜ (12)	½ (16)	¾ (21)	1 (27)	1¼ (35)	1½ (41)	2 (53)	2½ (63)	3 (78)	3½ (91)	4 (103)	5 (129)	6 (155)
CONDUCTORS														
RHH, RHW, RHW-2	14	2	4	7	11	20	27	45	—	—	—	—	—	—
	12	1	3	6	9	16	22	37	—	—	—	—	—	—
	10	1	2	4	7	13	18	30	—	—	—	—	—	—
	8	1	1	2	4	7	9	15	—	—	—	—	—	—
	6	0	1	1	3	5	7	12	—	—	—	—	—	—
	4	0	1	1	2	4	6	10	—	—	—	—	—	—
	3	0	1	1	1	4	5	8	—	—	—	—	—	—
	2	0	0	1	1	3	4	7	—	—	—	—	—	—
	1	0	0	1	1	1	3	5	—	—	—	—	—	—
	1/0	0	0	0	1	1	2	4	—	—	—	—	—	—
	2/0	0	0	0	1	1	1	3	—	—	—	—	—	—
	3/0	0	0	0	1	1	1	3	—	—	—	—	—	—
	4/0	0	0	0	0	1	1	2	—	—	—	—	—	—
	250	0	0	0	0	1	1	1	—	—	—	—	—	—
	300	0	0	0	0	1	1	1	—	—	—	—	—	—
	350	0	0	0	0	0	1	1	—	—	—	—	—	—
	400	0	0	0	0	0	1	1	—	—	—	—	—	—
	500	0	0	0	0	0	1	1	—	—	—	—	—	—
	600	0	0	0	0	0	0	1	—	—	—	—	—	—
	700	0	0	0	0	0	0	1	—	—	—	—	—	—
	750	0	0	0	0	0	0	1	—	—	—	—	—	—
	800	0	0	0	0	0	0	1	—	—	—	—	—	—
	900	0	0	0	0	0	0	1	—	—	—	—	—	—
	1000	0	0	0	0	0	0	0	—	—	—	—	—	—
	1250	0	0	0	0	0	0	0	—	—	—	—	—	—
	1500	0	0	0	0	0	0	0	—	—	—	—	—	—
	1750	0	0	0	0	0	0	0	—	—	—	—	—	—
	2000	0	0	0	0	0	0	0	—	—	—	—	—	—
TW, THHW, THW, THW-2	14	5	8	15	24	42	56	94	—	—	—	—	—	—
	12	4	6	11	18	32	43	72	—	—	—	—	—	—
	10	3	5	8	13	24	32	54	—	—	—	—	—	—
	8	1	2	4	7	13	18	30	—	—	—	—	—	—
RHH*, RHW*, RHW-2*	14	2	5	10	16	28	37	63	—	—	—	—	—	—
	12	2	4	8	13	22	30	50	—	—	—	—	—	—
	10	1	3	6	10	17	23	39	—	—	—	—	—	—
	8	1	1	3	6	10	14	23	—	—	—	—	—	—

(continues)

Δ   **TABLE C.7** *Continued*

Type	Conductor Size (AWG/ kcmil)	⅜ (12)	½ (16)	¾ (21)	1 (27)	1¼ (35)	1½ (41)	2 (53)	2½ (63)	3 (78)	3½ (91)	4 (103)	5 (129)	6 (155)
TW, THW,	6	1	1	3	4	8	11	18	—	—	—	—	—	—
THHW,	4	1	1	1	3	6	8	13	—	—	—	—	—	—
THW-2, RHH*,	3	0	1	1	3	5	7	11	—	—	—	—	—	—
RHW*,	2	0	1	1	2	4	6	10	—	—	—	—	—	—
RHW-2	1	0	0	1	1	3	4	7	—	—	—	—	—	—
	1/0	0	0	1	1	2	3	6	—	—	—	—	—	—
	2/0	0	0	1	1	1	3	5	—	—	—	—	—	—
	3/0	0	0	0	1	1	2	4	—	—	—	—	—	—
	4/0	0	0	0	1	1	1	3	—	—	—	—	—	—
	250	0	0	0	0	1	1	3	—	—	—	—	—	—
	300	0	0	0	0	1	1	2	—	—	—	—	—	—
	350	0	0	0	0	1	1	1	—	—	—	—	—	—
	400	0	0	0	0	1	1	1	—	—	—	—	—	—
	500	0	0	0	0	0	1	1	—	—	—	—	—	—
	600	0	0	0	0	0	1	1	—	—	—	—	—	—
	700	0	0	0	0	0	0	1	—	—	—	—	—	—
	750	0	0	0	0	0	0	1	—	—	—	—	—	—
	800	0	0	0	0	0	0	1	—	—	—	—	—	—
	900	0	0	0	0	0	0	1	—	—	—	—	—	—
	1000	0	0	0	0	0	0	1	—	—	—	—	—	—
	1250	0	0	0	0	0	0	0	—	—	—	—	—	—
	1500	0	0	0	0	0	0	0	—	—	—	—	—	—
	1750	0	0	0	0	0	0	0	—	—	—	—	—	—
	2000	0	0	0	0	0	0	0	—	—	—	—	—	—
THHW, THWN,	14	7	12	21	34	61	81	135	—	—	—	—	—	—
THWN-2	12	5	9	15	25	44	59	98	—	—	—	—	—	—
	10	3	5	10	15	28	37	62	—	—	—	—	—	—
	8	1	3	5	9	16	21	36	—	—	—	—	—	—
	6	1	2	4	6	11	15	26	—	—	—	—	—	—
	4	1	1	2	4	7	9	16	—	—	—	—	—	—
	3	0	1	1	3	6	8	13	—	—	—	—	—	—
	2	0	1	1	3	5	7	11	—	—	—	—	—	—
	1	0	1	1	1	3	5	8	—	—	—	—	—	—
	1/0	0	1	1	1	3	4	7	—	—	—	—	—	—
	2/0	0	0	1	1	2	3	6	—	—	—	—	—	—
	3/0	0	0	1	1	1	3	5	—	—	—	—	—	—
	4/0	0	0	1	1	1	2	4	—	—	—	—	—	—
	250	0	0	0	1	1	1	3	—	—	—	—	—	—
	300	0	0	0	1	1	1	3	—	—	—	—	—	—
	350	0	0	0	1	1	1	2	—	—	—	—	—	—
	400	0	0	0	0	1	1	1	—	—	—	—	—	—
	500	0	0	0	0	1	1	1	—	—	—	—	—	—
	600	0	0	0	0	1	1	1	—	—	—	—	—	—
	700	0	0	0	0	0	1	1	—	—	—	—	—	—
	750	0	0	0	0	0	1	1	—	—	—	—	—	—
	800	0	0	0	0	0	1	1	—	—	—	—	—	—
	900	0	0	0	0	0	1	1	—	—	—	—	—	—
	1000	0	0	0	0	0	0	1	—	—	—	—	—	—

Δ **TABLE C.7** *Continued*

Type	Conductor Size (AWG/kcmil)	⅜ (12)	½ (16)	¾ (21)	1 (27)	1¼ (35)	1½ (41)	2 (53)	2½ (63)	3 (78)	3½ (91)	4 (103)	5 (129)	6 (155)
FEP, FEPB, PFA, PFAH, TFE	14	7	12	21	33	59	79	131	—	—	—	—	—	—
	12	5	9	15	24	43	57	96	—	—	—	—	—	—
	10	4	6	11	17	31	41	68	—	—	—	—	—	—
	8	1	3	6	10	17	23	39	—	—	—	—	—	—
	6	1	2	4	7	12	17	28	—	—	—	—	—	—
	4	1	1	3	5	9	11	19	—	—	—	—	—	—
	3	1	1	2	4	7	10	16	—	—	—	—	—	—
	2	1	1	1	3	6	8	13	—	—	—	—	—	—
PFA, PFAH, TFE	1	0	1	1	2	4	5	9	—	—	—	—	—	—
PFA, PFAH, TFE, Z	1/0	0	1	1	1	3	4	8	—	—	—	—	—	—
	2/0	0	0	1	1	3	4	6	—	—	—	—	—	—
	3/0	0	0	1	1	2	3	5	—	—	—	—	—	—
	4/0	0	0	1	1	1	2	4	—	—	—	—	—	—
Z	14	9	14	25	40	71	95	158	—	—	—	—	—	—
	12	6	10	18	28	50	67	112	—	—	—	—	—	—
	10	4	6	11	17	31	41	68	—	—	—	—	—	—
	8	2	4	7	11	19	26	43	—	—	—	—	—	—
	6	1	3	5	7	13	18	30	—	—	—	—	—	—
	4	1	1	3	5	9	12	21	—	—	—	—	—	—
	3	1	1	2	4	7	9	15	—	—	—	—	—	—
	2	1	1	1	3	5	7	12	—	—	—	—	—	—
	1	0	1	1	2	4	6	10	—	—	—	—	—	—
XHHW, ZW, XHHW-2, XHH	14	5	8	15	24	42	56	94	—	—	—	—	—	—
	12	4	6	11	18	32	43	72	—	—	—	—	—	—
	10	3	5	8	13	24	32	54	—	—	—	—	—	—
	8	1	2	4	7	13	18	30	—	—	—	—	—	—
	6	1	1	3	5	10	13	22	—	—	—	—	—	—
	4	1	1	2	4	7	9	16	—	—	—	—	—	—
	3	1	1	1	3	6	8	13	—	—	—	—	—	—
	2	1	1	1	3	5	7	11	—	—	—	—	—	—
XHHW, XHHW-2, XHH	1	0	1	1	1	4	5	8	—	—	—	—	—	—
	1/0	0	1	1	1	3	4	7	—	—	—	—	—	—
	2/0	0	0	1	1	2	3	6	—	—	—	—	—	—
	3/0	0	0	1	1	2	3	5	—	—	—	—	—	—
	4/0	0	0	1	1	1	2	4	—	—	—	—	—	—
	250	0	0	0	1	1	1	3	—	—	—	—	—	—
	300	0	0	0	1	1	1	3	—	—	—	—	—	—
	350	0	0	0	1	1	1	2	—	—	—	—	—	—
	400	0	0	0	0	1	1	1	—	—	—	—	—	—
	500	0	0	0	0	1	1	1	—	—	—	—	—	—
	600	0	0	0	0	1	1	1	—	—	—	—	—	—
	700	0	0	0	0	0	1	1	—	—	—	—	—	—
	750	0	0	0	0	0	1	1	—	—	—	—	—	—
	800	0	0	0	0	0	1	1	—	—	—	—	—	—
	900	0	0	0	0	0	1	1	—	—	—	—	—	—
	1000	0	0	0	0	0	0	1	—	—	—	—	—	—
	1250	0	0	0	0	0	0	1	—	—	—	—	—	—
	1500	0	0	0	0	0	0	1	—	—	—	—	—	—
	1750	0	0	0	0	0	0	0	—	—	—	—	—	—
	2000	0	0	0	0	0	0	0	—	—	—	—	—	—

(continues)

Δ **TABLE C.7** *Continued*

Type	Conductor Size (AWG/ kcmil)	⅜ (12)	½ (16)	¾ (21)	1 (27)	1¼ (35)	1½ (41)	2 (53)	2½ (63)	3 (78)	3½ (91)	4 (103)	5 (129)	6 (155)
								Trade Size (Metric Designator)						
FIXTURE WIRES														
RFH-2, FFH-2, RFHH-2	18	5	8	14	23	40	54	90	—	—	—	—	—	—
	16	4	7	12	19	34	46	76	—	—	—	—	—	—
SF-2, SFF-2	18	6	10	18	29	51	68	114	—	—	—	—	—	—
	16	5	8	15	24	42	56	94	—	—	—	—	—	—
	14	4	7	12	19	34	46	76	—	—	—	—	—	—
SF-1, SFF-1	18	11	18	32	51	90	121	202	—	—	—	—	—	—
RFH-1, TF, TFF, XF, XFF	18	8	13	23	38	67	89	149	—	—	—	—	—	—
	16	6	11	19	30	54	72	120	—	—	—	—	—	—
XF, XFF	14	5	8	15	24	42	56	94	—	—	—	—	—	—
TFN, TFFN	18	13	22	38	60	107	143	239	—	—	—	—	—	—
	16	10	17	29	46	82	109	182	—	—	—	—	—	—
PF, PFF, PGF, PGFF, PAF, PTF, PTFF, PAFF	18	12	21	36	57	101	136	226	—	—	—	—	—	—
	16	10	16	28	44	78	105	175	—	—	—	—	—	—
	14	7	12	21	33	59	79	131	—	—	—	—	—	—
ZF, ZFF, ZHF	18	16	27	46	74	131	175	292	—	—	—	—	—	—
	16	12	20	34	54	96	129	215	—	—	—	—	—	—
	14	9	14	25	40	71	95	131	—	—	—	—	—	—
KF-2, KFF-2	18	24	40	69	111	196	263	438	—	—	—	—	—	—
	16	17	28	48	77	137	183	305	—	—	—	—	—	—
	14	11	19	32	52	92	123	205	—	—	—	—	—	—
	12	8	13	22	36	64	85	142	—	—	—	—	—	—
	10	5	8	15	24	42	56	94	—	—	—	—	—	—
KF-1, KFF-1	18	28	46	80	128	227	303	505	—	—	—	—	—	—
	16	20	32	56	90	159	213	355	—	—	—	—	—	—
	14	13	22	38	60	107	143	239	—	—	—	—	—	—
	12	9	14	25	40	71	95	158	—	—	—	—	—	—
	10	6	9	16	26	46	62	103	—	—	—	—	—	—
XF, XFF	12	3	4	8	13	22	30	50	—	—	—	—	—	—
	10	1	3	6	10	17	23	39	—	—	—	—	—	—

Notes:

1. This table is for concentric stranded conductors only. For compact stranded conductors, Table C.5(A) should be used.

2. Two-hour fire-rated RHH cable has ceramifiable insulation, which has larger diameters than other RHH wires.
Consult manufacturer's conduit fill tables.

*Types RHH, RHW, and RHW-2 without outer covering.

Δ **TABLE C.7(A)** *Maximum Number of Conductors of Fixture Wires in Liquidtight Flexible Nonmetallic Conduit (LFNC-C)* *(Based on Chapter 9: Table 1, Table 4, and Table 5A)*

Type	Conductor Size (AWG/ kcmil)	⅜ (12)	½ (16)	¾ (21)	1 (27)	1¼ (35)	1½ (41)	2 (53)	2½ (63)	3 (78)	3½ (91)	4 (103)	5 (129)	6 (155)
COMPACT CONDUCTORS														
THW, THW-2, THHW	8	1	2	4	6	11	15	25	—	—	—	—	—	—
	6	1	1	3	5	9	12	20	—	—	—	—	—	—
	4	1	1	2	3	6	9	15	—	—	—	—	—	—
	2	1	1	1	2	5	6	11	—	—	—	—	—	—
	1	0	1	1	1	3	4	7	—	—	—	—	—	—
	1/0	0	1	1	1	3	4	6	—	—	—	—	—	—
	2/0	0	0	1	1	2	3	5	—	—	—	—	—	—
	3/0	0	0	1	1	1	3	5	—	—	—	—	—	—
	4/0	0	0	1	1	1	2	4	—	—	—	—	—	—
	250	0	0	0	1	1	1	3	—	—	—	—	—	—
	300	0	0	0	1	1	1	2	—	—	—	—	—	—
	350	0	0	0	1	1	1	2	—	—	—	—	—	—
	400	0	0	0	0	1	1	1	—	—	—	—	—	—
	500	0	0	0	0	1	1	1	—	—	—	—	—	—
	600	0	0	0	0	1	1	1	—	—	—	—	—	—
	700	0	0	0	0	1	1	1	—	—	—	—	—	—
	750	0	0	0	0	0	1	1	—	—	—	—	—	—
	900	0	0	0	0	0	1	1	—	—	—	—	—	—
	1000	0	0	0	0	0	1	1	—	—	—	—	—	—
THHN, THWN, THWN-2	8	—	—	—	—	—	—	—	—	—	—	—	—	—
	6	1	2	4	7	13	17	29	—	—	—	—	—	—
	4	1	1	3	4	8	11	18	—	—	—	—	—	—
	2	1	1	1	3	6	7	13	—	—	—	—	—	—
	1	0	1	1	2	4	6	9	—	—	—	—	—	—
	1/0	0	1	1	1	3	5	8	—	—	—	—	—	—
	2/0	0	1	1	1	3	4	7	—	—	—	—	—	—
	3/0	0	0	1	1	2	3	5	—	—	—	—	—	—
	4/0	0	0	1	1	1	3	4	—	—	—	—	—	—
	250	0	0	0	1	1	1	3	—	—	—	—	—	—
	300	0	0	0	1	1	1	3	—	—	—	—	—	—
	350	0	0	0	1	1	1	3	—	—	—	—	—	—
	400	0	0	0	1	1	1	2	—	—	—	—	—	—
	500	0	0	0	0	1	1	1	—	—	—	—	—	—
	600	0	0	0	0	1	1	1	—	—	—	—	—	—
	700	0	0	0	0	1	1	1	—	—	—	—	—	—
	750	0	0	0	0	1	1	1	—	—	—	—	—	—
	900	0	0	0	0	0	1	1	—	—	—	—	—	—
	1000	0	0	0	0	0	1	1	—	—	—	—	—	—

(continues)

Δ **TABLE C.7(A)** *Continued*

Type	Conductor Size (AWG/kcmil)	Trade Size (Metric Designator)												
		⅜ (12)	½ (16)	¾ (21)	1 (27)	1¼ (35)	1½ (41)	2 (53)	2½ (63)	3 (78)	3½ (91)	4 (103)	5 (129)	6 (155)
XHHW, XHHW-2	8	1	3	5	8	15	20	33	—	—	—	—	—	—
	6	1	1	4	6	11	15	24	—	—	—	—	—	—
	4	1	1	3	4	8	11	18	—	—	—	—	—	—
	2	1	1	1	3	6	7	13	—	—	—	—	—	—
	1	0	1	1	2	4	6	9	—	—	—	—	—	—
	1/0	0	1	1	1	3	5	8	—	—	—	—	—	—
	2/0	0	1	1	1	3	4	7	—	—	—	—	—	—
	3/0	0	0	1	1	2	3	5	—	—	—	—	—	—
	4/0	0	0	1	1	1	3	5	—	—	—	—	—	—
	250	0	0	1	1	1	1	4	—	—	—	—	—	—
	300	0	0	0	1	1	1	3	—	—	—	—	—	—
	350	0	0	0	1	1	1	3	—	—	—	—	—	—
	400	0	0	0	1	1	1	2	—	—	—	—	—	—
	500	0	0	0	0	1	1	1	—	—	—	—	—	—
	600	0	0	0	0	1	1	1	—	—	—	—	—	—
	700	0	0	0	0	1	1	1	—	—	—	—	—	—
	750	0	0	0	0	1	1	1	—	—	—	—	—	—
	900	0	0	0	0	0	1	1	—	—	—	—	—	—
	1000	0	0	0	0	0	1	1	—	—	—	—	—	—

Definition: Compact stranding is the result of a manufacturing process where the stranded conductor is compressed to the extent that the interstices (voids between stranded wires) are virtually eliminated.

Δ **TABLE C.8** *Maximum Number of Conductors or Fixture Wires in Liquidtight Flexible Metal Conduit (LFMC)*
(Based on Chapter 9: Table 1, Table 4, and Table 5)

Type	Conductor Size (AWG/ kcmil)	Trade Size (Metric Designator)												
		⅜ (12)	½ (16)	¾ (21)	1 (27)	1¼ (35)	1½ (41)	2 (53)	2½ (63)	3 (78)	3½ (91)	4 (103)	5 (129)	6 (155)
CONDUCTORS														
RHH, RHW, RHW-2	14	2	4	7	12	21	27	44	66	102	133	173	—	—
	12	1	3	6	10	17	22	36	55	84	110	144	—	—
	10	1	3	5	8	14	18	29	44	68	89	116	—	—
	8	1	1	2	4	7	9	15	23	36	46	61	—	—
	6	1	1	1	3	6	7	12	18	28	37	48	—	—
	4	0	1	1	2	4	6	9	14	22	29	38	—	—
	3	0	1	1	1	4	5	8	13	19	25	33	—	—
	2	0	1	1	1	3	4	7	11	17	22	29	—	—
	1	0	0	1	1	1	3	5	7	11	14	19	—	—
	1/0	0	0	1	1	1	2	4	6	10	13	16	—	—
	2/0	0	0	1	1	1	1	3	5	8	11	14	—	—
	3/0	0	0	0	1	1	1	3	4	7	9	12	—	—
	4/0	0	0	0	1	1	1	2	4	6	8	10	—	—
	250	0	0	0	0	1	1	1	3	4	6	8	—	—
	300	0	0	0	0	1	1	1	2	4	5	7	—	—
	350	0	0	0	0	1	1	1	2	3	5	6	—	—
	400	0	0	0	0	1	1	1	1	3	4	6	—	—
	500	0	0	0	0	1	1	1	1	3	4	5	—	—
	600	0	0	0	0	0	1	1	1	2	3	4	—	—
	700	0	0	0	0	0	0	1	1	1	3	3	—	—
	750	0	0	0	0	0	0	1	1	1	2	3	—	—
	800	0	0	0	0	0	0	1	1	1	2	3	—	—
	900	0	0	0	0	0	0	1	1	1	2	3	—	—
	1000	0	0	0	0	0	0	1	1	1	1	3	—	—
	1250	0	0	0	0	0	0	0	1	1	1	1	—	—
	1500	0	0	0	0	0	0	0	1	1	1	1	—	—
	1750	0	0	0	0	0	0	0	1	1	1	1	—	—
	2000	0	0	0	0	0	0	0	0	1	1	1	—	—
TW, THHW, THW, THW-2	14	5	9	15	25	44	57	93	140	215	280	365	—	—
	12	4	7	12	19	33	43	71	108	165	215	280	—	—
	10	3	5	9	14	25	32	53	80	123	160	209	—	—
	8	1	3	5	8	14	18	29	44	68	89	116	—	—
RHH*, RHW*, RHW-2*	14	3	6	10	16	29	38	62	93	143	186	243	—	—
	12	3	5	8	13	23	30	50	75	115	149	195	—	—
	10	1	3	6	10	18	23	39	58	89	117	152	—	—
	8	1	1	4	6	11	14	23	35	53	70	91	—	—
TW, THW, THHW, THW-2, RHH*, RHW*, RHW-2*	6	1	1	3	5	8	11	18	27	41	53	70	—	—
	4	1	1	1	3	6	8	13	20	30	40	52	—	—
	3	1	1	1	3	5	7	11	17	26	34	44	—	—
	2	0	1	1	2	4	6	9	14	22	29	38	—	—
	1	0	1	1	1	3	4	7	10	15	20	26	—	—
	1/0	0	0	1	1	2	3	6	8	13	17	23	—	—
	2/0	0	0	1	1	2	3	5	7	11	15	19	—	—
	3/0	0	0	1	1	1	2	4	6	9	12	16	—	—
	4/0	0	0	0	1	1	1	3	5	8	10	13	—	—

(continues)

Δ **TABLE C.8** *Continued*

Type	Conductor Size (AWG/kcmil)	3/8 (12)	1/2 (16)	3/4 (21)	1 (27)	1¼ (35)	1½ (41)	2 (53)	2½ (63)	3 (78)	3½ (91)	4 (103)	5 (129)	6 (155)
								Trade Size (Metric Designator)						
	250	0	0	0	1	1	1	3	4	6	8	11	—	—
	300	0	0	0	1	1	1	2	3	5	7	9	—	—
	350	0	0	0	0	1	1	1	3	5	6	8	—	—
	400	0	0	0	0	1	1	1	3	4	6	7	—	—
	500	0	0	0	0	1	1	1	2	3	5	6	—	—
	600	0	0	0	0	1	1	1	1	3	4	5	—	—
	700	0	0	0	0	0	1	1	1	2	3	4	—	—
	750	0	0	0	0	0	1	1	1	2	3	4	—	—
	800	0	0	0	0	0	1	1	1	2	3	4	—	—
	900	0	0	0	0	0	0	1	1	1	3	3	—	—
	1000	0	0	0	0	0	0	1	1	1	2	3	—	—
	1250	0	0	0	0	0	0	1	1	1	1	2	—	—
	1500	0	0	0	0	0	0	0	1	1	1	2	—	—
	1750	0	0	0	0	0	0	0	1	1	1	1	—	—
	2000	0	0	0	0	0	0	0	1	1	1	1	—	—
THHN, THWN, THWN-2	14	8	13	22	36	63	81	134	201	308	401	523	—	—
	12	5	9	16	26	46	59	97	146	225	292	381	—	—
	10	3	6	10	16	29	37	61	92	141	184	240	—	—
	8	1	3	6	9	16	21	35	53	81	106	138	—	—
	6	1	2	4	7	12	15	25	38	59	76	100	—	—
	4	1	1	2	4	7	9	15	23	36	47	61	—	—
	3	1	1	1	3	6	8	13	20	30	40	52	—	—
	2	1	1	1	3	5	7	11	17	26	33	44	—	—
	1	0	1	1	1	4	5	8	12	19	25	32	—	—
	1/0	0	1	1	1	3	4	7	10	16	21	27	—	—
	2/0	0	0	1	1	2	3	6	8	13	17	23	—	—
	3/0	0	0	1	1	1	3	5	7	11	14	19	—	—
	4/0	0	0	1	1	1	2	4	6	9	12	15	—	—
	250	0	0	0	1	1	1	3	5	7	10	12	—	—
	300	0	0	0	1	1	1	3	4	6	8	11	—	—
	350	0	0	0	1	1	1	2	3	5	7	9	—	—
	400	0	0	0	0	1	1	1	3	5	6	8	—	—
	500	0	0	0	0	1	1	1	2	4	5	7	—	—
	600	0	0	0	0	1	1	1	1	3	4	6	—	—
	700	0	0	0	0	1	1	1	1	3	4	5	—	—
	750	0	0	0	0	0	1	1	1	3	3	5	—	—
	800	0	0	0	0	0	1	1	1	2	3	4	—	—
	900	0	0	0	0	0	1	1	1	2	3	4	—	—
	1000	0	0	0	0	0	0	1	1	1	3	3	—	—
FEP, FEPB, PFA, PFAH, TFE	14	7	12	21	35	61	79	130	195	299	389	507	—	—
	12	5	9	15	25	44	58	94	142	218	284	370	—	—
	10	4	6	11	18	32	41	68	102	156	203	266	—	—
	8	1	3	6	10	18	23	39	58	89	117	152	—	—
	6	1	2	4	7	13	17	27	41	64	83	108	—	—
	4	1	1	3	5	9	12	19	29	44	58	75	—	—
	3	1	1	2	4	7	10	16	24	37	48	63	—	—
	2	1	1	1	3	6	8	13	20	30	40	52	—	—
PFA, PFAH, TFE	1	0	1	1	2	4	5	9	14	21	28	36	—	—

△ **TABLE C.8** *Continued*

Type	Conductor Size (AWG/ kcmil)	⅜ (12)	½ (16)	¾ (21)	1 (27)	1¼ (35)	1½ (41)	2 (53)	2½ (63)	3 (78)	3½ (91)	4 (103)	5 (129)	6 (155)
PFA, PFAH, TFE, Z	1/0	0	1	1	1	3	4	7	11	18	23	30	—	—
	2/0	0	1	1	1	3	4	6	9	14	19	25	—	—
	3/0	0	0	1	1	2	3	5	8	12	16	20	—	—
	4/0	0	0	1	1	1	2	4	6	10	13	17	—	—
Z	14	9	15	26	42	73	95	156	235	360	469	611	—	—
	12	6	10	18	30	52	67	111	167	255	332	434	—	—
	10	4	6	11	18	32	41	68	102	156	203	266	—	—
	8	2	4	7	11	20	26	43	64	99	129	168	—	—
	6	1	3	5	8	14	18	30	45	69	90	118	—	—
	4	1	1	3	5	9	12	20	31	48	62	81	—	—
	3	1	1	2	4	7	9	15	23	35	45	59	—	—
	2	1	1	1	3	6	7	12	19	29	38	49	—	—
	1	1	1	1	2	5	6	10	15	23	30	40	—	—
XHHW, ZW, XHHW-2, XHH	14	5	9	15	25	44	57	93	140	215	280	365	—	—
	12	4	7	12	19	33	43	71	108	165	215	280	—	—
	10	3	5	9	14	25	32	53	80	123	160	209	—	—
	8	1	3	5	8	14	18	29	44	68	89	116	—	—
	6	1	1	3	6	10	13	22	33	50	66	86	—	—
	4	1	1	2	4	7	9	16	24	36	48	62	—	—
	3	1	1	1	3	6	8	13	20	31	40	52	—	—
	2	1	1	1	3	5	7	11	17	26	34	44	—	—
XHHW, XHHW-2, XHH	1	0	1	1	1	4	5	8	12	19	25	33	—	—
	1/0	0	1	1	1	3	4	7	10	16	21	28	—	—
	2/0	0	0	1	1	2	3	6	9	13	17	23	—	—
	3/0	0	0	1	1	1	3	5	7	11	14	19	—	—
	4/0	0	0	1	1	1	2	4	6	9	12	16	—	—
	250	0	0	0	1	1	1	3	5	7	10	13	—	—
	300	0	0	0	1	1	1	3	4	6	8	11	—	—
	350	0	0	0	1	1	1	2	3	5	7	10	—	—
	400	0	0	0	1	1	1	1	3	5	6	8	—	—
	500	0	0	0	0	1	1	1	2	4	5	7	—	—
	600	0	0	0	0	1	1	1	1	3	4	6	—	—
	700	0	0	0	0	1	1	1	1	3	4	5	—	—
	750	0	0	0	0	0	1	1	1	3	3	5	—	—
	800	0	0	0	0	0	1	1	1	2	3	4	—	—
	900	0	0	0	0	0	1	1	1	2	3	4	—	—
	1000	0	0	0	0	0	0	1	1	1	3	3	—	—
	1250	0	0	0	0	0	0	1	1	1	1	3	—	—
	1500	0	0	0	0	0	0	1	1	1	1	2	—	—
	1750	0	0	0	0	0	0	0	1	1	1	1	—	—
	2000	0	0	0	0	0	0	0	1	1	1	1	—	—
FIXTURE WIRES														
RFH-2, FFH-2, RFHH-2	18	5	8	15	24	42	54	89	134	206	268	350	—	—
	16	4	7	12	20	35	46	75	113	174	226	295	—	—
SF-2, SFF-2	18	6	11	19	30	53	69	113	169	260	338	441	—	—
	16	5	9	15	25	44	57	93	140	215	280	365	—	—
	14	4	7	12	20	35	46	75	113	174	226	295	—	—
SF-1, SFF-1	18	12	19	33	53	94	122	199	300	460	599	781	—	—

(continues)

Δ TABLE C.8 *Continued*

Type	Conductor Size (AWG/ kcmil)	Trade Size (Metric Designator)												
		⅜ (12)	½ (16)	¾ (21)	1 (27)	1¼ (35)	1½ (41)	2 (53)	2½ (63)	3 (78)	3½ (91)	4 (103)	5 (129)	6 (155)
RFH-1, TF, TFF, XF, XFF	18	8	14	24	39	69	90	147	222	339	442	577	—	—
	16	7	11	20	32	56	72	119	179	274	357	465	—	—
XF, XFF	14	5	9	15	25	44	57	93	140	215	280	365	—	—
TFN, TFFN	18	14	23	39	63	111	144	236	355	543	707	923	—	—
	16	10	17	30	48	85	110	180	271	415	540	705	—	—
PF, PFF, PGF, PGFF, PAF, PTF, PTFF, PAFF	18	13	21	37	60	105	136	224	336	515	671	875	—	—
	16	10	16	29	46	81	105	173	260	398	519	677	—	—
	14	7	12	21	35	61	79	130	195	299	389	507	—	—
ZF, ZFF, ZHF	18	17	28	48	77	136	176	288	434	664	865	1128	—	—
	16	12	20	35	57	100	130	213	320	490	638	832	—	—
	14	9	15	26	42	73	95	156	235	360	469	611	—	—
KF-2, KFF-2	18	25	42	72	116	203	264	433	651	996	1297	1692	—	—
	16	18	29	50	81	142	184	302	454	695	905	1180	—	—
	14	12	19	34	54	95	124	203	305	467	608	793	—	—
	12	8	13	23	38	66	86	141	212	325	423	552	—	—
	10	5	9	15	25	44	57	93	140	215	280	365	—	—
KF-1, KFF-1	18	29	48	83	134	235	304	499	751	1150	1497	1952	—	—
	16	20	34	58	94	165	214	351	527	808	1052	1372	—	—
	14	14	23	39	63	111	144	236	355	543	707	923	—	—
	12	9	15	26	42	73	95	156	235	360	469	611	—	—
	10	6	10	17	27	48	62	102	153	235	306	399	—	—
XF, XFF	12	3	5	8	13	23	30	50	75	115	149	195	—	—
	10	1	3	6	10	18	23	39	58	89	117	152	—	—

Notes:

1. This table is for concentric stranded conductors only. For compact stranded conductors, Table C.7(A) should be used.

2. Two-hour fire-rated RHH cable has ceramifiable insulation, which has much larger diameters than other RHH wires. Consult manufacturer's conduit fill tables.

*Types RHH, RHW, and RHW-2 without outer covering.

TABLE C.8(A) *Maximum Number of Conductors or Fixture Wires in Liquidtight Flexible Metal Conduit (LFMC)*
(Based on Chapter 9: Table 1, Table 4, and Table 5A)

Type	Conductor Size (AWG/ kcmil)	⅜ (12)	½ (16)	¾ (21)	1 (27)	1¼ (35)	1½ (41)	2 (53)	2½ (63)	3 (78)	3½ (91)	4 (103)	5 (129)	6 (155)
						Trade Size (Metric Designator)								
							COMPACT CONDUCTORS								
THW, THW-2, THHW	8	1	2	4	7	12	15	25	38	58	76	99	—	—
	6	1	1	3	5	9	12	19	29	45	59	77	—	—
	4	1	1	2	4	7	9	14	22	34	44	57	—	—
	2	1	1	1	3	5	6	11	16	25	32	42	—	—
	1	0	1	1	1	3	4	7	11	17	23	30	—	—
	1/0	0	1	1	1	3	4	6	10	15	20	26	—	—
	2/0	0	0	1	1	2	3	5	8	13	16	21	—	—
	3/0	0	0	1	1	1	3	4	7	11	14	18	—	—
	4/0	0	0	1	1	1	2	4	6	9	12	15	—	—
	250	0	0	0	1	1	1	3	4	7	9	12	—	—
	300	0	0	0	1	1	1	2	4	6	8	10	—	—
	350	0	0	0	1	1	1	2	3	5	7	9	—	—
	400	0	0	0	0	1	1	1	3	5	6	8	—	—
	500	0	0	0	0	1	1	1	3	4	5	7	—	—
	600	0	0	0	0	1	1	1	1	3	4	6	—	—
	700	0	0	0	0	1	1	1	1	3	4	5	—	—
	750	0	0	0	0	0	1	1	1	3	3	5	—	—
	900	0	0	0	0	0	1	1	1	2	3	4	—	—
	1000	0	0	0	0	0	1	1	1	1	3	4	—	—
THHN, THWN, THWN-2	8	—	—	—	—	—	—	—	—	—	—	—	—	—
	6	1	2	4	7	13	17	28	43	66	86	112	—	—
	4	1	1	3	4	8	11	17	26	41	53	69	—	—
	2	1	1	1	3	6	7	12	19	29	38	50	—	—
	1	0	1	1	2	4	6	9	14	22	28	37	—	—
	1/0	0	1	1	1	4	5	8	12	19	24	32	—	—
	2/0	0	1	1	1	3	4	6	10	15	20	26	—	—
	3/0	0	0	1	1	2	3	5	8	13	17	22	—	—
	4/0	0	0	1	1	1	3	4	7	10	14	18	—	—
	250	0	0	1	1	1	1	3	5	8	11	14	—	—
	300	0	0	0	1	1	1	3	4	7	9	12	—	—
	350	0	0	0	1	1	1	2	4	6	8	11	—	—
	400	0	0	0	1	1	1	2	3	5	7	9	—	—
	500	0	0	0	0	1	1	1	3	5	6	8	—	—
	600	0	0	0	0	1	1	1	2	4	5	6	—	—
	700	0	0	0	0	1	1	1	1	3	4	6	—	—
	750	0	0	0	0	1	1	1	1	3	4	5	—	—
	900	0	0	0	0	0	1	1	1	2	3	4	—	—
	1000	0	0	0	0	0	1	1	1	2	3	4	—	—

(continues)

TABLE C.8(A) *Continued*

Type	Conductor Size (AWG/ kcmil)	Trade Size (Metric Designator)												
		⅜ (12)	½ (16)	¾ (21)	1 (27)	1¼ (35)	1½ (41)	2 (53)	2½ (63)	3 (78)	3½ (91)	4 (103)	5 (129)	6 (155)
XHHW, XHHW-2	8	1	3	5	9	15	20	33	49	76	98	129	—	—
	6	1	2	4	6	11	15	24	37	56	73	95	—	—
	4	1	1	3	4	8	11	17	26	41	53	69	—	—
	2	1	1	1	3	6	7	12	19	29	38	50	—	—
	1	0	1	1	2	4	6	9	14	22	28	37	—	—
	1/0	0	1	1	1	4	5	8	12	19	24	32	—	—
	2/0	0	1	1	1	3	4	7	10	16	20	27	—	—
	3/0	0	0	1	1	2	3	5	8	13	17	22	—	—
	4/0	0	0	1	1	1	3	4	7	11	14	18	—	—
	250	0	0	1	1	1	1	3	5	8	11	15	—	—
	300	0	0	0	1	1	1	3	5	7	9	12	—	—
	350	0	0	0	1	1	1	3	4	6	8	11	—	—
	400	0	0	0	1	1	1	2	4	6	7	10	—	—
	500	0	0	0	0	1	1	1	3	5	6	8	—	—
	600	0	0	0	0	1	1	1	2	4	5	6	—	—
	700	0	0	0	0	1	1	1	1	3	4	6	—	—
	750	0	0	0	0	1	1	1	1	3	4	5	—	—
	900	0	0	0	0	0	1	1	1	2	3	4	—	—
	1000	0	0	0	0	0	1	1	1	2	3	4	—	—

Definition: *Compact stranding* is the result of a manufacturing process where the stranded conductor is compressed to the extent that the interstices (voids between strand wires) are virtually eliminated.

Δ **TABLE C.9** *Maximum Number of Conductors or Fixture Wires in Rigid Metal Conduit (RMC)*
(Based on Chapter 9: Table 1, Table 4, and Table 5)

Type	Conductor Size (AWG/ kcmil)	⅜ (12)	½ (16)	¾ (21)	1 (27)	1¼ (35)	1½ (41)	2 (53)	2½ (63)	3 (78)	3½ (91)	4 (103)	5 (129)	6 (155)
						Trade Size (Metric Designator)								
							CONDUCTORS							
RHH, RHW, RHW-2	14	—	4	7	12	21	28	46	66	102	136	176	276	398
	12	—	3	6	10	17	23	38	55	85	113	146	229	330
	10	—	3	5	8	14	19	31	44	68	91	118	185	267
	8	—	1	2	4	7	10	16	23	36	48	61	97	139
	6	—	1	1	3	6	8	13	18	29	38	49	77	112
	4	—	1	1	2	4	6	10	14	22	30	38	60	87
	3	—	1	1	2	4	5	9	12	19	26	34	53	76
	2	—	1	1	1	3	4	7	11	17	23	29	46	66
	1	—	0	1	1	1	3	5	7	11	15	19	30	44
	1/0	—	0	1	1	1	2	4	6	10	13	17	26	38
	2/0	—	0	1	1	1	2	4	5	8	11	14	23	33
	3/0	—	0	0	1	1	1	3	4	7	10	12	20	28
	4/0	—	0	0	1	1	1	3	4	6	8	11	17	24
	250	—	0	0	0	1	1	1	3	4	6	8	13	18
	300	—	0	0	0	1	1	1	2	4	5	7	11	16
	350	—	0	0	0	1	1	1	2	4	5	6	10	15
	400	—	0	0	0	1	1	1	1	3	4	6	9	13
	500	—	0	0	0	1	1	1	1	3	4	5	8	11
	600	—	0	0	0	0	1	1	1	2	3	4	6	9
	700	—	0	0	0	0	1	1	1	1	3	3	6	8
	750	—	0	0	0	0	0	1	1	1	3	3	5	8
	800	—	0	0	0	0	0	1	1	1	2	3	5	7
	900	—	0	0	0	0	0	1	1	1	2	3	5	7
	1000	—	0	0	0	0	0	1	1	1	1	3	4	6
	1250	—	0	0	0	0	0	0	1	1	1	1	3	5
	1500	—	0	0	0	0	0	0	1	1	1	1	3	4
	1750	—	0	0	0	0	0	0	1	1	1	1	2	4
	2000	—	0	0	0	0	0	0	0	1	1	1	2	3
TW, THHW, THW, THW-2	14	—	9	15	25	44	59	98	140	215	288	370	581	839
	12	—	7	12	19	33	45	75	107	165	221	284	446	644
	10	—	5	9	14	25	34	56	80	123	164	212	332	480
	8	—	3	5	8	14	19	31	44	68	91	118	185	267
RHH*, RHW*, RHW-2*	14	—	6	10	17	29	39	65	93	143	191	246	387	558
	12	—	5	8	13	23	32	52	75	115	154	198	311	448
	10	—	3	6	10	18	25	41	58	90	120	154	242	350
	8	—	1	4	6	11	15	24	35	54	72	92	145	209
TW, THW, THHW, THW-2, RHH*, RHW*, RHW-2*	6	—	1	3	5	8	11	18	27	41	55	71	111	160
	4	—	1	1	3	6	8	14	20	31	41	53	83	120
	3	—	1	1	3	5	7	12	17	26	35	45	71	103
	2	—	1	1	2	4	6	10	14	22	30	38	60	87
	1	—	1	1	1	3	4	7	10	15	21	27	42	61

(continues)

Δ **TABLE C.9** *Continued*

Type	Conductor Size (AWG/ kcmil)	⅜ (12)	½ (16)	¾ (21)	1 (27)	1¼ (35)	1½ (41)	2 (53)	2½ (63)	3 (78)	3½ (91)	4 (103)	5 (129)	6 (155)
									Trade Size (Metric Designator)					
	1/0	—	0	1	1	2	3	6	8	13	18	23	36	52
	2/0	—	0	1	1	2	3	5	7	11	15	19	31	44
	3/0	—	0	1	1	1	2	4	6	9	13	16	26	37
	4/0	—	0	0	1	1	1	3	5	8	10	14	21	31
	250	—	0	0	1	1	1	3	4	6	8	11	17	25
	300	—	0	0	1	1	1	2	3	5	7	9	15	22
	350	—	0	0	0	1	1	1	3	5	6	8	13	19
	400	—	0	0	0	1	1	1	3	4	6	7	12	17
	500	—	0	0	0	1	1	1	2	3	5	6	10	14
	600	—	0	0	0	1	1	1	1	3	4	5	8	12
	700	—	0	0	0	0	1	1	1	2	3	4	7	10
	750	—	0	0	0	0	1	1	1	2	3	4	7	10
	800	—	0	0	0	0	1	1	1	2	3	4	6	9
	900	—	0	0	0	0	1	1	1	1	3	3	6	8
	1000	—	0	0	0	0	0	1	1	1	2	3	5	8
	1250	—	0	0	0	0	0	1	1	1	1	2	4	6
	1500	—	0	0	0	0	0	1	1	1	1	2	3	5
	1750	—	0	0	0	0	0	0	1	1	1	1	3	4
	2000	—	0	0	0	0	0	0	1	1	1	1	3	4
THHN, THWN, THWN-2	14	—	13	22	36	63	85	140	200	309	412	531	833	1202
	12	—	9	16	26	46	62	102	146	225	301	387	608	877
	10	—	6	10	17	29	39	64	92	142	189	244	383	552
	8	—	3	6	9	16	22	37	53	82	109	140	221	318
	6	—	2	4	7	12	16	27	38	59	79	101	159	230
	4	—	1	2	4	7	10	16	23	36	48	62	98	141
	3	—	1	1	3	6	8	14	20	31	41	53	83	120
	2	—	1	1	3	5	7	11	17	26	34	44	70	100
	1	—	1	1	1	4	5	8	12	19	25	33	51	74
	1/0	—	1	1	1	3	4	7	10	16	21	27	43	63
	2/0	—	0	1	1	2	3	6	8	13	18	23	36	52
	3/0	—	0	1	1	1	3	5	7	11	15	19	30	43
	4/0	—	0	1	1	1	2	4	6	9	12	16	25	36
	250	—	0	0	1	1	1	3	5	7	10	13	20	29
	300	—	0	0	1	1	1	3	4	6	8	11	17	25
	350	—	0	0	1	1	1	2	3	5	7	10	15	22
	400	—	0	0	1	1	1	2	3	5	7	8	13	20
	500		0	0	0	1	1	1	2	4	5	7	11	16
	600	—	0	0	0	1	1	1	1	3	4	6	9	13
	700	—	0	0	0	1	1	1	1	3	4	5	8	11
	750	—	0	0	0	0	1	1	1	3	4	5	7	11
	800	—	0	0	0	0	1	1	1	2	3	4	7	10
	900	—	0	0	0	0	1	1	1	2	3	4	6	9
	1000	—	0	0	0	0	1	1	1	1	3	4	6	8
FEP, FEPB, PFA, PFAH, TFE	14	—	12	22	35	61	83	136	194	300	400	515	808	1166
	12	—	9	16	26	44	60	99	142	219	292	376	590	851
	10	—	6	11	18	32	43	71	102	157	209	269	423	610
	8	—	3	6	10	18	25	41	58	90	120	154	242	350
	6	—	2	4	7	13	17	29	41	64	85	110	172	249
	4	—	1	3	5	9	12	20	29	44	59	77	120	174
	3	—	1	2	4	7	10	17	24	37	50	64	100	145
	2	—	1	1	3	6	8	14	20	31	41	53	83	120

Δ **TABLE C.9** *Continued*

Type	Conductor Size (AWG/ kcmil)	⅜ (12)	½ (16)	¾ (21)	1 (27)	1¼ (35)	1½ (41)	2 (53)	2½ (63)	3 (78)	3½ (91)	4 (103)	5 (129)	6 (155)
PFA, PFAH, TFE	1	—	1	1	2	4	6	9	14	21	28	37	57	83
PFA, PFAH, TFE, Z	1/0	—	1	1	1	3	5	8	11	18	24	30	48	69
	2/0	—	1	1	1	3	4	6	9	14	19	25	40	57
	3/0	—	0	1	1	2	3	5	8	12	16	21	33	47
	4/0	—	0	1	1	1	2	4	6	10	13	17	27	39
Z	14	—	15	26	42	73	100	164	234	361	482	621	974	1405
	12	—	10	18	30	52	71	116	166	256	342	440	691	997
	10	—	6	11	18	32	43	71	102	157	209	269	423	610
	8	—	4	7	11	20	27	45	64	99	132	170	267	386
	6	—	3	5	8	14	19	31	45	69	93	120	188	271
	4	—	1	3	5	9	13	22	31	48	64	82	129	186
	3	—	1	2	4	7	9	16	22	35	47	60	94	136
	2	—	1	1	3	6	8	13	19	29	39	50	78	113
	1	—	1	1	2	5	6	10	15	23	31	40	63	92
XHHW, ZW. XHHW-2, XHH	14	—	9	15	25	44	59	98	140	215	288	370	581	839
	12	—	7	12	19	33	45	75	107	165	221	284	446	644
	10	—	5	9	14	25	34	56	80	123	164	212	332	480
	8	—	3	5	8	14	19	31	44	68	91	118	185	267
	6	—	1	3	6	10	14	23	33	51	68	87	137	197
	4	—	1	2	4	7	10	16	24	37	49	63	99	143
	3	—	1	1	3	6	8	14	20	31	41	53	84	121
	2	—	1	1	3	5	7	12	17	26	35	45	70	101
XHHW, XHHW-2, XHH	1	—	1	1	1	4	5	9	12	19	26	33	52	76
	1/0	—	1	1	1	3	4	7	10	16	22	28	44	64
	2/0	—	0	1	1	2	3	6	9	13	18	23	37	53
	3/0	—	0	1	1	1	3	5	7	11	15	19	30	44
	4/0	—	0	1	1	1	2	4	6	9	12	16	25	36
	250	—	0	0	1	1	1	3	5	7	10	13	20	30
	300	—	0	0	1	1	1	3	4	6	9	11	18	25
	350	—	0	0	1	1	1	2	3	6	7	10	15	22
	400	—	0	0	1	1	1	2	3	5	7	9	14	20
	500	—	0	0	0	1	1	1	2	4	5	7	11	16
	600	—	0	0	0	1	1	1	1	3	4	6	9	13
	700	—	0	0	0	1	1	1	1	3	4	5	8	11
	750	—	0	0	0	0	1	1	1	3	4	5	7	11
	800	—	0	0	0	0	1	1	1	2	3	4	7	10
	900	—	0	0	0	0	1	1	1	2	3	4	6	9
	1000	—	0	0	0	0	1	1	1	1	3	4	6	8
	1250	—	0	0	0	0	0	1	1	1	2	3	4	6
	1500	—	0	0	0	0	0	1	1	1	1	2	4	5
	1750	—	0	0	0	0	0	0	1	1	1	1	3	5
	2000	—	0	0	0	0	0	0	1	1	1	1	3	4

FIXTURE WIRES

Type	Conductor Size (AWG/ kcmil)	⅜ (12)	½ (16)	¾ (21)	1 (27)	1¼ (35)	1½ (41)	2 (53)	2½ (63)	3 (78)	3½ (91)	4 (103)	5 (129)	6 (155)
RFH-2, FFH-2, RFHH-2	18	—	8	15	24	42	57	94	134	207	276	355	557	804
	16	—	7	12	20	35	48	79	113	174	232	299	470	678
SF-2, SFF-2	18	—	11	19	31	53	72	118	169	261	348	448	703	1014
	16	—	9	15	25	44	59	98	140	215	288	370	581	839
	14	—	7	12	20	35	48	79	113	174	232	299	470	678
SF-1, SFF-1	18	—	19	33	54	94	127	209	299	461	616	792	1244	1794

(continues)

Δ **TABLE C.9** *Continued*

	Conductor Size (AWG/ kcmil)	Trade Size (Metric Designator)												
Type		⅜ (12)	½ (16)	¾ (21)	1 (27)	1¼ (35)	1½ (41)	2 (53)	2½ (63)	3 (78)	3½ (91)	4 (103)	5 (129)	6 (155)
RFH-1, TF, TFF, XF, XFF	18	—	14	25	40	69	94	155	221	341	455	585	918	1325
	16	—	11	20	32	56	76	125	178	275	367	472	741	1070
XF, XFF	14	—	9	15	25	44	59	98	140	215	288	370	581	839
TFN, TFFN	18	—	23	40	64	111	150	248	354	545	728	937	1470	2120
	16	—	17	30	49	84	115	189	270	416	556	715	1123	1620
PF, PFF, PGF, PGFF, PAF, PTF, PTFF, PAFF	18	—	21	38	61	105	143	235	335	517	690	888	1394	2011
	16	—	16	29	47	81	110	181	259	400	534	687	1078	1555
	14	—	12	22	35	61	83	136	194	300	400	515	808	1166
ZF, ZFF, ZHF	18	—	28	49	79	135	184	303	432	666	889	1145	1796	2592
	16	—	20	36	58	100	136	223	319	491	656	844	1325	1912
	14	—	15	26	42	73	100	164	234	361	482	621	974	1405
KF-2, KFF-2	18	—	42	73	118	203	276	454	648	1000	1334	1717	2695	3887
	16	—	29	51	82	142	192	317	452	697	931	1198	1880	2712
	14	—	19	34	55	95	129	213	304	468	625	805	1263	1822
	12	—	13	24	38	66	90	148	211	326	435	560	878	1267
	10	—	9	15	25	44	59	98	140	215	288	370	581	839
KF-1, KFF-1	18	—	48	84	136	234	318	524	748	1153	1540	1982	3109	4486
	16	—	34	59	96	165	224	368	526	810	1082	1392	2185	3152
	14	—	23	40	64	111	150	248	354	545	728	937	1470	2120
	12	—	15	26	42	73	100	164	234	361	482	621	974	1405
	10	—	10	17	28	48	65	107	153	236	315	405	636	918
XF, XFF	12	—	5	8	13	23	32	52	75	115	154	198	311	448
	10	—	3	6	10	18	25	41	58	90	120	154	242	350

Notes:

1. This table is for concentric stranded conductors only. For compact stranded conductors, Table C.8(A) should be used.

2. Two-hour fire-rated RHH cable has ceramifiable insulation, which has much larger diameters than other RHH wires. Consult manufacturer's conduit fill tables.

*Types RHH, RHW, and RHW-2 without outer covering.

TABLE C.9(A) *Maximum Number of Conductors or Fixture Wires in Rigid Metal Conduit (RMC)*
(Based on Chapter 9: Table 1, Table 4, and Table 5A)

Type	Conductor Size (AWG/kcmil)	⅜ (12)	½ (16)	¾ (21)	1 (27)	1¼ (35)	1½ (41)	2 (53)	2½ (63)	3 (78)	3½ (91)	4 (103)	5 (129)	6 (155)
COMPACT CONDUCTORS														
THW, THW-2, THHW	8	—	2	4	7	12	16	26	38	59	78	101	158	228
	6	—	1	3	5	9	12	20	29	45	60	78	122	176
	4	—	1	2	4	7	9	15	22	34	45	58	91	132
	2	—	1	1	3	5	7	11	16	25	33	43	67	97
	1	—	1	1	1	3	5	8	11	17	23	30	47	68
	1/0	—	1	1	1	3	4	7	10	15	20	26	41	59
	2/0	—	0	1	1	2	3	6	8	13	17	22	34	50
	3/0	—	0	1	1	1	3	5	7	11	14	19	29	42
	4/0	—	0	1	1	1	2	4	6	9	12	15	24	35
	250	—	0	0	1	1	1	3	4	7	9	12	19	28
	300	—	0	0	1	1	1	3	4	6	8	11	17	24
	350	—	0	0	1	1	1	2	3	5	7	9	15	22
	400	—	0	0	1	1	1	1	3	5	7	8	13	20
	500	—	0	0	0	1	1	1	3	4	5	7	11	17
	600	—	0	0	0	1	1	1	1	3	4	6	9	13
	700	—	0	0	0	1	1	1	1	3	4	5	8	12
	750	—	0	0	0	0	1	1	1	3	4	5	7	11
	900	—	0	0	0	0	1	1	1	2	3	4	7	10
	1000	—	0	0	0	0	1	1	1	1	3	4	6	9
THHN, THWN, THWN-2	8	—	—	—	—	—	—	—	—	—	—	—	—	—
	6	—	2	5	8	13	18	30	43	66	88	114	179	258
	4	—	1	3	5	8	11	18	26	41	55	70	110	159
	2	—	1	1	3	6	8	13	19	29	39	50	79	114
	1	—	1	1	2	4	6	10	14	22	29	38	59	86
	1/0	—	1	1	1	4	5	8	12	19	25	32	51	73
	2/0	—	1	1	1	3	4	7	10	15	21	26	42	60
	3/0	—	0	1	1	2	3	6	8	13	17	22	35	51
	4/0	—	0	1	1	1	3	5	7	10	14	18	29	42
	250	—	0	1	1	1	2	4	5	8	11	14	23	33
	300	—	0	0	1	1	1	3	4	7	10	12	20	28
	350	—	0	0	1	1	1	3	4	6	8	11	17	25
	400	—	0	0	1	1	1	2	3	5	7	10	15	22
	500	—	0	0	0	1	1	1	3	5	6	8	13	19
	600	—	0	0	0	1	1	1	2	4	5	6	10	15
	700	—	0	0	0	1	1	1	1	3	4	6	9	13
	750	—	0	0	0	1	1	1	1	3	4	5	9	13
	900	—	0	0	0	0	1	1	1	2	3	4	7	10
	1000	—	0	0	0	0	1	1	1	2	3	4	6	9

(continues)

TABLE C.9A *Continued*

Type	Conductor Size (AWG/ kcmil)	⅜ (12)	½ (16)	¾ (21)	1 (27)	1¼ (35)	1½ (41)	2 (53)	2½ (63)	3 (78)	3½ (91)	4 (103)	5 (129)	6 (155)
							Trade Size (Metric Designator)							
XHHW, XHHW-2	8	—	3	5	9	15	21	34	49	76	101	130	205	296
	6	—	2	4	6	11	15	25	36	56	75	97	152	220
	4	—	1	3	5	8	11	18	26	41	55	70	110	159
	2	—	1	1	3	6	8	13	19	29	39	50	79	114
	1	—	1	1	2	4	6	10	14	22	29	38	59	86
	1/0	—	1	1	1	4	5	8	12	19	25	32	51	73
	2/0	—	1	1	1	3	4	7	10	16	21	27	43	62
	3/0	—	0	1	1	2	3	6	8	13	17	22	35	51
	4/0	—	0	1	1	1	3	5	7	11	14	19	29	42
	250	—	0	1	1	1	2	4	5	8	11	15	23	34
	300	—	0	0	1	1	1	3	5	7	10	13	20	29
	350	—	0	0	1	1	1	3	4	6	9	11	18	25
	400	—	0	0	1	1	1	2	4	6	8	10	16	23
	500	—	0	0	0	1	1	1	3	5	6	8	13	19
	600	—	0	0	0	1	1	1	2	4	5	7	10	15
	700	—	0	0	0	1	1	1	1	3	4	6	9	13
	750	—	0	0	0	1	1	1	1	3	4	5	8	12
	900	—	0	0	0	0	1	1	1	2	3	5	7	11
	1000	—	0	0	0	0	1	1	1	2	3	4	7	10

Definition: *Compact stranding* is the result of a manufacturing process where the stranded conductor is compressed to the extent that the interstices (voids between strand wires) are virtually eliminated.

Δ **TABLE C.10** *Maximum Number of Conductors or Fixture Wires in Rigid PVC Conduit, Schedule 80*
(Based on Chapter 9: Table 1, Table 4, and Table 5)

Type	Conductor Size (AWG/ kcmil)	Trade Size (Metric Designator)												
		3/8 (12)	1/2 (16)	3/4 (21)	1 (27)	1 1/4 (35)	1 1/2 (41)	2 (53)	2 1/2 (63)	3 (78)	3 1/2 (91)	4 (103)	5 (129)	6 (155)
CONDUCTORS														
RHH, RHW, RHW-2	14	—	3	5	9	17	23	39	56	88	118	153	243	349
	12	—	2	4	7	14	19	32	46	73	98	127	202	290
	10	—	1	3	6	11	15	26	37	59	79	103	163	234
	8	—	1	1	3	6	8	13	19	31	41	54	85	122
	6	—	1	1	2	4	6	11	16	24	33	43	68	98
	4	—	1	1	1	3	5	8	12	19	26	33	53	77
	3	—	0	1	1	3	4	7	11	17	23	29	47	67
	2	—	0	1	1	3	4	6	9	14	20	25	41	58
	1	—	0	1	1	1	2	4	6	9	13	17	27	38
	1/0	—	0	0	1	1	1	3	5	8	11	15	23	33
	2/0	—	0	0	1	1	1	3	4	7	10	13	20	29
	3/0	—	0	0	1	1	1	3	4	6	8	11	17	25
	4/0	—	0	0	0	1	1	2	3	5	7	9	15	21
	250	—	0	0	0	1	1	1	2	4	5	7	11	16
	300	—	0	0	0	1	1	1	2	3	5	6	10	14
	350	—	0	0	0	1	1	1	1	3	4	5	9	13
	400	—	0	0	0	0	1	1	1	3	4	5	8	12
	500	—	0	0	0	0	1	1	1	2	3	4	7	10
	600	—	0	0	0	0	0	1	1	1	3	3	6	8
	700	—	0	0	0	0	0	1	1	1	2	3	5	7
	750	—	0	0	0	0	0	1	1	1	2	3	5	7
	800	—	0	0	0	0	0	1	1	1	2	3	4	7
	900	—	0	0	0	0	0	1	1	1	1	2	4	6
	1000	—	0	0	0	0	0	1	1	1	1	2	4	5
	1250	—	0	0	0	0	0	0	1	1	1	1	3	4
	1500	—	0	0	0	0	0	0	1	1	1	1	2	4
	1750	—	0	0	0	0	0	0	0	1	1	1	2	3
	2000	—	0	0	0	0	0	0	0	1	1	1	1	3
TW, THHW, THW, THW-2	14	—	6	11	19	35	49	82	118	185	250	324	514	736
	12	—	4	9	15	27	38	63	91	142	192	248	394	565
	10	—	3	6	11	20	28	47	68	106	143	185	294	421
	8	—	1	3	6	11	15	26	37	59	79	103	163	234
RHH*, RHW*, RHW-2*	14	—	4	8	13	23	32	55	79	123	166	215	341	490
	12	—	3	6	10	19	26	44	63	99	133	173	274	394
	10	—	2	5	8	15	20	34	49	77	104	135	214	307
	8	—	1	3	5	9	12	20	29	46	62	81	128	184

(continues)

Δ **TABLE C.10** *Continued*

Type	Conductor Size (AWG/kcmil)	⅜ (12)	½ (16)	¾ (21)	1 (27)	1¼ (35)	1½ (41)	2 (53)	2½ (63)	3 (78)	3½ (91)	4 (103)	5 (129)	6 (155)
TW, THW, THHW, THW-2, RHH*, RHW*, RHW-2*	6	—	1	1	3	7	9	16	22	35	48	62	98	141
	4	—	1	1	3	5	7	12	17	26	35	46	73	105
	3	—	1	1	2	4	6	10	14	22	30	39	63	90
	2	—	1	1	1	3	5	8	12	19	26	33	53	77
	1	—	0	1	1	2	3	6	8	13	18	23	37	54
	1/0	—	0	1	1	1	3	5	7	11	15	20	32	46
	2/0	—	0	1	1	1	2	4	6	10	13	17	27	39
	3/0	—	0	0	1	1	1	3	5	8	11	14	23	33
	4/0	—	0	0	1	1	1	3	4	7	9	12	19	27
	250	—	0	0	0	1	1	2	3	5	7	9	15	22
	300	—	0	0	0	1	1	1	3	5	6	8	13	19
	350	—	0	0	0	1	1	1	2	4	6	7	12	17
	400	—	0	0	0	1	1	1	2	4	5	7	10	15
	500	—	0	0	0	1	1	1	1	3	4	5	9	13
	600	—	0	0	0	0	1	1	1	2	3	4	7	10
	700	—	0	0	0	0	1	1	1	2	3	4	6	9
	750	—	0	0	0	0	0	1	1	1	3	4	6	8
	800	—	0	0	0	0	0	1	1	1	3	3	6	8
	900	—	0	0	0	0	0	1	1	1	2	3	5	7
	1000	—	0	0	0	0	0	1	1	1	2	3	5	7
	1250	—	0	0	0	0	0	1	1	1	1	2	4	5
	1500	—	0	0	0	0	0	0	1	1	1	1	3	4
	1750	—	0	0	0	0	0	0	1	1	1	1	3	4
	2000	—	0	0	0	0	0	0	0	1	1	1	2	3
THHN, THWN, THWN-2	14	—	9	17	28	51	70	118	170	265	358	464	736	1055
	12	—	6	12	20	37	51	86	124	193	261	338	537	770
	10	—	4	7	13	23	32	54	78	122	164	213	338	485
	8	—	2	4	7	13	18	31	45	70	95	123	195	279
	6	—	1	3	5	9	13	22	32	51	68	89	141	202
	4	—	1	1	3	6	8	14	20	31	42	54	86	124
	3	—	1	1	3	5	7	12	17	26	35	46	73	105
	2	—	1	1	2	4	6	10	14	22	30	39	61	88
	1	—	0	1	1	3	4	7	10	16	22	29	45	65
	1/0	—	0	1	1	2	3	6	9	14	18	24	38	55
	2/0	—	0	1	1	1	3	5	7	11	15	20	32	46
	3/0	—	0	1	1	1	2	4	6	9	13	17	26	38
	4/0	—	0	0	1	1	1	3	5	8	10	14	22	31
	250	—	0	0	1	1	1	3	4	6	8	11	18	25
	300	—	0	0	0	1	1	2	3	5	7	9	15	22
	350	—	0	0	0	1	1	1	3	5	6	8	13	19
	400	—	0	0	0	1	1	1	3	4	6	7	12	17
	500	—	0	0	0	1	1	1	2	3	5	6	10	14
	600	—	0	0	0	0	1	1	1	3	4	5	8	12
	700	—	0	0	0	0	1	1	1	2	3	4	7	10
	750	—	0	0	0	0	1	1	1	2	3	4	7	9
	800	—	0	0	0	0	1	1	1	2	3	4	6	9
	900	—	0	0	0	0	0	1	1	1	3	3	6	8
	1000	—	0	0	0	0	0	1	1	1	2	3	5	7

△ **TABLE C.10** *Continued*

Type	Conductor Size (AWG/ kcmil)	Trade Size (Metric Designator)												
		⅜ (12)	½ (16)	¾ (21)	1 (27)	1¼ (35)	1½ (41)	2 (53)	2½ (63)	3 (78)	3½ (91)	4 (103)	5 (129)	6 (155)
FEP, FEPB, PFA, PFAH, TFE	14	—	8	16	27	49	68	115	164	257	347	450	714	1024
	12	—	6	12	20	36	50	84	120	188	253	328	521	747
	10	—	4	8	14	26	36	60	86	135	182	235	374	536
	8	—	2	5	8	15	20	34	49	77	104	135	214	307
	6	—	1	3	6	10	14	24	35	55	74	96	152	218
	4	—	1	2	4	7	10	17	24	38	52	67	106	153
	3	—	1	1	3	6	8	14	20	32	43	56	89	127
	2	—	1	1	3	5	7	12	17	26	35	46	73	105
PFA, PFAH, TFE	1	—	1	1	1	3	5	8	11	18	25	32	51	73
PFA, PFAH, TFE, Z	1/0	—	0	1	1	3	4	7	10	15	20	27	42	61
	2/0	—	0	1	1	2	3	5	8	12	17	22	35	50
	3/0	—	0	1	1	1	2	4	6	10	14	18	29	41
	4/0	—	0	0	1	1	1	4	5	8	11	15	24	34
Z	14	—	10	19	33	59	82	138	198	310	418	542	860	1233
	12	—	7	14	23	42	58	98	141	220	297	385	610	875
	10	—	4	8	14	26	36	60	86	135	182	235	374	536
	8	—	3	5	9	16	22	38	54	85	115	149	236	339
	6	—	1	4	6	11	16	26	38	60	81	104	166	238
	4	—	1	2	4	8	11	18	26	41	55	72	114	164
	3	—	1	1	3	5	8	13	19	30	40	52	83	119
	2	—	1	1	2	5	6	11	16	25	33	43	69	99
	1	—	1	1	1	4	5	9	13	20	27	35	56	80
XHHW, ZW, XHHW-2, XHH	14	—	6	11	19	35	49	82	118	185	250	324	514	736
	12	—	4	9	15	27	38	63	91	142	192	248	394	565
	10	—	3	6	11	20	28	47	68	106	143	185	294	421
	8	—	1	3	6	11	15	26	37	59	79	103	163	234
	6	—	1	2	4	8	11	19	28	43	59	76	121	173
	4	—	1	1	3	6	8	14	20	31	42	55	87	125
	3	—	1	1	3	5	7	12	17	26	36	47	74	106
	2	—	1	1	2	4	6	10	14	22	30	39	62	89
XHHW, XHHW-2, XHH	1	—	0	1	1	3	4	7	10	16	22	29	46	66
	1/0	—	0	1	1	2	3	6	9	14	19	24	39	56
	2/0	—	0	1	1	1	3	5	7	11	16	20	32	46
	3/0	—	0	1	1	1	2	4	6	9	13	17	27	38
	4/0	—	0	0	1	1	1	3	5	8	11	14	22	32
	250	—	0	0	1	1	1	3	4	6	9	11	18	26
	300	—	0	0	1	1	1	2	3	5	7	10	15	22
	350	—	0	0	0	1	1	1	3	5	6	8	14	20
	400	—	0	0	0	1	1	1	3	4	6	7	12	17
	500	—	0	0	0	1	1	1	2	3	5	6	10	14
	600	—	0	0	0	0	1	1	1	3	4	5	8	11
	700	—	0	0	0	0	1	1	1	2	3	4	7	10
	750	—	0	0	0	0	1	1	1	2	3	4	6	9
	800	—	0	0	0	0	1	1	1	1	3	4	6	9
	900	—	0	0	0	0	0	1	1	1	3	3	5	8
	1000	—	0	0	0	0	0	1	1	1	2	3	5	7
	1250	—	0	0	0	0	0	1	1	1	1	2	4	6
	1500	—	0	0	0	0	0	0	1	1	1	1	3	5
	1750	—	0	0	0	0	0	0	1	1	1	1	3	4
	2000	—	0	0	0	0	0	0	1	1	1	1	2	4

(continues)

△ TABLE C.10 *Continued*

Type	Conductor Size (AWG/ kcmil)	Trade Size (Metric Designator)												
		⅜ (12)	½ (16)	¾ (21)	1 (27)	1¼ (35)	1½ (41)	2 (53)	2½ (63)	3 (78)	3½ (91)	4 (103)	5 (129)	6 (155)
FIXTURE WIRES														
RFH-2, FFH-2, RFHH-2	18	—	6	11	19	34	47	79	113	177	239	310	492	706
	16	—	5	9	16	28	39	67	95	150	202	262	415	595
SF-2, SFF-2	18	—	7	14	24	43	59	100	143	224	302	391	621	890
	16	—	6	11	19	35	49	82	118	185	250	324	514	736
	14	—	5	9	16	28	39	67	95	150	202	262	415	595
SF-1, SFF-1	18	—	13	25	42	76	105	177	253	396	534	692	1098	1575
RFH-1, TF, TFF, XF, XFF	18	—	10	18	31	56	77	130	187	293	395	511	811	1163
	16	—	8	15	25	45	62	105	151	236	319	413	655	939
XF, XFF	14	—	6	11	19	35	49	82	118	185	250	324	514	736
TFN, TFFN	18	—	15	29	50	90	124	209	299	468	632	818	1298	1861
	16	—	12	22	38	68	95	159	229	358	482	625	992	1422
PF, PFF, PGF, PGFF, PAF, PTF, PTFF, PAFF	18	—	15	28	47	85	118	198	284	444	599	776	1231	1765
	16	—	11	22	36	66	91	153	219	343	463	600	952	1365
	14	—	8	16	27	49	68	115	164	257	347	450	714	1024
ZF, ZFF, ZHF	18	—	19	36	61	110	152	255	366	572	772	1000	1587	2275
	16	—	14	27	45	81	112	188	270	422	569	738	1171	1678
	14	—	10	19	33	59	82	138	198	310	418	542	860	1233
KF-2, KFF-2	18	—	29	54	91	165	228	383	549	859	1158	1501	2380	3413
	16	—	20	38	64	115	159	267	383	599	808	1047	1661	2381
	14	—	13	25	43	77	107	179	257	402	543	703	1116	1600
	12	—	9	17	30	53	74	125	179	280	377	489	776	1113
	10	—	6	11	19	35	49	82	118	185	250	324	514	736
KF-1, KFF-1	18	—	33	63	106	190	263	442	633	991	1336	1732	2747	3938
	16	—	23	44	74	133	185	310	445	696	939	1217	1930	2767
	14	—	15	29	50	90	124	209	299	468	632	818	1298	1861
	12	—	10	19	33	59	82	138	198	310	418	542	860	1233
	10	—	7	13	21	39	54	90	129	203	273	354	562	806
XF, XFF	12	—	3	6	10	19	26	44	63	99	133	173	274	394
	10	—	2	5	8	15	20	34	49	77	104	135	214	307

Notes:

1. This table is for concentric stranded conductors only. For compact stranded conductors, Table C.9(A) should be used.

2. Two-hour fire-rated RHH cable has ceramifiable insulation, which has much larger diameters than other RHH wires. Consult manufacturer's conduit fill tables.

*Types RHH, RHW, and RHW-2 without outer covering.

TABLE C.10(A) *Maximum Number of Conductors or Fixture Wires in Rigid PVC Conduit, Schedule 80 (Based on Chapter 9: Table 1, Table 4, and Table 5A)*

Type	Conductor Size (AWG/ kcmil)	⅜ (12)	½ (16)	¾ (21)	1 (27)	1¼ (35)	1½ (41)	2 (53)	2½ (63)	3 (78)	3½ (91)	4 (103)	5 (129)	6 (155)
						COMPACT CONDUCTORS								
THW, THW-2, THHW	8	—	1	3	5	9	13	22	32	50	68	88	140	200
	6	—	1	2	4	7	10	17	25	39	52	68	108	155
	4	—	1	1	3	5	7	13	18	29	39	51	81	116
	2	—	1	1	1	4	5	9	13	21	29	37	60	85
	1	—	0	1	1	3	4	6	9	15	20	26	42	60
	1/0	—	0	1	1	2	3	6	8	13	17	23	36	52
	2/0	—	0	1	1	1	3	5	7	11	15	19	30	44
	3/0	—	0	0	1	1	2	4	6	9	12	16	26	37
	4/0	—	0	0	1	1	1	3	5	8	10	13	22	31
	250	—	0	0	1	1	1	2	4	6	8	11	17	25
	300	—	0	0	0	1	1	2	3	5	7	9	15	21
	350	—	0	0	0	1	1	1	3	5	6	8	13	19
	400	—	0	0	0	1	1	1	3	4	6	7	12	17
	500	—	0	0	0	1	1	1	2	3	5	6	10	14
	600	—	0	0	0	0	1	1	1	3	4	5	8	12
	700	—	0	0	0	0	1	1	1	2	3	4	7	10
	750	—	0	0	0	0	1	1	1	2	3	4	7	10
	900	—	0	0	0	0	0	1	1	1	3	4	6	8
	1000	—	0	0	0	0	0	1	1	1	2	3	5	8
THHN, THWN, THWN-2	8	—	—	—	—	—	—	—	—	—	—	—	—	—
	6	—	1	3	6	11	15	25	36	57	77	99	158	226
	4	—	1	1	3	6	9	15	22	35	47	61	98	140
	2	—	1	1	2	5	6	11	16	25	34	44	70	100
	1	—	1	1	1	3	5	8	12	19	25	33	53	75
	1/0	—	0	1	1	3	4	7	10	16	22	28	45	64
	2/0	—	0	1	1	2	3	6	8	13	18	23	37	53
	3/0	—	0	1	1	1	3	5	7	11	15	19	31	44
	4/0	—	0	0	1	1	2	4	6	9	12	16	25	37
	250	—	0	0	1	1	1	3	4	7	10	12	20	29
	300	—	0	0	1	1	1	3	4	6	8	11	17	25
	350	—	0	0	0	1	1	2	3	5	7	9	15	22
	400	—	0	0	0	1	1	1	3	5	6	8	13	19
	500	—	0	0	0	1	1	1	2	4	5	7	11	16
	600	—	0	0	0	1	1	1	1	3	4	6	9	13
	700	—	0	0	0	0	1	1	1	3	4	5	8	12
	750	—	0	0	0	0	1	1	1	3	4	5	8	11
	900	—	0	0	0	0	1	1	1	1	3	4	6	9
	1000	—	0	0	0	0	0	1	1	1	3	3	5	8

(continues)

TABLE C.(10A) *Continued*

Type	Conductor Size (AWG/ kcmil)	⅜ (12)	½ (16)	¾ (21)	1 (27)	1¼ (35)	1½ (41)	2 (53)	2½ (63)	3 (78)	3½ (91)	4 (103)	5 (129)	6 (155)
XHHW, XHHW-2	8	—	1	4	7	12	17	29	42	65	88	114	181	260
	6	—	1	3	5	9	13	21	31	48	65	85	134	193
	4	—	1	1	3	6	9	15	22	35	47	61	98	140
	2	—	1	1	2	5	6	11	16	25	34	44	70	100
	1	—	1	1	1	3	5	8	12	19	25	33	53	75
	1/0	—	0	1	1	3	4	7	10	16	22	28	45	64
	2/0	—	0	1	1	2	3	6	8	13	18	24	38	54
	3/0	—	0	1	1	1	3	5	7	11	15	19	31	44
	4/0	—	0	0	1	1	2	4	6	9	12	16	26	37
	250	—	0	0	1	1	1	3	5	7	10	13	21	30
	300	—	0	0	1	1	1	3	4	6	8	11	17	25
	350	—	0	0	1	1	1	2	3	5	7	10	15	22
	400	—	0	0	0	1	1	1	3	5	7	9	14	20
	500	—	0	0	0	1	1	1	2	4	5	7	11	17
	600	—	0	0	0	1	1	1	1	3	4	6	9	13
	700	—	0	0	0	0	1	1	1	3	4	5	8	12
	750	—	0	0	0	0	1	1	1	2	3	5	7	11
	900	—	0	0	0	0	1	1	1	2	3	4	6	9
	1000	—	0	0	0	0	0	1	1	1	3	3	6	8

Definition: *Compact stranding* is the result of a manufacturing process where the stranded conductor is compressed to the extent that the interstices (voids between strand wires) are virtually eliminated.

Δ **TABLE C.11** *Maximum Number of Conductors or Fixture Wires in Rigid PVC Conduit, Schedule 40 and HDPE Conduit (Based on Chapter 9: Table 1, Table 4, and Table 5)*

Type	Conductor Size (AWG/ kcmil)	Trade Size (Metric Designator)												
		⅜ (12)	½ (16)	¾ (21)	1 (27)	1¼ (35)	1½ (41)	2 (53)	2½ (63)	3 (78)	3½ (91)	4 (103)	5 (129)	6 (155)
CONDUCTORS														
RHH, RHW, RHW-2	14	—	4	7	11	20	27	45	64	99	133	171	269	390
	12	—	3	5	9	16	22	37	53	82	110	142	224	323
	10	—	2	4	7	13	18	30	43	66	89	115	181	261
	8	—	1	2	4	7	9	15	22	35	46	60	94	137
	6	—	1	1	3	5	7	12	18	28	37	48	76	109
	4	—	1	1	2	4	6	10	14	22	29	37	59	85
	3	—	1	1	1	4	5	8	12	19	25	33	52	75
	2	—	1	1	1	3	4	7	10	16	22	28	45	65
	1	—	0	1	1	1	3	5	7	11	14	19	29	43
	1/0	—	0	1	1	1	2	4	6	9	13	16	26	37
	2/0	—	0	0	1	1	1	3	5	8	11	14	22	32
	3/0	—	0	0	1	1	1	3	4	7	9	12	19	28
	4/0	—	0	0	1	1	1	2	4	6	8	10	16	24
	250	—	0	0	0	1	1	1	3	4	6	8	12	18
	300	—	0	0	0	1	1	1	2	4	5	7	11	16
	350	—	0	0	0	1	1	1	2	3	5	6	10	14
	400	—	0	0	0	1	1	1	1	3	4	6	9	13
	500	—	0	0	0	0	1	1	1	3	4	5	8	11
	600	—	0	0	0	0	1	1	1	2	3	4	6	9
	700	—	0	0	0	0	0	1	1	1	3	3	6	8
	750	—	0	0	0	0	0	1	1	1	2	3	5	8
	800	—	0	0	0	0	0	1	1	1	2	3	5	7
	900	—	0	0	0	0	0	1	1	1	2	3	5	7
	1000	—	0	0	0	0	0	1	1	1	1	3	4	6
	1250	—	0	0	0	0	0	0	1	1	1	1	3	5
	1500	—	0	0	0	0	0	0	1	1	1	1	3	4
	1750	—	0	0	0	0	0	0	1	1	1	1	2	3
	2000	—	0	0	0	0	0	0	0	1	1	1	2	3
TW, THHW, THW, THW-2	14	—	8	14	24	42	57	94	135	209	280	361	568	822
	12	—	6	11	18	32	44	72	103	160	215	277	436	631
	10	—	4	8	13	24	32	54	77	119	160	206	325	470
	8	—	2	4	7	13	18	30	43	66	89	115	181	261
RHH*, RHW*, RHW-2*	14	—	5	9	16	28	38	63	90	139	186	240	378	546
	12	—	4	8	13	22	30	50	72	112	150	193	304	439
	10	—	3	6	10	17	24	39	56	87	117	150	237	343
	8	—	1	3	6	10	14	23	33	52	70	90	142	205

(continues)

Δ **TABLE C.11** *Continued*

Type	Conductor Size (AWG/kcmil)	Trade Size (Metric Designator)												
		⅜ (12)	½ (16)	¾ (21)	1 (27)	1¼ (35)	1½ (41)	2 (53)	2½ (63)	3 (78)	3½ (91)	4 (103)	5 (129)	6 (155)
TW, THW, THHW, THW-2, RHH*, RHW*, RHW-2*	6	—	1	2	4	8	11	18	26	40	53	69	109	157
	4	—	1	1	3	6	8	13	19	30	40	51	81	117
	3	—	1	1	3	5	7	11	16	25	34	44	69	100
	2	—	1	1	2	4	6	10	14	22	29	37	59	85
	1	—	0	1	1	3	4	7	10	15	20	26	41	60
	1/0	—	0	1	1	2	3	6	8	13	17	22	35	51
	2/0	—	0	1	1	1	3	5	7	11	15	19	30	43
	3/0	—	0	1	1	1	2	4	6	9	12	16	25	36
	4/0	—	0	0	1	1	1	3	5	8	10	13	21	30
	250	—	0	0	1	1	1	3	4	6	8	11	17	25
	300	—	0	0	1	1	1	2	3	5	7	9	15	21
	350	—	0	0	0	1	1	1	3	5	6	8	13	19
	400	—	0	0	0	1	1	1	3	4	6	7	12	17
	500	—	0	0	0	1	1	1	2	3	5	6	10	14
	600	—	0	0	0	0	1	1	1	3	4	5	8	11
	700	—	0	0	0	0	1	1	1	2	3	4	7	10
	750	—	0	0	0	0	1	1	1	2	3	4	6	10
	800	—	0	0	0	0	1	1	1	2	3	4	6	9
	900	—	0	0	0	0	0	1	1	1	3	3	6	8
	1000	—	0	0	0	0	0	1	1	1	2	3	5	7
	1250	—	0	0	0	0	0	1	1	1	1	2	4	6
	1500	—	0	0	0	0	0	1	1	1	1	1	3	5
	1750	—	0	0	0	0	0	0	1	1	1	1	3	4
	2000	—	0	0	0	0	0	0	1	1	1	1	3	4
THHN, THWN, THWN-2	14	—	11	21	34	60	82	135	193	299	401	517	815	1178
	12	—	8	15	25	43	59	99	141	218	293	377	594	859
	10	—	5	9	15	27	37	62	89	137	184	238	374	541
	8	—	3	5	9	16	21	36	51	79	106	137	216	312
	6	—	1	4	6	11	15	26	37	57	77	99	156	225
	4	—	1	2	4	7	9	16	22	35	47	61	96	138
	3	—	1	1	3	6	8	13	19	30	40	51	81	117
	2	—	1	1	3	5	7	11	16	25	33	43	68	98
	1	—	1	1	1	3	5	8	12	18	25	32	50	73
	1/0	—	1	1	1	3	4	7	10	15	21	27	42	61
	2/0	—	0	1	1	2	3	6	8	13	17	22	35	51
	3/0	—	0	1	1	1	3	5	7	11	14	18	29	42
	4/0	—	0	1	1	1	2	4	6	9	12	15	24	35
	250	—	0	0	1	1	1	3	4	7	10	12	20	28
	300	—	0	0	1	1	1	3	4	6	8	11	17	24
	350	—	0	0	1	1	1	2	3	5	7	9	15	21
	400	—	0	0	0	1	1	1	3	5	6	8	13	19
	500	—	0	0	0	1	1	1	2	4	5	7	11	16
	600	—	0	0	0	1	1	1	1	3	4	5	9	13
	700	—	0	0	0	0	1	1	1	3	4	5	8	11
	750	—	0	0	0	0	1	1	1	2	3	4	7	11
	800	—	0	0	0	0	1	1	1	2	3	4	7	10
	900	—	0	0	0	0	1	1	1	2	3	4	6	9
	1000	—	0	0	0	0	0	1	1	1	3	3	6	8

Δ **TABLE C.11** *Continued*

Type	Conductor Size (AWG/ kcmil)	3/8 (12)	1/2 (16)	3/4 (21)	1 (27)	1¼ (35)	1½ (41)	2 (53)	2½ (63)	3 (78)	3½ (91)	4 (103)	5 (129)	6 (155)
								Trade Size (Metric Designator)						
FEP, FEPB, PFA, PFAH, TFE	14	—	11	20	33	58	79	131	188	290	389	502	790	1142
	12	—	8	15	24	42	58	96	137	212	284	366	577	834
	10	—	6	10	17	30	41	69	98	152	204	263	414	598
	8	—	3	6	10	17	24	39	56	87	117	150	237	343
	6	—	2	4	7	12	17	28	40	62	83	107	169	244
	4	—	1	3	5	8	12	19	28	43	58	75	118	170
	3	—	1	2	4	7	10	16	23	36	48	62	98	142
	2	—	1	1	3	6	8	13	19	30	40	51	81	117
PFA, PFAH, TFE	1	—	1	1	2	4	5	9	13	20	28	36	56	81
PFA, PFAH, TFE, Z	1/0	—	1	1	1	3	4	8	11	17	23	30	47	68
	2/0	—	0	1	1	3	4	6	9	14	19	24	39	56
	3/0	—	0	1	1	2	3	5	7	12	16	20	32	46
	4/0	—	0	1	1	1	2	4	6	9	13	16	26	38
Z	14	—	13	24	40	70	95	158	226	350	469	605	952	1376
	12	—	9	17	28	49	68	112	160	248	333	429	675	976
	10	—	6	10	17	30	41	69	98	152	204	263	414	598
	8	—	3	6	11	19	26	43	62	96	129	166	261	378
	6	—	2	4	7	13	18	30	43	67	90	116	184	265
	4	—	1	3	5	9	12	21	30	46	62	80	126	183
	3	—	1	2	4	6	9	15	22	34	45	58	92	133
	2	—	1	1	3	5	7	12	18	28	38	49	77	111
	1	—	1	1	2	4	6	10	14	23	30	39	62	90
XHHW, ZW, XHHW-2, XHH	14	—	8	14	24	42	57	94	135	209	280	361	568	822
	12	—	6	11	18	32	44	72	103	160	215	277	436	631
	10	—	4	8	13	24	32	54	77	119	160	206	325	470
	8	—	2	4	7	13	18	30	43	66	89	115	181	261
	6	—	1	3	5	10	13	22	32	49	66	85	134	193
	4	—	1	2	4	7	9	16	23	35	48	61	97	140
	3	—	1	1	3	6	8	13	19	30	40	52	82	118
	2	—	1	1	3	5	7	11	16	25	34	44	69	99
XHHW, XHHW-2, XHH	1	—	1	1	1	3	5	8	12	19	25	32	51	74
	1/0	—	1	1	1	3	4	7	10	16	21	27	43	62
	2/0	—	0	1	1	2	3	6	8	13	17	23	36	52
	3/0	—	0	1	1	1	3	5	7	11	14	19	30	43
	4/0	—	0	1	1	1	2	4	6	9	12	15	24	35
	250	—	0	0	1	1	1	3	5	7	10	13	20	29
	300	—	0	0	1	1	1	3	4	6	8	11	17	25
	350	—	0	0	1	1	1	2	3	5	7	9	15	22
	400	—	0	0	0	1	1	1	3	5	6	8	13	19
	500	—	0	0	0	1	1	1	2	4	5	7	11	16
	600	—	0	0	0	1	1	1	1	3	4	5	9	13
	700	—	0	0	0	0	1	1	1	3	4	5	8	11
	750	—	0	0	0	0	1	1	1	2	3	4	7	11
	800	—	0	0	0	0	1	1	1	2	3	4	7	10
	900	—	0	0	0	0	1	1	1	2	3	4	6	9
	1000	—	0	0	0	0	0	1	1	1	3	3	6	8
	1250	—	0	0	0	0	0	1	1	1	1	3	4	6
	1500	—	0	0	0	0	0	1	1	1	1	2	4	5
	1750	—	0	0	0	0	0	0	1	1	1	1	3	5
	2000	—	0	0	0	0	0	0	1	1	1	1	3	4

(continues)

△ **TABLE C.11** *Continued*

Type	Conductor Size (AWG/ kcmil)	⅜ (12)	½ (16)	¾ (21)	1 (27)	1¼ (35)	1½ (41)	2 (53)	2½ (63)	3 (78)	3½ (91)	4 (103)	5 (129)	6 (155)
							Trade Size (Metric Designator)							
FIXTURE WIRES														
RFH-2, FFH-2,	18	—	8	14	23	40	54	90	129	200	268	346	545	788
RFHH-2	16	—	6	12	19	33	46	76	109	169	226	292	459	664
SF-2, SFF-2	18	—	10	17	29	50	69	114	163	253	338	436	687	993
	16	—	8	14	24	42	57	94	135	209	280	361	568	822
	14	—	6	12	19	33	46	76	109	169	226	292	459	664
SF-1, SFF-1	18	—	17	31	51	89	122	202	289	447	599	772	1216	1758
RFH-1, TF, TFF,	18	—	13	23	38	66	90	149	213	330	442	570	898	1298
XF, XFF	16	—	10	18	30	53	73	120	172	266	357	460	725	1048
XF, XFF	14	—	8	14	24	42	57	94	135	209	280	361	568	822
TFN, TFFN	18	—	20	37	60	105	144	239	341	528	708	913	1437	2077
	16	—	16	28	46	80	110	183	261	403	541	697	1098	1587
PF, PFF, PGF,	18	—	19	35	57	100	137	227	323	501	671	865	1363	1970
PGFF, PAF, PTF,	16	—	15	27	44	77	106	175	250	387	519	669	1054	1523
PTFF, PAFF	14	—	11	20	33	58	79	131	188	290	389	502	790	1142
ZF, ZFF, ZHF	18	—	25	45	74	129	176	292	417	646	865	1116	1756	2539
	16	—	18	33	54	95	130	216	308	476	638	823	1296	1873
	14	—	13	24	40	70	95	158	226	350	469	605	952	1376
KF-2, KFF-2	18	—	38	67	111	193	265	439	626	969	1298	1674	2634	3809
	16	—	26	47	77	135	184	306	436	676	905	1168	1838	2657
	14	—	18	31	52	91	124	205	293	454	608	784	1235	1785
	12	—	12	22	36	63	86	143	204	316	423	546	859	1242
	10	—	8	14	24	42	57	94	135	209	280	361	568	822
KF-1, KFF-1	18	—	44	78	128	223	305	506	722	1118	1498	1931	3040	4395
	16	—	31	55	90	157	214	355	507	785	1052	1357	2136	3088
	14	—	20	37	60	105	144	239	341	528	708	913	1437	2077
	12	—	13	24	40	70	95	158	226	350	469	605	952	1376
	10	—	9	16	26	45	62	103	148	229	306	395	622	899
XF, XFF	12	—	4	8	13	22	30	50	72	112	150	193	304	439
	10	—	3	6	10	17	24	39	56	87	117	150	237	343

Notes:

1. This table is for concentric stranded conductors only. For compact stranded conductors, Table C.10(A) should be used.

2. Two-hour fire-rated RHH cable has ceramifiable insulation, which has much larger diameters than other RHH wires. Consult manufacturer's conduit fill tables.

*Types RHH, RHW, and RHW-2 without outer covering.

TABLE C.11(A) *Maximum Number of Conductors or Fixture Wires in Rigid PVC Conduit, Schedule 40 and HDPE Conduit (Based on Chapter 9: Table 1, Table 4, and Table 5A)*

Type	Conductor Size (AWG/ kcmil)	Trade Size (Metric Designator)												
		⅜ (12)	½ (16)	¾ (21)	1 (27)	1¼ (35)	1½ (41)	2 (53)	2½ (63)	3 (78)	3½ (91)	4 (103)	5 (129)	6 (155)
COMPACT CONDUCTORS														
THW, THW-2, THHW	8	—	1	4	6	11	15	26	37	57	76	98	155	224
	6	—	1	3	5	9	12	20	28	44	59	76	119	173
	4	—	1	1	3	6	9	15	21	33	44	57	89	129
	2	—	1	1	2	5	6	11	15	24	32	42	66	95
	1	—	1	1	1	3	4	7	11	17	23	29	46	67
	1/0	—	0	1	1	3	4	6	9	15	20	25	40	58
	2/0	—	0	1	1	2	3	5	8	12	16	21	34	49
	3/0	—	0	1	1	1	3	5	7	10	14	18	29	42
	4/0	—	0	1	1	1	2	4	5	9	12	15	24	35
	250	—	0	0	1	1	1	3	4	7	9	12	19	27
	300	—	0	0	1	1	1	2	4	6	8	10	16	24
	350	—	0	0	1	1	1	2	3	5	7	9	15	21
	400	—	0	0	0	1	1	1	3	5	6	8	13	19
	500	—	0	0	0	1	1	1	2	4	5	7	11	16
	600	—	0	0	0	1	1	1	1	3	4	5	9	13
	700	—	0	0	0	0	1	1	1	3	4	5	8	12
	750	—	0	0	0	0	1	1	1	2	3	5	7	11
	900	—	0	0	0	0	1	1	1	2	3	4	6	9
	1000	—	0	0	0	0	1	1	1	1	3	4	6	9
THHN, THWN, THWN-2	8	—	—	—	—	—	—	—	—	—	—	—	—	—
	6	—	2	4	7	13	17	29	41	64	86	111	175	253
	4	—	1	2	4	8	11	18	25	40	53	68	108	156
	2	—	1	1	3	5	8	13	18	28	38	49	77	112
	1	—	1	1	2	4	6	9	14	21	29	37	58	84
	1/0	—	1	1	1	3	5	8	12	18	24	31	49	72
	2/0	—	0	1	1	3	4	7	9	15	20	26	41	59
	3/0	—	0	1	1	2	3	5	8	12	17	22	34	50
	4/0	—	0	1	1	1	3	4	6	10	14	18	28	41
	250	—	0	0	1	1	1	3	5	8	11	14	22	32
	300	—	0	0	1	1	1	3	4	7	9	12	19	28
	350	—	0	0	1	1	1	3	4	6	8	10	17	24
	400	—	0	0	1	1	1	2	3	5	7	9	15	22
	500	—	0	0	0	1	1	1	3	4	6	8	13	18
	600	—	0	0	0	1	1	1	2	4	5	6	10	15
	700	—	0	0	0	1	1	1	1	3	4	5	9	13
	750	—	0	0	0	1	1	1	1	3	4	5	8	12
	900	—	0	0	0	0	1	1	1	2	3	4	7	10
	1000	—	0	0	0	0	1	1	1	2	3	4	6	9

(continues)

TABLE C.11(A) *Continued*

Type	Conductor Size (AWG/ kcmil)	⅜ (12)	½ (16)	¾ (21)	1 (27)	1¼ (35)	1½ (41)	2 (53)	2½ (63)	3 (78)	3½ (91)	4 (103)	5 (129)	6 (155)
						Trade Size (Metric Designator)								
XHHW, XHHW-2	8	—	3	5	8	14	20	33	47	73	99	127	200	290
	6	—	1	4	6	11	15	25	35	55	73	94	149	215
	4	—	1	2	4	8	11	18	25	40	53	68	108	156
	2	—	1	1	3	5	8	13	18	28	38	49	77	112
	1	—	1	1	2	4	6	9	14	21	29	37	58	84
	1/0	—	1	1	1	3	5	8	12	18	24	31	49	72
	2/0	—	1	1	1	3	4	7	10	15	20	26	42	60
	3/0	—	0	1	1	2	3	5	8	12	17	22	34	50
	4/0	—	0	1	1	1	3	5	7	10	14	18	29	42
	250	—	0	0	1	1	1	4	5	8	11	14	23	33
	300	—	0	0	1	1	1	3	4	7	9	12	19	28
	350	—	0	0	1	1	1	3	4	6	8	11	17	25
	400	—	0	0	1	1	1	2	3	5	7	10	15	22
	500	—	0	0	0	1	1	1	3	4	6	8	13	18
	600	—	0	0	0	1	1	1	2	4	5	6	10	15
	700	—	0	0	0	1	1	1	1	3	4	5	9	13
	750	—	0	0	0	1	1	1	1	3	4	5	8	12
	900	—	0	0	0	0	1	1	1	2	3	4	7	10
	1000	—	0	0	0	0	1	1	1	2	3	4	6	9

Definition: *Compact stranding* is the result of a manufacturing process where the stranded conductor is compressed to the extent that the interstices (voids between strand wires) are virtually eliminated.

Δ **TABLE C.12** *Maximum Number of Conductors or Fixture Wires in Type A, Rigid PVC Conduit (Based on Chapter 9: Table 1, Table 4, and Table 5)*

Type	Conductor Size (AWG/ kcmil)	⅜ (12)	½ (16)	¾ (21)	1 (27)	1¼ (35)	1½ (41)	2 (53)	2½ (63)	3 (78)	3½ (91)	4 (103)	5 (129)	6 (155)

CONDUCTORS

Type	Conductor Size	(12)	(16)	(21)	(27)	(35)	(41)	(53)	(63)	(78)	(91)	(103)	(129)	(155)
RHH, RHW, RHW-2	14	—	5	9	14	24	31	49	74	112	146	187	—	—
	12	—	4	7	12	20	26	41	61	93	121	155	—	—
	10	—	3	6	10	16	21	33	50	75	98	125	—	—
	8	—	1	3	5	8	11	17	26	39	51	65	—	—
	6	—	1	2	4	6	9	14	21	31	41	52	—	—
	4	—	1	1	3	5	7	11	16	24	32	41	—	—
	3	—	1	1	3	4	6	9	14	21	28	36	—	—
	2	—	1	1	2	4	5	8	12	18	24	31	—	—
	1	—	0	1	1	2	3	5	8	12	16	20	—	—
	1/0	—	0	1	1	2	3	5	7	10	14	18	—	—
	2/0	—	0	1	1	1	2	4	6	9	12	15	—	—
	3/0	—	0	1	1	1	1	3	5	8	10	13	—	—
	4/0	—	0	0	1	1	1	3	4	7	9	11	—	—
	250	—	0	0	1	1	1	1	3	5	6	8	—	—
	300	—	0	0	1	1	1	1	3	4	6	7	—	—
	350	—	0	0	0	1	1	1	2	4	5	7	—	—
	400	—	0	0	0	1	1	1	2	3	5	6	—	—
	500	—	0	0	0	1	1	1	1	3	4	5	—	—
	600	—	0	0	0	0	1	1	1	2	3	4	—	—
	700	—	0	0	0	0	1	1	1	2	3	4	—	—
	750	—	0	0	0	0	1	1	1	1	3	4	—	—
	800	—	0	0	0	0	1	1	1	1	3	3	—	—
	900	—	0	0	0	0	0	1	1	1	2	3	—	—
	1000	—	0	0	0	0	0	1	1	1	2	3	—	—
	1250	—	0	0	0	0	0	1	1	1	1	2	—	—
	1500	—	0	0	0	0	0	0	1	1	1	1	—	—
	1750	—	0	0	0	0	0	0	1	1	1	1	—	—
	2000	—	0	0	0	0	0	0	1	1	1	1	—	—
TW, THHW, THW, THW-2	14	—	11	18	31	51	67	105	157	235	307	395	—	—
	12	—	8	14	24	39	51	80	120	181	236	303	—	—
	10	—	6	10	18	29	38	60	89	135	176	226	—	—
	8	—	3	6	10	16	21	33	50	75	98	125	—	—
RHH*, RHW*, RHW-2*	14	—	7	12	20	34	44	69	104	157	204	262	—	—
	12	—	6	10	16	27	35	56	84	126	164	211	—	—
	10	—	4	8	13	21	28	44	65	98	128	165	—	—
	8	—	2	4	7	12	16	26	39	59	77	98	—	—

(continues)

Δ **TABLE C.12** *Continued*

Type	Conductor Size (AWG/ kcmil)	Trade Size (Metric Designator)												
		⅜ (12)	½ (16)	¾ (21)	1 (27)	1¼ (35)	1½ (41)	2 (53)	2½ (63)	3 (78)	3½ (91)	4 (103)	5 (129)	6 (155)
TW, THW, THHW, THW-2, RHH*, RHW*, RHW-2*	6	—	1	3	6	9	13	20	30	45	59	75	—	—
	4	—	1	2	4	7	9	15	22	33	44	56	—	—
	3	—	1	1	4	6	8	13	19	29	37	48	—	—
	2	—	1	1	3	5	7	11	16	24	32	41	—	—
	1	—	1	1	1	3	5	7	11	17	22	29	—	—
	1/0	—	1	1	1	3	4	6	10	14	19	24	—	—
	2/0	—	0	1	1	2	3	5	8	12	16	21	—	—
	3/0	—	0	1	1	1	3	4	7	10	13	17	—	—
	4/0	—	0	1	1	1	2	4	6	9	11	14	—	—
	250	—	0	0	1	1	1	3	4	7	9	12	—	—
	300	—	0	0	1	1	1	2	4	6	8	10	—	—
	350	—	0	0	1	1	1	2	3	5	7	9	—	—
	400	—	0	0	1	1	1	1	3	5	6	8	—	—
	500	—	0	0	0	1	1	1	2	4	5	7	—	—
	600	—	0	0	0	1	1	1	1	3	4	5	—	—
	700	—	0	0	0	1	1	1	1	3	4	5	—	—
	750	—	0	0	0	1	1	1	1	3	3	4	—	—
	800	—	0	0	0	0	1	1	1	2	3	4	—	—
	900	—	0	0	0	0	1	1	1	2	3	4	—	—
	1000	—	0	0	0	0	1	1	1	1	3	3	—	—
	1250	—	0	0	0	0	0	1	1	1	1	3	—	—
	1500	—	0	0	0	0	0	1	1	1	1	2	—	—
	1750	—	0	0	0	0	0	0	1	1	1	1	—	—
	2000	—	0	0	0	0	0	0	1	1	1	1	—	—
THHN, THWN, THWN-2	14	—	16	27	44	73	96	150	225	338	441	566	—	—
	12	—	11	19	32	53	70	109	164	246	321	412	—	—
	10	—	7	12	20	33	44	69	103	155	202	260	—	—
	8	—	4	7	12	19	25	40	59	89	117	150	—	—
	6	—	3	5	8	14	18	28	43	64	84	108	—	—
	4	—	1	3	5	8	11	17	26	39	52	66	—	—
	3	—	1	2	4	7	9	15	22	33	44	56	—	—
	2	—	1	1	3	6	8	12	19	28	37	47	—	—
	1	—	1	1	2	4	6	9	14	21	27	35	—	—
	1/0	—	1	1	2	4	5	8	11	17	23	29	—	—
	2/0	—	1	1	1	3	4	6	10	14	19	24	—	—
	3/0	—	0	1	1	2	3	5	8	12	16	20	—	—
	4/0	—	0	1	1	1	3	4	6	10	13	17	—	—
	250	—	0	1	1	1	2	3	5	8	10	14	—	—
	300	—	0	0	1	1	1	3	4	7	9	12	—	—
	350	—	0	0	1	1	1	2	4	6	8	10	—	—
	400	—	0	0	1	1	1	2	3	5	7	9	—	—
	500	—	0	0	1	1	1	1	3	4	6	7	—	—
	600	—	0	0	0	1	1	1	2	3	5	6	—	—
	700	—	0	0	0	1	1	1	1	3	4	5	—	—
	750	—	0	0	0	1	1	1	1	3	4	5	—	—
	800	—	0	0	0	1	1	1	1	3	4	5	—	—
	900	—	0	0	0	0	1	1	1	2	3	4	—	—
	1000	—	0	0	0	0	1	1	1	2	3	4	—	—

Δ **TABLE C.12** *Continued*

Type	Conductor Size (AWG/kcmil)	3/8 (12)	1/2 (16)	3/4 (21)	1 (27)	1¼ (35)	1½ (41)	2 (53)	2½ (63)	3 (78)	3½ (91)	4 (103)	5 (129)	6 (155)
FEP, FEPB, PFA, PFAH, TFE	14	—	15	26	43	70	93	146	218	327	427	549	—	—
	12	—	11	19	31	51	68	106	159	239	312	400	—	—
	10	—	8	13	22	37	48	76	114	171	224	287	—	—
	8	—	4	8	13	21	28	44	65	98	128	165	—	—
	6	—	3	5	9	15	20	31	46	70	91	117	—	—
	4	—	1	4	6	10	14	21	32	49	64	82	—	—
	3	—	1	3	5	8	11	18	27	40	53	68	—	—
	2	—	1	2	4	7	9	15	22	33	44	56	—	—
PFA, PFAH, TFE	1	—	1	1	3	5	6	10	15	23	30	39	—	—
PFA, PFAH, TFE, Z	1/0	—	1	1	2	4	5	8	13	19	25	32	—	—
	2/0	—	1	1	1	3	4	7	10	16	21	27	—	—
	3/0	—	1	1	1	3	3	6	9	13	17	22	—	—
	4/0	—	0	1	1	2	3	5	7	11	14	18	—	—
Z	14	—	18	31	52	85	112	175	262	395	515	661	—	—
	12	—	13	22	37	60	79	124	186	280	365	469	—	—
	10	—	8	13	22	37	48	76	114	171	224	287	—	—
	8	—	5	8	14	23	30	48	72	108	141	181	—	—
	6	—	3	6	10	16	21	34	50	76	99	127	—	—
	4	—	2	4	7	11	15	23	35	52	68	88	—	—
	3	—	1	3	5	8	11	17	25	38	50	64	—	—
	2	—	1	2	4	7	9	14	21	32	41	53	—	—
	1	—	1	1	3	5	7	11	17	26	33	43	—	—
XHHW, ZW, XHHW-2, XHH	14	—	11	18	31	51	67	105	157	235	307	395	—	—
	12	—	8	14	24	39	51	80	120	181	236	303	—	—
	10	—	6	10	18	29	38	60	89	135	176	226	—	—
	8	—	3	6	10	16	21	33	50	75	98	125	—	—
	6	—	2	4	7	12	15	24	37	55	72	93	—	—
	4	—	1	3	5	8	11	18	26	40	52	67	—	—
	3	—	1	2	4	7	9	15	22	34	44	57	—	—
	2	—	1	1	3	6	8	12	19	28	37	48	—	—
XHHW, XHHW-2, XHH	1	—	1	1	3	4	6	9	14	21	28	35	—	—
	1/0	—	1	1	2	4	5	8	12	18	23	30	—	—
	2/0	—	1	1	1	3	4	6	10	15	19	25	—	—
	3/0	—	0	1	1	2	3	5	8	12	16	20	—	—
	4/0	—	0	1	1	1	3	4	7	10	13	17	—	—
	250	—	0	1	1	1	2	3	5	8	11	14	—	—
	300	—	0	0	1	1	1	3	5	7	9	12	—	—
	350	—	0	0	1	1	1	3	4	6	8	10	—	—
	400	—	0	0	1	1	1	2	3	5	7	9	—	—
	500	—	0	0	1	1	1	1	3	4	6	8	—	—
	600	—	0	0	0	1	1	1	2	3	5	6	—	—
	700	—	0	0	0	1	1	1	1	3	4	5	—	—
	750	—	0	0	0	1	1	1	1	3	4	5	—	—
	800	—	0	0	0	1	1	1	1	3	4	5	—	—
	900	—	0	0	0	0	1	1	1	2	3	4	—	—
	1000	—	0	0	0	0	1	1	1	2	3	4	—	—
	1250	—	0	0	0	0	0	1	1	1	2	3	—	—
	1500	—	0	0	0	0	0	1	1	1	1	2	—	—
	1750	—	0	0	0	0	0	1	1	1	1	2	—	—
	2000	—	0	0	0	0	0	0	1	1	1	1	—	—

(continues)

Δ   **TABLE C.12** *Continued*

Type	Conductor Size (AWG/ kcmil)	⅜ (12)	½ (16)	¾ (21)	1 (27)	1¼ (35)	1½ (41)	2 (53)	2½ (63)	3 (78)	3½ (91)	4 (103)	5 (129)	6 (155)
						Trade Size (Metric Designator)								

FIXTURE WIRES

Type	Conductor Size (AWG/ kcmil)	⅜ (12)	½ (16)	¾ (21)	1 (27)	1¼ (35)	1½ (41)	2 (53)	2½ (63)	3 (78)	3½ (91)	4 (103)	5 (129)	6 (155)
RFH-2, FFH-2, RFHH-2	18	—	10	18	30	48	64	100	150	226	295	378	—	—
	16	—	9	15	25	41	54	85	127	190	248	319	—	—
SF-2, SFF-2	18	—	13	22	37	61	81	127	189	285	372	477	—	—
	16	—	11	18	31	51	67	105	157	235	307	395	—	—
	14	—	9	15	25	41	54	85	127	190	248	319	—	—
SF-1, SFF-1	18	—	23	40	66	108	143	224	335	504	658	844	—	—
RFH-1, TF, TFF, XF, XFF	18	—	17	29	49	80	105	165	248	372	486	623	—	—
	16	—	14	24	39	65	85	134	200	300	392	503	—	—
XF, XFF	14	—	11	18	31	51	67	105	157	235	307	395	—	—
TFN, TFFN	18	—	28	47	79	128	169	265	396	596	777	998	—	—
	16	—	21	36	60	98	129	202	303	455	594	762	—	—
PF, PFF, PGF, PGFF, PAF, PTF, PTFF, PAFF	18	—	26	45	74	122	160	251	376	565	737	946	—	—
	16	—	20	34	58	94	124	194	291	437	570	732	—	—
	14	—	15	26	43	70	93	146	218	327	427	549	—	—
ZF, ZFF, ZHF	18	—	34	57	96	157	206	324	484	728	950	1220	—	—
	16	—	25	42	71	116	152	239	357	537	701	900	—	—
	14	—	18	31	52	85	112	175	262	395	515	661	—	—
KF-2, KFF-2	18	—	51	86	144	235	310	486	727	1092	1426	1829	—	—
	16	—	36	60	101	164	216	339	507	762	994	1276	—	—
	14	—	24	40	67	110	145	228	341	512	668	857	—	—
	12	—	16	28	47	77	101	158	237	356	465	596	—	—
	10	—	11	18	31	51	67	105	157	235	307	395	—	—
KF-1, KFF-1	18	—	59	100	166	272	357	561	839	1260	1645	2111	—	—
	16	—	41	70	117	191	251	394	589	886	1156	1483	—	—
	14	—	28	47	79	128	169	265	396	596	777	998	—	—
	12	—	18	31	52	85	112	175	262	395	515	661	—	—
	10	—	12	20	34	55	73	115	171	258	337	432	—	—
XF, XFF	12	—	6	10	16	27	35	56	84	126	164	211	—	—
	10	—	4	8	13	21	28	44	65	98	128	165	—	—

Notes:

1. This table is for concentric stranded conductors only. For compact stranded conductors, Table C.11(A) should be used.

2. Two-hour fire-rated RHH cable has ceramifiable insulation, which has much larger diameters than other RHH wires. Consult manufacturer's conduit fill tables.

*Types RHH, RHW, and RHW-2 without outer covering.

TABLE C.12(A) *Maximum Number of Conductors or Fixture Wires in Type A, Rigid PVC Conduit (Based on Chapter 9: Table 1, Table 4, and Table 5A)*

Type	Conductor Size (AWG/kcmil)	Trade Size (Metric Designator)												
		⅜ (12)	½ (16)	¾ (21)	1 (27)	1¼ (35)	1½ (41)	2 (53)	2½ (63)	3 (78)	3½ (91)	4 (103)	5 (129)	6 (155)
COMPACT CONDUCTORS														
THW, THW-2, THHW	8	—	3	5	8	14	18	28	42	64	84	107	—	—
	6	—	2	4	6	10	14	22	33	49	65	83	—	—
	4	—	1	3	5	8	10	16	24	37	48	62	—	—
	2	—	1	1	3	6	7	12	18	27	36	46	—	—
	1	—	1	1	2	4	5	8	13	19	25	32	—	—
	1/0	—	1	1	1	3	4	7	11	16	21	28	—	—
	2/0	—	1	1	1	3	4	6	9	14	18	23	—	—
	3/0	—	0	1	1	2	3	5	8	12	15	20	—	—
	4/0	—	0	1	1	1	3	4	6	10	13	17	—	—
	250	—	0	1	1	1	1	3	5	8	10	13	—	—
	300	—	0	0	1	1	1	3	4	7	9	11	—	—
	350	—	0	0	1	1	1	2	4	6	8	10	—	—
	400	—	0	0	1	1	1	2	3	5	7	9	—	—
	500	—	0	0	1	1	1	1	3	4	6	8	—	—
	600	—	0	0	0	1	1	1	2	3	5	6	—	—
	700	—	0	0	0	1	1	1	1	3	4	5	—	—
	750	—	0	0	0	1	1	1	1	3	4	5	—	—
	900	—	0	0	0	0	1	1	1	2	3	4	—	—
	1000	—	0	0	0	0	1	1	1	2	3	4	—	—
THHN, THWN, THWN-2	8	—	—	—	—	—	—	—	—	—	—	—	—	—
	6	—	3	5	9	15	20	32	48	72	94	121	—	—
	4	—	1	3	6	9	12	20	30	45	58	75	—	—
	2	—	1	2	4	7	9	14	21	32	42	54	—	—
	1	—	1	1	3	5	7	10	16	24	31	40	—	—
	1/0	—	1	1	2	4	6	9	13	20	27	34	—	—
	2/0	—	1	1	1	3	5	7	11	17	22	28	—	—
	3/0	—	1	1	1	3	4	6	9	14	18	24	—	—
	4/0	—	0	1	1	2	3	5	8	11	15	19	—	—
	250	—	0	1	1	1	2	4	6	9	12	15	—	—
	300	—	0	1	1	1	1	3	5	8	10	13	—	—
	350	—	0	0	1	1	1	3	4	7	9	11	—	—
	400	—	0	0	1	1	1	2	4	6	8	10	—	—
	500	—	0	0	1	1	1	2	3	5	7	9	—	—
	600	—	0	0	0	1	1	1	3	4	5	7	—	—
	700	—	0	0	0	1	1	1	2	3	5	6	—	—
	750	—	0	0	0	1	1	1	2	3	4	6	—	—
	900	—	0	0	0	1	1	1	1	3	4	5	—	—
	1000	—	0	0	0	0	1	1	1	2	3	4	—	—

(continues)

TABLE C.12(A) *Continued*

Type	Conductor Size (AWG/ kcmil)	⅜ (12)	½ (16)	¾ (21)	1 (27)	1¼ (35)	1½ (41)	2 (53)	2½ (63)	3 (78)	3½ (91)	4 (103)	5 (129)	6 (155)
XHHW, XHHW-2	8	—	4	6	11	18	23	37	55	83	108	139	—	—
	6	—	3	5	8	13	17	27	41	62	80	103	—	—
	4	—	1	3	6	9	12	20	30	45	58	75	—	—
	2	—	1	2	4	7	9	14	21	32	42	54	—	—
	1	—	1	1	3	5	7	10	16	24	31	40	—	—
	1/0	—	1	1	2	4	6	9	13	20	27	34	—	—
	2/0	—	1	1	1	3	5	7	11	17	22	29	—	—
	3/0	—	1	1	1	3	4	6	9	14	18	24	—	—
	4/0	—	0	1	1	2	3	5	8	12	15	20	—	—
	250	—	0	1	1	1	2	4	6	9	12	16	—	—
	300	—	0	1	1	1	1	3	5	8	10	13	—	—
	350	—	0	0	1	1	1	3	5	7	9	12	—	—
	400	—	0	0	1	1	1	3	4	6	8	11	—	—
	500	—	0	0	1	1	1	2	3	5	7	9	—	—
	600	—	0	0	0	1	1	1	3	4	5	7	—	—
	700	—	0	0	0	1	1	1	2	3	5	6	—	—
	750	—	0	0	0	1	1	1	2	3	4	6	—	—
	900	—	0	0	0	1	1	1	1	3	4	5	—	—
	1000	—	0	0	0	0	1	1	1	2	3	4	—	—

Definition: *Compact stranding* is the result of a manufacturing process where the stranded conductor is compressed to the extent that the interstices (voids between strand wires) are virtually eliminated.

Δ **TABLE C.13** *Maximum Number of Conductors or Fixture Wires in Type EB, PVC Conduit (Based on Chapter 9: Table 1, Table 4, and Table 5)*

Type	Conductor Size (AWG/kcmil)	Trade Size (Metric Designator)												
		⅜ (12)	½ (16)	¾ (21)	1 (27)	1¼ (35)	1½ (41)	2 (53)	2½ (63)	3 (78)	3½ (91)	4 (103)	5 (129)	6 (155)
						CONDUCTORS								
RHH, RHW, RHW-2	14	—	—	—	—	—	—	53	—	119	155	197	303	430
	12	—	—	—	—	—	—	44	—	98	128	163	251	357
	10	—	—	—	—	—	—	35	—	79	104	132	203	288
	8	—	—	—	—	—	—	18	—	41	54	69	106	151
	6	—	—	—	—	—	—	15	—	33	43	55	85	121
	4	—	—	—	—	—	—	11	—	26	34	43	66	94
	3	—	—	—	—	—	—	10	—	23	30	38	58	83
	2	—	—	—	—	—	—	9	—	20	26	33	50	72
	1	—	—	—	—	—	—	6	—	13	17	21	33	47
	1/0	—	—	—	—	—	—	5	—	11	15	19	29	41
	2/0	—	—	—	—	—	—	4	—	10	13	16	25	36
	3/0	—	—	—	—	—	—	4	—	8	11	14	22	31
	4/0	—	—	—	—	—	—	3	—	7	9	12	18	26
	250	—	—	—	—	—	—	2	—	5	7	9	14	20
	300	—	—	—	—	—	—	1	—	5	6	8	12	17
	350	—	—	—	—	—	—	1	—	4	5	7	11	16
	400	—	—	—	—	—	—	1	—	4	5	6	10	14
	500	—	—	—	—	—	—	1	—	3	4	5	9	12
	600	—	—	—	—	—	—	1	—	3	3	4	7	10
	700	—	—	—	—	—	—	1	—	2	3	4	6	9
	750	—	—	—	—	—	—	1	—	2	3	4	6	9
	800	—	—	—	—	—	—	1	—	2	3	4	6	8
	900	—	—	—	—	—	—	1	—	1	2	3	5	7
	1000	—	—	—	—	—	—	1	—	1	2	3	5	7
	1250	—	—	—	—	—	—	1	—	1	1	2	3	5
	1500	—	—	—	—	—	—	0	—	1	1	1	3	4
	1750	—	—	—	—	—	—	0	—	1	1	1	3	4
	2000	—	—	—	—	—	—	0	—	1	1	1	2	3
TW, THHW, THW, THW-2	14	—	—	—	—	—	—	111	—	250	327	415	638	907
	12	—	—	—	—	—	—	85	—	192	251	319	490	696
	10	—	—	—	—	—	—	63	—	143	187	238	365	519
	8	—	—	—	—	—	—	35	—	79	104	132	203	288
RHH*, RHW*, RHW-2*	14	—	—	—	—	—	—	74	—	166	217	276	424	603
	12	—	—	—	—	—	—	59	—	134	175	222	341	485
	10	—	—	—	—	—	—	46	—	104	136	173	266	378
	8	—	—	—	—	—	—	28	—	62	81	104	159	227
TW, THW, THHW, THW-2, RHH*, RHW*, RHW-2*	6	—	—	—	—	—	—	21	—	48	62	79	122	173
	4	—	—	—	—	—	—	16	—	36	46	59	91	129
	3	—	—	—	—	—	—	13	—	30	40	51	78	111
	2	—	—	—	—	—	—	11	—	26	34	43	66	94
	1	—	—	—	—	—	—	8	—	18	24	30	46	66

(continues)

Δ **TABLE C.13** *Continued*

Type	Conductor Size (AWG/ kcmil)	⅜ (12)	½ (16)	¾ (21)	1 (27)	1¼ (35)	1½ (41)	2 (53)	2½ (63)	3 (78)	3½ (91)	4 (103)	5 (129)	6 (155)
	1/0	—	—	—	—	—	—	7	—	15	20	26	40	56
	2/0	—	—	—	—	—	—	6	—	13	17	22	34	48
	3/0	—	—	—	—	—	—	5	—	11	14	18	28	40
	4/0	—	—	—	—	—	—	4	—	9	12	15	24	34
	250	—	—	—	—	—	—	3	—	7	10	12	19	27
	300	—	—	—	—	—	—	3	—	6	8	11	17	24
	350	—	—	—	—	—	—	2	—	6	7	9	15	21
	400	—	—	—	—	—	—	2	—	5	7	8	13	19
	500	—	—	—	—	—	—	1	—	4	5	7	11	16
	600	—	—	—	—	—	—	1	—	3	4	6	9	13
	700	—	—	—	—	—	—	1	—	3	4	5	8	11
	750	—	—	—	—	—	—	1	—	3	4	5	7	11
	800	—	—	—	—	—	—	1	—	3	3	4	7	10
	900	—	—	—	—	—	—	1	—	2	3	4	6	9
	1000	—	—	—	—	—	—	1	—	2	3	4	6	8
	1250	—	—	—	—	—	—	1	—	1	2	3	4	6
	1500	—	—	—	—	—	—	1	—	1	1	2	4	6
	1750	—	—	—	—	—	—	1	—	1	1	2	3	5
	2000	—	—	—	—	—	—	0	—	1	1	1	3	4
THHN, THWN, THWN-2	14	—	—	—	—	—	—	159	—	359	468	595	915	1300
	12	—	—	—	—	—	—	116	—	262	342	434	667	948
	10	—	—	—	—	—	—	73	—	165	215	274	420	597
	8	—	—	—	—	—	—	42	—	95	124	158	242	344
	6	—	—	—	—	—	—	30	—	68	89	114	175	248
	4	—	—	—	—	—	—	19	—	42	55	70	107	153
	3	—	—	—	—	—	—	16	—	36	46	59	91	129
	2	—	—	—	—	—	—	13	—	30	39	50	76	109
	1	—	—	—	—	—	—	10	—	22	29	37	57	80
	1/0	—	—	—	—	—	—	8	—	18	24	31	48	68
	2/0	—	—	—	—	—	—	7	—	15	20	26	40	56
	3/0	—	—	—	—	—	—	5	—	13	17	21	33	47
	4/0	—	—	—	—	—	—	4	—	10	14	18	27	39
	250	—	—	—	—	—	—	4	—	8	11	14	22	31
	300	—	—	—	—	—	—	3	—	7	10	12	19	27
	350	—	—	—	—	—	—	3	—	6	8	11	17	24
	400	—	—	—	—	—	—	2	—	6	7	10	15	21
	500	—	—	—	—	—	—	1	—	5	6	8	12	18
	600	—	—	—	—	—	—	1	—	4	5	6	10	14
	700	—	—	—	—	—	—	1	—	3	4	6	9	12
	750	—	—	—	—	—	—	1	—	3	4	5	8	12
	800	—	—	—	—	—	—	1	—	3	4	5	8	11
	900	—	—	—	—	—	—	1	—	3	3	4	7	10
	1000	—	—	—	—	—	—	1	—	2	3	4	6	9
FEP, FEPB, PFA, PFAH, TFE	14	—	—	—	—	—	—	155	—	348	454	578	887	1261
	12	—	—	—	—	—	—	113	—	254	332	422	648	920
	10	—	—	—	—	—	—	81	—	182	238	302	465	660
	8	—	—	—	—	—	—	46	—	104	136	173	266	378
	6	—	—	—	—	—	—	33	—	74	97	123	189	269
	4	—	—	—	—	—	—	23	—	52	68	86	132	188
	3	—	—	—	—	—	—	19	—	43	56	72	110	157
	2	—	—	—	—	—	—	16	—	36	46	59	91	129

Δ **TABLE C.13** *Continued*

Type	Conductor Size (AWG/ kcmil)	⅜ (12)	½ (16)	¾ (21)	1 (27)	1¼ (35)	1½ (41)	2 (53)	2½ (63)	3 (78)	3½ (91)	4 (103)	5 (129)	6 (155)
PFA, PFAH, TFE	1	—	—	—	—	—	—	11	—	25	32	41	63	90
PFA, PFAH, TFE, Z	1/0	—	—	—	—	—	—	9	—	20	27	34	53	75
	2/0	—	—	—	—	—	—	7	—	17	22	28	43	62
	3/0	—	—	—	—	—	—	6	—	14	18	23	36	51
	4/0	—	—	—	—	—	—	5	—	11	15	19	29	42
Z	14	—	—	—	—	—	—	186	—	419	547	696	1069	1519
	12	—	—	—	—	—	—	132	—	297	388	494	759	1078
	10	—	—	—	—	—	—	81	—	182	238	302	465	660
	8	—	—	—	—	—	—	51	—	115	150	191	294	417
	6	—	—	—	—	—	—	36	—	81	105	134	206	293
	4	—	—	—	—	—	—	24	—	55	72	92	142	201
	3	—	—	—	—	—	—	18	—	40	53	67	104	147
	2	—	—	—	—	—	—	15	—	34	44	56	86	122
	1	—	—	—	—	—	—	12	—	27	36	45	70	99
XHHW, ZW, XHHW-2, XHH	14	—	—	—	—	—	—	111	—	250	327	415	638	907
	12	—	—	—	—	—	—	85	—	192	251	319	490	696
	10	—	—	—	—	—	—	63	—	143	187	238	365	519
	8	—	—	—	—	—	—	35	—	79	104	132	203	288
	6	—	—	—	—	—	—	26	—	59	77	98	150	213
	4	—	—	—	—	—	—	19	—	42	56	71	109	155
	3	—	—	—	—	—	—	16	—	36	47	60	92	131
	2	—	—	—	—	—	—	13	—	30	39	50	77	110
XHHW, XHHW-2, XHH	1	—	—	—	—	—	—	10	—	22	29	37	58	82
	1/0	—	—	—	—	—	—	8	—	19	25	31	48	69
	2/0	—	—	—	—	—	—	7	—	16	20	26	40	57
	3/0	—	—	—	—	—	—	6	—	13	17	22	33	47
	4/0	—	—	—	—	—	—	5	—	11	14	18	27	39
	250	—	—	—	—	—	—	4	—	9	11	15	22	32
	300	—	—	—	—	—	—	3	—	7	10	12	19	28
	350	—	—	—	—	—	—	3	—	6	8	11	17	24
	400	—	—	—	—	—	—	2	—	6	8	10	15	22
	500	—	—	—	—	—	—	1	—	5	6	8	12	18
	600	—	—	—	—	—	—	1	—	4	5	6	10	14
	700	—	—	—	—	—	—	1	—	3	4	6	9	12
	750	—	—	—	—	—	—	1	—	3	4	5	8	12
	800	—	—	—	—	—	—	1	—	3	4	5	8	11
	900	—	—	—	—	—	—	1	—	3	3	4	7	10
	1000	—	—	—	—	—	—	1	—	2	3	4	6	9
	1250	—	—	—	—	—	—	1	—	1	2	3	5	7
	1500	—	—	—	—	—	—	1	—	1	1	3	4	6
	1750	—	—	—	—	—	—	1	—	1	1	2	4	5
	2000	—	—	—	—	—	—	0	—	1	1	1	3	5
FIXTURE WIRES														
RFH-2, FFH-2, RFHH-2	18	—	—	—	—	—	—	107	—	240	313	398	612	869
	16	—	—	—	—	—	—	90	—	202	264	336	516	733
SF-2, SFF-2	18	—	—	—	—	—	—	134	—	303	395	502	772	1096
	16	—	—	—	—	—	—	111	—	250	327	415	638	907
	14	—	—	—	—	—	—	90	—	202	264	336	516	733
SF-1, SFF-1	18	—	—	—	—	—	—	238	—	536	699	889	1366	1940

(continues)

Δ **TABLE C.13** *Continued*

Type	Conductor Size (AWG/ kcmil)	⅜ (12)	½ (16)	¾ (21)	1 (27)	1¼ (35)	1½ (41)	2 (53)	2½ (63)	3 (78)	3½ (91)	4 (103)	5 (129)	6 (155)
RFH-1, TF, TFF,	18	—	—	—	—	—	—	176	—	396	516	656	1009	1433
XF, XFF	16	—	—	—	—	—	—	142	—	319	417	530	814	1157
XF, XFF	14	—	—	—	—	—	—	111	—	250	327	415	638	907
TFN, TFFN	18	—	—	—	—	—	—	281	—	633	826	1050	1614	2293
	16	—	—	—	—	—	—	215	—	484	631	802	1233	1751
PF, PFF, PGF,	18	—	—	—	—	—	—	267	—	600	783	996	1530	2174
PGFF, PAF, PTF,	16	—	—	—	—	—	—	206	—	464	606	770	1183	1681
PTFF, PAFF	14	—	—	—	—	—	—	155	—	348	454	578	887	1261
ZF, ZFF, ZHF	18	—	—	—	—	—	—	344	—	774	1010	1284	1973	2802
	16	—	—	—	—	—	—	254	—	571	745	947	1455	2067
	14	—	—	—	—	—	—	186	—	419	547	696	1069	1519
KF-2, KFF-2	18	—	—	—	—	—	—	516	—	1161	1515	1926	2959	4204
	16	—	—	—	—	—	—	360	—	810	1057	1344	2064	2933
	14	—	—	—	—	—	—	242	—	544	710	903	1387	1970
	12	—	—	—	—	—	—	168	—	378	494	628	965	1371
	10	—	—	—	—	—	—	111	—	250	327	415	638	907
KF-1, KFF-1	18	—	—	—	—	—	—	596	—	1340	1748	2222	3414	4850
	16	—	—	—	—	—	—	419	—	941	1228	1562	2399	3408
	14	—	—	—	—	—	—	281	—	633	826	1050	1614	2293
	12	—	—	—	—	—	—	186	—	419	547	696	1069	1519
	10	—	—	—	—	—	—	122	—	274	358	455	699	993
XF, XFF	12	—	—	—	—	—	—	59	—	134	175	222	341	485
	10	—	—	—	—	—	—	46	—	104	136	173	266	378

Notes:

1. This table is for concentric stranded conductors only. For compact stranded conductors, Table C.12(A) should be used.

2. Two-hour fire-rated RHH cable has ceramifiable insulation, which has much larger diameters than other RHH wires. Consult manufacturer's conduit fill tables.

*Types RHH, RHW, and RHW-2 without outer covering.

TABLE C.13(A) *Maximum Number of Conductors or Fixture Wires in Type EB, PVC Conduit (Based on Chapter 9: Table 1, Table 4, and Table 5A)*

Type	Conductor Size (AWG/ kcmil)	Trade Size (Metric Designator)												
		⅜ (12)	½ (16)	¾ (21)	1 (27)	1¼ (35)	1½ (41)	2 (53)	2½ (63)	3 (78)	3½ (91)	4 (103)	5 (129)	6 (155)
COMPACT CONDUCTORS														
THW, THW-2, THHW	8	—	—	—	—	—	—	30	—	68	89	113	174	247
	6	—	—	—	—	—	—	23	—	52	69	87	134	191
	4	—	—	—	—	—	—	17	—	39	51	65	100	143
	2	—	—	—	—	—	—	13	—	29	38	48	74	105
	1	—	—	—	—	—	—	9	—	20	26	34	52	74
	1/0	—	—	—	—	—	—	8	—	17	23	29	45	64
	2/0	—	—	—	—	—	—	6	—	15	19	24	38	54
	3/0	—	—	—	—	—	—	5	—	12	16	21	32	46
	4/0	—	—	—	—	—	—	4	—	10	14	17	27	38
	250	—	—	—	—	—	—	3	—	8	11	14	21	30
	300	—	—	—	—	—	—	3	—	7	9	12	19	26
	350	—	—	—	—	—	—	3	—	6	8	11	17	24
	400	—	—	—	—	—	—	2	—	6	7	10	15	21
	500	—	—	—	—	—	—	1	—	5	6	8	12	18
	600	—	—	—	—	—	—	1	—	4	5	6	10	14
	700	—	—	—	—	—	—	1	—	3	4	6	9	13
	750	—	—	—	—	—	—	1	—	3	4	5	8	12
	900	—	—	—	—	—	—	1	—	3	4	5	7	10
	1000	—	—	—	—	—	—	1	—	2	3	4	7	9
THHN, THWN, THWN-2	8	—	—	—	—	—	—	—	—	—	—	—	—	—
	6	—	—	—	—	—	—	34	—	77	100	128	196	279
	4	—	—	—	—	—	—	21	—	47	62	79	121	172
	2	—	—	—	—	—	—	15	—	34	44	57	87	124
	1	—	—	—	—	—	—	11	—	25	33	42	65	93
	1/0	—	—	—	—	—	—	9	—	22	28	36	56	79
	2/0	—	—	—	—	—	—	8	—	18	23	30	46	65
	3/0	—	—	—	—	—	—	6	—	15	20	25	38	55
	4/0	—	—	—	—	—	—	5	—	12	16	20	32	45
	250	—	—	—	—	—	—	4	—	10	13	16	25	35
	300	—	—	—	—	—	—	4	—	8	11	14	22	31
	350	—	—	—	—	—	—	3	—	7	9	12	19	27
	400	—	—	—	—	—	—	3	—	6	8	11	17	24
	500	—	—	—	—	—	—	2	—	5	7	9	14	20
	600	—	—	—	—	—	—	1	—	4	6	7	11	16
	700	—	—	—	—	—	—	1	—	4	5	6	10	14
	750	—	—	—	—	—	—	1	—	4	5	6	9	14
	900	—	—	—	—	—	—	1	—	3	4	5	8	11
	1000	—	—	—	—	—	—	1	—	3	3	4	7	10
XHHW, XHHW-2	8	—	—	—	—	—	—	39	—	88	115	146	225	320
	6	—	—	—	—	—	—	29	—	65	85	109	167	238
	4	—	—	—	—	—	—	21	—	47	62	79	121	172
	2	—	—	—	—	—	—	15	—	34	44	57	87	124
	1	—	—	—	—	—	—	11	—	25	33	42	65	93

(continues)

TABLE C.13(A) *Continued*

Type	Conductor Size (AWG/ kcmil)	Trade Size (Metric Designator)												
		⅜ (12)	½ (16)	¾ (21)	1 (27)	1¼ (35)	1½ (41)	2 (53)	2½ (63)	3 (78)	3½ (91)	4 (103)	5 (129)	6 (155)
	1/0	—	—	—	—	—	—	9	—	22	28	36	56	79
	2/0	—	—	—	—	—	—	8	—	18	24	30	47	67
	3/0	—	—	—	—	—	—	6	—	15	20	25	38	55
	4/0	—	—	—	—	—	—	5	—	12	16	21	32	46
	250	—	—	—	—	—	—	4	—	10	13	17	26	37
	300	—	—	—	—	—	—	4	—	8	11	14	22	31
	350	—	—	—	—	—	—	3	—	7	10	12	19	28
	400	—	—	—	—	—	—	3	—	7	9	11	17	25
	500	—	—	—	—	—	—	2	—	5	7	9	14	20
	600	—	—	—	—	—	—	1	—	4	6	7	11	16
	700	—	—	—	—	—	—	1	—	4	5	6	10	14
	750	—	—	—	—	—	—	1	—	3	5	6	9	13
	900	—	—	—	—	—	—	1	—	3	4	5	8	11
	1000	—	—	—	—	—	—	1	—	3	4	5	7	10

Definition: *Compact stranding* is the result of a manufacturing process where the stranded conductor is compressed to the extent that the interstices (voids between strand wires) are virtually eliminated.

Δ **TABLE C.14** *Number of Type MC Cables Permitted in Cable Tray (3C Multiconductor MC Cable Non-Jacketed Assembly) (Based on fill in accordance with 392.22, Table 392.22(A), column 1, ampacity in accordance with 392.80)*

Conductor Insulation Type	Conductor Size (AWG/ kcmil)	Ventilated Tray Width [mm (in.)]											Cable Diameter
		50 (2)	100 (4)	150 (6)	200 (8)	300 (12)	400 (16)	450 (18)	500 (20)	600 (24)	750 (30)	900 (36)	
THHN	14	13	27	41	55	82	110	124	138	165	206	248	0.46
	12	10	20	31	41	62	83	93	104	124	160	192	0.53
	10	7	15	23	31	47	62	70	78	94	119	149	0.61
	8	6	12	18	25	37	50	56	63	75	96	116	0.68
	6	4	8	13	17	26	34	39	43	52	66	79	0.82
	4	2	5	8	11	17	23	26	29	35	45	55	0.99
	3	2	5	7	10	15	21	23	26	31	40	48	1.05
	2	2	4	6	9	13	18	20	22	27	34	41	1.13
	1	2	4	6	8	12	16	18	20	24	30	36	1.2
	1/0	1	3	5	7	11	14	16	18	22	28	34	1.25
	2/0	1	3	4	6	9	13	14	16	19	24	29	1.34
	3/0	1	2	3	5	7	10	11	13	15	20	24	1.49
	4/0	1	2	3	5	7	10	11	12	15	19	22	1.57
	250	1	2	3	4	6	9	10	11	13	17	20	1.74
	300	1	2	3	4	6	8	9	10	12	16	19	1.86
	350	1	2	3	4	6	8	9	10	12	15	18	1.96
	400	0	1	2	3	5	7	8	9	11	14	17	2.11
	500	0	1	2	3	5	7	7	8	10	13	16	2.24
	600	0	1	2	3	4	6	7	8	9	12	15	2.38
	700	0	1	2	3	4	6	7	7	9	11	14	2.52
	750	0	1	2	2	4	5	6	7	8	11	13	2.67
	800	0	1	2	2	4	5	6	6	8	10	12	2.85
	900	0	1	1	2	3	5	5	6	7	10	12	2.99
	1000	0	1	1	2	3	4	5	6	7	9	11	3.25

Note: Single conductor diameters were obtained from Chapter 9, Table 5.

TABLE C.15 *Number of Type MC Cables Permitted in Cable Tray (4C Multiconductor MC Cable Non-Jacketed Assembly)*
(Based on fill in accordance with 392.22, Table 392.22(A), column 1, ampacity in accordance with 392.80)

| Conductor Insulation Type | Conductor Size (AWG/ kcmil) | Ventilated Tray Width [mm (in.)] | | | | | | | | | | | | Cable Diameter |
		50 (2)	100 (4)	150 (6)	200 (8)	225 (9)	300 (12)	400 (16)	450 (18)	500 (20)	600 (24)	750 (30)	900 (36)	
THHN	14	10	23	37	47	55	73	94	110	120	146	183	219	0.494
	12	8	18	28	37	43	57	73	85	93	114	142	171	0.56
	10	6	14	21	27	32	42	54	63	69	84	105	127	0.65
	8	5	11	17	22	25	33	43	50	55	67	84	100	0.73
	6	3	7	12	15	17	23	30	35	38	46	58	69	0.88
	4	2	5	8	10	11	15	19	23	25	30	38	45	1.09
	3	2	4	7	9	10	13	17	20	22	27	33	40	1.155
	2	2	3	5	7	8	11	14	16	18	21	27	32	1.29
	1	1	3	5	6	7	10	12	15	16	19	24	29	1.355
	1/0	1	3	5	6	7	9	12	14	15	19	24	28	1.375
	2/0	1	2	3	4	5	7	9	10	11	14	17	21	1.608
	3/0	1	2	3	4	4	6	7	9	10	12	15	17	1.75
	4/0	1	2	3	4	4	6	8	9	10	12	15	18	1.97
	250	0	1	2	3	4	5	7	8	9	11	14	17	2.01
	300	0	1	2	3	3	5	7	7	8	10	13	15	2.255
	350	0	1	2	3	3	5	6	7	8	10	12	15	2.39
	400	0	1	2	3	3	4	6	7	8	9	12	14	2.46
	500	0	1	2	2	3	4	5	6	7	8	11	13	2.71
	600	0	1	2	2	3	4	5	6	6	8	10	12	2.92
	700	0	1	2	2	3	4	5	6	6	8	10	12	2.98
	750	0	1	1	2	2	3	4	5	5	7	8	10	3.34
	800	0	1	1	2	2	3	4	4	5	6	8	9	3.71
	900	0	1	1	2	2	3	4	4	5	6	7	9	3.98
	1000	0	0	1	1	2	2	3	4	4	5	7	8	4.15

TABLE C.16 *Number of Type TC Cables Permitted in Cable Tray (3C Multiconductor TC Cable Assembly)*
(Based on fill in accordance with 392.22, Table 392.22(A), column 1, ampacity in accordance with 392.80)

| Conductor Insulation Type | Conductor Size (AWG/ kcmil) | Ventilated Tray Width [mm (in.)] | | | | | | | | | | | | Cable Diameter |
		50 (2)	100 (4)	150 (6)	200 (8)	225 (9)	300 (12)	400 (16)	450 (18)	500 (20)	600 (24)	750 (30)	900 (36)	
THHN	14	12	26	41	52	61	81	104	122	133	162	203	243	0.469
	12	9	20	32	41	47	63	81	95	104	126	158	189	0.532
	10	7	15	24	30	35	47	60	71	77	94	118	141	0.616
	8	5	12	19	24	29	38	49	57	62	76	95	114	0.685
	6	4	9	13	17	20	26	34	40	43	53	66	79	0.821
	4	4	8	13	16	19	26	33	38	42	51	64	77	0.834
	3	3	7	11	14	16	22	28	32	35	43	54	65	0.910
	2	3	6	9	12	14	18	23	27	30	36	45	55	0.990
	1	2	4	6	8	10	13	17	19	21	26	32	39	1.175
	1/0	2	4	6	8	9	12	15	18	20	24	30	36	1.220
	2/0	1	3	5	7	8	10	13	16	17	21	26	31	1.310
	3/0	1	3	4	6	7	9	12	13	15	18	22	27	1.410
	4/0	1	2	3	5	5	7	10	11	12	15	19	23	1.558
	250	1	2	3	4	5	6	9	10	11	13	17	20	1.720
	300	1	2	3	4	4	6	8	9	10	12	15	18	1.912
	350	1	2	3	4	4	6	8	9	10	12	15	18	1.953
	400	0	1	2	3	4	5	7	8	9	11	14	17	2.099
	500	0	1	2	3	4	5	7	8	8	10	13	16	2.239
	600	0	1	2	3	3	4	6	7	8	9	12	14	2.433
	700	0	1	2	3	3	4	6	6	7	9	11	13	2.661
	750	0	1	2	2	3	4	5	6	7	8	10	13	2.769
	800	0	1	2	2	3	4	5	6	6	8	10	12	2.988
	900	0	1	1	2	2	3	5	5	6	7	9	11	3.010
	1000	0	1	1	2	2	3	4	5	6	7	9	10	3.273

TABLE C.17 *Number of Type TC Cables Permitted in Cable Tray (4C Multiconductor TC Cable Assembly)*
(Based on fill in accordance with 392.22, Table 392.22(A), column 1, ampacity in accordance with 392.80)

| Conductor Insulation Type | Conductor Size (AWG/ kcmil) | Ventilated Tray Width [mm (in.)] | | | | | | | | | | | | Cable Diameter |
		50 (2)	100 (4)	150 (6)	200 (8)	225 (9)	300 (12)	400 (16)	450 (18)	500 (20)	600 (24)	750 (30)	900 (36)	
THHN	14	11	24	37	48	56	74	95	111	122	148	186	223	0.49
	12	8	18	28	37	43	57	73	85	93	114	142	171	0.56
	10	6	14	21	27	32	42	54	63	69	84	105	127	0.65
	8	5	11	17	22	25	33	43	50	55	67	84	100	0.73
	6	3	7	12	15	17	23	30	35	38	46	58	69	0.88
	4	3	6	10	12	15	19	25	29	32	39	48	58	0.96
	3	2	5	8	11	12	16	21	25	27	33	41	49	1.041
	2	2	4	7	9	10	14	18	21	23	28	34	41	1.137
	1	1	3	5	7	8	10	13	15	17	21	26	31	1.315
	1/0	1	3	5	6	7	10	12	14	16	19	24	29	1.365
	2/0	1	3	4	5	6	8	10	12	13	16	20	24	1.49
	3/0	1	2	3	4	5	7	9	10	11	14	17	21	1.598
	4/0	2	2	3	4	5	6	9	10	11	13	17	20	1.75
	250	1	2	3	4	4	6	8	9	10	12	15	18	1.91
	300	0	1	2	3	4	5	7	8	9	11	14	17	2.11
	350	0	1	2	3	4	5	7	8	9	11	13	16	2.16
	400	0	1	2	3	3	5	6	7	8	10	12	15	2.35
	500	0	1	2	3	3	4	6	7	7	9	11	14	2.52
	600	0	1	2	2	3	4	5	6	7	8	11	13	2.71
	700	0	1	2	2	3	4	5	6	6	8	10	12	2.91
	750	0	1	1	2	2	3	5	5	6	7	9	11	3.02
	800	0	1	1	2	2	3	4	5	6	7	9	10	3.33
	900	0	1	1	2	2	3	4	4	5	6	8	9	3.69
	1000	0	0	1	1	2	2	3	4	4	5	7	8	4.02

TABLE C.18 *Number of Single Conductor Cables Permitted in Cable Tray*
(Based on fill in accordance with 392.22, Table 392.22(A), column 1, ampacity in accordance with 392.80)

| Conductor Insulation Type | Conductor Size (AWG/ kcmil) | Ventilated Tray Width [mm (in.)] | | | | | | | | | | | | Cable Diameter |
		50 (2)	100 (4)	150 (6)	200 (8)	225 (9)	300 (12)	400 (16)	450 (18)	500 (20)	600 (24)	750 (30)	900 (36)	
THHN	1/0	4	8	12	16	18	24	32	36	40	48	61	74	0.486
	2/0	3	7	11	14	16	22	29	33	37	44	56	67	0.532
	3/0	3	6	10	13	15	20	26	30	33	40	51	61	0.584
	4/0	3	6	9	12	14	18	24	27	30	36	46	56	0.642
	250	5	10	16	21	23	32	42	48	53	64	81	98	0.711
	300	4	9	13	18	20	27	37	41	46	55	70	84	0.766
	350	4	8	12	16	18	24	32	36	40	48	61	74	0.817
	400	3	7	10	14	16	21	29	32	36	43	55	66	0.864
	500	3	6	9	12	13	18	24	27	30	36	45	55	0.949
	600	2	4	7	9	10	14	19	22	24	29	37	44	1.051
	700	2	4	6	8	9	12	17	19	21	25	32	39	1.122
	750	2	4	6	8	9	12	16	18	20	24	30	37	1.156
	800	1	3	5	7	8	11	15	17	19	23	29	35	1.188
	900	1	3	5	6	7	10	13	15	17	20	26	31	1.252
	1000	1	3	4	6	7	9	12	13	15	18	22	27	1.31

TABLE C.19 *Number of Single Conductor Cables Permitted in Cable Tray*
(Based on fill in accordance with 392.22, Table 392.22(A), column 1, ampacity in accordance with 392.80)

Conductor Insulation Type	Conductor Size (AWG/ kcmil)	Ventilated Tray Width [mm (in.)]												Cable Diameter
		50 (2)	100 (4)	150 (6)	200 (8)	225 (9)	300 (12)	400 (16)	450 (18)	500 (20)	600 (24)	750 (30)	900 (36)	
XHHW	1/0	4	8	12	16	18	24	32	36	40	49	62	74	0.482
	2/0	3	7	11	14	17	22	29	33	37	44	56	68	0.528
	3/0	3	6	10	13	15	20	27	30	33	40	51	62	0.58
	4/0	3	6	9	12	12	18	24	27	30	37	47	56	0.638
	250	5	10	16	21	24	32	43	49	54	65	83	98	0.705
	300	4	9	14	18	20	28	37	42	47	56	71	85	0.76
	350	4	8	12	16	18	24	33	37	41	49	62	75	0.811
	400	3	7	11	14	16	22	29	33	36	44	56	67	0.858
	500	3	6	9	12	13	18	24	27	30	36	46	55	0.943
	600	2	4	7	9	10	14	19	22	24	29	37	44	1.053
	700	2	4	6	8	9	12	17	19	21	25	32	39	1.124
	750	2	4	6	8	9	12	16	18	20	24	30	37	1.158
	800	1	3	5	7	8	11	15	17	19	23	29	35	1.19
	900	1	3	5	6	7	10	13	15	17	20	26	31	1.254
	1000	1	3	4	6	7	9	12	13	15	18	22	27	1.312

TABLE C.20 *Number of Single Conductor Cables Permitted in Cable Tray*
(Based on fill in accordance with 392.22, Table 392.22(A), column 1, ampacity in accordance with 392.80)

Conductor Insulation Type	Conductor Size (AWG/ kcmil)	Ventilated Tray Width [mm (in.)]												Cable Diameter
		50 (2)	100 (4)	150 (6)	200 (8)	225 (9)	300 (12)	400 (16)	450 (18)	500 (20)	600 (24)	750 (30)	900 (36)	
RHW	1/0	3	7	11	14	16	22	29	33	37	44	56	67	0.532
	2/0	3	6	10	13	15	20	27	30	34	40	51	62	0.578
	3/0	3	6	9	12	14	18	24	28	31	37	47	57	0.63
	4/0	2	5	8	11	13	17	22	25	28	34	43	52	0.688
	250	4	9	13	18	20	27	37	41	46	55	70	84	0.765
	300	4	8	12	16	17	24	32	36	40	48	61	73	0.82
	350	3	7	10	14	15	21	28	32	35	42	54	65	0.871
	400	3	6	9	12	14	19	25	28	32	38	49	58	0.918
	500	2	5	8	10	12	16	21	24	26	32	41	49	1.003
	600	2	4	6	8	9	13	17	19	21	26	33	40	1.113
	700	1	3	5	7	8	11	15	17	19	23	29	35	1.184
	750	1	3	5	7	8	10	14	16	18	21	27	33	1.218
	800	1	3	5	6	7	10	13	15	17	20	26	31	1.25
	900	1	3	4	6	7	9	12	14	15	18	23	28	1.314
	1000	1	2	4	5	6	8	11	12	14	17	21	26	1.372

Examples

Selection of Conductors. In the following examples, the results are generally expressed in amperes (A). To select conductor sizes, refer to the 0 through 2000 volt (V) ampacity tables of Article 310 and the rules of 310.14 that pertain to these tables.

Voltage. For uniform application of Articles 210, 215, and 220, a nominal voltage of 120, 120/240, 240, and 208Y/120 V is used in calculating the ampere load on the conductor.

Fractions of an Ampere. Except where the calculations result in a major fraction of an ampere (0.5 or larger), such fractions are permitted to be dropped.

Power Factor. Calculations in the following examples are based, for convenience, on the assumption that all loads have the same power factor (PF).

Ranges. For the calculation of the range loads in these examples, Column C of Table 220.55 has been used. For optional methods, see Columns A and B of Table 220.55. Except where the calculations result in a major fraction of a kilowatt (0.5 or larger), such fractions are permitted to be dropped.

SI Units. For metric conversions, $0.093 \text{ m}^2 = 1 \text{ ft}^2$ and $0.3048 \text{ m} = 1 \text{ ft}$.

In the Informative Annex D examples, loads are assumed to be properly balanced on the system. If loads are not properly balanced, additional feeder capacity may be required. Article 220 provides two methods to calculate feeder and service loads, known in the industry as the "standard" and "optional" calculations. Part III of Article 220 provides the rules for performing the standard calculation, and Part IV provides the rules for performing the optional calculation. In both, it is necessary to use the general requirements in Part I and the requirements in Part II covering branch-circuit load calculations. The service and feeder calculation requirements in Part III apply to all occupancies, while the requirements in Part IV cover dwelling units, multifamily dwellings, schools, and new restaurants. Both Parts III and IV provide requirements on how to calculate loads added to existing buildings and structures.

Example D1(a) One-Family Dwelling

The dwelling has a floor area of 1500 ft², exclusive of an unfinished cellar not adaptable for future use, unfinished attic, and open porches. Appliances are a 12-kW range and a 5.5-kW, 240-V dryer. Assume range and dryer kW ratings equivalent to kVA ratings in accordance with 220.54 and 220.55.

The general lighting and general-use receptacle load is computed from the outside dimensions of the building, apartment, or other area involved. For a dwelling unit, the computed floor area does not include open porches, garages, or, as stated in the opening paragraph of Example D1(a), "an unfinished cellar not adaptable for future use." A point to consider regarding this statement is that many of today's homes with basements or cellars do have space that is suitable for conversion to family rooms, bedrooms, home offices, or other habitable areas. In such instances, the basement or cellar space suitable for conversion needs to be included in the general lighting load calculation. See 220.41 for requirements on how to calculate the lighting and general-use receptacle load for this occupancy.

The two-story dwelling measures 30 ft × 30 ft for the first floor and 30 ft × 20 ft for the second floor.

First-floor area: 30 ft × 30 ft = 900 ft²
Second-floor area: 30 ft × 20 ft = 600 ft²
Total area = 1500 ft²

Calculated Load *(see 220.40)*

General Lighting Load 1500 ft² at 3 VA/ft² = 4500 VA

Minimum Number of Branch Circuits Required *[see 210.11(A)]*

General Lighting Load: 4500 VA ÷ 120 V = 38 A

This requires three 15-A, 2-wire or two 20-A, 2-wire circuits.
Small-Appliance Load: Two 2-wire, 20-A circuits *[see 210.11(C)(1)]*
Laundry Load: One 2-wire, 20-A circuit *[see 210.11(C)(2)]*
Bathroom Branch Circuit: One 2-wire, 20-A circuit (no additional load calculation is required for this circuit) *[see 210.11(C)(3)]*

Minimum Size Feeder Required [see 220.40]

General Lighting		4,500 VA
Small Appliance		3,000 VA
Laundry		1,500 VA
	Total	9,000 VA
3000 VA at 100%		3,000 VA
9000 VA – 3000 VA = 6000 VA at 35%		2,100 VA
	Net Load	5,100 VA
Range (see Table 220.55)		8,000 VA
Dryer Load (see Table 220.54)		5,500 VA
Net Calculated Load		18,600 VA

Net Calculated Load for 120/240-V, 3-wire, single-phase service or feeder

$$18{,}600 \text{ VA} \div 240 \text{ V} = 78 \text{ A}$$

Sections 230.42(B) and 230.79 require service conductors and disconnecting means rated not less than 100 amperes.

Calculation for Neutral for Feeder and Service

Lighting and Small-Appliance Load		5,100 VA
Range: 8000 VA at 70% (see 220.61)		5,600 VA
Dryer: 5500 VA at 70% (see 220.61)		3,850 VA
	Total	14,550 VA

Calculated Load for Neutral

$$14{,}550 \text{ VA} \div 240 \text{ V} = 61 \text{ A}$$

Example D1(b) One-Family Dwelling

Assume same conditions as Example No. D1(a), plus addition of one 6-A, 230-V, room air-conditioning unit and one 12-A, 115-V, room air-conditioning unit,* one 8-A, 115-V, rated waste disposer, and one 10-A, 120-V, rated dishwasher. See Article 430 for general motors and Article 440, Part VII, for air-conditioning equipment. Motors have nameplate ratings of 115 V and 230 V for use on 120-V and 240-V nominal voltage systems.

*(For feeder neutral, use larger of the two appliances for unbalance.)

From Example D1(a), feeder current is 78 A (3-wire, 240 V).

	Line A	Neutral	Line B
Amperes from Example D1(a)	78	61	78
One 230-V air conditioner	6	—	6
One 115-V air conditioner and 120-V dishwasher	12	12	10
One 115-V disposer	—	8	8
25% of air-conditioner (see 440.33)	3	3	2
Total amperes per conductor	99	84	104

Therefore, the service would be rated 110 A.

The air-conditioning load is calculated at 100 percent and is calculated separately to comply with the requirements of 220.14(C) and 220.50.

Example D2(a) Optional Calculation for One-Family Dwelling, Heating Larger Than Air Conditioning
(see 220.82)

The dwelling has a floor area of 1500 ft², exclusive of an unfinished cellar not adaptable for future use, unfinished attic, and open porches. It has a 12-kW range, a 2.5-kW water heater, a 1.2-kW dishwasher, 9 kW of electric space heating installed in five rooms, a 5-kW clothes dryer, and a 6-A, 230-V, room air-conditioning unit. Assume range, water heater, dishwasher, space heating, and clothes dryer kW ratings equivalent to kVA.

Air Conditioner kVA Calculation

$$6 \text{ A} \times 230 \text{ V} \div 1000 = 1.38 \text{ kVA}$$

This 1.38 kVA [item 1 from 220.82(C)] is less than 40% of 9 kVA of separately controlled electric heat [item 6 from 220.82(C)], so the 1.38 kVA need not be included in the service calculation.

General Load

1500 ft² at 3 VA		4,500 VA
Two 20-A appliance outlet circuits at 1500 VA each		3,000 VA
Laundry circuit		1,500 VA
Range (at nameplate rating)		12,000 VA
Water heater		2,500 VA
Dishwasher		1,200 VA
Clothes dryer		5,000 VA
	Total	29,700 VA

Application of Demand Factor [see 220.82(B)]

First 10 kVA of general load at 100%		10,000 VA
Remainder of general load at 40% (19.7 kVA × 0.4)		7,880 VA
	Total of general load	17,880 VA
9 kVA of heat at 40% (9000 VA × 0.4) =		3,600 VA
	Total	21,480 VA

Calculated Load for Service Size

$$21.48 \text{ kVA} = 21{,}480 \text{ VA}$$
$$21{,}480 \text{ VA} \div 240 \text{ V} = 90 \text{ A}$$

Therefore, the minimum service rating would be 100 A in accordance with 230.42 and 230.79.

Feeder Neutral Load in Accordance with 220.61

1500 ft² at 3 VA		4,500 VA
Three 20-A circuits at 1500 VA		4,500 VA
	Total	9,000 VA
3000 VA at 100%		3,000 VA
9000 VA - 3000 VA = 6000 VA at 35%		2,100 VA
	Subtotal	5,100 VA
Range: 8 kVA at 70%		5,600 VA
Clothes dryer: 5 kVA at 70%		3,500 VA
Dishwasher		1,200 VA
	Total	15,400 VA

Calculated Load for Neutral

$$15{,}400 \text{ VA} \div 240 \text{ V} = 64 \text{ A}$$

Example D2(b) Optional Calculation for One-Family Dwelling, Air Conditioning Larger Than Heating
[see 220.82(A) and 220.82(C)]

The dwelling has a floor area of 1500 ft², exclusive of an unfinished cellar not adaptable for future use, unfinished attic, and open porches. It has two 20-A small appliance circuits, one 20-A laundry circuit, two 4-kW wall-mounted ovens, one 5.1-kW counter-mounted cooking unit, a 4.5-kW water heater, a 1.2-kW dishwasher, a 5-kW combination clothes washer and dryer, six 7-A, 230-V room air-conditioning units, and a 1.5-kW permanently installed bathroom space heater. Assume wall-mounted ovens, counter-mounted cooking unit, water heater, dishwasher, and combination clothes washer and dryer kW ratings equivalent to kVA.

Air Conditioning kVA Calculation

$$\text{Total amperes} = 6 \text{ units} \times 7 \text{ A} = 42 \text{ A}$$
$$42 \text{ A} \times 240 \text{ V} \div 1000 = 10.08 \text{ kVA (assume PF} = 1.0)$$

Load Included at 100%

Air Conditioning: Included below *[see item 1 in 220.82(C)]*

Space Heater: Omit *[see item 5 in 220.82(C)]*

General Load

1500 ft² at 3 VA	4,500 VA
Two 20-A small-appliance circuits at 1500 VA each	3,000 VA
Laundry circuit	1,500 VA
Two ovens	8,000 VA
One cooking unit	5,100 VA
Water heater	4,500 VA
Dishwasher	1,200 VA
Washer/dryer	5,000 VA
Total general load	32,800 VA
First 10 kVA at 100%	10,000 VA
Remainder at 40% (22.8 kVA × 0.4 × 1000)	9,120 VA
Subtotal general load	19,120 VA
Air conditioning	10,080 VA
Total	29,200 VA

Calculated Load for Service

$$29{,}200 \text{ VA} \div 240 \text{ V} = 122 \text{ A (service rating)}$$

Feeder Neutral Load, in accordance with 220.61

Assume that the two 4-kVA wall-mounted ovens are supplied by one branch circuit, the 5.1-kVA counter-mounted cooking unit by a separate circuit.

1500 ft² at 3 VA	4,500 VA
Three 20-A circuits at 1500 VA	4,500 VA
Subtotal	9,000 VA
3000 VA at 100%	3,000 VA
9000 VA - 3000 VA = 6000 VA at 35%	2,100 VA
Subtotal	5,100 VA

Two 4-kVA ovens plus one 5.1-kVA cooking unit = 13.1 kVA. Table 220.55 permits 55% demand factor or 13.1 kVA × 0.55 = 7.2 kVA feeder capacity.

Subtotal from above	5,100 VA
Ovens and cooking unit: 7200 VA × 70% for neutral load	5,040 VA
Clothes washer/dryer: 5 kVA × 70% for neutral load	3,500 VA
Dishwasher	1,200 VA
Total	14,840 VA

Calculated Load for Neutral

$$14{,}840 \text{ VA} \div 240 \text{ V} = 62$$

Example D2(c) Optional Calculation for One-Family Dwelling with Heat Pump (Single-Phase, 240/120-Volt Service) *(see 220.82)*

The dwelling has a floor area of 2000 ft², exclusive of an unfinished cellar not adaptable for future use, unfinished attic, and open porches. It has a 12-kW range, a 4.5-kW water heater, a 1.2-kW dishwasher, a 5-kW clothes dryer, and a 2½-ton (24-A) heat pump with 15 kW of backup heat.

Heat Pump kVA Calculation

$$24 \text{ A} \times 240 \text{ V} \div 1000 = 5.76 \text{ kVA}$$

This 5.76 kVA is less than 15 kVA of the backup heat; therefore, the heat pump load need not be included in the service calculation *[see 220.82(C)]*.

General Load

2000 ft² at 3 VA	6,000 VA
Two 20-A appliance outlet circuits at 1500 VA each	3,000 VA
Laundry circuit	1,500 VA
Range (at nameplate rating)	12,000 VA
Water heater	4,500 VA
Dishwasher	1,200 VA
Clothes dryer	5,000 VA
Subtotal general load	33,200 VA
First 10 kVA at 100%	10,000 VA
Remainder of general load at 40% (23,200 VA × 0.4)	9,280 VA
Total net general load	19,280 VA

Heat Pump and Supplementary Heat*

$$240 \text{ V} \times 24 \text{ A} = 5760 \text{ VA}$$

15 kW Electric Heat:

$$5760 \text{ VA} + (15{,}000 \text{ VA} \times 65\%) = 5.76 \text{ kVA} + 9.75 \text{ kVA}$$
$$= 15.51 \text{ kVA}$$

*If supplementary heat is not on at same time as heat pump, heat pump kVA need not be added to total.

Totals

Net general load		19,280 VA
Heat pump and supplementary heat		15,510 VA
	Total	34,790 VA

Calculated Load for Service

$$34.79 \text{ kVA} \times 1000 \div 240 \text{ V} = 145 \text{ A}$$

Therefore, this dwelling unit would be permitted to be served by a 150-A service.

Examples D2(a) through D2(c) are sample calculations for one-family dwellings as defined in Article 100 under *dwelling, one-family (one-family dwelling)*. The optional method specified in 220.82 is the basis for these sample calculations. Although the samples were for one-family dwellings (also known in common nomenclature as single-family homes), this calculation can be used for the individual dwelling units of two-family and multifamily dwellings.

However, unlike the standard method, the optional calculation can be used only where the ungrounded service or feeder conductors have an ampacity of not less than 100 amperes regardless of the calculated load. The load can be less than 100 amperes, but the ungrounded conductors cannot have an ampacity of less than 100 amperes. The rationale behind this minimum conductor ampacity requirement is that with higher loads comes more diversity and that the significant demands permitted by 220.82 are appropriate only with service and feeder conductors having an ampacity not less than 100 amperes.

In Examples D4(a), D4(b), D5(a), and D5(b), the calculated load for each dwelling unit is less than 100 amperes; therefore, the standard calculation is used so that an ungrounded conductor having an ampacity of less than 100 amperes can be installed as a feeder conductor for the individual dwelling units.

The three-wire service or feeder conductors can be supplied from a single-phase 120/240-volt or three-phase 208Y/120-volt supply.

Example D3 Store Building

A store 80 ft by 60 ft, or 4,800 ft^2, has 30 ft of show window. There are a total of 80 duplex receptacles. The service is 120/240 V, single phase 3-wire service. Actual connected lighting load is 7,000 VA, all of which for this example is considered continuous. All calculations are rounded up or down as permitted in 220.5(B).

Calculated Load (see 220.40)

Δ **Noncontinuous Loads**

Receptacle Load *(see 220.47)*	
80 receptacles at 180 VA	14,400 VA
10,000 VA at 100%	10,000 VA
14,400 VA - 10,000 VA = 4,400 VA at 50%	2,200 VA
Subtotal	12,200 VA

Continuous Loads

General Lighting*	
4,800 ft^2 at 1.9 VA/ft^2	9,120 VA
Show Window Lighting Load	
30 ft at 200 VA/ft *[see 220.14(G)]*	6,000 VA
Outside Sign Circuit *[see 220.14(F)]*	1,200 VA
Subtotal	16,320 VA
Subtotal from noncontinuous	12,200 VA
Total noncontinuous loads + continuous loads =	28,520 VA

*In the example, the actual connected lighting load at 125% (7,000 VA × 1.25 VA) is less than the load from Table 220.42(A), so the required minimum lighting load from Table 220.42(A) is used in the calculation. Had the actual lighting load × 125% been greater than the value calculated from Table 220.42(A), the actual connected lighting load would have been used.

Minimum Number of Branch Circuits Required

General Lighting: Branch circuits need only be installed to supply the actual connected load *[see 210.11(B)]*.

$$7,000 \text{ VA} \times 1.25 = 8,750 \text{ VA}$$
$$8,750 \text{ VA} \div 240 \text{ V} = 36.45 \text{ A for 3-wire, 120/240 V}$$
$$8,750 \text{ VA} \div 120 \text{ V} = 72.92 \text{ A}$$

The lighting load would be permitted to be served by 2-wire or 3-wire, 15- or 20-A circuits with combined capacity equal to 36 A or greater for 3-wire circuits or 73 A or greater for 2-wire circuits. The feeder capacity as well as the number of branch-circuit positions available for lighting circuits in the panelboard must reflect the full calculated load of 9,120 VA. Lighting loads from Table 220.42(A) already include 125% for continuous load. See note at bottom of Table 220.42(A).

Show Window

$$6,000 \text{ VA} \times 1.25 = 7,500 \text{ VA}$$
$$7,500 \text{ VA} \div 240 \text{ V} = 31.25 \text{ A for 3-wire, 120/240 V}$$
$$7,500 \text{ VA} \div 120 \text{ V} = 62.5 \text{ A for 2-wire, 120 V}$$

The show window lighting is permitted to be served by 2-wire or 3-wire circuits with a capacity equal to 31 A or greater for 3-wire circuits or 63 A or greater for 2-wire circuits.

Receptacles required by 210.62 are assumed to be included in the receptacle load above if these receptacles do not supply the show window lighting load.

Receptacles

Receptacle Load:

$$14,400 \text{ VA} \div 240 \text{ V} = 60 \text{ A for 3-wire, 120/240 V}$$
$$14,400 \text{ VA} \div 120 \text{ V} = 120 \text{ A for 2-wire, 120 V}$$

The receptacle load would be permitted to be served by 2-wire or 3-wire circuits with a capacity equal to 60 A or greater for 3-wire circuits or 120 A or greater for 2-wire circuits.

Minimum Size Feeder (or Service) Overcurrent Protection
(see 215.3 or 230.90)

△ Subtotal noncontinuous loads	12,200 VA
Subtotal continuous loads not from Table 220.42(A) at 125% (7,200 VA × 1.25) (sign and show window)	9,000 VA
Subtotal of calculated continuous loads with 125% already included	9,120 VA
Total	30,320 VA

$$30,320 \text{ VA} \div 240 \text{ V} = 126 \text{ A}$$

The next higher standard size is 150 A *(see 240.6)*.

Minimum Size Feeders (or Service Conductors) Required *[see 215.2, 230.42(A)]*
For 120/240 V, 3-wire system,

$$30,320 \text{ VA} \div 240 \text{ V} = 126 \text{ A}$$

Service or feeder conductor is 1 AWG Cu in accordance with 215.3 and Table 310.16 (with 75°C terminations).

Example D3(a) Industrial Feeders in a Common Raceway

An industrial multi-building facility has its service at the rear of its main building, and then provides 480Y/277-volt feeders to additional buildings behind the main building in order to segregate certain processes. The facility supplies its remote buildings through a partially enclosed access corridor that extends from the main switchboard rearward along a path that provides convenient access to services within 15 m (50 ft) of each additional building supplied. Two building feeders share a common raceway for approximately 45 m (150 ft) and run in the access corridor along with process steam and control and communications cabling. The steam raises the ambient temperature around the power raceway to as much as 35°C. At a tee fitting, the individual building feeders then run to each of the two buildings involved. The feeder neutrals are not connected to the equipment grounding conductors in the remote buildings. All distribution equipment terminations are listed as being suitable for 75°C connections.

Each of the two buildings has the following loads:

Lighting, 11,600 VA, comprised of electric-discharge luminaires connected at 277 V

Receptacles, 22 125-volt, 20-ampere receptacles on general-purpose branch circuits, supplied by separately derived systems in each of the buildings

1 Air compressor, 460 volt, three phase, 5 hp

1 Grinder, 460 volt, three phase, 1.5 hp

3 Welders, AC transformer type (nameplate: 23 amperes, 480 volts, 60 percent duty cycle)

3 Industrial Process Dryers, 480 volt, three phase, 15 kW each (assume continuous use throughout certain shifts)

Determine the overcurrent protection and conductor size for the feeders in the common raceway, assuming the use of XHHW-2 insulation (90°C):

Calculated Load {Note: For reasonable precision, volt-ampere calculations are carried to three significant figures only; where loads are converted to amperes, the results are rounded to the nearest ampere *[see 220.5(B)]*}.

△ **Noncontinuous Loads**
Receptacle Load *(see 220.47)*

22 receptacles at 180 VA	3,960 VA

Welder Load *[see 630.11(A), Table 630.11(A)]*
Each welder: 480V × 23A × 0.78 = 8,610 VA
All 3 welders *[see 630.11(B)]* (demand factors 100%, 100%, 85% respectively)

8,610 VA + 8,610 VA + 7,320 VA =	24,500 VA
Subtotal, Noncontinuous Loads	**28,500 VA**

Motor Loads *(see 430.24, Table 430.250)*

Air compressor: 7.6 A × 480 V × √3 =	6,310 VA
Grinder: 3 A × 480 V × √3 =	2,490 VA
Largest motor, additional 25%:	1,580 VA
Subtotal, Motor Loads	**10,400 VA**

By using 430.24, the motor loads and the noncontinuous loads can be combined for the remaining calculation.

Subtotal for load calculations, Noncontinuous Loads	**38,900 VA**

Continuous Loads

General Lighting		11,600 VA
3 Industrial Process Dryers	15 kW each	45,000 VA
Subtotal, Continuous Loads:		**56,600 VA**

Overcurrent protection *(see 215.3)*
The overcurrent protective device must accommodate 125% of the continuous load, plus the noncontinuous load:

Continuous load	56,600 VA
Noncontinuous load	38,900 VA
Subtotal, actual load [actual load in amperes]	**95,500 VA**
[99,000 VA ÷ (480V × √3) = 119 A]	
(25% of 56,600 VA) *(See 215.3)*	**14,200 VA**
Total VA	**109,700 VA**

Conversion to amperes using three significant figures:
    109,700 VA / (480V × √3) = 132 A

Minimum size overcurrent protective device: 132 A

Minimum standard size overcurrent protective device *(see 240.6)*:
    150 amperes

Where the overcurrent protective device and its assembly are listed for operation at 100 percent of its rating, a 125 ampere overcurrent protective device would be permitted. However, overcurrent protective device assemblies listed for 100 percent of their rating are typically not available at the 125-ampere rating. *(See 215.3 Exception.)*

△ **Ungrounded Feeder Conductors**
The conductors must independently meet requirements for (1) terminations, and (2) conditions of use throughout the raceway run.

Minimum size conductor at the overcurrent device termination *[see 110.14(C) and 215.2(A), using 75°C ampacity column in Table 310.16]*: 1/0 AWG.

Minimum size conductors in the raceway based on actual load *[see Article 100, Ampacity, and 310.15(C)(1) and correction factors to Table 310.16]*:

$$95,500 \text{ VA} \div 0.7 \div 0.96 = 142,000 \text{ VA}$$

[70% = 310.15(C)(1)] & [0.96 = Correction factors to Table 310.16]

Conversion to amperes:

$$142,000 \text{ VA} \div (480 \text{ V} \times \sqrt{3}) = 171 \text{ A}$$

Note that the neutral conductors are counted as current-carrying conductors *[see 310.15(E)(3)]* in this example because the discharge lighting has substantial nonlinear content. This requires a 2/0 AWG conductor based on the 90°C column of Table 310.16. Therefore, the worst case is given by the raceway conditions, and 2/0 AWG conductors must be used. If the utility corridor were at normal temperatures [30°C (86°F)], and if the lighting at each building were supplied from the local separately derived system (thus requiring no neutrals in the supply feeders), the raceway result (95,500 VA ÷ 0.8 = 119,000 VA; 119,000 VA ÷ (480 V × √3) = 143 A, or a 1 AWG conductor @ 90°C) could not be used, because the termination result (1/0 AWG) based on the 75°C column of Table 310.16 would become the worst case, requiring the larger conductor.

In every case, the overcurrent protective device shall provide overcurrent protection for the feeder conductors in accordance with their ampacity as provided by this *Code (see 240.4)*. A 90°C 2/0 AWG conductor has a Table 310.16 ampacity of 195 amperes. Adjusting for the conditions of use (35°C ambient temperature, 8 current-carrying conductors in the common raceway),

$$195 \text{ A} \times 0.96 \times 0.7 = 131 \text{ A}$$

The 150-ampere circuit breaker protects the 2/0 AWG feeder conductors, because 240.4(B) permits the use of the next higher standard size overcurrent protective device. Note that the feeder layout precludes the application of 310.14(A)(2) Exception.

Feeder Neutral Conductor *(see 220.61)*

Because 210.11(B) does not apply to these buildings, the load cannot be assumed to be evenly distributed across phases. Therefore the maximum imbalance must be assumed to be the full lighting load in this case, or 11,600 VA. (11,600 VA ÷ 277 V = 42 A.) The ability of the neutral-to-return fault current *[see 250.32(B) Exception No. 2]* is not a factor in this calculation.

Because the neutral runs between the main switchboard and the building panelboard, likely terminating on a busbar at both locations, and not on overcurrent devices, the effects of continuous loading can be disregarded in evaluating its terminations *[see 215.2(A)(1) Exception No. 3]*. That calculation is (11,600 VA ÷ 277 V = 42 A), to be evaluated under the 75°C column of Table 310.16. The minimum size of the neutral might seem to be 8 AWG, but that size would not be sufficient to be depended upon in the event of a line-to-neutral fault *[see 215.2(B), second paragraph]*. Therefore, because the minimum size equipment grounding conductor for a 150 ampere circuit wired with 2/0

AWG conductors, as covered in Table 250.122, is 6 AWG, that is the minimum neutral size required for this feeder.

Example D4(a) Multifamily Dwelling

A multifamily dwelling has 40 dwelling units.

Meters are in two banks of 20 each with individual feeders to each dwelling unit.

One-half of the dwelling units are equipped with electric ranges not exceeding 12 kW each. Assume range kW rating equivalent to kVA rating in accordance with 220.55. Other half of ranges are gas ranges.

Area of each dwelling unit is 840 ft².

Laundry facilities on premises are available to all tenants. Add no circuit to individual dwelling unit.

Calculated Load for Each Dwelling Unit *(see Article 220)*

General Lighting: 840 ft² at 3 VA/ft² = 2520 VA

Special Appliance: Electric range *(see 220.55)* = 8000 VA

Minimum Number of Branch Circuits Required for Each Dwelling Unit *[see 210.11(A)]*

General Lighting Load: 2520 VA ÷ 120 V = 21 A or two 15-A, 2-wire circuits; or two 20-A, 2-wire circuits

Small-Appliance Load: Two 2-wire circuits of 12 AWG wire *[see 210.11(C)(1)]*

Range Circuit: 8000 VA ÷ 240 V = 33 A or a circuit of two 8 AWG conductors and one 10 AWG conductor in accordance with 210.19(C)

Minimum Size Feeder Required for Each Dwelling Unit *(see 215.2)*

Calculated Load *(see Article 220)*:

General Lighting	2,520 VA
Small Appliance (two 20-ampere circuits)	3,000 VA
Subtotal Calculated Load (without ranges)	5,520 VA

Application of Demand Factor *(see Table 220.45)*

First 3000 VA at 100%	3,000 VA
5520 VA - 3000 VA = 2520 VA at 35%	882 VA
Net Calculated Load (without ranges)	3,882 VA
Range Load	8,000 VA
Net Calculated Load (with ranges)	11,882 VA

Size of Each Feeder *(see 215.2)*

For 120/240-V, 3-wire system (without ranges)
    Net calculated load of 3882 VA ÷ 240 V = 16 A
For 120/240-V, 3-wire system (with ranges)
    Net calculated load, 11,882 VA ÷ 240 V = 50 A

D.8.6 Feeder Neutral

Lighting and Small-Appliance Load	3,882 VA
Range Load: 8000 VA at 70% *(see 220.61)*	5,600 VA
(only for apartments with electric range)	5,600 VA
Net Calculated Load (neutral)	9,482 VA

Calculated Load for Neutral

$$9482 \text{ VA} \div 240 \text{ V} = 39.5 \text{ A}$$

Minimum Size Feeders Required from Service Equipment to Meter Bank (For 20 Dwelling Units — 10 with Ranges)

Total Calculated Load:

Lighting and Small Appliance	
20 units × 5520 V	110,400 VA

Application of Demand Factor

First 3000 VA at 100%	3,000 VA
110,400 VA - 3000 VA = 107,400 VA at 35%	37,590 VA
Net Calculated Load	40,590 VA
Range Load: 10 ranges (not over 12 kVA)	
(see Col. C, Table 220.55, 25 kW)	25,000 VA
Net Calculated Load (with ranges)	65,590 VA

Net calculated load for 120/240-V, 3-wire system,

$$65,590 \text{ VA} \div 240 \text{ V} = 273 \text{ A}$$

Feeder Neutral

Lighting and Small-Appliance Load	40,590 VA
Range Load: 25,000 VA at 70% *[see 220.61(B)]*	17,500 VA
Calculated Load (neutral)	58,090 VA

Calculated Load for Neutral

$$58,090 \text{ VA} \div 240 \text{ V} = 242 \text{ A}$$

Further Demand Factor *[220.61(B)]*

200 A at 100%	200 A
242 A - 200 A = 42 A at 70%	29 A
Net Calculated Load (neutral)	229 A

Minimum Size Main Feeders (or Service Conductors) Required (Less House Load) (For 40 Dwelling Units — 20 with Ranges)

Total Calculated Load:

Lighting and Small-Appliance Load	
40 units × 5520 V	220,800 VA

Application of Demand Factor *(from Table 220.45)*

First 3000 VA at 100%	3,000 VA
Next 120,000 VA - 3000 VA = 117,000 VA at 35%	40,950 VA
Remainder 220,800 VA - 120,000 VA = 100,800 VA at 25%	25,200 VA
Net Calculated Load	69,150 VA
Range Load: 20 ranges (less than 12 kVA)	
(see Col. C, Table 220.55)	35,000 VA
Net Calculated Load	104,150 VA

For 120/240-V, 3-wire system

Net calculated load of $104,150 \text{ VA} \div 240 \text{ V} = 434 \text{ A}$

Feeder Neutral

Lighting and Small-Appliance Load	69,150 VA
Range: 35,000 VA at 70% *[see 220.61(B)]*	24,500 VA
Calculated Load (neutral)	93,650 VA

$$93,650 \text{ VA} \div 240 \text{ V} = 390 \text{ A}$$

Further Demand Factor *[see 220.61(B)]*

200 A at 100%	200 A
390 A − 200 A = 190 A at 70%	133 A
Net Calculated Load (neutral)	333 A

[See Table 310.16 through Table 310.21, and 310.15(B), (C), and (E).]

Example D4(b) Optional Calculation for Multifamily Dwelling

A multifamily dwelling equipped with electric cooking and space heating or air conditioning has 40 dwelling units.

Meters are in two banks of 20 each plus house metering and individual feeders to each dwelling unit.

Each dwelling unit is equipped with an electric range of 8-kW nameplate rating, four 1.5-kW separately controlled 240-V electric space heaters, and a 2.5-kW, 240-V electric water heater. Assume range, space heater, and water heater kW ratings equivalent to kVA. Calculate the load for the individual dwelling unit by the standard calculation (Part III of Article 220).

A common laundry facility is available to all tenants *[see 210.52(F), Exception No. 1]*.

Area of each dwelling unit is 840 ft^2.

Calculated Load for Each Dwelling Unit *(see Part II and Part III of Article 220)*

General Lighting Load:

840 ft^2 at 3 VA/ft^2	2,520 VA
Electric range	8,000 VA
Electric heat: 6 kVA (or air conditioning if larger)	6,000 VA
Electric water heater	2,500 VA

Minimum Number of Branch Circuits Required for Each Dwelling Unit

General Lighting Load: 2520 VA ÷ 120 V = 21 A or two 15-A, 2-wire circuits, or two 20-A, 2-wire circuits

Small-Appliance Load: Two 2-wire circuits of 12 AWG *[see 210.11(C)(1)]*

Range Circuit *(See Table 220.55, Column B)*:

8000 VA × 80% ÷ 240 V = 27 A on a circuit of three 10 AWG conductors in accordance with 210.19(C)

Space Heating: 6000 VA ÷ 240 V = 25 A Number of circuits *(see 210.11)*

Minimum Size Feeder Required for Each Dwelling Unit *(see 215.2)*

Calculated Load *(see Article 220)*:

General Lighting	2,520 VA
Small Appliance (two 20-A circuits)	3,000 VA
Subtotal Calculated Load (without range and space heating)	5,520 VA

Application of Demand Factor

First 3000 VA at 100%	3,000 VA
5520 VA - 3000 VA = 2520 VA at 35%	882 VA
Net Calculated Load	3,882 VA
(without range and space heating)	
Range	6,400 VA
Space Heating *(see 220.51)*	6,000 VA
Water Heater	2,500 VA
Net Calculated Load (for individual dwelling unit)	18,782 VA

Size of Each Feeder

For 120/240-V, 3-wire system,

Net calculated load of 18,782 VA ÷ 240 V = 78 A

Feeder Neutral *(see 220.61)*

Lighting and Small Appliance	3,882 VA
Range Load: 6400 VA at 70% *[see 220.61(B)]*	4,480 VA
Space and Water Heating (no neutral): 240 V	0 VA
Net Calculated Load (neutral)	8,362 VA

Calculated Load for Neutral

8362 VA ÷ 240 V = 35 A

Minimum Size Feeder Required from Service Equipment to Meter Bank (For 20 Dwelling Units)

Δ Total Calculated Load:

Lighting and Small-Appliance Load	
20 units × 5520 V	110,400 VA
Water and Space Heating Load	
20 units × 8500 V	170,000 VA
Range Load: 20 × 8000 V	160,000 VA
Net Calculated Load (20 dwelling units)	440,400 VA
Net Calculated Load Using Optional Calculation *(see Table 220.84(B))*	
440,400 VA × 0.38	167,352 VA

167,352 VA ÷ 240 V = 697 A

Minimum Size Main Feeder Required (Less House Load) (For 40 Dwelling Units)

Calculated Load:

Lighting and Small-Appliance Load	
40 units × 5520 V	220,800 VA
Water and Space Heating Load	
40 units × 8500 V	340,000 VA
Range: 40 ranges × 8000 V	320,000 VA
Net Calculated Load (40 dwelling units)	880,800 VA

Net Calculated Load Using Optional Calculation *(see Table 220.84(B))*

880,800 VA × 0.28 = 246,624 VA
246,624 VA ÷ 240 V = 1028 A

Feeder Neutral Load for Feeder from Service Equipment to Meter Bank (For 20 Dwelling Units)

Lighting and Small-Appliance Load	
20 units × 5520 V	110,400 VA
First 3000 VA at 100%	3,000 VA
110,400 VA - 3000 VA = 107,400 VA at 35%	37,590 VA
Net Calculated Load	40,590 VA
20 ranges: 35,000 VA at 70% *[see Table 220.55 and 220.61(B)]*	24,500 VA
Total	65,090 VA

65,090 VA ÷ 240 V = 271 A

Further Demand Factor *[see 220.61(B)]*

First 200 A at 100%	200 A
Balance: 271 A - 200 A = 71 A at 70%	50 A
Total	250 A

Feeder Neutral Load of Main Feeder (Less House Load) (For 40 Dwelling Units)

Lighting and Small-Appliance Load	
40 units × 5520 V	220,800 VA
First 3000 VA at 100%	3,000 VA
Next 120,000 VA - 3000 VA = 117,000 VA at 35%	40,950 VA
Remainder 220,800 VA - 120,000 VA = 100,800 VA at 25%	25,200 VA
Net Calculated Load	69,150 VA
40 ranges: 55,000 VA at 70% *[see Table 220.55 and 220.61(B)]*	38,500 VA
Total	107,650 VA

107,650 VA ÷ 240 V = 449 A

Further Demand Factor *[see 220.61(B)]*

First 200 A at 100%	200 A
Balance: 449 - 200 A = 249 A at 70%	174 A
Total	374 A

Example D5(a) Multifamily Dwelling Served at 208Y/120 Volts, Three Phase

All conditions and calculations are the same as for the multifamily dwelling [Example D4(a)] served at 120/240 V, single phase except as follows:

Service to each dwelling unit would be two phase legs and neutral.

Minimum Number of Branch Circuits Required for Each Dwelling Unit *(see 210.11)*

Range Circuit: 8000 VA ÷ 208 V = 38 A or a circuit of two 8 AWG conductors and one 10 AWG conductor in accordance with 210.19(C)

Minimum Size Feeder Required for Each Dwelling Unit *(see 215.2)*

For 120/208-V, 3-wire system (without ranges),

Net calculated load of 3882 VA ÷ 2 legs ÷ 120 V/leg = 16 A

For 120/208-V, 3-wire system (with ranges),

Net calculated load (range) of 8000 VA ÷ 208 V = 39 A

Total load (range + lighting) = 39 A + 16 A = 55 A

Reducing the neutral load on the feeder to each dwelling unit is not permitted *[see 220.61(C)(1)]*.

Minimum Size Feeders Required from Service Equipment to Meter Bank (For 20 Dwelling Units — 10 with Ranges)

For 208Y/120-V, 3-phase, 4-wire system,

Ranges: Maximum number between any two phase legs = 4
$$2 \times 4 = 8.$$

Table 220.55 demand = 23,000 VA

Per phase demand = 23,000 VA ÷ 2 = 11,500 VA

Equivalent 3-phase load = 34,500 VA

Net Calculated Load (total):

$$40,590 \text{ VA} + 34,500 \text{ VA} = 75,090 \text{ VA}$$
$$75,090 \text{ VA} \div (208 \text{ V} \times 1.732) = 208 \text{ A}$$

Feeder Neutral Size
Net Calculated Lighting and Appliance Load & Equivalent Range Load:

$$40,590 \text{ VA} + (34,500 \text{ VA at } 70\%) = 64,700 \text{ VA}$$

Net Calculated Neutral Load:

$$64,700 \text{ VA} \div (208 \text{ V} \times 1.732) = 180 \text{ A}$$

Minimum Size Main Feeder (Less House Load) (For 40 Dwelling Units — 20 with Ranges)

For 208Y/120-V, 3-phase, 4-wire system,

Ranges:

Maximum number between any two phase legs = 7
$$2 \times 7 = 14.$$

Table 220.55 demand = 29,000 VA

Per phase demand = 29,000 VA ÷ 2 = 14,500 VA

Equivalent 3-phase load = 43,500 VA

Net Calculated Load (total):

$$69,150 \text{ VA} + 43,500 \text{ VA} = 112,650 \text{ VA}$$
$$112,650 \text{ VA} \div (208 \text{ V} \times 1.732) = 313 \text{ A}$$

Main Feeder Neutral Size:

$$69,150 \text{ VA} + (43,500 \text{ VA at } 70\%) = 99,600 \text{ VA}$$
$$99,600 \text{ VA} \div (208 \text{ V} \times 1.732) = 277 \text{ A}$$

Further Demand Factor *(see 220.61)*

200 A at 100%	200.0 A
277 A - 200 A = 77 A at 70%	54 A
Net Calculated Load (neutral)	254 A

Example D5(b) Optional Calculation for Multifamily Dwelling Served at 208Y/120 Volts, Three Phase

All conditions and calculations are the same as for Optional Calculation for the Multifamily Dwelling [Example D4(b)] served at 120/240 V, single phase except as follows:

Service to each dwelling unit would be two phase legs and neutral.

Minimum Number of Branch Circuits Required for Each Dwelling Unit *(see 210.11)*

Range Circuit *(see Table 220.55, Column B):* 8000 VA at 80% ÷ 208 V = 31 A or a circuit of two 8 AWG conductors and one 10 AWG conductor in accordance with 210.19(C)

Space Heating: 6000 VA ÷ 208 V = 29 A

Two 20-ampere, 2-pole circuits required, 12 AWG conductors

Minimum Size Feeder Required for Each Dwelling Unit
120/208-V, 3-wire circuit

Net calculated load of 18,782 VA ÷ 208 V = 90 A

Net calculated load (lighting line to neutral):

3882 VA ÷ 2 legs ÷ 120 V per leg = 16.2 A

Line to line = 14,900 VA ÷ 208 V = 71.6 A

Total load = 16.2 A + 71.6 A = 88 A

Minimum Size Feeder Required for Service Equipment to Meter Bank *(for 20 Dwelling Units)*

Net Calculated Load

$$167,352 \text{ VA} \div (208 \text{ V} \times 1.732) = 465 \text{ A}$$

Feeder Neutral

Load for Feeder from Service Equipment to Meter Bank (for 20 Dwelling Units)

Lighting and Small-Appliance Load

$$20 \text{ units} \times 5520 \text{ VA} = 110,400 \text{ VA first}$$
$$3000 \text{ VA at } 100\% = 3000 \text{ VA}$$

$$110,400 \text{ VA} - 3000 \text{ VA} = 107,400 \text{ VA at } 35\% = 37,590 \text{ VA}$$

Net Calculated Load: 40,590 VA

Minimum Size Main Feeder (Less House Load) (for 20 Dwelling Units – 20 Ranges) for 208Y/120-V, 3-Phase, 4-Wire System

Ranges:

Maximum number between any two phase legs = 7

$$2 \times 7 = 14$$

Table 220.55 demand = 29,000 VA

Per phase demand = 29,000 VA ÷ 2 = 14,500 VA

Equivalent 3-phase load = 43,500 VA

Net Calculated Neutral Load (total): 40,590 VA + 43,500 VA = 84,090 VA

$$84,090 \text{ VA} \div (208 \text{ V} \times 1.732) = 234 \text{ A}$$

Minimum Size Service Required (Less House Load) (for 40 Dwelling Units) (Assume Balanced Load)

Net Calculated Load:

$$246,624 \text{ VA} \div (208 \text{ V} \times 1.732) = 685 \text{ A}$$

Feeder Neutral Load for Feeder from Service Equipment to Meter Bank (for 40 Dwelling Units)

Lighting and Small-Appliance Load

$$40 \text{ units} \times 5520 \text{ VA} = 220,800 \text{ VA}$$

First 3000 VA at 100% = 3000 VA

Next 120,000 VA − 3000 VA = 117,000 VA

$$117,000 \text{ VA at } 35\% = 40,950 \text{ VA}$$

Remainder 220,800 VA − 120,000 VA = 100,800 VA at 25% = 25,200 VA

Net Calculated Load: 69,150 VA

Minimum Size Main Feeder (Less House Load) (for 40 Dwelling Units — 40 with Ranges) for 208Y/120-V, 3-Phase, 4-Wire System

Ranges:

Maximum number between any two phase legs = 14

$$2 \times 14 = 28$$

Table 220.55 demand = 15,000 VA + (1000 VA × 28) = 43,000 VA per phase

Demand = 43,000 VA ÷ 2 = 21,500 VA

Equivalent 3-phase load = 21,500 VA × 3 = 64,500 VA

Neutral Load = 64,500 VA at 70% [See Table 220.55 and 220.61(B)] Neutral Load = 45,150 VA

Net Calculated Neutral Load (total): 69,150 VA + 45,150 VA = 114,300 VA

$$114,300 \text{ VA} \div (208 \text{ V} \times 1.732) = 317 \text{ A}$$

Further Demand Factor *[see 220.61(B)]*

200 A at 100%	200 A
317 A − 200 A = 117 A at 70%	82 A
Net Calculated Load (neutral)	282 A

Example D6 Maximum Demand for Range Loads

Table 220.55, Column C, applies to ranges not over 12 kW. The application of Note 1 to ranges over 12 kW (and not over 27 kW) and Note 2 to ranges over 8¾ kW (and not over 27 kW) is illustrated in the following two examples.

A. Ranges All the Same Rating *(see Table 220.55, Note 1)*
Assume 24 ranges, each rated 16 kW.

From Table 220.55, Column C, the maximum demand for 24 ranges of 12-kW rating is 39 kW. 16 kW exceeds 12 kW by 4.

5% × 4 = 20% (5% increase for each kW in excess of 12)

39 kW × 20% = 7.8 kW increase

39 + 7.8 = 46.8 kW (value to be used in selection of feeders)

B. Ranges of Unequal Rating *(see Table 220.55, Note 2)*
Assume 5 ranges, each rated 11 kW; 2 ranges, each rated 12 kW; 20 ranges, each rated 13.5 kW; 3 ranges, each rated 18 kW.

5 ranges	× 12 kW =	60 kW (use 12 kW for range rated less than 12)
2 ranges	× 12 kW =	24 kW
20 ranges	× 13.5 kW =	270 kW
3 ranges	× 18 kW =	54 kW
30 ranges, Total kW =		408 kW

408 ÷ 30 ranges = 13.6 kW (average to be used for calculation)

From Table 220.55, Column C, the demand for 30 ranges of 12-kW rating is 15 kW + 30 (1 kW × 30 ranges) = 45 kW. 13.6 kW exceeds 12 kW by 1.6 kW (use 2 kW).

5% × 2 = 10% (5% increase for each kW in excess of 12 kW)

45 kW × 10% = 4.5 kW increase

45 kW + 4.5 kW = 49.5 kW (value to be used in selection of feeders)

Example D7 Sizing of Service Conductors for Dwelling(s)

Service conductors and feeders for certain dwellings are permitted to be sized in accordance with 310.12.

With No Required Adjustment or Correction Factors. If a 175-ampere service rating is selected, a service conductor is then sized as follows:

175 amperes × 0.83 = 145.25 amperes per 310.12.

If no other adjustments or corrections are required for the installation, then, in accordance with Table 310.16, a 1/0 AWG Cu or a 3/0 AWG Al meets this rating at 75°C (167°F).

With Required Temperature Correction Factor. If a 175-ampere service rating is selected, a service conductor is then

175 amperes × 0.83 = 145.25 amperes per 310.12.

If the conductors are installed in an ambient temperature of 38°C (100°F), the conductor ampacity must be multiplied by the appropriate correction factor in Table 310.15(B)(1)(1). In this case, we will use an XHHW-2 conductor, so we use a correction factor of 0.91 to find the minimum conductor ampacity and size:

145.25/.91 = 159.6 amperes

In accordance with Table 310.16, a 2/0 AWG Cu or a 4/0 AWG Al would be required.

If no temperature correction or ampacity adjustment factors are required, the following table includes conductor sizes calculated using the requirements in 310.12. This table is based on 75°C terminations and without any adjustment or correction factors.

| Service or Feeder Rating (Amperes) | Conductor (AWG or kcmil) | |
	Copper	Aluminum or Copper-Clad Aluminum
100	4	2
110	3	1
125	2	1/0
150	1	2/0
175	1/0	3/0
200	2/0	4/0
225	3/0	250
250	4/0	300
300	250	350
350	350	500
400	400	600

Example D8 Motor Circuit Conductors, Overload Protection, and Short-Circuit and Ground-Fault Protection
(see 240.6, 430.6, 430.22, 430.23, 430.24, 430.32, 430.52, and 430.62, Table 430.52(C)(1), and Table 430.250)

Determine the minimum required conductor ampacity, the motor overload protection, the branch-circuit short-circuit and ground-fault protection, and the feeder protection, for three induction-type motors on a 480-V, 3-phase feeder, as follows:

(a) One 25-hp, 460-V, 3-phase, squirrel-cage motor, nameplate full-load current 32 A, Design B, Service Factor 1.15

(b) Two 30-hp, 460-V, 3-phase, wound-rotor motors, nameplate primary full-load current 38 A, nameplate secondary full-load current 65 A, 40°C rise

Conductor Ampacity The full-load current value used to determine the minimum required conductor ampacity is obtained from Table 430.250 *[see 430.6(A)]* for the squirrel-cage motor and the primary of the wound-rotor motors. To obtain the minimum required conductor ampacity, the full-load current is multiplied by 1.25 *[see 430.22 and 430.23(A)]*.

For the 25-hp motor,

$$34 \text{ A} \times 1.25 = 43 \text{ A}$$

For the 30-horsepower motors,

$$40 \text{ A} \times 1.25 = 50 \text{ A}$$
$$65 \text{ A} \times 1.25 = 81 \text{ A}$$

Motor Overload Protection Where protected by a separate overload device, the motors are required to have overload protection rated or set to trip at not more than 125% of the nameplate full-load current *[see 430.6(A) and 430.32(A)(1)]*.

For the 25-hp motor,

$$32 \text{ A} \times 1.25 = 40.0 \text{ A}$$

For the 30-hp motors,

$$38 \text{ A} \times 1.25 = 48 \text{ A}$$

Where the separate overload device is an overload relay (not a fuse or circuit breaker), and the overload device selected at 125% is not sufficient to start the motor or carry the load, the trip setting is permitted to be increased in accordance with 430.32(C).

Branch-Circuit Short-Circuit and Ground-Fault Protection The selection of the rating of the protective device depends on the type of motor and the protective device selected, in accordance with 430.52 and Table 430.52(C)(1). The following is for the 25-hp squirrel-cage motor:

(a) Nontime-Delay Fuse: The fuse rating is 300% × 34 A = 102 A. The next larger standard fuse is 110 A *[see 240.6 and 430.52(C)(1), Exception No. 1]*. If the motor will not start with a 110-A nontime-delay fuse, the fuse rating is permitted to be increased to 125 A because this rating does not exceed 400% *[see 430.52(C)(1), Exception No. 2(1)]*.

(b) Time-Delay Fuse: The fuse rating is 175% × 34 A = 59.5 A. The next larger standard fuse is 60 A *[see 240.6 and 430.52(C)(1), Exception No. 1]*. If the motor will not start with a 60-A time-delay fuse, the fuse rating is permitted to be increased to 70 A because this rating does not exceed 225% *[see 430.52(C)(1), Exception No. 2(2)]*.

The following is for the 30-hp wound-rotor motors:

(a) Nontime-Delay Fuse: The fuse rating is 150% × 40 A = 60 A. If the motor will not start with a 60-A nontime-delay fuse, the fuse rating is permitted to be increased to 150 A because this rating does not exceed 400% *[see 430.52(C)(1), Exception No. 2(1)]*.

(b) Time-Delay Fuse: The fuse rating is 150% × 40 A = 60 A. If the motor will not start with a 60-A time-delay fuse, the fuse rating is permitted to be increased to 90 A because this rating does not exceed 225% *[see 430.52(C)(1), Exception No. 2(2)]*.

(c) Inverse Time Circuit Breaker: The breaker rating is 150% × 40 A = 60 A. If the motor will not start with a 60-A breaker, the breaker rating is permitted to be increased to 150 A because this rating does not exceed 400% *[see 430.52(C)(1), Exception No. 2(3) for motor full-load currents of 100 A or less]*.

Feeder Short-Circuit and Ground-Fault Protection

(a) Example using nontime-delay fuse. The rating of the feeder protective device is based on the sum of the largest branch-circuit protective device for the specific type of device protecting the feeder. In the previous step above, the calculation for the 25 hp squirrel-cage motor results in the largest branch-circuit protective device: 300% × 34 A = 102 A (therefore the next largest standard size, 110 A, would be used) plus the sum of the full-load currents of the other motors, or 110 A + 40 A + 40 A = 190 A. The nearest standard fuse that does not exceed this value is 175 A *[see 240.6, Table 430.52(C)(1), and 430.62(A)]*.

(b) Example using inverse time circuit breaker. The largest branch-circuit protective device for the specific type of device protecting the feeder. The calculation for the 25-hp squirrel-cage motor results in the largest branch-circuit protective device, 250% × 34 A = 85. The next larger standard size is 90 A, plus the sum of the full-load currents of the other motors, or 90 A + 40 A + 40 A = 170 A. The nearest standard inverse time circuit breaker that does not exceed this value is 150 A *[see 240.6, Table 430.52(C)(1), and 430.62(A)]*.

Example D9 Feeder Ampacity Determination for Generator Field Control
[see 215.2, 430.24, 430.24 Exception No. 1, 620.13, 620.14, 620.61, and Table 430.22(E) and 620.14]

Determine the conductor ampacity for a 460-V 3-phase, 60-Hz ac feeder supplying a group of six elevators. The 460-V ac drive motor nameplate rating of the largest MG set for one elevator is 40 hp and 52 A, and the remaining elevators each have a 30-hp, 40-A, ac drive motor rating for their MG sets. In addition to a motor controller, each elevator has a separate motion/operation controller rated 10 A continuous to operate microprocessors, relays, power supplies, and the elevator car door operator. The MG sets are rated continuous.

Conductor Ampacity Conductor ampacity is determined as follows:

(a) In accordance with 620.13(D) and 620.61(B)(1), use Table 430.22(E), for intermittent duty (elevators). For intermittent duty using a continuous rated motor, the percentage of nameplate current rating to be used is 140%.

(b) For the 30-hp ac drive motor,

$$140\% \times 40 \text{ A} = 56 \text{ A}$$

(c) For the 40-hp ac drive motor,

$$140\% \times 52 \text{ A} = 73 \text{ A(I)}$$

(d) The total conductor ampacity is the sum of all the motor currents:

$$(1 \text{ motor} \times 73 \text{ A}) + (5 \text{ motors} \times 56 \text{ A}) = 353 \text{ A}$$

(e) In accordance with 620.14 and Table 620.14, the conductor (feeder) ampacity would be permitted to be reduced by the use of a demand factor. Constant loads are not included *(see 620.14, Informational Note)*. For six elevators, the demand factor is 0.79. The feeder diverse ampacity is, therefore, 0.79 × 353 A = 279 A.

(f) In accordance with 430.24 and 215.3, the controller continuous current is 125% × 10 A = 13 A.

(g) The total feeder ampacity is the sum of the diverse current and all the controller continuous current.

$$I_{total} = 279 \text{ A} + (6 \text{ elevators} \times 12.5 \text{ A}) = 354 \text{ A}$$

(h) This ampacity would be permitted to be used to select the wire size.

See Figure D9.

Example D10 Feeder Ampacity Determination for Adjustable Speed Drive Control
[see 215.2, 430.24, 620.13, 620.14, 620.61, and Table 430.22(E)]

Determine the conductor ampacity for a 460-V, 3-phase, 60-Hz ac feeder supplying a group of six identical elevators. The system is adjustable-speed SCR dc drive. The power transformers are external to the drive (motor controller) cabinet. Each elevator has a separate motion/operation controller connected to the load side of the main line disconnect switch rated 10 A continuous to operate microprocessors, relays, power supplies, and the elevator car door operator. Each transformer is rated 95 kVA with an efficiency of 90%.

Conductor Ampacity
Conductor ampacity is determined as follows:

(a) Calculate the nameplate rating of the transformer:

$$I = \frac{95 \text{ kVA} \times 1000}{\sqrt{3} \times 460 \text{ V} \times 0.90_{eff.}} = 133 \text{ A} \qquad \textbf{[D10]}$$

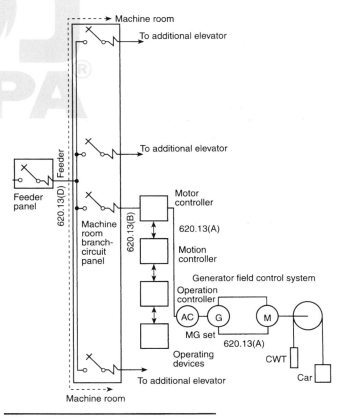

FIGURE D9 *Generator Field Control.*

(b) In accordance with 620.13(D), for six elevators, the total conductor ampacity is the sum of all the currents.

$$6 \text{ elevators} \times 133 \text{ A} = 798 \text{ A}$$

(c) In accordance with 620.14 and Table 620.14, the conductor (feeder) ampacity would be permitted to be reduced by the use of a demand factor. Constant loads are not included *(see 620.13, Informational Note No. 2)*. For six elevators, the demand factor is 0.79. The feeder diverse ampacity is, therefore, 0.79 × 798 A = 630 A.

(d) In accordance with 430.24 and 215.3, the controller continuous current is 125% × 10 A = 13 A.

(e) The total feeder ampacity is the sum of the diverse current and all the controller constant current.

$$I_{\text{total}} = 630 \text{ A} + (6 \text{ elevators} \times 12.5 \text{ A}) = 705 \text{ A}$$

(f) This ampacity would be permitted to be used to select the wire size.

See Figure D10.

Example D11 Mobile Home *(see 550.18)*

A mobile home floor is 70 ft by 10 ft and has two small appliance circuits; a 1000-VA, 240-V heater; a 200-VA, 120-V exhaust fan; a 400-VA, 120-V dishwasher; and a 7000-VA electric range.

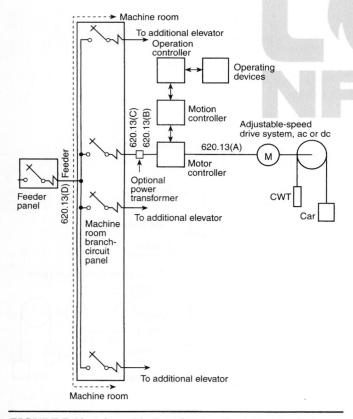

FIGURE D10 *Adjustable Speed Drive Control.*

Lighting and Small-Appliance Load

Lighting (70 ft × 10 ft × 3 VA per ft²)	2,100 VA
Small-appliance (1500 VA × 2 circuits)	3,000 VA
Laundry (1500 VA × 1 circuit)	1,500 VA
Subtotal	6,600 VA
First 3000 VA at 100%	3,000 VA
Remainder (6600 VA – 3000 VA = 3600 VA) × 35%	1,260 VA
Total	4,260 VA

$$4260 \text{ VA} \div 240 \text{ V} = 17.75 \text{ A per leg}$$

Amperes per Leg	Leg A	Leg B
Lighting and appliances	18	18
Heater (1000 VA ÷ 240 V)	4	4
Fan (200 VA × 125% ÷ 120 V)	2	—
Dishwasher (400 VA ÷ 120 V)	—	3
Range (7000 VA × 0.8 ÷ 240 V)	23	23
Total amperes per leg	47	48

Based on the higher current calculated for either leg, a minimum 50-A supply cord would be required.

For SI units, 0.093 m² = 1 ft² and 0.3048 m = 1 ft.

Example D12 Park Trailer *(see 552.47)*

A park trailer floor is 40 ft by 10 ft and has two small appliance circuits, a 1000-VA, 240-V heater, a 200-VA, 120-V exhaust fan, a 400-VA, 120-V dishwasher, and a 7000-VA electric range.

Lighting and Small-Appliance Load

Lighting (40 ft × 10 ft × 3 VA per ft²)	1,200 VA
Small-appliance (1500 VA × 2 circuits)	3,000 VA
Laundry (1500 VA × 1 circuit)	1,500 VA
Subtotal	5,700 VA
First 3000 VA at 100%	3,000 VA
Remainder (5700 VA – 3000 VA = 2700 VA) × 35%	945 VA
Total	3,945 VA

$$3945 \text{ VA} \div 240 \text{ V} = 16.44 \text{ A per leg}$$

Amperes per Leg	Leg A	Leg B
Lighting and appliances	16	16
Heater (1000 VA ÷ 240 V)	4	4
Fan (200 VA × 125% ÷ 120 V)	2	—
Dishwasher (400 VA ÷ 120 V)	—	3
Range (7000 VA × 0.8 ÷ 240 V)	23	23
Totals	45	46

Based on the higher current calculated for either leg, a minimum 50-A supply cord would be required.

For SI units, 0.093 m² = 1 ft² and 0.3048 m = 1 ft.

Example D13 Cable Tray Calculations
(see Article 392)

D.13(a) Multiconductor Cables 4/0 AWG and Larger

Use: *NEC* 392.22(A)(1)(a)

Cable tray must have an inside width equal to or greater than the sum of the diameters (Sd) of the cables, which must be installed in a single layer.

Example: Cable tray width is obtained as follows:

Cable Size Being Used	(OD) Cable Outside Diameters (in.)	(N) Number of Cables	SD = (OD) × (N) (Sum of the Cable Diameters) (in.)
3–conductor Type MC cable — 4/0 AWG	1.57	12	18.84

The sum of the diameters (Sd) of all cables = 18.84 in., therefore a cable tray with an inside width of at least 18.84 in. is required.

Note: Cable outside diameter is a nominal diameter from catalog data.

D.13(b) Multiconductor Cables Smaller Than 4/0 AWG
Use: *NEC*392.22(A)(1)(b)

The sum of the cross-sectional areas of all the cables to be installed in the cable tray must be equal to or less than the allowable cable area for the tray width, as indicated in Table 392.22(A)(1), Column 1.

TABLE D13(b) *from Table 392.22(A), Column 1*

Inside Width of Cable Tray (in.)	Allowable Cable Area (in.^{2})
6	7.0
9	10.5
12	14.0
18	21.0
24	28.0
30	35.0
36	42.0

Example: Cable tray width is obtained as follows:

Cable Size Being Used	(A) Cable Cross-Sectional Area (in.^{2})	(N) Number of Cables	Multiply (A) × (N) (Which Is a Total Cable Cross-Sectional Area in in.^{2})
4-conductor Type MC cable — 1 AWG	1.1350	9	12.15

The total cable cross-sectional area is 12.15 in.^{2}. Using Table D13(b) above, the next higher allowable cable area must be used, which is 14.0 in.^{2}. The table specifies that the cable tray inside width for an allowable cable area of 14.0 in.^{2} is 12 in.

Note: Cable cross-sectional area is a nominal area from catalog data.

D.13(c) Single Conductor Cables 1/0 AWG through 4/0 AWG
Use: *NEC*392.22(B)(1)(d)

Cable tray must have an inside width equal to or greater than the sum of the diameters (Sd) of the cables. The cables must be evenly distributed across the cable tray.

Example: Cable tray width is obtained as follows:

Single Conductor Cable Size Being Used	(OD) Cable Outside Diameters (in.)	(N) Number of Cables	Sd = (OD) × (N) (Sum of the Cable Diameters) (in.)
THHN — 4/0 AWG	0.642	18	11.556

The sum of the diameters (Sd) of all cables = 11.56 in., therefore, a cable tray with an inside width of at least 11.56 in. is required.

Note: Cable outside diameter from Chapter 9, Table 5.

D.13(d) Single Conductor Cables 250 kcmil through 900 kcmil
Use: *NEC*392.22(B)(1)(b)

The sum of the cross-sectional areas of all the cables to be installed in the cable tray must be equal to or less than the allowable cable area for the tray width, as indicated in Table 392.22(B)(1), Column 1.

TABLE D13(d) *from Table 392.22(B)(1), Column 1*

Inside Width of Cable Tray (in.)	Allowable Cable Area (in.^{2})
6	6.5
9	9.5
12	13.0
18	19.5
24	26.0
30	32.5
36	39.0

Example: Cable tray width is obtained as follows:

Cable Size Being Used	(A) Cable Cross-Sectional Area (in.^{2})	(N) Number of Cables	Multiply (A) × (N) (Which Is a Total Cable Cross-Sectional Area in in.^{2})
THHN — 500 kcmil	0.707	9	6.36

The total cable cross-sectional area is 6.36 in.^{2}. Using Table D13(d), the next higher allowable cable area must be used, which is 6.5 in.^{2}. The table specifies that the cable tray inside width for an allowable cable area of 6.5 in.^{2} is 6 in.

Note: Single-conductor cable cross-sectional area from Chapter 9, Table 5.

Types of Construction

This informative annex is not a part of the requirements of this NFPA document but is included for informational purposes only.

Δ **TABLE E.1** *Fire Resistance Ratings for Type I Through Type V Construction (hr)*

	Type I		Type II			Type III		Type IV	Type V	
	442	332	222	111	000	211	200	2HH	111	000
Exterior Bearing Walls[a]										
Supporting more than one floor, columns, or other bearing walls	4	3	2	1	0[b]	2	2	2	1	0[b]
Supporting one floor only	4	3	2	1	0[b]	2	2	2	1	0[b]
Supporting a roof only	4	3	1	1	0[b]	2	2	2	1	0[b]
Interior Bearing Walls										
Supporting more than one floor, columns, or other bearing walls	4	3	2	1	0	1	0	2	1	0
Supporting one floor only	3	2	2	1	0	1	0	1	1	0
Supporting roofs only	3	2	1	1	0	1	0	1	1	0
Columns										
Supporting more than one floor, columns, or other bearing walls	4	3	2	1	0	1	0	H	1	0
Supporting one floor only	3	2	2	1	0	1	0	H	1	0
Supporting roofs only	3	2	1	1	0	1	0	H	1	0
Beams, Girders, Trusses, and Arches										
Supporting more than one floor, columns, or other bearing walls	4	3	2	1	0	1	0	H	1	0
Supporting one floor only	2	2	2	1	0	1	0	H	1	0
Supporting roofs only	2	2	1	1	0	1	0	H	1	0
Floor/Ceiling Assemblies	2	2	2	1	0	1	0	H	1	0
Roof/Ceiling Assemblies	2	1½	1	1	0	1	0	H	1	0
Interior Nonbearing Walls	0	0	0	0	0	0	0	0	0	0
Exterior Nonbearing Walls[c]	0[b]	0[b]	0[b]	0[b]	0[b]	0[b]	0[b]	0[b]	0[b]	0[b]

Source: Table 7.2.1.1 from *NFPA 5000, Building Construction and Safety Code*, 2021 edition.

H: Heavy timber members.

[a]See 7.3.2.1 in *NFPA 5000*.

[b]See Section 7.3 in *NFPA 5000*.

[c]See 7.2.3.2.12, 7.2.4.2.3, and 7.2.5.6.8 in *NFPA 5000*.

Table E.1 contains the fire resistance rating, in hours, for Types I through V construction. The five different types of construction can be summarized briefly as follows (see also Table E.2):

Type I is a fire-resistive construction type. All structural elements and most interior elements are required to be noncombustible. Interior, nonbearing partitions are permitted to be 1- or 2-hour rated. For nearly all occupancy types, Type I construction can be of unlimited height.

Type II construction has three categories: fire-resistive, one-hour rated, and nonrated. The number of stories permitted for multifamily dwellings varies from two for nonrated and four for one-hour rated to 12 for fire-resistive construction.

Type III construction has two categories: one-hour rated and nonrated. Both categories require the structural framework and exterior walls to be of noncombustible material. One-hour rated construction requires all interior partitions to be one-hour rated. Nonrated construction allows nonbearing interior partitions to be of nonrated construction. The maximum permitted number of stories for multifamily dwellings and other structures is two for nonrated and four for one-hour rated.

Type IV construction includes traditional heavy timber construction and mass timber construction. In heavy timber construction, the structural framework and the exterior walls are required to be noncombustible except that wood members of certain minimum sizes are allowed. In mass timber construction, structural elements of cross-laminated timber (CLT) are permitted. Allowable building height for mass timber is much higher than for heavy timber.

Type V construction has two categories: one-hour rated and nonrated. One-hour rated construction requires a minimum of one-hour rated construction throughout the building. Nonrated construction allows nonrated interior partitions with certain restrictions. The maximum permitted number of stories for multifamily dwellings and other structures is two for nonrated and three for one-hour rated.

In Table E.1 the system of designating types of construction also includes a specific breakdown of the types of construction through the use of arabic numbers. These arabic numbers follow the roman numeral notation where identifying a type of construction [for example, Type I(442), Type II(111), Type III(200)] and indicate the fire resistance rating requirements for certain structural elements as follows:

(1) First arabic number — exterior bearing walls
(2) Second arabic number — columns, beams, girders, trusses and arches, supporting bearing walls, columns, or loads from more than one floor
(3) Third arabic number — floor construction

Table E.3 provides a comparison of the types of construction for various model building codes. [*5000*: A.7.2.1.1]

Table E.3 is reproduced from *NFPA 5000®*, *Building Construction and Safety Code®*. This table cross-references the building construction types described in NFPA 220, *Standard on Types of Building Construction*, to the construction types described in four other model building codes. For AHJs in a municipality where one of those model building codes is used, this table provides helpful information to assist in the proper application of 334.10(2), 334.10(3), 334.10(4), and 334.10(5) covering the permitted use of Type NM cable based on building construction type. The types of construction are based on NFPA 220, and Table E.3 facilitates assimilation of the 334.10 and 334.12 requirements with the construction types contained within the building code that is adopted by a jurisdiction.

The following is a description of the model building code acronyms contained in Table E.3:

UBC — *Uniform Building Code*, International Conference of Building Officials, Whittier, CA

B/NBC — *BOCA National Building Code*, Building Officials and Code Administrators International, Inc., Country Club Hills, IL

SBC — *Standard Building Code*, Southern Building Code Congress International, Inc., Birmingham, AL

IBC — *International Building Code*, International Code Council, Inc., Falls Church, VA

The information for the UBC, B/NBC, and SBC is included as reference for jurisdictions using those building codes. The publishers of those three codes founded the International Code Council in 1994 and combined the regional codes into a single document, the *International Building Code (IBC)*.

Δ **TABLE E.2** *Maximum Number of Stories for Types V, IV, and III Construction*

Construction Type	Maximum Number of Stories Permitted
V Nonrated	2
V Nonrated, Sprinklered	3
V One-Hour Rated	3
V One-Hour Rated, Sprinklered	4
IV Heavy Timber	4
IV Heavy Timber, Sprinklered	5
IV Mass Timber	12
III Nonrated	2
III Nonrated, Sprinklered	3
III One-Hour Rated	4
III One-Hour Rated, Sprinklered	5

Δ **TABLE E.3** *Cross-Reference of Building Construction Types*

NFPA 5000	I (442)	I (332)	II (222)	II (111)	II (000)	III (211)	III (200)	IV (2HH)	V (111)	V (000)
UBC	—	I FR	II FR	II 1 hr	II N	III 1 hr	III N	IV HT	V 1 hr	V N
B/NBC	1A	1B	2A	2B	2C	3A	3B	4	5A	5B
SBC	I	II	—	IV 1 hr	IV UNP	V 1 hr	V UNP	III	VI 1 hr	VI UNP
IBC	—	IA	IB	IIA	IIB	IIIA	IIIB	IV*	VA	VB

*Mass timber in the IBC is Type IV A, IV B, and IV C

Source: Table A.7.2.1.1 from *NFPA 5000, Building Construction and Safety Code*, 2021 edition.

UBC: *Uniform Building Code*.

FR: Fire rated.

N: Nonsprinklered.

HT: Heavy timber.

B/NBC: *National Building Code*.

SBC: *Standard Building Code*.

UNP: Unprotected.

IBC: *International Building Code*.

Availability and Reliability for Critical Operations Power Systems; and Development and Implementation of Functional Performance Tests (FPTs) for Critical Operations Power Systems

INFORMATIVE ANNEX F

This informative annex is not a part of the requirements of this NFPA document but is included for informational purposes only.

Δ **I. Availability and Reliability for Critical Operations Power Systems.** Critical operations power systems may support facilities with a variety of objectives that are vital to public safety. Often these objectives are of such critical importance that system downtime is costly in terms of economic losses, loss of security, or loss of mission. For those reasons, the availability of the critical operations power system, the percentage of time that the system is in service, is important to those facilities. Given a specified level of availability, the reliability and maintainability requirements are then derived based on that availability requirement.

Availability. Availability is defined as the percentage of time that a system is available to perform its function(s). Availability is measured in a variety of ways, including the following:

$$\text{Availability} = \frac{MTBF}{MTTF + MTTR} \quad [F.1]$$

where:

$MTBF$ = mean time between failures
$MTTF$ = mean time to failure
$MTTR$ = mean time to repair

See Table F.1 for an example of how to establish required availability for critical operation power systems:

TABLE F.1 *Availability for Critical Operation Power Systems*

Availability	Hours of Downtime
0.9	876
0.99	87.6
0.999	8.76
0.9999	0.876
0.99999	0.0876
0.999999	0.00876
0.9999999	0.000876

Note: Based on a year of 8760 hours.

Availability of a system in actual operations is determined by the following:

(1) The frequency of occurrence of failures. Failures may prevent the system from performing its function or may cause a degraded effect on system operation. Frequency of failures is directly related to the system's level of reliability.
(2) The time required to restore operations following a system failure or the time required to perform maintenance to prevent a failure. These times are determined in part by the system's level of maintainability.
(3) The logistics provided to support maintenance of the system. The number and availability of spares, maintenance personnel, and other logistics resources (refueling, etc.) combined with the system's level of maintainability determine the total downtime following a system failure.

Reliability. Reliability is concerned with the probability and frequency of failures (or lack of failures). A commonly used measure of reliability for repairable systems is *MTBF*. The equivalent measure for nonrepairable items is *MTTF*. Reliability is more accurately expressed as a probability over a given duration of time, cycles, or other parameter. For example, the reliability of a power plant might be stated as 95 percent probability of no failure over a 1000-hour operating period while generating a certain level of power. Reliability is usually defined in two ways (the electrical power industry has historically not used these definitions):

(1) The duration or probability of failure-free performance under stated conditions
(2) The probability that an item can perform its intended function for a specified interval under stated conditions [For nonredundant items, this is equivalent to the preceding definition (1). For redundant items, this is equivalent to the definition of mission reliability.]

Maintainability. Maintainability is a measure of how quickly and economically failures can be prevented through preventive maintenance, or system operation can be restored following failure through corrective maintenance. A commonly used measure of maintainability in terms of corrective maintenance is the mean time to repair (*MTTR*). Maintainability is not the same thing as maintenance. It is a design parameter, while maintenance consists of actions to correct or prevent a failure event.

Improving Availability. The appropriate methods to use for improving availability depend on whether the facility is being designed or is already in use. For both cases, a reliability/availability analysis should be performed to determine the availability of the old system or proposed new system in order to ascertain the hours of downtime (see the preceding table). The AHJ or government agency should dictate how much downtime is acceptable.

Existing facilities: For a facility that is being operated, two basic methods are available for improving availability when the current level of availability is unacceptable: (1) Selectively adding redundant units (e.g., generators, chillers, fuel supply) to eliminate sources of single-point failure, and (2) optimizing maintenance using a reliability-centered maintenance (RCM) approach to minimize downtime. (Refer to NFPA 70B-2019, *Recommended Practice for Electrical Equipment Maintenance.*) A combination of the previous two methods can also be implemented. A third very expensive method is to redesign subsystems or to replace components and subsystems with higher reliability items. *(Refer to NFPA 70B.)*

New facilities: The opportunity for high availability and reliability is greatest when designing a new facility. By applying an effective reliability strategy, designing for maintainability, and ensuring that manufacturing and commissioning do not negatively affect the inherent levels of reliability and maintainability, a highly available facility will result. The approach should be as follows:

(1) *Develop and determine a reliability strategy* (establish goals, develop a system model, design for reliability, conduct reliability development testing, conduct reliability acceptance testing, design system delivery, maintain design reliability, maintain design reliability in operation).

(2) *Develop a reliability program.* This is the application of the reliability strategy to a specific system, process, or function. Each step in the preceding strategy requires the selection and use of specific methods and tools. For example, various tools can be used to develop requirements or evaluate potential failures. To derive requirements, analytical models can be used, for example, quality function development (a technique for deriving more detailed, lower-level requirements from one level to another, beginning with mission requirements, i.e., customer needs). This model was developed as part of the total quality management movement. Parametric models can also be used to derive design values of reliability from operational values and vice versa. Analytical methods include but are not limited to things such as thermal analysis, durability analysis, and predictions. Finally, one should evaluate possible failures. A failure modes and effects criticality analysis (FMECA) and fault tree analysis (FTA) are two methods for evaluating possible failures. The mission facility engineer should determine which method to use or whether to use both.

(3) *Identify reliability requirements.* The entire effort for designing for reliability begins with identifying the mission critical facility's reliability requirements. These requirements are stated in a variety of ways, depending on the customer and the specific system. For a mission-critical facility, it would be the mission success probability.

Informational Note: For information regarding power system reliability, see IEEE 3006.5-2014, *Recommended Practice for the Use of Probability Methods for Conducting a Reliability Analysis of Industrial and Commercial Power Systems.*

II. Development and Implementation of Functional Performance Tests (FPTs) for Critical Operations Power Systems Development of FPT.

(1) Submit Functional Performance Tests (FPTs). System/component tests or FPTs are developed from submitted drawings, systems operating documents (SODs), and systems operation and maintenance manuals (SOMMs), including large component testing (i.e., transformers, cable, generators, UPS), and how components operate as part of the total system. The commissioning authority develops the test and cannot be the installation contractor (or subcontractor).

As the equipment/components/systems are installed, quality assurance procedures are administered to verify that components are installed in accordance with minimum manufacturers' recommendations, safety codes, and acceptable installation practices. Quality assurance discrepancies are then identified and added to a "commissioning action list" that must be rectified as part of the commissioning program. These items would usually be discussed during commissioning meetings. Discrepancies are usually identified initially by visual inspection.

(2) Review FPTs. The tests must be reviewed by the customer, electrical contractors, quality assurance personnel, maintenance personnel, and other key personnel (the commissioning team). Areas of concern include, among others, all functions of the system being tested, all major components included, whether the tests reflect the system operating documents, and verification that the tests make sense.

(3) Make Changes to FPTs as Required. The commissioning authority then implements the corrections, questions answered, and additions.

(4) FPTs Approval. After the changes are made to the FPTs, they are submitted to the commissioning team. When it is acceptable, the customer or the designated approval authority approves the FPTs. It should be noted that even though the FPT is approved, problems that arise during the test (or areas not covered) must be addressed.

Testing Implementation for FPTs. The final step in the successful commissioning plan is testing and proper execution of system-integrated tests.

(1) Systems Ready to Operate. The FPTs can be implemented as various systems become operative (i.e., test for the generator system) or when the entire system is installed. However, the final "pull the plug" test is performed only after all systems are completely installed. If the electrical contractor (or subcontractor) implements the FPTs, a witness must initial each step of the test. The electrical contractor cannot employ the witness directly or indirectly.

(2) Perform Tests (FPTs). If the system fails the test, the problem must be resolved and the equipment or system retested or the testing requirements re-analyzed until successful tests are witnessed. Once the system or equipment passes testing, it is verified by a designated commissioning official.

(3) Customer Receives System. After all tests are completed (including the "pull the plug" test), the system is turned over to the customer.

> Informational Note: For information regarding reliability of critical operations power systems, see IEEE 3006.2-2016, *Recommended Practice for Evaluating the Reliability of Existing Industrial and Commercial Power System*s.

INFORMATIVE ANNEX

G

Supervisory Control and Data Acquisition (SCADA)

This informative annex is not a part of the requirements of this NFPA document, but is included for informational purposes only.

(A) General. Where provided, the general requirements in the following shall apply to SCADA systems:

(1) The SCADA system for the COPS loads shall be separate from the building management SCADA system.

(2) No single point failure shall be able to disable the SCADA system.

(3) The SCADA system shall be permitted to provide control and monitor electrical and mechanical utility systems related to mission critical loads, including, but not limited to, the following:

 a. The fire alarm system
 b. The security system
 c. Power distribution
 d. Power generation
 e. HVAC and ventilation (damper position, airflow speed and direction)
 f. Load shedding
 g. Fuel levels or hours of operation

(4) Before installing or employing a SCADA system, an operations and maintenance analysis and risk assessment shall be performed to provide the maintenance parameter data

(5) A redundant system shall be provided in either warm or hot standby.

(6) The controller shall be a programmable logic controller (PLC).

(7) The SCADA system shall utilize open, not proprietary, protocols.

(8) The SCADA system shall be able to assess the damage and determine system integrity after the "event."

(9) The monitor display shall provide graphical user interface for all major components monitored and controlled by the SCADA system, with color schemes readily recognized by the typical user.

(10) The SCADA system shall have the capability to provide storage of critical system parameters at a 15-minute rate or more often when out-of-limit conditions exist.

(11) The SCADA system shall have a separate data storage facility not located in the same vicinity.

(B) Power Supply. The SCADA system power supply shall comply with the following:

(1) The power supply shall be provided with a direct-current station battery system, rated between 24 and 125 volts dc, with a 72-hour capacity.

(2) The batteries of the SCADA system shall be separate from the batteries for other electrical systems.

(3) The power supply shall be provided with a listed surge-protective device (SPD). The SPD shall be installed at the line-side terminals of the power supply in accordance with Part II of Article 242, with a direct low-impedance path to ground. Protected and unprotected circuits shall be physically separated to prevent coupling.

(C) Security Against Hazards. Security against hazards shall be provided in accordance with the following:

(1) Controlled physical access by authorized personnel to only the system operational controls and software shall be provided.

(2) The SCADA system shall be protected against dust, dirt, water, and other contaminants by specifying enclosures appropriate for the environment.

(3) Conduit and tubing shall not violate the integrity of the SCADA system enclosure.

(4) The SCADA system shall be located in the same secure locations as the secured systems that they monitor and control.

(5) The SCADA system shall be provided with dry agent fire protection systems or double interlocked preaction sprinkler systems using cross-zoned detection, to minimize the

threat of accidental water discharge into unprotected equipment. The fire protection systems shall be monitored by the fire alarm system in accordance with *NFPA 72*-2019, *National Fire Alarm and Signaling Code*.

(6) The SCADA system shall not be connected to other network communications outside the secure locations without encryption or use of fiber optics.

△ **(D) Maintenance and Testing.** SCADA systems shall be maintained and tested in accordance with D(1) and D(2).

(1) Maintenance. The maintenance program for SCADA systems shall consist of the following components:

(1) A documented preventive maintenance program
(2) Concurrent maintenance capabilities, to allow the testing, troubleshooting, repair, and/or replacement of a component or subsystem while redundant component(s) or subsystem(s) are serving the load
(3) Retention of operational data — the deleted material goes well beyond requirements to ensure proper maintenance and operation

(2) Testing. SCADA systems shall be tested periodically under actual or simulated contingency conditions.

> Informational Note No. 1: Periodic system testing procedures can duplicate or be derived from the recommended functional performance testing procedures of individual components, as provided by the manufacturers.
>
> Informational Note No. 2: See NFPA 70B-2019, *Recommended Practice for Electrical Equipment Maintenance*, for more information on maintenance and testing of SCADA.

Administration and Enforcement

Informative Annex H is not a part of the requirements of this NFPA document and is included for informational purposes only. Informative Annex H is intended to provide a template and sample language for local jurisdictions adopting the National Electrical Code®.

Informative Annex H is a model set of rules for adoption by jurisdictions, in whole or in part, to administer an electrical inspection program. Unless specifically adopted by a governmental or other entity charged with enforcing electrical installation requirements, the material is advisory only. This annex can serve as a template that can be used verbatim or with amendments tailored to fit the jurisdiction's needs. It provides an inspection process covering plan review, issuance of permits, number and types of inspections, and connection to or disconnection from the electrical supply system. Processes to file and adjudicate appeals and to adopt the current edition of the *NEC®* are also provided. In addition, requirements for qualified electrical inspection personnel and for establishing an electrical board are provided.

Informative Annex H is based on, and replaces, NFPA 70L, *Model State Law, Inspection of Electrical Installations,* which was adopted by NFPA on May 15, 1973, and a second edition of which was approved on March 27, 1987.

For most political subdivisions, adoption of the *NEC* can occur in two ways. It can be incorporated in a law, or a law can be enacted authorizing a governmental agency or board to adopt it. Rule-making and legislative processes in a particular jurisdiction determine which alternative is more appropriate.

80.1 Scope. The following functions are covered:

(1) The inspection of electrical installations as covered by 90.2
(2) The investigation of fires caused by electrical installations
(3) The review of construction plans, drawings, and specifications for electrical systems
(4) The design, alteration, modification, construction, maintenance, and testing of electrical systems and equipment
(5) The regulation and control of electrical installations at special events including but not limited to exhibits, trade shows, amusement parks, and other similar special occupancies

Δ **80.2 Definitions.**

Authority Having Jurisdiction. The organization, office, or individual responsible for enforcing the requirements of a code or standard, or for approving equipment, materials, an installation, or a procedure.

Chief Electrical Inspector. An electrical inspector who either is the authority having jurisdiction or is designated by the authority having jurisdiction and is responsible for administering the requirements of this *Code.*

Electrical Inspector. An individual meeting the requirements of 80.27 and authorized to perform electrical inspections.

80.3 Purpose. The purpose of this article shall be to provide requirements for administration and enforcement of the *National Electrical Code.*

80.5 Adoption. Article 80 shall not apply unless specifically adopted by the local jurisdiction adopting the *National Electrical Code.*

80.7 Title. The title of this *Code* shall be *NFPA 70®, National Electrical Code®,* of the National Fire Protection Association. The short title of this *Code* shall be the *NEC®.*

Δ **80.9 Application.**

(A) New Installations. This *Code* applies to new installations. Buildings with construction permits dated after adoption of this *Code* shall comply with its requirements.

(B) Existing Installations. Existing electrical installations that do not comply with the provisions of this *Code* shall be permitted to be continued in use unless the authority having jurisdiction determines that the lack of conformity with this *Code* presents an imminent danger to occupants. Where changes are required for correction of hazards, a reasonable amount of time shall be given for compliance, depending on the degree of the hazard.

(C) Additions, Alterations, or Repairs. Additions, alterations, or repairs to any building, structure, or premises shall conform to that required of a new building without requiring the existing building to comply with all the requirements of this *Code*. Additions, alterations, installations, or repairs shall not cause an existing building to become unsafe or to adversely affect the performance of the building as determined by the authority having jurisdiction. Electrical wiring added to an existing service, feeder, or branch circuit shall not result in an installation that violates the provisions of the *Code* in force at the time the additions are made.

△ **80.11 Occupancy of Building or Structure.**

(A) New Construction. No newly constructed building shall be occupied in whole or in part in violation of the provisions of this *Code*.

(B) Existing Buildings. Existing buildings that are occupied at the time of adoption of this *Code* shall be permitted to remain in use provided the following conditions apply:

(1) The occupancy classification remains unchanged.
(2) There exists no condition deemed hazardous to life or property that would constitute an imminent danger.

80.13 Authority. Where used in this article, the term *authority having jurisdiction* shall include the chief electrical inspector or other individuals designated by the governing body. This *Code* shall be administered and enforced by the authority having jurisdiction designated by the governing authority as follows:

(1) The authority having jurisdiction shall be permitted to render interpretations of this *Code* in order to provide clarification to its requirements, as permitted by 90.4.
(2) When the use of any electrical equipment or its installations is found to be dangerous to human life or property, the authority having jurisdiction shall be empowered to have the premises disconnected from its source of electric supply, as established by the Board. When such equipment or installation has been so condemned or disconnected, a notice shall be placed thereon listing the causes for the condemnation, the disconnection, or both, and the penalty under 80.23 for the unlawful use thereof. Written notice of such condemnation or disconnection and the causes therefor shall be given within 24 hours to the owners, the occupant, or both, of such building, structure, or premises. It shall be unlawful for any person to remove said notice, to reconnect the electrical equipment to its source of electric supply, or to use or permit to be used electric power in any such electrical equipment until such causes for the condemnation or disconnection have been remedied to the satisfaction of the inspection authorities.
(3) The authority having jurisdiction shall be permitted to delegate to other qualified individuals such powers as necessary for the proper administration and enforcement of this *Code*.
(4) Police, fire, and other enforcement agencies shall have authority to render necessary assistance in the enforcement of this *Code* when requested to do so by the authority having jurisdiction.

(5) The authority having jurisdiction shall be authorized to inspect, at all reasonable times, any building or premises for dangerous or hazardous conditions or equipment as set forth in this *Code*. The authority having jurisdiction shall be permitted to order any person(s) to remove or remedy such dangerous or hazardous condition or equipment. Any person(s) failing to comply with such order shall be in violation of this *Code*.
(6) Where the authority having jurisdiction deems that conditions hazardous to life and property exist, he or she shall be permitted to require that such hazardous conditions in violation of this *Code* be corrected.
(7) To the full extent permitted by law, any authority having jurisdiction engaged in inspection work shall be authorized at all reasonable times to enter and examine any building, structure, or premises for the purpose of making electrical inspections. Before entering a premises, the authority having jurisdiction shall obtain the consent of the occupant thereof or obtain a court warrant authorizing entry for the purpose of inspection except in those instances where an emergency exists. As used in this section, *emergency* means circumstances that the authority having jurisdiction knows, or has reason to believe, exist and that reasonably can constitute immediate danger to persons or property.
(8) Persons authorized to enter and inspect buildings, structures, and premises as herein set forth shall be identified by proper credentials issued by this governing authority.
(9) Persons shall not interfere with an authority having jurisdiction carrying out any duties or functions prescribed by this *Code*.
(10) Persons shall not use a badge, uniform, or other credentials to impersonate the authority having jurisdiction.
(11) The authority having jurisdiction shall be permitted to investigate the cause, origin, and circumstances of any fire, explosion, or other hazardous condition.
(12) The authority having jurisdiction shall be permitted to require plans and specifications to ensure compliance with this *Code*.
(13) Whenever any installation subject to inspection prior to use is covered or concealed without having first been inspected, the authority having jurisdiction shall be permitted to require that such work be exposed for inspection. The authority having jurisdiction shall be notified when the installation is ready for inspection and shall conduct the inspection within ___ days.
(14) The authority having jurisdiction shall be permitted to order the immediate evacuation of any occupied building deemed unsafe when such building has hazardous conditions that present imminent danger to building occupants.
(15) The authority having jurisdiction shall be permitted to waive specific requirements in this *Code* or permit alternative methods where it is assured that equivalent objectives can be achieved by establishing and maintaining effective safety. Technical documentation shall be submitted to the

authority having jurisdiction to demonstrate equivalency and that the system, method, or device is approved for the intended purpose.

(16) Each application for a waiver of a specific electrical requirement shall be filed with the authority having jurisdiction and shall be accompanied by such evidence, letters, statements, results of tests, or other supporting information as required to justify the request. The authority having jurisdiction shall keep a record of actions on such applications, and a signed copy of the authority having jurisdiction's decision shall be provided for the applicant.

Δ 80.15 Electrical Board.

(A) Creation of the Electrical Board. There is hereby created the Electrical Board of the _____ of _____, hereinafter designated as the Board.

(B) Appointments. Board members shall be appointed by the Governor with the advice and consent of the Senate (or by the Mayor with the advice and consent of the Council, or the equivalent).

(1) Members of the Board shall be chosen in a manner to reflect a balanced representation of individuals or organizations. The Chair of the Board shall be elected by the Board membership.

(2) The Chief Electrical Inspector in the jurisdiction adopting this Article authorized in 80.15(B)(3)(a) shall be the nonvoting secretary of the Board. Where the Chief Electrical Inspector of a local municipality serves a Board at a state level, he or she shall be permitted to serve as a voting member of the Board.

(3) The board shall consist of not fewer than five voting members. Board members shall be selected from the following:

a. Chief Electrical Inspector from a local government (for State Board only)
b. An electrical contractor operating in the jurisdiction
c. A licensed professional engineer engaged primarily in the design or maintenance of electrical installations
d. A journeyman electrician

(4) Additional membership shall be selected from the following:

a. A master (supervising) electrician
b. The Fire Marshal (or Fire Chief)
c. A representative of the property/casualty insurance industry
d. A representative of an electric power utility operating in the jurisdiction
e. A representative of electrical manufacturers primarily and actively engaged in producing materials, fittings, devices, appliances, luminaires, or apparatus used as part of or in connection with electrical installations
f. A member of the labor organization that represents the primary electrical workforce

g. A member from the public who is not affiliated with any other designated group
h. A representative of a telecommunications utility operating in the jurisdiction

(C) Terms. Of the members first appointed, _____ shall be appointed for a term of 1 year, _____ for a term of 2 years, _____ for a term of 3 years, and _____ for a term of 4 years, and thereafter each appointment shall be for a term of 4 years or until a successor is appointed. The Chair of the Board shall be appointed for a term not to exceed ____ years.

(D) Compensation. Each appointed member shall receive the sum of _____dollars ($_____) for each day during which the member attends a meeting of the Board and, in addition thereto, shall be reimbursed for direct lodging, travel, and meal expenses as covered by policies and procedures established by the jurisdiction.

(E) Quorum. A quorum as established by the Board operating procedures shall be required to conduct Board business. The Board shall hold such meetings as necessary to carry out the purposes of Article 80. The Chair or a majority of the members of the Board shall have the authority to call meetings of the Board.

(F) Duties. It shall be the duty of the Board to perform the following:

(1) Adopt the necessary rules and regulations to administer and enforce Article 80.
(2) Establish qualifications of electrical inspectors.
(3) Revoke or suspend the recognition of any inspector's certificate for the jurisdiction.
(4) After advance notice of the public hearings and the execution of such hearings, as established by law, the Board is authorized to establish and update the provisions for the safety of electrical installations to conform to the current edition of the *National Electrical Code* (NFPA 70) and other nationally recognized safety standards for electrical installations.
(5) Establish procedures for recognition of electrical safety standards and acceptance of equipment conforming to these standards.

(G) Appeals.

(1) *Review of Decisions.* Any person, firm, or corporation may register an appeal with the Board for a review of any decision of the Chief Electrical Inspector or of any Electrical Inspector, provided that such appeal is made in writing within fifteen (15) days after such person, firm, or corporation shall have been notified. Upon receipt of such appeal, said Board shall, if requested by the person making the appeal, hold a public hearing and proceed to determine whether the action of the Board, or of the Chief Electrical Inspector, or of the Electrical Inspector complies with this law and, within fifteen (15) days after receipt of the

appeal or after holding the hearing, shall make a decision in accordance with its findings.

(2) *Conditions.* Any person shall be permitted to appeal a decision of the authority having jurisdiction to the Board when it is claimed that any one or more of the following conditions exist:

    a. The true intent of the codes or ordinances described in this *Code* has been incorrectly interpreted.

    b. The provisions of the codes or ordinances do not fully apply.

    c. A decision is unreasonable or arbitrary as it applies to alternatives or new materials.

(3) *Submission of Appeals.* A written appeal, outlining the *Code* provision from which relief is sought and the remedy proposed, shall be submitted to the authority having jurisdiction within 15 calendar days of notification of violation.

(H) Meetings and Records. Meetings and records of the Board shall conform to the following:

(1) Meetings of the Board shall be open to the public as required by law.

(2) Records of meetings of the Board shall be available for review during normal business hours, as required by law.

80.17 Records and Reports. The authority having jurisdiction shall retain records in accordance with 80.17(A) and (B).

(A) Retention. The authority having jurisdiction shall keep a record of all electrical inspections, including the date of such inspections and a summary of any violations found to exist, the date of the services of notices, and a record of the final disposition of all violations. All required records shall be maintained until their usefulness has been served or as otherwise required by law.

(B) Availability. A record of examinations, approvals, and variances granted shall be maintained by the authority having jurisdiction and shall be available for public review as prescribed by law during normal business hours.

80.19 Permits and Approvals. Permits and approvals shall conform to 80.19(A) through (H).

(A) Application.

(1) Activity authorized by a permit issued under this *Code* shall be conducted by the permittee or the permittee's agents or employees in compliance with all requirements of this *Code* applicable thereto and in accordance with the approved plans and specifications. No permit issued under this *Code* shall be interpreted to justify a violation of any provision of this *Code* or any other applicable law or regulation. Any addition or alteration of approved plans or specifications shall be approved in advance by the authority having jurisdiction, as evidenced by the issuance of a new or amended permit.

(2) A copy of the permit shall be posted or otherwise readily accessible at each work site or carried by the permit holder as specified by the authority having jurisdiction.

(B) Content. Permits shall be issued by the authority having jurisdiction and shall bear the name and signature of the authority having jurisdiction or that of the authority having jurisdiction's designated representative. In addition, the permit shall indicate the following:

(1) Operation or activities for which the permit is issued

(2) Address or location where the operation or activity is to be conducted

(3) Name and address of the permittee

(4) Permit number and date of issuance

(5) Period of validity of the permit

(6) Inspection requirements

(C) Issuance of Permits. The authority having jurisdiction shall be authorized to establish and issue permits, certificates, notices, and approvals, or orders pertaining to electrical safety hazards pursuant to 80.23, except that no permit shall be required to execute any of the classes of electrical work specified in the following:

(1) Installation or replacement of equipment such as lamps and of electric utilization equipment approved for connection to suitable permanently installed receptacles

(2) Replacement of flush or snap switches, fuses, lamp sockets, and receptacles, and other minor maintenance and repair work, such as replacing worn cords and tightening connections on a wiring device

(3) The process of manufacturing, testing, servicing, or repairing electrical equipment or apparatus

(D) Annual Permits. In lieu of an individual permit for each installation or alteration, an annual permit shall, upon application, be issued to any person, firm, or corporation regularly employing one or more employees for the installation, alteration, and maintenance of electrical equipment in or on buildings or premises owned or occupied by the applicant for the permit. Upon application, an electrical contractor as agent for the owner or tenant shall be issued an annual permit. The applicant shall keep records of all work done, and the records shall be transmitted periodically to the electrical inspector.

(E) Fees. Any political subdivision that has been provided for electrical inspection in accordance with the provisions of Article 80 may establish fees that shall be paid by the applicant for a permit before the permit is issued.

(F) Inspection and Approvals.

(1) Upon the completion of any installation of electrical equipment that has been made under a permit other than an annual permit, it shall be the duty of the person, firm, or corporation making the installation to notify the Electrical

Inspector having jurisdiction, who shall inspect the work within a reasonable time.

(2) Where the Inspector finds the installation to be in conformity with the statutes of all applicable local ordinances and all rules and regulations, the Inspector shall issue to the person, firm, or corporation making the installation a certificate of approval, with duplicate copy for delivery to the owner, authorizing the connection to the supply of electricity and shall send written notice of such authorization to the supplier of electric service. When a certificate of temporary approval is issued authorizing the connection of an installation, such certificates shall be issued to expire at a time to be stated therein and shall be revocable by the Electrical Inspector for cause.

(3) When any portion of the electrical installation within the jurisdiction of an Electrical Inspector is to be hidden from view by the permanent placement of parts of the building, the person, firm, or corporation installing the equipment shall notify the Electrical Inspector, and the equipment shall not be concealed until it has been approved by the Electrical Inspector or until _____ days have elapsed from the time of such notification, provided that on large installations, where the concealment of equipment proceeds continuously, the person, firm, or corporation installing the equipment shall give the Electrical Inspector due notice in advance, and inspections shall be made periodically during the progress of the work.

(4) At regular intervals, the Electrical Inspector having jurisdiction shall visit all buildings and premises where work may be done under annual permits and shall inspect all electrical equipment installed under such permits since the date of the previous inspection. The Electrical Inspector shall issue a certificate of approval for such work as is found to be in conformity with the provisions of Article 80 and all applicable ordinances, orders, rules, and regulations, after payments of all required fees.

(5) If, upon inspection, any installation is found not to be fully in conformity with the provisions of Article 80, and all applicable ordinances, rules, and regulations, the Inspector making the inspection shall at once forward to the person, firm, or corporation making the installation a written notice stating the defects that have been found to exist.

(G) Revocation of Permits. Revocation of permits shall conform to the following:

(1) The authority having jurisdiction shall be permitted to revoke a permit or approval issued if any violation of this *Code* is found upon inspection or in case there have been any false statements or misrepresentations submitted in the application or plans on which the permit or approval was based.

(2) Any attempt to defraud or otherwise deliberately or knowingly design, install, service, maintain, operate, sell, represent for sale, falsify records, reports, or applications, or other related activity in violation of the requirements prescribed by this *Code* shall be a violation of this *Code*. Such violations shall be cause for immediate suspension or revocation of any related licenses, certificates, or permits issued by this jurisdiction. In addition, any such violation shall be subject to any other criminal or civil penalties as available by the laws of this jurisdiction.

(3) Revocation shall be constituted when the permittee is duly notified by the authority having jurisdiction.

(4) Any person who engages in any business, operation, or occupation, or uses any premises, after the permit issued therefor has been suspended or revoked pursuant to the provisions of this *Code*, and before such suspended permit has been reinstated or a new permit issued, shall be in violation of this *Code*.

(5) A permit shall be predicated upon compliance with the requirements of this *Code* and shall constitute written authority issued by the authority having jurisdiction to install electrical equipment. Any permit issued under this *Code* shall not take the place of any other license or permit required by other regulations or laws of this jurisdiction.

(6) The authority having jurisdiction shall be permitted to require an inspection prior to the issuance of a permit.

(7) A permit issued under this *Code* shall continue until revoked or for the period of time designated on the permit. The permit shall be issued to one person or business only and for the location or purpose described in the permit. Any change that affects any of the conditions of the permit shall require a new or amended permit.

(H) Applications and Extensions. Applications and extensions of permits shall conform to the following:

(1) The authority having jurisdiction shall be permitted to grant an extension of the permit time period upon presentation by the permittee of a satisfactory reason for failure to start or complete the work or activity authorized by the permit.

(2) Applications for permits shall be made to the authority having jurisdiction on forms provided by the jurisdiction and shall include the applicant's answers in full to inquiries set forth on such forms. Applications for permits shall be accompanied by such data as required by the authority having jurisdiction, such as plans and specifications, location, and so forth. Fees shall be determined as required by local laws.

(3) The authority having jurisdiction shall review all applications submitted and issue permits as required. If an application for a permit is rejected by the authority having jurisdiction, the applicant shall be advised of the reasons for such rejection. Permits for activities requiring evidence of financial responsibility by the jurisdiction shall not be issued unless proof of required financial responsibility is furnished.

80.21 Plans Review. Review of plans and specifications shall conform to 80.21(A) through (C).

(A) Authority. For new construction, modification, or rehabilitation, the authority having jurisdiction shall be permitted to review construction documents and drawings.

(B) Responsibility of the Applicant. It shall be the responsibility of the applicant to ensure the following:

(1) The construction documents include all of the electrical requirements.
(2) The construction documents and drawings are correct and in compliance with the applicable codes and standards.

(C) Responsibility of the Authority Having Jurisdiction. It shall be the responsibility of the authority having jurisdiction to promulgate rules that cover the following:

(1) Review of construction documents and drawings shall be completed within established time frames for the purpose of acceptance or to provide reasons for nonacceptance.
(2) Review and approval by the authority having jurisdiction shall not relieve the applicant of the responsibility of compliance with this *Code*.
(3) Where field conditions necessitate any substantial change from the approved plan, the authority having jurisdiction shall be permitted to require that the corrected plans be submitted for approval.

80.23 Notice of Violations, Penalties. Notice of violations and penalties shall conform to 80.23(A) and (B).

(A) Violations.

(1) Whenever the authority having jurisdiction determines that there are violations of this *Code*, a written notice shall be issued to confirm such findings.
(2) Any order or notice issued pursuant to this *Code* shall be served upon the owner, operator, occupant, or other person responsible for the condition or violation, either by personal service or mail or by delivering the same to, and leaving it with, some person of responsibility upon the premises. For unattended or abandoned locations, a copy of such order or notice shall be posted on the premises in a conspicuous place at or near the entrance to such premises and the order or notice shall be mailed by registered or certified mail, with return receipt requested, to the last known address of the owner, occupant, or both.

(B) Penalties.

(1) Any person who fails to comply with the provisions of this *Code* or who fails to carry out an order made pursuant to this *Code* or violates any condition attached to a permit, approval, or certificate shall be subject to the penalties established by this jurisdiction.

(2) Failure to comply with the time limits of an abatement notice or other corrective notice issued by the authority having jurisdiction shall result in each day that such violation continues being regarded as a new and separate offense.
(3) Any person, firm, or corporation who shall willfully violate any of the applicable provisions of this article shall be guilty of a misdemeanor and, upon conviction thereof, shall be punished by a fine of not less than _____dollars ($_____) or more than _____dollars ($_____) for each offense, together with the costs of prosecution, imprisonment, or both, for not less than _____ (_____) days or more than _____ (_____) days.

80.25 Connection to Electricity Supply. Connections to the electric supply shall conform to 80.25(A) through (E).

(A) Authorization. Except where work is done under an annual permit and except as otherwise provided in 80.25, it shall be unlawful for any person, firm, or corporation to make connection to a supply of electricity or to supply electricity to any electrical equipment installation for which a permit is required or that has been disconnected or ordered to be disconnected.

(B) Special Consideration. By special permission of the authority having jurisdiction, temporary power shall be permitted to be supplied to the premises for specific needs of the construction project. The Board shall determine what needs are permitted under this provision.

(C) Notification. If, within _____ business days after the Electrical Inspector is notified of the completion of an installation of electric equipment, other than a temporary approval installation, the Electrical Inspector has neither authorized connection nor disapproved the installation, the supplier of electricity is authorized to make connections and supply electricity to such installation.

(D) Other Territories. If an installation or electric equipment is located in any territory where an Electrical Inspector has not been authorized or is not required to make inspections, the supplier of electricity is authorized to make connections and supply electricity to such installations.

(E) Disconnection. Where a connection is made to an installation that has not been inspected, as outlined in the preceding paragraphs of this section, the supplier of electricity shall immediately report such connection to the Chief Electrical Inspector. If, upon subsequent inspection, it is found that the installation is not in conformity with the provisions of Article 80, the Chief Electrical Inspector shall notify the person, firm, or corporation making the installation to rectify the defects and, if such work is not completed within fifteen (15) business days or a longer period as may be specified by the Board, the Board shall have the authority to cause the disconnection of that portion of the installation that is not in conformity.

△ 80.27 Inspector's Qualifications.

(A) Certificate. All electrical inspectors shall be certified by a nationally recognized inspector certification program accepted by the Board. The certification program shall specifically qualify the inspector in electrical inspections. No person shall be employed as an Electrical Inspector unless that person is the holder of an Electrical Inspector's certificate of qualification issued by the Board, except that any person who on the date on which this law went into effect was serving as a legally appointed Electrical Inspector of _____ shall, upon application and payment of the prescribed fee and without examination, be issued a special certificate permitting him or her to continue to serve as an Electrical Inspector in the same territory.

(B) Experience. Electrical inspector applicants shall demonstrate the following:

(1) Have a demonstrated knowledge of the standard materials and methods used in the installation of electric equipment
(2) Be well versed in the approved methods of construction for safety to persons and property
(3) Be well versed in the statutes of _____ relating to electrical work and the *National Electrical Code*, as approved by the American National Standards Institute
(4) Have had at least _____ years' experience as an Electrical Inspector or _____ years in the installation of electrical equipment. In lieu of such experience, the applicant shall be a graduate in electrical engineering or of a similar curriculum of a college or university considered by the Board as having suitable requirements for graduation and shall have had two years' practical electrical experience.

(C) Recertification. Electrical inspectors shall be recertified as established by provisions of the applicable certification program.

(D) Revocation and Suspension of Authority. The Board shall have the authority to revoke an inspector's authority to conduct inspections within a jurisdiction.

80.29 Liability for Damages. Article 80 shall not be construed to affect the responsibility or liability of any party owning, designing, operating, controlling, or installing any electrical equipment for damages to persons or property caused by a defect therein, nor shall the _____ or any of its employees be held as assuming any such liability by reason of the inspection, reinspection, or other examination authorized.

80.31 Validity. If any section, subsection, sentence, clause, or phrase of Article 80 is for any reason held to be unconstitutional, such decision shall not affect the validity of the remaining portions of Article 80.

80.33 Repeal of Conflicting Acts. All acts or parts of acts in conflict with the provisions of Article 80 are hereby repealed.

80.35 Effective Date. Article 80 shall take effect _____ (_____) days after its passage and publication.

Recommended Tightening Torque Tables from UL Standard 486A-486B

This informative annex is not a part of the requirements of this NFPA document, but is included for informational purposes only.

In the absence of connector or equipment manufacturer's recommended torque values, Table I.1, Table I.2, and Table I.3 may be used to correctly tighten screw-type connections for power and lighting circuits*. Control and signal circuits may require different torque values, and the manufacturer should be contacted for guidance.

*For proper termination of conductors, it is very important that field connections be properly tightened. In the absence of manufacturer's instructions on the equipment, the torque values given in these tables are recommended. Because it is normal for some relaxation to occur in service, checking torque values sometime after installation is not a reliable means of determining the values of torque applied at installation.

Δ **TABLE I.1** *Tightening Torque for Screws*

Test Conductor Installed in Connector		Tightening Torque, N-m (lbf-in.)							
		Slotted head No. 10 and larger[a]							
AWG or kcmil	mm²	Slot width 1.2 mm (0.047 in.) or less and slot length 6.4 mm (¼ in.) or less		Slot width over 1.2 mm (0.047 in.) or slot length over 6.4 mm (¼ in.)		Split-bolt connectors		Other connectors	
30–10	0.05–5.3	2.3	(20)	4.0	(35)	9.0	(80)	8.5	(75)
8	8.4	2.8	(25)	4.5	(40)	9.0	(80)	8.5	(75)
6–4	13.2–21.2	4.0	(35)	5.1	(45)	18.5	(165)	12.4	(110)
3	26.7	4.0	(35)	5.6	(50)	31.1	(275)	16.9	(150)
2	33.6	4.5	(40)	5.6	(50)	31.1	(275)	16.9	(150)
1	42.4		—	5.6	(50)	31.1	(275)	16.9	(150)
1/0–2/0	53.5–67.4		—	5.6	(50)	43.5	(385)	20.3	(180)
3/0–4/0	85.0–107.2		—	5.6	(50)	56.5	(500)	28.2	(250)
250–350	127–177		—	5.6	(50)	73.4	(650)	36.7	(325)
400	203		—	5.6	(50)	93.2	(825)	36.7	(325)
500	253		—	5.6	(50)	93.2	(825)	42.4	(375)
600–750	304–380		—	5.6	(50)	113.0	(1000)	42.4	(375)
800–1000	405–508		—	5.6	(50)	124.3	(1100)	56.5	(500)
1250–2000	635–1010		—		—	124.3	(1100)	67.8	(600)

[a]For values of slot width or length not corresponding to those specified, select the largest torque value associated with the conductor size. Slot width is the nominal design value. Slot length shall be measured at the bottom of the slot.

TABLE I.2 *Tightening Torque for Slotted Head Screws Smaller Than No. 10 Intended for Use with 8 AWG (8.4 mm²) or Smaller Conductors*

Slot Length of Screw[a]		Tightening Torque, N-m (lbf-in.)	
mm	in.	Slot width of screw smaller than 1.2 mm (0.047 in.)[b]	Slot width of screw 1.2 mm (0.047 in.) and larger[b]
Less than 4	Less than 5/32	0.79 (7)	1.0 (9)
4	5/32	0.79 (7)	1.4 (12)
4.8	3/16	0.79 (7)	1.4 (12)
5.5	7/32	0.79 (7)	1.4 (12)
6.4	1/4	1.0 (9)	1.4 (12)
7.1	9/32		1.7 (15)
Above 7.1	Above 9/32		2.3 (20)

[a]For slot lengths of intermediate values, select torques pertaining to next shorter slot lengths. Also, see 9.1.9.6 of UL 486A-486B-2013, *Wire Connectors*, for screws with multiple tightening means. Slot length shall be measured at the bottom of the slot.
[b]Slot width is the nominal design value.

TABLE I.3 *Tightening Torque for Screws with Recessed Allen or Square Drives*

Socket Width Across Flats[a]		Tightening Torque, N-m (lbf-in.)	
mm	in.		
3.2	1/8	5.1	(45)
4.0	5/32	11.3	(100)
4.8	3/16	13.5	(120)
5.5	7/32	16.9	(150)
6.4	1/4	22.5	(200)
7.9	5/16	31.1	(275)
9.5	3/8	42.4	(375)
12.7	1/2	56.5	(500)
14.3	9/16	67.8	(600)

[a]See 9.1.9.6 of UL 486A-2003, *Wire Connectors and Soldering Lugs for Use with Copper Conductors*, for screws with multiple tightening means.

With the permission of Underwriters Laboratories Inc., material is reproduced from UL 486A-486B-2013, *Wire Connectors*, which is copyrighted by Underwriters Laboratories Inc., Northbrook, Illinois. While use of this material has been authorized, UL shall not be responsible for the manner in which the information is presented, nor for any interpretations thereof. For more information on UL, or to purchase standards, visit their website at www.comm-2000.com or call 1-888-853-3503.

ADA Standards for Accessible Design

This informative annex is not a part of the requirements of this NFPA document, but is included for informational purposes only.

The provisions cited in Informative Annex J are intended to assist the users of the *Code* in properly considering the various electrical design constraints of other building systems and are part of the 2010 ADA Standards for Accessible Design. They are the same provisions as those found in ANSI/ICC A117.1-2009, *Accessible and Usable Buildings and Facilities.*

J.1 Protruding Objects. Protruding objects shall comply with Section J.2.

J.2 Protrusion Limits. Objects with leading edges more than 685 mm (27 in.) and not more than 2030 mm (80 in.) above the finish floor or ground shall protrude a maximum of 100 mm (4 in.) horizontally into the circulation path. *(See Figure J.2.)*

Exception: Handrails shall be permitted to protrude 115 mm (4½ in.) maximum.

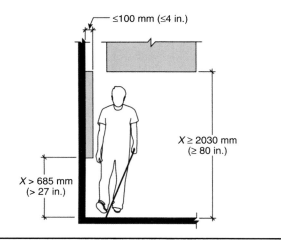

FIGURE J.2 *Limits of Protruding Objects.*

J.3 Post-Mounted Objects. Freestanding objects mounted on posts or pylons shall overhang circulation paths 305 mm (12 in.) maximum where located 685 mm (27 in.) minimum and 2030 mm (80 in.) maximum above the finish floor or ground. Where a sign or other obstruction is mounted between posts or pylons, and the clear distance between the posts or pylons is greater than 305 mm (12 in.), the lowest edge of such sign or obstruction shall be 685 mm (27 in.) maximum or 2030 mm (80 in.) minimum above the finish floor or ground. *(See Figure J.3.)*

Exception: The sloping portions of handrails serving stairs and ramps shall not be required to comply with Section J.3.

J.4 Vertical Clearance. Vertical clearance shall be 2030 mm (80 in.) high minimum. Guardrails or other barriers shall be provided where the vertical clearance is less than 2030 mm (80 in.) high. The leading edge of such guardrail or barrier shall be located 685 mm (27 in.) maximum above the finish floor or ground. *(See Figure J.4.)*

Exception: Door closers and door stops shall be permitted to be 1980 mm (78 in.) minimum above the finish floor or ground.

J.5 Required Clear Width. Protruding objects shall not reduce the clear width required for accessible routes.

J.6 Forward Reach.

J.6.1 Unobstructed. Where a forward reach is unobstructed, the high forward reach shall be 1220 mm (48 in.) maximum, and the low forward reach shall be 380 mm (15 in.) minimum above the finish floor or ground. *(See Figure J.6.1.)*

J.6.2 Obstructed High Reach. Where a high forward reach is over an obstruction, the clear floor space shall extend beneath the element for a distance not less than the required reach depth over the obstruction. The high forward reach shall be 1220 mm (48 in.) maximum where the reach depth is 510 mm (20 in.) maximum.

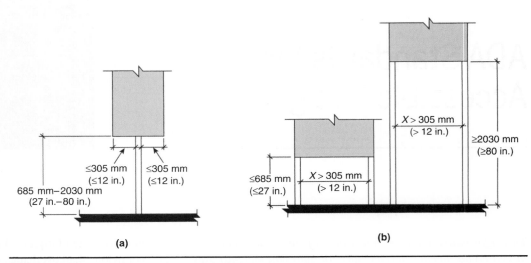

FIGURE J.3 *Post-Mounted Protruding Objects.*

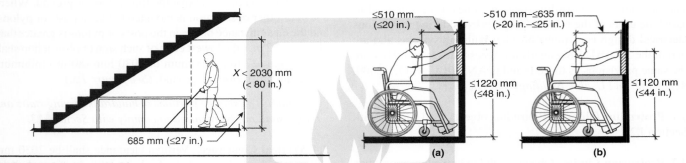

FIGURE J.4 *Vertical Clearance.*

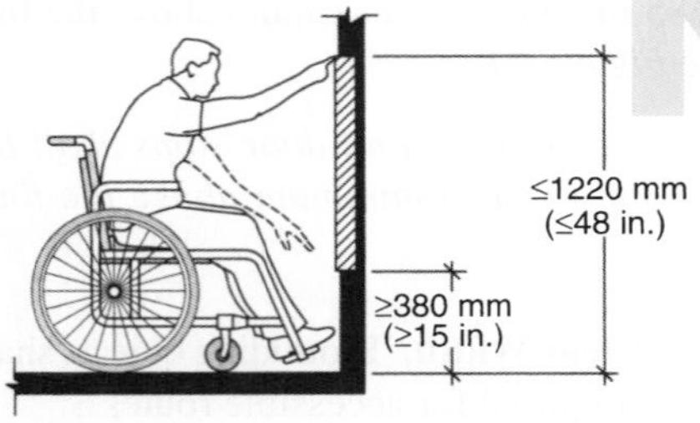

FIGURE J.6.2 *Obstructed High Forward Reach.*

FIGURE J.6.1 *Unobstructed Forward Reach.*

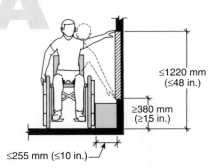

FIGURE J.7.1 *Unobstructed Side Reach.*

Where the reach depth exceeds 510 mm (20 in.), the high forward reach shall be 1120 mm (44 in.) maximum, and the reach depth shall be 635 mm (25 in.) maximum. (*See Figure J.6.2.*)

J.7 Side Reach.

J.7.1 Unobstructed. Where a clear floor or ground space allows a parallel approach to an element, and the side reach is unobstructed, the high side reach shall be 1220 mm (48 in.) maximum, and the low side reach shall be 380 mm (15 in.) minimum above the finish floor or ground. (*See Figure J.7.1.*)

Exception No. 1: An obstruction shall be permitted between the clear floor or ground space and the element where the depth of the obstruction is 255 mm (10 in.) maximum.

Exception No. 2: Operable parts of fuel dispensers shall be permitted to be 1370 mm (54 in.) maximum, measured from the surface of the vehicular way where fuel dispensers are installed on existing curbs.

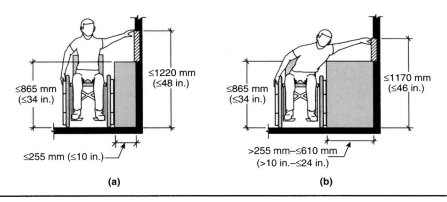

FIGURE J.7.2 *Obstructed High Side Reach.*

Δ **J.7.2 Obstructed High Reach.** Where a clear floor or ground space allows a parallel approach to an element and the high side reach is over an obstruction, the height of the obstruction shall be 865 mm (34 in.) maximum, and the depth of the obstruction shall be 610 mm (24 in.) maximum. The high side reach shall be 1220 mm (48 in.) maximum for a reach depth of 255 mm (10 in.) maximum. Where the reach depth exceeds 255 mm (10 in.), the high side reach shall be 1170 mm (46 in.) maximum for a reach depth of 610 mm (24 in.) maximum. *(See Figure J.7.2.)*

Exception No. 1: The top of washing machines and clothes dryers shall be permitted to be 915 mm (36 in.) maximum above the finish floor.

Exception No. 2: Operable parts of fuel dispensers shall be permitted to be 1370 mm (54 in.) maximum, measured from the surface of the vehicular way where fuel dispensers are installed on existing curbs.

Use of Medical Electrical Equipment in Dwellings and Residential Board-and-Care Occupancies

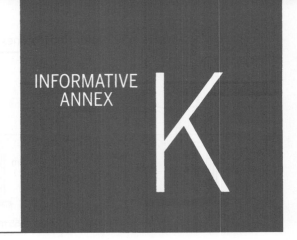

INFORMATIVE ANNEX

K

This informative annex is not a part of the requirements of this NFPA document, but is included for informational purposes only.

During the 2023 NEC® cycle, Informative Annex K, Use of Medical Electrical Equipment in Dwellings and Residential Board-and-Care Occupancies, was added. With increased medical care in homes, as well as other care occupancies, causing growth in the utilization of medical electrical equipment (MEE), the *NEC* has been updated to assist with safety in those installations. Informative Annex K covers the assessment and installation of MEE in dwellings and residential board-and-care occupancies, including assisted living residences.

N In recent years there has been a significant increase in home or remote patient use of electrical dependent medical equipment, and it is widely considered that this trend will continue in the coming years. Because of this trend, an investigation of the appropriate interaction of this critical equipment with *NFPA 70, National Electrical Code (NEC)* installations, both existing and new, should occur.

Medical electrical equipment (MEE) is equipment that has an applied part that transfers or detects energy to and from a patient. This equipment is provided with a single connection to electrical power and is intended for use by the manufacturer, either by marking on equipment or in instructions, to be used for diagnosis, treatment, or monitoring of a patient or for compensation or alleviation of disease, injury, or disability. The environment for intended use can be best described as a place where the patient lives or is present without continuous supervision or aid by professional workers. The Social Security Administration for Medicare provides a definition for durable medical equipment (DME) that is broader than electrical: equipment that is primarily used to serve a medical purpose and is appropriate for use in the home.

MEE is generally divided into classes relating to means of protection (MOP). With Class I equipment, protection against electric shock relies on bonding of equipment with an equipment grounding conductor. Class II equipment does not rely on bonding of equipment as an MOP but instead relies on double or reinforced insulation as an MOP against electric shock. Class II equipment does not have provisions for bonding of equipment or reliance upon installation conditions. For home use MEE, Class I equipment includes an equipment grounding conductor and must be permanently installed.

MEE evaluated for compliance with the ES 60601-1 series of standards are typically Class II equipment unless they are permanently connected to the building power. While there are varying applications and intended uses of medical equipment, one of the most critical is life support medical equipment. This equipment is intended to actively keep alive or resuscitate a patient. Due to the critical intent of this equipment, interaction with an *NEC* installation is critical. Reliable supply of power and understanding of availability of electrical power should there be an outage are key parameters to consider. This type of medical equipment is often supplied with backup power appropriate for the intended use and critical nature of its function. Life support equipment should be on a circuit with limited or no other loads to prevent overloading and unintended removal of power. It is recommended to supply this equipment with an individual branch circuit. If this is not a feasible option due to current installation conditions, it is recommended to conduct the following analysis and labeling:

(1) Conduct an analysis of the circuit intended to supply the life support equipment including all lighting or other outlets that are on the circuit.

(2) Follow the rules of 210.23(B)(2) limiting MEE loads on this circuit to 50 percent or less.

(3) Determine that adequate loading is available for the reliable supply of power to the life support equipment.

(4) Conduct an analysis around the need for backup power given the availability of the patient to access an alternate supply source should they lose primary power in the intended location of the equipment. This will be affected by the distance to the next available option for electrical power, the mobility of the patient, or access to others able to assist. If there is a concern in

this area, then backup power at the primary location is suggested.

(5) In the absence of an alarm integral to the MEE, provide an audible alarm that monitors the circuit supplying power to the equipment and sets off an alarm when power is lost at the outlet supplying the MEE.

(6) Investigate electrical devices and components in the premises wiring system to ensure that remote control or switching are not allowed. Confirm continuity of power by energizing equipment, and run through a normal cycle of functions to ensure reliable supply of power.

(7) Label all receptacles available to supply power to other loads on the circuit. The labeling should read as follows or similar language with the same intent:

WARNING — Power loss risk to
life-support and medical equipment on same circuit.
DO NOT OVERLOAD

When using medical equipment, it is critical to understand the conditions and environment in which it will be used. Locating the equipment in wet or damp locations or near other systems (e.g., water, gas, oxygen, sparks) can present hazards that need to be addressed in the installation. For wet and damp locations, MEE will be marked for use in these locations with an ingress protection IPXX (e.g., IP22) rating on the equipment. In the absence of IP21 or higher markings, the equipment should not be used in wet or damp locations. If the equipment is marked with an Umbrella (Keep Dry) symbol, it is limited to dry locations only.

IMPORTANT NOTICES AND DISCLAIMERS CONCERNING NFPA® STANDARDS

NOTICE AND DISCLAIMER OF LIABILITY CONCERNING THE USE OF NFPA STANDARDS

NFPA® codes, standards, recommended practices, and guides ("NFPA Standards"), of which the document contained herein is one, are developed through a consensus standards development process approved by the American National Standards Institute. This process brings together volunteers representing varied viewpoints and interests to achieve consensus on fire and other safety issues. While the NFPA administers the process and establishes rules to promote fairness in the development of consensus, it does not independently test, evaluate, or verify the accuracy of any information or the soundness of any judgments contained in NFPA Standards.

The NFPA disclaims liability for any personal injury, property or other damages of any nature whatsoever, whether special, indirect, consequential or compensatory, directly or indirectly resulting from the publication, use of, or reliance on NFPA Standards. The NFPA also makes no guaranty or warranty as to the accuracy or completeness of any information published herein.

In issuing and making NFPA Standards available, the NFPA is not undertaking to render professional or other services for or on behalf of any person or entity. Nor is the NFPA undertaking to perform any duty owed by any person or entity to someone else. Anyone using this document should rely on his or her own independent judgment or, as appropriate, seek the advice of a competent professional in determining the exercise of reasonable care in any given circumstances.

The NFPA has no power, nor does it undertake, to police or enforce compliance with the contents of NFPA Standards. Nor does the NFPA list, certify, test, or inspect products, designs, or installations for compliance with this document. Any certification or other statement of compliance with the requirements of this document shall not be attributable to the NFPA and is solely the responsibility of the certifier or maker of the statement.

ADDITIONAL NOTICES AND DISCLAIMERS

Updating of NFPA Standards
Users of NFPA codes, standards, recommended practices, and guides ("NFPA Standards") should be aware that these documents may be superseded at any time by the issuance of new editions or may be amended from time to time through the issuance of Tentative Interim Amendments or corrected by Errata. An official NFPA Standard at any point in time consists of the current edition of the document together with any Tentative Interim Amendments and any Errata then in effect. In order to determine whether a given document is the current edition and whether it has been amended through the issuance of Tentative Interim Amendments or corrected through the issuance of Errata, consult appropriate NFPA publications such as the National Fire Codes® Subscription Service, visit the NFPA website at www.nfpa.org, or contact the NFPA at the address listed below.

Interpretations of NFPA Standards
A statement, written or oral, that is not processed in accordance with Section 6 of the Regulations Governing the Development of NFPA Standards shall not be considered the official position of NFPA or any of its Committees and shall not be considered to be, nor be relied upon as, a Formal Interpretation.

Patents
The NFPA does not take any position with respect to the validity of any patent rights referenced in, related to, or asserted in connection with an NFPA Standard. The users of NFPA Standards bear the sole responsibility for determining the validity of any such patent rights, as well as the risk of infringement of such rights, and the NFPA disclaims liability for the infringement of any patent resulting from the use of or reliance on NFPA Standards.

NFPA adheres to the policy of the American National Standards Institute (ANSI) regarding the inclusion of patents in American National Standards ("the ANSI Patent Policy"), and hereby gives the following notice pursuant to that policy:

> NOTICE: The user's attention is called to the possibility that compliance with an NFPA Standard may require use of an invention covered by patent rights. NFPA takes no position as to the validity of any such patent rights or as to whether such patent rights constitute or include essential patent claims under the ANSI Patent Policy. If, in connection with the ANSI Patent Policy, a patent holder has filed a statement of willingness to grant licenses under these rights on reasonable and nondiscriminatory terms and conditions to applicants desiring to obtain such a license, copies of such filed statements can be obtained, on request, from NFPA. For further information, contact the NFPA at the address listed below.

Law and Regulations
Users of NFPA Standards should consult applicable federal, state, and local laws and regulations. NFPA does not, by the publication of its codes, standards, recommended practices, and guides, intend to urge action that is not in compliance with applicable laws, and these documents may not be construed as doing so.

Copyrights
NFPA Standards are copyrighted. They are made available for a wide variety of both public and private uses. These include both use, by reference, in laws and regulations, and use in private self-regulation, standardization, and the promotion of safe practices and methods. By making these documents available for use and adoption by public authorities and private users, the NFPA does not waive any rights in copyright to these documents.

Use of NFPA Standards for regulatory purposes should be accomplished through adoption by reference. The term "adoption by reference" means the citing of title, edition, and publishing information only. Any deletions, additions, and changes desired by the adopting authority should be noted separately in the adopting instrument. In order to assist NFPA in following the uses made of its documents, adopting authorities are requested to notify the NFPA (Attention: Secretary, Standards Council) in writing of such use. For technical assistance and questions concerning adoption of NFPA Standards, contact NFPA at the address below.

For Further Information
All questions or other communications relating to NFPA Standards and all requests for information on NFPA procedures governing its codes and standards development process, including information on the procedures for requesting Formal Interpretations, for proposing Tentative Interim Amendments, and for proposing revisions to NFPA standards during regular revision cycles, should be sent to NFPA headquarters, addressed to the attention of the Secretary, Standards Council, NFPA, 1 Batterymarch Park, P.O. Box 9101, Quincy, MA 02269-9101; email: stds_admin@nfpa.org

For more information about NFPA, visit the NFPA website at www.nfpa.org. All NFPA codes and standards can be viewed at no cost at www.nfpa.org/docinfo.

Abbreviations of Terms Used in the NEC®

AFCI	arc-fault circuit interrupter
AHJ	authority having jurisdiction
Al	aluminum
ANSI	American National Standards Institute
ATS	automatic transfer switch
AWG	American Wire Gage
BCSC	branch-circuit selection current
CATV	community antenna television
CB	circuit breaker
cmil	circular mil
CO/ALR	copper-aluminum, revised
COPS	critical operations power systems
CSA	Canadian Standards Association
Cu	copper
DCOA	designated critical operations areas
EGC	equipment grounding conductor
EMF	electromotive force
EMS	emergency management system
EMT	electrical metallic tubing
ENT	electrical nonmetallic tubing
EPL	equipment protection level
EPS	emergency power supply
EPSS	emergency power supply system
ESS	energy storage system
EV	electric vehicle
EVPE	electric vehicle power export equipment
EVSE	electric vehicle supply equipment
FEB	field evaluation body
FLC	full-load current
FMC	flexible metal conduit
FMP	fault-managed power
FMT	flexible metallic tubing
GEC	grounding electrode conductor
GFCI	ground-fault circuit interrupter
GFDI	ground-fault detector-interrupter
GFPE	ground-fault protection of equipment
GND	ground
HACR	heating, air conditioning, and refrigeration
HDCI	heat detecting circuit interrupter
HDPE	high density polyethylene
HID	high intensity discharge
hp	horsepower
HVAC	heating, ventilation, and air conditioning
Hz	hertz (cycle per second)
IAEI	International Association of Electrical Inspectors
IBP	insulated bus pipe
IBT	intersystem bonding termination
IEC	International Electrotechnical Commission
IEEE	Institute of Electrical and Electronics Engineers
IMC	intermediate metal conduit
IS	intrinsically safe
ITE	information technology equipment
kcmil	one thousand circular mils
kVA	kilovolt-amperes
kW	kilowatts
LCDI	leakage-current detector-interrupter
LED	light-emitting diode
LFMC	liquidtight flexible metal conduit
LFNC	liquidtight flexible nonmetallic conduit
LNG	liquefied natural gas
LP-Gas	liquefied petroleum gas
LRC	locked-rotor current
MCS	microgrid control system
MDC	modular data center
MG	motor-generator
MIC	minimum igniting current
MID	microgrid interconnect device
MIDI	musical instrument digital interface
MVA	mega-volt ampere
MW	megawatts
NECA	National Electrical Contractors Association
NEMA	National Electrical Manufacturers Association
NESC	National Electrical Safety Code
NIU	network interface unit
NPLFA	non-power-limited fire alarm circuit
NPT	National (American) Standard Pipe Taper
NUCC	nonmetallic underground conduit with conductors
OCPD	overcurrent protective device
ONT	optical network terminal
PCS	power control system
PCU	power-conditioning unit
PE	polyethylene
PF	power factor
PHEV	plug-in hybrid electric vehicle
PLC	programmable logic controller
PLFA	power-limited fire alarm circuit
PoE	Power over Ethernet
PV	photovoltaic
RMC	rigid metal conduit
RMS	root-mean-square
RNC	rigid nonmetallic conduit
RPM	revolutions per minute
RTRC	reinforced thermosetting resin conduit
SCADA	supervisory control and data acquisition
SCR	silicon controlled rectifier
SI	International System of Units
SPD	surge-protective device
SPGFCI	special purpose ground-fault circuit interrupter
SWD	switching duty
TCC	tubular covered conductor
TEFC	totally enclosed fan-cooled
TR	tamper-resistant
TRU	transport refrigerated unit
UPS	uninterruptible power supply
USB	universal serial bus
V ac	volts alternating current
V dc	volts direct current
VA	volt-ampere
VAM	valve actuator motor
VAr	volt-ampere reactive
W-hr	watt-hour
WPT	wireless power transfer
WPTE	wireless power transfer equipment
WR	weather resistant
WSAF	weight-supporting attachment fitting
WSCR	weight-supporting ceiling receptacle
Ze	effective impedance